# HANDBOOK OF INDUSTRIAL MIXING

# HANDBOOK OF INDUSTRIAL MIXING
## SCIENCE AND PRACTICE

Edited by

**Edward L. Paul**
Merck & Co., Inc.
Rahway, New Jersey

**Victor A. Atiemo-Obeng**
The Dow Chemical Company
Midland, Michigan

**Suzanne M. Kresta**
University of Alberta
Edmonton, Canada

Sponsored by the North American Mixing Forum

A JOHN WILEY & SONS, INC., PUBLICATION

*Cover*: The jet image is courtesy of Chiharu Fukushima and Jerry Westerweel, of the Laboratory for Aero and Hydrodynamics, Delft University of Technology, The Netherlands.

Copyright © 2004 by John Wiley & Sons, Inc. All rights reserved.

Published by John Wiley & Sons, Inc., Hoboken, New Jersey.
Published simultaneously in Canada.

No part of this publication may be reproduced, stored in a retrieval system, or transmitted in any form or by any means, electronic, mechanical, photocopying, recording, scanning, or otherwise, except as permitted under Section 107 or 108 of the 1976 United States Copyright Act, without either the prior written permission of the Publisher, or authorization through payment of the appropriate per-copy fee to the Copyright Clearance Center, Inc., 222 Rosewood Drive, Danvers, MA 01923, 978-750-8400, fax 978-750-4470, or on the web at www.copyright.com. Requests to the Publisher for permission should be addressed to the Permissions Department, John Wiley & Sons, Inc., 111 River Street, Hoboken, NJ 07030, (201) 748-6011, fax (201) 748-6008, e-mail: permreq@wiley.com.

Limit of Liability/Disclaimer of Warranty: While the publisher and author have used their best efforts in preparing this book, they make no representations or warranties with respect to the accuracy or completeness of the contents of this book and specifically disclaim any implied warranties of merchantability or fitness for a particular purpose. No warranty may be created or extended by sales representatives or written sales materials. The advice and strategies contained herein may not be suitable for your situation. You should consult with a professional where appropriate. Neither the publisher nor author shall be liable for any loss of profit or any other commercial damages, including but not limited to special, incidental, consequential, or other damages.

For general information on our other products and services please contact our Customer Care Department within the U.S. at 877-762-2974, outside the U.S. at 317-572-3993 or fax 317-572-4002.

Wiley also publishes its books in a variety of electronic formats. Some content that appears in print, however, may not be available in electronic format.

*Library of Congress Cataloging-in-Publication Data:*

Paul, Edward L.
  Handbook of industrial mixing : science and practice / Edward L. Paul,
Victor A. Atiemo-Obeng, Suzanne M. Kresta
      p. cm.
"Sponsored by the North American Mixing Forum."
Includes bibliographical references and index.
  ISBN 0-471-26919-0 (cloth : alk. paper)
  1. Mixing—Handbooks, manuals, etc. I. Atiemo-Obeng, Victor A. II. Kresta, Suzanne M. III. Title.

TP156,M5K74 2003
660'.284292—dc21
                                                        2003007731

Printed in the United States of America.

10 9 8 7 6 5 4 3 2 1

# CONTENTS

| | |
|---|---|
| **Contributors** | xxix |
| **Introduction** | xxxiii |
| *Edward L. Paul, Victor A. Atiemo-Obeng, and Suzanne M. Kresta* | |
| Mixing in Perspective | xxxiv |
|    Scope of Mixing Operations | xxxvi |
|    Residence Time Distributions: Chapter 1 | xxxvii |
|    Mixing Fundamentals: Chapters 1–5 | xxxix |
|    Mixing Equipment: Chapters 6, 7, 8, and 21 | xxxix |
|    Miscible Liquid Blending: Chapters 3, 7, 9, and 16 | xl |
|    Solid–Liquid Suspension: Chapters 10, 17, and 18 | xl |
|    Gas–Liquid Contacting: Chapter 11 | xli |
|    Liquid–Liquid Mixing: Chapter 12 | xlii |
|    Mixing and Chemical Reactions/Reactor Design: Chapters 13 and 17 | xlii |
|    Heat Transfer and Mixing: Chapter 14 | xliii |
|    Specialized Topics for Various Industries: Chapters 15–20 | xliii |
| Conversations Overheard in a Chemical Plant | xliv |
|    The Problem | xliv |
|    Competitive-Consecutive Reaction | xlv |
|    Gas–Liquid Reaction | xlvi |
|    Solid–Liquid Reaction | xlvi |
|    Liquid–Liquid Reaction | xlvii |
|    Crystallization | xlvii |
| Using the Handbook | xlix |
|    Diagnostic Charts | l |
|    Mixing Nomenclature and Unit Conversions | lv |
| Acknowledgments | lix |
| References | lx |

# Contents

**1 Residence Time Distributions** — 1
*E. Bruce Nauman*

- 1-1 Introduction — 1
- 1-2 Measurements and Distribution Functions — 2
- 1-3 Residence Time Models of Flow Systems — 5
  - 1-3.1 Ideal Flow Systems — 5
  - 1-3.2 Hydrodynamic Models — 6
  - 1-3.3 Recycle Models — 7
- 1-4 Uses of Residence Time Distributions — 9
  - 1-4.1 Diagnosis of Pathological Behavior — 9
  - 1-4.2 Damping of Feed Fluctuations — 9
  - 1-4.3 Yield Prediction — 10
  - 1-4.4 Use with Computational Fluid Dynamic Calculations — 14
- 1-5 Extensions of Residence Time Theory — 15
- Nomenclature — 16
- References — 16

**2 Turbulence in Mixing Applications** — 19
*Suzanne M. Kresta and Robert S. Brodkey*

- 2-1 Introduction — 19
- 2-2 Background — 20
  - 2-2.1 Definitions — 20
  - 2-2.2 Length and Time Scales in the Context of Turbulent Mixing — 24
  - 2-2.3 Relative Rates of Mixing and Reaction: The Damkoehler Number — 32
- 2-3 Classical Measures of Turbulence — 38
  - 2-3.1 Phenomenological Description of Turbulence — 39
  - 2-3.2 Turbulence Spectrum: Quantifying Length Scales — 45
  - 2-3.3 Scaling Arguments and the Energy Budget: Relating Turbulence Characteristics to Operating Variables — 53
- 2-4 Dynamics and Averages: Reducing the Dimensionality of the Problem — 61
  - 2-4.1 Time Averaging of the Flow Field: The Eulerian Approach — 62
  - 2-4.2 Useful Approximations — 63

|  |  | 2-4.3 | Tracking of Fluid Particles: The Lagrangian Approach | 69 |
|---|---|---|---|---|
|  |  | 2-4.4 | Experimental Measurements | 71 |
|  | 2-5 | Modeling the Turbulent Transport | | 72 |
|  |  | 2-5.1 | Time-Resolved Simulations: The Full Solution | 74 |
|  |  | 2-5.2 | Reynolds Averaged Navier–Stokes Equations: An Engineering Approximation | 78 |
|  |  | 2-5.3 | Limitations of Current Modeling: Coupling between Velocity, Concentration, Temperature, and Reaction Kinetics | 81 |
|  | 2-6 | What Have We Learned? | | 81 |
|  |  | Nomenclature | | 82 |
|  |  | References | | 83 |
| 3 | **Laminar Mixing: A Dynamical Systems Approach** | | | **89** |
|  | *Edit S. Szalai, Mario M. Alvarez, and Fernando J. Muzzio* | | | |
|  | 3-1 | Introduction | | 89 |
|  | 3-2 | Background | | 90 |
|  |  | 3-2.1 | Simple Mixing Mechanism: Flow Reorientation | 90 |
|  |  | 3-2.2 | Distinctive Properties of Chaotic Systems | 92 |
|  |  | 3-2.3 | Chaos and Mixing: Some Key Contributions | 94 |
|  | 3-3 | How to Evaluate Mixing Performance | | 96 |
|  |  | 3-3.1 | Traditional Approach and Its Problems | 96 |
|  |  | 3-3.2 | Measuring Microstructural Properties of a Mixture | 99 |
|  |  | 3-3.3 | Study of Microstructure: A Brief Review | 102 |
|  | 3-4 | Physics of Chaotic Flows Applied to Laminar Mixing | | 103 |
|  |  | 3-4.1 | Simple Model Chaotic System: The Sine Flow | 103 |
|  |  | 3-4.2 | Evolution of Material Lines: The Stretching Field | 108 |
|  |  | 3-4.3 | Short-Term Mixing Structures | 108 |
|  |  | 3-4.4 | Direct Simulation of Material Interfaces | 110 |
|  |  | 3-4.5 | Asymptotic Directionality in Chaotic Flows | 110 |
|  |  | 3-4.6 | Rates of Interface Growth | 112 |
|  |  | 3-4.7 | Intermaterial Area Density Calculation | 114 |
|  |  | 3-4.8 | Calculation of Striation Thickness Distributions | 116 |
|  |  | 3-4.9 | Prediction of Striation Thickness Distributions | 117 |
|  | 3-5 | Applications to Physically Realizable Chaotic Flows | | 119 |
|  |  | 3-5.1 | Common 3D Chaotic System: The Kenics Static Mixer | 119 |

|       |       | 3-5.2 | Short-Term Mixing Structures | 120 |
|---|---|---|---|---|
|       |       | 3-5.3 | Asymptotic Directionality in the Kenics Mixer | 120 |
|       |       | 3-5.4 | Computation of the Stretching Field | 123 |
|       |       | 3-5.5 | Rates of Interface Growth | 124 |
|       |       | 3-5.6 | Intermaterial Area Density Calculation | 125 |
|       |       | 3-5.7 | Prediction of Striation Thickness Distributions in Realistic 3D Systems | 128 |
|       | 3-6   | Reactive Chaotic Flows | | 130 |
|       |       | 3-6.1 | Reactions in 3D Laminar Systems | 134 |
|       | 3-7   | Summary | | 138 |
|       | 3-8   | Conclusions | | 139 |
|       |       | Nomenclature | | 140 |
|       |       | References | | 141 |

## 4 Experimental Methods — 145

### Part A: Measuring Tools and Techniques for Mixing and Flow Visualization Studies — 145
*David A. R. Brown, Pip N. Jones, and John C. Middleton*

|       | 4-1   | Introduction | | 145 |
|---|---|---|---|---|
|       |       | 4-1.1 | Preliminary Considerations | 146 |
|       | 4-2   | Mixing Laboratory | | 147 |
|       |       | 4-2.1 | Safety | 147 |
|       |       | 4-2.2 | Fluids: Rheology and Model Fluids | 148 |
|       |       | 4-2.3 | Scale of Operation | 154 |
|       |       | 4-2.4 | Basic Instrumentation Considerations | 155 |
|       |       | 4-2.5 | Materials of Construction | 156 |
|       |       | 4-2.6 | Lab Scale Mixing in Stirred Tanks | 156 |
|       |       | 4-2.7 | Lab Scale Mixing in Pipelines | 160 |
|       | 4-3   | Power Draw Or Torque Measurement | | 161 |
|       |       | 4-3.1 | Strain Gauges | 162 |
|       |       | 4-3.2 | Air Bearing with Load Cell | 164 |
|       |       | 4-3.3 | Shaft Power Measurement Using a Modified Rheometer | 164 |
|       |       | 4-3.4 | Measurement of Motor Power | 164 |
|       | 4-4   | Single-Phase Blending | | 164 |
|       |       | 4-4.1 | Flow Visualization | 165 |
|       |       | 4-4.2 | Selection of Probe Location | 167 |
|       |       | 4-4.3 | Approximate Mixing Time Measurement with Colorimetric Methods | 167 |
|       |       | 4-4.4 | Quantitative Measurement of the Mixing Time | 169 |

|  |  | 4-4.5 RTD for CSTR | 174 |
|---|---|---|---|
|  |  | 4-4.6 Local Mixedness: CoV, Reaction, and LIF | 174 |
| 4-5 | Solid–Liquid Mixing |  | 177 |
|  |  | 4-5.1 Solids Distribution | 177 |
|  |  | 4-5.2 Solids Suspension: Measurement of $N_{js}$ | 182 |
| 4-6 | Liquid–Liquid Dispersion |  | 187 |
|  |  | 4-6.1 Cleaning a Liquid–Liquid System | 187 |
|  |  | 4-6.2 Measuring Interfacial Tension | 188 |
|  |  | 4-6.3 $N_{jd}$ for Liquid–Liquid Systems | 189 |
|  |  | 4-6.4 Distribution of the Dispersed Phase | 189 |
|  |  | 4-6.5 Phase Inversion | 190 |
|  |  | 4-6.6 Droplet Sizing | 190 |
| 4-7 | Gas–Liquid Mixing |  | 194 |
|  |  | 4-7.1 Detecting the Gassing Regime | 194 |
|  |  | 4-7.2 Cavity Type | 194 |
|  |  | 4-7.3 Power Measurement | 196 |
|  |  | 4-7.4 Gas Volume Fraction (Hold-up) | 196 |
|  |  | 4-7.5 Volumetric Mass Transfer Coefficient, $k_L a$ | 196 |
|  |  | 4-7.6 Bubble Size and Specific Interfacial Area | 199 |
|  |  | 4-7.7 Coalescence | 199 |
|  |  | 4-7.8 Gas-Phase RTD | 200 |
|  |  | 4-7.9 Liquid-Phase RTD | 200 |
|  |  | 4-7.10 Liquid-Phase Blending Time | 200 |
|  |  | 4-7.11 Surface Aeration | 200 |
| 4-8 | Other Techniques |  | 201 |
|  |  | 4-8.1 Tomography | 201 |

**Part B: Fundamental Flow Measurement** — 202
*George Papadopoulos and Engin B. Arik*

| 4-9 | Scope of Fundamental Flow Measurement Techniques |  | 202 |
|---|---|---|---|
|  |  | 4-9.1 Point versus Full Field Velocity Measurement Techniques: Advantages and Limitations | 203 |
|  |  | 4-9.2 Nonintrusive Measurement Techniques | 206 |
| 4-10 | Laser Doppler Anemometry |  | 207 |
|  |  | 4-10.1 Characteristics of LDA | 208 |
|  |  | 4-10.2 Principles of LDA | 208 |
|  |  | 4-10.3 LDA Implementation | 212 |
|  |  | 4-10.4 Making Measurements | 220 |
|  |  | 4-10.5 LDA Applications in Mixing | 224 |

|   |   |   |   | |
|---|---|---|---|---|
| | 4-11 | Phase Doppler Anemometry | | 226 |
| | | 4-11.1 Principles and Equations for PDA | | 226 |
| | | 4-11.2 Sensitivity and Range of PDA | | 230 |
| | | 4-11.3 Implementation of PDA | | 233 |
| | 4-12 | Particle Image Velocimetry | | 237 |
| | | 4-12.1 Principles of PIV | | 237 |
| | | 4-12.2 Image Processing | | 239 |
| | | 4-12.3 Implementation of PIV | | 243 |
| | | 4-12.4 PIV Data Processing | | 246 |
| | | 4-12.5 Stereoscopic (3D) PIV | | 247 |
| | | 4-12.6 PIV Applications in Mixing | | 249 |
| | | Nomenclature | | 250 |
| | | References | | 250 |
| **5** | **Computational Fluid Mixing** | | | **257** |
| | *Elizabeth Marden Marshall and André Bakker* | | | |
| | 5-1 | Introduction | | 257 |
| | 5-2 | Computational Fluid Dynamics | | 259 |
| | | 5-2.1 Conservation Equations | | 259 |
| | | 5-2.2 Auxiliary Models: Reaction, Multiphase, and Viscosity | | 268 |
| | 5-3 | Numerical Methods | | 273 |
| | | 5-3.1 Discretization of the Domain: Grid Generation | | 273 |
| | | 5-3.2 Discretization of the Equations | | 277 |
| | | 5-3.3 Solution Methods | | 281 |
| | | 5-3.4 Parallel Processing | | 284 |
| | 5-4 | Stirred Tank Modeling Using Experimental Data | | 285 |
| | | 5-4.1 Impeller Modeling with Velocity Data | | 285 |
| | | 5-4.2 Using Experimental Data | | 289 |
| | | 5-4.3 Treatment of Baffles in 2D Simulations | | 289 |
| | | 5-4.4 Combining the Velocity Data Model with Other Physical Models | | 290 |
| | 5-5 | Stirred Tank Modeling Using the Actual Impeller Geometry | | 292 |
| | | 5-5.1 Rotating Frame Model | | 292 |
| | | 5-5.2 Multiple Reference Frames Model | | 292 |
| | | 5-5.3 Sliding Mesh Model | | 295 |
| | | 5-5.4 Snapshot Model | | 300 |
| | | 5-5.5 Combining the Geometric Impeller Models with Other Physical Models | | 300 |

|   |       |                                                              |      |
|---|-------|--------------------------------------------------------------|------|
|   | 5-6   | Evaluating Mixing from Flow Field Results                    | 302  |
|   |       | 5-6.1 Graphics of the Solution Domain                        | 303  |
|   |       | 5-6.2 Graphics of the Flow Field Solution                    | 304  |
|   |       | 5-6.3 Other Useful Solution Variables                        | 310  |
|   |       | 5-6.4 Mixing Parameters                                      | 313  |
|   | 5-7   | Applications                                                 | 315  |
|   |       | 5-7.1 Blending in a Stirred Tank Reactor                     | 315  |
|   |       | 5-7.2 Chemical Reaction in a Stirred Tank                    | 316  |
|   |       | 5-7.3 Solids Suspension Vessel                               | 318  |
|   |       | 5-7.4 Fermenter                                              | 319  |
|   |       | 5-7.5 Industrial Paper Pulp Chests                           | 321  |
|   |       | 5-7.6 Twin-Screw Extruders                                   | 322  |
|   |       | 5-7.7 Intermeshing Impellers                                 | 323  |
|   |       | 5-7.8 Kenics Static Mixer                                    | 325  |
|   |       | 5-7.9 HEV Static Mixer                                       | 326  |
|   |       | 5-7.10 LDPE Autoclave Reactor                                | 328  |
|   |       | 5-7.11 Impeller Design Optimization                          | 330  |
|   |       | 5-7.12 Helical Ribbon Impeller                               | 332  |
|   |       | 5-7.13 Stirred Tank Modeling Using LES                       | 333  |
|   | 5-8   | Closing Remarks                                              | 336  |
|   |       | 5-8.1 Additional Resources                                   | 336  |
|   |       | 5-8.2 Hardware Needs                                         | 336  |
|   |       | 5-8.3 Learning Curve                                         | 337  |
|   |       | 5-8.4 Common Pitfalls and Benefits                           | 337  |
|   |       | Acknowledgments                                              | 338  |
|   |       | Nomenclature                                                 | 339  |
|   |       | References                                                   | 341  |
| 6 | **Mechanically Stirred Vessels**                                 |      | **345** |

*Ramesh R. Hemrajani and Gary B. Tatterson*

|   |       |                                                              |      |
|---|-------|--------------------------------------------------------------|------|
|   | 6-1   | Introduction                                                 | 345  |
|   | 6-2   | Key Design Parameters                                        | 346  |
|   |       | 6-2.1 Geometry                                               | 347  |
|   |       | 6-2.2 Impeller Selection                                     | 354  |
|   |       | 6-2.3 Impeller Characteristics: Pumping and Power            | 358  |
|   | 6-3   | Flow Characteristics                                         | 364  |
|   |       | 6-3.1 Flow Patterns                                          | 366  |
|   |       | 6-3.2 Shear                                                  | 368  |
|   |       | 6-3.3 Impeller Clearance and Spacing                         | 371  |
|   |       | 6-3.4 Multistage Agitated Tanks                              | 372  |

|  |  |  | |
|---|---|---|---|
| | | 6-3.5 Feed Pipe Backmixing | 375 |
| | | 6-3.6 Bottom Drainage Port | 376 |
| | 6-4 | Scale-up | 376 |
| | 6-5 | Performance Characteristics and Ranges of Application | 378 |
| | | 6-5.1 Liquid Blending | 379 |
| | | 6-5.2 Solids Suspension | 380 |
| | | 6-5.3 Immiscible Liquid–Liquid Mixing | 381 |
| | | 6-5.4 Gas–Liquid Dispersion | 382 |
| | 6-6 | Laminar Mixing in Mechanically Stirred Vessels | 383 |
| | | 6-6.1 Close-Clearance Impellers | 385 |
| | | Nomenclature | 388 |
| | | References | 389 |

## 7 Mixing in Pipelines — 391
*Arthur W. Etchells III and Chris F. Meyer*

|  |  |  | |
|---|---|---|---|
| | 7-1 | Introduction | 391 |
| | 7-2 | Fluid Dynamic Modes: Flow Regimes | 393 |
| | | 7-2.1 Reynolds Experiments in Pipeline Flow | 393 |
| | | 7-2.2 Reynolds Number and Friction Factor | 394 |
| | 7-3 | Overview of Pipeline Device Options by Flow Regime | 396 |
| | | 7-3.1 Turbulent Single-Phase Flow | 398 |
| | | 7-3.2 Turbulent Multiphase Flow | 399 |
| | | 7-3.3 Laminar Flow | 401 |
| | 7-4 | Applications | 404 |
| | | 7-4.1 Process Results | 404 |
| | | 7-4.2 Pipeline Mixing Applications | 405 |
| | | 7-4.3 Applications Engineering | 405 |
| | | 7-4.4 Sample of Industrial Applications | 407 |
| | 7-5 | Blending and Radial Mixing in Pipeline Flow | 409 |
| | | 7-5.1 Definition of Desired Process Result | 410 |
| | | 7-5.2 Importance of Physical Properties | 417 |
| | 7-6 | Tee Mixers | 419 |
| | 7-7 | Static Or Motionless Mixing Equipment | 422 |
| | | 7-7.1 Types of Static Mixers | 426 |
| | | 7-7.2 Static Mixer Design Options by Flow Regime and Application | 429 |
| | | 7-7.3 Selecting the Correct Static Mixer Design | 429 |

| | | | |
|---|---|---|---|
| 7-8 | Static Mixer Design Fundamentals | | 429 |
| | 7-8.1 Pressure Drop | | 429 |
| | 7-8.2 Blending Correlations for Laminar and Turbulent Flow | | 432 |
| | 7-8.3 Which In-line Mixer to Use | | 437 |
| | 7-8.4 Examples | | 438 |
| 7-9 | Multiphase Flow in Motionless Mixers and Pipes | | 441 |
| | 7-9.1 Physical Properties and Drop Size | | 441 |
| | 7-9.2 Dispersion of Particulate Solids: Laminar Flow | | 450 |
| | 7-9.3 Pressure Drop in Multiphase Flow | | 451 |
| | 7-9.4 Dispersion versus Blending | | 452 |
| | 7-9.5 Examples | | 452 |
| 7-10 | Transitional Flow | | 459 |
| 7-11 | Motionless Mixers: Other Considerations | | 460 |
| | 7-11.1 Mixer Orientation | | 460 |
| | 7-11.2 Tailpipe/Downstream Effects | | 460 |
| | 7-11.3 Effect of Inlet Position | | 462 |
| | 7-11.4 Scale-up for Motionless Mixers | | 462 |
| 7-12 | In-line Mechanical Mixers | | 463 |
| | 7-12.1 Rotor–Stator | | 464 |
| | 7-12.2 Extruders | | 464 |
| 7-13 | Other Process Results | | 465 |
| | 7-13.1 Heat Transfer | | 465 |
| | 7-13.2 Mass Transfer | | 470 |
| 7-14 | Summary and Future Developments | | 473 |
| | Acknowledgments | | 473 |
| | Nomenclature | | 473 |
| | References | | 475 |

# 8 Rotor–Stator Mixing Devices  479

*Victor A. Atiemo-Obeng and Richard V. Calabrese*

| | | | |
|---|---|---|---|
| 8-1 | Introduction | | 479 |
| | 8-1.1 Characteristics of Rotor–Stator Mixers | | 479 |
| | 8-1.2 Applications of Rotor–Stator Mixers | | 480 |
| | 8-1.3 Summary of Current Knowledge | | 480 |
| 8-2 | Geometry and Design Configurations | | 482 |
| | 8-2.1 Colloid Mills and Toothed Devices | | 482 |

|     |       | 8-2.2 | Radial Discharge Impeller | 482 |
|---|---|---|---|---|
|     |       | 8-2.3 | Axial Discharge Impeller | 483 |
|     |       | 8-2.4 | Mode of Operation | 485 |
|     | 8-3   | Hydrodynamics of Rotor–Stator Mixers | | 489 |
|     |       | 8-3.1 | Power Draw in Batch Mixers | 489 |
|     |       | 8-3.2 | Pumping Capacity | 491 |
|     |       | 8-3.3 | Velocity Field Information | 491 |
|     |       | 8-3.4 | Summary and Guidelines | 496 |
|     | 8-4   | Process Scale-up and Design Considerations | | 496 |
|     |       | 8-4.1 | Liquid–Liquid Dispersion | 498 |
|     |       | 8-4.2 | Solids and Powder Dispersion Operations | 501 |
|     |       | 8-4.3 | Chemical Reactions | 501 |
|     |       | 8-4.4 | Additional Considerations for Scale-up and Comparative Sizing of Rotor–Stator Mixers | 502 |
|     | 8-5   | Mechanical Design Considerations | | 503 |
|     | 8-6   | Rotor–Stator Mixing Equipment Suppliers | | 504 |
|     |       | Nomenclature | | 505 |
|     |       | References | | 505 |
| **9** | **Blending of Miscible Liquids** | | | **507** |

*Richard K. Grenville and Alvin W. Nienow*

|     | 9-1   | Introduction | | 507 |
|---|---|---|---|---|
|     | 9-2   | Blending of Newtonian Fluids in the Turbulent and Transitional Regimes | | 508 |
|     |       | 9-2.1 | Literature Survey | 508 |
|     |       | 9-2.2 | Development of the Design Correlation | 508 |
|     |       | 9-2.3 | Use of the Design Correlation | 510 |
|     |       | 9-2.4 | Impeller Efficiency | 511 |
|     |       | 9-2.5 | Shaft Torque, Critical Speed, and Retrofitting | 512 |
|     |       | 9-2.6 | Nonstandard Geometries: Aspect Ratios Greater Than 1 and Multiple Impellers | 513 |
|     |       | 9-2.7 | Other Degrees of Homogeneity | 513 |
|     |       | 9-2.8 | Examples | 514 |
|     | 9-3   | Blending of Non-Newtonian, Shear-Thinning Fluids in the Turbulent and Transitional Regimes | | 516 |
|     |       | 9-3.1 | Shear-Thinning Fluids | 516 |
|     |       | 9-3.2 | Literature Survey | 517 |
|     |       | 9-3.3 | Modifying the Newtonian Relationships for Shear-Thinning Fluids | 518 |
|     |       | 9-3.4 | Use of the Design Correlation | 520 |
|     |       | 9-3.5 | Impeller Efficiency | 520 |

|  |  | 9-3.6 | Cavern Formation and Size in Yield Stress Fluids | 521 |
|---|---|---|---|---|
|  |  | 9-3.7 | Examples | 522 |
|  | 9-4 | Blending in the Laminar Regime | | 527 |
|  |  | 9-4.1 | Identifying the Operating Regime for Viscous Blending | 528 |
|  |  | 9-4.2 | Impeller Selection | 529 |
|  |  | 9-4.3 | Estimation of Power Draw | 529 |
|  |  | 9-4.4 | Estimation of Blend Time | 530 |
|  |  | 9-4.5 | Effect of Shear-Thinning Behavior | 530 |
|  |  | 9-4.6 | Design Example | 530 |
|  | 9-5 | Jet Mixing in Tanks | | 531 |
|  |  | 9-5.1 | Literature Review | 532 |
|  |  | 9-5.2 | Jet Mixer Design Method | 533 |
|  |  | 9-5.3 | Jet Mixer Design Steps | 535 |
|  |  | 9-5.4 | Design Examples | 536 |
|  | Nomenclature | | | 538 |
|  | References | | | 539 |

**10 Solid–Liquid Mixing** — 543
*Victor A. Atiemo-Obeng, W. Roy Penney, and Piero Armenante*

|  | 10-1 | Introduction | | 543 |
|---|---|---|---|---|
|  |  | 10-1.1 | Scope of Solid–Liquid Mixing | 544 |
|  |  | 10-1.2 | Unit Operations Involving Solid–Liquid Mixing | 544 |
|  |  | 10-1.3 | Process Considerations for Solid–Liquid Mixing Operations | 545 |
|  | 10-2 | Hydrodynamics of Solid Suspension and Distribution | | 548 |
|  |  | 10-2.1 | Settling Velocity and Drag Coefficient | 550 |
|  |  | 10-2.2 | States of Solid Suspension and Distribution | 556 |
|  | 10-3 | Measurements and Correlations for Solid Suspension and Distribution | | 557 |
|  |  | 10-3.1 | Just Suspended Speed in Stirred Tanks | 558 |
|  |  | 10-3.2 | Cloud Height and Solids Distribution | 562 |
|  |  | 10-3.3 | Suspension of Solids with Gas Dispersion | 562 |
|  |  | 10-3.4 | Suspension of Solids in Liquid-Jet Stirred Vessels | 563 |
|  |  | 10-3.5 | Dispersion of Floating Solids | 564 |
|  | 10-4 | Mass Transfer in Agitated Solid–Liquid Systems | | 565 |
|  |  | 10-4.1 | Mass Transfer Regimes in Mechanically Agitated Solid–Liquid Systems | 565 |

xvi    CONTENTS

|  |  | 10-4.2 | Effect of Impeller Speed on Solid–Liquid Mass Transfer | 568 |
|---|---|---|---|---|
|  |  | 10-4.3 | Correlations for the Solid–Liquid Mass Transfer | 569 |
|  | 10-5 | Selection, Scale-up, and Design Issues for Solid–Liquid Mixing Equipment |  | 573 |
|  |  | 10-5.1 | Process Definition | 573 |
|  |  | 10-5.2 | Process Scale-up | 574 |
|  |  | 10-5.3 | Laboratory or Pilot Plant Experiments | 575 |
|  |  | 10-5.4 | Tips for Laboratory or Pilot Plant Experimentation | 576 |
|  |  | 10-5.5 | Recommendations for Solid–Liquid Mixing Equipment | 577 |
|  |  | 10-5.6 | Baffles | 579 |
|  |  | 10-5.7 | Selection and Design of Impeller | 579 |
|  |  | 10-5.8 | Impeller Speed and Power | 580 |
|  |  | 10-5.9 | Shaft, Hub, and Drive | 580 |
|  | Nomenclature |  |  | 581 |
|  | References |  |  | 582 |

| 11 | **Gas–Liquid Mixing in Turbulent Systems** |  |  | **585** |
|---|---|---|---|---|
|  | *John C. Middleton and John M. Smith* |  |  |  |
|  | 11-1 | Introduction |  | 585 |
|  |  | 11-1.1 | New Approaches and New Developments | 586 |
|  |  | 11-1.2 | Scope of the Chapter | 586 |
|  |  | 11-1.3 | Gas–Liquid Mixing Process Objectives and Mechanisms | 589 |
|  | 11-2 | Selection and Configuration of Gas–Liquid Equipment |  | 591 |
|  |  | 11-2.1 | Sparged Systems | 595 |
|  |  | 11-2.2 | Self-Inducers | 595 |
|  |  | 11-2.3 | Recommendations for Agitated Vessels | 596 |
|  | 11-3 | Flow Patterns and Operating Regimes |  | 599 |
|  |  | 11-3.1 | Stirred Vessels: Gas Flow Patterns | 599 |
|  |  | 11-3.2 | Stirred Vessels: Liquid Mixing Time | 605 |
|  | 11-4 | Power |  | 607 |
|  |  | 11-4.1 | Static Mixers | 607 |
|  |  | 11-4.2 | Gassed Agitated Vessels, Nonboiling | 607 |
|  |  | 11-4.3 | Agitated Vessels, Boiling, Nongassed | 612 |
|  |  | 11-4.4 | Agitated Vessels, Hot Gassed Systems | 617 |
|  |  | 11-4.5 | Prediction of Power by CFD | 619 |
|  | 11-5 | Gas Hold-up or Retained Gas Fraction |  | 620 |
|  |  | 11-5.1 | In-line Mixers | 620 |

|  |  |  |  |
|---|---|---|---|
|  | 11-5.2 | (Cold) Agitated Vessels, Nonboiling | 620 |
|  | 11-5.3 | Agitated Vessels, Boiling (Nongassed) | 622 |
|  | 11-5.4 | Hold-up in Hot Sparged Reactors | 623 |
| 11-6 | Gas–Liquid Mass Transfer |  | 626 |
|  | 11-6.1 | Agitated Vessels | 627 |
|  | 11-6.2 | In-line Mixers | 630 |
|  | 11-6.3 | Gas–Liquid Mass Transfer with Reaction | 631 |
| 11-7 | Bubble Size |  | 632 |
| 11-8 | Consequences of Scale-up |  | 633 |
|  | Nomenclature |  | 634 |
|  | References |  | 635 |

## 12 Immiscible Liquid–Liquid Systems — 639
*Douglas E. Leng and Richard V. Calabrese*

|  |  |  |  |
|---|---|---|---|
| 12-1 | Introduction |  | 639 |
|  | 12-1.1 | Definition of Liquid–Liquid Systems | 639 |
|  | 12-1.2 | Practical Relevance | 640 |
|  | 12-1.3 | Fundamentals: Breakup, Coalescence, Phase Inversion, and Drop Size Distribution | 641 |
|  | 12-1.4 | Process Complexities in Scale-up | 646 |
|  | 12-1.5 | Classification by Flow Regime and Liquid Concentration | 647 |
|  | 12-1.6 | Scope and Approach | 649 |
| 12-2 | Liquid–Liquid Dispersion |  | 649 |
|  | 12-2.1 | Introduction | 649 |
|  | 12-2.2 | Breakup Mechanism and Daughter Drop Production in Laminar Flow | 651 |
|  | 12-2.3 | Drop Dispersion in Turbulent Flow | 656 |
|  | 12-2.4 | Time to Equilibrium and Transient Drop Size in Turbulent Flow | 668 |
|  | 12-2.5 | Summary | 679 |
| 12-3 | Drop Coalescence |  | 679 |
|  | 12-3.1 | Introduction | 679 |
|  | 12-3.2 | Detailed Studies for Single or Colliding Drops | 687 |
|  | 12-3.3 | Coalescence Frequency in Turbulent Flow | 692 |
|  | 12-3.4 | Conclusions, Summary, and State of Knowledge | 696 |
| 12-4 | Population Balances |  | 697 |
|  | 12-4.1 | Introduction | 697 |
|  | 12-4.2 | History and Literature | 698 |
|  | 12-4.3 | Population Balance Equations | 698 |

|     |       |                                                                 |     |
| --- | ----- | --------------------------------------------------------------- | --- |
|     | 12-4.4 | Application of PBEs to Liquid–Liquid Systems                   | 700 |
|     | 12-4.5 | Prospects and Limitations                                      | 700 |
| 12-5 | More Concentrated Systems                                              | 704 |
|     | 12-5.1 | Introduction                                                    | 704 |
|     | 12-5.2 | Differences from Low Concentration Systems                      | 705 |
|     | 12-5.3 | Viscous Emulsions                                               | 706 |
|     | 12-5.4 | Phase Inversion                                                 | 707 |
| 12-6 | Other Considerations                                                   | 710 |
|     | 12-6.1 | Introduction                                                    | 710 |
|     | 12-6.2 | Suspension of Drops                                             | 711 |
|     | 12-6.3 | Interrelationship between Suspension, Dispersion, and Coalescence | 713 |
|     | 12-6.4 | Practical Aspects of Dispersion Formation                       | 714 |
|     | 12-6.5 | Surfactants and Suspending Agents                               | 715 |
|     | 12-6.6 | Oswald Ripening                                                 | 717 |
|     | 12-6.7 | Heat and Mass Transfer                                          | 717 |
|     | 12-6.8 | Presence of a Solid Phase                                       | 718 |
|     | 12-6.9 | Effect of a Gas Phase                                           | 719 |
| 12-7 | Equipment Selection for Liquid–Liquid Operations                       | 719 |
|     | 12-7.1 | Introduction                                                    | 719 |
|     | 12-7.2 | Impeller Selection and Vessel Design                            | 719 |
|     | 12-7.3 | Power Requirements                                              | 727 |
|     | 12-7.4 | Other Considerations                                            | 727 |
|     | 12-7.5 | Recommendations                                                 | 729 |
| 12-8 | Scale-up of Liquid–Liquid Systems                                      | 730 |
|     | 12-8.1 | Introduction                                                    | 730 |
|     | 12-8.2 | Scale-up Rules for Dilute Systems                               | 731 |
|     | 12-8.3 | Scale-up of Concentrated, Noncoalescing Dispersions             | 732 |
|     | 12-8.4 | Scale-up of Coalescing Systems of All Concentrations            | 735 |
|     | 12-8.5 | Dispersion Time                                                 | 735 |
|     | 12-8.6 | Design Criteria and Guidelines                                  | 736 |
| 12-9 | Industrial Applications                                                | 737 |
|     | 12-9.1 | Introduction                                                    | 737 |
|     | 12-9.2 | Industrial Applications                                         | 737 |
|     | 12-9.3 | Summary                                                         | 742 |
|     | Nomenclature                                                            | 742 |
|     | References                                                              | 746 |

## 13  Mixing and Chemical Reactions                                            755
*Gary K. Patterson, Edward L. Paul, Suzanne M. Kresta,
and Arthur W. Etchells III*

    13-1   Introduction                                                        755
            13-1.1   How Mixing Can Cause Problems                             757
            13-1.2   Reaction Schemes of Interest                              758
            13-1.3   Relating Mixing and Reaction Time Scales:
                      The Mixing Damkoehler Number                             761
            13-1.4   Definitions                                               764

    13-2   Principles of Reactor Design for Mixing-Sensitive Systems           766
            13-2.1   Mixing Time Scales: Calculation of the
                      Damkoehler Number                                         766
            13-2.2   How Mixing Affects Reaction in Common
                      Reactor Geometries                                        778
            13-2.3   Mixing Issues Associated with Batch,
                      Semibatch, and Continuous Operation                       780
            13-2.4   Effects of Feed Point, Feed Injection Velocity,
                      and Diameter                                              782
            13-2.5   Mixing-Sensitive Homogeneous Reactions                    785
            13-2.6   Simple Guidelines                                         790

    13-3   Mixing and Transport Effects in Heterogeneous Chemical
          Reactors                                                            790
            13-3.1   Classification of Reactivity in Heterogeneous
                      Reactions                                                 794
            13-3.2   Homogeneous versus Heterogeneous Selectivity              795
            13-3.3   Heterogeneous Reactions with Parallel
                      Homogeneous Reactions                                     800
            13-3.4   Gas Sparged Reactors                                      800
            13-3.5   Liquid–Liquid Reactions                                   809
            13-3.6   Liquid–Solid Reactions                                    818

    13-4   Scale-up and Scale-down of Mixing-Sensitive Systems                 821
            13-4.1   General Mixing Considerations                             822
            13-4.2   Scale-up of Two-Phase Reactions                           824
            13-4.3   Scale-up Protocols                                        826

    13-5   Simulation of Mixing and Chemical Reaction                          833
            13-5.1   General Balance Equations                                 834
            13-5.2   Closure Equations for the Correlation Terms in
                      the Balance Equations                                     836
            13-5.3   Assumed Turbulent Plug Flow with Simplified
                      Closure                                                   839

|  |  | 13-5.4 | Blending or Mesomixing Control of Turbulently Mixed Chemical Reactions | 843 |
|---|---|---|---|---|
|  |  | 13-5.5 | Lamellar Mixing Simulation Using the Engulfment Model | 846 |
|  |  | 13-5.6 | Monte Carlo Coalescence–Dispersion Simulation of Mixing | 848 |
|  |  | 13-5.7 | Paired-Interaction Closure for Multiple Chemical Reactions | 850 |
|  |  | 13-5.8 | Closure Using β-PFD Simulation of Mixing | 853 |
|  |  | 13-5.9 | Simulation of Stirred Reactors with Highly Exothermic Reactions | 854 |
|  |  | 13-5.10 | Comments on the Use of Simulation for Scale-up and Reactor Performance Studies | 856 |
|  | 13-6 | Conclusions |  | 857 |
|  |  | Nomenclature |  | 859 |
|  |  | References |  | 861 |

## 14 Heat Transfer     869
*W. Roy Penney and Victor A. Atiemo-Obeng*

|  | 14-1 | Introduction |  | 869 |
|---|---|---|---|---|
|  | 14-2 | Fundamentals |  | 870 |
|  | 14-3 | Most Cost-Effective Heat Transfer Geometry |  | 873 |
|  |  | 14-3.1 | Mechanical Agitators | 874 |
|  |  | 14-3.2 | Gas Sparging | 874 |
|  |  | 14-3.3 | Vessel Internals | 874 |
|  | 14-4 | Heat Transfer Coefficient Correlations |  | 878 |
|  |  | 14-4.1 | Correlations for the Vessel Wall | 880 |
|  |  | 14-4.2 | Correlations for the Bottom Head | 880 |
|  |  | 14-4.3 | Correlations for Helical Coils | 881 |
|  |  | 14-4.4 | Correlations for Vertical Baffle Coils | 881 |
|  |  | 14-4.5 | Correlations for Plate Coils | 881 |
|  |  | 14-4.6 | Correlations for Anchors and Helical Ribbons | 881 |
|  | 14-5 | Examples |  | 882 |
|  |  | Nomenclature |  | 883 |
|  |  | References |  | 884 |

## 15 Solids Mixing     887

### Part A: Fundamentals of Solids Mixing     887
*Fernando J. Muzzio, Albert Alexander, Chris Goodridge, Elizabeth Shen, and Troy Shinbrot*

|  | 15-1 | Introduction | 887 |
|---|---|---|---|

| | | | |
|---|---|---|---|
| 15-2 | Characterization of Powder Mixtures | | 888 |
| | 15-2.1 Ideal Mixtures versus Real Mixtures | | 888 |
| | 15-2.2 Powder Sampling | | 891 |
| | 15-2.3 Scale of Scrutiny | | 895 |
| | 15-2.4 Quantification of Solids Mixing: Statistical Methods | | 896 |
| 15-3 | Theoretical Treatment of Granular Mixing | | 898 |
| | 15-3.1 Definition of the Granular State | | 899 |
| | 15-3.2 Mechanisms of Mixing: Freely-Flowing Materials | | 901 |
| | 15-3.3 Mechanisms of Mixing: Weakly Cohesive Material | | 904 |
| | 15-3.4 De-mixing | | 906 |
| 15-4 | Batch Mixers and Mechanisms | | 909 |
| | 15-4.1 Tumbling Mixers | | 909 |
| | 15-4.2 Convective Mixers | | 912 |
| 15-5 | Selection and Scale-up of Solids Batch Mixing Equipment | | 917 |
| | 15-5.1 Scaling Rules for Tumbling Blenders | | 917 |
| | 15-5.2 Final Scale-up and Scale-down Considerations | | 922 |
| 15-6 | Conclusions | | 923 |
| | Acknowledgments | | 923 |

### Part B: Mixing of Particulate Solids in the Process Industries — 924
*Konanur Manjunath, Shrikant Dhodapkar, and Karl Jacob*

| | | | |
|---|---|---|---|
| 15-7 | Introduction | | 924 |
| | 15-7.1 Scope of Solid–Solid Mixing Tasks | | 925 |
| | 15-7.2 Key Process Questions | | 925 |
| 15-8 | Mixture Characterization and Sampling | | 926 |
| | 15-8.1 Type of Mixtures | | 926 |
| | 15-8.2 Statistics of Random Mixing | | 928 |
| | 15-8.3 Interpretation of Measured Variance | | 931 |
| | 15-8.4 Sampling | | 931 |
| 15-9 | Selection of Batch and Continuous Mixers | | 933 |
| | 15-9.1 Batch Mixing | | 934 |
| | 15-9.2 Continuous Mixing | | 934 |
| | 15-9.3 Comparison between Batch and Continuous Mixing | | 934 |
| | 15-9.4 Selection of Mixers | | 936 |

| | | | |
|---|---|---|---|
| 15-10 | Fundamentals and Mechanics of Mixer Operation | | 936 |
| | 15-10.1 Mixing Mechanisms | | 936 |
| | 15-10.2 Segregation Mechanisms | | 939 |
| | 15-10.3 Mixer Classification | | 940 |
| 15-11 | Continuous Mixing of Solids | | 965 |
| | 15-11.1 Types of Continuous Mixers | | 967 |
| 15-12 | Scale-up and Testing of Mixers | | 968 |
| | 15-12.1 Principle of Similarity | | 969 |
| | 15-12.2 Scale-up of Agitated Centrifugal Mixers | | 969 |
| | 15-12.3 Scale-up of Ribbon Mixers | | 972 |
| | 15-12.4 Scale-up of Conical Screw Mixers (Nauta Mixers) | | 973 |
| | 15-12.5 Scaling of Silo Blenders | | 974 |
| | 15-12.6 Specifying a Mixer | | 974 |
| | 15-12.7 Testing a Mixer | | 975 |
| | 15-12.8 Testing a Batch Mixer | | 977 |
| | 15-12.9 Testing a Continuous Mixer | | 977 |
| | 15-12.10 Process Safety in Solids Mixing, Handling, and Processing | | 977 |
| | Nomenclature | | 981 |
| | References | | 982 |

# 16 Mixing of Highly Viscous Fluids, Polymers, and Pastes — 987
*David B. Todd*

| | | |
|---|---|---|
| 16-1 | Introduction | 987 |
| 16-2 | Viscous Mixing Fundamentals | 987 |
| | 16-2.1 Challenges of High Viscosity Mixing | 987 |
| | 16-2.2 Dispersive and Distributive Mixing | 988 |
| | 16-2.3 Elongation and Shear Flows | 989 |
| | 16-2.4 Power and Heat Transfer Aspects | 992 |
| 16-3 | Equipment for Viscous Mixing | 994 |
| | 16-3.1 Batch Mixers | 994 |
| | 16-3.2 Continuous Mixers | 1000 |
| | 16-3.3 Special Mixers | 1017 |
| 16-4 | Equipment Selection | 1020 |
| 16-5 | Summary | 1022 |
| | Nomenclature | 1023 |
| | References | 1024 |

## 17 Mixing in the Fine Chemicals and Pharmaceutical Industries 1027
*Edward L. Paul, Michael Midler, and Yongkui Sun*

| | | |
|---|---|---|
| 17-1 | Introduction | 1027 |
| 17-2 | General Considerations | 1028 |
| | 17-2.1 Batch and Semibatch Reactors | 1029 |
| | 17-2.2 Batch and Semibatch Vessel Design and Mixing | 1030 |
| | 17-2.3 Multipurpose Design | 1032 |
| | 17-2.4 Batch and Semibatch Scale-up Methods | 1035 |
| | 17-2.5 Continuous Reactors | 1035 |
| | 17-2.6 Reaction Calorimetry | 1036 |
| 17-3 | Homogeneous Reactions | 1038 |
| | 17-3.1 Mixing-Sensitive Reactions | 1039 |
| | 17-3.2 Scale-up of Homogeneous Reactions | 1042 |
| | 17-3.3 Reactor Design for Mixing-Sensitive Homogeneous Reactions | 1043 |
| 17-4 | Heterogeneous Reactions | 1044 |
| | 17-4.1 Laboratory Scale Development | 1045 |
| | 17-4.2 Gas–Liquid and Gas–Liquid–Solid Reactions | 1045 |
| | 17-4.3 Liquid–Liquid Dispersed Phase Reactions | 1050 |
| | 17-4.4 Solid–Liquid Systems | 1052 |
| 17-5 | Mixing and Crystallization | 1057 |
| | 17-5.1 Aspects of Crystallization that Are Subject to Mixing Effects | 1059 |
| | 17-5.2 Mixing Scale-up in Crystallization Operations | 1062 |
| | References | 1064 |

## 18 Mixing in the Fermentation and Cell Culture Industries 1071
*Ashraf Amanullah, Barry C. Buckland, and Alvin W. Nienow*

| | | |
|---|---|---|
| 18-1 | Introduction | 1071 |
| 18-2 | Scale-up/Scale-down of Fermentation Processes | 1073 |
| | 18-2.1 Interaction between Liquid Hydrodynamics and Biological Performance | 1073 |
| | 18-2.2 Fluid Dynamic Effects of Different Scale-up Rules | 1076 |
| | 18-2.3 Influence of Agitator Design | 1089 |
| | 18-2.4 Mixing and Circulation Time Studies | 1090 |
| | 18-2.5 Scale-down Approach | 1094 |
| | 18-2.6 Regime Analysis | 1095 |

18-2.7 Effects of Fluctuating Environmental Conditions on Microorganisms ... 1096
18-2.8 Required Characteristics of a Model Culture for Scale-down Studies ... 1103
18-2.9 Use of *Bacillus subtilis* as an Oxygen- and pH-Sensitive Model Culture ... 1104
18-2.10 Experimental Simulations of Dissolved Oxygen Gradients Using *Bacillus subtilis* ... 1104
18-2.11 Experimental Simulations of pH Gradients Using *Bacillus subtilis* ... 1110

18-3 Polysaccharide Fermentations ... 1113
18-3.1 Rheological Characterization of Xanthan Gum ... 1114
18-3.2 Effects of Agitation Speed and Dissolved Oxygen in Xanthan Fermentations ... 1115
18-3.3 Prediction of Cavern Sizes in Xanthan Fermentations Using Yield Stress and Fluid Velocity Models ... 1116
18-3.4 Influence of Impeller Type and Bulk Mixing on Xanthan Fermentation Performance ... 1119
18-3.5 Factors Affecting the Biopolymer Quality in Xanthan and Other Polysaccharide Fermentations ... 1123

18-4 Mycelial Fermentations ... 1124
18-4.1 Energy Dissipation/Circulation Function as a Correlator of Mycelial Fragmentation ... 1127
18-4.2 Dynamics of Mycelial Aggregation ... 1132
18-4.3 Effects of Agitation Intensity on Hyphal Morphology and Product Formation ... 1133
18-4.4 Impeller Retrofitting in Large Scale Fungal Fermentations ... 1137

18-5 *Escherichia coli* Fermentations ... 1137
18-5.1 Effects of Agitation Intensity in *E. coli* Fermentations ... 1138

18-6 Cell Culture ... 1139
18-6.1 Shear Damage and Kolmogorov's Theory of Isotropic Turbulence ... 1139
18-6.2 Cell Damage Due to Agitation Intensity in Suspension Cell Cultures ... 1141
18-6.3 Bubble-Induced Cell Damage in Sparged Suspension Cultures ... 1144
18-6.4 Use of Surfactants to Reduce Cell Damage Due to Bubble Aeration in Suspension Culture ... 1146

|  |  | 18-6.5 | Cell Damage Due to Agitation Intensity in Microcarrier Cultures | 1148 |
|---|---|---|---|---|
|  |  | 18-6.6 | Physical and Chemical Environment | 1149 |
|  | 18-7 | Plant Cell Cultures | | 1152 |
|  |  | Nomenclature | | 1154 |
|  |  | References | | 1157 |

## 19 Fluid Mixing Technology in the Petroleum Industry — 1171
*Ramesh R. Hemrajani*

|  |  |  |  |
|---|---|---|---|
| 19-1 | Introduction | | 1171 |
| 19-2 | Shear-Thickening Fluid for Oil Drilling Wells | | 1173 |
| 19-3 | Gas Treating for $CO_2$ Reduction | | 1174 |
| 19-4 | Homogenization of Water in Crude Oil Transfer Lines | | 1175 |
|  | 19-4.1 | Fixed Geometry Static Mixers | 1176 |
|  | 19-4.2 | Variable Geometry In-line Mixer | 1177 |
|  | 19-4.3 | Rotary In-line Blender | 1178 |
|  | 19-4.4 | Recirculating Jet Mixer | 1179 |
| 19-5 | Sludge Control in Crude Oil Storage Tanks | | 1179 |
|  | 19-5.1 | Side-Entering Mixers | 1180 |
|  | 19-5.2 | Rotating Submerged Jet Nozzle | 1181 |
| 19-6 | Desalting | | 1183 |
| 19-7 | Alkylation | | 1185 |
| 19-8 | Other Applications | | 1185 |
|  | Nomenclature | | 1186 |
|  | References | | 1186 |

## 20 Mixing in the Pulp and Paper Industry — 1187
*Chad P. J. Bennington*

|  |  |  |  |
|---|---|---|---|
| 20-1 | Introduction | | 1187 |
| 20-2 | Selected Mixing Applications in Pulp and Paper Processes: Nonfibrous Systems | | 1189 |
|  | 20-2.1 | Liquid–Liquid Mixing | 1189 |
|  | 20-2.2 | Gas–Liquid Mixing | 1189 |
|  | 20-2.3 | Solid–Liquid Mixing | 1192 |
|  | 20-2.4 | Gas–Solid–Liquid Mixing | 1194 |
| 20-3 | Pulp Fiber Suspensions | | 1196 |
|  | 20-3.1 | Pulp Suspension Mixing | 1196 |
|  | 20-3.2 | Characterization of Pulp Suspensions | 1196 |
|  | 20-3.3 | Suspension Yield Stress | 1199 |
|  | 20-3.4 | Turbulent Behavior of Pulp Suspensions | 1201 |
|  | 20-3.5 | Turbulence Suppression in Pulp Suspensions | 1203 |
|  | 20-3.6 | Gas in Suspension | 1204 |

|  |  |  |  |
|---|---|---|---|
| 20-4 | Scales of Mixing in Pulp Suspensions | | 1206 |
| 20-5 | Macroscale Mixing/Pulp Blending Operations | | 1206 |
| | 20-5.1 | Homogenization and Blending | 1206 |
| | 20-5.2 | Repulping | 1210 |
| | 20-5.3 | Lumen Loading | 1213 |
| 20-6 | Mixing in Pulp Bleaching Operations | | 1214 |
| | 20-6.1 | Pulp Bleaching Process | 1214 |
| | 20-6.2 | Mixing Equipment in Pulp Bleaching Objectives | 1221 |
| | 20-6.3 | Mixing Assessment in Pulp Suspensions | 1231 |
| | 20-6.4 | Benefits of Improved Mixing | 1237 |
| 20-7 | Conclusions | | 1238 |
| | Nomenclature | | 1238 |
| | References | | 1240 |

## 21 Mechanical Design of Mixing Equipment — 1247
*David S. Dickey and Julian B. Fasano*

|  |  |  |  |
|---|---|---|---|
| 21-1 | Introduction | | 1247 |
| 21-2 | Mechanical Features and Components of Mixers | | 1248 |
| | 21-2.1 | Impeller-Type Mixing Equipment | 1249 |
| | 21-2.2 | Other Types of Mixers | 1254 |
| 21-3 | Motors | | 1258 |
| | 21-3.1 | Electric Motors | 1258 |
| | 21-3.2 | Air Motors | 1267 |
| | 21-3.3 | Hydraulic Motors | 1267 |
| 21-4 | Speed Reducers | | 1267 |
| | 21-4.1 | Gear Reducers | 1268 |
| | 21-4.2 | Belt Drives | 1277 |
| 21-5 | Shaft Seals | | 1278 |
| | 21-5.1 | Stuffing Box Seals | 1278 |
| | 21-5.2 | Mechanical Seals | 1280 |
| | 21-5.3 | Lip Seals | 1285 |
| | 21-5.4 | Hydraulic Seals | 1285 |
| | 21-5.5 | Magnetic Drives | 1286 |
| 21-6 | Shaft Design | | 1287 |
| | 21-6.1 | Designing an Appropriate Shaft | 1287 |
| | 21-6.2 | Shaft Design for Strength | 1289 |
| | 21-6.3 | Hollow Shaft | 1292 |
| | 21-6.4 | Natural Frequency | 1293 |
| 21-7 | Impeller Features and Design | | 1308 |
| | 21-7.1 | Impeller Blade Thickness | 1309 |
| | 21-7.2 | Impeller Hub Design | 1310 |

|   |       |         |                                                              |      |
|---|-------|---------|--------------------------------------------------------------|------|
|   | 21-8  |         | Tanks and Mixer Supports                                     | 1310 |
|   |       | 21-8.1  | Beam Mounting                                                | 1311 |
|   |       | 21-8.2  | Nozzle Mounting                                              | 1313 |
|   |       | 21-8.3  | Other Structural Support Mounting                            | 1317 |
|   | 21-9  |         | Wetted Materials of Construction                             | 1318 |
|   |       | 21-9.1  | Selection Process                                            | 1318 |
|   |       | 21-9.2  | Selecting Potential Candidates                               | 1319 |
|   |       | 21-9.3  | Corrosion–Fatigue                                            | 1320 |
|   |       | 21-9.4  | Coatings and Coverings                                       | 1327 |
|   |       |         | Nomenclature                                                 | 1329 |
|   |       |         | References                                                   | 1330 |

## 22  Role of the Mixing Equipment Supplier   1333
*Ronald J. Weetman*

|   |       |         |                                                              |      |
|---|-------|---------|--------------------------------------------------------------|------|
|   | 22-1  |         | Introduction                                                 | 1333 |
|   | 22-2  |         | Vendor Experience                                            | 1334 |
|   |       | 22-2.1  | Equipment Selection and Sizing                               | 1334 |
|   |       | 22-2.2  | Scale-up                                                     | 1337 |
|   | 22-3  |         | Options                                                      | 1338 |
|   |       | 22-3.1  | Impeller Types                                               | 1338 |
|   |       | 22-3.2  | Capital versus Operating Costs: Torque versus Power          | 1343 |
|   | 22-4  |         | Testing                                                      | 1343 |
|   |       | 22-4.1  | Customer Sample Testing                                      | 1343 |
|   |       | 22-4.2  | Witness Testing                                              | 1344 |
|   |       | 22-4.3  | Laser Doppler Velocimetry                                    | 1345 |
|   |       | 22-4.4  | Computational Fluid Dynamics                                 | 1345 |
|   | 22-5  |         | Mechanical Reliability                                       | 1347 |
|   |       | 22-5.1  | Applied Loads Due to Fluid Forces                            | 1347 |
|   |       | 22-5.2  | Manufacturing Technologies                                   | 1348 |
|   | 22-6  |         | Service                                                      | 1349 |
|   |       | 22-6.1  | Changing Process Requirements                                | 1349 |
|   |       | 22-6.2  | Aftermarket and Worldwide Support                            | 1350 |
|   | 22-7  |         | Key Points                                                   | 1351 |
|   |       |         | References                                                   | 1352 |

**Index**   1353

# CONTRIBUTORS

**Albert Alexander**, Department of Chemical and Biochemical Engineering, Rutgers University, 98 Brett Road, Piscataway, NJ 08854-3058

**Mario M. Alvarez**, Department of Biochemical Engineering, Ave. Eugenio Garza Sada 2501 Sur, C. P. 64849, Monterrey, N.L. Mexico; e-mail: mario.alvarez@itesm.mx

**Ashraf Amanullah**, Merck Research Laboratories, Merck & Co., Inc., WP26C-1 101, 770 Sumneytown Pike, West Point, PA 19438; e-mail: ashraf_amanullah@merck.com

**Engin B. Arik**, VioSense Corporation, 36 S. Chester Ave., Pasadena, CA 91106-3105; e-mail: arik@viosense.com

**Piero M. Armenante**, Otto H. York Department of Chemical Engineering, New Jersey Institute of Technology, University Heights, Newark, NJ 07102-1982; e-mail: piero.armenante@njit.edu

**Victor A. Atiemo-Obeng**, The Dow Chemical Company, Building 1776, Midland, MI 48674; e-mail: vatiemoobeng@dow.com

**André Bakker**, Fluent, Inc., 10 Cavendish Court, Lebanon, NH 03766; e-mail: ab@fluent.com

**Chad P. J. Bennington**, Department of Chemical and Biological Engineering, Pulp and Paper Centre, University of British Columbia, 2385 East Mall, Vancouver, BC, Canada V6T 1Z4; e-mail: cpjb@chml.ubc.ca

**Robert S. Brodkey**, Department of Chemical Engineering, Ohio State University, 140 West 19th Avenue, Columbus, OH 43214-1180; e-mail: brodkey1@osu.edu

**David A. R. Brown**, BHR Group Ltd., Fluid Engineering Centre, Cranfield, Bedfordshire MK43 0AJ, United Kingdom; e-mail: dbrown@bhrgroup.co.uk

**Barry C. Buckland**, Merck Research Laboratories, Merck & Co., Inc., WP26C-1 101, 770 Sumneytown Pike, West Point, PA 19438; e-mail: barry_buckland@merck.com

**Richard V. Calabrese**, Department of Chemical Engineering, Building 090, Room 2113, University of Maryland, College Park, MD 20742-2111; e-mail: rvc@eng.umd.edu

**Shrikant Dhodapkar**, Solids Processing Laboratory, Engineering Sciences and Market Development, The Dow Chemical Company, Freeport, TX 77541; e-mail: sdhodapkar@dow.com

**David S. Dickey**, Mix Tech, Inc., 454 Ramsgate Drive, Dayton, OH 45430-2097; e-mail: d.dickey@mixtech.com

**Arthur W. Etchells III**, The DuPont Company, DuPont Engineering Technology (retired); 315 South 6th Street, Philadelphia, PA 19106; e-mail: etchells3@aol.com

**Julian B. Fasano**, Chemineer, Inc., P.O. Box 1123, Dayton, OH 45401; e-mail: jfasano@chemineer.com

**Chris Goodridge**, Department of Chemical and Biochemical Engineering, Rutgers University, 98 Brett Road, Piscataway, NJ 08854-3058

**Richard K. Grenville**, The DuPont Company, DuPont Engineering Technology, 1007 Market Street, Wilmington, DE 19898; e-mail: richard.k.grenville@usa.dupont.com

**Ramesh R. Hemrajani**, ExxonMobil Research and Engineering Company, Room 7A-2130, 3225 Gallows Road, Fairfax, VA 22037-0001; e-mail: ramesh.r.hemrajani@Exxonmobil.com

**Karl Jacob**, The Dow Chemical Company, Building 1319, Midland, MI 48674; e-mail: jacobkv@dow.com

**Pip N. Jones**, BHR Group Ltd., Fluid Engineering Centre, Cranfield, Bedfordshire MK43 0AJ, United Kingdom; e-mail: pjones@bhrgroup.com

**Suzanne M. Kresta**, Department of Chemical and Materials Engineering, University of Alberta, Edmonton, AB, Canada T6G 2G6; e-mail: suzanne.kresta@ualberta.ca

**Douglas E. Leng**, Leng Associates, 1714 Sylvan Lane, Midland, MI 48640-2538; e-mail: deleng@chartermi.net

**Konanur Manjunath**, Global Process Engineering/Solids Processing, The Dow Chemical Company, APB/1624, Freeport, TX 77541; e-mail: kmanjunath@dow.com

**Elizabeth Marden Marshall**, Fluent, Inc., 10 Cavendish Court, Lebanon, NH 03766; e-mail: emm@fluent.com

**Chris F. Meyer**, Sulzer Chemtech USA, Inc., 312-D Reichelt Road, New Milford, NJ 07646; e-mail: chris.meyer@sulzer.com

**John C. Middleton**, BHR Group Ltd., Fluid Engineering Centre, Cranfield, Bedfordshire, MK43 0AJ, United Kingdom; e-mail: jmiddleton@bhrgroup.co.uk

**Michael Midler**, Merck & Co., Inc., RY818-C312, 126 East Lincoln Avenue, Rahway, NJ 07065; ; e-mail: midler@merck.com

**Fernando J. Muzzio**, Department of Chemical and Biological Engineering, Rutgers University, 98 Brett Road, Piscataway, NJ 08854-3058; e-mail: muzzio@soemail.rutgers.edu

**E. Bruce Nauman**, Department of Chemical Engineering, Rensselaer Polytechnic Institute, Ricketts Building, 110 8th Street, Troy, NY 12180-3590; e-mail: nauman@rpi.edu

**Alvin W. Nienow**, Department of Chemical Engineering, School of Engineering, University of Birmingham, Edgbaston, Birmingham B15 2JJ, United Kingdom; e-mail: a.w.nienow@bham.ac.uk

**George Papadopoulos**, Dantec Dynamics, Inc., 777 Corporate Drive, Mahwah, NJ 07430; e-mail: george.papadopoulos@dantecdynamics.com

**Gary K. Patterson**, Department of Chemical Engineering, University of Missouri–Rolla, Rolla, MO 65401; e-mail: garyp@umr.edu

**Edward L. Paul**, Merck & Co., Inc. (retired); 308 Brooklyn Boulevard, Sea Girt, NJ 08750; e-mail: elpaul@verizon.net

**W. Roy Penney**, Department of Chemical Engineering, University of Arkansas, 3202 Bell Engineering Center, Fayetteville, AR 72701; e-mail: rpenny@engr.uark.edu

**Elizabeth Shen**, Department of Chemical and Biological Engineering, Rutgers University, 98 Brett Road, Piscataway, NJ 08854-3058; e-mail: eshen@rci.rutgers.edu

**Troy Shinbrot**, Department of Chemical and Biochemical Engineering, Rutgers University, 98 Brett Road, Piscataway, NJ 08854-3058; e-mail: shinbrot@sol.rutgers.edu

**John M. Smith**, University of Surrey, 28 Copse Edge, Cranleigh, Surrey GU6 7DU, United Kingdom; e-mail: jsmith@surrey.ac.uk

**Yongkui Sun**, Merck & Co., Inc., 126 East Lincoln Avenue, Rahway, NJ 07065; e-mail: yongkui-sun@merck.com

**Edit S. Szalai**, Schering-Plough Research Institute, 200 Galloping Hill Road, Mailstop F31A, Kenilworth, NJ 07033; e-mail: edit-szalai@yahoo.com

**Gary B. Tatterson**, Department of Chemical Engineering, North Carolina A&T State University, Greensboro, NC 27282; e-mail: gbt@ncat.edu

**David B. Todd**, New Jersey Institute of Technology, 35-H Chicopee Drive, Princeton, NJ 08540; e-mail: dbtodd@aol.com

**Ronald J. Weetman**, 185 Orchard Drive, Rochester, NY 14618; e-mail: ron@rjweetman.com

# INTRODUCTION

EDWARD L. PAUL
*Merck & Co. Inc.*

VICTOR A. ATIEMO-OBENG
*The Dow Chemical Company*

SUZANNE M. KRESTA
*University of Alberta*

Mixing as a discipline has evolved from foundations that were laid in the 1950s, culminating in the publication of works by Uhl and Gray (1966) and Nagata (1975). Over the last 30 years, many engineering design principles have been developed, and design of mixing equipment for a desired process objective has become possible. This handbook is a compilation of the experience and findings of those who have been most active in these developments. Together, the authors' experience extends over more than 1000 years of research, development, and consulting work.

This book is written for the practicing engineer who needs to both identify and solve mixing problems. In addition to a focus on industrial design and operation of mixing equipment, it contains summaries of the foundations on which these applications are based. To accomplish this, most chapters have paired an industrialist and an academic as coauthors. Discussions of theoretical background are necessarily concise, and applications contain many illustrative examples. To complement the discussions, a CD ROM is included which contains over 50 video clips and animations of mixing processes. These clips are accompanied by explanatory text. Internal cross-referencing and external references are used extensively to provide the reader with a comprehensive presentation of the core topics that constitute current mixing practice.

**The core mixing design topics are:**

- Homogeneous blending in tanks and in-line mixers
- Dispersion of gases in liquids with subsequent mass transfer
- Suspension and distribution of solids in liquids

- Liquid–liquid dispersions
- Heat transfer
- Reactions: both homogeneous and heterogeneous

**Underlying principles are presented in chapters on:**

- Residence time distribution
- Turbulence
- Laminar blending and flow

**Additional information is provided on ways of investigating mixing performance:**

- Experimental measurement techniques
- Computational fluid dynamics

**These topics are augmented by chapters on specific industrial mixing topics:**

- Solid–solid blending
- Polymer processing
- Fine chemical and pharmaceutical processes
- Fermentation and cell culture
- Petroleum
- Pulp and paper
- Mixing equipment: vessels, rotor–stators, and pipeline mixers
- Mechanical aspects of mixing equipment
- The vendor's role

At the end of this introduction, a set of charts is provided for the initial assessment of mixing related problems. These charts are designed to assist the reader who is meeting a mixing problem for the first time, and is unsure of where to start. They are not meant to replace the senior engineer or mixing specialist, who will typically be able to quickly evaluate the key issues in mixing-sensitive processes.

## MIXING IN PERSPECTIVE

***What is mixing?*** We define *mixing* as the reduction of inhomogeneity in order to achieve a desired process result. The inhomogeneity can be one of concentration, phase, or temperature. Secondary effects, such as mass transfer, reaction, and product properties are usually the critical objectives.

***What constitutes a mixing problem?*** Process objectives are critical to the successful manufacturing of a product. If the mixing scale-up fails to produce the required product yield, quality, or physical attributes, the costs of manufacturing may be increased significantly, and perhaps more important, marketing of the product may be delayed or even canceled in view of the cost and time required to correct the mixing problem.

Although there are many industrial operations in which mixing requirements are readily scaled-up from established correlations, many operations require more thorough evaluation. In addition to presenting the state of the art on the traditional topics, this book presents methods for recognition of more complex problems and alternative mixing designs for critical applications.

Failure to provide the necessary mixing may result in severe manufacturing problems on scale-up, ranging from costly corrections in the plant to complete failure of a process. The costs associated with these problems are far greater than the cost of adequately evaluating and solving the mixing issues during process development. Conversely, the economic potential of improved mixing performance is substantial. Consider the following numbers:

- *Chemical industry.* In 1989, the cost of poor mixing was estimated at $1 billion to $10 billion in the U.S. chemical industry alone. In one large multinational chemical company, lost value due to poor mixing was estimated at $100 million per year in 1993. Yield losses of 5% due to poor mixing are typical.
- *Pharmaceutical industry.* Three categories should be considered: costs due to lower yield (on the order of $100 million); costs due to problems in scale-up and process development (on the order of $500 million); and costs due to lost opportunity, where mixing problems prevent new products from ever reaching the market (a very large number).
- *Pulp and paper industry.* Following the introduction of medium consistency mixer technology in the 1980s, a CPPA survey documented chemical savings averaging 10 to 15% (Berry, 1990). Mills that took advantage of the improved mixing technology saw their capital investment returned in as little as three months.

From these numbers, the motivation for this handbook and for the research efforts that it documents becomes clear. The reader will almost certainly profit from the time invested in improved understanding of the design of mixing equipment. *Mixing equipment design must go beyond mechanical and costing considerations, with the primary consideration being how best to achieve the key mixing process objectives. Mixing solutions focus on critical issues in process performance.*

**How much mixing is enough, and when could overmixing be damaging to yield or quality?** These critical issues depend on the process and the sensitivity of selectivity, physical attributes, separations, and/or product stability to mixing intensity and time. The nonideality of residence time distribution effects combined with local mixing issues can have a profound effect on continuous processes.

Useful methods for mixing process development effort have been evolving in academic and industrial laboratories over the past several decades. They include improvements to traditional correlations as well as increasingly effective methods both for experiments and for simulation and modeling of complex operations. The combination of these approaches is providing industry with greatly improved tools for development of scalable operations. This handbook provides the reader with all the information required to evaluate and use these technologies effectively in process development and scale-up.

***How should new mixing problems be solved?*** Solutions for new mixing problems require answers to the question "Why?" as well as the very pressing question "How?" This question is best addressed with a good understanding of both the process and the underlying fundamentals. This requires discussion with both operations and developmental chemists. It is often well served by reposing the question "How can we scale this up?" as "How can we scale down the process equipment to closely replicate plant conditions in the lab?" The importance of this question should never be underestimated, as it often opens the door for discussions of geometric similarity and matching of mixing conditions. Good experimental design based on an understanding of mixing mechanisms is critical to obtaining useful data and robust solutions. Engineers who ignore the fundamentals always do so at their own peril. It is our hope in writing this book that mixing fundamentals will become accessible to a much wider audience of engineers, chemists, and operators whose processes are affected by mixing issues.

## Scope of Mixing Operations

Mixing plays a key role in a wide range of industries:

- Fine chemicals, agrichemicals, and pharmaceuticals
- Petrochemicals
- Biotechnology
- Polymer processing
- Paints and automotive finishes
- Cosmetics and consumer products
- Food
- Drinking water and wastewater treatment
- Pulp and paper
- Mineral processing

In all of these industries, the components of mixing problems can be reduced to some fundamental concepts and tools. The key variables to identify in any mixing problem are the time available to accomplish mixing (the time scale) and the required scale of homogeneity (the length scale of mixing). In the remainder of this section we briefly summarize the key mixing issues, the time and length scales of interest, from the perspective of key mixing objectives. We begin with residence time distributions, since this is typically the only area of mixing covered in the undergraduate curriculum.

## Residence Time Distributions: Chapter 1

Classical reactor analysis and design usually assume one of two idealized flow patterns: plug flow or completely backmixed flow. Real reactors may approach one of these; however, it is often the nonidealities and their interaction with chemical kinetics that lead to poor reactor design and performance (Levenspiel, 1998). Nonidealities include channeling, bypassing, and dead zones, among others.

A well-known method for assessing the nonideality of continuous process equipment is the determination of fluid residence time distributions. Residence time distribution (RTD) is a concept first developed by Danckwerts is his classic 1953 paper. In RTD analysis, a tracer is injected into the flow and the concentration of tracer in the outlet line is recorded over time (see Chapter 4). From the concentration history, the distribution of fluid residence times in the vessel can be extracted.

The limits of RTD analysis are the ideal plug flow of a pulse of tracer and a perfectly mixed pulse of tracer. In plug flow a pulse that is completely isolated from the rest of the reactor volume travels through the vessel in exactly the mean residence time. In a perfectly mixed stirred tank, the pulse of tracer is immediately mixed with the full volume of the reactor, leaving the vessel with an exponential decay of concentration as the volume is diluted with fresh feed. These two ideal limits provide us with a great deal of information about the bulk flow pattern or macromixing. When the mixing is ideal or close to ideal and the reaction kinetics are known, the RTD can be used to obtain explicit solutions for the reactor yield [see Levenspiel's classic introductory discussion (1972), Baldyga and Bourne's summary of the key cases (1999, Chap. 2), and Nauman's comprehensive treatment (2002)]. For many industrially important applications, the ideal and close-to-ideal models work very well.

The chief weakness of RTD analysis is that from the diagnostic perspective, an RTD study can identify whether the mixing is ideal or nonideal, but it is not able to uniquely determine the nature of the nonideality. Many different nonideal flow models can lead to exactly the same tracer response or RTD. The sequence in which a reacting fluid interacts with the nonideal zones in a reactor affects the conversion and yield for all reactions with other than first-order kinetics. This is one limitation of RTD analysis. Another limitation is that RTD analysis is based on the injection of a single tracer feed, whereas real reactors often employ the injection of multiple feed streams. In real reactors the mixing of separate feed streams can have a profound influence on the reaction. A third limitation is that RTD analysis is incapable of providing insight into the nature of micromixing. RTD studies and analyses deal primarily with bulk flow or macroscopic mixing phenomena.

***Where do the ideal models fail?*** For flow in a pipe, the ideal model is plug flow. This is a good assumption for fully turbulent flow with a uniform distribution of feed. There are two important cases where nonideal mixing must be addressed. If the second component is added from a small feed pipe rather than as a slug, radial dispersion of the feed must be considered. This case is discussed in Chapter 7. If the flow is laminar rather than turbulent, the velocity profile is parabolic (not flat), so the fluid in the center of the pipe will exit much sooner than the fluid close to the walls. This is the laminar axial dispersion problem which has been studied very extensively. The animation of flow in a Kenics mixer (CD ROM) illustrates this concept, showing axial dispersion of tracer particles for laminar flow in a static mixer. Ways to avoid this problem are also discussed in Chapter 7. For turbulent flow the problem of axial dispersion is less severe. A third practical consideration is partial plugging or fouling of a line. In this case the apparent residence time will be much shorter than expected because the effective volume of the vessel is less than the design volume.

For well-designed stirred tanks with simple reaction schemes and kinetics which are slow relative to the mixing time, the perfectly backmixed CSTR model works well. The most critical factor for design of a CSTR is placement of the feed and outlet locations. If a line drawn from the feed pipe to the outlet passes through the impeller, short circuiting is not likely to be a problem. If, however, the feed and the outlet are both located near the top of the vessel, short circuiting will almost certainly occur. Baffles may be used to reduce or eliminate this problem. The second characteristic of a well-designed CSTR is that the volume and mixing must be balanced with the feed rate. The volume must be big enough to allow 10 batch blend times to occur over the mean residence time (see Chapter 6). Alternatively, the primary impeller pumping capacity (see Chapter 6) should be 10 to 16 times the volumetric feed rate $q/Q = 10$ to 16 (Nauman, 2002, Chap. 8). These numbers are very conservative but are the best design standards currently available.

Residence time distributions, discussed in Chapter 1, represent the first generation of mixing models. The ideal cases of plug flow and perfectly mixed tanks provide solutions for most standard problems. Where the kinetics are more complex, are faster than the mixing time, or require a segregated feed strategy, the local mixing concepts discussed in this book and the zone-based models developed over the last 20 years have proved invaluable. The third generation of modeling will see coupling of computational fluid dynamics (Chapter 5) with reaction kinetics and heat transfer to obtain explicit and localized models for the most difficult mixing problems. Early reports of successes in this area include the production of adipic acid in the laminar flow regime in a stirred tank, modeling of crystallization reactions, and evaluation of the disinfection capabilities of ultraviolet treatment reactors in the water and wastewater treatment industries.

Residence time distributions are the first characteristic of mixing, but because they treat the vessel as a black box, they cannot address local mixing issues, which are the focus of much of this book. The characteristic time scale for a residence time distribution is the mean residence time of the vessel. The characteristic length scale is the vessel diameter, or volume. Many of the key process objectives of interest require more local information.

### Mixing Fundamentals: Chapters 1–5

There is a set of fundamental topics which, while not leading directly to design of mixing equipment, must be understood to address difficult mixing problems. Residence time distribution theory and modeling constitute the classical approach to mixing and were discussed earlier. Turbulent and laminar mixing theory is covered in Chapters 2 and 3. Laminar mixing theory springs from dynamical systems theory, or chaos theory. A number of topics are addressed, but perhaps most useful is the idea that well-designed laminar mixing devices repeat the stretching and folding patterns in the flow, thus producing repeating structures of mixing on ever smaller scales. Turbulent mixing theory is concerned primarily with two questions: "What is the range of time and length scales in the flow?" and the analog to this question, "Where is the energy dissipated?" The points of highest energy dissipation are the points of most intense mixing, or of the smallest time and length scales. Chapters 4 and 5 discuss the two principal tools used to investigate mixing phenomena and evaluate mixing equipment: laboratory experiments and computational fluid dynamics. There is a wide range of experimental and computational tools available with a wide range of experimental or computational difficulty and a wide range of detail in the results. Perhaps the most difficult question for the engineer is to understand the problem well enough to define a well-posed question. Once the question is defined, an appropriate tool can be selected relatively easily, and useful results can usually be obtained. These five fundamental topics provide the key tools needed to tackle new problems and to understand much of the theory underlying mixing design.

### Mixing Equipment: Chapters 6, 7, 8, and 21

A wide range of mixing equipment is now available, with the current generation of equipment typically designed for a specific process result. Chapter 6 covers traditional stirred tanks, baffling, the full range of impellers, and other tank internals and configurations. Chapter 7 provides information on equipment and design for pipeline mixing. Chapter 8 focuses on rotor–stators, which have been used for many years but have been investigated on a more fundamental level only in the last decade. Chapter 21 covers the mechanical aspects of mixing equipment design, providing a welcome primer on the vocabulary of mechanical engineering as well as important design information. Chapter 22 focuses on the vendor: what expertise can be offered and what information is needed for accurate specification of mixing equipment. Additional specialized equipment is discussed in

Chapters 15 (powder blending), 16 (high viscosity), 19 (petroleum), and 20 (pulp and paper industry). Key design concepts for equipment selection are:

- Selection of tanks versus in-line mixers and use of backmixed flow versus plug flow
- Selection of residence times are required: long residence times are well served by tanks, short residence times can be accomplished in pipes
- Design requirements: robust and flexible (typically stirred tanks) versus tight and specific (pipeline mixers and other specialized equipment)
- Mechanical design considerations: seals, dynamic loads, rotating shafts, and critical speed
- Classical and modern impeller design; the function and importance of baffles
- Characteristics of in-line mixing equipment, including static mixers and rotor–stators

### Miscible Liquid Blending: Chapters 3, 7, 9, and 16

Miscible liquid blending is the easiest mixing task. The reader is cautioned that miscible blending requires two things: The streams must be mutually soluble, and there must be no resistance to dissolution at the fluid interface. Chapters 7 and 9 present well-developed correlations for prediction of mixing time in this simplest case, and corrections for density and viscosity differences. Although laminar and non-Newtonian fluids are more difficult to handle, the current recommendations on these issues are also included in Chapters 7 and 9.

Chapter 3 provides a careful discussion of how we characterize and measure mixing scales. These concepts are combined with dynamical systems, or chaos theory, to identify similarities of scale in laminar mixing applications. This is a key theoretical concept that will allow rigorous advances in mixing design in the future. In Chapter 16, current polymer and high viscosity blending equipment is discussed. In these cases the blending objective must be combined with the heat transfer and high pressures required to produce polymer melts. For pastes, the fluids are typically non-Newtonian, so further specialized equipment is required.

### Solid–Liquid Suspension: Chapters 10, 17, and 18

Design methods for solid–liquid suspension were some of the first to be established (Zwietering, 1958), and this early work has withstood the test of time virtually unchanged. Solid–liquid mixing is discussed in Chapter 10, with design guidelines for:

- Mixing requirements for achieving and maintaining off-bottom suspension of solids (the just suspended speed, $N_{js}$)

- Requirements for achieving and maintaining uniform solids concentration throughout the tank—of interest particularly for slurry catalyst reactors and for feeding downstream equipment (e.g., centrifuges, continuous stirred tank reactors, fluid bed coaters)
- Mass transfer correlations for solids dissolution
- Maintaining the required slurry composition on discharge
- Tank draining with solids present: avoiding plugged nozzles

Difficult design problems that have not yet been resolved involve nonwetting, clumping, or floating solids. The key qualitative aspects of these problems can be identified and useful heuristic solutions are provided. Other mixing effects involving solids in suspension include clumping, agglomeration, fouling, and scaling. These problems can be reduced with good mixing designs, but a full discussion lies outside the scope of this book.

Reactions involving solids are discussed extensively in Chapters 13 and 17. Where solids are involved in reactions, there are two steps in the kinetics. The first, solids dissolution, is dominated by the particle size and the mixing conditions. The apparent reaction kinetics and even the reaction products can change depending on the mixing conditions. Key solids reaction topics include:

- Solids dissolution with reaction (Chapters 13 and 17)
- Potential for impeller damage to solids in suspension, including crystals (Chapter 17), cells (Chapter 18), and resin beads
- Mixing effects on nucleation and growth in crystallization (Chapter 17)

### Gas–Liquid Contacting: Chapter 11

Gas–liquid mixing has one key objective: the dispersion of gas in liquid with the maximum surface area for mass transfer. As with many multiphase systems, this objective is complicated by the difficulties of multiphase flow. The gas can flood the impeller, dramatically reducing its effectiveness; surface properties determine whether the system is coalescing or noncoalescing, and thus whether the surface area created is stable; boiling systems require completely different treatment; and gas–liquid reactions require consideration of local concentrations of gas. Chapter 11 includes the traditional topics:

- Correlations for prediction of $k_L$ a, including fermentation applications (also discussed in Chapter 18)
- Discussion of operating regimes: interaction of power and gassing rate to produce stable operation or flooding of the impeller
- Recommendations for sparger design and placement
- Design for sufficient gas phase residence time
- Gas–liquid reactions (also discussed in Chapter 13)

New discussions are provided on:

- The new generation of impellers designed for efficient gas dispersion
- Boiling systems

The reader should beware of conditions in the headspace, particularly for high viscosity and/or foaming systems. This is potentially detrimental for several types of operations. Excessive foaming can lead to interference with mass transfer. Gas entrained into high viscosity systems can be difficult to remove and severely affect product quality.

## Liquid–Liquid Mixing: Chapter 12

Liquid–liquid mixing is one of the most difficult and least understood mixing problems, despite extensive literature on both the mechanical agitation side of the problem and the surface science side of the problem. In spite of this, a number of important lessons emerge from the discussion in Chapter 12:

- Impurities, surface-active agents, and small changes in chemical composition can be critical in determining drop size distribution. Performance can change dramatically due to small changes in composition, even at the parts per million level, particularly for reactions, separations, and preparation of stable emulsions.
- Both the mixing system and duration of mixing can have an important effect on drop size distribution, drop breakup, and coalescence.
- Addition strategy can determine which phase is continuous.
- Phase inversion can play an important role in extraction and reaction.
- Overmixing can result in a stable emulsion or an overreacted product.
- Inadequate mixing can result in incomplete phase transfer or slow reaction.

## Mixing and Chemical Reactions/Reactor Design: Chapters 13 and 17

When mixing rates and chemical reaction rates occur on similar time scales, or when mixing is slower than chemical reaction, mixing effects can be very important. On the small scale, blend times and mixing time scales are typically very short and mixing effects may not be apparent. When reactions are scaled up, however, the chemical kinetics stay the same while mixing times get longer. Mixing effects are always worse on scale-up. These issues are discussed in some detail in Chapters 13 and 17. The key points are:

- How and when mixing effects can influence the yield and selectivity of complex homogeneous and heterogeneous chemical reactions.

- Yield and/or selectivity of homogeneous consecutive-competitive reactions that are subject to mixing effects can be lower on scale-up if proper precautions are not taken for mixing the reagents—mesomixing problems get worse on scale-up and blend times increase.
- Feeding at the impeller, or the region of most intense turbulence, is recommended for consecutive or competitive reactions to avoid reduced yield/selectivity on scale-up. It is better to feed at the impeller when this is not actually required than to feed on the surface when subsurface feed was in fact necessary.
- Mixing effects in heterogeneous reactions are often complex because of local effects in dispersed phase films and global mixing effects when competitive reaction(s) occur in the continuous phase.
- The yield and/or selectivity of heterogeneous complex reactions may in some cases be improved by the presence of the second or third phase.

### Heat Transfer and Mixing: Chapter 14

The principles of heat transfer in stirred tanks are discussed in Chapter 14, with a full set of design correlations for heat transfer coefficients. The key heat transfer concepts to keep in mind are as follows:

- Limitations in heat transfer normally result from surface area availability rather than from the mixing system.
- Limitations in heat transfer can sometimes be overcome by evaporative cooling: for example, during polymerization and other exothermic reactions.
- Good mixing can often reduce or prevent scaling and the resulting severe losses in heat transfer performance.
- Process modifications are sometimes needed to provide alternative solutions to limitations in heat transfer capability.

### Specialized Topics for Various Industries: Chapters 15–20

Mixing issues in several specialized industries are discussed in Chapters 15 to 20. In these chapters, the approach taken varies from author to author, depending on the state of knowledge in the industry. Powder blending and polymer or high viscosity blending both suffer from the difficulty of even characterizing the material of interest, making fully predictive design and scale-up nearly impossible. The fine chemicals industry typically uses equipment for a wide range of products, so the mixing must be both versatile and well understood. Reactions are often multiphase, and crystallization is a core competency with its own specialized mixing issues. Chapter 17 can be regarded as a more specialized reactions chapter. Biological processes, discussed in Chapter 18, are highly dependent on gas dispersion but must also consider the special requirements of living systems. The petroleum and pulp and paper industries have a range of key applications.

These applications are the focus of Chapters 19 and 20 and have significant value for extending one's understanding of mixing design fundamentals.

## CONVERSATIONS OVERHEARD IN A CHEMICAL PLANT

One Monday morning in an R&D center in Illinois, Marco's phone rings. "Hello—Marco? It's Bill from the Texas plant." Marco detects a mixture of excitement and concern in Bill's voice, "What's up, Bill?" "Well—ah—Marco, remember that mixing vessel you designed for us? Listen, we moved the reaction in Step 5 of the process—you remember the liquid–liquid phase transfer reaction step—to another vessel. The conversion and selectivity are both lower. We have checked the usual suspects, compound III from the previous step, temperature calibration, and charge meter calibrations, but everything looks OK. Any chance the change in vessel could be giving us a mixing problem?" Marco shakes his head and replies, "Very good chance, Bill. These liquid–liquid fast reactions can be devils to scale up. Let's get together and see what we can do."

In the chapters that follow, the varied roles of mixing in industrial operations are discussed by authors from both academic and industrial viewpoints, combining the fundamentals of mixing technology with industrial experience. Many examples are included, providing both illustrative calculations and more qualitative industrial mixing problems. In this section we follow Marco's journey as he works through the mixing issues that must be considered in development of a new process and its translation to manufacturing.

### The Problem

Marco first heard about this project when his boss, the director of chemical engineering R&D, called with the news that a new process was coming out of research with good potential to go through development and into manufacturing as the company's next product. "Talk to the head chemist, Lenny, and find out what the process looks like. Determine what engineering issues it may involve on scale-up as well as potential areas for process improvement."

This type of assignment had come to Marco many times before and always caused him concern. Achieving a successful scale-up always requires development of multiple steps that are easy to operate in the laboratory but can be very difficult to translate to manufacturing. In many cases a direct scale-up of the chemists' procedure is possible, but in others, mixing differences between large- and small scale equipment result in reduced yield and selectivity. One of Marco's first objectives was to determine the scalability of the new process in each of the reaction, purification, and isolation steps. In order to focus on the mixing issues of interest, the aspects of this process that require feasibility and optimization studies are not addressed in this discussion.

On initial review of the chemists' procedure, Marco notes that there are potential mixing issues in four of the chemical reactions as well as in the crystallization steps. The other steps in the process did not appear to pose significant mixing issues beyond prudent scale-up of blending, solids suspension, and so on. This initial diagnosis is critical to the success of the development program that Marco will put together in consultation with his colleagues. Marco is well aware that mixing problems can be difficult to forecast from a laboratory procedure and that careful modeling, engineering laboratory, and pilot plant studies will be essential to the success of the ultimate plant design.

**Competitive-Consecutive Reaction**

The first reaction is a bromination that is designed to add one bromine to an aromatic ring but can overreact to add a second bromine. The possibility of reduced selectivity caused by differences in micro- and mesomixing on scale-up must be investigated.

Marco calls one of his associates, Roger, who has looked into this issue in previous developmental studies. Roger advises that the first issue is to decide if the reaction could be influenced by mixing or is reaction-rate controlled. Despite some grumbling that "Mixing cannot affect homogeneous reactions, so why are we doing this?" a geometrically similar 4 liter reactor is set up to study high and low levels of mixing with two feed strategies: addition at the impeller for the high-level case and addition at the surface for the low-level case. No significant difference is observed in the amount of overbromination. Everyone's first reaction is to jump to the conclusion that there are no mixing issues for this reaction, as was predicted by the grumbling skeptics.

Roger warns that in some cases, a laboratory experiment might not reveal a significant difference because the blend time and micromixing on this scale could be sufficient to mask the problem. He proceeds to recommend further experiments: a reverse addition in which A is added to a solution of the brominating reagent instead of the brominating reagent being added to A, as the chemist's procedure specifies. The skeptics again object—even more forcefully—because it is obvious that this is no way to run the reaction. Marco asks Roger to explain his reasoning. "This extreme change will almost certainly reveal whether or not there is potential for overbromination since A is added into a sea of bromine reagent. This will exaggerate any overreaction." Roger notes that the 1 : 1 mole ratio will be maintained once the addition is completed. He also reminds Marco that even a small amount of dibromo (<1%) could be a problem since scale-up will increase the overreaction unless precautions for adequate mixing are recognized and taken. After some consideration, Marco decides that it is easier and cheaper to run the reverse addition experiment than to run the risk of overbromination.

The reverse addition revealed overreaction, indicating to Roger that further testing on a pilot plant scale would be required. After determination of the reaction rate ratio, $k_1/k_2$, Roger was concerned that successful scale-up in a stirred vessel might not be feasible because the consecutive overbromination reaction was relatively fast. A high-energy pipeline mixer reactor might be required for manufacturing. The skeptics were still not impressed, figuring that Roger was making a big deal out of a small problem.

For a discussion of mixing issues for this type of reaction, and more on who was right, Roger or the skeptics, the reader is referred to Chapters 13 and 17 on reactor design as well as Chapter 2 on homogeneous turbulence and Chapter 9 on blending in tanks. This particular issue, and a vindication of Roger's position, is detailed in Example 13-1.

**Gas–Liquid Reaction**

Moving along with his analysis, Marco notes that the second reaction is a hydrogenation. The chemist was running the hydrogenation in a 1 liter autoclave over a period of 8 hours (or sometimes overnight, thereby avoiding several evenings in the laboratory). Marco immediately suspects a mixing limitation on the gas–liquid contact even though the chemist adamantly maintained that it was just a slow reaction.

Mary, his expert in these types of reactions, has seen a similar issue before. She concurs that the mixing conditions in the autoclave will lead to inadequate hydrogen absorption. Surface reincorporation is not being achieved because of the design of the autoclave—high $z/T$ ratio and full baffles—a common but poor laboratory autoclave mixing system. Mary sets up a modified mixing system for the autoclave by cutting the baffles to create a vortex to the top pitched blade turbine in order to achieve surface reincorporation of hydrogen that has not reacted after sparging but has escaped to the headspace. By escaping to the headspace and building up pressure, this hydrogen effectively reduces or stops hydrogen flow into the vessel.

In Mary's modified autoclave, the reaction takes off and is complete in 30 minutes, much to the chemist's surprise. In making such a reactor modification, it is wise to prepare for more rapid heat evolution than was experienced in the improperly mixed original autoclave. Mary was prepared and had no difficulty controlling the temperature. It is also necessary to provide good bearing support for the shaft to counteract the increased vibration associated with vortexing.

The next issue for Marco to consider is how to scale-up this reaction for the pilot plant and production plant. Discussion of this type of reaction and the associated scale-up issues may be found in Chapters 13 and 17. Gas–liquid mixing issues are discussed in Chapter 11.

**Solid–Liquid Reaction**

The next reaction in the process is an alkylation using powdered potassium carbonate as a base to react with an organic acid reagent (in solution) to form the potassium salt (in solution). This reaction appears to be very straightforward and is transferred directly to the pilot plant. After the first pilot plant run, Marco gets an e-mail: the reaction was slower than expected and resulted in incomplete conversion. In addition to this, the operators spent the rest of the shift getting the batch out of the vessel because the bottom outlet was clogged with solids. Marco immediately suspects the culprit—inadequate off-bottom suspension of the powdered potassium carbonate. Marco explains that in a dissolving reagent reaction the solid dissolution step can be rate controlling. When the reaction time is extended to compensate for this, simultaneous decomposition of the product (in solution) can be enhanced. Another possibility is decomposition caused by contact of the base with the product in the high-pH liquid film around the particles. Marco recommends that the first issue to address is that all of the powder did not even get to react because it was mounded on the bottom of the vessel. To fix this problem, adequate mixing for off-bottom suspension is essential. In addition, improved mixing will favor rapid dissolution of the highly insoluble $K_2CO_3$. Marco cautions that overmixing could cause foaming in this type of fine-solid–liquid suspension. This may cause the particles to be coated with inert gas (nitrogen) and therefore have reduced dissolution characteristics. The particle size of the $K_2CO_3$ also has a significant effect on the dissolution time, so a change in supplier or grade of solids could be the problem, although in this case it is not.

Marco sets up a 4 liter reaction flask with a pitched blade turbine impeller and full baffles. Experiments are run to determine the effects of $K_2CO_3$ particle size and impeller speed on reaction rate over wide ranges. The sensitivity of reaction rate to both particle size and impeller speed is readily established. The loss of product due to decomposition is also demonstrated.

These results illustrate the critical nature of solids dissolution of reagents in chemical reactions. The reader is referred to Chapters 13 and 17 for discussion and further examples as well as to Chapter 10 for calculation of solids suspension requirements.

**Liquid–Liquid Reaction**

Six months later, Marco and his group have solved the three reaction problems outlined above by experimentation and by studying the appropriate literature references, including applicable parts of this handbook. The liquid–liquid reaction issues for the fifth reaction in the synthesis have also been solved in the laboratory and pilot plant and successfully scaled-up to manufacturing. This brings us to the phone call from Bill that we overheard at the beginning of this story. Manufacturing moved the reaction to a different vessel, causing a drop in conversion rate and selectivity in Step 5. As often happens, there are compelling reasons that prevent Bill from solving his problem by going back to the original, successful reactor, just as there were compelling reasons to change reactors in the first place. It may become obvious, however, that the compelling reasons are not as compelling as they may seem when a vessel transfer that appears straightforward turns into a nightmare for the plant.

On the phone, Bill indicates that the mixing still looks good in the new vessel but does acknowledge that the new vessel has a different impeller design and speed. Before leaving, Marco grabs a coffee with Vijay, who did some work in this area in graduate school. They agree to follow the mixing trail despite Bill's visual characterization of the mixing as good. As he sits on the plane to Texas, Marco wonders, "What could be causing slower reaction and increased by-product formation? A change in the mixing characteristics between the two reactors?" Marco remembers his anxiety during development of this liquid–liquid reaction because he was well aware of the potential for difficulty. He was very pleased that it had been piloted and scaled up to manufacturing successfully—at least until the change of vessel.

Liquid–liquid complex reactions have been classified as one of the most difficult—if not *the* most difficult—reaction scale-up mixing problem (Leng, 1997). The complexities of drop formation and coalescence both change with scale. They both depend on the location in a vessel and on very subtle changes in the composition of the fluids. These variations can cause problems like Bill's when a complex reaction occurs between reagents in separate liquid phases.

When Marco arrives at the plant, Bill suggests that the most readily achievable "fix" is to increase impeller speed. He can accomplish this with a change in drive gears, although a higher-power motor will be required. Should the increase in rpm be based on equal P/V, on equal tip speed, or on something else? Marco and Vijay know that an increase in rpm with a different impeller might not work since geometric similarity cannot be maintained. Marco was also harboring the disturbing thought that overmixing, by providing too much power while improving reaction rate, could actually reduce selectivity by exposing product in the droplet films to high concentrations of reagent in the aqueous phase. Is this possible? Has it ever been experienced?

One dilemma in answering these questions is that laboratory scale experimentation may not be able to provide a suitable model for scale-up. Bill, Vijay, and Marco may have to make a decision on the fix without quantitative information. Fortunately, most mixing problems can be addressed with more certainty than those involving fast, complex reactions in multiple phases. These issues are discussed in Chapters 13 and 17 as well as in Chapter 12 (liquid–liquid mixing). In addition, comparisons between impellers and general information on the components of stirred vessels may be found in Chapter 6, and the help that can be provided by mixing equipment suppliers is discussed in Chapter 22.

**Crystallization**

At the outset of this project, Marco noted that the seven-step synthesis includes four crystallization operations. One of these is a final reactive crystallization that will determine the physical attributes of the product. Any of these steps could produce crystals that are difficult to filter, wash, and dry because the particle size distribution could change on scale-up due to mixing effects during crystallization. Excess nucleation and/or crystal fracture are both expected to be more severe in plant operation than in the laboratory or even the pilot plant. From a process design point of view, elimination of one or more of the crystallization steps will yield large savings in both capital and operating costs. However, discussion of this type of development initiative is not included in the scope of this book because mixing issues would not be primary considerations in developing these strategies.

The reactive crystallization that is the last step of this complex process could present critical mixing issues because mixing can affect both the reaction and subsequent crystallization. The physical attributes and chemical purity of the final product will be determined by the success or failure of the scale-up. To add to this complexity, intense mixing may be required for the fast reaction, whereas modest blending may be simultaneously required to prevent crystal attrition. These mutually exclusive requirements require a compromise to achieve the best result possible.

Initially, Marco was at a bit of a loss because his group had always had difficulty with reactive crystallizations and had not developed a successful strategy for overcoming the basic issues inherent to this type of crystallization (also termed *precipitation*). Since these operations are almost always carried out at high supersaturation, they are nucleation based and therefore tend to produce small crystals, typically 5 to 10 μm in size, with many in the 1 μm range. Both occlusion of impurities and unacceptable physical attributes can result. Marco assigned Carol, a new engineer, to work with Joe, a veteran of many crystallization developments, who remained hopeful that this dilemma of precipitation could be solved.

Carol and Joe succeeded in balancing the reaction requirements with the crystallization parameters required to achieve a growth-dominated process. In doing so, they had to choose a mixing system that would achieve micromixing effectively for the fast reaction but which was compatible with crystal growth. Mixing issues in this and other types of crystallization operations are discussed in more detail in Chapter 17.

## USING THE HANDBOOK

This book is not meant to be read from beginning to end. It is designed as a reference, with extensive cross-referencing and indexing. The book is divided into three sections: fundamentals, design, and applications. Many examples are included to aid the reader in understanding the fundamentals as well as some case histories of mixing issues in industrial practice. Authorship of most of the chapters includes both academic and industrial contributors, for the purpose of providing a broad perspective on each topic. Also included is a CD ROM to aid in visualization of some specific mixing issues and examples. The sections in this introduction should help the reader new to the field of mixing in identifying what is meant by a mixing problem. We have summarized key issues (Mixing in Perspective) and discussed a process containing examples of many reactive mixing problems (Conversations Overheard in a Chemical Plant), and diagnostic charts follow this summary. All of these sections provide the reader with references to relevant chapters in the handbook.

The text and examples include guidance in troubleshooting mixing problems based on understanding the fundamental issues, aided by drawing on the experiences cited. It is often assumed that mixing scale-up is accomplished by direct scaling to a larger pot. This approach may work in some cases but is doomed to failure in others. The key question is the determination of process requirements for which direct scale-up will be inadequate. Another overall concern is

# I INTRODUCTION

to beware of the fact that multiple process objectives often must be realized in a single piece of mixing equipment, thereby requiring selection of a design basis compatible with the most critical scale-up issue(s).

## Diagnostic Charts

Figures I-1 through I-6 summarize some of the key symptoms and causes of mixing problems in the plant. They are in no way exhaustive and should not be used to replace the expertise of an experienced process specialist. They can be used to guide you through some of the implicit steps used in evaluating mixing problems, and may help to focus your reading.

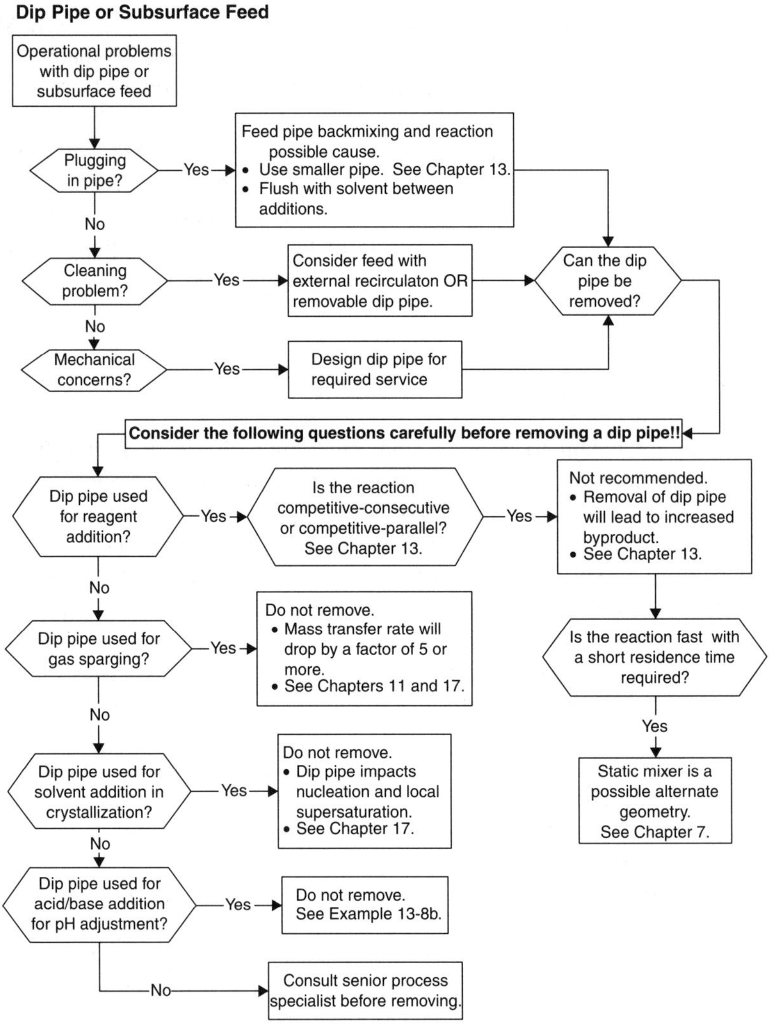

**Figure I-1** Dip pipe or subsurface feed.

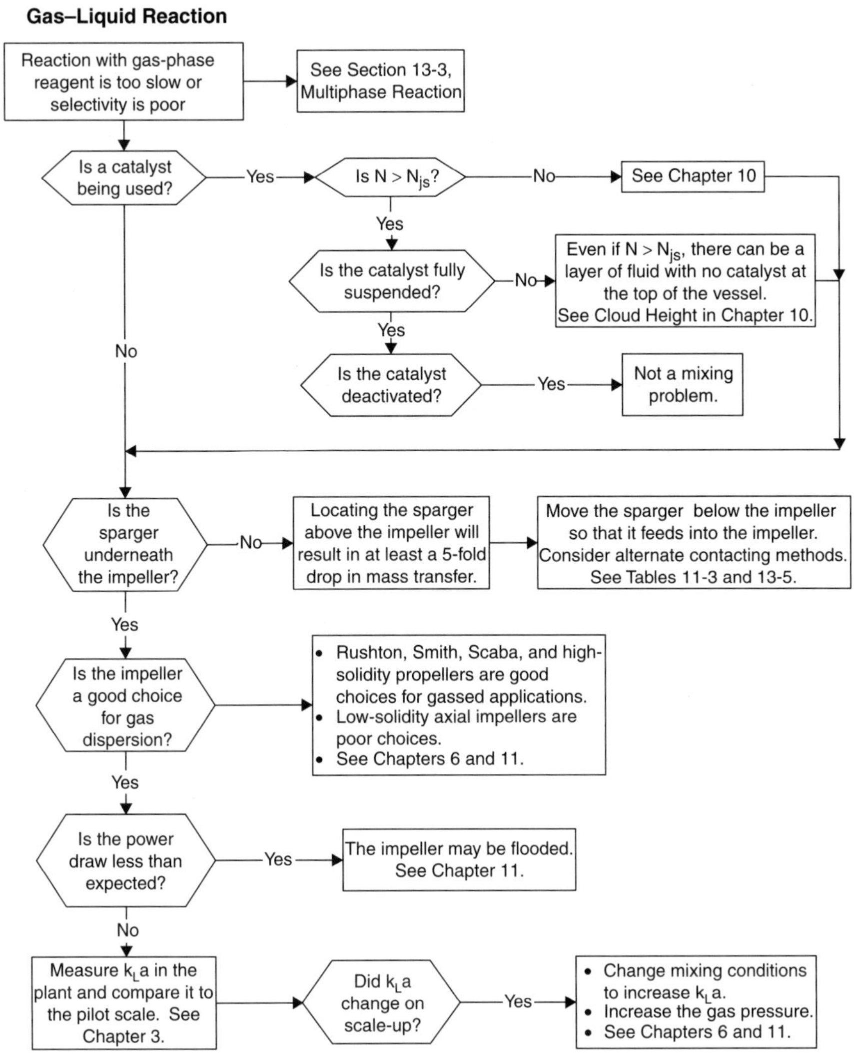

INTRODUCTION  li

## Gas–Liquid Reaction

```
Reaction with gas-phase
reagent is too slow or          → See Section 13-3,
selectivity is poor                Multiphase Reaction
         │
         ▼
  Is a catalyst  ──Yes──▶  Is N > N_js?  ──No──▶  See Chapter 10
  being used?                    │
         │                      Yes
         │                       ▼
         │             Is the catalyst fully  ──No──▶  Even if N > N_js, there can be a
         │                suspended?                    layer of fluid with no catalyst at
         │                       │                      the top of the vessel.
         │                      Yes                     See Cloud Height in Chapter 10.
         │                       ▼
         No            Is the catalyst   ──Yes──▶  Not a mixing problem.
         │              deactivated?
         │                       │
         │                       No
         ▼◀──────────────────────┘
  Is the sparger
  underneath     ──No──▶ Locating the sparger    ──▶ Move the sparger below the impeller
  the impeller?          above the impeller will     so that it feeds into the impeller.
         │               result in at least a 5-fold  Consider alternate contacting methods.
        Yes              drop in mass transfer.       See Tables 11-3 and 13-5.
         ▼
  Is the impeller
  a good choice  ──▶ • Rushton, Smith, Scaba, and high-
  for gas              solidity propellers are good
  dispersion?          choices for gassed applications.
         │           • Low-solidity axial impellers are
        Yes            poor choices.
         ▼           • See Chapters 6 and 11.
  Is the power
  draw less than  ──Yes──▶ The impeller may be flooded.
  expected?                See Chapter 11.
         │
         No
         ▼
  Measure k_La in the
  plant and compare it to  ──▶  Did k_La    ──Yes──▶ • Change mixing conditions
  the pilot scale. See           change on             to increase k_La.
  Chapter 3.                     scale-up?           • Increase the gas pressure.
                                                     • See Chapters 6 and 11.
```

**Figure I-2** Gas–liquid reaction.

lii    INTRODUCTION

**Batch Liquid–Liquid Extraction**

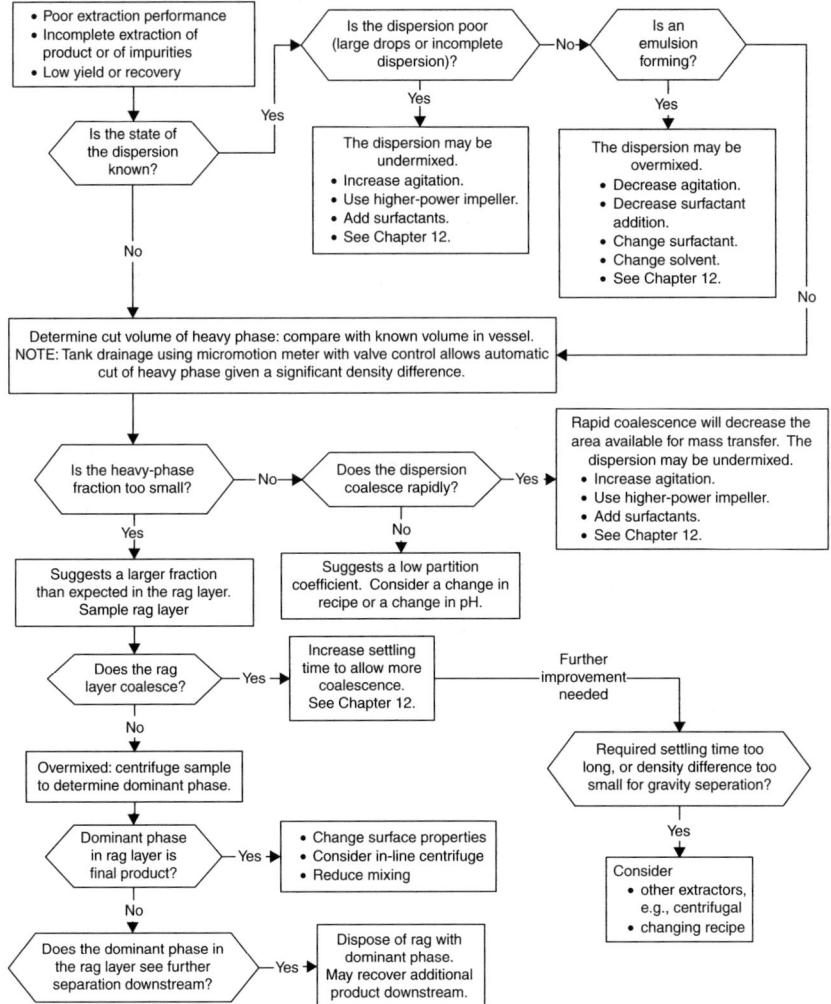

**Figure I-3**  Batch liquid–liquid extraction.

INTRODUCTION    liii

**Reaction in Liquid–Liquid Dispersion**

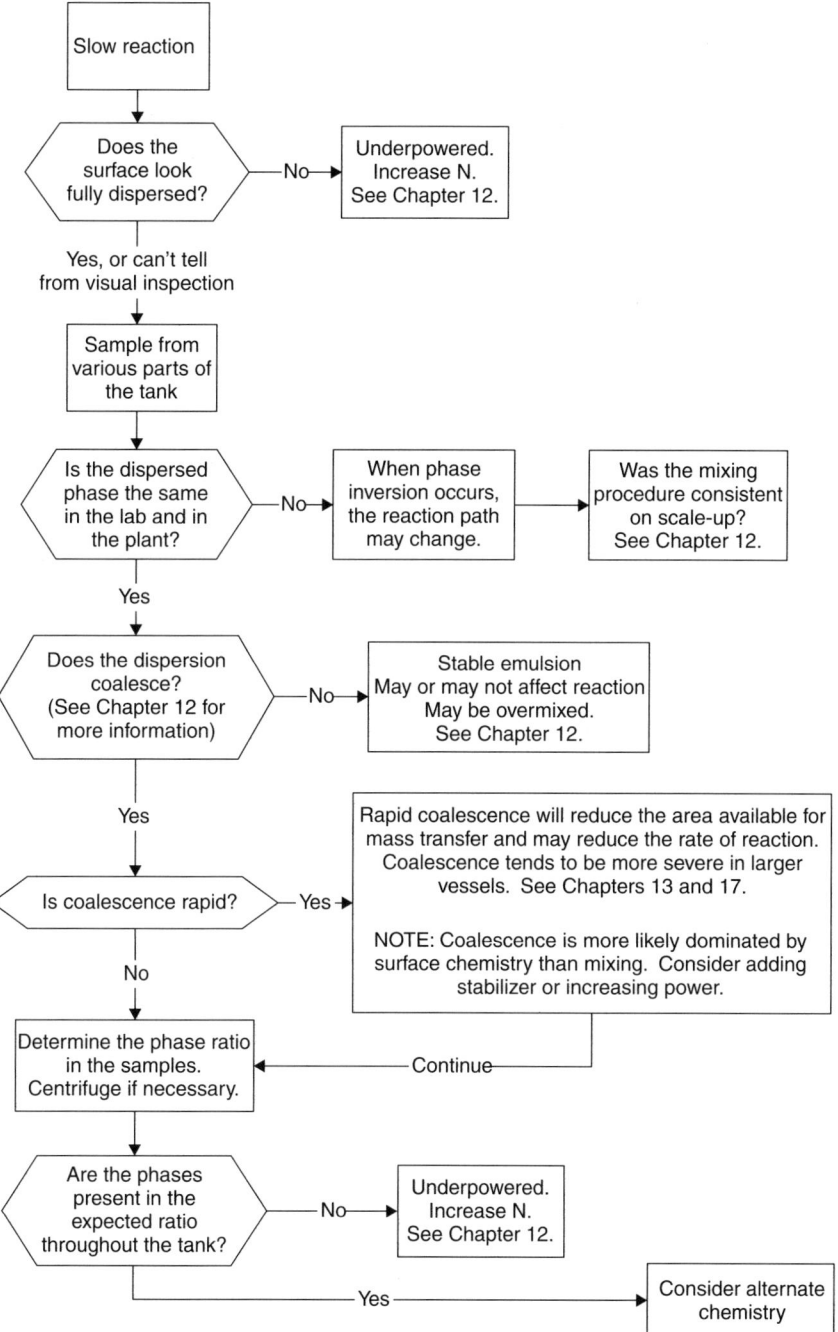

**Figure I-4**  Reaction in liquid–liquid dispersion.

liv    INTRODUCTION

**Solids Withdrawal from Stirred Tanks**

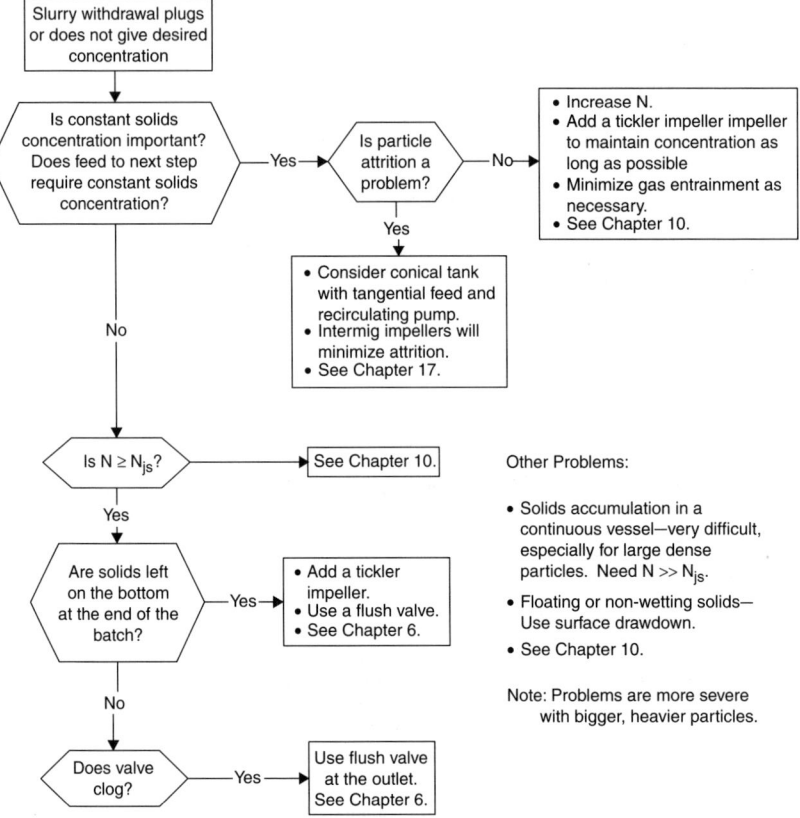

**Figure I-5**   Solids withdrawal from stirred tanks.

INTRODUCTION  lv

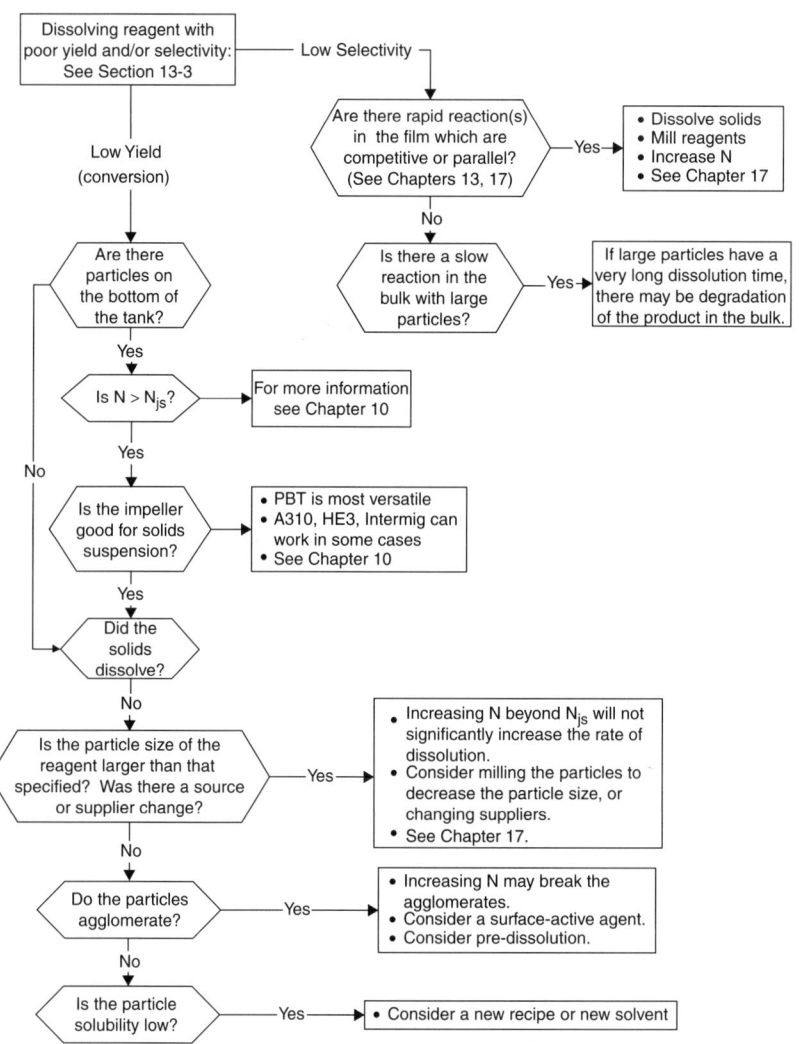

**Figure I-6** Solid–liquid reaction.

## Mixing Nomenclature and Unit Conversions

Table I-1 includes the common nomenclature used in mixing correlations and calculations. Many of the chapters in this book have more detailed lists of nomenclature for specific topics. Table I-1 is intended for general reference. Where a symbol is used for more than one purpose, the common multiple uses are given. Conversions are given in Tables I-2 and I-3.

The nomenclature follows that outlined by Oldshue (1977), Buck (1978), and the AIChE Equipment Testing Procedure for Mixing Equipment (2001) of symbols for use in the SI system. There are a few exceptions that are commonly

**Table I-1** Mixing Nomenclature

| Common Symbol | Quantity | Units |
|---|---|---|
| A, B, R, S | Reactants | |
| A, B, R, S | Reactant concentrations, $C_A$, $C_B$, etc. | mol/m$^3$ |
| B | Baffle width | m |
| C | Impeller off-bottom clearance | m |
| C | Reaction conversion, $(A_o - A)/A_o$ | % |
| $C_p$ | Specific heat | J/kg · K |
| D | Impeller diameter | m |
| $D_{AB}$ | Diffusivity | m$^2$/s |
| Da | Damkoehler number (see Chapter 13) | (−) |
| $d_{32}$ | Sauter mean diameter | m |
| Fr | Froude number, $N^2D/g$ | (−) |
| $g_c$ | Gravitational correction for British units, 32.2 lb$_m$/lb$_f$ × ft/s$^2$ | |
| H or Z | Liquid height | m |
| $k_1$, $k_2$, ... | Reaction rate constants | (mol/m$^3$)$^{1-n}$/s |
| k | Thermal conductivity | W/m · K |
| $k_g$, $k_l$ | Mass transfer coefficient | m/s |
| L | Length scale | m |
| N | Impeller rotational speed | rps or rpm |
| $N_c$ | Impeller critical rotational speed | rps or rpm |
| $N_{js}$ | Just suspended rotational speed | rps or rpm |
| $N_{min}$ | Just suspended speed for liquid drops | rps or rpm |
| Nu | Nusselt number, hT/k | (−) |
| $N_p$ or Po | Power number, $Pg_c/\rho N^3 D^5$ | (−) |
| P | Power, $N_p \rho N^3 D^5/g_c$ | W |
| P | Pressure | Pa |
| Pr | Prandtl number, $C_p \mu/k$ | (−) |
| Q | Heat transfer rate, $UA\Delta T_m$ | W |
| $Q_L$ | Pumping rate of impeller, $\alpha ND^3$ | m$^3$/s |
| R | Gas constant | J/mol · K |
| R | Impeller radius | m |
| Re | Reynolds number, $\rho DV/\mu$ | (−) |
| Re | Impeller Reynolds number, $\rho D^2 N/\mu$ | (−) |
| S | Reaction selectivity (see Chapter 13) | |
| Sc | Schmidt number, $\mu/D_{AB}\rho$ | (−) |
| T | Tank diameter | m |

INTRODUCTION lvii

**Table I-1** (*continued*)

| Common Symbol | Quantity | Units |
|---|---|---|
| T | Temperature | K,C |
| t | Time | s |
| $T_o$ or $T_Q$ | Torque, $\alpha \rho N^2 D^5$ | W/s |
| U | Overall heat transfer coeff, $Q/(A \Delta T_m)$ | $J/m^2 \cdot s \cdot K$ |
| $u'$ | Fluctuating velocity | m/s |
| $V_{impeller}$ | Impeller swept volume | $m^3$ |
| V | Volume | $m^3$ |
| v | Velocity | m/s |
| $W_b$ | Baffle width | m |
| $X_s$ | Impurity selectivity, $2S/R + 2S$ | % |
| Y | Reaction yield, $R/A_o$ | % |
| Z and H | Liquid height | m |

*Greek Symbols*

| | | |
|---|---|---|
| $\alpha$ | Blade angle | ° |
| $\gamma$ | Shear rate | $s^{-1}$ |
| $\delta$ | Width of shear gap, rotor and stator | m |
| $\varepsilon$ | Void fraction | (—) |
| $\varepsilon$ | Local rate of dissipation of turbulent kinetic energy per unit mass | $m^2/s^3$ |
| $\bar{\varepsilon}$ | Power input per mass of fluid in the tank, power per volume, $P/\rho V_{tank}$ | $m^2/s^3$ |
| $\bar{\varepsilon}_I$ | Power input per mass of fluid in the impeller swept volume, $P/\rho V_{Impeller}$ | $m^2/s^3$ |
| $\eta$ (also $\lambda_K$) | Kolmogorov scale, $(\nu^3/\varepsilon)^{1/4}$ | m |
| $\theta_B$, $t_{blend}$ | Blend time | s |
| $\theta$ | Angle of impeller blade with axis of rotation | ° |
| $\lambda$ | Taylor microscale of turbulence | m |
| $\lambda$ | Wavelength | m |
| $\lambda_B$ | Bachelor length scale, $(\nu D_{AB}^2/\varepsilon)^{1/4}$ | m |
| $\lambda_K$ (also $\eta$) | Kolmogorov scale, $(\nu^3/\varepsilon)^{1/4}$ | m |
| $\mu$ | Viscosity | $Pa \cdot s$ |
| $\nu$ | Kinematic viscosity, $\mu/\rho$ | $m^2/s$ |
| $\rho$ | Density | $kg/m^3$ |
| $\sigma$ | Interfacial tension | N/m |
| $\tau_M$ | Mixing time constant | s |
| $\tau_D$ | Diffusion time constant | s |
| $\tau_R$ | Reaction time constant | s |
| $\tau$ | Shear stress | Pa |
| $\tau$ (also $T_Q$) | Torque | $N \cdot m$ |
| $\phi$ | Volume fraction of dispersed phase | (—) |
| $\phi$ | Particle shape factor | (—) |

**Table I-2** Conversion from British to SI Units

| Non-SI Unit | Quantity | To Convert to SI Unit: | Multiply by: |
|---|---|---|---|
| Btu | Heat | Joule (J) | 1.0551 E +03 |
| Btu/lb$_m$ · °F | Heat capacity | J/kg · K | 4.1868 E +3 |
| Btu/hr | Heat flux | Watt (W) | 2.9307 E −01 |
| Btu/hr · ft$^2$ · °F | Heat transfer coefficient | W/m$^2$ · K | 5.6782 E +00 |
| Btu/ft · hr · °F | Thermal conductivity | W/m · K | 1.7307 E +00 |
| cal | Calorie | Joule (J) | 4.1868 E +00 |
| centipoise | Viscosity | Pa · s | 1.0000 E −03 |
| centistoke | Kinematic viscosity | m$^2$/s | 1.0000 E −06 |
| °F | Temperature | C | (°F − 32)(5/9) |
| dyne | | Newton (N) | 1.0000 E −05 |
| erg | | Joule (J) | 1.0000 E −07 |
| foot | | meter (m) | 3.0480 E −01 |
| foot$^2$ | | m$^2$ | 9.2990 E −02 |
| foot$^3$ | | m$^3$ | 2.8316 E −02 |
| ft · lb$_f$ | | Joule (J) | 1.3558 E +00 |
| gallon | U.S. liquid | m$^3$ | 3.7854 E −03 |
| horsepower | 550 ft-lb$_f$/sec | Watt (W) | 7.4570 E +02 |
| inch | | meter (m) | 2.5400 E −02 |
| inches Hg (60°F) | | Pascal (Pa) | 3.3768 E +03 |
| inches H$_2$0 (60°F) | | Pascal (Pa) | 2.4884 E +02 |
| kilocalorie | | Joule (J) | 4.1868 E +03 |
| micron | | meter (m) | 1.0000 E −06 |
| mmHg (0°C) | Pressure | Pascal (Pa) | 1.3332 E +02 |
| poise | Absolute viscosity | Pa · s | 1.0000 E −01 |
| lb$_f$ | | Newton (N) | 4.4482 E −00 |
| lb$_m$ | | kilogram (kg) | 4.5359 E −01 |
| lb$_m$/ft$^3$ | Density | kg/m$^3$ | 1.6018 E +01 |
| lb$_m$/ft-sec | Viscosity | Pa · s | 1.4882 E +00 |
| psi | Pressure | Pascal (Pa) | 6.8948 E +03 |
| Stoke | Kinematic viscosity | m$^2$/s | 1.0000 E −04 |
| tonne (long, 2240 lb$_m$) | | kilogram (kg) | 1.0160 E +03 |
| ton (short, 2000 lb$_m$) | | kilogram (kg) | 9.0718 E +02 |
| torr (mmHg, 0°C) | Pressure | Pascal (Pa) | 1.3332 E +02 |
| Watt | | Watt (W) | 1.0002 E +00 |
| Watt-h | | Joule (J) | 3.6000 E +03 |

**Table I-3** Conversion of SI Units

| SI Unit | To Convert to | Multiply by: |
|---|---|---|
| Joule (J) | Btu | 9 E −4 |
|  | ft-lb$_f$/sec | 0.7375 |
| Watt (W) | Btu/hr | 3.436 |
| Volume (m$^3$) | ft$^3$ | 35.32 |
|  | liter | 1000 |
|  | gallon | 264.2 |
| Meter (m) | angstrom | 1.000 E +10 |
|  | micron (μm) | 1.000 E +6 |
| Viscosity (Pa · s) | centipoises | 1.000 E +3 |
| Power (W) | horsepower | 0.0013 |
| Pressure (Pa) | inch Hg | 0.2953 E −3 |
|  | psi | 0.1451 E −3 |
|  | torr (mmHg at 0 K) | 7.5006 E −3 |

used in mixing terminology. The European Federation of Chemical Engineering Working Party on Mixing Terms, Symbols, and Units has also published a comprehensive list of nomenclature (Fort et al., 2000).

## ACKNOWLEDGMENTS

The editors would first like to acknowledge the contributions of the many authors whose efforts in writing their respective chapters have made publication of this handbook possible. Readers will appreciate the difficulty of finding time in very full professional lives to write authoritative chapters on fundamental ideas: chapters that required both reflection and compilation of the vast quantities of information in the technical literature. In addition, we would like to acknowledge the reviewers whose careful evaluations have been instrumental in helping the authors and editors to evaluate the technical content and relevance of each chapter:

Gary Anthieren, University of Alberta
Richard Berry, Paprican
Sujit Bhattacharya, University of Alberta
John Bourne, ETH Zurich (retired)
Clive Davies, Industrial Research Limited
David Dickey, MixTech
Arthur W. Etchells, DuPont
Rodney Fox, Iowa State University
Enrique Galindo, Instituto de Biotecnologia, Mexico
Alan Hall, Syngenta
Brian Johnson, Princeton University and Merck

Kishore Kar, Dow
Richard Kerekes, University of British Columbia
Roger King, BHRG (retired)
Mike Midler, Merck
Vesselina Roussinova, University of Alberta
Marco Salinas, Domtar
U. Sundararaj, University of Alberta
Phillipe Tanguy, École Polytechnique, Montreal
Larry Tavlarides, Syracuse University
Koutchen Tsai, Dow
Peitr Veenstra, Shell

For the authors and reviewers, the editors have drawn extensively on the considerable resources of the North American Mixing Forum. The handbook would not have been possible without the unwavering support of this rather remarkable organization. All royalties from the sale of this book will be returned to NAMF for the promotion of mixing research and education. The concept for this book originated in the Executive Council of NAMF and was first proposed by Arthur W. Etchells. The encouragement of the first president of NAMF, James Y. Oldshue, and indeed of all the presidents of NAMF, both past and current, is also gratefully acknowledged. Finally, the work of our editorial assistant, Kathy van Denderen, was indispensable, as were the efforts of the production team at Wiley.

The objective of this book is to provide mixing practitioners with a summary of the current state of mixing knowledge, both in terms of fundamentals and from the perspective of industrial practice and experience. We have found many of the chapters absolutely definitive in their area. We hope that readers find as much to stimulate and fascinate them in these pages as we have found during their editing.

As we complete the final pages of this manuscript, we would like to thank our families for their unwavering faith and encouragement during its preparation. At times the book appeared to be a fantasy of our own creation, and at times the work has been very real and seemingly unending. Through it all, they have shared our road less traveled by; and that made all the difference.

## REFERENCES

Baldyga, J., and J. R. Bourne (1999). *Turbulent Mixing and Chemical Reactions*, Wiley, Chichester, West Sussex, England.

Berry, R. (1990). High intensity mixers in chlorination and chlorine dioxide stages: survey results and evaluation, *Pulp Paper Can*, **91**(4), T151–T159.

Buck, E. (1978). Letter symbols for chemical engineering, *Chem. Eng. Prog.*, Oct., pp. 73–80.

Dickey, D.S. et al. (2001). *Mixing Equipment: AIChE Testing Procedure*, 3rd ed., AIChE, New York.

Fort, I., P. Ditl, and W. Tausher (2000). Working party on mixing of the European Federation of Chemical Engineering: terms, symbols, units, *Chem. Biochem. Eng. Q.*, **14**(2), 69–82.

Leng, D. E. (1997). Plenary talk, presented at Mixing XVI, Williamsburg, VA.

Levenspiel, O. (1972). *Chemical Reaction Engineering*, 2nd ed., Wiley, New York.

Levenspiel, O. (1998). *Chemical Reaction Engineering*, 3rd ed., Wiley, New York.

Nagata, S. (1975). *Mixing: Principles and Applications*, Wiley, New York.

Nauman, E. B. (2002). *Chemical Reactor Design, Optimization and Scaleup*, McGraw-Hill, New York.

Oldshue, J. Y. (1977). AIChE goes metric, *Chem. Eng. Prog.*, Aug., pp. 135–138.

Uhl, V. W., and J. B. Gray, eds. (1966). *Mixing Theory and Practice*, Vols. I and II, Academic Press, New York.

Zwietering, T. N. (1958). Suspending of solid particles in liquid by agitators, *Chem. Eng. Sci.*, **8**, 244–253.

# CHAPTER 1

# Residence Time Distributions

E. BRUCE NAUMAN

*Rensselaer Polytechnic Institute*

## 1-1 INTRODUCTION

The concept of residence time distribution (RTD) and its importance in flow processes first developed by Danckwerts (1953) was a seminal contribution to the emergence of chemical engineering science. An introduction to RTD theory is now included in standard texts on chemical reaction engineering. There is also an extensive literature on the measurement, theory, and application of residence time distributions. A literature search returns nearly 5000 references containing the concept of residence time distribution and some 30 000 references dealing with residence time in general. This chapter necessarily provides only a brief introduction; the references provide more comprehensive treatments.

The residence time distribution measures features of ideal or nonideal flows associated with the bulk flow patterns or *macromixing* in a reactor or other process vessel. The term *micromixing*, as used in this chapter, applies to spatial mixing at the molecular scale that is bounded but not determined uniquely by the residence time distribution. The bounds are extreme conditions known as *complete segregation* and *maximum mixedness*. They represent, respectively, the least and most molecular-level mixing that is possible for a given residence time distribution.

Most of this handbook treats *spatial mixing*. Suppose that a sample of fluid is collected and analyzed. One may ask: Is it homogeneous? Standard measures of homogeneity such as the striation thickness in laminar flow or the coefficient of variation in turbulent flow can be used to answer this question quantitatively. In this chapter we look at a different question that is important for continuous flow systems: When did the particles, typically molecules but sometimes larger particles, enter the system, and how long did they stay? This question

---

*Handbook of Industrial Mixing: Science and Practice*, Edited by Edward L. Paul,
Victor A. Atiemo-Obeng, and Suzanne M. Kresta
ISBN 0-471-26919-0    Copyright © 2004 John Wiley & Sons, Inc.

**2** RESIDENCE TIME DISTRIBUTIONS

involves *temporal mixing*, and its quantitative answer is provided by the RTD (Danckwerts, 1953).

To distinguish between spatial and temporal mixing, suppose that a flow system is fed from separate black and white streams. If the effluent emerges uniformly gray, there is good spatial mixing. For the case of a pipe, the uniform grayness corresponds to good mixing in the radial direction. Now suppose that the pipe is fed from a single stream that varies in shade or grayness. The effluent will also vary in shade unless there is good temporal mixing. In the context of a pipe, spatial mixing is equivalent to *radial mixing*, and temporal mixing is equivalent to *axial mixing*.

In a batch reactor, all molecules enter and leave together. If the system is isothermal, reaction yields depend only on the elapsed time and on the initial composition. The situation in flow systems is more complicated but not impossibly so. The counterpart of the batch reaction time is the age of a molecule. Aging begins when a molecule enters the reactor and ceases when it leaves. The total time spent within the boundaries of the reactor is known as the *exit age*, or *residence time*, t. In real flow systems, molecules leaving the system will have a variety of residence times. The distribution of residence times provides considerable information about homogeneous isothermal reactions. For single first-order reactions, knowledge of the RTD allows the yield to be calculated exactly, even in flow systems of arbitrary complexity. For other reaction orders, it is usually possible to calculate fairly tight limits, within which the yield must lie (Zwietering, 1959). If the system is nonisothermal or heterogeneous, the RTD cannot predict reaction yield directly, but it still provides a general description of the flow that is not easily obtained by velocity measurements.

Residence time experiments have been used to explore the hydrodynamics of many chemical processes. Examples include fixed and fluidized bed reactors, chromatography columns, two-phase stirred tanks, distillation and absorption columns, and trickle bed reactors.

## 1-2 MEASUREMENTS AND DISTRIBUTION FUNCTIONS

Transient experiments with inert tracers are used to determine residence time distributions. In real systems, they will be actual experiments. In theoretical studies, the experiments are mathematical and are applied to a dynamic model of the system. Table 1-1 lists the types of RTDs that can be measured using tracer experiments. The simplest case is a *negative step change*. Suppose that an inert tracer has been fed to the system for an extended period, giving $C_{in} = C_{out} = C_0$ for t < 0. At time t = 0, the tracer supply is suddenly stopped so that $C_{in} = 0$ for t > 0. Then the tracer concentration at the reactor outlet will decrease with time, eventually approaching zero as the tracer is washed out of the system. This response to a negative step change defines the *washout function*, W(t). The responses to other standard inputs are shown in Table 1-1. Relationships between the various functions are shown in Table 1-2.

**Table 1-1** Residence Time Distribution Functions

| Name | Symbol | Input Signal | Output Signal | Physical Interpretation | Properties |
|---|---|---|---|---|---|
| Washout function | $W(t)$ | Negative step change in tracer, concentration from an initial value of $C_0$ to a final value of 0 | $W(t) = C_{out}(t)/C_0$ | $W(t)$ is the fraction of particles that remained in the system for a time greater than t. | $W(0) = 1$<br>$W(\infty) = 0$<br>$dW/dt \leq 0$ |
| Cumulative distribution function | $F(t)$ | Positive step change in tracer concentration from an initial value of 0 to a final value of $C_\infty$ | $F(t) = C_{out}(t)/C_\infty$ | $F(t)$ is the fraction of particles that remained in the system for a time less than t. | $F(0) = 0$<br>$F(\infty) = 1$<br>$dF/dt \geq 0$ |
| Differential distribution function | $f(t)$ or $E(t)$ | Sharp impulse of tracer | $f(t) = \dfrac{C_{out}(t)}{\int_0^\infty C_{out}(t)\,dt}$ | $f(t)\,dt$ is the fraction of particles that remained in the system for a time between t and $t + dt$. | $f(t) \geq 0$<br>$\int_0^\infty f(t)\,dt = 1$ |
| Convolution integral | $C_{out}(t)$ | Any time-varying tracer concentration | $C_{out}(t) = \int_{-\infty}^{t} C_{in}(\theta)f(t-\theta)\,d\theta$<br>$= \int_0^\infty C_{in}(t-\theta)f(\theta)\,d\theta$ | The output signal is a damped response that reflects the entire history of inputs. | $[C_{out}]_{max} \leq [C_{in}]_{max}$ |

**4** RESIDENCE TIME DISTRIBUTIONS

**Table 1-2** Relationships between the Functions and Moments of the RTD

| Definition | Mathematical Formulation |
|---|---|
| Relations between the distribution functions | $f(t) = \dfrac{dF}{dt} = -\dfrac{dW}{dt}$ <br><br> $F(t) = \displaystyle\int_0^t f(t')dt'$ <br><br> $W(t) = \displaystyle\int_t^\infty f(t')dt'$ |
| Moments about the origin | $\mu_n = \displaystyle\int_0^\infty t^n f(t)\,dt = n\int_0^\infty t^{n-1} W(t)\,dt$ |
| First moment = mean residence time | $\bar{t} = \displaystyle\int_0^\infty t f(t)\,dt = \int_0^\infty W(t)\,dt$ |
| Moments about the mean | $\mu'_n = \displaystyle\int_0^\infty (t-\bar{t})^n f(t)\,dt = n\int_0^\infty (t-\bar{t})^{n-1} W(t)\,dt + (-\bar{t})^n$ |
| Dimensionless variance of the RTD | $\sigma^2 = \dfrac{\mu'_2}{\bar{t}^2} = \dfrac{\displaystyle\int_0^\infty (t-\bar{t})^2 f(t)\,dt}{\bar{t}^2} = \dfrac{2\displaystyle\int_0^\infty t W(t)\,dt}{\bar{t}^2} - 1$ |

A good input signal, usually a negative step change, must be made at the reactor inlet. The mixing-cup average concentration of tracer molecules must be accurately measured at the outlet. If the tracer has a background concentration, it is subtracted from the experimental measurements. The flow properties of the tracer molecules must be similar to those of the reactant molecules, and the change in total flow rate must be insignificant. It is usually possible to meet these requirements in practice. The major theoretical requirement is that the inlet and outlet streams have unidirectional flows, so that once the molecules enter the system they stay in until they exit, never to return. Systems with unidirectional inlet and outlet streams are *closed* so that a molecule enters the system only once and leaves only once. Most systems of chemical engineering importance are closed to a reasonable approximation.

Among RTD experiments, washout experiments are generally preferred since $W(\infty) = 0$ will be known a priori but $F(\infty) = C_0$ must usually be measured. The positive step experiment will also be subject to errors caused by changes in $C_0$ during the course of the experiment. However, the positive step change experiment requires a smaller amount of tracer since the experiment will be terminated before the outlet concentration fully reaches $C_0$. Impulse response experiments that measure $f(t)$ use still smaller amounts.

The RTD can be characterized by its moments as indicated in Table 1-2. The most important moment is the first moment about the mean, known as the *mean residence time* and usually denoted as $\bar{t}$:

$$\bar{t} = \int_0^\infty tf(t)\,dt = \int_0^\infty W(t)\,dt = \frac{\text{mass inventory in the system}}{\text{mass flow rate through the system}} = \frac{\text{hold-up}}{\text{throughput}} \quad (1\text{-}1)$$

Thus $\bar{t}$ can be found from inert tracer experiments. It can also be found from measurements of the system inventory and throughput. Agreement of the $\bar{t}$'s calculated by these two methods provides a good check on experimental accuracy. Occasionally, eq. (1-1) is used to determine an unknown volume or an unknown density from inert tracer data.

Roughly speaking, the first moment, $\bar{t}$, measures the size of an RTD, while higher moments measure its shape. One common measure of shape is the dimensionless second moment about the mean, also known as the *dimensionless variance*, $\sigma^2$ (see Table 1-2). In piston flow, all particles have the same residence time, so $\sigma^2 = 0$. This case is approximated by highly turbulent flow in a pipe. In an ideal continuous flow stirred tank reaction, $\sigma^2 = 1$. Well-designed reactors in turbulent flow have a $\sigma^2$ value between 0 and 1, but laminar flow reactors can have $\sigma^2 > 1$.

Note that either $W(t)$ or $f(t)$ can be used to calculate the moments. Use the one that was obtained directly from an experiment. If moments of the highest possible accuracy are desired, the experiment should be a negative step change to get $W(t)$ directly.

## 1-3 RESIDENCE TIME MODELS OF FLOW SYSTEMS

Figure 1-1 shows the washout functions for some flow systems. The time scale in this figure has been converted to dimensionless time, $t/\bar{t}$. This means that the integrals of the various washout functions all have unit mean so that the various flow systems can be compared independent of system size.

### 1-3.1 Ideal Flow Systems

The ideal cases are the *piston flow reactor* (PFR), also known as a *plug flow reactor*, and the *continuous flow stirred tank reactor* (CSTR). A third kind of ideal reactor, the *completely segregated CSTR*, has the same distribution of residence times as a normal, perfectly mixed CSTR. The washout function for a CSTR has the simple exponential form

$$W(t) = e^{-t/\bar{t}} \quad (1\text{-}2)$$

A CSTR is said to have an *exponential distribution* of residence times. The washout function for a PFR is a negative step change occurring at time $\bar{t}$:

$$W(t) = \begin{cases} 1 & t < \bar{t} \\ 0 & t > \bar{t} \end{cases} \quad (1\text{-}3)$$

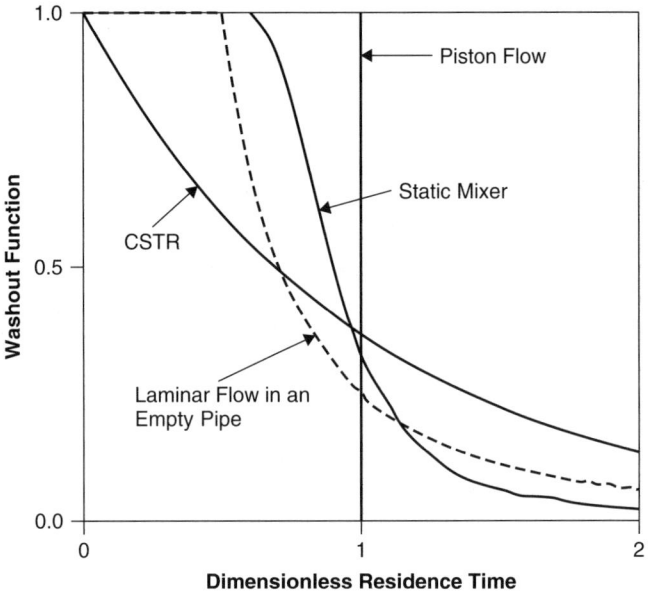

**Figure 1-1** Residence time washout functions for various flow systems.

The derivative of a step change is a delta function, and $f(t) = \delta(t - \bar{t})$. Thus, a piston flow reactor is said to have a *delta distribution* of residence times. The variances for these ideal cases are $\sigma^2 = 1$ for a CSTR and $\sigma^2 = 0$ for a PFR, which are extremes for well-designed reactors in turbulent flow. Poorly designed reactors and laminar flow reactors with little molecular diffusion can have $\sigma^2$ values greater than 1.

### 1-3.2 Hydrodynamic Models

The curve for laminar flow in Figure 1-1 was derived for a parabolic velocity profile in a circular tube. The washout function is

$$W(t) = \begin{cases} 1 & t < \bar{t}/2 \\ \dfrac{\bar{t}^2}{4t^2} & t > \bar{t}/2 \end{cases} \qquad (1\text{-}4)$$

Equation (1-4) is a theoretical result calculated from a hydrodynamic model, albeit a very simple one. It has a sharp *first appearance time*, $t_{\text{first}}$, where the washout function first falls below 1.0. Real systems, such as that for the static mixer illustrated in Figure 1-1, may have a fuzzy first appearance time. For the fuzzy case, a 5% response time [i.e., $W(t) = 0.95$] is used instead. Table 1-3 shows first appearance times for some laminar flow systems.

**Table 1-3** First Appearance Times in Laminar Flow Systems

| Geometry | $t_{first}/\bar{t}$ |
| --- | --- |
| Equilateral–triangular ducts | 0.450 |
| Square ducts | 0.477 |
| Straight, circular tubes | 0.500 |
| Straight, circular tubes (5% response) | 0.513 |
| 16 element Kenics mixer (5% response) | 0.598 |
| Helically coiled tubes | 0.613 |
| Annular flow | 0.500–0.667 |
| Parabolic flow between flat plates | 0.667 |
| 40 element Kenics mixer (5% response) | 0.676 |
| Single-screw extruder | 0.750 |
| Helical coils with changes in the direction of centrifugal force | >0.85 |

Flow patterns in the Kenics static mixer are too complicated to determine the residence time distribution analytically. Instead, experimental measurements were fit to a simple model. The model used for the Kenics mixer in Table 1-3 assumes regions of undisturbed laminar flow separated by planes of complete radial mixing, there being one mixing plane for every four Kenics elements. Simpler models are useful for systems in turbulent flow.

A system with a sharp first appearance time and $\sigma^2 < 1$ can be approximated as a PFR in series with a CSTR. This model is used for residence times in a fluidized bed reactor. If the system has a fuzzy first appearance time and $\sigma^2 \approx 1$, the tanks-in-series model or the axial dispersion model can be used. These models are used for tubular reactors in turbulent flow. The tanks-in-series is also used when the physical system consists of CSTRs in series, and it may be a good approximation for a single CSTR with dual Rushton turbines.

Tubular polymerization reactors frequently show large deviations from the parabolic velocity profile of constant viscosity laminar flow. The velocity profile of a polymerizing mixture can be calculated by combining the equations of motion with the convective diffusion equations for heat and mass, but direct experimental verification of the calculations is difficult. One way of testing the results is to compare an experimental residence time distribution to the calculated distribution. There is a one-to-one correspondence between velocity profile and RTD for well-developed diffusion-free flows in tubes. See Nauman and Buffham (1983) for details.

### 1-3.3 Recycle Models

High rates of external recycle have the same effect on the RTD as high rates of internal recycle in a stirred tank. The recycle system in Figure 1-2a can represent a loop reactor or it can be a model for a stirred tank. The once-through RTD must be known. In principle, it can be measured by applying a step change at the reactor inlet, measuring the outlet response, and then destroying the tracer before it has

**8**   RESIDENCE TIME DISTRIBUTIONS

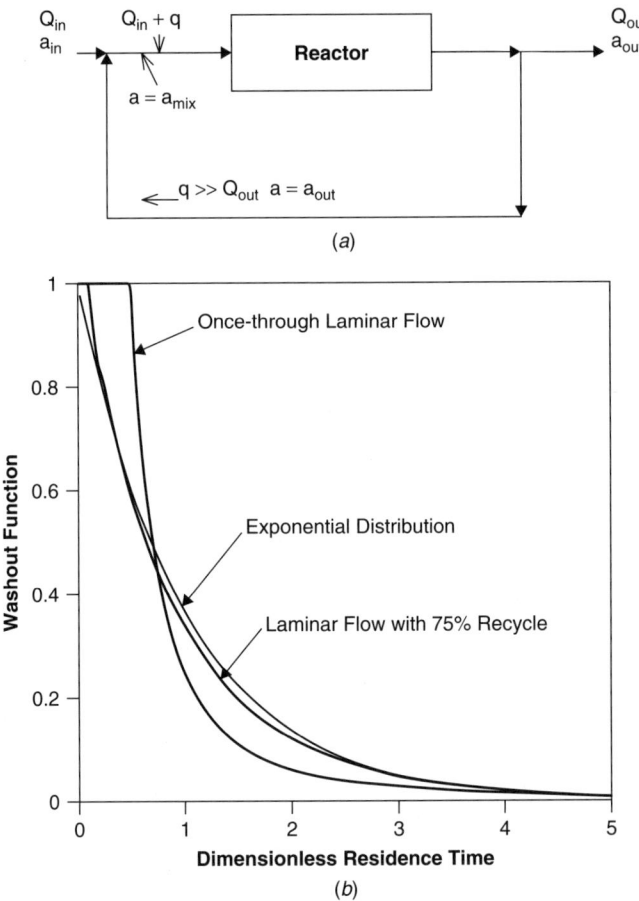

**Figure 1-2** Recycle reactor: (*a*) flow diagram; (*b*) washout function for a 3:1 recycle ratio.

a chance to recycle. A more elaborate analysis allows its estimation from tracer experiments performed on the entire system. In practice, mathematical models for the once-through distribution are generally used. The easiest way of generating the composite distribution is by simulation. As a specific example, suppose that the reactor in Figure 1-2*a* is a tube in laminar flow so that the once-through distribution is given by eq. (1-4). Results of a simulation for a recycle ratio of $q/Q = 3$ are shown in Figure 1-2*b*. This first appearance time for a reactor in a recycle loop is the first appearance time for the once-through distribution divided by $q/Q + 1$. It is thus 0.125 in Figure 1-2*b* and declines rather slowly as the recycle ratio is increased. However, even at $q/Q = 3$, the washout function is remarkably close to the exponential distribution of a CSTR. More conservative estimates for the recycle ratio necessary to approach the behavior of a CSTR range from 6 to 100. The ratio selected, of course, depends on the application.

## 1-4 USES OF RESIDENCE TIME DISTRIBUTIONS

The most important use of residence time theory is its application to equipment that is already built and operating. It is usually possible to find a tracer together with injection and detection methods that will be acceptable to a plant manager. The RTD is measured and then analyzed to understand system performance. In this section we focus on such uses. The washout function is assumed to have an experimental basis. Calculations using it will be numerical in nature or will be analytical procedures applied to a model that reproduces the data accurately. Data fitting is best done by nonlinear least squares using untransformed experimental measurements of $W(t)$, $F(t)$, or $f(t)$ versus time, $t$. Eddy diffusion in a turbulent system justifies exponential extrapolation of the integrals that define the moments in Table 1-2. For laminar flow systems, washout experiments should be continued until at least five times the estimated value for $\bar{t}$. The dimensionless variance has limited usefulness in laminar flow systems.

### 1-4.1 Diagnosis of Pathological Behavior

An important use of residence time measurements is to diagnose abnormalities in flow. The first test is whether or not $\bar{t}$ has its expected value (i.e., as the ratio of inventory to throughput). A lower-than-expected value suggests fouling or stagnancy. A higher value is more likely to be caused by experimental error.

The second test supposes that $\bar{t}$ is reasonable and compares the experimental washout curve to what would be expected for the physical design. Suppose that the experimental curve is initially lower than expected; then the system exhibits *bypassing*. If the tail of the distribution is higher than expected, the system exhibits *stagnancy*. Bypassing and stagnancy often occur together. If an experimental washout function initially declines faster than expected, it must eventually decline more slowly since the integrals under the experimental and model curves must both be $\bar{t}$. Bypassing and stagnancy are most easily distinguished when the system is near piston flow and the idealized model is a step change. They are harder to distinguish in stirred tanks because the comparison is made to an exponential curve. When a stirred tank exhibits either bypassing or stagnancy, $\sigma^2 > 1$. Extreme stagnancy will give a mean residence time less than that calculated as the ratio of inventory to throughput. Bypassing or stagnancy can be modeled as vessels in parallel. A stirred tank might be modeled using large and small tanks in parallel. To model bypassing, the small tank would have a residence time lower than that of the large tank. To model stagnancy, the small tank would have a longer residence time. The *side capacity model* shown in Figure 1-3 can also be used and is physically more realistic than a parallel connection of two isolated tanks.

### 1-4.2 Damping of Feed Fluctuations

One generally beneficial consequence of temporal mixing is that fluctuations in component concentrations will be damped. The extent of the damping depends on

# 10  RESIDENCE TIME DISTRIBUTIONS

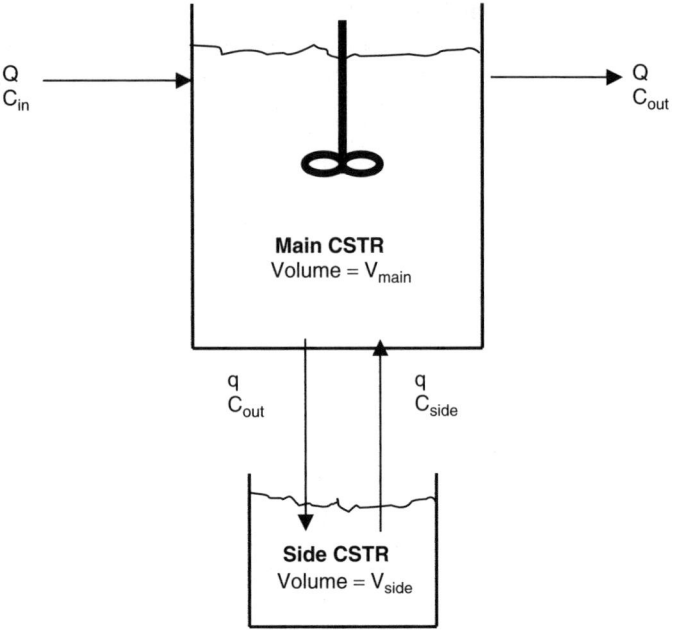

**Figure 1-3** Side capacity model for bypassing or stagnancy in a CSTR.

the nature of the input signal and the residence time distribution. The following pair of convolution integrals applies to an inert tracer that enters the system with time-varying concentration $C_{in}(t)$:

$$C_{out}(t) = \int_0^\infty C_{in}(t - t')f(t')\,dt' = \int_{-\infty}^t C_{in}(t)f(t - t')\,dt' \qquad (1\text{-}5)$$

A piston flow reactor causes pure dead time: a time delay of $\bar{t}$ and no damping. A CSTR acts as an exponential filter and provides good damping provided that the period of the disturbance is less than $\bar{t}$. If the input is sinusoidal with frequency $\omega$, the output will also be sinusoidal, but the magnitude or amplitude of the ripple will be divided by $\sqrt{1 + (\omega\bar{t})^2}$. Damping performance is not sensitive to small changes in the RTD. The true CSTR, the recycle reactor shown in Figure 1-3, and a recently designed axial static mixer give substantially the same damping performance (Nauman et al., 2002).

## 1-4.3  Yield Prediction

In this section we outline the use of RTDs to predict the yield of homogeneous isothermal reactions, based on the pioneering treatments of Danckwerts (1953) and Zwietering (1959) and a proof of optimality due to Chauhan et al. (1972). If there are multiple reactants, the feed stream is assumed to be premixed.

### 1-4.3.1 First-Order Reactions.
Suppose that the reaction is isothermal, homogeneous, and first order with rate constant k. Then knowledge of the RTD allows the reaction yield to be calculated. The result, expressed as the fraction unreacted, is

$$\frac{a_{out}}{a_{in}} = \int_0^\infty e^{-kt} f(t)\, dt = 1 - k \int_0^\infty e^{-kt} W(t)\, dt \tag{1-6}$$

Here, $a_{in}$ and $a_{out}$ are the inlet and outlet concentrations of a reactive component, A, that reacts according to A → products. Use the version of eq. (1-6) that contains the residence time function actually measured, W(t) or f(t).

Equation (1-6) provides a unique estimate of reaction yields because the first-order reaction extent depends only on the time that the molecule has spent in the system and not on interactions or mixing with other molecules. Reactions other than first order give more ambiguous results because the RTD does not measure spatial mixing between molecules that can affect reaction yields.

### 1-4.3.2 Complete Segregation.
A simple generalization of eq. (1-6) is

$$a_{out} = \int_0^\infty a_{batch}(t) f(t)\, dt = 1 - k \int_0^\infty a_{batch}(t) W(t)\, dt \tag{1-7}$$

where $a_{batch}(t)$ is the concentration in a batch reactor that had initial concentration $a_{in}$. This equation can be used to calculate the conversion of any reaction. It assumes an extreme level of local segregation; there is no mixing at all between molecules that entered the system at different times. Molecules that enter together leave together and remain in segregated packets while in the system. Figure 1-4a illustrates this possibility for a completely segregated CSTR.

### 1-4.3.3 Maximum Mixedness.
The micromixing extreme opposite to complete segregation is maximum mixedness and is the highest amount of molecular level mixing that is possible with a fixed residence time distribution. The conversion of a unimolecular but otherwise arbitrary reaction in a maximum mixedness reactor is found by solving Zwietering's differential equation (Zwietering, 1959):

$$\frac{da}{d\lambda} + \frac{f(\lambda)}{W(\lambda)}[a_{in} - a(\lambda)] + R_A = 0 \tag{1-8}$$

where $R_A = R_A(a)$ is the reaction rate. The boundary condition is that a must be bounded for all $\lambda > 0$. The outlet concentration, $a_{out}$, is found by evaluating the solution at $\lambda = 0$. For the special case of an exponential distribution, the solution of eq. (1-8) reduces to that obtained from a steady-state material balance on a perfectly mixed CSTR. A maximally mixed CSTR is the classic CSTR of reaction engineering. In the case of a delta distribution, eqs. (1-7) and (1-8) give the same answer. Reactors in which the flow is piston flow or near piston flow are insensitive to micromixing.

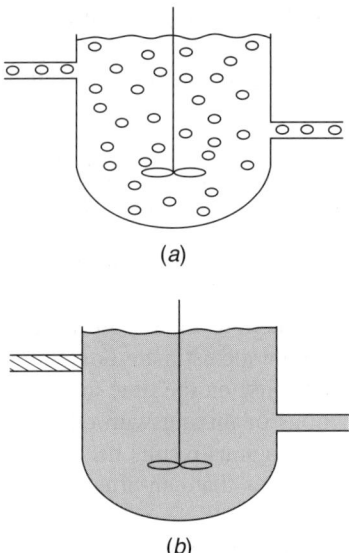

**Figure 1-4** Extremes of micromixing in a stirred tank reactor: (*a*) Ping-Pong balls circulating in an agitated vessel, the completely segregated stirred tank reactor; (*b*) molecular homogeneity, the perfectly mixed CSTR.

***1-4.3.4 Yield Limits.*** Equations (1-7) and (1-8) provide absolute limits on the conversion of most unimolecular reactions and many reactions involving multiple reactants, provided that the feed is premixed. There are three ideal reactors: piston flow, the perfectly mixed CSTR, and the completely segregated CSTR. Calculate the yields for all three types and the yield for a real system will usually lie within the limits of these yields. Measure the residence time distribution and eqs. (1-7) and (1-8) will provide closer limits. This is illustrated in the worked example that follows. A unique calculation of yield for any reaction other than first order is impossible based only on residence time data. It requires a micromixing model such as those developed by Bourne and co-workers (Baldyga et al., 1997). Such models are needed especially when the feed is unmixed or when there is a complex reaction with one or more fast steps. A CSTR cannot be considered well mixed unless the (internal) recycle ratio is very high and molecular-level mixing by molecular diffusion is rapid.

***Example 1-1.*** You have been asked to improve the performance of an existing polymerization reactor. Initially, you know only that it operates at an input flow rate of 10 000 lb/hr, gives a conversion of $62 \pm 1\%$ at a nominal operating temperature of 140°C, and reportedly once gave a higher conversion. The reactor drawings show a complicated arrangement of stirring paddles and cooling coils. The design intent was to approximate piston flow, but a detailed hydrodynamic analysis would be impractical. The drawings do show the working volume of the reactor, and you calculate that the fluid inventory should be about 12 500 lb. Thus you estimate $\bar{t} = 1.25$ hr.

The company library contains the original kinetics study for the polymerization, and it seems to have been done well. The major reaction is a self-condensation with rate eq. $R_A = -ka^2$, where $a_{in}k = 4$ hr$^{-1}$ at 140°C. The fraction unreacted in an isothermal batch reactor at t would be

$$\frac{a_{out}}{a_{in}} = \frac{1}{1 + a_{in}kt}$$

assuming that piston flow in the plant reactor gives $a_{out}/a_{in} = 0.167$, just like the batch reactor.

For a CSTR at maximum mixedness, $a_{out}/a_{in} = 0.358$. In principle, this result is found by solving eq. (1-8), but the result is the same as for a perfectly mixed CSTR.

For a segregated stirred tank, $a_{out}/a_{in} = 0.299$. This result is found by solving eq. (1-7) subject to an exponential distribution of residence times. The measured result, $a_{out}/a_{in} = 0.38$, is worse than any of the ideal reactors! There are several possibilities:

1. The RTD lies outside the normal region. In particular, there may be by-passing.
2. The laboratory kinetics are wrong.
3. The kinetics are right, but the calculated value for $a_{in}k\bar{t}$ is too high. This in turn leads to two main possibilities: (a) The actual temperature is lower than the measured temperature; or (b) the estimated value of $\bar{t}$ is too high.

The good engineer will consider all these possibilities and a few more. Temperature errors are very common, particularly in viscous, low thermal conductivity systems typical of polymers; and they lead to sizable errors in concentration. However, measured temperatures are usually lower than actual rather than higher.

Suppose you decide that the original kinetic study was sound, that there are no apparent changes in the process chemistry, and that the analytical techniques are accurate. This makes flow distribution or mixing a likely culprit. Besides, you would like to see just how that strange agitation/cooling system performs from a flow viewpoint.

Suppose you find an inert hydrocarbon that is not normally present in the system, which is easily detected by gas chromatography and can be tolerated in the product stream. You arrange for the tracer injection port and the product sampling ports to be installed during a maintenance shutdown. It is important that the tracer be well mixed in the inlet stream. Otherwise, it might channel though the system and give nonrepresentative results. You accomplish this by injecting the tracer at the suction side of the transfer pump that is feeding the reactor. You also dissolve a little polymer in the tracer stream to match its viscosity more closely to that of the reactor feed. Having carefully prepared, you perform

**14**  RESIDENCE TIME DISTRIBUTIONS

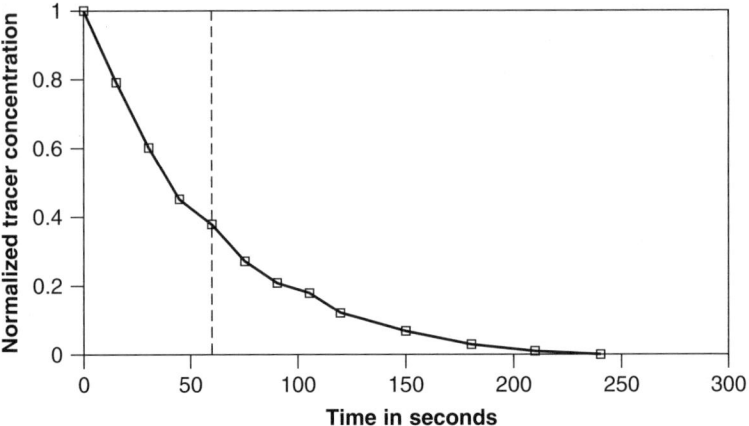

**Figure 1-5**  Experimental RTD data for Example 1-1.

a tracer washout experiment and obtain the results shown in Figure 1-5. The mean residence time is determined by integrating under the experimental washout curve and gives $\bar{t} = 59$ s. This is much less than the calculated value of 1.25 hr. You arrange for the reactor to be opened and find that it is partially filled with cross-linked polymer. When this is removed, the conversion increases to 74%: $a_{out}/a_{in} = 0.26$. A new residence time experiment gives $\bar{t} = 1.25$ hr as expected, and shows that the washout curve closely matches that for two stirred tanks in series:

$$f(t) = \frac{4t \exp(-2t/\bar{t})}{\bar{t}}$$

Now eqs. (1-7) and (1-8) can be used to calculate more precise limits on reactor performance. The results are $a_{out}/a_{in} = 0.290$ for complete segregation and $a_{out}/a_{in} = 0.287$ for maximum mixedness. Thus, as is typical of most industrial reactions, the extremes of micromixing provide tight limits on conversion. Since the actual result is outside these limits, something else is wrong. Quite likely it is the measured temperature that now seems too low.

### 1-4.4  Use with Computational Fluid Dynamic Calculations

Although they are increasingly popular, computational fluid dynamic (CFD) calculations are notoriously difficult to validate: Model equations may be available to the user, but the source code is typically proprietary, experimental data for comparison may be impossible to obtain, and the sheer volume of data available from the simulations makes complete and meaningful validations extremely difficult. Velocity measurements are difficult. Pressure drop measurements are easy but insensitive to the details of the flow. The RTD is a more sensitive test, but it is not unique since the RTD is derived from a flow-averaged velocity profile rather than the spatially resolved velocities that are predicted by CFD. Further, an experimental RTD will include effects of eddy or molecular diffusion that are not reliability captured by current CFD codes. Most CFD codes use convergence acceleration techniques that cause numerical diffusion that is an artifact of the computation. Numerical diffusion mimics molecular or eddy diffusion, although to an indeterminate extent.

Modern CFD codes are used routinely to calculate residence time distributions in complex flow systems such as static mixers. Care must be taken to sample according to flow rate rather than spatial position, and the number of particles must be surprisingly large for accurate results, particularly for the chaotic flow fields found in motionless mixers. The simulation of the recycle curve in Figure 1-2$b$ used $2^{18}$ tracer particles. The tail of the washout functions provides a demanding test for freedom from numerical diffusion. In the complete absence of diffusion, residence time distributions in laminar flow have slowly decreasing tails that give infinite variances. Specifically, they have algebraic tails for which $W(t)$ decreases as $t^{-2}$ so that all moments higher than the first diverge. Diffusion will cause the distributions to have rapidly decreasing exponential tails. The conclusion is that improvements in CFD codes and still faster computers are needed for accurate design calculations in complex geometries. Residence time calculations will be a useful tool for their validation. The situation becomes even more difficult when the equations of motion are combined with convective diffusion equations to estimate reactions yields and heat transfer. We anticipate significant near-term improvements in CFD codes, but they are now at the cutting edge of technology and have not yet become everyday tools for the practicing engineer.

## 1-5  EXTENSIONS OF RESIDENCE TIME THEORY

Residence time measurements are easiest in single-phase systems having one inlet and one outlet, but extensions to more complex cases are discussed in the General References. The RTD can be measured by component on an overall basis. Individual RTD's per inlet, per outlet, and per phase can also be measured. Most of the concepts discussed in this chapter can be applied to unsteady-state systems. The material leaving the systems at any time will have a time-dependent distribution of residence time. Analytical and numerical solutions are possible for a variable-volume CSTR, allowing calculation of time-dependent RTDs and reaction yields in a system subject to fluctuations in flow rate. For isothermal, solid-catalyzed reactions, the contact time distribution is the analog of the residence time distribution. It can be measured using adsorbable tracers. The results can be used to predict reaction yields or the upper and lower bounds of reaction yields. The thermal time distribution applies to nonisothermal homogeneous systems. It is a conceptual tool useful for optimizing the performance of nonisothermal tubular reactors and extruder reactors. Improved CFD codes will allow its calculation in static mixers and other complex geometries used for simultaneous heat transfer and reactor.

## NOMENCLATURE

### Roman Symbols

| | |
|---|---|
| $a$ | concentration of component A |
| $a_{batch}$ | concentration of component A in a batch reactor |
| $a_{in}$ | inlet reactant concentration |
| $a_{mix}$ | reactant concentration after the mixing point in a recycle reactor |
| $a_{out}$ | outlet reactant concentration |
| $C$ | concentration of inert tracer |
| $C_{in}$ | inlet tracer concentration |
| $C_{out}$ | outlet tracer concentration |
| $f$ | differential distribution function of residence times |
| $F$ | cumulative distribution function of residence times |
| $k$ | reaction rate constant |
| $q$ | internal flow rate or recycle flow rate |
| $Q$ | volumetric flow rate through the system |
| $R_A$ | reaction rate of component A |
| $t$ | residence time |
| $t_{first}$ | first appearance time |
| $\bar{t}$ | mean residence time |
| $V$ | volume |
| $W$ | residence time washout function |

### Greek Symbols

| | |
|---|---|
| $\lambda$ | residual life, the time variable in Zwietering's differential equation |
| $\mu_n$ | $n^{th}$ moment of the residence time distribution |
| $\sigma^2$ | dimensionless variance or residence times |
| $\omega$ | frequency of input disturbance |
| $\Theta$ | dummy variable of integration |

## REFERENCES

Baldyga, J., J. R. Bourne, and S. J. Hearn (1997). Interaction between chemical reactions and mixing on various scales, *Chem. Eng. Sci.*, **52**, 458–466.

Chauhan, S. P., J. P. Bell, and R. J. Adler (1972). On optimal mixing in continuous homogeneous reactors, *Chem. Eng. Sci.*, **27**, 585–591.

Danckwerts, P. V. (1953). Continuous flow systems: distribution of residence times, *Chem. Eng. Sci.*, **2**, 1–13.

Nauman, E. B., D. Kothari, and K. D. P. Nigam (2002). Static mixers to promote axial mixing, *Chem. Eng. Res. Des.*, **80**(A6), 681–685.

Zwietering, T. N. (1959). The degree of mixing in continuous flow systems, *Chem. Eng. Sci.*, **11**, 1–15.

In addition to the references above, the concepts introduced in this chapter are discussed at length in:

Nauman, E. B., and B. A. Buffham (1983). *Mixing in Continuous Flow Systems*, Wiley, New York.

Much of the material is also available in:

Nauman, E. B. (1981). Invited review: residence time distributions and micromixing, *Chem. Eng. Commun.*, **8**, 53.

Scale-up issues related to RTDs are discussed in:

Nauman, E. B. (2002). *Chemical Reactor Design, Optimization and Scaleup*, McGraw-Hill, New York.

# CHAPTER 2

# Turbulence in Mixing Applications

SUZANNE M. KRESTA
University of Alberta

ROBERT S. BRODKEY
Ohio State University

## 2-1 INTRODUCTION

Turbulence is central to much of liquid mixing technology and all of the typical processes (reaction, mass transfer, heat transfer, liquid–liquid dispersion, gas dispersion, solids suspension, and fluid blending) are dramatically affected by its presence. An understanding of the nature of turbulence is needed to deal with the interactions between turbulent fluctuations and mixing processes. Without an understanding of these basic physical phenomena, reliable predictions of performance can be difficult to achieve. Simple scale-up rules can be hopelessly inadequate. Unfortunately, the physics of turbulence still evades a general mechanistic description; and the flow in a stirred tank is complicated further by recirculation, strong geometric effects, and instabilities on several scales of motion. In this chapter we focus on providing a physical understanding of both turbulence and the tools that we use to understand its effects on process results.

The primary objective is to translate our current understanding of turbulence into an engineering context, providing the reader with a set of tools that can be used to solve practical mixing problems. In each section we begin with discussion of a central concept in turbulence and follow this with application of the idea to a practical problem, putting the concept into a practical context. Several facets of the turbulence problem are examined, in order to:

- Provide an engineering description of turbulence in terms of length and time scales.

---

*Handbook of Industrial Mixing: Science and Practice*, Edited by Edward L. Paul,
Victor A. Atiemo-Obeng, and Suzanne M. Kresta
ISBN 0-471-26919-0 Copyright © 2004 John Wiley & Sons, Inc.

- Illustrate the implications of these length and time scales for industrial mixing operations.
- Review the implications of isotropy and other approximate theoretical treatments.
- Consider the nature and implications of various experimental measures of the flow.
- Summarize the strengths and limitations of turbulence models and computational fluid dynamics (CFD) in general in the context of the design of mixing equipment.

In this chapter the topic of turbulence is broken down into four sections. First, in Section 2-2, the application of turbulence scaling principles to reactor design is discussed, to clarify for the reader the role played by the turbulent motions. In Section 2-3 we dig deeper into the description of turbulence, considering the various time and length scales involved in the description of turbulent flow, the scaling arguments that are used for engineering estimates, and how these estimates are related to the flow field. The information that is lost in the time averages and scaling arguments is revisited from the perspective of experimental and theoretical approximations of the flow in Section 2-4. Finally, the mathematical approach to the problem, the modeling of turbulence, is discussed in Section 2-5. The text is aimed at readers with no advanced training in fluid mechanics, and explanations of theoretical concepts are liberally interspersed with examples. Those with more experience will find summaries at the end of each section; they may also find a review of the more subtle concepts useful. Although the chapter can be read from beginning to end, it is also designed for independent reference to a specific subtopic. We begin by clarifying the initial definitions that we will need to discuss turbulence and the mixing operation.

## 2-2 BACKGROUND

### 2-2.1 Definitions

*These definitions are provided for the readers' reference. The case of B being mixed into a continuous A is used for the purpose of illustration.*

***2-2.1.1 Turbulence.*** An exact mechanistic definition of turbulence is limited by our understanding of its nature. Indeed, there can be no exact definition until we have exact understanding. However, as engineers, we need a working definition to ensure that we are all talking about the same thing.

We first look at the history of this moving target. The first historical phase was phenomenological theories where turbulence was defined by specific mechanistic concepts developed by researchers such as Prandtl. This led, for example, to the Prandtl mixing length. Taylor then suggested that statistical theory be applied to develop a more general view of turbulence. He proposed that the mechanism of turbulence is so complex that we cannot formulate a general model on which to base an analysis unless we restrict the meaning of turbulent motion to an irregular fluctuation about a mean value (Brodkey, 1967, pp. 260–261). Any motion that might have a regular periodicity (e.g., that from an impeller in a mixer) could not be considered as part of the turbulent motion. Within this context the eddy cascade picture of turbulence and time-averaged models like the k–ε model emerged. Frustration with this view has lead to the current concept of *coherent structures* in turbulence.

The coherent structure approach to turbulence is diametrically opposite to the statistical approach. Such coherent structures (e.g., ejections, sweeps, hairpin vortices, etc.) are to be distinguished from large scale organized motions that are forced upon the system externally. Today, coherent structures concepts are being extended to incorporate periodic structures generated by forcing or by geometry in the system. These forced structures can be generated, they evolve, and they interact with the natural turbulent coherent structures. In this context, coherent, regular structures are a feature of many turbulent flows, including mixing layers, and shed vortices can be included as part of our definition. To capture these structures, however, we are forced to adopt a more direct approach to the modeling, such as direct numerical simulation or large eddy simulation. Praturi and Brodkey (1978) offered the following commentary: "A mechanistic picture of turbulence cannot be treated on the average since such flows are dynamic. Many models can satisfy a long time-average picture. Emerging from this approach is the conclusion that turbulence can only be described as an evolving dynamic system. Reliable mechanistic models of turbulent shear flows that will enable reasonable predictions to be made should then be possible."

From the modeling perspective all flows, whether laminar, transitional, or turbulent, can be fully described by the Navier–Stokes equations with or without time dependency and with the restrictions of imposed geometry and appropriate boundary conditions. This suggests that there is no mechanistic difference between the flow regimes. Turbulent flow is simply a very complicated manifestation of the same physics that drives laminar flow.

Throughout this book, strong distinctions are made between laminar and turbulent mixing. The operation and design of mixing equipment in these two flow regimes are, in fact, quite different. Why? The flow for a given geometry and set of boundary conditions is a continuous development from very low Reynolds numbers (laminar operation) to very high ones (fully turbulent operation). At a low Reynolds number (Re), viscosity dominates, infinitesimal disturbances are damped out, and we have laminar flow. At a very high Re, inertial forces dominate, changes in viscosity have no effect on process results, and infinitesimal disturbances grow into a myriad of complex interacting structures so complex that we call it turbulence. With this complexity of interactions comes rapid dispersion and mixing. Somewhere between the extremes is a transitional region where both inertial and viscous forces play a role. Although our understanding of laminar mixing is imperfect and our understanding of turbulent mixing limited, our understanding of transitional flow and mixing is restricted to the simplest of cases.

**22**  TURBULENCE IN MIXING APPLICATIONS

Within this context, we offer the following working definitions of turbulence:

- *Turbulence* is a state of fluid motion where the velocity fluctuates in time and in all three directions in space. These fluctuations reflect the complex layering and interactions of large and small structural elements, such as vortices, sheets, ejections, and sweeps of a variety of shapes and sizes. In turbulent flows, scalar fields are rapidly dispersed compared to their laminar counterparts. At the time of writing, there is no completely acceptable way to model complex turbulent flow.
- *Fully turbulent flow* is an asymptotic state at very large Reynolds numbers. In fully turbulent flow, the velocity fluctuations are so intense that inertial forces overwhelm viscous forces. At all but the smallest scales of motion, viscous forces (and molecular diffusivity) become negligible. In fully turbulent flow, drag coefficients (e.g., friction factors and power numbers) and dimensionless blend times approach constant values. As is also the case in laminar flow, velocity profiles scale exactly with characteristic length and velocity scales. These conditions allow significant simplifications in modeling and design.

### 2-2.1.2  Mixing Mechanisms

- *Dispersion* or *diffusion* is the act of spreading out (B is dispersed in A).
- *Molecular diffusion* is diffusion caused by relative molecular motion and is characterized by the molecular diffusivity $D_{AB}$.
- *Eddy diffusion* or *turbulent diffusion* is dispersion in turbulent flows caused by the motions of large groups of molecules called *eddies*; this motion is measured as the turbulent velocity fluctuations. The turbulent diffusivity, $D_t$, is a conceptual analogy to $D_{AB}$ but is a property of the local flow rather than of the fluid.
- *Convection* (sometimes called *bulk diffusion*) is dispersion caused by bulk motion.
- *Taylor dispersion* is a special case of convection, where the dispersion is caused by a mean velocity gradient. It is most often referred to in the case of laminar pipe flow, where axial dispersion arises due to the parabolic velocity gradient in the pipe.

### 2-2.1.3  Measures of Mixedness

- *Scale of segregation* is a measure of the large scale breakup process (bulk and eddy diffusivity) without the action of diffusion, shown in Figure 2-1a. It is the size of the packets of B that can be distinguished from the surrounding fluid A.
- *Intensity of segregation* is a measure of the difference in concentration between the purest concentration of B and the purest concentration of A in

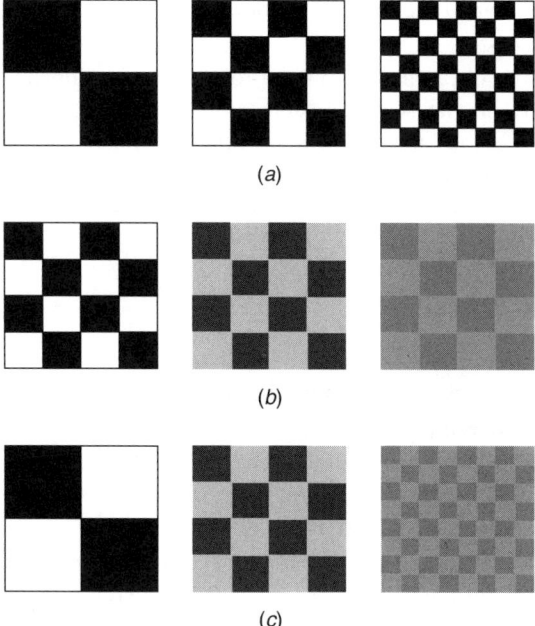

**Figure 2-1** Intensity and scale of segregation: (*a*) reduction in scale of segregation; (*b*) reduction in intensity of segregation; (*c*) simultaneous reduction of intensity and scale of segregation.

the surrounding fluid[1] shown in Figure 2-1*b*. Molecular diffusion is needed to reduce the intensity of segregation, as even the smallest turbulent eddies have a very large diameter relative to the size of a molecule.

A reduction in intensity of segregation can occur with or without turbulence; however, turbulence can help speed the process by reducing the scale of segregation, thus allowing more interfacial area for molecular diffusion. The scale of segregation is typically reduced by eddy motion while molecular diffusion simultaneously reduces the intensity of segregation, as shown in Figure 2-1*c*. When diffusion has reduced the intensity of segregation to zero, the system is considered completely mixed.[2] Two examples illustrate the importance of the scale of segregation:

1. In a jet injection reactor with liquid or gaseous feeds and a solid product, the solid is formed at the interface between A and B. The final particle

---

[1] The term *intensity of segregation* is also used by Danckwerts as a measure of the age of a fluid at a point (i.e., the backmixing or residence time distribution problem).

[2] A more careful consideration of the completely mixed condition would have to consider the scale of the probe volume relative to the scale of the molecules, or the largest acceptable striation in the fluid.

size is a strong function of the rate of reduction of segregation in the reaction zone.
2. In the mixing of pigment into paint for automotive finishes, the color quality depends on the scale of segregation of the pigment. If the scale of segregation is too large, the color is uneven, but if the scale and intensity of segregation are too small, the color loses its brightness and becomes muddy. This result is perhaps surprising; it is due to the reduced ability of an individual pigment particle to scatter light.

### *2-2.1.4 Scales of Mixing*

- *Macromixing* is mixing driven by the largest scales of motion in the fluid. Macromixing is characterized by the blend time in a batch system.
- *Mesomixing* is mixing on a scale smaller than the bulk circulation (or the tank diameter) but larger than the micromixing scales, where molecular and viscous diffusion become important. Mesomixing is most frequently evident at the feed pipe scale of semibatch reactors.
- *Micromixing* is mixing on the smallest scales of motion (the *Kolmogorov scale*) and at the final scales of molecular diffusivity (the *Batchelor scale*). Micromixing is the limiting step in the progress of fast reactions, because micromixing dramatically accelerates the rate of production of interfacial area available for diffusion.[3] This is the easiest way to speed up contact at the molecular level, since the molecular diffusivity is more or less fixed.[4]

We now proceed to our first exploration of turbulence in mixing applications: an evaluation of the time and length scales that are important for reactor design.

### 2-2.2 Length and Time Scales in the Context of Turbulent Mixing

For reactor design we would like to know how molecular diffusion and turbulent motions interact to bring molecules together. Turbulence can be used to break up fluid elements, reducing the scale of segregation. Energy is required for the generation of new surface area; so the limiting scale of segregation is associated with the smallest energy-containing eddies. These eddies are several times larger than the Kolmogorov scale,[5] $\eta$, and even the smallest scales of turbulence are much larger than a single molecule. As a result, even the smallest eddies will contain pockets of pure components A and B. Depending on the scale of observation, the fluid may appear well mixed; however, reaction requires submicroscopic

---

[3] The rate of diffusion is most frequently expressed as $k_{oL}a$. $k_{oL}$ is essentially determined by physical properties of the fluids; $a$ is increased by micromixing.

[4] For most liquids and gases the viscosity is greater than or equal to the molecular diffusivity ($Sc = \nu/D_{AB} \geq 1$), so it is easier to spread motion than molecules. Molten metals are a notable exception to this rule.

[5] See Section 2-3.

homogeneity, where molecules are uniformly distributed over the field. Molecules must be in contact to react. Turbulence alone cannot provide this degree of mixing. Molecular diffusion will always play an important role. Molecular diffusion, however, is very slow,[6] so the mixing process is critically dependent on both bulk mixing and turbulent diffusion to reduce the scales over which molecular diffusion must act. *To accomplish chemical reactions, we need the initial bulk mixing, efficient turbulence, and molecular diffusion for the final molecular contact.* Example 2-1 illustrates the impact of turbulence and molecular diffusion on mixing and reaction using a simplified physical model.

When mixing involves a chemical reaction, there are added complexities that depend on how the reactants are introduced into the mixing system. When a single stream is introduced and mixing occurs between fresh elements and older elements of the fluid, the mixing occurs in time and is called *self-mixing* or *backmixing*. When two streams enter a reactor and mixing occurs between the streams, two cases must be considered. If the reactants are all in one stream (*premixed* or *initially together*), the second stream acts as a diluent. With no mixing between the streams, the reaction proceeds as given by the kinetics. If, however, mixing dilutes the reactant concentrations and the order of the reaction is greater than 1, the dilution will depress the reaction rate. If the reactants are in separate streams (*unmixed* or *initially segregated*), molecular diffusion must take place for reaction to occur. In the last case, turbulence, molecular diffusion, and kinetics all interact to establish the course of the reaction. This is the critical turbulent mixing and kinetics problem that has received so much attention in the literature.

For chemical reactions in a known mixing field, the critical time scale depends on the relative rates of mixing and reaction. The limits of fast, slow, and intermediate reaction rates determine the relative importance of mixing and kinetics, as shown in Figure 2-2. Fast chemical reactions proceed as quickly as turbulence and molecular diffusion can bring the components together. The mixing rate dominates. Slow chemical reactions proceed much more slowly than any of the mixing time scales and are governed solely by reaction kinetics. For the important group of intermediate reaction rates, the reaction, diffusion, and mixing rates interact, and modeling is required. Example 2-2 illustrates these limits.

***Example 2-1a: Y-Tube—Identifying the Role of Various Mixing Mechanisms.***
A mixing Y-tube configuration (Figure 2-3a) was suggested for mixing two gaseous streams. The original reaction was studied in a $\frac{1}{4}$ in. bench scale reactor. The experimental results showed excellent selectivity and excellent conversion. The final plant design was to be a 12 in. tubular reactor, requiring a 48 : 1 scale-up. Because of the large scale-up factor, it was decided that a pilot scale would be tested. For this, a 2 in. reactor was designed and built (Figure 2-3b), which amounted to an 8 : 1 scale-up.

A series of experiments in the pilot unit revealed that the pilot reactor performed poorly both in selectivity and in overall conversion. The flow rates were

---
[6] This applies for $Sc \geq 1$. See note 4.

**26** TURBULENCE IN MIXING APPLICATIONS

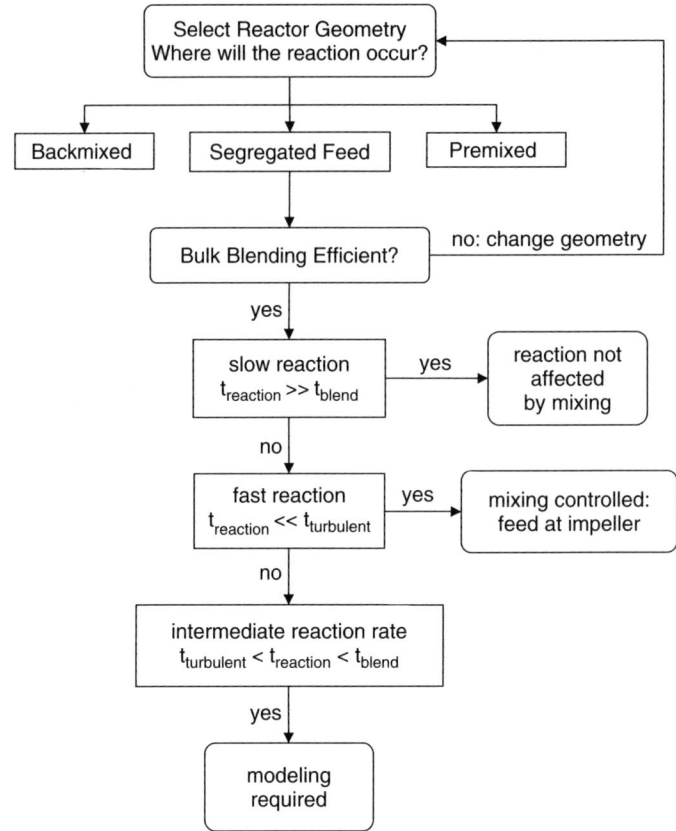

**Figure 2-2** Relationship between mixing and reaction time scales for equipment design.

such that the flow was turbulent. Recognizing the need for turbulence in a turbulent mixing system, the experimenter modified the pilot plant unit by adding screens near the entry to promote mixing via turbulent generation, as shown in Figure 2-3c. Contrary to expectations, the conversion and selectivity were further reduced by the turbulence-promoting screens. We now revisit the problem, considering not just the *amount* of turbulence, but also the *spectrum* of turbulence length scales.

Chemical reaction carried out in combination with mixing of the reactants has, as a first requirement, the large scale bulk dispersion of A into B. Only after this occurs can the finer scale mixing and molecular diffusion occur at a reasonable rate. In the bench scale reactor, the two incoming streams interacted vigorously and a grossly uniform mixture was obtained (Figure 2-3a). In sharp contrast, the results for the 2 in. reactor (Figure 2-3b) showed material segregation. The reaction occurred only on the interacting surface between the two streams. When the turbulence generation screens were added (Figure 2-3c), the

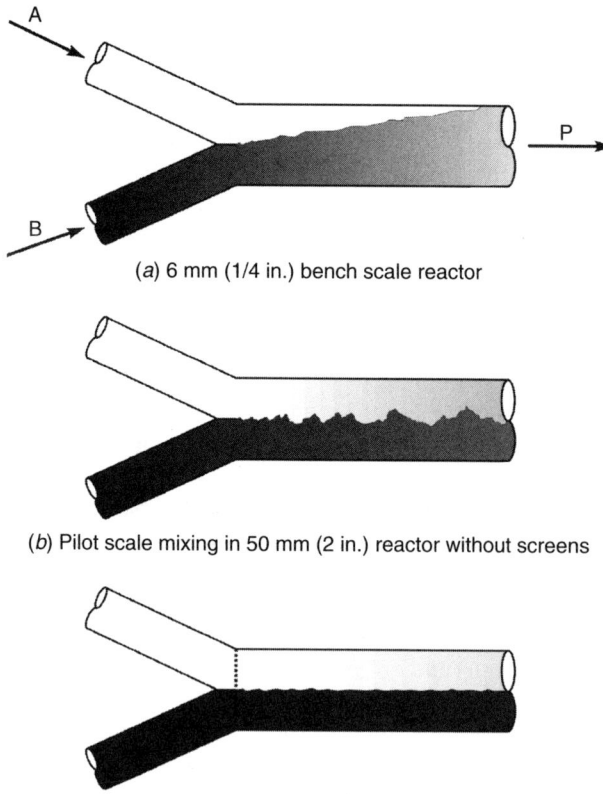

(a) 6 mm (1/4 in.) bench scale reactor

(b) Pilot scale mixing in 50 mm (2 in.) reactor without screens

(c) Pilot plant mixing in 50 mm (2 in.) reactor with screens

**Figure 2-3** Mixing and reaction carried out in a Y-tube.

screens eliminated the large scale interactions, providing a less contorted surface for reaction and thus further reduction of conversion. As a consequence of the screens, the pilot scale reactor provided extremely well-mixed material A and extremely well-mixed material B, but failed to bring A and B into contact. The Y-tube configuration has poor bulk mixing and thus fails the first test in Figure 2-2. An alternative geometry, such as a static mixer, T-junction, or stirred tank must be used for large scale dispersion (see, e.g., Monclova and Forney, 1995, or Wei and Garside, 1997). The key criterion for success in bulk mixing in a pipe is that the largest scale of segregation in the feed must be smaller than the largest scale of motion in the mixing geometry (Hansen et al., 2000).

This simple geometry provides us with an opportunity to explore the importance of various mixing mechanisms. We have already seen the disastrous impact of poor bulk mixing. Now, taking the length scale of the pipe, the problem is broken down into simplified models of pure molecular diffusion and pure eddy dispersion to clarify the interactions between these mixing mechanisms.

## 28  TURBULENCE IN MIXING APPLICATIONS

***Example 2-1b: Y-Tube—Limiting Case with No Eddy or Bulk Diffusion.*** What time would be required to achieve 99% mixing in the Y-tube if the only active mechanism was molecular diffusion? In this case there is poor bulk mixing between the layers, as observed above, and there is no turbulent enhancement of any mixing that does occur. Since we seek only the relative time scales, we reduce the problem to the case of plug flow with no turbulent fluctuations. This is clearly nonphysical,[7] but it will isolate the scales of mixing due to pure molecular diffusion. Since the materials are gaseous, we will assume equal-molal counter diffusion with equal diffusivities of the two components.

The diffusion is one dimensional in the x-direction; thus the rate of diffusion of A is written

$$\frac{\partial C_A}{\partial t} = D_{AB} \frac{\partial^2 C_A}{\partial x^2} \tag{2-1}$$

where $C_A$ is the concentration of species A. This is a common problem in chemical engineering, and solutions can be found in Brodkey and Hershey (1988). Probably the easiest approach for our purposes is to use the generalized chart solutions based on the original work by Gurney and Lurie (1923) and further improved upon by Heisler (1947). For our problem, each component must diffuse across a half width (from the centerline to the wall). To create the conditions required for 99% conversion of the reactants, we can take the "unaccomplished change" as 0.01, the centerline or half-width position (n = 1), and no resistance to transfer at the interface (m = 0). From the chart in Brodkey and Hershey (1988, pp. 670–672), the dimensionless time for these conditions is

$$\tau_D = \frac{D_{AB} t}{L^2} = 2.0 \tag{2-2}$$

We have been told that the diffusivity of our gases is about 25% greater than that of $CO_2$ in air; thus, we use $D_{AB} = 2 \times 10^{-5}$ m²/s. We use as the scale, L, the half-width of the system. For the $\frac{1}{4}$ in. bench unit, the diffusion time would be about 4 s. This would increase to a little over 1 min for the 2 in. diameter and to nearly 40 min for the 12 in. diameter commercial unit. At a Reynolds number of 2000, these times would correspond to pipe lengths of 5, 40, and 230 m, respectively. Clearly, mixing by pure molecular diffusion is an upper limit and would result in very long reactors.

If the mixing involves liquids rather than gases, the effect is more pronounced. The molecular diffusion for liquids would be very much lower (i.e., $D_{AB} = 1 \times 10^{-9}$ m²/s. Since t is inversely proportional to $D_{AB}$, the time required will increase by a factor of $2 \times 10^4$. The velocity in the liquid system will be lower

---

[7] If there are no fluctuations, the flow is laminar and the velocity profile is parabolic. The parabolic velocity profile will alter the diffusion characteristics, so this analysis is limited to a thought experiment. The superficial velocity from plug flow is needed to convert the required diffusion time to a distance down the reactor. To simplify the geometry without affecting the length scales significantly, we consider two dimensional plug flow between parallel plates, rather than the tubular geometry in the pipe.

**Table 2-1** Limiting Case of Equimolar Counterdiffusion

| Slab Thickness | 6 mm ($\frac{1}{4}$ in.) | 50 mm (2 in.) | 305 mm (12 in.) |
|---|---|---|---|
| *Air:* $D_{AB} = \mathbf{O}(2.00 \times 10^{-5})\,m^2/s,\ v = 1.50 \times 10^{-5}\,m/s^2,\ Re = 2000,\ Sc = \mathbf{O}(1)$ ||||
| Time to 99% diffused | 1 s | 1.1 min | 39 min |
| Velocity in plug flow | 4.7 m/s | 0.6 m/s | 0.1 m/s |
| Length of reactor | 4.8 m | 38 m | 230 m |
| *Water:* $D_{AB} = \mathbf{O}(1.00 \times 10^{-9})\,m^2/s,\ v = 1.00 \times 10^{-6}\,m/s^2,\ Re = 2000,\ Sc = \mathbf{O}(1000)$ ||||
| Time to 99% diffused | 5.6 hours | 15 days | 1.5 years |
| Velocity in plug flow | 0.3 m/s | 4 cm/s | 7 mm/s |
| Length of reactor | 6.3 km | 50.8 km | 305 km |

by a factor of 15, because the kinematic viscosity of water is 15 times lower than that for air at the same Re. This does not make up for the slower diffusion and would result in a bench scale unit over 6 km long! The results are summarized in Table 2-1.

***Example 2-1c: Y-Tube—Limiting Case with No Molecular Diffusion But Very Small Turbulent Eddies.*** Components A and B must come into contact on a molecular scale to react. If there is *no diffusion*, only a very thin monolayer of the product will form at the interface in Figure 2-3c. Once formed, it blocks any further reaction, since A and B cannot diffuse across the monolayer boundary. The familiar organic chemistry experiment where nylon is formed at the interface between sebacyl-chloride in tetrachloro-ethylene and an aqueous solution of hexamethylene diamine is a practical example of such a system. It is a simple geometric problem to obtain the interfacial area available per unit of length in our model reactor. Taking the molecular thickness to be ($d_m = 1$ Å) with a monolayer of reacted molecules at the interface gives a conversion of $8 \times 10^{-10}$ in the full scale reactor. This is an extremely small fraction of the molecules present!

Let us allow instead extremely effective turbulence, and suggest that the smallest scale of turbulence in our gas system is about ($l_t = 0.1$ mm). You might call this the smallest energy-containing eddy, if we knew what an eddy was. We could also assume that this "eddy" is spherical, and that the volume of A inside the eddies equals the volume of B outside the eddies. In our simplified physical model, we assume that the reduction in scale of segregation happens immediately on entering the pipe. It will quickly become clear that the length of pipe required to accomplish the reduction in scale is not the limiting factor for our "perfectly turbulent" nondiffusing reactor.

To calculate the maximum conversion in the perfectly turbulent reactor, we calculate the volume of product on the surface of the eddies. The volume of product is

$$V_{product} = \frac{\pi}{6}[l_t^3 - (l_t - d_m)^3] = 1.57 \times 10^{-18}\,m^3$$

and the remaining volume of A inside the eddy is

$$V_A = \frac{\pi}{6}(l_t - d_m)^3 = 5.24 \times 10^{-13} \text{ m}^3$$

which leaves a total remaining volume of reactants

$$V_{A+B} = 2V_A$$

Taking the ratio of volumes gives the conversion (assuming all molecules are approximately the same size)

$$\text{conversion} = \frac{V_{\text{product}}}{V_{A+B}} = 1.5 \times 10^{-6}$$

Diffusion and other physical properties are not a factor in this estimate of conversion. As long as the same Reynolds numbers and relative geometries are maintained, it will not matter if the system is a gas or liquid.

This conversion is clearly not good enough, so we decide to increase the turbulence and decrease the length scale, $l_t$, by a factor of 10 ($l_t = 10$ μm). This improves the conversion by a factor of 10, to $1.5 \times 10^{-5}$. The increase in power consumption, however, increases with $l_t^4$, so the power requirement per unit mass jumps from 33.75 W/kg to 0.3 MW/kg if our fluid is a gas.[8] If our fluid is a liquid, things are somewhat better because of the lower viscosity. In this case the power consumption jumps from 0.01 W/kg to 100 W/kg. While pure molecular diffusion was too slow, pure reduction of length scales gives disastrously low conversions, regardless of the length of the reactor.

Before moving on, we need to remind ourselves that we have used a simplified model of the physics. Our estimate ignores the fact that the packing of the molecules in the volume and on the surface may be different. One expects that this would be a small factor and not important in order-of-magnitude estimates. It also assumes an eddy diameter that is very small and is near the lower limit of turbulence, the Kolmogorov scale. Another possible estimate could use an *average* eddy diameter. On the other hand, the dynamic nature of turbulence will provide mixing *between* eddies, which can increase the effective surface area by several orders of magnitude. Despite these approximations, this estimate shows that very little reaction will occur without molecular diffusion. Now we consider the case where initial bulk mixing, efficient turbulence, and molecular diffusion for the final molecular contact are all present.

### *Example 2-1d: Y-Tube—Molecular Diffusion with Very Small Static Eddies.*
This time, let us assume that diffusion can occur in the very small eddies formed in Example 2-1c. Let us again assume that the smallest scale of turbulence is

---

[8] This calculation is based on η as discussed in Section 2-3, with the same fluid properties as those used in Example 2-1.

BACKGROUND 31

($l_t = 0.1$ mm) and that the eddy is spherical. The spherical surface containing $A + B$ becomes thicker as a result of molecular diffusion. We assume that all molecules within this thick surface will react.

For this example we combine the approaches used in Examples 2-1b and 2-1c, neglecting turbulent dispersion (see Section 2-3). Since the eddies are all assumed to be at their minimum size, all we need to determine is the time needed for the diffusion across an eddy radius ($l_t/2 = 0.05$ mm) for 99% diffusion. If the turbulence in the various test and commercial units does not change, the calculation will be the same for all cases, as it is based on a fixed eddy size, not on the system size. Of course, the total power will increase with the volume of the system. The only real difference from Example 2-1b is that we need to consider a sphere rather than a slab. The value of $D_{AB}t/L^2$ drops from 2.0 to 0.56 (see Brodkey and Hershey, 1988, p. 680), giving a diffusion time of

$$\tau = \frac{D_{AB}t}{L^2} = \frac{D_{AB}t}{(l_t/2)^2} = 0.56$$

$$t = \frac{(0.56)(1 \times 10^{-4} \text{ m}/2)^2}{2 \times 10^{-5} \text{ m}^2/\text{s}} = 7 \times 10^{-5} \text{ s}$$

on all scales of operation. This is, of course, a limiting estimate, which assumes that the same thing happens in all eddies at the same rate. For any practical gas reactor problem, this suggests that the combination of very efficient turbulence with molecular diffusion on the smallest scales will provide a very efficient reactor. Even if a more conservative eddy diameter of 1 mm is used, the time needed for the gaseous system is 0.001 s, still small enough for any practical reactor.

For a liquid system the time needed to reach the mixing conditions for 99% conversion is 1.4 s for the 0.1 mm diameter eddy and well over 2 min for an *average* eddy size of 1 mm. Although the reduction in scale due to the simple static model of turbulence has dramatically reduced the time needed to reach 99% diffusion, the time required is still long relative to the time scale of a fast reaction. Although this model contains dramatic simplifications of the physics for the purposes of a thought experiment, better models of the turbulence based on scaling arguments can be implemented successfully for simple geometries (Forney and Nafia, 2000). More realistic models of the turbulence are needed for complex reactor design, and these are discussed in later sections. Before moving to this discussion, we consider the impact of reaction kinetics on the problem, given good bulk blending.

### 2-2.2.1 *Interaction of Mixing Mechanisms: Summary of Example 2-1*

- In the case of segregated feed of reactants A and B to a reactor, the bulk mixing of the system needs to be addressed. The reactants need to be dispersed rapidly across the system and over a range of scales from the scale of the equipment to the point where individual molecules come into contact. Localized fine scale mixing of streams that remain segregated on the large scale contributes nothing to the overall mixing. (Example 2-1a)

- Diffusion alone, even in gas systems, is almost infinitely slow. (Example 2-1b)
- No diffusion results in essentially no reaction, even with a very dramatic reduction in the scale of segregation. (Example 2-1c)
- Adding molecular diffusion without eddy diffusion allows a crude estimate of the combined effects of (static) turbulence and molecular diffusion. The reduction in the time required for mixing on the molecular scale over previous cases is dramatic. For a gas system, this model is fast enough to reach practical limits. For liquids, the improvement is large, but not large enough to be realistic. (Example 2-1d)
- Although this example gives a dramatic illustration of the importance of all three mechanisms (bulk mixing, turbulent reduction of the scale of segregation, and molecular diffusion) to efficient mixing, a more realistic model of the turbulence is needed for accurate analysis.

### 2-2.3 Relative Rates of Mixing and Reaction: The Damkoehler Number

The outcome of a chemical reaction will depend on the rate of mixing compared to the rate of reaction. Figure 2-2 shows the interaction of the process and the key points to be considered. When the rate of reaction is slow compared to the mixing time, the reaction is not affected by mixing because the mixing is complete by the time significant reaction occurs (Example 2-2a). When the rate of reaction is fast compared to the rate of mixing, the kinetics are mixing limited, and the kinetics observed are effectively the mixing kinetics (Example 2-2b). Where the rate of reaction is similar to the rate of mixing, there will be strong interactions between the two rates (Example 2-2c). The relevant mixing time scales and reaction time scales are needed to determine the importance of mixing for a given reaction. In this chapter, only singular bimolecular reactions are considered. Also of considerable interest are bimolecular reactions that are either parallel-competitive (A + B → R, A + C → S) or series-competitive (A + B → R, B + R → S). These cases are discussed further in Chapter 13. It should be noted that what is said for the present single bimolecular case will apply equally well to the first reaction of the more complex cases.

*Example 2-2: Relative Rates of Mixing and Reaction.* To illustrate the role played by the turbulent scales across many different reactions, Toor (1969) and Mao and Toor (1971) obtained experimental conversion data in two different pipe flow reactors for a series of bimolecular reactions. Their results are combined with velocity measurements made in identical reactors by McKelvey et al. (1975). This allows us to compare various definitions of mixing time scales for pipe flow.

The defining number for this discussion is the Damkoehler number (Da), the ratio of mixing time to reaction time:

$$\text{Da} = \frac{\text{mixing time}}{\text{reaction time}} = \frac{\text{reaction rate}}{\text{scalar dissipation rate}} = k_r C_{B0} \frac{L_{1/2}}{u} = k_r C_{B0} t_{mixing} \quad (2\text{-}3)$$

In the case of a bimolecular reaction with the concentration of A in large excess, the term $k_r C_{B0}$ is the reciprocal time required for the fraction of B remaining to fall to one-half of the initial concentration. To obtain Da = 1 when the mixing time is just equal to the reaction time, Mao and Toor (1971) defined the mixing length as equal to $L_{1/2}$ and the mixing time as equal to $L_{1/2}/U$, where U is the superficial velocity in the pipe. Their $L_{1/2}$ must be determined from mixing studies or from the equivalent fast reaction measurements.

It would be more convenient to use a mixing time that is not geometry specific. A number of such times and Damkoehler numbers were compared by Brodkey and Kresta (1999) using various local turbulence scales in the Toor reactor. All of these times use local turbulence parameters or characteristic times. The position at which these are evaluated for the two multitube reactors is at the point of coalescence of the feed jets. It turns out that this is very close to Mao and Toor's (1971) characteristic half mixing length.

Independent of the turbulent time scale chosen, two distinct dividing points appeared in the Damkoehler number, allowing the identification of the two limiting cases of interest. Two of these measures are presented in Table 2-2 and Figure 2-4. The first measure is the microscale time, given by $t_\lambda = (\lambda^2/\varepsilon)^{1/3}$, and the second is the eddy dissipation time, given by $t_e = k/\varepsilon$. The eddy dissipation time (Spalding, 1971) is often used in reaction models (e.g., Forney and Nafia, 1998). Results such as these are a clear indication of the general value of these time constants, which in turn are based on turbulence scaling arguments. Turbulence scaling arguments are addressed in more detail in Section 2-3. For now, we accept these as given and focus on the three categories of reaction rate.

*(a) Slow reactions.* This is the case where the reaction time is much longer than the time needed to blend the reactants. There is plenty of time to complete the mixing before the reaction makes any significant progress. Vassilatos and Toor (1965) measured the progress of a slow reaction and were able to predict the results accurately by assuming a homogeneous concentration field and applying only the reaction kinetics. McKelvey et al. (1975) compared homogeneous calculations with calculations made using a known turbulent field. Their comparisons showed that the effect of turbulence was indeed negligible. Mao and Toor (1971) expressed the results in terms of a Damkoehler number based on the pipe diameter (Da = $k_r C_{B0} D_p/u$) such that for Da below 0.016, slow reaction conditions will apply. For all of the Damkoehler numbers defined in Table 2-2, the lower limits are on the order of 0.01. In this limit, turbulence is not important; the reactor is truly well mixed and a homogeneous kinetic calculation is sufficient.

**Table 2-2** Damkoehler Numbers Based on Different Mixing Times

| Mixing Time Scale | Mao and Toor | Taylor | Eddy Dissipation |
|---|---|---|---|
| *Definitions* | | | |
| Reaction rate (s$^{-1}$) | $k_r C_{B0}$ | $k_r C_{B0}$ | $k_r C_{B0}$ |
| Mixing time (s) | $t_M = L_{1/2}/u$ | $t_\lambda = (\lambda^2/\varepsilon)^{1/3}$ | $t_e = k/\varepsilon$ |
| Da | $Da_M = k_r C_{B0} \dfrac{L_{1/2}}{u}$ | $Da_\lambda = k_r C_{B0} t_\lambda$ | $Da_k = k_r C_{B0} t_e$ |
| *Limits Based on Experimental Results* | | | |
| Da$_u$ (fast reaction limit) | 100 | 30 | 150 |
| Da$_l$ (slow reaction limit) | 0.02 | 0.009 | 0.01 |
| *Experimental Results from Toor (1969) and Mao and Toor (1971), References Therein* | | | |
| Reaction    $k_r$ (L/mol·s) | Da$_M$ | Da$_\lambda$ | Da$_k$ |
| *Fast Reactions (Diffusion Controlled)* | | | |
| HCl–NaOH    $1.4 \times 10^{11}$ | $1.7 \times 10^7$ $-8.8 \times 10^7$ | $5.2 \times 10^6$ $-1.2 \times 10^7$ | $3.2 \times 10^7$ $-7.2 \times 10^7$ |
| Maleic acid–OH$^-$    $3 \times 10^8$ | $3.3 \times 10^4$ | $1.0 \times 10^4$ | $6.2 \times 10^4$ |
| Nitrilotriacetic acid–OH$^-$    $1.4 \times 10^7$ | $2.2 \times 10^3$ | $6.7 \times 10^2$ | $4.1 \times 10^3$ |
| *Intermediate Reactions* | | | |
| CO$_2$–2NaOH    $8.32 \times 10^3$ | 3.05–6.94 | 0.93 | 5.7 |
| CO$_2$–nNH$_3$    $5.85 \times 10^2$ | ~0.1 | 0.023 | 0.030 |
| *Slow Reaction (Kinetics Controlled)* | | | |
| HCOOCH$_3$–NaOH    $4.7 \times 10^1$ | ~0.01 | 0.0023 | 0.0030 |

*(b) Fast reactions.* When reactions are extremely fast, the time needed for a reaction to occur is much smaller than the time needed to blend the reactants. If two molecules can be brought together, they will react instantaneously. The controlling mechanism is the mixing due to both turbulence and diffusion. If the reactants are fed in stoichiometric balance, Toor (1962) has shown that the extent of the reaction is a direct measure of the mixing. The upper limit for Mao and Toor's (1971) definition of the Damkoehler number (Da $= k_r C_{B0} D_p/u$) is on the order of 100. For the definitions of Da in Table 2-2, the upper limit ranges from 30 to 150. Above this limit, fast reaction conditions apply. Acid-base reactions, which fall in this category, are often used as a means of measuring mixing times.

In addition to considering the limit of fast reactions, there is an effect of stoichiometry on the results, assuming that reactants are fed in stoichiometric ratio. Keeler et al. (1965) measured both mixing and fast reaction in the wake of a grid over a range of stoichiometric feed ratios. The effect of stoichiometry was also examined by Vassilatos and Toor (1965), who first assumed that mixing and very fast reaction results were equivalent for a stoichiometry of unity and then predicted the reaction at other stoichiometric ratios.

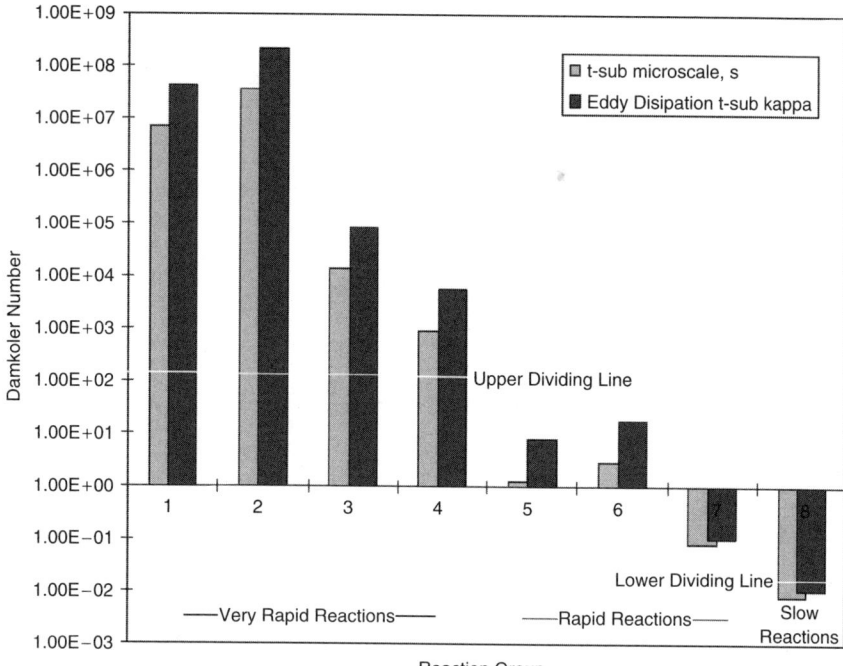

**Figure 2-4** Comparison of limits of Damkoehler number based on two different time scales.

(c) *Intermediate rates of reaction.* In the outer limits of very slow and very fast reactions, either mixing or kinetics becomes controlling and the other part of the physics can be ignored in the model. This simplifies the problem dramatically. In many real processing problems, however, both mixing and kinetics influence the course of reaction. In this example, the relatively simple case of a reaction in a pipe is used to illustrate our needs for the study of turbulent mixing.

Models reported in the literature have tended to focus on one of two parts of the problem. Simple reaction models (e.g., reacting slabs, random coalescence-dispersion, or multienvironment models) are designed to fit overall reaction data. More complex theoretical approaches require a model of the turbulence (see Section 2-5). The problem here is the adequacy of the turbulence model over a wide range of flow conditions. There still is no theory that takes into account structural aspects of turbulence with or without superimposed chemical reaction, although steady progress is being made in this direction [some current approaches are discussed by Fox (1998)].

As an example of what is needed, consider the models developed by McKelvey et al. (1975) using kinetic data from Toor (1969) and Mao and Toor (1971). McKelvey et al. (1975) measured the velocity field and mixing characteristics in exactly the same multinozzle pipe reactor that was used by Toor and co-workers to measure the kinetics. McKelvey et al. had two objectives. The first was to

## 36　TURBULENCE IN MIXING APPLICATIONS

establish the velocity and concentration fields in reactors used by Vassilatos and Toor (1965) to verify that turbulent mixing could be predicted from knowledge of the turbulent field. The second was to model the progress of a single second-order irreversible reaction where there is a significant impact of the dynamics of mixing on the observed reaction kinetics.

McKelvey et al. first consider the mass balance equation for a second-order reaction between species A and B [A + nB → (n + 1)P]. The equation for an individual species A in a differential control volume is

*accumulation of A + net bulk convection of A*

$= $ *net diffusion of A − disappearance of A due to reaction*

$$\frac{\partial C_A}{\partial t} + (\mathbf{U} \cdot \nabla)C_A = D_{AB}\nabla^2 C_A - k_r C_A C_B \tag{2-4}$$

$C_A$ and $C_B$ are the concentrations of species A and B. $D_{AB}$, the diffusion coefficient, and $k_r$, the reaction rate, are constant. The system is assumed incompressible and isothermal, and the scalar field has no effect on the velocity field (variations in concentration, for example, do not induce velocity gradients). Reynolds decomposition is used to separate the velocity and concentration fields into an average and fluctuating part. When these terms are substituted into eq. (2-4), the resulting equation can be averaged[9] to give

*accumulation of A + mean convection due to mean gradients of A*

*+ convection due to cross-fluctuations of velocity and concentration*

$= $ *bulk diffusion − reaction due to (mean field + fluctuating field)*

$$\frac{\partial \overline{C_A}}{\partial t} + (\overline{\mathbf{U}} \cdot \nabla)\overline{C_A} + (\nabla \cdot \overline{\mathbf{u}a}) = D_{AB}\nabla^2 \overline{C_A} - k_r(\overline{C_A}\,\overline{C_B} + \overline{ab}) \tag{2-5}$$

McKelvey et al. reduced this to a simplified form for the one dimensional experimental reactor

*mean convection in the x-direction = molecular diffusion in the x-direction*

*− reaction due to (mean field + fluctuating field)*

$$\overline{U_x}\frac{d\overline{C_A}}{dx} = D_{AB}\frac{d^2\overline{C_A}}{dx^2} - k_r(\overline{C_A}\,\overline{C_B} + \overline{ab}) \tag{2-6}$$

The term $\overline{ab}$ is the fluctuating component of the concentration field. It is related to the intensity of segregation and thus depends directly on the turbulent mixing field. The axial change in $C_A$ cannot be determined without $\overline{ab}$; however, if $\overline{ab}$ could be estimated, the equation could be numerically integrated. Toor

---

[9] See Brodkey and Hershey (1988, pp. 214–223) for a detailed presentation of the Reynolds averaging procedure.

(1969) showed that $\overline{ab}$ was the same for very slow reactions and very fast, stoichiometrically fed, second-order reactions. The former is dominated by kinetics, and the latter by mixing. With this as background, Toor hypothesized that "$\overline{ab}$ is independent of the speed of the reaction when the reactants are fed in stoichiometric proportion." He pointed out that this could not be exactly true for nonstoichiometric mixtures.

Based on this hypothesis, the measured intensity of segregation ($I_s$, discussed further in Section 2-3) was used for $\overline{ab}$ and the equation numerically integrated by McKelvey et al. (1975). They examined three of five experiments performed by Vassilatos and Toor (1965). The measured velocity and intensity of segregation were used for $\overline{U_x}$ and $I_s$ in the integration. The stoichiometric ratio varied from 1.0 to nearly 3.9. The agreement between the computations and the conversion experiments, shown in Figure 2-5, is remarkable, so for the simple second-order homogeneous reaction where both mixing and kinetics are of importance, the hypothesis of Toor clearly allows adequate predictions to be made. Mao and Toor's (1971) Damkoehler number should be between (0.02 < Da < 100) for the intermediate reaction conditions to apply.

For more complex reactions that are consecutive in nature, a fully adequate analysis is still lacking, but progress is being made. To adequately model the progress of a reaction, the terms equivalent to $\overline{ab}$ must be determined, and these depend in turn on the turbulent mixing. Where reactions depend on highly localized concentrations, the time average fluctuations, $\overline{ab}$, need to be modeled in

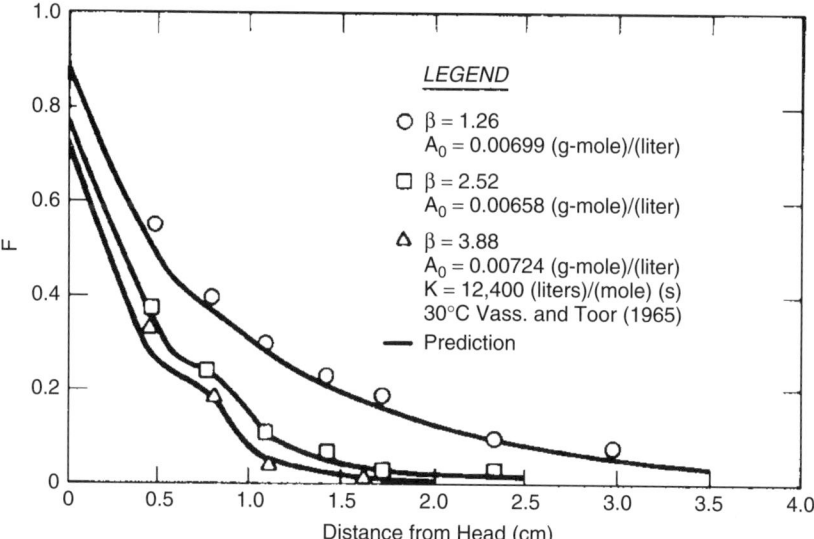

**Figure 2-5** Predicted conversion of an intermediate reaction where the turbulent mixing field is known. (From McKelvey et al., 1975.)

terms of probability density functions (PDF's) and other properties of the fluctuations. The simplified model used for the thought experiment in Example 2-1d is hopelessly inadequate because it neglects both the dynamics and the range of length scales present in a turbulent flow.

### 2-2.3.1 Importance of Turbulence in Modeling Reactions: Summary of Example 2-2

- In very fast reactions, the measured rate of reaction is wholly dependent on the rate of mixing. In this case, the mixing time is much longer than the reaction time. (Example 2-2b)
- In very slow reactions, the mixing time has no effect on the kinetics of reaction. The reaction takes much longer than the mixing. (Example 2-2a)
- Various Damkoehler numbers can be defined, preferably based on turbulence characteristics rather than on geometry-dependent variables. Regardless of the definition selected, there are definite limits of Da for the two limiting cases of fast and slow reactions. (Example 2-2)
- For intermediate reaction rates, modeling of the turbulence is essential, as the local concentration field is a function of the velocity field. The concentration field will change, and the velocity field may change, as the reaction proceeds. The fluctuating concentration field, in concert with the reaction kinetics, determines the progress of the reaction. (Example 2-2c)

Turbulent mixing covers a broad spectrum of applications beyond the field of reactions and reactor design, all of which are affected by the turbulence. Drop breakup, off-bottom solids suspension, gas dispersion, bulk blending, and heat transfer are all affected by the turbulent field. Without a better understanding of this part of the physics, it is difficult to make progress in these areas. With this broader objective clearly in mind, we now move forward to describe the key characteristics of turbulent flow.

## 2-3 CLASSICAL MEASURES OF TURBULENCE

In this section we review the classical approaches to the problem of turbulence and how they are applied in the field of mixing. For more detailed treatments, see Brodkey (1967, Chap. 14), Tennekes and Lumley (1972), Hinze (1975), Baldyga and Bourne (focusing on reactions in turbulent flows, 1999), Mathieu and Scott (2000), and Pope (2000). We begin with a description of turbulence, building up the model from the simple reduction of scale explored in Example 2-1c to something that is more realistic. This realistic picture is very difficult to model in its full complexity, so various ways of reducing the full physical complexity to a manageable scale of difficulty must be considered. First, the idea of the turbulence spectrum is discussed. This is a fingerprint of the scales of motion which are present in a flow. The turbulence spectrum provides us with a simplified image of the flow[10] and allows us to observe some general characteristics of turbulent transport. These simplifications, in turn, lead to some special cases of turbulence (e.g., homogeneous, isotropic, or locally isotropic) which underlie most turbulence modeling approaches used in computational fluid dynamics (CFD), and many experimental approaches as well. Finally, scaling arguments are developed and applied. The characteristic length and time scales that arise from scaling arguments depend on the physics outlined earlier in the chapter. The simplicity of the equations belies the challenges involved in successful scale-up and scale-down: our objective is to provide some physical understanding of scaling principles and some ground rules for their application. Now, we begin at the beginning with a physical model of turbulence.

## 2-3.1 Phenomenological Description of Turbulence

Development of an understanding of turbulence requires consideration of the details of turbulent motion. Much of our intuitive sense of fluid flow is based on what we can observe with the naked eye, and much of this intuitive sense can be applied to an understanding of turbulence, if we proceed with some care. We begin with the classical definition of simple shear flow, as shown in Figure 2-6. In this figure a Newtonian fluid is placed between two flat plates. The top plate moves with velocity $V_x$, requiring a force per unit area of plate surface (F/A) to maintain the motion. The force required is in proportion to the fluid viscosity,

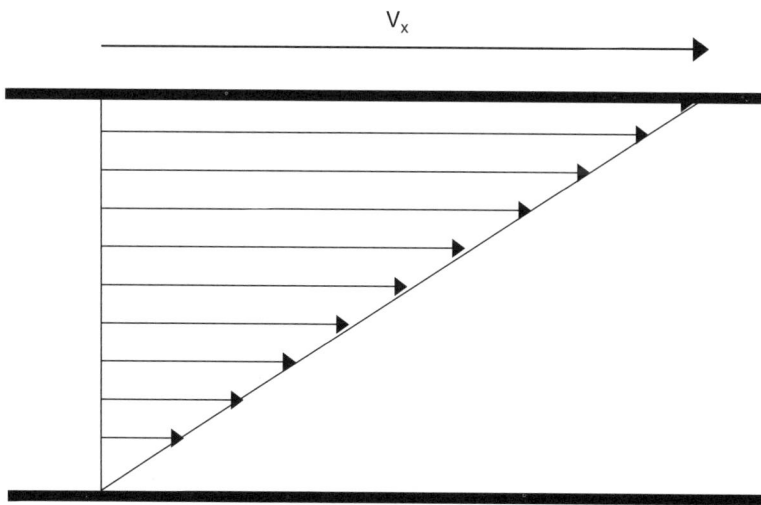

**Figure 2-6** Simple shear flow.

---

[10] See Brodkey (1967, pp. 273–278) for a full development of the meaning of the spectrum.

**40** TURBULENCE IN MIXING APPLICATIONS

returning Newton's law of viscosity:

$$\tau_{yx} = \frac{F}{A} = \mu \frac{\partial U}{\partial y} = \mu \frac{\Delta U}{\Delta y} \tag{2-7}$$

For the simple laminar shear flow between two flat plates, a probe placed anywhere in the flow will register a velocity that is constant in time (Figure 2-7). A probe placed in a stationary laminar recirculation zone will return the same result.

Now consider a stationary particle held in position by a stream flowing upward at just the terminal velocity of the particle, as shown in Figure 2-8. The fluid far away from the particle is in laminar flow, but in the wake of the particle,[11] eddies form. A two dimensional slice of the flow provides a picture that is similar, in our intuitive context, to the surface currents behind a rock in a flowing stream. The eddies are relatively stationary in space and are easy to observe. They are typically round (or elliptical) in cross-section and maintain their size, which invites our intuition to jump to the idea of a coherent[12] spherical (or ellipsoidal) eddy. We need to examine a general turbulent flow more carefully before making that assumption.

The trademark of a turbulent or transitional flow is that the velocity fluctuates in time,[13] as shown in Figure 2-7. These fluctuations occur as eddies change or

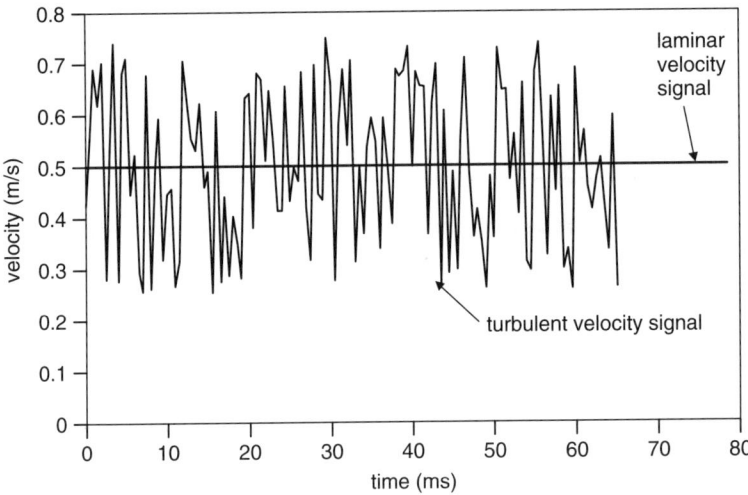

**Figure 2-7** Velocity as a function of time in laminar and turbulent flow.

---

[11] The wake is the downstream side of the particle, where the fluid flow is affected by the presence of the particle.

[12] A coherent structure in a flow field is one that maintains its shape but may evolve over time. If the structure grows, the velocity will decay as a requirement of the conservation of momentum. Stable coherent structures that maintain their shape, size, and velocity are often observed in lower Reynolds number flows.

[13] We defer the definition of transitional flow to Section 2-5.

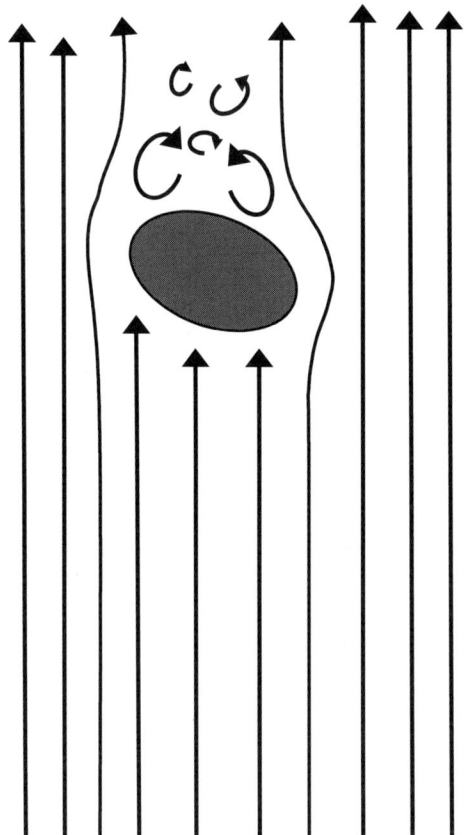

**Figure 2-8** Turbulence in the wake of a particle: two dimensional cross-section.

move past the probe. Returning to our observations in a flowing stream or in a pipe, a constant superficial velocity or local mean velocity can also be defined. At any point in the flow, the signal can be averaged to give a repeatable mean, although the details of the velocity signal fluctuations in any record are unique. We might represent this situation as a series of rotating simple shear flows of varying size, as shown in Figure 2-9, which are convected along with a velocity $U_c$. In the figure, a very limited size range is shown, and the velocity profiles are all linear. In a more realistic turbulent flow, a much broader range of eddy sizes is observed, and the velocity profiles take on various nonlinear shapes in response to the surrounding eddies. This image is left to the reader's imagination.

The idea of many miniature shear flows that are rotating in space and being convected across a probe seems like a useful one. This may allow us to make a link between the three flow regimes (laminar, transitional, and turbulent), but before adopting it, let's clarify the assumptions that underlie this model. First, we've assumed that the eddies do not change as they are convected along in the

**42**  TURBULENCE IN MIXING APPLICATIONS

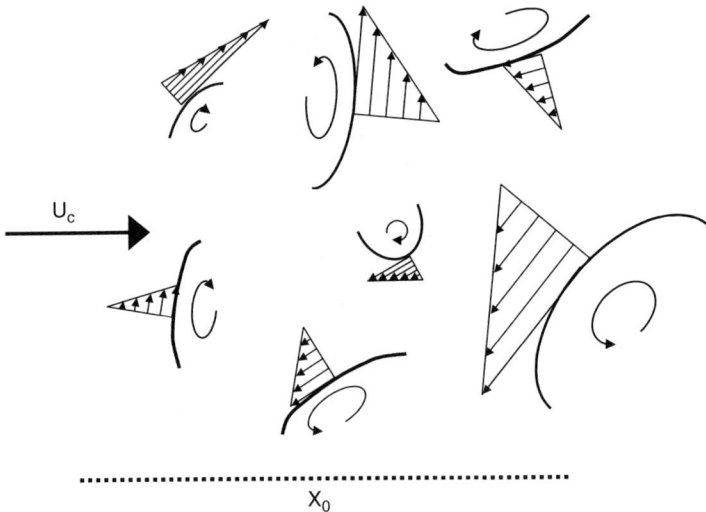

**Figure 2-9** Eddies of various sizes and velocities, each containing a component of simple shear. The convective velocity, $U_c$, and the integral length scale, $X_0$, are arbitrary.

flow. If this were true, we could take a set of signals collected at $X_0$ and back out an exact picture of the flow at an instant in time. We could then calculate instantaneous mean shear and deformation rates over the cross-section. This is known as *Taylor's hypothesis* (1921):

$$\frac{\partial u_i}{\partial x} = -\frac{1}{U_c}\frac{\partial u_i}{\partial t} \tag{2-8}$$

On average, Taylor's hypothesis will turn out to be quite useful, but it is a dangerous one for the development of our intuition. Taylor's hypothesis locks the turbulent eddies into two dimensional symmetrical shapes, which stay in the same place relative to each other as they rotate through space. To expand on our intuitive images and understand the dynamic three dimensional component of turbulence, we need to observe eddies in clouds on a windy day, or in stack plumes and car exhausts on a cold day. The eddies are highly three dimensional. If one eddy is observed as it is convected along at $U_c$, it rotates, changes shape, changes size, and exchanges material with the surrounding fluid as it moves downstream. Its life cycle is extremely dynamic. These characteristics are critical for turbulent mixing, as they allow much more rapid cutting, folding, and incorporation of new material than does Taylor's image of frozen turbulence.[14]

The critical characteristics in our discussion of turbulence so far are that it contains three dimensional eddies which have a wide range of sizes and shapes

---

[14] To be fair to G. I. Taylor, he clearly limited this hypothesis to very short sampling times, thus minimizing some of these problems. His hypothesis is often applied as the best available approximation over times that exceed the valid limits.

and which change dramatically over time. To complete the discussion, we need to investigate the three dimensional aspect of the problem a bit more closely.

Return to the particle suspended in upward flow, but now instead of a particle, consider a cylinder that is very long, placed perpendicular to $U_c$. A pragmatic example of this is a dip pipe or cross-flow heat exchanger. At low Reynolds numbers, the two dimensional wake of the cylinder will look the same in cross-section as the particle wake in Figure 2-8, but the three dimensional eddy is quite different from the deformed ellipsoid behind a particle. Now it is a long unstable tube behind a pipe. The tubular eddy has a diameter similar to that of the pipe, but it is very long in the third dimension. As the Reynolds number increases, the wake will shed eddies of various sizes, many of them long and skinny. If the velocity fluctuations are measured in the streamwise direction, the signal will be similar to that in the particle wake, but in the transverse direction, the signal will be affected by the long dimension of the eddies parallel to the pipe.

Another example of a three dimensional eddy arises in the boundary layer close to the wall. In this layer, horseshoe eddies, turbulent spots, and turbulent bursts separate from the wall and are swept into the bulk flow. Many of these eddies have highly distorted dimensions, as shown in the simple example in Figure 2-10. It is easy to imagine that for the single illustrated eddy, at least three distinct

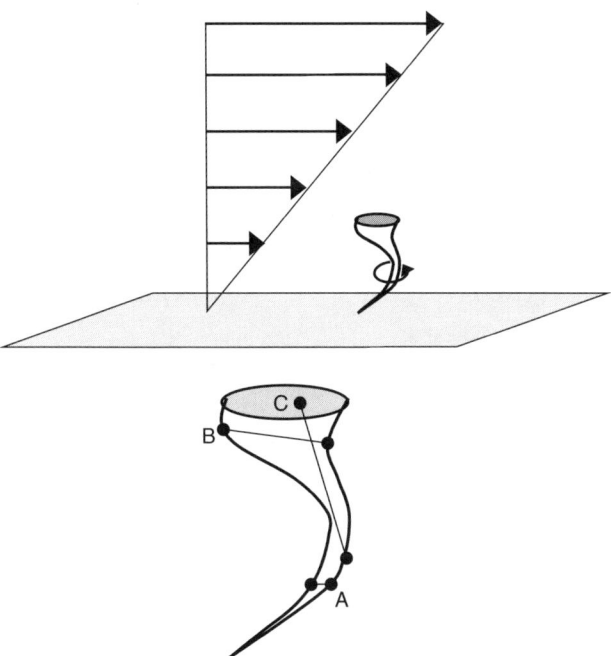

**Figure 2-10** Vortex in a boundary layer showing the different length scales (A, B, and C) or wavenumbers contained in a single extended eddy: the wavelength ($1/k = U_c/2\pi f$) is a single arbitrary dimension of a three dimensional time-varying structure.

length scales (also measured as frequencies, or wavenumbers) could be observed for different one dimensional slices through the flow at A, B, and C.

Next consider what will happen to a drop of immiscible fluid that is injected into a turbulent velocity field. At the beginning, the drop may be spherical or ellipsoidal, but it will quickly respond to the velocity field, and we assume that the fluid and surface properties are such that it may deform very quickly in response to the motion of surrounding eddies. When the drop is embedded in an "eddy" that is much larger than its own characteristic diameter, it will be convected along in the eddy with very little deformation (Figure 2-11a). However; when the drop encounters eddies close to its own size, it will be deformed due to the interactions between the drop and the eddies. It may be either torn apart by two co-rotating eddies (Figure 2-11b) or elongated as it is squeezed between two counterrotating eddies, (Figure 2-11c). Finally, when the drop encounters small eddies, packets of its volume are torn away to mix with the surrounding fluid (Figure 2-11d). If there is no molecular diffusion, the minimum drop sizes will be limited to the scale of the Kolmogorov eddies and the drop fragments generated on breakup. Some of the drop fragments can be much smaller than the Kolmogorov scale (Zhou and Kresta, 1998). The drop size distribution characterizes discrete drops of fluid and reaches an equilibrium distribution after some (long) time. If, on

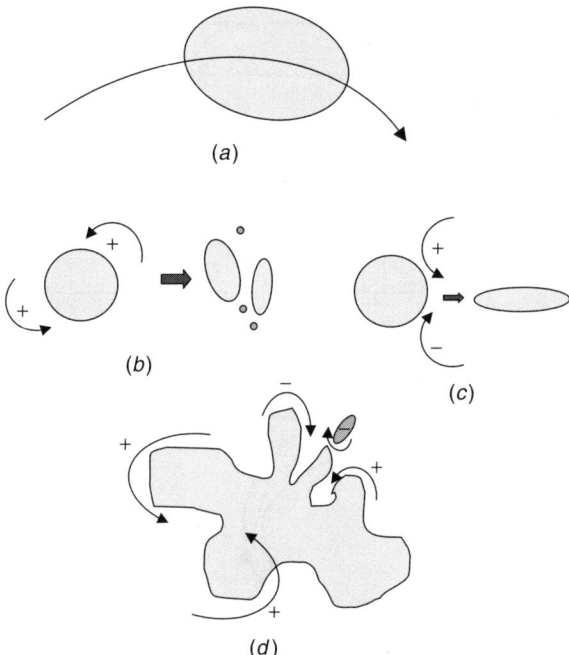

**Figure 2-11** Scalar deformation in a turbulent field with surface tension between the two phases: (a) convection by large eddies; (b) erosion by co-rotating eddies; (c) elongation by counterrotating eddies; (d) multiple scales of deformation.

the other hand, there is significant molecular diffusion and no surface tension between the two fluids, as is the case for blending of miscible liquids, the blob of scalar fluid will deform continuously without breaking and the edges of the blob will be smoothed out by molecular diffusion at the same time as turbulent eddies deform and break the blob. After a fairly short time, no discrete blobs will be observed. Both of these cases are illustrated in the drop breakup and blending videos on the Visual Mixing CD affixed to the back cover of the book.

In the same way that relative length scales of eddies and blobs affect the breakup of blobs, in multiphase flows the relative response times of particles and eddies determine how particles interact with eddies (Tang et al., 1992). Although we do not discuss this issue in detail, it is important to recognize two things: (1) the relevant length scales for multiphase flows can be much more difficult to scale accurately because of the complicated interactions between turbulence and particles; and (2) where tracer particles are used in experiments, the scales of motion that can be observed are a function of the particle size and characteristic response time.

### 2-3.1.1 Nature of Turbulence: Summary of Section 2-3.1

- Turbulence is a dynamic three dimensional multiscaled phenomenon.
- Eddies *are not spherical*, and turbulence length scales are not characteristic dimensions in the usual sense.
- Turbulence interacts with both scalars (dye, reactants) and dispersed phases (bubbles, drops, and particles) according to the relative length and time scales involved.
- The smaller the turbulent length scales, the finer the scale of micromixing, and the faster the rate of mixing at the smallest scales of motion.

If we can hold onto some sense of the three dimensional dynamic character of turbulent eddies, including the range of sizes that appear and how quickly the eddies change in time, we can start to extract the questions we must ask for applications of turbulence in mixing. In the next section we examine more formal ways of characterizing turbulence.

### 2-3.2 Turbulence Spectrum: Quantifying Length Scales

In Section 2-2, we used Figure 2-7 to illustrate the instantaneous velocity versus time signal and the mean velocity. A third velocity used widely for turbulent flows is the root-mean-square (RMS) velocity, or the standard deviation of the instantaneous velocity signal. Because the average fluctuation is zero by definition, the RMS velocity gives us an important measure of the amount or intensity of turbulence, but many different signals can return the same mean velocity and RMS fluctuating velocity,[15] so more information is needed to characterize the turbulence.

---

[15] The RMS velocity is exactly equivalent to the statistical measure known as the standard deviation.

If the velocity record is mean centered and transformed from the time domain to the frequency domain using a Fourier transform, we obtain the energy spectrum. The energy spectrum is a measure of the amount of energy present at each scale of motion. This allows us to take a fingerprint of the dominant frequencies in the flow in terms of their energy content (E), or power spectral density (PSD), as a function of wavenumber (k, in m$^{-1}$) or frequency (f, in Hz or s$^{-1}$). The spectrum gives the energy contained at each wavenumber, so the integral of the three dimensional spectrum returns the turbulent kinetic energy (2k or q), and the integral of the one dimensional spectrum (in the j-direction) over all wavenumbers returns the (j component of) RMS velocity, $u_j^2$.

The length scales of turbulence are contained in the measured frequencies, but the frequencies are a function of the mean velocity as well as of the rate of fluctuation. To obtain a more scalable picture of the length scales of turbulence, the measured frequencies are converted to wavenumbers using the mean convective velocity and Taylor's hypothesis:

$$k = \frac{2\pi f}{\overline{U_c}} \tag{2-9}$$

This application of Taylor's hypothesis is not quite the same as the faulty example given earlier, where we collected velocity versus time at $X_0$ and used it to back out the flow field at one instant in time. Now we are quantifying the time-averaged conditions in the flow, using Taylor's hypothesis to scale the spectrum with the mean convective velocity. The turbulence spectra, which develop at different mean velocities, can now be compared in terms of the wavenumbers (length scales) that are present in the turbulent part of the flow. We only have to assume that the eddies are coherent long enough to convect the largest length scale across $X_0$ in some *repeatable time-averaged* sense to justify using Taylor's hypothesis for scaling the spectrum.

To get a sense of the meaning of a wavenumber, consider a perfectly spherical eddy. As the eddy is convected past a probe at a constant velocity, it will give many different wavenumbers (frequencies), depending on where the sensor cuts through the sphere. The measured spectrum of lengths for a sphere will range from close to zero up to the diameter of the sphere. The three dimensional irregular dynamic multiscaled eddies present in fully turbulent flows will produce an analogous range of results. Wavenumbers are not physical lengths in the way we are used to thinking about them, but they will prove very useful as a means of scaling turbulence.

A typical spectrum for a stirred tank is shown in Figure 2-12b (Michelet, 1998). This spectrum is measured close to a Rushton turbine, where there are strong fluctuations in the velocity due to the moving blades and the trailing vortices, shown in Figure 2-12a (Yianneskis et al., 1987). The spectrum in Figure 2-12b is scaled with $f_p$, the blade passage frequency, so the blade passage frequency and its harmonics are evident as sharp peaks in the spectrum at 1 and at 2. The blade passages have a strong directional preference and are often

CLASSICAL MEASURES OF TURBULENCE 47

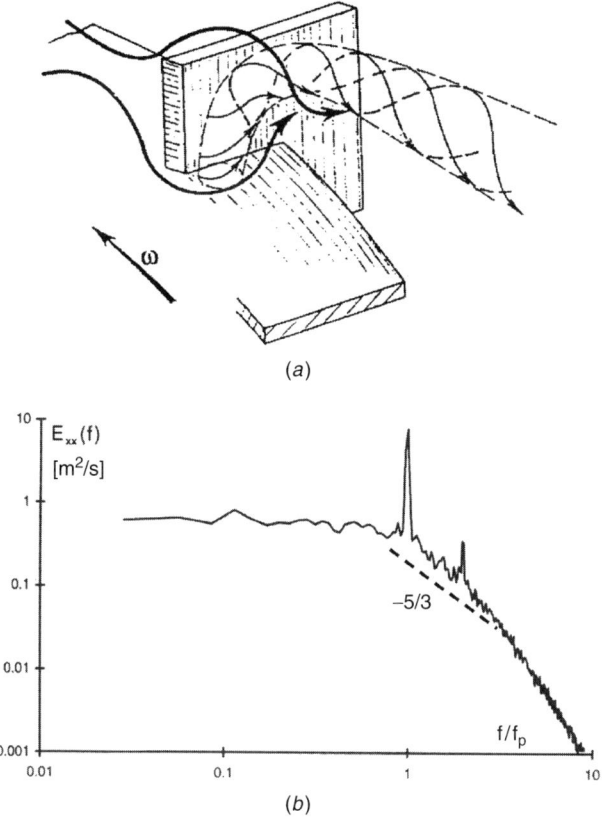

**Figure 2-12** Trailing vortex behind the blade of a Rushton turbine shown in (*a*) (Yianneskis et al., 1993) has a mirror image on the lower side of the blade. The frequency spectrum in (*b*) is normalized with the blade passage frequency and shows two peaks due to the blade passages. The trailing vortices are shown in motion on the Visual Mixing CD affixed to the back cover of the book.

removed from the signal before analysis of the turbulence. The reasons for this are discussed in detail in Example 2-5. In some configurations, an additional lower frequency is present at some fraction of the impeller speed, which must also be removed before the turbulence can be accurately quantified (Roussinova et al., 2000). Note that the range of *length scales* in the tank extends from T (the largest dimension) over at least three orders of magnitude (a factor of 1000) to $\eta$, and the measured *frequencies* in Figure 2-12*b* extend over a similar range. Using log scales allows us to cover the wide range of both power and wavenumbers.

The slope of the spectrum at frequencies higher than the blade passage frequency gives information about the distribution of energy across the turbulent scales of motion. If the energy distribution is in equilibrium, all the energy that enters the turbulent motion at large scales (i.e., in the form of low frequencies at the impeller) is dissipated at the same rate at the smallest scales of motion, where viscous dissipation is most effective. Where equilibrium exists, the slope in the equilibrium region must be $-\frac{5}{3}$ on a log-log scale.[16] It is evident from the figure that this criterion is satisfied only over a small range of frequencies close to the blades of a Rushton turbine.

The smallest scales of motion, or the smallest eddy dimensions, are characterized using the *Kolmogorov length scale*:

$$\eta = \left(\frac{v^3}{\varepsilon}\right)^{1/4} \quad (2\text{-}10)$$

At this length scale, the viscous forces in the eddy are approximately equal to the inertial forces due to turbulent velocity fluctuations. Somewhere close to this length scale, the dissipation of energy becomes rapid, and the slope of the spectrum increases dramatically, as shown in Figure 2-13. The Kolmogorov length, $\eta$, is a *defined* length equal to the inverse of the Kolmogorov wavenumber. It is used as a point of reference so that various conditions can be compared in a consistent way, but it is only one of a whole range of turbulent length scales which are present in any turbulent flow.

So what is the characteristic length scale of turbulence? This question is analogous to asking, "What is the diameter of an elephant?" We might say that the Kolmogorov scale is analogous to the diameter of the elephant's tail. Many

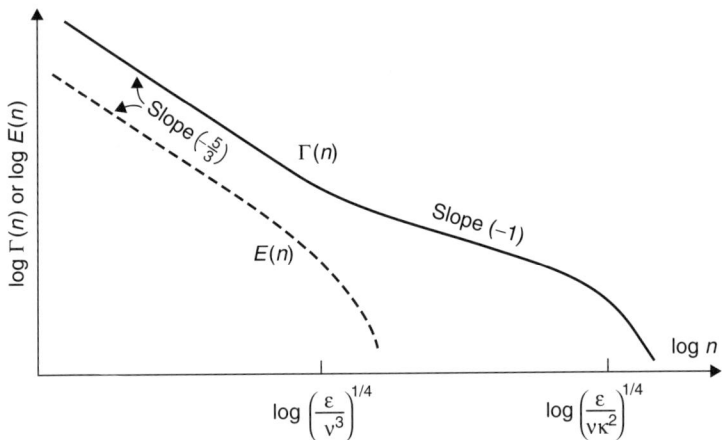

**Figure 2-13** Spectrum of velocity [E(n)] and temperature or concentration [Γ(n)] fluctuation wavenumbers (m$^{-1}$) in the equilibrium range of homogeneous isotropic turbulence for the case of large Sc or Pr (modified from Batchelor, 1959). In the Batchelor scale, κ is either the thermal diffusivity (k/ρC$_p$) or the molecular diffusivity (D$_{AB}$).

---

[16] Note, however, that the converse is not true: A $-\frac{5}{3}$ slope is not proof that equilibrium, or isotropy, exists.

other equally valid turbulent length scales have been defined, and we could say that they are analogous to the size of the elephant's trunk, ears, and legs. All of these length scales are related to each other and to the size of the elephant (the total energy in the spectrum) in a consistent and highly correlated way from one elephant to the next. Despite this, no single length scale can uniquely define the diameter of the elephant. In many cases of practical interest, the entire range of length scales is important to the process, and apparently small changes in the spectrum can make large changes in the process. An elephant without a tail is a perfectly good elephant, unless the elephant's main objective is to swat flies!

### 2-3.2.1 Spectral Arguments for Scalar Mixing and Mass Transfer.
Batchelor (1959) used scaling arguments to determine the size of a pure sphere of dye that will diffuse in exactly the time it takes the energy in an eddy of size $\eta$ to dissipate. This is called the *Batchelor scale:*

$$\lambda_B = \left(\frac{\nu D_{AB}^2}{\varepsilon}\right)^{1/4} \quad \text{and} \quad \frac{\eta}{\lambda_B} = Sc^{1/2} = \left(\frac{\nu}{D_{AB}}\right)^{1/2} \qquad (2\text{-}11)$$

This analysis is limited to cases where the molecular diffusivity is slow relative to the momentum diffusivity (kinematic viscosity), so that $Sc \geq 1$. In the same way that the Kolmogorov scale provides a limit where turbulent stresses are balanced by viscous stresses, the Batchelor scale provides a limiting length scale where the rate of molecular diffusion is equal to the rate of dissipation of turbulent kinetic energy. Below this scale, distinct packets of dye will quickly be absorbed into the bulk fluid by molecular diffusion, where our meaning of "quickly" is now consistent between the energy dissipation and molecular diffusion.

Figure 2-13 shows the gross characteristics of the velocity and concentration spectra. For a low viscosity liquid, Sc can be on the order of 1000, so the Batchelor scale can be 30 times smaller than the Kolmogorov scale. The ultimate scale of mixing needed for reaction is the size of a molecule, so in liquid-phase reactions, molecular diffusion is critically important for the final reduction in scale. For a gas, Sc is closer to 1, so the ratio is closer to 1, and the competition between the turbulent reduction in scale and molecular diffusion occurs at the same range of wavenumbers. The various length scales shown in Figure 2-13 are also summarized in Table 2-3

So far, our discussion of length scales has focused on the smallest scales of motion. At the larger scales of motion, Taylor (1921) considered the turbulent dispersion of fluid particles by homogeneous isotropic turbulence in the absence of molecular diffusion. In his model, each fluid particle leaving a point source in a uniform velocity field is expected to deviate from the linear mean path in a random manner, depending on the local nature of the turbulence. The RMS deviation of the particle paths is observed as a continued divergence, spread, or dispersion as the particles are carried downstream. *This eddy motion occurs even*

**Table 2-3** Useful Time and Length Scales Arising from Spectral Arguments

| Usual Notation | Name | Physical Meaning |
|---|---|---|
| $\eta = \left(\dfrac{v^3}{\varepsilon}\right)^{1/4}$ | Kolmogorov length scale | Eddy size at which the viscous forces are equal to the inertial forces. Viscous dissipation becomes important. Some authors place the dissipation limit at $5\eta$, where the viscous forces reach 20% of the inertial forces of turbulence. |
| $t_K = \left(\dfrac{v}{\varepsilon}\right)^{1/2}$ | Kolmogorov time scale | Time it takes to dissipate the energy contained in the smallest ($\eta$-sized) eddies. |
| $t_B = t_K \propto \dfrac{\eta^2}{D_{AB}}$ | Batchelor time scale | Time required for a pure scalar blob of A to diffuse into pure B if the blob diameter is $\eta$. |
| $\lambda_B = \left(\dfrac{D_{AB}^2 v}{\varepsilon}\right)^{1/4}$ | Batchelor length scale for mass transfer where Sc is large | Size of a pure scalar blob that will diffuse into pure surrounding fluid in exactly $t_K$. |

*in the absence of molecular diffusion.* The spread of the plume size, L, for large times (relative to the smallest scales of turbulence) can be approximated by

$$\frac{d(L^2)}{dt} = 2u^2 \tau_E \approx \frac{k^2}{\varepsilon} \quad (2\text{-}12)$$

where $\tau_E$ is the integral of the particle velocity autocorrelation function (constant for large integration times, but varying locally in the flow), and u, k, and $\varepsilon$ are local values that must be integrated over the path of the plume. The use of $k^2/\varepsilon$ in this context should be considered a scaling approximation. It assumes that the turbulence in the plume is at least locally isotropic and that it is uniform across the plume at any constant distance from the source. Inside the spreading plume, whether or not molecular diffusion plays an important role, the time-averaged concentration distribution will be Gaussian, making the reaction kinetics aspect of the problem more complicated.[17] This model is often the

---

[17] A colleague working in the area of pollutant dispersion notes an important weakness of the time-averaged approach: If the maximum concentration is a critical parameter, the average concentration is not a useful result. Take, for example, the dispersion of $H_2S$: If the local instantaneous concentration exceeds a toxic limit, people on the ground will die. "Alive on the average" is not a useful result, even if the average is very accurately determined! The necessary details of the distribution of concentrations can be extracted from statistical or PDF models which account for the time-varying characteristics of the concentration fluctuations.

best we can do with the available data, but it is clearly a simplified view of the physics.[18]

Taking a somewhat different approach, Corrsin (1957, 1964) considered the overall decay of concentration fluctuations, c(t), in a homogeneous turbulent field at high Re. In this case, the RMS fluctuations follow an exponential decay of the form

$$\overline{c^2(t)} = \overline{c^2(0)} \exp\left(\frac{-t}{\tau}\right) \qquad (2\text{-}13)$$

$\tau$ can be related to the physical properties of the fluid, $\nu$ and Sc, and the largest scales of concentration fluctuations, $L_s$, as well as the rate of dissipation of turbulent kinetic energy, $\varepsilon$. Corrsin integrated the approximate spectrum from the wavenumber corresponding to the size of the largest blobs of pure A ($k_0$) to beyond $1/\lambda_B$ to obtain an estimate of the mixing time constant. The resulting equations for both low and high Sc are given in Table 2-4. For liquids, Sc is large, $\lambda_B$ is smaller than $\eta$, and diffusion is very slow. The time constant increases due to the action of diffusion (second term in the expression for $\tau$), but this second term is often small when compared to the magnitude of the first term, particularly for low viscosity fluids. When Sc is near unity, as for gases, $\lambda_B$ is approximately equal to $\eta$, and the approximate spectrum is integrated from $k_0$ to beyond $\eta$. For the resulting equation to apply, Sc needs to be in the vicinity of 1; otherwise, the equation will predict infinite mixing time.

Corrsin's analysis was developed for an isotropic homogeneous turbulent field but has been very successfully applied in pipe flow, both in terms of the shape of the spectrum and in terms of the overall mixing time (see Example 2-1e). Others have applied Corrsin's scaling arguments in mixing tanks to determine the correct dimensionless groups to apply for blend time correlations (see Example 2-3). The key concept to understand from Table 2-4 is the relationship between the concentration fluctuation field and the velocity fluctuation field. This relationship is different for gases and liquids. The mixing time estimates in Table 2-4 allow us to make some useful arguments about the length scales that are retained on scale-up, and about the mass transfer time scales compared to the reaction time scales in cases where micromixing dominates the process.

---

[18] Going one step further, the combination of molecular diffusion with Taylor dispersion has also been treated in Brodkey (1967, p. 326). Turbulent eddies carry what has to be mixed from one part of the fluid to another, which accelerates the breakdown of blobs of pure A. At the same time, molecular diffusion is enhanced by the increase in surface area and the steep gradients of concentration that occur due to the action of the turbulent eddies. Using a statistical approach, the enhancement of mixing due to turbulent dispersion can be described with a simple first-order solution:

*mean-squared displacement of the interface = turbulent dispersion*

*+ result of interaction between turbulence and molecular diffusion*

$$L^2 = L_0^2 + 2D_{AB}t$$

This equation describes the mean spread of a plume in uniform (plug) flow and is most relevant for cases involving the dispersion of gases.

**Table 2-4** Effect of Schmidt Number on Concentration Length Scales and on Blend Time

| Schmidt Number $Sc = \nu/D_{AB}$ | Relative Length Scales $Sc^{1/2} = \eta/\lambda_B$ | Time Constant for Decay of Concentration Fluctuations |
|---|---|---|
| $Sc \ll 1$ Molecular diffusivity faster than momentum diffusivity | $\eta < \lambda_B$ Smallest length scales are in the velocity field; not realizable | |
| $Sc = 1$ or small Equal diffusivities of mass and momentum: typical of gases | Equal length scales | Governed by turbulence: $\tau = \left(\dfrac{5}{\pi}\right)^{2/3} \dfrac{2}{3-Sc^2}\left(\dfrac{L_s^2}{\varepsilon}\right)^{1/3}$ |
| $Sc \gg 1$ Molecular diffusion slow: typical of liquids | $\eta > \lambda_B$ Smallest length scales are in the concentration field | Mixing is slowed down by the effects of molecular diffusion: $\tau = \dfrac{1}{2}\left[3\left(\dfrac{5}{\pi}\right)^{2/3}\left(\dfrac{L_s^2}{\varepsilon}\right)^{1/3} + \left(\dfrac{\nu}{\varepsilon}\right)^{1/2}\ln(Sc)\right]$ The effect of Sc is usually small, particularly for low viscosity liquids |

*Example 2-1e: Realistic Models of Turbulent Mixing.* Now that we have some additional understanding of turbulent mixing of scalars, we can return to Example 2-1 to consider a more realistic analysis. Example 2-1d considered molecular diffusion across a single frozen eddy length scale. When we treat the turbulent eddies as a fixed reduction in the scale of segregation, the dynamic multiscaled nature of the turbulence is neglected. What we really need to know is how eddies across the entire spectrum of length scales interact dynamically with the concentration field in space and time. Because little is known about modeling these dynamics directly, the problem is formulated in terms of statistical averages in the theories developed by Taylor and by Corrsin.

Corrsin's (1957, 1964) theory considers the time scale, $\tau$, required for the decay of the concentration fluctuations. This can be expressed in terms of the intensity of segregation:

$$I_s = \frac{\overline{c'^2}}{\overline{c_0'^2}} = e^{-t/\tau} \quad (2\text{-}14)$$

The intensity of segregation is a measure of the mixing accomplished. The intensity is 1 when the components are unmixed and zero when they are fully mixed (zero fluctuations). This intensity is a point measurement, not an average over the entire vessel. The time constants, $\tau$, for low and high Sc are taken from Table 2-4.

The macroscale of mixing, $L_s$, is not well established, so estimates for this, and for $\varepsilon$, are needed. Brodkey (1967, p. 351) suggests

$$\left(\frac{5}{\pi}\right)^{2/3} \left(\frac{L_s^2}{\varepsilon}\right)^{1/3} = 0.341 \frac{r_0}{u} \tag{2-15}$$

for a pipe, where $r_0$ is the radius of the feed pipe and u is the streamwise RMS velocity. We will return to discuss the scaling arguments related to this expression in the next section, but first we would like to test the estimate using experimental evidence. McKelvey et al. (1975) established the velocity and concentration fields in the same pipe reactor as that used by Vassilatos and Toor (1965; Example 2-2c). Their objective was to test the time constants for mixing as predicted from velocity field measurements. This involved showing that there was an equivalence of the mixing and very fast reaction rates in Toor's reactor.[19] Figure 2-14 shows extremely good agreement between the mixing model based on velocity measurements and the reaction rate results from Vassilatos and Toor.

The general equivalence of time constants estimated over a wide variety of experiments involving gases and liquids and a variety of geometries was shown in a review by Brodkey (1975). He showed that nearly a 10 000 fold range in mixing times could be adequately estimated if one has some idea as to the proper value of $r_0$ (the characteristic dimension) to be used. This lends some credibility to Corrsin's theory and motivates further examination of $L_s$ and $\varepsilon$.

### 2-3.3 Scaling Arguments and the Energy Budget: Relating Turbulence Characteristics to Operating Variables

To relate the Kolmogorov scale, $\eta$, to operating variables, we need to get a measure of the rate of dissipation of turbulence kinetic energy per unit mass, $\varepsilon$. The easiest way to do this is via scaling arguments and the use of characteristic length and velocity scales. These scales are an important tool in engineering fluid mechanics and deserve some explanation.

In fully turbulent flow, viscous forces become negligible relative to turbulent stresses and can be neglected (except for their action at the dissipative scales of motion). This has an important implication: above a certain Reynolds number, *all velocities* will scale with the tip speed of the impeller, and the flow characteristics can be reduced to a single set of dimensionless information, *regardless of the fluid viscosity*. One experiment in the fully turbulent regime[20] can be applied for all tanks that are exactly geometrically similar to the model, at all Reynolds numbers

---
[19] Toor (1962) hypothesized that mixing and very fast reaction rates are equivalent when the reactants are stoichiometrically fed, and their stoichiometric ratio is 1.
[20] See the discussion of whether the tank is fully turbulent when the impeller region is fully turbulent in Section 2-5.

**54** TURBULENCE IN MIXING APPLICATIONS

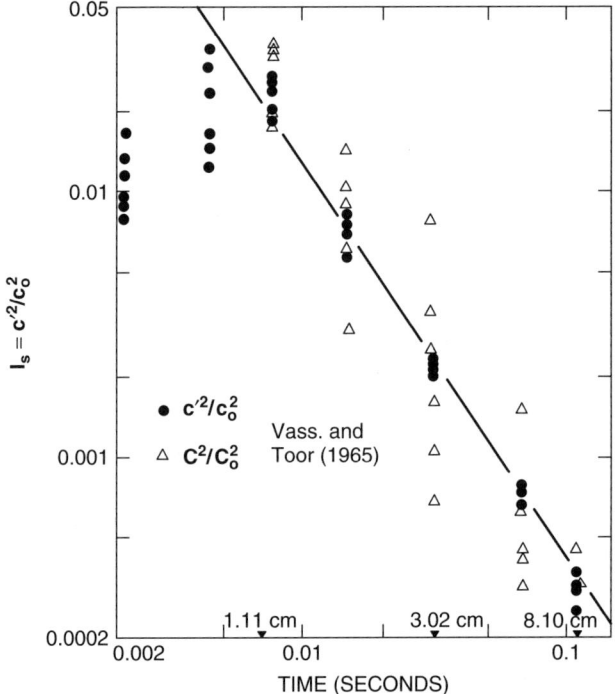

**Figure 2-14** Predicted intensity of segregation-based Corrsin-type analysis and Toor's hypothesis. (Data from McKelvey et al., 1975.)

in the fully turbulent regime, and for the full range of Newtonian working fluids. For the stirred tank, the characteristic turbulent velocity and length scales are

$$u_c = C'_u V_{TIP} = C_u ND$$
$$L_c = C_L D \tag{2-16}$$

For now, the characteristic length scale $L_c$ is assumed to scale with the impeller diameter, not the tank diameter.[21] If geometric similarity is observed and all impeller dimensions are scaled with the impeller diameter (including details such as blade thickness), the characteristic length scale ($C_L D$) will scale any of the impeller dimensions equally well; only $C_L$ will change. The constants $C_u$ and $C_L$ are a function of the impeller and tank geometry selected. For now, however, we retain them as constants.

The dissipation, $\varepsilon$, is the rate of dissipation of turbulent kinetic energy. The turbulent kinetic energy must scale with $u_c^2$. The rate of dissipation of energy is

---

[21] As long as strict geometric similarity is maintained, the only difference between D and T is a constant. See Example 2-3 for further discussion.

taken to scale with $u_c/L_c$, the characteristic time scale of the flow. This gives

$$\varepsilon \propto \frac{u_c^3}{L_c} = \frac{C_u^3 N^3 D^2}{C_L}$$

$$= A \frac{C_u^3 N^3 D^2}{C_L} \tag{2-17}$$

Note the large sensitivity of $\varepsilon$ to $C_u$, relative to its sensitivity to $C_L$ and A! When the dissipation is estimated from experimental data, A is taken to be equal to 1, $u_c$ is measured, and $L_c$ is either determined from an integral energy balance, or is estimated as some fraction of the impeller diameter. Direct measurements of the dissipation are extremely difficult [see review by Kresta (1998)].

A second estimate of turbulence characteristics, which avoids the need for $C_u$, is the power per unit mass of fluid in the tank. If the liquid depth, H, is equal to the tank diameter, T:

$$\frac{P}{\rho V_{tank}} = \frac{4 N_p \rho N^3 D^5}{\rho \pi T^2 H} \propto N_p N^3 D^2 \left(\frac{D}{T}\right)^3 \tag{2-18}$$

This scaling, however, introduces a factor of $(D/T)^3$. This may work well where the bulk characteristics of the flow dominate, but it is not an accurate measure of turbulence if local characteristics are needed. For the same power input per unit tank volume, or holding eq. (2-18) constant with variations in impeller type, diameter, and off-bottom clearance, Zhou and Kresta (1996a) provided an extensive set of data and showed that the local dissipation can vary by up to a factor of 100. This is illustrated for the Intermig on the Visual Mixing CD affixed to the back cover of the book. The best order-of-magnitude estimate of the maximum dissipation uses the swept volume of the impeller instead of the total tank volume:

$$\frac{P}{\rho V_{impeller}} \propto \frac{N_p \rho N^3 D^5}{\rho D^3} = N_p N^3 D^2 \tag{2-19}$$

and gives the same scaling with N and D as the original estimate of the dissipation. Note that this scaling suggests that the effect of $C_u^3/C_L$ is characterized by the power number, and that some fraction of the total energy is dissipated in the impeller swept volume. This fraction depends on the impeller geometry (Zhou and Kresta, 1996b). Figure 2-15 applies this scaling approach to measured estimates of $\varepsilon_{max}$ for various tank geometries, showing that this scaling estimate is accurate within a factor of 2 for four different impellers with power numbers ranging from 0.3 to 6. The importance of using the swept diameter in calculations, particularly for a PBT, is illustrated in the example below from Weetman (2002).

*Example 2-3: Swept Diameter Calculation.* In this example we consider two PBT's. The first is a standard geometry (W/D = 0.2, 45° blade angle or tip chord angle). The second is a PBT with W/D = 0.5 and a tip chord angle (TCA) of

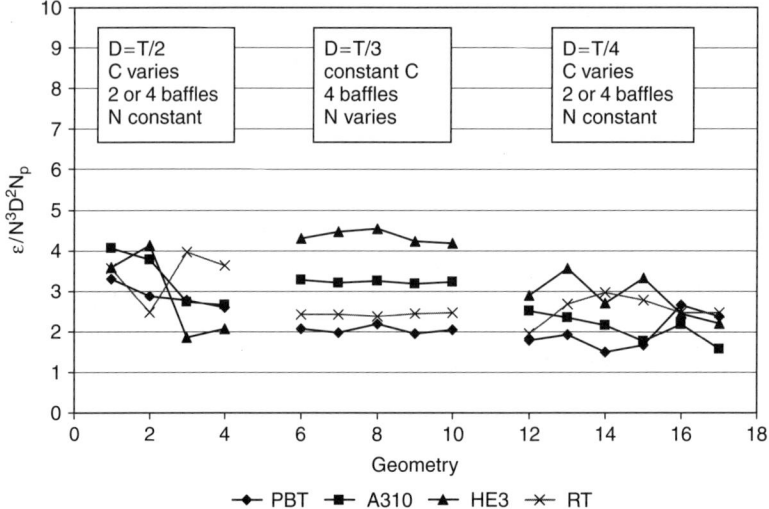

**Figure 2-15** Scaling of maximum local dissipation with the power per impeller swept volume across a range of geometries. Use of the power per tank volume with exact geometric similarity will give a similar result; however, when the geometry is varied, values of the local dissipation can vary dramatically from one tank to another, even at the same power per tank volume. (Modified from Zhou and Kresta, 1996b.)

30° to the horizontal. The blade length plus the hub radius is the perpendicular dimension, but the swept diameter must be corrected for the projection of the tip of the blade beyond the perpendicular radius.

*Standard impeller* ($W/D = 0.2$; TCA $= 45°$; blade thickness $= t_b = 0.01D$):

$$\frac{D_{swept}}{D} = \frac{1}{\{1 - [W\cos(TCA)/D + t_b\sin(TCA)/D]^2\}^{0.5}} = 1.0112$$

$$\frac{P_{swept}}{P} = \frac{N_p \rho N^3 D_{swept}^5}{N_p \rho N^3 D^5} = \left(\frac{D_{swept}}{D}\right)^5 = 1.057$$

$$\frac{\varepsilon_{swept}}{\varepsilon} = \frac{N_p N^3 D_{swept}^2}{N_p N^3 D^2} = \left(\frac{D_{swept}}{D}\right)^2 = 1.022$$

*30°, Wide-blade impeller* ($W/D = 0.5$; TCA $= 30°$; $t_b = 0.01D$):

$$\frac{D_{swept}}{D} = \frac{1}{\{1 - [0.5\cos(30°) + 0.01\sin(30°)]^2\}^{0.5}} = 1.112$$

$$\frac{P_{swept}}{P} = (1.112)^5 = 1.703$$

$$\frac{\varepsilon_{swept}}{\varepsilon} = (1.112)^2 = 1.237$$

For a standard PBT the error is on the order of 5% if the perpendicular distance is used. For the large-bladed impeller with a shallower angle, the errors are up to 70%! This D also makes sense when measuring the primary flow with a laser Doppler velocimeter for determination of the flow number (see Chapter 6). It is very important in a mixing installation when one has to be concerned with clearances from the tips of the blades.

Where does this leave us? We have three ways to estimate the dissipation and the Kolmogorov length scale: The first requires experimental information for $C_u$, $C_L$, and A; the second uses the power number and the impeller swept volume to get an estimate of the maximum local dissipation; the third uses the total volume of the tank to get an estimate of the gross average dissipation and introduces a factor of $(D/T)^3$ into the equation. More recent detailed studies on the Rushton turbine in particular (Michelet, 1998; Escudier, 2001) have shown that these estimates are reasonably accurate over some portion of the impeller discharge stream. All three methods will allow us to assess trends on scale-up, where physical properties often remain constant, but dimensions and rotational speeds change. The power per impeller swept volume is recommended as the best practice estimate.

***Example 2-4a: Blend Time.*** Now that we have ways to estimate $\varepsilon$ and the characteristic length scale $L_s$, we return to Corrsin's equations in Table 2-4. Probably the *most important practical point* is that the time constant of mixing scales with $(L_s^2/\varepsilon)^{1/3}$. All the rest of the terms in the equation for (Sc $\gg$ 1) are either constants or relatively minor effects of the Schmidt number. For mixing in a pipe, we take the radius of the feed pipe, $r_0$, as the initial integral length scale, and the fluctuating velocity, u, as a measure of the turbulent energy. Thus we can write

$$\varepsilon \propto \frac{u^3}{r_0}$$

$$\tau \propto \left(\frac{L_s^2}{\varepsilon}\right)^{1/3} \propto \left(\frac{r_0^2 r_0}{u^3}\right)^{1/3} = \frac{r_0}{u}$$

What are the practical implications of this result? The time constant goes up (longer mixing times are needed) directly with an increase in the size of the system and down with an increase in the turbulent RMS fluctuations. Stated in dimensionless terms, the mixing length, L/D, depends on the turbulence intensity in the pipe, U/u.

What about our underlying assumptions? We know from experimental measurements that u/U is a weak function of Reynolds number. We can assume that it is approximately constant. On scale-up at constant Reynolds number, the dimension increases and U decreases; thus u must decrease also. For the same scale-up, the largest concentration scales must also increase, as they scale with $r_0$. Substituting this back into the equation for $\tau$, we see that the mixing time will scale with the characteristic dimension squared. The mixing length is not quite

**58** TURBULENCE IN MIXING APPLICATIONS

so bad, as it will scale with τU, where U decreases on scale-up at constant Re. This means that the mixing length will scale with the characteristic dimension, $L/D =$ constant. This is essentially a requirement of geometric similarity. The only way to maintain a constant mixing time on scale-up is to increase the turbulence. Keeping our eye on the important scale-up parameters certainly helps us to understand mixing better.

Applying the same scaling arguments in a stirred tank, $L_s$ is equal to some fraction of D and ε is estimated using the power per impeller swept volume. This gives

$$\tau \propto \left(\frac{L_s^2}{\varepsilon}\right)^{1/3} \propto \left(\frac{D^2}{N_p N^3 D^2}\right)^{1/3} = \frac{1}{N_p^{1/3} N}$$

Compare this with the general form of correlation for blend time in the tank from Chapter 9, where the exponent n is 2 for axial impellers:

$$\theta_B \propto \frac{1}{N_p^{1/3} N} \left(\frac{T}{D}\right)^n \qquad (2\text{-}20)$$

The correct dependence of $\theta_B$ on $N_p$ and N is suggested by the scaling arguments. The effect of T/D can be extracted if we use a minimum dissipation instead of the maximum dissipation, and set the integral length scale equal to the tank diameter at the fully mixed conditions in the bulk:

$$u_{c,max} \propto (\varepsilon_{max} D)^{1/3} \propto (N_p N^3 D^2 D)^{1/3} = N_p^{1/3} ND$$

$$u_{c,min} \propto N_p^{1/3} ND \frac{D}{T} \quad \text{due to jet decay, so} \quad \varepsilon_{min} \propto \frac{u_{c,min}^3}{T} \propto \frac{N_p N^3 D^6}{T^4}$$

$$\tau_{max} \propto \left(\frac{L_s^2}{\varepsilon_{min}}\right)^{1/3} \propto \left[\frac{T^2}{N_p N^3 D^2 (D/T)^4}\right]^{1/3} = \frac{1}{N_p^{1/3} N} \left(\frac{T}{D}\right)^2$$

Where T/D is held constant on scale-up, this result reduces to $1/(N_p^{1/3} N)$. There are many different ways to make the scaling arguments (see, e.g., Grenville et al., 1995; Grenville and Tilton, 1996, 1997; or Nienow, 1997). The point is that the end result agrees well with Corrsin's approach. The most important thing to recognize is that $L_s^2/\varepsilon$, *however it is estimated, must be constant on scale-up to maintain constant blend time.* If the dissipation (ε) is held constant on scale-up, the blend time will always increase.

*Example 2-4b: Scale-up with Exact Geometric Similarity.* In this example we consider the relationship between the spectrum of velocity fluctuations and the micromixing scales. At the lab scale, a $T = 0.25$ m vessel is used to formulate a homogeneous reaction in an aqueous phase. The fully baffled vessel is equipped with a Rushton turbine impeller of $D = T/2$ at $C = T/3$ with $N_p = 5.0$. The

## CLASSICAL MEASURES OF TURBULENCE

reaction proceeds as desired at N = 240 rpm. Scale-up to the plant vessel follows exact geometric similarity, with T = 2 m. What is the appropriate N to use in the plant?

We could calculate the bulk blend time in the lab and in the plant, but in this case the process result requires a reaction. The reaction kinetics and molecular diffusivity are constant on scale-up, so we must ensure that the Batchelor scale is also preserved. The Batchelor scale can be defined using an estimate for the dissipation:

$$\varepsilon \propto \frac{(\pi ND)^3}{L_c} \propto N_p N^3 D^2$$

$$\lambda_B = \left( \frac{\nu D_{AB}^2}{N_p N^3 D^2} \right)^{1/4}$$

Setting the Batchelor scale equal in the lab and the plant gives

$$N^3 D^2 = \text{constant} = \left( \frac{240}{60 \text{ s}} \right)^3 \left( \frac{0.25 \text{ m}}{2} \right)^2 = 1.0 \text{ m}^2/\text{s}^3$$

$$N_{plant} = \left[ \frac{1.0 \text{ m}^2/\text{s}^3}{(2 \text{ m}/2)^2} \right]^{1/3} \left( 60 \frac{\text{s}}{\text{min}} \right) = 60 \text{ rpm}$$

So the use of N = 60 rpm (or higher) and exact geometric similarity will ensure that the Batchelor length scale for scalar mixing is preserved on scale-up.

Now check the Reynolds number and power consumption:

$$Re_{plant} = \frac{ND^2}{\nu} = \frac{(1 \text{ s}^{-1})(1 \text{ m})^2}{1 \times 10^{-6} \text{ m}^2/\text{s}} = 10^6$$

$$Re_{lab} = \frac{(4 \text{ s}^{-1})(0.125 \text{ m})^2}{1 \times 10^{-6} \text{ m}^2/\text{s}} = 6.25 \times 10^4$$

Both vessels are in the fully turbulent regime, so the scaling rules will hold. Thus,

$$\left( \frac{P}{V_{tank}} \right)_{plant} = \frac{4 N_p \rho N^3 D^5}{\pi T^2 H} = \frac{(4)(5.0)(1000 \text{ kg/m}^3)(1 \text{ s}^{-1})^3 (1 \text{ m})^5}{(3.14)(2 \text{ m})^2 (2 \text{ m})}$$

$$= 796 \text{ W/m}^3$$

$$\left( \frac{P}{V_{tank}} \right)_{lab} = \frac{(4)(5.0)(1000 \text{ kg/m}^3)(4 \text{ s}^{-1})^3 (0.125 \text{ m})^5}{(3.14)(0.25 \text{ m})^2 (0.25 \text{ m})} = 796 \text{ W/m}^3$$

Since we required a constant D/T and $\varepsilon$ on scale-up, the power per unit volume is also forced to remain constant. The power consumption provides what is considered intense agitation in both vessels.

## 60 TURBULENCE IN MIXING APPLICATIONS

Notice that the value for the molecular diffusivity was never used in this problem, because the physical properties were retained on scale-up!

***Example 2-4c: Scale-up where Exact Geometric Similarity Is Not Maintained.***
A more difficult case is one where geometric similarity is not maintained on scale-up. In this case the lab scale vessel is a round-bottomed flask with a magnetic stirrer, and an existing vessel with a PBT (T = 1 m, D = T/4, C = T/4, four baffles, $N_p = 1.2$) is to be used in the plant. The initial operating conditions set N at 45 rpm (Re = $4.7 \times 10^4$), but there is excessive formation of by-product. The chemists agree to run some scale-down experiments. The first experiment uses exact geometric similarity and the same scaling principles as outlined in Example 2-4b. The resulting product distribution matches the one obtained in the plant. The new conditions in the lab are T = 160 mm, D = 40 mm, C = 40 mm, and N = 152 rpm.

On increasing N to 400 rpm in the lab, the desired product distribution is obtained. *This is an indication that there is interaction between the reaction kinetics and the mixing.* To keep $N^3D^2$ constant, N in the plant must be 118 rpm. Unfortunately, the plant mixer has a fixed rpm. To keep a constant microscale, we decide to change the impeller diameter:

$$N_p N^3 D^2 = \text{constant} \qquad N = 45 \text{ rpm}$$
$$(400 \text{ rpm})^3 (0.04 \text{ m})^2 = (45 \text{ rpm})^3 D^2 \qquad D = 1.06 \text{ m}$$

This is larger than the existing tank diameter, so it is necessary to change the impeller geometry to something with a larger power number. Selecting a Rushton turbine (RT), $N_p$ is taken equal to 5.0 (conservative), so

$$(1.2)(400 \text{ rpm})^3 (0.04 \text{ m})^2 = 5.0(45 \text{ rpm})^3 D^2 \qquad D = 0.52 \text{ m}$$

This impeller diameter will fit in the existing tank. Now consider the relative blend times:

$$\theta_{B,\text{lab}} = \frac{5.2}{(400/60 \text{ s})(1.2)^{1/3}} \left(\frac{0.16 \text{ m}}{0.04 \text{ m}}\right)^2 = 11.7 \text{ s}$$

$$\theta_{B,\text{plant}} = \frac{5.2}{(45/60 \text{ s})(5.0)^{1/3}} \left(\frac{1.0 \text{ m}}{0.52 \text{ m}}\right)^2 = 15 \text{ s}$$

We expect to see a longer blend time in the plant, so this is probably acceptable. Both Reynolds numbers are in the turbulent regime and the fluids are the same, so this looks like a feasible design. One remaining problem is that we have moved from an axial impeller to a radial impeller, so the circulation patterns will change dramatically. It will be much cheaper to test the effect of this change in a scaled-down geometry than on the full plant scale! To complete the problem, we need to check the torque for the RT design versus the current operating conditions and make sure that the equipment can support the increased load.

### 2-3.3.1 Summary of Scaling Arguments

- In applying Corrsin's theory to real problems, we find that $L_s^2/\varepsilon$, however it is estimated, must be constant on scale-up. For a pipe, this requires scaling with $r_o/u$; for a tank where geometric similarity is preserved, it requires scaling with $1/N_p^{1/3}N$. (Example 2-4a)
- Scale-up with exact geometric similarity (or scale-down) requires very little empirical information. (Example 2-4b)
- Changing geometry on scale-up is a very complex undertaking that should be avoided wherever possible. (Example 2-4c)
- The crux of any problem is to determine the critical length scales and then to scale them correctly. (Example 2-4)

## 2-4 DYNAMICS AND AVERAGES: REDUCING THE DIMENSIONALITY OF THE PROBLEM

In turbulent flow, mixing is to a large extent controlled by the turbulence. Consequently, an understanding of turbulence per se is necessary before we can analyze transport phenomena. Recalling our phenomenological description from Section 2-3.1, turbulence is three dimensional, dynamic, and multiscaled, even in its most ideal form. In a stirred tank, the picture is further complicated by

- Chaotic macroinstabilities on the scale of the tank turnover time
- Anisotropic, coherent trailing vortices on the scale of the blade width
- The potential lack of fully turbulent flow (failure of Reynolds number scaling) in regions distant from the impeller
- The presence of internals and either gas or solid phases in the tank, which further complicate the generation and dissipation of turbulence

These additional variables make the stirred tank extremely versatile, but also make generalizations both difficult and dangerous. In this section we discuss various ways of simplifying our descriptions of the flow and the turbulence and illustrate where these simplifications have been applied successfully.

***Example 2-5: Solids Suspension versus Uniform Distribution.*** It is sometimes difficult to sort out exactly how each of the various length and time scales can dominate a process. To investigate this idea, consider solids suspension versus solids distribution in a tank. In the first case, our main interest is in making sure that all the solids are suspended. This is the constraint, for example, in solids dissolution, or leaching. In the second case it is important to have uniform solids distribution throughout the tank. This would be the constraint for a slurry catalyst or for continuous operation with slurry withdrawal at one point in the tank.

Consider the results for off-bottom solids suspension first. In 1958, Zwietering developed a correlation for the just suspended speed ($N_{js}$) of solids in a stirred tank. Despite numerous attempts to improve on the correlation, the result remains substantially unchanged. In 1978, Baldi et al. redeveloped the equation starting from an analysis of the fluctuating velocities in the boundary layer at the bottom of the tank. They argued that only the turbulent fluctuations can lift the solids off the bottom so that they can be convected into the main flow. The close agreement between their equation and Zwietering indicates that *the governing mechanism for off-bottom suspension is the scaling of turbulent fluctuations in the boundary layer at the bottom of the tank.*

A related problem is that of *uniform solids distribution* in the tank. Even when the $N_{js}$ criterion is met, solids are often not uniformly distributed throughout the tank. The vertical distribution of solids is still not well understood. In some cases, a sharp, stable interface forms above which there are few solids. The slip velocity between the particles and the fluid will certainly play a role in solids distribution, as will the upward velocity at the tank wall. To resolve this problem, a better understanding of the *vertical flow and macroinstabilities at the wall* is needed.

A third case is the rate of solids dissolution: once the solids are fully suspended ($N > N_{js}$), the rate of dissolution does not change significantly even if N is increased. Why? The mass transfer at the surface of the particle is determined by the boundary layer on the particle. The relative velocity between the particle and the fluid is the slip velocity, and this is not strongly affected by the fluid velocity. Once the particles are suspended, the slip velocity is approximately constant and no significant further gains can be made. *The governing mechanism for solids dissolution is the slip velocity between the particle and the fluid.*

These three cases illustrate the importance of considering the correct governing mechanism when trying to determine the most useful simplification of the flow.

### 2-4.1 Time Averaging of the Flow Field: The Eulerian Approach

Before the advent of fast computers, the time-averaged approach to the flow field was the only reasonable way to approach turbulent flows. In this approach, data taken at a single point are averaged over a sampling time long enough to provide a repeatable mean and RMS result. The only information available about transient behavior is the statistics of the signal (rms velocity) and the frequency spectrum. As long as the time scale of the process is longer than the time scale of the averaging, this approach is likely to be successful. In some other limiting cases (see Example 2-2b) the kinetics of the process are so fast that the mean mixing rate is the governing rate, and once again progress can be made.

For a basic analysis of the problem, we can use the Reynolds equations, which are the time-averaged form of the Navier–Stokes equations (see Section 2-5 and Chapter 5). The major problem is to simplify the equations and obtain additional relations between the unknowns. One idea to provide simplification is to assume that turbulent fluctuations are random in nature and can therefore be treated by means of statistics. Thus we approach the problem from a rigorous statistical theory into which we can introduce certain simplifying assumptions that will allow us to reduce the equations and solve for some of the variables of interest. The most important of these assumptions are defined and discussed in this section. The models of turbulence that result are discussed in Section 2-5.

## 2-4.2 Useful Approximations

A necessary objective in turbulence analysis is to define a limited number of simplifying assumptions that will simplify the problem while introducing only small errors in the solution. Any assumption is permissible as long as the limitations of the assumption are understood and taken into account. Let us begin by assuming that eddies range continuously in size from the very smallest to the largest, which are typically the same scale as the equipment. In the most ideal case, the boundaries influence only the large eddies and transfer energy to or from them. The larger eddies transfer their energy to the smaller eddies, and so on, until the energy is transferred to the smallest of eddies. These smallest eddies lose their energy by viscous dissipation. The five most useful assumptions required to build and work with this model are:

1. *Fully turbulent flow*. At very high Reynolds numbers, the inertial forces due to fluctuating velocities overwhelm the viscous forces, so the flow field becomes independent of fluid viscosity. Mean velocity profiles scale with a characteristic velocity and length scale, and drag coefficients (e.g., the power number) become independent of Reynolds number.
2. *Homogeneous turbulence*. The turbulence is completely random and is independent of position (i.e., RMS of u, v, w are constant over the field). The three fluctuating components u, v, and w are not necessarily equal.
3. *Full isotropy*. The fluctuations have no directional preference at any scale of motion. No gradients exist in the mean velocity.
4. *Local isotropy*. This assumption can be applied over a limited range of frequencies or eddy sizes (not a limited volume of space). Over this restricted range of eddy sizes, isotropy prevails. Eddies outside this range can be highly anisotropic, and mean velocity gradients are permitted.
5. *Turbulent shear flow*. This flow is a modification of completely homogeneous flow to allow for shear stresses and for well-defined mean velocity gradients, such as those found in a jet, a mixing layer, or a boundary layer. Usually, one or two of the Reynolds shearing stresses (Section 2-5) are zero.

The term *homogeneous turbulence* implies that the statistical characteristics of the turbulent velocity fluctuations are independent of position. We can further restrict the homogeneous system by assuming that the velocity fluctuations are independent of the axis of reference (i.e., invariant to axis rotation and reflection).

This is equivalent to saying that there is no directional preference in the fluctuating field. This restriction leads to *isotropic turbulence*, which by its definition is always homogeneous. To illustrate the difference between the two types of turbulence, consider the RMS velocity fluctuations. In homogeneous turbulence, the three components of the RMS velocity can all be different, but each value must be constant over the entire turbulent field. In isotropic turbulence, spherical symmetry requires that the fluctuations be independent of the direction of reference, or that all the RMS values be equal. A bowl of peanuts or pretzels is isotropic in a two dimensional sense: It is the same no matter how you look at it or where you place the reference axis. The same is true (in three dimensions) for isotropic turbulence. The branches and leaves on a tree, on the other hand, have a specific arrangement, so moving the axis changes the image. The tree is highly anisotropic.

True *isotropic homogeneous flow* requires that there be no directional preference in the three dimensional flow. There can be no mean velocity gradients, thus no shearing stresses. All three normal stresses must be equal, and all nonnormal stresses ($\overline{uv}$, $\overline{vw}$) must be equal to zero. If the flow has no directional preference and no coherent organized structures, there can be no correlation between components of the fluctuating velocity. The normal components ($\overline{uu}$, $\overline{vv}$, $\overline{ww}$), on the other hand, will always be positive because they are squared terms. Experimentally, such a flow can be obtained approximately in the turbulence developed behind a properly designed grid. This restriction excludes consideration of the trailing vortices in mixing vessels, which have a clearly defined orientation; it also excludes flow anywhere in the tank where velocity gradients exist. There is no possibility of seeing truly isotropic turbulence in a stirred tank.

While the fully isotropic assumption is not a good match to physical reality, the implications of isotropy are profound for turbulence modeling and measurements. Isotropy allows the entire turbulent spectrum to be defined from one component of fluctuating velocity, because the flow is perfectly without directional preference. It allows simplification of the equations to include only the normal stresses. It also allows one to make spectral arguments to simplify the measurement of the dissipation. This assumption is so powerful that it is often invoked *in the hope that it will be good enough for a first approximation*, despite the fact that it is a poor match for the full physical reality.

The area of turbulent study that holds the greatest interest for engineers is *turbulent shear flow*. This flow is a modification of completely homogeneous flow to allow for shear stresses and mean velocity gradients. Usually, one or two of the Reynolds shearing stresses are zero. Turbulent shear flow in turn may be divided into flows that are *nearly homogeneous* in the direction of flow and those that are *inhomogeneous* in the direction of flow. It has been found experimentally that the nearly homogeneous flows are those that are bounded, as in pipe flow, while the inhomogeneous shear flows are unrestricted systems, such as jets. Longitudinal homogeneity (or homogeneity in the direction of flow), arises from the fact that in pipe flow, turbulence is generated along the wall and there is no decay. Longitudinal decay arises from the dispersion of momentum and the decay of streamwise velocity, as is observed in jets. One flow of importance that has characteristics of both confined and free shear flows, depending on the location of study, is boundary layer flow. The area near the wall is nearly homogeneous in the direction of flow, and that near the bulk of the fluid is inhomogeneous and spreads as the boundary layer grows. Turbulent shear flow cannot be fully isotropic, but it may be locally isotropic.

Many misunderstandings have arisen due to a lack of care in distinguishing the *locally isotropic* assumption from the *isotropic* or fully isotropic assumption. The restriction of *local isotropy* can be applied over a limited range of frequencies or eddy sizes (not a limited volume of space). The conditions for local isotropy state that if the local Reynolds number (based on the turbulent length scale and the fluctuating velocity, not on the equipment length scale D and the mean velocity) is high enough, there may be a range of eddy sizes over which the turbulence energy cascade is in equilibrium. Under these conditions, energy enters at the top of the locally isotropic range of eddy sizes and is dissipated at the smallest locally isotropic scales of motion with no losses of energy at the intermediate scales. Over this range of eddy sizes, no memory of the oriented large scale motions (i.e., the trailing vortices) remains, and there is no directional preference in the flow. This condition extends up the cascade to some large eddy size $l$. Below this limiting length scale, the flow can be treated as locally isotropic. Eddies larger than $l$ may still be highly anisotropic. It should be understood that any conclusions that are valid for locally isotropic turbulence, are also valid for fully isotropic turbulence over the same range of wavenumbers.

This discussion of the basic simplifying assumptions used to describe different types of turbulence prepares the way for a better understanding and further interpretation of turbulence in mixing vessels. The next example involves applications of the five simplifying assumptions.

## *Example 2-6: Applications of the Simplifying Assumptions*

(a) *Homogeneous turbulence*. Is $\varepsilon = P/\rho V_{tank}$? In the early days of mixing research, there were very few data on the flow field, and some initial scaling variables were needed. Based on the first law of thermodynamics, the energy put into the tank can only be dissipated, since there is no energy out. Taking the power input at the shaft and dividing it by the mass of fluid in the tank ($P/\rho V_{tank}$) returns the same units as the rate of dissipation of turbulent kinetic energy per unit mass. It is not a big leap to abbreviate $P/\rho V_{tank}$ to $\varepsilon$, but is this physically meaningful? Is it a useful representation of the turbulence?

When the jump is made from $P/\rho V_{tank}$ to $\varepsilon$, an assumption that the turbulence is *homogeneous* is implied. This assumption is clearly a poor one in a stirred tank, where the levels of turbulence can vary by a factor of 100 from the impeller to the bulk. Generation and dissipation are vastly different between the impeller region and the regions away from the impeller. For shear-sensitive materials such as cells, their survival depends more on the maximum shear they see than on the average. In such cases, using $P/\rho V_{tank}$ as some kind of average dissipation is

**66** TURBULENCE IN MIXING APPLICATIONS

about as informative as saying that the average velocity in the tank is zero. So why does $P/\rho V_{tank}$ work so well as a correlating variable?

The success of $P/\rho V_{tank}$ is actually restricted to cases where *exact geometric similarity* is maintained. If this restriction is satisfied, $P/\rho V_{tank}$ *is really a scaling basis, not an average dissipation*. The local dissipation roughly scales with $N_p N^3 D^2$, or the power per impeller swept volume, which differs by a factor of $(D/T)^3$ from $P/\rho V_{tank}$. If D/T is constant, the two approaches are equivalent. There are many other good reasons for maintaining geometric similarity on scale-down, so this is not a bad restriction to keep—we just have to be careful of the basis for the argument.

Where the objective is to uncover the governing physics in the problem, the effects of the local dissipation must be separated from the effects of other variables. To accomplish this, geometric similarity will often not be maintained, and the best available scaling for the local dissipation is $N_p N^3 D^2$, or the power input per unit of *impeller* swept volume.

(b) *Fully turbulent flow*. Scaling variables work when the flow is *fully turbulent* and exact geometric similarity is maintained. When these two conditions are true, the effect of fluid viscosity is negligible. The flow field can be made dimensionless using a characteristic length scale and a characteristic velocity scale. Once fully developed turbulence is satisfied, dimensionless velocities scale exactly with the characteristic velocity. In a stirred tank, this velocity is the tip speed of the impeller. Figure 2-16a shows the radial velocity profile in the discharge stream of a Rushton turbine, scaled with the tip speed of the impeller. The velocity profile is measured at three different rotational speeds and in three different fluids. All of the data collapse onto one line. In Figure 2-16b, the local dissipation, $\varepsilon$, below an Lightnin A310 impeller is scaled in the same way. Note that the last place to attain this scaling in the impeller discharge is the velocity peak at the tip of the impeller blades.

In Figure 6-14 the power number is constant and independent of Re for Re $> 2 \times 10^4$. Similarly, the blend time scales exactly with N above the fully turbulent Re (see Chapter 9). Viscosity no longer has any effect on the velocity field or on the power draw. These dramatic simplifications are true only where the flow is fully turbulent. *Fully turbulent* does *not* mean that the turbulence will be fully homogeneous and the same everywhere. Processes that depend on local conditions, such as cell survival and apparent chemical kinetics, will be affected by the local variations that exist in mixing systems. Average quantities will work only if they reflect the distribution of the quantity as well as the average quantity (i.e., an average tank dissipation may be a valid parameter if when the average is doubled, the maximum is also doubled). For this reason, maintaining exact geometric similarity on scale-down is often critical.

(c) *Local isotropy*. Consider pipe flow at some relatively high Reynolds number. Throughout the pipe, the viscous forces along the wall provide the conditions necessary for turbulence formation. Rotation, very large vortices, or large eddies arise from the interaction of the mean flow with the boundary. In a mixing vessel,

DYNAMICS AND AVERAGES: REDUCING THE DIMENSIONALITY OF THE PROBLEM  67

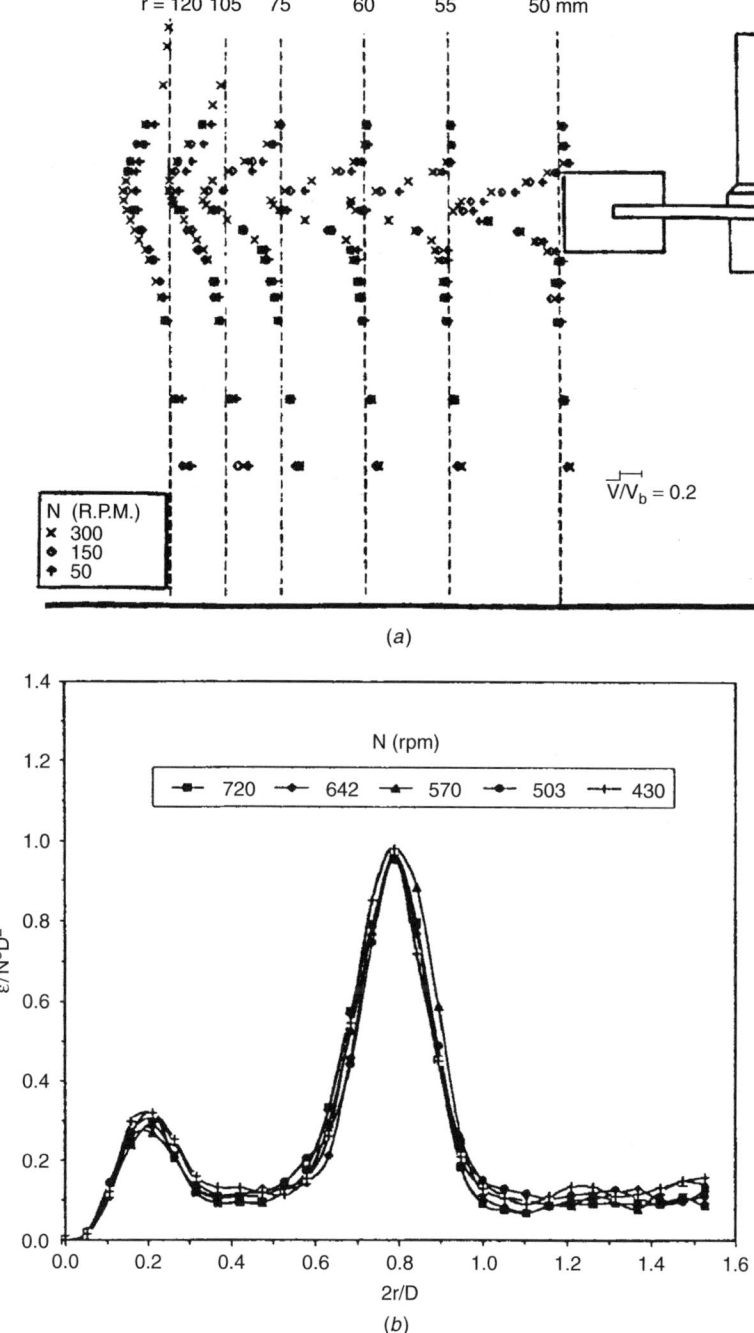

**Figure 2-16** Scaling of flow characteristics. (*a*) Scaling of velocity profiles with tip speed in fully turbulent flow. (From Nouri et al., 1987.) (*b*) Scaling of dissipation with $N^3D^2$ for the Lightnin A310 impeller, $D = 0.475T$. (From Zhou and Kresta, 1996b.)

the equivalent viscous forces and large eddies are generated at the impeller and baffles. The scale of these large eddies would be comparable to the pipe diameter, the impeller diameter, or the tank diameter. In the earlier section on locally isotropic turbulence, a model was proposed involving a cascade of energy from large to progressively smaller eddies. Now consider that the walls affect the largest fluid structures most strongly and lose their effect as the process moves down the chain. At very high wavenumbers or small eddy sizes, the effect of the boundaries is lost completely or is negligible. The small eddies are considered independent of the boundaries or mean flow. Even though the system may be inhomogeneous on the large scale, it may well be locally isotropic on the small scale, and thus an equilibrium range and inertial subrange might still be found. *Over this range*, and at the same local Re, the characteristics of the turbulence in the pipe and the turbulence in the stirred tank should be indistinguishable.

Several indicators are used to assess whether the assumption of local isotropy may be applied: the first is a high local Reynolds number, the second is a $-\frac{5}{3}$ slope in the frequency spectrum of the velocity signal, as tested in Figure 2-12, and the third is equality of the three RMS components of velocity. The final rigorous test of local isotropy is to transform the one dimensional energy spectrum measured for one component of velocity (xx) to another direction (yy or zz), and compare the results with the spectrum measured in that (yy or zz) direction. Michelet (1998) performed the first test of this condition for the flow in a stirred tank. Partial results from his work are shown in Figure 2-17. When applied to the flow closer to the impeller, as shown in Figure 2-17a and c, local isotropy must be considered an engineering approximation over a limited range of frequencies. As the probe is moved out into the discharge stream in Figure 2-17b and d, however, agreement quickly becomes very good. A similar growth in the extent of the $-\frac{5}{3}$ region was shown by Lee and Yianneskis (1998).

(d) *Turbulent shear flow.* As a first step, a very brief contemporary picture of turbulence in boundary layers and wall regions is provided. The flow can be divided into a wall region, outer region or regions away from walls, and the interactions that occur between the two regions. In wall regions (which would include impellers and baffles), the production of turbulent kinetic energy occurs. There are extensive studies of this for a variety of geometries. Often, there are intermittent periods when the Reynolds stresses are high. This is associated with an ordered sequence of events of *ejections* of low momentum fluid outward from the boundary, *interaction events*, and *sweeps* of high momentum fluid toward the area. For pipe and boundary layer flows, the entire sequence has been called a *burst phenomenon*. The outer region is characterized by the overall flow. For large systems, where boundary layers can form, these are the features that determine the turbulent/nonturbulent interfaces. The highly three dimensional bulges along the interface of boundary layers are vortical motions. Extensive measurements have been made of the turbulence characteristics inside these structures. Studies on jets and on the plane turbulent mixing layer have helped to uncover the basic features of the large scale structures in these flows. Much of the flow in the stirred tank is similar to jets, and there are valuable analogies to be made between this model

DYNAMICS AND AVERAGES: REDUCING THE DIMENSIONALITY OF THE PROBLEM    69

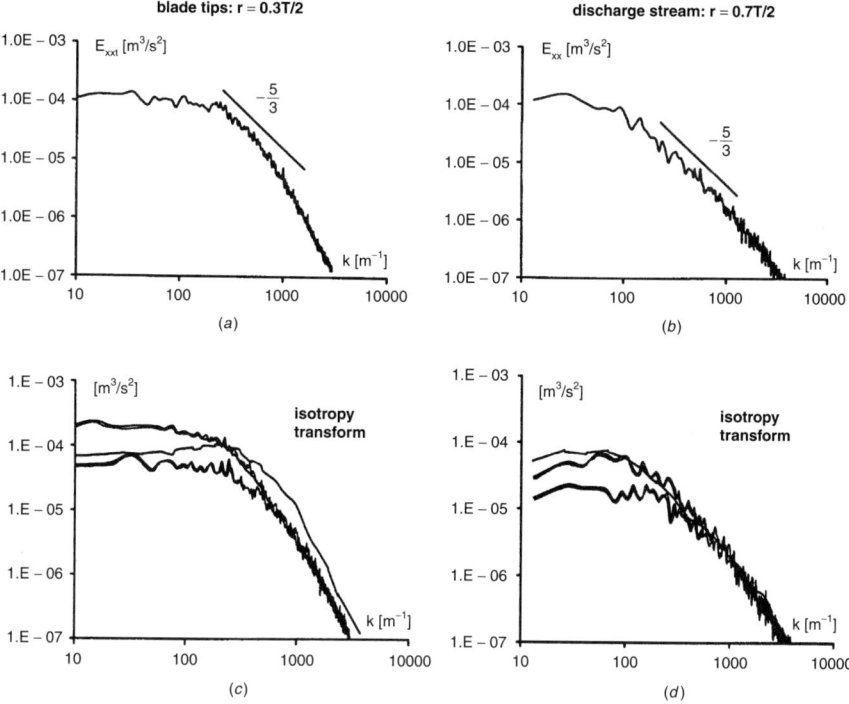

**Figure 2-17** Typical wavenumber spectra for a D = T/3 Rushton turbine with the blade passages removed. Parts (*a*) and (*c*) are taken at the tip of the impeller blades (r = 0.3T/2). Figures (*b*) and (*d*) are in the discharge stream (0.7T/2). Parts (*a*) and (*b*) show only one component of the wavenumber spectrum, $E_{xx}$. Parts (*c*) and (*d*) show the transformation of the xx spectrum (smoother line) onto the measured yy and zz spectra. Local isotropy quickly penetrates to high wavenumbers. (From Michelet, 1998.)

flow and the complicated recirculating flow in the tank (Fort, 1986; Bittorf and Kresta, 2001; Bhattacharya and Kresta, 2002; Kresta et al., 2002), as illustrated in Figure 2-18, and discussed in Section 10-3.2 in Chapter 10.

### 2-4.3  Tracking of Fluid Particles: The Lagrangian Approach

The Eulerian approach fails when there are significant variations of temperature or concentration in the tank that affect the process kinetics. One example of this is bioreactors, where cells may experience severe oxygen deprivation over large parts of the tank, changing their growth kinetics (Yegneswaran et al., 1991). A second example is crystallization (also discussed in Chapter 17), where the supersaturation varies significantly from the feed zone to the bulk. The local supersaturation determines growth and nucleation rates and thus the final particle size distribution and morphology (Baldyga et al., 1995; Wei and Garside,

**70** TURBULENCE IN MIXING APPLICATIONS

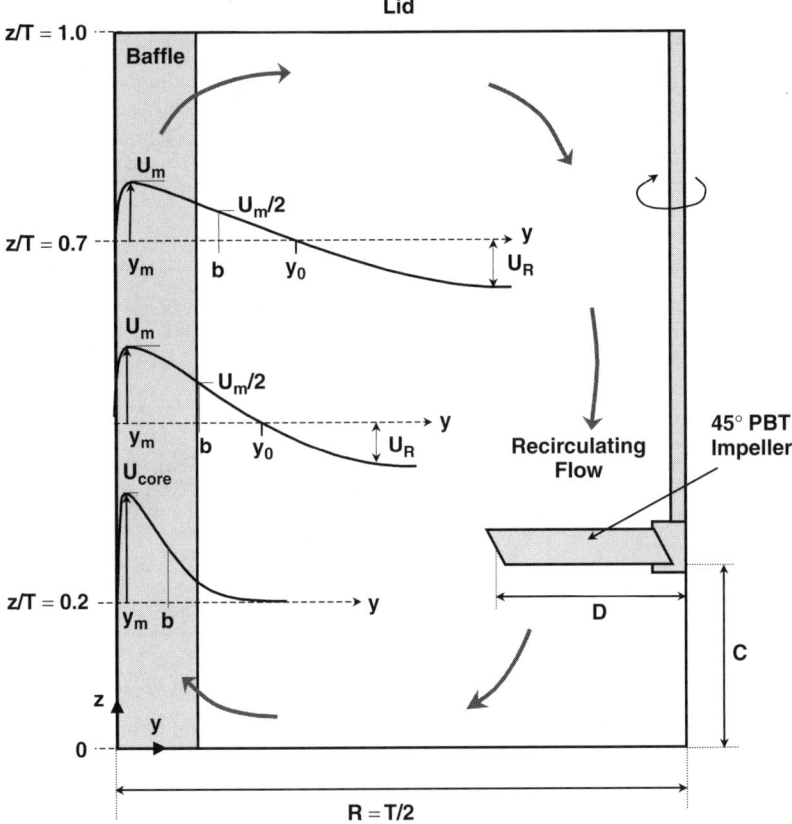

**Figure 2-18** Wall jet driven by axial impellers. The velocity profile at the wall scales with $U_m$.

1997). To model the process results accurately in both these cases, the Lagrangian experience of a fluid particle must be considered.

The Lagrangian approach follows a fluid particle over time as it moves through the flow field. Simulated or experimental particles are injected into the field at an arbitrary time and location. The particles are then tracked as they move under the influence of the velocity field. The injected particles can be neutrally buoyant or given a different density than of the fluid. In their most precise form, computed particle paths should follow experimental full-field time-resolved velocity vector data. Zhao and Brodkey (1998a) have illustrated the importance of using time-resolved data for the opposed jet system. If the process time is long, the mean concentration gradients are large, and/or the transient data are not important, significant progress can be made using a time-averaged velocity field with simple turbulent dispersion models (Bourne and Yu, 1994; Vivaldo-Lima et al., 1998). A third approach is to use circulation time distributions (Yegneswaran, 1991; Roberts et al., 1995) with stochastic modeling to incorporate the effect of variations in the particle path. Two of the greatest difficulties lie in defining accurate, reliable models of turbulent dispersion at intermediate (anisotropic) scales and in modeling complex higher-order kinetics. Both of these phenomena may well be present in cases where detailed modeling based on Lagrangian particle paths is warranted.

### 2-4.4 Experimental Measurements

The full resolution of a turbulent mixing problem would require full field measurements of three instantaneous velocity components over time [u(x,y,z,t), v(x,y,z,t), w(x,y,z,t)], plus full field concentration(s) for each component [c(x,y,z,t)]. This five dimensional space is not easily attainable with current methods, and the postprocessing requirements of this quantity of data suggest that some averaging will be required. In Section 2-3.4.1, we consider the various common experimental methods and what dimensions of this problem they measure.

#### 2-4.4.1 *Information Contained in Experimental Measurements*

- *Pointwise velocity as a function of time* [u(t) or v(t) or w(t)]. *Laser Doppler velocimetry* (LDV) is a single-point time series measurement, typically of one or two velocity components. From these data we can extract mean and RMS velocities, spectral information, and in the case of a two-component instrument, a single Reynolds stress ($u_i u_j$). We cannot obtain much information about the shape of large structures, or macroinstabilities in the flow, because only one spatial location can be measured at a time.

- *Pointwise velocity relative to the impeller blade* [u(θ) or v(θ) or w(θ)]. *Angle- or phase-resolved* LDV is still a single-point measurement but with the addition of a shaft encoder, which records the shaft angle versus time. The velocity versus time data are then sorted by angular position to give the velocity *relative to the impeller blade*. These data can be used to uncover cyclically appearing structures, such as the trailing vortices, and to define angle-resolved values of the RMS velocity and (again if two components are available) a single Reynolds stress. This information suggests that the peak levels of turbulence are rotating with the blades in a very small area behind the blades. Understanding this is important if we are to address the mechanisms of drop breakup and cell destruction vis-a-vis the instantaneous turbulence field.

- *Two components of velocity as a function of time over a full plane of the flow* [u(x,y,t), v(x,y,t)]. Full-plane *particle image velocimetry* (PIV) provides a full plane of velocity data with two components of the velocity at once. The measuring volume is thin in the direction normal to the plane. This is a problem if the component normal to the plane is large because the particles will not stay in the illuminated plane long enough to register a velocity. There are ways around this if the plane can be oriented to match the direction of the streamwise velocity, but in the highly three dimensional stirred tank, this requires significant insight into the flow. A newer extension of PIV can use two simultaneous views at two angles (stereoscopic imaging) to give the third component of the velocity, but still within a narrow plane.

- *Concentration as a function of time over a full plane of the flow* [c(x,y,t)]. *Laser-induced fluorescence* (LIF) provides a full plane of instantaneous concentration data and can be very valuable where the intermittency of concentration at the visible scales of motion must be understood. It has been applied successfully to several low Reynolds number mixing devices to elucidate mixing structures. Examples are given in Chapter 3. Quantitative analysis of the images can be done by converting light intensity to dye concentration at each pixel of data.

- *Three components of velocity as a function of position in three dimensional space* [u(x,y,z,t), v(x,y,z,t), w(x,y,z,t)]. *Particle tracking velocimetry* (PTV) tracks the image of several (up to 1000 (Guezennec et al., 1994; Zhao and Brodkey, 1998b)) particles in a three dimensional volume over time, giving the location of the particles over time. From the position records, three components of velocity can be extracted for each particle at each time step. If data are taken for a long enough time, the full time-averaged three dimensional velocity field can be extracted with all six Reynolds stresses. The time that is "long enough" can be very long if small numbers of particles are used, because at one instant in time only 2 views in the tank are measured, even though the full volume is recorded in the image. The success of PTV requires extensive image analysis, efficient tracking algorithms, and stereomatching techniques. The spatial resolution of this method is still low compared to PIV methods.

- *Three components of velocity as a function of position in three dimensional space* [u(x,y,z), v(x,y,z), w(x,y,z)]. *Scanning* PIV is a three dimensional extension of planar PIV at a higher spatial resolution than is possible with PTV. Time resolution in this method is more difficult than that for PIV, because a finite time is needed to scan the tank before the light sheet returns to the initial position. At least four of the six Reynolds stresses can be resolved with this approach. Another approach to this measurement is holographic PIV. None of these methods are commercially available at the time of writing.

## 2-5 MODELING THE TURBULENT TRANSPORT

Modeling can prove to be far less costly to use in the long run than actual mixing experiments, and may provide much more detailed information than is available from experiments, so there is a large incentive to develop reliable models of mixing processes. If the computed results do not adequately model the real physical system, they will not be of much use, so any useful model must be quantitatively validated.

A key part of any mixing process model will be the turbulence model, and an entire range of turbulence models has been developed in an effort to address this problem (Table 2-5). In these models there is a clear trade-off between complexity and representation of the underlying physics. The various theoretical approaches can be formulated in wavenumber space or physical space; can use long time averages, averages over specific structures, or no averages at all; and will usually involve some closure approximation based on statistical reasoning,

**Table 2-5** Summary of Approaches to Turbulence Modeling[a]

| Model | Physical Basis of the Model | Drawbacks |
|---|---|---|
| *Boussinesq approximation* <br> • One equation with one adjustable parameter <br> • Averaged over time and all length scales | • One length scale <br> • Based on analogy to laminar transport and apparent viscosity | • Oversimplification of the physics <br> • The apparent viscosity is a function of the flow field and of position |
| *Prandtl mixing length* <br> • One equation with two adjustable parameters <br> • Averaged over time and all length scales | • One length scale <br> • Based on analogy to mean free path in the kinetic theory of gases | • Oversimplification of the physics <br> • The assumption of a linear velocity profile does not match physical reality; however, the results are surprisingly good for the log-law region of a pipe |
| *Two-equation models* (taking the $k-\varepsilon$ model as an example) <br> • Two partial differential equations with five model constants <br> • Averaged over time, with models for two locally varying turbulent quantities (k and $\varepsilon$) | • Assumes that the three normal stresses are equal and that all cross-correlated stresses are zero (cross stresses may be estimated after the fact) <br> • Based on turbulent kinetic energy balance (k-equation) and a model for the rate of dissipation of turbulent kinetic energy ($\varepsilon$-equation) <br> • Variations have been developed to model subclasses of flows | • Five model constants have been determined for simplified flows <br> • Two-equation models cannot accurately model the effects of anisotropy on the large scale, although the form of the model may be useful for the locally isotropic range of turbulence <br> • $k-\varepsilon$ model tends to be overly diffusive |

(*continued overleaf*)

# 74  TURBULENCE IN MIXING APPLICATIONS

**Table 2-5**  (*continued*)

| Model | Physical Basis of the Model | Drawbacks |
|---|---|---|
| *Full Reynolds stress models (ASM, RSM, or DSM)*<br>• Model all six Reynolds stresses<br>• Averaged over time | • Treat anisotropy in the flow by modeling all six Reynolds stresses in their time-averaged form | • Computationally difficult<br>• Subject to problems with convergence<br>• Grid independence is difficult to attain |
| *Large eddy simulations (LES)*<br>• Model large scales and small scales separately<br>• Average small scales over time<br>• Allow transient (direct) simulation of large scales | • Model the larger, anisotropic scales of turbulence using a DNS approach, following their motion directly as it varies in time<br>• Treat the subgrid scales of turbulence as isotropic and in equilibrium: model these scales using a two-equation model of turbulence | • Requirements for data storage and data processing are outside the range of most users<br>• No consensus has emerged on subgrid modeling requirements<br>• Boundary conditions at solid surfaces are problematic |
| *Direct numerical simulations (DNS)*<br>• Solve the full time varying Navier–Stokes equations for the three dimensional field of fluctuating velocities<br>• No averaging required | • Using only the instantaneous form of the Navier–Stokes equations, solve the flow field at each instant in time, storing full three dimensional records of the fluctuating velocity<br>• Sometimes called a "numerical experiment" | • Computationally intensive<br>• Huge storage requirements; restricted to low Re<br>• Commercial versions are unlikely |

[a]Every time a new problem is attempted, model results must be validated. The level of complexity required in the model depends heavily on the level of accuracy and detail required in the results.

dimensional analysis, experimental evidence, or simplified conceptual modeling. Many facets of the physics need to be addressed to accurately represent any process of industrial importance. First, accurate models of the physics based on fundamental understanding are needed, and second, the inherent dynamics of turbulence, mixing, and reaction must be addressed.

## 2-5.1  Time-Resolved Simulations: The Full Solution

Since turbulence is by definition a time-varying phenomenon, the best hope for full resolution of the physics is in transient, or time-resolved, simulations. Both direct numerical simulation (DNS) and large eddy simulations (LES) use the governing equations directly without time averaging. These equations are the Navier–Stokes equations, the continuity equation, the individual species balance equations, and the energy balance equations. In such an approach there are as many equations as unknowns, so the problem is deterministic and the equations are, in principle, closed. However, the partial differential equations are nonlinear, higher order, and coupled. Problems in numerical resolution can be extreme, especially when DNS calculations are used.

The problem is complicated by the large range of length scales which are relevant to the process results and by the highly three dimensional nature of the stirred tank flow field, so simulation results that are grid- and time step–independent can be extremely difficult to attain. At the time of writing, time-resolved simulations are still in the province of the expert user. Despite this, a good deal of insight into modeling issues can be gained from a brief explanation of this approach to turbulence modeling.

**2-5.1.1 Direct Numerical Simulation.** The Navier–Stokes equations describe a momentum balance on a differential control volume at any instant in time. They are exactly correct, at any instant in time, so in principle all that is needed to solve turbulent flow is a transient solution of the Navier–Stokes equations with appropriate initial and boundary conditions. This is the approach used in DNS.

In a high Reynolds number turbulent flow, the changes with time can be very rapid, and the range of scales is extreme. In Example 2-1c, the smallest eddy was taken as 0.1 mm in a system that could be as large as 30 cm overall. The range of length scales in this simple geometry is 1 : 3000. A full computational domain would be $3000^3 = 2.7 \times 10^{10}$ cells big, a number that is far too large for present computers. The task is even more impressive when one realizes that the simulation must be transient with adequate resolution in time. It takes a very large computer indeed to do such modeling, even at low turbulent Reynolds numbers. Present computations can only be applied to low Reynolds numbers in somewhat simple geometries. Despite the lack of ability to do extremely detailed space and time resolution calculations, calculations in more modest grid structures (still fine when compared to LES) can be of use. In particular, when the geometry is complex and local conditions (as discussed earlier) are not as critical, such calculations could be very helpful in design.

In an ideal world, one could use DNS to reproduce the experimental flow field that controls mixing, then obtain measures of the individual terms in the Navier–Stokes equations on scales down to a small multiple of the grid size. These terms determine the coupling between mixing (and of course kinetics and heat transfer) with the instantaneous flow field. The results of these detailed, fully coupled calculations could then be used to test and develop models for subgrid scales in LES, and for other computational fluid dynamics (CFD) calculations where average forms of the equations are used.

**2-5.1.2 Large Eddy Simulation.** The second approach is to use large eddy simulations. The limitations of this method are much less severe. The large scale motions are computed in a manner similar to DNS but on a much coarser grid. The scale might be as coarse as 1 : 30. The computational domain would then be $30^3 = 2.7 \times 10^4$, which would not be difficult with current machines. A grid several times as fine as this would not be out of the question, and initial LES simulations in stirred tanks have recently been reported (Bakker et al., 1998; Revstedt et al., 1998; Derksen and van den Akker, 1999; Roussinova et al., 2001).

This modeling approach computes the larger scales of turbulence directly as they vary in time and models the finer scales of turbulence. The LES modeling technique has few assumptions, all of which can be modified to provide a match between the experimental statistical measures and the more detailed large scale results. The advantage of LES is that it is far less computationally demanding than DNS, so that the computations can be pushed to higher Reynolds number flows. The problem is to decide which, if any, of the subgrid models and filtering techniques are adequate to represent the data. As in the DNS effort, one cannot expect to match the data on an instantaneous basis, since any instantaneous velocity record is expected to be unique; however, by tracking the statistics that are important to the mixing process, the critical information can (in principle) be extracted. Initial results are promising, showing excellent agreement between experiment and simulation for the trailing vortices associated with a Rushton turbine (Derksen and van den Akker, 1999) and macroinstabilities associated with a pitched blade impeller in its resonant geometry ($D = T/2$, $C/D = 0.5$, $f = 0.186/N$; Roussinova et al., 2001).

***Example 2-7: Physical Implications of Large Scale Effects.*** Which eddies are large eddies for mixing processes? Are the additional resources required to resolve this level of detail, and to process detailed transient results, warranted? Bakker et al. (1996) did a comparative study between PIV and time-averaged CFD for the pitched blade impeller. While they were able to show good agreement between the time-averaged flow fields, as shown in Figure 2-19*a* and *b*, the instantaneous PIV results in Figure 2-19*c* show that the overall flow field does not resemble the time-averaged result. Roussinova et al. (2003) showed that there is a single dominant low frequency in the resonant geometry. Figure 2-19*d* shows the scaling of the macroinstability frequency for the resonant tank geometry (PBT, $D = T/2$, $C = T/4$, $St = f_{MI}/N = 0.186$). When the off-bottom clearance is changed, the frequency persists, but other frequencies may also appear. These macroinstabilities can induce strong vibrations of the tank and in some cases can cause breakage of vessel internals such as baffles, coupling bolts, and impeller shafts. Recent LES animation results from Roussinova et al. (2003), included on the Visual Mixing CD show the full complexity of this large scale variation. If the desired process result responds on a longer time scale than the scale of the time averaging (on the order of 10 s), as would be the case for a slow reaction or for bulk blending, the additional details are averaged into the result. If, however, the time scale of the process is shorter than the lifetime of these large eddies

MODELING THE TURBULENT TRANSPORT 77

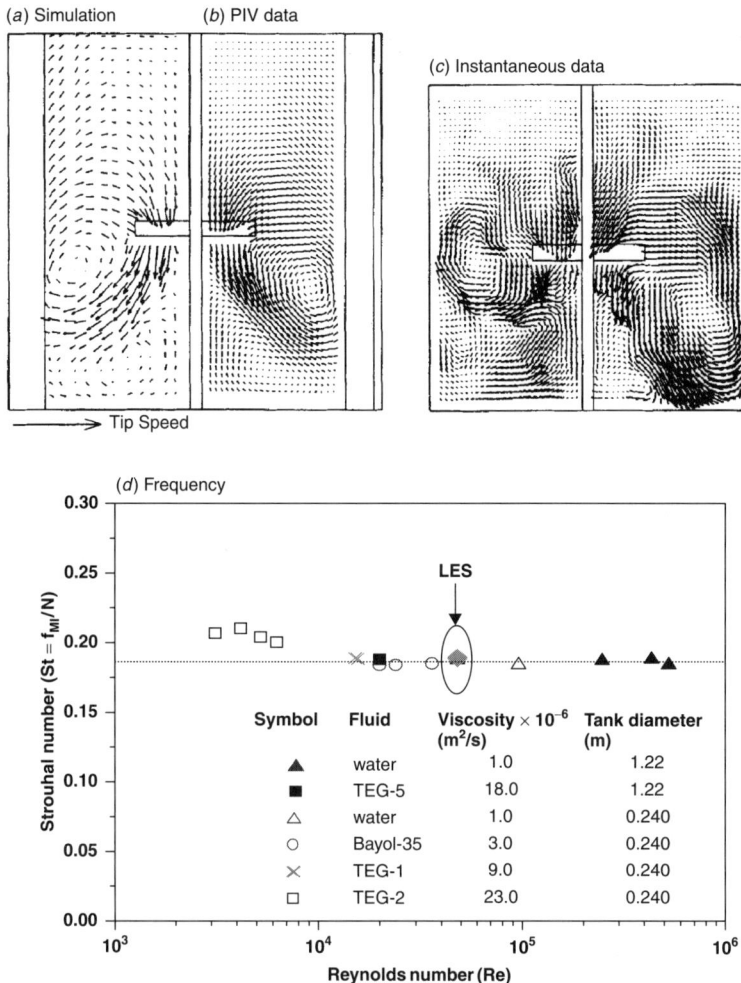

**Figure 2-19** Comparison of (*a*) time-averaged simulation with (*b*) time-averaged PIV data and (*c*) instantaneous PIV data. (From Bakker et al., 1996.) (*d*) Scaling of the frequency of the macroinstability for the resonant geometry (PBT impeller: D = T/2, C/D = 0.5, four baffles, $f_{MI} = 0.186N$; Roussinova et al., 2003). The Strouhal number St = $f_{MI}/N = 0.186$. An animation of the macroinstability is included on the Visual Mixing CD affixed to the back cover of the book.

but longer than the smallest scales of turbulence (e.g., intermediate reaction rates with higher-order kinetics), the process result may be affected by the mesoscales and it will be necessary to characterize these scales in order to make progress. Feed stream jet intermittency, a good example of how mesomixing is affected by large scale flow instabilities, is discussed further by Jo et al. (1994), Baldyga et al. (1997), and Houcine et al. (1999).

**78**   TURBULENCE IN MIXING APPLICATIONS

The next smallest scale of motion is the trailing vortices, shown in Figure 2-12 and animated on the Visual Mixing CD affixed to the back cover of the book. These are well predicted by explicit impeller modeling, sliding mesh, and DNS methods (Derksen and van den Akker, 1999). These coherent structures have stimulated ongoing debate about allowable ways to model turbulence in stirred tanks and have motivated many of the efforts to push this field forward. While the long time scales involved in the transient breakup of liquid–liquid dispersions point to the importance of these vortices and their presence certainly affects the analysis of turbulence for the Rushton turbine, their physical implications for other process results remain largely unexplored. For impellers other than the Rushton turbine, these vortices are much weaker, or even nonexistent (Roussinova et al., 2000).

The smallest intermediate scales that there is strong motivation to examine with LES (or DNS) are the larger inertial or mesomixing scales. These feed into the probability density functions used extensively by Fox (1998) to model the interactions between turbulence and chemical reactions.

## 2-5.2 Reynolds Averaged Navier–Stokes Equations: An Engineering Approximation

To reduce the modeling problem to a single steady solution, Reynolds formulated time-averaging rules. Application of these rules yields a time-averaged form of the Navier–Stokes and other equations, known as the Reynolds averaged, or RANS, equations. These equations now relate time-averaged quantities, not instantaneous time-dependent values. For this simplification, we pay a dear price in that there are now more unknowns than equations.

The additional unknowns are the six Reynolds stresses, which are the normal or mean-squared values (autocorrelations) and cross-correlations of the three components of fluctuating velocity:

$$\text{Reynolds stresses} = \rho \begin{bmatrix} \overline{uu} & \overline{uv} & \overline{uw} \\ \overline{vu} & \overline{vv} & \overline{vw} \\ \overline{wu} & \overline{wv} & \overline{ww} \end{bmatrix} \quad (2\text{-}21)$$

The terms on the diagonal are the normal stresses or variances, and these squared terms will always be positive. In an idealized flow with no directional preferences, they will all be equal. The off-diagonal elements are symmetric ($\overline{uv} = \overline{vu}$), so only three of them are unique. If the turbulence has no directional preference and there are no velocity gradients in the flow, the individual fluctuations will be completely random and the covariances will be equal to zero. This assumption of "no directional preference" or "isotropic turbulence" is an important concept for understanding the different classes of time-averaged turbulence models.[22] With these two conditions, the six unknowns can be reduced to a single unknown:

$$k = \tfrac{1}{2}(\overline{uu} + \overline{vv} + \overline{ww}) \quad (2\text{-}22)$$

---
[22] See a more complete discussion of isotropy in Section 2-4.3.

It turns out that this degree of simplification is too severe, and some way of treating the cross-correlations must also be considered. Although complete texts (Pope, 2000), and regular review articles (see, e.g., Launder, 1995) are written on the subject of turbulence modeling, the reader will benefit from understanding two important subsets of models. The simplest approach makes an initial assumption that the Reynolds stresses can be modelled using k and its rate of dissipation: these are the two-equation isotropic models, including the k–ε model. A more general, but more complex approach models each of the Reynolds stresses separately, allowing the development of anisotropy, or orientation of eddies, in the flow.

***2-5.2.1 Two-Equation Models of Turbulence.*** On application of Reynolds time averaging, six new unknowns (the Reynolds stresses) appear in the momentum equations. There are now more unknowns than equations, so the system of equations is no longer closed. This is the closure problem of turbulence. Physical flow models for the Reynolds stresses are needed to close the equations. Many logical closure schemes have been proposed and have met with some success for certain classes of flows, but there is no standard, fully validated approach to the modeling of Reynolds stresses.

A stable starting point for the kinds of flows encountered in a stirred tank is the k–ε model. This model assumes that the normal stresses are roughly equal and are adequately represented by k. Two differential equations are used to model the production, distribution, and dissipation of turbulent kinetic energy: the k-equation, and the ε-equation. These equations were developed for free shear flows, and experimentally determined constants are established for the model parameters. One of these constants is used to relate local values of k and ε to an estimate of $(\overline{uv})$ using a modified turbulent viscosity approach:

$$\overline{uv} = v_t \frac{\partial \overline{U}}{\partial y} = 0.09 \frac{k^2}{\varepsilon} \frac{\partial \overline{U}}{\partial y} \qquad (2\text{-}23)$$

Many variations on the k–ε model have been proposed and used, with varying degrees of success. Some of them are designed for the prediction of separation points, others incorporate some degree of anisotropy for cases where the flow is highly swirling (e.g., cyclones), and still others are being developed for application in multiphase flows. When a fully converged simulation using the k–ε equation does not predict the physical phenomena of interest to the desired degree of accuracy, other models should be considered.

One school of thought maintains that if the results do not have the desired degree of accuracy, the model constants should be tuned to improve agreement with experimental data. If the physical basis for the constants is considered carefully, and the adjustments based on an identifiable physical reason, the new constants might have some hope of general usefulness. On the other hand, when model constants are used as fitting parameters, the physical meaning of the turbulence model is reduced and the objective of the simulations (hopefully, one of validating the models to allow prediction of the flow field under new conditions) should be reassessed.

***Example 2-8: Prediction of Gross Circulation Patterns Using CFD.*** If the main objective of CFD modeling is determination of the mean flow patterns in the tank or of macroscopic quantities such as the power number, RANS simulations can provide good indications of the effects of changes in tank geometry and impeller geometry on the time-averaged results. Agreement for laminar flow is very good (Jaworski et al., 1998; Lamberto et al., 1999), while for fully turbulent flow the reported results vary, with the quality of the results dependent partially on the turbulence model and partially on the details of the grid and computational techniques. In general, one may expect good qualitative prediction of experimental trends where *accurate* experimental boundary conditions are used to model the impeller (Kresta and Wood, 1991; Fokema et al., 1994; Bakker et al., 1996; Coy et al., 1996; Harris et al., 1996; Armenante et al.,1997; Jaworski et al., 1998); where the impeller is simulated directly using multiple reference frames (Harris et al., 1996; Harvey and Rogers, 1996; Ranade and Dommeti, 1996; Ranade, 1997; Bhattacharya and Kresta, 2002); and where a sliding mesh is used to obtain transient solutions (Jaworski et al., 1998; Micale et al., 1999).

Several conditions are needed for accurate RANS simulation of gross circulation patterns:

- There must be fully turbulent flow at the impeller; $Re > 2 \times 10^4$.
- If impeller boundary conditions are used, they should be obtained for exactly the same geometry as is used in the simulation (Fokema et al., 1994).
- If a sliding mesh simulation is used, 20 or more rotations of the impeller are needed for convergence (Jaworski et al., 1998).
- Even with a good preprocessor, the user must pay careful attention to the layout of the grid. This is the single biggest factor affecting both convergence and accuracy of the results. The bottom line is that more cells are needed where large gradients are expected, usually close to the impeller and close to the baffles. Each impeller modeling method has its own gridding constraints in addition to the computational constraints listed above.

***2-5.2.2 Full Reynolds Stress Models.*** Full Reynolds stress modeling retains all six Reynolds stresses throughout the solution of the balance equations. The equations for these stresses are highly coupled and convergence is difficult. The advantage of this approach is that all of the stresses are available to play a role in the development of the flow field, and the transport of energy between components can develop strong directional preferences and coherent structures. This level of complexity in modeling is essential for very difficult, highly anisotropic flows, such as those found in a cyclone.

## 2-5.3 Limitations of Current Modeling: Coupling between Velocity, Concentration, Temperature, and Reaction Kinetics

Even with the rapid progress currently underway in the modeling of velocity fields for fully turbulent flow, the real objective remains the process result. The critical physics lies in interactions between equations of motion and scalar transport and the kinetics of reactions, crystal precipitation and growth, and other core processes. These are coupled higher order sets of equations that need to be solved simultaneously in a truly rigorous solution.

The alternative to this full solution is to take detailed velocity field calculations and extract critical information that can be applied over simplified zones. The reacting fluid particle is then tracked as it moves through the time-averaged (Eulerian) flow field. This Eulerian–Lagrangian approach has been followed by several authors (Bourne and Yu, 1994; Wei and Garside, 1997), with impressive results. The reader is referred to the review by Baldyga and Pohorecki (1995) the text by Baldyga and Bourne (1999), and Chapter 13 for more discussion and information about coupling reaction kinetics information to flow characteristics.

## 2-6 WHAT HAVE WE LEARNED?

- Turbulent blobs and their scalar counterparts are three dimensional, time-varying structures of arbitrary shape. They are represented by the wavenumber spectrum. Various portions of the spectrum, but not the whole spectrum, can be retained on scale-up.
- Models that account for all of the physics of turbulence cannot presently be solved for problems of practical interest. Turbulence models that can be solved do not contain all the physics needed to accurately predict all aspects of the velocity and turbulence fields.
- The effect of turbulence on scalars in the flow (c, T, reaction kinetics) is strong, and is sensitive to the details of the velocity and turbulence fields. Models that have been formulated to solve the combination of velocity and scalar fields have not yet accounted for the multiplicity of interactions between the fields, especially when complex reaction kinetics exist. Steady progress continues in the application of full PDF models to these problems.
- With a good phenomenological understanding of turbulence, many of the gross problems in design and operations can be addressed, despite our incomplete understanding of the physics. As engineers, it is often enough to have a good understanding of the process. Once the crucial issues have been identified, simpler scaling arguments can often provide a satisfactory engineering solution to the problem.

**82** TURBULENCE IN MIXING APPLICATIONS

## NOMENCLATURE

| | |
|---|---|
| a | fluctuating concentration of A (mol/L) |
| A | proportionality constant, 1.0 for isotropic turbulence (−) |
| b | fluctuating concentration of B (mol/L) |
| c | concentration fluctuation (mol/L) |
| $C_A$ | mean concentration of A (mol/L) |
| $C_{B0}$ | concentration of B at time 0 (mol/L) |
| $C_L$ | length scale proportionality constant (−) |
| $C_u$ | velocity scale proportionality constant (−) |
| $d_m$ | molecular diameter (m) |
| D | impeller diameter (m) |
| Da | Damkoehler number (−) |
| $D_{AB}$ | molecular diffusivity of A in B (m$^2$/s) |
| $D_p$ | pipe diameter (m) |
| $D_t$ | turbulent diffusivity (m$^2$/s) |
| E | energy content, or PSD power spectral density |
| f | frequency (s$^{-1}$) |
| H | liquid depth (m) |
| $I_s$ | intensity of segregation (−) |
| k | turbulent kinetic energy per unit mass (m$^2$/s$^2$) |
| k | wavenumber [$2\pi f/U_c$ in eq. (2-9)] (m$^{-1}$) |
| $k_0$ | wavenumber corresponding to largest scale of concentration (m$^{-1}$) |
| $k_r$ | reaction rate constant (units vary) |
| $l_t$ | smallest turbulent scale (Example 2-1c) (m) |
| L | length scale (m) |
| $L_c$ | characteristic length scale (m) |
| $L_s$ | Corrsin integral length scale (m) |
| $L_{1/2}$ | Mao and Toor mixing length (m) |
| N | impeller rotational speed (rps) |
| $N_{js}$ | just suspended speed, solids (rps) |
| $N_p$ | power number (−) |
| r | distance in the radial direction (m) |
| $r_0$ | feed pipe radius (m) |
| Re | Reynolds number (−) |
| Sc | Schmidt number, $\nu/D_{AB}$ (−) |
| t | time (s) |
| $t_e$ | eddy dissipation time scale, $k/\varepsilon$ (s) |
| $t_k$ | Kolmogorov time scale, $(\nu/\varepsilon)^{1/2}$ (s) |
| $t_\lambda$ | time scale based on Taylor microscale, $(\lambda^2/\varepsilon)^{1/3}$ (s) |
| T | tank diameter (m) |
| u | streamwise fluctuating velocity component (m/s) |
| $u_c$ | characteristic turbulent velocity scale (m/s) |
| U | mean velocity in the streamwise direction (m/s) |
| $U_c$ | convective velocity (m/s) |

| | |
|---|---|
| v | cross-stream fluctuating velocity component (m/s) |
| $V_{TIP}$ | impeller tip speed, $\pi ND$ (m/s) |
| V | volume (m$^3$) |
| $V_{impeller}$ | impeller swept volume (m$^3$) |
| $V_{tank}$ | tank volume (m$^3$) |
| w | cross-stream fluctuating velocity component (m/s) |
| x | distance in the x-direction (m) |
| y | distance in the y-direction (m) |
| z | distance in the z-direction (m) |

*Greek Symbols*

| | |
|---|---|
| $\varepsilon$ | rate of dissipation of turbulent kinetic energy per unit mass (m$^2$/s$^3$) |
| $\eta$ | Kolmogorov scale, $(\nu^3/\varepsilon)^{1/4}$ (m) |
| $\theta_B$ | blend time (s) |
| $\lambda$ | Taylor microscale of turbulence (m) |
| $\lambda_B$ | Batchelor length scale, $(\nu D_{AB}^2/\varepsilon)^{1/4}$ (m) |
| $\mu$ | absolute viscosity (kg/m·s) |
| $\nu$ | kinematic viscosity (m$^2$/s) |
| $\rho$ | density (kg/m$^3$) |
| $\tau$ | mixing time constant (s) |
| $\tau_D$ | dimensionless time for unsteady mass transfer [eq. (2-2)] |
| $\tau_{yx}$ | shear stress on the y-plane in the x-direction (Pa) |

# REFERENCES

Armenante, P. M., C. Luo, C. Chou, I. Fort, and J. Medek (1997). Velocity profiles in a closed, unbaffled vessel: comparison between experimental LDV data and numerical CFD predictions, *Chem. Eng. Sci.*, **52**, 3483–3492.

Bakker, A., K. J. Myers, R. W. Ward, and C. K. Lee (1996). The laminar and turbulent flow pattern of a pitched blade turbine, *Trans. Inst. Chem. Eng.*, **74A**, 485–491.

Bakker, A., H. Haidari, and E. Marshall (1998). Numerical modeling of mixing processes—What can LES offer? *Paper 238j*, presented at the AIChE Annual Meeting, Miami Beach, FL, Nov. 15–20.

Baldi, G., R. Conti, and E. Alaria (1978). Complete suspension of particles in mechanically agitated vessels, *Chem. Eng. Sci.*, **42**, 2949–2956.

Baldyga, J., and J. R. Bourne (1999). *Turbulent Mixing and Chemical Reactions*, Wiley, Chichester, West Sussex, England.

Baldyga, J., and R. Pohorecki (1995). Turbulent micromixing in chemical reactors: a review, *Chem. Eng. J.*, **58**, 183–195.

Baldyga, J., W. Podgorska, and R. Pohorecki (1995). Mixing-precipitation model with application to double feed semibatch precipitation, *Chem. Eng. Sci.*, **50**, 1281–1300.

Baldyga, J., J. R. Bourne, and S. J. Hearne (1997). Interaction between chemical reactions and mixing on various scales, *Chem. Eng. Sci.*, **52**, 457–466.

Batchelor, G. K. (1959). Small scale variation of convected quantities like temperature in turbulent fluid: Discussion and the case of small conductivity, *J. Fluid Mech.*, **5**, 113–133.

Bhattacharya, S., and S. M. Kresta (2002). CFD simulations of three dimensional wall jets in stirred tanks, *Can. J. Chem. Eng.*, **80**(4), 695–709.

Bittorf, K. J., and S. M. Kresta (2001). Three dimensional wall jets: axial flow in a stirred tank, *AIChE J.*, **47**, 1277–1284.

Bourne, J. R., and S. Yu (1994). Investigation of micromixing in stirred tank reactors using parallel reactions, *Ind. Eng. Chem. Res.*, **33**, 41–55.

Brodkey, R. S. (1967). *The Phenomena of Fluid Motions*, Dover, Mineola, NY.

Brodkey, R. S. (1975). Mixing in turbulent fields, in *Turbulence in Mixing Operations: Theory and Applications to Mixing and Reaction*, R. S. Brodkey, ed., Academic Press, New York, pp. 49–119.

Brodkey, R. S., and H. C. Hershey (1988). *Transport Phenomena: A Unified Approach*, McGraw-Hill, New York.

Brodkey, R. S., and S. M. Kresta (1999). Turbulent mixing and chemical reactions in an ideal tubular reactor, presented at Mixing XVII, Banff, Alberta, Canada, Aug. 20–25.

Corrsin, S. (1957). Simple theory of an idealized turbulent mixer, *AIChE J.*, **3**, 329–330.

Corrsin, S. (1964). The isotropic turbulent mixer: II. Arbitrary Schmidt number, *AIChE J.*, **10**, 870–877.

Coy, D., R. LaRoche, and S. Kresta (1996). Use of sliding mesh simulations to predict circulation patterns in stirred tanks, presented at AIChE Annual Meeting, Chicago, Nov. 10–15.

Derksen, J., and H. E. A. van den Akker (1999). Large eddy simulations of the flow driven by a Rushton turbine, *AIChE J.*, **45**, 209–221.

Escudier, R. (2001). Structure locale de l'hydrodynamique générer par une turbine de Rushton, Ph.D. dissertation, INSA, Toulouse, France.

Fokema, M. D., S. M. Kresta, and P. E. Wood (1994). Importance of using the correct impeller boundary conditions for CFD simulations of stirred tanks, *Can. J. Chem. Eng.*, **72**, 177–183.

Forney, L. J., and N. Nafia (1998). Turbulent jet reactors: mixing time scales, *Trans. Inst. Chem. Eng.*, **76A**, 728–736.

Forney, L. J., and N. Nafia (2000). Eddy contact model: CFD simulations of liquid reactions in nearly homogeneous turbulence, *Chem. Eng. Sci.*, **55**, 6049–6058.

Fort, I. (1986). Flow and turbulence in agitated vessels, Chapter 14 in *Mixing: Theory and Practice*, V. W. Uhl and J. B. Gray, eds., Academic Press, Toronto.

Fox, R. O. (1998). On the relationship between Lagrangian micromixing models and computational fluid dynamics, *Chem. Eng. Process.*, **6**, 521–535.

Grenville, R. K., and J. N. Tilton (1996). A new theory improves the correlation of blend time data from turbulent jet mixed vessels, *Trans. Inst. Chem. Eng.*, **74A**, 390–396.

Grenville, R. K., and J. N. Tilton (1997). Turbulence or flow as a predictor of blend time in turbulent jet mixed vessels, *Recent Prog. Genie Proc., Paris*, **11**, 67–74.

Grenville, R., S. Ruszkowski, and E. Garred (1995). *Blending of miscible liquids in the turbulent and transitional regimes*, presented at Mixing XV, Banff, Alberta, Canada.

Guezennec, Y., R. S. Brodkey, N. Trigui, and J. C. Kent (1994). Algorithms for fully automated three-dimensional particle tracking velocimetry, *Exp. Fluids*, **17**, 209–219.

Gurney, H. P., and J. Lurie (1923). Charts for estimating temperature distributions in heating or cooling solid shapes, *Ind. Eng. Chem.*, **15**, 1170–1172.

Hansen, L., J. E. Guilkey, P. A. McMurtry, and J. C. Klewicki (2000). The use of photoactivatable fluorophores in the study of turbulent pipe mixing: effects of inlet geometry, *Meas. Sci. Tech.*, **11**, 1235–1250.

Harris, C. K., D. Roekaerts, F. J. J. Rosendal, F. G. J. Buitendijk, Ph. Dakopoulos, A. J. N. Vreenegoor, and H. Wang (1996). CFD for chemical reactor engineering, *Chem. Eng. Sci.*, **51**, 1569–1594.

Harvey, A. D., and S. E. Rogers (1996). Steady and unsteady computation of impeller stirred reactors, *AIChE J.*, **42**, 2701–2712.

Heisler, M. P. (1947). Temperature charts for induction and constant temperature; heating, *Trans. ASME*, **69**, 227–236.

Hinze, J. O. (1975). *Turbulence*, 2nd ed., McGraw-Hill, Toronto.

Houcine, I., E. Plasari, R. David, and J. Villermaux (1999). Feedstream jet intermittency phenomenon in a continuous stirred tank reactor, *Chem. Eng. J.*, **72**, 19–30.

Jaworski, Z., M. L. Wyszynski, K. N. Dyster, V. P. Mishra, and A. W. Nienow (1998). A study of an up and a down pumping wide blade hydrofoil impeller: II. CFD analysis, *Can. J. Chem. Eng.*, **76**, 866–876.

Jo, M. C., W. R. Penny, and J. B. Fasano (1994). Backmixing into reactor feedpipes caused by turbulence in an agitated vessel, in *Industrial Mixing Technology: Chemical and Biological Applications*, G. B. Tatterson, ed., AIChE Symp. Ser., **90**, 41–49.

Keeler, R. N., et al. (1965). Mixing and chemical reaction in turbulent flow reactors, *AIChE J.*, **11**, 221–227.

Kresta, S. M. (1998). Turbulence in stirred tanks, anisotropic, approximate, and applied, *Can. J. Chem. Eng.*, **76**, 563–576.

Kresta, S. M., and P. E. Wood (1991). Prediction of the three dimensional turbulent flow in stirred tanks, *AIChE J.*, **37**, 448–460.

Kresta, S. M., K. J. Bittorf, and D. J. Wilson (2002). Internal annular wall jets: radial flow in a stirred tank, *AIChE J.*, **47**, 2390–2401.

Lamberto, D. J., M. M. Alvarez, and F. J. Muzzio (1999). Experimental and computational investigation of the laminar flow structure in a stirred tank, *Chem. Eng. Sci.*, **54**, 919–942.

Launder, B. E. (1995). Modeling the formation and dispersal of streamwise vortices in turbulent flow, 35th Lanchester Lecture, *Aeronaut. J.*, **99**(990), 419–431.

Lee, K. C., and M. Yianneskis (1998). Turbulence properties of the impeller stream of a Rushton turbine, *AIChE J.*, **44**, 13–24.

Mao, K. W., and H. L. Toor (1971). A diffusion model for reactions with turbulent mixing, *AIChE J.*, **16**, 49–52.

Mathieu, J., and J. Scott (2000). *An Introduction to Turbulent Flow*, Cambridge University Press, New York.

McKelvey, K. N., H. Yieh, S. Zakanycz, and R. S. Brodkey (1975). Turbulent motion, mixing and kinetics in a chemical reactor configuration, *AIChE J.*, **21**, 1165–1176.

Micale, G., A. Brucato, F. Grisafi, and M. Ciofalo (1999). Prediction of flow fields in a dual impeller stirred vessel, *AIChE J.*, **45**, 445–464.

Michelet, S. (1998). Turbulence et dissipation au sein d'un reacteur agité par une turbine Rushton: vélocimètrie laser Doppler a deux volume de mesure, Ph.D. dissertation, Institut National Polytechnique de Lorraine, France.

Monclova, L. A., and L. J. Forney (1995). Numerical simulation of a pipeline tee mixer, *Ind. Eng. Chem. Res.*, **34**, 1488–1493.

Nienow, A. W. (1997). On impeller circulation and mixing effectiveness in the turbulent flow regime, *Chem. Eng. Sci.*, **52**, 2557–2565.

Nouri, J. M., J. H. Whitelaw, and M. Yianneskis (1987). The scaling of the flow field with impeller size and rotational speed in a stirred reactor, presented at the 2nd International Conference on Laser Anemometry, Advances and Applications, Strathclyde, Scotland, Sept. 21–23.

Pope, S. B. (2000). *Turbulent Flows*, Cambridge University Press, New York.

Praturi, A. K., and R. S. Brodkey (1978). A stereoscopic visual study of coherent structures in turbulent shear flow, *J. Fluid Mech.*, **89**, 251–272.

Ranade, V. V. (1997). An efficient computational model for simulating flow in stirred vessels: a case of Rushton turbine, *Chem. Eng. Sci.*, **52**, 4473–4484.

Ranade, V. V., and S. M. S. Dommeti (1996). Computational snapshot of flow generated by axial impellers in baffled stirred vessels, *Trans. Inst. Chem. Eng.*, **74A**, 476–484.

Revstedt, J., L. Fuchs, and C. Tragardh (1998). Large eddy simulations of the turbulent flow in a stirred reactor, *Chem. Eng. Sci.*, **53**, 4041–4053.

Roberts, R. M., M. R. Gray, B. Thompson, and S. M. Kresta (1995). The effect of impeller and tank geometry on circulation time distributions in stirred tanks, *Chem. Eng. Res. Des.*, **73A**, 78–86.

Roussinova, V., B. Grgic, and S. M. Kresta (2000). Study of macro-instabilities in stirred tanks using a velocity decomposition technique, *Chem. Eng. Res. Des.*, **78**, 1040–1052.

Roussinova, V. T., S. M. Kresta, and R. Weetman (2003). Low frequency macroinstabilities in a stirred tank: scale-up and prediction based on large eddy simulations, *Chem. Eng. Sci.* **58**, 2297–2311.

Spalding, D. B (1971). Mixing and chemical reaction in confined turbulent flames, presented at the 13th International Symposium on Combustion, Combustion Institute, Pittsburgh, PA, pp. 649–657.

Tang, L., F. Wen, Y. Yang, C. T. Crowe, J. N. Chung, and T. R. Troutt (1992). Self-organizing particle dispersion mechanism in a plane wake, *Phys. Fluids A*, **4**, 2244–2251.

Taylor, G. I. (1921). Diffusion by discontinuous movements, *Proc. London Math. Soc.*, **20**, 196–212.

Tennekes, H., and J. L. Lumley (1972). *A First Course in Turbulence*, MIT Press, Cambridge, MA.

Toor, H. L. (1962). Mass transfer in dilute turbulent and nonturbulent systems with rapid irreversible reactions and equal diffusivities, *AIChE J.*, **8**, 70–78.

Toor, H. L. (1969). Turbulent mixing of two species with and without chemical reaction, *Ind. Eng. Chem. Fundam.*, **8**, 655–659.

Vassilatos, G., and H. L. Toor (1965). Second order chemical reactions in a nonhomogeneous turbulent fluid, *AIChE J.*, **2**, 666.

Vivaldo-Lima, E., P. E. Wood, A. E. Hamielec, and A. Penlidis (1998). Calculation of the PSD in suspension polymerization using a compartment mixing model, *Can. J. Chem. Eng.*, **76**, 495–505.

Wei, H., and J. Garside (1997). Application of CFD modeling to precipitation systems, *Trans. Inst. Chem. Eng.*, **75A**, 219–227.

Weetman, R. W. (2002). Personal correspondence.

Yegneswaran, P. K., B. G. Thompson, and M. R. Gray (1991). Experimental simulation of dissolved oxygen fluctuations in large fermentors: effect on Streptomyces clavuligerus, *Biotechnol. Bioeng.*, **38**, 1203–1209.

Yianneskis, M., Z. Popiolek, and J. H. Whitelaw (1987). An experimental study of the steady and unsteady flow characteristics of stirred reactors, *J. Fluid Mech.*, **175**, 537–555.

Zhao, Y., and R. S. Brodkey (1998a). Averaged and time resolved, full-field (three-dimensional), measurements of unsteady opposed jets, *Can. J. Chem. Eng.*, **76**, 536–545.

Zhao, Y., and R. S. Brodkey (1998b). Particle paths in three-dimensional flow fields as a means of study: opposing jet mixing system, *Powder Technol.*, **100**, 161–165.

Zhou, G., and S. M. Kresta (1996a). Distribution of energy between convective and turbulent flow for three frequently used impellers, *Chem. Eng. Res. Des.*, **74A**, 379–389.

Zhou, G., and S. M. Kresta (1996b). Impact of geometry on the maximum turbulence energy dissipation rate for various impellers, *AIChE J.*, **42**, 2476–2490.

Zhou, G., and S. M. Kresta (1998). Evolution of drop size distribution in liquid-liquid dispersions for various impellers, *Chem. Eng. Sci.*, **53**, 2099–2113.

Zwietering, T. N. (1958). Suspending of solid particles in liquid by agitators, *Chem. Eng. Sci.*, **8**, 244–253.

# CHAPTER 3

# Laminar Mixing: A Dynamical Systems Approach

EDIT S. SZALAI, MARIO M. ALVAREZ, and FERNANDO J. MUZZIO

*Rutgers University*

## 3-1 INTRODUCTION

Laminar mixing has been subject to formal study only recently, and it is safe to say that it is currently an "art" rather than a fully developed scientific discipline. The fundamentals evolved mainly from empiricism to a semiqualitative level, and a unifying framework that describes the dynamics of laminar mixing processes has only begun to emerge in the past decade. Our limited understanding of mixing fundamentals is a chief limiting factor in effective design of mixing equipment. Mixing, as an individual subject, is absent from most existing chemical engineering curricula. From a fundamental viewpoint, one of the main motivations to study of laminar mixing is that the underlying physics is amenable to direct analysis in a rigorous framework to develop meaningful theory. Additionally, mixing problems in laminar environments tend to be very difficult. In the context of pharmaceutical, food, polymer, and biotechnological processes, liquid mixing applications are frequently carried out at low speeds or involve high viscosity substances, such as detergents, ointments, creams, suspensions, antibiotic fermentations, and food emulsions. During the past few years there has been growing awareness of problems related to incomplete or inefficient mixing at various stages of these manufacturing processes. Some examples include dispersing particles or releasing bubbles from viscous liquids, oxygen or substrate limitation in high viscosity fermentation broths, preparation of sugar solutions for tablet or candy coatings, and difficulties controlling pH in batches of detergent (Muhr, 1995; Amanullah, 1998).

A full set of color figures for this chapter is provided on the Visual Mixing CD affixed to the back cover of the book.

---

*Handbook of Industrial Mixing: Science and Practice*, Edited by Edward L. Paul,
Victor A. Atiemo-Obeng, and Suzanne M. Kresta
ISBN 0-471-26919-0    Copyright © 2004 John Wiley & Sons, Inc.

In other instances, the concern for mechanical damage constrains mixing operations to the low-speed regime. Biological processes using shear-sensitive cell cultures (mammalian cells, plant cells, or mycelium) are prime examples where adequate mixing must be accomplished at low speeds to avoid compromising the metabolic and physical integrity of shear-sensitive cells and molecules. Increases in batch viscosity are also a common characteristic of these applications (e.g., production of antibiotics using mycelial fungus). As the biomass concentration increases at continued shearing, the fluid properties often become highly complex and strongly non-Newtonian, imposing new challenges on effective mixing.

## 3-2 BACKGROUND

### 3-2.1 Simple Mixing Mechanism: Flow Reorientation

Turbulent flows are intrinsically time-dependent. The velocity field is nonsteady, which implies a continuous reorientation of fluid particles along Lagrangian trajectories. However, steady laminar flows are often encountered when dealing with low fluid velocities or high viscosity materials. These types of processes can be poor mixing environments, because fluid motion is dominated by linear, viscous forces instead of nonlinear inertial forces. If the forcing is time-independent, fluid particles can follow concentric, closed streamlines. From a mixing standpoint, the lack of local time dependence often has a cost: poor performance.

Consider flow in a section of straight pipe, a typical example of steady, nonchaotic flows (often referred to as a regular flow). This problem is two dimensional, since the geometry possesses complete angular symmetry. It is well established that laminar flow of a Newtonian fluid produces a steady parabolic velocity profile within the pipe. To describe mixing in this flow, dye is injected at the inlet at different radial locations, as shown in Figure 3-1a. Consider how the dye streams behave as they are convected by the flow, if the flow regime is indeed steady. In the absence of diffusion, the material remains confined within streamlines that are forever parallel. Different streams do not mix with each other, and the surface separating representative streams (also known as the intermaterial contact area between two fluids) grows at most linearly in time. The pipe flow stretches the separating distance between neighboring particles at a constant rate, without any reorientation to the direction. The action of diffusion in viscous materials is minimal, because diffusivity is usually very low and the mixing time to achieve homogeneity via diffusion in most applications is prohibitively long. Should we conclude, then, that it is impossible to mix efficiently in laminar flows?

Quite the opposite is true, as mixing equipment such as stirred tanks (certainly the most widely used apparatus), static mixers, roller bottles, extruders, and so on, are employed commonly in laminar mixing applications with varying degrees of success. Such equipment, with the right choice of operational parameters and design, can disrupt continuous particle trajectories and create chaos (see Figure 3-1b). How are these systems different from our previous example, the straight pipe?

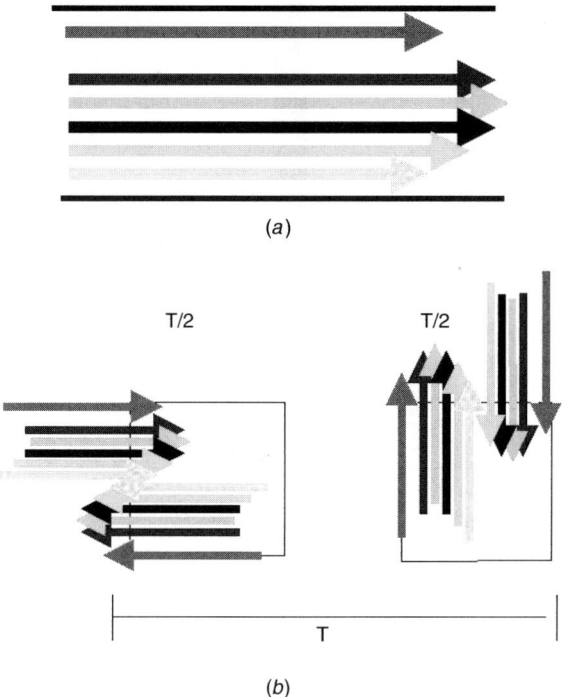

**Figure 3-1** Steady flows mix poorly. In a simple 2D flow, such as flow in a pipe (*a*), the intermaterial area between the colored streams does not grow in time. However, if two steady velocity fields are applied periodically in alternating directions to a stream of fluid (*b*), the interfaces between the dye streams grow exponentially in time.

These laminar mixing systems are effective because they make the mixing process time-dependent in a Lagrangian sense. The mechanism in all the equipment mentioned above is quite similar: a "periodic" forcing of the fluid. In the case of a stirred tank operated at a constant speed, each passage of the impeller blades disturbs the fluid periodically. In the case of a Kenics static mixer (Middlemann, 1977; Pahl and Muschelknautz, 1982; Ling and Zhang, 1995; Hobbs et al., 1997; Jaffer and Wood, 1998), the geometry of the system imposes spatial periodicity: Each element is a repetition of the preceding one twisting in the opposite direction, with a 90° rotation in between, forcing fluid elements to reorient in these transitional regions. Chaotic motion generated by periodic flows represents an important class of chaotic flows in general. Chaotic mixing is characterized by an exponential rate of stretching (as opposed to linear stretching in a nonchaotic flow) of fluid elements. As a fluid element travels through a chaotic flow, it is not only stretched, but also reoriented due to the repeated change in the direction of the flow field that acts on it. Reorientation leads to folding of material lines. The repetition of stretching and folding cycles increases the intermaterial area exponentially and reduces correspondingly the scale of segregation of the

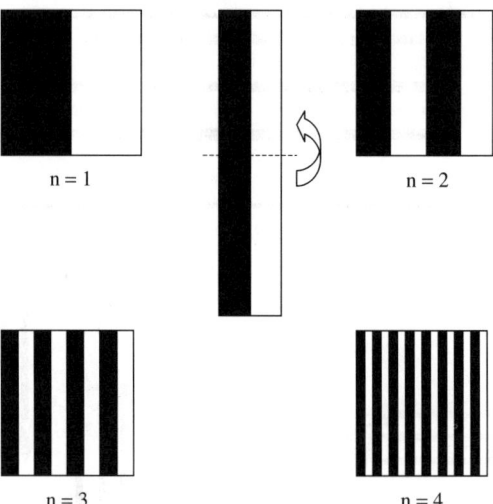

**Figure 3-2** Two components of chaotic mixing, stretching and folding, illustrated by a simple model. Baker's map defines a mixing protocol that stretches fluid elements to double length and folds them in each unit time n. The amount of intermaterial contact area (the interface between the light and the dark regions) grows at an exponential rate as the recipe is applied repeatedly.

system, also at an exponential rate (see Figure 3-2), following a general iterative "horseshoe" mechanism.

### 3-2.2 Distinctive Properties of Chaotic Systems

Although we have suggested a connection between chaos and laminar mixing, we have not established it formally. Two important properties of chaotic systems make them excellent candidates to mix in laminar conditions:

#### 3-2.2.1 Exponential Divergence of Nearby Particle Trajectories.
In a chaotic system, the distance separating two fluid particles initially located very close to one another will diverge exponentially in time. Considering that the objective of any mixing operation is to disperse clusters of material, exponential divergence of clusters of material that are initially close to each other is extremely desirable for mixing applications. Figure 3-3 illustrates some features of complex chaotic flows. Let us represent the initial distance between two particles as an infinitesimal vector of length $l_0$. Gradients in the velocity field continuously reorient and elongate the small material filament as it visits different regions in the flow. Its length at a later time $t_n$ is denoted by $l_n$. In a chaotic flow, the ratio $l_n/l_0$ (formally known as the *stretching* of a fluid element, denoted by $\lambda$) grows as

$$\lambda = \frac{l_n}{l_0} = e^{\Lambda t} \tag{3-1}$$

BACKGROUND 93

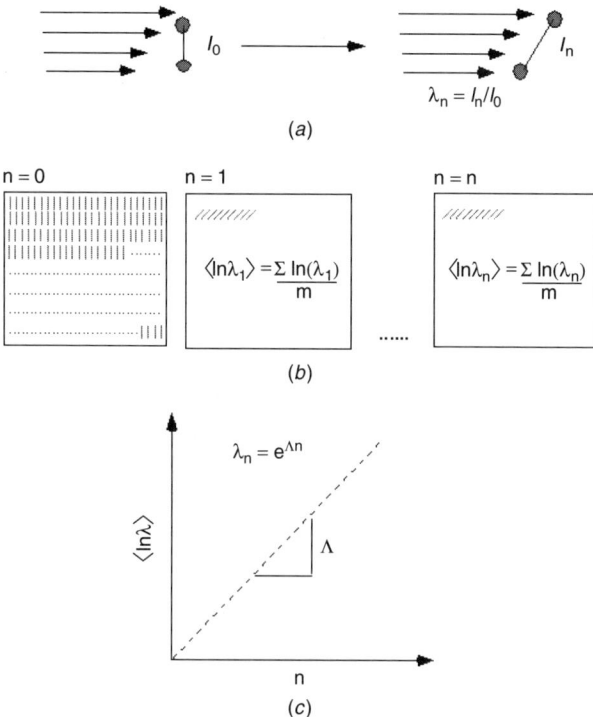

**Figure 3-3** Calculating the stretching field and the Lyapunov exponent. In (*a*) a small material filament, represented by a vector $l_0$, is convected by a flow. As a consequence, its length increases from the initial $l_0$ to $l_n$. The stretching ($\lambda$) experienced by the material line after each period n is the ratio $l_0/l_n$. in (*b*) an array of small vectors is placed in the flow, and the stretching of each is measured and an average $\lambda$ can be calculated. In (*c*) a chaotic flow, $\lambda$ grows exponentially, and the exponent characterizing the growth rate ($\Lambda$, the Lyapunov exponent) can be calculated from the slope of the curve $\langle \ln \lambda \rangle$ versus n.

where $l_n$ is the separation distance between the two particles at time $t_n$. The stretching $\lambda$ represents the intensity of the mixing process experienced by the material element. The spectrum of mixing intensities in the flow can be measured using a large number of vectors to represent a collection of tiny material filaments populating the entire flow domain. The exponential rate of average growth of vector length is usually represented by the constant $\Lambda$, called the *Lyapunov exponent* of the flow (Oseledek, 1968), which is a volume-averaged measure of the average rate of stretching at a given flow period. Strictly speaking, the Lyapunov exponent is the asymptotic limit on stretching as time approaches infinity and $l_o$ goes to zero. Chaotic flows are recognized by having positive Lyapunov exponents.

### 3-2.2.2 Frequency Distribution of Stretching.

A single particle traveling along a chaotic trajectory can explore an entire chaotic region densely and completely uniformly as time approaches infinity (i.e., in the asymptotic time limit). In a mixing context, this property of chaotic flows assures that every particle will eventually visit all areas in the chaotic region. The short-time frequency with which particles visit a particular region of the flow depends on their initial spatial position. The spatial position determines the amount of stretching and reorientation that a small fluid element experiences at that location. Eventually, the continuous stretching and reorientation process leads to the distribution of materials throughout the chaotic region.

The area visited by particles in a chaotic flow can be illustrated by the use of plots known as *Poincaré sections* or *maps*, a very common tool in dynamical systems theory. These plots show the long-time behavior of a mixing system by revealing whether the entire flow is chaotic or contains slow-mixing, segregated regions (*islands*). When preparing a Poincaré plot of a three dimensional dynamical system, such as a stirred tank or a static mixer, a two dimensional plane perpendicular to the main flow direction is typically chosen. This plane is intersected periodically by tracer particles following the flow. Some fluid tracer particles are marked in the flow and particle positions are recorded every time they cross the 2D plane. All these positions are superimposed on a single plot after many time periods.

An analogous definition of Poincaré sections can be applied to 2D time-dependent periodic systems. After some particles are marked in the flow, instantaneous snapshots of the system are taken at periodic intervals of time and overlapped on a single plot (Aref, 1984; Franjione and Ottino, 1987; Leong and Ottino, 1989; Muzzio and Swanson, 1991; Kusch and Ottino, 1992; Alvarez et al., 1997). The time period is usually chosen to be the period of the driving force that creates motion in the system. Poincaré plots simplify the analysis of a dynamical system, because the original system dimension is lowered by one, yet the characteristic dynamics are retained. In a Poincaré section, regions of chaotic motion appear as a cloud of points that will eventually fill the entire chaotic domain. Regions of regular motion (also known as islands or isolated regions) appear either as empty regions (if no particles were originally placed within) or as sets of closed curve (see Figure 3-4 for an example). The boundaries between the regular and chaotic regions are known as *KAM surfaces* [after Kolmogorov, Arnold, and Moser (Kolmogorov, 1954; Moser, 1962; Arnold, 1963)] and appear as closed curves in Poincaré sections. These surfaces pose a significant barrier to transport, because material exchange can occur across these boundaries only by diffusive mechanisms.

### 3-2.3 Chaos and Mixing: Some Key Contributions

The notion of complex, chaotic motion linked to simple dynamical systems is certainly not new. Aref introduced a set of ideas that led to major research efforts now spanning almost two decades (Aref, 1984). His main contribution was to incorporate concepts of dynamical systems theory to fluid mechanics, specifically in the context of laminar mixing. He was able to demonstrate that

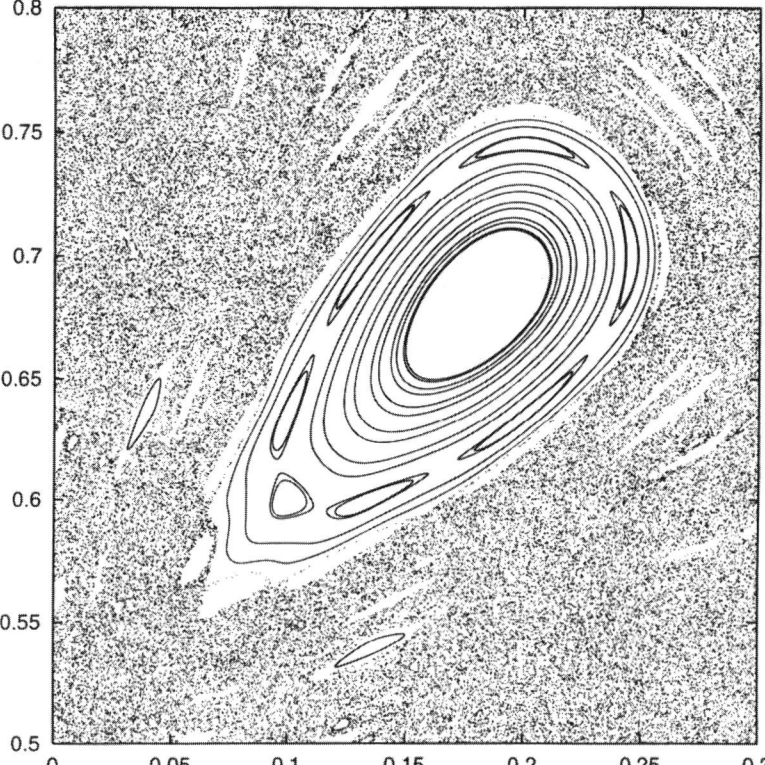

**Figure 3-4** Poincaré section of a model flow (the sine flow). Areas of chaotic motion appear as a random cloud of points, whereas segregated regions (i.e., islands) appear as empty regions, if no particles were originally placed inside. Tracers particles that are placed inside islands appear as sets of closed curves or KAM surfaces in quasi-periodic systems, and as points in a periodic system.

even very simple systems can exhibit chaos when operated in a time-periodic fashion. Working with a model 2D map, Aref's computations demonstrated that some time-periodic recipes could lead to chaotic motion almost everywhere in a flow domain.

The first experimental evidence of chaos in a 2D time-periodic flow was provided by Chien et al. (1986). The experimental system was a chamber filled with fluid, where two opposing surfaces could be moved independently. The shearing motion in the cavity causes flow and mixing in the system, otherwise referred to as *cavity flow*. If the direction and duration of motion for each surface is chosen carefully, the system can be fully chaotic or contain areas of regular flow (i.e., islands). Chaiken et al. (1986) provided strong evidence of chaotic motion in the journal bearing flow, a time-periodic flow between two eccentric cylinders. These three 2D systems—Aref's blinking vortex, Chien's cavity flow, and Chaiken's journal bearing—have since been investigated by many other authors (Muzzio and Ottino, 1988; Leong and Ottino, 1989; Swanson and Ottino, 1990; Metcalfe and Ottino, 1994; Souvaliotis et al., 1995). As a result, a solid theoretical framework for chaotic mixing processes has been established for 2D systems, where the dynamics governing fluid and particle motion are simpler than in realistic 3D flows.

The idea that chaotic motion can exist in three dimensional (3D) systems was known among dynamical systems theorists when Aref began his studies on 2D maps. Poincaré was perhaps the first investigator to formally observe that deterministic systems can exhibit "erratic" trajectories, later called *chaotic trajectories*. Kusch presented the first formal experimental study of chaotic mixing in laminar regime in 3D systems (Kusch and Ottino, 1992). Using two different geometries, the eccentric helical annular mixer and the partitioned pipe mixer, he demonstrated the existence of isolated regions coexisting with chaotic regions in 3D physically realizable systems (although these systems are not used in industrial applications). It is now well established that chaotic motion is the cause of effective laminar mixing in industrially relevant 3D systems such as stirred tanks and static mixers (Dong et al., 1994; Lamberto et al., 1996; Hobbs et al., 1997; Hobbs and Muzzio, 1998). Due to chaotic fluid motion in these mixing devices, fluid elements elongate exponentially fast while the diffusive length scale reduces at the same rate.

## 3-3 HOW TO EVALUATE MIXING PERFORMANCE

### 3-3.1 Traditional Approach and Its Problems

A long-standing issue in mixing theory (not only in the context of laminar mixing) has been how to characterize the state of a mixture. A largely debated issue is exactly what to measure and how to measure it. Resolving this problem is not a trivial task.

In Section 3-2 we described a common tool, Poincaré sections, which greatly simplify analysis of mixing in complex periodic systems since they reduce the dimensionality of the flow by one. Visualization of three dimensional mixing patterns becomes easier in two dimensional cross-sections, and some asymptotic characteristics such as permanently segregated areas in the flow can easily be revealed. However, important dynamics are lost when the evolution of mixing patterns is not considered on a shorter time scale. An inherent weakness of Poincaré sections is the loss of structure within the chaotic region of a flow. The fast, exponential separation between neighboring fluid particles severely limits the use of discrete points for representing continuous lamellar structures. Even if new points are added as mixing patterns develop, the complex spatial structure created by the flow is not retained with point tracers, because continuous material lines soon become a featureless, random cloud of points. See Figure 3-4, where large areas of the chaotic flow domain do not seem to possess any structure.

Some shortcomings of Poincaré sections are overcome when another measure, stretching, is used to analyze chaotic systems and exploit flow topology. However, this tool has not been the preferred approach among practitioners, probably

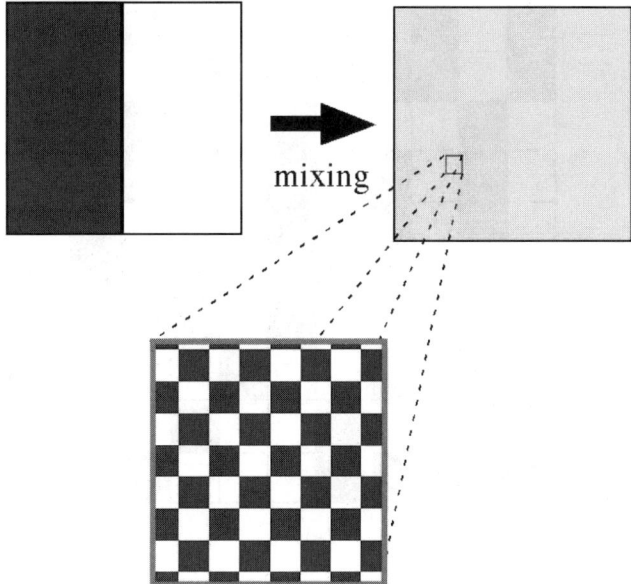

**Figure 3-5** Mixing is a process that increases the homogeneity of a system.

because it is not the most intuitive. When a mixing operation is performed, the ultimate objective is to achieve a target level of homogeneity within the mixture, and to do it in the fastest, cheapest, and if possible, the most elegant way. A simple mixing process is depicted in Figure 3-5, where the entire process volume is represented with a square domain. Starting from a highly segregated condition, dark fluid in one side of the square domain and light in the other, the mixing process generates a state at which the two colors are indistinguishable on the length scale of the system.

If the mixture is sampled in smaller quantities and examined on a smaller scale, there is always a level at which inhomogeneities can be detected. A first and obvious proposal to assess the state of a mixture would be to measure a property of interest throughout the system (e.g., concentration of a key component) and to determine the magnitude of deviations in the sample values from the target value desired for the entire mixture. On one hand, this is probably the most common approach to characterize mixtures both conceptually and in practice. On the other hand, this technique can lead to inconsistencies when trying to use it as an effective mixing measure. The definition of the size and location of samples and the total number taken are essential to establish the validity of any mixing measure. Thus, the placement and size of sampling probes must be matched if different experiments are to be compared. Consider sampling the mixture as depicted in Figure 3-6a. If different combinations of samples are analyzed from the same mixture the concentration estimation can deviate significantly from the "true" value of 0.5.

(a)

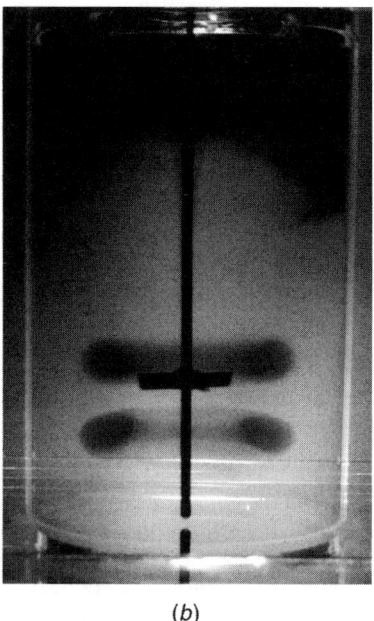

(b)

**Figure 3-6** Problems associated with some statistical methods of mixture characterization. If a different set of samples is selected from (a) than the one enclosed in the highlighted box, a different overall concentration is calculated each time. Real mixing systems, such as stirred tanks with laminar flow (b), exhibit such heterogeneity. If measured by a set of electrodes, the concentration of $H^+$ depends strongly on the placement of the electrodes in the tank.

Clearly, a measure that is so dependent on the scale and mechanism of application is not sufficiently robust. An even more significant aspect of this approach is not its reliability or its reproducibility[1] but its physical significance. Laminar mixing processes can show systematic effects, such as the presence of islands that can bring into question the practice of repeatedly sampling at discrete locations. Consider measuring the mixing time in a stirred tank using a neutralization reaction when operated in the laminar regime. In principle, if a set of pH electrodes is inserted at different positions in the tank, pH measurements can be taken as the mixing process evolves to indicate the state of the reaction. At the starting condition, all electrodes will read the same value within an experimental uncertainty of about 1%. It is expected that the global variance will decrease to a near-zero value, and at that point, the mixing time can be established. In reality, the measurement does not give such a clear indicative answer (see Figure 3-6b). Some electrode readings approach the target value much faster than others: as soon as 20 min for some probes, or only after several hours for others.

A simple mixing visualization technique can tell us why. If we perform the same neutralization in a transparent vessel, pH indicator can be added to the viscous media to indicate the local $H^+$ concentration. In acid environments, the liquid turns yellow, whereas in basic environments it is blue. A picture of an initially basic system taken after mixing for 10 min is presented in Figure 3-6b. A laboratory scale mixing vessel, equipped with three standard Rushton impellers, was initially filled with viscous fluid containing NaOH base and pH indicator. The entire tank appeared initially blue due to the amount of base premixed with the process fluid. The impellers were then set in motion at a constant speed, and acidic solution was injected into the tank near the blades shortly afterward. As the concentrated acidic injection spreads through the vessel, the instantaneous neutralization reaction between the acid and base marks areas of contact between the two components. The change in pH causes a color change in well-mixed areas. The picture after 10 min shows two distinct colors, indicating well-mixed and poorly mixed areas in the tank. The islands, or segregated regions, remain blue, while all regions where the neutralization reaction goes to completion become yellow. The existence of large, blue doughnut-shaped toroidal regions below and above each impeller in the vessel is obvious even to the naked eye. These structures remain in the flow for extended periods of time and are destroyed only by slow diffusion after hours of processing. Obviously, any "average" mixing time calculated based on a few randomly placed, dispersed electrode readings would be meaningless for this system, because it misses such important behavior.

### 3-3.2  Measuring Microstructural Properties of a Mixture

Even more serious inconsistencies can occur when mixture quality is assessed exclusively based on concentration measurements. It is possible that the estimated concentration distribution may be the same for two mixtures that are very different

---

[1] For example, we can improve the statistical significance of the measured concentration estimate by increasing the number of samples.

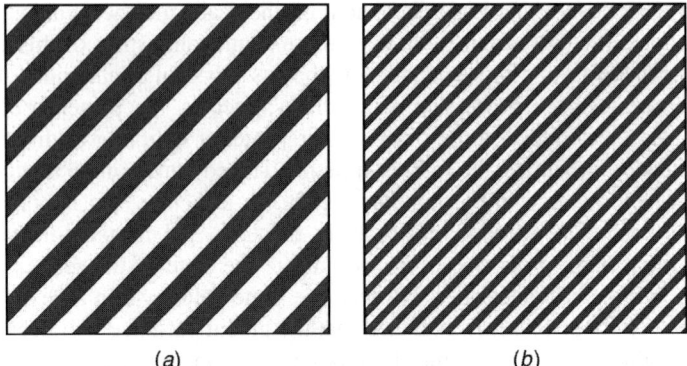

**Figure 3-7** Two mixtures with the same concentration (50% light and 50% dark component), but in (b) the intimacy of contact between the two materials is greater and the diffusional length scale is shorter. More surface area is available for transport in (b) than in (a). If a chemical reaction occurs in each of these systems, the observed reaction rates would be very different.

in nature. The two regions in Figure 3-7a and b have equal amounts of yellow material, 50%. However, the intermaterial contact area between the yellow and the blue constituents is higher in case (b). The parameters affecting mass transfer, the area available for transport and the diffusional distance, are both affected by the topology of the mixture. The area available for transport is greater and the diffusional distance is shorter in case (b). If this were a reactive mixture, these differences would result in a faster overall reaction rate in case (b). Similar situations occur in industrial mixtures (e.g., a polymerization reaction), which exhibit micromixing behavior that is strongly position-dependent.

Mixing processes involving miscible liquids (laminar and turbulent) generate complex striation patterns that are highly heterogeneous where mixing is much more intimate in some regions than in others. This topology is crucial to predicting the outcome of mass transfer and reactions. As an example of the complexity of lamellar structures generated by chaotic flows, refer to Figure 3-8. These photographs were taken at two different instances in time as two fluorescent dye steams are blended with a viscous liquid in a stirred tank. The shaft holding the impeller is visible at the center as a black line. The tan was equipped with a standard Rushton turbine, and green dye was injected above the impeller blades at the start of the experiment. Some time later, the red stream was added to follow the time evolution of the mixing structure in the tank. Within a small area, wide and thin striations coexist. Additionally, the intermaterial area density varies substantially between different regions of the system. This implies that the mixing quality—by any meaningful definition—is substantially different in different regions of the system.

Techniques for characterizing the evolving structural features in a mixing system need to be nonintrusive and robust. Some methods normally used (e.g.,

HOW TO EVALUATE MIXING PERFORMANCE    **101**

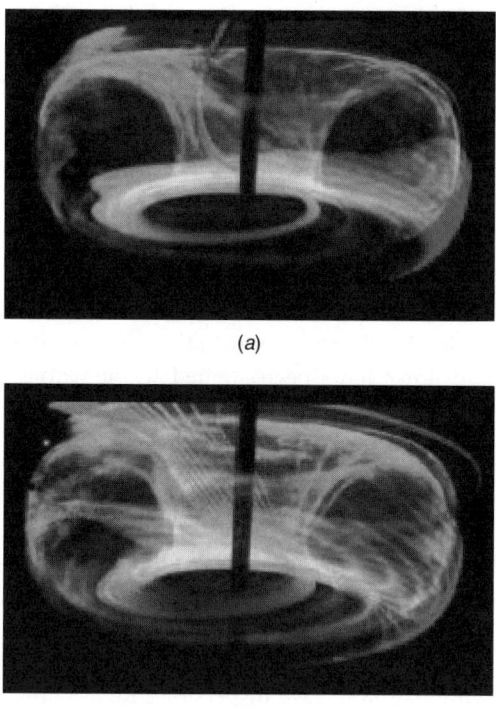

(a)

(b)

**Figure 3-8** The complex mixing patterns formed by chaotic flows are highly heterogeneous, and understanding the emerging structure is crucial to predicting heat and mass transfer in these systems. Here, some experimental pictures are shown of the mixing pattern formed by colored fluorescent dye in a stirred tank. The time evolution of the emerging structure can be monitored if a series of snapshots are taken in time, as in (a) and (b). See insert for a color representation of this figure.

probes) alter the structure that they intend to characterize. Optical methods are usually nonintrusive, but most of the mixing devices widely used in industry are opaque, which limits the applicability of these methods. Even if an optical method can be used, resolution of the most sophisticated optical equipment available today is not enough to resolve fine structural details.[2] As an alternative, recently developed numerical techniques can give insight to the statistics of flow and mixing in physical or model mixing systems. Simple 2D models are a valuable tool to represent and understand the behavior of industrially relevant 3D mixing systems. The analysis can be greatly simplified yet provide significant insight into the physics of realistic flows. It is well known that some characteristic features of real 3D chaotic systems are closely captured, at least qualitatively, by 2D time-periodic chaotic model flows. Namely, the coexistence of regular and chaotic

---

[2] As we will show in a later section, the distribution of length scales characteristic of chaotic flows spans more than four orders of magnitude.

regions, the shape of stretching distributions, and the scaling properties of those distributions are indicative of the similarity between 2D models and 3D flows.

Some methodologies of dynamical systems theory developed from model flows that can be extended to real 3D flows are introduced next. These new tools—stretching calculations, striation thickness distributions, and so on—are essential if one is to obtain detailed statistical information regarding the microstructure emerging in a mixing process and the dynamics of a mixing operation.

### 3-3.3 Study of Microstructure: A Brief Review

As many examples in the literature illustrate, the properties of the structure generated by chaotic flows have been studied by numerous methods in a list of different scientific disciplines. Mandelbrot introduced the concept of fractals, and with it, the notion of universality in the structure generated by chaotic systems (Mandelbrot, 1982). The structure generated by chaotic maps has been the subject of extensive analysis in mixing since dynamical system techniques have been applied to study the topology of mixing structures created by time-periodic stirring protocols (Aref, 1984). Many studies illustrating the complexity of these structures followed. Some numerical studies considering mixing structures in periodic flows include agglomeration in the blinking vortex flow (Muzzio and Ottino, 1988), or the evolution of mixing structure in time-periodic cavity flows (Leong and Ottino, 1989). Dye dispersion in the journal bearing flow was studied both computationally and experimentally (Swanson and Ottino, 1990). These studies recognized the creation of microstructure as an intrinsic characteristic of chaotic flows; nevertheless, the descriptions were primarily qualitative. Stretching and folding (Figure 3-2) were identified as the basic mechanisms that generate the complex structures observed in mixing experiments and simulations.

Since the flow topology governs local phenomena such as mass transfer and chemical reactions, its detailed characterization is the obvious next step. It is necessary to determine the distribution of length scales (or striation thicknesses) created during the stretching and folding process in a flow. Although the conceptual methods were already clear 10 years ago (Muzzio and Swanson, 1991), the limited resolution of image analysis techniques and the erosive effect of diffusion make experimental measurements of striation thickness distributions (STDs) nearly impossible. Currently, quantitative STDs cannot be resolved accurately using photographic techniques.

Until recently, numerical tools and resources were also underdeveloped relative to the magnitude of the striation measurement problem; Poincaré sections and tracer dispersion simulations were the primary techniques used to characterize mixing. To reconstruct striation patterns successfully by computational methods, it is necessary to track continuous material lines (i.e., dye blobs) injected in chaotic flows. The difficulty of such a numerical experiment is hidden in one word of the previous sentence: continuous. The feasibility of tracking material lines or surfaces numerically was explored by Franjione and Ottino, (Franjione and Ottino, 1987). These authors' estimations of time and disk space demands showed that with the computational resources existing at that time, accurate simulation of this type was impossible even in the simplest chaotic flows. As a result, following the evolution of mixing patterns by means of direct tracking of interfaces was abandoned for an entire decade. Several papers examined the evolution of an assumed lamellar microstructure under the effects of diffusion and fast chemical reactions using simplified models of flow topology (Muzzio and Ottino, 1989, 1990; Sokolov and Blumen, 1991).

In parallel, the strong experimental evidence of self-similar geometric properties of mixing structures began to be explored using a statistical approach. The repetitive nature of time- or space-periodic flows was suspected to generate structures endowed with statistical self-similarity. Evidence of the validity of this hypothesis was provided by Muzzio et al. for the drop size distribution produced by breakup in chaotic flows (Muzzio et al., 1991). Since neither experimental nor direct numerical characterization of the striation thickness distribution could be achieved, the computation of stretching, which is related to the increase in interfacial area available for mass and energy transport (feasible for even 3D flows), was suggested as an alternative route to characterize microstructure (Muzzio and Swanson, 1991).

In the remaining sections of this chapter we discuss the characterization of mixing and the implementation of these statistical and computational techniques using a simple time-periodic flow as an example. Since the implementation of the algorithms used here is mostly sequential, it is natural to describe them in the order they are applied.

## 3-4 PHYSICS OF CHAOTIC FLOWS APPLIED TO LAMINAR MIXING

The physics of laminar mixing may be interpreted from several perspectives, each of which is discussed in this section. The key results are summarized in Table 3-1.

### 3-4.1 Simple Model Chaotic System: The Sine Flow

The *sine flow*, a two dimensional model of a chaotic flow field, consists of two sinusoidal velocity fields which alternate for half a period (T/2) carrying fluid in the X and Y directions sequentially:

$$\underline{v} = (V_x, V_y) = \begin{cases} (\sin 2\pi y, 0); & \text{for } nT < t \leq (n+1/2)T \quad (3\text{-}2) \\ (0, \sin 2\pi x); & \text{for } (n+1/2)T < t \leq (n+1)T \quad (3\text{-}3) \end{cases}$$

A sketch of the velocity field is shown in Figure 3-9. This cycle can be repeated many times, where n denotes the number of periods. For most of the examples presented here, the sine flow is applied on a 2D domain. Periodic boundary conditions complete the definition of the system. Tracer particles that leave the domain through one side reenter the domain at the opposite side.

**Table 3-1** Measures of Mixing

| Method of Analysis | Type | Brief Description | Advantages/Limitations | Section |
|---|---|---|---|---|
| Mixing patterns by particle tracking (Poincaré sections) | Experimental/ computational | Colored fluid tracer particles are injected in a flow and their location is tracked during the mixing process by computational or experimental visualization methods. The efficiency of the mixing process is described by how fast and evenly the particles become dispersed | Segregated and slow-mixing regions can easily be identified. / Mixing patterns become indistinguishable as particles disperse in the flow. | 3-4.1 eq. (3-4) |
| Stretching field ($\lambda$) | Computational | The elongation of fluid filaments in each region of a flow is measured by the stretching of small vectors, which are attached to fluid tracer particles. | Effective injection locations and fast-mixing areas can easily be identified. / The effect of diffusion is not measured by the stretching field. | 3-4.2 eq. (3-1), eq. (3-7) |
| Lyapunov exponent ($\Lambda$) | Computational | This measure is the average stretching of fluid filaments after an infinite amount of time. The larger the Lyapunov exponent, the more efficient a mixing process is. | This measure can be used as a single number to compare the long-time efficiency of mixing processes. / $\Lambda$ does not give any indications about mixing patterns and dead zones in the flow. | 3-4.6 eq. (3-11) |

| | | | |
|---|---|---|---|
| Topological entropy ($\Theta$) | Computational | This measure is the average stretching of fluid filaments in a finite amount of time. The larger the topological entropy, the more efficient a mixing process is. | This measure is appropriate to compare average mixing rates for shorter times. / $\Theta$ does not give any indications about mixing patterns and dead zones in the flow. | 3-4.6 eq. (3-12) |
| Intermaterial area density ($\rho$) | Computational | This measure is the amount of contact area between mixture components in each region of a flow. Even and dense distribution of intermaterial area is desired for an efficient mixing process. | This measure can be used to characterize micromixing efficiency at all spatial locations. / $\rho$ cannot be obtained directly in complex flows, but it is predictable from the stretching field. | 3-4.7 eq. (3-13), eq. (3-15) |
| Striation thickness distribution (STD) | Computational | The thickness of fluid filaments (i.e., striations) measures the diffusive length scale. The effectiveness of a mixing process is determined by how fast the striation thickness is reduced in each region of a flow. Uniform distribution of striations is desired everywhere for most effective mixing. | This measure can be used to characterize the diffusional length scale at all spatial locations. / STD cannot be obtained directly in complex flows, but it is predictable from the stretching field. | 3-4.9 eq. (3-15), eq. (3-16) |

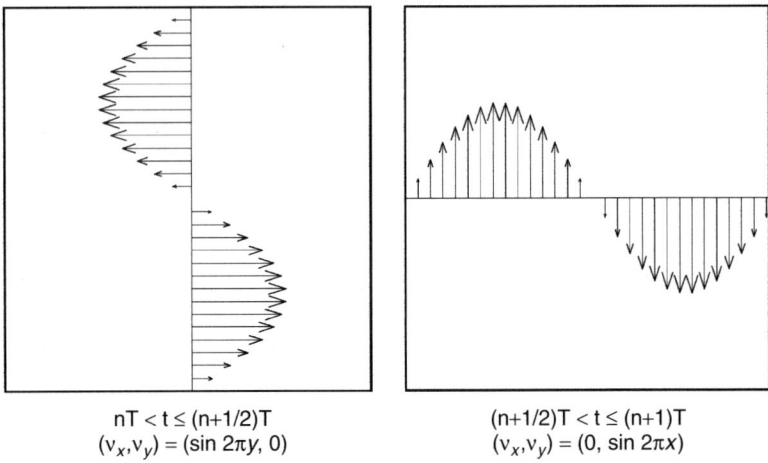

| $nT < t \leq (n+1/2)T$ | $(n+1/2)T < t \leq (n+1)T$ |
| $(v_x, v_y) = (\sin 2\pi y, 0)$ | $(v_x, v_y) = (0, \sin 2\pi x)$ |

**Figure 3-9** The sine flow is a 2D model flow that shows many characteristics of physically realizable 3D chaotic flows. The flow is defined by sequentially applying two steady sinusoidal velocity fields to the unit square flow domain. The duration of time that each velocity field is applied is known as the flow period, T. It is the only parameter that controls the types of mixing behavior that can be observed in this system.

Even though the sine flow is not physically realizable, it is a suitable model to study chaotic flows, for several reasons. First, tracking tracer particles in time (to follow their position, $\underline{X} = (x, y)$, as a function of time) becomes very easy computationally. This is done by integrating the differential equation

$$\frac{d\underline{X}}{dt} = \underline{v} \tag{3-4}$$

With the analytical expression for the velocity field $\underline{v}$ [eqs. (3-2) and (3-3)], there are no errors due to numerical approximations, and a closed-form solution for particle motion is attainable:

$$x_{n+1} = x_n + \frac{T}{2}[\sin(2\pi y_n)] \tag{3-5}$$

$$y_{n+1} = y_n + \frac{T}{2}[\sin(2\pi x_{n+1})] \tag{3-6}$$

The sinusoidal velocity profile and the periodic boundary conditions assure nth-order continuity, such that a particle leaving on one side of the system reenters on the opposite side with the same velocity. Using these expressions, the position of any tracer particle within the system can be calculated forward (or backward) for any time simply by knowing its initial (or final) position. Particle trajectories are a function of a single parameter (T), the period of the flow. The simplicity of the solution [eqs. (3-5) and (3-6)] embodies the beautiful paradox of chaotic flows.

PHYSICS OF CHAOTIC FLOWS APPLIED TO LAMINAR MIXING    **107**

Although the application of these equations for one flow period is a very simple exercise, it cannot be used to obtain an explicit prediction of particle positions over many periods in a deterministic manner. The algebraic complexity grows exponentially with the number of flow periods as the current particle locations are taken into account.

The sine flow reveals different spatial distributions of chaotic and nonchaotic regions for different values of the parameter T, similar to real mixing flows. Figure 3-10 shows Poincaré sections for three parameter values: T = 0.8, T = 1.2, and T = 1.6. Each section is computed by plotting the position of a single particle initially placed at (0.5,0.5) after 10 000 iterations of the flow. The segregated regions initially contain no particles, so they appear as empty islands in the flow. At T = 0.8, four such large regions and many smaller ones are

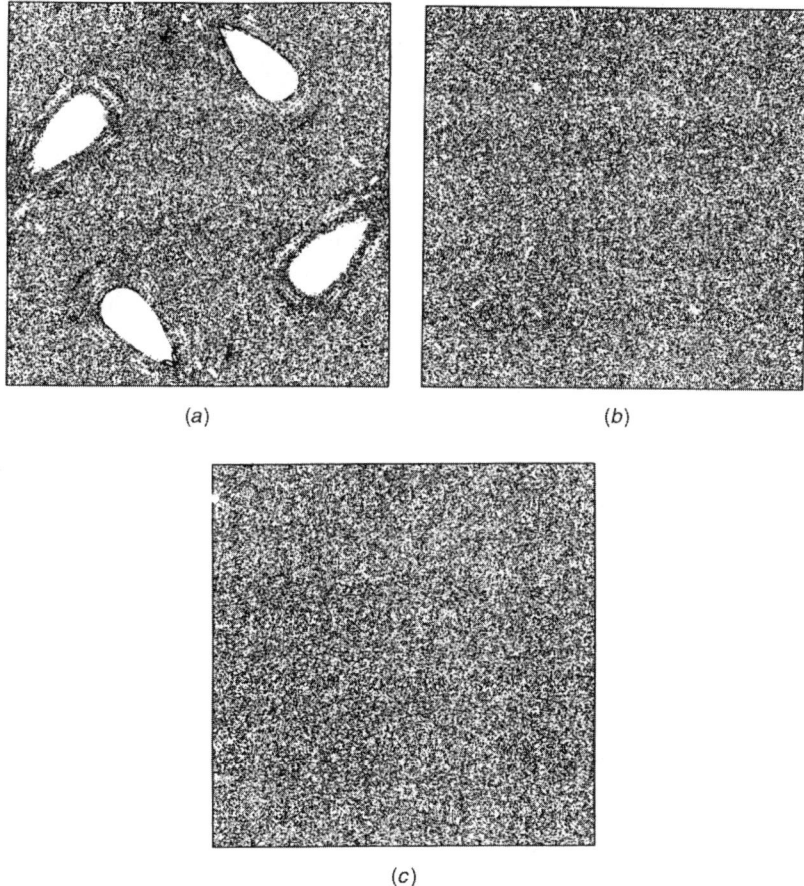

**Figure 3-10** Poincaré sections for three conditions in the sine flow: (*a*) T = 0.8, where large islands are visible; (*b*) T = 1.2, where only small islands are present; (*c*) T = 1.6, where particle motion is chaotic almost everywhere.

noticeable within the chaotic subdomain. (For a close-up, see Figure 3-4.) Islands significantly decrease in size for T = 1.2, and they nearly disappear at T = 1.6, thanks to the fact that this system is chaotic almost everywhere. These figures emphasize that the Poincaré analysis is valuable as an asymptotic diagnostic tool but not as a dynamical one. That is, the section shows the size and shape of the segregated regions in the flow, but it does not show the evolution to that asymptotic state, since the dynamic component (time dependence) is lost. Moreover, it creates an impression of "random, featureless mixing" that is entirely misleading and for many years deflected attention from an essential issue: the strong nonuniformity of mixing intimacy that is a prevailing feature of chaotic flows.

### 3-4.2 Evolution of Material Lines: The Stretching Field

More detailed information obtained by tracking the evolution of an arbitrary material line in a flow can be described by the stretching of a small vector attached to any particle in the flow:

$$\frac{dl}{dt} = (\nabla \underline{v})^T \cdot l \qquad l_{t=0} = l_0 \tag{3-7}$$

where $l_{t=0}$ is the initial stretching vector with a magnitude of 1 and random orientation in the flow field. Since filaments deform under the influence of the velocity gradient, $\nabla \underline{v}$, this equation is coupled with eq. (3-4) and must be integrated simultaneously.

Since the sine flow is continuous everywhere, it is differentiable within the entire flow domain. This property leads to a piecewise analytical expression for the stretching field in the flow domain:

$$l_{x,n+1} = l_{x,n} + 2\pi T[\cos(2\pi y_n)]l_{y,n} \tag{3-8}$$

$$l_{y,n+1} = l_{y,n} + 2\pi T[\cos(2\pi x_{n+1})]l_{y,n+1} \tag{3-9}$$

Here $l_x$ and $l_y$ are the two components of the small stretching vector $l_{n+1}$. Assuming that $|l_0| = 1$, the magnitude of the length stretch is given by

$$l_{n+1} = (l_{x,n+1}^2 + l_{y,n+1}^2)^{1/2} \tag{3-10}$$

The fact that the tracking of fluid particles and the calculation of stretching of material lines can be performed in closed form means that accurate calculations for the mixing microstructure are feasible.

### 3-4.3 Short-Term Mixing Structures

In practical applications we are often interested in the early stages of a mixing process, before the process becomes controlled by diffusion (i.e., when striations reach the length scale of diffusion). Short-term mixing dynamics have been extensively studied in the past by tracer dispersion simulations. Consider the sine flow

PHYSICS OF CHAOTIC FLOWS APPLIED TO LAMINAR MIXING **109**

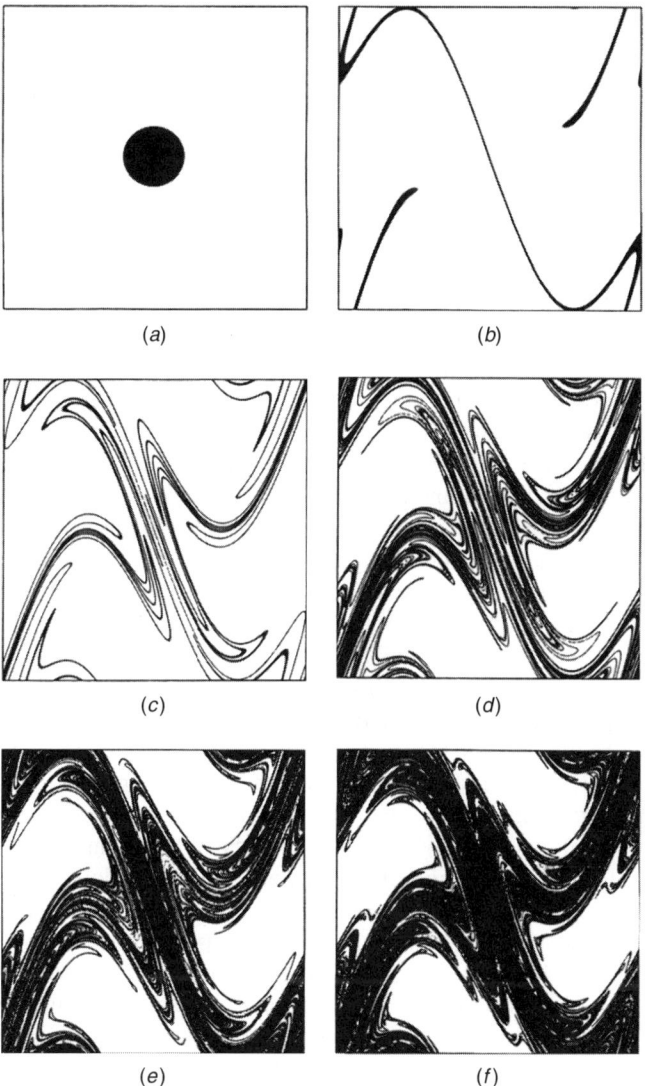

**Figure 3-11** Dispersion of a dye blob in the sine flow with T = 0.8: (*a*) Initially, the dye is at the center of the square flow domain and is spread throughout the chaotic region at times (*b*) n = 2; (*c*) n = 4; (*d*) n = 6; (*e*) n = 8; (*f*) n = 10.

with T = 0.8, where a circular blob of dye is placed at the center of the domain in Figure 3-11. The dye is represented as a collection of 100 000 massless tracer particles, and the position of each particle is plotted for n = 2, 4, 6, 8, and 10 flow periods. As the blob stretches and folds, it invades the flow domain at different rates in different areas, resulting in a highly nonuniform lamellar structure. However, an ability to resolve this structure is lost after a few iterations; as time advances, the points that compose the blob are dispersed and the fine detail of the partially mixed structure fades. After just 10 flow iterations, it is no longer possible to distinguish striations in the center of the domain (see Figure 3-11$f$). As the material is dispersed throughout the entire domain in time, it becomes equivalent to a Poincarè structure. A specialized algorithm is needed to preserve continuity along the material interface by introducing new tracers whenever the distance between adjacent tracers exceeds a predefined limit.

### 3-4.4 Direct Simulation of Material Interfaces

Material filaments are stretched and folded in a chaotic flow, but due to the Hamiltonian nature of incompressible flows (Aref, 1984), they never intersect. Thus, a dye blob can be represented by its surface in 3D or its perimeter in 2D to capture its structure and as a result, the computational effort to simulate the evolution of interfaces can be reduced significantly. The position and stretching of each point along the perimeter of the dye blob was calculated via eqs. (3-4) and (3-7), and the structure can be reconstructed after multiple iterations of the flow. The magnitude of the stretching experienced by the fluid along the original spatial distribution of points revealed continuous partially mixed structures such as the one presented in Figure 3-12. The dye blob was represented only by points tracing its perimeter initially, as in Figure 3-12$a$. This type of analysis reveals details that are "invisible" through conventional analysis of tracking point tracers. Compare the mixing structure that was computed via direct filament tracking (in Figure 3-12$b-d$) to the corresponding tracer spreading simulations in Figure 3-11$b-d$). The structure of the filament looks smooth and well dispersed in the particle tracking simulations, but it appears as a complex and nonuniform collection of material lines when the dynamics of the continuous filament is considered. The chaotic region at the center of the unit domain, packed very densely with material striations, attains a local orientation for all filaments.

Due to the nature of stretching in chaotic flows, an exponentially growing number of points is needed to sustain the continuity of material lines during the simulations, which makes these calculations extremely resource consuming. For example, for the case T = 1.2, after 10 iterations, $10^8$ points are needed to properly reconstruct the filament. Performing such calculations for real 3D mixing systems is considerably beyond the realm of today's computational power. A much more feasible alternative is to correlate measures of mixing intimacy, such as striation thickness distributions and intermaterial area density, to the stretching field. This correlation provides the necessary link to extend analysis to complex systems.

### 3-4.5 Asymptotic Directionality in Chaotic Flows

After only a few flow periods in a chaotic flow, a remarkably symmetric and intricate lamellar structure composed of thousands of striations emerges. This structure is self-similar in time when recorded at a fixed frequency (conveniently

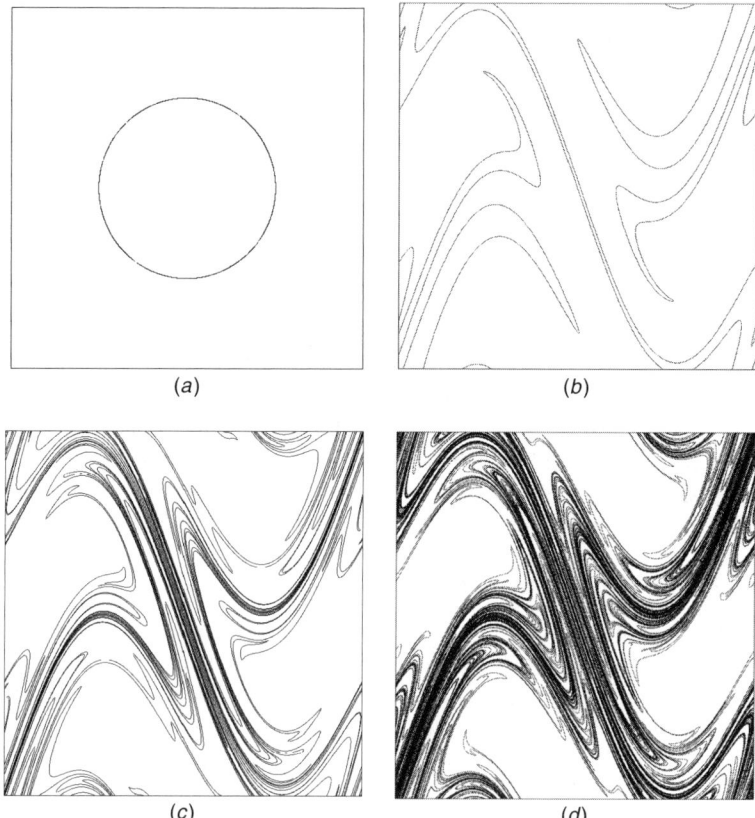

**Figure 3-12** Evolution of a continuous material filament in the sine flow for T = 0.8. The filament was initially placed at the center, and it is stretched throughout the chaotic region at times (a) initial filament; (b) n = 2; (c) n = 4; (d) n = 6. Dynamics of continuous material lines reveal details about the structure of the evolving mixing patterns that are "invisible" to the conventional analysis of tracking discrete points. Compare Figure 12b–d with Figure 11b–d.

after each flow period). As time evolves, the interface aligns to a template dictated by the flow, and more detail is added to the existing folds after each flow period. However, even though new striations appear at an exponential rate, the overall appearance of the structure remains the same. Asymptotically, as the material filament adapts to an invariant field of orientations in the flow, its structure becomes time-invariant (Muzzio et al., 2000).

The orientation process is very fast; the dependence on initial injection location is lost exponentially fast. After tracer fluid is injected at a specific location, the dye blob is stretched throughout the chaotic region of the flow; and dye filaments form an intricate and unique mixing pattern. This pattern is unique to the flow and becomes independent of the injection location after a short initial period.

Comparing the orientation of material lines that appear at the same location at different times in the flow field demonstrates that once a material line reaches a particular location of the flow, it adopts a characteristic local direction associated with that position in the flow. In other words, the orientation of a material filament in a periodic chaotic flow is determined by its instantaneous position, not by time.

The existence of an invariant field of orientations has important implications for the evolution of mixing patterns in chaotic flows. At each point in the chaotic domain there is a well-defined local orientation that is adopted by all material filaments in that region. The presence of an invariant field of orientations is referred to as asymptotic directionality in chaotic flows (Muzzio et al., 2000). In simple terms, when any portion of interface is in a particular region of the flow, it aligns to an orientation characteristic of that instantaneous spatial position. Thus, the evolution of mixing structures in chaotic flows is not a function of time but a function of instantaneous location. Asymptotic directionality (AD) explains the creation if self-similar mixing patterns in chaotic flows and provides a link to other properties of periodic chaotic flows in a general sense. For more details on AD property, see Giona et al. (1998, 1999), and Cerbelli (2000).

### 3-4.6 Rates of Interface Growth

Being able to measure or predict the rate of growth of interface and its area coverage is an important step toward understanding reactive processes. Intimate contact between mixture components is necessary to allow a chemical reaction to proceed. Although the spatial distribution of intermaterial area is unique in every chaotic flow, for all cases the interface grows exponentially fast in time and the rate of growth is known explicitly. In Figure 3-13 the natural logarithm of the total length of the interface is plotted versus number of iterations. One would expect that the rate of filament growth to be dictated by the Lyapunov exponent of the flow.

The Lyapunov exponent is the geometric average of the local stretching rates, which assigns equal weight to all local stretching values in the domain:

$$\Lambda \sim \lim_{n \to \infty} \left( \frac{1}{n} \langle \ln \lambda_i^n \rangle \right) = \frac{1}{n N_P} \sum_{i=1}^{N_P} \ln(\lambda_i^n) \qquad (3\text{-}11)$$

However, if we examine the three cases of sine flow shown in Figure 3-13, it is evident that the length of the interface is consistently underpredicted by the Lyapunov exponent. Because the filament does not sample the low and high stretching regions with equal probability, high stretching regions contribute more to the integral sum of the geometric average. Recall that the sine flow with $T = 1.6$ is a globally chaotic flow, whereas $T = 0.8$ and $T = 1.2$ are cases with mixed regimes containing both chaotic and regular flow regions. The deviation between the actual rate of elongation (along the filament) and the growth rate of the Lyapunov exponent is largest for $T = 0.8$. For this value of the period, the flow domain has large isolated regions of regular flow, which are not invaded by the stretching filament for long times.

PHYSICS OF CHAOTIC FLOWS APPLIED TO LAMINAR MIXING    113

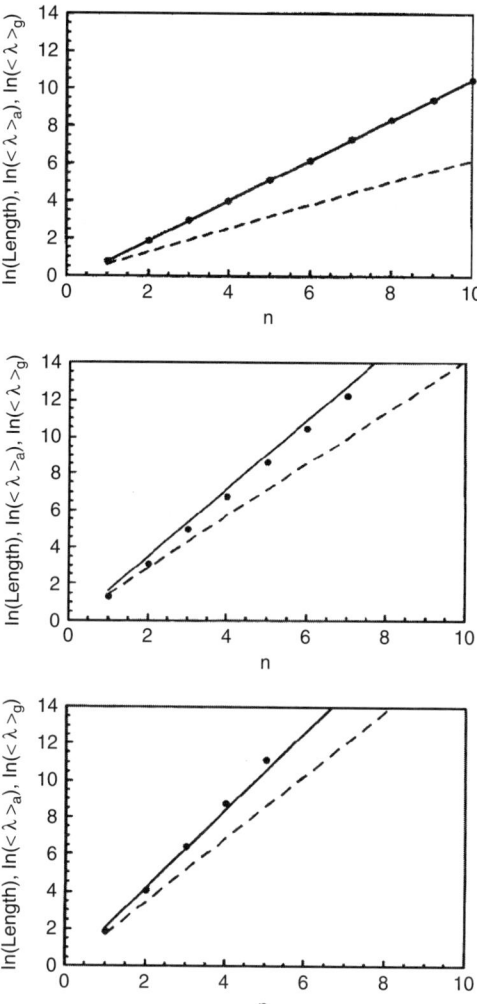

**Figure 3-13** Growth rate of material lines as measured by the Lyapunov exponent ($\Lambda$) (dashed lines) and topological entropy ($\Theta$) (solid lines). The three pictures represent three different flow conditions in the sine flow: (*a*) T = 0.8; (*b*) T = 1.2; (*c*) T = 1.6. In each case, the topological entropy closely predicts the length increase of the interface between mixture components (dots). The Lyapunov exponent underpredicts the mixing rate before the asymptotic time limit is reached.

The topological entropy, $\Theta$, is calculated from the logarithm of the arithmetic average of stretching:

$$\Theta \sim \lim_{n \to \infty} \left( \frac{1}{n} \ln \langle \lambda_i^n \rangle \right) = \frac{1}{nN_P} \ln \sum_{i=1}^{N_P} \lambda_i^n \qquad (3\text{-}12)$$

This measure is a better predictor of the rate of elongation of interfaces because it incorporates the finding that high stretching regions are populated more densely; and it can capture the biased dynamics of the stretching process. The topological entropy, $\Theta$, provides a basis for comparison of different mixing protocols using a single parameter, but still overlooks an important aspect of mixing processes. Spatial distribution of the interface in a chaotic flow (whether it is globally chaotic or not) is highly nonuniform, and this inhomogeneity is a permanent feature of the flow. Next, we incorporate the topology of the flow into the global analysis of mixing.

### 3-4.7 Intermaterial Area Density Calculation

Consider quantifying the amount of interface covering different areas of the chaotic domain by a simple box-counting technique. To do this, we subdivide the entire domain into identical-sized boxes and measure the amount of material filament that falls in each box. Ergodicity assures that the same amount of material will eventually fill every box, but the time to approach such macroscopic homogeneity is dependent on the box size. The larger the box, the shorter the time needed to form a seemingly uniform mixture. The intermaterial area density ($\rho$) is the length of filament in each cell ($L_i$) divided by the area of the cell ($area_i$). The average density in all boxes $\langle\rho\rangle$ is used to normalize each value to get a fraction of the total interface in each box. The spatial distribution of $\rho/\langle\rho\rangle$ is displayed in Figure 3-14 for $T = 1.6$, after seven periods of the sine flow. Since the spectrum of intermaterial area densities spans more than five orders of magnitude, results are described on the logarithmic scale, $\log(\rho/\langle\rho\rangle)$. It is evident that while the system is homogeneous from a macroscopic standpoint (i.e., nearly the same amount of tracer is present in each box), micromixing is much more intense in some regions of the flow than in others. Furthermore, regions of high intermaterial area density correspond closely to high stretching regions in the flow. This is a key link to measuring striation thickness distributions in physically realizable flows.

In the statistical domain, the probability density function of $\rho$ (PDF) is the quantitative characterization of this phenomenon. The frequency of $\log \rho$ for different flow periods is

$$H(\log \rho) = \frac{1}{N_\rho} \frac{dN(\log \rho)}{d\rho} \tag{3-13}$$

where $N_\rho$ is the total number of boxes ($512 \times 512$ grid) the flow region is divided into, and $dN(\log \rho)$ is the number of boxes with densities between $\rho$ and $\rho + d\rho$. Another manifestation of self-similarity of mixing structures generated by chaotic flows is that the normalized distributions of $\log \rho$ can be collapsed onto a self-similar set of curves.

Figure 3-15 shows the time evolution of the intermaterial area density field, revealing that the nonuniform distribution of contact area is a permanent property of chaotic flows. Once the filament has sampled the flow domain for a few

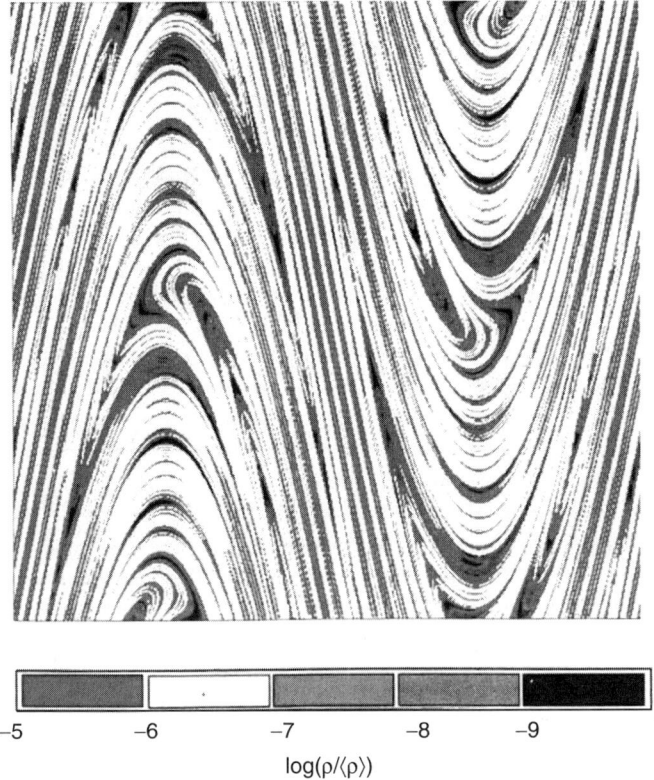

**Figure 3-14** Spatial distribution of intermaterial area density ($\rho$) in the sine flow without visible islands (T = 1.6). The shades of gray represent values on the logarithmic scale, so in the black regions the material interface is four orders of magnitude denser than in the white regions. See a color version of the figure on the Visual Mixing CD.

periods, it sufficiently approaches the characteristic invariant spatial distribution and then evolves everywhere at the same rate. The overall length of the interface $\langle \rho \rangle$ increases by several orders of magnitude, but its spatial distribution is preserved. Thus, the time evolution of micromixing intensity is expressed by a single value $\langle \rho \rangle \approx e^{n\Theta}$. It is important to realize that the scale $e^{n\Theta}$ is not valid for arbitrarily long times. Small striations are erased by diffusion as they are reduced to the molecular scale. At this state, the model based on $e^{n\Theta}$ both overpredicts the intensity of segregation and underpredicts the scale of segregation. Much of the interesting phenomena in practical laminar mixing applications occurs for much shorter times, before diffusion can have a significant influence on the overall rate of intermaterial area generation. Measuring intermaterial area density brings us closer to assessment of the rate of mass transfer and chemical reactions in laminar flows, since these processes occur through an interface. Still, $\rho$ is a coarse-grained average quantity, not truly a local one.

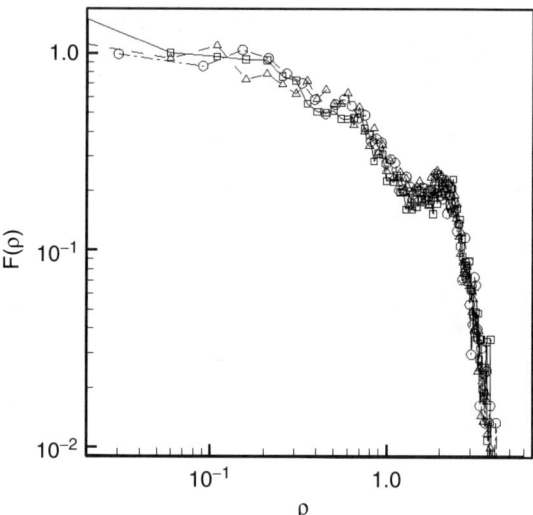

**Figure 3-15** Intermaterial area density distribution for two cases of the sine flow at T = 1.6 (almost globally chaotic). The figure compares the distribution computed from the coarse-grained stretching field to the distribution computed from direct tracking of a continuous material filament. Although the latter method cannot be applied to most chaotic flows, the first case is a fairly straightforward computation in both model and real chaotic flows. The distributions of $\rho$ are shown as a function of initial position (square), as a function of final position (circles), and as computed from direct filament tracking (triangles). All three curves collapse onto a single distribution.

### 3-4.8 Calculation of Striation Thickness Distributions

The local micromixing intensity in chaotic flows is usually characterized by the distribution of striation thicknesses. This scale measures the thickness of material striations in the lamellar structure generated by chaotic flows along an arbitrary reference line. Practical implementation of this idea is not trivial, because chaotic flows generate a wide spectrum of length scales (striation thickness values). Consider that after only a few iterations of the sine flow map, values as low as $10^{-12}$ are observed. In a physical context, if the length of the unit domain were 1 m, the striations are reaching molecular dimensions. Under such conditions, extreme care is necessary to prevent numerical errors from distorting the results; models need to be further developed to consider diffusion.

The striation thickness (s) at a given flow period is based on taking a cross-sectional cut along a straight line in the flow domain and computing the intersections of the material filament with that reference line. The frequency distribution of log s is defined as

$$H(\log s) = \frac{1}{N_s} \frac{dN_s(\log s)}{d(\log s)} \qquad (3\text{-}14)$$

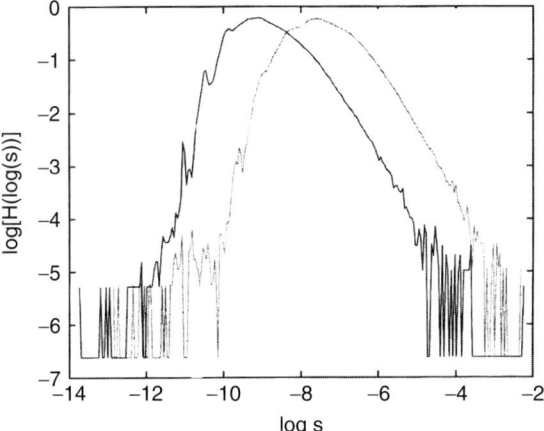

**Figure 3-16** Striation thickness distributions after nine time units in the sine flow, T = 1.2. The light- and dark-color curve is the distribution formed by the light and the dark mixture component. The horizontal shift in the mean is due to different amounts of light and dark material present in the mixture.

where $N_s$ is the total number of intersections and $dN_s(\log s)$ is the number of values of striation thickness between $\log s$ and $(\log s + d \log s)$. Figure 3-16 shows the striation thickness distribution (STD) generated by the sine flow for T = 1.2 after nine periods. The initial condition was similar to the one shown in Figure 3-11a. The distribution of the light-colored component in the binary mixture is the lighter curve; the distribution of the dark-colored component is the darker, continuous curve. The two are practically identical in shape. There is a difference in the mean of the two distributions (a horizontal shift along the axis), which is due to starting the mixing process with an excess amount of light-colored fluid. It is irrelevant where the reference line is placed in the chaotic flow domain, because after only a few periods the STDs along different reference lines are identical (Alvarez et al., 1997).

### 3-4.9 Prediction of Striation Thickness Distributions

In the preceding two sections, it was shown how to measure local micromixing intensity using intermaterial area distribution [$H(\log \rho)$] and length scale distributions [$H(\log s)$]. These tools are not directly applicable to 3D flows, because the resolution of the smallest length scales in 3D mixture structures is, to date, computationally prohibitive. In this section we present a predictive method for striation thickness distributions applicable to either 2D or 3D chaotic flows.

For incompressible 2D chaotic flows, the distribution of striation thicknesses was computed from the stretching distribution (Muzzio et al., 2000) using the idea that material filaments are stretched in one direction and simultaneously compressed in another direction at the same rate. Formally stated,

$$s \sim \frac{1}{\rho} \sim \frac{1}{\lambda} \tag{3-15}$$

The time evolution, shape, and scaling of stretching distributions in globally chaotic flows is almost one-to-one with the same properties of striation thickness distributions. The proportionality is the intermaterial area density ($\rho$) that links the stretching field with the number of striations influenced by each value of $\lambda$. If we consider that the *number* of stretching values is constant at each flow period, while the number of striations increases (i.e., $N_s \sim \langle \rho \rangle \sim \langle \lambda \rangle$), the striation thickness distribution as predicted from the stretching field is

$$H(\log s) = \frac{\lambda}{\langle \lambda \rangle} H(\log \lambda) \tag{3-16}$$

The real power of this relationship is that while the distribution on the left is unattainable directly in most flows, the second is obtainable from a fairly straightforward computation. Figure 3-17 provides evidence that $\rho \sim \lambda$, where the calculated and predicted intermaterial area density distributions are shown for the sine flow (T = 1.2). With the use of eq. (3-16), an extremely accurate prediction of STD is obtained based on computing the stretching field alone. It is important to note here that this method applies to flows devoid of large segregated regions, which does not limit its applicability for a practical application, since mixtures with macroscopic segregations are almost always undesired.

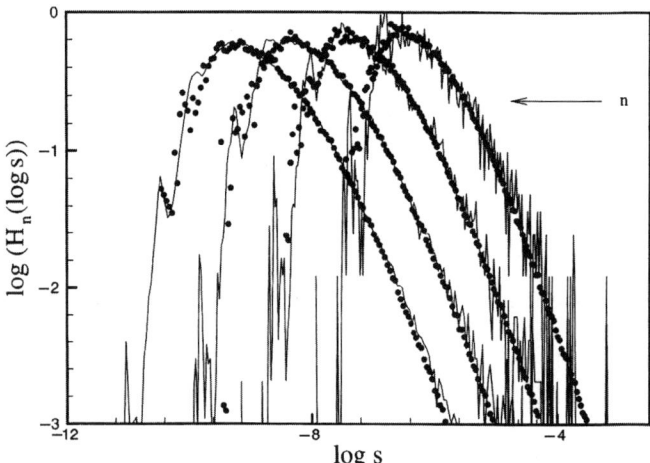

**Figure 3-17** Comparison of the striation thickness distribution calculated directly from simulating the evolution of material interfaces (as in Figure 3-12) and as predicted from the stretching field.

## 3-5 APPLICATIONS TO PHYSICALLY REALIZABLE CHAOTIC FLOWS

### 3-5.1 Common 3D Chaotic System: The Kenics Static Mixer

In the preceding sections we summarized and described some common tools to characterize mixing in complex flows using a 2D model system. However, these tools are universal and are applicable to industrially relevant 3D chaotic mixing systems as well. Chaotic flows have been shown to exist in stirred tanks operated in the laminar regime, static mixers, or extruders (Kim and Kwon, 1996; Lamberto et al., 1996; Hobbs et al., 1997). The use and design of static mixers is reviewed in detail in Chapter 7. Only a brief description of the important characteristics of these continuous flow devices is given here, to illustrate the application of concepts from dynamical systems theory to physically realizable flows using the Kenics static mixer as an example.

Static mixers are often employed in industry for viscous mixing applications, as heat and mass transfer promoters, or even as chemical reactors in a variety of applications. The designs commonly consist of an empty tube with mixer elements inserted to perturb the flow and mix streams of material. Since they contain no moving parts, the energy for flow is derived from the pressure drop across the mixer. As an example of a 3D chaotic flow, we focus on flow in the Kenics static mixer. A schematic drawing of a six-element mixer in Figure 3-18 illustrates the geometry and important parameters in the Kenics. Different flow regimes are usually distinguished on the basis of the open-tube Reynolds number:

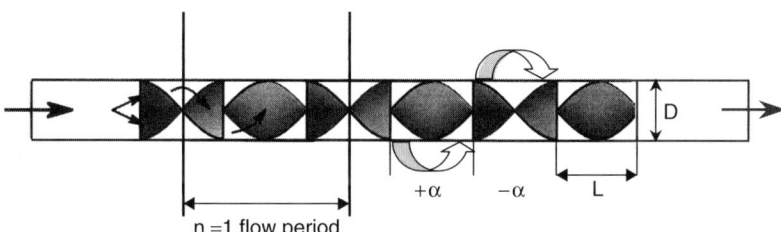

**Important Features:**

1. Spatial periodicity:
    2 Kenics elements = 1 flow period
2. Plane of symmetry: at center of each element
3. Parameters:
    Element length to diameter ratio = L/D
    Twist angle per element = $\alpha$
    Pitch per element = twist per unit length

**Figure 3-18** Geometry of a six-element Kenics static mixer. The element length-to-diameter ratio (L/D) is 1.5 and the twist angle is 180° in the standard design.

$Re = (\rho_f \langle v \rangle D)/\mu_f$, where $\langle v \rangle$ is the average axial velocity, $\rho_{fluid}$ and $\mu_{fluid}$ are the fluid density and viscosity, and D is the diameter of the mixer housing.

### 3-5.2 Short-Term Mixing Structures

Mixing patterns in the Kenics are examined by simulating 10% (by volume) injection of a tracer with equal density and viscosity as the main fluid. A closed-form solution for particle trajectories, such as eqs. (3-5) and (3-6), does not exist for 3D flows. Tracer mixing patterns can be captured experimentally by dye injection patterns, or computationally by integrating along the velocity field. The trajectory of 200 000 tracer particles is computed according to eq. (3-4) after they have been injected 1 mm upstream of the first element at the center of the cross-sectional area. The software technology and computational algorithms have been validated by experimental data available in the literature (Zalc et al., 2001; Szalai and Muzzio, 2002). The tracer mixing patterns for $Re = 10(^3\sqrt{10})$ are revealed after 8, 12, 16, and 20 mixer elements in Figure 3-19a–d. Inertial effects on the flow at this Re are small. This figure reveals very similar mixing behavior in the Kenics that we have shown in the 2D model system in Figure 3-11. It is evident that the tracer, injected initially at the center of the pipe, does not spread uniformly throughout the flow domain. Quite the contrary—as it is convected, stretched, and folded by the flow, it gives rise to a lamellar system composed of thousands of striations with a wide distribution of local length scales. The structure continues evolving as the number of flow periods increases; the lamellae become increasingly thinner as the result of the iterative stretching and folding process. As the fluid travels through an increasing number of elements, the mixture displays increasingly finer striations organized in a self-preserving topology and a self-similar process dominates the evolution of the structure.

Figure 3-19 illustrates the intrinsic self-similarity in the mixing structure. As time increases, the chaotic flow produces a partially mixed structure that is essentially identical to the structure recorded a period earlier, except that a larger number of thinner striations is found in each region. It is important to remark here that such self-similar structures are independent of initial conditions. When the mixing structure is recorded at periodic intervals, it "always looks the same" regardless of the initial location of the dyed fluid in experiments or the location of the filament in simulations (unless tracer is injected in an island). Independence of initial conditions in flows that create self-similar mixing structures has been described qualitatively in a number of 2D and 3D chaotic mixing flows (Leong and Ottino, 1989; Swanson and Ottino, 1990; Muzzio et al., 1991; Lamberto et al., 1996; Hobbs and Muzzio, 1998; Gollub and Cross, 2000).

### 3-5.3 Asymptotic Directionality in the Kenics Mixer

The explanation of self-similarity in chaotic flows was given earlier, when the asymptotic directionality (AD) property was introduced through examples of the sine flow system. Recall that AD is a local property of chaotic flows that creates

APPLICATIONS TO PHYSICALLY REALIZABLE CHAOTIC FLOWS   **121**

**Figure 3-19** Injecting an equal viscosity additive (10% by volume) just upstream of the inlet region reveals nonuniform mixing in the standard Kenics. Tracer mixing patterns for $Re = 10(10)^{1/3}$ are shown after (*a*) 8, (*b*) 12, (*c*) 16, and (*d*) 20 mixer elements. An animation of this mixing geometry is included on the Visual Mixing CD affixed to the back cover of the book.

an invariant spatial template of the mixture at different times. A material line that visits a certain region in the flow adopts an orientation that is characteristic of that position and independent of time. Therefore, periodic snapshots of a cross-section downstream look the same qualitatively, with more and more detail revealed in each picture.

This property, which controls the topology of chaotic flows, is readily observed in 3D industrially relevant flows. The technique used to reveal AD in the Kenics flow is similar to the method of stretching computations with a slight modification. If a small material element is represented as an infinitesimal vector attached to a fluid element in the flow, its position and stretching can be calculated from eqs. (3-4) and (3-7). However, instead of attaching only one such vector to a fluid particle, take a set of three orthogonal unit vectors (Figure 3-20). The evolution of the three vector components attached to the same trajectory reveals how the local spatial position affects the orientation of fluid filaments in a particular region. The three vectors could be thought of as material lines in close proximity to one another having different initial orientations. The relative direction of each vector in the ensemble can be monitored by measuring the cosine of the angle between the vectors as they are convected by the flow. We randomly distributed the "vector tripods" at the inlet. The cosine of the angle between any two components is 0.0 initially, and the gradual alignment of the vectors along the same trajectory is expressed as the enclosed angle ($\alpha_n$) decreases and the cosine tends to 1.0.

The reorientation and rapid alignment of vectors reveals how fluid filaments in a particular region tend to orient along a predetermined direction. Many sets of three vectors were dispersed at the inlet in the Kenics mixer, and the enclosed angle between the components of each set was recorded at consecutive downstream distances. The mean of the cosines ($\cos \alpha_n$), which is displayed in the main panel of Figure 3-20, starts from 0.0 at the inlet and rapidly approaches 1.0, indicating convergence to a single stretching direction at each spatial position in the flow. The direction to which all three vectors adapt is a function of the particular spatial position, and it is different for each set. Since the flow is spatially periodic, the field of orientations at a given phase angle is a stationary property of the system. Before the flow passes through a complete Kenics mixer element, the angle between the vector components is close to 0.0. This is an indication of strong directionality at all spatial locations in the flow. Material filaments, regardless of their previous orientation, arrive at a spatial position in the chaotic flow and they all adopt the local stretching direction. The probability density function of $\cos \alpha_n$ is indicated in the inset of Figure 3-20. The initial distribution is a Dirac delta function centered at zero [$\delta(0)$] since all sets of vectors initially have right angles between them. Each curve in the plot is the

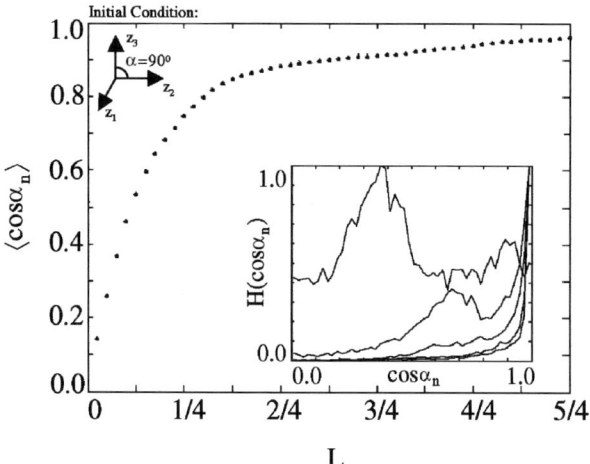

**Figure 3-20** Main panel: The mean of $\cos \alpha_n$ converges to 1 as the initially orthogonal vectors become increasingly aligned. $\alpha$ represents the angle enclosed by two of the vectors. Inset: The probability density function of $\cos \alpha_n$ in the standard Kenics mixer for one-tenth of the total average residence time.

PDF($\cos \alpha_n$) at consecutive time units in the flow. The important feature to note here is the rapid shift in the distribution from 0 to 1, showing the decrease in the enclosed angle from 90° toward 0°.

### 3-5.4 Computation of the Stretching Field

As stated earlier, the intensity of mixing and the intimacy between mixture components can be measured by computing the stretching of small vectors attached to fluid tracers. For 2D chaotic flows, closed-form solutions such as eqs. (3-8) and (3-9) can be obtained for the stretching field. Life is more complicated in 3D, where fluid filaments stretch and contract along multiple directions. The elongation (or stretching) of small vectors as they are convected by the flow is computed for 200 000 tracer elements in the Kenics. Numerical methods are used to solve eqs. (3-4) and (3-7) simultaneously in the entire 3D flow domain, and then the stretching magnitude is computed using eq. (3-10) with the third, z, component added. The accumulated stretching at several downstream distances is shown in Figure 3-21. The cross-sectional view of the mixer at each downstream distance is represented by circles, and the mixer element appears as a white line across the diameter. The initial orientation of the stretching vectors rapidly changes and they realign with the principal stretching direction of the flow within the first couple of mixer elements. After this realignment the initial vector orientations are no longer important.

The accumulated stretching of each tracer is plotted as a function of initial position and color-coded according to magnitude. Areas that experience high stretching correspond to high mixing intensities, where materials injected to the flow will spread rapidly over the domain. On the other hand, low-stretching areas correspond to poor mixing regions that exchange material slowly with the rest of the flow. Coherent structures that appear at every cross-section in the mixer are segregated regions, or islands, that are only destroyed by the slow action of diffusion. Two such large areas are noticeable at this flow condition (Re = 21.5) in the Kenics, one on each side of the mixer element near the center of the tube (see Figure 3-21$d$). A higher flow rate in this mixer does not necessarily lead to better mixing. It has been documented that these segregated regions exist at Re = 100 and cover nearly 10% of the total flow area. These figures reveal that these structures appear in the device at much lower flow rates and span a wider range of flow conditions than believed earlier.

It is important to point out that the range of stretching values is different in each cross-section, gradually increasing as the flow passes through more and more mixer elements. In Figure 3-21$a$ after two mixer elements, the minimum stretching magnitude is $4.80 \times 10^{-2}$, the maximum is $4.62 \times 10^8$, and the arithmetic average accumulated stretching per tracer is $4.84 \times 10^4$. By the time the fluid elements passes through 22 elements, the range is much broader, spanning 17 orders of magnitude with a minimum at $2.41 \times 10^{-1}$, the maximum at $2.01 \times 10^{15}$, and the average at $1.20 \times 10^{10}$.

Asymptotic directionality and self-similarity, the cause and effect of universal dynamical behavior in chaotic flows, is immediately apparent in each set of three

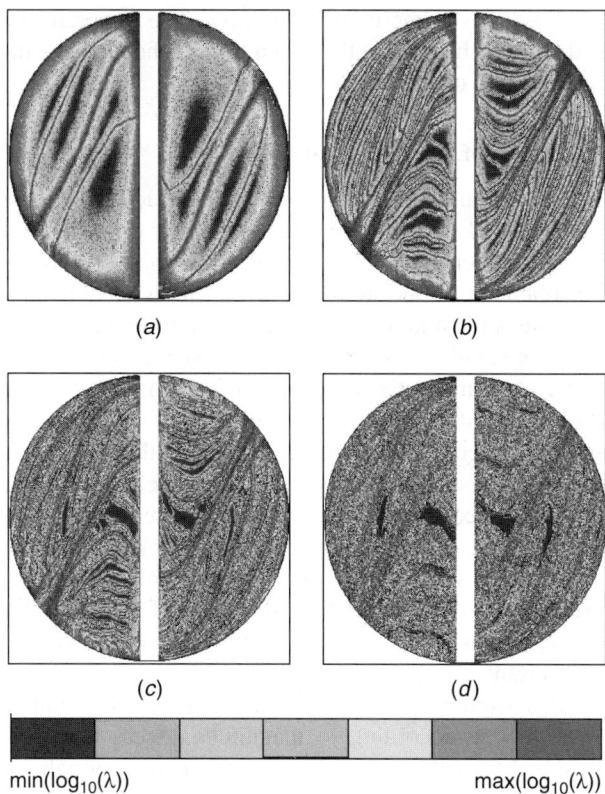

**Figure 3-21** Contours of the stretching field in the standard Kenics mixer at $Re = 10(10)^{1/3}$. The cross-sectional planes correspond to axial distances after (a) 2, (b) 6, (c) 10, and (d) 22 mixer elements. See insert for a color representation of this figure.

snapshots. Some features of the mixing structure appear early, after the fluid passes through just two mixer elements. New folds appear within existing ones in subsequent sections. The general features of each mixing pattern after 2, 6, or 10 elements (Figures 3-21a–c) are the same with more details added at each element. It should be pointed out at this point that dye mixtures in physical experiments would show the same pattern as observed in the stretching plots.

### 3-5.5 Rates of Interface Growth

We have shown in Figure 3-13 that intermaterial area density is created at an exponential rate in 2D chaotic flows. Figure 3-22 demonstrates that the same applies to 3D flows, where the average of the natural logarithm of stretching, $\langle \log \lambda \rangle$, is plotted as a function of downstream distance in the Kenics mixer for two flow rates. After some entrance effects, the mixing rate shows a uniform monotonic increase on a logarithmic scale, and the Lyapunov exponent

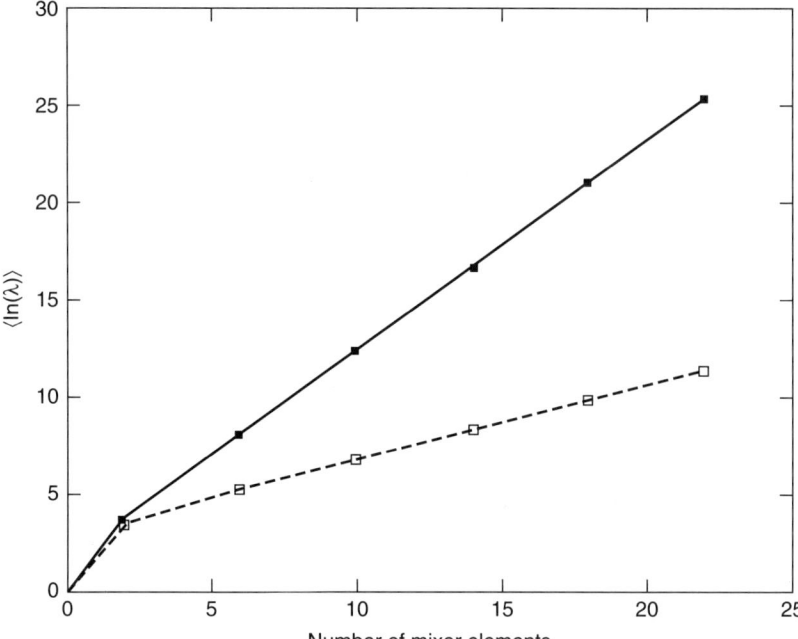

**Figure 3-22** Stretching in the standard Kenics mixer increases at an exponential rate, which is characteristic of chaotic flows. The rate of increase is higher at a higher flow rate Re = 1000 (dark squares) than at Re = $10(10)^{1/3}$ (light squares) but the energy cost of maintaining such high average throughput is not reflected in $\lambda$ alone.

is equal to the slope of the linear function. The rate of increase is higher at a higher flow rate Re = 1000 than at Re = $10(10)^{1/3}$. The different initial rate of increase (i.e., a kink in the slope after two elements) is due to the initial realignment of stretching vectors to the principal stretching direction during the first flow period.

### 3-5.6 Intermaterial Area Density Calculation

The real power of using stretching computations to characterize chaotic flows lies in the fact that stretching is the link between the macro- and micromixing intensities in laminar mixing flows. In this section we describe the method for computing striation thickness distribution in our 3D example, the Kenics mixer.

We used the direct relationship between intermaterial area density ($\rho$) and stretching ($\lambda$) stated in eq. (3-15) to compute the spatial structure of $\rho$ in the Kenics flow. Approximately $4 \times 10^6$ tracer particles were used in the stretching computations to assure statistically significant results. The specific method is as follows: First, the accumulated stretching of all the tracer filaments was recorded after 2, 6, 10, and 22 Kenics mixer elements. Then a uniform lattice of 225 × 225 boxes was overlaid each mixer cross-section, and all the boxes that were fully or partially outside the flow area were discarded. This includes boxes that were placed on the tube perimeter or on the tube diameter along solid surfaces. A total of 37 130 boxes were retained from the original 50 625. Finally, the accumulated stretching of tracer particles in each box was summed and divided by the overall average $\langle \lambda \rangle$.

Figure 3-23 shows the spatial distribution of intermaterial contact area, scaled by the overall length of the interface $\langle \rho \rangle$, at each axial position in the mixer. As discussed earlier in the chapter, the distribution of intermaterial area is related directly to the stretching field through the relationship $\lambda \sim \rho$. Qualitatively, this relationship can be confirmed if one compares the stretching field in Figure 3-21 to the field of intermaterial area densities in Figure 3-23. Quantitative proof was given in Figures 3-14 and 3-15, where the direct computation and prediction of $\rho$ were compared for the sine flow. Once the intermaterial area in each of the 37 130 cells is normalized by the overall average, $\langle \rho \rangle$, the distribution and scale are identical for all four cross-sectional positions. In other words, the function $\hat{\rho} = \rho / \langle \rho \rangle$ is invariant and describes the intermaterial area density at each period everywhere in the domain (i.e., each color represents the same range of $\rho$ in

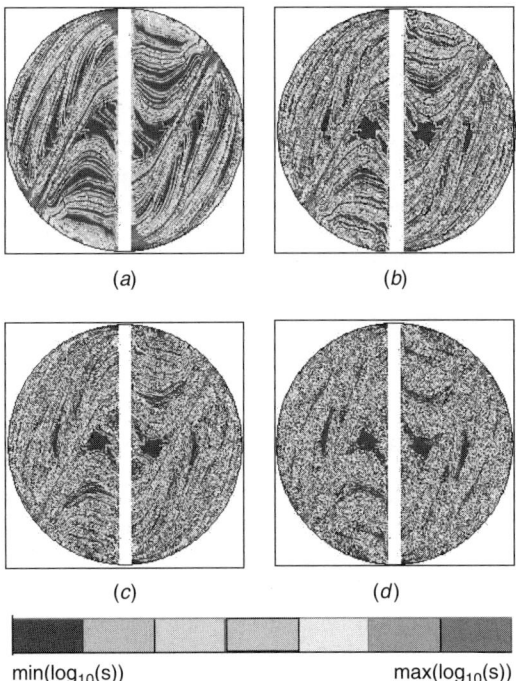

**Figure 3-23** Intermaterial area density ($\rho$) from coarse-grained stretching average in the standard Kenics mixer at Re $= 10(10)^{1/3}$ after (*a*) 8, (*b*) 12, (*c*) 16, and (*d*) 22 mixer elements. See insert for a color representation of this figure.

Figure 3-23a–d). This figure provides qualitative proof that the spatial structure of intermaterial area densities at different downstream distances is time-invariant. The spatial distribution of ρ exhibits strong fluctuations that are a *permanent* feature of the mixing process (i.e., the picture "looks the same" for the four flow periods). As shown in Figure 3-23, when the spatial distribution of ρ is computed after additional periods of the flow, although ⟨ρ⟩ increases by several orders of magnitude, its spatial distribution remains essentially unchanged.

The time invariance in the spatial distribution of ρ is another manifestation of the self-similarity of the structures generated by time- and spatially periodic chaotic flows. Such invariant statistical properties can be demonstrated by computing the probability density H(log ρ). If the probability density function of the scaled variable (ρ/⟨ρ⟩) is computed according to eq. (3-13), the distributions are scaled automatically by the mean density.

The distribution H(log(ρ(⟨ρ⟩))) is shown in Figure 3-24 after $n = 4, 6, 8$, and 11 on a logarithmic scale. $Re = 10(10)^{1/3}$ corresponds to a moderate flow rate of 1.29 m$^3$/h. From Figure 3-24 we see that the distributions approach an asymptotic shape characterized by a power law decay in the low-ρ region as the number of elements increases, followed by a much steeper, possibly exponential decay in the high-ρ region. The effective collapse of the scaled distribution over several orders of magnitude in ρ indicates that the above-mentioned nonuniformities in ρ are a permanent feature of periodic chaotic flows. As the number of elements increases, the intermaterial area density increases everywhere while preserving

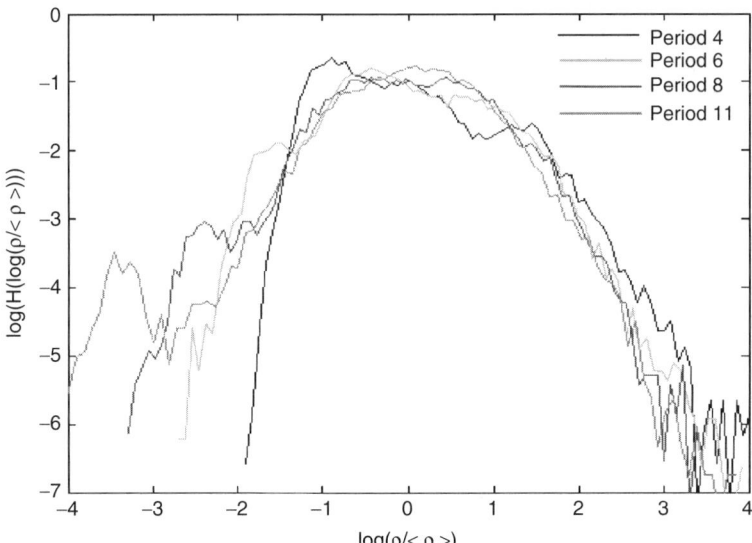

**Figure 3-24** The distribution of intermaterial area density (ρ) in the standard Kenics mixer at four axial cross-sections at $Re = 10(10)^{1/3}$. The probability density function of ρ/⟨ρ⟩ reaches an invariant shape as the mixing process evolves.

the same distribution, meaning that the number of low- and high-density locations, and the relative magnitude of the density on such regions, remain unchanged. This behavior is not unique to flows with low or moderate inertia, nor is it limited to static mixers. The evolution of mixing intensities and microstructure shows similar characteristics in cases where the flow is on the verge of turbulence (Szalai and Muzzio, 2003) for both static mixers and stirred tanks.

The observed scaling of $\rho$ has an important physical interpretation: Once $\rho$ approaches the characteristic invariant and statistical distributions generated by the global invariant manifold, it then evolves everywhere at the same rate as the mean density. In other words, if the mean intermaterial area density is doubled, the *local* density is doubled *everywhere*. This is important, because it means that the time evolution of $\langle \rho \rangle \approx e^{n\theta}$ determines the time evolution of $\rho$ at all locations of the chaotic flow (i.e., intimacy of mixing improves everywhere by the same factor). Similarly, the striation thickness $\langle s \rangle \approx \langle \rho \rangle^{-1}$ both locally and globally: The local average striation thickness decays everywhere at the same rate, as predicted by $\langle s \rangle \approx e^{-n\theta}$. Therefore, the mixing rate in periodic, chaotic mixing processes can be characterized by a single quantity, the growth of the average intermaterial area density $\langle \rho \rangle$. Moreover, $\Theta$, the topological entropy exponent, can be regarded as a mixing rate. While this observation is not true in nonchaotic regions, this distinction is immaterial in practical applications because flows with large nonchaotic regions are likely to generate such processing problems that in all likelihood such flows should be avoided in well-designed mixing applications.

As is clear from Figure 3-24, $\rho$ in some regions is several orders of magnitude higher than in other regions. This observation means that even if the system is homogeneous from a macromixing standpoint, in some regions the flow achieves much more intimate mixing (much more intense micromixing) than in other regions. Such fluctuations in $\rho$ can have huge effects on practical applications. For example, for systems that require dispersive mixing of a pigment or an additive (a common case in polymer processing applications), differences in $\rho$ directly affect the optical quality and material properties of the finished product. For diffusion-controlled reactions, the local reaction rate depends both on the amount of contact area and the average intensity of the concentration gradient, which is usually proportional to the intermaterial area density. For systems with multiple reactions, the value of $\rho$ determines the ratio of desired/undesired product generated in a given region of a reactor. Thus, strong fluctuations in interface density would have an enormous effect on reactive systems, resulting in different reaction rates and in different product distributions at each location. The micromixing properties of 3D chaotic flows are exploited in more detail in the next section.

### 3-5.7 Prediction of Striation Thickness Distributions in Realistic 3D Systems

The same iterative stretching-and-folding process that generates self-similar density distributions drives the evolution of striations and therefore controls the

APPLICATIONS TO PHYSICALLY REALIZABLE CHAOTIC FLOWS    **129**

dynamics of striation thickness distributions (STDs). STDs could be computed from the distributions of stretching values by realizing that as a portion of fluid is stretched, it generates striations with a thickness inversely proportional to the amount of stretching applied to the fluid. In other words, eqs. (3-15) and (3-16) can be extended to 3D systems such as the Kenics flow. The validity of eq. (3-16) was demonstrated in Figure 3-17, where more than $10^9$ tracer particles and over $10^6$ stretching values were required to compute the STD directly [by using eq. (3-14)] and were compared with the prediction of eq. (3-16). Such a level of numerical resolution is usually impossible and almost always impractical when dealing with realistic 3D flows. Equation (3-14) cannot be applied directly to 3D systems such as the example studied here.

Fortunately, additional theoretical developments allow one to circumvent the need for direct computation. Computing the distribution of stretching intensities and the overall stretching rate, as seen in Figures 3-21 and 3-22, is feasible in any deterministic chaotic flow. Subsequently, $H(\log s)$ is predicted by eq. (3-16) in Figure 3-25 for two cases, after 6 and 22 elements, in the Kenics. $H(\log s)$ versus $(-\log s)$ calculated by assuming that $s_o \approx 1$ (i.e., the initial unmixed region is assumed to be of similar size as the entire system) is shown in Figure 3-25 for Re = $10(10)^{1/3}$. Interestingly, the high stretching side of $H(\log \lambda)$ generates essentially the entire distribution $H(\log s)$, and the mode of $H(\log s)$ corresponds to stretching values far along the right tail of $H(\log \lambda)$. This observation has deep physical meaning. Figure 3-25 demonstrates that the overwhelming majority of

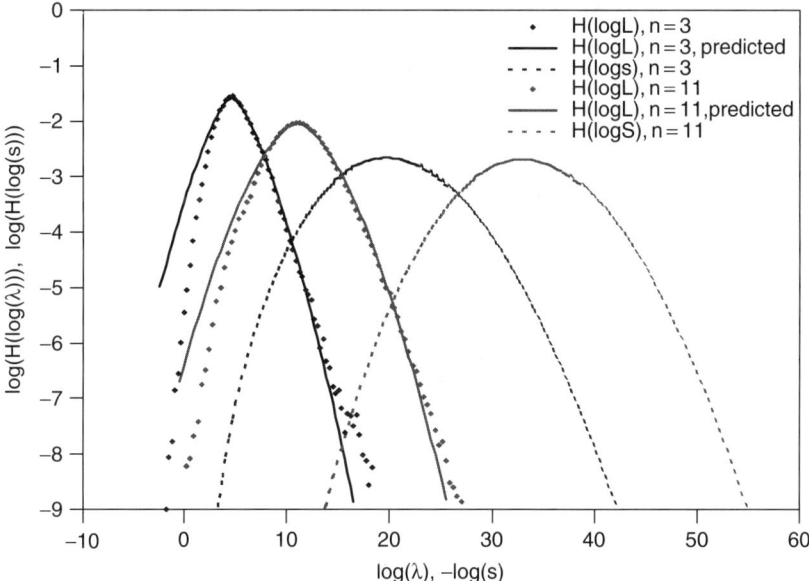

**Figure 3-25** The distribution of striation thicknesses, $H(\log s)$, is predicted from the right branch of the stretching distribution, $H(\log \lambda)$, for Re = $10(10)^{1/3}$.

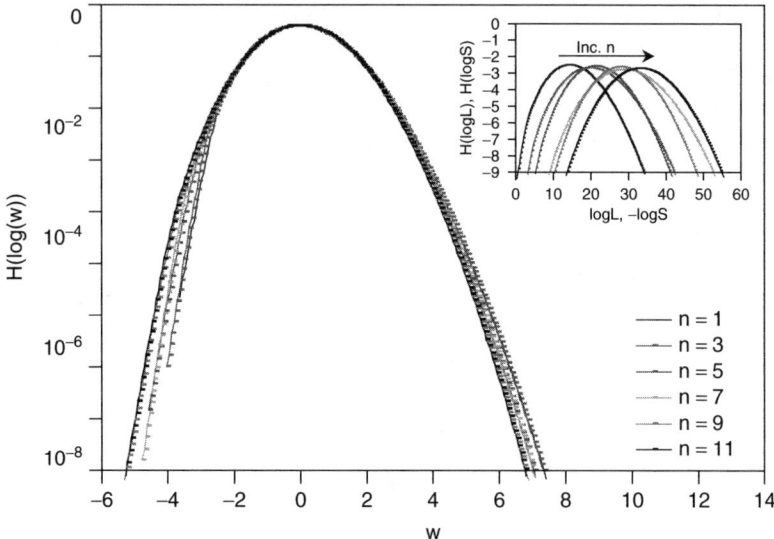

**Figure 3-26** H(log s), which represents the spectrum of diffusive length scales, is shown for n = 1–11 for Re = $10(10)^{1/3}$. The scale distributions in the main panel of each figure approach an invariant distribution as the period increases. The inset reveals the broadening distributions as striations are created by the flow.

striations, corresponding to most of the interfacial contact area in the system, is generated by a set of very rare but very intense stretching events (Szalai and Muzzio, 2003).

The evolution of H(log s), and its scaling properties, are illustrated in Figure 3-26 for Re = $10(10)^{1/3}$ for six cross-sections in the Kenics. The inset shows H(log s) after 2, 6, 10, 14, 18, and 22 elements. In the main panel the same data are shown after the distributions have been re-scaled by substracting the mean and dividing by the standard deviation. As revealed in Figure 3-26, the shape of the distribution gradually approaches a self-similar distribution where the curves increasingly overlap one another as n increases. Surprisingly, the convergence to a self-similar STD is much faster and much more complete for Re = 1000 (figure not shown), where the distributions overlap over three decades of probability after the first few elements (Szalai and Muzzio, 2003).

## 3-6 REACTIVE CHAOTIC FLOWS

The effects of flow and mixing on reactive systems have been recognized for many decades, yet the interplay of convection and diffusion has traditionally been treated with overly simplistic approaches. The complexity of applications where both diffusion and mechanical stirring occur simultaneously with reactions has been overwhelming to the point that perhaps in despair, critical aspects of convective flow have been altogether neglected. The classic approach of drawing a black box around the reactor and examining only the intrinsic kinetics of the reactive process is only meaningful when macroscopic inhomogeneities can no longer be detected. This method treats the intrinsic kinetics properly but can only give insight for processes with slow reactions. It is bound to fail when reactions occur faster than the time scale for homogenization in the system. In such cases we run the risk of overestimating the selectivity and conversion during the design process (a typical problem during reactor scale-up) but completely ignore that a "well-mixed" homogeneous state may never be reached during the entire life of the process (recall Figure 3-6).

Chaotic flows are the only effective route to destroy segregation rapidly in viscous mixing applications, which are particularly prone to remaining inhomogeneous for long times (Lamberto et al., 1996; Avalosse and Crochet, 1997; Hobbs et al., 1997; Unger and Muzzio, 1999). In the past two decades, the effect of chaotic mixing without reactions has been examined in a variety of increasingly complex mixer geometries on a case-by-case basis. However, without a deep understanding of how fluid motion couples with diffusion and molecular kinetics in reactive flows, we cannot yet take the leap to include all transport mechanisms in a realistic geometry. Simplifications can be made by considering 2D model chaotic flows, where the convection–diffusion–reaction mass balance can be solved directly (Muzzio and Ottino, 1988; Metcalfe and Ottino, 1994; Muzzio and Liu, 1996; Zalc and Muzzio, 1999; Neufeld, 2001; Giona et al., 2002).

The sine flow captures many features of 3D chaotic mixing processes accurately. The only parameter that controls the behavior of the system is the flow period T. As seen in Figure 3-11, this simple sinusoidal velocity field with alternating directions creates complex, layered structures that appear to be a collection of different-sized filaments filling the chaotic region. The two controlling mechanisms of the convective mixing process, stretching and folding, occur similarly in more complex, 3D systems, such as the Kenics mixer (Figure 3-19) or stirred tanks. To understand the effect of hydrodynamics and mixing on reactive flows, in the next section we explore how convective mixing, when coupled with diffusion and reactions, affects the evolution of chemical reactions.

Consider the unit square domain of the sine flow, where the right side is filled with reactant A and the left side contains only B. The initial interface between the two components is a vertical line along the center and one edge of the box. At $t = 0$ convective mixing is turned on in the model simultaneously with diffusion and reaction. The species balance equation in a dimensionless form for the flow with reaction, diffusion, and convection is

$$\frac{\partial c_i}{\partial t} + \bar{v} \cdot \nabla c_i = \frac{1}{Pe} \nabla^2 c_i + r \qquad (3\text{-}17)$$

where Pe, the Peclet number, measures the relative magnitude of the characteristic time for convection to diffusion. Applications usually have values of Pe in the range $10^2$ (laminar flames) to $10^{10}$ (turbulent reactive flows, polymerizations). For problems where convective mixing plays an important role, it is of interest to choose Pe as high as possible. If the two time scales were on the same order (i.e., Pe $\cong$ 1), mixing effects would be unimportant because diffusion would erase heterogeneities as quickly as convection creates them. At the other end of the spectrum, where Pe is very high, diffusion is very slow compared to the convective mixing process. In this case, very fine, partially mixed structures remain intact for long times before diffusion has an effect. The spatial resolution (i.e., maximum distance between neighboring nodes) that is necessary to capture the details of the mixing structure determines the upper bound of the Pe for practical applications. For the results presented in this chapter, given available computer resources, Pe = 2000 is the highest practical value.

We consider a bimolecular reaction: A + B → 2P and define $\varphi = c_A - c_B$. For infinitely fast reactions, since A and B cannot coexist, a single equation describes both the reactant and product concentrations:

$$\frac{\partial \varphi}{\partial t} + \overline{v} \cdot \nabla \varphi = \frac{1}{Pe} \nabla^2 \varphi \qquad (3\text{-}18)$$

To describe the evolution of $\varphi$, standard finite-difference methods were employed to solve eq. (3-17) at 1024 × 1024 nodes in the flow. Initially, half of the flow domain was filled with A and the other half contained only B. The reactant and product concentrations are computed from $\varphi$ according to eq. (3-18).

Once the product concentration is known as a function of time, the local reaction rate can be determined from mass conservation:

$$r = \frac{1}{2}\left(\frac{\partial c_P}{\partial t} + v \cdot \nabla c_P - D_f \nabla^2 c_P\right) \qquad (3\text{-}19)$$

Equation (3-19) in some sense captures the fact that three processes—the local reactivity, diffusive transport, and convective mixing—control the effective reaction rate. We now examine how each of these three components affects the overall evolution of a reactive mixing process.

The reaction rate after the first three periods of the sine flow with T = 1.6 is illustrated on the right-hand side of Figure 3-27. The three figures represent r after n = 1, 2, and 3, so its time evolution can be followed from the top to the bottom of the figure. We use a logarithmic scale to illustrate the spatial variations of r. The colors are relative to the maximum rate after each period, which is 69.82 after n = 1, 25.70 after n = 2, and 4.81 after n = 3. The darkest lines represent the hot zones in the flow, where the instantaneous reaction occurs. The rapid decrease in the maximum rate shows how fast reactants are consumed by the reaction. During the first three periods, the reaction proceeds very rapidly: The conversion is 49.3% at n = 1, 87.7% at n = 2, and reaches 96.5% at n = 3.

Compare the left and right columns of Figure 3-27, where the purely convective flow is compared with the reactive/diffusive case. The hot zones, where the reactive interface is, closely agree with the locations of highest filament density. The landscape of the nonreactive mixture (left side of Figure 3-27) looks almost identical to the landscape of the reactive flow (right side of Figure 3-27).

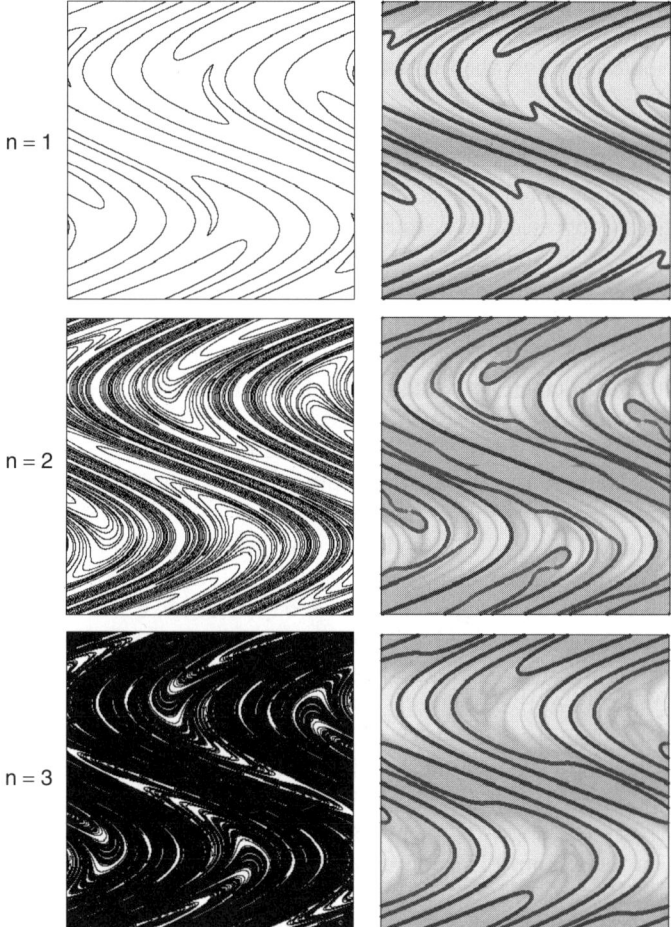

**Figure 3-27** The left-hand side of the figure indicates the distribution of intermaterial contact area in the sine flow with T = 1.6 after n = 1, 2, and 3. The right-hand side shows the rate of an instantaneous bimolecular reaction (A + B → 2P) in the same flow. The darkest areas represent the hot zone where the reaction proceeds very rapidly. Its location is almost identical to the landscape of intermaterial contact area.

This similarity is not coincidental; the two sets of figures are highly correlated. The stretching process is the sole determinant of where interface is created and establishes where intimate contact is achieved between fluids A and B. Once the playing field is set by convective stretching, diffusion and reaction follows: The landscape of the reactive zone closely outlines the location of the fluid–fluid interface, which is determined purely by convection. Diffusive transport in each region is proportional to $\rho$, the amount of interface at that location. The areas in the flow where striations contain only A or only B remain inactive, because no reaction takes place there. However, in the areas that contain neighboring striations of A and B (i.e., areas where the A–B interface is), diffusion immediately transports material across the interface, the reaction proceeds, and product begins to accumulate.

There are some differences between the nonreactive and the reactive landscapes (in the left and right columns of Figure 3-27). The thinnest striations contain the least amount of reactants, so they are consumed first by the reaction. As time progresses, these striations become exhausted and the reaction front advances. Early on, convection creates new striations and more interface, so the reaction zone stretches. Then it quickly reaches a balance with diffusion, and reaction proceeds until the area begins to run out of reactant. As a result, the reaction extinguishes and the hot zone migrates. The interplay of two mechanisms—creation of intermaterial area and diffusion—determines how quickly striations are consumed by the reaction. The exponential stretching process controls both of these mechanisms (Szalai et al., 2003).

Next, we examine the spatial distribution of reactants and products as they evolve in time. The left-hand column of Figure 3-28 shows the spatial distribution of reactants after n = 1, 2, and 3. The right-hand column of Figure 3-28 is the corresponding product distribution in the flow. Similar to previous contour plots, the colors of each figure are based on the relative maximum values of reactants and products, respectively. On the left side, striations of A (red) interpenetrate striations of B (blue) and regions near the reactive interface quickly become exhausted. At the same time, product appears at these locations, following the same landscape as in Figure 3-27. The initially thin product-rich filaments become increasingly thicker until isolated areas of low product concentration remain dispersed in the flow. These islands of low product concentration contain exclusively reactant A or B, as seen in the corresponding left-hand picture. They remain intact until diffusion transports reactants to these regions from the surrounding striations.

Diffusion increasingly smears the boundaries of striations in Figure 3-28. However, the difference is primarily cosmetic. As shown in Figure 3-29, the product concentration field remains statistically self-similar. The product concentration after each period is scaled using standardization. The probability density function of $c_P$ remains invariant over three flow periods after the mean ($\langle \log c_P \rangle$) is subtracted from each measurement and divided by the standard deviation ($\sigma_{\log c_P}$). Note here that the conversion at the end of the fourth period, corresponding to the pink curve on the plot, is over 96%. By this time the reaction is close to completion; in fact, most areas have become exhausted of reactants, thus demonstrating that self-similar dynamics control the entire evolution of the reactive mixing process even at Pe as low as Pe = 2000.

### 3-6.1 Reactions in 3D Laminar Systems

Convective mixing not only affects the progression of reactions in chaotic flows but has governing control on their evolution. The idea that it is sufficient to

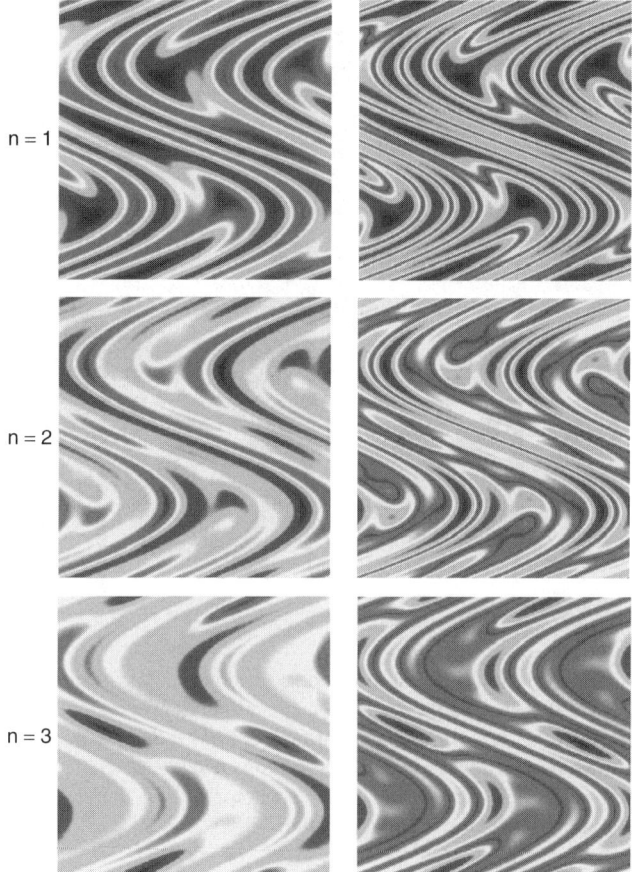

**Figure 3-28** The left-hand side of the figure shows the concentration of reactants A and B after the first three flow periods in the sine flow with $T = 1.6$. The right-hand side is the corresponding product concentration. See insert for a color representation of this figure.

have the end result in mind when designing a laminar reactive mixing process must be abandoned. The only route to fast mixing in laminar applications is via chaotic flows, and the mixing intensity in these systems is inherently nonuniform. In such systems, a wide spectrum of stretching intensities distributes material interface along a predefined, invariant template. It establishes intimate contact between mixture components (whether reactive or not) and controls the local rate of diffusion through the intermaterial area density and striation thickness distribution. All chaotic mixing flows share these characteristics.

Understanding the interplay of the three mechanisms—reaction, diffusion, and convection—is essential. In light of what happens in 2D reactive chaotic flows, we can attempt to extend the analysis to 3D reactive mixing applications. Let us take a fresh look at a common 3D example, a mixing tank stirred with three

**136** LAMINAR MIXING: A DYNAMICAL SYSTEMS APPROACH

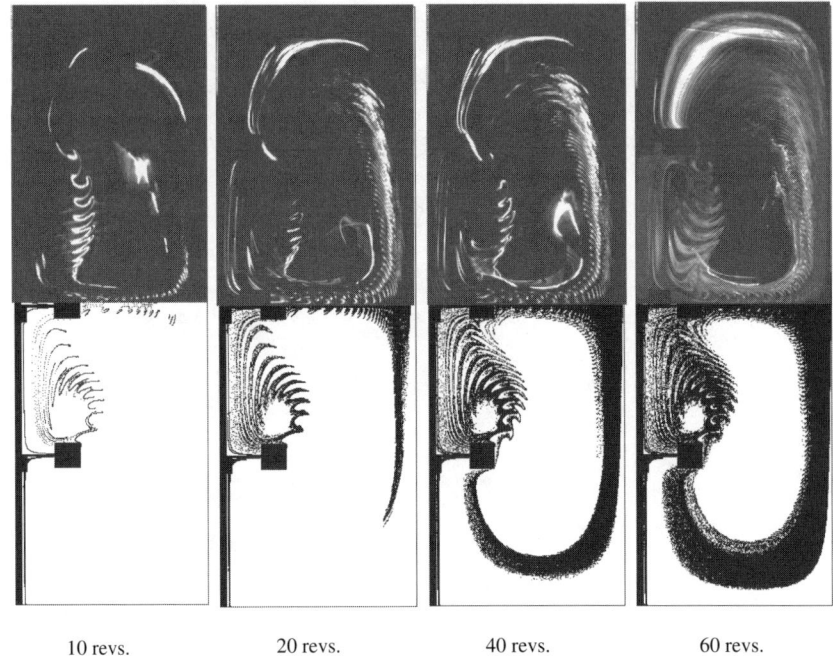

10 revs.　　　20 revs.　　　40 revs.　　　60 revs.

**Figure 3-29** The evolution of a fast reaction in a mixing tank stirred with three Rushton impellers at Re = 20. The reactive zones in a stirred tank are identical to the location of the intermaterial contact area in a mixing after 10, 20, 40, and 60 impeller revolutions. Each figure shows half of the vertical cross-section, where the shaft and three impeller blades are seen on the left. The upper half is a photograph of the reactive flow and the bottom is the computed chaotic mixing structure.

Rushton impellers. Analytical solutions for stretching and particle positions are not available for 3D flows, so we need to rely on numerical methods to compute the stretching and the filament structure. Figure 3-29 shows a dye filament, initially placed vertically near the impeller shaft, as it is stretched and folded into a complex mixing structure that eventually invades the entire chaotic region. The four pictures in the figure indicate four computational snapshots of the chaotic mixing process at consecutive times. We computed the structure of the filament by placing 200 000 tracer particles in a line along the shaft.

As time increases and convective mixing takes place, the material filament gradually invades portions of the chaotic region. A convoluted, partially mixed structure develops after 60 revolutions, and the location of the filament marks the interface between the mixture components, as indicated in Figure 3–29. An experiment using a pH-sensitive dye is compared to Figure 3–29, where dye was initially injected near the impeller blades. A photograph after 60 impeller revolutions indicates that the filament acquires an identical structure in the experiments as predicted by the computed mixing structure. Since the dye is pH sensitive, an instantaneous acid–base reaction occurs in the tank at the interface between the mixture components. The mechanism that controls the evolution of the reactive mixing process is largely the same as in the sine flow described earlier. The fluorescent, pH-sensitive dye gives a visual indication of the reaction zones, where the acid–base reaction occurs in the tank. The product-rich areas glow with high illumination and their shape unmistakably coincides with the mixing structure created by the convective flow.

We can go further than qualitative comparisons and measure the probability density function of the reactant and product concentrations in the tank. To measure the product concentrations experimentally, image analysis is used to quantify the level of luminescence in the experimental photograph. The time evolution of the reactive process is monitored by taking a series of pictures similar to those in Figure 3-29. Product concentration is quantified based on the pixel values in each photograph. The entire vessel is divided to equal-sized cells, and the amount of fluorescent dye is measured in each cell. The probability density function of the concentration is computed by counting the number of cells that have concentrations between values of c and $(c + \Delta c)$. The normalized values (w) are computed by subtracting the mean and dividing each by the standard deviation on a logarithmic scale. Figure 3-30 indicates the distribution of $c_P$ in the three-impeller stirred tank at three different times. The evolution of the acid–base

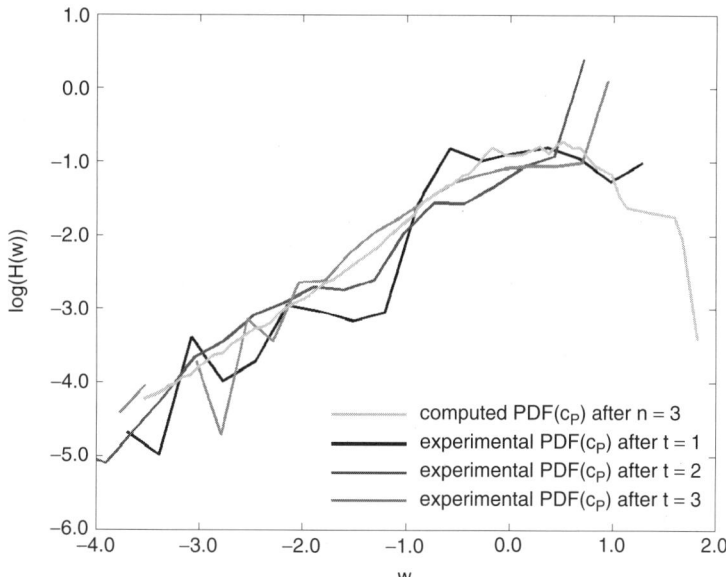

**Figure 3-30** The probability density function of the product concentration [PDF($c_P$)] in the tank is calculated from the experimental images after t = 1, 2, and 3 min. The experimental distributions in the 3D chaotic flow are invariant after scaling and match the concentration distribution of the model system. It is shown by overlaying PDF($c_P$) in the tank and in the sine flow at n = 3.

reaction 1, 2, and 3 min after injecting the dye is captured during the experiments. In light of the self-similarity that occurs in 2D reactive chaotic processes (recall Figure 3-27), we expect the scaled product concentration profiles to show an invariant distribution. Figure 3-30 confirms this expectation for the 3D reactive flow. The distribution of the scaled concentration profiles after 1, 2, and 3 min fall right on top of one another. Remarkably, they also agree closely with the scaled distribution obtained from the 2D model system. The agreement is shown in Figure 3-30 by overlaying the curve from n = 3 from the sine flow with the experimental data. This preliminary result suggests that self-similarity also appears in 3D reactive flows, and implies that the chaotic stretching and orientation process governs the evolution of reactions in 3D systems, indicating a clear conceptual direction pregnant with promise for future progress.

## 3-7 SUMMARY

Fluid mixing takes place by a combination of mechanisms: convection (stirring or macromixing), stretching, folding, and diffusion. Effective mixing processes, whether we think about turbulent flows or chaotic flows, all promote global uniformity by redistributing initially segregated components in space. In the laminar regime, the only efficient route to mix is through chaotic flows. The focus of this chapter was to discuss the nature of these flows through a set of experimental and computational observations. In a general sense, these can be summarized in a few key points:

- Defined geometric patterns and robust statistical properties are evident in chaotic mixing flows. The naive interpretation of chaos, as synonymous with disorder, is misleading. A defined mixing structure is created by each chaotic flow, and the generic geometric features these flows possess all have similar characteristics. We demonstrated this phenomenon on a 2D model flow (the sine flow) as well as on the flow field of an industrial mixer (the Kenics static mixer).
- The highly symmetric, self-similar nature of structures generated by chaotic mixing processes is connected to an intrinsic invariant field of orientations, generated by an invariant manifold structure in these flows. At each location in the domain, this predetermined orientation does not depend on time.
- The evolution of material filaments in chaotic flows can be described by the average intermaterial area density $\langle \rho \rangle$ at any period in the flow. This single measure, obtained from the elementary stretching of massless tracers, expresses the global time evolution of microstructure in chaotic flows: $\langle \rho \rangle \sim e^{n\Theta}$. The accumulation of interface in different areas of the flow domain is highly nonuniform. In some regions the intermaterial area density ($\rho$) is five orders of magnitude higher than in others. However, since it increases everywhere at the same rate $\Theta$, this parameter is a true rate for the overall mixing process.

- Due to the nonuniformity in ρ, the local rate of transport-controlled reactions (i.e., "fast" reactions) will be much faster in some areas of the flow than in others. In exothermic reactions, this can lead to the creation of "hot spots."
- The microstructure generated by chaotic periodic flows is very robust. Once the shape of the distribution of striations is determined and the average time evolution is known, the distribution of characteristic lengths can be predicted for any future time, because the striation thickness (s) is inversely proportional to ρ. Striation thickness distributions can be accurately predicted from stretching distributions, and their statistical properties are identical. Therefore, the stretching field provides convenient means of examining the length scales generated by chaotic mixing flows.
- Since stretching controls the location and density of the intermaterial area between mixture components, it has a direct impact on the evolution of reactions. The location of the reactive zones is practically identical to the mixing structure determined only by the chaotic stretching process. Furthermore, since stretching is a self-similar process, the distribution of product concentrations is self-similar as well and can be rendered invariant by statistical methods.

## 3-8 CONCLUSIONS

In this chapter we have discussed methods from dynamical systems theory for characterization of microstructure and topology generated by chaotic periodic flows. In the first half of the chapter, a simple two dimensional model flow, the sine flow, was introduced as an example to illustrate the application of some common and newly developed tools for mixing characterization. In the second half of the chapter we showed how these techniques can be extended to real, three dimensional chaotic flows, and our practical example was the Kenics static mixer. An important shortcoming of methods such as Poincaré sections is that the use of point tracers does not preserve the evolving mixing structure. Thus, understanding and controlling the dynamics of mixing, the very goal in mind when designing a process or equipment, becomes an impossible task.

Convective mechanisms in chaotic flows (stretching and folding) transform portions of materials into elongated striations and reorient these filaments with respect to the direction of deformation. Continuous reorientation creates a sustained exponential rate of increase in intermaterial contact area in chaotic flows. These mechanisms are captured by nontraditional tools, such as computing the stretching field, computing striation thickness distributions directly, or predicting them from stretching.

Another important component of mixing processes, diffusion, remains to be included in the discussion. The ability to measure or predict striation thickness distributions in chaotic flows provides the link between mixing on the macroscopic and microscopic scales. Diffusion and stretching are intimately coupled because as material filaments are stretched by the flow, the rate of diffusion is increased simultaneously. The increase is due partially to an increase in the interfacial area available for transport (as striation boundaries are increased) and also to a reduction in diffusional length scales (striation thicknesses are decreased everywhere at the same rate). Since stretching is spatially nonuniform in chaotic flows, fluctuations in local diffusional fluxes will also occur. This, in turn, will lead to spatial fluctuations in reaction rate, product concentration, and waste concentration. Continued efforts to incorporate all of the important mechanisms (convection, diffusion, and reactions) within a unified mixing theory is necessary before we can fully understand and predict the behavior of complex real-world industrial systems.

## NOMENCLATURE

| | |
|---|---|
| $c_i$ | concentration of a reactant, where i = A, B, or P |
| D | diameter of the housing for a continuous flow mixer |
| $D_f$ | diffusion coefficient |
| $l$ | length of an infinitesimal vector at a later time |
| $l_0$ | initial length of an infinitesimal vector representing a fluid filament |
| $l_x$ | x component of the stretching vector $l$ having initial length of $l_{xo}$ |
| $l_y$ | y component of the stretching vector $l$ having initial length of $l_{yo}$ |
| L | length of a static mixer element |
| L/D | element length-to diameter ratio, defined as the aspect ratio |
| n | integer representing the periodicity of the flow |
| $N_\rho$ | total number of cells the flow region is divided into |
| $N_P$ | number of fluid tracer particles in a flow |
| $N_s$ | number of fluid striations in a selected region of a flow |
| Pe | dimensionless Péclet number |
| r | local rate of a chemical reaction |
| Re | Reynolds number |
| s | thickness of a fluid striation |
| $\langle s \rangle$ | arithmetic average of the striation thickness, s |
| t | time |
| $t_n$ | time after n flow periods |
| T | period of a periodic chaotic flow |
| $\underline{v}$ | velocity field created by a fluid flow |
| $\langle \underline{v} \rangle$ | mean velocity in the main flow direction |
| $V_x$ | x-component of a velocity field |
| $V_y$ | y-component of a velocity field |
| w | scaled variable defined as $w = (\log c_P - \langle \log c_P \rangle)/\sigma_{\log c_P}$ |
| $\underline{X}$ | position vector of a fluid tracer particle, equal to (x,y) |
| x | x-position of a fluid tracer particle |
| y | y-position of a fluid tracer particle |
| $x_n$ | x-position of a fluid tracer particle after n flow periods |
| $y_n$ | y-position of a fluid tracer particle after n flow periods |

*Greek Symbols*

$\alpha_n$      enclosed angle between two vector components at time n
$\Theta$      topological entropy of a chaotic flow
$\lambda$      ratio of $l/l_0$ is defined as the stretching of a fluid filament
$\Lambda$      Lyapunov exponent of a chaotic flow
$\mu_F$      fluid viscosity
$\rho$      intermaterial area density created by a fluid flow
$\langle\rho\rangle$      average intermaterial area density
$\rho_F$      fluid density
$\sigma_{\log c_P}$      standard deviation of log $c_P$
$\varphi$      variable defined as the difference of the reactant concentrations, $c_A - c_B$

## REFERENCES

Alvarez, M. M., F. J. Muzzio, S. Cerbelli, and A. Adrover (1997). Self-similar spatiotemporal structure of material filaments in chaotic flows, *Fractals Eng.*, **1**, 323–335.

Amanullah, A., L. Serrano-Carreon, B. Castro, E. Galindo, and A. W. Nienow (1998). The influence of impeller type in pilot scale Xanthan fermentations, *Biotech. & Bioeng.*, **57**(1), 95–108.

Aref, H. (1984). Stirring by chaotic advection, *J. Fluid Mech.*, **143**, 1–21.

Arnold, V. I. (1963). Small denominators and problems of stability in classical and celestial mechanics, *Russ. Math Surv.*, **18**(6), 85–191.

Avalosse, Th., and M. J. Crochet (1997). Finite-element simulation of mixing: 2. Three-dimensional flow through a kenics mixer, *AIChE J.*, **43**(3), 588–597.

Cerbelli, S. (2000). *The topology of chaotic two-dimensional systems*, Ph.D. dissertation, Graduate School of New Brunswick, Rutgers University, New Brunswick, NJ.

Chaiken, J., R. Chevray, M. Tabor, and Q. M. Tan (1986). Experimental study of Lagrangian turbulence in Stokes flow, *Proc. R. Soc. London*, **A408**, 165–174.

Chien, W. L., H. Rising, and J. M. Ottino (1986). Laminar mixing and chaotic mixing in several cavity flows, *J. Fluid Mech.*, **170**, 355–377.

Dong, L., S. T. Johansen, and T. A. Engh (1994). Flow induced by an impeller in an unbaffled tank: I. Experimental, *Chem. Eng. Sci.*, **49**(4), 549–560.

Franjione, J. G., and J. M. Ottino (1987). Feasibility of numerical tracking of material lines and surfaces in chaotic flows, *Phys. Fluids*, **30**, 3641–3643.

Giona, M., S. Cerbelli, F. J. Muzzio, and A. Adrover (1998). Non-uniform stationary measure properties of chaotic area-preserving dynamical systems, *Physica A*, **254**, 451–465.

Giona, M., A. Adrover, F. J. Muzzio, S. Cerbelli, and M. M. Alvarez (1999). The geometry of mixing in time-periodic chaotic flows: I. Asymptotic directionality in physically realizable flows and global invariant properties, *Physica D*, **132**, 298–324.

Giona, M., S. Cerbelli, and A. Adrover (2002). Geometry of reaction interfaces in chaotic flows, *Phys. Rev. Lett.*, **88**(2), 024501-01.

Gollub, J. P., and M. C. Cross (2000). Chaos in space and time, *Nature*, **404**, 710–711.

Hobbs, D. M., and F. J. Muzzio (1998). Optimization of a static mixer using dynamical systems techniques, *Chem. Eng. Sci.*, **53**(18), 3199–3213.

Hobbs, D. M., M. M. Alvarez, and F. J. Muzzio (1997). Mixing in globally chaotic flows: a self-similar process, *Fractals*, **5**(3).

Jaffer, S. A., and E. P. Wood (1998). Quantification of laminar mixing in the Kenics static mixer: an experimental study, *Can. J. Chem. Eng.*, **76**, June, pp. 516–521.

Kim, S. J., and T. H. Kwon (1996). Enhancement of mixing performance of single-screw extrusion process via chaotic flows: II. Numerical study, *Adv. Polym. Technol.*, **15**(1), 55–69.

Kolmogorov, A. N. (1954). On conservation of conditionally periodic motions under small perturbations of the Hamiltonian, *Dokl. Akad. Nauk. SSSR*, **98**, 527–530.

Kusch, H. A., and J. M. Ottino (1992). Experiments on mixing in continuous chaotic flows, *J. Fluid Mech.*, **236**, 319–348.

Lamberto, D. J., F. J. Muzzio, and P. D. Swanson (1996). Using time-dependent rpm to enhance mixing in stirred vessels, *Chem. Eng. Sci.*, **51**(5), 733–741.

Leong, C. W., and J. M. Ottino (1989). Experiments on mixing due to chaotic advection in a cavity, *J. Fluid Mech.*, **209**, 463–499.

Ling, F. H., and X. Zhang (1995). A numerical study on mixing in the Kenics static mixer, *Chem. Eng. Commun.*, **136**, 119–141.

Mandelbrot, B. (1982). *The Fractal Geometry of Nature*, W.H. Freeman, San Francisco.

Metcalfe, G., and J. M. Ottino (1994). Autocatalytic processes in mixing flows, *Phys. Rev. Lett.*, **72**(18), 2875–2878.

Middlemann, S. (1977). Mixing, in *Fundamentals of Polymer Processing*, McGraw-Hill, New York.

Moser, J. (1962). On invariant curves of area-preserving mappings of an annulus, *Nachr. Akad. Wiss. Gottingen Math. Phys. Kl.*, **II**, 1–20.

Muhr, H., R. David, J. Villermaux, and P. H. Jezequel (1995). Crystallization and precipitation engineering—V. Simulation of the precipitation of silver bromide octahedral crystals in a double jet semi-batch reactor, *Chem. Eng. Sci.*, **50**(2), 345–355.

Muzzio, F. J., and M. Liu (1996). Chemical reactions in chaotic flows, *Chem. Eng. J.*, **64**, 117–127.

Muzzio, F. J., and J. M. Ottino (1988). Coagulation in chaotic flows, *Phys. Rev. A*, **38**, Sept., pp. 2516–2524.

Muzzio, F. J., and J. M. Ottino (1989). Dynamics of a lamellar system with diffusion and reaction: scaling analysis and global kinetics, *Phys. Rev. A*, **40**(12), 7182–7192.

Muzzio, F. J., and J. M. Ottino (1990). Diffusion and reaction in a lamellar system: self-similarity with finite rates of reaction, *Phys. Rev. A*, **42**(10), 5873–5884.

Muzzio, F. J., and P. D. Swanson (1991). The statistics of stretching and stirring in chaotic flows, *Phys. Fluids A*, **3**(5), 822–834.

Muzzio, F. J., M. Tjahjadi, and J. M. Ottino (1991). Self-similar drop size distributions produced by breakup in chaotic flows, *Phys. Rev. Lett.*, **67**(1), 54–57.

Muzzio, F. J., M. M. Alvarez, S. Cerbelli, M. Giona, and A. Adrover (2000). The intermaterial area density generated by time- and spatially periodic 2D chaotic flows, *Chem. Eng. Sci.*, **55**, 1497–1508.

Neufeld, Z. (2001). Excitable media in a chaotic flow, *Phys. Rev. Lett.*, **87**(10), 108301-1.

Oseledek, V. I. (1968). A multiplicative ergodic theorem: characteristic Lyapunov exponents of dynamical systems, *Trudy Mosk. Mat.* (Obsc.), **19**, 179–210.

Pahl, M. H., and E. Muschelknautz (1982). Static mixers and their applications, *Int. Chem. Eng.*, **22**(2), 197–205.

Sokolov, I. M., and A. Blumen (1991). Diffusion-controlled reactions in lamellar systems, *Phys. Rev. A*, **43**(6), 2714–2719.

Souvaliotis, A., S. C. Jana, and J. M. Ottino (1995). Potentialities and limitations of mixing simulations, *AIChE J.*, **41**(7), 1605–1621.

Swanson, P. D., and J. M. Ottino (1990). A comparative computational and experimental study of chaotic mixing of viscous fluids, *J. Fluid Mech.*, **213**, 227–249.

Szalai, E. S., and F. J. Muzzio (2002). A fundamental approach to the design and optimization of static mixers, submitted for publication in *AIChE J*.

Szalai, E. S., J. Kukura, P. Arratia, and F. J. Muzzio (2002). The effect of hydrodynamics on reactive mixing applications, submitted for publication in *AIChE J*.

Szalai, E. S., and F. J. Muzzio (2003). A fundamental approach to the design and optimization of static mixers, to appear in *AIChE J*.

Szalai, E. S., and F. J. Muzzio (2003). Predicting mixing microstructure in three dimensional systems, to appear in *Phys. Fluids*.

Szalai, E. S., J. Kukura, P. Arratia, and F. J. Muzzio (2003). Effect of hydrodynamics on reactive mixing in laminar flows, *AIChE J.*, **49**(1), 168–179.

Unger, D. R., and F. J. Muzzio (1998). Experimental and numerical characterization of viscous flow in an impinging jet contactor, *Can. J. Chem. Eng.*, **76**(6), 546–555.

Zalc, J. M., and F. J. Muzzio (1999). Parallel-competitive reactions in a two-dimensional chaotic flow, *Chem. Eng. Sci.*, **54**, 1053–1069.

Zalc, J. M., M. M. Alvarez, B. E. Arik, and F. J. Muzzio (2001). Extensive validation of computed laminar flow in a stirred tank with three Rushton turbines, *AIChE J.*, **47**(10), 2144–2154.

# CHAPTER 4

# Experimental Methods

DAVID A. R. BROWN, PIP N. JONES, and JOHN C. MIDDLETON

*BHR Group Ltd*

GEORGE PAPADOPOULOS

*Dantec Dynamics, Inc.*

ENGIN B. ARIK*

*VioSense Corporation*

## Part A: Measuring Tools and Techniques for Mixing and Flow Visualization Studies

*David A. R. Brown, Pip N. Jones, and John C. Middleton*

### 4-1 INTRODUCTION

The aim of this chapter is to provide information on some of the tools and techniques available to make qualitative and quantitative measurements of mixing processes. The list of techniques discussed here is not exhaustive. A large number of experimental methods have been employed over the years to quantify mixing processes, and many of those have proved to be unreliable, or reliable only if performed under very specific conditions. Emphasis is placed on providing practical information on techniques that are both reliable and repeatable.

The equipment, instrumentation, and protocols required to perform the various techniques are discussed, with emphasis on the potential errors, inaccuracies, inconsistencies, and limitations of the techniques, as well as any expected

---

* Formerly with Dantec Dynamics, Inc.

---

*Handbook of Industrial Mixing: Science and Practice*, Edited by Edward L. Paul, Victor A. Atiemo-Obeng, and Suzanne M. Kresta
ISBN 0-471-26919-0   Copyright © 2004 John Wiley & Sons, Inc.

deviations between laboratory models and industrial scale operation. Most of the techniques described in this chapter can be applied to stirred tank, jet, and in-line mixed systems with little modification.

### 4-1.1 Preliminary Considerations

Before beginning any experimentation, a careful assessment should be made of the problem that is being investigated and of the situation in which measurements are to be made. This will help to prevent an inappropriate technique being used for the measurement, and in many cases will lead to a rapid choice of the most suitable experimental method.

Experimental techniques for measuring mixing can be broadly divided into two categories: those that are performed in a laboratory, and those measurements that are performed in actual process plant equipment. The instrumentation and techniques used in each type of measurement are often different, although they can be based on very similar principles.

In the laboratory, experiments are most often carried out on a small scale using transparent vessels or pipe-work. Real process fluids should be used where possible; simulant fluids may be used if the process fluids are difficult to work with. It is straightforward to change the vessel and process configurations in the laboratory, making it possible to investigate a wide range of parameters relatively quickly and easily. The instrumentation that is used will require careful setup and expert operation, but with such treatment will yield precise and accurate data.

Process plant measurements are constrained by the nature of the process plant being used. The vessels are usually much larger than those found in a mixing laboratory, access to the vessels is often difficult, and visual observations of the mixing process are usually very difficult. Making alterations to the plant is costly and time consuming, particularly as a frequent requirement is that production must not be interfered with. In most cases the actual process fluids must be used, which may be difficult or dangerous to handle. The instrumentation used must be extremely robust and must be able to tolerate possible mishandling under the plant conditions. Measurements may have to be made at elevated temperatures or pressures, further increasing the cost and difficulty. The penalty for this robustness is usually a greatly reduced accuracy of measurement.

As a general rule, any extensive investigative or process development work should be performed at the laboratory scale, where a large number of parameters can be investigated easily, leading to rapid process optimization. Scale-up should be investigated using several different sizes of vessel at the laboratory (and pilot) scale, before moving up to the full production plant scale. Experiments with process plant should be carried out very selectively, as confirmation of laboratory work, or to investigate problems with existing process plant.

Before embarking on any series of mixing experiments, it is essential that the relevant published literature be checked to ensure full awareness of the methods and findings of other workers, avoiding unnecessarily repetition of what can be time consuming and costly experiments.

## 4-2 MIXING LABORATORY

In this section we outline the main requirements and considerations in setting up a laboratory with which to perform mixing experiments. Mixing experiments have been performed in the laboratory for many years and a vast array of measurement techniques is available to us. A large proportion of the measurement techniques described have not improved in any fundamental way over the last 10 years. The main improvements have probably come in the following areas:

- Increased data transfer rates, allowing the capture of more data
- Increased data storage capacity
- Increased computer processing power, allowing faster (and sometimes online) signal processing

This means that although we can now generate huge volumes of data, we still need to make sure that we analyze them properly and develop sensible conclusions.

Some other more recent advances that have had a significant effect on the measurements we can make are:

- Digital (or electronic analogue) video imaging and recording, allowing easy visualization of experiments
- Lasers, allowing the use of techniques such as laser Doppler anemometry (LDA) and particle image velocimetry (PIV), described later in this chapter

### 4-2.1 Safety

Safety is of paramount importance in the mixing laboratory. All rigs produced should undergo a safety assessment before use, including (where deemed necessary) a full HAZOP study. The safety assessment should result in a full set of standard operating procedures for the rig, a copy of which should be distributed to all operators. These procedures should include clear instructions is to the rig emergency shutdown procedure. A copy of the procedure should be displayed prominently on the rig. All rigs that include a rotating agitator shaft should have appropriate guards and a prominently displayed emergency stop button.

It is essential that a "Tank Contents" label be attached to each mixing vessel (or indeed, in-line mixing rig), along with relevant Control of Substances Hazardous to Health (COSHH) sheets or Material Safety Data Sheets (MSDSs) for each material present. This ensures that anybody in the laboratory encountering a problem with a rig is aware of the materials present and appropriate safety precautions.

As with all laboratories, it is vital that cleanliness and tidiness be maintained. The mixing laboratory is often subject to spillages of various fluids and particulate

**148** EXPERIMENTAL METHODS

solids, all of which can lead to slip hazards. All mixing vessels and rigs should be placed in a bounding wall or dike to contain any spillage.

## 4-2.2 Fluids

It is often not practical to use actual process fluids in the mixing laboratory, as this can involve the use of expensive and obstructive safety precautions as well as inconvenient temperatures and pressures. To avoid these problems, suitable simulant fluids must be found that will behave in a manner representative of the process fluid in the laboratory mixer. It should be noted that the simulant fluid must have the correct rheological properties *for the scale at which the measurements are to be made*. This is not necessarily the same as simply having the same properties as the fluid in the process: for example, if it is non-Newtonian.

### 4-2.2.1 Fluid Rheology[1].
It is important to model the fluid rheology correctly, particularly when non-Newtonian rheologies are concerned. The rheological properties of liquids handled in industrial processes can often change during the course of the process.

*Definitions.* Considering a small element of a liquid undergoing steady motion between the wall of a stationary cylindrical vessel and that of a rotating coaxial cylinder (Figure 4-1), the stress tensor ($\tau_{ij}$) can be defined by nine components:

$$\tau_{ij} = \begin{bmatrix} \tau_{xx} & \tau_{xy} & \tau_{xz} \\ \tau_{yx} & \tau_{yy} & \tau_{yz} \\ \tau_{zx} & \tau_{zy} & \tau_{zz} \end{bmatrix}$$

$\tau_{xy}, \tau_{xz}, \tau_{yz}, \tau_{yx}, \tau_{zx}$, and $\tau_{zy}$ are the shearing stresses, and for this configuration

$$\tau_{xz} = \tau_{zx} = 0, \tau_{yz} = \tau_{zy} = 0$$

Since the relative motion between the cylinders is in the x-direction, $\tau_{xy} = \tau_{yx}$. $\tau_{xx}, \tau_{yy}$, and $\tau_{zz}$ are the normal stresses. $\tau_{yx}$, the shear stress, varies in the y-direction and in most texts is denoted simply by $\tau$.

The flow curve of a fluid (Figure 4-2) is obtained by plotting the shear stress ($\tau$, Pa) as a function of the shear rate ($\dot\gamma$, s$^{-1}$; i.e., the change of shear strain per unit time). Shear stress is related to the shear rate by

$$\tau = \mu\dot\gamma$$

where $\mu$ is the liquid viscosity (Pa · s).

---

[1] Thanks to Gül Özcan-Taşkin for permission to reproduce this section from her Ph.D. thesis (Özcan-Taşkin, 1993).

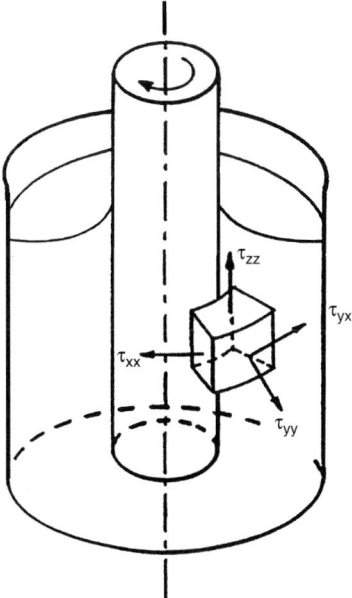

**Figure 4-1** Stress components around a rotating coaxial cylinder.

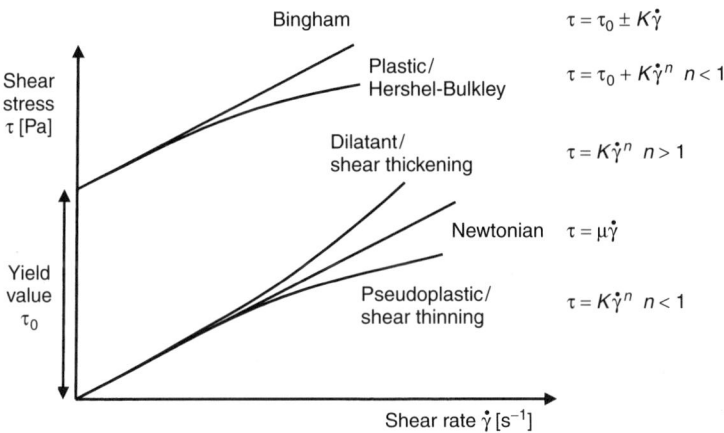

**Figure 4-2** Flow curves of Newtonian and non-Newtonian fluids.

*Newtonian Flow Behavior.* The viscosity of a Newtonian liquid depends only on temperature and pressure. At constant temperature and pressure, Newtonian behavior is characterized as follows (Barnes et al., 1989):

1. The shear stress is the only stress generated in simple shear flow.
2. The shear viscosity is independent of the shear rate.

3. The viscosity does not vary with the time of shearing and the stress in the liquid falls to zero immediately after the shearing is stopped. The viscosity from any subsequent shearing—regardless of the period of time between measurements—is as before.
4. The viscosities measured in different types of deformation (e.g., uniaxial extensional flow and simple shear flow) are always in simple proportion to one another.

Water, oil, glycerol, and sugar solutions are examples of Newtonian liquids. Dilute sludges such as unconcentrated activated and trickling-filter sludges also exhibit Newtonian behavior.

*Non-Newtonian Flow Behavior.* All liquids showing deviations from the behavior above are non-Newtonian. Many liquids encountered in industrial practice, such as paints, emulsions, most mineral slurries, latex, paper pulp, plastic melts, liquid foods, polymeric liquids, and concentrated wastewater sludge, are non-Newtonian.

Some examples of non-Newtonian flow behavior are shown in Figure 4-2. The viscosity of a non-Newtonian fluid is not a coefficient of the shear rate but a function of it and is called the *apparent viscosity* ($\mu_a$, Pa · s):

$$\mu_a = \frac{\tau}{\dot{\gamma}}$$

Non-Newtonian fluids can usefully be classified into three main categories (Skelland, 1967) even though sharp distinction between these groups is difficult:

1. *Time-independent fluids.* The shear stress at any point is dependent only on the instantaneous shear rate at that point.
2. *Time-dependent fluids.* The shear stress, and hence the viscosity, either decreases or increases with the duration of the shearing. These changes may be reversible or irreversible.
3. *Viscoelastic fluids.* These fluids have both elastic solid and viscous fluid characteristics. They exhibit partial elastic recovery after a deforming shear stress is removed.

Figure 4-2 shows typical flow curves of Newtonian and non-Newtonian fluids. Figure 4-3 illustrates shear viscosity as a function of shear rate for a pseudoplastic fluid, showing regions of low-shear limiting viscosity ($\dot{\gamma}_0$), high-shear limiting viscosity ($\dot{\gamma}_\infty$) and power law behavior.

*Time-Independent Fluids.* The majority of non-Newtonian fluids in this category are found to be pseudoplastic (Figures 4-2 and 4-3). The viscosity of pseudoplastic fluids decreases with the increase of shear rate over a large range of shear rates. Therefore, they are also called *shear-thinning liquids*. The characteristic of the

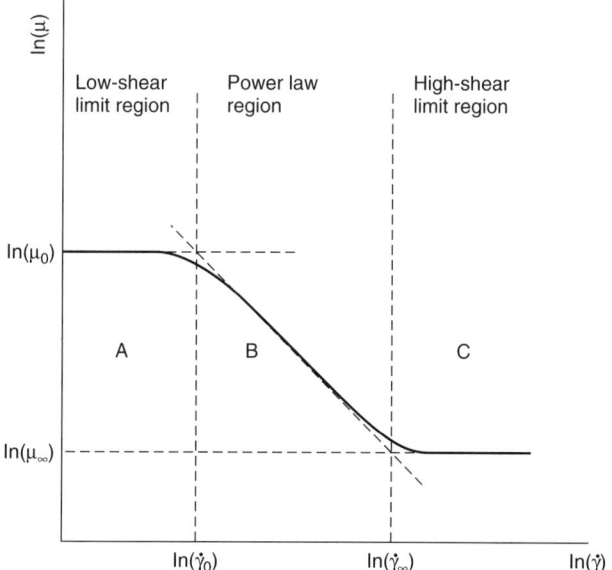

**Figure 4-3** Shear viscosity as a function of shear rate for a typical pseudoplastic fluid.

flow curve (Figure 4-3) is the linearity of the flow curve at very low and very high shear rates where the measured viscosity is constant ($\mu_0$ and $\mu_\infty$). Most industrial processes take place in the range of shear rates where the viscosity decreases with the shear rate. Therefore, the power law model of Oswald de Waele is widely used to characterize the shear-thinning behavior:

$$\tau = K\dot{\gamma}^n$$

where K (kg/m$^{-1}$ · s$^{n-2}$) is the consistency index and n(–) is the flow behavior index. The deviation of n from unity is a measure of the departure from Newtonian behavior. For shear-thinning fluids, $0 < n < 1$. The expression for the apparent viscosity is then

$$\mu_a = K\dot{\gamma}^{n-1}$$

where the apparent viscosity is linearly proportional to the shear rate on a logarithmic scale (section B in Figure 4-3). Most polymer solutions exhibit pseudoplastic flow behavior.

For *dilatant fluids*, $n > 1$. Rheological dilatancy refers to increasing viscosity with increasing shear rate. Therefore, these fluids are also called *shear-thickening*. Examples include whipped cream and starch slurries. They are rare in industrial practice.

Pseudoplastic and dilatant fluids begin to flow as soon as a stress is applied. For *plastic fluids*, a yield value ($\tau_y$) has to be exceeded before flow occurs (Figure 4-2). Two types of yield stress liquids are Bingham plastic and viscoplastic fluids

(Figure 4-2). Toothpaste, greases, certain mineral slurries such as ground limestone, tomato sauce, and wastewater sludge are some examples of plastic fluids. The Herschel–Bulkley model is used to characterize the behavior of these fluids:

$$\tau = \tau_y + K\dot{\gamma}^n$$

$$\mu = \frac{\tau_y + K\dot{\gamma}^n}{\dot{\gamma}}$$

where $n = 1$ for Bingham plastics.

*Time-Dependent Fluid Properties.* This category includes thixotropic and rheopectic fluids. When a shear stress is applied to a thixotropic fluid, the viscosity decreases gradually, and when the stress is removed, a gradual recovery is observed. The opposite behavior is observed with rheopectic fluids: a gradual increase of viscosity during shear followed by recovery on the removal of shear-inducing stress.

*Viscoelastic Fluids.* These materials exhibit elastic and viscous properties simultaneously. Under simple shear flow, normal stresses are generated as shown in Figure 4-1. The magnitude of these normal forces varies with the shear rate. It is usual to work in terms of normal stress differences rather than normal stresses themselves:

$$N_1 = A\dot{\gamma}^b$$

where $N_1$ (Pa) is the first normal stress difference. Polymer solutions exhibit viscoelastic behavior. While mixing viscoelastic fluids, forces generated normal to the plane of shear can lead to flow reversal and the Weissenberg effect where liquid climbs up a rotating shaft (see also Chapter 6).

**4-2.2.2 Simulant Fluids.** When choosing simulant fluids with which to perform experiments, the following must be taken into consideration:

- Is the fluid safe to use in a laboratory?
- Are the physical properties of the simulant fluid correct for the scale at which the experiments are being performed?
- Can the fluids be disposed of safely and economically at the laboratory scale?
- Can the fluid dissolve any ionic or surface active tracer materials?
- Is the fluid rheology sensitive to temperature?
- Will the fluid absorb moisture from the atmosphere?
- How long will the fluid last before decomposing?
- How expensive is the material?
- How must the material be disposed of?

Some common simulant materials are described below.

*Water.* Being cheap and readily available, water is an ideal simulant fluid for Newtonian mixing in the turbulent regime. Its density (and conductivity) can readily be modified by the addition of various salts, although care is required in the study of gas–liquid or liquid–liquid systems that the presence of dissolved salts does not alter the bubble or droplet coalescence properties of the system. Disposal of water is cheap, provided that the concentrations of any dissolved materials are within local discharge regulations.

*Glucose Syrup and Corn Syrup.* Undiluted glucose syrup is an extremely viscous transparent Newtonian material with a viscosity of about 150 Pa · s at room temperature. Corn syrup has similar properties and has a viscosity of about 18 Pa · s in undiluted form. At high viscosities both are extremely stable and will keep for a long period of time (months even). Once diluted with water (to about 1000 cP) biocide and fungicide should be added if solutions are to be kept for more than a day, as fermentation will take place, rendering the fluid unusable and producing an extremely unpleasant odor.

The extremely high viscosity and density of undiluted glucose make it quite difficult to perform the initial dilution required to make solutions of the required viscosity. The best method is to add the glucose slowly to a mixing vessel containing the required amount of water. Where this is not possible, mixing using a high-torque-rated mixer with lots of patience may be the only answer. Glucose affects the polarization of laser beams, so is not suitable for LDA measurements.

*Glycerol.* Glycerol is a viscous Newtonian fluid (99.9% glycerol is about 1.6 Pa · s at room temperature). It is extremely hygroscopic (it absorbs water from the atmosphere), so any free surfaces must be protected from the moist atmosphere to maintain the viscosity. Glycerol is even more transparent than glucose syrup and is good for transitional and laminar flow Newtonian LDA work.

*Silicone Oils.* Silicone oils (Dow Corning 200 fluids generally) are clear Newtonian fluids available in a wide range of viscosities. They are largely used in liquid–liquid dispersion experiments as the dispersed phase (in an aqueous bulk) as they have effectively no solubility in water or aqueous solutions. They are also largely inert and otherwise safe to use in the laboratory (the main danger being if some is spilled on the floor, as it is very difficult to remove and can be slippery). Care must be taken to ensure that any silicone oil purchased for the purpose of liquid–liquid dispersion experiments comes from a "pure" batch. Sometimes different viscosity grades are blended to produce the required final viscosity grade. Some workers have found that these blended oils do not produce consistent droplet size distribution results.

*CMC.* CMC (sodium carboxymethylcellulose) solutions need to be made up from the appropriate grade of powdered CMC. Although specialist equipment can be obtained to perform this mixing duty, a laboratory mixing vessel can also be used for this process. Care must be taken to feed the powder slowly, with adequate mixing intensity to give reproducible properties and avoid the formation of lumps. Once all the powder has been incorporated, the impeller should be stopped to allow any air ingested to escape before the fluid viscosity has had time to increase. Once the air has escaped, mixing should resume at a lower impeller speed, avoiding air entrainment. It can take up to 12 h of mixing to ensure full hydration of the substance. Newtonian (short-chain) or shear-thinning (long-chain) grades of CMC are available.

CMC solutions are particularly useful for performing mixing experiments because they are inexpensive and the viscosity is relatively insensitive to small changes in temperature and dilution. Significant quantities of salt can be dissolved in a CMC solution without greatly affecting its rheological properties; however, CMC solutions do have a significant background conductivity which can make them unsuitable for experiments using conductivity probes (e.g., blend-time measurements). Once diluted, fungicides must be added; otherwise, the fluids quickly degrade into a foul-smelling fluid with a rheology close to that of water.

CMC solutions tend to be quite cloudy, making them unsuitable for some flow visualization experiments, and useless for LDA and PIV experiments.

*Carbopol.* Carbopol is provided as powder in a manner similar to CMC. It can be obtained in a large number of grades, some of which will produce yield stress fluids. The rheology of most Carbopol solutions is very sensitive to pH.

*Natrosol.* Natrosol is also obtained as powder. It can be obtained in various grades, most of which produce fluids that have shear thinning behavior.

*Versicol.* Versicol (which comes in liquid form) produces acidic solutions with viscosities up to 5 Pa · s. It is little affected by temperature or slight dilution, but rapidly degenerates into an unpleasant-smelling liquid when dilute. Being acidic, it is highly conductive. Care should be taken in its use because its acidic nature means that it can dissolve mild steel and even concrete.

### 4-2.3 Scale of Operation

When a mixing experiment is designed to define a scale-up criterion, experiments should be performed at as large a range of scales as possible. The mechanisms controlling the mixing operation of interest at all scales can then be established with some confidence. It is of vital importance that the mixing system used at laboratory scale is a suitably accurate model of the large scale system. Dimensions must be rigorously scaled according to geometric similarity with the large scale system, even down to details like the thickness of metal sheet used to manufacture small scale impellers or static mixers. The power draw of such devices can be extremely sensitive to such details. The dimensions of baffles and vessel base shapes should also be subject to the same rigorous scaling.

In general, the guidelines in this chapter are applicable to vessels between about 0.2 and 2 m in diameter, although some techniques can be applied at much larger scales. It is generally not recommended to perform stirred tank-based mixing experiments in a vessel where the diameter is less than about 0.3 m, as maintaining geometric similarity becomes increasingly difficult, as does the manufacture of probes (conductivity, video, etc.) small enough to make only insignificant changes to the flow patterns. Also, the size of drops, particles, or bubbles may become comparable with those of impellers or baffles if the scale is too small. Typical laboratory measurements of in-line systems are made in pipes with diameters between 15 and 100 mm, although measurements in larger systems are not unknown.

Traditionally, mixing experiments in the laboratory are performed at a smaller scale than in a full production system. With a modern process-intensified plant, however, and the increased manufacture of fine chemicals and pharmaceuticals in smaller batches, it is possible to perform laboratory experiments at scales equal to or larger than some production scale facilities.

Care must be taken when developing mixing strategies at a small scale, as it is very easy to use excessively high specific power inputs, which cannot be replicated economically in large scale production systems.

### 4-2.4 Basic Instrumentation Considerations

In addition to the specialist mixing equipment described later in this section, the following common instruments are extremely useful for a variety of tasks in the mixing laboratory:

- Handheld optical tachometer used to calibrate and check the digital speed readout on most agitators.
- Averaging voltmeters used to measure signals from strain gauges, etc.
- Digital thermometers for monitoring temperature of fluids.
- Temperature-controlled water bath and recirculation system for temperature control of small vessels, usually recirculating the fluid in the surrounding viewing box.
- Digital (or other high-quality) video camera (and tripod) to record visualization experiments.
- Digital camera to photograph experiments.
- Distilled or deionized water supply to provide "clean" water of repeatable quality and low conductivity.
- Electronic balance accurate to 0.001 g for weighing of tracer materials, indicator solutions, density measurements, etc.
- Large balance for weighing large quantities of sand, etc.
- Sieves (and shaker) for crude sizing of solid particles and to help eliminate particular size ranges from a sample.
- *Rheometers.* No mixing laboratory should be without one or more reliable rheometers. There are a large number of different types of rheometer on the market, and one should be specified such that all the relevant rheological

properties can be identified and quantified. Care should be taken that rheology measurements can be and are made at shear rates appropriate to those that will be found in the mixing system being investigated. The instrument should also allow the sensitivity of fluid rheology to temperature (over the range to be experienced in the laboratory) to be investigated. Rheometers should be serviced and calibrated regularly.

### 4-2.5 Materials of Construction

The materials of construction of the vessel or pipe and all internals (impeller or mixers, shaft, baffles, bearings, seals, and lid) must be compatible with the fluids and any solids that are to be mixed. For most mixing experiments the ability to see clearly what is happening inside the mixer is extremely useful, if not essential. Vessels or pipes made of a transparent material such as polymethylmethacrylate (PMMA; Perspex or Lucite) or glass are ideal, provided that measurements are to be made at temperatures and pressures within the modest limits allowed by such materials (the glue used to seal PMMA vessels tends to fail at about 50°C). PMMA vessels up to about 1 m in diameter can be manufactured relatively simply.

At larger scales, or in situations where elevated or reduced pressures or temperatures are required, the use of completely transparent mixing vessels is unlikely or impossible. For such situations metals such as steel or other metallic alloys or plastics such as polythene can be used. In these situations the addition of windows in the side of the vessel can be extremely useful.

### 4-2.6 Lab Scale Mixing in Stirred Tanks

The design of mixing vessels to be used in performing experiments depends greatly on the scale of the vessel required. Small vessels can be of relatively simple construction, while larger vessels should undergo the same rigorous design process as production vessels. No matter what the scale of operation, the model should be of suitably accurate geometric scaling in all aspects that affect the mixing process vessel being investigated. When it is not clear if a particular design feature is significant, it is better to err on the side of caution and ensure full geometric similarity.

Small scale PMMA vessels up to about 400 mm can be manufactured by using the appropriate length of PMMA pipe to make the cylindrical section and attaching a base of PMMA that has been formed into the desired shape. Care should be taken to ensure that the base (whether it be flat or dished) is formed according to the relevant standards. A strengthening flange or lip should be added to the top rim of the vessel to increase rigidity.

Larger PMMA vessels (up to 1 m in diameter) are better manufactured in two or more sections (Figure 4-4). One or more cylindrical sections, flanged at the top and bottom to provide strength, can be bolted together to create vessels with a range of aspect ratios. Several base sections of different designs can then be

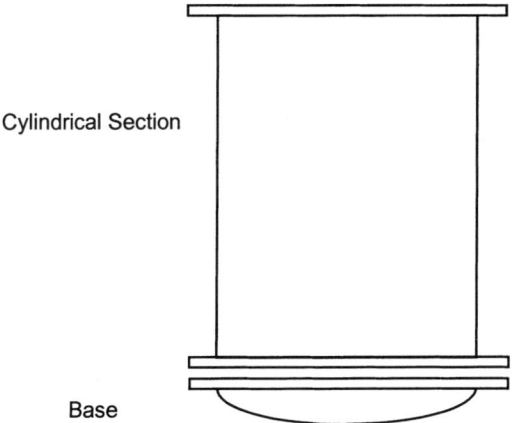

**Figure 4-4** PMMA mixing vessel made from a cylindrical section with a torrispherical base.

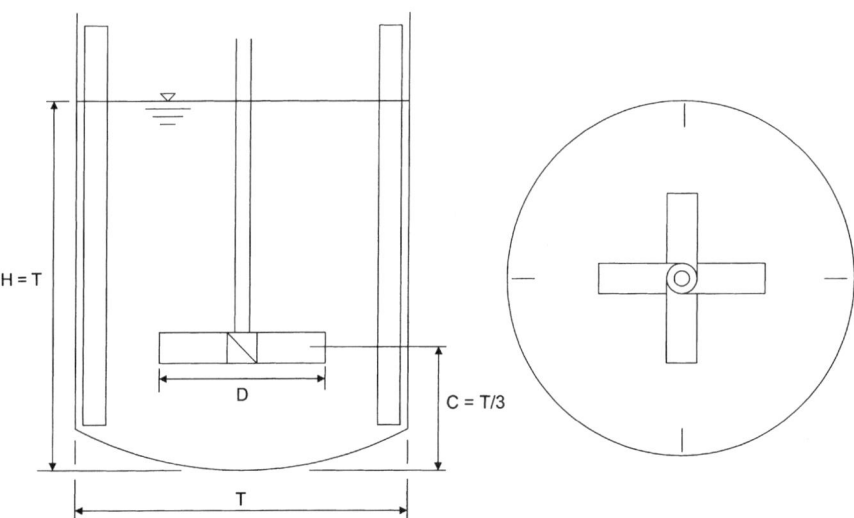

**Figure 4-5** Standard mixing vessel.

manufactured; all of which will fit to the cylindrical sections, allowing a large variety of different vessels to be produced. Double O-ring seals should be used between the flanges.

The standard mixing vessel (Figure 4-5) has a liquid fill depth equal to the tank diameter. To avoid splashing and to cope with a reasonable degree of gas hold-up, it is recommended that the vessel height be at least 30% greater than the fill depth. Laboratory mixers are usually designed without a tightly fitting lid, as they tend not to be used with particularly hazardous or volatile chemicals. However, when splashing is a problem, a lid made from two sections that fit around the central shaft can be used.

Small to medium scale vessels (up to about 0.6 m in diameter) can be placed in a viewing box as illustrated in Figure 4-6a. This is a rectangular box, with walls usually made of PMMA or glass and filled with the same fluid as in the vessel. Viewing boxes are also useful, as the fluid surrounding the vessel can be used to maintain the vessel contents at a constant temperature, or alternatively, can be used to modify the temperature. It is not practical to place larger vessels inside full viewing boxes, but small viewing boxes can be attached to transparent vessels (Figure 4-6b) to aid visualization. Where mixing experiments are to include heat transfer, the same rigor should be applied to the design of the heating/cooling coils and/or jacket as to the other aspects of the design.

It may seem convenient to place an outlet and drain valve in the middle of the base of a vessel; however, caution should be applied, as this can significantly affect any solid suspension measurements, and viscous materials added to the vessel may collect in the drain pipe or nozzle. In general, it is better that the vessel contents are removed by pumping through a piece of flexible hose, as this reduces the number of bosses and fittings on the vessel.

Wherever possible, mixing vessels should be supported above the laboratory floor (or benchtop for small vessels), as this provides access for observations of the vessel base. In all situations, the vessel, drive, bounding box, and any other equipment should be mounted in a strong frame that provides support and prevents excessive vibration.

Mixer drives are usually electric, although air motors are used in the presence of hazardous fluids. DC motors with variable speed and belt drive are popular, and direct-drive stepper motors are also recommended. It is generally simplest

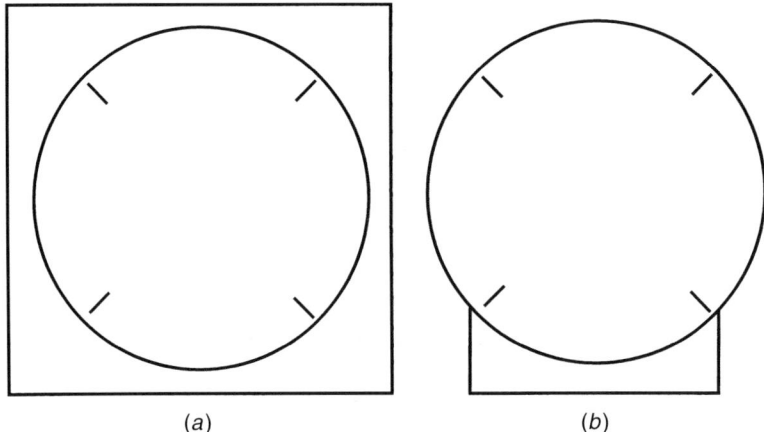

**Figure 4-6** Viewing boxes (a) and windows (b) to aid observations in transparent vessels.

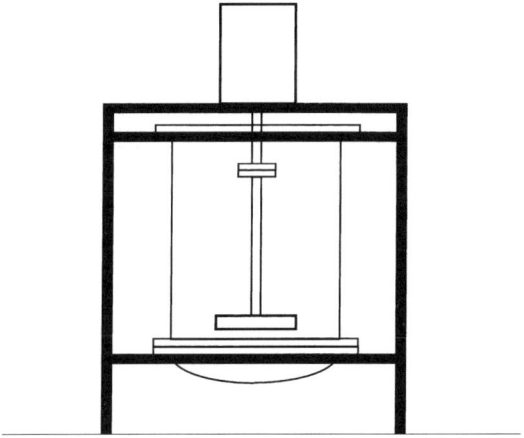

**Figure 4-7** Supported mixing vessel with overhung split shaft.

to have an overhead drive and gearbox in the mixing lab, which means that the shaft is supported from above. This saves dealing with awkward bottom-entry sealing problems and means that with a suitably split shaft (see Figure 4-7), the vessel or impeller can be removed easily.

Care should always be taken to check that the system is running at a speed lower than the first critical oscillation speed of the mixer shaft. The critical speed can be calculated or measured directly (see Chapter 21). In situations where operation close to the critical speed of the unsupported shaft is required, a bottom bearing must be used.

Impellers for use in mixing experiments should be rigorously geometrically scaled relative to the large scale impellers they mimic. Particular care should be made with the following aspects, which are often difficult to reproduce successfully at the small scale:

- *Hub diameter.* Shafts in small vessels tend to be overdesigned compared with large vessels; consequently, the hubs fitted to small impellers can be excessively large. This can sometimes be important, particularly when investigating the flow patterns generated by the impeller. However, due to the $D^5$ term in the power draw expression, the effect on the impeller power number may not be significant.
- *Blade edges.* The straightness and sharpness of the blade edges in small impellers must be as good (to scale) as their large scale equivalents.

The correct type of baffles should always be used, again, geometrically scaled down from the full scale and including the correct gap between the baffles and the vessel wall. Baffles are essential to most turbulent and transitional mixing processes with the majority of impellers as they convert the tangential motion into axial motion. Experiments with no baffling should only be performed if this is what is required at production scale, and with impeller systems specifically designed for use without baffles.

In later sections of this chapter we discuss in some detail the large number of types of probes that may be used to measure some quantity in a mixing vessel (or pipe). When mounting probes in a vessel, it is always desirable to place the probe in the correct position for the measurement while causing as little impact on the flow patterns in the vessel as possible. Consequently, the probe volume should be minimized as far as possible and care should be taken with the probe orientation and mounting supports to reduce disruption to the flow. Large probes designed for large vessels should not be squeezed into small vessels. In situations where measurement probes could cause significant interference with the flows, nonintrusive measurement techniques should be considered, although this may prove to be significantly more expensive. In situations where intrusion effects cannot be eliminated from scale-up experiments, probe size should be scaled geometrically with the vessel.

Probes can be mounted in vessels or pipes using a variety of methods. One common method is to insert the probes through bosses on the side of vessels, although this can, over time, lead to vessels that have so many patches on them that they are no longer transparent. Other good mounting techniques for probes are to attach them to baffles or to the rim of the vessel using rods and clamps. Extreme care must be taken when mounting probes in this way because once the impeller is rotating, the probe may be subject to quite extreme flow-induced vibrations. The various wires and fiber-optic cables commonly attached to probes should also be firmly held in place inside the vessel; otherwise, they may break or get tangled up in the impeller. When probes are mounted toward the base of a vessel but must be inserted into the tank from the top, it is wise to secure supporting rods and wires down behind a baffle, as this protects the probes while minimizing their effect on the flow.

In some situations it is possible to mount probes on the impeller or shaft itself. In these situations, the signal from the probe has to be transmitted either by radio-telemetry equipment or through slip rings on the shaft.

### 4-2.7 Lab Scale Mixing in Pipelines

Similar equipment and techniques can be employed for continuous flow experiments using in-line mixing systems. Again, PMMA is an ideal material for the construction of transparent test sections. Most measurements with in-line mixers require accurate measurements of pressure drop and flow rate. Calibrated gear pumps can be useful for small scale work.

Some other general considerations on in-line mixing experiments are:

- Leave enough straight pipe upstream of the measurement position to get fully developed flow if that is what is required for the experiment.
- Make sure that static mixer elements are manufactured correctly at small scale. Sometimes the wrong gauge of sheet or the wrong number of element layers can be used for small scale mixers.

- Often, small scale static mixers do not fit correctly into a circular pipe. Consideration must be made as to how significant this is for the mixing process being investigated.
- It is wise to use a size of static mixer for which manufacturer's friction factor data are available and to ensure confidence in any data of this type before using them in any further correlation. Whenever possible, the friction factor should be measured during the experiment.

## 4-3 POWER DRAW OR TORQUE MEASUREMENT

In a stirred tank mixing system, perhaps the most fundamental measurement one can make is that of the power draw of the system, as many scale-up rules depend heavily on the specific power input. The power draw, P, of an impeller is characterized by its power number, Po (also denoted $N_p$). The following expressions can be used to calculate the power draw of an impeller:

$$P = Po \cdot \rho N^3 D^5 \qquad (4\text{-}1)$$

$$P = 2\pi N T_q \qquad (4\text{-}2)$$

where $\rho$ is the density, N the impeller speed, D the impeller diameter, and $T_q$ the torque.

All the energy supplied to the fluid by the agitation system must eventually be dissipated as heat, so one possible method of measuring the power draw is to insulate the system and measure the temperature rise over time. In practice, this is extremely difficult to do accurately, due to problems with the effectiveness of the insulation and with the fluid physical properties being a function of temperature. Consequently, the method is rarely used, and methods based on measuring the reaction torque of the system are preferred.

As eq. (4-2) shows, the power input, P, into the system can be calculated if the impeller speed, N, and torque, $T_q$, are known. If the impeller diameter, D, and fluid density, $\rho$, are also known, the power number, Po, for an impeller can be calculated.

There are many methods available to measure the reaction torque. Some allow measurement of the individual torque contribution for each impeller (or any bearings, etc., in the system), while others can only be used to measure the entire system torque. Care should be taken with all methods to ensure that the true impeller torque can be isolated from any frictional torque loads applied by bearings and so on.

It should be remembered that some techniques are practical only over limited ranges of torque and hence scale or impeller speed. No matter what technique is used, extreme care must be taken to ensure that the torque measurement device is well calibrated over the range of torque values being measured, free of errors caused by friction, and compensated for any temperature effects. The calibration should ideally be checked both before and after the experiments, as some techniques are remarkably sensitive.

### 4-3.1 Strain Gauges

Strain gauges mounted on the mixer shaft (see Figure 4-8) are a popular and reliable method of measuring torque. Generally, strain gauges are used in pairs, with the alignment of the gauges being extremely important in ensuring reliable readings. With careful design and choice of shaft material they can be used to measure a very wide range of strains. With multiple-impeller systems, or where the contribution from a bottom bearing needs to be removed, multiple sets of gauges can be used on the same shaft. Strain gauges generally produce extremely low voltage signals, which must be amplified before the signal can be logged and recorded either in a PC-based data logging system or using an averaging voltmeter. The amplifier must be mounted on the mixer shaft above the water level, with the amplified signal passed to the recording device through either slip rings or a radio-telemetry system.

Extremely careful calibration of the strain gauges is required. Known torques must be applied to the shaft and the response measured. Note that no bending loads should be applied during calibration. If a single strain gauge is to be mounted on the shaft, it should be placed toward the top of the shaft where it will not be submerged and above where any impellers are to be mounted. Where

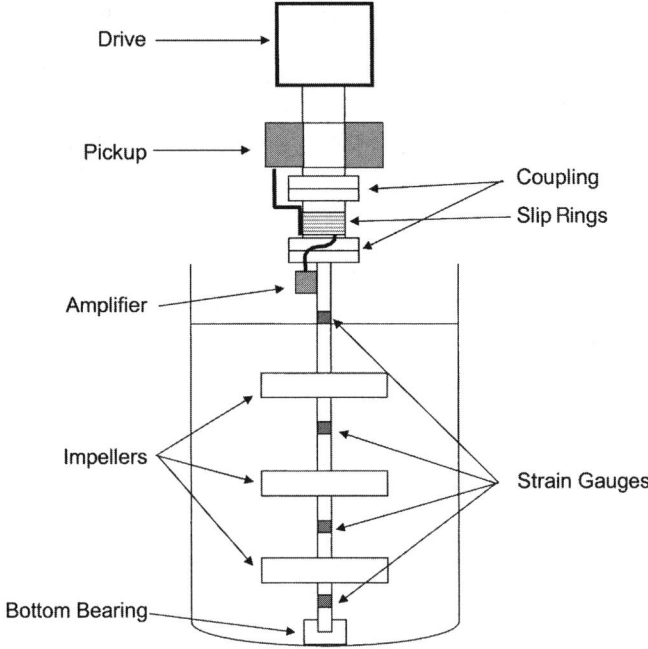

**Figure 4-8** Torque measurement using strain gauges and slip rings.

multiple strain gauges are to be mounted on the shaft, allowing the measurement of torque contributions from individual impellers and bearings, the gauges should be placed in positions that allow reasonable flexibility for impeller position (Figure 4-8).

The gauges are usually placed in turned-down sections of the shaft (which also increases sensitivity), and the connecting wires are best run along a keyway in the shaft or down the centre of a hollow shaft. If very small torques are to be measured, the gauges can be fitted to a special hollow plastic section of the shaft, which will deform more than a steel section, giving greater sensitivity. Gauges of this type are extremely delicate and must be handled and calibrated with extreme care.

Great care must be taken in the waterproofing of any gauges fitted to the shaft. The waterproofing usually consists of several layers of heat-treated wax and adhesive built-up until the surface of the shaft is smooth. Great care should be taken when using a strain-gauged shaft to ensure that the waterproof seal is not ruptured: for example, when the shaft is being removed from or positioned in the tank, or when the impellers are being moved.

Torque transducer units are available that contain strain-gauged sections of shaft that can be fitted in line with the mixer shaft. They must be suitably isolated from any bending moments and axial loads (Figure 4-9). They are generally subject to the same calibration and care during use precautions as are sensitive strain gauges (Chapple et al., 2002).

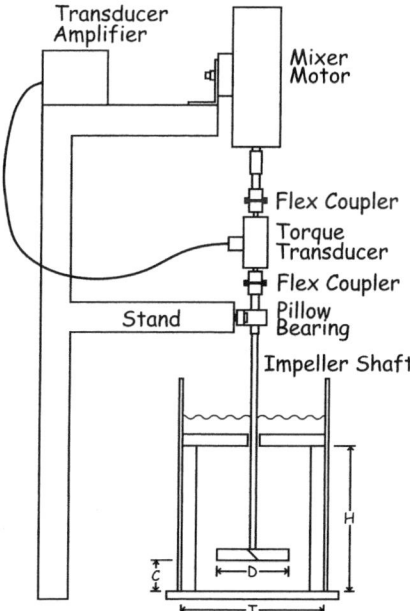

**Figure 4-9** Torque transducer. (From Chapple et al., 2002).

### 4-3.2 Air Bearing with Load Cell

An air bearing can be used to support either the mixing vessel or the motor, allowing rotation without friction. In either case only the total torque can be measured, by a suitably mounted load cell (Figure 4-10). Particular care must be taken when using an air bearing measurement with impellers of small D/T ratio, as the inertia of the tank or motor can become more significant. Air bearings can only be used in practice on vessels up to about 0.6 m (2 ft) diameter. Again, averaging of the signal from the load cell is required.

### 4-3.3 Shaft Power Measurement Using a Modified Rheometer

Some rheometers can be modified to provide accurate torque readings as they are designed to measure shear stresses at known shear rates. This can be a very good method to use in small scale vessels.

### 4-3.4 Measurement of Motor Power

Measurements of the electrical power consumed by the motor can be used to estimate the power delivered to the fluid, provided that the efficiency curves for the motor are known, along with knowledge of the losses in the gearbox and any bearings. In most situations the accuracy of the information on the efficiency and various losses is so poor that this technique is not recommended, particularly at small scale, where the losses tend to be greater than the power delivered to the fluid. Some mixer manufacturers sell small bench scale mixers that have digital torque readout. These almost always use this method and are generally very unreliable—certainly unsuitable for any scientific investigation.

## 4-4 SINGLE-PHASE BLENDING

The main aim of measurements of the single-phase mixing behavior in a stirred tank is to obtain a mixing time for the system under investigation. The mixing

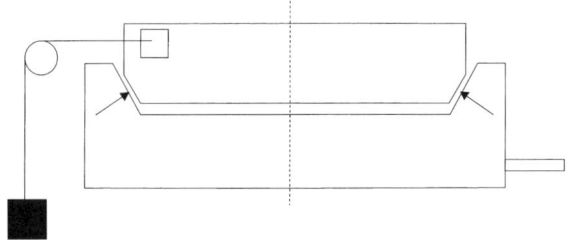

**Figure 4-10** Air bearing used to measure reaction torque.

time is the time taken for a volume of fluid added to a fluid in mixing vessel to blend throughout the rest of the mixing vessel to a pre-chosen degree of uniformity (see Chapter 9 for additional information on mixing time or blend time). Knowledge of the mixing time from experiments, from extrapolation from small scale tests, or from correlations provides confidence that:

- Any sample of product drawn from the vessel will have known chemical concentrations.
- An added species will become well distributed within a certain length of time.
- Any thermal gradients due to chemical reaction or differences in bulk and feed temperatures will be eliminated.
- All regions of the mixing vessel are moving and mixing with all other regions.
- Measurement of the bulk temperature or the concentration of a chemical species is representative of the entire vessel contents.

In an in-line mixing system, one is usually interested in quantifying how well mixed the fluids are at a particular location, or in a manner similar to the batch mixing time, how many mixer elements are required to achieve a given degree of mixedness.

## 4-4.1 Flow Visualization

Before making measurements of the mixing time for a given process, it is necessary to perform a number of flow visualization studies. These tests are designed not only to help decide how the mixing time should be measured, but also to provide information on flow and to highlight regions of poor fluid motion and flow compartmentalization within the mixing system. The techniques described below are discussed with respect to stirred tank systems; however, it should be made clear that all the techniques can be applied to in-line mixing systems with only a little modification.

The flow patterns of single-phase liquids in tanks agitated by various types of impellers have been widely reported in the literature. The simplest (and cheapest) technique for examining flow patterns within a mixing system is light sheet visualization (Figure 4-11). A narrow light sheet is shone through the mixing vessel, illuminating reflective tracer particles in the bulk fluid. Streak photography or, more commonly, video images of the reflections from the particles show the bulk flow patterns within the stirred tank. The technique can be used in any transparent laboratory mixing vessel, and the light sheet can be either vertical, to provide axial and radial flow information, or horizontal, with a video camera pointing through the bottom of the mixing vessel to provide radial and tangential flow information. Semiquantitative velocity component data can be obtained if the shutter speed or strobe duration is known by measuring the streak lengths. Further refinements to this technique include the use of a laser with a rod prism

**Figure 4-11** Flow visualization using streak photography.

to give a more intense and uniform light source, and using a polarizing filter to reduce glare from the vessel itself. It is important to minimize the amount of light from other sources within the laboratory to obtain the best possible images.

Video images from this technique have been used to show fluctuations in the discharge angle of impellers, the degree of flow compartmentalization and impeller interaction in multiple impeller systems, and to show the cavern size when mixing yield stress or highly shear thinning liquids.

Other techniques can provide information on local fluid velocities across the whole or parts of the vessel (e.g., PIV) or can measure fluid velocities at a single point with a high degree of accuracy (e.g., LDA/LDV or hot-wire anemometry). These techniques are discussed in Part B of this chapter. It should be stressed, however, that a simple light source and a video camera are substantially cheaper than these other techniques and can often be used as the first step in examining a mixing configuration. It is always useful to look at the flow field as a complement and guide to the use of more quantitative and expensive techniques.

For examination of mixing behavior, light sheet visualization is important, particularly in multiple-impeller systems, to help the experimentalist think about suitable points of addition to study a mixing system, possible choice of feed location, and an initial estimate of suitable probe locations for mixing time experiments. Further tests should be performed to provide more detailed information on suitable choices of probe location.

### 4-4.2 Selection of Probe Location

Mixing time techniques all work on the principle of adding material to the vessel which has different properties from the bulk. Measurements are then made (usually in a controlled volume within the vessel) that show the presence of the added material. The decay of material property fluctuations is used to measure the mixing time for the system. Many of the devices used for measurement of the added liquid are intrusive, and the experimenter must try to minimize the number of probes installed in the vessel in order to:

- Minimize the perturbation of the bulk flows by the probes and the probe supports.
- Reduce the amount of associated hardware required.
- Minimize the amount of data that need to be collected, stored, and processed.

However, it is important that the probes in the vessel be in the correct positions to provide as much relevant mixing information as possible. There will be certain points in the mixing vessel that will be less intensely agitated and where mixing rates are lower than in the bulk of the tank. These points will be the last points to become mixed and will control and limit the mixing rate for the entire mixing vessel. So if probes are used, one or more should be positioned in these poorly mixed regions to obtain the correct mixing time. Other probe locations should be chosen to indicate whether the bulk of the tank is well mixed. Extra probes could be distributed axially to provide information on flow compartmentalization, particularly in multiple-impeller systems.

### 4-4.3 Approximate Mixing Time Measurement with Colorimetric Methods

The simplest test of the rate of mixing of the bulk would be to add a dye and see how the dye moves throughout the fluid (Figure 4-13a). However, if a colored tracer is added to an agitated vessel, it is not possible to identify the last point of mixing because any dye in front of or behind the poorly mixed region will mask the last mixed pocket (Figure 4-12a). What is required is a decolorization technique where the last point to be mixed remains colored or marked while the rest of the tank is stripped of its color.

The technique most commonly used is known as (dye) decolorization. The entire contents of the vessel are colored using one chemical, and then a second chemical is added that removes the color. A poorly mixed region stands out as a pocket of color after the rest of the vessel has cleared (Figure 4-12b).

Two of the most common reactions used are pH change with an appropriate indicator (Figure 4-13), and an iodine–iodide change in the presence of starch. The experimenter should check for chemical compatibility when choosing an appropriate decolorization technique. It may be necessary to use a simulant fluid rather than the actual bulk chemical.

**168**  EXPERIMENTAL METHODS

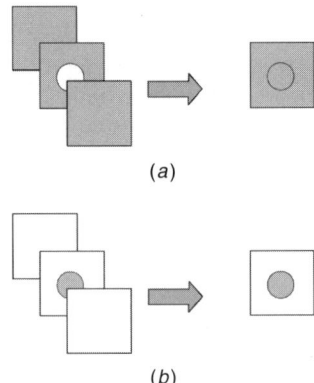

**Figure 4-12** Schematic representation of dye addition and dye decolorization for mixing visualization in three dimensions. (*a*) A poorly mixed pocket is uncolored, and cannot be seen in two dimensions. (*b*) A poorly mixed pocket is colored, and can clearly be seen in two dimensions.

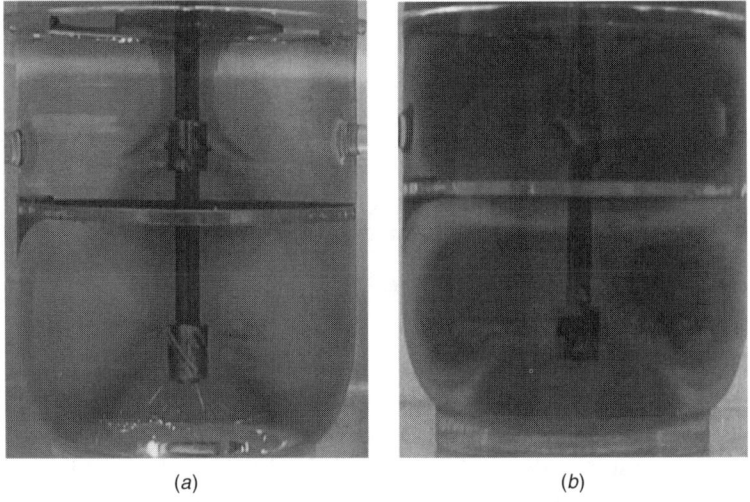

**Figure 4-13** Red dye tracer addition (*a*) and acid–base decolorization using bromophenol blue (*b*). (From Clark and Özcan-Taşkin, 2001.) See insert for a color representation of this figure; for a video clip, see the Visual Mixing CD affixed to the back cover of the book.

***4-4.3.1 Acid–Base with Indicator.*** An indicator such as bromophenol blue (blue to yellow, Figure 4-13*b*) or phenolphthalein (pink to colorless) is added to the bulk liquid. Small additions of acid and base can be used to color the entire mixing vessel and then strip out the color. A small excess of the acid–base used for the color change is typically used. The last point to remain colored is easily observed and selected as a suitable probe position. For standard geometries, the last point of mixing in the transitional regime is generally behind the baffles, where the fluid is protected from the bulk flows by the baffle itself.

Strong acids should always be used to avoid any buffering effects in the presence of salts from the neutralized acids and bases. With this technique, the bulk liquid can be reused repeatedly by re-adding more of the initial reactant.

**4-4.3.2 Iodine–Thiosulfate.** The reaction between sodium thiosulfate and iodine in the presence of starch is a most satisfactory decolorization reaction, with a strong color change from deep blue to clear.

*Preparation of Chemicals Required*

- *2 M iodine.* Dissolve 400 g of iodate-free potassium iodide in 0.5 L of distilled water in a 1 L volumetric flask. Add 254 g iodine to the flask and agitate until all the iodine has dissolved. Cool to room temperature, and make up to 1 L. Store in a dark place to prevent deterioration.
- *1 M sodium thiosulfate.* Weigh 248 g of sodium thiosulfate pentahydrate into a 1 L volumetric flask and make up to 1 L with distilled water. Add a small quantity of sodium carbonate (about 0.2 g) to aid preservation.
- *Starch indicator.* Manufacture a paste containing 10 g of soluble starch in a little water. Add this dropwise to a beaker containing 1 L of very hot (but not quite boiling) water while stirring constantly until a clear liquid is obtained. Cool the liquid to ambient temperature and dissolve it in 20 g of potassium iodide. This solution should be protected from the air.

*Procedure for Use*

1. Add about 50 mL of starch indicator to the vessel.
2. Add sufficient iodine solution to the vessel to produce an intense black color. Note the quantity used.
3. With the agitator running at the desired speed, quickly incorporate sodium thiosulfate into the vessel at a rate approximately 2.5 times the quantity of iodine solution used. As mixing progresses, the liquid will clarify.
4. After each test, slowly add iodine solution until the neutral point is again reached, demonstrated by a hint of light blue appearing. The test can then be repeated.

It should be noted that this technique must be used with care with solutions of CMC, which have a tendency to decolorize iodine solutions over time, albeit slowly.

### 4-4.4 Quantitative Measurement of the Mixing Time

After performing the decolorization technique to provide qualitative understanding of the mixing behavior, the method of mixing time measurement must be

**170** EXPERIMENTAL METHODS

selected. All mixing time techniques are based on addition of a chemical with a different property to the bulk, followed by measurement of the system to infer the presence and uniformity of this marked liquid.

***4-4.4.1 Off-line Sampling.*** If an off-line analysis technique is used, a chemical marker such as a particular salt, dye, or acid is added to the mixing vessel, and samples are removed regularly. The concentration of the marker in each sample is measured, and the degree of uniformity is inferred from these measurements. Installation of a suitable sampling system can be difficult, and this technique is not suitable if the mixing time is very short, since there will generally be a finite sampling time.

***4-4.4.2 Schlieren Effect-Based Mixing Measurements.*** The Schlieren-based technique relies on the light scattering that occurs when two liquids with different refractive indices are mixed. Light shone through the mixing vessel is scattered by the layers of different liquids and the tank appears cloudy. When the liquid is fully blended, the tank becomes clear once more, giving a mixing time (Van de Vusse, 1955). This technique does, however, require transparent bulk and added liquids, and the liquids must have different refractive indices. The liquids are also, therefore, likely to have different physical properties.

***4-4.4.3 Thermocouple-Based Mixing Time Measurements.*** A thermocouple-based mixing time test can be performed by adding a liquid that has a different temperature from the bulk. The temperature at different points in the mixing vessel is monitored over time, and the probe outputs are used to calculate the mixing time. This technique can be used with opaque and/or nonconducting liquids. A disadvantage of this technique is that it may not be suitable if the bulk liquid physical properties are very sensitive to changes in temperature, since the viscosity would then be a function of the temperature (and concentration) of the added liquid.

***4-4.4.4 Conductivity Probe Technique.*** The conductivity probe mixing time technique uses an electrolyte in the added liquid as the marker. Conductivity probes monitor the local conductivity as a function of time. If the electrolyte concentration is low, concentration is directly proportional to conductivity. The probe outputs are processed to calculate the mixing time for the system under consideration. This technique is not suitable for measurements in nonconducting systems and cannot be used in systems where the rheological properties of the bulk are sensitive to changes in salt concentration: for example, with certain gums and carrageenans. The technique is, however, cheap and easy to use. Conductivity probes can give very rapid response times, allowing measurements in mixing systems with short mixing times. Further details on probe designs suitable for conductivity-based mixing time measurements are available in Khang and Fitzgerald (1975). The probe itself is made by embedding pieces of platinum

SINGLE-PHASE BLENDING 171

**Figure 4-14** Liquid-phase conductivity probe.

or stainless steel wire, soldered onto the screened signal lead, in a "bullet" of epoxy resin. The diameter of the bullet is approximately 6 mm (Figure 4-14).

All probes have the outer electrode coiled around the probe tip in the shape of a cone. The electrode acts both as an earth for the measuring electrode and as a screen to prevent any interference from other electrodes or earthed objects in the vessel. The probes must be made so that the measuring volume is as small as practically possible, to minimize averaging effects when measuring the fluid conductivity. When placed within the vessel, the probes will affect local mixing by their presence. The position of the probes within the vessel is determined by the point at which a measurement is required and it is not possible to change a probe's position to minimize the effects on local mixing. The only way in which the effect on local mixing can be minimized is by making the probes as small as possible and arranging the supports in such a way that they intrude as little as possible into the vessel. To this end, the probe supports must, wherever possible, be placed behind baffles.

The typical probe positions used for a single-impeller mixing system are shown in Figure 4-15. Three probes are used and are sited to give as much information as possible about the mixing in the vessel. One probe is sited in the impeller stream, which is the most intensely mixed region of the vessel. The second probe is sited

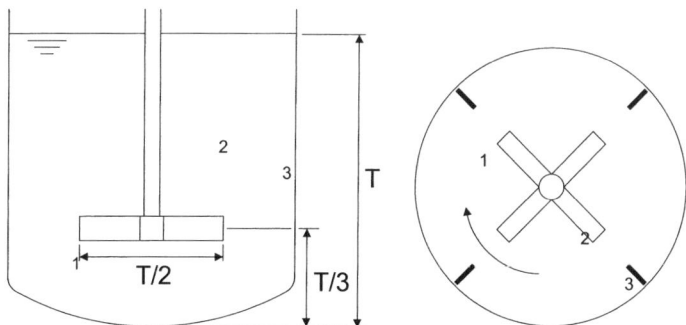

**Figure 4-15** Typical mixing time measurement probe positions.

close to a baffle, which is likely to be a poorly mixed region. The third probe is sited in the middle of the tank approximately halfway around a circulation loop. The position of this probe is particularly important in the transitional regime. It should be checked for new geometries using a decolorization test. The probes are mounted at different heights within the vessel, at different radial distances from the mixer shaft, and on different sides of the tank. These positions are selected so that any deficiencies in vertical convection or tangential convection within the tank will show up in the measured concentration records.

Great care must be taken to prevent unwanted chemical reactions occurring in the mixing vessel. The use of reactive tracers such as nitric acid and sodium hydroxide, rather than common salt, causes some problems. Materials that may normally be regarded as inert, since they react so slowly as not to cause significant drift over the time of the experiments, may start to react more rapidly. Materials that are initially inert may also start to react after a period of time when protective oxide films are broken down.

Care must also be taken to prevent metal objects in the vessel from inadvertently becoming charged. This will lead to electrochemical cells being set up, causing drifting conductivity and a noisy signal. This will also happen if two dissimilar metals, both dipping into the fluid, are allowed to come into contact.

If there is a drain hole in the bottom of the vessel, it should be plugged during experiments. This is done to prevent small amounts of fluid with a different conductivity from the vessel contents, occasionally mixing with the fluid in the vessel. The conductivity meters connected to each probe should use an alternating current-based technique to eliminate polarization or chemical reaction at the probe itself. If measurements are made using more than two probes, each probe should be operated with a different electrical frequency to eliminate probe crosstalk. The output from the conductivity meters is collected on a computer using an analog-to-digital converter card and can then be processed using dedicated software or using a spreadsheet program.

### 4-4.4.5 Processing Mixing Time Data.

Data collected by the conductivity, thermocouple, or pH techniques must be processed to obtain a characteristic mixing time for the system under investigation. The analysis is described below in terms of conductivity readings, but is equally applicable to the other data types.

The data must first be normalized to eliminate the effect of different probe gains. The data are normalized between an initial zero value measured before the addition of tracer, and a final stable value measured after the test is complete. These values are typically obtained by measuring the probe outputs over 30 s.

The normalized output is obtained by

$$C'_i = \frac{C_i - C_0}{C_\infty - C_0} \qquad (4\text{-}3)$$

where $C'_i$ is the normalized probe output. The normalized responses for three probes in a stirred tank are shown in Figure 4-16.

SINGLE-PHASE BLENDING  173

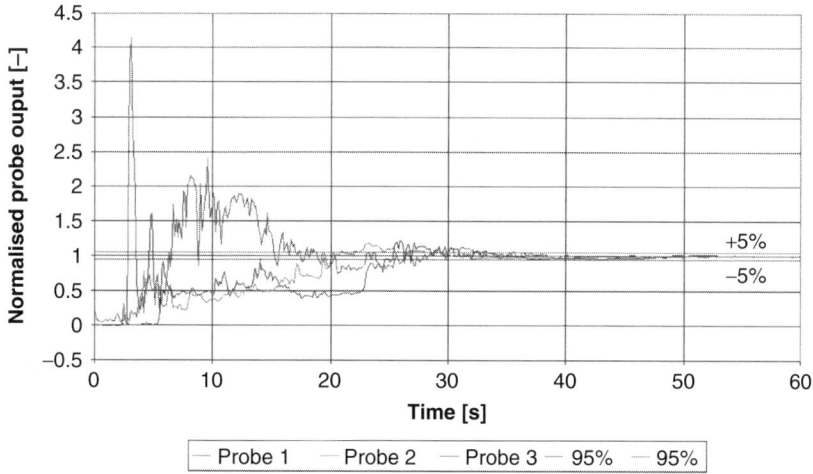

**Figure 4-16** Normalized conductivity probe responses for three probes.

The mixing time is defined as the time required for the normalized probe output to reach and remain between 95 and 105% (±5%) of the final equilibrium value. This value is called the 95% *mixing time*. It can be difficult to identify the 95% mixing time accurately from the normalized probe outputs (Figure 4-16) because of the fine scale around the endpoint. Because the probe fluctuations decay exponentially, the data can be conveniently replotted in terms of a probe log variance as a function of time (Figure 4-17):

$$\log \sigma^2 = \log(C'_t - 1)^2 \qquad (4\text{-}4)$$

This graph is more accurate for obtaining the 95% mixing time. The 95% mixedness line corresponds to the point where $C' = 0.95$. To obtain an overall mixing time for the system, the three probe responses must be combined and must be weighted toward the probe showing the largest concentration deviation to ensure that all regions of the vessel are mixed. This is achieved using an RMS variance and is plotted in Figure 4-17:

$$\log \sigma^2_{\text{RMS}} = \log \left\{ \tfrac{1}{3} \left[ (C'_{t,1} - 1)^2 + (C'_{t,2} - 1)^2 + (C'_{t,3} - 1)^2 \right] \right\} \qquad (4\text{-}5)$$

More stringent criteria of 99% or 99.9% mixing time are used in certain applications where a higher degree of uniformity is required (e.g., paint manufacture or pharmaceutical production). These values can also be obtained from the *log variance graph*. The linear response of log variance as a function of time is due to the exponential decay in concentration fluctuations. This relationship also allows one to calculate an arbitrary degree of uniformity from the 95% mixing time:

$$\frac{\theta_n}{\theta_{95}} = \frac{\ln[1.0 - (n/100)]}{\ln(1.0 - 0.95)} \qquad (4\text{-}6)$$

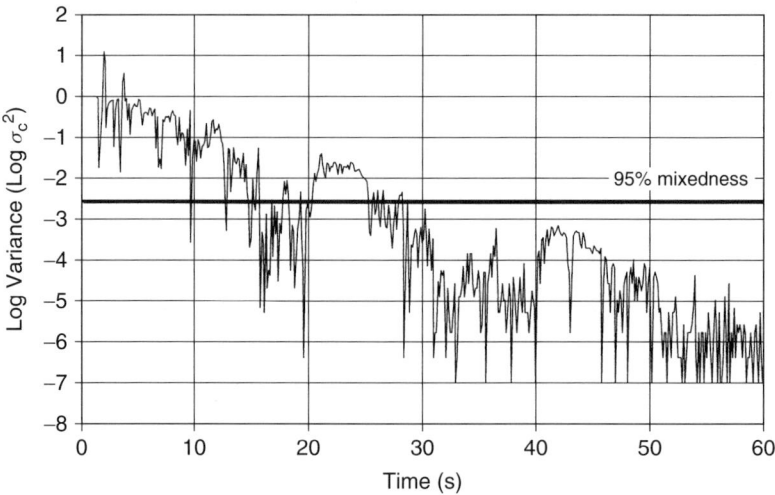

**Figure 4-17** Probe log variance.

where n represents the desired degree of mixedness. For example, a 99% mixing time can be calculated using

$$\frac{\theta_{99}}{\theta_{95}} = \frac{\ln(1.0 - 0.99)}{\ln(1.0 - 0.95)} = 1.54$$

Papers in the open literature (e.g., Harnby et al., 1992) often refer to a *dimensionless mixing time*, $N\theta$. This is the product of the 95% mixing time, $\theta$, and the impeller speed, N. The dimensionless mixing time is equivalent to the number of impeller revolutions required to achieve the desired degree of uniformity.

### 4-4.5 RTD for CSTR

The conductivity probe technique can also be used to measure the residence time distribution of continuous flow systems by installing probes at the inlet and outlet of the mixing vessel. The probe response can be normalized and interpreted as outlined by Levenspiel (1972) or as discussed in Chapter 1 of this book. Care should be taken to ensure that the data are collected over a sufficiently long time, because the tail can have a large effect on the measured mean residence time and the derived variance.

### 4-4.6 Local Mixedness

*4-4.6.1 Coefficient of Variation.* The coefficient of variation (CoV) and other measures used to describe mixedness are defined in Chapter 3, where details of the analysis are discussed, and reviewed in the context of pipeline mixing in Chapter 7. CoV can be measured using a variety of techniques, including laser-induced fluorescence, conductivity probes, and temperature probes.

**4-4.6.2 *Reactive Mixing Experiments*.** When two reactive fluids are brought together, reaction cannot proceed until the reactive molecules are mixed intimately on a molecular level. If there are several competing reactions occurring simultaneously whose reaction lifetimes are approximately the same as the time scale of the mixing process, the relative progress of the reactions will be governed by the mixing process(es). Generally speaking, the faster (usually desired) reactions will be favored by higher mixing rates and the slower (usually undesired by-product) reactions by slower mixing rates. The product distribution from the reactions will therefore reflect the mixing history; and with the aid of suitable mixing models, these distributions can be used to back-calculate mixing rates.

The overall mixing process occurs within a flow field continuum that covers a wide range of length scales. It is possible to decompose this continuum into several mixing subprocesses or mechanisms, each of which has distinct characteristics. If one of these subprocesses is much slower than the others, it is said to be *rate limiting*. Furthermore, if a system is sensitive to mixing rates (e.g., competitive fast reactions), it is the rate of the slowest mixing step that will influence the process outcome. By focusing on the most important mixing scale, the mixing characterization can be simplified greatly.

Mixing between two miscible fluids can be broken down into the following steps, where the length scales given are for turbulent mixing of liquids (high Sc):

1. Convection of an additive by mean velocities (typical length scale $10^{-1}-10^{-2}$ m)
2. Turbulent dispersion by fluctuating velocities in large eddies ($10^{-2}-10^{-3}$ m)
3. Reduction of segregation length scale by breakdown of large eddies ($10^{-3}-10^{-4}$ m)
4. Fluid engulfment in small eddies ($10^{-4}-10^{-5}$ m)
5. Lamellar stretching with molecular diffusion ($10^{-6}-10^{-7}$ m)

These subprocesses occur both in series and (to some extent) in parallel. For the purposes of this discussion, the five processes above can be spilt into three broader categories: steps 1 and 2 are responsible for *macromixing*, step 3 is responsible for *mesomixing*, and steps 4 and 5 are responsible for *micromixing*. The energy to drive these processes comes from the mean flow at the expense of pressure drop or shaft work. This energy is extracted from the flow at large scales (in the form of wakes, vortices, and mean radial velocities) and is dissipated at small scales, where viscosity becomes important. All of this is discussed in far greater detail in Chapter 2 in the context of turbulence and in Chapter 13 in the context of reactive mixing.

*Bourne Azo-Coupling Reactions*. The second Bourne reaction scheme (see Table 13-4) is the most commonly used reactive mixing characterization tool. It is a fairly robust scheme, with clean, well-defined kinetics that can be made sensitive to mixing rates over a wide range of conditions. The scheme involves the reaction of diazotized sulfanilic acid with 1- and 2-naphthol. The naphthols compete for the sulfanilic acid to produce a range of different-colored dye products according to the following scheme:

$$A + B \rightarrow R \quad k_{1-p} = 12\,238 \pm 446 \text{ m}^3/\text{mol} \cdot \text{s}$$
$$A + B \rightarrow T \quad k_{1-o} = 921 \pm 31 \text{ m}^3/\text{mol} \cdot \text{s}$$
$$R + B \rightarrow S \quad k_{2-o} = 1.835 \pm 0.018 \text{ m}^3/\text{mol} \cdot \text{s}$$
$$T + B \rightarrow S \quad k_{2-p} = 22.35 \pm 0.25 \text{ m}^3/\text{mol} \cdot \text{s}$$
$$C + B \rightarrow Q \quad k_3 = 124.5 \pm 1.0 \text{ m}^3/\text{mol} \cdot \text{s}$$

where A is 1-naphthol, B is diazotized sulfanilic acid, C is 2-naphthol, R and T are two mono-azo isomers, S is a bis-azo dye, and Q is another mono-azo dye.

The naphthols are generally used in large excess (to ensure that the reaction goes to completion in the mixer) with 2-naphthol present at a greater concentration than the 1-naphthol. The concentrations and relative volumes of the reagent solutions used are varied so that the reaction product distribution is sensitive to mixing intensity in the device used.

Typical concentrations used are $C_{A0} = 0.03 \text{ mol/m}^3$, $C_{C0} = 0.12 \text{ mol/m}^3$, $C_{B0} = 50.0 \text{ mol/m}^3$, $Q_{A,C}/Q_B = 3000$.

The product distribution is quantified in terms of product yield relative to the limiting reagent B.

$$X_Q = \frac{[Q]}{[R] + [T] + [Q] + 2[S]}$$
$$X_S = \frac{2[S]}{[R] + [T] + [Q] + 2[S]} \quad (4\text{-}7)$$

When relatively high mixing intensity conditions are used, very little of the bisubstituted product S is formed. The chemistry then reduces to a competitive-parallel scheme with 1-naphthol (A) competing with 2-naphthol (C) for sulfanilic acid (B), producing R and Q, respectively.

If the B solution was distributed perfectly in the A and C solution at a molecular level before any reaction occurred ("perfect mixing"), the product concentrations would be determined solely by the kinetics of the reactions. At the other extreme, if the mixing is extremely poor, any A and C entering the reaction zone will be consumed before fresh material can be incorporated. The ratio of the products will then be the same as the ratio of the feed reagents (ignoring the consecutive reactions). The actual values of $X_Q$ will depend on the initial concentration and volume of the B solution.

The Bourne schemes are dealt with in more detail in Chapter 13 and in Table 13-4.

### 4-4.6.3 LIF/PLIF. 
Laser-induced fluorescence (LIF) techniques are very promising but are still primarily a research tool requiring expert users. They are not discussed in this chapter.

## 4-5 SOLID–LIQUID MIXING

There are a large number of techniques available that can provide both qualitative and quantitative information on the suspension and distribution of solids in a liquid-filled stirred tank. Other techniques can be used to determine the "just suspended" speed (or "just drawn-down" speed with floating solids).

### 4-5.1 Solids Distribution

There is no simple way to describe the distribution of solids in a stirred vessel. The techniques described here either give accurate quantitative data on local concentrations of solids (conductivity probe, optical probe, sampling for external analysis, or process tomography), or provide general observations and semi-quantitative information (e.g., cloud height) on the distribution of solids (visual observation, process tomography). If several local measurements are made using the conductivity, optical, or sampling techniques described below, the relative standard deviation of the data can be used to give a single measure of how well the solids are distributed. The reader is cautioned that this quantity is extremely sensitive to the number and location of the data points used and is of little use without reference to visual observations and a full record of the local data measurement locations used to calculate it.

Two types of probe have commonly been used: conductivity and optical. Both types of probes and the associated instrumentation are commercially available, but care should be taken to ensure that the probe and instrumentation are suitable for a particular application. The dimensions of the measuring volume should be at least an order of magnitude greater than the dimensions of the solid particles, and the probe should allow free flow through the measuring volume. If these criteria cannot be met, inaccurate measurements will result, and a custom-built probe may have to be used.

Attempts to close a solids mass balance using these approaches generally fail unless the solids are very well distributed throughout the vessel. It is very difficult to measure accurately the quantity of any unsuspended solids on the vessel base. This problem is most acute if visual observations are not possible. The number of locations in the vessel at which measurements are made will usually be limited for practical reasons. This lack of spatial resolution will lead to inaccurate measurements being made in regions where there are high solids concentration gradients.

***4-5.1.1 Visual Observations.*** Visual observations are most commonly used to determine the minimum speed for suspension of solids ($N_{js}$) as discussed in Section 4-5.2. Visual observations can also be an extremely useful tool to give a rough estimate of the degree of homogeneity in a mixing vessel. Where a range of geometries is being optimized, this enables a rapid choice to be made of the two or three most efficient geometries. At this point further testing may be carried out. Observations of the multiphase flow are necessary in order to choose

**178** EXPERIMENTAL METHODS

suitable positions for mounting instruments in the vessel and can be used to rapidly identify problem areas such as stagnant areas where solids may collect. Such observations of the flow patterns also help in choosing sampling points in the vessel, and cloud height observations will help in selecting vessel outlet locations and in estimating how well mixed the vessel is. Visual observations also help to identify any unusual phenomena that may cause problems with the process (e.g., drop-out of solids at intermediate impeller speeds) and help in interpreting the data obtained by other means.

At high solids concentrations, visual observations become difficult: The solids screen most of the vessel from view. If a process with a high solids concentration is being investigated, observations at lower solids concentrations will aid in understanding the mixing mechanisms. However, observations should still be carried out at high solids concentration since the flow patterns in the vessel often vary considerably with solids concentration.

The addition of a colored dye (food dye works very well in aqueous systems) to the vessel, which is taken up preferentially by either the liquid phase or the solid phase, may assist with visualizing the flows. The addition of a small quantity of solids of a contrasting color (with the other relevant physical properties held constant) may help flow visualization, as will the use of a high intensity spotlight to allow light to penetrate as far into the vessel as possible. Where photographs or video are required, a polarizing filter can be extremely helpful in removing unwanted reflections. A ruler or some other scale attached to the side of the vessel can be very helpful for providing some quantitative information on cloud height observations.

### 4-5.1.2 Conductivity Probe.

If the suspending fluid is conducting (e.g., water) and the suspended solids nonconducting, a conductivity probe (Figure 4-18) can be used to measure the local solids concentration precisely and accurately. The conductivity probe will only measure an overall solids concentration in the measurement volume, and cannot discriminate between particles of different sizes. Any probe placed within the vessel has the disadvantage of being intrusive and hence altering the flows and possibly the solids concentration at the measuring point. The probes can be designed to minimize their effect on local flows, but some level of intrusiveness will have to be tolerated. Since the probe is intrusive, it is desirable to check for separation of the solids from the liquid near the probe. This may be done by rotating the probe. If no separation is taking place, the measured solids concentration will not change. If problems are occurring due to flow separation, the measured value of solids concentration will generally be too low, so the highest value of solids concentration should be taken as the true value.

The measured electrical conductivity of the probe volume will vary as the solids concentration in the probe volume changes. Generally, the measured conductivity is linearly related to the volume fraction of solids over a wide range of solids concentrations:

$$\frac{\kappa}{\kappa_O} = 1 - A\varepsilon_S \tag{4-8}$$

SOLID–LIQUID MIXING    179

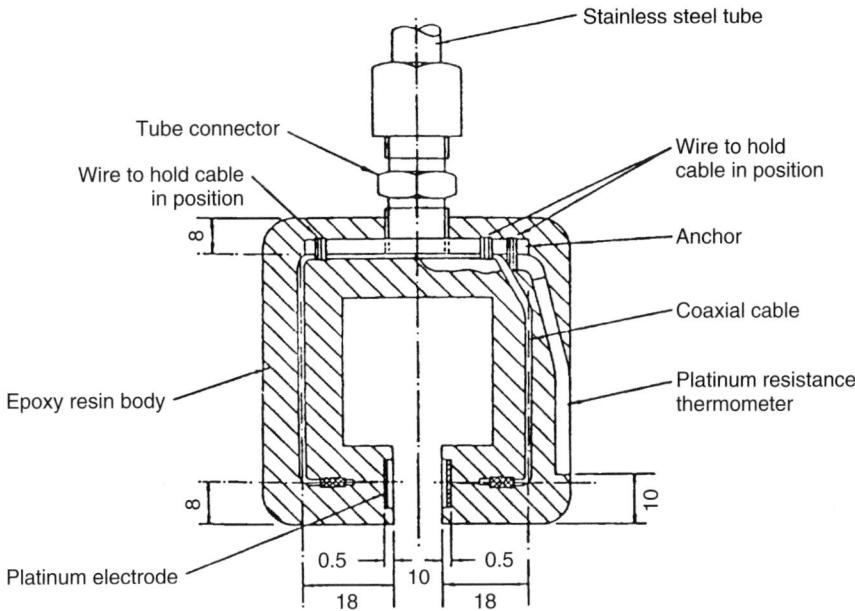

**Figure 4-18** Conductivity probe for measuring solid concentration.

where κ is the measured conductivity, $\kappa_O$ is the conductivity with no solids present, $\varepsilon_S$ is the volume fraction of solids, and A is a calibration factor.

The probes require calibration with the liquid and solids that are to be measured. For solids that are easily suspended, calibration may be carried out in a small vessel fitted with a laboratory stirrer. The suspension is agitated sufficiently to ensure homogeneity and allow measurement of the solids concentration with the probe being calibrated. Solids that are difficult to suspend may be calibrated using a fluidized bed, although it will be necessary to use solids of a very narrow size distribution for accurate calibration. Calibration using a laboratory stirrer is generally limited to low concentrations, as it is difficult to maintain homogeneity at high solids concentrations. Calibrations using a fluidized bed are generally limited to high solids concentrations, since the bed becomes unstable at low solids concentrations. It may be necessary to use a combination of the two techniques. There is some evidence that the calibration factor is a weak function of particle size and shape, but this can be ignored in most practical applications.

The fluid used for suspending the solids must be conducting but must not contain any electrolytes. Any electrolytes will cause polarization at the electrodes, leading to problems with noisy and drifting signals. Most commercially available probes have a platinum-black coating to prevent polarization. Care should be taken, as the solids present in the vessel often rapidly erode this coating. Before making any measurements, checks should be made to ensure that the solids contain no soluble impurities, which may cause the conductivity signal to drift, and that all the vessel internals are inert. Solids containing soluble impurities that may affect the conductivity must be washed thoroughly before use to remove the impurities.

The measured conductivity may change not only as a result of solids concentration, but also due to changes in the fluid temperature and to impurities dissolving in the fluid. During use it is essential to measure temperature accurately and to correct for any changes. The following expression can be used:

$$\kappa_C = \kappa_M \frac{\kappa_{OR}}{\kappa_{OM}} \quad (4\text{-}9)$$

where $\kappa_C$ is the conductivity corrected to the reference temperature, $\kappa_M$ the measured conductivity, $\kappa_{OR}$ the fluid conductivity at the reference temperature, and $\kappa_{OM}$ the fluid conductivity at the measured temperature.

Where more than one probe is used to make measurements, care should be taken that no crosstalk occurs between probes. A conductivity meter applies an AC voltage (approximately 1 to 2 kHz) across the probe, and the voltage applied to one probe may be detected by another probe. To prevent this, the probes must either be multiplexed, so that only one probe at a time has a voltage applied to it, or different AC frequencies, with bandpass filtering, must be used for each probe.

The conductivity probe is best suited to work under laboratory conditions, although if exceptional precautions are taken, it may be used in industrial process plant. Care must be taken to ensure that the probe(s) are mounted securely in the vessel. Measurements made with probes of this type are typically averages of several minutes' worth of data.

**4-5.1.3 Optical Probe.** Optical probes can also be used to measure solid concentration and are generally of two types: light absorption or light scattering. In both cases the amount of light absorbed or scattered is related to the solids concentration. For absorption, *Beer's law* applies:

$$\log \frac{I_0}{I} = E \varepsilon_S \ell \quad (4\text{-}10)$$

where $I_0$ is the intensity of the incident light, $I$ the intensity of the light after absorption, $E$ the extinction coefficient for the system, $\varepsilon_S$ the volume fraction of solids, and $\ell$ the optical path length.

The extinction coefficient is a function of particle size and light wavelength. The relative amounts of light absorbed and scattered will depend on both the particle size and light wavelength used, and it is essential that a calibration is carried out with the same solids and same light source which will be used in the mixing vessel. Calibration is carried out in a similar manner to the conductivity probes mentioned earlier.

The probe response becomes nonlinear as the solids concentration is increased, and measurements are extremely difficult at solids concentrations greater than 2 to 3 vol %. by volume. The upper limit on the instrument range is also dependent on the particle size of the solids. Reducing the gap between the light source and light receiver may extend the range, but this may cause problems with poor flow through the measuring volume.

Although the probe response is a function of both particle size and solids concentration, it is not possible to distinguish between the two effects. If the particle size distribution in the mixing vessel changes during the course of the experiment due to degradation of the solids, the probe response will drift. This problem is especially acute where the solids contain friable particles such as clay, which degrade easily to give a turbid suspension. Problems will also occur if solids with a very wide size distribution are used. The particle size distribution will vary with position in the vessel, and hence the probe response will become a function of position in the vessel.

Optical probes are insensitive to changes in temperature and to impurities that do not absorb or scatter light. An optical probe will give an accurate and precise measure of the concentration of solids for which a prior calibration has been carried out. The probes are generally less susceptible to interference than conductivity probes. They should be used under laboratory conditions, although in some cases they may be used in industrial process plant. Optical probes can normally be used only to measure low solids concentrations (up to 2 to 3 vol %). Care should be taken to avoid errors due to changes in particle size distribution during measurement.

*4-5.1.4 Sampling*. In many situations probes may be unsuitable for measuring local solids concentration, particularly if information on the particle size distribution is required. In such cases a sample must be taken from the vessel while it is in operation, and analyzed using standard laboratory methods. The laboratory analytical techniques are usually straightforward (e.g., sieve analysis), and generally no major problems are encountered with their use. The chief difficulty lies in obtaining a sample that is representative of the local conditions at the sampling point.

Two means are generally available for removing a sample from a mixing vessel: pumping the sample out through a pipe, or taking a grab sample. In both cases the sampling process may cause classification of the solids, or separation of the solids and liquid. When removing a sample through a pipe, the arrangement used must ensure that the sample is removed isokinetically; that is, the velocity at which the sample enters the pipe must match the local fluid velocity, and the pipe must be aligned in the local direction of flow. In a mixing vessel with its complex three dimensional turbulent flows, this is very difficult, but with care an approach can be made. A sharp-edged pipe should be used to prevent classification of the solid, and care should be taken to ensure that the solids do not separate out in the pipe. In practice there are few locations in the vessel where the flows are steady enough for the flow velocity and direction to be matched. Some flow visualization experiments must be carried out before sampling to ensure optimum matching of sampling flows with local flows in the vessel.

Grab sampling involves lowering a container fitted with a lid into the vessel, remotely opening the lid to fill the container, closing the lid and withdrawing the sample. Open-topped containers are unsuitable for this purpose, since they will tend to collect a sample from the surface of the vessel as they are lowered in. The measured solids concentration will also be a function of time unless the container allows a completely free flow of fluid while it is open.

Sampling is the most robust and most generally applicable technique for measuring solids concentration and is the only practical technique that will give information on particle size distribution [except, perhaps, phase Doppler anemometry (PDA), but that works only at very low concentrations]. Great care must be taken when sampling to ensure that representative samples are taken. The lack of commercially available equipment means that equipment will usually have to be designed and built for each job.

***4-5.1.5 Tomography.*** Electrical resistance tomography is very much an emerging technology (see Section 4-8.1) Measurement of solids distribution is one area where the technique is likely to be extremely useful.

### 4-5.2 Solids Suspension: Measurement of $N_{js}$

$N_{js}$ is defined as the minimum impeller speed at which all the solids in the vessel are suspended. This is the speed at which the surface area of all the solids in the vessel are in complete contact with the liquid and hence is an optimal operating point for mass transfer rate in the vessel (Figure 4-19). Solid suspension measurements are very sensitive to the precise shape of the vessel base. Measurements should not be made in a vessel with a drain in the middle

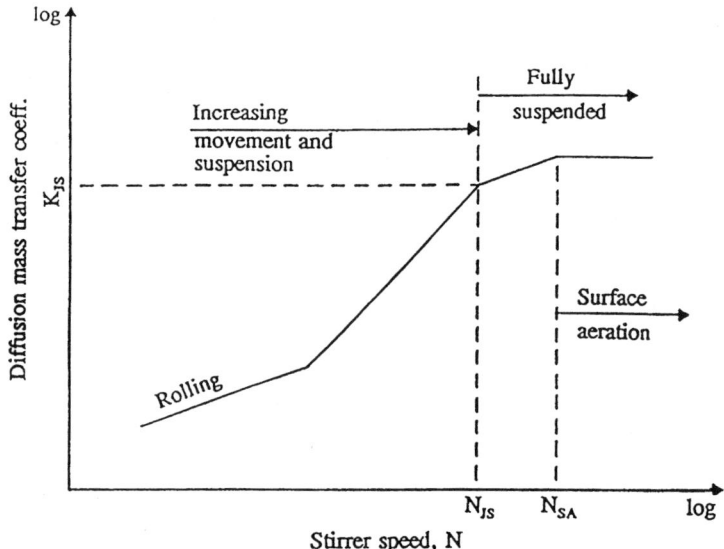

**Figure 4-19** Solid–liquid mass transfer coefficient over a range of impeller speeds.

of the vessel base unless they are done to model a particular production facility where this occurs.

The measurement (indeed, the existence of) of the "just suspension" speed is subject to the assumption that there exists a speed where there is equilibrium between the particles lifted from the vessel base and those settling back onto the base. This assumption is reasonable with fast settling particles but becomes more and more dubious as the settling velocity of the particles decreases.

**4-5.2.1 Visual Technique**. The most common way of determining $N_{js}$ is by making visual observations through the base of the vessel (provided that the vessel is equipped with a transparent base). This is a difficult technique, as there is generally no sharp endpoint at $N_{js}$, but a gradual transition from the partly suspended to the just suspended condition. This transition becomes more gradual as the solids concentration is increased. The problem is still further compounded, since many workers have used different criteria for defining the just suspended state. At various times, criteria such as "no solids remain stationary on the vessel base for more than 1 to 2 seconds," "no solids remain stationary on the vessel base for more than 2 to 4 seconds," and "no solids remain on the vessel base for more than 1 to 2 seconds" have been used. It is important to select an $N_{js}$ criterion relevant to the process under investigation, and one that is consistent with any other data that are being referenced.

Different workers, even when working with the same system and using the same suspension criterion, often obtain different values of $N_{js}$. Where possible, the entire set of measurements should be made by one person. When this is not possible, care should be taken to ensure that each operator trained in the technique produces consistent results.

On a practical level, transparent-based vessels designed for $N_{js}$ measurements should be mounted on a sturdy frame with the base raised several feet above the ground (Figure 4-7). In this way an operator can either lie on his or her back underneath the vessel, or can view a reflection of the base in a mirror. Care should be taken to ensure that a large enough section of the vessel base can be observed. With small vessels (T < 0.6 m) it should be possible to observe the entire base at once, while with larger vessels this may not be possible and a section (say, one quadrant) of the vessel base will have to be observed. If the decision is made to observe only a section of the vessel base, it should be noted that the eye does tend to wander, and it may help to hide the other sections from view.

The operator should have access to the impeller speed controller and a digital readout from the position in which he or she makes the measurements. Several repeat measurements should be made by slowly stepping up the impeller speed and observing the vessel base for stagnant particles. Sufficient time should be allowed for steady state to be reached before checking to see if the chosen "just suspension" criterion has been reached. If not, the impeller speed should be stepped up and the observation repeated. With fast-settling particles in a reasonably small vessel (say T < 1.0 m), the time required to reach steady state is not generally longer than about 1 min, although a longer period of measurement is required to test if the criterion has been reached. With much slower settling particles (either very small particles and/or a viscous liquid phase) the time to reach steady state and the measurement time may be large compared to the time it takes for a particle to settle back onto the vessel base. In these situations the entire measurement is meaningless, as the system is not in equilibrium and the concentration at the base is changing with time.

When making visual $N_{js}$ measurements, it is important to note not only the speed required to meet the chosen just suspended criterion but also the location of the point of last suspension. Some impeller and vessel geometry combinations tend to produce a point of last suspension in the middle of the base of the vessel; other combinations tend to leave a ring of settled solids around the base next to the vessel walls.

Similar visual techniques can be used to measure the equivalent speed for systems with floating particles, the "just drawn-down" criterion. The visual observations are subject to exactly the same issues and problems as the just suspended measurements; however, access to see the particles on the free surface of the fluid is generally easier to obtain than at the base of the vessel, as no transparent vessel parts are required.

In many situations, measurement of $N_{js}$ by direct observation is impractical. Other means must be resorted to, and two methods are available: detecting the movement of solids by means of an ultrasonic flowmeter, and inferring $N_{js}$ from measurements of solids concentration. Both techniques are difficult to use, and neither should be used without making visual observations in a geometrically similar vessel. It is in these situations that the knowledge of the location of the last point of suspension is vital.

**4-5.2.2 Ultrasonic Doppler Flowmeter Probe.** The ultrasonic Doppler flowmeter (UDF) instrument measures the velocity of a fluid by measuring the Doppler shift in an ultrasonic signal reflected from particles being carried by the fluid (Figure 4-20). This type of meter is normally used for measuring flows in pipes, but when mounted beneath the vessel base can be used to measure

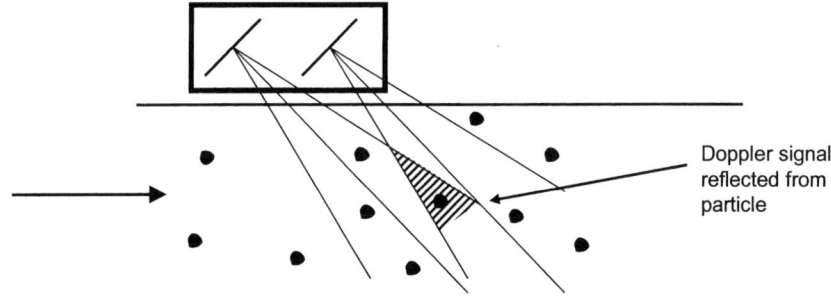

**Figure 4-20** Ultrasonic Doppler flowmeter: principles of operation.

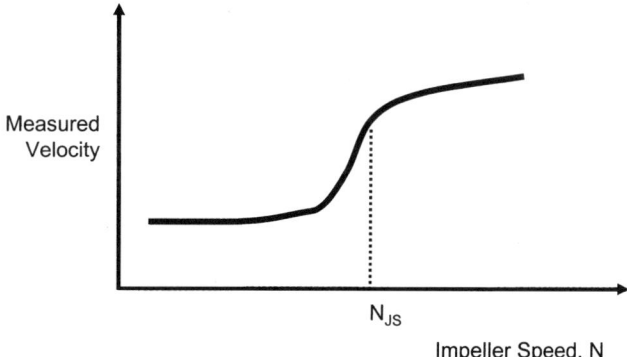

**Figure 4-21** Ultrasonic Doppler flowmeter: output signal.

$N_{js}$. At low impeller speeds the meter will indicate a zero velocity, since there will be an immobile bed of solids on the vessel base. As the impeller speed is increased, there will be a large increase in the indicated velocity at the point where the bed of solids becomes fully mobile. As impeller speed is increased beyond $N_{js}$, the indicated velocity will continue to increase, but at a much slower rate (Figure 4-21).

The meter does not require accurate calibration, since it is only necessary to measure relative velocities at different impeller speeds, and the position of the endpoint is taken as an indication of $N_{js}$. For successful operation a fairly high solids concentration is needed, at least 2 to 3 vol %. At lower concentrations the instrument can "see" right through the bed of solids and does not detect an endpoint. Several different types of sensing heads are available which measure velocities at various distances from the vessel wall. A head with as small a measuring distance as possible should be used to prevent this problem.

It is extremely important that the sensing head is positioned underneath the vessel base in the place where suspension occurs last. If visual observations are not possible in the vessel being used, some preliminary work must be carried out in a geometrically scaled transparent vessel. Placing the sensing head in an inappropriate position will result in erroneous and misleading measurements being made. Some correlation of meter response against visual observation must be made.

The ultrasonic Doppler flowmeter has the advantage of being mounted externally on the vessel and is capable of measuring $N_{js}$ where visual measurements are impossible. This makes it especially suitable for use in a process plant, where the process fluids may be aggressive, access to the vessel may be difficult, and visual observations cannot usually be made. The instrument is, however, unusable at low solids concentrations and still requires some visual observations to be made before installation.

*4-5.2.3 Conductivity Probe.* As described earlier, a probe that responds to changes in solids concentration may be placed at a point inside a vessel to

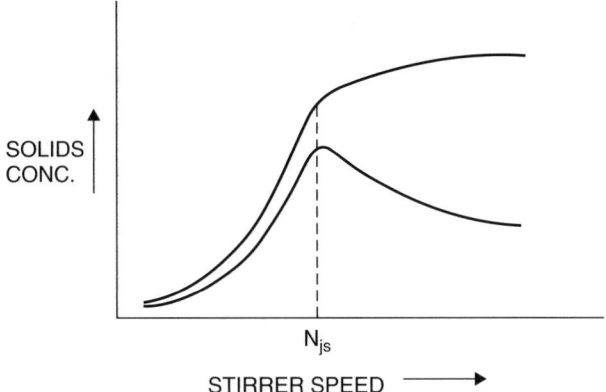

**Figure 4-22** Determination of $N_{js}$ from solids concentration measurement.

obtain information on the local solids concentration. From measurements of local solids concentration, $N_{js}$ may also sometimes be inferred. To infer $N_{js}$ from solids concentration measurements, there must be some feature of the solids concentration versus speed curve that can be correlated with $N_{js}$ (Figure 4-22).

The assumed mechanism here is that as the impeller speed is increased while the solids are partly suspended, the solids concentration near the vessel base will increase due to the increase in the quantity of solids removed from the immobile bed on the vessel base. Once $N_{js}$ has been reached, the solids concentration will start to fall, since the solids near the base are being redistributed to other regions of the vessel, and no more solids are available on the base to take their place.

In many cases such an idealized behavior has not been reported, and workers have observed only a kink in the solids concentration curve at $N_{js}$. It is extremely difficult to get reliable, reproducible results with this technique. The shape of the solids concentration versus speed curve is very sensitive to changes in the position of the measuring point. Often, there are no salient features that can be correlated with $N_{js}$, or worse still, there is a feature that is located at a speed far below $N_{js}$—hence the need for parallel visual observations. The technique does have the advantage that it is not prone to subjective interpretations by different observers. However, it should not he attempted unless there is a great deal of time available to develop the experimental method further.

Visual observations of the flows in the vessel, or in a geometrically similar small scale vessel, should be made to assist in probe positioning. Even in a well-mixed vessel the solids concentration can vary greatly between different parts of the vessel. A misleading interpretation of the results may occur if the solids in the vessel are incompletely suspended and measurements are made in regions of the vessel with relatively high solids concentrations.

## 4-6 LIQUID–LIQUID DISPERSION

### 4-6.1 Cleaning a Liquid–Liquid System

All experimentalists who work with immiscible liquid–liquid systems must be concerned about the cleanliness of the system. Liquid–liquid experiments, particularly drop size measurements, are notoriously sensitive to changes in the cleanliness of the fluids, the vessel, and the mixers. The concern is that any trace of impurity, be it surfactant or greasy fingerprint, can alter the interfacial tension, alter the dynamics of the interface (repressing coalescence), or if preferentially wetted by the dispersed phase, act as a center for coalescence.

Two approaches may be taken to cleanliness. The one that is more popular with industrial workers is to swamp the interface with a known contaminant. This gives a reproducible system that is completely stabilized, or *noncoalescing*, and this method is thus suitable only for study of such systems. The other technique, more popular with academic researchers, is to strive for a base case with a very high level of cleanliness. For this approach a range of extremely draconian, though very necessary cleaning regimes have been developed to ensure that vessels and pipes are at least uniformly clean. Most of these procedures involve the repeated scrubbing of all wetted mixer parts in copious quantities of hot water and surfactant, followed by several rinses with deionized water. Some procedures also stipulate flushing with chromic acid, but the present authors do not recommend the use of this substance. Further details of the cleaning procedures are presented in Chapter 12.

When moving to larger scale systems, things may get easier. At larger scales there is less surface area of mixer and vessel per unit volume available to contaminate, and at higher dispersed phase concentrations there is more interfacial area of dispersion per volume of contaminant. However, performing large scale experiments (anything at $T = 0.6$ m or over is considered large scale) is extremely difficult because of the high cost of materials and cleaning and the scarcity of vessels that can be cleaned to the degree required to give reproducible results.

The reader is strongly cautioned that a number of issues other than cleanliness may contribute to the variability often seen in liquid–liquid dispersion experiments. These include:

- *Materials of construction.* The fluids have different tendencies to wet the surfaces of different materials of construction, and this can lead to coalescence of one liquid at a surface, which distorts the results of the experiment. Glass and steel vessels are hydrophilic, while Perspex vessels are hydrophobic. These properties can be modified if surfaces become coated (typically with oil) over time.
- *Temperature.* The material properties, particularly interface tension, viscosity, and density, that influence liquid–liquid dispersions are functions of temperature.

**188**  EXPERIMENTAL METHODS

- *Slight geometry changes.* Small differences in impeller or vessel geometry can affect the power number of the impeller and hence the breakup and coalescence of drops. It may also be postulated that details such as blade sharpness or baffle thickness affect drop size in some cases.

### 4-6.2  Measuring Interfacial Tension

One of the key variables governing droplet breakup is the interfacial tension between phases. The pendant drop apparatus can be used to measure either the surface tension between a liquid and air or the interfacial tension between two liquids. The pendant drop method should be used in preference to the du Nüoy ring, as it is easier to use and gives more reliable results.

The apparatus consists of a glass microsyringe with a calibrated volume scale. The end of the syringe has a stainless steel hypodermic needle attached. The entire apparatus is clamped to a stand to eliminate vibrations (Figure 4-23). For interfacial tension measurements, a small PMMA or glass cell is positioned beneath the needle tip. The syringe is charged with the more dense fluid, and the cell is charged with the less dense fluid. The tip of the syringe needle is lowered to a position below the surface of the less dense fluid and a droplet is slowly squeezed out through the needle until it detaches and sinks to the base of the cell.

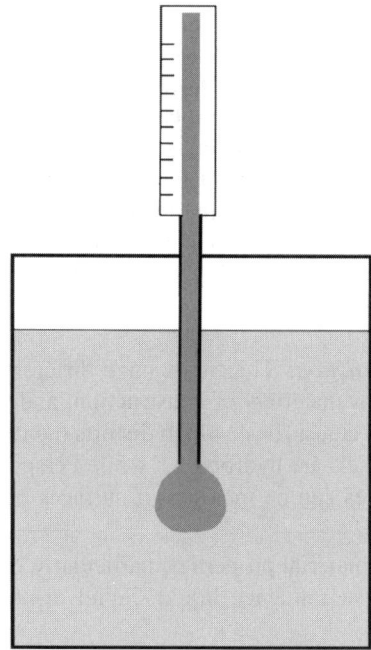

**Figure 4-23**  Pendant drop technique for interfacial tension measurement.

The method works on the principle that the volume that the droplet reaches within another liquid is related to the interfacial tension between the two liquids by the following expression:

$$\sigma = \frac{\theta V_d \Delta \rho g}{2\pi r} \quad (4\text{-}11)$$

where $\sigma$ is the interfacial tension, $V_d$ the volume of the detached drop, $\Delta \rho$ the difference in density between the two phases, g the acceleration due to gravity, and r the radius of the hypodermic needle. $\theta$ is a correction factor that allows for the amount of liquid left inside the needle tip after the drop has detached.

Great care must be taken when making these measurements to ensure that all equipment is scrupulously clean and that the temperature of the fluids and equipment is monitored and constant. The interfacial tension should be calculated based on the average droplet size of a large number of droplets. If repeatable measurements are not obtainable and suitable care has been taken in the performance of the experiment, it is possible that some form of contamination has occurred.

### 4-6.3  $N_{jd}$ for Liquid–Liquid Systems

Various techniques have been employed to try to measure the speed required to disperse successfully, to the conditions of "just drawdown," one immiscible phase in another (Skelland and Seksaria, 1978; Penney et al., 1999; Yamamura and Takahashi, 1999). These have included several variations on the use of conductivity probes (zero conductivity behind the baffle, indicating the presence of a pocket of organic phase), along with visual measurements. Measurements of this type are fraught with difficulty and it is extremely difficult to relate the results obtained using one technique with those obtained using another technique. There is a large difference between "just dispersed," where no continuous layer of the dispersed phase remains in the vessel, and "completely dispersed," where the liquid–liquid dispersion is distributed uniformly throughout the vessel. In a liquid–liquid dispersion system this can be quite difficult to detect. Figure 4-24 shows an immiscible liquid–liquid stirred tank system at impeller speeds below the full drawn-down condition.

### 4-6.4  Distribution of the Dispersed Phase

Most measurements on the distribution of a dispersed phase in a liquid–liquid system have been made using the "local" techniques of measuring the conductivity or light transmittance. The probes used for these measurements are practically identical to those discussed in Sections 4-5.1.2 and 4-5.1.3 with reference to solid–liquid mixing. The conductivity-type probe can, however, suffer from the added complication of the electrode(s) becoming completely coated with a layer of the nonconductive (organic) phase, which causes the instrument to fail. Sampling techniques are subject to the same problems discussed in Section 4-5.1.4.

**Figure 4-24** Immiscible liquid–liquid system at a speed below that required for complete dispersion. See insert for a color representation of this figure.

### 4-6.5 Phase Inversion

The point of phase inversion (where the continuous phase becomes dispersed and the dispersed phase becomes continuous) is generally best detected using a conductivity technique, provided that one phase is a good conductor (e.g., an aqueous solution) and the other is not (most organic liquids).

### 4-6.6 Droplet Sizing

Many efforts have been made to measure the size of liquid dispersed droplets. The techniques can be divided into those that require a sample to be withdrawn from the vessel (or pipe) and those that take the measurement in situ. Pulling a sample is generally considered to be acceptable if the system has been suitably stabilized using a surfactant. Significant care still needs to be taken to ensure that the resulting droplet size distribution is not affected by the sampling method itself (i.e., no droplet breakup, coalescence, or preferential sampling occurs).

The other main consideration when selecting a measurement technique is whether the system is to be dilute or high concentration. Large numbers of investigations into droplet size distributions (or simply $d_{32}$) have focused on extremely dilute dispersions, where it is assumed that coalescence can be eliminated. "High concentration" can actually mean anything from about 1 vol % to the point of phase inversion. Great care should be taken with any droplet size measurement system to ensure that a suitably large number of droplets have been sampled to provide a meaningful result. This means at least 500 drops for a unimodal distribution or for each peak of a multimodal distribution.

*4-6.6.1 Microscope Analysis.* Once a sample has been taken, perhaps the simplest way to measure the droplet size distribution is to place the sample on a microscope slide and take photographs of it. The photographs can either be taken digitally or conventionally, but in either case suitable calibration images will have to be collected so that the droplet sizes can be determined. Microscope slide cover slips should not be used, as they can "squash" the droplets flat, making them appear larger than they actually are.

***4-6.6.2 Laser Diffraction.*** There are now a large number of laser diffraction instruments on the market. This is an excellent technique, provided that a representative sample can be obtained and placed in the instrument. A key difficulty is sampling successfully (i.e., without altering the droplet size) and representatively, particularly at high concentrations when stabilization and dilution are often required.

***4-6.6.3 Video Probe Methods.*** Various commercial models of video probes are now available, and the components can easily be bought to produce a tailor-made system. A diagram of a typical probe (in situ in a stirred tank) is shown in Figure 4-25. A photograph of the same probe is shown in Figure 4-26. All probe surfaces that come into contact with the dispersion are made from stainless steel or glass. Other designs of probe are available. A typical image produced by probes of this type is shown in Figure 4-27. Video records are provided on the Visual Mixing CD affixed to the back cover of the book.

Some comments about the components of the video system:

- *Lens.* A compact lens with a suitable magnification should be chosen. The size of the droplets to be measured is the governing factor. The measurement

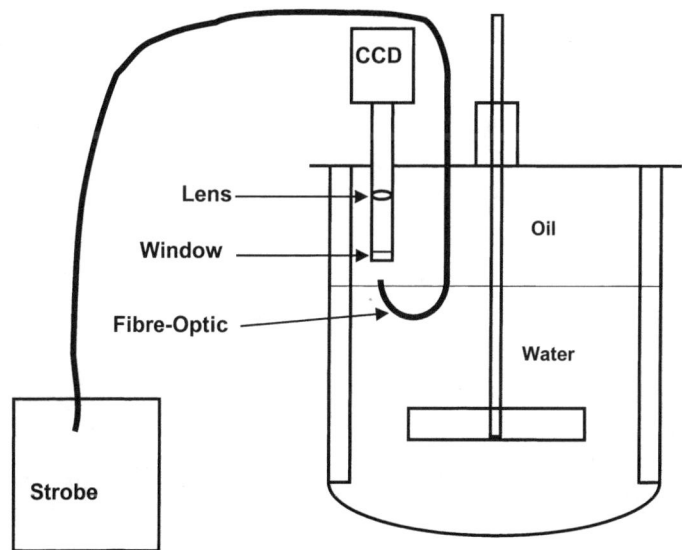

**Figure 4-25** Liquid–liquid drop size video probe in stirred tank.

**192** EXPERIMENTAL METHODS

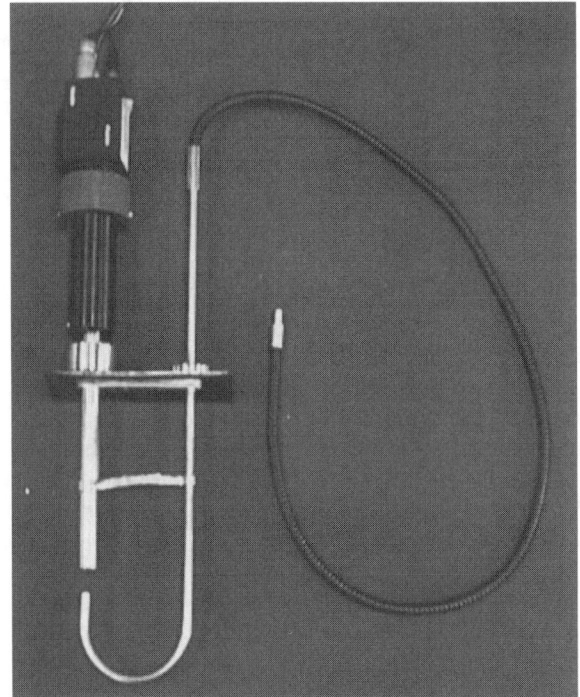

**Figure 4-26** Video probe for use in a STR.

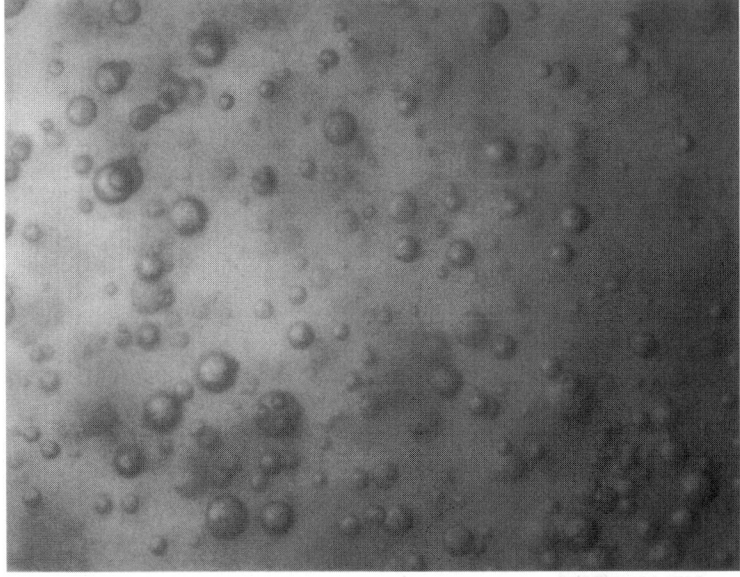

**Figure 4-27** Typical liquid–liquid droplet image.

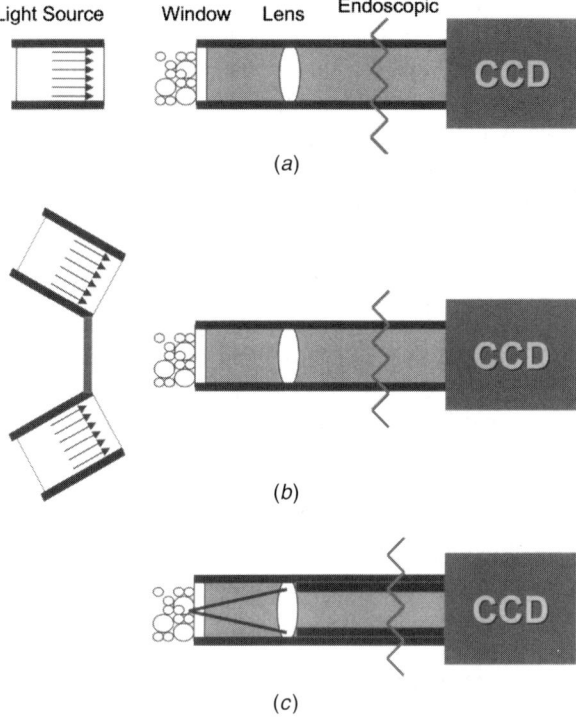

**Figure 4-28** Video probe lighting methods: (*a*) back lighting; (*b*) dark field; (*c*) ring lighting.

of a droplet that has a diameter less than 1/100 the image width is not recommended.
- *Lighting.* Backlighting produces the best images, although application of dark-field illumination and front lighting can prove useful in some situations (Figure 4-28). Great care is required in optimizing the lighting for a given measurement condition, as the lighting is critical in obtaining good-quality images but is sensitive to variables such as droplet size and dispersed phase fraction.
- *Shutter.* Most cameras have a controllable shutter. When a constant light source is used, the shutter should be set to as fast a speed as possible without the images appearing too dark, in order to give the clearest images with "stopped" droplets. A high-speed strobe (one with a short pulse width) running at the frequency of the camera can be used as an alternative to an electronic shutter. If the strobe is fast enough, it will "stop" the images of the droplets.
- *Capture.* Image capture uses either a digital video system, or an analog CCD camera, fed into a computer with an image capture card.

- *Analysis of images.* Various commercial software packages are now available that analyze images automatically and measure the size of the droplets. The quality of the results from these packages is variable and depends on a number of variables, including the quality of the initial images. Custom software can be written to incorporate previously validated image processing algorithms. Software is also available that allows the "manual" measurement of droplet images on computer. This process is extremely time consuming but generally considered essential for producing accurate results as well as checking the performance of automatic software.

***4-6.6.4 Chord Length Measurement Using Laser and Rotating Mirror.***
A device is available that uses a rotating mirror to pass a laser beam across a measurement area. The reflected light of the laser can be converted into chord lengths of droplets, and from this the droplet size distribution can be calculated, assuming that the droplets are spherical.

***4-6.6.5 Phase Doppler Anemometry.*** Phase Doppler anemometry (PDA) is discussed in Section 4-11. At the time of writing, it is only useful for very low dispersed phase concentrations.

## 4-7 GAS–LIQUID MIXING

### 4-7.1 Detecting the Gassing Regime

The regime describes the degree of dispersion of the gas and its flow patterns around the vessel (see Chapter 11; and Smith et al., 1987). Detecting the regime is best done by visual observation through the side of a transparent vessel using the real system or a good model system. If this is not possible, the appearance of the top surface gives an indication of the most important condition: If the gas is seen to be arriving at the center of the surface in large bubbles, this indicates that it is not being dispersed. Some clue to the regime may also be obtained by measuring the variation of local gas fraction between locations using a local void fraction probe (see Section 4-7.4).

### 4-7.2 Cavity Type

Direct observation is the best way to observe the type of blade cavity (refer to Chapter 11 for a description of cavities), although, of course, this requires that the vessel have a transparent base or bottom window. One method of capturing an image of the rotating cavity system involves observing or videoing through a derotational prism (Dove prism) mounted underneath the vessel (Figure 4-29). Another method involves a video camera mounted on a turntable underneath the vessel and rotated synchronously with the impeller. Early work (e.g., Warmoeskerken et al., 1984) used slip rings to transmit the signal to a stationary

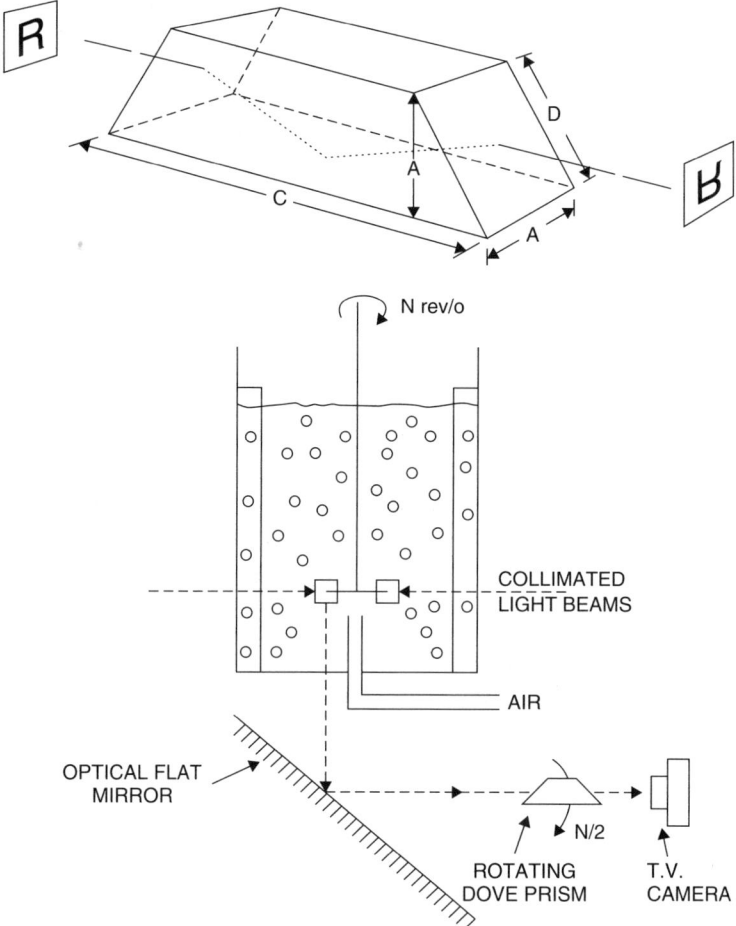

**Figure 4-29** Derotational prism. (Modified from Kuboi et al., 1983.)

recorder, which limited the speed and reliability. Self-contained camcorders are now available. Care is required (as always) with the lighting.

If a sufficiently powerful stroboscopic lamp is available, this can be used, flashing at N times the number of blades, with a stationary video camera (Nienow and Wisdom, 1974). For better-quality single shots, a high-speed flash can be used with a still camera.

The vane probe (Smith, 1985) has been used with success to detect cavity type. A probe with a small vane fitted with strain gauges is mounted in the turbine discharge stream, and the different cavity types are indicated by characteristic frequency spectra of the strain gauge signals. Cavity type can also be implied from power draw measurements if the ungassed power number of the impeller and the power losses in the motor and drive are known.

## 4-7.3 Power Measurement

The gassed power draw should be measured using the techniques described in Section 4-3.

## 4-7.4 Gas Volume Fraction (Hold-up)

Either local gas volume fraction or the vessel average fraction (*gas hold-up*) may be measured. Where the amount of surface motion in the gas–liquid system is not excessive, an estimate of the vessel average gas hold-up may be made by visual observation of the level change (possibly aided by video analysis).

A more reliable technique, and one that can be used where there are larger surface fluctuations, is to use an ultrasonic radar probe (Machon et al., 1991). Such probes are commercially available, from, for example, Endress and Hauser. The probe is mounted above a representative part of the fluctuating surface and measures the distance to the surface. Care must be taken in the calibration (especially if any foam is present) to ensure that the true surface is detected and that an adequate range of fill levels is covered. Some foams will not be penetrated by the ultrasonic beam. Calibration with a moving liquid surface is recommended.

A cruder but effective method for overall gas hold-up in batch vessels is the spillover technique. The vessel is filled to the overflow before gassing; then the gas displaces its own volume of liquid, which is collected and measured.

Local gas fraction is measured using conductivity probes or optical probes. These conductivity probes typically measure the conductivity across two parallel plates about 10 mm square and 10 mm apart. The optical probes are the same as those used in solid–liquid work. Both types require careful calibration with the liquid under study.

## 4-7.5 Volumetric Mass Transfer Coefficient, $k_L a$

Gas–liquid mass transfer and $k_L a$ are described in Chapter 11.

Dynamic absorption (usually of oxygen into aqueous liquids) has been a popular method for measuring $k_L a$, although it suffers from the need for rapid response, robust dissolved oxygen probes, which although being available (the polarographic type with thin membrane and small electrolyte path length) were not always used. The probes are temperature sensitive, and their response lag must be deconvoluted from the overall response signal. More important, the dynamic methods are subject to gross errors with systems containing small bubbles, which become exhausted quickly in oxygen-lean zones then act as an extra oxygen sink in oxygen-rich zones (Heijnen et al., 1980). For these reasons, steady-state methods are preferred.

Among the steady-state methods, the hydrogen peroxide–catalase method of Hickman (1988a) is the most commonly used. It can be used at any scale with aqueous systems at pH around 7, but not with ionic solutes such as NaCl, KCl, or $K_2SO_4$ (sometimes used to render the liquid noncoalescing). Air is the usual gas feed, and the liquid contains the enzyme catalase in excess. Hydrogen peroxide is fed at a steady rate such that dissolved oxygen concentration remains constant while the excess catalase decomposes the peroxide to water and oxygen. Thus at a steady state the oxygen transfer rate is obtained from the peroxide feed rate. Some precautions are necessary (Cooke et al., 1991):

- The catalase concentration must be in excess of stoichiometric by a factor of at least 15.
- For $k_L a > 0.5$ s$^{-1}$, catalase may deposit on surfaces and may affect coalescence properties.
- The pH must be maintained near 7 to minimize degradation of the catalase.
- The dissolved oxygen level must not exceed 300% saturation (otherwise, extraneous bubbles may be formed) but should be greater than 130% saturation to avoid large errors arising from small driving forces.
- The dissolved oxygen probe membrane must be resistant to peroxide attack (check by constancy of repeating measurements at different peroxide addition rates).
- Constant volume should be maintained by using concentrated peroxide feed and, for long experiments, withdrawing some liquid.
- Liquid mixedness should be checked by using several dissolved oxygen probes at various locations in the vessel.
- A high concentration of catalase will minimize the standing concentration of peroxide (thus minimize the risk of membrane damage) but will enhance spatial variations of concentration. Therefore, the catalase concentration must be optimized for the degree of mixing in each case.
- The decomposition of peroxide is exothermic, so temperatures must be checked before and after the measurement.
- The water produced could affect the viscosity of some model fluids.

One disadvantage of the catalase method for larger scales is the cost of the enzyme. A similar method involving manganese dioxide in place of catalase (Muller and Davidson, 1992; Martin et al., 1994; Vasconcelos et al., 1997) alleviates this problem and allows operation with noncoalescing electrolyte solutions. Although fine (20 μm) particles are involved, the concentration is low (0.8 g/l) and it is found (Vasconcelos et al., 1997) that they do not affect the coalescence properties of the system or the result obtained. However, manganese dioxide can absorb and degrade some solutes, such as carbohydrates (corn syrup, glucose, etc.) or polyhydroxy compounds (such as polypropylene glycol) often used in experiments.

A number of steady-state methods involving oxygen absorption with an oxygen sink dissolved in the liquid have been proposed. Among the oxygen sinks, sodium sulfite was used widely in older work but is very sensitive to conditions and catalysis (usually, copper or cobalt) and renders the solution noncoalescing. Active yeast has also been used as an oxygen sink (Hickman and Nienow, 1986) for work with fermenters.

Another method that has been suggested for plant vessels (especially, e.g., fermenters) where no additives are tolerated is the *pressure jump method* (Hickman, 1988b). The vessel pressure is rapidly dropped by a small amount, and the dissolved gas concentration in the liquid is monitored against time.

Chemical methods can also be used to measure $k_L a$. Robinson and Wilke (1974) describe an ingenious method to obtain $k_L a$ and $a$ simultaneously, by desorption of oxygen (rate $\propto k_L a$) and reaction-enhanced absorption of carbon dioxide (regime III; see Chapter 11) into potassium hydroxide solution (rate $\propto a$).

All the methods above require the correct gas flow pattern (plug flow, well back-mixed, or intermediate) to convert the transfer rate to a correct $k_L a$ value unless the degree of depletion of the gas phase is very low. This can be very important, as discussed in Chapter 11. Gas flow patterns can be determined from measurements of the gas residence time distribution using tracer gas (see Section 4-7.8). Two dynamic methods avoid this problem: the double response method (Chapman et al., 1982), in which the dynamic responses of both liquid and gas phases are measured, and the initial response method (Gibilaro et al., 1985).

All methods also assume that the liquid is well mixed. This can be checked by comparing the 90% mixing time (measured by the methods described earlier, but see Section 4-7.10) to the measured 90% mass transfer time ($= 2.3/k_L a$). If the liquid is not well mixed, an interlinked zone model (see, e.g., Figure 4-30

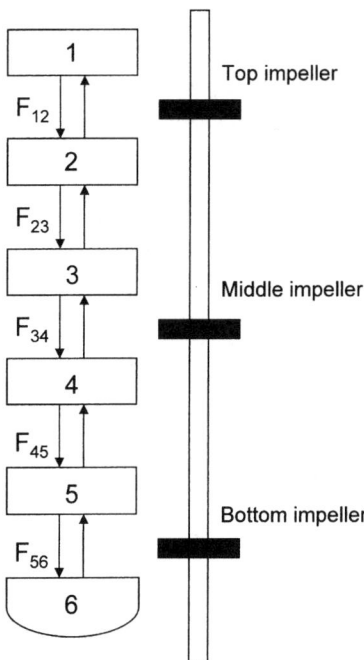

**Figure 4-30** Network of zones model.

and Whitton et al., 1997) can be used to describe and scale-up mass transfer, with the fluid volume divided into zones, chosen with regard to the flow patterns so as to be well mixed. The interzone flow rates to be used in the model are fitted from liquid residence time distribution measurements (see Section 4-7.9) or from zone-to-zone dye tracing experiments using the area between the normalized response traces of probes in neighboring zones.

### 4-7.6 Bubble Size and Specific Interfacial Area

Surface volume mean bubble size, d, and specific interfacial area, a, are linked via the local gas fraction, $\phi$:

$$a = \frac{6\phi}{d} \tag{4-12}$$

Thus, measurements of any two will give the third. Direct photography, either through the walls (Machon et al., 1997) (which may not give representative samples) or with an optical probe (see Section 4-6.6.3) with image analysis is the most direct method, yielding local size distributions and gas fractions.

Other probes have been used for measuring bubble size, often using arrays of optical fibers (Cochrane and Burgess, 1985; Frijlink, 1987) or electrodes [e.g., the Burgess–Calderbank five-point probe (Burgess and Calderbank, 1975), or the radio-frequency probe of Abuaf et al. (1979)] to register the gas–liquid interfaces as they pass over or between (Cochrane and Burgess, 1985) the probe tips. An array of probes enables shape and velocity to be detected also. Such probes are somewhat fragile and cannot measure bubbles smaller than about 1 mm in diameter.

Light transmittance methods for interfacial area have been widely employed in gas–liquid systems. Sridhar and Potter (1978) describe one of the more successful versions. The techniques are described in more detail in the liquid–liquid section. Chemical methods can be used to measure interfacial area, as described above (Section 4-7.5; Robinson and Wilke, 1974). Sampling methods are not suitable since it is impossible to withdraw a sample isokinetically. Generally, the results from chemical and physical methods do not agree. Chemical methods tend to have a bias toward the smaller bubbles of the distribution, whereas these may be missed by physical methods.

The method of dynamic gas disengagement (Sriram and Mann, 1977; Patel et al., 1989) to obtain an estimate of bubble size distribution is worthy of mention since it is convenient to use, sometimes even in real systems, especially for lower gas fractions. The impeller is stopped and the trend in the level measured against time. This trend indicates the bubble size distribution if terminal rise velocities are known and if coalescence is negligible.

### 4-7.7 Coalescence

The best test of whether a system is coalescing or noncoalescing is to set it up in a vessel of standard configuration and impeller speed, in which overall gas fraction can be measured. The gas fraction is compared with those of known coalescing and noncoalescing liquids and solutions.

Reith and Beek (1970) measured net coalescence rates in a vessel by injecting two gases separately and measuring concentrations in the leaving bubbles using a sampling probe. They also used a chemical method based on absorption in cobalt-catalyzed sulfite solution. The reaction, being second order in oxygen, proceeds at different rates for separate or mixed bubbles.

Several more fundamental measurements of quasi-static two-bubble coalescence times are reported; Machon et al. (1997) give a summary.

### 4-7.8 Gas-Phase RTD

Commonly, helium is used as a tracer gas in air or nitrogen, in the usual method for measuring RTD with a step change or pulse in the input gas stream. The tracer gas must exhibit the following properties:

- Very low solubility
- Safe to use
- Nonreactive
- Inexpensive
- Easy to sample
- Easily measurable at low concentration
- Easy to obtain

The method is described in Section 4-4.5. It is important to ensure that the dead spaces (gas volumes between the liquid surface and the exit gas concentration detector) are deconvoluted from the measured response. Gas-phase RTD has been measured by Hanhart et al. (1963) and Gal-or and Resnick (1966) and is often in between the ideal limits of plug flow and perfectly backmixed.

### 4-7.9 Liquid-Phase RTD

The method is the same as described in Section 4-4.5. The probes must be mounted or shielded such that the gas bubbles do not interfere with the readings.

### 4-7.10 Liquid-Phase Blending Time

Liquid-phase mixing time measurements can be made using the techniques described in the single-phase mixing section. Again the probes must be protected from interference from the gas bubbles.

### 4-7.11 Surface Aeration

Mass transfer by surface aeration is measured in the same ways as described in Section 4-7.5. If it is necessary to measure the gas induction rate in a closed batch vessel (e.g., with a surface aerator or self-inducing agitator), this can be done directly by measuring the gas flow rate into the sealed vessel required to maintain the head-space pressure. This gas flow can be corrected to compensate for the pressure drop of the entry into the vessel. Another method (Chapman et al., 1980) involves sampling and analyzing gas bubbles within the dispersion. This method was extended to measure surface aeration in the presence of sparged gas.

## 4-8 OTHER TECHNIQUES

### 4-8.1 Tomography

Tomography is the nonobtrusive localized measurement of velocity, density, concentration, and so on, in three dimensions. An image is built up from the responses of an array of sensors around the periphery of the domain. A variety of signals have been used in mixing vessels, including light, x-ray, γ-ray, ultrasonic, positron, magnetic resonance, and electrical resistance and capacitance.

The simplest form of tomography uses crossed light beams or sheets and cameras, and has been used to measure (plane-by-plane) velocities in the impeller stream (Takashima and Mochizuki, 1971), and concentration (pH) fields in acid-base-indicator mixing experiments (All-Saeedi, 1995). Lübbe (1982) used elaborate holographic interferometry equipment to measure the real-time dynamics of temperature fields in stirred vessels. Mewes and Ostendorf (1986) review this and other methods such as X-ray and γ-ray tomography. Ultrasonic tomography basically responds to density (and elastic modulus) differences and has been used successfully in gas–liquid pipe flow systems (including metal pipes). In theory it has a spatial resolution of about 1 mm and a temporal resolution of 16 ms (for a 200 mm pipe with multiple transmitters), but the stream velocity for imaging is limited to a maximum of about 0.5 m/s (Hoyle, 1996). Positron emission tomography (Parker and McNeil, 1996) provides high resolution but requires long times for data acquisition (e.g., 30 min). It has been used (McKee et al., 1994) for measuring solid particle concentration fields, and (Fangary et al., 1992) for velocity measurements (<0.5 m/s) in stirred vessels.

Electrical tomography (Dickin et al., 1992; McKee et al., 1994) is the most widely explored for stirred vessels, and probably the lowest-cost method, in cases where modest resolution is sufficient. Resistance tomography is used for conducting liquids and capacitance for nonconducting liquids. A typical capacitance system has a temporal resolution of 10 to 100 frames per second and a spatial resolution (radially) of $\frac{1}{20}$ of the vessel diameter. Several (typically eight) rings, each of typically 16 electrodes, are positioned on the vessel wall, and readings from all pairs on a ring are taken in turn and fed to an image reconstruction algorithm. A typical probe cage is shown in Figure 4-31. It can detect solids, gases, immiscible liquids, or ionic tracer solutions, and has been used in mixing studies in pilot scale stirred vessels (2.3 m$^3$) (Stanley et al., 2002). Resolution very near the base and surface is poor and deteriorates toward the axis of the

**Figure 4-31** Tomography cage. (From Cooke et al., 2001.)

vessel unless a central electrode is added. It is possible to use the method with metal vessels (Wang et al., 1993) using a modified reconstruction algorithm.

# Part B: Fundamental Flow Measurement

*George Papadopoulos and Engin B. Arik*

### 4-9 SCOPE OF FUNDAMENTAL FLOW MEASUREMENT TECHNIQUES

Early emphasis in turbulence research was on the statistical analysis and description of turbulent flow fields. Theoretical advances in the field have been complemented by experiments that included pressure measurements and by the point measurement technique of hot wire anemometry (HWA). The intrusive nature of this latter technique has precluded its use in some experiments, whereas in others, corrections have been introduced to the measurement results.

Optical diagnostic techniques are desirable for fluid flow measurements due to their nonintrusive nature. Soon after the invention of the laser in the 1960s, the technique of laser Doppler anemometry (LDA) was developed. During the last three decades, the LDA technique has witnessed significant advancements. Three-component fiber optic-based LDA systems with frequency-domain signal processors are currently the state of the art and are used in numerous facilities. The addition of a second photodetector to the first component LDA receiving optics gives the system the capability of size measurement, in addition to velocity, through phase difference analysis of the scattered light. The particle dynamics analyzer or phase Doppler anemometer (PDA) is an extension of the LDA and is a valuable tool for size determination of spherical particles.

Since the early 1960s, the investigation of three dimensional coherent structures has been of significant interest for turbulence researchers. Flow visualization techniques have been around since the days of Prandtl. Flow markers, such as seeding particles, dyes, smoke, and so on, are typically used, and the techniques have been improved, with new ones being developed over the years. The evolution of these whole field flow visualization techniques has led to quantification of the visualized results, especially during the last 10 years with the advent of digital imaging and fast growth of computational power. Particle image velocimetry (PIV) has evolved to be a highly powerful technique for 2D and 3D whole field velocity measurements, while planar laser-induced fluorescence (PLIF) is also becoming a powerful technique in the mixing research community for quantitative concentration measurements.

### 4-9.1 Point versus Full Field Velocity Measurement Techniques: Advantages and Limitations

Hot wire anemometry (HWA) or constant temperature anemometry (CTA), laser Doppler anemometry (LDA) or laser Doppler velocimetry (LDV), and particle image velocimetry (PIV) are currently the most commonly used and commercially available diagnostic techniques to measure fluid flow velocity. The great majority of the HWA systems in use employ the constant temperature anemometry (CTA) implementation. A quick comparison of the key transducer properties of each technique is shown in Table 4-1, with expanded details on spatial resolution, temporal resolution, and calibration provided in the following sections.

*4-9.1.1 Spatial Resolution.* High spatial resolution is a must for any advanced flow diagnostic tool. In particular, the spatial resolution of a sensor should be small compared to the flow scale, or eddy size, of interest. For turbulent flows, accurate measurement of turbulence requires that scales as small as two to three times the Kolmogorov scale be resolved. Typical CTA sensors are a few micrometers in diameter, and a few millimeters in length, providing sufficiently high spatial resolution for most applications. Their small size and fast response make them the diagnostic of choice for turbulence measurements.

The LDA measurement volume is defined as the fringe pattern formed at the crossing point of two focused laser beams. Typical dimensions are 100 μm for the diameter and 1 mm for the length. Smaller measurement volumes can be achieved by using beam expansion, larger beam separation on the front lens, and shorter focal length lenses. However, fewer fringes in the measurement volume increase the uncertainty of the measurement.

**Table 4-1** Transducer Comparison of Commonly Used Velocity Measurement Diagnostics

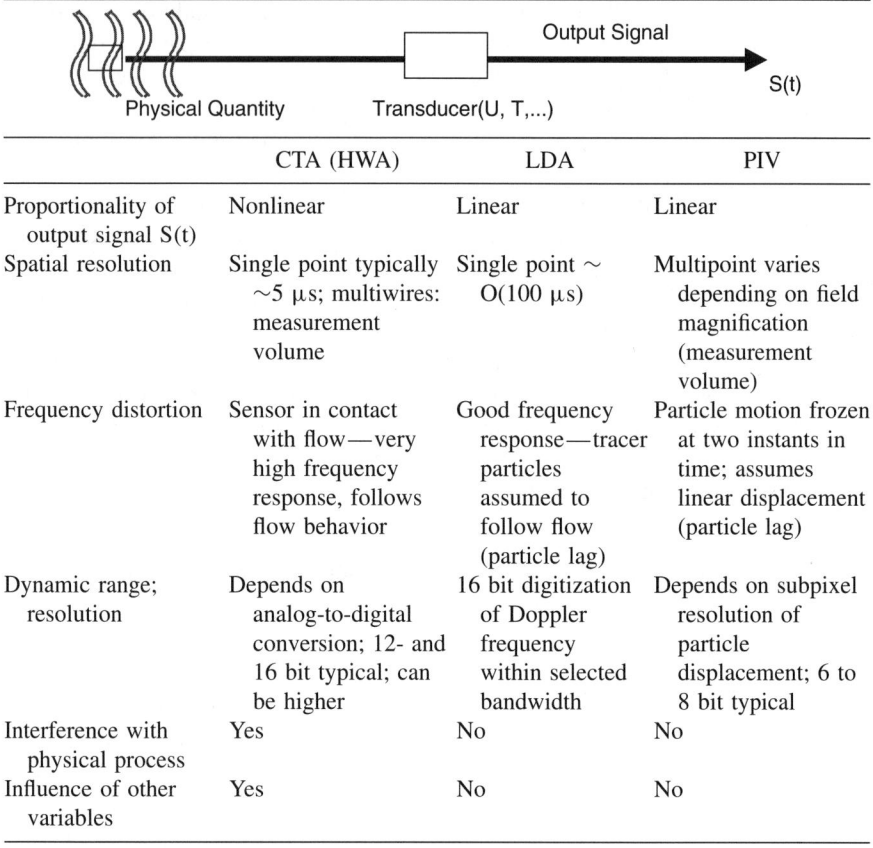

|  | CTA (HWA) | LDA | PIV |
|---|---|---|---|
| Proportionality of output signal S(t) | Nonlinear | Linear | Linear |
| Spatial resolution | Single point typically ~5 μs; multiwires: measurement volume | Single point ~ O(100 μs) | Multipoint varies depending on field magnification (measurement volume) |
| Frequency distortion | Sensor in contact with flow—very high frequency response, follows flow behavior | Good frequency response—tracer particles assumed to follow flow (particle lag) | Particle motion frozen at two instants in time; assumes linear displacement (particle lag) |
| Dynamic range; resolution | Depends on analog-to-digital conversion; 12- and 16 bit typical; can be higher | 16 bit digitization of Doppler frequency within selected bandwidth | Depends on subpixel resolution of particle displacement; 6 to 8 bit typical |
| Interference with physical process | Yes | No | No |
| Influence of other variables | Yes | No | No |

A PIV sensor is the subsection of the image, called an *interrogation region*. Typical dimensions are $32 \times 32$ pixels, which would correspond to a sensor having dimensions of $3 \times 3$ mm by the light sheet thickness (~1 mm) when an area $10 \times 10$ cm is imaged using a digital camera with a pixel format of $1000 \times 1000$. Spatial resolution on the order of a few micrometers has been reported by Meinhart et al. (1999), who have developed a micrometer resolution PIV system using an oil immersion microscopic lens. What makes PIV most interesting is the ability of the technique to measure hundreds or thousands of flow vectors simultaneously.

*4-9.1.2 Temporal Resolution.* Due to the high-gain amplifiers incorporated into the Wheatstone bridge, CTA systems offer a very high frequency response, reaching into the range of hundreds of kilohertz. This makes CTA an ideal instrument for the measurement of spectral content in most flows. A CTA sensor provides an analog signal, which is sampled using analog-to-digital converters at the appropriate rate obeying the Nyquist sampling criterion.

Commercial LDA signal processors can deal with data rates in the range of hundreds of kilohertz, although in practice, due to measurement volume size and seeding concentration requirements, validated data rates are typically in the 10 kHz or kHz range. This update rate of velocity information is sufficient to recover the frequency content of many flows.

The PIV sensor, however, is quite limited temporally, due to the framing rate of the cameras and pulsing frequency of the light sources used. Most common cross-correlation video cameras in use today operate at 30 Hz. These are used with dual-cavity Nd:Yag lasers, with each laser cavity pulsing at 15 Hz. Hence, these systems sample the images at 30 Hz and the velocity field at 15 Hz. High-framing-rate CCD (charge-coupled device) cameras are available that have framing rates in the 10 kHz range, albeit with lower pixel resolution. Copper vapor lasers offer pulsing rates up to the range 50 kHz, with energy per pulse around a fraction of 1 mJ. Hence, in principle, a high-framing-rate PIV system is possible with this camera and laser combination. But in practice, due to the low laser energy and limited spatial resolution of the camera, such a system is suitable for a limited range of applications. Recently, however, CMOS (complementary metal-oxide semiconductor)-based digital cameras have been commercialized with framing rates of 1 kHz and a pixel resolution of 1 K × 1 K. Combined with Nd:Yag lasers capable of pulsing at several kHz with energies of 10 to 20 mJ/pulse, the latter system brings us one step closer to measuring complex, 3D turbulent flow fields globally with high spatial and temporal resolution. The vast amount of data acquired using a high-framing-rate PIV system, however, still limits its common use due to the computational resources needed in processing the image information in nearly real time, as is the case with current low-framing-rate commercial PIV systems.

*4-9.1.3 Calibration.* The CTA voltage output, E, has a nonlinear relation with the input cooling velocity impinging on the sensor, U. Even though analytical treatment of the heat balance on the wire sensor shows that the transfer function has a power law relationship of the form $E^2 = A + BU^n$ (King's law), with constants A, B, and n, it is most often modeled using a fourth-order polynomial relation. Moreover, any flow variable that affects the heat transfer from the heated CTA sensor, such as the fluid density and temperature, affects the sensor response. Hence, CTA sensors need to be calibrated for their velocity response before use. In many cases, they also need to be calibrated for angular response. A common case such as gradually increasing ambient temperature can be dealt with either by performing a range of velocity calibrations for the range of ambient temperatures, or by performing analytical temperature corrections for the measured voltages.

The LDA measurement principle is given by the relation $V_x = d_f \cdot f_d$, where $V_x$ is the component of velocity in the plane of the laser beams and perpendicular to their bisector, $d_f$ is the distance between fringes, and $f_d$ is the Doppler frequency. The fringe spacing is a function of the distance between the two beams on the front lens and the focal length of the lens, given by the relation $d_f = \lambda/[2\sin(\theta/2)]$, where $\lambda$ is the laser wavelength and $\theta$ is the beam crossing angle. Since $d_f$ is a constant for a given optical system, there is a linear relation between the Doppler frequency and velocity. The calibration factor (i.e., the fringe spacing) is constant, calculable from the optical parameters, and mostly unaffected by other changing variables in the experiment. Hence, the LDA requires no physical calibration prior to use.

The PIV measurement is based on the simple relation $V = d/\Delta t$. Seeding particle velocities, which approximate the flow field velocity, are given by the particle displacement, d, obtained from particle images in at least two consecutive times, divided by the time interval between the images, $\Delta t$. Hence, the PIV technique also has a linear calibration response between the primary measured quantity (i.e., particle displacement) and particle velocity. Since the displacement is calculated from images, commonly using correlation techniques, PIV calibration involves measuring the magnification factor for the images. In the case of 3D stereoscopic PIV, calibration includes documenting the perspective distortion of target images obtained in different vertical locations by the two cameras situated off normal to the target.

**4-9.1.4 Summary of Transducer Comparison.** In summary, the point measurement techniques of CTA and LDA can offer good spatial and temporal response. This makes them ideal for measurements of both time-independent flow statistics, such as moments of velocity (mean, RMS, etc.) and time-dependent flow statistics such as flow spectra and correlation functions at a point. Although rakes of these sensors can be built, multipoint measurements are limited due primarily to cost.

The primary strength of the global PIV technique is its ability to measure flow velocity at many locations simultaneously, making it a unique diagnostic tool to measure 3D flow structures and transient phenomenon. However, since the temporal sampling rate is typically 15 Hz with today's commonly used 30 Hz cross-correlation cameras, the PIV technique is normally used to measure instantaneous velocity fields from which time-independent statistical information can be derived. Cost and processing speed are the main limiting factors that influence the temporal sampling rate of PIV, but such limitations are quickly disappearing.

## 4-9.2 Nonintrusive Measurement Techniques

Most emphasis in recent times has been in the development of nonintrusive flow measurement techniques for measuring vector as well as scalar quantities in the flow. These techniques have been mostly optically based, but when fluid opaqueness prohibits access, other techniques are available. A quick overview of several of these nonintrusive measurement techniques is given for completeness in the next few sections. More extensive discussion on these techniques can be found in the references cited.

### 4-9.2.1 Particle Tracking Velocimetry and Laser Speckle Velocimetry.

Just like PIV, PTV and LSV measure instantaneous flow fields by recording images of suspended seeding particles in flows at successive instants in time. An important difference among the three techniques comes from the typical seeding densities that can be dealt with by each technique. PTV is appropriate with low seeding density experiments, PIV with medium seeding density, and LSV with high seeding density. The issue of flow seeding is discussed later in the chapter.

Historically, LSV and PIV techniques have evolved separately from the PTV technique. In LSV and PIV, fluid velocity information at an interrogation region is obtained from many tracer particles, and it is obtained as the most probable statistical value. The results are obtained and presented in a regularly spaced grid. In PIV, a typical interrogation region may contain images of 10 to 20 particles. In LSV, larger numbers of particles in the interrogation region scatter light, which interferes to form speckles. Correlation of either particle images or speckles can be done using identical techniques and result in the local displacement of the fluid. Hence, LSV and PIV are essentially the same technique, used with different seeding density of particles. In the rest of the chapter the acronym PIV is used to refer to either technique.

In PTV, the acquired data provide a time sequence of individual tracer particles in the flow. To be able to track individual particles from frame to frame, the seeding density needs to be small. Unlike PIV, the PTV results in sparse velocity information located in random locations. Guezennec et al. (1994) have developed an automated three dimensional particle tracking velocimetry system that provides time-resolved measurements in a volume.

### 4-9.2.2 Image Correlation Velocimetry.

Tokumaru and Dimotakis (1995) introduced image correlation velocimetry (ICV) for the purpose of measuring imaged fluid motions without the requirement for discrete particles in the flow. Schlieren-based image correlation velocimetry was recently implemented by Kegrise and Settles (2000) to measure the mean velocity field of an axisymmetric turbulent free-convection jet. Papadopoulos (2000) demonstrated a shadow image velocimetry (SIV) technique which combined shadowgraphy with PIV to determine the temperature field of a flickering diffusion flame. Image correlation was also used by Bivolaru et al. (1999) to improve on the quantitative evaluation of Mie and Rayleigh scattering signal obtained from a supersonic jet using a Fabry–Perot interferometer. Although such developments are novel, we are still far from being able to fully characterize a flow by complete simultaneous measurements of density, temperature, pressure, and flow velocity.

## 4-10 LASER DOPPLER ANEMOMETRY

Laser Doppler anemometry is a nonintrusive technique used to measure the velocity of particles suspended in a flow. If these particles are small, in the order of micrometers, they can be assumed to be good flow tracers following the flow with their velocity corresponding to the fluid velocity. The LDA technique has some important characteristics that make it an ideal tool for dynamic flow measurements and turbulence characterization.

### 4-10.1 Characteristics of LDA

Laser anemometers offer unique advantages in comparison with other fluid flow instrumentation:

- *Noncontact optical measurement.* Laser anemometers probe the flow with focused laser beams and can sense the velocity without disturbing the flow in the measuring volume. The only necessary conditions are a transparent medium with a suitable concentration of tracer particles (or seeding) and optical access to the flow through windows or via a submerged optical probe. In the latter case the submerged probe will to some extent, of course, disturb the flow, but since the measurement takes place some distance away from the probe itself, this disturbance can normally be ignored.
- *No calibration—no drift.* The laser anemometer has a unique intrinsic response to fluid velocity—absolute linearity. The measurement is based on the stability and linearity of optical electromagnetic waves, which for most practical purposes can be considered unaffected by other physical parameters such as temperature and pressure.
- *Well-defined directional response.* The quantity measured by the laser Doppler method is the projection of the velocity vector on the measuring direction defined by the optical system (a true cosine response). The angular response is thus defined unambiguously.
- *High spatial and temporal resolution.* The optics of the laser anemometer are able to define a very small measuring volume and thus provides good spatial resolution and yields a local measurement of Eulerian velocity. The small measuring volume, in combination with fast signal processing electronics, also permits high-bandwidth time-resolved measurements of fluctuating velocities, providing excellent temporal resolution. Usually, the temporal resolution is limited by the concentration of seeding rather than by the measuring equipment itself.
- *Multicomponent bidirectional measurements.* Combinations of laser anemometer systems with component separation based on color, polarization, or frequency shift allow one-, two-, or three-component LDA systems to be put together based on common optical modules. Acoustooptical frequency shift allows measurement of reversing flow velocities.

### 4-10.2 Principles of LDA

*4-10.2.1 Laser Beam.* The special properties of the gas laser that make it so well suited for the measurement of many mechanical properties are spatial and temporal coherence. At all cross-sections along the laser beam, the intensity has

LASER DOPPLER ANEMOMETRY    209

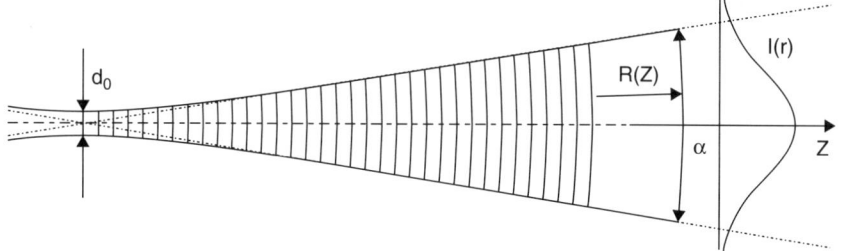

**Figure 4-32** Laser beam with Gaussian intensity distribution.*

a Gaussian distribution and the width of the beam is usually defined by the edge intensity being $1/e^2 = 13\%$ of the core intensity. At one point the cross-section attains its smallest value, and the laser beam is uniquely described by the size and position of this *beam waist*.

With a known wavelength $\lambda$ of the laser light, the laser beam is described uniquely by the size $d_0$ and position of the beam waist, as shown in Figure 4-32. With z describing the distance from the beam waist, the following formulas apply:

$$\text{beam divergence:} \quad \alpha = \frac{4\lambda}{\pi d_0} \tag{4-13}$$

$$\text{beam diameter:} \quad d(z) = d_0 \sqrt{1 + \left(\frac{4\lambda z}{\pi d_0^2}\right)^2} \rightarrow \alpha z \;\; \text{for } z \to \infty \tag{4-14}$$

$$\text{wavefront radius:} \quad R(z) = z\left[1 + \left(\frac{\pi d_0^2}{4\lambda z}\right)^2\right] \begin{cases} \to \infty & \text{for } z \to 0 \\ \to z & \text{for } z \to \infty \end{cases} \tag{4-15}$$

The beam divergence $\alpha$ is much smaller than indicated in Figure 4-32, and visually the laser beam appears to be straight and of constant thickness. It is important, however, to understand that this is not the case since measurements should take place in the beam waist to get optimal performance from any LDA equipment. This is due to the wavefronts being straight in the beam waist and curved elsewhere. According to the previous equations, the wavefront radius approaches infinity for z approaching zero, meaning that the wavefronts are approximately straight in the immediate vicinity of the beam waist, thus letting us apply the theory of plane waves and greatly simplify calculations.

### 4-10.2.2 Doppler Effect.
Laser Doppler anemometry utilizes the Doppler effect to measure instantaneous particle velocities. When particles suspended in a flow are illuminated with a laser beam, the frequency of the light scattered (and/or refracted) from the particles is different from that of the incident beam.

---
*Figure schematics and photographs in Part B of this chapter are courtesy of Dantec Dynamics A/S, unless otherwise stated.

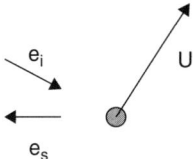

**Figure 4-33** Light scattering from a moving seeding particle.

This difference in frequency, called the *Doppler shift*, is linearly proportional to the particle velocity.

The principle is illustrated in Figure 4-33, where the vector **U** represents the particle velocity, and the unit vectors $\mathbf{e_i}$ and $\mathbf{e_s}$ describe the direction of incoming and scattered light, respectively. According to the Lorenz–Mie scattering theory, the light is scattered in all directions at once, but we consider only the light reflected in the direction of the LDA receiver. The incoming light has the velocity c and the frequency $f_i$, but due to the particle movement, the seeding particle "sees" a different frequency, $f_p$, which is scattered toward the receiver. From the receiver's point of view, the seeding particle acts as a moving transmitter, and the movement introduces additional Doppler shift in the frequency of the light reaching the receiver. Using Doppler theory, the frequency of the light reaching the receiver can be calculated as

$$f_s = f_i \frac{1 - \mathbf{e_i} \cdot (\mathbf{U}/c)}{1 - \mathbf{e_s} \cdot (\mathbf{U}/c)} \tag{4-16}$$

Even for supersonic flows the seeding particle velocity $|\mathbf{U}|$ is much lower than the speed of light, meaning that $|\mathbf{U}/c| \ll 1$. Taking advantage of this, eq. (4-16) can be linearized to

$$f_s \cong f_i \left[ 1 + \frac{\mathbf{U}}{c} \cdot (\mathbf{e_s} - \mathbf{e_i}) \right] = f_i + \frac{f_i}{c} \mathbf{U} \cdot (\mathbf{e_s} - \mathbf{e_i}) = f_i + \Delta f \tag{4-17}$$

With the particle velocity **U** being the only unknown parameter, then in principle the particle velocity can be determined from measurements of the Doppler shift $\Delta f$.

In practice, this frequency change can only be measured directly for very high particle velocities (using a Fabry–Perot interferometer). This is why in the commonly employed fringe mode, the LDA is implemented by splitting a laser beam to have two beams intersect at a common point so that light scattered from two intersecting laser beams is mixed, as illustrated in Figure 4-34. In this way both incoming laser beams are scattered toward the receiver, but with slightly different frequencies due to the different angles of the two laser beams. When two wavetrains of slightly different frequency are superimposed, we get the well-known phenomenon of a beat frequency due to the two waves intermittently interfering with each other constructively and destructively. The beat frequency

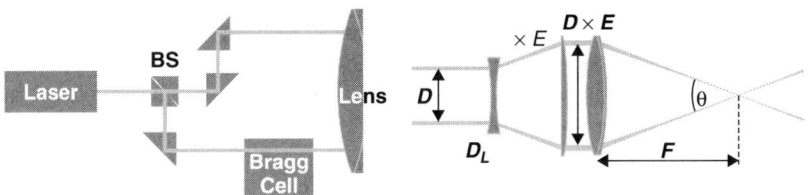

**Figure 4-34** LDA setup: left schematic shows beam splitter (BS) arrangement for creating two separate beams; right schematic shows the use of a beam expander to increase beam separation prior to focusing at a common point.

corresponds to the difference between the two wave frequencies, and since the two incoming waves originate from the same laser, they also have the same frequency, $f_1 = f_2 = f_I$, where the subscript I refers to the incident light:

$$\begin{aligned} f_D &= f_{s,2} - f_{s,1} \\ &= f_2 \left[ 1 + \frac{U}{c} \cdot (e_s - e_2) \right] - f_1 \left[ 1 + \frac{U}{c} \cdot (e_s - e_1) \right] \\ &= f_I \left[ \frac{U}{c} \cdot (e_1 - e_2) \right] \\ &= \frac{f_I}{c} (|e_1 - e_2| \cdot |U| \cdot \cos \varphi) \\ &= \frac{1}{\lambda} \cdot 2 \sin(\theta/2) \cdot u_x = \frac{2 \sin(\theta/2)}{\lambda} u_x \end{aligned} \quad (4\text{-}18)$$

where $\theta$ is the angle between the incoming laser beams and $\varphi$ is the angle between the velocity vector **U** and the direction of measurement. Note that the unit vector $e_s$ has dropped out of the calculation, meaning that the position of the receiver has no direct influence on the frequency measured. (According to the Lorenz–Mie light scattering theory, the position of the receiver will, however, have considerable influence on signal strength.) The beat frequency, also called the Doppler frequency $f_D$, is much lower than the frequency of the light itself, and it can be measured as fluctuations in the intensity of the light reflected from the seeding particle. As shown in eq. (4-18), the x-component of the particle velocity is directly proportional to the Doppler frequency and thus can be calculated directly from $f_D$:

$$u_x = \frac{\lambda}{2 \sin(\theta/2)} f_D \quad (4\text{-}19)$$

Further discussion on LDA theory and different modes of operation may be found in the classic texts of Durst et al. (1976) and Watrasiewics and Rudd (1976).

## 4-10.3 LDA Implementation

**4-10.3.1 Fringe Model.** Although the description of LDA above is accurate, it may be intuitively difficult to quantify. To handle this, the fringe model is commonly used in LDA as a reasonably simple visualization producing the correct results. When two coherent laser beams intersect, they will interfere in the volume of the intersection. If the beams intersect in their respective beam waists, the wave fronts are approximately plane, and consequently, the interference will produce parallel planes of light and darkness as shown in Figure 4-35. The interference planes are known as *fringes*, and the distance, $\delta_f$, between them depends on the wavelength and the angle between the incident beams:

$$\delta_f = \frac{\lambda}{2\sin(\theta/2)} \qquad (4\text{-}20)$$

The fringes are oriented normal to the x-axis, so the intensity of light reflected from a particle moving through the measuring volume will vary with a frequency proportional to the x-component, $u_x$, of the particle velocity:

$$f_D = \frac{u_x}{\delta_f} = \frac{2\sin(\theta/2)}{\lambda} u_x \qquad (4\text{-}21)$$

If the two laser beams do not intersect at the beam waists but elsewhere in the beams, the wavefronts will be curved rather than plane, and as a result the fringe spacing will not be constant but depend on the position within the intersection volume. As a consequence, the measured Doppler frequency will also depend on the particle position, and as such it will no longer be directly proportional to the particle velocity, hence resulting in a velocity bias.

**4-10.3.2 Measuring Volume.** Measurements take place in the intersection between the two incident laser beams, and the measuring volume is defined as the volume within which the modulation depth is higher than $e^{-2}$ times the peak core value. Due to the Gaussian intensity distribution in the beams, the measuring

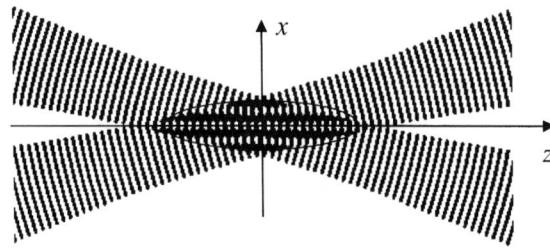

**Figure 4-35** Fringes at the point of intersection of two coherent beams.

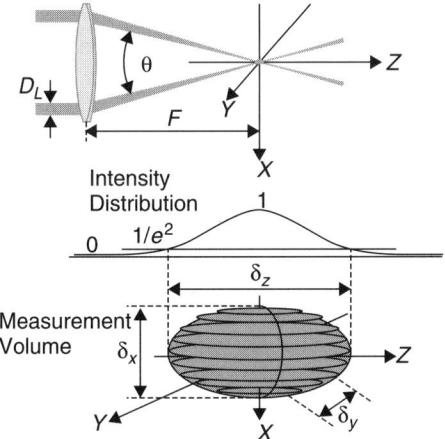

**Figure 4-36** LDA measurement volume.

volume is an ellipsoid, as indicated in Figure 4-36. Its dimensions are as follows:

$$\text{length:} \quad \delta_z = \frac{4F\lambda}{\pi ED_L \sin(\theta/2)} \quad (4\text{-}22)$$

$$\text{width:} \quad \delta_y = \frac{4F\lambda}{\pi ED_L} \quad (4\text{-}23)$$

$$\text{height:} \quad \delta_x = \frac{4F\lambda}{\pi ED_L \cos(\theta/2)} \quad (4\text{-}24)$$

where F is the focal length of the lens, E the beam expansion (see Figure 4-34), and $D_L$ the initial beam thickness ($e^{-2}$).

As important are the fringe separation and number of fringes in the measurement volume. These are given by:

$$\text{fringe separation:} \quad \delta_f = \frac{\lambda}{2\sin(\theta/2)} \quad (4\text{-}25)$$

$$\text{Number of fringes:} \quad N_f = \frac{8F\tan(\theta/2)}{\pi ED_L} \quad (4\text{-}26)$$

This number of fringes applies for a seeding particle moving straight through the center of the measuring volume along the x-axis. If the particle passes through the outskirts of the measuring volume, it will pass fewer fringes, and consequently there will be fewer periods in the recorded signal from which to estimate the Doppler frequency. To get good results from the LDA equipment, one should ensure a sufficiently high number of fringes in the measuring volume. Typical LDA setups produce between 10 and 100 fringes, but in some cases reasonable results may be obtained with less, depending on the electronics or technique

**214** EXPERIMENTAL METHODS

used to determine the frequency. The key issue here is the number of periods produced in the oscillating intensity of the reflected light, and while modern processors using FFT technology can estimate particle velocity from as little as one period, the accuracy will improve with more periods.

***4-10.3.3 Backscatter versus Forward Scatter.*** A typical LDA setup in the *forward scatter mode* is shown in Figure 4-37. The figure also shows the important components of a modern commercial LDA system. The majority of light from commonly used seeding particles is scattered in a direction away from the transmitting laser, and in the early days of LDA, forward scattering was commonly used, meaning that the receiving optics was positioned opposite the transmitting aperture [consult the text by Hulst (1981) for a discussion of light scattering].

A much smaller amount of light is scattered back towards the transmitter, but advances in technology have made it possible to make reliable measurements even on these faint signals, and today backward scatter is the usual choice in LDA. This *backscatter LDA* allows for the integration of transmitting and receiving optics in a common housing (as seen in Figure 4-38), saving the user a lot of tedious and time-consuming work aligning separate units.

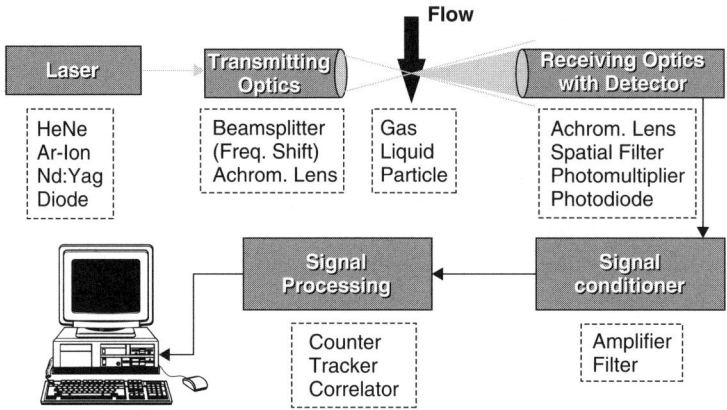

**Figure 4-37** Schematic of components for a typical LDA system.

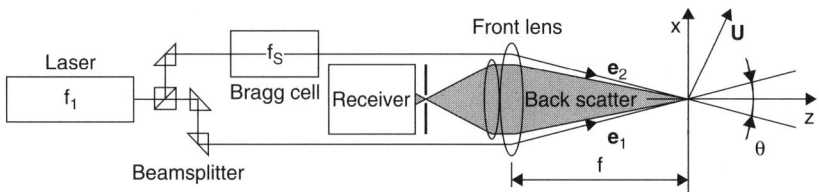

**Figure 4-38** Principles of a backscatter LDA system.

## LASER DOPPLER ANEMOMETRY

Forward scattering LDA is not completely obsolete, however, since in some cases its improved signal-to-noise ratio makes it the only way to obtain measurements at all. Experiments requiring forward scatter might include:

- High-speed flows, requiring very small seeding particles, which stay in the measuring volume for a very short time, and thus receive and scatter a very limited number of photons.
- Transient phenomena which require high data rates in order to collect a reasonable amount of data over a very short period of time.
- Very low turbulence intensities, where the turbulent fluctuations might drown in noise if measured with backscatter LDA.

Forward and backscattering is identified by the position of the receiving aperture relative to the transmitting optics. Another option is *off-axis scattering*, where the receiver is looking at the measuring volume at an angle. Like forward scattering, this approach requires a separate receiver and thus involves careful alignment of the different units, but it helps to mitigate an intrinsic problem present in both forward and backscatter LDA. As indicated in Figure 4-36, the measuring volume is an ellipsoid, and usually the major axis $\delta_z$ is much bigger than the two minor axes $\delta_x$ and $\delta_y$, rendering the measuring volume more or less cigar-shaped. This makes forward and backscattering LDA sensitive to velocity gradients within the measuring volume, and in many cases also disturbs measurements near surfaces due to reflection of the laser beams.

Figure 4-39 illustrates how off-axis scattering reduces the effective size of the measuring volume. Seeding particles passing through either end of the measuring volume will be ignored since they are out of focus, and as such contribute to background noise rather than to the actual signal. This reduces the sensitivity to velocity gradients within the measuring volume, and the off-axis position of the receiver automatically reduces problems with reflection. These properties make off-axis scattering LDA very efficient, for example, in boundary layer or near-surface measurements.

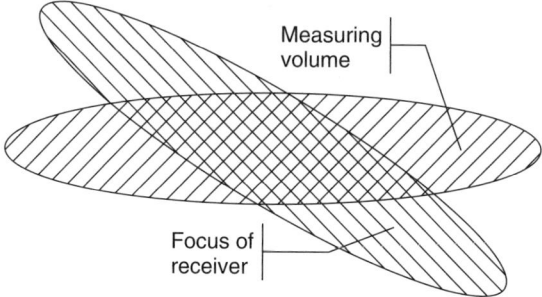

**Figure 4-39** Off-axis scattering.

**216**  EXPERIMENTAL METHODS

***4-10.3.4 Optics.*** In modern LDA equipment the light from the beamsplitter and the Bragg cell is sent through optical fibers, as is the light scattered back from seeding particles. This reduces the size and the weight of the probe itself, making the equipment flexible and easier to use in practical measurements. A photograph of a pair of commercially available LDA probes is shown in Figure 4-40. The laser, beamsplitter, Bragg cell, and photodetector (receiver) can be installed stationary and out of the way, while the LDA probe can be traversed between different measuring positions.

It is normally desired to make the measuring volume as small as possible. According to the formulas above, this means that the beam waist,

$$d_f = \frac{4F\lambda}{\pi E D_L} \qquad (4\text{-}27)$$

should be small. The laser wavelength $\lambda$ is a fixed parameter, and focal length F is normally limited by the geometry of the model being investigated. Some lasers allow for adjustment of the beam waist position, but the beam waist diameter $D_L$ is normally fixed. This leaves beam expansion as the only remaining way to reduce the size of the measuring volume. When no beam expander is installed, $E = 1$.

A beam expander is a combination of lenses in front of or replacing the front lens of a conventional LDA system. It converts the beams exiting the optical system to beams of greater width. At the same time, the spacing between the two laser beams is increased, since the beam expander also increases the aperture. Provided that the focal length F remains unchanged, the larger beam spacing will thus increase the angle $\theta$ between the two beams. According to the formulas in Section 4-10.3.2, this will further reduce the size of the measuring volume.

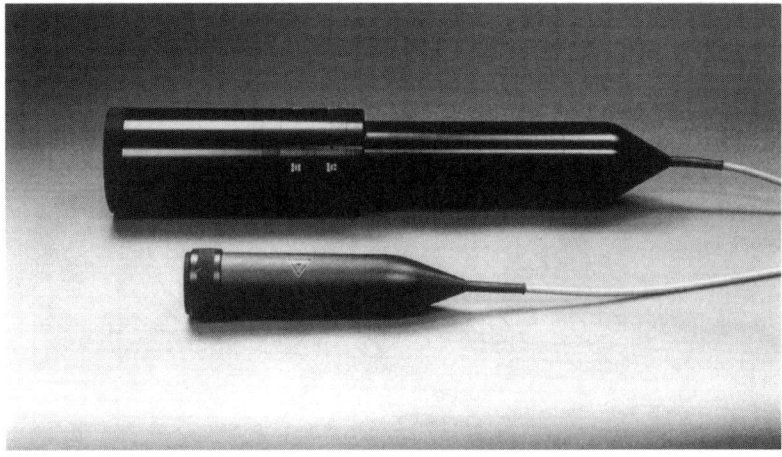

**Figure 4-40**  Modern commercial fiber optic-based LDA probes.

In agreement with the fundamental principles of wave theory, a larger aperture is able to focus a beam to a smaller spot size and hence generate greater light intensity from the scattering particles. At the same time the greater receiver aperture is able to pick up more of the reflected light. As a result, the benefits of the beam expander are threefold:

- Reduce the size of the measuring volume at a given measuring distance.
- Improve the signal-to-noise ratio at a given measuring distance, or
- Reach greater measuring distances without sacrificing the signal-to-noise ratio.

**4-10.3.5 Frequency Shift.** A drawback of the LDA technique described so far is that negative velocities $u_x < 0$ will produce negative frequencies $f_D < 0$. However, the receiver cannot distinguish between positive and negative frequencies, and as such, there will be a directional ambiguity in the measured velocities. To handle this problem, a Bragg cell is introduced in the path of one of the laser beams (as shown in Figure 4-34). The Bragg cell shown in Figure 4-41 is a block of glass. On one side, an electromechanical transducer driven by an oscillator produces an acoustic wave propagating through the block generating a periodic moving pattern of high and low density. The opposite side of the block is shaped to minimize reflection of the acoustic wave and is attached to a material absorbing the acoustic energy.

The incident light beam hits a series of traveling wavefronts that act as a thick diffraction grating. Interference of the light scattered by each acoustic wavefront causes intensity maxima to be emitted in a series of directions. By adjusting the acoustic signal intensity and the tilt angle, $\theta_B$, of the Bragg cell, the intensity balance between the direct beam and the first order of diffraction can be adjusted. In modern LDA equipment this is exploited, using the Bragg cell itself as the

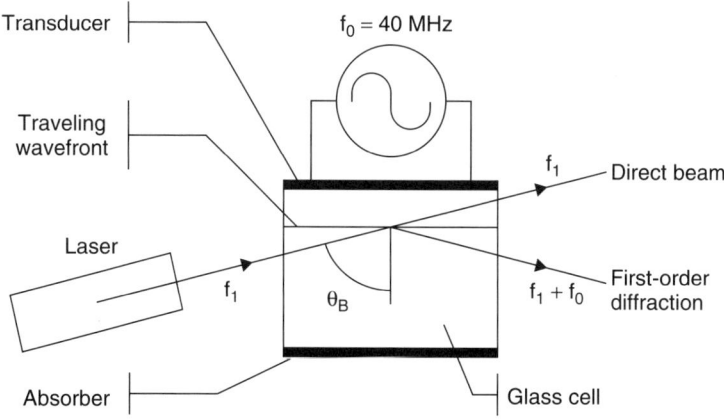

**Figure 4-41** Principles of operation of a Bragg cell.

beamsplitter. Not only does this eliminate the need for a separate beamsplitter, but it also improves the overall efficiency of the light-transmitting optics, since more than 90% of the lasing energy can be made to reach the measuring volume, effectively increasing the signal strength.

The Bragg cell adds a fixed frequency shift $f_0$ to the diffracted beam, which then results in a measured frequency off a moving particle of

$$f_D \simeq f_0 + \frac{2\sin(\theta/2)}{\lambda} u_x \qquad (4\text{-}28)$$

and as long as the particle velocity does not introduce a negative frequency shift numerically larger than $f_0$, the Bragg cell with thus ensure a measurable positive Doppler frequency $f_D$. In other words, the frequency shift $f_0$ allows measurement of velocities down to

$$u_x > -\frac{\lambda f_0}{2\sin(\theta/2)} \qquad (4\text{-}29)$$

without directional ambiguity (Figure 4-42). Typical values might be $\lambda = 500$ nm, $f_0 = 40$ MHz, $\theta = 20°$, allowing of negative velocity components down to $u_x > -57.6$ m/s. Upward, the maximum measurable velocity is limited only by the response time of the photomultiplier and the signal-conditioning electronics. In modern commercial LDA equipment, such a maximum is well into the supersonic velocity regime.

**4-10.3.6 Signal Processing.** The primary result of a laser anemometer measurement is a current pulse from the photodetector. This current contains the frequency information relating to the velocity to be measured. The photocurrent also contains noise, with sources for this noise being:

- Photodetection shot noise
- Secondary electronic noise
- Thermal noise from preamplifier circuit
- Higher-order laser modes (optical noise)
- Light scattered from outside the measurement volume, dirt, scratched windows, ambient light, multiple particles, etc.
- Unwanted reflections (windows, lenses, mirrors, etc.)

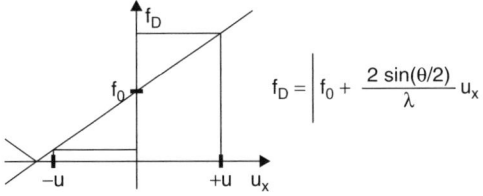

**Figure 4-42** Resolving directional ambiguity using frequency shift.

The primary source of noise is the photodetection shot noise, which is a fundamental property of the detection process. The interaction between the optical field and the photo-sensitive material is a quantum process, which unavoidably impresses a certain amount of fluctuation on the mean photocurrent. In addition, there is mean photocurrent and shot noise from undesired light reaching the photodetector. Much of the design effort for the optical system is aimed at reducing the amount of unwanted reflected laser light or ambient light reaching the detector.

A laser anemometer is most advantageously operated under such circumstances that the shot noise in the signal is the predominant noise source. This shot noise-limited performance can be obtained by proper selection of laser power, seeding particle size, and optical system parameters. In addition, noise should be minimized by selecting only the minimum bandwidth needed for measuring the desired velocity range by setting low- and high-pass filters in the signal processor input. Very important for the quality of the signal, and the performance of the signal processor, is the number of seeding particles present simultaneously in the measuring volume. If on average much less than one particle is present in the volume, we speak of a burst-type Doppler signal. Typical Doppler burst signals are shown in Figure 4-43. Figure 4-43a shows the filtered signal which is actually input to the signal processor. The DC part, which was removed by the high-pass filter, is known as the *Doppler pedestal*, and it is often used as a trigger signal, which starts sampling of an assumed burst signal. The envelope of the Doppler-modulated current reflects the Gaussian intensity distribution in the measuring volume.

If more particles are present in the measuring volume simultaneously, we speak of a multiparticle signal. The detector current is the sum of the current bursts from each individual particle within the illuminated region. Since the particles are located randomly in space, the individual current contributions are added with random phases, and the resulting Doppler signal envelope and phase will fluctuate. Most LDA processors are designed for single-particle bursts, and with a multiparticle signal, they will normally estimate the velocity as a weighted average of the particles within the measuring volume. One should be aware, however, that the random-phase fluctuations of the multiparticle LDA signal

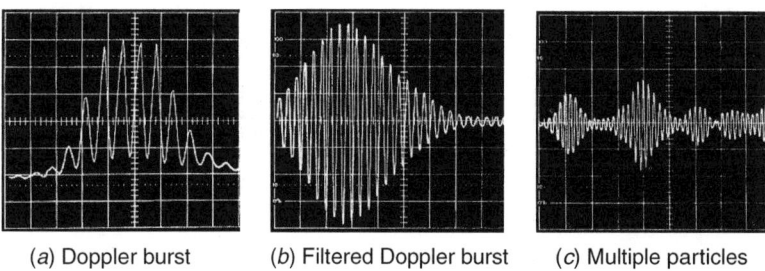

(a) Doppler burst     (b) Filtered Doppler burst     (c) Multiple particles

**Figure 4-43** Typical single- and multiple-particle Doppler bursts.

## 220 EXPERIMENTAL METHODS

add a phase noise to the detected Doppler frequency which is very difficult to remove.

To better estimate the Doppler frequency of noisy signals, frequency-domain processing techniques are used. With the advent of fast digital electronics, the fast Fourier transform of digitized Doppler signals can now be performed at a very high rate (hundreds of kHz). The power spectrum S of a discretized Doppler signal x is given by

$$S_k = \left| \sum_{n=0}^{N-1} x_n \exp\left(\frac{-j2\pi kn}{2N}\right) \right|^2 \qquad (4\text{-}30)$$

where N is the number of discrete samples, and $K = -N, -N+1, \ldots, N-1$. The Doppler frequency is given by the peak of the spectrum.

**4-10.3.7 Data Analysis.** In LDA there are two major problems faced when making a statistical analysis of the measurement data: velocity bias and the random arrival of seeding particles to the measuring volume. Although velocity bias is the predominant problem for simple statistics, such as mean and RMS values, the random sampling is the main problem for statistical quantities that depend on the timing of events, such as spectrum and correlation functions (see Tropea, 1995).

Figure 4-44 illustrates the calculation of moments, correlation, and spectra on the basis of measurements received from the processor. The velocity data coming from the processor consist of N validated bursts, collected during the time T, in a flow with the integral time scale $\tau_I$. For each burst the arrival time $a_i$ and the transit time $t_i$ of the seeding particle is recorded along with the non-Cartesian velocity components ($u_i$, $v_i$, $w_i$). The different topics involved in the analysis are described in more detail in the open literature and will be touched upon briefly in the following section.

### 4-10.4 Making Measurements

**4-10.4.1 Dealing with Multiple Probes (3D Setup).** The non-Cartesian velocity components ($u_1$, $u_2$, $u_3$) are transformed to Cartesian coordinates (u,v,w) using the transformation matrix **C**:

$$\begin{Bmatrix} u \\ v \\ w \end{Bmatrix} = \begin{Bmatrix} C_{11} & C_{12} & C_{13} \\ C_{21} & C_{22} & C_{23} \\ C_{31} & C_{32} & C_{33} \end{Bmatrix} \cdot \begin{Bmatrix} u_1 \\ u_2 \\ u_3 \end{Bmatrix} \qquad (4\text{-}31)$$

A typical 3D LDA setup requiring coordinate transformation is depicted in Figure 4-45, where 3D velocity measurements are performed with a 2D probe

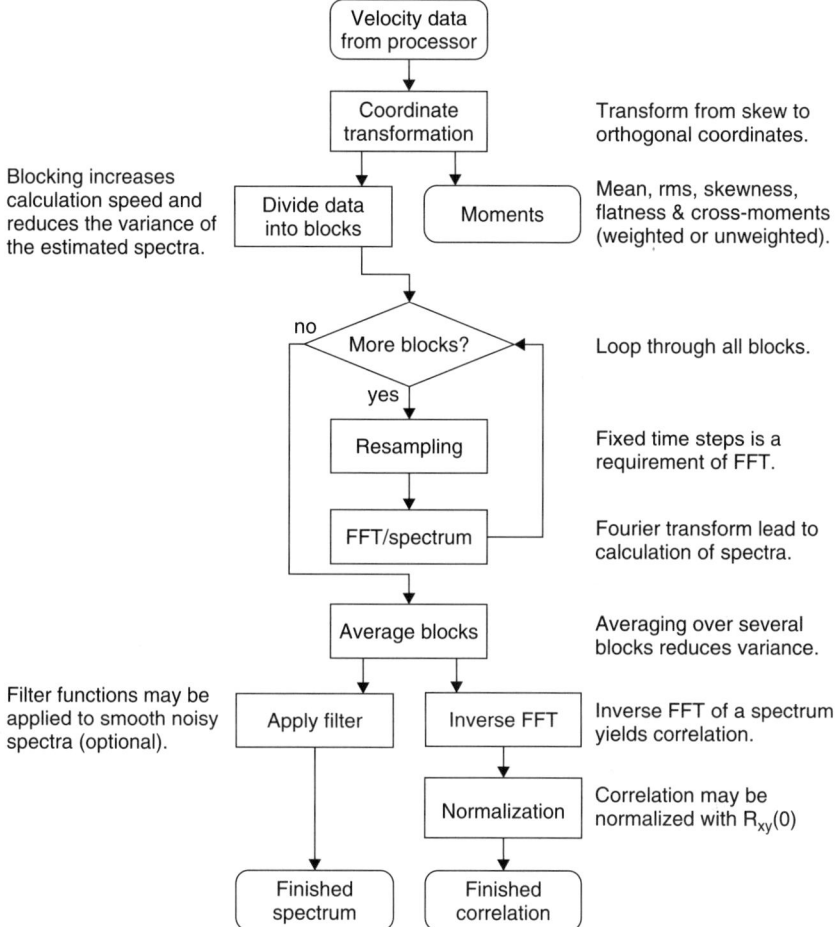

**Figure 4-44** Block diagram of data analysis and calculation from velocity estimate received from LDA processor.

positioned at off-axis angle $\alpha_1$ and a 1D probe positioned at off-axis angle $\alpha_2$. The transformation for this case is

$$\left\{\begin{array}{c} u \\ v \\ w \end{array}\right\} = \left\{\begin{array}{ccc} 1 & 0 & 0 \\ 0 & -\dfrac{\sin\alpha_2}{\sin(\alpha_1 - \alpha_2)} & \dfrac{\sin\alpha_1}{\sin(\alpha_1 - \alpha_2)} \\ 0 & \dfrac{\cos\alpha_2}{\sin(\alpha_1 - \alpha_2)} & \dfrac{\cos\alpha_1}{\sin(\alpha_1 - \alpha_2)} \end{array}\right\} \cdot \left\{\begin{array}{c} u_1 \\ u_2 \\ u_3 \end{array}\right\} \qquad (4\text{-}32)$$

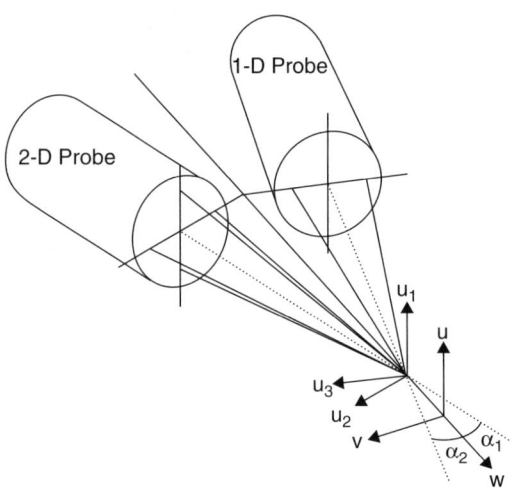

**Figure 4-45** Typical configuration of 3D LDA system.

***4-10.4.2 Calculating Moments.*** Moments are the simplest form of statistics that can be calculated for a set of data. The calculations are based on individual samples, and the possible relations between samples are ignored, as is the timing of events. This leads to moments sometimes being referred to as one-time statistics, since samples are treated one at a time.

Table 4-2 lists the formulas used to estimate the moments. The table operates with velocity components $x_i$ and $y_i$, but this is just examples, and could of course be any velocity component, Cartesian or not. It could even be samples of an external signal representing pressure, temperature, or something else. George (1978) gives a good account of the basic uncertainty principles governing the statistics of correlated time series with emphasis on the differences between equal-time and Poisson sampling (LDA measurements fall in the latter). Statistics treated by George include the mean, variance, autocorrelation, and power spectra. A recent publication by Benedict and Gould (1996) gives methods of determining uncertainties for higher-order moments.

***4-10.4.3 Velocity Bias and Weighting Factor.*** Even for incompressible flows where the seeding particles are statistically uniformly distributed, the sampling process is not independent of the process being sampled (i.e., the velocity field). Measurements have shown that the particle arrival rate and the flow field are strongly correlated (McLaughlin and Tiedermann, 1973; Erdmann and Gellert, 1976). During periods of higher velocity, a larger volume of fluid is swept through the measuring volume, and consequently, a greater number of velocity samples will be recorded. As a direct result, an attempt to evaluate the statistics of the flow field using arithmetic averaging will bias the results in favor of the higher velocities.

**Table 4-2** Definition of Statistical Measures Often Used for Turbulence Characterization

| Mean | $\bar{u} = \sum_{i=0}^{N-1} \eta_i u_i$ |
|---|---|
| Variance | $\sigma^2 = \sum_{i=0}^{N-1} \eta_i (u_i - \bar{u})^2$ |
| RMS | $\sigma = \sqrt{\sigma^2}$ |
| Turbulence | $Tu = \dfrac{\sigma}{\bar{u}} \cdot 100\%$ |
| Skewness | $S = \dfrac{1}{\sigma^3} \sum_{i=0}^{N-1} \eta_i (u_i - \bar{u})^3$ |
| Flatness | $F = \dfrac{1}{\sigma^4} \sum_{i=0}^{N-1} \eta_i (u_i - \bar{u})^4$ |
| Cross-moments | $\overline{uv} - \bar{u}\bar{v} = \sum_{i=0}^{N-1} \eta_i (u_i - \bar{u})(v_i - \bar{v})$ |

There are several ways to deal with this issue:

- *Ensure statistically independent samples.* The time between bursts must exceed the integral time scale of the flow field at least by a factor of 2. Then the weighting factor corresponds to the arithmetic mean, $\eta_i = 1/N$. Statistically independent samples can be accomplished by using very low concentration of seeding particles in the fluid.
- *Use dead-time mode.* The dead time is a specified period of time after each detected Doppler burst, during which further bursts will be ignored. Setting the dead time equal to two times the integral time scale will ensure statistically independent samples, while the integral time scale itself can be estimated from a previous series of velocity samples, recorded with the dead-time feature switched off.
- *Use bias correction.* If one plans to calculate correlations and spectra on the basis of measurements performed, the resolution achievable will be greatly reduced by the low data rates required to ensure statistically independent samples. To improve the resolution of the spectra, a higher data rate is needed, which, as explained above, will bias the estimated average velocity. To correct this velocity bias, a nonuniform weighting factor is introduced:

$$\eta_i = \frac{t_i}{\sum_{j=0}^{N-1} t_j} \quad (4\text{-}33)$$

The bias-free method of performing the statistical averages on individual realizations uses the transit time, $t_i$, weighting (see George, 1976). Additional information on the transit time weighting method can be found in George (1978), Buchhave et al. (1979), Buchhave (1979), and Benedict and Gould (1999). In the literature, transit time is sometimes referred to as *residence time*.

### 4-10.5 LDA Applications in Mixing

***4-10.5.1 Experimental Considerations.*** There are several important experimental issues specific to the application of LDA to mixing experiments, particularly in stirred tanks.

*Curvature.* As the laser beams travel through curved surfaces, they are deflected due to changes in the refractive index. This produces a lensing effect that moves the measuring volume away from its expected location. The problem can be addressed in two ways: by refractive index matching of the fluid with the vessel, or by performing beam tracing calculations to correct for the displacement (see Kresta and Wood, 1993).

*Use of Vessel Symmetry to Obtain Three Velocity Components.* The velocity component measured is perpendicular to the interference fringes shown in Figure 4-35. When the plane defined by the beams is vertical and the beam angle bisector is horizontal, the vertical velocity component (z-component) is measured. When the beams are in a horizontal plane, two velocity components can be measured, depending on the location of the beams in the tank. If the beam angle bisector intercepts the shaft, the angular velocity (θ-component) is measured. If a traverse at 90° to this is used, the radial velocity (r-component) is measured. Looking into the tank from the laser, moving forward and backward in front of the shaft axis (toward the six o'clock position from the top of the tank) will give the angular or tangential velocity. Moving sideways (toward the three o'clock or the nine o'clock position) will give the radial velocity. This assumes 90° symmetry of the tank, an assumption that has been validated for several tank geometries.

*Reflections from the Shaft and Impeller.* Signal quality close to the shaft or impeller can be improved by minimizing reflections. This can be done with black matte paint or by anodizing the parts after fabrication.

*Impeller Geometry.* The impeller geometry must be exactly scaled down from full scale dimensions. Thickness of blades, hub size, hub thickness, and placement of blades can all have significant effects on power draw, velocity profiles, and turbulence characteristics.

*Shaft Encoding.* LDV measurements in stirred vessels are typically combined with shaft encoding so that the velocity can be analyzed either as a long-time average at a fixed position in the tank, or as ensemble-averaged data relative to a specific angular position of the impeller. This is particularly important if the trailing vortices leaving the impeller blades are the feature of greatest interest. Many papers have been published that use this method (e.g., Yianneskis et al., 1987; Schaffer et al., 1997).

### 4-10.5.2 Uses of LDV Data

*Impeller Characterization.* The first stage of characterization of a new impeller will typically involve measurement of the mean velocity and RMS velocity profiles very close to the impeller. The position of the traverses relative to the impeller is crucial, as the velocity gradients are very steep in this region. The velocity drops quickly as one moves away from the impeller. These velocity traverses are used to define the flow number and the momentum of the fluid leaving the impeller and to characterize the turbulence intensity close to the impeller. The reader should note that streak photography is typically undertaken before LDV. This qualitative method will show the average user a great deal about the flow field without the full expense of LDV measurements. Quantitative full field measurements can be obtained using particle image velocimetry as an alternative to LDV.

*CFD Validation.* LDV measurements are widely used for validation of CFD codes. Velocity measurements can be compared directly with the results of simulations. There are some subtleties here as the simulations become more complex. When steady simulations are performed using velocity boundary conditions around the impeller, mean velocity measurements are used to validate circulation patterns and the RMS velocity components are combined for comparison with the simulated turbulent kinetic energy k. If trailing vortices dominate, as they do for the Rushton turbine, shaft-encoded LDV data are compared with the results of sliding mesh simulations, which are ensemble averaged. If macroinstabilities are of interest, large eddy simulations offer the next step of resolution in the time-varying flow. The definition of turbulence quantities for the second and third cases is still a topic of active research. When considering CFD validation, the reader will benefit from careful consideration of how best to characterize the flow: What are the key characteristics to be replicated, and how will they be validated?

*Turbulence Characterization.* LDV has been widely used to characterize the turbulence close to impellers. This work requires careful reflection and detailed experimental work. Approaches range from estimates (Zhou and Kresta, 1996) to full turbulent energy balances (Escudie, 2001) and direct measurements of the dissipation (Michelet, 1998). The best way to treat regular frequencies due to blade passages and macroinstabilities is still a topic of active investigation. Further discussion and references are given in Chapter 2 and by Kresta (1998).

*Application of Results.* It has often been noted that velocity fields, no matter how detailed, do not provide any direct information about process performance.

**226**   EXPERIMENTAL METHODS

This is true. The power of LDV and other velocity measurement methods lies in the interpretation of the results for the improvement of process performance. Velocity field measurements close to the base of vessels can be used to clarify mechanisms of solids suspension which depend on the shape of the bottom. Velocity measurements close to the top of the vessel can be used to probe dead zones and examine the effect of changing geometry on the size of these dead zones. These measurements are very targeted and can provide very powerful information to the user. This must be balanced with consideration of the costs of purchasing and supporting LDV equipment.

### 4-11  PHASE DOPPLER ANEMOMETRY

#### 4-11.1  Principles and Equations for PDA

In the LDA system discussed previously there was only one photodetector. If one considers the situation shown in Figure 4-46, two photodetectors spaced a certain distance apart will both receive the light scattered from the surface of a reflecting spherical particle. However, the difference in the optical path length for the reflections from the two incident beams changes with the position of the photodetector. This means that when the particle passes through the measuring volume, both photodetectors receive a Doppler burst of the same frequency, but the phases of the two bursts vary with the angular position of the detectors. This phenomenon was first utilized as an indication of the size of a particle by Durst and Zaré (1975).

Again it is convenient to introduce the fringe model as a first order of approximation. Figure 4-47 illustrates the intensity fluctuation in each of the photodetectors and the time lag, $\Delta t$, separating the wavefronts reaching the two

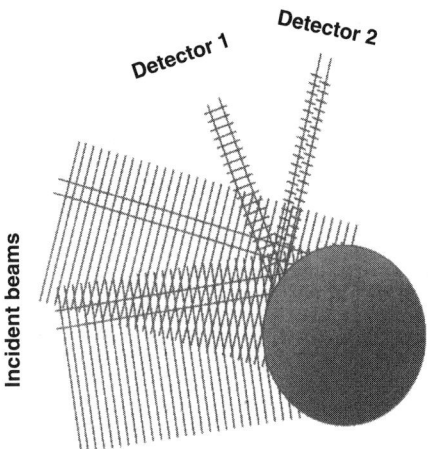

**Figure 4-46**  Interference patterns at two photodetectors will differ by a certain phase.

PHASE DOPPLER ANEMOMETRY    **227**

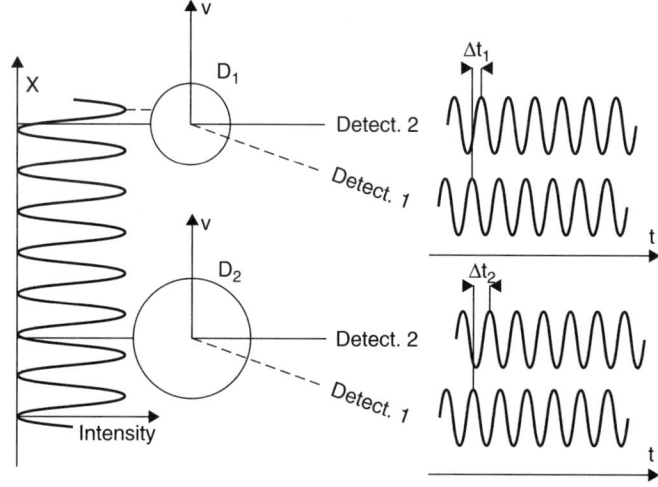

**Figure 4-47** Increasing phase difference with increasing particle diameter.

photodetectors. The corresponding phase difference for reflection would be

$$\Phi = \frac{2\pi d_p}{\lambda} \frac{\sin\theta \sin\psi}{\sqrt{2(1 - \cos\theta \cos\psi \cos\phi)}} \quad (4\text{-}34)$$

given the angular position of the detectors as illustrated in Figure 4-48. The property that is of foremost importance is that the phase difference between the two Doppler bursts depends on the size of the particle, provided that all other geometric parameters of the optics remain constant. Figure 4-47 shows

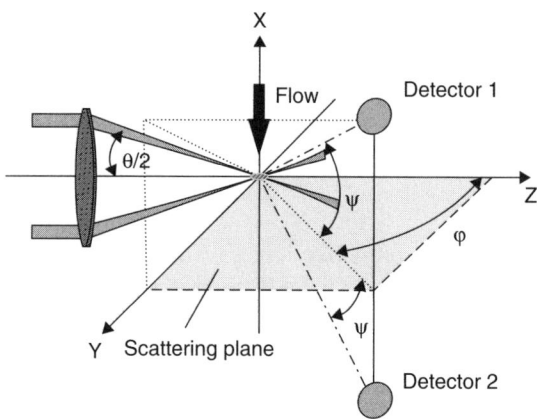

**Figure 4-48** Detector setup and general coordinate system used for PDA.

two particles of different size, illustrating how the phase difference between the Doppler bursts from the large particle exceeds that of the smaller particle.

Mathematically, the phase of a Doppler burst received at detector i can be expressed as

$$\Phi_i = \alpha \cdot \beta_i \tag{4-35}$$

where the size parameter is

$$\alpha = \pi \frac{n_1}{\lambda} d_P \tag{4-36}$$

and where $n_1$ is the refractive index of the scattering medium, $\lambda$ the laser wavelength in vacuum, and $d_P$ the particle diameter.

Thus, a linear relationship between particle size and phase exists. The geometrical factor, $\beta_i$, depends on the scattering mode and the three angles, $\theta$, $\varphi_i$, and $\psi_i$. The full intersection angle between the two incident beams, $\theta$, determines the fringe separation, while $\varphi_i$ and $\psi_i$ define the direction toward the (centroid of the) photodetector from the measuring volume. The angle of intersection between the two incident beams is determined by the beam separation and the focal length of the transmitting lens. The scattering angle, $\varphi_i$, is measured from the axis of the transmitting optics (the bisector of the two incident beams; the Z-axis), while the azimuth angle, $\psi_i$, gives the rotational position about the Z-axis.

The factor $\beta_i$ between particle diameter and phase shift also depends on the scattering mode. This is illustrated in Figure 4-49, in which ray tracing has been used to depict how incident light will scatter from a spherical particle. Three main contributions are included in the representation; reflection from the outer surface of the particle, refraction through the particle (first-order refraction), and refraction with one internal reflection (second-order refraction). The formulas expressing the geometrical factor are given below for reflection and (first-order) refraction.

*Reflection:*

$$\beta_i = \sqrt{2}\left(\sqrt{1 + \sin\frac{\theta}{2}\sin\varphi_i\sin\psi_i - \cos\frac{\theta}{2}\cos\varphi_i} \right.$$

$$\left. -\sqrt{1 - \sin\frac{\theta}{2}\sin\varphi_i\sin\psi_i - \cos\frac{\theta}{2}\cos\varphi_i}\right) \tag{4-37}$$

Note in eq. (4-37) that the refractive index of the particle, $n_2$, does not appear. Hence, in practice this means that the reflection mode is a useful scattering mode to employ in situations where the exact value of the refractive index is not known.

*First-order refraction:*

$$\beta_i = 2\left(\sqrt{1 + n_{rel}^2 - \sqrt{2}\,n_{rel}\sqrt{f_{i+}}} - \sqrt{1 + n_{rel}^2 - \sqrt{2}\,n_{rel}\sqrt{f_{i-}}}\right) \tag{4-38}$$

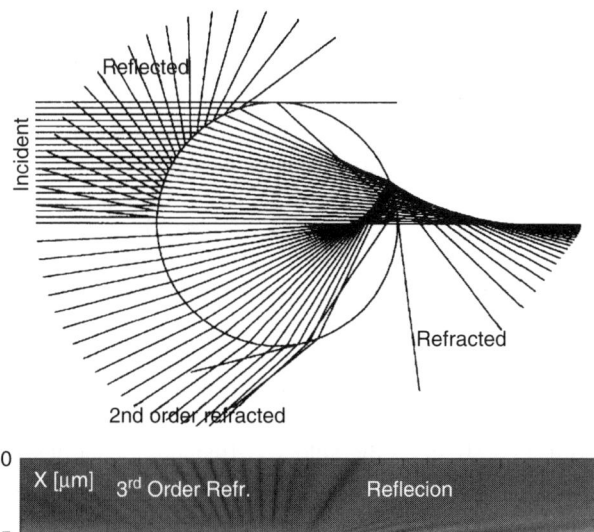

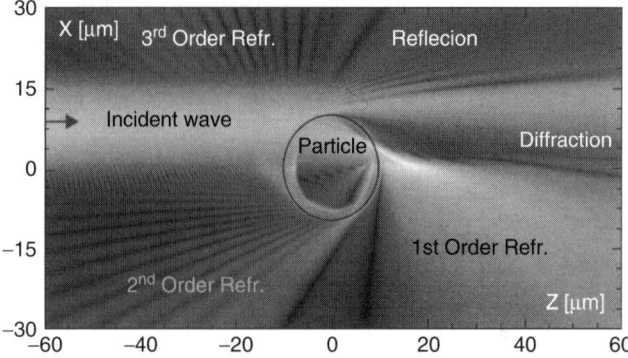

**Figure 4-49** Ray traces indicating the three significant modes of scattering—reflection and first- and second-order refraction—for a water droplet (top); light scattering of a Gaussian beam from a water droplet, simulated using the Fourier Lorenz–Mie theory (bottom; courtesy of C. Tropea and N. Damaschke, Technische Universität Darmstadt, Germany).

where

$$n_{rel} = \frac{n_2}{n_1} \tag{4-39}$$

$n_2$ is the particle refractive index, and

$$f_{i\pm} = 1 \pm \sin\frac{\theta}{2} \sin\varphi_i \sin\psi_i + \cos\frac{\theta}{2}\cos\varphi_i \tag{4-40}$$

For second-order refraction, $\beta_i$ cannot be given as a closed-form solution but must be solved for numerically by an iterative process. In this mode, small inaccuracies in the value of the refractive index of the particle may result to large errors in its diameter estimation; hence care should be used when employing this scattering mode.

**230**   EXPERIMENTAL METHODS

### 4-11.2  Sensitivity and Range of PDA

The geometric factor, and hence the sensitivity and range of the PDA, can be altered by changing any of the angles, $\theta$, $\varphi_i$, or $\psi_i$. However, in practice the three angles cannot be chosen freely. Typically, the selection of the scattering angle is quite restricted, either to ensure a particular mode of scattering (see Figure 4-50) or a sufficient signal-to-noise ratio, or from practical considerations of the measurement situation. The required working distance to the measurement point also affects the possible range of $\theta$ and $\psi_i$.

Figure 4-51 illustrates the increase in the slope of the diameter–phase relationship when the angular separation between the photodetectors is increased (i.e., increasing $\psi_{12}$ in the middle), and when the fringe separation is reduced by increasing the angle $\theta$ between the incident beams. Changing $\psi_{12}$ only affects the slope of the diameter–phase relationship (i.e., the sensitivity and range of the sizing) and has no effect on the velocity–frequency relationship. Changing $\theta$ affects both the slope of the diameter–phase curve and the velocity–frequency relationship. This is done in two ways: (1) by changing the focal length of the front lens of the transmitting optics, and (2) by changing the beam separation.

***4-11.2.1  Handling the $2\pi$ Ambiguity.*** Figure 4-52 shows the phase difference for three different particles of increasing size. While the phase difference for the first two particles is within $2\pi (= 360°)$, the third particle falls beyond this range. Thus, from measuring the phase difference between the Doppler bursts

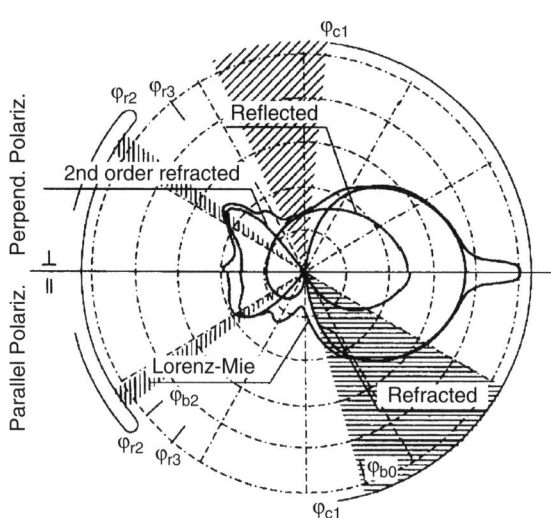

**Figure 4-50**  Light intensity distribution of a water droplet (log scale, one decade between each circle) for each of the scattering modes is shown in a polar plot for scattering angles from 0 to 180° and for two polarizations, the upper half at 90° to the scattering plane and the lower half parallel to the scattering plane.

PHASE DOPPLER ANEMOMETRY   **231**

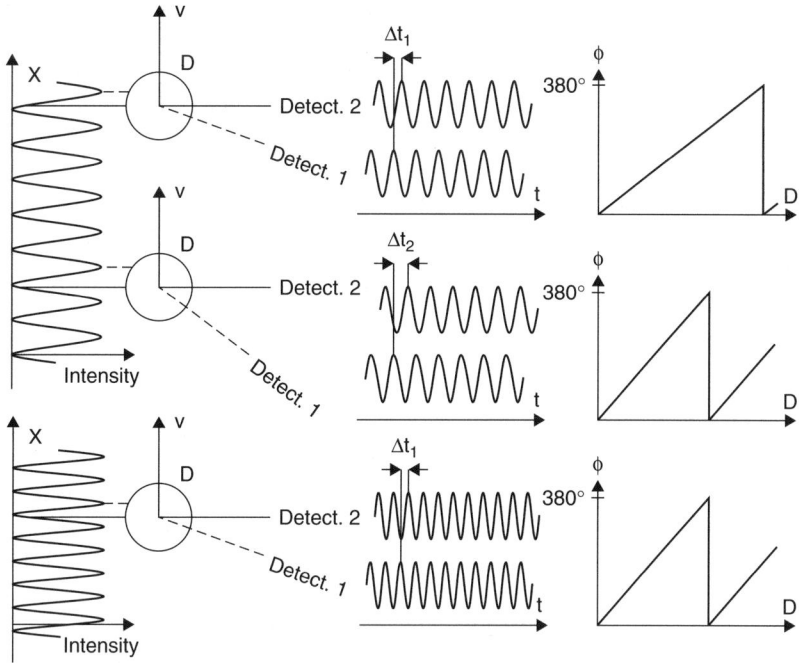

**Figure 4-51** Effect of changing the azimuth angle, $\psi$, and the angle between the incident beams, $\theta$, on the slope of the diameter–phase relationship.

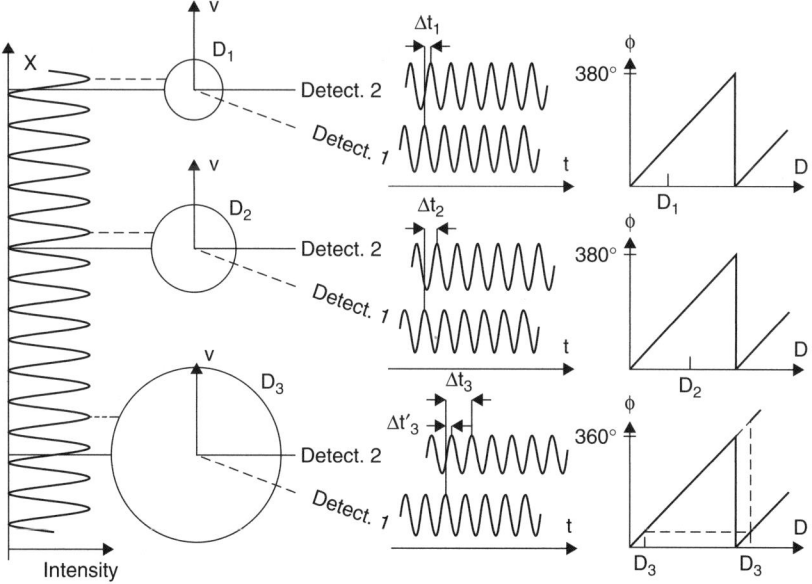

**Figure 4-52** The $2\pi$ ambiguity in PDA.

received by the two detectors alone there is no way to tell whether the diameter is $D_3$ or $D'_e$ (corresponding to the phase difference of $-2\pi$). This uncertainty is referred to as the $2\pi$ ambiguity in PDA. Under such conditions, a compromise is necessary between either high sensitivity and small measurement size range, or a larger measurement size range at the expense of sensitivity.

The solution to this problem is to use additional detectors. In conventional PDA a third detector is introduced so that the three are symmetrically positioned. Two detectors, U1 and U2, form the more distant pair giving the greater slope of the diameter–phase relationship and hence higher resolution and smaller working range. The detectors U1 and U3 form another pair, less separated and therefore giving a smaller slope to the diameter–phase relationship. This corresponds to a larger measurement size range but lower resolution. By comparing the phase differences from the two detector pairs, one can achieve at the same time, the high resolution and the large measurement range (see Figure 4-53).

### 4-11.2.2 Particle Sphericity.
The two-detector-pair arrangement has another useful feature, and that is to give information with regard to the curvature over a certain arc of the particle surface. If the curvature measured at two different locations on the surface (phase difference) is identical, the particle is said to be *spherical*. If the two local curvatures differ, $\Phi_{12}$ and $\Phi_{13}$ will point at diameter values differing by $\Delta D$. Consequently, a measure of the deviation from sphericity is available, and if $\Delta D$ exceeds a certain limit set by the user, the particle is said to be invalid. The underlying equations of size determination using the PDA technique assume that the particle is spherical, and hence any deviation from this assumption will introduce errors in the absolute determination of the particle size.

Ideally, the sum of the phase differences, $\Phi_{12}$, $\Phi_{23}$, and $\Phi_{31}$, is zero. However, due to the uncertainty of the phase measurements, this measured sum will deviate from zero. The absolute value of this deviation is the closed-loop phase error, which must not be greater than a certain value (to be set by the user in the software) for the particle to be accepted. This value should be typically set in the range 10 to 15°.

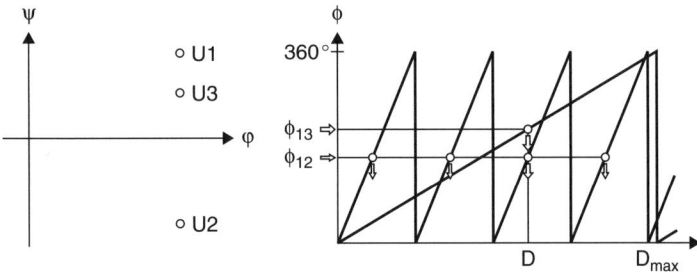

**Figure 4-53** Different slopes of the diameter–phase relation obtained in a conventional PDA setup with two pairs of photodetectors at different separations.

### 4-11.3 Implementation of PDA

There are three different optical configurations that can be used for size measurements using a PDA. These configurations are determined by the different modes of light scattering (see Figure 4-49):

- (First-order) refraction
- Reflection
- Second-order refraction

The majority of PDA applications make use of light scattered by first-order refraction. However, there can be several reasons for choosing one of the other two modes of light scattering. In any case, you should make sure that the selected mode of light scattering is dominating (i.e., the intensity of light scattered by one of the other modes is at least one order of magnitude smaller than the mode selected. Utilization of light scattering analysis codes is generally recommended for this purpose. A typical result of a light-scattering analysis computation is shown in Figure 4-50. Table 4-3 summarizes the application of the three modes.

*4-11.3.1 PDA Receiver Optics.* As with the LDA probes, commercial PDA equipment implements a fiber optic design to integrate all detectors in a single portable unit, as shown in Figure 4-54. Such units feature the flexibility of adjustable azimuthal positions (by means of exchangeable aperture plates) in combination with the convenience of alignment through a single common front lens. The front lens works as a collimator creating a beam of parallel light. This beam passes through an aperture plate, which divides the parallel light beam into three segments corresponding to the photodetectors U1, U2, and segmented lens then focuses each beam onto one of three slit-shaped spatial filters in front of an optical fiber that transmits the scattered light to the photodetectors.

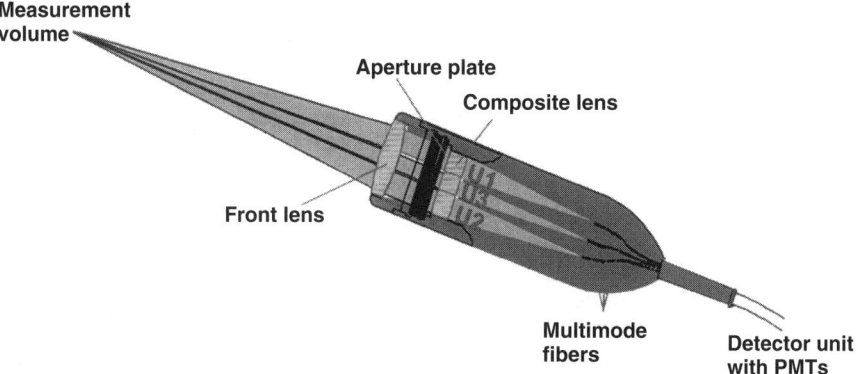

**Figure 4-54** Basic layout of a typical fiber optic-based PDA receiving probe.

**234** EXPERIMENTAL METHODS

**Table 4-3** Summary of PDA Measuring Modes[a]

*First-order refraction*
Transparent droplets: water or oil. Moderately great scattering angle (around 70°) is favored since the measured size will be independent of small changes of refractive index. Smaller angle (30°) is good when the signal-to-noise ratio is poor, due to higher scattered light intensity.

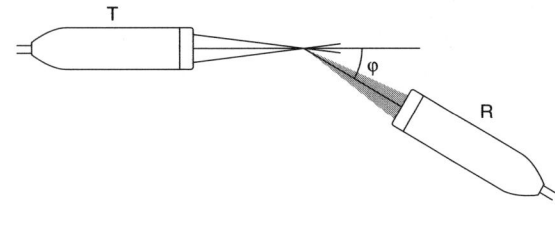

*Reflection*
Reflective particles or bubbles ($n_{rel} < 1$). For bubbles, scattering angle should be slightly less than 90°. Useful when the exact value of the relative refractive index is not known.

*Second-order refraction*
Transparent droplets. Scattering angle should be chosen carefully to yield a linear diameter–phase relationship. Accurate knowledge of the refractive index is required. Should be used only when physical restrictions, such as limited optical access, make it mandatory.

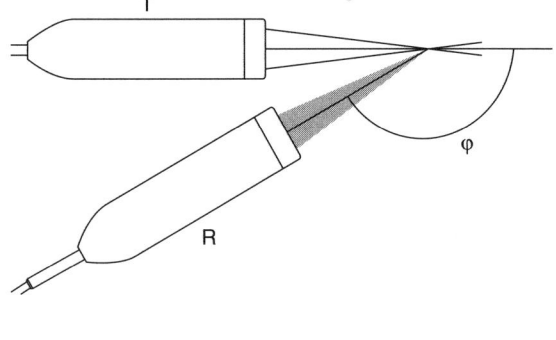

[a] Schematics show configuration with T And R labeling transmitting and receiving optics, respectively (the direction of flow is perpendicular to the plane of the illustrations). (Courtesy of Dantec Dynamics A/S.)

The measurable size range is determined by the following parameters:

- Beam intersection angle of the transmitting optics (defined by the beam separation and the front lens focal length)
- Focal length of the receiving front lens
- Aperture plate (controls azimuth angle)
- Selected scattering mode
- Scattering angle

All these parameters can be varied independently from one another to achieve a variety of size ranges that can be measured. As mentioned earlier, the working distance and optical access may fixate many of the parameters and thus limit implementation variability. One should be aware of the size range under investigation in order to best optimize the system. If not, setting parameters such as to achieve a large size range is a good first step toward optimizing the system configuration for the best measurement scenario. Typically, a PDA system can measure particles sizes of less than 1 μm micron to several millimeters.

**4-11.3.2 PDA Configurations.** The PDA principle is implemented using one of three configurations (Figure 4-55):

- *Conventional PDA:* three-detector system with detectors located off the scattering plane
- *Planar PDA:* three-detector system with detectors located in the scattering plane
- *Dual PDA:* four-detector system that combines a conventional two-detector PDA with a planar two-detector PDA

While the conventional and the planar PDA systems perform similar measurements, the dual PDA combines both configurations to make two simultaneous independent measurements of size to eliminate two common effects when measuring in refraction mode that potentially lead to incorrect size measurements when particles are more than one-third the size of the measurement volume. This error in size can translate to significant bias of the volume-weighted flux and concentration, since the volume is proportional to the third power of the diameter. The two effects are:

- The trajectory effect (or Gaussian beam effect)
- The slit effect

Both effects arise when the PDA system is set up to receive refractively scattered light, as is typical in spray applications, but instead receives reflectively scattered light. These signals are then processed with the phase–diameter relationship based on refraction and thus lead to incorrect size measurements.

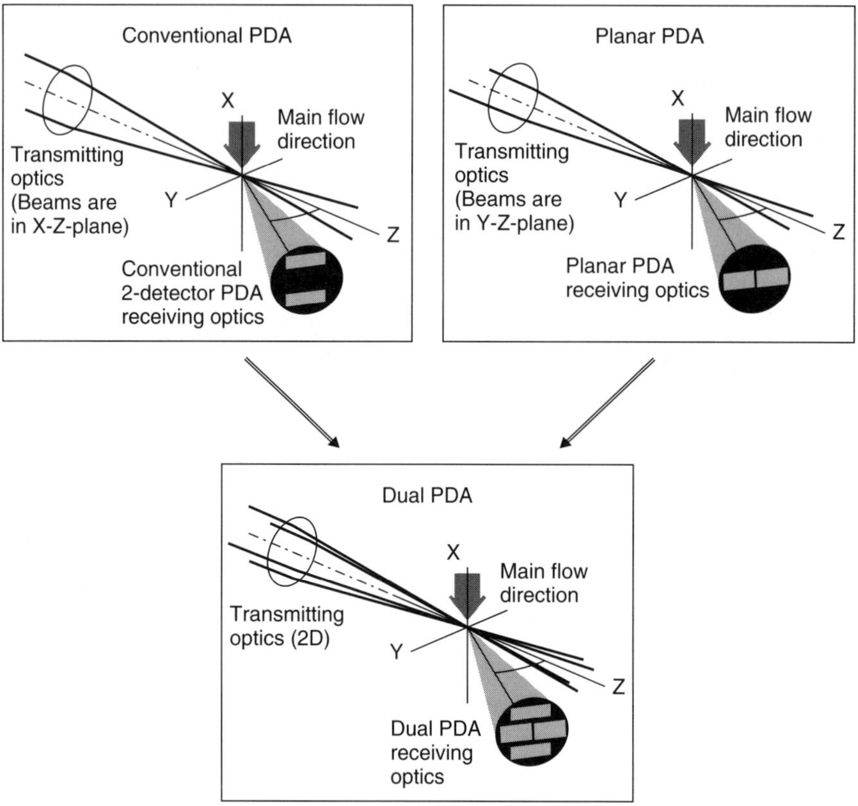

**Figure 4-55** PDA probe configurations.

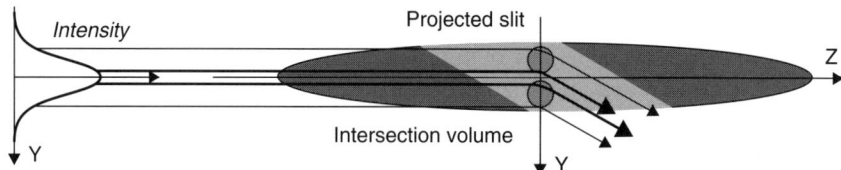

**Figure 4-56** Trajectory effect expressed in terms of geometric optics.

The two effects are illustrated in Figures 4-56 and 4-57, respectively. As indicated in Figure 4-56, there exist particle positions within the volume (trajectories) where reflection may become the dominant scattering mode due to the much higher intensity of the incident light, in particular on the negative Y-axis. On the other hand, in Figure 4-57 the scattering light from particles in certain positions may be suppressed because they lie outside the slit aperture, and hence unwanted reflections will overwhelm the signal.

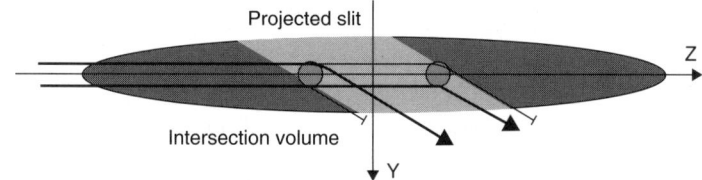

**Figure 4-57** Suppression of scattered light due to the slit aperture (slit effect).

By making two independent measurements, one using conventional PDA and the other using planar PDA, the same result regarding size is obtained only if the light received in both configurations is refractively scattered and the particle is spherical. Thus, both these effects that may result to significant errors in size and especially in volume estimation are eliminated using this redundancy.

*4-11.3.3 Optimizing Measurement Conditions in PDA.* In setting up any phase Doppler anemometer system for making reliable size measurements, some basic understanding of a few scattering phenomena is of great help. All three modes of scattering can be used with the PDA. There is, however, one crucial point that it is imperative to keep in mind. The PDA estimates the particle size from the phase differences of the Doppler bursts received by the three photodetectors in the receiving optics. Since the three modes of scattering give rise to different conversion factors (phase factors), receiving light from more than one scattering mode is likely to give rise to errors. Therefore, one should always set up a phase Doppler system so that only one mode of scattering dominates the light received by the receiving optics.

To assist in achieving a linear relationship between phase and diameter, Table 4-4 lists some typical scenarios and optical configurations that work best.

## 4-12 PARTICLE IMAGE VELOCIMETRY

### 4-12.1 Principles of PIV

PIV is a velocimetry technique based on images of tracer "seeding" particles suspended in the flow. In an ideal situation, these particles should be perfect flow tracers, homogeneously distributed in the flow, and their presence should not alter the flow properties. In that case, local fluid velocity can be measured by measuring the fluid displacement (see Figure 4-58) from multiple particle images and dividing that displacement by the time interval between the exposures. To get an accurate instantaneous flow velocity, the time between exposures should be small compared to the time scales in the flow; and the spatial resolution of the PIV sensor should be small compared to the length scales in the flow.

Principles of PIV have been covered in many papers, including Lourenco et al. (1989), Adrian (1991), and Willert and Gharib (1991). A more recent book by Raffel et al. (1998) is an excellent source of information on various aspects of PIV. The principal layout of a modern PIV system is shown in Figure 4-59.

**Table 4-4** Effect of Particle Type on Scattering Angle for PDA Measurements

| Particle Type | Scattering Angle | Polarization Relative to Scattering Plane |
|---|---|---|
| $n_{rel} > 1$ (bubbles) | Near forward scatter | Parallel or perpendicular |
| $n_{rel} > 1$ (droplets) | Near forward scatter | Must be parallel |
|  | Side scatter | Must be perpendicular |
|  | Near backscatter | Perpendicular (best) or parallel |
| Reflecting particles | Near forward or backscatter | Parallel or perpendicular |
|  | Side scatter | Must be perpendicular |

| Particle Type | Scattering Type | Scattering Angle |
|---|---|---|
| Totally reflecting | Reflection | Any angle except forward diffraction region; angle must be larger than arcsin $(1.4/D_{min})$ |
| Air bubbles in liquid | Reflection | Optimum near $\phi_\infty - 10°$; in liquid can be used from $\phi_\infty - 15°$ to $\phi_\infty - 5°$ |
| Air bubbles in water | Reflection | Optimum near $+70°$; in water can be used from $+65°$ to $+5°$ |
| Liquid droplets in air | Refraction | Optimum at $\phi_b$ second-order refraction can be used at $\phi_r + 5°$ to $\phi_r + 10°$ |
| Water droplets in air | Refraction | Optimum at Brewster's angle $= 73.7°$; can be used from 37 to 10°; second-order refraction can be used from 145 to 150° |
| Two-phase flow |  |  |
| $u_{rel} < 1$ | Reflection | Optimum near $\phi_\infty - 10°$; can be used from $\phi_\infty - 15°$ to $\phi_\infty - 5°$ |
| $u_{rel} > 1$ | Refraction | Optimum at Brewster's angle |

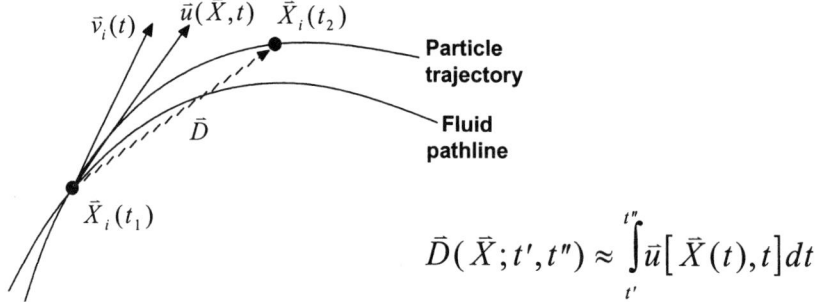

**Figure 4-58** Determination of particle displacement and its relationship to actual fluid velocity. (From Keane and Adrian, 1993.)

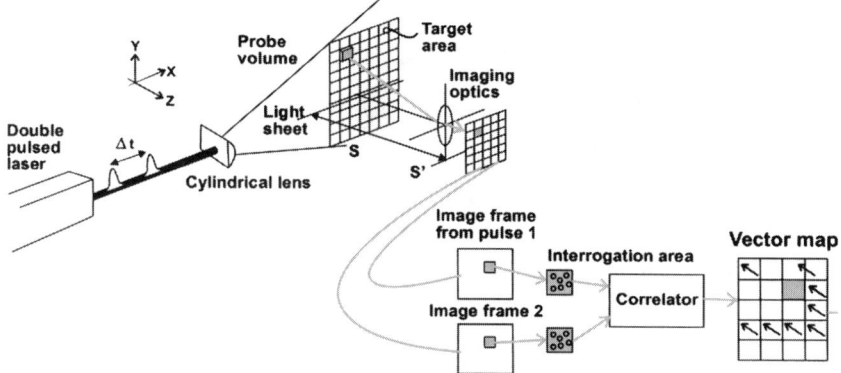

**Figure 4-59** Principal of operation for a PIV system.

The PIV measurement includes illuminating a cross-section of the seeded flow field, typically by a pulsing light sheet, recording multiple images of the seeding particles in the flow using a camera located perpendicular to the light sheet, and analyzing the images for displacement information.

The recorded images are divided into small subregions called *interrogation regions*, the dimensions of which determine the spatial resolution of the measurement. The interrogation regions can be adjacent to each other, or more commonly, have partial overlap with their neighbors. The shape of the interrogation regions can deviate from square to better accommodate flow gradients. In addition, interrogation areas A and B, corresponding to two different exposures, may be shifted by several pixels to remove a mean dominant flow direction (DC offset) and thus improve the evaluation of small fluctuating velocity components about the mean.

The peak of the correlation function gives the displacement information. For double or multiple exposed single images, an autocorrelation analysis is performed. For single exposed double images, a cross-correlation analysis gives the displacement information.

### 4-12.2  Image Processing

If multiple images of the seeding particles are captured on a single frame (as seen in the photograph of Figure 4-60), the displacements can be calculated by autocorrelation analysis. This analysis technique has been developed for photography-based PIV, since it is not possible to advance the film fast enough between the two exposures. The autocorrelation function of a double-exposed image has a central peak and two symmetric side peaks, as shown in Figure 4-61. This poses two problems: (1) although the particle displacement is known, there is an ambiguity in the flow direction, and (2) for very small displacements, the side peaks can partially overlap with the central peak, limiting the measurable velocity range. To overcome the directional ambiguity problem, image shifting techniques using

**240** EXPERIMENTAL METHODS

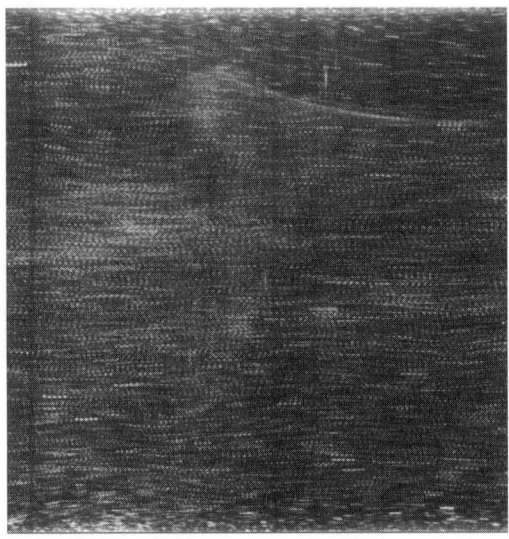

**Figure 4-60** Multiple-exposure image captured for PIV autocorrelation analysis. (From Stanislas et al., 2000, p. 65.)

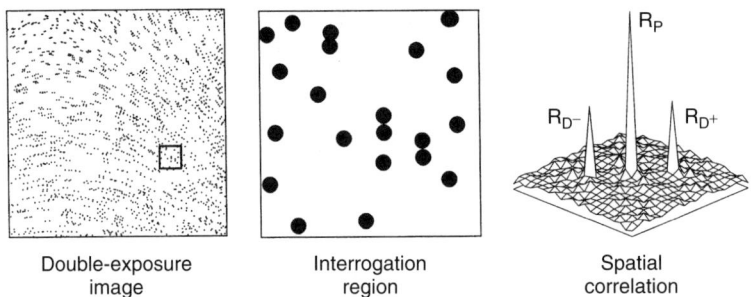

**Figure 4-61** Autocorrelation analysis of a double-exposed image. (From Westerweel, 1993.)

rotating mirrors (Landreth et al., 1988) and electrooptical techniques (Landreth and Adrian, 1988b; Lourenco, 1993) have been developed. To leave enough room for the added image shift, larger interrogation regions are used for autocorrelation analysis. By displacing the second image at least as much as the largest negative displacement, the directional ambiguity is removed. This is analogous to frequency shifting in LDA systems to make them directionally sensitive.

The preferred method in PIV is to capture two images on two separate frames and perform cross-correlation analysis. This cross-correlation function has a single peak,

$$R(\vec{s}) = \int W_1(\vec{x})I_1(\vec{x})W_2(\vec{x}+\vec{s})I_2(\vec{x}+\vec{s})\,d\vec{x} \qquad (4\text{-}41)$$

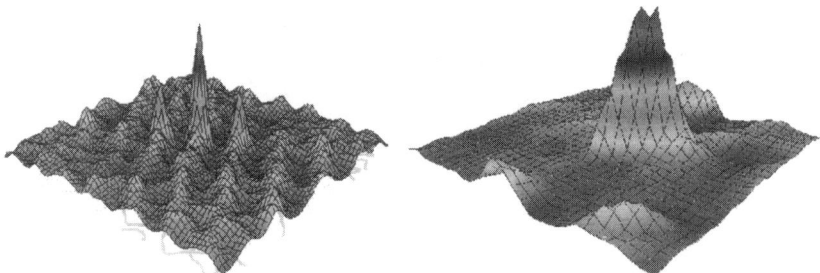

**Figure 4-62** Autocorrelation (left) versus cross-correlation (right) analysis result.

providing the magnitude and direction of the flow without ambiguity (Figure 4-62).

Common particles need to exist in the interrogation regions, which are being correlated; otherwise, only random correlation or noise will exist. The PIV measurement accuracy and dynamic range increase with increasing time difference $\Delta t$ between the pulses. However, as $\Delta t$ increases, the likelihood of having common particles in the interrogation region decreases and the measurement noise goes up. A good rule of thumb is to ensure that within time $\Delta t$, the in-plane components of velocity $V_x$ and $V_y$ carry the particles no more than a third of the interrogation region dimensions, and the out-of-plane component of velocity $V_z$ carries the particles no more than a third of the light sheet thickness.

Most commonly used interrogation region dimensions are $64 \times 64$ pixels for autocorrelation, and $32 \times 32$ pixels for cross-correlation analysis. Since the maximum particle displacement is about a third of these dimensions, to achieve reasonable accuracy and dynamic range for PIV measurements, it is necessary to be able to measure the particle displacements with subpixel accuracy.

Fast Fourier transform (FFT) techniques are used for the calculation of the correlation functions. Since the images are digitized, the correlation values are found for integral pixel values, with an uncertainty of $\pm 0.5$ pixel. Different techniques, such as centroids, Gaussian, and parabolic fits, have been used to estimate the location of the correlation peak. Using 8 bit digital cameras, peak estimation accuracy of 0.1 to 0.01 pixel can be obtained. For subpixel interpolation techniques to work properly, it is necessary for particle images to occupy multiple (2 to 4) pixels. If the particle images are too small (i.e., around 1 pixel), these subpixel estimators do not work properly, since the neighboring values are noisy. In such a case, slight defocusing of the image improves the accuracy.

Digital windowing and filtering techniques can be used in PIV systems to improve the results. Top-hat windows with zero padding, or Gaussian windows, applied to interrogation regions are effective in reducing the cyclical noise, which is inherent to FFT calculation. A Gaussian window also improves measurement accuracy in cases where particle images straddle the boundaries of the interrogation regions. A spatial frequency high-pass filter can reduce the effect of low-frequency distortions from optics, cameras, or background light variations.

**242** EXPERIMENTAL METHODS

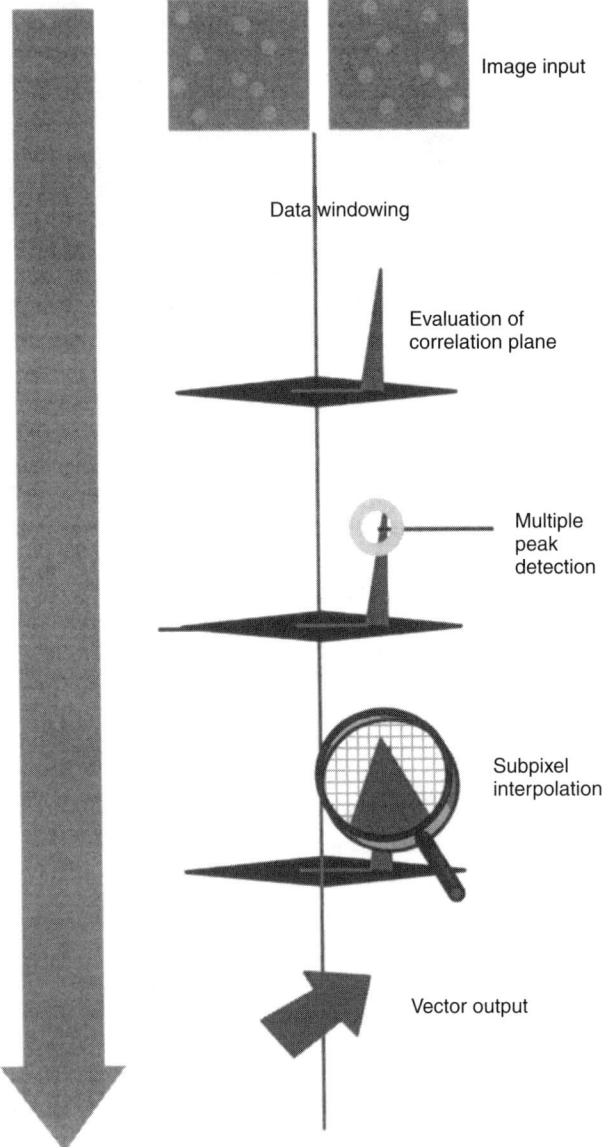

**Figure 4-63** Image to vector processing sequence.

A spatial frequency low-pass filter can reduce the high-frequency noise generated by the camera, and ensure that the subpixel interpolation algorithm can still work in cases where the particles in the image map are less than 2 pixels in diameter. Typical image to vector processing sequence is shown in Figure 4-63.

The spatial resolution of PIV can be increased using multipass correlation approaches. Offsetting the interrogation region by a value equal to the local integer displacement, a higher signal-to-noise ratio can be achieved in the correlation function, since the probability of matching particle pairs is maximized. This idea has led to implementations such as adaptive correlation *Dantec*, 2000), superresolution PIV (Keane et al., 1995), and hybrid PTV (Cowen and Monosmith, 1997).

### 4-12.3 Implementation of PIV

The vast majority of modern PIV systems today consist of dual-cavity Nd:Yag lasers, cross-correlation cameras, and processing using software or dedicated FFT-based correlation hardware. In this section we review the typical implementation of these components to perform a measurement process (as shown in the flowchart of Figure 4-64) and address concerns regarding proper seeding, light delivery, and imaging.

*4-12.3.1 Seeding Particles.* Rather than relying on naturally existing particles, it is common practice to add particles to the flow to have control over their size, distribution, and concentration. In general, these particles should be small enough to be good flow tracers and large enough to scatter sufficient light for imaging. They should also be nontoxic, noncorrosive, and chemically inert, if possible. Melling (1997) reviews a wide variety of tracer materials that have been used in liquid and gas PIV experiments. Methods of generating seeding particles and introducing them into the flow are also discussed.

Possibilities for liquid applications include silver-coated hollow glass spheres, polystyrene, polymers, titanium dioxide ($TiO_2$), aluminum oxide ($Al_2O_3$), conifer pollen, and hydrogen bubbles. Furthermore, fluorescent dies are used in conjunction with polystyrene or polymer particles to generate particles that will absorb the incident laser radiation and emit at another wavelength band. A common dye for Nd:Yag lasers operating in the 532 nm spectral range is Rhodamine-B, which when

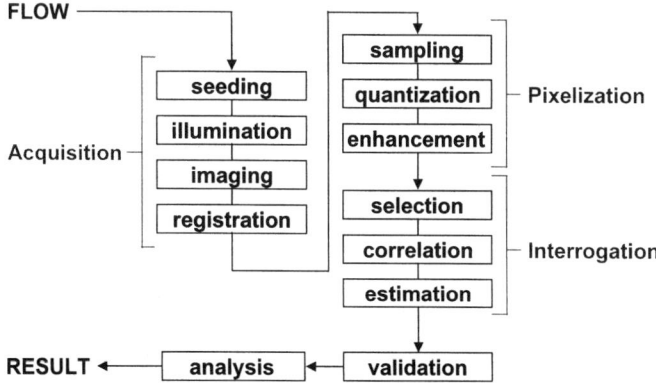

**Figure 4-64** Flowchart of the PIV measurement and analysis process. (From Westerweel, 1993.)

**244** EXPERIMENTAL METHODS

excited emits at wavelengths above 560 nm. Hence, for applications where many reflections exist from geometric boundaries (e.g., stirred tanks), the use of fluorescent particles is greatly advantageous (see Hammad and Papadopoulos, 2001).

For gas flow applications, theatrical smoke, different kinds of atomized oil, $TiO_2$, and $Al_2O_3$ have been used. Typical theatrical smoke generators are inexpensive, and they generate plenty of particles. Oil can be atomized using devices such as a Laskin nozzle, generating particles in the micrometer to submicrometer range, which are particularly useful for high-speed applications. $TiO_2$ and $Al_2O_3$ are useful for high-temperature applications such as combustion and flame measurements.

***4-12.3.2 Light Sources and Delivery.*** In PIV, lasers are used only as a source of bright illumination and are not a requirement. Flash lamps and other white light sources can also be used. Some facilities prefer these nonlaser light sources because of safety issues. However, white light cannot be collimated as well as coherent laser light, and their use in PIV is not widespread.

PIV image acquisition should be completed using short light pulses to prevent streaking. Hence, pulsed lasers are naturally well suited for PIV work. However, since many labs already have existing continuous-wave (CW) lasers, these lasers have been adapted for PIV use, especially for liquid applications.

Dual-cavity Nd:Yag lasers, also called PIV lasers, are the standard laser configuration for modern PIV systems. Nd:Yag lasers emit infrared radiation whose frequency can be doubled to give 532 nm green wavelength. Minilasers are available with power up to 200 mJ/pulse, and larger lasers provide up to 1 J/pulse. Typical pulse duration is around 10 ns, and the pulse frequency is typically 5 to 30 Hz. To achieve a wide range of pulse separations, two laser cavities are used to generate a combined beam. Control signals required for PIV Nd:Yag lasers are shown in Figure 4-65.

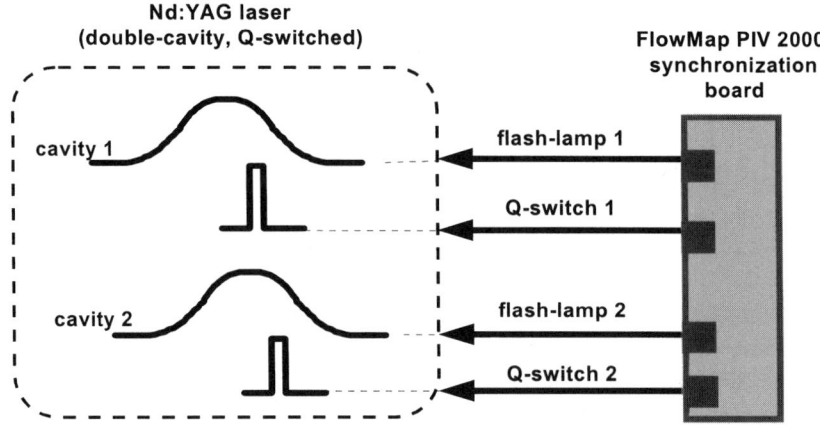

**Figure 4-65** Control signals required for PIV Nd:Yag lasers.

Argon-ion lasers are CW gas lasers whose emission is composed of multiple wavelengths in the green–blue–violet range. Air-cooled models emitting up to 300 mW and water-cooled models emitting up to 10 W are quite common in labs for LDA use. They can also be used for PIV experiments in low-speed liquid applications, in conjunction with shutters or rotating mirrors.

Copper-vapor lasers are pulsed metal vapor lasers that emit green (510 nm) and yellow (578 nm). Since the repetition rates can reach up to 50 kHz, energy per pulse is a few millijoules or less. They are used with high-framing-rate cameras for flow visualization and some PIV applications. High pulse rate Nd:Yag lasers have also been recently used to make time resolved PIV measurements (see Papadopoulos and Hammad, 2003).

Ruby lasers have been used in PIV because of their high-energy output. But their 694 nm wavelength is at the end of the visible range where typical CCD (charge-coupled diode) chips and photographic film are not very sensitive.

Fiber optics are commonly used for delivering Ar-ion beams conveniently and safely. Single-mode polarization-preserving fibers can be used for delivering up to 1 W of input power, whereas multimode fibers can accept up to 10 W. Although use of multimode fibers produces nonuniform intensity in the light sheet, they have been used in some PIV applications.

The short-duration high-power beams from pulsed Nd:Yag lasers can instantly damage optical fibers. Hence, alternative light guiding mechanisms have been developed, consisting of a series of interconnecting hollow tubes and flexible joints where high-power mirrors are mounted. Light sheet optics located at the end of the arm can be oriented at any angle and extended up to 1.8 m. Such a mechanism can transmit up to 500 mJ of pulsed laser radiation with 90% transmission efficiency at 532 nm, offering a unique solution for safe delivery of high-powered pulsed laser beams.

The main component of light sheet optics is a cylindrical lens. To generate a light sheet from a laser beam with small diameter and divergence, such as one from an Ar-ion laser, using a single cylindrical lens can be sufficient. For Nd:Yag lasers, one or more additional cylindrical lenses are used, to focus the light sheet to a desired thickness and height. For light sheet optics designed for high-power lasers, a diverging lens with a negative focal length is used first to avoid focal lines.

*4-12.3.3 Image Recorders.* Cross-correlation cameras are the preferred method of sampling data for PIV. The cross-correlation cameras use high-performance progressive-scan interline CCD chips. Such chips include m × n light-sensitive cells and an equal number of storage cells (blind cells). Figure 4-66 is a schematic illustration of the light-sensitive pixels and storage cell layout for these cameras.

The first laser pulse is timed to expose the first frame, which is transferred from the light-sensitive cells to the storage cells immediately after the laser pulse. The second laser pulse is then fired to expose the second frame. The storage cells now contain the first camera frame of the pair with information about the initial positions of seeding particles. The light-sensitive cells contain the second camera

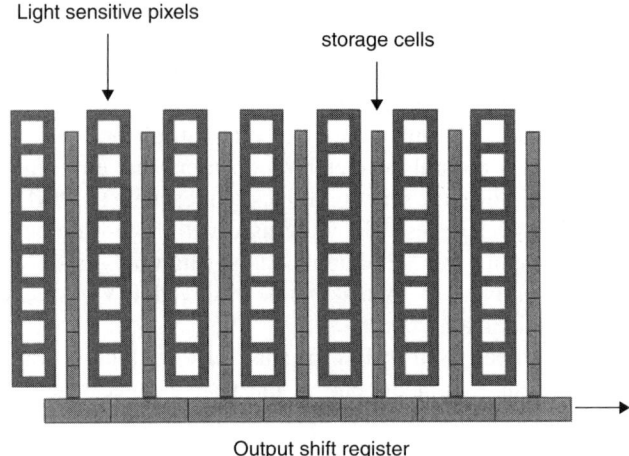

**Figure 4-66** Schematic illustration of light-sensitive pixels and storage cell layout of cross-correlation PIV cameras.

frame, which has information on the final positions of the seeding particles. These two frames are then transferred sequentially to the camera outputs for acquisition and cross-correlation processing.

Cross-correlation cameras are available with resolutions up to $2\,k \times 2\,k$ pixels, and framing rates up to 30 Hz; 8 bit cameras are sufficient for most purposes. However, 12 bit cameras are becoming common, especially for applications such as planar laser-induced fluorescence (PLIF), where extra sensitivity and dynamic range are required.

Flow fields with velocities ranging from micrometers per second to supersonic speeds can be studied since interframe time separations down to few hundred nanoseconds can be obtained. One interesting feature of the cameras is that they can be reset asynchronously. This is particularly useful in conjunction with the special triggering options for synchronizing the measurements to external events, such as rotating machinery.

### 4-12.4 PIV Data Processing

Typically, the raw PIV data obtained from the cross-correlation of two images need to be validated and optionally smoothed before statistical values are calculated or various derived quantities are computed.

The following are various data validation techniques that are commonly used:

- *Correlation peak-height validation* works based on the height of the peaks in the correlation plane. If $P_1$ is the highest peak and $P_2$ is the second highest peak, the most common approach, called the *detectability criterion*, validates vectors for which $P_1/P_2 \geq k$, where k is typically around 1.2 (Keane and Adrian, 1992).

- *Velocity-range validation* rejects vectors whose magnitude or components are outside a given range. Normally, the user has an idea about the range of velocities in the flow. This information is used as validation criteria.

$$V_{min} \leq |V| \leq V_{max} \quad \text{(length)}$$
$$V_{x,min} \leq |V_x| \leq V_{x,max} \quad \text{(x-component)}$$
$$V_{y,min} \leq |V_y| \leq V_{y,max} \quad \text{(y-component)}$$

  Hence, if the vector does not satisfy the required relations above, it is rejected.

- *Moving-average validation* is a special case of the general class of iterative filtered validation, described by Host-Madsen and McCluskey (1994). Since the vector field is oversampled by the PIV technique, there is a correlation between neighboring vectors, and there is not too much change from one vector to its neighbor. If a vector deviates too much from its neighbors, it must be an outlier. Hence, in this technique, the average of the vectors neighboring a given vector is calculated and compared with the vector. If the difference is larger than a certain acceptance factor, that vector is rejected. The rejected vector may be substituted by a local average of its neighbors.

- *Moving-average filter* substitutes each vector with the uniformly weighted average of the vectors in a neighborhood of a specified size m × n. Here, m and n are an odd number of vector cells symmetrically located around each vector. This filter takes out the high-frequency jitter in the PIV results.

After validation and optional filtering, the user is ready to calculate derived quantities and statistical values. Following are the commonly calculated derived quantities from PIV data:

- *Vorticity* is the curl of the 3D velocity vector. From 2D PIV calculations, the normal component of the vorticity vector can be calculated.
- *Streamlines* are curves parallel to the direction of the flow. They are defined by the equation $v_x \, dy = v_y \, dx$. They represent the path that a particle would follow if the flow field were constant with time. Hence, the streamlines calculated from PIV measurements are correct only if the flow is 2D in the plane of the light sheet.

The commonly employed statistical properties that are calculated from PIV measurements include the mean of each velocity component, the standard deviation of the mean, and the covariance coefficient.

## 4-12.5 Stereoscopic (3D) PIV

Conventional 2D planar PIV technique measures projections of 3D velocity vectors onto the 2D plane defined by the light sheet. It is not capable of measuring

**248** EXPERIMENTAL METHODS

the third component of the flow normal to the light sheet. In fact, if that normal component is large, the planar PIV technique can give wrong results even for the in-plane components of velocity, due to parallax error. This error gets increasingly large from the center to the edges of the image. In these situations, the problem is normally minimized by having a large-focal-length lens so that the distance from the camera to the image is large compared to the image area.

Since there are many applications where it is important to measure the third component of velocity normal to the light sheet, various approaches have been proposed to recover that third component. The most common technique, *stereoscopic PIV*, involves using an additional camera and viewing the flow from two different angles. It is based on the same principle as human stereo eyesight. The two eyes see slightly different images of the objects around us. The differences of the images are compared in the brain and interpreted as the 3D perception.

*4-12.5.1 Stereoscopic Imaging Basics: Scheimpflug Condition.* Stereoscopic PIV is a planar PIV technique for measuring all three components of velocity. Instead of having a single camera normal to the light sheet, two cameras are used, each looking at the same flow field at different angles. Due to different orientations, each camera records a different image. 3D displacements and hence velocities on the plane can be derived by proper calibration of the camera views of a target, and combining 2D results from each camera.

The principle of stereo PIV is indicated in Figure 4-67. When each camera views the measurement volume illuminated by the light sheet at an angle, the CCD chip in each camera needs to be tilted so that the entire field of view

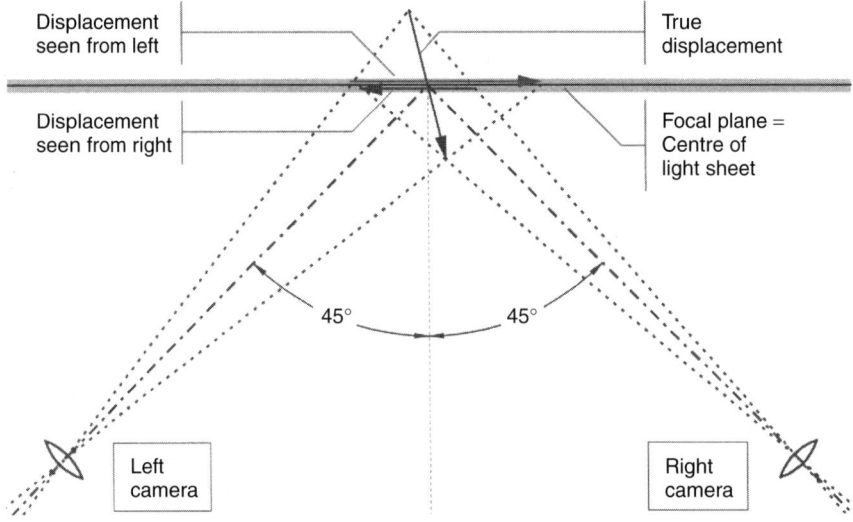

**Figure 4-67** Stereo vision.

of the camera can be focused. In fact, for each camera to be focused properly, the object (light sheet), camera lens, and image (CCD chip) planes should all intersect along a common line (Prasad and Jensen, 1995). This is called the *Scheimpflug condition*. When the Scheimpflug condition is satisfied, a perspective distortion is introduced into the two images as a side effect. Hence, the magnification factor is not constant across the image any more and needs to be evaluated via calibration.

### 4-12.5.2 Calibration and 3D Reconstruction.
To reconstruct the true 3D (X,Y,Z) displacements from two 2D (x,y) displacements as observed by the two cameras, a numerical model is necessary that describes how each of the two cameras image the flow field onto their CCD chips. Using the camera imaging models, four equations (which may be linear or nonlinear) with three unknowns are obtained.

Instead of a theoretical model that requires careful measurements of distances, angles, and so on, an experimental calibration approach is preferred. The experimental calibration estimates the model parameters based on the images of a calibration target as recorded by each camera. A linear imaging model that works well for most cases, the *pinhole camera model*, is based on geometrical optics. This leads to the following direct linear transform equations, where x,y are image coordinates, and X,Y,Z are object coordinates. This physics-based model cannot describe nonlinear phenomena such as lens distortions.

$$\begin{bmatrix} kx \\ ky \\ k \end{bmatrix} = \begin{bmatrix} A_{11} & A_{12} & A_{13} & A_{14} \\ A_{21} & A_{22} & A_{23} & A_{24} \\ A_{31} & A_{32} & A_{33} & A_{34} \end{bmatrix} \cdot \begin{bmatrix} X \\ Y \\ Z \\ 1 \end{bmatrix} \qquad (4\text{-}42)$$

In experiments involving significant lens distortion, refraction, and so on, higher-order nonlinear imaging models can be used (Soloff et al., 1997). These models are based not on a physical mapping of the geometry but on a least-squares fitting of the image–object pairs using adjustable parameters. Imaging parameters such as image magnification and focal length do not need to be determined, and higher-order terms can compensate for nonlinear effects.

### 4-12.6 PIV Applications in Mixing

*Stirred tanks* are used commonly in various mixing industries. These dynamic devices typically consist of a circular tank with a rotating shaft containing impellers of various geometries. There are also *static mixers*, where the mixing occurs as the fluid is forced through a complex geometry in a conduit. These devices have been adapted for use in such industries as pharmaceutical, chemical, personal care products, food and beverages, biotechnology, polymer, plastic, paper and pulp, oil, rubber, and waste disposal.

The use of computational fluid dynamics (CFD) to predict the flow fields in various industrial mixers has increased substantially during the last few years.

Time-dependent mixing flows, coupled with complex geometry, bring uncertainty to the CFD predictions, especially for turbulent flows. This has also increased the interest in experimental verification of the simulations.

Hence, in addition to the traditional experimentalist who obtains the mixing information experimentally, more and more CFD users are turning to PIV to measure the initial and boundary conditions in their mixer and to verify the results of their simulations.

Although the flow fields in stirred tanks have been measured by the LDA technique by many researchers since the mid-1970s, the use of PIV is more recent. Bakker et al. (1996) were early users of PIV to study the 2D flow pattern along the center plane of a stirred vessel. Sheng et al. (1998) have investigated methods for validating CFD simulations against PIV measurements. They have extracted the mean velocity, turbulent kinetic energy, Reynolds stresses, and dissipation rate from the PIV data and investigated the effect of boundary conditions on CFD simulation results using PIV and LDA data. More recently, Zalc (2000) has developed CFD tools to investigate mixing in a static mixer and a three–Rushton turbine stirred tank, where PIV was used to validate the stirred tank simulations.

## NOMENCLATURE

$D_{min}$    minimum particle diameter to be measured ($\mu$m)
$n_1$    index of refraction of external medium
$n_2$    index of refraction of particle
$n_{rel}$    relative index of refraction, $n_1/n_2$

*Greek Symbols*

$\phi_\infty$    critical angle for particles with $n_{rel} < 1$ (the largest scattering angle for reflection is 2 arccos $n_{rel}$)
$\phi_b$    Brewster's angle for particles with $n_{rel} > 1$ [the angle at which reflection is zero for polarization parallel to the scattering plane is $2 \arctan(1/n_{rel})$]
$\phi_{cl}$    critical angle for particles with $n_{rel} > 1$ [the largest scattering angle for refraction is $2 \arccos(1/n_{rel})$]
$\phi_r$    rainbow angle for particles with $n_{rel} > 1$ (the angle where second-order refraction is strongest is $4 \arccos[\cos(\gamma)n/_{rel}] - 2\gamma$, where
$$\gamma = \arcsin\sqrt{(n_{rel}^2 - 1)/3}$$

## REFERENCES

Abuaf, N., T. P. Felerabend, G. A. Zimmer, and O. C. Jones (1979). Radio-frequency probe for bubble size and velocity measurements, *Rev. Sci. Instrum.*, **50**(10), 1260–1263.

Adrian, R. J. (1991). Particle imaging techniques for experimental fluid mechanics, *Annu. Rev. Fluid Mech.*, **23**, 261–304.

All-Saeedi, J. N., M. O. Kirkpatrick, and R. W. Pike (1995). Optical tomography for measurement of concentration distributions in a stirred tank, *Paper 11.5*, presented at Mixing XVI, Banff, Alberta, Canada.

Bakker, A. K., J. Myers, R. W. Ward, and C. K. Lee (1996). The laminar and turbulence flow pattern of a pitched blade turbine, *Trans. Inst. Chem. Eng.*, **74**, 485–491.

Barnes, H. A., J. F. Hutton, and K. Walters (1989). *An Introduction to Rheology*, Elsevier, New York.

Benedict, L. H., and R. D. Gould (1996). Towards better uncertainty estimates for turbulence statistics, *Exp. Fluids*, **22**, 129–136.

Benedict, L. H., and R. D. Gould (1999). Understanding biases in the near-field region of LDA two-point correlation measurements, *Exp. Fluids*, **26**, 381–388.

Bivolaru, D., M. V. Ötügen, A. Tzes, and G. Papadopoulos (1999). Image processing for interferometric Mie and Rayleigh scattering velocity measurements, *AIAA J.*, **37**(6), 688–694.

Buchhave, P. (1979). The measurement of turbulence with the burst-like laser-Doppler-anemometer-errors and correction methods, *Tech. Rep. TRL-106*, State University of New York–Buffalo.

Buchhave, P., W. K. George, and J. L. Lumley (1979). The measurement of turbulence with the laser-Doppler anemometer, *Annu. Rev. Fluid Mech.*, **11**, 443–503.

Burgess, J. M., and P. H. Calderbank (1975). The measurement of bubble parameters in two-phase dispersions: I, *Chem. Eng. Sci.*, **30**, 743–750. [See also Raper, J. A., D. C. Dixon, and C. J. D. Fell (1978). Limitations of Burgess–Calderbank probe technique for characterization of gas liquid dispersions on sieve trays, *Chem. Eng. Sci.*, **33**, 1405–1406; and J. M. Burgess and P. H. Calderbank, Limitations of Burgess–Calderbank probe technique for characterization of gas liquid dispersions on sieve trays, authors' reply, p. 1407.]

Chapman, C. M., A. W. Nienow, and J. C. Middleton (1980). Surface aeration in a small agitated and sparged vessel, *Biotechnol Bioeng.*, **22**, 981–993.

Chapman, C. M., L. G. Gibilaro, and A. W. Nienow (1982). A dynamic response technique for the estimation of gas–liquid mass transfer coefficients in a stirred vessel, *Chem. Eng. Sci.*, **37**(6), 891–896.

Chapple, D., S. M. Kresta, A. Wall, and A. Afacan (2002). The effect of impeller and tank geometry on power number for a pitched blade turbine, *Trans. Inst. Chem. Eng. A (Chem. Eng. Res. Des.)*, **80**, 364–372.

Clark, H. and Özcan-Taşkin, N. G. (2001). A study of drop break up and deformation in the laminar regime, *AIChE Meeting*, November, Reno, NV.

Cochrane, D. R., and J. M. Burgess (1985). A new method for the measurement of the properties of two-phase gas–liquid dispersions, *Proc. Chemeca'85 Conference, Paper C1A*, pp. 263–268.

Cooke, M., M. K. Dawson, A. W. Nienow, G. Moody, and M. J. Whitton (1991). Mass transfer in aerated agitated vessels: assessment of the NEL/Hickman steady state method, *Proc. 7th European Conference on Mixing, KVIV*, Bruges, Belgium, pp. 409–418.

Cooke, M., G. T. Bolton, D. H. Jones, and D. Housley (2001). Demonstration of a novel retrofit tomography baffle cage for gas–liquid mixing studies under intense operating conditions, *Proc. 2nd World Congress on Industrial Process Tomography*, Hanover, Germany, Aug. 29–31.

Cowen, E. A., and S. G. Monosmith (1997). A hybrid digital particle tracking velocimetry technique, *Exp. Fluids*, **22**, 199–211.

Dantec (2000). Adaptive correlation, *Dantec Product Information*, Dantec Dynamics, Inc., Mahwah, NJ.

Dickin, F. J., T. Dyakowski, R. A. Williams, R. C. Waterfall, C. G. Xie, and M. S. Beck (1992). Process tomography for improving the design and control of multiphase systems: its current status and future prospects, presented at the ECAPT Conference, Manchester, Lancashire, England.

Durst, F., and M. Zaré (1975). Laser-Doppler measurements in two-phase flows, *Proc. LDA Symposium*, Copenhagen, pp. 403–429.

Durst, F., A. Melling, and J. H. Whitelaw (1976). *Principles and Practice of Laser-Doppler Anemometry*, Academic Press, London.

Erdmann, J. C., and R. I. Gellert (1976). Particle arrival statistics in laser anemometry of turbulent flow, *Appl. Phys. Lett.*, **29**, 408–411.

Escudie, R. (2001). Ph.D. dissertation, INSA, Toulouse, France.

Fangary, Y. S., J. P.K. Seville, and M. Barigou (1992). Flow studies in stirred tanks by positron emission particle tracking (PEPT), *Proc. Fluid Mixing 6 Conference*, Bradford, Yorkshire, England. *Inst. Chem. Eng. Symp. Ser.*, **146**, 23–34.

Frijlink, J. J. (1987). *Physical aspects of gassed suspension reactors*, Ph.D. dissertation, Technical University of Delft, The Netherlands.

Gal-or, B., and W. Resnick (1966). Gas residence time in agitated gas–liquid contactor: experimental test of mass transfer model, *Ind. Eng. Chem. Process. Des. Dev.*, **5**(1), 15–19.

George, W. K. (1976). Limitations to measuring accuracy inherent in the laser Doppler signal, in *The Accuracy of Flow Measurements by Laser Doppler Methods: Proceedings of the LDA–Symposium Copenhagen 1975*, P. Buchhave, J. M. Delhaye, F. Durst, W. K. George, K. Refslund, and J. H. Whitelaw (eds.), Hemisphere Publishing Corporation, pp. 20–63.

George, W. K. (1978). Processing of random signals, *Proc. Dynamic Flow Conference*, Skovlunde, Denmark.

Gibilaro, L. G., S. N. Davies, M. Cooke, P. M. Lynch, and J. C. Middleton (1985). Initial response analysis of mass transfer in a gas sparged stirred vessel, *Chem. Eng. Sci.*, **40**(10), 1811–1816.

Guezennec, Y. G., R. S. Brodkey, N. T. Trigue, and J. C. Kent (1994). Algorithms for fully automated three-dimensional particle tracking velocimetry, *Exp. Fluids*, **17**, 209–219.

Hammad, K. J., and G. Papadopoulos (2001). Phase-resolved PIV measurements within a triple impeller stirred-tank, *ASME 2001 Fluids Engineering Division Summer Meeting*, FEDSM2001-18224, May 29–June 1, New Orleans, LA.

Hanhart, J., H. Kramers, and K. R. Westerterp (1963). The residence time distribution of the gas in an agitated gas–liquid contactor, *Chem. Eng. Sci.*, **18**, 503–509.

Harnby, N., M. F. Edwards, and A. W. Nienow (Eds.) (1992). *Mixing in the Process Industries*, Butterworth-Heinemann, Wolburn, MA.

Heijnen, J. J., K. van't Riet, and A. J. Wolthuis (1980). Influence of small bubbles on the dynamic $k_L$ a measurement in viscous gas–liquid systems, *Biotechnol. Bioeng.*, **22**, 1945–1956.

Hickman, A. D. (1988a). Gas–liquid oxygen transfer and scale-up: a novel experimental technique with results for mass transfer in aerated mixing vessels, *Proc. 6th European Conference on Mixing, Pavia, Italy*, p. 369.

Hickman, A. D. (1988b). The measurement of oxygen mass transfer coefficients using a simple novel technique, Proc. 3rd Bioreactor Project Research Symposium, NEL, East Kilbride, Glasgow, Scotland, May 4.

Hickman, A. D., and A. W. Nienow (1986). Mass transfer and hold-up in an agitated simulated fermentation broth as a function of viscosity, *Proc. First International Conference on Bioreactor Fluid Dynamics*, BHRA, Cambridge, Paper 22, p. 301.

Host-Madsen, A., and D. R. McCluskey (1994). On the accuracy and reliability of PIV measurements, *Proc. 7th International Symposium on Applications of Laser Techniques to Flow Measurements*, Lisbon.

Hoyle, B. S. (1996). Process tomography using ultrasonic sensors, *Meas. Sci. Technol.*, **7**, 272–280.

Hulst, H. C. (1981). *Light Scattering by Small Particles*, Dover, New York.

Keane, R. D., and R. J. Adrian (1992). Theory of cross-correlation analysis of PIV images, *Appl. Sci. Res.*, **49**, 191–215.

Keane, R. D., and R. J. Adrian (1993). *Flow Visualization and Image Analysis*, Kluwer, Norwell, MA, pp. 1–25.

Keane, R. D., R. J. Adrian, and Y. Zhang (1995). Super-resolution particle image velocimetry, *Meas. Sci. Technol.*, **6**, 754–768.

Kegrise, M. A., and G. S. Settles (2000). Schlieren image-correlation velocimetry and its application to free-convection flows, *Proc. 9th International Symposium on Flow Visualization*, G. M. Carlomagno and I. Grant, eds., Henriot-Watt University, Edinburgh, pp. 380: 1–13.

Khang, S. J., and T. J. Fitzgerald (1975). A new probe and circuit for measuring electrolyte conductivity, *Ind. Eng. Chem. Fundam.*, **14**(3), 208–211.

Kresta, S. M. (1998). Turbulence in stirred tanks: anisotropic, approximate, and applied, *Can. J. Chem. Eng.*, **76**, 563–576.

Kresta, S. M., and P. E. Wood, 1993. The flow field produced by a 45 degree pitched blade turbine: changes in the circulation pattern due to off bottom clearance, *Can. J. Chem. Eng.*, **71**, 42–53.

Kuboi, R., A. W. Nienow, and K. V. Allsford (1983). Using a derotational prism for studying gas dispersion processes, *Chem. Eng. Commun.*, **22**, 29.

Landreth, C. C., and R. J. Adrian (1988). Electro-optical image shifting for particle image velocimetry, *Appl. Opt.*, **27**, 4216–4220.

Landreth, C. C., R. J. Adrian, and C. S. Yao (1988). Double-pulsed particle image velocimeter with directional resolution for complex flows, *Exp. Fluids*, **6**, 119–128.

Levenspiel, O. (1972). *Chemical Reaction Engineering*, 2nd ed., Wiley, New York.

Lourenco, L. M. (1993). Velocity bias technique for particle image velocimetry measurements of high speed flows, *Appl. Opt.*, **32**, 2159–2162.

Lourenco, L. M., A. Krothopalli, and C. A. Smith (1989). Particle image velocimetry, in *Advances in Fluid Mechanics Measurements*, Springer-Verlag, Berlin, p. 127.

Lübbe, D. (1982). Thesis, University of Hanover, Germany.

Machon, V., C. M. McFarlane, and A. W. Nienow (1991). Power input and gas hold-up in gas liquid dispersions agitated by axial flow impellers, *Proc. 7th European Conference on Mixing, KVIV*, Bruges, Belgium, p. 242.

Machon, V., A. W. Pacek, and A. W. Nienow (1997). Bubble sizes in electrolyte and alcohol solutions in a turbulent stirred vessel, *Trans. Inst. Chem. Eng.*, **75A**, 339–348.

Martin, T., C. M. McFarlane, and A. W. Nienow (1994). The influence of liquid properties and impeller type on bubble coalescence behaviour and mass transfer in sparged, agitated reactors, *Proc. 8th European Conference on Mixing, Chem. Eng. Symp. Sers.*, **136**, 57–64.

McKee, S. L., R. A. Williams, F. J. Dickin, R. Mann, J. Brinkel, P. Ying, A. Boxman, and G. McGrath (1994). Measurement of concentration profiles and mixing kinetics in stirred tanks using a resistance tomography technique, *Proc. 8th European Conference on Mixing, Inst. Chem. Eng. Symp. Ser.*, **136**, 9–16.

McLaughlin, D. K., and W. G. Tiedermann, Jr. (1973). Biasing correction for individual realization of laser anemometer measurements in turbulent flow, *Phy. Fluids*, **16**(12), 2082–2088.

Meinhart, C. D., S. T. Werely, and J. G. Santiago (1999). PIV measurements of a microchannel flow, *Exp. Fluids*, **27**, 414–419.

Melling, A. (1997). Tracer particles and seeding for particle image velocimetry, *Meas. Sci. and Technol.* **8**(12), 1406–1416.

Mewes, D., and W. Ostendorf (1986). Application of tomographic measurement techniques for process engineering studies, *Int. Chem. Eng.*, **26**(1), 11–21.

Michelet, St. (1998). Ph.D. dissertation, INPL, CNRS, Nancy, France.

Muller, F. L., and J. F. Davidson (1992). On the contribution of small bubbles to mass transfer in bubble columns containing highly viscous liquids, *Chem. Eng. Sci.*, **47**(13/14), 3525–3532.

Nienow, A. W., and D. J. Wisdom (1974). Flow over disc turbine blades, *Chem. Eng. Sci.*, **29**, 1994–1997.

Özcan-Taşkin, N. G. (1993). On the effects of viscoelasticity in stirred tanks, Ph.D Thesis, Birmingham University.

Papadopoulos, G. (2000). Inferring temperature by means of a novel shadow image velocimetry technique, *J. Thermophys. Heat Transfer*, **14**(4), 593–600.

Papadopoulos, G., and K. J. Hammad (2003). Time-resolved PIV measurements within a triple impeller stirred-tank. *ASME/JSME Joint Fluids Engineering Summer Meeting*, FEDSM2003-45295, July 6–11, Honolulu, HI.

Parker, D. J., and P. A. McNeil (1996). Positron emission tomography for process applications, *Meas. Sci. Technol.*, **7**(3), 287–296.

Patel, S. A., J. Daly, and D. B. Bukur (1989). Holdup and interfacial area measurements using dynamic gas disengagement, *AIChE J.*, **35**(6), 931–942.

Penney, W. R., M. Myrick, P. Popejoy and R. Goff (1999). Liquid–liquid suspension using six-blade disk impellers in agitated vessels: an experimental and correlational study, presented at the AIChE Meeting, Dallas, TX.

Prasad, A. K., and K. Jensen (1995). Scheimpflug stereo camera for particle image velocimetry in liquid flows; *Appl. Opt.*, **34**(30), 7092–7099.

Raffel, M., M. Willert, and J. Kompenhans (1998). *Particle Image Velocimetry: A Practical Guide*, Springer-Verlag, Berlin.

Reith, T., and W. J. Beek (1970). Bubble coalescence rates in a stirred contactor, *Trans. Inst. Chem. Eng.*, **48**, T63–T68, T76.

Robinson, C. W., and C. R. Wilke (1974). Simultaneous measurement of interfacial area and mass transfer coefficients for a well-mixed gas dispersion in aqueous electrolyte solutions, *AIChE J.*, **20**(2), 285–294.

Schaffer, M., M. Hofken, and F. Durst (1997). Detailed LDV measurements for visualization of the flow field within a stirred-tank reactor equipped with a Rushton turbine, *Chem. Eng. Res. Des.*, **75**, 729–736.

Sheng, J., H. Meng, and R. O. Fox (1998). V*alidation of CFD simulations of a stirred tank using particle image velocimetry data*, *Can. J. Chem. Eng.*, **76**, 611–625.

Skelland, A. H. P. (1967). *Non-Newtonian Flow and Heat Transfer*, Wiley, New York.

Skelland, A., and R. Seksaria (1978). Minimum impeller speeds for liquid–liquid dispersion in baffled vessels, *Ind. Eng. Chem. Process. Des. Dev.*, **17**, 56–61.

Smith, J. M. (1985). Dispersion of gases in liquids, *Chapter 5 in Mixing of Liquids by Mechanical Agitation*, J. J. Ulbrecht, and G. K. Patterson, eds., Gordon & Breach, New York..

Smith, J. M., M. M.C. G. Warmoeskerken, and E. Zeef (1987). Flow conditions in vessels dispersing gases in liquids, *Biotechnol. Prog. AIChE*, **3**, 107–115.

Soloff, S. M., R. J. Adrian, and Z. C. Liu (1997). Distortion compensation for generalized stereoscopic particle image velocimetry, *Meas. Sci. and Technol.*, **8**, 1441–1454.

Sridhar, R., and O. E. Potter (1978). Interfacial area measurements in gas–liquid agitated vessels: comparison of techniques, *Chem. Eng. Sci.*, **33**, 1347–1353.

Sriram, K., and R. Mann (1977). Dynamic gas disengagement: a new technique for assessing the behaviour of bubble columns, *Chem. Eng. Sci.*, **32**, 571–580.

Stanislas, M., J. Kompenhans, and J. Westerweel (eds.) (2000). *Particle Image Velocimetry: Progress Toward Industrial Applications*, Kluwer, Dordrecht, The Netherlands.

Stanley, S. J., R. Mann, and K. Primrose (2002). Tomographic imaging in three dimensions for single-feed semi-batch operation of a stirred vessel, *Trans. Inst. Chem. Eng. A (Chem. Eng. Res. Des.)*, **80**, 903–909.

Takashima, I., and M. Mochizuki (1971). Tomographic observations of the flow around an agitator impeller, *J. Chem. Eng. Jpn*, **4**(1), 66–72.

Tokumaru, P. T., and P. E. Dimotakis (1995) Image correlation velocimetry, *Exp. Fluids*, **19**(1), 1–15.

Tropea, C. (1995). Laser Doppler anemometry: recent developments and future challenges, *Meas. Sci. Technol.*, **6**, 605–619.

Van de Vusse, J. G. (1955). Mixing by agitation of miscible liquids. *Chem. Eng. Sci.*, **4**, 178–200.

Van den Hulst, H. C. (1981). *Light Scattering by Small Particles*, Dover, New York.

Vasconcelos, J. M.T., A. W. Nienow, T. Martin, S. S. Alves, and C. M. McFarlane (1997). Alternative ways of applying the hydrogen peroxide steady state method of $K_L a$ measurement, *Trans. Inst. Chem. Eng.*, **75A**, 467–472.

Wang, M., F. J. Dickin, R. A. Williams, R. C. Waterfall, and M. S. Beck (1993). Electrical resistance tomography on metal walled vessels, presented at the ECAPT Conference, Karlsruhe, Germany.

Warmoeskerken, M. M. C. G., J. Speur, and J. M. Smith (1984). Gas–liquid dispersion with pitched-blade turbines, *Chem. Eng. Commun.*, **25**, 11–29.

Watrasiewics, B. M., and M. J. Rudd (1976). *Laser Doppler Measurements*, Butterworth, London.

Westerweel, J. (1993). *Digital Particle Image Velocimetry—Theory and Application*, Delft University Press, Delft, The Netherlands.

Whitton, M. J., S. Cropper, and N. G. Özcan-Taşkin (1997). Mixing of liquid phase in large vessels equipped with multiple impellers, Proc. 4th International Conference on Bioreactor and Bioprocess Fluid Dynamics, BHRG, Edinburgh, Scotland, pp. 277–294.

Willert, C. E., and M. Gharib (1991). Digital particle image velocimetry, *Exp. Fluids*, **10**, 181–193.

Yamamura, H., and K. Takahashi (1999). Minimum impeller speeds for complete liquid–liquid dispersion in a baffled vessel, *J. Chem. Eng. Jpn.* **32**(4), 395–401.

Yianneskis, M., Z. Popiolek, and J. H. Whitelaw (1987). An experimental study of the steady and unsteady-flow characteristics of stirred reactors, *J. Fluid Mech.*, **175**, 537–555.

Zalc, J. M. (2000). Computational fluid dynamic tools for investigating flow and mixing in industrial systems: the koch–Glitsch SMX static mixer and a three Rushton turbine stirred tank, Ph.D. dissertation, Rutgers, The State University of New Jersey.

Zhou, G., and S. M. Kresta (1996). Impact of geometry on the maximum turbulence energy dissipation rate for various impellers, *AIChE J.*, **42**, 2476–2490.

# CHAPTER 5

# Computational Fluid Mixing

ELIZABETH MARDEN MARSHALL and ANDRÉ BAKKER

*Fluent, Inc.*

## 5-1 INTRODUCTION

Mixing processes can be based on a number of mechanisms, from agitation to sparging to static flow manipulation. Agitation in a stirred tank is one of the most common operations, yet presents one of the greatest challenges in the area of computer simulation. Stirred tanks typically contain an impeller mounted on a shaft, and optionally can contain baffles and other internals, such as spargers, coils, and draft tubes. Modeling a stirred tank using computational fluid dynamics (CFD) requires consideration of many aspects of the process. First, any computational model requires that the domain of interest, in this case the volume occupied by the fluid inside the vessel, be described by a computational grid, a collection of small subdomains or cells. It is in these cells that problem-specific variables are computed and stored. The computational grid must fit the contours of the vessel and its internals, even if the components are geometrically complex. Second, the motion of the impeller in the tank must be treated in a special way, especially if the tank contains baffles or other internals. The special treatment employed affects both the construction of the computational grid and the solution method used to obtain the flow field numerically. In this chapter the process of modeling the flow inside a stirred tank is examined, and these special considerations are discussed at length.

In Section 5-2, an introduction to the field of computational fluid dynamics is given, with an emphasis on the fundamental equations that are used to describe processes that are common in mixing applications. An overview of the numerical methods used to solve these equations is presented in Section 5-3. Numerical simulations of stirred tanks are normally done in either two or three dimensions. In two dimensional (2D) simulations, the geometry and flow field are assumed to

*Handbook of Industrial Mixing: Science and Practice*, Edited by Edward L. Paul, Victor A. Atiemo-Obeng, and Suzanne M. Kresta
ISBN 0-471-26919-0 Copyright © 2004 John Wiley & Sons, Inc.

**258** COMPUTATIONAL FLUID MIXING

be axisymmetric or independent of the angular dimension. The solution domain extends from the axis of the vessel out to the vessel wall. Approximations are required for elements that do have angular dependence, such as the impellers and baffles. These approximate methods are discussed in Section 5-4. In three dimensional (3D) simulations, the impellers, baffles, and other internals can be modeled using their exact geometry. The challenge in these simulations is to incorporate the motion of the impeller in the presence of the stationary tank and internals. Methods for performing 3D simulations are discussed in Section 5-5. Section 5-6 illustrates how CFD results can be interpreted for mixing analysis. Several application examples are presented in Section 5-7, and closing remarks, including a review of some of the common pitfalls to success, are given in Section 5-8.

Figure 5-1*a* shows the outline of a simple baffled stirred tank containing a Rushton turbine on a centrally mounted shaft. The tank has diameter T. The impeller has diameter D and is located a distance C off the bottom of the tank. These symbols are used throughout the chapter.

In addition, references will be made to the computational grid that is necessary for computing a numerical solution for the flow field in a stirred tank when the impeller is operational. This grid can take on many forms, as discussed in Section 5-3. One example of a computational grid for the vessel of Figure 5-1*a* is shown in Figure 5-1*b*.

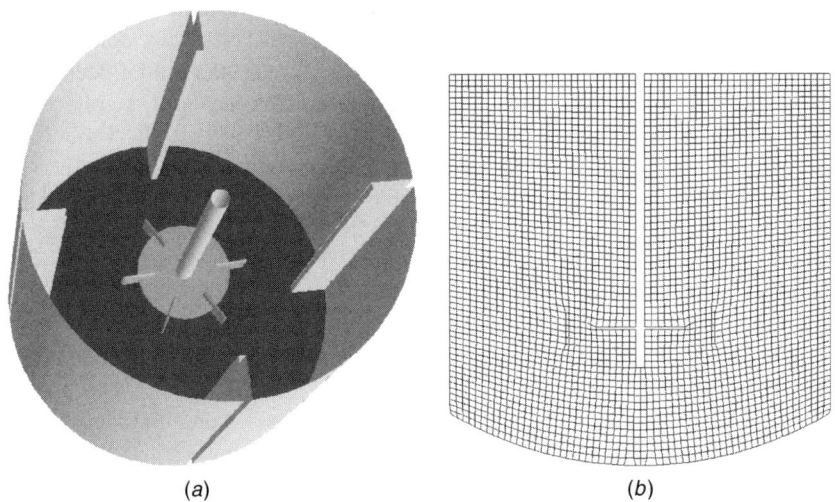

(a)          (b)

**Figure 5-1** (*a*) Mixing vessel showing a Rushton turbine on a central shaft and baffles. (*b*) Example of a computational grid that can be used for solution of the flow field in this vessel.

## 5-2 COMPUTATIONAL FLUID DYNAMICS

Computational fluid dynamics (CFD) is the numerical simulation of fluid motion. While the motion of fluids in mixing is an obvious application of CFD, there are hundreds of others, ranging from blood flow through arteries, to supersonic flow over an airfoil, to the extrusion of rubber in the manufacture of automotive parts. Numerous models and solution techniques have been developed over the years to help describe a wide variety of fluid motion. In this section, the fundamental equations for fluid flow are presented.

Although the primary focus is on specific models that are relevant to the analysis of mixing processes, a number of advanced models for more complex flows are also discussed.

### 5-2.1 Conservation Equations

If a small volume, or element of fluid in motion is considered, two changes to the element will probably take place: (1) the fluid element will translate and possibly rotate in space, and (2) it will become distorted, either by a simple stretching along one or more axes or by an angular distortion that causes it to change shape. The process of translation is often referred to as *convection*, and the process of distortion is related to the presence of gradients in the velocity field and a process called *diffusion*. In the simplest case, these processes govern the evolution of the fluid from one state to another. In more complicated systems, sources can also be present that give rise to additional changes in the fluid. Many more phenomena can also contribute to the way a fluid element changes with time. Heat can cause a gas to expand, and chemical reactions can cause the viscosity to change, for example. Many of the processes such as those that are involved in the description of generalized fluid motion are described by a set of conservation or transport equations. These equations track, over time, changes in the fluid that result from convection, diffusion, and sources or sinks of the conserved or transported quantity. Furthermore, these equations are coupled, meaning that changes in one variable (say, the temperature) can give rise to changes in other variables (say, the pressure). The equations discussed below describe many of these coupled phenomena, with an emphasis on those processes that are typical in mixing applications.

***5-2.1.1 Continuity.*** The continuity equation is a statement of conservation of mass. To understand its origin, consider the flow of a fluid of density $\rho$ through the six faces of a rectangular block, as shown in Figure 5-2. The block has sides of length $\Delta x_1$, $\Delta x_2$, and $\Delta x_3$ and velocity components $U_1$, $U_2$, and $U_3$ in each of the three coordinate directions. To ensure conservation of mass, the sum of

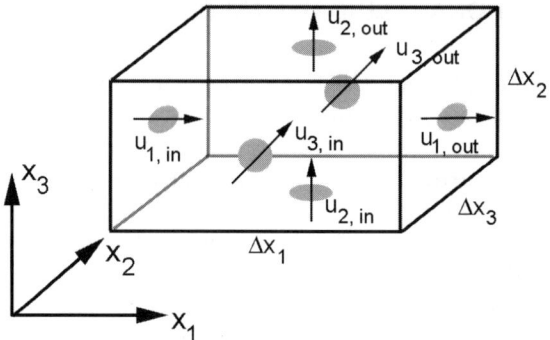

**Figure 5-2** A rectangular volume with inflow and outflow can be used to illustrate a conservation equation.

the mass flowing through all six faces must be zero:

$$\rho(U_{1,out} - U_{1,in})(\Delta x_2 \Delta x_3) + \rho(U_{2,out} - U_{2,in})(\Delta x_1 \Delta x_3) \\ + \rho(U_{3,out} - U_{3,in})(\Delta x_1 \Delta x_2) = 0 \quad (5\text{-}1)$$

Dividing through by $(\Delta x_1 \Delta x_2 \Delta x_3)$ the equation can be written as

$$\rho \frac{\Delta U_1}{\Delta x_1} + \rho \frac{\Delta U_2}{\Delta x_2} + \rho \frac{\Delta U_3}{\Delta x_3} = 0 \quad (5\text{-}2)$$

or, in differential form,

$$\rho \frac{\partial U_1}{\partial x_1} + \rho \frac{\partial U_2}{\partial x_2} + \rho \frac{\partial U_3}{\partial x_3} = 0 \quad (5\text{-}3)$$

A more compact way to write eq. (5-3) is through the use of Einstein notation:

$$\rho \frac{\partial U_i}{\partial x_i} = 0 \quad (5\text{-}4)$$

With this notation, whenever repeated indices occur in a term, the assumption is that there is a sum over all indices. Here, and elsewhere in this chapter, $U_i$ is the ith component of the fluid velocity, and partial derivatives with respect to $x_i$ are assumed to correspond to one of the three coordinate directions. For more general cases, the density can vary in time and in space, and the continuity equation takes on the more familiar form

$$\frac{\partial \rho}{\partial t} + \frac{\partial}{\partial x_i}(\rho U_i) = 0 \quad (5\text{-}5)$$

### 5-2.1.2 Momentum.
The momentum equation is a statement of conservation of momentum in each of the three component directions. The three momentum equations are collectively called the *Navier–Stokes equations*. In addition to momentum transport by convection and diffusion, several momentum sources are also involved:

$$\frac{\partial(\rho U_i)}{\partial t} + \frac{\partial}{\partial x_j}(\rho U_i U_j) = -\frac{\partial p}{\partial x_i} + \frac{\partial}{\partial x_j}\left[\mu\left(\frac{\partial U_i}{\partial x_j} + \frac{\partial U_j}{\partial x_i} - \frac{2}{3}\frac{\partial U_k}{\partial x_k}\delta_{ij}\right)\right] + \rho g_i + F_i \quad (5\text{-}6)$$

In eq. (5-6) the convection terms are on the left. The terms on the right-hand side are the pressure gradient, a source term; the divergence of the stress tensor, which is responsible for the diffusion of momentum; the gravitational force, another source term; and other generalized forces (source terms), respectively.

### 5-2.1.3 Turbulence.
A number of dimensionless parameters have been developed for the study of fluid dynamics that are used to categorize different flow regimes. These parameters, or numbers, are used to classify fluids as well as flow characteristics. One of the most common of these is the *Reynolds number*, defined as the ratio of inertial forces, or those that give rise to motion of the fluid, to frictional forces, or those that tend to slow the fluid down. In geometrically similar domains, two fluids with the same Reynolds number should behave in the same manner. For simple pipe flow, the Reynolds number is defined as

$$\text{Re} = \frac{\rho U d}{\mu} \quad (5\text{-}7)$$

where $\rho$ is the fluid density, U the axial velocity in the pipe, d the pipe diameter, and $\mu$ the molecular or dynamic viscosity of the fluid. For mixing tanks, a modified definition is used:

$$\text{Re} = \frac{ND^2\rho}{\mu} \quad (5\text{-}8)$$

where N is the impeller speed, in rev/s, and D is the impeller diameter. Based on the value of the Reynolds number, flows fall into either the laminar regime, with small Reynolds numbers, or the turbulent regime, with high Reynolds numbers. The transition between laminar and turbulent regimes occurs throughout a range of Reynolds numbers rather than at a single value. For pipe flow, transition occurs in the vicinity of Re = 2000 to 4000, while in mixing tanks, it usually occurs somewhere between Re = 50 and 5000, depending on the power number of the impeller. In the turbulent regime, fluctuations in the mean velocity and other variables occur, and for the model to be able to provide meaningful results, their effect needs to be incorporated into the CFD model. This is done through the use of a turbulence model.

Several methods are available for including turbulence in the Navier–Stokes equations. Most of these involve a process of time averaging the conservation equations. When turbulence is included, the transported quantity, say velocity, is assumed to be the sum of an equilibrium and a fluctuating component, $U_i + u'_i$. After time averaging over many cycles of the fluctuation, terms containing factors of the fluctuating component average to zero. The only term that remains positive definite is one containing the product of two fluctuating terms. The remaining terms are identical to those in eq. (5-6). Thus, the *Reynolds-averaged Navier–Stokes* (RANS) *equation* for momentum is

$$\frac{\partial (\rho U_i)}{\partial t} + \frac{\partial}{\partial x_j}(\rho U_i U_j) = -\frac{\partial p}{\partial x_i} + \frac{\partial}{\partial x_j}\left[\mu\left(\frac{\partial U_i}{\partial x_j} + \frac{\partial U_j}{\partial x_i} - \frac{2}{3}\frac{\partial U_k}{\partial x_k}\delta_{ij}\right)\right]$$
$$+ \frac{\partial}{\partial x_j}(-\overline{\rho u'_i u'_j}) + \rho g_i + F_i \tag{5-9}$$

The new terms involving $\overline{u'_i u'_j}$ are called the *Reynolds stresses*. The overbar indicates that these terms represent time-averaged values. Reynolds stresses contribute new unknowns to the RANS equations and need to be related to the other variables. This is done through various models, collectively known as *turbulence models*.

*Boussinesq Hypothesis.* The Boussinesq hypothesis makes the assumption that the Reynolds stresses can be expressed in terms of mean velocity gradients. The following statement of the hypothesis shows the introduction of a new constant that is dimensionally equivalent to viscosity:

$$\overline{\rho u'_i u'_j} = \frac{2}{3}\rho k \delta_{ij} + \left[\mu_t\left(\frac{\partial U_i}{\partial x_j} + \frac{\partial U_j}{\partial x_i}\right)\right] \tag{5-10}$$

The new constant, $\mu_t$, is the turbulent or eddy viscosity. It can be seen that when eq. (5-10) is substituted into eq. (5-9), the terms containing the partial derivatives can be combined and a new quantity, the effective viscosity, can be introduced:

$$\mu_{eff} = \mu + \mu_t \tag{5-11}$$

The hypothesis also introduces another term involving a new variable, k, the kinetic energy of turbulence. This quantity is defined in terms of the velocity fluctuations u', v', and w' in each of the three coordinate directions:

$$k = \tfrac{1}{2}(\overline{u'^2} + \overline{v'^2} + \overline{w'^2}) \tag{5-12}$$

It is the job of the turbulence model to compute the Reynolds stresses for substitution into eq. (5-9). In some cases, this is done by computing the parameters k and $\mu_t$ (or k and $\mu_{eff}$) for substitution into eq. (5-10) and ultimately, eq. (5-9). All turbulence models use some level of approximation to accomplish this goal, and it is the nature of the flow conditions in each specific application that determines which set of approximations is acceptable for use. A brief summary of some of the popular turbulence models in use today for industrial applications is given below.

*k–ε Model.* The k–ε model is one of a family of two-equation models for which two additional transport equations must be solved to compute the Reynolds stresses. (Zero- and one-equation models also exist but are not commonly used in mixing applications.) It is a robust model, meaning that it is computationally stable, even in the presence of other, more complex physics. It is applicable to a wide variety of turbulent flows and has served the fluid modeling community for many years. It is semiempirical, based in large part on observations of high-Reynolds-number flows. The two transport equations that need to be solved for this model are for the kinetic energy of turbulence, k, and the rate of dissipation of turbulence, ε:

$$\frac{\partial (\rho k)}{\partial t} + \frac{\partial}{\partial x_i}(\rho U_i k) = \frac{\partial}{\partial x_i}\left(\mu + \frac{\mu_t}{\sigma_k}\right)\frac{\partial k}{\partial x_i} + G_k - \rho\varepsilon \qquad (5\text{-}13)$$

$$\frac{\partial (\rho \varepsilon)}{\partial t} + \frac{\partial}{\partial x_i}(\rho U_i \varepsilon) = \frac{\partial}{\partial x_i}\left(\mu + \frac{\mu_t}{\sigma_\varepsilon}\right)\frac{\partial \varepsilon}{\partial x_i} + C_1\frac{\varepsilon}{k}G_k + C_2\rho\frac{\varepsilon^2}{k} \qquad (5\text{-}14)$$

The quantities $C_1$, $C_2$, $\sigma_k$, and $\sigma_\varepsilon$ are empirical constants. The quantity $G_k$ appearing in both equations is a generation term for turbulence. It contains products of velocity gradients and also depends on the turbulent viscosity:

$$G_k = \mu_t \left(\frac{\partial U_i}{\partial x_j} + \frac{\partial U_j}{\partial x_i}\right)\frac{\partial U_j}{\partial x_i} \qquad (5\text{-}15)$$

Other source terms can be added to eqs. (5-13) and (5-14) to include other physical effects, such as swirl, buoyancy, or compressibility, for example. The turbulent viscosity is derived from both k and ε and involves a constant taken from experimental data, $C_\mu$, which has a value of 0.09:

$$\mu_t = \rho C_\mu \frac{k^2}{\varepsilon} \qquad (5\text{-}16)$$

To summarize the solution process for the k–ε model, transport equations are solved for the turbulent kinetic energy and dissipation rate. The solutions for k and ε are used to compute the turbulent viscosity, $\mu_t$. Using the results for $\mu_t$ and k, the Reynolds stresses can be computed from the Boussinesq hypothesis for substitution into the momentum equations. Once the momentum equations have been solved, the new velocity components are used to update the turbulence generation term, $G_k$, and the process is repeated.

*RNG k–ε Model.* The renormalization group (RNG) model (Yakhot and Orszag, 1986) was developed in response to the empirical nature of the standard k–ε model. Rather than being based on observed fluid behavior, it is derived using statistical methods used in the field of RNG theory. It is similar in form to the standard k–ε model but contains modifications in the dissipation equation to better describe flows with regions of high strain, such as the flow around a bend or reattachment following a recirculation zone. In addition, a differential equation is solved for the turbulent viscosity. When the solution of this differential equation is evaluated in the high Reynolds number limit, eq. (5-16) is returned with a coefficient, $C_\mu$, of 0.0845, within 7% of the empirical value of 0.09. While the RNG model works well for high Reynolds number flows, it also works well for transitional flows, where the Reynolds number is in the low turbulent range.

*Realizable k–ε Model.* The realizable k–ε model (Shih et al., 1995) is a fairly recent addition to the family of two-equation models. It differs from the standard k–ε model in two ways. First, the turbulent viscosity is computed in a different manner, making use of eq. (5-16) but using a variable for the quantity $C_\mu$. This is motivated by the fact that in the limit of highly strained flow, some of the normal Reynolds stresses, $\overline{u_i'^2}$, can become negative in the k–ε formulation, which is unphysical, or unrealizable. The variable form of the constant $C_\mu$ is a function of the local strain rate and rotation of the fluid and is designed to prevent unphysical values of the normal stresses from developing.

The second difference is that the realizable k–ε model uses different source and sink terms in the transport equation for eddy dissipation. The resulting equation is considerably different from the one used for both the standard and RNG k–ε models. The modified prediction of ε, along with the modified calculation for $\mu_t$, makes this turbulence model superior to the other k–ε models for a number of applications. In particular, the model does better in predicting the spreading rate of round jets, such as those emitted from a rotating impeller blade.

*RSM Model.* The Reynolds stress model (RSM) does not use the Boussinesq hypothesis. Rather than assume that the turbulent viscosity is isotropic, having one value as in the k–ε model, the Reynolds stress model computes the stresses, $\overline{u_i' u_j'}$, individually. For 2D models, this amounts to four additional transport equations. For 3D models, six additional transport equations are required. Along with the transport equation for ε, which must also be solved in the RSM model, the full effect of turbulence can be represented in the momentum equations with greater accuracy than can be obtained from the k–ε models. Flows for which the assumption of isotropic turbulent viscosity breaks down include those with high swirl, rapid changes in strain rate, or substantial streamline curvature. As computer power and speed have increased during the past several years, the use of the Reynolds stress turbulence model has become more widespread, giving rise to improved accuracy over other RANS-based turbulence models when compared to experimental results for a number of applications, such as the flow in unbaffled stirred vessels.

*LES Model.* A fairly recent entry to the group of commercially available turbulence models is the large eddy simulation (LES) model. This approach recognizes that turbulent eddies occur on many scales in a flow field. Large eddies are often sized according to the extents of the physical domain. Small eddies, however, are assumed to have similar properties and behavior for all problem domains, independent of their overall size or purpose. With the LES model, the continuity and momentum equations are filtered prior to being solved in a transient fashion. The filtering process isolates the medium and large scale eddies from those that are smaller than a typical cell size. The effects of the small eddies are included in the filtered equations through the use of a subgrid scale model. The transient simulation is then free to capture the random fluctuations that develop on medium and large scales. Despite the fact that a transient simulation is needed for this turbulence model, it has proven to be worth the effort. Simulations to date have predicted unstable behavior successfully in jets, flames, and both static mixers and stirred tanks. See, for example, Section 5-7.11, where LES is used to simulate the flow in an HEV static mixer. An overview of the turbulence models discussed in this section, including the primary advantages and disadvantages of each, is provided in Table 5-1.

**5-2.1.4 Species.** The species equation is a statement of conservation of a single species. Multiple-species equations can be used to represent fluids in a mixture with different physical properties. Solution of the species equations can predict how different fluids mix, but not how they will separate. Separation is the result of different body forces acting on the fluids, such as gravity acting on fluids of different density. To model separation, separate momentum equations are required for each of the fluids so that the body forces can act on the fluids independently (see Section 5-2.2.2). Species transport is nevertheless a very useful tool for predicting blending times or chemical reaction. For the species $i'$, the conservation equation is for the mass fraction of that species, $m_{i'}$, and has the following form:

$$\frac{\partial(\rho m_{i'})}{\partial t} + \frac{\partial}{\partial x_i}(\rho U_i m_{i'}) = -\frac{\partial}{\partial x_i} J_{i',i} + R_{i'} + S_{i'} \quad (5\text{-}17)$$

In eq. (5-17), $J_{i',i}$ is the i component of the diffusion flux of species $i'$ in the mixture. For laminar flows, $J_{i',i}$ is related to the diffusion coefficient for the species and local concentration gradients (Fick's law of diffusion). For turbulent flows, $J_{i',i}$ includes a turbulent diffusion term, which is a function of the turbulent Schmidt number. $R_{i'}$ is the rate at which the species is either consumed or produced in one or more reactions, and $S_{i'}$ is a general source term for species. The general source term

**Table 5-1** Summary of Turbulence Models

| Turbulence Model | Description, Advantages, and Disadvantages |
| --- | --- |
| Standard k–ε | The most widely used model, it is robust, economical, and has served the engineering community well for many years. Its main advantages are a rapid, stable calculation, and reasonable results for many flows, especially those with high Reynolds number. It is not recommended for highly swirling flows, round jets or for flows with strong flow separation. |
| RNG k–ε | A modified version of the k–ε model, this model yields improved results for swirling flows and flow separation. It is not well suited for round jets and is not as stable as the standard k–ε model. |
| Realizable k–ε | Another modified version of the k–ε model, the realizable k–ε model correctly predicts the flow in round jets and is also well suited for swirling flows and flows involving separation. |
| RSM | The full Reynolds stress model provides good predictions for all types of flows, including swirl, separation, and round and planar jets. Because it solves transport equations for the Reynolds stresses directly, longer calculation times are required than for the k–ε models. |
| LES | Large eddy simulation is a transient formulation that provides excellent results for all flow systems. It solves the Navier–Stokes equations for large scale turbulent fluctuations and models only the small scale fluctuations (smaller than a computational cell). Because it is a transient formulation, the required computational resources are considerably larger than those required for the RSM and k–ε style models. In addition, a finer grid is needed to gain the maximum benefit from the model and to accurately capture the turbulence in the smallest, subgrid scale eddies. Analysis of LES data usually requires some degree of advance planning. |

can be used for nonreacting sources, such as the evaporated vapor from a heated droplet, for example. When two or more species are present, the sum of the mass fractions in each cell must add to 1.0. For this reason, if there are n species involved in a simulation, only n − 1 species equations need to be solved. The mass fraction of the $n$th species can be computed from the required condition:

$$\sum_{i'}^{n} m_{i'} = 1.0 \tag{5-18}$$

More details about reacting flow are presented in Section 5-2.2.1.

### 5-2.1.5 Heat Transfer.
Heat transfer is often expressed as an equation for the conservation of energy, typically in the form of static or total enthalpy. Heat can be generated (or extracted) through many mechanisms, such as wall heating (in a jacketed reactor), cooling through the use of coils, and chemical reaction. In addition, fluids of different temperatures may mix in a vessel, and the time for the mixture to come to equilibrium may be of interest. The equation for conservation of energy (total enthalpy) is

$$\frac{\partial(\rho E)}{\partial t} + \frac{\partial}{\partial x_i}[U_i(\rho E + p)] = \frac{\partial}{\partial x_i}\left[k_{eff}\frac{\partial T}{\partial x_i} - \sum_{j'} h_{j'} J_{j',i} + U_j(\tau_{ij})_{eff}\right] + S_h \quad (5\text{-}19)$$

In this equation, the energy, E, is related to the static enthalpy, h, through the following relationship involving the pressure, p, and velocity magnitude, U:

$$E = h - \frac{p}{\rho} + \frac{U^2}{2} \quad (5\text{-}20)$$

For incompressible flows with species mixing, the static enthalpy is defined in terms of the mass fractions, $m_{j'}$, and enthalpies, $h_{j'}$, of the individual species:

$$h = \sum_{j'} m_{j'} h_{j'} + \frac{p}{\rho} \quad (5\text{-}21)$$

The enthalpy for the individual species j' is a temperature-dependent function of the specific heat of that species:

$$h_{j'} = \int_{T,ref}^{T} c_{p,j'}\, dT \quad (5\text{-}22)$$

Once the enthalpy has been determined from the relationships shown above, the temperature can be extracted using eq. (5-22). This process is not straightforward because the temperature is the integrating variable. One technique for extracting the temperature involves the construction of a look-up table at the start of the calculation, using the known or anticipated limits for the temperature range. This table can subsequently be used to obtain temperature values for corresponding enthalpies obtained at any time during the solution.

The first term on the right-hand side of eq. (5-19) represents heat transfer due to conduction, or the diffusion of heat, where the effective conductivity, $k_{eff}$, contains a correction for turbulent simulations. The second term represents heat transfer due to the diffusion of species, where $J_{j',i}$ is the diffusion flux defined in Section 5-2.1.4. The third term involves the stress tensor, $(\tau_{ij})_{eff}$, a collection of velocity gradients, and represents heat loss through viscous dissipation. The fourth term is a general source term that can include heat sources due to reactions, radiation, or other processes.

### 5-2.2 Auxiliary Models

While a wide range of applications can be modeled using the basic transport equations described above, others involve more complex physics and require additional modeling capabilities. Some of these models are discussed below.

**5-2.2.1 Chemical Reaction.** Chemically reacting flows are those in which the chemical composition, properties, and temperature change as the result of a simple or complex chain of reactions in the fluid. Depending on the implementation, reacting flows can require the solution of multiple conservation equations for species, some of which describe reactants, and others of which describe products. To balance the mass transfer from one species to another, reaction rates are used in each species conservation equation, and have as factors the molecular weights, concentrations, and stoichiometries for that species in all reactions.

Consider, for example, the single-step first-order reaction $A + B \rightarrow R$, for which the reaction rate is given by

$$R_i \propto C_A C_B + \overline{c_A c_B} \qquad (5\text{-}23)$$

Here $C_A$ and $C_B$ denote the mean molar concentrations of reactants A and B, while $c_A$ and $c_B$ denote the local concentration fluctuations that result from turbulence. When the species are perfectly mixed, the second term on the right-hand side, containing the correlation of the concentration fluctuations, will approach zero. If the species are not perfectly mixed, this term will be negative and will reduce the reaction rate. The estimation of this correlation term is not straightforward, and numerous models are available (Hannon, 1992) for this purpose. Its presence suggests, however, that the reaction rate should incorporate not only the mean concentrations of the reactant species but the turbulent fluctuations of the reactant species as well, since the latter gives an indication of the degree to which these species are mixed.

One popular method for computing the reaction rates as a function of both mean concentrations and turbulence levels is through the Magnussen model (Magnussen and Hjertager, 1976). Originally developed for combustion, it can also be used for liquid reactions by tuning some of the model parameters. The model consists of rates calculated by two primary means. An Arrhenius, or kinetic rate, $R_{K\_i',k}$, for species $i'$ in reaction $k$, is governed by the local mean species concentrations and temperature in the following manner:

$$R_{K\_i',k} = -v_{i',k} M_{i'} A_k T^{\beta_k} \exp\left(-\frac{E_k}{RT}\right) \prod_{j'=1}^{N} [C_{j'}]^{\eta_{j',k}} = K_{i',k} M_{i'} \prod_{j'=1}^{N} [C_{j'}]^{\eta_{j',k}} \qquad (5\text{-}24)$$

This expression describes the rate at which species $i'$ is consumed in reaction k. The constants $A_k$ and $E_k$, the Arrhenius preexponential factor and activation energy, respectively, are adjusted for specific reactions, often as the result of experimental measurements. The stoichiometry for species $i'$ in reaction k is represented by the factor $v_{i',k}$, and is positive or negative, depending upon whether the species serves as a product or reactant. The molecular weight of the species $i'$ appears as the factor $M_{i'}$. The temperature, T, appears in the exponential term and also as a factor in the rate expression, with an optional exponent, $\beta_k$. Concentrations of other species, $j'$, involved in the reaction, $[C_{j'}]$, appear as factors with optional exponents associated with each. Other factors and terms, not appearing in eq. (5-24), can be added to include effects such as the presence of nonreacting species in the rate equation. Such third-body reactions are typical of the effect of a catalyst on a reaction, for example. Many of the factors appearing in eq. (5-24) are often collected into a single rate constant, $K_{i',k}$.

In addition to the Arrhenius rate, two mixing rates are computed that depend on the local turbulent kinetic energy and dissipation rate. One rate, $R_{M1,i',k}$, involves the mass fraction of the reactant in reaction k, $m_R$, with returns the smallest rate:

$$R_{M1,i',k} = v_{i',k} M_{i'} A \rho \frac{\varepsilon}{k} \frac{m_R}{v_{R,k} M_R} \tag{5-25}$$

where the subscript R refers only to the reactant species, $i' = R$. The other mixing rate, $R_{M2,i',k}$, involves the sum-over-product species mass fractions, $m_P$, and product stoichiometries, $v'_{j',k}$:

$$R_{M2,i',k} = v_{i',k} M_{i'} A B \rho \frac{\varepsilon}{k} \frac{\sum_P m_P}{\sum_{j'}^N v'_{j',k} M_{j'}} \tag{5-26}$$

In the mixing rate expressions, the values 4.0 and 0.5 are often used for the constants A and B, respectively, when the model is used for gaseous combustion. These values can be adjusted, however, for different types of reactions, such as those involving liquids.

After the rates in eqs. (5-24), (5-25), and (5-26) are computed, the smallest, or slowest, is used as a source term in the species transport equations for all species involved in any given reaction. The basic idea behind the Magnussen model is that in regions with high turbulence levels, the eddy lifetime, $k/\varepsilon$, is short, mixing is fast, and as a result the reaction rate is not limited by small scale mixing. In this limit, the kinetic rate usually has the smallest value. On the other hand, in regions with low turbulence levels, small scale mixing may be slow and limit the reaction rate. In this limit, the mixing rates are more important.

The Magnussen model was initially developed for simple, one- or two-step reaction sets, in which all reaction rates are fast relative to the small scale mixing, even though it has found use for more complex systems. Recently, for more complex reaction sets, a new model has been developed (Gran and Magnussen, 1996) called the *eddy dissipation concept* (EDC) *model*. This model assumes that reaction occurs in small turbulent structures, called the *fine scales*. A volume fraction of the small scales is calculated, which depends on the kinematic viscosity of the fluid, the eddy dissipation rate, and the turbulent kinetic energy. Reactions are assumed to occur in the fine turbulent structures, over a time scale that depends on the kinematic viscosity and the energy dissipation rate. A source term for each chemical species is then calculated that depends on the volume fraction of the fine scales, the time scale, and the difference in species concentrations between the fine scale structures and the surrounding fluid. This extension of the Magnussen model provides improved accuracy for complex, multistep reaction sets in which not all reactions are fast relative to the rate at which small scale mixing occurs.

Numerous other reaction models exist that can be coupled to a CFD calculation. The probability density function (PDF) modeling approach (also known as the mixture fraction approach) is one that is based on the assumptions of infinitely fast reactions and chemical equilibrium at all times. In this model a collection of reacting species is described by a mixture fraction, which, under certain circumstances, is a conserved quantity. For the turbulent combustion of fuel and oxygen, for example, the mixture fraction is the elemental mass fraction in the incoming fuel stream. In turbulent conditions, fluctuations in the mixture fraction exist, along with fluctuations in the velocities and other variables. A PDF is used to describe these fluctuations. Choices are available for the assumed shape of the PDF, with the beta PDF being the most popular for engineering applications, since it offers the best agreement with experiment. This two-parameter function depends on the mean and variance of the mixture fraction. These variables are tracked by transport equations that are solved alongside the fluid equations. Based on the values of the mixture fraction (mean and variance) and enthalpy, the mass fractions of all reactant and product species can be obtained. Thus, whereas the kinetic rate expression uses time-averaged values for species mass fractions, the PDF model allows for fluctuations in these quantities. Although this model has many benefits for gaseous combustion systems, it is not the best choice for liquid reactions that are typical of mixing applications. This is because reacting liquid mixtures are not always characterized by chemical equilibrium at all times, and reaction rates can range from being very fast to being very slow compared to typical mixing rates.

Another reaction modeling approach incorporates the methodology used to describe micromixing, or mixing on the smallest scales (Bourne et al., 1981; Hannon, 1992; Fox, 1998). In the context of a CFD calculation, micromixing is on a scale that is smaller than a typical computational cell. Macromixing, on the other hand, is responsible for large scale blending, and mesomixing is in between these limits. The identification of these mixing regimes is drawn from assumptions at the core of turbulence modeling theory: namely, that turbulence energy is generated in large eddies within a domain and cascades to successively smaller eddies before being dissipated at the smallest scales. This cascade of turbulence is associated with a cascade of mixing, from macromixing on the large scales, to mesomixing throughout the midscales, to micromixing on the subgrid scales. One motivation for the interest in micromixing in liquid reactions is that micromixing must occur before reactions can take place. It therefore plays an important role when the reaction times are on the same order as the mixing times. Micromixing models typically use a mixture fraction approach and use a PDF formulation for the turbulence–chemistry interaction. The micromixing models are incorporated through the calculation of the variance of the mixture fraction.

**5-2.2.2 Multiphase Flows.** When multiple fluids are involved in a flow, representing them by multiple species equations works only if the fluids are mixing and not separating. Any separation caused by the action of body forces, such as gravity or centrifugal force, can only be captured by treating the fluids with a multiphase model. When such a model is used, each of the fluids is assigned a separate set of properties, including density. Because different densities are used, forces of different magnitude can act on the fluids, enabling the prediction of separation. Five of the most popular multiphase models that are in wide use in commercial software today are described below.

*Dispersed Or Discrete Phase Model.* The dispersed phase model uses the Navier–Stokes equations to describe a continuous fluid phase and a Lagrangian particle tracking method to describe a dispersed phase consisting of particles, droplets, or bubbles. Heat, mass, and momentum exchange are permitted between the dispersed and fluid phases. Thus, gas bubbles can rise in a liquid, sand particles can settle, and water droplets can evaporate or boil, releasing steam to a background of warm gas, for example. The model is widely used for coal and liquid fuel combustion, bubble columns, and gas spargers in stirred tanks. It is best when the dispersed phase does not exceed 10% of the volume of the mixture in any region.

*VOF Model.* The volume of fluid (VOF) model is designed for two or more immiscible fluids. Because the fluids do not mix, each computational cell is filled with purely one fluid, purely another fluid, or the interface between two (or more) fluids. Because of this unique set of conditions, only a single set of Navier–Stokes equations is required. Each fluid is allowed to have a separate set of properties. The properties used are those of the fluid filling the control volume. If the interface lies inside the control volume, special treatment is used to track its position and slope in both the control volume and neighboring cells as the calculation progresses. This model is used to track free surface flows or the rise of large bubbles in a liquid, for example.

*Eulerian Multiphase Model.* The Eulerian multiphase model is designed for systems containing two or more interpenetrating fluids. The fluids can be in the form of liquids, gases, or solids. Whereas the dispersed phase model works best for low-volume fraction mixtures (<10%), the Eulerian multiphase model is general enough that any volume fraction of any phase is allowed. Separate sets of momentum and continuity equations are used to describe each fluid. Momentum transfer between the phases is incorporated through the use of exchange terms in the momentum equation. When heat and mass transfer between phases occurs, exchange terms are used in the energy and continuity equations as well. The volume fractions of the phases are tracked, with the condition that the sum of the volume fractions for all phases is identically 1.0 at all times in all control volumes. Separate equations can also be used for turbulence and species transport for each phase. Although momentum, mass, heat, and species transfer between phases may be well understood, the same cannot be said for the coupling of the turbulence equations. This is an area that is currently undergoing active research at a number of institutions worldwide.

*Eulerian Granular Multiphase Model.* When the primary phase is a liquid or a gas and the secondary phase consists of solid particles, a modified form of the Eulerian multiphase model can be used. The Eulerian granular multiphase (EGM) model uses kinetic theory to describe the behavior of the granular or particulate phase, which is different in many ways from that of a fluid phase (see, e.g., Ogawa et al., 1980; Ding and Gidaspow, 1990; Syamlal et al., 1993). In particular, the viscosity of the granular phase undergoes a discontinuous change as the granular material transforms from a packed bed at rest to a fluid in motion, and this can only be captured by the special treatment at the heart of the EGM model. Also unique to the model is a solids pressure, which arises in part from inelastic collisions between particles. As is typical of a gas described by kinetic theory, a Maxwellian velocity distribution can be assumed for the granular phase. The width of this distribution, or spread in velocity fluctuations about the mean value, is related to the granular temperature, a parameter that can contribute to several other phenomena in granular multiphase flows. The maximum volume fraction that the granular phase can occupy is always less than 1.0 (typically, 0.6), owing to the void that is always present between the particles. These and other issues are addressed by the EGM model, allowing it to simulate a wide array of granular flow applications, from solids suspension in stirred tanks to fluidized bed flow patterns to flow in a riser.

*Algebraic Slip Mixture Model.* As with the Eulerian multiphase model, the algebraic slip mixture, or ASM model, is designed for use with two interpenetrating fluids. A full set of Navier–Stokes equations is solved for the primary fluid. Rather than solve a complete set for the secondary fluid, however, an algebraic equation for the slip velocity between the fluids is solved instead. The slip velocity is derived from the fluid properties and local flow conditions and is used to compute the velocity of the secondary phase. The ASM model is best when used for liquid–liquid or gas–liquid mixtures. It can also be used for lightly loaded granular mixtures, where the physics associated with the granular phase as it approaches the packing limit are not as important.

**5-2.2.3 *Non-Newtonian Viscosity*.** For Newtonian fluids, the viscosity often varies weakly with the temperature, by an amount that depends on the temperature range in use. Many fluids do not fit this simple pattern, however, and have viscosities that also depend on the shear rate in the fluid. The viscosity of these non-Newtonian fluids can be described by one of a number of laws that involve the local shear rate of the fluid in one way or another. The dependence can be in the form of a power law (the shear rate raised to some power) and can involve a discontinuous transition after a minimum yield stress has been exceeded. In some cases, a fluid will transition from non-Newtonian to Newtonian behavior after a threshold stress has been exceeded. In general, shear-thinning fluids exhibit a drop in viscosity in regions of high shear, while shear-thickening fluids exhibit an increase in viscosity in these regions. For computational fluid dynamics, the consequence of non-Newtonian flow modeling is that the viscosity, a fluid property, becomes coupled to the fluid motion, making the equation set more difficult to solve if the viscosity is strongly varying within the limits of the flow field conditions.

Some non-Newtonian fluids are also described by a property called *viscoelasticity*. As for Newtonian fluids, these fluids deform when a shearing force is applied, but they have a partial memory of their state prior to the application of the force. Thus, when the force is withdrawn, they return, to a greater or lesser degree, to their previous state. Specialty CFD codes exist that have comprehensive models for both non-Newtonian and viscoelastic fluids. These codes are used for certain laminar mixing processes in stirred tanks and extruders.

## 5-3 NUMERICAL METHODS

The differential equations presented in Section 5-2 describe the continuous movement of a fluid in space and time. To be able to solve those equations numerically, all aspects of the process need to be discretized, or changed from a continuous to a discontinuous formulation. For example, the region where the fluid flows needs to be described by a series of connected control volumes, or computational cells. The equations themselves need to be written in an algebraic form. Advancement in time and space needs to be described by small, finite steps rather than the infinitesimal steps that are so familiar to students of calculus. All of these processes are collectively referred to as *discretization*. In this section, discretization of the domain, or grid generation, and discretization of the equations are described. A section on solution methods and one on parallel processing are also included.

### 5-3.1 Discretization of the Domain: Grid Generation

To break the domain into a set of discrete subdomains, or computational cells, or control volumes, a *grid* is used. Also called a *mesh*, the grid can contain elements of many shapes and sizes. In 2D domains, for example, the elements are usually either quadrilaterals or triangles. In 3D domains (Figure 5-3), they can be tetrahedra (with four sides), prisms (five sides), pyramids (five sides), or hexahedra (six sides). A series of line segments (2D) or planar faces (3D) connecting the boundaries of the domain are used to generate the elements.

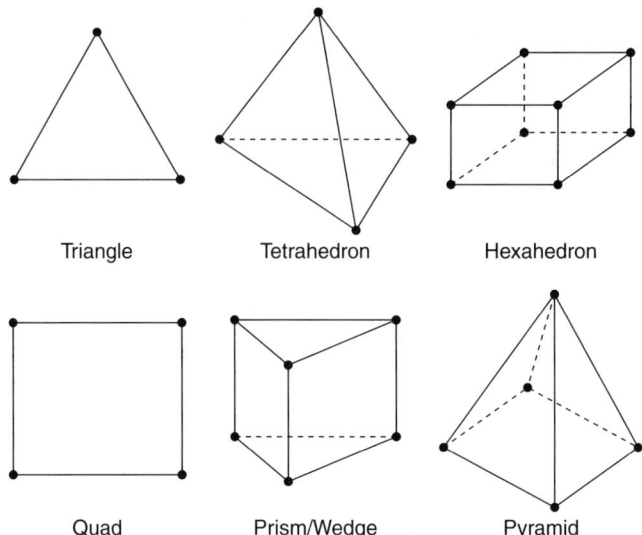

**Figure 5-3** Element types that can be used in computational grids.

Structured grids are always quadrilateral (2D) or hexahedral (3D) and are such that every element has a unique address in I, J, K space, where I, J, and K are indices used to number the elements in each of the three computational directions (Figure 5-4). The I, J, and K directions can, but need not be, aligned with the coordinate directions x, y, and z. Unstructured grids do not follow this addressing rule (Figure 5-5). Hybrid meshes are unstructured meshes that make use of different types of elements (e.g., triangles and quadrilaterals, as in Figure 5-5b). Block structured meshes use quadrilateral (2D) or hexahedral (3D) elements and have I, J, K structures in multicell blocks rather than across the entire domain. The top section of Figure 5-5b is an example of a block structured grid, although the grid as a whole (including the bottom section) is unstructured.

In general, the density of cells in a computational grid needs to be fine enough to capture the flow details, but not so fine that the overall number of cells in the domain is excessively large, since problems described by large numbers of cells require more time to solve. Nonuniform grids of any topology can be used to focus the grid density in regions where it is needed and to allow for expansion in other regions.

In laminar flows, the grid near boundaries should be refined to allow the solution to capture the boundary layer flow detail. A boundary layer grid should contain quadrilateral elements in 2D and hexahedral or prism elements in 3D, and should have at least five layers of cells. For turbulent flows, it is customary to use a wall function in the near-wall regions. This is due to the fact that the transport equation for the eddy dissipation has a singularity at the wall, where k [in the denominator in the source terms in eq. (5-14)] is zero. Thus, the equation for $\varepsilon$ must be treated in an alternative manner. Wall functions rely on the fact

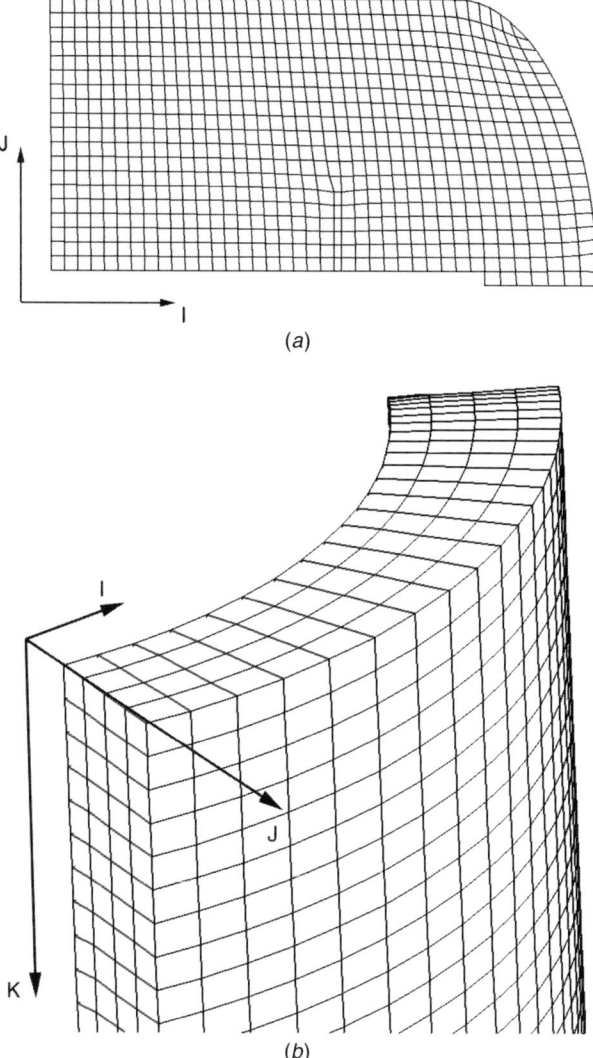

**Figure 5-4** Structured grids in (a) 2D and (b) 3D showing the I, J, and K directions.

that the flow in a turbulent boundary layer consists of a narrow viscous sublayer and a broad, fully turbulent, or *log-law* layer in which the behavior is well documented. In particular, the shear stress due to the wall can be extracted from a linear relationship involving the log of the perpendicular distance to the wall. Guidelines exist so that the placement of the cell center in the cell nearest the wall lies outside the viscous sublayer and inside the log-law layer. If these guidelines are followed, the wall shear stress will be captured correctly, resulting in the best possible predictions for pressure drop and heat transfer in the simulation.

**276** COMPUTATIONAL FLUID MIXING

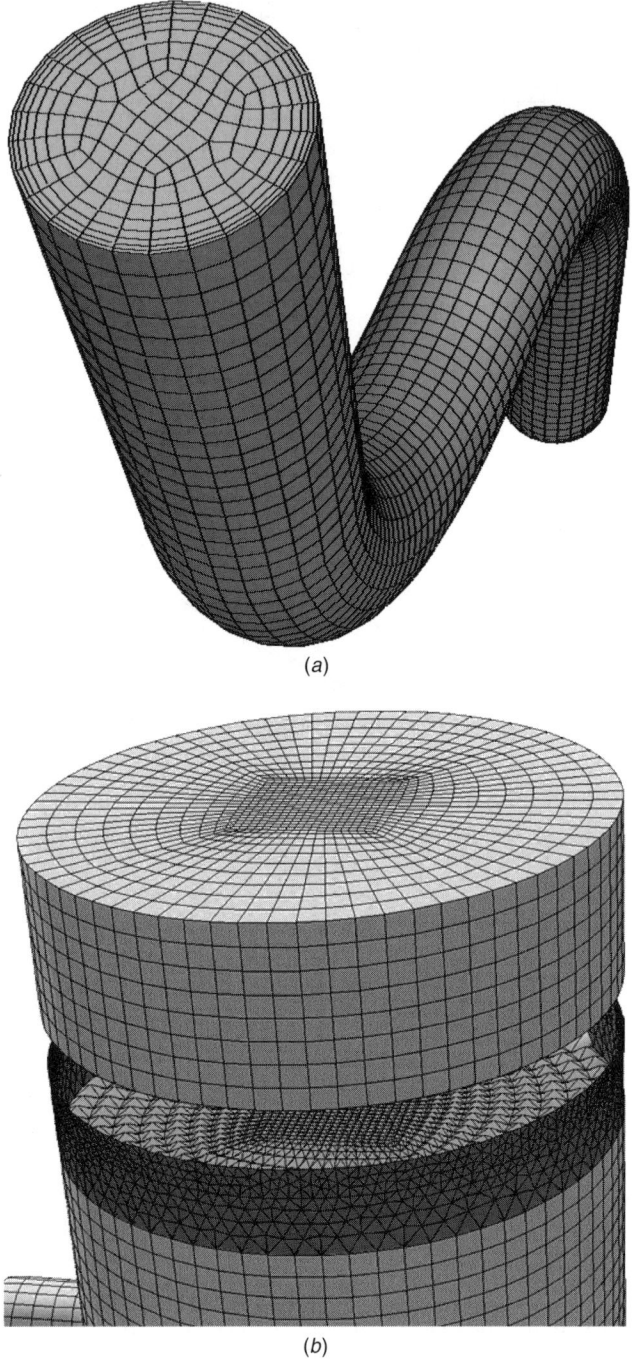

**Figure 5-5** (*a*) Unstructured grid using hexahedral elements. (*b*) Unstructured grid using a mixture of elements.

## 5-3.2 Discretization of the Equations

Several methods have been employed over the years to solve the Navier–Stokes equations numerically, including the finite difference, finite element, spectral element, and finite volume methods. The focus of this chapter is on the finite volume method, which is described in detail below. Once the method and terminology have been presented, the other methods are discussed briefly in Section 5-3.2.3.

To illustrate the discretization of a typical transport equation using the finite-volume formulation (Patankar, 1980; Versteeg and Malalasekera, 1995), a generalized scalar equation can be used with the rectangular control volume shown in Figure 5-6a. The scalar equation has the form

$$\frac{\partial(\rho\phi)}{\partial t} + \frac{\partial}{\partial x_i}(\rho U_i \phi) = \frac{\partial}{\partial x_i}\left(\Gamma \frac{\partial \phi}{\partial x_i}\right) + S' \tag{5-27}$$

The parameter $\Gamma$ is used to represent the diffusion coefficient for the scalar $\phi$. If $\phi$ is one of the components of velocity, for example, $\Gamma$ would represent the viscosity. All sources are collected in the term $S'$. Again, if $\phi$ is one of the components of velocity, $S'$ would be the sum of the pressure gradient, the gravitational force, and any other additional forces that are present. The control volume has a node, P, at its center where all problem variables are stored. The transport equation describes the flow of the scalar $\phi$ into and out of the cell through the cell faces. To keep track of the inflow and outflow, the four faces are labeled with lowercase letters representing the east, west, north, and south borders. The neighboring cells also have nodes at their centers, and these are

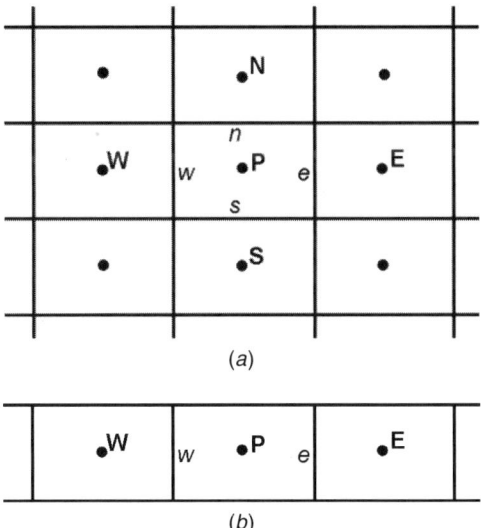

**Figure 5-6** (a) Simple 2D domain showing the cell centers and faces. (b) 1D rectangular simplification of the 2D domain.

labeled with the capital letters E, W, N, and S. For the purpose of this example, flow in the one dimensional row of cells shown in Figure 5-6b is considered.

The first step in the discretization of the transport equation is an integration over the control volume. The volume integral can be converted to a surface integral by applying the divergence theorem. Using a velocity in the positive x-direction, neglecting time dependence, and assuming that the faces e and w have area A, the integrated transport equation takes the following form:

$$(\rho_e U_e \phi_e - \rho_w U_w \phi_w) A = \left( \Gamma_e \left[ \frac{d\phi}{dx} \right]_e - \Gamma_w \left[ \frac{d\phi}{dx} \right]_w \right) A + S \qquad (5\text{-}28)$$

where S is the volume integral of the source terms contained in S'. This expression contains four terms that are evaluated at the cell faces. To obtain the face values of these terms as a function of values that are stored at the cell centers, a discretization scheme is required.

### 5-3.2.1 Discretization Schemes
Since all of the problem variables are stored at the cell center, the face values (e.g., the derivatives) need to be expressed in terms of cell center values. To do this, consider a steady-state conservation equation in one dimension without source terms:

$$\frac{d}{dx}(\rho U \phi) = \frac{d}{dx}\left( \Gamma \frac{\partial \phi}{\partial x} \right) \qquad (5\text{-}29)$$

This equation can be solved exactly. On a linear domain that extends from $x = 0$ to $x = L$, corresponding to the locations of two adjacent cell nodes, with $\phi = \phi_0$ at $x = 0$ and $\phi = \phi_L$ at $x = L$, the solution for $\phi$ at any intermediate location (such as the face) has the form

$$\phi = \phi_0 + (\phi_L - \phi_0) \frac{\exp[Pe(x/L) - 1]}{\exp(Pe - 1)} \qquad (5\text{-}30)$$

The Péclet number, Pe, appearing in this equation is the ratio of the influence of convection to that of diffusion on the flow field:

$$Pe = \frac{\rho U L}{\Gamma} \qquad (5\text{-}31)$$

Depending on the value of the Péclet number, different limiting behavior exists for the variation of $\phi$ between $x = 0$ and $x = L$. These limiting cases are discussed below, along with some more rigorous discretization or differencing schemes that are in popular use today.

*Central Differencing Scheme.* For $Pe = 0$ (i.e., $U = 0$), there is no convection, and the solution is purely diffusive. This would correspond to heat transfer due to pure conduction, for example. In this case, the variable $\phi$ varies linearly from cell center to cell center, so the value at the cell face can be found from linear interpolation. When linear interpolation is used in general, i.e., when both convection and diffusion are present, the discretization scheme is called *central differencing*. When used in this manner, as a general purpose discretization scheme, it can lead to errors and loss of accuracy in the solution. One way to reduce these errors is to use a refined grid, but the best way is to use another differencing scheme. There is one exception to this rule. Central differencing is the preferred discretization scheme when the LES turbulence model is used.

*Upwind Differencing Schemes.* For Pe $\gg$ 1, convection dominates, and the value at the cell face can be assumed to be identical to the upstream or upwind value (i.e., $\phi_w = \phi_W$). When the value at the upwind node is used at the face, independent of the flow conditions, the process is called *first-order upwind differencing*. A modified version of first-order upwind differencing makes use of multidimensional gradients in the upstream variable, based on the upwind neighbor and its neighbors. This scheme, which makes use of a Taylor series expansion to describe the upwind gradients, is called *second-order upwind differencing*. It offers greater accuracy than the first-order upwind method, but requires additional computational effort.

*Power Law Differencing Scheme.* For intermediate values of the Péclet number, $0 \leq Pe \leq 10$, the face value can be computed as a function of the local Péclet number, as shown in eq. (5-30). This expression can be approximated by one that does not use exponentials, involving the Péclet number raised to an integral power. It is from this approximate form that the power law differencing scheme draws its name. This first-order scheme is identical to the first-order upwind differencing scheme in the limit of strong convection, but offers slightly improved accuracy for the range of Péclet numbers mentioned above.

*QUICK Differencing Scheme.* The QUICK differencing scheme (Leonard and Mokhtari, 1990) is similar to the second-order upwind differencing scheme, with modifications that restrict its use to quadrilateral or hexahedral meshes. In addition to the value of the variable at the upwind cell center, the value from the next neighbor upwind is also used. Along with the value at the node P, a quadratic function is fitted to the variable at these three points and used to compute the face value. This scheme can offer improvements over the second-order upwind differencing scheme for some flows with high swirl.

*Choosing a Differencing Scheme.* If the flow is aligned with the grid, first-order differencing schemes such as upwind and power law differencing are acceptable. Flow in a straight pipe modeled with a hexahedral grid is one example where these schemes would be sufficient. However, since flow patterns in both static and stirred mixers do not, in general, satisfy this condition, especially if unstructured grids are used, second-order differencing is recommended to reduce the numerical errors in the final solution. In general, first-order schemes allow the error to be

**Table 5-2** Summary of Discretization Schemes

| Discretization Scheme | Description, Advantages, and Disadvantages |
|---|---|
| Central | Good when diffusion dominates. Assumes that there is no convection and that variables vary linearly from cell center to cell center. For convective flows, errors can be reduced by the use of a refined grid. This scheme is recommended for LES simulations. |
| First-order upwind | Good when convection dominates and the flow is aligned with the grid. Assumes that the face value for each variable is equal to the upstream cell center value. Stable, and a good way to start off a calculation. A switch to a higher-order scheme is usually recommended once the solution has partially converged. |
| Second-order upwind | Good for full range of Peclet numbers. Computes the face value for each variable from gradients involving the upwind neighbor and its neighbors. |
| Power law | Good for intermediate values of Peclet number. Computes the face value for each variable from gradients expressed in the form of a power law function. For high Péclet numbers, results are equivalent to first-order upwind. |
| QUICK | Good for full range of Péclet numbers. Similar to second-order upwind, but restricted to quadrilateral and hexahedral meshes. |

reduced linearly with the grid spacing, while second-order schemes allow the error to be reduced as the square of the grid spacing. A common practice in CFD is to obtain a partially converged solution using one of the first-order schemes and then switch to a higher-order scheme to obtain the final converged result. The discretization schemes discussed above are summarized in Table 5-2.

**5-3.2.2 Final Discretized Equation.** Once the face values have been computed using one of the above differencing schemes, terms multiplying the unknown variable at each of the cell centers can be collected. Large coefficients multiply each of these terms. These coefficients contain information that includes the properties, local flow conditions, and results from previous iterations at each node. In terms of these coefficients, $A_i$, the discretized equation has the following form for the simple 2D grid shown in Figure 5-6:

$$A_P \phi_P = A_N \phi_N + A_S \phi_S + A_E \phi_E + A_W \phi_W = \sum_{i, \text{neighbors}} A_i \phi_i \qquad (5\text{-}32)$$

For a complex, or even a simple flow simulation, there will be one equation of this form for each variable solved, in each cell in the domain. Furthermore, the equations are coupled, since for example, the solution of the momentum equations will affect the transport of every other scalar quantity. It is the job of the solver to solve these equations collectively with the most accuracy in the least amount of time.

**5-3.2.3 Alternative Numerical Techniques.** As mentioned earlier, other methods for solving the Navier–Stokes equations exist. Two of these are described briefly below.

*Finite Difference Method.* The finite difference, or Taylor series formulation replaces the derivatives in eq. (5-27) with finite differences evaluated at the variable storage sites (cell centers) using a truncated Taylor series expansion. The differences for each variable are computed using the cell value and/or the adjacent neighbor values, depending on the order of the derivative. The variation of the variable between storage sites is ignored during the solution process. Although this is an acceptable method to solve for some simply varying functions, it is not the best choice for general purpose CFD analysis because the method is limited to simple grids and does not conserve mass on coarse grids.

*Finite Element Method.* The finite element method uses piecewise linear or quadratic functions to describe the variation of the variable $\phi$ within a cell. By substituting the selected function into the conservation equation for each cell and applying the boundary conditions, a linear system of coupled equations is obtained. These equations are then solved (iteratively) for the unknown variable at all storage sites.

This method is popular for use with structural analysis codes and some CFD codes. In the early days of CFD, when structured orthogonal grids were used for most applications of the finite volume method, the finite element method offered the luxury of unstructured meshes with nonorthogonal elements of various shapes. Now that the use of unstructured meshes is common among finite volume solvers, the finite element method has been used primarily for certain focused CFD application areas. In particular, it is popular for flows that are neither compressible nor highly turbulent, and for laminar flows involving Newtonian and non-Newtonian fluids, especially those with elastic properties.

## 5-3.3 Solution Methods

The result of the discretization process is a finite set of coupled algebraic equations that need to be solved simultaneously in every cell in the solution domain. Because of the nonlinearity of the equations that govern the fluid flow and related processes, an iterative solution procedure is required. Two methods are commonly used. A segregated solution approach is one where one variable at a time is solved throughout the entire domain. Thus, the x-component of the velocity is solved on the entire domain, then the y-component is solved, and so on. One iteration of the solution is complete only after each variable has been solved in this manner. A coupled solution approach, on the other hand, is one where all variables, or at a minimum, momentum and continuity, are solved simultaneously in a single cell before the solver moves to the next cell, where the process is repeated. The segregated solution approach is popular for incompressible flows with complex physics, typical of those found in mixing applications.

Typically, the solution of a single equation in the segregated solver is carried out on a subset of cells, using a Gauss–Seidel linear equation solver. In some cases the solution time can be improved (i.e., reduced) through the use of an algebraic multigrid correction scheme. Independent of the method used, however, the equations must be solved over and over again until the collective error is reduced to a value that is below a preset minimum value. At this point, the solution is considered converged, and the results are most meaningful. Converged solutions should demonstrate overall balances in all computed variables, including mass, momentum, heat, and species, for example. Some of the terminology used to describe the important aspects of the solution process is defined below.

**5-3.3.1 SIMPLE Algorithm.** For 3D simulations, the three equations of motion [eq. (5-6)] and the equation of continuity [eq. (5-5)] combine to form four equations for four unknowns: the pressure and the three velocity components. Because there is no explicit equation for the pressure, special techniques have been devised to extract it in an alternative manner. The best known of these techniques is the SIMPLE algorithm, semi-implicit method for pressure-linked equations (Patankar, 1980). Indeed, a family of algorithms has been derived from this basic one, each of which has a small modification that makes it well suited to one application or another.

The essence of the algorithm is as follows. A guessed pressure field is used in the solution of the momentum equations. (For all but the first iteration, the guessed pressure field is simply the last updated one.) The new velocities are computed, but these will not, in general, satisfy the continuity equation, so corrections to the velocities are determined. Based on the velocity corrections, a pressure correction is computed which when added to the original guessed pressure, results in an updated pressure. Following the solution of the remaining problem variables, the iteration is complete and the entire process is repeated.

**5-3.3.2 Residuals.** If the algebraic form of a conservation equation in any control volume [eq. (5-32)] could be solved exactly, it would be written as

$$A_P \Phi_P - \sum_{i,\text{neighbors}} A_i \Phi_i = 0 \tag{5-33}$$

Since the solution of each equation at any step in an iterative calculation is based on inexact information, originating from initial guessed values and refined through repeated iterations, the right-hand side of eq. (5-33) is always nonzero. This nonzero value represents the error or residual in the solution of the equation in the control volume:

$$A_P \Phi_P - \sum_{i,\text{neighbors}} A_i \Phi_i = R_P \tag{5-34}$$

The total residual is the sum over all cells in the computational domain of the residuals in each cell:

$$\sum_{P,cells} R_P = R \tag{5-35}$$

Since the total residual, R, defined in this manner, depends on the magnitude of the variable being solved, it is customary either to normalize or to scale the total residual to gauge its changing value during the solution process. Although normalization and scaling can be done in a number of ways, it is the change in the normalized or scaled residuals that is important in evaluating the rate and level of convergence of the solution.

**5-3.3.3 Convergence Criteria.** The convergence criteria are preset conditions for the (usually normalized or scaled) residuals that determine when an iterative solution is converged. One convergence criterion might be that the total normalized residual for the pressure equation drop below $1 \times 10^{-3}$. Another might be that the total scaled residual for a species equation drop below $1 \times 10^{-6}$. Alternatively, it could be that the sum of all normalized residuals drop below $1 \times 10^{-4}$. For any set of convergence criteria, the assumption is that the solution is no longer changing when the condition is reached and that there is an overall mass balance throughout the domain. When additional scalars are being solved (e.g., heat and species), there should be overall balances in these scalars as well. Whereas the convergence criteria indicate that overall balances probably exist, it is the wise engineer who will examine reports to verify that indeed they do.

**5-3.3.4 Underrelaxation.** The solution of a single differential equation, solved iteratively, makes use of information from the preceding iteration. If $\phi_n$ is the value of the variable from the preceding iteration and $\phi_{n+1}$ is the new value, some small difference or change in the variable brings the variable from the old value to the new one:

$$\phi_{n+1} = \phi_n + \Delta\phi \tag{5-36}$$

Rather than use the full computed change in the variable, $\Delta\phi$, it is often necessary to use a fraction of the computed change when several coupled equations are involved:

$$\phi_{n+1} = \phi_n + f\Delta\phi \tag{5-37}$$

This process is called *underrelaxation*, and underrelaxation factors, f, typically range from 0.1 to 1.0, depending on the complexity of the flow physics (e.g., laminar flow or turbulent reacting flow), the variable being solved (pressure or momentum), the solution method being used, and the state of the solution (during the first few iterations or near convergence). Underrelaxation makes the convergence process stable, but slower. Guidelines exist for the optimum choices for underrelaxation factors for a variety of conditions. As the solution converges, the underrelaxation factors should be gradually raised to ensure convergence that is both rapid and stable at all times.

### 5-3.3.5 Numerical Diffusion.

Numerical diffusion is a source of error that is always present in finite volume CFD, owing to the fact that approximations are made during the process of discretization of the equations. It is so named because it presents itself as equivalent to an increase in the diffusion coefficient. Thus, in the solution of the momentum equation, the fluid will appear more viscous; in the solution of the energy equation, the solution will appear to have a higher conductivity; in the solution of the species equation, it will appear that the species diffusion coefficient is larger than in actual fact. These errors are most noticeable when diffusion is small in the actual problem definition.

To minimize numerical diffusion, two steps can be taken. First, a higher-order discretization scheme can be used, such as the QUICK or second-order upwinding schemes discussed earlier. Second, the grid can be built so as to minimize the effect. In general, numerical diffusion is more of a problem on coarse grids, so it is wise to plan ahead and avoid coarse meshes in regions where the most accuracy is sought. Numerical diffusion is usually less of a problem with quadrilateral or hexahedral meshes, provided that the flow is aligned with the mesh. Unfortunately, the flow is rarely aligned with the mesh throughout the entire flow field, so some degree of numerical diffusion is unavoidable.

### 5-3.3.6 Time-Dependent Solutions.

To solve a time-dependent problem, the time derivative appearing in eq. (5-27) must be discretized. If $F(\phi)$ is the spatially discretized part of eq. (5-27), the time derivative can be approximated to first order as

$$\frac{\phi^{n+1} - \phi^n}{\Delta t} = F(\phi) \qquad (5\text{-}38)$$

In this expression, $\phi^n$ is the solution at time t and $\phi^{n+1}$ is the solution at time $t + \Delta t$. While certain flow conditions, such as compressible flow, are best suited to an explicit method for the solution of eq. (5-38), an implicit method is usually the most robust and stable choice for a wide variety of applications, including mixing. The major difference between the explicit and implicit methods is whether the right-hand side of eq. (5-38) is evaluated at the current time $[F(\phi) = F(\phi)^n]$ or at the new time $[F(\phi) = F(\phi^{n+1})]$. The implicit method uses the latter:

$$\phi^{n+1} = \phi^n + \Delta t F(\phi^{n+1}) \qquad (5\text{-}39)$$

The assumption at the core of this quasi-steady approach is that the new value of the variable $\phi$ prevails throughout the entire time step, which takes the solution from time t to time $t + \Delta t$.

## 5-3.4 Parallel Processing

Parallel processing is a procedure in which a large calculation can be performed on two or more processors working in parallel. The processors can reside on the same (multiprocessor) computer or can be on a network of computers. For the calculation to run on the processors in a parallel fashion, the calculation domain (the computational grid) must be divided into partitions, or subdomains. The equations in each partition are solved simultaneously on the multiple processors (using the segregated or coupled approach), and the results at the boundaries of the partitions are communicated to the neighbor partitions on a regular basis. As the number of nodes increases, the computation time for each node decreases, and the communication between partitions increases. In this limit, the efficiency of parallel computing decreases. Recent advances in parallel algorithms have pushed back this limiting behavior, however.

## 5-4 STIRRED TANK MODELING USING EXPERIMENTAL DATA

Stirred tanks typically contain one or more impellers mounted on a shaft, and optionally, baffles and other internals. Although it is a straightforward matter to build a 3D mesh to contour to the space between these elements, the mesh must be built so that the solution of the flow field incorporates the motion of the impeller. This can be done in two ways. First, the impeller geometry can be modeled directly, or explicitly, and the grid and solution method chosen so as to incorporate the motion of the impeller using either a steady-state or time-dependent technique. This approach is discussed in detail in Section 5-5. Second, the motion of the impeller can be modeled implicitly, using time-averaged experimental velocity data to represent the impeller motion. The second approach is the subject of this section.

### 5-4.1 Impeller Modeling with Velocity Data

When modeling the impeller using velocity data, the time-averaged velocities in the outflow of the impeller are prescribed, and the CFD solver calculates the flow in the remainder of the vessel. An illustration of this process is shown in Figure 5-7 for a radial flow impeller. The parabolic velocity profile in the impeller outflow region is prescribed as a boundary condition in the simulation, and the well-known radial flow pattern with circulation loops above and below the impeller results from the CFD calculation. It is important to note that the volume swept by the impeller is also part of the model but that other than for the fixed velocities in the outflow region, it is treated as part of the fluid domain by the CFD solver. Figure 5-7 also illustrates the fact that for this particular case it is indeed sufficient to prescribe the velocities in the impeller outflow only to obtain a good flow field prediction. Kresta and Wood (1991) and Bakker and Van den Akker (1994) presented quantitative validations for this particular case, and other authors have presented similar validations for other cases.

Over the years, practical experience has demonstrated that it is usually sufficient to prescribe the velocity data only along the edges of the impeller where the flow exits. One or two edges of the impeller are typically needed for this purpose. For an impeller that creates a purely radial flow pattern, such as the radial flow impeller of Figure 5-7, prescribing the velocities on the side of the impeller is sufficient, since flow is drawn into the impeller at the top and bottom edges.

**286** COMPUTATIONAL FLUID MIXING

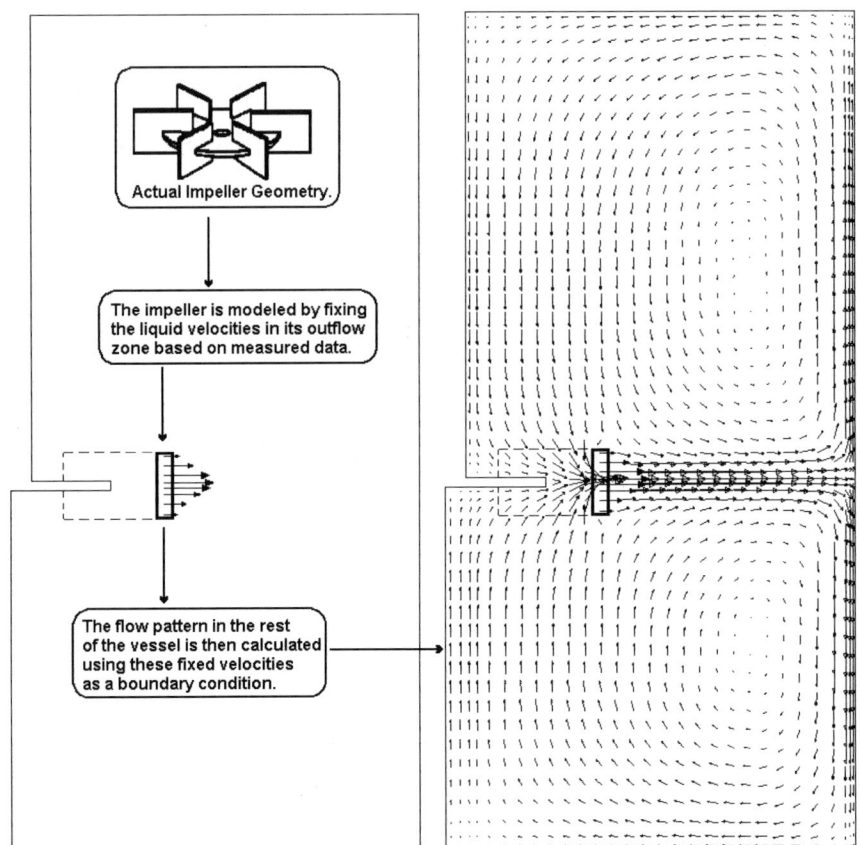

**Figure 5-7** Velocity data measured radially outside a radial flow impeller are applied to a 2D CFD simulation, resulting in the well-known double-loop flow pattern.

In general for all impeller types, all three velocity components should be prescribed in the discharge region. For turbulent flow it is also recommended that values for the turbulent kinetic energy, k, and dissipation rate, ε, be prescribed. The turbulent kinetic energy can be computed from measured fluctuations in the velocity components using eq. (5-12). Using k, the eddy dissipation can be calculated using

$$\varepsilon = \frac{k^{3/2}}{L_t} \quad (5\text{-}40)$$

where $L_t$ is a characteristic turbulent length scale in the outflow of the impeller. The authors often use $L_t = W_b/4$, where $W_b$ is the width of the impeller blade. Note, however, that there is some debate in the literature about the exact value of the factor relating $L_t$ and $W_b$.

Figure 5-8 shows where to prescribe the velocity data for various cases, including the previously discussed radial flow impeller (Figure 5-8a). For a

STIRRED TANK MODELING USING EXPERIMENTAL DATA    **287**

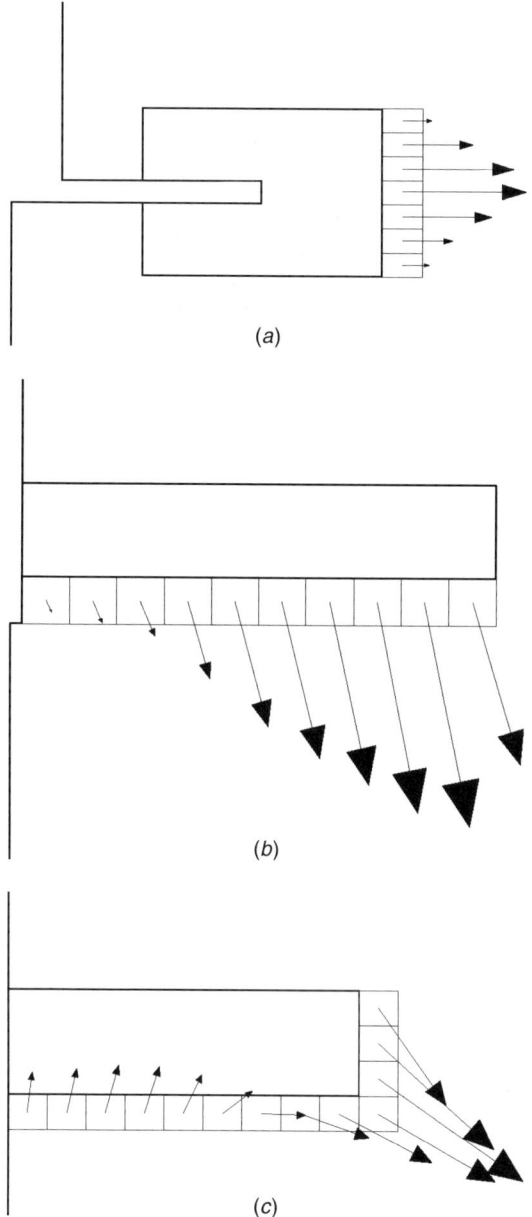

**Figure 5-8** Suggested locations for prescribing impeller boundary conditions for (*a*) a radial flow impeller in the turbulent flow regime, (*b*) an axial flow impeller in the turbulent flow regime, (*c*) an axial flow impeller operating in the laminar flow regime. (*Continued*)

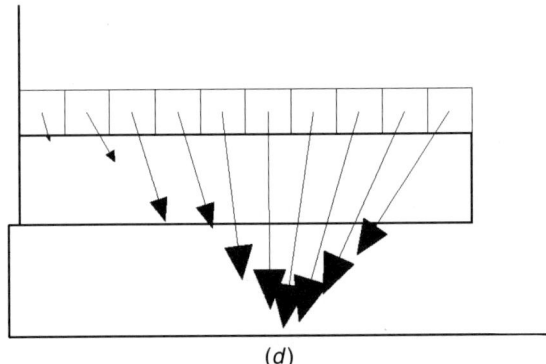

**Figure 5-8** (*d*) An axial flow impeller close to the vessel bottom (e.g., C/T < 0.1).

down-pumping impeller that creates a purely axial flow pattern (Figure 5-8*b*), liquid will enter the impeller from the top and the side and exit the impeller on the bottom. In such a case it is sufficient to prescribe the liquid velocities along the bottom edge only. For an up-pumping impeller under the same conditions, the velocities would be prescribed along the top edge. When an axial flow impeller operates in the laminar flow regime, however, it will have a combined axial–radial flow (Figure 5-8*c*). On the bottom of the impeller, flow both enters and exits depending on the radial location. Furthermore, flow exits the impeller on the side. On the top of the impeller, flow enters but does not exit. Therefore, for this situation the proper modeling method is to prescribe complete velocity profiles on both the bottom and the side edges. Although it is not in general recommended to prescribe all velocity components on the top of the impeller as well, for laminar flow conditions the prediction of the swirling flow pattern in the top of the vessel can be improved by prescribing the tangential velocity component only along this edge, in addition to the prescriptions along the side and bottom. For an up-pumping impeller, the velocities should be prescribed along the top and the side edges, with the swirl (optionally) prescribed along the lower edge for laminar flows only.

In general, caution should be used when applying velocity data in circumstances where the impeller discharge has a strong interaction with the tank. This behavior has been examined experimentally and through the use of CFD by Fokema et al. (1994) for a pitched blade turbine. Indeed, when a down-pumping axial flow impeller is mounted very close to the vessel bottom (or an up-pumping impeller close to the liquid surface), velocities should not be prescribed in the impeller discharge region. Such cases present several difficulties. On the experimental side, measuring velocities in regions close to walls can be difficult and may result in inaccuracies. In the CFD simulation, there may be only a few computational cells between the vessel bottom and the impeller. In these circumstances, good results can often still be obtained if the velocities are prescribed at the top inflow of the impeller (Figure 5-8*d*).

## 5-4.2 Using Experimental Data

Several experimental methods are available for measuring the velocities imparted to the fluid by a working impeller. These include laser Doppler velocimetry (LDV) and particle image velocimetry (PIV). These methods are discussed in Chapter 4. Under ideal circumstances, the velocity data prescribed for a simulation would have been obtained from measurements made on an identical system. In practice, however, this is rarely the case. The experimental data that are available were probably obtained for conditions that are different from the system being modeled. Nonetheless, several scaling rules can be applied to the existing data so that appropriate velocity profiles for the case at hand can be generated.

The first step involves normalization of the available data. Typically, the measured liquid velocities are normalized by the impeller tip speed, $U_{tip}$, used during the experiment. The turbulent kinetic energy is usually normalized by $U_{tip}^2$. The eddy dissipation can be normalized by $U_{tip}^3/D$, with a possible constant of proportionality. Radial measurement locations are typically normalized by the impeller radius, R, and axial locations by the impeller blade height, z, measured from the impeller centerline. To perform the simulation, profiles for the liquid velocities, k and ε, are obtained by multiplying the normalized profiles by the $U_{tip}$, $U_{tip}^2$, and $U_{tip}^3/D$ used in the simulation, respectively. The locations at which the velocity data are available are calculated by multiplying the normalized measurement locations by the actual impeller radius or blade height.

When prescribing the velocity data above or below the impellers, it is recommended that the computational grid be constructed such that the center of the cells where the velocities are prescribed fall within a quarter-cell height of the normalized axial measurement locations. Similarly, when prescribing data at the side of the impeller, it is recommended that the cell centers are within a quarter-cell width of the normalized radial measurement locations. For both cases, interpolation can then be used to determine the velocity values at the radial and axial grid locations of the individual cell centers, respectively.

The exact shape of the velocity profile in the outflow of an impeller does not depend solely on the impeller. It is also affected by such variables as the impeller Reynolds number, impeller off-bottom distance C/T, and impeller diameter D/T. If the flow is fully turbulent (i.e., $Re > 10^4$), the impeller outflow profiles are typically independent of Reynolds number. If the flow is transitional or laminar, however, care should be taken so that the velocity profiles used were either measured at a similar Reynolds number, or that the prescribed velocities are being interpolated from data sets measured over a range of Reynolds numbers. Similarly, for impeller off-bottom clearance and diameter, if data for various C/T and D/T values are available, interpolations can be used to obtain the prescribed velocities for the actual conditions.

## 5-4.3 Treatment of Baffles in 2D Simulations

As mentioned earlier, the time-averaging method used to record velocity data for an impeller makes the data useful for 2D simulations in the radial–axial plane, where angular—and therefore time—dependence of the geometry and flow field is ignored. While ignoring the angular dependence of the impeller motion can be done in this manner, the angular dependence of the baffles needs to be addressed as well. Baffles are used to reduce the swirl introduced by the rotating impeller. One way of including this effect in a 2D simulation is to omit the swirling component of the velocity data in the numerical simulation, using the radial and axial components instead. Another way to model baffles is to set a boundary condition of zero swirl in the baffle region in the 2D simulation. By setting the boundary condition on the swirl only, the axial and radial velocities can be computed in the baffle region as they are in the remainder of the vessel.

### 5-4.4 Combining the Velocity Data Model with Other Physical Models

The steady-state implicit impeller model, which uses time-averaged experimental data, can be used to model other steady-state and time-dependent processes, as described below. Because of its simplicity, it has no effect on other scalar transport in the domain. The models that do require special consideration are those involving multiple phases, with separate sets of momentum boundary conditions, as described below. Species blending is also discussed, because it is a calculation that is commonly performed in conjunction with the implicit impeller model.

*5-4.4.1 Volume of Fluid Model.* In stirred tank applications, the volume of fluid (VOF) free surface model is useful for tracking the shape of the liquid surface during operation. This includes the transition to a parabolic shape during startup, which can lead to the (undesired) drawdown of air. The velocity data model can be used in 2D or 3D for simulations of this type. The VOF model can have a steady or time-dependent implementation, and both are fully compatible with this steady-state treatment of the impellers.

If air drawdown does occur, caution is needed. If air passes through cells where large momentum sources exist, resulting from the velocity data boundary conditions, the liquid–air interface will be broken, resulting in many small bubbles that will mix with the liquid. The VOF model is not equipped to handle this condition accurately, so the simulation should be terminated at this point. Thus, whereas the model can be used to predict if drawdown will occur, it should not be used to predict the flow conditions afterward.

*5-4.4.2 Multiphase Model.* Both solids suspension and gas sparging can be simulated using an experimental data model for the impeller. The manner in which the multiphase parameters are input depends on the multiphase model being used. For solids suspension, an Eulerian granular multiphase model is recommended, and separate sets of momentum equations are used for the liquid and solids phases. This model, run in a time-dependent fashion, is fully compatible with the time-averaged representation of the impellers. Experimental velocity data are set as a boundary condition independently for each of the phases. Note, however, that there is usually some degree of slip between the fluid and granular phases, a value that increases with the density difference between the phases. Thus the velocities used to represent the impeller for a pure liquid need to be adjusted somewhat for the granular phase. This can be accomplished by estimating the slip velocity between the two phases. The measured data can be used to represent the impeller for the fluid phase, and a corrected set of data, obtained by subtracting the slip velocity from the experimental data, can be used to represent the impeller for the solids phase.

Gas sparging can be modeled using the Eulerian multiphase model or the algebraic slip mixture model. For the Eulerian multiphase model, two sets of momentum equations are used, and the same comments regarding the slip velocity between phases apply, although the issue is not as critical. That is, the velocity data used for the gas phase could be corrected slightly from the liquid-phase velocities but need not be because the gas phase has so little inertia compared to the liquid phase. When the algebraic slip mixture model is used, separate boundary conditions are not required for the individual phases, so a correction of the velocity data is not required.

Another consideration in the case of gas–liquid mixtures is the impact of the impeller on gas bubble size. In an actual stirred tank, the momentum of the rotating impeller often acts to break up gas bubbles as they pass through the region. This reduces the bubble size and can lead to an increase in the gas hold-up as well as a change in the momentum exchange term (drag) between the phases. When experimental data are used, this phenomenon is missing from the formulation but can often be incorporated into the calculation if subroutines, written by the user, are available to modify the model in the commercial software.

*5-4.4.3 Turbulence.* The use of a transient turbulence model, such as the large eddy simulation model, is inconsistent with the experimental data formulation because the latter is intrinsically steady-state. All of the RANS models, however, are fully compatible with the velocity data approach.

*5-4.4.4 Species Blending.* When a neutrally buoyant tracer, one with the same fluid properties, is added to the liquid in a vessel, a simplified approach to predicting the mixing time can be used. Rather than model the complete set of transport equations in a transient manner, the steady-state flow field can be computed first, including the inflow and outflow for the anticipated tracer and resulting mixture, respectively. Prior to beginning the transient species calculation for the tracer, the calculation of the flow field variables (pressure, momentum, and turbulence) can be disabled, since the overall properties of the mixture will not change. Thus, the dispersion of the tracer species can be tracked by solving only a single scalar transport equation. (The same technique can be used for heat transfer if the properties are the same and not temperature-dependent.) This method for computing species blending is fully compatible with the experimental data representation of the impellers.

## 5-5 STIRRED TANK MODELING USING THE ACTUAL IMPELLER GEOMETRY

To model the geometry of the impeller exactly, a 3D simulation must be performed. A number of solution approaches are available to incorporate the motion of the impeller, and the computational grid used must be able to adapt to the solver method employed. The models in popular use today are reviewed in the following sections. Particular attention is paid to the sliding mesh model, the most rigorous of them all. The solver methods described are all designed to capture the motion of a rotating impeller in a stationary tank, but they vary in accuracy. Three of the models are steady-state and one is time-dependent.

### 5-5.1 Rotating Frame Model

The rotating frame model solves the momentum equations for the entire domain in a rotating frame. The Coriolis force is included in the process. Problems solved in a rotating frame typically use the angular velocity of the primary rotating component, $\Omega$, as the angular velocity of the frame. In stirred tanks, the impeller serves this purpose, so the frame is assumed to rotate with the impeller. Thus, the impeller is at rest in the rotating frame. The tank, however, rotates in the opposite direction, so must have a rotational boundary condition of $-\Omega$. If baffles exist, they would need to rotate into the fluid with the same angular velocity, $-\Omega$. Unfortunately, this simple steady-state model is not equipped to handle the motion of elements such as baffles into or through the fluid. The approach is therefore only useful for unbaffled tanks with smooth tank walls that are geometrically equivalent to a perfect surface of revolution. Thus an unbaffled cylindrical tank with an axisymmetric bottom shape and no angular-dependent internals could be simulated in this manner. Vessels with baffles, dip tubes, or inflow–outflow ports could not.

### 5-5.2 Multiple Reference Frames Model

A modification of the rotating frame model is the multiple reference frames (MRF) model (Luo et al., 1994). The modification is that more than one rotating (or nonrotating) reference frame can be used in a simulation. This steady-state approach allows for the modeling of baffled stirred tanks and tanks with other complex (rotating or stationary) internals. A rotating frame is used for the region containing the rotating components while a stationary frame is used for regions that are stationary (Figure 5-9). In the rotating frame containing an impeller, the impeller is at rest. In the stationary frame containing the tank walls and baffles, the walls and baffles are at rest. The fact that multiple reference frames can be used means that multiple impeller shafts in a rectangular tank can each be modeled with separate rotating frames (with separate rotation frequencies) while the remaining space can be modeled with a stationary frame.

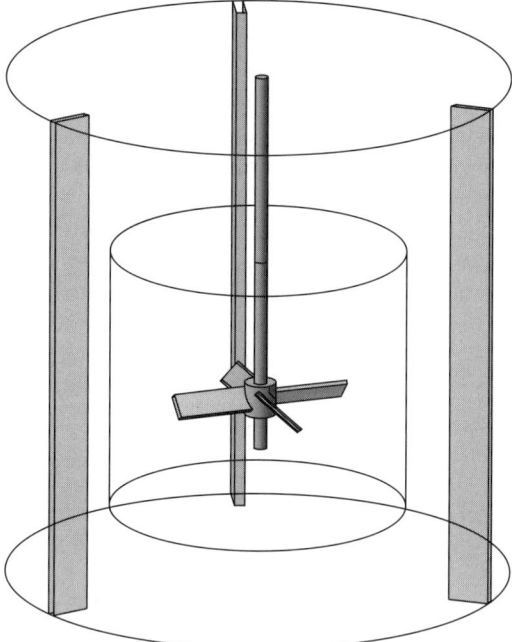

**Figure 5-9** Cylindrical mixing tank with an MRF boundary surrounding the impeller.

The grid used for an MRF solution must have a perfect surface of revolution surrounding each rotating frame. The momentum equations inside the rotating frame are solved in the frame of the enclosed impeller while those outside the rotating frame are solved in the stationary frame. A steady transfer of information is made at the MRF interface as the solution progresses. While the solution of the flow field in the rotating frame in the region surrounding the impeller imparts the impeller rotation to the region outside this frame, the impeller itself does not move during this type of calculation. Its position is static. If the impeller is mounted on a central shaft in a baffled tank, this means that the orientation of the impeller blades relative to the baffles does not change during the solution. If the interaction between the impeller and baffles is weak, the relative orientation of the impeller and baffles does not matter. If the interaction is strong, however, the solution with the impeller in one position relative to the baffles will be different from that with the impeller in a different position. The model is therefore recommended for simulations in which the impeller–baffle interaction is weak. Note, however, that if the solution is to be used to obtain spatially averaged macroscopic properties of the flow field, such as power draw, the orientation of the impeller relative to the baffle may not matter. The careful engineer will perform two solutions with the impeller in two different locations and use both results (e.g., averaging them) rather than just one.

A modified version of the MRF model is the mixing plane model, in which the variables at the MRF boundary are spatially averaged in the circumferential direction prior to being passed from one side to the other. After the averaging process, all angular dependence on the boundary is eliminated, so the variables are functions of radial and axial position only. This approach is popular for turbomachinery, where many closely spaced rotors and stators are in relative motion. It has not had widespread use in the mixing community, however, owing in part to asymmetries in the flow field that are common in stirred tanks. For example, a tracer species introduced through a single dip tube on the side of the vessel would appear to be uniformly distributed on the interface shortly after reaching it, which is clearly unphysical. As another example, any stirred tank with inflow and outflow ports could have flow through the MRF interface that is not unidirectional. When the averaging process is done, this condition could also result in unphysical results. The mixing plane approach is therefore not recommended for most stirred tank applications.

### 5-5.2.1 Validation of the MRF Model.

To validate the MRF model, a Lightnin A310, operating in a baffled vessel ($Re = 4.6 \times 10^5$), was simulated using a number of turbulence models (Marshall et al., 1999). Results for the velocity field, power number, and flow number were compared to measurements performed by Weetman (1997). The vessel used for the simulation had a diameter $T = 1.22$ m (Figure 5-10a) and three baffles. The A310 impeller (with surface grid shown in Figure 5-10b) had a diameter and off-bottom clearance of $D/T = C/T = 0.352$. A $120°$ sector of the domain was modeled using a grid of approximately 150 000 hexahedral cells.

Figure 5-11 shows a comparison of the velocity data from the LDV measurements with the velocities in a nonbaffle plane computed by the MRF model, using RSM for turbulence. The CFD calculation picks up the features of the flow field correctly. In Table 5-3, the results for flow number, $N_Q$ (Section 5-6.4.2), and power number, $N_P$ (Section 5-6.4.1), show good agreement for all turbulence models. The power drawn by the impeller was computed by integrating the pressure force over the impeller blades to obtain the torque. The flow rate was computed by integrating the flow through a circular discharge area below the impeller.

**Table 5-3** Results of the MRF Impeller Model with Several Turbulence Models as Compared to Experiment

| Turbulence Model | $N_Q$ | $N_P$ |
| --- | --- | --- |
| Experiment[a] | 0.56 | 0.30 |
| Standard k–ε | 0.50 | 0.30 |
| RNG k–ε | 0.53 | 0.28 |
| Realizable k–ε | 0.52 | 0.29 |
| RSM | 0.51 | 0.29 |

[a] Experimental data provided by the impeller manufacturer.

STIRRED TANK MODELING USING THE ACTUAL IMPELLER GEOMETRY **295**

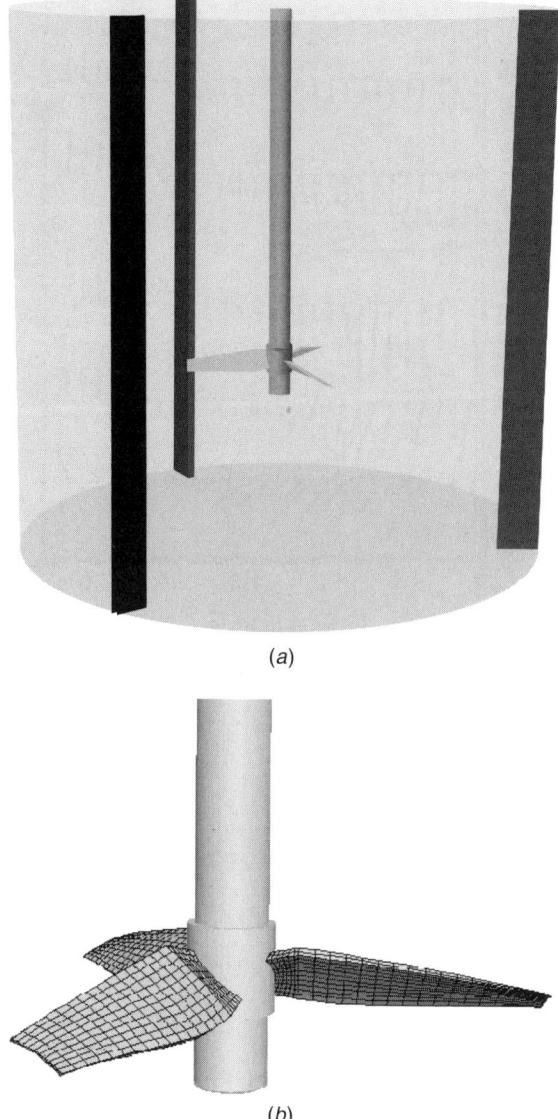

**Figure 5-10** (*a*) Tank containing a Lightnin A310 impeller. (*b*) Grid detail for the impeller surface.

## 5-5.3 Sliding Mesh Model

The sliding mesh model is a time-dependent solution approach in which the grid surrounding the rotating component(s) physically moves during the solution (Figure 5-12). The velocity of the impeller and shaft relative to the moving mesh region is zero, as is the velocity of the tank, baffles, and other internals in

**296** COMPUTATIONAL FLUID MIXING

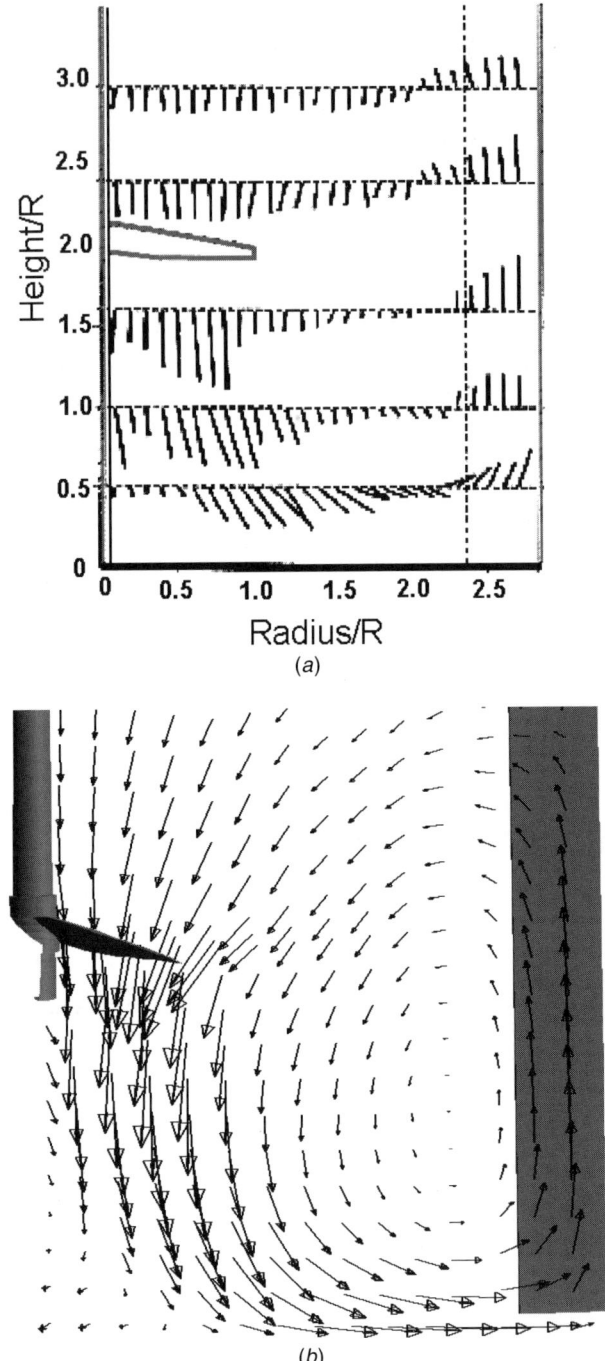

**Figure 5-11** (*a*) Experimental data from Weetman (1997). (*b*) CFD solution using the MRF model for the impeller and RSM for turbulence.

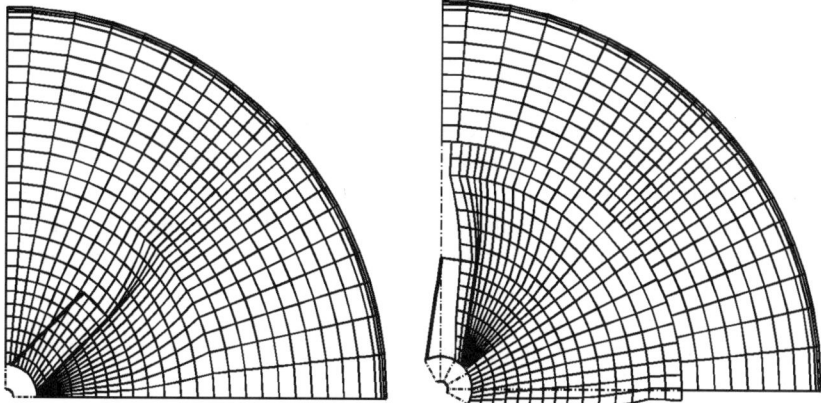

**Figure 5-12** Sliding mesh in two orientations (shown in 2D).

the stationary mesh region. The motion of the impeller is realistically modeled because the grid surrounding it moves as well, giving rise to a time-accurate simulation of the impeller–baffle interaction. The motion of the grid is not continuous. Rather, it is in small, discrete steps. After each such motion, the set of conservation equations is solved in an iterative process until convergence is reached. The grid moves again, and convergence is once again obtained from an iterative calculation. During each of these quasi-steady calculations, information is passed through the interface from the rotating to the stationary regions and back again.

In order to rotate one mesh relative to another, the boundary between the meshes needs to be a surface of revolution. When in its initial (unrotated) position, the grid on this boundary must have two superimposed surfaces. During the solution, one will remain with the rotating mesh region, and the other will remain with the stationary mesh region. At any time during the rotation, the cells will not (necessarily) line up exactly, or conform to each other. When information is passed between the rotating and stationary grid regions, interpolation is required to match each cell with its many neighbors across the interface.

The sliding mesh model is the most rigorous and informative solution method for stirred tank simulations. Transient simulations using this model can capture low-frequency (well below the blade passing frequency) oscillations in the flow field (Bakker et al., 2000; Roussinova et al., 2000) in addition to those that result from the periodic impeller–baffle interaction.

*5-5.3.1 Solution Procedures*. Because this is a transient model involving the motion of the impeller, starting the simulation with the impeller at rest is analogous to modeling startup conditions. After a period of time the flow field reaches periodic steady state, but this period of time may correspond to dozens of revolutions. If the goal of the simulation is to study the periodic steady-state conditions, minimizing the time spent reaching this state is desirable.

One way to pass through the startup conditions rapidly is to move the impeller by large increments each time step in the early stage of the calculation. If the model is a 90° sector, for example, the first few revolutions of the impeller can be modeled using a coarse time step that corresponds to a 30° displacement. The time step can then be refined to correspond to a 10° displacement, and refined again (and again) until the desired temporal and spatial accuracy is achieved. The solutions during these initial coarse time steps do not need to be converged perfectly, provided that the simulation involves a single fluid phase and there are no inflow and outflow boundaries. In these instances, improved convergence can be obtained in the later stages of the calculation.

An alternative way to bypass calculation of the startup period is to solve for a steady-state solution first using the MRF model. The MRF model (Section 5-5.2) provides a solution for the moving impeller at a fixed orientation relative to the baffles. Tools are available in commercial codes to use the solution data from the MRF simulation and apply it to the sliding mesh simulation as an initial condition. A moderately coarse time step can be used initially (say, corresponding to a 10° rotation, as in the example above) and reduced at a quicker rate than would otherwise be advisable. This approach can also be used if inflow and outflow boundaries are present or if a multiphase calculation is to be performed. In the case of multiphase flows, however, care must be taken to wait until the periodic steady-state condition has been reached before introducing the secondary phase.

*5-5.3.2 Validation of the Sliding Mesh Model.* One validation of the sliding mesh model was presented in a paper by Bakker et al. (1997). A pitched blade turbine was operated in a baffled vessel with diameter $T = 0.3$ m under laminar conditions (Re = 40). The impeller diameter and off-bottom clearance were such that D/T = C/T = 1/3. A 90° sector of the stirred tank was modeled using approximately 50 000 cells.

Figure 5-13 shows a comparison between LDV data on the left and CFD results on a midbaffle plane on the right. Because the impeller was operating at a low rotational speed, its discharge was more radial than axial. This structure is captured by the CFD model, in agreement with the experimental data, where circulation loops above and below the impeller can be seen. Calculations for this system operating at higher Reynolds numbers through transition and into the turbulent regime were also performed. Results for the flow number (Section 5-6.4.2), computed throughout both laminar and turbulent regimes, are in excellent agreement with values based on LDV measurements, as shown in Figure 5-14.

*5-5.3.3 Unstable Flows.* In recent years, much attention has been paid to instabilities that are observed in stirred tanks. These instabilities typically have frequencies that are low compared to the impeller frequency and involve the slow asymmetric wobble of material or momentum from one side of the vessel to the other. Instabilities of this type can be predicted using the sliding mesh technique on a 360° model of a stirred tank, particularly if the LES turbulence model is

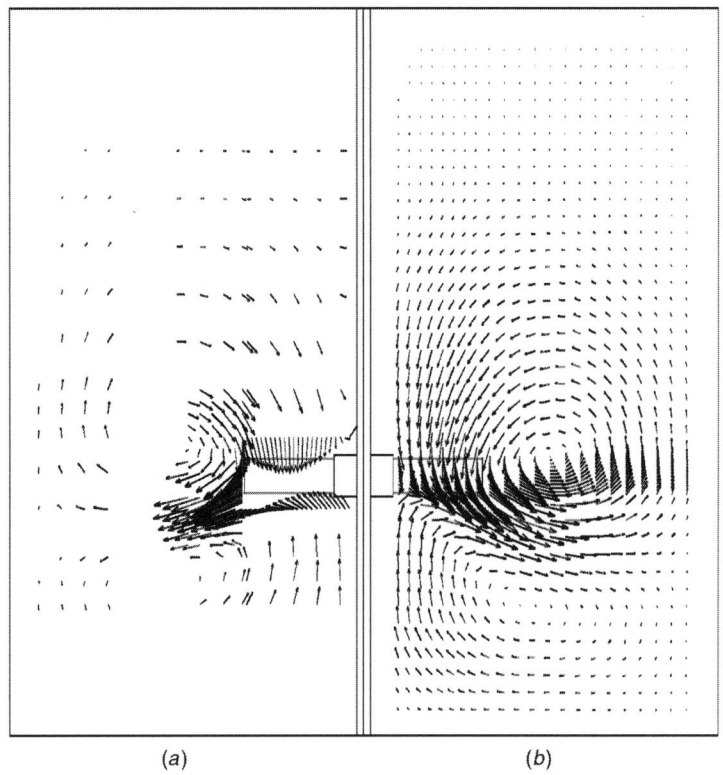

**Figure 5-13** Comparison of (*a*) LDV data and (*b*) CFD sliding mesh results.

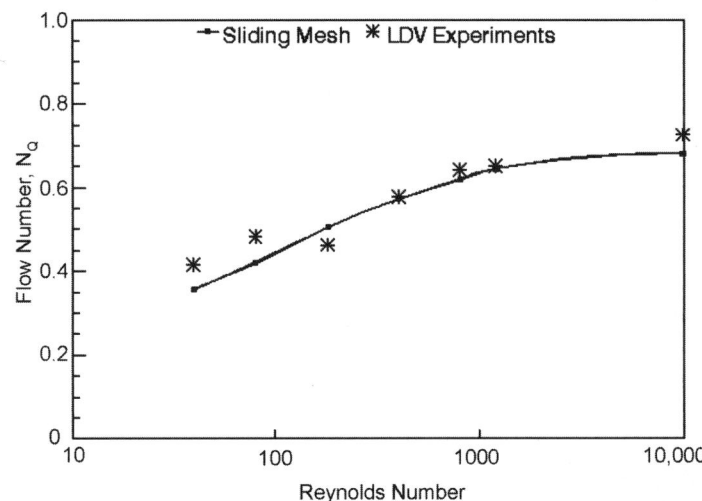

**Figure 5-14** Flow number based on experimental measurements and computed by the sliding mesh model.

**300** COMPUTATIONAL FLUID MIXING

used (Bakker et al., 2000, 2001). Roussinova et al. (2001) showed exact agreement between the frequency predicted by LES simulations and low frequencies observed for a PBT impeller (see Chapter 2).

### 5-5.4 Snapshot Model

The snapshot model (Ranade and Dommeti, 1996) is a steady-state approach that captures the flow field at a single instant in time, when the impeller position relative to the baffles is fixed. When the impeller is rotating, the leading face of the blade exerts a force on the fluid in front of it and acts to push the fluid away. Behind the rotating blade, there is a void of low pressure, which acts to pull the surrounding fluid in. These two complementary functions can be represented as balanced mass sources in front of and in back of the impeller blade, and this premise is the basis of the snapshot model. A grid is built with the impeller in one position relative to the baffles, and a steady-state solution is performed. Mass sources in front of and behind the impeller blade are used to simulate the action of the impeller as if it were rotating. The flow field is therefore characteristic of a fully developed flow for a rotating impeller but is limited to a snapshot of the motion when the impeller is in the single position described by the model. Because the results are highly dependent on the accuracy of the source terms used, this approach, while offering 3D effects, has drawbacks similar to those of the velocity data model.

### 5-5.5 Combining the Geometric Impeller Models with Other Physical Models

The geometric impeller models described above can be used to model both steady-state and time-dependent processes, but attention must be paid to the timescales, where appropriate, and other special requirements of each. In this section, some of these considerations are reviewed for the two most popular of the geometric formulations: the MRF and sliding mesh models.

*5-5.5.1 VOF Model.* In stirred tank applications, the VOF model is useful for tracking the shape of the liquid surface during operation. Even though the steady-state shape of the surface is usually of interest, a transient VOF formulation is usually the best way to obtain it. With this in mind, either the steady-state MRF or transient sliding mesh model can be used for this purpose. If the MRF model is used, the gradual change in the free surface can be predicted using the VOF method. Note, however, that because the orientation of the impeller relative to the baffles is fixed, any irregularities in the free surface that result from the impeller rotation will not be captured. If these details are important, the sliding mesh model should be used.

When the VOF model is solved in conjunction with the sliding mesh model, the smallest required time step for the two models must be used. Since a smaller time step is often required for the VOF calculation than for the sliding mesh calculation, this means that the motion of the impeller will advance in time at a slower rate than is necessary for a calculation involving the sliding mesh model alone. One way to circumvent this problem is to use the sliding mesh model to obtain periodic steady-state conditions using a single fluid first, and then introduce the second fluid with the VOF model and continue the transient calculation until a new periodic steady state is reached. For simple cases in which the free surface is axisymmetric, an implicit impeller model (using fixed velocity data) (Section 5-4) may be preferable for use with the VOF calculation.

*5-5.5.2 Multiphase Model.* Gas–liquid or liquid–solids mixtures can be solved using the Eulerian multiphase or ASM model in conjunction with either the sliding mesh or MRF model. Whereas the common goal of free surface modeling using VOF is to obtain the steady-state shape of the liquid interface, the goal of multiphase modeling can be to examine the unsteady behavior of the mixture as well as to predict the final settling of solids or final gas hold-up. The advantage of using the MRF model is that its steady-state basis can be combined with the time-stepping needed for complex multiphase flows. The disadvantage, however, is that the fixed orientation of the impeller blade with the baffles introduces error in the transient behavior by ignoring the impact of the impeller-baffle interaction on the flow.

For cases in which the transient behavior of the process is of interest, the sliding mesh model should be used instead. Here, the same issues apply that are important for VOF modeling. Generally speaking, a smaller time step is required for the Eulerian multiphase model than for the sliding mesh calculation. It is good practice to obtain a periodic steady-state solution of the single-phase liquid first using the sliding mesh model, prior to introducing the additional phases. The development of the solids suspension or gas hold-up can then be computed most accurately in the presence of the rotating impeller.

*5-5.5.3 Turbulence.* As discussed in Section 5-2, there are several steady-state turbulence models in widespread use today. These so-called RANS models address a time-averaged state of the fluid such that all turbulent fluctuations are represented by averaged values. The RANS models are often used with both the MRF and sliding mesh models, as well as with many other transient models used in CFD analysis. This practice is justified in part because the time scales of turbulence fluctuations are assumed small compared to those of the other processes being modeled, such as the blade passing time in a stirred tank. It has also been justified because until recently, other more rigorous treatments have not been available in commercial software or solvable in a realistic time on the computers of the day.

The large eddy simulation (LES) model (Section 5-2.1.3) is a fairly recent model to appear in commercial software. It offers considerably more rigor than the RANS models. It makes use of a steady-state model for the smallest turbulent eddies, but treats the large scale eddies in a transient manner. The use of LES is inconsistent with the use of the MRF modeling approach, because the approximation introduced with the MRF model is on a longer time scale than the detail offered by the LES calculation. The use of LES with the sliding mesh model, on the other hand, is a powerful combination that has demonstrated great potential for capturing not just small scale fluctuations but large scale fluctuations as well, including instabilities with frequencies that are several times larger than the impeller rotation frequency (Bakker, 2001; Roussinova et al., 2001).

***5-5.5.4 Species Transport.*** When the sliding mesh model is used, species blending can be tracked along with the transient motion of the impeller. The species is normally introduced after the system has reached periodic steady state, but need not be. If an inflow boundary is to be used for species calculations after periodic steady state has been reached, it should be assigned the velocity of the species jet (using the background fluid) during the startup period. One method that can be used to hasten the calculation during the startup period is to start with a solution based on the MRF model, as discussed in Section 5-5.3.1.

When the MRF model is used, transient species transport should be done with great care or avoided altogether. This is due to the fact that the velocities in the rotating frame, whether stored in the local or absolute frame, will give rise to erroneous behavior when they are used to convect a scalar in a transient manner. Graphical displays of the species distribution are suspect, even though the method can accurately capture the average species concentration as a function of time in the vessel as a whole.

***5-5.5.5 Dispersed Phase Particle Tracking.*** The dispersed phase model, discussed in Section 5-2.2.2, allows for the coupled motion of a particle, bubble, or droplet stream with the fluid phase. When used with the sliding mesh model, the trajectories are computed in segments, with one segment per time step. The solver must ensure that the total time of each trajectory segment does not exceed the duration of the time step. If this condition is met, the particles can cross the sliding mesh interface without any incompatibility in the assumptions of either model. When combined with the MRF model, however, the implementation must be able to incorporate the particle motion in the rotating frame as well. Although there are techniques for doing so, it is not clear that the results are meaningful in all reference frames. This combination of models should therefore be avoided.

## 5-6 EVALUATING MIXING FROM FLOW FIELD RESULTS

Although there are numerous options for simulating the fluid flow inside a stirred tank, the goal of the simulation is to learn about the various aspects of the flow field. On a simple level, this might include velocity vectors in one or more regions, path lines followed by infinitesimal fluid elements as they wind their way through the vessel, or the distribution of a tracer species after some period of time has passed, for example. On another level, the analyst might want to understand the power requirements for the motor, the time required to achieve adequate blending, or the fate of vortices trailing from the edges of the impeller blades. This type of information and more can generally be extracted from the CFD results or can be obtained from auxiliary CFD calculations based on those results. To illustrate how, we present summaries and examples in the following sections, designed to provide an overview of several of the methods used to make CFD analysis of mixing a meaningful endeavor.

### 5-6.1 Graphics of the Solution Domain

A stirred tank can be displayed in a number of ways to illustrate the relevant features of the vessel and its internals. These are described below.

***5-6.1.1 Geometry Outline.*** Perhaps the simplest method for displaying the vessel is to draw an outline of the geometry. An outline consists of the features of the tank and internals, but little else. For 2D simulations, either a side view or dotted lines (or both) can be used to represent the impeller and the location where the experimental data are applied to represent it. For 3D simulations modeled using the explicit geometry, all edges are shown.

***5-6.1.2 Surfaces.*** In addition to the features shown in an outline, the surfaces can also be drawn. If solid surfaces are used for the tank, the viewer cannot see inside unless the viewpoint is through an opening in the side or the top (Figure 5-15). Alternatively, solid surfaces can be used for the internals, and translucent surfaces can be used for the vessel walls (Figure 5-10a). When displayed with lighting, the image can accurately convey the 3D nature of the entire geometry.

***5-6.1.3 Grids.*** For 2D simulations, a display of the grid (Figure 5-1b) is an excellent way to illustrate the potential level of accuracy in the solution. Despite having a deeply converged solution, a coarse grid cannot deliver accuracy on a scale any finer than the grid itself. A fine grid, however, has the potential to deliver a much better resolved flow field, assuming that the solution is converged adequately. Most grids are nonuniform, with fine and coarse grid regions that show the areas where the most (and least) accurate details can be expected. For 3D simulations, displays of the grid are more difficult to do in a meaningful way. When the grid is structured, a single grid plane can be displayed. In addition to showing the distortion in the grid (if the grid plane is distorted), this type of display can also show fine and coarse grid regions. For unstructured grids, single grid planes do not exist. A cut through the solution domain on, say, a surface of constant x-value, shows the cross-section through a number of cells and is not necessarily helpful. A more common approach is a display of the surface grid in 3D simulations (Figure 5-10b). If the surface grid is fine (or coarse) in a region,

**304** COMPUTATIONAL FLUID MIXING

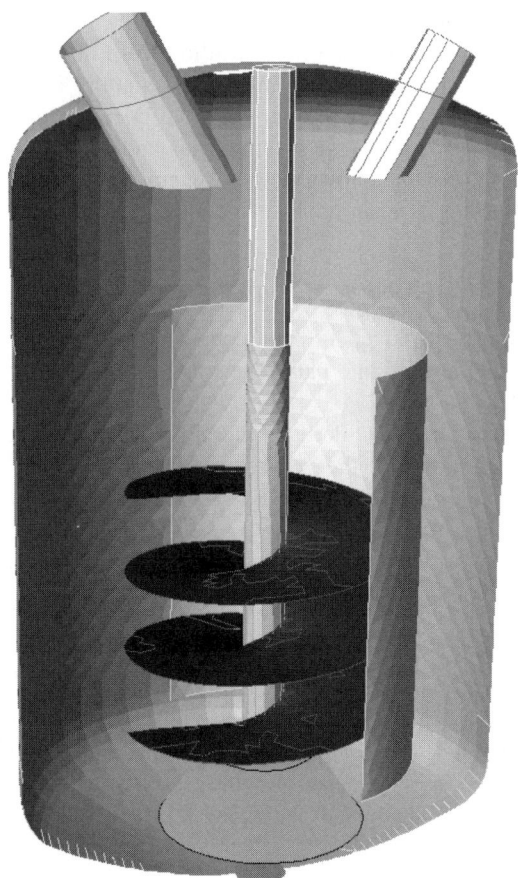

**Figure 5-15** Geometry using solid surfaces with a cut in the wall and top to look inside.

the chances are good that the volumetric mesh in that region is fine (or coarse) as well.

### 5-6.2 Graphics of the Flow Field Solution

There are many ways to examine the flow field results, some of which are described below.

*5-6.2.1 Velocity Vectors.* Velocity vectors can be used to illustrate the magnitude and direction of the flow field throughout the solution domain. For 2D simulations, a plot of all velocity vectors gives an overall picture of the fluid behavior. For 3D simulations, a plot of all vectors in the domain is too crowded to be useful. Vectors need to be plotted on one or more planes or surfaces instead, as shown in Figure 5-11*b* and again in Figure 5-13. Note that the planes can be single grid planes (e.g., J = 10) or Cartesian grid planes (x = 3.5 m). Surfaces can be planar or nonplanar, such as a surface of constant temperature or a surface of constant radius. The important point is that for vector plots to be meaningful, the vectors (with length and orientation) need to be clearly visible, so the surfaces or planes used to plot them need be chosen accordingly.

### 5-6.2.2 Streamlines.
In 2D simulations, a quantity called the *stream function*, $\psi$, is defined in terms of the density and gradients of the x- and y-components of the velocity, U and V. In terms of cylindrical coordinates, which are most appropriate for axisymmetric stirred tank models, the definition takes the form

$$\rho U = \frac{1}{r}\frac{\partial \psi}{\partial r} \quad \text{and} \quad \rho V = -\frac{1}{r}\frac{\partial \psi}{\partial x} \tag{5-41}$$

where U and V are the axial and radial components of velocity. The stream function is constant along a streamline, a line that is everywhere tangent to the velocity field. When defined in the manner above, $\psi$ incorporates a statement of conservation of mass. The difference between the stream function defined on any two streamlines is equal to the mass flow rate between the streamlines. Thus when a pair of streamlines has close spacing, the implication is that the velocity is greater than when the same pair has wide spacing, since the same amount of mass must pass through the space between the lines. Streamlines therefore have the ability to convey not only the relative movement of the flow, but the relative speed as well. In Figure 5-16, streamlines in a 2D simulation of a stirred tank

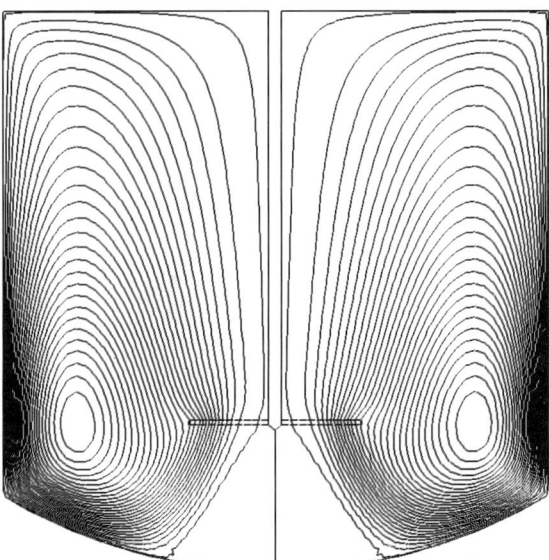

**Figure 5-16** Streamlines for a 2D simulation of a pitched blade impeller with a single recirculation zone showing high- and low-speed regions.

**306**    COMPUTATIONAL FLUID MIXING

are close as they pass through the impeller, where the boundary conditions are imposed and the flow speed is high. They are also close along the outer wall, but are more widely spaced elsewhere, where the flow recirculates in a larger area at a much slower speed.

*5-6.2.3 Path Lines.* Since the stream function is defined only for 2D flows, an alternative method is needed to visualize 3D flows in the same manner. Path lines can be used for this purpose. Path lines follow the trajectories that would be followed by massless particles seeded at any location within the domain. These particles move with the flow field and leave behind tracks in one form or another that allow the flow field to be visualized. In Figure 5-17, path lines are used to illustrate the flow through a static mixer. Path lines can be drawn as simple lines or as tubes, ribbons, or a series of dots. They can usually be colored by problem variables, such as temperature. When colored by time, they give information on residence time if inflow and outflow of fluid are involved.

*5-6.2.4 Contours.* Contours are lines where a chosen variable has a constant value. The streamlines illustrated in Figure 5-16 are actually contours of stream function, since $\psi$ is constant on each of the lines shown. In addition to line contours, filled contours, plotted on an entire 2D domain or on a surface in a 3D domain, are also very useful for showing the maximum and minimum values as well as local gradients. In Figure 5-18, contours of a tracer species are shown on a cross-section through a 3D domain.

*5-6.2.5 Isosurfaces.* Isosurfaces in 3D flow fields are analogous to contour lines in a 2D flow field. These 3D surfaces are constructed in such a way that a particular variable has a constant value everywhere on it. If the isosurface has a constant value of the Cartesian coordinate x, for example, it is planar. If it

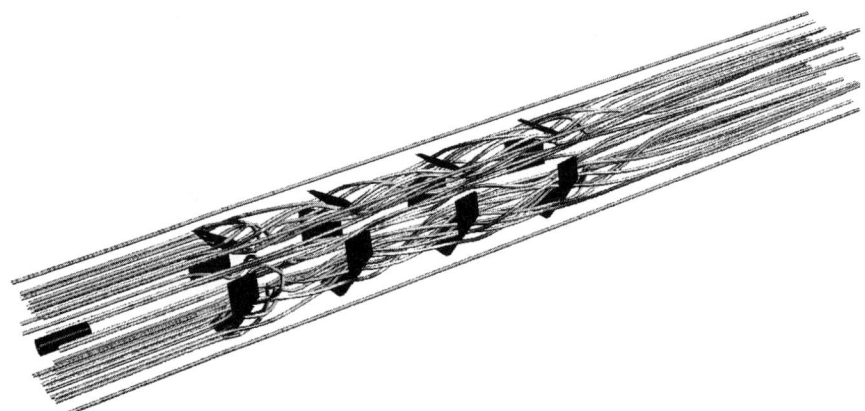

**Figure 5-17**   Path lines, colored by velocity magnitude, illustrating the flow through an HEV static mixer.

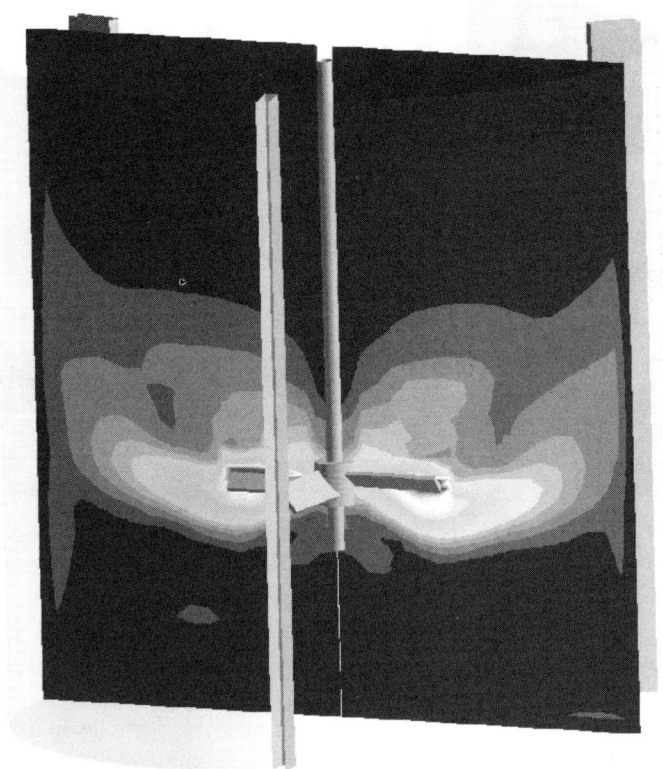

**Figure 5-18** Filled contours of a tracer species shown on a planar surface in a 3D domain.

has a constant value of velocity in a stirred tank, it is complex in shape and can have several disconnected regions. Isosurfaces of this type can be plotted as solid surfaces with lighting (Figure 5-19), to convey the 3D nature of the variable distribution. They can also be used to plot contours, showing how one variable changes as another one is held fixed.

**5-6.2.6 Particle Tracks.** Whenever the discrete phase model is used (Section 5-2.2.2), particle tracks can be used to illustrate the trajectories of the particles, bubbles, or droplets. Trajectories can usually be displayed in a number of ways. For example, lines can be colored by the time of the trajectory or temperature of the particle itself. In addition to lines, ribbons and tubes can generally be used. The tracks can be computed and displayed using the mean fluid velocities, or in the case of turbulent flows, using random fluctuations in the mean fluid velocities as well. These *stochastic tracks* often give a more realistic picture of the extent to which the particles reach all corners of the solution domain than do tracks computed from the mean velocities alone.

**Figure 5-19** Dispersion of a tracer in a stirred tank. A blob of tracer is injected at time zero, and its dispersion is shown after $\frac{1}{4}, \frac{1}{2}, \frac{3}{4}, 1, 1\frac{1}{4}, 1\frac{1}{2}, 1\frac{3}{4}$, and 2 impeller revolutions, respectively.

*5-6.2.7 Animations.* Animations can be created from groups of image files that follow a process from beginning to end, or during some period of operation. They can also be used to follow the motion of massless particles in a steady-state flow field. Numerous postprocessing packages are commercially available for the creation of animations, and many CFD packages have built-in functionality to do so as well.

*Types of Animations.* Some examples of how animations can be used for displaying flow field results are described below. In general, anywhere from 20 to hundreds of images or frames can be created and concatenated, or joined together to form the animation. The content of these images depends on whether the simulation is steady-state or time-dependent, and what the display goals are intended to be. In general when creating animations, care should be taken to avoid incorporating too much information into a single image, since some of this information will inevitably be lost on the viewer.

*Time-Dependent Simulations.* For time-dependent flow fields, images should be made at uniform time intervals for the purpose of creating a meaningful animation. Examples of time-dependent animations include

- Contours of tracer concentration on a single plane during blending
- Velocity vectors on a plane during a turbulent simulation modeled using large eddy simulation
- Gas from a sparger filling a stirred tank or a bubble column

- Lifting and suspension of solids off the vessel floor in a stirred tank
- Isosurfaces of vorticity trailing from a rotating impeller in a sliding mesh model

*Path Lines.* Path lines are normally created by a simultaneous calculation and display of trajectories, using the problem geometry and flow field data. To generate an animation of evolving path lines, frames of the trajectories at intermediate stages need to be created and stored. To do this, a total time for the animation needs to be determined along with a number of frames to be made. Tools are available in most visualization packages to generate the intermediate frames based on these inputs, using dots, lines, or other geometric entities. The intermediate frames can be written to files in one of a number of available formats. When played in succession, the concatenated frames will mimic the display that is generated by the original visualization software.

*Moving Slice Planes.* One method of illustrating the change in a variable throughout a 3D domain is through the use of animated slices. For example, in a stirred tank, the velocity field at different angular locations—from one impeller blade to the next or from one baffle to the next—might be of interest. Planar slices at equal angle intervals can be used for each frame, on which either contours or in-plane velocity vectors are displayed. A series of axial slices is another useful way to examine the change in a variable from one end of a mixer to another. This type of animation is particularly useful for static mixers.

*Moving Isosurfaces.* When injecting a tracer, one method of following its evolution is by animating isosurfaces of the tracer mass fraction. The animation is made most effective if the same numerical value is chosen for each frame. The value should be small in magnitude so that the expanding surface at later times can be captured. If the data for all times exist prior to the creation of the images, the data for the first and last times should be used to plan the best isosurface value to track for the duration of the process. When plotted as a solid surface with lighting, the 3D nature of the isosurface is easy to discern and the effect makes for an exciting and informative animation. The frames shown in Figure 5-19 are taken from an animation of this type.

*Moving Impeller Blades.* In stirred tank animations, it is always helpful if the motion of the impeller can be animated as well. This is possible in sliding mesh simulations, where the changing position of the impeller can be captured in successive frames. Some animation software can extract this motion from MRF simulation data, where the rotation speed of the impeller is known. Based on the time interval between frames, the impeller is advanced by a computed angle in each display created. When the frames are animated, a continuous motion of the impeller can be seen, along with other animated variables, such as path lines or changing contours on a stationary surface.

**310** COMPUTATIONAL FLUID MIXING

*Moving Viewpoint.* For steady-state external flows, animations based on a moving viewpoint are popular. These animations can also be used to illustrate the complex geometry of a system, such as a stirred tank and its internals. Beginning with a distant view, the camera can approach the object and peer inside to get close-up views of the components. For the Kenics static mixer described in Section 5-7.10, a sequence of frames representing equally spaced axial slices (Figure 5-29) can be used to illustrate the progression of mixing as the fluid moves through the helical elements.

*Creating Animations from a Collection of Images.* Numerous commercial software packages are available for creating an animation from a collection of images. Different image file formats are available for this purpose. Once the images have been concatenated to form the animation, tools are available in most animation packages to set the speed of the animation. A choice of about 0.05 s between frames usually results in a smoothly playing animation, but this also depends on the number of frames and the capabilities of the computer. It should be noted that the time interval mentioned here refers to the playing time, not the physical time between the data used for each frame display.

### 5-6.3 Other Useful Solution Variables

In Section 5-6.2, methods of plotting several common solution variables, such as velocity, stream function, and species concentration, were discussed. Plots of turbulent kinetic energy and dissipation are also of interest in turbulent flows, especially if other processes, such as chemical reactions, are to take place. In multiphase flows, the volume fraction of the phases is the most useful tool to assess the distribution of the phases in the vessel. In this section, three additional quantities are reviewed that are derived from the velocity field. These can provide a deeper understanding of the flow field than can plots of the velocity alone.

**5-6.3.1** *Vorticity.* Vorticity, a vector quantity, is a measure of the rotation of the fluid. In terms of a fluid element, a nonzero vorticity implies that the element is rotating as it moves. The vorticity is defined as the curl of the velocity vector, **U**:

$$\xi = \nabla \times \mathbf{U} \qquad (5\text{-}42)$$

Vorticity can be defined in both 2D and 3D flows. In 2D flows, the direction is normal to the plane of the simulation. This means that for a 2D axisymmetric simulation of flow in a stirred tank, the vorticity is always in the circumferential direction:

$$\xi_\theta = \frac{\partial U_x}{\partial r} - \frac{\partial U_r}{\partial x} \qquad (5\text{-}43)$$

In 2D simulations, positive values indicate counterclockwise rotations, while negative values indicate clockwise rotation. In a 3D simulation, vorticity can take on any direction, and plots of vorticity magnitude, rather than the individual components, are often the most helpful. The units of vorticity are s$^{-1}$, the same as those used for shear rate. In Figure 5-20a, contours of vorticity are shown for a 2D flow in a stirred tank with velocity vectors superimposed on the display. White regions (near the impeller) have a maximum positive value, and black regions (near the walls) have a maximum negative value. These regions are those where steep normal gradients occur in the velocity. The fact that the vorticity is positive near the impeller and negative near the wall indicates simply that the direction of the curl is opposite in these two regions. In Figure 5-20b, isosurfaces of constant vorticity magnitude in a 3D simulation show the trailing vortices behind a Rushton impeller. The simulation was performed using the LES turbulence model.

**5-6.3.2 Helicity.** The helicity is defined as the dot product of the velocity vector with the vorticity vector:

$$\mathbf{H} = \mathbf{U} \cdot \boldsymbol{\xi} = \mathbf{U} \cdot (\nabla \times \mathbf{U}) \tag{5-44}$$

Clearly, the helicity has a value of zero in 2D simulations. In 3D simulations, it gives an indication of how well the local rotation of a fluid element is aligned with the velocity of the element. It is useful for illustrating longitudinal vortices, or spiral motion, as is often found in vortex cores. In Figure 5-21, isosurfaces of helicity are used to depict the longitudinal vortices generated in the Kenics static mixer described in Section 5-7.10.

**5-6.3.3 Rate of Deformation.** The rate of deformation or strain rate tensor is a collection of terms that together describe the complete deformation of a fluid element in motion. The deformation can be the result of linear strain, which gives rise to a linear deformation or stretching of the element, and shear strain, which gives rise to an angular deformation or change in shape of the element. The symmetric tensor has components of the generalized form

$$S_{ij} = \frac{1}{2}\left(\frac{\partial U_i}{\partial x_j} + \frac{\partial U_j}{\partial x_i}\right) = S_{ji} \tag{5-45}$$

Although the tensor components themselves offer little insight into the behavior of the flow field, functions of the tensor components often do. In terms of the Cartesian coordinates x, y, and z, the diagonal terms are

$$S_{xx} = \frac{\partial U_x}{\partial x} \quad S_{yy} = \frac{\partial U_y}{\partial y} \quad S_{zz} = \frac{\partial U_z}{\partial z} \tag{5-46}$$

Each of these terms represents a linear strain rate or rate of elongation of the fluid element in each of the three coordinate directions. The sum of these diagonal terms is the *trace* or first invariant of the tensor. For incompressible fluids, this quantity is always zero, since the volume of the fluid element must be conserved.

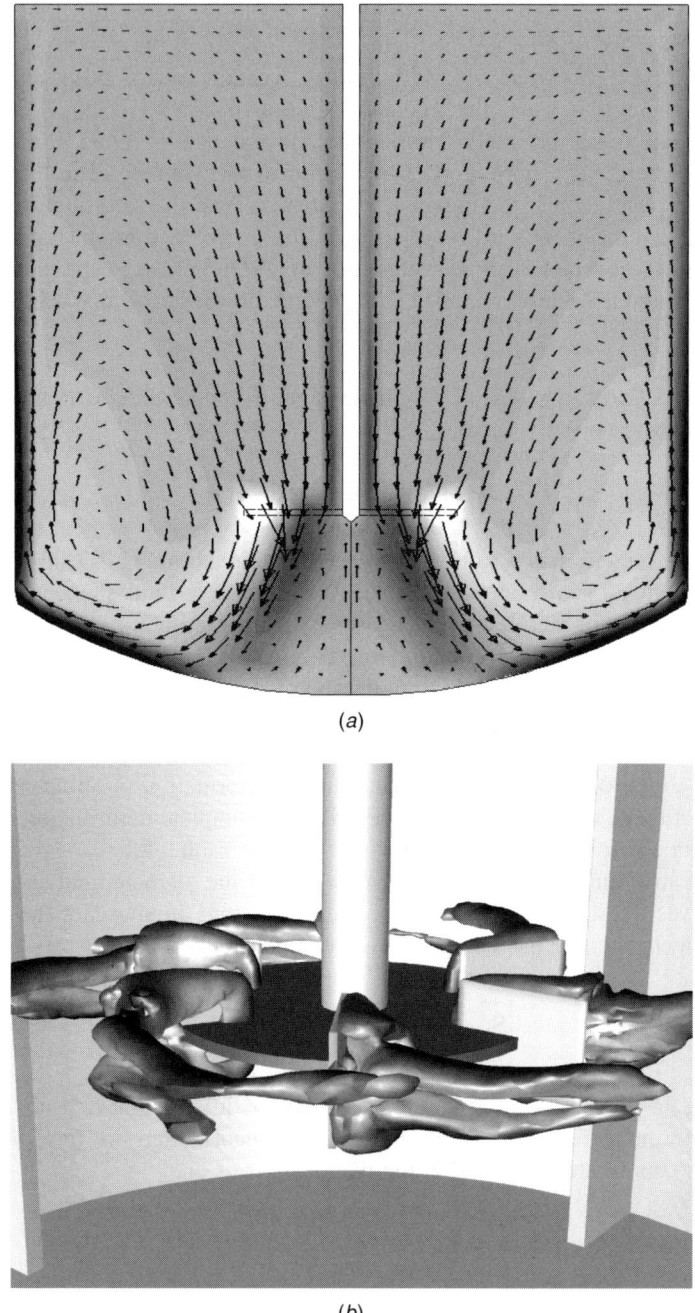

**Figure 5-20** (*a*) Contours of vorticity in a 2D simulation with superimposed velocity vectors. (*b*) Isosurfaces of vorticity magnitude behind a Rushton turbine.

EVALUATING MIXING FROM FLOW FIELD RESULTS    313

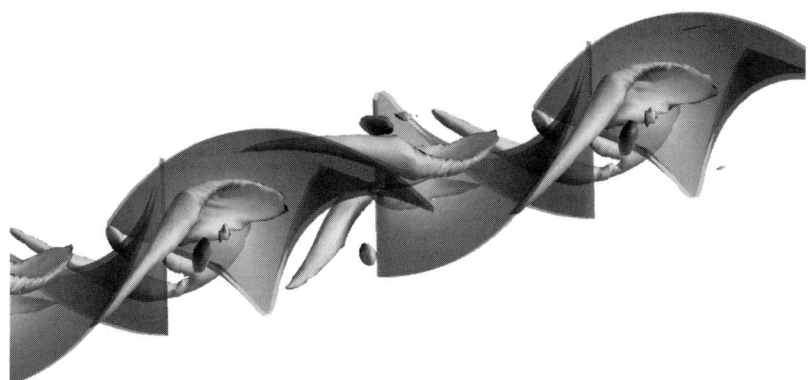

**Figure 5-21**  Isosurfaces of helicity are used to show the longitudinal vortices in a Kenics static mixer.

In addition to the trace, another quantity, often referred to simply as the *strain rate*, is of interest. The strain rate, taken from the modulus of the tensor, is a positive-definite representation of all possible components of the strain rate tensor. It is used to determine the viscosity in strain-dependent non-Newtonian fluids and is also helpful as a reporting tool for mixing applications. In particular, regions with a high strain rate play an important role in liquid dispersion.

### 5-6.4  Mixing Parameters

Parameters that are used to characterize stirred tank flows and mixing processes in general can be computed by correlations that can be found in the literature. In many cases, these parameters can also be computed from the CFD results. Examples of how to compute some of these parameters are given below.

***5-6.4.1  Power Number.***  The *power number* is a dimensionless parameter that provides a measure of the power requirements for the operation of an impeller. It is defined as

$$N_P = \frac{P}{\rho N^3 D^5} \qquad (5\text{-}47)$$

In eq. (5-47) P is the power applied to the impeller of diameter D, $\rho$ the density, and N the impeller rotation speed in Hertz. Correlations are available that provide the dependence of $N_P$ on the Reynolds number. Thus, if CFD is not available, the power requirements can generally be obtained from one of these correlations. The correlations can break down, however, if they do not address the D/T or C/T ratios of single impellers or the presence and spacing of multiple impellers. In such cases, CFD results can be used to compute $N_P$, or simply, the power requirements.

**314** COMPUTATIONAL FLUID MIXING

The power delivered to the fluid is the product of the impeller speed, $2\pi N$, in rad/s, and torque, $\tau$, which is obtained by integration of the pressure on the impeller blade:

$$P = 2\pi N\tau \qquad (5\text{-}48)$$

Reports are usually available for the torque delivered to the fluid by the impeller. In some cases, reports of power or even power number can be obtained from the software.

*Integration of the Dissipation.* In principle, the power delivered to the mixer is equivalent to that lost or dissipated in the fluid. An integration of both the viscous and turbulent dissipation throughout the volume should, therefore, be an acceptable way to compute the power draw. The dissipation rate predicted by the various turbulence models can vary significantly, however, and there is no guarantee that the turbulence model that gives the best flow pattern prediction also gives the best dissipation rate prediction. For laminar flows, even with a refined mesh near the impeller blades, CFD can have difficulty predicting viscous dissipation in a satisfactory manner. For this reason, the best method for extracting the power drawn by the impeller is by calculation of the torque on the blade surfaces.

**5-6.4.2 Flow Number.** The *flow number* is a measure of the pumping capacity of an impeller. Different measures for pumping capacity exist, but the flow number is used widely. It is defined as

$$N_Q = \frac{Q_\ell}{ND^3} \qquad (5\text{-}49)$$

In this expression, $Q_\ell$ is the flow rate produced by the impeller. The subscript is used to ensure that the flow rate for the liquid phase alone is used in the calculation. To compute $Q_\ell$ for an impeller, a surface needs to be created for the discharge region. This surface would be circular for an axial flow impeller and a section of cylinder wall for a radial flow impeller. By integrating the total outflow through this surface, the flow rate, $Q_\ell$, and subsequently the flow number, $N_Q$, can be obtained.

**5-6.4.3 Evaluating Mixing Time.** A transient blending calculation is the best method for determining the time required to achieve a certain level of blending. When a tracer is added to a fluid in a mixing tank, the transient calculation can be made exclusive of the flow field calculation if the properties of the tracer and background liquid are identical. When this is the case, a steady-state calculation can be performed for the background liquid using either experimental data or the MRF method, although care should be exercised when using the latter, as discussed in Section 5-5.5.4. If inflow and outflow ports are to be used, the simulation of the background liquid alone should include the inflow boundary conditions for velocity that will ultimately be used for the tracer. Once the flow field for the background fluid is satisfactorily converged, the tracer can be introduced. Since the mixture fluid properties will not change with the addition of the tracer, the transport equations for momentum, continuity, and turbulence can be disabled while the transient species calculation takes place. The transient solution of this single scalar equation will be robust (since it is not coupled to other variables that are in a state of change) and economical, advancing rapidly with few iterations required each time step. Averages of the tracer concentration, along with standard deviations, can be computed throughout the vessel to determine when the tracer has become fully blended.

There are two exceptions to the use of the method described above, in which the flow field calculation can be disabled during the species calculation. First, if the sliding mesh model is used, the flow field data are required for each time step, so it is not possible to disable the flow field calculation to perform the species transport calculation. Second, if the tracer is to be added through an inlet or dip tube for a finite period of time, after which the inlet flow is disabled, calculation of the flow field should resume at that time, especially if the inlet delivers a jet of significant momentum to the vessel.

*5-6.4.4 Information from LES Simulations*. Large eddy simulations are transient simulations designed to capture the fluctuations that are the result of turbulent eddies. For this reason, LES images and animations have the potential to capture small and large scale activity that would otherwise be averaged to zero with a RANS turbulence model. Some of the small scale activity includes the birth and death of eddies or small vortices. Some of the large scale activity includes low-frequency instabilities in stirred tanks. A common way to visualize the turbulent structure present in LES simulations of mixers is by animating vectors or isosurfaces of vorticity magnitude.

## 5-7 APPLICATIONS

To illustrate the successful application of CFD to many types of process equipment, a number of examples are presented in this section. Unless otherwise noted, these simulations were performed with software from Fluent Inc.

### 5-7.1 Blending in a Stirred Tank Reactor

Mixing time correlations for stirred tank reactors are available, but these are often difficult to extend outside the experimentally studied parameter range. One advantage of CFD is that it can be used to evaluate industrial sized equipment or equipment for which no correlations are available. A comprehensive evaluation of the accuracy of mixing time predictions using CFD was presented by Oshinowo et al. (1999). The main conclusion drawn was that although unsteady tracer dispersion predictions based on a steady-state flow field are acceptable, the accuracy of the predicted mixing time is greatest when the mixing simulation is based on a time-dependent calculation, using the sliding mesh model. For the latter method, either the LES model or a standard turbulence model such as RSM may be used.

Figure 5-19 shows an example of the dispersion of a chemical tracer in a stirred tank. A standard pitched blade turbine is used to mix two waterlike materials. The neutrally buoyant tracer is injected at time zero as a blob above the impeller, as shown on the top left in the figure. The flow field is calculated using the sliding mesh and LES models, and the dispersion of the tracer is derived from the flow field. The blob is stretched and the chemical is mixed with the rest of the fluid over time. It is interesting to see that despite the fact that there are four impeller blades and four baffles, the concentration field is not symmetric because of the off-axis injection. The consequence is that the full tank needs to be modeled instead of a 90° section. Bakker and Fasano (1993b) presented a successful comparison between blend time predicted by CFD and calculated from experimental correlations.

### 5-7.2 Chemical Reaction in a Stirred Tank

The blending of chemical reactants is a common operation in the chemical process industries. When a competitive side reaction is present, the final product distribution is often unknown until the reactor is built. This is partly because the effects of the position of the feed stream on the reaction by-products are difficult to predict. In this example from Bakker and Fasano (1993), the product distribution for a pair of competing chemical reactions is calculated with CFD and compared with experimental data from the literature. The model used here is a slightly modified version of the standard Magnussen model discussed in Section 5-2.2.1.

The following competitive-consecutive reaction system was studied:

$$A + B \xrightarrow{K_1} R$$

$$B + R \xrightarrow{K_2} S \qquad (5\text{-}50)$$

This is the reaction system used by Bourne et al. (1981) and Middleton et al. (1986). The first reaction is much faster than the second reaction: $K_1 = 7300$ m$^3$/mol · s versus $K_2 = 3.5$ m$^3$/mol · s. The experimental data published by Middleton et al. were used to determine the Magnussen model constants. Two reactors were studied, a 30 L reactor equipped with a $D/T = \frac{1}{2}$ Rushton turbine and a 600 L reactor with a $D/T = \frac{1}{3}$ Rushton turbine. In the CFD analysis, a converged flow field was computed first for each reactor, using experimental data for the impeller boundary conditions. The reactants A and B were then introduced to the tank on an equimolar basis. The reactant A was assigned a weak but uniform concentration throughout the vessel. The reactant B was added in a high concentration in a small region. The calculation of the flow field variables was disabled after the addition of the reactants, and the species calculations alone were performed. Once the solution converged, the product distribution $X_S$ was calculated using

$$X_S = \frac{2C_S}{C_R + 2C_S} \quad (5\text{-}51)$$

In the reaction model used here it was assumed that small scale mixing affected only the first reaction and that once this reaction had occurred, the species were locally well mixed. As a result, small scale turbulent mixing did not affect the second reaction. This was achieved by using different values of the Magnussen model constants for the two reactions.

Figure 5-22 shows a comparison between the experimental data from Middleton et al. and the CFD predictions for both reactors. The product distribution, $X_S$, is plotted as a function of impeller speed, in rpm. This graph shows that the model predicts the effects of scale and impeller rotational speed correctly and is usually within 10% of the experimental results. The effect of the inlet position of the feed stream on the formation of the by-product, S, was also studied. Figure 5-23 shows values of $X_S$ for various feed locations. $X_S$ varies only slightly when the inlet is located in the fluid bulk. However, when the feed is injected directly above the impeller, such that the feed stream passes immediately through the highly turbulent impeller zone, local mixing is much faster and does not limit the rate of the first reaction. As a result there is less reaction by-product, S, and the final $X_S$ is only 50% of what it would be if the feed were located away from the impeller. This qualitatively agrees with the experimental results of Tipnis et al. (1993), who used a different set of reactions and tank geometries but also found that injection near the impeller resulted in a lower $X_S$ value than

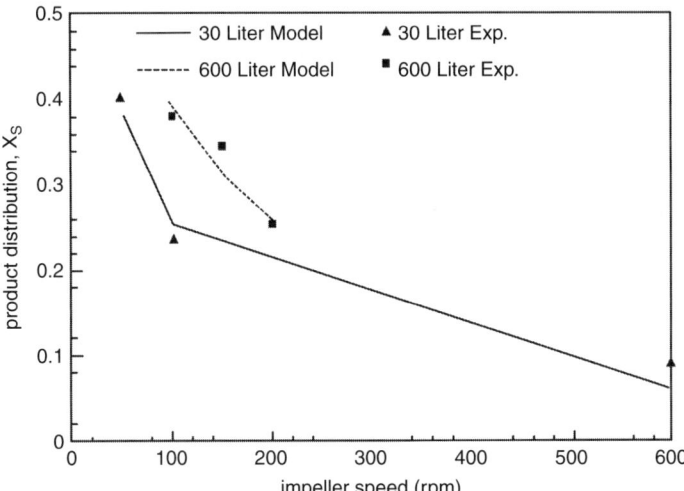

**Figure 5-22** Product distribution, $X_S$ as a function of impeller speed (rpm) for two vessels of different size, with the second reactant being added in the outflow of the impeller. Model predictions are compared with data from Middleton et al. (1986).

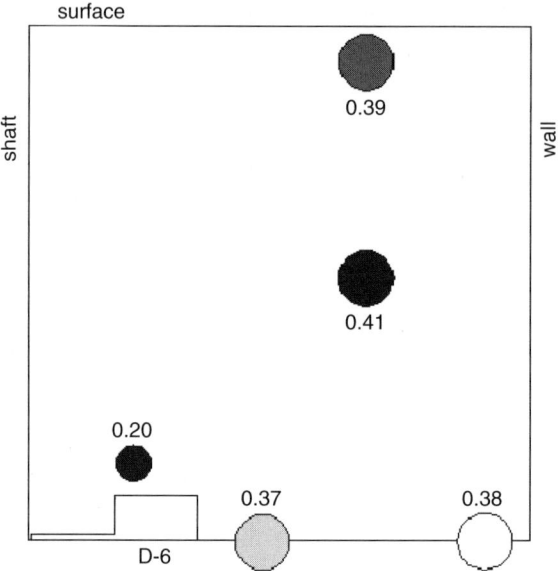

**Figure 5-23** Product distribution $X_S$ as a function of feed location for a 600 L vessel with a Rushton turbine operating at 100 rpm. The product distribution is reduced by about a factor of 2 when the feed is positioned directly above the impeller.

injection farther away from it. The relative differences found by Tipnis et al. are similar to those shown in this example. The effect of mixing on reaction is discussed further in Chapters 13, 17, and 2.

### 5-7.3 Solids Suspension Vessel

Stirred tanks for solids suspension applications have traditionally been designed using the just suspended impeller rotational speed, $N_{JS}$. Although much work about solids suspension has been published, most of it concentrates on providing correlations for the just suspended speed. Attempts to develop mathematical models for the solids suspension process are often based on the total power draw of the impeller, or the average liquid velocity in the tank, without taking local effects into account. The effect of the flow pattern on the spatial distribution of the solids has received relatively little attention. It is now known that the solids spatial distribution is strongly affected by the number of impellers, their location, and certain flow transitions. When either the D/T or C/T ratios are too large, a flow transition with reversed flow at the vessel base may occur. This results in an undesired increase in the power needed to suspend the solids, or more simply, $N_{JS}$.

Adding a second impeller typically has a very small effect on the just-suspended speed. In multiple-impeller systems, zoning occurs when the impeller separation is too large. The most efficient solids mixing occurs just before the

**Figure 5-24** Solids suspension in a tall vessel. The solids distribution with a single impeller is shown on the left, and with a dual-impeller system is shown on the right.

flow between the impellers separates. Unfortunately, designing on the basis of the just suspended speed or on the basis of power consumption does not necessarily lead to an optimum multiple-impeller system. Figure 5-24 shows a comparison of the solids distribution for a single (left)- and a dual (right)-impeller system in a tall stirred tank, modeled using experimental data for the impellers and the Eulerian granular multiphase model for the solids suspension (Oshinowo et al., 2000). The results on the left show that in a tall tank equipped with a single impeller, the solids do not move up higher than about half the liquid level. When a second impeller is added, however, such that one long flow loop is formed, the solids reach the level of the second impeller, as shown on the right. When the second impeller is placed too far above the first impeller and zoning occurs, the solids do not reach the upper impeller (not shown; see Bakker et al., 1994b). From the differences between the solids suspension performance of these two configurations it can be concluded that consideration of the just suspended speed or power draw alone does not necessarily lead to the best design. The impeller system has to be designed so that it provides the optimum flow pattern for the suspension duty to be performed. To design such a system, the effects of the flow pattern on the solids distribution must be taken into account. Computer simulation provides an excellent tool for this purpose.

### 5-7.4 Fermenter

Large scale fermenters are used to make such products as yeast, vitamin C, xanthan gum, citric acid, and penicillin, for example. Fermentations are usually carried out in tall vessels with multiple-impeller systems. Air is sparged in at the bottom to provide the microorganisms in the vessel with a supply of oxygen. It is important that the mixer disperse the gas into fine bubbles, a condition that is required to ensure good mass transfer from the air to the broth. See Chapter 11

**320** COMPUTATIONAL FLUID MIXING

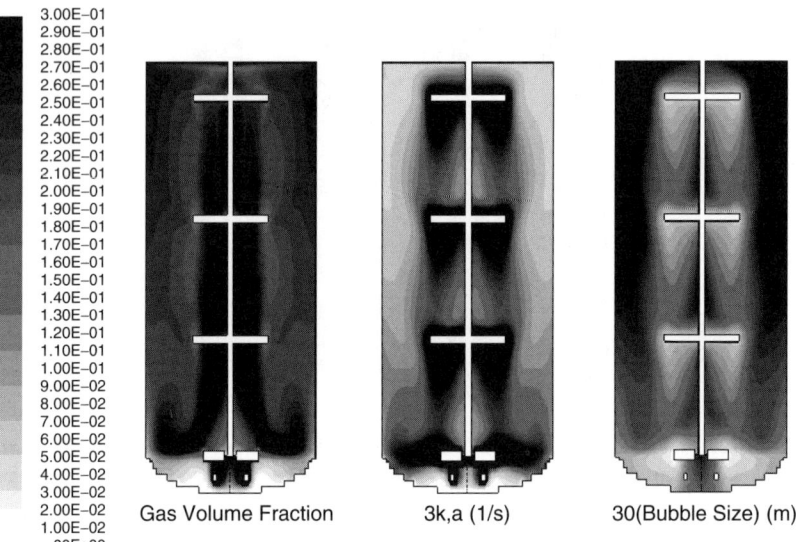

**Figure 5-25** The local gas volume fraction (left), the local mass transfer coefficient $k_l a$ (center), and the local bubble size (right). The bubble size is smallest near the impellers (white) and increases away from the impellers, due to coalescence. The mass transfer coefficient is highest near the impellers (black) because this is where the bubble size is small (leading to a large interfacial area) and where the turbulence intensity is high (leading to fast surface renewal around the bubbles).

for further discussion of gas dispersion and mass transfer, and Chapter 18 for a discussion of mixing in biological applications.

Figure 5-25 shows the results of a gas dispersion simulation of a fermenter. The fermenter is equipped with a radial flow CD-6 impeller with concave blades at the bottom, and three down-pumping HE-3 impellers on top. The vessel has no baffles but is equipped with 12 sets of eight cooling coils, which also act as swirl suppressors. Flow field simulations can be performed to design the impeller system such that there is sufficient liquid movement around these coils.

The gas–liquid simulations shown here were performed with software developed by Bakker (1992), which contains models for gas dispersion, bubble coalescence and breakup, and interphase mass transfer. The local gas volume fraction is shown on the left. The local mass transfer coefficient $k_l a$ (with values multiplied by 3) is shown in the middle, and the local bubble size (with values multiplied by 30) is shown on the right. All figures share the same scale from 0 to 0.3 (which is why the mass transfer coefficient and local bubble size distributions are multiplied by a factor). The bubble size is smallest near the impellers and increases away from them, due to coalescence. The mass transfer coefficient is highest near the impellers, where the bubble size is smallest (leading to a large interfacial area) and where the turbulence intensity is highest (leading to fast surface renewal around the bubbles). The stair-stepped representation of the curved vessel bottom was necessary using the software available at the time of this simulation. Rectangular cells have become obsolete with the introduction of boundary-fitted cells and unstructured grids.

### 5-7.5 Industrial Paper Pulp Chests

One example of a difficult mixing problem is found in the paper industry. Paper pulp, which is a suspension of thin, flexible fibers, exhibits a very complex rheology. As a result, multiple flow regimes are found in paper pulp storage tanks, or chests, which can be rectangular or cylindrical in shape. Laminar flow is common in some parts of the chest, while turbulent flow is common in others. The bottom of the chest is usually filleted, and either sloped, curved, or both. Although paper pulp chests are sometimes equipped with top-entering agitators, the preference in the paper industry is to use side-entering agitators.

The rheological properties of fiber suspensions are discussed in a paper by Gullichsen (1985). The fiber suspension initially behaves as a non-Newtonian fluid with a yield stress $\tau_y$. Above $\tau_y$ the paper pulp behavior is non-Newtonian. When the shear stress exceeds a second threshold value, $\tau_d$, the fiber network structure is disrupted and the suspension behavior is similar to that of a turbulent Newtonian fluid. As a result of this rheological behavior, fiber suspensions are extremely difficult to agitate. To provide motion through the whole tank, the shear stress has to exceed the yield stress everywhere in the fluid. Since gradients in the shear stresses can be expected, there will be regions in the fluid where the fiber network structure is disrupted and the flow is turbulent. At the same time the flow may be laminar or even stagnant in other parts of the chest. This combination of turbulent flow and laminar flow of a non-Newtonian fluid makes paper pulp storage chests difficult to model with CFD.

In an effort to address this problem, Bakker and Fasano (1993a) developed a model for the flow of paper pulp. To model the complex fiber suspension, the following method was used. For every computational cell, the computations are first performed as if the flow were turbulent. A check is then done to see if the total shear stress is indeed larger than $\tau_d$. If this condition is not met, the calculations for that particular cell are repeated as if the flow were laminar. The local apparent viscosity is then calculated from the experimental shear stress versus shear rate curves and the local shear rate. The model has since been used successfully to predict the flow patterns in large industrial chests where zones with turbulent mixing, laminar mixing, and stagnant regions can easily be located. See Chapter 20 for additional discussion of mixing effects in the pulp and paper industry.

Figure 5-26 shows the flow pattern in one example of a stock chest for mixing and storage of paper pulp. The agitator is modeled using experimental data. The flow pattern with a solution of 1% pulp is shown in Figure 5-26 part (*a*); part (*b*) shows how the flow pattern changes when the concentration is increased to 5% and the same impeller speed is used. The results show that more power must be applied to maintain adequate flow conditions when the pulp concentration is

**322** COMPUTATIONAL FLUID MIXING

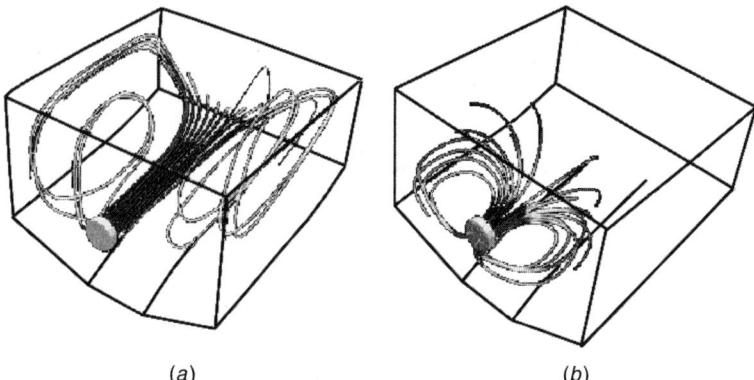

(a)  (b)

**Figure 5-26** Flow pattern in a stock chest for the mixing and storage of paper pulp with (*a*) a solution of 1% pulp and (*b*) a solution of 5% pulp.

increased. The model is an excellent tool for the optimization of agitators for large industrial storage chests and has been used successfully over the years for many different paper pulp applications.

### 5-7.6 Twin-Screw Extruders

The twin-screw extruder is one of the most widely used tools, not only in the plastics and rubber industry but also in other areas, such as food processing. Single- and twin-screw extruders are used to melt, convey, compress, and mix the different compounds involved in any given process, and these steps can considerably affect the quality of the final product. This explains the large interest in screw analysis and, more specifically, the numerous attempts to model twin-screw extruders through numerical simulations. The challenges involved in such simulations (e.g., moving parts, thermal behavior, difficult meshing and remeshing tasks, and partial filling) often lead to many simplifications of the actual problem.

To ease the setup of a three dimensional unsteady twin-screw extruder, a technique referred to as *mesh superposition* (MST) has been developed (Avalosse and Rubin, 1999). This robust technique greatly simplifies the meshing of the geometric entities and does not present the complexities and limitations of other commonly used techniques. The transient algorithm was developed for 2D and 3D nonisothermal, generalized Newtonian fluids. It is designed to work with a finite element solver. A mesh is generated for each part of the flow simulation: one for the flow domain and one for each screw. The screws are assumed to be rigid and their motion is a combination of translation and rotation. At each time step the screw meshes are moved to a new position, overlapping the flow mesh. For each node of this new domain that lies within a given screw, a special formulation is used that imposes a velocity that matches the rotation speed of the screw. The movement of the screws imparts momentum to the surrounding fluid. The flow is calculated in this manner for a set of successive screw positions at constant angular displacement. The history of the flow pattern is thus obtained and stored for further analysis. This application is discussed further in Chapter 16.

Figure 5-27a shows the grid for a typical twin-screw extruder. The grid in the screw regions is shown on the surfaces of the elements. The black lines show the outline of the region containing the fluid. Figure 5-27b shows the shear rate on a planar surface through the extruder. High shear rates are found near the tips of the extruder elements, as expected. This information is relevant when dealing with shear-sensitive materials. Other quantities of interest, such as residence time distributions, material thermal history, and stretching rates, for example, can also be obtained. This allows for a detailed comparison between alternative designs. For example, using this technique it was found that an extruder in which conveying elements were alternated with kneading elements provided 25% better mixing per unit length than a standard extruder that contained only conveying elements. The residence time distribution was narrower, however, with the standard design. Being able to obtain such detailed performance information without experimentation allows process engineers to design advanced and more efficient process equipment with confidence.

### 5-7.7 Intermeshing Impellers

The mesh superposition technique (MST) can also be used to model the flow in vessels equipped with multiple impellers whose swept volumes overlap. In this example, the mixing in such a system (a planetary mixer) operating at a very low Reynolds number ($1 \times 10^{-4}$) is considered. Figure 5-28 shows two anchor impellers mounted on separate shafts. The impellers are set at a 90° angle relative to each other. Although the impellers do not touch each other, there is a volume that is swept by both impellers. Such a system cannot be modeled using the sliding mesh models implemented in most commercial CFD programs. The main benefits of using the mesh superposition technique for such as system are that each part can be meshed separately and that these intermeshing parts can rotate freely without having to be remeshed.

To create the mixer geometry, a cylindrical mesh is generated for the tank. Two other, completely independent meshes are defined for the blades. The three meshes are then combined into one. As the blades rotate, the transient flow pattern in the tank can be calculated and illustrated by the dispersion of tracer particles, as shown in the figure. As the total number of rotations increases, the tracer becomes more uniformly distributed. After six rotations, the dispersion of the tracer particles in the horizontal plane is satisfactory. Note, however, that the particles have moved little in the vertical direction. This is because the anchor impellers in use impart little or no axial momentum to the fluid. Twisted blades, which also impose an axial motion on the flow, might perform better to distribute the tracer throughout the vessel. The mesh superposition technique is well suited to study such systems. For other examples of flow in planetary mixers, see Tanguy et al. (1999) and Zhou et al. (2000).

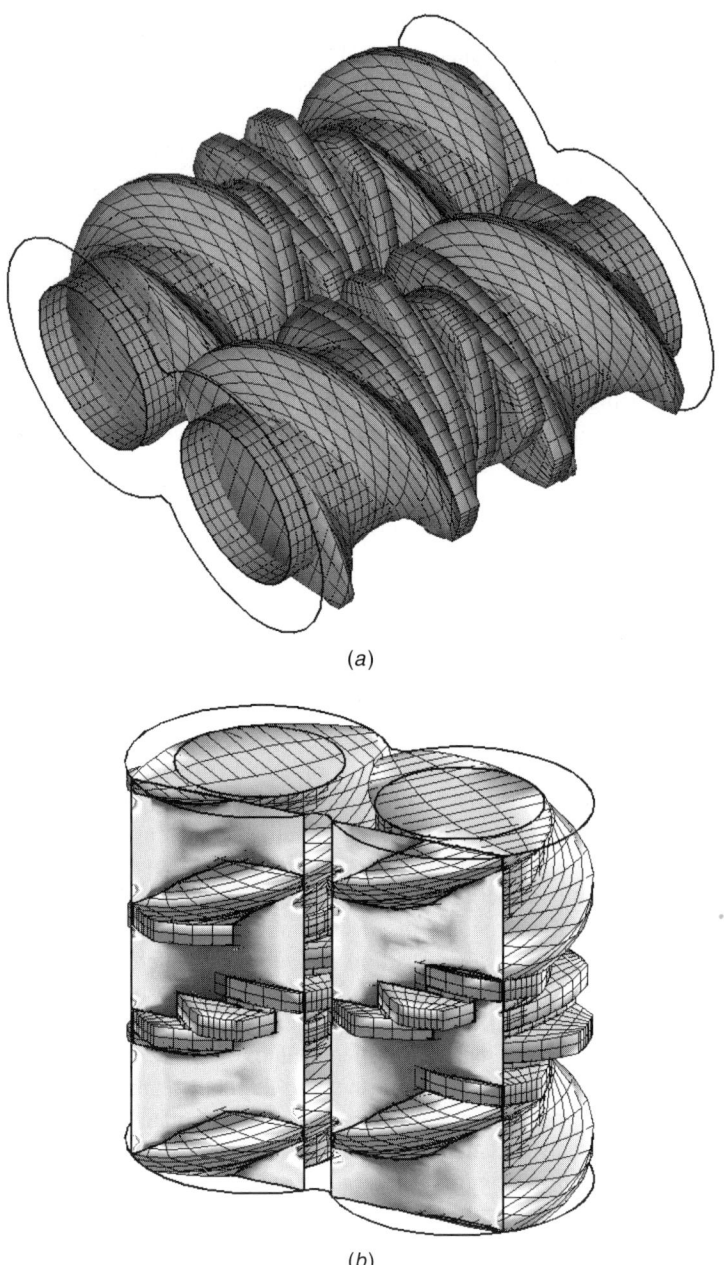

**Figure 5-27** (*a*) Surface grid for the screws in a twin-screw extruder. (*b*) Local shear rate on a planar slice through the twin-screw extruder, with white denoting regions of high shear rate and black denoting regions of low shear rate. Three meshes were used for this configuration, one for each screw and one for the flow domain.

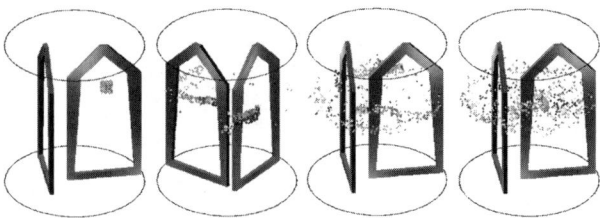

**Figure 5-28** Dispersion of a particle tracer in a vessel equipped with two intermeshing anchor impellers, calculated using the mesh superposition technique. After six full rotations, the particles are well dispersed on the horizontal plane where they were released.

## 5-7.8 Kenics Static Mixer

Static mixers are used widely in the process industries. Static mixers consist of motionless elements mounted in a pipe, which create flow patterns that cause fluids to mix as they are pumped through the pipeline. Most of the experimental work on static mixers has concentrated on establishing design guidelines and pressure drop correlations. The number of investigations into the flow and mixing mechanisms is limited, probably due to difficulties encountered in obtaining meaningful experimental measurements.

The Kenics in-line mixer consists of a number of elements of alternating right- and left-hand 180° helices. The elements are positioned such that the leading edge of each element is perpendicular to the trailing edge of the preceding element. The length of the elements is typically one and a half tube diameters. This type of static mixer is used for mixing under laminar flow conditions, such as the mixing of polymers or food products like peanut butter and chocolate. To evaluate the mixing mechanism of the Kenics mixer, Bakker and Marshall (1992) and Bakker and LaRoche (1993) calculated the transport of two chemical species through a six-element device. The center of the inlet was 100% of one species, designated by white in Figure 5-29. The outside of the inlet was 100% of the other species, shown as black. The results are presented as a series of contour plots, showing the concentration fields of the chemical species at various axial positions along the tubes. The concentration fields after 18°, 54°, 90°, 126°, and 162° of rotation in each of the six Kenics mixing elements are shown. In the first element, the white core coming from the inlet is split into two white islands. These islands are stretched and move outward. The black, which was initially on the outside, is split into two semicircular filaments, which move toward the inside. Similar stretching and folding processes occur in the next several elements. At the inlet of the third element the black species is now on the inside, meaning that the concentration field has basically flipped inside out. This process of splitting, stretching, folding, and flipping inside out repeats itself every two elements, until the fluids are mixed. The number of elements can be adjusted to the requirements of the process, but typically varies between six and 18, depending on the Reynolds number. See, for example, Hobbs and Muzzio (1997), Hobbs et al. (1998), and Zalc et al. (2002).

**326** COMPUTATIONAL FLUID MIXING

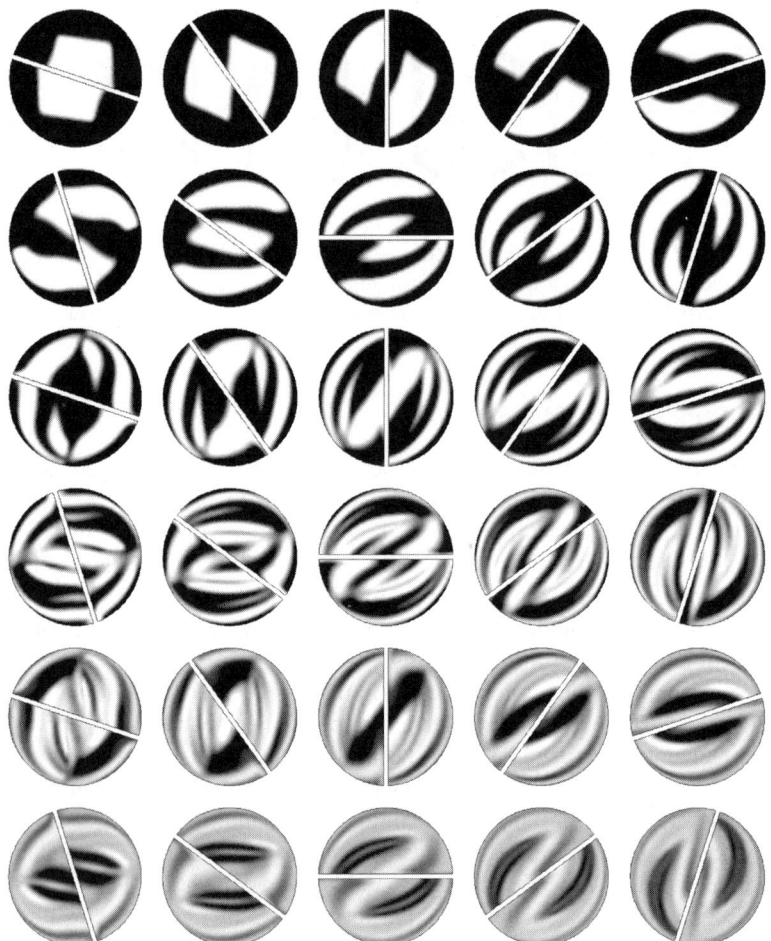

**Figure 5-29** Concentration profiles in a Kenics static mixer. Rows 1 to 6 show the concentration in elements 1 to 6, respectively. Columns 1 to 5 show the concentration profiles at 18°, 54°, 90°, 126°, and 162°, respectively.

### 5-7.9 HEV Static Mixer

The traditional helical mixing element is used primarily for in-line blending under laminar and transitional flow conditions. The high efficiency vortex (HEV) mixer is used for turbulent blending of gases or miscible liquids. It consists of a series of tab arrays, which are placed along a length of pipe. The advantages of this design are that it is easily adapted to both cylindrical and square pipe cross-sections and that it has a relatively low pressure drop. HEV mixers have been in use in the process industries for several years now, for both liquid–liquid and gas–gas mixing. Applications include wastewater treatment, burners, exhaust stacks, beverage manufacturing, and many others. The wide range of applications and scales in which the HEV mixer is used requires a technique to analyze custom applications on demand. Gretta (1990) investigated the flow pattern generated by the tabs using a combination of hot wire anemometry, hydrogen bubble visualization, and dye visualization and found that the tabs not only generate a pair of counterrotating longitudinal vortices but also shed *hairpin vortices*. The smaller hairpin vortices, generated in a transient manner, move downstream with the larger longitudinal vortices.

Bakker et al. (1994a) modeled the flow pattern generated by an HEV mixer using the Reynolds stress model for turbulence. This steady-state model correctly predicted the formation of the longitudinal vortices, but the hairpin vortices only showed up in the results as regions of high turbulence intensity at the edges of the tabs. Due to the steady-state nature of that model and the assumption of eightfold symmetry made for the purpose of the calculation, the mixing of fluids near the center of the pipe was underpredicted compared to what was known from operational experience and laboratory studies.

Because of the shortcomings of the RANS turbulence models in predicting the hairpin vortices, the HEV mixer was selected as a good candidate for the LES turbulence model. In the LES model, no symmetry assumptions were made, meaning that the full 360° pipe was modeled. The advantage of modeling the full pipe is that periodic interactions between the vortices that form behind the different tabs are not restrained. The simulation was started with a steady-state calculation based on the $k-\varepsilon$ turbulence model. After partial convergence, the LES model was enabled. As hoped, the transient results showed the periodic shedding of hairpin vortices off the back sides of the tabs. Figure 5-30 shows

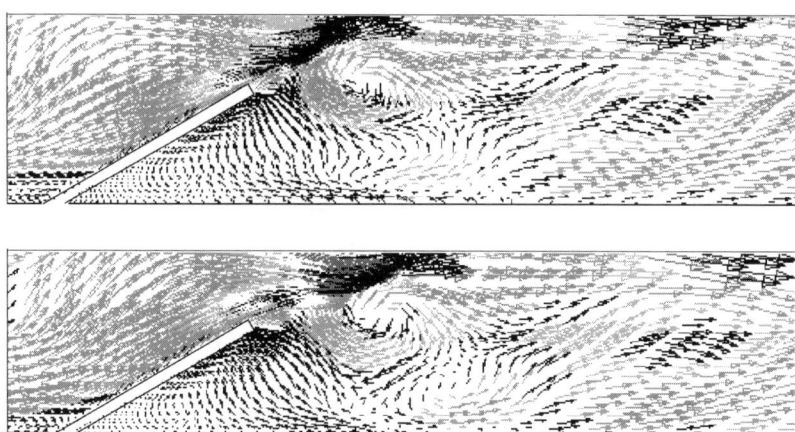

**Figure 5-30** The hairpin vortex (in cross-section) that forms behind the tab in an HEV mixer at two different instances in time is shown. Vortices such as these are shed in a time-dependent fashion. The LES model was used for this simulation. A similar HEV mixer, solved using the steady-state Reynolds stress turbulence model, failed to capture this flow detail.

these vortices at two different instances in time. It is clear that the hairpin vortex forming around the tab in the top image has shifted downstream during the 0.06 s that separates the two flow pattern snapshots. This shows that the LES model is well suited to capture complex time-dependent vortex systems such as these.

### 5-7.10 LDPE Autoclave Reactor

Low density polyethylene (LDPE) reactors are used to manufacture polymer products. The reactors are typically of the tubular or autoclave variety. To make the (multimolecule chain) polymer, a minute amount of initiator is added to a (single-molecule) monomer. Several reaction steps take place in which the monomer is transformed to intermediate polymers, or radicals, and finally to a polymer product with a range of chain lengths (corresponding to a range of molecular weights). Heat is released in many of the reactions, and one goal of LDPE reactor design is to prevent hot spots that give rise to a condition called *thermal runaway*, which is characterized by an undesired product distribution. In this example, the nearly infinite set of reactions in the chain is approximated by six finite rate reactions using the method of moments (Kiparissides et al., 1997). These reactions are solved using the finite rate reaction model with the help of user-defined functions. As a consequence of the method of moments, quantities that describe the product distribution can also be computed. These include the molecular weight distribution, which, if narrow, indicates a high-quality (uniform) product.

A hybrid mesh of 166 000 cells, shown in Figure 5-31, is used for the simulation. The reactor contains both paddle and twisted blade impellers, whose

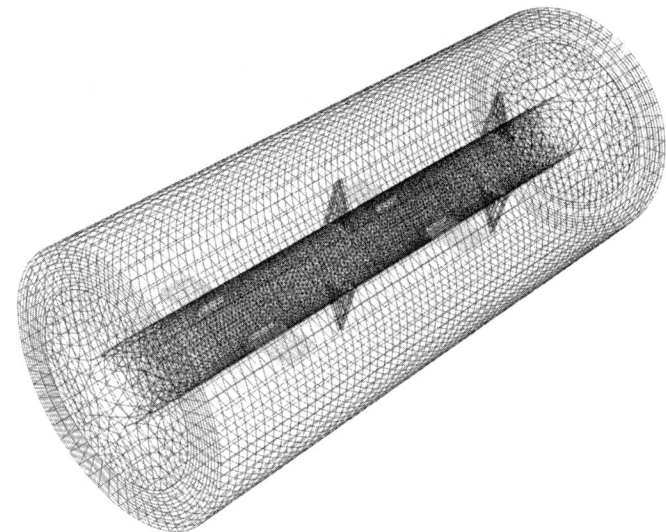

**Figure 5-31** Surface mesh used for the LDPE reactor.

rotation is modeled using a sliding mesh. The monomer and initiator used are ethylene and DTBP, respectively. The initiator is premixed with the monomer and injected into the reactor through an annular ring at one end of the vessel. The mixture leaves the device through an annular exit at the opposite end. The flow field is characterized by high swirl, which is induced by the rapidly rotating impellers in the unbaffled vessel. The RNG $k-\varepsilon$ model is used to account for turbulence in the highly swirling flow.

Four axial slices are used in the next two figures to show the progression of two problem variables as the mixture advances through the reactor. In these figures, the inlet annulus is at the top of the figure and the outflow annulus is at the bottom. In Figure 5-32 the conversion of the monomer (to both radicals and product polymers) is shown to increase gradually to about 7% as the flow passes through the vessel, in reasonably good agreement with published data (Read et al., 1997). (Higher values are shown in dark gray.) Contours of the molecular weight distribution (Figure 5-33) vary from 41 500 to 41 900, or by about 1%.

**Figure 5-32** Conversion of monomer increases to about 7% as the material moves through the reactor from top to bottom. The conversion is highest at the bottom of the reactor, where the contours are darkest.

**Figure 5-33** Contours of molecular weight distribution are used to assess the range of molecular weights in the product.

As the mixture moves through the reactor, the spread in the distribution narrows, indicating a product of high quality. The solution also indicates that the molecular viscosity increases as the chains of radicals grow, consistent with expectations.

### 5-7.11 Impeller Design Optimization

Ever since the 1950s the Rushton turbine has been the standard impeller for gas dispersion applications. It features six flat blades mounted on a disk. As shown in Figure 5-20*b*, the flow behind the impeller blades separates and trailing vortices form. On gassing, gas accumulates in the low-pressure regions behind the blades and cavities form. This leads to a significant drop in power draw and loss of gas dispersion ability. During the late 1980s and early 1990s, modified Rushton turbines with semicircular blades became standard. These models reduce flow separation and cavity formation behind the blades but do not eliminate them completely.

To date, the disk-style gas dispersion impellers studied in the literature have blades that are symmetric with respect to the plane of the disk. This is not necessarily optimal, since the gas usually enters from the bottom, causing a distinctly asymmetric flow pattern. In this example, the operation of the Chemineer BT6 gas dispersion impeller is reviewed (Bakker, 1998; Myers et al., 1999). The BT6 impeller, with vertically asymmetric blades, is designed to accommodate the various flow conditions above and below the impeller disk. The turbulent flow pattern created by the BT6 was modeled using a fully unstructured tetrahedral mesh with approximately 500 000 cells. The MRF approach and RNG $k-\varepsilon$ turbulence model were used. Second-order upwind differencing was used for the momentum and turbulence equations. The flow pattern was converged using the SIMPLEC pressure–velocity coupling method, which allows for the use of high underrelaxation factors, resulting in fast convergence.

The triangular mesh on the impeller blade is shown in Figure 5-34. The blades have a concave shape, which consists of three curves of different radii and length. The top part of the blade is longer than the bottom part. The back side of the blade is rounded. After the flow field was converged, the torque on the impeller was calculated by integrating the pressure on the impeller blade surfaces. From the torque, the impeller power number, based on the nominal diameter at the impeller disk level, was calculated to be 2.3, which is in excellent agreement with experiments.

Figure 5-35 shows the velocity field around the impeller blades. The velocity vectors are drawn in the frame of reference of the impeller. It is clear that no flow separation occurs behind the impeller blades. This means that cavity formation under gassed conditions will be reduced. Indeed, visualization studies have shown that gas is captured under the top overhang and dispersed from a deep vortex on the inside of the blade. No large gas-filled cavities have been observed behind the blade. As a result, the BT6 has a gassed power curve that is flatter than that of

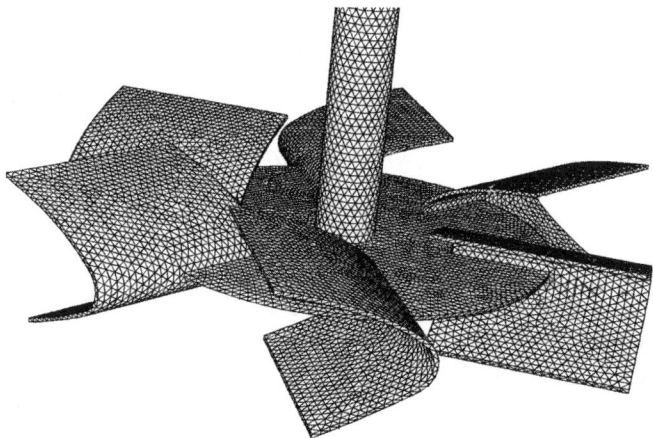

**Figure 5-34** Triangular surface mesh on a Chemineer BT6 impeller.

**Figure 5-35** Velocity field around the blades of a BT6 impeller. No flow separation occurs behind the blades.

other impellers. It can disperse more gas before flooding than the impellers with symmetric semicircular blades and is less affected by changes in liquid viscosity.

### 5-7.12 Helical Ribbon Impeller

High viscosity mixing applications occur in most chemical process industry plants. For instance, the polymer industries must blend high viscosity reaction masses to thermal and chemical uniformity. This industry must also blend small amounts of low viscosity antioxidants and colorants into polymer streams. The personal-care products industry encounters many high viscosity mixing applications in the preparation of creams, lotions, pastes, and drugs. Other high viscosity applications occur in the production of food, paint, drilling mud, and greases, to name a few. Viscosities can be in a range from about 1 Pa · s all the way up to 25 000 Pa · s in some extreme cases. The quality of the final mixed product in these applications can be very important economically.

Low viscosity mixing applications can usually be handled efficiently with impeller systems consisting of one or more turbines. To obtain adequate mixing under the laminar flow conditions encountered in high viscosity applications, on the other hand, close-clearance impellers such as anchors and helical ribbons are required. These impellers sweep the whole wall surface of the vessel and agitate most of the fluid batch through physical contact. Helical ribbon impellers are typically used for industrial applications where the viscosity is in the range 20 000 to 25 000 Pa · s. Wall scrapers can be mounted on the impeller blades to improve heat transfer.

APPLICATIONS  333

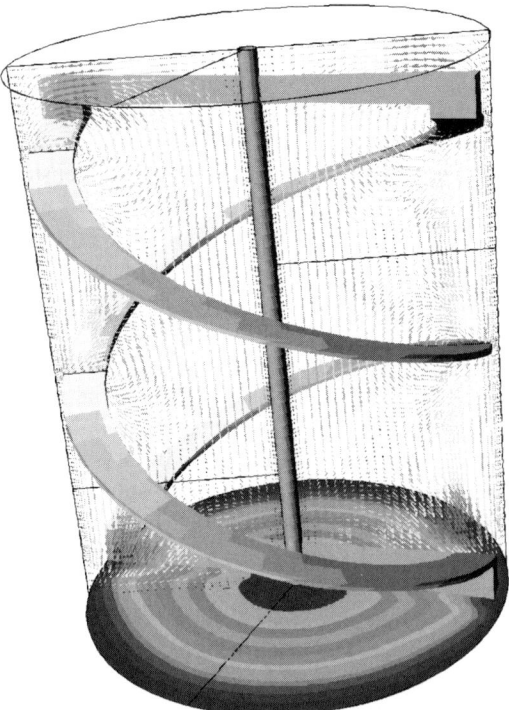

**Figure 5-36** Flow field in a vessel equipped with a helical ribbon impeller. Velocity vectors in a vertical plane are shown. The bottom of the vessel is colored by velocity magnitude.

Figure 5-36 shows the flow pattern in the vertical plane of a vessel equipped with a helical ribbon. A fully structured hexahedral mesh with approximately 100 000 cells was used. The structured 3D mesh was created by extruding and twisting a 2D planar mesh. The fluid is viscous and the impeller Reynolds number is approximately 10. The velocity vectors show that the impeller pumps down at the wall and up in the center. Contours of velocity magnitude on the tank bottom show that there are low velocities in the center and higher velocities near the outside wall. Small circulation loops form between the impeller blades and the vessel wall, as discussed in the general literature. These indicate the need for an even larger D/T or the use of wall scrapers if optimum heat transfer is to be obtained.

### 5-7.13 Stirred Tank Modeling Using LES

In turbulent flows, large scale eddies with coherent structures are primarily responsible for the mixing of passive scalars. The large scale eddies embody themselves in the form of identifiable and organized distributions of vorticity. In addition, the mixing process involves all mechanisms typically found in vortex dynamics, such as stretching, breakup, concatenation, and self-induction of vortices. Recent experimental work (Bakker and Van den Akker, 1994) suggests that large scale time-dependent structures with periods much longer than the time of an impeller revolution are involved in many of the fundamental hydrodynamic processes in stirred vessels. For example, local velocity data histograms may be bimodal or trimodal, even though they are being analyzed as having only one mode in most laser Doppler experiments. In solids suspension processes, solids can be swept from one side of the vessel to the other in an oscillating pattern, even in dilute suspensions. Digital particle image velocimetry experiments have shown that large scale asymmetries with periods of up to several minutes exist in stirred vessels equipped with axial flow impellers.

The advantage of large eddy simulation (LES) over other turbulence models is that it explicitly resolves the large eddies, which are responsible for much of the mass, energy, and momentum transport. Only the small eddies are represented by a time-averaged subgrid scale model. In mixing tank simulations, the LES turbulence model is typically combined with a sliding mesh model for the impeller so that the most rigorous time-accurate solution can be obtained. One parameter that is pivotal to the success of an LES simulation is the density of the grid throughout the domain. To determine an optimum grid size, the following, straightforward method is recommended. First a steady-state, three dimensional calculation is performed that uses the standard k–ε turbulence model and the MRF model for the impeller. From the converged flow field, volume averages for the following three turbulent length scales are calculated:

- Integral length scale: $L_t = k^{3/2}/\varepsilon$
- Taylor length scale: $L_a = (15 \nu u'^2/\varepsilon)^{0.5}$
- Kolmogorov scale: $L_k = (\nu^3/\varepsilon)^{1/4}$

The integral length scale is a measure of the large scale turbulence. The Kolmogorov length scale is a measure of the smallest scale eddies at which dissipation occurs. The Taylor length scale is an intermediate length scale that can be used as a guide to determine the grid size required for LES simulations. For a typical turbulent small scale vessel, $L_t/T \sim 10^{-1}$, $L_a/T \sim 10^{-2}$, and $L_k/T \sim 10^{-3}$. Based on the Taylor length scale, a suitable grid size for an LES simulation would be on the order of $10^{-2}T$, which would result in a grid on the order of $10^6$ cells. The large number of cells, along with the transient solution method (one that requires a small time step), contribute to the increased calculation time required by the LES model as compared with RANS models. Figure 5-37 shows how the CPU time and required grid size for the LES model compare with other turbulence modeling options.

In this example, the use of LES and the sliding mesh model to predict large scale chaotic structures in stirred tanks is demonstrated for a single high efficiency impeller. A full hexahedral mesh was used for the simulation. The vessel diameter is 0.29 m, and the impeller rotates at 60 rpm, resulting in a Reynolds number of 13 000. The central differencing scheme for the momentum equations was used along with a time step of 0.01 s. The RNG modification of the Smagorinsky

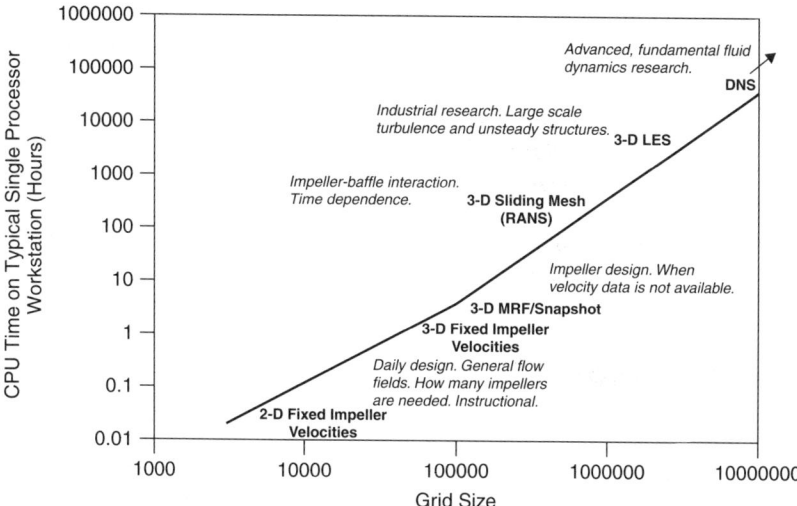

**Figure 5-37** CPU time and grid size requirements for various impeller modeling options.

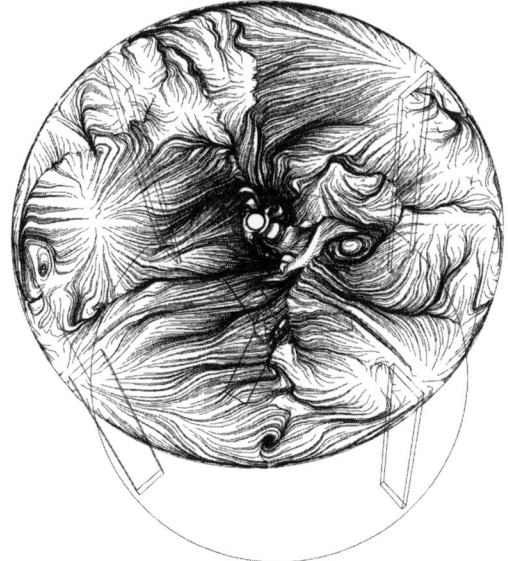

**Figure 5-38** Flow pattern at the surface of a vessel equipped with a high efficiency impeller calculated using an LES turbulence model.

model was used for the subgrid scale turbulence. A period of approximately 40 s was simulated. The results show that the flow pattern indeed exhibits large scale unsteady motion, similar to what has been reported from experimental data in the literature. Figure 5-38 shows the flow field at the liquid surface at one instant in time, using oil flow lines, which are path lines that are confined to the surface from which the flow followers are released. The turbulent structure of the flow is clearly visible.

When performing such LES calculations, it is advised to visualize the results by creating flow field images after every time step. These can then be used to create animations. Similarly, statistical data can be obtained by creating monitor points or lines in the domain and saving important variables in these locations. The time series that are obtained in this manner can be analyzed further using standard statistical and signal analysis techniques.

## 5-8 CLOSING REMARKS

It could be said that what comes out of a CFD simulation is only as good as what goes in. Although this is true in part, there are many other considerations that can lead to the success—or lack thereof—of CFD. One is based on the choice of software. Many commercial packages are available today, and resources to help find and evaluate them are given in Section 5-8.1. Comments on basic hardware requirements for CFD codes, which are computationally intensive, are given in Section 5-8.2. Issues regarding the learning curve, or the time required for an engineer to "come up to speed" and be successful with CFD, are discussed in Section 5-8.3. Once the proper software, proper hardware, and trained user are in place, there are still some common pitfalls to be avoided. These, along with some of the benefits of CFD, are discussed in Section 5-8.4.

### 5-8.1 Additional Resources

Many commercial and even some freeware or shareware CFD codes are available, each with different capabilities, special physical models, numerical methods, geometric flexibility, and user interfaces. Specialized pre- and postprocessing programs are also available for generation of the geometry and grid, input of model parameters, and viewing of results. Excellent overviews of these products can be found on the Web (see, e.g., CEWES, Christopher, Larsson, and Wyman).

### 5-8.2 Hardware Needs

In the past, CFD use was often associated with the realm of high-powered computer systems. But much of today's modeling work can be accomplished on low-end Unix workstations or high-end personal computers (PCs). A typical PC configuration might be a one- or two-processor system, running Windows or Linux. Unix workstations with one, two, or more processors are also commonly used. These systems are more than adequate for moderately sized, steady-state or time-dependent analyses. For complicated models, or those using a large number of computational cells ($>1$ million), multiprocessor workstations are often used. Although supercomputers are still employed for high-end research and development work, they are not commonly needed for typical engineering design applications. Another recent trend involves the clustering of multiple inexpensive PCs into a parallel- or cluster-computing network. Such systems provide supercomputing power at a fraction of the cost.

### 5-8.3 Learning Curve

The user friendliness of CFD software has also increased significantly during recent years. In the past, CFD software was characterized by text- or command-file-based interfaces and difficult- to-configure solvers that made fluid flow analysis the exclusive domain of highly trained experts. However, the latest generation of commercial CFD software has been developed with graphical user interfaces. They have much more stable and robust solvers and allow easy geometry exchange between CAD programs and the CFD solver. This has allowed engineers who are not experts in fluid dynamics to make efficient use of CFD and use this technology on a day-to-day basis in their design and optimization work. Most commercial CFD companies provide training and ongoing technical support with a software license. The average engineer typically requires one week of training to get started using one of these modern CFD packages.

### 5-8.4 Common Pitfalls and Benefits

Despite the increased user friendliness of modern CFD software, there are still a number of potential pitfalls that can beset the analyst. Some of the mistakes made most commonly when using CFD are listed below.

- *Use of a low-quality, coarse grid.* Details that are smaller than the cell size cannot be resolved. Often, small flow features in one region need to be resolved in great detail in order to predict large flow features accurately in other regions. For example, a jet penetrating to a vessel will appear to diffuse more rapidly than in actual fact if a coarse grid is used in the jet region. Satisfying grid needs such as this may lead to a finer grid containing far more cells than estimated initially.
- *Use of unconverged results.* CFD solvers are iterative and it is often tempting to cut a calculation short when deadlines are approaching or the coffee break is over. However, the analyst should always ensure that proper convergence has been obtained before using the results from any CFD solver.
- *Use of the wrong physical property data.* This is not as trivial as it sounds. For example, viscosity curves may have been determined in one temperature and shear rate range, but if the actual shear rates or temperatures in the flow domain are outside this range, the curves may no longer be valid and incorrect results may be obtained. As another example, accurate average particle size and density are needed to best predict solids suspension behavior.

Fortunately, none of these problems is fundamental to the CFD technology itself. A coarse grid may be refined, unconverged calculations continued, and accurate physical constants may be measured. These easily avoided pitfalls are far outweighed by the following benefits:

- CFD can be used to augment design correlations and experimental data.
- CFD provides comprehensive data that are not easily obtainable from experimental tests.
- CFD reduces scale-up problems, because the models are based on fundamental physics and are scale independent. Models of the actual unit can be simulated just as easily as models of lab scale versions, so predictions, and indeed optimization of the actual unit, can be achieved.
- When evaluating plant problems, CFD can often be used to help understand the root cause of a problem, not just the effect.
- CFD can be used to complement physical modeling. Some design engineers actually use CFD to analyze new systems before deciding which and how many validation tests need to be performed.
- Many "what if" scenarios can often be analyzed in less time than experimental tests would take.

In summary, if the CFD analyst is careful when addressing the issues of problem setup and solution convergence, the potential benefits that can be extracted from the simulation are numerous. Furthermore, the computational resources available today, in terms of both speed and power, should encourage engineers to make use of high density grids and complex models so as to achieve results of the best possible quality.

## ACKNOWLEDGMENTS

The authors gratefully acknowledge the contributions of the following people: Lanre M. Oshinowo for numerous discussions and his assistance with the blending and stirred vessel solids suspension simulations; Richard D. LaRoche for his conceptual contributions and his cooperation on the static mixer simulations; Ahmad H. Haidari for sharing his many ideas; Thierry Avalosse and Yves Rubin for the twin-screw extruder simulations; and Bernard Alsteens for the intermeshing impeller simulations. Furthermore, Liz Marshall wishes to thank Ronald J. Weetman for many helpful discussions over the years on mixing processes and analysis; and André Bakker wishes to thank Kevin J. Myers, Julian B. Fasano, Mark F. Reeder, Lewis E. Gates, John M. Smith, Robert F. Mudde, Jaap J. Frijlink, Marijn M. C. G. Warmoeskerken, Ivo Bouwmans, and Harrie E. A. van den Akker for many fruitful discussions and contributions.

# NOMENCLATURE

| | |
|---|---|
| A | magnussen mixing rate constant (−) |
| $A_k$ | Arrhenius constant for reaction k (variable units) |
| B | Magnussen mixing rate constant (−) |
| $c_A$ | fluctuation in the concentration of species A (mol/m$^3$) |
| C | off-bottom clearance (m) |
| $C_1$ | turbulence model constant (−) |
| $C_2$ | turbulence model constant (−) |
| $C_A$ | concentration of species A (mol/m$^3$) |
| $C_{j'}$ | concentration of species j' (mol/m$^3$) |
| D | impeller diameter (m) |
| E | total enthalpy (J) |
| $E_k$ | activation energy for reaction k (J/mol) |
| f | underrelaxation factor (−) |
| F(φ) | spatially discretized transport equation |
| $F_i$ | net force in the i direction (N) |
| g | gravitational acceleration (m/s$^2$) |
| $G_k$ | generation term for turbulence (kg/m · s$^3$) |
| h | static enthalpy (J) |
| $h_{j'}$ | enthalpy for the species j' (J) |
| $J_{i',i}$ | diffusion flux of species i' in direction I (kg/m$^2$ · s$^1$) |
| k | turbulent kinetic energy (m$^2$/s$^{-2}$) |
| $k_{eff}$ | effective conductivity (W/m · K) |
| $K_{i',k}$ | reaction rate of species i' in reaction k (variable units) |
| L | length of domain in definite integral over coordinate x (m) |
| $L_a$ | Taylor length scale (m) |
| $L_k$ | Kolmogorov scale (m) |
| $L_t$ | integral length scale (m) |
| $m_{i'}$ | mass fraction of species i' (−) |
| $M_{i'}$ | molecular weight of species i' (kg/kg-mol) |
| N | impeller rotational speed (s$^{-1}$) |
| $N_P$ | power number (−) |
| $N_Q$ | flow number (−) |
| p | pressure (Pa) |
| P | power drawn by an impeller (W) |
| Pe | Péclet number |
| $Q_l$ | liquid flow rate (m$^3$/s) |
| r | spatial coordinate in the radial direction (m) |
| R | universal gas constant (J/mol · K) |
| R | impeller radius (m) |
| Re | Reynolds number (−) |

| | |
|---|---|
| $R_{i'}$ | generalized source term for reactions in the species i′ transport equation (kg/m$^3 \cdot$ s) |
| $R_{K\_i',k}$ | kinetic reaction rate for species i′ in reaction k (kg/m$^3 \cdot$ s$^{-1}$) |
| $R_{M1-i',k}$ | mixing limited reaction rate for the reactant species i′ in reaction k (kg/m$^3 \cdot$ s) |
| $R_{M2-i',k}$ | mixing limited reaction rate for the product species i′ in reaction k (kg/m$^3 \cdot$ s) |
| $S_h$ | generalized source term for the enthalpy equation (W/m$^3$) |
| $S_{i'}$ | net species source term in the species i′ transport equation (kg/m$^3 \cdot$ s) |
| t | time (s) |
| T | tank diameter (m) |
| T | temperature (K) |
| $T_{ref}$ | reference temperature for formation enthalpy (K) |
| U | velocity vector (m/s) |
| $u'_i$ | fluctuating velocity component (due to turbulence) in the direction i (m/s) |
| $U_i$ | velocity in the direction i (m/s) |
| $U_{tip}$ | impeller tip speed (m/s) |
| $W_b$ | width of impeller blade (m) |
| $x_i$ | spatial coordinate in direction i (m) |
| $X_s$ | product distribution (−) |
| z | impeller blade height (m) |

*Greek Symbols*

| | |
|---|---|
| $\beta_k$ | temperature exponent in Arrhenius rate expression (−) |
| $\Gamma$ | generalized diffusion coefficient (variable units) |
| $\delta_{ij}$ | Kronecker delta (−) |
| $\varepsilon$ | turbulent kinetic energy dissipation rate (m$^2$/s$^3$) |
| $\eta_{j',k}$ | exponent for concentration of species j′ in reaction k (−) |
| $\mu$ | molecular viscosity (kg/m $\cdot$ s) |
| $\mu_{eff}$ | effective viscosity (kg/m $\cdot$ s) |
| $\mu_t$ | turbulent viscosity (kg/m $\cdot$ s) |
| $\nu$ | kinematic viscosity (m$^2$/s) |
| $\nu_{i'}$ | stoichiometry of species i′ (−) |
| $\xi$ | vorticity (s$^{-1}$) |
| $\rho$ | liquid density (kg/m$^3$) |
| $\sigma_k$ | turbulence model constant (−) |
| $\sigma_\varepsilon$ | turbulence model constant (−) |
| $\sigma_\mu$ | turbulence model constant (−) |
| $\tau$ | shear stress (Pa) |
| $\tau_d$ | disruptive shear stress (Pa) |
| $\tau_y$ | yield stress (Pa) |
| $\phi$ | generalized conserved quantity (variable units) |
| $\Omega$ | angular speed (rad/s) |

# REFERENCES

Avalosse, T., and Y. Rubin (1999). Analysis of mixing in co-rotating twin screw extruders through numerical simulation, *Proc. 15th Polymer Society Conference*, Hertogenbosch, The Netherlands.

Bakker, A. (1992). Hydrodynamics of stirred gas–liquid dispersions, Ph.D. dissertation, Delft University of Technology, The Netherlands.

Bakker A. (1998). Impeller assembly with asymmetric concave blades, U.S. patent 5,791,780.

Bakker A., and J. B. Fasano (1993a). A computational study of the flow pattern in an industrial paper pulp chest with a side entering impeller, presented at the annual AIChE meeting, Nov. 1992; *AIChE Symp. Ser. 293*, **89**, 118–124.

Bakker A., and J. B. Fasano (1993b). Time dependent, turbulent mixing and chemical reaction in stirred tanks, presented at the annual AIChE meeting, St. Louis, MO, Nov. *AIChE Symp. Ser. 299*, **90**, 71–78.

Bakker A. and R. LaRoche (1993). *Flow and mixing with Kenics static mixers*, Cray Channels, Volume **15**(3), p. 25–28.

Bakker, A., and E. M. Marshall (1992). Laminar mixing with Kenics in-line mixers, *Fluent User's Group Meeting Proc.*, Burlington, VT, Oct. 13–15, pp. 126–146.

Bakker, A., and H. E. A. Van den Akker (1994). Single-phase flow in stirred reactors, *Chem. Eng. Res. and Des., Trans. Inst. Chem. Eng.*, **72**, 583–593.

Bakker A., N. Cathie, and R. LaRoche (1994a). Modeling of the flow and mixing in HEV static mixers, presented at the 8th European Conference on Mixing, Cambridge, Sept. 21–23; *Inst. Chem. Eng. Symp. Ser.*, 136, 533–540.

Bakker A., J. B. Fasano, and K. J. Myers (1994b). Effects of flow pattern on the solids distribution in a stirred tank, *Proc. 8th European Conference on Mixing*, Cambridge, Sept. 21–23; *Inst. Chem. Eng. Symp. Ser.*, **136**, 1–8.

Bakker, A., R. D. LaRoche, M. H. Wang, and R. V. Calabrese (1997). Sliding mesh simulation of laminar flow in stirred reactors, *Trans. Inst. Chem. Eng.*, **75A**, Jan.

Bakker, A., L. Oshinowo, and E. Marshall (2000). The use of large eddy simulation to study stirred vessel hydrodynamics, *Proc. 10th European Conference on Mixing*, Delft, The Netherlands, pp. 247–254.

Bakker A., A. Haidari, and E. M. Marshall (2001). Modeling stirred vessels using large eddy simulation, presented at the 18th Biennial North American Mixing Conference, Pocono Manor, PA.

Bourne, J. R., F. Kozicki, and P. Rys (1981). Mixing and fast chemical reaction: I. Test reactions to determine segregation, *Chem. Eng. Sci.*, **36**, 1643.

CEWES MSRC (n.d.). Computational fluid dynamics software data log, http://phase.go.jp/nhse/rib/repositories/cewes_cfd/catalog/index.html.

Christopher, W. (n.d.). CFD codes list, http://www.icemcfd.com/cfd/CFD_codes.html.

Ding, J., and D. Gidaspow (1990). A bubbling fluidization model using kinetic theory of granular flow, *AIChE J.*, **36**, 523–538.

Fluent (1998). Fluent 5 User's Guide, Fluent, Inc., Lebanon, NH.

Fokema, M. D., S. M. Kresta, and P. E. Wood (1994). Importance of using the correct impeller boundary conditions for CFD simulations of stirred tanks, *Can. J. Chem. Eng.*, **72**, 177–183.

Fox, R. O. (1998). On the relationship between Lagrangian micromixing models and computational fluid dynamics, *Chem. Eng. Process.* **37**, 521–535.

Gidaspow, D., M. Syamlal, and Y. C. Seo (1986). Hydrodynamics of fluidization: supercomputer generated vs. experimental bubbles, *J. Powder Bulk Solids Technol.*, **10**, 19–23.

Gran, I. R., and B. F. Magnussen (1996). A numerical study of a bluff-body stabilized diffusion flame: 2. Influence of combustion modeling and finite-rate chemistry, *Combust. Sci. Technol.*, **119**, 119–191.

Gretta W. J., and C. R. Smith (1993). The flow structure and statistics of a passive mixing tab, *J. Fluids Eng.* **115**, 255–263.

Gullichsen J. (1985). Medium consistency processing: Fundamentals, *Bleach Plant Operations/TAPPI Seminar Notes*, pp. 135–142.

Hannon, J. (1992). Mixing and chemical reaction in tubular reactors and stirred tanks, Ph.D. dissertation, Cranfield Institute of Technology, Cranfield, Bedfordshire, England.

Hobbs, D. M., and F. J. Muzzio (1997). The Kenics static mixer: a three-dimensional chaotic flow, *Chem. Eng. J.*, **67**(3), 153–166.

Hobbs, D. M., P. D. Swanson, and F. J. Muzzio (1998). Numerical characterization of low Reynolds number flow in the Kenics static mixer, *Chem. Eng. Sci.* **53**(8), 1565.

Kiparissides, C., D. S. Achilias, and E. Sidiropoulou (1997). Dynamical simulation of industrial poly(vinyl chloride) batch suspension polymerization reactors, *Ind. Eng. Chem. Res.*, **36**, 1253.

Kresta, S. M., and P. E. Wood (1991). Prediction of the three dimensional turbulent flow in stirred tanks, *AIChE J.*, **37**, 448–460.

Larsson, J. (n.d.). CFD online, http://www.cfd-online.com.

Leonard, B. P., and S. Mokhtari (1990). ULTRA-SHARP nonoscillatory convection schemes for high-speed steady multidimensional flow, NASA TM 1-2568 (ICOMP-90-12), NASA Lewis Research Center.

Luo, J. Y., R. I. Issa, and A. D. Gosman (1994). *Prediction of impeller induced flows in mixing vessels using multiple frames of reference*, Inst. Chem. Eng. Symp. Ser. **136**, 549–556.

Magnussen, B. F., and B. H. Hjertager (1976). On mathematical models of turbulent combustion with special emphasis on soot formation and combustion, *Proc. 16th International Symposium on Combustion*, Combustion Institute, Pittsburgh, PA.

Marshall, E. M., Y. Tayalia, L. Oshinowo, and R. Weetman (1999). *Comparison of turbulence models in CFD predictions of flow number and power draw in stirred tanks*, presented at Mixing XVII, Banff, Alberta, Canada.

Middleton, J. C., F. Pierce, and P. M. Lynch (1986). Computations of flow fields and complex reaction yield in turbulent stirred reactors and comparison with experimental data, *Chem. Eng. Res. Des.*, **64**, 18–21.

Myers, K. J., A. J. Thomas, A. Bakker, and M. F. Reeder (1999). Performance of a gas dispersion impeller with vertically asymmetric blades, *Trans. Inst. Chem. Eng.*, **77**, 728–730.

Ogawa, S., A. Umemura, and N. Oshima (1980). On the equation of fully fluidized granular materials, *J. Appl. Math. Phys.*, **31**, 483.

Oldshue, J. Y., and N. R. Herbst (1992). *A Guide to Fluid Mixing*, Lightnin, Rochester, NY.

Oshinowo L., A. Bakker, and E. M. Marshall (1999). *Mixing time: a CFD approach*, presented at Mixing XVII, Banff, Alberta, Canada.

Oshinowo L. M., E. M. Marshall, A. Bakker, and A. Haidari (2000). Benefits of CFD in modeling solids suspension in stirred vessels, presented at the AIChE Annual Meeting, Los Angeles.

Patankar, S. V. (1980). *Numerical Heat Transfer and Fluid Flow*, Hemisphere, Washington, DC.

Ranade, V. V., and S. M. S. Dommeti (1996). Computational snapshot of flow generated by axial impellers in baffled stirred vessels, *Trans. Inst. Chem. Eng.*, **74**.

Read, N. K., S. X. Zhang, and W. H. Ray (1997). Simulations of a LDPE reactor using computational fluid dynamics, *AIChE J.*, **43**, 104–117.

Roussinova, V. T., B. Grgic, and S. M. Kresta (2000). Study of macro-instabilities in stirred tanks using a velocity decomposition technique, *Chem. Eng. Res. Des.*, **78**, 1040–1052.

Roussinova, V., S. M. Kresta, and R. J. Weetman 2001. Low frequency macroinstabilities in a stirred tank: scale-up and prediction based on large eddy simulations, presented at the 18th Biennial North American Mixing Conference, Pocono Manor, PA, June.

Shih, T.-H., W. W. Liou, A. Shabbir, and J. Zhu (1995). A new k–ε eddy-viscosity model for high Reynolds number turbulent flows: model development and validation, *Comput. Fluids*, **24**, 227–238.

Syamlal, M., W. Rogers, and T. J. O'Brien (1993). *MIFX Documentation*, Vol. 1, *Theory Guide*, DOE/METC-9411004, NTIS/DE9400087, National Technical Information Service, Springfield, VA.

Tanguy, P. A., F. Thibault, C. Dubois, and A. Ait-Kadi (1999). Mixing hydrodynamics in a double planetary mixer, *Chem. Eng. Res. Des.*, **77**(4), 318–324.

Tipnis, S. K., W. R. Penney, and J. B. Fasano (1993). An experimental investigation to determine a scale-up method for fast competitive parallel reactions in agitated vessels, presented at the AIChE Annual Meeting, St. Louis, MO.

Versteeg, H. K., and W. Malalasekera (1995). *An Introduction to Computational Fluid Dynamics: The Finite Volume Method*, Longman Scientific & Technical, Harlow, Essex, England.

Weetman, R. J. (1997). Automated sliding Mesh CFD computations for fluidfoil impellers, *Proc. 9th European Conference on Mixing*, Paris.

Wyman, N. (n.d.). CFD review, http://www.cfdreview.com.

Yakhot, V., and S. A. Orszag (1986). Renormalization group analysis of turbulence: I. Basic theory, *J. Sci. Comput.*, **1**, 1–51.

Zalc, J. M., E. S. Szalai, F. J. Muzzio, and S. Jaffer (2002). Characterization of flow and mixing in an SMX static mixer, *AIChE J.* **48**(3), 427–436.

Zhou, G., P. A. Tanguy, and C. Dubois (2000) Power consumption in a double planetary mixer with non-newtonian and viscoelastic materials, *Chem. Eng. Res.* **78**(3), 445–453.

# CHAPTER 6

# Mechanically Stirred Vessels

RAMESH R. HEMRAJANI
*ExxonMobil Research and Engineering Company*

GARY B. TATTERSON
*North Carolina A&T State University*

## 6-1 INTRODUCTION

There are a number of ways to perform mixing in vessels. Mechanical agitation, gas sparging, and jets are often used. Due to the variety of processing needs and process objectives, a number of different mixer geometries have been developed. This chapter is intended to introduce some of the more prominent geometries used for mechanical agitation in vessels. Blending in-line in pipes and in stirred vessels are topics of Chapters 7 and 9, respectively.

Mixing and contacting in agitated tanks can be accomplished in continuous, batch, or fed-batch mode. A good mixing result is important for minimizing investment and operating costs, providing high yields when mass transfer is limiting, and thus enhancing profitability.

Processing with mechanical mixers occurs under either laminar or turbulent flow conditions, depending on the impeller Reynolds number, defined as Re = $\rho ND^2/\mu$. For Reynolds numbers below about 10, the process is laminar, also called *creeping flow*. Fully turbulent conditions are achieved at Reynolds numbers higher than about $10^4$, and the flow is considered transitional between these two regimes.

Fluid mixing is carried out in mechanically stirred vessels for a variety of objectives, including for homogenizing single or multiple phases in terms of concentration of components, physical properties, and temperature. The fundamental mechanism involves physical movement of material between various parts of the entire mass using rotating impeller blades. Over 50% of the world's

---

*Handbook of Industrial Mixing: Science and Practice*, Edited by Edward L. Paul,
Victor A. Atiemo-Obeng, and Suzanne M. Kresta
ISBN 0-471-26919-0 Copyright © 2004 John Wiley & Sons, Inc.

**346** MECHANICALLY STIRRED VESSELS

chemical production involve these stirred vessels for manufacturing high-added-value products. These vessels are commonly used for:

- Blending of homogeneous liquids such as lube oils, gasoline additives, dilution, and a variety of chemicals
- Suspending solids in crystallizers, polymerization reactors, solvent extraction, etc.
- Blending and emulsification of liquids for hydrolysis/neutralization reactions, extraction, suspension polymerization, cosmetics, food products, etc.
- Dispersing gas in liquid for absorption, stripping, oxidation, hydrogenation, ozonation, chlorination, fermentation, etc.
- Homogenizing viscous complex liquids for polymer blending, paints, solution polymerization, food products, etc.
- Transferring heat through a jacket and/or internal coils for heating or cooling

An optimum approach to designing these mixing systems consists of the following steps:

- Define the process mixing requirements, such as blending quality, drop sizes, degree of solids suspension, mass transfer rates, etc.
- A suitable impeller type must then be chosen based on the type of fluid system and mixing requirements.
- The overall mixing system can then be designed, which involves determining the appropriate number of impellers, sizing the impeller, determining mixer speed, and estimating energy requirements.
- Other components, such as baffles, must also be specified based on desired flow patterns.
- One must design the mechanical components, such as shaft diameter, impeller blade thickness, baffles and supports, bearings, seals, etc. (see Chapter 21).

## 6-2  KEY DESIGN PARAMETERS

To design an effective stirred tank, an efficient impeller should be chosen for the process duty. More than one impeller may be needed for tanks with high aspect ratio ($Z/T > 1.5$). Sizing of the impeller is done in conjunction with mixer speed to achieve the desired process result. The appropriate size and type of wall baffles must be selected to create an effective flow pattern. The mixer power is then estimated from available data on impeller characteristics, and the drive size is determined. The mixer design is finalized with mechanical design of the shaft, impeller blade thickness, baffle thickness and supports, inlet/outlet nozzles, bearings, seals, gearbox, and support structures.

## 6-2.1 Geometry

A conventional stirred tank consists of a vessel equipped with a rotating mixer. The vessel is generally a vertical cylindrical tank. Nonstandard vessels such as those with square or rectangular cross-section, or horizontal cylinder vessels are sometimes used. The rotating mixer has several components: an impeller, shaft, shaft seal, gearbox, and a motor drive. Wall baffles are generally installed for transitional and turbulent mixing to prevent *solid body rotation* (sometimes called *fluid swirl*) and cause axial mixing between the top and bottom of the tank. This is illustrated in a video clip recorded on the Visual Mixing CD affixed to the back cover of the book. A conventional vertical cylindrical stirred tank with a top-entering mixer is shown in Figure 6-1. Occasionally, a small impeller, called a *tickler* or *kicker*, is installed close to the tank bottom to maintain agitation when the liquid level drops below the main impeller.

In tall tanks, the mixer may be installed from the bottom (Figure 6-2) to reduce the shaft length and provide mechanical stability. The mixers can be side entering (Figure 6-3) for large product storage and blending tanks or inserted from the top at an angle (Figure 6-4) for nonbaffled small tanks. The flows generated with side entering and angled mixers are asymmetric, and therefore wall baffles are no longer needed. In horizontal cylindrical tanks, the mixer can be installed on the side or from top, as shown in Figures 6-5 and 6-6, respectively.

#### 6-2.1.1 Impeller Types.
The typical impellers used in transitional and turbulent mixing are listed in Table 6-1. These have been divided into different general classes, based on flow patterns, applications, and special geometries. The classifications also define application types for which these impellers are used. For example, axial flow impellers are efficient for liquid blending and solids suspension, while radial flow impellers are best used for gas dispersion. Up/down impellers can be disks and plates, are considered low-shear impellers, and are commonly used in extraction columns. The pitched blade turbine, although classified as an axial flow impeller, is sometimes referred to as a mixed flow impeller, due to the flow generated in both axial and radial directions. Above a D/T ratio of 0.55, pitched blade turbines become radial flow impellers (see the Visual Mixing CD for an illustrative video). Further details of these applications and impeller selection criteria are given later in the chapter.

**Table 6-1** Impeller Classes and Specific Types

| | |
|---|---|
| Axial flow | Propeller, pitched blade turbine, hydrofoils |
| Radial flow | Flat-blade impeller, disk turbine (Rushton), hollow-blade turbine (Smith) |
| High shear | Cowles, disk, bar, pointed blade impeller |
| Specialty | Retreat curve impeller, sweptback impeller, spring impeller, glass-lined turbines |
| Up/down | Disks, plate, circles |

**348** MECHANICALLY STIRRED VESSELS

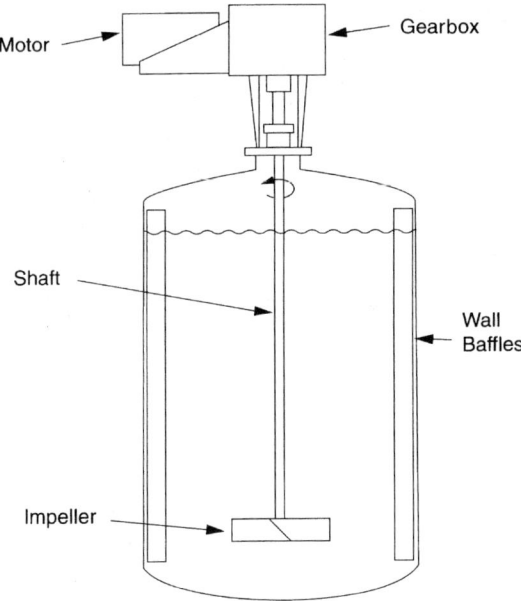

**Figure 6-1**  Conventional stirred tank with top-entering agitator.

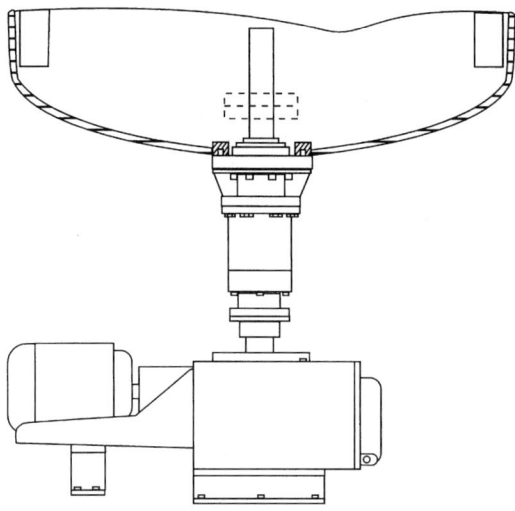

**Figure 6-2**  Bottom-entering agitator.

***6-2.1.2 Wall Baffles.*** Baffles are generally used in transitional and turbulent mixing, except in severe fouling systems, which require frequent cleaning of tank internals. For laminar mixing of viscous fluids, baffles are not needed. In square and rectangular tanks, the corners break up the tangential flow pattern and thus

KEY DESIGN PARAMETERS **349**

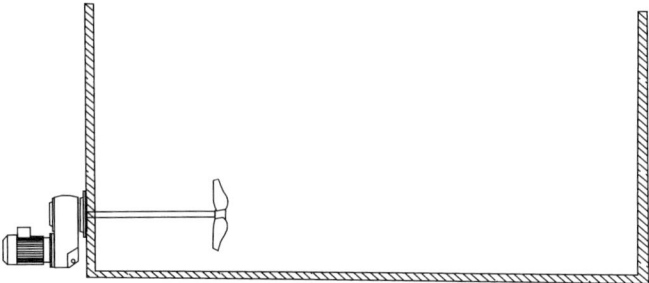

**Figure 6-3** Side-entering mixer for large product storage and blending tanks.

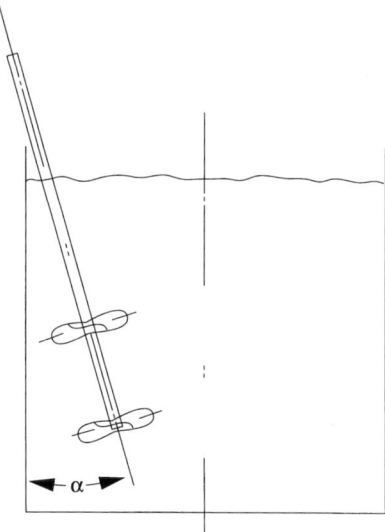

**Figure 6-4** Angular top-entering mixer for small tanks with portable mixers.

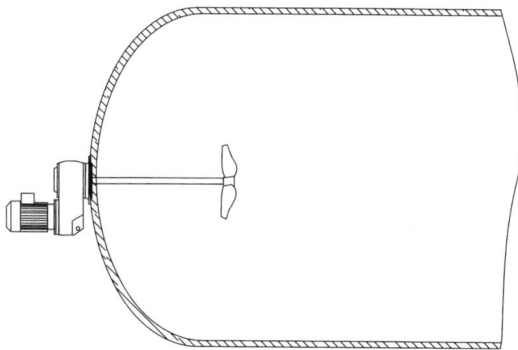

**Figure 6-5** Side-entering mixer for horizontal cylindrical vessel.

**350** MECHANICALLY STIRRED VESSELS

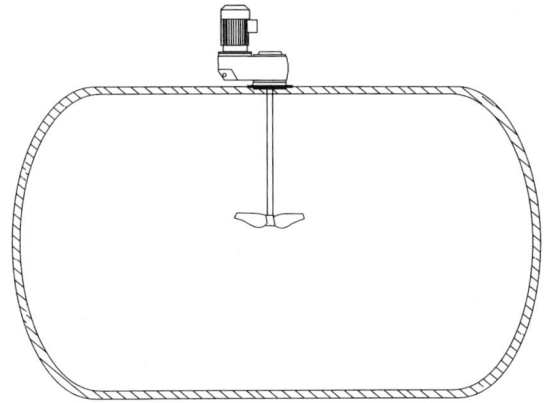

**Figure 6-6** Top-entering mixer for horizontal cylindrical vessel.

provide a baffling effect, and wall baffles may not be needed. Baffles are also not used for side-entering mixers in large product tanks and angled mixers in small agitated tanks.

Wall baffles typically consist of solid surfaces positioned in the path of tangential flows generated by a rotating impeller. Wall baffling has a significant influence on the flow behavior and resulting mixing quality. In the absence of baffles, the flow created by impeller rotation is two dimensional and causes swirling action, i.e., solid body rotation. Wall baffles transform tangential flows to vertical flows, provide top-to-bottom mixing without swirl, and minimize air entrainment. Baffles increase the drag and power draw of the impeller.

A standard baffle configuration consists of four vertical plates having width equal to 8 to 10% (T/12 to T/10) of the tank diameter. Narrower baffles are sometimes used for high viscosity systems, buoyant particle entrainment (width = 2% of T), or when a small vortex is desired. A small spacing between baffles and the tank wall (1.5% of T) is allowed to minimize dead zones particularly in solid–liquid systems. Wall baffles increase the power consumption of the mixer and generally enhance the process result.

For glass-lined vessels and retreat curve blade impellers and glass-lined turbines, five different types of baffles (shown in Figure 6-7) are commonly used: finger, flattened pipe, h style, concave baffle, and fin. These baffles can be conveniently supported in the vessel heads of glass-lined reactors. Of these, the fin baffle has become a more standard choice.

Other types of baffling (e.g., surface baffles, retractable baffles, twisted baffles, and partition baffles) are also used for satisfying specific process needs. Surface baffles can prevent gas entrainment from the vapor head. Retractable baffles are used in systems where rheology changes during the process, and baffles must be removed at low Reynolds numbers. Partition baffles are used for staging of tall vessels. The selection, sizing, and location of baffles depend on the process requirements and mixing regime.

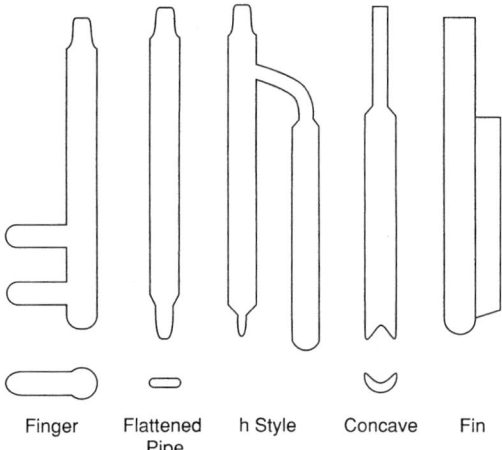

**Figure 6-7**  Common glass-lined baffle types.

***6-2.1.3 Tank Bottoms***. The conventional stirred vessel uses a cylindrical tank with a flat or dished bottom. Dished bottom heads can be 1 : 2 ellipsoidal, ASME dish, or hemispherical. Information on the geometry of pressure vessel heads can be found in Dimoplon (1974). The flow patterns below the impeller can be different with different heads and result in different mixing efficiencies. For solids suspension in flat-bottomed tanks, solids tend to accumulate in the corners. Dished bottoms are preferred to maximize suspension quality. Large tanks are constructed with flat bottoms or with a shallow cone or inverted cone, as in crude oil storage tanks. Deep cone bottoms are used in tanks when inventories and heels must be minimized. Such a geometry may require a special impeller shaped to conform to the cone geometry and placed near the bottom of the cone to provide agitation at low liquid levels.

***6-2.1.4 Draft Tubes***. A draft tube is a tube installed centrally within the vessel. Axial flow impellers located inside a draft tube are used to provide an efficient top-to-bottom circulation pattern, which is important for flow-controlled processes. Draft tubes reduce the standard deviations in process variables such as concentration, density, and viscosity. They are also useful in tanks with a high ratio of height to diameter.

***6-2.1.5 Motor/Gearbox***. The motor and gearbox constitute the drive system of the mixer. The motor can be electric (induction or DC), or driven by air pressure, hydraulic fluid, steam turbine, or diesel and gas engine. Typical power ratings of commercially available electric motors are given in Table 6-2. A gearbox is used to obtain the desired mixer shaft speed from the motor speed; the speed is fixed based on the frequency of the power supply, typically 1750 rpm at 60 Hz electric power. Depending on the desired mixer shaft speed, a gearbox can

**352** MECHANICALLY STIRRED VESSELS

**Table 6-2** Standard Motor Power and Mixer Speeds

| Motor Power (hp) | | | | | | Mixer Speed (rpm) | | | | |
|---|---|---|---|---|---|---|---|---|---|---|
| $\frac{1}{4}$ | $\frac{1}{2}$ | 1 | $1\frac{1}{2}$ | 2 | 3 | 4 | 5 | 6 | $7\frac{1}{2}$ | 9 | 11 |
| 5 | $7\frac{1}{2}$ | 10 | 15 | 20 | 25 | $13\frac{1}{2}$ | $16\frac{1}{2}$ | 20 | 25 | 30 | 37 |
| 30 | 40 | 50 | 60 | 75 | 100 | 45 | 56 | 68 | 84 | 100 | 125 |
| 125 | 150 | 200 | 250 | 300 | 350 | 155 | 190 | 230 | 280 | 350 | |
| 400 | 500 | 600 | | | | | | | | | |

have a two- or three-step gear reduction. Although the gearbox can be fabricated to provide any gear ratio, there are standard gear ratios to provide mixer shaft speeds given in Table 6-2 and represented by Rautzen et al. (1976) as a chart of the available motor horsepowers and the corresponding available shaft speeds for each given horsepower.

**6-2.1.6  Inlets/Outlets.** The location and design of inlets and outlets are based on the process, type of feed, and sensitivity of the process result to the rate of feed dispersion. For slow batch processes, the feed inlet can be from the top. It should be pointed at an active surface away from the tank wall and the impeller shaft. For processes requiring quick dispersion of the feed, the inlet nozzle should be located in a highly turbulent region such as the suction or discharge of the impeller, as discussed extensively in Chapter 13. The inlet nozzle should be sized to prevent backmixing of the tank contents into the inlet pipe, where lack of mixing may cause poor process results. Specific guidance is available from Jo et al. (1994). When feeding solids into a liquid, the feed rate must be controlled to closely match the rate of solid wetting, incorporation, and dispersion by the mixer.

The outlet is generally located on the side near the tank bottom or in the bottom head if the vessel needs to be drained completely. When solids are present, this bottom outlet can get plugged and can cause poor contacting of liquid and solids unless fitted with a flush-bottomed valve. A small impeller, installed very close to the tank bottom, also helps to eliminate this problem and provides mixing at low liquid levels. In continuously operated agitated tanks, the outlet must be located far from the inlet to minimize short-circuiting of the feed.

**6-2.1.7  Heat Transfer Surfaces.** When the process requires heat addition to or removal from the process fluid, the mixing tank must be equipped with appropriate heat transfer surfaces. Liquid motion supplied by the mixer enhances the heat transfer coefficient. Commonly used heat transfer surfaces, shown in Figure 6-8, include jackets, internal helical coils, and internal baffle coils. A jacket can be a tank outside the main tank, baffled, half-pipe, or dimpled. Each of these heat transfer surfaces can also be used in combination with a single coil or multiple heating coils installed within the space between the impeller and the tank wall. A suitable heat transfer fluid must be supplied on the service side of the heat transfer surfaces.

Positioning of internal coils should be such that they are not placed in the discharge flow of the impeller. Since the impeller discharge flow is typically

KEY DESIGN PARAMETERS   353

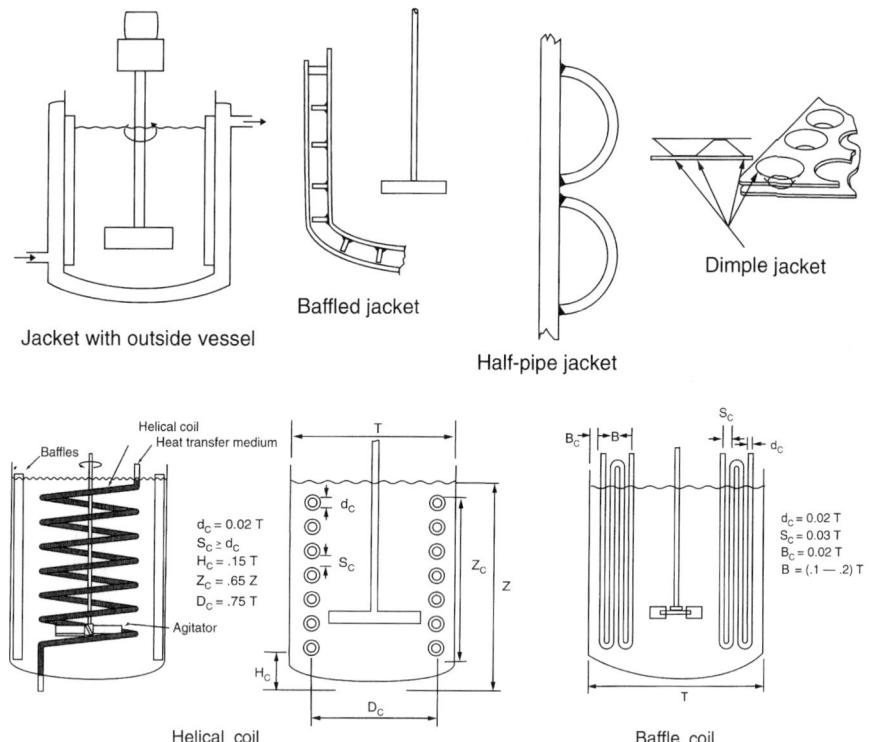

**Figure 6-8**  Heat transfer surfaces for stirred tanks.

pulsating, the coils and supports may suffer excessive fatigue and wear. Helical coils can be sized to act as a draft tube and enhance internal circulation and mixing. (See Chapter 14 for additional information on heat transfer and heat transfer correlations.)

*6-2.1.8  Gas Sparger.* A gas sparger is used when a gas is introduced into the liquid for efficient gas–liquid contacting for mass transfer and/or reaction. While the mixer design and operation control the gas–liquid interfacial area, a well-designed and well-located sparger can enhance the gas–liquid process result by maximizing contacting and eliminating maldistribution. A commonly used sparger configuration consists of a ring with equally spaced sparger holes positioned below the impeller. The sparger diameter should be less than the impeller diameter, typically 0.8 times the impeller diameter. Other shapes can also be effective as long as the sparger holes are well distributed across the tank cross-section and are active. A common problem with gas spargers in a gas–liquid–solids system is that the sparger can be sanded in quickly with solids. Under such conditions, maldistribution of gas can be significant. Each sparger hole may have to be individually controlled. The reader is referred to Chapter 11 for additional information on gas–liquid mixing and to Chapter 18 for special considerations in sparger design for biological reactors.

### 6-2.2 Impeller Selection

There are literally hundreds of impeller types in commercial use. Determination of the most effective impeller should be based on the understanding of process requirements and knowledge of physical properties. Impellers can be grouped as turbines for low to medium viscosity fluids and close-clearance impellers for high viscosity fluids. Turbine impellers are further characterized, based on flow patterns, as axial flow and radial flow. Recent developments in the impeller technology have been focused on increasing axial flow at reduced shear. These impellers use a hydrofoil blade profile for efficient and more streamlined pumping. There are also many specialty impeller designs developed for specific process needs.

In this section we describe turbine impellers used in transitional and turbulent flow applications. High viscosity applications and appropriate impeller types are discussed in Section 6-6. Further discussions of impellers may be found in Chapter 21 and in Dickey et al. (2001). A number of video clips illustrating the effects of impeller selection are included on the Visual Mixing CD.

There are four types of turbine impellers, which are characterized by the flow patterns and level of shear they create: axial flow, radial flow, hydrofoil, and high-shear impellers. They have the widest use in low and medium viscosity liquid applications, solids suspension, liquid–liquid emulsification, and gas dispersion. Turbine impellers can have blades varying from 2 to 12 in number. Two blades are normally unstable mechanically, while it is difficult to install more than six blades on a hub. Axial flow impellers generally have three or four blades, and radial flow impellers are designed with six blades.

#### 6-2.2.1 Axial Flow Impellers.
Axial flow impellers (Figure 6-9) are used for blending, solids suspension, solids incorporation or draw down, gas inducement, and heat transfer. The oldest axial flow impeller design is the marine propeller, which is often used as a side-entering mixer in large tanks and as a top-entering mixer in small tanks. It can be designed with a different pitch to change the combination of pumping rate and thrust. Due to its fabrication by casting, a propeller becomes too heavy when large. It is not generally used as a top-entering impeller for tank sizes larger than 5 ft.

A pitched blade turbine consists of a hub with an even number of blades bolted and tack-welded on it. It is lighter in weight than a propeller of the same diameter. The blades can be at any angle between 10 and 90° from the horizontal, but the most common blade angle is 45°. The flow discharge from a pitched blade impeller has components of both axial and radial flow velocity in low to medium viscosity liquids, and is considered to be a mixed-flow impeller. Most

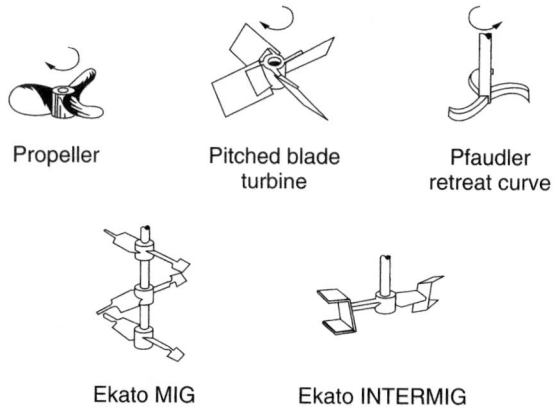

**Figure 6-9** Axial flow impellers.

applications require the impeller rotation to direct the flow toward the bottom head or down-pumping. However, in some situations, such as gas dispersion and floating solids mixing, up-pumping may be more effective.

The retreat blade impeller was developed by the Pfaudler Company specifically for glass-lined reactors used for highly corrosive fluids. This was the only impeller geometry at that time which could hold a glass covering. However, current technology allows glassing of very complex impeller geometries for several impeller types, including the ability to combine two impellers on the same shaft. The retreat blade impeller is now being phased out in favor of the more effective and scalable new generation of impellers made possible by major advances in glass-lined technology. Further information on these impellers may be found in Chapter 17. In glass-lined tanks, glassed baffles (see Figure 6-7) are not installed on the wall but are supported through nozzles in the top and/or bottom head. However, one manufacturer (DeDietrich) has recently developed a three-baffle system in which the baffles are integral with the vessel wall.

The Ekato Company developed two two-blade axial flow impellers, the Mig and the Intermig, mainly for high viscosity liquids. However, they can be effective for low to medium viscosity liquids as well. These impellers are designed at high impeller/tank diameter ratio (D/T) and have two sections of blades at opposite angles. If the inner blade pumps down, the outer blade pumps up to enhance the liquid circulation. The outer blade section of Intermig has two staggered sections designed for minimizing local form drag losses, which results in more distinct axial flow and a lower power number. Three Mig impellers are recommended for a liquid height/tank diameter ratio (H/T) of 1.0, while two Intermig impellers are adequate for the same configuration. Both impellers are sized at $D/T = 0.7$ for turbulent conditions and require wall baffles. For laminar conditions, $D/T > 0.7$ is used without wall baffles. These impellers have been found to be excellent for crystallization operations because they combine low shear with good circulation.

*6-2.2.2 Radial Flow Impellers.* Like axial flow turbine impellers, radial flow impellers (Figure 6-10) are commonly used for low to medium viscosity fluids. Although they can be used for any type of single- and multiple-phase mixing duty, they are most effective for gas–liquid and liquid–liquid dispersion. Compared to axial flow impellers, they provide higher shear and turbulence levels with lower pumping. Radial flow impellers discharge fluid radially outward to the vessel wall. With suitable baffles these flows are converted to strong top-to-bottom flows both above and below the impeller.

Radial flow impellers may either have a disk (Rushton turbine) or be open (FBT) and may have either flat or curved blades (backswept turbine). Impellers without the disk do not normally pump in a true radial direction since there is pressure difference between each side of the impeller. This is also true when the impellers are positioned in the tank at different off-bottom clearances. They can pump upward or downward while discharging radially. Radial discharge flow patterns can cause stratification or compartmentalization in the mixing tank. Disk-type radial impellers provide more uniform radial flow pattern and draw more power than open impellers. The disk is a baffle on the impeller, which prevents gas from rising along the mixer shaft. In addition, it allows the addition of a large number of impeller blades. Such blade addition cannot be done easily on a hub. A disk can also be used with a pitched blade turbine for use in gas–liquid mixing.

The Rushton turbine is constructed with six vertical blades on the disk. Standard relative dimensions consist of blade length of D/4, blade width of D/5, and the disk diameters of 0.66 and 0.75D. The backswept turbine has six curved blades with a power number 20% lower than the Rushton turbine. The backswept nature of the blades prevents material buildup on the blades. It is also less susceptible to erosion. Typical applications include general waste and fiber processing in pulp and paper industries.

The recently developed hollow-blade impellers (e.g., Scaba SRGT, Chemineer CD6, and the Smith impeller) provide better gas dispersion and higher gas-holding

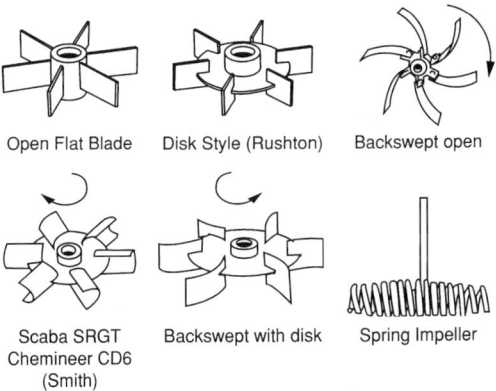

Figure 6-10  Radial flow impellers.

capacity than the Rushton turbine. The impeller blades are semicircular or parabolic in cross-section. This general shape allows for much higher power levels to be obtained in the process than that obtained by the Rushton turbine during gas dispersion. Gas dispersion is discussed in Chapters 11 and 18.

The coil or spring impeller was developed for systems where solids frequently settle to the tank bottom. When buried in stiff solids, a spring impeller is able to dig itself out of the solids without breaking an impeller blade.

*6-2.2.3 Hydrofoil Impellers.* Hydrofoil impellers (Figure 6-11) were developed for applications where axial flow is important and low shear is desired. They have three or four tapering twisted blades, which are cambered and sometimes manufactured with rounded leading edges. The blade angle at the tip is shallower than at the hub, which causes a nearly constant pitch across the blade length. This produces a more uniform velocity across the entire discharge area. This blade shape results in a lower power number and higher flow per unit power than with a pitched blade turbine. The flow is more streamlined in the direction of pumping, and the vortex systems of the impeller are not nearly as strong as those of the pitched blade turbine.

Lightnin A310, Chemineer HE3, and EMI Rotofoil are characterized by a low solidity ratio, defined by a projected area of impeller blades divided by the impeller horizontal cross-sectional area. They are very efficient impellers for liquid blending and solids suspension.

Hydrofoil impellers with a high solidity ratio include the Lightnin A315 and Prochem Maxflo. This feature makes them effective for gas dispersion in viscous systems in addition to liquid blending and solids suspension. The Maxflo impeller is constructed with a larger drum-type hub with three or five trapezoidal and cambered blades. The two-bladed Ekato Interprop is designed with a high angle of attack of the blade and with an additional leading blade wing. This configuration provides an improved lift/resistance ratio and more intense axial impulse compared to other hydrofoils. Interprop is, therefore, used effectively for dispersion applications in addition to blending and solids suspension.

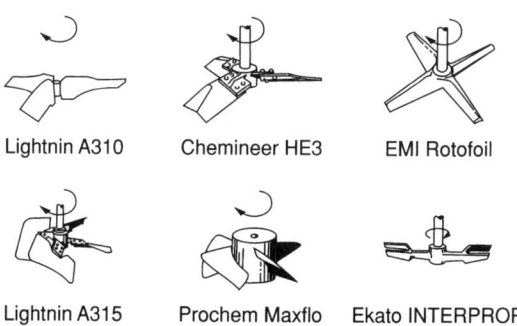

**Figure 6-11** Hydrofoil impellers.

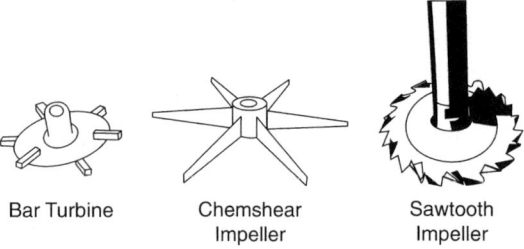

**Figure 6-12**  High-shear impellers.

***6-2.2.4 High-Shear Impellers.*** High-shear impellers (Figure 6-12) are operated at high speeds and are used for the addition of a second phase (e.g., gas, liquid, solid, powder) in grinding, dispersing pigments, and making emulsions. These dispersing impellers are low pumping and therefore are often used along with axial flow impellers for providing both high-shear and homogeneous distribution. At the lower end of the high-shear range is the bar turbine, which has square cross-section bars welded to a disk. The Chemshear impeller has tapered blades and provides intermediate shear levels. A very high-shear producing sawtooth impeller consists of a disk with serrations around its circumference. It provides reasonably high intensity of turbulence in the vicinity of the impeller.

### 6-2.3  Impeller Characteristics: Pumping and Power

Power numbers, pumping numbers, shear levels, and flow patterns characterize the various impellers described above. All the power applied to the mixing system produces circulating capacity, Q, and velocity head, H, given by

$$Q \propto ND^3 \qquad (6\text{-}1)$$

$$H \propto N^2 D^2 \qquad (6\text{-}2)$$

Q represents internal circulation and H provides the shear in mixing. In a sense, the velocity head, H, provides the kinetic energy that generates shear through the jet or pulsating motion of the fluid. Both expressions have not included the effects of the number of blades and blade width. Head results in shear and is dissipated by turbulence. Equation (6-1) can be rewritten as

$$Q = N_Q ND^3 \qquad (6\text{-}3)$$

where $N_Q$ is the pumping number, which depends on the impeller type, the D/T ratio, and impeller Reynolds number, defined as

$$Re = \frac{\rho ND^2}{\mu} \qquad (6\text{-}4)$$

**6-2.3.1 Pumping and Pumping Number.** Pumping is the amount of material discharged by the rotating impeller. The values of $N_Q$ under turbulent conditions are known for the commonly used impellers and are given in Table 6-3. As can be seen, the values of pumping number for most commonly used impellers vary in the range 0.4 to 0.8. As a result, all standard impellers will pump at about the same rate for a given diameter and mixer speed.

Figure 6-13 shows the relationship of $N_Q$ with Re and D/T for a 45° pitched blade turbine (PBT). As evidenced in this figure, $N_Q$ increases as Re increases up to Re of 10 000 and becomes constant at higher Re. Also, smaller diameter impellers have higher pumping numbers. Figure 6-13 also shows that these impellers should not be used below a Reynolds number of 1000 if high pumping efficiency is desired. Similar plots of $N_Q$ are available for a variety of impellers from the respective vendors.

**Table 6-3** Pumping Number, $N_Q$, under Turbulent Conditions for Various Impellers

| Impeller Type | $N_Q$ |
|---|---|
| Propeller | 0.4–0.6 |
| Pitched blade turbine | 0.79 |
| Hydrofoil impellers | 0.55–0.73 |
| Retreat curve blade | 0.3 |
| Flat-blade turbine | 0.7 |
| Disk flat-blade turbine (Rushton) | 0.72 |
| Hollow-blade turbine (Smith) | 0.76 |

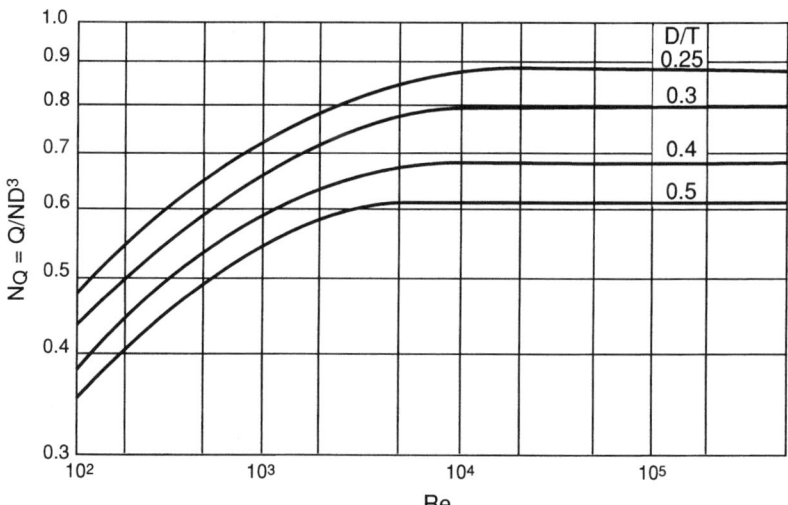

**Figure 6-13** Pumping number versus impeller Reynolds number for pitched blade turbine.

**360**   MECHANICALLY STIRRED VESSELS

Pumping of an impeller changes with the changes in impeller geometry and batch size. For example, pumping is dependent on number of blades and the blade width. Limited data are available in the literature and from vendors on these effects. Pumping also changes with varying liquid level. Unfortunately, quantification of this effect is not available.

It is important to note that some impellers and mixer configurations do not pump well. For example, the retreat curve blade impeller in an unbaffled vessel creates solid body rotation and poor pumping. Pumping with close-clearance impellers such as anchors and helical ribbon can be very high or, sometimes, very poor, depending on conditions and the materials being pumped. Turbulent impellers in laminar applications only pump locally. Often, the rest of the tank goes unmixed.

### 6-2.3.2 Power and Power Number.

The power consumed by a mixer can be obtained by multiplying pumping, Q, and head, H, and is given by

$$P = \frac{N_p \rho N^3 D^5}{g_c} \quad (6\text{-}5)$$

where $N_p$ is the power number and depends on impeller type and impeller Reynolds number.

Using another viewpoint, power, generated by an individual section of an impeller, is equal to the drag, F, multiplied by the impeller velocity, V, for that section or

$$P = FV \quad (6\text{-}6)$$

This is then summed over the entire impeller to obtain the total power. Form and skin drag in the turbulent regime are represented by

$$F = 0.5 C_d \rho V^2 A_p \quad (6\text{-}7)$$

where $C_d$ is drag coefficient, $\rho$ the density of fluid around the impeller, and $A_p$ the projected area of the impeller blade.

Substituting eq. (6-7) into (6-6) yields

$$P = 0.5 C_d \rho V^3 A_p \quad (6\text{-}8)$$

Since all velocities in a mixing tank are proportional to the tip speed ($= \pi N D$) and the impeller projected area is proportional to $D^2$, the power can be represented by

$$P \propto C_d \rho N^3 D^5 \quad (6\text{-}9)$$

Comparing eqs. (6-5) and (6-9), the power number $N_p$ can be considered similar to a drag coefficient. Just as the drag coefficient under turbulent flow is a function

KEY DESIGN PARAMETERS 361

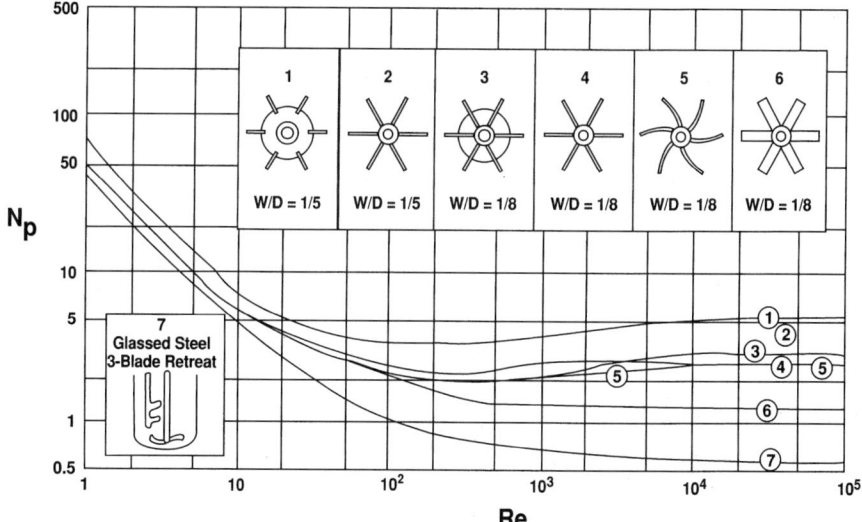

**Figure 6-14** Power number versus impeller Reynolds number for seven different impellers. (Modified from Rushton et al., 1950.)

of geometry and independent of Reynolds number, $N_p$ also is constant at high Re for a given impeller geometry.

The power number, $N_p$, also is a function of impeller blade width, number of blades, blade angle, D/T, baffle configuration and impeller elevation. Figure 6-14 shows the relationship between $N_p$ and Re for seven impellers. It is important to recognize that at Re < 100, the conditions become laminar flow; and mixing quality, obtained using these impellers, becomes extremely poor. Under such conditions, impellers designed for laminar flow conditions are recommended.

While Figure 6-14 provides the power number data in a wide range of Re, the information should not be used below an impeller Reynolds number of 1000. The flow regimes are (1) laminar flow below a Reynolds number of 10, (2) transition between Reynolds numbers of 10 and $10^4$, and (3) turbulent above a Reynolds number of $10^4$. The functionality between $N_p$ and Re can be described as follows:

- $N_p \propto Re^{-1}$ in the laminar regime and power depends greatly on viscosity.
- $N_p$ = constant in turbulent regime (Re > 10 000) and is independent of liquid viscosity.
- $N_p$ changes slightly in the transitional regime (100 < Re < 10 000).
- $N_p$ for turbine impellers varies with blade width as follows:
  For a six-bladed Rushton

$$N_p \propto (W/D)^{1.45} \qquad (6\text{-}10)$$

For a four-bladed 45° pitched blade

$$N_p \propto (W/D)^{0.65} \qquad (6\text{-}11)$$

- The functionality with number of blades is given by:
  For three to six blades
  $$N_p \propto (n/D)^{0.8} \tag{6-12}$$
  For six to twelve blades
  $$N_p \propto (n/D)^{0.7} \tag{6-13}$$
- For turbines with four to eight curved blades, eq. (6-12) is valid.
- For pitched blade turbines, changing the blade angle $\theta$ changes the power number by
  $$N_p \propto (\sin\theta)^{2.6} \tag{6-14}$$

The scale or size of a turbine impeller has a very small effect on its power number in the commonly used range $D/T = 0.33$ to $0.5$, when standard baffles are used.

The number of baffles ($N_b$) and their width (B) have a significant effect on $N_p$. As the parameter $N_b B$ increases, $N_p$ increases (Figure 6-15) up to the power number of the conventional configuration, with four baffles having width equal to T/10. At higher $N_b B$ values, the power number is constant at a level which depends on D/T.

The effect of impeller elevation C on the power number is small for turbine impellers, especially the radial flow impellers. Figure 6-16 shows this effect for pitched blade (PBT), flat-blade (FBT), and disk flat-blade (DFBT) turbines with two different width ratios.

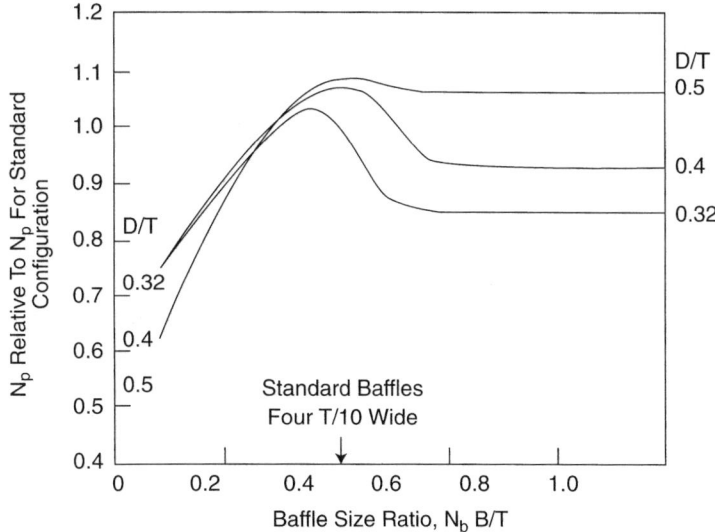

**Figure 6-15** Effect of baffling and D/T on power number.

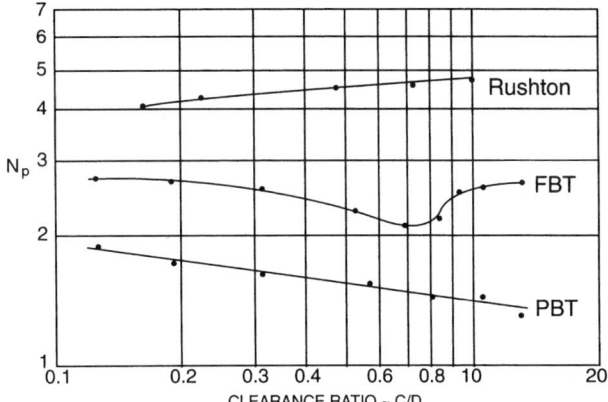

**Figure 6-16** Effect of turbine clearance on power number for PBT, FBT, and DFBT with $N_b B/T = 0.33$.

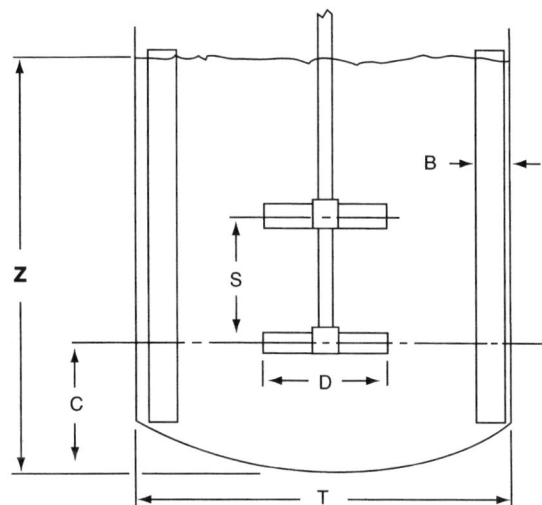

**Figure 6-17** Two impellers on a shaft of top-entering mixer.

For PBTs the power number correlates with C as

$$N_p \propto (C/D)^{-0.25} \qquad (6\text{-}15)$$

When multiple impellers are used on the same shaft (Figure 6-17), the combined power number may or may not be additive of individual power numbers. The power number for such a system depends on impeller type and spacing between the impellers (S/D) as shown in Figure 6-18.

Typical spacing between impellers is one impeller diameter. If impellers are placed closer than this, there is considerable interaction between them. In the

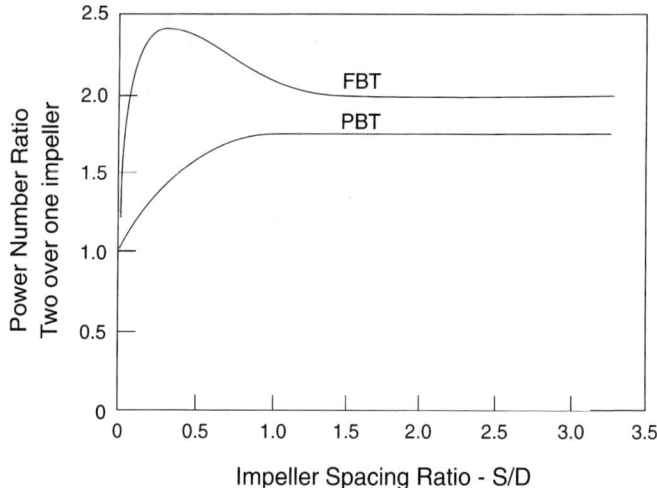

**Figure 6-18** Effect of dual turbine spacing on power number for FBT and PBT.

case of an axial flow impeller such as the pitched blade turbine, the combined power is significantly less than twice the single-impeller power. With a flat-blade turbine, however, the total power may exceed twice the single-impeller power, depending on the impeller spacing. If the impellers are too close to each other, the total power is reduced.

The power number of side-entering propellers depends on the impeller Reynolds number and the pitch. The propeller pitch is defined as the distance traversed by the propeller in one revolution divided by the diameter. For a square pitch (pitch = 1.0), this distance is equal to the diameter. As shown in Figure 6-19, $N_p$ is higher for higher-pitch propellers at all values of Re.

It is important to recognize that at Re < 100, the flow conditions would approach laminar flow and mixing quality with these propellers would be poor. The power number functionality with propeller pitch between 1.0 and 2.0 at Re > 1000 can be approximated by

$$N_p \propto (p/D)^{1.5} \qquad (6\text{-}16)$$

The power numbers of several other commonly used impellers under turbulent conditions are given in Table 6-4.

## 6-3 FLOW CHARACTERISTICS

Flow characteristics for an impeller can be divided into:

- Flow patterns
- Pumping
- Shear

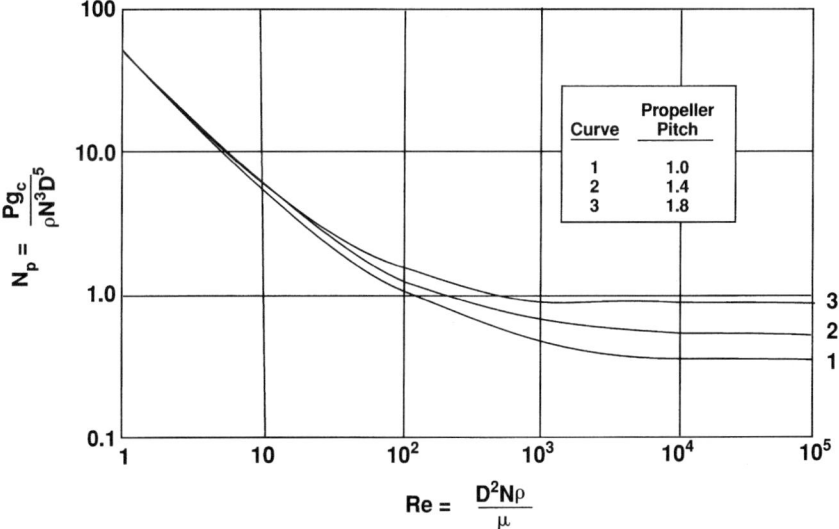

**Figure 6-19** Power number of side-entering propellers versus Reynolds number.

**Table 6-4** Power Numbers of Various Impellers under Turbulent Conditions with Four Standard Baffles

| Impeller Type | $N_p$ |
|---|---|
| Concave- or hollow-blade turbine | 4.1 |
| Ekato MIG—3 impellers, $D/T = 0.7$ | 0.55 |
| Ekato Intermig—2 impellers, $D/T = 0.7$ | 0.61 |
| High-shear disk at Re = 10 000 | 0.2 |
|  | (lower for lower Re) |
| Lightnin A310 | 0.3 |
| Chemineer HE3 | 0.3 |
| The following are all for $D = T/3$, $C = T/3$, and blade width $W = D/5$: | |
| 45°PBT; 4 blades | 1.27 |
| 45°PBT; 6 blades | 1.64 |
| Marine propeller (1.0 pitch) | 0.34 |
| Marine propeller (1.5 pitch) | 0.62 |
| Smith or concave- or hollow-blade with 6 blades | 4.4 |

All impellers generate some sort of flow pattern. These flow patterns, coupled with the flow regime, determine relative levels of pumping and shear. All impellers can therefore be categorized by variations in their pumping and shear capabilities. For example, axial flow hydrofoils are mostly pumping and low-shear impellers. Radial flow impellers, on the other hand, provide high shear but low pumping.

## 6-3.1 Flow Patterns

The mixing process result is highly influenced by the impeller flow patterns. There are mainly two types of flow patterns with top-entering mixers, axial and radial, depending on the impeller type (Figure 6-20). Axial flow impellers, including propellers, pitched blade turbines, and hydrofoils, produce a flow pattern throughout the entire tank volume as a single stage, as shown in Figure 6-20b and c. The pitched blade turbine (PBT) has a good balance of pumping and shear capabilities and therefore is considered to be a general-purpose impeller. The hydrofoils produce about the same pumping but at lower shear and turbulence levels than a PBT. The discharge from hydrofoils is more streamlined compared to a PBT, which gives a small reverse loop underneath.

Radial flow impellers, on the other hand, produce two circulating loops, one below and one above the impeller (Figure 6-20a). Mixing occurs between the two loops but less intensely than within each loop. This is an example of compartmentalization mentioned earlier. A true axial flow is usually created with hydrofoil impellers, which provide a confined flow similar to that created in a draft tube. These differences in flow patterns can cause variations in distribution of shear rate and energy dissipation rate within the mixing tank. Depending on the process requirements, a suitable impeller can be chosen based on the flow patterns and resulting shear rates. For example, liquid blending can be achieved

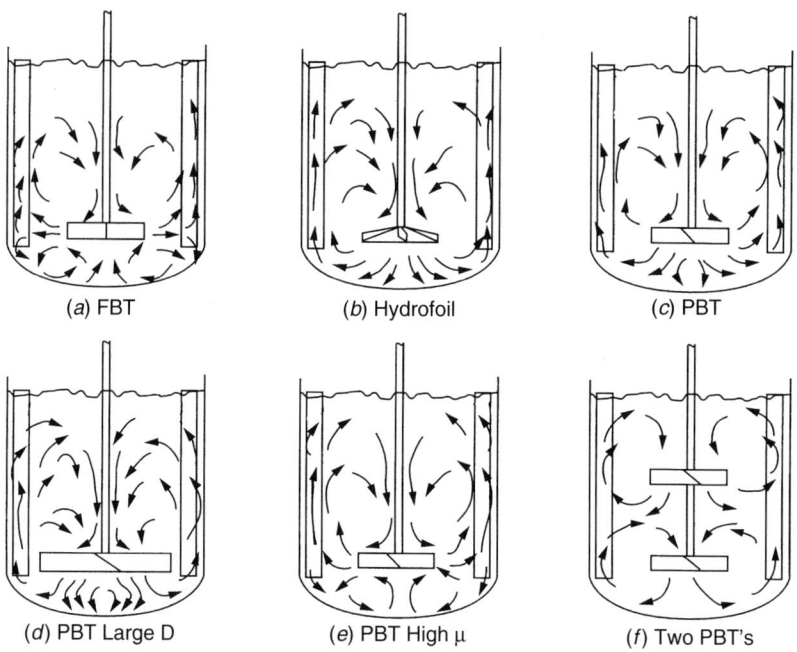

**Figure 6-20** Flow patterns with different impellers, impeller diameter, and liquid viscosity.

efficiently through a single circulation loop from axial flow impellers, whereas dispersion of gas bubbles is better obtained with dual circulation loops using radial flow disk impellers.

The flow patterns with a given impeller are altered by parameters such as impeller diameter, liquid viscosity, and use of multiple impellers. For example, the flow pattern with a PBT becomes closer to radial as the impeller diameter is increased (Figure 6-20*d*) or liquid viscosity is increased (Figure 6-20*e*). Multiple impellers are used when liquid depth/tank diameter ratio is higher than 1.0. In that case, more circulation loops are formed (e.g., two loops with PBT; Figure 6-20*f*). Radial flow impellers give two circulation loops with each impeller.

For suspension of sinking solids, it is important to provide liquid velocities directed to the tank floor for an effective sweeping action. Hydrofoils perform well in this duty. However, if the solids have any tendency to be sticky and cling to the blades, the effectiveness of a hydrofoil can be reduced. This can reduce its versatility as a multipurpose impeller.

In addition to suspending solids off the tank bottom, a process may require homogeneous suspension throughout the bulk. An additional axial flow impeller, perhaps an up-pumping one, may be needed at a higher level for this purpose. Radial flow impellers can be designed to suspend solids, especially if placed on the bottom of the tank, but are less efficient and provide relatively poor solids homogeneity in the bulk.

When using axial flow impellers, the mixer rotation can be reversed to create up-pumping action. This pumping mode can be effective for some systems, such as entrainment of floating solids and gas dispersion. The up-pumping flow provides an effective mechanism for incorporating lighter solids on the liquid surface near the wall. This avoids the need for creating a vortex, which can cause air entrainment and mechanical vibrations (see Chapter 10). For gas dispersion, the up-pumping impeller is generally used at the bottom along with a down-pumping impeller at the top. Such a configuration can provide good gas-holding capacity and prevent mechanical vibrations caused by opposite flows resulting from a down-pumping impeller at the bottom. Up-pumping applications are relatively new and require careful testing and study before use (see Chapter 11).

For effective blending of liquids in large tanks using side-entering propeller mixers, it is important to create the flow patterns shown in Figure 6-21. This is achieved by positioning the horizontal mixer 10 degrees to the left of tank centerline, assuming that the propeller rotates clockwise looking from the motor side. If the mixer is positioned along the tank centerline, the jet flow can cause vortexing on the surface. This vortexing can reduce blending efficiency and possibly entrain air into the liquid product.

Variations in flow patterns for axial flow impellers can be generated to advantage by changing the impeller position or by baffling. For example, axial flow impellers provide radial flows when placed near the tank bottom. For submergence of floating solids, the tank is often unbaffled in the top half to create a controlled vortex useful for pulling down solids. The same method can also be used for surface inducement of gas from vapor space (Oldshue, 1983) and

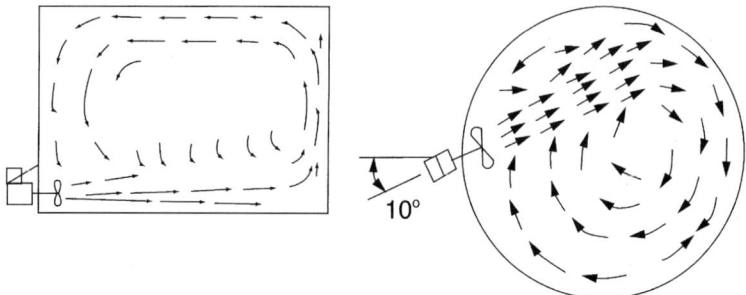

**Figure 6-21** Flow patterns with side-entering propeller mixers.

has been found to be very effective in hydrogenation applications. Care must be taken, however, to avoid gas overloading that could cause mechanical vibrations and damage to seals and bearings.

Flow patterns of high-shear impellers, such as the bar turbine, Chemshear, and sawtooth impeller, are similar to those of radial flow impellers. The major difference is in lower pumping at higher shear. Backswept turbine and spring impeller also have similar radial flow patterns. It is important to understand the flow patterns around the impeller blades, where dispersion and attrition processes occur. Changing the blade geometry changes these flow patterns and alters the shear.

High-speed flows and vortices occur behind the impeller blades and remain coherent as the flow moves into the bulk. The velocities in these flows can be higher than the impeller tip velocity. Vortices are low-pressure regions that can coalesce lower-density materials and sometimes form gas pockets and cavities. Strong vortices are important for dispersion processes, while high velocity flows are desired for liquid blending and solids suspension.

While velocities and shear are high near the impeller, velocities and shear are generally low away from the impeller, particularly in corner areas and at the liquid surface. Processes such as coalescence, agglomeration, and flocculation occur in these regions where energy levels are low. For maximizing mixing efficiency, it is important to add materials away from these regions. See Chapters 13 and 17 for discussions of the impact of feed position on fast reactions.

## 6-3.2 Shear

Whenever there is relative motion of liquid layers, shearing forces exist that are related to the flow velocities. These forces, represented by shear stress, carry out the mixing process and are responsible for producing fluid intermixing, dispersing gas bubbles, and stretching/breaking liquid drops. The shear stress is a complex function of shear rate defined by the velocity gradients, impeller blade pressure drop, turbulence level, and viscosity. These velocity gradients represent velocity differences between adjacent portions of materials, which are therefore sheared and dispersed. By measuring time-averaged velocities near the impeller blade

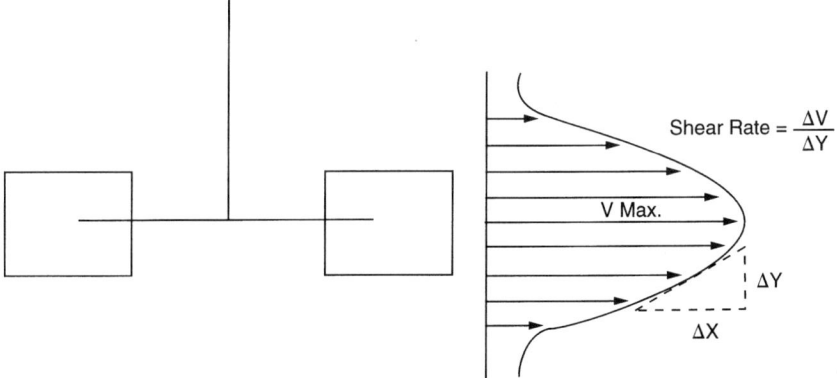

**Figure 6-22** Vertical velocity profile near impeller blade.

(Figure 6-22), macroflow velocity gradient and shear rate can be obtained by taking the slope.

Shear rate, with reciprocal time as the unit, can be viewed as a time constant. If a process has a shear rate of 1000 s$^{-1}$, the events in the flow occur on the order of 1 ms. Such high shear rates are generated in the immediate vicinity of the impeller. However, the volume of this region is relatively small and, therefore, a very small amount of the material experiences these shear rates. The conditions in the vortices are similar, with high shear rate but small volume. The overall mixing process is defined by the combination of shear rate and the volume. Detailed information on the distribution of shear rates and respective volumes is difficult to obtain experimentally. Computational fluid dynamics can be used to extract such information for given mixing conditions.

The local shear rate in the flow for a disk turbine in Newtonian and shear thinning fluids is proportional to mixer speed, as indicated by data of Metzner and Taylor (1960) shown in Figure 6-23:

$$\gamma = KN \qquad (6\text{-}17)$$

where the proportionality constant K decreases rapidly with distance from the impeller blade tip. Equation (6-17) is often referred to as the *Metzner–Otto relationship*. The validity of eq. (6-17) has been demonstrated for laminar flow, transitional flows, and a portion of the turbulent regime. The values of proportionality factor for various impellers are given in Table 6-5.

The Metzner–Otto relationship, eq. (6-17), does not apply for other non-Newtonian fluids, such as shear thickening fluids, Bingham plastics, and false body fluids. In these fluids, shear rates are highly localized around the impeller blade, with the rest of the tank stagnant. The relationship does not apply in highly turbulent flow as well.

As discussed earlier, shear rates are different in different parts of a mixing tank. Therefore, there are several types of shear rate:

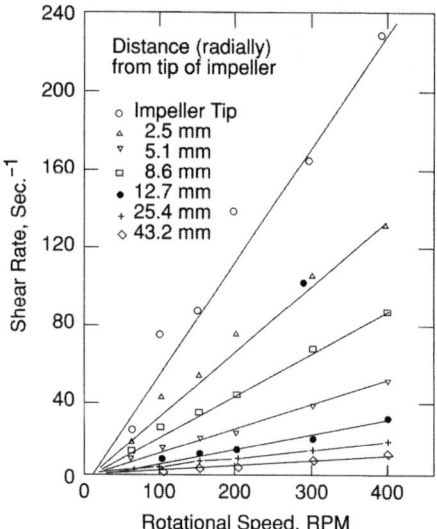

**Figure 6-23** Shear rate distributions at different distances from the impeller in Karo Syrup. (Data from Metzner and Taylor, 1960.)

**Table 6-5** Metzner–Otto Constant for Shear Rate versus Mixer Speed

| Impeller | Propeller | Rushton | Helical Ribbon | Anchor |
|---|---|---|---|---|
| K | 10 | 12 | 30 | 25 |

1. *Maximum shear rate on the impeller blade is:*

$$\gamma \sim 2000 \, N \qquad (6\text{-}18)$$

2. *Maximum shear rate in the flow has two interpretations:*

$$\gamma \sim 150 \, N \qquad (6\text{-}19a)$$

$$\gamma \propto \text{Tip Speed}(= \pi N D) \qquad (6\text{-}19b)$$

with a dimensional proportionality constant. It occurs near the blade tip in high-speed jets and vortices.

3. *Average shear rate in the impeller region is:*

$$\gamma \sim KN \qquad (6\text{-}20)$$

where the proportionality constant varies between 5 and 40 for most impellers as listed in Table 6-5.

4. *Average shear rate in the entire tank:* about one order of magnitude less than definition 3.

5. *Minimum shear rate:* about 25% of definition 4 and is near the liquid surface.

Understanding the magnitude and location of shear in an agitated tank has significant implications for design. For example, if the power of the feed jet is substantial, feed nozzles should be located in low-shear regions. The feed jet would do the mixing in such regions and prevent dead zones. If the feed jet has relatively little power, feeding into a dead zone is probably not advantageous to processing. Very little material actually experiences the very high shear rates on the blade in an agitated tank. Some material experiences the maximum shear rates. Maximum shear in the flow may need to be limited for shear-sensitive materials: for example, crystals (Chapter 17) and biological materials (Chapter 18).

At high Reynolds numbers, the concepts of shear, mean shear, and the impeller rotational speed, N, become unimportant relative to the impeller tip velocity, ND. Viscosity is no longer the mechanism by which momentum is transferred.

Dispersion processes are correlated with Weber number to $-0.6$ power and sometimes impeller tip speed. Weber number is typically defined as $\rho N^2 D^3 / \sigma$, where $\sigma$ is the surface or interfacial tension.

When scaling-up agitated tanks, mixer speed generally decreases and tip speed increases; resulting in lower average shear and higher maximum shear rate at the blade tip. If scale-up is based on constant power per unit volume (P/V), the shear rate away from the impeller becomes lower in the larger vessel. This results in a wider distribution of shear rates on scale-up. The change in shear rate distributions on scale-up affects the mixer performance significantly. For example, feed nozzle location near the impeller blade tip for single-phase systems is more important in large vessels then in laboratory scale vessels. For liquid–liquid mixing, the dispersed phase drop size distribution can be wider in the large agitated tank. For crystallization, the average particle size can be larger and size distribution wider in the large tank.

### 6-3.3 Impeller Clearance and Spacing

Impeller clearance from the tank bottom and impeller spacing for multiple-impeller systems can have a significant impact on the power number. In addition, these parameters influence the flow patterns and the process result. The extent of these impacts depends on the type of mixing system and mixing requirements. If an impeller is located very close to the tank bottom, down-pumping axial flow impellers provide flow patterns similar to radial flow impellers. This can result in reduced pumping and higher shear. For suspension of solids, this condition may be superior for keeping the tank bottom clear of solids but at a cost of reduced bulk homogeneity.

In tall agitated tanks, multiple impellers are often used to improve circulation and narrow the distribution of shear and energy dissipation. Generally, these impellers are spaced away from each other by a distance equal to one impeller diameter. However, to avoid splashing and the formation of a vortex, the top

**Table 6-6** Recommended Impeller Clearance and Spacing

| Mixing System | Maximum Liquid Height, Z/T | Number of Impellers | Impeller Elevation from Tank Bottom Bottom | Top |
|---|---|---|---|---|
| Liquid blending | 1.4 | 1 | Z/3 | — |
|  | 2.1 | 2 | T/3 | 2 Z/3 |
| Solids suspension | 1.2 | 1 | Z/4 | — |
|  | 1.8 | 2 | T/4 | 2 Z/3 |
| Gas dispersion | 1.0 | 1 | T/6 | — |
|  | 1.8 | 2 | T/6 | 2 Z/3 |

impeller should not be located too close to the liquid surface. A deep vortex can cause air or vapor entrainment and dispersion. If impeller spacing of less than D is used, higher shear is generated between the impellers. General guidelines for impeller clearance and impeller spacing for different mixing processes are given in Table 6-6.

The corollary of impeller clearance is impeller submergence, also represented by the height of liquid above the impeller, or the top impeller in a multiple-impeller system. Adequate submergence is necessary to avoid excessive vortexing and entrainment of headspace gas. The minimum submergence requirement depends on the impeller type, pumping direction, and baffle width. For example, a submergence greater than D/2 is sufficient for down-pumping axial flow impellers and conventional baffles (width = T/12), while a minimum of D is required for narrow baffles (width = T/50). When a deep vortex is formed with low impeller submergence, excessive vibrations can occur along with incorporation of vapor bubbles and loss of mixing. Such conditions cause the mixing performance in an agitated tank to deteriorate. Alternative gas–liquid contacting methods are discussed in Chapter 11.

### 6-3.4 Multistage Agitated Tanks

When near-plug flow conditions along with high intensity mixing are desired, multistage mixing tanks provide effective and economical designs. The alternatives, such as a series of continuously stirred tanks, can be highly cumbersome, requiring high maintenance. Compartmentalized horizontal cylindrical tanks require the use of several mixers each with a motor/gearbox/shaft/impeller set. Plug flow can also be achieved with in-line mixers but at a cost of high pressure drop at short residence times. A multistage mixing tank (Figure 6-24), consisting of a vertical column divided by horizontal donut baffles and multiple mixers, can provide staged mixing for a variety of fluids and can be sized for the desired residence time. These mixing tanks are commonly used for blending with chemical reaction, gas absorption, extraction, dissolution, crystallization, caustic treatment, water wash, polymerization, and alkylation.

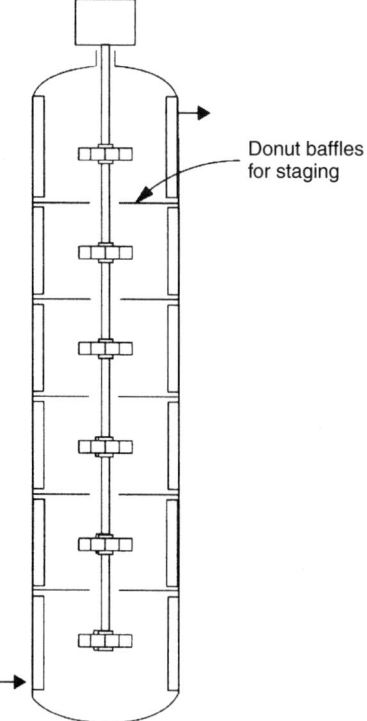

**Figure 6-24** Multistage mixing tank.

Plug flow is important for continuous reacting and nonreacting processes to achieve high yield, selectivity, and mass transfer rates. As shown in Figure 6-25, the required volume of a CSTR can be two orders of magnitude higher than the volume of a plug flow reactor, depending on reaction order and conversion level.

Key issues in designing multistage mixing tanks include tank volume, height/diameter ratio, mixer design, interstage baffle opening size, and inlet–outlet locations. The tank volume is based on the desired mean residence time. The aspect ratio is determined on the basis of number of stages. Although more stages approach plug flow, generally four to six stages are satisfactory. The mixer is designed on the basis of process mixing requirements such as blend time, or quality of solids suspension, liquid–liquid emulsification, and gas dispersion. The mixer design includes type and number of impellers, diameter, mixer speed, and driver power. The interstage baffle opening must be sized based on acceptable level of exchange flows between stages. A larger opening increases exchange flow and backmixing, and as a result leads to lower stage efficiency. Draft tubes are sometimes used at these openings to reduce backmixing.

The inlet and outlet should be located at opposite ends of the mixing tank for best plug flow conditions. Several models are available in the literature for predicting backmixing and can be used to evaluate reactor performance. The

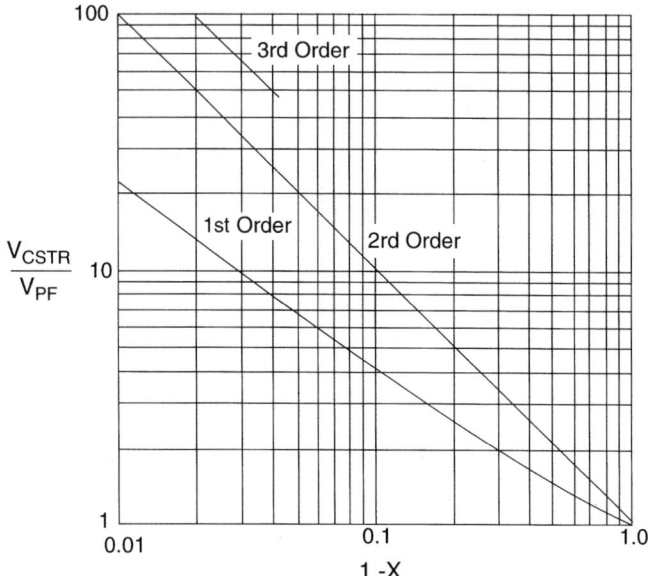

**Figure 6-25** Ratio of volumes of CSTR/plug flow for chemical reactions.

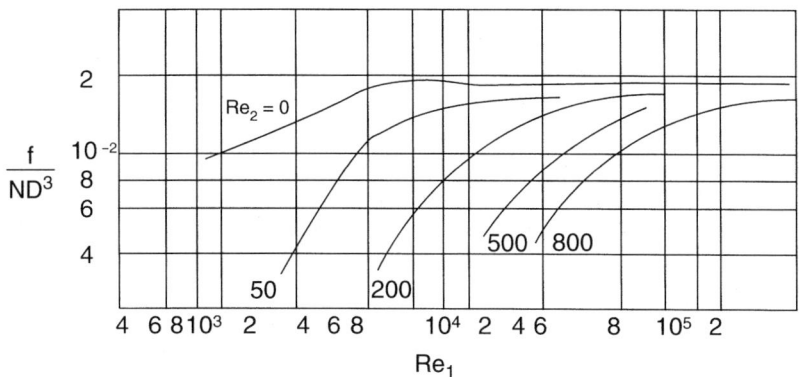

**Figure 6-26** Backflow rate in a multistage mixing tank.

interstage back flow rate, f, is a function of impeller Reynolds number $Re_1 = ND^2/v$ and the flow Reynolds number $Re_2 = q/vD$, as shown in Figure 6-26 for an interstage opening of 33%. As evidenced by these data, backflow rate decreases as flow Reynolds number is increased.

The effect of forward flow velocity on backflow velocity was demonstrated by Xu et al. (1993). Their data, shown in Figure 6-27, indicate that backmixing can be reduced significantly by increasing the forward flow rate. The backflow velocity can be reduced significantly at high forward flow velocity or the reactor

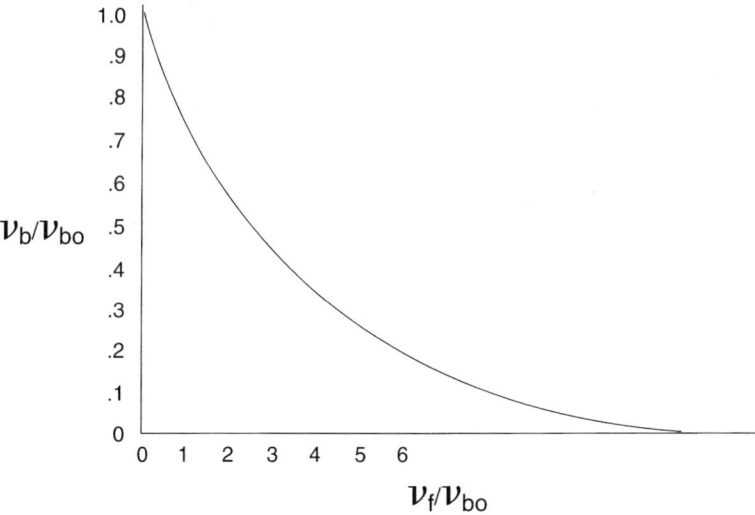

**Figure 6-27** Backflow velocity as a function of forward flow velocity in a multistage mixing tank where $v_b$ is the backflow velocity, $v_{bo}$ the backflow velocity at zero throughput, and $v_f$ the forward flow velocity.

throughput rate. It was also found that backflow velocity is proportional to a/A for a small opening of a/A < 0.25. If an opening at the wall is made available, backflow can increase considerably compared to an equal-sized opening in the center.

In multiphase reacting and nonreacting systems such as liquid–liquid and gas–liquid, very small openings of 1 to 2% of cross-sectional area may be required for high mass transfer. Small openings also minimize bypassing of the dispersed phase.

### 6-3.5 Feed Pipe Backmixing

For fast competitive chemical reactions, improper feed blending can cause the formation of undesired by-products. This occurs due to the high local concentrations of a feed component and the nature of the kinetics. To avoid such a problem, the feed nozzle should be located near the impeller, either in the flow of material directly to the impeller or near the exit flow of the impeller. If the feed nozzle velocity is too low or designed improperly, the tank contents may penetrate into the feed pipe and react. If the feed velocity is too high, the portions of the feed jet can pass through the impeller volume. Both situations can lead to poor reaction yield and selectivity. Backmixing in the feed pipe under poor mixing and high concentration of the fed component permits some undesirable reactions to occur, which result in poor selectivity. Similar results can occur with an excessively high velocity feed jet. For optimum design of the feed nozzle, the criteria listed in Table 6-7 are recommended based on the ratio of feed pipe velocity to impeller tip speed ($v_f/v_t$) for two impeller types (see also Chapter 13).

**Table 6-7** Recommended $v_f/v_t$ Values for Two Impeller Geometries for Turbulent Flow in Feed Pipe

| Impeller Type | Feed Location | G/D[a] | Recommended $v_f/v_t$ |
|---|---|---|---|
| Rushton turbine | Radial | 0.1 | 1.9 |
|  | Above impeller | 0.55 | 0.25 |
| Chemineer HE3 | Radial | 0.1 | 0.1 |
|  | Above impeller | 0.55 | 0.15 |

[a] G = Vertical distance between feed nozzle and impeller tip, D/T = 0.53

### 6-3.6 Bottom Drainage Port

Like any storage tanks, agitated tanks must be emptied frequently, especially in batch systems. Therefore, a bottom drain must be installed to minimize or eliminate any heel. These drain ports, even though closed during the mixer operation, can get plugged when sediments or solids are present in the liquid. These inventories can lead to product contamination and hinder mechanical operation of the drain valve. Pneumatically activated valves with stems designed to be flush with the inside wall can be used to avoid such problems. Small "kicker" blades (D/T < 0.2) positioned at a low level (C/Z < 0.1) can be very effective and consume little power because of the small diameter. The lower turbine of the newer glass-lined impeller combinations or as a single impeller in smaller vessels discussed in Section 6-2.2.2 can be positioned within a few inches of the bottom and can also be effective in solids discharge.

## 6-4 SCALE-UP

The main objective of scale-up is to design a large scale mixing system that will achieve the same mixing quality as in a laboratory tank. Since the distributions of shear rate and energy dissipation widen as the volume is increased, the mixer design must be adjusted to obtain the same process result. Therefore, it is important to understand the impact of these differences on the process. The scale-up criteria depend strongly on the process type and requirements. While the details are discussed in other chapters (liquid blending, Chapter 9; solids suspension, Chapter 10; gas dispersion, Chapter 11; immiscible liquid mixing, Chapter 12; and chemical reactions, Chapter 13), a few commonly used scale-up methods are discussed here. Other references on scale-up include Dickey et al. (2001) and Tatterson (1994).

Some scale-up methods emphasize geometric similarity. This refers to holding constant the impeller geometry, the impeller dimensional ratios (such as D/T, W/D, C/T), the liquid height/tank diameter ratio, and baffling. There are many situations when complete geometric similarity is not feasible: for example, when the aspect ratio of commercial scale tanks needs to be larger than the laboratory tank.

There are two commonly used scale-up criteria based on holding power per unit volume (P/V) or torque per unit volume ($T_Q/V$) constant on scale-up.

SCALE-UP    377

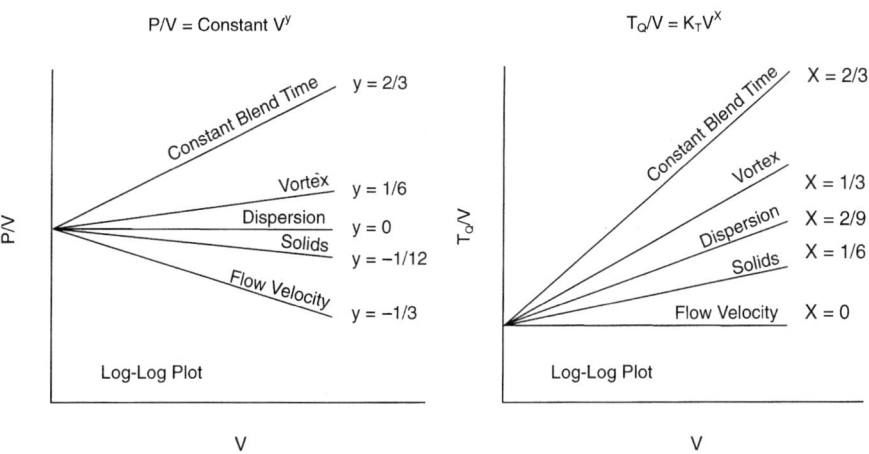

**Figure 6-28**  Scale-up methods for different process types and requirements.

Figure 6-28 shows changes in these parameters as the vessel volume is increased for several processes. The exponents x and y in Figure 6-28 should be determined experimentally or verified even for the processes listed in these plots. Some mixing equipment vendors prefer to use the $T_Q/V$ criterion because it has a direct impact on the overall size and cost of the mixer, including the gearbox.

When choosing a scale-up method, one must consider changes in other flow and power parameters and their impact on the process result. Table 6-8 shows how these important parameters change on scale-up to 10 times the diameter and 1000 times the volume of laboratory mixing tank. Scale-up methods based on constant blend time require the mixer speed in the commercial vessel to be the same as in the laboratory vessel. This, however, results in a very large increase in the motor power. Such a demanding criterion is necessary for very fast to instantaneous reactions where the reaction lifetime may be a few seconds. Commercial reactors for such systems are, therefore, relatively small in size. Using constant P/V, the mixer speed decreases by 78%, but the blend time increases by a factor of 4.6. If constant P/V is used in scaling up a reacting system, the reactors may need to be sized for longer residence time than the laboratory reactor because of the increase in blend time. It should be noted that the Reynolds number increases by a factor of 21.5, and therefore,

**Table 6-8**  Most Important Changes in Mixing Parameters on Scale-up by a Factor of 10 in Diameter and 1000 in Volume for Geometrically Similar Systems

| Quantity | N | Q/V | Tip Speed | Re | $T_Q/V$ | We | P/V | P |
|---|---|---|---|---|---|---|---|---|
| Changes | 1 | 1 | 10 | 100 | 100 | 1000 | 100 | $10^5$ |
| in | 0.1 | 0.1 | 1 | 10 | 1 | 10 | 0.1 | 100 |
| parameters | 0.22 | 0.22 | 2.2 | 21.5 | 4.8 | 48.4 | 1 | 1000 |

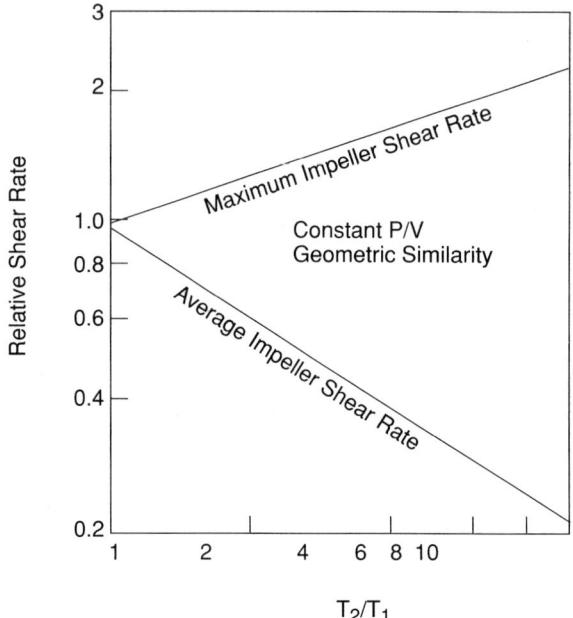

**Figure 6-29** Changes in maximum and average shear rate on scale-up.

the flow regime may significantly change and affect the mixing quality. Also, the Weber (We) number increases by a factor of 48.4, which may decrease the dispersed phase drop size on scale-up of an immiscible liquid system. Constant tip speed and equal $T_Q/V$ are some other scale-up criteria and are used only when flow velocities in the impeller region need to be the same as in the laboratory tank.

It must be recognized that rotational speed and shear rate change significantly on scale-up at constant P/V. Based on eqs. (6-19) and (6-20) average shear rate in the impeller region decreases while the maximum shear rate increases on scale-up. These changes are shown clearly in Figure 6-29.

## 6-5 PERFORMANCE CHARACTERISTICS AND RANGES OF APPLICATION

The performance of stirred vessels is characterized on the basis of the process result for the mixing operation. Detailed mixing mechanisms and design criteria are discussed in respective chapters for these mixing operations. In this section we cover only the general principles and mixer selection for different ranges of fluid properties.

## 6-5.1 Liquid Blending

Mutually soluble liquids are blended to provide a desired degree of uniformity in an acceptable mixing time. An efficient mixer design is important for good product quality at a high production rate. The critical issues that need to be addressed include number of liquids and their volumes, tank configuration, batch mixing times or residence time distribution in a continuous system, and physical properties.

Mixing system designs fall into two main categories based on liquid viscosity. Low to medium viscosity liquids up to 10 000 cP can be blended effectively by internal pumping action from turbine impellers. For higher viscosity liquids, discharge stream velocity from a turbine dissipates rapidly, resulting in poor homogeneity and very long blend times. Such systems require the use of close-clearance impellers, such as helical ribbons.

Mixing of low to medium viscosity liquids occurs at two levels: macromixing and micromixing. Macromixing is established by the mean convective flow divided in different circulation loops between which the material is exchanged. Micromixing occurs because of turbulent diffusion between small cells in the fluid causing intermingling of molecules.

The selection of mixer type for blending depends on the tank size and configuration. For large blending tanks, typically larger than 30 ft in diameter, side-entering propeller (SEP) mixers are recommended. Small tanks can be equipped with top-entering mixers with one or more impellers on the shaft. SEP mixers are useful for homogenizing two or more liquids in terms of temperature and physical properties. Such product tanks generally take hours to achieve the desired homogeneity in the range 80 to 95%. Typical applications include crude oil and chemical products blending processes. The design of SEP mixers involves sizing the propeller and drive on the basis of required pumping rate using eq. (6-3). Desired blend time and number of turnovers dictate the pumping rate. The number of turnovers to achieve 95% homogeneity is a function of liquid viscosity. General guidelines for the required number of turnovers are given in Table 6-9. For low viscosity liquids, eq. (6-3) sometimes can lead to an undersized SEP mixer. It is recommended that these mixers be sized at a minimum P/V of 0.25 hp per kilobarrel (kbbl; 1 barrel = 42 gal = 0.159 m$^3$) required for creating a full tank recirculation.

Top-entering mixers can perform a wide range of mixing duties, from gentle blending for product homogeneity to concentration homogenization in reacting

**Table 6-9** Number of Tank Turnovers for 95% Homogeneity

| Liquid viscosity (cP) | <100 | 100–1000 | 1000–5000 | >5000 |
|---|---|---|---|---|
| Number of turnovers for 95% homogeneity | 3 | 10 | 50 | >100 |

[a] See also Chapter 9.

systems. Mixing times can be minutes to a few seconds, and often a high degree of homogeneity, better than 99.9%, is required. Since pumping liquid causes homogenization, axial flow impellers especially hydrofoil type, are more efficient than radial flow impellers. The impeller sizing and mixer speed requirements are determined from the tank turnover method or blend time estimation. These design methods are discussed in Chapter 9 for stirred vessels.

### 6-5.2 Solids Suspension

Stirred tanks are commonly used for suspending both types of solids, sinking and floating. Suspending solid particles in a turbulent liquid can be considered as balancing of energy supplied by a rotating impeller and energy needed to lift and suspend solids. Industrial applications requiring adequate mixing of solids in liquids include coal slurries, catalyst–polymer systems, solids dissolution, crystallization, paper pulp, ore slurrying for leaching, and so on. Axial flow impellers with high pumping efficiencies are most suitable for solids suspension. These impellers generate a flow pattern which sweeps the tank bottom and suspends the solids.

For suspension of sinking solids, the mixer must be designed for a variety of mixing conditions. At the minimum, good motion of solids on the tank bottom is needed, although this condition is rarely sufficient. For most applications, off-bottom suspension is necessary. For enhanced solid–liquid contacting and mass transfer, complete uniformity of particles may be sought. The design guidelines for these different degrees of mixing are discussed in Chapter 10. An effective mixing system consists of a tank with dished or ellipsoidal bottom head, a down-pumping axial flow impeller, four wall baffles having width equal to T/12, a baffle wall clearance of about 1.5% T, and an impeller bottom clearance equal to T/4. A variety of other mixing objectives may also affect the design of the mixer.

For entrainment of floating solids, a mixer must be designed to provide downward pulling drag force to offset the upward buoyancy force. There are three types of floating solids encountered in the industry: solids lighter than the liquid, difficult-to-wet solids, and solids with low bulk density. Two mechanisms are used for mixing these solids; one uses a central controlled vortex on the surface, and the other uses a recirculation loop to entrain floating solids near the wall at the liquid surface. The vortex is formed using a down-pumping axial flow impeller and narrow baffles (width = T/50) at the wall. An effective circulation loop can be generated with an up-pumping axial flow impeller and standard baffles having width = T/12. In this case, the solids with upward circulation loop move to the wall on a liquid surface where they are incorporated.

A variety of mixing issues need to be addressed when mixing floating solids. The mixing requirements vary from just dispersion to complete slurry homogeneity, solids wetting, shearing, and breakup of agglomerates. In continuous processes it is important to achieve an entrainment rate exceeding the solids feed rate. If the solids are sticky, such as polymers, they can agglomerate and accumulate on the impellers, baffles, and supports. When mixing with narrow baffles, vortex formation can also result in vapor entrainment and mechanical vibrations.

Testing in the lab or pilot plant will help define the appropriate design and scale-up requirements. The reader is referred to Chapter 10 on solid–liquid mixing and Chapter 13 on reacting solids.

### 6-5.3 Immiscible Liquid–Liquid Mixing

Intermixing of mutually insoluble liquids can be achieved in stirred tanks with turbine impellers for the purpose of creating large enhancements in interfacial area. This significantly boosts the rate of mass transfer and reaction. These operations are frequently encountered in industries such as chemical, petroleum, pharmaceutical, cosmetics, food, and mining. Several reacting and nonreacting systems include extraction, alkylation, suspension polymerization, emulsification, and phase transfer catalysis. Energy spent in maximizing the liquid–liquid interfacial area is generally cheaper than the improved process result. However, optimization of mixing energy is necessary because too much energy can create undesirable process results. It can create highly stable emulsions and generate excessive heat, which may have adverse impact on product quality. In addition, it can cause foaming and vapor entrainment from the headspace through the vortex, which can seriously affect liquid–liquid dispersion.

To design an optimum mixer, it is important to define the process needs (e.g., homogenization of phases, fine dispersions for fast reactions, dispersion with narrow drop sizes, minimize diffusional resistance in the continuous phase, induce convection within the drops, generate very fine stable emulsions, etc.). Turbine impellers provide the desired mixing conditions for contacting of immiscible liquids. Even with high viscosity liquids, the shear needed for emulsification must be supplied with turbines and not with close-clearance impellers. Typically, low shear hydrofoils can be used for coarse dispersions. Axial and radial flow impellers are effective for fine emulsions. High-shear impellers are necessary for preparing stable emulsions.

Several correlations have been published in the literature for predicting average drop size and drop size distribution based on mixer design parameters and liquid physical properties. These correlations, discussed in Chapter 12, are based on balancing the rates of drop breakup and coalescence. Dispersed drops break up due to shearing action near the impeller as they are circulated, and then coalesce when they reach low shear zones away from the impeller. The time required to reach an equilibrium drop size distribution depends on system properties and can sometime be longer than the process time.

The criterion of maintaining equal power per unit volume has commonly been used for duplicating dispersion qualities on scale-up and scale-down. However, this criterion would be conservative if only gentle homogeneity of the two phases is desired. Other scale-up criteria may be needed for different processes and should be developed through pilot plant testing.

The mixing conditions in a stirred tank can be modified to cause a phenomenon called *phase inversion*. This involves interchange of dispersed and continuous phases. The phase, which will become dispersed, depends on the position of the impeller, liquid volumes and their physical properties, the feeding conditions, and the dynamic characteristics of the mixing process. There is always a range of volume fractions throughout which either component would remain dispersed, and this is called the *range of ambivalence*. The limits of this range are influenced by the size and shape of the vessel, mixer speed, physical properties of the liquids, and presence of contaminants. Depending on specific process requirements, phase inversion may be desired for product quality or avoided to maintain high mass transfer and reaction.

### 6-5.4 Gas–Liquid Dispersion

Mechanically agitated gas–liquid contactors are widely used in industrial processes for absorption, stripping, oxidation, hydrogenation, chlorination, carbonylation, fermentation, and so on. They are also used for carrying out biochemical processes such as aerobic fermentation, manufacture of protein, and wastewater treatment. The fractional hold-up of gas ($\phi$) in these contactors is a basic measure of their efficiency. The hold-up in conjunction with Sauter mean bubble diameter ($d_{32}$) determines the interfacial area (i.e., $a = 6\phi/d_{32}$) and hence the mass transfer rate. Knowledge of $\phi$ also gives the residence time of each phase.

Disk turbine impellers are the most suitable type for gas–liquid dispersion. The disk is useful in forcing the sparged gas bubbles to move through high shear zones near the impeller blade tip. Recently, many concave-blade impellers (e.g., SRGT, CD-6, in Figure 6-10) have been developed to obtain even higher gas holding capability. This impeller characteristic reflects the maximum rate of gas sparging before the mixer approaches the flooding regime. The flooding regime represents excessive bypassing of gas bubbles along the shaft and large reduction in power consumption. A mixer under flooding conditions loses its capability of providing adequate gas hold-up and liquid pumping, and thus gives poor process result.

Although all turbine impellers can be used to disperse gas, axial flow impellers are inferior to radial flow impellers. In addition, down-pumping axial flow impellers can create an unstable hydrodynamic regime due to opposite and out-of-phase frequencies of liquid pumping and bubble rise. This can cause severe torque fluctuations and mechanical vibrations. Large vessels such as fermenters can use high solidity-ratio hydrofoils to achieve high circulation throughout the vessel, which is especially critical in high viscosity biological operations to prevent local oxygen starvation (see Chapters 11 and 18).

When the gas used in the process is hazardous and/or expensive, it is desirable to recycle it from the vapor space in the stirred tank. This can be achieved by using gas-inducing mixing systems. Typical applications include hydrogenation, chlorination, carbonylation, and phosgenation processes. There are three types of gas-inducing mixing systems: a hollow shaft/impeller, axial flow impellers with narrow baffles, and Praxair AGR system (Figure 6-30).

A hollow shaft/impeller system uses the acceleration of the liquid over the blades to reduce the pressure locally at an orifice and induce the gas flow through

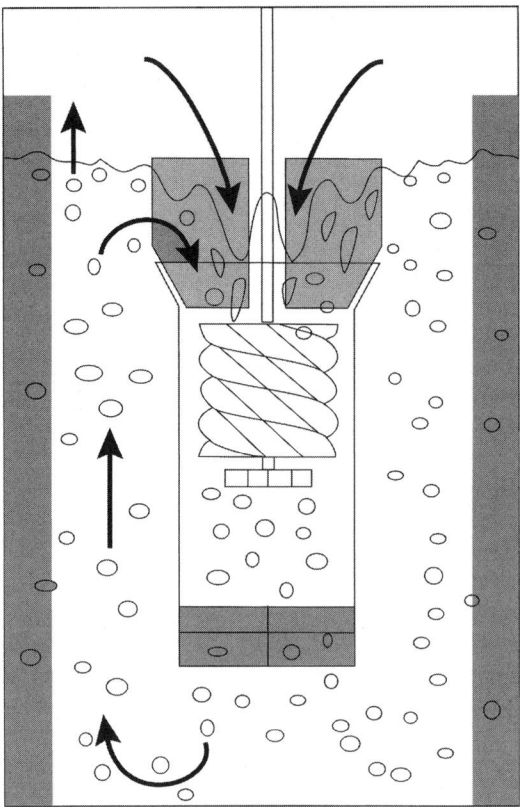

**Figure 6-30** Advanced gas reactor by Praxair.

a hollow shaft. An axial flow impeller with narrow baffles creates a surface vortex through which vapors are entrained and dispersed away. The AGR system uses a combination of high-speed helical screw impeller and draft tube to entrain vapors with a flat-blade impeller rotating just below the draft tube.

## 6-6 LAMINAR MIXING IN MECHANICALLY STIRRED VESSELS

A laminar mixing regime occurs when the impeller Reynolds number drops below 10, due primarily to high fluid viscosity rather than low impeller rotational speed. If turbine impellers are used with highly viscous liquids, flow velocities rapidly decay to low values away from the impeller. This results in formation of a cavern around the impeller. Mixing can be good inside the cavern and poor outside. Flow patterns flatten out and axial flow impellers produce radial flow. These flow changes significantly diminish the blending quality. Turbine impellers are therefore not recommended for use in the laminar regime. For such conditions, close-clearance impellers such as anchors and helical ribbons are commonly used.

**384**   MECHANICALLY STIRRED VESSELS

There are a variety of viscous materials that are mixed in the laminar regime, including polymer solutions, pastes, gums, and semisolids. As the viscosity of the material increases, the material undergoes different fluid motions, including (1) slipping over itself, (2) fracturing, (3) stretching and relaxing back, (4) agglomerating, and (5) clinging to walls or impeller blades. As a result, mixing is poor. If an additional phase is present, mixing in processes such as reactions, adsorption, melting, dissolution, polymerization, dispersion, and contacting can become very difficult (see also Chapter 16).

The mechanism of laminar mixing involves reorientation and redistribution of the viscous material. This is achieved by cutting, dicing, chopping, and so on, and then restacking the sectioned material. The stacked material is then sheared or normally elongated and then redistributed by folding for further reorientation. As the number of reorientations and redistributions increases, the interfacial area increases. This large interfacial area eventually allows diffusion to homogenize the material.

Power for laminar mixing can be derived based on Stokes' drag and written as

$$P \propto \frac{\mu N^2 D^3}{g_c} \qquad (6\text{-}21)$$

This correlation does not include the effects of blade number and blade width as expected from Stokes' drag. Using the definition of Reynolds number in eq. (6-4) and turbulent power number $N_p$ in eq. (6-5), this power expression can be rearranged to

$$N_p Re = B \qquad (6\text{-}22)$$

where B is a constant dependent on the mixer geometry. Typically, B has an average value of 300 and can range between 10 and 40 000 for a variety of impellers. A number of relationships are available for calculating power input under laminar flow conditions in Tatterson (1991).

The power draw can be very high in laminar mixing compared to turbulent mixing. In addition, these mixers are operated at low speeds and the torque on the shaft can be extremely high. The mixer drives designed for high torque require high investment costs. Since most of the power consumed by the mixer is dissipated into heat, removal of heat may be required to avoid possible adverse effects on the process and product quality.

Pumping numbers for the helical ribbon and screw impellers are available in the literature. They range from 0.04 to 0.5 and are highly dependent on the geometry. The anchor impeller only pumps along its radial arms, and pumping numbers are not readily available for this impeller.

Quite often, the fluid viscosity may change during processing and the mixing system design may be based on average process viscosity. This can lead to a very large increase in power inputs unless the mixer speed is dropped as viscosity increases. Use of a variable speed mixer becomes necessary for such systems. Since power going to the mixing is proportional to viscosity, the fluid viscosity can be estimated through power measurements.

A number of mixing system designs are used for mixing of viscous fluids in laminar regime. They include (1) close-clearance impellers, (2) planetary impellers that move throughout the tank, and (3) fixed impellers in tanks that move to expose the material to the impeller. All of these systems generate the necessary three dimensional flows required for mixing.

## 6-6.1 Close-Clearance Impellers

These impellers are designed to physically turnover the fluids because viscous fluids are difficult to pump. These impellers are typically large in size, nearly the same size as the tank diameter, and provide gentle macroscale blending of liquids at low shear. The most common designs are the anchor and the helical ribbon, shown in Figure 6-31.

Anchors are used for liquid viscosities between 5000 and 50 000 cP because at low viscosities there is not enough viscous drag at the wall to provide pumping. Above 50 000 cP, especially with non-Newtonian fluids, the pumping capacity of the anchor declines and the impeller slips in the liquid. When heat transfer through a jacket is desired along with good mixing, the anchor blades are designed with wipers for scraping the wall. A typical wall scraper design is shown in Figure 6-32. Mixing with an anchor can be complemented by adding inside turbines and/or using a draft tube.

Helical ribbon impellers provide top-to-bottom physical movement of the liquid. In addition to one outer helix, they can be designed with an inner helix pumping in the opposite direction. This is particularly needed for direct-action mixing for high viscosity materials. These impellers can also have two outer helixes. The most commonly used pitch for the helixes is 0.5. A higher pitch

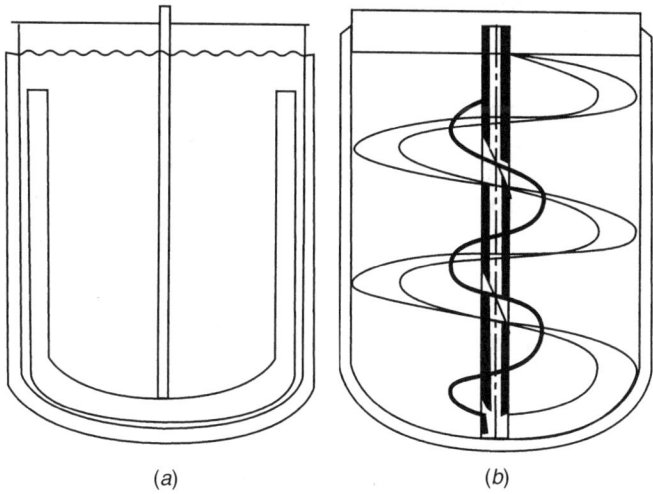

(a)  (b)

**Figure 6-31** Close-clearance impellers: (*a*) anchor; (*b*) helical ribbon. The differences in flow are illustrated on the Visual Mixing CD affixed to the back cover of the book.

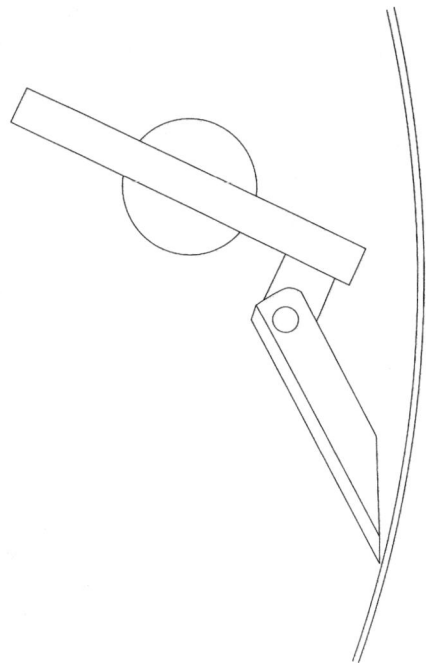

**Figure 6-32** Wiper attachment for anchor impeller blade.

reduces top-to-bottom mixing, while a lower pitch causes excess friction and energy consumption. For enhancing heat transfer through a jacket, wipers and scrapers can be attached to the blades. Helical ribbons are also designed at $D/T = 0.7$ with a draft tube to provide top-to-bottom recirculation.

The power number of anchor agitators depends on the wall clearance, in addition to the impeller Reynolds number. This relationship is shown in Figure 6-33 for two anchor geometries: flat and round blade anchors, at two values of wall clearance.

The effect of wall clearance can also be expressed mathematically for different flow regimes. For laminar flow conditions, $Re < 30$:

$$N_p \propto (e/T)^{-0.5} \tag{6-23}$$

For transitional conditions, $30 < Re < 1000$:

$$N_p \propto (e/T)^{-0.25} \tag{6-24}$$

The height of anchor arm (h) also changes the power number as

$$N_p \propto [0.89(h/D) + 0.11] \tag{6-25}$$

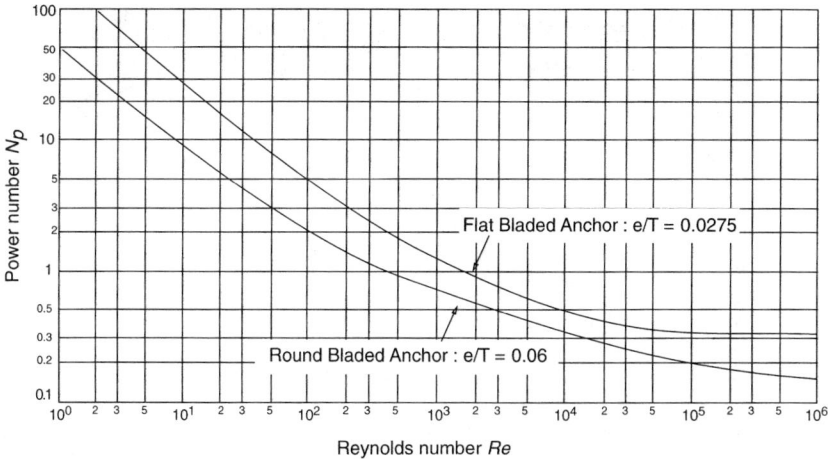

**Figure 6-33** Power number curves for anchor agitators at two wall clearances.

The power number for helical ribbon mixers depends on Reynolds number; blade/wall clearance, e; height, h; pitch of the helix, p (height of one turn around the helix); blade width, w; and number of helical flights, n. The relationship is described as

$$N_p = \frac{150}{Re} \frac{h}{D} \sqrt{\frac{n}{\frac{p}{D}\left(\frac{e}{w}\right)^{0.67}}} \qquad (6\text{-}26)$$

Anchor and helical ribbon impellers can also be used in turbulent applications where high shear is not necessary. This is an advantage for processes in which the fluid viscosity changes significantly and both laminar and turbulent conditions occur at different stages. Although laminar impellers do mix well in turbulent applications, they are not generally recommended, due to their high investment costs.

Direct-action impellers are needed for extremely high viscosity liquids and plastic masses. Such materials include bread dough, battery paste, saltwater taffy, carbon black mixed in rubber, and so on. Suitable mixers for such systems include the co-kneader, extruders, and the Banbury mixer. These are described in Chapter 16.

For specifically difficult applications, special combinations of anchors/helixes and anchor/turbines can be used. For example, a helical ribbon impeller can be supported on an anchor for providing both top-to-bottom material movement and folding action. For dispersing powder into a viscous liquid, a high-speed dispersing disk is used in combination with the helix. Intermeshing cone helical ribbon impellers also exist for self-cleaning action and for viscous plastic masses, which may accumulate on the impeller blades.

The time it takes to achieve the desired mixing quality in laminar mixing depends on the mixer geometry and mixer speed. For most processes it takes

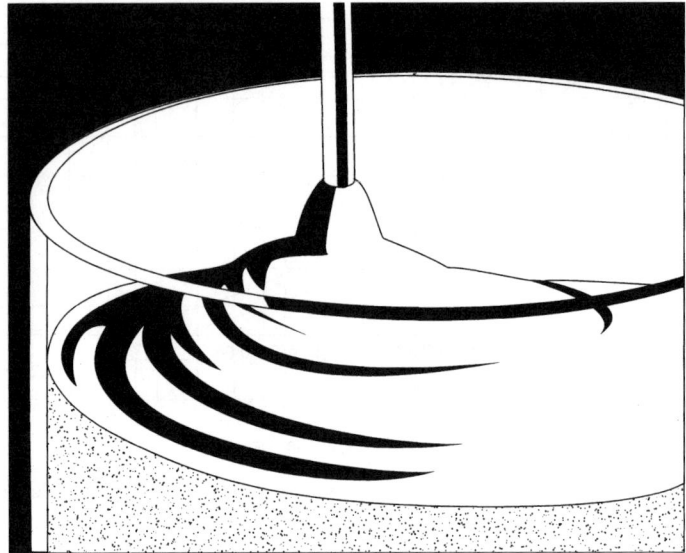

**Figure 6-34** Weissenberg effect with mixing of viscoelastic liquid.

between 15 and 300 revolutions with well-designed geometries. This mixing time and mixer speed should be determined through pilot plant testing. The data for scale-up should be obtained for conditions that ensure the elimination of dead zones. It is also important to study the effect of location and rate of material addition.

For viscoelastic liquids, both turbine and close-clearance impellers are used, depending on the liquid viscosity ranges. Due to viscoelasticity, normal stresses are created in addition to the usual tangential stresses during impeller rotation. These stresses give rise to the Weissenberg effect, which causes the fluid to climb up a rotating shaft (Figure 6-34). As a result, mixing quality deteriorates and blend times become longer, and the impeller power number increases in the laminar regime. This reduced mixer performance is more pronounced in the laboratory scale than in the commercial scale.

## NOMENCLATURE

| | |
|---|---|
| a | area of opening, interfacial area |
| A | tank or column cross-sectional area |
| $A_p$ | projected area |
| B | baffle width, constant, function of geometry |
| C | clearance |
| $C_d$ | drag coefficient |
| $d_{32}$ | Sauter mean diameter |
| D | impeller diameter |

| | |
|---|---|
| e | blade/wall clearance |
| f | interstage back flow rate |
| F | force, drag force |
| $g_c$ | gravitation constant |
| G | distance between feed nozzle and impeller tip |
| h | blade height |
| H | velocity head |
| K | constant |
| n | number of blades, number of flights |
| N | impeller rotational speed |
| $N_b$ | number of baffles |
| $N_p$ | power number |
| $N_Q$ | impeller pumping number |
| p | blade pitch, height of one turn around the helix |
| P | power |
| q | forward flow rate |
| Q | impeller pumping capacity |
| Re | impeller Reynolds number |
| S | impeller separation or spacing |
| T | tank diameter |
| $T_Q$ | torque |
| $v_b$ | backflow velocity |
| $v_{bo}$ | backflow velocity at zero throughput |
| $v_f$ | forward flow velocity, feed pipe velocity |
| $v_t$ | impeller tip velocity |
| V | velocity, volume |
| W,w | blade width |
| Z | liquid height |

*Greek Symbols*

| | |
|---|---|
| $\gamma$ | shear rate |
| $\theta$ | blade angle |
| $\mu$ | viscosity |
| $\nu$ | kinematic viscosity |
| $\rho$ | density |
| $\sigma$ | surface or interfacial tension |
| $\phi$ | hold-up fraction |

## REFERENCES

Dickey, D. S., et al. (2001). *Mixing Equipment (Impeller Type): AIChE Equipment Testing Procedure*, 3rd ed., AIChE, New York.

Dimoplon, W. (1974). How to determine the geometry of pressure vessel heads, *Hydrocarbon Process*, Aug., 71–74.

Jo, M. C., W. R. Penney, and J. B. Fasano (1994). Backmixing into reactor feedpipes caused by turbulence in an agitated vessel, *Industrial Mixing Technology*, G. B. Tatterson, R. V. Calabrese, and W. R. Penney, eds., *AIChE Symp. Ser.: Chemical and Biological Applications*, **90**, #299, p. 41.

Metzner, A. B., and J. S. Taylor (1960). Flow patterns in agitated tanks, *AIChEJ.*, **6**, 109.

Oldshue, J. Y. (1983). *Fluid Mixing Technology*, McGraw-Hill, New York.

Rautzen, R. R., R. R. Corpstein, and D. S. Dickey (1976). How to use scale-up methods for turbine agitators, *Chem. Eng.*, Oct. 25, 119.

Rushton, J. H., E. W. Costich, and H. J. Everett (1950). Power characteristics of mixing impellers, *Chem. Eng. Prog.*, **46**(8), 395–476.

Tatterson, G. B. (1991). *Fluid Mixing and Gas Dispersion in Agitated Tanks*, McGraw-Hill, New York.

Tatterson, G. B. (1994). *Scaleup and Design of Industrial Mixing Processes*, McGraw-Hill, New York.

Xu, B. C., W. R. Penney, and J. B. Fasano (1993). Private communication, Mixing XIV, Santa Barbara, CA, June 20–25.

# CHAPTER 7

# Mixing in Pipelines

ARTHUR W. ETCHELLS III
*The DuPont Company (retired)*

CHRIS F. MEYER
*Sulzer Chemtech USA, Inc.*

## 7-1 INTRODUCTION

Most industrial mixing processes take place in tanks or vessels. They are ubiquitous in the process industries. However, mixing can and often does take place in the pipes connecting these process vessels, and when this is the case, the pipelines themselves actually serve as process vessels. In many cases the pipe, especially when equipped with static mixing internals, is a better place to mix and more economical than a vessel. This is often true when fast blending is required or when long hold-ups associated with vessels are not desirable: for instance, when dealing with molten polymers that degrade with time. There are a number of pipeline devices in mixing related service throughout the chemical and hydrocarbon processing industries:

- Static mixer
- Tee mixer
- Impinging jet mixer
- Spray nozzle
- Empty pipe or duct, elbows, etc.
- In-line mechanical mixer

Pipeline mixing is most useful when:

- Process is continuous versus batch

---

*Handbook of Industrial Mixing: Science and Practice,* Edited by Edward L. Paul,
Victor A. Atiemo-Obeng, and Suzanne M. Kresta
ISBN 0-471-26919-0  Copyright © 2004 John Wiley & Sons, Inc.

- Component feed rates are uniform
- Plug flow is preferred to backmixing
- Short residence time is desirable (long residence times are special consideration involving slow reactions)
- With solids of consistent concentration and usually small particle size
- Gas phase continuous (agitated tanks not applicable)
- High pressure (seal concerns)
- Limited space available, limited access–low maintenance desirable

See Myers et al. (1997) for additional selection criteria.

Pipeline or in-line mixing has evolved to play a well-established role in process engineering. Equipment design and use in both the turbulent and laminar flow regimes are well documented. Application reports by industrial users, academic papers, patents, and the literature of equipment manufacturers present a wealth of technology. It is very clear that the value of pipeline mixing technology in the process industry far exceeds the equipment capital cost. Investment in pipeline equipment is small compared to that of in-tank dynamic agitators and other mechanical mixing devices, but is increasing. This growth results largely from static mixers having proven their capability, not only in bulk blending and mixing, but also in applications involving the dispersion of immiscible fluids, heat transfer, interphase mass transfer, and establishing plug flow in tubular reactors.

Static in-line mixers are continuous radial mixing devices, characterized by an effective degree of plug flow, depending on the specific design and application. Unlike dynamic mix tanks, large recirculation flows are not required to achieve desired results. Since there is little backmixing, residence times can be very short, and consequently, many commercial scale static mixers are compact relative to the scale of fluid flow being processed. Since they have short residence times and little backmixing, proper dosing of the feed components with no fluctuation in time is a prerequisite for good performance. When backmixing is required, static mixers are incorporated into pump-around loops. Like agitated vessels they provide a single stage of contacting for interphase mass transfer. Since static mixers have no moving parts, they are low maintenance and sealing problems are nonexistent.

Static mixers range in size from a few millimeters in diameter to units with equivalent diameters exceeding 3 m and volumes exceeding 100 $m^3$. They exist in both round and nonround cross-section. Small static mixers, for example, are found in laboratory scale processes for mixing and fast reaction, used in meter–mix–dispense systems to combine reactive components at the point of application, and in synthetic fiber production for flow homogenization prior to spinning. Large static mixers result from requirements to handle large flow rates, to hold a process stream for a long residence time, or to provide a large surface area for heat transfer. Mixer applications involving very high volumetric flow include gas mixing in utility scale power/incineration plants and additives blending in municipal water treatment facilities. Mixers required to provide large fluid hold-up typically involve slow reactions, as, for example polymerization, hydrolysis, and the catalysis of soluble enzyme reactions, or are mass transfer limited, as in some absorption and extraction processes. Heat transfer applications involve both heating and cooling of viscous and/or heat-sensitive materials. Examples are the rapid preheating of a polymer solution prior to devolatilization and the cooling of viscous products prior to packaging. Applications, device selection criteria, and design fundamentals are presented in the chapter. It is our intent to provide the reader with both a general overview of pipeline mixing technology and an ability to select and size equipment for a variety of applications.

## 7-2 FLUID DYNAMIC MODES: FLOW REGIMES

The flow regime, laminar or turbulent, sets the mechanisms and the relations used in the selection and detailed design of in-line mixing equipment. An early first step in the understanding of a pipeline mixing application is the identification of the fluid dynamic mode or flow regime in which the process operates. The determinates are the fluid flow rate and physical properties. Flow regime can vary with processing rate. Additionally, it must be recognized that fluid properties can change with time during the mixing process, which in the case of pipeline mixing devices means fluid properties and possibly flow regime can vary along the length of the mixer. An understanding of fluid dynamic flow regime is required to calculate degree of mixing, energy expenditure (pressure drop), heat transfer, and drop size in the case of multiphase processes. The design correlations are often valid in only one flow regime. Different correlations are required as flow changes from laminar to turbulent, and vice versa.

### 7-2.1 Reynolds Experiments in Pipeline Flow

The Reynolds number characterizes turbulence in any given pipeline flow or mixing device. It is instructive to consider first the empty or unpacked pipe and look at the classic experiment by Osborne Reynolds (1883). His demonstration consisted of flowing water through a clear glass tube with capability to vary the water flow rate to achieve a broad range of fluid velocity. At the center of the tube a fine jet of water-soluble dye is introduced through a capillary tube so that a thin filament of dye injected coaxially into the stream of water has a velocity equal to that of the water at the point of introduction. Figure 7-1a shows that at low water velocity the dye filament retains its identity in the water stream, tending to widen very slightly during the downstream passage because of molecular diffusion of the dye into the water. At a slightly higher mean velocity as shown in Figure 7-1b, the dye filament breaks up into finite large eddies. Further downstream the eddies break up further, and the dye that has been introduced tends to become homogeneously dispersed or mixed with the water. At much higher mean velocity (Figure 7-1c) the eddy activity becomes extremely violent, and the region of homogeneous dye color approaches the point of dye entry. From visual observation it is evident that the eddies in normal pipe

**394**  MIXING IN PIPELINES

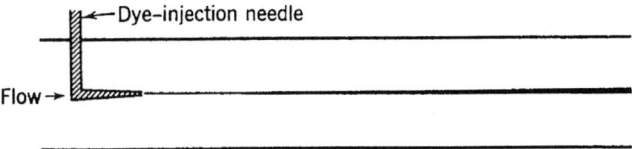

(a) Flow pattern at low mean velocity with dye injection.

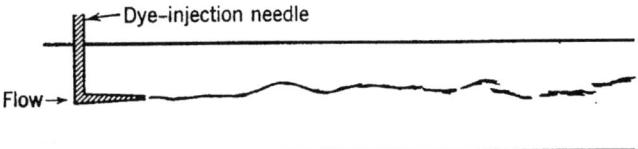

(b) Flow pattern at higher velocity with dye injection.

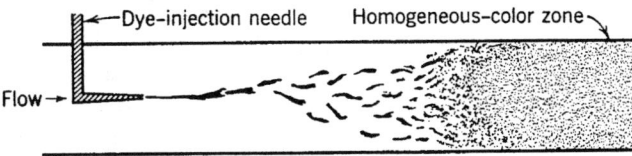

(c) Flow pattern at high velocity with dye injection.

**Figure 7-1** Reynolds experiments.

flow were on the order of one-tenth the pipe diameter and move in completely random patterns. Subsequent experiments showed further that eddy formation was influenced by system factors such as pipe wall finish, vibration, dissolved gases, and other factors. Abnormal or metastable flow aside, it was shown that an upper limit of viscous flow and a lower limit of turbulent flow seemed to exist and that the limits were separated by a transition region.

The conclusions that can be drawn from the Reynolds experiment are:

1. Above a certain mean velocity for a given system, relatively large eddies form that flow cross stream in some random behavior.
2. These eddies are larger and more abundant at the center of the tube.
3. An increase in mean velocity of the fluid widens the turbulent core until the tube is essentially filled with the core of eddy activity.

### 7-2.2 Reynolds Number and Friction Factor

The *Reynolds number*, defined as

$$\text{Re} = \frac{\rho DV}{\mu} \tag{7-1}$$

was proposed to delineate the flow regime in ducts and pipes. Various forms were later proposed and used for systems other than circular pipes and for pipes containing structures such as static mixers. All are the dimensionless ratio of momentum transferred by eddy mechanisms to momentum transferred by molecular transport. Two systems that operate at the same Reynolds number are dynamically similar with respect to forces associated with momentum transfer.

A friction factor, f, was also derived to express the ratio of total momentum transferred to momentum transferred by turbulent mechanisms. Friction factor is a function of Reynolds number, $f = \Phi(Re)$.

The *Fanning friction factor*, the form used in this chapter, is defined as follows:

$$f = \frac{\Delta P}{\rho V^2} \frac{D}{2L} \tag{7-2}$$

Another definition of friction factor is that used in the Darcy equation for pressure drop. The *Darcy friction factor*, f', is defined as

$$f' = \frac{2\Delta P}{\rho V^2} \frac{D}{L} \tag{7-3}$$

and thus

$$f = \frac{f'}{4} \tag{7-4}$$

The European literature often uses the term *Newton number* for the friction factor. The Newton number is defined as

$$Ne = \frac{\Delta P}{\rho V^2} \frac{D}{L} \tag{7-5}$$

and is one-half the Darcy friction factor or twice the Fanning friction factor.

The definition of friction factor should be noted carefully when using friction factor–Reynolds number plots contained in academic and vendor literature. Also, the influence of pipe roughness should be taken into account when designing for turbulent flow. Publications dealing with pipeline mixing are typically based on measurements taken on clean commercial pipe and plate surfaces. If you contemplate using something else, a roughness factor should be considered.

A plot of the Darcy friction factor versus Reynolds number for flow in open circular pipe of various roughnesses from Moody (1944) is shown in Figure 7-2. The plot shows three distinct flow regimes:

- Re < 2100          Laminar
- 2100 < Re < 10 000     Transition (the range from 2100 to 3500 is especially unstable)
- Re > 10 000         Fully turbulent

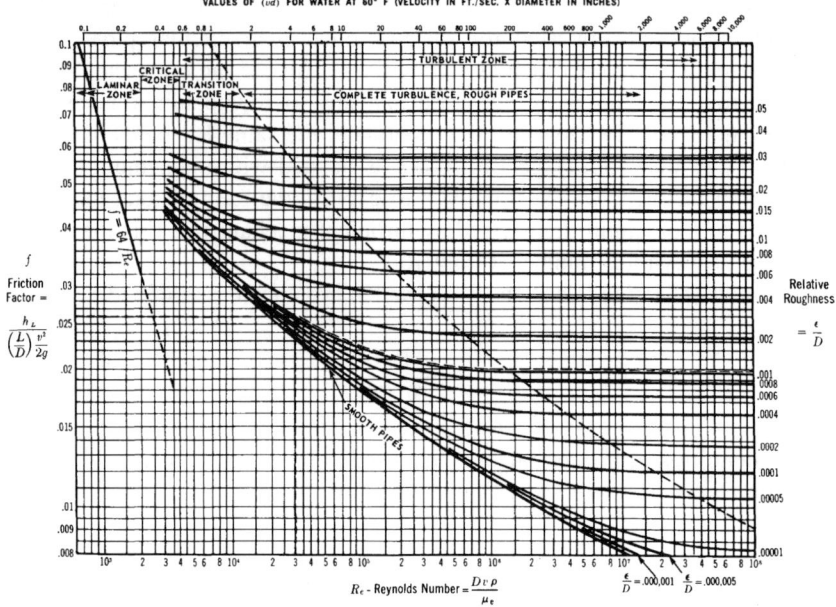

**Figure 7-2** Friction factor (Darcy) versus Reynolds number.

These regimes change when inserts such as static mixers are installed, due to their shape, which changes the hydraulic diameter of the mixing channel. Vendor literature should always be consulted.

Later studies have supplemented the early work of Reynolds, providing a better understanding of the nature of turbulence, and more important for those in the process industries, how to apply the information to fluid processing. With an understanding of the fundamentals, equipment can be engineered to meet specific process requirements efficiently. Pipeline mixing is applied over the entire range of fluid flow regimes. It is an attractive processing option for bulk blending, additive/reactant mixing, multiphase dispersion and contacting, tubular reactors, and heat transfer.

In each of these applications, turbulence strongly influences the mixing process and the ability to achieve desired results. These applications are further discussed later in the chapter. Fundamentals of turbulence and the influence on mixing are developed further elsewhere in the book. For more information, see Chapter 2.

## 7-3 OVERVIEW OF PIPELINE DEVICE OPTIONS BY FLOW REGIME

There are a broad variety of method and equipment options for the continuous processing of fluids in pipelines to achieve objectives in mixing, dispersion, heat transfer, and reaction. The fluid flow regime is a main determinate for equipment selection. Additionally, the available pressure in both the main stream and additive stream are important in the selection criteria.

Fluid flow in the pipe itself may generate adequate turbulence to accomplish simple mixing and dispersion processes. This option is often used successfully in highly turbulent flow where mixing length and time are not important.

Tee mixers, impingement jet mixers, and spray nozzles (for liquid into gas) are also often used, especially when adequate pressure energy is available or can be made available in the additive stream. These sidestream additive injection devices are sometimes used in combination with static mixers to optimize design and performance. The design and application of tee mixers are described in detail in the following section.

Static mixers are the dominant design choice for motionless pipeline mixing. They are essential in the laminar flow regime. They are well established in turbulent processes, both single and multiphase, due to their simplicity, compactness, and energy efficiency. Properly designed static mixers offer predictable performance and operate over a broad range of flow conditions with high reliability. Static mixer design options and basic design principles are described in the following sections.

In-line mechanical mixers are primarily the rotor–stator type and extruder design. They are mentioned because of their importance in pipeline mixing processes but are really beyond the scope of this chapter. They are relatively high energy devices capable of applying high shear stresses to fluids. A typical application is the dispersion of solids into a liquid matrix. High shear stress can prevent the formation of solid agglomerates, and break them up if formed in the initial stage of the process. Extruders are also used where the heat generated by mechanical shear stress is used for melting materials and for providing heat in devolatilization processes. High-speed rotor–stator mixers are very attractive where extremely short contact time is required.

Table 7-1 provides a rough method for initially selecting among the various pipeline equipment options. Subsequent detailed design is used to select the optimum device where several options are possible.

In summary:

- Energy is required for achieving the desired result.
- In laminar flow, static mixers are typically required. The energy for mixing must be made available as pressure that will be dissipated in the process. Line pressure drop will be a design criterion.
- In turbulent flow, if there are no time or length restrictions, the simple pipeline uses the minimum energy and is often the best choice for blending applications. If there are energy limitations, the energy for mixing must be supplied by either the main stream or the side stream.
- Simple mixing or blending can typically be done at low energy expenditure if time is available. With miscible components, diffusion will eventually result in a homogeneous mix. The in-line mixer device merely accelerates the process by bringing the components into more intimate contact.

**398**   MIXING IN PIPELINES

**Table 7-1**   Pipeline Equipment Options

| Flow Regime | Pipe | Tee Mixer | Impingement Jet Mixer | Spray Nozzle | Static Mixer | Inline Mechanical |
|---|---|---|---|---|---|---|
| Laminar regime | | | | | | |
| Mixing/blending | | | | | × | × |
| Dispersion | | | | | × | × |
| Heat transfer | | | | | × | × |
| Reaction | | | | | × | × |
| Plug flow | | | | | × | |
| Turbulent regime | | | | | | |
| Mixing/blending | × | × | × | × | × | |
| Dispersion | × | × | × | × | × | |
| Heat transfer | × | | | × | | |
| Reaction | | | × | × | × | × |

- Dispersion involves the creation of surface area and is more energy demanding. With immiscible components the end result with time is phase separation. The in-line mixer must overcome this to create a dispersion. Significant energy input may be required, depending on the fluid properties.

### 7-3.1   Turbulent Single-Phase Flow

When the flow is highly turbulent single phase, there are many design options, including:

- Empty pipe or pipe works without any special internals
- Valves, nozzles, and orifice plates
- Tee mixers and jet mixers
- Static or motionless mixers

Special features of these design options are described below.

***7-3.1.1   Empty Pipe or Pipe Works without Special Internals.*** Blending or simple mixing is achieved as a result of naturally occurring turbulent eddies in straight pipe and additional turbulence resulting from changes in flow direction, as, for example, in 90° elbows, pipe restrictions and expansions, and so on. Typically, a very long flow path length is required to achieve homogeneity, especially in large pipes and flow conduits. The main advantage is that cost is low and pressure drop increase is minimal or nothing if the pipe works already exists or is being built for fluid transport. Empty pipe is used when residence time and residence time distribution are not design issues.

***7-3.1.2   Valves, Nozzles, and Orifice Plates.*** These devices provide some degree of control of the mixing process versus empty pipe. Mixing length can be reduced, but only at the expense of increased pressure drop. They are sometimes also used in multiphase flow to create dispersions and mists, but it is difficult to predict the drop/bubble size and size distribution. The drop size distribution is typically very broad.

***7-3.1.3 Tee Mixers and Jet Mixers.*** Both of these pipeline devices can be designed to achieve rapid turbulent mixing in a short length of pipe. Since the energy required to achieve mixing is in the additive or side stream, the process side pressure drop required to achieve homogeneity is very low. These designs are most workable when the additive can be supplied at a pressure significantly above the mainline pressure and can be injected into the main stream at a single point. This design option becomes less attractive when dealing with either very small or very large volumetric flows. Small additive flows necessitate small injector size and are subject to plugging. Large flows in large diameter pipe necessitate multiple injection points, increasing complexity and concerns about feed distribution to the various injection points.

***7-3.1.4 Static or Motionless Mixers.*** These mixing devices are readily available and highly engineered for continuous operation. Static mixers achieve predictable mixing performance through a definable pressure drop. A high degree of homogeneity can be achieved in a very short length of pipe. The most attractive designs for single-phase flow components at Reynolds numbers greater than 10 000 are based on vortex generating principles—large scale vortex flow is initiated at a mixer blade or blades with bulk flow mixing immediately downstream. Mixers of about five pipe diameters of total length, including the built-in empty pipe sections, are capable of achieving variation coefficients below 0.05 for moderate amounts of additive. A value of 0.05 (5%) is considered thoroughly mixed in most industrial applications (see Section 7-5.2). Shorter mixing lengths are possible with mixers built of structured plate or bars and well-designed inlet injectors. These designs more aggressively direct the flow, using the increased turbulent energy to achieve mixing. Mixers of corrugated plate design are often used in large pipes and ducts, where length is limited. In many cases involving large flow conduits (both round, square, and rectangular cross-section), the mixer internals represent significantly lower installed cost versus empty conduit based on achieving equivalent mixing. Unlike jet mixers and tee mixers, most of the energy cost is with the main process steam, and higher than the empty pipe option.

## 7-3.2 Turbulent Multiphase Flow

When the flow is highly turbulent multiphase, there are only two practical design options:

- Static or motionless mixers
- Valves, nozzles, and orifice plates

**400**  MIXING IN PIPELINES

*7-3.2.1 Static or Motionless Mixers.* Static mixers are well established in multiphase turbulent flow and meet industrial requirements for absorption, reaction, extraction, and heat transfer/phase change. Designs are engineered to achieve specific results at minimum cost and energy expenditure. The mixers are very compact, making them very attractive for single-stage contacting applications versus countercurrent flow options such as packed, tray, or mechanically driven towers.

Static mixers are recommended for multiphase flow applications with a continuous liquid phase and a dispersed gas or immiscible liquid phase. Turbulent shear is applied efficiently to the additive liquid or gas to create a dispersion or droplets or bubbles. The mean drop size depends on the energy expenditure. Also important is the drop or bubble size distribution. Static mixers are designed specifically for the application to create uniform drop size distributions with the interfacial surface area required for reaction or extraction. Uniform size distribution also facilitates downstream separation of the phases in some type of gravity or inertial separator. In addition to creating interfacial surface area, the static mixer performs bulk homogenization, ensuring that all flow components are distributed uniformly in the cross-section and exposed to similar levels of turbulent energy dissipation in the fluid surrounding the droplet or bubble. The required mixer pressure drop or energy dissipation depends on the amount of interfacial surface area required for mass transfer-limited and reaction rate-limited applications as well as the required residence time when reaction rate is limiting. Surface area generation varies with power input per unit mass, and consequently, there are turndown limitations that must be considered when designing static mixer processes for multiphase applications. Scale-up criteria are well established for the static mixer designs that are used in turbulent multiphase flows. This is a very important consideration since many processes are lab scale or pilot scale tested prior to commercialization.

Static mixers in multiphase applications where the gas is continuous are typically highly structured designs, providing large surface area per unit volume. Surface area is needed for absorption of gas phase components, stripping of components from the liquid, condensation, or vaporization. The properly selected mixer is a compact, highly efficient phase contactor. Turbulent flow energy is used to break up the liquid feed, achieving some equilibrium droplet size and corresponding total surface area. Flow turbulence is maintained uniformly over the pipe cross-section in individual interconnected flow channels. Liquid that wets the mixer surfaces is continuously stripped off and redispersed in the gas stream. Flow stability is maintained over a greater range of gas–liquid flows versus what would occur in an empty pipe, an important factor considering that liquid- and gas phase mass and volume flow rates could change significantly during the process as a result of phase change of all or part of the streams. As with all multiphase processes the initial drop (or bubble in the continuous liquid analog) size is an important design factor. Spray nozzles (with or without an atomizing fluid) are often used to create the initial drop size distribution utilizing additive stream energy and designing for relatively low mixer pressure drop.

*7-3.2.2 Valves, Nozzles, and Orifice Plates.* These devices are used in continuous liquid-phase processes, but less and less so as static mixing technology has evolved to dominate multiphase continuous liquid processes. In continuous liquid applications these devices are less efficient and often require a higher energy expenditure than that required by a properly designed static mixer. Also, as a result of highly concentrated energy dissipation, the drop size distribution is less controlled and typically much broader than that which can be achieved in a static mixer. Broad drop size distributions have a negative impact on equipment size and performance of downstream phase separation equipment as described above for static mixing. Valve designs do have the benefit of providing a degree of adjustment to tune the process and deal with turndown, but have proved difficult with respect to continuously achieving design performance. They are rarely used in continuous gas processes, except for two fluid nozzles, where the gas component can be used to atomize the liquid stream. Design uncertainty and potential for flow instability are typical concerns.

## 7-3.3 Laminar Flow

When the flow is laminar, either single or multiphase, there is only one design class option: static or motionless mixers. Other pipeline mixing devices described for turbulent flow are not usable for even the simplest mixing applications in the laminar regime. All rely on turbulence and cannot function at low Reynolds numbers. The only alternative technology is in-line dynamic mixers, which include extruders, rotor–stator mixers, and a variety of rotating screw devices. None of these has the benefits of simplicity and the little or no maintenance characteristic of static mixers. In-line mechanical mixers are discussed briefly later in the chapter.

*7-3.3.1 Static or Motionless Mixers.* Static mixers are proven in a broad range of laminar flow processes involving both Newtonian and shear thinning fluids. Some processes are more complicated than others. Very often, commercial installations follow laboratory or pilot scale evaluations, and success is dependent on proper scale-up. Scale-up methodology is well established for the predominant static mixer designs used in the laminar flow regime. In addition to mixing applications there is value in the use of static mixer packings to enhance laminar flow heat transfer and for creating plug flow in laminar tubular reactors.

*7-3.3.2 Blending of Fluids with Similar Viscosity.* There exist a broad range of static mixer design options for blending or distributive mixing. Selected static or motionless mixers are designed specifically for operation in laminar flow. At a minimum they operate by cutting and dividing the feed into substreams, distributing the substreams across the pipe diameter and recombining them in a continuous manner creating fluid layers. The layers are stretched, reducing layer thickness, and recombined as they are folded into each other. The number of layers is increased exponentially as fluids flow through the mixer. Layer thickness is decreased until thickness is so small that differences in composition or temperature are indistinguishable on a macroscopic level. Final micromixing occurs by diffusion or conduction made possible in the laminar region at very small distances between layers. The mixer is designed to achieve the desired degree of homogeneity. The time required for diffusion may influence the mixer design, but in most cases it is not an issue since the time normally required to create the striations is quite significant, at least relative to what can be accomplished in short time when operating in the turbulent regime. This process of division and recombination of the process stream as fluids flow through the static mixer represents what is called *simple* or *distributive mixing*. Many static mixer designs can achieve the desired result, but there are significant differences in mixer residence time, length, and pressure drop. The optimum static mixer from the perspective of equipment design would be the one most compact and operating at the lowest pressure drop. The optimum mixer design would be the one that is the most efficient generator of fluid surface, or a high division rate device.

*7-3.3.3 Blending of Fluids with High Viscosity Ratio*. When laminar distributive mixing is complicated by viscosity differences, static mixer design options are limited. Elongational flow static mixers are required. The description of the simple distributive mixing process above applies to cases where the component fluids are not only miscible, but in addition can be infinitely divided into each other. In other words, resistance at the interface between flow streams is minimal or nonexistent. There is, however, a phenomenon in viscous laminar flow called the *miscible interface* which serves to block the mixing process. It comes into play, for example, when processing polymeric material, where there are significant differences in molecular weights and/or crystallinity, and most important, when there is a significant viscosity difference between the two components. Mixer design options are now more limited than for simple blending applications.

When there is a resistance at the interface, the static mixer must be one that operates at uniform shear stress. Additionally, when there is some degree of immiscibility and a significant difference in viscosity, elongational shear has been demonstrated to be more effective then rotational or simple shear (Grace, 1982). Miscible materials actually behave like they are immiscible when there is a large difference in viscosity, and this must be taken into account when designing the mixer. Channeling of the low viscosity component at the wall must also be prevented for a successful mixer operation. The structured X or cross-bar design has many of the characteristics necessary for these difficult applications and at this time is the only proven performer in many industrial processes.

*7-3.3.4 Liquid–Liquid and Gas–Liquid Dispersion*. When the flow is laminar and multiphase, elongational flow static mixers are required to mix and disperse additives into viscous bulk streams. The dispersion of a low viscosity immiscible additive into a viscous mainstream is a common and very difficult static mixing application in laminar flow. As mentioned above, the uniform application of shear is required to prevent channeling of the low viscous additive. Mixing efficiency is strongly dependent on elongational flows within the mixer structure. Mixing of higher viscosity material into a lower viscosity but laminar stream is not as common but equally difficult, requiring controlled elongational flows. Shear stressing of the additive gas or liquid results in it being extended to the point where it becomes unstable and breaks to smaller size. This process continues until the droplet or bubble is reduced to a size that is stable under mixer flow conditions. In addition to creating this dispersion, the mixer must also distribute the additive phase uniformly over the pipe cross-section. The structured X or cross-bar design is at this point in time, the only significant commercially available design for this very difficult application.

*7-3.3.5 Heat Transfer Enhancement.* When the application is laminar flow with heat transfer, static mixers can enhance the heat transfer and provide a more gentle product treatment than what can be achieved in empty or unpacked tubes. Heat transfer enhancement is achieved by the use of static mixer heat transfer packings (mixing elements) in jacketed pipes and the tubes of multitube heat exchangers. The internals mix the fluid during the heat transfer process, continuously exchanging material at the wall with material at the core. Disruption of the laminar boundary layer at the wall results in increased heat transfer capability along the length of the pipe. In laminar flow this is usually the controlling resistance. Thermal start conditions are continually reestablished as material flows through the heat exchanger. In addition to improving the process side heat transfer coefficient, the static mixer packing provides a more gentle thermal processing. Baking on the wall is reduced in heating, and skinning or precipitation minimized in cooling. The fluid flowing through a static mixer packed heat exchanger pipe has a more uniform thermal profile, uniform shear history, and narrower residence time distribution compared to fluid flowing through an empty (unfilled) pipe.

In addition to designs with static elements serving as inserts in pipe and tubes, there exists commercially available designs where the static mixing element itself is made out of hollow tube through which heat transfer fluid flows. The entire mixing element is an active heat transfer surface with process fluid flowing externally. Large diameter monotube designs are possible, and very attractive in many processes, especially when the process fluid viscosity increases dramatically, as in polymerization reactions and some cooling applications.

It is important to note that in most cases the value static mixing brings to heat transfer does not translate to turbulent flow, where cost and pressure drop cannot be justified by the heat transfer enhancement. Turbulent flow heat transfer processes are best handled by empty tubes and tubes with spiral wrapped cores or tubes containing twisted tapes. These other pipeline devices, though not discussed here, are nevertheless important in industry. The reader is encouraged to seek other literature if interested (see Burmeister, 1983a,b).

*7-3.3.6 Tubular Plug Flow Reaction.* When the application is a laminar flow tubular reactor, static mixing internals can provide great benefits in terms of performance, control, and reliability. Static mixer reactor packings are specially designed to create plug flow in tubular reactors. The parabolic velocity profile of laminar flow in empty pipe is flattened as material flows through the static mixer. Reactors can be designed to achieve near-plug-flow conditions even in pipes with large diameters and short lengths. Static mixers are applied in laminar flow reactors processing both low and high viscosity materials. Although the designs are significantly different, depending on viscosity, plug flow is readily achieved in both cases. Density-driven internal recirculations are eliminated by proper mixer selection and detailed design. When a heat load exists simultaneously with a reaction, a static mixer with temperature-controlled surfaces may be required. Lab scale and/or pilot testing is most often required to establish exact conditions.

## 7-4 APPLICATIONS

Very many commercial scale applications are efficiently handled with in-line static mixing equipment. Other in-line devices are also used, but the range of applications is less broad. For the purpose of the applications discussed in this section, we are referring primarily to static mixing applications, but recognizing that in some specific cases, other in-line devices may also be employed.

### 7-4.1 Process Results

Processes carried out with in-line mixing equipment are very similar to those with stirred tanks. They include the following three main classes:

- Blending of miscible fluids or distribution
  - Same physical properties
  - Different physical properties
  - For homogeneous chemical reactions
- Area generation or dispersion
  - Liquid in a liquid
  - Gas in a liquid
  - Solid in a liquid
- Heat transfer

Additionally, there are applications where there is no analogy with stirred tanks, for example:

- Blending of gases
- Contacting of gases with liquids (scrubbing, vaporization, desuperheating)
- Creating plug flow in a pipe

**Table 7-2** Applications of Pipeline Devices

| Application | Laminar Flow | Turbulent Flow | Measurement Criteria |
|---|---|---|---|
| Blending of components | × | × | Variation coefficient |
| Temperature or thermal homogenization | × | × | Variation coefficient |
| Liquid–liquid dispersion extraction/reaction | × | × | Drop size, drop size distribution, mass transfer |
| Gas–liquid dispersion with gas continuous | – | × | Drop size, total surface area |
| stripping/vaporization absorption | – | × | Mass transfer, reaction |
| Gas–liquid dispersion with liquid continuous |   |   |   |
| Reaction | × | × | Bubble size, bubble size distribution |
| Absorption | × | × | Mass transfer, reaction |
| Heat transfer enhancement | × | – | Heat transfer coefficient, heat duty |
| Fast reaction | – | × | Mix time, variation coefficient |
| Slow reaction (plug flow) | × | – | Plug flow characteristics |
| Gas–solid fluid bed | – | × | Contacting efficiency |

### 7-4.2 Pipeline Mixing Applications

Pipeline devices are used in a broad spectrum of applications (Table 7-2). Performance criteria are well established.

### 7-4.3 Applications Engineering

Static mixers differ widely in their construction and performance characteristics. Technical criteria should be used to determine the best design for each specific application. The process requirements should dictate the static mixer design or design options. There are three fundamental steps in the thought process to select the correct mixing design for a given application:

1. Determine if pipeline mixing is applicable.
2. If applicable, determine what type of pipeline equipment is best for the application.
3. Complete a detailed design after selection of equipment type.

A nine-step design procedure is recommended:

1. Clearly identify the application (blending, dispersion, heat transfer, reaction).
2. Fully define process flow conditions (stream flow rates, densities, viscosity, etc.).

3. Identify constraints (e.g., space limitation, available pressure drop).
4. Specify desired process results and measurement criteria.
5. Pick candidate designs (several will probably achieve the desired process result).
6. Identify secondary requirements.
7. Evaluate candidates on secondary requirements (includes cost, length, etc.).
8. Select an optimum design.
9. Design the mixer.

Note that the mixer is chosen on secondary requirements. All candidate mixers must achieve the process result. Thus "better" or optimum design will be based on secondary considerations. Typical secondary considerations are cost, length, pressure drop, delivery, and past experience.

Design of the static mixer is possible only after a thorough specification of the application. Numerous product brochures and technical publications are available to assist the designer. Additionally, the major static mixer manufacturers offer application and design assistance as part of their proposal process. Secondary, but also important factors that need to be considered are startup, turndown, upset conditions, and mechanical requirements. Most static mixer designs are available in a broad selection of materials of construction, both metals and plastics, to meet plant/process requirements.

Many applications have become routine, and equipment can be sized following well-established design procedures. However, as the technology evolves, so does interest in more difficult applications, many of which involve large commercial scale production rates, complicated chemistry and physical interaction between components, unique products not already characterized at flow conditions, and continuous tubular systems involving simultaneous mixing, heat transfer, and reaction. These should be laboratory and/or pilot tested prior to attempting full scale design. Design fundamentals and the ability to scale-up a design then become very important. Development of new static mixing processes should start with analysis of the full scale system. After final requirements are identified, laboratory and pilot scale mixers can be selected based on their ability to perform and their suitability for scale-up.

It is interesting to note that it is sometimes advantageous to change the mixing design along its length to optimize results at any given point in the mixing process. For example:

- Follow a high intensity jet mixer, utilizing side-stream energy and operating at high local turbulence, with a low pressure drop static mixer to distribute or bulk mix the mixer products throughout the pipe cross-section.
- Provide a length of empty pipe downstream of a vortex-generating device to achieve bulk mixing.
- Change pipe diameter to maintain turbulence as volumetric rate changes, as, for example, during gas absorption (rate decreasing) or conversely during

vaporization of a liquid or gas being generated as a result of a reaction (rate increasing).
- Arrange mixers of different design (diameter) in parallel to handle extreme process changes and turndown.
- Change mixer design or diameter to account for a change in fluid viscosity, as, for example, during cooling of a viscous product, or reactions building molecular weight, both of which lead to increasing viscosity and reduction in turbulence. This is analogous to polymerization reactions conducted in a series of agitated tanks, each designed with an agitator best suited for process viscosity, mixing, and heat transfer requirements.

### 7-4.4 Sample of Industrial Applications

*Chemicals*

- Mixing miscible/dispersing immiscible reactants
- Dissolving gases (e.g., chlorination processes)
- Providing plug flow and controlled-reaction conditions in tubular reactors with low or high viscous fluids
- Dispersing liquids in extraction and washing processes
- Mixing gases in front of catalytic reactors (e.g., the production of styrene, nitric acid, maleic anhydride)
- Vaporizing liquids in front of oxidation reactors (e.g., xylene in phthalic anhydride plants)
- Co-current scrubbing acid process gas components
- Homogenizing process and product streams for representative sampling
- Controlled heating and cooling of slurries in catalyst production
- Neutralizing or pH adjustment/control of process streams with caustic or acid

*Cosmetics and Detergents*

- Saponifying greases with caustic soda
- Sulfonating fatty alcohols with oleum
- Mixing components of toothpaste, lotions, shampoo, soaps, or detergents
- Diluting surfactants

*Energy*

- Mixing blast furnace and coke oven gas
- Reheating flue gas in desulfurization plants
- Blending emulsifier for water into fuel, dispersing/emulsifying water, and fuel
- Blending fuel gases with air before combustion

## Environmental Protection

- Scrubbing $H_2S$ from exhaust gas with caustic
- Oxidizing sulfite-laden scrubber blow over with air
- Vaporizing ammonia solution and mixing it with exhaust gas before the catalyst bed in a SCR DeNox installation

## Foods

- Dissolving $CO_2$ into beer, fruit juice, or wine
- Heating coffee extract before spray drying of flash evaporation
- Heating and cooling chocolate mixtures
- Heating and cooling starch slurries under plug flow conditions
- Mixing enzymes and chemicals into starch suspensions
- Diluting concentrated juices and admixing flavorings
- Mixing fruits and flavors into yogurt and ice cream
- Diluting molasses and sugars
- Mixing color and flavor into pet food

## Natural Gas

- Scrubbing $H_2S$ from natural gas with caustic or specialty chemicals
- Dehydrating natural gas with glycols
- Adjusting the Btu content of natural gas with propane
- Cooling natural gas in the compressor loop of LNG terminals by injection and vaporizing LNG

## Polymers, Plastics, and Textile Fibers

- Mixing additives, catalyst, and inhibitors into polymer melts and solutions
- Providing plug flow in polymerization reactors (e.g., polystyrene, PA6, silicone, and many others)
- Dispersing a low viscosity stripping agent into polymer solutions prior to devolatilization (e.g., water into polystyrene)
- Rapid uniform heating of polymers prior to flash devolatilization
- Mixing additives (e.g., mineral oil, pigments, ultraviolet stabilizer, antioxidants into polymer prior to pelletization)
- Homogenization of temperature and colorants in polymer melts in extruder and injection molding machines
- Cooling polymer melts before processing and removing heat of polymerization

## Petrochemicals

- Chlorinating hydrocarbons (e.g., ethylene to EDC)

- Mixing ethyl benzene with stream before the first dehydrogenation reactor in styrene plants
- Scrubbing acid components with caustic from exhaust gas during catalyst regeneration

*Pulp and Paper*
- Mixing bleaching chemicals with pulp stock
- Blending stocks and diluting stock with water for consistency control
- Mixing retention agents in front of the head box
- Admixing caustic or acid for pH control
- Admixing chlorine and chlorine dioxide to stocks for bleaching
- Steam injection for heating

*Refining*
- Homogenizing crude for representative BS&W measurement, custody transfer
- Desalting crude with water
- Dispersing sulfuric acid or HF in hydrocarbons in alkylation processes
- Neutralizing/washing hydrocarbon streams with caustic and water
- Sweetening kerosene and gasoline with caustic
- Blending different hydrocarbon streams and additives into gasoline
- Scrubbing acidic components with water from hydrogen gas
- Establishing vapor–liquid equilibrium in hydrocarbon streams
- Adjusting the viscosity of heavy oil with gas oil and other additives
- Contacting steam with catalyst in the FCC catalyst stripper

*Water and Wastewater*
- Aerating water (e.g., for improvement of oxygen level or oxidation of iron and/or manganese)
- Neutralization, adjusting pH/conditioning of water with acid, caustic, lime solution, or by dissolving $CO_2$
- Diluting flocculants (e.g., polyacrylamide)
- Mixing flocculants such as $FeCl_3$ or $Al_2(SO_4)_3$ into water, wastewater, or sludge
- Stripping excess $CO_2$ with air for deacidification
- Dissolving ozone

## 7-5 BLENDING AND RADIAL MIXING IN PIPELINE FLOW

Blending in a pipeline can be radial or axial. The best designs create a high degree of plug flow, achieving radial mixing while minimizing backmixing. This is an important characteristic of in-line devices, especially static mixers. Backmixing, however, often occurs at the point of additive injection, something that must be considered in the overall system design. The amount of mixing that can be achieved in any given pipeline mixer and energy expenditure is strongly dependent on the flow regime, laminar or turbulent. If a high degree of backmixing is required, as, for example, to even out time fluctuations in the feed, an agitated tank may be a better design option.

### 7-5.1 Definition of Desired Process Result

The process result of heat transfer is a heat transfer coefficient. For dispersion it is a drop or particle size and size distribution. For blending in tanks it is blend time to achieve a certain degree of mixing. The equivalent for mixing in pipeline flow is not as clear. Alloca and Streiff (1980) proposed using a radial coefficient of variation, and this concept is now widely accepted. Since it is unique in the process industries to pipeline flow, it merits some extended discussion.

***7-5.1.1 Coefficient of Variation as a Measure of Homogeneity.*** Consider the cross-section of a pipe to which a small amount of material has been added. Initially, it is completely segregated into two areas, one occupied by each fluid. As mixing occurs, the areas intermingle. Figure 7-3, showing the mixing of two fluids in laminar flow in a motionless mixer (type SMX), presents the process graphically. How can the differences in this set of pictures be described? If we now superimpose a grid of squares over the cross-section, we can describe the process by estimating how much of each color is in each individual square. The overall average concentration will stay the same, but the individual boxes will start very segregated and approach the average with continued mixing length. Alternatively, we could sample at many points in the cross-section at axial positions downstream of the mixer and use these point values as a measure of segregation. Of course, the sampler should not interfere with or promote

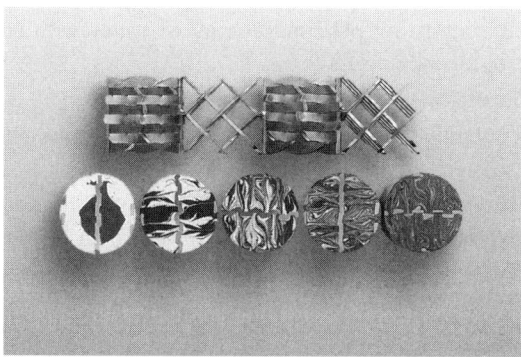

**Figure 7-3** Homogeneity with laminar mixing of two fluids (shown with an SMX mixer). (Courtesy of Koch-Glitsch, LP.)

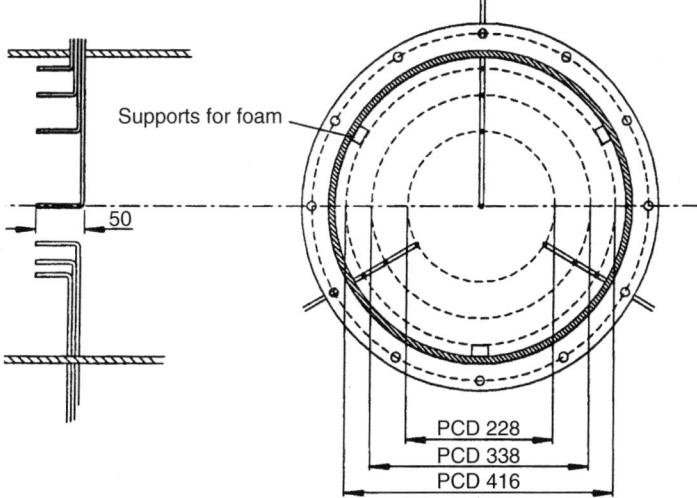

**Figure 7-4** Sampling array arrangement: 10 points. (Courtesy of BHRG.)

the mixing process. This can be done with temperature probes or by extracting individual samples at points located in the cross-section. Other methods can be worked out. The results are a set of numbers at defined spacing which with time approach the average. Given this set of numbers, statistics can be used to produce a measure of uniformity or mixedness. Table 7-3 gives some typical experimental data obtained by BHRG for a mixer study for DuPont. A mixture of air and air with $CO_2$ was used for the two streams. Ten points were sampled using a gas chromatograph. The sampling array is shown in Figure 7-4. Measurements were made with time and with position. The time variation was due to fluctuations in the feed system. This sets the minimum degree of uniformity that can be obtained. From the 10 discrete readings, an average and standard deviation are calculated. The average should not change, of course. This is just an internal check of the data quality. The standard deviation is normalized by dividing it by the average, giving the function called the *coefficient of variation* (CoV = standard deviation of concentration measurements/mean concentration). This is a useful concept, as coefficient of variation (often reported as a percent) is easy for laypeople to comprehend. This is often also called the *intensity of mixing* or *degree of segregation*. From statistics for a normal or Gaussian distribution of data, two-thirds of the data will lie within ±1 standard deviation, 95% within ±2 standard deviations and >99% within ±3 standard deviations. Thus, in practice, one can say that ±2 or 2.5 CoV is the spread of all the data. One can talk about everything within 10, 5, or 1% of the average. Often, the process will indicate what is an acceptable coefficient of variation. For example, in a typical industrial mixing process, an additive might be considered well mixed at 5% CoV, while in a more critical application such as the addition of color to an extruded sheet, product might require 0.5% CoV in order to escape the strong discriminating

**Table 7-3** Experimental Data from Radial Mixing Study

ANALYSIS OF MIXING DATA
0.315 m³/s air     0.069 m³/sec side stream with $CO_2$
DATA IS IN PPM OF CO2

| TIME STEP \ POSITION | 1 | 2 | 3 | 4 | 5 | 6 | 7 | 8 | 9 | 10 | ave. | position s.d. | CoV |
|---|---|---|---|---|---|---|---|---|---|---|---|---|---|
| 1 | 3000 | 2590 | 2790 | 2560 | 3380 | 3470 | 3190 | 2490 | 2590 | 2660 | 2872 | 363.1896 | 0.126459 |
| 2 | 2920 | 2730 | 2720 | 2490 | 3280 | 3300 | 2480 | 2770 | 2710 | 2710 | 2811 | 282.9389 | 0.100654 |
| 3 | 2770 | 2950 | 2570 | 2840 | 3280 | 3260 | 3030 | 2650 | 2590 | 2720 | 2866 | 258.6804 | 0.090258 |
| 4 | 2820 | 2830 | 2770 | 2620 | 3480 | 3110 | 3100 | 2790 | 2480 | 2680 | 2868 | 289.3978 | 0.100906 |
| 5 | 2640 | 2390 | 2780 | 2830 | 3230 | 3260 | 3150 | 2840 | 2600 | 2510 | 2823 | 304.4868 | 0.107859 |
| ave | 2830 | 2698 | 2726 | 2668 | 3330 | 3280 | 2990 | 2708 | 2594 | 2656 | 2848 | | |
| s.d. time | 138.5641 | 217.0714 | 91.26883 | 159.2796 | 100 | 128.6468 | 291.2902 | 140.4279 | 81.42481 | 85.02941 | 28.69669 | | |
| CoV | 0.048963 | 0.080456 | 0.033481 | 0.0597 | 0.03003 | 0.039222 | 0.097421 | 0.051857 | 0.03139 | 0.032014 | 0.010076 | | |

each time step is about 15 seconds

overall average is    2848 compared to a theoretical of 2603 ppm

average time CoV    0.050453

average position CoV    0.105227

CoV initial    2.09

ability of the human eye. Note that the final CoV is usually independent of the amount to be mixed. The mixer length required to achieve a given CoV depends on the amount to be mixed because of the initial state of unmixedness. It is of interest to look at the original state of mixing in similar terms. The coefficient of variation for an unmixed sample (CoV$_0$) is given by statistical theory to be based on the unmixed volume fraction, $C_v$:

$$\text{CoV}_0 = \left(\frac{1 - C_v}{C_v}\right)^{0.5} \tag{7-6}$$

Thus, the initial degree of "unmixedness," CoV$_0$, depends on how much needs to be mixed. The following table is illustrative.

| Initial Additive (Volume Fraction), $C_v$ | CoV$_0$ |
|---|---|
| 0.5 | 1.0 |
| 0.1 | 3 |
| 0.01 | 10 |
| 0.001 | 33 |

Thus, the smaller the amount added, the greater the initial CoV or state of unmixedness. We define a mixing task as reducing from an initial coefficient of variation (CoV$_0$) to a final chosen CoV which is independent of the initial coefficient of variation (CoV$_0$). It is believed that the performance of a motionless mixer is set by this mixing task and is independent of the magnitude of the initial or final values. Some recent unpublished data challenge this, but until better data are obtained, we will presume that the mixer performs as a reducer of variance. This is the assumption of most vendors.

This, of course, neglects molecular diffusion effects, which tend to shorten the process. The motionless mixer can then be thought of as a transfer function which reduces the CoV from an initial value to a lower final value. Thus, CoV reduction as a function of length is a measure of quality of mixing of a motionless mixer. The smaller the volumetric flow of additive (mean concentration, or % addition), the longer the mixer. Variation coefficient versus mixer length for two mixer designs (SMX and SMXL) operating in laminar flow at 0.1, 1, 10, and 50% additive rates is shown in Figure 7-5. This is also shown in turbulent flow for a KVM vortex mixer (L/D = 5) at 0.1, 1, and 10% additive rates in Figure 7-6 and for the SMV mixer at 0.1, 1, 10, and 50% additive rates in Figure 7-7. Note by comparison with Figure 7-5 that as one would expect, in turbulent flow homogeneity is achieved much more quickly than in laminar flow and that there is additional mixing achieved after the mixer due to the turbulence in the tailpipe section.

**414** MIXING IN PIPELINES

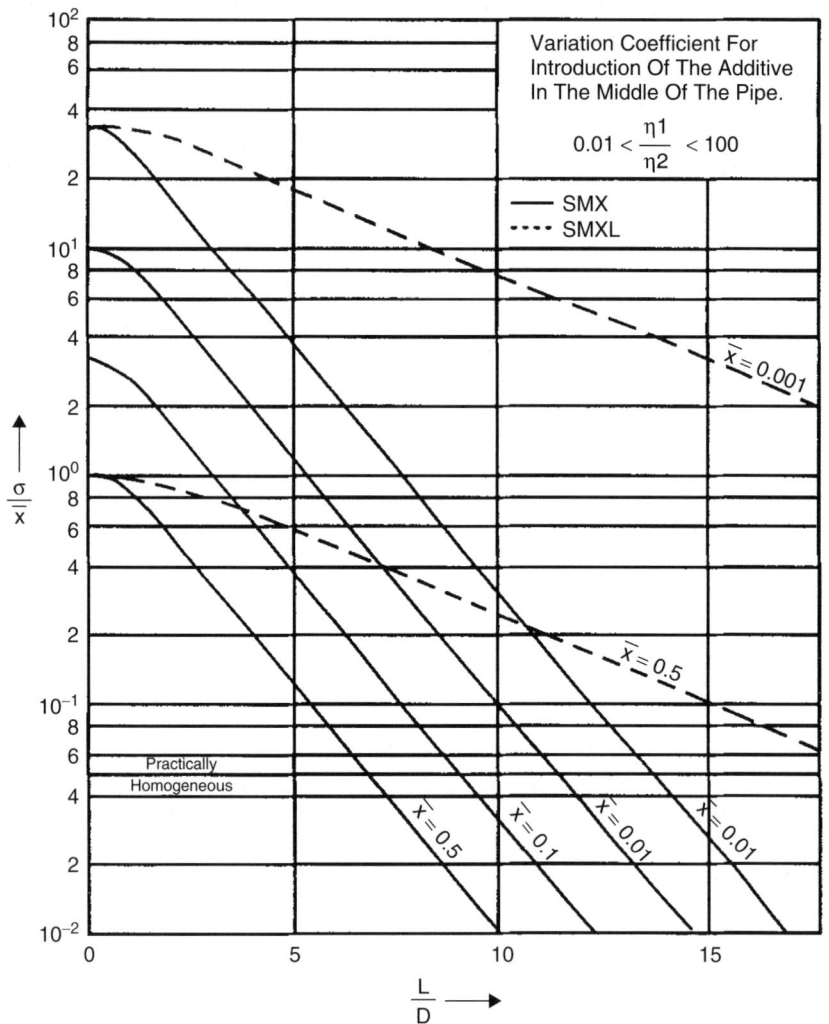

**Figure 7-5** Homogeneity expressed as variation coefficient versus mixer length for SMX and SMXL static mixers operating in laminar flow. (From Schneider, 1981.)

**7-5.1.2 *Other Characterization Measures*.** Many other mixing measures have been developed. Several of them have been taken from work on mixing of particulate solids. However, in miscible liquid blending and gas mixing there will always be diffusion at the lowest and final scales to finish off the process. There is also no mechanism for de-mixing as can occur with solids due to particle mass differences.

In recent years the advances in computational fluid mechanics (CFD, computational fluid dynamics) and its application to mixing (CFM, computational

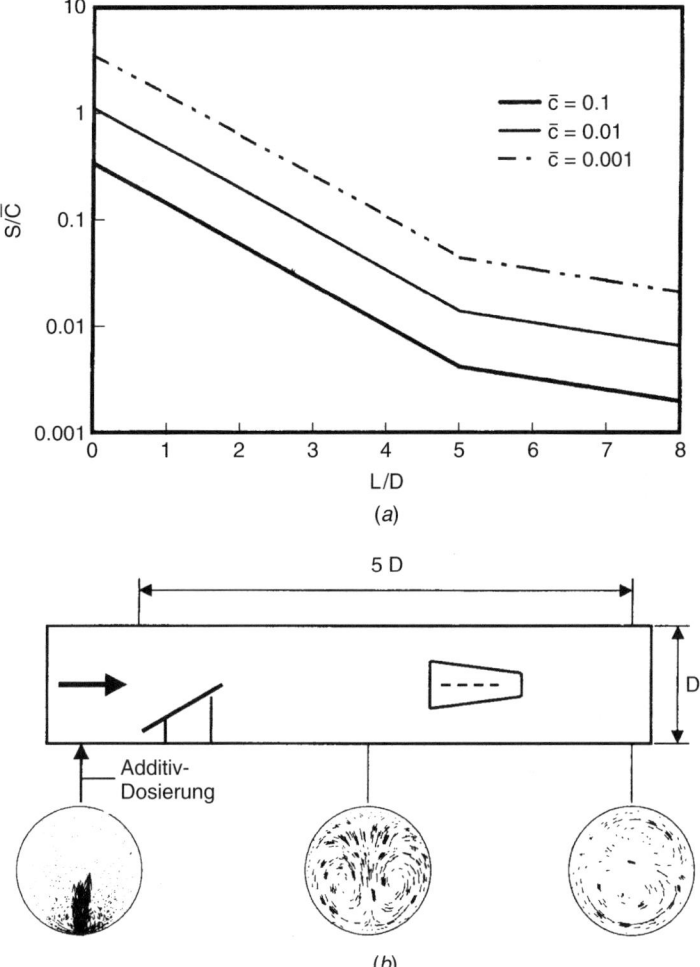

**Figure 7-6** (*a*) Homogeneity expressed as variation coefficient versus mixer length for a vortex-type static mixer type (KVM at L/D = 5) in turbulent flow. (*b*) KVM mixer layout and computational velocity vectors shown at three points. (From Streiff et al., 1999.)

fluid mixing) have allowed more detailed analysis to be performed of the state of mixing. Early computational studies showed more rapid mixing than experimental studies. This was due to a phenomena called *numerical diffusion*, where with coarse grids numerical rounding errors cause a smoothing of concentration gradients. With finer grids and the tracking of large number of massless particle tracking, the accuracy of the calculation matches well with the experimental measurements. This has been demonstrated in laminar flow for both the Koch SMX and the Kenics KMS motionless mixers by Zalc et al. (2003) and for turbulent mixing with Kenics HEVs by Bakker and LaRoche (1993).

**416** MIXING IN PIPELINES

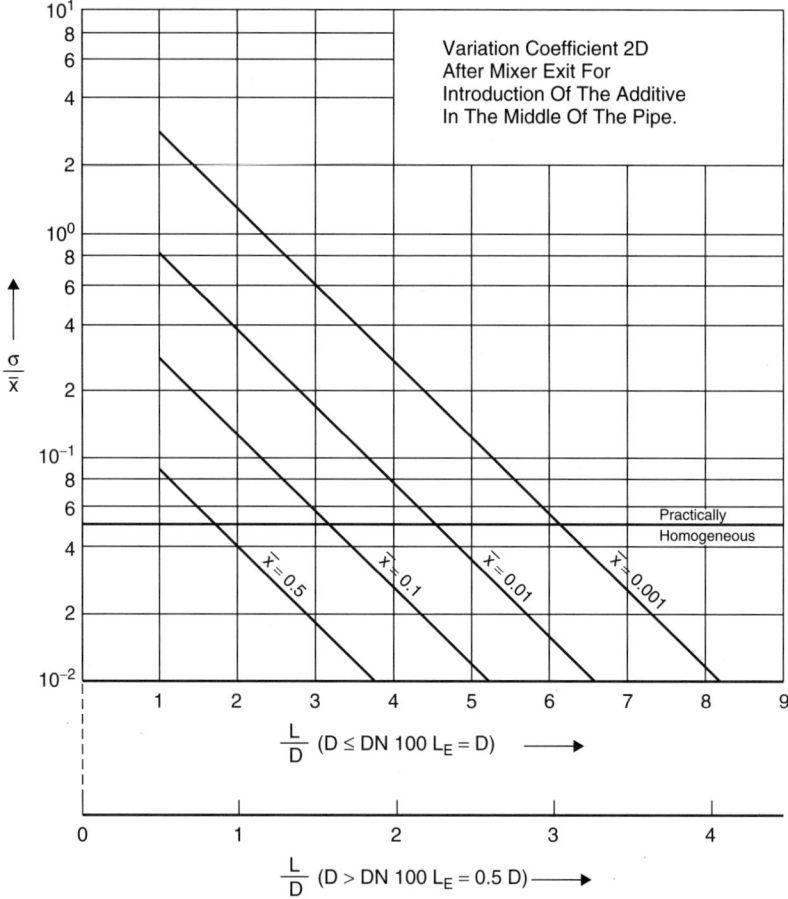

**Figure 7-7** Homogeneity expressed as a variation coefficient versus mixer length for an SMV static mixer operating in turbulent flow. (From Schneider, 1981.)

Experimentally, Etchells et al. (1995) calculated both CoV and Danckwerts mixing length and found similar relations in laminar flow. The Danckwerts length has the dimension of distance and may be thought of as laminae or diffusional thickness in laminar flow.

In the previous examples, color and composition were used; however, concentration or temperature can be used as well. In some cases not enough data exist for calculation of a coefficient of variation. In that case, another measure, such as spread, can be used. This is most useful in dealing with temperature problems, where the spread can be estimated coming in and the desired spread specified going out. Note that the mixing concepts above are independent of flow regime. The concept of coefficient of variation to describe mixing in pipeline devices is valid in both the laminar and turbulent flow regime and is used routinely in process design to describe or specify homogeneity for blending applications. In Chapter 2, several other concepts are discussed in detail, using the motionless mixer as an example.

### 7-5.2 Importance of Physical Properties

The two key properties in single-phase flow are the fluid density and the viscosity. The density is quite straightforward; it is the mass per unit volume. In turbulent flow, pressure drop is directly proportional to density, so that the accuracy of the density is the accuracy of the pressure drop prediction. It is easy to get better than 1% accuracy on such values. Viscosity, on the other hand, is a more complex measurement. Low viscosity systems usually run in turbulent flow, where the viscosity has little or no effect on mixing or pressure drop. For low viscosity material the prime use of the viscosity is in calculating a Reynolds number to determine if the flow is laminar or turbulent. If turbulent, little accuracy is needed. An error in viscosity of a factor of 2 will have negligible effect. In laminar flow, however, the viscosity becomes all important and pressure drop is directly proportional to it, so that an accuracy of 10% or less is often required. For laminar processing a complete relation of stress versus strain or shear rate versus shear stress is required. See Chapter 4 for the means and type of data required.

*7-5.2.1 Laminar Flow Regime.* In laminar pipeline flow the velocity vectors are parallel and there is no radial mixing. Because of the parabolic velocity distribution the velocity across the pipe is nonuniform. This results in a residence time distribution that is not plug flow (see Figure 7-33). The presence of this residence time distribution does not indicate backmixing, but rather, de-mixing, in that it works against radial mixing. The age distribution of fluid elements is being increased and this may be undesirable. For all practical purposes there is no radial mixing in laminar pipeline flow. Pipeline laminar flow often introduces gradients in age and temperature that must be removed by mixing. This mixing task is often referred to as *simple blending* or *low homogenization* since there are no additives being introduced to the bulk stream. Contents of the flow stream are merely being blended with themselves to eliminate gradients.

Static or motionless mixers are the only effective pipeline devices in the laminar flow regime. Flow inverters are related devices since they take annular fluid flowing at the pipe wall and interchange it with material at the center of the pipe. The purpose of this flow inversion is to equalize residence time to prevent degradation effects in sensitive processes such as melt fiber spinning. Static mixers are good flow inverters and, in addition, accomplish mixing. Flow inverters are, however, not good mixers—in principle, a second ideal flow inverter device in series with another merely undoes any mixing achieved by the first.

*7-5.2.2 Turbulent Flow Regime.* With turbulent flow there is mass interchange in both the radial and axial directions, due to turbulent eddies. Radial mixing has been summarized extensively by Gray (1986). The study by Ger and

Holley (1976) is most useful. They looked at single-point coaxial addition with equal main and injection velocities. They determined a radial diffusion coefficient in terms of a friction factor and an average velocity. They showed how an initial centerline injection spread out to the walls of the pipe as the fluid traveled along. Because of the linear relation between time and diffusivity, an increase in velocity increases turbulent diffusivity but reduces contact time so that the critical parameter become the length/diameter ratio (L/D) and friction factor, which was relatively insensitive to velocity. Thus, the empirical observation by many early workers that radial mixing in an empty pipe took between 50 and 100 diameters was verified by this work. In addition, the work showed that side injection took about twice as long as centerline, due to the longer diffusion path. Ger and Holleys's relation is

$$\frac{L}{D}_{mix} = 20.5 Re^{0.10} \left(\frac{f_s}{f}\right)^{0.5} \log(2.40 \text{ CoV}) \quad (7\text{-}7)$$

Unfortunately, Ger and Holley do not include the initial degree of unmixedness (CoV$_0$) in their correlation. It is interesting to note that the length required for mixing increases with increasing Reynolds number. See Figure 7-8 for measurements across a 4 in. diameter pipe at various distances downstream of a centerline injection of radioactive tracer into water at 1% concentration with a Reynolds number of 77 000. The publications also present results achieved with additive injection at the wall.

There is also axial dispersion. Again this is not mixing but a mechanism that introduces a residence time distribution. If a pulse is added to a turbulent pipeline, it will gradually lengthen with time. The best discussion of this is in Levenspiel's book (1967), where the work of Levenspiel and Bischoff is discussed. Mixing time, an important concept for reactive mixing, is given by

$$t_{mix} = 50 \frac{D}{V} \quad (7\text{-}8)$$

Thus, small pipes had short mixing times, and larger pipes take proportionally longer. This could present a scale-up problem from semiworks to full scale plant. In many plant cases, involving large pipe, the mixing length and time are too long. In many control and measurement situations, shorter time and distances are needed or used. This requirement to reduce mixing length and time has led to the concept of the multicoaxial mixer. There has been much academic work (Toor, 1975) on using parallel arrays of tubes for coaxial injection. This adjusts the initial scale of turbulence to a lower value (higher turbulence) than an empty pipe. While of continuing academic interest, they are not commonly used in industry; they are too complex and relatively ineffective. Tee mixers and static mixers are the two designs routinely employed to accomplish distributive mixing in the turbulent flow regime. Occasionally, they are used together. They are described in detail in subsequent sections of this chapter.

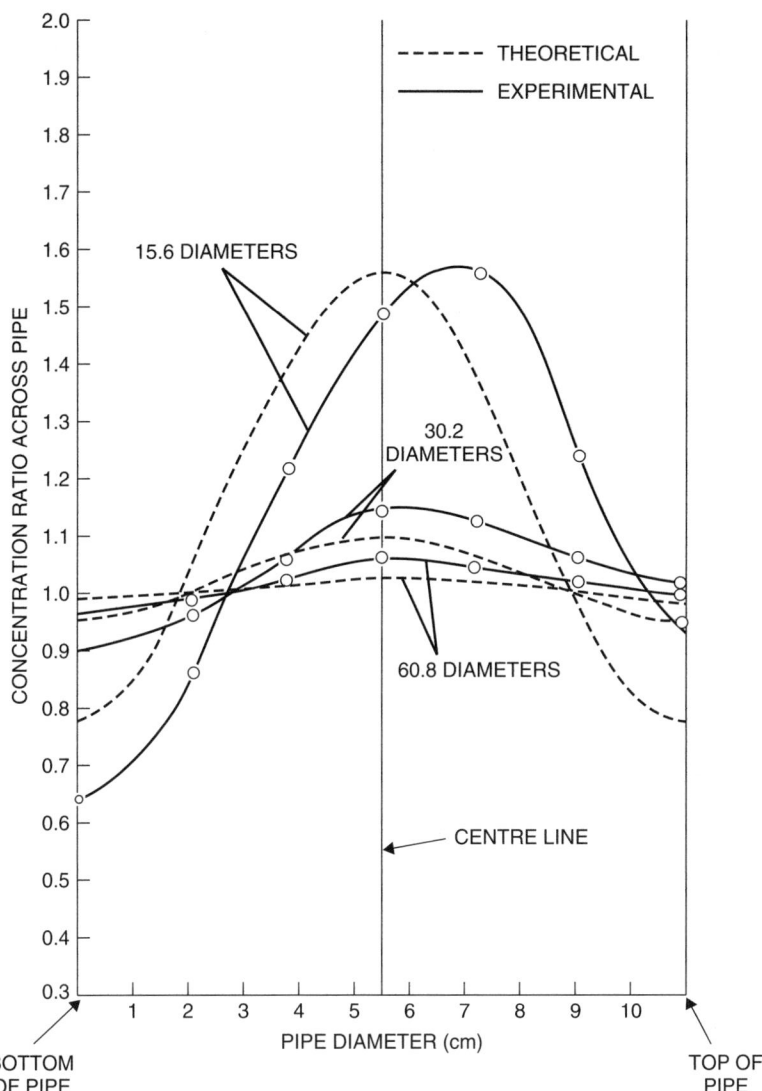

**Figure 7-8** Variation in concentration across a round pipe at different lengths downstream of a central injection with RE = 77 000. (From Clayton, 1979.)

## 7-6 TEE MIXERS

One of the two most popular approaches to pipeline mixing in turbulent flow involves the use of side injection tees. The other is the use of motionless static mixers (which are discussed later). The most complete work on side tees is by Forney and Lee (1982). They and others have found that the momentum of the side stream must be high enough to mix across the pipe with the bulk stream for shortest mixing length. When the momentum is low, the side stream will be deflected and become a sidewall injection, and the mixing length will be about 50 to 100 diameters. Therefore, much of the work has been centered around finding the optimum relation between main and side flow rate and velocity or diameter ratios. For low side stream momentum, the side stream is plastered along the near wall and mixing rates are similar to that with a pipeline. For too high a side-stream momentum, the incoming jet plasters against the far wall and backmixing occurs, which can be undesirable in the case of reactions. Figure 7-9 provides a diagram of the flow variables and shows the turbulent jet created by the properly designed tee. Tee mixing is the prime technology used in many reaction injection molding (RIM) systems for making polyurethanes. Both side tee and opposed tees are used. Angled jets that cause swirl reduce the amount of mixing and lengthen mixing length. The following rules from Forney and Lee (1982) give a set of optimum relations:

$$\frac{v}{V} = \frac{1.0}{(d/D)^{0.5}} \tag{7-9}$$

or

$$\frac{q}{Q} = \left(\frac{d}{D}\right)^{1.5} \tag{7-10}$$

since

$$\frac{q}{Q} = \frac{v}{V}\left(\frac{d}{D}\right)^{2} \tag{7-11}$$

Slight adjustments would be required to account for differences in densities between the two streams.

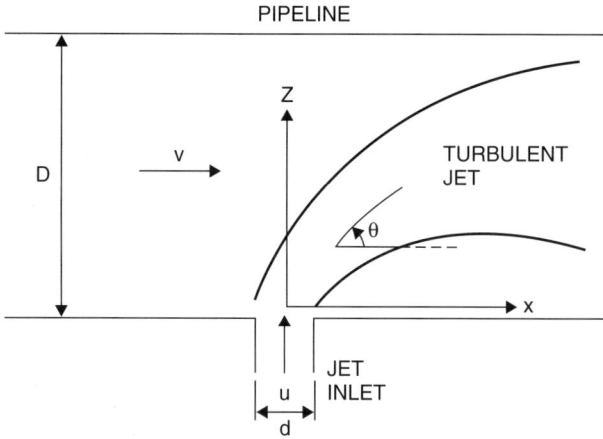

**Figure 7-9** Turbulent jet created by Tee mixer. (Modified from Forney, 1986.)

The energy for mixing comes mainly from the higher flow velocity in the side tee. Because there is a higher flow velocity, there must be a higher pressure to drive the side stream. This pressure loss is typically defined in terms of velocity head loss and depends on the design of the nozzle, which sets the head loss coefficient.

$$\Delta P = \frac{Kv^2}{2} \qquad (7\text{-}12)$$

where $\Delta P$ is the side-stream pressure loss, K the head loss coefficient, and v the side-stream velocity. K can vary between 1.0 for a very smooth inlet to 2.5 for an orifice, with 1.5 being very typical.

For many practical situations the side stream has to come in between two and five times the velocity of the main stream. As the volumetric amount of side stream becomes smaller, this ratio grows higher, the inlet hole gets very small, and the pressure drop gets very large. Tee mixers become less attractive for adding a small amount of one material into a large amount of a main stream.

Multijets have been studied, and four arranged circumferentially around the pipe wall is somewhat better than one. For an optimum design, the quality of mixing in terms of coefficient of variation is given by Forney et al. (2001) as

$$\text{CoV}^2 \left(\frac{x}{D}\right)^E = \frac{0.32}{B^{0.86}} \qquad (7\text{-}13)$$

where

$$B = n^2 R^2 \left(\frac{d}{D}\right)^2 \qquad (7\text{-}14)$$

Here n is the number of jets, R the velocity ratio (jet over main), d the side-stream diameter, and D the main stream diameter, x is the tailpipe length up to 5D and E is an empirically determined constant depending on B and geometry:

$E = 1.33$      for $B < 0.7$ and $n = 1$

$E = 1/33 + 0.95 \ln \dfrac{B}{0.7}$      for $B > 0.7$ and $n = 1$

$E = 1.97$      for $B < 2.0$ and $n = 4$

$E = 1.97 + 0.95 \ln \left(\dfrac{B}{2}\right)^{2.0}$      for $B > 2$ and $n = 4$

Equations (7-13) and (7-14) is applicable only to a mixing length of five pipe diameters for systems turbulent in the tailpipe. Equations (7-13) and (7-14) have been developed for low viscosity fluids. As viscosity increases, the flow approaches transitional. Due to the high energy dissipation in the tee, the flow can be turbulent even at Reynolds numbers down to about 1000, depending on design and flows. However, once outside the mixer, the flow rapidly becomes laminar in the tailpipe. Equations (7-13) and (7-14) should be used with great care with such systems.

## 7-7 STATIC OR MOTIONLESS MIXING EQUIPMENT

In the early 1950s a number of devices were developed in industry to handle thermal nonhomogeneous regions in polymer piping (transfer lines). Molten polymers usually are in laminar flow and have no radial exchange when flowing. Thus temperature gradients can form and be propagated. Heating and cooling of these materials through the wall is very difficult because of the residence time distribution associated with the laminar velocity profile. Material at the center moves much faster than the material at the wall, and in addition, has less contact time because of the poor conductive heat transfer at the center. These devices would reduce radial thermal gradients that occur in polymer processing. They were called *thermal homogenizers* and *flow inverters*. One of the first commercial units was the Kenics device. In the Kenics, a set of twisted elements with left- and right-hand twists caused the material to move from the wall to the center and from the center to the wall (see Visual Mixing CD). After traveling through a number of these elements, the fluid is homogenized with respect to age, composition, and temperature. These devices were called *motionless mixers* or *static mixers* because the mixer did not move, although the liquid did. The term *static mixer* was originally copyrighted by Kenics Corporation, but the term is now commonly used for all such in-line motionless mixers.

Over the years a large number of companies have produced motionless mixers all based on the principle of moving the streams radially by a series of metal baffles. These baffles may consist of twists of metal, corrugated sheets, parallel bars, small-diameter passages, or tabs sticking out from the wall. They are essentially plug flow devices with some small degree of backmixing, depending on the exact design. Two common types are the twisted-ribbon mixer (Kenics KMS; see Figure 7-10) and the structured-packing mixer, one of which makes use of layers of crisscrossed corrugations (Koch-Sulzer SMV; see Figure 7-11). Another structured packing static mixer is the overlapping lattice type (Koch-Sulzer SMX and SMXL; see Figure 7-12). The lattice members in the SMX are all oriented at 45° to the direction of flow and at 30° for the SMXL. A simplification of the lattice type which generates a mixing flow somewhat similar to the twisted ribbon is made of crossed elliptical plates whose flat surfaces are at 45° to the

**Figure 7-10**  Spiral static mixer. (Courtesy of Chemineer, Inc.)

STATIC OR MOTIONLESS MIXING EQUIPMENT    **423**

**Figure 7-11**   Corrugated plate static mixer (SMV). (Courtesy of Koch-Glitsch, LP.)

direction of flow (Koch SMXL-B). A version of this has triangular plates connecting the straight sides of the ellipses, making each element resemblea crude twisted ribbon. Another incorporates a flat to divide the ellipse at the centerline (Komax mixer). An example of a radial mixer with tabs extending from the pipe wall is the Koch-Sulzer SMF, a low intensity mixer with wide-open structure used in highly plugging service.

A recent static mixer innovation for application exclusively in highly turbulent flow is the use of small tabs projecting from the wall of the pipe into the core region of a turbulent flow (Kenics HEV; see Visual Mixing CD). Another design (Koch KVM; Figure 7-13) utilizes a single larger tab mounted off the tube wall to create large counterrotating vortices for mixing. The industry found a number of years later that even in turbulent flow where there is radial turbulent mixing, this mixing can be enhanced by using motionless mixers. The mechanism was different in detail, but the effect was the same. Some of the improved radial mixing came from increased radial turbulent diffusion. In the Ger and Holley formulation for empty pipe there is a friction factor. In motionless mixers the friction factor is many times larger than for empty pipe (i.e., the pressure drop is higher). This in itself would increase mixing and reduce mixing length. Also, there is in some motionless mixers a bulk radial flow. Etchells and Short (1988) took some limited data on SMV motionless mixers and showed that the improved mixing rate over an empty pipe was due almost entirely to the increased friction factor. Subsequent data on the HEV, however, do not fit that model. It is now believed that only a portion of the pressure drop energy expended goes into radial mixing and that the rest is lost in skin friction. Thus, the newer motionless mixers for application in turbulent flow rely on vortex generation away from surfaces to mix and take less pressure drop to get equivalent blending results.

**Figure 7-12** SMX static mixer shown with two mixing elements. (Courtesy of Koch-Glitsch, LP.)

Plate-type mixers (SMV) are, however, still very attractive design options for turbulent flow applications in large diameter ducts and pipe where mixing length is limited. The method for introduction of the additive stream becomes a very important part of design optimization. An example of this is found in the selective catalytic reduction process for the removal of nitrogen oxides from combustion flue gas (DeNox), where a small amount of ammonia is added and mixed with the flue gas prior to the catalyst bed. A large SMV static mixer element for this application is shown in Figure 7-14. A high degree of mixing is achieved in only two pipe diameters using sparger designs that introduce the additive to each mixing cell of an SMV mixing element (Fleischli and Streiff, 1995). Pressure drop is low since some of the mixing is accomplished downstream of the individual mixing elements by utilizing the swirl flow, which is induced in the wake of the mixer hardware. Figure 7-15 compares the mixing achieved using a plate mixer (SMV) with that of a vortex design (KVM). Note that spacers are included between the mixing elements to take advantage of mixing in the tailpipe

STATIC OR MOTIONLESS MIXING EQUIPMENT 425

**Figure 7-13** KVM vortex static mixer shown in flanged housing. (Courtesy of Koch-Glitsch, LP.)

**Figure 7-14** Large duct static mixer (SMV) for application in mixing ammonia with hot flue gas selective catalytic reduction DeNox process. (Courtesy of Koch-Glitsch, LP.)

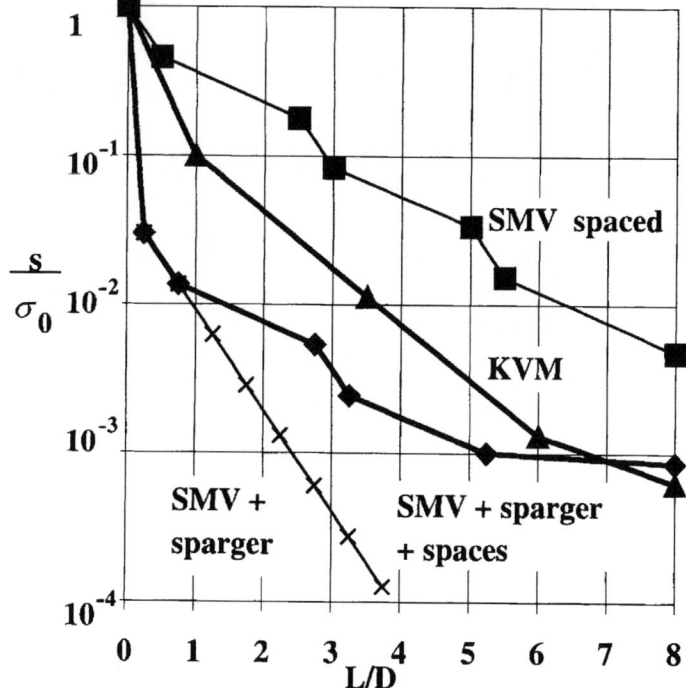

**Figure 7-15** Homogeneity for plate (SMV) and vortex (KVM) static mixers in turbulent flow at additive concentration = 0.01 (1%). (Modified from Streiff et al., 1999.)

downstream of each element. Also note the contribution of a properly designed sparger to the total mixing achieved. Modern mixer design for turbulent flow involves optimization of the sparger or injector, the static mixing element, and properly located empty pipe spaces.

### 7-7.1 Types of Static Mixers

Examples of the most commercially significant static mixers are, by manufacturer:

*Chemineer, Inc. (Kenics)*

- *KMS:* twisted ribbon or bowtie type, with alternating left- and right-hand twists. An element is 1.5 or 1.0 diameter in length. The KME variation is edge sealed to the tube wall. See Visual Mixing CD.
- *KMX:* a series of inclined retreat curve rods forming an X lattice; alternating in direction every diameter an element is one diameter in length.
- *HEV:* a series of four tabs spaced around the pipe. An element consists of four tabs symmetrically placed. Axially, the tabs are about 1.5 diameters apart. See Visual Mixing CD.

### Koch-Glitsch, LP

- *SMV:* several stacked sheets of corrugated metal running at 30 or 45° to the pipe axis. Each element is 0.5 to 1.0 diameter in length and adjacent elements are rotated 90° relative to each other. Mixer hydraulic diameter is determined by the height of the corrugation or the number of stacked corrugated sheets.
- *SMX:* guide vanes are intersecting bars at 45° to the pipe axis. Each mixing element is 1.0 diameter length. Adjacent elements are rotated 90°. See Visual Mixing CD.
- *SMXL:* similar to the SMX but with intersecting bars at 30° to the pipe axis. Typically, fewer bars per element, and the element length is variable, depending on application.
- *SMR:* guide vanes are hollow tubes through which heat transfer fluid circulates. The tubular bundle is arranged similar to the shape of the SMX design.
- *KVM:* single inclined tab mounted off the tube wall. Axially, tabs are about 2.5 diameters apart.
- *KHT:* twisted ribbon with alternating right- and left-handed twists.
- *SMF:* three guide vanes project from the tube wall so as not to contact each other. This is a special design for high plugging applications. Element length is approximately 1.0 diameter.
- *KFBE:* special version of the SMX/SMXL design with guide bars for exclusive application in gas fluidization of solid particles.

### Komax Systems, Inc.

- *Komax mixer:* crossed elliptical plates with a flat at the centerline. Adjacent mixing elements are rotated 90°.

### Charles Ross & Son Company

- *ISG:* solid tube inserts with shaped ends so that adjacent elements form a tetrahedral chamber, each with four holes drilled at oblique angles.

### Sulzer Chemtech

- *SMV, SMX, SMXL, SMF, and SMR:* as described above. These products, initially developed jointly by Koch and Sulzer under a licensing agreement, now expired.

See Figure 7-16 for an illustration of some of these static mixers. There are many more types, and new ones are being developed constantly. Most of the above are characterized by good vendor technical information. Omission of a type from the list above, however, does not necessarily indicate an inferior product, just a lack of quantitative information. Many manufacturers have copied the basic

**Figure 7-16** Static mixer design options. From left: vortex mixer (type KVM), corrugated plate (type SMV), wall-mounted vanes (type SMF), cross-bar (type SMX), helical twist (type KHT), cross-bar (type SMXL). (Courtesy of Koch-Glitsch, LP.)

**Table 7-4** Rough Guidelines for Applications in the Laminar and Turbulent Flow Regimes[a]

| | Static Mixer Design | | | | | | | | | |
|---|---|---|---|---|---|---|---|---|---|---|
| Flow Regime | KMS | KMX | HEV | SMV | SMX | SMXL | SMR | KVM | SMF | ISG |
| Laminar | | | | | | | | | | |
| Mixing/blending | c | a | | | c | c | | | a | a |
| High–low viscosity | | a | | | c | a | | | | a |
| Dispersion | a | a | | | c | a | | | | a |
| Heat transfer | c | | | | b | c | c | | | |
| Plug flow | b | | | | c | b | c* | | | |
| Turbulent | | | | | | | | | | |
| Mixing/blending | | | | | | | | | | |
| High turbulence | a | | c | c[†] | | | | c | | |
| Low turbulence | c | | | c | a | a | | | a | |
| Dispersion | | | | | | | | | | |
| Liquid–liquid | c | | | c | a | a | c* | | a | |
| Gas in liquid | c | | | c | a | a | a* | | a | |
| Liquid in gas | a | | | c | a | | | | | |
| Fluidized beds | | | | | c[‡] | | | | | |

[a] a, Applicable; b, typically applied; c, best design choice. *, Where temperature control is required; [†], especially for very large diameters and nonround cross-sections; [‡], gas fluidized solid particles, specialized design (Koch-type KFBE).

helical mixing element, Kenics type KMS/KME, and have competing products on the market.

### 7-7.2 Static Mixer Design Options by Flow Regime and Application

Table 7-4 provides rough guidelines for applications in the laminar and turbulent flow regimes. Equipment selection and sizing should be based on application engineering to meet specific process requirements.

### 7-7.3 Selecting the Correct Static Mixer Design

See Figure 7-17.

## 7-8 STATIC MIXER DESIGN FUNDAMENTALS

It is appropriate to start this section with a word of caution. The equations and design constants given here and in the tables for the various types of commercial motionless mixers come from the open literature and the vendors' literature. All calculations based on them should be confirmed by the vendors. There are differences in design and construction in different pipe sizes which can significantly affect pressure drop and energy dissipation. For example, high pressure drop designs tend to be made of heavier construction (thicker sheets or bars) so that the actual velocities are somewhat higher than the superficial or empty pipe velocity usually calculated. Additionally, new mixers are constantly being developed, and the parameters of those are not given.

### 7-8.1 Pressure Drop

In both laminar and turbulent cases, the addition of the baffles of motionless mixers increases the pressure drop, and therefore extra energy is required to get the additional mixing effect. Twisted-ribbon and structured packing static mixers will increase pressure drop per unit pipe length over standard open pipe by as much as a factor of 7 to several hundred, depending on the Reynolds number. Vortex-generating designs operate with less flow resistance, but the pressure drop is still significantly higher than it is for the same pipe size without elements. A fundamental fact is that pressure energy is required to mix in pipeline flow. Reduced mix time (shorter mixer length) requires higher-energy dissipation rates.

The pressure drop in a static mixer of fixed geometry is expressed as the ratio of the pressure drop through the mixer to the pressure drop through the same diameter and length of open pipe KL for laminar and KT for turbulent flow.

$$\Delta P_{sm} = \begin{cases} K_L \ \Delta P_{pipe} & \text{when laminar} \quad (7\text{-}15) \\ K_T \ \Delta P_{pipe} & \text{when turbulent} \quad (7\text{-}16) \end{cases}$$

**430** MIXING IN PIPELINES

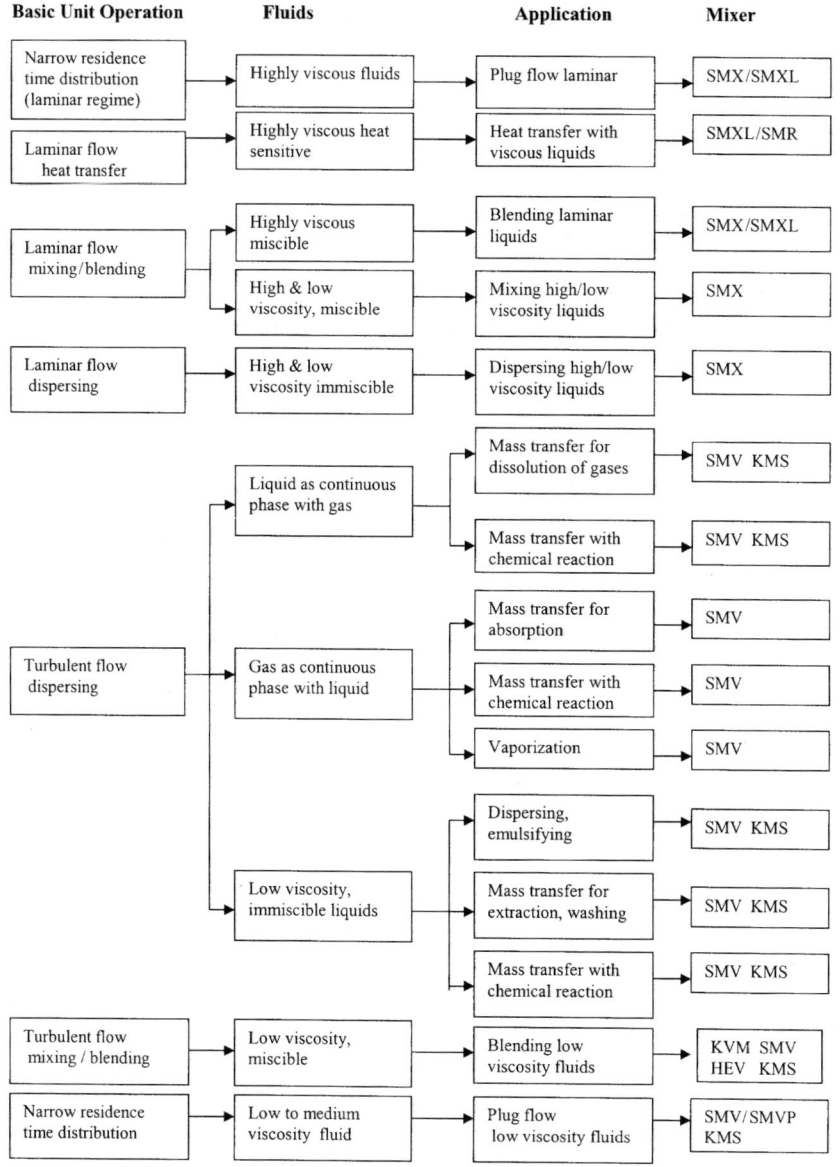

**Figure 7-17** Correct static mixer design: applications.

In Tables 7-5 and 7-6, values of $K_L$ and $K_T$ are given. These values are considered good to about 15%.

For completeness the standard pressure-drop equation for open pipe is

$$\Delta P = 4f \frac{L}{D} \rho \frac{V^2}{2} \qquad (7\text{-}17)$$

**Table 7-5** Laminar Blending and Pressure Drop Parameters for Motionless Mixers

| Device | KL | KiL |
|---|---|---|
| Empty pipe | 1 | — |
| KMS | 6.9 | 0.87 |
| SMX | 37.5 | 0.63 |
| SMXL | 7.8 | 0.85 |
| SMF | 5.6 | 0.83 |
| SMR | 46.9 | 0.81 |

*Source*: Streiff et al. (1999).

**Table 7-6** Turbulent Blending and Pressure Drop Parameters for Motionless Mixers

| Device | Ne[a] | KT | KiT |
|---|---|---|---|
| Empty pipe | 0.01 | 1 | 0.95 |
| KMS | 1.5 | 150 | 0.50 |
| KVM | 0.24 | 24 | 0.42 |
| SMX | 5 | 500 | 0.46 |
| SMXL | 1 | 100 | 0.87 |
| SMV | 1–2 | 100–200 | 0.21–0.46 |
| SMF | 1.3 | 130 | 0.40 |

[a] Ne is the Newton number, equivalent to 2f, twice the Fanning friction factor.

*Source*: Streiff et al. (1999).

where f is the Fanning friction factor introduced in Section 7-2.2. It is correlated empirically for turbulent flow in smooth pipes by the Blasius equation, given by

$$f = \frac{0.079}{Re^{0.25}} \quad (7\text{-}18)$$

for Reynolds numbers between 4000 and 100 000, and in laminar flow by

$$f = \frac{16}{Re} \quad (7\text{-}19)$$

for Reynolds numbers below 2000.

Most vendors have more accurate correlations that take into account a slight Reynolds number effect in transitional and turbulent flow, and the volume fraction occupied by the mixer, which varies with mixer diameter and pressure rating. A more detailed approach is necessary for some designs that have the option for variable but similar geometry. For the most accurate pressure drop predictions, the manufacturer should always be consulted.

## 7-8.2 Blending Correlations for Laminar and Turbulent Flow

The results for blending for motionless mixers can be correlated by plotting coefficient of variation reduction CoV$_r$ versus L/D. In laminar flow there is no effect of viscosity, flow rate or initial CoV on these correlations. CoV$_r$ is usually found to correlate with the L/D in an exponential form,

$$\text{CoV}_r = K_i^{L/D} \tag{7-20}$$

where K$_i$ depends on the mixer type. Tables 7-5 and 7-6 give typical values for both the blending coefficient (KiL for laminar, KiT for turbulent) and pressure drop coefficient (KL for laminar, KT for turbulent).

At low L/D there is some deviation as the flow develops, but this is usually neglected because there is also little mixing. The Kenics HEV shows a weak Reynolds number dependence (Figure 7-18), along with a length/number of element dependence. This vortex-generating mixer design is typically applied at a Reynolds numbers above 10 000.

***7-8.2.1 Laminar Flow: Effective Shear Rate.*** In laminar flow, the fluids are often shear thinning (i.e., the viscosity decreases with increasing shear rate). The apparent or effective shear rate in an empty pipe with Newtonian fluids is expressed as

$$G' = 8\frac{V}{D} \tag{7-21}$$

For motionless mixers in laminar flow, the shear rate is higher. This and the extra surface area are what contribute to the higher pressure drop. Table 7-7 gives some

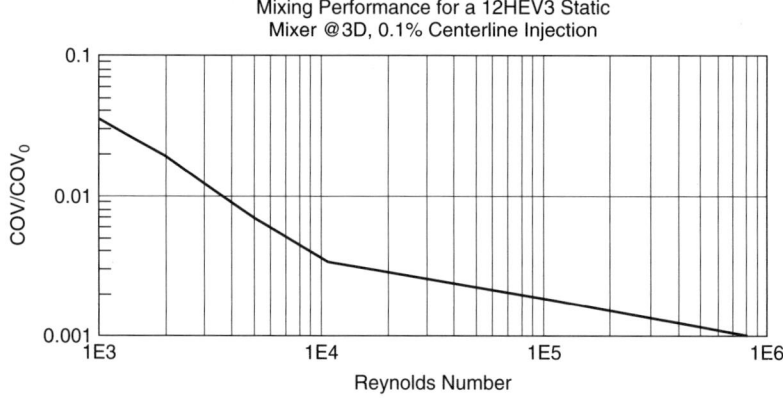

**Figure 7-18** Coefficient of variation reduction versus Reynolds number for the HEV mixer. (Courtesy of Chemineer, Inc.)

## STATIC MIXER DESIGN FUNDAMENTALS

**Table 7-7** Effective Shear Rate in Motionless Mixers $K_G = G'/(V/D)$

| Device | KG |
|---|---|
| Empty pipe | 8 |
| KMS | 28 |
| SMX | 64 |
| SMXL | 30 |
| SMV | 50 |
| SMF | 25 |
| SMR | 60 |

*Source*: Streiff et al. (1999).

estimated effective shear rates in a variety of mixers, based on

$$K_G = \frac{G'}{V/D} \qquad (7\text{-}22)$$

With this apparent or effective shear rate a rheogram relating effective viscosity to shear rate can be used to calculate an effective viscosity for pressure drop calculations. For another experimental approach, see Jaffer and Wood (1998).

***7-8.2.2 Laminar Flow: Layer Generation.*** In laminar flow, mixing of miscible components with similar viscosity and nonelastic behavior is achieved by the formation of layers as the materials are stretched and deformed into each other. All static mixers employ the principle of dividing the flow into substreams, distributing the substreams radially, and recombining them in a reordered sequence. The number of layers is increased and layer thickness is reduced by each successive mixing element. The process is represented schematically in Figure 7-19 and shown for real mixing processes in Figure 7-20 (KMS cross-cuts, cutting flow into two substreams) and Figure 7-3 (SMX cross-cuts, cutting flow into eight

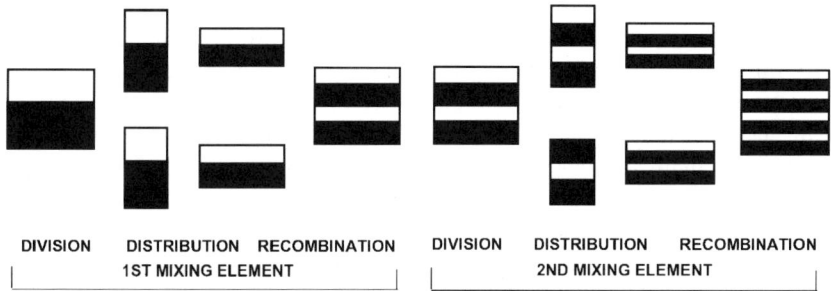

**Figure 7-19** Generalized portrayal of simple mixing in a static mixing device (each element dividing the fluid flow into two substreams).

**434** MIXING IN PIPELINES

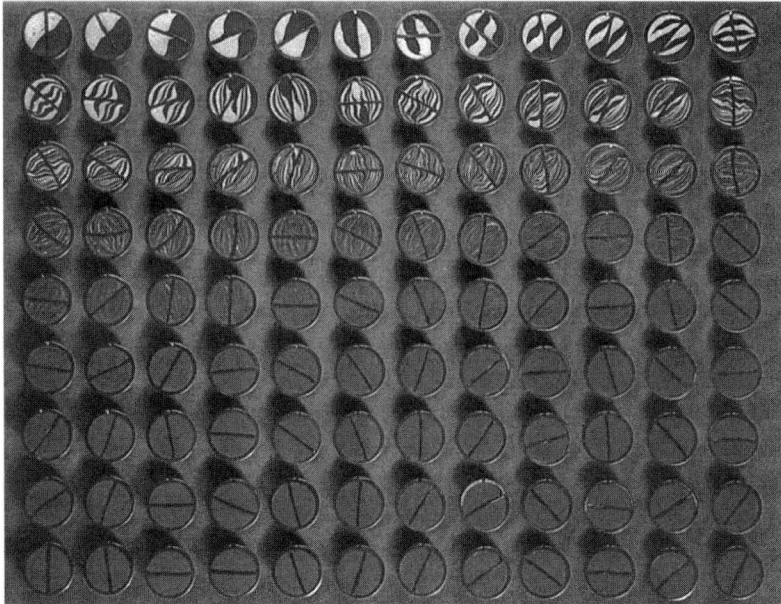

**Figure 7-20** Laminar mixing of fluids by division and recombination (KMS mixer). Cross-sections of the mixer are shown in sequence from left to right, top to bottom. (Courtesy of Chemineer, Inc.) See insert for a color representation of this figure; for the animation, see the Visual Mixing CD affixed to the back cover of the book.

substreams). It is interesting to note that the rate of layer generation differs with different designs.

***7-8.2.3 Comparison of Several Static Mixers in Laminar Flow.*** Several static mixer designs were studied (Alloca and Streiff, 1980) for the purpose of comparing their performance as distributive mixers operating in laminar flow. The conductivity tracer method was used to measure the degree of mixing achieved at 10% additive fraction. Testing was performed using concentrated glucose solutions with a viscosity in the range 7 to 9 Pa · s. The additive and bulk stream were equal in composition except for a small amount of tracer in the additive stream. Additionally, they were of equal viscosity to eliminate the effect of viscosity differences at the interface. Variation coefficient as a function of relative mixer length (L/D) is plotted in Figure 7-21. The difference in the degree of mixing per unit mixing length is shown clearly. Table 7-8 provides a comparison of mixer diameter, length and volume, fluid hold-up, and pressure drop to achieve the same degree of mixing, in this case a variation coefficient of 0.05 (5%). There is a significant degree of design flexibility in laminar distributive mixing. Mixers can be optimized to minimize pressure drop, length requirement, or residence time, or some combination. Of the mixers shown in Figure 7-21, the PMR design is of unknown manufacture and the Lightnin mixer tested most often used in turbulent flow.

STATIC MIXER DESIGN FUNDAMENTALS    435

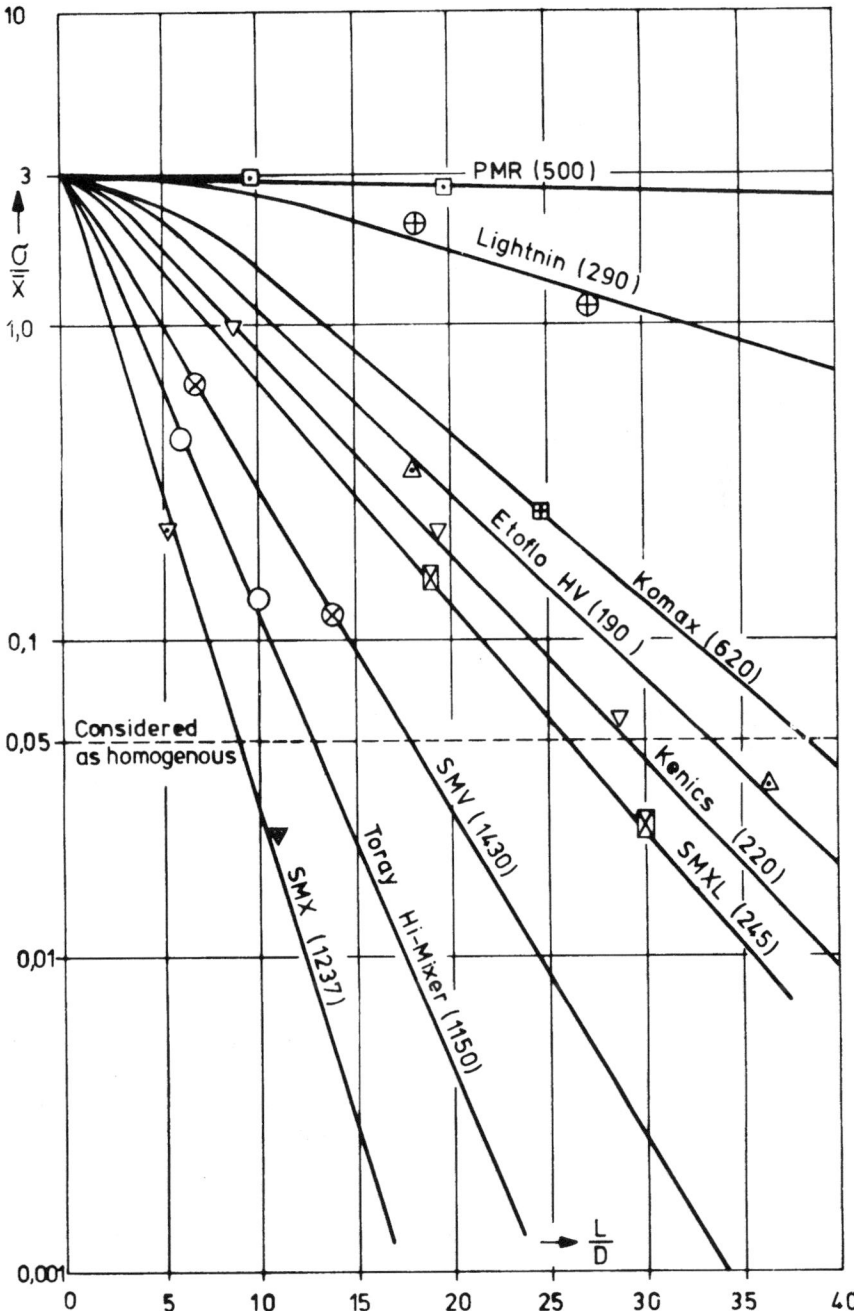

**Figure 7-21** Coefficient of variation (CoV) versus relative mixer length for several static mixer designs operating in laminar regime at 10% additive and equal viscosity. (From Alloca and Streiff, 1980.)

**Table 7-8** Comparison of Static Mixers for Equivalent Homogeneity in Laminar Flow

|  | Measured Values | | Comparisons | | | | |
|---|---|---|---|---|---|---|---|
| Mixing Unit | L/D for $\sigma/\bar{x} = 0.05$ | $Ne \cdot Re_D$ | Volume[a] | Holdup[a] | Diameter[a] | Length[a] | Pressure Drop[b] |
| SMX | 9 | 1237 | 1 | 1 | 1 | 1 | 1 |
| SMXL | 26 | 245 | 1.8 | 1.8 | 0.84 | 2.4 | 0.6 |
| SMV | 18 | 1430 | 4.6 | 4.5 | 1.3 | 2.7 | 2.3 |
| Kenics | 29 | 220 | 1.9 | 1.8 | 0.84 | 2.7 | 0.6 |
| Etoflo HV | 32 | 190 | 2 | 2 | 0.84 | 2.7 | 0.6 |
| Komax | 38 | 620 | 8.9 | 8.2 | 1.3 | 5.4 | 2.1 |
| Lightnin | 100 | 290 | 29 | 27 | 1.4 | 15.3 | 2.6 |
| PMR | 320 | 500 | 511 | 460 | 2.4 | 86 | 14.5 |
| Cunningham | | | | No mixing | | | |
| Toray | 13 | 1150 | 1.94 | 0.88 | 1.1 | 1.6 | 1.35 |
| N-Form | 29 | 544 | 4.5 | 3.8 | 1.1 | 3.6 | 1.40 |
| Ross ISG | 10 | 9600 | 9.6 | 3.4 | 2.1 | 2.3 | 8.6 |

[a] Multiple of volume, fluid holdup, diameter and length as compared to the SMX design for equal volumetric flow, viscosity, pressure drop, and variation coefficient (CoV = 0.05).
[b] Multiple of pressure drop as compared to the SMX design for equal volumetric flow, viscosity, pipe diameter, and variation coefficient (CoV = 0.05).

### 7-8.2.4 Effect of Physical Property Differences in Blending.

Two fluids may be miscible but not have the same physical properties. For example, a high viscosity soluble dispersant may be added to water. A small amount of solvent may be added to a highly viscous polymer stream. Two polymers of different molecular weight and therefore different viscosity may need to be blended. The correlations given above are for materials that are miscible and of the same physical properties. The effect of density is usually not large but can be significant (see Section 7-11.1 on orientation of motionless mixers for more on density effects). On the other hand, viscosity can differ by orders of magnitude, and the materials are still miscible.

In turbulent mixing systems the trick is to keep the more viscous materials (almost always the additive phase) from getting into a low-turbulence area. There is interesting work on stirred tanks by Smith and Schoenmakers (1988) which is equally applicable to static mixers. They found that if the high viscosity additive is allowed to touch the wall of the vessel, it takes a long time to dissolve. If added into the turbulent zone, mixing time is the same as with low viscosity material. For motionless mixers this suggests that the additive should be added not at the inlet but between the elements, where high levels of turbulence occur. When added at the inlet, the additive stream can drift into a low-turbulence area.

In viscous systems the additive viscosity (dispersed phase, $\mu_d$) is usually of lower viscosity than the bulk stream (continuous phase, $\mu_c$). With such systems the low viscosity additive slips between the areas of high shear rate and shear stress and the flow is segregated. Mixing length is much greater to reach a desired CoV.

The following empirical relation seems to describe the situation (Streiff et al., 1988):

$$\left(\frac{L}{D}\right)_{unequal} = \left(\frac{L}{D}\right)_{equal} + K \log \frac{\mu_c}{\mu_d} \qquad (7\text{-}23)$$

This equation applies for long mixers where the outlet CoV is low. The number of values for K is limited. For the SMX design it has been determined experimentally to be 1.0. For other designs, values between 2 and 10 are probably realistic. Note that for large viscosity ratios $\mu_c/\mu_d$ such as 10 000 : 1, which are not uncommon in the polymer industry, the mixer length can be 1.5 to 3 times longer than it would be if the streams were of equal viscosity. Design selection is limited when mixing low viscosity additives into viscous bulk streams.

For miscible additives of higher viscosity with a $\mu_c/\mu_d$ ratio of 1 : 10 000, Streiff (1999) claims that no mixing is possible using static mixers operating in laminar flow. Diffusion will still occur and this suggests that multiple inlets will be helpful.

A special case is one where materials are miscible but because of molecular differences have different surface tensions, and so when initially contacted there is a finite interfacial tension. Such materials act initially as immiscible, but as mass transfer takes place the interfacial forces disappear and the system acts totally miscible. Extra mixer length is required while this transformation takes place.

### 7-8.3 Which In-line Mixer to Use

***7-8.3.1 Turbulent Blending.*** In all cases energy is required to mix. For in-line mixers that energy comes from pressure drop. Motionless mixers usually do not take very high pressure drops compared to the total for liquid systems. For gas systems, however, the pressure drop, while low, is often significant compared with that of the whole system. Special high-efficiency designs have been developed which are most useful for gas systems.

If there are no time or length limitations, the simple pipeline uses the minimum energy. If there are limitations, then:

- If the main stream has sufficient pressure, a static or motionless mixer should be used.
- If the main stream does not have sufficient pressure but the additive side stream does, a Tee mixer (or spray nozzle for liquids into gas) should be used.
- If neither the main stream nor side stream has adequate pressure for mixing, an in-line mechanical device where the power can be supplied externally should be considered.

***7-8.3.2 Laminar Flow.*** In laminar flow there is no radial mixing without a motionless mixer, mechanical in-line mixer, or a stirred tank. The choice is

**438** MIXING IN PIPELINES

between various motionless mixers. On a simple pressure drop basis it is a trade-off between length needed to get to a certain quality of mixing versus the required pressure drop. This balance can be determined by using the data in the earlier tables. Usually, the device with the smallest KL is the one with the lowest pressure drop but the longest length. Then the investment must be looked at. Longer mixers cost more. In some cases the extra length required to obtain a certain CoV is not possible due to space limitations, and shorter mixers operating with higher pressure drop are preferred. When the viscosity ratio of the main stream to the additive stream is large (greater than 100), only the SMX and KMX have been demonstrated to be effective.

### 7-8.4 Examples

*Example 7-1: Gas–Gas Blending—Turbulent Blending.* Two gases are to be mixed prior to entering a reactor with a catalytic bed. Prior to contacting the catalyst, the gases are inert. Since there is little radial mixing in the catalyst bed, a high degree of uniformity of stoichiometry is required in the feed gas. To this end a mixer has been proposed. Three types will be evaluated: a tee mixer, an SMV motionless mixer, and a HEV motionless mixer.

The main pipe is 762 mm in diameter. The main stream flow is 11.7 m³/s and the side stream 2.74 m³/s. Densities are 1.79 and 1.77 kg/m³, respectively; viscosities are 0.014 and 0.020 mPa · s.

*Physical Properties:*

|  | Major Flow | Minor Flow | Total |
|---|---|---|---|
| Density (kg/m³) | 1.79 | 1.77 | 1.79 |
| Viscosity (Pa·s) | $1.4 \times 10^{-5}$ | $2.0 \times 10^{-5}$ | $1.4 \times 10^{-5}$ |
| Flow (m³/s) | 11.7 | 2.74 | 14.44 |

*Dimensions:*

$$D = 0.762 \text{ m} \qquad \text{area} = \frac{\pi}{4}(0.762)^2 = 0.46 \text{ m}^2$$

*Velocity:*

$$14.44/0.46 = 31.4 \text{ m/s}$$

*Reynolds number:*

$$\frac{\rho D V}{\mu} = 0.762 \times 31.4 \times \frac{1.79}{1.4 \times 10^{-5}} = 3 \times 10^6$$

The flow is turbulent. The density difference is negligible.

The initial degree of unmixedness, from eq. (7-6):

$$C_v = \frac{2.74}{14.44} = 0.19$$

$$CoV_0 = \left(\frac{1 - 0.19}{0.19}\right)^{0.5} = 2.06$$

Specify the side tee:

$$V = \frac{11.7}{0.46} = 25.43 \text{ m/s}$$

$$\frac{q}{Q} = \frac{2.74}{11.7} = 0.23$$

From eq. (7-10),

$$\frac{q}{Q} = 0.23 = \left(\frac{d}{D}\right)^{1.5}; \quad \text{then} \quad \frac{D}{d} = 2.66, \quad v = 45.67 \text{ m/s}$$

The side tee nozzle diameter is 0.29 m with a velocity of 45.67 m/s.
This is the optimum design. The degree of mixing is given by eqs. (7-13) and (7-14):

$$B = n^2 R^2 \left(\frac{d}{D}\right)^2 = (1)\left(\frac{45.67}{25.43}\right)^2 \left(\frac{0.29}{0.762}\right)^2 = 0.47$$

for B = 0.47, E = 1.33. Use the maximum x/D of 5.0 for the mix length:

$$x = 5 \times 0.762 = 3.81 \text{ m}$$

$$CoV^2 \times 5^{1.33} = \frac{0.32}{0.47^{0.86}} = 0.62$$

$$CoV = \left(\frac{0.62}{5^{1.33}}\right)^{0.5} = 0.27$$

Thus, at the end of 3.81 m, the CoV has been reduced from 2.06 to 0.27. That is still a high variability and probably not good enough for this application. A long tailpipe would still be required.

For the HEV, use Figure 7-18 at a Reynolds number of $3 \times 10^6$ the coefficient of variation reduction, $CoV_r = CoV/CoV_0 = 0.001$ for three sets of HEV tabs as measured three diameters downstream. This means a CoV of 0.002 at 2.2 m downstream for the last tab. The three sets of HEV tabs will take up about another three diameters, for a total length of 4.4 m. The pressure drop must be estimated by the vendor.

For the SMV motionless mixer, use eq. (7-20) and Table 7-6.

**440** MIXING IN PIPELINES

For the SMV the friction factor is 1-2 and $K_iT$ is between 0.21 and 0.46, depending on exact design. We will use an average 0.33.
From eq. (7-20), $CoV_r = 0.33^{L/D}$
To get the same $CoV_r$ as with the HEV,

$$\frac{L}{D} = \frac{\ln 0.0010}{\ln 0.33} = 6.23$$

The mixer length would have to be 4.74 m.

An alternative would be to use Figure 7-7; however, it does not go out to such a low $CoV_r$ value. The vendor would probably optimize this further for pressure drop and length by employing a multipoint sparger and spacers between the elements, both subjects beyond the scope of this chapter.

**7-8.4.2 Thermal Homogenization—Laminar Blending.** A polymer solution exiting a simple shell-and-tube heat exchange shows erratic behavior in downstream processing. Contact temperature measurements around the exit pipe gave temperatures of 113, 93, 115, and 126°C. We will design a motionless mixer to get the temperature spread down to ±0.5°C. The polymer solution has a density of 1100 kg/m³ and a viscosity of 650 mPa·s at process conditions. The flow rate is 15 060 lb/hr (6845 kg/h). The exit pipe is 50 mm inner diameter.

*Physical properties:*

$$\rho = 1100 \text{ kg/m}^3 \qquad \mu = 0.650 \text{ Pa} \cdot \text{s}$$
$$\text{flow} = 6485 \text{ kg/h} = 1.6 \times 10^{-3} \text{m}^3/\text{s}$$
$$\text{pipe size} = 2 \text{ in.} = 50 \text{ mm} = 0.05 \text{ m} \qquad \text{area} = 0.002 \text{ m}^2$$
$$\text{velocity} = 0.88 \text{ m/s} = V$$
$$\text{Re} = \frac{\rho DV}{\mu} = 0.05 \times 0.88 \times \frac{1100}{0.650} = 74.5$$

Flow is laminar.

$$\text{maximum temperature difference} = 126 - 93 = 33°C$$
$$\text{desired temperature difference} = 1°C$$
$$\text{desired reduction } 1/33 = CoV_r = 0.033$$

From Table 7-2, for the Kenics KMS,

$$K_i = 0.87 \qquad K_L = 6.9$$
$$\frac{L}{D} = \frac{\ln 0.033}{\ln 0.87} = \frac{3.41}{0.14} = 24.36$$
$$L = 1.22 \text{ m}$$

For the SMX,

$$K_i = 0.63 \quad K_L = 37.5$$
$$\frac{L}{D} = \frac{\ell n(0.033)}{\ell n(0.63)} = \frac{3.41}{0.46} = 7.41$$
$$L = 0.37 \text{ m}$$

For pressure drop in an empty pipe, use eq. (7-17):

$$\Delta P = 4f \frac{L}{D} \rho \frac{V^2}{2}$$

For 1 m of pipe without mixers,

$$f = \frac{16}{Re} = \frac{16}{74} = 0.215$$

$$\Delta P = 4 \times 0.215 \times \frac{1.0}{0.05} \times 1100 \times \frac{0.2^2}{2} = 7324 \text{ Pa or } 0.07 \text{ bar}$$

For the KMS, the length is 1.22 m:

$$\Delta P = 1.2 \times 6.9 \times 7324 = 63\,444 \text{ Pa or } 0.63 \text{ bar}$$

For the SMX, the length is 0.37 m:

$$\Delta P = 0.37 \times 37.5 \times 7324 = 103\,928 \text{ Pa or } 1.04 \text{ bar}$$

Thus, the SMX takes more pressure drop but is shorter and has less volume, while the KMS takes less pressure drop but has more volume and length. The final best choice of mixer for this application would probably depend on what the secondary criteria were, since the pressure drops are not high in either case.

## 7-9 MULTIPHASE FLOW IN MOTIONLESS MIXERS AND PIPES

Dispersive multiphase mixing is distinguished from simple blending or distributive mixing in that the additive phase breaks up into discrete drops or bubbles which are surrounded by the other phase. A basic comparison is shown in Figure 7-22.

### 7-9.1 Physical Properties and Drop Size

In the process result of area generation or dispersion, the most important physical property for low viscosity fluids is the interfacial tension. This is the force at an interface between immiscible fluids: liquids with liquid or liquids with gas

**442** MIXING IN PIPELINES

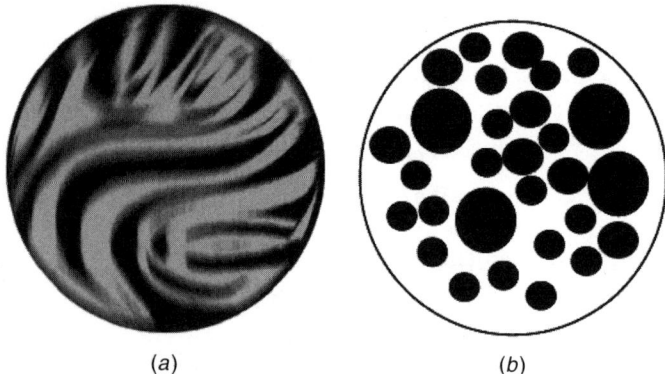

(a)　　　　　　　　　　　(b)

**Figure 7-22** Blending (simple distributive mixing) and dispersive mixing: (*a*) blending: flow streams are interleaved; (*b*) dispersive mixing: one phase is discontinuous.

that resists elongation or deformation. For single liquids this is called the *surface tension*. There is no surface tension for gases, as they form no interfaces. However, for liquids it is a physical property that can and often is measured. For two totally immiscible fluids, the force at the surface is called the *interfacial tension*. It is the difference in their individual surface tensions. However, if there is mutual solubility of the components in each other, the interfacial tension is less than this ideal defined value. Interfacial tensions can be lowered further by the presence of surface-active chemicals in even trace amounts in the liquids. These surface-active agents reside at the interface and change the surface energy of the drops. They affect not only the interfacial tension but also the tendency for interfaces to break or coalesce.

With two immiscible fluids it is also necessary to know which is the continuous phase and which is the dispersed phase. For liquid and gases, all drop size/bubble correlations are for gas dispersed in liquid. In theory, at high enough gas rates it should be possible to make a spray in a motionless mixer, but such a phenomenon has not been reported. Even at inlet volume ratios of eight parts of gas to one of liquid, the liquid phase is continuous, with gas holdups well over 50%. In a two-phase gas–liquid flow system there are several possible flow regimes. These include bubble flow, wavy flow, slug flow, annular flow, dispersed flow (also referred to as spray or mist flow), and a few other variations. A good description of these regimes for horizontal flow in empty pipe is provided by Lockhart–Martinelli (Govier and Aziz, 1972). When it is desirable to operate in the dispersed flow regime, liquid volumetric flow should be kept below 10% of the gas volumetric flow. Baker and Rogers (1989) report that the mixer/contactor geometry has a strong impact on the point at which dispersed flow is achieved. The SMV mixer will induce dispersed flow at much lower gas velocity than that required in an empty pipe. Flow regime (and operating pressure drop) at any gas and liquid flow rate will be influenced by mixer orientation and flow direction (horizontal, vertical up or down).

When dealing with the contacting of liquid in continuous gas flow in static mixers, one thinks in terms of liquid film flow on the mixer surfaces and droplets being formed as the liquid film is sheared off the mixer blades. The total interfacial surface is a combination of film plus droplet surface. Film flow is well defined and equal to the mixer plate surface area and is often used as the contact surface area for calculation of mass and heat transfer capacity. Mixers with high surface/volume ratios are most often the preferred design.

For two liquids the continuous phase is determined by physical and surface properties in a way not thoroughly understood in either agitated tanks or pipeline mixers. A rough guideline is that the liquid with volume fraction below 30% is dispersed. Between 30 and 70% is an ambiguous region where there are no hard rules, although it should be noted that water tends to be the continuous phase even above 50% organic (see Pacek et al., 1994). With static mixers providing a large surface area, the phase that best wets the mixer surface will tend to be the continuous phase, even if the minor component is within limits in the ambiguous region. Thus, water is the continuous phase in metal and glass equipment and would tend to be the dispersed phase in polyethylene, PVC, and polytetrafluoroethylene. This has little effect on the drop size, which is most strongly affected by interfacial tension. In turbulent flow, which phase is dispersed has little effect. However, it may be important for downstream operation such as decanters or at high volume fractions where coalescence is important. In laminar flow, where viscosity of the continuous phase is important, the determination of which phase is continuous is important even in dispersion.

### 7-9.1.1 Turbulent Flow: Dispersed Phase Drop or Bubble Size. 

The breakup of drops and bubbles in turbulent fields in agitated tanks is discussed elsewhere in the book. Similar theories apply to motionless mixers as uniform turbulence generators and similar equations can be developed. In pipeline flow the equations for dispersed gas and dispersed liquid are identical. The physical property differences are handled in the physical property corrections. Little or no coalescence is assumed. The equations apply in vertical flow and in horizontal flow when the criterion in Chapter 12 is met. Unlike stirred tanks, the motionless mixer system reaches equilibrium droplet size very quickly, in a few pipe diameters of mixer length. The most popular equation form (Middleman, 1974; Streiff et al., 1999) is

$$d_{max} = k_1 \left(\frac{\sigma}{\rho_c}\right)^{0.6} \left(\frac{\rho_c}{\rho_d}\right)^{0.2} \varepsilon^{-0.4} \qquad (7\text{-}24)$$

$k_1$ is on the order 1.0, and $d_{max}$ is about 1.5 times $d_{32}$.

An alternative is the Weber number form,

$$\frac{d_{32}}{D} = \frac{K}{We^{0.6}} \qquad (7\text{-}25)$$

Here K is different for various mixers and the form does not include all the density, viscosity, and concentration effects. For fluids with a low density

**444** MIXING IN PIPELINES

ratio and low viscosity ratio in turbulent flows, K = 0.49 for the Kenics KMS mixer (Calabrese and Berkman, 1988). For other mixers the drop size is inversely proportional to the friction factor to the 0.4 power.

Additional terms are often included to take into account dispersed phase viscosity and coalescence due to holdup. They usually depend on the volume fraction of dispersed phase:

$$1 + kC_v \tag{7-26}$$

where k is 3 to 5 and $C_v$ is the volume fraction of dispersed phase. Streiff et al. (1997) report that this k depends on flow orientation and varies from 1.7 to 3.4 for upflow versus downflow for air in demineralized water. This effect of dispersed phase volume fraction on drop size is probably a coalescence rather than a turbulence dampening effect as suggested by some. Several authors also include a density ratio to get both gas and liquid dispersed phase data in the same correlation. This ratio is continuous over dispersed and the exponent varies from 0.1 to 0.5.

Notice that this relation predicts that gas bubbles will be larger than liquid droplets at the same energy dissipation. This is observed experimentally. A similar correlation is used for gas-driven sprays where the gas is the continuous phase, and this predicts that spray drops would be finer than drops in a motionless mixer. This is also observed.

For a viscous dispersed phase the derivation of the drop size equations are modified (Calabrese and Berkman, 1988; Streiff et al., 1997) with an extra term representing the viscous resistance to drop breakup. This adds a new term:

$$d_{max} = K_1 \left(\frac{\sigma}{\rho_c}\right)^{0.6} \left(\frac{\rho_c}{\rho_d}\right)^{0.2} \varepsilon^{-0.4}(1 + Vi) \tag{7-27}$$

where

$$Vi = \frac{\mu_d(\varepsilon d_{max})^{0.333}}{\sigma(\rho_c/\rho_d)^{0.5}} \tag{7-28}$$

or

$$Vi = \mu_d \frac{V}{\sigma} \left(\frac{\rho_c}{\rho_d}\right)^{0.5} \tag{7-29}$$

Sometimes the Ohnesorge number (Oh) is used to show the breakpoint between viscosity-controlled and surface tension–controlled breakup. Unfortunately, it is depends on the drop size and is of limited usefulness.

It is, however, used extensively in the spray literature (Lefebvre, 1989):

$$Oh = \frac{(We)^{0.5}}{Re} = \frac{\mu_d}{(\rho_c \sigma d)^{0.5}} \tag{7-30}$$

It thus is the ratio of the resisting viscous force to the surface force.

Though seldom explicitly mentioned, the dispersed phase viscosity in the drop breakup equations should be the elongational viscosity. For simple Newtonian fluids it is three times the shear viscosity. For elastic fluids the elongational viscosity can be much larger than the shear viscosity. Few data exist for the breakup of elastic drops.

Streiff et al. (1999) propose an empirical equation for low viscosity dispersed phase, similar to eq. (7-16) but allowing some of the dimensionless constants to float:

$$d_{max} = 0.93 \left(\frac{\sigma}{\rho_c}\right)^{0.6} \left(\frac{\rho_c}{\rho_d}\right)^{0.1} \varepsilon^{-0.4} \tag{7-31}$$

and for all the data (Streiff et al., 1997),

$$d_{max} = 1(1 + 1.7C_v) \left(\frac{\sigma}{\rho_c}\right)^{0.6} \left(\frac{\rho_c}{\rho_d}\right)^{0.1} \varepsilon^{-0.4} 0.9^{0.6} (1 + 0.3Vi)^{0.6} \tag{7-32}$$

Figure 7-23 is a comparison of the measured drop size for different mixer types with the drop size calculated according to this equation [see Streiff et al. (1997) for more information].

In addition, Chandavimol et al. (1991a,b) have estimated the kinetic rate at which the bubbles go from initial size to the maximum equilibrium size as a function of energy dissipation. The rate of dispersion was found to be approximately proportional to energy dissipation rate. [See Figure 7-24 for a comparison of bubble breakup rate between vortex (HEV) and spiral (KMS type) static mixers.] In general, the equilibrium drop size is reached in a few pipe diameters. However, the drop size distribution is narrowed as the simultaneous processes of drop breakup and coalescence are continued, depending on the mixer design and fluid properties. See also Hesketh et al. (1987, 1991).

In the correlations above, $\varepsilon$ for a motionless mixer is given by

$$\varepsilon = \frac{\Delta PQ}{\rho AL} = \frac{\Delta P}{\rho L} V \tag{7-33}$$

Most theories and analysis of the dispersion process neglect the coexisting coalescence process. This is probably valid at very low concentration or when anticoalescing (stabilizing) chemicals are present. For example, in agitated mix tanks, if a dispersion is made at one speed and then agitated at a lower speed, the drops will grow via coalescence. Similarly, with static mixers there is an equilibrium drop size that depends on fluid properties and specific energy dissipation within the device. If the energy input is reduced, as for example in the downstream pipeline device to provide residence time, the drop will grow to a new equilibrium value.

Droplet breakup and coalescence occur in parallel in most industrial processes. An example is found in the mining industry, where metals are extracted from leach solutions by contacting with an organic phase. In both agitated vessels and static mixers, there are two stages of contacting. The first is with high specific

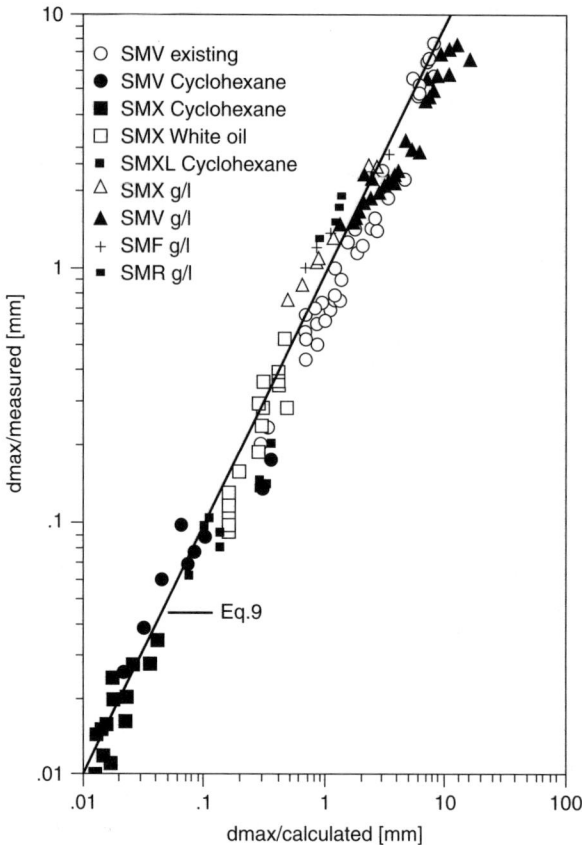

**Figure 7-23** Comparison of measured to calculated drop size for different static mixers. (From Streiff et al., 1997.)

energy dissipation, creating small droplets and a high interfacial surface area for mass transfer and reaction. This is followed by a second stage, operating at a lower energy input, providing residence time and maintaining a dispersion with increasing drop size due to coalescence. It is also interesting to note that in these processes, which operate at a high dispersed phase fraction, the mass transfer rate between phases is dependent on which phase is continuous and which is dispersed.

There have been a few studies of drop size distributions, and they appear to be similar. Calabrese et al. (1988) give

$$f_v(x) = 0.5 \left[ 1 + \mathrm{erf}\left( \frac{X - X_{ave}}{1.414\sigma} \right) \right] \qquad (7\text{-}34)$$

where $f_v(X)$ is the volume frequency distribution, erf the error function, $X_{ave}$ is the mean of the drop size distribution, and $\sigma$ is the volume-weighted standard

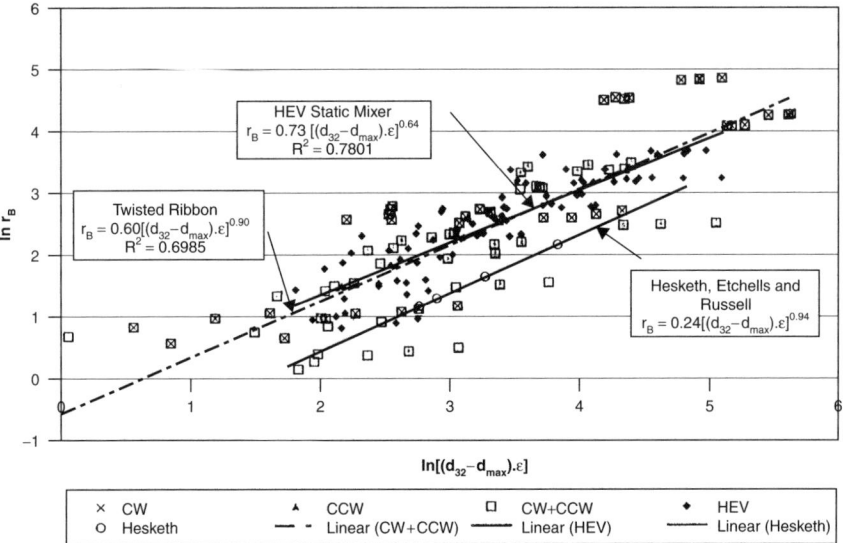

**Figure 7-24** Comparison of overall correlation of bubble breakup rate for vortex (HEV) and spiral (KMS) static mixers. (Data from Chandavimol et al., 1991b.)

deviation. $X = d/d_{32}$, $X_{ave}$ and $\sigma$ are the two constants that define the distribution. Calabrese and Berkman (1988) found that $X_{ave} = 1.12$ and $\sigma = 0.31$ fit both tank and motionless mixer data. Middleman (1974) got slightly narrower parameters, of $X_{ave} = 1.06$ and $\sigma = 0.25$. Figure 7-25 shows the distribution obtained by Calabrese and Berkman (1988).

Also of use is the table from Streiff et al. (1997) for calculating the characteristic drop size of interest: $d_{min}$, $d_{10}$, $d_{50}$, $d_{32}$, $d_{90}$, and so on. For example, $d_{50}$ means that 50% of the drop swarm volume is in drops below this diameter. The drop diameter $d_{50}$ shown below is 60% of $d_{max}$. For mass transfer the surface/volume mean diameter, $d_{32}$, is used. It is the drop diameter that will give the same mass transfer surface area as the swarm.

$$d_{min} = 0.2 d_{max} \qquad d_{10} = 0.35 d_{max} \qquad d_{50} = 0.6 d_{max}$$
$$d_{32} = 0.65 d_{max} \qquad d_{90} = 0.85 d_{max}$$

### 7-9.1.2 Laminar Flow: Dispersed Phase Drop or Bubble Size.

The laminar mechanism for dispersed phase breakup is discussed in detail elsewhere. Again there is the balance between the forces holding the dispersed phase together and those generated by the flow through the mixer. Figure 7-26 illustrates how dispersive mixing occurs in a static mixer. The drop will break up until the force holding the drop together is larger than the continuous phase force. Thus, at

**448**  MIXING IN PIPELINES

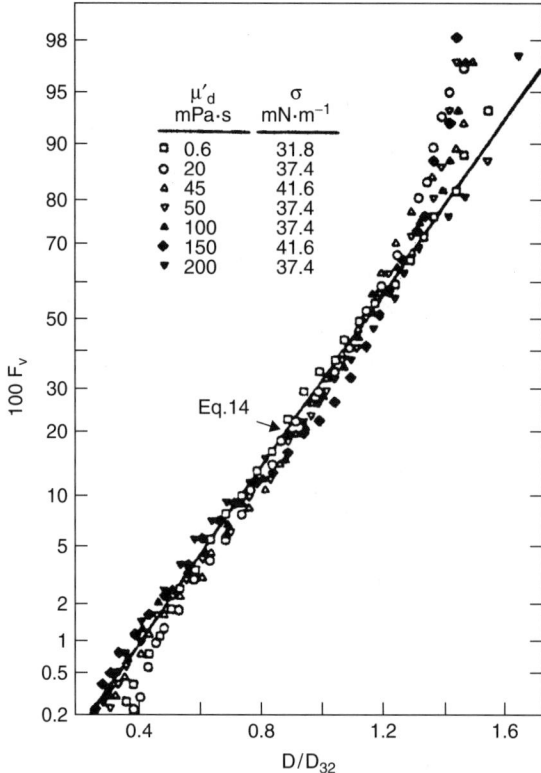

**Figure 7-25** Normalized volume distribution for constant conditions of agitation at Re = 18 000. (From Calabrese and Berkman, 1988.)

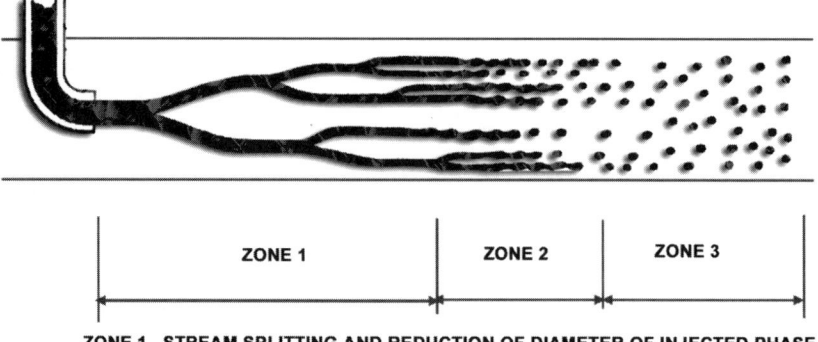

ZONE 1.. STREAM SPLITTING AND REDUCTION OF DIAMETER OF INJECTED PHASE
ZONE 2.. STREAM INSTABILTY AND BREAK-UP INTO DROPLETS
ZONE 3.. DISTRIBUTIVE MIXING

**Figure 7-26** Simplified portrayal of dispersive mixing in a static mixing device. A video clip of this process is provided on the Visual Mixing CD affixed to the back cover of the book.

MULTIPHASE FLOW IN MOTIONLESS MIXERS AND PIPES   449

equilibrium the forces are about equal.

$$\frac{\sigma}{d} = \mu_c G' \tag{7-35}$$

This leads to the relation (sometimes called a *capillary number*)

$$\text{Ca} = \frac{d\mu G'}{\sigma} \tag{7-36}$$

which can be considered as a drop size. One would expect this to depend only on the flow regime (elongational or shearing) and the viscosity ratio. Grace (1982) showed that such a relationship exists for drops broken up in laminar flow in a four-roll mill. Mutsakis et al. (1986) showed that a similar relation could be obtained for a motionless mixer. This relation and work are shown in Figure 7-27. Note that the breakup of drops only seems to occur at dispersed/continuous viscosity ratios of less than about 4. At this and higher ratios the droplets do not seem to break in shear flow. At low viscosity ratios the exponent on the viscosity ratio is between 0.33 and 0.5. Some authors give a theoretically derived viscosity ratio effect,

$$\frac{q+16}{q+19} \tag{7-37}$$

where q is the viscosity ratio, dispersed to continuous. For a range of q from 1 to 0.0, this value varies from 16/19 to 17/20. This is a very minor effect and does not accurately predict the effects measured.

From the above the maximum stable drop size can be estimated. There will be smaller drops, but in theory no drops larger than this. No data on distribution as yet exist for laminar breakup. Figure 7-28 compares drop size by laminar mechanisms with those calculated for turbulent flow. Smaller droplets are expected for laminar versus turbulent flow at the same energy dissipation rate.

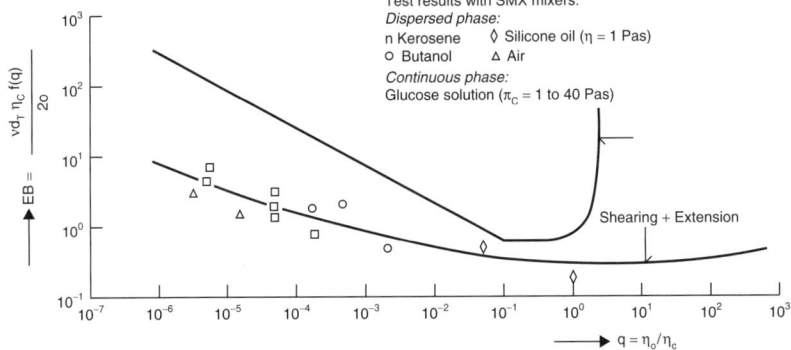

**Figure 7-27** Shear stressing, causing drop disintegration. (From Mutsakis et al., 1986.)

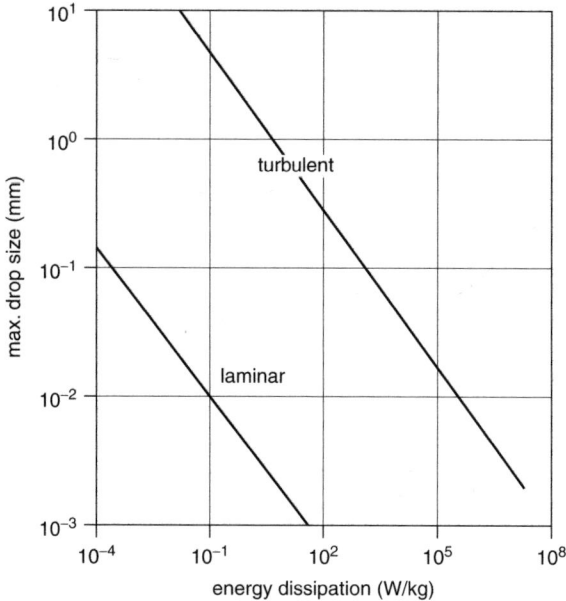

**Figure 7-28** Maximum drop size versus energy dissipation for laminar and turbulent flow. (From Streiff et al., 1999.)

Data from Grace indicate that elongational flow is more efficient at breakage than is simple shearing flow. The flow in most motionless mixers is a combination of shear and elongation. An interesting comparison of several commercially significant static mixers utilizing flow numerical analysis is provided by Rauline et al. (1998). The investigation compares mixers using criteria of extensional efficiency, stretching, mean shear rate, and intensity of segregation. Highest efficiency-creating dispersions with static mixers are achieved by designs with high extensional efficiency throughout the mixer volume. If the region of high extensional efficiency is segregated within the mixer volume, there is no guarantee that the drop to be dispersed will pass through it, and the mixer efficiency is reduced. Two designs of commercial significance in this area of application are the SMX and ISG. Use of the ISG is limited due to its high pressure drop versus that of the SMX and other designs.

Note that after the dispersion is created, distributive mixing of the discrete drops is often required to achieve a uniform mix. This can be done at a lower energy expenditure than that required to create the dispersion initially.

### 7-9.2 Dispersion of Particulate Solids: Laminar Flow

By analogy to liquid–liquid drop breakup, we can determine the mechanism for the breakup of agglomerates of fine particles. This is a common phenomenon in polymer processing, where various solid pigments and additives need to be deagglomerated and then dispersed into a viscous fluid. Tadmor and Manas-Zloczower (1994) give for the cohesive strength of an agglomerate,

$$\tau = \frac{9}{8}\left(\frac{1-\varepsilon}{\varepsilon}\right)\frac{F}{d^2} \quad (7\text{-}38)$$

where

$$F = \frac{Ad}{24z^2} \quad (7\text{-}39)$$

$\varepsilon$ is the porosity, F the inter particle force, d the primary particle size, A the Hamaker constant, and z the physical adsorption distance.

From this we see that densely packed agglomerates of fine particles will be quite strong. Breakup will be by attrition, and it will take many breakups to get to all primary particles. For example, titanium dioxide has a primary particle size of 0.25 $\mu$m but exists as a powder as agglomerates of 10 to 100 $\mu$m. Similarly, carbon black has a primary particle size of 0.03 $\mu$m, some aggregates of 0.15 $\mu$m, and agglomerates of 100 $\mu$m. Again there is a tendency for primary particles broken off agglomerates to re-form unless there are chemicals added to prevent reagglomeration (see Section 7-11.2.3).

Results of tests with the SMX mixing element in high-concentration slurries are given by Furling et al. (2000). The tests demonstrate that it is possible to prepare high solid content slurries with in-line static mixers. The slurry product is reported to have the same quality as those prepared with in-tank agitators operating at significantly higher power input. Unfortunately, a method to wet out particles using a motionless mixer has not been developed, and a tank and agitator to make the initial slurry are usually required.

### 7-9.3 Pressure Drop in Multiphase Flow

In both laminar and turbulent flow it is assumed that the mixture is pseudohomogeneous with respect to density; that is, a volume average density is used. The viscosity of the continuous phase is used. Note that for very small dispersed phase drops or bubbles (under 10 $\mu$m), the viscosity may even be higher and non-Newtonian, such as in foams and emulsions. In such cases direct measurements are required of a well-dispersed sample. Avoid correlations that average viscosities. For gas–liquid systems, the method of Lockhart and Martinelli (see Govier and Aziz, 1972) for turbulent flow is very successful and more accurate then the pseudohomogeneous method. The pressure drop for each phase flowing alone is calculated. The liquid/gas pressure drop ratio is made. This is used with an empirical correlation to get an enhancement factor for the liquid-alone pressure drop,

$$\left(\frac{\Delta P_{\text{liquid}}}{\Delta P_{\text{gas}}}\right)^{0.5} = X \quad (7\text{-}40)$$

and the two-phase pressure drop is given by

$$\Delta P_{tp} = \left(1 + \frac{20}{X} + \frac{1}{X^2}\right) \Delta P_{liquid} \qquad (7\text{-}41a)$$

or

$$\Delta P_{tp} = (1 + 20X + X^2) \Delta P_{gas} \qquad (7\text{-}41b)$$

### 7-9.4 Dispersion versus Blending

Note that the mechanisms for blending and dispersion are very different and lead to very different relations. In simple blending (without extreme viscosity differences) the degree of mixing depends on the length of the mixer and is independent of the flow rate, shear rate, and continuous phase fluid properties. Turndown does not change the blending CoV. Extra length improves mixing. The cost of mixing (i.e., pressure drop) does depend on the flow and physical properties. In dispersion the physical properties have a strong influence. Flow rate sets the drop size, a function of specific energy input. The length to achieve equilibrium drop size is short, and more length beyond that does not affect the process result of droplet size.

### 7-9.5 Examples

***Example 7-3: Liquid–Liquid Contacting—Turbulent Dispersion.*** A stream from a reactor is be contacted with an immiscible solvent to extract the product. A motionless mixer is planed. After the mixer the two streams will enter a decanter (Table 7-9). The cut size of the decanter is 125 μm, so a goal drop size for the mixer is 500 μm. Choose a Kenics KMS mixer based on past experience. No line size is given.

Typical line velocities are 2 to 3 m/s. Use 2 m/s as a first guess. The line size initial estimate: $0.0075/2 = 0.0037$ m² $\Rightarrow 0.069$ m $= 69$ mm; use 3 in. pipe with an inside diameter of 76.2 mm. The velocity is then 1.63 m/s.

**Table 7-9** Process Stream Data

|  | Main Stream | Secondary Stream |
|---|---|---|
| Density (kg/m³) | 1154 | 982 |
| Viscosity (mPa · s) | 1.07 | 0.475 |
| Flow (kg/h) | 21 755 | 8149 |
| Flow (m³/s) | 0.0052 | 0.0023 |
|  | Combined Streams |  |
| Total flow | 0.0075 m³/s |  |
| Interfacial tension | 12 mN · m |  |
| Volume fraction | 0.30 |  |

MULTIPHASE FLOW IN MOTIONLESS MIXERS AND PIPES    **453**

Do a rough estimate of drop size using the simpler but less exact Weber number from eq. (7-25):

$$We = D\frac{\rho V^2}{\sigma} = \frac{0.0762 \times 1154 \times 1.63^2}{12 \times 10^{-3}} = 19\,283$$

$$d_{32} = \frac{KD}{We^{0.6}} = 0.49 \times \frac{0.076}{(19\,283)^{0.6}} = 100 \; \mu m$$

Then the average drop size from eq. (7-25) is 100 μm.

This is too small; we must go to larger pipe—use 4 in. pipe with a diameter of 0.100 m and a velocity of 0.95 m/s:

$$We = \frac{0.100 \times 0.95^2 \times 1154}{12 \times 10^{-3}} = 8596$$

$$d_{32} = 0.49 \times \frac{0.100}{(8596)^{0.6}} = 213 \; \mu m \qquad \text{looks good}$$

Check the Reynolds number:

$$Re = \frac{\rho DV}{\mu} = \frac{0.1 \times 0.85 \times 1154}{1.07 \times 10^{-3}} = 91\,672 \qquad \text{turbulent}$$

Check for orientation using eq. (7-42) to calculate the Froude number:

$$Fr = \frac{\rho V^2}{\Delta \rho \, Dg}$$

$$= \frac{1154 \times 0.95^2}{(1154 - 982) \times 0.1 \times 9.8} = 6.2$$

A vertical installation is required since the Froude number is less than 20.

For the exact droplet size calculation, the energy dissipation is obtained from eqs. (7-33) and (7-17):

$$\varepsilon = \frac{\Delta P}{L}\frac{V}{\rho} = \frac{4f\,V^3}{D\,2} = 2f\frac{V^3}{D}$$

The friction factor for the KMS is given in Table 7-6 as $0.5 \times 1.5 = 0.75$:

$$\varepsilon = \frac{2 \times 0.75 \times 0.95^3}{0.1}$$

$$\frac{\sigma}{\rho} = \frac{15 \times 10^{-3}}{1154} = 1.3 \times 10^{-5}$$

Using eq. (7-31) yields

$$d_{max} = \frac{0.93 \times (1.3 \times 10^{-5})^{0.6}}{(1154/982)^{0.1}(12.86)^{0.4}}$$

To check the viscosity effect, use eq. (7-28):

$$Vi = \mu_d \frac{(\varepsilon d_{max})^{0.33}}{\sigma} \left(\frac{\rho_c}{\rho_d}\right)^{0.5}$$

$$= 0.475 \times 10^{-3} \frac{(12.86 \times 5 \times 10^{-4})^{0.33}}{15 \times 10^{-3}} \left(\frac{1154}{982}\right)^{0.5} = 0.0058$$

so the effect of viscosity is to increase the droplet size by 1.056, which is negligible.

To check the density effect:

$$\frac{\rho_c}{\rho_d} = \frac{1154}{982} = 1.1752^{0.1} = 1.01 \quad \text{negligible}$$

To check the concentration effect, use eq. (7-26):

$$1 + 0.3 \times 5 = 2.5 \quad \text{significant}$$

We can use a 4 in. pipe with a KMS mixer located vertically. Use about 10 diameters of mixer for a residence time of about 1 s.

Using Figure 7-25, and eq. (7-34) with $d/d_{32} = 125/500$ gives a volume fraction less than 0.2% under 125 μm, and this would be the approximate carryover in the decanter with a cut size of 125 μm.

*Example 7.4: Blending and Dispersion—Large Viscosity Ratios.* A molten polymer is to have several materials added. One is a soluble antioxidant and the other is an immiscible silicone oil slip agent (Table 7-10). The pipe size is 50 mm. We will work out a motionless mixer design.

$$\text{flow} = \frac{1500/3600}{900} = 4.62 \times 10^{-4} \text{ m}^3/\text{s}$$

$$\text{area} = \frac{3.1416}{4 \times 0.05^2} = 1.96 \times 10^{-3} \text{ m}^2$$

$$\text{miscible volume fraction } C_v = \frac{15/600}{15/600 + 1500/900} = 0.015$$

**Table 7-10**

|  | Polymer | Miscible | Immiscible |
|---|---|---|---|
| Flow (kg/h) | 1500 | 15 | 10 |
| Density (kg/m³) | 900 | 600 | 1010 |
| Viscosity (Pa · s) | Shear-thinning index of 0.7 1200 at 1 s⁻¹ | 100 | 0.5 |
| Interfacial tension (m · N/m) |  | 12 |  |

## MULTIPHASE FLOW IN MOTIONLESS MIXERS AND PIPES 455

velocity = 0.2357 m/s

empty pipe shear rate $G' = 8\dfrac{V}{D} = 8 \times \dfrac{0.2357}{0.05} = 37 \text{ s}^{-1}$

viscosity ratio miscible = $\dfrac{0.5}{1200 \times 1000} = 4 \times 10^{-7}$

use an SMX because of the viscosity ratio.

shear rate in mixer (Table 7-7) $G' = 64 \times \dfrac{0.2357}{0.05} = 301 \text{ s}^{-1}$

effective viscosity $\mu = \dfrac{1200}{G'^{(1-0.07)}} = 216 \text{ Pa} \cdot \text{s}$

viscosity ratio in mixer $\dfrac{0.5/216}{1000} = 2.3 \times 10^{-6}$

Reynolds number $\text{Re} = \dfrac{\rho D V}{\mu} = \dfrac{0.05 \times 0.2357 \times 900}{216} = 0.0491$

friction factor $f = \dfrac{16}{\text{Re}} = 325$

pressure drop empty pipe per meter = $\dfrac{\Delta P}{L} = \dfrac{4f \rho V^2}{D \cdot 2}$

$= \dfrac{4(325)}{0.050} \times 900 \times \dfrac{0.2357^2}{2} = 649\,987 \text{ Pa/m or 6.5 bar/m}$

which is rather high.

From Table 7-5, the pressure drop for the SMX is 37.5 times larger. What length is required to get to an outlet coefficient of 5%? The starting CoV from the volume fraction is, from eq. (7-6),

$$\text{CoV} = \left(\dfrac{0.985}{0.015}\right)^{0.5} = 8.1$$

$$\text{CoV}_r = \dfrac{0.05}{8.1} = 0.0062$$

For equal viscosities, using Table 7-5 we have

$$\text{CoV}_r = 0.63^{L/D}$$

$$\ln 0.0062 = \dfrac{L}{D} + \ln 0.63 = 4.62$$

Add a length for the viscosity ratio, eq. (7-23):

$1.0 \log(2.3 \times 10^6) = 6.36$ additional length

total length = $4.62 + 6.36$ for $11 \times 50$ mm = 0.549 m

For the immiscible system, from Figure 7-27 and eq. (7-36):

$$\frac{\mu_d}{\mu_c} = \frac{100}{260 \times 1000} = 4 \times 10^4$$

The capillary number is

$$1.0 = d\frac{\mu_c G'}{\sigma} = d \times \frac{260 \times 301}{12 \times 10^{-3}}$$
$$= 0.15 \; \mu m$$

Because of the length needed for blending the system should reach this value. The final pressure drop is $6.5 \times 0.549 \times 37.5 = 133$ bar. We may decide that the pressure drop is too high and go to larger pipe, say 75 mm.

The mixing length for blending will not change in terms of L/D because there is no effect on blending of flow rate. The actual length will now be

$$11 \times 75 \text{ mm} = 0.825 \text{ m}$$

The new velocity, Reynolds number, and shear rate will be lower:

$$V = 0.1053 \text{ m/s}$$
$$G' = 64 \times \frac{0.1053}{0.075} = 89 \text{ s}^{-1}$$
$$\mu = \frac{1200}{89^{0.3}} = 312 \text{ Pa} \cdot \text{s}$$
$$Re = 0.075 \times 1053 \times \frac{900}{312} = 0.0228$$

The new pressure drop is

$$f = \frac{16}{Re} = 702$$
$$\frac{\Delta P}{L} = \frac{4f \; \rho V^2}{D \; 2} = 4 \times \frac{702}{0.075} \times 900 \times \frac{0.1053^2}{2}$$
$$= 1.87 \text{ bar per meter of empty pipe}$$

For the static mixer, $\Delta P = 0.825 \times 1.87 \times 37.5 = 58$ bar about one-half the pressure drop.

## MULTIPHASE FLOW IN MOTIONLESS MIXERS AND PIPES  457

**Table 7-11**  Process Stream Data

|  | Main Steam: Sour Gas | Side Stream: Dilute Caustic |
| --- | --- | --- |
| Flow rate | 280 000 scfh | 44 gpm |
| Density (kg/m$^3$) | 4.97 | 1057 |
| Viscosity (mPa · s) | 0.02 | 1.8 |
| Pressure (psig) | 73 | as required |
| Temperature (°F) | 105 | 100 |

*Example 7-5: Gas Continuous Multiphase—Turbulent Dispersion.* A refinery sour gas stream containing 1.21 mol % hydrogen sulfide is to be scrubbed to reduce H$_2$S to below 50 vppm. Since the flow rates are relatively low, it is decided that a dilute caustic solution is the most economical chemistry (Table 7-11). With fast chemical reaction, a single stage of contacting is all that is required to achieve desired contaminate removal. This is provided by spray nozzle feeding the caustic solution into a static mixer where the contacting/absorption and chemical reaction take place. Downstream of the static mixer a mist eliminator separates entrained liquid from the cleaned product gas. A 6 in. diameter line size and horizontal installation are preferred to best fit the available space. A pressure drop of 10 psi is available for the static mixer. A SMV design mixer consisting of 6 diameters packed length is determined to be required for the process. We are to determine the suitability of the 6 in. diameter mixer and the pressure drop.

The Froude number will be evaluated to determine the suitability of horizontal flow. The pressure drop will then be calculated for both the gas phase and liquid phase flowing separately in the empty pipe. The pressure drop for the combined flow streams will then be determined using the method of Lockhart and Martinelli. Finally, the mixer pressure drop will be determined using the multiplier of the empty pipe pressure drop reported for the SMV mixer specified.

Convert from standard conditions to flow at actual operating temperature and pressure using the ideal gas law. Standard conditions are 60°F and 1 atm in the gas process industry.

$$Q_{actual} = Q_{standard} \frac{P_{standard}}{P_{actual}} \frac{T_{actual}}{T_{standard}}$$

$$= 280\,000 \times \frac{14.7}{14.7 + 73} \times \frac{460 + 105}{520} = 50\,992 \text{ acfm}$$

or in metric, the actual gas flow rate is 0.4012 m$^3$/s.

Six-inch schedule 40 pipe specified: ID = 6.06 in. or 0.154 m, and an open cross-section of 0.0186 m$^2$. The gas velocity in open pipe flow is (0.4012 m$^3$/s) 0.0186 m$^2$ = 21.57 m/s. Calculate the Reynolds number using eq. (7-1) to confirm the flow regime:

$$Re = \frac{\rho DV}{\mu} = \frac{0.154 \times 21.57 \times 4.97}{2.0 \times 10^{-5}} = 825\,462 \quad \text{the flow is turbulent}$$

Check the Froude number using eq. (7-42) to see if stratification will be a problem in horizontal flow:

$$Fr = \frac{\rho V^2}{\Delta \rho \, Dg}$$

$$= \frac{4.97 \times 21.57^2}{(1150 - 4.97) \times 0.154 \times 9.81}$$

$$= 1.34 \quad (<20 \text{ and too low for horizontal flow})$$

Since the Froude number is below 20, the mixer should be set in vertical orientation with flow down. The horizontal installation requested would be risky, especially at startup and turndown operation.

The friction factor is required to calculate the pressure drop. Use eq. (7-18) for turbulent flow:

$$f = \frac{0.079}{Re^{0.25}} = \frac{0.079}{825\,462^{0.25}} = 0.00259$$

The pressure drop for gas flow only in the empty pipe is then determined using eq. (7-17):

$$\Delta P_{gas} = 4f \frac{\rho V^2}{2} \frac{L}{D}$$

$$= \frac{4(0.00259)(4.97)(21.57)^2(0.924)}{2 \times 0.154} = 71.87 \text{ Pa or } 0.00072 \text{ bar}$$

Now in similar fashion, calculate the pressure drop for the liquid phase only. At a feed rate of 44 gal/min, or 0.00278 m³/s:

$$\text{velocity, } V = \frac{0.00278}{0.0186} = 0.149 \text{ m/s}$$

$$\text{Reynolds number, } Re = \frac{1150 \times 0.154 \times 0.149}{0.0018} = 14\,660$$

$$\text{friction factor, } f = \frac{0.079}{14\,660^{0.25}} = 0.0072$$

$$\text{pressure drop, } \Delta P_{liquid} = \frac{4(0.0072)(1150)(0.149)^2(0.924)}{2 \times 0.154}$$

$$= 2.2 \text{ Pa or } 0.000022 \text{ bar}$$

For the combined stream pressure drop in empty pipe, we use eqs. (7-40) and (7-41):

$$X = \left(\frac{\Delta P_{liq}}{\Delta P_{gas}}\right)^{0.5} = \left(\frac{2.2}{71.87}\right)^{0.5} = 0.175$$

$$\Delta P_{total} = \left(1 + \frac{20}{X} + \frac{1}{X^2}\right) \times \Delta P_{liq} \quad \text{[using eq. (7-41a), liquid predominant]}$$

$$= 1 + \frac{20}{0.175} + \frac{1}{0.175^2} \times 0.000022 = 0.0032 \text{ bar in empty pipe}$$

or

$$\Delta P_{total} = (1 + 20X + X^2) \times \Delta P_{gas} \quad \text{[using eq. (7-41b), gas predominant]}$$

$$= (1 + 20 \times 0.175 + 0.175^2) \times 0.00072 = 0.0033 \text{ bar in empty pipe}$$

In this case there is good agreement between the values calculated for total pressure drop. Since gas flow is predominant, the total empty pipe pressure drop calculated from the gas-only pressure drop using eq. (7-41b) should be used. It should be noted that this Lockhart–Martinelli correlation is considered to be conservative when used in vertical downward flow. The original work was all in horizontal flow.

Now for the SMV mixer in turbulent flow, we use eq. (7-16) and Table 7-6:

$$\Delta P_{sm} = K_T \, \Delta P_{pipe} = K_T(0.0033)$$

With $K_T$ given as ranging from 100 to 200, the pressure drop is expected to be 0.33 to 0.66 bar, or in English units, 4.8 to 9.6 psi. This is less than the maximum 10 psi allowed, and the design is acceptable based on this preliminary calculation. The static mixer vendor should be consulted for more exact determination of pressure drop based on the specific SMV mixing element being used.

There are other comments to be made about this type of application. The liquid spray nozzle should supply liquid to the face of the mixing elements without appreciably wetting the vessel wall. A 30° full cone nozzle is typically used. Ideally, the spray should consist of droplets in the range 1000 to 2000 μm range. A quick review of spay nozzle literature indicates that the appropriate spray nozzle for 44 gal/min of alkaline water would operate at about 100 psi pressure drop. Multiple nozzles could be used if a lower liquid-side pressure drop is desired. Fine atomization spray nozzles should be avoided since fine spray drops are difficult to separate in downstream mist eliminator equipment. A filter or strainer should be installed on the liquid feed to prevent plugging the feed nozzle, especially if the nozzle orifice size is small.

## 7-10 TRANSITIONAL FLOW

The previous discussion was devoted to processes that were either laminar or turbulent. The transition between these flow regimes is set by the Reynolds number based on the pipe diameter. For a simple pipe the traditional number is 2100, but there is a large transition range. For motionless mixers the Reynolds number is usually much lower based on pressure drop. This is due to the much higher rates of energy dissipation due to the internals. The exact value depends on the mixer design but is in the 500 range for many. However, this transition is based on the change measured by pressure drop. The eddy structure starts to change at higher Reynolds numbers. There is very limited test work showing that the quality of turbulent mixing is poorer at low Reynolds number, due to these turbulence changes. Care should be taken in this region.

The Kenics HEV mixer, which consists of tabs, shows a transition in mixing performance at a very high Reynolds number. This is believed due to the change in vortex structure off the tabs at a specific tab Reynolds number rather than a pipe Reynolds number. Since the tab/diameter ratio is kept constant, this occurs at a higher pipe Reynolds number.

## 7-11 MOTIONLESS MIXERS: OTHER CONSIDERATIONS

### 7-11.1 Mixer Orientation

*7-11.1.1 Density Ratio Effects in Blending.* If the mixer is located vertically, there is little or no effect of density ratio between the added and main streams. If the mixer is horizontal, a density ratio could cause separation before mixing. This is most important in plain pipe, where a 10% density difference is reported to increase the mixing length tenfold. Vertical orientation is recommended by some manufacturers when the densiometric Froude number is less than 20 in turbulent flow situations, and the ratio of Froude to Reynolds number should be is less than 1.0 in laminar systems. Criteria for orientation in the turbulent flow regime:

$$Fr' = \frac{\rho V^2}{\Delta \rho\, D_h g} < 20 \qquad (7\text{-}42)$$

Criteria for orientation in the laminar flow regime:

$$\frac{Fr'}{Re} = \frac{\mu V}{\Delta \rho g D_h^2} < 1 \qquad (7\text{-}43)$$

*7-11.1.2 Density Ratio Effects in Dispersion.* Again, the manufacturers recommend vertical arrangements. Horizontal arrangements will cause separation and increase coalescence, which will cut down on the interfacial area. The same criteria as used above in blending are therefore recommended.

### 7-11.2 Tailpipe/Downstream Effects

*7-11.2.1 Turbulent Blending.* The length downstream of the mixer is often referred to as the *tailpipe*. In laminar flow, no further mixing occurs in this region.

In turbulent flow, however, mixing continues as the extra turbulence generated by the mixing elements dies out. This effect lasts one or two diameters. For the HEV an extra factor of 2 in CoV reduction has been observed for SMVs; the range is from 1.5 to 2, depending on the number of plates in the mixer (Tauscher and Streiff, 1979). This observation suggests that the overall pressure drop can be minimized for a given mixing task by spacing mixers and empty pipe.

**7-11.2.2 Turbulent Dispersion: Coalescence.** After the dispersed phase leaves the motionless mixer, it will tend to coalesce to an equilibrium drop or bubble size characteristic of the shear field in the downstream piece of pipe. This coalescence is not just a phenomenon of the downstream tailpipe but is a process happening in parallel with dispersion. It is not as well understood. We do know that just like dispersion, coalescence is affected by volume concentration and is promoted by turbulence. Coalescence is strongly affected by surface chemistry effects. The role of many chemicals added to stabilize dispersions is to slow down the coalescence rate.

**7-11.2.3 Laminar Dispersion: De-mixing.** Often, motionless mixers are used to mix particulate solids into fluids flowing in the laminar flow regime. A high concentration slurry is mixed into a main stream and the mixers are to distribute the material across the diameter of the pipe (e.g., a pigment concentrate is added to a polymer line). This is a common and successful application of a mixer. However, because of the nature of laminar flow, discrete particles may move to agglomerate due to the velocity gradients in laminar flow. There is a radial diffusivity that causes particles to move away from the wall. This is usually not a problem [but for more information see the work of Acrivos and Leighton (1987)]. The parabolic profile has been a problem in some cases. Because of the differential velocity, particles on one streamline can catch up with those on another. This can lead to agglomerates, which can adversely affect downstream processing. This is often termed *Smolachowksi agglomeration*. Agarwal et al. (1998) and Chimmili et al. (1998) have shown that in laminar flow the key variable can be combined as

$$\frac{d}{d_0} = \exp\left(\frac{-8fC_vG't}{3\pi}\right) \qquad (7\text{-}44)$$

or

$$\frac{d}{d_0} = \exp\left(\frac{-8fC_vL}{3\pi D}\right) \qquad (7\text{-}45)$$

where f is the efficiency of collision. Thus, this is a problem at high concentrations and with long pipes in laminar flow. It has been suggested that this value be kept lower than 1 (see Agarwal et al., 1998). To prevent such de-mixing agglomeration, extra motionless mixers are installed along long transfer lines.

## 7-11.3 Effect of Inlet Position

In laminar flow, vendors usually recommend coaxial centerline injection, often at the edge of an element. An interesting study was done by Hobbs et al. (1997, 1998) on the effect of injection position on degree of blending in laminar flow. It found that the initial injection position affected quality of mixing and that this effect was equivalent to several lengths to diameters of mixing, depending on mixer type. For a twisted mixer (KMS), nonoptimum addition could add the equivalent of four diameters of mixer to get the same CoV. For a cross-member mixer (SMX), nonoptimum costs about two diameters of length. For short mixers, both of these are significant reductions in mixing capability. Staged addition along the mixer axis may be required when adding large volumetric rates (>25%) of low viscosity fluids to high viscosity base materials in laminar flow. If the low viscosity phase should become continuous, the shear stress will be reduced and mixing will be reduced. When adding immiscible additives, consideration must be made to ensure that the mixer hydraulic diameter is of adequate size to prevent *flooding*, a term used to describe the low viscosity additive phase becoming continuous. Staged mixers with decreasing hydraulic diameter may be required to avoid flooding and also to achieve desired drop size. In turbulent flow evidence exists that injection is also very important, as the number of mixing elements are very low. Off-center injection and bends in front of turbulent mixers can drastically reduce the effect of the first element and thereby significantly reduce overall performance. Vendors' guidelines should be followed. If they cannot be followed, extra diameters of mixers should be added.

## 7-11.4 Scale-up for Motionless Mixers

In most cases motionless mixers can be designed based on fluid physical properties and process understanding without the need to run experiments on a small scale. The equations given above and vendor correlations will allow a large percentage of mixers to be designed without any scale-up or scale-down testing. However, in some cases data exist only on a small scale and the desire is to scale up to a larger processing capacity, achieving the same results as demonstrated on the smaller scale. An example of where small scale laboratory or pilot testing is required is when the fluid physical properties or the rate constants are unknown. Another is when the exact process result is unclear for example attrition or agglomeration or dispersion. Scale-up in heat transfer applications is a special case discussed in Chapter 14. In such cases the basic understanding discussed above gives guidance for scale-up.

We will use the flow rate ratio between big and small scale, $R$, as a scaling parameter and small and capital letters for the various variables:

$$R = \frac{Q}{q} = \frac{V}{v}\left(\frac{D}{d}\right)^2 \qquad (7\text{-}46)$$

For operation in a laminar flow regime, we get the same time effects when we keep the residence time the same:

$$\frac{L}{V} = \frac{\ell}{v} \tag{7-47}$$

If we keep the shear rate the same in both scales, we also keep the shear stresses the same:

$$\frac{V}{D} = \frac{v}{d} \tag{7-48}$$

This leads immediately to keeping the length/diameter ratio the same, and

$$D = dR^{0.33} \tag{7-49}$$

and the pressure drops will be

$$\Delta P_{large} = \Delta P_{small} \tag{7-50}$$

since the pressure drop is proportional to velocity and length and inversely to diameter squared.

For the turbulent flow regime, the residence time is again held the same, and now the energy dissipation is held constant for any reactive or two-phase effects. This leads to keeping

$$\frac{L}{V} \text{ and } \frac{V^3}{D} \text{ constant} \tag{7-51}$$

Then

$$D = d^{3/7} \tag{7-52}$$

The length/diameter ratio is no longer constant.

In some cases there may be a flow regime change so that the Reynolds number, which always increases on scale-up, must always be compared on both scales. Scale-up is most reliable when both large and small scale systems are operating in the same flow regime. As mentioned in Section 7-10, care should be taken when operating in the transition regime.

## 7-12  IN-LINE MECHANICAL MIXERS

There are a number of cases in which mechanical mixers are put in line to promote mixing. Many years ago, small tanks (often made of pipe) and mechanical agitators were manufactured to be put in line. The residence times were low and the agitators of modest power with high speed. These devices could give very short blend times or high local energy dissipations. The internal flow is high compared to the through flow to avoid bypassing. Sometimes, staged vessels were used with horizontal baffles. In most cases these devices have been replaced by motionless mixers when the pressure drop is available. The chief incentive was removing a rotational piece of equipment with its shaft and seals from a process. Maintenance is reduced and leakage eliminated with static mixers. This movement away from in-line mechanical mixers to static mixers is given momentum by the movement toward process chemistries operating at very high pressure, 1000 psi and higher.

### 7-12.1 Rotor–Stator

Still very popular for in-line dispersion are the class of rotor–stator mixers. These devices look more like pumps than like stirred tanks. Volumes are small but rotational speeds and powers are high, giving high local energy dissipation. They are often staged with several rotors separated by stators that reduce bypassing.

The literature is small on rotor–stator devices, as discussed in Chapter 8. [They are also discussed in Chapter 12, since this is the process result (liquid–liquid and solid–liquid) that these devices are most used for.] It should be mentioned that because of their very high speed they can produce turbulent motion in some rather high viscosity fluids. Also see Cohen (1998), Dietsche (1998), and Myers et al. (1999) for general information about the industrial application of rotor–stator mixers versus other mixing options.

### 7-12.2 Extruders

Extruders are sometimes used as mixers. But the primary purpose of most extruders is to melt polymer pellets and to increase the pressure in the melt (i.e., to pump). Neither of these steps requires a lot of radial mixing. What radial mixing may occur is secondary. In particular, single-screw extruders generate little radial mixing. Some radial mixing occurs in the melting zone, but this is very hard to predict. If material is injected into the single-screw flights, very little radial mixing is found. Twin-screw extruders, however, can show some radial mixing, due to their configuration, which promotes radial mixing. Attempts to quantify this type of mixing have not been as successful as with motionless mixers. For more information, consult a good polymer processing text such as Tadmor and Manas-Zloczower (1994). Specially designed twin-screw extruders (both co-rotating and counterrotating) are used for compounding (mixing), cooling (heat transfer), reaction (mixing and plug flow), and devolatilization (heating). See Biesenberger (1983) for more on the subject of devolatilization. In all cases, static mixers are attractive alternatives. For the topic of *compounding*, a fancy name for mixing and dispersion in polymers, see Tadmor and Manas-Zloczower (1994). Many mechanical dispersion processes are in commercial operation, and others are continually being developed. Manufacturers of static mixing equipment have varying degrees of experience in these areas of application in polymer processing and should be interviewed if there is interest in this growing area of activity.

## 7-13 OTHER PROCESS RESULTS

### 7-13.1 Heat Transfer

***7-13.1.1 Turbulent Flow Heat Transfer in Pipes.*** In turbulent flows in pipes, the relations for heat transfer to or from the fluid and the wall are well known and are the basis of many heat exchanger designs. The overall heat transfer resistance consists of a resistance on the cooling or heating side, a wall resistance, and a fluid resistance inside the pipe. The latter depends on the nature of the fluid and the flow regime and is the one of primary interest in this section. As is common practice we will use the individual heat transfer concept to describe resistances. For a detailed discussion, see any heat transfer text (e.g., Burmeister, 1983a,b). Static mixers and other pipe inserts are not typically justified in turbulent flow, since the cost and added pressure drop are high relative to the benefit achieved. However, in some cases, such as condensing, flow turbulence promoters in the form of long, spiral internals are added to promote film forming (Burmeister, 1983a,b). A general form of the equations for predicting inside or process heat transfer coefficient is given by

$$\mathrm{Nu} = A \cdot \mathrm{Re}^a \cdot \mathrm{Pr}^b \left(\frac{\mu}{\mu_w}\right)^c \tag{7-53}$$

where, typically, $A = 1$ to $2$, $a = 0.66$, $b = 0.33$, and $c = 0.14$. Note that even though the flow is turbulent, there is still a significant viscosity effect. The addition of the viscosity ratio is empirical and based on the observation that when cooling, the heat transfer coefficients are lower than when heating. This adjustment is needed because heat transfer takes place in a wall film and the conditions of this film are better characterized by local properties at the local temperature. In general, the viscosity is much more sensitive to the temperature than are any of the other physical properties.

***7-13.1.2 Laminar Flow Heat Transfer in Pipes.*** In laminar flow the heat transfer rates are greatly reduced. There is no radial flow and temperature gradients build up, reducing the heat transfer rate. In addition, the parabolic velocity distribution causes the center to spend little time in the heating zone and the fluid near the walls to spend more time. The overall effect is that the local heat exchange coefficient becomes a function of length and can actually approach zero at long lengths. In other words, heating in laminar flow is very poor and uniform. For example, when heating in an empty pipe, the flow can be viewed as a series of streamlines, as shown in Figure 7-29. Material near the center of the pipe flows fast and heats up slowly compared to material flowing near the pipe wall. At the outlet the product is a mixture of material with drastically different time, temperature, and shear histories. There exists a thermally overloaded zone near the wall and a thermally underloaded zone near the center.

**466**  MIXING IN PIPELINES

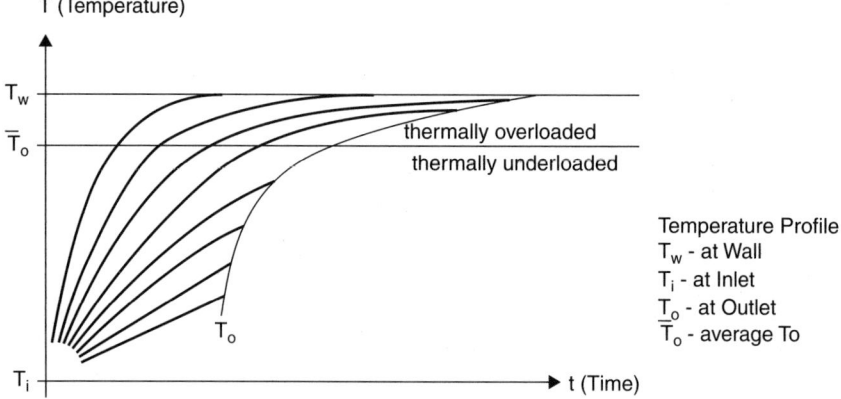

**Figure 7-29** Thermal loading profile for laminar flow heat transfer in empty pipe. (From Heierle, 1989.)

Performance problems related to maldistribution also exist in cooling applications, especially where viscosity increases as temperature is lowered. At worst, case equipment could become inoperable, due to plugging of all but the center of the flow channel. This condition can be eliminated by the use of static mixer internals, discussed later. Heat transfer coefficients for laminar flow in empty pipe are correlated by

$$\text{Nu} = A \cdot \text{Re}^a \cdot \text{Pr}^b \left(\frac{\mu}{\mu_w}\right)^c \left(\frac{D}{L}\right)^d \qquad (7\text{-}54)$$

where $A = 1.61$, $a = 0.33$, $b = 0.33$, $c = 0$ to $0.14$, and $d = 0.33$. This equation has a limit at a Nusselt number of about 3.7. Higher Nusselt numbers are achieved only at short lengths, providing only limited heat transfer surface.

In laminar flow where pressure drops are high, there is viscous heating, which is on the order of 1°C per 10 bar pressure drop for most polymers. This temperature rise is quite nonuniform, causing radial temperature gradients of several degrees to appear. In addition, extruders and pumps also generate thermal nonuniformity and radial temperature gradients. Even multiflow heat exchangers show large radial temperature gradients. Radial temperature gradients up to 10° or more are not uncommon.

Since the static or motionless mixer promotes radial flow of both momentum and heat, it will significantly enhance the heat transfer rate. Figure 7-30 shows static mixer heat transfer packing installed in the tubes of a multitube heat exchanger.

The same form of the heat transfer equation is used:

$$\text{Nu} = A(\text{Re} \cdot \text{Pr})^a \left(\frac{\mu}{\mu_w}\right)^b \left(\frac{D}{L}\right)^c \qquad (7\text{-}55)$$

OTHER PROCESS RESULTS **467**

**Figure 7-30** Static mixer for heat transfer enhancement in a multitube heat exchanger (SMXL). (Courtesy of Koch-Glitsch, LP.)

where A depends on vendor correlation, a = depends on vendor correlation, b = 0.14, and c = depends on vendor correlation. Note that there is now no effect of viscosity. The mechanism is one of surface renewal and differs from purely convective flow in that it is unaffected by film effects. It is also debatable whether there is an actual D/L effect, as there is continual surface renewal. There is few experimental data, and it appears that the laminar form was used by some vendors with a fixed L/D. Most vendors do not include length in their correlation for laminar flow heat transfer. Table 7-12 gives some typical values for constants A and a used in eq. (7-55) (Streiff, 1986). Note also that sometimes the constant $a$ is set to 0.33 and A is restated for comparison of data from different sources.

The increase in heat transfer resulting from surface renewal related to radial mixing is significant. Figure 7-31 compares the laminar flow heat transfer coefficient achieved with the type SMX and SMXL mixing element with that of an empty pipe of the same diameter (Heierle, 1988). A four to tenfold increase in

**Table 7-12** Heat Transfer Coefficients for Established Flow

|  | A | $a^a$ |
|---|---|---|
| Kenics KMS |  |  |
|   Edge seal | 1.5 | 0.33 |
|   No edge seal | 2.25 | 0.33 |
| SMX | 2.6 | 0.35 |
| SMXL | 0.98 | 0.38 |

[a] See eq. (7-55).
*Source*: Streiff (1986).

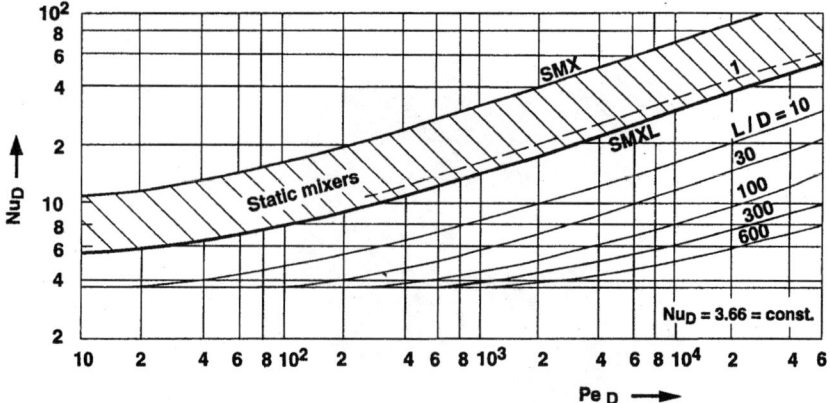

**Figure 7-31** Heat transfer to the inner tube wall with laminar flow in empty pipe and static mixers. (From Heierle, 1989.)

heat transfer is shown. A slight added benefit could be obtained by conduction of heat through the element blade (called the *fin effect*), but only if the mixing element is securely sealed to the pipe wall. Also note that for SMX and SMXL mixing elements, unlike empty pipe and other heat transfer enhancements, the heat transfer coefficient is not a function of the diameter/length ratio.

An interesting motionless mixer heat exchanger is the SMR, where the cross members that provide the radial mixing are actually heat exchanger tubes with heat transfer medium flowing in them. This increases the heat exchange area per unit length five to tenfold. Heat transfer surface per unit volume can be maintained as equipment volume increases, making the design very attractive for scale-up of reactor processes requiring precise temperature control. See Figure 7-32 for a picture of the SMR mixer-heat exchanger-reactor. The radial mixing is good, giving high heat transfer coefficients (see Table 7-2). SMR mixers do not produce radial temperature uniformity on a fine scale. They are often followed by a short length of smaller conventional mixer to give fine scale radial temperature uniformity.

### 7-13.1.3 Notes Regarding the Scale-up of Heat Exchangers Containing Static Mixer Heat Transfer Packing.

Conventional static mixers such as the SMX, SMXL, and KMS types accomplish heating and cooling by the transfer of heat through the vessel wall. This limits scale-up as a single tube design. The heat transfer coefficient decreases in inverse proportion, while product volume and total heat load increase as the second power with increasing pipe diameter. Therefore, the volume-related heat transfer capacity is very small with larger diameters. Pipe diameters can be increased only up to about a nominal 100 mm (4 in.) maximum. With high throughputs and/or heat load, the required pipe length, and consequently pressure drop, would become excessive.

Multitube designs offer a partial solution to the limitation above. Tube diameters can be maintained small and high volume related heat transfer capacity

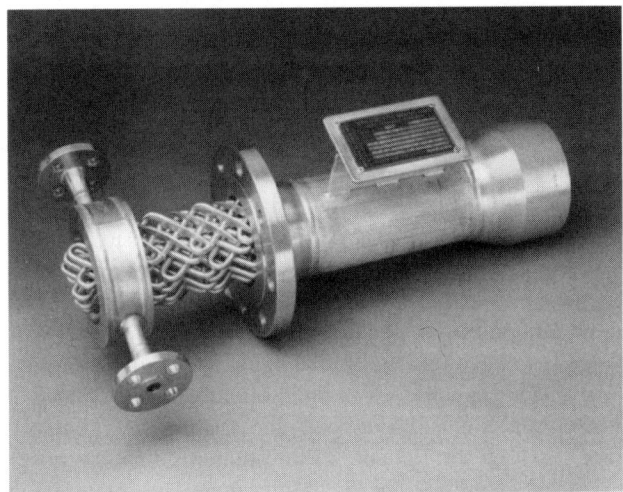

**Figure 7-32** Static mixer with internal heat transfer surface (SMR). (Courtesy of Koch-Glitsch, LP.)

achieved at high throughput when the flow is divided into a large number of parallel flow paths. High capacity and uniform product treatment are achieved, but only when flow is divided equally to all tubes. It is important to recognize that the mixing elements achieve their effect only within the individual tubes treating partial flows. With isolated flow paths, there is no radial mixing effect over the entire product stream as with the monotube design. This is proved not to be a problem in heating applications where the viscosity decreases as temperature is increased. At a moderate pressure drop, a properly designed multitube viscous heat exchanger is self-correcting under conditions of decreasing viscosity. Size (number of parallel tubes) can be increased to handle whatever maximum heat load and throughput are required.

Heat exchanger scale-up is much more difficult for cases where the viscosity increases during the process. This is the case in many viscous cooling applications and for polymerization reactions where viscosity can increase by several orders of magnitude. The *maldistribution effect* in multitube designs occurs when one or several of the parallel tubes behave a little differently than the others. There are numerous causes for this. Maybe there is a maldistribution in the feed stream; or resistance to flow may be minimally different due to slightly different surface roughness or slightly different diameter within tolerance limits; or heat transfer may be slightly different due to minor differences in coolant flow around the tube. If for some reason an individual tube cools a little faster than the others, its effective viscosity will increase at a greater rate, causing higher resistance to flow. Since the pressure drop must be the same over all parallel tubes, the flow through the offending tube will decrease. This results in a longer residence time, an even lower temperature, and still higher viscosity. At greatly diminished flow, the tube could eventually reach the temperature of the cooling fluid, and if

**470** MIXING IN PIPELINES

viscosity at this temperature is high enough, the tube is essentially plugged. Meanwhile, at constant overall volumetric throughput flow has been diverted to the unaffected tubes, resulting in higher velocities, less heat transfer, and shorter residence time. Multiple steady states are established. The situation always results in reduced cooling performance, a very wide residence time spectrum, and in many cases, a higher pressure drop than calculated originally. The design of multitube equipment for viscous cooling or polymerization reaction involving increasing viscosity requires thorough analysis of the rate of viscosity increase versus the flow rate. In many cases, very high pressure drop becomes necessary to avoid creation of a multiple steady-state situation. A single flow channel is highly desirable for heat transfer/reaction processes with steeply increasing viscosity. Isolated flow paths are avoided and radial exchange of material ensures uniform product treatment. This is achieved in the monotube SMX, SMXL, or KMS heat exchanger, but only at small throughput, due to limited availability of the heat transfer surface. At high throughput, special features such as those provided by the SMR design are required. The SMR eliminates the requirement to transfer heat through the wall of the containing vessel. Heat transfer occurs throughout the structure of the mixing element, thus providing a very high volume-related heat transfer capacity (Streiff, 1986), as shown in Figure 7-33. The SMR structure achieves mixing, boosts the viscous heat transfer coefficient, and achieves plug flow similar to that of SMX and SMXL mixing elements. A very low temperature driving force and low pressure drop operation are possible. The SMR is recommended for scale-up when viscosities are steeply increasing and when extremely uniform product treatment is required.

### 7-13.2  Mass Transfer

Given the drop or bubble size in laminar or turbulent flow, the mass transfer coefficient can be estimated. For turbulent flow, the techniques in Chapter 11 will be applicable. Experience shows that in such cases mass transfer is very fast and that the equivalent of an equilibrium stage (90 to 99% of equilibrium) is reached in less than 10 diameters. Static mixers are beneficial not only in creating the drop dispersion but also in maintaining the drop size distribution and a high degree of turbulence, promoting exchange at the interface.

*7-13.2.1 Chemical Reactor.* Motionless mixers and pipelines are often used as chemical reactors, particularly for fast reactions. Chapter 13 is recommended for a discussion of these aspects. Of particular note is the work of Baldyga et al. (1997), in which different mixers were found to give different energy efficiencies for reactive mixing.

*7-13.2.2 Plug Flow Characteristics.* In turbulent flow the residence time distribution (RTD) is close to plug, but in laminar flow the RTD results from the parabolic velocity distribution are quite skewed. Middleman (1977) gives for the RTD in laminar flow

$$F(t) = 1 - 0.25 \left(\frac{t}{t_{ave}}\right)^{-2} \qquad (7\text{-}56)$$

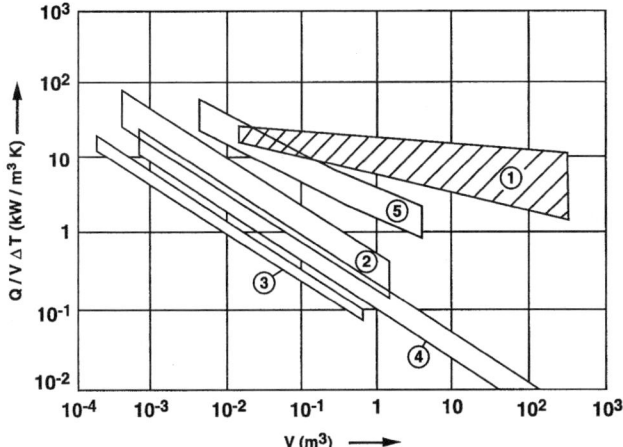

**Figure 7-33** Heat transfer capacity per unit volume of SMR and other chemical reactors: (1) SMR; (2) static mixer; (3) empty pipe; (4) stirred tank; (5) extruder.

where F(t) is the fraction of material leaving the pipe before a given time t. Notice that no fluid leaves a pipe before one-half the residence time and that at the average residence time half the material has left. F(t) can also be though of as the well-mixed normalized concentration exiting the pipe. In actuality the flow leaving a pipe is radially distributed, but F(t) as a normalized concentration is the concentration it would be if there was instantaneous mixing at the end of the pipe. This distribution can cause many problems in treatment and reactions in laminar flow. In most cases, plug flow where all parts of the fluid spend the same time in the pipe is preferred.

Static mixing technology is very applicable to slow chemical reactions requiring plug flow for long duration (see Streiff and Rogers, 1994). Plug flow is achievable with SMX and SMXL static mixers, and with the SMR when precise control of temperature during reaction is also important. Mixing and heat transfer criteria have already been discussed. Reaction presents additional requirements and constraints on the design. First laboratory, and then pilot testing is almost always necessary to establish full scale reactor design.

The static mixer reactor is preferably a monotube design. Multitube configuration presents high potential for reactor instability and nonperformance due to tube–tube variation, as described previously. This is especially true when attempting to control high viscosity exothermic polymerization reactions (Nguyen et al., 1984b; Cusack, 1999).

Commercial scale tubular plug flow reactors utilizing SMX and SMXL static mixer internals have now been operating for more than 20 years. These static mixer designs have demonstrated their ability to mix and achieve plug flow in large scale equipment. The residence time distribution and shear stress–temperature history are very uniform. The Bodenstein number (also sometimes known as a Péclet number) is a measure of the width of the RTD:

$$\text{Bo} = \frac{LV}{D_{ax}} \tag{7-57}$$

The Bodenstein number (Bo) is equal to zero for an ideal continuously stirred tank reactor (CSTR, backmixed reactor) and infinity for perfect plug flow as shown in Figure 7-34. For a cascade of j continuously stirred tanks in series, Bo = 2j (Levenspiel, 1967). Measurements on the SMX and SMXL static mixers in nominal 100 mm (4 in.) diameter pipe at a Reynolds number of 0.1 gave values of the Bodenstein number ranging from 50 to 100 per meter length, corresponding to a very high degree of plug flow. According to Streiff et al. (1999), well-designed static plug flow tubular reactors have residence times corresponding to greater than 30 CSTRs (Bo > 60). Tracer studies conducted during the development of a tubular reactor design for the continuous bulk polymerization of polystyrene showed flow through an SMX packed reactor to be equivalent to a large number of continuously stirred tanks in series (Nguyen et al., 1984a).

Additionally, plug flow requires the reduction or elimination of both localized and large scale recirculation. These phenomena can originate within an improperly designed reactor as a result of temperature and density gradients occurring in the feed or generated during the reaction. Reactor performance is affected by both fluid viscosity and velocity. Parametric testing during pilot scale evaluation should be designed to identify a range of performance and points of instability. Design symmetry with regard to both the shape and size of the hydraulic flow channel should be maintained as best possible on scale-up. With large units, inlet and outlet conditions are also important. Flow simulation models are now starting to be used to evaluate these requirements.

When heat transfer is required for initiation and/or control of a viscous reaction, the SMR mixer-reactor is used. It offers, for example, a unique solution to the difficult problem of laminar flow heat transfer during exothermic polymerization reactions. As described previously, it provides a large area of heat

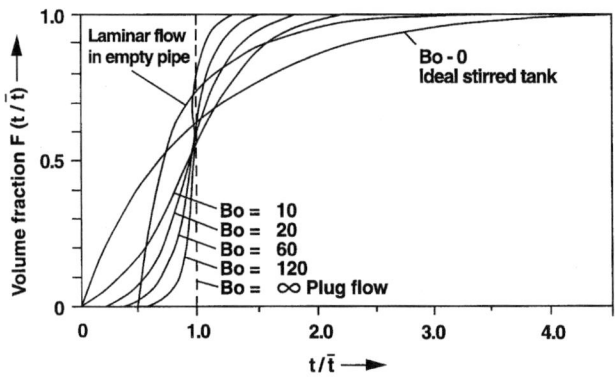

**Figure 7-34** Residence time distribution.

transfer surface throughout the reacting mass as well as accomplishing mixing and maintaining plug flow. Reaction temperatures can be controlled with very small transverse temperature gradients. A near-isothermal reaction process is possible (Craig, 1987).

## 7-14 SUMMARY AND FUTURE DEVELOPMENTS

From the information presented in this chapter it can be seen that for a limited but significant number of cases, an in-line mixer and particularly a static mixer offers a way to achieve many process results. In addition, there are a large number of theory-based empirical relations based on academic and vendor correlation for calculating designs to achieve blending, dispersion, and heat transfer process results. In many cases the methods are more accurate than the available basic data. Future vendor designs will target new applications and be aimed at achieving a given process result with less pressure drop or length.

Uncertainties in the current correlations for in-line mixers are similar to those in other fields of mixing. They involve systems where the physical properties of the streams to be mixed are very different and the understanding of the physics is poor. For example:

- The role of coalescence in gas–liquid and liquid–liquid drop size prediction
- Rate processes in making dispersions
- Blending of miscible systems with large viscosity ratio

Recent unpublished work suggests that the effect of feed ratio in turbulent flow is not as strong as predicted by the coefficient of variation method. It is hoped that computational fluid mechanics will increase our knowledge in this and other areas of pipeline mixing.

As more of the fundamental physics and physical chemistry of such processes becomes understood it will be incorporated into the rather large existing knowledge base. Work is just starting to appear on the effect of non-Newtonian fluids on pressure drop and blending. Because of the simple geometry and the uniformity of the shear stresses, both laminar and turbulent and the controlled limitation of bypassing and tight RTD, the motionless static mixer may be a better device to study some of these phenomena than the conventional stirred tank with its wide variability of shear stress time history.

## ACKNOWLEDGMENTS

The authors gratefully acknowledge Chris Wolfe for her work on the examples.

## NOMENCLATURE

| | |
|---|---|
| a | exponent in heat transfer correlations |
| A | preexponential term in heat transfer calculations |

## 474 MIXING IN PIPELINES

| | |
|---|---|
| b | exponent in heat transfer correlations |
| Bo | Bodenstein number, $Bo = LV/D_{ax}$ |
| c | exponent in heat transfer correlations |
| Ca | capillary number, $Ca = d\mu G'/\sigma$ |
| CoV | coefficient of variation |
| $CoV_0$ | initial coefficient of variation |
| $CoV_r$ | relative CoV reduction |
| $C_p$ | specific heat |
| $C_v$ | volumetric concentration of dispersed phase |
| d | exponent in heat transfer correlations |
| d | drop or bubble diameter |
| $d_{max}$ | maximum drop or bubble diameter |
| D, d | pipe diameter |
| $D_{ax}$ | axial dispersion coefficient |
| f, f' | friction factor |
| $f_s$ | friction factor for smooth pipe |
| g | gravitational acceleration |
| G' | shear rate |
| h | heat transfer coefficient |
| k | thermal conductivity |
| K, k | arbitrary constants |
| $K_L$, $K_T$ | pressure drop ratios for motionless mixers |
| KiT, KiL | mixing rate coefficient for blending |
| $K_G$ | shear rate constant |
| L, l | length variable |
| Ne | Newton number, $Ne = (\Delta P/\rho V^2)(D/L)$ |
| Nu | Nusselt number, $Nu = hD/k$ |
| Oh | Ohnesorge number, $Oh = (We)/Re^{0.5} = \mu_d/(\rho_c \sigma d)^{0.5}$ |
| $\Delta P$ | pressure drop |
| $\Delta P_{tp}$ | pressure drop two-phase (gas–liquid) flow |
| Pr | Prandtl number, $Pr = C_p\mu/k$ |
| q | viscosity ratio—dispersed to continuous |
| Q | volumetric flow rate |
| Re | Reynolds number, $Re = \rho DV/\mu$ |
| V, v | velocity |
| Vi | viscosity number, $Vi = \mu_d(V/\sigma)(\rho_c/\rho_d)^{0.5}$ |
| We | Weber number, $We = D(\rho V^2/\sigma)$ |

### *Greek Symbols*

| | |
|---|---|
| $\varepsilon$ | energy intensity or power per unit mass |
| $\mu$ | viscosity |
| $\mu_c$ | continuous phase viscosity |
| $\mu_d$ | dispersed phase viscosity |
| $\mu_w$ | viscosity calculated at wall temperature |
| $\sigma$ | interfacial or surface tension |

| | |
|---|---|
| ρ | density of the fluid |
| ρ_c | continuous phase density |
| ρ_d | dispersed phase density |
| Δρ | density difference |

## REFERENCES

Acrivos, A., and D. Leighton (1987). The shear induced migration of particles in concentrated suspensions, *J. Fluid Mech.*, **181**, 415–439.

Agarwal, S., D. Doraiswamy, and R. K. Gupta (1998). Demixing effects in laminar shearing flows: a practically useful approach, *Can. J. Chem. Eng.*, **76**, 511.

Alloca, P. T. (1982). Mixing efficiency of static mixing units in laminar flow, *Fiber Producer*, 12–19.

Alloca, P., and F. A. Streiff (1980). presented at the AIChE Annual Meeting, Chicago.

Baker, J. R., and J. A. Rogers (1989). High efficiency co-current contactors for gas conditioning operations, *Paper AF-2*, presented at the Laurence Reid Gas Conditioning Conference, Norman, OK.

Bakker, A., and R. LaRoche (1993). Flow and mixing with Kenics static mixers, *Cray Channels*, **15**, 25–28.

Baldyga, J., J. R. Bourne, and S. J. Hearn (1997). Interaction between chemical reactions and mixing on various scales, *Chem. Eng. Sci.*, **52**, 457–466.

Biesenberger, J. A. (1983). Devolatilization of Polymers: Fundamentals–Equipment–Applications, Hanser, Munich.

Burmeister, L. C. (1983a). Laminar heat transfer in ducts, Chapter 6 in *Convective Heat Transfer*, Wiley, New York.

Burmeister, L. C. (1983b). Condensation, Chapter 10 in *Convective Heat Transfer*, Wiley, New York.

Calabrese, R. V., and P. D. Berkman (1988). Dispersion of viscous liquids by turbulent flow in a static mixer, *AIChE J.*, **34**, 602–609.

Chandavimol, M. (1991a). Gas dispersion in liquid in static mixers in turbulent flow, M.S. thesis, University of Missouri–Rolla.

Chandavimol, M., D. Randall, M. Vielhaber, and G. K. Patterson (1991b). Gas dispersion in a liquid in twisted ribbon and HEV static mixers, presented at the AIChE Annual Meeting, Dallas.

Chimmili S., D. Doraiswamy, and R. K. Gupta (1998). Shear induced agglomeration of particulate suspensions, *I&EC Res.*, **37**, 2073.

Clayton, C. G. (1979). Dispersion of mixing during turbulent flow of water in a circular pipe, AERE-R5569. Isotope Research Division, Wantage Research Laboratory, Wantage, Berkshire, England.

Cohen, D. (1998). How to select rotor–stator mixers, *Chem. Eng.*, Aug.

Craig, T. O. (1987). Heat transfer during polymerization in motionless mixers, *Polym. Eng. Sci.*, **27**, Oct.

Cusack, R. W. (1999). A fresh look at reaction engineering, *Chem. Eng.*, Oct.

Dietsche, W. (1998). Choose the Best Mixers Every Time, *Chem. Eng.*, Aug.

Etchells, A. W., and D. G. R. Short (1988). Pipeline mixing: a user's view; turbulent blending, *Proc. 6th European Mixing Conference*, Pavia, Italy, pp. 539–544.

Etchells, A. W., J. Fasano, D. Doraiswamy, and B. Rubin (1995). Characteristics of radial mixing in motionless mixers in laminar flow, presented at Mixing XVI, Banff, Alberta, Canada.

Fleischli, M., and F. A. Streiff (1995). U.S. patent 5,380,088.

Forney, L. J. (1986). Chapter 32 in *Encyclopedia of Fluid Mechanics*, N. Cheremissinoff, ed., Vol. 1, Gulf Publishing, Houston, TX.

Forney, L. J., and H. C. Lee (1982). Optimum dimensions for pipeline mixing at T junctions, *AIChE J.*, **28**, 980.

Forney, L. J., and L. A. Monclova (1993). Numerical simulation of a pipeline tee mixer: comparison with data, presented at the AIChE Meeting, St. Louis, MO.

Forney, L. J., A. T. G. Giorges, and X. Wang (2001). Numerical study of multi-jet mixing, *Trans. Inst. Chem. Eng.*, **79**, July.

Furling, O., P. A. Tanguy, L. Choplin, and H. Z. Li (2000). Solid liquid mixing at high concentrations with SMX static mixers, *Proc. 10th European Conference on Mixing*, Delft, The Netherlands, July.

Ger, A. M., and E. R. Holley (1976). Turbulent jets in cross flow, *J. Hydraul. Div., Proc. ASCE*, **102**, 731.

Govier, G. W., and K. Aziz (1972). Chapter 10 in *The Flow of Complex Mixtures in Pipes: Horizontal Flow of Gas–Liquid and Liquid–Liquid Gas Mixtures*, Van Nostrand, New York.

Grace, H. P. (1982). Dispersion phenomena in high viscosity immiscible fluid systems and application of static mixers as dispersion devices in such systems, *Chem. Eng. Commun.*, **14**, 225–227.

Gray, J. B. (1986). Turbulent mixing in pipes, Chapter 13 in *Mixing*, Vol. III, V. W. Uhl and J. B. Gray, eds., Academic Press, New York.

Heierle, A. (1989). Static mixer-heat exchangers, *Chem. Plants Process.*, Nov., pp. 30–35.

Hesketh, R. P. A., W. Etchells, and T. W. F. Russell (1987). Bubble sizes in horizontal pipelines, *AIChE J.*, **33**, 663–667.

Hesketh, R. P. A., W. Etchells, and T. W. F. Russell (1991). Experimental observations of bubble breakage in turbulent flow, *Ind. Eng. Chem. Res.*, **29**.

Hobbs, D. M., and F. J. Muzzio (1997). Effects of injection location, flow ratio and geometry on Kenics mixer performance, *AIChE J.*, **43**, 3121–3132.

Hobbs, D. M., P. D. Swanson, and F. J. Muzzio (1998). Numerical characterization of low Reynolds number flow in the Kenics static mixer, *Chem. Eng. Sci.*, **53**, 1565–1584.

Jaffer, S. A., and P. E. Wood (1998). Quantification of laminar mixing in a Kenics static mixer: an experimental study, *Can. J. Chem. Eng.*, **76**, 516–521.

LaRoche, R. D. (2000). CFD in the process industries, *Chemical Reaction Engineering VIII, Computational Fluid Dynamics*, Quebec City, Quebec, Canada, Aug.

Lefebvre, A. H. (1989). *Atomization and Sprays*, Taylor & Francis, New York.

Levenspiel, O. (1967). *Reaction Engineering*, Wiley, New York.

Middleman, S. (1974). Drop size distributions produced by turbulent pipe flow of immiscible fluids through a static mixer, *Ind. Eng. Chem. Prod. Des. Dev.*, **13**, 78.

Middleman, S. (1977). *Fundamentals of Polymer Processing*, McGraw-Hill, New York, pp. 301–306 (for RTD) and 86–88 (for Poisseuille flow).

Moody, L. F. (1944). Friction factors for pipe flow, *Trans. ASME*, **66**, 671–678.

Mutsakis, M., F. A. Streiff, and G. Schneider (1986). Advances in static mixing technology, *Chem. Eng. Prog.*, July, pp. 42–48.

Myers, K. J., A. Bakker, and D. Ryan (1997). Avoid agitation by selecting static mixers, *Chem. Eng. Prog.*, June, pp. 28–38.

Myers, K. J., M. F. Reeder, D. Ryan, and G. Daly (1999). Get a fix on high shear mixing, *Chem. Eng. Prog.*, Nov., pp. 33–42.

Nguyen, K. T., et al. (1984a). Motionless mixers for the design of multi-tubular polymerization reactors, presented at the AIChE Meeting, San Francisco, Nov. 25–30.

Nguyen, K. T., et al. (1984b). Static mixers as tubular reactors for highly viscous reaction media, *Swiss Chem.*, **6**(12a), 45–48.

Pacek, A. W., A. W. Nienow, I. P. T. Moore, and J. Homer (1994). Fundamental studies of phase inversion in a stirred vessel, *Proc. 8th European Conference on Mixing*, pp. 171–178.

Rauline, D, P. A., J. Tanguy, J. LeBlevec, and J. Bousquet (1998). Numerical investigation of the performance of several static mixers, *Can. J. Chem. Eng.*, **76**, 527–535.

Reynolds, O. (1883). An experimental investigation of whether the motion of water in parallel channels shall be direct or sinuous and of the law of resistance in parallel channels, *Philos. Trans. R. Soc. London, Ser. A*, **174**, 935.

Riley, C. D., and A. B. Pandit (1988). The mixing of Newtonian liquids with large density and viscosity differences in mechanically agitated contactors, *Proc. 6th European Mixing Conference*, Pavia, Italy, pp. 69–77.

Schneider, G. (1981). Continuous mixing of liquids using static mixing units, *Process. Chem. Eng., Aust.*, June, pp. 32–42.

Smith, J. M., and A. W. Schoenmakers (1988). Blending liquids of differing viscosity, *Trans. Inst. Chem. Eng.*, **66A**, 16–21.

Streiff, F. A. (1986). Statische Wärmeubertragungsaggregate, in *Wärmeubertragung bei der Kunstoff-aufbereititung*, VDI, Dusseldorf, Germany.

Streiff, F. A., and J. A. Rogers (1994). Don't overlook static mixer reactors, *Chem. Eng.*, June.

Streiff, F. A., P. Mathys, and T. U. Fischer (1997). New fundamentals for liquid–liquid dispersion in static mixers, *Proc. Mixing IX*, Paris, pp. 307–314.

Streiff, F. A., S. Jaffer, and G. Schneider (1999). Design and application of motionless mixer technology, *Proc. ISMIP3*, Osaka, pp. 107–114.

Tadmor, Z., and I. Manas-Zloczower (1994). *Mixing and Compounding of Polymers*, Hanser, New York.

Tauscher, W., and F. A. Streiff (1979). Static mixing of gases, *Chem. Eng. Prog.*, 61–65.

Toor, H. L. (1975). Chapter 3 in *Turbulence in Mixing Operations*, Academic Press, New York, p. 133.

Wang, C. Y., and R. V. Calabrese (1986). Drop breakup in turbulent stirred tank contactors, *AIChE J.*, **32**, 667.

Zalc, J. M., E. S. Szalai, and F. J. Muzzio (2003). Mixing dynamics in the SMX static mixer as a function of injection location and flow ratio, *Polymer Eng. Sci.*, **43**(4), 875–890.

# CHAPTER 8

# Rotor–Stator Mixing Devices

VICTOR A. ATIEMO-OBENG

*The Dow Chemical Company*

RICHARD V. CALABRESE

*University of Maryland*

## 8-1 INTRODUCTION

### 8-1.1 Characteristics of Rotor–Stator Mixers

The distinguishing feature of a rotor–stator mixer is a high-speed rotor (the driven mixing element) in close proximity to a stator (the fixed mixing element). Typical rotor tip speeds range from 10 to 50 m/s. They are also called *high-shear devices* because the local energy dissipation and shear rates generated in these devices are much higher than in a conventional mechanically stirred vessel. In a rotor–stator mixer the shear rate ranges from 20 000 to 100 000 $s^{-1}$. The local energy dissipation may be three orders of magnitude greater than in a conventional mechanically agitated vessel. High speed, high shear, and higher power are the main characteristics of rotor–stator mixers.

The action of the rotor and stator together generates the mixing energy, shear and elongational stresses, turbulence, and cavitation (in various proportions depending on the speed, viscosity, and other fluid flow parameters), which provide the mixing or size reduction. Thus, the rotor–stator assembly is often called a *generator*.

Commercially available rotor–stator mixers range in size from small lab units to large production units capable of flow rates of 1000 gal/min or more, driven by drives with power greater than 100 hp. There are many geometric variations in rotor–stator generator design. They may, however, be classified into three main geometric groups whose features are described in Section 8-2.

---

*Handbook of Industrial Mixing: Science and Practice*, Edited by Edward L. Paul,
Victor A. Atiemo-Obeng, and Suzanne M. Kresta
ISBN 0-471-26919-0 Copyright © 2004 John Wiley & Sons, Inc.

Rotor–stator generators may be assembled or configured for either batch, semibatch, or continuous operation. They may also be combined in various ways and with conventional mechanical agitation for applications requiring high shear and high local energy dissipation. A few of these configurations are also described in Section 8-2.

### 8-1.2 Applications of Rotor–Stator Mixers

Rotor–stator mixers are used in the chemical, biochemical, agricultural, cosmetics, and food-processing industries. They are employed in many process operations that involve:

- Homogenization
- Dispersion
- Emulsification
- Grinding
- Dissolving
- Chemical reaction
- Cell disruption
- Coagulation (due to shear)

Their major use includes the production of latexes, adhesives, personal care and cleaning products, dispersion and microdispersions of chemicals, and agricultural pesticide formulations. These dispersions, in general, have a viscosity less than 150 Pa · s (150 000 cP). When the viscosity of the fluids or dispersion is much greater than this, extruders are employed instead of rotor–stator mixers.

A single rotor–stator mixer can be used to accomplish several of the above operations within a single process. This makes the rotor–stator an invaluable piece of processing equipment. An example is the grinding and subsequent dissolution of "whole bales of rubber into a solvent" using the Silverson BE2500 Disintegrating/Dissolver Plant (Silverson Machines, 2002). Here, the grinding and mixing needed to enhance dissolution occur simultaneously. These mixers can save time due to the highly localized energy and high shear rate that they input to the system. Although the ability to accomplish multiple operations is a benefit, it also makes rotor–stator mixers difficult to understand. Each task may be governed by a different hydrodynamic or operational variable that may be competitive with respect to scale-up and operation.

### 8-1.3 Summary of Current Knowledge

Despite their widespread use, the current understanding of rotor–stator devices has almost no fundamental basis. There are few theories by which to predict, or systematic experimental protocols by which to assess, the performance of these mixers. In fact, there are very few archival publications on rotor–stator processing. Furthermore, the equipment is developed and manufactured by small, highly competitive and highly specialized companies in an environment that is not conducive to the development of a generalized knowledge base. Consequently, process development, scale-up, and operation are done mostly through engineering judgment and trial and error instead of through sound engineering principles. This leads to a multitude of problems during the startup of a process, which inevitably result in lost time to market as well as increased costs. Often, extensive qualification programs are needed to ensure that pilot and manufacturing scale processes make the same product. In recent years there has been some activity of a fundamental nature, some of which is discussed below.

*8-1.3.1 Key Issues and Current Research Efforts.* Key questions relating to the use of rotor–stator mixers include:

- When should one consider a rotor–stator mixer?
- What design(s) and/or configuration(s) should be selected, and why?
- What criteria should be used for scale-up and/or design?

There are no straightforward answers to these questions at the moment. Users currently depend mostly on manufacturers or suppliers and their publications for information on the performance characteristics, scale-up, and design of these devices. Determining what is an appropriate rotor–stator design and operating conditions for a particular process application requires extensive lab, pilot, or plant trials. Information gathered may then be used for scale-up and design. For some processes, especially in the food and cosmetic industries, prior experience may point to a specific rotor–stator design, but the relative efficiency to other designs may still be unknown.

The need for a fundamental characterization of the performance of these devices and the criteria for scale-up and design are well recognized. In response to this need, two consortia, the British Hydromechanics Research Group (BHRG) and the High Shear Mixing Research Program at the University of Maryland, have embarked upon systematic study and characterization of rotor–stator mixers.

BHRG began work in this area as part of the HILINE program in the late 1980s. Most of the research has focused on power-draw measurements, residence time distribution tests, and reactive mixing studies in a rotor–stator device using low viscosity fluids. Very little work has been done using multiphase and high viscosity fluids. Work done so far has shown that:

1. Only a small fraction of the total power input is effective in mixing.
2. Both the shaft power and power losses increase with increasing rotor speed, larger flow rate, higher viscosity, and smaller shear gap.
3. The effect of the width of the shear gap on power draw is weak.
4. A preliminary test with a set of fast competitive reactions in a single-stage toothed rotor–stator mixer revealed significant fluid bypassing of the

region of high turbulence intensity, making the mixer ineffective as a high intensity reactor. However, the extent to which the use of multiple rows of rotor and stator teeth would change the situation was not considered.

Most of the BHRG results are contained in proprietary reports that are available only to consortium members.

The High Shear Mixing Research Program (HSMRP) at the University of Maryland has focused on two fundamental aspects. The first is developing a mechanistic understanding of the governing fluid dynamics occurring in a continuous in-line rotor–stator mixer with a single row of rotor and stator teeth. Both computational fluid dynamics simulations and velocity field measurements via laser Doppler anemometry have been performed. The second involves monitoring dilute liquid–liquid dispersions in batch mixers for insights into the physics of drop breakup in this type of rotor–stator design. These studies have examined the effect of rotor speed, gap width, power draw, and the geometry of the openings in the stator head on the drop size distribution (DSD). Since one of the coauthors of this chapter heads the HSMRP consortium, some of the results are reported below and in Chapter 12.

## 8-2 GEOMETRY AND DESIGN CONFIGURATIONS

Numerous companies design and supply rotor–stator mixers. Many of the available designs often differ only slightly in geometry, although their suppliers make vastly different claims of performance. In this section the main rotor–stator geometries are described briefly and illustrated with figures. Several configurations for batch or continuous operation are also described and illustrated.

### 8-2.1 Colloid Mills and Toothed Devices

Figure 8-1*a* represents the simplest rotor–stator geometry. It is a conical couette device, called a *colloid mill*. There are many design variations of this basic couette geometry. Figure 8-1*b* represents a slightly more complex variation, with slots or teeth built into both the rotor and stator pieces to provide multiple channels for the flow of the fluid through the device. A schematic representation of the hydrodynamics is shown in Figure 8-1*c*. A commercial colloid mill with groves in the rotor and stator is shown in Figure 8-2*a*. Commercial toothed designs often have multistage assemblies of two or more sets of rotor–stator teeth in a mixing head, as shown in Figure 8-2*b*.

### 8-2.2 Radial Discharge Impeller

Another popular rotor–stator design is the Silverson Machines or Ross type. The rotor is a radial impeller that rotates inside a stationary housing with slots as shown in Figure 8-3*a*. The rotor moves the fluid radially out of the mixer head through the slots or holes in the stator. Superimposed on the radial flow is a

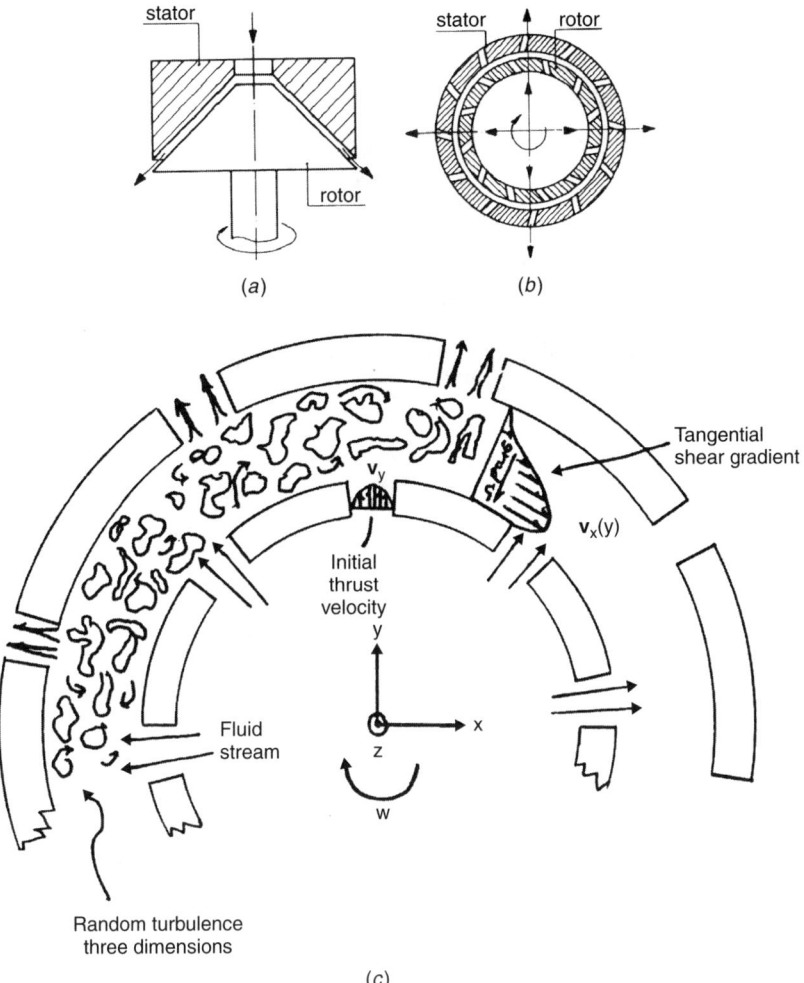

**Figure 8-1** Couette rotor–stator geometry, single-stage design: (a) simple Couette; (b) toothed rotor–stator; (c) schematic representation of hydrodynamics; (d) commercial example of a toothed rotor–stator device.

tangential shear flow inside the stator. Both Silverson Machines and Ross offer a variety of slot shapes and sizes, as shown in Figure 8-3b–d.

### 8-2.3 Axial Discharge Impeller

Chemineer Greerco offers a different rotor–stator geometry. In this case, as shown in Figure 8-4a and b, the rotor is an axial impeller that pushes the fluid axially through holes bored into the stator. The rotating action of the impeller, however, creates some tangential shear flow inside the stator. There is a two-stage version called the *tandem shear pipeline mixer*. This consists of a primary disperser

**Figure 8-2** Commercial devices: (*a*) groved colloid mill; (*b*) IKA Works multistage rotor–stator design.

GEOMETRY AND DESIGN CONFIGURATIONS    **485**

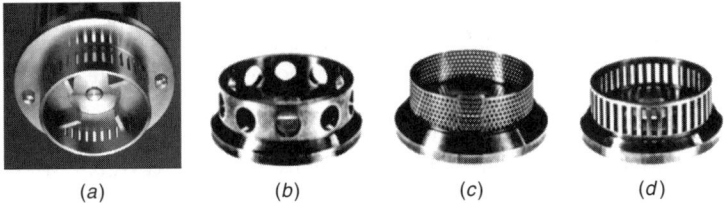

**Figure 8-3** Silverson Machines or Ross rotor–stator design: (*a*) head assembly; (*b*) general purpose disintegrating head; (*c*) high-shear screen; (*d*) slotted-screen disintegrating head.

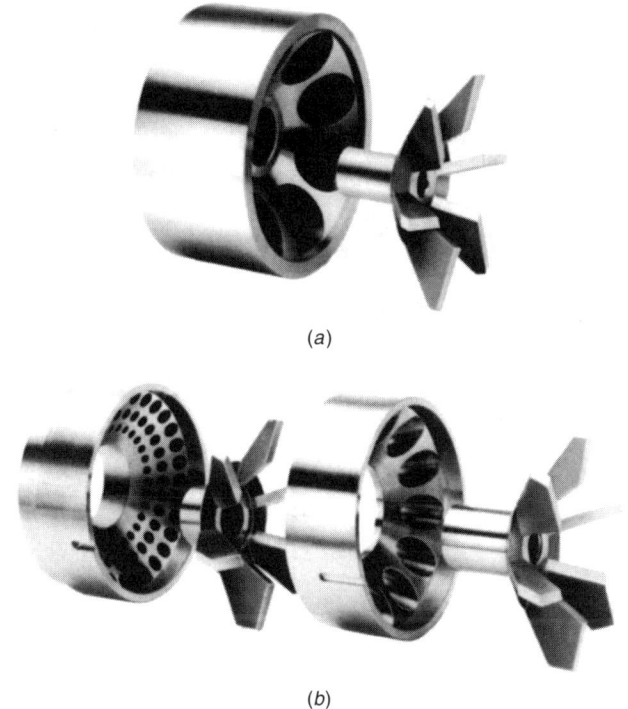

**Figure 8-4** Chemineer Greerco rotor–stator design: (*a*) single-stage design; (*b*) two-stage design.

with larger holes followed by a second rotor–stator head with smaller holes in the stator. Another notable feature of this design is that the gap width between the rotor and stator may be adjusted in the field by installing washers or bushings.

### 8-2.4  Mode of Operation

Any of the mixing heads described above can be assembled and configured for batch semibatch or continuous operation. For a batch or semibatch operation, the mixing head is attached to a long drive shaft and supporting rods, as shown in Figure 8-5a and b. The shaft and supporting rods may also be designed with appropriate seals for mounting onto a nonatmospheric vessel, as illustrated in Figure 8-5c.

Because these mixers are small in diameter, they do not provide much circulation flow in batch operations. Often, especially for large vessels (>0.04 m$^3$ or 10 gal), the vessel may be equipped with an auxiliary impeller whose function is to create circulation flow, as illustrated in Figure 8-5c. For continuous operation, the head is mounted inside a casing equipped with an inlet and an outlet. Usually, the inlet is located so that the fluid enters the center or "eye" of the rotor. Figure 8-6 shows rotor–stator mixers configured for continuous operation.

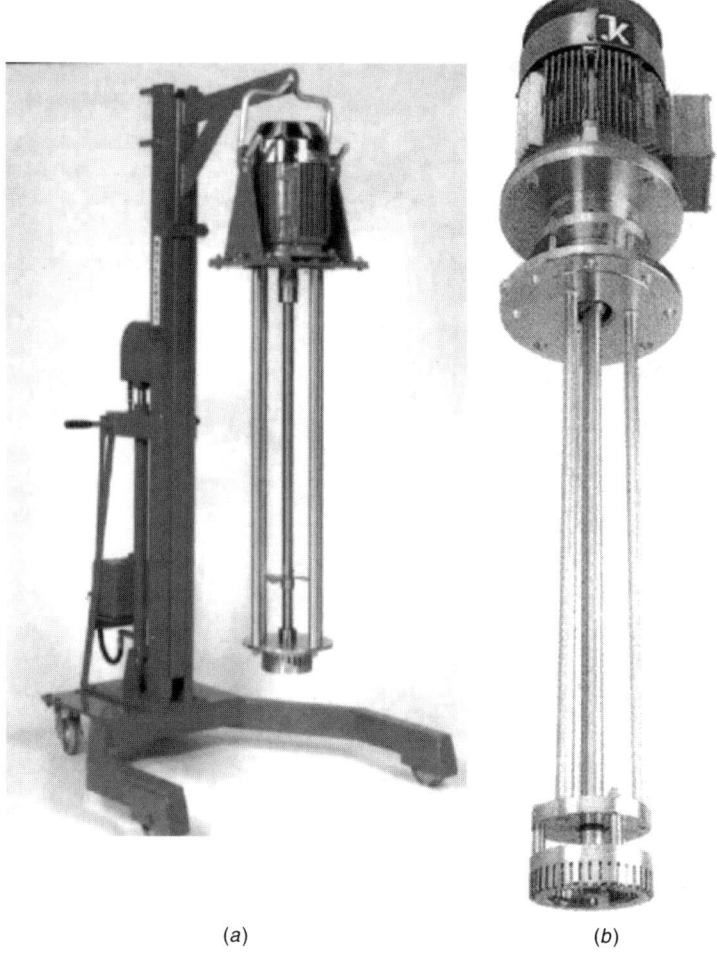

(a)          (b)

**Figure 8-5** Batch or semibatch rotor–stator assemblies: (a) Silverson Machines batch mixer on hydraulic double-lift stand; (b) IKA Works batch mixer. (Myers, 1999, reproduced with permission of AIChE © 1999.)

GEOMETRY AND DESIGN CONFIGURATIONS **487**

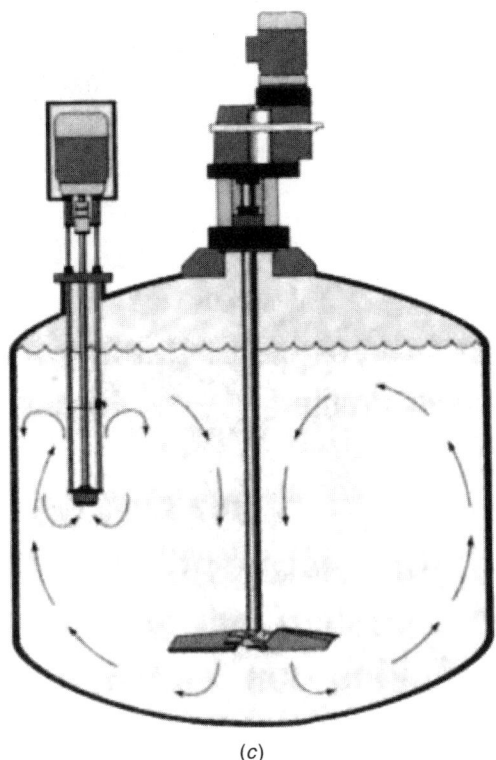

(c)

**Figure 8-5** (c) Vessel equipped with rotor–stator and impeller. (Myers, 1999, reproduced with permission of AIChE © 1999.)

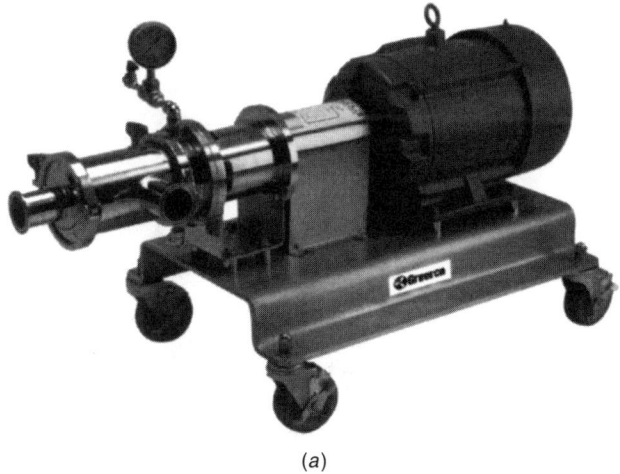

(a)

**Figure 8-6** Rotor–stator mixers for continuous operation: (a) low capacity Chemineer Greerco in-line mixer. (Myers, 1999, reproduced with permission of AIChE © 1999.)

(*Continued*)

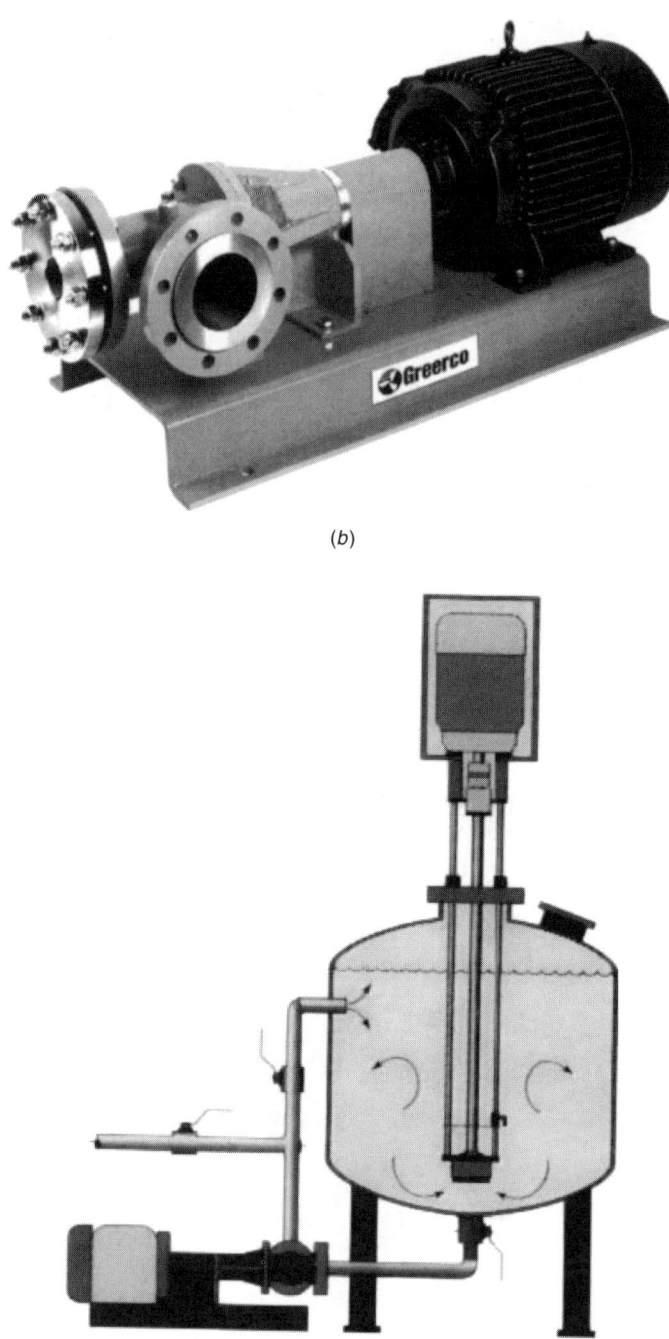

**Figure 8-6** (*b*) High capacity Chemineer Greerco in-line mixer; (*c*) Chemineer Greerco in-line mixer in a circulation loop around a batch mixer. (Myers, 1999, reproduced with permission of AIChE © 1999.)

## 8-3 HYDRODYNAMICS OF ROTOR–STATOR MIXERS

There are few published data for power draw and pumping capacity in either batch or in-line rotor–stator mixers. Even less is known about the velocity fields in these devices, so there is little hard evidence to support proposed mechanisms for dispersion and emulsification. As a result it is often necessary to rely on equipment vendors for scale-up rules. Although many vendors have facilities for customer trials, few have well-equipped laboratories for acquisition of basic data for performance characterization. In reality, it is difficult to know how many vendor data are available, since many consider the information to be proprietary. Until recently, there has been little academic interest in high-shear mixers. This work is only starting to appear in the open literature, and it is important for the practitioner to stay informed as a body of knowledge evolves.

### 8-3.1 Power Draw in Batch Mixers

To the authors' knowledge there are no published power draw data for in-line rotor–stator mixers. The few available data sets for batch mixers are limited to devices with a single rotor with blades (not teeth) surrounded by a single stator. There are several characteristic lengths for these devices, including rotor diameter, D, and the clearance between the rotor and stator or gap width, $\delta$. If the stator has openings through which fluid jets exit the mixing head, an additional length scale exists. For conventional stirred tanks the primary length scale is the impeller (rotor) diameter, so the Power number and Reynolds number are defined by $Po = P/\rho N^3 D^5$ and $Re = \rho N D^2/\mu$, respectively. In the laminar regime Po is inversely proportional to Re. In the turbulent regime Po is often constant and varies from about 2 to 6. The transition from laminar to turbulent flow occurs around $Re \sim 10^4$.

For the limited studies to date for batch rotor–stator mixers, it has been found that there was no advantage to using a different definition of Po or Re. However, for a given geometry only a single-size mixing head was studied. As a result, the Po versus Re behavior was quite similar to that of a stirred tank, displaying about the same range of Po. Myers et al. (2001) reported results for a Greerco $1\frac{1}{2}$ HR Homomixer ($D_{max} = 20$ cm, $\delta = 0.25$ mm), which is an axial flow device with a conical impeller swept volume. The mixing head can be operated to pump either upward or downward and has a solid circular baffle plate above it and an annular plate below it that serve to control the flow direction. The use and position of the plates affect the power draw. The mixing head was centered in the tank, so baffling of the fluid in the vessel also played an important role. However, there was little effect of off-bottom clearance. Power numbers were similar to that for an axial impeller in a stirred tank, but the actual power draw is higher, due to increased rotational speed.

Padron (2001) acquired power draw data for a Greerco Homomixer ($D_{max} = 4.3$ cm, $\delta = 0.5$ mm) and found that this much smaller mixer exhibited behavior that was both similar to and different from that of the larger mixer studied by

Myers et al. (2001). The presence of baffles had little effect in upward pumping but yielded a greater power draw for downward pumping. A major difference was that this mixer could not draw power at lower speeds for low viscosity materials, resulting in an unusually large transition region not seen at the pilot scale. As a result, the onset of fully turbulent flow at constant mixer size depended strongly on fluid viscosity. The plates that are located above and below the mixer head cause flow patterns to vary widely with rotor speed and pumping direction.

Padron (2001) also acquired power draw data for the Ross ME 100LC (D = 3.4 cm, δ = 0.5 mm) and the Silverson L4R (D = 2.8 cm, δ = 0.2 mm) bench scale mixers. These are radial flow mixers with a four-blade rotor and replaceable stators of various geometry. The Silverson mixer head is shown in Figure 8-7. The Ross design is slightly different but has an equally large variety of stator geometry. Baffles were not used, so the mixers were placed off-center in the tanks to ensure good top-to-bottom mixing. In the laminar regime Po is somewhat independent of stator geometry, and the Silverson mixer draws slightly higher power at constant Re. This is due in part to its smaller gap width. Fully turbulent conditions occur above Re $\sim 10^4$, but the Ross mixer has a smaller transition region, possibly due to its larger size. In the turbulent regime the Ross mixer draws more power than the Silverson device, indicating that energy dissipation

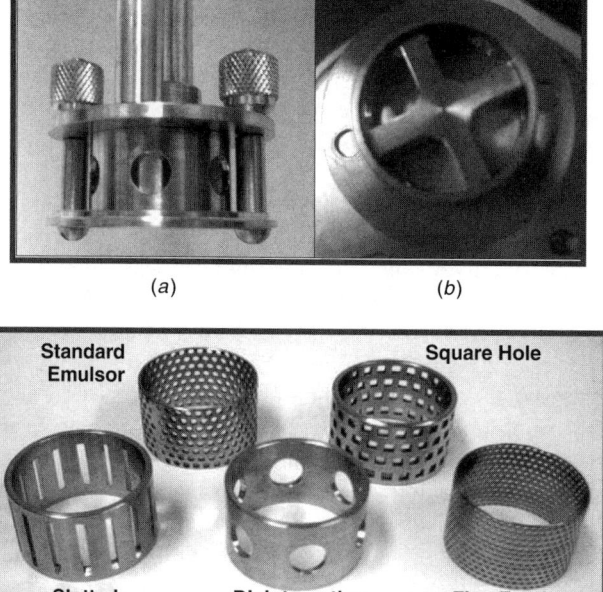

**Figure 8-7** Rotor–stator head geometry for the Silverson L4R batch rotor–stator mixer: (*a*) side view; (*b*) bottom view showing rotor; (*c*) range of stator geometries.

in the shear gap is not controlling. The constant turbulent power number was dependent on stator geometry and varied from 2.4 to 3.0 for the Ross mixer head and 1.7 to 2.3 for the Silverson head. In general, for a given geometry, the power number increased with the number of openings in the stator, indicating that energy dissipation was controlled by fluid impingement on stator slot surfaces or turbulence in the jets emanating from the stator slots. For instance, for the slotted head, the power number ratio for the Ross head (Po = 3.0) to the Silverson head (Po = 2.1) is 1.43. The slots themselves are quite similar for both devices, but the Ross has 1.56 times as many slots, indicating that the power number per stator slot is the same.

### 8-3.2 Pumping Capacity

To obtain the pumping capacity of a batch rotor–stator mixer, it is necessary to measure the velocity field entering and/or emanating from the mixer head. Since this is often tedious and requires sophisticated instrumentation, few pumping capacity data exist for batch devices. One must usually rely on vendor information or trial-and-error experimentation to estimate batch time. Rotor–stator mixers are often operated off-center in unbaffled vessels to promote good bulk mixing. As an alternative, the vessel may be equipped with a standard axial impeller to provide bulk mixing (see Figure 8-5c). Mixer location affects both pumping capacity and blend time.

It is much easier to measure the pumping capacity of an in-line rotor–stator mixer, so vendor data are much more reliable. Units that have a rotor with blades, such as the Silverson in-line series, can simultaneously pump and emulsify/disperse material. However, many designs, such as those with multiple rows of rotor and stator teeth, have marginal pumping capacity and may even cause a pressure drop. While rotor rotation acts to create pressure and pumping, the resulting tangential velocities are redirected radially as the fluid passes through the stator slots to counteract the pumping action of the rotor. However, it is the dissipation of this pressure energy that promotes good dispersion/emulsification. This is discussed in more detail below. It is often necessary to feed the mixer with a separate pump.

### 8-3.3 Velocity Field Information

Much can be learned about the mechanisms of dispersion and emulsification in a rotor–stator mixer by studying the detailed velocity and deformation fields for a pure liquid passing through the mixer. However, this can only be accomplished using advanced computational fluid dynamics (CFD) techniques or using sophisticated measurement techniques such as laser Doppler anemometry (LDA) or particle image velocimetry (PIV). The best approach is to use a first-generation simulation to design an experimental program, to use the acquired experimental data to develop a more sophisticated model, and so on. Although still the subject of long-range academic research, rapid advancements in computational resources and instrumentation make this an attractive means to quickly increase our basic understanding of rotor–stator devices, as well as many other little studied and geometrically complex mixers and dispersion devices.

Detailed hydrodynamic studies have been performed for pump-fed in-line mixers with rotors and stators comprised of teeth with slots between them. LeClair (1995) reported an early attempt for a Kady mixer. Although the simulation was quite simplistic and considered only a small section with assumed perfect symmetry, the results revealed a complex circulation pattern in the stator slot.

Calabrese and co-workers have made a more comprehensive study, consisting of modeling and measurement. Calabrese (1999) reported preliminary results. Kevala (2001) and Calabrese et al. (2002) have reported more recent results. To allow acquisition of detailed information while maintaining realism, they considered a simplified prototype of an IKA Works, Inc. mill containing a single set of rotor and stator teeth. A schematic diagram is shown in Figure 8-8. Sliding mesh simulations with the Fluent CFD code were performed at a rotational speed of 30 rps and a throughput of 2.86 L/s (45.4 gal/min). The working fluid was water, so the flow was turbulent even in the narrow rotor–stator gap. Therefore, the RANS equations were solved with the standard $k-\varepsilon$ turbulence model. Corresponding velocity measurements were made using a two-color Dantec LDA system. The front of the volute was made of Plexiglas (or Perspex) to allow laser beam access.

Because the tooth depth was small compared to the mixer diameter, it was initially believed that the flow field was two dimensional. Figure 8-9 shows mean velocity vectors, resulting from a 2D simulation, in the quadrant closest to the exit pipe. These results are angularly resolved in that the flow field changes as the rotor passes the stator. The direction of rotor rotation is clockwise and an extremely complex flow pattern is revealed in the stator slots and volute. Circulation cells in the stator slots allow reentrainment of volute fluid back into

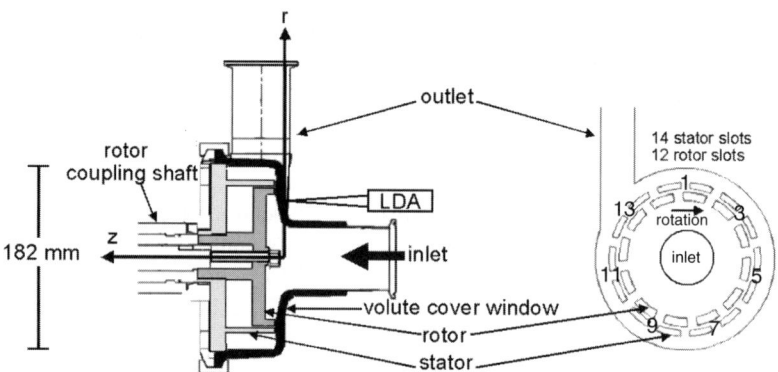

**Figure 8-8** Schematic diagram of prototype IKA mixer with 12 rotor and 14 stator teeth. Inner diameter of rotor is 11.8 cm; outer diameter of stator is 15.4 cm. The rotor tooth depth is 1 cm and the rotor–stator gap width is 0.5 mm. Approximately drawn to scale, except for gap width.

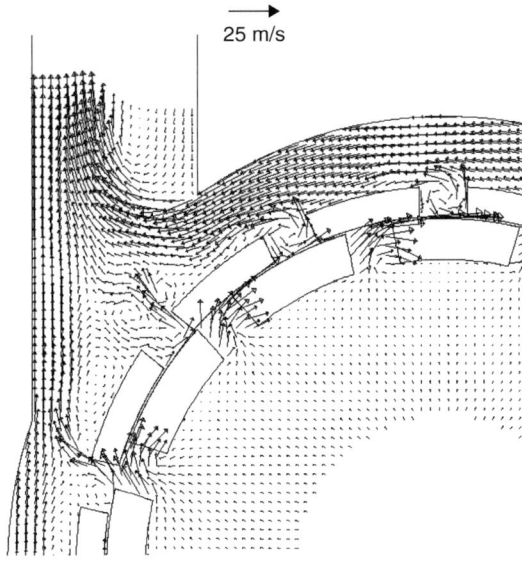

**Figure 8-9** Mean velocity vector data (m/s) acquired via LDA for IKA prototype mixer. Single-time snapshot.

the shear gap. Below the exit the flow in the volute is in the direction of rotation. To the right of the exit the volute flow is counterclockwise against the direction of rotation. As the fluid enters a stator slot it is redirected radially as it impinges on the downstream stator tooth and seeks the path of least resistance to the exit. The flow in other quadrants is equally complex. Each stator slot has a different circulation and reentrainment pattern. At the point farthest from the exit, there are recirculation zones in the volute that divide the clockwise and counterclockwise zones. Figure 8-10 shows the measured mean velocity field in the same quadrant as Figure 8-9. These data are averaged over all rotor positions. That is, LDA data are acquired in the stator slot without regard to the position of the rotor, so all angularly correlated information is lost. The angularly correlated LDA data reveal similar trends. The data were acquired at middepth in the z-direction indicated in Figure 8-8 or halfway between the front and back of the teeth. The measurements show a stronger jet emanating from the stator slots into the volute, as a result of fluid impingement on the leading edge of the downstream stator tooth, than does the 2D simulation. The recirculation zones in the stator slot are more focused and reentrainment into the shear gap is stronger. Nevertheless, it is seen that much physical insight can be gained from the more detailed and less labor-intensive simulation.

Figure 8-11 shows the total mass flow rate exiting each of stator slots 1 to 7 (refer to Figure 8-8) predicted by the 2D simulation, as a function of rotor tooth position. As the rotor tooth blocks the stator slot, the flow rate drops rapidly and does not recover until the rotor tooth passes the slot completely. The maximum

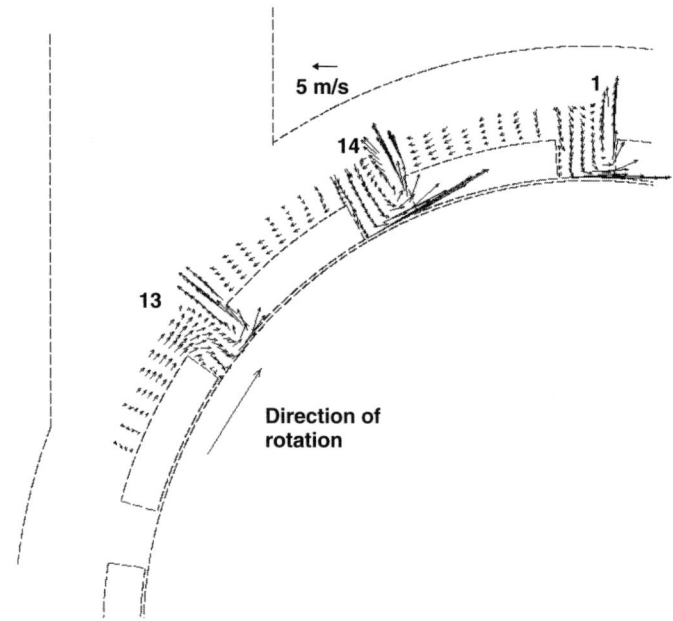

**Figure 8-10** Mean velocity vector data (m/s) acquired via LDA for IKA prototype mixer. Velocity field is angularly averaged over all rotor positions.

flow rate does not occur when the rotor and stator slots are aligned but when the stator slot is about half blocked by the rotor tooth. Stator slots closest to the exit have higher mass flow rates than those far from the exit. The LDA measurements are somewhat different. While the exit flow rate drops rapidly as the rotor tooth blocks the stator slot, it partially recovers well before the rotor tooth has passed the stator slot. Furthermore, all slots have about the same flow rate. The data reveal that the flow field is highly three dimensional and that when the rotor tooth blocks the stator slot, there is leakage flow over the top of the rotor tooth into the stator slot. As with extruder screw flights, there must be a clearance between the rotor and the front volute cover. This clearance cannot be much smaller than the shear gap itself. Preliminary 3D simulations by Calabrese and co-workers indicate that the results are much more in line with the LDA data.

Figure 8-12 shows the turbulent kinetic energy (TKE) predicted by the 2D simulation. In simple flow fields the TKE can be related to the energy dissipation rate, which is, in turn, a measure of emulsification or dispersion capacity. The LDA data are somewhat different, but the same physical insights emerge. When rotor and stator slots are aligned, the TKE is low. As the rotor tooth blocks the stator slot, the TKE builds on the leading edge of the stagnated downstream stator tooth. The most intense TKE field is created just as the stator slot is blocked and as the flow rate begins to fall. As a result, much of the fluid leaving the stator slots does not experience an intense deformation field. It is because of this "bypassing"

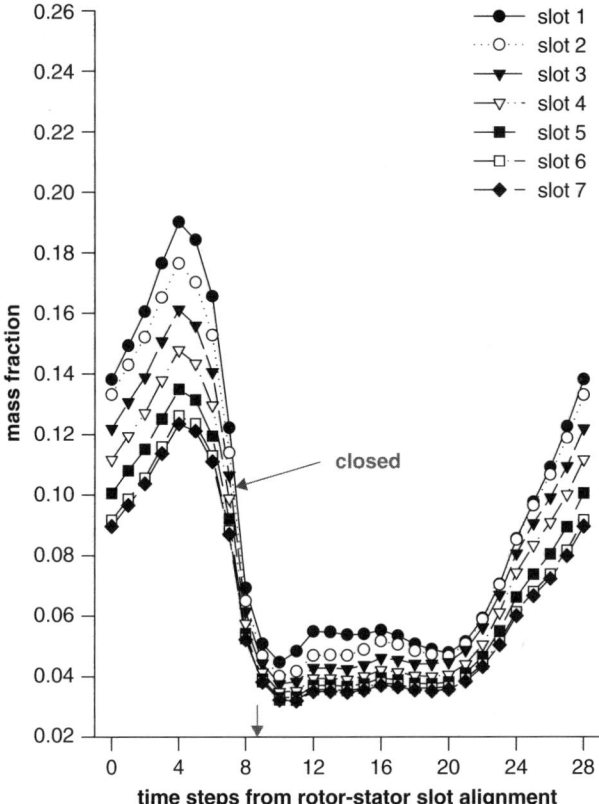

**Figure 8-11** Mass flow rate exiting stator slots 1 to 7 as a function of time or rotor position. The rotor tooth completely blocks the stator slot at time step 9 and the slot begins to open again at time step 21. Results of 2D CFD simulation. Stator slot numbers are defined in Figure 8-8.

that devices with multiple rows of rotor and stator teeth are more effective for emulsification and dispersion. It is often necessary to run multiple passes through the mixer to achieve the desired product.

For turbulent flow through rotor–stator devices with teeth, the aforementioned velocity field results indicate that flow stagnation on the leading edge of the downstream stator teeth provides a major energy field for emulsification and dispersion. It is not clear from these results what role is played by flow in the shear gap. The simulations indicate that the flow in the rotor–stator gap is not a simple shear flow but is more like a classical turbulent shear flow. Use of nominal shear rate may not be useful in scale-up.

Simulations were performed and some LDA data were acquired for a similar device that has an enlarged shear gap of 4 mm rather than the standard 0.5 mm gap. The results indicate that there is much less stagnation on the stator teeth, so

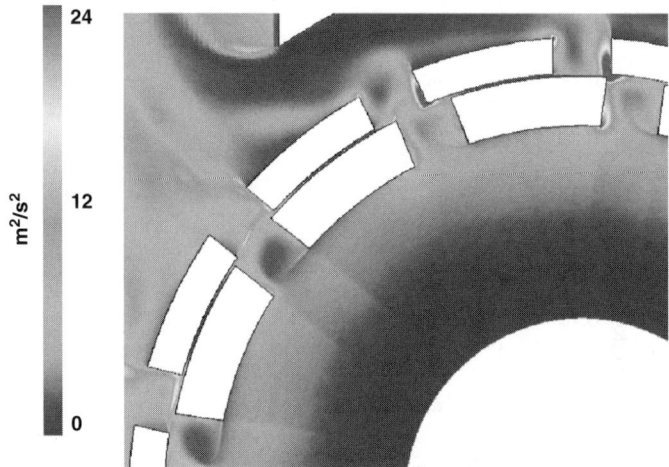

**Figure 8-12** Turbulent kinetic energy (m$^2$/s$^2$) from 2D CFD simulation for IKA prototype mixer. Single-time snapshot. See insert for a color representation of this figure.

that high levels of TKE are not seen. When the shear gap is too large, too much fluid remains in the gap rather than being forced onto the leading edge of the stator tooth, thereby exiting the stator slot. As a result, it is necessary to have a narrow shear gap even if the shear in this gap is not a major contributor to the dispersion process.

### 8-3.4 Summary and Guidelines

As stated above, there are few fundamental data to allow design and scale-up of rotor–stator mixers available in the open literature. Academic activity is increasing and it is important to remain aware of new information. It is important to note that there is activity in this field that may be documented in non-English publications that has not been discussed above. The results discussed above were general in nature. Additional information may be available in industry-specific publications, such as those for food processing, paints and pigments, and so on.

## 8-4 PROCESS SCALE-UP AND DESIGN CONSIDERATIONS

There are many important process and mixer variables to consider in the process selection, scale-up, and operation of rotor–stator mixers. Key variables among them include:

- Chemistry and physics of the process and requirements for macro- or micro-mixing
- Density and rheological properties of the feed and product streams

- Coalescence behavior with or without surface-active agents (surfactants, dispersants, stabilizers, etc.)
- Role of the materials of construction of fluid-wetted parts on the chemistry or physics of the process
- Number of feed streams and the need for staging or premixing
- Desired throughput
- Control of the flow ratios of feed streams
- Location and order of addition of reagents to different stages of mixer heads
- Temperature of operation and the need for heating or cooling for temperature control
- Effect of shear on the degradation or coagulation, and so on, of the feed and product streams
- Operating mode: batch, semibatch, or continuous operation
- Rotor–stator design and operation
- Speed of the rotor
- Gap between rotor and stator
- Diameter of the generator or rotor–stator head or assembly
- Design or geometry of the rotor and stator, including number of teeth, tooth to tooth spacing, number of stages, percent open area, and so on.

The fluid properties at the desired operating temperature must be known. These properties determine, in large measure, the required mixing operation and how difficult it will be to achieve the desired process result. Fluids can heat up substantially while being processed in a rotor–stator mixer due to the high-energy input and small fluid volumes. As the viscosity increases, so does the viscous heating effect. Therefore, provisions for cooling, or in a few cases heating, must be considered. Jacketed units are not able to add much cooling (or heating) because of the limited surface area and short contact time. Jackets are used mainly to maintain temperature during downtime or to prevent polymerization during periods of idle operation.

For dispersion processes, the role of coalescence on the mean particle size and particle size distribution should not be overlooked. The type and amounts of surface-active agents added, as well as the material of construction of the rotor and stator, all affect coalescence, as discussed in greater detail in Chapter 12.

The appropriate operating mode (batch, semibatch, or continuous operation) will depend on the mixing process and the process result desired. Batch and semibatch operations are best suited for processes that require long time scales, greater than a few seconds.

For batch or semibatch applications, the vessel size and shape, location of the mixing head (i.e., mounted centrally or off-center), and the presence of baffles are additional items for consideration. Commercial rotor–stator mixers are matched to specific batch volumes.

For continuous applications, additional issues for consideration include:

- Flow rate through the unit and control of multiple feed streams
- Pumping capacity or need for a separate feed pump
- Premixing of feed streams and location in the mixing head where the feed is introduced
- Residence time
- Operating pressure and back pressure (for some systems), especially those that are sensitive to compression or flashing

The precise control of feed rates and ratios is vital for continuously operated rotor–stator mixers, since the extent of backmixing to even out variations in feed rates is unknown. Some processes may require premixing of the feed streams upstream of the rotor–stator mixer. For very fast reactive processes, premixing must be avoided. Others may benefit from staged addition of the streams, especially in multistaged mixers. When mixing streams of widely different viscosities, it is instructive to explore the effects of switching feed locations on the process result.

Although rotor–stator mixers can pump to some extent, it is preferred to use a pump to control the feed rate to the mixer. This way, one does not need to vary the rotor speed to control the flow rate. Instead, the rotor speed can be varied to control the energy input, turbulent kinetic energy, and shear rates in the device, independent of the flow rate. To prevent equipment failure, it is also important to ensure that the unit is "fully flooded" and not starved during operation.

Feed streams must be introduced into the mixing head so as to minimize bypassing or short-circuiting to the outlet of the mixer. An important consideration is the unit's orientation (vertical or horizontal) for the correct delivery of the material to the shear zone. Streams must flow through the shear gaps and regions of the stator exit slots where most of the energy is dissipated. The residence time must be related to the volume in the vicinity of the rotor and stator where the mixing action occurs, not the entire volume of the device. For the device to be effective, the residence time must match the required mixing time for the processes.

The rheology of the dispersed or homogenized stream can be very different from the feed streams. This greatly influences the ability to move the fluids though the system and can affect the backpressure on the unit. Some emulsions will not form unless there is enough backpressure for the materials to stay in intimate contact with each other in the mixing zones. However, once formed, thick emulsions may need to be diluted quickly to facilitate fluid flow.

### 8-4.1 Liquid–Liquid Dispersion

Rotor–stator mixers are widely used in the chemical process and allied industries to produce liquid–liquid dispersions and emulsions. Although production of dispersions and emulsions in rotor–stator mixers is often highlighted in industry-specific publications, there are few studies that relate drop size to energy input or other variables related to agitation intensity. The authors are not aware of generic studies for in-line rotor–stator mixers. Such experiments are difficult and require disposal of large amounts of material, making the approach of Section 8-3, to infer mechanisms, attractive. In this section the discussion is limited to batch rotor–stator mixers, where drop size data are more easily acquired and the results more readily correlated.

Whereas power numbers for batch rotor–stator mixers are of the same magnitude as for stirred tank impellers, the power per unit volume is much higher, due to high rotational speed. Dispersions and emulsions of practical significance are usually of high dispersed phase volume fraction. As a result, surfactants or other stabilizers must be present to prevent coalescence. Unfortunately, the roles played by the surfactant and mechanical forces in determining ultimate drop size are difficult to separate and poorly understood. To circumvent this problem, Calabrese et al. (2000) made drop size measurements for dilute dispersions of low viscosity organic liquids in aqueous solutions, for which surfactants were not required, due to negligible coalescence rate. The experiments were performed in Ross ME 100 LC and Silverson L4R mixers (Figure 8-7). These are the same mixers as those used by Padron (2001) for the power number studies discussed in Section 8-3.1. The flow was turbulent.

The results showed that when the rotor–stator shear gap width was increased at constant rotor speed smaller drops were produced even though the nominal gap shear rate had decreased. Furthermore, the Ross and Silverson mixers produced the same drop size when the slotted stator head was used, since the power per stator slot was the same (Section 8-3.1). Even though the power numbers were similar, the slotted head produced smaller drops than the disintegrating head. LDA measurements in the jets emanating from the stator openings indicated that the slotted head focused the energy better and produced higher levels of TKE. These observations lead the authors to suggest that drop breakup occurred on stator surfaces or in the jets emanating from the stator rather than in the shear gap.

The smallest scale of turbulence (Kolmogorov microscale) was estimated from the power draw data and found to be slightly smaller than the drop size. That is, at high-energy input the turbulent microscale is decreased so that the resulting drop size is still governed by turbulent stresses acting on the drops. As a result, the mean drop size data could be correlated in the same way as data for stirred tank and other turbulent liquid–liquid dispersion devices. These correlations are discussed in Chapter 12. Figure 8-13 shows that for low viscosity drops of constant physical properties, the maximum drop size correlates with local energy dissipation rate (power per unit mass). The rotor–stator data fall below those for stirred tanks and above those for liquid whistles.

*8-4.1.1 Mean Drop Size and Drop Size Distribution.* Estimates of drop size that are achievable by various mixing devices for dispersing immiscible liquid–liquid systems are also shown in Table 8-1. In the table, the specific

# 500  ROTOR–STATOR MIXING DEVICES

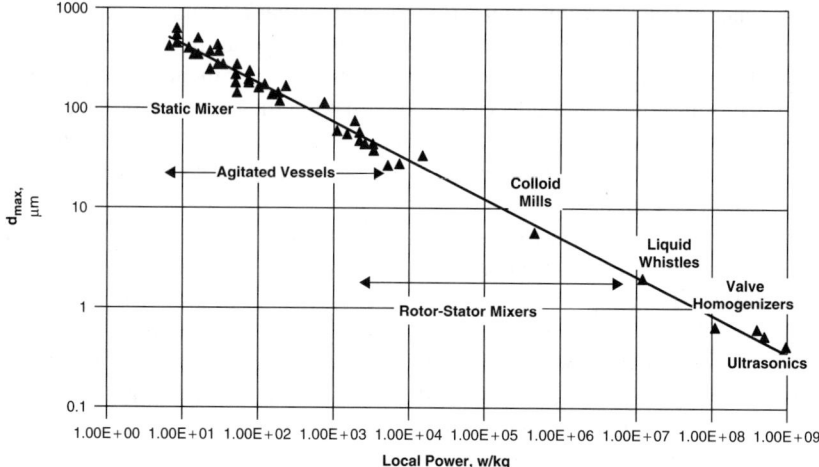

**Figure 8-13** Maximum drop size $d_{max}$ versus local power draw for dilute oil-in-water dispersions. The local power per mass of fluid is the total power input divided by the mass of fluid in the high intensity dispersion region of the mixer. (After Davies, 1987.)

**Table 8-1** Performance Features of Various Dispersion Devices

| Type of Device | Energy Dissipation Range (W/kg) or (m²/s³) | Typical Size Range (μm) | Comments |
|---|---|---|---|
| Static mixers | 10–1000 | 50–1000 | Narrower DSD than agitated vessel |
| Agitated vessel | 0.1–100 | 20–500 | With Rushton turbine in a fully baffled vessel; usually, broad DSD |
| High speed rotor–stator | 1000–100 000 | 0.5–100 | Can be smaller with the correct chemistry |
| Valve homogenizer | ~$10^8$ | 0.5–1 | Requires high pressure, 5000 to 10 000 psi |
| Ultrasonics | ~$10^9$ | 0.2–0.5 | Sonification devices |

energy input increases as one moves down the "Type of Device" column. In general, as local energy input is increased, a smaller droplet size is produced, as illustrated in Figure 8-13. This does not mean that the dispersion will be stable, as coalescence can occur outside the device. Surfactants are used to stabilize against coalescence. Batch systems inherently produce a wider drop size distribution than continuous systems.

When choosing a disperser or homogenizer for evaluation, important considerations include the turn-down ratio (maximum flow/minimum flow) and the ease with which equipment components can be changed to control the DSD.

## 8-4.2 Solids and Powder Dispersion Operations

Rotor–stator mixers are an important technological option for wet mixing and for dispersion of hard-to-wet powders into a liquid. They find use in the preparation of "fish-eye" free solutions of powders (thickeners, stabilizers, flour, starches, caseinates, powdered milk, clays, etc.). Suppliers have various names for the systems used for such applications, but in principle, they include the same basic components: an in-line rotor–stator mixer, a high-pressure centrifugal pump, a venturi assembly, and a powder feed hopper system. The centrifugal pump conveys the fluid through a venturi with a TEE connection to the powder feed line. The resulting low pressure created in the throat of the venturi by the flowing fluid causes the powder to be sucked into the flowing liquid, which immediately enters the rotor–stator mixer. Here, the intense mechanical and hydraulic shear in the mixer head quickly disperses the powder into the liquid.

The authors are not aware of any published fundamental studies of solid dispersion in rotor–stator mixers but have seen videos supplied by vendors that show preparation of "fish-eye" free solutions of hard-to-wet powders such as Carbopol, with apparently remarkable ease. Scale-up and design of these systems would be based on operational information obtained with vendor equipment. This is highly recommended.

Although rotor–stator devices are not considered highly energy intensive relative to some types of solids size-reduction devices, they are sometimes employed when more controlled size reduction is desired, especially when it is desirable to minimize the production of fines. One example of such use is the conditioning of seed particles for crystallization operations when the seed is a portion of the previous batch ("heel" crystallization).

## 8-4.3 Chemical Reactions

The authors are not aware of any nonproprietary industrial application of rotor–stator mixers for carrying out fast or mixing controlled chemical reactions, even though they appear suitable for the purpose. Indeed for this purpose, Bourne and co-workers (Bourne and Garcia-Rosas, 1986; Bourne and Studer, 1992) evaluated the suitability of a commercially available rotor–stator mixer using the fast competitive azo-coupling reactions of 1-naphthol and diazotized sulfanilic acid. The rationale was to accelerate micromixing by exploiting the general characteristics of rotor–stator mixers, that is, they generate a locally intense turbulence in a small volume with a short residence time.

Their results indicate that it is necessary to feed at or near the shear gap to take advantage of the high-energy dissipation rate in the gap. This was difficult to do. They estimated the turbulent energy dissipation in the shear gap to be on the order of 1000 W/kg. By comparing the product distribution at two scales, they concluded that scale-up on the basis of tip speed was better than on the basis of power per volume.

## 8-4.4 Additional Considerations for Scale-up and Comparative Sizing of Rotor–Stator Mixers

As noted earlier, a fundamental understanding of the underlying principles and of the important variables that affect the performance of rotor–stator mixers is only now being developed and disseminated. However, most suppliers possess proprietary know-how and experience acquired over many years with their specific equipment and can make recommendations for the selection and scale-up for specific process applications. To do this effectively, the suppliers will require information on the process and fluid properties. Often, they will insist on testing their lab or pilot equipment on actual process fluids before making recommendations. Suppliers will treat process information shared with them, or acquired with their test equipment, as confidential. For additional protection, however, consider secrecy agreements before sharing proprietary process information. Most suppliers may not have facilities to handle hazardous chemicals and high viscosity fluids. For these situations the client can instead suggest model fluids or rent the equipment for installation and testing in-house.

Vendors often design and scale-up rotor–stator mixers based on equal rotor tip speed, $V_{tip} = \pi ND$, where N is the rotational speed of the rotor and D is the rotor diameter. This criterion is equivalent to equal nominal shear rate in the rotor–stator gap, $\dot{\gamma}$. In most industrial rotor–stator mixers, the shear gap width $\delta$, remains the same on scale-up, making the two criteria equivalent.

The nominal shear rate in the rotor–stator gaps is calculated as follows:

$$\dot{\gamma} = \frac{\pi ND}{\delta} \tag{8-1}$$

It is important to recall from the discussion above that for turbulent flow the actual shear rate in the rotor–stator gap varies substantially from $\dot{\gamma}$, and that gap shear rate does not directly control power draw and dispersion. However, tip speed may control turbulence characteristics, especially if the spacing between stator elements (teeth or stator openings) as well as the shear gap width do not change on scale-up.

Usually, different rotor–stator devices are compared on the basis of the throughput, the amount of fluid that one can move through a specific unit. The throughput listed in catalogs is based on the volumetric flow rate of water as the fluid pumped and mixed under specified conditions of rotor speed and gap width. When viscous fluids are processed, the actual flow rate will be less and the unit's capacity will be lower. Modifications to the rotor may be needed to handle greater throughput and/or higher viscosity fluids. These modifications can enhance throughput but may reduce the mixing performance. They often add significantly to the cost of the unit.

Units may also be compared on the basis of the residence time in the mixing zone. It is important to determine the true volume of the mixing zone for this type of comparison to be meaningful.

## 8-5 MECHANICAL DESIGN CONSIDERATIONS

As illustrated in Figure 8-14, rotor–stator mixers are complex mechanical devices with many potentially wearable components. Their reliability requires careful attention to important mechanical details, including:

- Proper selection of the motor and design for variable speeds
- Gearbox selection and design
- Seal design
- Pressure rating of the unit
- Bearing design and tolerances, especially for a larger mixer with a cantilevered shaft
- Use of appropriate materials of construction

Most rotor–stator devices are equipped with variable speed drives to allow operations at different rotor speeds. There is sufficient know-how about variable speed drives. However, one must pay careful attention to:

- Operations at the low end of the speed range because of the required higher torque and associated mechanical instability
- Mechanical limits specified by the supplier, including shaft and seal design, to be sure the design torque and speed are not exceeded

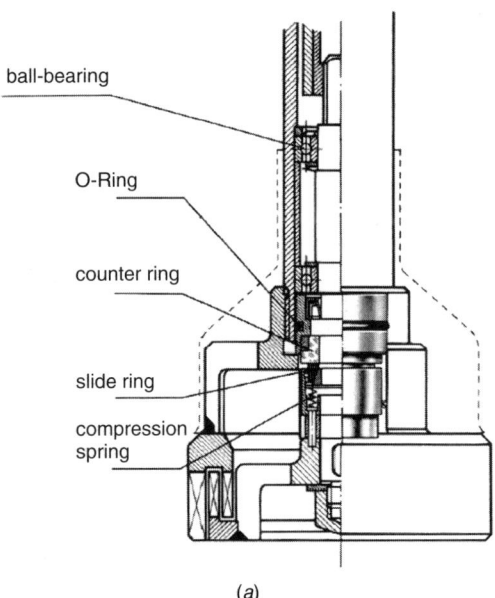

(a)

**Figure 8-14** Mechanical complexity of rotor–stator mixers: (*a*) seal of IKA Ultra Turrax batch rotor–stator. (*Continued*)

**504**  ROTOR–STATOR MIXING DEVICES

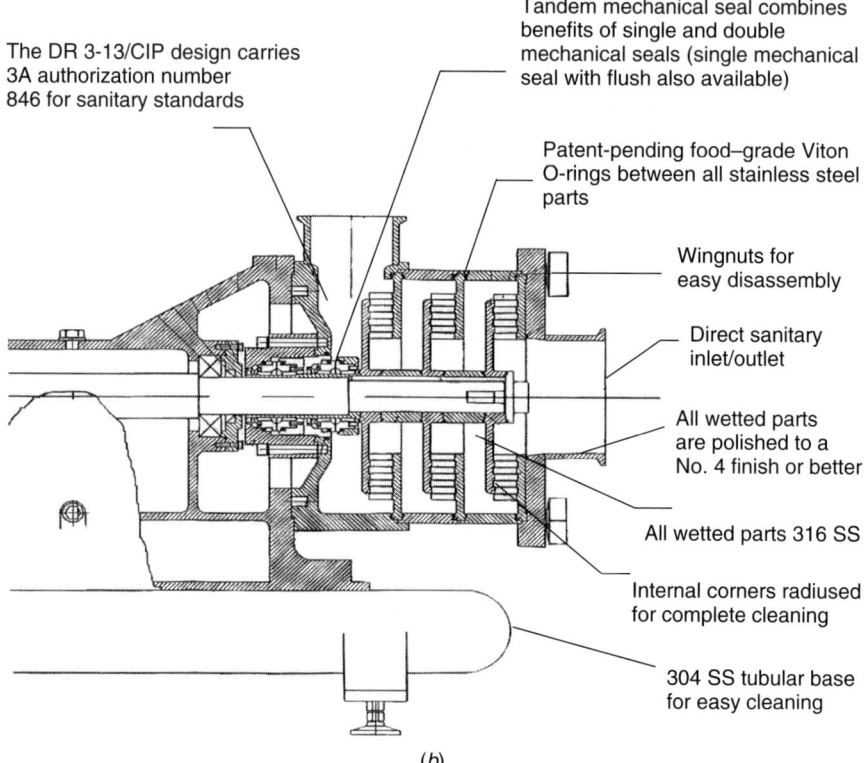

**Figure 8-14** (*b*) Multistage mixing head.

It is important to verify, a priori, that the specific equipment has been used at the specified conditions without any known problems. Because of the close tolerances or small gaps between the rotor and stator parts, one must confirm that the gap tolerances will be maintained as the components heat up during processing.

Details of mechanical design, fabrication, and assembly must be entrusted to the suppliers of the mixer. Users must resist the temptation to mechanically redesign or modify standard mixers without a supplier's recommendation and approval.

## 8-6  ROTOR–STATOR MIXING EQUIPMENT SUPPLIERS

There are many vendors of rotor–stator mixers. The reader may refer to current editions of *Thomas's Register, Chemical Engineering Buyers' Guide*, and so on, for an up-to-date list of major suppliers. There are a number of other suppliers whose equipment is used to produce dispersions or emulsions but which does not fit the definition of having a rotor and stator. These include high-pressure devices such as valve homogenizers, whistles, and cavitation-based devices.

## NOMENCLATURE

- $d_{max}$    maximum stable drop size (μm)
- D    rotor diameter (m)
- $D_{max}$    maximum diameter of a Greerco rotor (m)
- N    rotor rotational speed (rps)
- P    power (W)
- Po    power number, $\dfrac{P}{\rho N^3 D^5}$
- Re    Reynolds number, $\dfrac{\rho N D^2}{\mu}$
- $V_{tip}$    rotor tip speed (m/s)

### Greek Symbols

- $\dot{\gamma}$    nominal shear rate ($s^{-1}$)
- $\delta$    width of shear gap between rotor and stator (m)
- $\mu$    viscosity (Pa · s)
- $\rho$    density (kg/m$^3$)

## REFERENCES

Bourne, J. R., and J. Garcia-Rosas (1986). Rotor stator mixers for rapid micromixing, *Chem. Eng. Res. Des.*, **64**, 11–17.

Bourne, J. R., and M. Studer (1992). Fast reactions in rotor stator mixers of different size, *Chem. Eng. Process.*, **31**, 285–296.

Calabrese, R. V. (1999). Plenary lecture: assessment of rotor–stator mixing devices, *Paper 5.1*, presented at Mixing XVII, 17th Biennial North American Mixing Conference, Banff, Alberta, Canada, Aug.

Calabrese, R. V., M. K. Francis, V. P. Mishra, and S. Phongikaroon (2000). Measurement and analysis of drop size in a batch rotor–stator mixer, *Proc. 10th European Conference on Mixing*, H. E. A. van den Akker and J. J. Derksen, eds., Elsevier Science, Amsterdam, pp. 149–156.

Calabrese, R. V., M. K. Francis, K. R. Kevala, V. P. Mishra, G. A. Padron, and S. Phongikaroon (2002). Fluid dynamics and emulsification in high shear mixers, *Proc. 3rd World Congress on Emulsions*, Lyon, France, Sept.

Davies, J. T. (1987). A physical interpretation of drop sizes in homogenizers and agitated tanks, including the dispersion of viscous oils, *Chem. Eng. Sci.*, **42**, 1671–1676.

Kevala, K. R. (2001). Sliding mesh simulation of a wide and narrow gap inline rotor–stator mixer, M.S. thesis, University of Maryland, College Park, MD, May.

LeClair, M. L. (1995). Optimize rotor–stator performance using computational fluid dynamics, *Paint Coat. Ind.*, **1**(4), 46–48.

Myers, K. J., M. F. Reeder, and D. Ryan (2001). Power draw of a high-shear homogenizer, *Can. J. Chem. Eng.*, **79**, 94–99.

Padron, G. A. (2001). Measurement and comparison of power draw in batch rotor–stator mixers, M.S. thesis, University of Maryland, College Park, MD, Oct.

Silverson Machines, Inc. (2002). High speed preparation of rubber solutions, Application report No. C7CA1, East Longmeadow, MA.

# CHAPTER 9

# Blending of Miscible Liquids

RICHARD K. GRENVILLE

*The DuPont Company*

ALVIN W. NIENOW

*University of Birmingham*

## 9-1 INTRODUCTION

Blending is a common mixing operation in the chemical and process industries. The objective is to take two or more miscible fluids and blend them to a predetermined degree of homogeneity. The time taken to reach this degree of homogeneity is the *blend time*. This is also known as the *macroscale mixing time* since it is the time scale associated with mixing the contents of a vessel.

Blending operations are carried out for low viscosity fluids in the turbulent regime, moderately viscous fluids in the transitional regime, and highly viscous fluids in the laminar regime. In most cases, the viscous fluids will be non-Newtonian and generally shear thinning. This must be taken into account in the design of appropriate mixing equipment. Occasionally, the fluids may exhibit a yield stress and/or viscoelasticity, and these complex behaviors also need to be considered, although viscoelasticity is not going to be considered here (see Section 4-2.2.1).

Agitated and jet mixed vessels are used for blending duties. The choice of equipment will depend on the viscosity of the fluid, the desired blend time, and the size of the vessel. The chapter is divided into four sections that cover design rules for:

- Blending of Newtonian fluids in the turbulent and transitional regimes (Re > 200) for turbine and hydrofoil impellers

---

*Handbook of Industrial Mixing: Science and Practice*, Edited by Edward L. Paul, Victor A. Atiemo-Obeng, and Suzanne M. Kresta
ISBN 0-471-26919-0  Copyright © 2004 John Wiley & Sons, Inc.

- Blending of shear-thinning fluids in the turbulent and transitional regimes (Re > 200) for turbine and hydrofoil impellers
- Agitation of yield stress fluids
- Blending of Newtonian and shear-thinning fluids in the laminar regime with helical ribbon impellers
- Jet mixing of low viscosity fluids in the turbulent regime

## 9-2 BLENDING OF NEWTONIAN FLUIDS IN THE TURBULENT AND TRANSITIONAL REGIMES

### 9-2.1 Literature Survey

Blend times have been measured in agitated vessels using a variety of techniques; conductivity, temperature, or pH (using an indicator for color change, as discussed in Section 4-4 and illustrated on the Visual Mixing CD). The results are presented as a relationship between the dimensionless blend time, which is the product of the measured blend time and the impeller rotational speed, dimensionless geometrical ratios, and in some cases, Reynolds and Froude numbers.

***9-2.1.1 Turbulent Regime.*** The majority of references in the literature report that in the turbulent regime, the dimensionless blend time is a constant, independent of Reynolds and Froude numbers. The value of the constant is dependent on the impeller type and diameter relative to the vessel. These references include Kramers et al. (1953), Procházka and Landau (1961), Hoogendoorn and den Hartog (1967), Khang and Levenspiel (1976), Sano and Usui (1987), and others. There is a smaller group of references that report a weak dependence on Reynolds and Froude numbers in the turbulent regimes, and these include Fox and Gex (1956) and Norwood and Metzner (1960).

### 9-2.2 Development of the Design Correlation

The recommended correlations for design of agitators for blending in the turbulent and transitional regimes were developed at the Fluid Mixing Processes consortium at Cranfield in the U.K. The work is discussed in detail in Grenville (1992). Briefly, blend times were measured using a conductivity technique in vessels 0.30, 0.61, 1.83, and 2.97 m in diameter. The vessels had a standard torispherical base and were all fitted with standard baffles. The correlation is based on experiments carried out with one impeller located one-third of the liquid depth above the vessel base.

A variety of impellers were tested, including hydrofoils, pitched and flat blade turbines, and disk turbines, and their diameters ranged from one-third to one-half of the vessel diameter. Three conductivity probes were placed in the vessel in regions of differing agitation intensity (see Chapter 4):

- *Beneath the impeller:* T/50 below the impeller, T/8 from the shaft axis

- *Halfway between the agitator shaft and the vessel wall:* T/4.5 below the liquid surface, T/4.7 from the shaft axis
- *Behind a baffle:* T/3 below the liquid surface, T/2.2 from the shaft axis

In the turbulent regime, the local blend times were the same. In the transitional regime, the blend time measured beneath the impeller did not change significantly, but behind the baffle, the blend time increased. Ultimately, the local blend time behind the baffle controlled the blend time for the entire vessel.

**9-2.2.1 Turbulent Regime.** After rounding off the exponents obtained from regression of the data, the correlation for blend time to reach 95% homogeneity for all the impellers at all the scales tested by Grenville (1992) is

$$Po^{1/3} N\theta_{95} \frac{D^2}{T^{1.5} H^{0.5}} = 5.20 \tag{9-1}$$

The standard deviation of the constant is ±10.0%.

In a vessel where the liquid depth is equal to the vessel diameter,

$$Po^{1/3} N\theta_{95} \left(\frac{D}{T}\right)^2 = 5.20 \tag{9-2}$$

Since the impeller's power number is constant in a baffled vessel operating in the turbulent regime, $N\theta$ is a constant and independent of Reynolds number.

The equation can be rearranged to

$$Po^{1/3} ND^2 = 5.20 \frac{T^2}{\theta_{95}} \tag{9-3}$$

Multiplying both sides by $\rho/\mu$ yields

$$Po^{1/3} \frac{\rho ND^2}{\mu} = 5.20 \frac{\rho T^2}{\mu \theta_{95}} \tag{9-4}$$

$$Po^{1/3} Re = \frac{5.20}{Fo} \tag{9-5}$$

The dimensionless groups on the left-hand side are the power and Reynolds numbers of the impeller. The dimensionless group on the right-hand side is the Fourier number that is used in analysis of unsteady transfer processes. Hoogendoorn and den Hartog (1967) used it in their work but called it the *vessel Reynolds number*.

In an area where many studies have been made leading to different correlations, it is very valuable if independent corroborating work is available. Nienow (1997) found that blend time data already published with his co-workers or available from them fitted eq. (9-2) very well. This additional work covered further impeller types and a wider range of energy dissipation rates (down to 0.01 W/kg found in fermenters containing animal cell cultures), although still in the turbulent regime. He also gave some theoretical justification for the relationships based on fundamental turbulence concepts from Corrsin (1964). The critical assumptions in the analysis were that the macro scale of turbulence was related to the diameter of the vessel, and the critical local energy dissipation rate was that at the wall.

### 9-2.2.2 Transitional Regime.
The data taken in the transitional regime were correlated by performing a regression of $Po^{1/3}Re$ on $1/Fo$:

$$Po^{1/3}Re = \frac{183}{\sqrt{Fo}} \tag{9-6}$$

The standard deviation on the constant is $\pm 17.4\%$. This equation can be expanded to give

$$N\theta_{95} = \frac{183^2}{Po^{2/3}Re} \left(\frac{T}{D}\right)^2 \tag{9-7}$$

Since the power number is roughly constant in the transitional regime (the variation with Re is much less than the 1/Re dependence observed in the laminar regime), the dimensionless blend time is inversely proportional to Reynolds number, as other workers found.

Solving the two correlations for $Po^{1/3}Re$ on $1/Fo$ gives the values of these two dimensionless groups at the boundary between the turbulent and transitional regimes:

$$Po^{1/3}Re_{TT} = 6370 \quad \text{and} \quad \frac{1}{Fo_{TT}} = 1225$$

## 9-2.3 Use of the Design Correlation

There are two ways in which an engineer may have to use the blend time correlation; the first is to design a new agitator, and the second is to rate an existing agitator for a new process. Expressing the correlation in terms of $Po^{1/3}Re$ and $1/Fo$ makes this easier to do. When designing a new process, the vessel size, fluid physical properties, and desired blend time will be specified and $1/Fo$ can be calculated. Immediately, the regime in which the impeller will operate can be identified. The appropriate correlation can be used to calculate $Po^{1/3}Re$. The impeller type and diameter must be chosen such that so that the rotational speed necessary to achieve the desired blend time can be calculated.

When rating an existing agitator/vessel, the impeller type, speed, and diameter are known with the fluid physical properties. Now $Po^{1/3}Re$ can be calculated and the operating regime identified. Then $1/Fo$ can be calculated and the blend time calculated as the final step. Also, the standard deviation of the constants can be used to give a level of confidence to be included in the design procedure.

Approximately 67% of observations will lie within ±1 standard deviation. Similarly, 95% lie within ±2 standard deviations and 99% lie within ±3 standard deviations. So the level of confidence can be incorporated into the design process by defining the correlation constant to be used as:

$$\text{turbulent regime:} \quad 5.20 + 0.52s$$
$$\text{transitional regime:} \quad 183 + 31.1s$$

where s = 1 for 67% confidence level, s = 2 for 95% confidence level, and s = 3 for 99% confidence level. Examples of the correlation's use are given at the end of this section.

### 9-2.4 Impeller Efficiency

The question of which impeller is the most efficient for blending can be answered by rearranging the blend time correlations. In the turbulent regime, for a vessel where H = T,

$$Po^{1/3} N\theta_{95} \left(\frac{D}{T}\right)^2 = 5.20 \tag{9-2}$$

$$\theta_{95} \propto \left(\frac{T^3}{Po \cdot N^3 D^5}\right)^{1/3} \left(\frac{T}{D}\right)^{1/3} T^{2/3} \tag{9-8}$$

$$\theta_{95} \propto \left(\frac{1}{\bar{\varepsilon}}\right)^{1/3} \left(\frac{T}{D}\right)^{1/3} T^{2/3} \tag{9-9}$$

This analysis shows that:

- All impellers of the same diameter are equally energy efficient (i.e., achieve the same blend time at the same power per unit mass of fluid, $\bar{\varepsilon}$).
- A larger impeller diameter will achieve a shorter blend time for the same power input per unit mass.
- Blend time is independent of the fluid's physical properties in the turbulent regime.
- When scaling-up at constant power per unit mass and geometry, blend time will increase by the scale factor raised to the two-thirds power.

These conclusions are strongly supported by the theoretical analysis and experimental results of Nienow (1997) and Langheinrich et al. (1998). Most surprising is the conclusion concerning the equivalence of different impellers, which is counter-intuitive and contrary to what many vendors claim.

## 512 BLENDING OF MISCIBLE LIQUIDS

A similar analysis for the transitional regime leads to some similar conclusions but some which are quite different:

$$N\theta_{95} = \frac{183^2}{Po^{2/3} Re} \left(\frac{T}{D}\right)^2 \tag{9-7}$$

$$\theta_{95} \propto \left(\frac{T^3}{Po \cdot N^3 D^5}\right)^{2/3} \left(\frac{\mu}{\rho}\right) D^{-2/3} \tag{9-10}$$

$$\theta_{95} \propto \left(\frac{T^3}{Po \cdot N^3 D^5}\right)^{2/3} \left(\frac{\mu}{\rho}\right) \left(\frac{T}{D}\right)^{2/3} T^{-2/3} \tag{9-11}$$

$$\theta_{95} \propto \left(\frac{1}{\bar{\varepsilon}}\right)^{2/3} \left(\frac{\mu}{\rho}\right) \left(\frac{T}{D}\right)^{2/3} T^{-2/3} \tag{9-12}$$

This analysis shows that:

- All impellers of the same diameter are equally energy efficient (i.e., achieve the same blend time at the same power per unit mass of fluid, $\bar{\varepsilon}$).
- A larger impeller diameter will achieve a shorter blend time for the same power input per unit mass.
- Blend time is proportional to the fluid viscosity and inversely proportional to the density.
- When scaling-up at constant power per unit mass and geometry, blend time will *decrease* by the scale factor raised to the two-thirds power.

The first two conclusions are the same as for turbulent operation, but the last two are different. The final one in particular, that blend time will decrease on scale-up, may seem odd. However, it is correct, and the reason for this is that scaling-up at constant power input per unit mass, the Reynolds number will increase and the dimensionless blend time is inversely proportional to Reynolds number. This shows that care needs to be taken when scaling-up from lab or pilot scale vessels, which may be operating in the transitional regime, to plant scale, which may be operating in the turbulent regime. Using the design methods described above will take care of this scaling issue.

### 9-2.5 Shaft Torque, Critical Speed, and Retrofitting

Another consideration is the shaft torque. Since impellers of the same diameter require the same power input to achieve the same blend time, an impeller with a lower power number will have to operate at a higher speed and hence will have a lower shaft torque ($P = 2\pi N\Lambda$). Since the size of the shaft and gearbox are related to the torque, reducing the torque may reduce the agitator size and reduce the cost.

One factor to be aware of when making this decision is the critical speed of the shaft and impeller assembly. Running at a higher speed may reduce the torque, but the agitator may require a larger shaft anyway because the operating speed is now too close to the first critical speed. On the other hand, a high-power-number agitator may be replaced by a lower Po agitator while the speed is maintained. If the diameter is increased so that the power and torque stay the same, critical speed problems are generally avoided, while the mixing time is shortened. This is a modestly effective retrofitting strategy.

### 9-2.6 Nonstandard Geometries: Aspect Ratios Greater Than 1 and Multiple Impellers

Cooke et al. (1988) reported mixing times for a range of multiple-impeller systems. They found the time increased very significantly but did not distinguish whether the increase was due particularly to the increase in height or the extra number of impellers. The equation that they gave for multiple Rushton turbines is similar to eq. (9-2):

$$\theta_m = \frac{3.3}{Po^{1/3} N} \left(\frac{H}{D}\right)^{2.43} \tag{9-13}$$

Cronin et al. (1994) showed by a decolorization technique the staged mixing associated with radial flow Rushton turbines, and Otomo et al. (1993) showed similar results with radial flow hollow-blade turbines. The latter also used conductivity probes and found mixing times about twice as long as with a single impeller, with values close to the predictions of the equation of Cooke et al. (1988).

When using two down-pumping wide-blade Lightnin A315 hydrofoil impellers in a vessel containing liquid with an aspect ratio of 2, Otomo et al. (1995) found that staged mixing or zoning was largely eliminated. As a result, the mixing time was significantly reduced compared to that of two radial flow impellers, typically by about 50% at the same specific energy dissipation rate. At the same aspect ratio, Hari-Prajitno et al. (1998) found a similar reduction with two up-pumping wide-blade hydrofoils (40%) and an even greater reduction with an up-pumping hydrofoil below a down-pumping (60%). The use of a radial flow impeller beneath either up-pumping (Vrabel et al., 2000) or down-pumping hydrofoils (Manikowski et al., 1994) is also quite effective as a means of lowering the mixing time and reducing zoning compared to radial flow impellers.

Clearly, once multiple impellers are employed, particularly with aspect ratios significantly different from 1, the system becomes considerably more complex. Mixing times can greatly increase and the choice of impellers now can be very significant.

### 9-2.7 Other Degrees of Homogeneity

The design correlation is based on experiments in which the blend time required to reach 95% homogeneity were measured. The blend time required to reach another degree of homogeneity can be calculated because the blending process is first order (i.e., the concentration fluctuations which are being "smoothed" as

**514** BLENDING OF MISCIBLE LIQUIDS

the blending progresses decay exponentially):

$$\frac{dc'}{dt} = -kc' \tag{9-14}$$

$$\int_1^x \frac{dc'}{c'} = -k \int_0^\theta dt \tag{9-15}$$

$$[\ln c']_1^x = -k[t]_0^\theta \tag{9-16}$$

$$\ln(1-x) = -k\theta \tag{9-17}$$

Here x is the relative magnitude of the concentration fluctuations and equals 1 at time t = 0 (i.e., 0% homogeneity). For 95% homogeneity, x = 0.05. So the equation to adjust the blend time for a degree of homogeneity other than 95% is

$$\theta_z = \theta_{95} \frac{\ln[(100 - \% \text{ homogeneity})/100]}{\ln 0.05} \tag{9-18}$$

If the blend time for 99% homogeneity is required,

$$\theta_{99} = \theta_{95} \frac{\ln[(100 - 99)/100]}{\ln 0.05} \tag{9-19}$$

$$\theta_{99} = \theta_{95} \frac{\ln 0.01}{\ln 0.05} = 1.537 \theta_{95} \tag{9-20}$$

### 9-2.8 Examples

***Example 9-1: Designing a New Agitator.*** A new process is to be carried out in a baffled vessel that is 6 ft (1.83 m) in diameter. The liquid depth will be 6 ft (1.83 m). At the end of the process an inhibitor is added to stop the reaction and it must be blended to 99.5% homogeneity within 0.5 min to prevent "overreaction" and production of a product with too high a molecular weight. At this point in the process the fluid has a specific gravity of 1.02 and a viscosity of 18 cP.

SOLUTION

1. Determine the operating regime by calculating 1/Fo. In SI units:

    viscosity: $\mu = 0.018$ Pa · s
    density: $\rho = 1020$ kg/m$^3$
    desired blend time: $\theta = 30$ s
    vessel diameter: $T = 1.83$ m

$$\frac{1}{\text{Fo}} = \frac{\rho T^2}{\mu \theta} = \frac{1020 \text{ kg/m}^3 \times (1.83 \text{ m})^2}{0.018 \text{ Pa} \cdot \text{s} \times 30 \text{ s}} = 6326$$

The process will operate in the turbulent regime since 1/Fo > 1225.

BLENDING OF NEWTONIAN FLUIDS IN THE TURBULENT AND TRANSITIONAL REGIMES    515

2. Determine the multiplier to convert 95% to 99.5% blend times:

$$\theta_{99.5} = \theta_{95} \frac{\ln[(100-99.5)/100]}{\ln 0.05}$$

$$\theta_{99} = \theta_{95} \frac{\ln 0.005}{\ln 0.05} = 1.768 \theta_{95}$$

So the constant in the calculation will be 9.19 (i.e., 5.20 × 1.77).

3. Calculate $Po^{1/3}Re$ using the turbulent correlation with the adjusted constant:

$$Po^{1/3}Re = \frac{9.19}{Fo} = 9.19 \times 6326 = 58\,136$$

4. Choose the impeller type and power number and calculate the Reynolds number. Use a pitched blade turbine with a power number of 1.80.

$$Re = \frac{58\,136}{Po^{1/3}} = \frac{58\,136}{1.8^{1/3}} = 47\,792$$

5. Choose the impeller diameter and calculate the impeller speed. An impeller diameter of 50% the vessel diameter will be most energy efficient, so choose $D = 0.915$ m.

$$N = \frac{Re \cdot \mu}{\rho D^2} = \frac{47\,792 \times 0.018}{1020 \times 0.915^2} = 1.01 \text{ rps}$$

6. The calculated speed is 1 rps, or 60 rpm, but this is not a standard gearbox output speed. The closest standard speeds are 56 and 68 rpm (see Table 6-2). With the same diameter, running at 56 rpm will increase the blend time by 7% while running at 68 rpm will decrease the blend time by 12%. Alternatively, the speed and diameter can be changed to give the desired blend time. This will be a judgment that the engineer has to make.

7. Design the agitator to run at 56 rpm accepting the slightly longer blend time; calculate the power input by the impeller and choose the motor size:

$$P = Po \cdot \rho N^3 D^5 = 1.8 \times 1020 \text{ kg/m}^3 \times \left(\frac{56}{60 \text{ s}}\right)^3 \times (0.915 \text{ m})^5 = 957 \text{ W}$$

The power input by the impeller is 957 W, or 1.28 hp. The next highest standard motor power would be 1.5 hp (see Table 6-2). This is acceptable since the power drawn by the impeller is roughly 85% of the available motor power.

8. The design is complete. The agitator will require a 1.5 hp motor with an output speed of 56 rpm. The impeller will be a pitched blade turbine 36 in. in diameter.

**Example 9-2: Rating an Existing Agitator.** An existing vessel and agitator are being considered for a new process. The vessel is 3 m in diameter and the liquid

# 516  BLENDING OF MISCIBLE LIQUIDS

depth will be 2.5 m. The fluid will have a viscosity of 500 mPa · s and a density of 980 kg/m³. The impeller is a hydrofoil, with a power number of 0.33, 1.0 m in diameter and operating at 125 rpm. What will the blend time be?

SOLUTION

1. Determine the operating regime by calculating $Po^{1/3}Re$:

$$Po^{1/3}Re = Po^{1/3}\frac{\rho ND^2}{\mu} = 0.33^{1/3} \times \frac{980 \text{ kg/m}^3 \times (125/60 \text{ s}) \times (1.0 \text{ m})^2}{0.5 \text{ Pa} \cdot \text{s}}$$

$$= 2822$$

The process will operate in the transitional regime since $Po^{1/3}Re < 6370$.

2. Calculate 1/Fo using the transitional correlation:

$$Po^{1/3}Re = \frac{183}{\sqrt{Fo}}$$

$$\frac{1}{\sqrt{Fo}} = \frac{Po^{1/3}Re}{183} = \frac{2822}{183} = 15.42$$

$$\frac{1}{Fo} = 237.8$$

3. Calculate the blend time:

$$\frac{1}{Fo} = \frac{\rho T^{1.5}H^{0.5}}{\mu\theta}$$

$$\theta = \frac{\rho T^{1.5}H^{0.5}}{\mu(1/Fo)} = \frac{980 \text{ kg/m}^3 \times (3.0 \text{ m})^{1.5} \times (2.5 \text{ m})^{0.5}}{0.5 \text{ Pa} \cdot \text{s} \times 237.8} = 67.7 \text{ s}$$

4. The blend time for 95% homogeneity will be 68 s. If a higher, or lower, degree of homogeneity is required, the appropriate correction factor can be calculated.

## 9-3  BLENDING OF NON-NEWTONIAN, SHEAR-THINNING FLUIDS IN THE TURBULENT AND TRANSITIONAL REGIMES

### 9-3.1  Shear-Thinning Fluids

Methods for designing agitators to blend Newtonian fluids were discussed in Section 9-2. Unfortunately, the vast majority of viscous fluids in the "real world" are non-Newtonian, and the Newtonian design rules must be modified to take account this fact. The most common type of non-Newtonian fluid exhibits shear-thinning behavior.

The behavior of a shear-thinning fluid can be described mathematically by the *power law*, which relates the shear stress in the fluid to the shear rate being exerted on it:

$$\tau = K\dot{\gamma}^n \tag{9-21}$$

so that from the definition of dynamic viscosity,

$$\mu_A = \frac{\tau}{\dot{\gamma}} = K\dot{\gamma}^{n-1} \tag{9-22}$$

where $\mu_A$ is the apparent viscosity of the fluid, and K and n are the consistency and flow behavior indices, respectively ($n < 1$ for a shear-thinning fluid). For this case quantitative relationships are available and these are discussed in this section.

The shear rate in an agitated vessel will vary with position, being highest near the impeller where the velocity gradients are steepest and low near the walls and surface. For a shear-thinning fluid this variation means that the apparent viscosity near the impeller is low, and near the wall, it is high. To estimate the blend time for a non-Newtonian fluid, the appropriate shear rate must be identified. This will then be used to estimate a value for the apparent viscosity of the fluid. Once this has been done, the Newtonian correlations can be used to estimate the blend time.

## 9-3.2 Literature Survey

Metzner and Otto (1957) developed the best-known definition of shear rate in an agitated vessel. They measured the power number for a variety of impellers in the laminar regime in Newtonian fluids and then repeated the measurements with shear-thinning fluids. They assumed that the power number was unaffected by the fluid's non-Newtonian behavior and that the Newtonian viscosity and shear-thinning apparent viscosity were equal for equal power number and Reynolds number. Once an estimate of the apparent viscosity is made, eq. (9-22) can be rearranged and the shear rate can be calculated from the power law model:

$$\dot{\gamma} = \left(\frac{\mu_A}{K}\right)^{1/(n-1)} = k_s N \tag{9-23}$$

Metzner and Otto concluded that the shear rate is proportional to the impeller speed with the constant of proportionality, $k_S$, taking a value between 10 and 15 for turbine impellers and 25 to 30 for close-clearance impellers. With this approach, the impeller Reynolds number can be modified to give

$$\text{Re} = \frac{\rho N D^2}{\mu_A} = \frac{\rho N^{(2-n)} D^2}{K k_S^{(n-1)}} \tag{9-24}$$

Strictly, this method is valid only in the laminar regime, where power number is inversely proportional to Reynolds number. Nagata et al. (1971) repeated these experiments over an extended range of Reynolds numbers for a helical ribbon, anchor, flat blade, and Rushton turbine. They concluded that the Metzner and Otto shear rate did work in the laminar regime but failed in the transitional regime for the two turbine impellers (flat blade and Rushton). It did work for the two close-clearance impellers.

The Metzner and Otto method gives an estimate of the shear rate at the impeller based on the power measurement and the apparent viscosity at the impeller. Other processes in shear-thinning fluids, especially heat transfer, have also been studied where the shear rate at the impeller is not important. In this case the apparent viscosity and shear rate at the heat transfer surface are controlling and the Metzner–Otto method no longer applies.

Pollard and Kantyka (1969), Bourne et al. (1981), and Wang and Yu (1989) all concluded that in the laminar regime, the shear rate is proportional to the impeller speed, but in the transitional regime, the dependence on speed is more complicated. Generally, they found that the shear rate in the transitional regime is proportional to the square root of the power input per unit mass, indicating that turbulence is contributing to the generation of shear.

Wichterle et al. (1984) used an electrochemical technique to measure the shear rates on the surface of a Rushton turbine's blades for Newtonian and non-Newtonian fluids with a Reynolds number that varied between 1 and 10 000. They correlated their data by

$$\dot{\gamma}_m = (1 + 5.3n)^{1/n} \mathrm{Re}_m^{1/(n+1)} N \quad (9\text{-}25)$$

where

$$\mathrm{Re}_m = \frac{\rho N^{2-n} D^2}{K} \quad (9\text{-}26)$$

At low Reynolds numbers, the shear rate is proportional to impeller speed as reported by Metzner and Otto (1957). As Reynolds number increases, the exponent on the impeller speed increases from 1.0 to 1.5 (in a Newtonian fluid). This is due to the presence of shear stresses resulting from the turbulent fluctuating velocities that will start to appear. Note that eq. (9-26) does not contain the Metzner–Otto constant, $k_S$, because only one impeller type was used in the study.

### 9-3.3 Modifying the Newtonian Relationships for Shear-Thinning Fluids

The correlation for design of agitators for blending shear-thinning fluids in the turbulent and transitional regimes was developed at the fluid mixing processes consortium at Cranfield in the U.K. and is discussed in detail in Grenville (1992). The equipment and experimental technique described in Section 9-2.2 were used.

The Newtonian experiments had shown that in turbulent regime, the local blend times were the same throughout the vessel. As the viscosity increased and the Reynolds number decreased, the blend time measured behind the baffle increased significantly while those measured beneath the impeller and in the middle of the vessel increased slightly compared with the turbulent values. The local blend time measured behind the baffle was controlling the blend time for the entire vessel. The approach that Grenville (1992) took to analyze the shear-thinning data was to determine the apparent viscosity of the fluid at the wall and use this value to calculate the values of Reynolds and Fourier numbers.

Bird et al. (1960) give an equation for the shear rate (tangential velocity gradient) on the wall of a baffled vessel and the pressure exerted on the baffles as a function of the torque on the agitator shaft:

$$\Lambda = \mu \iint_S R\left(\frac{\partial v_\theta}{\partial r}\right)_W dS + \iint_A R p_{baff}\, dA \qquad (9\text{-}27)$$

Assuming that the shear rate is constant on the surfaces of the vessel wall and base, eq. (9-27) can be rewritten in terms of the shear stress at the vessel wall:

$$\Lambda = \tau_W \iint_S R\, dS + \iint_A R p_{baff}\, dA \qquad (9\text{-}28)$$

The pressure exerted by the fluid on the baffles was estimated as

$$p_{baff} = \frac{\rho(\Delta v)^2}{2} \qquad (9\text{-}29)$$

where $\Delta v$ is the change in tangential velocity as the fluid impinges on the baffle.

Applying the appropriate integration limits, the shear stress at the wall in a vessel where $H = T$ with standard baffles and a torispherical bottom can be estimated from

$$\tau_W = \frac{1}{1.622}\left[\frac{\Lambda}{T^3} - 0.0638\rho(\Delta v)^2\right] \qquad (9\text{-}30)$$

In order to use eq. (9-30), an estimate of the fluid velocity impinging on the baffle should be made, but the contribution of the pressure is small compared to the torque and, for engineering calculations, can be ignored. Thus, the estimated shear stress at the wall is

$$\tau_W = \frac{1}{1.622}\left(\frac{\Lambda}{T^3}\right) \qquad (9\text{-}31)$$

The power law can then be used to determine the shear rate at the vessel wall and the apparent viscosity:

$$\tau_W = K \dot\gamma_W^n \qquad (9\text{-}32)$$

$$\dot\gamma_W = \left(\frac{\tau_W}{K}\right)^{1/n} \qquad (9\text{-}33)$$

$$\mu_W = K \dot\gamma_W^{n-1} \qquad (9\text{-}34)$$

Once an estimate of the viscosity at the wall of the vessel has been made, Reynolds and Fourier numbers can be calculated and the method used for Newtonian fluids can be followed. It is important to remember that if any change

**520** BLENDING OF MISCIBLE LIQUIDS

is made to the agitator's operation, the apparent viscosity at the wall must be recalculated.

### 9-3.4 Use of the Design Correlation

Again, there are two ways in which an engineer may have to use the blend time correlation; the first is to design a new agitator and the second is to rate an existing agitator for a new process. In both cases the extra step of estimating the fluid's apparent viscosity will be necessary.

The procedure for rating an existing agitator/vessel is relatively straightforward since the impeller type, speed, and diameter are known with the fluid physical properties. The torque on the agitator shaft can be calculated followed by the fluid's apparent viscosity at the wall. Then $Po^{1/3}Re_W$ can be calculated and the operating regime identified, and then $1/Fo_W$ can be calculated using the appropriate correlation. Finally, the blend time can be calculated.

The procedure for designing a new process is more complicated because the fluid's apparent viscosity at the wall is determined by the impeller type, diameter, and operating speed, which determine the shear stress and shear rate at the wall. The vessel size, fluid density, and desired blend time can be specified, but the viscosity is required in order to calculate $1/Fo_W$.

Since the apparent viscosity is a function of the impeller properties and $1/Fo_W$ cannot be calculated immediately, an iterative procedure must be made. This can be simplified because there are a limited number of possible gearbox output speeds. Once the impeller type and diameter have been chosen, the torque at each speed can be calculated followed by the shear stress and shear rate at the wall. Then the viscosity and $1/Fo_W$ can be calculated and the regime in which the impeller would operate can be identified. The appropriate correlation can then be used to calculate $Po^{1/3}Re_W$, and this can be rearranged to solve for the impeller speed. The condition where the output speed from rearranging $Po^{1/3}Re_W$ is just less than the input speed used to calculate $1/Fo_W$ is the one on which the design will be based. An example of this method is given in Section 9-3.7.

### 9-3.5 Impeller Efficiency

Again, the question of which impeller is the most efficient for blending can be answered by rearranging the blend time correlations. In the turbulent regime there is no dependence of blend time on viscosity, so the conclusions drawn for Newtonian fluids apply to non-Newtonian, shear-thinning fluids.

In the transitional regime the blend time is proportional to the fluid's apparent viscosity:

$$N\theta_{95} = \frac{183^2}{Po^{2/3}Re_W}\left(\frac{T}{D}\right)^2 \tag{9-35}$$

$$\theta_{95} \propto \left(\frac{T^3}{PoN^3D^5}\right)^{2/3} \frac{\mu_W}{\rho}D^{-2/3} \tag{9-36}$$

or

$$\theta_{95} \propto \left(\frac{T^3}{P_o N^3 D^5}\right)^{2/3} \frac{\mu_W}{\rho} \left(\frac{T}{D}\right)^{2/3} T^{-2/3} \quad (9\text{-}37)$$

$$\theta_{95} \propto \left(\frac{1}{\varepsilon}\right)^{2/3} \frac{\mu_W}{\rho} \left(\frac{T}{D}\right)^{2/3} T^{-2/3} \quad (9\text{-}38)$$

Now the blend time is proportional to the viscosity of the fluid at the wall of the vessel. This in turn is dependent on the torque on the agitator's shaft; the higher the torque, the higher the shear stress and shear rate and the lower the fluid's apparent viscosity. So if two impellers of the same diameter are compared at the same power input per unit volume, the one with the lower power number, running at the higher speed, will give the longer blend time.

## 9-3.6 Cavern Formation and Size in Yield Stress Fluids

In very viscous, highly shear-thinning fluids (with n values on the order of 0.3 or less) whatever the physical reason for these particular rheological properties (e.g., mycelial fermentation broths, yogurt, high concentration, fine solid suspensions, emulsions, polymer solutions), agitation tends to cause cavern formation. A streak photograph of a cavern is shown in Figure 18-13. Thus, regions of liquid mixing and motion around the impeller are found, outside which the fluid is stagnant in dead zones or nearly so (Jaworski et al., 1994). In addition, there is no exchange of material (other than by diffusion) between the cavern and the bulk (Solomon et al., 1981). For Rushton turbines the cavern is usually cylindrical, centered on the agitator (Nienow and Elson, 1988) and of height/diameter ratio of 0.4. The shape is similar with the pitched blade turbine (Elson, 1988), the Scaba 6SRGT (Galindo and Nienow, 1993), and the Lightnin A315 (Galindo and Nienow, 1992), although with the latter the aspect ratio is a little higher ($\sim 0.6$).

The boundary of the cavern can be defined as the surface where the local shear stress equals the fluid yield stress. If it is assumed that the predominant flow in the cavern is tangential [and LDA studies suggest that this is a reasonable approximation (Hirata et al., 1994)] and that the cavern shape, fluid yield stress, and impeller power number are known, the cavern size may be determined. A right circular cylinder of height $H_c$ and diameter $D_c$ centered on the impeller is a good model for the cavern shape, which allows for the effect of different impellers (Elson et al., 1986). Thus,

$$\left(\frac{D_C}{D}\right)^3 = \frac{P_o \cdot \rho N^2 D^2}{\tau_y} \frac{1}{(H_c/D_c + \frac{1}{3})\pi^2} \quad (9\text{-}39)$$

Since the ratio of cavern height to diameter is typically 0.4, eq. (9-39) can be simplified to give

$$\left(\frac{D_C}{D}\right)^3 = \frac{1.36 \, P_o \cdot \rho N^2 D^2}{\pi^2 \, \tau_y} \quad (9\text{-}40)$$

This yield stress cavern model has been used by industrialists with some success (Etchells et al., 1987; Carpenter et al., 1993).

To use eq. (9-40) for agitator design, the cavern diameter must be set equal to the vessel diameter (i.e., the edge of the cavern must reach the vessel wall):

$$\left(\frac{T}{D}\right)^3 = \frac{1.36 \, Po \cdot \rho N_C^2 D^2}{\pi^2 \, \tau_y} \tag{9-41}$$

Equation (9-41) can be rearranged to give

$$\frac{Po \cdot \rho N_C^2 D^5}{T^3} = \tau_y \frac{\pi^2}{1.36} \tag{9-42}$$

where $N_C$ is the impeller speed when the cavern reached the vessel wall.

The term on the left-hand side of eq. (9-42) is the agitator shaft torque per unit volume. It does not matter what impeller type, diameter, and speed are chosen for the design, the torque must reach a lower limiting value in order for the cavern to reach the vessel wall. An agitator with a low power requirement can be designed by choosing a large diameter impeller ($\sim T/2$) with a higher power number. This will run at a lower speed, and since power is the product of torque and speed, a lower speed will result in less power required.

Once the cavern reaches the wall, it continues to rise up the vessel with increasing speed, but only rather slowly. Thus, very significant increases in energy dissipation rate are required to achieve motion everywhere. The most efficient way to ensure such motion is to use two impellers with D/T values of about 0.5 to 0.6 in a vessel of H/T and with one impeller being placed at $C = 0.25H$ and one at 0.75H. In this way, the two caverns approximately fill the vessel when either one produces a cavern that reaches the wall. The power requirement is just twice that required for a single impeller to reach the wall.

Recently, because of the difficulty of accurately determining the yield stress, a new model has been developed (Amanullah et al., 1998) based on assuming that a power law model with a low n value fits the flow curve. It also defines the cavern size by a minimum speed at its edge as the motion/no-motion boundary. As yet, an independent report has not been published confirming the effectiveness of this new approach. Finally, it should be noted that although there has been significant work dedicated to defining the size of the zones of motion in yield stress fluids, work has not been done to determine the blend time of the fluid inside the cavern.

### 9-3.7 Examples

***Example 9-3: Designing a New Agitator.*** A new process is to be carried out in a baffled vessel that is 2 m in diameter operating with a liquid depth of 2 m. The fluid has a density of 995 kg/m$^3$ and a shear-thinning rheology. The consistency index, K, has a value of 5.25 Pa · s$^n$ and a flow behavior index, n, of 0.654.

## SOLUTION

1. Choose the impeller type and diameter. A large diameter impeller with a high power number will be best suited for blending a shear-thinning fluid. Choose a pitched blade turbine with Po = 1.75 and D = 1.0 m (or T/2).
2. Calculate the torque at a range of operating speeds. The torque is calculated from:

$$\Lambda = \frac{Po \cdot \rho N^2 D^5}{2\pi} = \frac{1.75 \times 995 \text{ kg/m}^3 \times N^2 \times (1.0 \text{ m})^5}{2\pi} = 277.1 N^2 \text{ N} \cdot \text{m}$$

So for standard operating speeds between 30 and 155 rpm the torque will be as shown in Table 9-1.

**Table 9-1**

| N (rpm) | $\Lambda$ (N·m) | $\tau_W$ (Pa) | $\dot{\gamma}_W$ (s$^{-1}$) |
|---|---|---|---|
| 30  | 69   | 5.34   | 1.03   |
| 37  | 105  | 8.12   | 1.95   |
| 45  | 156  | 12.01  | 3.55   |
| 56  | 241  | 18.60  | 6.92   |
| 68  | 356  | 27.43  | 12.53  |
| 84  | 543  | 41.86  | 23.91  |
| 100 | 770  | 59.33  | 40.76  |
| 125 | 1203 | 92.70  | 80.64  |
| 155 | 1849 | 142.53 | 155.69 |

3. Calculate the shear stress and shear rate at the wall for each condition using

$$\tau_W = \frac{1}{1.622} \frac{\Lambda}{T^3}$$

$$\dot{\gamma}_W = \left(\frac{\tau_W}{K}\right)^{(1/n)}$$

Again, the results are given in Table 9.1.

4. Calculate the viscosity of the fluid at the wall and 1/Fo$_W$ using

$$\mu_W = K \dot{\gamma}_W^{(n-1)}$$

$$\frac{1}{Fo_W} = \frac{\rho T^2}{\mu_W \theta}$$

as given in Table 9-2.

**524** BLENDING OF MISCIBLE LIQUIDS

**Table 9-2**

| N (rpm) | $\Lambda$ (N·m) | $\tau_W$ (Pa) | $\gamma_W$ (s$^{-1}$) | $\mu_W$ (Pa·s) | $1/Fo_W$ |
|---|---|---|---|---|---|
| 30 | 69 | 5.34 | 1.03 | 5.20 | 15.30 |
| 37 | 105 | 8.12 | 1.95 | 4.17 | 19.10 |
| 45 | 156 | 12.01 | 3.55 | 3.39 | 23.49 |
| 56 | 241 | 18.60 | 6.92 | 2.69 | 29.61 |
| 68 | 356 | 27.43 | 12.53 | 2.19 | 36.36 |
| 84 | 543 | 41.86 | 23.91 | 1.75 | 45.47 |
| 100 | 770 | 59.33 | 40.76 | 1.46 | 54.69 |
| 125 | 1203 | 92.70 | 80.64 | 1.15 | 69.25 |
| 155 | 1849 | 142.53 | 155.69 | 0.92 | 86.95 |

5. Identify the regime in which the impeller would operate and calculate $Po^{1/3}Re_W$. In each case, $1/Fo_W$ is less than 1225, so the impeller operates in the transitional regime. Calculate $Po^{1/3}Re_W$ using

$$Po^{1/3}Re_W = \frac{183}{\sqrt{Fo_W}}$$

and Table 9-3.

**Table 9-3**

| N (rpm) | $\mu_W$ (Pa·s) | $1/Fo_W$ | $Po^{1/3}Re_W$ |
|---|---|---|---|
| 30 | 5.20 | 15.30 | 715.75 |
| 37 | 4.17 | 19.10 | 799.74 |
| 45 | 3.39 | 23.49 | 887.00 |
| 56 | 2.69 | 29.61 | 995.80 |
| 68 | 2.19 | 36.36 | 1103.52 |
| 84 | 1.75 | 45.47 | 1234.05 |
| 100 | 1.46 | 54.69 | 1353.30 |
| 125 | 1.15 | 69.25 | 1522.88 |
| 155 | 0.92 | 86.95 | 1706.44 |

6. For each case, calculate the impeller speed from

$$N = \frac{(Po^{1/3}Re_W)\mu_W}{Po^{1/3}\rho D^2}$$

as given in Table 9-4. The gearbox output speed at which the solution converges is 100 rpm, so this must be used to size the power required to run the agitator.

BLENDING OF NON-NEWTONIAN, SHEAR-THINNING FLUIDS    525

**Table 9-4**

| N (rpm) | $\mu_W$ (Pa·s) | $1/F_{OW}$ | $Po^{1/3}Re_W$ | N (rpm) |
|---|---|---|---|---|
| 30 | 5.20 | 15.30 | 715.75 | 186 |
| 37 | 4.17 | 19.10 | 799.74 | 167 |
| 45 | 3.39 | 23.49 | 887.00 | 150 |
| 56 | 2.69 | 29.61 | 995.80 | 134 |
| 68 | 2.19 | 36.36 | 1103.52 | 121 |
| 84 | 1.75 | 45.47 | 1234.05 | 108 |
| 100 | 1.46 | 54.69 | 1353.30 | 98 |
| 125 | 1.15 | 69.25 | 1522.88 | 87 |
| 155 | 0.92 | 86.95 | 1706.44 | 78 |

7. Calculate the power drawn by the impeller and choose the appropriate motor size:

$$P = Po \cdot \rho N^3 D^5 = 1.75 \times 995 \text{ kg/m}^3 \times \left(\frac{100}{60 \text{ s}}\right)^3 \times (1.0 \text{ m})^5$$

$$= 8061 \text{ W (or } 10.80 \text{ hp)}$$

The next standard motor size is 15 hp (see Table 6-2).

8. The design is complete. The agitator will require a 15 hp motor with an output speed of 100 rpm. The impeller will be a pitched blade turbine 1.0 m in diameter.

**Example 9-4: Rating an Existing Agitator.** An existing vessel and agitator are being considered for a new process. The vessel is 3 m in diameter and the liquid depth will be 3 m. The fluid is shear-thinning with a power law constant of $K = 8.98$ Pa · s$^n$, a power law index of $n = 0.467$, and a density of 1050 kg/m$^3$. The impeller is a hydrofoil, with a power number of 0.33, 1.5 m in diameter, and operating at 125 rpm. What will the blend time be?

SOLUTION

1. Calculate the torque on the agitator shaft:

$$\Lambda = \frac{Po \cdot \rho N^2 D^5}{2\pi} = \frac{0.33 \times 1050 \text{ kg/m}^3 \times (125/60 \text{ s})^2 \times (1.5 \text{ m})^5}{2\pi}$$

$$= 1817.6 \text{ N} \cdot \text{m}$$

2. Calculate the shear stress at the wall:

$$\tau_W = \frac{1}{1.622}\left(\frac{\Lambda}{T^3}\right) = \frac{1}{1.622} \times \frac{1817.6 \text{ N} \cdot \text{m}}{(3 \text{ m})^3} = 41.50 \text{ Pa}$$

## 526 BLENDING OF MISCIBLE LIQUIDS

3. Determine the shear rate at the wall:

$$\dot{\gamma}_W = \left(\frac{\tau_W}{K}\right)^{1/n} = \left(\frac{41.50}{8.98}\right)^{1/0.467} = 26.51 \text{ s}^{-1}$$

4. Determine the apparent viscosity at the wall:

$$\mu_W = K\dot{\gamma}_W^{n-1} = 8.98 \times 26.51^{0.467-1} = 1.565 \text{ Pa} \cdot \text{s}$$

5. Determine the operating regime by calculating $Po^{1/3}Re_W$:

$$Po^{1/3}Re_W = Po^{1/3}\frac{\rho ND^2}{\mu_W}$$

$$= 0.33^{1/3} \times \frac{1050 \text{ kg/m}^3 \times (125/60 \text{ s}) \times (1.5 \text{ m})^2}{1.565 \text{ Pa} \cdot \text{s}} = 2173$$

The process will operate in the transitional regime since $Po^{1/3}Re_W < 6370$.

6. Calculate $1/Fo_W$ using the transitional correlation

$$Po^{1/3}Re_W = \frac{183}{\sqrt{Fo_W}}$$

$$\frac{1}{\sqrt{Fo_W}} = \frac{Po^{1/3}Re_W}{183} = \frac{2176}{183} = 11.89$$

$$\frac{1}{Fo_W} = 141.4$$

7. Calculate the blend time:

$$\frac{1}{Fo_W} = \frac{\rho T^2}{\mu_W \theta}$$

$$\theta = \frac{\rho T^2}{\mu_W (1/Fo)} = \frac{1050 \text{ kg/m}^3 \times (3.0 \text{ m})^2}{1.565 \text{ Pa} \cdot \text{s} \times 141.1} = 42.8 \text{ s}$$

The blend time for 95% homogeneity will be 43 s.

**Example 9-5: Minimum Speed for Agitation of a Yield Stress Fluid.** A fluid with a density of 1560 kg/m³ and exhibiting a yield stress of 18 Pa is to be stored in a vessel 3 m in diameter with a maximum operating depth of 2.3 m. Design an agitator that will eliminate stagnant zones in the vessel.

SOLUTION

1. Choose the impeller type and diameter. A large diameter impeller ($\sim T/2$) with a higher power number will operate at a lower speed and power. Choose a pitched blade turbine (Po = 1.8), 1.5 m in diameter.

2. Calculate the minimum speed, $N_C$, for the cavern to reach the vessel wall from

$$\left(\frac{T}{D}\right)^3 = \frac{1.36}{\pi^2}\left(\frac{Po \cdot \rho N_C^2 D^2}{\tau_y}\right)$$

$$N_C = \left(\tau_y \frac{\pi^2}{1.36} \frac{T^3}{Po \cdot \rho D^5}\right)^{1/2}$$

$$= \left(18 \text{ Pa} \times \frac{\pi^2}{1.36} \times \frac{(3.0 \text{ m})^3}{1.8 \times 1560 \text{ kg/m}^3 \times (1.5 \text{ m})^5}\right)^{1/2} = 0.407 \text{ rps}$$

The minimum speed will be 24.4 rpm; from Table 6-2 30 rpm is the next highest standard speed.

3. Determine the number and location of the impeller(s). The height of the cavern will be approximately 40% of its diameter, in this case 1.2 m. Two impellers located at a clearance off the vessel base of 0.6 and 1.8 m will produce two intersecting caverns to a total height of 2.4 m which is higher than the maximum operating level.

4. Calculate the power drawn by the two impellers and choose the motor size. The power drawn will be calculated from

$$P = 2Po \cdot \rho N^3 D^5 = 2 \times 1.8 \times 1560 \text{ kg/m}^3 \times \left(\frac{30}{60 \text{ s}}\right)^3 \times (1.5 \text{ m})^5$$

$$= 5331 \text{ W (or 7.14 hp)}$$

The next standard motor size is 7.5 hp, but this would mean that the impeller power draw will be 95% of the motor power. Choose a 10 hp motor.

5. The design is complete. The agitator will require a 10 hp motor with an output speed of 30 rpm. The impellers will be two pitched blade turbines 1.5 m in diameter located 0.6 and 1.8 m above the vessel base.

## 9-4 BLENDING IN THE LAMINAR REGIME

Turbine and hydrofoil impellers operating in the turbulent and transitional regimes rely on entrainment to move fluid from the impeller region to the vessel walls and surface. As the viscosity of the fluid increases, primary flow generated by the impeller and level of entrainment are reduced until the regions away from the impeller become "stagnant."

In the transitional regime the dimensionless blend time, $N\theta$, is inversely proportional to Reynolds number. As the fluid's viscosity increases (and Reynolds number decreases) a value is reached where the dimensionless blend time becomes more sensitive to changes in viscosity. At this value of Reynolds number, the decision of whether to use a turbine impeller or to change to an impeller better suited to operation in the laminar regime must be made. The differences between these impellers are illustrated on the Visual Mixing CD. At very low Reynolds numbers, one of these impellers will be used. In this section we cover the calculations that can be made to determine if the process operates in the laminar regime, and if it does, what impeller to use and how to design it.

### 9-4.1 Identifying the Operating Regime for Viscous Blending

Two methods have been used to identify the boundary between transitional and laminar blending, and they give similar results for the value of Reynolds number at the boundary. Wichterle and Wein (1981) made visualization studies of the flow in vessels agitated by Rushton and pitched blade turbines. They defined two Reynolds numbers: the value when motion first appears and the value when all stagnant zones disappear. The second definition is used here:

$$\mathrm{Re_{TL}} = \left(\frac{1.8\,\mathrm{T}}{\mathrm{aD}}\right)^2 \tag{9-43}$$

The value of a can be calculated from

$$\mathrm{a} = 0.375\mathrm{Po}^{1/3} \tag{9-44}$$

So, for example, a pitched blade turbine with a power number of 1.8 and diameter equal to T/2 will have a value of $\mathrm{Re_{TL}}$ of 62.

Hoogendoorn and den Hartog (1967) measured blend times for a variety of impellers. They found that when the dimensionless blend time data were plotted, the exponent on Reynolds number changed from $-1$ to $-10$ at a value of Reynolds number of 170 for a Rushton turbine (i.e., the blend time became highly sensitive to the value of viscosity for Re < 170). Johnson (1967) found that the exponent was $-13$.

Hoogendoorn and den Hartog proposed that the boundary between the laminar and transitional regimes could be estimated for all impellers by

$$\frac{1}{\mathrm{Fo}} = 1 \text{ at } \mathrm{Re_{TL}} \tag{9-45}$$

This agrees well with the conclusions of Zlokarnik (1967), who concluded that the boundary occurs at a value of 0.25. Substituting the value of 1/Fo from eq. (9-45) into (9-6) and rearranging gives

$$\mathrm{Re_{TL}} = \frac{183}{\mathrm{Po}^{1/3}} \frac{1}{\sqrt{\mathrm{Fo}}} = \frac{183}{\mathrm{Po}^{1/3}} \tag{9-46}$$

For the pitched blade turbine with a power number of 1.8, eq. (9-46) predicts that $\mathrm{Re_{TL}}$ will be 150 (Hoogendoorn and den Hartog, 1967) or 75 (Zlokarnik, 1967). Use of an impeller specifically designed for laminar operation must be considered when the Reynolds number is in the range 100 to 200. At lower Reynolds numbers ($< \sim 50$) it becomes easier to make the decision.

## 9-4.2 Impeller Selection

The most commonly used impeller for laminar blending applications is the helical ribbon. Other impeller types have been studied, including anchors and helical screws, but the helical ribbon is most effective. A helical ribbon impeller will have a large diameter, typically 90 to 95% of the vessel diameter. This ensures that the fluid is "positively displaced" by the ribbons. This is important because there is no mixing due to entrainment by eddies in the laminar regime.

A typical helical ribbon is shown in Figure 6-31. Although different numbers of ribbons can be supplied, it is usual for the impeller to have two. This ensures that the hydraulic forces exerted on the shaft are balanced. The pitch (the ribbon height of one 360° turn) is usually equal to the impeller diameter, and the width of a ribbon blade is typically 10% of the impeller diameter. A tighter pitch and wider ribbon will increase the power draw.

## 9-4.3 Estimation of Power Draw

The power drawn by an impeller operating in the laminar regime is calculated from

$$Po = \frac{K_P}{Re} \tag{9-47}$$

The power drawn by an impeller in the laminar regime can be calculated from

$$P = Po \cdot \rho N^3 D^5 = \frac{K_P \mu}{\rho N D^2} \rho N^3 D^5 = K_P \mu N^2 D^3 \tag{9-48}$$

The power drawn by an impeller is proportional to the fluid viscosity.

The constant $K_P$ is a function of the impeller's geometry and a variety of correlations have been produced to relate its value to the geometrical ratios of a helical ribbon impeller. There are a number of correlations available in the literature for estimating the value of $K_P$. For example, Shamlou and Edwards (1985) correlated their data by

$$K_P = 150 \frac{h}{D} \left(\frac{p}{D}\right)^{-0.50} \left(\frac{c}{w}\right)^{0.33} n_b^{0.50} \tag{9-49}$$

Brito-de la Fuente et al. (1997) did not vary D/T and as a consequence, c/D, and found that

$$K_P = 173.1 \left(\frac{p}{D}\right)^{-0.72} \left(\frac{w}{D}\right)^{0.14} \tag{9-50}$$

Rieger et al. (1988) reported that

$$K_P = 82.8 \frac{h}{D} \left(\frac{c}{D}\right)^{-0.38} \left(\frac{p}{D}\right)^{-0.35} \left(\frac{w}{D}\right)^{0.20} n_b^{0.78} \tag{9-51}$$

**530** BLENDING OF MISCIBLE LIQUIDS

Their correlation included data from eight geometries that they measured themselves and 69 others that were available in the literature. Given the similarities between the various correlations, using this version will give a good estimate for $K_P$ and will account for all possible geometrical variations. It is also based on total of 77 experimental observations, which increases the level of confidence in its use.

### 9-4.4 Estimation of Blend Time

The dimensionless blend time, $N\theta$, is a constant for a helical ribbon operating in the laminar regime (see, e.g., Hoogendoorn and den Hartog, 1967; Johnson, 1967; Rieger et al., 1986). This means that the blend time is independent of Reynolds number and the fluid viscosity, so that even if the fluid is shear-thinning the blend time will not be affected by the rheological behavior. This is not true for visco-elastic behavior.

Grenville et al. (2001) took the data of Rieger et al. (1986) and found that the dimensionless blend time could be correlated with the constant $K_P$ by

$$N\theta = 896 \times 10^3 K_P^{-1.69} \qquad (9\text{-}52)$$

The standard error for the constant is $\pm 17\%$.

The higher the value of $K_P$, the lower the dimensionless blend time. Grenville et al. (2001) also found that an impeller with a high $K_P$ value was the most energy efficient geometry (i.e., gave the shortest blend time for a given power input).

### 9-4.5 Effect of Shear-Thinning Behavior

The power drawn by any impeller in the laminar regime is proportional to the fluid viscosity, so an estimate of the apparent viscosity must be made for a shear-thinning fluid. Since the impeller is operating in the laminar regime, the Metzner and Otto approach to estimating the shear rate is valid, and for helical ribbons, the constant $k_S$ has a value of 30. Shamlou and Edwards (1985) found that there is a weak effect of the gap between the ribbon and the vessel wall on this value, but for engineering calculations, the value of 30 is accurate enough.

### 9-4.6 Design Example

*Example 9-6.* A small volume of liquid is to be added to a large volume of viscous fluid in a vessel that is 2 m in diameter. The depth will also be 2 m. The fluid has a density of 990 kg/m$^3$ and is shear-thinning with a consistency index of 1450 Pa · s$^n$ and a flow behavior index of 0.45. A sample will be taken after 10 min to check that the fluid is homogeneous. Design an agitator for this process.

## SOLUTION

Start by assuming a standard helical ribbon impeller with two blades, $p/D = h/D = 1$; $D/T = 0.95$; $w/D = 0.1$. From eq. (9-51), $K_p = 357$. From eq. (9-52)

$$N = \frac{8.96 \times 10^5}{(357)^{1.69}(600 \text{ s})} = 0.07 \text{ s}^{-1} = 4 \text{ rpm}$$

The closest standard speed is 16.5 rpm (Table 6-2). This is much higher than the 4 rpm required and will result in a higher-than-necessary power consumption. Decrease the impeller diameter to $D/T = 0.9$, keeping everything else the same. The new $N = 7$ rpm is much closer to the smallest available speed of 16.5 rpm (0.275 rps).

The next step is to calculate the Reynolds number using the apparent viscosity and the Metzner–Otto equation. For helical ribbon impellers, $k_s = 30$:

$$\dot{\gamma} = 30N = 8.25 \text{ s}^{-1}$$

$$\mu_{app} = 1450\dot{\gamma}^{-0.55} = 454 \text{ Pa} \cdot \text{s}$$

$$\text{Re} = \frac{\rho N D^2}{\mu_{app}} = \frac{990 \text{ kg/m}^3 \times 0.275 \text{ s}^{-1} \times (1.8 \text{ m})^2}{454 \text{ Pa} \cdot \text{s}} = 1.94$$

This is far into the laminar regime; check the $\text{Re}_{TL}$. From eq. (9-46), $\text{Re}_{TL} = 35$, so the helical ribbon is a good choice. Because a helical ribbon impeller was selected, caverns are not a concern in this application. The power draw will be

$$P = \text{Po} \cdot \rho N^3 D^5 = 53.9 \text{ kW} = 72 \text{ hp}$$

The closest standard motor size is 75 hp, and the next largest is 100 hp (Table 6-2). A slight further reduction in the impeller diameter to 0.88T reduces the power draw to 62 hp, which is a better match for the motor size. The blend time is still well below the requirement of 10 min.

## 9-5 JET MIXING IN TANKS

Mixing of fluids requires the input of mechanical energy to achieve a process result, and previous sections in this chapter have dealt with equipment that consists of an impeller, or impellers, attached to a rotating shaft. An alternative method for getting energy into the fluid is to generate a high velocity jet of fluid in the vessel. Vertically oriented jets are illustrated on the Visual Mixing CD. The jet entrains and mixes the surrounding fluid and the mechanical energy is supplied from a pump. The rules for designing jet mixers for use in low viscosity turbulent applications are very well defined and can be used with a great deal of confidence.

Jet mixers are commonly used in large storage tanks, where the contents must be homogenized, but the required blend time can be on the order of hours rather

**532** BLENDING OF MISCIBLE LIQUIDS

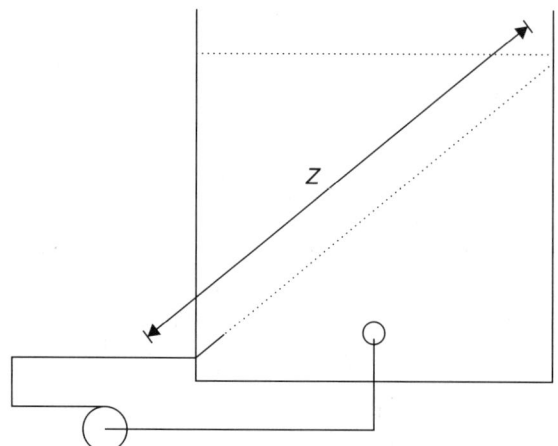

**Figure 9-1** Jet mixer configuration for blending operations.

than minutes or seconds. This will be the main area covered in this chapter. When used in large storage tanks the jet usually enters from the side of the vessel close to the base and is directed toward the opposite top corner (see Figure 9-1).

Jet mixers are driven by pumps that can be located on the ground next to the vessel, giving easy access for maintenance. The vessel will often need a pump for filling and emptying, and this pump can also be used for the jet mixer, thus reducing the capital investment needed, especially if an agitator is being considered.

### 9-5.1 Literature Review

A number of studies have been done over the years measuring blend time in jet mixed vessels. During World War II, Fossett and Prosser (1949) examined the blending of tetraethyl lead (TEL) into aviation fuel in underground storage tanks. Their main concern was to ensure that the dense TEL stream was well mixed with the fuel, but they proposed a correlation for estimating the blend time as a function of the vessel diameter, nozzle diameter, and jet velocity:

$$\theta \propto \frac{T^2}{UD} \tag{9-53}$$

or in dimensionless terms,

$$\frac{U\theta}{D} \propto \left(\frac{T}{D}\right)^2 \tag{9-54}$$

This correlation predicts that in the turbulent regime, the dimensionless blend time is independent of the jet Reynolds number.

Fox and Gex (1956) also measured blend times in jet mixed tanks and proposed a correlation that included an effect of the jet Reynolds and Froude numbers:

$$\frac{U\theta}{D} \propto \left(\frac{Fr}{Re}\right)^{1/6} \frac{TH^{1/2}}{D^{3/2}} \quad (9\text{-}55)$$

Van de Vusse (1959) measured blend times in a 12 000 m$^3$ vessel and concluded that the Fossett and Prosser form of correlation best fit his data and that the Fox and Gex correlation underpredicted the measured blend times. Several other workers have measured blend times in jet-mixed vessels, with most finding that the Fossett and Prosser correlation fit their data. These include Okita and Oyama (1963) and Ràcz and Wassink (1973).

Grenville et al. (1992) measured blend times in three jet-mixed vessels 0.61, 1.68, and 3.98 m in diameter and force-fit the data into the two correlations that had been proposed in the literature. They found that the Fossett and Prosser correlation fit the data with a standard deviation of 8.15%, whereas for the Fox and Gex correlation the standard deviation was 18.9%. Also, a regression of the data showed that the Fossett and Prosser correlation was, in fact, the best fit for the data.

### 9-5.2 Jet Mixer Design Method

The correlations developed for estimating blend time in jet-mixed vessels were based on regression analysis of experimental data but without any physical understanding of the phenomena that control the process. Grenville and Tilton (1996) proposed that the overall blend time in a jet-mixed vessel would be determined from the mixing in the region of the vessel where the local mixing rate was slowest. The mixing rate at the end of the jet path could be estimated and compared with the mixing rate for the entire vessel.

The mixing rate can be estimated using the Corrsin (1964) time scale (see Chapter 2), where the turbulent energy dissipation rate is calculated at the end of the jet path and the appropriate length scale is the jet's free path, Z:

$$\theta = K_Z \left(\frac{\varepsilon_Z}{Z^2}\right)^x \quad (9\text{-}56)$$

where $K_Z$ is a dimensionless constant. If the data fit the Corrsin model, the exponent, x, would be $-\frac{1}{3}$.

The turbulent kinetic energy dissipation rate at the end of the jet is estimated from the jet's centerline velocity and diameter at the end of its free path:

$$\varepsilon_Z = A_Z \frac{U_Z^3}{D_Z} \quad (9\text{-}57)$$

The velocity on the centerline of a turbulent jet can be estimated from Rajaratnam (1986):

$$U_Z = 6\frac{UD}{Z} \quad (9\text{-}58)$$

## 534 BLENDING OF MISCIBLE LIQUIDS

As the jet moves away from the nozzle it expands and slows down as it entrains the surrounding fluid. Its momentum is conserved. The relationship between the jet velocity at the nozzle, the nozzle diameter, and the jet's velocity and diameter at any distance along its path will be

$$UD = U_Z D_Z \tag{9-59}$$

Substituting eqs. (9-58) and (9-59) into (9-57), it can be shown that

$$\varepsilon_Z \propto \frac{(UD)^3}{Z^4} \tag{9-60}$$

and thus:

$$\theta \propto \left[\frac{(UD)^3}{Z^6}\right]^x \tag{9-61}$$

So the turbulent kinetic energy dissipation rate at the end of the jet can be calculated from quantities that are known: the jet velocity at the nozzle, the nozzle diameter, and the path length.

Blend time data measured in three scales of vessel were fitted to this relationship and the regression showed that the exponent was $-\frac{1}{3}$ as expected. Equation (9-61) is rearranged using $x = -\frac{1}{3}$ to give

$$\theta = K_Z \frac{Z^2}{UD} \tag{9-62}$$

for the conditions:

- Re > 10 000 (turbulent flow)
- 0.2 < H/T < 2.0
- 0.178 < V < 1200 m$^3$
- 1.32 × 10$^{-2}$ < (UD/Z) < 0.137 m/s
- 86 < Z/D < 753

The constant $K_Z$ has a value of 3.00 with a standard deviation of ±11.0%.

Equation (9-62) can be rearranged into two dimensionless groups:

$$UD = K_Z \frac{Z^2}{\theta} \tag{9-63}$$

Multiplying both sides of eq. (9-63) by $\rho/\mu$ yields

$$\frac{\rho UD}{\mu} = K_Z \frac{\rho Z^2}{\mu \theta} \tag{9-64}$$

The dimensionless group on the left-hand side of eq. (9-64) is the jet Reynolds number and the group on the right-hand side is the reciprocal of Fourier number, as used for agitator design and rating calculations.

$$\text{Re} = \frac{K_Z}{\text{Fo}} \tag{9-65}$$

Expressing the correlation in this way makes it very useful for design since the properties of the jet are separated from the properties of the mixing duty (the blend time, vessel size, and liquid physical properties).

The correlation can be used in two ways; the first is simply to use the value $K_Z = 3.00$ with no account taken of the standard deviation of the experimental results. The second, taking account of the standard deviation, allows a level of confidence to be included in the design procedure. Approximately 67% of observations will lie within ±1 standard deviation. Similarly, 95% lie within ±2 standard deviations and 99% lie within ±3 standard deviations. So if the second approach is to be taken, including a level of confidence, the constant $K_Z$ in eq. (9-65) can be defined as $3.00 + 0.33\,s$, where $s = 1$ for 67% confidence, $s = 2$ for 95% confidence, and $s = 3$ for 99% confidence. Examples of the correlation's use are given in Section 9-5.3.

## 9-5.3 Jet Mixer Design Steps

There are two ways in which the correlation for blend times can be used:

1. Designing a new vessel and jet mixer
2. Rating an existing vessel and jet mixer

***9-5.3.1 Designing a New Jet Mixer.*** For a new application, the vessel dimensions and required blend time will be defined. The Fourier number can be calculated immediately, followed by the required jet Reynolds number. A jet velocity needs to be chosen at this point and a typical value would be 10 m/s. Once this is done, the nozzle diameter can be calculated from the jet Re, followed by the pressure drop and the pump flow rate. The jet nozzle should be constructed from standard pipe, and sizes are given in Perry and Green (1984). Choose the next larger standard pipe above the calculated diameter and recalculate the pressure drop and flow rate. This will give a shorter blend time, so it will be possible to use a lower jet velocity with this standard pipe size. If the tank operates in continuous mode, the flow through the vessel may be used to drive the jet mixer (see Example 9-8).

***9-5.3.2 Pump Sizing.*** Using the design correlation to size a jet mixer will determine what the required flow rate through the nozzle has to be to achieve the desired blend time. In order to specify the pump, it is necessary to know the pressure drop through the system. It is quite likely that the actual operating point on the pump curve will not give exactly the flow rate specified, so the pump curve and the mixing curve must be combined to find the operating conditions for the system. The pump curve for a centrifugal pump can be fitted to a quadratic equation with the head on the y-axis and flow rate on the x-axis. The mixing time correlation can be expressed in terms of the flow rate and the head loss through the piping system and mixer nozzle.

## 9-5.4 Design Examples

***Example 9-7: Design of a New Jet Mixer.*** A monomer storage vessel is 10 m in diameter with a straight-side height of 8 m. Thirty minutes after delivery of a fresh shipment of monomer, the vessel contents are sampled and analyzed. A jet mixer will be installed in the vessel to blend the new shipment with the existing fluid. The monomer has a density of 850 kg/m$^3$ and viscosity of 1.2 mPa · s. Design the jet mixer.

SOLUTION

1. Calculate the jet path length. For an optimum jet geometry with maximized jet path length,

$$Z = \sqrt{H^2 + T^2} = \sqrt{8^2 + 10^2} = 12.81 \text{ m}$$

2. The Fourier number can be calculated immediately since the required blend time has been defined as 30 min:

$$\text{Fo} = \frac{\mu\theta}{\rho Z^2} = \frac{1.2 \times 10^{-3} \text{ Pa} \cdot \text{s} \times (30 \times 60 \text{ s})}{850 \text{ kg/m}^3 \times (12.81 \text{ m})^2} = 1.55 \times 10^{-5}$$

3. Now calculate the required jet Reynolds number using $K = 3.00$.

$$\text{Re} = \frac{K_Z}{\text{Fo}} = \frac{3.00}{1.55 \times 10^{-5}} = 1.935 \times 10^5$$

4. Setting the jet velocity equal to 10 m/s, calculate the required nozzle diameter:

$$D = \frac{\text{Re} \cdot \mu}{\rho U} = \frac{(1.935 \times 10^5) \times 1.2 \times 10^{-3} \text{ Pa} \cdot \text{s}}{850 \text{ kg/m}^3 \times 10 \text{ m/s}} = 0.027 \text{ m}$$

The next larger standard pipe size is 0.035 m (1.25 in. schedule 40 pipe). The jet path/nozzle diameter ratio $Z/D = 366$, which is acceptable.

5. Calculate the required flow rate:

$$Q = \frac{\pi}{4} U D^2 = \frac{\pi}{4} \times 10 \text{ m/s} \times (0.035 \text{ m})^2 = 9.62 \times 10^{-3} \text{ m}^3/\text{s}$$

(or 152 US gal/min).

6. Finally, estimate the head loss through the piping and jet nozzle:

$$h_L = 2.5 \frac{U^2}{2g} = 2.5 \times \frac{(10 \text{ m/s})^2}{2 \times 9.81 \text{ m/s}^2} = 12.75 \text{ m of fluid}$$

Once the piping has been laid out, a more rigorous pressure drop calculation can be made to size the pump.

JET MIXING IN TANKS    537

If a 95% confidence level is to be applied to this design, the constant $K_Z$ in step 3 would be 3.66 (i.e., $3 + 2 \times 0.33$).

***Example 9-8: Design of a Jet Mixer for an Existing Process.*** An effluent stream is pumped to a large vessel prior to treatment in aerobic digesters. The flow rate can range between 4 and 7 m³/min. The vessel is 36 m in diameter and has an operating volume of 8000 m³, giving a residence time of about 1 day. To improve the operation of the digesters, it has been decided that the contents of the vessel must be blended to prevent spikes in effluent concentration from reaching the microorganisms. Can the pump provide enough flow to blend the vessel contents in a short enough time?

SOLUTION

1. Calculate the nozzle diameter. Choose a jet velocity of 10 m/s at the highest flow rate as a starting point for the calculation:

$$D = \sqrt{\frac{Q}{(\pi/4)U}} = \sqrt{\frac{7 \text{ m}^3/60 \text{ s}}{(\pi/4) \times 10 \text{ m/s}}} = 0.122 \text{ m}$$

2. Calculate the Reynolds number. The fluid is water with a density of 1000 kg/m³ and a viscosity of 1.0 mPa·s:

$$\text{Re} = \frac{\rho U D}{\mu} = \frac{1000 \text{ kg/m}^3 \times 10 \text{ m/s} \times 0.122 \text{ m}}{0.001 \text{ Pa} \cdot \text{s}} = 1.22 \times 10^6$$

3. Calculate the Fourier number. The jet is turbulent, so $K_Z = 3.00$ and

$$\text{Fo} = \frac{3.00}{\text{Re}} = \frac{3.00}{1.22 \times 10^6} = 2.46 \times 10^{-6}$$

4. Calculate the jet path length. First, the liquid level must be calculated:

$$H = \frac{V}{(\pi/4)T^2} = \frac{8000 \text{ m}^3}{(\pi/4) \times (36 \text{ m})^2} = 7.86 \text{ m}$$

The jet path length is

$$Z = \sqrt{H^2 + T^2} = \sqrt{(7.86 \text{ m})^2 + (36 \text{ m})^2} = 36.85 \text{ m}$$

5. Calculate the blend time:

$$\theta = \frac{\text{Fo} \cdot \rho Z^2}{\mu} = 3340 \text{ s} = 56 \text{ min}$$

This is much less than the residence time, so the jet mixer will be effective at the high flow rate. Repeating the calculations for the low flow rate gives

$$U = \frac{4}{60(\pi/4)(0.122)^2} = 5.7 \text{ m/s}$$

$$\text{Re} = 7 \times 10^5 \quad \text{turbulent since Re} > 10\,000$$

$$\text{Fo} = 4.3 \times 10^{-6}$$

$$\theta = 5840 \text{ s} = 97 \text{ min}$$

This is still much less than 10% of the mean residence time, so the jet mixer will be sufficient.

## NOMENCLATURE

| | |
|---|---|
| A | area of baffle (m$^2$) |
| A$_Z$ | dimensionless constant |
| c | helical ribbon impeller wall clearance (m) |
| c' | concentration fluctuation (concentration units) |
| C | impeller off-bottom clearance (m) |
| D | impeller diameter (m) |
| D | jet diameter at the nozzle (m) |
| D$_c$ | cavern diameter (m) |
| D$_z$ | jet diameter at the end of the jet path (m) |
| Fo | Fourier number, $\mu\theta/\rho T^2$ |
| Fo$_{TT}$ | transition to turbulent Fourier number |
| Fo$_w$ | Fourier number at the wall |
| F$_r$ | jet Froude number |
| h | helical ribbon impeller height (m) |
| h$_L$ | head loss through the jet piping (m) |
| H | fluid height (m) |
| H$_c$ | cavern height (m) |
| k | blending rate constant (s$^{-1}$) |
| k$_s$ | Metzner–Otto constant |
| K | power law constant or consistency index (Pa · s$^n$) |
| K$_p$ | P$_o$ · R$_e$ in the laminar regime |
| K$_Z$ | constant for jet mixing time correlation (= 3.00) |
| n | power law exponent or flow behavior index |
| n$_b$ | number of blades, helical ribbon impeller |
| N | impeller rotational speed (rps) |
| N$_c$ | impeller speed at which the cavern reaches the wall (rps) |
| p | helical ribbon impeller pitch (m per 360° rotation) |
| p$_{baff}$ | pressure on the baffles (Pa) |
| Po | power number, P/$\rho$N$^3$D$^5$ |

| | |
|---|---|
| Q | volumetric flow rate through the nozzle (m³/s) |
| r | radius (m) |
| R | tank radius (m) |
| Re | impeller Reynolds number, $ND^2/\nu$ |
| $Re_{TL}$ | transition to laminar Reynolds number |
| $Re_{TT}$ | transition to turbulent Reynolds number |
| $Re_w$ | Reynolds number at the wall |
| S | wall area (m²) |
| t | time (s) |
| T | tank diameter (m) |
| U | jet velocity at the nozzle (m/s) |
| $U_z$ | velocity at the end of the jet (m/s) |
| V | vessel volume (m³) |
| v | tangential velocity (m/s) |
| w | helical ribbon impeller blade width (m) |
| x | relative magnitude of the concentration fluctuation |
| Z | jet path length in a jet mixer (m) |

*Greek Symbols*

| | |
|---|---|
| $\dot{\gamma}_w$ | wall shear rate (s⁻¹) |
| $\dot{\gamma}$ | shear rate (s⁻¹) |
| $\bar{\varepsilon}$ | power dissipated per unit mass, $Po \cdot N^3 D^5 / V$ (m²/s³) |
| $\varepsilon_Z$ | turbulent energy dissipation rate at the end of the jet (m²/s³) |
| $\theta_{95}$ | blend time to 95% reduction in variance (s) |
| $\theta_M$ | blend time to 95% reduction in variance for multiple impellers (s) |
| $\Lambda$ | torque on the shaft (N · m) |
| $\mu$ | dynamic viscosity (Pa · s or kg/m · s) |
| $\mu_A$ | apparent viscosity (Pa · s) |
| $\mu_w$ | viscosity at the wall (Pa · s) |
| $\rho$ | fluid density (kg/m³) |
| $\tau$ | shear stress (N/m² or Pa) |
| $\tau_W$ | shear stress at the wall (N/m² or Pa) |
| $\tau_Y$ | yield stress (Pa) |

## REFERENCES

Amanullah, A., S. J. Hjorth, and A. W. Nienow (1998). A new mathematical model to predict cavern diameters in highly shear thinning, power law liquids using axial flow impellers, *Chem. Eng. Sci.*, **53**, 455–469.

Bird, R. B., W. E. Stewart, and E. N. Lightfoot (1960). *Transport Phenomena*, Wiley, New York, Chap. 6, p. 205.

Bourne, J. R., M. Buerli, and W. Regenass (1981). Power and heat transfer to agitated suspensions: use of heat flow calorimetry, *Chem. Eng. Sci.*, **36**, 782–784.

Brito-de la Fuente, E., L. Choplin, and P. A. Tanguy (1997). Mixing with helical ribbons: effect of highly shear-thinning behaviour and impeller selection, *Trans. Inst. Chem. Eng.*, **75**, 45–52.

Carpenter, K. J., P. C. Lines, R. Aldington, A. Keron, and W. A. J. Hindson (1993). Two examples of designing full scale reactors for multi-stage synthesis involving non-Newtonian mixtures, presented at the AIChE Annual Meeting, St. Louis, MO.

Cooke, M., J. C. Middleton, and J. R. Bush (1988). Mixing and mass transfer in filamentous fermentations, *Proc. 2nd International Conference on Bioreactor Fluid Dynamics*, pp. 37–64.

Corrsin, S. (1964). The isotropic turbulent mixer: II. Arbitrary Schmidt number, *AIChE J.*, **9**, 870–877.

Cronin, D. G., A. W. Nienow, and G. W. Moody (1994). An experimental study of the mixing in a proto-fermenter agitated by dual Rushton turbines, *Food Bioprod. Process. (Trans. Inst. Chem. Eng. C)*, **72**, 35–40.

Elson, T. P. (1988). Mixing of fluids possessing a yield stress, *Proc 6th European Conference on Mixing*, pp. 485–492.

Elson, T. P., D. J. Cheesman, and A. W. Nienow (1986). X-ray studies of cavern sizes and mixing performance with fluids possessing a yield stress, *Chem. Eng. Sci.*, **41**, 2555–2562.

Etchells, A. W., W. N. Ford, and D. G. R. Short (1987). *Fluid Mixing 3, Inst. Chem. Eng. Symp. Ser.*, **108**, pp. 1–10.

Fossett, H., and L. E. Prosser (1949). The application of free jets to mixing of fluids in bulk, *Proc. Inst. Mech. Eng.*, **160**, 224–232.

Fox, E. A., and V. E. Gex (1956). Single phase blending of liquids, *AIChE J.*, **2**, 539–544.

Galindo, E., and A. W. Nienow (1992). Mixing of highly viscous simulated xanthan fermentation broths with the Lightnin A315 impeller, *Biotechnol. Prog.*, **8**, 233–239.

Galindo, E., and A. W. Nienow (1993). The performance of the Scaba 6SRGT agitator in the mixing of simulated xanthan gum broths, *Chem. Eng. Technol.*, **16**, 102–108.

Grenville, R. K. (1992). Blending of viscous Newtonian and pseudo-plastic fluids, Ph.D. dissertation, Cranfield Institute of Technology, Cranfield, Bedfordshire, England.

Grenville, R. K., and J. N. Tilton (1996). A new theory improves the correlation of blend time data from turbulent jet mixed vessels, *Trans. Inst. Chem. Eng.*, **74**, 390–396.

Grenville, R. K., A. T. C. Mak, and S. W. Ruszkowski (1992). Blending of fluids in mixing tanks by re-circulating turbulent jets, *Proc. 1992 Institution of Chemical Engineers Research Event*, University of Manchester Institute of Science and Technology, Manchester, Lancashire, England, pp. 128–130.

Grenville, R. K., T. M. Hutchinson, and R. W. Higbee (2001). Optimisation of helical ribbon geometry for blending in the laminar regime, presented at MIXING XVIII, NAMF.

Hari-Prajitno, D., V. P. Mishra, K. Takenaka, W. Bujalski, A. W. Nienow, and J. McKemmie (1998). Gas–liquid mixing studies with multiple up- and down-pumping hydrofoil impellers: power characteristics and mixing time, *Can. J. Chem. Eng.*, **76**, 1056–1068.

Hirata, Y., A. W. Nienow, and I. P.T. Moore (1994). Estimation of cavern sizes in a shear-thinning plastic fluid agitated by a Rushton turbine based on LDA measurements, *J. Chem. Eng. Jpn.*, **27**, 235–237.

Hoogendoorn, C. J., and A. P. den Hartog (1967). Model studies on mixers in the viscous flow region, *Chem. Eng. Sci.*, **22**, 1689–1699.

Jaworski, Z., A. W. Pacek, and A. W Nienow (1994). On the flow close to cavern boundaries in yield stress fluids, *Chem. Eng. Sci.*, **49**, 3321–3324.

Johnson, R. T. (1967). Batch mixing of viscous liquids, *IEC Proc. Des. Dev.*, **6**, 340–345.

Khang, S. J., and O. Levenspiel (1976). New scale-up and design method for stirrer agitated batch mixing vessels, *Chem. Eng. Sci.*, **31**, 569–577.

Kramers, H., G. M. Baars, and W. H. Knoll (1953). A comparative study of the rate of mixing in stirred tanks, *Chem. Eng. Sci.*, **2**, 35–42.

Langheinrich, C., T. Eddleston, N. C. Stevenson, A. N. Emery, T. M. Clayton, N. K. H. Slater, and A. W. Nienow (1998). Liquid homogenisation studies in animal cell bioreactors of up to 8 m$^3$ in volume, *Food Bioprod. Process. (Trans. Inst. Chem. Eng. C)*, **76**, 107–116.

Manikowski, M., S. Bodemeier, A. Lübbert, W. Bujalski, and A. W. Nienow (1994). Measurement of gas and liquid flows in stirred tank reactors with multiple agitators, *Can. J. Chem. Eng.*, **72**, 769–781.

Metzner, A. B., and R. E. Otto (1957). Agitation of non-Newtonian fluids, *AIChE J.*, **3**, 3–10.

Nagata, S., M. Nishikawa, H. Tada, and S. Gotoh (1971). Power consumption of mixing impellers in pseudo-plastic liquids, *J. Chem. Eng. Jpn.*, **4**, 72–76.

Nienow, A. W. (1997). On impeller circulation and mixing effectiveness in the turbulent flow regime, *Chem. Eng. Sci.*, **52**, 2557–2565.

Nienow, A. W., and T. P. Elson (1988). Aspects of mixing rheologically complex fluids, *Chem. Eng. Res. Des.*, **66**, 5–15.

Norwood, K. W., and A. B. Metzner (1960). Flow patterns and mixing rates in agitated vessels, *AIChE J.*, **6**, 432–437.

Okita, N., and Y. Oyama (1963). Mixing characteristics in jet mixing, *Jpn. Chem. Eng.*, **1**, 92–101.

Otomo, N., W. Bujalski, and A. W. Nienow (1993). Mixing time measurements for an aerated, single- and double-impeller stirred vessel by using a conductivity technique, *Proc. 1993 Institution of Chemical Engineers Research Event*, Birmingham, Jan., pp. 669–671.

Otomo, N., W. Bujalski, and A. W. Nienow (1995). An application of a compartment model to a vessel stirred with either dual radial or dual axial flow impellers, *Proc. 1995 Institution of Chemical Engineers Research Event*, Edinburgh, Jan. pp. 829–831.

Perry, R. H., and D. Green (eds.) (1984). *The Chemical Engineers' Handbook, 6th ed.*, Mcgraw-Hill, New York, Sec. 6.

Pollard, J., and T. A. Kantyka (1969). Heat transfer to agitated non-Newtonian fluids, *Trans. Inst. Chem. Eng.*, **47**, T21–T27.

Procházka, J., and J. Landau (1961). Studies on mixing: XII. Homogenisation of liquids in the turbulent regime, *Coll. Czech. Chem. Commun.*, **26**, 2961–2973.

Ràcz, I., and J. G. Wassink (1974). Strömungsverlauf und Mischzeiten in axialen Strahlmischern, *Chem. Ing. Tech.*, **46**, 261.

Rajaratnam, N. (1986). In *The Encyclopaedia of Fluid Mechanics*, Vol. 2, N. Cheremisnoff, ed., Gulf Publishing, Houston, TX, Chap. 15.

Rieger, F., V. Novák, and D. Havelková (1986). Homogenization efficiency of helical ribbon agitators, *Chem. Eng. J.*, **33**, 143–150.

Rieger, F., V. Novák, and D. Havelková (1988). The influence of the geometrical shape on the power requirements of ribbon impellers, *Int. Chem. Eng.*, **28**, 376–383.

Sano, Y., and H. Usui (1987). Effects of paddle dimensions and baffle conditions on the interrelations among discharge flow rate, mixing power and mixing time in mixing vessels, *J. Chem. Eng. Jpn.*, **20**, 399–404.

Shamlou, P. A., and M. F. Edwards (1985). Power consumption of helical ribbon mixers in viscous Newtonian and non-Newtonian fluids, *Chem. Eng. Sci.*, **40**, 1773–1781.

Solomon, J., A. W. Nienow, and G. W. Pace (1981). Flow patterns in agitated plastic and pseudo-plastic fluids, in *Fluid Mixing*, *Inst. Chem. Eng. Symp. Ser.*, **64**, A1–A13.

Van de Vusse, J. G. (1959). Vergleichende Rührversuche zum mischen löslicher Flüssigkeiten 12000 m$^3$ Behälter, *Chem. Ing. Tech.*, **31**, 583–587.

Vrabel, P., R. G. J. M. van der Lans, K. Ch. A. M. Luyben, L. A. Boon, and A. W. Nienow (2000). Mixing in large scale vessels stirred with multiple radial or radial and axial up-pumping impellers: modelling and measurements, *Chem. Eng. Sci.*, **55**, 5881–5896.

Wang, K., and S. Yu (1989). Heat transfer and power consumption of non-Newtonian fluids in agitated vessels, *Chem. Eng. Sci.*, **44**, 33–40.

Wichterle, K., and O. Wein (1981). Threshold of mixing non-Newtonian fluids, *Int. Chem. Eng.*, **21**, 116–120.

Wichterle, K., M. Kadlec, L. Žák, and P. Mitchka (1984). Shear rates on turbine impeller blades, *Chem. Eng. Commun.*, **26**, 25–32.

Zlokarnik (1967). Eigung von Rühren zum homogenisieren von Flüssigkeitenmischen, *Chem. Ing. Tech.*, **39**(9/10), 539–548.

# CHAPTER 10

# Solid–Liquid Mixing

VICTOR A. ATIEMO-OBENG

*The Dow Chemical Company*

W. ROY PENNEY

*University of Arkansas*

PIERO ARMENANTE

*New Jersey Institute of Technology*

## 10-1 INTRODUCTION

In this chapter the focus is on mixing operations involving, primarily, solid and liquid phases carried out in agitated or stirred vessels. Fundamental aspects of the hydrodynamics and mass transfer as well as practical design issues for solid–liquid mixing of both settling and floating solids in ungassed or gassed suspensions are discussed. Settling solid particles have a higher density than the liquid and will settle without agitation. Solids that float without agitation include solids that are less dense than the liquid, dense solids with trapped gas, and solids that are difficult to wet. Often, solid–liquid mixing operations are carried out in the presence of gas bubbles. These are known as *gassed suspensions*, in contrast to ungassed suspensions in the absence of gas bubbles. The gas bubbles may be introduced, directly as in solid-catalyzed hydrogenation reactions, entrained inadvertently or deliberately from the headspace, or evolved as in an evaporative crystallization or as a gaseous reaction product.

Solid suspensions are typically carried out in mechanically agitated or stirred vessels. Pumped liquid jets have also been used to suspend low concentrations of relatively slow settling solids. Although static mixers have been used to disperse fine solids into polymers, application of the technology is limited and beyond the scope of the present discussion.

---

*Handbook of Industrial Mixing: Science and Practice*, Edited by Edward L. Paul, Victor A. Atiemo-Obeng, and Suzanne M. Kresta
ISBN 0-471-26919-0 Copyright © 2004 John Wiley & Sons, Inc.

**544**  SOLID–LIQUID MIXING

Not included in this chapter are several solid–liquid contacting operations, such as:

1. Dispersion of very fine particles in liquids where interfacial phenomena dominate both the dispersion process and the rheology of the suspension. An application of this technology is in the preparation of a stable solid suspension such as an agricultural "flowable" formulation by the addition of suspending aids, stabilizers, and so on. The book by Parfitt (1973) discusses this technology.
2. Liquid or gas fluidized beds.
3. Liquid–solid contacting in fixed bed systems.
Froment and Bischoff (1990) discuss both fixed bed and fluidized bed systems.

### 10-1.1 Scope of Solid–Liquid Mixing

The primary objectives of solid–liquid mixing are to create and maintain a slurry and/or to promote and enhance the rate of mass transfer between the solid and liquid phases. The mixing operation promotes the

- Suspension of solids
- Resuspension of settled solids
- Incorporation of floating solids
- Dispersion of solid aggregates or control of particle size from the action of fluid shear as well as any abrasion due to particle–particle and impeller–particle impacts
- Mass transfer across the solid–liquid interface

### 10-1.2 Unit Operations Involving Solid–Liquid Mixing

Solid–liquid mixing is a key aspect of common unit operations in the chemical industry, including:

1. Dispersion of solids
2. Dissolution and leaching
3. Crystallization and precipitation
4. Adsorption, desorption, and ion exchange
5. Solid-catalyzed reaction
6. Suspension polymerization

These unit operations, with the exception of dispersion, involve mass transfer between the solid and liquid phases.

*Dispersion* of solids is a physical process where solid particles or aggregates are suspended and dispersed by the action of an agitator in a fluid to achieve a uniform suspension or slurry. Applications include the preparation of a slurry of solid reactants or catalyst to feed a reactor as well as dispersion of solid pigments and other materials into a liquid.

*Dissolution* is a mass transfer unit operation during which the solid particle decreases in size and ultimately disappears as it is incorporated as solute in the liquid. In *leaching*, a soluble component of the solid dissolves, usually leaving a particle of different size, density, and/or porosity. For some rubber or plastic materials, the particles may actually swell initially. The density and viscosity of the resulting liquid may differ considerably from the original liquid for some systems. The process goal here is to achieve the desired rate of dissolution or leaching by agitation.

*Crystallization* and *precipitation* start with a solid-free liquid phase if unseeded. The solid particles form during the crystallization or precipitation operation. The solids grow in size as well as in population. The viscosity and density of the slurry thus formed usually increase. The process goals include control of the rate of nucleation and growth of the particles as well as the minimization of particle breakage or attrition. Both the average size and the particle size distribution are important properties. Liquid-phase mixing to achieve uniformity of supersaturation or to avoid local high concentration regions is important in achieving particle size control. Crystallization is discussed further in Chapter 17.

In *adsorption*, *desorption*, and *ion exchange*, there is mass transfer between the solid and the solution. Mass transfer is from the liquid phase into the solid in adsorption and from the solid into the liquid phase for desorption. In ion-exchange operations there is an exchange of ions between the solid and the liquid.

*Solid-catalyzed reactions* usually involve adsorption of reactants onto the surfaces of the catalyst particles where the reactions take place, followed by the desorption of the reaction products from the surface. A uniform suspension of catalyst particles ensures a uniform concentration of reactants and reaction products throughout the vessel. In addition, agitation reduces the diffusional mass transfer boundary layer, thus enhancing the solid–liquid mass transfer.

*Suspension polymerization* starts with the creation of a stabilized dispersion of monomer droplets. As polymerization proceeds, the monomer droplets polymerize, usually passing through a sticky phase. The protective coating of suspending agents (surfactants, etc.) and agitation conditions keep the droplets from coalescing. They also control particle size and size distribution. The mixing objective here is to produce and maintain, by agitation, a dispersion of uniform size drops and suspension of both monomer drops and eventually, polymer particles. The dispersion of monomer droplets and emulsion polymerization is discussed further in Chapter 12.

## 10-1.3 Process Considerations for Solid–Liquid Mixing Operations

The desired process results for solid–liquid mixing vary from process to process as indicated above in the brief discussion of several unit operations. It is the responsibility of the process researcher and/or process engineer to determine the pertinent and specific process needs. Sometimes, results associated with other mixing operations—blending, gas–liquid, liquid–liquid, heat transfer, and so on—may be more important. Therefore, it is essential to consider and understand, early in the process development stage, all the physical and chemical phenomena necessary to achieve the desired process results. In particular, how these phenomena are influenced by the process chemistry, the properties of the solid and liquid phases, and the operational variables of mixing must be understood. The key considerations include the:

1. *Mode of process operations:* batch, semibatch [continuous addition to batch (con-add)], or continuous
2. Phases—solid, liquid, and/or gas phases—that are present or occur from the beginning to the end of the process
3. Properties of the solid and liquid phases, including stickiness and tendency to agglomerate
4. Unit operations involved from the beginning to the end of the process
5. Vessel geometry and internals
6. *Mixing parameters:* local or average fluid velocity or flow, local or average shear rates, blend time, power input, and so on.

### 10-1.3.1 Key Process Questions for Solid–Liquid Mixing.
For each mixing operation, several key process-related issues must be addressed before scale-up and design. For solid–liquid mixing operations, key process questions include the following.

- *What is the process mode of operation: batch, semibatch, or continuous?* Whether a process is best run as a batch, semibatch, or continuous operation depends on the unit operation, upstream and/or downstream operations, and the volume of materials processed. For example, in a single stirred tank, a solid–liquid mixing operation requiring complete solid dissolution or complete reaction of the solid must, of necessity, be batch or semibatch. The solid–liquid mixing operations where a slurry is the end product can be batch, semibatch or continuous. For batch operations, the mixing requirements often change during the batch as a result of changes in physical and chemical properties and/or changes in the mixing volume for semibatch operations. It is therefore important to determine all the physical and chemical phenomena taking place during the entire duration of the batch. For continuous operations, the physical and chemical phenomena occurring during startup and shutdown must also be determined.
- *What phases are present or occur during the process?* The type of mixing operation to study, and the degree of difficulty in achieving the desired process result, depend on the phases present. The presence of solid and liquid phases only suggests that the mixing problem of interest is one of

solid–liquid mixing operation. For example, the mixing problem is blending rather than solid–liquid mixing if the settling velocity is less than about 0.5 ft/min or 0.0025 m/s. This condition occurs if the viscosity of the suspending liquid is very high, the solid particles are so small, and/or the density difference between the solid particle and the liquid is small. The presence of gas bubbles and/or immiscible liquids can significantly influence the ability to suspend the solids.

- *Is there a chemical reaction of the solid with the liquid?* Solid–liquid mixing operations involving chemical reactions often require a high relative velocity between the solid particle and the liquid—high local shear rate or agitation intensity—to minimize the thickness of the boundary layer for mass transfer. This is also true for the dissolution of a sparingly soluble solid, as discussed further in Chapter 13.

- *What are the physical properties of the solid and liquid phases present?* The degree of difficulty in solid suspension depends on several properties of the fluid and solid particles discussed in Section 10-2. The properties of interest include the relative density of the solid and liquid phases, the viscosity of the liquid, the wetting characteristics of the solid, the shape of the solid particles, and the mass or volume ratio of solids to liquid. Large and dense solids are more difficult to suspend than small light ones; spherical particles are also more difficult to suspend than thin flat disks. The impact of these properties on solid–liquid mixing must be studied and understood early in process research and development.

- *What degree or level of suspension is required?* The required degree or level of suspension depends on the desired process result and the unit operations involved. (Levels of suspension are discussed in Section 10-2.2.) For example, a higher degree of suspension is required in a crystallizer or slurry feed vessel than in a vessel for the dissolution of a highly soluble solid.

- *What is the minimum agitator speed to suspend the solids?* In stirred tanks, there is always an impeller speed below which settling solids will tend to accumulate on the bottom of the vessel. This speed is different for different types of impellers and for identical impellers located at different clearances from the bottom of the vessel. It also depends on the properties of the solid and liquid phases. The minimum speed may be estimated for certain impeller and tank geometries using the Zwietering correlation. It is advisable, however, to determine this value experimentally for processes where solid–liquid mixing is deemed critical. See Section 10-2.2 for details.

- *What happens to the suspension when agitation is decreased or interrupted?* Obviously, solids will settle or float depending on the properties of the solid relative to the liquid phase. The more important issues are whether the solids agglomerate and/or cake as they settle or how easy it is to resuspend them when agitation is increased or restored. This information is crucial for the proper mechanical design as well as instrumentation and control of the agitation. See Sections 10-2.2 and 10-5.9.

- *What happens to the suspension when agitation is increased?* Most solid–liquid mixing operations operate above the minimum speed for suspension. A higher agitation speed improves the degree of suspension and enhances mass transfer rates. The higher speed also translates into higher turbulence as well as local and average shear rates, which for some processes may cause undesirable particle attrition. Obviously, there is also a practical economic limit on the maximum speed of agitation.
- *What effect does vessel geometry have on the process?* The geometry of the vessel, in particular the shape of the vessel base, affects the location of dead zones or regions where solids tend to congregate. It also influences the minimum agitation speed required to suspend all particles from the bottom of the vessel. In flat-bottomed vessels, dead zones and thus "fillet formation" tend to occur in the corner between the tank base and the tank wall, whereas in dished heads the solids tend to settle beneath the impeller or midway between the center and the periphery of the base. The minimum agitation speed is typically 10 to 20% higher in a flat-bottomed vessel than in one with a dished head. Both the minimum agitation speed and the extent of fillet formation are also a function of impeller type, ratio of impeller diameter to tank diameter, and location of the impeller from the vessel bottom. In general, a dished-head vessel is preferred to a flat-bottomed vessel for solid–liquid mixing operations. There is little or no difference between ASME dished, elliptical, or even hemispherical dished heads as far as solid–liquid mixing is concerned. However, elliptical heads are preferred for higher-pressure applications.
- *What is the appropriate material of construction for the process vessel?* The main issue here is that, for steel or alloy vessels, the standard four wall-mounted baffles provide a better environment for solid–liquid mixing. The standard glass-lined vessels are usually underbaffled because of a deficiency of nozzles from which to mount baffles.

## 10-2 HYDRODYNAMICS OF SOLID SUSPENSION AND DISTRIBUTION

Solid suspension requires the input of mechanical energy into the fluid–solid system by some mode of agitation. The input energy creates a turbulent flow field in which solid particles are lifted from the vessel base and subsequently dispersed and distributed throughout the liquid. Nienow (1985) discusses in some detail the complex hydrodynamic interactions between solid particles and the fluid in mechanically agitated vessels. Recent measurements (Guiraud et al., 1997; Pettersson and Rasmuson, 1998) of the 3D velocity of both the fluid and the suspension confirm the complexity.

Solids pickup from the vessel base is achieved by a combination of the drag and lift forces of the moving fluid on the solid particles and the bursts of turbulent eddies originating from the bulk flow in the vessel. This is clearly evident in

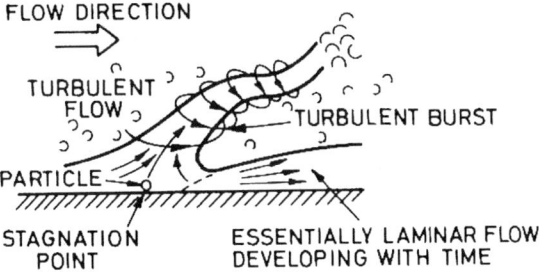

**Figure 10-1** Sudden pickup of solids by turbulent burst (Cleaver and Yates, 1973).

visual observations of agitated solid suspensions as in the video clip included on the accompanying CD ROM. Solids settled at the vessel base mostly swirl and roll around there, but occasionally, particles are suddenly and intermittently lifted up as a tornado might lift an object from the ground. An illustration of sudden pickup by turbulent bursts is shown in Figure 10-1.

The distribution and magnitude of the mean fluid velocities and large anisotropic turbulent eddies generated by a given agitator determine to what degree solid suspension may be achieved. Thus, different agitator designs achieve different degrees of suspensions at similar energy input. Also for any given impeller the degree of suspension will vary with D/T as well as C/T at constant power input. One of the video clips on the accompanying CD ROM shows the effect of D/T on solid suspension for a pitched blade impeller at constant power input.

For small solid particles whose density is approximately equal to that of the liquid, once suspended they continue to move with the liquid. The suspension behaves like a single-phase liquid at low solid concentrations; the mixing operation is more like blending than solid suspension. For heavier solid particles, their velocities will be different from that of the liquid. The drag force on the particles caused by the liquid motion must be sufficient and directed upward to counteract the tendency of the particles to settle by the action of gravity.

The properties of both the liquid and the solid particles influence the fluid–particle hydrodynamics and thus the suspension. Also important are vessel geometry and agitation parameters. The important fluid and solid properties and operational parameters include:

1. Physical properties of the liquid, such as:
   a. Liquid density, $\rho_l$ (lb/ft$^3$ or kg/m$^3$)
   b. Density difference, $\rho_s - \rho_l$ (lb/ft$^3$ or kg/m$^3$)
   c. Liquid viscosity, $\mu_l$ (cP or Pa · s)
2. Physical properties of the solid, such as:
   a. Solid density, $\rho_s$ (lb/ft$^3$ or kg/m$^3$)
   b. Particle size, $d_p$ (ft or m)

c. Particle shape or sphericity, ψ (dimensionless factor defined by the ratio of surface area of a spherical particle of the same volume to that of a nonspherical particle)
d. Wetting characteristics of the solid
e. Tendency to entrap air or headspace gas
f. Agglomerating tendencies of the solid
g. Hardness and friability characteristics of the solid
3. Process operating conditions, such as:
a. Liquid depth in vessel, Z (ft or m)
b. Solids concentration, X (lb solid/lb liquid or kg solid/kg liquid)
c. Volume fraction of solid, φ
d. Presence or absence of gas bubbles
4. Geometric parameters, such as:
a. Vessel diameter, T (ft or m)
b. Bottom head geometry: flat, dished, or cone-shaped
c. Impeller type and geometry
d. Impeller diameter, D (ft or m)
e. Impeller clearance from the bottom of the vessel, C (ft or m)
f. Liquid coverage above the impeller, CV (ft or m)
g. Baffle type and geometry and number of baffles
5 Agitation conditions, such as:
a. Impeller speed, N (rps)
b. Impeller power, P (hp or W)
c. Impeller tip speed (ft/s or m/s)
d. Level of suspension achieved
e. Liquid flow pattern
f. Distribution of turbulence intensity in the vessel

## 10-2.1 Settling Velocity and Drag Coefficient

A dense solid particle placed in a quiescent fluid will accelerate to a steady-state settling velocity. This velocity, often called the *free* or *still-fluid settling velocity*, occurs when the drag force balances the buoyancy and gravitational force of the fluid on the particle. In an agitated solid suspension, because of the complex turbulent hydrodynamic field, including solid–solid interactions, it is difficult to clearly define and/or measure a particle settling velocity. However, the particle settling velocity in an agitated solid suspension is a function of the free settling velocity and is always less than the free settling velocity (Guiraud et al., 1997).

The magnitude of the free settling velocity has proven useful in characterizing solid suspension problems into easy, moderate, or difficult categories (see Table 10-2). It is also used in solid–liquid mixing correlations, as described below.

Correlations for the free settling velocity have been derived for spherical particles. In Newtonian fluids, the free settling velocity, $V_t$, is calculated by the expression (Perry and Green, 1984).

$$V_t = \left(\frac{4g_c d_p (\rho_s - \rho_l)}{3 C_D \rho_l}\right)^{1/2} \quad (10\text{-}1)$$

where $g_c$ is the gravitational constant (32.17 ft/sec$^2$ or 9.81 m/s$^2$) and the drag coefficient, $C_D$, is a function of the particle Reynolds number, $Re_p$, and particle shape (see Figure 10-2):

$$Re_p = \frac{\rho_l V_t d_p}{\mu} \quad (10\text{-}2)$$

In Figure 10-2, the flow is assumed normal to the flat side of the disk and normal to the axis of the cylinder. The cylinder is assumed to have an "infinite" aspect ratio—length/diameter ratio.

The correlation for $C_D$ (like the friction factor and the impeller power number, $N_p$) covers several hydrodynamic regimes. The corresponding ranges for $Re_p$ and the correlating expression for $C_D$ are shown in Table 10-1 for three hydrodynamic regimes.

When the expressions for $C_D$ are substituted in eq. (10-1), the resulting expressions for the free settling velocity, $V_t$ are, respectively:

- For the Stokes' law (laminar) regime, $Re_p < 0.3$:

$$V_t = \frac{g_c d_p^2 (\rho_s - \rho_l)}{18\mu} \quad (10\text{-}3)$$

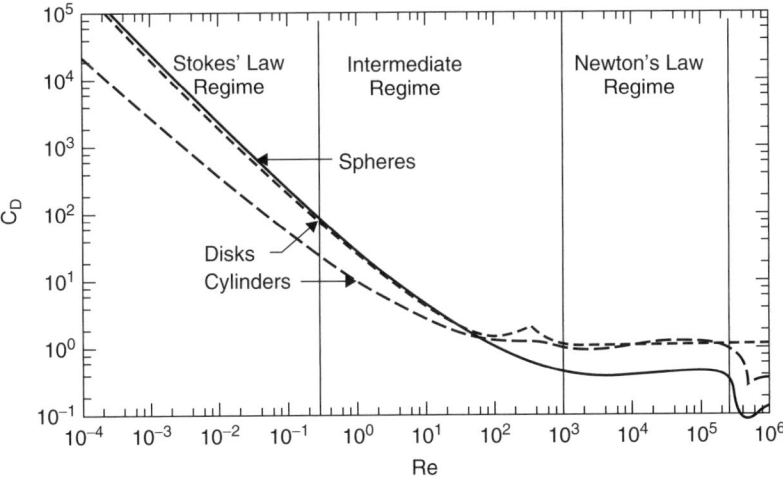

**Figure 10-2** Drag coefficient as a function of Reynolds number.

**Table 10-1** Hydrodynamic Regimes for Settling Particles

| Regime | Reynolds Number | $C_D$ Expression |
| --- | --- | --- |
| Stokes' law (laminar) | $Re_p < 0.3$ | $C_D = 24/Re_p$ |
| Intermediate law | $0.3 < Re_p < 1000$ | $C_D = 18.5/Re_p^{3/5}$ |
| Newton's law (turbulent) | $1000 < Re_p < 35 \times 10^4$ | $C_D = 0.445$ |

- For the Newton's law (turbulent) regime, $1000 < Re_p < 35 \times 10^4$:

$$V_t = 1.73 \left[ \frac{g_c d_p (\rho_s - \rho_l)}{\rho_l} \right]^{1/2} \qquad (10\text{-}4)$$

Figure 10-3 is a chart for estimating the free settling velocity for particles settling in water at ambient conditions.

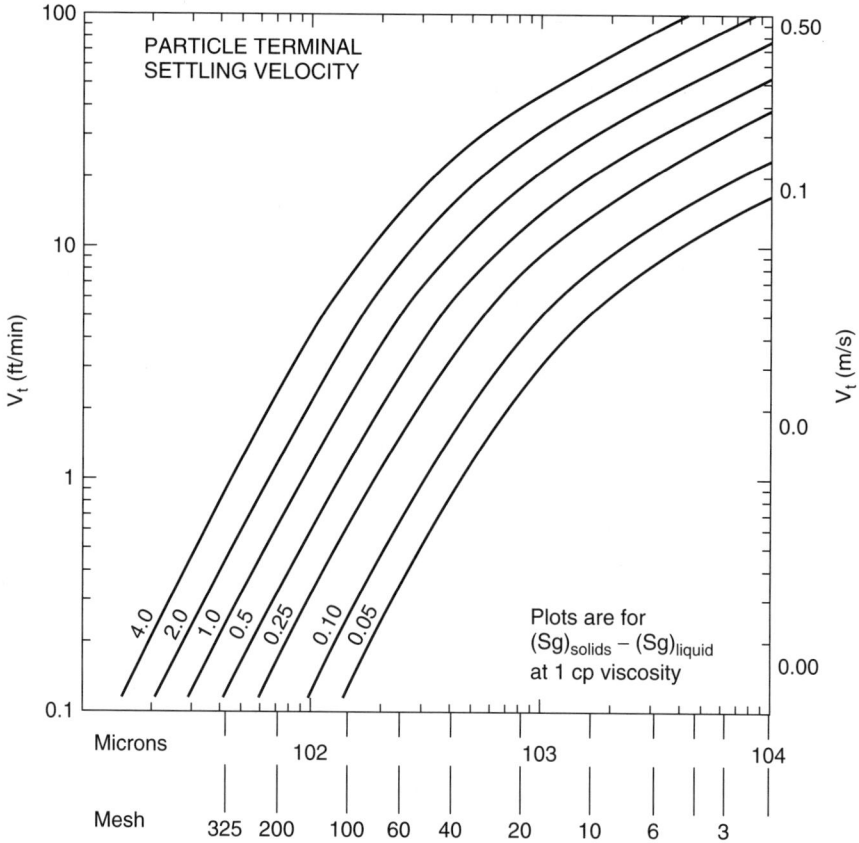

**Figure 10-3** Free settling velocity, $V_t$, as a function of particle size and density difference.

### 10-2.1.1 Effect of Solids Particle Size and Distribution.
Solids particles encountered in industrial applications usually have a distribution of sizes. Larger particles settle faster than smaller ones. Studies by Baldi et al. (1978) suggest that for a distribution of particle sizes, the appropriate particle diameter to use in the expressions above is the mass-mean diameter, $(d_p)_{43}$. This is calculated from size distribution data by

$$(d_p)_{43} = \frac{\sum_{i=1}^{N} n_i d_i^4}{\sum_{i=1}^{N} n_i d_i^3} \qquad (10\text{-}5)$$

where $d_i$ is the mean particle diameter of the ith size class and $n_i$ is the number of particles in the ith size class. The value of $n_i$ is calculated from the weight percent data by the expression

$$n_i = \frac{\text{mass of solids in the ith size class}}{\text{mass of particle of diameter } d_i}$$

However, in practice, the process engineer selects the largest particle size that must be suspended to achieve the desired process result.

### 10-2.1.2 Effect of Particle Shape and Orientation to Flow.
As indicated by Figure 10-2, the shape of the particle, and particularly its orientation to flow, affects the settling velocity. Particle shape is often quantified by the sphericity, $\psi$, which is the ratio of the surface area of a spherical particle of the same volume to that of the nonspherical particle. Chapman et al. (1983) reported that for particles with sphericity between 0.7 and 1, it is sufficient to use eqs (10-3) and (10-4) and replace the particle diameter, $d_p$, with the diameter of a sphere of equal volume. For particles with sphericity less than 0.7, the estimation of the settling velocity is complicated by the fact that the orientation to flow is a function of the Reynolds number. The effect of shape on the settling of such particles must be evaluated experimentally. Correlations presented by Pettyjohn (1948) and Becker (1959) are recommended only for preliminary estimates.

### 10-2.1.3 Effect of Solids Concentration.
The settling velocity expression above is based on the hydrodynamics of a single settling particle. The presence of other particles lowers the value of $V_t$. *Hindered settling* occurs because of the (1) interactions with surrounding particles, (2) interactions with the upward flow of fluid created by the downward settling of particles, and (3) increase in the apparent suspension viscosity and density. An empirical correlation for hindered settling in monodispersed suspensions is reported by Maude (1958) as

$$V_{ts} = V_t(1 - \chi)^n \qquad (10\text{-}6)$$

where $V_{ts}$ is the hindered settling velocity, $V_t$ the free settling velocity, $\chi$ the volume fraction of solids in the suspension, and n is a function of the particle

**554** SOLID–LIQUID MIXING

Reynolds number, Re$_p$, as follows: n = 4.65 for Re$_p$ < 0.3, n = 4.375 Re$_p^{-0.0875}$ for 0.3 < Re$_p$ < 1000, and n = 2.33 for Re$_p$ > 1000. This expression is recommended for preliminary estimates of the effect of solid concentration on the settling velocity. Davis and Gecol (1994) have reviewed hindered settling functions at low particle Reynolds numbers for mono- and poly-dispersed systems.

***Example 10-1: Calculation of Settling Velocity.*** Calculate the free settling velocity for AlCl$_3$ crystals in methylene chloride using Figure 10-3 and also eqs (10-3) and (10-4). The solid and liquid properties are:

| | |
|---|---|
| Particle size of AlCl$_3$ (d$_p$) | 4–14 mesh (5000–1000 10$^{-6}$ m) |
| Particle density of AlCl$_3$ ($\rho_s$) | 2.44 g/mL (2440 kg/m$^3$) |
| Density of MeCl$_2$ ($\rho_l$) | 1.326 g/mL at 20°C (1326 kg/m$^3$) |
| Viscosity of MeCl$_2$ ($\mu$) | 1 cP at 20°C (0.001 Pa · s or kg/m · s) |

SOLUTION: Calculate ($\rho_s - \rho_l$) = 2.44 − 1.326 = 1.114 and read the value of the free settling velocity from Figure 10-3. The free settling velocity for the solids is approximately:

1. For particles of 5000 µm, V$_t$ = 55 ft/min.
2. For particles of 1000 µm, V$_t$ = 22 ft/min.

Note that using eqs. (10-3) and (10-4) require an iterative calculation since the value of the Reynolds number determines the flow regime and thus which equation to use. On the other hand, to evaluate the Reynolds number, one needs the value of V$_t$. Such problems are easily solved with an equation solver such as TK Solver software from Universal Technical Systems, Inc. or Microsoft Excel software.

1. For the Stokes' law (laminar) regime, Re$_p$ < 0.3:

$$V_t = \frac{g_c d_p^2 (\rho_s - \rho_l)}{18\mu} \tag{10-7}$$

$$V_t = \frac{9.81(5000 \times 10^{-6})^2 (2.44 - 1.326)10^3}{18 \times 0.001}$$

= 15.2 m/s or 49.8 ft/s or 3000 ft/min   which seems impractical

Checking the particle Reynolds number, Re$_p$, yields

$$Re_p = \frac{\rho_l V_t d_p}{\mu} \tag{10-8}$$

$$= \frac{(1.326 \times 10^3 \text{ Kg/m}^3)(0.351 \text{ m/s})(5000 \times 10^{-6} \text{ m})}{0.001 \text{ kg/m} \cdot \text{s}}$$

= 100 776

The particle Reynolds number is outside the Stokes' law regime; therefore, we discard the calculated settling velocity.

2. For the Newton's law (turbulent) regime, $1000 < \text{Re}_p < 35 \times 10^4$:

$$V_t = 1.73\sqrt{\frac{g_c d_p (\rho_s - \rho_l)}{\rho_l}} \quad (10\text{-}9)$$

$$= 1.73\sqrt{\frac{9.81 \times 5000 \times 10^{-6}(2.44 - 1.326)}{1.326}}$$

$$= 0.35 \text{ m/s or } 1.15 \text{ ft/s or } 69 \text{ ft/min}$$

Checking the particle Reynolds number, $\text{Re}_p$, yields

$$\text{Re}_p = \frac{(1.326 \times 10^3 \text{ kg/m}^3)(0.351 \text{ m/s})(5000 \times 10^{-6} \text{ m})}{0.001 \text{ kg/m} \cdot \text{s}}$$

$$= 2327$$

Since this is within the Newton's law limits, we accept the velocity calculated.

Repeating the calculations for the 1000 μm particle size yields the following results:

1. For the Stokes' law (laminar) regime, $\text{Re}_p < 0.3$:

$$V_t = 0.608 \text{ m/s}$$

$$\text{Re}_p = 806$$

The particle Reynolds number is outside the Stokes' law limits; therefore, we discard the settling velocity calculated.

2. For the Newton's law (turbulent) regime, $1000 < \text{Re}_p < 35 \times 10^4$:

$$V_t = 0.157 \text{ m/s}$$

$$\text{Re}_p = 208$$

The particle Reynolds number is outside the Newton law limits, therefore, we discard the settling velocity calculated.

3. For the intermediate law regime, $0.3 < \text{Re}_p < 1000$:

$$V_t = 0.107 \text{ m/s or } 19.3 \text{ ft/min}$$

$$\text{Re}_p = 141.6$$

Since the particle Reynolds number is within the intermediate law limits, we accept the velocity calculated.

## 10-2.2 States of Solid Suspension and Distribution

In agitated vessels, the degree of solids suspension is generally classified into three levels: on-bottom motion, complete off-bottom suspension, and uniform suspension. These are illustrated in Figure 10-4.

**10-2.2.1 On-Bottom Motion or Partial Suspension.** This state is characterized by the visual observation of the complete motion of all particles around the bottom of the vessel. It excludes the formation of *fillets*, a loose aggregation of particles in corners or other parts of the tank bottom. Since particles are in constant contact with the base of the vessel, not all the surface area of particles is available for chemical reaction or mass or heat transfer. On-bottom motion conditions are sufficient for the dissolution of highly soluble solids.

**10-2.2.2 Off-Bottom or Complete Suspension.** The state of suspension known as off-bottom or complete suspension is characterized by the complete motion of all particles, with no particle remaining on the base of the vessel for more than 1 to 2 s. This condition is known as the *Zwietering criterion*. Under this condition, the maximum surface area of the particles is exposed to the fluid for chemical reaction or mass or heat transfer. The "just suspended" condition refers to the minimum agitation conditions at which all particles attain complete suspension.

In mechanically agitated vessels, the minimum agitation speed for the just suspended state, $N_{js}$, has been the subject of many experimental and theoretical analyses (Nienow, 1985). The pioneering study by Zwietering (1958) covered by far the widest range variables. The resulting correlation is discussed below.

**10-2.2.3 Uniform Suspension.** Uniform suspension corresponds to the state of suspension at which particle concentration and particle size distribution are practically uniform throughout the vessel; any further increase in agitation speed or power does not appreciably enhance the solids distribution in the fluid. A

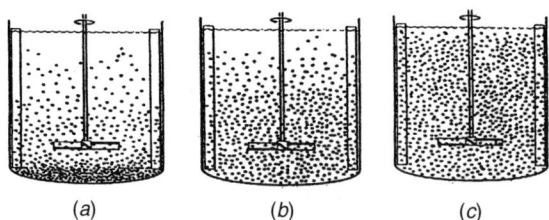

(a)         (b)         (c)

**Figure 10-4** Degrees of suspension. (*a*) Partial suspension: some solids rest on the bottom of the tank for short periods; useful condition only for dissolution of very soluble solids. (*b*) Complete suspension: all solids are off the bottom of the vessel; minimum desired condition for most solid–liquid systems. (*c*) Uniform suspension: solids suspended uniformly throughout the vessel; required condition for crystallization, solid catalyzed reaction. See Visual Mixing CD affixed to the back cover of the book for several illustrative videos.

**Table 10-2** Impact of Desired Result on Mixer Design[a]

|  |  | Power Ratio at Settling Velocity (ft/min) |  |  |
|---|---|---|---|---|
| Suspension Criteria | Speed Ratio | 16–60 Difficult | 4–8 Moderate | 0.1–0.6 Easy |
| On-bottom motion | 1 | 1 | 1 | 1 |
| Complete off-bottom suspension | 1.7 | 5 | 3 | 2 |
| Total uniformity | 2.9 | 25 | 9 | 4 |

*Source*: Oldshue (1983).
[a] Power and speed depend on mixing criteria and settling velocity.

coefficient of variation of the solid concentration of about 0.05, or a uniformity of 95%, is often considered adequate for most process applications. A uniformity of 100% is impractical because there is always a fluid layer a few inches thick at the surface where particle concentration is lower because the axial lift velocity is small near the fluid surface.

Uniform suspension is often the desired process result for process operations where a representative sample of solids is required or a uniform concentration of solids must be achieved. For example, in crystallization, nonuniform solids concentration may lead to unacceptably high local supersaturation levels and subsequent nonuniformity in crystal growth. Also, in as practical a way as possible, a slurry must be fed at a uniform solids concentration to a continuous reactor or to a centrifuge for uniform buildup of solids required for proper filtration and washing of the solid cake.

As illustrated in Table 10-2, it requires increasing energy input to progress from on-bottom motion through complete suspension to the level of uniform suspension. For particles with a free settling velocity of 0.1 to 6.0 ft/min, the power required to achieve complete suspension and total uniformity is two and four times, to respectively, that required for on-bottom motion. For particles with a free settling velocity of 4 to 8 ft/min, the power ratios are 3 and 9 for complete suspension and total uniformity, respectively. For very fast settling particles, with a free settling velocity of 16 to 60 ft/min, the power ratios are 5 and 25 for complete suspension and total uniformity, respectively.

## 10-3 MEASUREMENTS AND CORRELATIONS FOR SOLID SUSPENSION AND DISTRIBUTION

Techniques for measuring the speed required for the condition for "just suspension" are discussed in Chapter 4 of this book and by Choudhury (1997). Also discussed are key aspects of the criteria, techniques, and precautions that one must take to obtain reliable data for solids suspension correlations. The Zwietering criterion of no particle remaining at the base of the vessel for more than 1 to 2 s is the basis for most of the published studies.

## 10-3.1 Just Suspended Speed in Stirred Tanks

There have been many experimental studies and theoretical analyses, with the pioneering work of Zwietering (1958) as the earliest known. He derived the following correlation from dimensional analysis and estimated the exponents by fitting to data for just suspended impeller speed, $N_{js}$:

$$\text{Re}_{\text{imp}}^{0.1} \, \text{Fr}^{0.45} \left(\frac{D}{d_p}\right)^{0.2} X^{0.13} = S \qquad (10\text{-}10)$$

The correlation is often expressed in dimensional form as

$$N_{js} = S\nu^{0.1} \left[\frac{g_c(\rho_s - \rho_l)}{\rho_l}\right]^{0.45} X^{0.13} d_p^{0.2} D^{-0.85} \qquad (10\text{-}11)$$

where $\text{Re}_{\text{imp}}$ is the impeller Reynolds number, $\text{Re}_{\text{imp}} = N_{js}D^2/\nu$; Fr the Froude number, $\text{Fr} = \frac{[\rho_l]}{(\rho_s-\rho_l)} N_{js}^2 D/g_c$; D the impeller diameter (m); $d_p$ the mass-mean particle diameter, $(d_p)_{43}$ (m); X the mass ratio of suspended solids to liquid × 100 (kg solid/kg liquid); S the dimensionless number which is a function of impeller type, as well as of D/T and C/T; $N_{js}$ the impeller speed for "just suspended" (rps); $\nu$ the kinematic viscosity of the liquid (m²/s); $g_c$ the gravitational acceleration constant, 9.81 m/s²; $\rho_s$ and $\rho_l$ the density of particle and the density of liquid (kg/m³).

With the exception of the density difference, the influence of fluid and particle properties on $N_{js}$ is not large, as indicated by the small exponents on the kinematic viscosity, $\nu$, the particle diameter, $d_p$, and the solid loading parameter, X, in eq. (10-11). The density difference is the property with the largest influence on $N_{js}$. Its exponent reflects the effect of the terminal settling velocity of the particles. The exponent on the impeller diameter, D, represents the effect of scale. Note that an exponent of $-0.67$ on D would imply a scaling rule based on power per volume.

More recent studies (Nienow, 1968; Baldi et al., 1978; Rao et al., 1988; Mak, 1992; Choudhury, 1997) generally corroborate Zwietering's original findings. Choudhury (1997) has pointed out regions of interest where Zwietering's correlation is not as reliable. They include solids loading below 2 vol %, high $d_p$/T values, and high solids loading (greater than 15 vol %).

### 10-3.1.1 Effect of Fluid Viscosity.
Most studies and applications of solid suspension occur in the turbulent regime, so the small effect of viscosity is expected. In fact, published values of the viscosity exponent range from 0 to 0.2 for experimental studies (Zwietering, 1958; Chapman et al., 1983; Ibrahim and Nienow, 1994; Rieger and Ditl, 1994). This suggests that the true hydrodynamic mechanism for the just suspended condition remains fuzzy. There may actually exist a hydrodynamic regime where there is little or no influence of viscosity and another where the influence is reflected in a positive value of the exponent.

The highest viscosity tested in the studies cited is only 100 mPa · s (Ibrahim and Nienow, 1994). What happens when the fluid viscosity is even higher remains to be determined. With a more viscous fluid, or as the transitional flow regime is approached, the hydrodynamics near the vessel base may change and make it more difficult for solids to be picked up, even though the bulk of the fluid remains turbulent.

*10-3.1.2 Effect of Solid Loading.* Zwietering chose to represent the effect of solid loading with the parameter X defined above. The exponent on this parameter fits experimental data reliably for values of X from about 5 to 170, which corresponds to about 2 to 40 vol% by volume for sand at a solid density of 2600 kg/m$^3$.

Choudhury (1997) and Choudhury et al. (1995) questioned the appropriateness of the use by Zwietering of the X parameter to correlate the effect of solid loading. They preferred the use of the volume fraction as a percent, %V, because a designer can specify it directly. The following expressions are useful for converting between various measures of solid loading in a slurry. To convert from volume percent, vol %, use

$$X = 100 \frac{\rho_s \text{vol}\%}{\rho_l(100 - \text{vol}\%)}$$

In terms of weight percent of solids, wt%, the corresponding expression is

$$X = 100 \frac{\text{wt}\%}{100 - \text{wt}\%}$$

When converting from slurry density, $\rho_{av}$, the expression is

$$X = 100 \frac{M_s}{\rho_{av} V - M_s}$$

where V is the total slurry volume and $M_s$ is the mass of solids.

*10-3.1.3 Effect of Fluid Particle Size.* Several studies (Zolfagharian, 1990; Choudhury et al., 1995; Choudhury, 1997) indicate that the effect of particle diameter is not as simple as the Zwietering correlation suggests, particularly at solid loading less than about 5 wt%. The exponent reported by Zwietering appears to be an average value for $d_p$ between 0.20 and 1 mm. For particles greater than about 1 mm in diameter, $N_{js}$ appears to be unaffected by the particle size. Choudhury reported this critical particle in terms of $d_p/D$ at a value of about 0.01. On the other hand, for particles smaller than 0.20 mm, the average value of the exponent was about 0.5.

*10-3.1.4 Effect of Vessel and Impeller Geometry and Scale.* The effects of the geometry of the impeller, vessel, and its internals are subsumed in the

**Table 10-3** Parameters for Solids Suspension in Dished Vessels

| Impeller Geometry and Location | Zwietering Constant, S |
|---|---|
| A-310 (T/2.4) | |
| C = T/4 | 6.9 |
| A-310 (T/2) | |
| C = T/4 | 7.1 |
| 30° PBT (T/3, D/2.5) | |
| C = T/4 | 6.4 |
| C = T/6 | 7.1 |
| C = T/8 | 7.2 |
| 45° PBT (T/3.3, D/2.1) | |
| C = T/4 | 4.5 |
| C = T/8 | 4.3 |
| 45 PBT (T/3, D/3.5) | |
| C = T/4 | 4.8 |
| C = T/6 | 4.6 |
| C = T/8 | 4.2 |
| 45° PBT (T/2.5, D/2.8) | |
| C = T/4 | 4.7 |
| C = T/8 | 3.4 |
| 45° PBT (T/2, D/3.5) | |
| C = T/4 | 5.2 |
| C = T/6 | 4.2 |
| C = T/8 | 3.7 |
| 45° PBT (T/2, D/6) | |
| C = T/4 | 5.5 |
| C = T/8 | |
| 45° PBT (T/1.7, D/3.5) | |
| C = T/4 | 6.7 |
| C = T/6 | 5.1 |
| C = T/8 | 4.4 |
| 45° PBT (T/1.7, D/4.3) | |
| C = T/4 | 6.8 |
| C = T/8 | 3.8 |
| 45° PBT (T/1.4, D/5.0) | |
| C = T/4 | 5.4 |
| C = T/8 | 4.5 |
| 45° PBT (T/3, D/4) | |
| C = T/4 | 4.4 |
| C = T/6 | 4.1 |
| C = T/8 | 3.7 |
| 90° PBT (T/3, D/5) | |
| C = T/4 | 4.4 |
| C = T/6 | 4.1 |
| C = T/8 | 4.1 |

*Source*: Mak (1992).

**MEASUREMENTS AND CORRELATIONS FOR SOLID SUSPENSION AND DISTRIBUTION** **561**

S parameter. Representative values of the impeller-specific Zwietering constant, S, are listed in Table 10-3 for a variety of impellers. Note that the value of S varies with D/T and C/T. It is smaller at smaller C/T (i.e., an impeller mounted closer to the vessel bottom), and larger D/T (i.e., a larger-diameter impeller). Obviously, there are practical as well as performance limits on these dimensions. For example, it is clearly evident in the solid suspension video clip on the accompanying CD ROM that a large-diameter pitched blade turbine (D/T = 0.75) is poor at solid suspension because of the resulting flow patterns. When the power number of the impellers is taken into account, it becomes clear that axial flow impellers (e.g., Lightnin A-310, Chemineer HE-3) are able to achieve a just suspended state at a lower rotational speed than can a pitched blade or disk turbine. The resulting axial flow developed by high efficiency impellers is higher at the vessel base than for radial flow impellers. They are also more effective at higher clearances from the vessel base (i.e., larger values of C/T).

Zwietering provided plots of S as a function of D/T and C/T. Armenante et al. (1998) and others [see references cited in Armenante and Nagamine (1998)] have sought simple mathematical expressions to describe the effects of geometry (D/T and C/T) to facilitate the calculation of $N_{js}$. Their results are yet to be validated with data from large scale tests and for vessels with dished bottoms. Published data (Guerci et al., 1986) indicate that the just suspended condition is more easily achieved in dish-bottomed vessels than in flat-bottomed ones. Just suspension is impractical with conical bottoms.

It must be emphasized that studies of the minimum agitation speed for the just suspended state, $N_{js}$, address primarily hydrodynamic mechanisms associated with particle pickup from the vessel base and not necessarily the distribution of particles. Therefore, it is not expected that the use of multiple impellers would significantly affect $N_{js}$.

*Example 10-2: Calculation of the Impeller Speed for Just Suspension.* Calculate the just suspension impeller speed for suspending $AlCl_3$ crystals in methylene chloride. The solid and liquid properties are given in Example 10-1. Other data are as follows: Ratio of solid to liquid, X: 0.4. Kinematic viscosity of the liquid, $\nu$: (0.001 kg/m · s)/1326 kg/m³ or $7.541 \times 10^{-7}$ m²/s.

The impeller is a 45° pitched blade with a D/T value of 1/3 and blade width of D/4 located at C/T value of $\frac{1}{8}$ in a vessel with a diameter, T, of 48 in. The impeller diameter is 28.5 in. or 0.724 m; the S value from Table 10-3 is 3.7.

SOLUTION: We use the Zwietering correlation. For 5000 μm particles,

$$N_{js} = 3.7 \times (7.541 \times 10^{-7})^{0.1} \left[\frac{9.81(2.44 - 1.326)}{1.326}\right]^{0.45}$$
$$\times 0.4^{0.13}(5 \times 10^{-3})^{0.2}(0.724^{-0.85})$$
$$= 0.95 \text{ rps or } 57 \text{ rpm}$$

For 1000 μm particles, use the fact that $N_{js} \propto d_p^{0.2}$, to obtain $N_{js} = 0.69$ rps or 41 rpm.

### 10-3.2 Cloud Height and Solids Distribution

In solid suspensions there is a distinct level to which most of the solids are lifted within the fluid even at speeds above $N_{js}$. The distance from the bottom of the vessel to this level is called the *cloud height*. The liquid below this height is solid-rich, while above it there is only an occasional visit by a few small solids. Hicks et al. (1993, 1997) and Bujalski et al. (1999) have reported extensive data on cloud height and solid distribution. Bujalski et al. (1999) also reported that the blending between the solid-rich and solid-free portions is rather poor, and can result in a blend time as much as 20 times longer in the solid-free region than in the solid-rich volume.

The data of Hicks et al. (1993) for single impellers showed that the cloud height increases with increasing impeller D/T at $N_{js}$. They reported a cloud height at $N_{js}$ to be at about 70% of the slurry height for a single four-bladed 45° pitched blade turbine or a Chemineer HE-3 impeller with D/T = 0.35 located at C/T = 0.25 in a fluid with Z/T = 1. The cloud height was greater than 95% of the slurry height at impeller speeds of 1.5 times $N_{js}$. When the slurry height was increased to Z/T = 1.75, the cloud height was only about 40% of the slurry height at $N_{js}$ and never got above 70%, even at three times $N_{js}$. They also reported that the cloud height improves with the addition of a second impeller. The best separation distance between impellers was three impeller diameters S/D = 3). Bakker et al. (1994) showed that at this separation the dual impellers generate one large flow loop. However, when S/D is increased to 3.7, two separated flow loops are formed and the cloud height drops to the same level as for the single impeller.

Bittorf and Kresta (2002) have applied a wall jet model successfully to predict the cloud height data of Hicks et al. (1997) and Bujalski et al. (1999). The proposed model for purely axial impellers (i.e., A310 or HE3) is

$$CH = \frac{N}{N_{js}} \left[ 0.84 - 1.05 \frac{C}{T} + 0.7 \frac{(D/T)^2}{1 - (D/T)^2} \right] \qquad (10\text{-}12)$$

where CH is the cloud height made dimensionless with T. The model is good for $0.154 < D/T < 0.52$ for solids with a terminal settling velocity less than 0.143 m/s. The model agrees with the data of Hicks et al. (1993, 1997) and predicts that an impeller with a larger D/T located at small D/T results in a higher cloud height.

### 10-3.3 Suspension of Solids with Gas Dispersion

Three-phase (gas–liquid–solid) systems such as gaseous slurry reactions in stirred vessels are common in the chemical industry. They present special mixing challenges. The presence of gas tends to disturb the liquid flow patterns established by the rotating impellers. Sometimes the gas is entrapped by solid agglomerates increasing their tendency to float. In general, laboratory or pilot testing is a must for reliable scale-up and design of three-phase slurry systems.

In a study of gassed solid suspension in an agitated vessel, Chapman (1981) found that for small-diameter (D = T/4) 45° pitched blade impellers, a sudden collapse of the suspension occurs at some critical gas rate. This is when the flow pattern becomes dominated by gas flow as opposed to impeller flow. The gas flow decreases the eddies and the upward velocities that maintain the suspension.

A theoretical correlation for $N_{js}$ by Baldi et al. (1978) implies that $N_{js}$ for gassed slurry systems is higher than for ungassed systems. This has been confirmed (Chapman et al., 1983) in experiments performed in 0.56 m-diameter vessels using particles of size greater than 80 mm and particle density greater than 1.2 g/cm$^3$ in distilled water. Chapman found that as the gas rate is increased, substantial increases in $N_{js}$ are required to achieve a complete suspension of the solids. He also found that the impeller speed required for the just suspended state is always higher than that required for a complete dispersion of the gas bubbles. At low gas rates (volume of gas per minute per volume of liquid, vvm, less than 0.75), he found 45° pitched blade impellers to be more efficient than disk or Rushton turbines for solid suspension.

### 10-3.4 Suspension of Solids in Liquid-Jet Stirred Vessels

Jet mixers (see Chapter 9) are not normally used for solid suspension. However, it may be more economical to use liquid jets to suspend incidental solids in a vessel not initially intended for a solid–liquid mixing application. For example, a vessel designed normally for liquid storage may occasionally see sludge or solids accumulate from process upsets, including changes in concentration or temperature or failure of an upstream filter. It may be more difficult or expensive to retrofit such a vessel with a mechanical agitator than to install a jet mixer using the existing loading or unloading pump and piping system.

Shamlou and Zolfagharian (1990) have studied liquid-jet stirred suspension and found the mechanism of suspension to be similar to impeller-stirred suspension. The preferred design consists of a downward-pointing feeder nozzle centrally mounted with the tip fully submerged in the slurry. They found that to achieve an acceptable cloud height, the tip of the nozzle should be below half the slurry height. They showed also that:

1. There is no significant effect of the jet clearance—the distance between the tip of the jet and the vessel bottom—on the minimum jet velocity for solid suspension. The recommendation is to use the smallest practical jet clearance, but greater than eight jet nozzle diameters, to avoid erosion of the tank base.

2. The minimum jet velocity for off-bottom suspension, $V_{js}$, may be estimated using the following dimensional correlation:

$$V_{js} = 2\left(\frac{\rho_s - \rho_l}{\rho_l}\right)^{2.08} \frac{v^{0.16} g^{0.42} T^{1.16} d_p^{0.1} C_w^{0.24}}{D_j} \quad (10\text{-}13)$$

where $V_{js}$ is the minimum jet velocity for off-bottom suspension or "just suspended" (m/s); $d_p$ the mass-mean particle diameter, $(d_p)_{4,3}$ (m); $C_w$ the percentage weight fraction of solids; $D_j$ the jet diameter, (m). T the vessel diameter (m); It is worth noting the similarity between the proposed equation for $V_{js}$ and the Zwietering correlation.

## 10-3.5  Dispersion of Floating Solids

Without adequate agitation, solid particles less dense than the liquid will float. Also, fine solids such as flour or powders may entrap large amounts of air, which reduces the effective density, causing them to float. Sometimes the solids are difficult to wet with the liquid and may form large clumps with entrapped air. Below is a brief summary of work reported in the open literature. This is offered as a guide but with the advise that in almost all cases, lab and pilot testing will be required for meaningful scale-up and design.

Studies using 10 wt% polyethylene in tap water and others using cork or polypropylene particles in water or corn syrup solutions (Joosten et al., 1977; Hemrajani et al., 1988; Thring, 1990; Siddiqui, 1993) indicate that formation of a controlled vortex is the key to achieving a complete dispersion and suspension of floating solids. The controlled vortex is obtained by using various partial baffles in the vessel rather than no baffles at all. All these studies indicate that dispersion of floating solids requires more energy than for settling solids.

The Froude number, $N_{Fr}$, is a predominant correlating parameter in these systems, where liquid surface behavior is so important. Joosten et al. (1977) have developed a correlation that has been used successfully to design a commercial mixing system for suspending floating solids in a 50 m³ vessel. The correlation is

$$N_{Fr} = 3.6 \times 10^{-2} \left(\frac{D}{T}\right)^{-3.65} \left(\frac{\rho_l - \rho_s}{\rho_l}\right)^{0.42} \quad (10\text{-}14)$$

where

$$N_{Fr} = \frac{N^2 D}{g_c} \quad (10\text{-}15)$$

Joosten et al. (1977) recommend a down-pumping 45° pitched blade impeller in a vessel with a single baffle whose width is one-fifth the impeller diameter submerged to a depth of one-third the impeller diameter to produce a noncentral vortex. Hemrajani et al. (1988) recommend a down-pumping 45° pitched blade impeller in a vessel with four baffles that are $\frac{1}{50}$ the tank diameter. Siddiqui also recommends a down-pumping 45° pitched blade impeller but in a vessel with three partially immersed standard baffles 90° apart immersed to different depths but with one or two of the baffles extending to the top impeller. Siddiqui (1993) found this design to be more effective than Hemrajani's for either standard vessels or for a tall vessel with a liquid height/tank diameter greater than 1.2. The variety of recommendations by these researchers is indicative of the complexities involved in suspending floating solids. Reliable scale-up and design will require careful experimental studies.

## 10-4 MASS TRANSFER IN AGITATED SOLID-LIQUID SYSTEMS

As noted earlier, with the exception of the purely physical process of producing a slurry, unit operations involving solid-liquid mixing are mass transfer processes. These include:

- Leaching
- Dissolution of solids with or without chemical reaction
- Precipitation
- Crystallization—nucleation and crystal growth
- Adsorption
- Desorption
- Ion exchange
- Solid-catalyzed reactions
- Suspension polymerization

Mass transfer between a solid and the liquid is discussed in great detail by Doraiswamy and Sharma (1984) and in other books devoted to a particular mass transfer operation, such as crystallization (Mullins, 1993). In the following sections we highlight several important aspects.

### 10-4.1 Mass Transfer Regimes in Mechanically Agitated Solid-Liquid Systems

In solid-liquid mass transfer processes, the rate-controlling steps are:

1. Diffusion in the liquid film surrounding the solid particles (film diffusion)
2. Diffusion within the particles—in pores or through the solid phase itself (particle diffusion), as in ion exchange
3. Chemical reaction at the surface of the particle (surface reaction)

Agitation affects only the film diffusion controlled process.

**566** SOLID–LIQUID MIXING

The rate of diffusional mass transfer, M, is defined as a product of the diffusional mass transfer coefficient, $k_{SL}$, the interfacial area for mass transfer, $a_p$, and the concentration driving force, $[A^*] - [A]$:

$$M = k_{SL} a_p ([A^*] - [A]) \quad (10\text{-}16)$$

The variables $[A^*]$ and $[A]$ are the concentration of the solid material, A, at the solid surface and in the bulk of the liquid, respectively. The interfacial area per unit volume is

$$a_p = 6 \frac{\phi}{\rho_s d_p} \quad (10\text{-}17)$$

where $\phi$ is the solid loading with units of g/cm$^3$ solid-free liquid.

In a reactive diffusion system the dissolved solid undergoes a reaction in the bulk liquid or at the solid–liquid interface. The reaction rate may be expressed as a product of the reaction rate constant, $k_r$, and concentration to some power, n:

$$M = k_r [A]^n \quad (10\text{-}18)$$

The constants, $k_{SL}$, $k_r$, and n are to be determined from experimental data or from available correlations.

A key issue in solid–liquid reactions is determination of the controlling process regime: chemical reaction in the bulk liquid phase or mass transfer in the liquid film surrounding the solid particle. Experimentally, this is done by checking the effect of agitator speed on the observed process rate.

The controlling regime depends on the relative values of $k_{SL}$ and $k_r$ for first-order reactions as follows:

1. Chemical reaction controls when $k_r/k_{SL} \leq 0.001$.
2. Diffusional mass transfer controls when $k_r/k_{SL} \geq 100$.

For a reaction such as A + B → products, where A is the solid and B is a liquid phase reagent, Figure 10-5 shows schematically the concentration gradients for four different regimes that can occur: bulk reaction (regime 1), film diffusion (regime 2), film kinetics (regime 3), and instantaneous reaction (regime 4). Note that when the process is mass transfer controlled, there are three possible regimes (regimes 2, 3, or 4), depending on the kinetics of the reaction:

- In regime 1, as noted above, the reaction is so slow or the solubility of the solid is so high that the concentration of the solid species is essentially equal to the equilibrium conditions at the solid–liquid interface. Bulk liquid-phase reaction governs the overall process.
- In regime 2, the reaction is fast enough to keep the bulk liquid-phase concentration of the solid essentially zero but not fast enough to occur substantially

MASS TRANSFER IN AGITATED SOLID–LIQUID SYSTEMS 567

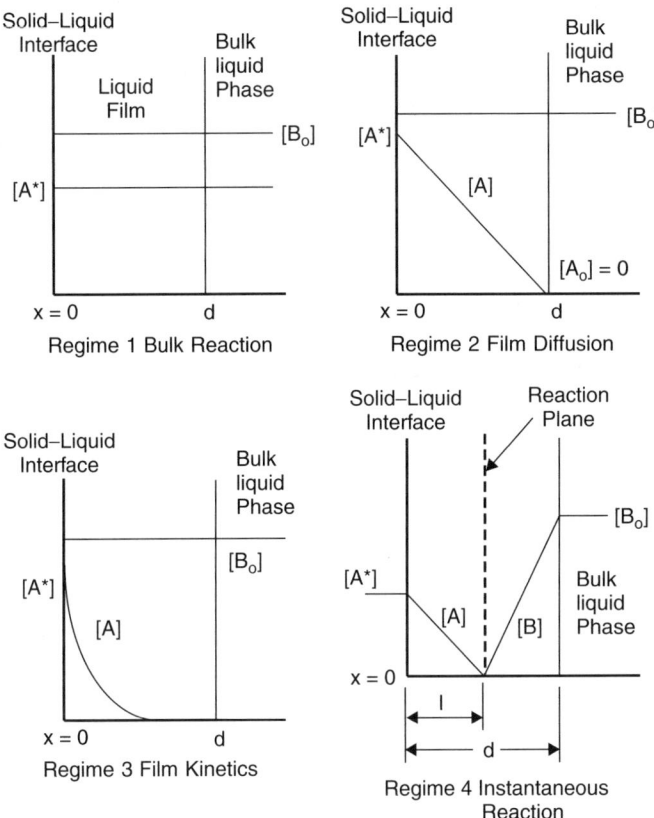

**Figure 10-5** Schematic diagram of concentration gradients for solid–liquid reactions.

in the liquid film. There is no enhancement of mass transfer due to reaction. Diffusion and reaction take place in series.

- In regime 3, the reaction is sufficiently fast to consume the dissolved solid reactant completely in the liquid film. Diffusion and reaction occur simultaneously in a parallel fashion in the liquid film. The mass transfer coefficient has no effect on the overall process rate.
- In regime 4, the reaction is so fast (virtually instantaneous) that the reactant (A) and the liquid-phase reactant (B) cannot coexist. Diffusion of A from the solid–liquid interface and diffusion of B from the bulk liquid toward the reaction plane control the overall process.

Intermediate conditions between regimes are also possible. These correspond to cases where the concentration of species in the liquid film remains finite instead of going to zero. Doraiswamy and Sharma (1984) discuss these cases in some detail.

It is important to know the regime of a particular reaction system, since the equipment choice and the effect of design and operating variables on the process performance depend on the regime. A lack of this fundamental understanding leads to many apparent discrepancies between different scales of operations and sometimes to scale-up failures. For instance, in the lab the process may be operating in regime 1, while in production scale it could be in regime 2. Alternatively, one may design equipment for minimal mass transfer requirements based on the confirmation of regime 1 on lab scale only to find much lower process rates. The effect of temperature is minimal if the system is in regimes 2 and 4, substantial in regime 3 (apparent activation energy is half of the true activation energy), and maximum in regime 1 (apparent activation energy is equal to the true activation energy). Solid–liquid reactions are discussed further in Chapters 13 and 17.

### 10-4.2 Effect of Impeller Speed on Solid–Liquid Mass Transfer

Many authors (Nienow, 1975; Nienow and Miles, 1978; Chaudhari, 1980; Conti and Sicardi, 1982) have reported the effect of agitation on the diffusional mass transfer coefficient, $k_{SL}a_p$. It is sufficient to say that the diffusional mass transfer rate is affected primarily by the impact of agitation on the hydrodynamic environment near the surface of the particle, in particular the thickness of the diffusional boundary layer surrounding the solid. The hydrodynamic environment near the particle surface depends on the properties of the fluid properties as well as those of the particles. The specific variables were introduced in Section 10-2.1.1. In addition to these, the diffusivity, $D_A$, also influences the diffusional mass transfer.

The important hydrodynamic variables are the relative velocity, $V_s$, between the solids and the liquid (also know as *slip velocity*) and the rate of renewal of the liquid layer near the solid surface. The relative velocity, $V_s$, obviously varies from point to point within the vessel, and the average value is difficult to estimate. So, in practice, the relative velocity, $V_s$, is assumed equal to the free settling velocity, $V_t$. The renewal of the boundary layer depends on the intensity of turbulence around the solid particle as well as the convective velocity distribution in the vessel.

The observed effect of agitation is depicted in Figure 10-6. As the stirrer speed increases, the volumetric mass transfer coefficient, $k_{SL}a_p$, increases. If the process is mass transfer controlled, the observed rate of reaction increases with increasing impeller speed. However, beyond the just suspended or complete suspension state the observed rate may not increase much with increasing rpm or mixing intensity, indicating that the overall process is bulk reaction controlled. For extremely slow reactions of highly soluble solids, on-bottom motion to prevent stagnant pockets may be all that is needed.

In general, the specific impact of agitation must be determined experimentally for each system. The correlations discussed below are presented to provide a guide and insight into the expected effects of various variables on solid–liquid mass transfer.

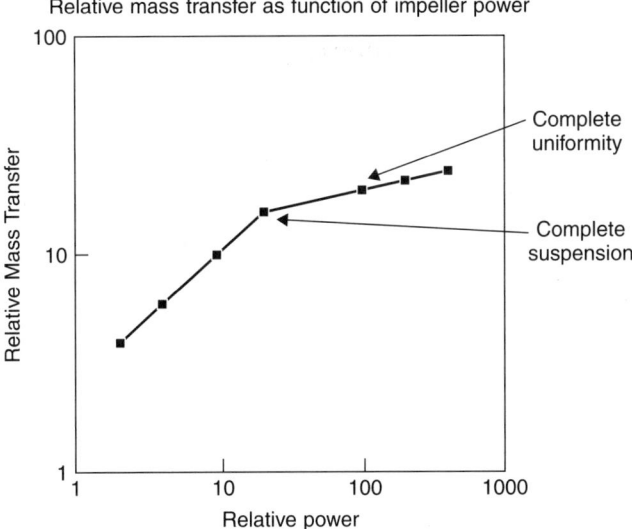

**Figure 10-6** The mass transfer increases sharply up to the point of complete suspension and at a much lower rate to complete uniformity.

### 10-4.3 Correlations for the Solid–Liquid Mass Transfer

Several correlations for $k_{SL}$ have been reported in the literature. The Froessling type equation developed by Nienow and Miles (1978) based on the theory of slip velocity between the liquid and solid particles: namely,

$$\text{Sh} = 2 + 0.44 \text{Re}_p^{1/2} \text{Sc}^{0.38} \tag{10-19}$$

has proven useful for estimating $k_{SL}$ or establishing the effect of solid and fluid properties as well as agitation parameters. In this equation, the Sherwood number, Sh, the particle Reynolds number, $\text{Re}_p$, and the Schmidt number, Sc, are defined in terms of the particle diameter, $d_p$, the liquid density, $\rho_l$, liquid viscosity, $\mu_l$, terminal velocity, $V_t$, and diffusivity, $D_A$, as

$$\text{Sh} = \frac{k_{SL} d_p}{D_A} \tag{10-20}$$

$$\text{Re}_p = \frac{\rho_l V_t d_p}{\mu_l} \tag{10-21}$$

$$\text{Sc} = \frac{\mu_l}{\rho_l D_A} \tag{10-22}$$

The Froessling correlation is not applicable to solid–liquid systems where the settling velocity or slip velocity is small, $\ll 0.1$ ft/min or 0.0005 m/s. Figure 10-3 can be used to estimate the combination of the range of particle sizes and density

**570** SOLID–LIQUID MIXING

difference, $(\rho_s - \rho_l)$ that lead to small values of the settling velocity. For such systems, the correlation

$$\text{Sh} = 2 + 0.47 \text{Re}_p^{0.62} \text{Sc}^{0.36} \left(\frac{D}{T}\right)^{0.17} \quad (10\text{-}23)$$

developed by Levins and Glastonbury (1972a,b), based on Kolmogoroff's theory of isotropic turbulence, is recommended (Nienow, 1975). (See Chapter 2 for a discussion of isotropic turbulence.) In this correlation the particle Reynolds number, $\text{Re}_p$, is defined in terms of the power input per unit mass of solid, $\varepsilon_p$, as follows:

$$\text{Re}_p = \frac{\rho_l \varepsilon_p^{1/3} d_p^{4/3}}{\mu_l} \quad (10\text{-}24)$$

*Key Points*

1. Experiments show that the measured value of $k_{SL}$ can be significantly different from that estimated with the correlations above (Nienow, 1975). Therefore, for reliable scale-up or design, laboratory- or pilot-plant experimentation to measure the rate of mass transfer is a must for systems where mass transfer is important.

2. Experiments indicate that solid–liquid mass transfer rate increases relatively rapidly with increasing impeller speeds up to the just suspended state, $N_{js}$. This is a result of increases in both the interfacial area per volume, $a_p$, and the mass transfer coefficient, $k_{SL}$. Beyond $N_{js}$, $a_p$ is independent of agitation because all the solid surface available for mass transfer is now exposed, but the mass transfer coefficient, $k_{SL}$, continues to increase, although at a much lower rate. The overall effect is illustrated in Figure 10-6.

3. At impeller speeds corresponding to $N_{js}$, the value of $k_{SL}$ is independent of the geometry of the vessel, impeller design, or the specific power consumption (Doraiswamy and Sharma 1984).

4. The functional relationship between $k_{SL}$ and the speed of agitation depends on the hydrodynamic regime of agitation. In the turbulent regime, where the impeller Reynolds number is greater than 1000, the value of $k_{SL}$ is independent of particle size and practically independent of the density difference (Doraiswamy and Sharma, 1984).

5. At impeller speeds near $N_{js}$, the value of $k_{SL}$ will be a strong function of the density difference between the particle and the liquid (Doraiswamy and Sharma, 1984).

*Example 10-3: Calculation of Solid–Liquid Mass Transfer Coefficient.* (Adapted from Doraiswamy and Sharma, 1984) It is desired to prepare a 25°C aqueous solution of potassium sulfate containing 0.09 g $K_2SO_4$/g solution in an agitated 48 in. diameter stainless steel reactor. Calculate:

(a) The solid–liquid mass transfer coefficient at $N_{js}$, the minimum impeller speed required to suspend the potassium particles completely.
(b) The rate of dissolution of the solids at $N_{js}$.

The required data for solving this problem include:

| | |
|---|---|
| Solid loading | 0.05 g/cm³ of solid free liquid |
| Solution viscosity | 1.01 cP or 0.00101 kg/m · s |
| Solution density | 1.08 g/cm³ or 1080 kg/m³ |
| $K_2SO_4$ density | 2.66 g/cm³ or 2660 kg/m³ |
| $K_2SO_4$ particles size | 324 μm or 0.000324 m |
| Solubility of $K_2SO_4$ | 0.12 g/g of solution |
| Bulk concentration | 0.09 g $K_2SO_4$/g solution |
| Diffusivity of $K_2SO_4$ in water | $9.9 \times 10^{-6}$ cm²/s |

SOLUTION: (a) For this simple dissolution a high efficiency impeller will be used (see below for the rationale). From Table 10-3 select an A-310 impeller with diameter, D, equal to the half the vessel diameter, T (i.e., $D = T/2$, 24 in. or 0.61 m and located at T/4 from the vessel bottom). First, calculate $N_{js}$ using the Zwietering correlation,

$$N_{js} = S\upsilon^{0.1} \left[\frac{g_c(\rho_s - \rho_l)}{\rho_l}\right]^{0.45} X^{0.13} d_p^{0.2} D^{-0.85} \quad (10\text{-}25)$$

The value of S from Table 10-3 for a T/2 A-310 located at a clearance of T/4 from the vessel bottom is 7.1.

$$N_{js} = 7.1 \left(\frac{0.00101}{1080}\right)^{0.1} \left[\frac{9.81(2.66 - 1.08)}{1.08}\right]^{0.45} \left(\frac{0.09 \times 100}{1 - 0.09}\right)^{0.13}$$
$$\times (0.000324)^{0.2}(0.61^{-0.85})$$
$$= 2.4 \text{ rps or } 144 \text{ rpm}$$

Second, calculate the Sherwood number, Sh, using the correlation developed by Nienow and Miles (1978):

$$Sh = 2 + 0.44 \, Re_p^{1/2} Sc^{0.38} \quad (10\text{-}26)$$

The Schmidt and Reynolds numbers for this system are

$$Sc = \frac{1.01 \times 10^{-2}}{1.08 \times 9.9 \times 10^{-6}}$$
$$= 945$$

Calculate $V_t$. As pointed out earlier, the calculation is an iterative one since the equation to use depends on the value of $Re_p$, which in turn depends on $V_t$. An

**572**  SOLID–LIQUID MIXING

equation solver such as TK Solver can be used to quickly perform the required iterative solution. For this system it turns out that the intermediate law is what applies, as shown below. So we use eq. (10-3) and the appropriate expression for $C_D$ from Table 10-1:

$$V_t = \sqrt{\frac{4 g_c d_p (\rho_s - \rho_l)}{3 C_D \rho_l}} \qquad (10\text{-}27)$$

where $C_D$ is given by
$C_D = 18.5/\text{Re}_p^{3/5}$ and $\text{Re}_p$ by

$$\text{Re}_p = \frac{\rho_l V_t d_p}{\mu} \qquad (10\text{-}28)$$

Substituting values in eqs. (10-3) and (10-4), we obtain

$$V_t = \sqrt{\frac{(4 \times 9.81 \times 0.000324)(2.66 - 1.08)}{3 \times 1.08 C_D}}$$

$$\text{Re}_p = \frac{(1.08 \times 10^3 \text{ kg/m}^3)(V_t \text{ m/s})(0.000324 \text{ m})}{0.00101 \text{ Kg/m} \cdot \text{s}}$$

Solving these iteratively with an equation solver, we obtain

$$V_t = 0.04 \text{ m/s or } 0.13 \text{ ft/sec} \quad \text{and} \quad \text{Re}_p = 14$$

The value of $\text{Re}_p$ is within the intermediate law regime; therefore, we accept the settling velocity calculated.

Substituting the values for $\text{Re}_p$, and Sc into eq. (10-23) gives

$$\text{Sh} \equiv \frac{k_{SL} d_p}{D_A} = 2 + 0.44 \times 14^{1/2}(945^{0.38})$$

$$= 24$$

and therefore

$$k_{SL} = \text{Sh}\frac{D_A}{d_p} = 24 \times \frac{9.9 \times 10^6}{0.0324}$$

$$= 7.4 \times 10^{-3} \text{ cm/s}$$

(b) The initial dissolution rate corresponds to the case where $[A] = 0$ and can now be calculated with eq. (10-17) as follows.

$$a_p = 6 \frac{5 \times 10^{-2}}{2.66 \times 3.24 \times 10^{-2}} = 3.5 \text{ cm}^2/\text{cm}^3$$

The initial rate of dissolution using eq. (10-16) is

$$M = (7.4 \times 10^{-3} \times 10^{-3} \times 3.5)(0.12 \times 1.08 - 0.0) \text{ g/cm}^3 \cdot \text{s}$$
$$= 3.36 \times 10^{-3} \text{ g/cm}^3 \cdot \text{s}$$

Note that as the particles dissolve: (1) the particle size, $d_p$, decreases; (2) the bulk concentration increases, thus decreasing the driving force, and (3) $k_{SL}$ increases. These time-dependent changes have to be accounted for to obtain the final dissolution rate and how long it takes to dissolve all the particles completely.

## 10-5 SELECTION, SCALE-UP, AND DESIGN ISSUES FOR SOLID–LIQUID MIXING EQUIPMENT

The selection, scale-up, and design of the components that make up the mixing system are based on the fundamental and experimental descriptions of the hydrodynamics and mass transfer aspects of solids suspension discussed earlier. The following issues must be addressed:

1. Process needs assessment, including:
    a. Phases—solid, liquid, and gas—present or occurring during the process
    b. Mixing operations and the desired process results
    c. Unit operations of interest
    d. Quantities and properties of solid and liquid phases
2. Vessel design and internals, including:
    a. Bottom head design
    b. Size and dimensions
    c. Baffles and other internals
3. Selection and design of the agitator or mixer components, including:
    a. Impeller type, number, and dimensions
    b. Impeller location in the vessel
    c. Impeller speed and power
    d. Shaft diameter and length
    e. Drive and seal system

### 10-5.1 Process Definition

The first task in analyzing a mixing problem, determining experiments to perform for mixer scale-up, or designing a mixing system is to define the process needs. It is important to consider carefully the potential impact of mixing on all the physical and chemical phenomena necessary to achieve the desired process result. Invariably, one of these phenomena will be the critical operation on which to base the selection, scale-up, or design of the mixing system.

The definition should include:

- A list of all the phases of matter (gas, liquid, solid) involved or that can occur, even by accident, from start to end of the process; in particular, instances where two or more phases coexist must be noted.
- A list of all the mixing operations (blending, solids suspension, gas dispersion, immiscible liquids dispersion, etc.) involved in the process or carried out in the same vessel.
- A statement of the purpose and duty of the mixing operations, including the desired process result. For solids suspension, one must choose from among the applicable process objectives as well as the desired degree of suspension. The selection must be based on knowledge of the process determined experimentally or by comparison with a similar process.
- The quantities of solid and liquid phases involved as well as the properties of the solid and liquid to assess how difficult it might be to achieve the aforementioned desired results.

### 10-5.2 Process Scale-up

Scale-up is an effort to understand the fundamental phenomena occurring in a process in order to predict the performance in larger scale equipment. It begins with process research at the bench scale, often in small glassware, through pilot scale studies to full production. The value of scale-up is captured in the following comment attributed to L. H. Baekland, the father of plastics: "Commit your blunders on a small scale and make your profits on a large scale."

In solid–liquid mixing applications, the purpose of scale-up is to determine the operating conditions at different scales at which mixing yields equivalent process results. The tasks involve:

1. Definition of the appropriate desired process result, such as level of uniformity of the solid distribution in a vessel, the time to achieve complete dissolution, the rate of reaction between a solid and a liquid reactant, and so on.
2. Developing reliable correlations that describe the effects of key process properties, mixer design, and operating variables on the desired process result by either experimentation or mathematical analysis of the physicochemical phenomena
3. Determining and confirming the key controlling physicochemical phenomena and the associated correlating parameters, preferably in dimensionless form
4. Applying the key correlations to predict the process performance at different scales

Occasionally, heuristics based on extensive experience with similar processes are sufficient. Often, especially for processes involving multiple phases or fast reactions, it is necessary to perform several experiments at two or more different scales, where the vessel size based on diameter is varied by at least a factor of 2.

## 10-5.3 Laboratory or Pilot Plant Experiments

Simple laboratory or pilot plant experiments carried out in transparent vessels, such as glassware, where one can observe the behavior of the various phases during agitation often provides great insight and understanding of the mixing challenges and opportunities. Often, these are augmented with pilot scale tests to determine or evaluate pertinent scale-up requirements. The lab experiments should be designed to answer specific process-related questions such as those discussed. Ultimately, the tests should provide information including:

1. The desired level of suspension required by the process
2. The properties of the solids and liquids required to estimate the necessary solid–liquid mixing parameters, including:
   a. Settling velocity, $V_t$
   b. Minimum speed for suspension, $N_{js}$
   c. Solid–liquid mass transfer coefficient, $k_{SL}$
   d. Materials of construction

In the various correlations presented earlier, the magnitude and sign of the exponents on the variables establish their parametric effects and may be used as a guide for selection of the more sensitive parameters to explore in a laboratory or pilot plant.

Typical lab experiments must include evaluation of the following effects:

1. Impeller speed to establish the effect, if any, on the process result as well as the speed beyond which there is no further significant gain in or deterioration of the desired process results
2. Particle size to determine the effect on reaction rates for solid-catalyzed reactions: in particular, the particle size at which mass transfer effects are negligible
3. Addition rate of solids and/or liquid, as well as the ratio of solids to liquid to determine their effects on rheology, suspension level, reaction, or other mass transfer rate
4. Impeller design and geometry to explore the relative effects of flow and shear distribution in the vessel for particle size control, micromixing for fast kinetics, and so on. Geometric ratios of importance include:
   a. Ratio of the impeller to tank diameter, D/T, to determine the effect of the ratio of overall pumping capacity to fluid shear
   b. Blade width to impeller diameter, W/D, to evaluate the relative effects of microscale and macroscale mixing processes and also fluid shear rates

5. Number and location of the impeller to explore the effect of liquid coverage on headspace gas entrainment, uniformity of solids distribution, and so on. Parameters of interest include:
    a. Ratio of the impeller clearance from vessel bottom to tank diameter, C/T
    b. Ratio of liquid coverage above impeller to tank diameter, CV/T
6. Baffle design and location to explore effects of vortex formation for entrainment of floating solids, and so on.

### 10-5.4  Tips for Laboratory or Pilot Plant Experimentation

In any laboratory or pilot plant tests, the first thing to vary is the impeller speed. This changes pumping capacity, blend time, and shear rates.

- On-bottom motion or partial suspension is rarely a useful desired mixing result except, perhaps, for the dissolution of very soluble solids.
- Complete suspension is the minimum desired mixing goal for most solid–liquid mixing operations involving settling solids. The equivalent condition for floating solids is complete incorporation and dispersion of the floating solids.
- Uniform suspension is required for crystallization, solid-catalyzed reactions, and suspension polymerization where high local concentrations may lead to poor yields of the desired product. Also, as practical as possible, crystallization slurries must be fed to a centrifuge at a uniform solids concentration for the proper cake buildup required for effective filtration and washing of the solid cake.
- Specified mass transfer rate such as dissolution rate, reaction rate, and so on, may be the desired process result to achieve a given production capacity.
- Particle size control may be the desired result in certain formulation operations.
- The measurement of power on a full or pilot scale vessel is best accomplished with a wattmeter. Ammeter readings, at best, must be ratioed to the full-load nameplate amperage, which varies with voltage, power factor, and motor type.
- For the fractional-horsepower motor used in the laboratory or pilot plant, power draw is best determined by calculation using the defining equation for the power number. This requires power number versus Reynolds number data or correlation.
- To estimate the viscosity of complex non-Newtonian slurries, Oldshue and Sprague (1974) recommend the use of a mixing viscometer that mimics the hydrodynamic environment likely to be encountered in an agitated vessel.

## 10-5.5 Recommendations for Solid–Liquid Mixing Equipment

Solids suspension is usually carried out in mechanically agitated vessels with or without draft tubes. A schematic representation of a typical mechanically agitated vessel is shown in Figure 10-7. A mechanically agitated vessel with a draft tube employed for certain crystallization operations is shown in Figure 10-8.

In the following sections we provide several design guidelines and examples of the selection, design, and operation of equipment for solid–liquid mixing.

### 10-5.5.1 Vessel Geometry and Vessel Nozzles.
The vessel design, in particular, the bottom head design, can have a profound effect on the agitation requirements for a given desired result. The bottom head geometry influences the flow patterns responsible for lifting solids up from the vessel bottom.

**Design Tip.** Dished heads (ASME dished, elliptical, or torispherical heads) are the preferred design. To achieve complete suspensions, flat-bottomed heads require 10 to 20% higher impeller speeds than for dished heads (Mak, 1992). Conical bottoms must be avoided.

The aspect ratio of the vessel—actually, the ratio of liquid depth, H, to vessel diameter, T (see Figures 10-7 and 10-8)—is an important determinant of the number of impellers to be used. The fluid velocities decrease with increasing distance from the impeller region and may not be sufficient to counteract the tendency of the solid to settle. Also, impellers mounted far above the vessel base

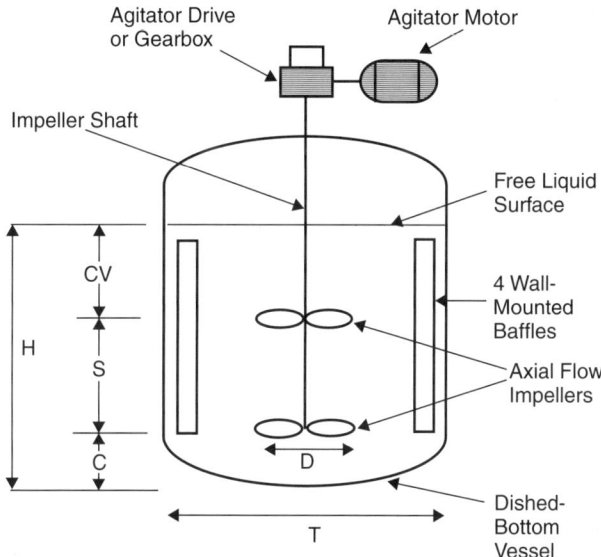

**Figure 10-7** Schematic representation of a typical mechanically agitated vessel.

**578**  SOLID–LIQUID MIXING

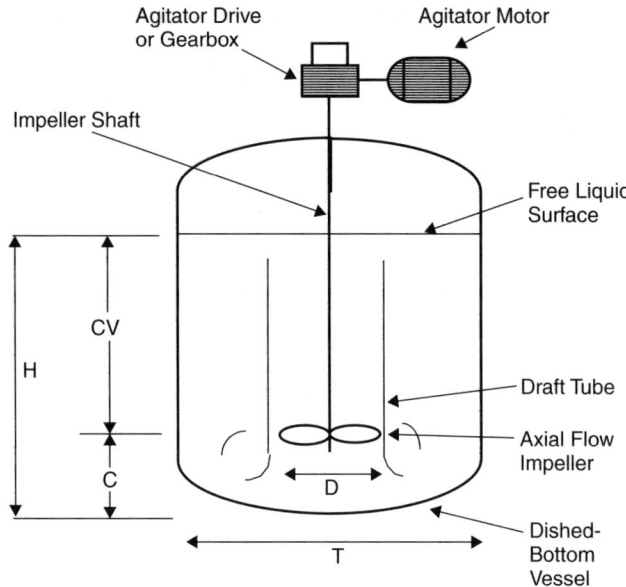

**Figure 10-8**  Mechanically agitated vessel with a draft tube.

may not generate enough turbulent velocity at the base of the vessel to lift any settled solids.

### Design Tips

- A single impeller is usually sufficient for off-bottom suspension in vessels with dished heads, $H/T < 1.3$.
- Dual impellers are recommended for vessels with $1.3 < H/T < 2.5$, used for uniform suspension of fast-settling solids.
- Three impellers may be required if $2.5 < H/T$. A vessel with such a high aspect ratio is a poor choice for solid suspension.
- Vessel nozzles should be located and oriented to avoid or minimize any interference with the mixing system's performance.

### Nozzle Design Tips

- Nozzles and dip pipes for liquid addition should not allow the liquid jet to impinge directly on the impeller. At too high a liquid jet velocity, the jet force will contribute to higher shaft deflections.
- Dip pipes and other probes must be supported—usually by attaching to wall-mounted baffles—or stiff enough to withstand the bending moments imposed by the fluid forces. Discuss with your local mechanical engineer.

- Install grating or screen on nozzles for solid addition to keep very large solid chunks or foreign matter from the liquid.
- Bottom nozzles should be as short as practical and be installed with flush-bottom valves to prevent solids from collecting.

### 10-5.6 Baffles

Baffles are highly recommended for solids suspension operations involving solids that are heavier than the liquid. They convert the swirling motion into top-down or axial fluid motion that helps to lift and suspend the solids (see Visual Mixing CD for an illustrative video). For floating solids, consider the use of submerged or partial baffles to achieve a controlled vortex to draw down the floating solids as recommended by Joosten et al. (1977), Hemrajani et al. (1988), Thring (1990), and Siddiqui (1993).

**Baffle Design and Installation Tips**

- In steel or alloy vessels, the recommended baffle design for solid suspension of settling solids is four flat blade baffles, each with width, B, equal to T/12 at a wall clearance of at least T/72. The baffles should extend to the lower edge of the lower impeller or to the lower tangent line.
- In glass-lined equipment, the recommended baffles are either fin or beaver-tail type (see Chapter 17). A minimum of two baffles is recommended. These baffles are generally less effective than the standard four flat blade baffles.
- Fin baffles must be installed with the edge of the fin pointing toward the vessel wall; the flat face must be perpendicular to the tangential flow.

### 10-5.7 Selection and Design of Impeller

Solids suspension and solids distribution is governed primarily by the bulk or convective flows in a vessel. *High efficiency impellers* (e.g., Lightnin A310 and A320, Chemineer HE3, APV LE20, Ekato Viscoprop), whose discharge is flow dominated as well as axially directed, are more efficient than others in achieving solids suspension. However, high efficiency impellers may be a poor choice when the solid suspension is accompanied by other mixing duties, such as liquid–liquid dispersion or gas dispersion. For these cases a multiple-impeller system consisting of a high efficiency impeller in combination with a 45° pitched blade impeller should be evaluated in pilot plant studies.

Small pitched blade impellers with diameter $D < T/2.5$, located nearer the vessel base ($C < T/4$), are good for solid suspension (see Table 10-3). They also aid in the discharge of the solids during slurry transfer. Typical values for impeller clearance are T/4 for hydrofoils and T/3 for pitched blade turbines.

For glass-lined vessels, one is no longer limited to the Pfaudler "crowfoot," also known as the *retreat blade* or *retreat curve impeller* (RCI). Most impeller

**580** SOLID–LIQUID MIXING

designs can now be obtained with a glass lining. Removable glassed impeller designs are preferred over the integral glassed shaft-impeller design (see Chapters 6 and 17).

### 10-5.8 Impeller Speed and Power

The impeller speed recommended will in general be higher than $N_{js}$, the speed required for the just suspended state estimated by the Zwietering correlation. The speed required should be based on experimental data. For quick estimates of the speed and power requirements for complete uniformity, the ratios in Table 10-2 may be applied to the estimated value of $N_{js}$.

**Design Tip**. For multiprocess batch reactors, mixers equipped with variable speed drives permit the mixer to be operated at different impeller speeds to accommodate the different mixing needs of the various steps in the process.

### 10-5.9 Shaft, Hub, and Drive

In the design of the shaft and drive system (see Chapter 21), careful consideration should be given to issues, including the need for:

- Startup of the mixer in settled solids.
- Filling and emptying while the mixer is running—the fluid forces on the impeller and shaft are amplified significantly when the liquid surface runs through the impeller, causing severe shaft deflections and vibrations.
- Ensuring that the suspension is maintained during emptying of the vessel to very low levels—for top-mounted agitators, a longer shaft fitted with a smaller-diameter impeller; a tickler, located at the lowest possible clearance from the base of the vessel, is required.
- Employing the same mixer for multiple mixing operations in the same process or for different processes.

**Design Tip**. The need for startup of a mixer in settled solids will require a larger shaft. This should be stated clearly in any mixer specification or request for quotation. The American Gear Manufacturing Association (AGMA) service rating for the gearbox will be higher. The shaft and gearbox design should be based on a minimum service rating factor of 2. An experienced mechanical engineer should be consulted for help in specifying the mixer or in reviewing any vendor proposals or quotations.

Mixing equipment suppliers have calculational tools to size the shaft to minimize shaft deflections.

**Design Tip**. Sizing mixers to handle startup in settled solids requires measuring torque under test conditions with actual settled solids. In the absence of such a measurement, any design for such conditions can only be a "wild" guess. Use other means, such as air sparging, lancing with high-pressure liquid, heating to melt or dissolve the solid, and so on, to loosen the settled solid first. Before attempting to start the agitator drive, check and confirm by hand-turning the shaft that the impeller is indeed free.

## NOMENCLATURE

### Dimensional Variables and Parameters

| | |
|---|---|
| $a_p$ | interfacial area for mass transfer per volume of fluid (ft$^2$/ft$^3$, m$^2$/m$^3$) |
| [A*]–[A] | concentration driving force (mol/ft$^3$, mol/m$^3$) |
| C | impeller clearance from the bottom of the vessel (ft, m) |
| CH | cloud height (–) |
| CV | liquid coverage above the impeller (ft, m) |
| D | Impeller diameter (ft, m) |
| $D_A$ | diffusivity (ft$^2$/h, m$^2$/s) |
| $(d_p)_{43}$ | mass-mean diameter (ft, m) |
| $d_i$ | mean particle diameter of the ith size (ft, m) |
| $d_p$ | particle size or diameter (ft, or m) |
| $g_c$ | gravitational constant (32.17 ft/sec$^2$ or 9.81 m/sec$^2$) |
| $k_{SL}$ | diffusional mass transfer coefficient |
| M | rate of diffusional mass transfer |
| N | impeller speed (rps) |
| $n_i$ | number of particles in the ith size class |
| $N_{js}$ | impeller speed for "just suspended" state of particles (rps) |
| P | impeller power (hp, W) |
| T | vessel diameter (ft, m) |
| $V_t$ | particle-free settling velocity (ft/s, or m/s) |
| $V_{ts}$ | particle-hindered settling velocity (ft/s, or m/s) |
| X | mass ratio of suspended solids to liquid time 100 (kg solid/kg liquid)×100 |
| Z | liquid depth in vessel (ft, m) |

### Dimensionless Parameters

| | |
|---|---|
| $C_D$ | drag coefficient |
| $Fr = \left(\dfrac{\rho_l}{\rho_s - \rho_l}\right) N_{ts}^2 D/g_c$ | Froude number |
| $N_{Fr}$ | Froude number |
| $N_p$ | impeller power number |
| $Re_p$ | particle Reynolds number |
| $Re_{imp} = N_{js}D^2/\nu$ | impeller Reynolds number |

| | |
|---|---|
| S | Zwietering constant, dimensionless number which is a function of impeller type, as well as D/T, C/T |
| Sc | Schmidt number |
| Sh | Sherwood number |

*Greek Symbols*

| | |
|---|---|
| $\mu_l$ | liquid viscosity (cP or Pa · s) |
| $\nu$ | kinematic viscosity of the liquid (m$^2$/sec) |
| $\rho_l$ | liquid density (lb/ft$^3$ or kg/m$^3$) |
| $\rho_s$ | solid or particle density (lb/ft$^3$, kg/m$^3$) |
| $\phi$ | volume fraction of solid |
| $\phi$ | solid loading (g/cm$^3$ solid-free liquid) |
| $\psi$ | particle shape or sphericity, (dimensionless factor defined by the ratio of surface area of a spherical particle of the same volume to that of a nonspherical particle) |
| $\chi$ | volume fraction of solids in suspension |

## REFERENCES

Armenante, P. M., and E. U. Nagamine (1998). Effect of low off-bottom impeller clearance on minimum agitation speed for complete suspension of solids in stirred tanks, *Chem. Eng. Sci.*, **53**(9), 1757–1775.

Armenante, P. M., E. U. Nagamine, and J. Susanto (1998). Determination of correlations to predict the minimum agitation speed for complete solid suspension in agitated vessels, *Can. J. Chem. Eng.*, **76**, 413–419.

Bakker, A., J. B. Fasano, and K. J. Myers (1994). Effect of flow pattern on solids distribution in a stirred tank, *Inst. Chem. Eng. Symp. Ser.*, **136**, 65–72.

Baldi, G., R. Conti, and E. Alaria (1978). Complete suspension of particles in mechanically agitated vessels, *Chem. Eng. Sci.*, **33**, 21.

Becker, H. A. (1959). The effects of shape and Reynolds number on drag in the motion of a freely oriented body in an infinite fluid, *Can. J. Chem. Eng.*, **37**, 85–91.

Bittorf, K. J., and S. M. Kresta (2002). Prediction of cloud height for solid suspension in stirred tanks, *CHISA Conference Proc.*, Prague, Aug. 25–29.

Bujalski, W. K., et al. (1999). Suspension and liquid homogenisation in high solids concentration stirred chemical reactors, *Trans. Inst. Chem. Eng.*, **77**, 241–247.

Chapman, C. M. (1981). Studies of gas–liquid–particle mixing in stirred vessels, Ph.D. dissertation, University of London.

Chapman, C. M., A. W. Nienow, and M. Cooke (1983). Particle–gas–liquid mixing in stirred vessels: 1. Particle–liquid mixing, *Chem. Eng. Res. Dev.*, **61**, 71–81.

Chaudhari, R. V. (1980). Three phase slurry reactors, *AIChE J.*, **26**, 179.

Choudhury, N. H. (1997). Improved predictive methods for solids suspension in agitated vessels at high solids loadings, Ph.D. dissertation, University of Arkansas, Fayetteville, AR.

# REFERENCES

Choudhury, N. H., W. R. Penney, K. Meyers, and J. B. Fasano (1995). An experimental investigation of solids suspension at high solids loadings in mechanically agitated vessels, *AIChE Symp. Ser.*, **305**(91), 131–138.

Cleaver, J. W., and B. Yates (1973). Mechanism of detachment of colloidal particles from a flat substrate in turbulent flow, *J. Colloid Interface Sci.*, **44**, 464.

Conti, R., and S. Sicardi (1982). Mass transfer from freely-suspended particles in stirred tanks, *Chem. Eng. Commun.*, **14**, 91.

Davis, R. H., and H. Gecol (1994). Hindered settling functions with no empirical parameters for polydisperse suspensions, *AIChE J.*, **40**, 570–575.

Doraiswamy, L. K., and M. M. Sharma (1984). *Heterogeneous Reactions: Analysis, Examples and Reactor Design*, Vol. 2, *Fluid–Fluid–Solid Reactions*, Wiley, New York, pp. 233–316.

Froment, G. F., and K. B. Bischoff (1990). *Chemical Reactor Analysis*, Wiley, New York.

Guerci, D., R. Conti, and S. Sicardi (1986). *Proc. International Colloquium on Mechanical Agitation,* ENSIGC, Toulouse, France, pp. 3–8 to 3–24.

Guiraud, P., J. Costes, and J. Bertrand (1997). Local measurements of fluid and particle velocities in a stirred suspension, *Chem. Eng. J.*, **68**, 75–86.

Hemrajani, R. R., et al. (1988). Suspending floating solids in stirred tanks: mixer design, scale-up and optimization, *Proc. 6th European Conference on Mixing*, Pavia Italy, May 24–26, pp. 259–265.

Hicks M. T., et al. (1993). Cloud height, fillet volume, and the effect of multiple impellers in solid suspension, presented at Mixing XIV, Santa Barbara, CA, June 20–25.

Hicks M. T., K. J. Myers, and A. Bakker (1997). Cloud height in solids suspension agitation, *Chem. Eng. Commun.*, **160**, 137–155.

Ibrahim, S. B., and A. W. Nienow (1994). The effect of viscosity on mixing pattern and solid suspension in stirred vessels, *Inst. Chem. Eng. Symp. Ser.*, **136**, 25–36.

Joosten, G. E. H., J. G. M. Schilder, and A. M. Broere (1977). The suspension of floating solids in stirred vessels, *Trans. Inst. Chem. Eng.*, **55**, 220.

Levins, D. M., and J. Glastonbury (1972a). Application of Kolmogoroff's theory to particle–liquid mass transfer in agitated vessels, *Chem. Eng. Sci.*, **27**, 537–542.

Levins, D. M., and J. Glastonbury (1972a). Particle–liquid hydrodynamics and mass transfer in a stirred vessel, *Trans. Inst. Chem Eng.*, **50**, 132–146.

Mak, A. T. C. (1992). Solid–liquid mixing in a mechanically agitated vessel, Ph.D. dissertation, University College–London.

Maude (1958). Cited in *Chemical Engineers' Handbook*, R. H. Perry and D. Green, eds., McGraw-Hill, New York, 1984, pp. 5–66.

Mullins, J. W. (1993). *Crystallization*, 3rd ed., Butterworth, London.

Nienow, A. W. (1968). Suspension of solid particles in turbine-agitated baffled vessels, *Chem. Eng. Sci.*, **23**, 1453.

Nienow, A. W. (1975). Agitated vessel particle–liquid mass transfer: a comparison between theories and data, *Chem. Eng. J.*, **9**, 153.

Nienow, A. W. (1985). The dispersion of solids in liquids, in *Mixing of Liquids by Mechanical Agitation*, J. J. Ulbrecht and G. K. Patterson, eds., Gordon & Breach, New York, pp. 273–307.

Nienow, A. W., and D. Miles (1978). The effect of impeller/tank configurations on fluid–particle mass transfer, *Chem. Eng. J.*, **15**, 13.

Oldshue, J. Y. (1983). Fluid mixing technology and practice, *Chem. Eng.*, June 13, pp. 83–108.

Oldshue, J. Y., and J. Sprague (1974). Theory of mixing, *Paint Varnish Prod.*, **3**, 19–28.

Parfitt, G. D. (1973). *Dispersion of Powders in Liquids*, 2nd ed., Applied Science Publishers, London.

Perry, R. H., and D. Green (1984). In *Chemical Engineers' Handbook*, R. H. Perry and D. Green, eds., McGraw-Hill, New York, pp. 5–63, 5–68.

Pettersson, M., and A. C. Rasmuson (1998). Hydrodynamics of suspensions agitated by a pitched-blade turbine, *AIChE J.*, **44**(3), 513–527.

Pettyjohn (1948). Cited in *Chemical Engineers' Handbook*, R. H. Perry and D. Green, eds., McGraw-Hill, New York, 1984, pp. 5–64.

Rao, K. S. M. S. R., V. B. Rewatkar, and J. B. Joshi (1988). Critical impeller speed for solid suspension in mechanically agitated contactors, *AIChE J.*, **34**(8), 1332.

Rieger, F., and P. Ditl (1994). Suspension of Solid Particles, *Chem. Eng. Sci.*, **49**(14), 2219–2227.

Shamlou, P. A., and A. Zolfagharian (1990). Suspension of solids in liquid-jet stirred vessels, *Fluid Mixing IV*, H. Benkreira, ed., Hemisphere Publishing, Washington, DC, pp. 365–377.

Siddiqui, H. (1993). Mixing technology for buoyant solids in a non-standard vessel, *AIChE J.*, **39**(3), 505.

Thring, R. W. (1990). An experimental investigation into the complete suspension of floating particles, *Ind. Eng. Chem.*, **29**, 676.

Zolfagharian, A. (1990). Solid suspension in rotary-stirred and in liquid-jet stirred vessels, Ph.D. dissertation, University College–London.

Zwietering, T. N. (1958). Suspending of solid particles in liquid by agitators, *Chem. Eng. Sci.*, **8**, 244.

# CHAPTER 11

# Gas–Liquid Mixing in Turbulent Systems

JOHN C. MIDDLETON
*BHR Group Ltd.*

JOHN M. SMITH
*University of Surrey*

## 11-1 INTRODUCTION

There are many processes in which gas–liquid contacting is important. Gas must be *effectively* and *efficiently* contacted with liquid to provide mass transfer (absorption or desorption; absorption of gas into liquid to produce a chemical reaction is often a particularly critical duty). Sometimes the gas merely provides energy (via buoyancy, level rise, bubble wakes, bubble coalescence, or gas expansion) for mixing the liquid.

Different contexts bring different challenges. Fermentations and effluent treatment can be at very large scale but the product value and workup tend to be comparatively low, so mixer capital and energy are important, whereas mass transfer requirements can be modest (fortunate if the microorganisms are shear sensitive). Gas–liquid reactions in low viscosity liquids

- Are often also at large scale
- Have reaction selectivity issues involving the dissolved gas concentration
- Have rapid reactions with large exotherms
- Involve subsequent processing producing comparatively valuable products

So for these, scale-up, liquid mixedness, and mass and heat transfer are important but impeller capital and energy cost are not. Chlorinations and sulfonations

---

*Handbook of Industrial Mixing: Science and Practice*, Edited by Edward L. Paul,
Victor A. Atiemo-Obeng, and Suzanne M. Kresta
ISBN 0-471-26919-0   Copyright © 2004 John Wiley & Sons, Inc.

**586** GAS–LIQUID MIXING IN TURBULENT SYSTEMS

tend to be fast reactions with soluble gases, so high mass transfer intensity with short contact time is efficient. With oxidations the gas is less soluble, but selectivity is often critical. Hydrogenations involve longer contact times, often with gas recycling (compression safety issues!) and solid particles to be kept in suspension.

### 11-1.1 New Approaches and New Developments

How is this chapter different from previous texts on gas–liquid mixing? First, it takes the viewpoint of a practitioner with the task of designing or scaling-up a process vessel, so the spectrum of information, conflicts, and priorities is always in view. There is mention of the more academic side when it helps to provide understanding and therefore confidence in the design methods. Second, there are reviews of some newer features, such as:

- The behavior of high-vapor-pressure systems, which may be either boiling or hot sparged
- The behavior at "high" (>0.08 m/s) superficial gas velocity (often found in industry, yet very little researched and very different from the usual regime reported in the literature)
- The extended range of impellers, including concave blade designs and up-pumping wide-blade hydrofoils
- The correlation of gas recirculation ratio and its value in calculating mass transfer driving force correctly

### 11-1.2 Scope of the Chapter

Table 11-1 lists many of the process considerations that will influence the selection of equipment for gas–liquid contacting operations. The equipment possibilities are

**Table 11-1** Process Factors Controlling the Selection of Gas–Liquid Contacting Equipment

---

Required residence time for either phase
Allowable pressure drop
Relative flow rates of gas and liquid
Need for countercurrent contact
Local mass transfer performance (dispersion size and turbulent mass transfer)
Need to supply or remove heat
Corrosion considerations
Presence of solid particles
Foaming behavior and phase separation
Relative importance of micromixing
Flow pattern requirements of reaction scheme
Interaction of reaction with mass transfer
Rheological behavior in laminar and transitional flow regimes

**Table 11-2**  General Classification of Gas–Liquid Reactors

---

Contactors in which the liquid flows as a thin film
  Packed columns
  Trickle bed reactors
  Thin-film reactors
  Rotating disk reactors
Contactors in which gas is dispersed into the liquid phase
  Plate columns (including control cycle reactors)
  Mechanically agitated reactors (principally stirred tanks)
  Bubble columns
  Packed bubble columns
  Sectionalized bubble columns
  Two-phase horizontal contactors
  Co-current pipeline reactors
  Coiled reactors
  Plunging jet reactors, ejectors
  Vortex reactors
Contactors in which liquid is dispersed in the gas phase
  Spray columns
  Venturi scrubbers

---

outlined in Table 11-2, with their main operational characteristics (at least when arising in air–water systems) presented in Table 11-3.

The emphasis is on providing practical advice, underpinned as much as possible by analysis of the basic mechanisms involved. We consider turbulent systems, concentrating on stirred vessels with "high-speed" agitators (i.e., not anchors or helical ribbons) and certain static mixers. Stirred vessels are very commonly used for gas–liquid reactions on account of their flexibility and good performance for mass and heat transfer, so much of this chapter is concerned with them.

Static mixers operating in turbulent flow can be useful where plug flow and/or higher intensity of mass transfer are required and their short contact time is acceptable. For some cases, other equipment is more suitable (see Figure 11-3); for example:

- Bubble columns (cheaper than stirred vessels; if modest mass transfer performance is acceptable) (Deckwer, 1992) and gas-lift recirculating columns
- Ejectors (Nagel et al., 1973; Zlokarnik, 1979) and plunging jets (van de Sande and Smith, 1973, 1974; Bin and Smith, 1982)
- Sprays (low liquid hold-up)
- Packed towers and plate columns (countercurrent flow)
- In-line rotor–stator mixers (for high viscosity liquids)

The topics covered in this chapter include mass transfer, liquid mixedness, liquid and gas flow patterns and residence time distribution, gas fraction ("gas hold-up"), and impeller power demand. Bubble size is important in all these

**Table 11-3** Characteristics of Gas–Liquid Contacting Equipment[a]

| Type of Absorber | Typical Gas Velocity $\times 10^2$ (m/s) | Residence Time Distribution Gas | Residence Time Distribution Liquid | Residence Time of Liquid | Fractional Liquid Holdup | $k_L \times 10^4$ (m/s) | $a$ (m$^2$/m$^3$) | $k_L a \times 10^2$ (s$^{-1}$) |
|---|---|---|---|---|---|---|---|---|
| Film type: packed column and trickle bed reactors | 10–100 | Plug | Plug | Very low | 0.05–0.1 | 0.3–2 | 2–35 | 0.06–7 |
| With gas dispersed as bubbles in liquids | | | | | | | | |
| Bubble columns | 1–30 | Plug | Mixed | Unlimited | 0.6–0.8 | 1–4 | 2.5–100 | 0.25–40 |
| Packed bubble columns | 1–20 | Plug | Mixed | Unlimited | 0.5–0.7 | 1–4 | 10–30 | 1–12 |
| Bubble cap plate columns | 50–200 | Plug | Mixed | Unlimited | 0.7–07 | 1–4 | 10–40 | 1–16 |
| Plate columns without downcomers | 50–300 | Plug | Mixed | Limited variation | 0.5–0.5 | 1–4 | 10–20 | 1–8 |
| Mechanically agitated contactors | 0.1–2 | Mixed | Mixed | Unlimited | 0.5–0.8 | 1–5 | 20–100 | 2–50 |
| Horizontal pipeline contactors | 5–300 | Plug | Plug | Low | 0.1–0.8 | 2–6 | 10–40 | 2–24 |
| Static mixers | 0.05–20 | Plug | Plug | Low | 0.01–0.99 | 1–20 | 10–100 | 10–200 |
| With liquid dispersed in gas | | | | | | | | |
| Spray columns | 5–300 | Mixed | Plug | Very low | — | 0.5–1.5 | 2–15 | 0.1–2.25 |
| Sieve plate in spray regime | 100–300 | Plug | Mixed | Unlimited | — | 1–3 | 5–20 | 0.5–6 |

[a] These are comparative values only, based on air and water. See also Table 13-9.

aspects of gas–liquid mixing, so some remarks on breakup and coalescence are also included, partly to illustrate the difficulty of providing accurate design correlations. Heat transfer is often important, and although in many cases the agitation required for gas dispersion is more than adequate to satisfy the demands of heat transfer, there is insufficient information on the effects of gassing to include a worthwhile discussion in this chapter.

This chapter covers only processes with low viscosity liquids: those in which turbulent or near-turbulent flow is achievable in practice. For stirred vessels this implies an impeller Reynolds number $ND^2\rho/\mu > \sim 10^4$, or for static mixers and the like, a Reynolds number $UD\rho/\mu > \sim 3000$. The dispersion of gases into viscous liquids is a different problem and largely outside the scope of this chapter. In such fluids the dispersion action is best achieved by elongating and folding the gas into the liquid, a principle that is exploited in a variety of beaters and rollers exemplified by those empirically developed for the food-processing industry over the last 2000 years. Rotor stator devices can also be used as a means of bringing these high viscosity fluids into turbulent motion, but the high viscosities and small clearances involved make this difficult to achieve. Various static mixers can be reasonably successful in achieving dispersion, notably the various Sulzer or Koch SMX designs.

It is worth mentioning that in many processes the avoidance of air entrainment and/or the removal of bubbles from viscous liquid is a greater problem. Lowering the pressure to increase bubble volume and reducing the liquid viscosity by heating and/or spreading the liquid into thin films are probably the most generally used techniques to de-gas viscous fluids.

Computational fluid dynamics (CFD) is now quite well established as a tool for modeling mixing processes with single-phase systems, but its success in predicting multiphase coalescing or dispersing flows has hitherto been limited. A brief overview in the context of the modeling of gas–liquid systems has been included in Section 11-3.1.

### 11-1.3 Gas–Liquid Mixing Process Objectives and Mechanisms

*11-1.3.1 Turbulent Mechanisms.* The processes of liquid mixing, generation of interface area, and gas–liquid mass transfer in turbulent systems are controlled primarily by the *power* dissipated in the fluids and the *gas volume fraction* $\phi$. The power (together with the fluid properties) influences the *bubble size*. The gas is broken up into a dispersion of bubbles in a high-shear zone such as at the discharge from the sparger holes in a bubble column, the impeller tips in an agitated vessel, or the gas inlet and wall-shear zones in a static mixer. It is the power dissipated in that zone which controls the bubble breakup process. However, with agitated vessels the design correlations are commonly based on the average energy dissipation per unit mass in the vessel, $P/\rho V$. The power in this expression is the sum of the shaft power and the (principally potential) energy introduced as a result of injecting the gas at depth (Middleton et al., 1994). It may be noted that the ratio of local to average energy dissipation rates can be large and will differ between impeller types.

**590** GAS–LIQUID MIXING IN TURBULENT SYSTEMS

The bubbles may or may not subsequently recoalesce to some extent, depending on the local fluid dynamics and the interfacial behavior. The unpredictability of this rules out a priori prediction of bubble size and interface area in general, so design via scale-up from experiments is preferred. The *gas fraction* in an agitated vessel is determined by the bubble size and the degree of bubble recirculation [itself a function of agitation, bubble size, and scale (Middleton, 1997)]. For a static mixer, ϕ is largely set by the ratio of the average gas flow to liquid flow, but with corrections for bubble "slip," which depend on flow orientation and the bubble size.

***11-1.3.2 Factors Influencing the Power.*** In a given baffled agitated vessel, with given fluid properties, the independent variables controlling P and ϕ are the impeller type, impeller diameter, impeller speed, and gas rate. However, the gas rate for a process is usually set by the process flow sheet, that is, by the stoichiometry and the required inlet and outlet gas compositions (or absorption efficiency), so the contribution of gas buoyancy to the total energy dissipation rate is fixed. Calculation of the other (usually main) contribution, being the impeller power input per unit mass, $P/\rho V$, is well established for single-phase systems. For some (unfortunately, still common) impeller types such as the Rushton disk turbine and the downflow pitched blade turbine, the impeller power draw is greatly reduced when gas is introduced. The power draw is affected by the degree of gas recirculation and to some extent by the detailed geometry of the equipment. Modern gas–liquid impellers, such as the concave-blade disk turbines (Scaba SRGT, Chemineer BT6 and CD6 impellers, and the Lightnin R130; sample shown in Figure 11-5) and the up-pumping wide-blade hydrofoils (Lightnin A345, Prochem MaxfloW, APV B6; sample shown in Figure 11-5), maintain more than 70% of their ungassed power draw on gassing.

For an inline mixer, a value for the specific power $(P/\rho V)$ can easily be estimated from the manufacturer's (or measured) friction factor, adjusted for the gas–liquid ratio using the correction of Lockhart and Martinelli (1944). Although this may not be rigorously applicable, some success has been achieved by applying the approach to static mixers using the laminar gas–turbulent liquid regime factors.

***11-1.3.3 Liquid Mixing.*** The bulk circulation is the rate-determining step for *liquid mixing* (*blending*) in stirred vessels. The turbulence ensures that mixing on smaller scales (mesomixing and micromixing) is comparatively fast. (Note, however, that extremely fast reactions can be even faster than the micromixing.) Again, the gas affects this. At modest gas rates, the gas affects the intensity of liquid mixing because of its effect on the impeller power, and its location because of changes to the flow field. At high gas fractions, presumably, the gas buoyancy must contribute.

***11-1.3.4 Gas–Liquid Mass Transfer.*** Good *mass transfer* performance requires large interface area between gas and liquid (resulting directly from small bubble size and high gas fraction, given the fixed gas rate) and a high mass transfer coefficient (associated with local levels of turbulence). A high gas fraction is not always desirable since the profitability of a reactor is largely controlled by the quantity of liquid it contains. Excessive gas retention may also lead to overreaction. It is only necessary to allow enough time for the required mass transfer.

**11-1.3.5 Heat Transfer**. *Heat transfer* in the turbulent regime is essentially a macromixing process. Heat transfer coefficients are controlled by the turbulence levels (hence boundary layer thickness) near the heat transfer surfaces. In many cases the process demands of suspension or dispersion and mass transfer are more than sufficient to ensure adequate heat transfer.

**11-1.3.6 Solid Particles**. *Particle suspension* from the base and *drawdown* from the surface are often required in gas–liquid agitated vessels and are influenced in a complex manner by gassing. There are no well-established correlations for the influence of gas. Particle suspension is probably controlled by the energy and frequency of turbulent bursts, and drawdown by details of local flow patterns and vorticity at the surface, both of which could be expected to be affected by the presence of gas bubbles.

**11-1.3.7 Flow Patterns**. Flow patterns can be important. A "slow" reaction scheme (occurring in the bulk liquid) with competing steps may exhibit selectivity dependent on the local concentration of a liquid or dissolved gas reactant. In this case the liquid flow pattern (i.e., whether the liquid undergoes backmixing or plug flow or, as is almost always the case, somewhere in between) is important. A "fast" reaction scheme (occurring mainly near the gas–liquid interface) with dependence of selectivity on local dissolved gas concentration, will be sensitive to the history of gas concentration in the bubbles as they travel through the reactor. In other words, the selectivity will be sensitive to the degree of backmixing of the gas phase, and therefore to the bubble flow pattern. Even for simple gas–liquid mass transfer, the gas flow pattern is critical unless a very small proportion of the dissolvable gas is absorbed per pass. For example, if 95% of the inlet dissolvable gas is absorbed, its mean concentration in the gas phase (and hence its mean transfer rate to the liquid), if in plug flow, is 5.17 times that for an perfectly backmixed gas phase.

Here a conflict can arise in an agitated vessel. High power input per unit mass is required to enhance mass transfer area and heat transfer coefficient, but this will result in a high degree of gas recirculation, reducing the mean gas phase concentration "driving force" for mass transfer. Local shear rates will also increase with power input. The balance will vary with scale.

## 11-2 SELECTION AND CONFIGURATION OF GAS–LIQUID EQUIPMENT

Tables 11-1, 11-2, and 11-3 give an indication of the aspects to be considered in this section which gives a procedure for defining the components of gas–liquid mixing equipment. The procedure applies only to low viscosity liquids in which turbulent flow can be achieved. If it is not clear whether it is practical to achieve turbulent flow, an outline design will be useful. For example, if an agitated vessel is to be used (see below), take a typical power number (e.g., 0.8 if a Lightnin A345 upflow hydrofoil is to be used, or 5.0 for a six-blade Rushton turbine) and an impeller diameter of 0.4T (or 0.33T), where T is the intended vessel diameter, and calculate the speed N required to provide a specific power input of, say, 2 kW m$^{-3}$ (see Section 11-1.4.2). The Reynolds number can then be calculated and compared with that required to give turbulent mixing. If the fluid is non-Newtonian, an appropriate viscosity will be that at a shear rate of about 10 times N, the agitator speed (Metzner and Otto, 1957; see Section 9-3). Skelland (1967) gives a table of the constant for a number of impeller types.

First the *gas entry method* can be decided. With a vessel, the gas is preferably *sparged* in through a dip pipe discharging (preferably via a sparge

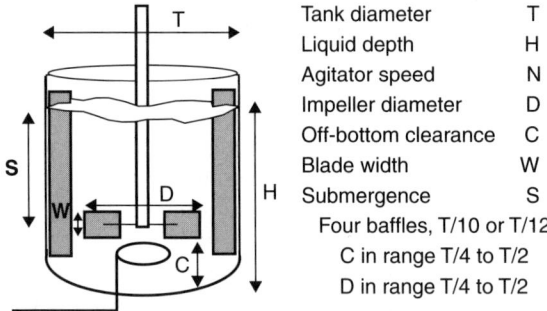

**Figure 11-1** Standard vessel geometry (single impeller, H ~ T).

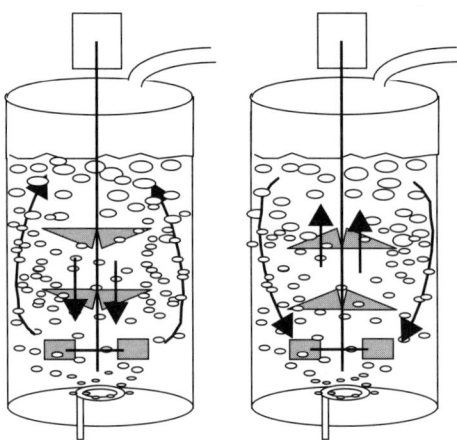

**Figure 11-2** Multiple-impeller agitators, down- and up-pumping hydrofoils above a radial dispersing impeller.

ring of diameter less than the impeller diameter) underneath the impeller (see Figures 11-1 and 11-2). This ensures that the gas has a good chance of being dispersed into fine bubbles by the impeller, providing a high gas–liquid contact area. For in-line mixing, gas will generally be fed to the inlet of a static mixer (see Figure 11-3), preferably via an axially positioned feed pipe or, with larger mixer diameters, via a multipoint distributor.

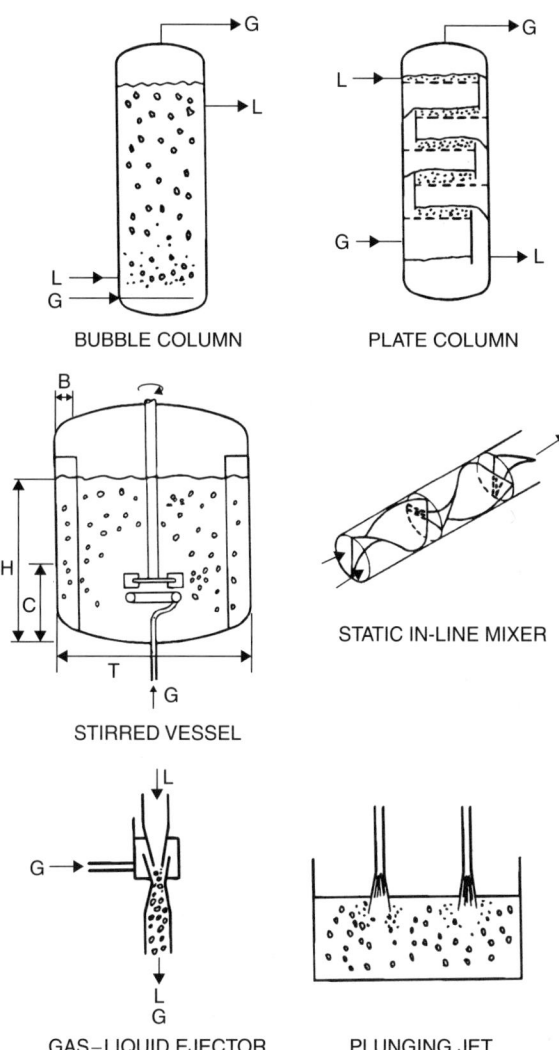

**Figure 11-3** Gas–liquid contacting equipment for low viscosity liquids. (From Middleton, 1997; reproduced by permission of Butterworth–Heinemann.) An illustration of gas–liquid contacting is included on the Visual Mixing CD affixed to the back cover of the book.

**594**   GAS–LIQUID MIXING IN TURBULENT SYSTEMS

In some cases, sufficient gas pressure may not be available (e.g., if avoiding the dangers of compressing hydrogen) and the gas can be drawn in by means of the energy in the liquid flow. In a vessel, gas is drawn down from the headspace using, preferably, a proprietary *self-inducing agitator*, which draws gas down a hollow shaft to the impeller (see Figure 11-4). An impeller near the surface is sometimes used to draw in gas, although this arrangement can be unstable and very sensitive to small changes of level. For in-line mixing, an *ejector*, in which gas is sucked in and dispersed by entrainment into a liquid jet, may well be chosen (see Figure 11-3).

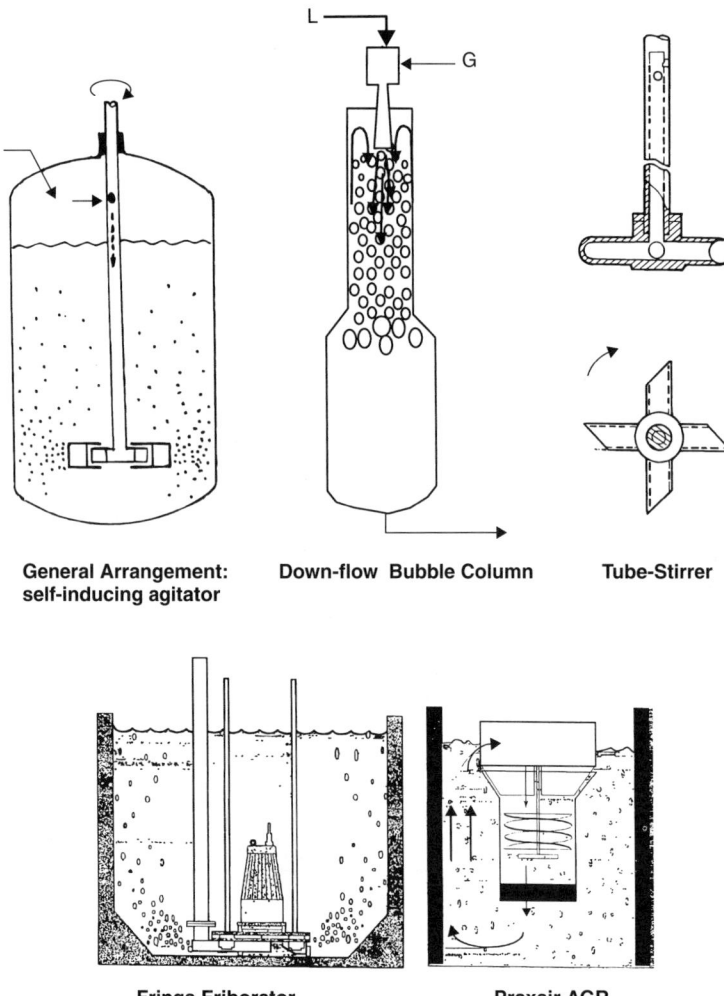

**Figure 11-4**  Self-inducing gas–liquid equipment. (Part from Middleton, 1997; reproduced by permission of Butterworth–Heinemann.)

## 11-2.1 Sparged Systems

The next choice concerns the intensity of mass transfer and turbulence required. For a first selection, three levels can be defined [see Middleton (1997, Sec. 15.1) for more detail, and also Section 11-6.3 for reacting systems]:

1. *Low intensity:* $k_L a$ values (air–water equivalent) of order $0.005$ s$^{-1}$; for slow reactions, without a severe particle suspension or heat transfer duty. Large liquid volume is required since the reaction occurs throughout the liquid phase. Here a *bubble column* should be considered: possibly with packing to enhance the plug flow characteristics of the gas. Where it is appropriate to enhance the driving force for mass transfer by using countercurrent flow, or if the liquid needs to be nearer plug flow, a *plate column* may be selected. To meet low cost and intensity requirements when liquid flow pattern is not an issue, *plunging jets* could be considered. See Figure 11-3 and Table 11-3.

2. *Moderate intensity:* $k_L a$ of order $0.05$ s$^{-1}$; for fast reactions with other slower steps; where particle suspension and/or heat transfer require enhancement. *Agitated vessels* are useful here, and indeed are often selected where the intensity needs are uncertain, or may vary widely (as in general-purpose reactors). The larger top surface area per unit volume than can be achieved with bubble columns allows higher exit gas flow rates without liquid entrainment and carryover.

3. *High intensity:* $k_L a$ of order $0.5$ s$^{-1}$; for very fast reactions and short residence times: *Static mixers in turbulent flow* offer plug flow in both phases. *Thin-film contactors* such as wiped-film columns or spinning disks offer large surface per unit volume, giving very rapid mass transfer and evaporative flux.

## 11-2.2 Self-Inducers

A variety of *surface aerators* are available that entrain gas into a liquid surface, but these are generally applicable only in the wastewater treatment area. The simplest self-inducer for an *agitated vessel* is an impeller located near the surface, sometimes with the upper part of the baffles removed so as to encourage the formation of a surface vortex. This is, however, a sensitive and unstable arrangement. It is better, although probably more expensive, to use a self-inducing impeller system in which gas is drawn down a hollow shaft to the low-pressure region behind the blades of a suitable, often shrouded impeller (see Figure 11-4). Various proprietary designs are available, such as the Ekato gasjet Praxair AGR and the Frings Friborator (see Figure 11-4).

Self-inducing impellers are not generally successful for drawing gas down to depths greater than about 2.5 m. Success of scale-up while changing to an undersparged system will be uncertain. With limited pressure differences across the orifices there is a potential danger of plugging when operating in systems liable to cause reactor fouling.

In either case the achievable gas flow rates and gas penetration depths are limited, so large scale units may not be very successful. Scale-up will normally be on the basis of maintaining a given impeller Froude number, and as the equipment becomes larger, this will inevitably result in operation at very high specific power input levels. Performance can be sufficient for some fermentations and hydrogenations but is generally insufficient to satisfy the demands of higher intensity reactions.

Higher intensity self-induction can be achieved by an *ejector* (or *eductor*), in which a liquid stream (either the feed stream or the circulating stream of a loop reactor) is used to draw in gas and disperse it with high $k_L a$. The loop also usually contains a pump, heat exchanger, and a gas disengagement space. In the special case of total absorption, where there is no exit gas, a *downflow bubble column* may be suitable (Figure 11-4): Gas and liquid flow in at the top and the gas is dispersed, perhaps using an ejector. The bubbles are held in the downflow liquid stream until they disappear.

### 11-2.3 Recommendations for Agitated Vessels

Since agitated vessels are so common, it is worth noting some points arising from recent work that lead to recommended designs for turbulent systems. Most of this work has been with sparged systems, but the remarks on impeller blade shapes may also apply to self-inducers.

*11-2.3.1 Sparged Stirred Vessel Geometry.* As mentioned above, the gas should be fed beneath the impeller such that the impeller will "capture" the rising gas plume. With radial or upward flow impellers it is sufficient to use a sparger that has a smaller diameter than the impeller itself (a ring sparger of diameter about 0.75D is recommended). To provide the maximum gas contact time, the impeller should be near the base of the vessel but not so near as to inhibit its liquid pumping action: a clearance of T/4 is recommended. The bubble breakup mechanism relies on a high relative velocity between the blades and the liquid, so wall baffles are necessary to restrict the circumferential motion of the liquid. They also enhance the vertical motion of the liquid and hence the mixing of the liquid bulk and the recirculation of liquid and gas back to the impeller, increasing the gas hold-up. For any single impeller this recirculation is favored by an aspect ratio liquid height/vessel diameter $\equiv H/T$ of about 1. All of these factors lead to a recommended geometry, which is illustrated in Figure 11-1.

A vessel of larger aspect ratio may be required, for example, to:

- Obtain more wall surface for heat transfer
- Provide a longer contact time for the gas
- Give a staged countercurrent system
- Circumvent a mechanical limitation on available vessel diameter

In this case, more than one impeller will be required (see Section 11-2.3.3).

## SELECTION AND CONFIGURATION OF GAS–LIQUID EQUIPMENT 597

***11-2.3.2 Impeller Type.*** An impeller that approximately maintains the ungassed power level when gas is introduced will give more stable operation and minimal scale-up difficulties. Recommended types (Figure 11-5) include, for radial flow, hollow-blade designs such as the Scaba SRGT, Chemineer CD6 or BT6, Lightnin R130, or for axial flow, an upward-pumping wide-blade hydrofoil such as the Lightnin A345 or A340 or the Prochem-Chemineer MaxfloW. Downflow hydrofoils or pitched blade turbines may be unstable during gas–liquid operation (Chapman et al., 1983; Nienow et al., 1986; Hari-Prajitno et al., 1998). The liquid flow induced by a downpumping impeller is opposed to the natural tendency of buoyant gas to rise. With a single impeller this is evidenced in the transition between indirect and direct loading that occurs as the gas flow is increased (Warmoeskerken et al., 1984). At certain impeller speeds there may be an accumulation of gas below the impeller plane which can become hydrodynamically unstable. These physical phenomena, which are independent of scale, have been found, within the authors' experience, to lead to an unpredictable loading of the impeller and a source of mechanical problems (see Section 11-4.2).

A single upflow hydrofoil may not be optimum in a vessel with H = T, if the D/T ratio is larger than say 0.5 (which may occur if high P/$\rho$V is required), since recirculation will be localized and zones of high local gas fraction will be formed.[1]

***11-2.3.3 Multiple Impellers.*** In vessels taller than H/T = 1.2, or when Reynolds numbers are below about 5000, additional impellers may be required. These would improve the liquid mixing, but also, especially in the heterogeneous

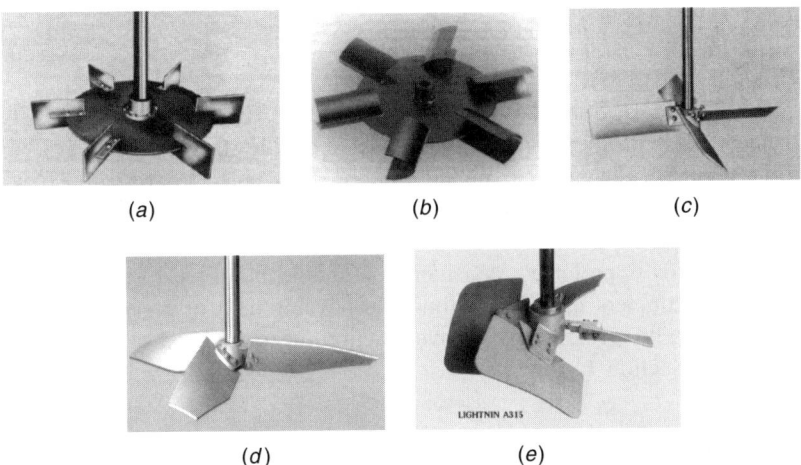

**Figure 11-5** Various impellers: (*a*) Rushton disk turbine; (*b*) hollow-blade turbine; (*c*) pitched blade turbine; (*d*) narrow-blade hydrofoil; (*e*) wide-blade hydrofoil.

---

[1] *Editors' note:* The question of up-pumping versus down-pumping axial impellers for gas–liquid operation is still under active investigation.

regime or at high gas velocities, will help to redisperse and redistribute gas from the large bubbles which otherwise tend to bypass the impellers. Generally, spacing between impellers should be larger than their diameter D; otherwise, the flow patterns will interact and the power dissipated by the combined impellers will be less than the sum of the individuals. Multiple radial impellers tend to generate zoned or compartmentalized flow fields, in contrast with the better top-to-bottom circulation generated by multiple axial flow configurations. A combination of a radial flow impeller to produce dispersion together with one or more axial flow impellers is often recommended. Many operators use upward-pumping wide-blade hydrofoils (D/T approximately 0.6) even though there is a tendency for these to develop regions of very high gas fraction in the upper part of the vessel (Smith et al., 2001b).

### 11-2.3.4 High Gas Velocities.
In high gas velocity systems (superficial gas velocity >0.02 to 0.03 m/s, the lower value referring to lower N), gas fraction and mass transfer do not increase with impeller power as might be expected, and much of the gas flows through as large bubbles (Gezork et al., 2000). This is the heterogeneous regime (see Section 11-3.1).

### 11-2.3.5 Boiling (Nonsparged) Systems.
Although purely boiling systems are not very common, they do arise in certain polymerizations (e.g., propylene), liquid-phase exothermic reactions, and evaporative crystallization (e.g., sugar, salt). To avoid cavitation and maintain known impeller performance, impellers such as the axial flow A315u or the radial flow BT6 should be selected. These are suitable for single impeller installations as well as for the uppermost impeller of multiple impeller agitators (see Section 11-4.3).

### 11-2.3.6 Near-Boiling Gas Sparged Systems.
Gas sparged or gas evolving hot systems pose different problems. Ventilated cavities (see Section 11-3.1) will almost inevitably develop, so impellers should be selected from those which maintain the power input level on gassing. Again, deep hollow-blade radial flow impellers or upward-pumping wide-blade hydrofoils are suitable. If a multiple-impeller agitator is preferred, consideration should be given to using impellers of differing diameters in order to limit the development of zones of very high void fraction, which might lead to overreaction, near the level of the uppermost impeller (see Section 11-4.4).

### 11-2.3.7 Other Points.
In a three-phase reactor it is necessary to ensure that the requirements of solid suspension and gas dispersion are separately satisfied. Liquid macromixing may be as much a limitation as gas–liquid mass transfer, especially in larger gas–liquid reactors. A model comparing the kinetics of the uptake of the dissolved gas by the reaction with the supply rate via the liquid from a bubble will be useful.

## 11-3 FLOW PATTERNS AND OPERATING REGIMES

Characterization of the flow pattern of either phase is often limited to the ideals of perfect plug flow or fully backmixed flow (see Chapter 1). In practice, it is necessary to consider degrees in between: many in-line mixers such as ejectors and static mixers in turbulent flow achieve a close approximation to plug flow for both phases, but in industrial agitated vessels a close approach to complete backmixing is rare for either phase. If gas–liquid mass transfer is the process rate-controlling step, the flow pattern of the gas is important: Typically, it has a very great effect on the rate of mass transfer, as illustrated in Section 11-3.1. If the limiting step is reaction in the bulk liquid phase, the liquid-phase flow pattern (residence time distribution if continuous flow) may be important (see Section 11-1.3.7).

For batch systems a stirred vessel or loop reactor with an in-line mixer is used. Where plug flow is required, for long residence times a cascade of stirred vessels or loop reactors is commonly used, and for short residence times the choice will often be a static mixer or ejectors. For continuous flow systems requiring an approach to backmixed flow, stirred vessels or loop reactors are indicated.

### 11-3.1 Stirred Vessels: Gas Flow Patterns

In the *homogeneous regime* in an agitated vessel, the superficial gas velocity, $v_S < 0.02$ to $0.03$ m/s (lower value for lower N), and the bubbles have a monomodal size distribution with a small mean size, generally between 0.5 and 4 mm. Here, the impeller controls the flow pattern and bubble size. At higher gas superficial velocities, the *heterogeneous regime* occurs (Gezork et al., 2000), in which the bubble size distribution is bimodal, with some large bubbles (say 10 mm or greater), and is controlled more by the gas velocity (possibly void fraction) than by the agitator. In this regime the influences of impeller speed and gas rate are different from those in the homogeneous regime, as will be seen in Sections 11-4 and 11-5.

Gas flow pattern is important. It controls the degree of recirculation and backmixing of the gas phase, which in turn determines the mean concentration driving force for mass transfer. It can also profoundly affect the liquid-phase macrocirculation and homogenization. One way to quantify the gas backmixing is to use the *recirculation ratio*, α (van't Riet, 1976), defined as the ratio of the gas flow recirculated to the impeller to that sparged. Since in the homogeneous regime gas is mixed with other gas only at the impeller, α represents the degree of backmixing of the gas. This implies that there is little coalescence in the bulk of the two-phase mixture in the reactor. In large scale equipment (larger than about 1 m³) liquid velocities are usually less than in small scale vessels, so even when the gas distribution is described as homogeneous (e.g., monomodal in size distribution), it is unusual for much gas to be recirculated below the level of the (bottom) impeller.

For a standard baffled agitated vessel with $H = T$ and a single *six-flat-blade disk turbine* of $D/T = 0.3$ to $0.5$ operating within the range $PT/V = 500$ to

5000 Wm$^{-2}$ and $v_S = 0.005$ to 0.04 ms$^{-1}$ (i.e., in the homogeneous regime), a correlation for the degree of gas recirculation, α, in terms of power per unit volume is proposed (Middleton, 1997):

$$\alpha = c \left(\frac{PT}{V}\right)^{1.42} \quad (11\text{-}1)$$

where c is a constant equal to $18 \times 10^{-6}$ for water (a *coalescing* system) or $21 \times 10^{-6}$ for ionic solutions (*noncoalescing*), with P in watts, T in meters, and V in cubic meters.

[Note that this is empirical and is not dimensionally consistent, but it will give a guide for other systems, using the water value for liquids without surface active solutes. It covers the useful regimes above the loading point (see below).] No correlations are available for other impeller types and vessel configurations, but CFD may be used to calculate gas recirculation, using a suitable estimate or measured value for the mean bubble size.

α is used in mass transfer calculations to estimate the overall mean concentration driving force, as follows: If ΔC is the mean mass transfer driving force (C* − C$_L$), where C* is the equilibrium dissolved gas concentration at the gas–liquid interface and C$_L$ is the bulk dissolved gas concentration, the mean driving force for the vessel is given approximately by

$$\Delta C = \frac{\Delta C_{IN} - \Delta C_{OUT}}{(\alpha + 1) \ln[(\Delta C_{IN} + \alpha \Delta C_{OUT})/(\alpha + 1)\Delta C_{OUT}]} \quad (11\text{-}2)$$

(For α in the range 0.1 to 10, this gives values about 10 to 20% low.)

The flow pattern of the gas depends on the regime of gas–impeller interaction. For six-blade disk-turbine impellers, three regimes of flow in the vessel can be defined, as shown in Figure 11-6:

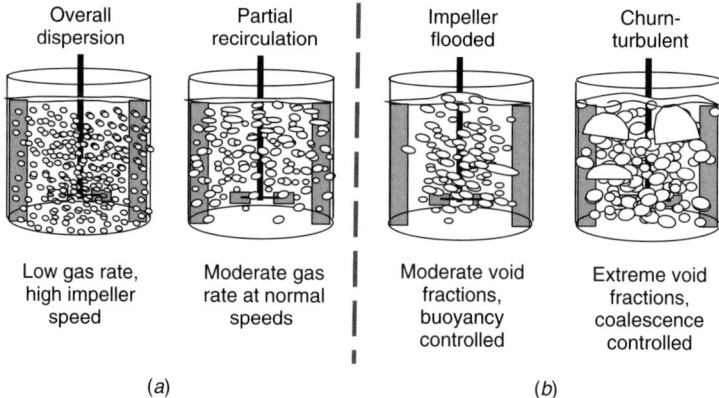

**Figure 11-6** Typical void fraction distributions in vessels with a single impeller: (*a*) impeller-controlled regimes; (*b*) void fraction-controlled regime.

1. *Flooding* in which the impeller is overwhelmed by gas and gas–liquid contact; mixing, and so on, are very poor
2. *Loading* in which the impeller disperses the gas through the upper part of the vessel
3. *Complete dispersion* in which gas bubbles are distributed throughout the vessel and significant gas is recirculated back to the impeller

These are closely related to the regimes of gas–impeller interaction: As more gas is fed to the impeller (or speed diminishes), there is more tendency for gas to be accumulated in the low-pressure regions behind the blades, forming ventilated "cavities." When these are large they can cause a profound reduction in the power number of the impeller (related to their obstruction of the liquid discharge from the impeller) (see Figure 11-7 and Section 11-4.2) and hence in its performance for mixing, mass, and heat transfer. This is particularly important for flat-blade turbines with four, six, or eight blades. For six-blade disk turbines the cavity regime is best obtained from the flow regime maps of Warmoeskerken and Smith (1986) (Figure 11-8) [also summarized in Middleton (1997)] since they are dimensionless and tested for several scales.

However, it should be noted that the published maps for disk turbines refer only to impellers with $D = 0.4T$; for other ratios the regime boundaries should be adjusted using the appropriate correlations given below. The transitions between the various regimes generated by a gassed Rushton turbine can be characterized with the main dimensionless numbers, the gas flow number ($Fl_G = Q_G/ND^3$), the impeller Froude number ($Fr = N^2D/g$), and the geometry ($D/T$) (Smith et al., 1987):

1. Below a certain minimum speed, the impeller has no discernible action. This is approximately when

$$Fr < 0.04 \qquad (11\text{-}3)$$

2. The gas flow will swamp the impeller (flooding) if

$$Fl_G > 30 Fr \left(\frac{D}{T}\right)^{3.5} \qquad (11\text{-}4)$$

3. Large cavities are developed by a Rushton turbine when

$$Fl_G > \sim 0.025 \left(\frac{D}{T}\right)^{-0.5} \qquad (11\text{-}5)$$

The constant in this expression has a weak dependence (to the power of about 0.2) on the scale of the equipment.

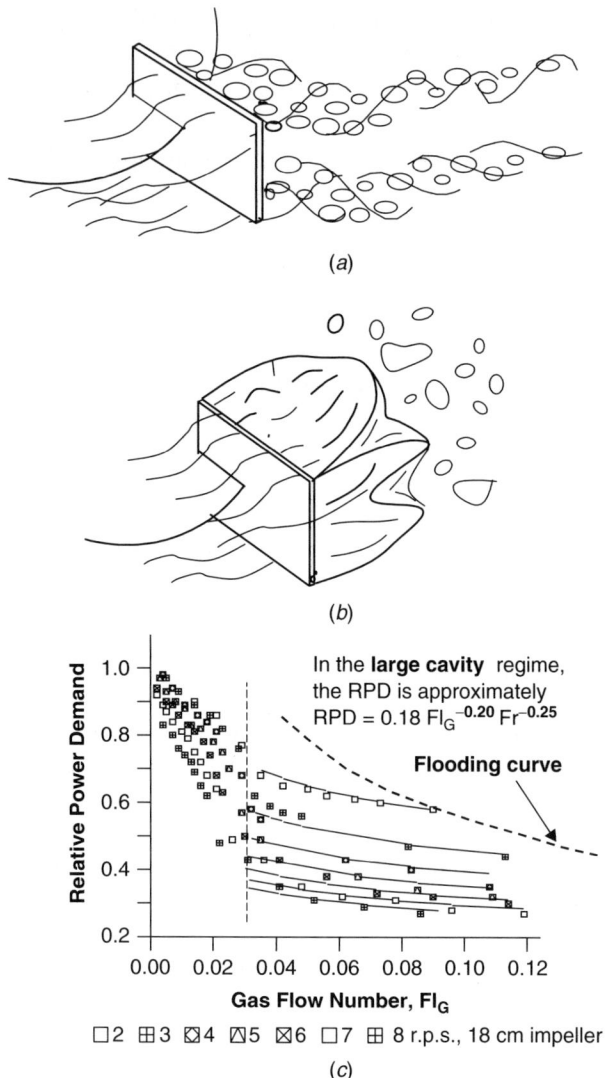

**Figure 11-7** (a) and (b) ventilated gas cavity forms [(a) vortex cavities; (b) large cavity] on turbine blades and (c) relative power demand for a gassed Rushton turbine (D/T = 0.4). (Data from Warmoeskerken et al., 1982.)

4. Nienow et al. (1977) developed a relationship for the speed of a Rushton turbine that would recirculate a given gas rate which can be reformulated and expressed as

$$Fl_G < 13 Fr^2 \left(\frac{D}{T}\right)^{5.0} \tag{11-6}$$

# FLOW PATTERNS AND OPERATING REGIMES

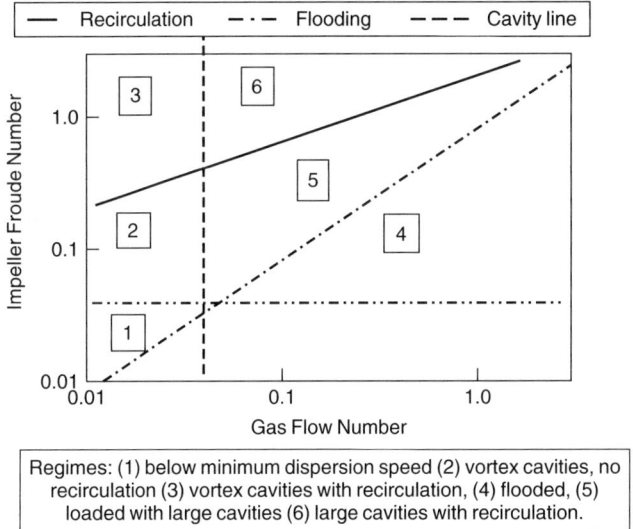

Regimes: (1) below minimum dispersion speed (2) vortex cavities, no recirculation (3) vortex cavities with recirculation, (4) flooded, (5) loaded with large cavities (6) large cavities with recirculation.

**Figure 11-8** Flow map for single Rushton turbine ($T/D = 2.5$).

These equations allow us to predict the operating conditions in any equipment. In the large-cavity regime of gassed aqueous systems, a good approximation for the gassed power of a single Rushton turbine ($D/T = 0.4$) is given by

$$\text{RPD} = \frac{P_G}{P_U} = 0.18 \text{Fl}_G^{-0.20} \text{Fr}^{-0.25} \qquad (11\text{-}7)$$

and lines corresponding to this equation can easily be added to the flow map. A similar map has been produced for a concave-blade impeller similar to the CD6 (Warmoeskerken, and Smith, 1989). It should be pointed out that non-Newtonian systems behave differently at transitional Reynolds numbers [see Middleton (1997) for a brief summary].

With axial flow impellers in down-pumping mode, two important regimes are identified: direct and indirect loading (Warmoeskerken et al., 1984) (Figure 11-9). At lower gas rates and higher impeller speeds, the downflow from the impeller dominates and gas enters the impeller from above; this is known as *indirect loading*. If the gas buoyancy dominates, the gas loads the impeller directly, and the impeller now pumps radially with much diminished power number (see Figure 11-10 and Section 11-4.2). Operation near the transition is to be avoided since the regime can flip unstably, giving rise to serious mechanical and operational problems. It is preferable to avoid this possibility altogether by operating in upward-pumping mode. Here the gas and liquid flows are not in conflict, and the power curve with gassing is stable and much flatter (Figure 11-11).

**604** GAS–LIQUID MIXING IN TURBULENT SYSTEMS

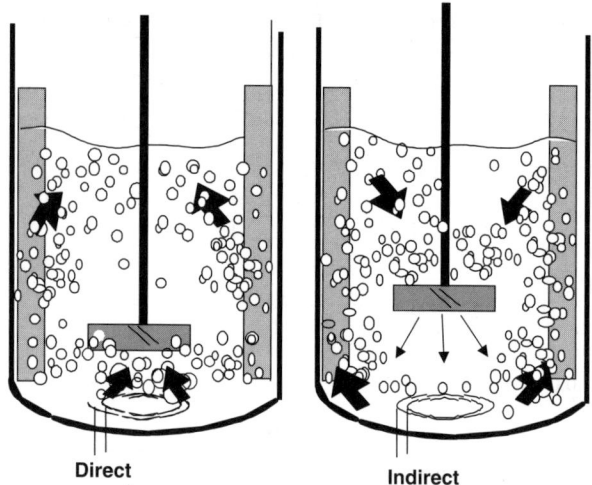

**Figure 11-9** Direct and indirect loading of a downward-pumping axial flow impeller.

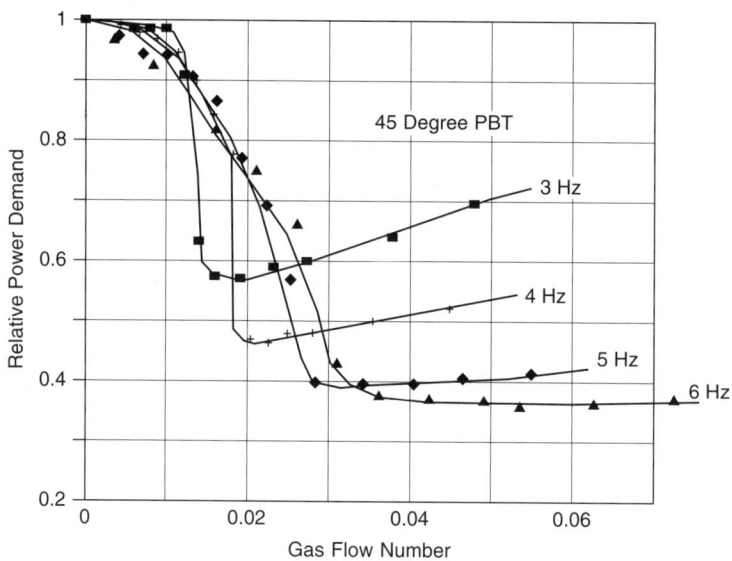

**Figure 11-10** RPD For a down-pumping 45° pitched blade turbine. (From Warmoeskerken et al., 1984.)

*11-3.1.1 Flow Computation.* As was remarked in Section 11-1.2, CFD is now quite well established as a tool for modeling mixing processes in single-phase systems, although the currently popular Reynolds-averaging models using the k–ε turbulence model are not appropriate for the local turbulence conditions

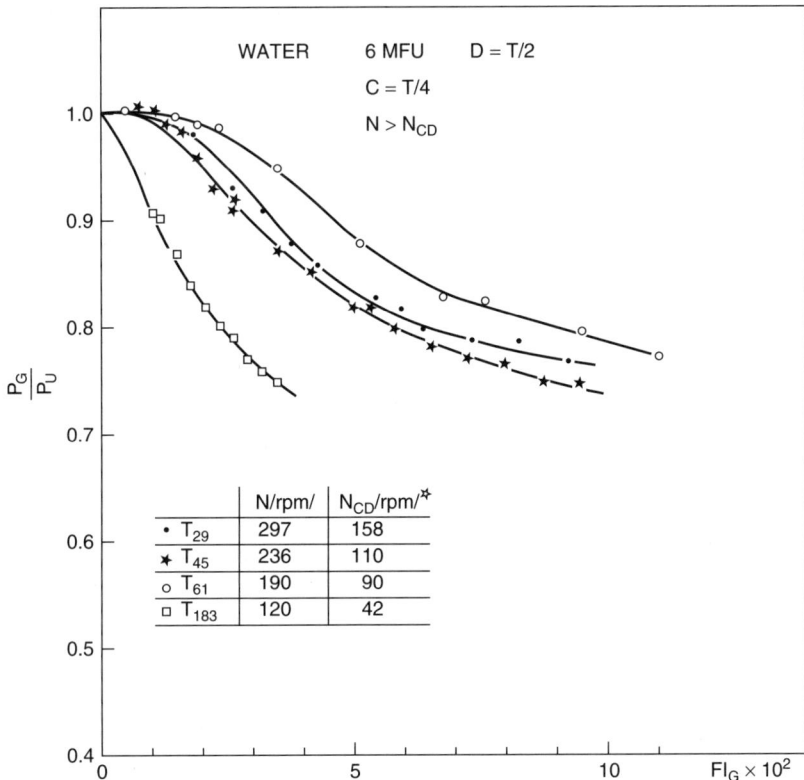

**Figure 11-11** Power curves for typical upflow pitched blade turbines. $T_{29}$ data in a 29 cm diameter vessel, etc. (From Nienow et al., 1987.)

around the impeller. Most CFD packages now offer a version of two-phase treatment, generally either particle tracking or a full Eulerian solution for each phase. In the former a selection of bubbles can be tracked as "particles," but a bubble size has to be assumed, and it is also assumed that the liquid flow patterns are unchanged by the gas, so that this is appropriate only for low gas fractions of small bubbles. In due course it is to be expected that a full two-phase treatment will account for interactions between the phases, bubble breakup, and coalescence, but development of these is in the early stages. Computational meshes can now be generated that are sufficiently fine to model the vortices behind the blades, but the remarks above concerning poor prediction of local turbulence still apply, and the gas cavities, if present, have still to be adequately modeled.

## 11-3.2 Stirred Vessels: Liquid Mixing Time

There is some conflict in the literature as to the effect of gassing on liquid mixing time in the homogeneous regime. However, the effects can all be related to the reduction in power number caused by gassing, as described above. Cooke et al. (1988) found some success by substituting $(P/V)^{1/3}$ for N in the usual expression for turbulent mixing ($Nt_M$ = constant): This correlated single phase and gassed cases (for subsurface addition) when the total of the gassed impeller shaft power together with the gas buoyancy power was used.

Recent work (Gao et al., 2000; Zhao et al., 2001) has compared liquid mixing times in ungassed, cold, and hot sparged and boiling conditions. It was shown that in an aerated "standard" tank, the mixing time correlates with the specific power input (W kg$^{-1}$) and superficial gas velocity (m s$^{-1}$), provided that both the possible changes in the relative power demand of the impeller and the potential energy added by the sparged gas (the saturated volume in hot operation) are taken into consideration. The surprising result was that mixing in a truly boiling system is significantly faster than would be expected on this basis, although, of course, with most of the gas being released near the liquid surface in a truly boiling liquid, the potential energy term is difficult to evaluate. Addition at the boiling liquid surface gave more rapid overall liquid blending than addition near the impeller—this is the only situation in which this has been found to be the case. It should be noted that recent (to date unpublished) work implies that this result does not necessarily apply when a combination of a radial flow impellers surmounted by strongly pumping axial flow impellers is used, although there are still advantages from surface addition in a compartmentalized reactor mixed with a multiple-impeller agitator.

As stated above, at high superficial gas velocities, in the heterogeneous regime, large bubbles are formed which rise faster than the liquid and take some liquid with them within their wakes. Since only a few of these will be recirculated, there is a net upflow of liquid produced by the large bubbles. Presumably this will enhance liquid mixing and reduce the mixing time, but this awaits quantification and correlations (Gezork et al., 2000, 2001).

When multiple impellers are used, care must be taken in their selection for gas–liquid systems. For example, a vessel of H = 3T with three radial flow Rushton turbines gives rise to "compartmentalization" of the flow with poor overall top-to-bottom mixing: mixing times can be very much longer than with similar specific power input in a tank with H = T mixed by a single impeller (Cooke et al., 1988). Mixing times for a combination of one to three radial flow impellers for Re > 4400 were well correlated by

$$t_{M90\%}[Po(RPD)]^{0.33}N\left(\frac{D}{T}\right)^{2.4}\left(\frac{T}{H}\right)^{2.4} = 3.3 \tag{11-8}$$

The same vessel with a Rushton turbine at the bottom surmounted by two down-flow axial impellers gave less compartmentalization and a mixing time of seven times the H = T single-Rushton value. The best option in this case is probably three upflow hydrofoils (such as the Lightnin A345). Even these give some localized circulation loops, and these dominate at transitional Reynolds numbers. Overall mixing time is not always the full story. There are generally some comparatively dead zones (e.g., in the bottom corners or near the surface) in an agitated vessel, and these may be a problem with some processes, with solids deposition or fouling, for example.

CFD (as described above) has been used to predict mixing times in liquid-phase systems. In the authors' experience, once the tracer input condition has been carefully modeled to match an experiment, reasonable agreement with experimental values has been obtained for axial flow impellers. For radial flow impellers, the predicted mixing times were longer than the measured values: this appears to be caused by inadequate description of the vertical transfer between the blade vortices in the impeller discharge stream. The effects of gassing have, however, not been explored.

## 11-4 POWER

This section deals only with turbulent flow conditions. In this regime power dissipation is the controlling factor for mixing and phase dispersion. For in-line mixers the power is derived from the flow energy of the fluid, and for stirred vessels it is obtained from the impeller and, where density differences occur, from buoyancy forces.

### 11-4.1 Static Mixers

Noting that power = volumetric flow rate × pressure drop, the overall power per unit mass of liquid is straightforward to calculate for single-phase systems given the friction factors and voidage fraction in the mixer as supplied by mixer manufacturers or measured in the laboratory. For gas–liquid systems the volume of fluid in the mixer must be multiplied by $(1 - \phi)$ to obtain the liquid volume, so the gas fraction $\phi$ must be known (see Section 11-5). It has been found that the Lockhart–Martinelli (1944) correction for the effect of the gas phase on pressure drop in pipe flow can be applied to static mixers with reasonable accuracy ($\pm 20\%$).

### 11-4.2 Gassed Agitated Vessels, Nonboiling

*11-4.2.1 Single Impellers*. The well-known equation for impeller power is often modified for gas–liquid systems to give

$$P = Po(RPD)\rho N^3 D^5 \qquad (11\text{-}9)$$

where RPD is the relative power demand or gassing (or K) factor ($P_G/P_U$), which depends on the blade shape, $Q_G$, N, and D. It generally decreases with increased dimensionless gas rate [or gas flow number ($Fl_G = Q_G/ND^3$)]. The value of RPD is particularly important for six-blade disk turbines and for downflow pitched blade turbines and hydrofoils, since it can easily fall as low as 0.4, as shown in Figure 11-7. For the recommended impellers with parabolic concave blades, such as the Scaba SRGT or Chemineer BT6, it falls to only about 0.9 (and only then at high flow numbers); with the semicircular blades of the Chemineer CD6, it falls to about 0.7. Where higher power numbers are required, flat-blade turbines with more than six blades (preferably 12 or 16) have been used, for which RPD eventually drops to about 0.4 but not until much higher flow numbers than for six flat blades (Figure 11-7). The RPD of up-pumping wide-blade hydrofoils remains close to 1.0, as shown in Figure 11-11.

This behavior has been shown (Bruijn et al., 1974; Warmoeskerken and Smith, 1982) to be related to the buildup of cavities of gas behind the blades, as described in Section 11-3.1. The flatter the blade, the larger the cavities that can form. These act as though they obstruct the passage of liquid through the impeller, and it is this that most directly reduces the effective power number [a summary of cavity formation and its effect on power can be found in Middleton (1997)]. The best way to predict the gassing effect (RPD) is first to predict the cavity regime, then obtain the value of RPD for that regime. The results in Bruijn et al. (1974) may be interpreted to relate RPD to the cavity regime to within engineering tolerance:

vortex − clinging cavities:   RPD ∼ 0.9

three clinging + three large cavities:

$$RPD \approx 0.18 Fl_G^{-0.20} Fr^{-0.25} \qquad (11\text{-}10)$$

six large cavities:   RPD ∼ 0.5 → 0.4

Intermediate conditions are less distinct.

Where axial flow impellers are preferred, they should be operated in the upflow direction, when they are stable and suffer only modest power drop on gassing (e.g., RPD for an upflow pitched blade turbine or a Lightnin A345 falls only to about 0.75, even at high gas rates). Downflow axial flow impellers, especially pitched blade turbines and narrow-blade hydrofoils, have a seriously unstable operating regime in gassed systems and suffer a drastically sharp fall in RPD under particular conditions (the direct−indirect loading transition) with dire consequences, such as fluctuating process performance, rapid seal and bearing wear, and high risk of shaft failure. However, wide-blade hydrofoils can be quite effective, especially as the upper impellers in multiple-impeller agitators.

With pitched blade impellers, cavities form in an analogous way to their development behind Rushton turbine blades. It is a convenient approximation to assume that indirect loading produces vortex cavities and direct-loading large cavities, although in reality the transition may occur at slightly different loadings. The RPD curves for downward-pumping pitched blade turbines are more complex than those for radial flow impellers since the liquid discharge is acting against the gas rising from the sparger. As was the case with radial flow impellers, the gassed RPD of a pitched blade impeller depends on both the gas flow number and the Froude number. The curves shown in Figure 11-10 are for a down-pumping 0.18 m (∼7 in.) diameter impeller with four 45° blades in a 0.44 m (18 in.) tank. The 45° impeller in down-pumping mode is fairly unstable, especially at low speeds. Even at the highest speed used (6 s$^{-1}$), the rate of power drop is much steeper than that found with a Rushton turbine (Figure 11-7).

For all down-pumping PBT impellers, the RPD lines cross over and do not follow the orderly progression found with radial flow turbines. General sensitivity to the geometry of down-pumping two-phase hydrodynamics, particularly with respect to the transition between direct and indirect loading, has discouraged the construction of flow maps for these turbines.

Design for mass transfer entails producing a given impeller power, so it is the product of power number Po and the gassing factor (RPD) that is of importance. A summary of typical values for popular impellers at high gas flow number (say, 0.1) may be useful for guidance (Table 11-4). An example of the calculation of power for an agitated reactor is given in Example 11-1.

A very popular basis for predicting gassed power is the equation proposed by Michel and Miller (1962), which arrives at a value of the gassed power in terms of the product of the square of the ungassed power draw, the impeller pumping, $ND^3$, and the gas rate raised to the arbitrary power of 0.56: $[P_U^2 ND^3/Q_G^{0.56}]^{0.45}$. Unfortunately, as Nienow et al. pointed out in 1977, this equation is specious, effectively depending on a plot of $N^3 D^5$ against $N^7 D^{13}$. To emphasize this point, Figure 11-12 shows the all-too-plausible correlation of gassed power ($P_G$) against $P_U^2 ND^3/Q_G^{0.56}$ based on allocating random numbers to N and D in the ranges 2 to 9 and 0.2 to 1.5, respectively, and random numbers for the parameters that really matter, RPD (in the range 0.35 to 1.0) and $Q_G$ (10 to 1000 L per minute). A plot of the values of RPD versus $Q_G$ actually used to generate these "data" is shown in Figure 11-13.

Table 11-4 Comparative Gassed Power for Various Impellers

| Impeller Type | Po | (RPD)$_{Fl=0.1}$ |
|---|---|---|
| Radial flow | | |
| 6 blade disk turbine[a] D = T/3 | 5 | 0.4 |
| 12 blade disk turbine, D = T/3 | 10 | 0.6 |
| 18 blade disk turbine, D = T/3 | 12 | 0.7 |
| Chemineer CD6 | 2.3 | 0.8 |
| Chemineer BT6 | 2.0 | 0.9 |
| Scaba 6SRGT | 1.5 | 0.9 |
| Axial upflow | | |
| 4 pitched blade turbine, D = T/3, C = T/3 | 1.3 | 0.75 |
| 6 pitched blade turbine, D = T/3, C = T/3 | 1.7 | 0.75 |
| Lightnin A345, D = 0.4T | 0.8 | 0.75 |
| Axial downflow | | |
| 4 pitched blade turbine, D = T/3, C = T/3 | 1.3 | 0.3 |
| 6 pitched blade turbine, D = T/3, C = T/3 | 1.7 | 0.4 |
| Prochem MaxfloW 5, D = 0.45T | 1.3 | 0.7 |
| Lightnin A315, D = 0.4T | 0.8 | 0.7 |

[a] This is actually a function of scale (see Bujalski et al., 1987).

**610** GAS–LIQUID MIXING IN TURBULENT SYSTEMS

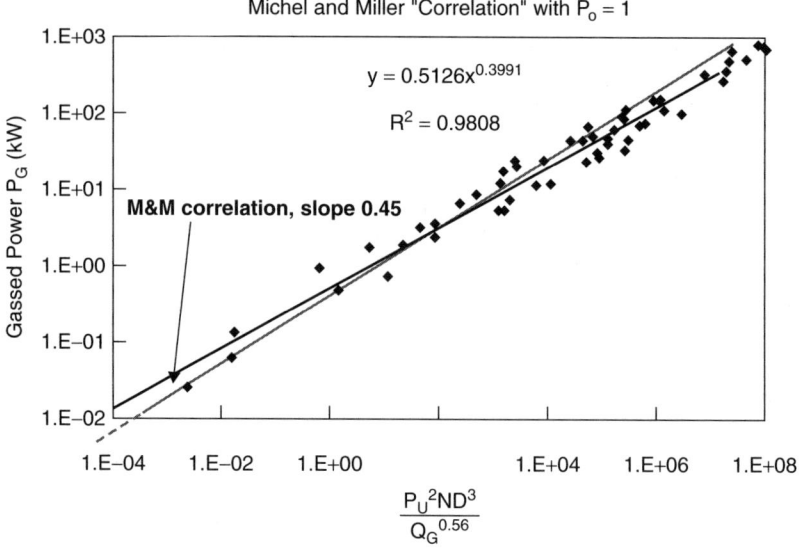

**Figure 11-12** Michel–Miller relationship.

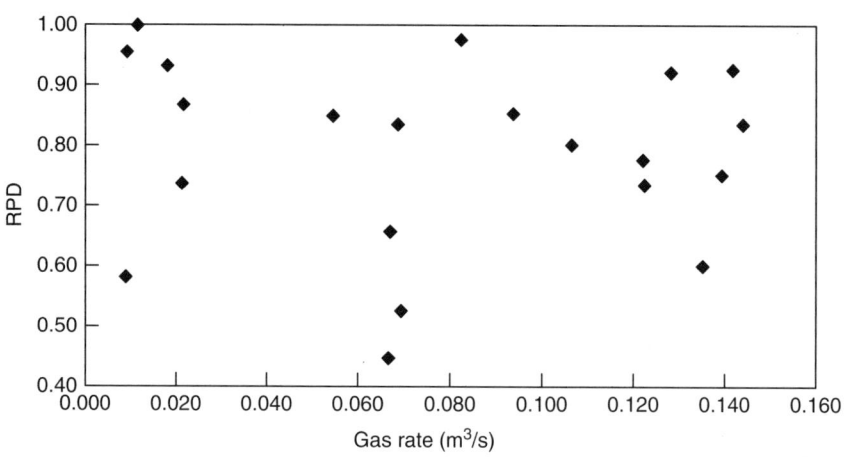

**Figure 11-13** Random numbers used in generating Figure 11-12.

***Example 11-1: Power Draw of an Agitated Reactor.*** A 5 m$^3$ vessel (177 ft$^3$ or 1320 gal) has an impeller 0.52 m (1.7 ft) in diameter and an ungassed power number of 5.0 driven at 42 rpm in water (density 62.4 lb/ft$^3$ and viscosity 1 cP). What is the ungassed power draw of this impeller? (*Ans.* 65.2 W.) If the gas flow number is 0.04, what is the gassed power demand? (*Ans.* 55.7 W.)

**Table 11-5**

| Name | SI Value | SI Unit | U.S. Engg. Value | U.S. Engg. Unit |
|---|---|---|---|---|
| N | 0.7 | s$^{-1}$ | 42 | rpm |
| D | 0.52 | m | 1.71 | ft |
| T | 1.3 | m | 4.27 | ft |
| H | 3.77 | m | 12.4 | ft |
| V | 5 | m$^3$ | 1320 | gal |
| Po | 5 | | 5 | |
| Fl$_G$ | 0.04 | | 0.04 | |
| Q$_G$ | 0.00394 | m$^3$/s | 0.14 | ft$^3$/sec |
| Fr | 0.026 | | 0.026 | |
| RPD | 0.854 | | 0.854 | |
| Re | 189 000 | | 189 000 | |
| P$_U$ | 65.2 | W | 0.0874 | hp |
| P$_G$ | 55.7 | W | 0.0746 | hp |
| g | 9.81 | m/s$^2$ | 32.2 | ft/sec$^2$ |
| $\rho_L$ | 1000 | kg/m$^3$ | 62.4 | lb/ft$^3$ |
| $\mu_L$ | 0.001 | Pa·s | 0.00067 | lb$_m$/ft-sec |

See Table 11-5 for the calculations. The relevant equations, which are solved using TK Solver or a similar program, are

$$Fl_G = \frac{Q_G}{ND^3} \qquad Fr = \frac{N^2 D}{g} \qquad P_U = Po \cdot N^3 D^5 \rho_L$$

$$RPD = 0.18 Fr^{-0.25} Fl_G^{-0.20} \qquad P_G = RPD \cdot P_U$$

**11-4.2.2 Multiple Impellers.** Assuming that the lowest impeller is used for the primary gas dispersion, the upper impellers are not loaded by all the gas entering through the sparger (Smith et al., 1987). It can be assumed for the purpose of power demand estimation that upper impellers experience about half the total gas rate. This can be illustrated on a flow regime map (Figure 11-14).

*Example 11-2: Power Demand of a Large Fermenter.* The agitator in a 20 m$^3$ fermenter agitator is to have a lower dispersing impeller (Po$_2$ = 5.0) surmounted by a wide-blade hydrofoil, (Po$_1$ = 1.0). The fermenter height is twice the tank diameter and the hydrofoil impeller is to be 40% of the tank diameter. When aerated, the lower impeller is expected to have an RPD of 0.7 and the upper impeller an RPD of 0.9. It is desired that the same energy should be transferred to the liquid from each impeller. What should be the diameter of the dispersing turbine? At what speed can the assembly be driven if the specific gassed power input is to be limited to 0.6 kW/m$^3$ (3 hp per 1000 gal)? See Example 11-1 for physical properties.

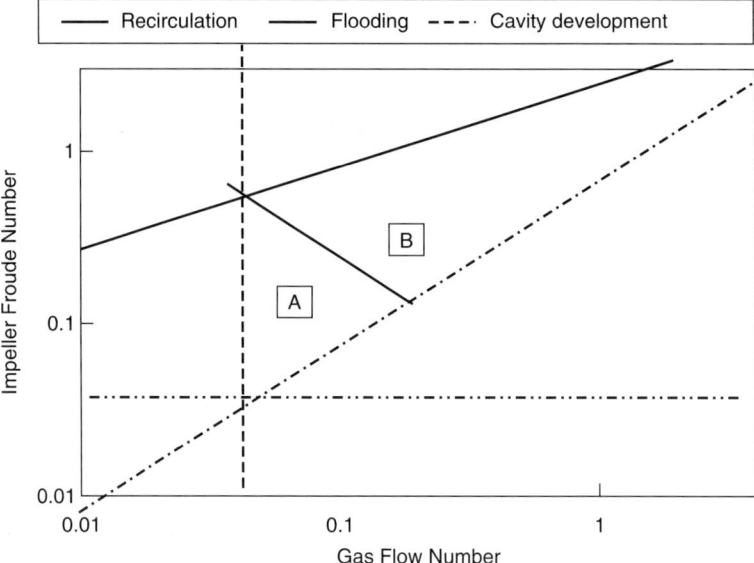

**Figure 11-14** Flow map for triple Rushton turbines (T/D = 2.5). Regimes as for Figure 11-8, except that in region A there are large cavities on the lowest impeller only; in region B large cavities are present on all three impellers.

See Table 11-6 for the calculations. The relevant equations, which are solved using TK Solver or a similar program, are

$$P_U = Po \cdot N^3 D^5 \rho_L \qquad P_G = RPD \cdot P_U$$

### 11-4.3 Agitated Vessels, Boiling, Nongassed

Early studies (Breber, 1986; Smith and Verbeek, 1988; Smith and Smit, 1988) demonstrated the general similarities between the ventilated cavities formed during the dispersion of gases with agitators and those developed in unsparged boiling systems. However, there are major differences in performance between boiling and gas–liquid systems. During boiling, the RPD is essentially independent of the boil-up rate (Smith and Katsanevakis, 1993). Figure 11-15 illustrates results obtained with a 0.18 m diameter Rushton turbine. It is clear that neither the total boil-up rate nor changes in the flow field when vapor from the immersion heaters is directed into or away from the impeller have any effect on the relationship between impeller speed and the RPD. This implies that vapor does not load the impeller in the same manner as does noncondensable gas, and that generation in the low-pressure regions behind the impeller blades is limited.

In boiling systems the processes of the initiation and further development or collapse of vapor cavities are crucial. Conditions in the vapor cavities behind the

**Table 11-6**

| Name | SI Value | SI Unit | U.S. Engg. Value | U.S. Engg. Unit |
|---|---|---|---|---|
| V | 20 | m$^3$ | 5 280 | gal |
| T | 2.34 | m | 7.66 | ft |
| H | 4.67 | m | 15.33 | ft |
| $D_1$ | 0.934 | m | 3.06 | ft |
| $D_2$ | 0.712 | m | 2.34 | ft |
| N | 2.11 | s$^{-1}$ or rps | 127 | rpm |
| $Po_1$ | 1 | | 1 | |
| $Po_2$ | 5 | | 5 | |
| $Pu_1$ | 6 670 | W | 8.94 | hp |
| $Pu_2$ | 8 570 | W | 11.5 | hp |
| $RPD_1$ | 0.9 | | 0.9 | |
| $RPD_2$ | 0.7 | | 0.7 | |
| $Pg_1$ | 6 000 | W | 8.05 | hp |
| $Pg_2$ | 6 000 | W | 8.05 | hp |
| $P_{tot}$ | 12 000 | W | 16.1 | hp |
| Specific power | 0.6 | W/kg | 3.05 | hp/gal |

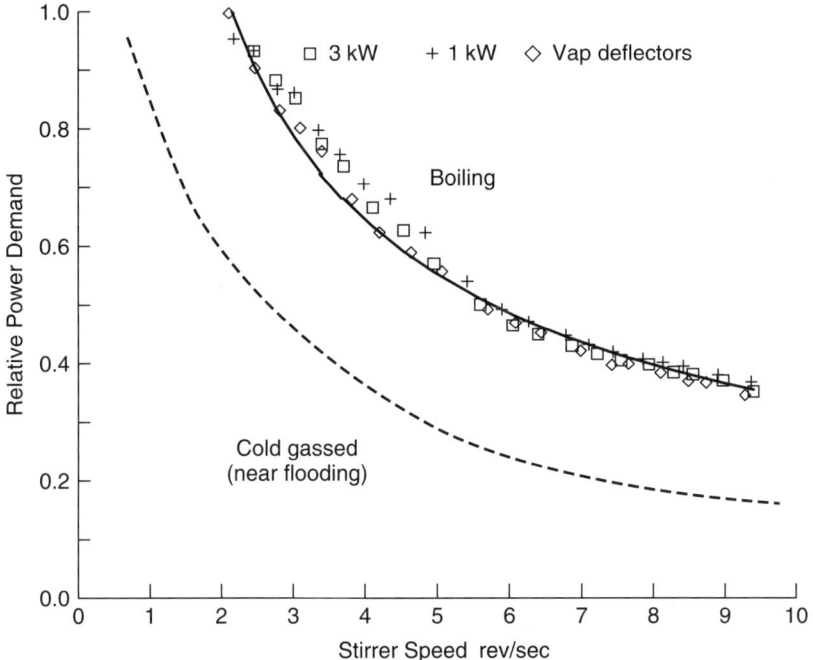

**Figure 11-15** Power demand of a 0.18 m Rushton turbine with different boil-up rates and vapor flow arrangements. (From Smith and Katsanevakis, 1993.)

# 614 GAS–LIQUID MIXING IN TURBULENT SYSTEMS

impeller blades can be represented by the ratio between the nominal stagnation pressure on the front of the impeller blade, near the tip, $\frac{1}{2}\rho v_t^2$, and the difference between the pressure within the cavity and that at the free liquid surface. At a submergence S, measured to the midplane of the impeller, the latter pressure difference is approximately that due to the nominal hydrostatic head ($\rho g S$), so that we can define an *agitation cavitation number*, $C_{Ag}$, now sometimes referred to as the *Smith number* (Sm):

$$\text{Sm} = C_{Ag} = \frac{2gS}{v_t^2} = \frac{2}{\pi^2}\left(\frac{S}{D}\right)\frac{1}{\text{Fr}} \qquad (11\text{-}11)$$

This is similar to a traditional cavitation number except that the pressure within the vapor cavity is strongly affected by both local fluid mechanics and thermal factors. It was shown by Smith and Katsanevakis (1993) that the RPD in a boiling agitated system can be described adequately by relationships of the form

$$\text{RPD} = \frac{P_B}{P_U} = A\left(\frac{2Sg}{v_t^2}\right)^B = AC_{Ag}^B = A \cdot \text{Sm}^B \qquad (11\text{-}12)$$

The constant A, which is often about unity, depends on the impeller type. (see Table 11-7). Impellers with a high gas-handling capacity, such as hollow-blade disk designs, have the highest values. As will also be seen in Table 11-7, the exponent B varies considerably with impeller type but is about 0.4 for Rushton and pitched blade turbines. Figure 11-16 reproduces some results for a six-blade Rushton turbine working at various submergences and boil-up rates. In this case the constant A in eq. (11-12) is 0.74.

A critical $(C_{Ag})_{crit}$ (or $\text{Sm}_{crit}$) can be defined as that value above which the power draw is essentially the same as when this impeller, is ungassed. For a Rushton impeller, this value is about 2.1 and values for other impellers are given

**Table 11-7** Impeller Constants for Unsparged Boiling

| Impeller | Constant A | Exponent B | Critical $C_{Ag}$ |
|---|---|---|---|
| Rushton turbine | 0.69 | 0.4 | 2.15 |
| PBT$_D$ (down-pumping) | 0.74 | 0.4 | 2.10 |
| PBT$_U$ (up-pumping) | 0.90 | 0.4 | 1.30 |
| Chemineer CD-6 | 1.17 | 0.2 | 0.46 |
| Chemineer BT-6 | 1.16 | 0.1 | 0.23 |
| Chemineer MaxfloW$_D$ | 1.03 | 0.4 | 0.93 |
| Chemineer MaxfloW$_U$ | 1.61 | 0.4 | 0.30 |
| Lightnin A315$_D$ | 1.16 | 0.4 | 0.69 |
| Lightnin A340$_D$ | 1.07 | 0.4 | 0.84 |
| Lightnin A340$_U$ | 1.12 | 0.2 | 0.57 |

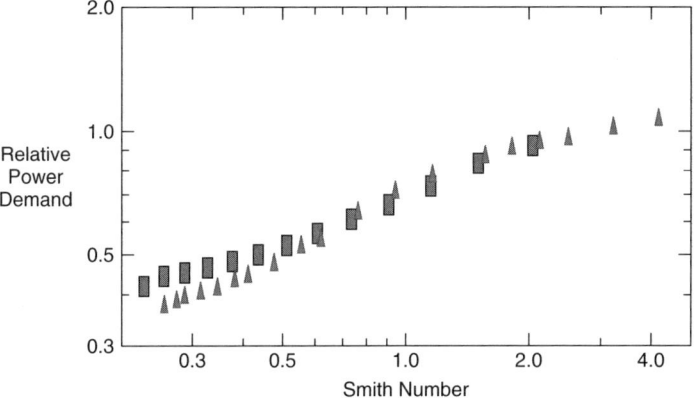

**Figure 11-16** Boiling power demand, Rushton turbine. (From Gao et al., 2001a.)

in Table 11-7. Using this critical value, we can write

$$\text{RPD} = \left[\frac{C_{Ag}}{(C_{Ag})_{crit}}\right]^B = \left(\frac{\text{Sm}}{\text{Sm}_{crit}}\right)^B \quad (11\text{-}13)$$

The values do not appear to be very sensitive either to D/T or to absolute scale.

Figures 11-17 and 11-18 show data from Smith et al. (2001a) relating to various hollow-blade and hydrofoil impellers from which the values in Table 11-7 have been derived. Modern hydrofoil impellers, which are designed to have good gas-handling characteristics, almost maintain their cold ungassed power levels when up-pumping. In this respect they behave almost as if cavitation does not occur. This will be a very desirable feature of these impellers when used in evaporative crystallizers. Later work (Smith and Tarry, 1994) confirmed that the

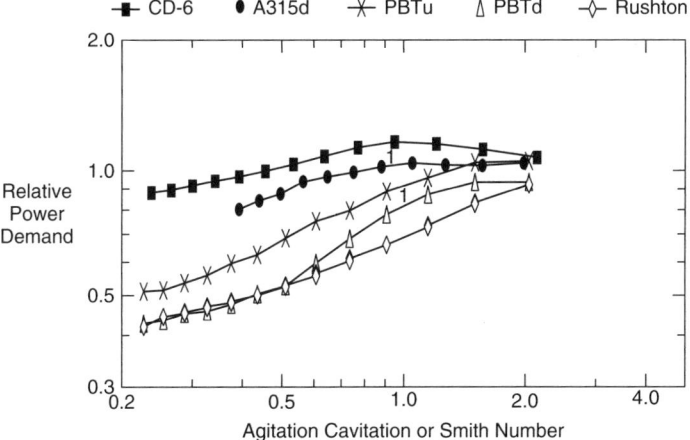

**Figure 11-17** Boiling RPD for common impellers. (From Gao et al., 2001a.)

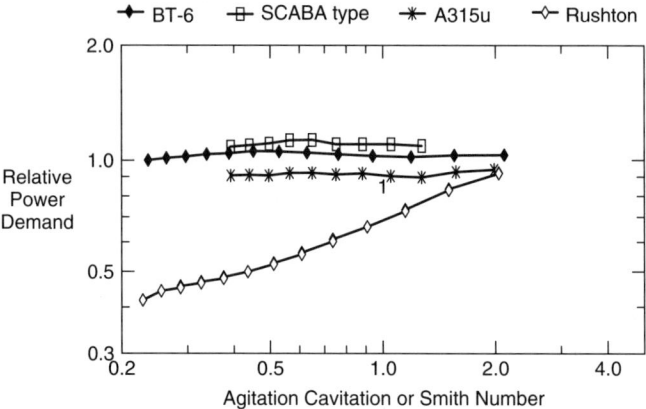

**Figure 11-18** Boiling RPD for modern gas-dispersing impellers. (From Gao et al., 2001.)

identical relationship is valid with boiling solutions in which the elevation of boiling point would have the same effect as that of significant increases in impeller submergence. It is also unlikely that liquid viscosity will have a significant influence as long as the Reynolds number is high. These two facts encourage the conclusion that the results will be valid for all low viscosity liquids.

The locus of the limiting power appropriate to flooding an 0.18 m diameter Rushton impeller at a submergence of 0.3 m is also shown in Figure 11-15. The much higher relative power demand of an impeller in rapidly boiling liquid compared with that in the near-flooded, cold-sparged condition at the same shaft speed (i.e., of Fr or Sm) is evident. These boiling cavitation and (cold) gas flooding lines represent limits between which a sparged boiling reactor might be expected to operate.

*Example 11-3: Impeller Power in a Boiling Crystallizer.* An upward-pumping pitched blade impeller of 0.6 m in diameter is to be specified for a boiling crystallizer in which it is submerged by 0.7 m. If the critical Smith number for this impeller is 1.3 with RPD obeying a $Sm^{0.4}$ law, and the RPD is not to be lower than 60% of the ungassed value, what is the maximum speed at which the impeller should be driven?

See Table 11-8 for the calculations. The relevant equations, which are solved using TK Solver or a similar program, are

$$Fl_G = \frac{Q_G}{ND^3} \qquad Fr = \frac{N^2 D}{g} \qquad P_U = Po \cdot N^3 D^5 \rho_L$$

$$P_G = RPD \cdot P_U \qquad Sm = \frac{2gS}{v_t^2} = \frac{2}{\pi^2}\left(\frac{S}{D}\right)\frac{1}{Fr} \qquad v_t = \pi ND$$

$$RPD = \left(\frac{Sm}{Sm_{crit}}\right)^B \qquad \text{where } B = 0.4$$

**Table 11-8**

| Name | SI Value | SI Unit | U.S. Engg. Value | U.S. Engg. Unit |
|---|---|---|---|---|
| RPD | 0.6 | | 0.6 | |
| Sm | 0.36 | | 0.36 | |
| $Sm_{crit}$ | 1.3 | | 1.3 | |
| S | 0.7 | m | 2.30 | ft |
| g | 9.81 | m/s$^2$ | 32.2 | ft/sec$^2$ |
| $v_t$ | 6.15 | m/s | 20 | ft/sec |
| D | 0.6 | m | 1.97 | ft |
| N | 3.26 | s$^{-1}$ or rps | 196 | rpm |

## 11-4.4 Agitated Vessels, Hot Gassed Systems

Unsparged boiling and cold sparging generate two quite different sets of conditions with large differences between the physical properties of the liquid and gas phases. The cavitation line for Figure 11-15 is based on liquid at its boiling-point generating vapor, while the flooding correlation refers to cold systems with little further vaporization. It has been shown that at a given speed [i.e., a fixed Froude or agitation cavitation (Smith) number] most impellers draw more power in a boiling system than in cold, preflooding, gassed conditions. The interactions of gas rate and impeller operation need to be understood so that the transitional hot sparged gas case can be quantified for industrially important conditions.

When an inert gas is passed through a boiling liquid, there is a change in the thermodynamic equilibrium. Since the bubbles consist of a mixture of vapor and inert gas, the partial vapor pressure of the condensable components is less than the total pressure at which the liquid was previously boiling. The liquid is therefore superheated relative to the mixed gas phase and there will be an immediate increase in the evaporation rate so that latent heat can remove the excess energy. The liquid temperature will fall until the energy supply and removal rates are in balance. A steady equilibrium temperature will be established when there are constant net heat input and gas throughput rates.

Saturation is rapid. Figure 11-19 shows the results of a simplified calculation suggesting that bubbles leaving a sparger are brought to within 90% of saturation in about 500 ms (Gao et al., 2001). This suggests that calculations based on complete saturation are accurate enough for most design purposes. The evaporation of the liquid into the bubbles removes heat from the system. Saturated bubbles will contain vapor to a partial pressure that will correspond to the temperature of the liquid and is less than the total pressure of the system. It follows that even when there is a heat source, no continuously sparged liquid can be at its true boiling point.

The equilibrium temperature is sensitive to the sparged gas and heat supply rates but is independent of the impeller speed, a result confirmed by experiment (Figure 11-20). This simplifies the analysis since a constant vapor pressure can

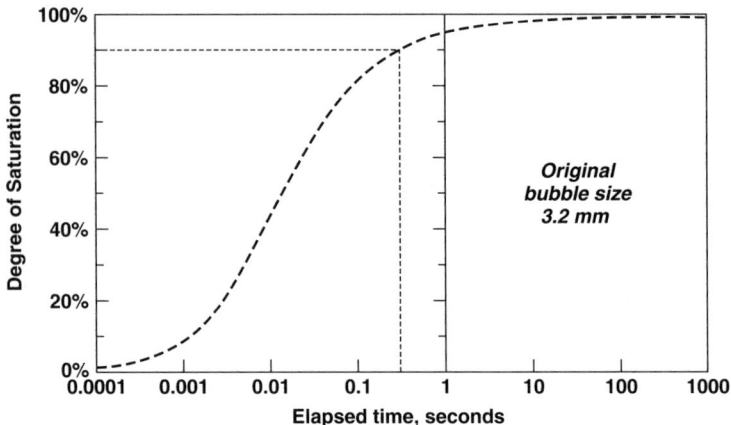

**Figure 11-19** Saturation of an air bubble introduced into boiling water.

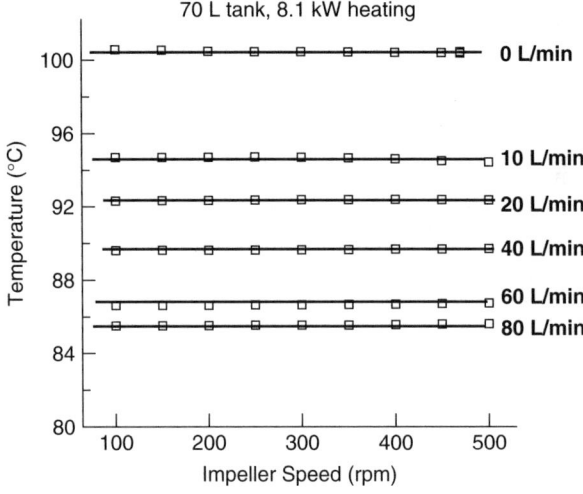

**Figure 11-20** Temperature of heated sparged water showing the independence of impeller speed.

be assumed for a given gas rate, and this allows reasonable estimates to be made of the combined gas and vapor flow loading the impeller.

When an existing boiling reactor is sparged, the sparged gas rate can be corrected using the vapor pressure of the liquid at the temperature measured. Assuming that the partial pressure of the vapor $p_v$ is then known, the total volumetric rate, $Q_{GV}$, is given by

$$\frac{Q_{GV}}{Q_G} = \frac{p_0}{p_0 - p_v} \tag{11-14}$$

When the liquid vapor pressure is known in terms of the usual relationship, $p_v = Ae^{b/\theta}$, the correction can be expressed as

$$\frac{Q_{GV}}{Q_G} = \frac{1}{1 - e^{b(1/\theta_1 - 1/\theta_0)}} \qquad (11\text{-}15)$$

where $\theta_0$ is the boiling point at the ambient pressure $p_0$ and $\theta_1$ is the temperature measured during sparged operation. This relationship will always be true even if there has not been sufficient time for the equilibrium conditions consistent with the heat balance to be established.

Some RPD results from experiments in a dish-bottomed vessel of 0.44 m diameter with three 1.2 kW heaters and a 0.18 m Rushton turbine are shown in Figure 11-21 using the log(RPD) versus log(Sm) format (after Smith and Millington, 1996). Most of the values fall between the pool boiling cavitation and gas flooding lines, with higher gas supply rates corresponding to lower values of the RPD. As can be seen from the figure, within the accuracy of the measurements the relative power demand of a sparged "boiling" system is independent of the impeller speed until the impeller speed becomes low enough for the data points to concatenate onto the cold flooding line. Many other experiments have confirmed this behavior.

### 11-4.5 Prediction of Power by CFD

If the methods referred to earlier are used with care, CFD can predict the power number of an impeller in a single-phase system to within 20%. A well-chosen grid with local refinement around the impeller and at least 200 000 cells is required. The only successful method seems to be to integrate the torque on the impeller. Summation of energy dissipation, $\varepsilon$, over the vessel does not give the

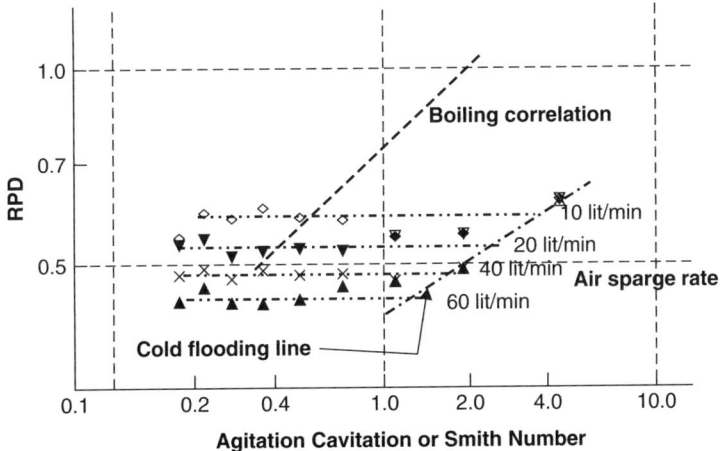

**Figure 11-21** Relative power demand in aerated hot 70 L reactor.

correct answer, probably because of the shortcomings of the turbulence models; this is illustrated by the underprediction of $\varepsilon$ in the discharge region of disk turbine impellers (Montante et al., 2001). CFD methods have not yet been developed to the point that they will predict the correct effect of gassing on power demand.

## 11-5 GAS HOLD-UP OR RETAINED GAS FRACTION

### 11-5.1 In-line Mixers

As remarked earlier, because in-line static mixers are plug flow devices, the gas fraction is comparatively easy to determine from the ratio of mean gas flow rate to total flow rate, with adjustment for bubble "slip" if the flow orientation is nonhorizontal. Often, vertical downflow is preferred, since the gas–buoyancy leads to the bubble velocity being less than the liquid velocity, so the gas fraction (and hence the gas–liquid interface area) is greater than for other configurations. While there is much literature on bubble slip velocities, the predictions are said to be unreliable (Zuber and Findlay, 1965) and it is usually preferred to use empirical correlations of the gas fraction based on measurements [such as those in Middleton (1978)], although so far these all seem to be for air–water systems with negligible depletion of bubble size, so may need adjustment for other systems.

When gases are dispersed in liquids of high vapor pressure, there are significant effects due to vaporization or condensation of the liquid. For example, if the pressure surrounding an air bubble in water at around 97°C, which has a vapor pressure of about 0.9 bar, is reduced from 1.2 bar to 1 bar, the volume of the bubbles will increase threefold, not by the 20% or so that would be the case at room temperature (see Figure 11-23). This effect will be particularly important in changing the phase ratio of a two-phase flow through a static mixers operating with a large overall pressure drop.

### 11-5.2 (Cold) Agitated Vessels, Nonboiling

Gas fraction in agitated vessel is difficult to predict a priori, but in the homogeneous regime, scale-up can be made reasonably accurately using empirical correlations. These are best expressed in the form

$$\phi = \alpha' \left(\frac{P}{\rho V}\right)^{\beta'} (v_s)^{\gamma'} \qquad (11\text{-}16)$$

where the constants $\alpha'$, $\beta'$, and $\gamma'$ are independent of scale. Although such equations are unsatisfactory in principle both because the $P/\rho V$ and $v_s$ terms are often mutually dependent and because of the need for $\alpha'$ to have noninteger dimensions in order to provide dimensional consistency, they have been more successful than alternative formulations. The implication is the rather counterintuitive result that the impeller design or configuration is only of secondary importance, provided that the energy is transferred to the liquid.

The value of α' depends on the physical properties of the liquid, in a way that is in general difficult to predict (hence the recommendation to obtain at least one measurement at semitech or pilot scale during process development, and use the correlation for scale-up only). The published data are for aqueous systems, in which the addition of any solute that exhibits surface activity (this includes electrolytes and alcohols as well as surfactants) has a large impact on the gas fraction; for example, a system in which water gives, say, φ = 0.1 and d = 4 mm may give φ = 0.25 and d = 0.5 mm with a solution of a simple electrolyte (above a plateau concentration). The considerable literature on this effect currently aligns observations with a reduction of bubble coalescence caused by the solute via gradients of surface tension repressing drainage of the liquid film between approaching bubbles. Such effects could also occur with small concentrations of water in organic liquids or with small particles caught at the interface. For engineering purposes, the situation has been simplified to cover (for the homogeneous regime) two "extreme" classes of liquid system—coalescing and noncoalescing systems—with separate correlations for gas fraction and mass transfer; but there is no guarantee that all industrial systems fit between these classes.

Values of β' and γ' vary in the literature between 0.2 and 0.7, but generally, β' = 0.48 and γ' = 0.4 are quite reliable (Smith et al., 1977). More recent work (Gao et al., 2001) has led to the equation (expressed in W, m, s units)

$$\phi = 0.9 \left(\frac{P}{\rho V}\right)^{0.20} (v_s)^{0.55} \tag{11-17}$$

for hold-up in vessels with multi-impeller agitators dispersing air in water at ambient temperature. It would be better to have separate correlations for each flow regime; for example, as the Reynolds number is decreased into the transitional region, β' tends to fall and γ' to rise (Cooke et al., 1988). The same trends occur in β' and γ' as gas superficial velocity $v_s$ rises into the heterogeneous regime; eventually (above $v_s = 0.08$ m s$^{-1}$ and $P/\rho V = 1$ W/kg), the total gas fraction actually decreases slightly with increased $P/\rho V$, and the fraction of small bubbles remains constant, with the large bubble fraction increasing as $v_s$ is increased (Gezork et al., 2000). The latter work was carried out with one liquid system (air–polypropylene glycol solution) which gives very high gas fractions (up to 0.55), and no general correlations for this regime are yet available. The presence of large bubbles implies that it may not be an optimal regime for mass transfer.

These equations are for operation at ambient temperature. In the fully turbulent regime there is a dependence of void fraction on temperature which is discussed below. This gives $\phi \propto \mu^{0.55}$. Measurements of the void fraction distribution in gas-sparged vessels clearly show a region of high gas fraction in the violently agitated regions near the impeller plane. There may also be gas accumulation in the liquid downflow centrally above a radial pumping impeller and near the walls below the impeller plane.

### 11-5.3 Agitated Vessels, Boiling (Nongassed)

Bubbles can survive in a hot liquid only if the temperature is high enough that the vapor pressure of the liquid matches the local pressure. This implies that vapor generation in a well-mixed liquid is limited to boiling near the free surface, possibly in low-pressure regions behind impeller blades and in any superheated liquid that may be near heat sources. In the case of boiling water at 1 bar, a superheat of 1 K will sustain a bubble at a depth of about 35 cm, so this is the maximum depth that bubbles can exist in a tank in which the temperature is as uniform as that. Visual observation in pilot scale rigs confirms that vapor generation is limited to the topmost few centimeters of the vessel, and the vertical distribution shown in Figure 11-22 is typical.

The situation is rather different at 10 bar. Because of the steeply rising vapor pressure, a water temperature of 1 K superheat will support bubbles to a depth of about 7.5 m. Since large vessels, with their greater likelihood of temperature inhomogeneity, frequently operate at high pressure, the void distribution in them can be expected to be distributed much more uniformly and to approach those more typical of gassed systems.

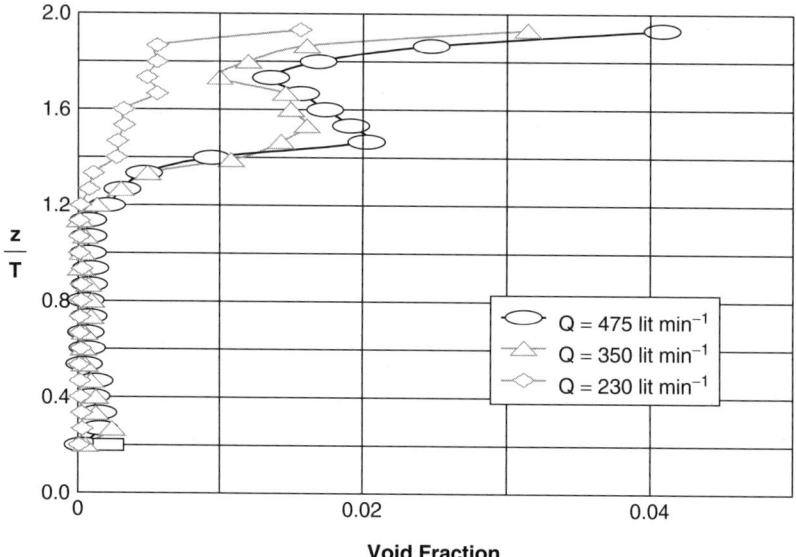

**Figure 11-22** Vertical void fraction distributions at three boil-off rates in a boiling reactor with twin radial pumping 18 cm CD6 impellers at 240 rpm.

## 11-5.4 Hold-up in Hot Sparged Reactors

When gas is sparged into a hot liquid, there is an immediate change in the thermodynamic status as the liquid vaporizes into the bubbles. As noted above, this process continues until the latent heat required removes all that is available. In continuous operation the liquid will settle at a temperature below its nominal boiling at a value determined by the rate of supply of sparge gas and heat. Any sparged or evolved gas will produce this effect. The vapor dilutes the sparged gas, so reduces the driving force for mass transfer.

Although the effects of pressure are less spectacular than in purely boiling conditions, they cannot be neglected. Figure 11-23 illustrates the difference when a small air bubble is released into open tanks of hot and cold water at a depth of 2 m: at ambient temperature the bubble expands by about 20%, whereas at 97°C the expansion is 300%. The effect will be less marked at high pressure, and again, the closer the liquid is to its boiling point at the operating pressure, the greater the effect, so that void distributions in purely boiling liquids at high pressure can be expected to be closer to those in sparged systems. Sparged hold-up measured in hot systems differs markedly from that at room temperature. Experimental measurements suggest that in an air–water system around 80°C, void fractions are at least 30% lower than at room temperature.

Overall void fraction measurements (made by a radar probe detecting the surface level averaged over several seconds) are shown in Figure 11-24. Similar data confirm the lower gas holdup in heated systems. In this figure the sharp fall-away in void fraction at low shaft power (i.e., at low speeds) seen at room temperature is visually correlated with the loss of radial pumping action by the asymmetric BT-6 impeller.

Extensive work with configurations involving up-pumping hydrofoils, which have become a generally favored arrangement for large gas–liquid reactors, has led to a correlation for overall gas retention that is a function of the absolute temperature. Specifically, in an air–water system with multiple impellers, the

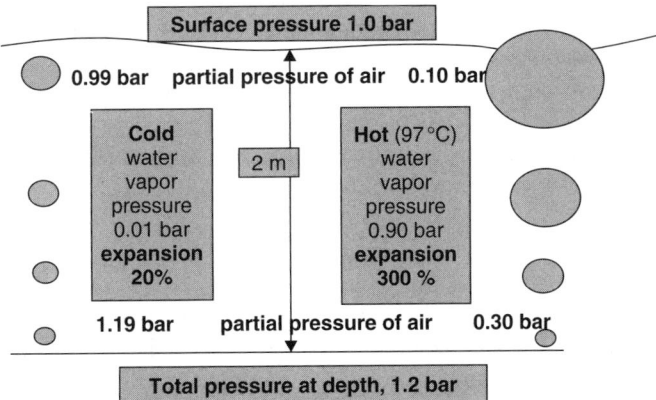

**Figure 11-23** Expansion of air bubbles rising in cold and hot water.

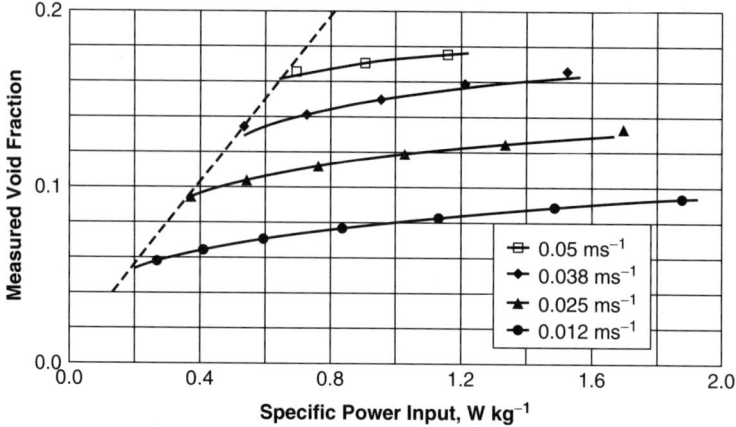

**Figure 11-24** Overall void fractions with a multiple impeller agitator CD6 + 2MFU.

average void fraction

$$\phi = 70 \times 10^6 \left(\frac{P}{\rho V}\right)^{0.20} (v_s)^{0.55} \theta^{-3.2}$$

where $P/\rho V$ is the specific power input (W/kg), $v_s$ the superficial gas velocity (m/s), and $\theta$ the absolute temperature (K). This equation is consistent with that by Gao et al. (2001) given above for ambient holdup data.[2] In a vessel with a single impeller agitator, the void fraction will be lower, about 65% of this predicted value. Since vapor pressure and liquid-phase viscosity have similar dependence on temperature, there is not enough evidence to decide which is controlling, but broadly similar behavior can be expected whatever the composition of the liquid phase.

***Example 11-4: Void Fraction in a Gas–Liquid Reactor.*** A void fraction of 7% is measured in an aerated reactor containing water at 20°C. What will be the void fraction if the reactor is operated at the same specific power input and superficial gas velocity (after allowing for the contribution of water vapor) at 90°C?

SOLUTION: Void fraction varies with absolute temperature, $\theta^{-3.2}$. For this example

$$7\left(\frac{363.2 \text{ K}}{293.2 \text{ K}}\right)^{-3.2} = 3.5\%$$

The mean residence time of gas passing through a reactor at 20°C (when the partial pressure of water is negligible) is 1 min [i.e., the sparge rate is 1 vvm (1 volume of gas per volume of liquid per minute)], at which flow rate the void

---

[2] This conclusion has not been confirmed in a system of very high purity (Shaper et al., 2002).

fraction is 7%. What will be the mean residence time if the temperature is raised to 90°C (when the partial pressure of water is 0.9 bar)?

The oxygen content will be reduced from 21% to 2.1% and the residence time reduced from 1 min to 31 s. Mass transfer might be expected to be about 20 times as difficult except that diffusion coefficients will be increased (by about $\theta^{3/2}$, i.e., ≈40%) at the higher temperature.

### 11-5.4.1 Void Fraction Profiles with Multiple-Impeller Agitators.

Figure 11-25 shows data obtained with two 18 cm radial flow CD6 impellers in tank of T = 44 cm, H = 2T. The highest void fraction occurs just above the level of the uppermost impeller, with a peak value in cold operation that is about 40% higher than that just above the lower impeller. In hot operation (generating about 250 L/min of steam into about 130 L/min of sparged air), although the void profile has an overall shape which is generally similar to that at room temperature, the gas fraction is clearly considerably less at all levels. In hot conditions the void fraction near the upper impeller is nearly twice that at the lower impeller. When the liquid is boiling, with a vapor generation rate giving a similar off-gas volume, the voids are limited to the top few centimeters of the reactor since efficient liquid mixing eliminates significant superheat.

Figure 11-26 shows comparable data when the agitator is a combination of a radial impeller with two up-pumping wide blade hydrofoils. This provides a rather different picture from that with twin radial impellers. The highest void fraction again develops just above the plane of the uppermost impeller, whatever the operating temperature. The maximum void fraction is spectacular, approaching 50% at room temperature. Again, hot sparged conditions generate similar void fraction profiles, but over the entire reactor height the gas fraction is lower than at room temperature. The contribution of the middle impeller to gas retention

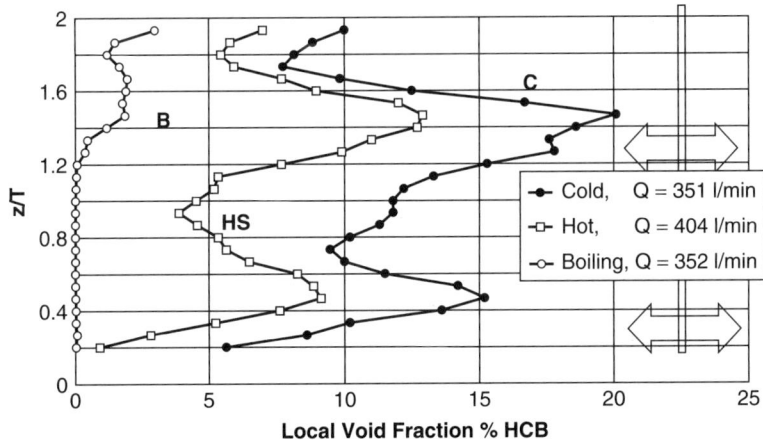

**Figure 11-25** Void profiles in cold and hot sparged and boiling conditions with twin CD6 impellers. (Data from Smith et al., 2001a.)

**626** GAS–LIQUID MIXING IN TURBULENT SYSTEMS

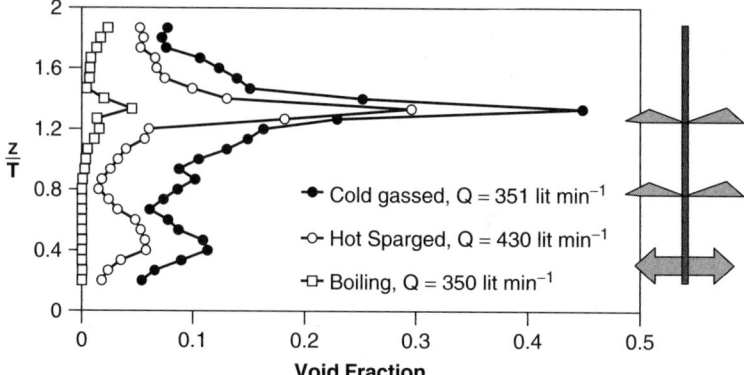

**Figure 11-26** Void distribution with a multiple-impeller agitator (CD6 with two MaxfloW up-pumping hydrofoils). Note the high local void fraction just above the upper impeller.

is slight, but the effect on liquid circulation almost certainly remains important. Truly boiling conditions again have very low void fractions throughout the tank, with some evidence of vapor bubbles being released from the topmost impeller.

The very strong liquid circulation induced by the hydrofoils forces gas through the bottom (radial) impeller to the extent that the discharge from the impeller has a strong upward component. The conditions differ from the usual buoyancy-induced flooding in that the dispersing action of the impeller appears not to be badly affected. This combination with up-pumping hydrofoils is currently popular as a means of ensuring good top-to-bottom mixing in tall reactors. Again the large peak in void fraction is seen just above the level of the uppermost impeller. The profiles depend on temperature, with significantly less gas retained in a hot system and very few vapor bubbles being found below the liquid surface in boiling conditions.

## 11-6 GAS–LIQUID MASS TRANSFER

This section is concerned mainly with predicting or scaling-up the mass transfer rate between gas and liquid, in which the controlling factor is film diffusion on the liquid side of the interface, as described by the mass transfer coefficient, $k_L$. Ideally, perhaps, this should be done from a basis of predicting local bubble sizes and gas fractions, using perhaps CFD, but this is not established within the realms of process engineering. The traditional method is (as for gas fraction) to use empirical correlations for the mass transfer factor $k_L a$, and to use this in mass balance equations:

$$\text{overall transfer rate} = k_L a . V . (C^* - C_L)_{\text{mean}} \qquad (11\text{-}18)$$

This has the advantage of not requiring knowledge of bubble sizes, but also has some inherent disadvantages which are set out later in this section. Evidently, it will also be necessary to use an appropriate value of the mean for $(C^* - C_L)$, which, as discussed in Section 11-3, will in general be between those for the ideal backmixed and plug flow cases. It should be noted that this is important also for the extraction of $k_L a$ values from laboratory concentration measurements and may not have been observed correctly in the derivation of some older correlations.

## 11-6.1 Agitated Vessels

The homogeneous region correlations for $k_L a$ (again like those for gas fraction) for the turbulent regime are best expressed in the form

$$k_L a = \alpha'' \left(\frac{P}{\rho V}\right)^{\beta''} (v_s)^{\gamma''} \qquad (11\text{-}19)$$

where P includes shaft power and gas buoyancy power $[QHg(\rho_L - \rho_G)]$ but not gas kinetic energy (Middleton et al., 1994). Typical values for the air–water system at 20°C are $\alpha'' = 1.2$, $\beta'' = 0.7$, and $\gamma'' = 0.6$, with P in watts, V in m³, $v_s$ in m/s, $k_L a$ in s⁻¹, and $\alpha''$ dimensioned appropriately (Middleton, 1997). However, it has been found (e.g., by Smith et al., 1977) that whereas the indices $\beta''$ and $\gamma''$ do not change with liquid type, impeller type, or scale, $\alpha''$ is a strong function of liquid type and properties, the noncoalescing value being about twice that for coalescing systems. Thus such correlations can be used for scale-up purposes but not for general prediction. However, two concerns remain: one is the need for fractionally dimensioned constants, and the other is that [as shown in Smith et al. (1977)] the correlations are actually composed of smaller, very nonlinear curves, so should not be extrapolated outside their $v_s$ range (in this case, 0.004 to 0.02 m/s). It should also be noted (also for gas fraction), especially for disk turbines, that $P/\rho V$ is itself a function of $v_s$, so the variables in the correlation are not independent. This may explain why the indices $\beta''$ and $\gamma''$ vary between workers and data sets even when the $k_L a$ values may be similar. It is therefore recommended to use only those correlations that cover the relevant ranges of $P/\rho V$ and $v_s$, and not to extrapolate.

For extension into transitional Reynolds numbers (range 100 to 10⁶), Cooke et al. (1988) obtained

$$k_L a \propto \left(\frac{P}{\rho V}\right)^{0.5} (v_s)^{0.3} \mu_{app}^{-1} \qquad (11\text{-}20)$$

to ±30% for aqueous suspensions of fibers and several combinations of impellers at scales of 20 to 60 L, with H = T, and with a different value of the constant, with H = 3T. The P term in these correlations includes the contribution of gas buoyancy $[Q_G Hg(\rho_L - \rho_G)]$.

Although it is commonly assumed that when agitation conditions are sufficiently intense for effective gas–liquid dispersion, the liquid mixedness will be

**628**   GAS–LIQUID MIXING IN TURBULENT SYSTEMS

good. It is worth checking this, particularly for large vessels [noting that, on scaling up at constant $P/\rho V$ in the turbulent regime, N will decrease in proportion to (scale)$^{-2/3}$ and mixing time with $1/N$]. If it turns out that mixing time is longer than mass transfer time (90% mass transfer time $= 2.3/k_L a$), preferably the liquid mixing should be improved; otherwise, a more complex design calculation with interlinked zones of different driving force (and even perhaps local values for $k_L a$) will be necessary.

***Example 11-5: Impeller Size and Speed for Mass Transfer.*** Assume that 0.2 mol/s of gas A is to be absorbed into a coalescing type of aqueous solution of B in a baffled vessel of 2 m³ liquid capacity with a DIN torispherical base. What is the required design if 99% of gas A is to be absorbed and reacted?

The temperature $\theta$ is 300 K; the pressure at the sparger is 1.5 bar abs., and the inlet concentration of A in gas, $y_{A0}$, is 0.1 mol/mol. Henry's constant He $= 10^{-8}$ mol fr./Pa; molar volume of liquid $M_V = 50\,000$ g-mol/m³.

SOLUTION: Calculate the gas flow rate from the mass balance and absorption efficiency, $\eta$:

$$N_0 y_{A0} \eta = k_L a V (\Delta C)_{mean} = J$$

where $N_0$ is the inlet molar flow rate of gas and $y_{A0}$ is the concentration of A at the inlet.

$$N_0 = \frac{J}{y_{A0} \eta} = \frac{0.2 \text{ mol/s}}{0.1 \times 0.99} = 2.0 \text{ mol/s}$$

$P_0 Q_0 = N_0 R \theta$   ideal gas law applied at the inlet

$$Q_0 = \frac{N_0 R \theta}{P_0} = \frac{2.0 \text{ mol/s} \times 8.314 \text{ m}^3 \cdot \text{Pa/mol} \cdot \text{K} \times 300 \text{ K}}{152\,000 \text{ Pa}} = 0.0332 \text{ m}^3/\text{s}$$

Calculate the vessel dimensions for a DIN torispherical base, specifying that $H = T$:

$$H = T = \left(\frac{V}{0.7320}\right)^{1/3} = \left(\frac{2}{0.7320}\right)^{1/3} = 1.40 \text{ m}$$

Calculate the gas superficial velocity at the inlet:

$$v_{s0} = \frac{4 Q_{G0}}{\pi T^2} = \frac{4 \times 0.033 \text{ m}^3/\text{s}}{\pi \times (1.40 \text{ m})^2} = 0.022 \text{ m/s}$$

Calculate the pressure at the surface. This is given approximately by

$$P_1 = P_0 - \rho g H = 152\,000 \text{ Pa} - 1000 \text{ kg/m}^3 \times 9.81 \text{ m/s}^2$$
$$\times 1.40 \text{ m} = 138\,000 \text{ Pa}$$

The partial pressure of the reactant at the base is given by

$$p_{A0} = y_{A0}P_0 = 0.1 \times 152\,000 = 15\,200 \text{ Pa}$$

The partial pressure of the reactant at the surface is given by:

$$p_{A1} = y_{A1}P_1 = 0.001 \times 138\,000 \text{ Pa} = 138 \text{ Pa}$$

The saturation concentrations of the reactant at the base and the surface are given by

$$x_{A0}^* = \text{He} \cdot p_{A0} = 1 \times 10^{-8} \times 15\,200 = 1.52 \times 10^{-4} \text{ mol A/mol liquid}$$
$$x_{A1}^* = \text{He} \cdot p_{A1} = 1 \times 10^{-8} \times 132 = 1.32 \times 10^{-6} \text{ mol A/mol liquid}$$

Calculate the mean concentration driving force. Assume that the reaction is rapid such that the concentration of A in the bulk liquid phase is approximately zero. Assume also that the gas recirculation ratio α is low, approaching plug flow.

$$(\Delta C)'_{\text{mean}} = \frac{\Delta C_0 - \Delta C_1}{\ln(\Delta C_0/\Delta C_1)} = \frac{(1.52 \times 10^{-4} - 0) - (1.38 \times 10^{-6} - 0)}{\ln[(1.52 \times 10^{-4} - 0)/(1.38 \times 10^{-6} - 0)]}$$
$$= 3.20 \times 10^{-5} \text{ mol A/mol liquid}$$

Converting units yields

$$(\Delta C)_{\text{mean}} = (\Delta C)'_{\text{mean}} \times M_V = 3.20 \times 10^{-5} \times 50\,000 = 1.60 \text{ mol/m}^3$$

Calculate the mass transfer coefficient required:

$$k_L a V (\Delta C)_{\text{mean}} = J$$

$$k_L a = \frac{J}{V(\Delta C)_{\text{mean}}} = \frac{0.2 \text{ mol/s}}{2 \text{ m}^3 \times 1.60 \text{ mol/m}^3} = 0.063 \text{ s}^{-1}$$

To calculate the shaft power required; the correlation chosen for $k_L a$ is

$$k_L a = 1.2 \left(\frac{P}{\rho V}\right)^{0.7} (v_s)^{0.6}$$

(Note that preferably the constant in the $k_L a$ correlation is confirmed from the results of semitech scale tests.) Therefore;

$$\frac{P}{\rho V} = \left[\frac{k_L a}{1.2(v_s)^{0.6}}\right]^{1/0.7} = 0.39 \text{ W/kg}$$

The shaft power required will therefore be

$$P = 780 \text{ W}$$
$$P = Po \cdot RPD \cdot \rho N^3 D^5$$

Specify a BT-6 impeller with a ring sparger (Po = 2), with D = 0.4, T = 0.56 m. Assume that RPD is approximately 0.9 (this will be checked later). The required impeller speed is therefore

$$N = 2.0 \text{ s}^{-1}$$

Calculate the Reynolds number:

$$\text{Re} = \frac{\rho N D^2}{\mu} = \frac{1000 \times 2.0 \times 0.56^2}{10^{-3}} = 6.3 \times 10^5$$

confirming that the impeller flow is turbulent. Confirm the value of RPD:

$$\text{gas flow number } \text{Fl}_G = \frac{Q_G}{ND^3} = \frac{0.0332}{2.0 \times 0.56^3} = 0.094$$

Hence RPD = 0.9 is correct (see Table 11-4).

Confirm the estimated gas recirculation ratio $\alpha$: If the impeller had been a Rushton disk turbine, the correlation of Section 11-6.1 would have applied, giving a value of 0.14 for $\alpha$, confirming the approximation to plug flow of the gas assumed above. No correlation is available for the BT6 impeller, but the flow pattern is similar.

### 11-6.2  In-line Mixers

With static mixers in the turbulent regime, scale-up can be made using a correlation of almost the same form as that for vessels:

$$k_L a = \alpha''' \left(\frac{P}{\rho V}\right)^{\beta'''} (v_s)^{\gamma'''} \tag{11-21}$$

and $\beta''' = 0.42$ and $\gamma''' = 0.42$ have been found to fit data for coalescing and non-coalescing liquids for several mixer types and scales [in agreement with Holmes and Chen (1981)]. For air–water systems at 20°C, a value for $\alpha'''$ of 0.38 is obtained (with P in watts, V in m³, $k_L a$ in s$^{-1}$): however, $\alpha'''$ seems to vary slightly with liquid and mixer type and possibly scale, so fitting it from small scale tests (during process development) with the actual process fluids is advised. Note that $k_L a$ values are very high and that except for the fastest reactions, it can usually be assumed that equilibrium is achieved with a few elements.

## 11-6.3 Gas–Liquid Mass Transfer with Reaction[3]

When the reaction rate is comparable to that of the mass transfer through the diffusion film, interactions must be taken into account. The interactions can be delineated as five regimes, as shown in Figure 11-27. These are identified by the

| | REGIME | CONDITIONS | IMPORTANT VARIABLES | CONCENTRATION PROFILES |
|---|---|---|---|---|
| I | Kinetic control<br>Slow reaction | $\sqrt{\dfrac{t_D}{t_R}} < 0.02$ | Rate $\alpha\ \epsilon_L$<br>$\alpha\ k_{nm}$<br>$\alpha\ (C_{AL}^*)^n$<br>$\alpha\ (C_{BL})^m$<br>Independent of $a$<br>(if $a$ adequate)<br>Independent of $k_L$ | 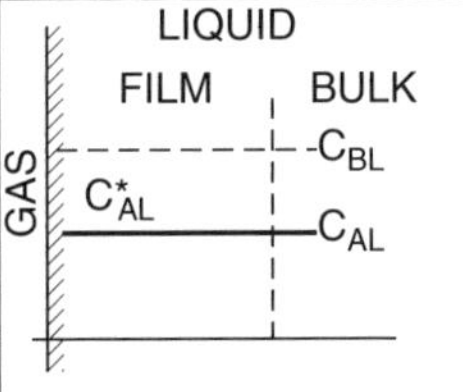 LIQUID<br>FILM \| BULK<br>$C_{AL}^*$ ---- $C_{BL}$<br>$C_{AL}$ |
| II | Diffusion control<br>Moderately fast reaction in bulk of liquid<br>$C_{AL} \approx 0$ | $0.02 < \sqrt{\dfrac{t_D}{t_R}} < 2$<br>Design so that<br>$\dfrac{\epsilon_L}{a} > 100\ \dfrac{D_{AL}}{k_L}$ | Rate $\alpha\ a$<br>$\alpha\ k_L$<br>$\alpha\ C_{AL}^*$<br>Independent of $k_{nm}$<br>Independent of $\epsilon_L$<br>(if $\epsilon_L$ adequate) | 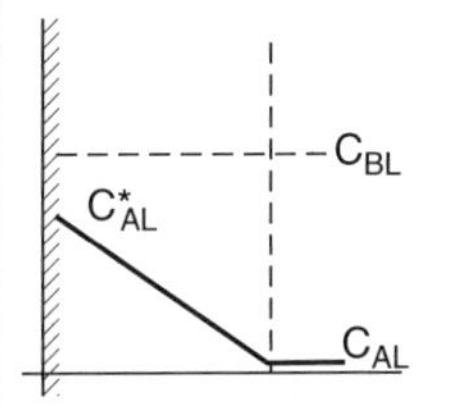 $C_{AL}^*$ ---- $C_{BL}$<br>$C_{AL}$ |
| III | Fast reaction<br>Reaction in film<br>$C_{AL} \approx 0$<br>(pseudo first order in A′) | $2 < \sqrt{\dfrac{t_D}{t_R}} < \dfrac{C_{BL}}{qC_{AL}^*}$<br>$C_{BL} \gg C_{AL}^*$ | Rate $\alpha\ a$<br>$\alpha\ \sqrt{k_{nm}}$<br>$\alpha\ (C_{AL}^*)^{(n+1)/2}$<br>Independent of $k_L$<br>Independent of $\epsilon_L$ | 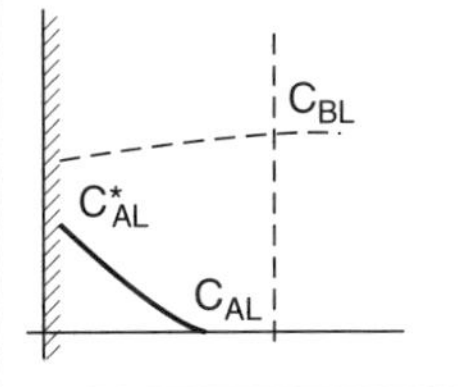 $C_{BL}$<br>$C_{AL}^*$<br>$C_{AL}$ |
| IV | Very fast reaction<br>General case of III | $2 < \sqrt{\dfrac{t_D}{t_R}}$<br>$C_{BL} \sim C_{AL}^*$ | Rate $\alpha\ a$<br>depends on<br>$k_L,\ k_{nm},\ C_{AL}^*,\ C_{BL}$<br>Independent of $\epsilon_L$ | 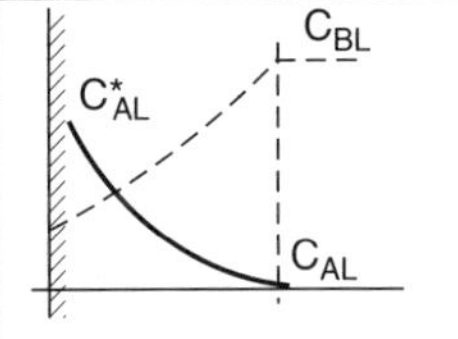 $C_{BL}$<br>$C_{AL}^*$<br>$C_{AL}$ |
| V | Instantaneous reaction<br>Reaction 'at interface'.<br>Controlled by transfer of B to interface from bulk.<br>$J \alpha\ k_L a$ | $\sqrt{\dfrac{t_D}{t_R}} \gg \dfrac{C_{BL}}{qC_{AL}^*}$ | Rate $\alpha\ a$<br>$\alpha\ k_L$<br>Independent of $C_{AL}^*$<br>Independent of $k_{nm}$<br>Independent of $\epsilon_L$ | 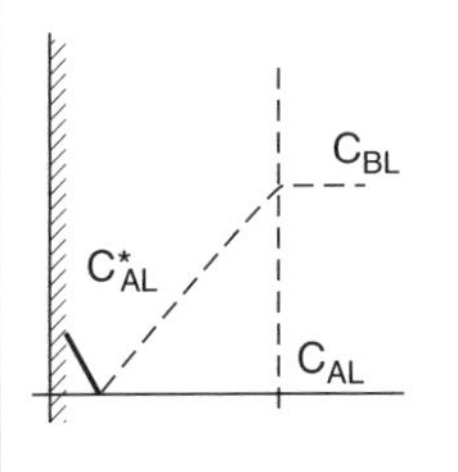 $C_{BL}$<br>$C_{AL}^*$<br>$C_{AL}$ |

**Figure 11-27** Regimes of gas–liquid mass transfer with reaction. (From Middleton, 1997; reproduced by permission of Butterworth-Heinemann.)

---
[3] This material is taken from Middleton (1997) by permission of Butterworth-Heinemann.

## GAS-LIQUID MIXING IN TURBULENT SYSTEMS

value of the *Hatta number*, Ha, which is defined as the square root of the ratio of the diffusion time, $t_D$, to the reaction time, $t_R$. For a reaction of the type

$$nA + mB \rightarrow products$$

these are defined as follows:

$$t_D = \frac{D_{AB}}{k_L^2} \tag{11-22}$$

$$t_R = \frac{n+1}{2k_{mn}C_{LA}^{n-1}C_{LB}^m} \tag{11-23}$$

For regimes III and IV the reaction effectively enhances the mass transfer rate, and an "enhanced" effective value of $k_L$ is often used, defined as $k_L^*$:

$$k_L^* = \left[ \frac{2D_{AB}k_{nm}(C_{LA}^* - C_{LA})^{n-1}C_{LB}^m}{n+1} \right]^{0.5} \tag{11-24}$$

Note that this is now a function of the reaction rate, not the hydrodynamics. If heat of reaction is significant, this expression must be modified to allow for the effects of local temperature on gas solubility and reaction rate (Mann and Moyes, 1977).

The regime dictates the choice of reactor. From Figure 11-27, the following choice of equipment for each regime can be inferred:

- *Regime I:* reaction in bulk, modest $k_La$: bubble column
- *Regimes II, IV, and V:* high a and $k_La$: stirred vessel
- *Regime III:* all reaction in film, high a: thin-film reactor (packed column or spinning disk)

It should also be noted that as the reaction rate increases, it becomes more likely that the gas-side resistance will become important, so this should be checked if the gas phase is multicomponent.

## 11-7 BUBBLE SIZE

The apparent success of $P/\rho V$ as a correlating parameter for $\phi$ and $k_La$ in the turbulent regime implies that it is strongly linked to bubble size as well as to liquid circulation. Indeed, $P/\rho V$ can be equated to the vessel-average value of $\varepsilon$, the turbulent energy dissipation rate at the smallest scales of turbulence, if it is assumed (classically) that all the power eventually dissipates at these scales. Several workers have postulated that bubble breakup occurs (finally) by impact of turbulent eddies at this smallest (Kolmogorov) scale, presumably via pressure fluctuations distorting the bubble sufficiently to disrupt it. Hinze (1979) balanced this external force with the restoring surface tension to obtain a critical Weber number (We = $\tau d/\sigma$) above which breakup will occur, with for turbulent breakup, $\tau = 2\rho(\varepsilon d)^{2/3}$; thus, d is the maximum bubble size to survive. There are, however, some conceptual problems with applying this. First, breakup occurs only in the regions of highest stress (in the impeller vortices of an agitated vessel or the wall shear layers in static mixers), and the ratio of maximum to mean $\varepsilon$ (i.e., $P/\rho V$) differs between impeller types, but the same correlations for $k_L a$ apply to different impeller types. Second, the Kolmogorov scale of turbulence in the discharge of typical impellers ($\varepsilon \approx 10$ W/kg) is $(\nu^3/\varepsilon)^{1/4} \approx 0.02$ mm, which is considerably smaller than the final bubble size, so a mechanism whereby sufficiently large bubble distortions are produced by this mechanism, and for sufficiently long time scales for breakup to occur, is difficult to imagine. Breakup has been observed to occur only in the blade and blade vortex region, and several possible mechanisms have been postulated [see, e.g., Kumar et al. (1991) for liquid droplets] involving bubble (or drop) stretching by shear and elongational flows. Rationalization of the correlations cited above with these observations is still awaited. The bimodal bubble size distributions found in the heterogeneous regime also await fundamental explanation.

## 11-8 CONSEQUENCES OF SCALE-UP

It is evident that having made a choice of scale-up relationship, other factors will be affected in different ways; there is often no way to scale up all the significant factors together, so priorities have to be chosen. An example of this is given in Middleton (1997), to which reference should be made for full details. A summary is given here.

In the example a gas–liquid reaction with particulate solids (e.g., a catalyst) operating in regime II in a stirred reactor with a Rushton turbine is to be scaled up. The primary process requirement is for the same degree of reaction conversion at each scale, which means the same number of moles of gas transferred per mole of liquid fed:

$$\frac{k_L a V (C_A^* - C_A)}{C_{LBfeed} Q_L} = \text{constant} \qquad (11\text{-}25)$$

Assume for simplicity that $C_A = 0$ (a good approximation for regime II) and that the degree of gas backmixing is the same at all scales (this should be checked at the end of the calculation and reiterations performed if necessary). Given a constant feed concentration at all scales,

$$k_L a V C_L \propto Q_L \qquad (11\text{-}26)$$

Sometimes it is necessary for the outlet gas concentration to be constant (e.g., with hazardous gases); then from the mass balance this becomes

$$k_L a V \propto Q_G \qquad (11\text{-}27)$$

# GAS–LIQUID MIXING IN TURBULENT SYSTEMS

Substituting a suitable correlation for $k_L a$, for example

$$k_L a \propto \left(\frac{P}{V}\right)^{0.7} (v_S)^{0.6} \tag{11-28}$$

and a curve fit for the gassed power curve, such as

$$P = N^{3.3} D^{6.3} Q_G^{-0.4} \quad \text{(not necessarily reliable!)} \tag{11-29}$$

an expression such as

$$N^{3.4} T^{6.0} \propto Q_L \tag{11-30}$$

results.

Another constraint will then fix the design. In this example maintaining $N > N_{JS}$ for the suspension of the catalyst particles is important, so $NT^{0.76} =$ constant could be added (although not strictly applicable to gassed systems), giving

$$Q_G \propto Q_L \propto T^{3.4} \tag{11-31}$$

This scale-up method has the effects, on increasing the scale, of:

- Increasing $v_S$, so foaming and entrainment become more likely
- Decreasing P/V
- Decreasing the heat transfer flux per unit throughput
- Nearer approach to poor gas dispersion
- Longer liquid mixing time

## NOMENCLATURE

| | |
|---|---|
| a | gas–liquid interfacial area per unit volume of liquid (m²/m³) |
| c | constant in eq. (11-1) |
| C | off-bottom clearance of impeller (m) |
| $C_L$ | saturation concentration of solute gas in bulk liquid (mol/m³) |
| C* | concentration of solute gas in liquid at interface (mol/m³) |
| $C_{ag}$ | agitation cavitation number, eq. (11-11) |
| d | surface mean bubble size (m) |
| D | impeller diameter (m) |
| $D_{AB}$ | molecular diffusivity of A in B (m²/s) |
| $Fl_G$ | gas flow number, $Q_G/ND^3$ (–) |
| Fr | Froude number, $N^2 D/g$ (–) |
| H | liquid height (m) |
| J | gas flux (mol/s) |
| $k_L$ | mass transfer coefficient (m²/s) |
| N | impeller speed (rps) |
| $N_{js}$ | just suspended speed for solids suspension (rps) |
| P | power draw (W) |
| $P_G$ | gassed power draw (W) |

| | |
|---|---|
| $P_U$ | ungassed power draw (W) |
| Po | power number (−) |
| q | mass transfer enhancement coefficient due to reaction (−) |
| $Q_L$ | liquid volumetric flowrate (m$^3$/s) |
| $Q_G$ | mean gas volumetric flowrate (m$^3$/s) |
| RPD | relative power demand, $P_G/P_U$ (−) |
| S | submergence of the impeller below the liquid surface (m) |
| $S_m$ | Smith number $2gS/v_t^2$ |
| $t_D$ | diffusion time (s) |
| $t_M$ | mixing time (s) |
| $t_R$ | reaction time (s) |
| T | vessel diameter (m) |
| U | superficial velocity (m/s) |
| $v_s$ | gas superficial velocity (m/s) |
| $v_t$ | impeller tip speed (m/s) |
| V | volume of liquid (m$^3$) |
| W | blade width (m) |
| We | Weber number $\tau d/\sigma$ (−) |

*Greek Symbols*

| | |
|---|---|
| α | gas recirculation ratio (−) |
| ε | turbulent energy dissipation rate (m$^2$/s$^3$) |
| $\varepsilon_L$ | liquid hold-up (−) |
| φ | gas volume fraction |
| μ | viscosity (kg/ms) |
| $\mu_{APP}$ | apparent viscosity (kg/ms) |
| ν | kinematic viscosity (m$^2$/s) |
| θ | absolute temperature (K) |
| ρ | density (kg/m$^3$) |
| σ | surface tension (N/m) |
| τ | local shear stress (N/m$^2$) |

## REFERENCES

Bin, A. K., and J. M. Smith (1982). Mass transfer in a plunging jet absorber, *Chem. Eng. Commun.*, **15**, 367–383.

Breber, G. (1986). The decrease in power number of impellers due to cavitation phenomena, Paper 77c. AIChE Meeting, Miami Beach, FL.

Bruijn, W., K. van't Riet, and J. M. Smith (1974). Power consumption with aerated Rushton turbines, *Trans. Inst. Chem. Eng.*, **52**, 88–104.

Bujalski, W., A. W. Nienow, S. Chatwin, and M. Cooke (1987). The effect of scale on power number for Rushton disc turbines, *Chem. Eng. Sci.*, **42**, 317–326.

Chapman, C. M., A. W. Nienow, J. C. Middleton, and M. Cooke (1983). Particle–gas–liquid mixing in stirred vessels, *Chem. Eng. Res. Des. (Trans. Inst. Chem. Eng. A)*, **61**, 71–82, 167–182.

Cooke, M., J. C. Middleton, and J. R. Bush (1988). Mixing and mass transfer in filamentous fermentations, *Paper B1, Proc. 2nd International Conference on Bioreactors,* pp. 37–64.

Deckwer, W.-D. (1992). *Bubble Column Reactors,* Wiley, Chichester, West Sussex, England.

Gao, Z., J. M. Smith, D. Zhao, and H. Müller-Steinhagen (2000). Void fraction and mixing in sparged and boiling reactors, *Proc. 10th European Mixing Conference,* Delft, The Netherlands, Elsevier, Amsterdam, pp. 213–220.

Gao, Z., J. M. Smith, and H. Müller-Steinhagen (2001). The effect of temperature on the void fraction in gas–liquid reactors, *Proc. 5th Symposium on Gas–Liquid–Solid Systems,* Melbourne, Australia.

Gezork, K. M., W. Bujalski, M. Cooke, and A. W. Nienow (2000). The transition from homogeneous to heterogeneous flow in a gassed stirred vessel, *Chem. Eng. Res. Des. (Trans. Inst. Chem. Eng. A),* **78A**, 363–370.

Gezork, K. M., W. Bujalski, M. Cooke, and A. W. Nienow (2001). Mass transfer and hold-up characteristics in a gassed, stirred vessel at intensified operating conditions, *Proc. ISMIP4,* Toulouse, France [submitted to *Chem. Eng. Res. Des. (Trans. Inst. Chem. Eng. A)*].

Hari-Prajitno, H., V. I. Mishra, K. Takemaka, W. Bujalski, A. W. Nienow, and J. McKemmie (1998). Gas–liquid mixing studies with multiple up and down pumping hydrofoil impellers: power characteristics and mixing times, *Can. J. Chem. Eng.,* **76**, 1056–1068.

Hinze, J. O. (1979). *Turbulence,* McGraw-Hill, New York.

Holmes, T. L., and T. L. Chen (1981). Gas–liquid contacting with horizontal static mixing systems, presented at the AIChE Annual Meeting, New Orleans, LA.

Kumar, S., R. Kumar, and K. S. Gandhi (1991). Alternative mechanisms of drop breakage in stirred vessels, *Chem. Eng. Sci.,* **46**, 2483–2489.

Lockhart, R. W., and R. C. Martinelli (1944). Proposed correlation of data for isothermal two phase, two component flow in pipes, *Chem. Eng. Prog.,* **45**, 39–48.

Mann, R., and M. Moyes (1977). Exothermic gas absorption with chemical reaction, *AIChE J.,* **23**, 17–23.

Metzner, A. B., and R. E. Otto (1957). Agitation of non-Newtonian fluids, *AIChE J.,* **3**, 3–10.

Michel, B. J., and S. A. Miller (1962). Power requirements of gas–liquid agitated systems, *AIChE J.,* **8**, 264–266.

Middleton, J. C. (1978). Motionless mixers as gas–liquid contacting devices, *Paper 74e,* presented at the AIChE Annual Meeting, Miami Beach, FL.

Middleton, J. C. (1997). Gas–liquid dispersion and mixing, in *Mixing in the Process Industries,* N. Harnby, A. W. Nienow, and M. F. Edwards, eds., Butterworth—Heinemann, Oxford.

Middleton, J. C., M. Cooke, and L. Litherland (1994). The role of kinetic energy in gas–liquid dispersion, *Proc. 8th European Mixing Conference, Inst. Chem. Eng. Symp. Ser.,* **136**, 595–602.

Montante, G., K. C. Lee, A. Brucato, and M. Yianneskis (2001). Numerical simulations of the dependency of flow pattern on impeller clearance in stirred vessels, *Chem. Eng. Sci.,* **56**, 3751–3770.

Nagel, O., H. Kuerten, and B. Hegner (1973). The interfacial area in gas–liquid contact apparatus, *Chem. Ing. Tech.,* **45**, 913–920.

Nienow, A. W., D. J. Wisdom, and J. C. Middleton (1977). The effect of scale and geometry on flooding, recirculation and power in gassed stirred vessels, *Proc. 2nd European Mixing Conference*, Cambridge, pp. 17–34.

Nienow, A. W., M. Konno, and W. Bujalski (1986). Studies on three phase mixing: a review and recent results, *Chem. Eng. Res. Des. (Trans. Inst. Chem. Eng. A)*, **64**, 35–42.

Schaper, R., A. B. de Haan, and J. M. Smith (2002). Temperature effects on the hold-up in agitated vessels, *Chem. Eng. Res. Des. (Trans. Inst. Chem. Eng. A*, **80**, 887–892.

Skelland, A. H. P. (1967). *Non-Newtonian Flow and Heat Transfer*, Wiley, New York.

Smith, J. M., and A. Katsanevakis (1993). Impeller power demand in mechanically agitated boiling systems, *Chem. Eng. Res. Des. (Trans. Inst. Chem. Eng. A)*, **71A**, 145–152, 466.

Smith, J. M., and C. A. Millington (1996). Boil-off and power demand in gas–liquid reactors, *Chem. Eng. Res. Des. (Trans. Inst. Chem. Eng. A)*, **74A**, 424–430.

Smith, J. M., and L. Smit (1988). Impeller hydrodynamics in boiling reactors, *Proc. 6th European Conference on Mixing*, Pavia, Italy, pp. 297–304.

Smith, J. M., and K. Tarry (1994). Impeller power demand in boiling solutions, *Chem. Eng. Res. Des. (Trans. Inst. Chem. Eng. A)*, **72A**, 739–740.

Smith, J. M., and D. G. F. Verbeek (1988). Impeller cavity development in nearly boiling liquids, *Trans. Inst. Chem. Eng.*, **66**, 39–46.

Smith, J. M., K. van't Riet, and J. C. Middleton (1977). Scale up of agitated gas–liquid reactors for mass transfer, *Proc. 2nd European Conference on Mixing*, pp. F4-51 to F4-66.

Smith, J. M., M. M. C. G. Warmoeskerken, and E. Zeef (1987). Flow conditions in vessels dispersing gases in liquids with multiple impellers, in *Biotechnology Processes*, C. S. Ho and J. Y. Oldshue, eds., AIChE, New York, pp. 107–115.

Smith, J. M., Z. Gao, and J. C. Middleton (2001a). The unsparged power demand of modern gas dispersing impellers in boiling liquids, *Chem. Eng. J.*, **84**, 15–22.

Smith, J. M., Z. Gao, and H. Müller-Steinhagen (2001b). Void fraction distributions in sparged and boiling reactors with modern impeller configurations, *Chem. Eng. Process.*, **40**, 489–497.

Van de Sande, E., and J. M. Smith (1973). Surface entertainment of air by high velocity water jets, *Chem. Eng. Sci.*, **28**, 1161–1168.

Van de Sande, E., and J. M. Smith (1974). Mass transfer with plunging water jets, in *Multi-phase Flow Systems, Inst. Chem. Eng. Symp. Ser.*, **38**(J3), 1–11.

Van't Riet, K. (1976). Turbine agitator hydrodynamics and dispersion performance, Ph.D. dissertation, University of Delft, The Netherlands.

Warmoeskerken, M. M. C. G., and J. M. Smith (1982). Description of the power curves of turbine stirred gas dispersions, *Proc. 4th European Conference on Mixing*, Noordwijkerhout, The Netherlands, pp. 237–246.

Warmoeskerken, M. M. C. G., and J. M. Smith (1986). Flow regime maps for Rushton turbines, *Paper W-624*, presented at the 3rd World Congress of Chemical Engineering, Tokyo.

Warmoeskerken, M. M. C. G., and J. M. Smith (1989). Hollow blade turbine impellers for gas dispersion and mass transfer, *Chem. Eng. Res. Des. (Trans. Inst. Chem. Eng. A)*, **67A**, 193–198.

Warmoeskerken, M. M. C. G., J. Speur, and J. M. Smith (1984). Gas–liquid dispersion with pitched blade turbines, *Chem. Eng. Commun.*, **25**, 11–29.

Zhao, D., Z. Gao, H. Müller-Steinhagen, and J. M. Smith (2001). Liquid-phase mixing times in sparged and boiling agitated reactors with high gas loading, *Ind. Eng. Chem. Res.*, **40**, 1482–1497.

Zlokarnik, M. (1979). Sorption characteristics of slot injectors and their dependency on the coalescence behaviour of the system, *Chem. Eng. Sci.*, **34**, 1265–1271.

Zuber, N., and J. A. Findlay (1965). Average volumetric concentration in two-phase flow systems, *J. Heat Transfer ASME*, **87C**, 453–468.

# CHAPTER 12

# Immiscible Liquid–Liquid Systems

DOUGLAS E. LENG
*Leng Associates*

RICHARD V. CALABRESE
*University of Maryland*

## 12-1 INTRODUCTION

### 12-1.1 Definition of Liquid–Liquid Systems

In this chapter we describe the use of agitated vessels and other equipment to create immiscible liquid–liquid dispersions. The primary purpose is to help the reader become acquainted with the physical and interfacial phenomena involved with coalescence and dispersion, and how to use these phenomena in practice. The goal is to predict mean drop size and drop size distribution for a given design, set of properties, and operating conditions. The subject matter is complex, often failing to predict accurate information. Unforeseen impurities, interfacial "scum," phase inversions, and poorly defined objectives complicate reliable predictions for the practitioner.

The term *immiscible liquid–liquid system* refers to two or more mutually insoluble liquids present as separate phases. These phases are referred to as the *dispersed* or *drop phase* and the *continuous* or *matrix phase* and are given subscripts of d and c, respectively. The dispersed phase is usually smaller in volume than the continuous phase, but under certain highly formulated conditions, it can represent up to 99% of the total volume of the system. Immiscible liquid–liquid systems can also contain additional liquid, solid, or gas phases.

Agitation plays a controlling role in the liquid–liquid systems considered herein. It controls the breakup of drops, referred to as *dispersion*; the combining of drops, known as *coalescence*; and the suspension of drops within the

---

*Handbook of Industrial Mixing: Science and Practice,* Edited by Edward L. Paul, Victor A. Atiemo-Obeng, and Suzanne M. Kresta
ISBN 0-471-26919-0  Copyright © 2004 John Wiley & Sons, Inc.

system. The magnitude and direction of convective flows produced by an agitator affect distribution and uniformity throughout the vessel as well as the kinetics of dispersion. Agitation intensity is also important. Intense turbulence found near the impeller leads to drop dispersion, not coalescence. Lower turbulence or laminar/transitional conditions found elsewhere in the vessel promote coalescence by enabling drops to remain in contact long enough for them to coalesce. Laminar shear also leads to drop dispersion. If a drop is stretched beyond the point of critical elongation, it breaks. If not, it returns to its prestressed state as it enters a more quiescent region.

### 12-1.2 Practical Relevance

*12-1.2.1 Industrial Applications.* Immiscible liquid–liquid systems are found extensively throughout the chemical, petroleum, and pharmaceutical industries. The rate of chemical reactions is often mass transfer controlled and affected by interfacial area. Examples include nitration, sulfonation, alkylation, hydrogenation, and halogenation. For example, the nitration of aromatic compounds involves use of a continuous phase of concentrated mixed acids ($HNO_3 + H_2SO_4$) and a dispersed organic phase to be nitrated. Dispersion, coalescence, and suspension are all involved, along with heat and mass transfer. The nitronium ion from the continuous phase is transported to the drop surface, where reaction occurs. Water, a by-product of the reaction, transfers to the continuous phase. Nitration reactions are exothermic, and reaction rates and temperatures are controlled by interfacial area, created by agitation. Failure to suspend drops adequately can lead to catastrophic results, as described in Section 12-9.

The petroleum industry depends on efficient coalescence processing to remove aqueous brine drops in crude refinery feed streams to prevent severe corrosion of processing equipment. Control of mean drop size and drop size distribution (DSD) is vital to emulsification and suspension polymerization applications. Extraction processes depend on repeated drop coalescence and dispersion to accomplish the required mass transfer.

Coalescence, dispersion, and suspension phenomena are complex and scale dependent. Nevertheless, some industrial processes can be simplified, as suggested in Table 12-3, if they are either noncoalescing or slowly coalescing. This simplifies design and scale-up. Coalescence can usually be neglected, for practical purposes, in applications where the volume fraction of dispersed phase, $\phi \leq 0.1$. This is particularly true if surfactants and/or interfacial contaminants are present.

*12-1.2.2 Design Scope.* Stirred vessels, rotor–stator mixers, static mixers, decanters, settlers, centrifuges, homogenizers, extraction columns, and electrostatic coalescers are examples of industrial process equipment used to contact liquid–liquid systems. Although this chapter emphasizes stirred vessels, the fundamentals of phase behavior are applicable to a broad range of other equipment types. Immiscible liquid–liquid systems are processed in batch, continuous, and semicontinuous modes.

In the case of stirred vessels, the resulting mean drop size and drop size distribution depend on the selection, placement, and operational speed of the agitator. Excessive speed leads to hard-to-separate emulsions. Inadequate speed can cause phase separation. Coalescence and dispersion are both fluid motion-dependent rate processes. Drop sizes depend on flow, shear, turbulence, and dispersion time as well as on physical and interfacial system properties.

### 12-1.3 Fundamentals

An agitated liquid–liquid process involves many simultaneous, interdependent phenomena, such as dispersion, coalescence, suspension, heat and mass transfer, and chemical reaction. Previously described nitration requires control of the interfacial area rather than specific drop size, but some processes require precise control of drop size. For example, equipment for suspension polymerization processes must be capable of producing uniform beads of specified size range as well as providing for heat transfer and drop suspension.

Flow patterns and turbulence in stirred vessels are complex phenomena that can often be better appreciated using modern tools such as laser Doppler velocimetry (LDV), particle image velocimetry (PIV), and computational fluid dynamics (CFD). All show turbulence consisting of high-energy eddies near the impeller, and lower-energy eddies located farther away. Turbulence intensities near the impeller can be ≈40 times greater than the mean for the entire vessel. Turbulence intensities are very low in regions close to the wall and at the top and bottom of the vessel. Flow patterns are sensitive to impeller geometry, the number of impellers, and their position in the vessel. The reader is referred to Chapters 2 and 6 for more detail. Since drop size depends on flow-dependent dispersion and coalescence phenomena, it can be concluded that certain regions of the vessel are dominated by dispersion while others are dominated by coalescence. When a drop is contained within a larger eddy, it rotates within that eddy and does not break up. However, if it encounters an eddy of its size or smaller, it can be deformed and dispersed. When drops suspended in gentle flows collide, they often remain in contact long enough to coalesce.

*12-1.3.1 Breakup, Coalescence, and Phase Inversion.* Drop deformation is caused by mechanical forces induced by the surrounding fluid and is resisted by surface and internal viscous forces. Drop breakage occurs when fluid forces exceed the combined resistance force. Figure 12-1 shows different types of drop deformation due to different disruptive forces. Impact drop collisions (walls, impeller blades, and baffles) lead to lenticular deformation, uniform shear leads to cigar-shaped deformation, and turbulent conditions lead to bulgy deformation. It is common practice to refer to all fluid dynamical forces that cause drop deformation as *shear forces* regardless of the controlling mechanisms. These include shear and extension in laminar flow, and pressure fluctuations in turbulent flow.

An elongated drop does not necessarily break. In simple shear flows, differences in surface drag establish an internal rotation or circulation within the drop that helps stabilize it. This circulation does not develop for the case of bulgy deformation.

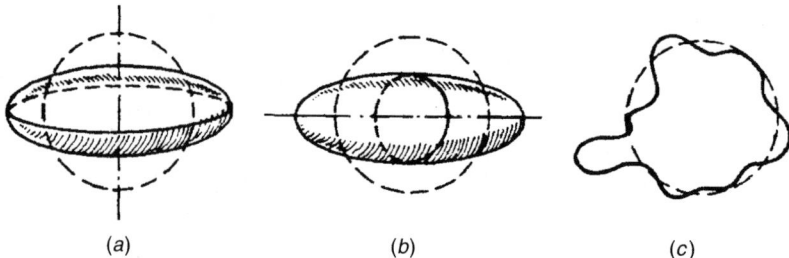

**Figure 12-1** Basic types of globule deformation: (*a*) lenticular; (*b*) cigar-shaped; (*c*) bulgy. (Reproduced from Hinze, 1955.)

Surface forces due to interfacial tension attempt to minimize surface area by forcing the elongated drop to return to its original spherical shape. Breakage does not occur unless a critical deformation is reached during stretching. The drop either breaks or reverts to a condition of lower deformation as it passes to a region of lower shear rate. Dispersion also occurs by collisions with solid surfaces such as impeller blades, baffles, and vessel walls. Impeller selection and tank geometry are important in preventing this undesirable, uncontrolled form of dispersion. Fluid shear forces are mostly responsible for drop dispersion in stirred tanks, but impingement can be important in static mixers and rotor-stator machines. This topic is discussed in Section 12-2.

Coalescence is the combining of two or more drops, or a drop with a coalesced layer. The two-step process involves collision followed by film drainage. The drainage step depends on the magnitude and duration of the force acting on the drop(s), to squeeze out the separating film to a critical thickness, believed to be in the range $\approx 50$ Å. In the case of a drop coalescing to a settled layer, the force is gravitational. The rate of film thinning also depends on the interfacial tension and the viscosity of the phases. Collision frequency depends on both agitation rate and the volume fraction of the dispersed phase. Not all collisions result in coalescence. If the contact is of short duration, critical thickness is not reached during contact, and the drops separate. The coalescence rate is the product of the collision rate and the coalescence efficiency. The mobility of the liquid–liquid interface also affects the film drainage rate. Clean, mobile interfaces promote efficient film drainage and lead to higher coalescence probability. As drops collide, a flattened disk forms at the leading drop surfaces. The diameter of this disk is important. If the system has a low interfacial tension, a large disk forms and more continuous phase fluid is trapped. This increases the task of drainage and reduces coalescence probability. A viscous continuous phase lowers drainage rates and therefore coalescence probabilities. Coalescence probabilities have been correlated in terms of the ratio of the contact time to drainage time. Section 12-3 deals with this subject quantitatively.

Phase inversion is the transitioning of water dispersed in oil (w/o) to oil dispersed in water (o/w), or vice versa. It can occur in more concentrated systems as a result of changes in stabilization, physical properties, or phase proportions.

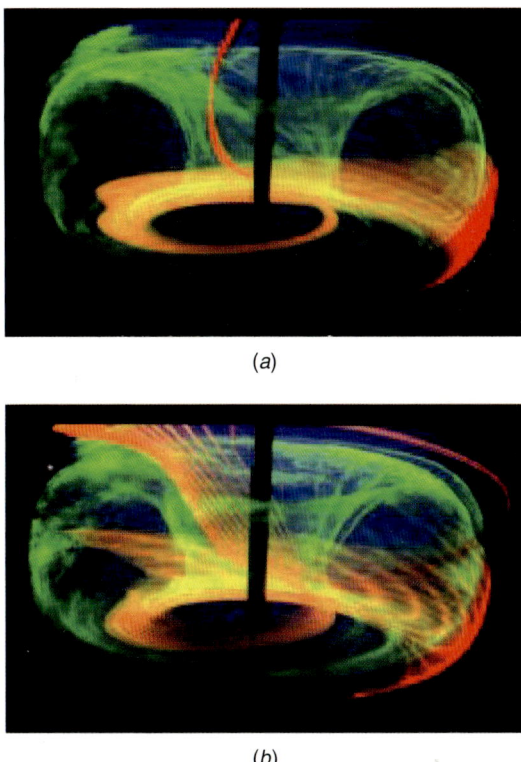

(a)

(b)

**Figure 3-8** The complex mixing patterns formed by chaotic flows are highly heterogeneous, and understanding the emerging structure is crucial to predicting heat and mass transfer in these systems. Here, some experimental pictures are shown of the mixing pattern formed by colored fluorescent dye in a stirred tank. The time evolution of the emerging structure can be monitored if a series of snapshots are taken in time, as in (a) and (b).

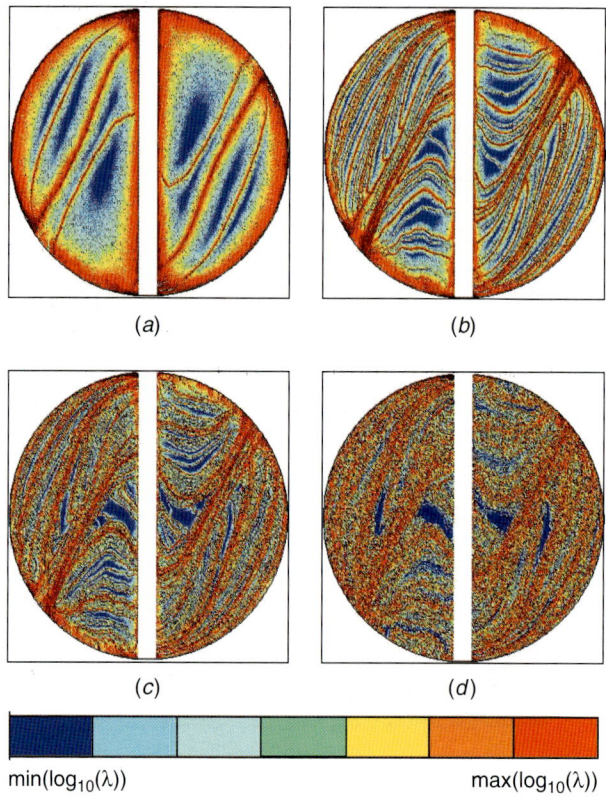

**Figure 3-21** Contours of the stretching field in the standard Kenics mixer at $Re = 10(10)^{1/3}$. The cross-sectional planes correspond to axial distances after (a) 2, (b) 6, (c) 10, and (d) 22 mixer elements.

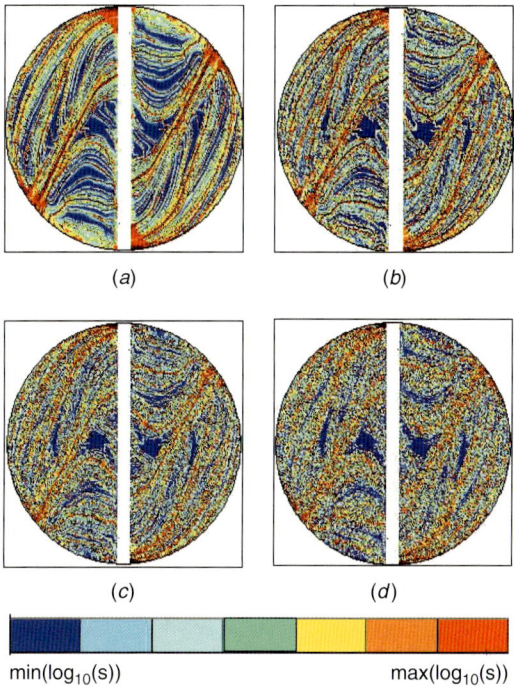

**Figure 3-23** Intermaterial area density ($\rho$) from coarse-grained stretching average in the standard Kenics mixer at Re = $10(10)^{1/3}$ after (a) 8, (b) 12, (c) 16, and (d) 22 mixer elements.

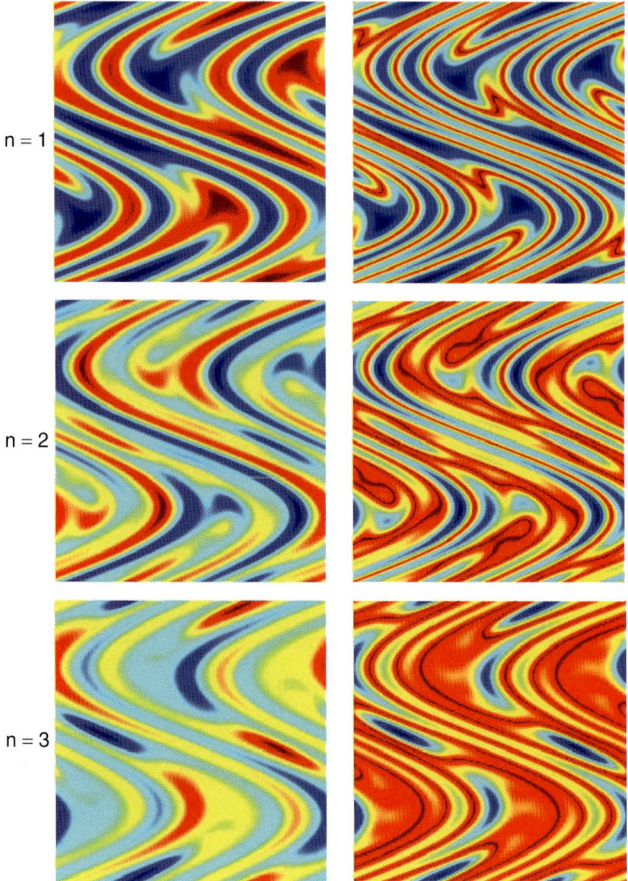

**Figure 3-28** The left-hand side of the figure shows the concentration of reactants A and B after the first three flow periods in the sine flow with T = 1.6. The right-hand side is the corresponding product concentration.

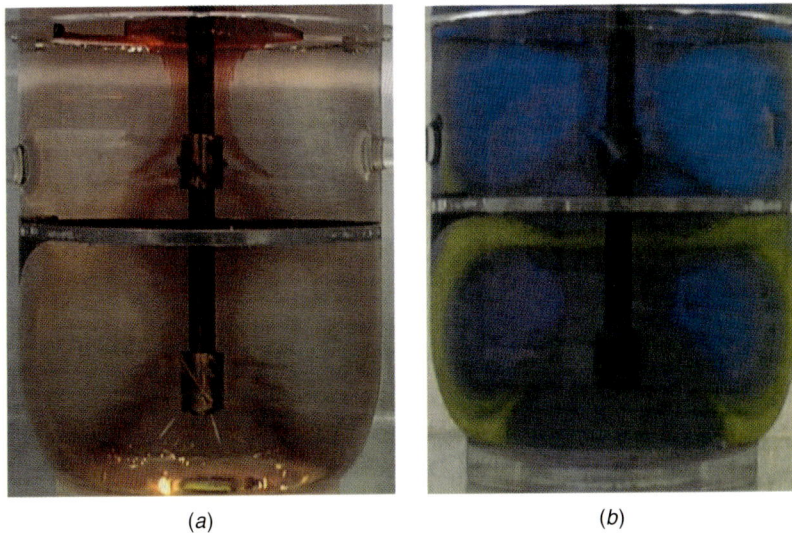

**Figure 4-13** Red dye tracer addition (*a*) and acid–base decolorization using bromophenol blue (*b*). (From Clark and Özcan-Taşkin, 2001.)

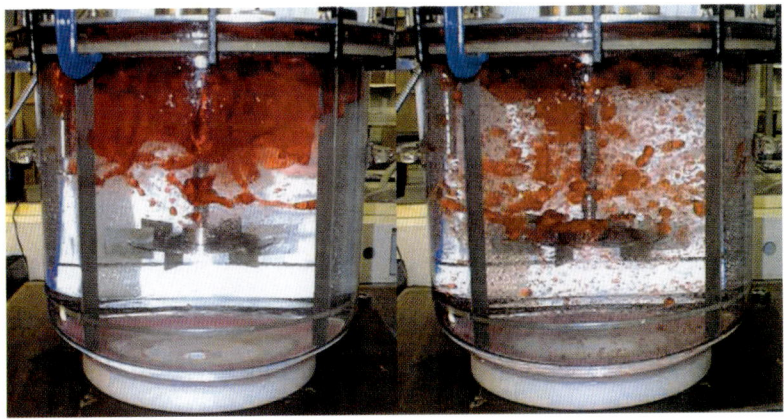

**Figure 4-24** Immiscible liquid–liquid system at a speed below that required for complete dispersion.

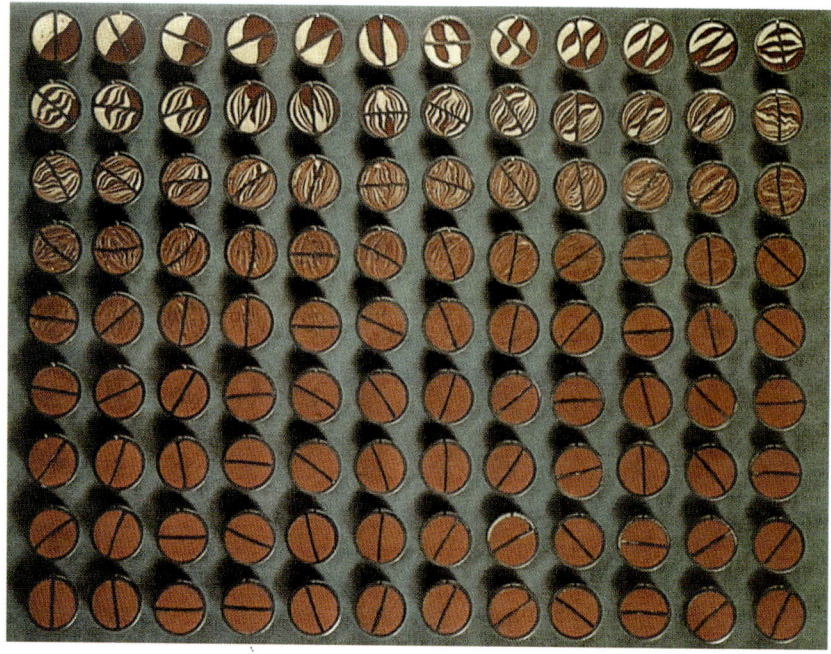

**Figure 7-20** Laminar mixing of fluids by division and recombination (KMS mixer). Cross-sections of the mixer are shown in sequence from left to right, top to bottom. (Courtesy of Chemineer, Inc.)

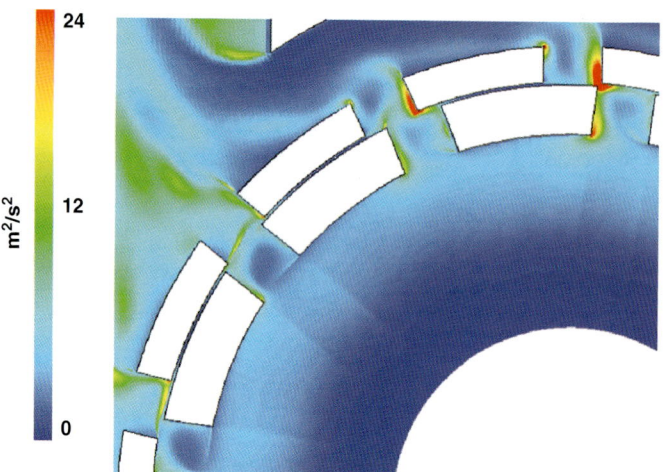

**Figure 8-12** Turbulent kinetic energy (m$^2$/s$^2$) from 2D CFD simulation for IKA prototype mixer. Single-time snapshot.

**Figure 20-4** Photographs of fibers and suspensions: (a) photomicrograph of softwood trachieds ($l_w$ = 2.37 mm); (b) light transmission through a $C_m$ = 0.005 kraft pulp suspension showing mass nonuniformities (flocs). The line in the image is 5 cm in length; (c) hands hold a medium consistency ($C_m$ = 0.10) kraft pulp suspension.

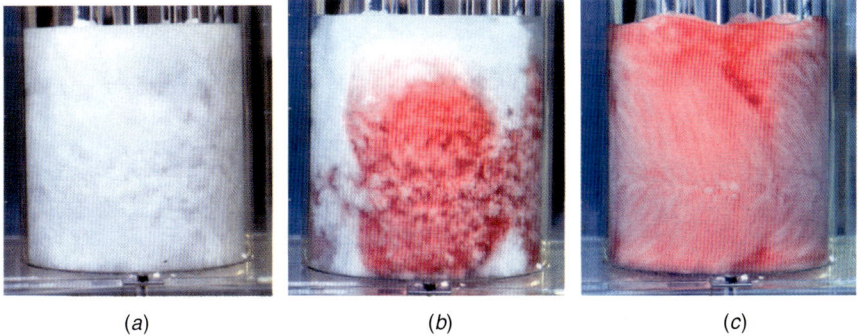

**Figure 20-6** Effect of yield stress on suspension motion in a stirred tank. $C_m$ = 0.02 FBK suspension. The vessel is 30 cm in diameter with the suspension height set at 30 cm. A D = 10 cm diameter Rushton turbine was located 10 cm from the vessel floor. Impeller speeds are N = (a) 4, (b) 7, and (c) 14 rps. The red dye shows regions of suspension motion. In image (a) the cavern has not reached the vessel wall.

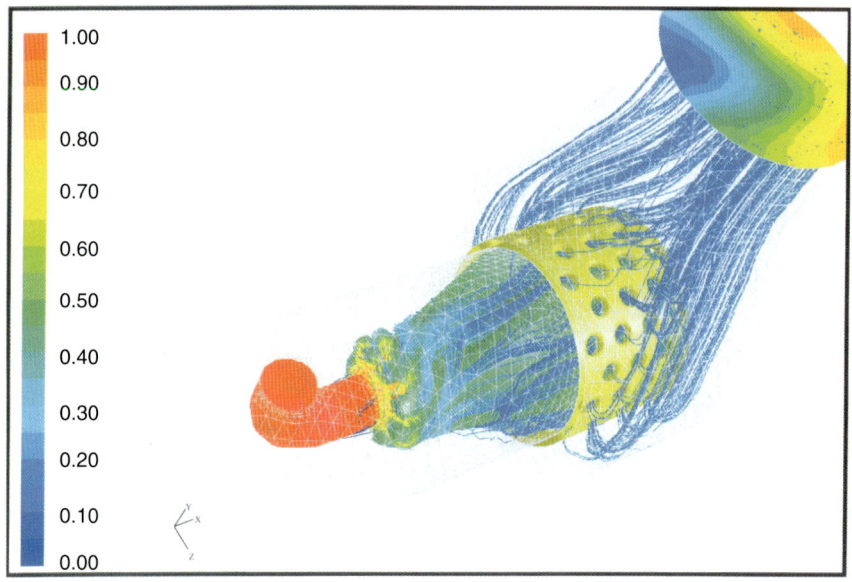

**Figure 20-19** CFD solution for flow in an early tri-phase mixer (GL&V) prior to design modifications. Note the spatial distribution of chemical (chlorine dioxide) in the pipe exiting the mixer.

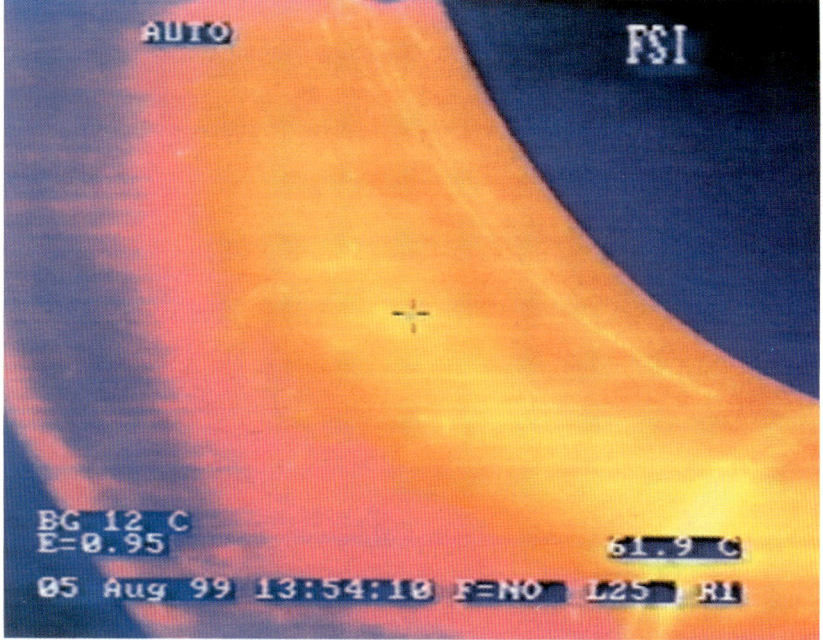

**Figure 20-23** Thermograph of exit piping following a high-shear chlorine dioxide mixer in $D_1$ service. Elbow immediately following mixer discharge. The temperature is 12°C higher along the top of the pipe.

# INTRODUCTION

For example, if chemical changes result in one phase becoming more viscous, that phase will tend to become the continuous phase. If this phase was originally the continuous phase, no inversion occurs, but if it was the dispersed phase, inversion is likely to occur. This phenomenon is discussed in Section 12-5.

The initial dispersion of two settled layers can create either (o/w) or (w/o) systems, often both temporarily. This is shown in Figure 12-34. However, the continued addition of one phase normally makes that phase the continuous phase. Surface-active materials also influence which phase ultimately becomes the dispersed phase.

### 12-1.3.2 Terms Used to Represent Mean Drop Size and Drop Size Distribution.
The following expressions describe the common drop size notation used in this chapter. The volume fraction of dispersed phase is $\phi$, the total interfacial area per unit volume of mixed phases is $a_v$, and $d_{max}$ is the maximum drop size. The Sauter mean diameter, $d_{32}$, is defined by

$$d_{32} = \frac{\sum_{i=1}^{i=m} n_i d_i^3}{\sum_{i=1}^{i=m} n_i d_i^2} \qquad (12\text{-}1)$$

where m is the number of size classes describing the DSD, $n_i$ the number of drops, and $d_i$ the nominal diameter of drops in size class i. The subscripts indicate that $d_{32}$ is formed from the ratio of the third to second moments of the DSD.

The mean diameter of choice is often $d_{32}$, since it is directly related to $\phi$ and $a_v$ by

$$d_{32} = \frac{6\phi}{a_v} \qquad (12\text{-}2)$$

Another commonly used mean drop diameter is the mass mean diameter where $d_{43}$ is the ratio of the fourth to third moments of the DSD. Since drop mass is proportional to the cube of diameter, eq. (12-3) represents a mass-weighted average.

$$d_{43} = \frac{\sum_{i=1}^{i=m} n_i d_i^4}{\sum_{i=1}^{i=m} n_i d_i^3} \qquad (12\text{-}3)$$

The number mean diameter is given by

$$d_n = \frac{\sum_{i=1}^{i=m} n_i d_i}{\sum_{i=1}^{i=m} n_i} \qquad (12\text{-}4)$$

For consistency, the number mean diameter should be referred to as $d_{10}$, since it represents the ratio of the first to zero moments of the DSD. Although eq. (12-4) is the most common statistical definition of the mean, it is seldom used in the analysis of liquid–liquid dispersions since it provides little useful practical information.

We define $d_{10}$ as 10% by volume of all drops smaller than $d_{10}$, $d_{50}$ is defined as 50% by volume of all drops smaller than $d_{50}$, and $d_{90}$ as 90% by volume of all drops smaller than $d_{90}$. These drop diameters are determined from plots of size distribution data in terms of the cumulative volume frequency, defined below. In practice, $d_{50}$ and $d_{32}$ are close in value and are often used interchangeably. Overall mass transfer coefficients are commonly reported as $k_m a_v$ where $k_m$ is the mass transfer coefficient and $a_v$ is the interfacial area per unit volume, defined by eq. (12-2).

Drop sizes depend on many factors that are discussed throughout this chapter. For any given system, drop sizes are never uniform; rather, they exist in a continuous size spectrum. The large end of the drop size spectrum is controlled by agitation intensity, and the small end by the physics of drop breakage events. The DSD is sometimes bimodal or trimodal. Multimodal distributions are usually a result of multiple breakage mechanisms and unusual breakage patterns, such as those that result when viscous and/or viscoelastic drops are dispersed. Certain coalescence events can also lead to bimodal drop size distributions.

The DSD is usually represented in a discrete or histogram form in terms of number frequency, $f_n(d_i)$, or volume frequency, $f_v(d_i)$, given by

$$f_n(d_i) = \frac{n_i}{\sum_{j=1}^{m} n_j} \quad \text{and} \quad f_v(d_i) = \frac{n_i d_i^3}{\sum_{j=1}^{m} n_j d_j^3} \tag{12-5}$$

The DSD can also be described in a continuous or cumulative form (e.g., fraction up to size d). The cumulative number frequency $F_n(d_k)$, is defined by

$$F_n(d_k) = \frac{\sum_{i=1}^{k} n_i d_i}{\sum_{j=1}^{m} n_j d_j} = \int_0^{d_k} P_n(d') \, d\,d' \tag{12-6}$$

where $d_k$ is the size of drops in the $k^{th}$ size class, $d'$ is a dummy variable of integration, and $P_n(d)$, a continuous function, is the number probability density for drops of diameter d. The discrete and continuous distribution functions are related by $f_n(d_i) = P_n(d_i)\Delta d_i$, where $d_i$ is the nominal diameter and $\Delta d_i$ is the bin width for size class i.

An industrially important quantity is the cumulative volume frequency $F_v(d_k)$. For example, it relates to the yield of suspension polymerization products as defined by product specifications. It is defined by

$$F_v(d_k) = \frac{\sum_{i=1}^{k} n_i d_i^3}{\sum_{j=1}^{m} n_j d_j^3} = \int_0^{d_k} P_v(d') \, d\,d' \tag{12-7}$$

where $P_v(d)$ is the volume probability density function and $f_v(d_i) = P_v(d_i)\Delta d_i$.

Cumulative drop size distributions can be plotted conveniently on linear or log probability paper. A straight line on linear or normal probability paper means that the drop sizes follow a *normal* or *Gaussian distribution*. If data form a straight line on log probability paper, the distribution is referred to as *lognormal*.

# INTRODUCTION

The probability density functions for the normal and lognormal distributions are given in eqs. (12-8) and (12-9), respectively:

$$P_x(d) = \frac{1}{(2\pi)^{1/2}\sigma_{SD}} \exp\left[-\frac{1}{2}\left(\frac{d-\bar{d}}{\sigma_{SD}}\right)^2\right] \quad (12\text{-}8)$$

$$P_x(d) = \frac{1}{(2\pi)^{1/2}\sigma_{SD}}\frac{1}{d} \exp\left\{-\frac{1}{2}\left[\frac{\ln(d/\bar{d})}{\sigma_{SD}}\right]^2\right\} \quad (12\text{-}9)$$

For the number distribution, the dummy subscript x equals n, and the mean, $\bar{d}$, and standard deviation, $\sigma_{SD}$, are number-averaged quantities. For the volume distribution, x equals v, so $\bar{d}$ and $\sigma_{SD}$ are volume-averaged quantities.

Figures 12-10 and 12-12 are examples of cumulative frequency plots for distributions that are normally distributed in volume. Values of $d_{10}$, $d_{50}$, and $d_{90}$ (defined above) are readily determined for $100F_v$ equal to 10, 50, and 90, respectively. The slopes of the curves are a measure of the breadth of the distribution. A steeper slope means a narrower size distribution.

A commonly used measure of the breadth of a size distribution is the coefficient of variation, CoV. This can be determined easily from normal or lognormal plots of cumulative frequency data. The smaller the value of the CoV, the narrower the drop size distribution:

$$\text{CoV} = \frac{d_{16} - d_{84}}{2\,d_{50}} \quad (12\text{-}10)$$

where $d_{16}$ is the drop diameter in the spectrum where 16% of drops are smaller than $d_{16}$. Similarly, $d_{84}$ is the size where 84% are smaller, and $d_{50}$ is the midpoint.

Empirical relations such as the Schwarz–Bezemer equation, given by eq. (12-11), are used to relate the Sauter mean diameter, $d_{32}$, to the maximum drop size, $d_{max}$. $a^*$ is an empirical constant.

$$d_{32} = \frac{a^*}{1 + a^*/d_{max}} \quad (12\text{-}11)$$

All batch agitation-formed dispersions show transient behavior. Initially, the distribution is broad, due to incomplete dispersion. With continued agitation the distribution becomes narrower as large drops continue to disperse. This is described in more detail in Section 12-2.

In chemical processing, typical dispersion drop sizes range from 1000 μm $\geq d_{32} \geq$ 50 μm. Certain products, such as paint, personal care, and pharmaceutical products, require submicron sizes for reasons of shelf-life stability. Microdispersions are liquid–liquid systems where $d_{32}$ lies between 0.5 and 50 μm.

**Table 12-1** Drop Size Classification of Immiscible Liquid–Liquid Systems

| Drop Size | Comments | Equipment/Agents |
|---|---|---|
| <0.5 μm | Stabilized by Brownian motion; nonsettling | Emulsifiers, ultrasonic devices, rotor–stator mixers, high-pressure homogenizers; surface agents usually required |
| 0.5–3.0 μm | Marginally stable; can cream and separate | Rotor–stator and impingement mixers, static mixers |
| >3.0 μm | Usually unstable; coalescence and phase separation common when agitation ceases | Static mixers, in-line mixers, and stirred vessels |

Emulsions are liquid–liquid systems where $d_{32}$ is less than 0.5 μm. Table 12-1 gives characteristics of liquid–liquid systems based on drop size.

### 12-1.4 Process Complexities in Scale-up

Successful scale-up means that larger scale operations are fully anticipated and understood. Usually, the performance will be poorer than witnessed on a smaller scale. Scale-up must address several interdependent, flow-sensitive physical processes occurring simultaneously. These are dispersion, dispersion kinetics, coalescence, and drop suspension, as mentioned previously.

Scale-up is system dependent. For example, the scale-up of a dilute, neutral-density, noncoalescing system is a matter of balancing shear with dispersion time. However, the scale-up of a concentrated coalescing liquid–liquid system is much more complex. For this case, scale differences in fluid flow in the vessels result in different proportions of the vessel causing coalescence and dispersion. The small vessel tends to be dominated by dispersion, and the large one, by coalescence. This is due to coalescence being promoted by gentle shear leading to soft, long-duration collisions, while dispersion requires unsteady intense shear. In turbulent flow, rates of drop deformation, collision, and film drainage are governed by the small scale turbulence structure, which is somewhat insensitive to tank size. However, the amount of time that drops spend in the high-shear and quiescent regions depends on mean circulation time, governed by impeller pumping rate and macroscale turbulence phenomena. These are strongly influenced by tank size, so the balance of these rates is not readily scaled. These widely different conditions exist in all stirred vessels. In light of these complexities, scaling up liquid–liquid systems using "rules" such as constant tip speed or power per volume can lead to failures. Scale-up practices are discussed in Section 12-8.

*12-1.4.1 Drop Suspension.* A completely suspended condition is necessary to control and ensure a steady and predictable DSD. Segregation and layering in all cases lead to inferior results. The ease with which a suspension forms depends on phase density differences, agitation rate, impeller type/size, and its location within the vessel. The formation of a suspension starting from separated, settled phases is determined by empirical equations such as those developed by Pavlushenko and Yanishevskii (1958), Nagata (1975), and Skelland and Seksaria (1978). These are discussed in Section 12-6.

***12-1.4.2 Role of Surfactants, Solid Particles, and Other Materials***. Surfactants, dispersants, surface-active colloids, and very fine solids are all used to control drop size by stabilizing the system against coalescence. They are present at low concentrations, usually less than 1% based on the continuous phase. Composition and functionality are varied. Surfactants and suspending agents reduce the interfacial tension and drop size, and stop or reduce coalescence by affecting interfacial mobility. Fine solids act as structures preventing drop surfaces from touching. A commercialized process, known as limited coalescence, using solid suspending agents, is described in Section 12-9. Polymeric compounds that accumulate at drop surfaces are also used as suspending agents. When adsorbed, they can totally prevent coalescence.

## 12-1.5 Classification by Flow Regime and Liquid Concentration

***12-1.5.1 Flow Regimes: Laminar, Transition, and Turbulent***. Flow regimes are separated by the value of the Reynolds number, Re, the ratio of inertial to viscous forces. The impeller Reynolds number is

$$\text{Re} = \frac{D^2 N \bar{\rho}}{\bar{\mu}} \quad (12\text{-}12)$$

where $\bar{\rho}$ and $\bar{\mu}$ are the bulk density and viscosity of the mixed phases, respectively. For dilute dispersions (defined below) they are equal to those for the continuous phase. Laminar conditions exist when $0 \leq \text{Re} \leq 10$, transition flow occurs when $10 \leq \text{Re} \leq 10^4$, and fully turbulent flow occurs when $\text{Re} > 10^4$. Despite this generalization, it is common to find turbulent conditions near the impeller and transitional or laminar conditions elsewhere in the stirred vessel. This is particularly true for non-Newtonian fluids.

***12-1.5.2 Dispersed Phase Concentration***. The dispersed phase concentration is usually expressed as a volume fraction, $\phi$. Coalescence, dispersion, and settling are all affected by dispersed phase concentration. For example, coalescence rates increase with increasing $\phi$. This is due to both an increase in collision frequency and to rheological changes that enable longer contact intervals to be obtained. A high dispersed phase concentration also affects small scale turbulent eddies, reducing their intensity and making them less able to disperse drops. Therefore, the amount of information available and the means by which we approach the design process depend significantly on drop phase concentration. It is useful to categorize liquid–liquid systems with respect to their dispersed phase concentration as defined below.

*Dilute Systems:* $\phi < 0.01$. Ideally, a dilute system is one in which dispersion is affected only by hydrodynamics, and each drop is a single entity experiencing continuous phase fluid forces. Coalescence is neglected because few collisions occur. These simplifications enable a fairly fundamental treatment of dispersion to be made. Coalescence can become significant for clean systems at $\phi \geq 0.05$.

*Moderately Concentrated Systems:* $\phi < 0.2$. The behavior and technical treatment of systems in this concentration range depend on coalescence behavior. Ideal dilute dispersion theories may still apply, particularly if the system is noncoalescing. A simple test to detect coalescence is to agitate or shake a sample for 5 min and then watch it settle and coalesce. If only a trace of coalesced layer appears on the surface after 5 min, the system can be considered to be stable. The system is considered to be strongly coalescing if complete separation occurs in less than 30 s. Obviously, many results fall between these limits. More details coalescence tests are given in Section 12-3.1.5.

Even in the presence of coalescence, it is possible to predict the DSD for moderately concentrated systems. For $\phi < 0.2$, the drop phase does not appreciably affect the structure of the continuous phase flow field above the scale of the drop size. This allows single-phase flow concepts that describe the mechanical forces causing drop deformation, collisions, and film drainage to be used.

*More Concentrated Systems:* $\phi > 0.2$. This range is common in industry. Fast coalescence is probable for clean systems. Sprow (1967b) found that with coalescing systems, drop sizes were position dependent within the vessel. This behavior is very complex and extremely difficult to scale-up, since coalescence and dispersion dominate in different regions of the vessel, as described earlier. A special case is suspension polymerization. It is typically a concentrated system where $\phi \approx 0.5$ and coalescence is prevented by the use of polymeric suspending agents. This enables theories based on dilute systems to be used. This is described fully in Section 12-8. Overall, it is more difficult to predict mean drop size and DSD in systems of high dispersed phase concentration.

*Other Considerations.* The presence of a third phase can affect liquid–liquid dispersion and coalescence. Fine solids have little effect on drop dispersion but often affect coalescence. Gas bubbles affect dispersion by reducing the effective continuous phase viscosity and lead to a loss in momentum transport, hence dispersion capability. Tiny gas bubbles reduce probability of coalescence by interfering with film drainage rates between colliding drops. This subject is complex and is best studied experimentally at different scales.

Mass transfer to and from drops affects coalescence. Mass transfer creates concentration gradients in the region of the thinning film. Depending on the interfacial tension–concentration characteristics of the system, this can lead to Marangoni effects, causing surface flows and internal circulation within the drops. Such movement accelerates film drainage and increases the probability of coalescence.

### 12-1.6 Scope and Approach

Liquid–liquid dispersion is among the most complex of all mixing operations. It is virtually impossible to make dispersions of uniform drop size, because of the wide range of properties and flow conditions. Our chapter provides a fundamental framework for analysis and understanding of dispersion and coalescence, based often on idealized experiments and theories. This framework can be applied to more complicated processes, including scale-up. Throughout the chapter, references are made to state-of-the-art information, often not yet proven in practice. The chapter concludes with commercialization advice and recommendations.

Section 12-2 deals with liquid–liquid dispersion, while in Section 12-3 we discuss coalescence. Section 12-4 gives an introduction to the methods used for population balance models, along with references for further reading. In Section 12-5 we describe more concentrated dispersed phase systems, including phase inversion. Section 12-6 deals with other considerations, such as suspension, mass transfer, and other complexities, Section 12-7 with equipment used in liquid–liquid operations, Section 12-8 with scale-up, and Section 12-9 provides industrial examples. Nomenclature and references then follow. Although every attempt has been made to make this a stand-alone chapter, space limitations occasionally make it necessary to refer to other chapters in the book.

## 12-2 LIQUID–LIQUID DISPERSION

### 12-2.1 Introduction

*12-2.1.1 Breakup of Single Drops in Laminar and Turbulent Flow.* The breakup of a single drop in laminar and turbulent flow fields forms the starting point for this section. Although not industrially relevant, it shows in the simplest possible form what occurs when fluid forces act on a drop, and thus provides important insight. The progeny of a single breakage event may be few and orderly or may be many drops of broad size distribution.

Simple laminar shear or extension flow produces orderly dispersion since the flow field surrounding the drop is constant and continuous. In contrast, simple turbulent flows produce more random breakup events, due to the time-dependent nature of fluid–drop interactions. The effect of breakage mechanism on the resulting DSD is sometimes counterintuitive.

Simple theories are described in which breakup results when disruptive forces in the surrounding fluid exceed cohesive forces, due to interfacial tension and drop viscosity. The results for a single drop are then extended to dilute dispersions in order to predict and correlate data for the DSD. The methodology is extended to more concentrated noncoalescing systems of wider practical importance as well as other dispersion devices. The scope includes a broad range of factors. Although most of the section is devoted to the development of the equilibrium mean drop size and DSD, dispersion kinetics and the time evolution of the DSD are included.

**12-2.1.2 Description of Forces Causing Breakup.** The forces acting to deform a drop in simple laminar flow can be characterized by the shear or extension rate (velocity gradient) in the surrounding fluid. In turbulent flows, these forces are best characterized in terms of the energy dissipation rate, since it is not practical to resolve the instantaneous velocity gradients. The two approaches are consistent, since in general, the energy dissipation rate is the product of the stress and velocity gradient tensors.

In stirred vessels, the forces causing drop dispersion are extremely nonuniform. Velocity gradients or deformation rates are highest near the impeller and diminish rapidly with distance from the impeller. Turbulent energy dissipation rates per volume of fluid in the impeller region are often ≈40 times greater than the average or power draw per unit volume for the tank. Some regions are apt to be turbulent, while others can be laminar. From the point of view of the drop, it seems obvious that it matters little how the force or disruptive energy is produced. This allows for a more general application of the dispersion process.

In laminar flow, the spatially dependent flow field is time periodic to the stationary observer but steady in time with respect to the rotating stirrer blades. However, the flow field as seen by a drop as it moves through regions of varying shear appears time dependent. This transient nature of the deformation process is important. Once deformed, a drop passing to a less intense region tends to return to a spherical shape. However, if it has already reached a critical state of deformation, it will become unstable and break up. In some instances, a deformed drop will remain stable in a steady force field and will not break until the force is relaxed. This is because internal circulation stabilizes it. For a drop exceeding the critical deformation, if internal circulation stops before the drop begins to return to its spherical state, it is likely to disperse. Newton's law of viscosity, or an appropriate non-Newtonian constitutive equation, can be used to describe the forces acting on the drop in laminar flow.

Turbulent flows contain a spectrum of eddies of different size, intensity, and lifetime. However, each eddy has an element of simple shear or extension, and creates forces that lead to drop deformation. The drop sees a time-dependent deformation field even if the Reynolds-averaged velocity field does not vary in space. This is illustrated in Figure 12-2 and is explained more fully in Chapter 2. In reality, forces in turbulent stirred vessels arise from both spatial and temporal velocity fluctuations. These arise from mean velocity gradients, interacting turbulent eddies and impingement of jetlike flows on walls, baffles, and impeller blades.

Figure 12-3 contrasts the time-averaged and instantaneous velocity fields acquired in a turbulent stirred tank. The probability of drop dispersion in such a transient flow field depends on two time scales. One characterizes the turbulent stretching force and the other the restorative surface force. It should be noted that drop viscosity opposes both deformation and relaxation. Observe the nature of flow in the data shown in Figure 12-3. The top view shows the time-averaged velocity field acquired by both LDV (laser Doppler velocimetry) and PIV (particle image velocimetry). These data show regions of high and low liquid velocity and

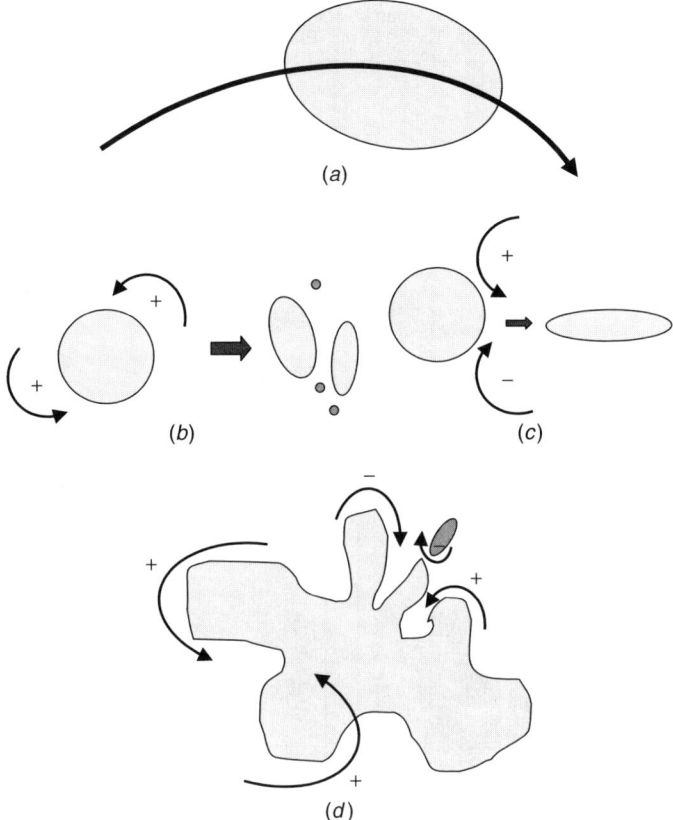

**Figure 12-2** Scalar deformation in a turbulent field: (*a*) convection by large eddies; (*b*) erosion by co-rotating eddies; (*c*) elongation by counterrotating eddies; (*d*) multiple scales of turbulent deformation. (From Kresta and Brodkey, Chapter 2, this volume.)

are useful to predict overall convective or bulk mixing. The lower two pictures, acquired by PIV, show instantaneous transient velocity fields. These transient fields create forces leading to drop breakage. However, even if a large quantity of these data were available, a detailed analysis of drop dynamics is not currently possible. Therefore, it is more practical to employ mechanistic theories that relate drop deformation to local energy dissipation rates.

### 12-2.2 Breakup Mechanism and Daughter Drop Production in Laminar Flow

To the authors' knowledge, there are practically no data or fundamental analysis for drop dispersion in stirred tanks under laminar flow conditions. There are several reasons for this somewhat surprising occurrence. Viscous formulations are often produced in highly specialized equipment and exhibit complex and varied

**652** IMMISCIBLE LIQUID–LIQUID SYSTEMS

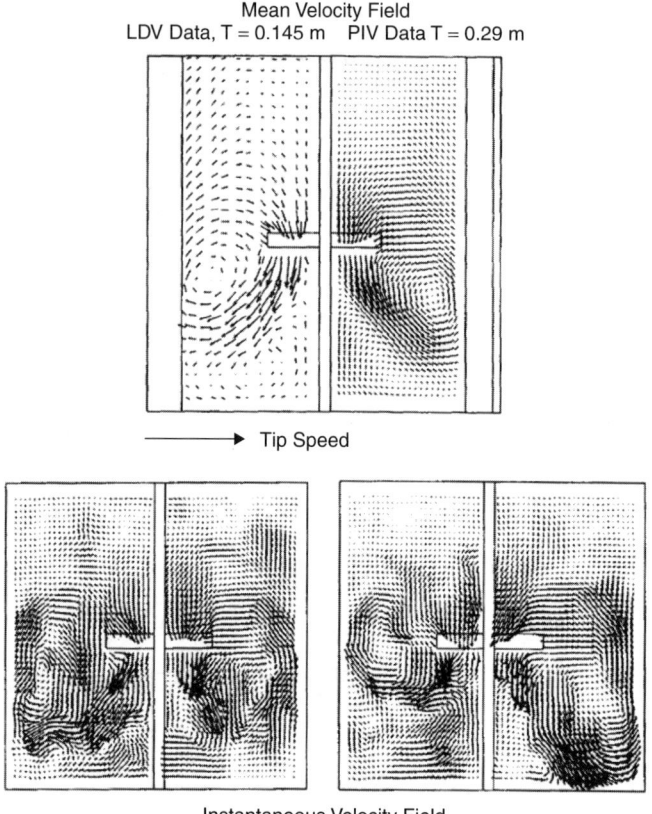

**Figure 12-3** Comparison of time-averaged and instantaneous velocity fields in a turbine stirred vessel. (Reproduced from Bakker et al., 1996.)

rheological behavior, so that results are not readily generalized. Drop size data are difficult to acquire, due to limited measurement techniques and numerous handling and disposal issues. Despite this, dispersion does certainly take place under laminar conditions, and it is important. For example, the continuous addition of low viscosity monomer to a stirred mass polymerization system results in monomer dispersion in a viscous matrix phase. Drops are formed long before they dissolve. Product quality often depends on how rapidly the monomer can be made available to growing polymer chains. Another example is given in Section 12-5 where Figure 12-24 shows a steady rotational shear flow and the initial creation of a water-in-oil dispersion prior to phase inversion.

From an analytical viewpoint, the flow fields in laminar devices are highly dependent on geometry, and individual drops experience varied deformation paths of long time scale that are difficult to analyze. Even if Lagrangian tracking of deformation and breakup history of many drops were possible, it would be difficult to apply this information to real-life systems. Therefore, most studies have focused on single drops in highly idealized flow fields such as simple shear and/or extension. These studies have led to a better understanding of drop dispersion and form a basis for process design and scale-up by judicious application of this fundamental information.

It is not our purpose to provide a complete discussion of drop deformation and breakup in idealized laminar flow fields. There have been numerous studies that have been reviewed by Rallison (1984), Stone (1994), and others. Only the most practically relevant studies are discussed below. Of central importance is to predict and/or correlate the size above which a parent drop of known physical properties (that is subjected to an imposed deformation) will become unstable and break up into smaller drops. This size is referred to as the *critical* or *maximum stable drop size*, $d_{max}$.

If the breakup of a single drop in an idealized laminar flow is confined to low Reynolds number (creeping flow), inertial forces can be neglected. Nondimensionalization of the resulting Stokes equations reveals that drop size data can be correlated in terms of a capillary number, $Ca = \mu_c G a/\sigma$, and a viscosity ratio, $\mu_d/\mu_c$. G is the deformation rate (shear or extension rate). The capillary number is the ratio of the viscous force acting to deform the drop to the surface force opposing deformation. This is illustrated in Figure 12-4, which shows the critical stability curve for Newtonian fluids in simple shear flow (SSF) and simple extensional flow (SEF). A drop at conditions above the curve is unstable and will break. The drop is stable at conditions below the curve. Consider a drop of known physical properties at the critical capillary number. Then $a$ is the radius of the largest drop that exists for a deformation rate G; or G is the smallest deformation rate required to break a drop of radius $a$. Note that the shape of the curves for shear and extensional flows are quite different.

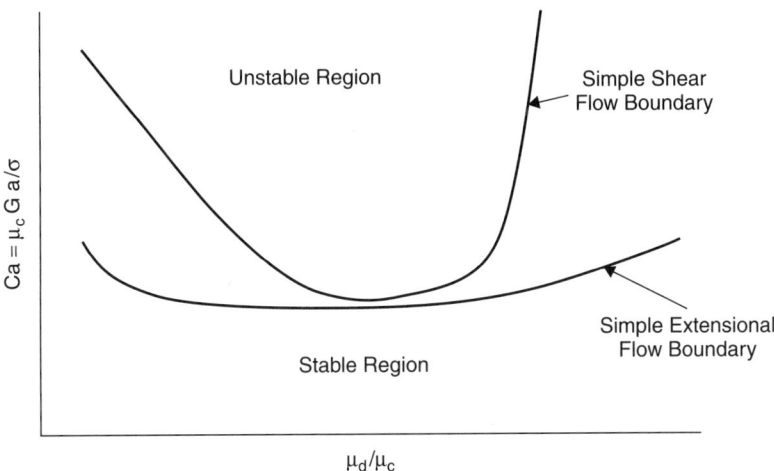

**Figure 12-4** Critical stability curves for simple shear (SSF) and simple extensional (SEF) flow.

In SSF or Couette flow, it is not possible to break a drop if the viscosity ratio is greater than about 3. The deformed drop shape is stabilized by internal circulation. It has been concluded that extension is more effective than shear at breaking drops. With respect to practical flows, it is important not to interpret this statement too literally since in practical applications, a single steady shear gradient rarely exists. Bear in mind that the physical definitions of shear and extension depend on the environment seen by the drop along its trajectory, while the mathematical definitions are related to the choice of coordinate system.

Taylor (1934) was first to establish an analytical relationship between the degree of deformation of a drop and the deformation rate. For SSF this is given by

$$D_{crit} = \frac{L_d - B_d}{L_d + B_d} = \frac{G a \mu_c}{\sigma} \frac{1.19 (\mu_d/\mu_c) + 1}{(\mu_d/\mu_c) + 1} = Ca \cdot f(\mu_d/\mu_c) \qquad (12.13)$$

where $L_d$ and $B_d$ are the length and breath of the deformed drop and $D_{crit}$ is the critical deformation for breakage. Since then there has been considerable effort, both analytically and computationally, to determine the critical or maximum stable drop size in a variety of idealized laminar flow situations. The reader is again referred to the reviews referenced above.

Karam and Bellinger (1968) and Grace (1982) studied drop deformation and breakup in simple shear flow. The conditions for breakage were observed within a glass-walled Couette apparatus consisting of independently controlled, counterrotating concentric glass cylinders. When the rotational speeds were about equal, the centroid of a drop would remain stationary, enabling deformation and breakage information to be observed and recorded. Drop data from two breakage

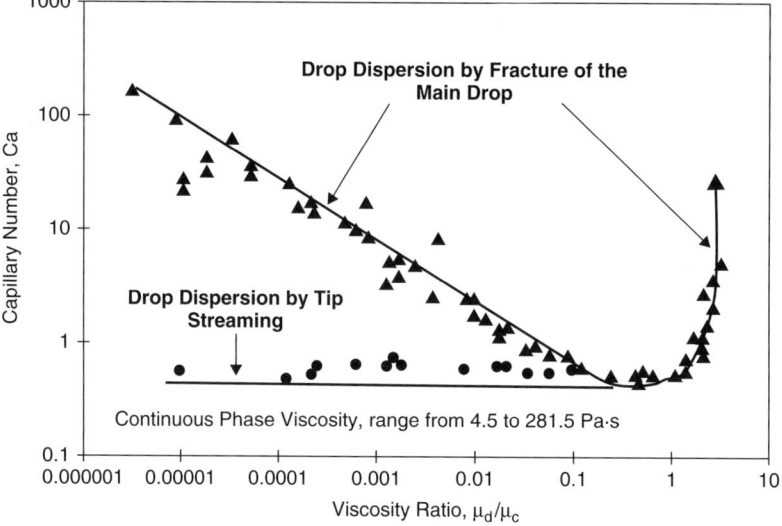

**Figure 12-5** Drop stability data for simple shear flow. (Data of Grace, 1982.)

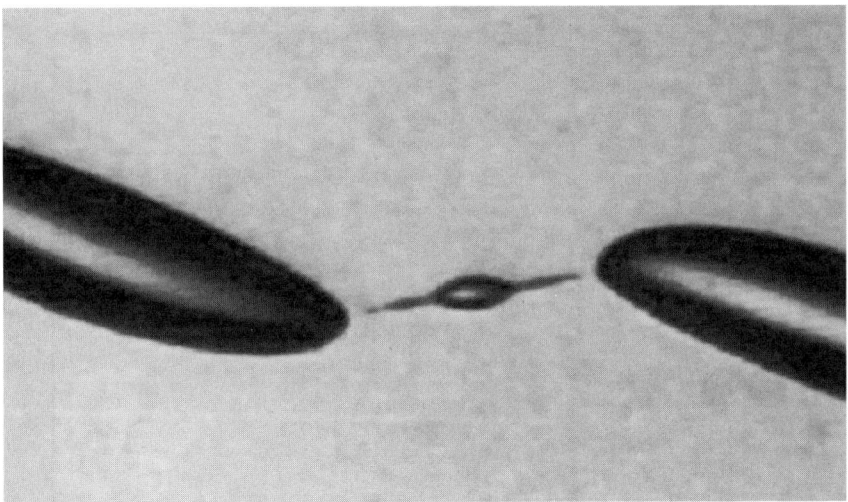

**Figure 12-6** Breakup of a drop in simple shear flow. (Photo from Grace's archives; courtesy of E.I. DuPont de Nemours & Co.)

modes are shown in Figure 12-5. Typical breakage patterns showing the breakage of the main drop into orderly size daughter drops and the shedding of smaller drops from the tip ends of the main drop are given in Figure 12-6.

Bentley and Leal (1986) and Stone et al. (1986) studied deformation and breakup of a drop in a four-roll mill that allowed specification of idealized flows with various degrees of shear and extension. In addition to determining the critical deformation rate under steady conditions, the following experiments were performed. Drops were deformed to a steady nonspherical shape. After stopping the flow motion, the stability of the drop was monitored as it relaxed to a sphere. This enabled the conditions leading to breakup to be determined. Tjahjadi and Ottino (1991) studied the breakup of drops subjected to both stretching and folding.

There have been fewer studies to observe drop breakup and the resulting daughter drop size distribution. Figure 12-6 shows a typical breakage pattern for fracture of the main drop in SEF. Observe there are three distinct drop diameters that form upon breakup of the parent drop. As the imposed deformation rate exceeds the critical value required to just break the drop, a larger number of daughter drops are formed. Figure 12-7 correlates daughter drop production with the ratio of imposed to critical shear rate ($G/G_{crit}$). It shows that tens to thousands of daughters can result. Tjahjadi et al. (1992) measured relaxation of a stretched drop in SSF. They included details of satellite drop production as the drop broke up while trying to regain its initial spherical shape.

In a more recent study, Marks (1998) observed the deformation and breakup of a drop in SSF that was exposed to a steady shear rate greater than $G_{max}$. His results show that the breakage mechanism and breadth of the daughter DSD

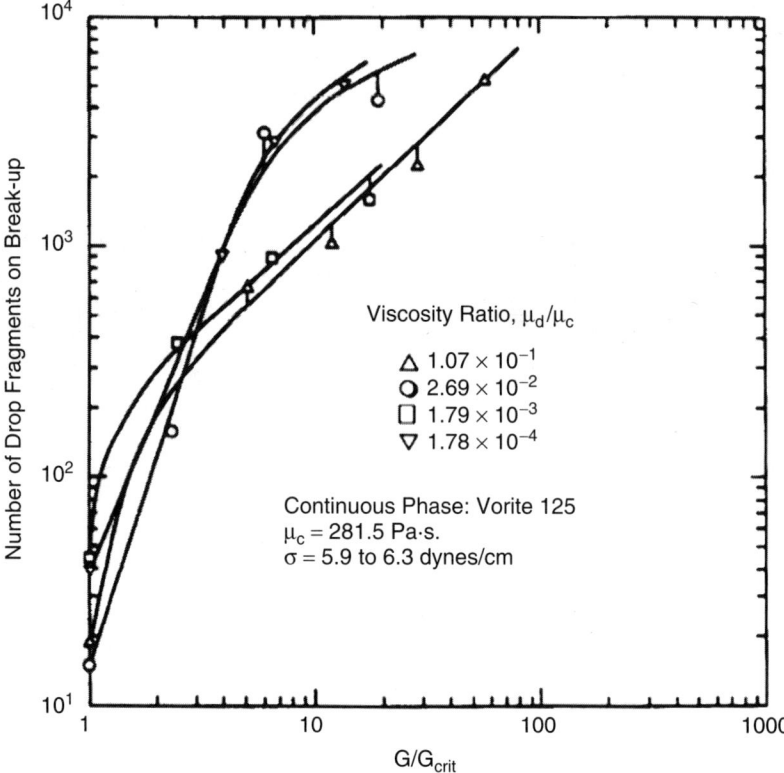

**Figure 12-7** Number of drop fragments from breakup of a single drop in SSF as a function of the ratio of imposed to critical shear rate, $G/G_{crit}$. (Data of Grace, 1982.)

depend uniquely on $G/G_{crit}$. Furthermore, there was no further breakage of the largest fragments formed. If an application requires a narrow size distribution, it is important to operate as close to the critical shear rate as possible and to provide for a uniform deformation field. Alternatively, one can ramp up the deformation rate temporally or spatially to continue the break up the largest existing fragments.

In the absence of data for practical flows, the engineer must make use of the insights gained from these idealized studies. The literature for blending of immiscible polymers in extruders may also provide useful insights.

### 12-2.3  Drop Dispersion in Turbulent Flow

In contrast to laminar flow, there are numerous studies of drop dispersion in practical turbulent flows, particularly for dilute systems, when coalescence can be neglected. Data are relatively easy to acquire, since water and other nontoxic Newtonian fluids serve as the continuous phase and waste disposal issues are minimized in dilute systems. With respect to mixing flows, most of the studies have been conducted on a bench scale in fully baffled batch stirred vessels equipped with a single Rushton (RDT) impeller. Fortunately, the small scale turbulence structure that determines ultimate drop size is independent of geometry, and turbulent time scales are such that statistically repeatable results can be obtained. This allows the development of mechanistic analysis coupled with similarity arguments to develop correlations for mean drop size and DSD that when applied carefully, perform adequately under extrapolation to larger scale. On the other hand, there are few observations of the breakup of single drops in practical turbulent flows, since these experiments are quite difficult to perform.

We begin by considering mechanistic theories that allow correlation of equilibrium mean drop size in dilute systems. An example of their application is given. Drop size distributions are then discussed. The predictive approach is extended to other contacting devices and to moderately concentrated noncoalescing systems. Some additional factors are considered, followed by a discussion of transient effects and time to achieve equilibrium.

### 12-2.3.1 Mechanistic Models and Correlation of Mean Drop Size.
Mechanistic models for maximum stable drop size in turbulent flow are based on arguments put forth by Kolmogoroff (1949) and Hinze (1955). The stress acting to deform a drop of size d is given by

$$\tau_c = \rho_c \overline{v'(d)^2} = \rho_c \int_{1/d}^{\infty} E(k)\, dk \qquad (12\text{-}14)$$

where $\overline{v'(d)^2}$ is the mean-square velocity difference across the surface of the drop of diameter d, E(k) the energy spectral density function, and k the wavenumber or inverse eddy length. Only energy contained in eddies of scale smaller than $k = 1/d$ is considered, since larger eddies carry rather than deform the drop.

For energy dissipation rates that commonly occur in stirred vessels, final drop sizes are small compared to the turbulence macroscale but large compared to the Kolmogoroff microscale, defined by

$$\eta = \left(\frac{\nu_c}{\varepsilon}\right)^{1/4} \qquad (12\text{-}15)$$

Therefore, eddies that interact with the drops to determine the ultimate DSD fall within the inertial subrange of turbulence. These eddies are locally isotropic and E(k) can be described by Kolmogoroff's (1941a,b) theory of local isotropy:

$$E(k) = \beta_K\, \varepsilon^{2/3}\, k^{-5/3} \qquad (L_T \gg d \gg \eta) \qquad (12\text{-}16)$$

$\beta_K \sim 3/2$ is the Kolmogoroff constant. When eq. (12-16) is used in (12-14), the result is

$$\tau_c \approx \rho_c\, \varepsilon^{2/3}\, d^{2/3} \qquad (L_T \gg d \gg \eta) \qquad (12\text{-}17)$$

It should be noted that $\varepsilon$ is the local energy dissipation rate, which varies widely throughout stirred tanks and other contacting devices.

Cohesive forces due to interfacial tension and drop viscosity oppose drop deformation. The surface force per unit area is given by

$$\tau_s \approx \frac{\sigma}{d} \tag{12-18}$$

According to Hinze (1955), the viscous stress within the drop is

$$\tau_d \approx \mu_d \frac{(\tau_c/\rho_d)^{1/2}}{d} \tag{12-19}$$

This is Newton's law of viscosity, with the characteristic velocity within the drop, $(\tau_c/\rho_d)^{1/2}$, related to the turbulent stress on the surface.

An examination of eqs. (12-17) to (12-19) reveals that there exists a maximum stable drop size, $d_{max}$, above which the disruptive forces are sufficient to break the drop, and below which the drop is stabilized by surface and internal viscous forces. For $d = d_{max}$, the disruptive force exactly balances the cohesive forces, so that

$$\tau_c = \tau_s + \tau_d \tag{12-20}$$

*Low Viscosity Dispersed Phase.* If the drop is inviscid, $\tau_d$ is negligible, and only the surface force contributes to drop stability. According to eq. (12.20), we can then equate (12-17) and (12-18) and rearrange to obtain

$$d_{max} = C_1 \left(\frac{\sigma}{\rho_c}\right)^{3/5} \varepsilon_{max}^{-2/5} \tag{12-21}$$

where the constant $C_1$ must be determined empirically. Given the broad spatial distribution in energy dissipation rate in a stirred vessel, the maximum stable drop size will not be achieved until all dispersed phase globules experience the highest energy region of the flow. Therefore, $d_{max}$ is determined by the maximum energy dissipation rate. Hence $\varepsilon$ is replaced by $\varepsilon_{max}$ in eq. (12-21). Furthermore, it will take a large number of impeller passes before equilibrium is achieved. Time to complete dispersion is discussed later in this section.

For geometrically similar turbulent systems, $\varepsilon_{max} \propto \varepsilon_{avg}$, where $\varepsilon_{avg}$ is the power draw per unit mass ($P/\rho_c V$) of fluid. For constant power number this gives $\varepsilon_{max} \sim N^3 D^2$. For a dilute system, the equilibrium DSD will consist of drops of size $d_{max}$ and smaller. There is considerable experimental evidence that $d_{max}$ is proportional to $d_{32}$. This relationship has also been argued mechanistically. Therefore, for geometrically similar systems, eq. (12-21) is equivalent to

$$\frac{d_{32}}{D} = C_2 \cdot We^{-3/5} \tag{12-22}$$

LIQUID–LIQUID DISPERSION    659

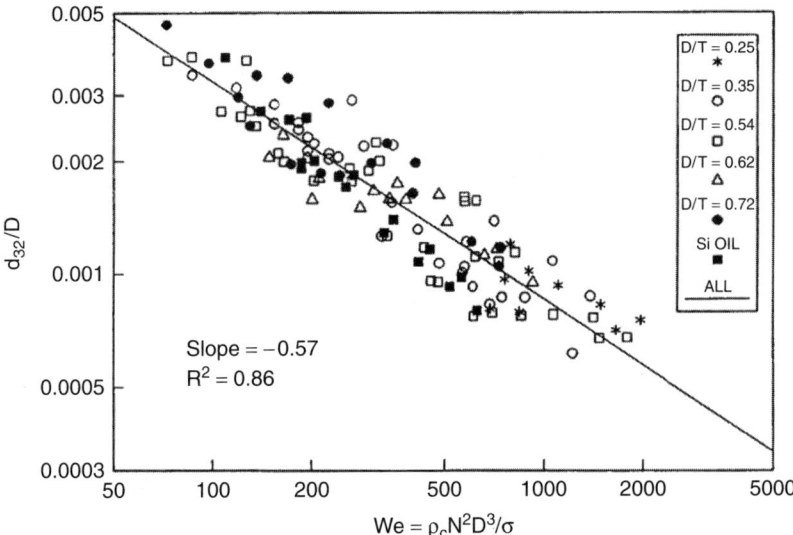

**Figure 12-8** Experimental data for 14 different liquid–liquid pairs. (Data of Chen and Middleman, 1967 for a RDT.)

where We = $\rho_c N^2 D^3/\sigma$ is the ratio of inertial (disruptive) to surface (cohesive) forces. This expression is the well-known Weber number theory, which has been derived and validated by Chen and Middleman (1967), among others. Their substantial data set for a Rushton turbine covered a broad range of physical properties and tank size and is shown in Figure 12-8. The data are best fit by eq. (12-22) with $C_2 = 0.053$.

Equations (12-21) and (12-22) show that dispersed phase systems created by turbulent flow scale-up by maintaining constant $\varepsilon_{max}$; or for practical industrial purposes, by constant P/V. Large Weber numbers result in small drops, and vice versa. These expressions are valid for dilute, noncoalescing systems of low $\mu_d$. It turns out that many stabilized or noncoalescing industrial systems with $\phi > 0.05$ can also be scaled by the constant P/V criterion.

***Example 12-1.*** It is proposed to recover a fermentation product by solvent extraction. The broth has a viscosity of $\mu_c = 0.3$ Pa·s (300 cP). While the broth is viscous, the drop phase is not. The bulk or mixture viscosity is $\bar{\mu} = 0.0386$ Pa·s (38.6 cP). The interfacial tension is $\sigma = 0.003$ N/m (3.0 dyn/cm). The broth has density $\rho_c = 1000$ kg/m³ (1.0 g/cm³), but the bulk or mixture density is 1100 kg/m³ (1.1 g/cm³). The vessel volume is 3.54 m³ (750 gal). The vessel has a diameter T = 1.524 m (5.0 ft) and is equipped with an RDT with D/T = 0.4. Laboratory studies have shown that acceptable extraction results are obtained if the mean drop size is $d_{32} = 50$ μm. Determine the required impeller speed and power draw.

SOLUTION: The solvent will disperse in the broth and the system will be slow to coalesce because of the high broth viscosity. As a result, eq. (12-22) will be used with $C_2 = 0.053$, even though it is not a dilute system. Substituting $d_{32} = 0.005$ cm, $D = 61$ cm, $\sigma = 3.0$ dyn/cm, $\rho_C = 1.0$ g/cm$^3$ into eq (12-22) and solving for N yields: $N = 48$ rpm ($0.8$ s$^{-1}$).

The Reynolds number, $Re = D^2 N \bar{\rho}/\bar{\mu}$, is 7700. Flow is nearly fully turbulent, so the use of eq. (12-22) is acceptable. For an RDT, an average power number is $N_p = P/\bar{\rho} N^3 D^5 = 5.0$. Using the bulk density, the power required is $P = N_p \bar{\rho} N^3 D^5 = 0.3$ hp.

Equations (12-21) and (12-22) are independent of the device used. However, $C_1$, $C_2$, and $\varepsilon_{max}/\varepsilon_{avg}$ do depend on impeller type and tank geometry. In principle, one can apply data for an RDT to other geometries from knowledge of their respective values of $\varepsilon_{max}$. These can be estimated from LDV measurements (see Chapter 3), as demonstrated by Zhou and Kresta (1998a), who successfully correlated drop size data for several impeller geometries with $\varepsilon_{max}$. Accurate DSD data are even more difficult to acquire than accurate LDV data. This makes measurements of $\varepsilon_{max}$ an efficient means to convert literature data for RDTs to other geometries.

A simple concept is to use the impeller swept volume as the dissipation volume to correlate data for different geometries in the absence of data for $\varepsilon_{max}$. The idea is to assume that all power is dissipated uniformly in the volume swept out by the impeller rather than throughout the tank volume. Then, according to eq. (12-21), drop size for different geometry should scale approximately with $N_p^{-2/5}$. McManamey (1979) correlated many systems with other types of impellers using

$$\frac{d_{32}}{D} = C_3 N_p^{-2/5} We^{-3/5} \tag{12-23}$$

In other words, if a turbine other than the RDT is used for dispersion (say turbine X), first calculate $d_{32}$ from eq. (12-22) for the RDT and then correct it by multiplying it with a *factor* represented by the ratios of the power numbers to the $\frac{2}{5}$ power. For example,

$$factor = \left( \frac{N_{p\,Rushton}}{N_{p\,Impeller\,X}} \right)^{2/5}$$

It cannot be overstated that the basis of the mechanistic theory and scale-up criteria discussed here assumes that there is no coalescence and that the drops are large compared to the Kolmogoroff microscale but small compared to the macroscale ($L_T \gg d \gg \eta$). Otherwise, eq. (12-17), and hence (12-21) and (12-22), are not valid. Correlations and scale-up for other criteria are discussed later in this section.

*Viscous Dispersed Phases.* If the drop is viscous, the internal viscous resistance to deformation cannot be ignored. Both interfacial tension and viscosity contribute to drop stability and the development of the preceding section can be extended in a straightforward manner. When all of eqs. (12-17) to (12-19) are substituted into (12-20), the result is

$$\frac{\rho_c \varepsilon_{max}^{2/3} d_{max}^{5/3}}{\sigma} = C_4 \left[ 1 + C_5 \left( \frac{\rho_c}{\rho_d} \right)^{1/2} \frac{\mu_d \varepsilon_{max}^{1/3} d_{max}^{1/3}}{\sigma} \right] \qquad (12\text{-}24)$$

In the limit as $\mu_d$ vanishes and/or $\sigma$ becomes large, the right-hand term in brackets becomes small with respect to unity, and eq. (12-24) reduces to (12-21). In the limit of large $\mu_d$ and/or small $\sigma$, internal viscous forces predominate over surface forces. The right-hand term in brackets becomes large with respect to unity, and eq. (12-24) reduces to

$$d_{max} = C_6 (\rho_c \rho_d)^{-3/8} \mu_d^{3/4} \varepsilon_{max}^{-1/4} \qquad (12\text{-}25)$$

For the case of geometrically similar systems with constant power number, eq. (12-24) yields

$$\frac{d_{32}}{D} = C_7 \cdot We^{-3/5} \left[ 1 + C_8 \cdot Vi \left( \frac{d_{32}}{D} \right)^{1/3} \right]^{3/5} \qquad (12\text{-}26)$$

The viscosity group, $Vi = (\rho_c/\rho_d)^{1/2} \mu_d N D/\sigma$ represents the ratio of viscous to surface forces stabilizing the drop. In the limit as $Vi \to 0$, eq. (12-26) yields (12-22). In the limit as $Vi \to \infty$, eq. (12-26) yields the counterpart to eq. (12-25):

$$\frac{d_{32}}{D} = C_9 \left( \frac{\rho_c}{\rho_d} \right)^{3/8} \left( \frac{\mu_d}{\mu_c} \right)^{3/4} Re^{-3/4} \qquad (12\text{-}27)$$

Equation (12-27) can be misleading. Since $Re = \rho_c N D^2/\mu_c$, there is actually no dependence on $\mu_c$. Calabrese et al. (1986a,b) and Wang and Calabrese (1986) extended the work of Chen and Middleman (1967) to dilute dispersions of viscous drops in turbulent stirred vessels equipped with Rushton turbines. They found that the mechanistic correlations were valid for $\mu_d \leq 500$ cP. Figure 12-9 is taken from their substantial data set and verifies that $d_{32} \sim \mu_d^{3/4}$ for large $\mu_d$. Based on their results and several other data sources (ca. 350 data sets), they found that $C_7 = 0.054$ and $C_8 = 4.42$. They also found that the following empirical equation was equally accurate for the RDT:

$$\frac{d_{32}}{D} = 0.053 \, We^{-3/5} (1 + 0.92 Vi^{0.84})^{3/5} \qquad (12\text{-}28)$$

Both correlations collapse to the Chen and Middleman result in the inviscid limit.

According to eqs. (12-21), (12-24), and (12-25), the dependency of drop size on $\varepsilon_{max}$ or P/V varies from the $-\frac{2}{5}$ to the $-\frac{1}{4}$ power as $\mu_d$ or $Vi$ increases. The ideas discussed earlier about scale-up and application to other impeller types still

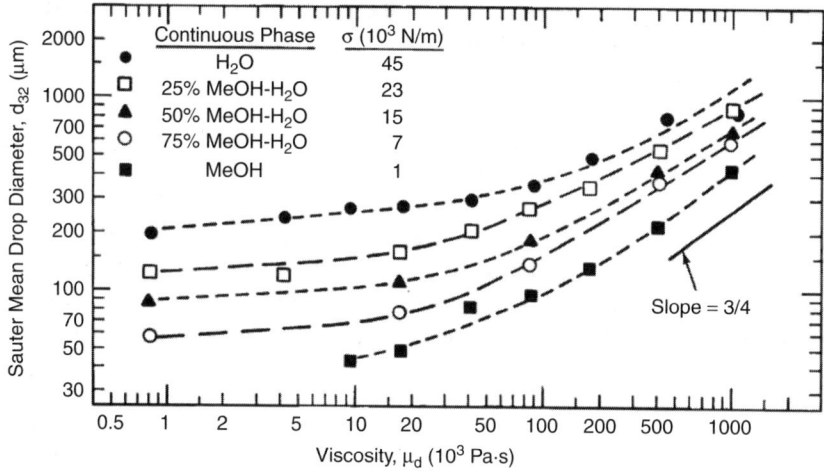

**Figure 12-9** Relative influence of $\mu_d$ and $\sigma$ on $d_{32}$ for constant conditions of agitation. Silicone oils dispersed in aqueous methanol solutions. RDT with N = 3.0 rps, D/T = 0.5, T = 0.2 m. (Reproduced from Wang and Calabrese, 1986.)

apply here, except that one additional complexity arises. The power to which you scale $\varepsilon_{max}$ or the $N_p$ ratio now varies, depending on the value of viscosity group. Vi is also scale dependent, so an approximate power dependency must be assumed.

For $\mu_d > 500$ cP, dispersion behavior and the dependency of $d_{32}$ on system variables is quite complex. The reader is referred to the original work of Calabrese et al. (1986a).

### 12-2.3.2 Equilibrium Drop Size Distribution.
Chen and Middleman (1967) found that for turbulent Rushton turbine stirred vessels, the equilibrium DSD for dilute inviscid dispersions was normally distributed in volume and therefore described by eq. (12-8). Wang and Calabrese (1986) found a similar result for low- to moderate-viscosity dispersed phases ($\mu_d \leq 500$ cP). Figure 12-10 shows that the cumulative volume frequency exhibits straight-line behavior on normal probability coordinates that is indicative of a Gaussian DSD. The distribution broadens with increasing drop viscosity, increasing interfacial tension, and decreasing impeller speed.

Both authors argued that for dynamically similar breakage mechanisms, the equilibrium DSD should only depend on the ratio of disruptive ($\tau_c$) to cohesive ($\tau_s$ and/or $\tau_d$) forces acting on the drops. Thus, the individual DSDs could be collapsed to a single correlation by normalization with $d_{32}$. Defining $X = d/d_{32}$, the volume probability density function becomes

$$P_V(X) = \frac{1}{\sqrt{2\pi}\,\sigma_V} \exp\left(-\frac{X-\overline{X}}{\sqrt{2}\,\sigma_V}\right)^2 \qquad (12\text{-}29)$$

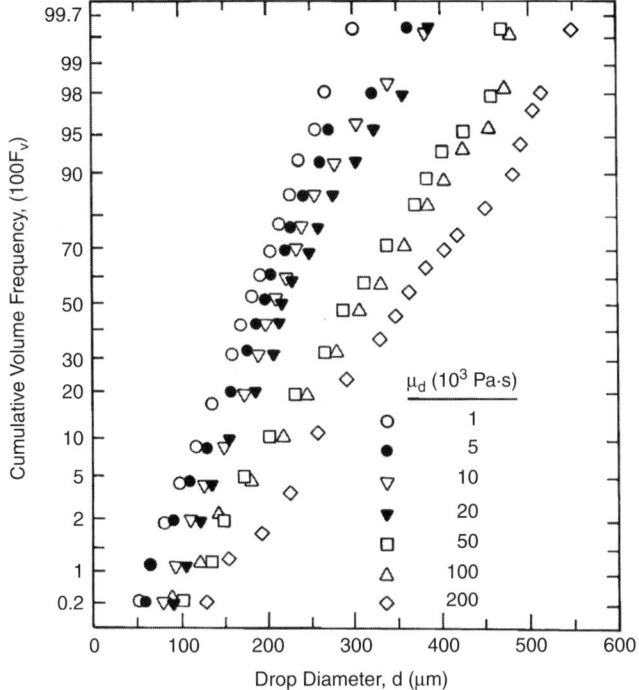

**Figure 12-10** Effect of $\mu_d$ on DSD. Silicone oils dispersed in water, $\sigma = 0.045$ N/m. RTD with N = 4.67 rps, D/T = 0.5, T = 0.15 m. (Reproduced from Wang and Calabrese, 1986.)

where $\overline{X}$ is the mean of $d/d_{32}$ and $\sigma_v$ is the volume standard deviation. Chen and Middleman found that their inviscid dispersed phase data were well correlated by $\overline{X} = 1.07$ and $\sigma_V = 0.24$. Wang and Calabrese found essentially the same result ($\overline{X} = 1.07$ and $\sigma_V = 0.23$) for viscous drops with $\mu_d \leq 500$ cP. Therefore, a single correlation can be used to include a broad range of physical properties.

In the absence of direct information, it is reasonable to assume that the functional form of the DSD, and its mean and standard deviation, are not a strong function of scale or geometry. Then, once $d_{32}$ has been estimated, the DSD is known. The solid line of Figure 12-13 (discussed later) is just eq. (12-29) in its cumulative form, with $\overline{X} = 1.07$ and $\sigma_V = 0.24$.

Dispersion behavior becomes more complex for $\mu_d > 500$ cP. The DSD broadens considerably and transitions to a lognormal distribution in volume, due to a shift in the breakage mechanism, resulting in the production of numerous small satellite drops. The reader is referred to the original work of Calabrese et al. (1986a).

### 12-2.3.3 Extension to Finite ϕ. 
The equations given in Sections 12-2.3.1 and 12-2.3.2 hold strictly only for dilute systems. By *dilute* it is meant that

neighboring drops do not alter the turbulence structure or interfere with the drop breakage forces themselves. Furthermore, coalescence is neglected even for coalescing systems, since collision rates are low. Dilute dispersion studies are typically made at dispersed phase volume fractions of $\phi < 0.01$. In more concentrated systems, say for $0.01 \leq \phi \leq 0.3$, that can be described as noncoalescing, many of the relationships given above can still be used if modified appropriately to account for the effect of phase fraction on turbulence forces. For this to be valid, the dispersion must be well stabilized against coalescence and the high phase fraction cannot alter the rheological behavior of the system. Under these conditions, the presence of droplets tends to suppress small scale turbulent fluctuations and thereby reduce the stress acting to break the drops.

For inviscid drops, eq. (12-22) has been modified both mechanistically and empirically to yield

$$\frac{d_{32}}{D} = C_{10}(1 + b\,\phi)We^{-3/5} \qquad (12\text{-}30)$$

Doulah (1975) argued mechanistically that if the only effect of the presence of drops was to alter the local energy dissipation rate, $C_{10} = C_2$ and $b = 3$. Brown and Pitt (1970) measured drop size at the tip of an RDT. Their data for $\phi < 0.3$ were well correlated by the Chen and Middleman (1967) correlation with the Doulah correction [eq. (12-30) with $C_{10} = 0.053$ and $b = 3.0$]. Furthermore, the normalized DSD was also well correlated by the Chen and Middleman correlation [eq. (12-29), with $\overline{X} = 1.07$ and $\sigma_V = 0.23$]. Calderbank (1958) and Mlynek and Resnick (1972) found similar correlations for $d_{32}$. Calabrese et al. (1986b) used the Doulah approach to correct eq. (12-26) and their correlation for viscous drops. In addition to the $(1 + 3\phi)$ term in front of We, they included a $(1 - 2.5\phi)$ term in front of Vi. They suggested that their modified correlation for $d_{32}$ and their original correlation for DSD applied to RDTs for $\mu_d \leq 500$ cP and $\phi < 0.3$. However, they offered no experimental validation. In the absence of additional information, these extensions allow application of the correlations and scale-up procedures discussed above for RDTs and other impellers to noncoalescing systems of higher phase fraction.

It should be noted that numerous researchers have used eq. (12-30) to correlate drop size data for coalescing systems. Both $C_{10}$ and b varied widely and were greater than the values reported above. Except in special circumstances, the use of such correlations is not recommended, since they do not mechanistically account for coalescence, making their performance under extrapolation questionable. This is discussed further in Section 12-3.

### 12-2.3.4 Extension to Other Devices.
Equations (12-17) to (12-27) and (12-30) were the result of mechanistic arguments that were independent of device geometry. They are based on the argument that the equilibrium DSD is such that $L_T \gg d \gg \eta$, so that the turbulent stress is derived from eddies in the inertial subrange of turbulence. The structure (isotropic) and energy content of these eddies do not depend on the large scale motion or how the power is introduced. Therefore, these equations apply to a variety of contactors, provided

that $L_T \gg d \gg \eta$. Scale-up and extension of device specific correlations to other geometries requires knowledge of the $\varepsilon_{max}/\varepsilon_{avg}$ ratio or the power number, or the friction factor in the case of continuous flow devices.

*Static Mixers.* Middleman (1974) studied the dispersion of dilute inviscid dispersed phases in turbulent flow in a Kenics static mixer. He found that equilibrium was achieved after 10 mixer elements. To ensure equilibrium, Berkman and Calabrese (1988) performed a similar study for viscous dispersed phases in a 24-element static mixer. For a Kenics mixer,

$$\varepsilon_{avg} = V'_s \frac{\Delta P}{\rho_c L_p} = \frac{2V'^3_s f}{D_p} \tag{12-31}$$

where $V'_s$ is the superficial velocity, $\Delta P$ the pressure drop, $L_p$ the mixer length, f the constant friction factor, and $D_p$ the pipe diameter. The latter authors found that both data sets were well correlated by eq. (12-26) with $C_7 = 0.49$ and $C_8 = 1.38$, with the impeller diameter replaced by the pipe diameter and the Weber number and viscosity groups now defined as $We = \rho_c V_s^2 D_p/\sigma$ and $Vi = (\rho_c/\rho_d)^{1/2} \mu_d V'_s/\sigma$. When compared on an equal power per unit mass basis, the RDT produces smaller drops than the static mixer. This is because the RDT focuses energy in the trailing vortices behind the impeller blades, while the static mixer dissipates energy more uniformly. That is, the ratio $\varepsilon_{max}/\varepsilon_{avg}$ is very different in the two devices. It is tempting to conclude that the RDT is more efficient than the static mixer, but this is not the case. The energy in a stirred tank is intensely focused, but the time to reach equilibrium is relatively long. Many drop paths do not pass through the high dispersion zone. This is not the case in a static mixer. All drops are exposed to fairly uniform shear as they pass through the mixer. Berkman and Calabrese (1988) also found that the DSD is well correlated by eq. (12-29), with $\overline{X} = 1.12$ and $\sigma_V = 0.31$. While the mean is almost the same as for the RDT, the distribution is broader. It is not clear if this is real or if improvements in photographic measurement techniques allowed for better capture of the smaller drops.

*Rotor–Stator Mixers.* Calabrese et al. (2000) studied dilute dispersions of inviscid drops in turbulent flow in Ross ME100LC and Silverson L4R batch rotor-stator mixers. These devices are discussed in Chapter 8. These machines have four blade rotors, are geometrically similar, and discharge the flow radially outward from the mixing head. Although the Power numbers are similar, in magnitude to those for stirred tank turbines, these devices operate at higher speed and energy input, producing smaller drops that are close in size to the Kolmorgoff microscale. Nevertheless, the data for a slotted stator head were well correlated using eq. (12-22) with $C_2 = 0.038$, making the correlation similar to that for an RDT. The authors also found that many smaller drops were produced, resulting in the volume probability density function being a lognormal rather than a Gaussian distribution function [see eq. (12-9)]. A reasonably good correlation for DSD could still be obtained by normalization with $d_{32}$. The lognormal distribution had log mean $\overline{X} = 1.01$ and volume log standard deviation $= 0.31$.

Phongikaroon (2001) extended the study of Calabrese et al. (2000) to include viscous dispersed phases produced in the Silverson mixer, with both slotted and disintegrating (round hole) stator heads. Since he studied a broader range of physical properties and rotor speeds, he was able to produce drops at low $\mu_d$ and $\sigma$ and at high N that were smaller than the Kolmorgoff scale. As a result, only the larger values of $d_{32}$ in his data set could be correlated using eq. (12-26). Correlations that result when the restriction $L_T \gg d \gg \eta$ is not valid are discussed below. Phongikaroon (2001) also found that the normalized probability density function was lognormally distributed in volume.

*Local Power Per Mass Approach.* Davies (1987) showed that values of $d_{max}$ for a wide variety of dispersion devices could be correlated with local power per mass if a rough estimate of $\varepsilon_{max}/\varepsilon_{avg}$ could be obtained. By extending the ideas of McManamey (1979), he argued that this could be accomplished by assuming that all the power is dissipated in a localized, device-specific volume. The results of his analysis of literature data for dilute inviscid dispersed phases, corrected for interfacial tension, are shown in Figure 12-11. The slope of the line bounding the data is $-\frac{2}{5}$, as predicted by eq. (12-21). The rotor–stator data discussed above would lie between the data for agitated vessels and liquid whistles.

Experimental evidence shows that the scale-up procedures discussed above can be applied to a broad range of dispersion geometries, provided that the criterion $L_T \gg d \gg \eta$ is met. Furthermore, a few comprehensive data sets can be used to design a variety of dilute dispersion processes when applied with good judgment

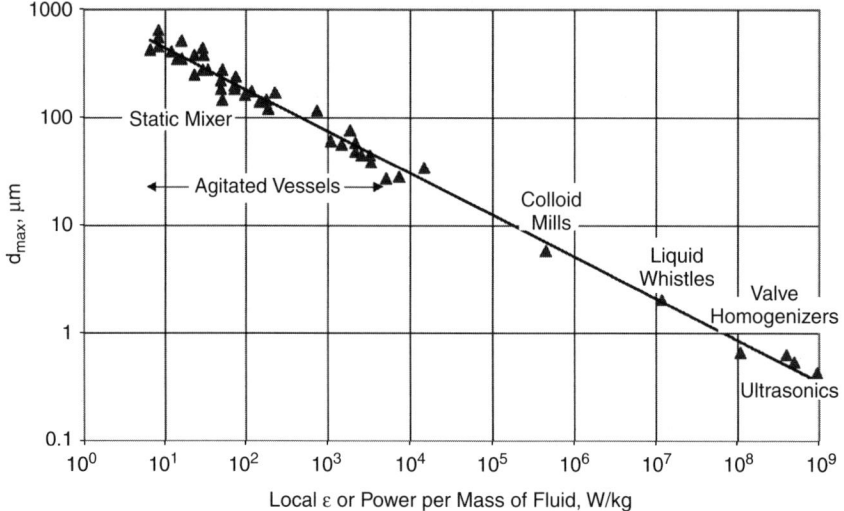

**Figure 12-11** Dependence of drop size on local power draw for various dispersion devices. (After Davies, 1987.)

by a skilled practitioner. Scale-up recommendations and examples are given in Section 12-8.

### 12-2.3.5 Additional Factors for Dilute Turbulent Dispersions

*Fine Scale Intermittency.* Baldyga and Bourne (1992) argued that the equilibrium drop size was ultimately determined by violent but relatively rare bursts of turbulent energy. Therefore, on long time scales the ultimate value of $\varepsilon_{max}$ in eq. (12-21), (12-24), and (12-25) is determined by the intermittent nature of the fine scale turbulence. They redeveloped the mechanistic theory of Section 12-2.3.1 to show that different dependencies on system parameters would result. For instance, for inviscid dispersed phases, the dependency of $d_{32}/D$ on Weber number, would be to a power less than the value $-\frac{3}{5}$ given by eq. (12-22) and could be as low as $-0.93$. Although the theory is well grounded, it is difficult to implement from a practical viewpoint. It is not clear how much time is needed to experience the ultimate turbulent burst or how system dependencies would vary at very long times. Other factors that complicate the interpretation are discussed below.

*More Sophisticated Models.* More sophisticated models have been developed to predict equilibrium mean drop size. For instance, Arai et al. (1977), Lagisetty et al. (1986), and Clark (1988) have used a Voigt (spring and dashpot) model to account for the interaction between interfacial and dispersed phase viscous forces rather than assume that they were additive, as in eq. (12-20). Although this approach is more realistic, it has not resulted in more reliable data correlation. Models have been developed for non-Newtonian drops (Lagisetty et al., 1986; Koshy et al., 1988b) and for drop breakup in the presence of drag reducing agents (Koshy et al., 1989). Unfortunately, they have only been weakly validated by data.

*Other Breakage Regimes.* The models and correlations discussed above are based on the Kolmogoroff (1949) theory for the inertial subrange of turbulence. That is, the stress acting to deform the drop is given by eq. (12-17), so the models apply only for $L_T \gg d \gg \eta$. Correlations can be developed for other breakup regimes by replacing eq. (12-17) with an appropriate model for $\tau_c$. For instance, Chen and Middleman (1967) developed a model for inviscid drops that applies when $d \ll \eta$. Shinnar (1961) and Baldyga and Bourne (1993) proposed expressions for $\tau_c$ that apply to both viscous and inertial disruptive forces when $d < \eta$. These models were not validated, since stirred vessels are usually not operated at a sufficiently high power draw to produce such small drops. Recently, Calabrese et al. (2000) have developed models for inviscid drops based on the Shinnar arguments to correlate data for $d_{32}$ of order $\eta$, produced in high-shear mixers. Phongikaroon (2001) extended these to viscous drops.

Turbulence in laboratory scale vessels may be entirely nonisotropic. That is, the drop size may be of the same order as the turbulent macroscale ($d \sim L_T$) and an equilibrium turbulence subrange may not exist. Konno et al. (1983), Pacek

**668** IMMISCIBLE LIQUID–LIQUID SYSTEMS

et al. (1999), and others have shown that in this limit, $d_{32}/D = C_{11} \cdot We^{-1}$ for inviscid drops. Therefore, the exponent on the Weber number varies from $-\frac{3}{5}$ to $-1$ as the impeller size and speed decrease. Using small tanks, Blount (1995) and others have shown this to be the case. Unfortunately, the fine scale intermittency argument above leads to a similar shift in the We exponent. These considerations illustrate how difficult it is to develop correlations for extrapolation, even for dilute dispersions.

*Effect of Surfactants.* For dilute dispersions, the presence of surfactants influences drop size only by reducing interfacial tension. To a first approximation, the drop size may be estimated within the framework developed above using the static interfacial tension in the presence of surfactant. However, drop stretching and breakup occur rapidly. As new interface is created, the rate at which surfactant diffuses to the surface may not be sufficient to maintain a constant interfacial tension. The dynamic $\sigma$ will vary from the static value in the presence of a surfactant to the value for a clean interface. Phongikaroon (2001) found that for this reason, drop sizes produced in a rotor–stator mixer with a surfactant-laden system of known static $\sigma$ were larger than those produced for a clean system of the same $\sigma$.

At high surfactant concentration, the resistance to deformation may be due solely to drop viscosity, and/or the ultimate size may be dictated by thermodynamic considerations. Koshy et al. (1988a) developed a model for drop breakup in the presence of surfactants. Unfortunately, there are few experimental data to support its implementation.

*Correlations for Sauter Mean Diameter.* Table 12-2 summarizes a large number of correlations for $d_{32}$ in stirred vessels reported before 1990. The table contains many of the studies discussed above. It also contains many studies that are largely empirical. Many apply to low viscosity drops and are based on eq. (12-30). As noted previously, this equation applies only to dispersion-dominated systems stabilized against coalescence. Yet many of the table entries are for coalescing systems at high dispersed phase fraction. As stated previously, the reader should exercise caution in extrapolating such correlations. Coalescence is discussed in Section 12-3. Most of the studies are for RDTs, demonstrating the lack of data for other impellers. Several measurement techniques, including light transmission, in situ photography, and sample withdrawal, are represented. Since these correlations were acquired for a broad spectrum of processing conditions, it is not surprising that the results are varied. Recently, Zhou and Kresta (1998a, b) and Pacek et al. (1999) have acquired data for several other impeller geometries.

### 12-2.4 Time to Equilibrium and Transient Drop Size in Turbulent Flow

Several investigators, including Chen and Middleman (1967), Arai et al. (1977), and Wang and Calabrese (1986), have reported that after introduction into the tank, several hours are required for a dilute dispersion to reach the equilibrium

**Table 12-2** Review of Correlations for Mean Drop Diameter in Liquid–Liquid Stirred Vessels[a]

| Authors | Correlation[a] | Physical Properties $\dfrac{\rho_d}{\rho_c}$ (g/cm³) | $\dfrac{\mu_d}{\mu_c}$ (cP) | $\sigma$ (dyn/cm) | D (cm) | T (cm) | $\phi$ | N (rps) | Impeller/Measurement Techniques and Comments |
|---|---|---|---|---|---|---|---|---|---|
| Vermeulen et al. (1955)[b] | $\dfrac{d_{32}}{D} = Bf_\phi We^{-0.6}$ | $\dfrac{0.693-1.595}{0.693-1.595}$ | $\dfrac{0.378-184}{1.81-65.4}$ | 3.1–55.1 | — | 25.4, 50.8 | 0.1–0.4 | 1.80–6.67 | 4-blade paddles/in situ (light transmittance) |
| Rodger et al. (1956) | $\dfrac{d_{32}}{D} = B(D/T)^{-b}We^{-0.36}$ | $\dfrac{0.761-1.101}{1.0}$ | $\dfrac{0.578-3.91}{1.0}$ | 2.1–49 | 5.1–30.0 | 15.5, 45.7 | 0.5 | 1–20 | 6-blade RT/in situ (photography, light transmittance) |
| Calderbank (1958) | $\dfrac{d_{32}}{D} = 0.06(1 + 3.75\phi)We^{-0.6}$ $\dfrac{d_{32}}{D} = 0.06(1 + 9\phi)We^{-0.6}$ | — | — | 35–40 | 5.8–25.4 | 17.8, 38.1 | 0–0.2 | — | 4-blade paddles/in situ (light transmittance); 6-blade RT/in situ (light transmittance) |
| Shinnar (1961) | $\dfrac{d_{32}}{D} = BWe^{-0.6}$ (breakage control) $\dfrac{d_{32}}{D} = B(\sigma D)^{-3/8}We^{-3/8}$ (coalescence control) | — | $\dfrac{22.5}{0.4}$ | — | 12.7 | 29.0 | 0.05 | 2.6–10.5 | No experiments; paddle turbine/sample withdrawal |

*(continued overleaf)*

**Table 12-2** (*continued*)

| Authors | Correlation[a] | Physical Properties | | | Operating Conditions/ Geometric Parameters | | | | Impeller/Measurement Techniques and Comments |
|---|---|---|---|---|---|---|---|---|---|
| | | $\dfrac{\rho_d}{\rho_c}$ (g/cm³) | $\dfrac{\mu_d}{\mu_c}$ (cP) | $\sigma$ (dyn/cm) | D (cm) | T (cm) | $\phi$ | N (rps) | |
| Chen and Middleman (1967) | $\dfrac{d_{32}}{D} = 0.053 \text{We}^{-0.6}$ | $\dfrac{0.703-1.101}{0.997-1.001}$ | $\dfrac{0.52-25.8}{0.890-1.270}$ | 4.75–48.3 | 5.1–15.2 | 10.0–45.7 | 0.001–0.005 | 1.33–16.7 | 6-blade RT/in situ (photograph) |
| Sprow (1967a) | $\dfrac{d_{32}}{D} = 0.0524 \text{We}^{-0.6}$ | $\dfrac{0.692}{1.005}$ | $\dfrac{0.51}{0.99}$ | 41.8 | 3.2–10.0 | 22.2, 30.5 | 0–0.015 | 4.2–33.4 | 6-blade RT, modified turbine/sample withdrawal (Coulter counter) |
| Brown and Pitt (1970) | $\dfrac{d_{32}}{D} = 0.051(1 + 3.14\phi)\text{We}^{-0.6}$ | $\dfrac{0.783-0.838}{0.972-0.998}$ | $\dfrac{0.59-3.30}{1.0-1.28}$ | 1.9–50.0 | 10 | 30 | 0.05–0.3 | 4.2–7.5 | 6-blade RT/in situ (photograph) |
| Van Heuven and Beek (1971) | $\dfrac{d_{32}}{D} = 0.047(1 + 2.5\phi)\text{We}^{-0.6}$ | $\dfrac{—}{0.998}$ | — | 8.5–49.5 | 3.75–40.0 | 12.5–120 | 0.04–0.35 | — | 6-blade RT/ encapsulation, sample withdrawal |
| Mlynek and Resnick (1972) | $\dfrac{d_{32}}{D} = 0.058(1 + 5.4\phi)\text{We}^{-0.6}$ | $\dfrac{1.055}{1.0}$ | $\dfrac{—}{1.0}$ | 41 | 10 | 29 | 0.025–0.34 | 2.3–8.3 | 6-blade RT/in situ (photograph) sample withdrawal (drop encapsulation) |

| Reference | Correlation | | | | $D$ (cm) | | | System |
|---|---|---|---|---|---|---|---|---|
| Weinstein and Treybal (1973)[c] | $d_{32} = 10^{(-2.316+0.672\bar{\phi})}\bar{v}_c^{0.0722}$ $\times \varepsilon^{-0.194}(\sigma g_c/\rho_c)^{0.196}$ (batch process) | $\dfrac{0.831-0.997}{0.831-0.997}$ | $\dfrac{0.722-7.43}{0.722-7.43}$ | 3.76–36.0 | 7.62–12.7 | 24.5, 37.2 | 0.079–0.593 | 2.5–10.33 | 6-blade RT, unbaffled tank/in situ (light transmittance) |
| | $d_{32} = 10^{(-2.066+0.732\bar{\phi})}\bar{v}_c^{0.047}$ $\times \varepsilon^{-0.204}(\sigma g_c/\rho_c)^{0.274}$ (continuous process) | | | | | | | | |
| Brown and Pitt (1974)[d] | $d_{32} = B(\sigma/\rho_c\varepsilon_{avg}t_{circ})^{0.6}$ | $\dfrac{0.783-0.838}{0.972-0.998}$ | $\dfrac{0.59-3.30}{1.0-1.28}$ | 1.9–50.0 | 10, 15 | 30 | 0.05 | 2.1–7.5 | 6-blade RT/in situ (light transmittance) |
| Coulaloglou and Tavlarides (1976) | $\dfrac{d_{32}}{D} = 0.081(1+4.47\phi)We^{-0.6}$ (continuous process) | $\dfrac{0.972}{1.0}$ | $\dfrac{1.3}{1.0}$ | 43 | 10.0 | 24.5 | 0.025–0.15 | 3.2–5.2 | 6-blade RT/in situ (photomicrography) |
| Godfrey and Grilc (1977) | $\dfrac{d_{32}}{D} = 0.058(1+3.6\phi)We^{-0.6}$ $d_{32} = 10^{(-3.18+0.74\phi)}$ $\times \varepsilon_{avg}^{-0.2755}(\sigma/\rho_c)^{0.1787}$ | $\dfrac{0.783-0.829}{0.986-0.997}$ | $\dfrac{2.05-8.6}{0.89-1.19}$ | 1.9–34.5 | 5.1 | 15.2 | 0.05–0.5 | 8.33–15.0 | 6-blade RT, unbaffled square tank/sample withdrawal |
| Arai et al. (1977)[e] | $\dfrac{d_{max}}{d_{max,0}} = (1+9Vi')^{3/5}$ | $\dfrac{0.879-0.922}{1.00}$ | $\dfrac{0.78-1500}{0.97}$ | 22 | Incorrect value of (D/10) is given in the paper | 12.7 | <0.003 | 2.5–13.7 | 6-blade RT/In situ (photograph) |

*(continued overleaf)*

**Table 12-2** (continued)

| Authors | Correlation[a] | Physical Properties | | | Operating Conditions/Geometric Parameters | | | | Impeller/Measurement Techniques and Comments |
|---|---|---|---|---|---|---|---|---|---|
| | | $\dfrac{\rho_d}{\rho_c}$ (g/cm$^3$) | $\dfrac{\mu_d}{\mu_c}$ (cP) | $\sigma$ (dyn/cm) | D (cm) | T (cm) | $\phi$ | N (rps) | |
| Lagisetty et al. (1986) | $\dfrac{d_{max}}{D} = 0.125(1 + 4.0\phi)^{1.2}\text{We}^{-0.6}$ | $\dfrac{0.88-1.47}{0.78, 1.0}$ | $\dfrac{\text{Non-Newtonian}}{1.0, 2.1}$ | 20, 45.2, 50 | 7.25 | 14.5 | 0.02 | 3.33–10 | 6-blade RT/sample withdrawal; correlation is based on Voigt model; limited experiments used for verification of the model |
| Calabrese et al. (1986a)[f] | $\dfrac{d_{32}}{d_0} = (1 + 11.5\text{Vi}'')^{5/3}$, $\text{Vi}'' < 1$ (moderate viscosity, $\mu_d$) | $\dfrac{0.960, 0.970}{0.997}$ | $\dfrac{96.0, 486}{0.893}$ | 37.8 | 7.1–19.6 | 14.2–39.1 | <0.0015 | 0.93–5.95 | 6-blade RT/In situ (photograph) |
| | $\dfrac{d_{32}}{D} = 2.1(\mu_d/\mu_c)^{3/8}\text{Re}^{-3/4}$ (high viscosity, $\mu_d$) | $\dfrac{0.971-0.975}{0.997}$ | $\dfrac{971-10\,510}{0.893}$ | 37.8 | 7.1–19.6 | 14.2–39.1 | <0.0015 | 0.93–5.95 | — |
| Wang and Calabrese (1986)[g] | $\dfrac{d_{32}}{D} = 0.053\text{We}^{-0.6} \times (1 + 0.97\text{Vi}^{0.79})^{3/5}$ | $\dfrac{0.834-0.986}{0.792-0.997}$ | $\dfrac{0.81-459}{0.52-0.89}$ | 0.21–47 | 7.1–15.6 | 14.2–31.2 | <0.002 | 1.4–4.7 | 6-blade RT/In situ (photograph) |

| Reference | Equation | | | | | | | | Notes |
|---|---|---|---|---|---|---|---|---|---|
| Calabrese et al. (1986b)[g] | $\dfrac{d_{32}}{D} = 0.053 We^{-0.6} \times (1 + 0.91 Vi^{0.84})^{3/5}$ | $\dfrac{0.692-1.101}{0.792-1.005}$ | $\dfrac{0.51-520}{0.52-1.27}$ | 0.21–48.3 | 7.1–19.6 | 14.2–39.1 | <0.005 | 0.93–33.4 | Mainly 6-blade RT/Correlate Sprow (1967b), Arai et al. (1977), Calabrese et al. (1986a), and Wang and Calabrese (1986) |
| Berkman and Calabrese (1988)[h] | $\dfrac{d_{32}}{D} = 0.49 We'^{-0.6} \times [1 + 1.38 Vi'''(d_{32}/D_p)^{1/3}]^{3/5}$ | $\dfrac{0.852-0.967}{\sim 1}$ | $\dfrac{0.63-204}{-}$ | 31.8–41.6 | — | 1.91 | 0.00057–0.001 | — | Kenics static mixer/In situ (photograph) |
| Nishikawa et al. (1987a) | $\dfrac{d_{32}}{D} = 0.095 N_p^{-2/5} We^{-0.6}$ $\times (1 + 2.5\phi^{2/3}(\mu_d/\mu_c)^{1/5})_d$ $\times (\mu_d/\mu_c)_c^{1/8}$ (breakup region) $\dfrac{d_{32}}{D} = 0.035 N_p^{-1/4} We^{-3/8} D^{-3/8}$ $\times (1 + 3.5\phi^{3/4}(\mu_d/\mu_c)^{1/5})_d$ $\times (\mu_d/\mu_c)_c^{1/8}$ (coalescence region) | $\dfrac{0.81}{0.972}$ | $\dfrac{17.0}{0.356}$ | 17.9 | 12.5 | 25.0 | 0.0045–0.36 | 1.3–5.0 | 6-blade RT/sample withdrawal; the suffix d or c outside the bracket ($\mu_d/\mu_c$) means to keep the viscosity of dispersed phase or the viscosity of continuous phase constant |

*(continued overleaf)*

**Table 12-2** (continued)

| Authors | Correlation[a] | Physical Properties | | | Operating Conditions/ Geometric Parameters | | | | Impeller/Measurement Techniques and Comments |
| --- | --- | --- | --- | --- | --- | --- | --- | --- | --- |
| | | $\dfrac{\rho_d}{\rho_c}$ (g/cm$^3$) | $\dfrac{\mu_d}{\mu_c}$ (cP) | $\sigma$ (dyn/cm) | D (cm) | T (cm) | $\phi$ | N (rps) | |
| Nishikawa et al. (1987b)[i] | $\dfrac{d_{32}}{D} = 0.095 N_p^{-2/5} We^{-0.6} (T/T_0)^{-2/5}$ $\times (1 + 2.5(T/T_0)^{1/2} \phi^{2/3})$ $\times (\mu_d/\mu_c)_d^{1/5} (\mu_d/\mu_c)_c^{1/8}$ (breakup region) $\dfrac{d_{32}}{D} = 0.035 N_p^{-1/4} We^{-3/8}$ $\times (T/T_0)^{-1/4}$ $\times (1 + 3.5(T/T_0)^{1/2} \phi^{3/4})$ $\times (\mu_d/\mu_c)_d^{1/5}$ $\times (\mu_d/\mu_c)_c^{1/8} D^{-3/8}$ (coalescence region) | $\dfrac{0.81}{0.972}$ | $\dfrac{17.0}{0.356}$ | 17.9 | T/2 | 12–50 | 0.005–0.36 | 1.0–10.0 | 6-blade RT/sample withdrawal; the suffix d or c outside the bracket ($\mu_d/\mu_c$) means to keep the viscosity of dispersed phase or the viscosity of continuous phase constant |

| Chatzi et al. (1989) | $\dfrac{d_{32}}{D} = 0.056(1 + 10.97\phi)We^{-0.6}$ | $\dfrac{0.8792, 0.9014 \ 0.4591, 0.7303}{0.9881, 0.9971 \ 0.5502, 0.9147}$ | 7.4, 11.5 | 7.5 | 15 | 0.01–0.03 | 2.50–5.00 | 4-blade turbine/sample withdrawal; suspending agent (polyvinyl alcohol added in the tank) |

Reproduced from Zhou and Kresta (1998a).

[a] B and b are constants.

[b] $f_\phi$ is the ratio of the actual mean diameter to that at $\phi = 0.1$.

[c] $\nu_c = \mu_c/\rho_c$ = continuous-phase kinematic viscosity. For batch process; $\varepsilon = [Pg_c/V(1-\overline{\phi})\rho_c]$ (V is the volume of the fluids). For continuous process: $\varepsilon = [(P - 6Q_d\sigma/\overline{d_{32}})g_c]/V(1-\overline{\phi})\rho_c$ ($Q_d$ is the flow rate of dispersed phase; $g_c$ is the gravitational constant).

[d] $t_{circ}$ is the circulation time.

[e] Based on a Voigt model; $d_{max,0}$ is the value of $d_{max}$ when $Vi' \to 0$; $Vi' = \mu_d \varepsilon_{avg}^{1/3} d_{max}^{1/3}/\sigma$.

[f] $d_0$ is the $d_{32}$ for an inviscid dispersed phase ($Vi'' \to 0$); $Vi'' = (\rho_c/\rho_d)^{1/2}(\mu_d \varepsilon_{avg}^{1/3} d_{32}^{1/3}/\sigma)$.

[g] $Vi = (\mu_d ND/\sigma)(\rho_c/\rho_d)^{1/2}$.

[h] $We' = \rho_c V_s^2 D_p/\sigma$ and $Vi''' = (\rho_c/\rho_d)^{1/2}(\mu_d V_s/\sigma)$. $D_p$ is the pipe diameter, and $V_s$ is the superficial velocity.

[i] $T_0$ is a reference tank diameter.

DSD. This implies that breakage rate slows considerably as d approaches $d_{max}$. The long time behavior of $d_{32}$ is uncertain. As drop sizes decrease with time, the energy required for them to disperse further increases continually. This amounts to shrinking the effective dispersion volume. Since all dispersing drops must pass through this shrinking volume, it explains the long times required to reach an equilibrium state. Lam et al. (1996) argued that their data supported the idea that turbulent intermittency caused $d_{32}$ to decrease without limit. Blount (1995) and others have found that $d_{32}$ actually increased at very long times. This could be due to very slow coalescence rates becoming important as breakage ceases; or possibly to Ostwald ripening or to redispersion of dispersed phase liquid collected on impeller, tank, and baffle surfaces. These considerations could have limited practical consequence, since the time to reach equilibrium decreases drastically for coalescing systems as the dispersed phase volume fraction increases. Hong and Lee (1985) measured times to equilibrium of less than 10 min for $0.05 < \phi < 0.2$. The time to achieve a dynamic equilibrium between breakage and coalescence appears to be much shorter than that to achieve inconsequential breakage in the absence of coalescence. In large scale vessels the time to reach equilibrium is longer than on the bench scale.

Using an intuitive approach, several researchers have proposed that the time for $d_{32}$ to reach equilibrium could be described by analogy to reaction kinetics:

$$\frac{d\Omega}{d\theta} = -\alpha_1 \Omega^{\alpha_2} \quad \text{where} \quad \Omega = \frac{d_{32}(t) - d_{32}^{\infty}}{d_{32}^{\infty}} \quad \text{and} \quad \theta = Nt \quad (12\text{-}32)$$

where N is the impeller speed, $d_{32}(t)$ the Sauter mean diameter at time t, and $d_{32}^{\infty}$ its value at equilibrium. The terms $\alpha_1$ and $\alpha_2$ are analogous to the reaction rate constant and reaction order, respectively. An implicit assumption is that the entire DSD evolves similarly. For $\alpha_2 = 1$, $d_{32}(t)$ decays exponentially. Hong and Lee (1985) found this to be the case for stirred tank systems undergoing simultaneous breakage and coalescence ($0.05 < \phi < 0.2$). Al Taweel and Walker (1997) argued that data for a dilute dispersion in a Lightnin static mixer were well correlated by $\alpha_2 = 2$, where t was the transit time through the mixer.

### 12-2.4.1 Prediction of Transient Drop Size Distribution.
The initial stage of forming a dilute dispersion shows a broad size distribution. This is illustrated in Figure 12-12. Narsimhan et al. (1980) and Sathyagal et al. (1996) have acquired similar data. As stirring continues, drops of all size continue to break to form smaller droplets. Both impeller speed and physical properties affect the dispersion time and breadth of the DSD. Faster speeds tend to hasten dispersion and give a narrower DSD. Increases in interfacial tension and drop viscosity result in longer dispersion time and broader distributions.

The aforementioned investigators found that after a relatively short time, the DSD became normally distributed in volume, like the equilibrium DSD discussed in Section 12-2.3.2. Both Narsimhan et al. (1980) and Chang (1990) found that the data from many experiments could be collapsed to a single curve

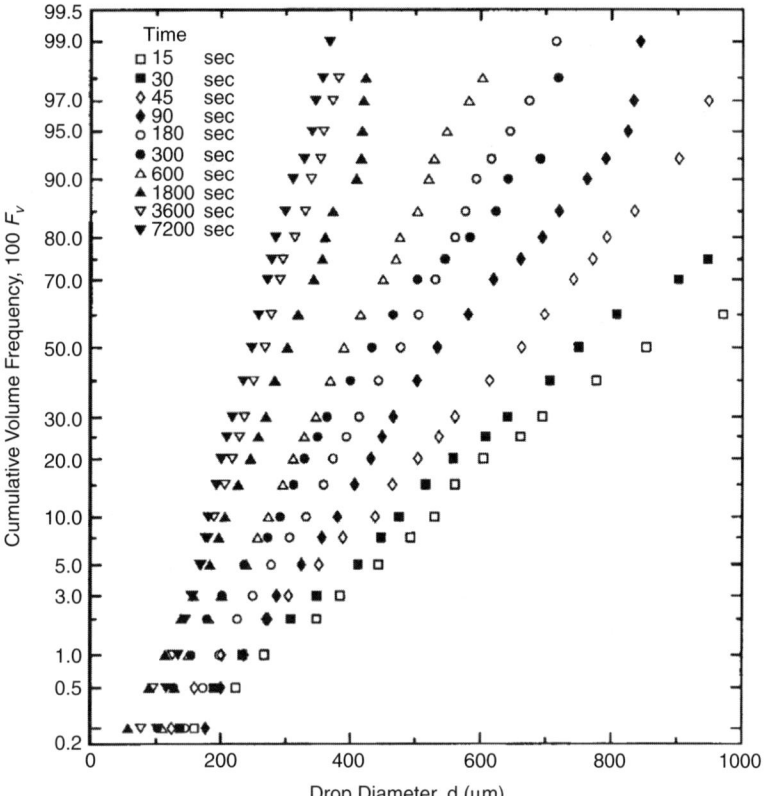

**Figure 12-12** Effect of stirring time on DSD for a paraffin oil dispersed in water, $\sigma = 0.048$ N/m, $\mu_d = 0.040$ Pa·s. For RDT with N = 4.67 rps, D/T = 0.5, T = 0.21 m. (Data of Chang, 1990.)

by normalization with the instantaneous $d_{32}(t)$. This is shown in Figure 12-13. Therefore, the instantaneous DSD can be described by eq. (12-29), where $X(t) = d/d_{32}(t)$ and the mean and the volume standard deviation are defined similarly. Chang found that his data for a RDT were well correlated by $\overline{X} = 1.07$ and $\sigma_V = 0.27$ for both inviscid and viscous drops with $\mu_d \leq 0.140$ Pa·s. This is essentially the same result as for the equilibrium DSD of Section 12-2.3.2, as should be expected. Therefore, for dilute systems, a single correlation describes the time evolution of the DSD provided that $d_{32}(t)$ is known. The correlation also fits the data of Narsimhan et al. for a flat-blade turbine.

Chang (1990) used a population balance framework (discussed in Section 12-4) to develop correlations for $d_{32}(t)$. For nonviscous oils dispersed in water, he obtained

$$\frac{d_{32}(t)}{d_{32}^{\infty}} = \left(\frac{3.8 \times 10^3}{Nt}\right)^{1/6} \quad \text{for } 100 < Nt < 3800 \quad (12\text{-}33)$$

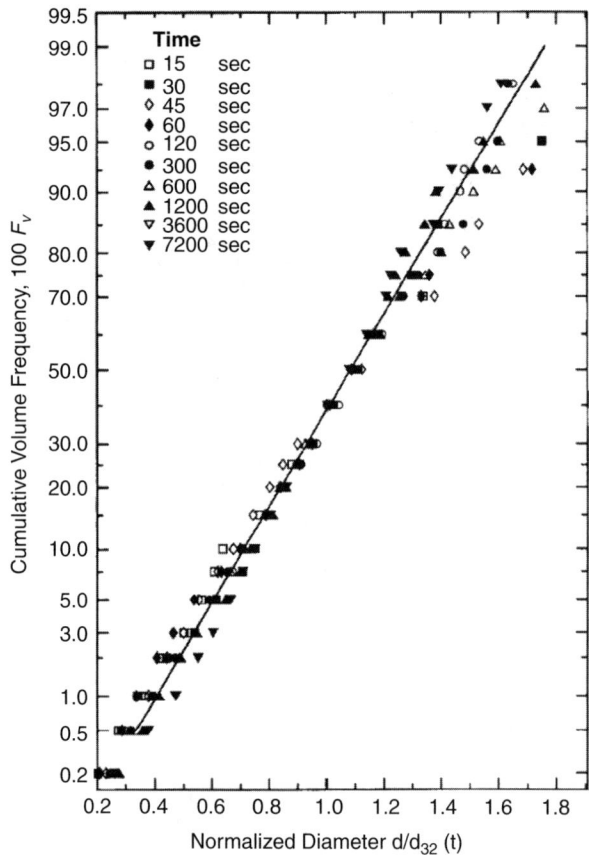

**Figure 12-13** Normalized transient DSD for a paraffin oil dispersed in water, $\sigma = 0.048$ N/m, $\mu_d = 0.140$ Pa · s. For RDT with N = 3.0 rps, D/T = 0.5, T = 0.21 m. (Data of Chang, 1990.)

For viscous oils ($0.040 < \mu_d \leq 0.140$ Pa · s) dispersed in water, he obtained

$$\frac{d_{32}(t)}{d_{32}^{\infty}} = \left(\frac{2.1 \times 10^4}{Nt}\right)^{1/6} \quad \text{for } 100 < Nt < 21\,000 \quad (12\text{-}34)$$

In his experiments with a single RDT, the tank mean circulation time was given by $N \bar{t}_{circ} = 3.8$. The inviscid drops reached equilibrium after about 1000 impeller passes. The viscous drops approached equilibrium more slowly, requiring about 5500 impeller passes.

Chang's results can be used to estimate the time evolution of the DSD as follows. The equilibrium Sauter mean diameter, now called $d_{32}^{\infty}$, can be estimated using the correlations developed in Section 12-2.3.1. The value of $d_{32}(t)$ can then be obtained from the more appropriate of eq. (12-33) and (12-34). The

DSD follows from eq. (12-29) with $\overline{X} = 1.07$ and $\sigma_v = 0.27$. Chang (1990) and Calabrese et al. (1992) have summarized the method.

For impellers other than the RDT, the estimation of $d_{32}^\infty$ was discussed above. The dimensionless DSD for the radial impeller studied by Narsimhan et al. (1980) is also fit by $\overline{X} = 1.07$ and $\sigma_v = 0.27$, so it is reasonable to apply these values to approximate the DSD for other impellers. The weak link is prediction of $d_{32}(t)$, since it is directly dependent on circulation time. If the circulation time is known relative to the RDT work of Chang (1990), it may be possible to guess the decay rate by reference to eq. (12-32) and (12-33), since it is the number of impeller passes that determine the time to achieve equilibrium. The reader is reminded that these methods apply only to dilute dispersions and will significantly overestimate the dispersion time in the presence of coalescence.

### 12-2.5 Summary

Estimations of mean drop size and drop size distribution is complex, even for non-coalescing systems. They depend on the completeness of the dispersion process, local turbulent intensities (which in turn depend on impeller selection), vessel and impeller design, and operating conditions. They also depend on physical and interfacial properties which are often affected by the presence of surfactants, suspending agents, and impurities. Furthermore, concentration of the dispersed phase plays an important role. Drop size distributions are also affected by the violence of drop breakage. When barely enough energy is available to cause breakage, the result is for relatively few daughter drops to form. If the breakage event is caused by excessive energy, often orders-of-magnitude more daughter drops form. We have tried to summarize the state of knowledge for dilute systems and how this information can be applied to more concentrated dispersions. Reliable data are available for relatively few process geometries, so engineering judgment is required to apply these data to other configurations. It is difficult to predict the time to achieve a steady DSD, since this depends on drop concentration and circulation/residence time. Even in the absence of detailed information, it is important to make a rough estimate of the dispersion time relative to the process time.

Most of the studies discussed above were carried out at a bench scale. If scale-up is involved, further complications can arise. These are discussed in Section 12-8.

## 12-3 DROP COALESCENCE

### 12-3.1 Introduction

*12-3.1.1 Basic Principles*. Coalescence is the process of combining two or more drops to form one or more larger drops. It occurs when drops, suspended in a moving fluid, collide with one another as shown in Figure 12-14. Coalescence also occurs when drops rise or settle due to gravity to a condensed layer, as in a

**680** IMMISCIBLE LIQUID–LIQUID SYSTEMS

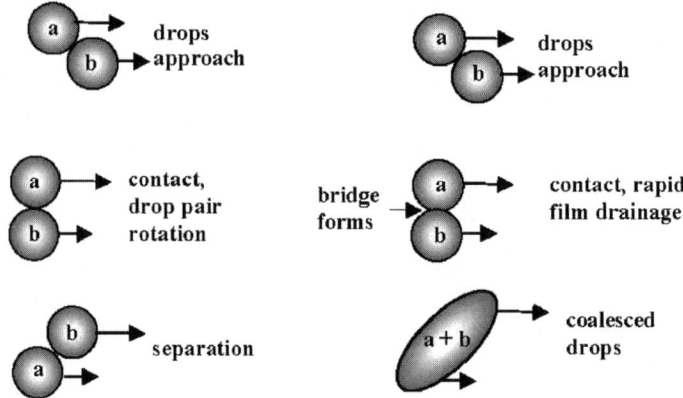

**Figure 12-14** Rebounding and coalescence of two drops in shear flow.

decanter. It is also caused by impacts, such as when drops collide with impeller blades, baffles, vessel walls, static mixer elements, or fibers in coalescers. *Coalescence efficiency*, defined as the probability of coalescence per collision, depends on the collision force, the cleanliness of the interface, and the time of contact.

From an industrial viewpoint, coalescence is undesirable for some processes and desirable for others. For example, coalescence during suspension polymerization is undesirable and leads to reactor setup, or buildup of polymer on vessel walls and agitation equipment. On the other hand, mass transfer processes, such as extraction, centrifugation, and decantation, depend on coalescence to achieve desirable rates of operation. Coalescence between drops leads to intimate mixing in the newly formed larger drop.

Coalescence depends on the collision rate, which increases with dispersed phase concentration. To quantify this process, it is convenient to define a collision frequency $\xi(d, d')$, between drops of diameter d and d', which is independent of concentration. The collision frequency depends on agitation rate and drop size. As shown in Figures 12-14 and 12-17, the collision of two drops does not ensure coalescence. As the drops approach each other, a film of continuous phase fluid keeps them apart. Coalescence depends on the rupture of this film. It must drain to a critical thickness before coalescence can occur. The critical drainage time is the time it takes for the film to thin sufficiently that rupture occurs; or in other words, coalescence occurs only if the collision interval, referred to as the *contact time*, exceeds the critical film drainage time. The probability that this will occur is called the *coalescence efficiency*, $\lambda(d,d')$. It depends on a different set of hydrodynamic factors as well as drop size and physicochemical variables. Because collision frequency and coalescence efficiency depend on different factors, their contributions to coalescence are treated separately. As a result, the coalescence frequency $\Gamma(d, d')$ between two drops of diameter d and d' is defined as

$$\Gamma(d, d') = \xi(d, d')\lambda(d, d') \qquad (12\text{-}35)$$

For fine aerosol particles, $\lambda \to 1.0$ and the agglomeration rate is the collision rate. However, for liquid–liquid systems, the coalescence efficiency is often small and rate limiting. Therefore, classical agglomeration theory (e.g., Smoluchowski equation) cannot be directly applied to liquid–liquid dispersions. Coalescence is known as a second-order process ($\sim n^2$) since the coalescence rate is proportional to $\Gamma(d, d')n(d)n(d')$, where $n(d)$ and $n(d')$ represent an appropriate measure of the number of drops of size $d$ and $d'$, respectively.

### 12-3.1.2 Empirical Approach for Turbulent Stirred Vessels.

In turbulent stirred vessels with small but finite $\phi$, drop size often varies linearly with dispersed phase concentration. For low viscosity drops, Figure 12-15 shows a linear relationship up to $\phi = 10\%$, between $\phi$ and drop size, expressed as $d_{32}(\phi)/d_{32}(0)$, presumably at the same agitation rate. $d_{32}(\phi)$ is the equilibrium Sauter mean diameter at dispersed phase fraction $\phi$, while $d_{32}(0)$ is its counterpart for a dilute dispersion ($\phi \to 0$). This ratio also shows dependence on impeller type and D/T. Larger D/T impellers (e.g., Intermigs in Figure 12-15), promote gentler agitation throughout the vessel, enhancing coalescence. In the earlier work of Vermeulen et al. (1955), the degree of coalescence was less than for the Todtenhaupt et al. (1991) data of Figure 12-15. This could be due to different physical properties and impurities. As discussed below, coalescence rates depend on many factors and are operation dependent. However, the Vermeulen et al. data did show that the dependence of $d_{32}$ on $\phi$ was nonlinear at high $\phi$.

For inviscid dispersed phases, $d_{32}(0)$ is given by eq. (12-22). Since the curves of Figure 12-15 are fit by an empirical equation of the form $d_{32}(\phi)/d_{32}(0) =$

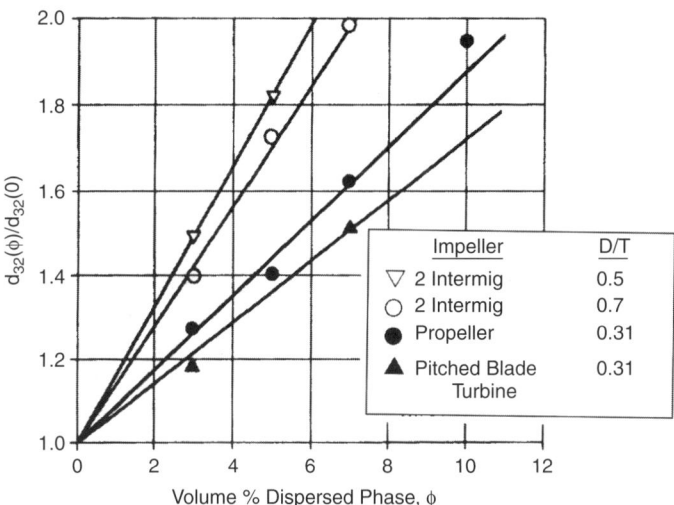

**Figure 12-15** Typical dependence of drop size on dispersed phase concentration for a coalescing system. $d_{32}(\phi)/d_{32}(0)$ increases with $\phi$ due to coalescence. (Reproduced from Todtenhaupt et al., 1991.)

**682** IMMISCIBLE LIQUID–LIQUID SYSTEMS

$1 + b \phi$, the data should be correlated by an equation of the form of eq. (12-30). Early investigators treated coalescence as an addendum to dispersion theory, where b ranged from 3 to 9, depending on the system and the investigator. As discussed previously, Table 12-2 summarizes much of the work done using this approach.

The use of eq. (12-30), a breakage equation, to correlate data for coalescing systems can be further rationalized by reference to eq. (12-35). Since the b $\phi$ term represents the coalescence frequency, the use of eq. (12-30) suggests that there is a constant coalescence efficiency represented by b and a constant collision rate that is proportional to $\phi$. Presently, there is no systematic way to relate b to the many factors governing coalescence, and there is no way to extend the approach to viscous drops. This empirical approach, although simple to use, lacks technical interpretation and is therefore risky to apply for scale-up work.

### 12-3.1.3 Factors Influencing Coalescence.

Drop coalescence is not as well understood as drop breakage, since the relevant physical mechanisms are more complex and data acquisition (sampling and analysis) becomes more difficult with increasing dispersed phase concentration. The collision frequency is determined largely by the dynamics of the continuous phase flow field, which determines the trajectories of the colliding drops. Calculations that account for the effect of drop deformation and other drop-surrounding fluid interactions on collision rate are difficult, and it is often assumed that the drops are rigid and behave as inertialess fluid points. For laminar flow, these calculations depend strongly on geometry and are tedious for realistic processing equipment.

For turbulent flows, the collision rate depends on the frequency at which eddies bring drops into contact. Since the drops are usually small compared to the macroscale ($L_T \gg d \gg \eta$, as in Section 12-2.3.1), isotropic turbulence theory can be used to model the collision frequency, the force with which two drops collide, and the time that they remain in contact before subsequent eddies carry them apart. These factors depend on the drop size and the magnitude of the energy dissipation rate, which depends on the impeller speed and diameter. For instance, Coulaloglou and Tavlarides (1977) show that for equal drop size and $L_T \gg d \gg \eta$, the collision frequency, $\xi(d, d)$, is given by

$$\xi(d, d) = C_{12}d^{7/3}\varepsilon^{1/3} = C_{13}d^{7/3}ND^{2/3} \qquad (12\text{-}36)$$

They show further that the approach force is given by $F \sim (d^8\varepsilon^2)^{1/3}$ and the contact time is given by $t_c \sim (d^2/\varepsilon)^{1/3}$. These quantities, derived from turbulence theory, are required inputs to models for the drainage rate of the laminar film and the coalescence efficiency, as described below.

The collision frequency and approach force increase with drop size and agitation rate. For monodisperse drops, the collision rate is of order $n^2(d)\xi(d, d)$. Contact times increase with drop size and decrease with agitation rate. Coulaloglou and Tavlarides (1977) have also modified these results to apply to unequal-sized drops.

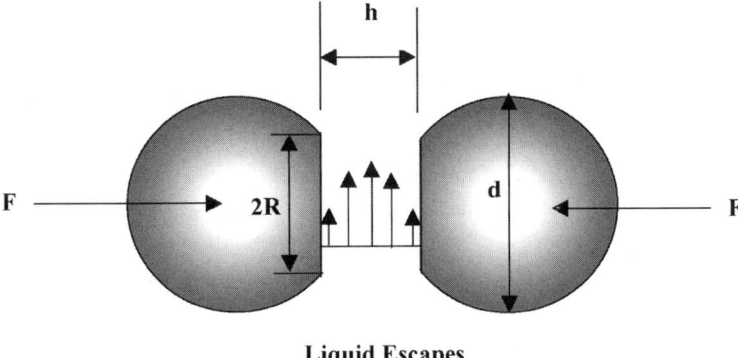

**Figure 12-16** Film drainage and thinning for deformable equal-sized colliding drops.

The collision efficiency is much more difficult to quantify. Consider the collision of two equal-sized drops as illustrated by Figure 12-16. The drops approach each other with a transient force F. This force squeezes out the film of continuous phase fluid, of thickness h, trapped between the drops. The contact time should be sufficiently long so that a critical thickness is achieved, whereupon film rupture and coalescence will take place. During the impact or contact period, the drops deform and flatten, thereby increasing the surface area of contact. The degree of flattening, characterized by disk radius R in Figure 12-16, affects the film drainage time since the amount of entrapped film and the resistance to drainage both increase with increasing contact area. The film-thinning rate also depends on the mobility of the interface between the drops and the draining film. If the interface is rigid, the drop fluid remains stationary and is not dragged in the direction of the draining film. A rigid interface offers the maximum resistance to film drainage due to a no-slip condition at the film–drop interface. A mobile interface is one in which the drop phase fluid is dragged in the direction of film drainage, so that the velocity is equal on both sides of the interface. A mobile interface offers the minimum resistance to film drainage, due to a "complete slip" condition at the film–drop interface. It was shown by Murdoch and Leng (1971) that when interfaces are immobile, the drainage flow develops a parabolic velocity profile instead of the plug flow profile that exists for mobile surfaces. Most interfaces are partially mobile, falling somewhere between the two limits. The film drainage time increases as the interface becomes less mobile.

Physicochemical factors affecting coalescence efficiency are complex and often difficult to quantify. A high drop viscosity promotes coalescence by increasing resistance to leading surface deformation during impact, but it inhibits coalescence by making film drainage more difficult. The latter factor is usually dominant. Suspension polymerizations go through a sticky stage. This is caused by the collision of partially polymerized drops having sticky surfaces.

Interfacial tension is an important physicochemical factor. Decreasing the interfacial tension inhibits coalescence since it leads to greater flattening for a given impact force. Surfactants, suspending agents, and certain impurities reduce coalescence by immobilizing drop-film interfaces and increasing disk size, due to lower interfacial tension. Small quantities of surface-active impurities can significantly reduce coalescence rates. Suspending agents and surfactants are designed to act in the same way. Suspending agents are adsorbed more slowly at the interface than surfactants, due to their higher molecular weights and thus slower diffusion rates. Typical molecular weights for surfactants ≈300 and for suspending agents ≈30 000. However, once polymeric suspending agents are adsorbed at the drop interface, they can form physically coherent "skins," which are a solid polymer network and prevent coalescence for days, even at stagnant conditions.

The film drainage time is also affected by the magnitude and duration of collision forces, which depend not only on external hydrodynamic conditions but also on the electrochemical state of the interface. It would appear at first glance that higher approach forces would lead to faster film drainage. But this is not necessarily the case, since increasing this force promotes more flattening, and excessive pressure buildup, resulting in rebounding of drop pairs. This is another reason why coalescence is promoted by "gentle collisions."

In certain cases when drops approach one another, repulsive forces begin to act. For instance, increasing the pH inhibits coalescence in water–organic systems due to increased surface adsorption of $OH^-$, causing stronger repulsive forces. Tobin and Ramkrishna (1992) found that absorption of $CO_2$ from the headspace in a stirred tank decreased the pH and caused an increased coalescence rate of organic drops in water. Ionic surfactants inhibit coalescence by increasing electrostatic repulsive forces. For such systems, increasing the ionic strength by the addition of electrolytes promotes coalescence by decreasing the effect of double-layer protection. In summary, electrical charges can create either a force of attraction or a force repulsion between drops. Coalescers employed by the petroleum industry use charged plates to promote coalescence of saltwater drops in crude petroleum fractions.

It is difficult to develop a single model for coalescence efficiency because of the numerous factors influencing the film drainage rate and therefore the coalescence frequency. Even if all impurities could be eliminated, it would still be difficult to interpret the most systematic experiments in surfactant-free and charge-neutral systems. For instance, in a turbulent stirred tank, increasing the agitation rate (N or $\varepsilon$) at constant $\phi$ increases the collision rate by increasing the collision frequency directly and by increasing the number of drops, due to increased dispersion. However, increased agitation decreases the coalescence efficiency by increasing the approach force and by decreasing the contact time. The decrease in drop size inhibits coalescence due to decreased collision frequency but will promote coalescence by reducing drop flattening. Uncontrolled impurities, unqualified electrical forces, and other interfacial phenomena will further complicate interpretation. As a result, considerable judgment must be exercised when scaling-up from lab scale studies or in using empirical correlations. Mechanistic models are discussed below.

Fundamental studies have focused on the more complex film drainage step, by precisely monitoring the coalescence of a single drop at a plane interface or the interaction between two colliding drops under precisely controlled conditions. These studies elucidate the complexities of the coalescence process.

### 12-3.1.4 *Coalescence Mechanisms in Mixing Flows*.

As explained previously, coalescence between colliding drops occurs when the film of continuous phase fluid separating them thins to a critical thickness during contact. Once the critical thickness is reached, a hole opens up which enlarges rapidly, resulting in coalescence and internal drop mixing. Sometimes the combination is so rapid that internal pressures cause satellite drops to be ejected from opposite ends of the newly formed oscillating drop. If the force holding drops in contact is brief and insufficient drainage occurs, coalesce will not take place.

The approach forces needed to bring about film drainage can be hydrodynamic, hydrostatic, or physicochemical. As discussed above, hydrodynamic forces are bought about by shear (laminar or turbulent) and are of finite duration. Such forces can be intense but are definitely not constant during drop contact. Forces can also be due to gravity acting on density differences between the drops and the continuous phase. Gravitational forces are constant and are of long duration. They control coalescence times for drops approaching settled layers. The film thinning mechanism still applies even though there is no critical time beyond which departure occurs. The time for an emulsion or suspension to settle completely can be quite long, particularly if the phases have similar densities or if interfaces contain surfactants, repulsive charges, or impurities.

Solid surfaces, particularly those easily wetted by the dispersed phase, can be major collectors of drops. In the case of a rotating impeller, drops collect and coalesce on blade surfaces to form a condensed film. As this film grows in thickness, it flows under centrifugal forces to the impeller tips and disperses into tiny drops. This process is similar to the breakup of a cylindrical liquid jet. A film of dispersed phase can also collect on free surfaces, baffles, tank walls, and the impeller shaft, where the surface vortex meets the shaft. In the case of emulsion and suspension polymerization, coalescence also leads to fouling of heat transfer surfaces.

Electrostatic forces are used in electrostatic precipitators to coalesce aqueous brine from crude oil. Fibrous beds are used to coalesce flowing drop suspensions where fibers are chosen that will be wetted by the dispersed phase. As the drop suspension is forced through the bed, drops coalesce and build up a wet layer on the fibers. This layer continues to thicken until drag forces caused by the flow result in break-off. The departing drops, however, are much larger than the incoming drops, so the device achieves its desired function. Centrifuges amplify gravitational forces. The cream separator is a good example.

### 12-3.1.5 *Practical Classification of Coalescing Systems*.

While it has been stated repeatedly that coalescence is highly complex and that scale-up is difficult, not all liquid–liquid systems are complex. A simple way to characterize systems is to measure the time for a dispersion to separate. The tested system is thoroughly agitated to form either a water-in-oil (w/o) or oil-in-water (o/w) dispersion, depending on the system under investigation. Following 3 to 5 min of vigorous agitation or shaking, the system is allowed to settle and the time to form two distinct layers is noted. Complete separation may not occur, but two distinct layers ought to be visible. Guidelines for scale-up, based on separation time, are given in Table 12-3. When applying this method, be aware that density differences affect both settling time and the forces acting on the drops that cause film drainage. If coalescence appears to be severe and undesirable, reverse the phases if possible and repeat the test. Finally, compare the times for coalescence to see which phase should be dispersed.

A more quantitative method is to use a baffled stirred vessel containing a light transition probe similar to the one described by Rodger et al. (1956). Record the probe output at moderately high levels of agitation. After a constant baseline is established, reduce the agitation to just maintain full suspension and observe changes in the recorded output. The probe can be calibrated to read interfacial area. The slope at the time just after speed transition is proportional to the rate of coalescence under dynamic as opposed to static conditions. This method is insensitive to effects of density difference and is described by Howarth (1967).

If the interfacial area appears to remain constant after decreasing the agitation, the system can be considered to be noncoalescing. If not, the steepness of the

**Table 12-3** Characterization of the Coalescibility of Immiscible Liquid–Liquid Systems

| Time to Separate | Characterization | Process Implication |
|---|---|---|
| <10 s | Very fast coalescence | Expect severe scale-up problems for agitated vessels, provide more dispersion opportunities. For example, use multiple impellers, provide for strong flow at the top and bottom of the vessel. Consider use of long static mixers. |
| <1 min | Fast coalescence | Scale-up problems can be managed by careful selection of mixing equipment. Use multiple impellers, eliminate unnecessary internals, and provide for complete circulation. |
| 2–3 min | Moderate coalescence | Problems are less severe, design for coalescence. Use large impellers for dispersion and flow. Maintain ample flow at the top/bottom surfaces. Often can treat this case as noncoalescing. |
| >5 min | Slow coalescence | Application can be treated as dispersion only. |

slope is a measure of the severity of coalescence. Care must be taken when choosing the slow speed to ensure that settling does not occur.

### 12-3.2 Detailed Studies for Single or Colliding Drops

#### 12-3.2.1 Coalescence of a Single Drop with a Plane Interface.
Numerous studies have dealt with the coalescence of a single drop at a plane interface created by a settled, coalesced layer. These studies involve measurement of the elapsed time from drop arrival at the interface to coalescence. Many factors influence the rest or film drainage time, including the age of the interface. Times are correlated using film drainage theory. The approach force acting on the drop is constant and caused by gravity (density difference). Although drop rest time studies are relatively simple compared to dynamic measurements, they yield useful information concerning film drainage rates and the critical film thickness necessary for coalescence to occur. The nearly static system permits in situ transient film thickness measurements to be made (e.g., by interferometry) during the thinning process. The earliest studies were reported by Gillespie and Rideal (1956), followed by Charles and Mason (1960), Allan and Mason (1962), MacKay and Mason (1963), Jeffreys and Hawksley (1965), Lang and Wilke (1971a, b), Hartland and Jeelani (1987), Hartland (1990), and others.

The simplest model for film drainage assumes that the conditions affecting the drainage rate are time invariant. By analogy to squeezing flow between parallel disks (lubrication approximation), the rate at which the film thins is given by

$$\frac{dh}{dt} = -\alpha_3 h^3 \qquad (12\text{-}37)$$

The interface is assumed to be mobile but motionless. The initial separation distance is $h_0$, and h is the separation distance after time t. The constant $\alpha_3$ accounts for all the factors that determine the drainage time. Integration of eq. (12-37), with initial condition $h = h_0$ at $t = 0$, leads to

$$\frac{1}{h^2} - \frac{1}{h_0^2} = \alpha_4 t \qquad (12\text{-}38)$$

Estimation of the initial film thickness $h_0$ is not critical, since initial thinning is fast. After a short time, $h^{-2} \gg h_0^{-2}$, allowing evaluation of the drainage rate constant $\alpha_4$, from precise measurements of film thickness versus time. Estimates for the film thickness at rupture from 25 to 500 Å have been reported. Studies involving mass transfer from drops show that in the presence of mass transfer, coalescence times are much shorter.

#### 12-3.2.2 Coalescence of Two Colliding Drops.
Refer again to Figure 12-16, which is a schematic diagram showing the collision between two drops of equal diameter d. The leading edges of both deformable drops become flattened on collision. This deformation creates a parallel disklike geometry.

Therefore, the dynamics of film drainage can be represented as a squeezing flow between two disks of radius R, separated by distance h, that approach each other due to force F. The relationship governing this process is given by eq. (12-39), which applies only to an immobile interface:

$$\frac{dh}{dt} = -\frac{2F}{3\pi\mu_c R^4} h^3 \qquad (12\text{-}39)$$

The reasoning is similar to that for drops resting at a flat liquid–liquid interface. Equation (12-37) is the same as eq. (12-39) with $\alpha_3 = 2F/3\pi\mu_c R^4$ equal to a constant. The rate of film thinning (dh/dt) depends, among other things, on the approach force F and the radius R of the disks. The approach force and disk radius are not independent, since $F \sim \pi R^2 \cdot (4\sigma/d)$. That is, the excess pressure in the film must be on the order of the Young–Laplace pressure. Using this result to substitute for R leads to

$$\frac{dh}{dt} = -\frac{32\pi\sigma^2}{3\mu_c d^2 F} h^3 \qquad (12\text{-}40)$$

Equation (12-40) shows that the film drainage rate is inversely proportional to the approach force, again demonstrating that coalescence is promoted by gentle collisions. Integration of eq. (12-40) with initial condition $h = h_0$ at $t = 0$ and final condition $h = h_c$ at $t = \tau$ leads to

$$\tau = \frac{3\mu_c d^2 F}{64\pi\sigma^2} \left( \frac{1}{h_c^2} - \frac{1}{h_0^2} \right) \qquad (12\text{-}41)$$

where $h_c$ is the critical thickness required for film rupture. The initial distance $h_0$ is usually much greater than $h_c$, so that $h_c^{-2} - h_0^{-2} \approx h_c^{-2}$. The time required for film rupture is $\tau$. Coalescence occurs only if the contact time $t_c$ is greater than $\tau$. There are several versions of this equation that reflect variable approach force, circulation in the drop, and the mobility of the drop interface. Further details can be found in Murdoch and Leng (1971), Scheele and Leng (1971), and Chesters (1991). Further discussion is given in Section 12-3.3.

Scheele and Leng (1971) and Murdoch and Leng (1971) investigated the coalescence behavior of colliding drop pairs. Anisole drops ($\Delta\rho = 0$, d = 3 mm) suspended in water were fired at one another from nozzles and their movement filmed at 1000 fps. Figures 12-17 and 12-18 show the interaction patterns for drop pairs that rebound (bouncing) and coalesce, respectively. As drops left the nozzle, an oscillation was established that affected the curvature of the leading edge at impact. Drops having more pointed leading edges at impact coalesced, as seen in Figure 12-18. Drops striking with a blunt leading edge usually bounced apart, as seen in Figure 12-17. As they traveled toward each other, the leading-edge shape oscillated between pointed and blunt. Therefore, changing the nozzle spacing changed the shape of the leading surface at impact. The measured coalescence efficiencies varied with separation distance from 25 to 100%, as determined

DROP COALESCENCE **689**

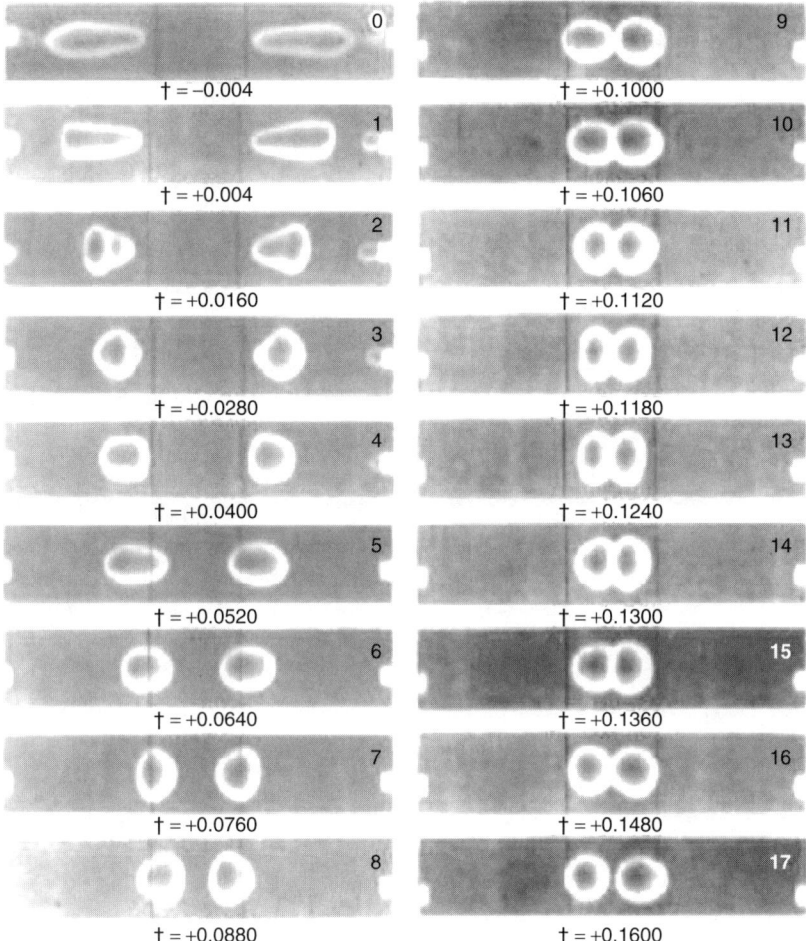

**Figure 12-17** Rebounding (bouncing) of colliding anisole drop pairs in water. Times are in seconds before (−) or after (+) drops become independent of the nozzle. (Reproduced from Scheele and Leng, 1971.)

by witnessing 100 events per distance setting. Figure 12-19 shows that coalescing pairs (upper half) had smaller disk radii during contact than bouncing pairs (lower half). The run numbers refer to a specific filmed experiment of single drop pair collisions.

There are many theories of how actual rupture occurs and at what thickness it happens. For example, a hypothesis by Vrij (1966) suggests that as hydrodynamic thinning proceeds, a point is reached where van der Waals attractive forces dominate over surface (interfacial tension) forces. Therefore, surface waves develop and become unstable, creating a hole where the film is thinnest. Film thickness at rupture was estimated to be in excess of 100 Å.

**690** IMMISCIBLE LIQUID–LIQUID SYSTEMS

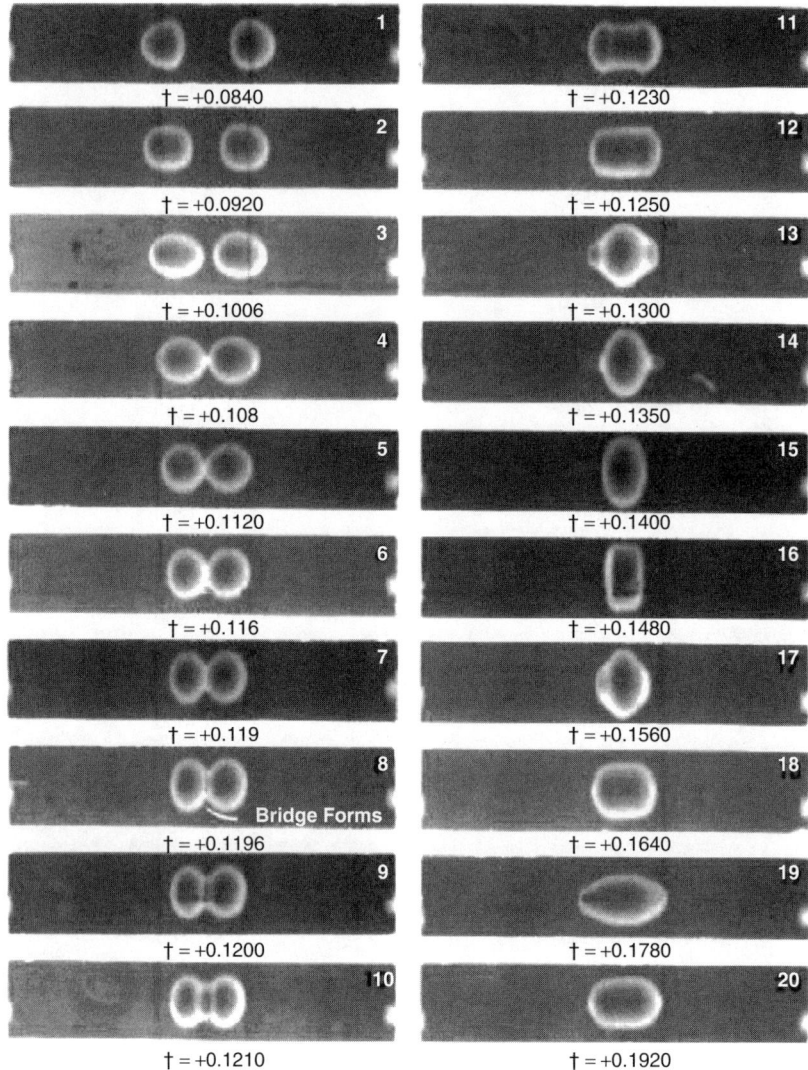

**Figure 12-18** Coalescence of colliding anisole drop pairs in water. Times are in seconds after (+) drops become independent of the nozzle. (Reproduced from Scheele and Leng, 1971.)

### 12-3.2.3 Practical Implications of Single Drop and Drop Pair Studies. 
The observations made from detailed single drop and drop pair studies have several practical implications, which complement the discussion of Section 12-3.1:

- If sufficient drainage occurs during the contact interval, a critical thickness is reached and the drops will coalesce. This requires that $\tau$ in eq. (12-41) be

## DROP COALESCENCE

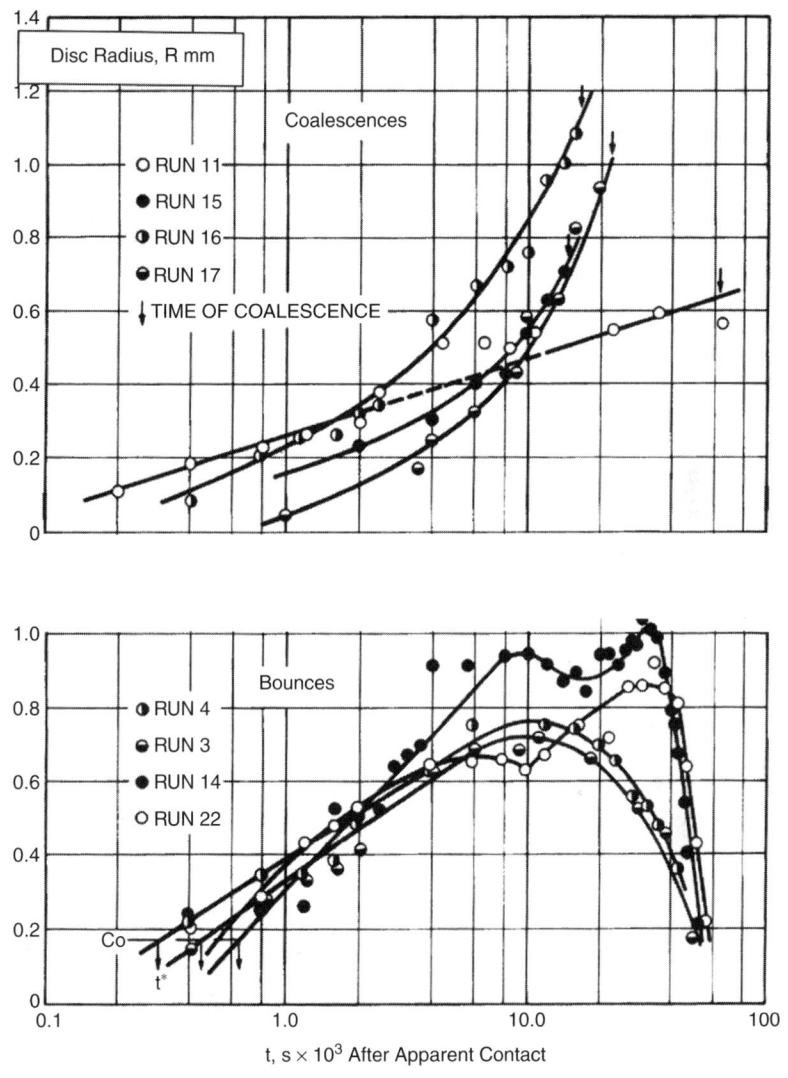

**Figure 12-19** Expansion/contraction of apparent contact radius with time for coalescing and bouncing drops. (Reproduced from Scheele and Leng, 1971.)

equal to or less than the contact time. If insufficient drainage occurs during the contact interval, the drops depart one another. A higher contact force, F, decreases drainage rates by creating larger disk radii, thereby increasing the time required for coalescence.

- Low interfacial tension leads to greater flattening upon contact, thereby trapping more continuous phase fluid. This increases the drainage time and decreases the likelihood of coalescence. Surfactants normally lower the interfacial tension, σ, and therefore reduce coalescence probability. Adsorbed surfactants also immobilize the drop–film interface. This also affects the slip velocity of the draining film, further reducing coalescence probability.

- A higher continuous phase viscosity increases the resistance to film drainage by partially immobilizing the drop–fluid interface. This reduces coalescence probability. If two similar volumes of immiscible liquids are dispersed, the fluid having the higher viscosity will normally become the continuous phase. The first attempts to produce suspension polymers used sugar to thicken the suspending phase and to retard coalescence.
- Solids trapped in the thinning film prevent critical thicknesses from being reached, and therefore reduce coalescence probability. Solid particles have been used as suspending agents in suspension polymerization processes.
- The argument put forth in Section 12-2, that P/V be maintained constant for scale-up in order to maintain equal drop size under turbulent conditions, does not hold true for scaling-up of coalescing systems.

### 12-3.3 Coalescence Frequency in Turbulent Flow

One of the earliest attempts to quantify coalescence frequencies was the work of Howarth (1967). A procedure was used that is similar to the one described in Section 12-3.1.5. A steady dispersion was established at a high agitation rate. The stirrer speed was then lowered so that only coalescence occurred, at least initially. Howarth defined a global or macroscopic coalescence frequency as the initial slope of a plot of interfacial area (related to $d_{32}$) versus time and demonstrated that systematic experiments could be conducted to determine the effect of various system variables on coalescence rate. Since the coalescence frequency depends strongly on drop diameter, most models are based on the approach discussed below.

The coalescence frequency, $\Gamma(d, d')$, is the product of the collision frequency, $\xi(d, d')$, and coalescence probability, $\lambda(d, d')$, as shown by eq. (12-35). A schematic diagram illustrating how models for $\Gamma(d, d')$ are developed for flow-driven collisions is given in Figure 12-20. The diagram follows the overview given by Chesters (1991). From a hydrodynamic viewpoint, two separate models are developed. The model for the external flow surrounding the drops produces the collision frequency, $\xi(d, d')$, approach force, F, and contact time, $t_c$. This model can be for laminar or turbulent flow, depending on the contacting equipment and process variables. The model for the internal flow yields the film drainage time. This model is that for a squeezing flow, driven by F and constrained by $t_c$ of the external model. Given the dimensions of the draining film, this is a viscous model, usually assumed to be a lubrication flow.

The coalescence efficiency is determined by comparing the time to reach critical thickness with the available contact time determined by the external flow model. The approach of Figure 12-20 allows development of a variety of models. Whereas the form of $\xi(d, d')$ depends on the process flow field, that for $\lambda(d, d')$

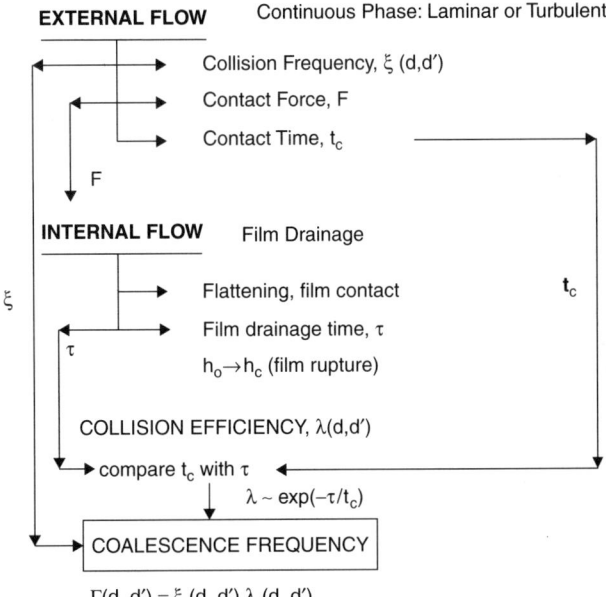

**Figure 12-20** Model for coalescence frequency.

depends on interface mobility and the physicochemical and electrostatic state of the interface. Chesters (1991) demonstrates how laminar and turbulent models for $\xi(d, d')$ are developed, as well as how models for $\lambda(d, d')$ that apply to rigid spheres and mobile, partially mobile, and immobile interfaces are developed. His review is excellent and does not need repeating here. Chesters gives an example by application of the method to simple shear flow. Here we provide one of the earliest examples of the approach in the form of the model developed by Coulaloglou and Tavlarides (1977) for turbulent stirred tank systems.

The model developed by Coulaloglou and Tavlarides (1977) for turbulent stirred tanks applies to drops whose collision rates are determined by interaction with eddies that fall within the inertial subrange of isotropic turbulence ($L_T \gg d \gg \eta$). For equal-sized drops, assuming uniform energy distribution throughout the vessel, the collision frequency is given by eq. (12-36). For unequal-size drops, these authors obtained

$$\xi(d, d') = C_{14}(d^2 + d'^2) \cdot (d^{2/3} + d'^{2/3})^{1/2} \varepsilon^{1/3} \qquad (12\text{-}42)$$

where $\varepsilon$ is the energy dissipation rate typical of the quiescent regions of the tank; $\varepsilon/\varepsilon_{avg} \leq 0.1$. Equation (12-42) was derived by assuming that the collision mechanism was similar to that for molecules in the kinetic theory of gases. As shown in Figure 12-20, the coalescence efficiency, $\lambda(d, d')$, is described in terms of the time, τ, required for sufficient film drainage to take place compared to the time that drops remain in contact with one another, $t_c$. If $t_c > τ$, coalescence occurs, and if $t_c < τ$, drops fail to coalesce. This is a simple concept, having a somewhat unrealistic yes or no criterion. Accordingly, coalescence efficiency for drop diameters d and d' is expressed as

$$\lambda(d, d') = C_{15}\, e^{-\tau/t_c} \tag{12-43}$$

For drops having immobile interfaces, Coulaloglou and Tavlarides (1977) show the drainage time for unequal size drops to be

$$\tau = \frac{3}{16}\frac{\mu_c F}{\pi\sigma^2}\left(\frac{1}{h_c^2} - \frac{1}{h_o^2}\right)\left(\frac{d\,d'}{d+d'}\right)^2 \tag{12-44}$$

Equation (12-44) reduces to (12-41) for $d = d'$. The drops are brought into contact by eddies whose size is of order $d + d'$. Consistent with eq. (12-14), $\overline{v'(d+d')^2}$ is the mean-square turbulent velocity difference across these eddies and is given by

$$\overline{v'(d+d')^2} \approx \varepsilon^{2/3}(d+d')^{2/3} \tag{12-45}$$

This is consistent with eq. (12-17). The authors argue that the average contact force, F, is given by

$$F \approx \rho_c \overline{v'(d+d')^2}\left(\frac{d\,d'}{d+d'}\right)^2 \tag{12-46}$$

Combining eq. (12-45) and (12-46) and substituting into eq. (12-44) yields

$$\tau = \frac{\mu_c \rho_c \varepsilon^{2/3}(d+d')}{\sigma^2}\left(\frac{1}{h_c^2} - \frac{1}{h_o^2}\right)\left(\frac{d\,d'}{d+d'}\right)^2 \tag{12-47}$$

The contact time, $t_c$, is proportional to the eddy arrival time for eddies of size $d + d'$:

$$t_c \approx \frac{(d+d')^{2/3}}{\varepsilon^{1/3}} \tag{12-48}$$

Since drop volumes are additive upon coalescence, it is convenient to write quantities in terms of drop volume, v, rather than diameter. Inserting eqs. (12-47) and (12-48), into (12-43) with $d \sim v^{1/3}$ gives the following result for coalescence efficiency when the interface is immobile:

$$\lambda(v, v') = C_{15} \exp\left[-\frac{C_{16}\rho_c\mu_c\varepsilon}{\sigma^2}\left(\frac{v^{1/3}\,v'^{1.3}}{v^{1/3} + v'^{1/3}}\right)^4\right] \tag{12-49}$$

Equation (12-49) expresses the coalescence efficiency in terms of drop volumes (v, v'), physical properties ($\mu_c$, $\rho_c$, $\sigma$), and the energy dissipation rate, ε, for quiescent regions of the vessel. It should be noted that for fully mobile and rigid interfaces, the drop viscosity does not play a role. For a partially mobile interface, the drop viscosity contributes to interface immobility.

It is thus possible to combine eqs. (12-42) and (12-49) using (12-35) to give the coalescence frequency under agitated conditions:

$$\Gamma(v, v') = C_{17}(v^{2/3} + v'^{2/3}) \cdot (v^{2/9} + v'^{2/9})^{1/2} \varepsilon^{1/3}$$

$$\times \exp\left[-\frac{C_{16}\rho_c\mu_c\varepsilon}{\sigma^2}\left(\frac{v^{1/3}v'^{1/3}}{v^{1/3} + v'^{1/3}}\right)^4\right] \quad (12\text{-}50)$$

For equal-sized drops, eq. (12-50) is reduced to

$$\Gamma(v, v') = C_{19}v^{7/9}\varepsilon^{1/3} \exp\left[-\frac{C_{18}\mu_c\rho_c\varepsilon v^{4/3}}{\sigma^2}\right] \quad (12\text{-}51)$$

As noted earlier, $\varepsilon$ is the local energy dissipation rate, so eq. (12-50 and 51) can be used for spatially dependent calculations. For constant power number and relatively uniform energy dissipation in the circulation region of the tank, $\varepsilon \sim N^3D^2$ and the dependency on impeller speed and diameter can be established. The terms in eq. (12-50) are consistent with practice and our discussion in Section 12-3.1.3. The coalescence frequency $\Gamma(v, v')$ is independent of the volume fraction of dispersed phase. The coalescence rate can be obtained from the coalescence frequency by accounting for the number of drops of size v and v'. This is best demonstrated by reference to the population balance equations discussed in Section 12-4.

Although difficult to apply in practice, models for coalescence rate provide an appreciation for the physical phenomena that govern coalescence. They also provide an appreciation for why it is difficult to interpret stirred tank data or even to define the appropriate experiment. For instance, it can be clearly seen from eq. (12-49) to (12-51) that the collision frequency increases with $\varepsilon$, whereas the coalescence efficiency decreases with $\varepsilon$. For constant phase fraction, the number of drops also increases with $\varepsilon$. The models for coalescence of equal-sized drops are quite useful to guide the interpretation of data that elucidate the time evolution of both mean diameter and drop size distribution during coalescence. To this end, Calabrese et al. (1993) extended the work of Coulaloglou and Tavlarides (1977) to include turbulent stirred tank models for rigid spheres and deformable drops with immobile and partially mobile interfaces. The later model accounts for the role of drop viscosity. In practice, models for unequal-sized drops are even more difficult to apply, but they do suggest that rates are size dependent. They are useful in the application of the population balance models discussed in Section 12-4.

Numerous authors have developed models for coalescence frequency. These include the models of Muralidhar and Ramkrishna (1986), Das et al. (1987), Muralidhar et al. (1988), Tsouris and Tavlarides (1994), and Wright and Ramkrishna (1994), for turbulent stirred tanks, as well as those of Davis et al. (1989),

Vinckier et al. (1998), and Lyu et al. (2002) for laminar flows and extruder applications. The models differ in how they describe the film drainage and/or drop collision process. Although the effect of surface charge and absorbed surfactant at the interface can be addressed in principle, these are rarely considered, due to their complexity. Although some models have been validated in a global sense, there has been little quantitative validation of dependencies on system variables.

### 12-3.4  Conclusions, Summary, and State of Knowledge

It is obvious that coalescence is a complex phenomenon. Here we have focused on creating an overall understanding rather than presenting an exhaustive literature review. We apologize for the omission of important studies not reported. We conclude this section by reiterating some important points and by providing additional practical observations.

- One must consider whether coalescence is desirable or undesirable for the application.
- Although it is difficult to apply the fundamental equations of this section, it is useful to use them to determine the effects that variables have on coalescence rates.
- There are regions close to the impeller where dispersion predominates. For scale-up under geometrically similar conditions, the effective dispersion volume shrinks with increasing vessel size.
- There are large regions in a vessel where coalescence can occur. Gentle agitation promotes coalescence because it provides for longer contact times enabling more film drainage to take place.
- Bench scale processes may occur at steady-state conditions, while larger scale industrial processes may not.
- Coalescence rates depend on both dispersed phase concentration and physicochemical factors. Except for strongly coalescing systems, coalescence effects are minimal at concentrations less than 5%.
- In practice, it is helpful to characterize coalescence rates by the simple methods presented in Section 12-3.1.5.
- Models for coalescence frequency show the importance of agitation rate, physicochemical phenomena, and interfacial properties on coalescence. This information is broadly useful for explaining the behavior of stirred vessels, decanters, extractors, and centrifuges, as well as how to prevent coalescence. It is also useful in the determination of which phase will tend to dominate as the continuous phase and in the interpretation of phase inversion phenomena.
- Scale-up is discussed in Section 12-8. Since different scale vessels have different proportions of drop time spent in coalescence and dispersion zones, it is a major challenge to design for duplicate results. One promising approach is to use CFD to create circulation time and energy dissipation rate profiles

at the various scales under consideration. Assuming that coalescence dominates in regions where $\varepsilon_{local}/\varepsilon_{avg} < 0.1$ and that dispersion dominates in regions where $\varepsilon_{local}/\varepsilon_{avg} \geq 10$. CFD enables one to see what effects design variables have on the size of and residence time in these regions. For example, using more or larger impellers in the larger vessel can be a way to increase the dispersion region and decrease the circulation time. CFD can guide the selection process.

- Surfactants, suspending agents, and other stabilizers can make a system totally noncoalescing. For such systems, scale-up becomes a dispersion and kinetics problem.

## 12-4 POPULATION BALANCES

### 12-4.1 Introduction

Population balances are a set of mathematical tools that enable one either to predict the time evolution of the DSD or to determine specific information, such as breakage frequency and daughter size distribution, or collision frequency and coalescence efficiency, from an analysis of time-variant drop size data. They were first developed by Valentas et al. (1966) and Valentas and Amundson (1966), as applied to liquid–liquid dispersions. These techniques have been used for both batch and continuous systems and for steady state as well as unsteady conditions.

Population balances are analogous to material balances, but instead of applying them to each chemical species, they are applied to each drop size class comprising the entire DSD. Therefore, accumulation and depletion terms are referred to as *birth* and *death rates* for a drop of specific diameter or volume. Figure 12-21 shows a general scheme for the events taking place. Within the enclosure, or control volume, are drops of volume v. The population of these drops is determined as follows. Drops of volume v enter by convection and because they are formed by the coalescence of smaller drops and the breakage of larger drops. Drops of volume v leave by convection and because they are depleted as they themselves break and/or coalesce. It is important to note that both breakup and coalescence produce a gain and a loss to the control volume or size class, as indicated by the arrows.

As discussed above, population balance models account for the influent and effluent of drops into the control volume. The control volume can be the entire tank or a particular region of the tank. The resulting equations are referred to as *integro-differential equations*. Analytical solutions of these equations exist only for unreasonably simplistic assumptions. Usually, the equations are solved by numerical methods, either by direct numerical integration or by a statistical simulation such as a Monte Carlo technique. Several authors opt for simplifying assumptions, such as imparting similarity conditions on one or more variables. One similarity argument can be illustrated by reference to Figure 12-13, which shows that the shape of the DSD is time independent. The solution methods are beyond the scope of this chapter.

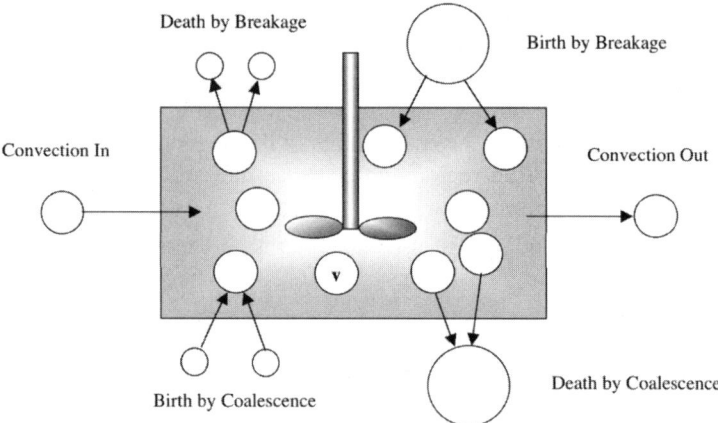

**Figure 12-21** Population balance events for drops of volume v.

In addition to liquid–liquid systems, the population balance equation (PBE) has been applied to crystallization, grinding, interphase heat and mass transfer, multiphase reactions, and floatation.

### 12-4.2 History and Literature

There are two principal ways in which population balances have been used in liquid–liquid systems. These involve using experimentally or phenomenologically derived models for the breakage frequency and resulting daughter size distribution (known as the breakage kernel), along with similarly derived models for collision frequency and coalescence efficiency, to compute the evolution of the drop size distribution. The other procedure, referred to as the *inverse problem*, is to use transient drop size distribution data to compute or infer the breakage frequency and kernel, or the collision frequency and coalescence efficiency. From a computational point of view, the latter is more complex. Tavlarides and Stamatoudis (1981) give an excellent review of population balance models for stirred vessels. These authors address reaction and mass transfer as well as coalescence and dispersion. The theory, solution, and general application of population balance equations are well described in a book by Ramkrishna (2001). Table 12-4 lists some of the important contributions to the literature. The table includes methods used, results, and some conclusions. It is meant to be representative rather than comprehensive.

### 12-4.3 Population Balance Equations

The most general form of the population balance equation, applicable for a flow system, can be written

$$\frac{\partial n_d}{\partial t} + \nabla \cdot (\overline{U} n_d) - \dot{B}_d + \dot{D}_d = 0 \tag{12-52}$$

For a given drop size, $n_d$ is the number of drops, $\dot{B}_d$ the birth rate, $\dot{D}_d$ the death rate, and $\overline{U}$ the velocity vector.

The complete population balance equation given by Coulaloglou and Tavlarides (1977) in the form of number density over drop volume for a CSTR is given by:

$$\frac{\partial}{\partial t}[N_T(t)A(v,t)] = \int_v^{v_{max}} \beta(v',v)\upsilon(v')g'(v')N_T(t)A(v',t)\,dv'$$
$$- g'(v)N_T(t)A(v,t) + \int_0^{v/2} \xi(v-v',v')\lambda(v-v',v')$$
$$\times N_T(t)A(v-v',t)N_T(t)A(v')\,dv' - N_T(t)A(v,t)$$
$$\times \int_0^{v_{max}-v} \xi(v,v')\lambda(v,v')N_T(t)A(v',t)\,dv'$$
$$+ N_{T0}(t)A_0(v,t) - N_T(t)A(v,t)f_e(v) \quad (12\text{-}53)$$

where $N_T(t)$ is the total number of drops in the vessel at time t; $A(v,t)$ the number probability density for drops of volume v at time t; $\beta(v',v)$ the breakage kernel or the number probability density of daughter drops of volume v formed by the breakup of a parent drop of volume $v'$; $\upsilon(v')$ the mean number of daughter drops resulting from breakage of a parent drop of volume $v'$; $g'(v')$ the breakage frequency of drops of volume $v'$; $\xi(v,v')$ the collision frequency of drops of volume v with drops of volume $v'$; $\lambda(v,v')$ the coalescence efficiency between drops of volume v and drops of volume $v'$; $N_{T0}(t)$ the number feed rate of drops at time t; $A_o(v,t)$ the number probability density of drops of volume v at time t in the feed; and $f_e(v)$ the escape frequency of drops of volume v in the product stream.

The first two terms on the right represent the addition (birth) and loss (death) of drops of volume v due to breakage. The next two terms deal with formation and loss due to coalescence, and the last two terms represent droplet flow into and out of the vessel. The last two terms are eliminated for batch operation. If the system is noncoalescing, the middle two terms are eliminated. If the system is purely coalescing (no breakage), the first two terms are eliminated. Purely coalescing systems exist, at least initially, when the impeller speed is decreased. At steady state, eq. (12-53) can be written as

$$N_{TS}A(v)[g'(v) + \gamma(v) + f_e(v)]$$
$$= N_0A_o(v) + \int_v^{v_{max}} \beta(v',v)\upsilon(v')g'(v')N_{TS}A(v')\,dv'$$
$$+ \int_0^{v/2} \xi(v-v',v')\lambda(v-v',v')N_{TS}A(v-v',v')N_{TS}A(v')\,dv' \quad (12\text{-}54)$$

where

$$\gamma(v) = \int_0^{v_{max}-v} \xi(v,v')\lambda(v,v')N_{TS}A(v')\,dv'$$

Here $N_T(t) = N_{TS}$ and $N_{T0}(t) = N_0$, since the number of drops in the vessel and in the feed remain constant. Furthermore, $A(v,t) = A(v)$ and $A_0(v,t) = A_0(v)$. Many variations of eq. (12-53) and (12-54) exist, and the nature of the problem dictates the selection of terms to be used. As stated previously, these equations are in the form of number density over drop volume. That is, number probability density functions are applied to a drop of specified volume. It is often more convenient to use volume probability density functions for a drop of specified diameter. This form, given for a batch, noncoalescing system in eq. (12-55) represents volume density over drop volume. It was applied successfully by Konno et al. (1983):

$$\frac{\partial P_v(d, t)}{\partial t} = \int_d^{d_{max}} P_v(d', t) g'(d') \beta'(d', d) \, dd' - P_v(d, t) g'(d) \qquad (12\text{-}55)$$

where $P_v(d, t)$ is the volume probability density for drops of size d at time t; $g'(d)$ the breakage frequency of a drop of size d; and $\beta'(d, d')$ the breakage kernel or the number of daughter drops of size d formed by the breakup of a parent drop of size d'. Note that $\beta'$ as defined in eq. (12-55) is equivalent to the product of $\upsilon$ and $\beta$ in eq. (12-53).

### 12-4.4 Application of PBEs to Liquid–Liquid Systems

To apply practically the equations given in Section 12-4.3, it is important to have experimental data. As mentioned earlier, there are two approaches shown in Table 12-4. The direct approach is to use phenomenological models for the breakage and coalescence terms in the appropriate PBE to solve for the DSD. Favorable comparison of experimental and computed DSDs leads to the confirmation of phenomenological expressions. An excellent review of models for the breakage terms, including direct comparison to breakage rate data, is given by Lasheras et al. (2002). The other approach, referred to as the *inverse method*, uses transient drop size distribution data as input, and the solution of the PBE yields quantitative breakage and coalescence information, such as $\xi(v, v')$, $\lambda(v, v')$, $\beta(v, v')$, and $g'(v)$. The experimental and numerical procedures used to determine these quantities are discussed by Ramkrishna (2001) and others listed in Table 12-4. They will not be repeated here.

The numerical solution of the PBE often leads to errors. Some of these include discretization errors, truncation errors, round-off errors, and propagated errors. Inverse problems are particularly stiff. Experimental errors include determining when steady state has been reached, noise in the tails of the DSD, sampling and analysis errors, and uncertainties that arise when in situ measurements cannot be made.

### 12-4.5 Prospects and Limitations

Despite the fact that PBE technology has been around since the 1960s, little practical industrial use has been made of it. Part of this is due to the formidable task of solving these equations, and part is due to the difficulty in obtaining quality

Table 12-4  Population Balance Studies in Stirred Vessels

| Reference | Year | Methods | Problem Type | Conclusions |
|---|---|---|---|---|
| Bajpai et al. (1976) | 1976 | — | Direct | Coalescence followed by immediate redispersion into drops of specified distribution. Numerical results are compared with data from the literature. |
| Coulaloglou and Tavlarides (1977) | 1977 | Iterative technique backward marching from largest size increment | Direct | Used phenomenological models to develop breakage and coalescence rate and distribution functions. Used these to predict the DSD, which compared favorably with CSTR data. |
| Narsimhan et al. (1980) | 1980 | Similarity hypothesis | Inverse problem | Experimental DSD data for dilute system used to extract breakage frequency; led to validation of similarity hypothesis. |
| Tavlarides and Stamatoudis (1981) | 1981 | — | — | A major review of population balance work, including mass transfer and chemical reactions. |
| Tavlarides (1981) | 1981 | Monte Carlo methods | Direct | Used phenomenological models describing coalescence and dispersion to predict DSD, with results compared with experiments. A review. |
| Rod and Misek (1982) | 1982 | Monte Carlo methods | Inverse problem | Compared Monte Carlo simulation with exact solution of PBEs. Concluded MC methods are adequate and efficient for calculating rate parameters in response to DSD data. |
| Bapat et al. (1983) | 1983 | Interval of quiescence Monte Carlo method | Direct | Used breakage and coalescence functions to predict spatially varying drop size distribution and mass transfer rates. |

(*continued overleaf*)

**Table 12-4** (continued)

| Reference | Year | Methods | Problem Type | Conclusions |
|---|---|---|---|---|
| Tavlarides and Bapat (1983) | 1983 | — | — | General review, including PBE modeling. |
| Narsimhan et al. (1984) | 1984 | Discretized drop volume intervals | Direct | Dilute dispersion data; observed similarity behavior for DSD. Compared DSD predicted by PBE with measurements; good agreement. |
| Rajamani et al. (1986) | 1986 | Monte Carlo methods | — | Compared time-driven with event-driven methods, applied to batch liquid–liquid dispersion. |
| Jeon and Lee (1986) | 1986 | — | Various | Showed applicability of PBEs to solving mass transfer problems. Using various assumptions, it was shown that all were in close agreement with experimental DSD and mass transfer data. |
| Laso et al. (1987a, b) | 1986 | Discrete model; size classes defined by geometric series | Direct | Developed a computationally efficient discrete model for breakage and coalescence; breakage and coalescence rates determined by optimizing the fit of experimental DSD data. |
| Muralidar and Ramkrishna (1988) | 1988 | Similarity hypothesis for PBE | Inverse problem; applicable if DSDs are self-preserving | 1. Transient DSD data for purely coalescing systems used to determine coalescence efficiencies. <br> 2. Mechanistic models developed for coalescence efficiency that include physical and turbulence information regarding drop motion, rest time, etc. |

| | | | | |
|---|---|---|---|---|
| Chang et al. (1991) | 1991 | Discrete size classes | Direct | Calculated DSD for different breakage frequencies. Size classes defined. |
| Calabrese et al. (1993) | 1993 | Discrete size classes | Both direct and inverse problem | Calculated DSD from described breakage model, and breakage frequencies from DSD data. |
| Alvarez et al. (1994) | 1994 | Quasi-steady state assumed for population dynamics | Direct | Applied PBE to suspension polymerization using breakage-coalescence parameters from the literature. Predicted DSD in agreement with published data. |
| Wright and Ramkrishna (1994) | 1994 | Similarity hypothesis | Inverse problem | Experimental data for purely coalescing system used with PBE to extract and develop correlation for coalescence frequency. |
| Chatzi and Kiparissides (1994) | 1994 | Numerical solutions | Direct | Steady-state distributions were calculated and found to be in generally good agreement with experimental data. Used to help determine effectiveness of polymeric suspending agent, and its performance vs. composition. |
| Sathyagal et al. (1995) | 1995 | Similarity transformation used | Inverse problem | Transient dilute dispersion DSD data used to extract breakage frequencies and kernels. |
| Ramkrishna (2001) | 2001 | — | — | Book covering all aspects, including theory and solution, of population balance equations. |

**704** IMMISCIBLE LIQUID–LIQUID SYSTEMS

data required for analysis. This is particularly true for coalescence phenomena, as discussed in Section 12-3. Recall that the number of drops, the collision rate, and the coalescence efficiency depend in a complicated and often competing way on agitation rate, drop diameter, and physicochemical variables, making validation of phenomenological models difficult.

Another limitation is that the breakage and coalescence kernels and frequency information tend to be specific to the equipment used to acquire the data. It is highly scale dependent; all quantities are flow dependent. Once information is obtained using PBEs, it cannot be used, with confidence, for scale-up work. At a specific scale, however, system information can prove useful. For example, the effect of surfactant concentration, stirring rate, impeller design, phase composition, and so on, could all be interpreted in terms of $\xi(v, v')$, $\lambda(v, v')$, $\beta(v, v')$, and $g'(v)$. This information could be used to improve and control product quality.

Vastly improved and faster computers can overcome the previously expensive task of solving the equations. In the past, simplifying assumptions have been used to shorten computation time. Today, and in the future, the most rigorous numerical techniques should be employed to eliminate the compromises of the past.

We expect that in the near future, CFD technology that has proven valuable in characterizing differences in flow behavior due to scale can be coupled with PBEs to give reasonably accurate drop size information, including scale effects, for estimating interfacial area and drop size uniformity.

## 12-5 MORE CONCENTRATED SYSTEMS

### 12-5.1 Introduction

Most industrial liquid–liquid applications fall into the category of being more concentrated systems. We identify more concentrated systems as $\phi > 0.20$ by volume fraction of dispersed phase. Industrial examples include suspension and emulsion polymerization, extraction, and separations, including decantation, centrifugation, and electrostatic precipitation. Because practice is as much an art as a science, much of the industrial experience on concentrated systems is proprietary and not published, contrasting the vast amount of academic work published for dilute and "clean" systems.

Concentrated liquid–liquid systems often involve dispersion and coalescence as well as rheological complexities. Data conflicts are common, often arising from the presence of impurities, sometimes unknown to investigators. There is also the challenge of obtaining representative samples and analyzing them. Describing the microscale interactions between the drops and the surrounding fluid, necessary for theoretical interpretation, is seldom a goal. That is, for concentrated systems, the small scale structure of the continuous phase turbulence is unknown and the drop–eddy interactions are undetermined. Salts, surface-active materials, and other impurities lead to system-specific behavior, complicating the development of industrial technology. As a result, many of the points discussed in this section are tied to specific process examples.

Dispersion, coalescence, and suspension phenomena are all important in concentrated liquid–liquid dispersions. Convective mixing patterns are also affected by the changes in rheology brought about by high dispersed phase concentrations. Heat transfer becomes more critical because of high concentrations of reactive materials often in the dispersed phase. For example, heat is managed in emulsion polymerization by controlling the addition rate of monomer fed to the reactor. Certain smaller scale processes can maintain temperature control through jacket cooling. For highly exothermic reactions, reflux condensers are used. If the end product is not shear sensitive, cooling by recirculation through an external heat exchanger is often used.

As discussed in previous sections, turbulent eddies are affected by high dispersed phase concentrations. Elasticlike behavior of deformable drops "cushion" eddies, reducing momentum transport. This means drop dispersion is limited to a smaller region closer to the impeller than for dilute systems.

Coalescence is also different in concentrated systems. Drop coalescence in dilute and moderate concentration systems was shown to originate from drop–drop collisions, contact with surfaces, or settling to a nondispersed, settled layer. Turbulence-induced collisions lead to brief contact intervals during which the separating film thins due to shear forces acting on the drop pairs. The total extent of thinning during contact determines coalescence probability, as shown in Section 12-3. Drops are closer together in concentrated systems (sometimes touching), and relative drop movement due to eddy fluctuations is less. This leads to longer contact intervals and a higher coalescence probability. In the case of highly concentrated systems, drops move relative to one another due to the local velocity gradient. Collisions, as we have described previously, are not likely to occur.

Gravitational effects are also different for concentrated systems. Quiescent settling of dilute dispersions leads to a gradient in both drop size and phase fraction. For $\rho_d < \rho_c$, the largest drops concentrate near the liquid surface and the smallest drops are closest to the lower cleared layer. Coalescence rates for unprotected drops are also accelerated due to the greater hydrostatic force on the settled drops, promoting faster film drainage. Dense drop populations lead to slower, hindered drop settling.

Surface-active materials are used to stabilize dispersions in industrial applications when coalescence must be prevented, as for suspension and emulsion polymerization processes. Concentrated dispersions are more likely to undergo phase inversion. This complex coalescence-dominated phenomenon is discussed later in this section.

### 12-5.2 Differences from Low Concentration Systems

Differences and similarities are illustrated by example. Suspension and emulsion polymerizations are examples of industrial processes having high drop concentrations, where coalescence is prevented by the use of suspending agents/emulsifiers. Polyvinyl alcohol (PVA) is typical of the aqueous suspending agents used. Concentrates of ≈2% of partially hydrolyzed PVA are diluted to ≈0.05 to 0.2%

for use in polymerization reactions. This concentration is usually sufficient to prevent coalescence once drop interfaces become sufficiently covered. The stabilizing efficiency depends on its chemical composition (degree of hydrolysis for PVA) and its molecular weight. The typical phase ratios are close to 1 : 1 or $\phi \sim 0.5$. Monomer containing an initiator is dispersed into water containing the suspending agent. Agitation continues at ambient temperatures to establish desired drop size and consistency. The temperature is then increased to the point where free radicals are formed form the initiator and polymerization begins. Each drop formed by agitation becomes a polymer particle of similar, but slightly smaller, size compared to the liquid drop. Heat transfer is seldom a problem, since drops have a large surface/volume ratio, and water, the suspending medium, provides good conduction and convection for heat transfer to the jacketed vessel walls. Suspension polymerization reactions are typically low viscosity operations. Vivaldo-Lima et al. (1997) have given an excellent review of suspension polymerization.

Leng and Quarderer (1982) show that for certain applications, dilute dispersion theories can be applied successfully to concentrated noncoalescing systems. It was shown that boundary layer shear on impeller surfaces controlled drop dispersion for drops in the size range 300 to 1000 μm and turbulence-controlled dispersion for smaller drops. The expressions given by eqs. (12-73) and (12-74) were supported by data from bench to production scale experiments. These results give encouragement that some industrially complex noncoalescing systems behave similarly to dilute systems. The result is not surprising since flow patterns were simple and independent of scale, rheology was close to Newtonian, and shear brought about drop dispersion. Additional details are given in Section 12-8.3.

### 12-5.3 Viscous Emulsions

#### 12-5.3.1 Emulsion Viscosity and Stability.
Drop sizes for emulsions are less than 0.1 μm, as distinguished from dispersions, which contain larger drops. Emulsions typically contain high concentrations of emulsifiers, and the dispersed phase volume fraction can be as high as 99%. Such high internal phase compositions often have unusually high viscosity and display complex rheological behavior. The apparent emulsion viscosity is much higher than single-component viscosities, and this is due both to large quantities of adsorbed surface-active materials and to large interfacial areas, causing internal flow resistance. In certain industrial applications, the viscosity of such systems has been found to be several hundred poise. Latex paints and similar products are strongly formulated to provide optimum film uniformity, durability, and adhesion. Balances of short-range forces stabilize these emulsions. These are electrostatic and steric repulsion forces and London–van der Waals attraction forces. The addition of an electrolyte reduces the repulsion forces and causes the emulsion to coalesce.

Emulsion polymerization involves simultaneous nucleation and growth phenomena. Monomer is first dispersed into drops enabling the aqueous phase to become saturated. Monomer moves by a convection–diffusion mechanism to growing micelles or suspended particles. Although nucleation and growth occur simultaneously, growth continues after nucleation stops. The growth phase stops when the monomer supply or free-radical generation is exhausted. Emulsion polymerization reactions are nearly always exothermic. Heat transfer is managed by a combination of controlled monomer feeding and the use of external heat transfer surfaces, such as reflux condensers or heat exchangers arranged in a circulation loop. Pumped circulation can be used only when dealing with shear stable products. The role of agitation is to disperse monomer into drops and to provide adequate movement for suspension and heat transfer. Despite the presence of stabilizers, many latex products are shear sensitive and prone to coagulation. Coagulum is undesirable and costly to remove. Agitation equipment should be chosen to minimize coagulum formation. A common design consists of a baffled jacketed glass-lined steel vessel equipped with a three-blade retreat curve impeller, shown in Figure 12-28. Use of glass-lined equipment helps prevent fouling and leads to higher product quality.

**12-5.3.2 Drop Dispersion.** Both turbulence and shear can break up drops in concentrated systems, but due to the dampening of eddies, it is likely that mean shear plays an important role in drop dispersion. This effect has been quantified by Coulaloglou and Tavlarides (1977) and shown by

$$N_{eff}^* = \frac{N^*}{1 + \phi} \qquad (12\text{-}56)$$

where $N_{eff}^*$ is the rotational speed necessary for equivalent dispersion for a volume fraction $\phi$, equivalent to that for a dilute system operating at a speed $N^*$. Drop dispersion occurs only near the impeller, and coalescence occurs throughout the rest of the vessel, similar to dilute dispersion. The high dispersed phase fraction leads to a higher collision rate.

## 12-5.4 Phase Inversion

**12-5.4.1 General Description.** Phase inversion is a commonly observed and practiced phenomenon in which the continuous phase becomes the dispersed phase, and vice versa. Coalescence is the fundamental phenomenon involved with phase inversion. Figure 12-22 shows schematically the steps occurring during phase inversion. The left column shows the preinverted condition. The middle column shows bridging and coalesce taking place, and the right column shows the inverted condition. The bottom row shows how irregular bridging (center) leads to drops in drops (right column), as rapid coalescence traps some of the

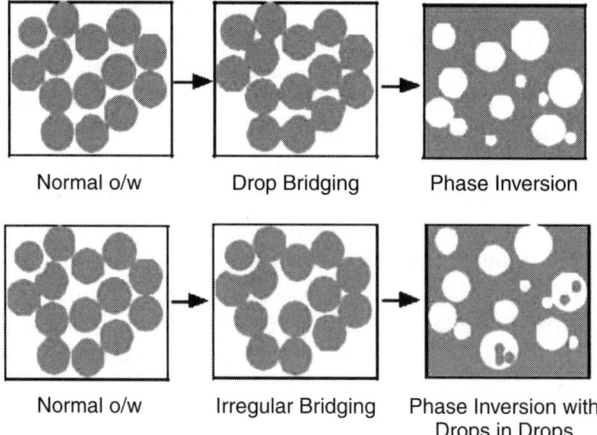

**Figure 12-22** Sequences in phase inversion.

continuous phase. The condition shown in the bottom right view is metastable, usually existing only temporarily, but not always.

Although conflicting information exists on the subject of phase inversion, the following conclusions can be made:

- Coalescence, not dispersion, dominates as the controlling mechanism in phase inversion. Factors discussed in Section 12-3 affecting film drainage rates, such as agitation rate, interfacial tension, interface mobility, $\mu_c$, and contact time, all apply.
- Inversion behavior is system specific.
- Surface-active agents play an important role, affecting film drainage rates.
- Every system has an operating region in which the oil phase is continuous, a region in which the aqueous phase is continuous and an ambivalent region where either phase can be continuous.
- The probability for phase inversion increases as drops get closer together. For uniform drops, the distance between drops is $s_d/d = (c_p/\phi)^{1/3} - 1$, where $s_d$ is the separation distance between drops, d the drop diameter, $\phi$ the volume fraction dispersed phase, and $c_p$ a packing parameter (0.7404 for face-centered cubic or hexagonal packing).
- The phase boundaries, or volume fractions at which phase inversion occurs, depend to some extent on initial conditions, path and agitation intensity, resulting in an *ambivalent region*. Beyond a certain point, phase inversion becomes independent of operating conditions.
- Several studies show metastable conditions of drops in drops, or water in oil in water.

*12-5.4.2 Physical Description.* Phase inversion is the transformation from o/w to w/o or from w/o to o/w. Sometimes phase inversion is initiated as a result of physical property changes brought about by chemical reaction. Both o/w and w/o phases usually coexist temporarily during the inversion process. For example, if the dispersed aqueous phase becomes viscous (as a result of polymerization) and coalescence occurs, it becomes the continuous phase as a result of inversion. However, if the continuous oil phase were to thicken, it would remain the continuous phase and no inversion would take place.

### 12-5.4.3 Phase Inversion Boundaries/Regime Map and Ambivalent Region.
Figure 12-23 is an example of a regime map. The region above the curves is where oil is always the continuous phase. Water is the continuous phase in the region below the curves. Between the two sets of lines is the ambivalent region, where either o/w or w/o systems can exist. The top arrow shows o/w going to w/o as more oil is added to the system. The bottom arrow shows the inversion of a w/o to an o/w system as water is added. Both the upper and lower boundary lines show a weak dependence on agitation rate, becoming even less dependent at higher levels of agitation.

Pacek et al. (1994a) have developed an effective video technique for concentrated liquid–liquid systems enabling phase inversion to be recorded in situ. Pacek et al. (1993, 1994b) found that when water was dispersed in oil at $\phi > 0.25$, water drops appeared in oil (drops in drops), but drops in drops did not appear when oil was dispersed in water.

Surfactant concentration can also be used to drive phase inversion. At high surfactant concentration, agitation and the method of addition may play a less

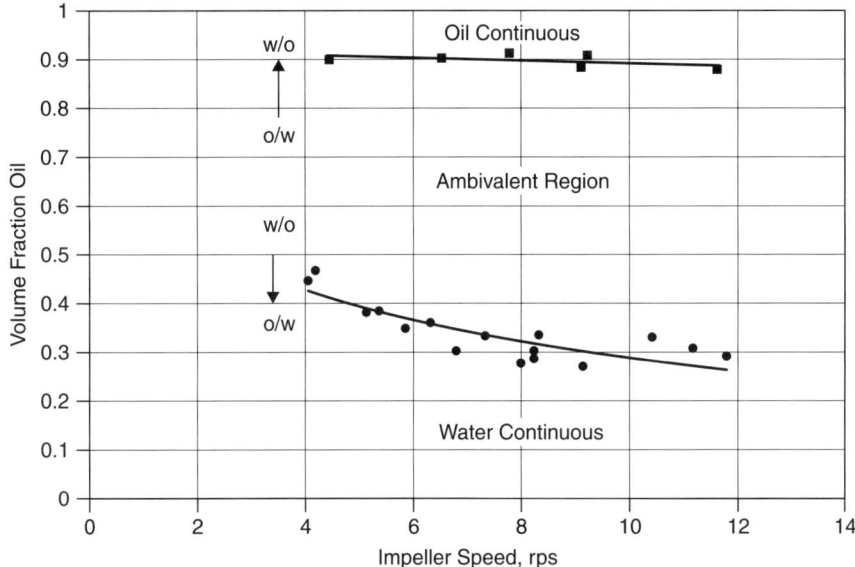

**Figure 12-23** Phase inversion boundaries for the kerosene–water system showing oil and water continuous regions and an ambivalent region. (Data of Kinugasa et al., 1997.)

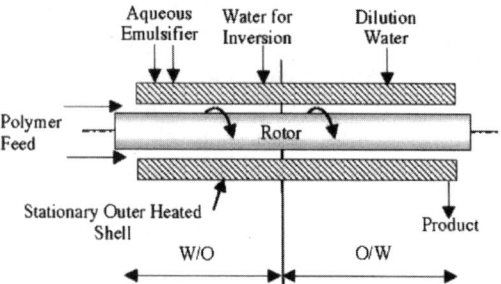

**Figure 12-24** Continuous Couette-type phase inversion emulsifier.

important role. Systematic studies on the effect of surfactant concentration and mixing on phase inversion and emulsion drop size have been carried out by Brooks and Richmond (1991, 1994a–c).

***12-5.4.4 Other Types of Phase Inversion.*** Synthetic emulsions were prepared by phase inversion at The Dow Chemical Company. A long Couette-like concentric cylinder apparatus was developed and is shown in Figure 12-24. All feed streams were precisely metered and controlled. Polymer in the form of either a melt or solution is fed in as shown on the left. Two aqueous streams are added to permit the gradual buildup of a w/o (polymer) phase. These aqueous streams contain significant quantities of surfactant. The third addition of water forces phase inversion, similar to that shown in Figure 12-23. A final water addition is for dilution to obtain the desired solids concentration. Typically, the final product contained 40 to 60% solids consisting of 0.1 to 1.0 μm particles in water. A wide variety of both heat- and solvent-plasticized feeds were demonstrated. The variables maintained constant for successful scale-up were shear rate and dispersion time. The process is more fully described in patents issued to Warner and Leng (1978) and Leng et al. (1985). The process was commercialized.

High capillary numbers were obtained as a result of high-shear rate (typically, 200 s$^{-1}$), high continuous phase viscosity, and low interfacial tension. Once steady-state conditions are established, cooling to the outer cylinder is applied to compensate for heat generation caused by viscous energy dissipation. Overheating leads to lower viscosity and in a reduction of shear stress required for dispersion. Different feed streams were used, requiring different feed preparation. Some polymers required use of solvents to adjust viscosity, whereas for others, simple heating was sufficient to pump in the feed. When solvents were used, they were removed by continuous stripping.

## 12-6 OTHER CONSIDERATIONS

### 12-6.1 Introduction

Section 12-6 provides a discussion of drop suspension, dispersion formation, and the interrelationships between dispersion, coalescence, and suspension. Additional topics include the role of surfactants and suspending agents, Oswald ripening, mass and heat transfer, and the effect of the presence of solids and gas bubbles on dispersion and coalescence.

## 12-6.2 Suspension of Drops

Settling and coalescence are common when the dispersed and continuous phases are of different density and when agitation provides only minimal circulation throughout the vessel. It is therefore important to determine the minimum speed for drop suspension. Most reported work is semiempirical and follows the approach of Zwietering (1958) for the just suspended state of solids in liquids.

There are analogies between the minimum impeller speed $N_{js}$ for solids suspension and $N_{min}$ for drop suspension. Both depend on density difference, continuous phase viscosity, and impeller diameter. However, $N_{js}$ depends directly on particle size, while $N_{min}$ depends instead on interfacial tension and the other physical properties that determine drop size. Skelland and Seksaria (1978) determined the minimum speed to form a liquid–liquid dispersion from two settled (separated) phases of different density and included the sensitivity to impeller location. The vessels used were fully baffled. They determined $N_{min}$ for systems of equal volumes of light and heavy phase. Studies included use of single impellers placed midway in the dense phase (C = H/4), at the o/w interface (C = H/2) and midway in the lighter phase (C = 3H/4). They also examined the use of dual impellers located midway in both phases. Several impeller types were tested, including a propeller (Prop), a 45° pitched blade turbine (PBT), a flat-blade turbine (FBT), and a curved-blade turbine (CBT). Their results are correlated by the following equation, which is dimensionless:

$$\frac{N_{min} D^{0.5}}{g^{0.5}} = C_{20} \left(\frac{T}{D}\right)^{\alpha_5} \left(\frac{\mu_c}{\mu_d}\right)^{1/9} \left(\frac{\Delta\rho}{\rho_c}\right)^{0.25} \left(\frac{\sigma}{D^2 \rho_c g}\right)^{0.3} \quad (12\text{-}57)$$

$\Delta\rho = |\rho_d - \rho_c|$. The magnitude of the constants $C_{20}$ and $\alpha_5$, given in the Table 12-5, are a measure of the ease of suspension formation. Low $C_{20}$ values indicate that dispersions are formed at low speeds. Large $C_{20}$ values (single impellers) suggest that higher speeds are required for minimum suspension. Turbines at the o/w interface require lower speed than in other locations. Radial flat-blade turbines placed in the light phase appear to be inefficient.

In an earlier study, Nagata (1975) determined minimum agitation conditions for forming a dispersion using a baffled cylindrical vessel and four-blade turbine impellers of $D/T = \frac{1}{3}$, placed at $C = T/2$. The following equation shows his dimensional correlation:

$$N_{min} = C_{21} T^{-2/3} \left(\frac{\mu_c}{\rho_c}\right)^{1/9} \left(\frac{\rho_c - \rho_d}{\rho_c}\right)^{0.26} \quad (12\text{-}58)$$

The value of $C_{21}$ is 750 for normal centered agitation and 610 for off-center agitation with eccentricity D/4. Units are $\rho_c$ (kg/m³), $\rho_d$ (kg/m³), $\mu_c$ (kg/m · s),

**Table 12-5** Constants for Use in Eq. (12-57)

| Type | Clearance | $C_{20}$ | $\alpha_5$ |
|---|---|---|---|
| Prop. | H/4 | 15.3 | 0.28 |
| Prop. | 3H/4 | 9.9 | 0.55 |
| Prop. | H/2 | 15.3 | 0.39 |
| Prop. | H/4 + 3H/4 | 5.2 | 0.92 |
| PBT | H/4 | 6.8 | 1.05 |
| PBT | 3H/4 | 6.2 | 0.82 |
| PBT | H/2 | 3.0 | 1.59 |
| PBT | H/4 + 3H/4 | 3.4 | 0.87 |
| FBT | H/4 | 3.2 | 1.62 |
| FBT | 3H/4 | [a] | [a] |
| FBT | H/2 | 4.0 | 0.88 |
| FBT | H/4 + 3H/4 | [a] | [a] |
| CBT | H/4 | 3.6 | 1.46 |
| CBT | 3H/4 | [a] | [a] |
| CBT | H/2 | 4.7 | 0.80 |
| CBT | H/4 + 3H/4 | 4.3 | 0.54 |

[a] Insufficient data for correlation purposes.

T (m), and $N_{min}$ (rpm). Off-center locations are seldom used, but the vortex due to eccentricity creates an efficient means to help form dispersions. The lack of dependency on $\mu_d$ indicates that only low viscosity dispersed phases were considered.

Pavlushenko and Yanishevskii (1958) determined the minimum speed for suspension in experiments that he conducted in a 0.3 m baffled vessel. The following equation gives his dimensional result, where SI units are used and N has units of rps.

$$N_{min} = \frac{5.67 \Delta \rho^{0.08} \mu_c^{0.06} \mu_d^{0.04} \sigma^{0.15} T^{0.92}}{\rho_c^{0.33} D^{1.87}} \quad (12\text{-}59)$$

Armenante and Tsai (1988) studied the effects of many variables on $N_{min}$. Their results for inviscid dispersed phases are given by

$$N_{min} = C_{22} (g \Delta \rho)^{5/12} \sigma^{1/12} \rho_c^{-0.5} D^{-2/3} \left(\frac{T}{D}\right)^{0.67} \left(\frac{H}{D}\right)^{0.33} N_P^{-1/3} \quad (12\text{-}60)$$

The results in terms of minimum Reynolds number are given by

$$Re_{min} = C_{23} \cdot Su^{1/12} Ar^{5/12} \left(\frac{T}{D}\right)^{0.67} \left(\frac{H}{D}\right)^{0.33} N_P^{-1/3} \quad (12\text{-}61)$$

where $Su = \rho_c \sigma D / \mu_c^2$ is the Suratman number, $Ar = g \rho_c \Delta \rho D^3 / \mu_c^2$ is the Archimedes number, and Np is the power number. The equation was found to be in good agreement with other work.

Armenante and Huang (1992) and Armenante et al. (1992) found practically no advantage in using multiple impellers for determining $N_{min}$. This is similar to the result for solid–liquid suspension. However, multiple impellers were useful in improving dispersed phase uniformity. Results agreed with the work of Skelland and Seksaria (1978).

We recommend use of eq. (12-57) in the absence of direct experimental data. It describes more specific impeller arrangements than the other work reported and is confirmed by the more recent work of Armenante and co-workers. These bench scale minimum-speed equations have not been validated by scale-up experiments, so caution is advised. For important applications, we recommend that scale-up experiments be conducted on a minimum of a fourfold volume scale using eq. (12-57) to guide in the variable selection and correlation.

### 12-6.3 Interrelationship between Suspension, Dispersion, and Coalescence

Church and Shinnar (1961) described the interrelationship between suspension, dispersion, and coalescence. Figure 12-25 shows drop size as a function of agitator speed in a turbulent process vessel. A stable region exists in the center area bounded by three lines representing dispersion, coalescence, and suspension phenomena. Consider constant impeller speed. If a large drop exists above the upper dispersion line, it will continue to break up until the dispersion line is reached. Breakage can result in some drops whose size lies below the lower coalescence line. These drops will continue to coalesce until the coalescence line is reached. Inside the bounded region, equilibrium is established between dispersion and coalescence.

A drop existing to the left of the suspension line will only be suspended when the speed is increased to the intersection of that drop size with the suspension line. If agitation speeds are to the right of the suspension line, the drops are always suspended. In the figure, the equations for the three lines apply to

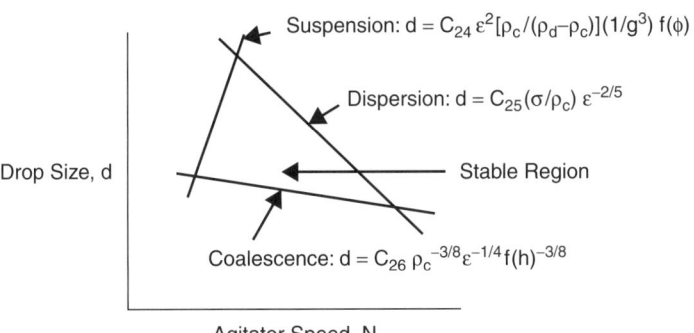

**Figure 12-25** Stable region concept for liquid–liquid processing in a stirred vessel. (After Church and Shinnar, 1961.)

inviscid dispersed phases. Symbols are defined in the "Nomenclature" section. The dispersion equation is analogous to eq. (12-21). Church and Shinnar (1961) derived the suspension and coalescence equations. Although somewhat simplistic compared to later work, they well illustrate the concept.

An extension of the Church and Shinnar concepts as they apply to suspension polymerization is as follows. In suspension polymerization, a conflict exists between suspension and dispersion since large uniform drops must be formed. Suspension of these large drops is often a problem, due to the phase density difference. The speed necessary for the prevention of "layering out" can produce smaller than desired beads. Figure 12-26 depicts the interaction between suspension and dispersion in the process vessel. For given properties and equipment, drop size decreases with increasing impeller speed, but the size of drops that can be suspended increases with speed. For a given system and reactor design, the largest practical drop size lies at the intersection of the two lines. Different agitation designs and suspending agents can shift the position of these lines, as suggested by the lighter lines on the figure, to meet bead size requirements.

### 12-6.4 Practical Aspects of Dispersion Formation

Placing a turbine (RDT) in the aqueous or lower phase, close to the interface, can make o/w dispersions. A central interfacial vortex forms with the commencement of impeller motion. This directs a stream of the lighter oil phase to the impeller, where it disperses. The volume of oil layer decreases with continued dispersion until it is exhausted. Placing the turbine in the oil, or upper phase, close to the interface can make w/o dispersions. A water-containing vortex forms, allowing water to be dispersed into the lighter oil phase.

Dispersions may also be formed by the continuous addition of one phase into another under agitation conditions. This method offers a safe procedure for handling exothermic reactions such as nitration and emulsion polymerization. The amount of phase addition will determine if phase inversion occurs as discussed in Section 12-5.4.

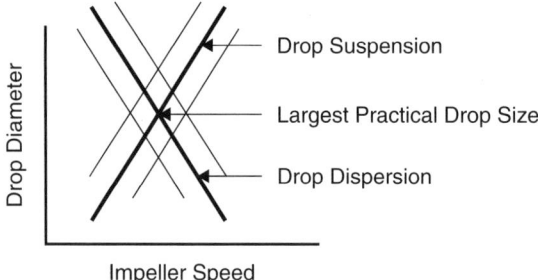

**Figure 12-26** Relationship between drop suspension and dispersion.

Listed below are some general recommendations for o/w and w/o systems:

- Use multiple turbines if the system is rapidly coalescing to provide additional dispersion capability. Axial flow turbines can also be used to achieve better uniformity in circulation.
- Avoid excessive dispersion in noncoalescing systems. Creation of tiny hard to coalesce drops can become a real problem if phase separation is required later. Test the system using bench scale equipment to see if and at what speed undesirably small drops form.
- Use at least one axial flow hydrofoil-type impeller of high D/T (i.e., $0.4 \leq D/T \leq 0.6$) in addition to the RDT for systems having large phase density differences.
- Interfacial tension controls the ease of drop breakage. Systems of low interfacial tension ($\sigma \leq 10$ dyn/cm or 0.01 N/m) require much lower power for dispersion than do those of high interfacial tension ($\sigma \geq 30$ dyn/cm or 0.03 N/m). We described this in more detail in Section 12-2.
- Baffling is always required for liquid–liquid dispersion, with the exception of suspension polymerization and certain highly shear-sensitive emulsion polymerizations.

### 12-6.5 Surfactants and Suspending Agents

Surfactants are organic compounds, often liquids, that have a hydrophobic and a hydrophilic portion of the molecule. Typical molecular weights range from 100 to 400. Suspending agents are usually polymeric in nature. They also have a hydrophobic portion, often the polymer backbone, and a hydrophilic group added to the backbone. Typical molecular weights range from 10 000 to 40 000. They are often only sparingly soluble in water. In practice, surfactant and/or suspending agents inhibit coalescence. This means that drop sizes are controlled by dispersion rather than by equilibrium between dispersion and coalescence, thus simplifying scale-up. The problem becomes one of dispersion kinetics and suspension. Suspending agent/surfactant concentrations are application dependent. However, typical concentrations are about 0.2 wt % for suspending agents and about 1 wt % for surfactants based on water content.

Surfactant/suspending agent molecules adsorb at liquid–liquid interfaces until equilibrium is reached between the adsorbed layer and the bulk fluid. The interfacial tension decreases with increasing bulk concentration until the critical micelle concentration (CMC) is reached. The interfacial tension remains constant beyond the CMC. Figure 12-27 shows a typical dependence of interfacial tension on surfactant concentration. Surface viscosity behavior is different. The viscosity remains practically constant up to the CMC and increases beyond it. The CMC is an equilibrium phenomenon. As surface area is created by agitation, surfactant molecules leave the CMC cluster, transfer to the aqueous phase, and then transfer to the liquid–liquid interface. Adsorption and protective action are not

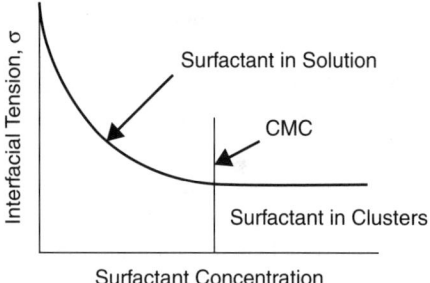

**Figure 12-27** Interfacial tension dependence on surfactant concentration.

instantaneous. The diffusion-dependent adsorption rate is faster for lower molecular weight materials. To illustrate this point, an attempt to produce uniform drops using a static mixer failed because newly made drops collided with one another and coalesced faster than they could be protected. The age of all drops produced in a static mixer is the same, whereas in a stirred tank, a large age distribution exists. If a protected drop collides with an unprotected drop, the pair does not coalesce; but the collision of two unprotected drops can result in coalescence.

Unlike surfactants, suspending agents usually create a viscous or semisolid skin over the surface of the drops. This makes coalescence impossible. Furthermore, dispersion is governed by viscous rather than interfacial resistance. Suspending agents used in suspension polymerization include materials such as polyvinyl alcohol and derivatized methylcellulose. In industry, the composition of effective suspending agents is closely guarded technology.

Approximations can be made to estimate how much surfactant is needed to maintain a desired dispersion. This is illustrated by example.

***Example 12.2.*** Suppose that a dispersion is to be 50% oil dispersed in water and consist of 50 μm drops. The surfactant molecular weight is 350. Assume that the molecular dimensions of the surfactant are 4 Å × 7 Å and that drop stability is obtained when surfaces are 50% covered. Estimate the surfactant requirement.

SOLUTION: Let $c_s$ be the surfactant concentration in g/L. The molar concentration is then $(c_s/350)$ g-mol/L. The number of surfactant molecules in solution can be obtained using Avogadro's number and is $(c_s/350)(6.023 \times 10^{23})$ molecules/L. The interfacial surface area is $(c_s/350)(6.023 \times 10^{23})(4 \times 7)$ Å$^2$/L. A 50% ($\phi = 0.5$) o/w dispersion having $d_{32} = 50$ μm has a specific surface area of $a_v = 6 \times 0.5/50 \times 10^4$ cm$^2$/cm$^3$ [refer to eq. (12-2)]. This converts to $6 \times 10^5$ cm$^2$/L or $6 \times 10^{21}$ Å$^2$/L. Since stability is reached with only 50% coverage, the surfactant needs to cover only $3 \times 10^{21}$ Å$^2$/L. Equating molecular area to drop surface area gives $c_s = 0.062$ g/L. Since the surfactant is supplied to the water phase (50% of total volume), the aqueous phase needs to contain 0.124 g/L or 0.0124 wt % surfactant.

## 12-6.6 Oswald Ripening

*Ostwald ripening* is a phenomenon resulting from slight differences in solubility due to differences in drop or crystal size. Small drops are slightly more soluble in the surrounding phase than large ones. This causes small drops, over time, to decrease in size and larger ones to get larger. The driving force for this phenomenon comes from consideration of the minimum surface free energy and is best explained in fundamental texts on phase equilibria.

Nyvlt et al. (1985) showed that for the case of a pure crystal of species A, the relationship between the bulk solubility of A in solution $c_{A\infty}$ and the solubility of a small particle of radius $\bar{r}$ in the same solution $c_{A\bar{r}}$ is given by

$$\bar{r} = \frac{\beta_S \hat{V}_A \hat{E}_S c_{A\infty}}{(c_{A\bar{r}} - c_{A\infty})(kT)_b} \quad (12\text{-}62)$$

where $\beta_S$ is a shape factor for the crystal, $\hat{V}_A$ the molecular volume of A, $\hat{E}_S$ the specific surface energy of the particle, and $(kT)_b$ the product of the Boltzmann constant and absolute temperature. As $\bar{r} \to \infty$, $c_{A\bar{r}} \to c_{A\infty}$. An alternative interpretation of eq. (12-62) is that for a solution at concentration $c_{A\bar{r}}$, $\bar{r}$ is a critical particle radius. Smaller particles will disappear due to their higher solubility, and larger particles will grow due to their lesser solubility. Ostwald ripening is diffusion controlled and is often important for long-term storage of emulsified or formulated products. A model for Ostwald ripening in emulsions has been developed by Yarranton and Masliyah (1997).

## 12-6.7 Heat and Mass Transfer

Many industrially important chemical reactions occur in liquid–liquid systems since heat and mass transfer can be very efficient in agitated heterogeneous stirred reactors. The reaction usually takes place in the dispersed phase. Transport rates depend on the slip velocity between the phases as shown in eqs. (12-63) and (12-64). They are applicable only to single drops that are larger than the turbulent macroscale and are presented for illustrative purposes only. A tank-specific correlation is given later. The heat transfer coefficient, $h_T$, for a single sphere is given by

$$\frac{h_T d}{k_{cf}} = 2.0 + 0.6(Re_\infty)^{1/2}(Pr_f)^{1/3} \quad (12\text{-}63)$$

The mass transfer coefficient, $k_m$, is given by

$$\frac{k_m d}{D_{AB}} = 2.0 + 0.6(Re_\infty)^{1/2}(Sc_f)^{1/3} \quad (12\text{-}64)$$

The Reynolds number $Re_\infty = d v_\infty \rho_f / \mu_f$, the Prandtl number $Pr_f = C_{pf} \mu_f / k_{cf}$, and the Schmidt number $Sc_f = \mu_f / \rho_f D_{AB}$ are based on the physical properties (density $\rho_f$, viscosity $\mu_f$, heat capacity $C_{pf}$, thermal conductivity $k_{cf}$, and mass diffusivity $D_{AB}$) of the surrounding fluid.

The Reynolds number includes $v_\infty$, the drop velocity relative to its surroundings or slip velocity. If drops move with the surrounding fluid, $v_\infty$ is negligible, and heat and mass transfer rates depend solely on conduction and diffusion, respectively. If drops are suspended as in fluidization, heat and mass transfer coefficients will increase due to increased slip velocity.

The mass transfer rate, $\dot{m}_A$, of species A into or out of a drop depends on the interfacial area, $\pi d^2$, the concentration driving force, $\Delta C_A$, and the mass transfer coefficient, $k_m$, as shown by

$$\dot{m}_A = k_m \pi d^2 \Delta C_A \qquad (12\text{-}65)$$

where $\Delta C_A$ is the difference in concentration of species A inside and outside the drop. The actual driving force for interphase mass transfer is the difference in chemical potential. Therefore, one of these concentrations must be adjusted using a partition coefficient, or equivalent, so that $\Delta C_A$ is defined relative to either the drop or continuous phase. Increasing agitation intensity increases mass transfer in two ways. Since drop size decreases, interfacial area is increased. Eddy motion increases, causing an increase in slip velocity.

Mass transfer can affect the rate of film thinning between drops and hence coalescence rate. When mass transfer is not uniform, surface concentration and interfacial tension gradients are established. This leads to a phenomenon known as the *Marangoni effect*. Differences in concentration result in differences in interfacial tension and surface pressure that cause surface flows that facilitate film drainage and coalescence. Coalescence affects mass transfer since the coalescing drops can have different composition.

Skelland and Moeti (1990) and Skelland and Xien (1990) measured mass transfer rates using an electrical conductivity probe for drops suspended in an agitated vessel. Results of 180 different systems were correlated by

$$\frac{k_m d}{D_m} = 1.237 \times 10^{-5} (Sc_c)^{1/3} Re^{2/3} Fr^{5/12} \left(\frac{D}{d}\right)^2 \left(\frac{d}{T}\right)^{1/2} \left(\frac{\rho_d d^2 g}{\sigma}\right)^{5/4} \phi^{-1/2} \qquad (12\text{-}66)$$

where $D_m$ is the mass diffusivity of the solute in the continuous phase, $Fr = N^2 D/g$ is the impeller Froude number, Re the impeller Reynolds number, and $Sc_c = \mu_c/\rho_c D_m$ is the Schmidt number.

## 12-6.8 Presence of a Solid Phase

Solids affect coalescence in some instances by slowing the rate of film drainage. They can also have the opposite effect of helping to bridge the film, thereby increasing the probability of coalescence. Dispersion is less sensitive to the presence of solids. At low solids concentration there is little effect. At continuous phase concentrations above 10 vol %, a higher average viscosity tends to reduce coalescence and create higher shear stresses. Therefore, drop sizes become smaller with increasing solids content. "Limited Coalescence" is a patented, high concentration dispersed phase process that utilizes solids to stabilize against coalescence. It is described fully in Section 12-9.2.3.

### 12-6.9 Effect of a Gas Phase

Gas bubbles play a complicating role in both dispersion and coalescence. The effects of gas bubbles are size dependent. Large bubbles (larger than drops) interfere with momentum transfer. This results in a loss of shear stress and the ability to transport momentum necessary for drop dispersion. Large bubbles often collect drops in their wake or in the trapped liquid between them. When bubbles are trapped in the liquid film, buoyancy forces create a squeezing flow that enhances drop coalescence. On the other hand, microbubbles, located in the film drainage region between drops, interfere with film drainage and thus reduce coalescence rates.

## 12-7 EQUIPMENT SELECTION FOR LIQUID–LIQUID OPERATIONS

### 12-7.1 Introduction

Any impeller in a vessel capable of pumping fluid and providing shear can produce liquid–liquid dispersions. The impellers commonly used for immiscible liquid–liquid systems include disk turbines, pitched blade turbines, propellers, hydrofoils, paddles, retreat curve impellers, and other proprietary designs. We showed in Section 12-2 that drop size depends on maximum energy dissipation rate. More specifically, eq. (12-23) shows that the power number of an impeller affects drop size. In this section we deal with equipment used for two common industrial applications: creating the maximum interfacial area and creating uniformly sized drops.

Most drop dispersion results from shear forces created by a rotating impeller. To a lesser extent, drop dispersion occurs by drops impinging on baffles and vessel walls, and by streaming from dispersed phase liquid collected on impeller blades and other surfaces. Dispersion in a static mixer involves both shear forces and drop impingement on the leading edges of mixer elements. Although the major emphasis of this section is on stirred vessels, other contacting equipment is also considered.

### 12-7.2 Impeller Selection and Vessel Design

*12-7.2.1 Impeller Selection.* Design for liquid–liquid contactors includes impeller geometry, number of impellers required, D/T ratio, and location in the vessel. Commonly used impellers are classified as producing shear or flow. If the application requires high interfacial area (small drop diameters), a high-shear impeller, such as the Rushton turbine shown on the left in Figure 12-28, is a

**720** IMMISCIBLE LIQUID–LIQUID SYSTEMS

good choice. These turbines are also known as radial disk turbines (RDT) and by other vendor designations, such as the Lightnin R100 or Chemineer D6. If moderate, yet gentle shear is required, such as for emulsion polymerization, the retreat curve impeller, shown in the center of Figure 12-28, is commonly chosen. When larger drops of a narrow size distribution are required, the loop impeller, shown in the right view of Figure 12-28, is a reasonable choice. Broad blade paddles are also used. Acceptable substitutes for the RDT include the Scaba and Chemineer's BT6 and CD6 impellers, commonly used for gas–liquid mixing.

RDTs produce strong radial flows and intense turbulence. When the impeller flow meets the vessel wall, it divides, forming two distinct circulation zones, as shown in Figure 12-29. Baffles increase dispersing power by increasing power draw and eliminating vortexing.

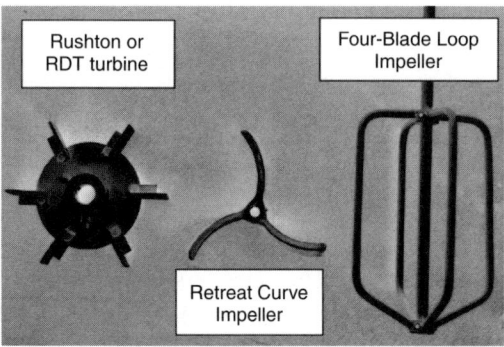

**Figure 12-28** Some impellers used for liquid–liquid dispersion.

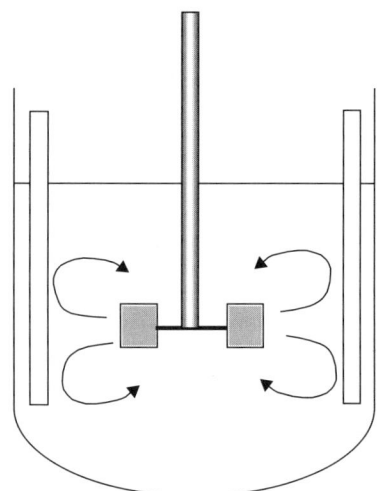

**Figure 12-29** Overall flow pattern for a radial disk turbine in a baffled vessel.

EQUIPMENT SELECTION FOR LIQUID–LIQUID OPERATIONS   721

Other high-shear impellers include the tapered blade ChemShear impeller and dispersing disks such as the Cowles impeller. These provide excellent shear, but far less flow than the RDT. They are used primarily in small scale batch applications where dispersion time is not critical. Pitched blade turbines (PBT) are used when large density differences could lead to a suspension problem. They require higher speed to create the same drop size as the RDT, since they have a lower power number. The flow discharge angle for PBTs varies with Reynolds number and blade angle.

Impeller size is conveniently specified in terms of the D/T ratio. This helps conceptualization and scale-up. This ratio varies from 0.25 to 0.40 for RDTs and from 0.4 to 0.6 for flow-type hydrofoils and propellers. D/T ratios for retreat curve, glassed steel impellers are larger, usually ranging from 0.5 to 0.8. Vertical placement of the impeller depends on vessel shape and application. For example, for dispersion by continuous addition of a dense phase fluid into a less dense fluid, the impeller should be placed fairly low in the vessel at a clearance C $\approx$ H/4 to H/5, where H is the liquid height. For dispersion of light liquids, it is good practice to place a single impeller between $0.2 \leq C/H \leq 0.5$. The subject of impeller type and location with respect to drop suspension was covered in Section 12-6.

The production of pharmaceuticals and specialty chemicals frequently requires the same vessel and agitation equipment be used for each processing step. Therefore, a gas–liquid dispersion step might require special impellers for that operation. If a liquid–liquid processing step is also required, the equipment chosen for the gas–liquid step will usually be well suited for liquid–liquid dispersion. For such multiuse applications, it is essential to use a variable speed drive. It is common to have to deal with slurries. Care must be taken to ensure adequate mixing during off-loading, so impellers are often located close to the bottom for such applications.

Multiple impellers are recommended if $H/T \gg 1.2$ or if $\Delta \rho > 150$ kg/m$^3$. Assuming a less dense dispersed phase, the second or top impeller often is a hydrofoil placed midway between the RDT and the surface of the liquid. This impeller produces high flow at low power, provides excellent circulation, and complements the flow pattern produced by the RDT. The diameter of the second impeller is usually greater than the RDT, typically $D/T \geq 0.45$. A good practice is to distribute the total power to $\approx$20% for the hydrofoil and $\approx$80% for the RDT. Since the power number, $N_p$, is known for each turbine, setting the power distribution enables the diameter of the hydrofoil to be determined. The vertical position of the upper turbine must ensure that fluid reaches the lower impeller, but must avoid gas entrainment that could occur if placement is too close to the liquid surface. Flow from a PBT does not complement that from a RDT and is therefore not recommended. Power requirements are discussed in Section 12-7.3. Table 12-6 lists equipment options for different drop sizing objectives (desired result). If $d_{32}$ must be less than 30 µm, the use of a stirred tank is not recommended, so other devices are also included in the table.

Mass transfer among drops is enhanced by repeated coalescence and redispersion. This is very important in liquid–liquid extraction. Disk turbines used in

**Table 12-6** Common Types of Equipment Used for Liquid–Liquid Dispersion

| Description | Impeller Types | Batch or Continuous | Desired Result | Comments |
|---|---|---|---|---|
| Stirred tanks; baffles | Flat, pitch, and disk type | Either | $30 \leq d_{32} \leq 300$ μm | General; mass transfer operations |
| Stirred tanks; baffles | Retreat curve | Either | $30 \leq d_{32} \leq 300$ μm | General; emulsion polymerization[a] |
| Stirred tanks; no baffles | Paddle, loop, special types | Batch | $100 \leq d_{32} \leq 1000$ μm | Suspension polymerization; suspending agent required |
| Static/in-line mixers | None | Continuous | $10 \leq d_{32} \leq 200$ μm | Dispersant or protective colloid needed |
| Rotor–stator mixers | Slotted ring or impeller, along with slotted stator | Either, often continuous | $1 \leq d_3 \leq 50$ μm | Sparse data for scale-up; need extensive testing |
| Impingement mixers | None | Continuous | $1 \leq d_{32} \leq 50$ μm | Sparse data; work with vendors |
| Valve homogenizers; ultrasonic mixers | None | Usually continuous | $0.1 \leq d_{32} \leq 10$ μm | Sparse data; work with vendors; feed is predispersed |

[a]Drop size refers to monomer drops. Latex products are much smaller particles, in the range 0.1 to 0.5 μm.

extraction are operated at moderately low speed to avoid over dispersing, thus forming hard to coalesce drops. Suspension polymerization applications require production of nearly monodispersed drops, since these become the final product. Figure 12-30 shows a loop impeller that creates low, uniform shear for suspension polymerization. It was described by Leng and Quarderer (1982) and is discussed in Section 12-8. Four long vertical arms produce regions of relatively uniform shear and provide wall movement for heat transfer. This design is not easily adaptable to systems of large phase density difference, due to weak axial flow. D/T ratios are between 0.6 and 0.8. Two- and four-blade backswept square paddles can also used. Baffling is kept to a minimum to minimize shear.

**12-7.2.2 Tank Geometry.** It is essential to avoid stagnant regions in liquid–liquid operations, regardless of the process. This means that use of flat- and cone-bottomed tanks and tall slender vessels should be avoided if possible. Placing baffles away from the wall, to permit flow between the wall and the baffle, prevents dispersed phase buildup on surfaces. Internal heating coils and ladders should also be avoided if possible. Optimum flow patterns normally

EQUIPMENT SELECTION FOR LIQUID–LIQUID OPERATIONS **723**

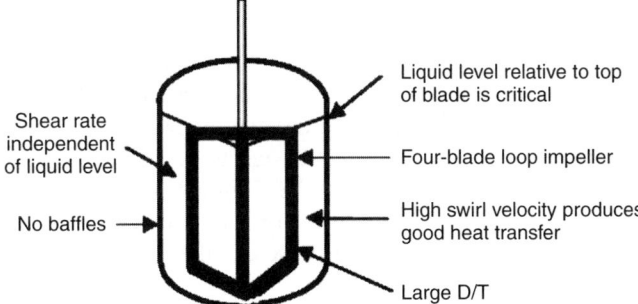

**Figure 12-30** Low-shear agitation for suspension polymerization.

develop when the overall vessel shape is $1 < H/T < 1.2$. It is certainly possible to operate successfully well beyond this range, as shown later in this section, but the design must provide for excellent flow throughout the vessel.

For mass transfer dependent reactions, agitation must promote dispersion, discourage coalescence, and prevent settling. Usually, a single impeller can accomplish these tasks for vessels of $H/T \leq 1.2$ and for $0.9 < \rho_d/\rho_c < 1.1$. However, additional impellers are used when $H/T \geq 1.2$ or when $\rho_d/\rho_c$ is outside the limits cited above. The selection of a second impeller was discussed in Section 12-7.2.1. Dispersions of 1 mm drops are easily suspended in square vessels ($H = T$) and normally do not require use of a second impeller.

High-pressure autoclaves are sometimes designed as tall, slender vessels to minimize construction cost due to wall thickness. Figure 12-31 shows such an

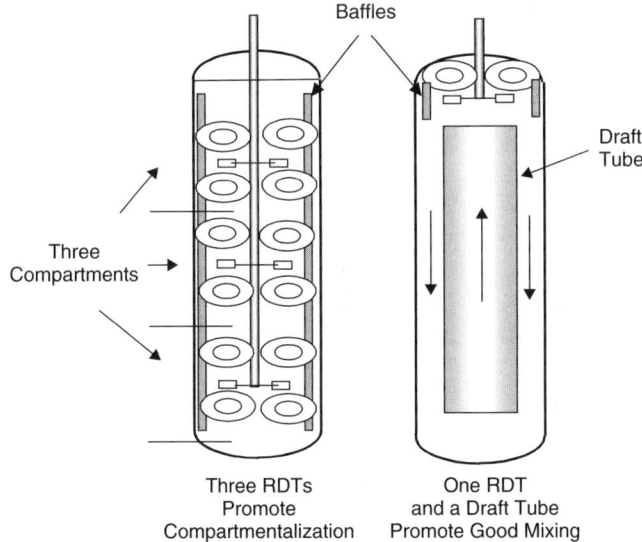

**Figure 12-31** Internal arrangements for tall vessels.

application. The slender shape complicates efficient top-to-bottom mixing. A solution is to use a draft tube, with a tube/tank diameter ratio of ~0.7, and a top entering Rushton turbine. This is shown in the right-hand view. This design avoids compartmentalization problems leading to poor circulation, shown by the design on the left. Multiple RDTs set up circulation cells around each impeller. Reaction modeling shows results for this design to be consistent with those for a multistage CSTR. The preferred design of Figure 12-31 was commercialized and operated for over 20 years, for a high-pressure reaction requiring both high shear and circulation. Its features are discussed further in Section 12-8.

**12-7.2.3 Forming Dispersions.** The initial condition is important in forming dispersions, as illustrated in Figures 12-32 and 12-33. In these examples, oil is the lighter or upper phase. If the lower phase is to be dispersed in the upper phase, the RDT is placed in the upper phase and an up-pumping axial flow turbine is placed in the lower phase. Figure 12-32 shows the suggested arrangement. When the upper oil layer is to be dispersed in the lower water layer, the arrangement shown in Figure 12-33 is recommended. Here the axial flow turbine pumps downward. Both figures show the use of a RDT for dispersion and a propeller to improve circulation. Single impellers can also be used. Often, both o/w and w/o regions initially coexist. The amount of each phase, and the relative rates of coalescence (o/w versus w/o) during transient conditions, determines whether the final system is o/w or w/o. Figure 12-34 shows the ideal location for a single turbine.

**12-7.2.4 Baffles and Baffle Placement.** Baffles increase the axial velocity component that promotes circulation and reduce the tangential or swirl velocity. This lower tangential velocity leads to a higher relative velocity and shear rate near the impeller. Higher rates of shear and circulation result in faster overall dispersion. Good surface movement helps prevent settled layers from forming. Poor surface movement can lead to surface coalescence and the formation of a condensed layer. Baffles help prevent this. However, suspension polymerization reactors use little or no baffles to help reduce shear and therefore produce larger drops.

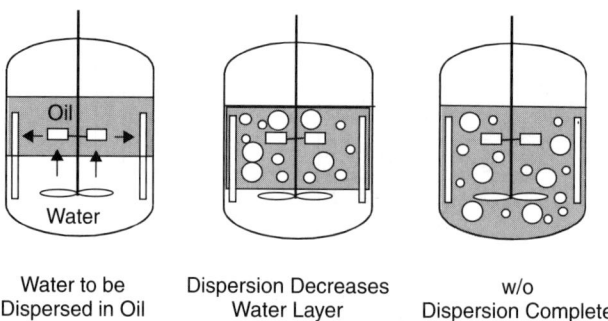

Water to be Dispersed in Oil    Dispersion Decreases Water Layer    w/o Dispersion Complete

**Figure 12-32** Dual impeller arrangement for water-in-oil dispersion. Propeller is upward pumping.

EQUIPMENT SELECTION FOR LIQUID–LIQUID OPERATIONS **725**

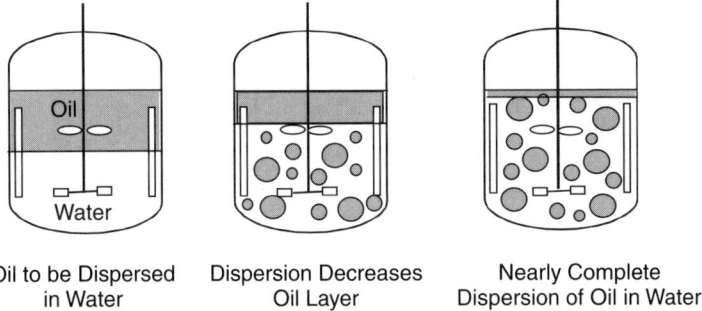

**Figure 12-33** Dual impeller arrangement for oil-in-water dispersion. Propeller is downward pumping.

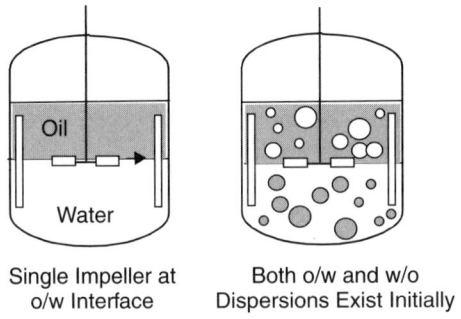

**Figure 12-34** Single RDT placed at oil–water interface.

Short baffles, H/3 in length and T/12 in width, located just below the liquid surface, can be used to promote improved axial flow while producing only a slight increase in effective shear rate. They are positioned well above the plane of the impeller and are able to convert tangential into axial momentum without significantly increasing shear rates. If phase density differences are great enough to require better overall circulation, narrow width baffle designs (<T/12) should be considered. Baffles used in glass-lined equipment (beavertail or "D" or finger designs) have proven beneficial, since the degree of baffling can be adjusted by changing the baffle angle relative to the flow. However, baffles can cause dispersed phase and polymer buildup, stagnation, and some loss of heat transfer through the wall, due to lower tangential velocities at the wall. A nonfouling design is to provide weak baffling by welding four 90° angle sections to the vessel walls, to create triangular fins. Baffles cause an increase in power supplied to the vessel and therefore reduce drop size.

As a general rule, four equally spaced baffles should be used. The baffle width should be T/10 to T/12 and should be located a minimum distance of T/72 from the wall. This enables liquid to pass between the baffle and the wall. Baffles should extend from just below the surface of the liquid to the lower end of the

**726**   IMMISCIBLE LIQUID–LIQUID SYSTEMS

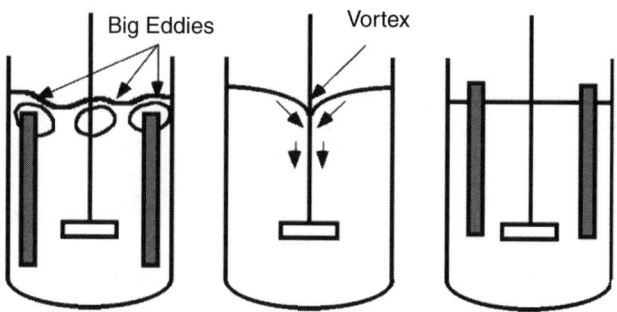

**Figure 12-35**   Importance of baffling to surface conditions.

straight wall, or in the case of dish-bottomed vessels, the lower tangent line. For good mixing, Nagata (1975) proposed suspending baffles from the top of the vessel a radial distance two-thirds out from the center and to submerge them to a depth of H/3. This arrangement is commonly found in glass-lined vessels, where baffles are suspended from the top head. However, for conventional vessels, top-mounted baffles are seldom used.

As noted above, correct baffle placement can improve surface flow. The location of the top edge of the baffles relative to the liquid surface is important in creating eddies that are helpful in facilitating drop suspension. When baffle tips are just below the surface, unrestricted eddy motion facilitates engulfment of surface materials into the bulk liquid. If baffles extend through the surface, they create local stagnation, causing slow surface engulfment and sometimes pooling. This is shown in Figure 12-35. The left-hand view shows how ideally placed baffles can aid in creating surface motion. The center view shows that a central vortex forms when no baffles are used. Although poor from a mixing point of view, the vortex can assist in the engulfment of feed streams. The right-hand view shows baffles extending through the surface and creating stagnation and poor surface mixing.

### 12-7.2.5 Location of Feed and Exit Streams.
When rapid initial mixing is required, direct feed injection through a dip pipe to the impeller is often used for nonplugging conditions. Other considerations, such as differences in phase densities, need to be considered. Low density liquids are introduced near the bottom and heavy liquids near the top of the vessel. Feed discharge onto the surface is not recommended.

Most batch processes are drained from the bottom of the vessel, but for continuous processes, removal can be from any well-mixed region. It is good practice to keep feed and exit locations as far apart as possible to prevent "short-circuiting."

## 12-7.3 Power Requirements

The questions to address when estimating power requirements are:

- How much power is needed for the desired result?
- Which impeller(s) size and speed will deliver that power?
- What vessel geometry, shape, and baffling are to be used?

As discussed in Section 12-2, the ultimate drop size is determined by $\varepsilon_{max}$, not $\varepsilon_{avg}$. However, most correlations for drop size use $\varepsilon_{avg}$, since data for $\varepsilon_{max}$ are not readily available. Many investigators, starting with Corrsin (1964), determined that $\varepsilon_{max}/\varepsilon_{avg} \simeq 40$. Once T, D/T, and $\varepsilon_{avg}$ have been selected, it is a straightforward task to calculate the operating speed, motor power, and torque. The power number, $N_p$, is needed for the calculation. Power numbers for different impellers are a function of impeller Reynolds number and are found in Chapters 6 and 9. Once $N_p$ is known, the hydraulic power is calculated from

$$P = \frac{N_p \bar{\rho} D^5 N^3}{f_{conv}} \qquad (12\text{-}67)$$

If the units are P in hp, D in ft, N in rpm, and $\bar{\rho}$ in lb/ft$^3$, the conversion factor is $f_{conv} = 17{,}710$. If the units are P in kW, D in meters, N in rps, and $\bar{\rho}$ in kg/m$^3$, then $f_{conv} = 0.001$. The vessel average power per unit mass, $\varepsilon_{avg} = P/V\bar{\rho}$.

## 12-7.4 Other Considerations

### 12-7.4.1 Time to Reach Equilibrium.
As discussed previously, studies over the past two decades have shown that large differences in turbulence energy and shear exist in different regions of stirred vessels. Turbulence is highest near the impeller surfaces and lowest near vessel walls and the free surface. As a result, the power input is not evenly dissipated throughout the tank, so that $d_{max}$ is achieved only when the last drop of size $d > d_{max}$ experiences the region of maximum energy dissipation, $\varepsilon_{max}$. For example, in tests witnessed by one of the authors, a 1000 gal suspension polymerizer took over 30 h to reach terminal dispersion conditions. A light transmission probe was used to measure the transient interfacial area. Most industrial processes using noncoalescing liquid–liquid systems operate at transient drop size conditions. Steady-state (equilibrium) conditions are reached more quickly in coalescing systems.

Equation (12-68) shows an empirical relationship developed by Hong and Lee (1983, 1985) for the time to reach equilibrium, $t_{eq}$. They conducted 181 experiments (representing five different liquid–liquid systems) and two scale sizes. The range of dispersed phase volume fraction was $0.05 < \phi < 0.20$.

**728** IMMISCIBLE LIQUID–LIQUID SYSTEMS

$$N t_{eq} = 1995.3 \left(\frac{D}{T}\right)^{-2.37} \left(\frac{We}{Re}\right)^{0.97} \frac{\mu_d}{\mu_c} Fr^{-0.66} \quad (12\text{-}68)$$

The time for a dilute system to reach equilibrium was discussed in Section 12-2.4.

### 12-7.4.2 Breakage and Coalescence Regimes. 
As stated previously, dispersion depends on maximum local energy, $\varepsilon_{max}$, and coalescence depends on gentle shear. For coalescing systems, drops will coalesce in the more quiescent regions of a stirred vessel and will disperse close to the impeller. This is illustrated by the work of Sprow (1967b), shown in Figure 12-36. Using a small baffled stirred vessel and a coalescing system consisting of methyl isobutyl ketone in water, Sprow found that drop size varied with agitator speed and with location in the vessel. Position C represents a location where dispersion controls. At this location he found that $d_{32} \sim N^{-1.5}$, which is a much stronger speed dependence than the $N^{-1.2}$ prediction of the Weber number theory given by eq. (12-22). Position D is well away from the impeller where gentle flow promotes coalescence. Drops in this region were less sensitive to agitation with $d_{32} \sim N^{-0.75}$. Coalescence dominates at position D but dispersion dominates at C. Since $\varepsilon \sim N^3 D^2$, then $\varepsilon^{-1/4} \sim N^{-0.75}$, and Sprow's data at position D are consistent with those of Church and Shinnar (1961) for the coalescence line in Figure 12-25. Sprow's results imply that coalescence and dispersion rates are as fast or faster than overall mixing and circulation rates. This may be the exception, not the rule.

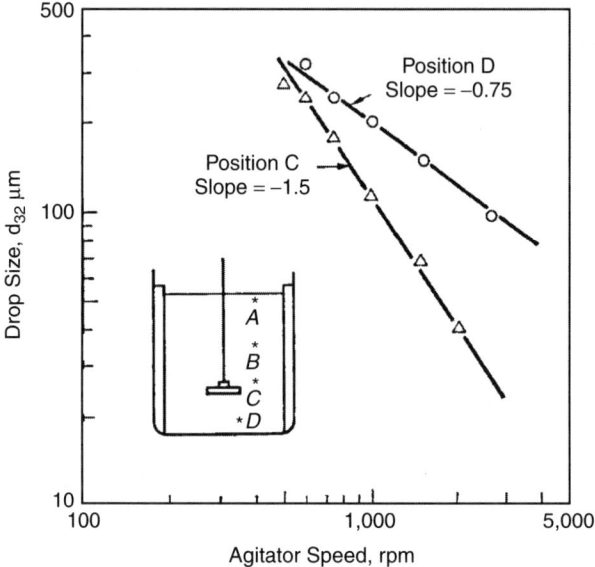

**Figure 12-36** Drop size dependence on impeller speed and spatial location for a coalescing system. (From Sprow, 1967b, reproduced with permission of AIChE © 1967.)

For example, Hong and Lee (1983, 1985) argued that Sprow's results were exaggerated. Using light transmission probes to measure interfacial area, they found only a small spatial variation in drop size. We can only conclude that spatial dependence can occur and is system dependent. It is less apt to be a problem, even for coalescing systems, if the vessel has short circulation times.

### 12-7.4.3 Circulation Time.
As stated previously, both time to equilibrium and the competition between coalescence and dispersion depend on circulation time. Holmes et al. (1964) performed bench scale experiments in a baffled flat-bottomed vessel with H = T and C/T = $\frac{1}{2}$. For turbulent flow their data were well correlated by

$$N \bar{t}_{circ} = C_{27} \left(\frac{T}{D}\right)^2 \qquad (12\text{-}69)$$

where $\bar{t}_{circ}$ is the mean circulation time and $C_{27} = 1.0$ for their RTD and vessel geometry. Middleton (1979) considered the effect of scale and performed experiments in three vessels of similar geometry, ranging from 0.61 m < T < 1.8 m. However, in his work C/T = $\frac{1}{3}$. His data were well correlated by

$$N \bar{t}_{circ} = 0.5 V^{0.3} \left(\frac{T}{D}\right)^3 \qquad (12\text{-}70)$$

There appears to be mechanistic arguments as well as further experimental evidence to support both correlations. It is apparent that other variables, such as C/T, H/T, bottom geometry, and so on, will significantly influence circulation time.

An intuitive approach is to assume that $V/Q_V$ gives the mean circulation time, where $Q_V$ is the volume flow from the impeller. The latter quantity is usually correlated in terms of flow number, $N_q = Q_V/ND^3$. Then

$$N \bar{t}_{circ} = \frac{V}{N_q \cdot D^3} \qquad (12\text{-}71)$$

Equations (12-69) and (12-70) both indicate that the intuitive approach of eq. (12-71) is too optimistic. However, it has been applied successfully in limited cases.

### 12-7.5 Recommendations

- Select RDTs for demanding applications. As discussed in Section 12-2, this is presently the only well-studied geometry.
- Use dished/elliptical-bottomed vessels of overall proportions of H/T = 1 to 1.2. These give better circulation and minimize creation of dead zones.
- Use multiple dispersing impellers, full baffles (either conventional or those proposed by Nagata), and larger D/T impellers if strongly coalescing systems are involved. Minimize circulation time through the use of

secondary large high-flow impellers. $\bar{t}_{circ}$ is reduced by a factor of 2 if two equal-sized RDTs are used in place of one.

## 12-8 SCALE-UP OF LIQUID–LIQUID SYSTEMS

### 12-8.1 Introduction

Scale-up of agitated immiscible liquid–liquid systems can be a challenge that should not be taken lightly. The problems arise from incomplete or inaccurate process information and few quantitative tools to deal with complex technology. In this section we describe some proven practices for scale-up and caution that liquid–liquid dispersion technology is highly system specific.

Most problems are not observed in glass bench scale equipment because unrealistically high rates of circulation mask coalescence and suspension problems. These problems usually surface at the time of scale-up. Throughout this chapter it has been emphasized that production scale vessels are dominated by coalescence, whereas small vessels are dominated by dispersion. As discussed previously, Sprow (1967b) worked with a coalescing system in a small bench scale vessel and found that different regions of the vessel responded differently to agitation. The technology to cope with these complex issues lags other mixing operations, such as blending and solids suspension. Often, all three flow-dependent phenomena—dispersion, coalescence, and drop suspension—must be dealt with simultaneously.

A successful scale-up does not mean that identical results are obtained at two different scales, but rather, that the scale-up results are predictable and acceptable. Problem correction at large scale is costly, time consuming, and sometimes not possible (see Section 12-9.2.2). Scale-up errors can lead to losses in capacity, quality, safety, and profits. For example, an explosion resulted from increasing agitation for an inadequately suspended mixed acid nitration. Faster agitation created a large increase in interfacial area at reaction temperatures and led to an uncontrolled exothermic reaction and property loss.

The scale-up of certain liquid–liquid processes can be straightforward. Dilute dispersions are the easiest processes to scale up. The most difficult ones involve simultaneous coalescence, dispersion, suspension, mass transfer, and chemical reaction. If multiple complex reactions are involved, inadequate mixing often leads to yield losses.

The first step is to understand the goals of the process and to acquire accurate data for all components, including physical, chemical, and interfacial properties as well as reaction kinetics. This also includes the influence of minor impurities. Differences in the quality of raw materials need to be considered.

It is important to undertake bench scale studies that simulate the poorer mixing conditions in the larger vessel. For example, simulate the large scale vessel circulation time. Although dispersion is apt to be unrealistic, coalescence and settling problems can be observed. Examination of the flow patterns in the proposed full scale vessel using CFD can help visualize potential problems related to design. Once the CFD model has been developed and validated, design and operating parameters can be compared to determine design sensitivities. One observation seems to hold universally—better results are always obtained in small equipment.

Identify applications by types likely to cause problems, and separate these from more trivial applications. For example, mixing is critical in the following applications:

- Chemical reactors/polymerizers in which reaction rates are equal to, or faster than, mixing rates
- Competing chemical reactions when yields depend on good mixing
- Mass transfer dependent reactions involving coalescence and dispersion

Less demanding tasks include:

- Heat transfer
- Reactors involved with slow chemical reactions

### 12-8.2 Scale-up Rules for Dilute Systems

Many processes have been scaled successfully using $ND^X = $ constant. This simple rule is based on years of industrial experience. To apply it, the tank Reynolds number must be greater than $10^4$ and vessels must be geometrically similar. Table 12-7 lists the rule and the application best suited to the rule. Other operations, such as blending and solids suspension, are included to provide the reader

**Table 12-7** "Rules" for Scale-up of Geometrically Similar Vessels at Turbulent Conditions, Based on $ND^X = $ Constant

| Value of X | Rule | Process Application |
|---|---|---|
| 1.0 | Constant tip speed, constant torque/volume | Same maximum shear; simple blending; shear-controlled drop size. |
| 0.85 | Off-bottom solids suspension | Used in Zwietering equation for $N_{js}$, for easily suspended solids; also applies to drop suspension (see Section 12-6.2). |
| 0.75 | Conditions for average suspension | Used for applications of average suspension difficulty. |
| 0.67 | Constant P/V | Used for turbulent drop dispersion; fast settling solids; reactions requiring micromixing; gas–liquid applications at constant mass transfer rate. |
| 0.5 | Constant Reynolds number | Similar heat transfer from jacket walls; equal viscous/inertial forces. |
| 0.0 | Constant speed | Equal mixing time; fast/competing reactions. |

# 732 IMMISCIBLE LIQUID–LIQUID SYSTEMS

with an overview of how the exponent on impeller diameter varies from operation to operation. One can see from the table that different scale-up rules apply for suspension, dispersion, heat transfer, and reaction, making it necessary to focus on the most important or limiting task. As mentioned earlier, the indiscriminate use of rules can lead to problems.

***Example 12.3.*** Consider scale-up of a process for a dilute (noncoalescing) liquid–liquid system. For inviscid drops,

$$d_{max} = C_1 \left(\frac{\sigma}{\rho_C}\right)^{0.6} \varepsilon_{max}^{-2/5} \qquad (12\text{-}21)$$

SOLUTION: Assume similar geometry, $Re > 10^4$, an equal ratio of $\varepsilon_{max}/\varepsilon_{avg}$ on both scales and identical physical properties. For $N_p =$ constant, $\varepsilon_{avg} \sim N^3 D^2$. Then eq. (12-21) for scale 1 and scale 2, with the condition $d_{max}(1) = d_{max}(2)$, can be written

$$\frac{d_{max}(1)}{d_{max}(2)} = 1 = \left[\frac{\varepsilon_{avg}(2)}{\varepsilon_{avg}(1)}\right]^{2/5} = \left[\frac{P/V(2)}{P/V(1)}\right]^{2/5} = \frac{N(2)^{1.2}D(2)^{0.8}}{N(1)^{1.2}D(1)^{0.8}} \qquad (12\text{-}72)$$

Since $(ND^{0.67})^{1.2} = N^{1.2}D^{0.8}$, eq. (12-72) is consistent with Table 12-7, row 4.

Equation (12-72) can be used to calculate the speed required for a $T = 3.0$ m vessel with $D/T = \frac{1}{3}$, to achieve the same drop size as a $T = 1.0$ m geometrically similar vessel operating at 200 rpm. Substituting $N(1) = 200$ rpm, $D(1) = 0.33$ m, and $D(2) = 1.0$ m into eq. (12-72) gives $N(2) = 95.5$ rpm.

## 12-8.3 Scale-up of Concentrated, Noncoalescing Dispersions

Dilute, low viscosity dispersions are nearly always controlled by turbulence. At high dispersed phase concentrations, small scale turbulent eddies are damped out by the drops and bulk viscosity increases. As a result, laminar shear forces can control drop dispersion in concentrated systems. Turbulence theories developed for dilute dispersions can sometimes apply to concentrated, noncoalescing systems. However, in other cases, they may not. This is illustrated, by example, below for the scale-up of a suspension polymerization application, described by Leng and Quarderer (1982).

The system consisted of free radical initiated styrene–divinylbenzene monomers dispersed in water containing 0.2% dissolved polyvinyl alcohol. The dispersed phase was 50 vol %. The process was to be carried out in a vessel containing a loop impeller (see Figure 12-28) operating at low-shear conditions. Bench scale studies showed important variables to be speed, impeller diameter, baffling, selection of the suspending agent, and continuous phase viscosity. Polymerization reactions were completed and bead size distributions were determined by sieve analysis.

Theories based on laminar and turbulent dispersion conditions were developed, and tested by comparing bead size against each specific variable. Results showed that beads of size greater than 300 μm were formed under laminar shear-controlled conditions, and smaller beads were formed under turbulence-controlled conditions.

Leng and Quarderer (1982) reasoned that dispersion occurred in the boundary layer adjacent to the loop impeller surfaces and that the impeller vertical elements could be approximated by cylinders moving through the suspension at the relative impeller tip speed. When laminar shear forces predominated, it was shown that

$$d_{max} = C_{28}\sigma \left(\frac{D_C}{\mu_c \rho_c}\right)^{1/2} \frac{1}{[ND(1-k_v)]^{3/2}} \left(\frac{(\mu_d/\mu_c)+1}{1.19(\mu_d/\mu_c)+1}\right) f(\mu_d/\mu_c) \quad (12\text{-}73)$$

where $D_C$ is the diameter of the cylinder and $k_v$ is the ratio of the tangential velocity at the impeller tip to the tip speed. All other variables follow earlier use.

The equation for turbulent dispersion was based on the classical development of Chen and Middleman (1967) (see Section 12-2), with the energy dissipation term calculated for drag on a cylinder. Two cases were assumed for the dissipation volume in the wake region behind the cylindrical impeller blade. The first was that an eddy length proportional to the cylinder diameter determined the dissipation volume. The second was that this volume was proportional to the velocity of the cylinder (tip speed) and a characteristic eddy decay time. Equation (12-74) results from the second case. It showed reasonable agreement with data taken at higher speeds.

$$d_{max} = C_{29} \left(\frac{\sigma}{\rho_c}\right)^{3/5} \frac{1}{(ND)^{4/5}(1-k_v)^{2/5}} \quad (12\text{-}74)$$

Typical low-speed laboratory results showing the effect of impeller speed and baffling are given in Figure 12-37. Using paddle impellers, Aiba (1958) found that $k_v = 0.6$ for unbaffled and 0.3 for baffled conditions. These values were used to correct for baffling effects. Uncorrected data fell on two parallel lines of the same slope. In Figure 12-37, $d_{max} \sim N^{-1.5}$, confirming the validity of eq. (12-73) based on the bench scale data. Other laboratory scale results are given in the paper.

Scale-up experiments were conducted in four larger scale vessels, ranging in volume from 0.082 to 15.1 m$^3$. The D/T ratio varied from 0.478 to 0.676. For each vessel, the impeller speed that gave $d_{max} \approx 1000$ μm was determined. Identical physical properties and chemical composition were used at all scales. Then, according to eq. (12-73), for $d_{max}$ to be the same on all scales, the quantity $D_C^{1/2}/(ND)^{3/2}$, based on the measured speed, must be the same on all scales. Table 12-8 shows values of this quantity. The numbers in the second column appear to be scale independent. This supported the hypothesis that dispersions were formed by laminar shear. Equation (12-74) did apply to runs made at higher impeller speeds.

**734** IMMISCIBLE LIQUID–LIQUID SYSTEMS

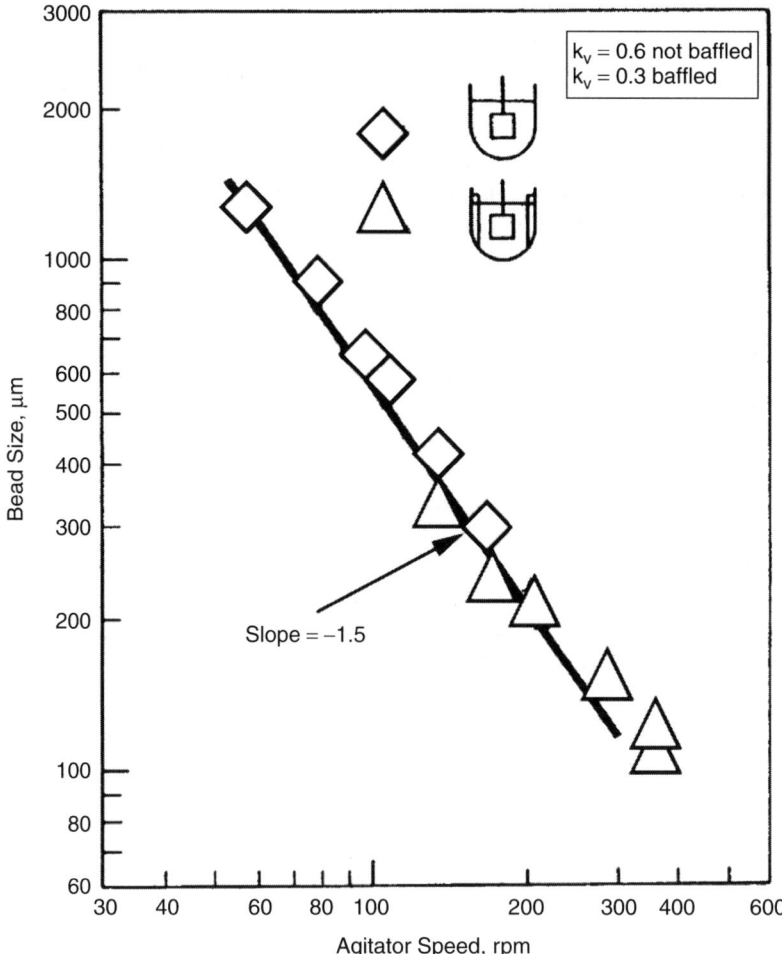

**Figure 12-37** Suspension bead size versus agitation rate for styrene/DVB. (Reproduced from Leng and Quarderer, 1982.)

**Table 12-8** Validation of Eq. (12-73) for Scale-up

| Reactor Volume (m$^3$) | $Dc^{1/2}/(ND)^{3/2}$ |
|---|---|
| 0.082 (laboratory) | $7.5 \times 10^{-4}$ |
| 0.1135 | $7.5 \times 10^{-4}$ |
| 0.330 | $6.0 \times 10^{-4}$ |
| 2.840 | $4.8 \times 10^{-4}$ |
| 15.15 (production) | $5.5 \times 10^{-4}$ |

*Source*: Data of Leng and Quarderer (1982).

### 12-8.4 Scale-up of Coalescing Systems of All Concentrations

No exact method exists to assure successful scale-up of strongly coalescing systems. The following considerations are offered.

- Does the process require coalescing or noncoalescing conditions? Extractions require coalescence; suspension and emulsion polymerization processes do not.
- Few industrial systems are rapidly coalescing. Impurities, salts, and residues often ensure slow coalescence.
- Coalescence rates can be characterized using either the static or dynamic method described in Section 12-3.1.5 and Table 12-3.
- Make the more viscous phase continuous if coalescence is to be minimized. Consider adding a thickener to the continuous phase.
- Suspending aids, such as polymeric suspending agents, detergents, or fine solids, reduce or stop coalescence.
- A static mixer in a recirculation loop can complement conventional agitation in the vessel.
- The use of multiple and larger diameter impellers can increase the "effective" dispersion zone.
- CFD can be used to examine flow field details for both the small scale and the proposed larger scale vessels, and to map out regions of constant energy dissipation rate. The dispersion volume can be approximated as the region in which $\varepsilon_{local}/\varepsilon_{avg} \geq 3.0$. Similarly, the coalescence region is where $\varepsilon_{local}/\varepsilon_{avg} \leq 0.1$. The probability of success upon scale-up will improve if the volume ratio of the dispersion to coalescence regions is scale independent.

### 12-8.5 Dispersion Time

Dispersion kinetics is discussed in Section 12-2.4 for dilute systems and in Section 12-7.4.1 for more concentrated systems. As stated previously, dispersion kinetics in turbulent stirred vessels follows a first-order rate process, and rate constants depend on interfacial tension, drop size, and flow conditions (Hong and Lee 1983, 1985). Figure 12-38 shows a typical drop size versus dispersion time relationship for a batch vessel. Upon introduction of the dispersed phase, the drop size falls off rapidly and approaches the ultimate size within a factor of 2 or so, at times that are often short compared to the process time. However, the decay to equilibrium size is quite slow. This is why equilibrium drop size correlations perform adequately despite the fact that the process time is often smaller than the time to equilibrium.

Dispersion time adds a complication to the scale-up of liquid–liquid systems. For a coalescing system, a small vessel reaches $d_{32}^{\infty}$ in a shorter time than the larger one. This is illustrated in Figure 12-39. A steady $d_{32}^{\infty}$ is reached at time

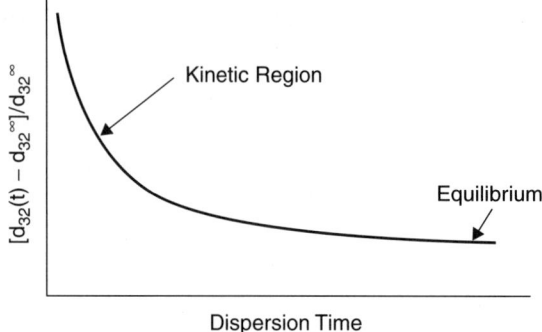

**Figure 12-38** Drop size as a function of dispersion time in a batch vessel.

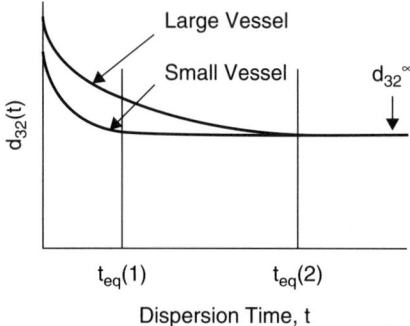

**Figure 12-39** Typical dispersion times in vessels of different size.

$t_{eq}(1)$ in the small vessel, but not until time $t_{eq}(2)$ in the large one. If the large vessel is required to have the same dispersion time as the bench scale, the agitation rate must be increased beyond the value for equal drop size. The mean drop size can be smaller and the drop size distribution may be affected.

## 12-8.6 Design Criteria and Guidelines

Table 12-9 gives a summary of practical guidelines for scale-up of coalescing and noncoalescing systems. Based on the static test for coalescibility, described in Section 12-3.1.5 and Table 12-3, a non/slowly coalescing system has a settling time that is greater than 5 min. A rapidly coalescing system has a settling time of less than 1 min. In Table 12-9, the scale-up limitation refers to the ratio of vessel volumes (large $V_L$ to small $V_S$) that should not be exceeded. That is, for non/slowly coalescing systems, it is safe to scale-up by a factor of 100 in volume, but for rapidly coalescing systems, scale-up should be limited to a 10 to 20 fold increase in volume.

**Table 12-9** Guidelines for Scale-up of General Purpose Liquid–Liquid Stirred Vessels

| Feature | Non/Slowly Coalescing System | Rapidly Coalescing System |
|---|---|---|
| Scale-up criterion | P/V = constant | Circulation time = constant |
| Scale-up limitation, $V_L/V_S$ | 100 : 1 | 10 : 1 to 20 : 1 |
| Baffles | Yes but not for suspension polymerization | Yes |
| Impellers | RDT and optional axial flow/hydrofoil impeller | Multiple RDTs and axial flow/hydrofoil impeller for better circulation |
| D/T | 0.3–0.5 | $\geq 0.5$ |
| Time to reach terminal drop size | Long times for large vessels | Short times under 30 min for most coalescing systems (all vessel sizes) |
| Geometric similarity | Maintain close similarity | Use more and larger turbines in larger vessel; do not try to maintain geometric similarity |
| Speed/drives | Variable or fixed speed | Variable speed capability is essential; consider overdesign to meet unpredicted performance |
| Risk | Low to moderate risk | High risk |

## 12-9 INDUSTRIAL APPLICATIONS

### 12-9.1 Introduction

Common problems encountered in the industrial applications of liquid–liquid systems include (1) failure to meet requirements for interfacial area, often due to effects of coalescence, (2) failure to meet requirements for drop size distribution, (3) failure to meet requirements for drop suspension and process heat transfer, and (4) failure to recognize problems caused by interfacial debris and tiny drops.

Every problem is unique. Sometimes, differences in quality of raw materials lead to unexpected by-products that prevent coalescence. When coalescence is required for separation, filters, centrifuges, and fibrous bed coalescers can sometimes alleviate these problems.

### 12-9.2 Industrial Applications

Several examples of industrial scale-up problems are given below. In some cases the problems were corrected. For others, less than ideal performance had to be accepted.

***12-9.2.1 Inverse Suspension Polymerization.*** *Inverse suspension polymerization* refers to the polymerization of an aqueous monomer dispersion in an organic continuous phase. For this application, the aqueous dispersed phase consisted of initiated monomer dissolved in water, and the continuous phase was xylene containing a dissolved polymeric suspending agent. On the production scale, polymerization was rapid relative to dispersion, and a viscoelastic dispersed phase was initially produced. Viscoelasticity proved to be a problem during the dispersion step. Deformation and drop breakage of the elastic drops was partly due to tip streaming, and that led to considerable quantities of undesirable fines, dusty particles that caused problems for the customer. Laboratory studies failed to reveal the problem, since dispersion was fast and complete prior to polymerization. In the production plant dispersion was incomplete at the time of initiation.

The problem was solved using initially fast agitation to establish the desired particle size, followed by slower agitation just prior to initiation. This way, dispersion was completed before the elasticity developed. Coalescence was not a problem due to the presence of the suspending agent.

***12-9.2.2 Pharmaceutical Process Scale-up.*** The second step in the synthesis of a pharmaceutical intermediate was to reduce an organic reactant, using powdered zinc and concentrated HCl as the reducing agent. This reduction reaction was mass transfer controlled. Studies in a 10 L glass reactor gave acceptable reaction rates that served as a basis for production goals. The large reactor was a typical 3500 gal glass-lined vessel, containing beavertail baffles and a single retreat curve impeller located at the bottom. Production results were unexpectedly poor. Low yields and reaction rates (17 times slower than expected) were observed. Laboratory tests in glass vessels showed that after stopping agitation, complete coalescence took place in seconds. This evidence suggested that coalescence was the cause of the problem. The large reactor was controlled by coalescence, not by dispersion, as was the case in the laboratory vessel. Loss of much needed interfacial area explained the results.

Agitation in the production scale equipment was changed to include dual glass-lined impellers of $D/T = 0.45$, consisting of a lower four-blade FBT and an upper four-blade 45° PBT. This was an attempt to increase the volume of the dispersion region and to improve circulation. The modified system did improve reaction rates, but not to the degree desired.

***12-9.2.3 Limited Coalescence.*** Limited coalescence is a commercialized process (Ballast et al., 1961) that produces uniform polymer particles. The principle involves providing the correct amount of very fine particles to interfere with film drainage, thereby suppressing coalescence. Inorganic materials, such as zinc oxide or silica, and organic materials, such as sulfonated polyvinyl toluene, have been used. A given number of particles support a given surface area. Since agitation creates more surface area by dispersion, fewer particles are available per unit area to protect against coalescence. Drop sizes then grow by coalescence, which then reduces the total interfacial area. Therefore, ultimate drop size is a result of a dynamic equilibrium process that depends on the number of particles present, not on agitation intensity. The process is fine-tuned by the use of wetting agents that control the position of the particles relative to the o/w interface. Solids partially wet by the oil phase move into the drop and are less able to prevent coalescence. Nonwetted particles are located at the drop surface, where they effectively prevent coalescence. While drop size is controlled by coalescence phenomena, vigorous agitation is required for good mixing and drop–drop interactions.

**12-9.2.4 Agricultural Intermediate.** The first step in producing an agricultural intermediate was to nitrate an aromatic feed. Nitration usually involves a sequence of reactions leading to mono-, di-, and trinitro products. In this case, only the mononitro compound was desired. The nitrating agent consists of an anhydrous mixture of $HNO_3$ and $H_2SO_4$. The nitronium ion becomes available to the dispersed organic phase by mass transfer from the continuous phase through the drop surfaces. Laboratory work in a batch CSTR gave favorable results. A continuous fed columnlike apparatus was used for scale-up. The design was a failure because it did not provide adequate suspension and dispersion, and displayed a predominant tailing residence time distribution pattern. Production was slower than expected, and large quantities of multinitrated products resulted from the undesirable residence time distribution.

**12-9.2.5 Largest Surviving Drop Applications.** Under the conditions described below, the maximum mixing intensity can be characterized by measuring the largest surviving drop size using a dilute, noncoalescing test system. The method consists of contacting an aqueous phase, typically containing 0.1 to 0.2 wt % polyvinyl alcohol (PVA) in water, and an oil phase with $\phi = 0.01$ to 0.02. The oil phase can be any nonpolar liquid such as monochlorobenzene, ethylbenzene, toluene, mineral oil, or silicone oils. Samples are withdrawn to allow measurement of drop size. Analysis is usually done photometrically or by using a Coulter counter. The largest drop diameter in the distribution characterizes maximum mixing intensity. In practice, $d_{90}$ is selected rather than $d_{max}$. It takes a long time to reach equilibrium in large equipment. It is important to dissolve the PVA completely and to disperse for a long enough time to ensure that all drops "see" the region of maximum shear. Three examples using this technique are given below.

*Emulsion Polymerization.* Emulsion polymerization processes are used to produce synthetic latexes. Changing product requirements dictate producing and testing many new formulations. Agitation disperses drops, provides mixing, and promotes heat and mass transfer. Latexes are usually shear sensitive and agglomerate if exposed to excessive agitation. With each new product there is the question of agitation optimization. Traditionally, optimal conditions were arrived at by trial and error. The surviving drop method was used to calibrate production scale vessels, identify maximum shear rate, and anticipate product quality from studies in smaller 1 to 5 gal scale equipment.

## 740 IMMISCIBLE LIQUID–LIQUID SYSTEMS

*High-Pressure Autoclave.* A high-pressure two-phase alkyl phenol reaction was to be scaled up from experimental data collected from sealed rocking bomb bench scale experiments. At any given rocking speed, mixing intensity varies with the amount of liquid in the bomb. For example, a half-full autoclave provides more mixing intensity than a nearly full one. Mixing intensities in the rocking bomb experiments were compared to agitation rates in a stirred autoclave reactor using the surviving drop method. Drop sizes were determined as a function of rocking bomb loading and compared to those versus speed in the stirred autoclave. Figure 12-40 shows the relationship found between the fill level in the bomb and the agitation speed in the stirred autoclave.

The laboratory stirred vessel was actually never used for reaction experiments. It was simply a scaled-down version of the commercial autoclave geometry and operating conditions. Scale-up/scale-down was accomplished using constant P/V and geometric similarity. Reactions in the commercial reactor proved to be identical to results obtained in the rocking bomb autoclave.

*High Pressure Reactor Design.* Diphenyloxide and orthophenylphenol were produced continuously as co-products in a high-temperature, high-pressure (410°C, 4000 psig) two-phase reactor by reacting sodium hydroxide with monochlorobenzene. Scale-up was a big challenge. Visual appreciation of mixing was impossible, due to reaction conditions. Operations were calculated to be close to supercritical conditions. Minimizing fabrication cost and providing for an 8 min mean residence time led to a design that was 96 in. tall by 18 in. internal diameter. The process demanded intense dispersive mixing, rapid circulation, and a narrow residence time distribution. These were difficult to obtain in the tall cylinder, shown in the right-hand view of Figure 12-31. An experimental $\frac{1}{3}$ scale Lucite vessel containing a top-entering six-blade RDT, a draft tube, and four wall baffles located in the impeller region was constructed. At the bench scale, continuous reactions (with an 8 min residence time) were carried out in a 1.0 L stirred autoclave reactor to determine the critical change over in impeller speed from mass transfer to reaction rate control.

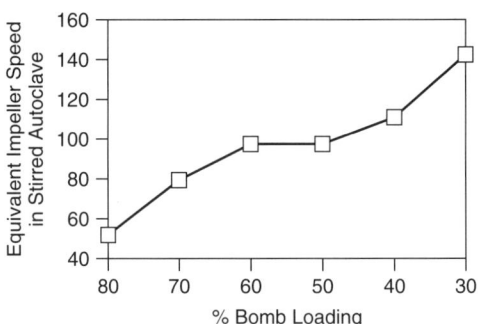

**Figure 12-40** Relationship between rocking bomb reactor and lab scale stirred autoclave using the maximum stable drop size as a calibration tool.

INDUSTRIAL APPLICATIONS    741

The mixing intensity at the critical crossover speed was characterized for the small reactor, employing the maximum stable drop method with mineral oil as the dispersed phase. Similar experiments were run in the $\frac{1}{3}$ scale Lucite prototype vessel. The goal was to find the impeller speed in the $\frac{1}{3}$ scale prototype that gave the same mixing intensity (maximum drop size) as the autoclave operating at the critical change over speed. This information made it possible to establish the speed and power requirements for the production scale vessel. Scale-up was accomplished using equal P/V and circulation time. The commercial scale reactor produced precisely the expected result, and a second identical reactor was installed. This plant operated successfully for over 20 years.

*Caution.* Example applications using the surviving drop method were for either noncoalescing or slowly coalescing systems. This technique should not be used if the application is a rapidly coalescing system.

### 12-9.2.6 Suspension Polymerization: Cross-Linked Polystyrene.

A suspension polymerization process was to be scaled up from a 1000 gal to a 4000 gal vessel. Attempts to do this failed because existing designs produced beads that were too small when suspension needs were met. Correctly sized beads could be made but not adequately suspended. Reactor setups were common. Comparative testing showed that a new impeller design, shown in Figures 12-28 (right view) and 12-30, seemed to meet both needs. It consisted of a four-blade loop-type impeller placed in a nonbaffled vessel. The design provided excellent surface mixing needed for drop suspension, while producing much larger drops of good uniformity. Long vertical arms provided uniform shear as well as good heat transfer to the wall. With reference to Figure 12-26, the new design raised the suspension line, enabling larger drops to be produced.

### 12-9.2.7 Suspension Polymerization: Vinyl Polymerization.

A well-established suspension polymerization process was being scaled from existing 3500 gal production reactors to new, more scale-efficient 10 000 gal vessels. The reactor functions consisted of: blending two dense monomers, mixing water with a suspending agent, mixing an initiator with the monomers, dispersing the monomers in the aqueous phase to form an o/w dispersion, and then carrying out the exothermic reaction isothermally. The two reactors (3500 and 10 000 gal) were geometrically similar, and no problems were expected. This was not to be the case. The first three batches in the larger vessel underwent mass polymerization (bulk polymerization), resulting in a difficult to remove mass of polymer. In these reactors, the single retreat curve impeller was located at the bottom of the vessel. Investigative laboratory tests showed that in the 10 000 gal vessel, a dispersion of water (the less dense phase) in the (more dense) mixed initiated monomer phase was formed rather than vice versa. Thus heating resulted in a mass, not a suspension polymerization.

Prior to forming the dispersion, two separated layers existed, with the impeller in the lower monomer phase. A large interfacial vortex formed on starting the

**742**  IMMISCIBLE LIQUID–LIQUID SYSTEMS

agitator, drawing the upper water layer down, like a tornado, into the impeller, where it was dispersed into the monomer. It is hard to explain why this did not happen in the 3500 gal vessels. Possibly, differences in wall drag had prevented the deep vortex from forming in the older, rough surfaced vessels. Laboratory simulations showed that a second impeller, located in the water phase, would inhibit interfacial vortex formation. This was adapted, and the production plant operated as expected with monomer dispersed in water.

### 12-9.3 Summary

The applications presented in this section serve to demonstrate that fundamental knowledge must be coupled with practical insight and engineering judgment to solve problems associated with real industrial applications. Apart from certain formulated products, liquid–liquid dispersion is rarely carried out for its own sake. It is usually accompanied by heat/mass transfer and chemical reaction, thereby complicating scale-up.

## NOMENCLATURE

| | |
|---|---|
| a | drop radius (m) |
| $a_v$ | interfacial area per unit volume ($m^{-1}$) |
| $a^*$ | constant in eq. (12-11) |
| A(v,t) | number probability density function for drops of volume v at time t |
| $A_o$(v,t) | value of A(v,t) in vessel feed stream |
| A(v), $A_o$(v) | steady-state values of A(v,t) and $A_o$(v,t), respectively |
| B | baffle width (m) |
| $B_d$ | breadth of deformed drop (m) |
| $\dot{B}_d$ | birth rate of drops of size d ($s^{-1}$) |
| $c_{A\bar{r}}$ | solubility of a particle of species A of radius $\bar{r}$ in solution (kg-mol/$m^3$) |
| $c_{A\infty}$ | bulk solubility of species A in solution (kg-mol/$m^3$) |
| $c_s$ | surfactant concentration (kg/$m^3$) |
| C | clearance from tank bottom (m) |
| CoV | coefficient of variation |
| $C_1 \cdots C_{29}$ | dimensionless empirical constants |
| $\Delta C_A$ | concentration driving force for mass transfer (kg/$m^3$) |
| $C_{pf}$ | heat capacity of fluid (J/kg · K) |
| d, d' | drop diameter (m) |
| $d_i$, $d_j$, $d_k$ | nominal diameter of drops in size class i, j, and k, respectively (m) |
| $d_{10}$, $d_{16}$, $d_{50}$, $d_{84}$, $d_{90}$ | drop diameters defined by cumulative volume frequencies of 0.1, 0.16, 0.5, 0.84, and 0.9, respectively (e.g., 50% of the volume is contained in drops of size $d_{50}$ and smaller) (m) |

# NOMENCLATURE

| | |
|---|---|
| $d_{32}$ | Sauter mean drop diameter, general use (m) |
| $d_{32}(t)$ | instantaneous Sauter mean diameter (at time t) (m) |
| $d_{32}^{\infty}$ | equilibrium Sauter mean diameter (m) |
| $d_{32}(0)$ | Sauter mean diameter for $\phi \to 0$ (m) |
| $d_{32}(\phi)$ | Sauter mean diameter for finite $\phi$ (m) |
| $d_{43}$ | mass mean drop diameter (m) |
| $d_{max}$ | maximum stable drop diameter (m) |
| $d_n$ | number mean drop diameter (m) |
| $\bar{d}$ | average drop diameter in eqs. (12-8) and (12-9) (m) |
| $\Delta d_i$ | bin width for size class i in DSD (m) |
| $D$ | impeller diameter (m) |
| $D_{AB}$ | mass diffusivity, general use (m$^2$/s) |
| $D_C$ | diameter of cylinder (m) |
| $D_{crit}$ | critical drop deformation |
| $D_m$ | mass diffusivity in continuous phase (m$^2$/s) |
| $D_p$ | diameter of static mixer pipe (m) |
| $\dot{D}_d$ | death rate of drops of size d (s$^{-1}$) |
| $E(k)$ | energy spectral density function for eddies of wavenumber k |
| $\hat{E}_S$ | specific surface energy of a particle (J/kg-mol) |
| $f$ | friction factor |
| $f(h)$ | energy necessary to separate two adhering drops separated by distance h (J) |
| $f(\phi)$ | function of dispersed to continuous phase volume fraction ratio |
| $f(\mu_d/\mu_c)$ | function of dispersed to continuous phase viscosity ratio |
| $f_e(v)$ | escape frequency of drops of volume v from vessel |
| $f_n(d_i)$ | number frequency of drops in size class i |
| $f_v(d_i)$ | volume frequency of drops in size class i |
| $F$ | approach force acting on drop pairs (N) |
| $F_n(d_k)$ | cumulative number frequency up to drop size $d_k$ |
| $F_v(d_k)$ | cumulative volume frequency up to drop size $d_k$ |
| $g$ | gravitational acceleration (m/s$^2$) |
| $g_c$ | gravitational constant (m/s$^2$) |
| $g(d), g(v)$ | breakage frequency of drops of diameter d and volume v, respectively |
| $G$ | deformation rate (shear or extension) (s$^{-1}$) |
| $G_{crit}$ | critical deformation rate (shear or extension) (s$^{-1}$) |
| $h$ | film thickness/separation distance between colliding drops (m) |
| $h_o$ | initial value of h (m) |
| $h_c$ | critical film thickness for coalescence to occur (m) |
| $h_T$ | heat transfer coefficient (W/m$^2 \cdot$ K) |
| $H$ | height of liquid in vessel (m) |
| $k$ | wavenumber of eddy (m$^{-1}$) |

| | |
|---|---|
| $k_{cf}$ | thermal conductivity of fluid (W/m · K) |
| $k_m$ | mass transfer coefficient (m/s) |
| $k_v$ | ratio of tangential velocity at blade tip to impeller tip speed |
| $(kT)_b$ | product of Boltzmann constant and absolute temperature (N · m) |
| $L_d$ | length of deformed drop (m) |
| $L_p$ | length of a static mixer (m) |
| $L_T$ | turbulent macro length scale (m) |
| m | number of size classes representing drop size distribution |
| $\dot{m}_A$ | mass transfer rate to/from drop (kg/s) |
| $n(d), n(d')$ | number of drops of size d and d', respectively |
| $n_d$ | number of drops of size d |
| $n_i, n_j$ | number of drops in size class i and j, respectively |
| N | impeller speed (rps) |
| $N_{js}$ | minimum impeller speed to just suspended solid particles in vessel (rps) |
| $N_{min}$ | minimum impeller speed to suspend liquid drops in vessel (rps) |
| $N_T(t)$ | total number of drops in vessel at time t |
| $N_{T0}(t)$ | total number of drops in vessel feed stream at time t |
| $N_{TS}, N_0$ | steady-state values of $N_T(t)$ and $N_{T0}(t)$, respectively |
| $N^*, N^*_{eff}$ | impeller speeds defined by eq. (12-56); $N^*$ applies to a dilute dispersion and $N^*_{eff}$ to a more concentrated dispersion (rps) |
| P | power (W) |
| $P_n(d)$ | number probability density function for drop size d |
| $P_n(d, t)$ | number probability density function for drop size d at time t |
| $P_v(d)$ | volume probability density function for drop size d |
| $P_V(X)$ | volume probability density function for dimensionless drop size X |
| $P_x(d)$ | probability density function for drop size d, where x = n or x = v |
| $\Delta P$ | pressure drop in static mixer (Pa) |
| $Q_V$ | impeller volumetric flow rate (m³/s) |
| $\bar{r}$ | radius of particle undergoing Ostwald ripening (m) |
| R | radius of disk formed on flattened drop during collision with another drop (m) |
| t | time (s) |
| $t_c$ | contact time between two colliding drops (s) |
| $\bar{t}_{circ}$ | mean circulation time in tank (s) |
| $t_{eq}$ | time to reach equilibrium (s) |
| T | tank diameter (m) |
| $\overline{U}$ | velocity vector (m/s) |

| | |
|---|---|
| $v, v'$ | volume of drop (m³) |
| $v_{max}$ | volume of largest drop, (m³) |
| $v_\infty$ | slip velocity of spherical particle (m/s) |
| $\overline{v'(d)^2}$ | root mean square turbulent velocity difference across drop surface (m/s) |
| $V$ | volume of tank (m³) |
| $V_L, V_S$ | volume of large (L) and small (S) scale tanks during scale-up (m³) |
| $\hat{V}_A$ | molar volume of species A (m³/kg-mol) |
| $V'_s$ | superficial velocity in static mixer (m/s) |
| $W$ | width of an impeller blade (m) |
| $X = d/d_{32},$ $X(t) = d/d_{32}(t)$ | dimensionless or normalized drop diameter |
| $\overline{X}$ | mean value of X in DSD |

*Greek Symbols*

| | |
|---|---|
| $\alpha_1 \ldots \alpha_5$ | constants |
| $\beta(v, v')$ | frequency of daughter drops of volume v resulting from breakage of a parent drop of volume v' |
| $\beta'(d, d')$ | number of daughter drops of size d resulting from breakage of a parent drop of size d' |
| $\beta_K$ | Kolmogoroff constant = 1.5 |
| $\beta_S$ | crystal shape factor |
| $\Gamma(d, d')$ | coalescence frequency between drops of diameter d and d' |
| $\Gamma(v, v')$ | coalescence frequency between drops of volume v and v' |
| $\varepsilon$ | local energy dissipation rate per mass of fluid (W/kg) |
| $\varepsilon_{avg}$ | average energy dissipation rate per mass of fluid or power draw per mass (W/kg) |
| $\varepsilon_{max}$ | maximum energy dissipation rate per mass of fluid (W/kg) |
| $\eta$ | Kolmogoroff microscale of turbulence (m) |
| $\theta = Nt$ | dimensionless time in vessel |
| $\lambda(d, d')$ | coalescence efficiency between drops of diameter d and d' |
| $\lambda(v, v')$ | coalescence efficiency between drops of volume v and v' |
| $\mu_c$ | viscosity of continuous phase (Pa · s) |
| $\mu_d$ | viscosity of dispersed phase (Pa · s) |
| $\mu_f$ | viscosity of fluid (Pa · s) |
| $\overline{\mu}$ | bulk viscosity of liquid–liquid mixture (Pa · s) |
| $\nu_c$ | kinematic viscosity of continuous phase, $\mu_c/\rho_c$ (m²/s) |
| $\xi(d, d')$ | collision frequency between drops of diameter d and d' |
| $\xi(v, v')$ | collision frequency between drops of volume v and v' |

| | |
|---|---|
| $\rho_c$ | density of continuous phase (kg/m³) |
| $\rho_d$ | density of dispersed phase (kg/m³) |
| $\rho_f$ | density of fluid (kg/m³) |
| $\bar{\rho}$ | bulk density of liquid–liquid mixture (kg/m³) |
| $\Delta\rho = |\rho_d - \rho_c|$ | density difference between phases (kg/m³) |
| $\sigma$ | interfacial tension (N/m) |
| $\sigma_{SD}$ | standard deviation, general (m) |
| $\sigma_V$ | volume standard deviation of normalized DSD |
| $\tau$ | time for the film between two coalescing drops to drain to a critical thickness (s) |
| $\tau_c$ | turbulent stress (force per area) acting on surface of drop (N/m²) |
| $\tau_d$ | internal viscous stress (force per area) resisting drop deformation (N/m²) |
| $\tau_s$ | stress (force per area) due to interfacial tension resisting drop deformation (N/m²) |
| $\upsilon(v), \upsilon(v')$ | number of daughter drops formed upon breakage of a parent drop of volume v and v', respectively |
| $\phi$ | volume fraction of dispersed phase |
| $\Omega = \dfrac{d_{32}(t) - d_{32}^\infty}{d_{32}^\infty}$ | dimensionless instantaneous Sauter mean diameter |

*Dimensionless Groups*

| | |
|---|---|
| Ar | Archimedes number, $g\rho_c\Delta\rho\, D^3/\mu_c^2$ |
| Ca | capillary number, $\mu_c Ga/\sigma$ |
| Fr | Froude number for stirred vessel, $N^2 d/g$ |
| $N_p$ | power number, $P/\rho_c N^3 D^5$ |
| $N_q$ | flow number, $Q_V/ND^3$ |
| $Pr_f$ | Prandtl number, $C_{pf}\mu_f/k_{cf}$ |
| Re | Reynolds number for stirred vessel (impeller), $\bar{\rho}ND^2/\bar{\mu}$ |
| $Re_\infty$ | Reynolds number for spherical particle, $\rho_f v_\infty d/\mu_f$ |
| $Sc_f$ | Schmidt number, $\mu_f/\rho_f D_{AB}$ |
| Su | Suratman number for stirred vessel, $\rho_c \sigma D/\mu_c^2$ |
| Vi | viscosity group for stirred vessel, $(\rho_c/\rho_d)^{1/2}\mu_d ND/\sigma$; for static mixer, $(\rho_c/\rho_d)^{1/2}\mu_d V'_s/\sigma$ |
| We | Weber number for stirred vessel, $\rho_c N^2 D^3/\sigma$; for static mixer, $\rho_c V'^2_s D_p/\sigma$ |

# REFERENCES

Aiba, S. (1958). Flow patterns of liquids in agitated vessels, *AIChE J.*, **4**, 485.

Al Taweel, A. M., and L. D Walker (1997). Dynamics of drop breakup in turbulent flow, presented at Mixing XIV, 14th Biennial North American Mixing Conference, Williamsburg, VA, June.

## REFERENCES

Allan, R. S., and S. G. Mason (1962). Particle motions in sheared suspensions: XIV. Coalescence of liquid drops in electric and shear fields, *J. Colloid Sci.*, **17**, 383–408.

Alvarez, J., J. Alvarez, and M. Hernandez (1994). A population balance approach for the description of particle size distribution in suspension polymerization reactors, *Chem. Eng. Sci.*, **49**, 99–113.

Arai, K., N. Konno, Y. Matunaga, and S. Saito (1977). The effect of dispersed phase viscosity on the maximum stable drop size for breakup in turbulent flow, *J. Chem. Eng. Jpn.*, **10**, 325–330.

Armenante, P. M., and Y.-T. Huang (1992). Experimental determination of the minimum agitation speed for complete liquid–liquid dispersion in mechanically agitated vessels, *Ind. Eng. Chem. Res.*, **31**, 1398–1406.

Armenante, P. M., and D. Tsai (1988). Agitation requirements for complete dispersion of emulsions, presented at the AIChE Annual Meeting, Washington, DC, Nov.

Armenante, P. M., Y.-T. Huang, and T. Li (1992). Determination of the minimum agitation speed to attain the just dispersed state in solid–liquid and liquid–liquid reactors provided with multiple impellers, *Chem. Eng. Sci.*, **47**, 2865–2870.

Bajpai, R. K., D. Ramkrishna, and A. Prokop (1976). Coalescence redispersion model for drop-size distributions in an agitated vessel, *Chem. Eng. Sci.*, **31**(10), 913–920.

Bakker, A. J., K. J. Myers, R. W. Ward, and C. K. Lee (1996). The laminar and turbulent flow pattern of a pitched blade turbine, *Trans. Inst. Chem. Eng.*, **74A**, 485–491.

Baldyga, J., and J. R. Bourne (1992). Some consequences for turbulent mixing of fine scale intermittency, *Chem. Eng. Sci.*, **47**, 3943–3958.

Baldyga, J., and J. R. Bourne (1993). Drop breakup in the viscous subrange: a source of possible confusion, *Chem. Eng. Sci.*, **49**, 1077–1078.

Ballast, D. E., S. I. Bates, and R. M. Wiley (1961). Formation of solid beads by congelation of suspended liquid drops, U.S. patent 2,968,066, The Dow Chemical Company.

Bapat, P. M., L. L. Tavlarides, and G. W. Smith (1983). Monte Carlo simulation of mass transfer in liquid–liquid dispersions, *Chem. Eng. Sci.*, **38**(12), 2003–2013.

Bentley, B. J., and L. G. Leal (1986). An experimental investigation of drop deformation and breakup in steady, two-dimensional linear flows, *J. Fluid Mech.*, **167**, 241–283.

Berkman, P. D., and R. V. Calabrese (1988). Dispersion of viscous liquids by turbulent flow in a static mixer, *AIChE J.*, **34**(4), 602–609.

Blount, J. M. (1995). Mechanisms of drop breakage in dilute, agitated liquid–liquid systems, M.S. thesis, University of Maryland, College Park, MD.

Brooks, B. W., and H. N. Richmond (1991). Dynamics of liquid–liquid phase inversion using non-ionic surfactants, *Colloids Surfaces*, **58**, 131–148.

Brooks, B. W., and H. N. Richmond (1994a). Phase inversion in non-ionic surfactant–oil–water systems: I. The effect of transitional inversion on emulsion drop sizes, *Chem. Eng. Sci.*, **49**, 1053–1064.

Brooks, B. W., and H. N. Richmond (1994b). Phase inversion in non-ionic surfactant–oil–water systems: II. Drop size studies in catastrophic inversion with turbulent mixing, *Chem. Eng. Sci.*, **49**, 1065–1075.

Brooks, B. W., and H. N. Richmond (1994c). Phase inversion in non-ionic surfactant–oil–water systems: III. The effect of oil phase viscosity on catastrophic inversion and the relationship between the drop size present before and after catastrophic inversion, *Chem. Eng. Sci.*, **49**, 1843–1853.

Brown, D. E., and K. Pitt (1970). Drop breakup in a stirred liquid–liquid contactor, *Proc. Chemeca '70*, Melbourne and Sydney, Australia.

Brown, D. E., and K. Pitt (1974). Effect of impeller geometry on drop break-up in a stirred liquid contactor, *Chem. Eng. Sci.*, **29**, 345–348.

Calabrese, R. V., T. P. K. Chang, and P. T. Dang (1986a). Drop breakup in turbulent stirred-tank contactors: I. Effect of dispersed phase viscosity, *AIChE J.*, **32**(4), 657–666.

Calabrese, R. V., C. Y. Wang, and N. P. Bryner (1986b). Drop breakup in turbulent stirred-tank contactors: III. Correlations for mean size and drop size distribution, *AIChE J.*, **32**(4), 677–681.

Calabrese, R. V., M. H. Wang, N. Zhang, and J. W. Gentry (1992). Simulation and analysis of particle breakage phenomena, *Trans. Inst. Chem. Eng.*, **70A**, 189–191.

Calabrese, R. V., A. W. Pacek, and A. W. Nienow (1993). Coalescence of viscous drops in a stirred dispersion, *Proc. 1993 Inst. Chem. Eng. Research Event*, January, pp. 651–653.

Calabrese, R. V., M. K. Francis, V. P. Mishra, and S. Phongikaroon (2000). Measurement and analysis of drop size in a batch rotor–stator mixer, *Proc. 10th European Conference on Mixing*, Delft, The Netherlands, Elsevier Science, Amsterdam, pp. 149–156.

Calderbank, P. H. (1958). Physical rate processes in industrial fermentation: I, *Trans. Inst. Chem. Eng.*, **36A**, 443–463.

Chang, K. C. (1990). Analysis of transient drop size distributions in dilute agitated liquid–liquid systems, Ph.D. dissertation, University of Maryland, College Park, MD.

Chang, Y. C., R. V. Calabrese, and J. W. Gentry (1991). An algorithm for determination of size-dependent breakage frequency of droplets, flocs and aggregates, *Part. Part. Syst. Charact.*, **8**, 315–322.

Charles, G. E., and S. G. Mason (1960). The mechanism of partial coalescence of drops at a liquid–liquid interface, *J. Colloid. Sci.*, **15**, 105–122.

Chatzi, E. G., and D. Kiparissides (1994). Drop size distributions in high hold-up fraction dispersed systems: effect of the degree of hydrolysis of PVA stabilizer, *Chem. Eng. Sci.*, **49**(24B), 5039–5052.

Chatzi, E. G., A. D. Gavrielides, and C. Kiparissides (1989). Generalized model for prediction of the steady-state drop size distributions in batch stirred vessels., *Ind. Eng. Chem. Res.*, **28**, 1704–1711.

Chen, H. T., and S. Middleman (1967). Drop size distribution in agitated liquid–liquid systems, *AIChE J.*, **13**(5), 989–995.

Chesters, A. K. (1991). The modelling of coalescence processes in fluid–liquid dispersions, *Trans. Inst. Chem. Eng.*, **69A**, 259–270.

Church, J. M., and R. Shinnar (1961). On the behavior of liquid dispersions in mixing vessels, *Ind. Eng. Chem.*, **53**, 479–484.

Clark, M. M. (1988). Drop breakup in turbulent flow: I. Conceptual and modeling considerations, *Chem. Eng. Sci.*, **43**, 671–679.

Corrsin, S. (1964). The isotropic turbulent mixer: II. Arbitrary Schmidt number, *AIChE J.*, **10**, 870–877.

Coulaloglou, C. A., and L. L. Tavlarides (1976). Drop size distribution and coalescence frequencies of liquid–liquid dispersions in flow vessels, *AIChE J.*, **22**, 289–297.

Coulaloglou, C. A., and L. L. Tavlarides (1977). Description of interaction processes in agitated liquid–liquid dispersions, *Chem. Eng. Sci.*, **32**, 1289–1297.

Das, P. K., R. Kumar, and D. Ramkrishna (1987). Coalescence of drops in stirred dispersion: a white noise model for coalescence, *Chem. Eng. Sci.*, **42**, 213–220.

Davies, J. T. (1987). A physical interpretation of drop sizes in homogenizers and agitated tanks, including the dispersion of viscous oils, *Chem. Eng. Sci.*, **42**, 1671–1676.

Davis, R. H., J. A. Schonberg, and J. M. Rollison (1989). The lubrication force between two viscous drops, *Phys. Fluids A*, **1**, 77–81.

Doulah, M. S. (1975). An effect of hold-up on drop sizes in liquid–liquid dispersion, *Ind. Eng. Chem. Fundam.*, **14**, 137–138.

Gillespie, T., and E. Rideal (1956). The coalescence of drops at an oil–water interface, *Trans. Faraday Soc.*, **52**, 173–183.

Godfrey, J. C., and V. Grilc (1977). Drop size and drop size distribution for liquid–liquid dispersions in agitated tanks of square cross-section, *Proc. 2nd European Conference on Mixing*, Cambridge.

Grace, H. P. (1982). Dispersion phenomena in high viscosity immiscible fluid systems and application of static mixers as dispersion devices in such systems, *Chem. Eng. Commun.*, **14**, 225–277.

Hartland, S. (1990). Coalescence in close packed gas–liquid and liquid–liquid dispersions, *Ber. Bunsenges. Phys. Chem.*, **85**(10), 851–863.

Hartland, S., and S. A. K. Jeelani (1987). Drainage in thin planar non-Newtonian films, *Can. J. Chem. Eng.*, **65**(3), 382–390.

Hinze, J. O. (1955). Fundamentals of the hydrodynamic mechanism of splitting in dispersion processes, *AIChE J.*, **1**, 289–295.

Holmes, D. B., R. M. Voncken, and J. A. Dekker (1964). Fluid flow in turbine stirred, baffled tanks: I. Circulation time, *Chem. Eng. Sci.* **19**, 201–208.

Hong, P. O., and J. M. Lee (1983). Unsteady-state liquid–liquid dispersions in agitated vessels, *Ind. Eng. Chem. Process. Des. Dev.*, **22**, 130–135.

Hong, P. O., and J. M. Lee (1985). Changes of average drop sizes during initial period of liquid–liquid dispersions in agitated vessels, *Ind. Eng. Chem. Process. Des. Dev.*, **24**, 868–872.

Howarth, W. J. (1967). Measurement of coalescence frequency in an agitated tank, *AIChE J.*, **13**, 1007–1013.

Jeffreys, G. V., and J. L. Hawksley (1965). Coalescence of liquid droplets in two-component–two-phase systems: I. Effect of physical properties on the rate of coalescence, *AIChE J.*, **11**, 413–417.

Jeon, Y., and W. K. Lee (1986). Drop population model for mass transfer in liquid–liquid dispersion: I. Simulation and its results, *Ind. Eng. Chem. Fundam.*, **25**(2), 293–300.

Karam, H. J., and J. C. Bellinger (1968). Deformation and breakup for liquid droplets in a simple shear field, *Ind. Eng. Chem. Fundam.*, **1**, 576–581.

Kinugasa, T., K. Watanabe, T. Sonove, and H. Takeuchi (1997). Phase inversion of stirred liquid–liquid dispersions, presented at the International Symposium on Liquid–Liquid Two Phase Flow and Transport Phenomena, Session 13, Antalya, Turkey.

Kolmogoroff, A. N. (1941a). The local structure of turbulence in incompressible viscous fluid for very large Reynolds numbers, *Compt. Rend. Acad. Sci. USSR*, **30**, 301–305.

Kolmogoroff, A. N. (1941b). Dissipation of energy in locally isotropic turbulence, *Compt. Rend. Acad. Sci. USSR*, **32**, 16–18.

Kolmogoroff, A. N. (1949). The breakup of droplets in a turbulent stream, *Dokl. Akad. Nauk.*, **66**, 825–828.

Konno, M., A. Aoki, and S. Saito (1983). Scale effect on breakup process in liquid–liquid agitated tanks, *J. Chem. Eng. Jpn.*, **16**, 312–319.

Konno, M., K. Kosaka, and S. Saito (1993). Correlation of transient drop sizes in breakup process in liquid–liquid agitation, *J. Chem. Eng. Jpn.*, **26**(1), 37–40.

Koshy, A., T. R. Das, and R. Kumar (1988a). Effect of surfactants on drop breakage in turbulent liquid dispersions, *Chem. Eng. Sci.*, **43**, 649–654.

Koshy, A., T. R. Das, R. Kumar, and K. S. Ghandi (1988b). Breakage of viscoelastic drops in turbulent stirred dispersions, *Chem. Eng. Sci.*, **43**, 2625–2631.

Koshy, A., R. Kumar, and K. S. Gandhi (1989). Effect of drag reducing agents on drop breakage in stirred dispersions, *Chem. Eng. Sci.*, **44**, 2113–2120.

Lagisetty, J. S., P. K. Das, R. Kumar, and K. S. Ghandi (1986). Breakage of viscous and non-Newtonian drops in stirred dispersions, *Chem. Eng. Sci.*, **41**, 65–72.

Lam, A., A. N. Sathyagal, S. Kumar, and D. Ramkrishna (1996). Maximum stable drop diameter in stirred dispersions, *AIChE J.*, **42**, 1547–1552.

Lang, S. B., and C. R. Wilke (1971a). A hydrodynamic mechanism for the coalescence of liquid drops: I. Theory of coalescence at a planar interface, *Ind. Eng. Chem. Fundam.*, **10**, 329–340.

Lang, S. B., and C. R. Wilke (1971b). A hydrodynamic mechanism for the coalescence of liquid drops: II. Experimental studies, *Ind. Eng. Chem. Fundam.*, **10**, 341–352.

Lasheras, J. C., C. Eastwood, C. Martinez-Bazan, and J. L. Montanes (2002). A review of statistical models for the break-up of an immiscible fluid immersed into a fully developed turbulent flow, *Int. J. Multiphase Flow*, **28**, 247–278.

Laso, M., L. Steiner, and S. Hartland (1987a). Dynamic simulation of liquid–liquid agitated dispersions: I. Derivation of a simplified model, *Chem. Eng. Sci.*, **42**, 2429–2436.

Laso, M., L. Steiner, and S. Hartland (1987b). Dynamic simulation of liquid–liquid agitated dispersions: II. Experimental determination of the breakage and coalescence rates in a stirred tank, *Chem. Eng. Sci.*, **42**, 2437–2445.

Leng, D. E., and G. J. Quarderer (1982). Drop dispersion in suspension polymerization, *Chem. Eng. Commun.*, **14**, 177–201.

Leng, D. E., W. L. Sigelko, and F. L. Saunders (1985). Aqueous dispersions of plasticized polymer particles, U.S. patent 4,502,888, The Dow Chemical Company.

Lyu, S. P., F. S. Bates, and C. W. Macosko (2002). Modeling of coalescence in polymer blends, *AIChE J.*, **48**(1), 7–14.

MacKay, G. D. M., and S. G. Mason (1963). The gravity approach and coalescence of fluid drops at liquid interfaces, *Can. J. Chem. Eng.*, **41**, 203–212.

Marks, C. R. (1998). Drop breakup and deformation in sudden onset strong flows, Ph.D. dissertation, University of Maryland, College Park, MD.

McManamey, W. J. (1979). Sauter mean and maximum drop diameters of liquid–liquid dispersions in turbulent agitated vessels at low dispersed phase hold-up, *Chem. Eng. Sci.*, **34**, 432–434.

Middleman, S. (1974). Drop size distributions produced by turbulent pipe flow of immiscible fluids through a static mixer, *Ind. Eng. Chem. Process. Des. Dev.*, **13**, 78–83.

Middleton, J. C. (1979). Measurement of circulation within large mixing vessels, *Proc. 3rd European Conference on Mixing*, York, Yorkshire, England, Vol. 1, pp. 15–36.

Mlynek, T., and W. Resnick (1972). Drop size in an agitated liquid–liquid system, *AIChE J.*, **18**, 122–127.

Muralidhar, R., and D. Ramkrishna (1986). Analysis of droplet coalescence in turbulent liquid–liquid dispersions, *Ind. Eng. Chem. Fundam.*, **25**, 554–560.

Muralidhar, R., and D. Ramkrishna (1988). Coalescence phenomena in stirred liquid–liquid dispersions, *Proc. 6th European Conference on Mixing*, Pavia, Italy.

Muralidhar, R., D. Ramkrishna, P. K. Das, and R. Kumar (1988). Coalescence of rigid droplets in a stirred dispersion: II. Band-limited force fluctuations, *Chem. Eng. Sci.*, **43**, 1559–1586.

Murdoch, P. G., and D. E. Leng (1971). The mathematical formulation of hydrodynamic film thinning and its application to colliding drops suspended in a second liquid: II, *Chem. Eng. Sci.*, **26**, 1881–1892.

Nagata, S. (1975). *Mixing Principles and Applications*, Halstead Press, Wiley, New York.

Narsimhan, G., D. Ramkrishna, and J. P. Gupta (1980). Analysis of drop size distributions in lean liquid–liquid dispersions, *AIChE J.*, **26**, 991–1000.

Narsimhan, G., G. Nejfelt, and D. Ramkrishna (1984). Breakage functions of droplets in agitated liquid–liquid dispersions, *AIChE J.*, **30**, 457–467.

Nishikawa, M., F. Mori, and S. Fujieda (1987a). Average drop size in a liquid–liquid phase mixing vessel, *J. Chem. Eng. Jpn.*, **20**, 82–88.

Nishikawa, M., F. Mori, S. Fujieda, and T. Kayama (1987b). Scale-up of liquid–liquid phase mixing vessel, *J. Chem. Eng. Jpn.*, **20**, 454–459.

Nyvlt, J., O. Sohnel, M. Matuchova, and M. Bruol (1985). *The Kinetics of Industrial Crystallization*, Elsevier, Amsterdam.

Pacek, A. W., I. P. T. Moore, R. V. Calabrese, and A. W. Nienow (1993). Evolution of drop size distributions and average drop diameters in liquid–liquid dispersions before and after phase inversion, *Trans. Inst. Chem. Eng.*, **71A**, 340–341.

Pacek, A., I. P. T. Moore, A. W. Nienow and R. V. Calabrese (1994a). A video technique for the measurement of the dynamics of liquid–liquid dispersions during phase inversion, *AIChE J.* **40**, 1940–1949.

Pacek, A. W., A. W. Nienow, and I. P. T. Moore (1994b). On the structure of turbulent liquid–liquid dispersed flows in an agitated vessel, *Chem. Eng. Sci.*, **49**(20), 3485–3498.

Pacek, A. W., S. Chamsart, A. W. Nienow, and A. Baker (1999). The influence of impeller type on mean drop size and drop size distribution in a agitated vessel, *Chem. Eng. Sci.*, **54**, 4211–4222.

Pavlushenko, I. S., and A. V. Yanishevskii (1958). Effective number of revolutions of a stirrer for the dispersion of two mutually immiscible liquids, *Zhur. Priklad. Khim.*, **31**, 1348–1354.

Phongikaroon, S. (2001). Drop size distribution for liquid–liquid dispersions produced by rotor–stator mixers, Ph.D. dissertation, University of Maryland, College Park, MD.

Rajamani, K., W. T. Pate, and D. J. Kinneberg (1986). Time-driven and event-driven Monte Carlo simulations of liquid–liquid dispersions: a comparison, *Ind. Eng. Chem. Fundam.*, **25**(4), 746–752.

Rallison, J. M. (1984). The deformation of small viscous drops and bubbles in shear flows, *Annu. Rev. Fluid Mech.*, **16**, 45–66.

Ramkrishna, D. (2001). *Population Balances*, Wiley, New York.

Rod, V., and T. Misek (1982). Stochastic modeling of dispersion formation in agitated liquid–liquid systems, *Trans. Inst. Chem. Eng.*, **60**(1), 48–53.

Rodger, W. A., V. G. Trice, and J. H. Rushton (1956). Effect of fluid motion on interfacial area of dispersions, *Chem. Eng. Prog.*, **52**, 515–520.

Sathyagal, A. N., D. Ramkrishna, and G. Narsimhan (1995). Solution of inverse problems in population balances: II. Particle break-up, *Comput. Chem. Eng.*, **19**(4), 437–451.

Sathyagal, A. N., D. Ramkrishna, and G. Narsimhan (1996). Droplet breakage in stirred dispersions: breakage functions from experimental drop-size distributions, *Chem. Eng. Sci.*, **51**, 1377–1391.

Scheele, G. F., and D. E. Leng (1971). An experimental study of factors which promote coalescence of two colliding drops suspended in water: I, *Chem. Eng. Sci.*, **26**, 1867–1879.

Shinnar, R. (1961). On the behaviour of liquid dispersions in mixing vessels, *J. Fluid Mech.*, **10**, 259–275.

Skelland, A. H. P., and L. T. Moeti (1990). Mechanisms of continuous-phase mass transfer in agitated liquid–liquid systems, *Ind. Eng. Chem. Res.*, **29**, 2258–2267.

Skelland, A. H. P., and R. Seksaria (1978). Minimum impeller speeds for liquid–liquid dispersion in baffled vessels, *Ind. Eng. Chem. Process. Des. Dev.*, **17**, 56–61.

Skelland, A. H. P., and H. Xien (1990). Dispersed-phase mass transfer in agitated liquid–liquid systems, *Ind. Eng. Chem. Res.*, **29**, 415–420.

Sprow, F. B. (1967a). Distribution of drop sizes produced in turbulent liquid–liquid dispersion, *Chem. Eng. Sci.*, **22**, 435–442.

Sprow, F. B. (1967b). Drop size distributions in strongly coalescing liquid–liquid systems, *AIChE J.*, **13**, 995–998.

Stone, H. A. (1994). Dynamics of drop deformation and breakup in viscous fluids, *Annu. Rev. Fluid Mech.*, **26**, 65–102.

Stone, H. A., B. J. Bentley, and L. G. Leal (1986). An experimental study of transient effects in the breakup of viscous drops, *J. Fluid Mech.*, **173**, 131–158.

Tavlarides, L. L. (1981). Modelling and scale-up of dispersed phase liquid–liquid reactors, *Chem. Eng. Commun.*, **8**, 133–164.

Tavlarides, L. L., and P. M. Bapat (1983). Models for the scale-up of dispersed phase liquid–liquid reactors, *AIChE Symp. Ser.*, **80**, 12–46.

Tavlarides, L. L., and M. Stamatoudis (1981). The analysis of interphase reaction and mass transfer in liquid–liquid dispersions, *Adv. Chem. Eng.*, **11**, 199–273.

Taylor, G. I. (1934). The formation of emulsions in definable fields of flow, *Proc. R. Soc.*, **A146**, 501–523.

Tjahjadi, M., and J. M. Ottino (1991). Stretching and breakup of drops in chaotic flows, *J. Fluid Mech.*, **232**, 191–219.

Tjahjadi, M., H. A. Stone, and J. M. Ottino (1992). Satellite and subsatellite formation in capillary breakup, *J. Fluid Mech.*, **243**, 297–317.

Tobin, T., and D. Ramkrishna (1992). Coalescence of charged droplets in agitated liquid–liquid dispersions, *AIChE J.*, **38**(8), 1199–1205.

Todtenhaupt, P., E. Todtenhaupt, and W. Muller (1991). Handbook of Mixing Technology, Ekato Ruhr und Mischtechnik, Schopfheim, Germany.

Tsouris, C., and L. L. Tavlarides (1994). Breakage and coalescence models for drops in turbulent dispersions, *AIChE J.*, **40**, 395–406.

Valentas, K. J., and N. R. Amundson (1966). Breakage and coalescence in dispersed phase systems, *Ind. Eng. Chem. Fundam.*, **5**, 533–542.

Valentas, K. J., O. Bilous, and N. R. Amundson (1966). Analysis of breakage in dispersed phase systems, *Ind. Eng. Chem. Fundam.*, **5**, 271–279.

Van Heuven, J. W., and W. J. Beek (1971). Power input, drop size and minimum stirrer speed for liquid-liquid dispersions in stirred vessels, *Proc. International Solvent Extraction Conference*, The Hague, The Netherlands, **1**, 70–81.

Vermeulen, T., G. M. Williams, and G. E. Langlois (1955). Interfacial area in liquid–liquid and gas–liquid agitation, *Chem. Eng. Prog.*, **51**, 85F–95F.

Vinckier, I., P. Moldenaers, A. M. Terracciano, and N. Grizzuti (1998). Drop size evolution during coalescence in semi-concentrated model blends, *AIChEJ*, **44**, 951–958.

Vivaldo-Lima, E., P. E. Wood, and A. E. Hamielec (1997). An updated review on suspension polymerization, *Ind. Eng. Chem. Res*, **36**, 939–965.

Vrij, A. (1966). Possible mechanism for the spontaneous rupture of thin free liquid films, *Disc. Faraday Soc.*, **42**, 3505–3516.

Wang, C. Y., and R. V. Calabrese (1986). Drop breakup in turbulent stirred-tank contactors: II. Relative influence of viscosity and interfacial tension, *AIChE J.*, **32**(4), 667–676.

Warner, G. L., and D. E. Leng (1978). Continuous process for preparing aqueous polymer microdispersions. U.S. patent 4,123,403, The Dow Chemical Company.

Weinstein, B., and R. E. Treybal (1973). Liquid–liquid contacting in unbaffled agitated vessels, *AIChE J.*, **19**, 304–312.

Wright, H., and D. Ramkrishna (1994). Factors affecting coalescence of droplets in a stirred liquid–liquid dispersion, *AIChE J.*, **40**(5), 767–776.

Yarranton, H. W., and J. H. Masliyah (1997). Numerical simulation of Ostwald ripening in emulsions, *J. Colloid Interface Sci.*, **196**, 157–169.

Zhou, G., and S. M. Kresta (1998a). Correlation of mean drop size with the turbulence energy dissipation and the flow in an agitated tank, *Chem. Eng. Sci.*, **53**, 2063–2079.

Zhou, G., and S. M. Kresta (1998b). Evolution of drop size distribution in liquid–liquid dispersions for various impellers, *Chem. Eng. Sci.*, **53**, 2099–2113.

Zwietering, T. N. (1958). Suspending of solid particles in liquid by agitators, *Chem. Eng. Sci.*, **8**, 244–253.

# CHAPTER 13

# Mixing and Chemical Reactions

GARY K. PATTERSON
University of Missouri–Rolla

EDWARD L. PAUL
Merck and Co., Inc.

SUZANNE M. KRESTA
University of Alberta

ARTHUR W. ETCHELLS III
The DuPont Company (retired)

## 13-1 INTRODUCTION

Mixing and chemical reaction are intimately entwined. The method of bringing together reactants that are to undergo reaction can have a significant impact on the course of the reaction. If the reaction can result in only one product, the mixing and mass transfer can influence only the reaction rate. If more than one product is possible, contacting can influence the product distribution as well. These considerations apply to both homogeneous and heterogeneous reaction systems. This issue was identified qualitatively by Danckwerts (1958) and Levenspiel (1962) and demonstrated experimentally by Paul and Treybal (1971). An early theoretical paper by Corrsin (1964) established the framework for modeling turbulent mixing in chemical reactors. Brodkey and co-workers (McKelvey et al., 1975) achieved experimental verification of the Corrsin theory. This topic was then expanded with development of test reaction systems and modeling by Bourne and co-workers as summarized in the comprehensive treatise by Baldyga and Bourne (1999). Many workers in this field have made valuable contributions, not all of which can be discussed in this chapter.

*Handbook of Industrial Mixing: Science and Practice*, Edited by Edward L. Paul,
Victor A. Atiemo-Obeng, and Suzanne M. Kresta
ISBN 0-471-26919-0 Copyright © 2004 John Wiley & Sons, Inc.

**756**  MIXING AND CHEMICAL REACTIONS

In this chapter we address the key conditions that determine whether mixing is important. The main objectives of the chapter are to answer the following questions:

- When are mixing effects important?
- What are the criteria for quantifying mixing and reaction?
- What mixing design will optimize yield and selectivity?

Our current understanding of these issues is discussed in the context of industrial applications.

To determine what conditions are required for mixing processes to affect reaction processes, we will use a number of concepts. *Most important is the comparison of time constants of the various processes.* The processes of interest are blending, mixing, mass transfer between phases, and chemical reaction. Some typical time constants are the blend time and reaction half-life. For simple exponential processes (first-order reactions), rates and characteristic times, such as reaction half-lifes, are related. The first-order rate equation is

$$\frac{dC_A}{dt} = -k_R C_A \tag{13-1}$$

where $k_R$ is the reaction rate constant and $1/k_R$ is a characteristic reaction time for a first-order reaction. We can also find the reaction half-life by integrating the rate expression to give

$$\frac{C_A}{C_{A_o}} = e^{-k_R t} \tag{13-2}$$

giving the time when $C_A$ has dropped to half of $C_{A_o}$:

$$t_{1/2} = \frac{-\ln(0.5)}{k_R} \tag{13-3}$$

Even for more complicated reactions, the linear half-life expression is a good approximation for short times. Second-order reactions have a characteristic time of $1/kC$, and a general time constant for higher-order reactions can be defined: $1/kC^{n-1}$. The concepts of rate and characteristic time are used interchangeably throughout the chapter.

Mixing effects in chemical reactions are complicated in that the easily formulated global time constants, such as blend time, are not the ones of interest, but rather, time constants based on local conditions in the reactor, such as local mixing time or local mass transfer rate. When the rates of reaction, mixing, and mass transfer approach one another, mixing will affect the outcome of the process. At the lab scale, mixing effects change the apparent kinetics of the reaction so that the measured kinetics are limited by the rate of mixing rather than by the rate of reaction.

## 13-1.1 How Mixing Can Cause Problems

Consider two beakers of reactive reagents. They are low viscosity miscible liquids that will react when combined. However, no reaction will take place until the liquids are brought into intimate contact by being mixed on the smallest scales. Thus, the processes of mixing and chemical reaction are linked; they operate in series initially, then in parallel. Now consider the case where the chemical reaction is slow, with a half-life of several minutes. If the mixing takes place quickly, say within seconds, the mixing is essentially finished before significant chemical reaction takes place. There is no effect of mixing on the slow chemical reaction, and the ideal mixed batch reactor analysis may be used. Now consider a very fast chemical reaction: for example, an acid–base neutralization with a half-life of 0.001 s. If the mixing again takes place in seconds, as before, the rate of the chemical reaction depends on the rate of the mixing, which is much slower. If the reaction rate were measured, the result would be the mixing rate, not the molecular chemical reaction rate. The result is the "apparent" reaction rate.

The rate of mixing for fast reactions can often be mistaken for the rate of chemical reaction. Tests in which the rate of mixing is varied (say, by varying mixer speed) must be used to determine the true reaction kinetics. The information presented in this chapter is aimed at solving the problem of fast chemical reactions where the mixing rate and the reaction rates are intertwined. When reactions are fast relative to the mixing rate, not only are the reaction rates affected but the entire time and temperature history of the reactions is affected, yielding different selectivities and yields depending on the intensity of the mixing. This leads to the scale-up problem, where yields of desirable products in a plant scale reactor are not as good as in the small scale reactor in the laboratory or the pilot plant. If the yield is poorer in the plant scale reactor, there is a mixing problem, assuming that other important variables are held constant, such as temperature, pressure, and composition.

The competition between reaction and mixing is well represented by a mixing Damkoehler number, $Da_M$, which is the ratio between the reaction rate and the local mixing rate, or conversely, the ratio of the characteristic mixing time, $\tau_M$, and the reaction time, $\tau_R$:

$$Da_M = \frac{\tau_M}{\tau_R} \qquad (13\text{-}4)$$

A smaller $Da_M$ indicates less effect of mixing; a larger $Da_M$ indicates that mixing will be a concern. Estimates of mixing rates and mass transfer rates can be made from existing information for several reactor configurations for both homogeneous and heterogeneous reactions. These estimates combined with an estimate of the magnitude of the reaction rate can give a rough but useful approximation of the conditions under which mixing effects may be critical to the course of a reaction system and in scale-up. This chapter is focused on the determination of those conditions. Several examples are included to illustrate reactor design problems and solutions for the major types of reacting systems. Example 13.1 shows how mixing affects selectivity in various reactor configurations.

**Example 13-1: How Mixing Conditions Affect Selectivity to a Desired Product.** A comparison of the selectivity of a competitive-consecutive chemical reaction under various mixing conditions is made (see Section 13-1.4 for a definition of selectivity). The chemical reactions are as follows:

$$A + B \rightarrow R \quad k_{R1} = 35 \text{ m}^3/\text{kmol} \cdot \text{s} \quad C_{Ao} = 0.2 \text{ kmol/m}^3$$

$$R + B \rightarrow S \quad k_{R2} = 3.8 \text{ m}^3/\text{kmol} \cdot \text{s} \quad C_{Bo} = 0.2 \text{ kmol/m}^3$$

R and S are not present in the feed solutions. The solutions containing A and B must mix in order for the reaction to proceed. The reactions are allowed to go to completion to obtain the final yield of R. Table 13-1 shows the selectivity for various reaction conditions. The nonideal results are taken from simulations discussed in Section 13-5.

From this example of a fast, competitive consecutive reaction scheme we can see that nonideal mixing can cause a decrease in selectivity in both continuous and semibatch reactors. Residence time distribution issues can cause a reduction in yield and selectivity for both slow and fast reactions (see Chapter 1), but for fast reactions, the decrease in selectivity and yield due to inefficient local mixing can be greater than that caused by RTD issues alone. In semibatch reactors, poor bulk mixing can also cause these reductions (see Example 13-3).

### 13-1.2 Reaction Schemes of Interest

Mixing effects on product distribution are of importance in multiple reactions because the impact on design and economics can be profound. In such reactions the desired product is one of two or more possible products. Economics are directly affected by yield of desired product, and both design and economics are affected by downstream separation requirements.

The effects of mixing on selectivity have been most carefully investigated for a competitive-consecutive reaction of the type

$$A + B \xrightarrow{k_{R1}} R$$

$$R + B \xrightarrow{k_{R2}} S$$

**Table 13-1** Selectivity for Ideal and Imperfectly Mixed Reactions

| Type of Reactor | Selectivity of $R = R/(A_o - A_{final})$ |
|---|---|
| Ideal plug flow with perfectly mixed feed | 0.861 |
| Ideal CSTR | 0.731 |
| Imperfect tubular reactor | 0.571 with turbulence parameters: $k = 0.008$ m$^2$/s$^2$, $\varepsilon = 0.03$ m$^2$/s$^3$, 10 s residence time |
| Imperfect stirred tank | 0.652 for 6280 L, Rushton turbine with N = 24.4 rpm, feed at impeller discharge |

B is added to A in the semibatch case. A and B are mixed continuously in the tubular reactor case. R is considered to be the desired product. The objective is to determine how mixing conditions can affect the yield of R. We are concerned with the time period from when the reactants are first contacted until they are completely mixed to a molecular scale. During this time, zones of local B concentration can vary from an upper limit equal to the feed concentration to a lower limit of essentially zero. This critical stage is depicted schematically in Figure 13-1, where B is added to A and B is the limiting reagent. The reaction of A with B to form the desired product, R, is occurring along with the normally undesired reaction of R with B to form S. In the first case, the mixing takes place before reaction occurs. A and B are intimately mixed and very little unwanted material S is formed. In the other case, there is a boundary between A and B. Although a lot of desirable product R is formed, it quickly reacts with high concentrations of B to form undesirable product S. While this reaction system has received the most attention, the course of any reaction that is influenced by concentration has the potential to be influenced by mixing. The effect can be on the reaction rate, the product distribution, or both (see Examples 13-3 and 13-4).

Competitive-parallel reactions can also be subject to mixing effects, as shown by Baldyga and Bourne (1990) and Paul et al. (1992). Many variations are possible, but the basic reactions of these systems are as follows:

$$A + B \xrightarrow{k_{R1}} R$$

$$A + C \xrightarrow{k_{R2}} U$$

where the first reaction is the desired one and the second is a simultaneous decomposition of A to undesired U (see Example 13-8a).

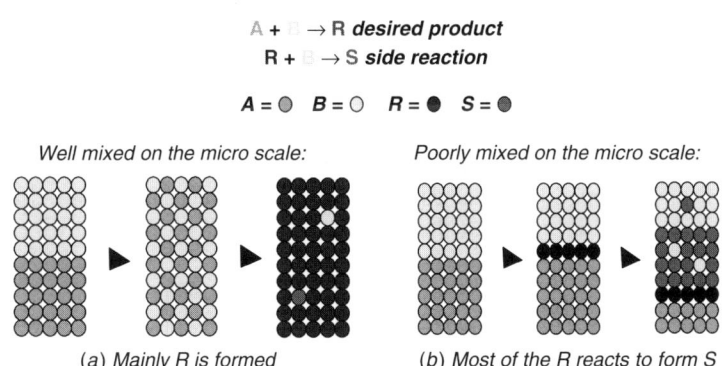

(a) Mainly R is formed    (b) Most of the R reacts to form S

**Figure 13-1** Diffusion and chemical reaction at an A–B mixing surface. In this competitive-consecutive reaction, the first reaction, which forms the desired product (R), is fast, and the consecutive reaction step, forming the undesired by-product (S), is slower. Local mixing conditions at the molecular scale determine the amount of undesired by-product (S) formed.

A very significant variation on this basic system is the decomposition of a product during pH adjustment as follows:

$$\text{acid} + \text{base} \rightarrow \text{salt}$$

When base is added to acid,

$$A + \text{base} \rightarrow U$$

When acid is added to base,

$$A + \text{acid} \rightarrow U$$

(Both A and the desired product, R, could be decomposed in this way.)

Although the acid–base neutralization is normally orders of magnitude faster than the decomposition of A, areas of extreme pH resulting from inadequate mixing can exist and a significant loss of A during seemingly straightforward pH adjustment operations can result in loss of product. This effect is particularly important on scale-up to large vessels, including fermenters, and provision must be made for adequate mixing (see Example 13-8b).

Returning to the basic competitive-consecutive reaction system, consider a semibatch operation. Reagent B is added over time and is instantaneously mixed to a molecular level with the vessel contents. The maximum selectivity for the desired product, R, is a function of the rate constants $k_{R1}$ and $k_{R2}$, the overall molar charge ratio of A to B, and the degree of conversion of A. The degree of conversion of A can depend on the charge ratio and the residence time. The discussion that follows is limited to the case of sufficient residence time such that all of the B charged will react, provided that B is not charged in excess of what is required for complete conversion of A to S. The maximum selectivity for R, in the absence of mixing effects, then becomes a function only of $k_{R1}/k_{R2}$ and the molar charge ratio. If we now fix $k_{R1}/k_{R2}$ and the molar charge ratio, the selectivity is fixed, as are the yield and the degree of conversion of A.

The term *expected (ideal) yield*, $Y_{exp}$, is used to denote the yield that would be obtained for a competitive-consecutive reaction under conditions of perfect mixing and complete conversion of the limiting reactant, as presented by Levenspiel (1972):

$$Y_{exp} = \frac{R}{A_o} = \frac{1}{1-\kappa}\left[\left(\frac{A}{A_o}\right)^\kappa - \frac{A}{A_o}\right] \tag{13-5}$$

where $\kappa = k_{R2}/k_{R1}$ and capital letters denote molar concentrations. This equation applies to both batch and semibatch operations, provided that both reaction rates depend on B in the same way (e.g., second order) and provided that B is added to A in the semibatch case and B is consumed completely. This equation is often used in flowsheeting programs to solve for A given a specified yield. There is

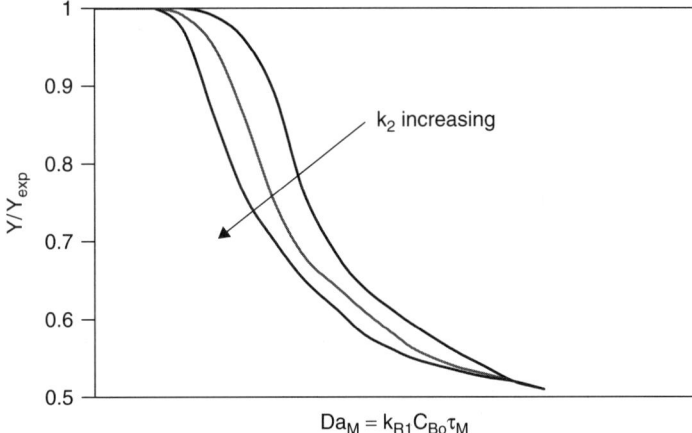

**Figure 13-2** Normalized yield, $Y/Y_{exp}$, as a function of Damkoehler number based on $k_1$. This is a qualitative conceptualization of the interaction between mixing rate as expressed by a local mixing time, $\tau_M$, reaction rate, $k_1 C_{Bo}$, and reaction yield. As the mixing improves (smaller $Da_M$), the yield increases. As the second reaction gets faster (increasing $k_2$), the mixing time must also drop, to maintain yield.

no guarantee that the desired yield, $Y_{exp}$, and A will be obtained—unless the equipment is carefully designed for good mixing conditions.

A useful way of visualizing the relationship between the magnitude of the primary reaction rate constant and its potential to affect yield is illustrated in Figure 13-2, in which normalized yield, $Y/Y_{exp}$, is plotted against a mixing Damkoehler number based on $k_{R1}$ [$Y_{exp}$ is the maximum yield as calculated using eq. (13-5)]. For low values of $k_{R1}$ the yield equals that expected from the chemical kinetics. As $k_{R1}$ increases, yield decreases because of mixing effects. The decline accelerates with increasing values of $k_{R2}$, as shown in Figure 13-2. These relationships can also be expressed as shown in Figure 13-3, where (Sharratt, 1997) $X_S$ is used to represent the amount of S formed where $X_S = 2S/(2S + R)$ and $k_{R2}$ to represent the undesired reaction kinetics.

Mixing effects for homogeneous reactions can only reduce yield below the expected (ideal) as calculated by eq. (13-5). The primary concern is the magnitude of the yield reduction attributable to deviation from instantaneous perfect mixing to the molecular level.

## 13-1.3 Relating Mixing and Reaction Time Scales: The Mixing Damkoehler Number

The final phase of mixing during which chemical reactions can occur and before complete molecular homogeneity is achieved may be visualized as the molecular diffusion-controlled mixing of the smallest eddies in the turbulence energy dissipation spectrum. The smallest eddy size can vary over several orders of

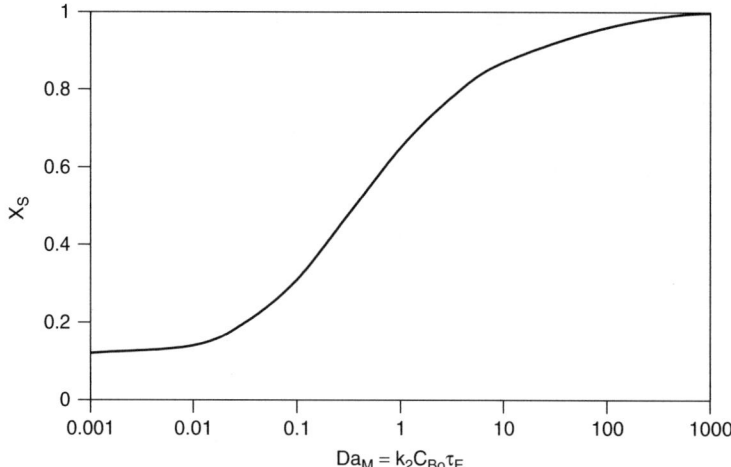

**Figure 13-3** By-product selectivity, $X_s$, as a function of Damkoehler number based on $k_2$. These data of Bourne in Sharratt (1997) show the increased by-product formation with increasing mixing time based on the engulfment model, $\tau_E$. As the reaction rate for the second reaction, $k_2 C_B$, increases, the mixing time must decrease to maintain yield.

magnitude, from ~1 μm in intense jet mixing to >100 μm in stirred tanks with low-shear impellers. The reader is referred to Chapter 2 for a discussion of the time and length scales of turbulence and small scale diffusion.

When fluids mix, the elements of the two fluids are stretched into striations or lamellae. In laminar flow, the average lamellar thickness, $\delta$, can be used to generate a mixing time, $\tau_L$, based on the molecular diffusivity, $D_{AB}$. This gives

$$\tau_L = \frac{\delta^2}{D_{AB}} \qquad (13\text{-}6)$$

The final stage of diffusion in turbulent flow, although conceptually identical to this model, is more complicated, and we defer definition of turbulent mixing time scales to Section 13-2.1.3. In the case of consecutive-competitive chemical reactions [A + B → R; R + B → S] product R must mix with reactant B for the second chemical reaction to proceed. At any location in the vessel and at any instant in time, local concentration gradients normal to stretching fluid lamella may appear as shown in Figure 13-4. This concentration pattern is repeated layer upon layer throughout the mixing fluid as the two mixing fluids diffuse together. The lamellae or striations are not flat: they twist, roll up, and are stretched thinner and thinner by the turbulent vortices in the flow.

The magnitude of yield reduction due to imperfect mixing is determined by the following major factors:

1. *Local mixing time:* a measure of the time from initial contact of the reactants to final homogeneity on a molecular scale at a given point. Any

# INTRODUCTION

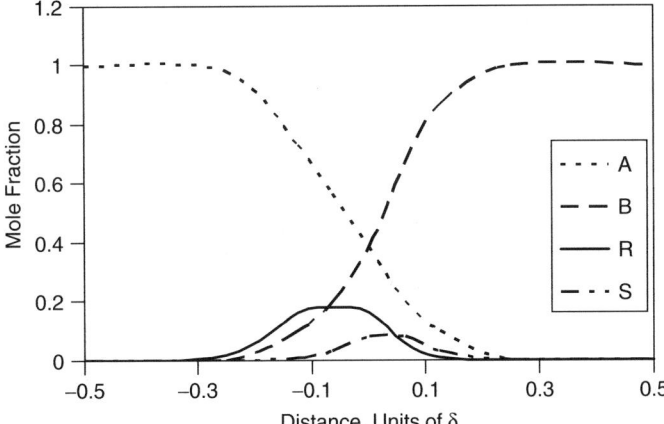

**Figure 13-4** Mole fraction profiles across a lamella or striation. The lamellar thickness is $\delta = 1.0$. At one edge, the mole fraction of A is 1.0; at the other edge the mole fraction of B is 1.0. The components diffuse across the layer, reacting to A and B. While Figure 13-1 showed the molecular scale at the interface, this figure shows the mole fraction across a full striation. Figure 13-12 shows the same phenomenon at the surface of a bubble or a drop.

overreaction of R to S must occur during this time because once the reactants are molecularly mixed, the relative amounts of R and S obtained are fixed by $k_{R1}/k_{R2}$ and the molar charge ratio $A_0$ to $B_0$ according to eq. (13-5) and the dependence of A on $B_0$. Estimating the local mixing time at a given point in a reactor is not easy and will be strongly affected by both the reactor configuration and the way the reagent is fed into the reactor. There are a number of models: Corrsin, Baldyga, and Bourne's micromixing, Baldyga and Bourne's engulfment, and Villermaux's interchange models. All of these try to predict how the reaction conditions at the addition point are affected by local mixing and by the subsequent history of the feed as it is dispersed throughout the mixing vessel. All the models depend on local turbulence conditions as measured by local energy dissipation per unit volume. Depending on the model, a scale of turbulence is often required. This will be an eddy scale ranging from the Corrsin integral length scale to the Kolmogorov scale. In some cases physical properties such as viscosity and molecular diffusivity are required. See Section 13-2.1 for further discussion.

2. *Chemical kinetics:* the absolute values of $k_{R1}$ and $k_{R2}$. The magnitude of the rate constant, $k_{R1}$, will determine how much A can be converted during the time required to achieve molecular mixing. The extent of the conversion will determine the amount of R that is subject to excess B concentration and hence overreaction to S as determined by $k_{R2}$. In some cases the kinetics can be determined by use of a stopped-flow reactor or similar device. For

the results to be valid, the response of the device must be much faster than the fastest reaction.
3. *The mixing Damkoehler number:* the ratio of rates of the first or second reaction and the local mixing rate.

R is converted to S, depending on the probability of a molecule of R reacting with a molecule of B. In a B-rich zone this probability is greater than in the perfectly mixed zone and the extent to which it occurs will depend on the rate at which R can diffuse out of the B-rich zone relative to the rate at which it reacts with B. For reactor design purposes, the key issues are (1) methods of determining and/or predicting which reactions are mixing sensitive, and (2) reactor design guidelines to minimize yield loss on scale-up in which the information discussed above is used to predict and describe mixing considerations in industrial reactors. These issues are considered further later in the chapter.

### 13-1.4 Definitions

To assure accurate and consistent interpretation of theories, models, and results, precise definitions of important terms must be established. The most important definitions used in this chapter are as follows:

- *Conversion:* ratio of moles of a key reactant reacted to moles charged, $(A_0 - A)/A_0$
- *Yield:* ratio of moles of the desired product to moles of a key reactant charged, $Y = R/A_0$
- *Selectivity:* ratio of moles of the desired reaction product to moles of key reactant consumed, $S = R/(A_0 - A)$

**Note:** *Some texts, including Levenspiel (1962), Fogler (1999), and Baldyga and Bourne (1999), use alternative definitions:*

- *Yield* (alternative)*:* $R/(A_0 - A)$ (same as selectivity above)
- *Selectivity* (alternative)*:* R/U, where U is an undesired product
- *Selectivity* (alternative)*:* selectivity as used by Baldyga and Bourne for the competitive-consecutive reaction scheme described in Section 13-1.2, $X_S = 2S/(2S + R)$

*The definitions for yield and selectivity used throughout this book are the first definitions since in many industrial reaction systems, the amounts of individual undesired reaction by-products may not be known. When another definition is used, it will be noted explicitly.*

- *Blending rate:* the rate that concentration differences are reduced by large scale circulation and convective flow down to a selected level of variation everywhere in the whole vessel.

PRINCIPLES OF REACTOR DESIGN FOR MIXING-SENSITIVE SYSTEMS  765

- *Blend time:* the reciprocal of the blending rate, typically the blending time constant, $\tau_B$, for reduction of concentration fluctuations by 95% according to eqs. (13-8) and (13-9).
- *Local mixing rate:* the reciprocal of the local mixing time defined below.
- *Local mixing time:* the time constant for local mixing to molecular scale, which depends on geometry, local shear rates, and physical properties (see Section 13-2).
- *Mixing Damkoehler number:* the ratio of mixing time to reaction time, $Da_M = \tau_M/\tau_R$. The mixing Damkoehler number may be referred to simply as the Damkoehler number. (Note that the traditional Damkoehler number is the vessel residence time divided by the reaction time.)
- *Reaction time:* the time constant for chemical reaction based on the molecular reaction rate constant as follows:

$$\tau_R = \begin{cases} 1/k & \text{for first-order reactions} \\ 1/kC & \text{for second-order reactions} \\ 1/kC^{n-1} & \text{for higher-order reactions} \end{cases}$$

- *Scale of segregation:* a measure of the large scale breakup process (bulk and eddy diffusivity) without the action of diffusion. It is the size of the packets of B that can be distinguished from the surrounding fluid A. See the discussion in Chapters 2 and 3.
- *Segregation:* a measure of the difference in concentration between the purest concentration of B and the purest concentration of A in the surrounding fluid. Molecular diffusion is needed to reduce the segregation, as even the smallest turbulent eddies have a very large diameter relative to the size of a molecule. Segregation can be defined mathematically as $s = \overline{c_i^2}$, where $c_i$ is the fluctuating concentration of component i, given by $c_i = C_i - \overline{C_i}$, and $\overline{C_i}$ is the average concentration. Intensity of segregation is the segregation divided by the product of the average concentrations of A and B.
- *Micromixing:* mixing at the scale of the smallest turbulent eddies and concentration striations (see Figure 13-5a).
- *Macromixing:* another term for blending to a degree of homogeneity throughout a vessel. For the blend time correlation, this degree is 95%. This is the largest scale reduction of concentration fluctuations (see Figure 13-5b).
- *Mesomixing:* all intermediate scales of mixing. Mesomixing effects most typically occur when the feed rate is greater than the local mixing rate, allowing a plume of higher concentration to spread from the feed point (see Figure 13-5c).

**766** MIXING AND CHEMICAL REACTIONS

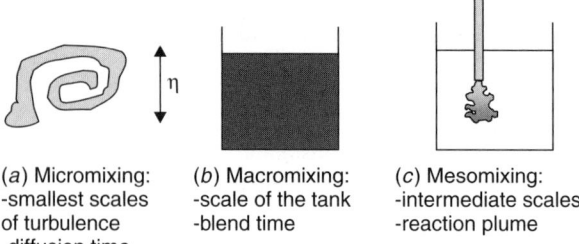

(a) Micromixing:
-smallest scales
of turbulence
-diffusion time

(b) Macromixing:
-scale of the tank
-blend time

(c) Mesomixing:
-intermediate scales
-reaction plume

**Figure 13-5** (a) Micromixing at the smallest scales; (b) macromixing at the largest scale; and (c) mesomixing at intermediate scales.

## 13-2 PRINCIPLES OF REACTOR DESIGN FOR MIXING-SENSITIVE SYSTEMS

The rate at which reactants are brought together is very important in many reactions. For very fast acid–base reactions, the time it takes to mix is the apparent reaction time. Reaction cannot take place before mixing, so the processes take place essentially in series. A reduction in apparent reaction rate is often not critical to the process result. However, with fast reactions, slower mixing results in high local reactant concentrations, which can allow an undesired consecutive or parallel reaction to proceed to a greater extent than predicted by the rate constant ratio, thereby decreasing selectivity. Mixing rates are frequently important in determining the yields of desired products in semibatch reactors, since the reaction rates may be fast relative to mixing rates. Scale-up from bench scale to commercial production scale can result in yield reductions of more than 10%, unless the mixing requirements are recognized in development and provided for on scale-up.

### 13-2.1 Mixing Time Scales: Calculation of the Damkoehler Number

There are several mixing or blending times that can be measured and observed in an agitated vessel. The bulk blending time is the time it takes to get all points in the tank within some arbitrary range of all other points. Local mixing time is the measure of how fast material at a given point losses its identity. Thus, the local mixing time varies with position, while the bulk blending time may vary with position of addition but not position of measurement. Bulk blending time is usually based on the longest time or the slowest rate of mixing in the vessel. Local mixing times depend on the local turbulence.

The Damkoehler number requires characteristic time scales for both mixing and reaction. Calculation of the reaction time scale is relatively straightforward, although the necessary data may be difficult to obtain. Many choices for the mixing time have been proposed, and data are available for many common semibatch geometries.

### 13-2.1.1 Characteristic Reaction Time.
As shown in Example 13-1, mixing can affect the selectivity of a reaction, not just the rate. Reactions that show selectivity are usually two-step reactions which are either consecutive or parallel. One reaction is usually so fast that it is mixing controlled. The second reaction has a characteristic time constant of the order of the local mixing time. The reaction time is usually given by

$$\tau_R = \frac{1}{k_{R2} C_{Bo}} \quad (13\text{-}7)$$

where $k_{R2}$ is the rate constant of the second undesirable reaction and $C_{Bo}$ is the initial concentration of B in the feed—*not* the well-mixed concentration. The component A is usually present in large excess, so its concentration is essentially constant and does not appear in the equation. The reaction half-life $[\tau_R = -\ln(0.5)/k_R]$ as given in eq. (13-3) is another characteristic reaction time, but it does not account for the effect of concentration. Use of the feed concentration, $C_{Bo}$ rather than the well mixed concentration gives the reaction time at the end of the feed pipe. This is the worst condition in the reactor and is the location where mixing must overcome kinetics in order to avoid the formation of undesirable by-products.

### 13-2.1.2 Blend Times.
Even though it is known that local mixing time is more relevant to yield effects for mixing rate-controlled reactions, blend times are a more common way to compare mixing and reaction time constants. The blend time is the time it takes after an input change to a stirred vessel for spatial variation of average concentration to drop to 5% of the original variation. Typically, changes in conductivity are used to make measurements of degree of blending (see Grenville, 1992; Nienow, 1997). The Grenville correlations for blend times are used extensively for design and scale-up. They are dependent on the Reynolds number range as follows:

$$N\tau_B = \begin{cases} \dfrac{5.4}{N_p^{1/3}} \left(\dfrac{T}{D}\right)^2 & \text{for } Re > 6400 \quad (13\text{-}8) \\[2mm] \dfrac{1}{Re} \dfrac{184.2}{N_p^{2/3}} \left(\dfrac{T}{D}\right)^2 & \text{for } 500 < Re < 6400 \quad (13\text{-}9) \end{cases}$$

Vessel blend times are typically about 2 s in a 1 L vessel and about 20 s in a 20 000 L vessel for low viscosity liquids.

Other blend time correlations were presented by Penney (1971), Khang and Levenspiel (1976), and Fasano and Penney (1991). Use of these correlation equations allows the estimation of blending times, which can be compared to molecular reaction times for all the reactions in the reactor. Even though local mixing time is the critical time for determining apparent reaction rate, blend time can be used in an approximate manner. If the characteristic molecular reaction time (e.g., the half-life) is much greater than the characteristic blend time (typically, 100 times), the chemical reactions occur under well-mixed conditions. If, on the other hand, the characteristic blend time is very long compared to the characteristic reaction time, there will be regions rich in some reactants that could lead to unwanted by-products and reduced yield of the desired products. The study of such yield effects as affected by mixing rate is frequently called *micromixing* because it deals with small characteristic times and small (local) scales of concentration fluctuation. It must be emphasized that the use of blend time is only approximate because it is the spectrum of local mixing times that actually determines how the mixing rate affects the yield. The reader is referred to Chapter 9 for more details on blending in tanks and to Chapter 7 for in-line blending.

### 13-2.1.3 Local Mixing Time Scales.
In dealing with mixing effects on reaction two topics are of interest. The first is the size of the additive blob or feed stream. The second is its rate of disappearance or the inverse local mixing time. For low viscosity liquids, very rapid mixing with local mixing time constants, $\tau_M$, as short as 0.01 s is easily obtained in liter-sized reactors, but due to mechanical limitations, local mixing times on the order of 0.1 s or longer typically occur in reactors of 10 000 or more liters. The size of the blob together with the local mixing time determines the amount of undesirable product that can be formed. There are many formulations for these two effects.

The discussion of local mixing time scales must begin with a definition of the turbulent scales which underlie many of the mixing time formulations. These scales are developed in Chapter 2, so only a brief summary is provided here. The range of turbulent length scales starts at the largest integral scales of motion, which is a dimension close to the blade width or the feed pipe diameter. The eddies cascade energy down through smaller and smaller scales until the turbulent energy is dissipated by viscosity at the smallest scales of motion. The Kolmogorov length scale is the size of the smallest turbulent eddy:

$$\eta = \left(\frac{\nu^3}{\varepsilon}\right)^{1/4} \tag{13-10}$$

At the Kolmogorov scale, the following statements apply:

$$\varepsilon \propto \frac{u_\eta'^3}{\eta}$$

where $\varepsilon$ is the rate of dissipation of turbulent kinetic energy per unit mass.

$$Re_\eta = 1.0 = \frac{\eta u_\eta'}{\nu}$$

where $Re_\eta$ is the local Reynolds number at the Kolmogorov scale. Thus,

$$u_\eta' = \frac{\nu}{\eta} = (\nu\varepsilon)^{1/4}$$

PRINCIPLES OF REACTOR DESIGN FOR MIXING-SENSITIVE SYSTEMS   769

so the time that it takes to dissipate a Kolmogorov sized eddy is

$$\tau_K = \frac{u_\eta'^2}{\varepsilon} = \frac{u_\eta'^2 \eta}{u_\eta'^3} = \frac{1}{(\nu\varepsilon)^{1/4}}\left(\frac{\nu^3}{\varepsilon}\right)^{1/4}$$

$$= \left(\frac{\nu}{\varepsilon}\right)^{1/2} \quad (13\text{-}11)$$

This the Kolmogorov time scale.

Batchelor (1959) developed an expression for the smallest concentration (or temperature) striation based on the argument that for diffusion time scales longer than the Kolmogorov scale, turbulence would continue to deform and stretch the blobs to smaller and smaller lamellae. Only once the lamellae could diffuse at the same rate as the viscous dissipation scale would the concentration striations disappear. The Batchelor length scale is the size of the smallest blob that can diffuse by molecular diffusion in one Kolmogorov time scale. Using the lamellar diffusion time from eq. (13-6) gives

$$\tau_B = \frac{\lambda_B^2}{D_{AB}}$$

If the Batchelor and Kolmogorov times are equal, $\tau_B = \tau_K$, the Batchelor length scale is

$$\lambda_B = \left(\frac{\nu D_{AB}^2}{\varepsilon}\right)^{1/4} = \left(\frac{D_{AB}}{\nu}\right)^{1/2}\left(\frac{\nu^3}{\varepsilon}\right)^{1/4} = \frac{\eta}{\sqrt{Sc}} \quad (13\text{-}12)$$

where $Sc = \nu/D_{AB}$ is the Schmidt number. Because the Batchelor and Kolmogorov time scales are equal, mixing times proportional to $\tau_K$ are referred to as Batchelor scale mixing (see Table 2-3 and related text). For liquids with Schmidt numbers much larger than 1, the smallest striation thicknesses are given by the Batchelor scale. For large Sc, say 1000, the Batchelor length scale can be 30 times smaller than the Kolmogorov length scale.

*Corrsin Mixing Time.* One of the first theoretical formulations of mixing time is due to Corrsin (1964). For isotropic homogeneous turbulence, he determined the time required for a reduction of scale from the largest scales of concentration fluctuations, $L_s$, through the full range of the inertial convective scales of turbulence to the Kolmogorov scale, $\eta$, and then through the viscous scales to the Batchelor scale, $\lambda_B$, by integrating the scalar (concentration) and turbulence spectra. This gives

$$\tau_M = \begin{cases} 2\left(\frac{L_s^2}{\varepsilon}\right)^{1/3} + \frac{1}{2}\left(\frac{\nu}{\varepsilon}\right)^{1/2}\ln(Sc) & \text{for liquids where } Sc \gg 1 \quad (13\text{-}13) \\ 1.36\left(\frac{L_s^2}{\varepsilon}\right)^{1/3} & \text{for gases where Sc is about } 1.0 \quad (13\text{-}14) \end{cases}$$

$L_S$ is the local scale of segregation or the average size of unmixed regions and $\varepsilon$ is the local rate of energy dissipation. The first term arises from describing the large inertial scales which contain most of the turbulent energy. The second term gives the time scales at the smallest scales of mixing. This is the time required to reduce the blob from the Kolmogorov length scale [eq. (13-10)] to the Batchelor length scale [eq. (13-12)] for large Sc, where molecular diffusion is much slower than the diffusion of momentum. Baldyga and Bourne have restated the second term in eq. (13-13) as asinh(0.05Sc) using a somewhat more rigorous derivation than Corrsin's. In both cases, this term will be vanishingly small most of the time.

*Micromixing.* Alternative ways of expressing the local mixing time constant have been developed by Bourne and co-workers, and they have dubbed this approach *micromixing*. In the micromixing analysis it is assumed that the amount added and the rate of addition are very small and that the scale of interest is set by the local turbulence. The earliest approach was to assume that added material did not do anything until the Kolmogorov scale was reached and the subsequent mixing took place by molecular diffusion. Using eqs. (13-6) and (13-10) yields

$$\tau_M = \frac{\eta^2}{D_{AB}} = \left(\frac{\nu^3}{\varepsilon D_{AB}^2}\right)^{1/2} = Sc\left(\frac{\nu}{\varepsilon}\right)^{1/2} \qquad (13\text{-}15)$$

This concept was replaced by the engulfment model, which is a more realistic way of treating the breakup of the added reactant. Here the engulfment rate is

$$E = 0.06\left(\frac{\varepsilon}{\nu}\right)^{1/2}$$

and thus

$$\tau_E = \frac{1}{E} = 17\left(\frac{\nu}{\varepsilon}\right)^{1/2} \qquad (13\text{-}16)$$

The differences in the approaches are small. Both include local energy dissipation, and both include the viscosity. The molecular diffusivity is important only when the viscosity is high (see Section 13-2.1.5). The similarity between $\tau_E$ and the last term in the Corrsin development is not surprising, but the implication of the much larger coefficient (17 instead of 0.5) is that viscosity may play a role at scales significantly larger than the Kolmogorov scale, and the effective micromixing rate for reactions must include these scales.

*Mesomixing.* This term is used to describe a set of phenomena between macromixing, which involves the whole vessel, and micromixing, which involves a small volume at the smallest eddy scales. Although the term *mesomixing* was first used by Bourne and co-workers, the first group to describe the phenomenon was that of Villermaux. His group was doing experiments similar to those of Bourne, but in their experiments with colored materials and precipitating materials they observed a plume near the point of addition. The size and rate of disappearance of this plume in semibatch experiments did not fit any of the models that Bourne and others had proposed. Villermaux and Devillon (1972) and Villermaux and David (1987) developed a semiempirical model called *IEM* (interaction by exchange with the mean) to describe the volume and the rate of exchange between the plume volume and the bulk. They assumed that the plume was at or near the composition of the inlet and that the bulk was well mixed. Turbulent mass transfer occurred across the boundary by turbulent interchange. Empirical relations were developed for the size of the plume and the rate of mixing. Thus, effects of inlet geometry velocity and flow rates were taken into account. Thoma (1989), Bourne and Thoma (1991), and Thoma et al. (1991) looked at the time of addition in semibatch operation and observed that when the addition time was very short there was an additional undesirable selectivity change. This effect is shown for a sample reaction system in Figure 13-6. It appeared that higher rates of addition were sufficient to overcome the local ability to take material away, and micromixing by turbulence of small packets was overwhelmed. They called this phenomenon *mesomixing*. It now seems that what Villermaux's group observed was very similar. A plume exists when the feed is added faster than the fine scale micromixing turbulence can take it away. This plume is clearly shown in concentration isoplots of mixing-reaction simulations. The processes governing mesomixing are not as well worked out as those for micromixing, but many useful thoughts come from understanding the concepts. Thoma et al. (1991) discussed the relationship of micromixing to macromixing in detail.

In terms of the Corrsin development, for mesomixing the initial scale is set by the inlet conditions (e.g., feed pipe diameter), not by the local turbulence. The first term of eq. (13-13) accounts for the mesomixing effect, and the second term is related to the micromixing effect: large values of the first term occur when an unmixed plume is evident.

Bourne and Thoma (1991) found that the critical addition time was inversely proportional to impeller speed. When running below the critical addition time, scale-up could be affected by absolute impeller speed in addition to local energy dissipation. For addition times longer than the critical mixing time, local turbulent energy dissipation alone governed selectivity. Mesomixing occurs mainly at intermediate scales of turbulence, which are not affected by viscosity. Micromixing occurs at scales smaller than the Kolmogorov scale, $\eta$, where there is a definite viscosity effect, as shown above. Bourne and Hilber (1990) showed that the number of addition points affected the critical feed time so that the following expression could be developed:

$$\tau_{crit} Nn = \text{constant} \qquad (13\text{-}17)$$

The constant is a function of local turbulence and chemistry, as shown in Figure 13-6. Attempts have also been made to define the mesomixing parameters from basic turbulence theory. For example, Baldyga and Bourne (1999) suggest

# 772  MIXING AND CHEMICAL REACTIONS

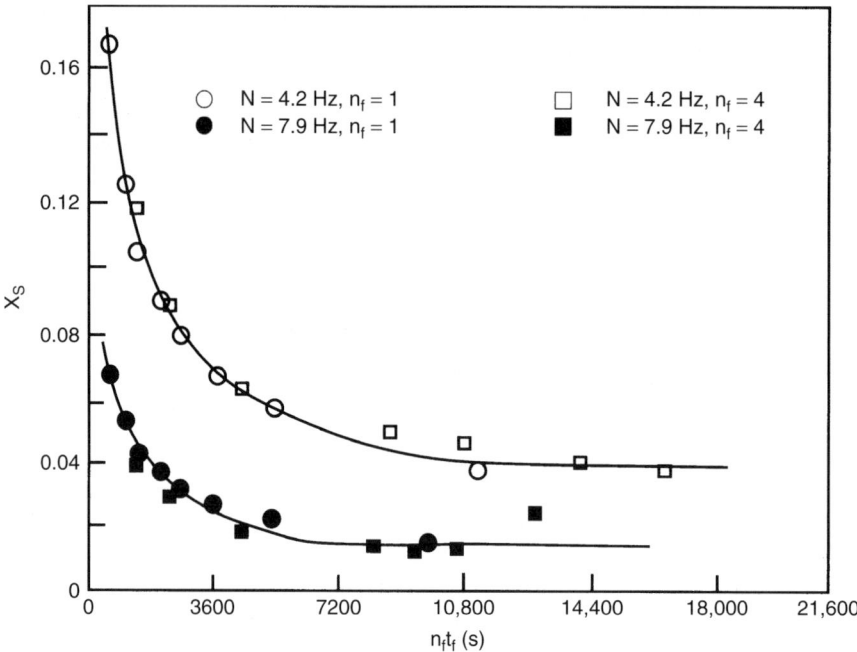

**Figure 13-6** Effect of addition time on selectivity in semibatch operation. Number of nozzles and feed time ($n_f t_f$) determine selectivity at constant N. If N decreases on scale-up, the minimum critical addition time ($n_f t_f$) must increase to achieve the same selectivity. Feed nozzles are in the impeller discharge region. (Data from Bourne and Hilber, 1990.)

that the mesomixing time is given either by

$$\tau_D = \frac{Q_B}{U D_t} \tag{13-18}$$

where $\tau_D$ is the mesomixing time for dispersion of feed, $Q_B$ the volumetric feed rate of B, U the local velocity in surrounding fluid at the feed point, and $D_t$ the local turbulent diffusivity ($D_t = 0.1 k^2/\varepsilon$) in the surrounding fluid; or by Corrsin's form for the mesomixing time scale:

$$\tau_S = A \left( \frac{L_S^2}{\varepsilon} \right)^{1/3} \propto \left( \frac{Q_B}{U_B \varepsilon} \right)^{1/3} \tag{13-19}$$

where $\tau_S$ is the mesomixing time for disintegration of large eddies, A is a constant between 1 and 2, $L_S$ is the concentration macroscale, and $\varepsilon$ is the local turbulent energy dissipation rate.

However, these expressions work to only a limited extent. For example, an inlet jet can be designed to develop a high local energy dissipation at the inlet and rapid mixing. Current theories cannot incorporate this effect. A simple ordering argument would suggest that if the energy dissipation from the jet given by the velocity cubed divided by the jet diameter is larger than the surrounding local energy dissipation per unit mass, that is the energy to use and the pipe diameter is the dimension. More detail on the experimental and theoretical foundation of these concepts is given in Baldyga and Bourne (1999, Chap. 12). Cases for jet mixers and motionless mixers are discussed.

An interesting sidelight of these mixing effects is the increasing importance of mixing on scale-up, as illustrated in Example 13-2.

***Example 13-2: Scale Effects on Mixing in Stirred Vessels.*** Determine whether the fast reaction from Example 13-1 will be affected by mixing on scale-up if the feed point is close to the impeller. Compute the values of the Corrsin mixing time, $\tau_M$, at the impeller tip and the blend time, $\tau_B$, for (a) 1 L and (b) 20 000 L vessels stirred by a disk turbine ($N_P = 6$) at power per unit volume of 0.36 kW/m³. Use properties of water: $\rho = 1000$ kg/m³; $\nu = 10^{-6}$ m²/s; $Sc = 2000$ for a typical solute.

SOLUTION: The reaction time scale is taken from the first reaction in Example 13.1: $\tau_{R1} = 1/k_{R1}C_{A0} = 0.14$ s. The Corrsin equation (Section 13-2.1.3) is

$$\tau_M = 2\left(\frac{L_S^2}{\varepsilon}\right)^{1/3} + \frac{1}{2}\left(\frac{\nu}{\varepsilon}\right)^{1/2}\ln(Sc)$$

where $L_s$ is the largest length scale of the scalar, often taken to be the feed pipe diameter. If this length scale is not known, Zipp and Patterson (1998) suggest using $L_S = 0.39 L_T$, where $L_T$ is the largest eddy size, proportional to $k^{3/2}/\varepsilon$. This gives a modified Corrsin equation based on k and $\varepsilon$:

$$\tau_M = 1.1\left(\frac{k}{\varepsilon}\right) + \frac{1}{2}\left(\frac{\nu}{\varepsilon}\right)^{1/2}\ln(Sc) \qquad (13\text{-}20)$$

From Wu and Patterson (1989) we know that the energy dissipation per unit mass at the impeller tip is 20 times the average for the tank and that the random turbulence energy per unit mass, k, is approximately $0.06 U_{tip}^2$. Our objective is to compare the mixing time scales with the reaction time scale for both the small and the large vessel. As long as the mixing time scales are shorter than the reaction time scale, the reaction will not be limited by mixing.

(a) For the 1 L vessel:

$$T = 0.108 \text{ m}$$
$$D = 0.036 \text{ m}$$
$$P = (360 \text{ W/m}^3)(0.001 \text{ m}^3)$$
$$= N_p \rho N^3 D^5 = (6)(1000 \text{ kg/m}^3)(N^3)(0.036 \text{ m})^5$$

774    MIXING AND CHEMICAL REACTIONS

$$N = 9.97 \text{ s}^{-1} \text{ or } 598 \text{ rpm}$$
$$U_{tip} = \pi ND = (\pi)(9.97 \text{ s}^{-1})(0.036 \text{ m}) = 1.13 \text{ m/s}$$

For this geometry, the random turbulence energy, k, is about $0.06 U_{tip}^2$, so

$$k = 0.06(1.13 \text{ m/s})^2 = 0.0766 \text{ m}^2/\text{s}^2 \text{ at the tip of the impeller}$$
$$\varepsilon = 20(P/V)/\rho = 20(360 \text{ W/m}^3)/(1000 \text{ kg/m}^3) = 7.2 \text{ m}^2/\text{s}^3$$

Therefore, for $\tau_M = 0.5[2.2 \, k/\varepsilon + (\nu/\varepsilon)^{1/2} \ln(\text{Sc})]$:

$$\tau_M = 0.5\{2.2(0.0766 \text{ m}^2/\text{s}^2)/(7.2 \text{ m}^2/\text{s}^3)$$
$$+ [(10^{-6} \text{ m}^2/\text{s})/(7.2 \text{ m}^2/\text{s}^3)]^{1/2} \ln(2000)\}$$
$$= 0.0117 \text{ s} + 0.0014 \text{ s} = 0.0131 \text{ s}$$

The inertial mixing (first term) is controlling since its time constant is about eight times the time constant of the Batchelor scale mixing. The Corrsin scale mixing is much faster than the reaction time constant.

$$\text{Re} = ND^2 \rho/\mu = (9.97 \text{ s}^{-1})(0.036 \text{ m})^2 (1000 \text{ kg/m}^3)/(0.001 \text{ kg/m} \cdot \text{s})$$
$$= 12\,921$$

Therefore,

$$N\tau_B = 5.4/N_p^{0.333}/(D/T)^2 \quad \text{for Re} > 6400$$
$$\tau_B = 5.4/6^{0.333}/(1/3)^2/9.97 \text{ s}^{-1} = 2.7 \text{ s}$$

about 206 times the local mixing time at the impeller tip, or 230 times the inertial term in the Corrsin equation. Bulk blend time is much slower than the reaction time, so it is important to feed into the zone of maximum dissipation, close to the tip of the impeller blades.

(b) For the 20 000 L vessel:

$$T = 2.94 \text{ m}$$
$$D = 0.98 \text{ m}$$
$$P = (360 \text{ W/m}^3)(20 \text{ m}^3) = 7200 \text{ W}$$
$$= N_p \rho N^3 D^5 = (6)(1000 \text{ kg/m}^3)(N^3)(0.98 \text{ m})^5$$

so $N = 1.1 \text{ s}^{-1}$ or 66 rpm.

$$U_{tip} = \pi ND = (\pi)(1.1 \text{ s}^{-1})(0.98 \text{ m}) = 3.39 \text{ m/s}$$
$$k \approx 0.06(3.39^2) = 0.688 \text{ m}^2/\text{s}^2$$

PRINCIPLES OF REACTOR DESIGN FOR MIXING-SENSITIVE SYSTEMS  775

at the tip of the impeller, assuming that $k/U_{tip}^2$ remains constant during scale-up.

$$\varepsilon = 20(P/V)/\rho = 20(360 \text{ W/m}^3)/(1000 \text{ kg/m}^3) = 7.2 \text{ m}^2/\text{s}^3$$

assuming that $\varepsilon/\varepsilon_{avg}$ remains constant during scale-up. Therefore, for $\tau_M = 0.5[2.2k/\varepsilon + (\nu/\varepsilon)^{1/2} \ln(\text{Sc})]$,

$$\tau_M = 0.5\{2.2(0.688 \text{ m}^2/\text{s}^2)/(7.2 \text{ m}^2/\text{s}^3)$$
$$+ [(10^{-6} \text{ m}^2/\text{s})/(7.2 \text{ m}^2/\text{s}^3)]^{1/2} \ln(2000)\}$$
$$= 0.1051 \text{ s} + 0.0014 \text{ s} = 0.1065 \text{ s}$$

The inertial mixing (first term) is controlling since its time constant is about 75 times the time constant of the Batchelor scale mixing. The contribution due to the Batchelor scale mixing is negligible when the vessel is scaled up. At the large scale, the Corrsin time scale is similar to the reaction time scale (0.14 s). This very fast reaction may be limited by mixing on scale-up, allowing the second reaction to produce additional undesired by-product.

$$\text{Re} = ND^2\rho/\mu = (0.98 \text{ m})^2(1.1 \text{ s}^{-1})(1000 \text{ kg/m}^3)/(0.001 \text{ kg/m} \cdot \text{s})$$
$$= 1\,056\,440$$

Therefore,

$$N\tau_B = 5.4/N_p^{0.333}/(D/T)^2 \quad \text{for Re} > 6400$$
$$\tau_B = 5.4/6^{0.333}/(1/3)^2/1.1 \text{ s}^{-1} = 24.3 \text{ s}$$

about 228 times the local mixing time at the impeller tip. Again, it is important to feed into the zone of maximum dissipation, close to the tip of the impeller blades.

*Message:* From the comparisons above, it is clear that $\tau_M$ and $\tau_B$ scale-up in the same way. It turns out that each is proportional to $1/N$ as long as geometry and the power per unit volume remain constant and the contribution due to the Batchelor scale mixing can be neglected. The key time scales for this problem are summarized in Table 13-2. Chemical reactions and their rates are scale-independent phenomena while the local mixing time is both scale and position dependent. Mixing effects get worse on scale-up.

### Summary of Key Time Constants

- Reaction: $\quad \tau_R = \dfrac{1}{k_{R2}C_{Bo}}$ $\hfill$ (13-7)

- Lamellar diffusion: $\quad \tau_L = \dfrac{\delta^2}{D_{AB}}$ $\hfill$ (13-6)

**Table 13-2  Summary of Time Scales in Example 13-2**

| Time | Small Scale | Large Scale |
|---|---|---|
| Reaction $\tau_R$ | 0.14 s | Same |
| Corrsin–Batchelor micromixing term | 0.0014 s | Same |
| Corrsin mesomixing term | 0.0117 s | 0.105 s |
| Corrsin mixing time $\tau_M$ | 0.013 s | 0.107 s |
| Blend time $\tau_B$ | 2.7 s | 24.3 s |
| Bourne engulfment $\tau_E$ | 0.006 s | Same |
| N | 9.97 rps | 1.1 rps |

- Kolmogorov or Batchelor time scale:  $\tau_K = \left(\dfrac{\nu}{\varepsilon}\right)^{1/2}$  (13-11)
- Bourne engulfment time scale for micromixing:  $\tau_E = 17\left(\dfrac{\nu}{\varepsilon}\right)^{1/2}$  (13-16)
- Baldyga and Bourne mesomixing time for dispersion of feed:

$$\tau_D = \dfrac{Q_B}{UD_t} \qquad (13\text{-}18)$$

- Corrsin mesomixing time for disintegration of large eddies:

$$\tau_M = 2\left(\dfrac{L_s^2}{\varepsilon}\right)^{1/3} + \dfrac{1}{2}\left(\dfrac{\nu}{\varepsilon}\right)^{1/2} \ln(Sc) \qquad (13\text{-}13)$$

### 13-2.1.4 Laminar Micromixing.
Looking at the simple case of laminar mixing the initial feed of reactant appears as a blob which is then stretched by the laminar mixing action. The blob is still at its inlet concentration. Molecular diffusivity starts to spread the reactant out and reaction takes place at the interface. With progressive mixing the interface stretches increasing transfer area and reducing diffusion distance. There may also be an interchange of streamlines if a mixer is present. The growth and redistribution of streamlines is discussed in Chapter 3 on laminar flow and in the work of Ottino (1980). The process continues until all the controlling reactant is used up. A good example of this technique applied to a copolymerization is the work of Tosun (1997).

### 13-2.1.5 Turbulent Micromixing: Effect of High Viscosity.
In a turbulent field a similar phenomenon happens when a blob of one reactant is distorted and diffusion and chemical reaction take place. The initial model of Bourne pictured a blob of reactant fluid that rapidly broke down to the smallest eddy size without much diffusion and reaction. The smallest eddy size is the Kolmogorov size, and at that size diffusion takes place via molecular diffusion [see eq. (13-15)]. Later, Bourne abandoned that model and went to an engulfment model based on

a concept of stretching lamellae and is believed to more accurately represent the turbulent process [eq. (13-16)].

Higher viscosity generally reduces the mixing rate at the same turbulence energy dissipation rate or power per unit mass. The mixing theory of Corrsin (1964) accounts for the effects of viscosity and molecular diffusivity on the time constant for local mixing, although the viscosity appears in the smaller term. The second term in his time constant equation is frequently an order of magnitude smaller that the first term. For instance, if viscosity is increased by a factor of 100, the impeller stream mixing time constant would be almost doubled for the 1 L vessel but would be little affected in the 20 000 L vessel, since in the latter, inertial mixing dominates completely.

A similar estimate of the effect of viscosity on local mixing rate can be obtained from the "engulfment" model of Baldyga and Bourne (1989). Their time constant, $\tau_E$, for the final step of mixing, engulfment of unmixed fluid, $\tau_E = 17(\mu/\rho\varepsilon)^{1/2}$, shows that the mixing time constant increases in proportion to the square root of viscosity if turbulence energy dissipation rate and density are constant and engulfment rate (Batchelor scale mixing) is controlling.

From these time scales, it is clear that both the viscosity and the diffusivity affect mixing at the smallest scales. The Schmidt number,

$$Sc = \frac{\mu}{\rho D_{AB}} = \frac{\nu}{D_{AB}} = \frac{\text{momentum diffusivity}}{\text{molecular diffusivity}} \qquad (13\text{-}21)$$

defines limits for the different mechanisms as discussed in Chapter 2. For Schmidt numbers smaller than 4000, the turbulent engulfment model works (see Baldyga and Bourne, 1999, p. 576), while for larger numbers the mixing is by viscous stretching. Note that for low viscosity liquids such as water, the Schmidt number is on the order of 1000; for gases it is on the order of 1; and for viscous liquids or feeds at 1000 mPa · s it is on the order of $10^6$. A large Sc value means that the smallest eddy dissipated by viscosity (the Kolmogorov scale eddies) will be much larger than the smallest concentration striations, which are dissipated by molecular diffusivity (the Batchelor scale striations).

### 13-2.1.6 Summary: Da$_M$.
Given an estimate of reaction time and an estimate of the appropriate local mixing time constant, one can calculate Da$_M$ and use Figure 13-2 and/or Figure 13-3 for an estimate of yield and selectivity. This is, however, based on the assumption that the local mixing time is constant as the initial blob moves away from the inlet position. In fact, there is a wide distribution of values of local mixing time constants, and the entering material moves through many different zones of varying energy dissipation. The distribution of energy dissipation in a reactor is thus very important. This explains the interest in laser Doppler anemometry and computational fluid dynamics.

This distribution of energy dissipation complicates any mathematical analysis immensely. It also explains why many modelers have gone to zone model analyses to predict the path of the reactants more accurately. In such models the

**778** MIXING AND CHEMICAL REACTIONS

vessel is divided into a number of zones of different energy intensity where local mixing varies. An example of such a model developed by Patterson (1975) is discussed in Section 13-5. Other examples of contributions in this area are the papers by Bourne and Yu (1994) and Baldyga et al. (1995).

Often, a full analysis is not possible because of the lack of full kinetic data for all the steps. In such cases the scale-up protocols in Section 13-4.3 can be very useful. Running small scale experiments in which key parts of the local mixing rate are varied, including position, number of feed points, and rate of addition, can greatly aid in understanding any choice of final reactor design. In some cases the final reactor design is known, and then the local mixing time for the large scale can be estimated and experiments under similar time scales run on the small scale. This process is often called *scaling down*.

### 13-2.2  How Mixing Affects Reaction in Common Reactor Geometries

Although there are many reactor geometries in practice, discussion here is limited to four geometries where mixing is of particular interest: the pipe, Tee mixer, static mixer, and stirred tank. Figure 13-7 illustrates these geometries. The full range of stirred tank geometries and impellers is the subject of Chapter 6. A brief description of each geometry and the mixing issues particular to each is given below.

*13-2.2.1  Pipes*. The simplest mixed chemical reactor is a pipe with reactant injectors at one end. The reactants mix as they flow toward the outlet, forming a *tubular reactor*. There are two measures of mixing in a pipe: (1) the degree of uniformity of the average concentration in the radial direction, and (2) the mean square of the level of concentration fluctuations (referred to as *segregation*) at various locations across the pipe.

The perfect reactor analysis assumes that there is no radial distribution of concentration and the reaction occurs with time along a length with no radial effect. This reduces the analysis to a simple differential equation with time or, at constant velocity, with length. This is the ideal plug flow assumption. When there is only a single reactant, radial concentration gradients are small if the velocity profile is nearly flat, as in high Reynolds number turbulent flow. For multiple reactants it takes a finite time for them to achieve radial uniformity at the molecular scale where chemical reactions occur, so the ideal plug flow assumption does not hold and mixing rates must be considered.

Pipe reactors can be operated in laminar or turbulent flow. In *laminar flow* radial diffusivity is molecular only, which is very slow, particularly if the viscosity is high. In turbulent flow the radial fluctuating velocity component produces the radial turbulent diffusivity which is much faster than molecular diffusivity. Many devices have been developed to promote fast radial mixing in laminar flow, such as static mixers, which are discussed below and in Chapter 7. Besides static mixers, a number of methods exist to promote faster radial mixing in *turbulent flow*, since even in turbulent flow it takes 50 to 100 pipe diameters to achieve

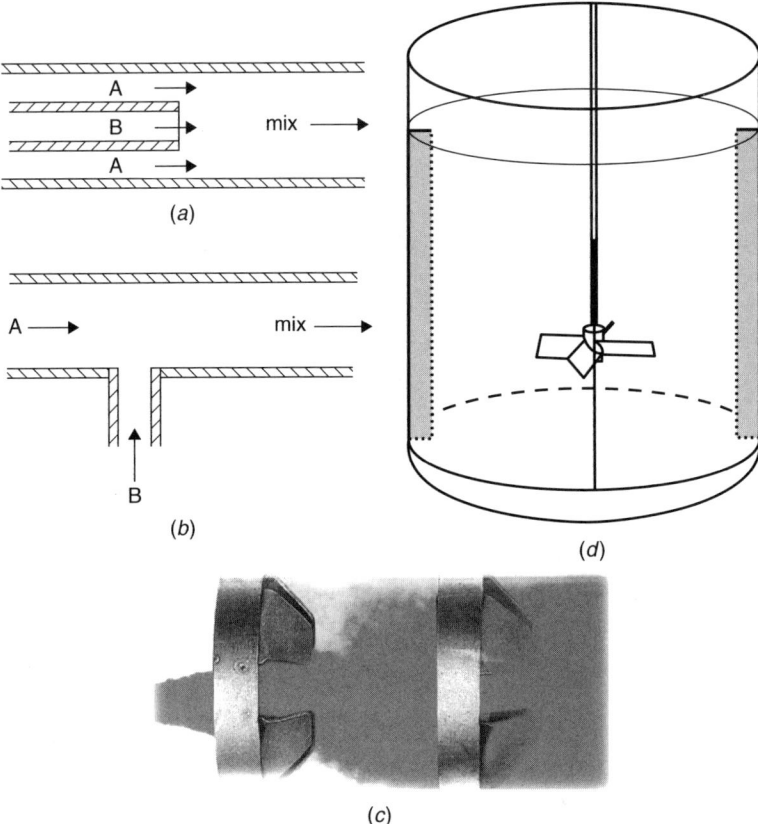

**Figure 13-7** Various mixing geometries used for chemical reactors: (*a*) co-axial jet and (*b*) Tee mixing in a pipe, (*c*) static mixer, and (*d*) stirred tank.

mixing to 95% uniformity if one reactant is injected into the centerline of the pipe. Mixing in pipes is discussed in more detail in Chapter 7.

***13-2.2.2 Tee Mixers.*** To shorten the mixing length of a pipe reactor, one variation is the tee mixer. Tee mixers can shorten the length for blending to 95% uniformity to three to five pipe diameters. The tee mixer is a simple version of the pipe reactor in which one reactant is injected into a flow of the other reactant by a side-entering flow. Care must be taken, however, to prevent the injected reactant from staying adjacent to the pipe wall and not mixing with the flowing stream as expected. This is particularly true for laminar flow and high viscosity, as shown by Forney et al. (1996).

Another type of tee mixer involves opposing flows of reactants with outflow through the side exit into a mixing pipe. One example is the impinging jet reactor used in reaction injection molding technology. The Tee mixer, which is frequently used for liquid-phase reactions, is easier to construct and maintain

**780**   MIXING AND CHEMICAL REACTIONS

than the multiport injector or the coaxial flow injector frequently used for rapid gas phase reactions such as combustion. The downstream mixing pipe for a tee mixer is again considered to be a tubular reactor.

*13-2.2.3 Static Mixers*. Strictly stated, pipe and tee mixers are static mixers since there are no moving parts within them. The term *static mixer*, however, is more frequently associated with pipes containing internal flow diverters and obstacles that promote mixing. Two common types are the twisted-ribbon mixer (Kenics KM) and the structured-packing mixer, one of which makes use of layers of criss-crossed corrugations (Koch-Sulzer SMV). Another structured packing static mixer is the overlapping lattice type (Koch-Sulzer SMX). Details of the construction and operation of various types of static mixers are given in Chapter 7 and additional examples of operation in Examples 13-3, 13-6, and 13-8a below.

*13-2.2.4 Stirred Tanks*. There is a large variety of stirred-vessel reactor types. They range from laminar regime mixing for bulk polymerization or fermentation to highly turbulent mixing for promotion of high yields in competitive reaction synthesis schemes. They cover batch, semibatch (or fed-batch), and continuous flow reactors. They can be single-phase liquid, liquid with suspended solids (usually catalysts, but sometimes reactants or products), liquid with sparged gas reactants, liquid with vaporizing products (boiling reactors), liquid with immiscible suspended liquid droplets, or three-phase reactors. The stirred-vessel reactors can be nearly isothermal or highly exothermic or endothermic. Stirred-vessel reactors can range in size from 0.004 m$^3$ to 10 m$^3$ (a gallon or two to thousands of gallons) may require glass or other special surface treatments to inhibit corrosion, and require a variety of impeller types to achieve success in all the uses above. Mixing issues for all of these types of contacting are discussed in the appropriate chapters of this book and in examples of reactions throughout this chapter. Typical values for the maximum local energy dissipation in selected geometries are given in Table 13-3.

### 13-2.3 Mixing Issues Associated with Batch, Semibatch, and Continuous Operation

*13-2.3.1 Batch Operation*. Pure batch operation is actually a rare application since something must trigger a chemical reaction to proceed, usually the addition of a reactant, catalyst, or heat. If the chemical reaction is fast enough to proceed during the addition of a chemical reactant or a catalyst, the mode is actually semibatch, and mixing effects may be present. If the chemical reaction is so slow that the times needed for the addition of reactants and/or catalyst and for mixing are negligible compared to reaction time, mixing rate is not likely to be a factor in chemical yield. If a threshold temperature must be reached through heating, mixing rate is again unlikely to be a controlling factor, because heating is usually much slower than mixing. Pure batch reactors are, therefore, not generally considered to be affected by mixing rate, with the exception of heterogeneous

**Table 13-3** Maximum Energy Dissipation in Various Geometries

### Stirred Tanks

| Impeller[a] | D/T | C/T | 2r/D (Axial) or 2z/W (RT) | $\varepsilon_{max}$ $P/\rho V_{tank}$ | $\varepsilon_{bulk}$ $P/\rho V_{tank}$ | Source[b] |
|---|---|---|---|---|---|---|
| RT | 0.33 | 0.5 | 0.50 | 21 | 0.9 | 1 |
| RT | 0.5 | 0.5 | 0.0 | 21 | 0.5 | 2 |
| RT | 0.33 | 0.33 | 0.0 | 48 | 0.7 | 2 |
| PBT-6U | 0.33 | – | – | 37 | 0.7 | 3 |
| PBT-4U | 0.5 | 0.5 | 1.15 | 18 | 0.6 | 2 |
| PBT-4U | 0.33 | 0.33 | 0.85 | 37 | 0.7 | 2 |
| A310 | 0.55 | 0.5 | 0.68 | 19 | 0.4 | 2 |
| A310 | 0.35 | 0.25 | 0.85 | 40 | 0.7 | 2 |
| HE3 | 0.5 | 0.5 | 0.85 | 27 | 0.3 | 2 |
| HE3 | 0.33 | 0.33 | 0.80 | 99 | 0.3 | 2 |

### Other Geometries

| | |
|---|---|
| Pipe mixing | $\bar{\varepsilon} = (\Delta P V_S)/\rho L$, where $V_S$ is the superficial velocity. See Chapter 7 for calculation of $\Delta P/L$ for static mixers and other pipeline devices |
| Gas–liquid devices | See Tables 13-7 and 11-3 |
| Liquid–liquid devices | See Figure 12-11 |

[a] Fully baffled with four rectangular T/10 baffles, flat-bottomed tanks at fully turbulent Reynolds numbers (Re > $2 \times 10^4$). Tank diameters are 0.24 m, with traverses taken within 2 mm of the impeller blades. All fluctuations are included unless otherwise noted. Data cited are a small selection of data available in the literature.

[b] 1, Wu and Patterson (1989); blade passage fluctuations removed; 2, Zhou and Kresta (1996); 3, Medek (1980).

reactions, including (1) dissolving solid reactants and (2) two liquid-phase reactions when the reaction rate could be limited by the dissolution rate and the interfacial area, respectively.

**13-2.3.2 Semibatch Operation.** Semibatch reactors, also referred to as fedbatch, are very common in the specialty chemical and pharmaceutical manufacturing industries. Semibatch operations are typically carried out in a more-or-less standard type of stirred mixing vessel in both homogeneous and heterogeneous applications, although special provision is often required for fast and/or heterogeneous reactions (see examples in Section 13-3). Their use is very flexible in that they can be quickly reconfigured for various types of chemical reactions needed in a series of chemical synthesis steps (see Chapter 17). The key step of blending in semibatch operation is addressed in Chapter 9. Often, by feeding a particularly reactive reagent later or slower and in a region of high energy dissipation, the reactions can be forced along a more desirable path, producing a better yield of desired products. This is particularly true of consecutive-competitive reactions as discussed below.

### 13-2.3.3 Continuous Flow Operation.

Continuous flow mixed reactors are most common in high-capacity processing operations. Continuous flow reactions may be carried out in all the geometries discussed above: pipe and tee mixers, static mixers and other types of in-line mixers, and many types of stirred vessel. Sizes of such reactors range from very small tee mixer reactors on the scale of 1 cm to stirred vessels holding thousands of liters of liquid with impellers in the range of 3 m or more in diameter. Yields for very small continuous reactors may be studied in a pilot unit and applied directly to the plant, but yields for the very large reactors represent a severe design problem if the reactions are very mixing-rate sensitive. These and the semibatch reactors discussed above bring major scale-up problems, even for single-phase chemical reactions. In both continuous flow and semibatch reactors, feed location and local turbulence intensity have a major effect on yield. Multiple-phase chemical reactions cause even more complex scale-up problems, particularly if the mass transfer rate effects are compounded by chemical reaction yield effects in the reaction phase.

Continuous and semicontinuous flow reactors are sometimes used in fine chemical and pharmaceutical applications, primarily because some reactions require the high intensity of mixing that can only be achieved in in-line mixers (see Examples 13-3, 13-7, and 13-8a and discussion in Chapter 17).

An important question for the design of continuous flow systems is: When can the classic perfectly mixed assumption (ideal CSTR) be used in a continuous flow stirred tank reactor? The blend time concept can be used here. If the blend time is small compared to the residence time in the reactor, the reactor can be considered to be well mixed. That is because the residence time is proportional to the characteristic chemical reaction time. A 1 : 10 ratio of blend time to reaction time is often used, but often, larger values result because the mixer must do other jobs, which lead to even smaller blend times. Frequently, residence time distributions are used to determine whether a reactor is well-mixed. It is usually easy to achieve well-mixed conditions in continuous flow, turbulent stirred vessels unless the reactions are very fast, such as acid–base neutralizations. Even in laminar systems the blend time can be made much less than the required residence time for the chemical reaction mainly because required residence times are so long for high viscosity reactants. For discussions of residence time distribution analysis, see Chapter 1, Levenspiel (1972), and Nauman (1982).

In Chapter 9 it is suggested that if the batch blending time is less than one-tenth the residence time and the inlet and outlet are separated in such a way that a line drawn from the inlet to the outlet passes through the impeller, fully back mixed conditions will be achieved. Even in the case of a perfectly backmixed vessel, mixing effects on selectivity must also be checked.

### 13-2.4 Effects of Feed Point, Feed Injection Velocity, and Diameter

One of the most important concepts that comes from the micromixing theory is the importance of addition position for selectivity in competitive-consecutive homogeneous reactions. In Chapter 2 it was shown that there is a wide range of turbulent length scales and intensities in a stirred tank. The effect of position on mixing selectivity has been shown by a number of researchers, and several methods have been used to demonstrate this effect. A CFD simulation of this effect is included on the Visual Mixing CD affixed to the back cover of the book.

Nienow and Inoue (1993) gave an interesting set of examples using a small tank and the semibatch barium sulfate method of Villermaux to demonstrate the importance of feed position, as shown in Figure 13-8. In all cases the mixer speed and rate of addition were held constant; only the position of addition was changed. All vessels were at about the same power per unit tank volume. The selectivity given is that of unwanted by-product. High numbers mean that more by-product was formed.

Tank 1 was agitated by a radial turbine. The turbulence levels in the tank vary with position, and so does the local mixing rate leading to different by-product selectivities for different feed locations. At the high intensity region just entering the flow through the impeller or near the impeller tips, the turbulence is high and the by-product formation is low. At the top surface where turbulence levels are very low, the by-product selectivity is very high. Similarly, at the vessel bottom the by-product selectivity is poor. The radial impeller located off bottom delivers little turbulence to the bottom.

In tank 2, an axial down-pumping impeller is used. Again, feed at the surface has the most by-product formation, and feed in the impeller gives the best result. For the axial impeller, the feed position at the bottom of the tank is not bad because the impeller is delivering turbulence at that position in contrast with the radial impeller where the bottom has very high by-product selectivity.

Tank 3 is unbaffled. Near the impeller there is high turbulence and low by-product selectivity results. The surface and throat of the vortex with its rotational motion and poor incorporation have very high by-product formation.

Generally, the fastest, most immediate mixing of feeds or of a feed with resident fluid occurs when the feed is introduced into the region with the shortest local mixing time constant, or the most intense turbulence, whether in a pipe

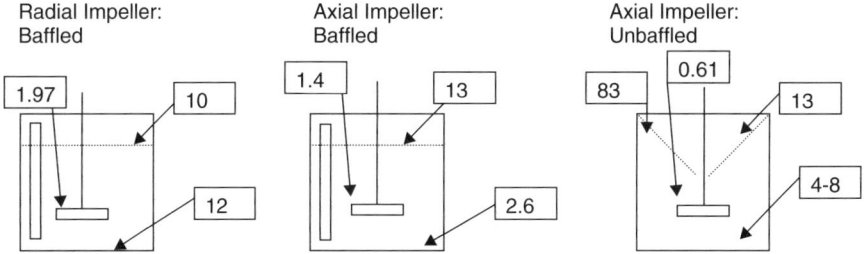

**Figure 13-8** Impact of different feed positions on the precipitation of barium sulfate. The selectivity to by-product as percent of reactant is shown for feed into zones of high and low turbulent energy dissipation. The impeller speed and reactant addition time were held constant. More by-product is formed at feed points where the local mixing is slow.

mixer, a static mixer, or a stirred tank. For instance, in a stirred tank the region with the fastest mixing is in or near the impeller discharge flow. The rate of turbulence energy dissipation is greatest and the scale of mixing is smallest in that region. For feed in other regions of a vessel, especially on the top liquid surface, the exposure of high fed reactant B concentrations at reduced mixing intensity can result in the most dramatic reduction in selectivity and increase in by-product formation.

Experimental measurements of yield and selectivity as a function of feed location have borne out the ideas expressed above (see Paul and Treybal, 1971; Bourne et al., 1981; Bourne and Rohani, 1983; Bourne and Dell'Ava, 1987; Baldyga and Bourne, 1990; Bourne and Yu, 1994; Baldyga et al., 1997).

Tipnis et al. (1994) investigated experimentally the scale-up of a competitive-parallel chemical reaction (the third Bourne reaction; see Section 13-2.5) in stirred vessels of 2.15, 20, 178, and 600 L. The vessels were geometrically similar, with two feed positions in each. The impellers were all six-blade disk turbines with $D/T = \frac{1}{3}$. The feed points were at two distances (G/T of 0.33 and 1.33) above the impellers near the impeller shaft. The chemical reaction rates at 298 K were as follows:

$$NaOH + HCl \rightarrow NaCl + H_2O \qquad k_{R1} = 1.3 \times 10^{11} \text{ L/g-mol} \cdot \text{s}$$

$$NaOH + CH_2ClCOOC_2H_5 \rightarrow CH_2ClCOONa$$
$$+ C_2H_5OH \qquad k_{R2} = 700 \text{ L/g-mol} \cdot \text{s}$$

For these conditions they found that equal blend time, which implies equal impeller rotational speed, N, gave nearly equal yields of $C_2H_5OH$ for all scales at each of the two feed locations. Plots of yield versus blend time up to 30 s for all scales gave nearly identical curves with little effect of feed concentration. Feed pipe backmixing (see Section 13-4.1.4) was not an issue since $v_f/v_t$ was relatively high. Both injection locations are in zones of relatively low turbulence, close to the impeller shaft. The implication is that tank blending rate ($\propto N$) is more important than local mixing rate $[\propto (\varepsilon/\nu)^{1/2}]$ for these competitive-parallel reactions with a nearly instantaneous first reaction. This is an example of mesomixing (see Section 13-2.1).

Constant blend time scale-up, however, leads to prohibitively high power requirements at large scale. Tipnis et al. (1994) recommended the use of a static mixer in a pump-around loop to reduce the total power requirement for these competitive-parallel reactions. This shows that high-energy intensities for a short time are a better way to distribute the energy than trying to generate a high intensity in a large tank. All the turbulence energy is focused on mixing a small volume in the confined space of the static mixer. See also Example 13-8a, in which essentially no significant yield of product could be achieved in a vessel, whereas the mixing capability of a static mixer resulted in satisfactory performance.

## 13-2.5 Mixing-Sensitive Homogeneous Reactions

Laboratory studies of mixing-sensitive homogeneous reactions have been done by many investigators using the four reactions that have become known as the *Bourne reactions*, developed through work by Bourne and co-workers (numerous references), and the reaction of iodine with tyrosine, first used by Paul and Treybal (1971). These studies and others fleshed out the previous theoretical predictions and established the experimental grounds for their confirmation. These experimental results have provided input for many modeling studies, some of which are discussed in Section 13-5. The reactions have been used by many investigators to study the mixing characteristics of stirred vessels in various configurations, including scale-up studies, as well as several types of in-line mixers.

The four Bourne reactions are as follows:

1. Diazo coupling between 1-naphthol and diazotized sulfanilic acid
2. Simultaneous diazo coupling between 1- and 2-naphthols and diazotized sulfanilic acid
3. Competitive neutralization of hydrochloric acid and alkaline hydrolysis of monochloroacetate esters with sodium hydroxide
4. Competitive neutralization of sodium hydroxide and acid hydrolysis of 2,2-dimethoxypropane with hydrochloric acid

Some of the key features of these reactions are summarized in Table 13-4. A case study from an industrial application follows.

*Example 13-3: Development of a Mixing-Sensitive Homogeneous Reaction.*
You are a process development engineer with laboratory and pilot plant facilities available for experimentation and gathering data for scale-up to manufacturing. Your current assignment is to develop a manufacturing process for a reaction that is now being run in flasks by chemists. The reaction is known to be competitive-consecutive:

$$A + B \rightarrow R + S1 + S2 + \text{higher MWs}$$

where R is the desired product and S1, S2, and higher MWs are overreaction products that both consume starting material, causing yield loss, and also react in the subsequent step to cause further yield loss. Separation before the subsequent steps is not feasible. The chemistry is shown in Figure 13-9.

*Q:* What do you want to know from the chemists before starting developmental studies and experiments?

*A:* Laboratory procedure and apparatus used; yield of R and analysis of other products.

*Procedure:* The chemists added B to A in a 1 L flask with good mixing with a paddle impeller. They made the addition in 1 min to minimize overreaction to S's. There is no significant exotherm. They obtained a yield to R of 90% using a B/A molar ratio of 1.0. Overreaction products were S1 = 5% and S2 = 2% with higher MWs not measurable. The A remaining was 5%.

**Table 13-4** Bourne Reactions

| Practical Range of $\tau_{rxn} = 1/k_2 C_{Bo}$ from 20 to 35°C, $\alpha = 1$ ms | Practical Range of Energy Dissipation, $\varepsilon =$ W/kg | Relative Ease of Use | Reagent Stability | Material Balance Close? | Method of Analysis | Oral Toxicity $LD_{50}$ (g/kg rat) | Reagents per m³ Solution at 350 ms | Waste Stream |
|---|---|---|---|---|---|---|---|---|
| *Competitive Consecutive* | | | | | | | | |
| *First: 1-naphthol with diazotized sulfanilic acid (Bourne et al., 1990)* | | | | | | | | |
| 65–5000 | Up to 400 | Difficult | Marginal | No (>95%) | UV multicomponent | 2 | 150 | <1 wt % dyes; water |
| *Competitive Consecutive and Competitive* | | | | | | | | |
| *Second: 1- and 2-naphthol with diazotized sulfanilic acid (Bourne et al., 1992)* | | | | | | | | |
| 30–5000 | Up to $10^5$ | Difficult | Marginal | No (>95%) | UV multiparameter | 2 | 150 | <1 wt % dyes; water |
| *Competitive (Parallel) Reaction* | | | | | | | | |
| *Third (a): Base hydrolysis of ethylchloroacetate vs. HCl (Bourne and Yu, 1994)* | | | | | | | | |
| 350–9000 | Up to 1 | Good | Marginal 4% in 30 min at 32°C and 130 mM starting soln. | Yes (>99%) | G.C. single component | 0.2 | 550 | <1% NaCl–acetate; <1% alcohol; salts; water |
| *Third (b): Base hydrolysis of methylchloroacetate vs. HCl (Bourne and Yu, 1994)* | | | | | | | | |
| 200–5000 | Up to 10 | Good | Marginal | Yes (>99%) | G.C. single component | 0.2 | 550 | <1% NaCl–acetate; <1% alcohol; salts; water |
| *Fourth: Acid hydrolysis of 2,2-dimethoxypropane vs. NaOH (Baldyga et al., 1998)* | | | | | | | | |
| 1–2000 | 1–$10^6$ | Excellent | Excellent <0.6%/day at 25°C and 200 mM DMP | Yes (>99%) | G.C. single parameter linear regression | 1 | 180 | 25% EtOH: <5% MeOH–acetone; salts; water |

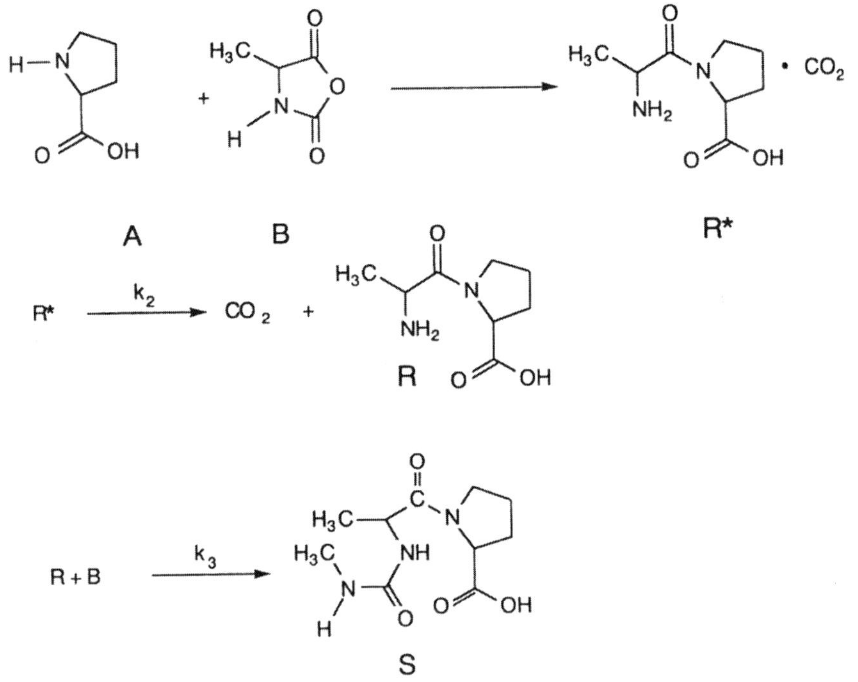

**Figure 13-9** Coupling reaction of L-alanyl-L-proline: chemistry of the consecutive-competitive reaction.

Q: *What experiments do you run before the pilot plant trials? What were the results?*
A: As time and starting material supplies permit, some runs in a 4 L cylindrical vessel with a fully baffled 6 cm. Rushton turbine at two speeds with two addition points and two rates of addition (Table 13-5).
Time did not permit completion of all the experiments. These results indicate mixing sensitivity, as noted by the improved yield at higher speed, matching the chemists' results. At this scale, however, it is noted that the yield differences are not large, so pilot scale operation can be started (material is needed for further studies).
Q: *What vessel and mixing system would you select for the pilot pant study? What results were obtained?*

Table 13-5

| Speed (rpm) | Addition Point | Feed Time (s) | k (m²/s²) | ε (m³/s³) | $\tau_M = 1.1k/\varepsilon$ (s) | P/V (W/m³) | R (%) | S (%) |
|---|---|---|---|---|---|---|---|---|
| 200 | Surface | 60 | 0.0052 | 0.031 | 0.18 | 38 | 82 | 12 |
| 400 | Impeller | 60 | 0.021 | 0.25 | 0.09 | 304 | 90 | 7 |

**Table 13-6**

| Speed (rpm) | Addition Point | Feed Time (s) | k (m²/s²) | ε (m²/s³) | $\tau_M$ (s) | P/V (W/m³) | R (%) | S (%) |
|---|---|---|---|---|---|---|---|---|
| 100 | Surface | 1000 | 0.0172 | 0.0527 | 0.347 | 77 | 80 | 15 |
| 150 | Impeller | 60 | 0.179 | 3.57 | 0.075 | 260 | 86 | 10 |
| 150 | Impeller | 1000 | 0.179 | 3.57 | 0.075 | 260 | 80 | 13 |

*A:* A fully baffled 200 L vessel with a 22 cm Rushton turbine. Addition points are on the surface and into the impeller discharge stream (Table 13-6).

These results continue to indicate mixing sensitivity, indicating that extreme caution must be taken on scale-up to manufacturing. The effect of addition time is not as expected for a classic consecutive-competitive reaction system, suggesting that the reaction pathway contains a step that requires maintaining short addition time on scale-up.

*Note:* This kind of unexplained difference in result is not uncommon, and in many cases, time does not permit finding the exact cause as long as the negative effect can be overcome by effective design.

*Q:* What vessel and mixing system would you specify for manufacturing that would achieve laboratory results?

*A:* The production requirements for this step of a multistep process require a 10 m³ (10 000 L) stirred tank. The sensitivity to mixing experienced in the experiments above indicates that scale-up to this size vessel may not be feasible to obtain expected yield. The complication of the unexpected sensitivity to addition time further indicates that an alternative design is indicated.

*Alternatives to be considered:*

- Multiple injection points in the tank
- Rapid recycle loop on a standard reactor for addition into the high-shear zone
- In-line mixer device in semicontinuous operation

In-line premixing of the reactants with a static mixer was selected since this method of operation was compatible with the overall process design, and the mixing intensity required could be expected at the size and throughput required.

*Solution:* A static mixer was developed successfully for production scale operation, as shown schematically in Figure 13-10. The mixer chosen was a static mixer with an L/D ratio of 4. The nominal residence time of the combined two-liquid-phase stream was 1 s. The Reynolds number in the mixer was 2000 based on empty tube diameter. The reactant mole ratio was 0.95 to 1.0 mol B/A.

Results for the static mixer in both laboratory scale 0.008 m (0.8 cm) and plant scale 0.0254 m (2.54 cm) operation were excellent. No change in selectivity or product distribution occurred over this scale-up. When there are compelling reasons to use a semibatch reactor instead of a semicontinuous system, the reactor

# PRINCIPLES OF REACTOR DESIGN FOR MIXING-SENSITIVE SYSTEMS 789

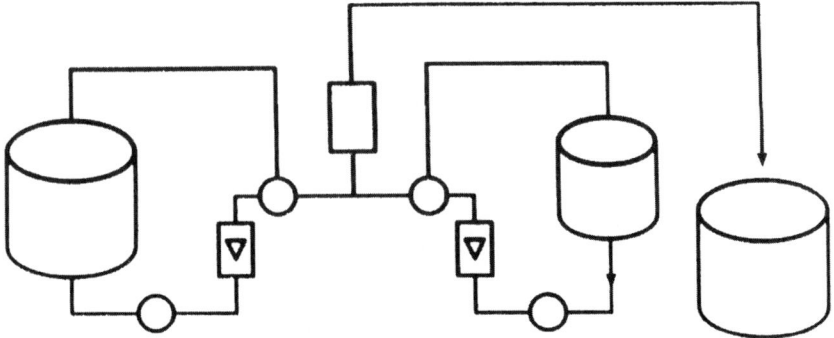

**Figure 13-10** In-line mixer for the L-alanyl-L-proline reaction: developed to maintain expected yield on scale-up to full scale. The pilot plant scale showed a drop in yield from the bench scale results. Intense local mixing is required.

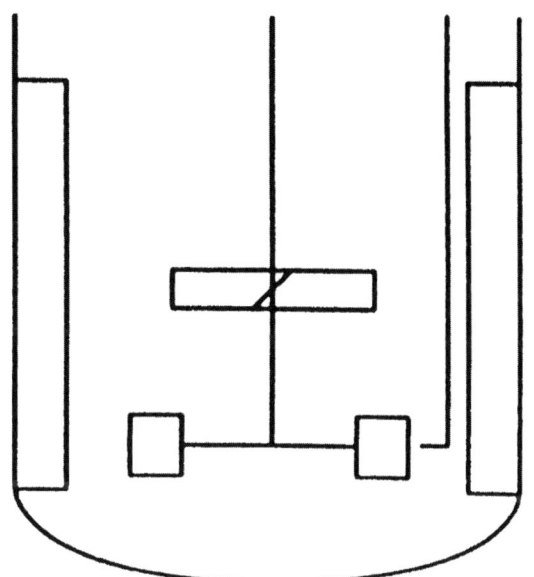

**Figure 13-11** Mixing configuration for semibatch reactors for mixing-sensitive reactions when in-line reaction systems are not viable. Note the feed directly into the region of highest turbulence and the second impeller used to maintain turbulence and flow in the top third of the tank.

shown in Figure 13-11 with dual turbines and a properly located subsurface addition line can provide the best scale-up opportunity to achieve expected selectivity for a fast, complex reaction.

*Message:* This example shows that such a reaction system requires feed addition directly into the fast-mixing zone to achieve maximum selectivity and to maintain that selectivity upon scale-up. The mixing intensity of a stirred vessel in semibatch operation was believed to be too low to maintain selectivity on scale-up. A static mixer was developed that achieved the required selectivity.

The increased local energy dissipation rate of the static mixer over the impeller region of a stirred tank is the key to this scale-up. Although not measured quantitatively in this system, the mixing literature (see Chapter 7 for static mixers, Chapter 12 and Table 8-1 for liquid–liquid dispersion devices, Table 13-9 for gas–liquid dispersion equipment, and Table 13-3 for stirred tanks) provides information that allows choices to be made. The effectiveness of the equipment can then be verified experimentally.

The change in selectivity on initial scale-up from the laboratory to a pilot plant vessel showed that mixing was a key issue without prior quantitative determination of rate constants, a step that is often not feasible in the time available.

### 13-2.6 Simple Guidelines

The concepts of micromixing and mesomixing can be reduced to a set of simple guidelines. For fast reactions having time scales of seconds and tens of seconds:

- Always add the ingredients to the point of highest turbulence. Avoid adding to the surface, a point of low turbulence.
- Scale-up and scale-down based on constant power per unit volume or mass. Even then there can be a loss of yield or by-product selectivity on scale-up.
- Consider using smaller reactors with higher energy dissipation rates, such as in-line mixers in recirculation loops.
- Consider diluting the incoming reagents.
- Question the size of the reactor.
- If experiments show a possibility of mixing reaction interactions and the rate of addition is important, consider multiple point injections. The feed time will have to be increased in large scale equipment.

## 13-3 MIXING AND TRANSPORT EFFECTS IN HETEROGENEOUS CHEMICAL REACTORS

When chemical reactors have more than one phase, the problem increases in complexity because the reaction and mass transfer processes interact. The interaction is governed by the relative rates of the reaction and mass transfer. In some cases, chemical reactions are mass transfer rate controlled (very fast chemical reactions) and in others they are reaction kinetics controlled (very slow chemical reactions); however, in reality very few reactions strictly fit this classification. To understand the complex interactions and the various variables involved, the following simplified discussion and equations may be useful in explaining certain topics of interest and the relations between key variables. Thorough discussions of this problem are given by Astarita (1967), Cichy and Russell (1969), and Schaftlein and Russell (1968).

For simplicity we assume a single reactant entering in one phase and reacting in the other. Examples are a gaseous reactant passing as bubbles through an agitated liquid in a tank, a solid powdered reactant being added to a liquid, and even two liquid phases, one dispersed in the other. The reactant that is being transferred we will call B. Initially, there is no B in the liquid. A further simplification will be to consider a gas phase as containing the reactant.

The general mass balance on B in the continuous liquid in units of moles per time is

$$\underbrace{k_L a' \Phi V (C_B^* - C_B)}_{\text{mass transfer rate}} \quad \underbrace{-rV}_{\text{reaction rate}} \quad \underbrace{-Q_L C_B}_{\text{flow out}} \quad = \underbrace{V \frac{dC_B}{dt}}_{\text{accumulation}} \quad (13\text{-}22)$$

In a steady-state batch liquid vessel, $dC_A/dt$ and $Q_L$ are zero and the mass transfer rate equals the reaction rate. All other cases are more complicated.

First consider the terms making up the mass transfer rate.

- **$k_L$** The first term is the overall mass transfer coefficient $k_L$. This is the reciprocal of the mass transfer resistance. This development is based on the film theory concept of mass transfer and consists of resistances in series: the resistance of the gas film (usually negligible), the resistance of the liquid film (the most important), and the resistance of any contaminant layer between the phases, such as solids or a surfactant. Again for simplicity we assume that all resistance is in the liquid phase. The variable $k_L$ is most dependent on the chemistry of the fluids. It can be estimated from surface renewal theory and related to molecular diffusivity and dispersed phase bubble or drop or particle size. In addition, the liquid film coefficient can be increased due to rapid reactions which effectively thin the diffusion layer. This is discussed below. $k_L$ estimates are often not reliable and are usually obtained by experiment in well-understood geometries.

- **$a'$** This is the area per unit volume of a bubble. Actually, it is an average of the total surface area over the total bubble volume. Thus, it can be written as $6/d$, where d is the average bubble or drop or particle size. In Chapter 11 there are several correlations for this value based on a large variety of experiments. The average drop size is the result of the combination of drop or bubble breakup and coalescence. Breakup is determined by fluid forces and the surface or interfacial tension-resisting force. Coalescence is controlled by various physiochemical effects, such as double layers and the presence of surface-active agents. For example, coalescing systems such as air and water will have a certain bubble size under a given set of agitation conditions. The addition of small amounts of salt will decrease coalescence and make smaller drops, giving larger holdups and increased mass transfer. Addition of surfactants will also reduce bubble size and increase holdup, but the extra film resistance often balances this effect, leading to no increase in mass transfer rate.

- **Φ** This is the holdup as a volume fraction of fluid in the vessel. It is the volume of dispersed phase (e.g., gas) in the vessel divided by the total volume. This is a variable strongly affected by the mixing conditions. In Chapter 11 there are several correlations for this variable. Holdup (Φ) times $a'$ gives total mass transfer surface area per unit vessel volume, which is often called $a$. Thus, one often sees correlations of $k_L a$ versus mixing parameters. It should always be remembered that this value contains implicitly the holdup and the bubble size. One way to think of holdup is as the ratio of the superficial gas velocity to the bubble rise velocity. This comes from a simplistic picture of the motion of the gas phase:

$$Q_G = A v_s = A \Phi v_r \tag{13-23}$$

where A is the cross-section of the vessel, $v_s$ is the superficial gas velocity, and $v_r$ is the effective rise velocity of the gas.

The advantage of this concept is that it shows the strong effect to be expected of the gas superficial velocity on mass transfer. This is certainly found experimentally. Assuming a typical rise velocity of gas bubbles of 0.3 m/s, it gives a crude estimate of holdup. As gas volume fraction increases, hindered rising and bubble swarms break down this simple relation. This relation also shows that a decrease in bubble size which leads to more surface area, and slower rise velocity results in more holdup.

Bubble size, holdup, rise velocity, and area per unit volume are all tied together in a complex way.

- **$C_B^*$** This is the saturated concentration of B in the liquid in equilibrium with the other phase (e.g., the gas phase). This relation is a thermodynamic property (such as a Henry's law coefficient) and is affected only by pressure and temperature, not by fluid dynamics or mixing.
- **$C_B$** This is the concentration of B in the liquid phase. This is the variable that is most affected by the transport and reaction rates. We return to this below.
- **r** This is the reaction rate and for a simple reaction could be expressed as $k_R C_B$. However, there is no necessity to use such simple forms. In most cases the reaction rate will depend on the concentration of the transferred ingredient in the liquid/continuous phase.
- **$Q_L$** This is the flow rate of liquid phase from the vessel. It can be zero in semibatch operation.

Now consider the effect of reaction rate. For a given design all the other variables are fixed except $C_B$. For a fast reaction rate the maximum mass transfer rate will occur when $C_B$ is zero. This condition is sometimes called *mass transfer* *control* because the reaction rate is fixed by mass transfer limitations. Such terminology often creates confusion. $C_B$ cannot be zero but only small compared to the saturation concentration $C_B^*$. It must be finite for there to be any reaction. One can think of the reaction rate being limited by the rate of mass transfer. In this case it is quite likely that the mass transfer rate will be enhanced by the reaction, and the mass transfer rate with reaction will be faster than without reaction.

If the reaction rate is very slow, the concentration difference between $C_B^*$ and $C_B$ grows closer. In the limit, $C_B$ is equal to $C_B^*$ and the maximum reaction rate is obtained at the saturation composition. It almost all cases it is assumed that the continuous or liquid phase is well mixed, so that no gradients exist. This is true in most equipment because the blend time is usually small compared to the mass transfer time. This means that $C_B$ is the same at all places in the vessel.

There is another equation to consider: the reactant balance in the gas (mol/time):

$$Q_G(Y_{Bo} - Y_{Bi})K_H - k_L aV(C_B^* - C_B) = \frac{\Phi V \, d(K_H Y_B)}{dt} \qquad (13\text{-}24)$$

where $Q_G$ is the volumetric flow rate of the gas (m³/s), or more generally of the dispersed phase; $Y_{Bo}$, and $Y_{Bi}$ are the outlet and inlet mole fractions of B in the gas phase; $\Phi V$ is the volume of the dispersed phase (e.g., the gas); $K_H$ is the Henry's law constant, relating $C_B$ and $Y_B$ [(mol/m³)/(mole fraction)]. Thus, the value of $C_B$ depends on what goes on in the second phase, but this equation shows a problem. With what concentration in the gas phase is $C_B^*$ associated? For a continuous gas flow ($dC_B/dt = 0.0$) there are several choices. If the gas is backmixed so that the volume $\Phi V$ is all of the same composition, $Y_B$ is given by the outlet composition $Y_{Bo}$. If the gas phase is not backmixed and has gradients in it, an average of the inlet and outlet concentrations needs to be used. This can often be a simple average. A log mean concentration difference can be used if the approach is close, as in heat transfer. Now we may need to know something about the residence time distribution of the gas or other second phase, even if it is dispersed.

In many mass transfer operations the effect of gas phase residence time distribution is neglected. In fermentations and in wastewater aeration systems, a 15% consumption of the oxygen from the inlet air is on the high side. This translates to going from 21% oxygen to 18% at the outlet. If the air is backmixed, the gas phase composition in equilibrium with the liquid would be based on 18% oxygen. If a simple plug flow assumption is made instead, the composition only rises to 19.5%. This is a minor effect.

When chemical reactions such as organic oxidations are present, more of the reactant is often removed from the gas phase. In organic oxidations it is not uncommon to have exit oxygen composition as low as 3%. This is because of the fast reactions and also for safety reasons. The well-mixed composition would then be 3% and the plug flow average would be 12%. This leads to a factor of 4 difference when estimating the composition in the liquid. Gas phase residence time distributions are thus of more interest when reaction and mass transfer interact. This is discussed in Section 13-3.4.

*Time Constants.* If the transport equation were a bit simpler, one could treat reaction with mass transfer as a process of rates in series, such as heat or mass transfer. One could compare half-times of reaction with inverse mass transfer coefficients and even include mixing times as a rate constant. Although an interesting thought, it is hardly ever done.

Some time constants that have been used are

- $\Phi V/Q_G$ is the gas residence time.
- $(1 - \Phi)V/Q_L$ is the liquid residence time.
- $C_B^* V/r$ is a measure of how long the reactor can coast without mass transfer. It applies when the full liquid volume is saturated, and then the gas is suddenly turned off.

For second phases other than gas, the relations are much simpler. For liquid–liquid and liquid–solid systems both phases are usually considered backmixed. Note that there is still holdup and it may not be the inlet volume fraction that is often assumed.

### 13-3.1 Classification of Reactivity in Heterogeneous Reactions

Astarita (1967), Levenspiel (1972), and Doraiswamy and Sharma (1984) describe an effective framework in which to evaluate the relative contributions of mass transfer and reaction kinetics in heterogeneous systems. This classification is as follows.

- *Regime 1: very slow reactions.* Reaction rates are much slower than the mass transfer rate, so that reaction follows homogeneous kinetics and the reactions are not affected by the mixing and mass transfer rates. The reactants are supplied to the reacting zone at the expected molar ratio, resulting in no departure in conversion or selectivity from that predicted by the reaction rate constants and their ratio. Mixing would not affect the reaction, assuming that the reagents are blended. Only very poor bulk or macromixing (i.e., solids settled on the bottom or large dispersed phase drop size) could result in slow conversion.
- *Regime 2: slow reactions.* The reaction rate is fast enough that significant reaction occurs in the film between the reactants, but the consecutive or competing reactions are slow relative to the primary reaction. In this case, the conversion rate would be slower than expected, but selectivity would be unaffected.
- *Regimes 3 and 4: fast and very fast reactions.* These regimes are combined for purposes of this simplified discussion, although Doraiswamy and Sharma (1984) treat several subsets within these regimes as a function of relative reaction and mass transfer rates. In all cases in these regimes, both

conversion rate and selectivity are affected by mixing and reactor design. In most cases, selectivity is reduced by a restricted supply of reagents in and through the films between the phases. However, selectivity can be improved significantly by manipulation of the interfacial conditions. Mixing design and scale-up are critical to successful performance in manufacturing for these regimes.

The classification system by Doraiswamy and Sharma was treated quantitatively for gas–liquid systems by Middleton (1992), as summarized in Chapter 11 and Section 13-3.4. Middleton describes five regimes instead of four, but in general the classifications are similar. Deviations in the case of homogeneous reactions are more amenable to quantitative analysis and can therefore be developed more completely. The same local considerations developed in Section 13-2 for homogeneous reactions apply for heterogeneous reactions where expected overall molar ratios between reactants cannot be maintained. In heterogeneous reactions, mass transfer limitations at phase boundaries as well as local mixing limitations may affect the reaction. For simple reactions overall reaction rates may be affected and usually decrease, but yield is unaffected given equal degrees of conversion. For complex reactions, the selectivity may be decreased, but unlike homogeneous systems, may also be increased under certain circumstances (see Example 13-8a).

The other key difference between homogeneous and heterogeneous reactions regarding selectivity is that significant selectivity effects can occur in heterogeneous systems at far lower absolute reaction rates because the mass transfer limitations can be very severe. In addition, these effects can be subject to considerable magnification on scale-up to plant operations. These effects can be visualized as changing the inherent $k_{R2}/k_{R1}$ ratio, as measured by independent determination of the rate constants, to an apparent value caused by mass transfer limitations.

In Section 13-3, examples are used to illustrate the mixing issues that can be significant for various types of heterogeneous systems. The analysis of deviations from ideal behavior in homogeneous systems applies in many of these cases. Homogeneous reactions are more amenable to quantitative analysis and can therefore be developed more completely. The extension of the principles to heterogeneous systems will be more qualitative because of the complexity of these systems.

Heterogeneous systems can in some cases be manipulated to achieve improved yields compared to a homogeneous system with the same reactions. There can, therefore, be a great advantage in running under heterogeneous conditions or in some cases to deliberately creating a heterogeneous system for the purpose of improving selectivity. See Example 13-8a for an illustration.

### 13-3.2 Homogeneous versus Heterogeneous Selectivity

The discussion of selectivity considerations in homogeneous reactions in Section 13-2 provides an introduction to the far more complex issues involving heterogeneous reactions. The continuity of theoretical and practical considerations between these different types of reacting systems is provided by the obvious

**796** MIXING AND CHEMICAL REACTIONS

fact that the course of reactions is determined by events at the molecular scale, whether or not the reactive molecules are in the liquid, solid, or gas phase when they enter the reaction zone. As in the case of homogeneous reactions, the course of a complex reaction will be determined by local molar ratios and chemical kinetics. The degree of deviation from expected kinetic behavior is determined by the reaction rate relative to the rates of mass transfer and mixing. The possible chemical interactions in the film around a dissolving reagent particle, a reactive gas bubble, or a dispersed liquid drop are illustrated in Figure 13-12 for a consecutive-competitive reaction. B is added to A, and B is the limiting reagent. The reaction of A with B to form the desired product, R, is occurring along with the normally undesired reaction of R with B to form S. In the heterogeneous case, there is a mass transfer boundary between A and B. Although a lot of desirable product R is formed, it quickly reacts with high concentrations of B to form undesirable product S. Differences in selectivity between the same reaction run under homogeneous conditions and heterogeneous conditions are illustrated in Example 13-4.

*Example 13-4: Competitive-Consecutive Reaction—Solid–Liquid Compared with Homogeneous (Homsi et al., 1993)*

- *Goal:* development of a solid–liquid competitive-consecutive reaction system

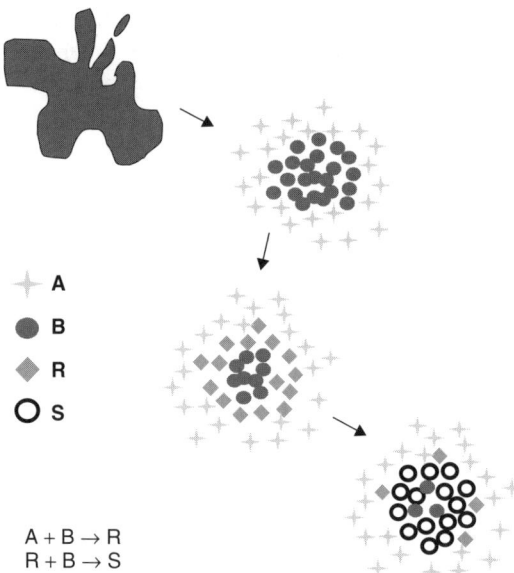

**Figure 13-12** Simultaneous mass transfer and reaction in the films around solid particles, gas bubbles, and liquid drops. For a heterogeneous competitive consecutive reaction, mass transfer rates, reaction rates, and mixing rates can all play a role.

- *Issue:* laboratory selectivity not reproducible in the pilot plant
- *Classical bromination:* homogeneous versus heterogeneous selectivity

This example compares a reaction run using reagent addition as a dissolving solid and the same reagent added in solution. The two reactions were run in the same pilot plant equipment with the same mixing conditions. The data for the product distribution in this consecutive-competitive reaction system allow direct comparison of product distributions obtained under homogeneous and heterogeneous conditions.

The reaction shown in Figure 13-13 is a classical competitive-consecutive bromination to mono- and dibromo-substituted products where the mono-substituted product is the desired product. (Both 3- and 5-bromo products are acceptable for the following steps.) Dibromo formation represents a yield loss both in this step and in the reaction steps to follow. The reaction is run in the semibatch mode in all cases. The dissolving reagent is $N$-bromosuccinimide (NBS), and the reaction solvent is acetone. The pilot plant conditions are shown in Table 13-7. The NBS is added over a 6 h period because the reaction is very exothermic. The actual reaction rate is not known, but the addition requires 6 h for heat removal. The impeller is a six-blade Rushton turbine.

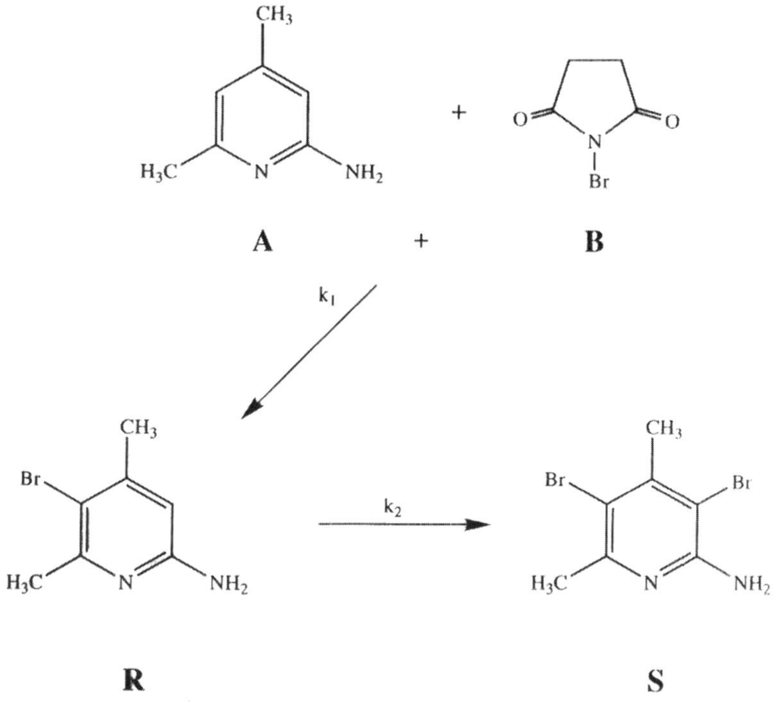

**Figure 13-13** Chemistry of a classical consecutive-competitive bromination reaction subject to mixing effects.

**798** MIXING AND CHEMICAL REACTIONS

**Table 13-7** Pilot Plant Conditions for Bromination with *N*-Bromosuccinimide

| Variable | Pilot Plant Condition |
|---|---|
| Vessel volume (m$^3$) | 0.75 |
| Vessel diameter, T (m) | 1 |
| Impeller diameter, D (m) | 0.4 |
| Impeller speed, N (rpm) | 175 |
| Reaction volume H/T | 0.5 |
| Power/volume (W/m$^3$) | 117 |
| Local power at point of solution addition (W/m$^3$) | 1170 |

*Results from powder addition of NBS:*

- *Laboratory:* 91% monobromo, 2% dibromo. This is an acceptable level of impurity. The relative rates of reaction and mixing are not known or suspected to be a problem. Scale-up to the pilot plant is attempted.
- *Pilot plant:* 83% monobromo, 8% dibromo. This is an unacceptable increase in overreaction to dibromo. The apparent rate constant ratio, $k_{R2}/k_{R1}$, for the two scales can be calculated from the product distributions [eq. (13-5)]. This apparent ratio increased from 0.02 to 0.08, resulting in a decrease in selectivity.

Mixing effects in the film around the dissolving NBS are the obvious reason—the reaction rate is fast enough to allow significant reaction in the film before the dissolved NBS can be mixed to the molecular level. This indicates that the mass transfer rate is slower than the reaction rate.

*Possible solutions:*

1. Reduce the particle size of the NBS by milling to reduce dissolution time.
2. Eliminate mass transfer limitations by predissolving NBS in the reaction solvent and running as a homogeneous reaction.

*Evaluation of alternative solutions:*

1. Reduction in particle size can reduce dissolution time, but its overall effect in reducing reaction in the films around the dissolving particles may not be sufficient. Also, milling of a noxious material such as NBS is not feasible.
2. For a soluble reagent, another alternative is to predissolve it and add it as a solution. The mixing time could be further decreased and could achieve a significant reduction in Da$_M$.

MIXING AND TRANSPORT EFFECTS IN HETEROGENEOUS CHEMICAL REACTORS   799

*Solution:* Option 2 was run in the laboratory and was shown to reduce dibromo below that obtained with a powder addition (<1%). The same reduction was achieved in the pilot plant under the conditions shown in Table 13-8.

*Message:* This example indicates that the homogeneous reaction environment is more selective than the film around a dissolving reagent for a consecutive reaction. The result can be represented as an increase in the apparent rate constant ratio, $k_{R2}/k_{R1}$, for the heterogeneous condition as indicated in Figure 13-14, where the loss in selectivity (increase in $X_S$) is plotted against $k_{R1}$.

**Table 13-8** Percent Yield of Dissolving Solid and Homogeneous Reactions under Identical Reactor Configurations[a]

| Compound | Lab Solid NBS Addition | Batch 1 Solid NBS Addition | Batch 2 Solution Addition |
|---|---|---|---|
| 5-Bromo | 87 | 75 | 82 |
| 3-Bromo | 4 | 8 | 7 |
| Dibromo | 2 | 8.4 | <1 |

[a] N-Bromosuccinimide is the reactant and 5- and 3-bromo are the desired products.

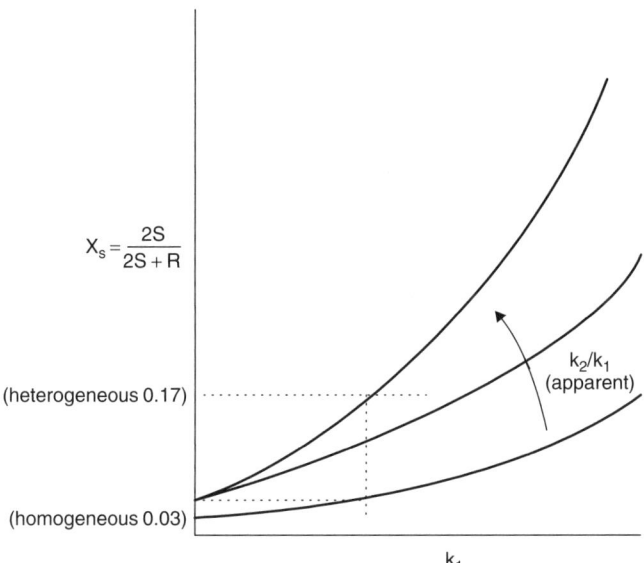

**Figure 13-14** Correlation of apparent rate constant ratio with reaction rate and impurity selectivity (Xs); the effect of the mass transfer limitation on a dissolving solid can be shown as an increase in the apparent $k_2/k_1$ ratio.

As in Example 13-3, time was not available to measure individual rate constants. However, the laboratory and initial poor pilot plant results showing the effect of scale-up on product distribution were sufficient to illustrate the mixing sensitivity of the reaction. The key to solving the problem was not to try to improve mixing but to eliminate the mass transfer and local effects of a dissolving powder by changing the process to use a solution addition. An improvement was realized at both the laboratory and pilot plant scales.

### 13-3.3 Heterogeneous Reactions with Parallel Homogeneous Reactions

The yield and selectivity of heterogeneous reactions can also be affected by mass transfer in extending the time for completion of a reaction during which a parallel reaction—possibly decomposition of A, B, or R—can be occurring in the bulk phase as well as in the films around the dispersed phase (or in the dispersed phase for liquid–liquid reactions). This problem can develop when the desired reaction rate can only be achieved at a temperature at which the starting materials, any intermediate, or the product can react or decompose during the reaction time. This reaction time can be longer than expected on scale-up if the mass transfer rates do not duplicate those in the laboratory or in piloting. Shorter overall reaction times can also be realized on scale-up when mass transfer rates are increased by improved mixing (e.g., for liquid–liquid or gas–liquid dispersion). Reasons for slower mass transfer and extended reaction time for each type of contact are as follows:

- *Gas–liquid:* lower $k_L a$ because of insufficient gas dispersion–holdup and surface area
- *Solid–liquid:* slower dissolution time because of variation in reagent particle size and mass transfer
- *Liquid–liquid:* larger dispersed phase drop size and higher coalescence rates than expected

All of these factors are mixing dependent and can contribute to scale-up difficulty if mass transfer rates are not reproduced successfully.

### 13-3.4 Gas Sparged Reactors

Gas sparged chemical reactors are designed and used in many different geometries. These reactors are usually continuous in gas, and batch or continuous in liquid. Some of the geometries in use are bubble columns, pipe and static mixer reactors, stirred vessels, packed columns, tray columns, spray columns, jet loop reactors, and venturi ejector reactors. Design equations for each geometry are based on correlations and simplifying assumptions, such as uniform $k_L a$ in the stirred vessel. Other gas–liquid reactors include spray columns and spray combustors.

**Table 13-9** Comparison of Different Gas–Liquid Contacting Devices[a]

| Device | $k_L a$ ($s^{-1}$) | V ($m^3$) | $k_L aV$ ($m^3/s$) | $a'$ ($m^2/m^3$) | $1-\Phi$ | Liquid Flow | Gas Flow | P/V ($kW/m^3$) |
|---|---|---|---|---|---|---|---|---|
| Baffled agitated tank | 0.02–0.2 | 0.002–100 | $10^{-4}$–20 | ~200 | 0.9 | ~Backmixed | Both | 0.5–10 |
| Bubble column | 0.005–0.01 | 0.002–300 | $10^{-5}$–3 | ~20 | 0.95 | ~Plug | Plug | 0.01–1 |
| Packed tower | 0.005–0.02 | 0.005–300 | $10^{-5}$–6 | ~200 | 0.05 | Plug | ~Plug | 0.01–0.2 |
| Plate tower | 0.01–0.05 | 0.005–300 | $10^{-5}$–15 | ~150 | 0.15 | Both | ~Plug | 0.01–0.2 |
| Static mixer | 0.1–2 | 0.001–10 | 1–20 | ~1000 | 0.5 | ~Plug | Plug | 10–500 |

[a] Approximate values for typical cases; not to be used for design of specific processes. Middleton (see Table 11-3) and Lee and Tsui (1990) found similar characteristics. Note that the characteristic mass transfer time is given by $1/k_L a$.

Typical performance values such as $k_L a$, $a'$, $\Phi$, and P/V for each geometry are given in Table 13-9. Of interest is that $k_L$ ranges from 0.0001 m/s for the packed, spray, and tray columns to 0.001 m/s for the stirred vessel and jet loop reactors. The pipe and static mixer reactors and the bubble columns are intermediate. The result of this observation is that for a given reactor, such as the stirred vessel, the local mass transfer rate is probably approximately proportional to the mass transfer area per unit volume, $a$, since the $k_L$ values are not very sensitive to hydrodynamics. Of course, fluid properties can also affect $k_L$ values, and they must be taken into account.

Reactions most commonly occur in the liquid phase in gas–liquid reactors. The most likely exception to this is spray combustors in which the reactions occur in the gas phase after or as the liquid droplets vaporize. Usually, in chemical reactors one reactant is transferred from the gas phase to the liquid phase, where the chemical reactions occur, as in chlorinations, oxidations, and hydrogenations.

If the time scale of a chemical reaction is short compared to the time scale of mass transfer, the mass transfer slows the chemical reaction but can also cause the concentration in the liquid-side mass transfer film to be decreased, resulting in an increased driving force and an enhanced mass transfer rate. Levenspiel (1999) and Middleton (1992) present diagrams for the interface concentration profiles likely to happen at the various reaction rates relative to the mass transfer rate. Those are shown in Table 13-10 along with estimates of the ranges of variables for the various regimes [similar to those of Doraiswamy and Sharma (1984)] and important variables for design and scale-up.

Levenspiel (1999) has estimated the effect of liquid-phase chemical reaction on the mass transfer coefficient, $k_L$. The modified coefficient is given here as $k_L^*$. For Middleton's regime V, where the chemical reaction is so fast that the reaction front is within the mass transfer film, the modified mass transfer coefficient for the gas phase component of the reaction is

$$k_L^* = k_L \left(1 + \frac{D_{AL} C_A}{D_{BL} C_B^*}\right) \quad (13\text{-}25)$$

**802**  MIXING AND CHEMICAL REACTIONS

**Table 13-10  Various Gas–Liquid Reaction Regimes and Parameters of Importance**

| Regime | Conditions | Important Variables | Concentration Profile |
|---|---|---|---|
| I. Kinetic control, slow reaction | $(t_D/t_R)^{1/2} < 0.02$ | Rate $\propto \theta_L$ <br> $\propto K_R C_{AL} C_{BL}$ <br> Independent of $a$ <br> Independent of $K_L$ | GAS \| LIQUID; film \| bulk; $C_{AL}^*$ — $C_{BL}$; $C_{AL}$ |
| II. Diffusional control moderately fast reaction in bulk of liquid, $C_{AL} \approx 0$ | $0.02 < (t_D/t_R)^{1/2} < 2$ <br> Design so that <br> $\Phi_L/a > 100 D_{AL}/K_L$ | Rate $\propto a$ <br> $\propto K_L$ <br> $\propto C_{AL}$ <br> Independent of $K_R$ <br> Independent of $\Phi_L$ <br> if $\Phi_L$ is adequate | $C_{AL}^*$ — $C_{BL}$; $C_{AL}$ |
| III. Fast reaction in film $C_{AL} \approx 0$ (pseudo first-order in A) | $2 < (t_D/t_R)^{1/2}$ <br> $< C_{BL}/q C_{AL}^*$ <br> $C_{BL} \gg C_{AL}^*$ | Rate $\propto a$ <br> $\propto K_R^{1/2}$ <br> $\propto (C_{AL}^*)^{(n+1)/2}$ <br> Independent of $K_L$ <br> Independent of $\Phi_L$ | $C_{AL}^*$ — $C_{BL}$; $C_{AL}$ |
| IV. Very fast reaction, general case for regime III | $2 < (t_D/t_R)^{1/2}$ <br> $C_{BL} \approx C_{AL}^*$ | Rate $\propto a$ <br> depends on <br> $K_L, K_R, C_{AL}^*, C_{BL}$ <br> Independent of $\Phi_L$ | $C_{AL}^*$ — $C_{BL}$; $C_{AL}$ |
| V. Instantaneous reaction at interface, controlled by transfer of B to interface from bulk, $J \propto K_L a$ | $(t_D/t_R)^{1/2} \gg C_{BL}/q C_{AL}^*$ | Rate $\propto a$ <br> $\propto K_L$ <br> Independent of $C_{AL}^*$ <br> Independent of $K_R$ <br> Independent of $\Phi_L$ | $C_{AL}^*$ — $C_{BL}$; $C_{AL}$ |

where $C_A$ is the bulk concentration of the liquid-phase reactant and $C_B^*$ is the concentration of the gas phase reactant in the liquid at the interface (the equilibrium concentration). For Middleton's regime IV, the very fast reaction regime, the modified mass transfer coefficient is

$$k_L^* = (D_{BL} k_R C_A)^{1/2} \qquad (13\text{-}26)$$

For both eqs. (13-25) and (13-26), the chemical reactions are for stoichiometric coefficients and kinetic rate exponents of one for both components. Middleton and Levenspiel give equations for general coefficients, but such cases are not common for fast reactions, since more than one reaction is usually involved in such cases. Note also that these equations assume that the gas phase mass transfer rate is very high relative to the liquid-phase rate. Levenspiel gives the equations for cases where the gas phase mass transfer rates could affect the value of $k_L^*$. A very general analysis of the interactions of chemical reaction and mass transfer was presented by Astarita (1967).

Middleton indicates that for his regime I (very slow reaction), where $k_L$ is little affected by the chemical reaction, the interface surface area per unit volume, $a$, is of little importance since the reaction takes place in the bulk liquid phase, so a bubble column is the typical reactor of choice. For Middleton's regimes II, IV, and V—diffusional control, very fast reaction, and instantaneous reaction, respectively—both high $a$ and $k_L^*$ are needed, so a stirred tank is the typical reactor recommended. In regime III—reaction in the mass transfer film—the most important variable is the interface area, so a packed column yielding much liquid surface area may be appropriate.

If a detailed simulation of the local mass transfer rates and reactions rates in the reactor is not to be done, a key question in attempting to design a gas–liquid reactor is the residence time distribution of the gas phase. In stirred reactors the liquid phase is usually well mixed because of the necessity to disperse the gas adequately, but the gas flows can range from plug flow to well mixed, depending on the gas rate and agitator design. This can have a significant effect on driving force for the mass transfer rates if the inlet and outlet concentrations of the reactant in the gas are significantly different, as shown in the introduction to Section 13-3. In many pure-gas mass transfer cases, the change in gas composition is insignificant and residence time distribution has little effect on mass transfer rates. With highly reactive systems, such as oxidizers, where almost all of the reactant is consumed leaving mostly inert gas components, the difference in residence time distributions can have a large effect. For example, if a reactor has an air feed at 21% oxygen but a gas outlet oxygen fraction of only 4%, whether the gas is well mixed in the vessel or travels through in an almost plug flow manner will affect the overall mass transfer rate. If the gas is well mixed, most of the bubbles are at 4% oxygen, which determines the driving force for the mass transfer. It the gas is in plug flow, an average of inlet and outlet concentrations is closer to determining the overall driving force, which would give about a threefold increase in mass transfer rate for a very fast reaction in the liquid if the same interfacial surface area is produced. Of course, the well-mixed case would probably have much greater surface area for mass transfer, so the problem is not so simple.

When competitive reactions exist in gas–liquid systems, the yield can be strongly affected by the rate of mass transfer. This is an area of continuing investigation. In some cases the rate at which products are removed from the interface can affect the yield if the second reaction is fast. In oxidations and chlorinations where overreaction can lead to undesirable by-products, the rate of mixing of the products with the bulk fluid can help reduce the overreaction effect. In such cases the placement of the sparger for the gas might not noticeably affect the rate of mass transfer from the gas, but could affect the level of by-product formation. In this case the liquid mixing is used to remove products from the bubble surface, not to achieve the micromixing, which is important for single-phase reactions. A case study of a chlorination reaction is given in Example 13-6.

### 13-3.4.1 Gas–Liquid and Gas–Liquid–Solid Reactions, Gas as Reagent.

With the exception of fermentation, which is the subject of Chapter 18, one of the most common gas–liquid reactions is hydrogenation. The intrinsic reaction rates of hydrogenation reactions vary over several orders of magnitude and can fall into any of the categories discussed above. Design of a hydrogenation system is generally focused on supplying a sufficient quantity of hydrogen so that hydrogen concentrations in the bulk or adsorbed on the catalyst will not be limiting. In many cases, this can be accomplished by suitable design of a subsurface sparger to accomplish absorption during transit from the sparger discharge to the vapor space. In some cases, however, the absorptivity or reaction rate is slow enough that reabsorption from the pressurized vapor space is required. This can be accomplished with alternative mixing systems that are described in Lee and Tsui (1999).

The Editors' Introduction to this book contains a discussion of the importance of mixing configuration in hydrogenation. In this example, a reaction in a laboratory autoclave with a large H/D ratio (>2) appears to be very slow when the limitation is actually ineffective sparging and lack of surface reincorporation. If this limitation is not recognized, translation to a well-mixed vessel with surface reincorporation can result in very rapid and unexpected hydrogen uptake. Heat removal could then become critical.

Surface reincorporation can be accomplished by modifications of the standard reactor configuration, including that of the Praxair AGR system, which is shown in Chapter 6. Other systems employing high recycle in loops with gas induction are also effective, as discussed in Chapter 11. Another alternative using surface reincorporation follows.

### *Example 13-5: Hydrogen Uptake as a Function of Vessel Mixing Configuration*

- *Goal:* determination of the cause of greatly reduced hydrogen uptake rate in a manufacturing scale hydrogenator compared to a laboratory autoclave and pilot plant reactor
- *Issue:* initial manufacturing scale reaction rate unacceptably slow

Scale-up of a Raney nickel-catalyzed reduction of a phenylazo-substituted pyrimidine to a triaminopyrimidine in a 6 m$^3$ (6000 L) vessel required careful configuration of the hydrogen sparger, turbine agitator, and baffles. The reaction is run in water at 130°C and 8 bar hydrogen pressure. The first batch run in the 6000 L vessel during startup of production facilities had an extremely slow hydrogen uptake rate compared with expected uptake based on pilot plant experience. The fully baffled single-turbine impeller and sparger configuration is shown in Figure 13-15a.

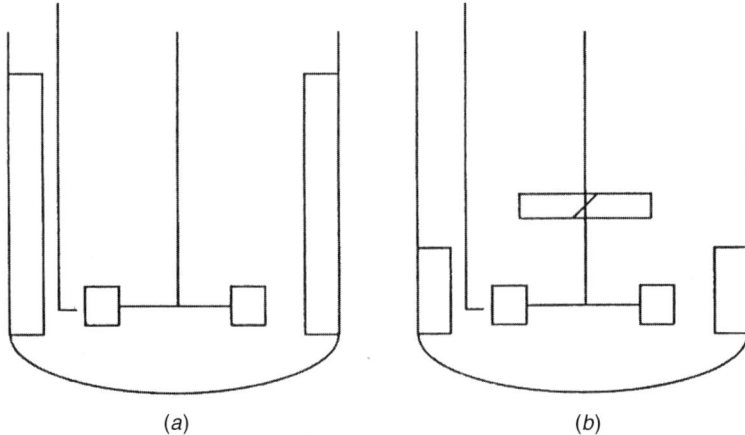

**Figure 13-15** Hydrogenation internal configurations for gas dispersion: (*a*) an ineffective sparger configuration compared to (*b*) a design for increasing surface reincorporation.

Modification of sparger location, baffles, and upper axial flow turbine size and location, as shown in Figure 13-15*b*, was successful in achieving (and exceeding) the expected hydrogen uptake rate. These modifications were based on the recommendations by Oldshue (1980), where vortexing at the upper surface is used to incorporate hydrogen from the vapor space to augment that injected by sparging. These recommendations include shorter baffles and an axial flow turbine near the top surface in addition to a flat-blade lower turbine near the sparger for efficient gas dispersion.

*Message:* The modified configuration achieved efficient vortexing, and effective surface re-incorporation and rapid hydrogenation was established. It is apparent from the dramatic improvement accomplished with surface reincorporation of hydrogen that very little hydrogen uptake was accomplished with the initial ineffective sparger placement and the fully baffled single turbine. Unreacted hydrogen broke through to the vapor space, which became pressurized. The vapor space pressure then increased to the feed pressure, effectively shutting off the flow of hydrogen that was being fed from an on-demand system.

Many gas–liquid reactions other than hydrogenations are run in the chemical industry. Reaction system design depends primarily on the solubility of the gas in the reaction mixture and on its rate of reaction. Soluble gases can often be added without sparging by vortexing. Pitched blade turbines with partial baffles are effective in this regard. However, care must be taken, as discussed in Chapters 6 and 11, to avoid impeller balance and vibration problems.

For complex reactions, the product distribution can be affected by mixing in direct analogy to the homogeneous case discussed earlier. Some of the first experiments in this area were conducted on chlorination of *n*-decane by van de

**806**   MIXING AND CHEMICAL REACTIONS

Vusse (1966), in which mixing was shown to affect the distribution of chlorinated products. The chlorination of acetone is also a mixing-sensitive gas–liquid reaction as described in Example 13-6.

## Example 13-6: Chlorination of Acetone: Mixing-Sensitive Gas–Liquid Reaction (Paul et al., 1981)

- *Goal:* determination of the cause of reduced selectivity in a manufacturing scale gas–liquid competitive-consecutive reaction and modification of the reactor to achieve target selectivity
- *Issue:* decrease in selectivity experienced on scale-up

This example presents reactor design problems experienced in the scale-up of a classical competitive-consecutive reaction from bench to manufacturing scale. Expected selectivity was not achieved initially, and a revised reactor was required.

The chlorination of acetone is a very fast reaction that produces both monochloroacetone and dichloroacetones as well as polychlorinated species. The product desired is monochloroacetone. The reactions are shown in Figure 13-16 and the product distribution in Figure 13-17. Elevated local concentrations will increase the local reaction rates.

As is often the case, the di- and polychlorinated species not only reduce yield but cause ongoing yield and purity problems in subsequent steps of a multistep synthesis because of their reactivity. The rate constant ratio between mono- and dichlorinated species is very unfavorable for making high-purity monochlorinated species, thereby requiring excellent mixing, a very high molar ratio of acetone to chlorine ($>10:1$), and subsequent acetone recovery.

*Laboratory results:* Semibatch addition of chlorine gas to liquid acetone in a 5 L flask; 98% monochloro, 2% overchlorinated products.

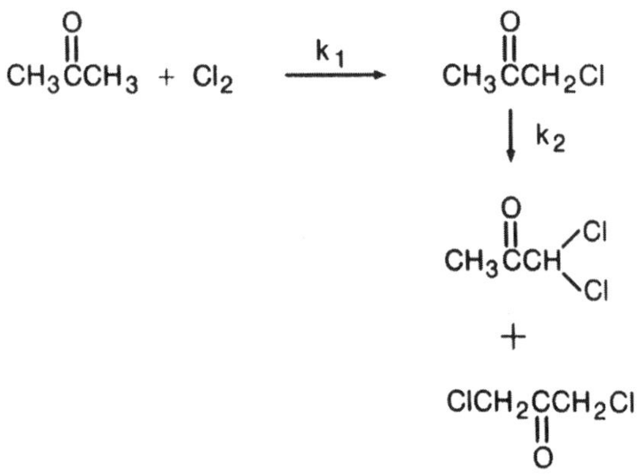

**Figure 13-16** Chemistry of the classical consecutive-competitive chlorination of acetone.

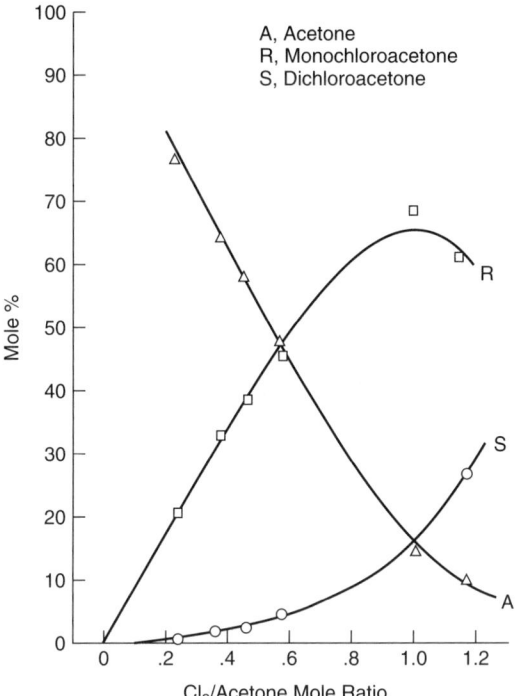

**Figure 13-17** Product distribution in the chlorination of acetone as a function of the chlorine/acetone mole ratio. The formation of overchlorinated products increases with increasing mole ratio.

*Manufacturing results:* The ratio of products achieved in the manufacturing scale continuous vapor-phase reactor was 92% monochloro, 8% overchlorinated products. The initial configuration is shown in Figure 13-18.

*Troubleshooting analysis:* Laboratory results are in agreement with expected selectivity based on measured $k_{R1}/k_{R2}$ ratio and chlorine/acetone molar charge ratio as calculated by eq. (13-5).

Manufacturing results indicate that monochloroacetone once formed is overreacting to a greater extent than indicated by the laboratory results. Figure 13-18 shows the gas phase reactor chosen for manufacturing. Both reactants are vaporized before entering the tubular reactor. Since there is no mass transfer resistance or reacting film in the gas phase, the expected selectivity should be achieved. (*Note:* Gas phase reactions are not subject to the mixing issues being discussed in this chapter because of the order-of-magnitude increase in molecular diffusivity compared to homogeneous liquid-phase or gas–liquid mixing.)

*Possible problem:* The reaction may not be going to completion (not consuming all of the chlorine) in the gas phase tubular reactor, thereby allowing gas phase chlorine to enter the fractionating column and react with refluxing acetone, as shown schematically in Figure 13-19.

**808** MIXING AND CHEMICAL REACTIONS

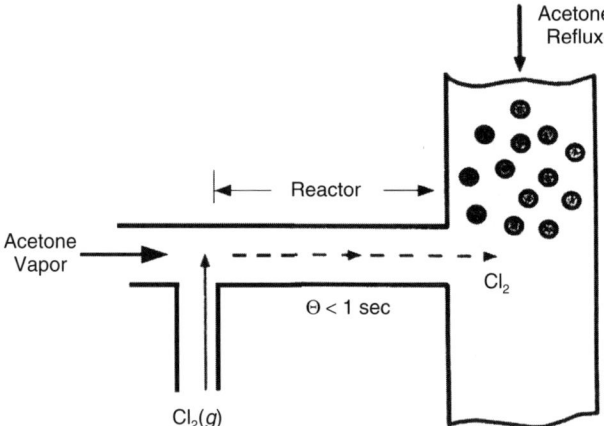

**Figure 13-18** Schematic drawing of the manufacturing scale gas phase tubular chlorination reactor and downstream (connected) fractionator for the recovery of unreacted acetone (molar feed ratio 10:1 acetone/chlorine).

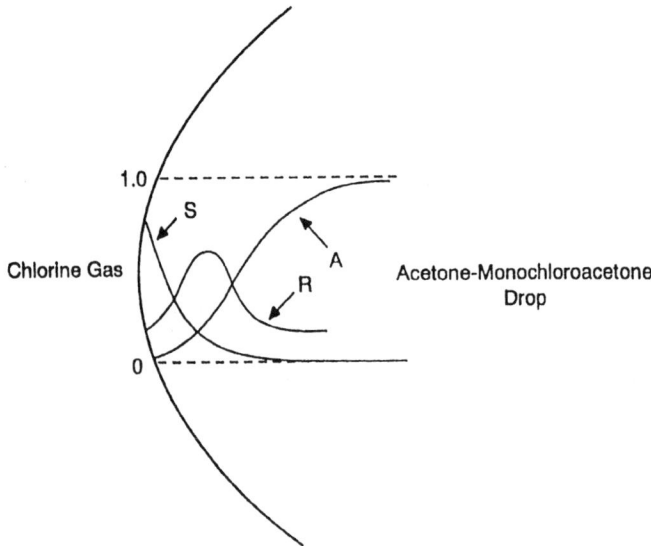

**Figure 13-19** Qualitative model for consecutive reaction in the film around a liquid drop of acetone surrounded by gas phase chlorine (see Figure 13-12).

*Possible solutions:* Extend the length of the tubular reactor or redesign the reactor as a gas–liquid contactor.

*Evaluation of alternative solutions:* Temperature rise measurements in the existing gas phase reactor indicated that the conversion of chlorine was very low, thereby indicating that the length of the reactor would have to be increased considerably. In addition, residence time distribution issues could allow unreacted chlorine to reach the fractionator.

A gas-liquid reactor with liquid acetone as the continuous phase could achieve the required conversion in a much smaller reactor.

*Resolution:* The solution to the problem was to mix liquid acetone and gaseous chlorine in an in-line reactor in which the reaction was completed before reaching the fractionator. The configuration is shown in Figure 13-20. The system was piloted to determine the design conditions that would allow complete reaction of chlorine. Successful scale-up was achieved in a turbulent 0.05 m (5 cm) mixer with very short residence time.

*Message:* As in Examples 13-3 and 13-4, the key to realization of a potential mixing issue was the knowledge of (and analytical confirmation of) consecutive reactions combined with the qualitative observation that the reactions are very fast. Large production quantities required that manufacturing operations be run in a continuous reactor, thereby ruling out a stirred vessel (which would not have been a good choice in any case because of mixing limitations). A single-point injection-line mixer was chosen. The high energy dissipation rate and gas–liquid dispersion capabilities of a static mixer would have been preferable, but these devices were not available at the time the work was done. The same conversion and product distribution were achieved as in a semibatch laboratory flask.

### 13-3.5 Liquid–Liquid Reactions

In liquid–liquid reactors both interphase mass transfer from the continuous phase to the dispersed phase (or vice versa) and dispersed phase mixing through coalescence and dispersion (CD) can occur. The reader is referred to the comprehensive discussion of liquid–liquid systems in Chapter 12.

As pointed out in the Editors' Introduction to this book, liquid–liquid reactions may present the most difficult scale-up challenge in heterogeneous reactions. They are very common and occur in all the regimes discussed in the classification of hetrogeneous reactivity. In regimes 1 and 2 (see Section 13-3.1), slow reactions combined with low solubility of the reactants in their respective phases,

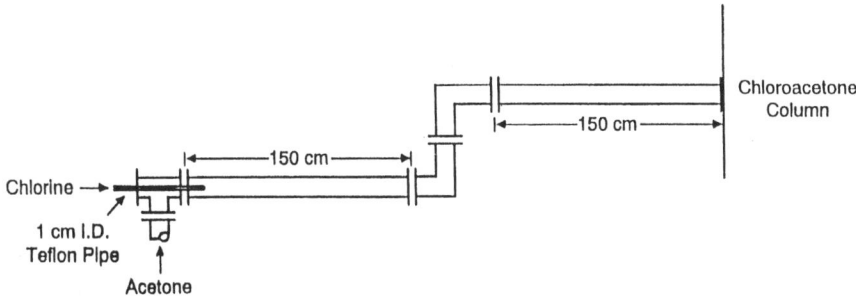

**Figure 13-20** In-line reactor for mixing gas phase chlorine with liquid-phase acetone to achieve complete consumption of chlorine before fractionation.

the actual conversion rate for even very well-mixed systems may be negligible. In these cases, a third solvent may be added to improve mutual solubility or a phase transfer catalyst may be added to transfer one reagent, usually ionic, from aqueous to solvent phase. Both types of additions add downstream separation operations, however, so their use is avoided if possible.

Other strategies to promote reactivity include (1) generation of large interfacial area by intense mixing and (2) removal of one of the phases by distillation of the more volatile solvent, thereby combining the reactants in the remaining phase. The last method may be complicated by the appearance of a solid phase (reagent or product becoming insoluble) but may still be preferable to an additive.

Two examples of liquid–liquid reactions are provided below to illustrate that selectivity can be a significant issue on scale-up.

## Example 13-7: Agitated Thin-Film Reactors and Tubular Reactors with Static Mixers for a Rapid Exothermic Multiple Reaction (Schutz, 1988)

- *Goal:* design of a scalable reaction system for a very fast, highly exothermic complex reaction.
- *Issue:* choice of a suitable reactor for manufacturing scale operation. This reaction scheme is:

$$A + B \to R$$
$$R + B \to S$$

with an enthalpy of reaction of $-440$ kJ/mol. The reactants are single-phase liquids, but the reaction mixture is two liquid phases. Optimum temperature is $-20$ to $-30°C$, and the adiabatic temperature rise at the operating concentrations is 80°C. Significantly lower yields were obtained in a 0.25 m$^3$ (250 L) stirred tank under semibatch conditions (30 min addition) than in the laboratory using a stirred tank.

Two alternative reactor configurations were then investigated in the laboratory; (1) agitated thin-film reactor and (2) tubular reactor with static mixers. The reaction time was found to be at most a few tenths of a second and yield increased with increasing agitator speed in the thin-film reactor and increasing flow rate in the tubular reactor. Semicommercial scale reactors of both types were assembled and tested.

The agitated thin-film reactor was an 0.08 m diameter wiped-film evaporator and was cooled by either convection or evaporation. Because of vacuum requirements for evaporation of the solvent, only convective cooling was utilized. Yields were found to be 10 to 15% higher than in the 250 L stirred tank. The increase in temperature affected the results far less than that which occurred in the stirred tank. Although no exact data on local energy dissipation rates in wiped-film evaporators was available for the unit, higher local energy dissipation rates occurred than in stirred tanks.

The tubular reactor with static mixers was chosen for the documented capability of static mixers to accomplish the following important functions for fast multiple reactions in turbulent flow: (1) homogeneity down to the molecular level can be achieved in a few tube diameters; (2) very short mixing time and narrow residence time distributions are required; and (3) high rates of energy dissipation are achievable (average energy dissipation rates can be calculated from pressure drop and local rates can be estimated).

The cooling capability of static mixers under conditions of extremely rapid heat generation is very low, leading to solvent evaporation in the mixing elements. This two-phase flow reduced the residence time further. The tubular reactor with static mixing elements discharged into a stirred tank, where evaporation and condensation removed the heat.

Yields in the tubular reactor were 15 to 25% better than in the stirred tank, depending on flow rate. This reactor configuration was also superior to the wiped-film evaporator, possibly for two reasons: (1) higher rates of turbulence energy dissipation achieved more rapid micromixing, and (2) superior mixing at the entrance to the static mixers compared to the entrance of the wiped-film evaporator, where some backmixing took place. The possible significance of backmixing in fast multiple reactors is underscored by this example and is also discussed by Bourne et al. (1981) and by Bourne and Garcia-Rosas (1984).

If the heat rise in the tubular reactor (unspecified) could not have been tolerated, the wiped-film evaporator with more effective convective heat removal (since more surface area to volume is achieved) would have been required.

*Resolution:* A static mixer reactor was superior to a wiped-film evaporator in yield. Heat removal was accomplished in a subsequent flash vessel.

*Message:* This example is illustrative of the necessity to meet multiple criteria, including liquid–liquid dispersion, short contact time, minimum backmixing, and high heat transfer rates, in a single reactor configuration. The heat transfer rate, however, was not achievable in an on-line mixer that could achieve the mixing criteria because of the extremely short contact time. Use of flash evaporative cooling at the reactor discharge is an excellent example of effective process integration.

### *13-3.5.1 Reactive Extraction.*

Enhancement of selectivity because of the presence of an immiscible phase is an important aspect of liquid–liquid systems. The improvement in selectivity is achieved by protection of the reactant(s) or product in a separate phase from an active reagent to reduce consecutive or competitive reaction to undesired by-products. Sharma (1988) discusses this subject and presents examples of very large increases in selectivity. An example from Wang (1984) is presented by Sharma in which isocyanates were prepared from amides or $N$-bromoamides by Hofmann rearrangement under phase catalysis conditions. Without a second phase the isocyanate overreacted under alkaline conditions in the aqueous phase. Addition of a carefully selected solvent achieves reaction and rapid extraction of the isocyanate, which can then be obtained in high yield. This route to isocyanates obviates the use of phosgene.

**812** MIXING AND CHEMICAL REACTIONS

Another example of reactive extraction is provided by King et al. (1985). In this case, an acid hydrolysis could be replaced by a highly advantageous change to alkaline hydrolysis to achieve improved selectivity, productivity, quality, and waste minimization. However, the decomposition rates of reagent and reaction product under aqueous alkaline conditions are prohibitive. By running under reactive-extractive conditions, the objectives were achieved. Conventional mixing in a vessel was not feasible because of the rapid decomposition. In-line mixing followed by rapid phase separation proved to be an extremely effective method to carry out this complex reaction, which is discussed further in Example 13-8a.

## *Example 13-8a: Reactive Extraction (King et al., 1985)*

- *Goal:* determination of the feasibility of process improvements requiring a reaction to be run under conditions of rapid simultaneous decomposition of substrate and product
- *Issue:* design of scalable reaction system for a classical parallel and consecutive reaction system

The chemistry of the hydrolysis of an intermediate in the synthesis of an antibiotic is shown in Figure 13-21. Although the chemistry for liquid–liquid acid catalyzed hydrolysis was satisfactory, several process advantages could result from a change to base hydrolysis. However, base hydrolysis was known to be unsatisfactory because of the simultaneous decomposition.

*Laboratory results:* Running the reaction in a standard laboratory semibatch mode adding aqueous NaOH to a solution of A at the pH required for hydrolysis (<10) resulted, as predicted by the chemists, in an unacceptable degree of decomposition of A before the hydrolysis was complete and the base could be neutralized.

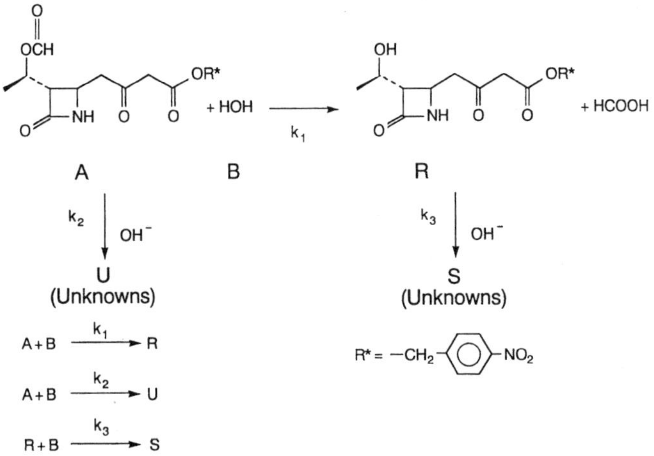

**Figure 13-21** Chemistry of hydrolysis reaction in the synthesis of an antibiotic.

MIXING AND TRANSPORT EFFECTS IN HETEROGENEOUS CHEMICAL REACTORS    **813**

*Challenge:*   Is there a reaction system that could accomplish the hydrolysis and neutralization fast enough to achieve acceptable yields of R without decomposition of A and R to unknowns? (Decomposition to unknowns occurs rapidly after the unstable four-membered ring is opened.)

*Possibility:*   A is soluble in the organic phase (methylene chloride). It hydrolyzes to the enolate salt in the aqueous-phase film at the liquid–liquid interface, as shown in Figure 13-22, and then becomes soluble in the aqueous phase as the enolate salt. A reaction system with simultaneous contacting of the two phases and extraction of the product could be feasible. The relative rates of hydrolysis and base decomposition would determine the feasibility of the system proposed. These rates were measured independently and the ratio was found to be sufficiently favorable ($k_{R1}/k_{R2} > 100$) to proceed with design.

*Reaction system:*   The reaction system with a static mixer followed by a centrifugal separator was assembled in the laboratory and was able to produce acceptable yields of R. A plant design was then developed as shown in Figure 13-23.

*Key components:*

1. Static mixer to achieve liquid–liquid contact and residence time sufficient to transfer A to the aqueous phase
2. Static mixer and extractor to provide residence time to complete the hydrolysis of A to R ($\sim 15$ s)
3. Limited residence time in extractor to minimize base decomposition to U
4. Separate phases to allow continuous transfer of the aqueous phase to a vessel containing aqueous acid for neutralization and crystallization of R

*Results:*

1. The final plant design is as shown in Figure 13-23. The static mixer achieved the required reaction conversion and subsequent separation was completed in a suitable time frame to minimize base-catalyzed decomposition.
2. The reactor–extractor chosen was a Podbielniak centrifugal extractor. These units are normally run with countercurrent feed for extraction and are capable of having two- or three-stage efficiency. The operation, including the

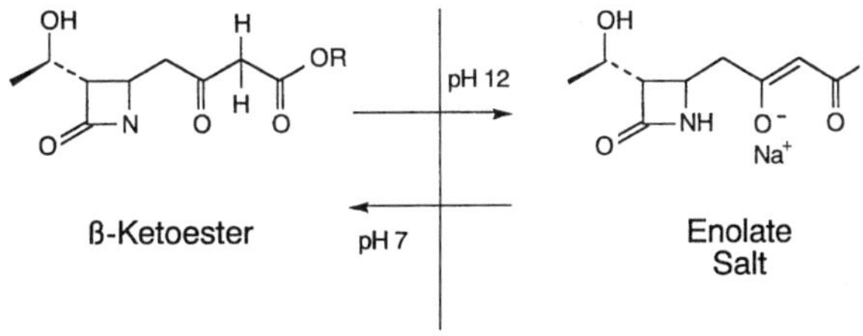

**Figure 13-22**   Conceptual procedure of enolization for hydrolysis under basic conditions.

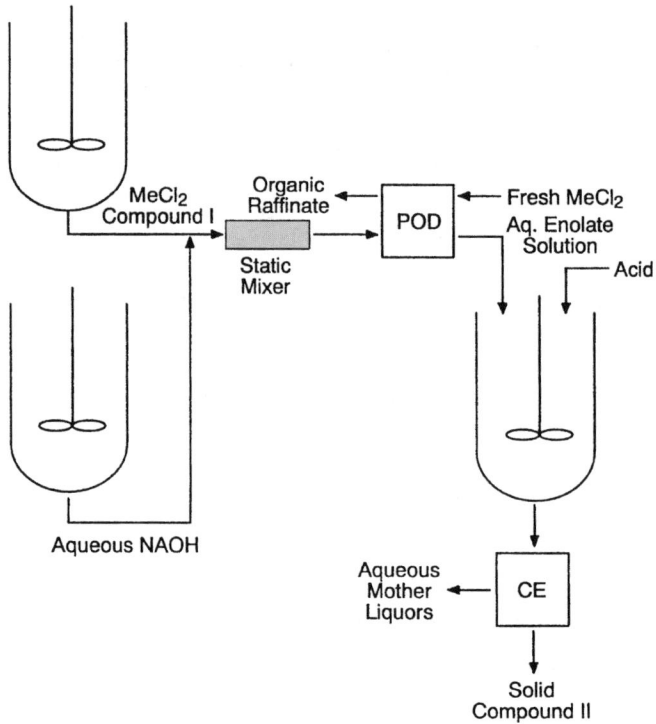

**Figure 13-23** Schematic diagram of static mixer for in-line reaction, centrifugal extractor, and crystallization train.

reaction zone, is shown in Figure 13-24 in a cutaway side view of the centrifugal rotor. The mixed phases leaving the static mixer are fed to the extractor as shown. The mixing–contacting zone between the organic phase and the aqueous base provided enough interfacial area and residence time to complete the reaction and extraction. By maintaining the principal interface, the two phases are separated as exit streams and the aqueous phase transferred directly into aqueous acid to stop formation of U and to crystallize R. Although successful operation could be achieved without premixing the feeds, the improvement with a backwash with solvent was chosen as shown. This utilized the counterflow capability of the extractor to remove impurities from the organic stream. The improvements realized by base hydrolysis compared to acid hydrolysis are as follows:

- *Yield:* 95% versus 81%
- *Impurity removal:* 20% versus 0%

*Message:* In this example, laboratory development was actually more difficult to characterize than plant operation because a small scale extractor to simulate Podbielniak performance was not available. This case is illustrative,

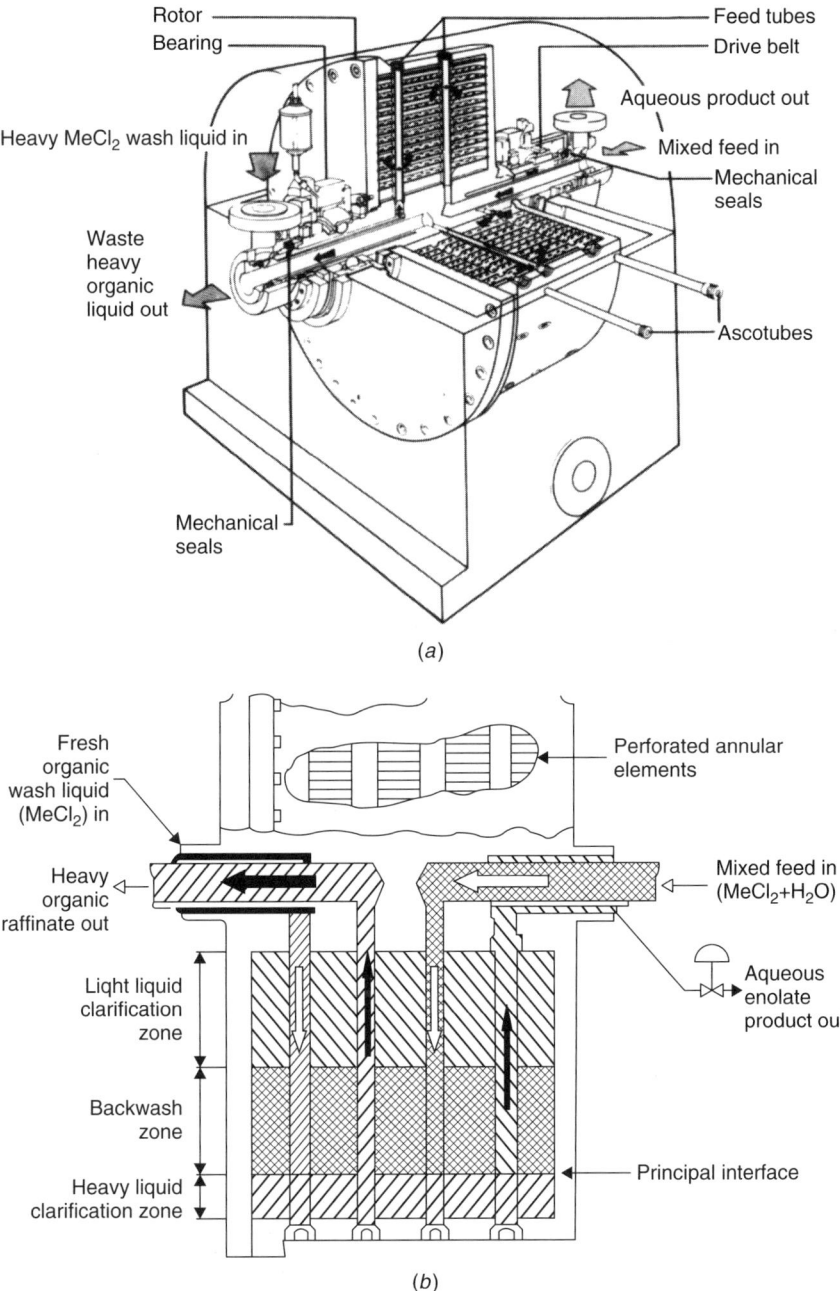

**Figure 13-24** (*a*) Continuous flow countercurrent extractor. (Modified from Thornton, 1992.) (*b*) Cutaway drawing of Podbielniak extractor with mixed feeds. The feed and backwash liquids enter at the principal interface. Centrifugal force separates the heavy phase from the light phase. The direction of rotation is out of the page.

therefore, of the need to conceptualize full scale performance and equipment design in the absence of an integrated laboratory model and to utilize separate laboratory reaction rate data on the various reactions to design the overall reaction system.

The final plant design illustrates a semicontinuous operation in which the run starts with the feed streams in separate vessels and ends with the reaction/extraction product in a third vessel, in this case a crystallizer. This system allows the reaction to be carried out under the same local conditions throughout the run and at a residence time consistent with the stability of the reactants and product. This reaction could not be carried out successfully in a semibatch mode in which the product would accumulate at high pH. The overall run time is, therefore, not a function of product stability but only of production requirements and equipment sizes. In this example, an overall run time of about 2 h was satisfactory.

This example illustrates the effectiveness of two liquid-phase reactions in protecting unstable reactants and/or products from a reactive aqueous phase. In this case the protecting solvent was present in the feed stream from a previous step. In some cases, the protecting solvent is added for that purpose.

A second example of a parallel reaction involving the starting material, A, is illustrative of how mixing sensitivity during neutralization with a strong acid or base can result in unwanted reaction and/or decomposition, as discussed in Section 13-1.2. This problem is highlighted in Example 13-8b.

## *Example 13-8b: Neutralization Involving Parallel Decomposition (Paul et al., 1992)*

- *Goal:* determination of cause of unexpected decomposition of an intermediate during pH adjustment
- *Issue:* manufacturing scale operation that resulted in 4 to 5% unexpected decomposition

This example outlines a case in which a simple pH adjustment with a strong base, sodium hydroxide, of a two-liquid-phase mixture resulted in some unexpected decomposition of the compound R (structure same as in Figure 13-21). This pH adjustment is in preparation for the hydrolysis reaction described in Example 13-8a. A is dissolved in a solvent, methylene chloride (SG = 1.4), and the pH adjusted from 2 to 7 with aqueous sodium hydroxide. No change in the concentration of A is expected, but a decrease is observed in manufacturing.

The reaction system is the classic parallel type:

$$A \text{ (acid)} + B \text{ (base)} \rightarrow P \text{ (water)}$$

$$B \text{ (base)} + C \text{ (formate)} \rightarrow Q \text{ (in this case R of Figure 13-21)}$$

where the base B is added to neutralize acid A in the presence of C (in Example 13-8a, C was A in a competitive-consecutive high-pH scheme). In this analysis, P is the water formed in the neutralization reaction and Q is the phenol R from Figure 13-21. Q is symbolic of a reaction by-product of the base B reacting with substrate C while neutralizing the acid.

In most cases, the rate of the neutralization reaction will be so much faster than the parallel decomposition reaction C to Q that no Q would be formed. However, as seen in Example 13-8a, C is sensitive to high pH, and in this case the local high concentration of sodium hydroxide during neutralization was the possible cause of the loss of C.

*Laboratory results:* A laboratory study was made of the neutralization step to determine whether A could be reacting because of insufficient mixing. The laboratory reactor was a 0.006 m³ (6 L) fully baffled vessel with a 0.072 m (7.2 cm) six-bladed Rushton turbine. The results are shown in Figure 13-25, where the amount of Q formed is shown to depend on turbine speed. The effect of changing feed position is shown in Figure 13-26, where the expected best results from feeding near the turbine were not observed. The cause is the differences in composition of the two-liquid phases within the vessel. When the base is added in a zone that is primarily aqueous, the base strength is reduced rapidly, whereas when added in a region of high solvent composition, the decomposition is accelerated because the base strength is not reduced as readily. The laboratory reactor was run at a low turbine speed, where phase dispersion varied with depth, to exaggerate the possible effects of poor mixing that could be experienced in the manufacturing vessel (12 m³).

*Power comparisons:* The local energy dissipation rates (as power per volume, $\rho\varepsilon$) at the impeller discharge in the laboratory compared to the manufacturing vessel are as follows:

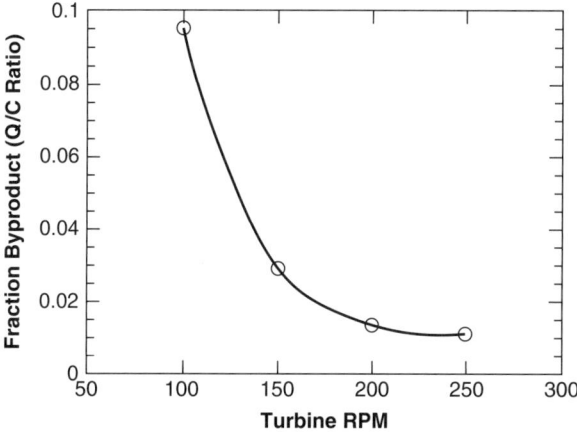

**Figure 13-25** By-product formation as a function of impeller speed for a pH adjustment with competitive decomposition of substrate at the laboratory scale.

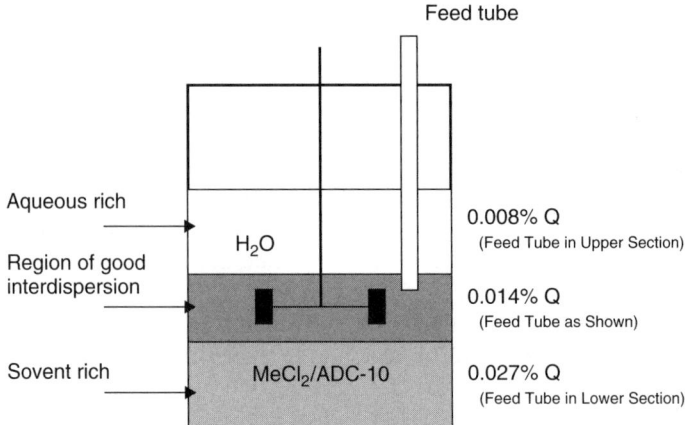

**Figure 13-26** Effect of feed tube location on decomposition during the pH adjustment at the laboratory scale. The caustic (feed) is quickly diluted in the aqueous-rich zone but is more concentrated for a longer time in the lower two zones.

Lab at 4 rps       185 W/m$^3$
Lab at 3 rps       78 W/m$^3$
Plant at 1.5 rps   340 W/m$^3$

*Manufacturing modification options:*

1. Increase impeller speed and/or change type.
2. Use sodium bicarbonate (NaHCO$_3$) in place of sodium hydroxide (NaOH) as the base, thereby reducing the maximum pH that could be experienced even by poor mixing from >12 to ~8.

*Solution:* Option 2 was chosen as a far less costly and less time-consuming solution.

*Message:* Local extremes of pH can occur in a pH adjustment because of imperfect mixing in homogeneous or heterogeneous systems, and these extremes can be expected to be more severe on scale-up. Care must be taken that possible parallel reactions are recognized and minimized by adequate mixing or reactor design. In some cases, the sensitivity may be sufficiently severe to require a high intensity line mixer for the addition point of the acid or base, possibly in a recycle system around the primary vessel.

### 13-3.6 Liquid–Solid Reactions

Solids in reacting systems can be either heterogeneous catalysts, dissolving reagents, precipitating products, or other reaction components, such as adsorption agents or ion-exchange resins. Reaction rates can fall in all regimes of the kinetic spectrum, as described at the beginning of this section. A discussion of solid–liquid mixing without reaction may be found in Chapter 10.

As with all other types of heterogeneous reactions, very slow reactions in the liquid phase (regime 1) are unaffected by mass transfer in the film surrounding dissolving reagents or adsorption agents, and mixing is required only to maintain solids suspension. However, in the case of precipitating or crystallizing products, mixing can affect the particle size of the product just as it would in a precipitation without chemical reaction. Therefore, an effect of mixing in regime 1 must be considered. Reactions in regimes 2, 3, and 4 are all sensitive to local conditions and the films around solids and are therefore subject to mixing effects.

### 13-3.6.1 Solids as Dissolving Reagents.

Organic and inorganic reagents are often incompletely soluble in reaction solvents for a variety of reaction types. The particle size of these reagents can be a major factor in reaction rate and/or selectivity. One objective of a laboratory development program is to determine the effect of particle size and to separate dissolution kinetics from chemical kinetics. An effective method of studying these reactions is to run the reaction under homogeneous conditions to measure true reaction kinetics. This can be accomplished by preparing a saturated solution of the reagent, even if the maximum concentration is very low, and determining the reaction rate as discussed in Section 13-4.3. Once the true chemical kinetics are established, the overall reaction rate can be evaluated for dissolution limitations.

A second effective but less quantitative method is to run the reaction with different particle size distributions of the insoluble reactant to determine effect on overall reaction rate. If no effect is measured, it could be concluded that regime 1 applies and chemical kinetics—not dissolution—controls. This method must be used with care, however, since other factors, such as surface coatings and incompletely characterized particle size distribution, can mask mixing effects and lead to erroneous conclusions.

An example of a reaction with dissolving solids was presented in Example 13-4 in which a direct comparison can be made with the same reaction run in the same pilot scale vessel under homogeneous and heterogeneous conditions. The selectivity is significantly lower for the heterogeneous conditions.

### 13-3.6.2 Solids as Precipitating/Crystallizing Products.[1]

Several studies have shown the effect of mixing on the precipitation of inorganic salts. Mixing intensity was shown to affect particle size for the instantaneous reaction to form $BaSO_4$ by Pohorecki and Baldyga (1988). Particle size was found to increase with increasing impeller speed in a segregated feed CSTR. Barthole et al. (1982) used a modification of the precipitation of $BaSO_4$ (modified to indicate the degree of micromixing) by characterizing product distribution of a $BaSO_4$ EDTA complex in alkaline medium under the influence of an acid.

---
[1] The distinction between precipitation and crystallization is not always clear. For purposes of this discussion, precipitation is the formation of a solid phase from solution by chemical reaction. Reactive crystallization applies to cases in which the solid is crystalline and not amorphous.

Garside and Tavare (1985) modeled the effect of micromixing limits on elementary chemical reaction and subsequent crystallization. Two limiting cases are analyzed, and although the conversions of the chemical reactions are the same, the crystal size distributions can be very different. These differences are caused by the nonuniformity of supersaturation profiles that can be experienced by different fluid elements within a tank, owing to micromixing as well as macromixing effects. This modeling work also explores the sensitivity of two mixing models to reaction rate constant and nucleation kinetic parameters.

Literature references to experimental work on the crystallization or precipitation of products of organic reaction are rare, even though this is a common reaction type. The difference between crystallization and precipitation is not well defined and is interpreted differently by different investigators. The interpretation that is used here is that crystallization generates a crystalline product, whereas precipitates form rapidly and can be crystalline or amorphous. The differences are often blurred, however, because many organics actually appear first as amorphous noncrystalline solids which later turn truly crystalline. In these cases nucleation is difficult to separate from precipitation of an amorphous solid. Mersmann and Kind (1988) present an excellent discussion on precipitation as it is affected by micromixing. Additional discussion of mixing effects in crystallization may be found in Chapter 17.

An experimental study by Marcant et al. (1991) of the crystallization of calcium oxalate concluded that the particle size distribution was significantly affected by impeller speed and other mixing variables. The particle size distribution increased, passed through a maximum, and then decreased as the impeller speed was increased. This result is interpreted as changes in the key factors controlling nucleation and growth as well as reaction. Other mixing variables, including reagent addition point, were also significant and affected particle size distribution in different ways. Another observation by these authors is that measurement of particle size distribution as a function of addition on the surface compared with at the turbine can indicate whether or not micromixing effects have any influence on crystallization or whether other factors, such as nucleation and/or growth rate, dominate. This work is also summarized by Baldyga and Bourne (1999), including pictorial representation of the importance of the addition point in determination of particle size distribution for this system.

The initial appearance of a solid that results from generation of supersaturation by a chemical reaction is a very complex series of events. The conditions affecting crystallization can be critical to the overall process result, for several possible reasons. Yield from a complex reaction can be a function of the rate of crystallization and degree of supersaturation since these factors determine the concentration of that reaction product in solution at any given time. When in solution, all of the factors affecting selectivity can be significant, as discussed above in Section 13-2.5. Delayed nucleation because of improper seeding, mixing conditions, or excessive impurity levels can result in significantly reduced selectivity.

The purity of the crystallization product can be affected by the parameters that control any crystallization as well as the presence of other chemical species, including the starting materials, that can be occluded from the reaction mixture. The particle size distribution can be affected by supersaturation, reaction rate, mixing, and other factors that affect crystallization in general. The degree to which control of the crystallization must be of concern obviously depends on downstream processing. In some cases, physical attributes may not be significant and the reaction can be optimized on the basis of chemical kinetics alone. In other cases, however, the requirements for maximum selectivity may be different than those for physical attributes, requiring a trade-off in actual system design.

An example of reaction-induced crystallization where the particle size and purity must be controlled is discussed in Chapter 17 (Example 17-3; Larson et al., 1995). In this case, mixing played a key role by balancing circulation with shear to achieve micromixing for the reaction but avoiding overmixing. This configuration achieves growth without shedding and/or crystal fracture. A key factor is the time of addition—a mesomixing issue as well as a means of regulating supersaturation.

Mixing can also play a key role in affecting the morphology of a crystalline product. This effect results from the complex interaction between the impeller and nucleation and growth. The reader is referred to the discussion on mixing and crystallization in Chapter 17.

## 13-4 SCALE-UP AND SCALE-DOWN OF MIXING-SENSITIVE SYSTEMS

When perfect mixing or plug flow cannot be assumed but it is not feasible to perform complete simulations of the flow for mixing and chemical reactions in the vessel, scale-up based on local mixing conditions is essential. For stirred reactors with multiple reactions and mixing effects on yield, the simplest approach is to hold constant the power per unit volume. This will work only if the feed locations are in the most turbulent location and geometric similarity is maintained. A more precise scale-up criterion is to hold the rate of turbulent energy dissipation per unit mass in the region of most intense mixing constant. This is particularly useful when the feed is into the impeller stream of a stirred vessel where the mixing is at the fastest rate. Indeed, when yield is an issue, the region of most intense turbulence is almost always the best location for the feed. For geometrically similar mixing vessels, the local turbulence energy dissipation rate per unit mass is generally proportional to the overall power per unit volume, so the two criteria are essentially the same. In some cases, as shown in Example 13-3, equal selectivity on scale-up, even to a pilot scale vessel, cannot be achieved, and an in-line mixer is required. Results such as this are controlled by the local intensity of turbulence.

Scale-down for development and scale-up to manufacturing will benefit from consideration of the key points related to reactions and mixing. The idea of a local

**822**   MIXING AND CHEMICAL REACTIONS

mixing rate, as described above, is the central point, but it is only one of several issues. Interactions between reactions are always of concern. The way in which the feed is added is also critical: Concentrations should be held constant, the addition time must be slow enough to allow mixing to proceed before reaction, and the possibility of feed pipe backmixing must be avoided. Finally, the heat transfer surface area per unit volume will decrease on scale-up, possibly leading to hot spots in the vessel. These issues apply for all reactions but may be fully understood in the context of a simple single-phase (i.e., homogeneous) reaction. Additional issues must be considered for heterogeneous reactions. Experimental protocols for scaled-down development of all cases are provided in Section 13-4.3.

## 13-4.1   General Mixing Considerations

When it is suspected that a reacting system is subject to mixing effects, scale-up can be particularly critical since most organic reactions have multiple by-products. Success or failure on scale-up could be determined not only by the selectivity of the desired product, as it affects costs, but also by the ability to maintain the ratio of other by-products constant to minimize effects on product quality and downstream processing. Increases in by-products as little as 0.1% can be a significant problem.

### 13-4.1.1   Effect of Concentration on Yield for Competitive Consecutive Reactions.
The molecular rates of the chemical reactions taking place in a mixed reactor are determined by rate constants, concentrations of reactants, and temperature. At a given temperature the rate of a second-order reaction in moles per unit volume per unit time depends on the product of the rate constant and the concentrations of the reactants. A characteristic time of reaction is the reciprocal of the product of the rate constant and a representative concentration (the geometric average or the resident reactant concentration in a semibatch reactor). As discussed in Section 13-1.3, the concentration at which the chemical reaction becomes faster than the mixing is the critical concentration at which conversion and yield will be affected by mixing. The correlation presented in Figures 13-2 and 13-3 illustrates the local concentration effect. Again, it is emphasized that the relative reaction rate at a point in the reactor is proportional to the product of the rate constant and the *local* concentration.

The data of Paul and Treybal (1971; see Figure 13-35), Middleton et al. (1986; see Figure 13-32), and Baldyga and Bourne (1992; see Figure 13-34), were analyzed to determine the approximate impeller rotation rate where the yield began to drop substantially: 160 rpm at 0.1 kW/m$^3$ for the 0.065 m$^3$ (65 L) Baldyga and Bourne vessel, 600 rpm at 5 kW/m$^3$ for the 0.03 m$^3$ (30 L) Middleton et al. vessel, and 400 rpm at 0.2 kW/m$^3$ for the 0.05 m$^3$ (5 L) Paul and Treybal vessel. At those rotation rates the mixing time constants were determined using the best available values of L$_S$ and ε. The time constants for the reactions were determined as k$_{R2}$C$_{B0}$, where k$_{R2}$ is the molecular rate constant of the second chemical reaction and C$_{B0}$ is the initial concentration of the resident reactant in the vessel.

These values were combined into a mixing Damkoehler number as the ratio of mixing time constant to reaction time constant:

$$Da_M = \frac{\tau_M}{\tau_R} = \left(\frac{L_s^2}{\varepsilon}\right)^{1/3} k_{R2} C_{B0}$$

The mixing Damkoehler numbers obtained were 0.050, 0.032, and 0.023, respectively. These are reasonably close for such radically different experimental conditions: vessel size, impeller rotation rate, feed location, and concentration.

Baldyga and Bourne (1992) have used a similar Damkoehler number approach to determine the relative importance of *microscale mixing*, the time constant of which is given by eq. (13-16) and *mesoscale mixing*, the time constant of which is given by the first term in the Corrsin equation [eq. (13-13)]. For the typical case where the mesoscale mixing is controlling (see Section 13-2.1), the biggest change in yield seems to occur between mixing Damkoehler numbers as defined above of 0.01 and 0.05, which corresponds well with the analyses of the chemical reactors above.

Generally, it is recommended that bench and pilot data for mixing sensitive reactions be obtained at the same concentrations as are to be used in the commercial plant. That eliminates concentration as a concern in scale-up.

### 13-4.1.2 Concentration Effects in Parallel Reactions: Product Degradation Due to High pH.
The effect of concentration may be very significant in the case of the use of a strong acid or base in pH adjustment. These pH adjustments actually are parallel reactions if the desired components can react with the acid or base. In addition to the neutralization reaction, a parallel reaction between the acid or base and the desired components (possible decomposition) may occur that is mixing dependent (see Example 13-8b). A common practice is the use of concentrated acid or base in production to minimize semibatch volume changes and to avoid the use of dilution equipment. An increase in unwanted decomposition could result on scale-up if the mixing effectiveness is not provided to overcome the increased feed concentration. In cases with a higher sensitivity to pH extremes, mixing intensity alone may be insufficient and dilution of the acid or base may be required to avoid yield loss and impurity generation. When appropriate, a weaker acid or base may be substituted as utilized by Paul et al. (1992) and Example 13-8b.

### 13-4.1.3 Effect of Feed Rate or Addition Time on Yield.
The importance of feed rate on yield for a mixing-sensitive reaction has been well demonstrated by Baldyga and Bourne (1992). The time of addition of a reagent in a semibatch reaction is often increased on scale-up to production equipment because of heat transfer limitations (Chapter 14). In the case of a reaction that is sensitive to mixing, the time of addition often is increased on scale-up to account for the increase in blend time of the reagent (A) in the vessel with the added reagent (B) to maintain expected molar ratio at the feed point. The minimum feed time

**824**   MIXING AND CHEMICAL REACTIONS

to achieve expected yield is, therefore, scale dependent. Shorter feed times will result in reduced yield. At feed times greater than the minimum for that scale, the yield becomes independent of feed time, assuming that there is no parallel decomposition reaction that continues to produce unwanted by-product with time, as discussed in Section 13-3.3. The minimum feed time for expected yield is a function primarily of the rate constant of the primary reaction and the mixing intensity at the point of feed introduction and is, therefore, a mesomixing issue.

*13-4.1.4  Feed Pipe Backmixing.* Backmixing into reactor feed pipes can also lower yield by causing a slower overall mixing rate of the reactants. Jo et al. (1994) have shown that feed pipe backmixing can have a significant effect on yield, and they developed recommendations for $v_f/v_t$, the feed pipe exit velocity divided by impeller blade tip velocity. For turbulent flow in the feedpipe, Table 13-11 gives minimum values of $v_f/v_t$. For feedpipe laminar flow, $v_f/v_{t,min}$ was always lower than for turbulent flow. A good rule of thumb for turbulent flow is to design for $v_f/v_t > 0.5$ except for case 1, where $v_f/v_t > 2$ is necessary.

*13-4.1.5  Hot Spots.* For exothermic reactions, yields may be substantially lower in a large vessel than in bench or pilot scale vessels, particularly if the activation energy for the reaction producing the unwanted product is very high. This effect can result from hot spots (high localized temperatures) in a reactor that can develop because the rate of heat removal by mixing is insufficient for the rate of addition (e.g., adding water to sulfuric acid). The high local temperatures overcome activation energies that are a barrier to reaction on the small scale. This mesomixing effect has been illustrated using the simulations by Randick (2000) [see also Patterson and Randick (2000) and Section 13-5]. Design for heat transfer requirements is discussed in Chapter 14.

### 13-4.2  Scale-up of Two-Phase Reactions

*13-4.2.1  Scale-up of Gas–Liquid Reactions.* The many types of gas–liquid reactions require different considerations on scale-up. In addition to the discussions

**Table 13-11**  Recommended Minimum $v_f/v_t$ For Selected Geometries for Turbulent Feed Pipe Flow Conditions

| Case | Impeller | Feed Position | D/T | G/D | $v_f/v_{t,min}$ |
|---|---|---|---|---|---|
| 1 | 6BD[a] | Radial/midplane[b] | 0.53 | 0.1 | 1.9 |
| 2 | 6BD[a] | Above/near shaft[c] | 0.53 | 0.55 | 0.25 |
| 3 | HE-3[d] | Radial/midplane[b] | 0.53 | 0.1 | 0.1 |
| 4 | HE-3[d] | Above/near shaft[c] | 0.53 | 0.55 | 0.15 |

*Source*: Jo et al. (1994), Table 5.
[a] Six-blade disk turbine.
[b] Injection radially inward toward the impeller at its midplane at a distance G/D.
[c] Injection downward into the impeller at about D/4 from the centerline of the impeller shaft and G/D above the impeller midplane.
[d] High efficiency three-blade down-pumping turbine.

in Section 13-3, the reader is referred to Chapters 7, 11, and 18 and to Examples 13-5 and 13-6.

### 13-4.2.2 Scale-up of Liquid–Liquid Reactions.
Despite the frequent need to run reactions in immiscible liquid systems, the reliability and applicability of correlations to predict drop size distribution and surface area of the dispersed phase, especially in the presence of reactions, is limited. The reader is referred to Chapter 12 and Section 13-3.5. This problem is due in part to effects that small changes in physical aspects such as small agitator blade width can have on dispersed phase drop size as well as on surfactant effects resulting from reacting substrates. It is sometimes even difficult to predict which phase will be continuous and which dispersed. Although this factor is normally a property of a given system, it can sometimes be reversed by the manner in which the phases are contacted (i.e., by mixing during addition as opposed to starting with both phases present).

Extreme care must be taken during laboratory and piloting studies to determine the extent to which interfacial differences are significant so that the impact of changes in dispersion that are very likely to occur on scale-up can be evaluated. In many cases these changes may not be significant because other aspects of the reacting system are controlling. However, phase dispersion can be critical to selectivity in some cases because of complex interfacial interactions. Selection of impellers and speeds to achieve the desired drop size distribution (which has a direct effect on settling rate) can also be critical to reactions that require subsequent phase separation.

The uncertainties inherent in scale-up of liquid–liquid systems, especially if selectivity is affected, require testing over a wide range of operating conditions in the laboratory and possibly the pilot plant to determine the sensitivity of each system to changes in dispersion characteristics. These studies should include mixing configurations and impeller speeds as well as system compositions. Despite these qualifications, much can be gained from applying scale-up correlations to specific problems to establish guidelines and limits for performance. As in the case of gas–liquid systems, the reader is referred to the texts of Oldshue (1983), Tatterson (1991), and Harnby et al. (1992).

Scale-up from laboratory data on the same system can be predicted to some extent. Constant power per unit volume is a good guide, but care must be taken with large tanks and density differences, as mentioned above. Two-phase mixing effects on chemical reactions generally result when the mass transfer rate required to bring the reactants together is much slower than the chemical reactions. This can occur for gas–liquid systems, where the chemical reaction occurs in the liquid phase or in solid–liquid systems, where reactants must diffuse to the solid surface to react. In liquid–liquid systems an interphase mass transfer effect, a droplet coalescence and dispersion effect, and an intraphase mixing effect can be present: for instance, in the case where internal circulation in droplets accounts for the mixing of the diffusing reactant with the droplet-resident reactant. (See Chapter 12 for mass transfer rates in these cases, and Example 13-8a.)

***13-4.2.3 Scale-up of Liquid–Solid Reactors.*** Fluid dynamic scale-up of liquid–solid suspensions has been well characterized by many studies. The reader is referred to Chapter 10 for a comprehensive discussion. For reacting systems, power and speed should in some cases be above the minimum for homogeneous suspension since energy consumption is generally a smaller contributor to cost than other aspects of scale-up uncertainty (conversion and selectivity). Even this recommendation must be qualified by the potentially negative aspects of overmixing, as discussed in Section 13-3.6.2. Reacting solids can also agglomerate and thereby require large increases in energy to maintain adequate dispersion.

These system-dependent properties are extremely difficult to characterize quantitatively and require specific scaling studies at extremes of possible operating ranges to determine sensitivity. Such systems are primary contributors to the case for built-in versatility. The more important consideration in reacting systems than solid suspension may be mass transfer rate since considerably more power and speed may be required to achieve expected reaction rate for reacting solids than that required for homogeneous suspension. As discussed in Section 13-3, selectivity can also be affected in complex reactions because of the potential overreaction in the diffusive film around the dissolving or precipitating particles. An excellent discussion of mass transfer and reaction is presented by Fogler (1999).

Another critical aspect determining the effectiveness of mass transfer correlations for prediction of coefficients in reacting systems is the very troublesome but all-too-common tendency for the surface of a reacting solid, catalyst, or precipitating product to become covered by another solid or second phase liquid, or by a gas in a three-phase mixture. The gas or vapor can also come from entrainment from the headspace. Such a heterogeneous film would obviously have a profound effect on the expected mass transfer coefficient and in many cases can cause a reaction to stop before the expected conversion is achieved. These films are obviously unique to each reacting system, thereby preventing any generalizations as to whether they are susceptible to chemical or physical manipulation. Chemical manipulation could be achieved by addition of a surfactant that would be able to modify surface properties to prevent or modify formation of the film.

Physical manipulation of such films may be possible through variation in mixing intensity, primarily by local shear. Such interactions would be very scale dependent and could readily be masked in smaller scale operations. The extent to which reactions can be affected by coating of particles is illustrated in an excellent example by Wiederkehr (1988). This study also includes other aspects of reaction system design, such as the choice of continuous smaller volume reactors over batch reactors to reduce the size (and potential energy) of the reacting mass as well as the criticality of residence time distribution in complex reactions.

## 13-4.3 Scale-up Protocols

The concepts embodied in the mixing Damkoehler number ($Da_M$) are extremely useful for initial evaluation of reaction conditions in which mixing effects must

be considered:
$$Da_M = \text{mixing time/reaction time} = \frac{\tau_M}{\tau_R}$$

These interactions are shown in Figure 13-2, in which $Y/Y_{exp}$ is plotted against an expression of $Da_M$ using $k_{R1}$ as a measure of reaction rate and in Figure 13-3, in which $X_S$, a measure of overreaction product, $2S/(2S + R)$, is plotted against an expression of $Da_M$ that uses $k_{R2}$ as a measure of the overreaction rate.

The reaction rate constant of the consecutive reaction, $k_{R2}$, can vary over several orders of magnitude and for a particular reaction, the magnitude of $k_{R2}$ can be estimated within two orders of magnitude or less. The mixing rate in vessels should not vary by more than two orders of magnitude. With these bracketed values, upper and lower limits on $Da_M$ can readily be estimated and used as a first measure of mixing sensitivity by using the estimates of Bourne (Sharratt, 1997) for three regions of mixing sensitivity as follows:

| | |
|---|---|
| $Da_M < 0.001$ | when reaction rate is much slower than mixing rate and chemical kinetics only determine selectivity |
| $Da_M > 1000$ | when reaction rate is much faster than mixing rate and the selectivity could approach asymptotic limit in the instantaneous reaction |
| $0.001 < Da_M < 1000$ | when reaction and mixing rates compete and both micromixing and chemical kinetics must be considered |

These concepts can be further utilized in a developmental program for a new chemical reaction as summarized in the following brief outline of an experimental protocol for a homogeneous reaction. Similar protocols for heterogeneous reactions are outlined in Sections 13-4.3.2, 13-4.3.3, and 13-4.3.4.

### 13-4.3.1 Scale-up Protocol for Homogeneous Reactions.
Chemists report a yield of R of 68% for a reaction in which they added reagent B to A in the ratio 1.05 A/B in a round-bottomed flask with paddle impeller over a 1 h period with cooling to control the temperature at 50°C. A and B are both dissolved in solvents that are miscible in all proportions. B is consumed completely. The amounts of unreacted A and by-product S in the final reaction mixture were determined analytically to be 19% and 14%, respectively. Evaluation of the effectiveness of mixing in the round-bottomed flask can be useful but is difficult to characterize, as the types of impellers often used provide good circulation but low shear. The small scales involved may mask mixing effects.

The development and scale-up of this reaction is now taken on by the chemical engineering group, who need to answer the following questions:

1. Is this the maximum yield that can be obtained in this reaction?
2. Was there an effect of mixing in the laboratory?

**828** MIXING AND CHEMICAL REACTIONS

3. Could there be an effect of mixing on scale-up?
4. What reactor design is most suitable for a large production requirement?

Experimental work and modeling/simulation are both to be utilized. A few key experiments are required at the outset that require setup of a scalable laboratory reactor (preferred minimum volume 0.004 m$^3$ (4 L) and materials that are compatible with the reacting materials—assuming in this case stainless steel, fully baffled flat or pitched blade impellers in standard configuration (see Chapter 6).

The apparent rate constant ratio can be calculated from the reported yield of 68% using eq. (13-5), resulting in $k_{R2}/k_{R1} = 0.14$. The question is whether or not the chemists' yield is less than the maximum for this reaction because the flask was not sufficiently mixed to achieve the conditions for perfect mixing and the maximum yield. This question can be answered by running the reaction with increasing mixing intensity to determine whether the yield is sensitive to mixing. Addition of B on the surface when compared to the optimum position at the impeller should be compared. The mixing rates at these two extremes of addition points can vary by a factor of 10 or more.

For this reaction system, at each increased mixing intensity, even with feed into the impeller, the yield continued to increase. This information can be used to evaluate a range for $Da_M$ in Figure 13-2 and/or 13-3 that would indicate mixing sensitivity by noting that concentration and $k_{R2}$ are constant but mixing rate is increasing, giving lower values of $Da_M$. By not reaching a constant minimum value of $Da_M$, it can also be concluded that mixing effects are still preventing achievement of the maximum yield, $Y_{exp}$. To determine the true $Y_{exp}$ value for this reaction system, the rate constants can be determined separately and their ratio used to calculate the true $Y_{exp}$ from eq. (13-5):

$$Y_{exp} = \frac{R}{C_{A_o}} = \frac{1}{1-\kappa}\left[\left(\frac{C_A}{C_{A_o}}\right)^\kappa - \frac{C_A}{C_{A_o}}\right]$$

*Note:* If this yield is still not being achieved, further work is necessary to determine the cause since further increases in mixing would not appear to be effective. A very high shear device such as a rotor-stator or Waring blender could be tested to determine if the reaction is still too fast ($Da_M$ too large) to realize the maximum possible yield (minimum $X_S$).

$Da_M$ can be further reduced by dilution. Experiments at 10× dilution of B and A were run at increasing mixing rates. The yield leveled off at $Y = 0.75$, beyond which further increases in mixing rate had no effect. From this result, the actual rate constant ratio can be calculated as $k_{R2}/k_{R1} = 0.11$. Using this value, the amount of residual A and by-product S can be calculated by material balance and checked against the experimental results to determine that the original chemists' results did not achieve the maximum yield possible and that there is an effect of mixing on all scales. These results answer questions 1 through 3.

The absolute value of the primary rate constant should be estimated to help in design of a manufacturing scale reactor. The one hour addition time used by the chemists to control the reaction exotherm does not provide any information on the magnitude of $k_{R1}$ since this addition time was actually controlled by heat transfer, not by reaction kinetics. Methods for the evaluation of $k_{R1}$ are available, including use of stopped-flow reaction techniques.

The mixing sensitivity found in the experiments above indicates that (1) the primary reaction can be described qualitatively as being very fast, and (2) the mixing rate achievable in a stirred vessel will not be sufficient to achieve $Y_{exp}$. This result indicates that an in-line mixer is a good choice since these devices can maximize the required local energy dissipation at the point of feed injection and achieve complete reaction in a short residence time. (Dilution of A and/or B is usually not feasible for productivity reasons even if a stirred vessel could then achieve the maximum yield, as determined above.) The absolute value of $k_{R1}$ can be used to predict the contact time required after in-line mixing to complete the reaction. The use of an in-line mixer would require provision for heat removal for this exothermic system and would limit choices to designs that would address both heat exchange requirements and complete initial mixing. Injection of B along the mixer length with heat exchange between injection points may be feasible. An impinging jet mixer could also be considered for manufacturing assuming that the heat generated can be adequately removed or tolerated from the adiabatic temperature rise.

If an in-line mixer is not feasible and a stirred vessel is to be used, the design shown in Figure 13-11 is recommended to provide (1) high shear and micromixing at the lower turbine with proper placement of the feed line in the impeller discharge at the point of maximum energy dissipation rate, and (2) good circulation from the upper pitched blade. Prediction of applicable feed pipe diameter and feed velocity must be evaluated by methods described in Section 13-4.1.4 and in Jo et al. (1994).

The addition time on scale-up may have to be increased to account for the slower bulk mixing time and mesomixing effects, as discussed by Bourne and Thoma (1991). A critical minimum addition time can be determined experimentally above which $X_S$ remains constant but below which $X_S$ can increase. Note that the local mixing rate discussed above may be held constant on scale-up, but the addition time may have to be increased because the bulk mixing time will increase and affect the rate at which reagent A is circulated throughout the reactor. This can change the local A/B mole ratio, giving rise to mesomixing effects.

During or after this experimental program, the reaction system can be modeled and the results used to check experimental results and/or predict performance. For example, the engulfment model developed by Baldyga and Bourne (1999) and described further in Sharratt (1997) can be used to calculate the $Da_M$ relationship to $X_S$ (as in Figure 13-3) for this reaction, thereby establishing the appropriate region of $Da_M$ and indicating the degree of mixing sensitivity. Full reaction simulation, discussed in Section 13-5 and summarized in Table 13-10, could be used for similar purposes, giving a more complete picture of scale-up requirements.

The actual value of $Da_M$ for this reaction system may also be determined by measurement of the absolute value of $k_{R2}$, from which $k_{R1}$ can be calculated

**830** MIXING AND CHEMICAL REACTIONS

from the estimate of $k_{R2}/k_{R1}$ that had been determined previously from $Y_{exp}$ using eq. (13-5). Using the calculated mixing rate for a particular reacting condition, a value of $Da_M$ can be calculated, a point on Figure 13-2 or 13-3 as they apply for the reaction of interest determined from this, and thus the $Y_{exp}$ or $X_S$ measured at this mixing condition. This point can be compared with the result from modeling and thereby provide excellent insight into the reactor design issues for the system.

The effects of higher and lower values of $Da_M$ on reactor design are as follows:

- If $X_S$ has reached a minimum value at a mixing rate that can be achieved on scale-up, a stirred vessel can be used to achieve $Y_{exp}$. At values of $Da_M < 0.001$, mixing is only necessary for blending and heat exchange, and the concerns about feed pipe placement and addition rate are not applicable. Caution must be used in reaching this conclusion, as even small increases in $X_S$ can cause downstream problems in separation.
- If the lowest value of $X_S$ that could be achieved was much larger than that predicted by $k_{R2}/k_{R1}$ [independently measured using eq. (13-5) and a material balance to calculate $X_S$ from R and S], a possibly severe yield loss can be expected, and new conditions for this step are required.

*Note:* This protocol is focused on mixing effects for the classic competitive-consecutive reaction system. Reaction systems may also include parallel reactions in which A, B, or R are reacting to form unwanted products that are not represented by the consecutive-competitive system as used to derive eq. (13-5). To keep these reactions from making more unwanted products on scale-up, the overall reaction (addition) time may have to be held constant. In this case, the mesomixing issue for the primary reactions, $A + B \rightarrow R$ and $R + B \rightarrow S$, would predict that more S would be formed. These issues may require selection of an alternative reactor, such as an in-line mixer, for successful scale-up.

### 13-4.3.2 Scale-up Protocol for Solid–Liquid Reactions.
Refer to Section 13-4.3.1 and change reagent B from being dissolved in a miscible solvent to being added as a fine powder, all other factors remaining unchanged.

A fifth question must be added to the four questions in the developmental strategy for homogeneous reactions:

5. Does the particle size and/or addition time of B affect yield (mesomixing)?

In addition to running the reaction with increased mixing intensity, the effect of particle size and addition time can be evaluated by running at two or three different particle sizes and addition times. If the yield continues to increase with decreasing particle size, increasing addition time, and increased mixing rate, mixing conditions are clearly demonstrated to be critical. The maximum possible yield may not have been achieved because these three factors can all affect overall reaction time, the degree to which by-products can form in the films, and the continuous phases in consecutive and parallel reactions. All experiments must be run at impeller speeds at or above $N_{js}$, as defined and discussed in Chapter 10, to be valid representations of solid–liquid mixing.

Unlike homogeneous reactions, even in the laboratory, overall reaction time can have an effect on yield that is caused by the effect of mixing on mass transfer rate if there are parallel reactions in the continuous phase or in the films between phases. With slower mass transfer, these reactions have longer to generate by-products—often decomposition products—so that time of reaction is important on all scales. Determination of this possibility must be included in the experimental plan. Increased amounts of S and other by-products for longer overall reaction times would indicate this sensitivity to mass transfer rate.

For a heterogeneous system, a possible method of determining the maximum possible yield is to find a solvent system in which the reactants and products are soluble and miscible, thereby creating a homogeneous environment. As in the previous outline, high dilution may be required. Again, assuming that these changes can be tested and a yield plateau can be reached, this may be the maximum yield possible. This conclusion can be verified by separate determinations of $k_{R1}$ and $k_{R2}$ and calculation of the maximum yield from eq. (13-5), as before. If the $k_{R1}$ and $k_{R2}$ predict a still higher maximum yield, the mixing effect has reached its asymptotic value, and other factors may be the cause; in this case the reactions are not a classic consecutive-competitive system.

The scale-up recommendations for in-line mixers with very fast homogeneous reactions must be modified for many cases for heterogeneous reactions because in-line reactors may not be feasible with, for example, high solids content from dissolving reactants or crystallizing products. Achievement of the required mass transfer rates in all types of heterogeneous systems may also be an issue, and the reader is referred to the appropriate chapters and examples for discussions of these factors. As cautioned above and discussed in Example 13-7, heat exchange requirements for fast reactions in in-line mixers may be limiting.

### 13-4.3.3 Scale-up Protocol for Gas–Liquid Reactions.

Refer to Section 13-4.3.1 and change reagent B from being dissolved in a miscible solvent to being added as a gas, all other factors remaining unchanged.

A fifth question must be added to the four questions in the developmental strategy for homogeneous reactions:

5. Does the gas–liquid mass transfer rate and its influence on addition time of the gas, reagent B, affect yield?

The reaction can be run under differing mass transfer rates by changes in impeller speed and system pressure. These changes can also affect the addition time necessary for completion of gas uptake. If the yield increases with increased mass transfer (higher impeller speed and/or higher system pressure), mixing conditions are clearly demonstrated to be critical. The maximum possible yield may not have been achieved because these factors can all affect overall reaction time and the degree to which by-products can form in the films and the continuous phases in consecutive and parallel reactions.

An additional influence on mass transfer rate can be the type and location of the gas sparger. In the case of ineffective sparging, the reaction may be very slow because the reagent gas is passing through the liquid before it can react, either because of small gas–liquid surface area or because of poor sparger location. Alternative gas–liquid contacting methods are discussed in Chapter 11 and Section 13-3.4. For exothermic reactions, caution must be taken when increasing mass transfer rate because of increased heat transfer requirements.

Unlike homogeneous reactions, even in the laboratory, overall reaction time can have an effect on yield that is caused by the effect of mixing on mass transfer rate if there are parallel reactions in the continuous phase or in the films between phases. With slower mass transfer, these reactions have longer to generate by-products—often decomposition products—so that time of reaction is important on all scales. Determination of this possibility must be included in the experimental plan. Increased amounts of S and other by-products for longer overall reaction times would indicate this sensitivity to mass transfer rate.

The scale-up recommendation on in-line mixers for very fast homogeneous reactions must be modified in many cases for heterogeneous reactions because in-line reactors may not be feasible with, for example, high gas–liquid ratios (see discussion in Chapter 7). As cautioned above and discussed in Example 13-7, heat exchange requirements for fast reactions in in-line mixers may be limiting.

### 13-4.3.4 Scale-up Protocol for Heterogeneous Liquid–Liquid Reactions.
Refer to Section 13-4.3.1 and change reagent B from being dissolved in a miscible solvent to being added in a solvent that is immiscible with the solvent containing dissolved A. All other factors remain unchanged.

A fifth question must be added to the four questions in the developmental strategy for homogeneous reactions:

5. Does the drop size distribution and/or addition time of the solution containing B affect yield?

To determine the effect of mixing on the reaction rate, the reaction should be run with increasing mixing intensity. If the reaction rate continues to increase, the effectiveness of mixing on drop size dispersion is clearly demonstrated to affect mass transfer rate. If there is no increase in reaction rate, the chemical kinetics may be controlling.

The addition time of B may not be an important variable since the reaction time may be determined by the mass transfer rate and it would be advantageous to add all of the B solution early to maximize this rate. One exception would be a fast reaction with high mass transfer rate, which could cause heat removal to be the limiting factor.

If it is shown that shorter reaction times result in improved selectivity, a parallel or consecutive reaction in the bulk or in the liquid–liquid films around the reacting drops could result in significant yield loss if the reaction time (mass transfer) is not duplicated on scale-up. In this case, scale-up of drop size distribution is critical. The reader is referred to Chapter 12 for a comprehensive discussion of this issue. As indicated in the Editors' Introduction, this is one of the more difficult reaction scale-up problems because drop size distributions in a large vessel with broader dispersion–coalescence rates than in small vessels can be very difficult to duplicate.

If reaction time is found to be critical, in-line mixers can be considered for liquid–liquid reactions because of their effectiveness in creating scalable drop size distributions and mass transfer rates (see Chapter 7 and Example 13-8a).

## 13-5 SIMULATION OF MIXING AND CHEMICAL REACTION

The methods for reactor design and scale-up described above are the usual approach to achieving a workable and economic reactor system when mixing and reaction interact. A better understanding of this interaction is needed than is available from scaling concepts alone: mixing and reaction may interact over a wide range of scales, particularly in the realm of mesomixing effects. Without the more detailed results available from simulations, these issues cannot be fully addressed. Detailed spatial simulation of the reactor using computational fluid mechanics (CFD; see Chapter 5) as the starting point is useful and can often be enlightening for some design and scale-up problems, where, for example, local concentrations and temperature are critical to the success of the process.

Efforts to link mixing and reaction rates to local flow and turbulence characteristics in combustion applications have proceeded independent of mixed reactor work in fine chemical applications. In combustion, the relationships between the degree of conversion and the degree of mixing usually depend on either a chemical equilibrium approximation or an instantaneous local mixing assumption [see the review by Patterson (1985)]. The rate of local mixing for the first approximation is almost always based on some variation of eq. (13-14), the Corrsin (1964) equation for gases, which makes use of the local rate of turbulence energy dissipation and the local concentration length scale.

Most early work on mixing effects in chemical reactors treated the reactor as a uniform field (box) with various processes (such as coalescence, reaction, and dispersion, called *C-D models*) occurring simultaneously or as a collection of environments, linked by flows, each of which had different mixing effects. Most of these models did not link the modeled effects directly to the local turbulence characteristics of the reactor, making them highly empirical. More recent models divide the reactor into zones, where accurate experimental data are available for the velocities and turbulence quantities. Although these models have provided some very useful results, significant process insight is required to develop them, and this is their main weakness. General models incorporated into CFD packages have the potential to overcome this limitation.

Models that couple the local reaction and mixing processes allow simulation of the spatial variations of concentrations due to mixing and diffusion, and thus the rates of chemical reaction. These coupled models usually use some type of computational fluid dynamics (CFD) computer program as a basis for the calculations, as discussed in Chapter 5. Simulation methods may be divided into those using the Lagrangian (fluid element following) coordinate frame and those using the Eulerian (fixed in space) coordinate frame for computation. The Lagrangian coordinate frame is easiest to implement in one dimensional flows but becomes quite complex in three dimensions, so Eulerian simulations are the most common. The main exception is particle or fluid element tracking simulations which use the C-D model to simulate local mixing and chemical reaction.

Table 13-12 summarizes the main simulation methods that have been or are in use. In the discussion that follows, Eulerian methods based on time-averaged (or Reynolds-averaged) balance equations for the component concentrations and segregation will be emphasized, but the Lagrangian-oriented engulfment model and Monte Carlo coalescence–dispersion models are also presented.

### 13-5.1 General Balance Equations

Simulation of turbulent fluid mechanics, mass transfer, mixing, and chemical reaction requires the use of one (typically differential) balance equation for each

**Table 13-12** Some of the Current Models Used for Determining Chemical Reaction Conversion and Yield in the Presence of Mixing

| Model Name[a] | Authors and Refs. | Model Type | Frame | Implementation |
|---|---|---|---|---|
| Turbulent plug flow* | Vassilatos and Toor (1965); Patterson (1973) | Simplified closure | Eulerian or Lagrangian | 1D flow |
| Blending controlled* | Middleton et al. (1986) | Null closure | Eulerian | CFD |
| Four-environment | Mehta and Tarbell (1983) | Mechanistic | Eulerian | Box or CFD |
| Spectral relaxation | Fox (1995) | Large eddy | Eulerian | CFD |
| Engulfment* | Baldyga and Bourne (1984, 1989); Baldyga et al. (1997) | Lamellar | Lagrangian | 1D flow |
| Random walk mixing | Heeb and Brodkey (1990) | C-D | Lagrangian | 1D flow |
| Mixing rate vs. reaction rate | Magnussen and Hjertager (1976) | Null closure | Eulerian | 3D flow |
| Monte Carlo mixing* | Canon et al. (1977); van den Akker (2001) | C-D | Lagrangian | 3D |
| Paired-interaction* | Patterson (1973, 1975, 1985) | Spiked PFD | Eulerian | CFD or 1D |
| β-PFD* | Baldyga (1994) | Continuous PFD | Eulerian | CFD |
| Direct numerical | Leonard et al. (1995) | No closure | Eulerian | CFD |

[a]Those with an asterisk are discussed in some detail in this chapter.

variable to be solved. Commercial CFD codes generally use Reynolds-averaged variables, although work is progressing on large eddy simulations, where time-varying solutions are obtained for all but the smallest scales of motion. In the case of a chemical reactor, transport equations must be solved simultaneously for each velocity component (momentum per unit mass) and for the concentration of each chemical component. If Reynolds-averaged transport equations are used, differential balance equations for turbulence energies (k, averaged velocity fluctuations squared), some variable to relate turbulence energies to eddy viscosity (usually, the turbulence energy dissipation rate per unit mass, $\varepsilon$, leading to the famous k–$\varepsilon$ model), and segregations ($s_i$) for each chemical component must also be solved. Closure equations are required to relate Reynolds stresses in the momentum equations to the turbulence energy and energy dissipation rate and segregation in the concentration equations to averaged-concentration fluctuation correlations. The Reynolds stresses and segregation terms arise from the Reynolds averaging process. The reader is referred to the book by Bird et al. (1960) for a full explanation of Reynolds averaging and formation of the differential balance equations. The basic equations for the k–$\varepsilon$ model are not presented here [refer to Chapter 5 for the basics and to the book by Launder and Spalding (1972)].

The equations for component mass (concentration) and segregation used in the modeling of a second-order chemical reaction are as follows: For component mass,

$$\frac{D\overline{C_i}}{Dt} = \underbrace{\frac{\partial}{\partial x_j}\left(\frac{v_t \partial \overline{C_i}}{\sigma_c \partial x_j}\right)}_{\text{I}} \underbrace{- k_R(\overline{C_i}\overline{C_k} + \overline{c_i c_k})}_{\text{II}} \qquad (13\text{-}27)$$

and for segregation ($\overline{s_i} = \overline{c_i^2}$),

$$\frac{D\overline{s_i}}{Dt} = \underbrace{\frac{\partial}{\partial x_j}\left(\frac{v_t \partial \overline{s_i}}{\sigma_s \partial x_j}\right)}_{\text{I}} + \underbrace{C_{g1} v_t \left(\frac{\partial \overline{C_i}}{\partial x_j}\right)^2}_{\text{III}} \underbrace{- \frac{2\overline{s_i}}{C_C(k/\varepsilon)}}_{\text{IV}} \underbrace{- k_R(\overline{C_i c_i c_k} + \overline{C_k} \overline{s_i} + \overline{s_i c_k})}_{\text{II}}$$

(13-28)

A note on index notation is in order for those unfamiliar with its use. Index notation is used to shorten the transport equations that are presented above. Each index (shown as a subscript) can represent one of the three Cartesian directions or one of the chemical components. The index that appears on both sides of the equation in all or most terms is the *equation index*. If it is x, the equation is a balance for quantities in the x-direction; if it is 1, the equation is for chemical component 1. The indices that appear only on one side of the equation in occasional terms are running indices and take on each of the three values x, y, and z or each of the component values 1, 2, 3, ... such that the term in which they appear will have several forms that are additive. In the case of component values, indices of product components may not be equal. To illustrate use of the index notation, the first component of eq. (13-27) is written out in full:

$$\frac{D\overline{C_1}}{Dt} = \frac{\partial}{\partial x_x}\left(\frac{v_t \partial \overline{C_1}}{\sigma_c \partial x_x}\right) + \frac{\partial}{\partial x_y}\left(\frac{v_t \partial \overline{C_1}}{\sigma_c \partial x_y}\right) + \frac{\partial}{\partial x_z}\left(\frac{v_t \partial \overline{C_1}}{\sigma_c \partial x_z}\right)$$
$$- k_R[\overline{C_1}(\overline{C_2}+\overline{C_3}) + \overline{c_1 c_2} + \overline{c_1 c_3}]$$

We now turn our attention to the physical meaning of each of the terms in eq. (13-27) and (13-28), as a discussion of the solution of the full differential equations is best treated elsewhere.

The left-hand sides of these equations are the substantial derivatives describing convective transport, or transport by bulk motion. Terms I are turbulent diffusion; terms II are the rates of decrease due to chemical reaction; term III is the Spalding (1971) rate of segregation production, where concentration fluctuations may increase with time as bulk mixing penetrates into previously uniform (but unmixed) portions of the vessel; term IV is the Corrsin (1964) rate of segregation decay or mixing without the Schmidt number term, which is usually small. The complete form of the Corrsin term is given in eq. (13-13), with the substitution of $k/\varepsilon$ for $(L_S^2/\varepsilon)^{1/3}$.

A number of values have been proposed for the constant $C_C$. The value used here, $C_C = 2.2$, is based on a best fit to the data of Vassilatos and Toor (1965) for turbulent mixing in a tubular reactor (Zipp and Patterson, 1998). Corrsin predicted a value of $C_C$ of about 4.0, but he used the term $(L_S^2/\varepsilon)^{1/3}$ instead of $k/\varepsilon$. Where the scales of the mixing lamellae are determined by the turbulent flow, that is, away from the influence of the feed jets, the term $k/\varepsilon$ is considered to be about twice as large as $(L_S^2/\varepsilon)^{1/3}$ (Pope, 1985), which probably leads to the smaller constant. A value of 4.0, instead of the 2.7 recommended by Spalding (1971) and the 2.8 recommended later by Elghobashi et al. (1977), was used by Zipp and Patterson for $C_{g1}$ because $C_{g1} = 4.0$ gave closer results for segregation production without reaction. This larger value does not correspond to Spalding's prescription of $C_{g1} = 2/\sigma_C$.

In following sections some methods for modeling the transport and reactions described by eqs. (13-27) and (13-28) are discussed and demonstrated. Particular attention is directed to terms containing the fluctuating concentration, c. All of these terms require modeling. This is the closure problem discussed at the beginning of the section. The objective is to determine the effect of mixing on the conversion and yield of competing chemical reactions.

### 13-5.2 Closure Equations for the Correlation Terms in the Balance Equations

Ever since Toor and co-workers (Vassilatos and Toor, 1965; Toor, 1969; Mao and Toor, 1971; Li and Toor, 1986) defined methods for relating reaction conversion for non-premixed reactants to their degree of mixing, workers in the field of mixed chemical reactors have attempted to build upon and refine their analysis, which was based on the use of an assumed probability density function (PDF) for reactant concentration (see Donaldson 1975; Brodkey and Lewalle, 1985; Kosaly, 1987; Baldyga, 1994; Baldyga and Henczka, 1995). Pope (1985) presented an extensive review of concentration PDF closure methods for mixed chemical reactions, but only the spiked and β-PDFs are presented here.

The *paired-interaction closure* (Patterson, 1975, 1985) is one of the simplest closures and depends on a spiked PDF shown in Figure 13-27, which represents the probabilities of zero, maximum, and mean concentrations for each chemical component. The paired-interaction closures for $\overline{c_i c_k}$ for the reaction terms in both equations and for $\overline{s_i c_k}$ in the segregation equation are as follows:

$$\overline{c_i c_k} = \frac{-\overline{s_i s_k}}{\overline{C_i C_k}} \tag{13-29}$$

$$\overline{s_i c_k} \simeq 0 \tag{13-30}$$

The assumption of $\overline{s_i c_k}$ equal to zero is based on the idea that since $s_i$ is always positive, its correlation with $c_k$ should be much smaller in magnitude than the correlation of $c_i$ with $c_k$.

A more representative PDF for the mixing process is the beta-probability density distribution (β-PDF), which has been used by Baldyga (1994) and Baldyga

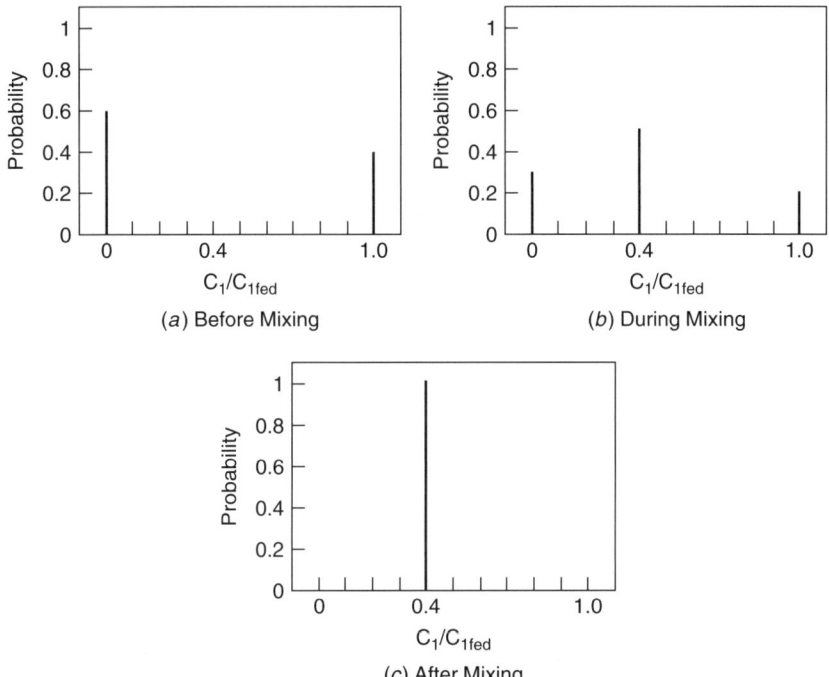

**Figure 13-27** Spiked PDF for paired-interaction closure.

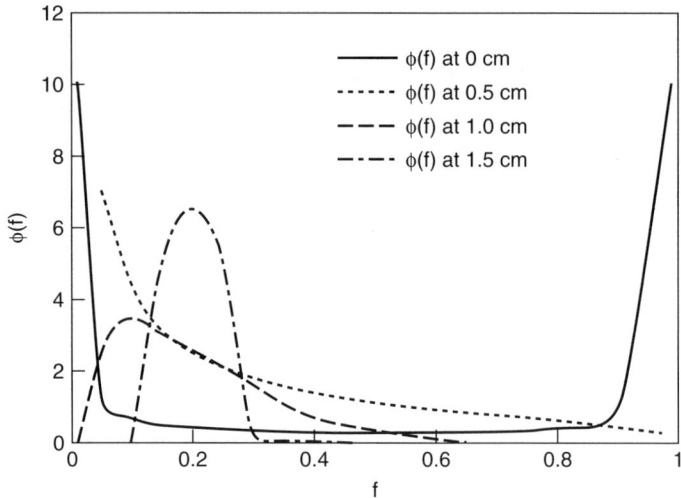

**Figure 13-28** β-PFD versus distance downstream in a mixing pipe. (Based on the data of Vassilatos and Toor, 1965 and Baldyga, 1994.)

and Henczka (1995, 1997) to simulate chemical reactions with turbulent mixing. The β-PDF is gradually transformed from a heterogeneous mixture of pure components to a homogeneous solution with a peak at the average concentration as shown in Figure 13-28. The equations that describe the β-PDF are as follows:

$$\phi(f) = \frac{f^{v-1}(1-f)^{w-1}}{\int_0^1 u^{v-1}(1-u)^{w-1}\,du} \tag{13-31}$$

where

$$v = f\left[\frac{f(1-f)}{s-1}\right] \quad \text{and} \quad w = (1-f)\left[\frac{f(1-f)}{s-1}\right]$$

The product of the β-PDF with the reactant concentrations is integrated at all points to obtain the mean of the product of instantaneous concentrations. When this is multiplied by the reaction rate constant, the last term of eq. (13-27) becomes

$$k_R(\overline{C_i C_k}) = k_R(\overline{C_i}\,\overline{C_k} + \overline{c_i c_k}) = k_R \int_0^1 C_i(f)C_k(f)\phi(f)\,df \tag{13-32}$$

In this set of equations f represents the concentration of a nonreacting (passive) scalar, which is depicted by the following equation:

$$f = \frac{C_i - C_k + C_{ko}}{C_{io} + C_{ko}} \tag{13-33}$$

In the application of this closure it is assumed that the rate of chemical reaction has no effect on the rate of mixing, which is, however, inherent in eq. (13-28). The current value of segregation, s, is computed with the Corrsin equation, eq. (13-13), modified as in eq. (13-28). Therefore, eqs. (13-13), (13-27), and (13-28) (omitting the last term) and eqs. (13-31) through (13-33) constitute the β-PDF closure. Typically, the closure is applied to the first reaction of a competitive-consecutive reaction scheme (A + B → R; R + B → S) and to both reactions of a competitive-parallel reaction scheme (A + B → R; A + C → S). Baldyga (1994) and Baldyga and Henczka (1995, 1997) demonstrated the use of this β-PDF closure for the plug flow pipe reactor [data by Vassilatos and Toor (1965)], an opposed jet reactor, and a concentric jet flow into a pipe reactor. Good results were shown for all geometries.

### 13-5.3 Assumed Turbulent Plug Flow with Simplified Closure

If the mixing of two fluids flowing downstream in a pipe mixer can be assumed to be occurring in a plug flow at a given turbulence energy and energy dissipation rate, the mixing rate and rate of chemical reaction can be computed. This approach is particularly applicable to the multiple-jet header issuing reactants into a pipe and the static mixer geometries. The following equations then apply for the concentration and segregation of each reacting component with the paired-interaction closure (see Section 13-5.2) used for the $c_i$ and $s_i$ terms. Other closures may be substituted.

$$\frac{d\overline{C_i}}{dx} = -\frac{k_R(\overline{C_iC_j} + \overline{c_ic_j})}{U_x} \tag{13-34}$$

$$\frac{d\overline{c^2}}{dx} = -\frac{k_R(\overline{c_i^2C_j} + \overline{C_ic_ic_j})}{U_x} - \frac{\overline{c_i^2}}{\tau_M U_x} \tag{13-35}$$

$$\overline{c_ic_j} = -\frac{\overline{c_i^2c_j^2}}{\overline{C_iC_j}} \tag{13-36}$$

Equation (13-13) is used to compute $\tau_M$ with $L_S = 0.39k^{3/2}/\varepsilon$ if the turbulence energy k instead of segregation scale $L_S$ is used. Initial values of $\overline{c_ic_j}$ should be set equal to the products of the initial concentrations of the reactants as if they were completely mixed. If the mixing time constant is very small, $\overline{c_ic_j}$ will quickly become zero and $d\overline{C_i}/dx = -k_R(\overline{C_iC_j})$, making the molecular kinetic rate equation valid. Note that no segregation production term is used, since no segregation production is expected beyond the maximum assumed at the injection point. This is because the scale of segregation is already comparable to the size of the grid at injection.

Use of this one dimensional method leads to results that compare well with experimental data taken in the pipe reactor of Vassilatos and Toor (1965), shown in Figure 13-29. The value of $U_x$ was 0.75 m/s. The chemical reaction was

**840** MIXING AND CHEMICAL REACTIONS

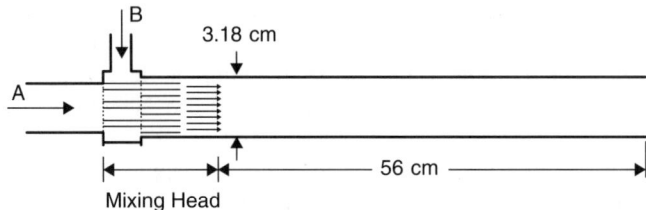

**Figure 13-29** Schematic of tubular reactor with multiple injector header used by Toor and co-workers and by Brodkey and co-workers.

an acid–base neutralization ($H_2CO_3 + NaOH$) with $k_R = 12\,400$ m$^3$/kmol · s. Injection concentrations were 0.025 and 0.031 kmol/m$^3$, giving a reactant ratio of 1.26. Values of $L_S$ and $\varepsilon$, as determined from experimental data obtained in the same geometry by McKelvey et al. (1975), varied from the injection point to the pipe outlet; $L_S$ increased from a low of 0.0005 m to level out at 0.0050 m, and $\varepsilon$ decreased from a high of 7 m$^2$/s$^3$ to level out at 0.2 m$^2$/s$^3$, all in a distance of 0.065 m. Beyond 0.065 m these values were nearly constant at the center of the pipe. Comparison of the simulation results with the experimental results for the pipe reactor is shown in Figure 13-30.

Data for a Kenics twisted-ribbon static mixer geometry obtained by Baldyga et al. (1997) is shown in Figure 13-31. In this case only final yields for a complex reaction were measured. The static mixer used by Baldyga et al. (1997) was 0.04 m in diameter. The method developed above was used to simulate the reactions in the static mixer. Even though it is not true in individual elements of the static mixer, plug flow overall was assumed. Also, in contrast with the

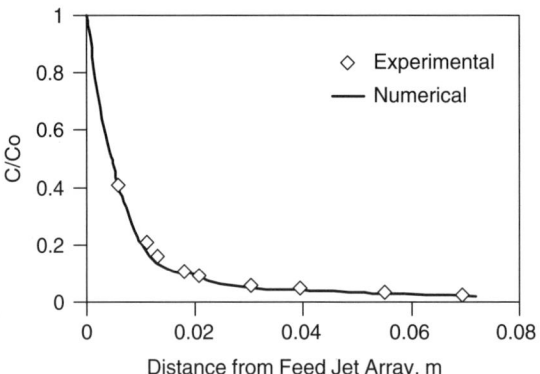

**Figure 13-30** Normalized concentration downstream of the feed jet array in the Toor tubular reactor for $k_r = 12\,400$ L/mol · s; reactant feed ratio of 1.26 and an average velocity of 0.75 m/s. The experimental values of Vassilatos and Toor (1965) are compared to simulation values using paired-interaction closure. The reaction was a single second-order acid–base neutralization.

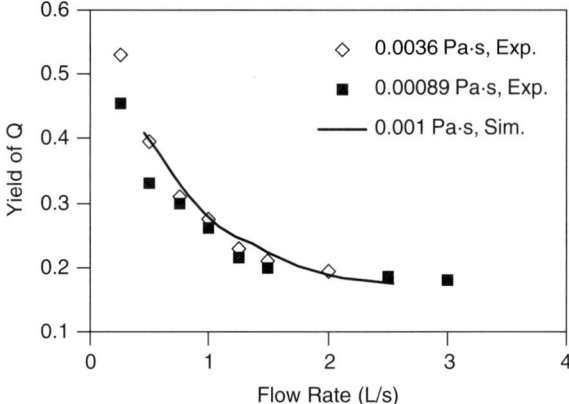

**Figure 13-31** Comparison of the Baldyga et al. data for mixed reaction in a static mixer with results of paired-interaction closure for the reaction A + B → p-R + o-R; p-R + o-R + B → S; AA + B → Q. (See Baldyga et al., 1997, for details.)

mixing-head pipe mixer discussed above, the values of $L_S$ and $\varepsilon$ were assumed constant since they were generated by the mixing elements throughout the static mixer. Details of this one dimensional simulation are given in Example 13-9. If significant radial blending must be accounted for in such a simulation, a two or three dimensional simulation using CFD may be necessary.

***Example 13-9: Yields from a Static Mixer Reactor Assumed to Be Plug Flow.*** The approach given in eqs. (13-34) to (13-36) may be used to compute the yield values for a static mixer reactor. Following Baldyga et al. (1997), the pipe radius divided by 2 was assumed as an approximation of the mixing scale, $L_S$. The method may easily be modified to compute conversions and yields as a function of distance downstream for any turbulent plug flow reactor and set of chemical reactions if realistic feed conditions can be given.

The chemical reactions with their respective rate constants in the Baldyga et al. case may be depicted as follows:

$$A + B \rightarrow \text{p-R} \quad k_{R1} = 12\,238 \text{ m}^3/\text{kmol} \cdot \text{s}$$
$$A + B \rightarrow \text{o-R} \quad k_{R2} = 921 \text{ m}^3/\text{kmol} \cdot \text{s}$$
$$\text{p-R} + B \rightarrow S \quad k_{R3} = 1.835 \text{ m}^3/\text{kmol} \cdot \text{s}$$
$$\text{o-R} + B \rightarrow S \quad k_{R4} = 22.25 \text{ m}^3/\text{kmol} \cdot \text{s}$$
$$AA + B \rightarrow Q \quad k_{R5} = 125 \text{ m}^3/\text{kmol} \cdot \text{s}$$

Since the static mixer has a diameter of 0.04 m, the value of $L_S$ is given as 0.01 m, one-half the radius, and is assumed to be constant, although that is a strong approximation. Normalized values of the reactants and products were

## 842 MIXING AND CHEMICAL REACTIONS

computed for comparison with the experimental results as A/A₀, B/B₀, p-R/B₀, S/B₀, Q/B₀, and AA/AA₀. The values of ε at various flow rates were calculated from pressure drop data given by the authors and were as follows:

| Q (m³/s) | ε (m²/s³) |
|---|---|
| 0.0005 | 1.38 |
| 0.0010 | 11.0 |
| 0.0015 | 37.1 |
| 0.0020 | 88.0 |
| 0.0025 | 171.9 |

The values of Sc and $\nu$ were taken to be 2000 and $0.89 \times 10^{-6}$ m²/s. The resulting yields for flow rates ranging from 0.0005 to 0.0025 m³/s are shown in Figure 13-31, where they are compared with the experimental data. The experimental data showed that the viscosity had some effect on the yield of Q at low flow rate, and therefore at low mixing rate, but little effect at the higher flow rates. The simulations focused on the effect of geometry, so viscosity was held constant.

The equations that were solved to obtain the simulated yields are as follows:

$$\frac{dC_A}{dt} = -k_{R1}(C_A C_B + c_A c_B)$$
$$- k_{R2}(C_A C_B + c_A c_B) \qquad C_{A0} = 0.02 \text{ kmol/m}^3$$

$$\frac{dC_B}{dt} = -k_{R1}(C_A C_B + c_A c_B)$$
$$- k_{R2}(C_A C_B + c_A c_B)$$
$$- k_{R3}(C_{p-R} C_B + c_{p-R} c_B)$$
$$- k_{R4}(C_{o-R} C_B + c_{o-R} c_B)$$
$$- k_{R5}(C_{AA} C_B + c_{AA} c_B) \qquad C_{B0} = 0.0166 \text{ kmol/m}^3$$

$$\frac{dC_{p-R}}{dt} = k_{R1}(C_A C_B + c_A c_B)$$
$$- k_{R3}(C_{p-R} C_B + c_{p-R} c_B) \qquad C_{p-R,0} = 10^{-6} \text{ kmol/m}^3$$

$$\frac{dC_{o-R}}{dt} = k_{R1}(C_A C_B + c_A c_B)$$
$$- k_{R4}(C_{o-R} C_B + c_{o-R} c_B) \qquad C_{o-R,0} = 10^{-6} \text{ kmol/m}^3$$

$$\frac{dC_{AA}}{dt} = -k_{R5}(C_{AA} C_B + c_{AA} c_B) \qquad C_{AA,0} = 0.08 \text{ kmol/m}^3$$

$$\frac{dC_S}{dt} = k_{R3}(C_{p-R} C_B + c_{p-R} c_B) \qquad C_{S0} = 0 \text{ kmol/m}^3$$
$$+ k_{R4}(C_{o-R} C_B + c_{o-R} c_B)$$

$$\frac{dC_Q}{dt} = k_{R5}(C_{AA}C_B + c_{AA}c_B) \qquad C_{Q0} = 0 \text{ kmol/m}^3$$

$$c_A c_B = -\frac{s_A s_B}{C_A C_B}$$

$$c_{p-R} c_B = -\frac{s_{p-R} s_B}{C_{p-R} C_B}$$

$$c_{o-R} c_B = -\frac{s_{o-R} s_B}{C_{o-R} C_B}$$

$$c_{AA} c_B = -\frac{s_{AA} s_B}{C_{AA} C_B}$$

$$\frac{ds_A}{dt} = -\frac{s_A}{\tau_M} \qquad s_A = 0.000332 \text{ (kmol/m}^3)^2$$

$$\frac{ds_B}{dt} = -\frac{s_B}{\tau_M} \qquad s_B = 0.000332 \text{ (kmol/m}^3)^2$$

$$\frac{ds_{p-R}}{dt} = -\frac{s_{p-R}}{\tau_M} \qquad s_{p-R0} = 0$$

$$\frac{ds_{AA}}{dt} = -\frac{s_{AA}}{\tau_M} \qquad s_{AA} = 0.00531 \text{ (kmol/m}^3)^2$$

$$\frac{ds_{o-R}}{dt} = -\frac{s_{o-R}}{\tau_M} \qquad s_{o-R0} = 0$$

$$\tau_M = 2.05 \left(\frac{L_S^2}{\varepsilon}\right)^{1/3} + 0.5 \left(\frac{\nu}{\varepsilon}\right)^{0.5} \ln(Sc)$$

Computations for this example were done using the program Polymath (distributed by the Cache Corp., an affiliate of AIChE), but any program that integrates sets of stiff differential equations (e.g., those using Gear methods) may be used. For this problem steady state was attained in the reactor after 0.5 s of integration for all flow rates used. This corresponds to reactor lengths which depend on the velocity of the feed stream(s).

### 13-5.4 Blending or Mesomixing Control of Turbulently Mixed Chemical Reactions

Middleton et al. (1986) measured yields of a competitive-consecutive chemical reaction under various stirred vessel conditions (size, impeller rotation rate) for semibatch reactors in which the added reactant was injected rapidly. They used the *Bourne reaction*, which is the reaction of 1-naphthol (component A) with diazotized sulfanilic acid (component B) to produce two products according to the scheme A + B → R; R + B → S (see Section 13-2.5). The reaction rate constants of these reactions are $7.3 \times 10^3$ and 3.5 L/mol · s, respectively. Reactant B at a concentration of 0.016 mol/L was added very rapidly at the top of the reactor vessel to reactant A at 0.0058 mol/L. This rapid feed injection produced a large unmixed cloud of reactant B. From this we would expect blending or mesomixing effects to dominate (see Section 13-2.1.3) and the resulting mixing to scale with N.

Middleton et al. (1986) compared their experimental results with the results of a simulation which assumed no effect of local concentration fluctuations. Local average concentrations were used with the kinetic rate equations, and it was assumed that segregation on the small scale was zero everywhere. There were concentration variations throughout the vessel which gradually diminished as blending and chemical reaction occurred. This is the assumption of large scale mesomixing control as opposed to small scale mesomixing or micromixing control.

Figure 13-32 shows the results of Middleton et al. (1986) plotted as yield versus power per unit volume, which would be consistent with micromixing control. It is clear that the results for the two reactor sizes do not coincide on such a plot. This work shows that constant power per unit volume (which corresponds to micromixing rate control) is inadequate for scale-up under mesomixing conditions. Yields for the large vessel were considerably lower than for the small vessel at equal power per unit volume.

If Figure 13-32 is replotted as yield versus N (impeller rotation rate), which is based on the assumption that mesomixing controls, the result is as shown in

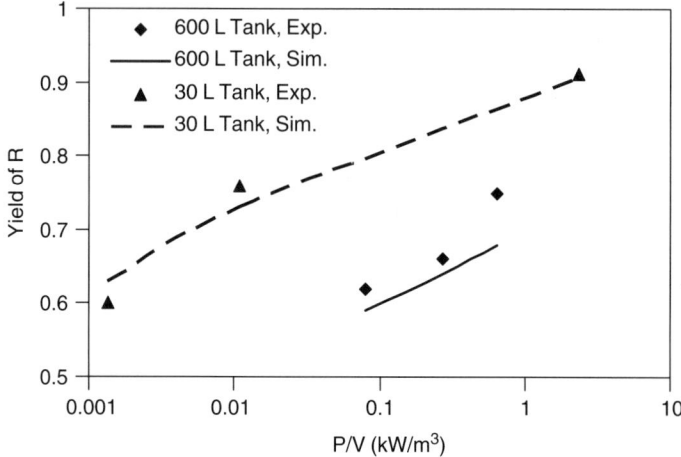

**Figure 13-32** Yield of R in the reaction A + B → R; R + B → S from experimental data of Middleton et al. (1986) and from their simulations, which assume no local mixing rate effect. Simulations using paired-interaction closure agree with the Middleton et al. simulations, showing that the controlling mixing rate is not micromixing.

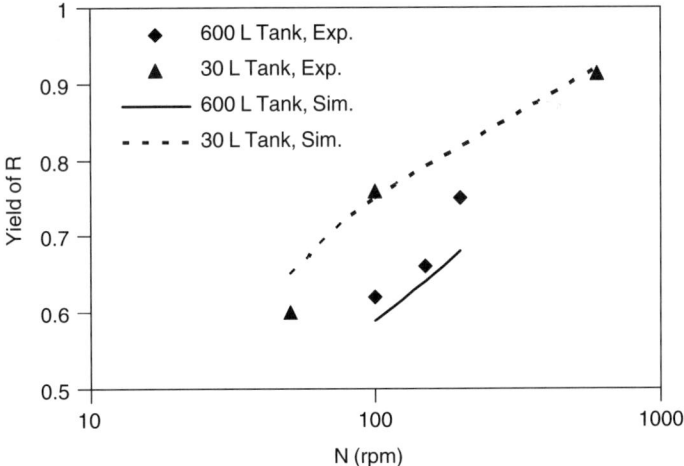

**Figure 13-33** Yield versus N for Middleton et al. (1986) data shown in Figure 13-32. The results are close to mesomixing control, as shown by the closer agreement between small and large scale results when scaling is based on N.

Figure 13-33. The yields for the two scales are closer together for equal N, but there is still significant deviation, suggesting that for this chemical reaction there is an effect of flow pattern on yield.

***Example 13-10: Use of Fluent to Simulate Blending Rate Control in a Stirred Reactor.*** Simulation of mesomixing controlled chemical reactions can be accomplished using any of the common commercial CFD codes. The experiments resulting in Figure 13-32 were simulated using the commercial code Fluent. The simulation was set up in the usual way to compute the flow patterns in the vessel with some attention given to the flow or flows of feed into the vessel. The reactor is semibatch, and since Fluent does not accommodate changing volume, an outflow far from the feed point(s) was specified. The volume fraction of feed was small compared to the volume of the resident liquid in the reactor, so the errors incurred in this strategy were small. The mixing of the reactants and products was accounted for by the mass balance capabilities of the program. To account for the very rapid feed injection of the Middleton et al. experiments, a starting condition with a cloud of unmixed component B was established. For the chemical reaction equations the kinetic rate constant option without any mixing effect was chosen. This means that at any point in the reactor, the reaction rate is determined by local average concentrations at the scale of the grid, as if all components are perfectly mixed at that point. The model then consists of eq. (13-27) with the term $\overline{c_i c_k}$ equal to zero. The results of the simulation are essentially the same as those obtained by Middleton et al. using their own computer code.

A somewhat related simulation method is the one proposed by Magnussen and Hjertager (1976) for combustion, which has been incorporated into the code

Fluent. That method computes reaction rate using the intrinsic kinetics if the mixing rate is fast and using the mixing rate if the intrinsic reaction rates are fast. The Magnussen–Hjertager method has found success for combustion simulations where diffusion is fast and mesomixing rate effects dominate. No account of the degree of mixing (segregation) is kept in either of these methods, in contrast with PDF-based methods (paired-interaction, β-PDF, etc.). A complete description with examples of the use of the Magnussen–Hjertager method in Fluent was given by Bakker et al. (2001).

### 13-5.5 Lamellar Mixing Simulation Using the Engulfment Model

Bourne (1983) realized the important effect of the mixing process on the yield and product distribution of series–parallel chemical reactions. He and co-workers used the first Bourne reaction, described in Section 13-2.5, to do experimental studies of the effects of mixing on yield. They made extensive use of this reaction system (Bourne et al., 1981; Bourne and Rohani, 1983; Angst et al., 1984; Bourne and Dell'Ava, 1987) and used the results to determine parameters in their models.

Alkaline hydrolysis of nitromethane was used by Klein et al. (1980). These results were used to determine the parameters in the interaction and exchange with mean (IEM) model of Villermaux and Zoulalian (1969), which uses the balance equation for component j in the injection region for component i:

$$\frac{dC_j}{dt} = I(\overline{C_j} - C_j) + r_j \qquad (13\text{-}37)$$

where I is the interaction rate, $\overline{C_j}$ the concentration of component j in the region surrounding the injection region (mixing-reaction zone), and $r_j$ the rate of appearance or disappearance in the injection region of component j based on molecular kinetics. This model is basically a Lagrangian model in which the progress of the chemical reaction is followed in time as reacting fluids flow downstream. The flowing stream can be the feed jet in a plug flow reactor or stirred tank. The surrounding fluid is drawn into the flowing jet by turbulent diffusion, causing expansion of the jet. The interaction parameter may be assumed constant or changing as the fluid flows downstream.

More recently, Baldyga and Bourne (1988, 1992) and Baldyga et al. (1993) have shown experimentally and through use of simulation the effects of mixing intensity, feed location, order of reactant addition, and relative molecular kinetics on yields in stirred vessels of competitive-consecutive and parallel-competitive reactions. The data of Baldyga and Bourne using the Bourne reaction are plotted in Figure 13-34 as the yield of R as a function of power per volume for long feed times at various feed points relative to disk turbine impellers. Under these conditions, scaling based on a micromixing mechanism was expected. The chemical reactions were carried out using semibatch addition of reactant B at a concentration of 11.8 mol/L into reactant A at 0.128 mol/L. They did simulations using

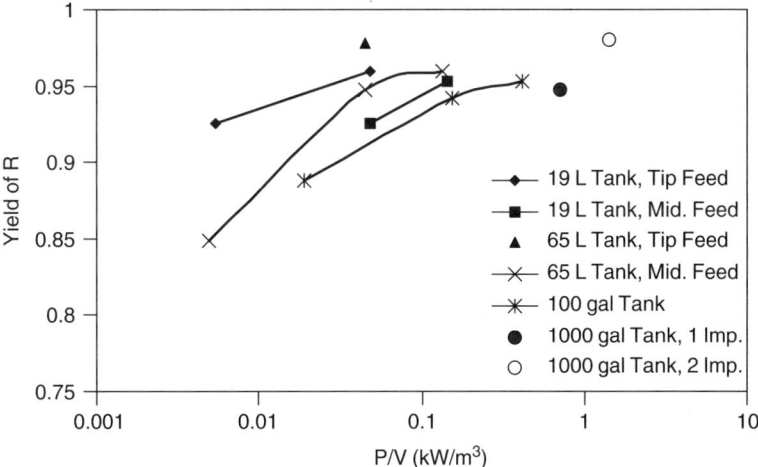

**Figure 13-34** Yields in four stirred vessel sizes. The 19 and 65 L tanks were studied by Baldyga and Bourne (1992): feed points were at the impeller tip and midway between the tip and the tank wall. In the 100 and 1000 gal (378 and 3780 L, respectively) tanks studied by Paul (1988), the feed points were all at the impeller blade tips. In both cases a Bourne reaction of the form $A + B \rightarrow R$; $R + B \rightarrow S$ was used.

the engulfment model for micromixing, which is a Lagrangian simulation of the expanding jet (zone) of mixing and reaction. The model assumes that the volumetric flow rate of completely mixed, B-rich fluid through the expanding volume of the mixing-reaction zone is proportional to distance from the feed nozzle, the fraction of B-rich fluid in the zone, and the turbulent diffusivity, which is related to the engulfment rate, E. Through manipulation of the basic balance equations, the following equation is reached for the rate of change of the fraction B-rich fluid in the mixing-reaction zone:

$$\frac{dX_B}{dt} = EX_A X_B - \frac{X_B}{t} \tag{13-38}$$

where $E = 1/\tau_E$, $X_B$ is the fraction of B-rich fluid, and $X_A$ is the fraction of A-rich fluid. Here $\tau_M$ is computed using eq. (13-16).

Equation (13-38) was combined with the equation for the engulfment model to give the rate of change of the concentration of any of the reactants or products:

$$\frac{d\overline{C_j}}{dt} = E(1 - X_B)(\overline{C_j} - C_j) + r_j \tag{13-39}$$

The model is similar to the interaction and exchange with the mean model, but the engulfment rate and the fraction B-rich fluid in the reaction zone determine the interaction rate, that is, $I = E(1 - X_B)$.

Note that contrary to the Middleton et al. results, the yields in Figure 13-34 are greater for the larger tank in the Baldyga and Bourne results when plotted as a function of power per unit volume. This may be caused by the slow semi-batch addition of reactant B into the high-turbulence region near the impeller, causing micromixing conditions to occur, rather than the very rapid injection near the top of the tank, which has very low turbulence and causes mesomixing conditions to occur. Similar experimental conditions were considered by Paul (1988) and are shown in Figure 13-35. The decrease in yield as power per unit volume decreases seems to begin at about the same level (0.1 to 1 kW/m$^3$) for all cases, but data do not agree in absolute values of yield. Differences in feed pipe velocities are suspected to be the major variant. This can cause variations in the balance between mesomixing and micromixing, even when micromixing is the dominant effect. See the discussions in Section 13-2.1.3 under "Mesomixing" and in Section 13-2.4 for more details on the effects of feed pipe velocities.

Baldyga et al. (1993) simulated the 19 and 65 L yield results of Figure 13-34 using eq. (13-39) and data on the flow rates of the impeller stream and the feed stream. The results were very close to the experimental data, indicating a dependence on Batchelor scale micromixing. The method for accomplishing the simulation is straightforward and was published as a TK Solver program by Penney et al. (1997; available from Penney).

### 13-5.6 Monte Carlo Coalescence–Dispersion Simulation of Mixing

Generally, yield of the component R in the reaction sequence $A + B \to R$; $R + B \to S$ increases with power dissipation (increased impeller rotation rate),

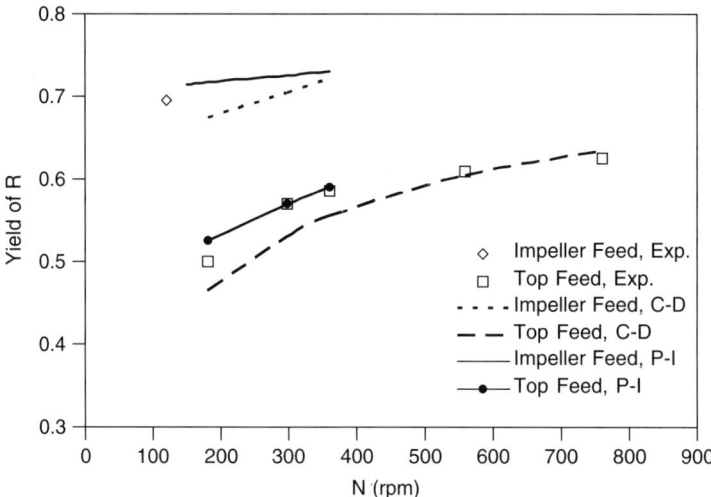

**Figure 13-35** Results from Paul and Treybal (1971) experiments compared with C-D simulation (Canon et al., 1977) and paired-interaction (P-I) simulation (Patterson and Randick, 2000).

## SIMULATION OF MIXING AND CHEMICAL REACTION   849

impeller size, and proximity of feed to the impeller. Feed component B is usually added to resident component A when component B participates in both reactions. Paul and Treybal (1971) made measurements of the final yield of one product of a competitive-consecutive reaction as a function of impeller speed in two small (T = 0.15 m and 0.29 m) semibatch reactors, showing how mixing rate affects yield. The data for the 0.29 m-diameter vessel are plotted in Figure 13-35. In the reaction sequence used by Paul and Treybal, A was tyrosine, B was iodine, R was monoiodated tyrosine, and S was diiodated tyrosine. Iodine was fed at a concentration of 2 mol/L into tyrosine at 0.2 mol/L.

Canon et al. (1977) simulated the flow, mixing, and reaction in the Paul and Treybal stirred reactor using a Monte Carlo coalescence and dispersion (C-D) method. In this method elements of the fluid are simulated by points that move according to the flow pattern in the vessel. These points have mass and composition representing some fraction of the fluid in the vessel. The points are caused to mix (coalesce), react, then disperse. The number of points undergoing C-D during each time increment is proportional to a C-D frequency. The local C-D frequency (coalescences/time/site) was found to be related to local turbulence as follows:

$$I_{CD} = 0.1 \left( \frac{\varepsilon}{L_S^2} \right)^{1/3} \equiv 0.186 \left( \frac{\varepsilon}{k} \right) \qquad (13\text{-}40)$$

The choice of points to undergo C-D is done by a Monte Carlo algorithm in which each of two points for each C-D event is chosen randomly within a flow zone.

As shown in Figure 13-35, the C-D results are reasonably close to the experimental ones. Similarly good results based on Monte Carlo methods have been obtained by van den Akker (2001).

**Example 13-11: Use of Monte Carlo C-D Simulation for a Mixed Chemical Reactor.** The McKelvey et al. (1975) study of mixing in the Toor geometry (see Figure 13-29) gives data for the segregation of mixing water solutions as a function of distance downstream of the injection nozzles. The data, summarized in Table 13-13, also give hydrodynamic data such as velocities, turbulence intensities, length scales, and rates of turbulence energy dissipation. The data for $L_S$ and $\varepsilon$ were discussed in Example 13-10. If one assumes that $I_{CD} = K_{CD}(\varepsilon/L_S^2)^{1/3} = -\frac{1}{2}(ds/s\,dt)$ based on the incremental effect of each coalescence on s, then $K_{CD}$ should be approximately 0.25 when the integral of this equation is compared with the integral of the Corrsin equation. A plot of the values of $-\frac{1}{2}(\Delta s/s\,\Delta t)$ versus $(\varepsilon/L_S^2)^{1/3}$ is shown in Figure 13-36, with the value of $K_{CD}$ given as the slope of the straight line. The zero value at the origin is fixed. The slope is $K_{CD} = 0.0944$. A more complete analysis using many data points and the chemical results of Vassilatos and Toor (1965) gives a value of $K_{CD} = 0.1$, as given in eq. (13-40).

**Table 13-13**

| $L_x$ (cm) | t(s) | $s/s_o$ | $\Delta s/s$ | $-\frac{1}{2}(\Delta s/s\Delta t)$ $(s^{-1})$ | $\varepsilon_{avg}$ $(cm^2/s^3)$ | $L_{S,avg}$ (cm) | $(\varepsilon/L_S^2)^{1/3}$ $(s^{-1})$ |
|---|---|---|---|---|---|---|---|
| 1.1 | 0.0147 | 0.015 | 0 | – | 0 | – | 0 |
| 3.0 | 0.040 | 0.0020 | 1.73 | 35.2 | 30 600 | 0.0175 | 378 |
| 5.0 | 0.067 | 0.00095 | 0.71 | 13.2 | 9675 | 0.025 | 132 |
| 8.1 | 0.108 | 0.00040 | 0.81 | 9.95 | 3050 | 0.038 | 96 |

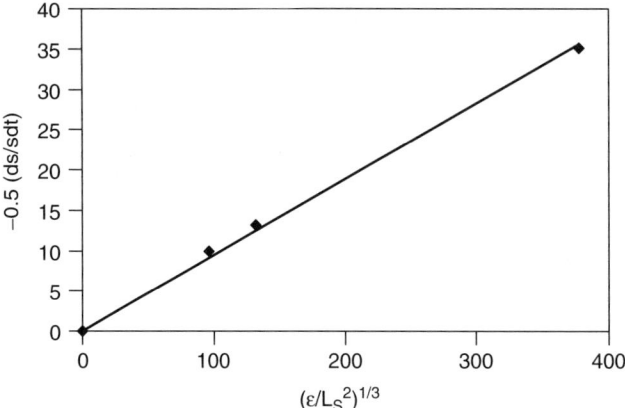

**Figure 13-36** Linear fit of $K_{CD}$ as a function of rate of segregation decay and the time constant of mixing. Best-fit line gives $K_{CD} = 0.0944$.

### 13-5.7 Paired-Interaction Closure for Multiple Chemical Reactions

Paired-interaction closure [eqs. (13-29) and (13-30)] may be used in three dimensional simulations of turbulent mixed reactors with multiple simultaneous chemical reactions. The assumptions of the closure are that (1) only one concentration probability need be used to capture the most important aspects of the interaction of mixing and chemical reaction, and (2) that the reactants need only be considered in pairs and that higher-order interactions of the reactants are not important. These assumptions give a closure that is very fast to compute and results very close to experimental results (see Zipp and Patterson, 1998; Randick, 2000).

The earliest attempts to simulate mixing and chemical reaction in stirred vessels were based on the use of connected zones within which mixing and reaction took place. In some of these simulations, the flow in and out, $L_S$ and $\varepsilon$, were based on experimental data (Patterson, 1975; Mann and Knysh, 1984). It is now possible using standard CFD codes to numerically compute spatial distributions of Reynolds-averaged variables of the turbulent flow and mixing of miscible fluids with similar viscosities and densities in almost any geometry (see Hutchings et al., 1989; Bakker and van den Akker, 1990; Ju et al., 1990; Bakker and Fasano, 1993; Perng and Murthy, 1993; Dilber and Rosenblat, 1995; Harvey et al., 1995; Fox, 1995; Armenante and Chou, 1996; Zipp and Patterson, 1998). Circulation patterns and distribution of turbulence intensities are generally good and predict the trends correctly, but the values of the turbulent quantities k and ε are not always correct. Turbulence energy and energy dissipation rate, which are the major parameters determining local mixing rate, can be distorted by the presence of trailing vortices (Wu and Patterson, 1989) and macroinstabilities (Roussinova et al., 2000). The usual problem is that the k and ε values are too low in regions where the vortices feed energy into the turbulence. Interestingly, in the case of the radial flow impeller, the ratio of k to ε, the important quantity for mesomixing rate computation, is still nearly correct (Zipp and Patterson, 1998). This has not been tested for pitched blade impellers or other axial flow impellers. Great care must therefore be exercised when simulating mixing and chemical reaction in stirred vessels. Ways to achieve an acceptable simulation are discussed in Example 13-12. Use of sliding mesh methods solves these problems only partially, as the vortices are very small and energetic, requiring a very fine mesh for simulation. Further improvement could be obtained by using large eddy simulation (LES), in which the large eddies are simulated on a time-dependent basis. Derksen and van den Akker (1999) have done pioneering work in using LES for stirred vessel simulations.

Simulation of single and multiple chemical reactions in stirred vessels may be done using a CFD computer code with added subroutines for mixing rates and chemical reaction rates. Validation of this simulation method has been done by simulation of the semibatch vessel and chemical reaction used by Paul and Treybal (1971), shown in Figure 13-35, and by comparisons of semi-batch measurements made by Doshi (2001) with corresponding simulations by Randick (2000) and Gross (2002) as discussed in Example 13-12.

***Example 13-12: Use of Paired-Interaction Closure for Multiple Chemical Reactions—Isothermal Case.*** An example of recent more detailed simulations are those made by Randick (2000) and Gross (2002) to determine the effects of impeller rotation rate, impeller type, feed location, chemical reaction rate constants, heat of reaction and activation energy, and vessel size on yield of a consecutive-competitive chemical reaction. The fluid dynamics code Fluent was used to simulate the flow patterns and turbulence in the vessel in the usual way. The outflow from the impellers was simulated by fixing the velocity of the fluid at the locus of points swept by the impeller edge at the fluid outflow. This is more efficient than using sliding mesh to simulate the impeller flows, but is not always feasible (see Chapter 5). When detailed impeller outflow and turbulence data are not available, a sliding mesh (or similar) method must be used. Complete experimental impeller flow data are necessary for the best accuracy, but for the Rushton turbine fixing only the angular velocity at the impeller tip gave good results. The resulting radial velocities, turbulence energy, and dissipation rates were close to those determined by fixing all the values at the blade tips.

**852** MIXING AND CHEMICAL REACTIONS

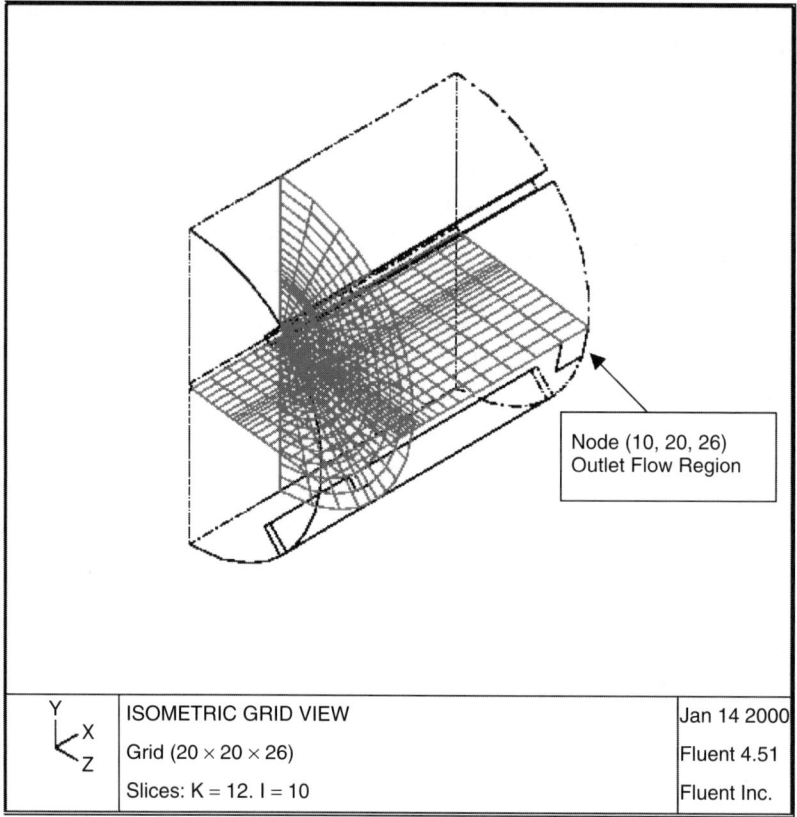

**Figure 13-37** Isometric view of grid used in simulations in a mixing tank.

The grid used in the simulations is shown in Figure 13-37. It consisted of 20 r-z grids unevenly spaced in the angular direction. They were closer together near the baffles. Each r-z grid was $20 \times 26$ nodes with node compression near the impeller tips in both directions. Grid refinement experiments showed that this grid spacing was adequate for the mixing effects being modeled. Grid refinement should always be done to determine whether any grid-size effects are influencing the results of the simulation.

In this example eqs. (13-27) to (13-30) (the paired-interaction closure) were incorporated into a subroutine called by Fluent to compute the rates of segregation growth and decay and the rates of the chemical reactions. The subroutine, called Pairin, is available from author Patterson. Pairin may easily be adapted to other fluid dynamics simulators if desired. Pairin may also be used as an example for development of new subroutines using more or less sophisticated closures. In addition, the subroutine developed by Baldyga and co-workers (see Sections 13-5.2 and 13-5.8) for use of the $\beta$-PDF may be used instead of the Pairin subroutine.

Two impeller types were simulated: the standard six-blade disk turbine and a 45° pitched blade turbine. The outflow from the disk turbine was simulated by fixing the tangential velocities at the blade tip locus (the FIX option in Fluent). The radial velocities and k/ε ratios generated were close to the values that have been measured by Wu and Patterson (1989). The outflow velocities and turbulence energy from the pitched blade turbine were fixed at the bottom locus of the impeller blades using the data of Fort et al. (1999). The resulting flow patterns were close to the data measured.

Doshi (2001) made measurements of the final yields for the tyrosine–iodine reaction used by Paul and Treybal (1971), but in vessels (0.785 and 19.1 L) with standard geometry (T = H = 3 D), using both Rushton or disk and pitched blade impellers. Four feed locations were used: the top surface of the liquid in the vessel, the center of the impeller (disk turbine only), into the impeller discharge very near the blade tips, and the inside edge of the baffles at the height of the impeller directed toward the shaft. A range of feed times were examined, showing a threshold of minimum feed time for micromixing similar to that obtained by Baldyga and Bourne (1992). These experimental data for feed into the impeller discharge very near the blade tips are compared in Figure 13-38 with the simulation results obtained by the method described above by Randick and Gross.

It is clear that for the conditions of these experiments the larger vessel gives slightly higher yields at the same power per volume. This is consistent with the data of Baldyga and Bourne (1992) for feed into the high-turbulence region near the impeller. The absolute volume of this region is larger in the larger vessel.

### 13-5.8  Closure Using β-PFD Simulation of Mixing

Baldyga (1994) gives several examples of conversion and yield calculations using the β-PFD closure for various reactor geometries where mixing is a factor. One

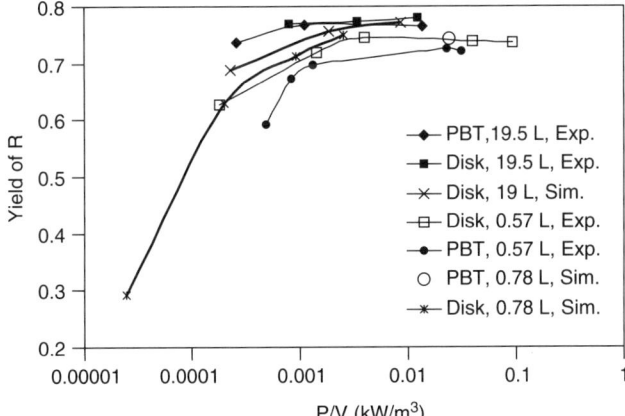

**Figure 13-38** Comparisons for the Doshi (2001) experimental data for fed-batch reaction (impeller feed) with the Randick (2000) and Gross (2002) simulations.

of these examples is the simulation of the yield results obtained experimentally by Li and Toor (1986). They measured yields at various flow rates (Reynolds numbers) in the multijet reactor illustrated in Figure 13-29 using the Bourne reaction of 1-naphthol with diazotized sulfanilic acid. Details of the calculation method were given in the Baldyga (1994) article. Their results show very good correspondence with the experiments, as shown in Figure 13-39. The volume feed ratio and reactant concentration ratio were both near 1.

### 13-5.9 Simulation of Stirred Reactors with Highly Exothermic Reactions

Of great importance because of the prevalence of semibatch reactors are the effects of temperature rise in semibatch reactors with exothermic chemical reactions. Several semibatch cases with exothermic reaction were simulated by Randick (2000) using Fluent with Pairin. The method used for the simulations was the same as that described in Example 13-12. The simulations frequently required 2 or more days on a fast computer. The reactions were $A + B \rightarrow C$ ($k_{R1}$) and $A + C \rightarrow D$ ($k_{R2}$). The conditions for the semibatch simulations were $k_{R1} = 36$ L/kmol·s, $k_{R2} = 3.6$ L/kmol·s, $\Delta H_{R1} = 0$, $\Delta H_{R2} = -1.0 \times 10^8$ J/kmol, $E_{R1} = 0$, $E_{R2} = 1.0 \times 10^7$ to $3.0 \times 10^7$ J/kmol, where $k_{Ri}$ are rate constants, $\Delta H_{Ri}$ are heats of reaction, and $E_{Ri}$ are activation energies. Figure 13-39 shows the results of simulations that were done in the semibatch mode.

The effect of activation energy on yield is clearly shown by Figure 13-40, where yield is plotted as a function of stirring power per unit volume for the various determining variables. Power per unit volume and tank size have a much stronger effect for the exothermic reactions than for the isothermal reactions, as shown. In both cases direct comparison of results in the 0.785 L tank with the 3785 L tank indicates a yield difference of about 25%. Figures 13-41 and 13-42 show how concentration and temperature change with time as the semibatch

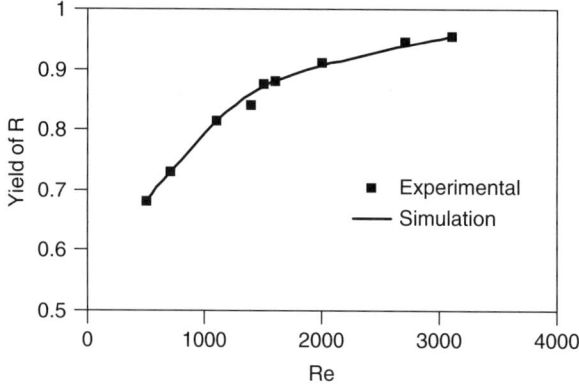

**Figure 13-39** Comparison of β-PDF simulation of yield from Baldyga (1994) as a function of Reynolds number with the data of Li and Toor (1986).

SIMULATION OF MIXING AND CHEMICAL REACTION    855

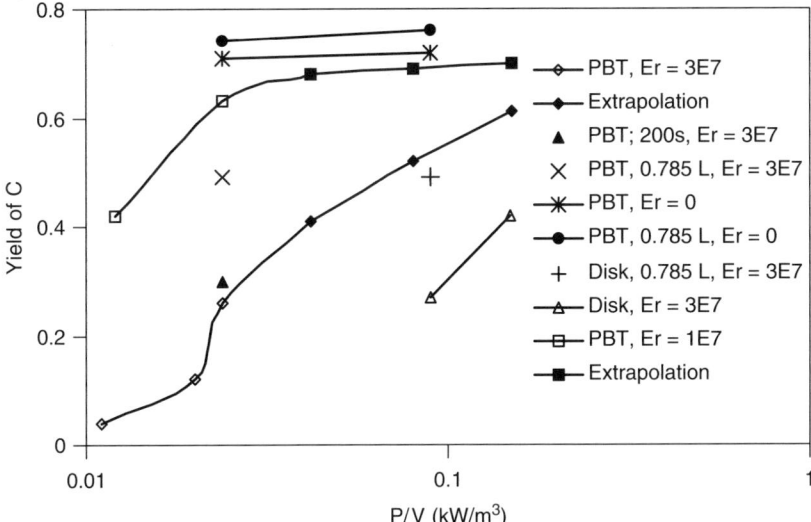

**Figure 13-40** Comparisons of yields for exothermic semibatch cases; $k_{R1}/k_{R2} = 10$, volume = 3785 L, and addition time = 50 s except where noted.

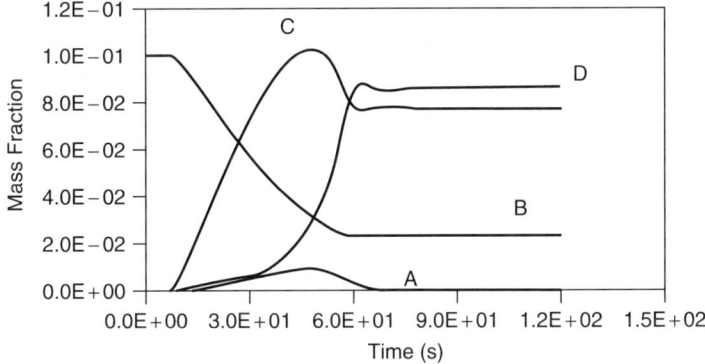

**Figure 13-41** Mole fraction versus time at a position near the top of the vessel for a fed-batch reaction (Randick's trial 10): $A + B \rightarrow C$; $A + C \rightarrow D$ (see Randick, 2000). The conditions in the reactor are: volume = 3785 L; pitched blade impeller; N = 100 rpm; $V_{feed} = 1.045$ m/s; $t_f = 50$ s; $\Delta H_{R1} = 0$; $\Delta H_{R2} = -10^8$ J/kmol; $E_{R1} = 0$; $E_{R2} = 3 \times 10^7$ J/kmol. Yield of C was 41.3%.

reactions proceed. Wall cooling with a heat transfer coefficient of 283 J/s · m² · K and a temperature of 323 K were assumed.

This preliminary research shows that prediction of yield on scale-up of a highly exothermic chemical reaction series can be particularly difficult. The heats of reaction and activation energies of each reaction must be well known. The

**856**  MIXING AND CHEMICAL REACTIONS

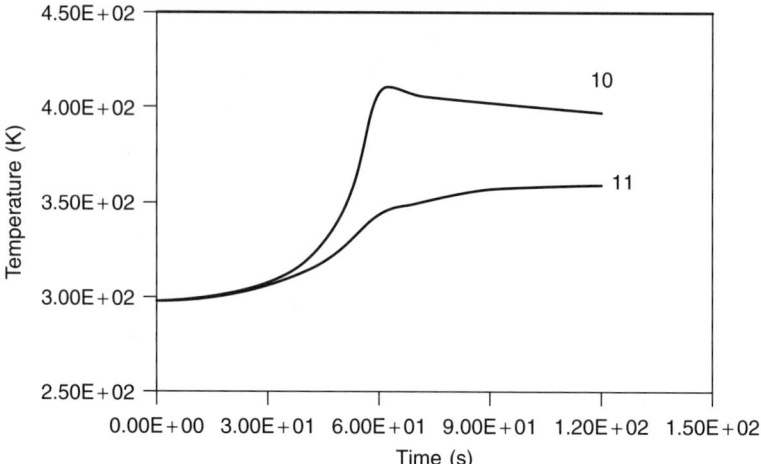

**Figure 13-42** Temperature response for two exothermic reactions measured at the same location as the concentrations in Figure 13-24. One plot is for Randick's trial 10 (see Figure 13-41) and the other is for a reaction with a lower reaction activation energy.

power per unit volume, feed point, and impeller type must also be carefully specified. Only a full simulation of the reaction system using CFD is capable of incorporating all these determining variables.

It must be emphasized that even though these results seem reasonable and that extensive validation of stirred-tank simulations has been done for various continuous flow and semibatch cases, only isothermal validation has been done, and no large ratio scale-up validations have been done. The feed time of 50 s for a 3.8 m$^3$ vessel is very short and exaggerates the heat effects. These aspects of the problem need to be studied before complete confidence may be placed on the CFD simulation of highly exothermic reactions.

### 13-5.10 Comments on the Use of Simulation for Scale-up and Reactor Performance Studies

Simulation has not yet reached the point that it is a replacement for experimental results. However, simulation has several very useful aspects:

1. Many results may be obtained in a relatively short time using simulation once the users become highly conversant with the methods.
2. Simulations may be cheaper per result than experiments if personnel are available who have expertise in the simulation methods equivalent to the expertise of the experimenters.
3. Simulations may be used profitably to study details of flow, turbulence, or mixing rate that cannot conveniently be done experimentally or possibly cannot be done at all. This is particularly true where simulations at the

same conditions as the experiments give results close to the experimental results. Significant design insights can sometimes come from such studies.
4. Simulation is based on the fundamental physics of the process and therefore always has the potential of giving more realistic information on the performance of the process than that of methods based on dimensional analysis, mechanistic approximations, or space-averaged (one-point) theories or correlations, *provided that the model equations are well defined.* The directional changes shown by the simulations should in general be the same as the experimental ones, and the magnitudes of those changes should also be similar.
5. Simulation of mixing effects in reactors requires good data for reaction kinetics; physical constants such as viscosity, density, and diffusivity; and knowledge of the exact geometrical configuration of the vessel, including feed locations.
6. The approximations involved in simulation are constantly improving, the main ones being spatial and time averaging (Reynolds averaging), the numerical methods used to reach convergence, and the closures needed to link the averages to provide a closed set of equations. Large-eddy simulation (LES) methods may well provide a way to reduce the severity of these approximations and lead to more realistic simulations. At this writing, objective comparison of LES results with time-resolved experimental results is extremely difficult, so validation is reduced to time averaging the LES results for comparison with time-averaged experimental results.

Simulations should always be used in conjunction with some experiments, the balance between the two being dictated by the level of uncertainty in the simulations being done. Uncertainty increases as reactors progress from homogeneous stirred tanks and pipes to the much more difficult two- and three-phase reactors where in addition to the chemical reactions, the following must be addressed: gas bubble and/or dispersed-liquid droplet sizes; gas, liquid, or solids distributions in the vessel; coupling of momentum between the phases; rates of breakup, agglomeration, and coalescence; surface effects; and mass transfer coefficients. Progress over the last 10 years has been significant, but many challenges remain to be resolved.

## 13-6 CONCLUSIONS

In this chapter we have shown that the mechanism and kinetics of a chemical reaction scheme can be combined with a rather detailed fine scale picture of the fluid dynamics in a simple reactor to predict yield and selectivity for fast homogeneous reactions. With a lot more information we can predict some of the reaction effects when another phase is present and interphase mass transfer becomes important. Heterogeneous analysis is severely limited because our knowledge of the phenomena and the descriptive equations dictating surface area creation and disappearance are only now being developed.

Industrially, the information needed about chemical mechanisms and intrinsic chemical kinetics is difficult to obtain. An experimental program to measure fast kinetics under industrial conditions (often, high temperatures and pressures) is very expensive. Frequently, the industrialist must deal with lumped rate expressions for a multistep reaction which are obtained from a poorly characterized lab reactor. Many industrialists and consultants have developed experimental protocols to determine mixing effects in laboratory scale chemical reactors. When combined with the concepts in this chapter, this allows us to scale up to full scale without all of the details of the kinetics, phase behavior, and full mathematical models. We emphasize for a final time that laboratory equipment tends to have shorter mixing times because of scale effects and higher turbulence levels. This can obscure rate-controlling kinetic steps in several important classes of reaction. The reader will note that there is no general summary table for reaction scale-up rules, there is no unique definition of $Da_M$, no unique correlation for critical feed time. This is because there are many things which are still not understood or are poorly quantified. The first of these is the wide spectrum of reaction mechanisms and kinetics. The second is the entire area of inertial scale or mesoscale mixing, combined with the infinite possible variations of industrial reactor geometries, leading to an equally wide range of local turbulence conditions.

The fluid flow and turbulence in many plant reactors is more complex than in vessels modeled by academics. CFD is useful for understanding the flow patterns in such reactors; however, at the time of writing CFD still does a poor job of predicting the local turbulence quantities vital to micromixing and mesomixing analysis. This is another area where greater understanding is needed.

There is an entire class of problems where the particle size and particle size distribution are critical characterization parameters in addition to yield or purity. Another set of equations describing nucleation and growth along with population balances must then be added to the models. Success in predicting these effects has so far been limited, possibly due to inaccurate or unrealistic kinetics. When the question of morphology is added to the mix, the problem becomes daunting. Again, experimental protocols based on an understanding of micromixing can help the engineer determine good and bad strategies for scale-up, but more understanding is needed.

There is a final class of phenomena associated with two-phase surface energy systems, such as liquid–liquid dispersions and emulsions, in which rapid physiochemical kinetics can be observed. The rate of mixing can have a profound effect on the properties of such systems. One example is the shocking of a stabilized dispersion where the addition of more continuous phase upsets the local surface-active-agent balance and can cause agglomeration or coagulation. Rapid addition with rapid dilution (mixing) can avoid such affects.

As this is written, there are undoubtedly more examples of complex reactions being developed in laboratories around the world. An understanding of how fluid motion and particularly, turbulence can affect the path of reactions is extremely useful for understanding many of the problems encountered in developing these reactions, and more important, in determining ways to avoid mixing problems on scale-up.

## NOMENCLATURE

| | |
|---|---|
| a | mass transfer area per unit volume of the total fluid (m$^2$/m$^3$) |
| a' | mass transfer area per unit bubble volume (m$^2$/m$^3$) |
| A | reactant or moles of reactant |
| A | cross-sectional area (m$^2$) |
| A$_0$ | initial moles of reactant A |
| B | baffle width (m) |
| B | moles of reactant B |
| $\overline{c_0^2} = \overline{s_0}$ | initial or maximum segregation (usually, $\overline{C_{10}} < \overline{C_{20}}$) (mol/m$^3$)$^2$ |
| c$_i$ | fluctuation of component i concentration about average (mol/m$^3$) |
| C$_{i0}$ | initial concentration of reactant i as if already mixed (mol/m$^3$) |
| C$_i$ | concentration of component i at any location (mol/m$^3$) |
| C$_i$* | saturated or equilibrium concentration of component i in liquid (mol/m$^3$) |
| C$_C$ | constant in Corrsin term for energy dissipation |
| C$_{g1}$ | constant in Spalding term for segregation production |
| d$_{jo}$ | initial jet diameter (m) |
| D | impeller diameter or pipe diameter (m) |
| D$_{AB}$ | molecular diffusivity of A in B (m$^2$/s) |
| Da$_M$ | mixing Damkoehler number |
| D$_t$ | turbulent diffusivity (m$^2$/s) |
| E | engulfment rate constant (s$^{-1}$) |
| E$_{Ri}$ | activation energy for reaction i |
| f | mixture fraction |
| G | ratio of distance from impeller to impeller diameter (−) |
| G | molar gas flow rate (mol/s) |
| H | vessel height (m) |
| H$_{Ri}$ | heat of reaction for reaction i (kJ/mol) |
| I | interaction rate in the IEM model |
| I$_{CD}$ | rate of coalescence and dispersion in CD simulation of mixing rate |
| k | turbulence energy (m$^2$/s$^2$) |
| k$_L$ | liquid-side mass transfer coefficient |
| k$_L^*$ | liquid-side mass transfer coefficient corrected for reaction rate effect |
| k$_R$ | reaction rate constant (depends on reaction order) |
| K | constant in the C-D equation |
| K$_H$ | Henry's law constant |
| L | pipe or static mixer length (m) |
| L$_S$ | concentration macroscale (m) |

**860** MIXING AND CHEMICAL REACTIONS

| | |
|---|---|
| $L_f$ | scale of the feed (m) |
| n | number of feed points () |
| n | reaction order () |
| N | impeller rotation rate (rps) |
| $N_p$ | power number for stirred vessels () |
| p | probability of a given concentration () |
| P | fluid mixing power (W) |
| q | lumped parameter (see Section 11-6.3) |
| $Q_B$ | volumetric feed rate of component B (m$^3$/s) |
| $Q_L$ | volumetric liquid flow rate (m$^3$/s) |
| $Q_G$ | gas feed rate (m$^3$/s) |
| r | radial distance from vessel axis (m) |
| $r_i$ | rate of production of component i by reaction (mol/s) |
| R | vessel radius, T/2 (m) |
| R | moles of product R (mol) |
| Re | Reynolds number (ND$^2$/$\nu$ for a stirred tank; DV/$\nu$ for a pipe) |
| $s_i$ | square of concentration fluctuation; its mean square is segregation (mol/m$^3$) |
| S | moles of product S (mol) |
| Sc | Schmidt number for molecular diffusion, $\nu/D_{AB}$ () |
| t | time (s) |
| $t_f$ | feed time (s) |
| $t_{1/2}$ | half-life (s) |
| T | vessel diameter (m) |
| u' | root-mean-square fluctuating velocity (m/s) |
| U | bulk-mean velocity in a pipe or ambient velocity near a feed location (m/s) |
| $U_{jo}$ | bulk-mean jet velocity (m/s) |
| $U_{tip}$ | impeller blade tip velocity (m/s) |
| $U_x$ | velocity in a plug flow reactor (m/s) |
| V | total volume of fluid in vessel (m$^3$) |
| $V_L$ | liquid-phase volume (m$^3$) |
| $v_f$ | feed pipe velocity (m/s) |
| $v_r$ | rise velocity of bubbles in vessel (m/s) |
| $v_s$ | superficial gas velocity (m/s) |
| $v_t$ | impeller blade tip velocity (m/s) |
| x | distance downstream (m) |
| x | mole fraction in liquid phase () |
| $x_i$ | Cartesian coordinate (m) |
| $X_A, X_B$ | fraction A- and B-rich fluids |
| $X_S$ | selectivity to product S |
| $Y_{exp}$ | ideal yield with perfect mixing |
| $Y_{Ao}, Y_{Ai}$ | outlet and inlet mole fractions of A in gas |
| $Y_R, Y_S$ | yield of R or S product |
| z | axial distance from impeller disk plane (m) |

*Greek Symbols*

| | |
|---|---|
| $\varepsilon$ | turbulence energy dissipation rate ($m^2/s^3$) |
| $\theta$ | tangential coordinate |
| $\lambda_B$ | Batchelor length scale (m) |
| $\mu$ | fluid viscosity (kg/ms) |
| $\nu$ | kinematic viscosity ($m^2/s$) |
| $\nu_t$ | turbulent momentum diffusivity ($m^2/s$) |
| $\rho$ | fluid density ($kg/m^3$) |
| $\sigma_c$ | Schmidt number for mass diffusion |
| $\sigma_s$ | Schmidt number for segregation diffusion |
| $\tau_B$ | blending time (s) |
| $\tau_{crit}$ | feed time beyond which there are no yield effects of feed rate (s) |
| $\tau_D$ | mesomixing time based on feed rate, ambient velocity, and turbulent diffusivity (s) |
| $\tau_M$ | Corrsin mixing time (s) |
| $\tau_R$ | reaction time constant ($1/k_R C$) for second-order reaction (s) |
| $\tau_S$ | mesomixing time based on the Corrsin formulation (s) |
| $\Phi$ | volume fraction gas or dispersed phase holdup |

## REFERENCES

Angst, W., J. R. Bourne, and P. Dell'Ava (1984). Mixing and fast chemical reaction: IX. Comparison between models and experiments, *Chem. Eng. Sci.*, **39**, 335–342.

Armenante, P. M., and C.-C. Chou (1996). Velocity profiles in a baffled vessel with single or double pitched-blade impellers, *AIChE J.*, **42**, 42–54.

Astarita, G. (1967). *Mass Transfer with Chemical Reaction*, Elsevier, New York.

Bakker, R. A., and J. B. Fasano (1993). Time dependent turbulent mixing and chemical reactions in stirred tanks, *Paper 70c*, presented at the AIChE Annual Meeting, St. Louis, MO.

Bakker, R. A., and H. E. van den Akker (1990). *A Computational Study on Dispersing Gas in a Stirred Reactor*, Kramers Laboratory, Delft University of Technology, Delft, The Netherlands.

Bakker, A., A. H. Haidari, and L. M. Marshall (2001). Design reactors via CFD, *Chem. Eng. Prog.*, **97**, Dec., pp. 31–39.

Baldyga, J. (1994). A closure model for homogeneous chemical reactions, *Chem. Eng. Sci.*, **49**, 1985–2003.

Baldyga, J., and J. R. Bourne (1984). A fluid mechanical approach to turbulent mixing and chemical reaction, *Chem. Eng. Commun.*, **28**, 231–278.

Baldyga, J., and J. R. Bourne (1988). Calculation of micromixing in inhomogeneous stirred tank reactors, *Chem. Eng. Res. Des.*, **66**, 33–38.

Baldyga, J., and J. R. Bourne (1989). Simplification of micromixing calculations: I. Derivation and application of new model, *Chem. Eng. Sci.*, **42**, 83–92.

Baldyga, J., and J. R. Bourne (1990). The effect of micromixing on parallel reactions, *Chem. Eng. Sci.*, **45**, 907–916.

Baldyga, J., and J. R. Bourne (1992). Interactions between mixing on various scales in stirred tank reactors, *Chem. Eng. Sci.*, **47**, 1839–1848.

Baldyga, J., and J. R. Bourne (1999). *Turbulent Mixing and Chemical Reactions*, Wiley, Chichester, West Sussex, England.

Baldyga, J., and M. Henczka (1995). Closure problem for parallel chemical reactions, *Chem. Eng. J.*, **58**, 161–173.

Baldyga, J., and M. Henczka (1997). Turbulent mixing and parallel chemical reactions in a pipe: application of a closure model, *Recent. Prog. Genie Precedes*, **11**, 341–348.

Baldyga, J., J. R. Bourne, and Y. Yang (1993). Influence of feed pipe diameter on meso-mixing in stirred tank reactors, *Chem. Eng. Sci.*, **48**, 3383–3390.

Baldyga, J., W. Podgorska, and R. Pohorecki (1995). Mixing-precipitation model with application to double feed semibatch precipitation, *Chem. Eng. Sci.*, **50**, 1281–1300.

Baldyga, J., J. R. Bourne, and S. J. Hearn (1997). Interaction between chemical reactions and mixing on various scales, *Chem. Eng. Sci.*, **52**, 457–466.

Baldyga, J., J. R. Bourne, and B. Walker (1998). Non-isothermal micromixing in turbulent liquids: theory and experiment, *Can. J. Chem. Eng.*, **76**, 641–649.

Barthole, J., R. David, and J. Villermaux (1982). *A New Chemical Method for the Study of Local Micromixing Conditions in Industrial Stirred Tanks*, ISCRE, Boston.

Batchelor, G. K. (1959). Small scale variation of convected quantities like temperature in turbulent fluid: I. Discussion and the case of small conductivity, *J. Fluid Mech.*, **5**, 113–133.

Bird, R. B., W. Stewart, and E. N. Lightfoot (1960). *Transport Phenomena*, Wiley, New York.

Bourne, J. R. (1983). Mixing on the molecular scale, *Chem. Eng. Sci.*, **38**, 5–8.

Bourne, J. R., and P. Dell'Ava (1987). Micro- and macro-mixing in stirred tank reactors of different sizes, *Chem. Eng. Res. Des.*, **65**, 180–186.

Bourne, J. R., and J. Garcia-Rosas (1984). Rapid mixing and reaction in a Ystral in-line dynamic mixer, *Paper 52c*, presented at the AIChE Annual Meeting, San Francisco.

Bourne, J. R., and C. P. Hilber (1990). The productivity of micro-mixing-controlled reactions: effect of feed distribution in stirred tanks, *Chem. Eng. Res. Des.*, **68**, 51–56.

Bourne, J. R., and S. Rohani (1983). Mixing and fast chemical reaction: VII. Determining reaction zone model for the CSTR, *Chem. Eng. Sci.*, **38**, 911–916.

Bourne, J. R., and S. Thoma (1991). Some factors determining the critical feed time of a semi-batch reactor, *Chem. Eng. Res. Des.*, **69**, 321–323.

Bourne, J. R., and S. Yu (1994). Investigation of micromixing in stirred tank reactors using parallel reactions, *Ind. Eng. Chem. Res.*, **33**, 41–55.

Bourne, J. R., F. Kozicki, U. Moergeli, and P. Rys (1981). Mixing and fast chemical reaction: III. Model–experiment comparisons, *Chem. Eng. Sci.*, **36**, 1655–1663.

Bourne, J. R., O. M. Kut, J. Lenzner, and H. Maire (1990). Kinetics of the diazo coupling between 1-naphthol and diazotized sulfanilic acid, *Ind. Eng. Chem. Res.*, **29**, 1761–1765.

Bourne, J. R., O. M. Kut, and J. Lenzner (1992). An improved reaction system to investigate micromixing in high-intensity mixers, *Ind. Eng. Chem. Res.*, **31**, 949–958.

Boussinesq, J. (1903). *Théorie analytique de la chaleur*, Vol. 2, Gauthier-Villars, Paris.

Brodkey, R. S., and J. Lewalle (1985). Reactor selectivity based on first-order closures of the turbulent concentration equations, *AIChE J.*, **31**, 111–118.

Canon, R. M., A. W. Smith, K. W. Wall, and G. K. Patterson (1977). Turbulence level significance of the coalescence–dispersion rate parameter, *Chem. Eng. Sci.*, **32**, 1349–1352.

Cichy, P. T., and T. W. F. Russell (1969). Two-phase reactor design tubular reactors: reactor model parameters, *Ind. Eng. Chem. Res.*, **61**, 15.

Corrsin, S. (1964). The isotropic turbulent mixer: II. Arbitrary Schmidt number, *AIChE J.*, **10**, 870–877.

Danckwerts, P. V. (1958). The effect of incomplete mixing on homogeneous reactions, *Chem. Eng. Sci.*, **8**, 93–99.

Derksen, J., and H. E. A. van den Akker (1999). Large eddy simulations of the flow driven by a Rushton turbine, *AIChE J.*, **45**, 209–221.

Dilber, I., and S. Rosenblat (1995). Anisotropic turbulence models for simulation of mixing processes, Paper 122f, presented at the AIChE Annual Meeting, Miami Beach, FL, Nov.

Donaldson, C. (1975). On the modeling of the scalar correlations necessary to construct a second-order closure description of turbulent reacting flows, in *Turbulent Mixing in Nonreactive and Reactive Flows*, S. N. B. Murthy, ed., Plenum Press, New York, pp. 131–162.

Doraiswamy, L. K., and M. M. Sharma (1984). *Heterogeneous Reactions: Analysis, Examples, and Reactor Design*, Wiley, New York.

Doshi, J. (2001). Effect of mixing on multiple chemical reactions, M.S. thesis, University of Missouri–Rolla.

Elghobashi, S. E., W. W. Pun, and D. B. Spalding (1977). Concentration fluctuations in isothermal, turbulent confined jets, *Chem. Eng. Sci.*, **32**, 161–166.

Fasano, J. B., and W. R. Penney (1991). Avoid blending mix-ups, *Chem. Eng. Prog.*, **87**(10), 56–63.

Fogler, H. S. (1999). *Elements of Chemical Reaction Engineering*, 3rd ed., Prentice Hall, Upper Saddle River, NJ.

Forney, L. J., N. Nafia, and H. X. Vo (1996). Optimum jet mixing in a tubular reactor, *AIChE J.*, **42**, 3113–3122.

Fort, I., P. Vortuba, and J. Medek (1999). Turbulent flow of liquid in mechanically agitated closed vessel, manuscript from I. Fort.

Fox, R. O. (1995). The spectral relaxation model of the scalar dissipation rate in homogeneous turbulence, *Phys. Fluids*, **7**, 1082–1094.

Garside, J., and N. S. Tavare (1985). Mixing, reaction and precipitation in an MSMPR crystallizer: effects of reaction kinetics on the limits of micromixing, in *Industrial Crystallization*, Vol. 84, S. Jancic and E. J. de Jong, eds., Elsevier, Amsterdam, pp. 131–136.

Grenville, R. K. (1992). Blending of viscous Newtonian and pseudo-plastic fluids, Ph.D. dissertation, Cranfield Institute of Technology, Cranfield, Bedfordshire, England.

Gross, J. (2002). Simulation of effects of impeller rotation rate, reactor size and density gradients on yield in a stirred, fed-batch reactor, M.S. thesis, University of Missouri–Rolla.

Harnby, N., M. F. Edwards, and A. W. Nienow, eds. (1992). *Mixing in the Process Industries*, Butterworth-Heinemann, Wolburn, MA.

Harvey, A. D., C. K. Lee, and S. E. Rogers (1995). Steady-state modeling and experimental measurement of a baffled impeller stirred tank, *AIChE J.*, **41**, 2177–2186.

Heeb, T. G., and R. S. Brodkey (1990). Turbulent mixing with multiple second-order chemical reactions, *AIChE J.*, **36**, 1457–1470.

Homsi, K. L., A. Thompson, and M. P. Thien (1993). A facile and selective large-scale bromination of aminotoluidine using N-bromosuccinimide/acetone solution addition, presented at the AIChE National Meeting, St. Louis, MO, Nov.

Hutchings, B. J., R. R. Patel, and R. J. Weetman (1989). Computation of flow fields in mixing tanks with experimental verification, presented at the ASME Annual Winter Meeting, San Francisco, Dec.

Jo, M. C., W. R. Penney, and J. B. Fasano (1994). Backmixing into reactor feedpipes caused by turbulence in an agitated vessel, *AIChE Symp. Ser. 299*, **90**, 41–49.

Ju, S. Y., T. M. Mulvahill, and R. W. Pike (1990). Three-dimensional turbulent flow in agitated vessels with a nonisotropic viscosity turbulence model, *Can. J. Chem. Eng.*, **68**, 3–16.

Khang, S. J., and O. Levenspiel (1976). New scale-up and design method for stirrer agitated batch mixing vessels, *Chem. Eng. Sci.*, **31**, 569–580, and *Chem. Eng.*, Oct. 11, pp. 141–148.

King, M. L., A. L. Forman, C. Orella, and S. H. Pines (1985). Extractive hydrolysis for pharmaceuticals, *Chem. Eng. Prog.*, **81**(5), 36–39.

Klein, J., R. David, and J. Villermaux (1980). Interpretation of experimental liquid phase micromixing phenomena in a continuous stirred tank reactor with short residence times, *Ind. Eng. Chem. Fundam.*, **19**, 373–379.

Kosaly, G. (1987). Premixed simple reaction in homogeneous turbulence, *AIChE J.*, **33**, 1998–2002.

Larson, K. A., M. Midler, and E. Paul (1995). Reactive crystallization: control of particle size and scale-up, presented at the Association for Crystallization Technology Meeting, Charlottesville, VA.

Launder, B. E., and D. B. Spalding (1972). *Mathematical Models of Turbulence*, Academic Press, New York.

Lee, S. Y., and Y. P. Tsui (1999). Succeed at gas/liquid contacting, *Chem. Eng. Prog.*, July, pp. 23–49.

Leonard, A. D., R. D. Hamlen, R. M. Kerr, and J. C. Hill (1995). Evaluation of closure models for turbulent reacting flows, *Ind. Eng. Chem. Res.*, **34**, 3640–3652.

Levenspiel, O. (1962). *Chemical Reaction Engineering*, Wiley, New York.

Levenspiel, O. (1972). *Chemical Reaction Engineering*, 2nd ed., Wiley, New York.

Levenspiel, O. (1999). *Chemical Reaction Engineering*, 3rd ed., Wiley, New York.

Li, K. T., and H. L. Toor (1986). Turbulent reactive mixing with a series–parallel reaction: effect of mixing on yield, *AIChE J.*, **32**, 1312–1320.

Magnussen, B. F., and B. H. Hjertager (1976). On mathematical models of turbulent combustion with special emphasis on soot formation and combustion, *Proc. 16th International Symposium on Combustion*, Combustion Institute, Pittsburgh, PA.

Mann, R., and P. Knysh (1984). Utility of networks of interconnected backmixed zones to represent mixing in a closed stirred vessel, *Inst. Chem. Eng. Symp. Ser.*, **89**, 127–145.

Mao, K. W., and H. L. Toor (1971). Second-order chemical reactions with turbulent mixing, *Ind. Eng. Chem. Fundam.*, **10**, 192–197.

Marcant, B., R. David, R. Mamourian, and J. Villermaux (1991). In *Industrial Crystallization,* Vol. 90, 205, A. Mersmann, ed., Germmish, Germany.

McKelvey, K. N., H.-C. Yieh, S. Zakanycz, and R. S. Brodkey (1975). Turbulent motion, mixing and kinetics in a chemical reactor configuration, *AIChE J.*, **21**, 1165–1176.

Medek, J. (1980). Power characteristics of agitators with flat inclined blades, *Int. Chem. Eng.*, **20**, 664–672.

Mehta, R. V., and J. M. Tarbell (1983). Four-environment model of mixing and chemical reaction, *AIChE J.*, **29**, 320–328.

Mersmann, A., and M. Kind (1988). Chemical engineering aspects of precipitation from solution, *Chem. Eng. Technol.*, **11**, 264–276.

Middleton, J. C. (1992). Gas–liquid dispersion and mixing, Chapter 15 in *Mixing in the Process Industries*, N. Harnby, M. F. Edwards, and A. W. Nienow, eds., Butterworth-Heinemann, Wolburn, MA.

Middleton, J. C., F. Pierce, and P. M. Lynch (1986). Computations of flow fields and complex reaction yield in turbulent stirred reactors and comparison with experimental data, *Chem. Eng. Res. Des.*, **64**, 20–22.

Nauman, E. B. (1982). Reactions and residence time distributions in motionless mixers, *Can. J. Chem. Eng.*, **60**, 136–140.

Nienow, A. W. (1997). On impeller circulation and mixing effectiveness in the turbulent flow regime, *Chem. Eng. Sci.*, **52**, 2557–2565.

Nienow, A. W., and K. Inoue (1993). A study of precipitation micromixing, macromixing, size distribution, and morphology, Paper 9.4, presented at CHISA, Prague, Czech Republic.

Oldshue, J. Y. (1980). Mixing in hydrogenation processes, *Chem. Eng. Prog.*, **76**(6), 60–64.

Oldshue, J. Y. (1983). *Fluid Mixing Technology*, McGraw-Hill, New York.

Ottino, J. M. (1980). Lamellar mixing models for structured chemical reactions and their relationship to statistical models, *Chem. Eng. Sci.*, **35**, 1377–1391.

Patterson, G. K. (1973). Model with no arbitrary parameters for mixing effects on second-order reaction with unmixed feed reactants, in *Fluid Mechanics of Mixing*, E. M. Uram, and V. W. Goldschmidt, eds., ASME, New York, pp. 31–38.

Patterson, G. K. (1975). Simulating turbulent-field mixers and reactors, in *Application of Turbulence Theory to Mixing Operations*, R. S. Brodkey, ed., Academic Press, New York, pp. 223–275.

Patterson, G. K. (1985). Modeling of turbulent reactors, in *Mixing of Liquids by Mechanical Agitation*, J. J. Ulbrecht, and G. K. Patterson, eds., Gordon & Breach, New York, pp. 59–91.

Patterson, G. K., and J. Randick (2000). Simulation with validation of mixing effects in fed-batch reactors, *Proc. 10th European Conference on Mixing*, H. E. A. van den Akker, and J. J. Derksen, eds., Delft, The Netherlands, July, pp. 53–60.

Paul, E. L. (1988). Design of reaction systems for specialty organic chemicals, *Chem. Eng. Sci.*, **43**, 1773–1782.

Paul, E. L., and R. E. Treybal (1971). Mixing and product distribution for a liquid-phase, second-order, competitive-consecutive reaction, *AIChE J.*, **17**, 718–731.

Paul, E. L., J. Aiena, and W. A. Sklarz (1981). Effect of mixing on the selectivity of the competitive-consecutive chlorination of acetone, presented at the AIChE National Meeting, New Orleans, LA.

Paul, E. L., H. Mahadevan, J. Foster, M. Kennedy, and M. Midler (1992). The effect of mixing on scale-up of a parallel reaction system, *Chem. Eng. Sci.*, **47**, 2837–2840.

Penney, R. W. (1971). Scale-up for mixing operations, *Chem. Eng.*, Mar., p. 22.

Penney, R. W., H. X. Vo, and G. K. Patterson (1997). Implementation of the Bourne engulfment model using TK-Solver, presented at the AIChE Annual Meeting, Los Angeles.

Perng, C. Y., and J. Y. Murthy (1993). A moving mesh technique for simulation of flow in mixing tanks, in *Process Mixing: Chemical and Biochemical Applications*, Part II, G. B. Tatterson, R. V. Calebrese, and W. R. Penney, eds., AIChE Symp. Ser.

Pohorecki, R., and J. Baldyga (1988). The effects of micromixing and the manner of reactant feeding on precipitation in stirred tank reactors, *Chem. Eng. Sci.*, **43**, 1949–1954.

Pope, S. B. (1985). PDF methods for turbulent reactive flows, *Prog. Energy Combust. Sci.*, **11**, 119–192.

Randick, J. J. (2000). Simulation of scale-up effects on stirred chemical reactors, M.S. thesis, University of Missouri–Rolla.

Roussinova, V., B. Grgic, and S. M. Kresta (2000). Study of macro-instabilities in stirred tanks using a velocity decomposition technique, *Chem. Eng. Res. Des.*, **78**, 1040–1052.

Schaftlein, R. W., and T. W. F. Russell (1968). Two-phase reactor design, tank-type reactors, *Ind. Eng. Chem.*, **60**, 12–20.

Schutz, J. (1988). Agitated thin-film reactors and tubular reactors with stator mixers for a rapid exothermic multiple reaction, *Chem. Eng. Sci.*, **43**, 1975–1980.

Sharma, M. M. (1988). Multiphase reactions in the manufacture of fine chemicals, *Chem. Eng. Sci.*, **43**, 1749–1750.

Sharratt, P. N., ed. (1997). *Handbook of Batch Process Design*, Blackie, London.

Spalding, D. B. (1971). Concentration fluctuations in a round turbulent free jet, *Chem. Eng. Sci.*, **26**, 95–108.

Tatterson, G. B. (1991). *Fluid Mixing and Gas Dispersion in Agitated Tanks*, McGraw-Hill, New York.

Thoma, S. (1989). Interactions between macro- and micro-mixing in stirred tank reactors, Dissertation 9012, ETH, Zurich.

Thoma, S., V. V. Ranade, and J. R. Bourne (1991). Interaction between macro- and micromixing during reactions in stirred tanks, *Can. J. Chem. Eng.*, **69**, 1135–1141.

Thornton, J. D. (1992). *Science and Practice of Liquid–Liquid Extraction*, Vol. 2, Oxford University Press, Oxford, pp. 290–291.

Tipnis, S. K., W. R. Penney, and J. Fasano (1994). An experimental investigation to determine a scale-up method for fast competitive-parallel reactions in agitated vessels, *AIChE Symp. Ser. 299*, **90**, 78–91.

Toor, H. L. (1969). Turbulent mixing of two species with and without chemical reaction, *Ind. Eng. Chem. Fundam.*, **8**, 655–659.

Tosun, G. (1997). A study of diffusion and reaction in unpremixed step growth copolymerization in a micro-segregated continuous stirred reactor, *Ind. Eng. Chem.*, **36**, 4075–4086.

Van de Vusse, J. G. (1966). Consecutive reactions in heterogeneous systems, *Chem. Eng. Sci.*, **21**, 631–643, 1239–1252.

Van den Akker, H. E. A. (2001). Poster presented at Mixing XVIII, Pocono Manor, PA, June.

Vassilatos, G., and H. L. Toor (1965). Second-order chemical reactions in a nonhomogeneous turbulent field, *AIChE J.*, **11**, 666–672.

Villermaux, J., and R. David (1987). Interpretation of micromixing effects on fast consecutive competitive reactions in semi-batch stirred tanks by a simple interaction model, *Chem. Eng. Commun.*, **54**, 333–352.

Villermaux, J., and J. C. Devillon (1972). Representation de coalescence de la redispersion des domaines de segregation, *Proc. 2nd International Symposium on Chemical Reaction Engineering*, Amsterdam, pp. B1–B13.

Villermaux, J., and Z. Zoulalian (1969). Etat du mélange du fluide dans in réacteur continu: a propos d'un modèle de Weinstein et Alder, *Chem. Eng. Sci.*, **24**, 1513–1518.

Wang, Y. D. (1984). Study on the Hofmann rearrangement in a two phase system, *Jilin Daxue Ziran Kexue Xuebao*, **2**, 89–93.

Wiederkehr, H. (1988). Examples of process improvements in the fine chemicals industry, *Chem. Eng. Sci.*, **43**, 1783–1791.

Wu, H., and G. K. Patterson (1989). Laser-Doppler measurements of turbulent flow parameters in a stirred mixer, *Chem. Eng. Sci.*, **44**, 2207–2221.

Zhou, G., and S. M. Kresta (1996). Impact of geometry on the maximum turbulence energy dissipation rate for various impellers, *AIChE J.*, **42**, 2476–2490.

Zipp, R. P., and G. K. Patterson (1998). Experimental measurements and simulation of mixing and chemical reaction in a stirred tank, *Can. J. Chem. Eng.*, **76**, 657–669.

# CHAPTER 14

# Heat Transfer

**W. ROY PENNEY**
*University of Arkansas*

**VICTOR A. ATIEMO-OBENG**
*The Dow Chemical Company*

## 14-1 INTRODUCTION

Heat transfer in agitated vessels, a common industrial practice, has been researched extensively. Experimental work started in the 40s with peak activity in the 50s and 60s. The 1959 paper by Brooks and Su is an excellent example of the work done in the 50s and 60s. Comprehensive coverage is beyond the scope of this work; only a summary of the most useful and general information is presented here. Books by Sterbacek and Tausk (1965), Holland and Chapman (1966), Uhl and Gray (1966), and Nagata (1975) present comprehensive coverage. Parker (1964), Jordan (1968), Edwards and Wilkinson (1972a,b), and Rase (1977) present less comprehensive coverage. Penney (1983) and Dream (1999) give summaries of the most useful correlations for heat transfer coefficients in agitated vessels. Fasano et al. (1994) give correlations for the vessel wall and for the vessel bottom head for various impellers. Haam et al. (1992, 1993) discuss an experimental technique based on surface calorimeters for measuring local heat flux.

The intent of this chapter is to provide sufficient information to enable the designer to design a heating or cooling system for the job at hand. Only the most commonly used agitator impellers and heat transfer surfaces are covered here. To determine the economic optimum system will often require going beyond the knowledge base of this chapter. The reader is advised to contact vendors and use their expertise for a more economical design than that of the "base case," which one can obtain by using the information in this chapter.

---

*Handbook of Industrial Mixing: Science and Practice*, Edited by Edward L. Paul,
Victor A. Atiemo-Obeng, and Suzanne M. Kresta
ISBN 0-471-26919-0  Copyright © 2004 John Wiley & Sons, Inc.

## 14-2 FUNDAMENTALS

A jacketed agitated vessel may be used for heating or cooling its fluid contents. The rate of heat transfer, Q, can be expressed by Newton's law of heat transfer as follows:

$$Q = UA\, \Delta T \qquad (14\text{-}1)$$

U, the overall heat transfer coefficient, depends on the fluid properties, the operating parameters of the mixer, and the system configuration. It is the key parameter that is affected by the operation of the mixer. The available area for heat transfer, A, depends on the geometry of the system. Usually, the area per volume decreases on vessel scale-up. The temperature driving force, $\Delta T$, depends on the operating conditions of both the process and the heating or cooling fluid.

Heat transfer between the jacket fluid and the vessel contents occurs by conduction and forced convection. The resistance to heat transfer is a composite of the resistances through the various sections indicated in Figure 14-1. Using the classical film theory and heat conduction through composite layers, the overall heat transfer coefficient can be expressed as

$$\frac{1}{U_{ps}} = \frac{1}{h_{htfs\ film}} \frac{A_{ps}}{A_{htfs}} + f_{htfs}\frac{A_{ps}}{A_{htfs}} + \frac{x_{wall}}{k_{wall}}\frac{A_{ps}}{A_{wall}} + \frac{x_{lining}}{k_{lining}}\frac{A_{ps}}{A_{lining}} + f_{ps} + \frac{1}{h_{ps\ film}} \qquad (14\text{-}2)$$

In a jacketed agitated vessel, mixing affects only the process-side film heat transfer coefficient, $h_{ps\ film}$. The largest resistance in the expression dominates the value of the overall heat transfer coefficient. The thermal resistance due to fouling on either the heat transfer fluid side, $f_{htfs}$, or the process fluid side, $h_{ps}$, can significantly affect heat transfer. For carbon steel vessels, the wall conductivity is normally high enough so that the conductive resistance is a minor fraction of the overall thermal resistance. The thermal conductivity is lower for stainless steel and glass lining and can affect the overall heat transfer coefficient significantly. Values of thermal conductivity for various materials are given in Table 14-1.

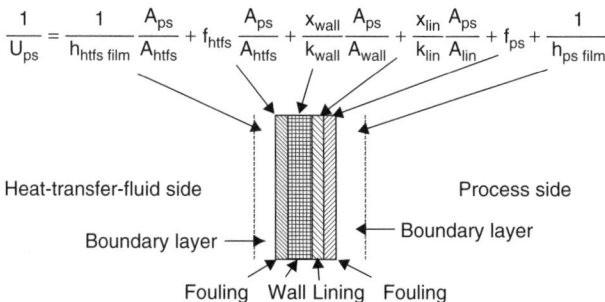

**Figure 14-1**  Resistances to heat transfer.

**Table 14-1** Physical Properties of Vessel Materials

| Material | Thermal Conductivity, k<br>Btu/(hr ft² °F/ft)<br>[W/m · K] | Specific Heat, $c_p$<br>Btu/lb °F<br>[J/kg · K] | Density, $\rho$<br>lb/ft³<br>[g/cm³] |
|---|---|---|---|
| Carbon steel | 30 [52] | 0.11 [460] | 484 [7.8] |
| Copper | 218 [377] | 0.092 [385] | 559 [9.0] |
| Cupro-nickel 90/10 | 30 [52] | 0.106 [444] | 541 [8.7] |
| Cupro-nickel 70/30 | 20 [35] | 0.106 [444] | 519 [8.3] |
| Glass | 0.67 [1.16] | 0.2 [835] | 155 [2.5] |
| Hastelloy C276 | 7.5 [11] | 0.092 [385] | 558 [8.9] |
| Incoloy 825 | 6.9 [12] |  | 508 [8.1] |
| Inconel 600 | 9.2 [16] | 0.106 [444] | 520 [8.3] |
| Monel (400, 404, R405, 411) | 15 [26] | 0.102 [427] | 551 [8.8] |
| Nickel (200, 201, 220, 225) | 38 [66] | 0.105 [440] | 555 [8.9] |
| Stainless steel (304, 316, 321, 347) | 9.8 [17] | 0.12 [502] | 481 [7.7] |
| Tantalum | 32 [54] | 0.036 [151] | 1036 [16.6] |
| Titanium | 11.5 [20] | 0.139 [582] | 283 [4.5] |

The following steps are required to design an agitated vessel to satisfy certain heat transfer requirements:

1. Select the agitator and vessel geometry.
2. Select the vessel internals.
3. Size the agitator and heat transfer surfaces.

The most important parameters affecting the design of an agitated vessel for heat transfer are:

1. The process results, other than heat transfer, to be obtained
2. The heat duty per unit of vessel volume
3. The fluid physical properties (primarily viscosity)
4. The vessel volume

For low to moderate viscosities ($\mu$ < 10 000 cP, i.e., 100 poise), in industrial-sized vessels (volume > 1 m³), high impeller pumping rates producing turbulent motion are possible, and nonproximity impellers (shown in Figure 14-2 and Chapter 2) are used. For fluids with higher viscosity where laminar flow patterns are likely, proximity impellers such as anchors and helical ribbons are used.

For low to moderate heat duties (in terms of heat duty per unit of vessel volume), a vessel jacket is usually adequate to provide the required heat transfer surface. As heat duty increases, internal heat transfer surfaces (helical coils,

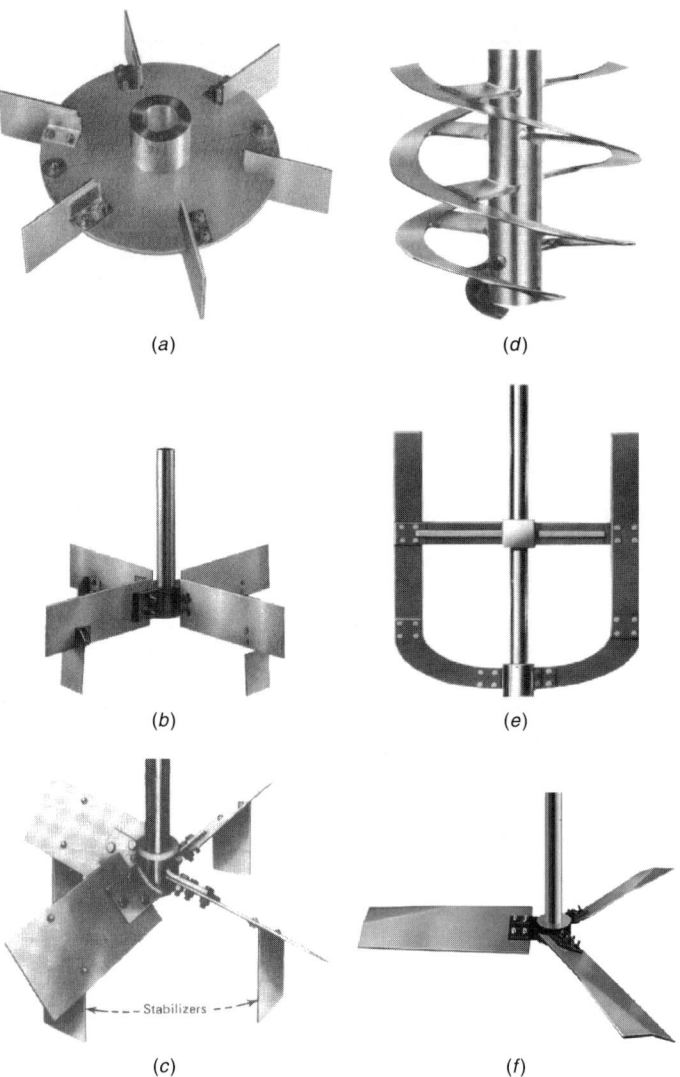

**Figure 14-2** Typical impellers for mechanically agitated vessels: (*a*) six-blade disk impeller (6BD) or Rushton turbine; (*b*) four-blade flat impeller (4BF); (*c*) four-blade pitched impeller (4BP); (*d*) helical ribbon impeller; (*e*) anchor impeller; (*f*) high efficiency turbine impeller.

baffle pipes, or plate coils) may be required. For some systems, with very high heat duties, adequate heat exchange is not possible even with the installation of internal heat transfer surfaces. For such systems additional heat transfer surface can readily be provided by using an external pumped-through heat exchanger.

For systems where a volatile component can be vaporized and condensed, evaporative cooling is often the most economical means of heat removal. The vaporized component is condensed and returned (i.e., refluxed) back to the vessel. For information on the selection and design of external heat exchangers and condensers, the reader should refer to other sources, such as Saunders (1988).

A special problem exists for heat removal applications. Because agitator power requirements (which always add heat to the vessel contents) are much more strongly dependent on agitator speed than are heat transfer coefficients, a maximum heat removal capability exists for any particular agitated vessel. This phenomenon is of particular importance for heat removal in high viscosity systems. Refer to Penney and Koopman (1971) for recommendations concerning optimum design. Penney and Koopman have given the following recommendations for the magnitude of agitator power input to obtain maximum net heat removal. For the laminar regime, the agitator power to obtain maximum net heat removal is about 20% of the net heat removed, and for the turbulent regime, the agitator power to obtain maximum net heat removal is in the range 5 to 20% of the net heat removal. A value of 20% for the turbulent regime is generally applicable where all thermal resistances other than the inside vessel fluid film are about 0.001 hr-ft$^2$-°F/Btu, and a value of 5% is applicable where they are about 0.01 hr-ft$^2$-°F/Btu for typical organic liquids in plant-sized vessels.

## 14-3  MOST COST-EFFECTIVE HEAT TRANSFER GEOMETRY

The first consideration regarding heat transfer needs in agitated vessels concerns which surface and which heat removal system to use. Which heat removal system is most cost-effective? In general the choices listed in order of least to most expensive are:

1. Vessel wall jacket
2. Vessel bottom head jacket
3. Internal surface (e.g., helical coils, plate coils, or vertical harp coils)
4. Reflux cooling by solvent evaporation and external condenser
5. External pumped-through heat exchangers

It is important to remember that for the most demanding heat removal requirements, one should always consider introducing a solvent that can be made to boil at the process temperature and pressure, vaporizing that solvent, condensing it, and returning (i.e., refluxing) the condensate to the vessel. This option is often overlooked when designing a heat removal system. There are myriad reflux cooling applications in industrial practice, and an excellent example is the reflux cooling used to remove the heat of polymerization from polystyrene reactors by evaporating and condensing the styrene monomer.

The geometry of internal coils needs to be selected carefully because all geometries are not equally effective. The most important geometrical considerations are summarized in Table 14-2. The vessel, vessel internals, and an agitation source are the components of the typical installation. Mechanical agitators (consisting of a drive, a drive shaft, and an impeller) are most often used to

**874** HEAT TRANSFER

**Table 14-2** Most Effective Geometry for Heat Transfer

| Geometrical Variable | Most Effective Value(s) (Best → Least) |
|---|---|
| Type of surface | + → Helical coils (see Figure 14-3). |
| | → Harp (i.e., vertical tube baffles) coils (see Figure 14-4). |
| | → Plate coils (see Figure 14-4). |
| Number of coils, plates, etc. | + For helical coils, use two maximum. |
| | + For harps and plate coils, up to 16 can be used effectively. |
| Position of surface in vessel | + Helical coils are placed inside and attached to baffles. |
| | + Harp coils and plate coils act as baffles and are positioned vertically along the vessel walls as baffles. |
| Distance between coil banks | + Minimum distance is twice the tube diameter. |
| Spacing of harps and plate coils | + Up to 16 are used; above about 8 the harps and plate coils are normally positioned at about 45° to the vessel diameter. |
| Spacing between tubes in harps and helical coils | + Minimum spacing is one tube diameter. |

provide agitation; however, gas sparging and liquid jets (i.e., jet mixing) entering the vessel (primarily from circulation through a pump) are also used to provide agitation. Quantitative design methods for gas sparging are available in the literature and the source literature references are given later in this chapter, but quantitative design methods are not available for jet mixing.

### 14-3.1 Mechanical Agitators

Figure 14-2 presents visuals of commonly used mechanical agitator impellers. Heat transfer correlations for these impellers are presented later. The reader is referred to Chapter 6 for information on their performance characteristics and guidelines for their selection and use.

### 14-3.2 Gas Sparging

For bubble columns with height/diameter $> 5$, a simple open pipe at the bottom of the column is often adequate. For height/diameter $< 5$, a ring or finger-style perforated pipe sparger is desirable to obtain uniform radial distribution of the gas and to prevent excessive channeling of the gas up the center of the vessel. For heat transfer in bubble agitated columns, see Hart (1976) and Tamari and Nishikawa (1976).

### 14-3.3 Vessel Internals

With nonproximity agitators, baffles are almost always used to prevent swirl and subsequent vortexing and to increase top-to-bottom motion and turbulence.

Helical coils, pipe baffles, and plate coil baffles are the most common heat transfer surfaces within the vessel.

***14-3.3.1 Wall Baffles.*** Figure 14-3 shows the recommended geometry; the baffle width is $\frac{1}{12}$ of the vessel diameter, and the baffles are most often positioned a short distance (about T/75) from the vessel wall. Sometimes vortexing is desirable (e.g., when wetting semibatch fed powders or dispersing small volumes of gas from the vessel headspace) (Deeth et al., 2000). For these applications partial baffling is recommended. The most commonly used partial baffling is half-baffles, which are normally $\frac{1}{12}$ the vessel diameter in width but extend only halfway up the liquid height up the vessel wall from the vessel bottom.

Vortex depth in nonbaffled agitated vessels has been investigated by Brennan (1976) and Rieger et al. (1979); both references present predictive methods. For

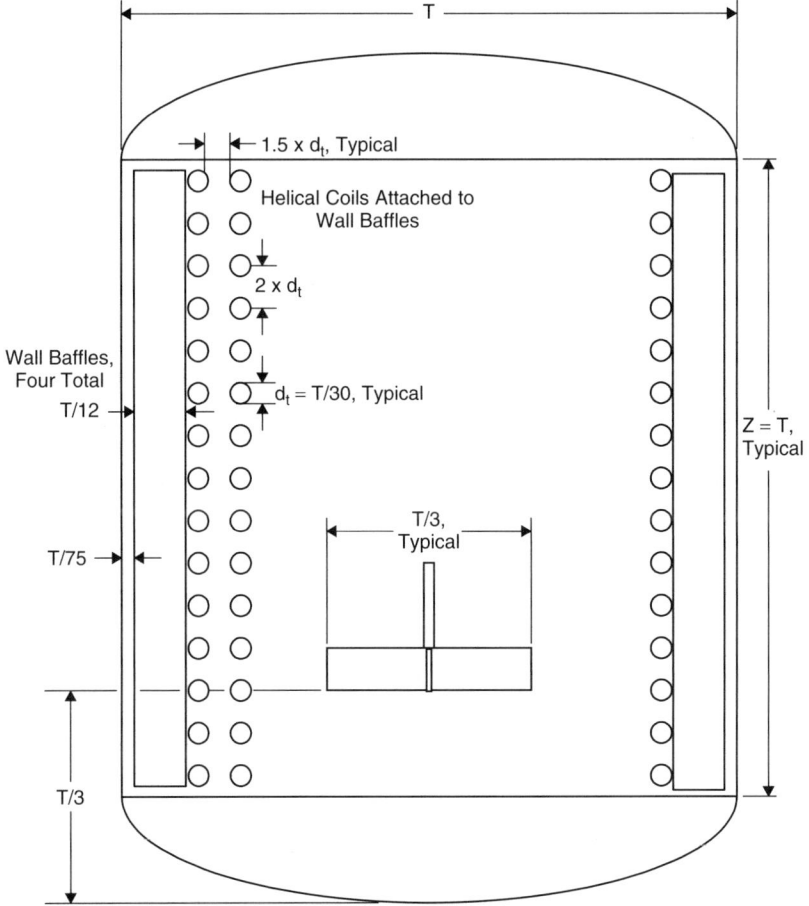

**Figure 14-3** Recommended geometries for wall baffles and helical coils.

thin liquids with turbulent conditions, for quick estimating purposes X/D $\simeq$ 4(Fr) [see Table 1 in Brennan (1976) and Figure 14-3 in Rieger et al. (1979)]. Often, half-baffles (i.e., B = T/12 and the baffles extend only halfway up the vessel wall) are used in the lower portion of the vessel to produce a significant vortex while reducing swirl sufficiently to provide sufficient shear to be effective in dispersing solids agglomerates or to entrain and disperse small volumes of gas from the vessel headspace.

Axial flow impellers (<5 kW) are often mounted angled, off-center (Rushton, 1947; Weber, 1963; Uhl and Gray, 1966) to prevent swirl and vortexing. For clockwise rotation (looking along the shaft from the drive to the impeller), Uhl and Gray (1966, pp. 153–156) recommend that the agitator shaft be moved off-center T/6, tilted back at an angle of 10°, and then moved back so that the impeller centerline is located on the centerline of the vessel.

**14-3.3.2 Helical Coils.** Figure 14-3 presents the recommended geometry. Rushton (1947), Parker (1964), and Hicks and Gates (1975) give recommendations concerning coil geometry. This geometry is not the economical optimum for all cases; for example, Rushton (1947, p. 653) says, "clearance between pipes in a helix need not be great and, if $\mu$ < 500 cP, the clearance between layers may be as little as one-third pipe diameter." Unfortunately, insufficient quantitative information is available to allow prediction of the heat transfer coefficient for a gap spacing of less than one pipe diameter; thus, the recommendation given in Figure 14-3 is a gap spacing of one pipe diameter. For cases where the geometry of Figure 14-3 does not provide sufficient heat duty, an equipment manufacturer should be contacted. They may be able to recommend more compact geometries which have greater heat transfer surface per unit volume than those recommended here. Manufacturers often have additional design information which allows selection of a more economical geometry. For example, for some situations a second and a third coil (Marshall and Yazdani, 1970; Hicks and Gates, 1975), may be economical, although the heat transfer coefficients will be reduced for the middle and outer coils in a three-coil band—to about 60 and 40%, respectively, of the inner coil—according to Marshall and Yazdani (1970).

**14-3.3.3 Baffle Pipes and Plate Coil Baffles.** The recommended geometries are presented in Figure 14-4. The baffle pipes can be placed either radially across the vessel or at an angle of 45° with the vessel radius. With radial positioning, four pipes per baffle are recommended; and with angled positioning, five pipes per baffle are recommended. The experimental work discussed later was done with four baffles. When using plate coils, more than four baffles are used; in fact, as many as 16 angled plate coils have been installed in agitated vessels. Probably six baffles would have little effect on the heat transfer coefficient; however, above some number, perhaps six, the heat transfer coefficient will be reduced as additional plate coils are added. For more than six baffles, a manufacturer of plate coils and/or a manufacturer of fluid mixing equipment should be consulted.

HEAT TRANSFER COEFFICIENT CORRELATIONS **877**

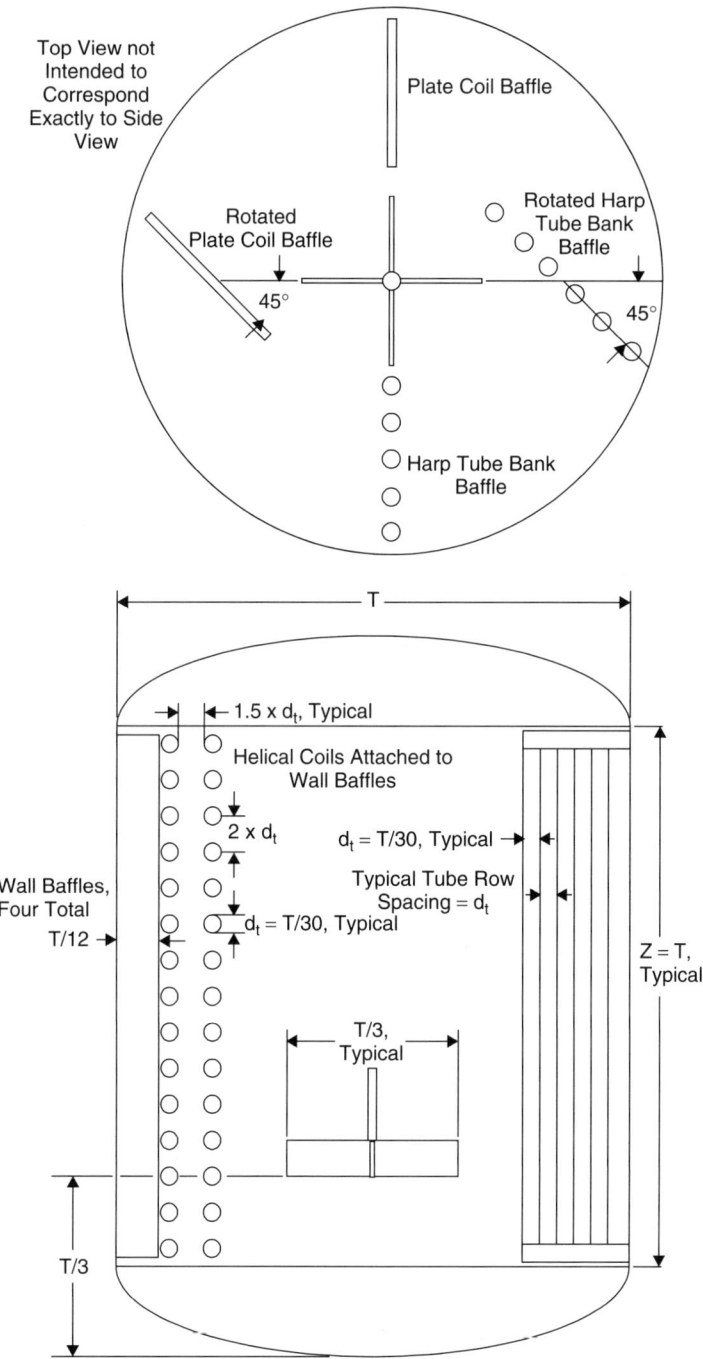

**Figure 14-4** Recommended geometries for vertical tube baffles (harp coils) and plate coils.

## 14-4  HEAT TRANSFER COEFFICIENT CORRELATIONS

Published correlations for the process-side heat transfer coefficient are all of the form

$$\text{Nu} = K \cdot \text{Re}^a \text{Pr}^b \mu_R^c G_c \tag{14-3}$$

where $G_c$ represents a geometry correction. Several correlations are presented in Table 14-3. The dimensionless numbers are explained below.

### Nusselt Number, Nu
For heat transfer to or from the vessel wall or bottom head in a jacketed vessel

$$\text{Nu} = \frac{hT}{k} \tag{14-4a}$$

For a harp or helical coil

$$\text{Nu} = \frac{hd_t}{k} \tag{14-4b}$$

For a plate coil

$$\text{Nu} = \frac{h(W_{PC}/4)}{k} \tag{14-4c}$$

The characteristic length of $W_{PC}/4$ is recommended for a plate coil because Petree and Small (1978) used the width of the plate coil divided by the number of utility fluid passes within the utility side of the plate coil as the characteristic length in the Nusselt number. It is unlikely that the number of passes on the utility side have any significant effect on the heat transfer coefficient. Thus, the only reason to use $W_{PC}/4$ as the characteristic length in the Nusselt number is that Petree and Small (1978) used a plate coil with four passes in their experimental apparatus.

### Prandtl Number, Pr

$$\text{Pr} = \frac{\mu_b C_p}{k} \tag{14-5}$$

where the subscript b refers to the bulk of fluid.

### Viscosity Ratio, $\mu_R$

$$\mu_R = \frac{\mu_b}{\mu_w} \tag{14-6}$$

where the subscript b refers to the bulk fluid and the subscript w refers to the vessel wall.

*Geometrical Corrections (i.e., $G_C$).* In some cases the exact forms of the geometric corrections have been changed from the original references, and in several

**Table 14-3** Heat Transfer Coefficient Correlations for Agitated, Baffled Vessels[a]

| Impeller | Surface | Re Range | K | a | B | c | Geometry Correction[a] | Reference |
|---|---|---|---|---|---|---|---|---|
| 6BD | Wall | >100 | 0.74 | 2/3 | 1/3 | 0.14 | $(1/[H/T])^{0.15}(L/L_S)^{0.2}$ | Fasano et al. (1994) |
| 4BF | Wall | >100 | 0.66 | 2/3 | 1/3 | 0.14 | $(1/[H/T])^{0.15}(L/L_S)^{0.2}$ | Fasano et al. (1994) |
| 4BP | Wall | >100 | 0.45 | 2/3 | 1/3 | 0.14 | $(1/[H/T])^{0.15}(L/L_S)^{0.2}$ | Fasano et al. (1994) |
| HE3 | Wall | >100 | 0.31 | 2/3 | 1/3 | 0.14 | $(1/[H/T])^{0.15}$ | Fasano et al. (1994) |
| PROP | Wall | >100 | 0.5 | 2/3 | 1/3 | 0.14 | $(1/[H/T])^{0.15}(1.29[P/D])/(0.29+[P/D])$ | Strek et al. (1963) |
| 6BD | BH[c] | >100 | 0.50 | 2/3 | 1/3 | 0.14 | $(1/[H/T])^{0.15}(L/L_S)^{0.2}$ | Fasano et al. (1994) |
| 4BF | BH | >100 | 0.40 | 2/3 | 1/3 | 0.14 | $(1/[H/T])^{0.15}(L/L_S)^{0.2}$ | Fasano et al. (1994) |
| 4BP | BH | >100 | 1.08 | 2/3 | 1/3 | 0.14 | $(1/[H/T])^{0.15}(L/L_S)^{0.2}$ | Fasano et al. (1994) |
| HE3 | BH | >100 | 0.9 | 2/3 | 1/3 | 0.14 | $(1/[H/T])^{0.15}$ | Fasano et al. (1994) |
| RCI[d] | Wall and BH | >100 | 0.54 | 2/3 | 1/3 | 0.14 | $(1/[H/T])^{0.15}$ | Fasano et al. (1994) |
| PROP | Helical coil | >100 | 0.016 | 0.67 | 0.37 | 0.14 | $([D/T]/[1/3])^{0.1} ([d/T]/0.04)^{0.5}$ | Ackley (1960) |
| 6BD | Helical coil | >100 | 0.03 | 2/3 | 1/3 | 0.14 | $(1/[H/T])^{0.15}(L/L_S)^{0.2}([D/T]/[1/3])^{0.1}([d/T]/0.04)^{0.5}(2/N_B)^{0.2}$ | Oldshue (1966) Oldshue and Gretton (1954) |
| 6BD | 45° harp coils | >100 | 0.021 | 0.67 | 0.4 | 0.27[e] | $(1/[H/T])^{0.15}(L/L_S)^{0.2}([D/T]/[1/3])^{0.33}([d/T]/0.04)^{0.5}$ | Gentry and Small (1978) |
| 4BF | 0° harp coils | >100 | 0.06 | 0.65 | 0.3 | 0.42[e] | $(1/[H/T])^{0.15}(L/L_S)^{0.2}([D/T]/[1/3])^{0.33}([d/T]/0.04)^{0.5}$ | Dunlap and Rushton (1953) |
| 6BD | Plate coils | >100 | 0.031 | 0.66 | 0.33 | 0.5[e] | $(1/[H/T])^{0.15}(L/L_S)^{0.2}$ | Petree and Small (1978) |
| Anchor |  | <1 |  |  |  | Not recommended! In the laminar regime; the anchor fails to give top-to-bottom fluid motion |  |  |
|  | Wall | >12, <100 | 0.69 | 1/2 | 1/3 | 0.14 |  | Harry and Uhl (1973) |
|  | Wall | >100 | 0.32 | 2/3 | 1/3 | 0.14 |  | Harry and Uhl (1973) |
| Helical | Wall | <13 | 0.94 | 1/3 | 1/3 | 0.14 |  | Ishibashi et al. (1979) |
| ribbon | Wall | >13, <210 | 0.61 | 1/2 | 1/3 | 0.14 |  | Ishibashi et al. (1979) |
|  | Wall | >210 | 0.25 | 2/3 | 1/3 | 0.14 |  | Ishibashi et al. (1979) |

[a] The correlations are of the following form: $Nu = K \cdot Re^a Pr^b \mu_R^c G_c$, where Nu (vessel wall), Nu (coil), Pr, and $\mu_R$ are as defined in eq. (14-2), (14-3), and (14-4).
[b] d, tube diameter in helical coil; D, impeller diameter; L, height of the impeller blade parallel with the axis of rotation; $L_S$, standard height of the impeller blade parallel to the axis of rotation (for 6BD, $L_S/D = 0.2$; for 4BF, $L_S/D = \frac{1}{5}$; for 4BP, $L_S/D = 0.17$); $N_B$, number of blades on the Impeller; P, pitch of a propeller or a helical ribbon impeller (forward motion of the impeller blade over 360° of rotation); T, tank diameter, H, height of the batch.
[c] BH bottom: head.
[d] RCI retreat curve impeller.
[e] These authors used the ratio to bulk to film viscosity. It is recommended that one use $c = 0.14$ and use the ratio of bulk to wall viscosity.

879

**880**  HEAT TRANSFER

cases, additional geometric corrections have been included. In some cases this procedure changed the value of K from the original reference, but the final correlation remains the same. The geometrical corrections have been consistently included as a ratio of a geometrical parameter to a standard geometrical parameter. The value for the standard value is normally the experimental value (or a value in the midpoint of the experimental data) used by various investigators. This practice, where the geometrical correction is unity for the standard value of a parameter, makes it much easier to compare various correlations directly.

For turbulent conditions, typical values for the exponents in eq. (14-3) are $\frac{2}{3}$, $\frac{1}{3}$, and 0.14, respectively, for a, b, and c. With these values for the exponents, eq. (14-3) can be written in terms of the specific energy input or energy dissipation, $\varepsilon$, for a given impeller and physical properties as follows:

$$h \propto \varepsilon^{2/9} \left(\frac{D}{T}\right)^{2/9} T^{-1/9} \qquad (14\text{-}7)$$

This form helps one appreciate the effect of energy input and impeller size on the process-side film heat transfer coefficient. To double h, $\varepsilon$ must increase by a factor of 23! Similarly, the effect of impeller size, represented by D/T, is also weak. A change of impeller diameter from 0.33T to 0.67T improves h by only 10%.

### 14-4.1 Correlations for the Vessel Wall

- *Flat, disk, pitched blade turbines and HE-3*. The correlations by Fasano et al. (1994) are recommended. Corrections for dimensionless batch height and dimensionless impeller height are given.
- *Propeller*. The correlation of Strek et al. (1965) is recommended. A correction for dimensionless batch height is given. The vessel is baffled with heat transfer at the wall.
- *Glass-coated three-bladed impeller (RCI)–one finger-style baffle*. The correlation of Ackley (1960) is recommended. A correction for dimensionless batch height is given.

### 14-4.2 Correlations for the Bottom Head

- *Flat, disk, pitched blade turbines and HE-3*. The correlations recommended by Fasano et al. (1994) are recommended. Corrections for dimensionless batch height and dimensionless impeller height are given.
- *Propeller*. The correlation of Strek et al. (1965), which was developed for the vessel wall, is recommended.
- *Glass-coated three-bladed impeller–one finger-style baffle*. The correlation of Ackley (1960), which was developed for the vessel wall, is recommended.

## 14-4.3 Correlations for Helical Coils

- *Flat, disk, and pitched blade turbines.* Oldshue and Gretton's (1954) correlation is recommended for the standard geometry. Corrections for dimensionless batch height and dimensionless impeller height are given.
- *Propeller.* Oldshue's (1966) correlation is recommended for the standard geometry. Corrections for dimensionless batch height and dimensionless impeller height are given.

## 14-4.4 Correlations for Vertical Baffle Coils (i.e., Vertical Baffle Pipes)

- *Four-blade disk turbines–vertical baffle coil.* Dunlap and Rushton's (1953) correlation is recommended with appropriate geometric corrections added. Table 14-3 gives K = 0.06, which seems high based on the findings of Gentry and Small (1978) for baffle pipes oriented at 45° to the vessel diameter. It may be prudent to use K = 0.04 rather than K = 0.06.
- *Two six-blade disk turbines–vertical tube baffles.* Gentry and Small's (1978) correlation is recommended with corrections added for dimensionless batch height and dimensionless impeller height.

## 14-4.5 Correlations for Plate Coils

- *Two six-blade disk turbines–vertical plate coils.* Petree and Small's (1978) correlation is recommended.

## 14-4.6 Correlations for Anchors and Helical Ribbons

Uhl (1970), Harry and Uhl (1973), and Nishikawa et al. (1975) have summarized all previous work. Ishibashi et al. (1979) and Rautenbach and Bollenrath (1979) have published the latest works. Coyle et al. (1970) have presented very useful experimental data. Nagata et al. (1970) and Mitsuishi and Miyairi (1973) are of interest. The correlations by Harry and Uhl (1973) and Ishibashi et al. (1979) are recommended. The recommended impeller geometries (Penney, 1983) are given in Table 14-4 and the applicable correlation parameters are given at the bottom of Table 14-3.

Table 14-4  Recommended Impeller Geometries for Anchors and Helical Ribbons

| Geometric Ratio | Anchor | Helical Ribbon |
| --- | --- | --- |
| P/D | ∞ | 1/2 |
| W/D | 0.082 | 0.082 |
| $C_w/D$ | 0.02 | 0.02 |
| D/T | 0.96 | 0.96 |

## 14-5 EXAMPLES

*Example 14-1: Turbine Impeller.* Determine the process-side heat transfer coefficient for Problem 15.9, page 460, from McCabe et al. (1993). A turbine-agitated (6BD) vessel of diameter T = 2 m contains 6233 kg of a dilute aqueous solution at 40°C. The agitator is a standard-geometry (thus $L = L_s$) six-blade disk impeller of diameter $D = \frac{2}{3}$ m and N = 140 rpm. Determine the vessel wall heat transfer coefficient.

SOLUTION

| | |
|---|---|
| Solution specific heat, $C_p$ | 4187 J/kg · K |
| Solution density, $\rho$ | 992 kg/m³ |
| Solution viscosity, $\mu$ | 0.000657 kg/m · s |
| Solution thermal conductivity, k | 0.63 W/m · K |

We first need to consider the system geometry. Let's calculate the batch height (i.e., H).

$$H = \frac{m}{\rho} \Big/ \frac{\pi D^2}{4} = \frac{(6233 \text{ kg})}{(992 \text{ kg/m}^3)} \Big/ \frac{\pi (2 \text{ m})^2}{4} = 2 \text{ m}$$

Thus, the batch is what we refer to as a "square" batch (i.e., H = T). We need to use the correlation from Table 14-3 for a 6BD and the vessel wall.

$$\text{Nu} = \frac{hT}{k} = 0.74 \text{Re}^{2/3} \text{Pr}^{1/3} \mu_R^{0.14} \left[ \frac{1}{(H/T)} \right]^{0.15} \left( \frac{L}{L_s} \right)^{0.2}$$

$$\text{Re} = \frac{ND^2 \rho}{\mu} = \frac{(140/60 \text{ rps})(2/3 \text{ m})^2 (992 \text{ kg/m}^3)}{0.000657 \text{ kg/m} \cdot \text{s}} = 1.57 \times 10^6$$

$$\text{Pr} = \frac{\mu C_p}{k} = \frac{(0.000657 \text{ kg/m} \cdot \text{s})(4187 \text{ J/kg} \cdot \text{K})}{0.63 \text{ J/m} \cdot \text{K}} = 4.37$$

Assume that $\mu_R \approx 1$.

$$\text{Nu} = \frac{hT}{k} = 0.74(1.57 \times 10^6)^{2/3}(4.37)^{1/3}(1)^{0.14} \left( \frac{1}{1} \right)^{0.15} (1)^{0.2}$$

$$= (0.74)(13\,500)(1.64)(1)(1)(1) = 16\,400$$

$$h = \frac{\text{Nu} \cdot k}{T} = \frac{16\,400(0.63 \text{ W/m} \cdot \text{K})}{(2 \text{ m})} = 5170 \text{ W/m}^2 \cdot \text{K}$$

*Example 14.2: Helical Ribbon Impeller.* Determine the process-side heat transfer coefficient for the tank blending design example for a helical ribbon impeller (Bakker and Gates, 1995):

## SOLUTION

| | |
|---|---|
| Tank diameter, T | 2.5 m |
| Impeller diameter, D | 0.96(T) = 0.96(2.5) = 2.4 m |
| D/T | 0.96 |
| H/T | 1 |
| Batch height, H | 2.5 m |
| Fluid viscosity, $\mu$ | 25 Pa·s = 25 kg/m·s |
| Fluid density, $\rho$ | 1200 kg/m$^3$ |
| Impeller speed, N | 16.4 rpm |
| Fluid thermal conductivity, k | 0.25 W/m·K |
| Fluid specific heat, C$_p$ | 2500 J/kg·K |

Calculate the Reynolds number:

$$\text{Re} = \frac{ND^2\rho}{\mu} = \frac{(16.4/60 \text{ rps})(2.4 \text{ m})^2(1200 \text{ kg/m}^3)}{(25 \text{ kg/m·s})} = 76$$

Select the appropriate heat transfer coefficient correlation. From Table 14-3 (for the standard helix pitch/impeller diameter ratio P/D = $\frac{1}{2}$, the appropriate heat transfer correlation is

$$\text{Nu} = 0.61\text{Re}^{1/2}\text{Pr}^{1/3}(\mu_B/\mu_w)^{0.14}$$

$$\text{Pr} = \frac{\mu_B C_P}{k} = \frac{(25 \text{ kg/m·s})(2500 \text{ J/kg·K})}{(0.25 \text{ W/m·K})} = 2.5 \times 10^5$$

$$\text{Nu} = \frac{hT}{k} = (0.61)(76)^{1/2}(2.5 \times 10^5)^{1/3}(\sim 1)^{0.14} = (0.61)(8.72)(63.4) = 337$$

$$h = \frac{337(0.25 \text{ W/m·K})}{(2.5 \text{ m})} = 34 \text{ W/m}^2\text{·K}$$

## NOMENCLATURE

| | |
|---|---|
| A | heat transfer area (m$^2$) |
| B | baffle width (m) |
| C$_p$ | specific heat of the fluid (J/kg·K) |
| C$_w$ | wall clearance for close clearance impellers (m) |
| d | outside diameter of the tube of which a coil is made (m) |
| D | impeller diameter (m) |
| d$_t$ | tube diameter (m) |
| $\Delta$T | temperature driving force (°C or K) |
| f | thermal resistance due to fouling $\left(\frac{m^2 K}{W}\right)$ |
| Fr | Froude number = $N^2D/g$ |

| | |
|---|---|
| $G_c$ | geometry correction factor |
| $h$ | process-side heat transfer coefficient (W/m² · K) |
| $h_U$ | utility-side heat transfer coefficient (W/m² · K) |
| $H$ | tank height (m) |
| $k$ | fluid thermal conductivity (W/m · K) |
| $K$ | precorrelation factor for Nu |
| $L$ | height of the impeller blade parallel with the axis of rotation (m) |
| $L_S$ | standard height of the impeller blade parallel to the axis of rotation (m) |
| $N$ | impeller rotational speed (rps or rpm) |
| $N_B$ | number of blades on the impeller |
| Nu | Nusselt number = hT/k for the vessel wall or bottom head; $N_u$ = hd/k for a coil |
| $P$ | pitch of a propeller or helical ribbon impeller (i.e., the distance along the axis of rotation which the impeller would move over 360° of rotation) (m) |
| Pr | Prandtl number = $C_p \mu / k$ |
| $Q$ | rate of heat transfer (W) |
| Re | impeller Reynolds number = $\dfrac{ND^2 \rho}{\mu}$ |
| $t$ | time to heat or cool the batch from $T_I$ to $T_F$ (s) |
| $T$ | tank diameter (m) |
| $T_F$ | final temperature after cooling or heating of the batch is complete (K) |
| $T_I$ | initial temperature of the batch before heating or cooling starts (K) |
| $T_U$ | utility fluid temperature (K) |
| $U$ | overall heat transfer coefficient (W/m² · K) |
| $W$ | blade width (m) |
| $x$ | Wall thickness (m) |
| $X$ | vortex depth below surface (m) |
| $Z$ | height of the batch (m) |

*Greek Symbols*

| | |
|---|---|
| $\rho$ | density $\left(\dfrac{kg}{m^3}\right)$ |
| $\theta$ | angle of the impeller blade with the axis of rotation |
| $\mu$ | fluid viscosity (kg/m · s) |
| $\mu_b$ | fluid viscosity at the bulk fluid temperature (kg/m · s) |
| $\mu_R$ | viscosity ratio: bulk viscosity/wall viscosity = $\mu_b/\mu_w$ |
| $\mu_w$ | fluid viscosity at the fluid wall temperature (kg/m · s) |

## REFERENCES

Ackley, E. J. (1960). Film coefficients of heat transfer for agitated process vessels, *Chem. Eng.*, Aug. 22, pp. 133–140.

Bakker, A., and L. E. Gates (1995). Properly choose mechanical agitators for viscous liquids, *Chem. Eng. Prog.*, Dec., pp. 25–33.

Brennan, D. J. (1976). Vortex geometry in unbaffled vessels with impeller agitation, *Trans. Inst. Chem. Eng.*, **54**, 209–217.

Brooks, G., and G. J. Su (1959). Heat transfer in agitated vessels, *Chem. Eng. Prog.*, **55**, 54–57.

Coyle, C. K., et al. (1970). Heat transfer to jackets with close clearance impellers in viscous liquids, *Can. J. Chem. Eng.*, **48**, 275–278.

Deeth, B. P., W. R. Penney, and M. F. Reeder (2000). Gas dispersion from vessel headspace: experimental scale-up studies to maintain $k_L a$ constant, *Paper f*, Session 178, Winter Annual AIChE Meeting, Los Angeles, Nov.

Dream, R. F. (1999). Heat transfer in agitated jacketed vessels, *Chem. Eng. Jen.*, pp. 90–96.

Dunlap, I. R., and J. H. Rushton (1953). Heat transfer coefficients in liquid mixing using vertical-tube baffles, *Chem. Eng. Prog. Symp. Ser.* 5, **49**, 137–151.

Edwards, M. F., and M. A. Wilkinson (1972a). Heat transfer in agitated vessels: I. Newtonian fluids, *Chem. Eng.*, Aug., pp. 310–319.

Edwards, M. F., and M. A. Wilkinson (1972b). Heat transfer in agitated vessels: II. Non-Newtonian fluids, *Chem. Eng.*, Sept., pp. 328–335.

Fasano, J. B., A. Bakker, and W. R. Penney (1994). Advanced impeller geometry boosts liquid agitation, *Chem. Eng.*, Aug., pp. 110–116.

Gentry, C. L., and W. M. Small (1978). Heat transfer and power consumption for agitated vessels having bare and finned, vertical tube baffles, *Proc. 6th International Heat Transfer Conference*, **4**, 13–18.

Haam, S. J., R. S. Brodkey, and J. B. Fasano (1992). Local heat transfer in a mixing vessel using heat flux sensors, *Ind. Eng. Chem. Res.*, **31**, 1384–1391.

Haam, S. J., R. S. Brodkey, and J. B. Fasano (1993). Local heat transfer in a mixing vessel using a high-efficiency impeller, *Ind. Eng. Chem. Res.*, **32**, 575–576.

Harry, F. P., and V. W. Uhl (1973). Heat transfer to viscous materials in a vessel with a helical ribbon impeller, presented at the 74th National AIChE Meeting, New Orleans, LA, Mar.

Hart, W. F. (1976). Heat transfer in bubble-agitated systems: a general correlation, *Ind. Eng. Chem. Process Des. Dev.*, **15**, 109–114.

Hicks, R. W., and L. E. Gates (1975). Fluid agitation in "Fluid agitation in polymer reactors", *Chem. Eng. Prog.*, **71**, Aug., pp. 74–79.

Holland, F. A., and F. S. Chapman (1966). *Liquid Mixing and Processing in Stirred Tanks*, Reinhold, New York.

Ishibashi, K., A. Yamanaka, and N. Mitsubishi (1979). Heat transfer in agitated vessels with special types of impellers, *J. Chem. Eng. Jpn.*, **12**, 230–235.

Jordan, D. G. (1968). *Chemical Process Development*, Interscience, New York, Pt. 1, Chap. 3, pp. 111–174.

Marshall, V. C., and N. Yazdani (1970). Design of agitated, coil-in-tank coolers, *Chem. Process. Eng.*, Apr., pp. 89–101.

McCabe W. L., J. C. Smith, and P. Harriott (1993). *Unit Operations of Chemical Engineering*, 5th cd., McGraw-Hill.

Mitsuishi, N., and Y. Miyairi (1973). Heat transfer to non-Newtonian fluids in an agitated vessel, *J. Chem. Eng. Jpn.*, **6**, 415–420.

Nagata, N. (1975). *Mixing: Principles and Applications*, Wiley, New York.

Nagata, S., M. Nishikawa, and T. Kayama (1972). Heat transfer to vessel wall by helical ribbon impeller in highly viscous liquids, *J. Chem. Eng. Jpn.*, **5**, 83–85.

Nishikawa, M., N. Kamata, and S. Nagata (1975). Heat transfer for highly viscous liquids in mixing vessels, *Heat Transfer Jpn. Res.*, **5**, 84–92.

Oldshue, J. Y. (1966). Fluid mixing, heat transfer and scale-up, *Chem. Eng. Prog.*, Apr., pp. 183–188.

Oldshue, J. Y., and A. T. Gretton (1954). Helical coil heat transfer in mixing vessels, *Chem. Eng. Prog.*, **50**, Dec., pp. 615–621.

Parker, N. H. (1964). Mixing: modern theory and practice on the universal operation, *Chem. Eng.*, June 8, pp. 165–220.

Penney, W. R. (1983). *Agitated vessels*, Chapter 14 in *Heat Exchanger Design Handbook*, Hemisphere, New York.

Penney, W. R. and R. N. Koopman (1971). Prediction of new heat removal for agitated vessels (and pumped-through heat exchangers), *AIChE Preprint 10*, 12th National Heat Transfer Conference, AIChE-ASME, Tulsa, OK.

Petree, D. K., and W. M. Small (1978). Heat transfer and power consumption for agitated vessels with vertical plate coils, *AIChE Symp. Ser. 174*, **74**, 53–59.

Rase, H. F. (1977). *Chemical Reactor Design for Process Plants*, Vol. 1, *Principles and Techniques*, Wiley, New York, pp. 331–392.

Rautenbach, R., and F. M. Bollenrath (1979). Heat transfer in stirred vessels to high-viscosity Newtonian and non-Newtonian substances, *Ger. Chem. Eng.*, **2**, 18–24.

Rieger, F., P. Ditl, and V. Novak (1979). Vortex depth in mixed unbaffled vessels, *Chem. Eng. Sci.*, **34**, 397–401.

Rushton, J. H. (1947). Design and utilization of internal fittings for mixing vessels, *Chem. Eng. Prog.*, **43**, Dec., pp. 649–657.

Saunders, E. A. D. (1988). *Heat Exchangers: Selection, Design and Construction*, Longman, Harlow, Essex, England.

Sterbacek, Z., and P. Tausk (1965). *Mixing in the Chemical Industry*, Pergamon Press, London.

Strek, F., S. Masiuk, G. Gawor, and R. Jagiello (1965). Heat transfer in mixers for liquids (studies of propeller agitators), *Int. Chem. Eng.*, **5**, 695–710.

Tamari, M., and K. Nishikawa (1976). The stirring effect of bubbles upon heat transfer to liquids, *Heat Transfer Jpn. Res.*, **5**, 31–44.

Uhl, V. W. (1970). Mechanically aided heat transfer to viscous materials, *Proc. Symposium on Augmentation of Convective Heat and Mass Transfer*, ASME Winter Annual Meeting, New York, Dec., pp. 109–117.

Uhl, V. W., and J. B. Gray, eds. (1966). *Mixing: Theory and Practice*, Vols. I and II, Academic Press, New York.

Weber, A. P. (1963). Selecting propeller mixers, *Chem. Eng.*, Sept. 2, pp. 91–98.

# CHAPTER 15

# Solids Mixing

FERNANDO J. MUZZIO, ALBERT ALEXANDER, CHRIS GOODRIDGE, ELIZABETH SHEN, and TROY SHINBROT

*Rutgers University*

KONANUR MANJUNATH, SHRIKANT DHODAPKAR, and KARL JACOB

*The Dow Chemical Company*

## Part A: Fundamentals of Solids Mixing

*Fernando J. Muzzio, Albert Alexander, Chris Goodridge, Elizabeth Shen, and Troy Shinbrot*

### 15-1 INTRODUCTION

Solids mixing is essential to many industries, including ceramics, metallurgy, chemicals, food, cosmetics, coal, plastics, and pharmaceuticals. To give an idea of the magnitude of applications involving granular processes, worldwide production annually accounts for over a trillion kilograms of granular and powdered products, much of which must be uniformly blended to meet quality and performance goals. In this chapter we present an example-oriented overview of current understanding of mixing and de-mixing mechanisms of importance to powder blending operations. We focus on blending in tumblers, which simultaneously comprises the bulk of solids blending operations and represents the greatest opportunity for future predictive modeling. We direct the reader to existing literature sources (e.g., Harnby, 1997) for more specialized blending equipment.

Numerous distinct mechanisms for both mixing and de-mixing of granular materials have been cataloged, including convection, diffusion, shear, and percolation, and in most applications several mechanisms act concurrently and interact in complex ways. For example, details of loading of powders into blenders of

---

*Handbook of Industrial Mixing: Science and Practice,* Edited by Edward L. Paul, Victor A. Atiemo-Obeng, and Suzanne M. Kresta
ISBN 0-471-26919-0 Copyright © 2004 John Wiley & Sons, Inc.

common design can alter the time needed to homogenize them by two orders of magnitude, and by the same token, given that a certain blender can be designed to deliver acceptable performance in the laboratory, we have no consistent a priori mechanism to scale the process up and achieve the same performance in blenders of industrial size. The opposite problem, lack of dynamical similarity during process scale-down, is also quite common, haunting practitioners who attempt to undertake benchtop product design or wish to reproduce manufacturing problems in the lab. Nevertheless, although comprehensive predictive understanding of practical blending problems remains a distant goal, it has recently become possible to define models that generate respectable agreement with observations in practical granular devices (e.g., 3D tumblers). Progress has been made to develop systematic techniques to analyze new products and equipment. Some of these advances are reviewed in this chapter, following a description of the current level of understanding of blending and segregation mechanisms in commonly used industrial devices.

## 15-2 CHARACTERIZATION OF POWDER MIXTURES

A prerequisite to meaningful evaluation and interpretation of mixing is the development of a reliable measure of mixing. Straightforward though this concept may seem, some care needs to be exercised in its implementation. Any mixing measure is obtained by first evaluating a relevant quantity, typically concentration, in specified sample regions. Ideally, for the samples to be representative, they should be taken uniformly from a flowing stream that is itself uniform in both space and time. In tumbling blenders, this is not practical, and sampling usually consists of extracting small samples from a static bed. We discuss techniques for extracting such samples shortly, but first it is worthwhile to review the description of ideal mixtures, for which particle distributions are known throughout the mixture.

### 15-2.1 Ideal Mixtures versus Real Mixtures

Mixing is so common an every day experience to both specialist and layperson that it is often taken for granted. Throughout the undergraduate curriculum in engineering, processes that are clearly mixing-dependent (such as chemical reaction, crystallization, die filling) are assumed to be homogeneous. This widespread preconception is also reflected in the common attitude toward powder mixtures, especially for relatively small particles that, due to their ability to scatter visible light, tend to look more uniform to the naked eye than is often warranted. Early conceptions of homogeneous particle assemblies assumed the particles to be distributed in a state of perfect homogeneity, such that any sample containing a large number of particles would have essentially the same composition. Three conceptual approaches to such blissful state—perfect, random, and ordered mixtures—are discussed below. Real mixtures, unfortunately, tend to show at least some degree of heterogeneity, obeying to one of three main causes: incomplete

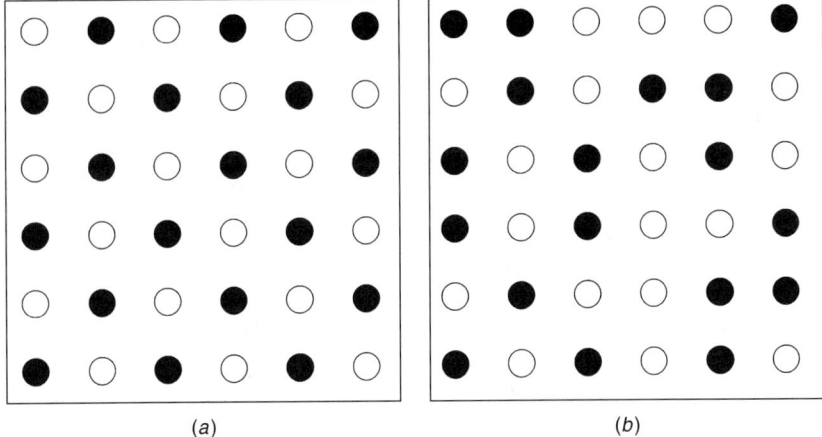

**Figure 15-1** Simulated mixtures: (*a*) perfect mixture; (*b*) random mixture. (From Williams, 1986.)

mixing, agglomeration, and segregation, resulting in different types of textures, also discussed below.

*15-2.1.1 Perfect Mixtures versus Random Mixtures.* The first and simplest conception of a homogeneous system is the perfectly uniform mixture, where particles alternate themselves along a lattice (Figure 15-1*a*), very much resembling the position of atoms of different species inside a perfect crystal. Samples taken from such a mixture are necessarily identical. This highly ordered state is never achieved unless painstakingly created by positioning particles one at a time. If the particles are freely moving and differing from one another by a property that does not affect their movement in any way (such as, perhaps, color for identically sized glass beads), the best achievable state is that of a random mixture (Figure 15-1*b*), rigorously defined as a mixture where the probability of a particle belonging to a certain moiety is statistically independent of the nature of its neighbors. Sample extracted from such a mixture follow a binomial (or multinomial) distribution.

*15-2.1.2 Ordered Mixtures.* For cohesive systems where the particles apply surface forces to one another, it is common to observe the formation of agglomerates. Depending on the relative magnitude of forces between like-particles and unlike-particles, it is possible to see agglomerates of a single species (the "guest"), as well as agglomerates where a small-size moiety essentially coats another, larger moiety (the "host"). This latter situation motivated the concept of an "ordered mixture" (which the reader should distinguish from the situation depicted in Figure 15-1*a*). In the ideal case, the same exact number of identical guest particles covers every identical host (Figure 15-2*a*). Samples taken from such a system would be, once again, identical, thus resulting in a higher degree of sample homogeneity than the random mixtures depicted in Figure 15-1*b*. In reality,

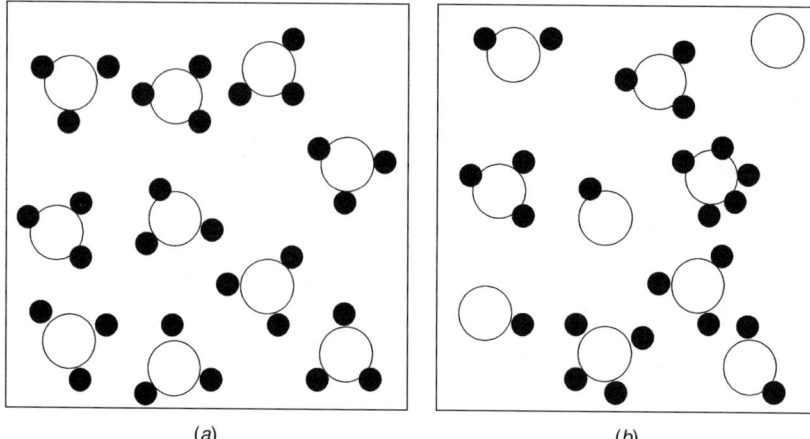

**Figure 15-2** Distributions of individual particles that form an (*a*) ideal random mixture and a (*b*) random mixture. Part (*b*) illustrates a less structured blend that is still well mixed but does not exhibit long-range order in the spatial distribution of particles. This distribution has been called the ideal random mixture, one for which the location of any particle has no influence on the particle (or particles) that are adjacent to it. In other words, a particle that is removed from any location in the mixture has an equivalent chance of being of either species type. In practical terms, this distribution is often the best attainable for a real system of interacting particles.

one observes a distribution in the number of guests on each host, as well as free (unassociated) guests, leading to a less homogeneous outcome (Figure 15-2*b*).

### 15-2.1.3 Textured (Segregated) Mixtures.
The most troublesome mixtures are those that exhibit long scale texture (i.e., segregation), complicating description of mixture distributions and characterization. Textured mixtures form when a characteristic of one or more particle species causes that component to separate into specific regions of the mixture, depending on the type of agitation applied to the bulk mixture. Also, dead zones or incomplete agitation of the powder can lead to segregated regions in blenders. In general, more free-flowing mixtures exhibit more extreme segregated states. Cohesivity acts to inhibit mixture segregation, as individual particles have trouble moving independently of the bulk mixture. Determining mixture quality of textured mixtures depends on accurately determining the size, location, and severity of the segregated regions.

Figure 15-3 shows two types of segregated mixtures, one for free-flowing materials and the other for a cohesive mixture. In the free-flowing case, if the particles do not differ in any particle characteristic other than color, long scale texturing of the mixture will not occur for a sufficiently agitated system. But when there are differences in particle characteristics (size, shape, density, etc.), a situation like that shown in Figure 15-3*a* can arise, where individual particle species are preferentially found in specific regions of the mixture (left to right in this case). For more cohesive systems, a partially randomized mixture can

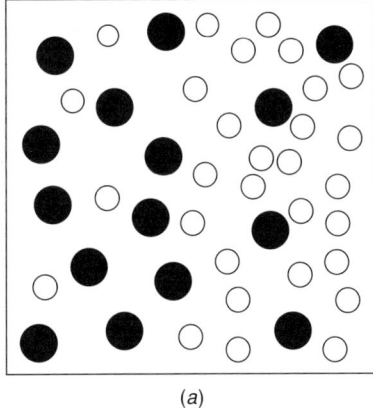 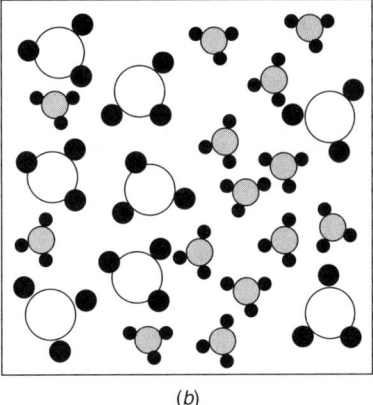

(a)            (b)

**Figure 15-3** Distribution of particles differing in size for (a) freely-flowing and (b) cohesive materials.

develop, as shown in Figure 15-3b. This drawing depicts segregation of ordered units with different-sized carrier particles, but segregation of ordered units with leftover adherent particles is also possible.

In any real mixture there will be areas that correlate closely to many of the ideal distributions discussed previously. Unfortunately, the characterization of mixture quality cannot currently be done by viewing particle distributions throughout the mixture. For real systems, samples are extracted from specific regions of the mixture and it is important to ensure that the sample size is representative.

### 15-2.2 Powder Sampling

Real systems do not yield complete and pristine data on the distribution of particulate species within the bed. Instead, it is necessary to extract a finite, typically small number of samples from the mixture. These samples often have important limitations and biases, as discussed here. The most common means for sampling powder constructs is through the use of sampling thieves. These devices are inserted into the bed and extract samples from the interior. When devising a sampling scheme, it is important to adequately sample all regions of the powder bed. As mentioned, granular materials can segregate spontaneously, and can mix very slowly (especially when dispersion is the major mixing mechanism). Hence, sampling at only a few locations can lead to significant undersampling as regions of poor mixing are completely missed or underrepresented. Furthermore, postprocessing of a powder mixture can cause a previously well-mixed sample to de-mix and adversely affect further applications.

***15-2.2.1 Physical Sampling Methods.*** The behaviors of two popular types of thief samplers are shown in Figures 15-4 and 15-5 (Muzzio et al., 1999). In Figure 15-4a we illustrate the bed disturbances that occur when using a *side-sampling thief*. This device consists of a tube with a slot in its side that can be

# 892 SOLIDS MIXING

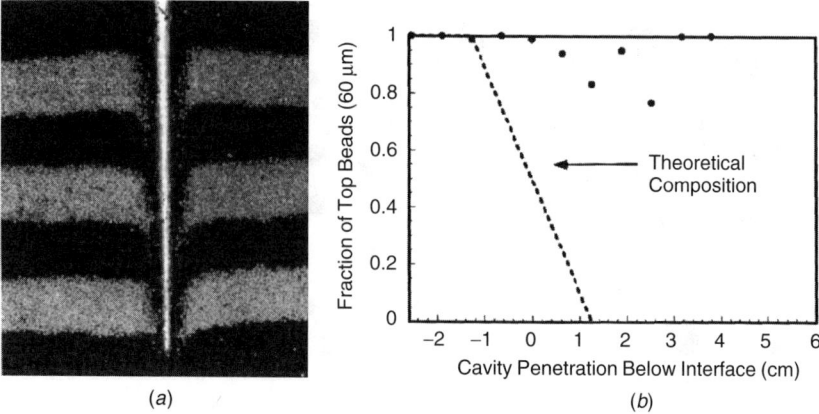

**Figure 15-4** Systematic sampling errors introduced by a side-sampling thief. (*a*) Initially layered configuration of large (light) and small (dark) particles are noticeably disturbed as the thief entrains particles during insertion. (*b*) This type of thief relies on free flow of particles to fill a cavity when a slot is opened in the side of the sampling tube. Consequently, fine and freely flowing particles are overrepresented by this probe, and fine particles are transported to regions where they were not placed originally.

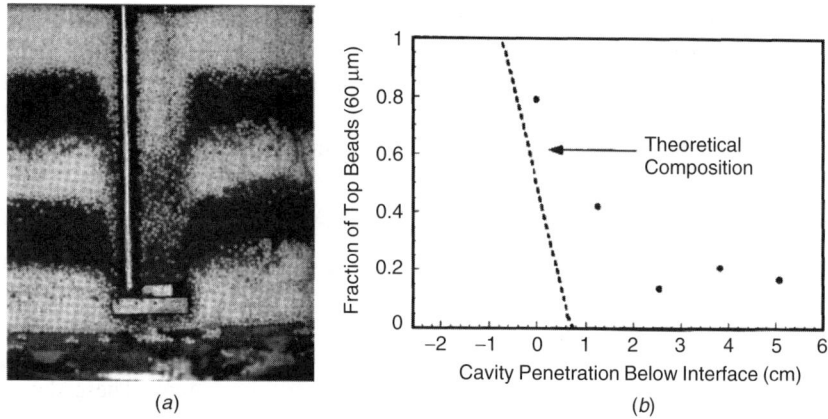

**Figure 15-5** Sampling errors introduced by an end-sampling thief differ from those introduced by a side-sampling thief, but persist nonetheless. In this type of thief, a window is opened at the bottom of the sampling tube, and particles are forced into a cavity by further insertion of the thief. This eliminates the bias toward particles that passively fill a cavity more easily than others, but on the other hand, (*a*) these thieves entrain more particles during insertion, and (*b*) their performance again suffers from substantial systematic error.

opened to allow particles to flow into a cavity, and closed to extract the sample. An initially layered system of light gray 200 and dark 60 μm particles is visibly disturbed by inserting the probe. Particles are entrained along the insertion route, causing local particle rearrangements that typically result in the bed appearing to be anomalously well mixed. It is also significant that side-sampling thieves rely on particle flow into the sampling cavity to obtain particles; consequently, free-flowing or smaller particles can flow into the sampling cavity more readily than more cohesive or larger particles. These observations are quantified in Figure 15-4b that shows the fraction of smaller beads in samples obtained using a side-sampling thief in separate experiments in which 60 μm particles are initially arranged in a single thick layer over a bed of 200 μm particles. The thief obtains samples almost entirely consisting of the smaller species, irrespective of the actual concentration at the sampling location.

Sampling problems that arise from differences in particle flow into the sampling cavity can be mitigated through the use of *end-sampling thieves*, such as the one shown in Figure 15-5. For these thieves, the sampling tube is inserted to a desired depth in the bed, an aperture at the distal end of the probe is opened, and then the probe is pushed deeper into the bed to capture the sample; closing the aperture allows extraction of the sample. Particles are actively forced into the cavity rather than passively flowing into it, as in side-sampling thieves. Thus, this device is relatively free of differential sampling problems caused by differences in particle flowability. However, Figure 15-5a demonstrates that these devices are typically bulky and consequently entrain and disturb considerable material during their insertion. For the case discussed here, the resulting sample concentration measurements (Figure 15-5b) are improved over those of the side-sampling thief but remain very inaccurate, as data consistently overestimates mixture quality.

An alternative that is nearly free of either entrainment (Figure 15-4a) or flow (Figure 15-4b) anomalies is the *core sampler*. This sampler extracts an entire contiguous core of particles throughout the depth of insertion. At its simplest, the probe consists of a thin-walled tube that is inserted into a granular bed, together with a mechanized extrusion apparatus to permit samples to be extracted in a last-in, first-out manner after the tube has been removed from the bed. For capturing free-flowing particles, which can flow out of the tube, an end cap that can be opened during insertion and then closed during extraction is added to the device. Unlike the end-sampling thief, the end-cap mechanism here is internal to the sampling tube, and an entire core is extruded from the bed. The behavior of this device is demonstrated in Figure 15-6. Using the end cap (shown closed in Figure 15-6a), the concentration data obtained compare favorably with other methods, as shown in Figure 15-6b. Importantly, in the core sampler the core extends through the depth of the sampling tube, allowing for precise determination of concentrations between different layers of the bed. Furthermore, sample size is completely variable and can easily be adjusted for different mixtures, core sampler diameters, or changes in process parameters.

By foregoing use of the end cap, core sampler performance is improved further. In Figure 15-7 we display core sampling results for three different inner-diameter sampling tubes using a two-layer bed of common pharmaceutical excipient powders: microcrystalline cellulose and lactose. For all sampler diameters, the experimental data are indistinguishable from ideal expected concentrations. In practice, we note that it is important that the walls of the sampling tubes be

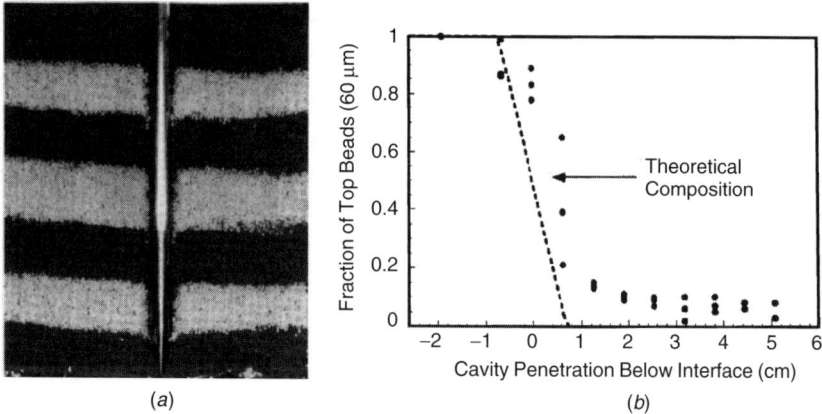

**Figure 15-6** Core sampler with end cap can be used for freely-flowing (e.g., granulated) materials that would escape from the sampling tube during removal from the bed without the end cap. (*a*) Very little entrainment is visible after insertion, and (*b*) systematic errors are reduced. (From Robinson, 1999.)

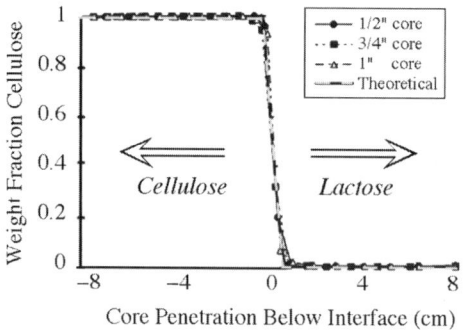

**Figure 15-7** Non-free-flowing powdered material can be extracted from a blend using a core sampler with an open end. This case shows a sampling experiment using a thick layer of microcrystalline cellulose above a bed of lactose, where measurement errors are virtually undetectable.

polished (to prevent excessive entrainment and difficulty filling the tube during insertion) and that a well-regulated extrusion device be employed.

Once samples have been obtained, one can use a variety of available chemical, optical, spectroscopic, chromatographic, or other assays to determine concentration. For example, data in Figure 15-7 were obtained using a calibrated densitometric technique in which one of the two species was colored in advance. Similar results have been obtained using other assay techniques, such as reflection near infrared spectroscopy to evaluate concentrations of magnesium stearate (a common pharmaceutical lubricant) or conductivity assays to evaluate the mixing of salt (NaCl, KCl) in anionic excipients (Avicel).

*15-2.2.2 Noninvasive Methods.* Other, more technologically complex techniques have also been developed for visualizing the interior of granular beds. These include:

- *Diffusing wave spectroscopy*, where statistics of fluctuations in relatively thin, Hele–Shaw configurations are measured
- *Positron emission tomography*, where a single radioactive particle is tracked during flow within a granular bed using an array of external photomultipliers
- *Magnetic resonance imaging*, where magnetic moments of hydrogenated particles are aligned in structured configurations (e.g., stripes) and these structures are tracked for short periods of time
- *X-ray tomography*, where a population of radiopaque particles are tracked in a flow of interest

These techniques are typically expensive and cumbersome to implement; nevertheless, they reveal flows within an optically opaque bed and provide valuable information not available otherwise. For example, in Figure 15-8, we display results of x-ray tomography experiments that show the evolution of the interior mixing structure within a double-cone blender using molybdenum-doped tracer particles (dark in Figure 15-8). Data of this kind reveal a complexity in flow and mixing evolution that simultaneously represents the cause of historical difficulty in understanding the subject and the opportunity for future developments. As these methods are improved, they will yield more quantitative information about mixture quality, leading to more robust methods for characterization of powder mixtures.

### 15-2.3  Scale of Scrutiny

When extracting samples from sampling thieves, it is necessary to specify the sample size. This determination is based on the necessary scale of scrutiny for a particular system. Typically, this scale can be determined from the end use of

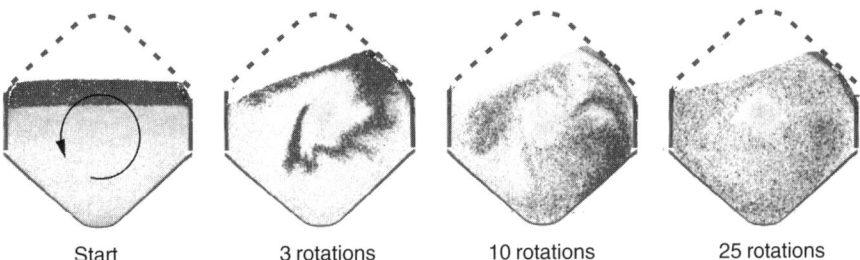

**Figure 15-8**  X-ray tomographic time series of blending of radiopaque grains in a double-cone blender is representative of several new techniques available for on-line and in situ assays of blending mechanisms. (From Chester et al., 1999.)

**896** SOLIDS MIXING

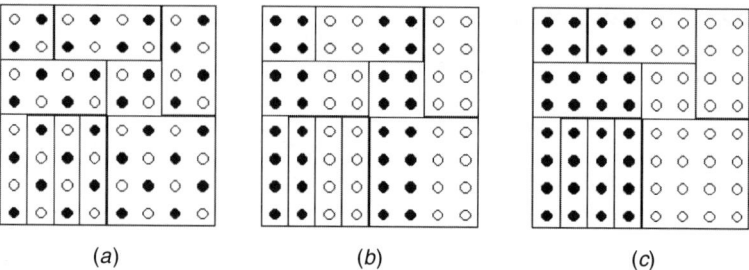

**Figure 15-9** Three ideal mixture distributions are shown, representing (*a*) well-mixed, (*b*) striated, and (*c*) completely segregated mixtures. A number of boxes are drawn over each distribution to represent possible sample size and orientations. These posited sample locations show how sample size and location can adversely affect the true measure of mixture quality.

the product (e.g., pharmaceutical tablet). Component variations within a single dose generally do not affect the chemical uniformity of the final product units (mechanical properties are a different matter altogether). However, relying on too large a scale of scrutiny can mask mixing problems within a system.

Three model systems are shown in Figure 15-9. A series of boxes of varying size drawn over these particle distribution schematics illustrate some of the problems than can arise from improperly defining the scale of scrutiny. If the sample size is chosen as either the entire domain or a single particle, all the distributions shown in Figure 15-9 would have the same measure of mixing, which is clearly incorrect if the finished product has an intermediate scale. For other box sizes, we see that box size is largely irrelevant to the well-mixed case (Figure 15-9*a*), but the box size and orientation can have major influences on the apparent mixture distributions for the striated (Figure 15-9*b*) and completely segregated (Figure 15-9*c*) mixtures. For these mixtures, a symmetric $2 \times 2$ box encompassing four particles would appear to be the minimum scale of scrutiny necessary to extract reliable mixing information. The upper limit is bounded by a $4 \times 4$ box size, as the $4 \times 4$ box clearly would not detect the striated mixture in Figure 15-9*b*.

A key point to gather from the textured mixtures above is that the orientation of sampling can have a large impact on mixing measures as well as the sample size. Sampling perpendicular to a striated mixture must be undertaken with care to ensure that the sample size is smaller than the striation size to obtain accurate measures of mixture quality (or lack thereof).

### 15-2.4 Quantification of Solids Mixing: Statistical Methods

Clearly, the sampling protocol and extraction technique can have a major impact on the accuracy of the mixing measure. Once samples have been obtained, it then becomes necessary to develop mixing measures that give an accurate representation of mixture quality.

One of the most useful measures of mixedness is the intensity of segregation, I. This is essentially a normalized variance of concentration measurements. Intrinsic to the use of intensity of segregation therefore is the presumption that the mixing distribution is, at least to a first approximation, Gaussian. This raises two issues: (1) it is not clear that granular mixing tends toward a Gaussian state, and (2) in many practical applications a Gaussian is not the desired outcome. Indeed, in pharmaceutical processing, if a blend were Gaussian rather than uniform, the unavoidable presence of exponential tails on a Gaussian distribution would guarantee that some small fraction of tablets made from the blend would be beyond any therapeutic range that one could specify. Moreover, the expectation of a Gaussian distribution provides the manufacturer of a regulated product an incentive for extracting as few samples as possible, since a larger number of samples increases the probability of detecting product out of specification. Fortunately, granular flows appear to scatter particles more uniformly than a simple Gaussian would predict, although the details and mechanisms for this behavior are not yet well understood.

With these caveats in mind, the intensity of segregation, I, is defined as

$$I = \frac{\sigma^2 - \sigma_r^2}{\sigma_0^2 - \sigma_r^2} \quad (15\text{-}1)$$

where $\sigma^2$ is the variance of sampled data, $\sigma_r^2$ is the variance of the same number of randomly chosen concentration data, and $\sigma_0^2$ is the variance of an initial, typically fully segregated state, again consisting of the same number of data points. Several forms of I appear in the literature; the form presented here is useful because it is normalized so that $I = 1$ and $I = 0$ correspond to completely segregated and randomly mixed states, respectively. In practice, values below 0.7 are rarely encountered, changing the range of expected values for good and poor mixed constructs.

Another mixing measure of importance to powder mixing is the relative standard deviation (RSD), also known as the coefficient of variance (CoV), defined to be

$$\text{RSD} = \text{CoV} = \frac{\sigma}{M} \quad (15\text{-}2)$$

where $\sigma$ is the standard deviation and M is the mean concentration over all samples taken. Often, in pharmaceutical applications, mixing protocols are written to ensure that a specified percentage of all extracted samples meet an RSD ceiling.

Related to the standard deviation of a mixture is the mixture variance. Variance measurements have the benefit of being additive, meaning that total variance can be subdivided into mixture variance plus sampling error, assay error, and so on. Using this quality, a more detailed analysis of bed variability can be obtained by separating the total variance measurement into separate dependent measurements. For example, for mixtures of cohesive and free-flowing components, it is very important to design stratified sampling schemes where multiple samples are taken from each of a series of predetermined sampling locations, allowing the experimenter to distinguish within-location and between-location variability. For tumbling blenders it is extremely useful to divide the measured variance into axial variance and radial variance components. Axial variance measures the differences in concentrations between sampling locations, while radial variance measures variance within the bed at a single location. Using a core sampler greatly aids these measurements, as concentration data from a single core and average values between different cores can be used separately. Formally, for each core j,

$$\overline{x}_j = \frac{\sum_i x_{ij}}{N_i} \quad (15\text{-}3)$$

where the core mean concentration is $\overline{x}_j$, $x_{ij}$ is a given sample concentration, and $N_i$ is the number of samples in that core. The standard definition of variance is

$$\sigma^2 = \sum_j \sum_i \frac{(x_{ij} - \overline{x})^2}{N} \quad (15\text{-}4)$$

where $\sigma^2$ is variance, N is the number of samples, and $\overline{x}$ is the mean composition. Substituting eq. (15-1) into eq. (15-2) and rearranging leads to

$$\sigma^2 = \frac{1}{N} \sum_j N_i (\overline{x}_j - \overline{x})^2 + \frac{1}{N} \sum_j \sum_i (x_{ij} - \overline{x}_j)^2 \quad (15\text{-}5)$$

In eq. (15-5) the first term is a measure of axial variance ($\sigma_A^2$) and the second term, radial variance ($\sigma_R^2$). These two measures give a more accurate description of mixing quality within a granular blend than can be achieved with any single measurement.

## 15-3 THEORETICAL TREATMENT OF GRANULAR MIXING

Research into granular flow and mixing can be roughly divided by chronology: prior to 1990, industrially usable results were mostly empirical (e.g., in experiments using a particular blend in a specific device), and fundamental research was largely analytic (e.g., using continuum approximations to the granular state applicable only to one phase of granular behavior). Although significant progress has been made into developing specialized engineering solutions as well as models of fundamental behaviors of ideal granular systems, little generally applicable knowledge was attainable using either approach. Over the past decade, computational and methodological advances have permitted quantitative evaluations of granular flow, transport, and mixing at a much greater level of detail. In this section we review progress on tumbling flow and blending phenomenology that has led to the development of the best existing predictive models.

### 15-3.1 Definition of the Granular State

A chief limitation, and the principal area of opportunity for the future, in developing predictive understanding of granular flows is the coexistence of multiple, history-dependent granular states. Within a device—be it a tumbler, a mill, a fluidized bed, or a high-shear intensifier—granular material can, and typically will, exhibit multiple rheologically different phases that vary nontrivially and often with profound consequences as a function of minor changes in material or operational parameters. This is a particular problem in the pharmaceutical industry, where products may be developed in dry northern latitudes, and produced in wet equatorial climates. Both hygroscopic excipients and actives behave very differently in these two environments, and blending regimens that work in one may well fail in the other. Moreover, even within a single well-controlled bench scale device, multiple phases are typically present. The tumbling blender is a case in point.

In Figure 15-10 we display a deceptively common outcome of an attempt to blend dissimilar materials, here grains differing only in size and color. In this transparent 4 L capacity[1] V-blender, we have tumbled equal volumes of smaller light-gray and larger black grains at 6 rpm for 200 revolutions. The visibly segregated state is one of several distinct segregated configurations that

**Figure 15-10** Left–right segregated state, here in a transparent V-blender, between larger (dark) and smaller (light) grains. This state occurs spontaneously at high fill levels and fast tumbling speeds in many tumbler designs.

---

[1] The reader should note that capacity customarily refers to a fraction (generally, 60%) of the total interior volume of a blender.

form spontaneously and reproducibly in all common blender geometries and scales. Once formed, these patterns persist despite the practitioner's best efforts at modification of process parameters. Developing cures for this type of problem demands a systematic understanding of why de-mixing occurs in the first place, so that the cause of segregation can be addressed directly. This understanding, in turn, requires an analysis of the different granular behaviors seen during the tumbling operation.

A first step in the analysis of granular behaviors is the characterization of the different granular phases that are inevitably present during flow. It has been recognized for over a century that grains, unlike common fluids, must dilate in order to flow—that is, grains in the static state are interlocked and cannot move without separating (see Figure 15-11a). The locations and timing of flow can be quite complex; nevertheless, far from flowing regions a granular bed remains static and solidlike, whereas near a shear interface, grains are fluidlike or gaslike as the shear rate increases. The modifier "like" is important to include, for a solidlike region is not truly elastic, as it transmits stress along irregular compressive chains, it undergoes slow creep and settling on time scales ranging from seconds to hours, and it can solidify into a rigid cake over time scales of days to months. Similarly, the fluidlike phase transmits shear discontinuously in both space and time and does not obey Navier–Stokes equations, and the gaslike phase is far from equilibrium and is not characterized by Maxwell–Boltzmann statistics. The selection of these three latter phases—glassy, fluidlike, or gaslike—in a specified location depends on details of the bed dynamics, including the rate of shear, the extent of compaction of the bed, and the geometry in which the bed is confined. It is the differences both between qualitative behaviors of different regions of a granular bed at different times and between any one of the behaviors and accepted models for flow and dispersion that make predictive understanding of even the simplest granular systems challenging.

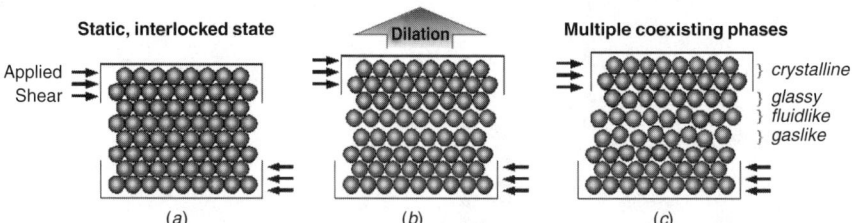

**Figure 15-11** Schematic of dilation mechanism that is a prerequisite for the flow of solids. (a) In an undisturbed state, grains are interlocked and behave much like an ordinary solid. (b) A granular bed dilates in response to applied shear and can then flow. (c) In the flowing state, the bed can form distinct crystalline, glassy, fluidlike, and gaslike phases. The crystalline phase is regular and ordered, the glassy phase is disordered but static, the fluidlike state flows but exhibits enduring contacts, and the gaslike state is characterized by rapid and brief interparticle contacts.

The ability of granular systems to coexist in rheologically distinct states has many consequences. It has prevented the scientific community from developing effective devices for measuring rheometric properties. Essentially all available techniques provide an average measure of strongly nonlinear, strongly variable behavior. Although accumulated experience can help the practitioner use this type of information within a narrow range of parameter variability, in practice a lack of rheometric techniques is perhaps the main reason for a lack of effective constitutive models. Models abound, but their experimental verification is somewhere between unfeasible and impossible. As a direct consequence, computer simulation techniques for granular flows are much less developed than those used for fluids, and our ability to develop granular flow systems *in silica* is largely a goal for future generations.

Nonetheless, despite intrinsic difficulties in developing all-encompassing models for granular flow, important blending problems of practical interest have been effectively solved using analytic, computational, and semiempirical means. We summarize the current understanding of granular blending and de-mixing in subsequent sections.

### 15-3.2 Mechanisms of Mixing: Freely-Flowing Materials

In tumbling applications, dilation and flow principally play out near the unconstrained upper surface of a granular bed, and except for solid-body rotation, the bulk of grains beneath are thought to remain nearly motionless during rotation of the blender. This simplified picture changes for some blenders (notably the V-blender, in which flow is strongly intermittent; see Moakher et al., 2000), but predictive models for blending in most common blending geometries can be derived by disregarding all transport beneath the free surface. In the sections following, we summarize the best existing models and methods and describe their application to common tumbler designs. A useful design choice for the purposes of illustration is the horizontal drum tumbler. The horizontal drum is used in many chemical, metallurgical, and pharmaceutical industries in the form of ball mills, dryers, rotary kilns, coating pans, and mixers. Flow in rotating drums with increasing tumbling speed has been described qualitatively in terms of regimes termed: slipping (or slumping), avalanching, rolling, cascading, cataracting, and centrifuging. These are defined as follows.

***15-3.2.1 Slipping.*** The slipping regime occurs when the granular bed undergoes solid body rotation and then slides, usually intermittently, against the rotating tumbler walls. This occurs most frequently in simple drums that are only partially filled and is typically counteracted by including baffles of various designs along the inner walls of the tumbler. While the slipping regime is not important for blending purposes per se, it is encountered even in effective blending systems, and an evaluation of the number of times a bed turns over per tumbler revolution will often reveal the presence of some slipping.

## 902 SOLIDS MIXING

**15-3.2.2 Avalanching.** A second regime seen at slow tumbling speeds is avalanching flow, also referred to as *slumping*. In this regime, flow consists of discrete avalanches that occur as a grouping of grains travel down the free surface and come to rest before a new grouping is released from above. The avalanching regime is not seen in tumblers larger than a few tens of centimeters in diameter, but it is an instructive case because a flow and mixing model can be derived in closed form for simplified drum geometries.

To analyze this problem, one needs only observe that if the angle of repose at the free surface immediately before an avalanche is $\theta_i$, and after an avalanche is $\theta_f$, the effect of the avalanche is to carry a wedge of material in the angle $\theta_f - \theta_i$, downhill, as sketched in Figure 15-12a for an idealized two dimensional disk blender. The same behavior occurs for all fill levels, and one can readily use this model to make several concrete predictions. First, mixing occurs during avalanches through two distinct mechanisms: (1) particles within a wedge rearrange during a single avalanche, and (2) particles rearrange globally between wedges during successive avalanches. Second, at 50% fill (Figure 15-12b) no two avalanching wedges intersect, so no global mixing between separated regions can exist, and mixing must slow. Third, since flow occurs only near the avalanching surface, at high fill levels a nonmixing core necessarily develops (Figure 15-12c). Although this model is oversimplified and neglects material variations, boundary effects, and other important phenomena, these conclusions carry over to more realistic tumbling systems.

**15-3.2.3 Rolling.** At higher tumbling speeds, discrete avalanches give way to continuous flow at the surface of a blend (Shinbrot et al., 1999a). Grains

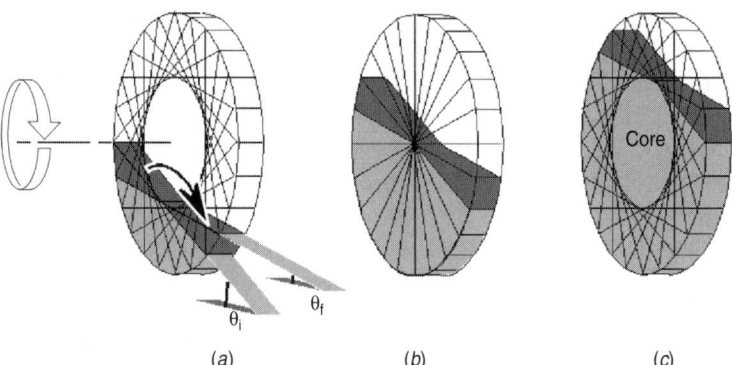

**Figure 15-12** (*a*) Avalanching flow in an idealized disk tumbler transports grains from an uphill wedge to a downhill wedge as the free surface relaxes from an initial angle, $\theta_i$, to a final angle, $\theta_f$. This implies that global mixing occurs in quadrilateral regions where grains within one wedge intersect with a second wedge. (*b*) Consequently, tumblers mix more efficiently at low fill levels than at high, and global mixing nearly stops at 50% fill. (*c*) At fill levels above 50%, a core develops that does not visit the avalanching surface and therefore does not experience transport or mixing.

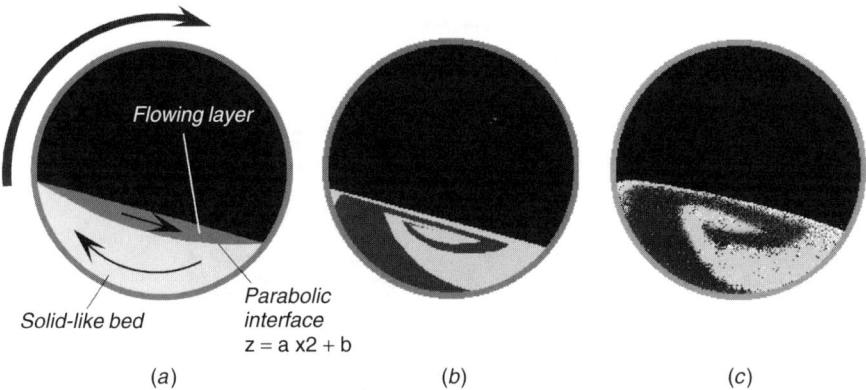

**Figure 15-13** (*a*) In the rolling regime, the blend separates into a flowing layer near the surface and a solidlike supporting bed. (*b*) By establishing simple conditions such as mass conservation, one can generate an analytic model for the flow, producing mixing patterns between initially separated and different-colored but otherwise identical grains. (*c*) Comparison with experimental mixing patterns using freely flowing grains in a small drum tumbler reveals substantial agreement. The snapshot in part (*c*) is obtained from the interior of the blend using a solidification technique.

beneath this surface flowing layer rotate nearly as a solid body with the blender until they reach the surface. One can solve for flow and transport subject to certain simplifying assumptions in this regime as well. For this solution, one assumes that the grains are so small as to be regarded as a continuum and one takes the free surface to be nearly flat, as sketched in Figure 15-13*a*. The interface between the flowing layer and the bed beneath has been determined experimentally and computationally to be roughly parabolic in shape, and by demanding mass conservation at this interface, one can construct continuum flow equations for this system. If one simulates the mixing in an idealized disk blender of mechanically identical grains initially separated by color to left and right of a vertical central plane, one obtains the results displayed in Figure 15-13*b* (for a particular fill level and flowing layer depth). Corresponding experimental results are shown in Figure 15-13*c*.

### 15-3.2.4 *Cascading, Cataracting, and Centrifuging*.

For larger tumblers, or for tumblers rotated at higher speeds, the surface is manifestly not flat, as shown in Figure 15-14 in a 1 m diameter disk tumbler. This flow, termed *cascading*, differs qualitatively from the rolling flow solution; here the flowing layer is thin, is nearly uniform in speed and thickness, and has been modeled as depth-averaged pluglike flow. As the rotation speed of the tumbler is increased, the surface becomes increasingly sigmoidal until grains become airborne, and at higher speeds yet, the grains centrifuge against the tumbler wall. These regimes are termed *cataracting* and *centrifuging*, respectively, and have not been well analyzed.

**904** SOLIDS MIXING

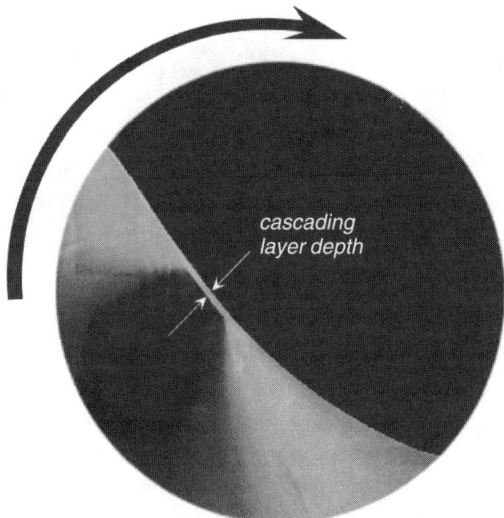

**Figure 15-14** Cascading flow occurs in large tumblers or during tumbling of fine but freely flowing grains. This snapshot shows a 1 m diameter 1 cm wide transparent disk tumbler partially filled with colored ∼500 μm irregular grains. Initially, light and dark grains were placed to the left and right of a central plane; this snapshot shows the mixing pattern at one-half revolution of the disk. This tumbler is thin, so grains are under the strong influence of wall effects; nevertheless, this example serves to illustrate that the free surface is manifestly not flat, and the cascading layer is thin and nearly uniform along the flowing surface.

### 15-3.3 Mechanisms of Mixing: Weakly Cohesive Material

Another mechanism of granular and powder mixing is associated with blending of weakly cohesive materials. Weakly cohesive materials (e.g., powders and fine grains in the size range 50 to 300 μm) exhibit stick-slip motion so that flow becomes intermittent rather than continuous. This is a situation of practical importance since most industrial applications use particles across a broad range of sizes and materials. As the size of grains diminishes or as interparticle cohesion grows, stick-slip flow transforms mixing interfaces from a smooth, regular patterns as shown in Figure 15-15 (500 or 700 μm cases) to a complex, irregular pattern, shown in Figure 15-15 (300 or 100 μm cases). In simple geometries this response to shear can be modeled accurately: If we assume that the flowing surface of a bed sticks and slips periodically, the mechanism displayed in Figure 15-13a can be embellished by allowing the shear band between flowing layer and bed to deform periodically (Brone et al., 1997). This produces mixing patterns between initially separated but identical grains that are substantially similar to experimentally observed ones, as shown at the bottom of Figure 15-15. This is important for blending because in smooth regular flow, adjacent particles remain nearby for long periods of time, while

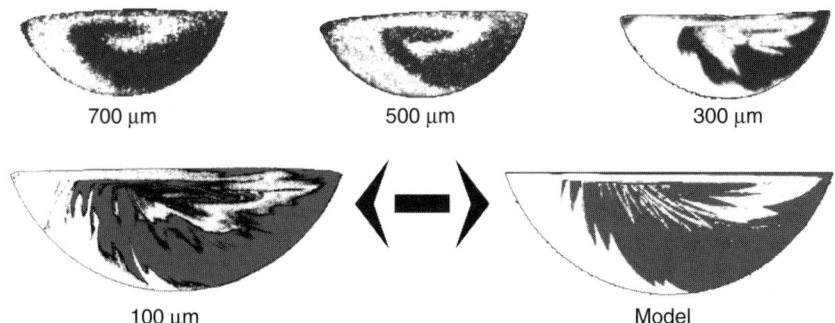

**Figure 15-15** Mixing patterns after one revolution in identical drum tumblers loaded with identical (except for color) grains in four experiments using successively finer grains as well as in a model simulation of idealized stick-slip flow. At 700 and 500 μm, the mixing interface remains smooth and regular; below about 300 μm, it becomes variegated due to intermittent slipping of the cascade. Each experimental snapshot shows a view from the interior of a blend using the solidification technique described by Wightman et al. (1995), and all cases began with light grains to the left of center and dark grains to the right.

in intermittent stick-slip flow, particles can rapidly relocate across the blender, resulting in an exponentially rapid growth of interfaces between separated regions of grains (Shinbrot et al., 1999b).

For particles smaller than about 100 μm, cohesive forces (believed to be due to van der Waals interactions for intimate contacts, and to surface tension of adsorbed water layers for lubricated contacts) between particles become comparable to particle weights, and small particles can stick to one another in relatively rigid aggregates. Unless such aggregates are destroyed, the system will behave as if it had an effective particle size much larger than the primary particle size.

For strongly cohesive materials, it is typically necessary to fragment agglomerates through the introduction of high-shear, intensification devices, such as impellers or mills that energetically deform grains on the finest scale. Many forms of intensification are used in industrial practice. Some common approaches include passing the blend through shaker sieves or through hammer or pin mills between blending stages, as well as using high-speed devices within blenders, such as intensifier bars in tumbling or choppers in high-shear granulator-style mixers.

Essentially no detailed systematic quantitative information is available concerning the effects of intensification on blend quality. We are aware of no studies investigating the micromixing quality as a function of intensity and duration of applied shear. Scale-up and design information provided by equipment vendors is largely limited to advising the user to keep intensifier tip speed and time of operation constant during scale-up. Although this guideline is reasonable in lieu of rigorous information, it is clear that in situ intensifiers apply shear only locally, and nonuniformly, to the mixture; the end result is almost guaranteed to be affected by the interplay of the intensity of the shear field, the residence time of particles in the shear field, and the global homogenization capabilities of the blender. At the present time, laboratory devices for applying shear uniformly and at a known rate are unavailable, making study of the problem even harder. Given a tendency across industry to deal with ever smaller, ever more cohesive materials, understanding the role of shear on blend quality is undoubtedly one of the areas in greatest need of attention by the scientific community.

### 15-3.4 De-mixing

Processing blends of dissimilar grains almost invariably promotes *de-mixing*, also referred to as *segregation*, characterized by the spontaneous emergence of regions of nonuniform composition. Segregation due to differences in particle size in a blend has drawn the greatest attention in the literature, including studies of fluidized beds, chutes, hoppers, vibrated beds, and tumbling blenders, but segregation due to differences in particle density, shape, and triboelectric order have also been recorded. As a practical matter, segregation manifests itself in granular mixing that characteristically improves over a brief initial period, while convection generates large scale mixing, and then degrades, often dramatically as slower segregational fluxes take over. De-mixing should not be confused with the phenomenon of overblending, which is also frequently encountered in blending applications. Overblending is associated with physical degradation of material properties, as occurs, for example, when a waxy lubricant is excessively deformed, causing it to coat pharmaceutical grains and reduce their bioavailability, or when coated granules are damaged through abrasion or fracture.

At the present time, mechanisms for segregation, even in the simple tumbling drum, remain obscure, and work on more complex and industrially common blender geometries is extremely limited. Three distinct types of de-mixing are moderately well characterized in tumblers: radial de-mixing, axial de-mixing, and competitive patterned de-mixing. We describe each of these in turn.

***15-3.4.1 Radial De-mixing.*** Segregation typically proceeds in two stages. First, large grains rapidly segregate radially, producing a central core of fine grains surrounded by larger grains, identified in Figure 15-16 for a simple drum tumbler. Unlike the core seen in overfilled tumblers, this core appears at fill levels under 50% and is associated exclusively with migration of fine grains toward the center of an overturning blend. Radial segregation is seen in both quasi-2D and fully 3D blenders of various geometries. In simpler 3D geometries, such as the drum, double-cone, or tote, the core is nearly always apparent when blending significantly dissimilar grains, while in more complicated geometries such as the V-blender or slant cone, the core becomes significantly distorted and may only be conspicuous for higher fill levels or in certain (e.g., upright) orientations of the blender. Even in the simplest case of the drum tumbler, however, the location and dynamics of the core remain somewhat enigmatic—for example, as shown

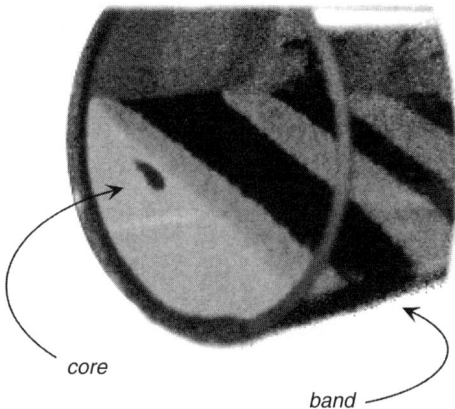

**Figure 15-16** Typical segregation pattern seen between fine (dark) and coarse (light) grains in a small transparent drum tumbler. A core of fines travels along the entire length of the tumbler, connecting the bands that emerge at the surface in a single bulging tube. The coarse grains are constrained to flow within the confines defined by this tube. This constraint is important for understanding mechanisms of de-mixing in more complex geometries, as summarized in Section 15-3.4.

in Figure 15-16, the core is actually located upstream of the geometric center of the granular cascade.

The core appears to form as a result of two cooperative influences. First, smaller grains percolate through the flowing layer to occupy successively lower strata each time the bed overturns. Second, once a sufficient volume of smaller grains has accumulated, the larger grains tend to roll increasingly freely over the (comparatively smooth) substrate of smaller grains. This higher-speed surface flow reinforces the segregated state by expelling remaining slower small grains. These mechanisms are very robust, and cores are almost invariably found in tumbling of freely flowing grains with diameter ratios between about 1 : 1.5 and 1 : 7. As the diameter ratio approaches unity, the core becomes more diffuse, while as the diameter ratio grows sufficiently large, fine grains can percolate increasingly freely through a matrix of larger grains or, if sufficiently fine, can coat the larger species.

*15-3.4.2 Axial De-mixing.* A second stage of segregation occurs in drum tumblers as grains in the core migrate *along* the tumbling axis. Numerical and experimental investigations have attributed this migration to conflicting causes (e.g., a secondary flow within the core leading to a bulging of the core toward the surface versus different angles of repose of fine, mixed, and coarse grains). Whatever the ultimate cause, the result of this axial migration is the formation of a series of bands as shown in Figure 15-16. In this final state, two pure phases of material are formed, divided by sharp boundaries with very little intermixing.

## 15-3.4.3 Competitive Patterned De-mixing.

In more complex, and more common tumbler geometries, several distinct segregation patterns have been observed. These patterns are believed to arise from a competition between surface segregation of coarse grains flowing over a radially segregated core of fine grains and interactions with the boundaries of the tumbler. Despite significant differences between common blender geometries, there is substantial commonality in the ultimate patterns seen. For example, mixing of large, light-gray, and small, dark-gray grains in a double cone and a V-blender generate similar patterns in both experiments and particle-dynamic simulations (described in Section 15-4), as shown in Figure 15-17.

As parameters such as fill level, tumbler speed, and concentrations of the different particle species are varied, the patterns observed change significantly. Importantly, there appear to be few dominant and recurring patterns that are seen in both experiments and simulations in all blender geometries. Notably at high fill levels and tumbling speeds, the left–right state shown in Figure 15-10 appears to dominate. This pattern and two other common variants are shown at the top of Figure 15-18 in top views of the surface of a double-cone blender. Each of these patterns appears reproducibly and spontaneously whenever different-sized grains are tumbled in any of several blender geometries. Simulations shown beneath the experimental figures in Figure 15-18 use a continuum model in which large

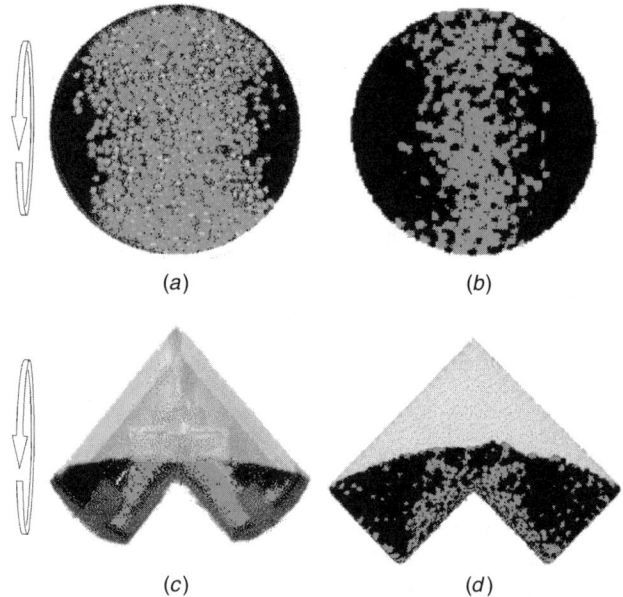

**Figure 15-17** Axial segregation in top views of double-cone blender from (*a*) experiment and (*b*) particle-dynamic simulation using large, light and small, dark spherical grains. Similar patterns are seen in other tumbler designs: for example, in the V-blender in (*c*) experiment and (*d*) simulation.

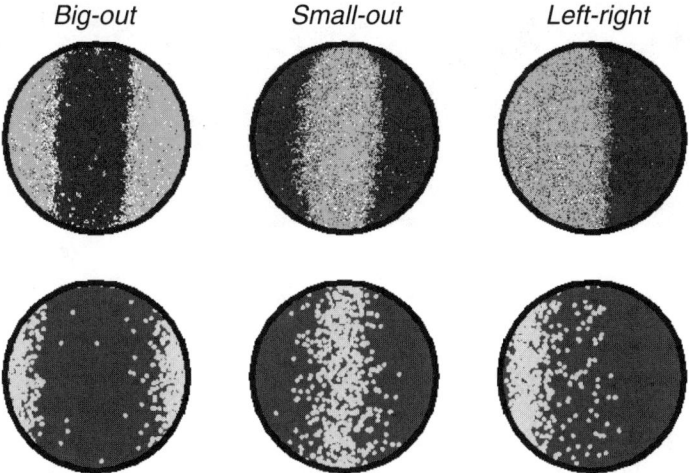

**Figure 15-18** Three common segregation patterns between large (light) and small (dark) grains seen in top views of a double-cone blender. Top: experimental snapshots; bottom: simplified continuum simulations.

particles are convected on the surface of an idealized convex bed of smaller grains. Container geometry is included by assuming that large particles rebound specularly when they reach the downstream boundary of the idealized blender. Correspondence between experimental data and this simulation indicates that ongoing improvements in modeling show promise for unveiling the underlying mechanisms of de-mixing and permitting eventual accurate modeling of practical granular processing systems.

## 15-4 BATCH MIXERS AND MECHANISMS

### 15-4.1 Tumbling Mixers

Although drum blenders represent a convenient paradigm for the purpose of categorizing granular behaviors, most blending operations occur in more complex tumbler geometries. Three of the most common geometries used in pharmaceutical operations are the double cone, the V-blender, and the bin blender, sketched in Figure 15-19. Each of these geometries possesses many variants; for example, symmetry can be broken to introduce cross-flow by slanting the double cone, by elongating one of the arms of the V-blender, or by inserting baffles in a bin.

To model flow and blending in complicated geometries, particle-dynamic simulations have been applied. In these simulations, particles are treated as individual entities with physical properties (e.g., size, static and dynamic friction coefficients, coefficient of restitution, etc.) appropriate to the problem of interest, and Newton's laws of motion are integrated for each particle. Particle-dynamic simulations are similar in concept to molecular-dynamic simulations but include

# 910 SOLIDS MIXING

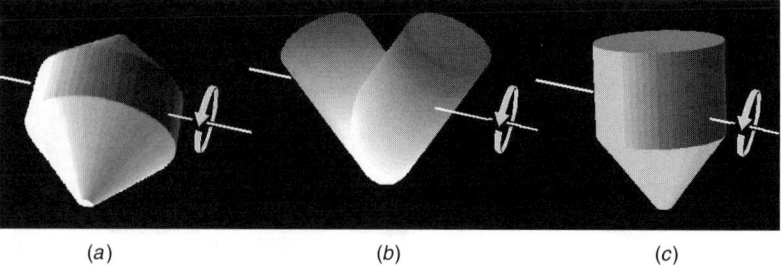

(a)    (b)    (c)

**Figure 15-19** Three common tumbler designs: (*a*) double cone, (*b*) V, and (*c*) tote or bin blenders. Video clips of these mixers are provided on the Visual Mixing CD affixed to the back cover of the book.

features of importance to the flow of macroscopic particles (e.g., static and dynamic friction models) in place of microscopic properties (e.g., bond strengths and chemical potentials). Particle-dynamic simulations come in many different types, depending on how they treat physical parameters, such as rolling friction and particle shape, or numerical issues, such as search algorithms and routines to maintain computational stability. As such, results of distinct computational simulations can differ, sometimes significantly, and the importance of experimental validation of numerical results cannot be overemphasized.

Two of the most common classes of particle-dynamic simulations are termed hard-particle and soft-particle methods. *Hard-particle methods* calculate particle trajectories in response to instantaneous, binary collisions between particles, and allow particles to follow ballistic trajectories between collisions. This class of simulation permits only instantaneous contacts and is consequently often used in rapid flow situations such as are found in chutes, fluidized beds, and energetically agitated systems. *Soft-particle methods*, on the other hand, allow each particle to deform elastoplastically and compute responses using standard models from elasticity and tribology theory. This approach permits enduring particle contacts and is therefore the method of choice for tumbler applications. The simulations described in this chapter use soft-particle methods and have been validated and found to agree in detail with experiments.

***15-4.1.1 V-Blender.*** Mixing in all tumbling blenders consists of a fast convective stage, driven by the mean velocity of many particles, followed by a much slower dispersive stage, caused by velocity fluctuations leading to rearrangements of individual particles. Convection in grains (as in fluids) is by far the faster and more efficient mixing mechanism, yet at the same time it suffers from the same mixing limitations known for fluids: convective flows can—and very often do—possess barriers to mixing (e.g., islands) that do not interact with surrounding material. Two pathologies are readily observed: overfilled mixers develop elliptic, nonchaotic islands that rotate as a unit in the center of the granular bed (discussed in Section 15-3.2.2), and symmetric blenders (seen in most standard designs) exhibit separatrices that divide the flow into noninteracting

BATCH MIXERS AND MECHANISMS   **911**

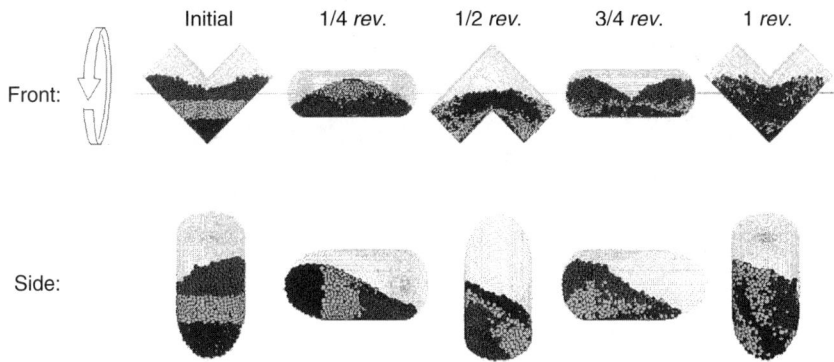

**Figure 15-20** Rapid convective flow seen in particle-dynamic simulation of identical but colored spheres in a V-blender. Top: Front view reveals that unlike in some designs, convection in this blender drives grains axially, alternately outward toward the tumbler arms and inward toward its center. This axial flow strongly influences mixing. Bottom: Side view indicates that transport is dominated by a spiraling flow, seen also in drums and other blenders (cf. Figure 15-13). The full simulation is included on the Visual Mixing CD affixed to the back cover of the book.

sectors. Beyond this, little is currently known of details of particle flow patterns and mixing barriers in practical, three dimensional blender geometries, although there is strong evidence indicating that flow bifurcations analogous to those seen in fluids may be present in granular tumblers.

Convection in the context of granular blenders refers to transport associated with flow driven by gravity (in tumbling blenders) or impellers (in intensified, ribbon, or other blenders). Convection is observed in all functioning blender geometries and can be visualized using particle-dynamic simulations. In Figure 15-20 we display successive front and side views taken a quarter revolution apart of 20 000 identical but colored spheres tumbled in a V-blender in the cascading regime. These snapshots illustrate the qualitative motion produced in this blender, which causes the bed to overturn from top to bottom. Mixing due to convective flow grows linearly with time insofar as the area of an interface (e.g., between differently colored layers in these snapshots or in Figure 15-13*b* and *c*) grows characteristically linearly with time. Similar qualitative behaviors are seen in all tumbler geometries, although the quantitative mixing seen can differ considerably between geometries (Moakher et al., 2000).

***15-4.1.2 Bin Blender.*** In contrast to convection, which can effectively intersperse grains in a tumbler within tens to hundreds of revolutions, is *dispersion*, or *diffusion*. Dispersion refers to the random relocation of individual grains due to collisions between adjacent particles and can take hundreds to thousands of revolutions to act. Thus, particles can only cross a plane separating the two arms of the V-blender (or an equivalent symmetry plane in many other blender geometries) as a result of occasional collisional happenstances and not

**912** SOLIDS MIXING

**Figure 15-21** Dispersive mixing is slow across the symmetry plane of a blender, here a bin design. After 10 revolutions, a front view reveals clear evidence of the initial left–right distribution of identical but colored spheres in this particle-dynamic simulation. The full simulation is included on the Visual Mixing CD affixed to the back cover of the book.

as a result of an overall mean flow. Various stratagems, including the use of baffles, asymmetric cross-flow designs (referred to earlier), irregular rotation protocols, and axial rocking, have been introduced to mitigate this limitation. Notwithstanding these improvements, dispersion is the rate-limiting mechanism for mixing, and there is much potential for improvement of dispersive mixing.

Although convection is typically orders of magnitude more rapid than dispersion, the relative contribution of each mechanism to blending is strongly influenced by the initial distribution of species in the mixer. Thus, ingredients loaded in horizontal layers (as in Figure 15-20) can be mixed relatively rapidly, while ingredients layered side by side, either intentionally (as in Figure 15-21) or inadvertently (as a result of careless loading of a tumbler), will typically mix enormously more slowly.

To visualize this effect, in Figure 15-21 we display dispersive mixing of 8000 identical but colored grains loaded side by side, in a bin blender. With each successive revolution, only a few particles cross the interface separating the two symmetric halves of the tumbler, and as a result, after 10 revolutions the original particle ordering is still unmistakable. Systematic assays obtained from experiments of blending of realistic pharmaceutical excipients and actives confirm that imperfectly loaded blends retain any initial asymmetry for many hundreds of tumbler revolutions.

### 15-4.2 Convective Mixers

The second class of blenders commonly used in industrial applications is convective blenders. In contrast to tumbling blenders, convective blenders primarily mix by transporting material throughout a mixing vessel by the motion of a stirring device. A typical convective blender consists of a stationary chamber swept out by stirring mechanisms, such as rotating impellers. Convective blenders have a broad range of applications and can be used to blend components that cannot be adequately combined in tumbling blenders, such as materials that are prone to segregate or agglomerate. Many convective blenders can be designed to accommodate continuous rather than batch processing, further adding to their utility.

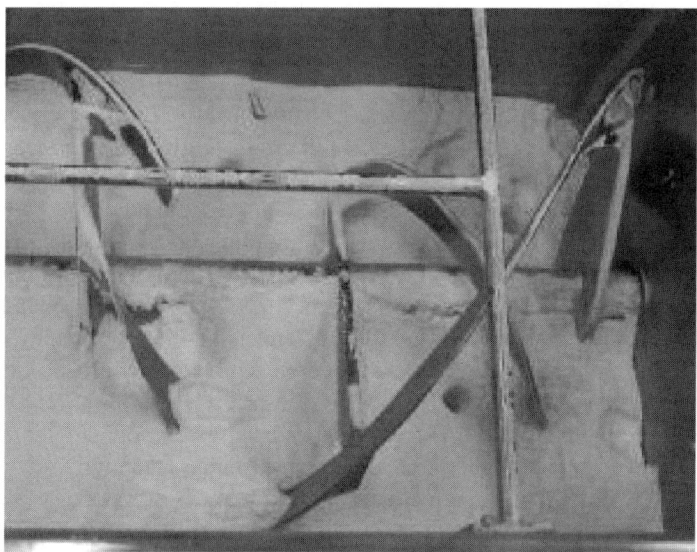

**Figure 15-22** Well-mixed powder bed in a five-spoke ribbon blender, with both sampled cores and regions blocked by mixing blades visible.

However, despite this versatility, we have only a very limited understanding of the dynamics and performance of convective blenders. This is due to both the inherent difficulty involved in characterizing powder and granular mixing and the fact that few researchers have investigated these devices (see, e.g., Masiuk, 1987). In fact, much of the limited quantitative analysis was performed over 30 years ago (Adams and Baker, 1956; Greathead and Simmons, 1957; Poole et al., 1964; Ashton and Valentin, 1966; Harnby, 1967; Williams and Khan, 1973) and suffers from many of the limitations and difficulties of powder bed and granular mixing characterization discussed elsewhere in this chapter. Convective blenders also pose special challenges to powder bed analysis, as the impellers may hinder or block acquisition of powder samples, rendering thorough and uniform sampling difficult. Figure 15-22 shows a blended powder bed in a ribbon blender, a common convective blender. Several cores have been extracted from this bed, but it is evident that certain locations were impossible to sample, due to blades obstructing that location. Process design and optimization utilizing convective blenders are therefore performed predominately on a case-by-case basis.

*15-4.2.1 Blending Mechanisms.* Powder blending in a convective mixer is accomplished primarily by convection and shear effects. The motion of the impeller initiates blending by transporting material from region to region within the mixing chamber. This motion also generates slip planes, which often results in nonnegligible shear mixing as well. Diffusive mixing typically has a minimal role in mixing in these blenders. This combination of mixing mechanisms produces superior results in many applications. Segregation of particles of different properties can be a major problem in tumbling mixers, potentially resulting in de-mixing or agglomeration (accretion of smaller particles into larger clumps). The motion of a convective mixing impeller both prevents the formation of and destroys any preexisting agglomerates. Convection is the mixing process least likely to result in pronounced segregation caused by difference in size or other physical properties. Convective mixers typically transfer relatively large amounts of material throughout the mixing vessel, allowing only limited opportunities for particles to segregate. This is in contrast to tumbling blenders, where the material has many opportunities to segregate during the tumbling/avalanching phase of the mixing cycle. In addition, mixing occurs throughout the chamber in a convective blender, whereas in a tumbling blender, mixing is typically confined to the surface of the powder bed, leaving large regions undisturbed during long periods of the mixing cycle.

While providing resistance to segregation and agglomeration, the mixing mechanisms employed by a convective blender can produce potentially adverse consequences. The motion of the impeller induce high levels or rates of shear, which may damage the material being mixed. Two other effects that may occur in a convective blender are attrition (grinding the powder into finer particles) or intense heating of the powder. Attrition is a common phenomenon in these blenders, as the shearing action of the impeller blades can cause rapid reduction in the individual particle size. This effect is sometimes desirable, for example, when designing a process to equalize the particle size of the material being processed. The motion of the impellers may aerate the mixture and cause the regions near the surface to fluidize. Convective blenders are typically operated at rotational speeds (15 to 60 rpm) and fill levels (>50%) similar to those of tumbling blenders. However, no systematic study of mixing performance in terms of mixing time, rotation rate, fill level, or loading patterns has been published to date for any convective blender.

*15-4.2.2 Applications and Types.* The design of convective blenders allows for the efficient processing of a wide range of material states, including powders, granular solids, slurries, liquids, pastes, and combinations of these. Consequently, these blenders are utilized in a wide range of industries, including construction, agriculture, chemicals, pharmaceuticals, and foods. Convective blenders come in a wide variety of designs, all following the basic design of a stationary vessel swept by a rotating impeller. The stationary mixing chamber may be conical, cylindrical, or trough-shaped, and impeller designs range from ribbons to paddles to screws. The impeller may sweep though the entire mixing chamber each rotation, or it may stir small regions progressively until the entire chamber is stirred. Common convective blenders include the ribbon blender (a cylindrical vessel with a helical ribbon impeller mounted to a horizontal shaft), the paddle blender (a modified ribbon blender with paddles instead of a helical ribbon), and the Nauta blender (a vertically oriented conical tank swept out by a rotating and precessing screw impeller). Other blenders used in industrial applications include the Forberg mixer (two paddle blender drives sweeping two connected troughs), the Z-blade blender (a cylindrical vessel swept out by a Z-shaped blade), and the Lodige (similar to a kitchen mixer, where plough-shaped shovels rotate within a cylindrical drum). Further discussion of these mixers is provided in Section 15-10.3.2.

*15-4.2.3 Ribbon Blender.* The ribbon blender is one of the most common general purpose mixers, as it is capable of effectively performing a wide range of mixing processes including liquid, solid, and liquid–solid blending. Common industrial applications of these blenders include mixing the powder components of pharmaceutical tablets, blending oils and shortenings into dry ingredients to form a cake batter, and combining gravel and asphalt. A batch ribbon blender is depicted in Figure 15-23. The motion of the ribbon blades near the vessel walls can result in *pinch points*, regions of high shear and compression which may damage fragile materials or cause attrition. The capacity of these blenders is set by the span of the ribbon, which must clear the top of the powder bed in order to mix the entire bed. As is true for many convective blenders, the intensity of shear can result in heating that can adversely affect the quality of the product.

During operation of a standard ribbon blender, two sets of helical ribbon blades transport material in opposite directions; the outer ribbons will transport material toward the center of the mixing vessel while the inner ribbons transport material toward the ends of the vessel (Figure 15-24a). Turbulent convective currents caused by these counterrotating elements act to blend the different components. Unlike many tumbling blenders, a ribbon blender is often not completely discharged by gravity, requiring additional blade rotation to complete this process. This can result in additional mixing, segregation, and attrition, which must be taken into account during process design.

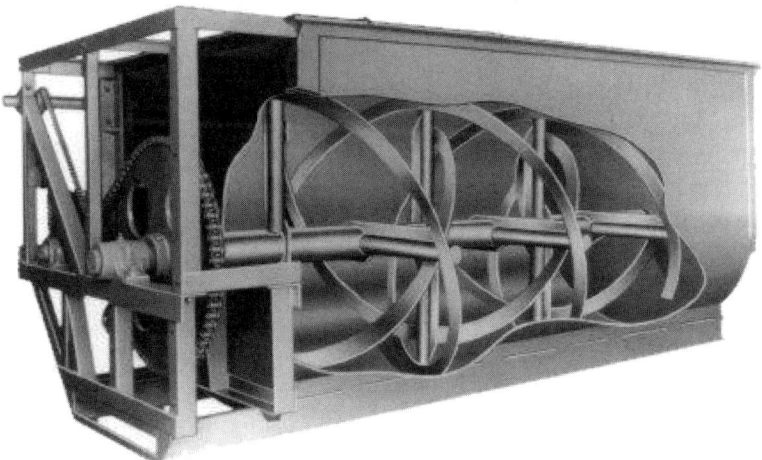

**Figure 15-23** Schematic of a ribbon blender, consisting of a cylindrical vessel swept out by a rotating helical impeller blade. (Courtesy of H.C. Davis Sons Manufacturing Co., Inc., Bonner Springs, KS, printed with permission.)

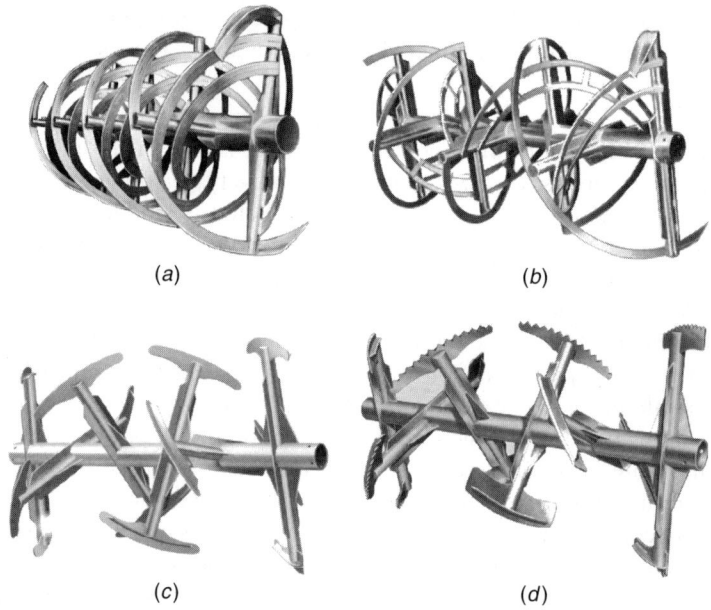

**Figure 15-24** Four convective blender impellers: (*a*) double ribbon agitator; (*b*) center-discharge ribbon agitator; (*c*) paddle agitator; (*d*) sawtooth paddle agitator. (Courtesy of H.C. Davis Sons Manufacturing Co., Inc., Bonner Springs, KS, printed with permission.)

Ribbon blenders are simple to modify for specific processes, and many refinements have been commercialized. The most common type of modification is to change the ribbon blade design; for example, Figure 15-24 shows some variations of the standard ribbon blade (Figure 15-24*a*). Figure 15-24*b* shows a common variation: the center-discharge ribbon. Here the two outer ribbons bring material to the center of the blender, while the inner ribbons force the ingredients outward to each end of the vessel. The paddle agitator (Figure 15-24*c*) contains both forward and reversing paddles in place of the smooth ribbon, constantly moving ingredients from one end to the other. To limit pinch points between the paddles and the blender surface, notches, or saw teeth, can be cut into the paddles (Figure 15-24*d*). Other types of modifications include creating a hybrid ribbon-paddle agitator or adding components to or removing sections of a ribbon to vary shear effects. Two agitators of the latter type are the cut-it-in ribbon agitator (a standard ribbon supplemented with cutting wires mounted on ribbon arms) used to cut thick materials (fats, oils, shortening) into powders (flour) and the cutout agitator (alternating sections of a standard ribbon are removed) used for heavier materials. Although there are many variations on the simple ribbon, these are all ad hoc, and there are few rigorous and usable scientific studies of these devices. See Section 15-10.3.2.3 for applications.

## 15-5 SELECTION AND SCALE-UP OF SOLIDS BATCH MIXING EQUIPMENT

As mentioned above, tumbling blenders can be grouped into two categories: convective blenders and tumbling blenders. Convective blenders rely on the action of impellers or paddles within a bowl, trough, cone, and so on, to move powders around and to generate a well-mixed product. Tumbling blenders consist of a hollow vessel attached to a rotating shaft; different blender types are identified by the geometry of the vessel. In either case, powders are mixed by rotating the blades or the vessel itself at a fixed rotation rate.

From a design and scale-up viewpoint, the major difference between the two types of blenders is the amount of shear imparted to the mixture during the blending process. In the absence of intensifier bars, tumbling blenders provide low-shear environments and are used when materials are shear sensitive or nonagglomerating. Convective blenders impart much more shear into the mixture and tend to be utilized for cohesive materials. Some tumbling blenders are equipped with a high-speed impeller, which can greatly increase the shear environment and allow for blending of some cohesive mixtures. From a manufacturing standpoint, tumbling blenders are often preferred because they come in a wide range of capacities and have shorter cleaning times. The choice of mixer often comes down to the properties of the mixture in question. Unfortunately, without well-established methods for measuring cohesion or agglomerating tendencies for different mixtures, it is impossible to develop a priori rules for blender choice based on the characteristics of the mixture materials.

In the remainder of this section we discuss recent findings regarding the scale-up of tumbling blenders, which have more easily classified flow fields and mixing mechanisms than convective blenders. As mentioned previously, the description of mixing mechanisms in convective blenders has not been the subject of considerable experimental investigation work, relegating scale-up considerations to trial and error.

### 15-5.1 Scaling Rules for Tumbling Blenders

The ensuing discussion will revolve around experiments run in 14, 56, and 300 L tote blenders using two mixtures: a free-flowing binary 50:50 w/w% mixture of 400 μm sand particles and a cohesive mixture of 3% micronized acetaminophen (~30 μm) in a 50:50 w/w% matrix of PH102 Avicel and Fast-Flo lactose. All experiments were run at 60% of blender capacity at a rotation rate of 10 rpm. In raw form the acetaminophen was extremely cohesive and agglomerates (up to $O(1\ cm)$ in diameter) formed in the bulk mixture. The effect of blender scale on the breakup of these agglomerates is an important consideration for scale-up of tumbling blenders.

Probably the most important rule governing the basic dynamics of mixing in tumbling blenders is that the loading protocol has the most direct impact on the mixing rate. When the blender is symmetrically loaded in a top/bottom fashion,

**918** SOLIDS MIXING

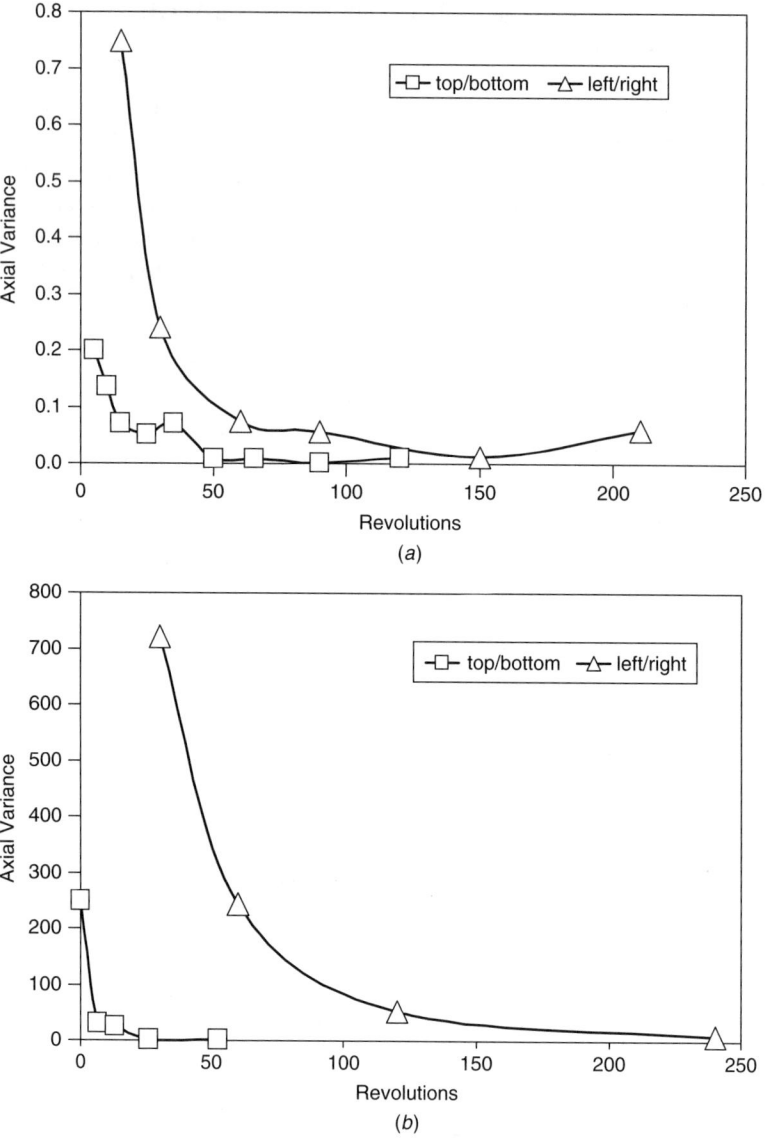

**Figure 15-25** Axial variance decrease for top/bottom- and left/right-loaded experiments in a 56 L tote blender with (*a*) a cohesive mixture and (*b*) a free-flowing mixture.

mixing rates can be more than an order of magnitude greater than when there are axial differences in the loading conditions. Figure 15-25 shows the decrease in axial variance (radial variance decrease, not shown, was nearly identical for both cases) for top/bottom- and left/right-loaded experiments using both the cohesive mixture and the free-flowing mixture.

Radial mixing rates (emphasized by top/bottom loading) are faster than axial mixing rates (emphasized by left/right loading) for both mixtures. However, the change in loading pattern has a much greater impact on mixing rates for the free-flowing mixture than the cohesive mixture. This disparity indicates that mixing mechanisms do not scale equivalently for the two mixtures, which can be a major consideration when changing vessel sizes.

*15-5.1.1 Scale-up of Axial and Radial Mixing Rates.* The effects of changing the vessel size on mixing rate is the predominant concern for the scale-up of manufacturing processes. Figures 15-26 and 15-27 compare the effect of changing blender size on the radial mixing rate (i.e., loaded top/bottom) and axial mixing rate (i.e., loaded left/right) for both cohesive and free-flowing mixtures.

For radial mixing rates, the change in vessel size has almost no impact on the observed mixing rate, regardless of mixture characteristics (variability in the cohesive data is caused by the capture of agglomerates). On the other hand, axial mixing rates are significantly different for the free-flowing mixture but nearly identical for the cohesive mixture. These differences in the scaling of the axial mixing process can be attributed to the difference in the way that these two mixtures flow in a tumbling blender.

Generally, when a free-flowing mixture is rotated in a tumbling blender, there is a regular flow, characterized by a nearly flat axial surface (i.e., there is little variability in bed height perpendicular to the mean flow), and particles travel along path lines nearly perpendicular to the axis of rotation. Cohesive mixture flow displays completely different behavior: flow is characterized by a series of dislocations that mark the onset of flow for a discrete portion of the mixture at various locations. These failures occur at seemingly random positions in the mixture and contain variable amounts of material. Thus, the surface of the cohesive mixture is marked by many hills and valleys, and flow down the cascade is rarely straight or perpendicular to the axis of rotation. Flow of cohesive material has an inherent axial component that greatly enhances axial mixing of the powder. Hence, the width (capacity) of the vessel does not play a significant role affecting mixing rates for cohesive mixtures, in stark contrast to free-flowing materials that rely on dispersion as the major mechanism for axial mixing. In this case, vessel width plays a dominant role.

*15-5.1.2 Shear Effects on Mixing.* Free-flowing mixtures do not exhibit significant particle–particle interactions, and the presence of shear, or lack thereof, has little or no effect on the mixing rate. It has been shown that mixing of free-flowing particles in tumbling blenders is independent of rotation rate and only a function of fill level and number of revolutions. However, for cohesive mixtures, the presence of shear can have a major impact on the homogeneity of the final blend.

Shear effects are demonstrated in Figure 15-28, which shows the evolution of the sample mean (mean value of all retrieved samples) from the cohesive mixture for top/bottom-loaded experiments in 14, 56, and 300 L blenders. To

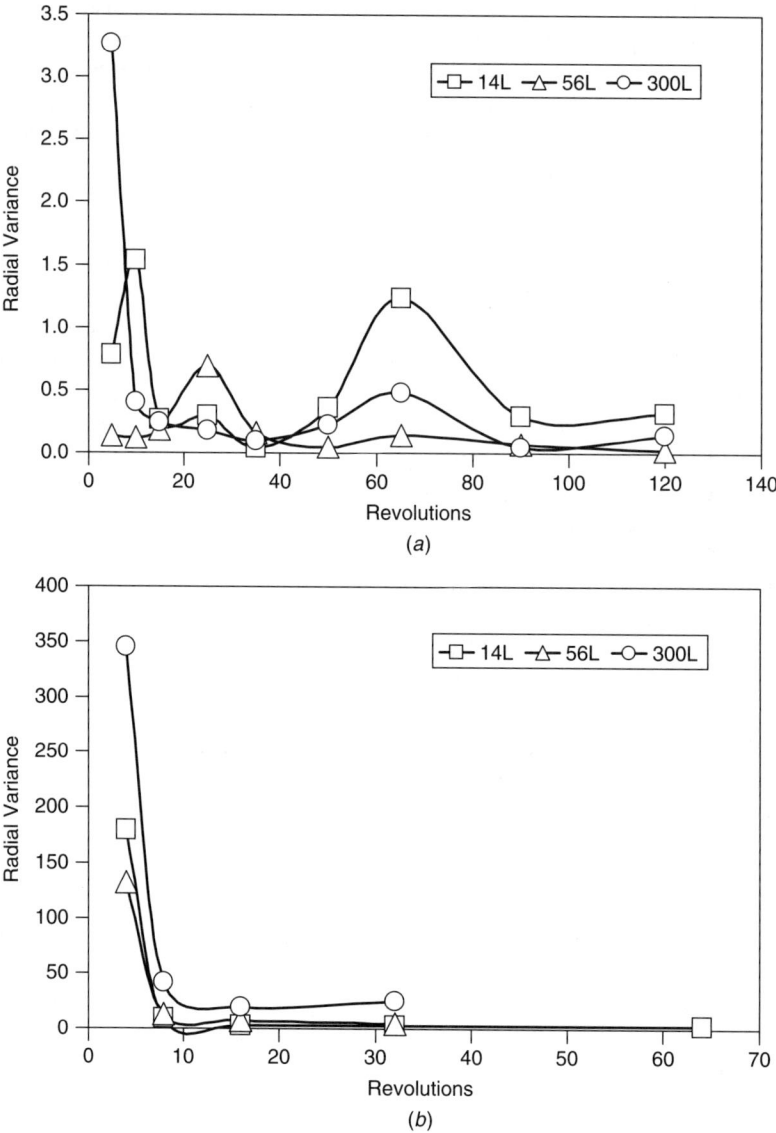

**Figure 15-26** Radial mixing rates for top/bottom-loaded (*a*) cohesive and (*b*) free-flowing systems in tote blenders of 14, 56, and 300 L capacity.

highlight the effect of shear, the acetaminophen was loaded into the blender without presieving, so that numerous agglomerates were initially present in the mixture. Sample means for the smallest blender (14 L) reach a maximum of ~1.8% acetaminophen, indicating that a significant amount of acetaminophen was still trapped in clumps too large to be sampled. For larger vessels, the sample

SELECTION AND SCALE-UP OF SOLIDS BATCH MIXING EQUIPMENT    **921**

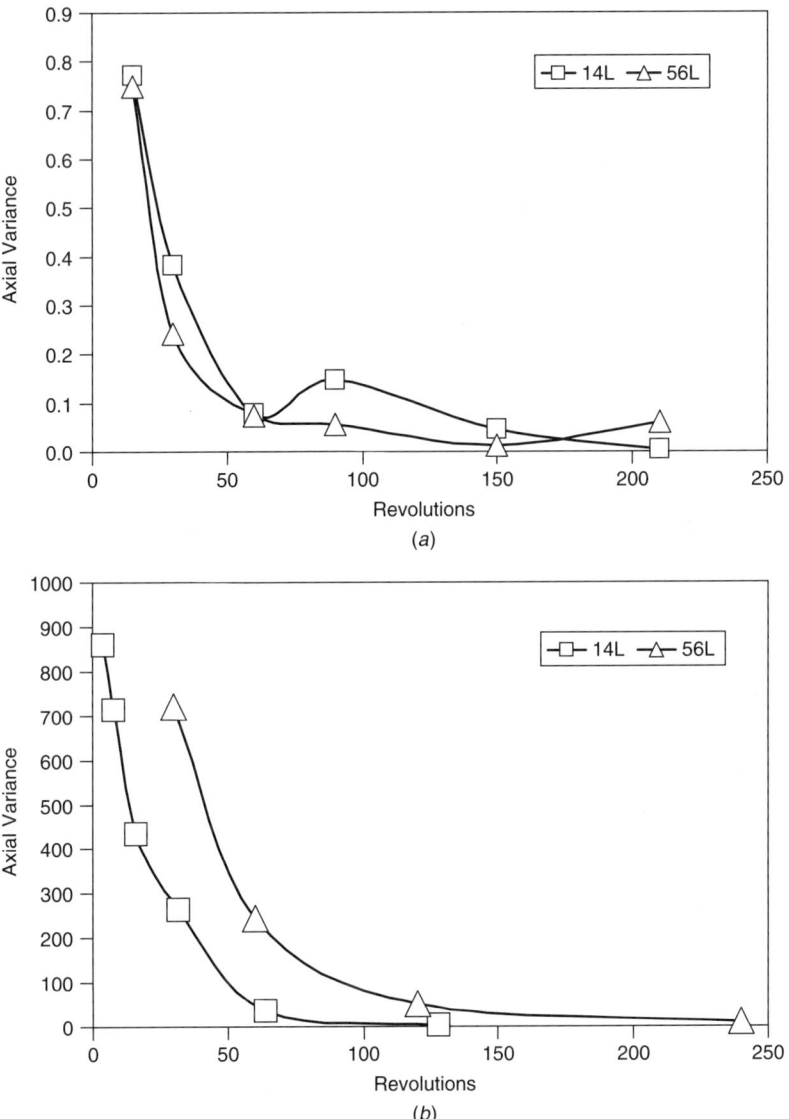

**Figure 15-27** Axial mixing rates for left/right-loaded (*a*) cohesive and (*b*) free-flowing systems in tote blenders of 14 and 56 L capacity.

mean approached the mixture mean more rapidly; after 120 revolutions in the 300 L blender, the sample mean was 2.9%. The only mechanism for agglomerate breakup was through shear-induced diminution in the cascading layer.

All experiments were run at the same rotation rate, making it appear that shear energy scales with vessel size at constant rotation rate. Particle velocities in the flowing layer have been shown to scale with the radial length of the blender

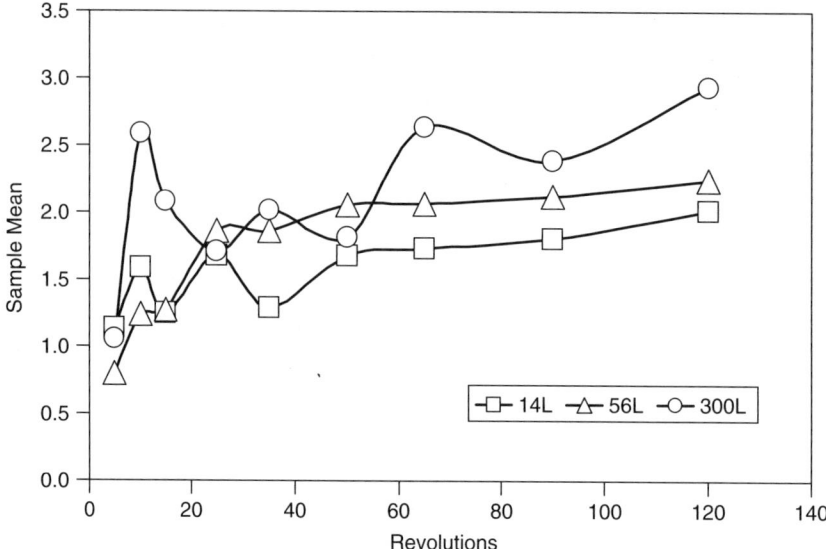

**Figure 15-28** Evolution of the mean of all retrieved samples taken from 14, 56, and 300 L tote blenders. Larger blenders come closer to the actual mean value of 3%.

at constant rotation rate. Using particle velocity as a rough estimate for shear rates supports the idea that increased vessel size at constant rotation rates leads to increased shear and hence increased acetaminophen agglomerate dispersion into the bulk mixture. This finding can be enormously important for scale-down of mixing processes involving cohesive mixtures because agglomerate breakup, which is necessary for the achievement of a well-mixed final product, might occur more slowly (or not at all) in a smaller vessel. Shear energy can also be affected by changes in rotation rate, fill level, and the use of high-speed impellers. To frame scale-up rules properly for cohesive mixtures, it will be necessary to devise a means for quantifying changes in shear with these operational variables.

## 15-5.2 Final Scale-up and Scale-down Considerations

Generally, mixture characteristics (rather than blender characteristics) play the determining role in scale-up of tumbling blenders. For free-flowing mixtures, changing vessel size has an enormous impact on axial mixing rates, making the loading conditions the most important variable affecting the change in mixing rates with changes in scale. For cohesive mixtures, the amount of shear in the mixing process has the greatest effect on the mixing rate because the breakup and dispersion of agglomerates is necessary for creating well-mixed blends. Thus, one must be cautious of *increases* in vessel size for free-flowing mixtures, but for cohesive mixtures it is *decreases* in vessel size that pose the most problems. It is clear that the processes that control mixing in these devices will not be fully understood until quantitative methods for measuring cohesion of powder mixtures are developed.

## 15-6 CONCLUSIONS

Despite sustained efforts during the past decade both at Rutgers and elsewhere, powder blending remains largely an "art," governed by empiricism and subject to frequent failure. In the opinion of the authors, the chief limitations in advancing the scientific understanding of powder mixing is a lack of effective experimental methods for measuring powder flow and powder constitutive behavior. Lacking hard data, constitutive modeling remains in its infancy, severely limiting our ability to achieve effective models for predicting powder flow and mixing from first principles. This situation is in stark contrast with the state of the art in fluid flow and mixing, which is discussed abundantly in the remainder of this book. For fluids, well-established methods for measuring constitutive behavior have greatly advanced our understanding of fluid rheology, which is a major building block of CFD models. Excellent EFD technology facilitates validation of CFD efforts. The net result is that fluid mixing systems can be designed with increasing reliability, and in many cases, entirely by computer. Although limitations exist, the fluids community is taking them by storm, one after another, at a rapid pace.

Thus, to advance beyond the current state of the art, we need to improve experimental techniques for measuring powder flow and powder constitutive behavior. Important efforts, mentioned earlier in this chapter, are under way using a variety of noninvasive technologies, but much remains to be done before powder blending processes can be designed and scaled-up reliably. The reader is advised to stay tuned, as the field is likely to evolve rapidly in coming years.

## ACKNOWLEDGMENTS

The work discussed here has been largely performed at Rutgers University with support from multiple sources, including the National Science Foundation, New Jersey Commission on Science and Technology, Kraft, IFPRI, Merck, Pfizer, Tor Pharm, Gallay Systems, Patterson-Kelley, Bristol-Myers-Squibb, Schering-Plough, M&M Mars, 3 M, ExxonMobil, Abbott Laboratories, Johnson & Johnson, DuPont, Dow, Ganes Chemicals, and National Starch and Chemical Company, among many others. Their support is gratefully acknowledged.

Many graduate students, postdoctoral researchers, and industrial collaborators have participated in this work in many ways over the years, and their contribution is acknowledged in the cited literature. The authors wish to express their special gratitude to dozens of Rutgers undergraduate students who tirelessly, and often anonymously, collected and analyzed the hundreds of thousands of samples that compose the data set and supports the observations and conclusions communicated in Part A of this chapter.

# Part B: Mixing of Particulate Solids in the Process Industries

*Konanur Manjunath, Shrikant Dhodapkar, and Karl Jacob*

## 15-7 INTRODUCTION

Solid–solid mixing is a ubiquitous unit operation in particulate processes where consistency and homogeneity of the product is a key requirement. Mixtures are seen in all different phases of processing, ranging from the mixture of raw materials, as is the case with glass batch, ceramics, and blast furnace charges to the final formulation of products such as cereal mixtures, cosmetics, and polymer master batches. The quality of mixing is often key to product performance. For example, good mixing of cement and aggregate, color concentrate and base polymer, active ingredient in tablet formulations, individual components of a fertilizer formulation, and various components of a cereal mix is key to successful application.

Mixing can also be done in combination with other unit operations, such as agglomeration, size reduction, particle coating, and chemical reaction, to name a few. Often, selection of an appropriate unit operation (e.g., grinding, agglomeration) can reduce the cost of the process by combining two unit operations.

There is a wide variety of solids mixing equipment on the market today. Equipment ranges in size from small mixers that hold a few hundred pounds to large silos for large scale blending operations. It can be seen that certain mixers have long-standing domination of certain market segments, and improved designs have resulted from close cooperation with the end user. Newer concepts, such as Forberg mixers, continue to be introduced and accepted as the applications gain ground.

Selection of an appropriate mixer begins with an understanding of process requirements (e.g., quantity of bulk solids to be mixed, desired degree of mixedness, the need for batch integrity, upstream/downstream process) and material properties (e.g., particle size distribution, cohesiveness, particle shape, abrasiveness). The final analysis must then combine these factors with the operating and purchase cost to arrive at an engineering decision.

In this part of the chapter we cover solid–solid mixing as routinely practiced in the industrial environment, from small scale batch blending to continuous homogenization of polymer pellets. An effort is made to compile typical operating ranges and practical guidelines from various sources and the authors' experience. The following text is a summary of currently acceptable industrial practices; however, it is possible that certain segments of the industry may have variant practices. The focus here is also primarily on mixing of solids in the "dry" state. The subject of mixing/wetting of powders to form suspensions, slurries, and pastes is not treated here. A detailed discussion of tumbling mixers and a fundamental treatment of mixing phenomena was covered in detail in Part A of this chapter.

## 15-7.1 Scope of Solid–Solid Mixing Tasks

A myriad of different mixing tasks are performed in the process industries today. Some common tasks are:

- Mixing of product for homogenization of quality or reduction of variance (e.g., blending of polymer pellets, blending of batches in a lot)
- Mixing of active ingredient onto a carrier material (e.g., formulation of insecticides or herbicides for household applications where the carrier particles are clay granules or other inerts)
- Mixing of multicomponent mixtures as a formulation (e.g., cereal mix, specialty polymers)
- Coating of a cohesive component onto a carrier particle (e.g., coating of antiblocking agents on polymer pellets or granules, formulation of agricultural products)
- Mixing of fine powders to create a homogeneous mixture at the particulate level (e.g., masterbatch preparation for medicinal drugs)
- Coating of liquid additives onto the base material (e.g., waxes or additives on polymer granules, food applications)

The nature of application and process requirements will dictate the selection and specification of mixing equipment.

## 15-7.2 Key Process Questions

There are numerous factors that govern the successful mixing of bulk solids. Although it is difficult to make a complete and absolutely comprehensive list, there are several key process questions that deserve attention prior to mixer selection.

- What are the desired mixture quality and acceptable variation? This is a key process issue. It is extremely important to determine the acceptable mixture quality and variation early in the process of selection of a new mixer or while troubleshooting an existing mixer.
- What quantity of material is to be mixed or homogenized?
- What is the nature of the process? Is the process more conducive to batch or continuous mixing?
- Are there other unit operations, such as grinding, granulation, and drying, which can be combined with the mixing operation?
- What are the consequences of product degradation and cross-contamination? Does the equipment need to be cleaned frequently?
- Is the production rate constant? What kind of turn-up or turn-down is needed?

- Does the mixing process need to be contained because of industrial hygiene and safety concerns?
- What are the safety issues related to rotating piece of equipment or fire and explosion hazards?
- Does the process need to have on-line control? What is the extent of automation required?
- What are the constraints from space and accessibility perspective?
- What is the cost to benefit analysis?

## 15-8 MIXTURE CHARACTERIZATION AND SAMPLING

The ability of particles to mix and their tendency to segregate depend on differences in their size, density, shape, elastic properties, surface characteristics, and magnitude of interparticle forces. The difference in particle size is probably the most important factor. Unlike immiscible liquid systems, the density differences play a relatively minor role in de-mixing or segregation of particulate mixtures. The large body of literature available on fluid mixing therefore cannot be used to predict or evaluate solids mixing applications.

### 15-8.1 Type of Mixtures

Terminology associated with mixture classification is mired in controversy (Egermann, 1980; Thiel, 1982; Nienow et al., 1985; Hersey, 1975). Previously accepted terms have come under scrutiny as our understanding of mixing mechanisms has improved over the years. There is a serious lack of homogeneity and consistency in the terms used in the literature. An effort is made here to present basic concepts and various viewpoints on mixture classification.

*15-8.1.1 Perfect Mixture.* A perfect mixture of two types of particles is one in which any sample randomly taken from the mixture will contain the same proportion of each particle as the proportions present in the mixture taken as a whole. As shown in Figure 15-29a, alternate arrangement of black and white particles will create a perfect mixture. Such perfect mixtures are rarely found in nature.

*15-8.1.2 Random or Stochastic Mixture.* When two noninteracting components (e.g., free-flowing pellets) with similar properties (size, shape, elasticity, etc.) are mixed in an ideal mixer, the quality of mixing reaches an asymptotic limit of random mixing (see Figure 15-29b). This is a statistical process or probabilistic process. Each square in Figure 15-29b has a 50:50 chance of being black or white. Extended mixing of random mixtures does not result in improvement of mixing quality. For particles with different physical properties, it is not always possible to achieve a random mix. A random mixture cannot be achieved in the presence of significant interparticle forces (e.g., van der Waals, electrostatic, cohesive).

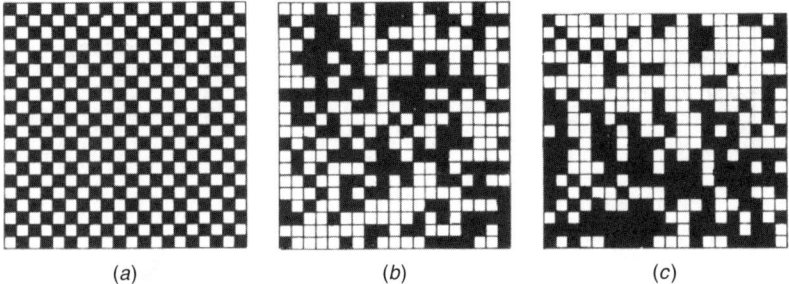

**Figure 15-29** Simulated mixtures: (*a*) perfect mixture; (*b*) random mixture; (*c*) segregated mixture. (From Williams, 1986.)

*15-8.1.3 Ordered Mixture.* When two interacting components are mixed together, a structure or order can build up into the mixture. The units ordered could be a result of agglomeration or cohesion of one component to the other or a mixture of the two. A perfectly ordered mixture can be obtained by:

1. Applying sufficient energy to break any agglomerate of the cohesive fraction (minor component) and distributing it on available sites on the carrier particles. The available sites should be sufficient to hold the cohesive fraction; otherwise, the cohesive particles can reagglomerate.
2. Ensuring complete randomization of carrier particles.

It should be noted that an ordered mixture can have a variance less than that of a random mixture. A perfectly ordered mixture will have zero variance.

*15-8.1.4 Partially Ordered Random Mixture.* Once the cohesive fraction or minor component saturates available sites on the carrier particles, the remaining fraction will agglomerate. In partially ordered random mixtures, these agglomerates and the ordered units are randomly mixed together.

*15-8.1.5 Pseudorandom Mixture.* When the ordered units contain different number of adherent particles and the carrier particles are randomly mixed, the mixture is called a *pseudorandom mixture*. The carrier particles are not saturated with the minor component, and there are no agglomerates in the mixture.

The following illustration (Stainforth, 1982) in Figure 15-30 is an excellent summary of various types of mixtures and the influence of surface forces.

*15-8.1.6 Alternative Definitions.* Egermann (1980, and Table 15-1) proposed that the nomenclature reflect the degree of homogeneity of the mixture rather than the underlying mechanisms. He proposed the following definitions:

- *Ideally ordered or perfect mixtures* are defined by a standard deviation of the sample composition equal to zero provided that the sample size is greater than one ordered unit.

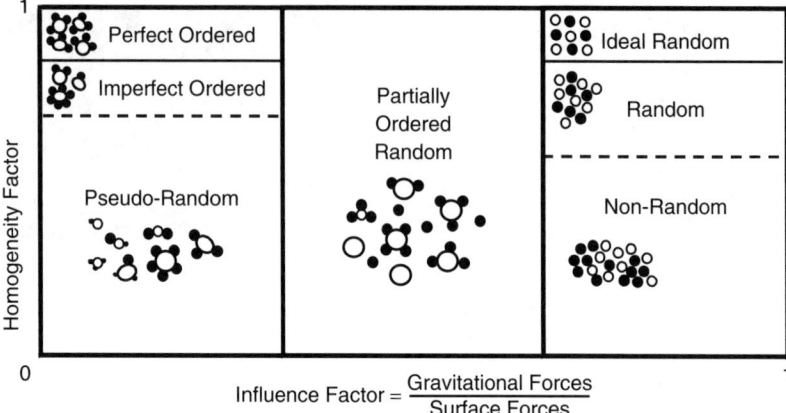

**Figure 15-30** Representation of relationship between various mixtures. (From Stainforth, 1982.)

**Table 15-1** Summary of Mixing Nomenclature

| Type of Mixture | Homogeneity | Some Degree of Order? |
|---|---|---|
| Ideally ordered (perfect) | $\sigma$ equal to 0 | Yes |
| Ordered | $\sigma$ smaller than $\sigma_R$ | Yes |
| Pseudorandom | $\sigma$ equal to $\sigma_R$ | Yes |
| Random | $\sigma$ equal to $\sigma_R$ | No |
| Incomplete | $\sigma$ higher than $\sigma_R$ | Yes |

*Source*: Egermann (1980).

- *Ordered mixtures* feature a higher degree of homogeneity than random mixtures. The standard deviation will be smaller than random standard deviation.
- *Random mixtures* show a completely disordered distribution of the individual particles in the absence of interparticle interactions.
- *Pseudorandom mixtures* show the degree of homogeneity but not the fully disordered texture of random mixtures.
- *Incomplete or segregated mixtures* are of poorer quality than random mixtures.

### 15-8.2 Statistics of Random Mixing

If random samples of N particles are taken from the known mixture of average composition q, where the fraction of the first component is P and second component is $1 - P$, the composition of the samples will be normally distributed with a standard deviation of

$$\sigma = \sqrt{\frac{P(1-P)}{N}} \tag{15-6}$$

As can be seen, as the sample size decreases, the variance or standard deviation increases. In a mixture of total mass W, if m random samples are analyzed for concentration of component 1 (concentration $c_1, c_2, c_3, c_4, \ldots, c_m$), the sample mean can be calculated as

$$\overline{X} = \frac{\sum_{i=1}^{m} c_i}{m} \tag{15-7}$$

The standard deviation of a set of samples is given by

$$s = \sqrt{\frac{(c_i - \overline{X})^2}{(m-1)}} \tag{15-8}$$

It is assumed here that the sample concentrations are normally distributed about the mean.

It is known that multiple samples taken from a bulk (population) will give a distribution of measurements. For completely random mixture, such as a mixture of two free-flowing granular materials of equal size and density, the variance can be calculated theoretically as

$$\sigma^2_{\text{random mixture}} = \frac{P(1-P)}{N} \tag{15-9}$$

where P is the fraction of one of the components in the mixture and N is the number of individual particles.

For nonrandom mixtures, Williams (1986) found that the relationship above no longer holds. The relationship between sample size and variance is

$$\sigma^2_{\text{nonrandom mixture}} = \left[L + \frac{P(1-P) - L}{N}\right]^2 \tag{15-10}$$

Figure 15-31 shows the effect of sample size on the standard deviation for a nonrandom mixture. L is a constant for a given mixture or state of mixedness. It can be determined experimentally if the value of $\sigma$ is known at one value of N. The condition of $L = 0$ corresponds to random mix:

$$\lim_{L \to 0} [\sigma_{\text{nonrandom}}] = \sigma_{\text{random}} \tag{15-11}$$

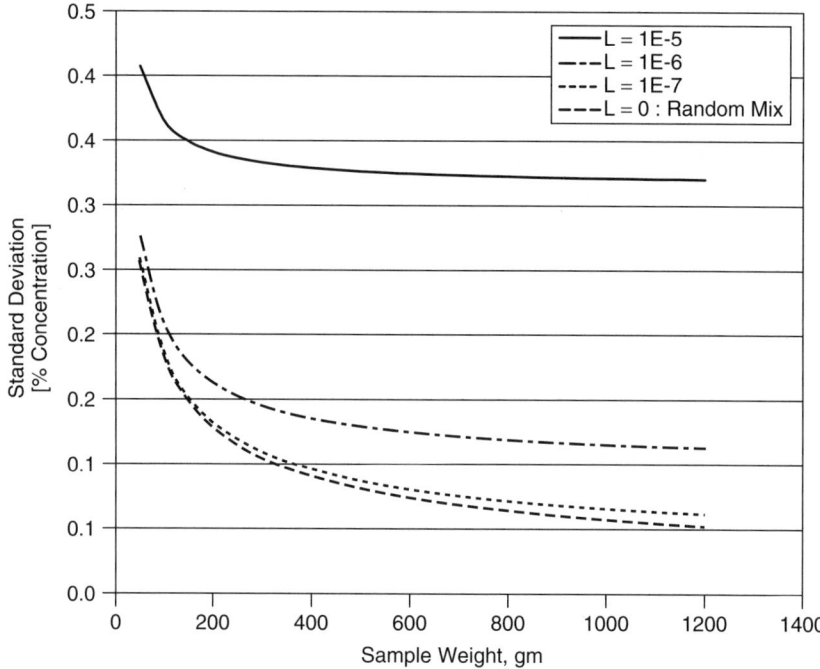

**Figure 15-31** Effect of sample size on mixture standard deviation for nonrandom mixtures.

For a system where the two components are completely unmixed, the variance $\sigma_o^2$ of the sample composition is

$$\sigma_o^2 = P(1 - P) \tag{15-12}$$

A completely unmixed system can be visualized as filling a jar first with component 1 and then topping it with component 2 without disturbing the layer of component 1. Note that the expression above is independent of the sample size. It is also assumed here that no sample straddles the boundary between the two components.

The equations discussed so far assume that all the particles in the mixture have the same size. However, if we have a binary random mixture in which each component has a particle size distribution, it is necessary to calculate the number of particles of each component independently. Stange's (1954) derivation has been used widely:

$$\sigma_R = \sqrt{\frac{P(1-P)}{W/[Pw_P(1+C_P^2) + (1-P)w_{1-P}(1+C_{1-P}^2)]}} \tag{15-13}$$

where W the total weight of the sample, P is the mass fraction of the first component corresponding to $\sigma_R$, $w_P$ the weight fraction of the first component, $w_{1-P}$ the weight fraction of the second component, and $C_P$ the coefficient of variation of the particle weight of the first component, $C_{1-P}$ the coefficient of variation of the particle weight of the second component.

### 15-8.3 Interpretation of Measured Variance

The total measured variance for a set of samples taken from bulk (population) is summation of contributions from various sources: namely,

$$\sigma^2_{measured} = \sigma^2_{mixture} + \sigma^2_{sampling} + \sigma^2_{analytical} \tag{15-14}$$

The sampling and analytical variances must be determined from independent measurements. It is well established that once the analytical uncertainty (standard deviation) is reduced to a third or less of the sampling uncertainty, further reduction in analytical uncertainty is of little importance. Therefore, if the uncertainty in sampling is very large, it may be beneficial to opt for an analytical method that is rapid even though it might have lower precision. This will permit more samples to be analyzed, thereby resulting in a better estimate of the mean value.

### 15-8.4 Sampling

To evaluate the state of mixedness of a mixture, a representative sample must be retrieved and analyzed. The result of this analysis combines errors due to sampling, analytical method, and uncertainty due to state of the mixture. Following certain guidelines and good practices can minimize the error due to sampling and analysis. The subject of sampling is very complex and detailed treatment can be found in the literature (Hersey, 1970; Kristensen, 1973; Sommer, 1986; Muzzio et al., 1997). The objective of this section is to outline important issues and provide practical guidance.

Some common questions pertaining to sampling encountered during mixer selection and performance evaluation are:

- Where should the samples be obtained?
- How should the optimal sample size be determined?
- How often should samples be collected?
- What apparatus is available for sampling from stationary material and from moving (process) streams? How should the appropriate unit be selected?
- How should the gross sample be reduced to analytical size with minimal bias?

***15-8.4.1 Sampling Location.*** The selection of sampling locations depends on the objective of the study and mode of mixer operation. The objective of the mixing study could be (1) selection of a mixer for a given process, or (2) evaluation of the performance of mixer in the process. A mixer could be operated in batch or continuous mode.

For selection of a mixer, it is important to understand the spatial and temporal variability of the characteristic property within the mixer. The entire volume of the mixer must then be spatially divided and samples randomly taken. Knowledge of flow patterns within the mixer is helpful to identify the location where material is likely to be stagnant.

For performance evaluation, it is recommended that samples be taken at the discharge spout. From a process perspective, it is important that the mixture coming out of the mixer is homogeneous and no segregation occurs during discharge. It is possible for a well-mixed sample to de-mix during discharge if proper care is not taken. Proper care implies that one is cognizant of the segregation mechanisms possible for the mixture and implementation of proper engineering controls to control it.

One must follow the two *Golden rules of sampling* as proposed by Allen (1981):

1. Sample a moving stream.
2. The whole of the stream should be taken for many short increments of time in preference to part of the stream being for the whole of the time.

***15-8.4.2 Selection of Sample Size.*** Ideally, the size of a sample should be equal to the scale of scrutiny of the mixture. The scale of scrutiny is the scale (or amount) of material at which homogeneity is desired. For instance, for certain pharmaceutical applications, the size of a single tablet is the scale of scrutiny where it is critical to ensure that the active ingredient is well mixed to the level of single tablet. Similarly, for polymer extrusion processes, the scale of scrutiny is the volume of the polymer mixing zone in the extruder. For agricultural chemicals, one or a number of bags of fertilizer could be considered as the appropriate scale. Determination of scale is independent of the property that is being scrutinized.

When it is not possible to obtain samples comparable to the ideal size, appropriate sample size reduction techniques must be employed to obtain representative sample for analysis. For random mixtures, the sample variance is inversely proportional to the sample size. For a certain composition of noninteracting particulate and a given sample size, there exists a minimum theoretical value of standard deviation that can be achieved through random mixing. If the chosen sample size is very small, one must take a large number of samples to reduce the uncertainty in determination of mean mixture property (Student's t-test).

***15-8.4.3 Number of Samples or Sampling Frequency.*** The value of sample variance approaches mixture variance as the number of samples becomes very large (following chi-square statistics). For batch mixers it is common practice to stop the mixer and sample the stationary bed at various locations. Since the mixture mean or standard deviation is not known a priori, historical data must be used as guideline. In continuous mixers, a sample can be obtained at the mixer outlet following the golden rules of sampling. Care must be taken to avoid any long-term cycling of the process variables. In practice, the number of samples, is limited by the capability of the analytical technique to process the samples. For online measurement devices, such as those that measure moisture, higher sampling frequency can be implemented.

### 15-8.4.4 Samplers and Their Selection.

Although it is highly recommended that the golden rules of sampling be followed, sampling situations from a stationary bed or pile are unavoidable. If the total quantity of a stationary bed or pile is not very large, a chute riffler or spinning riffler should be used to obtain a representative sample. If the stationary bed is large or while sampling from a mixer, one may use a sampling thief, pneumatic lance, or a scoop. Each of these methods results in biased sample and must be used with caution (Allen, 1981). See also Section 15-2.2.

As mentioned earlier, it is best practice to sample a mixer at the discharge location. A number of online samplers are available in the market: whole stream samplers, cross-cut samplers, and split-stream samplers. The following factors must be considered for selection of an appropriate sampling device:

1. *Flowability of the material* (cohesive versus free-flowing). The sample must be capable of flowing into the sampler.
2. *Maximum particle size.* The sampler must be sufficiently large to accommodate the maximum particle size.
3. *Friability of the material.* The sample must not be crushed in the sampler.
4. *Size of sample desired.* The sample size should match its intended use.
5. *Availability of space.* Because of their large size, some samplers cannot be fit into an existing process.

### 15-8.4.5 Sample Size Reduction.

The sample obtained from the sampler is generally larger than the sample required for analysis. If a small portion of this sample is arbitrarily scooped for analysis, it will introduce a bias. The sample size can be reduced to the analytical size using various sample reduction techniques: (1) spinning riffler, (2) chute riffler, (3) ICI method, and so on. All these methods follow the golden rules for sampling. For details on these sample reduction techniques, see Allen (1981).

## 15-9 SELECTION OF BATCH AND CONTINUOUS MIXERS

In this section the distinction between batch and continuous mixers is discussed so that appropriate selection can be made to suit both the process and mixture requirements [see also Brennen (1990) and Michael (1992)].

### 15-9.1 Batch Mixing

Batch mixing is mixing ingredients in any amount in individual batches in an individual mixer or a vessel. All ingredients are loaded into a mixer and agitated for a certain period until they are homogeneously distributed or mixed. The resulting mixture is then discharged out of the vessel. The critical parameters that influence the selection of such mixers is the mixing duration, the size and the geometry of the mixer, and the operating conditions.

### 15-9.2 Continuous Mixing

Continuous mixing is used to mix ingredients continuously in a mixer in a single pass. The ingredient quantity to be mixed may vary in any range; however, unlike batch mixing, care must be taken to feed the mixer in a controlled fashion. Mixing in batch often leads to variation in the mixing quality, which can be controlled or almost eliminated by continuous mixers. Even though continuous mixing is gaining popularity, selection of continuous mixers is not as straightforward as batch, which can be accomplished by running trials.

Continuous mixers are more compact than batch mixers. While discharging from the mixers, segregation can be reduced by positioning the discharge closer to packaging units or as an integral part of it. In continuous mixing, mixing has to be achieved in both the radial and axial (in the direction of conveying) directions.

### 15-9.3 Comparison between Batch and Continuous Mixing

*Advantages of Batch Mixing*

- Any type of powders either cohesive or free flowing can be well mixed.
- Loading either powder or liquid ingredients into the mixer is straightforward.
- Mixers are easier to maintain and clean.
- It is easier to identify a batch for further follow-up (batch integrity).

*Disadvantages of Batch Mixing*

- It is not suited for minor ingredients, particularly at very low loading, since the ingredient can get coated onto the vessel.
- Segregation is unavoidable and may be severe if the packaging is located far from the discharge.
- If changing ingredients, the mixer must be cleaned after every batch, work that is labor intensive.

## Advantages of Continuous Mixing

- *High capacity.* Compared to batch types, continuous mixers of small capacity and power can be used to produce large quantities of a mixture. Hence, for a given capacity, they are more compact than batch mixers.
- *Efficient dispersion of minor ingredients.* Mixing is very intense and thorough. Minor ingredient can be mixed more effectively.
- *Low hold-up.* The residence time in continuous mixers is much less and hence the holdup in continuous mixers is usually low.
- *Suitability for automatic control.* These systems are suitable for the application of online instrumentation and quality control. It is also possible to carry out formulation and rate changes to match the performance to the process requirements.
- *Minimum segregation.* Continuous mixing can cut down and control segregation of products because it can be located physically close to the next unit operation. If a batch mixer is selected for such a case, one could expect segregation during discharge of the product from the blender and subsequent handling of the mixture.
- *Low cost.* Continuous mixers tend to be cheaper than the equivalent batch mixer because they are compact and require less space. However, the cost for the feeders for metering the product into the mixer could be higher, especially if the number of components being mixed is large.
- *Minimum labor.* Since filling and emptying goes on automatically, minimal labor is required.

## Disadvantages of Continuous Mixing

- *Lack of flexibility.* The continuous mixer is designed for a particular application and it cannot easily be tailored to mix a number of different ingredients unless the necessary facilities are built into the system at the outset. The effective turndown for batch mixer is quite high, and it is easier to vary the production rate. Continuous mixers generally have feeders that are precisely controlled, and it is not easy to change these narrow limits around the feed rate. If a new ingredient is introduced, this calls for a change in the protocol, and the feeders have to be calibrated to suit the application in hand.
- *Equipment break down.* Continuous mixers depend on many other pieces of equipment, such as the metering feeders and the monitoring devices to function at the level desired. If any piece of equipment fails to operate reliably, the entire mixing process is affected. Once the feeders are set to deliver the desired mixture quality, the system must be left undisturbed except for the purpose of calibration.

- *Calibration and checking.* The feeding devices require careful calibration, particularly if the process is being operated within a narrow range. To have good mixture quality, all associated equipment must be properly maintained.
- *Batch integrity.* This is something well suited with batch mixing operations. In continuous mixing, bad batches of incoming products get mixed, and it can be very difficult to spot the "off-specification" product.
- *Good sampling procedures.* To keep track of the mixture quality, one has to have an inspection routine for the mixture and for the metered feeds to the mixer. Variation in the input quality and quantity will affect the final mix.
- *Component limitation.* Since the final mixture is a function of the feed products, one will have to specify feeds in a very tight range, and this will increase the cost of the final mixture. If multiple components must be fed, batch is mostly preferred.
- *Controlled feeding.* Controlled feeding of input materials is required, and this will increase the capital costs for such mixers. The throughput is more or less fixed and is difficult to vary without significant modification of the expensive feeding and controlling devices.

### 15-9.4 Selection of Mixers

***15-9.4.1 Selection of Batch Mixers.*** A selection chart for batch mixers is shown in Figure 15-32. The chart, however, excludes consideration of properties such as abrasiveness of the product, the buildup of static charge in the mixer, friability of ingredients, and so on.

***15-9.4.2 Selection of Continuous Mixers.*** A selection chart is shown in Figure 15-33 for deciding whether batch or continuous mixers are suitable, based on criteria proposed by van den Bergh.

### 15-10 FUNDAMENTALS AND MECHANICS OF MIXER OPERATION

See also discussions by Bridgewater (1976), Williams (1986), Fan et al. (1990), and Harnby et al. (1992).

### 15-10.1 Mixing Mechanisms

There are three underlying mechanisms for solids–solids mixing: diffusive mixing, shear mixing, and convective mixing. Diffusive mixing occurs due to small scale random motion of particles when they roll over a free surface. A high degree of particle mobility is required. When the material is sheared, either externally with a mechanical agitator or internally due to rotating motion of the shell, the shear zones within the bulk cause mixing by exchange of particles across the shear zones. Large scale mixing or convective mixing can be achieved by moving large portions of material from one location to the other within the bulk. The particles do not have much mobility, and therefore segregation is minimized.

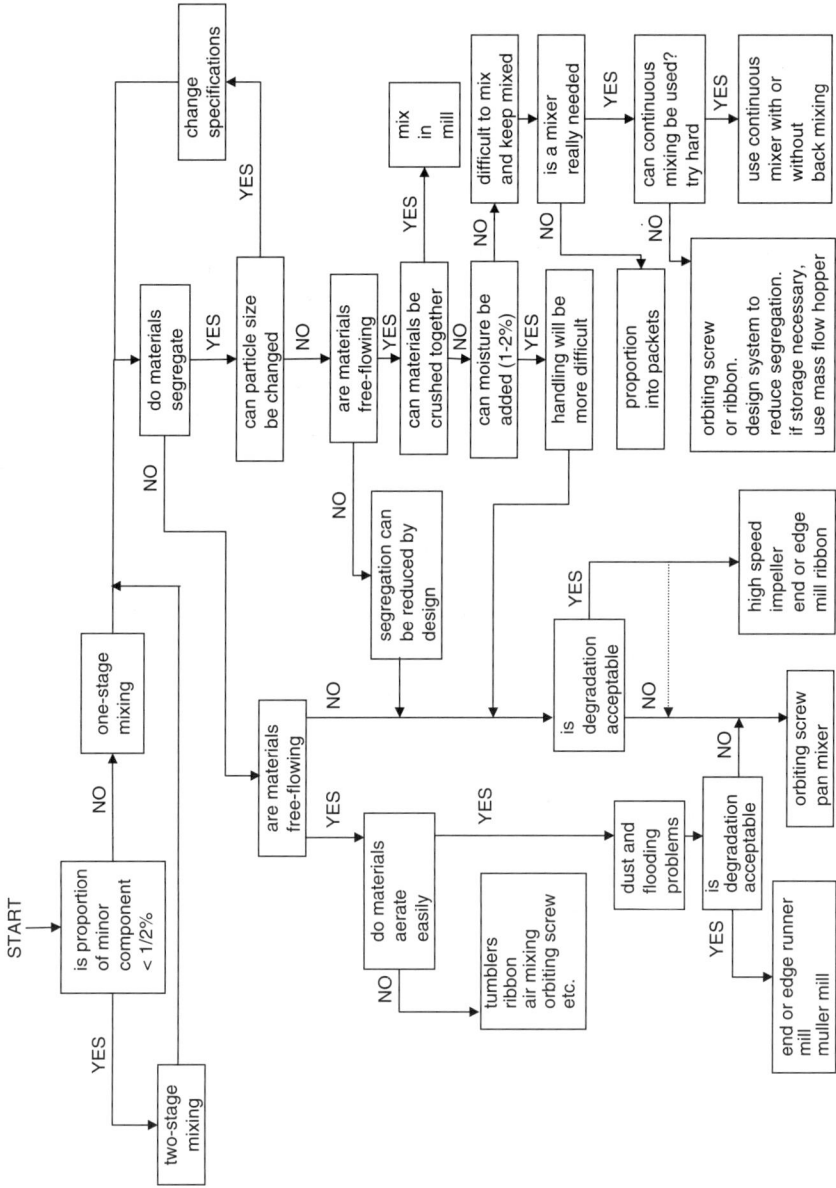

**Figure 15-32** Selection chart for a batch mixer. (Modified from Miles and Schofield, 1970a.)

**938** SOLIDS MIXING

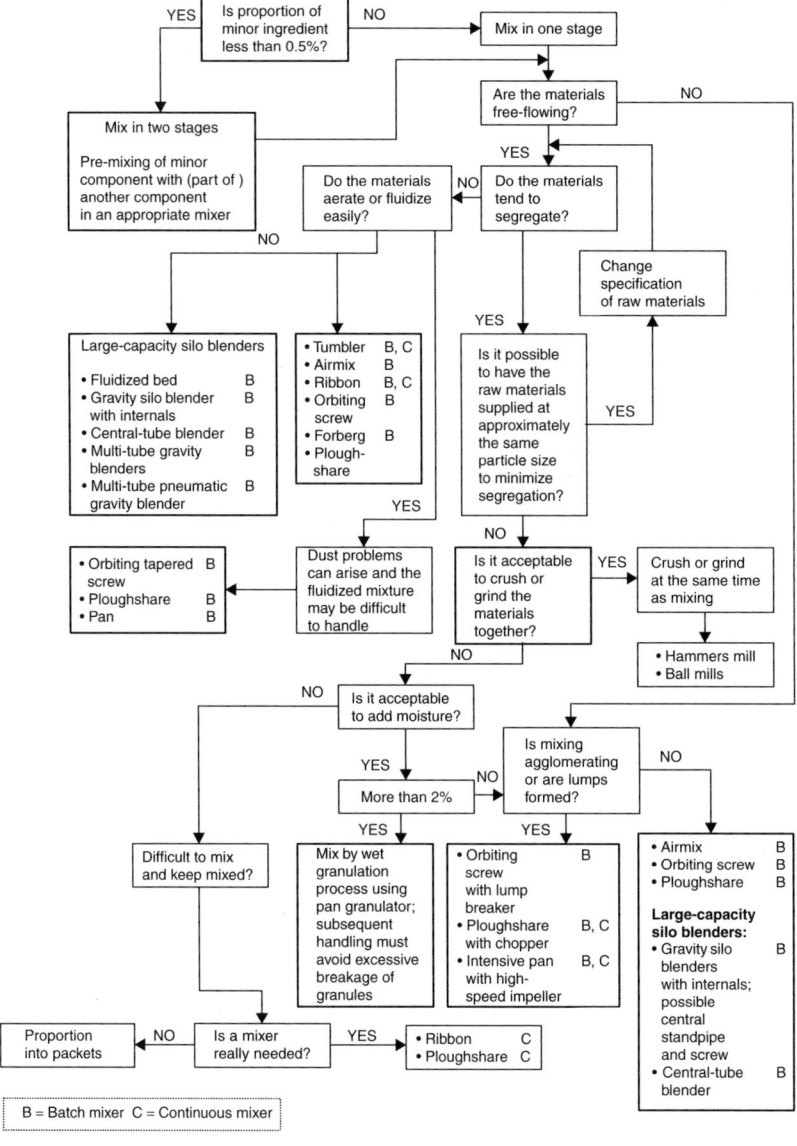

**Figure 15-33** Selection criteria for continuous and batch mixers. (Modified from van den Bergh, 1994.)

During mixing, irrespective of the type of mixers, all the three mechanisms will occur, and one will probably dominate. Some of the mixing mechanisms can also result in segregation. For instance, free motion of particles on the surface (diffusive mixing) can also result in size-based segregation. Extended mixing time may actually cause de-mixing and poor mixture quality. Therefore, an understanding of the potential segregation mechanisms is helpful while selecting a mixer. Similarly, a low-shear tumbling mixer can cause agglomeration of a fine fraction in a fine/coarse mixture.

### 15-10.2 Segregation Mechanisms

No discussion of process mixing can take place without a concurrent discussion on segregation of particles. It is often said that segregation (particularly in batch mixing processes) begins the moment the agitation is switched off! Since the process engineer/operator has typically taken great care to mix the components, segregation is usually an undesirable side effect that must be recognized, and either appropriate engineering controls must be installed or the mixture must be "designed" such that it cannot segregate.

There are five generally recognized mechanisms of segregation: momentum, trajectory, sifting, fluidization, and air currents. These are most significantly affected by one particle property, particle size. Although there are other mechanisms involving other particle properties (e.g., the authors have witnessed segregation induced by a difference in coefficient of restitution), these five predominate and are described below (Carson et al., 1986).

*15-10.2.1 Momentum.* This mechanism occurs when a flowing bulk solid impacts a pile that is formed below the bulk solid. Because of differences in the momentum and in some cases the cohesion between the coarse and fine fractions, the coarse particles tend to gather at the outer periphery of the pile while the fines are deposited directly below the point of impact. This mechanism is quite common and has been shown to occur when the ratio of particle diameters is as small as 1.3. This mechanism would probably occur when, for example, a batch mixer is discharged into a hopper below it. Fortunately, the use of mass flow hoppers can effectively combat this radial segregation.

*15-10.2.2 Trajectory.* When a stream of bulk solids is allowed to flow off a conveyor or an inclined chute, segregation of the coarse and fine fractions can occur. The fines tend to drop directly below the chute/conveyor, while the coarse fraction is flung away. In mixing operations, care must be exercised when discharging either a batch or continuous mixer through an inclined chute. If this chute empties into a hopper, mass flow hoppers can help to combat the segregation, but not as effectively as the case for radial segregation. This is because of the asymmetric radial distribution of sizes in the hopper. If the mixture is to be packaged directly upon emptying the mixer, it may be possible for segregation in the package.

*15-10.2.3 Sifting.* This mechanism occurs when fine particles sift (much like a screening operation) into the interstices between the coarse fraction. One common cause is the shearing flow of bulk solid, as in the case of emptying a funnel flow silo (centrally moving core of material with a nonflowing annular region).

Here fine particles sift in between the coarse as the bulk solid avalanches down toward the center of the silo. This usually manifests itself by showing an increase in fines content in the material exiting the silo at the end of the discharge of the silo. Another common cause of sifting is vibration. In this case, the finer fraction can sift into the interstices of the coarse. This occurs frequently during product shipment.

***15-10.2.4 Fluidization.*** When conveying material into a silo, it is not unusual for the fines to remain suspended in the headspace above the material for a long period of time, eventually settling as a layer on the material conveyed. This mechanism is likely to occur when mixtures containing slow settling particles (usually less than 100 μm) are conveyed.

***15-10.2.5 Air Currents.*** Air currents within a vessel can cause the fines and coarse to segregate. This is quite common when a polydisperse mixture is being conveyed into a silo or hopper. The resulting segregation pattern is difficult to predict, but since the powder is being spread across the entire cross-section of the silo, mass flow can aid in remixing the segregated material.

### 15-10.3 Mixer Classification

Most industrial mixers can be broadly classified into the following categories:

- Tumbling mixers (V-cone, double cone, etc.)
- Agitated mixers
  - Paddle and plow mixers
  - Ribbon mixers (vertical and horizontal)
  - Screw mixers (vertical and horizontal and orbiting types)
  - Sigma-blade and Z-blade mixer
  - Forberg mixer
- Gravity silo blenders
- Pneumatic blenders
- High intensity mixers
  - Henschel mixer
  - Paddle mixer
- High-intimacy or high-shear mixer
  - Muller mixer
  - Compaction rollers

A brief description, typical operating ranges, and practical application information for these mixers follow.

***15-10.3.1 Tumbling Mixers.*** As the name suggests, these mixers "tumble" the powder mass. Mixing is achieved predominantly by random motion when particles roll down a sloping surface. As the whole shell tumbles either on its own axis or eccentrically, the ingredients are bodily displaced, and mixing takes place in the radial direction. To achieve better dispersion, internal baffles and/or counterrotating impellers can be mounted and can be tailored for difficult-to-handle mixtures as well. Stationary spray nozzles for liquid coating can also be mounted along the axis of rotation.

Typical tumbling mixers are:

- Double-cone blender (batch); refer to Figure 15-19
- V and Y blenders (batch); refer to Figures 15-19 and 15-20
- Cylindrical blender (batch or continuous)

Tumbling mixers are discussed in detail in Section 15-4.1.

**15-10.3.2  Agitated Mixers.** Agitating mixers [see also discussions by Pahl (1986), Steiss (1995), Fuller (1998), Kent (2002), and Ramponi et al. (2002); discussion of convective mixers in Section 15-4.2] use mechanical means (e.g., paddles, plows, and ribbons) to create mixing action while keeping the shell stationary. A typical agitated mixer consists of a stationary shell (vertical or horizontal) with a single or twin shafts on which agitating devices are mounted. During mixing, particles are thrown randomly and the product is sheared or fluidized mechanically, depending on the tip speed of the paddles or plows. These mixers can handle a wide range of bulk solids from free-flowing to cohesive to pastes. While mixing is taking place, one can incorporate a liquid injection for further agglomeration and choppers or delumpers for breaking up the agglomeration, depending on the requirement.

In agitating mixers, the mixing is predominantly due to particles moving randomly from one point to the other, along with the bulk mass. So there is a combination of both shear and convection occurring within the mixer. Depending on the handling characteristic of powder, a certain degree of aeration will take place, and at higher tip speeds, the mass is capable of being fluidized. Once the material is aerated, the frictional effects including the interparticle forces are reduced, and sometimes eliminated, which enables the particles to move freely and randomly.

*Paddle and Plow Mixers.* The mixer typically has a single or double U-shaped trough with an impeller that consists of a single shaft or twin shafts mounted with plows/paddles at regular pitch in between, as shown in Figure 15-34. The plow helps to lift the solids creating chaotic motion causing shear in the powder mass that results in mixing. As shown in Figure 15-35, the motion of the powder in the mixer results in convective mixing whose intensity is proportional to the tip speed of the impeller.

At lower speeds, which is called *cascading*, the powder is carried by rotation and descends by rolling and/or sliding along the surfaces of the solids mass just as in tumbling mixers. At medium speeds, which is called *cataracting*, the powder is carried by the plow and drops either by sliding, rolling, or cascading. At higher speeds, which is called the *equilibrium regime*, the powder is mostly lifted by

## 942  SOLIDS MIXING

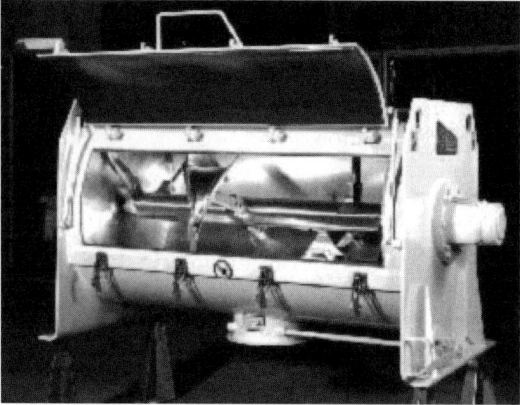

**Figure 15-34**  Plow mixer (Courtesy of Scott Equipment Company.)

**Figure 15-35**  Plow mixer agitator and solids flow direction. (Courtesy of Kemutec.)

the plow and slides off at the end. In the equilibrium regime, there is hardly any chance for rolling, let alone shearing, and the desired mixing level will not be promoted. Hence, the right speed for operating the plow mixer depends on the physical properties of the product and the rotational speed of the mixer.

PRACTICAL INFORMATION FOR PLOW MIXERS

- *Typical size range for ingredients:* up to 5 mm, suitable for free flowing to slightly cohesive type of powders, can handle varying densities. Not suitable for very cohesive solids.
- *Mode of operation:* batch or continuous.
- *Choice of internal configuration:* single and double shafts with plows placed at regular intervals.
- *Suitability for special unit operations:* spray nozzle for agglomeration, high-speed choppers for breaking loose the agglomerates in the feed or mixture.
- *Mixing time/order:* up to 5 min with random mixing.
- *Particle degradation/attrition:* negligible when operated properly.

- *Type of industry:* food, ceramics, chemicals, building, plastics, pharmaceuticals.
- *Typical problems:* complete emptying is a problem.
- *Mechanical issues:* if steam cleaning with high-pressure water cleaning is adopted, mechanical seals must be selected to withstand the temperature and pressure limits. This requires regular maintenance. Seals will require routine inspection and maintenance.
- *Energy consumption:* up to 150 kW/m$^3$.

The paddle mixer (Figure 15-36) is similar to the plow type but the paddles mounted on the shaft differ in design from the plow mixer. Paddles can be oriented so as to impart lateral/back and axial mixing. The paddles are generally operated at higher speed than the plows. The speed has to be determined from running trials, and the right speed can result in good mixing. Higher speeds can cause segregation resulting in heavier product thrown near the wall and lighter product in the center of the mixer. Both the plow and paddle types can have double shafts with two horizontal impellers. The paddles or blades overlap those on the other shaft. In operation, the impellers counterrotate, fluidizing and mixing the material.

The paddle mixer can also be used for kneading and for mixing pastes since the kneading action allows phase changes from dry to paste when liquid is added. To evaluate the suitability of such mixers, it is recommended that test trials are conducted on the mixture to judge the duration and the quality of mixing. Paddle mixers are so versatile, they are also used as dryers and coolers for bulk solids.

PRACTICAL INFORMATION FOR PADDLE MIXERS

- *Typical size range for ingredients:* suitable for cohesive powders.
- *Mode of operation:* batch or continuous.

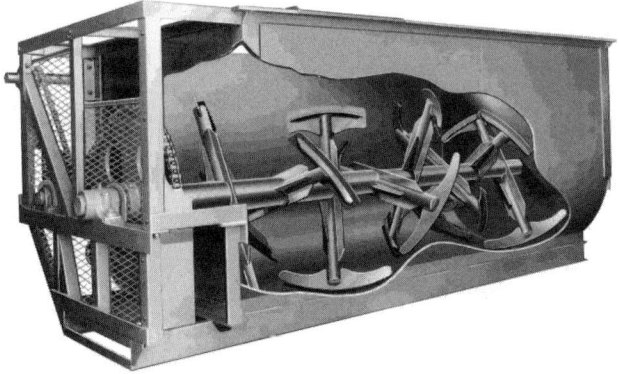

**Figure 15-36** Single-shaft paddle mixer. (Courtesy of H.C. Davies Sons Manufacturing Co.)

**944**   SOLIDS MIXING

- *Choice of internal configuration:* single and double shafts with paddles placed at regular intervals. These paddles can be oriented in such a way as to impart both axial and vertical mixing.
- *Suitability for special unit operations:* spray nozzle for agglomeration; high-speed choppers for breaking loose the agglomerates in the feed or mixture. Can be adopted for heating or cooling.
- *Mixing time/order:* up to 6 min with random mixing.
- *Particle degradation/attrition:* negligible. Higher paddle speeds can cause some attrition, but require quantification through trials.
- *Type of industry:* sewage treatment, dyes and pigments, animal feed, building, pharmaceuticals.
- *Typical problems:* clean up is easier. Complete emptying is not a problem.
- *Mechanical issues:* high speeds for the paddles calls for properly sized good-quality bearings and seals. Seals must be provided with air purge arrangements to insure that they stay dust free.
- *Energy consumption:* up to 150 kW/m$^3$.

*Fluidizing Paddle Mixer (Forberg Mixer).* A Forberg batch mixer [see also discussions by Forberg (1992) and Smith (1997)], shown in Figure 15-37, consists of paddles mounted on twin shafts in a twin trough. The ingredients to be mixed are fed from the top. The counterrotating paddles moves through the mixture throwing it in air, thus mechanically fluidizing the contents. Rapid fluidization is achieved. Discharge of the mixture take place through a large set of twin doors at the bottom of the mixer to minimize segregation. The peripheral speed of the paddle is about 1.5 m/s with gentle operation and very fast mixing of about 1 min (Forberg, 1992). Mixer volumes up to 50 m$^3$ are possible. Forberg mixers have been adapted to continuous mixing as well. Applications of Forberg

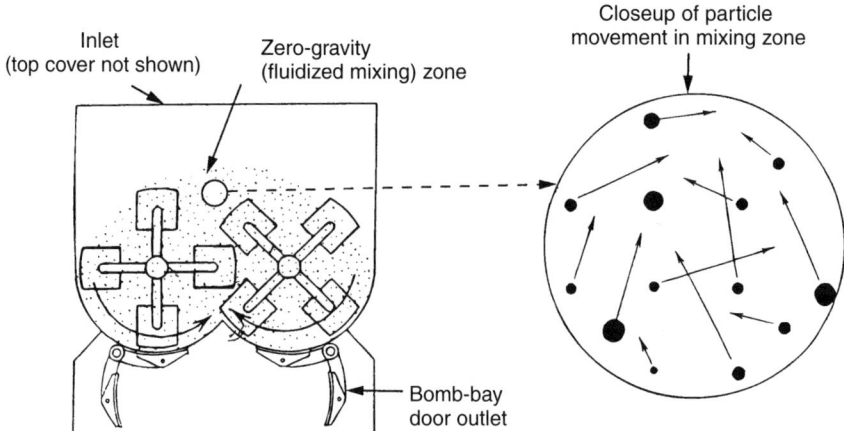

**Figure 15-37** Forberg mixer.

mixers are in industries such as feed, food, pharmaceutical, chemical, building, and environmental.

MIXING TIME WITH A FORBERG MIXER. The time for mixing is dependent on the cohesion of the powders being mixed. As the cohesion increases, the mixing duration increases. However, the chances for the mixed solid to segregate while discharging are reduced due to solid cohesion. Normally, one requires a test at the supplier to verify the degree of mixing. The scale-up must be based on factors such as mixing duration, tip speed, solids filling point, and the discharge options after the process. Results related only to mixing time without such supporting information are useless (Muller, 1982). A comparison of the duration of a Forberg mixer to a generic plow mixer is shown in Figure 15-38. The results show that the paddle mixer is much more efficient then the plow mixer in terms of both mixing quality and duration. This is expected, since in the case of plow mixers, mixing is achieved primarily by diffusion and shear of the mass.

## Ribbon Mixers

VERTICAL RIBBON MIXER. The vertical ribbon mixer [see also Pahl (1986), Steiss (1995), and Cavender (2000)] shown in Figure 15-39, is similar to a ribbon mixer turned upright. The mixer is designed to operate with 90% full of solids. During operation the ribbon rotates slowly. This action creates a shearing zone at the

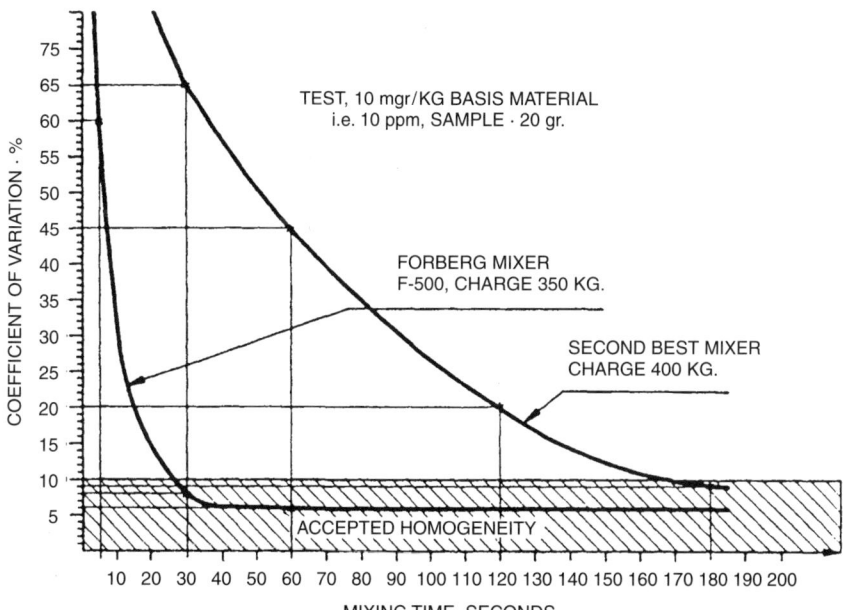

**Figure 15-38** Forberg/twin shaft paddle mixer performance. (From Forberg, 1992.)

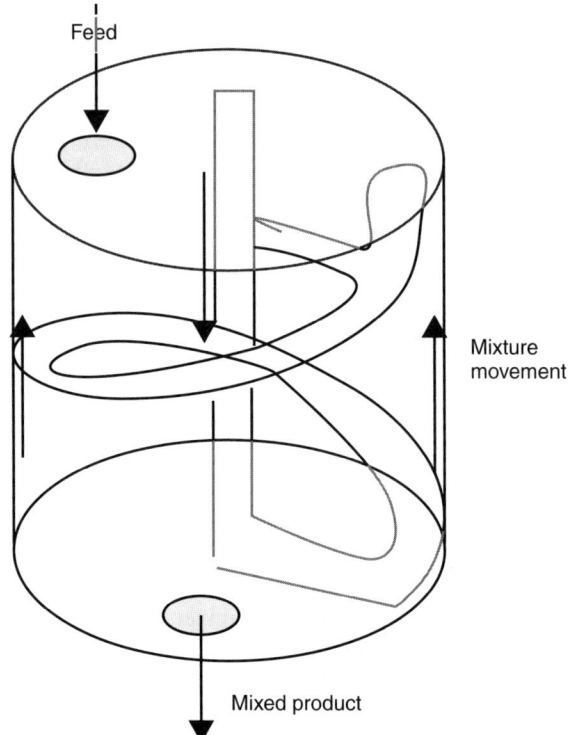

**Figure 15-39** Vertical ribbon mixer.

wall, where the material moves in a helical fashion upward and flows down centrally. The mixer may be designed with a single or double shaft. The mixer with a single shaft has a capacity of up to 30 m$^3$ and can be operated under pressure or vacuum from 50 to 250°C. The mixer can handle friable products such as cereals, plastics, pigments, and pharmaceutical powders.

HORIZONTAL RIBBON MIXER. The ribbon (see Fuller, 1998) mixes the product in a trough by pushing it along the axis in both directions and displacing it by centrifugal force. Segregation may occur, resulting in separation into the bottom of the trough. The mixer is suitable for free flowing to cohesive products and is not suitable for sticky products. Emptying the entire contents may be difficult due to the small clearance between the trough and the ribbon. These mixers can be very large and the power required may be as high as 6 kW/m$^3$. A ribbon mixer, shown in Figure 15-40, has a double shaft, double helix, and top cover. The shaft has welded radial supports for the ribbons. The clearance between the ribbon's outer edge and the trough wall ranges from 3 to 6 mm. An inlet is located in the trough's cover at the top, and a discharge opening is either located in the center of the trough bottom or at either end. Figure 15-41 shows the direction of solids movement in such a mixer.

## FUNDAMENTALS AND MECHANICS OF MIXER OPERATION 947

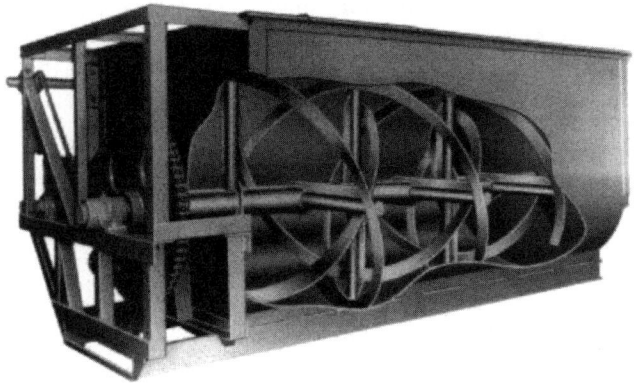

**Figure 15-40** Horizontal ribbon mixer. (Courtesy of H.C. Davies Sons Manufacturing Co.)

**Figure 15-41** Direction of solids movement in ribbon mixer. (Courtesy of Kemutec.)

The mixture can be agglomerated by a liquid spray mounted above the ribbons. To reduce agglomerates, the mixer wall can be fitted with high-speed choppers. In operation, 40 to 85% of the mixer capacity is filled. The shaft rotates at relatively slow speed and the moving ribbons push the material back and forth. The inner ribbons move the solids toward the trough end, and the outer ribbons push the solids in the opposite direction, toward the center discharge valve.

For most powders, mixing can take 15 to 20 min. However, if the powders are fibrous, it could take much longer (up to 2 h) depending on the mixture quality requirements. During discharge, complete emptying can be a problem since a certain amount of product remains due to the clearance of the ribbon to the wall. This has to be cleaned out if ingredient change is required. These mixers can handle products ranging from free-flowing to cohesive powders and even slurries. Typically, they are used for adding a small amount of the ingredient to a larger bulk for general-purpose mixing.

PRACTICAL INFORMATION FOR RIBBON OR SHAFTLESS SCREW MIXERS

- *Typical size range for ingredients:* up to 5 mm, suitable for free to cohesive type of powders; can handle varying densities.
- *Mode of operation:* batch or continuous.

- *Choice of internal configuration:* single and double shafts with ribbons; single and up to three pitches designed either left- or right-handed to allow backmixing and conveying. Cut and folded screws are used for mixing of fine solids such as clay.
- *Suitability for special unit operations:* high-speed choppers for breaking loose the agglomerates in the feed or mixture. Can be adopted for heating or cooling.
- *Mixing time/order:* variable to 30 min, but can take as much as 2 h for fibrous products.
- *Particle degradation/attrition:* negligible. Particles can be broken at the clearance between the ribbon and the wall.
- *Type of industry:* widespread use. Food, chemicals, environmental, dyes and pigments, animal feed, wood and paper industry, pharmaceuticals.
- *Typical problems:* cleanup can be difficult. Emptying is not a problem.
- *Advantages:* emptying is not a problem.
- *Mechanical issues:* the ribbons must be designed to withstand the load equivalent to the volume of the mixer. The drive selected must be robust and allow for easier stop and start during mixing. Chunks, if caught at the clearance or in between the ribbons, can destroy the orientation of the ribbon. Care must be exercised to delump the agglomerates before feeding. Bearings are often gas purged. Material of construction depends on the product mixed.
- *Specific power consumption:* up to 12 kW/m$^3$.

## Screw Mixers

VERTICAL ORBITING SCREW MIXERS. This type of mixer [see other discussions by Hixon and Ruschmann (1992), Pahl (1986), Steiss (1995), and Hosakawa Micron (1998)] consists of a hopper-shaped vessel and a screw feeder placed along the wall. The clearance varies from 3 to 6 mm, depending on the product. The screw orbits around the hopper but at the same time rotates in its own axis. This motion causes the mixture to lift and spread on a surface as shown in Figure 15-42. During this motion some shearing also occurs. It may not be suitable for products that are friable. Liquid can be introduced into the mixer through a suitable spray system. When two orbiting screws are used, one screw is normally shorter and is called a *satellite screw*. In some cases, two separate screws are used in two hopper vessels joined along the wall, as shown in Figure 15-42. Screw design can vary in flight design, pitch, and diameter. Recent designs incorporate varying diameter screws, which in fact reduces the ineffective area of mixing, thus increasing the residence time for particles.

PRACTICAL INFORMATION FOR VERTICAL OR ORBITING MIXERS

- *Typical size range for ingredients:* up to 500 μm, suitable for free to very cohesive type of powders; cannot handle varying densities.

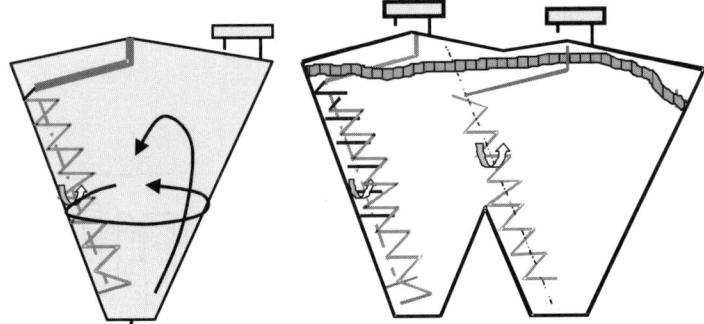

**Figure 15-42** Vertical orbiting screw mixer.

- *Mode of operation:* batch only.
- *Choice of internal configuration:* screws can be driven from either above or below. A short satellite screw with normally larger diameter can be provided for better and faster mixing; screw designs vary from ribbon to screws with varying diameters to reduce the ineffective area of mixing. Spray nozzles can be mounted for liquid addition.
- *Suitability for special unit operations:* can be adopted for heating or cooling.
- *Mixing time/order:* at least 10 min.
- *Particle degradation/attrition:* friable products may degrade.
- *Type of industry:* food, chemicals, environmental, plastics, pharmaceuticals.
- *Advantages:* cleanup is not easy when sticky solids are handled. Emptying is easy.
- *Mechanical issues:* mechanical seals are used (Kent, 2002). These can be quite expensive and require special lubricants. There may be problems if this lubricant contaminates the product. Special seals are required in such cases.
- *Specific power consumption:* up to 80 kW/m$^3$.

*Sigma-Blade and Z-Blade Mixers.* These mixers [see also Pahl (1986), Fuller (1998), and Harnby (2000)] consist of twin troughs each fitted with a rotating agitator, and each one of these agitators is a heavy-duty Z-shaped blade, as shown in Figure 15-43. The product is introduced from the top of the mixer. The same access is used for cleaning purposes. The mixer can be tilted for emptying the products. A spray bar can be mounted above the blades. While in operation, the product is loaded up to 40 to 65% of the mixer's capacity. The blades can overlap and rotate at the same speed or at variable speeds. The mixing duration can be 10 to 30 min with good homogeneity of up to 99%. No delumper is required in this mixer. In Z-blades, two counterrotating Z-blades fold and shear the material quite severely. These are seldom used for dry solids, but are used for producing doughs and thick viscous pastes.

*Comparison of Agitated Mixers.* A classification of the rotating mixers or mixers with rotating components is made according to Froude number, Fr (Rumpf

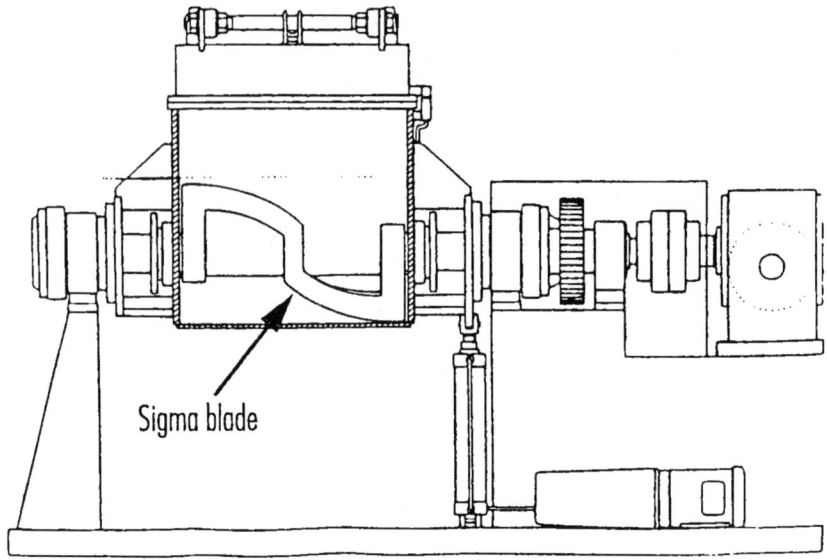

**Figure 15-43** Sigma-blade mixer.

**Table 15-2** Types of Equipment for Agitation of Solids by Rotating Inserts or Rotating Vessels

| Type of Mixer | Motion by: | $Fr = \dfrac{\overline{r\omega}^2}{g}$ | Capacity ($m^3$) | Power ($kW/m^3$) |
|---|---|---|---|---|
| Tumbling type with rotating vessels; cylinders, drums, cubes, V and Y types, double cones | Free fall | <1 | <2 | <1–2 |
| Double ribbon mixers, trough blenders, vertical screw types, orbiting type such as Nauta | Thrust | <1 | <30 | 3–10 |
| Centrifugal mixers with paddles plows | Thrust, centrifugal force | >1 | <30 | 20 |
| High intensity mixer Turbine and fluid mixers | Centrifugal force | ≫1 | <1.5 | <500, 20 |

*Source*: Rumpf and Muller (1962), Pahl (1986).

and Muller, 1962). Froude number is the ratio of the centrifugal force and acceleration due to gravity. Hence,

$$Fr = \frac{\overline{r\omega}^2}{g} \qquad (15\text{-}15)$$

where r is the radius of the mixer, ω is the angular velocity, and g is the acceleration due to gravity.

Tables 15-2 to 15-5 show different parameters of significance for agitator mixers. Table 15-2 shows different classes in batch mixers followed by the mode of operation and Froude number and respective capacity ranges and their power requirements. Note that only the high intensity mixers have $Fr \gg 1$, and centrifugal types with $Fr > 1$; otherwise, most of the mixers operate with $Fr < 1$. Sections to follow will treat each of the classes of mixers above.

This type of classification is one of the ways to characterize the mixer similar to liquid mixing; however, as Muller (Weinekotter and Gericke, 1999) points

**Table 15-3** Comparison of Agitated Mixers

| | Agitating | | | |
|---|---|---|---|---|
| Factor | Ribbon/Paddle | Plow | Fluidizing Paddle | Sigma-Blade |
| Material consistency | Powders/ granules | Powders/ granules | Powders/ granules | Pasty sticky gritty slurries up to $2 \times 10^6$ cP |
| Allowable fill level or batch size (% of total mixer capacity) | 40–85 | 30–70 | 40–140[a] | 40–65 |
| Liquid addition configuration | Spray bar above ribbons | Spray nozzles at mixer top | Spray bar above paddles | Spray bar above blade |
| Delumping agitator configuration | High-speed chopper blades at sides | High-speed chopper blades at sides | Pin mills above paddles | None |
| Mixing cycle length (minutes) | 15–20 | <5 | <1 | 10–30 |
| Final moisture homogeneity (% of complete homogeneity) | 90–95 or better | 95–98 or better | 98–99 or better | 99 or better |
| Rotating or stationary vessel | Stationary | Stationary | Stationary | Stationary |
| Degree of particle shear | Some | High | Slight | Very high |

[a] Percent fill more than 100% of the total capacity for another agitating batch mixer of equal volume.

**Table 15-4** Comparison of Various Agitated Mixers

| Property | Plow | Paddle | Ribbon | Shaftless Screw |
|---|---|---|---|---|
| Free-flowing powders, $50 < x < 500$ μm | Yes | Possible | Yes | Yes |
| Free-flowing granules, $200 < x < 5000$ μm | Yes | Possible | Yes | Yes |
| Cohesive powder | Possible | Yes | Yes | Yes |
| Energy transmitted to powder | Yes | Yes | No | Yes |
| Capacity (m³/h) | 1.8–1500 | <1000 | 10–50 | 1–50 |
| Size (m³) | <40 | <40 | <50 | <60 |
| Filling ratio (%) | <70 | <70 | <60 | <70 |
| Specific power | 10–150 | 10–150 | 3–12 | 8–12 |
| Mixing time (min) | 0.5–5 | 1–6 | 3–20 | 2–8 |
| Froude number | 1–9 | <9 | <1 | <1.5 |
| Cost/performance (U.S.$/ft³) | 0.18 | 0.14 | 0.09 | 0.14 |

*Source*: Adapted from Ramponi et al. (2002).

**Table 15-5** Typical Tip Speeds for Agitated Mixers

| Type of Mixer | Tip Speed (m/s) |
|---|---|
| Ribbon | 1.4 |
| Turbine | 3 |
| Paddle | 2–6 |
| Twin shell tumbler with: | — |
|    Pin-type intensifier | 8.6 |
|    Liquid feed bar | 17 |
| Twin rotor | Up to 6.6 |
| Single rotor | 30–45 |
| Mills of various types | 12–100 |

*Source*: Adapted from (Weidenbaum 1973, Fayed and Otten 1984).

out, it has not yet been possible to characterize the solids mixer like the one for liquids. There is no relationship between a powder parameter that might be comparable to viscosity. Mixing of solids is by no means a straightforward task, and must never be underestimated. To appreciate and understand the mixing, one must understand the mechanisms influencing such an operation, and this is dealt with in the sections to follow.

Let us consider the case of a cohesive powder. During mixing of such a product, the ingredient may form lumps. These lumps may just circulate on their own without taking part in mixing. If the product is free flowing, one has to restrain the movement of particles, and this is contrary to the approach taken for mixing a cohesive product.

## FUNDAMENTALS AND MECHANICS OF MIXER OPERATION

**15-10.3.3 Gravity Silo Blenders.** There are several instances in industry where the method of production or the nature of a process leads to variations in the quality of a particulate powder as a function of time. If these bulk solids are stored in silos before further processing or delivery to customers, these quality variations are propagated further in the process. In general, the contents of the silo are too large for eliminating these variations by the use of ordinary mixers, and homogenization has to be undertaken in situ.

Homogenization in silos can be undertaken using several techniques. The most common are fluidization, internal mechanical recirculation, and external recirculation with or without a hopper type of static mixing device. Since the variations in the quality of the stored powder occur as a function of time, the individual layers in the silo must be mixed with one another to obtain homogenization. Thus, one must design the hopper section of the silo with as large a hopper half-angle as possible to achieve the desired flow patterns. However, the flow patterns developed must be of mass flow to allow reliable bulk powder flow and to prevent segregation of the bulk powder upon discharge from the silo. Hence, the hopper half-angle should be designed on the mass flow/funnel flow limit to achieve the optimum blending efficiency of the silo.

The following classification along with Table 15-6 is helpful:

- Multitube blenders (refer to Figure 15-44)
    - Waeschle's gravity blender and combiflow blender
    - Phillips blender
    - Fuller blender
    - Mixing silo blender (Muller, 1982)
    - Zeppelin Centro blender

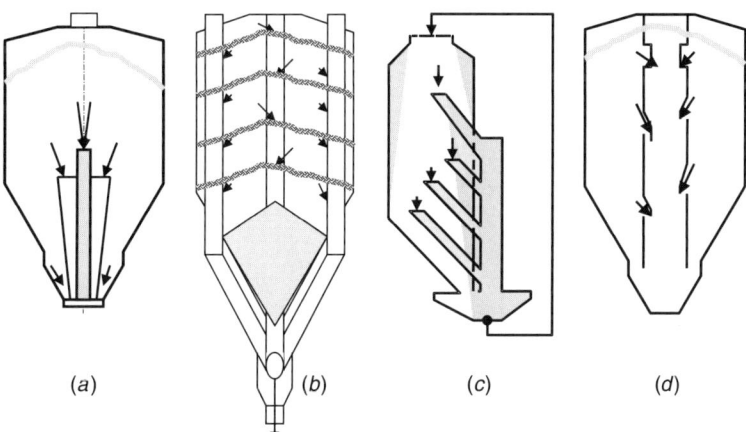

**Figure 15-44** Multitube gravity blenders: (*a*) Waeschle's gravity blender and combiflow blender; (*b*) Phillips blender; (*c*) Mixing silo blender; (*d*) Zeppelin Centro blender.

**Table 15-6** Survey of Blending Systems[a]

| | Fluid Bed | Air Merge Type | Vertical Screw | Nauta Type | Zeppelin Type | Binsert Type | Waeschle Type | Phillips Type |
|---|---|---|---|---|---|---|---|---|
| Free-flowing powders 50 μm < x < 500 μm | + | 0 | + | + | + | + | (+) | (+) |
| Free-flowing pellets 200 μm < x < 5000 μm | 0 | 0 | 0 | 0 | + | + | + | + |
| Cohesive powders | – | + | + | + | (+) | + | – | – |
| Maximum vessel size (m$^3$) | 1000–30 000 | 200 | 100 | 30–60 | 200 | 200 | 600 | 100 |
| Energy required | Pneumatic | Pneumatic | Mechanical | Mechanical | Gravity | Gravity | Pneumatic | Gravity |
| Specific energy (kWh/t) | 1–2 | 2–7 | 2–10 | 2–10 | 1–3 | 1–3 | Ca2 | <1 |

*Source*: Wilms (1992).

[a] +, Suitable; (+), Limited Suitably; –, Not Suitable; 0, Not Economical.

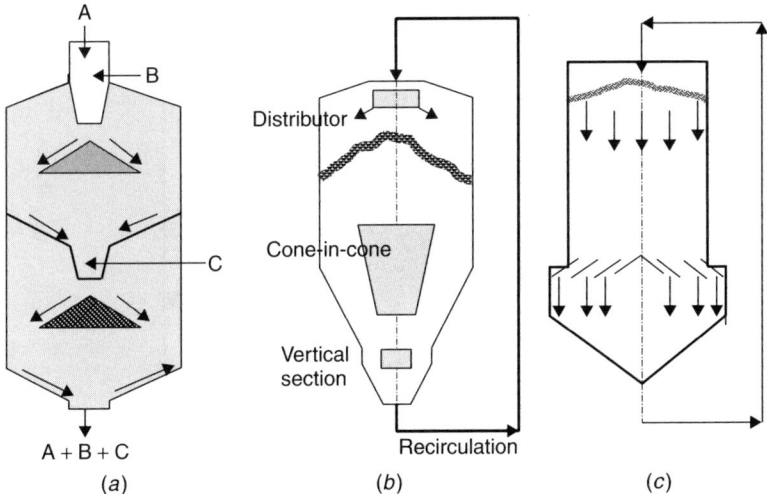

**Figure 15-45** Blenders with inserts: (*a*) Roth blender; (*b*) Binsert blender; (*c*) Peschl blender. (As referred by Manjunath et al., 1992.)

- Insert-type blenders (refer to Figure 15-45)
  - Roth blender
  - Binsert blender
  - Peschl universal blender (Peschl, 1996)
  - Johanson blender with rings along the length
- Mechanical blender (refer to Figure 15-46)
  - Dual flow blender
  - Orbital screw-type blender (refer to Figure 15-42)

Various techniques have been used for homogenization of the silo contents. Figures 15-44*b*, and 15-45*a* show what are essentially multisilo arrangements. The Phillips blender, shown in Figure 15-44*b*, utilizes the principle of simultaneously drawing down powders from different levels in the silo and further promotes homogenization by providing two blending stages. The Roth blender in Figure 15-45*a* is a type of static mixer, where three product streams are mixed simultaneously.

Both principles are suitable for nonsegregating, uniformly sized particles whose physical or chemical characteristics may be marginally different. Figure 15-44*d* shows a Zeppelin Centro blender, and Figure 15-44*c* shows a mixing silo described by Muller (1982). In these types of blenders it is important to ensure that the discharge capacity of the central tube is larger than the combined inlet capacity of the ports along it and that the ports allow roughly similar amounts of powder to enter the central tube. Another blender manufactured by Zeppelin is a multipipe blender, which allows not merely powders from various levels to be mixed on

**956** SOLIDS MIXING

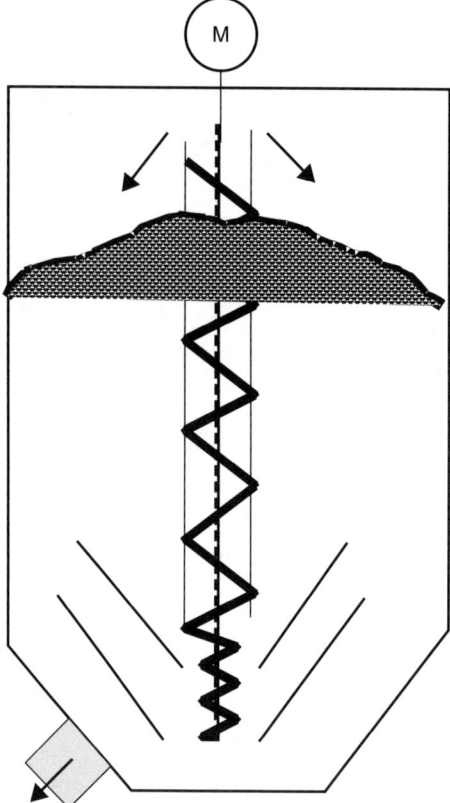

**Figure 15-46** Mechanical blender.

discharge but also ensures that powders from various zones are also mixed. The disadvantages of these systems are that they are relatively costly to manufacture, and it is difficult to ensure that the various ports have similar capacities.

Figure 15-45 shows systems that use mechanical activation in addition to gravity. The Peschl universal blender (Figure 15-45c) also ensures mixing of the zones plus layers by creating a velocity profile across the silo diameter. This is achieved by applying differential levels of vibration to the concentric activating rings. It should be noted that care must be taken (Figure 15-46) to ensure that there is a differential rate of ingress to the screw from the various (concentric) zones in the silo to achieve effective blending.

It is clear that what one wishes to achieve is a mixing of the various zones plus the various layers in the silo. Doing this by gravity alone is considerably cheaper and more convenient than resorting to mechanical means. To achieve this blending objective, one must create a marked velocity gradient over the diameter of the silo while ensuring that the entire contents of the silo are in motion during discharge. As is also the case with most of the blenders described in the foregoing, it may be necessary to recirculate the contents of the silo several times to achieve an homogeneous mixture.

An effective method in which to create a large velocity differential of the type desired for a gravity blending is the Binsert hopper-in-hopper axisymmetric silo, shown in Figure 15-45*b*. Bulk material flows through the inner as well as through the annulus between the two hoppers. Placing the inner hopper at a predetermined position controls the velocity of the material. The design limits for this configuration are chosen near the limits for mass flow, to obtain a high velocity differential between the center and the outside unit. The critical point with reference to this type of blender is the aspect ratio, H/D, for the cylindrical section of the silo. Johanson calculated the required aspect ratio versus the hopper half-angle for a certain bulk powder, as shown in Figure 15-47. It is seen that the H/D ratio of 1.5 seems to be a limit for effective blending. It should be noted that as the hopper half-angle is increased to more than 35°, the aspect ratio necessary for effective blending is reduced drastically, to approximately 0.5.

*Operation of Gravity Blenders.* It is of considerable advantage to be able to predict the degree of blending that will be achieved in a given silo without the need to resort to experimental procedures on pilot scale test rigs. Consider a layer of markers, or marked particles, placed on top of a bulk powder, which is to be homogenized as shown in Figure 15-48. When the powder is discharged, some blending will take place. This degree of blending can be determined by plotting a graph of the percentage of markers in a sample against the volume of the powder removed and recirculated. During the first pass, a certain volume, $V_1$, must be discharged before any of the marked particles emerge from the silo. Johanson assumes an idealized triangular distribution as depicted in Figure 15-48 (Johanson, 1970). The volume, $V_p$, corresponds to the discharge of

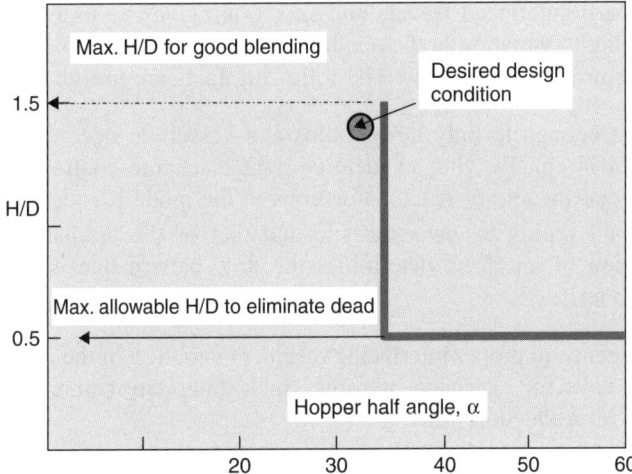

**Figure 15-47** Optimal design of gravity blenders. (From Johanson, 1970.)

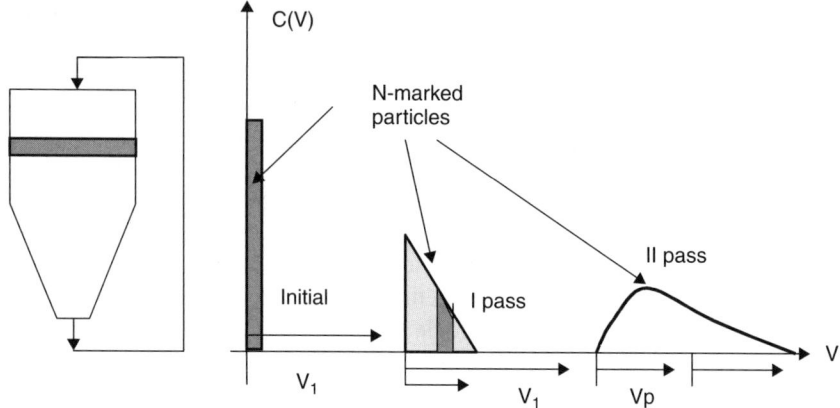

**Figure 15-48** In-bin blending process representation. (From Johanson, 1970 and Roberts, 1990.)

the markers as the silo is emptied. During the second pass, the marked particles will be discharged over the volume $2V_p$ (Figure 15-48). After several passes the markers will be evenly distributed in the total volume. Effective blending can be obtained when $V_1/V_p < 1$. This implies that a portion of the markers must discharge as rapidly as possible after flow is initiated and that there must be a large time lag between the first marked particle to discharge and the last to discharge, for each cycle.

*Silo–Feeder Interface (Manjunath et al., 1992).* The well-designed blenders are often mismatched in terms of interfacing with discharge control equipments such as screw, belts, or vibratory feeders. Mismatching occurs due to either negligence or due to lack of information on the flow of solids. The result could be costly, since improperly interfaced feeders and silos often give rise to asymmetric flow patterns leading to variation in the quality of the products. Proper interfacing will help to overcome such problems. The following facts are useful:

- It is not enough to only design silos and vessels to operate in mass flow mode; it is equally vital to identify right discharge control equipment to ensure that the silo or reactor functions in the mode it is designed for.
- The small region between the silo and that of the discharge equipment, the region of interface, determines the flow pattern that is established in the blender.

Any negligence to proper interfacing results in variation of the product quality, which again calls for expensive retrofits. Such things are common in industry, causing loss in production time.

### 15-10.3.4 Pneumatic Blenders
If powders exhibit expansion characteristics when aerated, they may be a good candidate for pneumatic blending. The

expansion nature of powders can be determined by passing air through a permeable membrane or fluidizing media. The particles rise due to the drag force of the gas. Further increase in the air velocity, called the *superficial air velocity*, causes agitation in the bed, resulting in the formation of bubbles, causing mixing to take place. Such an operation (van den Bergh, 1994) is used for cement blending (10 000 m$^3$) and for blending of pellets (1000 m$^3$) in the petrochemical sector; however, fluidizing blenders consume very high energy, so industry also has considered gravity blenders. Refer to Table 15-7 for blender power consumption.

The amount of air to fluidize a system, the minimum fluidization velocity, is a function of the particle size, particle density, bulk density, and gas density. Bulk solids can be classified into four types, based on the density difference and mean particle size, as in Figure 15-49 (Geldart, 1973):

**Table 15-7** Blender Power Consumption[a]

| Blender Type | Power Consumption (kWh/t) |
|---|---|
| Air fluidized system | 1–1.5 |
| Gravity system, single inverted cone | 0.25–0.5 |
| Gravity system, multioutlet | 0.1–0.13 |

[a] Power consumption Based on kWh/t of raw mill feed.

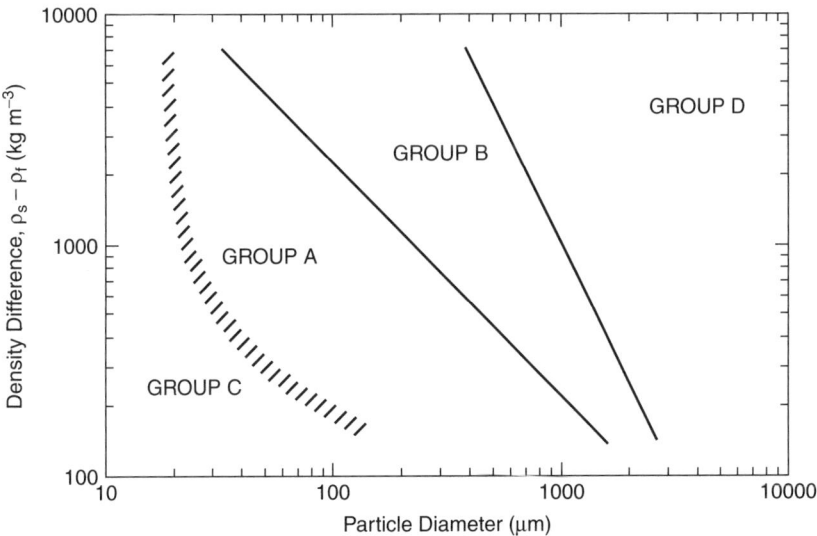

**Figure 15-49** Geldart classification.

- *Class A:* aeratable, such as alumina; bubbling starts when the gas velocities are higher than the minimum fluidization velocity.
- *Class B:* bubbly; starts to bubble when the minimum fluidization velocity is reached.
- *Class C:* cohesive, such as cement; due to interparticle forces such as van der Waals and cohesion; cannot be mixed in fluidized beds.
- *Class D:* spoutable, such as plastic pellets; the required gas velocities are too high and spoutable beds can be an answer, not the fluidized beds.

*Principle of Operation.* The mixing action in the blender can be achieved by fluidizing the contents. If performed at velocities higher than the minimum fluidization velocity, the fluidization creates bubbles in the bed, except for C-type powders. Bubbles are necessary for mixing, as they drag solids in their wake, promoting mixing. Refer to Figure 15-50. A bubble consists a dome-shaped void plus particles in its lower region called the *wake*. As the bubble rises up to the top, solids are frequently exchanged between the wake and the drift. Finally, the bubble bursts forth from the surface of the bed and the powder is ejected from the bed. This process is largely responsible for blending or mixing of solids. The size and shape of a bubble depend on the particle/powder properties. To effectively express the degree of homogenization, it is essential to know the *bed turnover time*, and this is analogous to the number of silo volume recirculation in the case of gravity blenders. To evaluate the bed turnover time, modeling tools must be developed for better understanding.

The following three models in fluid–solid systems suggest how mixing takes place:

- Diffusion model
- Diffusion–convection model
- Convection model

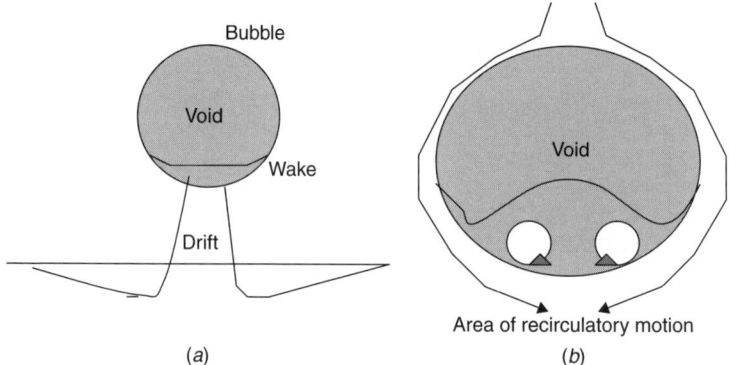

**Figure 15-50** Bubble formation in a fluid bed blender: (*a*) bubble consists of a wake, drift, and void space; (*b*) circulation of solids around the bubble.

These models are also based on the Fickian equation (Lacey, 1954; Fan et al., 1990). The difficult part is the description of the diffusion coefficient for the diffusion model, which is a function of bubble size, its axial velocity, the particle density, particle size, the viscosity and density of the fluid, and the minimum fluidization velocity. When describing these models, experimentation is required for verification of the above parameters. Axial dispersion takes place when the bubbles are rising in the bed, and horizontal dispersion when the bubbles burst. However, there exists evidence that horizontal dispersion also occurs as the bubbles form and move up the bed due to continuous displacement of the mass around the bubble.

*Comparison with Mechanical Blenders or Homogenizers.* Mechanical blenders normally have recirculation systems either within the blender or placed outside it. There are several types, but they have in common steep hopper half-angles with respect to vertical as compared to fluid bed blenders. This is required to allow all the mixed material to flow out of the blender. The silo geometry has to ensure mass flow conditions, which accounts for even withdraw of products, but at the same time, the angles have to be somewhat shallower to allow shear between the flowing layers of products to cause blending. If the outer hoppers are shallower, then normally, blenders are equipped with inserts, such as hopper-in-hopper types. These inserts can work in both axisymmetric and plane flow blenders, although the former types are more common.

*Types of Pneumatic Blenders.* The following types are available:

- Air mix blender
- Air merge blender
- Modified air merge blender
- Pneumatic blender (Krambrock, 1976)
- Entire hopper bottom fluidized

There are several configurations of mixers available. Figure 15-51*a* is called the *Air mix*. To achieve mixing, the system employs a blower or compressor to generate air flow, a control valve to vary air velocity, and downstream of the blender, gas cleaning systems are required to capture the fines. The Air mix can operate in a closed cycle, so that any gas can be employed to fluidize the contents. The gas is introduced through the mixing head as shown in the figure. It is claimed that such a design produces a swirling turbulent action within the blender in a matter of 15 to 30 s, and a similar period is required for the particles to resettle.

On the other hand, the *Air merge* blender employs a hopper divided into several segments, and each segment can be fluidized (Figure 15-51*b*). The fluidization of these quadrants or segments can be controlled. A modified version of the Air merge blender is the column blender shown in Figure 15-51*c*, which consists of a central draft tube through which powder can flow during mixing, thereby

**962** SOLIDS MIXING

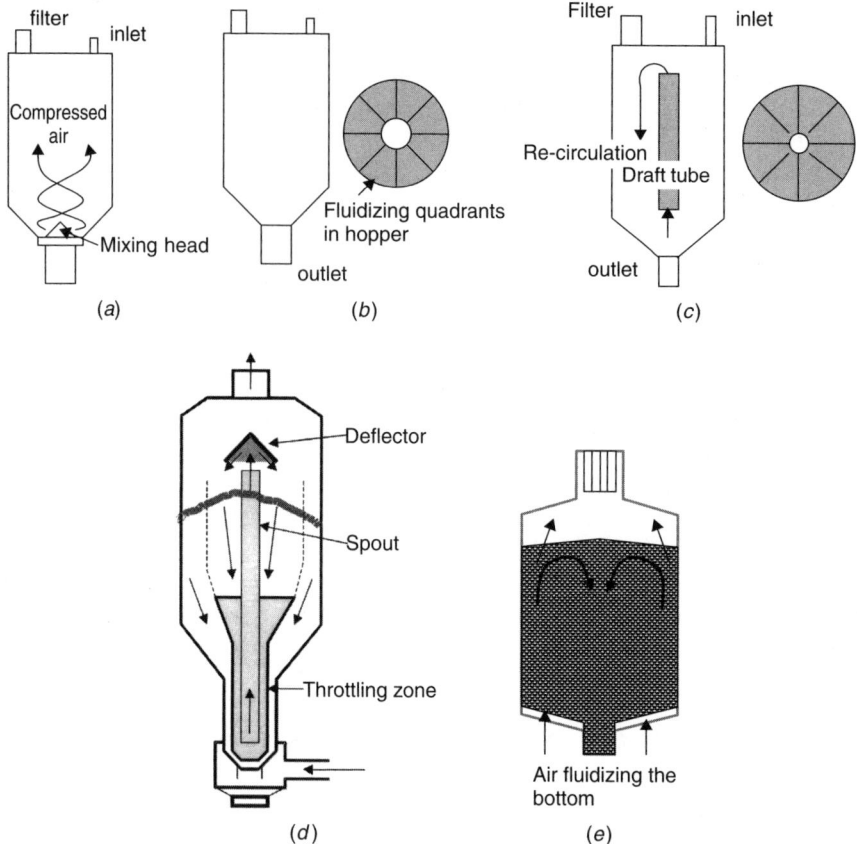

**Figure 15-51** Pneumatic and fluid bed blenders: (*a*) Air mix; (*b*) Air merge; (*c*) modified air merge; (*d*) pneumatic blender (Krambrock, 1976); (*e*) entire hopper bottom fluidized.

increasing effectiveness during the process. The blending area is located directly in the center of the silo. It is claimed that the powder flows up the draft tube, overflows at the top, and flows down the outside of the tube and into the bottom of the tube for recirculation. Certain blenders have hopper parts completely fluidized without sectoring the bottom, unlike the two types above. Fluidization of the mixture is important and causes the development of bubbles for achieving mixing. While discharging from such hoppers, a velocity below the minimum fluidization value, called *deaeration velocity*, must be used for better emptying of the contents, which are well mixed, else segregation of the mixture will result. Krambrock (Figure 15-51*d*) focused on the petrochemical industry to blend plastic granules in large quantities, particularly for obtaining uniform product quality, such as melt index, color values, and the number of additives during long production periods. Table 15-8 contains the design and operating data of some mixer sizes suitable for plastic granules. It is important to note that the mechanical aspects of fluidizing

**Table 15-8** Design Data for Homogenizers

| Volume (m$^3$) | Mixing Time, for 3 Recirculations (h) | Total Air (SCM/h) | Diameter of Blender, D (m) | Height of Mixer, H (m) | $\Delta P$ without Filter Piping (mbar) |
|---|---|---|---|---|---|
| 1 | 0.25 | 1100 ⎫ | | | |
|  | 0.5 | 550 ⎬ | 1 | 2.9 | 32 |
|  | 1.0 | 275 ⎭ | | | |
| 10 | 0.95 | 5400 ⎫ | | | |
|  | 1 | 2700 ⎬ | 2 | 6.3 | 45 |
|  | 2 | 1350 ⎭ | | | |
| 100 | 2 | 14 000 ⎫ | | | |
|  | 4 | 7000 ⎬ | 3 | 19.4 | 90 |
|  | 8 | 3500 ⎭ | | | |
| 200 | 4 | 14 000 ⎫ | | | |
|  | 8 | 7000 ⎬ | 3.5 | 28.2 | 120 |
|  | 16 | 3500 ⎭ | | | |

*Source*: Krambrock (1976).

silos must be carefully designed because of the possibility of development of hydrostatic pressures in the silo.

The pneumatic blender (Figure 15-51*d*) consists of a mixing silo, a central conveying tube, and a cone at the top for spreading the mixture. During filling the outlet is closed and the slide valve prevents entry of product into the fan. The cone prevents product from falling back into the central tube. Once the blender is filled, air is introduced through the central tube; the product is carried with it and spreads at the top as it hits the cone. This motion is carried out for a certain duration, for mixing. In this way, the inner and the outer product layers are displaced relative to one another in the axial direction, to achieve good mixing. The capacity or throughput is determined from the gap between the tube and the annular region; as the gap increases, so does the capacity. This also increases the pressure drop.

*15-10.3.5 High Intensity Mixers*. An impaction mixer (the Henschel mixer) is shown in Figure 15-52. Granules repeatedly break as they form and re-form. The shape of these mixers lends itself to easy cleanup and maintenance. The impaction mixer resembles a typical kitchen food processor. The blades rotate at speeds within the range 2000 to 3000 rpm, so these mixers require significant energy compared to other types with similar capacities (Harnby, 1992). The impaction mixer is used as a mixer–granulator.

*15-10.3.6 High-Shear Mixers*. Harnby states; "These are the alchemist's mortar and pestle and the miller's milestone for grinding of grain" (Harnby, 2000). As powder is pressed between two pressurized rolls (Figure 15-53*a*), any agglomerates will be pulverized. These mixers are commonly preceded by a

**964** SOLIDS MIXING

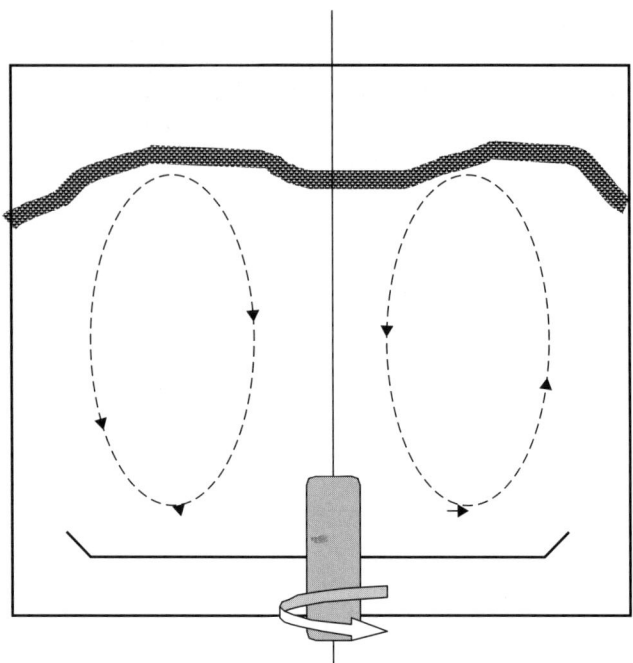

**Figure 15-52** Impaction mixer (Henschel mixer).

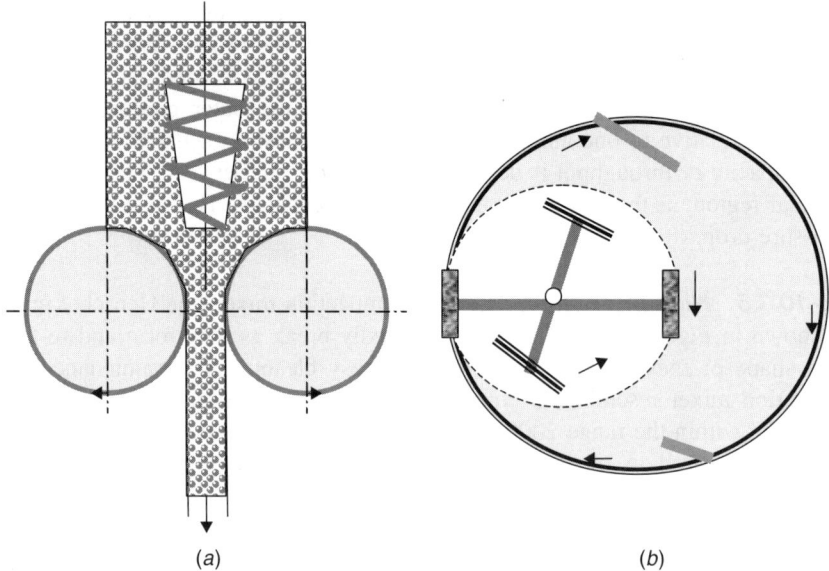

**Figure 15-53** (a) High-shear mixer (Harnby et al., 1992) and (b) Muller mixer (Weidenbaum, 1973; Fayed and Otten, 1984.)

convective tumbler mixer to provide a reasonable quality before the product is conditioned. The muller mixer is intended for finely ground powders and is available in both batch and continuous modes of operation. The turret and the pan can be designed to rotate in opposite directions. Some designs have a stationary pan with only the turret rotating. As shown in Figure 15-53b, the rollers grind the material into a very finely divided and well-mixed consistency by high shearing, and at the same time by folding and turning the mixture over in each turn, resulting in intimate mixing of ingredients. It is not easy to clean or to empty these mixers.

## 15-11 CONTINUOUS MIXING OF SOLIDS

Continuous mixing is the preferred option for processes where throughput is high, space is a constraint, storage of intermediates must be avoided, or the material has a tendency to segregate. A continuous mixer can be a low-cost and reliable option if the operating variables are appropriately controlled. Selection of batch versus continuous mixing has been discussed in Section 15-9.

Continuous mixing is achieved by homogenization of incoming component streams at a fixed rate and ratio. The time required to obtain a homogeneous mix determines the size of a continuous mixer. If the streams are being fed side by side in a mixer, it is the radial mixing that dictates the mixer efficiency. Radial mixing can be achieved by agitated internals (paddles or plows), by tumbling the shell (zigzag mixer) or by static internals in a static mixer. The size of a continuous mixer based on radial mixing alone would be extremely compact. However, it is very challenging to feed fine particulate consistently without fluctuations. This problem is magnified when a mixture contains multiple components.

Axial mixing can dampen the effect of feed stream fluctuations on the composition of the mixture at the outlet. The greater the axial mixing, the lower is the influence of fluctuation or inconsistencies of feed stream. Danckwerts (1953) proposed a variance reduction ratio (VRR) to quantify this characteristic.

$$\text{VRR} = \frac{\sigma^2_{\text{input}}}{\sigma^2_{\text{output}}} \qquad (15\text{-}16)$$

A good mixer will have a high VRR when $\sigma^2_{\text{output}}$ is low, even when $\sigma^2_{\text{input}}$ is high. The ratio above is a metric for the performance characteristics of a continuous mixer. For a given mixer, this ratio also depends on the operating parameters and the nature of input variation. To demonstrate this relationship, Weinekotter and Gericke (1999) carried out detailed investigations using the setup shown in Figure 15-54. Input disturbances of different time periods were imposed on the mixer. The output concentration was monitored using an optical probe. The results are summarized in Figure 15-55.

When the time period of input fluctuation (120 s) is greater than the residence time (44 s), the mixer is incapable of dampening out the fluctuations. The input

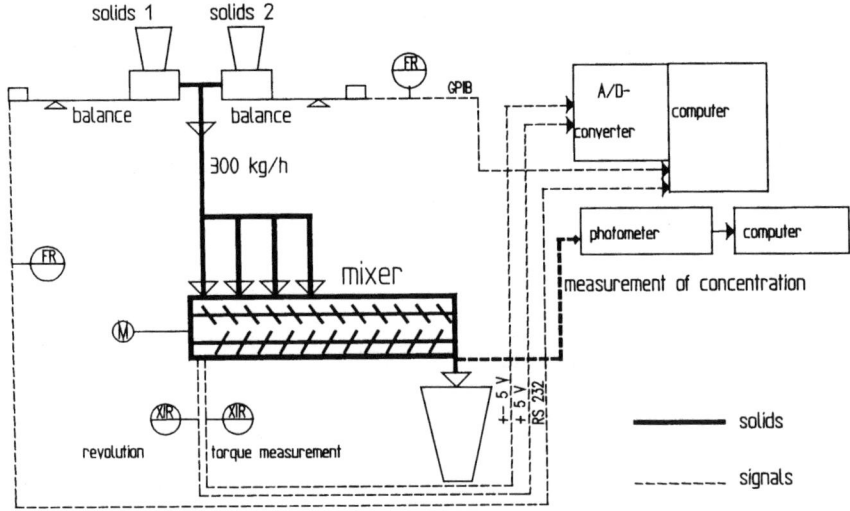

**Figure 15-54** Experimental test set for continuous mixers. (From Weinekotter and Gericke, 1999.)

variability is propagated to output concentration and results in a VRR of 1. On the other hand, when the time period of the input fluctuations is 30 s, a significant dampening is observed and the resulting VRR value is 82. In general, higher-frequency fluctuations in the feed stream are easier to dampen out than low-frequency fluctuations.

Therefore, one must consider the relationship between average residence time and possible cycle time of feeder fluctuations while specifying a continuous mixing system. The average residence time in a mixer is given by

$$\text{average residence time} = \frac{\text{retention volume in mixer}}{\text{mass flow rate/bulk density}} \qquad (15\text{-}17)$$

The retention volume refers to the volume of solids in the mixer at any instant.

Given the stochastic nature of the motion of particles in a mixer, it is common to observe a distribution of residence time. This is a direct result of axial mixing or dispersion. Axial dispersion/mixing will help dampen the feed fluctuations; however, it also broadens the residence time distribution. A broad residence time distribution implies a variable time history for particles passing through the mixer. If the mixer is used as a dryer, granulator, coater, or reactor, broad residence time distribution is not desirable. A balance between the process requirements (allowable residence time distribution) and process capability (feeder characteristics and mixer residence time) must be reached to design a robust continuous mixing system.

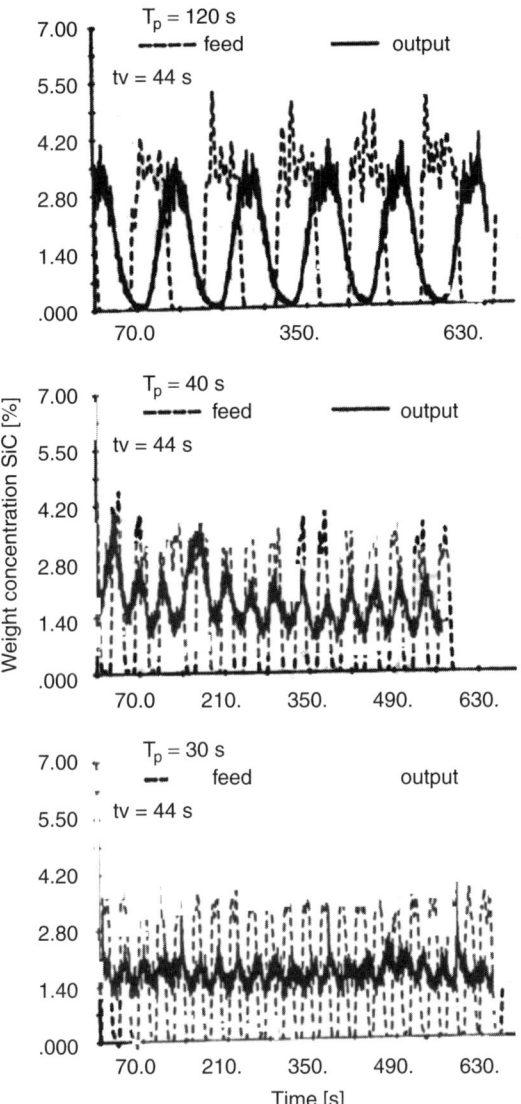

**Figure 15-55** Relationship between residence time and feed fluctuation reduction. (From Weinekotter and Gericke, 1999.)

### 15-11.1  Types of Continuous Mixers

*15-11.1.1  Mixing by Simultaneous Feed.* Mixing using two or more feeders represents a continuous mixing process. Metering each ingredient precisely and bringing the streams together blends the components. For free-flowing solids, a static mixer can be used. There is little chance for backmixing or axial mixing in such units. Static mixers must always be starve-fed to achieve proper randomization of feed streams.

### 15-11.1.2 Agitated Mixers as Continuous Mixers.
Ribbon and paddle mixers can be run in continuous mode. The material is fed at one end of the mixer while a weir on the opposite end provides retention control. The average residence time and residence time distribution depend on the mixer speed, geometry of agitators, and the rpm. These mixers are typically run 30 to 50% full during normal operation. Typical residence times are in the range 0.5 to 5 min. Some agitated mixers, such as the Nauta (vertical screw) and plow types, are not suitable for operating in continuous mode.

### 15-11.1.3 Tumbling Mixers as Continuous Mixers.
Rotating drum or zigzag mixers are commonly used as continuous mixers. These are designed with a high aspect ratio and operated in horizontal mode. Continuous drums are inclined slightly for natural movement of material along the length. These are particularly suitable for coating applications, where low intensity mixing and long residence time are required.

## 15-12  SCALE-UP AND TESTING OF MIXERS

In this section we outline the commonly accepted scale-up criteria used in industry for agitated mixers and silo blenders. Scale-up of tumbling mixers was addressed in Section 15-5. The complexity of interaction between physical properties of solids, mixer configuration and velocity, and stress profiles within a mixer makes it difficult to formulate generalized scale-up criteria. Various experimental investigations into scale-up, however, do provide useful guidance for scale-up. No concurrence on acceptable approach has been reached, and various manufacturers tend to follow their experience.

A general axiom for scale-up is: "Commit your blunders on a small scale, and make your profits on a large scale" (Zlokarnik, 1984). Although it is easy to conduct experiments on a small scale, one must also consider the following questions:

- How small can the pilot mixer be?
- What geometric, dynamic, and kinematic parameter ratios need to be kept constant to ensure the validity of data on a large scale?
- Are the material properties (e.g., particle size distribution, temperature, and moisture) comparable to those in a large scale process?
- Are there other processes (heating, cooling, granulation) that require scale-up along with the mixing process?

### 15-12.1 Principle of Similarity

Similarity principles are applied routinely for scaling up of industrial liquid–liquid mixers. Wang and Fan (1978) first proposed using geometric, kinematic, and dynamic similarity for scale-up of tumbling mixers. *Geometric similarity* involves keeping the ratio of linear dimensions of pilot and full scale mixers constant. *Kinematic similarity* requires the velocity ratio between corresponding points in the two (pilot and full scale) systems to be constant while maintaining geometric similarity. *Dynamic similarity* dictates that the ratio of forces at corresponding points in the two systems be constant. Sometimes, nondimensional force ratios (e.g., Froude number, Reynolds number) are used. When the governing equations are not known, such parameters can be derived from dimensional analysis.

An industrial practitioner is interested in the following questions:

- What are the power consumption and mixing time in a full scale mixer for a given quality of mixing?
- How do the mixing quality and power consumption change with mixer rpm?

### 15-12.2 Scale-up of Agitated Centrifugal Mixers

As mentioned earlier, it is nearly impossible to formulate generalized scale-up equations for solids mixers. However, extensive experimental investigations conducted by Muller (1982), Scheuber et al. (1980), and Merz and Holzmuller (1981) have resulted in the following usable criteria. As shown in Figure 15-56, two regions are demarcated at a Froude number of 3. The improvement in mixing coefficient for a given mixer at Fr > 3 is dramatic.

The coefficient of mixing, M, proposed by Muller, is a parameter used in his semiempirical one dimensional model of horizontal mixers. The mixing coefficient determines how quickly concentration equalization will occur in the mixer. A large mixing coefficient will result in a short mixing time for a given quality of mix. The mixing coefficient is assumed to remain constant at all points in the mixer for the duration of the mix. It should be noted that M depends on the type of mixer, geometry of the internals, and the operating conditions. It does not depend on the properties of the mixture components (e.g., size or density).

$$\frac{M}{D^2 n} = \text{constant} \quad (\text{Fr} < 3) \qquad (15\text{-}18)$$

$$\frac{M}{D^2 n} \approx \text{Fr}^2 \quad (\text{Fr} > 3) \qquad (15\text{-}19)$$

where M is the mixing coefficient, D the diameter of the mixer (m), n the mixer rpm, and Fr the Froude number, defined as

$$\text{Fr} = \frac{v^2}{gR} = \frac{R\omega^2}{g} \qquad (15\text{-}20)$$

# SOLIDS MIXING

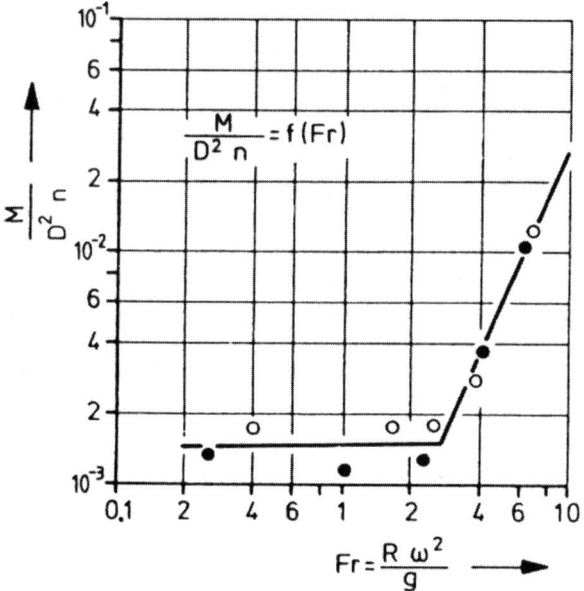

**Figure 15-56** Relationship between dimensionless mixing coefficient and Froude number. (From Muller, 1982.)

where v is the peripheral velocity of mixing element (plow, paddle) (m/s), R the mixer radius ($= D/2$) (m), and $\omega$ the angular velocity of the agitators (rad/s):

$$n = \frac{30\omega}{\pi} = \frac{30}{\pi}\left(\frac{v}{R}\right) \quad (15\text{-}21)$$

Two common approaches are used for scaling these mixers:

1. Keep the peripheral speed constant between the pilot mixer and the full scale mixer.
2. Keep the Froude number constant between the pilot mixer and the full scale mixer.

Note that each of these approaches is used by mixer equipment manufacturers, and this suggests that more research and development are required to increase our understanding of solids mixing processes.

- *Criterion 1: Keep v constant:*

$$\frac{n_{\text{pilot}}}{n_{\text{full scale}}} = \frac{R_{\text{full scale}}}{R_{\text{pilot}}} \quad (15\text{-}22)$$

assuming geometric similarity and the same quality of mixing.

- *Criterion 2: Keep Froude numbers constant:*

$$\frac{n_{pilot}}{n_{full\ scale}} = \sqrt{\frac{R_{full\ scale}}{R_{pilot}}} \qquad (15\text{-}23)$$

assuming geometric similarity and the same quality of mixing.

It is a common practice to use a Froude number of 7 for mixing nonfriable materials. For friable materials, the effect of agitator impact on breakage must be evaluated. Attrition is nonlinear with impact velocity, whereas it is linear with mixing time. Therefore, an optimum can be found through experimentation.

**15-12.2.1  Mix Time.** Rumpf and Muller (1962) have shown experimentally that the mixing coefficient can also be related to mixer length (L) if the mixer diameter (D) is kept constant:

$$\frac{Mt}{L^2} = \text{constant} \qquad (15\text{-}24)$$

where M is the mixing coefficient, t the mixing time (s), and L the mixer length (m).

For Froude numbers below 3 and for geometrically similar mixers operating at the same peripheral speed of agitator, the mixing time increases linearly with the mixer diameter,

$$t \approx \left(\frac{L}{D}\right)^2 \frac{D}{v} \qquad (15\text{-}25)$$

where D is the mixer diameter.

At higher Froude numbers (>3), the mixing time is linear with mixer volume (not the diameter). The effect of agitator speed is significant in this range.

$$t \approx \left(\frac{L}{D}\right)^2 \frac{D^3}{v^5} \qquad (15\text{-}26)$$

**15-12.2.2  Power Consumption.** The relationship between power consumption and Froude number for agitated centrifugal mixers is shown in Figure 15-57. The power consumption is expressed in a nondimensional form as the Newton number (Ne):

$$\text{Ne} = \frac{P}{\rho_s(1-\varepsilon)D^5 n^3 \left(\dfrac{L}{D}\right)} \qquad (15\text{-}27)$$

# 972 SOLIDS MIXING

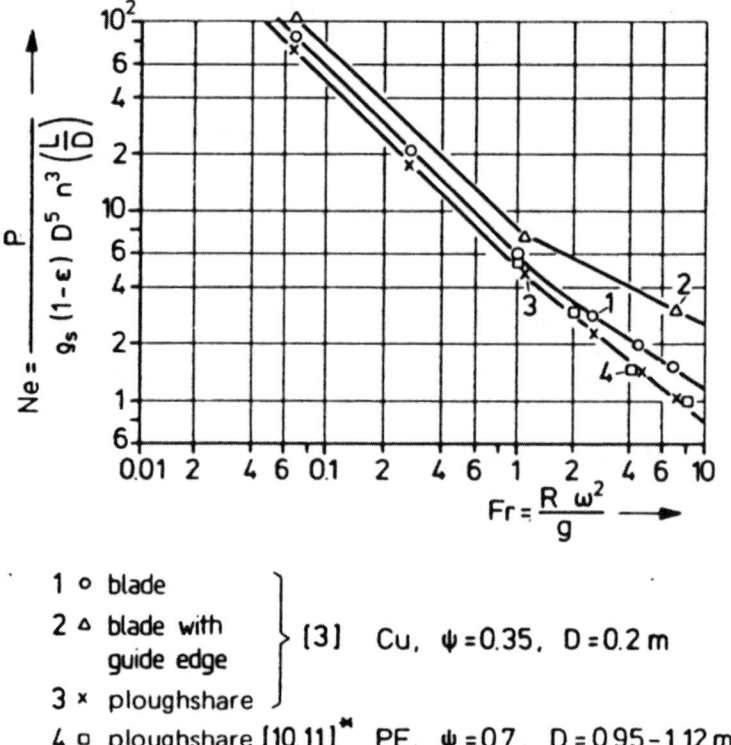

1 ○ blade
2 △ blade with guide edge
3 × ploughshare
} [3] Cu, $\psi = 0.35$, $D = 0.2$ m

4 □ ploughshare [10,11]* PE, $\psi = 0.7$, $D = 0.95 - 1.12$ m

**Figure 15-57** Relationship between power consumption and Froude number for centrifugal agitated mixers. (From Muller, 1982.)

For Fr < 1, where the acceleration forces are relatively small, the material is not fluidized and under plastic shear, the following relationship holds:

$$\text{Ne} \sim \text{Fr}^{-1} \qquad (15\text{-}28)$$

At higher Froude numbers, the configuration of paddles/agitators will have a significant influence on the shape of the curve (Figure 15-57). The surface roughness, shape, and size of the particles also have a significant influence on this curve.

In practice, the driver motor and shaft must be designed such that the mixer can be started with a normal load of material. The peak torque and power associated with the startup conditions far exceeds the operating conditions.

## 15-12.3  Scale-up of Ribbon Mixers

The literature on mixing in ribbon mixers is not extensive. The following rules can be extracted from current literature:

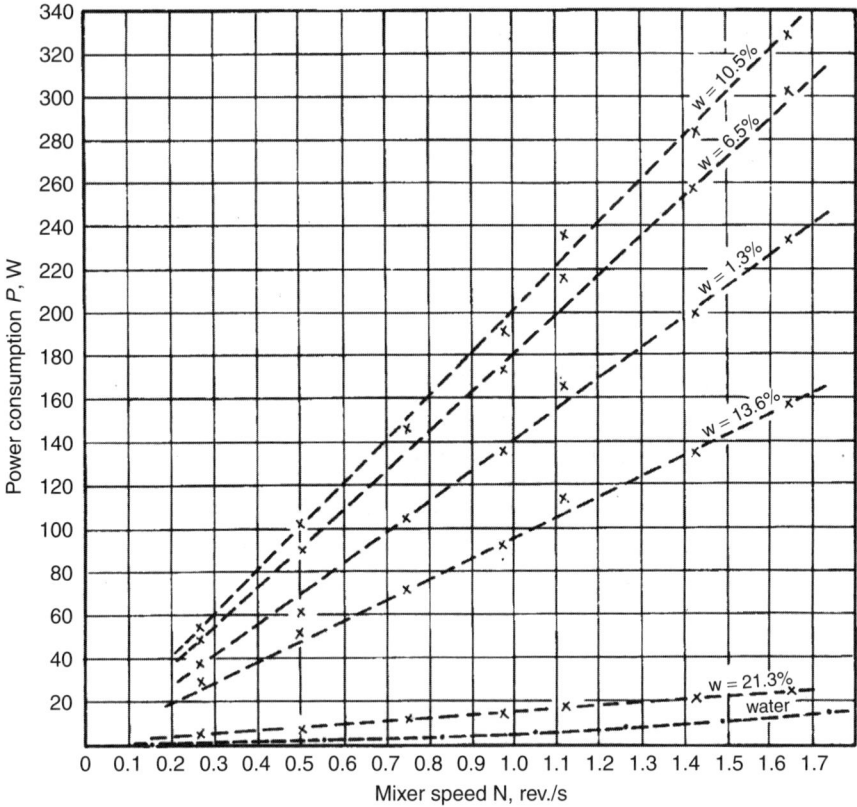

**Figure 15-58** Variation of power consumption with rpm. (From Masiuk, 1987.)

- Power consumption in a ribbon mixer increases linearly with rpm at a given level of fill in the mixer. The rate of increase in power consumption with rpm is greater at higher fill levels (see Figure 15-58).
- The power consumption depends largely on the pitch and diameter of the mixing ribbon rather than its width.

### 15-12.4 Scale-up of Conical Screw Mixers (Nauta Mixers)

Entrop (1978) conducted a wide range of experiments in conical screw mixers. The results can be summarized as follows:

#### 15-12.4.1 Power Consumption

$$\frac{P}{n_s \rho_s (1-\varepsilon) d_s^4 g} = k_1 \frac{n_s}{n_a} \left(\frac{1}{d_s}\right)^{1.7} \quad (15\text{-}29)$$

where P is the consumption (W); $n_s$ the screw rpm (min$^{-1}$); $\rho_s$ the true particle density (kg/m$^3$); $\varepsilon$ the packed bed voidage; $d_s$ the diameter of the orbiting screw (m); $n_a$ the arm rotational speed rpm (min$^{-1}$); l the immersion length of the screw (m).

### 15-12.4.2 Mix Time

$$t = \frac{k_2}{n_s}\left(\frac{1}{d_s}\right)^{1.93} \tag{15-30}$$

$k_1$ and $k_2$ are constants that must be determined experimentally on the pilot scale.

### 15-12.5 Scaling of Silo Blenders

The mixing action of silo blenders is due to velocity gradients generated during discharge and recirculation of material. The mixing behavior of blenders of two different sizes will be the same if they are geometrically similar (H/D ratio) and have similar velocity gradients. It has been shown in practice that homogeneity tests conducted on a 3 m$^3$ (volume) blender can be scaled to a 500 m$^3$ blender (Wilms, 1988). It should be noted that the blender internals do not always scale down with respect to blender diameter. The arching dimension (mechanical and cohesive) of a material dictates the dimensions of the internals for reliable flow.

Very few experimental data on silo blenders are available in the published literature. All manufacturers offer test facilities for pilot scale testing and use proprietary data for scale-up.

### 15-12.6 Specifying a Mixer

When specifying equipment for mixing of particulate solids, the following issues need to be considered:

*Performance*

- Metric for mixedness or expression for mixing index
- Scale of scrutiny
- Statistical limits for mixture concentration variations and acceptable frequency for concentration being outside the limits

*Mechanical Issues*

- *Wear:* effect of material abrasiveness on mixer internals
- *Attrition:* breakage of material during mixing process
- *Material of construction:* mixer body, seals, shafts, and gaskets
- *Headroom availability and cost:* accountability of mixer design for the cost of the structure required to house a mixer

- *Drives:* necessity for practical size of the drive and motor
- *Contamination:* material contamination due to bearings or internals
- *Cross-contamination:* need to clean the mixer between batches or product grades (especially critical for pharmaceutical processes)
- *Dust explosion:* potential for dust explosion and need to purge with inert gas
- *Hazardous materials:* need for dust or gastight containment
- *Design of internals and clearances:* depends on the potential of the particles to smear and degrade

*Configuration Issues*

- *Feed system:* accurate dosing system required for batch mixers, good feeder systems for continuous mixers
- *Discharge configuration:* importance of designing the system downstream of the mixer such that segregation is minimized (chutes, silos, and pneumatic transfer systems can de-mix the mixture and result in poor system performance)

After an initial selection of a suitable mixer has been made (see Figure 15-33), it is highly recommended that pilot scale testing be conducted to ensure acceptable performance upon installation.

### 15-12.7 Testing a Mixer

Many mixers can be operated in either batch or continuous mode. It is advised that the selection of batch versus continuous operation be made before any tests are conducted. The mixing tests should be conducted in appropriate mode. For instance, performance of a paddle mixer in continuous mode of operation cannot be reliably estimated from batch experiments.

In either case, we need to address the following questions regarding sampling and analysis:

1. Sample size
2. Location of sampling
3. Method of sampling
4. Number of samples

***15-12.7.1 Sample Size.*** An ideal sample size is equal to the scale of scrutiny or the scale at which the product specification has been developed. If the scale of scrutiny is much smaller than the minimum amount of sample that can reliably be obtained from a sampler, suitable sample reduction techniques must be utilized. The spinning riffler has been found to be the most reliable method of reducing the sample size (Allen, 1981). The sample size must be greater than the minimum amount required for the analytical technique.

***15-12.7.2 Location of Sampling.*** In practice, the most meaningful and reliable method of sampling is to take a full-stream sample for a short duration of time at the outlet of the mixer. Not only does it conform to the golden rules of sampling, this sample also accounts for any de-mixing that might occur during the discharge process.

For studying the mixing patterns within a batch mixer, the mixer must be stopped at various mixing times and sampled at various locations. These locations are usually chosen by dividing the space into equal regions. The start–stop transients during such experiments can affect the results; therefore, extended mixing without stopping must be performed for confirmation.

In the case of continuous mixers, samples must be acquired at the mixer outlet at regular frequency, starting at three times the residence time of the mixer. Usually, it is sufficient to sample up to 10 times the residence time. However, the data should also be checked for long-time scale patterns, in which case the total test time must be extended. Sampling within the mixer can be done by stopping the mixer and using sampling probes. Such data are useful in the determination of mixing length and identification of stagnant regions.

***15-12.7.3 Method of Sampling.*** Sampling thieves or probes are commonly used for taking samples from stationary mixtures. It has been shown that many samplers introduce a bias by disturbing the very mixture they are supposed to sample (Figures 15-4, 15-5, and 15-6). Special designs mitigate the sampler bias. Using a scoop sample from the top surface of the mixture is the most nonrepresentative sample that one can obtain. Bias can be checked by comparing the mean composition of the samples with the composition of the entire mixture using Student's $t$ test.

For continuous mixers, full-stream samplers at the mixer outlet are recommended. In-line or in situ probes to measure the physical/chemical property of interest, if available, can give useful insight into the mixing process. Before sampling, make sure that the system has reached a steady state and that at least three residence times have passed.

***15-12.7.4 Number of Samples.*** Determination of the number of samples depends largely on the confidence level in the estimate of population standard deviation from sample standard deviation. The precision of the estimate of standard deviation increases with the number of samples; however, so does the analytical cost. The analytical cost is generally a small price to pay to avoid selection of an improper mixer. It is a common practice to take 15 to 30 samples during testing. As a rule of thumb, the total amount of sample removed from a batch mixer should not exceed 5% of the charge. This will prevent sample bias. Repeat tests must be conducted if the total number of samples required for analysis exceeds 5% of the charge.

### 15-12.8 Testing a Batch Mixer

For reliable scale-up and extrapolation of test data, it important to maintain geometric, kinematic, and dynamic similarity between the test mixer and the full scale mixer. A batch mixer should have access for sampling within the mixer at various mixing times. Batch mixers require reliable dosing measurement, which can be accomplished by weighing the components individually.

A typical batch mixer test would include the following steps:

1. Load the components in predetermined sequence. The sequence of addition of components can affect the rate at which homogeneity is achieved.
2. Mix for a known period.
3. Stop the mixer and take 10 to 15 samples from various locations within the mixer. Refer to the sampling guidelines discussed earlier.
4. Start the mixer again and repeat steps 2 and 3 at least three more times. The subsequent mixing times can be chosen in a geometric progression (e.g., 2, 4, 8, and 16 min).

### 15-12.9 Testing a Continuous Mixer

A continuous mixer cannot be tested reliably without a reliable feed system for the components. The feed system should be capable of turndown ratios similar to those of the full scale system. The objective of the test is threefold:

1. To determine the quality of mixture at the discharge.
2. To assess the ability of the continuous mixer to dampen feeder-related fluctuations or variance in composition.
3. To determine the residence time distribution of the components in the mixer.

The mixer operating conditions must be selected carefully based on kinematic and dynamic similarity conditions. The following experimental considerations must be heeded:

- Allow the system to reach a steady state.
- Follow the golden rules of sampling.
- Do not start sampling until three residence times have elapsed.
- Sample until about 10 residence times have elapsed or longer if necessary.
- Study start–stop transients and transients associated with feed system instabilities.

### 15-12.10 Process Safety in Solids Mixing, Handling, and Processing

The dust arising from the mixing or processing of combustible solid materials can form explosive mixtures with air. In the design of plants to handle such materials, consideration must be given to the potential for dust explosions and actions to prevent or minimize dust explosions. A dust explosion hazard exists when a combustible dust with particle size distribution less than 420 μm is dispersed in air or other oxidant. In general, most organic solids and metals can form explosive dust clouds. For dust to explode, certain conditions have to exist:

- Particles of dust must be of suitable size.
- Concentration of dust in the air must fall within explosive limits.
- There must be a source of ignition energy.

*15-12.10.1 Important Parameters.* In any unit operation involving drying, mixing, milling, conveying, storage, and so on, the following major properties of the product will probably need to be known for safe operation.

- Melt/decomposition temperature
- Potential evolution of toxic or flammable gases upon heating/combustion
- Potential for spontaneous heating
- Minimum ignition energy (MIE)
- Limiting oxidant concentration (LOC)
- Resistivity (ability to generate and accumulate static electric charge)
- Explosiveness (rate of pressure rise in closed test apparatus)
- Smoldering characteristics

*15-12.10.2 General Design Safety.* In addition to taking specific explosion prevention and protection steps, the conditions that can result in a secondary explosion and the frequency of ignition must be minimized. The following are important:

- Good housekeeping
- Control of mechanical sparks and friction
- Control of cutting, welding, and other open ignition sources
- Electrical bonding and grounding
- Electrical classification
- Insulating hot surfaces

*15-12.10.3 Housekeeping.* Good housekeeping is required for dust explosion safety in rooms or areas where dust may escape and accumulate. This dust cloud provides fuel for the secondary explosion within the room or building. Reducing significant additional dust accumulation is therefore a major factor in reducing the hazard in areas where a dust hazard can exist. A dust layer 0.8 mm thick and covering the floor of a building is sufficient to produce a uniform dust cloud of optimum concentration 3 m high throughout the building. Therefore, good housekeeping is critical for safety.

## 15-12.10.4 Control of Ignition Sources

*Control of Mechanical Sparks and Friction.* Foreign materials (such as tramp metal) that are capable of igniting combustible material being processed should be removed from the process stream by magnetic separators, pneumatic separators, or grates or other separation devices.

*Control of Cutting and Welding and Open-Flame Ignition Sources.* Although cutting and welding are well recognized as an ignition sources for dust explosions, these operations continue to be ignition sources

*Control of Static Electricity.* Bonding is the process of electrically connecting adjacent conductive components so that they are at the same electrical potential and no sparks can occur between them. When one or more of these components is additionally connected to an electrical ground, all of the components are then grounded, and sparks cannot occur either between them or to other systems which are also grounded. Bonding and grounding of electrically conductive components of conveying, mixing, feeding, blending, and storage systems should be provided to dissipate electrostatic buildup below hazardous accumulations. The type or extent of bonding/grounding needed is a function of the ignition sensitivity and conductivity of the material being processed or handled.

### 15-12.10.5 Electrical Classification.
In those areas of the plant where a hazardous quantity of dust accumulates or is suspended in the air, the area should be classified and all electrical equipment and installations in those areas should comply with Article 502 or Article 503 of NFPA 70, the National Electrical Code, as applicable.

### 15-12.10.6 Control of Hot Surfaces.
It is generally accepted that it is good engineering practice to limit the surface temperature in dust-handling or dust-processing areas to less than 80% of the minimum ignition temperature of the dust layer. Areas requiring control of hot surfaces are generally identified as those requiring Division II Electrical Classification.

### 15-12.10.7 Explosion Prevention and Protection.
If the evaluation determines that dust explosion protection is required, dust explosions can either be prevented, or explosion venting provided, to protect against unacceptable consequences. Prevention methods are outlined in National Fire Protection Association's *Explosion Prevention Systems* (NFPA69, 2000). Explosion (deflagration) venting is outlined in *Guide for Venting of Deflagrations* (NFPA654, 2000). An outline of the general approach to protection (Eckhoff, 1991) is shown in Figure 15-59.

The following are methods of explosion protection by preventing or containing explosions:

1. Oxidant concentration reduction
2. Combustible concentration reduction

**980** SOLIDS MIXING

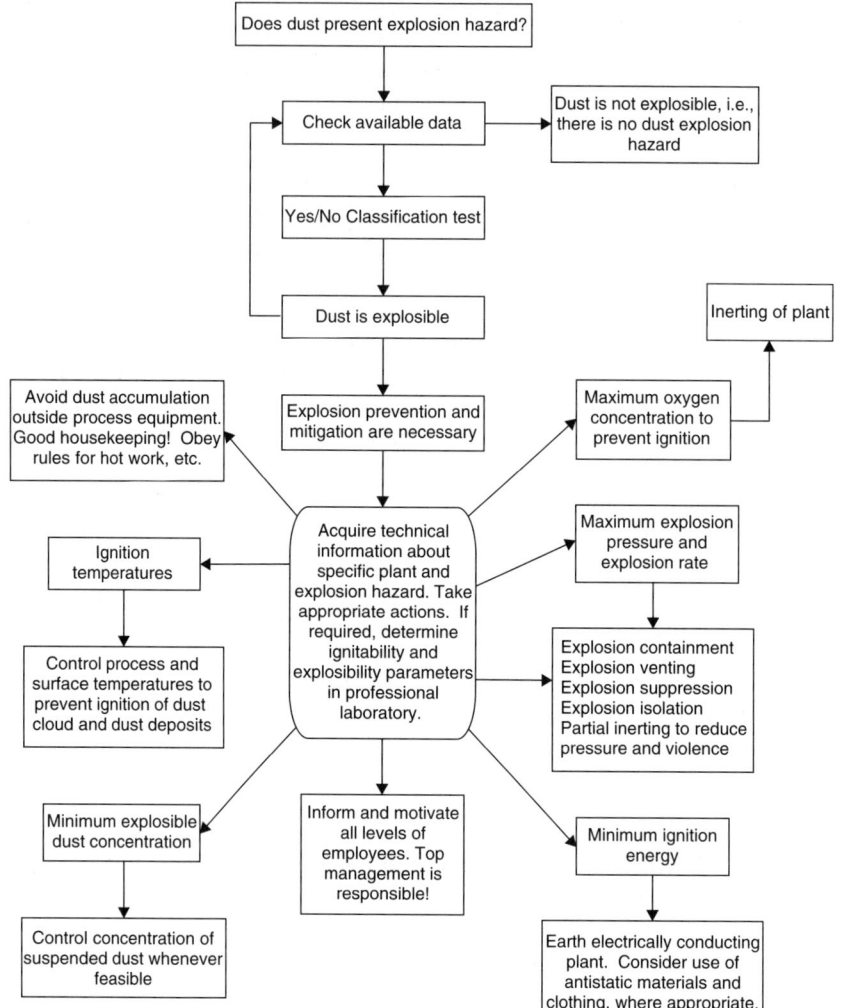

**Figure 15-59** Outline of general approach to practical dust explosion protection. (From Eckhoff, 1991, modified and extended version according to Field 1982.)

3. Explosion suppression
4. Deflagration pressure containment

Methods 1 and 2 prevent combustion at a rate sufficient to result in an explosion; methods 3 and 4 are listed as protection methods based on limiting or preventing damage. Spark extinguishing is also listed as an explosion prevention method by NFPA69 (2000) but is only applicable to ducts transporting combustible dusts and must be used in conjunction with one of the other explosion prevention methods or explosion venting for protection of the complete system.

# NOMENCLATURE

## Part A

| | |
|---|---|
| CoV | coefficient of variance (eq. 15-2) |
| I | intensity of segregation (eq. 15-1) |
| M | mean concentration |
| $N_j$ | number of samples contained in core j |
| N | number of samples |
| RSD | relative standard deviation (eq. 15-2) |
| $\bar{x}$ | mean composition |
| $x_{ij}$ | concentration of a sample |
| $\bar{x_j}$ | mean concentration of core $j$ |

## Greek Symbols

| | |
|---|---|
| $\sigma$ | standard deviation |
| $\sigma^2$ | variance of sampled data (eqs. 15-4 and 15-5) |
| $\sigma_A^2$ | axial variance |
| $\sigma_R^2$ | radial variance |
| $\sigma_r^2$ | variance of randomly chosen concentration data |
| $\sigma_0^2$ | initial variance of a fully segregated system |

## Part B

| | |
|---|---|
| $c_i$ | concentration of the $i^{th}$ sample |
| $C_i$ | coefficient of variation of the $i^{th}$ component |
| $d_s$ | diameter of orbiting screw (m) |
| D | diameter (m) |
| g | gravitational constant (m/s$^2$) |
| H | level of solids in hopper (m) |
| $k_1, k_2$ | constants in eqs. (15-29) and (15-30) |
| l | immersion length of a screw in a Nauta mixer (m) |
| L | constant |
| L | mixer length (m) |
| m | number of samples |
| M | coefficient of mixing (m$^2$/s) |
| n | mixer agitator speed (rps) |
| $n_a$ | screw rotation speed (min$^{-1}$) |
| $n_s$ | arm rotation speed in a Nauta mixer |
| N | number of particles |
| P | component fraction |
| r | mixer radius (m) |
| R | mixer radius (m) |
| s | standard deviation of sample |
| t | time (s) |
| v | peripheral velocity of the mixing element (m/s) |

V        volume (m$^3$)
$w_i$    weight fraction of the i$^{th}$ component
W        total mass (kg)
x        particle size ($\mu$m)
X        sample average
Fr       Froude number (eq. 15-20)
Ne       Newton number (eq. 15-27)
VRR      variance reduction ratio (eq. 15-14)

*Greek Symbols*

$\varepsilon$    packed bed voidage
$\rho_s$         true particle density (kg/m$^3$)
$\sigma$         standard deviation of population
$\omega$         angular velocity (rad/s)

## REFERENCES

Adams, J., and A. Baker (1956). An assessment of dry blending equipment, *Trans. Inst. Chem. Eng.*, **34**, 91.

Allen, T. (1981). *Particle Size Measurement*, 3rd ed., Chapman & Hall, London.

Ashton, M. D., and F. H. H. Valentin (1966). The mixture of powders and particles in industrial mixers, *Trans. Inst. Chem. Eng.*, **44**, T166.

Brennen, A. K., Jr. (1990). Selecting the right mixer: batch or continuous, *Powder Bulk Eng.*, **4**(1), 38–50.

Bridgwater, J. (1976). Fundamental powder mixing mechanisms, *Powder Technol.*, **15**, 215–236.

Brone, D., C. Wightman, K. Connor, A. W. Alexander, F. J. Muzzio, and P. Robinson (1997). Using flow perturbations to enhance mixing of dry powders in V-blenders, *Powder Technol.*, **91**, 165–172.

Carson, J. W., T. A. Royal, and D. J. Goodwill (1986). Understanding and eliminating particulate segregation mechanisms, *Bulk Solids Handl.*, **6**(1), 139–144.

Cavender, J. (2000). Quick, thorough and gentle blending with a vertical ribbon mixer, *Powder Bulk Eng.*, **14**(1), 46–57.

Chester, A. W., J. A. Kowalski, M. E. Coles, E. L. Muegge, F. J. Muzzio, and D. Brone (1999). Mixing dynamics in catalyst impregnation in double-cone blenders, *Powder Technol.*, **102**, 85–94.

Danckwerts, P. V. (1953). Theory of mixtures and mixing, *Chem. Eng. Res.*, **6**, 355.

Donald, M. B., and B. Roseman (1962). Mixing and de-mixing of solid particles: I. Mechanisms in a horizontal drum mixer, *Br. Chem. Eng.*, **7**, 749–753.

Eckhoff, R. K. (1991). *Dust Explosions in the Process Industries*, Butterworth-Heinemann, Oxford.

Egermann, H. (1980). Effects of adhesion on mixing homogeneity: I. Ordered adhesion–random adhesion, *Powder Technol.*, **27**, 203–206.

Entrop, W. (1978). *Proc. European Conference on Mixing in the Chemical and Allied Industries*, Mons, Belgium, D1, pp. 1–14.

Fan, L. T., Y.-M. Chen, and F. S. Lai (1990). Recent developments in solids mixing, *Powder Technol.*, **61**, 255–287.

Fayed, M. E., and L. Otten, eds. (1984). *Handbook of Powder Science and Technology*, Van Nostrand Reinhold, New York.

Field, P. (1982). *Dust Explosions*, Elsevier, Oxford.

Forberg, H. (1992). Short note on modern mixing: theory and practice, *Powder Handl. Process.*, **4**(3), 318–320.

Fuller, W. O. (1998). Mixing up a batch: batch mixer types and selection tips, *Powder Bulk Eng.*, **12**(1), 48–66.

Geldart, D. (1973). Types of gas fluidization, *Powder Technol.*, **7**, 285–292.

Greathead, J., and W. Simmons (1957). *Chem. Eng. Prog.*, **53**, 194.

Harnby, N. (1967). A comparison of the performance of industrial solids mixers using segregating materials, *Powder Technol.*, **1** (2), 94–102.

Harnby, N. (1997). The selection of powder mixers, in *Mixing in the Process Industries*, Butterworth-Heinemann, Oxford, pp. 42–61.

Harnby, N. (2000). An engineering view of pharmaceutical powder mixing, *Pharm. Sci. Technol. Today*, **3**(9), 303–309.

Harnby, N., M. F. Edwards, and A. W. Nienow (1992). *Mixing in the Process Industries*, 2nd ed., Butterworth-Heinemann, Oxford.

Hersey, J. A. (1970). Sampling and assessment of powder mixtures for cosmetics and pharmaceuticals, *J. Soc. of Cosmet. Chem.*, **31**, 259–269.

Hersey, J. A. (1975). Ordered mixing: a new concept in powder mixing practice, *Powder Technol.*, **11**, 41–44.

Hixon, L., and J. Ruschmann (1992). Using a conical screw mixer for more than mixing, *Powder Bulk Eng.*, **6**(1), 37–43.

Hosakawa (1994). *Mechanical and Thermal Processing Technology*, Hosakawa Micron, Doetinchem, The Netherlands.

Johanson, J. R. (1970). In-bin blending, *Chem. Eng. Prog.*, **66**(6), 50–55.

Kent, M. (2002). Mixer selection: an aid to those involved in the selection, specification and purchasing of mixing machines, *Powder Handl. Process.*, **14**(1), 41–43.

Krambrock, W. (1976). Mixing and homogenizing of granular materials in a pneumatic mixer unit, *Powder Technol.*, **15**, 199–206.

Kristensen, H. G. (1973). Statistical properties of random and non-random mixtures of dry solids: I and II, *Powder Technol.*, **7**, 249–257; **8**, 149–157.

Lacey, P. M. C. (1954). Developments in the theory of particle mixing, *J. Appl. Chem.*, **4**, 257–268.

Manjunath, K. S., S. R. de Silva, and A. W. Roberts (1992). Homogenization of bulk powders in plane symmetric silos, *Powder Handl. Process.*, **4**(3), 283–292.

Masiuk, S. (1987). Power consumption, mixing time and attrition action for solids mixing in a ribbon mixer, *Powder Technol.*, **51**, 217–229.

Merz, A., and R. Holzmuller (1981). Radionuclide investigations into the influence of mixing chamber geometry on the mass transport in a continuous plowshare mixer, *Proc.*

*Powtech Conference on Mixing of Particulate Solids*, Institution of Chemical Engineers, pp. S1/D/1–11.

Michael, B. J. (1992). Answers to 10 common questions about batch mixing, *Powder Bulk Eng.*, **6**(1), 44–49.

Moakher, M., T. Shinbrot, and F. J. Muzzio (2000). Experimentally validated computations of flow, mixing and segregation of non-cohesive grains in 3D tumbling blenders, *Powder Technol.*, **109**, 58–71.

Muller, W. (1982). Mixing of solids: methods and present state of design, *Germ. Chem. Eng.*, **5**, 263–277.

Muzzio, F. J., P. Robinson, C. Wightman, and D. Brone (1997). Sampling practices in powder blending, *Int. J. Pharm.*, **155**(2), 153–178.

Muzzio, F. J., M. Roddy, D. Brone, A. W. Alexander, and O. Sudah (1999). An improved powder-sampling tool, *Pharm. Technol.*, **23**, 92–110.

NFPA654 (2000). *Explosion Prevention Systems*, National Fire Protection Association, Quincy, MA.

NFPA69 (2000). *Guide for Venting of Deflagrations*, National Fire Protection Association, Quincy, MA.

Pahl, H. (1986). *Mischen beim herstellen und verarbeiten von Kunststoffen*, VDI-Verlag, Germany.

Peschl, I. A. S. Z. (1996). Universal blender: A blending and mixing system for cohesive and free flowing powders, *Bulk Solids Handl.*, **6**(3), 173–176.

Poole, K. R., R. F. Taylor, and G. P. Wall (1964). Mixing powders to fine scale homogeneity: studies of batch mixing, *Trans. Inst. Chem. Eng.*, **42**, T305.

Ramponi, S., D. Negrini, and M. Passerini (2002). Mixer selection for powders, *Powder Handl. Process.*, **14**(1), 30–39.

Roberts, A. W. (1990). Storage and discharge of bulk solids from silos with special reference to the use of inserts, POSTEC Research Report, Porsgrunn, Norway.

Robinson, P., F. J. Muzzio, C. Wightman, D. Brone, and E. K. Gleason (1999). End-sampling thief probe, U.S. patent 5, 996, 426.

Rumpf, H., and W. Muller (1962). An investigation into the mixing of powders in centrifugal mixers, *Proc. Symposium on the Handling of Solids, Institution of Chemical Engineers*, pp. 38–46.

Scheuber, G., Alt, Cr., and R. Leucke (1980). Untersuchung des Mischungverlaufs in Feststoffmischern unterscheidlicher Groesse, *Aufbereit.-Tech.*, **21**(2), 57–68.

Shinbrot, T., A. Alexander, M. Moakher, and F. J. Muzzio (1999a). Chaotic granular mixing, *CHAOS*, **9**, 611–620.

Shinbrot, T., A. Alexander, and F. J. Muzzio (1999b). Spontaneous chaotic granular mixing, *Nature*, **397**, 675–678.

Smith, J. L. (1997). Mechanically fluidized bed plow mixers: fast intense processing without high shear, *Powder Bulk Eng.*, **11**(1), 45–57.

Sommer, K. (1986). *Sampling of Powders and Bulk Materials*, Springer-Verlag, Berlin.

Stainforth, J. N. (1982). Determination and handling of total mixes in pharmaceutical systems, *Powder Technol.*, **33**(2), 147–159.

Stange, K. (1954). The mixture quality of a random mixture of two component mixtures, *Chem. Ing. Tech.*, **26**, 331–337.

Steiss, M. (1995). *Mechanische Verfahrenstechnik*, Vol. 1, 2nd ed., Springer-Verlag, Berlin, pp. 168–275.

Thiel, W. J. (1982). Further comments on "Suggestions on the nomenclature of powder mixtures," *Powder Technol.*, **33**, 287–288.

Van den Bergh, W. (1994). Removing the uncertainty in solids mixer selection, *Chem. Eng.*, Dec., 70–77.

Weidenbaum, S. S. (1973). Solid–solid mixing, in *Chemical Engineers Handbook*, 5th ed., J. H. Perry, ed., McGraw-Hill, New York, pp. 21–31 to 21–36.

Weidenbaum, S. S. (1984). Mixing of powders, in *Handbook of Powder Science and Technology*, M. E. Fayed and L. Otten, eds., Van Nostrand Reinhold, New York, pp. 345–364.

Weinekotter, R., and H. Gericke (1999). *Mixing of Solids*, Kluwer, Dordrecht, The Netherlands.

Wightman, C., P. R. Mort, F. J. Muzzio, R. E. Riman, and E. K. Gleason (1995). The structure of mixtures by particles generated by time-dependent flows, *Powder Technol.*, **84**, 231–240.

Williams, J. C. (1986). Mixing of particulate solids, in *Mixing: Theory and Practice*, Vol. III, V. W. Uhl and J. A. Von Essen, eds., Academic Press, New York, pp. 265–305.

Williams, J. C., and M. I. Khan (1973). The mixing and segregation of particulate solids of different particle size, *Chem. Eng. London*, **269**, Jan. 19.

Wilms, H. (1988). Blending and homogenizing of bulk solids with the Zeppelin multi-pipe gravity blender, *Bulk Solids Handl.*, **8**(6), 733–736.

Wilms, H. (1992). Blending silos: an overview, *Powder Handl. Process.*, **4**(3), 293–299.

Zlokarnik, M. (1984). Scale-up in process engineering, *Ger. Chem. Eng.*, **7**(3), 150–159.

# CHAPTER 16

# Mixing of Highly Viscous Fluids, Polymers, and Pastes

DAVID B. TODD

New Jersey Institute of Technology

## 16-1 INTRODUCTION

Viscous mixing involves the many applications in processes wherein the viscosity is sufficiently high (e.g., greater than 10 Pa · s) that turbulent mixing is usually unobtainable, or the dissipation of the viscous energy involved would result in an unacceptably high product temperature. From simple tasks such as stirring the morning oatmeal to sophisticated industrial processes involved in manufacturing today's elastomeric and plastic compounds, recourse to laminar mixing techniques must be employed. Many industrially important products, such as pastes, putties, chewing gum, soap, grease, solid propellant, and some foods, fall into this category.

The key features that distinguish viscous mixing from nonviscous (turbulent) mixing are described in Chapters 2 and 3. In mixers handling very viscous materials, it is necessary to promote both lateral and transverse motion, with the material(s) being pulled, sheared, compressed, kneaded, and folded by the action of rotor(s) against vessel walls, saddles, or projecting stators. The condition of the feed can be an important consideration in selection of the mixer for the task. For example, a mixer for producing a uniform rubber cement solution must first cope with masticating an entire bale of rubber before starting the gradual letdown and final homogeneous operation.

## 16-2 VISCOUS MIXING FUNDAMENTALS

### 16-2.1 Challenges of High Viscosity Mixing

Mixing is an operation whereby the nonuniformity within a mixture is reduced. Mixing in very viscous systems can be a formidable task. There are no turbulent

---

*Handbook of Industrial Mixing: Science and Practice*, Edited by Edward L. Paul,
Victor A. Atiemo-Obeng, and Suzanne M. Kresta
ISBN 0-471-26919-0 Copyright © 2004 John Wiley & Sons, Inc.

eddies to help distribute components. Because of the high matrix viscosity, diffusion coefficients for even very small molecules are exceedingly low. Most high viscosity fluids are also non-Newtonian; many are shear-thinning, some have a yield point. Viscous energy dissipation during mixing can cause significant temperature variation throughout a vessel, thus contributing to further viscosity nonuniformity, and possibly product degradation.

Heat transfer coefficients are also poor in very viscous systems, and the steps normally taken to improve heat transfer in cooling, such as increased agitator speed or greater temperature differences, can be counterproductive because of the heat generated from viscous energy dissipation.

The mixing task can be further complicated by the changes in state that may need to be accommodated during the process. For example, solution or homopolymerization may start out with watery thin liquids into which a very small amount of equally thin catalyst needs to be distributed uniformly, and turbines or propellers would be appropriate. However, as polymerization proceeds and viscosity starts to build, probably into the 10 to 50 Pa · s range, these simple agitators no longer suffice, and anchor or helical ribbon mixers are more suitable. Keeping the polymerization under control means not only avoiding local hot spots but also may involve reflux cooling, so the mixing system must allow for rapid blending of the thin reflux into the viscous matrix, as well as vapor disengagement.

## 16-2.2 Dispersive and Distributive Mixing

Mixing can be brought about in viscous systems only by mechanical action or by the forced shear or elongational flow of the matrix. Solids with a cohesive nature (such as agglomerated particles) or other immiscible fluids (drops with interfacial tension) will require intensive mechanical stress to achieve the required dimunition. *Dispersive mixing* is defined as the breakup of agglomerates or lumps to the desired ultimate grain size of the solid particulates or the domain size (drops) of other immiscible fluids. Thus, dispersive mixing is a consequence of the history of the fluid mechanical stresses imposed on the mixture.

When mixing thermodynamically miscible fluids, or mixing hot and cold segments of the same material, mixing is determined simply by the history of deformation imparted to the fluid (the strain). *Distributive mixing* is defined as providing spatial uniformity of all the components. The interrelationship between dispersive and distributive mixing is illustrated in Figure 16-1.

In general, viscous mixing operations require some combination of dispersive and distributive actions; *intensive dispersive* mixing to break up globs or agglomerates, and *extensive distributive mixing* to apportion the broken phase throughout the co-mixture. Dispersion may be either a continuing (chronic) erosion or an abrupt (acute) rupture after surpassing a critical stress level sufficient to overcome the cohesive strength.

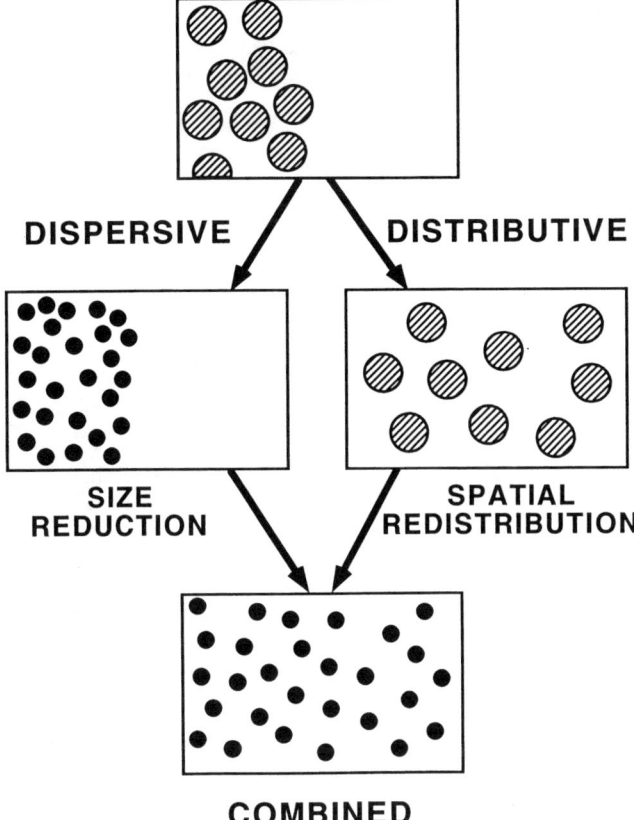

**Figure 16-1** Interrelationship between dispersive and distributive mixing.

### 16-2.3 Elongation and Shear Flows

Convective mixing can be achieved by imposing deformation on a system through laminar flow, which can be the consequence of shear, elongation (stretching), or squeezing (kneading). Laminar flow will orient the phases, so a critical aspect of viscous mixing is to provide frequent reorientation of dispersed elements.

For *miscible* fluids (no interfacial tension), a glob or layer of A in a B matrix can undergo stretching and folding to even thinner lamella thicknesses until diffusivity, even though very low, can achieve the desired homogenization (Figure 16-2).

With *immiscible* fluids, interfacial tension ($\sigma$) will resist the forces tending to tear the dispersed phases apart. If the rheologies and concentrations are similar, the mixture could end up co-continuous, as shown in Figure 16-3. With unequal viscosities and phase ratios, the restorative action of interfacial tension ($\sigma$) will cause the stretched-out regions to break into segments (Figure 16-4). In simple shear flow, Karam and Bellinger (1968) showed that there is a maximum viscosity

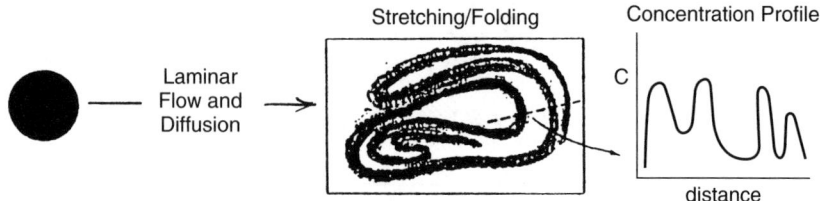

**Figure 16-2** Distributive laminar mixing and diffusion with miscible fluids (no interfacial tension and very low diffusivities).

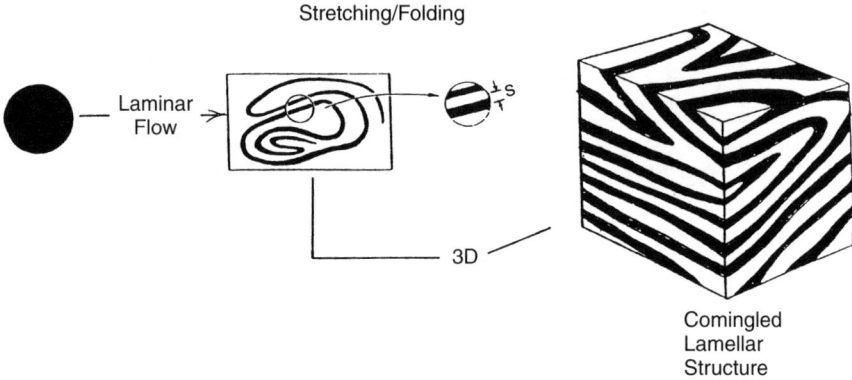

**Figure 16-3** Distributive laminar mixing of immiscible fluids with similar rheologies (s is the striation thickness).

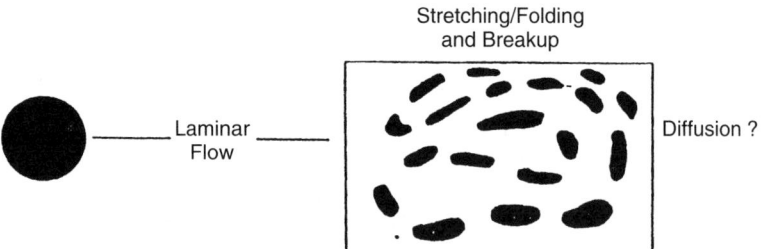

**Figure 16-4** Laminar flow dispersive mixing while undergoing high stresses.

ratio, p (p equals the ratio of the dispersed phase viscosity, $\mu_d$, to continuous phase viscosity, $\mu_c$), beyond which a liquid droplet cannot be broken up by shear alone. Grace (1982) indicated, however, that elongational flow did not have this limitation. Figure 16-5 represents the Weber or capillary number, or the ratio of viscous forces to restorative forces, $\gamma\mu_c r/\sigma$, plotted against the viscosity ratio, p ($\gamma$ is the shear rate, r is the drop radius). Based on Figure 16-5, the minimum

# VISCOUS MIXING FUNDAMENTALS

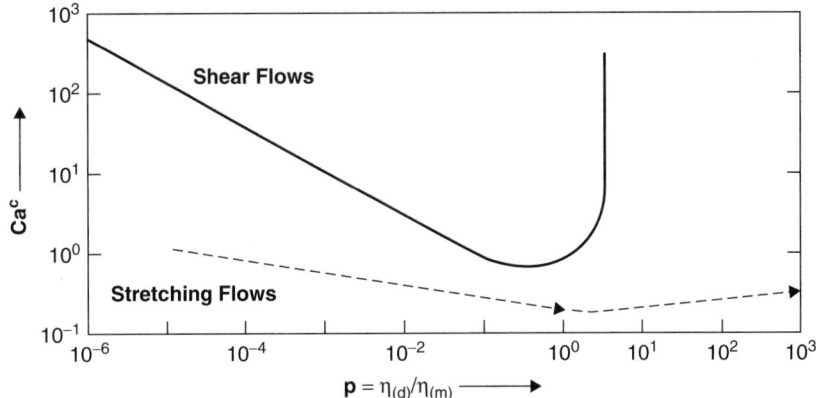

**Figure 16-5** Critical Weber (or capillary) number versus viscosity ratio.

dispersed phase drop radius can be achieved where the viscosity ratio p is close to unity, but dispersion by shear flow is not possible if p exceeds 4. This limit may be different for viscoelastic fluids.

When the viscosity of a mixture exceeds 10 Pa · s, simple mixing with a conventional impeller such as a turbine or propeller stirrer will not suffice. The high viscosity may arise from a high concentration of solids in a slurry, the high viscosity of the matrix fluid itself, or by interactions between ingredients. When the viscosity is high, the mixing Reynolds number (Re = $\rho D^2 N/\mu$) is probably less than 100. As such, mixing can occur only by viscous forces, and turbulence will play no part.

Most high viscosity mixers have a limited high-shear zone to minimize total power and heat buildup. The impellers are preferably designed to circulate all of the mixer contents past the localized high-shear zones. Particular attention must be paid to avoiding stagnant zones in batch equipment and to ensure that unmixed components are not carried through continuous mixers. Many high viscosity mixtures are shear thinning, so there may be a rapid fall-off of shear forces away from the shear-creating device. High velocity impellers may be completely ineffective since they may create an isolated cavity in the vessel without producing the required circulation (see Figure 18-13 and Chapter 9).

The basic requirement for accomplishing mixing in viscous systems is deformation of the mixture. Simple shear is inefficient and insufficient. Additional complexity must be incorporated into the system. The inefficient orientation of simple shear must be disrupted. When dispersion is also required, region(s) of intense deformation must be created by having flow forced through narrow passageways either through passive orifices (consuming pressure drop) or between walls which move with respect to each other, such as provided by a closely fitting impeller (consuming power for rotation).

The critical effect of orientation is shown in Figure 16-6, wherein the inner cylinder rotates with respect to the outer cylinder. If placed initially

**992** MIXING OF HIGHLY VISCOUS FLUIDS, POLYMERS, AND PASTES

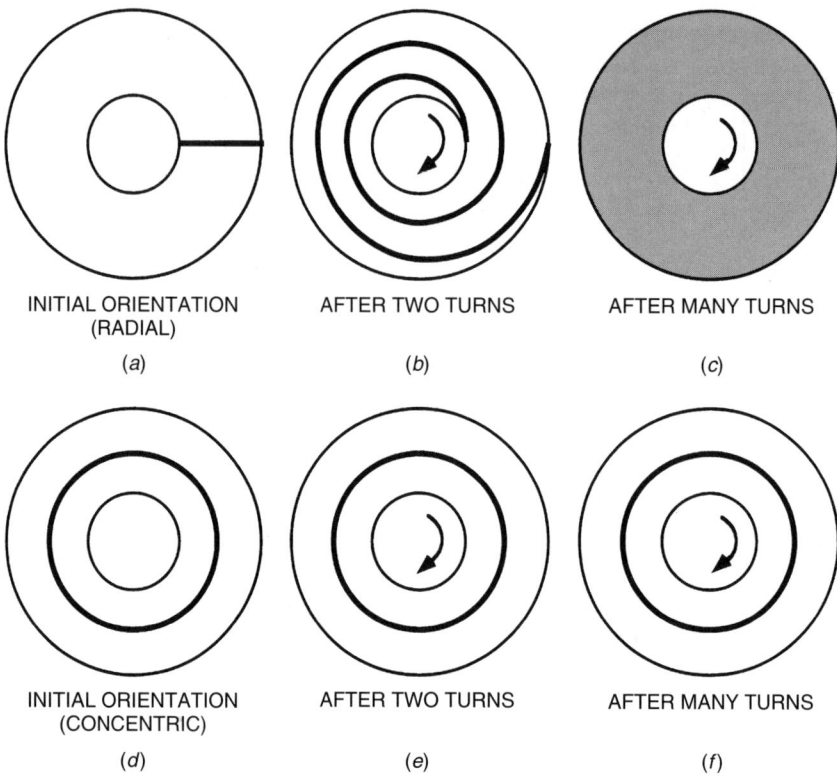

**Figure 16-6** Effect of orientation on mixing in a concentric cylinder mixer.

radially between the two cylinders, the interface, increases linearly with time (Figure 16-6$a$–$c$), but with an ever-decreasing striation thickness. If the interface is originally also concentric, as in Figure 16-6$d$, the striation thickness remains constant regardless of the amount of rotation. A video clip of this process is provided on the Visual Mixing CD affixed to the back cover of the book.

### 16-2.4 Power and Heat Transfer Aspects

**16-2.4.1 *Power*.** For equipment wherein the agitator sweeps near the vessel wall, the power drawn is due primarily to viscous drag rather than from the pumping required for circulation. The power for a Newtonian viscous mixer evolves thus:

shear rate  $= \alpha DN/t$
shear stress $=$ (shear rate) (viscosity) $= \alpha DN\mu/t$
shear area  $= \alpha$ DL           $= \alpha$ DL
force       $=$ (shear stress) (area) $= \alpha$ $D^2 LN$ $\mu/t$
torque      $=$ (force) (radius)      $= \alpha$ $D^3 LN$ $\mu/t$
power       $=$ (torque) (speed)      $= \alpha$ $D^3 LN^2 \mu/t$

On scale-up, with dimensional similarity of both blade length L and clearance t proportional to diameter,

$$\text{power} = \alpha \mu D^3 N^2 \qquad (16\text{-}1)$$

For very viscous mixtures, the viscosity is likely non-Newtonian. For shear-thinning fluids that can be represented by a power law exponent (n);

$$\text{power} = \alpha \mu D^3 N^{1+n} \qquad (16\text{-}2)$$

**16-2.4.2 Heat Transfer.** Correlations that have been developed for heat transfer in wiped vessels usually take the form of the dependence of Nusselt number (hD/k) on Reynolds number ($D^2 N \rho / \mu$), and Prandtl number ($C_p \mu / k$), and include a slight dependence on viscosity ratio ($\mu / \mu_w$), where $\mu_w$ is the viscosity at the wall:

$$\text{Nu} \propto \text{Re}^a \text{Pr}^b \left( \frac{\mu}{\mu_w} \right)^c$$

Figure 16-7 shows correlations for the dependence of heat transfer on Reynolds number for a variety of such vessels. As the viscosity increases, dependency on Reynolds number, exponent $a$, shifts from $\frac{2}{3}$ to $\frac{1}{3}$ (Figure 16-7). A selection of correlations for vessels is given in Chapter 14. In the Reynolds number range $10^{-3}$ to 10, Todd (1988) correlated data for twin-screw extruders as follows:

$$\text{Nu} = 0.94 \text{Re}^{0.33} \text{Pr}^{0.33} \left( \frac{\mu}{\mu_w} \right)^{0.14} \qquad (16\text{-}3)$$

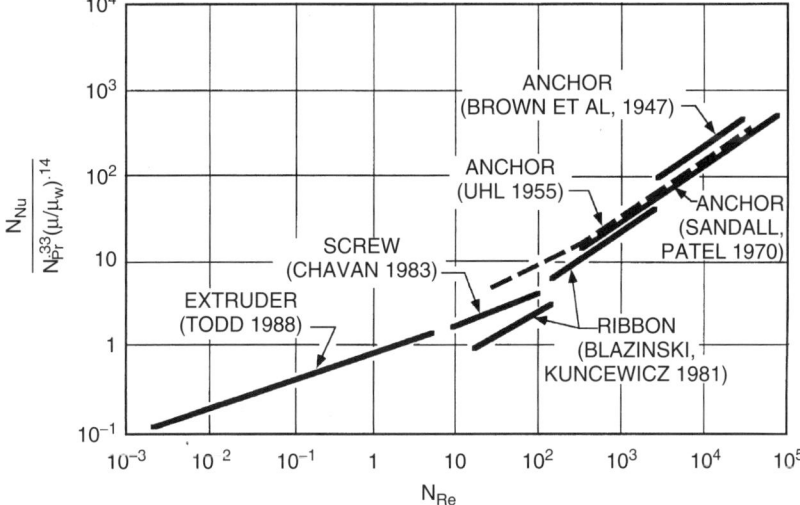

**Figure 16-7** Nusselt–Prandtl–Reynolds number correlations for heat transfer.

## 16-3 EQUIPMENT FOR VISCOUS MIXING

Characteristics of mixers for viscous mixing may include:

- Small clearances between impeller and vessel walls
- High power per unit volume
- Relatively small volume
- Slow impeller speeds to limit heat buildup
- Smearing blade profile if dispersion is important
- Scraping profile if heat transfer is critical
- Intermeshing blades or stators to keep the material from cylindering on the rotating impeller
- Special consideration for emptying

### 16-3.1 Batch Mixers

***16-3.1.1 Single-Stirrer Mixers.*** Viscous mixtures such as thick pastes or polymer solutions can be handled in a batch mixer as long as the agitator is in close proximity to the vessel wall. The two most common types are anchor blades (Figure 16-8) and helical ribbons (Figure 16-9). The latter are generally preferred because they provide for end-to-end axial turnover as well, and usually require less power than anchor mixers. Since very viscous mixtures are frequently shear thinning, the goal of achieving effective flow throughout the vessel usually precludes the use of propellers or turbine blade agitators, which may merely spin

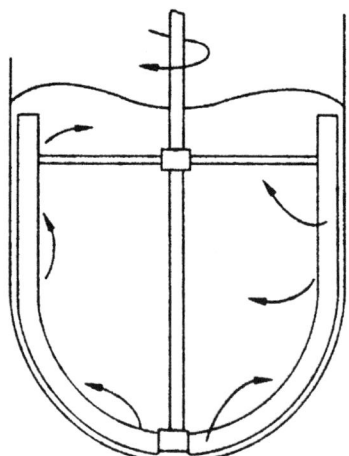

**Figure 16-8** Anchor mixer. A video clip of mixing with the anchor mixer is provided on the Visual Mixing CD affixed to the back cover of the book.

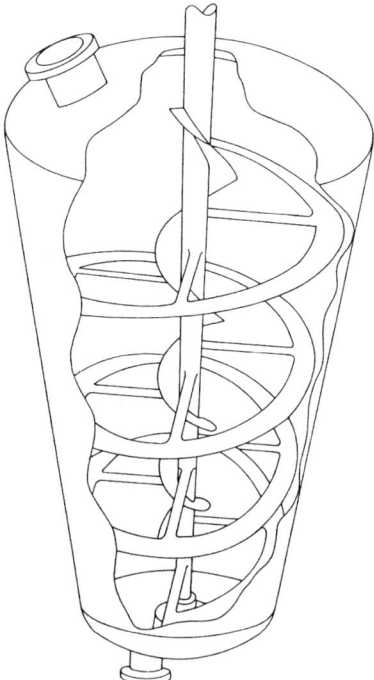

**Figure 16-9** Helical blade mixer. A video clip of blending with a helical ribbon is provided on the Visual Mixing CD affixed to the back cover of the book.

in a central zone without causing any motion at the vessel walls. This is discussed at more length in Chapter 9.

Bakker and Gates (1995) provide some guidelines for power and blending time, and specifically compare a helical ribbon against a turbine impeller for an intermediate viscosity (25 Pa · s) application (Re = 90). For the same blend time (4.5 min), the turbine impeller would require over four times as much power.

Carreau et al. (1993) determined power consumption for six helical ribbon–tank combinations differing in diameter, impeller pitch, and blade width, and with various test fluids. In the laminar region (Re < 50) the power number ($N_p$) for Newtonian fluids could be correlated by

$$N_p = \frac{K_p}{Re}$$

with $K_p$ ranging from 120 to 192 for the six combinations. Carreau et al. (1993) indicate that the effective shear rate [based on the Metzner and Otto (1957) correlations] is strongly dependent on the rheological characteristics of the fluid being mixed, but only weakly dependent on agitator geometry.

With very viscous mixers, the time to empty the vessel may also be an important consideration. Significant degradation may occur if fluid remains stuck to the tank walls and internals. Again, the helical ribbon mixer would be the preferred option for fixed installations.

**16-3.1.2 Change Can Mixers.** These mixers allow for separation of the mixing blade(s) from the mixing vessel, thereby providing more accurate weighing of ingredients prior to mixing, less batch-to-batch cross-contamination, easier cleaning, and less tie-up of the mixer while the tank is being emptied. Frequently, agitation is provided by planetary mixing blades (Figure 16-10). Change can mixers are available over a wide range of sizes from 1 L to 4 $m^3$.

**16-3.1.3 Double Arm Kneading Mixers.** As illustrated in Figure 16-11, these mixers have two horizontally mounted mixing blades, either tangential or slightly overlapping. The bottom of the trough consists of two half-cylinders that match the sweep of the mixing blades. Various blade shapes have evolved, as

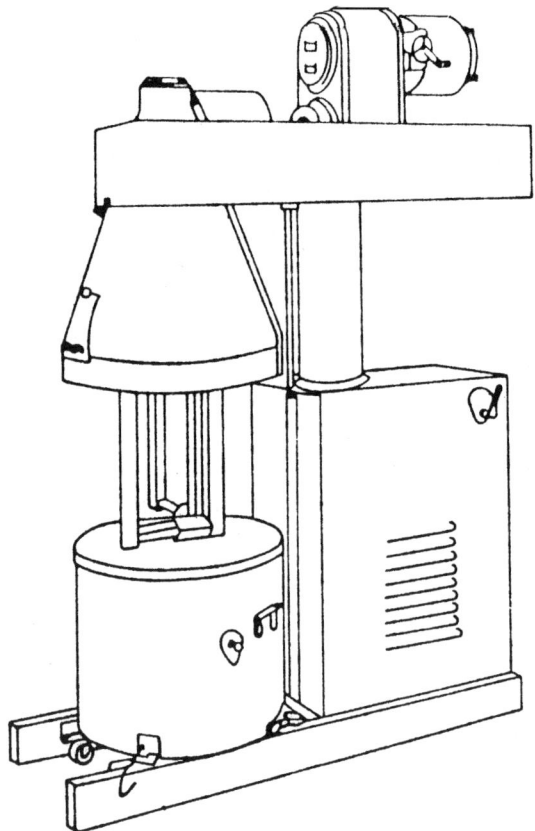

**Figure 16-10** Change can mixer.

EQUIPMENT FOR VISCOUS MIXING    **997**

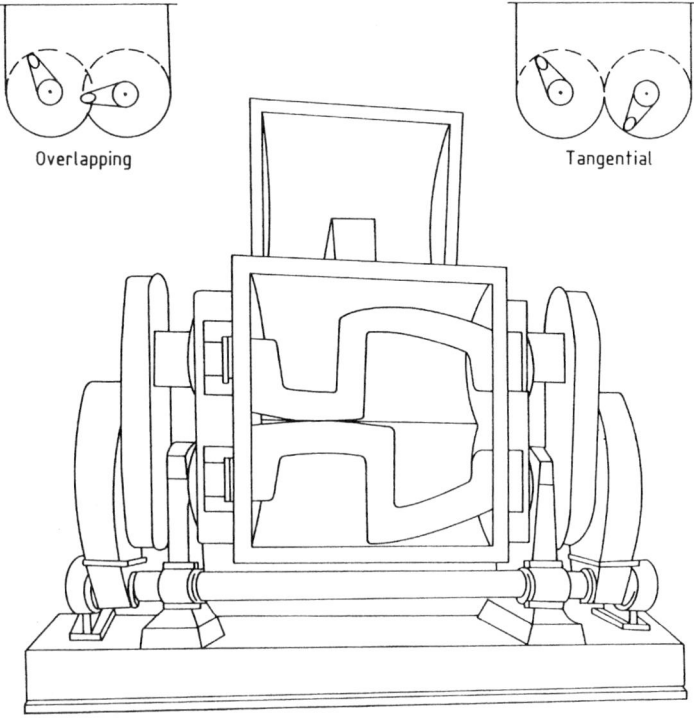

**Figure 16-11**   Double arm kneading mixer.

shown in Figure 16-12. The most common is the Sigma-blade (Figure 16-12*a*), with unequal wings to induce randomness.

With tangential blades, additional randomness in introduced by having the blades turning at different speeds. Using acid–base titration in a viscous non-Newtonian CMC (carboxymethyl cellulose) solution, the time for complete mixing was determined as a function of blade speed ratio, as shown in Figure 16-13, where equal rotational speeds are shown to take almost twice as long to reach molecular scale mixing as when the blades are turning at different speeds (e.g., a speed ratio > 1.2). For Newtonian and many non-Newtonian mixtures, the time to achieve molecular scale mixing is generally inversely proportional to average blade speed, as shown in Figure 16-14.

One of the problems in scale-up is that viscous energy dissipation is usually great with high viscosity fluids, yet the surface/volume ratio decreases with increasing size. At the same mixer speed, the torque and dissipation are higher in a high viscosity fluid. Scale-up at constant dissipation requires decreasing N as $N_L/N_S = (D_S/D_L)^{3/2}$, but the area per volume that is available for heat transfer decreases faster, with $D_S/D_L$. Consequently, most mixer manufacturers decrease blade speed and power/volume as size is increased, to prevent product degradation. This results in longer mixing times at the large scale.

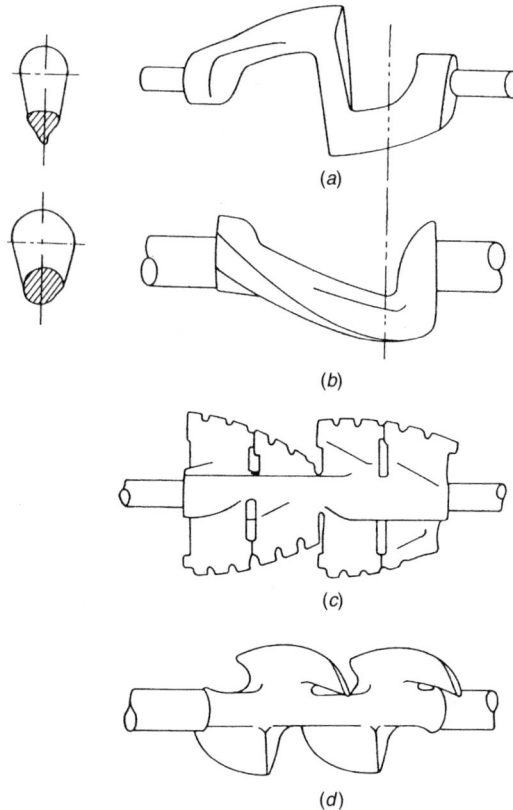

**Figure 16-12** Agitator blades for double arm kneaders: (*a*) Sigma; (*b*) dispersion; (*c*) multiwing overlap; (*d*) double Naben.

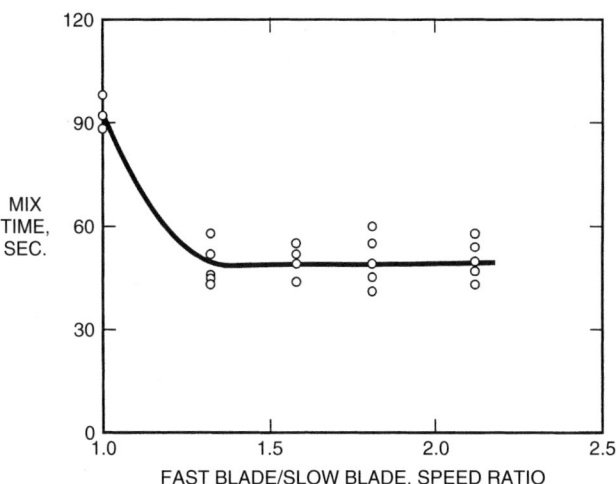

**Figure 16-13** Effect of blade speed ratio on mixing time.

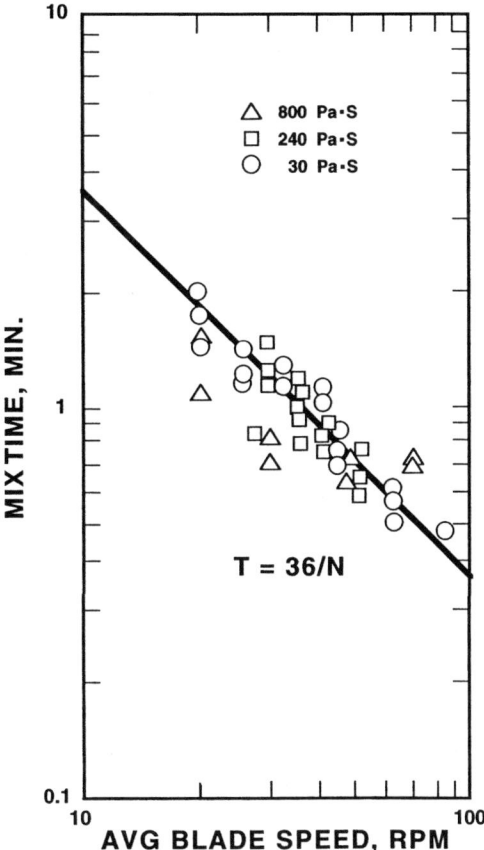

**Figure 16-14** Mixing time in a 9 L Sigma-blade batch mixer.

The Sigma-blade mixer may be tilted for discharge, be equipped with a bottom discharge door, or contain an auxiliary screw discharge, as shown in Figure 16-15. In the latter case, the screw is rotated inward during the mixing cycle to provide additional mixing action, and then reversed for discharge. Sigma-blade mixers range in size from 1 L to 5 m$^3$. Power inputs range from 0.02 to 0.5 kW/kg, as shown in Figure 16-16 for typical applications (Parker, 1965).

The Banbury mixer (Figure 16-17) is a very high power (up to 6 kW/kg) batch mixer equipped with a top ram to force material into the mixing zone, and with bottom discharge when the batch cycle is complete. Because of the short L/D and with the shafts supported at both ends, this mixer is frequently used for very viscous materials such as rubber.

Plow mixers (Figure 16-18), ribbon blenders (Figure 16-19), cone and screw mixers (Figure 16-20), and Mullers (Figure 16-21) are used for free-flowing paste mixing applications (as well as just for solids blending purposes) where the power requirements are not too high.

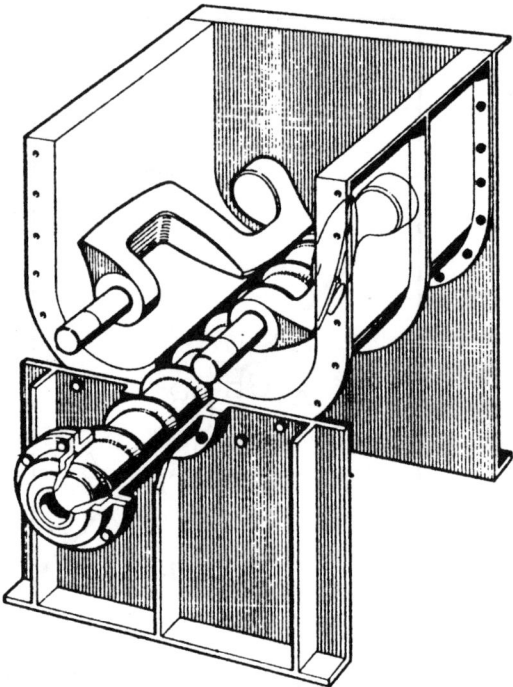

**Figure 16-15** Screw discharge batch mixer.

## 16-3.2 Continuous Mixers

Although batch mixers could perhaps be converted to continuous operation, they rarely are because the broad residence time distribution would lead to product nonuniformity. Continuous mixers require accurate metering of ingredients. Usually, one or two rotors operate in an open or closed trough, which may be jacketed. In larger units, the rotors may be cored to provide additional heat transfer area. The rotor(s) may have interrupted flights to interact with pins or baffles projecting inward from the trough wall to improve blending by the action of flow division and recombination. The rotors usually are in close proximity to the trough or barrel wall. A restriction may be placed at the discharge end to control holdup, residence time, and mixing energy.

***16-3.2.1 Single-Screw Extruders.*** A necessary step in most polymer processing is melting of the plastics to be able to homogenize mixtures incorporating stabilizing agents, color, fillers, and so on. Melting cannot be accomplished by direct heat transfer because of the inherently low thermal conductivities of most polymers, and too great a temperature difference driving force at the vessel wall will lead to scorching and product degradation.

Single-screw extruders (SSEs) convert mechanical energy of the drive into thermal energy, most of which is utilized in melting the feed polymer. The

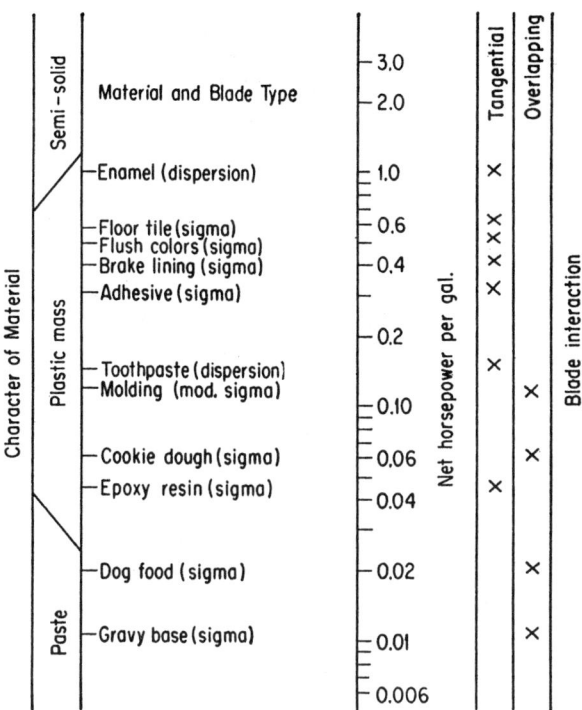

**Figure 16-16** Typical applications and power for double arm kneading mixers.

enthalpy of some common polymers is shown in Figure 16-22 relative to a datum at 20°C. SSEs enable the processor to combine melting, mixing, off-gassing, and developing pressure for product shaping (profile, film, sheet, pellets).

A basic SSE is shown in Figure 16-23. Solid feed, as powder, granules, or pellets, is generally flood-fed to the extruder, wherein the feed rate is controlled by the takeaway capacity of the rotating screw. An SSE generally has a continuous helical flight, typically with a lead, Z (length for 360° turn), about equal to the diameter, D, and a channel depth, h. The root diameter must be large enough to handle the torque.

Melting occurs by friction of the solids being moved forward by the screw against the close-fitting barrel wall as the channel depth decreases through a transition zone between the feed and the metering section. As melt forms, viscous energy dissipation becomes the predominant energy transfer mechanism.

The metering section acts as a melt pump, dragging the melt forward. Since sufficient pressure will need to be generated to overcome the resistance of any filter or shaping die at the discharge end of the extruder, the molten mixture will want to flow back upstream, offsetting some of the drag flow.

The net capacity of an extruder (Q) for pumping a viscous melt is expressed as

$$Q = Q_d - Q_p \qquad (16\text{-}4)$$

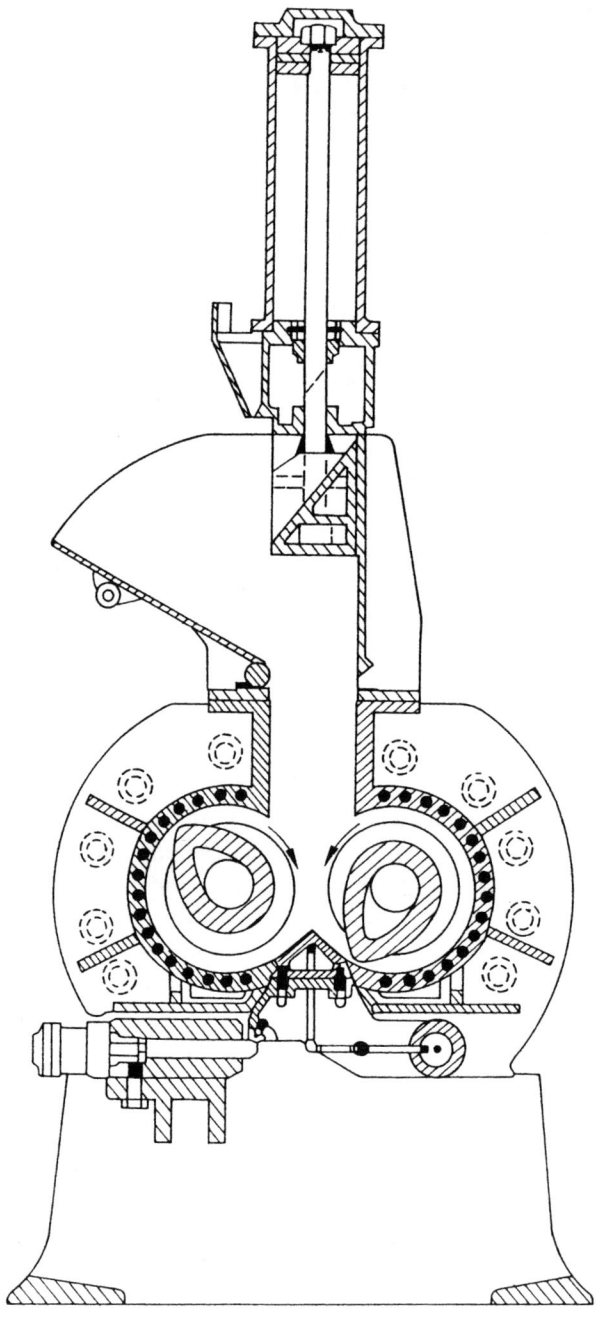

**Figure 16-17** Banbury mixer.

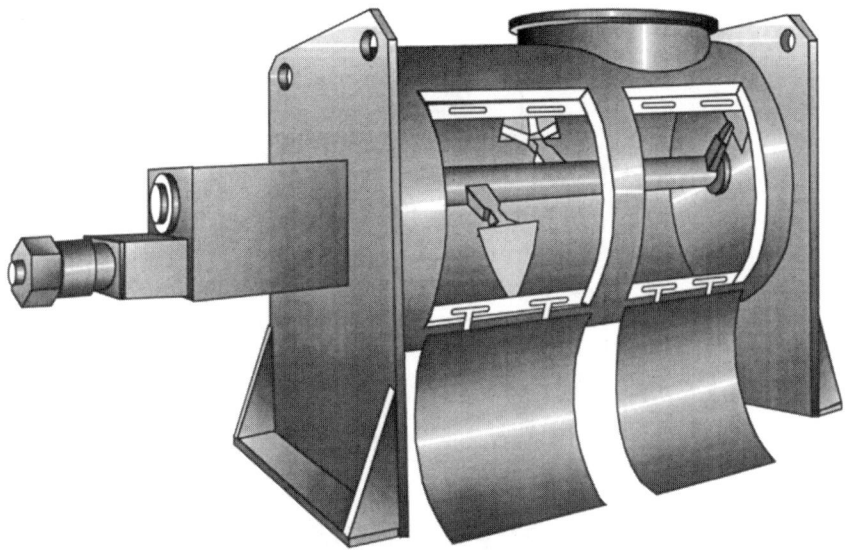

**Figure 16-18** Plow mixer.

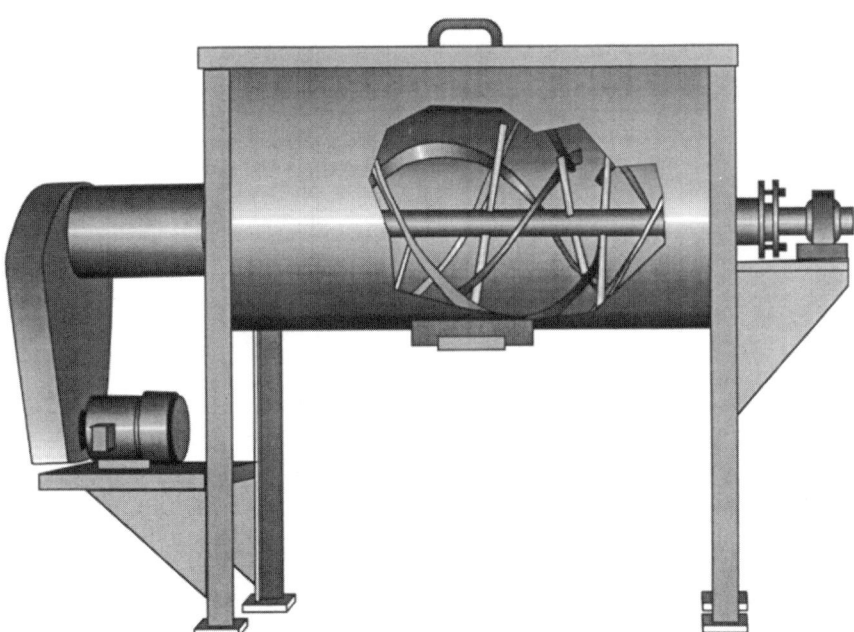

**Figure 16-19** Ribbon blender.

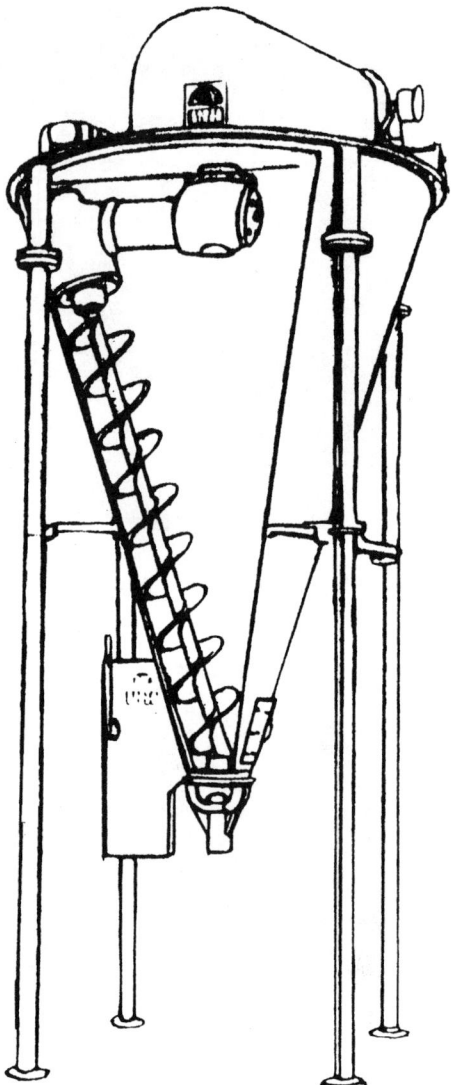

**Figure 16-20** Cone and screw mixer.

where $Q_d$ is the volumetric drag flow forward and $Q_p$ is the volumetric pressure flow back down the channel.

$$Q_d = aN \qquad (16\text{-}5)$$

$$Q_p = \frac{b\Delta P}{\mu L} \qquad (16\text{-}6)$$

EQUIPMENT FOR VISCOUS MIXING **1005**

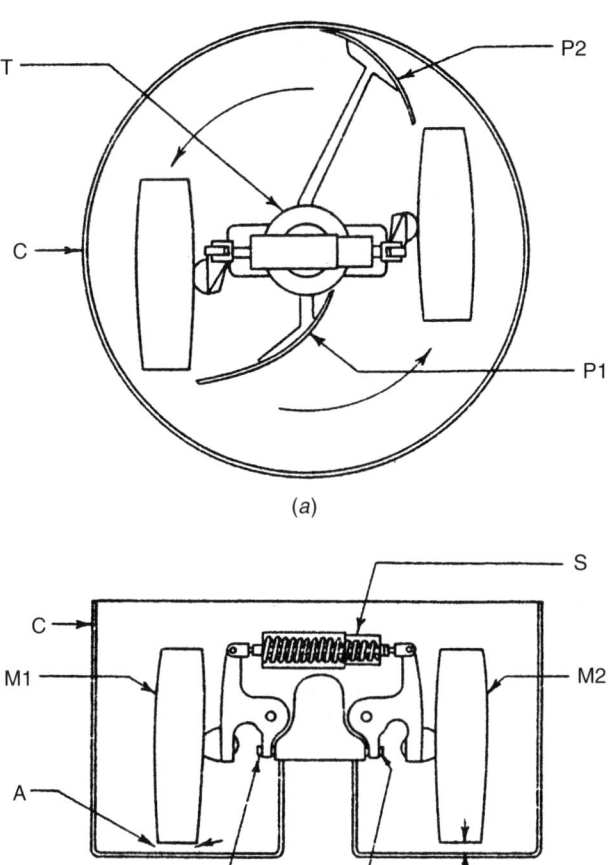

**Figure 16-21** Muller.

$$a = \frac{F_d \pi D W h \cos \theta}{2} \qquad (16\text{-}7)$$

$$b = \frac{F_p W h^3 \sin \theta}{2} - e \qquad (16\text{-}8)$$

$$W = \frac{\pi D \sin \theta}{2} - e \qquad (16\text{-}9)$$

where F is the shape factor dependent upon h/w, D is the screw diameter, W the channel width, h the channel depth, θ the helix angle, and e the flight width. Textbooks such as Tadmor and Gogos (1979), Rauwendaal (1990) and Chung (2000), provide many details on the workings of SSEs.

**1006** MIXING OF HIGHLY VISCOUS FLUIDS, POLYMERS, AND PASTES

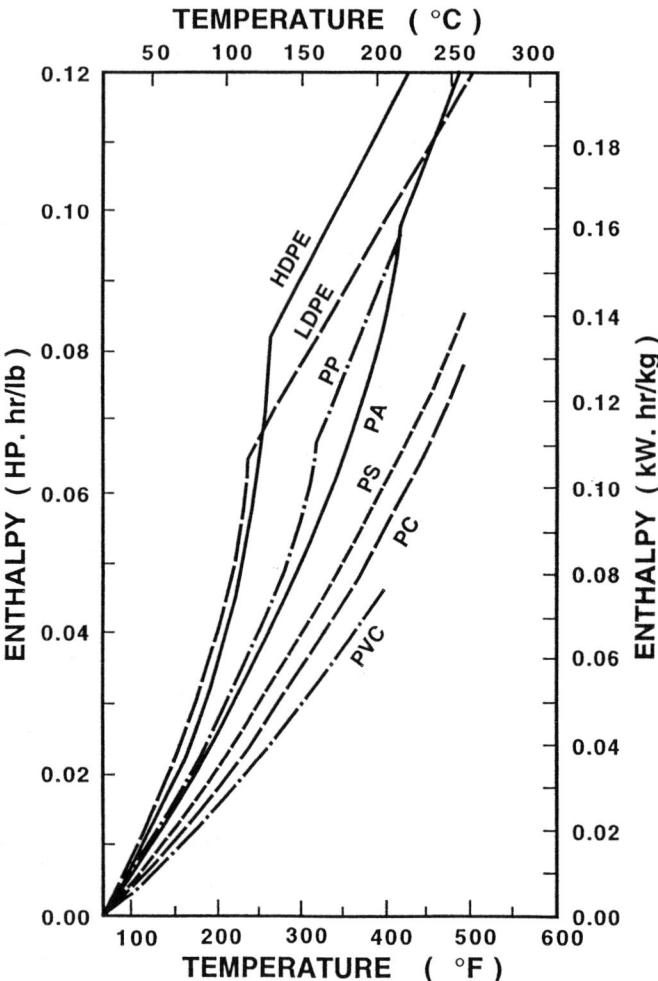

**Figure 16-22** Enthalpy of common polymers (enthalpy = 0 at 20°C).

Solid additives may be dry blended in the desired ratio in a ribbon blender or intensive mixer upstream of the extruders. Alternatively, it may be important to melt the polymer fully before solid fillers are added downstream, particularly if a large volume is to be incorporated. In the act of melting pellets in the presence of powdered fillers, compressive forces may be created that can actually cause agglomeration, akin to briquetting. The agglomerates so formed may then survive into the final product. Postponing filler addition until after the base resin is fully molten will also lessen barrel and screw wear. Downstream addition of filler will require proper venting of the air that accompanies introduction of low-bulk-density powders. The merits of alternative modes of incorporating solids are described by Todd (2000).

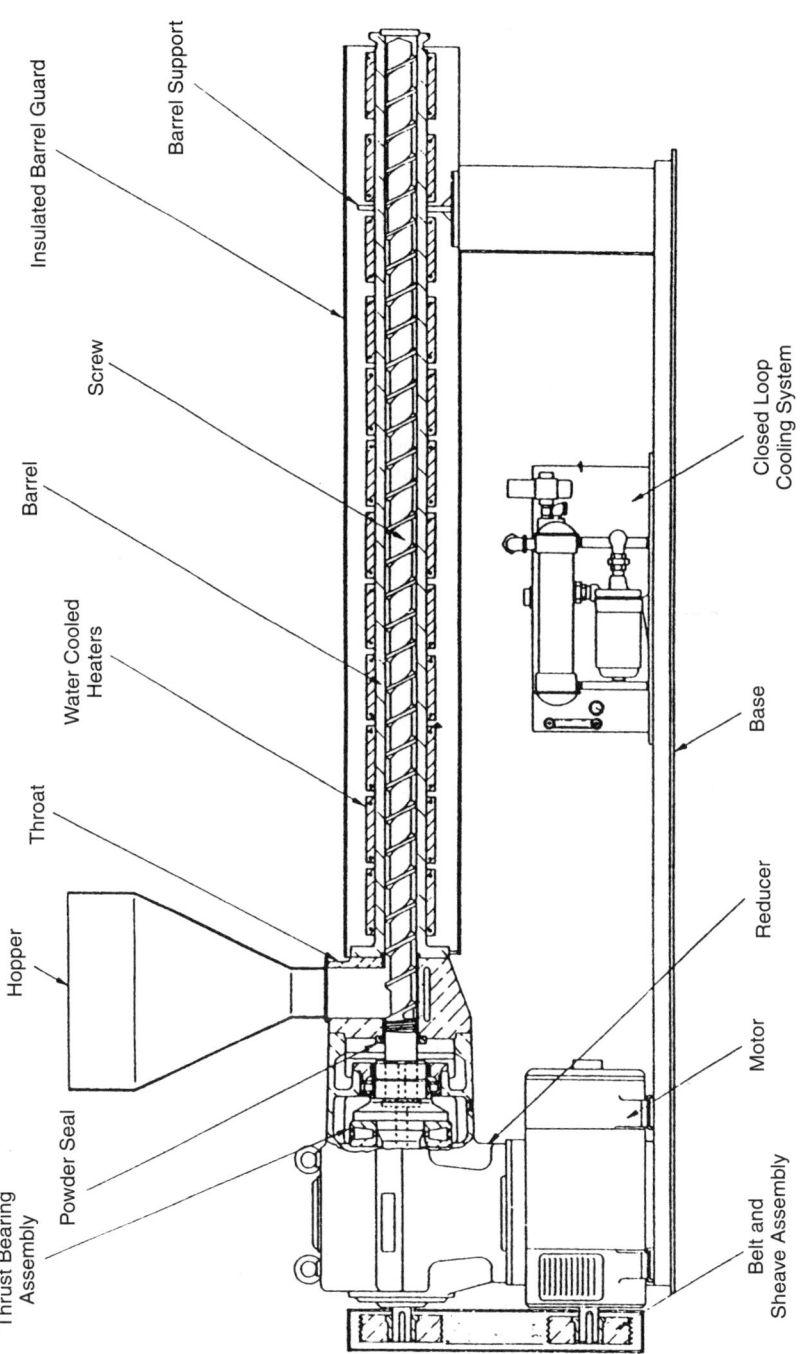

**Figure 16-23** Single-screw extruder.

**1008** MIXING OF HIGHLY VISCOUS FLUIDS, POLYMERS, AND PASTES

If venting is required, a two-stage extruder is used, with an intermediate deep channel zone following the first metering zone. A melt seal is created at the end of the metering zone to prevent sucking gases from the feed port. After transit through the vent zone, the channels are only partly filled and the volatile material is removed through the vent port in the barrel. The melt mixture passes through a second metering zone to develop the required discharge pressure.

**16-3.2.2 Mixing Enhancers.** Cross-channel flow occurs in the full flights of an extruder due to the combined actions of drag and pressure flows. Frequently, most of the mixing may be accomplished by the time that the last bit of polymer has been melted. Various devices have been employed to provide final homogenization before discharge, such as those illustrated in Figure 16-24.

The Maddock mixing section forces flow over a narrow clearance between inlet and outlet flutes. This device not only forces all the product through a high-shear zone but can also act as a crude filter to prevent the passage of gross agglomerates downstream. Most of the other mixing promoters shown depend on generating elongational flow patterns and multiple dividing and recombining of the split flows (as well as localized high-shear zones). As such, they also require an additional pressure driving force, but the latter involves less energy than the original melting.

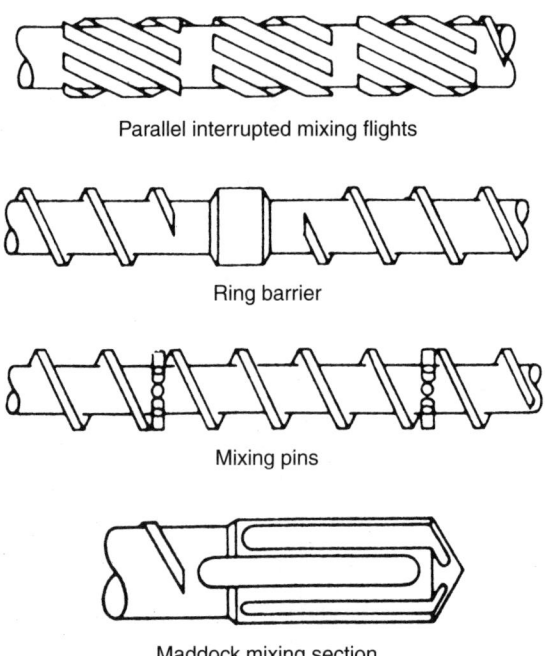

**Figure 16-24** Mixing enhancers for single-screw extruders.

Generating pressure in an extruder is not thermodynamically efficient, and extruders may be only about 10% efficient (E) as a pump. The adiabatic temperature rise ($\Delta T$) accompanying pressure rise ($\Delta P$) for an extruder is:

$$\Delta T = \frac{\Delta P}{E \rho C_p} \qquad (16\text{-}10)$$

With a typical specific gravity ($\rho$) equal to unity, a melt heat capacity $C_p$ of 0.5 cal/g · °C, and 10% efficiency (E):

$$\Delta T(°C) \approx \frac{\Delta P(\text{bar})}{2} \qquad (16\text{-}11)$$

For example, following venting in a two-stage extruder, the adiabatic temperature rise accompanying generation of a 1450 psi (100 bar) discharge pressure would be about an additional 50°C.

The Kokneader (Figure 16-25) is a special single-screw extruder with interrupted flights on the screw and mixing teeth projecting inward from the barrel wall. The screw reciprocates as well as rotates. Passage of the teeth through the channels creates multiple mixing actions, as described by Case (1998), as well as preventing adherence of the mixture within the screw channel.

### 16-3.2.3 Twin-Screw Extruders. 
Single-screw extruders take advantage of the interaction between screw and barrel. Twin-screw extruders (TSEs) capitalize on the interaction between the two screws. TSEs can be classified as being counterrotating or co-rotating, and tangential or intermeshing (accepting also varying degrees of intermesh). The three types of commercially available TSEs are shown in Figures 16-26 and 16-27 (corotating tangential designs are not offered). In polymer processing, TSEs perform the full gamut of solids feeding, melting, mixing, reacting, venting, and pressure development (viscous pumping). How the various functions are performed in representative types of commercially available equipment is fully described in a book edited by Todd (1998). We concern ourselves here primarily with the mixing function.

In all extruders being fed solids, some mixing may occur simultaneously with melting. However, more reliance is placed on creating flow fields within the screw channels to enhance both strain and elongational stresses conducive to dispersive and distributive mixing. For *nonintermeshing counterrotating TSEs* (Figure 16-26), the flow within each screw's channel is similar to that in SSEs, with additional reorienting as flow oscillates between the screws when in a staggered array. As with SSEs, the screw channels effectively achieve mixing only when full, so blister rings or reverse helix screws can be artfully placed to dictate the degree of fill. Nonintermeshing counterrotating screws can provide good distributional mixing, but are not particularly effective in dispersive mixing. An excellent summary of this type of TSE is given by Bash (1998).

Another type of nonintermeshing rotor mixer is the continuous mixer, such as that shown in Figure 16-28, which is in essence a continuous version of the

**Figure 16-25** Reciprocating single-screw extruder (Kokneader).

Banbury mixer (Figure 16-17). Again, the short L/D shafts are supported at each end, and very high intensity mixing is possible (Canedo and Valsamis, 1998). Discharge is radial, with a variable restriction to control the holdup. A separate melt extruder is used to develop pressure for pelletizing.

Traditional intermeshing counterrotating TSEs, shown in Figure 16-29, consist of a series of essentially closed off C-shaped chambers (Figure 16-30) that march down the barrel with little interchange between chambers except that which occurs by leakage, as shown by Janssen (1978). There is a milling effect, not unlike that of a two-roll mill, with good elongational flow and good dispersion

EQUIPMENT FOR VISCOUS MIXING  **1011**

Tangential
Counter-Rotating

Intermeshing
Counter-Rotating

Intermeshing
Co-Rotating

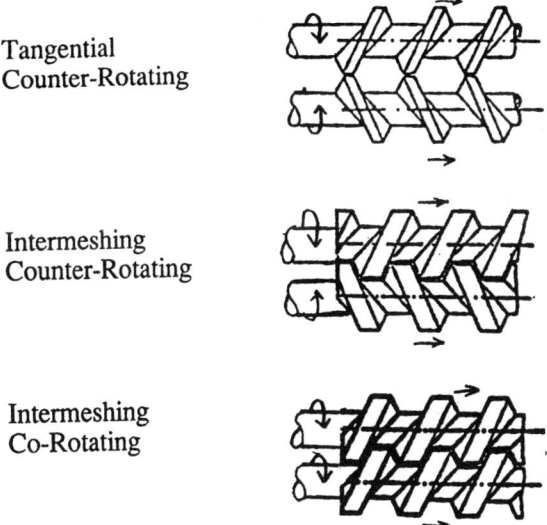

**Figure 16-26** Classification of twin-screw extruders.

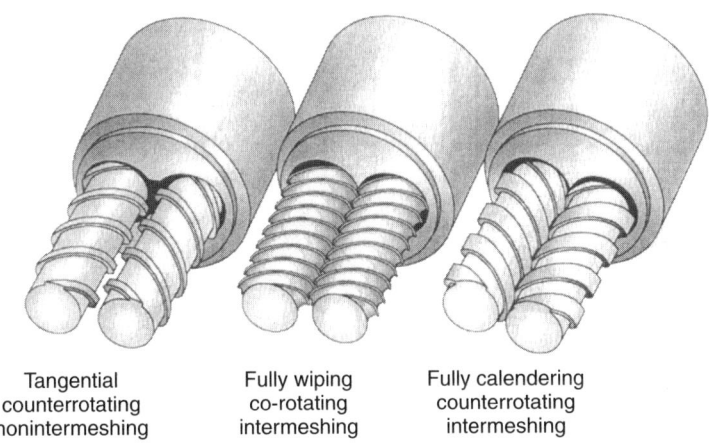

Tangential
counterrotating
nonintermeshing

Fully wiping
co-rotating
intermeshing

Fully calendering
counterrotating
intermeshing

**Figure 16-27** Classical formats of twin-screw compounding extruders.

between the flights of one screw in cooperation with the channels of the other. With almost half of each screw in close contact with the barrel, the conventional intermeshing counterrotating TSE has regions of low volume undergoing high shear and potential overheating, thereby limiting the screw speed and potential capacity of this type of TSE. Thiele (1998) has described the mixing action that can be achieved with lobe-type mixing elements, as depicted in Figure 16-31. These elements can be multilobed to produce a great variety of flow patterns conducive to enhanced mixing, both dispersive and distributive.

**1012** MIXING OF HIGHLY VISCOUS FLUIDS, POLYMERS, AND PASTES

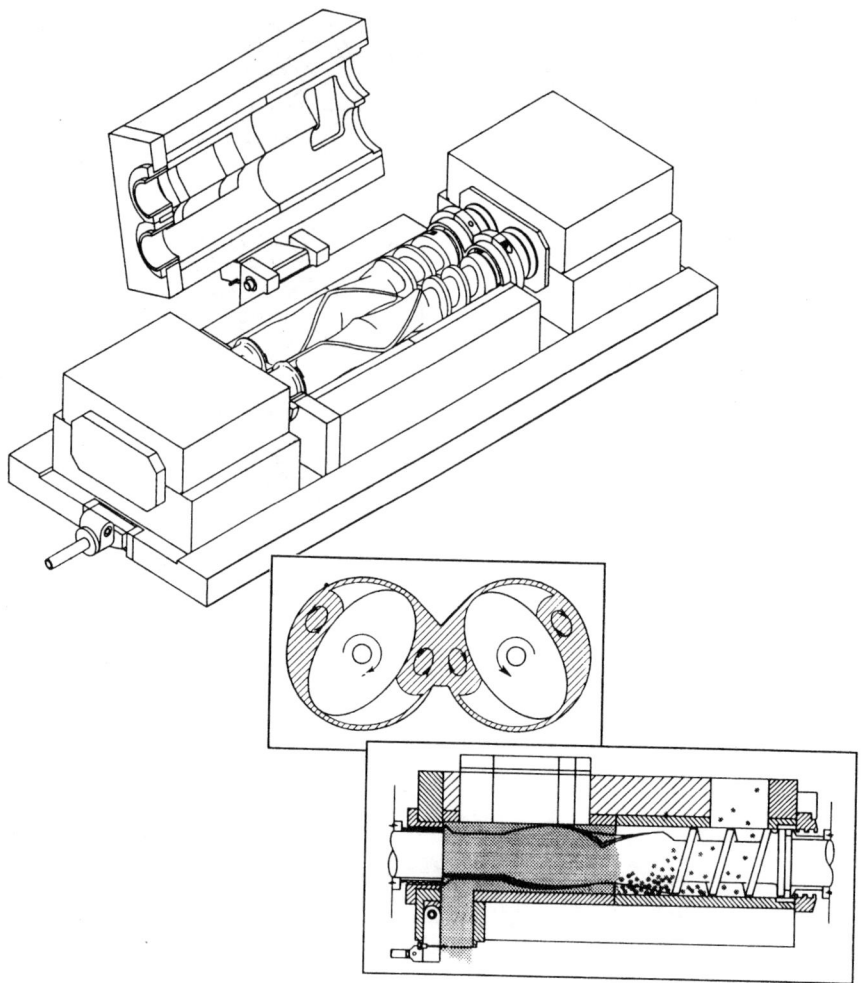

**Figure 16-28** Farrel continuous mixer.

The most common variety of TSE is the intermeshing corotating type, shown in Figure 16-32. Flight tips on one rotor intermesh and wipe the channel of the other. The main mixing action, however, arises from the use of kneading paddles, which can generate a mixing action not available in the other types of TSEs. The kneading paddle has the same cross-section as the screw, with the same self-wiping feature. The most common configuration for mixing is a bilobe design (Figure 16-33), but single- and trilobe designs also exist (Figure 16-33). The kneading paddles are frequently grouped in units of three to five for mechanical strength, to form kneading blocks of fixed angular offset and with a variety of axial lengths. The kneading blocks may have some of the conveying characteristics of screws, depending on paddle width and offset, as shown in Figure 16-34.

EQUIPMENT FOR VISCOUS MIXING  **1013**

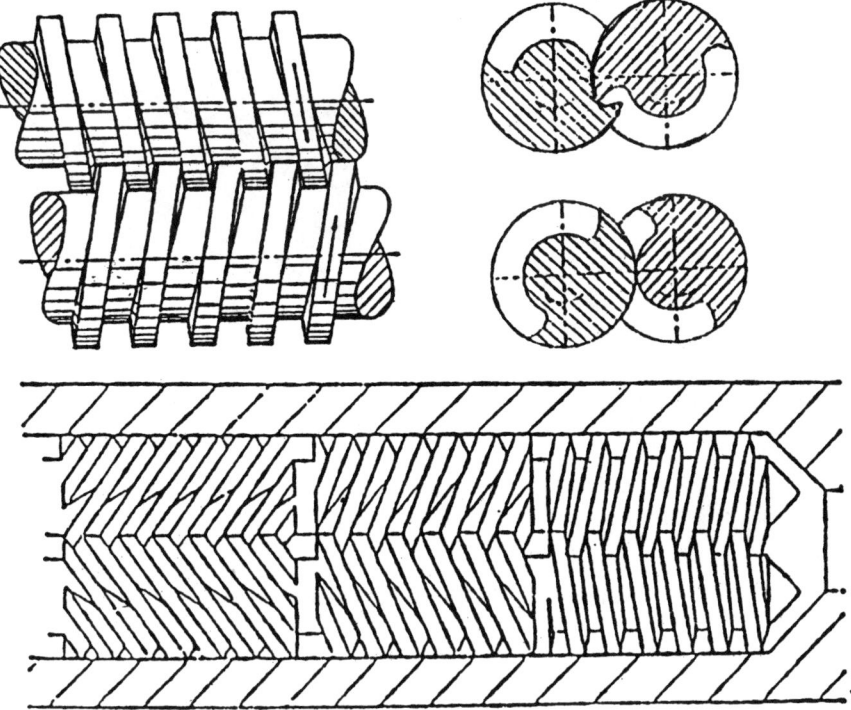

**Figure 16-29** Counterrotating intermeshing screws.

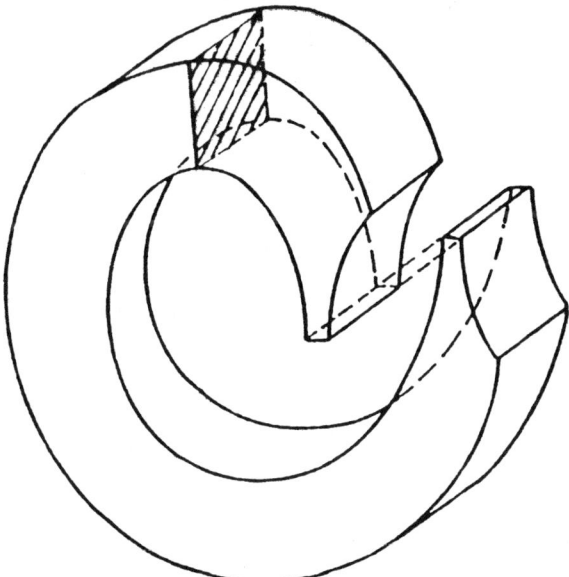

**Figure 16-30** C-shaped channel.

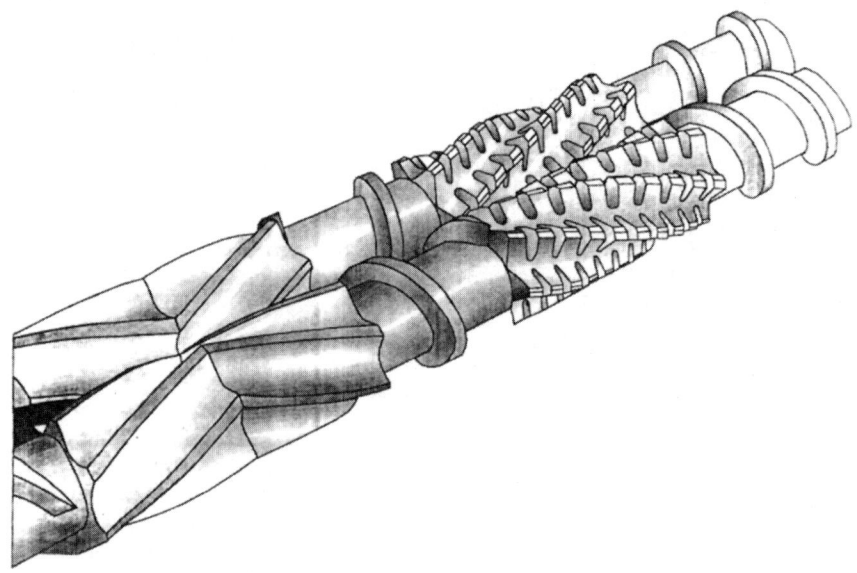

**Figure 16-31**  Hexalobal mixing screws.

The response to pressure generation is completely different, as backflow can easily occur through the gaps of the offset.

As with single-screw extruders [eqs. (16-4) to (16-6)], equations can be developed for drag and pressure flow terms for TSEs (Janssen, 1978; Todd, 1991). However, because of the great variety of screw and kneading paddle arrays, it is necessary to determine the relevant equations for each configuration in series.

It is also desirable to know the volume available for the various processing functions, such as reaction and devolatilization. For the common intermeshing bilobe configuration (Figure 16-33), the open cross-section (A) available for processing is

$$A = 3.08hD \qquad (16\text{-}12)$$

where h is channel depth and D is screw diameter.

The fluid-conveying capacity, or drag flow $Q_d$, can be approximated by

$$\frac{Q_d}{N} = \frac{AZ}{2} \qquad (16\text{-}13)$$

where N is the rotational speed and Z is the lead length (the axial distance required for 360° of a flight tip). Equation (16-19) is based on the observation that one-half of the material contained in one turn of the screw is conveyed forward per revolution. The degree of fill (f) in a barrel section where pressure

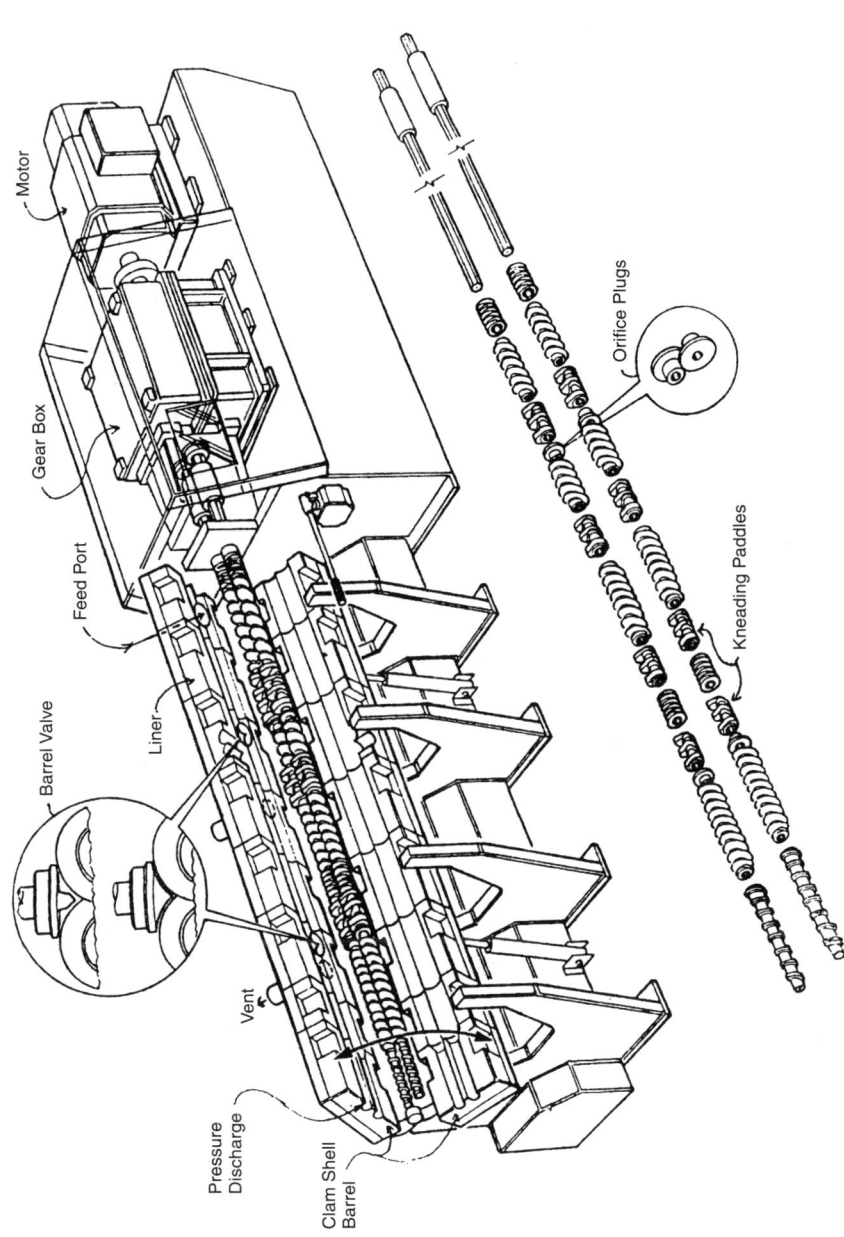

**Figure 16-32** Co-rotating intermeshing twin-screw extruder with clamshell barrel. (Courtesy B & P Process Equipment and Systems, Saginaw, MI.)

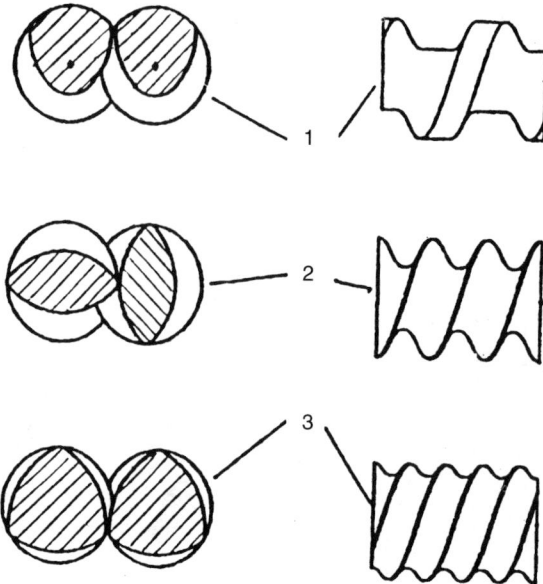

**Figure 16-33** One-, two-, and three-start screw profiles for intermeshing co-rotating twin-screw extruders.

is not being generated is merely

$$f = \frac{Q}{Q_d} \qquad (16\text{-}14)$$

where Q is the net flow.

The primary mixing action in intermeshing co-rotating TSEs is caused by multiple expansion/compression action as the rotors rotate. Figure 16-35 illustrates how the shaded material in one crescent section can receive material from upstream or downstream sections, and then be squeezed out a quarter-turn later. The elongational flow so produced is excellent for both dispersive and distributive mixing. This expansion/compression mixing action occurs only if that section of the barrel is completely full, so fill control is another necessary aspect of mixing in these TSEs (Todd, 1998). Additional mixing effects occur from the dispersive face of the kneading paddles (Figure 16-36) and the slicing action as the two rotors intermesh (Figure 16-37).

In addition to screw sections of differing helix angles, and kneading blocks of differing axial length, other special mixing elements are available, such as screws or blister rings with flight interruptions, as shown in Figure 16-38. The relative flow-pressure characteristics of such mixing elements have been reported by Brouwer et al. (2002). The screw devices are generally close to pressure neutral, as the slots in the flights are usually of opposite hand to the screw helix. The

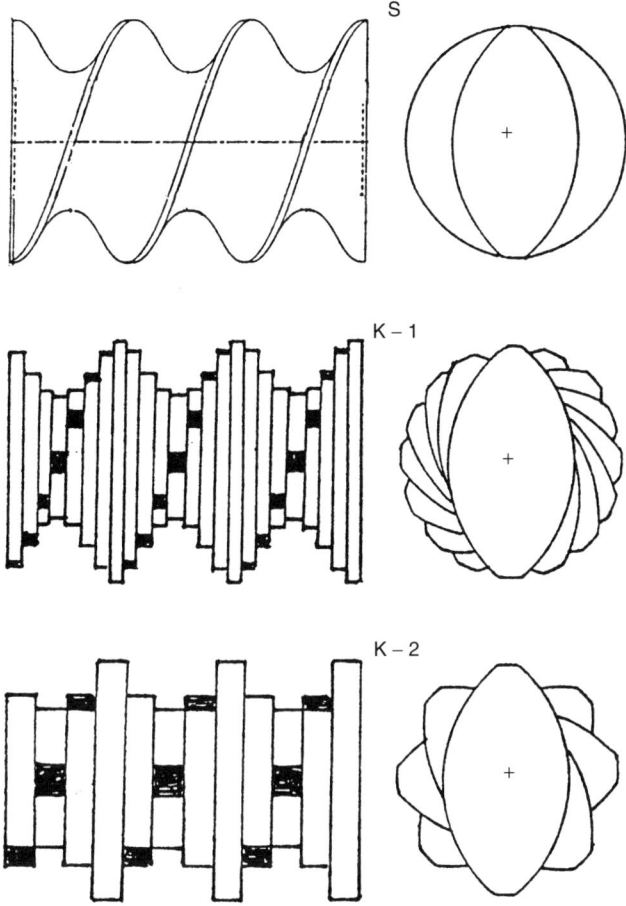

**Figure 16-34** Kneading paddles in arrays matching square-pitch screws.

gear mixers (Figure 16-38, TME) may have neutral or angular slots either forwarding or reversing. These interrupted flight mixing elements provide multiple splitting and recombining actions. Residence time tests (Brouwer et al., 2002) indicate that these devices increase radial homogenization, with only a minimum of axial mixing. Hrymak and Wood (1999) have utilized a transparent acrylic TSE with the same index of refraction as their Newtonian test fluid to validate computational fluid dynamics models. The comparisons for the kneading disk region of an intermeshing co-rotating TSE were done using particle image velocimetry experimental data (Jaffer et al., 2000).

### 16-3.3 Special Mixers

Because viscous mixing needs are so varied, equipment manufacturers have created a broad range of devices to fill specific needs. Where a single agitator cannot

**1018** MIXING OF HIGHLY VISCOUS FLUIDS, POLYMERS, AND PASTES

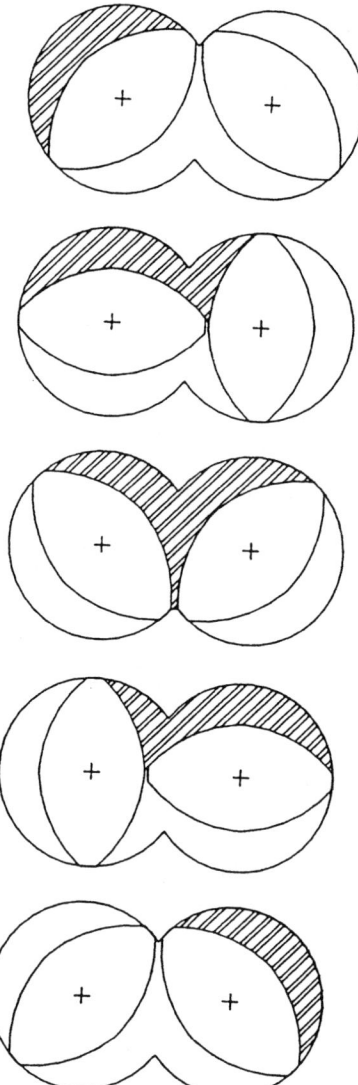

**Figure 16-35** Expansion/compression mixing cycle in kneading paddles.

provide circulation and high shear simultaneously, a second mixing blade may be added. Blenders such as the plow mixer (Figure 16-18) may also be equipped with a high-speed chopper to break up clumps. Similarly, a high-speed impeller can be supplied along with an anchor mixer or with a combination anchor/helical blade agitator (Figure 16-39).

Pasty and tacky products frequently may be plagued with adherence to the agitator(s). Single-shaft mixers such as depicted in Figure 16-40, with additional

EQUIPMENT FOR VISCOUS MIXING **1019**

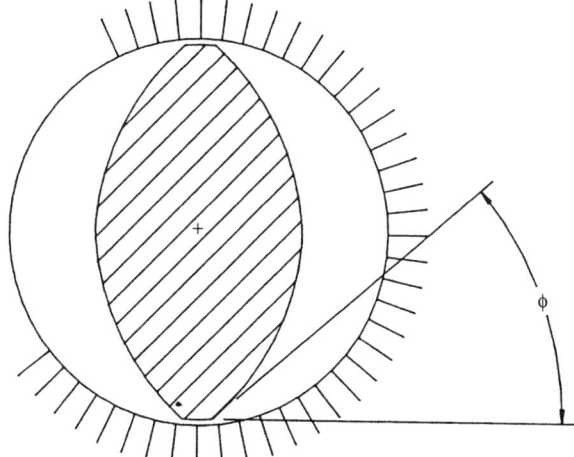

**Figure 16-36** Dispersion face of a kneading paddle.

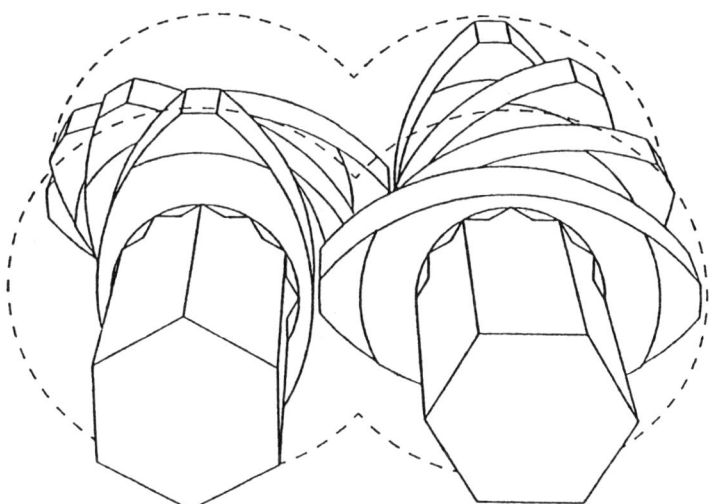

**Figure 16-37** Slicing action between opposed pairs of kneading paddles.

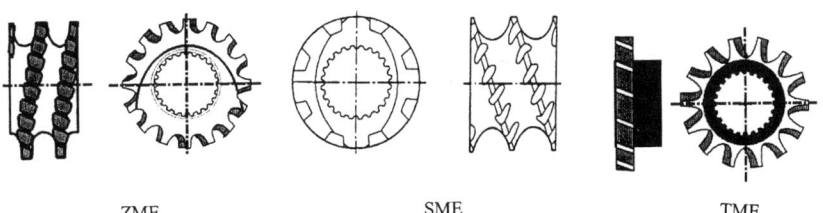

ZME　　　　　　　　　SME　　　　　　　　　TME

**Figure 16-38** Mixing enhancers for intermeshing co-rotating twin-screw extruders.

**1020** MIXING OF HIGHLY VISCOUS FLUIDS, POLYMERS, AND PASTES

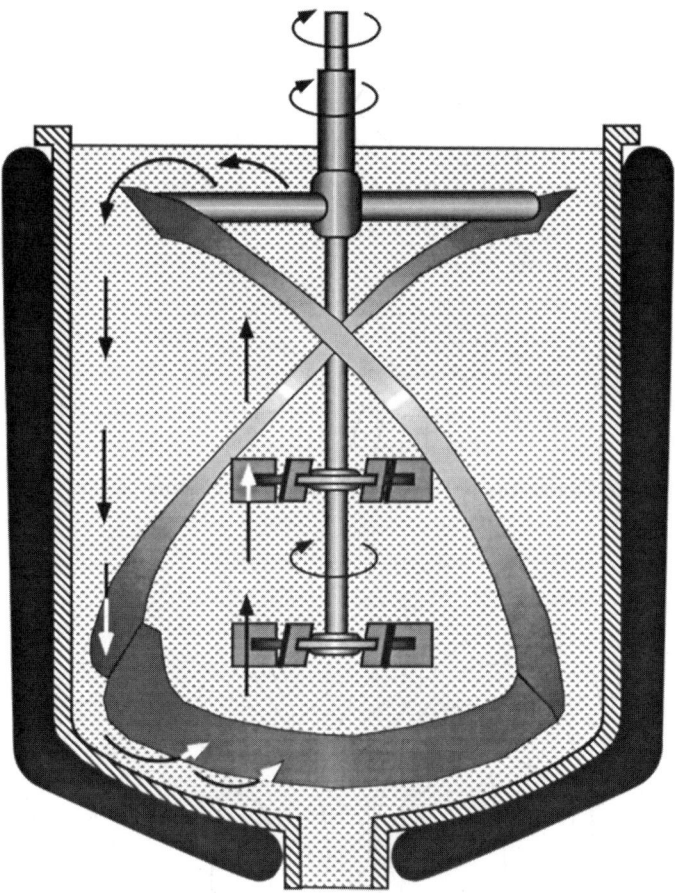

**Figure 16-39** Combination anchor/helical blade, turbine mixer.

rotary scrapers for the disk blades, can operate in batch or continuous mode. Twin-shaft mixers (Figure 16-41) also are particularly suitable for mechanical and thermal processing with pasty, highly viscous, and crust-forming products.

## 16-4 EQUIPMENT SELECTION

The trade journals *Chemical Engineering* and *Chemical Processing* frequently run feature articles on mixing equipment. *Chemical Engineering's* annual *Buyers' Guide* provides a convenient list of mixer suppliers classified by type. *Plastics Technology's* annual *Processing Handbook & Buyers' Guide* has sections on extrusion systems and compounding and mixing systems. *Modern Plastics'*

EQUIPMENT SELECTION 1021

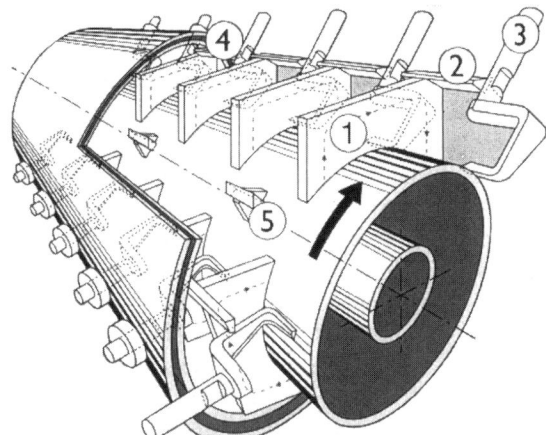

**Figure 16-40** Batch thermal processor with wiped blades: 1, disk elements; 2, mixing/kneading bars; 3, counter hooks; 4, kneading space; 5, finger bars. (Courtesy of List, Inc., Acton, MA.)

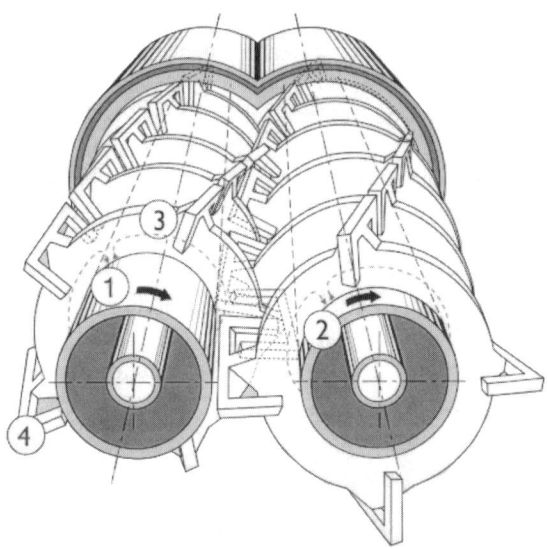

**Figure 16-41** Twin-shaft continuous mechanical and thermal processor: 1, main agitator; 2, cleaning shaft; 3, disk; 4, kneading bars. (Courtesy of List, Inc., Acton, MA.)

annual *World Encyclopedia* includes a classified product listing of single- and twin-screw extruder suppliers, and descriptive text of extrusion, compounding, and mixing.

Most equipment vendors offer rental and pilot-size testing services and will take responsibility for sizing appropriate to meet the client's expected production-size requirements. During pilot testing, the sensitivity of the process to changes in shear rate, temperature, residence time, specific energy (kWh/kg), and any other special process requirement can be evaluated. Additional discussion is provided in Chapter 22.

With single-screw extruders, maintaining geometric similarity on scale-up usually does not provide an equal process result because of the importance of barrel surface temperature on the melting and mixing steps and the decrease in surface/volume ratio as diameter is increased. Scale-ups are usually limited to scale ratios of 2.5:1 or less.

Large twin-screw extruders operate almost adiabatically (typically, <10% of the energy imparted to the product can be transmitted through the barrel walls). Consequently, it is important to conduct pilot tests as close to adiabatic conditions as possible. Adiabatic operation permits scale-up of geometrically similar twin-screw extruders with capacity proportional to screw speed and the cube of diameter. Even though not much energy may be transferred through the barrel into the process, barrel temperature can still play an important role in controlling the temperature of the barrel wall film layer, and thus the viscosity and consequent viscous energy dissipation.

Specific energy (kWh/kg) is frequently a useful guide for achieving the desired mixed quality in the product (Irving and Saxton, 1967). Poor design, however, can lead to overmixing in one region and undermixing in another because of poor circulation.

Temperature control may be the most difficult task on scale-up. Putting more heat in is easy—you can always run the agitator(s) faster. Taking heat out is the problem. Consequently, one may have to run a large mixer at a lower rotational speed than the pilot unit, and thus it will take longer to reach the same level of homogeneity.

Because of the viscosity changes that may be taking place during the mixing process and the greater sensitivity of Reynolds number in viscous processing, it may be very difficult to model the system. Thus, it is better to rely on the experience of the equipment vendor for scale-up guidance.

## 16-5 SUMMARY

Mixing of viscous materials is the intermingling of two or more components by mechanical action of smearing, folding, stretching, wiping, and kneading. With the exception of static mixers (Chapter 7), where the fluid is forced around baffles in a pipe, the required state of product uniformity is achieved by rotation of one or more rotors within the processing vessel. Operation may be either batch or continuous. Viscous energy dissipation and poor heat transfer may limit the operating speed of the rotor(s) and thus prolong the time required when potential product degradation imposes an upper temperature constraint. Possible changes of state, and lack of information regarding the changing properties of the constituents during the mixing process, make a priori prediction of mixing performance of specific equipment most difficult. Equipment suppliers can provide rental equipment or conduct pilot tests to demonstrate feasibility and provide data required for scale-up.

## NOMENCLATURE

| | |
|---|---|
| a | dimensionless constant |
| A | area |
| b | dimensionless constant |
| c | dimensionless constant |
| $C_p$ | heat capacity |
| D | diameter |
| e | flight width |
| E | screw pumping efficiency |
| F | shape factor |
| f | degree of fill |
| h | channel depth |
| h | heat transfer coefficient |
| k | thermal conductivity |
| $K_p$ | proportionality constant |
| L | filled screw length |
| n | power law exponent |
| N | rotational speed |
| $N_p$ | power number |
| Nu | Nusselt number |
| P | pressure |
| p | ratio of viscosity of dispersed phase to viscosity of continuous phase |
| Pr | Prandtl number |
| Q | volumetric flow rate |
| Re | Reynolds number |
| r | radius |
| T | temperature |
| t | clearance |
| W | channel width |
| Z | lead length of screw |

*Greek Symbols*

| | |
|---|---|
| $\alpha$ | dimensionless proportionality constant |
| $\gamma$ | shear rate |
| $\mu$ | viscosity |
| $\rho$ | density |
| $\sigma$ | interfacial tension |

## Subscripts

c  continuous
d  dispersed
d  drag
L  large
p  pressure
S  small
w  wall

## REFERENCES

Bakker, A., and L. E. Gates (1995). Properly choose mechanical agitators for viscous liquids, *Chem. Eng. Prog.*, **91**(12), 25.

Bash, T. F. (1998). Welding engineers CRNI twin-screw extruders, Chapter 8 in *Plastic Compounding: Equipment and Processing*, D. B. Todd, eds., Hanser, Munich.

Blazinski, H., and C. Kuncewicz (1981). Heat transfer during mixing of pseudoplastic fluids with ribbon agitators, *Int. Chem. Eng.*, **21**, 679.

Brouwer, T., D. B. Todd, and L. P.B. M Janssen (2002). Flow characteristics of screws and special mixing enhancers in a co-rotating twin screw extruder, *Int. Polym. Proc.*, **17**, 26.

Brown, R. W., M. A. Scott, and C. Toyne (1947). An investigation of heat transfer in agitated cast iron vessels, *Trans. Inst. Chem. Eng.*, **25**, 181.

Canedo, E. L., and L. N. Valsamis (1998). Farrel continuous mixers systems for plastics compounding, Chapter 9 in *Plastics Compounding: Equipment and Processing*, D. B. Todd, eds., Hanser, Munich.

Carreau, P. J., R. P. Chhabra, and J. Cheng (1993). Effect of rheological properties on power consumption with helical ribbon agitators, *AIChE J.*, **39**(9), 1421.

Case, C. (1998). The reciprocating single screw extruder and its applicability to compounding, Chapter 10 in *Plastics Compounding: Equipment and Processing*, D. B. Todd, eds., Hanser, Munich.

Chavan, V. V. (1983). Close-clearance helical impeller: a physical model for Newtonian liquids at low Reynolds numbers, *AIChE J.*, **29**(2), 177.

Chung, C. I. (2000). Extrusion of Polymers: Theory and Practice, Hanser, Munich.

Grace, H. P. (1982). Dispersion phenomena in high viscosity immiscible fluid systems and application of static mixers as dispersion devices in such systems, *Chem. Eng. Commun.*, **14**, 225.

Hrymak, A. N., and P. E. Wood (1999). Optical methods for flow measurement and mixing system validation, *Proc. 15th Polymer Society International Meeting*, Hertogenbosch, The Netherlands.

Irving, H. F., and R. L. Saxton (1967). *Mixing of high viscosity materials*, in *Mixing: Theory and Practice*, Vol. II, V. W. Uhl and J. B. Gray, eds., Academic Press, New York.

Jaffer, S. A., V. L. Bravo, P. E. Wood, A. N. Hrymak, and J. D. Wright (2000). Experimental validation of numerical simulations of the kneading disc section in a twin screw extruder, *Polym. Eng. Sci.*, **40**(4), 892.

Janssen, L. P. B. M. (1978). *Twin Screw Extrusion*, Elsevier, New York.

Karam, H. J., and J. C. Bellinger (1968). Deformation and breakup of liquid droplets in a simple shear field, *Ind. Eng. Chem. Fundam.*, **7**(4), 576.

Metzner, A. B., and J. C. Otto (1957). Agitation of non-Newtonian fluids, *AIChE J.*, **3**, 3.

Parker, N. H. (1965). How to select double-arm mixers, *Chem. Eng.*, **72**(18), 125.

Rauwendaal, C. (1990). *Polymer Extrusion*, Hanser, Munich.

Sandall, O. C., and K. G. Patel (1970). Heat transfer to non-Newtonian pseudoplastic fluids in agitated vessels, *Ind. Eng. Chem. Process. Des. Dev.*, **9**(1), 139.

Tadmor, Z., and C. G. Gogos (1979). *Principles of Polymer Processing*, Wiley, New York.

Thiele, W. C. (1998). Counter-rotating intermeshing twin-screw extruders, Chapter 3 in *Plastics Compounding: Equipment and Processing*, D. B. Todd, eds., Hanser, Munich.

Todd, D. B. (1988). Heat transfer in twin-screw extruders, *SPE ANTEC Papers* 34, 54.

Todd, D. B. (1991). Drag and pressure flow in twin screw extruders, *Int. Polym. Proc.*, **6**, 143.

Todd, D. B., ed. (1998). *Plastics Compounding: Equipment and Processing*, Hanser, Munich.

Todd, D. B. (2000). Improving incorporation of fillers in plastics, *Adv. Polym. Technol.*, **19**(1), 54.

Uhl, V. W. (1955). Heat transfer to viscous materials in jacketed agitated kettles, *Chem. Eng. Prog. Symp. Ser.*, **51**(17), 93.

# CHAPTER 17

# Mixing in the Fine Chemicals and Pharmaceutical Industries

EDWARD L. PAUL, MICHAEL MIDLER, and YONGKUI SUN

Merck & Co., Inc.

## 17-1 INTRODUCTION

This chapter is directed toward the unique mixing issues that may be encountered in the fine chemicals and pharmaceutical industries. Relevant mixing issues are the subjects of other chapters in this book, and extensive references to these chapters will be made. Mixing in chemical reactions is very important in many fine chemical and pharmaceutical processes. Because of this critical aspect, the reader is referred to the detailed discussion and examples in Chapter 13. Other chapters relevant to the pharmaceutical industry are Chapters 15 and 18. The mixing concerns in the fine chemicals and pharmaceutical industries that are addressed in this book include those associated with the synthesis of intermediates and active ingredients but do not include pharmaceutical operations such as granulation.

One of the unique aspects of the pharmaceutical and agricultural-chemical industries is the degree of regulatory control over processing, both in limiting operations to narrow ranges and in restricting process changes directed at improvement. Mixing issues are often a part of these regulatory concerns because of the sensitivity of some operations, to changes in mixing conditions either on scale-up or on changes in vessels in manufacturing operations. The operations that are among the most sensitive to these concerns are mixing sensitive reactions and crystallization. In this chapter we address many of the sensitivities of these operations, with the intent of providing guidelines for development and scale-up that can be helpful in the design of mixing systems to minimize regulatory issues.

---

*Handbook of Industrial Mixing: Science and Practice*, Edited by Edward L. Paul, Victor A. Atiemo-Obeng, and Suzanne M. Kresta
ISBN 0-471-26919-0    Copyright © 2004 John Wiley & Sons, Inc.

## 17-2 GENERAL CONSIDERATIONS

Although the subject of this chapter is general mixing in the fine chemical and pharmaceutical industries, much of the focus is on reactions and their associated operations, since they present a wide range of mixing challenges. In addition, multipurpose plants are designed with a series of vessels that are used for reactions as well as extraction, distillation, crystallization, and so on, and in some applications, these operations may all be performed in the same vessel. A major challenge for mixing in stirred vessels for the pharmaceutical and fine chemical industries is, therefore, the variety of functions that must be performed—often simultaneously, such as heat transfer, reaction, mixing and so on. To make matters worse for the development and design engineer, mixing that may be good for one aspect of an operation may be deleterious to another. A central question is: Can one set of operating conditions [e.g., impeller(s), rotational speed(s), power input, baffles, internals, control systems] cover all requirements?

Consider a reacting solid–liquid–gas system with mass transfer occurring among the phases. The operations involved may include chemical reaction and mass transfer for gas absorption. The solids are reagents that are dissolving as the reaction proceeds. The operations are solids suspension and solids dissolution. The products or by-products, whose physical and purity characteristics are important in downstream processing, are crystallizing out of solution. The operations are mass transfer and crystal growth. Overall, the system is highly complex. Furthermore, it should be apparent that no single set of design correlations can cover this range of simultaneous interactions, even though it might be possible to design adequate systems for each of the operations.

One simple solution to the problem would be to specify a mixing system with enough energy to exceed the requirements for the most difficult individual operation. Often, this is done. A pilot plant version of the process may even run satisfactorily using this design strategy. However, for scale-up to the manufacturing plant, this strategy may give detrimental results due to overmixing e.g., overmixing with regard to crystallization and possible production of excessive fines, emulsions, or intractable foam from vapor-phase incorporation. In some cases, it may not be possible to accomplish the most difficult individual operation on the scale required by the manufacturing facility. It may then be necessary to use a specialty design such as a continuous in-line device or other special mixing system.

This scale-up problem is posed to illustrate the interacting features of multiphase batch reactors and the importance of developing a comprehensive strategy for scale-up. It stresses the need for versatility that must be built into reactor systems, the effectiveness that this can achieve in plant operation, and the limitations in cases that exceed such capabilities. In the laboratory, there is no substitute for testing reaction sensitivity over a broad range of conditions to provide as much information on allowable ranges as possible. The reader is referred to Section 17-2.6 for a discussion of this important tool for this type of experimentation.

## 17-2.1 Batch and Semibatch Reactors

A large majority of reactions are run in the batch or semibatch mode, making the stirred vessel the mainstay of chemical reaction engineering in the pharmaceutical and specialty chemical industries. The reasons for this reactor choice include:

1. *Complete conversion.* Reactions are generally run to achieve complete conversion of the limiting reagent—controlled by time and not subject to differences in completeness of conversion because of residence time distribution in a continuous stirred tank reactor.
2. *Accuracy of charge.* Reagent quantities can be carefully controlled and procedures for overchecks of quantities actually utilized.
3. *Productivity.* Reactor volume is often consistent with the limited productivity requirements characteristic of this industry.
4. *Flexibility.* Batch reactors can process a large variety of homogeneous and heterogeneous reactions successfully with little modification of internals and can be used in dedicated or multipurpose facilities. The use of variable speed drives along with versatile impellers are key factors.

Another aspect of flexibility is with respect to process changes (e.g., reaction times can be changed in response to scale-up problems), often resulting from short lead times from lab to manufacture, ability to rework material, and so on. The *disadvantages* of batch reactors in processing should also be noted.

1. *Reaction environment.* There may exist nonuniformity of mixing intensity throughout the vessel that can lead to undesirable side reactions caused by variations in local concentration environments.
2. *Optimum conditions.* One aspect of a reaction system may require different conditions than another which may overlap in time (i.e., a reaction that results in precipitation of a product or by-product may require different mixing intensities for the reaction and precipitation).
3. *Heat transfer.* High rates of heat transfer are not achievable without external pumping through a heat exchanger or by utilization of unwieldy internal coils. In these systems, heat transfer is often achieved by operating at reflux and using an external condenser to remove the heat.
4. *Thermal hazards.* A large volume of a reacting system with highly exothermic reactions or decompositions can pose severe thermal hazards.

Given this conflicting set of advantages and disadvantages, the mixing system design challenge is to identify the requirements of each reaction and to recognize the need for alternative reaction systems when a fit is not practical or possible for scale-up. The latter aspect is discussed later in this section. For homogeneous reactions, blending in the required micro-time scale is required, as discussed at length in Section 13-2.5. For heterogeneous reactions, all combinations of liquid–liquid, gas–liquid, and solid–liquid conditions may be encountered, requiring evaluation of power requirements, impeller design, and vessel internals. As in the case of special reactor considerations, the role of the process development engineer is to determine when the standard batch reactor must be augmented in some way to carry out a particular reaction or reaction sequence successfully.

### 17-2.2 Batch and Semibatch Vessel Design and Mixing

The versatility of the glass-lined vessel in a large variety of chemical environments has made it the workhorse of the industry. These reactors range in size from 80 to 20 000 L and larger. One limitation in the use of glass-lined vessels that is related to mixing and heat transfer is that the limit of temperature difference between jacket and batch is about 125°C. (The manufacturer should be consulted for specific limitations for the type of glass lining and base metal in use.)

The retreat-blade and anchor impellers that have been widely used for many years are now being replaced by glass-lined turbines and other shapes that have been developed by manufacturers using sophisticated methods of applying the glass to more sharply angled shapes. These turbines are available for vessels as small as 80 L, although shafts with removable interchangeable impellers are not available for tank sizes smaller than about 1200 L. These new impellers, especially in multitier configuration, have greatly improved the mixing capabilities of the glass-lined reactor by providing increased shear and circulation. A number of glassed impeller types are now available, including curved, pitched, vertical blade, and gas-handling turbines, several types of hydrofoils, helical ribbon, and traditional anchor. Some examples of these are shown in Figure 17-1. The lower turbine can be positioned within about 10 cm of the vessel bottom. For single turbines in larger vessels, however, the low turbine position may not provide the desired overall circulation.

An additional consideration in the choice and testing of glassed mixing systems is that glassed turbines, because of at least minimal required rounding at the edges, do not provide exactly the same fluid dynamics (e.g., vortex shedding) as that of their metal counterparts, and this could affect some operations. The three manufacturers of glass-lined mixing systems use different methods of attachment of the blades of the impeller(s) to the shaft, Table (17-1).

Glassed baffle design changes have significantly improved mixing performance. However, most glass-lined vessel applications are limited to only a single baffle, to maximize the number of tank nozzles available for other purposes. Figure 17-2 shows baffling efficiency (power loading) of some traditional glassed baffle designs, relative to four wall baffles in a metal tank. For mixing applications that require more than one baffle, but for which additional nozzles cannot be spared, specialized multiple baffle configurations have been provided by several manufacturers.

The other commonly used batch reactor is the fully baffled turbine-agitated vessel, made from stainless steel and other alloys. This reactor can be used for

GENERAL CONSIDERATIONS **1031**

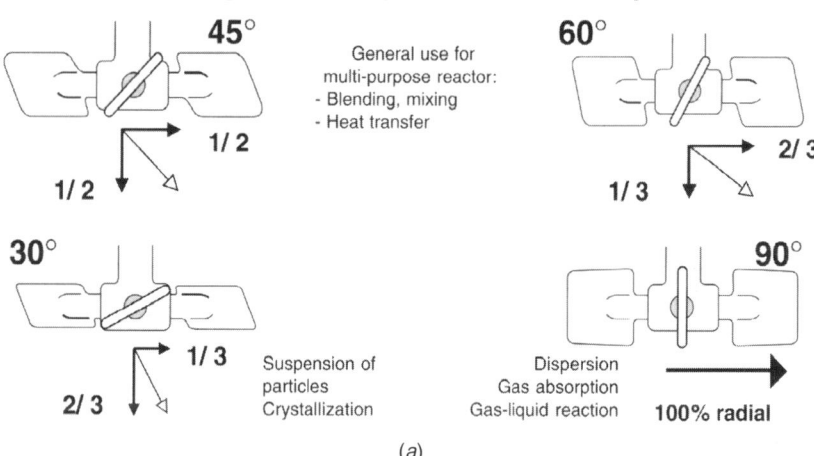

**Figure 17-1** Typical glass-coated impellers showing recently developed technology by the manufacturers to coat sharp edges. (*a*) GlasLock® glass-lined still impellers. (Courtesy of DeDietrich Process Systems.)

some chemical environments that are not compatible with glass (i.e., strong bases, hydrofluoric acid). This reactor also has several advantages in construction and internal configuration, because of the advantages and versatility of stainless steel and other alloys in fabrication. An additional advantage is that, in general, a wider variety of impellers can be used, and as noted above, they do not have the (at least minimally) rounded surfaces required for glass. The most common metal

**1032** MIXING IN THE FINE CHEMICALS AND PHARMACEUTICAL INDUSTRIES

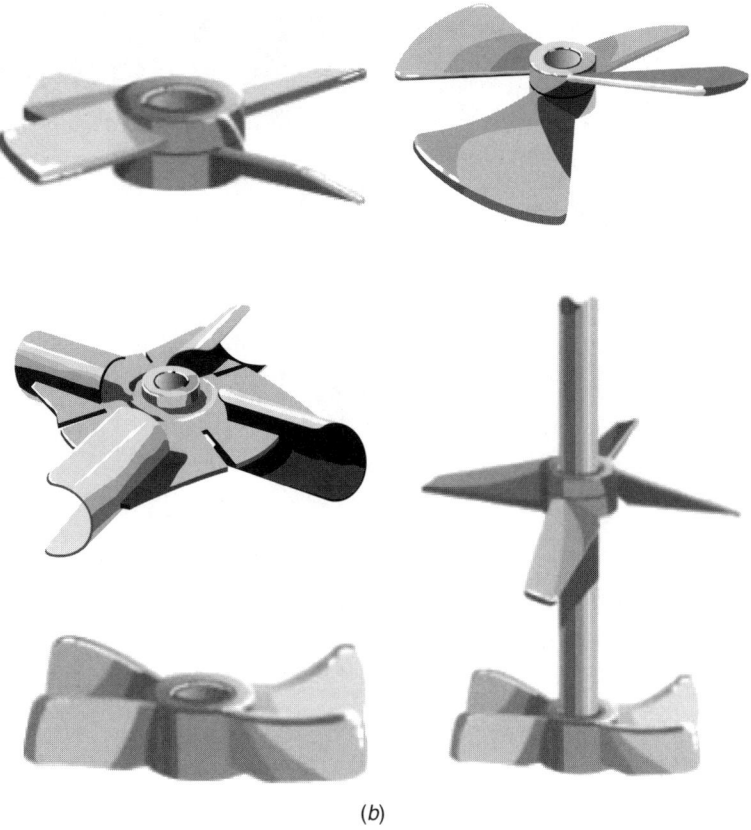

(b)

**Figure 17-1** (b) Cryo-Lock® impeller. (Courtesy of Pfaudler, Inc.)

impellers are pitched blade turbines and hydrofoils for axial flow and flat-blade (Rushton) turbines for increased dispersion. See Chapter 6 for specifics on stirred vessel components, and Chapter 21 for the mechanical aspects of stirred vessels.

### 17-2.3 Multipurpose Design

Mixing requirements of batch and semibatch reactors vary over the full range of the mixing spectrum, from simple blending to high-shear specialty designs. A well-designed mixing system using the impellers described above, equipped with a variable speed drive, can cover the majority of cases. Variable speed capability permits mixing to be adjusted for specific functions such as dissolution of solids, crystallization during reaction, addition of gases, and so on. Scale-up is also greatly facilitated when the speeds of both the pilot and production scale vessels can be varied, thereby decreasing dependence on the reliability of mixing scale-up correlations to achieve specific results. Although variable speed drives add additional capital expense ($\sim$5% of vessel cost) and may not be needed in many

GENERAL CONSIDERATIONS 1033

(c)

**Figure 17-1** (c) ElcoLock® and fixed impellers. (Courtesy of Tycon Technoglass, a Robbins & Myers company.)

cases, the versatility of a reactor is greatly enhanced for future applications or in multipurpose applications (see Chapter 21 for specifics on variable speed drives).

Mixing in semibatch vessels is also complicated by the range in volume that may be encountered during the course of a particular series of steps being carried out in the same vessel. Correlations normally characterize mixing at

**Table 17-1** Glass-Lined Impellers and Their Methods of Attachment

| | Blade Attachment | Removable? | Variable Pitch? |
|---|---|---|---|
| DeDietrich | GlasLock: friction fit of blades into holes in the hub | Yes | Yes (up to power limit of motor, gearbox, shaft) |
| Pfaudler | Cryo-Lock: liquid $N_2$ cooling to shrink the shaft | Yes | No (different pitches available) |
| Tycon | ElcoLock | Yes | No (different pitches available) |

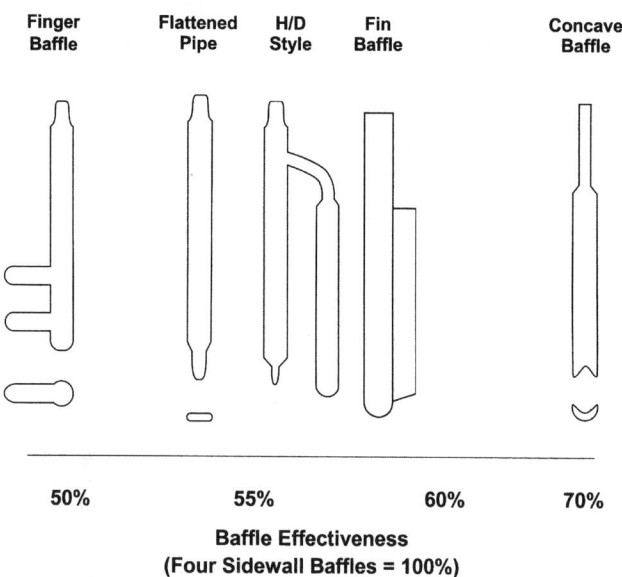

**Figure 17-2** Power-loading efficiency of some typical glass-coated baffles. (Data courtesy of Pfaudler, Inc.).

height/diameter ratios from 2 to 1. However, low volume ratios (<0.5) are often encountered at the beginning or end of an operation when reagents are added or solvents removed. Under these conditions, vapor may be entrained, causing foaming (may be prevented by reduced speed), or solids may not be suspended properly (may require provision of a special low-volume blade in some cases). In glass-lined vessels, the batch may actually be below the bottom of the baffle(s), leading to instabilities, especially with solids present.

Different types of impellers are used for a variety of mixing requirements. Although it is not possible to cover all cases, applications of the various impellers can be classified by their relative input of shear and circulation (see Chapter 6 for power numbers and mixing characteristics of various impellers). Velocity distributions and shear can be determined by laser Doppler anemometry and other methods (Chapter 4). More specific mixing requirements are discussed in the sections on homogeneous and heterogeneous reactions later in this chapter and in Chapter 13, as well as in Chapters 18 and 20. Chemical reaction engineering issues in the fine chemicals industry are discussed by Carpenter (1985, 2001).

### 17-2.4 Batch and Semibatch Scale-up Methods

Much of the literature on scale-up of reaction systems has focused on continuous systems. However, scale-up methods for batch and semibatch operations have been included in several books, including Oldshue (1983), Whitaker and Cassano (1986) Carberry and Varma (1987) Froment and Bischoff (1990), Tatterson (1991), Harnby et al. (1992), and Baldyga and Bourne (1999). Correlations for heat transfer, mass transfer, liquid–liquid dispersions, solids suspensions, and dissolution are available and are discussed in these references and in several chapters of this book. Mixing requirements for scale-up of homogeneous reactions are discussed in Chapter 13, including explanation of the limitations of the usual mixing scale-up parameter of equal power per unit volume. The reader is referred to the texts listed in the references, in which these correlations are well developed. These correlations are not reproduced in this chapter.

Characterization of the physical and chemical parameters of multiphase systems with complex reactants and interfacial phenomena is extremely difficult and may limit the usefulness of the correlations mentioned above. Achievement of a scalable microenvironment is also difficult but may be crucial to successful scale-up. These factors, combined with the multiplicity of uses for batch reactors, argues for maximizing the versatility of both pilot and production scale equipment to encompass a range of operating conditions for specific reactions, as well as to maximize the number of different reactions that can be run successfully. Methods of achieving this versatility are discussed in later sections and in Chapter 13. However, as wide as this range may be, there will be many reactions that cannot be scaled-up successfully without incorporation of reactor design alternatives. We discuss some of these in the next section.

### 17-2.5 Continuous Reactors

Continuous reactors in the pharmaceutical and specialty chemical industries may not only be needed for high productivity as in other segments of the chemical industry, but additionally to solve specific reactor design problems caused by limitations in batch operation. These limitations include heat transfer, mass transfer, and mixing. Continuous reactors are also used to minimize the reacting volume of thermally potent and/or noxious reactions and to decrease the potential and exposure for catastrophic failure of a vessel. Chemical industry reactor standards such as packed bed, fluid bed, and trickle bed reactors find limited utility since this type of phase contacting can usually be achieved in a slurry reactor, where residence time distribution variations, which can lead to changes in product distributions, are eliminated. Continuous stirred tank reactor operation is used only rarely to increase productivity. Operation in this mode can be subject to product selectivity problems.

Tubular reactors with high mixing efficiency and good control of reaction temperature profiles and contact times are used for fast consecutive reactions. An interesting and effective variation on a tubular line mixer is the addition of a centrifugal extractor in series to achieve rapid separation of phases (Example 13-8a, Chapter 13). The local mixing intensity of a well-designed in-line mixer can exceed that of a well-mixed reactor even in the zone of an impeller. The line mixer has the added advantage of allowing the reaction to proceed under more constant conditions of concentration and mixing intensity than can be achieved in a batch or semibatch reactor (Examples 13-3, 13-6, and 13-8a).

The design of continuous reactors has received extensive coverage in the literature. Operating parameters such as residence time, residence time distribution (Nauman and Buffham, 1983), and mixing requirements must be rigorously established for successful scale-up. Since most of the applications are for fast reactions, however, large scale-up factors are often not encountered. The limitation to experimentation may well be the amount of intermediates available for testing at the piloting stage since throughput of the smallest prototype may be relatively large. The combination of high heats of reaction, high reactant concentration, extremely fast reaction rate, and simultaneous or consecutive reactions to undesired products presents an extreme challenge to development and design engineers. Heat transfer and micromixing requirements must be satisfied simultaneously. Stirred tank, wiped film evaporator, and tubular reactors with static mixing elements are compared in an industrial study by Schutz (1988). This system is described in Example 13-7.

## 17-2.6 Reaction Calorimetry

Reaction calorimetry is hardly a new technique. One often thinks of thermodynamics when one thinks of calorimetry since calorimetry has long been used to determine thermodynamic properties of materials and chemical processes. However, with the advent of automated reaction calorimeters such as Mettler's RC1, calorimetry began to be employed as a powerful laboratory tool for studying kinetics and mechanisms as well as scale-up and mixing properties of chemicals and processes in process development for pharmaceutical applications (Landau et al., 1994, 1995; Sun et al., 1996a).

Accurate measurement of kinetics is essential for investigation of reaction mechanisms, since kinetics are a "reflection" of the reaction mechanism, as well as for successful process development and optimization. It is important to evaluate the impact of process parameters such as agitation rate, mixing intensity, and mass transfer on overall rate and selectivity of the chemical process. According to the types of data obtained, methods for measurement of kinetics may be classified into two categories. The first category provides information on concentrations and conversion, or *integral properties*, since they measure integration of variables such as rate from the beginning of the reaction to the point of measurement.

To obtain rate, one has to differentiate these integral results with respect to time. Obviously, an accurate determination of rate from these integral properties requires high rates of sampling.

The most commonly used integral method for monitoring kinetics of synthetic organic reactions is direct sampling from the reactor followed by chemical analysis. Although this method is valuable in providing concentration, or conversion information, and in particular, chemical identities of components in the reactor, it has a threefold deficiency insofar as a determination of kinetics is concerned. First, only integral properties (e.g., concentration) are measured. Second, in reality, only a limited number of samples may be taken per reaction, a factor that makes it highly inaccurate to determine instantaneous rate by differentiating the concentration data with respect to time. Finally, it is not an in situ method. The sample has to be taken away from the reacting atmosphere for workup and analysis, during which period the sample may undergo chemical changes, thus distorting the true profile of components in the reactor.

Other integral methods include in situ spectroscopic techniques, such as time-resolved infrared spectroscopy, which measures concentration or conversion as well as provides information on chemical identities of components in the reactor. In contrast to the chemical sampling method, the infrared (IR) spectrum of the reacting system may be collected at a much faster pace than that normally possible for the chemical sampling method. Consequently, rate data may be derived with good accuracy by differentiating the IR intensity data with respect to time.

The second category of methods is characterized by direct measurement of *differential properties* such as the instantaneous rate of the reaction. Reaction calorimetry, representing this category of methods, is a more powerful method for monitoring kinetics. Modern reaction calorimetry measures the *rate* of heat flow into or out of a reactor, typically 1 L in volume, during reaction while maintaining precise control of the temperature of the contents in the reactor. A schematic of a reaction calorimeter is shown in Figure 17-3. The characteristics of the reactor and the reaction mixture, i.e., UA, and $mC_p$, are determined through a simple and automatic sequence, thus allowing accurate determination of the heat flow, $q_r$, as a function of time during reaction. The heat flow measured under isothermal conditions is directly proportional to reaction rate, or to be more precise, a summation of the rate of each reaction step as weighed by heat of reaction $\Delta H_i$ of the corresponding step:

$$q_r = V_r \sum_i \Delta H_i \frac{dC_i}{dt}$$

where $V_r$ is the volume of the contents in the reactor and $\Delta H_i$ is the heat of reaction for the ith step. In addition, the reaction calorimeter conducts the measurements in an in situ, noninvasive, and continuous fashion, which is difficult, if not impossible, to achieve using the chemical sampling method. Although the chemical sampling method is most valuable in providing information on the

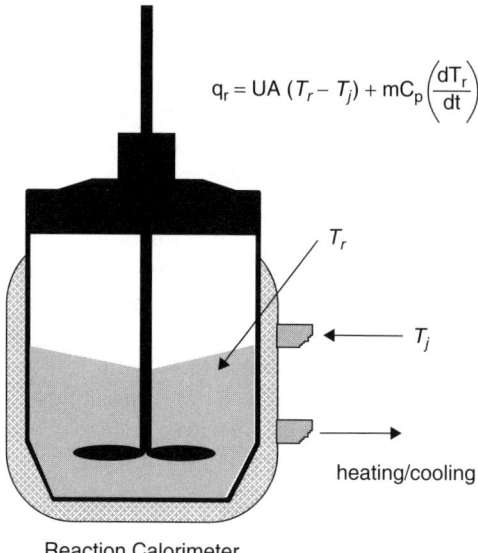

**Figure 17-3** Schematic of a reaction calorimeter.

chemical identities of components in the reactor, it is not an accurate kinetic tool. The salient advantages of the reaction calorimetry may make up for its obvious shortcoming in such a way that accurate measurements of kinetics may be achieved. Furthermore, when combined with other in situ compositional analysis tools, such as infrared spectroscopy as well as direct sampling, reaction calorimetry would enable one to obtain a complete picture of kinetics and reaction pathways associated with a chemical reaction. This integrated approach has been used successfully to elucidate the reaction pathways of a number of catalytic selective hydrogenation reactions (LeBlond et al., 1998), to evaluate the effects of mixing intensity and gas–liquid mass transfer on enantioselectivity of asymmetric hydrogenation (Sun et al., 1996b,c) and to determine whether a process is operating under mass diffusion limitations (Landau et al., 1995). Reaction calorimetry allows measurement of the intrinsic kinetics of a chemical process as well as a determination of the influence of process parameters, including mixing intensity, on the outcome of a chemical process, such as rate and selectivity, thus allowing one to predict process scale-up properties.

## 17-3 HOMOGENEOUS REACTIONS

The reader is referred to Chapter 13 for detailed discussion of this topic. A comprehensive treatise is that of Baldyga and Bourne (1999).

Reactions that are truly homogeneous throughout their entire course may or may not be affected by mixing or other scaling variables, depending on their reaction rates. Slow, homogeneous reactions can be scaled up directly by appropriate increase in volume, when blend time for reagents is rapid compared with reaction rates and when heat transfer rates are adequate to maintain any fixed temperature or temperature profile. The reaction(s) must be sufficiently slow that significant conversion occurs only after the reagents are homogeneously blended on a molecular scale (i.e., the meso- and micromixing conditions have been satisfied).

### 17-3.1 Mixing-Sensitive Reactions

Many examples of mixing-sensitive reactions have been identified in reactions that are homogeneous but whose reaction rates are sufficiently rapid to result in significant conversion during mixing of reagents prior to the achievement of complete homogeneity on a molecular scale. Many of these examples are from reaction types that are run in the fine chemicals and pharmaceutical industries, and their identification can be essential to successful development and scale-up. Homogeneity of the reaction systems makes them more accessible to analysis than the more frequently encountered heterogeneous reactions. However, the generalizations that are developed can be applied qualitatively, and in some cases quantitatively, to heterogeneous systems.

One of the most difficult aspects of the scale-up of both homogeneous and heterogeneous reactions is the prediction and control of by-product distribution. Mixing sensitivity is not only an issue through loss of yield of the major product by consecutive and/or parallel reactions but also by an increase in the amount of by-products formed. These by-products may be negligible on a laboratory or even a pilot plant scale, but may increase on scale-up to production. An increase of as little as 0.1 to 1.0% in the amount of a particular by-product may not be acceptable when it cannot be adequately removed by downstream processing. These impurities may affect physical form, particle size, downstream liquid–liquid separation, or foaming tendency. If the impurity level rises above about 0.1%, product registration may be affected.

This problem is difficult because the threshold reaction rates and rate constant ratios that are significant may be far lower than anticipated by laboratory experiments. In addition, the mixing scale-up issue with regard to a decrease in local mixing intensity and an increase in circulation time may result in an unexpected increase in by-products. The reader is referred to Chapter 13 for additional discussion and several examples.

*17-3.1.1 Laboratory Prediction of Mixing Sensitivity.* The challenge in development is to predict mixing-sensitive behavior so that the mixing and/or reaction system can be modified to circumvent the issue. The use of estimates of the Damkoehler number (Da), the ratio of the reaction rate to the mixing rate, is recommended as outlined in, Sections 13-2 and 13-4. As a first approximation, the following types of laboratory experiments can be recommended to attempt to establish whether or not a potential problem exists:

1. For consecutive reactions, the mixing sensitivity can be assessed by running the reaction in the reverse addition mode such that there is a large

excess of the reagent that can cause overreaction. One reason for no overreaction is that there is no pathway in the reaction system for a consecutive reaction. That information can be very helpful on scale-up by eliminating one possible pathway for failure.

2. For consecutive or parallel reactions, runs comparing very poor mixing and very good mixing can be helpful as a first indication of sensitivity. Any differences, however small, should be cause for further study. In some cases, laboratory mixing, even when very poor, can be sufficient to prevent differences in product distribution. In these cases, pilot scale experiments may be required.

3. A step-by-step recommended procedure for determining mixing sensitivity based on evaluation of the Damkoehler number may be found in Section 13-4.

4. A laboratory reaction calorimeter can be very helpful in this type of experimentation. Use of this device is described in Section 17-2.6.

5. Although there is no general rule or agreement on minimum size since there are critical dependency requirements of the specific operation, 4 L is considered a minimum, recognizing the probable limited amounts of materials available or practical to obtain. Larger vessels can obviously provide better scale-up data. Smaller vessels (1 L) can also be adequate provided that they follow a standardized configuration.

6. The reaction vessel should be cylindrical with a standard turbine, alloy or glass, agitation system, with baffles.

7. A laboratory favorite, the round-bottomed flask, must be avoided since the results cannot be deemed reliable for mixing information and even for accurate assessment of selectivity for fast reactions.

Additional methods of assessing sensitivity are available for heterogeneous systems and are discussed in Section 17-4 and in Chapter 13.

### 17-3.1.2 Previous Work on Mixing and Homogeneous Reactions.
The earliest work on the effect of mixing on homogeneous reactions as initially carried out by Danckwerts (1957, 1958) was focused on the effect of mixing on conversion rate resulting from incomplete blending in continuous reactors. Several papers appeared on this subject, including those of Toor (1962), Keeler, et al. (1965), Vassillatos and Toor (1965), Kattan and Adler (1967), and Harris and Srivastava (1968). Experimental and theoretical work in this area has continued with a primary focus on the study of micromixing. Acid–base neutralizations are essentially instantaneous and therefore mixing controlled. These and other reactions have been used in studies by Rice et al. (1964), Mao and Toor (1971), Klein et al. (1980), Li and Toor (1985), Shenoy and Toor (1989, 1990), to measure micromixing using indicators and other means to measure instantaneous reactions.

Kinetic problems with reaction systems in the pharmaceutical industry are more concerned with selectivity in complex systems than with conversion rate.

However, the fundamental mixing characteristics that affect mixing-controlled conversion are the same as those that can affect selectivity and yield in complex systems. The effect of mixing on selectivity was predicted qualitatively by Levenspiel (1962) in his classic text on chemical reaction engineering. Several studies on specific reactions have since appeared in the literature, including Paul and Treybal (1971), Truong and Methot (1976), Bourne and Kozicki (1977), Bourne et al. (1977a,b, 1981b), and Nabholz and Rys (1977).

In addition, excellent papers on the chemical and physical aspects of mixing-sensitive reactions, including modeling of the mixing effects, have been published, including Ott and Rys (1975), Canon et al. (1977), Belevi et al. (1981), Bourne et al. (1981a), Angst et al. (1982a,b), Bolzern and Bourne (1983), Baldyga and Bourne (1984a, 1988, 1989), Bourne (1984), Mann and Hamouz (1991), Wang and Mann (1990), Angst et al. (1982a), and Laufhutte and Mersmann (1987). A recent publication by Heeb and Brodkey (1990) presents a molecular-based statistical simulation model that was developed to study the covariance terms for mass transfer during mixing of reactants undergoing complex reactions. Finally, some publications on the effect of scale-up have appeared, including Bourne and Dell'ava (1987), Bourne and Hilber (1990), Paul (1990), Rice and Baud (1990), and Wang and Mann (1992). The subject is covered comprehensively in the treatise of Baldyga and Bourne (1999).

### 17-3.1.3 Mixing-Kinetic Problem.
The reaction scheme that has received the most attention in both theoretical and experimental investigations of the effects of mixing on selectivity is the competitive-consecutive reaction. In addition, the parallel reaction system is receiving attention for its importance in reactions and pH adjustments. These systems are discussed in Chapter 13 and highlighted here because of their fundamental importance in the fine chemicals and pharmaceutical industries. The reaction scheme is as follows:

$$A + B \xrightarrow{k_1} R$$

$$R + B \xrightarrow{k_2} S$$

with B added to A in the semibatch case or A and B mixed continuously in the tubular reactor case. R is considered as the desired product. The objective is to determine how mixing conditions can affect the yield of R. We are concerned with the time period during which the reactants are first contacted and when they are completely mixed to a molecular scale. During this time, zones of local B concentration can vary from an upper limit equal to the feed concentration to a lower limit of essentially zero. The course of any reaction that is influenced by concentration has the potential to be influenced by mixing. The effect can be on the reaction rate, the product distribution, or both (see Examples 13-3, 13-4, and 13-6).

Parallel reactions can also be subject to mixing effects, as shown by Baldyga and Bourne (1990), Paul et al. (1992), and Wang and Mann (1992), (see Example 13-8b). A critically important type of parallel reaction is the potential for decomposition of substrates during pH adjustments. Such reactions are very

**1042**  MIXING IN THE FINE CHEMICALS AND PHARMACEUTICAL INDUSTRIES

common in the fine chemical and pharmaceutical industries and in fermentation (see Chapter 18). Scale-up to large vessels can result in serious losses in selectivity.

Mixing effects are of far greater importance on product distribution in multiple reactions because the impact on design and economics is more profound. In such reactions the product desired is one of two or more possible products. The selectivity of a reacting system is defined as the ratio of the amount of desired product to the total amount of limiting reagent actually reacted. The yield, Y, is the ratio of the amount of limiting reagent reacting to produce the desired product to the total amount of limiting reagent charged.

***17-3.1.4 Selectivity in Homogeneous Reactions.*** In semibatch operations, provided that reactant B is mixed instantaneously to a molecular level with the vessel contents, the maximum selectivity in a competing-consecutive reaction system is a function of the rate constants $k_1$ and $k_2$, the overall molar charge ratio of A to B, and the degree of conversion of A. The degree of conversion of A can depend on the charge ratio and residence time. This discussion is limited to the case of sufficient residence time such that all of the B charge will react, provided that B is not charged in excess for complete reaction to S. The maximum selectivity of R, in the absence of mixing effects, then becomes a function only of $k_1/k_2$ and the molar charge ratio. For purposes of the present discussion, it is convenient to establish a fixed molar charge ratio. The ratio chosen is not necessarily intended to give the maximum selectivity since consumption of starting material may be a prime consideration. For a particular reaction system, (fixed $k_1/k_2$), a fixed molar charge ratio, and conditions of perfect mixing, the selectivity is also fixed, as is the yield and the degree of conversion of A.

At this point it is convenient to discuss the yield of R rather than the selectivity. The term *expected yield* is used to denote the yield that would be obtained under conditions of perfect mixing, as derived by van de Vusse (1966) and Levenspiel (1962).

$$Y_{exp} = \frac{R}{A_0} = \frac{1}{(1 - k_2/k_1)} \left[ \left( \frac{A}{A_0} \right)^{k_2/k_1} - \frac{A}{A_0} \right] \qquad (17\text{-}1)$$

where capital letters denote molar concentrations. This equation applies to both batch and semibatch operations, provided that both reaction rates depend on B in the same way and provided that B is added to A in the semibatch case.

Less than perfect mixing may reduce, but not increase, the yield in homogeneous systems. The primary concern is the magnitude of the yield reduction attributable to deviation from instantaneous, perfect mixing to a molecular level. The reader is referred to Chapter 13 for additional discussion of these key issues.

## 17-3.2  Scale-up of Homogeneous Reactions

Scale-up of reacting systems in the pharmaceutical industry is often considered to be a simple matter of an appropriate increase in size of the reaction vessel. This simplification is indeed applicable to homogeneous reactions in which the product distribution is a function of the kinetics only and thermal effects can be controlled by conventional methods. The majority of homogeneous reactions fall into this category or at least appear to. In the pharmaceutical industry this scale-up issue requires particular attention during process development. Even very small deviations in product distribution can result in significant separation problems that must be addressed to meet the stringent requirements for consistency of product purity between laboratory-, pilot-, and production scale operations. Reaction scale-up problems are thus better resolved by reactor design modification than by downstream purification modifications. The objective is to achieve constant product distribution on scale-up: changes in product distribution profiles as small as 0.1% in an impurity level may make the product unacceptable.

The primary function of the development chemist or engineer, therefore, is to determine for each reaction whether or not special design considerations are required. In some cases, laboratory work at different mixing levels will indicate mixing selectivity. In other cases, however, even pilot plant operation will not reveal subtle deviations from expected product distribution. In general, the absolute value of the primary reaction rate is the most reliable predictor of mixing sensitivity. If the reaction half-life is within one or two orders of magnitude of the blend time for the plant scale reactor, a mixing dependence might be expected, as can be determined by estimation of the Damkoehler number.

It is important to note that the reaction rate as represented by $k_1$ may apply either to the principal reaction or to addition of a substance that could react with any of the substrates in a parallel reaction. For example, if an acid or base is added to adjust pH, the local concentrations at the point of addition could cause reactions that would not be expected over the intended pH limits. Such unexpected reactions could lead to substrate decomposition. The reader is referred to Chapter 13 and Example 13-8b for more information on this effect.

Scale-up studies on mixing-sensitive homogeneous reactions were run using diazotization of 1-naphthol as the model system as reported by Bourne and Dell'ava (1987), Paul (1988), Bourne and Gablinger (1989), and Rice and Baud (1990). These studies showed that both micromixing and macromixing must be considered in larger vessels because the circulation time increases even when local micromixing at the point of addition can be maintained relatively constant. Increased circulation may therefore be required to maintain the mole ratio balance between reagents A and B in the mixing zone where the actual reaction is occurring. A local insufficiency in A will result in an overreaction to S regardless of local micromixing intensity. B may also be less available because of "engulfment" (mesomixing effect), discussed in Chapter 13.

### 17-3.3 Reactor Design for Mixing-Sensitive Homogeneous Reactions

*17-3.3.1 Semibatch Reactors*. When other considerations preclude the use of in-line mixers and large vessels must be used, two methods of minimizing or eliminating mixing deficiencies can be used. Both multiple turbines and multiple addition points have been shown to be effective in reducing overreaction. These methods are compatible and can both be used in the same reactor. The effectiveness of multiple turbines was shown by Paul (1988) and for multiple addition points by Bourne and Hilber (1990). A preferred configuration for multiple turbines with a single optimal addition point is shown in Figure 13-9. The addition of a multiple-point feed distributor would be expected to further enhance this effectiveness. Care must be taken, however, in the design of this type of addition line with regard to placement and nozzle discharge velocity. Furthermore, the time of addition must be sufficient to prevent macromixing and mesomixing from becoming factors.

*17-3.3.2 In-line Mixers*. The preferred reactor design for extremely fast or sensitive reactions is an in-line mixer of appropriate design. Various possibilities have been investigated, including centrifugal pumps (Bolzern and Bourne, 1985), rotor–stator mixers (Bourne and Garcia-Rosas, 1986), impinging thin liquid sheets (Demyanovich and Bourne, 1988, 1989), reaction injection molding (Lee et al., 1980), and vortex mixing (Bowe, 1990). The impinging jet design inherent in the technology of reaction injection molding and the vortex design seem to present high degrees of micromixing such that the blending of reagents should be completed to the molecular level in the minimum time. The advantage of in-line mixing devices is to eliminate the need for scale-up of macromixing and mesomixing parameters since the correct mole ratio can be maintained by the feed system and the mixer is required to provide adequate micromixing only. An example of a homogeneous consecutive reaction is presented in Example 13-3.

## 17-4 HETEROGENEOUS REACTIONS

Heterogeneous reaction systems are very common in the pharmaceutical and specialty chemical industries because of the limited and unusual solubility of many reagents and reaction products. Contributing factors to these characteristics are high-molecular-weight (300 to 1000) and multifunctional molecular structures. In addition, because heterogeneous systems can in some cases be manipulated to achieve improved yields compared to a homogeneous system with the same reactions, there can be an advantage in running under heterogeneous conditions or in some cases to deliberately create a heterogeneous system for the purpose of improving selectivity.

The discussion of selectivity considerations in homogeneous reactions in Section 17-3 is intended to provide an introduction to the far more complex issues involving heterogeneous reactions. The continuity of theoretical and practical considerations between these different types of reacting systems is provided by the obvious fact that the course of reactions is determined by events at the molecular scale whether or not the reactive molecules are in the liquid, solid, or gas phase when they enter the reaction zone. As in the case of homogeneous reactions, the course of a complex reaction will be determined by local molar ratios and kinetics. The degree of deviation from expected kinetic behavior is determined by the reaction rate relative to the rates of mass transfer and mixing. Differences in selectivity between the same reaction run under homogeneous conditions and heterogeneous conditions are illustrated in Example 13-4.

This example indicates that the homogeneous reaction environment with regard to minimizing a consecutive reaction is more selective than the film around a dissolving reagent. The result can be represented as an increase in the apparent rate constant ratio, $k_2/k_1$, for the heterogeneous condition as indicated in Figure 13-11, where the loss in selectivity [increase in X, where $X = 2S/(2S + R)$] is plotted against $k_1$. Deviations in the case of homogeneous reactions are more amenable to quantitative analysis and can therefore be developed more completely. The same local considerations apply in heterogeneous reactions where expected overall molar ratios between reactions cannot be maintained because of mass transfer limitations at phase boundaries. For simple reactions, overall reaction rates may be affected and usually decrease, but yield is unaffected, given equal degrees of conversion. For complex reactions, the selectivity may be decreased, but unlike homogeneous systems, can also be increased under certain circumstances.

The other key difference between homogeneous and heterogeneous reactions regarding selectivity is that significant effects can occur in heterogeneous systems at far lower absolute reaction rates because the mass transfer limitations can be very severe. In addition, these effects can be subject to considerable magnification on scale-up to plant operations.

### 17-4.1 Laboratory Scale Development

Laboratory experimentation is the primary approach to characterization of reaction systems. This work must be designed carefully to prevent diffusion and mixing limitations from appearing unimportant at this scale when they may be significant at the manufacturing scale. Extremes of mixing and/or reaction conditions must be explored to fully characterize the interacting responses of these complex physical and chemical factors (see Section 13-4 for a recommended laboratory program). One final consideration in the selection of equipment is the subsequent purification and isolation process. It may prove advantageous or even essential to integrate key elements, such as simultaneous extraction or crystallization with the reaction step. This consideration is developed further in Examples 13-8a and 17-2.

### 17-4.2 Gas–Liquid and Gas–Liquid–Solid Reactions

*17-4.2.1 Gas as Reagent.* With the exception of fermentation, which is the subject of Chapter 18, the most common gas–liquid reaction in the pharmaceutical industry is hydrogenation. The intrinsic reaction rates of hydrogenation reactions vary over several orders of magnitude and can fall into any of the reaction categories discussed in Table 13-8. Design of a hydrogenation system is generally focused on supplying a sufficient quantity of hydrogen so that hydrogen concentrations in the bulk or adsorbed on the catalyst will not be limiting. In many cases, this can be accomplished by suitable design of a subsurface sparger to accomplish absorption during transit from sparger discharge to vapor space. In some cases, however, the absorptivity or reaction rate is slow enough so that reabsorption from the pressurized vapor space is required. This can be accomplished with special gas–liquid mixing designs, such as the Praxair advanced gas reactor (Litz, 1985). Both cases are often seen in batch operations with high rates and in mass transfer control initially, changing to slow rates with kinetic control at the end. Liquid levels may vary, making reabsorption difficult to predict.

Many gas–liquid reactions other than hydrogenations are run in the pharmaceutical industry. Reaction system design depends primarily on the solubility of the gas in the reaction mixture and on its rate of reaction. Soluble gases can often be added without sparging via a vortex created by impeller design. Pitched blade turbines with partial baffles are effective in this regard. For complex reactions, the product distribution can be affected by mixing in direct analogy to the homogeneous case discussed earlier. Some of the first experiments in this area were conducted on chlorination of $n$-decane by van de Vusse (1966) in which mixing was shown to affect the distribution of chlorinated products. The chlorination of acetone (Example 13-6) also shows an effect of mixing on this reaction.

*17-4.2.2 Gas as By-product.* Removal of a volatile reaction product by fractionation to drive equilibrium reactions to completion (e.g., esterification) is well known and presents no mixing issues beyond heat transfer rate and suspension. When a volatile product reacts in competing or consecutive reactions to form an undesired side product, special design measures must be considered. The goal is to accomplish removal of the volatile product rapidly enough to maintain its concentration in the reaction mixture at a suitably low level. In most cases, the volatile product can be removed by simultaneous distillation provided that the distillation rate and volatility are sufficient to maintain the desired critical solution concentration. Poor mixing could cause reduced selectivity by causing a reduction in heat transfer rate, thereby reducing the effective primary reaction rate and extending the time available for the production of by-products. Foaming may often be a problem, especially due to impurities, fine solids, or second liquid phase.

In some cases, other methods for removing the volatile product are required. The distillation may be too slow, relative to the reaction rate, or low relative volatilities may preclude removal of the volatile product to a concentration low enough to prevent overreaction. This problem can be solved by the addition of adsorbents, such as ion-exchange resins or molecular sieves, for the in situ removal of the volatile product (e.g., water or HCl). Mass transfer considerations will govern the effectiveness of such solid adsorbents; surface area and particle size are the critical factors. In cases where the mass transfer limitation is still significant under all practical surface areas and particle sizes, it is sometimes possible to add a chemical scavenger that removes the by-product by homogeneous chemical reaction. These mixing and mass transfer issues are illustrated in Example 17-1.

*Example 17-1: Removal of By-product Gas by Adsorption* (Weinstock 1986). This example is of a reaction system in which a by-product (HCl) that is generated by the primary reaction would decompose both the desired product and the starting material to give essentially no product yield unless its concentration is controlled. The actual selectivity as well as the conversion rate is a function of the method and extent of this control. The chemistry is shown in Figures 17-4 and 17-5, and the reaction system is summarized in Figure 17-6. The method of controlling the concentration of HCl below a value that causes excessive decomposition while maintaining its concentration high enough for its participation in the required reactions is critical to the success of the overall scheme. The product, R, is made in relatively large volume. A feasible, commercially viable synthesis of this compound was essential for operation in a manufacturing environment.

There are two distinct reaction types taking place: (1) reactions to form imides and (2) HCl-promoted imide cleavages producing amides and an acid chloride. Consecutive decomposition by reaction with HCl always proceeds depending on the concentration of HCl. If no method of mediating the HCl concentration was applied, the concentration of HCl would increase to 0.1 $M$ and result in complete decomposition of R. It was determined that an optimum concentration of 0.004 $M$ is required for imide cleavage.

Molecular sieves (3A or 4A) were found to be very effective for this mediation under very well defined conditions. The HCl concentration in solution is affected

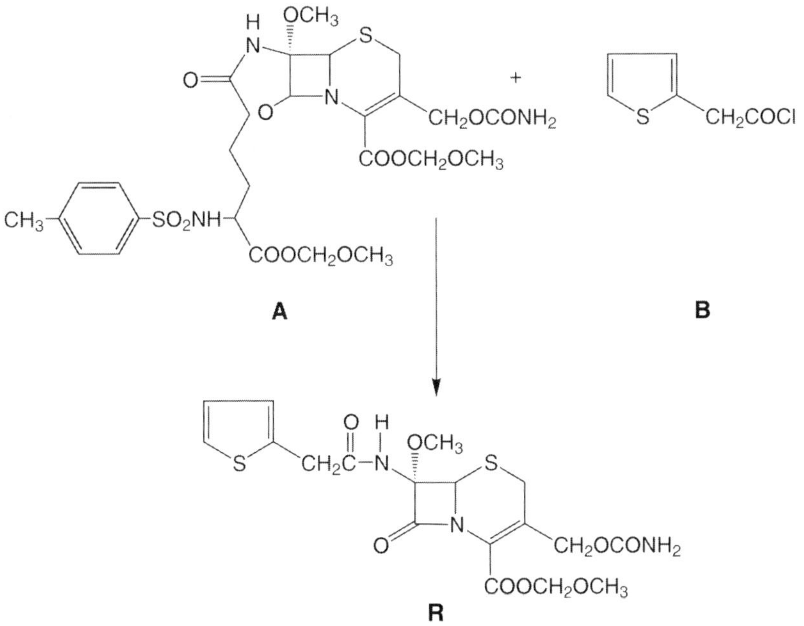

**Figure 17-4** Overall chemistry of the transacylation reaction in an antibiotic synthesis in which rapid removal of by-product HCl is essential for practical operation.

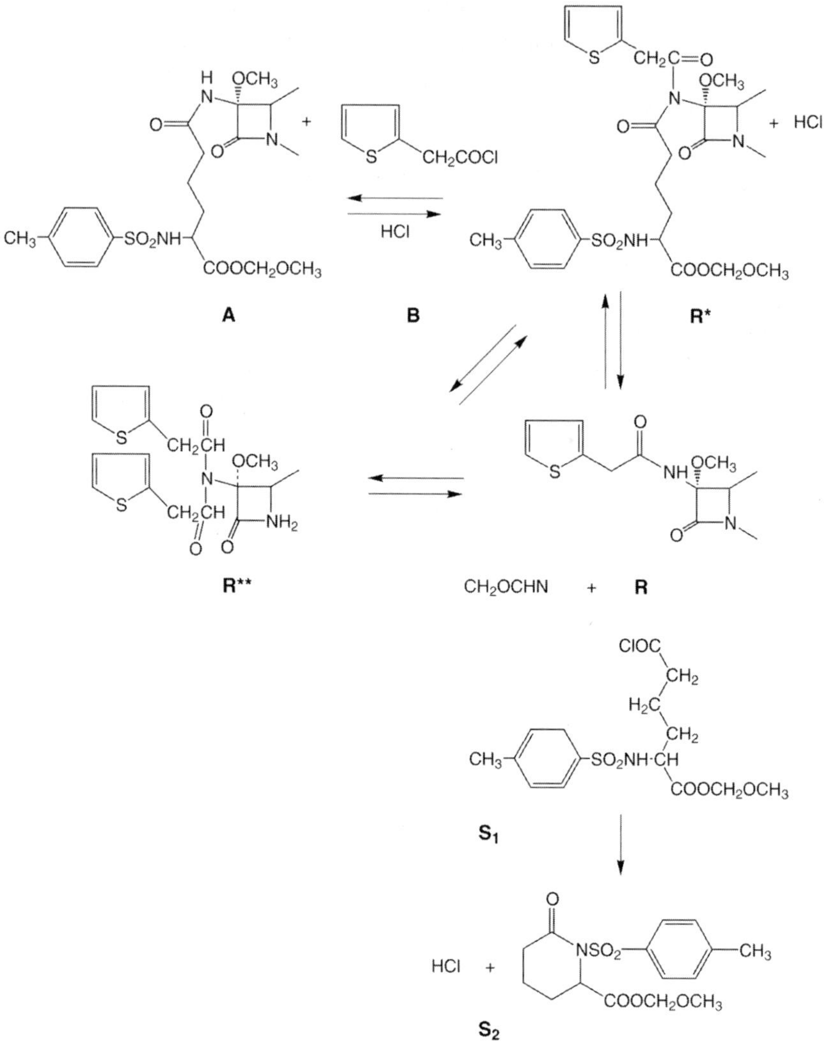

**Figure 17-5** Chemical intermediates in the transacylation reaction pathway.

$$A + B \rightleftharpoons R^* + HCl$$

$$R^* + B \rightleftharpoons R^{**} + S_1 + HCl$$

$$R^* + HCl \rightleftharpoons S_1 + R$$

$$R^{**} + HCl \rightleftharpoons R + B$$

$$S_1 \rightarrow S_2 + HCl$$

**Figure 17-6** Kinetic representation of the transacylation pathway.

**Table 17-2** Comparison of Reaction Selectivity for Different Methods of By-product Removal

| Method | Relative Selectivity |
| --- | --- |
| None | Essentially zero |
| Distillation | Essentially zero |
| Molecular sieves | + |
| Homogeneous scavenger | ++ |

by both the amount of sieves used as well as their external surface area. Sieve pellets (−400 μm) were not satisfactory because of rate-controlling diffusion in the pores, whereas powdered sieves (1 to 4 μm) were. The concentration and removal rate of HCl were critical. Improvements in selectivity were subsequently achieved through the development of a homogeneous HCl scavenger, trimethyl silyl methyl carbamate. Elimination of the mass transfer resistance at the sieve surface by the presence in solution of a reagent that reacts directly with HCl resulted in a significant yield increase. Comparison of selectivity of R by four different methods of HCl mediation is shown in Table 17-2. All of the reaction studies were carried out in the laboratory. Scale-up of sieve and homogeneous scavenger mediated reactions was relatively straightforward once the concentrations and reaction were defined in the laboratory. Successful plant scale operation required rapid heat-up and cool-down, however, to minimize time at other than optimum temperature.

Typical production scale reaction kinetic profiles, as determined by high-performance liquid chromatography, are shown in Figure 17-7. If the run had been allowed to continue, a significant yield loss would have been experienced, as indicated by the dashed lines.

***17-4.2.3 Scale-up.*** A great deal of work has been done to characterize gas–liquid mixing for prediction of power requirements to achieve equivalent mass transfer on scale-up. Much of this work concerns gas–liquid–solid where the solids are catalysts. The reader is referred to Chapter 11 and many excellent texts and literature articles, including Nagata (1975), Oldshue (1983), and Smith (1985). Oldshue (1985) pointed out the relation between the gas flow energy and mixer energy. For a radial flow impeller, the mixer energy must be about three times greater than the energy in the expanding gas stream or the mixer will not control the flow pattern. For axial systems, this ratio is about 10. Below this mixer energy level, the axial flow pattern is destroyed completely by the gas flow. In both cases, however, satisfactory performance may be possible if other aspects of mixing are not critical. Oldshue (1983) points out that although axial flow impellers are often specified for solid–liquid suspensions, the effect of gas sparging can disrupt the flow pattern to the extent that radial flow impellers could be a better choice. Scale-up requires consideration of the balance of these characteristics as well as consideration of the power per volume.

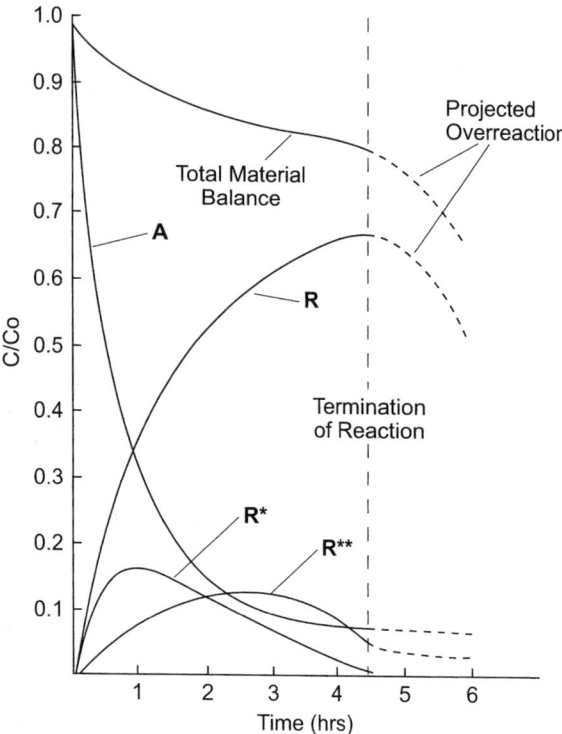

**Figure 17-7** Profile of a manufacturing scale reaction as determined by high-performance liquid chromatography.

### 17-4.3 Liquid–Liquid Dispersed Phase Reactions

*17-4.3.1 Reactivity.* Reactions in liquid–liquid dispersed phase systems are common, and the reaction rates span all the regimes of reactivity. For slow reactions of reactants with low solubility in their respective phases, the actual conversion rate for even very well-mixed systems may be negligible. In these cases, a third solvent may be added to improve mutual phase solubility, or a phase transfer catalyst may be added to transfer one reagent, usually ionic, from aqueous to solvent phase. Both additions add downstream separation operations, however, so their use is avoided if possible. Strategies to promote reactivity include (1) generation of large interfacial area by intense mixing, and (2) removal of one of the phases by distillation of the more volatile solvent, thereby combining the reactants in the remaining phase. The last method may be complicated by the formation of a solid phase (reagent or product becoming insoluble) but may still be preferable to an additive.

*17-4.3.2 Selectivity in Liquid–Liquid Dispersed Phase Reactions.* Enhancement of selectivity because of the presence of an immiscible phase is an important aspect of liquid–liquid systems. The improvement in selectivity is achieved by protection of the reactant(s) or product in a separate phase from an active reagent to reduce consecutive or competitive reaction to undesired by-products. Sharma (1988) discusses this subject and presents examples of very large increases in selectivity. Wang et al. (1984) present an example in which isocyanates were prepared from amides or N-bromoamides by Hofmann rearrangement under phase catalysis conditions. Without a second phase the isocyanate overreacted under alkaline conditions in the aqueous phase. Addition of a carefully selected solvent achieves reaction and rapid extraction of the isocyanate, which can then be obtained in high yield. This route to isocyanates obviates the use of phosgene.

Another example of reactive extraction is provided in Example 13-8a (King et al., 1985). In this case an acid hydrolysis could be replaced by a highly advantageous change to alkaline hydrolysis to achieve improved selectivity, productivity, quality, and waste minimization. However, the decomposition rates of reagent and reaction product under aqueous alkaline conditions are prohibitive. By running under reactive-extractive conditions, the objectives were achieved. Conventional mixing in a vessel was not feasible because of the rapid decomposition. Line mixing followed by rapid phase separation proved to be an extremely effective method to carry out this complex reaction.

*17-4.3.3 Scale-up.* Despite the frequent need to run reactions in immiscible liquid systems, the reliability and applicability of correlations to predict drop size distribution and surface area of the dispersed phase, especially in the presence of reactions, is limited. This is due, in part, to effects that small changes in geometry such as small agitator blade width can have on dispersed phase drop size as well as in surfactant effects resulting from reacting substrates. It is sometimes even difficult to predict which phase will be continuous and which dispersed. Although the continuous phase is normally a property of the system, the phases can sometimes be inverted by the manner in which they are contacted (e.g., by mixing during addition as opposed to starting with both phases present).

Care must be taken during laboratory and pilot plant studies to examine a wide range of interfacial areas (drop sizes). This allows the engineer to determine the extent to which interfacial area affects the conversion and selectivity. Changes in dispersion characteristics are very likely to occur on scale-up. In many cases these changes may not be significant because other aspects of the reacting system are controlling. However, phase dispersion can be critical to selectivity in some cases because of complex interfacial interactions. Selection of impellers and speeds to achieve the desired drop size distribution—which has a direct effect on settling rate—can also be critical to reaction processes that require subsequent phase separation. If when adding a reagent to the dispersed phase it is difficult to ensure good mixing of the drops, it can be concluded that concentration variations in dispersed phase can lead to problems.

Despite these qualifications, much can be gained from applying scale-up correlations to specific problems to establish guidelines and limits for performance.

As in the case of gas–liquid systems, the reader is referred to Chapter 12 and the text by Nagata (1975) for additional discussion. Scale-up from laboratory data on the same system can be predicted to some extent. Constant power per unit volume is a good guide, but care must be taken with large tanks and density differences, as mentioned above.

### 17-4.4 Solid–Liquid Systems

Solids in reacting systems can be either heterogeneous catalysts, dissolving reagents, precipitating products, or other reaction components, such as adsorption agents and ion-exchange resins. Reaction rates can fall in all regimes of the kinetic spectrum. The reader is referred to Chapter 10 for a discussion of solid–liquid mixing. As with all other types of heterogeneous reactions, very slow reactions in the liquid phase are unaffected by mass transfer in the film surrounding dissolving reagents or adsorption agents, and mixing is required only to maintain solids suspension. However, in the case of precipitating or crystallizing products, mixing can affect the particle size of the product just as it would in a precipitation without chemical reaction. Therefore, an effect of mixing must always be considered. Reactions in which reaction rates compete with mass transfer rates are all sensitive to local conditions and the films around solids and are therefore subject to mixing affects.

*17-4.4.1 Solids as Dissolving Reagents*. Both organic and inorganic reagents are often incompletely soluble in reaction solvents for a variety of reaction types. The particle size of these reagents can be a major factor in rate and/or selectivity. One objective of a laboratory development program is to determine the effect of particle size and to separate dissolution kinetics from chemical kinetics. An effective method of studying these reactions is to run the reaction under homogeneous conditions to measure true kinetics. This can be accomplished by preparing a saturated solution of the reagent, even if the maximum concentration is very low, and determining the reaction rate. Once having established true chemical kinetics, the overall reaction rate can be evaluated for dissolution limitations.

A second, effective but less quantitative method is to run the reaction with different particle size distributions of the insoluble reactant to determine the effect on overall reaction rate. If no effect is measured, it could be concluded that regime 1 applies and kinetics—not dissolution—controls. This method must be used with care, however, since other factors, such as surface coatings and incompletely characterized particle size distribution, can mask kinetic effects and lead to erroneous conclusions. An example of a reaction with dissolving solids was discussed previously (Example 13-4) in which a direct comparison can be made with the same reaction run in the same pilot scale vessel under homogeneous conditions. The effect on selectivity is significantly different between the heterogeneous and homogeneous conditions.

*17-4.4.2 Solids as Precipitating Products.* Studies have appeared on the effect of mixing on the precipitation of inorganic salts. Mixing intensity was shown by Pohorecki and Baldyga (1988) to affect particle size for the instantaneous reaction to form $BaSO_4$. Particle size was found to increase with increasing impeller speed in a segregated feed CSTR. Barthole et al. (1982) and Meyer et al. (1988) used a modification of the precipitation of $BaSO_4$ (modified to indicate the degree of micromixing) by characterizing product distribution of a $BaSO_4$ ethylenediaminetetracetic acid complex in alkaline medium under the influence of an acid.

Garside and Tavare (1985) modeled the effect of micromixing limits on elementary chemical reaction and subsequent crystallization. Two limiting cases are analyzed, and although the conversions of the chemical reactions are the same, the crystal size distributions can be very different. These differences are caused by the nonuniformity of supersaturation profiles that can be experienced by different fluid elements within a tank, owing to micromixing as well as macromixing effects. This modeling work also explores the sensitivity of two mixing models to reaction rate constant and nucleation kinetic parameters.

Literature references to experimental work on the crystallization or precipitation of products of organic reaction is rare even though this is a common reaction type. The difference between crystallization and precipitation is not well defined and is interpreted differently by different investigators. The interpretation that is implied here is that crystallization generates a crystalline product, whereas precipitates form rapidly and can be crystalline or amorphous. The differences are often blurred, however, because many organics actually appear first as amorphous noncrystalline solids which later turn truly crystalline. In these cases, nucleation is difficult to separate from precipitation of an amorphous solid. A further complication is that organics often separate first as oils or gums. This problem is very common when a reaction product is formed that is insoluble in the reaction system. Mersmann and Kind (1988) present an excellent discussion on precipitation as it is affected by micromixing.

An experimental study by Marcant and David (1991) of the crystallization of calcium oxalate concluded that the resulting particle size distribution was affected significantly by impeller speed and other mixing variables. The particle size distribution increased, passed through a maximum, and then decreased as the impeller speed was increased. This result is interpreted as changes in the key factors controlling nucleation and growth as well as reaction. Other mixing variables were also significant and affected particle size distribution in different ways.

The initial appearance of a solid that results from generation of supersaturation by a chemical reaction is a very complex series of events. As mentioned in the introduction to this section, the conditions affecting crystallization can be critical to the overall process result for several possible reasons. Selectivity of a complex reaction can be a function of the rate of crystallization and degree of supersaturation since these factors determine the concentration of that reaction product in solution at any given time. When in solution, all of the factors affecting selectivity can be significant, as discussed in Section 17-3. Delayed nucleation because of improper seeding or excessive impurity levels can result in significantly reduced selectivity.

The purity of the crystallization product can be affected by the parameters that control any crystallization as well as the presence of the other chemical species, including the starting materials, that can be occluded from the reaction mixture. The particle size distribution can be affected by supersaturation, reaction rate, mixing, and other factors that affect crystallization in general. The degree to which control of the crystallization must be of concern obviously depends on the downstream processing. In some cases, physical attributes may not be significant, and the reaction can be optimized on a kinetic basis only. In other cases, however, the requirements for maximum selectivity may be different from those for physical attributes requiring a trade-off in actual system design.

An example of reaction-induced crystallization where the particle size and purity must be controlled is discussed in Example 17-2. In this case, mixing played a key role by balancing circulation with shear to achieve sufficient micromixing and mesomixing for the reaction but avoiding overmixing to achieve growth without shedding and/or crystal fracture.

***Example 17-2: Reaction and Simultaneous Crystallization*** (Larson et al., 1995)

- *Goal:* scale-up of a reactive crystallization with crystal growth and impurity rejection
- *Issue:* determination of conditions to limit nucleation and promote growth

One of the steps in the preparation of a side chain in the synthesis of an antibiotic involves a reaction between two soluble reactants to form a product that crystallizes from the reaction solvent, toluene. The reaction itself is relatively straightforward, requiring no special consideration for scale-up once the reaction conditions are established in the laboratory. The problem on scale-up to the pilot plant proved to be the crystal size distribution of the product. The crystals were bimodal with fines mixed with very large crystals. The filtration rate proved to be impracticably slow, and severe occlusion of starting material was experienced.

Development of this reaction system was then focused on the crystallization. A laboratory experimental program established the following:

1. The crystallization was inhibited by a component of the reaction mixture (determined by crystallizing the product from the same solvent but without chemical reaction, in which case large, well-shaped crystals were obtained).
2. When crystallization occurred as the result of reaction, the system developed a high degree of supersaturation before self-nucleation, leading to the generation of small crystals with little growth.

These observations led to the conclusion that the reaction system had to be modified to achieve controlled nucleation and crystal growth while minimizing the concentration of nucleation-inhibiting compound (identity unknown but known to a component of the reaction mixture). The reaction was run under conditions in which one reagent was added last to the otherwise complete reaction mixture. The addition was sufficiently rapid to generate a high level of supersaturation, and crystallization did not initiate when the saturation concentration was reached by reaction.

The traditional approaches to inducing crystal growth were then explored, including seeding, control of supersaturation, and optimization of mixing to prevent crystal fracture. Control of supersaturation was achieved by control of reagent addition rate. The reagents were added simultaneously to a seeded mixture over a several-hour period to minimize concentrations of reactants and supersaturation of product. Crystallization was improved. However, crystal fracture of the improved crystals was then observed, which continued to cause slow filtration rates. Crystal fracture was reduced dramatically by a change in impeller to the Ekato Intermig (Chapter 6). This impeller was capable of providing sufficient blending for the reaction while providing the necessary low-shear environment for crystal growth. The resulting crystals from the slow addition (6 h) are shown in Figure 17-8*a* for comparison to those for more rapid addition (Figure 17-8*b*) (2 h), in which the bimodal distribution causing slow filtration is apparent. The key factor is control of low supersaturation and sufficient mixing for the fast reaction while avoiding shear damage to the crystals.

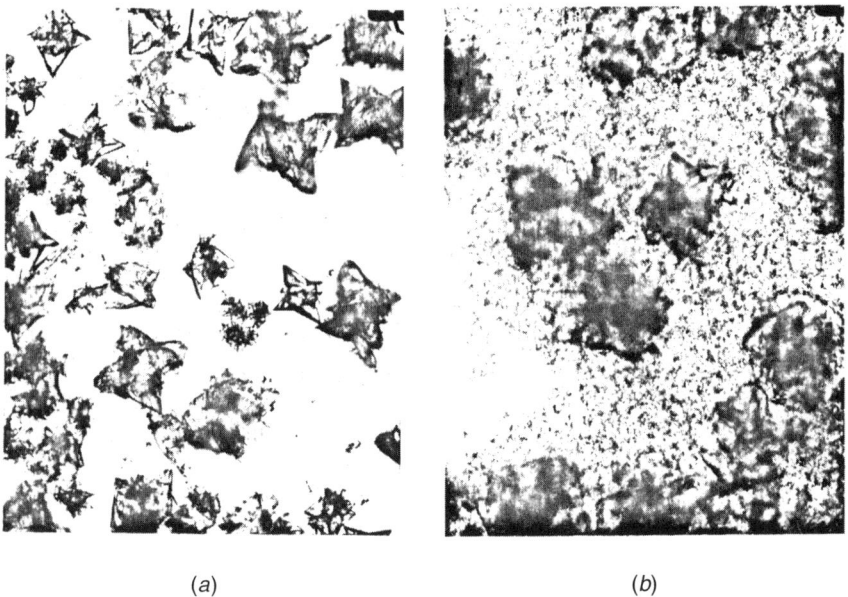

(a)          (b)

**Figure 17-8** Photomicrographs of crystals from manufacturing scale reactive crystallizations using two addition rates, showing (*a*) minimization of fines by essentially all-growth at low rates, and (*b*) fines formation and bimodal distribution at high rates caused by nucleation.

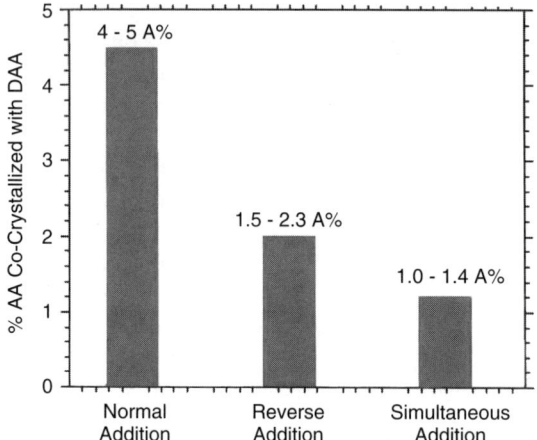

**Figure 17-9** Inclusion of impurities from three addition modes in pilot scale operation.

This method of controlling supersaturation was also effective in rejecting impurities from the growing crystals, as shown in Figure 17-9, where impurity occlusion for a faster addition is shown to be increased substantially.

*Resolution:* Crystal growth in this reactive crystallization can be controlled by limiting supersaturation by slow reagent addition, high level seeding, and low-shear, high-circulation mixing.

This example illustrates the complex role that mixing is often required to play in providing the necessary shear for reagent blending while simultaneously suspending a crystal slurry in a growth environment without causing overnucleation (small crystals) or crystal breakage. In cases involving these conflicting requirements, it is necessary to select the most critical criteria and to design the mixing system on that basis. In some cases, this will result in a compromise of the optimum for some of the criteria. In this example, all of the requirements could be met.

*17-4.4.3 Scale-up.* Scale-up of liquid–solid suspensions has been well characterized by many studies. The reader is referred to Chapter 10 for a comprehensive discussion. For reacting systems, power and speed should be above the minimum for homogeneous suspension since energy consumption is generally a smaller contributor to cost than other aspects of scale-up uncertainty (conversion and selectivity). Even this recommendation must be qualified by the potentially negative aspects of overmixing, as discussed in the introduction to Section 17-2. Reacting solids can also agglomerate and thereby require large increases in energy to maintain adequate dispersion.

These system-dependent properties are extremely difficult to characterize quantitatively and require specific scaling studies at extremes of possible operating ranges to determine sensitivity. Such systems are primary contributors to the case for built-in versatility. The more important consideration in reacting systems than solid suspension may be mass transfer, since considerably more power and speed may be required to achieve expected reaction rate for reacting solids than that required for homogeneous suspension. As developed in Section 17-4.3.2, selectivity can also be affected in complex reactions because of the potential overreaction in the diffusive film around the dissolving or precipitating particles. An excellent discussion of mass transfer and reaction is presented by Fogler (1986).

Another critical aspect in the effectiveness of mass transfer correlations to predict coefficients in reacting systems is the very troublesome but all-too-common tendency for the surface of a reacting solid, catalyst, or precipitating product to become covered by another solid or second-phase liquid, or by a gas in a three-phase mixture. Such a heterogeneous film would obviously have a profound effect on the expected mass transfer coefficient and in many cases can cause a reaction to stop before the expected conversion. These films are obviously unique to each reacting system, thereby preventing any generalizations as to whether they are susceptible to chemical or physical manipulation. Chemical manipulation could be achieved by addition of a surfactant that would be able to modify surface properties to prevent or modify formation of the film.

Physical manipulation of such films may be possible through variation in mixing intensity primarily by local shear. Such interactions would be very scale dependent and could readily be masked in smaller scale operations. The extent to which reactions can be affected by coating of particles is well illustrated in an excellent example by Wiederkehr (1988). Other aspects of reaction system design are included in this case study, such as the choice of continuous smaller-volume reactors over batch reactors to reduce the size (and potential energy) of the reacting mass as well as the criticality of residence time distribution in complex reactions.

Another example of the impact of dissolving solids on reaction performance on the industrial scale is provided by Yamazaki et al. (1989). This example highlights the difficulty of scale-up of systems in which the mass transfer rate influences the product distribution and selectivity of complex reactions and may control the overall reaction rate. In this example the rate-determining step is the dissolution of $K_2CO_3$ particles in dipolar solvents.

## 17-5 MIXING AND CRYSTALLIZATION

Interactions between mixing and crystallization are often ignored. They should not be. In many cases, these interactions can affect every aspect of a crystallization operation, including nucleation, growth, and maintenance of a crystal slurry. To complicate the problem further, mixing optimization for one aspect of an operation may require different parameters than for another aspect, even though both requirements must be satisfied simultaneously. In addition, these operations are often scale dependent. For these and other reasons discussed below, many would contend that crystallization is the most difficult operation to scale-up—successfully.

*Successful scale-up implies that both physical and chemical properties have been duplicated between pilot plant and plant operations. These rigid criteria are not always required but are for example, for final bulk active pharmaceutical products by biobatch regulations (duplication of physical attributes and chemical purity from pilot plant to plant scale operation). In all cases, however, it is prudent to apply these criteria in development, planning, and experimentation in order to reduce the risk of a dramatic failure (i.e., increased impurity levels or small crystal size causing drastically reduced filtration rates, poor washing, and slow-drying product). An even more drastic failure would be the inability to reproduce the required biobatch physical and chemical attributes.*

Successful operations depend on identifying the mixing parameters for the most critical aspects of the process and then evaluating whether those parameters will be satisfactory for the other aspects. Although this approach may be satisfactory in most cases, there will be crystallization procedures that require operation at conditions that are not optimum for mixing for some aspects of the operation, as discussed below. This discussion is limited to crystallization in stirred vessels by batch and semibatch operation. The crystallizers normally employed in the fine chemical and pharmaceutical industries are multipurpose vessels with various impeller and baffle configurations, as shown in Figures 17-1 and 17-2 and Chapter 6. The workhorse impeller is the pitched blade turbine because of its ability to create good circulation at relatively low shear. These attributes help reduce secondary nucleation and crystal breakage while achieving good suspension and circulation. The flat-blade turbine is less applicable because of high shear and less overall circulation. The Intermig (Ekato, Chapter 6) has proven to have superior performance in some crystallization operations because of its combination of excellent circulation with low shear. Baffles are required in all cases to prevent poor mixing due to swirling as well as entrainment of vapor that can provide nucleation sites. This discussion does not include more specific types of equipment, such as vessels with draft tubes or fines dissolution loops (see Mersmann, 2001; Mullin, 2001; Myerson, 2002).

Two important alternative types of crystallizers are fluidized beds and impinging jets. The fluidized bed is a very effective crystallizer for minimizing nucleation and promoting growth by providing very low shear, low energy, and minimum velocity impact between crystals. Both continuous and semicontinuous operation can be utilized. This principle has been applied successfully in the resolution of optical isomers in which nucleation must be minimized, preferably eliminated, to achieve isomer separation as described by Midler (1970, 1975, 1976). Impinging jet crystallization achieves an opposite extreme of mixing, high shear, and energy input, to promote nucleation by intense mixing. This principle has been utilized successfully in an industrial application to achieve a small average particle size (3 to 5 μm) and a narrow particle size distribution. Impinging jet crystallization is described by Midler et al. (1994) for an industrial application and by Mahajan and Kirwan (1996), Benet et al. (1999), and Condon (2001) in laboratory studies. This technology has been employed to produce nanoparticles stabilized by block copolymers (Johnson, 2003; Johnson and Prud'homme, 2003). Impinging jet crystallization with sonication (Lindrud et al. 2001) is another variant.

## 17-5.1 Aspects of Crystallization that Are Subject to Mixing Effects

The aspects of crystallization that may be affected by mixing are discussed below.

*17-5.1.1 Nucleation.* Primary nucleation may be induced by mixing. The effects of mixing on true primary nucleation are exceedingly complex. The overall result is a reduction in the width of the metastable region when the width for a static solution is compared to that for an agitated solution. Therefore, an unagitated solution can be cooled further before nucleation than can an agitated solution. Since an industrial system will always be agitated (except for operations such as melt crystallization), this has theoretical interest only. Since secondary nucleation becomes important as soon as nuclei appear, the nucleation mechanisms become virtually impossible to characterize. In addition, any seeded crystallization is by definition secondary even though some nuclei may form simultaneously by a primary mechanism. Therefore, the major part of this discussion will be on secondary nucleation.

Secondary nucleation is mixing dependent as follows:

- *Crystal–crystal impact:* a function of both the local micromixing environment and the overall macromixing circulation
- *Crystal–impeller and crystal–wall impact:* functions of the impeller speed, shape of blade, and material of construction
- *Adsorbed layer:* thickness decreased by increased mixing

These factors affect the rate of nucleation, which in turn determines the number of nuclei formed and their size. These events can then dominate the entire crystallization operation with respect to both physical and chemical purity attributes. Ultimate crystal size is a function of the number of nuclei generated, as shown in Table 17-3. Nominal dimensions of the resulting crystals for cubic particles (3D growth), flat plate (2D growth), and needle-shaped crystals (1D growth) are shown versus the quantity of nuclei (each "nucleus" assumed to be a 5 μm cube). It can be seen that the number of nuclei generated by the various causes of nucleation, including agitation, has an exponential effect, as expected from this purely geometrical relationship, on the ultimate size that can be achieved by growth subsequent to nucleation. Since the nucleation rate can often increase on scale-up because of local power dissipation differences, the average particle size on scale-up could be reduced. Other mixing factors that affect growth could increase the size distribution as well, as discussed below.

The effect of agitation on secondary nucleation has been reported in the literature and several references are discussed by Mullin (2001). This discussion highlights the complex nature and unpredictability of these interactions. Moreover, the critical nature of these interactions is the key factor in causing difficulty in scale-up of nucleation-dominated crystallization processes, even with small quantities of seed. The critical mixing factors are impeller speed and type and their influence on local turbulence and overall circulation. Since neither the localized turbulence distribution nor the overall circulation time can realistically be

**Table 17-3** Effect of Extent of Nucleation on Final Crystal Particle Size

|  | Product Particle Dimensions (μm) | | |
|---|---|---|---|
| % Nucleation | Length | Width | Thickness |
| a. Cubic Particles (3D Growth) | | | |
| 0.5 | 29 | 29 | 29 |
| 1 | 23 | 23 | 23 |
| 5 | 14 | 14 | 14 |
| 10 | 11 | 11 | 11 |
| b. Flat Plates (2D Growth)[a] | | | |
| 0.5 | 71 | 71 | 5 |
| 1 | 50 | 50 | 5 |
| 5 | 22 | 22 | 5 |
| 10 | 16 | 16 | 5 |
| c. Needles (1D Growth)[a] | | | |
| 0.5 | 1000 | 5 | 5 |
| 1 | 500 | 5 | 5 |
| 5 | 100 | 5 | 5 |
| 10 | 50 | 5 | 5 |

[a] Very thin particles unlikely to survive attrition in tank.

maintained constant on scale-up, the extent to which changes in the crystallizing environment will affect nucleation are extremely difficult to predict. To the mixing issue must be added the uncertainties caused by soluble and insoluble impurities that may be present in sufficiently different concentrations from batch to batch to cause variation in nucleation rate.

It is important to remember that low-level impurities can also have significant impact on crystal growth, usually by blocking growth sites on the growth surface, reemphasizing the importance of controlling reaction conditions with suitable local mixing. However, impurities can more easily disturb a molecular cluster trying to arrange itself into a critical sized nucleate than they can an already formed growing surface, so the effect is clearly more pronounced in nucleation In general, the greater the dependence on nucleation, the greater the difficulty in developing a stable process for scale-up and/or ongoing production.

If no process alternative is possible to avoid dependence on nucleation, mixing scale-up can be based on equal power per unit volume assuming that the same impeller type is used and geometric similarity is maintained. In most cases, however, this approach will result in changes in particle size distribution (PSD) on scale-up, which may or may not be acceptable. In general, the PSD will be broader and the average particle size smaller if the P/V scale-up criterion is used. This generalization is suggested by Nývlt (1971). As often experienced in crystallization scale-up, however, the opposite of this expectation can be realized, depending on the specific nucleation characteristics of the system. A further generalization may be proposed: that fast nucleating systems tend toward smaller size distribution on scale-up, whereas slow nucleation can give the opposite result. An alternative to equal power per unit volume was suggested by Nienow (1976) and is discussed further below. For excellent insight into this complex phenomenon of nucleation, the reader is referred to Mersmann (2001), Mullin (2001), and Myerson (2002).

*17-5.1.2 Growth*. Mixing can affect crystal growth in several ways, as summarized below:

- Mass transfer rate in the diffusion film around growing crystals
- Bulk turnover rate and its affect on minimizing differences in the supersaturation ratio throughout the vessel
- Heat transfer rate and wall film thickness
- The effect of shear on crystal breakage
- Dispersion of an antisolvent or reagent
- Growth rate dispersion
- Minimizing impurity concentration at the crystal surface

The need to maintain high mass transfer rates to minimize supersaturation gradients in the film around a growing crystal is one of the primary functions of mixing in a crystallization operation. As in other types of mass transfer operations, the coefficient increases with increased mixing, although at high Reynolds numbers, this increase becomes less significant. Additional factors that improve with increased mixing are (1) heat transfer, (2) bulk turnover, (3) dispersion of an additive such as an antisolvent or reagent, (4) uniformity of crystal suspension, (5) avoidance of settling and minimization of wall scale, and (6) minimization of impurity concentration at the crystallizing surface. However, these needs must be balanced against the possibly negative results of overmixing, which can cause crystal breakage and/or shedding of nuclei as well as increased secondary nucleation. Increased growth dispersion is also possible, since increased mixing can increase the growth rate of large crystals (assuming that the growth rate is dependent on mass transfer), but has little effect on small crystals ($<10$ $\mu$m) since these crystals are smaller than the turbulent eddies and have little relative movement. The latter effect may be a contributing factor in the increase in size distribution that is common on scale-up.

These concerns lead to the conclusion referred to above that it is often necessary to choose a mixing condition (impeller speed, type, etc.) that may not be optimum for every aspect of the crystallization and may actually not be optimum for any of them. In many cases, however, one end result [i.e., PSD, bulk density, uniformity of suspension, approach to equilibrium solubility (yield)] may dictate the choice of mixing conditions. In this case it becomes essential to determine if the negatively affected aspects can be tolerated.

All of these factors are properties of a given crystallization system, thereby requiring choices for each specific operation. Experimentation is required to determine the key responses to mixing for each system and could include determination of the following:

- *Effect of impeller speed and type on PSD at a minimum of two seed levels and two supersaturation ratios.* These results should indicate the sensitivity of the system to mixing. A small response could indicate that other system properties were controlling (i.e., inherent crystal growth rate or nucleation rate). A large response would indicate sensitivity to secondary nucleation and/or crystal cleavage and require additional experimentation and evaluation of scale-up requirements. The laboratory results should be evaluated relative to each other since scale-up can be expected to make additional changes in PSD, especially when a large response is experienced in these simple experiments.
- *Effect of impeller speed on crystallization rate and approach to equilibrium solubility (yield).* Failure to achieve equilibrium solubility may indicate accumulation of impurities at the crystallizing surfaces. An increase in impeller speed resulting in further reduction in solution concentration could indicate resumption in growth or additional nucleation (see Example 17-3).
- *Suspension requirements as indicated by the settling rate to achieve off-bottom suspension.* The impeller speed that is found necessary to achieve off-bottom suspension should be the minimum speed on which to base scale-up, as calculated for the appropriate scale of operation. Higher speeds may be required to satisfy other requirements, as indicated below.

For nucleation-dependent operations, it is recommended that additional information be obtained as follows:

- Effect of impeller speed on width of metastable region
- Effect of impeller speed and type on rate of nucleation

This experimentation is focused primarily on evaluation of mixing sensitivity.

### 17-5.2 Mixing Scale-up in Crystallization Operations

The compromises in mixing optimization that may be required on scale-up often result in use of the common mixing criterion of equal power per unit volume or in some cases equal tip speed. Both of these require utilization of the same impeller type as well as geometric similarity in order to have a reasonable chance of success. Preliminary laboratory evaluation of the mixing requirements for a crystallization operation should be carried out in a minimum 1 L vessel, with further evaluation at 100 to 1000 L—as much as is practical. The smaller scale operations will generally produce more uniform PSD and larger mean crystal size than the manufacturing scale (typically, 10 000 L) when using equal power per unit volume. These changes typically are caused by the local differences in impeller shear (an unavoidable result of the equal power per unit volume criterion) that cause increased nucleation, leading to a larger number of particles, an increased spread in PSD, and a smaller particle average diameter.

For transfer of a crystallized mixture to another operation, there may be a requirement to have a homogeneous dispersion. One example would be feeding a centrifuge or other solid–liquid separation device. A guideline that can be helpful in avoiding local overmixing, other than equal power per unit volume, was suggested by Nienow (1976). Using this guideline, the agitator speed at the manufacturing scale would be selected to be sufficient to just maintain off-bottom suspension, thereby resulting in reduced nucleation, fewer particles, and more growth. In general, this speed would be considerably less than equal power per unit volume. Limitations on this guideline would be cases of high-density crystals, which could require higher speeds to prevent excessive settling. In addition, antisolvent and reactive crystallization applications may require higher speeds to prevent local supersaturation at the point of addition. In the latter case, scale-up based on equal local energy dissipation at the point of addition may be necessary. This subject is discussed in many literature references, including Sohnel and Garside (1992) and Mersmann (2001). The requirements for particle off-bottom suspension are also discussed in Chapter 10. A further caution on reduced speed is a possible increase in encrustation caused by crystal contact with the bottom surface and potential for sticking.

The effect of mixing on PSD has been determined experimentally by Marcant and David (1991) for the reactive crystallization of calcium oxalate. This work is an excellent example of the multiple dependencies on mixing that can be experienced in a crystallization operation. The factors noted above that are mixing dependent are shown to have positive or negative influences on the resulting physical characteristics, thereby illustrating the necessity of selecting the most important result to be achieved. The effect of increasing agitator speed is shown initially to cause an increase in particle size, followed by passing through a maximum and then decreasing particle size. This result is attributed to changes in controlling factors resulting from the changes in mixing.

*Example 17-3: Slow Approach to Equilibrium.* Scale-up of the crystallization of an intermediate (MW ~850) in a multistep synthesis resulted in a slow approach to equilibrium solubility at the end of a combined cooling/antisolvent crystallization. Agitation rate was held at a minimum value for off-bottom suspension because of concern for shear damage of the crystals. Accumulated soluble impurities were known to be present, and it was suspected that their accumulation at the crystallizing surfaces could be a factor in stopping growth. An increase in agitation rate by about 20% late in the cooling cycle was successful in causing a further reduction in dissolved product solubility. The increased agitation was successful in resuming growth because of reduction in diffusion film thickness. (Another possibility is a resumption in nucleation resulting from the increased energy of the impeller.)

Incorporation of impurities in the crystals was a major concern that indicated higher impeller speeds but generated excessive fines, and the resulting poor filtration rates were a counterbalancing influence on the determination of impeller speed. The two-level agitation rate scheme was a balance between these conflicting factors, which gave passable purity, yield, and filtration rate. As in many high-impurity systems, however, the average particle size was small because of the need to operate at relatively high supersaturation to achieve practical growth rates, thereby incurring more nucleation than desired. The qualitative aspects of this difficult crystallization provide examples of the trade-offs that are often encountered in development and scale-up.

## REFERENCES

Angst, W., J. R. Bourne, and R. N. Sharma (1982a). Mixing and fast chemical reaction: V. Influence of diffusion within the reaction zone on selectivity, *Chem. Eng. Sci.*, **37**, 1259–1264.

Angst, W., J. R. Bourne, and R. N. Sharma (1982b). Mixing and fast chemical reaction: IV. The dimensions of the reaction zone, *Chem. Eng. Sci.*, **37**, 585–590.

Baldyga, J., and J. R. Bourne (1984a). Mixing and fast chemical reactions: VIII. Initial deformation of material elements in isotropic, homogeneous turbulence, *Chem. Eng. Sci.*, **39**, 329–334.

Baldyga, J., and J. R. Bourne (1988). Calculation of micromixing in inhomogeneous stirred tank reactors, *Chem. Eng. Res. Des.*, **66**, 33–38.

Baldyga, J., and J. R. Bourne (1989). Simplification of chemical engineering calculations, *Chem. Eng. J.*, **42**, 83–101.

Baldyga, J., and J. R. Bourne (1990). The effect of micromixing on parallel reactions, *Chem. Eng. Sci.*, **45**(4), 907–916.

Baldyga, J., and J. R. Bourne (1999). *Turbulent Mixing and Chemical Reactions*, Wiley, Chichester, West Sussex, England.

Barthole, J. P., R. David, and J. Villermaux (1982). A new chemical method for the study of local micromixing combinations in industrial stirred tanks, *ACS Symp. Ser.* **196U**, 545–554.

Belevi, H., J. R. Bourne, and P. Rys (1981). Mixing and fast chemical reaction: II. Diffusion reaction model for the CSTR, *Chem. Eng. Sci.*, **36**, 1649–1654.

Benet, N., L. Falk, H. Muhr, and E. Plasari (1999). Experimental study of a two-impinging-jet mixing device for application in precipitation processes, *Proc. 14th International Symposium on the Industrial Crystals (Computer Optical Disc)* pp. 1007–1016.

Bolzern, O., and J. R. Bourne (1983). Mixing and fast chemical reactions: VI. Extension of the reaction zone, *Chem. Eng. Sci.*, **38**, 999–1003.

Bolzern, O., and J. R. Bourne (1985). Rapid chemical reactions in a centrifugal pump, *Chem. Eng. Res. Des.*, **63**, 275–282.

Bourne, J. R. (1984). Micromixing revisited, *Proc. 8th International Symposium on Chemical Reaction Engineering, Symp. Ser. Ind. Chem. Eng. (London)*, **87**, 797–814.

Bourne, J. R., and P. Dell'ava (1987). Micro- and macromixing in stirred tank reactors of different sizes, *Chem. Eng. Res. Des.*, **65**(3), 180–186.

Bourne, J. R., and H. Gablinger (1989). Local pH gradients and the selectivity of fast reactions, *Chem. Eng. Sci.*, **44**(6), 1347–1352.

Bourne, J. R., and J. Garcia-Rosas (1986). Rotor–stator mixers for rapid micromixing, *Chem. Eng. Res. Des.*, **64**, 11–17.

Bourne, J. R., and C. P. Hilber (1990). The productivity of micromixing-controlled reactions: effect of feed distribution in stirred tanks, *Chem. Eng. Res. Des.*, **68**, 51–56.

Bourne, J. R., and F. Kozicki (1977). Mixing effects during the bromination of 1,3,5-trimethoxybenzene, *Chem. Eng. Sci.*, **36**, 1538–1539.

Bourne, J. R., E. Crivelli, and P. Rys (1977a). Chemical selectivities disguised by mass diffusion, *Helv. Chim. Acta*, **60**(8), 2944–2957.

Bourne, J. R., P. Rys, and R. Suter (1977b). Mixing effects in the bromination of resorcin, *Chem. Eng. Sci.*, **32**, 711–716.

Bourne, J. R., et al. (1981a). Mixing and fast chemical reaction: III. Model–experiment comparisons, *Chem. Eng. Sci.*, **36**, 1655–1663.

Bourne, J. R., F. Kozicki, and P. Rys (1981b). Mixing and fast chemical reaction: I. Test reactions to determine segregation, *Chem. Eng. Sci.*, **36**, 1643–1648.

Bowe, M. (1990). Fluidics puts mixing in a spin, *Processing*, Feb., pp. 43–44.

Canon, R. M., et al. (1977). Turbulence level significance of the coalescence-dispersion rate parameter, *Chem. Eng. Sci.*, **32**, 1349–1352.

Carberry, J. J., and A. Varma, eds., (1987). *Chemical Reaction and Reaction Engineering*, Marcel Dekker, New York.

Carpenter, K. J. (1985). A fine chemicals view of the mixing world, *Proc. 5th European Conference on Mixing*, BHRA, pp. 233–241.

Carpenter, K. J. (2001). Chemical reaction engineering aspects of fine chemicals manufacture, *Chem. Eng. Progr.*, **56**, 305–322.

Condon, J. M. (2001). Investigation of impinging jet crystallization for a calcium oxalate model system, Ph.D. dissertation, Rutgers University, New Brunswick, NJ.

Danckwerts, P. V. (1957). Measurement of molecular homogeneity in a mixture, *Chem. Eng. Sci.*, **7**(1), 116–117.

Danckwerts, P. V. (1958). The effect of incomplete mixing on homogeneous reactions, *Chem. Eng. Sci.*, **8**(1), 93–102.

Demyanovich, R. J., and J. R. Bourne (1988). A new method using thin liquid sheets, *Proc. 6th European Conference on Mixing*, BHRA, pp. 177–182.

Demyanovich, R. J., and J. R. Bourne (1989). Rapid micromixing by the impingement of thin liquid sheets, *Ind. Eng. Chem. Res.*, **28**(6), 825–839.

Fogler, H. S. (1986). *Elements of Chemical Reaction Engineering*, Prentice Hall, Englewood Cliffs, NJ.

Froment, G. F., and R. B. Bischoff (1990). *Chemical Reactor Analysis and Design*, Wiley, New York.

Garside, J., and N. S. Tavare (1985). Mixing, reaction, and precipitation: limits of micromixing in an MSMPR crystallizer, *Chem. Eng. Sci.*, **40**(8), 1485–1493.

Harris, I. J., and R. D. Srivastava (1968). The simulation of single phase tubular reactors with incomplete reactant mixing, *Can. J. Chem. Eng.*, **46**, 66–69.

Heeb, T. G., and R. S. Brodkey (1990). Turbulent mixing with multiple second-order chemical reactions, *AIChE J.*, **36**(10), 1457–1470.

Johnson, B. K. (2003). Flash nanoprecipitation of organic actives via confined micromixing and block copolymer stabilization, PhD Thesis, Princeton Univ.

Johnson, B. K., and R. J. Prud'homme (2003). Chemical processing and micromixing in confined impinging jets, *AIChE J.*, in press.

Kattan, A., and R. J. Adler (1967). A stochastic mixing model for homogeneous, turbulent, tubular reactors, *AIChE J.*, **13**(3), 580–585.

Keeler, R. N., E. E. Petersen, and J. M. Prausnitz (1965). Mixing and chemical reaction in turbulent flow reactors, *AIChE J.*, **11**(2), 221–227.

King, M. L., A. L. Forman, C. Orella, and S. H. Pines (1985). Extractive hydrolysis for pharmaceuticals, *Chem. Eng. Prog.*, **81**(5), 36–39.

Klein, J. P., R. David, and J. Villermaux (1980). Interpretation of experimental liquid phase micromixing phenomena in a continuous stirred reactor with short residence times, *Ind. Eng. Chem. Fundam.*, **19**, 373–379.

Landau, R. N., D. Blackmond, and H. H. Tung (1994). Calorimetric investigation of an exothermic reaction: kinetic and heat flow modelling, *Ind. Eng. Chem. Res.*, **33**, 814.

Landau, R. N., U. K. Singh, F. G. Gortsema, Y.-K. Sun, S. C. Gomolka, T. Lam, M. Futran, and D. G. Blackmond (1995). A reaction calorimetric investigation of the hydrogenation of a substituted pyrazine, *J. Catal.*, **157**, 201.

Larson, K. A., M. Midler, and E. L. Paul (1995). Reactive crystallization: control of particle size and scale-up, presented at the Association for Crystallization Technology 1995 Meeting, Charlottesville, VA.

Laufhutte, H. D., and A. Mersmann (1987). Local energy dissipation in agitated turbulent fluids and its significance for the design of stirring equipment, *Chem. Eng. Technol.*, **10**, 56–63.

LeBlond, C., J. Wang, R. Larsen, C. Orella, and Y.-K. Sun (1998). A combined approach to characterization of catalytic reactions using in situ kinetic probes, *Topics Catal.*, **5**, 149.

Lee, L. J., J. M. Ottino, W. E. Ranz, and C. W. Macosko (1980). Impingement mixing in reaction injection molding, *Polym. Eng. Sci.*, **20**, 868–874.

Levenspiel, O. (1972). *Chemical Reaction Engineering*, Wiley, New York.

Li, K. T., and H. L. Toor (1985). Chemical indicators as mixing probes: a possible way to measure micromixing time, *Ind. Eng. Chem. Fundam.*, **25**, 719–723.

Lindrud, M. D., S. Kim, and C. Wei (2001). Sonic impinging jet crystallization apparatus and process, U.S. Patent 6,302,958.

Litz, L. M. (1985). A novel gas–liquid stirred tank reactor, *Chem. Eng. Prog.*, **81**(11), 36–39.

Mahajan, A. J., and D. J. Kirwan (1996). Micromixing effects in a two-impinging-jets precipitator, *AIChE J.*, **42**(7), 1801–1814.

Mann, R., and A. E-Hamouz (1991). Effect of macromixing on a competitive consecutive reaction in a semi-batch stirred reactor: Paul's iodination experiments interpreted by networks of zones, presented at the 7th European Conference on Mixing, Brugge, Belgium, Sept. 18–20.

Mao, K. W., and H. L. Toor (1971). Second-order chemical reactions with turbulent mixing, *Ind. Eng. Chem. Fundam.*, **10**(2), 192–197.

Marcant, B. N., and R. David (1991). Experimental evidence for and prediction of micromixing effects in precipitation, *AIChE J.*, **37**(11), 1698–1710.

Mersmann, A. (2001). *Crystallization Technology Handbook*, 2nd ed., Marcel Dekker, New York.

Mersmann, A., and M. King (1988). Chemical engineering aspects of precipitation from solution, *Chem. Eng. Technol.*, **11**, 264–276.

Meyer, T., R. David, A. Renken, and J. Villermaux (1988). Micromixing in a stator mixer and an empty tube by a chemical method, *Chem. Eng. Sci.*, **43**, 1955–1960.

Midler, M. (1970). Production of crystals in a fluidized bed with ultrasonic vibrations, U.S. patent 3,510,266.

Midler, M. (1975). Process for production of crystals in fluidized bed crystallizers, U.S. patent 3,892,539.

Midler, M. (1976). Crystallization system and method using crystal fracturing external to a crystallizer column, U.S. patent 3,996,018.

Midler, M., E. Paul, E. Whittington, M. Futran, P. Liu, J. Hsu, and S. Pan (1994). Crystallization method to improve crystal structure and size", U.S. patent 5,314,506.

Mullin, J. W. (2001). *Crystallization*, 4th ed., Butterworth-Heinemann, Oxford.

Myerson, A. S., (ed.) (2002). *Handbook of Industrial Crystallization*, 2nd ed., Butterworth-Heinemann, Newton, MA.

Nabholz, F., and P. Rys (1977). Chemical selectivities disguised by mass diffusion, *Helv. Chim. Acta*, **60**(8), 2937–2943.

Nagata, S. (1975). *Mixing: Principles and Applications*, Wiley, New York.

Nauman, E. B., and B. A. Buffham (1983). *Mixing in Continuous Flow Systems*, Wiley, New York.

Nienow, A. W. (1976). The effect of agitation and scale-up on crystal growth rates and on secondary nucleation, *Trans. Inst. Chem. Eng.*, **54**, 205.

Nývlt, J. (1971). *Industrial Crystallization from Solutions*, Butterworth & Co. Ltd, London.

Oldshue, J. Y. (1983). *Fluid Mixing Technology*, McGraw-Hill, New York.

Oldshue, J. Y. (1985). Scale-up of unique industrial fluid mixing processes, *Proc. 5th European Conference on Mixing*, BHRA, pp. 35–42.

Ott, R. J., and P. Rys (1975). Chemical selectivities disguised by mass diffusion, *Helv. Chim. Acta*, **58**(7), 2074–2091.

Paul, E. L. (1988). Design of reaction systems for specialty organic chemicals, *Chem. Eng. Sci.*, **43**(8), 1773–1782.

Paul, E. L. (1990). Reaction systems for bulk pharmaceutical production, *Chem. Ind.*, **21**, May, pp. 320–325.

Paul, E. L., and R. E. Treybal (1971). Mixing and product distribution for a liquid-phase, second-order, competitive-consecutive reaction, *AIChE J.*, **17**(5), 718–724.

Paul, E. L., H. Mahadevan, J. Foster, M. Kennedy, and M. Midler (1992). The effect of mixing on scaleup of a parallel reaction system, *Chem. Eng. Sci.*, **47**, 2837–2840.

Pohorecki, R., and J. Baldyga (1988). The effects of micromixing and the manner of reactant feeding on precipitation in stirred tank reactors, *Chem. Eng. Sci.*, **43**, 1949–1954.

Rice, R. W., and R. E. Baud (1990). The role of micromixing in the scaleup of geometrically similar batch reactors, *AIChE J.*, **36**(2), 293–298.

Rice, A. W., H. L. Toor, and F. S. Manning (1964). Scale of mixing in stirred vessels, *AIChE J.*, **10**(1), 125–129.

Schutz, J. (1988). Agitated thin-film reactors and tubular reactors with stator mixers for a rapid exothermic multiple reaction, *Chem. Eng. Sci.*, **43**, 1975–1980.

Sharma, M. M. (1986). Intensification of heterogeneous reactions: theory and practice, *Proc. Indian Natl. Sci. Acad.*, **52A**, 449–475.

Sharma, M. M. (1988). Multiphase reactions in the manufacture of fine chemicals, *Chem. Eng. Sci.*, **43**(8), 1749–1758.

Shenoy, U. V., and H. L. Toor (1989). Turbulent micromixing parameters from reactive mixing measurements, *AIChE J.*, **35**(10), 1692–1700.

Shenoy, U. V., and H. L. Toor (1990). Unifying indicator and instantaneous reaction methods for measuring micromixing, *AIChE J.*, **36**(2), 227–232.

Smith, J. M., and M. M. G. G. Warmoeskerken (1985). The dispersion of gases in liquids with turbines, Proc. 5th European Conference on Mixing, BHRA, pp. 115–126.

Sohnel, O., and J. Garside, (1992). *Precipitation*, Butterworth-Heinemann, Oxford.

Sun, Y. -K., C. LeBlond, J. Wang, R. Larsen, C. J. Orella, A. Forman, R. N. Landau, J. Laquidara, J. R. Sowa, Jr., and D. G. Blackmond (1996a). Reaction calorimetry as an in situ kinetic tool for characterizing complex reactions, *Thermochim. Acta*, **289**, 189.

Sun, Y.-K., R. N. Landau, J. Wang, C. LeBlond, and D. G. Blackmond (1996b). A re-examination of pressure effects on enantioselectivity in asymmetric catalytic hydrogenation, *J. Am. Chem. Soc.*, **118**, 1348.

Sun, Y. -K., J. Wang, C. LeBlond, and D. G. Blackmond (1996c). Asymmetric hydrogenation of ethyl pyruvate: diffusion effects on enantioselectivity, *J. Catal.*, **161**, 759.

Tatterson, G. B. (1991). *Fluid Mixing and Gas Dispersion in Agitated Tanks*, McGraw-Hill, New York.

Toor, H. L. (1962). Mass transfer in dilute turbulent and non-turbulent systems with rapid irreversible reactions and equal diffusivities, *AIChE J.*, **8**(1), 71–78.

Truong, K. T., and J. C. Methot (1976). Segregation effects on consecutive competing reaction in a CSTR, *Can. J. Chem. Eng.*, **54**, 572–577.

van de Vusse, J. G. (1966). Consecutive reactions in heterogeneous systems, *Chem. Eng. Sci.*, **21**, 631–643, 1239–1252.

Vassillatos, G., and H. L. Toor (1965). Second-order chemical reactions in a nonhomogeneous turbulent fluid, *AIChE J.*, **11**(4), 666–673.

Wang, S., et al. (1984). Study on the Hofmann rearrangement in a two phase system, *Jilin Daxue Ziran Kexue Xuebao*, **2**, 89–93 (Chinese).

Wang, Y. D., and R. Mann (1990). Mixing in a stirred semi-batch reactor: partial segregation for a pair of competing reactions analyzed via networks-of-zones, *Inst. Chem. Eng. Symp. Ser.*, **121**(Fluid Mixing 4), 241–258.

Wang, Y. D., and R. Mann (1992). Partial segregation in stirred batch reactors: effect of scale-up on the yield of a pair of competing reactions, *Chem. Eng. Res. Des.*, **70**(A3), 282–290.

Weinstock, L. M. (1986). Evolution of the cefoxitin process, *Chem. Ind. (London)*, **86**(3), 86–90.

Whitaker, S., and A. E. Cassano, eds. (1986). *Concepts and Design of Chemical Reactors*. Gordon & Breach, New York.

Wiederkehr, H. (1988). Examples of process improvements in the fine chemicals industry, *Chem. Eng. Sci.*, **43**, 1783–1791.

Yamazaki, H., H. Yazawa, and K. Minima (1989). Effect of mixing on esterification of cepharosporic acid in a solid–liquid system, *Chem. Eng. Sci.*, **41**, 109–116.

# CHAPTER 18

# Mixing in the Fermentation and Cell Culture Industries

ASHRAF AMANULLAH and BARRY C. BUCKLAND
*Merck & Co., Inc.*

ALVIN W. NIENOW
*University of Birmingham*

## 18-1 INTRODUCTION

Given the importance of mixing and mass transfer in fermentation and cell culture processes and the potentially huge literature available on the subject, it is not possible to cover all aspects of this topic within the scope of this chapter. What are considered the most important subject matters are addressed here. In Section 18-2 we focus on the aspects of scaling up and scaling down fermentation processes. Although only microbial and fungal systems are considered, similar principles can be applied for cell culture processes. Scale-up of industrial fermentation processes occurs either when a new process is scaled up or when an existing process is subjected to modifications (e.g., media or strain improvements). Since scale-up is still largely performed using empirical knowledge and although scale-up can sometimes be successful, it is difficult to use it for optimization purposes. A more rapid, process-specific approach that is capable of predicting the performance with greater confidence on scale-up is desired. The limitations of traditional scale-up methods have been highlighted, and alternative methods using a scale-down approach are described in detail. Only stirred tank bioreactors have been considered in this discussion since the use of such systems is overwhelmingly dominant in the fermentation and cell culture industries. A significant part of this chapter is devoted to a description of studies that have measured spatial variations in dissolved oxygen, substrate, and pH at large scales of operation and

---

*Handbook of Industrial Mixing: Science and Practice*, Edited by Edward L. Paul,
Victor A. Atiemo-Obeng, and Suzanne M. Kresta
ISBN 0-471-26919-0 Copyright © 2004 John Wiley & Sons, Inc.

those that have investigated the effects of repeated exposure of microorganisms to the nonhomogeneous distribution of microenvironmental conditions. Such studies form the basis for the rational design of scale-down models used to simulate the microenvironment experienced by cells at the large scale. Particular emphasis is given to the practical design of scale-down models.

One of the most challenging tasks in the fermentation industry today is the design of bioreactors for highly shear thinning, viscous fermentation broths, including those for commercially important antibiotic and polysaccharide fermentations. In such fermentations, the maximum productivity, product concentration, and quality achievable depends primarily on bulk mixing and oxygen mass transfer, which in turn are governed by process operation, impeller type, and fluid properties. These generic problems inherent in viscous polysaccharide fermentations have been investigated and reported in Section 18-3 using xanthan gum as a model fermentation system. In addition, the effects of bulk mixing oxygen transfer on the quality of xanthan and other polysaccharide gums are also discussed.

Fungal strains for secondary metabolite, organic acid, and heterologous protein production are widely used industrially. In fungal fermentation, engineering variables such as agitation conditions require attention due its effect on the morphology, which in some cases can affect productivity. In many fungal fermentations, the high apparent viscosities and the non-Newtonian behavior of the broths necessitate the use of high agitation speeds to provide adequate mixing and oxygen transfer. However, mycelial damage at high stirrer speeds (or power input) can limit the acceptable range of speeds, and consequently, the oxygen transfer capability and volumetric productivity of the fermenter. The effects of hydrodynamic stress on fungal physiology are not always readily understood. An understanding of how agitation affects mycelial morphology and productivity ought to be valuable in optimizing the design and operation of large scale fungal fermentations for the production of secondary metabolites and recombinant proteins. The effects of agitation intensity on hyphal morphology and product formation in two commercially important fungal fermentations (*Penicillium chrysogenum* and *Aspergillus oryzae* for penicillin and recombinant protein production, respectively) are considered in Section 18-4.

Protein production by recombinant technology has been the subject of much industrial interest. However, production has been limited to a few well-known overexpression systems such as *Escherichia coli*, although knowledge of the process engineering variables on performance is still limited. The cultivation of *E. coli* in fed-batch mode using high-substrate feed concentrations to produce high cell densities is the preferred industrial method for increasing the volumetric productivity of bacterial derived products. Mixing is critical in such situations to ensure that addition of the concentrated feed is mixed as quickly as possible. However, information on the impact of intense mixing on bacterial physiology is very scarce. This subject is dealt with briefly in Section 18-5.

The commercial use of animal and insect cell culture at scales up to 20 m$^3$ (20 000 L) for the production of posttranslationally modified proteins using recombinant DNA techniques has made cell culture a cornerstone of modern biotechnology. Given that both suspension and microcarrier cell cultures are potentially more sensitive then microbial cells, to agitation and aeration in stirred tank bioreactors, proper design and operation of bioreactors in relation to agitation and aeration, including the use of surfactants to minimize cell damage, are critical for process optimization. These issues are described in Section 18-6. In addition, limitations on the use of Kolmogorov's theory of isotropic turbulence for the prediction of shear damage are discussed.

Issues related to mixing in plant cell cultures with reference to hydrodynamic shear are discussed in Section 18-7. Generally, plant cell damage mechanisms due to agitation have been difficult to identify given the diversity of cell lines, aggregate morphologies, culture age, and history. Greater understanding of these interactions is still required to for widespread commercialization of plant cell cultures.

## 18-2 SCALE-UP/SCALE-DOWN OF FERMENTATION PROCESSES

Successful scale-up often depends on the extent to which system characteristics resemble each other on the production and laboratory scales. One problem frequently encountered in the scale-up of bioprocesses is the nonideal or even unknown fluid flow behavior at large scale. Whereas the time constants of a biological reaction remain independent of vessel size, this is not true of many of the physical parameters involved. Mixing is sufficiently intense and uniform in laboratory scale fermenters that the microenvironment experienced by cells is effectively homogeneous throughout. With increasing fermenter dimensions, circulation times increase and the microenvironment experienced by the cells becomes a function of bulk flow, mixing, and turbulence. The behavior of such a system with its numerous interrelated processes is complex and difficult to predict, particularly when significant spatial variations exist within the bioreactor. It is in most cases the nonhomogeneous distributions of dissolved oxygen, substrate, pH, temperature, and dissolved carbon dioxide that are responsible for differences in performance at large scales of operation.

Although the effects of environmental extremes may be predictable in general terms, the overall effects of continually fluctuating conditions are not well understood. Whether the biological performance of the microorganisms is influenced by the changing environment in dissolved oxygen, substrate concentration, pH, and dissolved carbon dioxide depends on the magnitude of the characteristic times of the cell reactions. In this context, scale-down models can be used effectively to understand the effects of a nonhomogeneous microenvironment on cell metabolism and for process optimization.

### 18-2.1 Interaction between Liquid Hydrodynamics and Biological Performance

In aerobic fermentations, the most important consideration is often the adequate supply of oxygen to the cells. Oxygen is used continuously by growing cells and due to its low solubility in the liquid phase, a continuous supply is necessary from the gas phase. Oxygen gradients can occur as a result of the interaction between oxygen transfer, long circulation times, and microbial kinetics. Since oxygen consumption occurs in segregated fluid elements circulating in the fermenter, the time constant for oxygen consumption is often the same order of magnitude as the liquid circulation. Depletion of the oxygen may occur at long circulation times, whereas the oxygen concentration remains relatively high at short circulation times. The time scales for circulation in large scale bioreactors may be comparable to the time scales for certain metabolic processes and adjustments (Roels, 1982). Bailey and Ollis (1986) assumed that the circulation time in a stirred vessel is lognormally distributed and that oxygen consumption rate follows a zero-order reaction. They showed that the exposure of cells to starvation conditions increases as the mean circulation time and the standard deviation of circulation times increases. This may also indicate that the kinetic models developed under the very different mixing conditions in a small reactor may not apply when greater mixing times and environmental fluctuations are encountered at the large scale.

Several authors have indicated the presence of dissolved oxygen concentration gradients, resulting from insufficient mixing and mass transfer. Their existence has been inferred from regime analysis (see Section 18-2.6) (Oosterhuis, 1984; Sweere et al., 1986) and in some cases by physical measurement (Carilli et al., 1961; Steel and Maxon, 1966; Manfredini et al., 1983; Oosterhuis and Kossen, 1983, 1984). Oosterhuis and Kossen (1984) reported the existence of dissolved oxygen concentration gradients in a 25 $m^3$ production scale fermenter equipped with two Rushton impellers, using a low viscosity broth (refer to Figure 18-1). The values of dissolved oxygen were corrected for hydrostatic pressure. Relatively high concentrations were measured in the impeller region (8 to 15% of air saturation at ambient pressure) and low values (0 to 6% of saturation) away from the vicinity of the impeller region in the macromixed zones. A strong radial gradient was also observed at the level of the agitator. Therefore, cells circulate in agitated vessels from the well-mixed impeller region (active zone) to the relatively poorly mixed regions (quiescent zones) and will experience fluctuations in dissolved oxygen concentration.

In some fermentations, performance may be governed by the efficiency with which nutrients such as glucose are mixed. Gradients in glucose concentration and their effects on microbial metabolism are discussed further in Section 18-2.7.1. Inhomogeneities also occur when the addition or removal of a component in a system is made in a nonuniform manner. Thus, the addition of concentrated acid or base for pH control will raise the local pH to a high value that will persist for longer if the mixing rate is slow. Solution pH is a fundamental parameter in the regulation of cellular metabolism, and the effects of spatial variations in pH can be important for successful scale-up. The effect of pH on cell metabolism is discussed in Section 18-2.11.

The viscosity of the broth will influence the bulk mixing, air dispersion, and power draw by the agitator. The rheological behavior of fermentation broths have been reviewed (Metz, 1976; Charles, 1978; Riley et al., 2000) and will generally

SCALE-UP/SCALE-DOWN OF FERMENTATION PROCESSES  **1075**

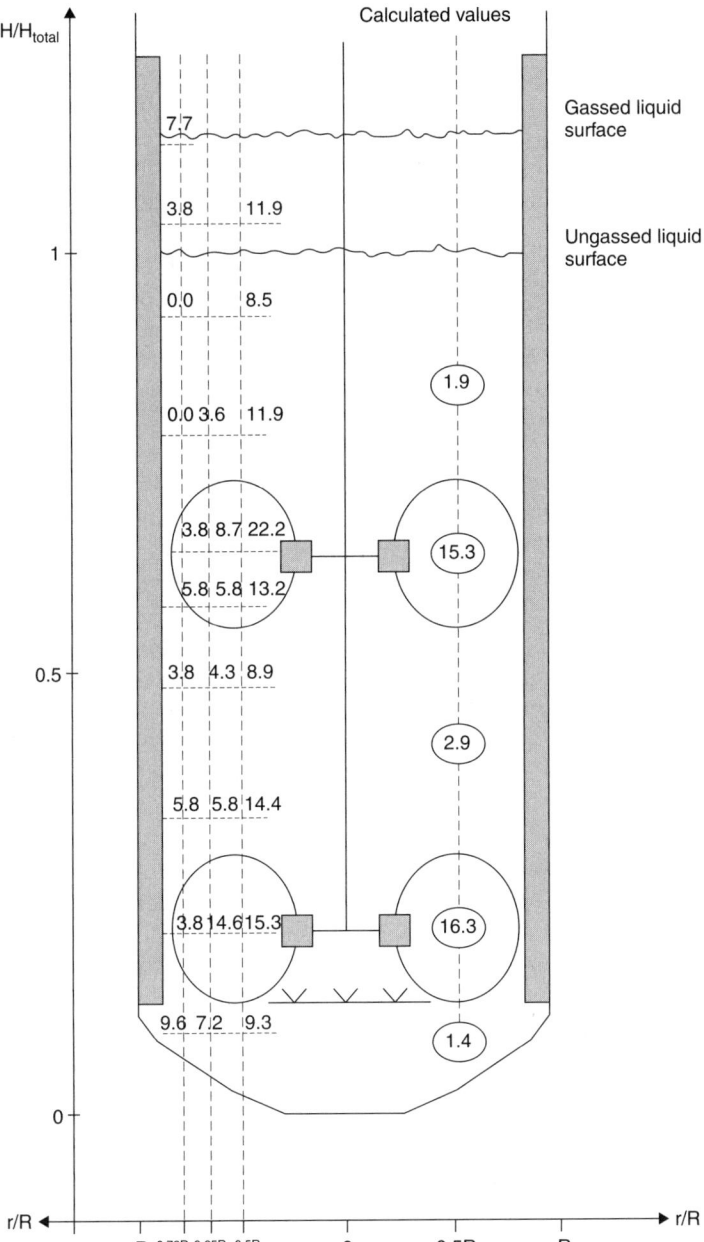

**Figure 18-1** Dissolved oxygen concentration profiles in a 25 m$^3$ (19 m$^3$ working volume) fermenter. Relatively high concentrations of DOT were measured in the well-mixed impeller regions, while low values were measured in the radial and axial planes away from the impellers. The calculated values of DOT (using a two-compartment model for oxygen transfer as shown in the right of the figure in circles) were in broad agreement with the measured values. (From Oosterhuis and Kossen, 1984.)

be influenced by the morphology of the microorganism and in some cases by the formation of extracellular products such as xanthan gum (Amanullah et al., 1998b) and gellan gum (Dreveton et al., 1996). In suspensions of filamentous microorganisms, such as *Streptomycetes, Penicillium,* or *Aspergillus* species, the mycelial hyphae readily become entangled. Together with a high biomass concentration, this can lead to very viscous non-Newtonian suspensions, many exhibiting a yield stress or shear thinning behavior. These properties will have serious implications for bulk mixing. The presence or absence of a yield stress (or very low shear rates) will dictate whether there is flow in regions of low shear stress in the vessel. The stagnant regions outside the cavern persist even in aerated suspensions, and hence adequate oxygen transfer may take place effectively only in the vicinity of the impeller. Low dissolved oxygen levels as a result of poor oxygen transfer and mixing may cause changes in microbial metabolism, productivity, and product quality. For instance, the level of dissolved oxygen can have a marked effect on the recombinant protein production in fungal cultures (Amanullah et al., 2002) as well as on the quality of microbial polysaccharides as determined by the molecular weight (Trujillo-Roldan et al., 2001). In some instances microorganisms such as *Streptomycetes* can aggregate to form pellets, mats, or flocs. This can give rise to diffusional limitations. Steel and Maxon (1966) suggested that the limiting factor in the *Streptomyces* fermentation, where the microorganism was in the form of clumps, was the transfer of oxygen within the clump and not, as in unicellular fermentations, the transfer of oxygen into the liquid phase.

Important interactions between the turbulence intensity at different scales and the morphology (and hence the metabolic state) of certain organisms can be expected. The interaction is most important for organisms that grow to a size scale comparable to the turbulent Kolmogorov eddy scales expected. These scales range from the largest eddies, on the scale of the height of a turbine blade of an agitator ($\approx 0.1$ m), to the smallest eddies, which are produced by the cascade of turbulent eddies. In agitated bioreactor systems, the smallest eddy size is on the order of 10 to 100 $\mu$m (Bailey and Ollis, 1986). Flocs of microorganisms, mycelial aggregates, and animal cells are intermediate in the size spectrum of turbulence and therefore may potentially be influenced by mixing intensity and the distribution of turbulence fields encountered in the reactor. On the other hand, unicellular bacteria and yeast are generally considered "shear" insensitive since their size is considerably smaller than the Kolmogorov eddy scale. The effects of hydrodynamic stress in fungal, bacterial, and animal cells are discussed in Sections 18-4.1, 18-5.1, and 18-6.1, respectively.

### 18-2.2 Fluid Dynamic Effects of Different Scale-up Rules

The scale-up of biotechnological processes developed in the laboratory often presents problems that owing to the complexity of multiple parameters do not permit a generalized solution. This section focuses on the empirical approach to scale-up and highlights the difficulties in maintaining kinematic similarity at different scales. In addition, it highlights the need to account for the biological response of cells to the effects of changing scale. Later, a more process-specific approach is described in which the biological response of microorganisms can be predicted on scale-up using a scale-down approach. The effects of using different scale-up criteria on mixing of aerated stirred vessels has been discussed in this chapter with the aim of understanding the physical phenomena that may affect the biological response of microorganisms. Traditionally, many methods of scale-up of aerated, stirred fermenters have been considered and reviewed (Hempel and Dziallas, 1999). These include the following criteria:

1. Equal specific energy dissipation rates
2. Maintaining geometric similarity
3. Equal impeller tip speeds
4. Constant mixing times
5. Equal volumetric mass transfer coefficients
6. Equal oxygen transfer rates
7. Extrapolation or interpolation of test data generally secured for two scales
8. Combination of more than one of the criteria above

In the following section, the effects of the most commonly used scale-up methods on the mixing process are discussed using mixing theory, with the aid of theoretical and empirical correlations. Particular emphasis has been given to the use of equal specific energy dissipation rates. Correlations to predict the energy dissipation rate are essential, and the problems with their use are also mentioned.

### 18-2.2.1 Scale-up at Equal Specific Energy Dissipation Rate.

Equal specific energy dissipation, P/V, is commonly used to scale-up fermentation and cell culture processes. In the following analysis of the use of equal P/V, only geometrically similar systems have been considered. Geometric similarity implies that all vessel dimensions have a common ratio (H/T, D/T, C/T, etc.). Furthermore, the power dissipated due to aeration is not considered, which may be significant compared to power input, due to agitation in cell culture processes (Langheinrich et al., 1998).

The ungassed power, Po, required by an impeller is given by

$$P = P_o \rho N D^3 \tag{18-1}$$

where P is the power input, $\rho$ the fluid density, N the impeller speed, and D the impeller diameter.

Correlations for the gassed power number are discussed in Section 18-2.2.3, but it is reasonable to assume that the gassed power input is proportional to the ungassed power input in the turbulent flow regime. Thus,

$$P \propto P_g \propto N^3 D^5 \quad \text{and} \quad V \propto D^3 \tag{18-2}$$

where $P_g$ is the gassed power input and V is the liquid volume.

Using $(P/V)_{large} = (P/V)_{small}$, where $(P/V)_{large}$ and $(P/V)_{small}$ represent P/V at the large and small scales, respectively, results in $P \propto D^3$, and substituting for P gives $N^3 D^2 = $ constant, resulting in

$$N \propto D^{-2/3} \tag{18-3}$$

Therefore, the impeller speed will decrease on scale-up for geometrically similar vessels at constant P/V.

Revill (1982) recommends an impeller flow number, Fl, defined as $Q/ND^3$, of 0.75 for Rushton turbines in a fully turbulent system. Thus the impeller pumping capacity, Q, will increase with scale:

$$Q \propto D^{-2/3} D^3 \propto D^{7/3} \tag{18-4}$$

However, the specific pumping capacity, Q/V, will decrease:

$$\frac{Q}{V} \propto D^{7/3} D^{-3} \propto D^{-2/3} \tag{18-5}$$

The impeller tip speed,

$$U_T \propto ND \propto D^{-2/3} D \propto D^{1/3} \tag{18-6}$$

Therefore, higher tip speeds are found on scale-up for geometrically similar vessels at constant P/V.

$$\text{Reynolds number, } Re \propto ND^2 \propto D^{-2/3} D^2 \propto D^{4/3} \tag{18-7}$$

$$\text{Froude number, } Fr \propto N^2 D \propto D^{-4/3} D \propto D^{-1/3} \tag{18-8}$$

Thus, higher Reynolds number and smaller Froude numbers are found on scale-up. The Froude number is usually important only in situations where gross vortexing exists and can be neglected if the $Re < 300$ (Harnby et al., 1997). For higher Reynolds numbers, the effects of the Froude number can be eliminated by the use of baffles.

$$\text{circulation time, } t_c = \frac{V}{Q} \quad \text{or} \quad t_c \propto N^{-1} \propto D^{2/3} \tag{18-9}$$

Since mixing time, $t_m \propto t_c$, therefore

$$t_m \propto D^{2/3} \tag{18-10}$$

This inherent increase in mixing time is one of the major problems in scale-up (see also eq. 9.9). Fermentation processes are often scaled up using constant

$P_g/V$ and volumetric flow of gas per liquid volume per minute (vvm) or constant $P_g/V$ and $v_s$ (superficial gas velocity) and the effects of their use on the mass transfer coefficient and gas holdup also can be illustrated. The vvm is defined as

$$\text{vvm} = \frac{60 Q_G}{(\pi/4)T^2 H} \tag{18-11}$$

where $Q_G$ is the gas flow rate, T, the tank diameter, and H the liquid height. The superficial gas velocity is given by

$$v_s = \frac{Q_G}{(\pi/4)T^2} \tag{18-12}$$

Substituting for $Q_G$ from eq. (18-11) into (18-12) results in

$$v_s = \text{vvm} \frac{H}{60} \tag{18-13}$$

and for $H = T$,

$$v_s \propto \text{vvm}(T) \tag{18-14}$$

This implies that higher superficial gas velocities will result from scaling up at constant vvm using geometrically similar systems.

To calculate $k_L a$ for noncoalescing salt solutions (typical of fermentation media), van't Riet (1979) suggests a correlation of the form

$$k_L a \propto \left(\frac{P_g}{V}\right)^{0.7} v_s^{0.2} \tag{18-15}$$

Therefore, scaling up at constant $P_g/V$ and $v_s$ results in $k_L a =$ constant (if the exponents on $P_g/V$ and $v_s$ are constant).

Scaling up at constant $P_g/V$ and vvm and substituting $v_s$ in eq. (18-15) using (18-14) gives

$$k_L a \propto (\text{vvm} \cdot T)^{0.2} \propto T^{0.2} \tag{18-16}$$

At constant $P_g/V$ and vvm, $k_L a$ will increase with scale for geometrically similar systems.

The effect on gas hold-up, $\varepsilon$, can be analyzed by using the correlation proposed by Smith et al. (1978):

$$\varepsilon \propto \left(\frac{P_g}{V}\right)^{0.48} v_s^{0.4} \tag{18-17}$$

By means of an analysis similar to that for $k_L a$, it can be shown that scaling up at constant $P_g/V$ and $v_s$ results in $\varepsilon =$ constant, and scaling up at constant $P_g/V$ and vvm and assuming that $H = T$ results in

$$\varepsilon \propto T^{0.4} \tag{18-18}$$

## 18-2.2.2 Effect of Viscosity on Scale-up Equal Specific Energy Rates.

High viscosity broths arise as a result of product formation in polysaccharide fermentations (refer to Section 18-3) such as the production of xanthan or pullulan gum or due to the growth of filamentous species such as *Penicillium* or *Aspergillus* (refer to section 18-4). For non-Newtonian fluids obeying the power law, the average shear rate concept of Metzner and Otto (1957) can be used to estimate Reynolds number. Aeration of these fluids in stirred tanks results in the formation of stable equi-sized cavities behind each impeller blade. Increases in the aeration rate do not change the cavity size significantly and hence the power draw (Nienow et al., 1983). In viscous shear thinning fermentation broths (whether a yield stress exists or not is debatable; see Amanullah et al., 1998a), cavern formation can occur. These are regions around the impeller where there is intense gas–liquid mixing and motion. Outside these regions, the fluid is stagnant, and in this situation, the shear stress at the cavern boundary equals the fluid yield stress. More recently, an alternative mathematical model based on a fluid velocity approach has been proposed to estimate cavern sizes and can be applied to radial as well as axial impellers (Amanullah et al., 1998a) (see Section 18-3.3). Elson et al. (1986) proposed a correlation to predict the size of the cavern relative to the impeller diameter for Rushton turbines (using fluids with a yield stress) and also demonstrated the implications for scale-up:

$$\left(\frac{D_c}{D}\right)^3 = \frac{1.36 Po}{\pi^2} \rho N^2 D^2 \tau_y \qquad (18\text{-}19)$$

where $D_c$ and $D$ are the cavern and impeller diameters, respectively, and $\tau_y$ is the fluid yield stress.

Expanding eq. (18-19) for constant fluid properties gives $D_c^3 \propto PoN^2D^5$. At constant power input (assuming that Po is constant) $N \propto D^{-5/3}$, and with Re > 30 gives

$$D_c \propto D^{5/9} \qquad (18\text{-}20)$$

Therefore, for a given power input, the cavern size will increase with larger impellers. Also,

$$\frac{D_c}{D} \propto N^2 D^2 \qquad (18\text{-}21)$$

Therefore, the size of the cavern relative to the size of the vessel increases on scale-up at constant P/V. Also, a constant impeller tip speed is required to maintain the same value of $D_c/D$ on scale-up (see also Section 9-3.6).

## 18-2.2.3 Correlations for Impeller Power Consumption under Gassed Conditions.

The gassed power consumption is one of the most important parameters in the successful design and scale-up of stirred tank bioreactors since it influences numerous mixing parameters whose interactions are complex. However, a major weakness of scale-up at constant $P_g/V$ is in the estimation of the gassed power consumption. Typically, the power consumption of Rushton turbines under aerated conditions is approximately 50% (or less) than the ungassed power, provided that the flow is turbulent. The nature of the gas cavities and bubble dynamics in the vessel affects the flow patterns and the power draw. Most of the correlations in the literature do not account for the flow regime. Establishing the flow regime is important since significant changes in power consumption can occur as a result of changing the flow regime. Studies by Nienow et al. (1985) and Warmoeskerken and Smith (1985) incorporating this concept have proposed different correlations for each type of gas cavity, which is determined essentially by the gas flow number and is discussed later. A selection of correlations from the literature to estimate gassed power is included below.

Calderbank (1958) proposed two correlations for $P_g/P$:

$$\frac{P_g}{P} = 1 - 12.6 \left(\frac{Q_G}{ND^3}\right) \qquad \text{for Fl}_G < 0.035 \qquad (18\text{-}22)$$

where $Fl_G$, is the gassed flow number, defined as $Q_G/ND^3$, and

$$\frac{P_g}{P} = 0.62 - 1.85 \frac{Q_G}{ND^3} \qquad \text{for Fl}_G > 0.035 \qquad (18\text{-}23)$$

Nagata (1975) proposed

$$\log \frac{P_g}{P} = -192 \left(\frac{D}{T}\right)^{4.38} \left(\frac{\rho D^2 N}{\mu}\right)^{0.115} Fr^{1.96(D/T)} Fl \qquad (18\text{-}24)$$

Luong and Volesky (1979) correlated $P_g/P$ by

$$\frac{P_g}{P} = 0.497 \left(\frac{Q_G}{ND^3}\right)^{-0.38} \left(\frac{\rho D^3 N^2}{\sigma}\right)^{-0.18} \qquad (18\text{-}25)$$

where $\sigma$ is the fluid surface tension.

Reuss et al. (1980) used dimensional analysis to obtain the following correlation:

$$\frac{P_g}{P} = 0.0312 Fr^{-0.16} Re^{0.064} Fl^{-0.38} \left(\frac{T}{D}\right)^{0.8} \qquad (18\text{-}26)$$

Hughmark (1980) reviewed the various correlations for the gassed power ratio and suggested

$$\frac{P_g}{P} = 0.1 \left(\frac{Q_G}{NV}\right)^{-0.25} \left(\frac{N^2 D^4}{gwV^{0.67}}\right)^{-0.2} \qquad (18\text{-}27)$$

where w is the blade width and g is the gravitational constant.

Greaves and Kobaccy (1981) proposed the following correlation to calculate the gassed power input in watts:

$$P_g = 1007 \left[ \frac{N^{3.33} D^{6.33}}{(\eta Q_G)^{0.404}} \right] \quad (18\text{-}28)$$

The efficiency index, $\eta$, was correlated depending on whether or not there was recirculation. For $N_F < N < N_R$, $\eta = 1$, where $N_R$ and $N_F$ represent the minimum impeller speed to prevent flooding and the speed at which the onset of recirculation occurs, respectively. To determine $N_F$ and $N_R$, they used

$$N_F = 1.52 \left( \frac{T^{0.2} Q_G^{0.29}}{D^{1.74}} \right) \quad (18\text{-}29)$$

$$N_R = 0.57 \left( \frac{T^{0.97} Q_G^{0.13}}{D^{2.34}} \right) \quad (18\text{-}30)$$

Warmoeskerken (1986) showed that plots of the power ratio $P_g/P$ versus the dimensionless flow number for different impeller speeds yield separate curves. This is due to increased gas recirculation at higher speeds. Most correlations for predicting the aerated power input do not take gas recirculation into account. Nienow et al. (1979) quantified the recirculation rate for a single Rushton turbine and found that it could be as high as three times the sparged rate. Van't Riet (1975) noted the presence of vortices behind the blades of Rushton impellers in a single-phase system. When the stirrer was operated in a gas–liquid system, gas was drawn to these regions (vortices) of low pressure and this led to the formation of gas-filled cavities. Van't Riet distinguished and defined three cavity forms, dependent on stirrer speed and gas flow rate; vortex, clinging, and large cavities. Warmoeskerken (1986) has identified flow regimes to relate the formation of these cavities to the power consumption of Rushton impellers and has proposed correlations, which take gas recirculation into account, to calculate the gassed power input for each type of cavity structure. Warmoeskerken (1986) combined the concepts of cavity formation and gas recirculation with empiricism to give:

- For vortex and clinging cavities, $0 < Fl_G < (Fl_G)_{3-3}$, where

$$(Fl_G)_{3-3} = 0.0038 \left( \frac{Re^2}{Fr} \right)^{0.07} \left( \frac{T}{D} \right)^{0.5} \quad (18\text{-}31)$$

$$\frac{P_g}{P} = 1 - 16.7(Fl_G)(Fr)^{0.35} \quad (18\text{-}32)$$

- For small 3–3 cavities, $(Fl_G)_{3-3} < Fl_G < 0.1$,

$$\frac{P_g}{P} = B - \left[ \frac{0.1(A - B)}{(Fl_G)_{3-3} - 0.1} \right] + \frac{(A - B)(Fl_G)}{(Fl_G)_{3-3} - 0.1} \quad (18\text{-}33)$$

where

$$A = 1 - 17(Fl_G)_{3-3}(Fr)^{0.35} \quad \text{and} \quad B = 0.27 + 0.022 Fr^{-1} \qquad (18\text{-}34)$$

- For large 3–3 cavities, $0.1 < (Fl_G) < (Fl_G)_F$,

$$\frac{P_g}{P} = 0.27 + 0.022 Fr^{-1} \qquad (18\text{-}35)$$

The correlations for predicting gassed power consumption of Rushton turbines mentioned earlier have been used to demonstrate the differences obtained in estimation of the gassed to ungassed power ratio, $P_g/P$, as a function of the flow number. For a fixed impeller size, the flow number can be varied by altering either the gassing rate or impeller speed. Thus, a low flow number can result from low gassing rates or a high impeller speed, and similarly, a high flow number can mean high gassing rates or low impeller speed. In the following example to demonstrate the effects of the use of different correlations to estimate $P_g/P$, the impeller speed has been held constant while the gassing rate has been varied from 0 to 3.5 vvm (Figure 18-2). The scale of the tank has been chosen such that it is within the range covered by the correlations. Considering a flat-bottomed tank equipped with a single Rushton impeller (unaerated power number = 5.5), with T = 0.3 m, H/T = 1, D/T = 0.33, D = 0.1 m, w = 0.02 m, containing water (V = 0.0213 m$^3$, $\rho$ = 1000 kg/m$^3$, $\sigma = 72 \times 10^{-3}$ J/m$^2$) and operated at a

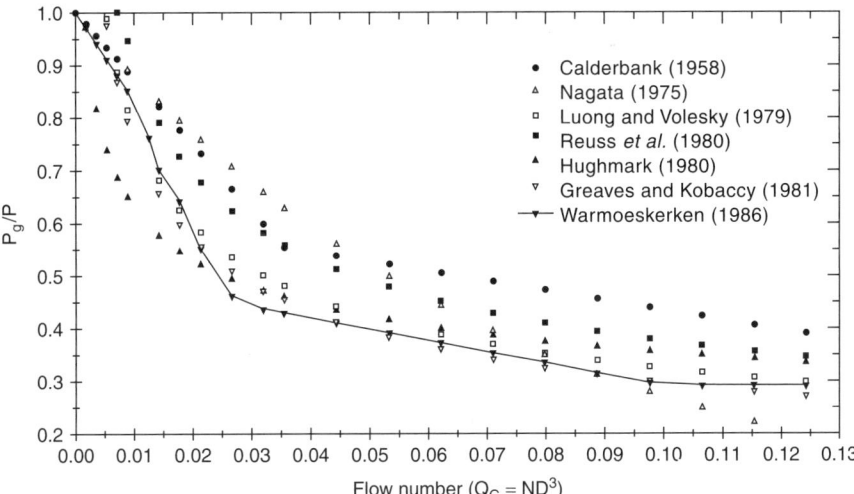

**Figure 18-2** Prediction of gassed power consumption using different correlations using a single Rushton turbine (unaerated power number = 5.5), with T = 0.3 m, H/T = 1, D/T = 0.33, D = 0.1 m, W = 0.02 m, containing water (V = 0.0213 m$^3$, $\rho$ = 1000 kg/m$^3$, $\sigma = 72 \times 10^{-3}$ J/m$^2$) and operated at a constant impeller speed of 600 rpm. Under these conditions, the ungassed power consumption is 2.6 kW/m$^3$.

constant impeller speed of 600 rpm. Under these conditions, the ungassed power consumption is given by

$$P = (5.5)(1000)(10)^3(0.1)^5 = 55 \text{ W or } 2.6 \text{ kW/m}^3$$

At constant impeller speed, the gassed/ungassed power ratio has been calculated as a function of the gassing rate (in $m^3$/s, unless otherwise stated) for the various correlations. The $P_g/P$ ratio as evaluated by the use of these correlations, as a function of the flow number, is shown in Figure 18-2. It is not surprising that differences in the prediction of $P_g/P$ for the same flow number arise since some of the correlations have been obtained using different scales of operation and geometry. Also, most of these correlations have been derived from experiments in small scale vessels, and the power measurements may not be accurate (especially in the earlier studies). Thus, the differences in the predicted gassed power may be expected to increase with scale. Despite a lot of research in agitated gas–liquid systems, no satisfactory method exists for accurately predicting the gassed power consumption. This is due primarily to the complexity of the hydrodynamics of stirred gas–liquid systems. It is therefore difficult to predict the power consumption by simple correlations based on either empirical data or dimensionless analysis. The inaccuracy of this prediction is likely to increase in the case of multiple-impeller systems, which are commonly used in industrial fermentations. Further complications can arise when considering the power input in rheologically complex fermentations, where the availability of power input correlations are limited and further compounded by time-varying rheological characteristics of the fermentation broth.

### 18-2.2.4 Scale-up by Maintaining Geometric Similarity.
Johnston and Thring (1957) have reviewed the principles of similarity for scale-up of processes and in particular for agitation applications. Although a useful purpose is served by the principle of similarity approach, it is seldom possible to apply it directly. Very few companies have geometrically similar bioreactors throughout their laboratory, pilot, and production scale facilities (Einsele, 1978). Geometric similarity may be maintained in going from bench to pilot scale tests. However, at the commercial scale, dimensions such as the H/T ratio may be changed from 1 to 2 or more to improve the efficiency of air utilization to reduce operating costs, and it may necessitate the use of multiple impellers to ensure adequate mixing. Most small scale fermenters are operated with Rushton turbines using a D/T ratio of 0.33. However, this ratio may be increased on scale-up since higher D/T ratios show advantages. Rushton impellers with large D/T ratios are more energy efficient for bulk blending in both low viscosity broths (Nienow, 1984) and for high viscosity shear thinning broths (Nienow and Ulbrecht, 1985).

The effect of the D/T ratio on gas dispersion can be demonstrated. Nienow et al. (1985) proposed the following equations for the flooding–loading transition (denoted by the subscript F):

$$(Fl_G)_F = 30 \left(\frac{D}{T}\right) Fr_F \qquad (18\text{-}36)$$

and the complete dispersion (denoted by the subscript CD) phenomena:

$$(Fl_G)_{CD} = 0.2 \left(\frac{D}{T}\right)^{0.5} Fr_{CD}^{0.5} \qquad (18\text{-}37)$$

Considering the flooding correlation, at constant vvm using geometrically similar systems, we have

$$\frac{1}{N_F} \propto N_F^2 D \quad \text{since} \quad \frac{Q_G}{D^3} = \text{constant} \quad \text{and therefore} \quad \left(\frac{P_g}{V}\right)_F \propto D \qquad (18\text{-}38)$$

Similarly, using the complete dispersion correlation, it can be shown that

$$N_{CD} \propto D^{-0.25} \quad \text{and} \quad \left(\frac{P_g}{V}\right)_{CD} \propto D^{1.25} \qquad (18\text{-}39)$$

The analysis above shows that higher specific power input is necessary both to prevent impeller flooding and to achieve complete dispersion at the large scale. Also, from the flooding correlation, for constant aeration rate and a fixed vessel size,

$$\frac{1}{N_F D^3} \propto D^{4.5} N_F^2 \quad \text{and therefore} \quad N_F \propto D^{-5/2} \qquad (18\text{-}40)$$

Since constant pumping capacity implies that $N \propto D^{-3}$, eq. (18-40) may be taken to imply that at a fixed scale of operation, a constant pumping rate is required to disperse a given flow of gas. Also, the use of large D/T ratios is more economical since

$$P_F \propto N_F^3 D^5 \text{ and replacing } N_F \text{ using eq. (18-40) results in} \quad P_F \propto D^{-5/2}$$
$$(18\text{-}41)$$

Therefore, lesser power is required to prevent flooding using large D/T ratios, or for the same power input, the gas-handling capacity of the impeller is increased. This would also imply that the drop in aerated power draw would be less when larger D/T ratios are used.

Similar analysis using the complete dispersion correlation results in

$$P_{CD} \propto D^{-1} \qquad (18\text{-}42)$$

The same conclusions as for flooding can be made, although the power dependence on the D/T ratio is reduced ($P_{CD} \propto D^{-1}$ compared to $P_F \propto D^{-2.5}$).

### 18-2.2.5 Scale-up at Equal Impeller Tip Speeds.

The impeller tip speed, $U_T$, has been used as a scale-up criterion for mycelial fermentations since it is often cited that the growth of the filamentous organisms is sensitive to the shear produced by the impeller. Typical values of $U_T$ employed are 5 m/s (Wang et al., 1979). The use of this scale-up criterion results in a higher power input ($P \propto D^2$), lower power per unit volume ($P/V \propto D^{-1}$), higher pumping capacity ($Q \propto D^2$), and longer circulation times ($t_c \propto D$). The effects of different scale-up rules, including tip speed and specific energy dissipation, on fragmentation of mycelial hyphae are considered in Section 18-4.1 and the use of $U_T$ as a scale-up criterion to correlate hyphal fragmentation in mycelial fermentations is shown to be unsatisfactory.

### 18-2.2.6 Scale-up at Constant Mixing Times.

Scaling up at constant mixing time or circulation time using geometrically similar systems is generally not acceptable since their use results in $P \propto D^5$ and $P/V \propto D^2$. It is rare to use a strategy that results in a higher P/V value at the large scale. If such a strategy is implemented, further benefits would have to be demonstrated to justify its use.

### 18-2.2.7 Scale-up at Equal Volumetric Mass Transfer Coefficients.

Scale-up at equal volumetric mass transfer coefficients ($k_L a$) has to rely on the use of correlations to enable the calculation of overall values of $k_L a$, which are of the form

$$k_L a = A \left(\frac{P_g}{V}\right)^a v_s^b \tag{18-43}$$

where A, a, and b are approximately constant for a given fermenter system (geometry and system), independent of agitator type. Van't Riet (1979) has proposed correlations to predict $k_L a$ within 20 to 40% accuracy for coalescing and non-coalescing salt solutions in stirred vessels. For water under coalescing conditions:

$$k_L a = 2.6 \times 10^{-2} \left(\frac{P_g}{V}\right)^{0.4} v_s^{0.5} \tag{18-44}$$

where $0.002 \leq V \leq 2.6$ m$^3$ and $500 \leq (P_g/V) \leq 10\,000$ W/m$^3$. For salt solutions (noncoalescing):

$$k_L a = 2.0 \times 10^{-3} \left(\frac{P_g}{V}\right)^{0.7} v_s^{0.2} \tag{18-45}$$

where $0.002 \leq V \leq 4.4$ m$^3$ and $500 \leq (P_g/V) \leq 10\,000$ W/m$^3$.

The effect of scaling up at equal $k_L a$ and vvm using geometrically similar systems can be demonstrated as follows: Using eq. (18-45) for noncoalescing solutions (typical of fermentation broths) and assuming that $H = T$ and replacing $v_s$ using eq. (18-14) gives

$$k_L a \propto \left(\frac{P_g}{V}\right)^{0.7} (\text{vvm} \cdot T)^{0.2} \tag{18-46}$$

Several authors (Humphrey, 1977; Moo-Young and Blanch, 1981; Oldshue, 1983; Charles 1985; Bailey and Ollis, 1986) recommend the use of equal $k_L a$ and vvm as a scale-up criterion together with the use of a correlation of the form of eq. (18-44) to calculate $k_L a$. The value of A in eq. (18-42) is sensitive to and is significantly reduced by antifoam (Martin et al., 1994). Humphrey (1977) reports that the exponents a and b vary with scale, and this is also in agreement with the observations of Bartholomew (1960). The apparently unspecific dependency of the exponents a and b in the correlation on a given fermenter system can lead to problems in using this approach for scale-up. For viscous fluids, a viscosity term, $\mu_a$, is introduced in eq. (18-44). Hickman and Nienow (1986) have shown that $k_L a \propto \mu_a^{-0.5}$.

### 18-2.2.8 Scale-up at Equal Oxygen Transfer Rates.

Often, the supply of oxygen is the factor limiting the productivity of large scale fermenters, especially in high-cell-density cultivations. The low solubility of oxygen in aqueous solutions necessitates the continual supply of oxygen from the gas phase. The lack of oxygen may result in the death of cells or may be responsible for diverting the metabolic pathways of some species. The oxygen transfer rate (OTR) can be calculated from

$$\text{OTR} = k_L a (C_L^* - C_L) \qquad (18\text{-}47)$$

$C_L^*$ and $C_L$ represent the dissolved oxygen concentration at air saturation and in the liquid phase.

The maximum oxygen uptake rate ($\text{OUR}_{max}$) is related to biomass concentration (x) by

$$\text{OUR}_{max} = (Q_{O_2\,max}) x \qquad (18\text{-}48)$$

where $Q_{O_2\,max}$ is the maximum specific oxygen uptake rate.

Online oxygen uptake rate (and carbon dioxide production rate) have been possible to measure and calculate routinely for some time now using mass spectrometry (Buckland et al., 1985). Coupled with online dissolved oxygen measurements, this technique can also be used to measure online $k_L a$. Mass spectrometers offer fast, reliable, and accurate measurements of these parameters and have proved to be invaluable for process monitoring, control, and scale-up. In fact, the data generated using online exhaust gas analysis forms the basis of this widely used scale-up rule using the maximum oxygen transfer capability of fermenters (which, in turn, dictates its biomass production capability) as a scale-up criterion.

Successful scale-up (from 0.005 to 57 m$^3$), using constant OTR, of penicillin and streptomycin and baker's yeast (from 0.019 to 114 m$^3$) fermentations have been reported (Hempel and Dzialas, 1999). However, examples have also been reported (Bartholomew, 1960) where in vitamin B$_{12}$ fermentations, the use of equal OTR as a scaling parameter led to an oversizing of the large scale fermenter. This may occur partly due to the fact that the measurement of $C_L$ by point-positioned dissolved oxygen probes may be unrepresentative of the global $C_L$ distribution at the large scale in which dissolved oxygen gradients may be severe.

**1088** MIXING IN THE FERMENTATION AND CELL CULTURE INDUSTRIES

This can be due to several reasons. First, the rate of transfer to the liquid phase, and hence $C_L$, increases with increasing hydrostatic head. This is exacerbated by the depletion of oxygen from the gas phase by bubbles rising through the liquid. Second, both oxygen transfer and oxygen uptake are position dependent, due to the combined effects of regions of different mixing intensities, resulting in inhomogeneities and the presence of a circulation time distribution.

### 18-2.2.9 Effects of Different Scale-up Criteria Using a Linear Scale-up Factor of 10.

The extrapolation or interpolation of test data generally secured for two scales is used extensively in chemical engineering for scale-up. However, there is a limited range in which the results can be used and caution has to be exercised if extrapolation is extensive. The effects of the various scale-up criteria discussed in previous sections on the mixing process have been evaluated and their consequence for large scale operation is shown in Table 18-1. This is based on an extension from an earlier analysis conducted by Oldshue (1966) for geometrically similar systems under unaerated conditions and a linear scale-up factor of 10, or a 1000 fold increase in volume. The present analysis has been conducted for aerated conditions where the gassed power input has been assumed to be proportional to the ungassed power input. It has also been extended to include the effects of scale-up at equal $k_L a$ and vvm and equal $k_L a$ and $v_s$ as well as the consequence for scale-up on the impeller pumping capacity, Froude number, and the circulation time. Correlations proposed by van't Riet (1979) for noncoalescing salt solutions have been used to calculate $k_L a$.

From Table 18-1 it can be seen that scale-up at equal power per unit volume for geometrically similar systems results in a lower impeller speed, higher tip speed, pumping capacity, $k_L a$ (at constant vvm), and circulation time. Scale-up at equal impeller speed or mixing time is unrealistic since the power input per

**Table 18-1** Effect of Different Scale-up Criteria Using a Linear Scale-up Factor of 10 and Maintaining Geometrical Similarity (Re > $10^4$)

| Large Scale/<br>Small Scale<br>Value | Scale-up Criteria ||||||
|---|---|---|---|---|---|---|
| | Equal P/V | Equal N | Equal $U_T$ | Equal Re | Equal $k_L a$ and vvm | Equal $k_L a$ and $v_s$ |
| $P \propto N^3 D^5$ | 1000 | $10^5$ | 100 | 0.1 | 829 | 1000 |
| $P/V \propto N^3 D^2$ | 1 | 100 | 0.1 | $10^{-4}$ | 0.8 | 1 |
| N or $T_m^{-1}$ | 0.22 | 1 | 0.1 | 0.01 | 0.3 | 0.22 |
| $U_T \propto ND$ | 2.2 | 10 | 1 | 0.1 | 2.7 | 2.2 |
| Re $\propto ND^2$ | 22 | 100 | 10 | 1 | 27.2 | 22 |
| $Q \propto ND^3$ | 220 | 1000 | 100 | 10 | 272 | 220 |
| Fr $\propto N^2 D$ | 0.48 | 10 | 0.1 | $10^{-3}$ | 0.5 | 0.48 |
| $T_c \propto N^{-1}$ | 4.55 | 1 | 10 | 100 | 9.4 | 4.55 |
| $k_L a$ at equal vvm | 1.59 | 39.8 | 0.32 | $2.5 \times 10^{-5}$ | 1 | — |
| $k_L a$ at equal $v_s$ | 1 | 25.1 | 0.20 | $1.6 \times 10^{-3}$ | — | 1 |

unit volume has to be increased substantially. Scale-up at constant Reynolds number is also not feasible since in this case the $P_g/V$ value is reduced by a factor of $10^4$ at the large scale. Equal impeller tip speed can be used, although the reduction in the $P_g/V$ value by a factor of 10 also results in lower $k_La$ values. Also, in this case there is a 10 fold increase in the circulation time. Equal $k_La$ and vvm and equal $k_La$ and $v_s$ are also commonly used for scale-up. One of the consequences of using the former scale-up criterion rather than the latter is that the power consumption is lower due to the higher superficial gas velocity, and this results in higher circulation times.

It is important to note that regardless of the choice of scale-up criterion (except scale-up at equal impeller speed or mixing time, both of which are economically unrealistic), there is an increase in the circulation time at the large scale. This increase, coupled with high oxygen demands, can cause severe oxygen gradients, and coupled with the addition of concentrated reagents for pH control and nutrient availability, can have a significant impact on fermentation yield. It is clear from Table 18-1 that different scale-up criteria result in entirely different process conditions at the larger scale. It is impossible to maintain similarity of all aspects of the microenvironment at different scales. The scale-up criteria are system specific and it is therefore necessary to select a scale-up basis depending on the transport property most critical to the performance of the bioprocess. Thus, if oxygen transfer is the limiting factor in a fermentation, scale-up at equal $P_g/V$ may be invoked, or if shear rates are significant, the energy dissipation/circulation function (see Section 18-4.1) may be employed. However, keeping one parameter constant also results in a change in other important variables. Therefore, the choice of scale-up rule is not easy given the potentially sensitive and diverse responses of cells to each of the transport phenomena influenced by impeller design, system geometry, scale, fluid properties, and operating parameters.

Kossen and Oosterhuis (1985) proposed two ways to solve the problem of scale-up of bioreactors: first, by acquiring more knowledge about the hydrodynamics and interaction of the hydrodynamics with other mechanisms in production scale fermenters, and second, by developing scale-up procedures that give an adequate estimation of the performance of production scale fermenters based on small scale investigations. This approach is discussed in detail in later sections.

### 18-2.3 Influence of Agitator Design

For many years, Rushton turbines of approximately one-third the fermenter diameter were considered as the optimum design for mixing of fermentation processes. These radial flow impellers induce high turbulence around the impeller region and thus promote good gas dispersion and bubble breakup. Bulk blending is considered to be poor, using such turbines due to their tendency to compartmentalize (Nienow and Ulbrecht, 1985) and can lead to broth inhomogeneities of either pH or oxygen. This may be expected to be amplified in viscous, non-Newtonian broths (Buckland et al., 1988a; Nienow et al., 1995; Amanullah et al., 1998b).

Bryant (1977) and Bajpai and Reuss (1982) have suggested that the critical factor that determines the overall effectiveness of oxygen uptake by microorganisms is the frequency at which cells are circulated through the highly oxygenated impeller region. However, the circulation capacity of standard Rushton impellers with D/T ratios of 0.33 may be insufficient to induce the necessary bulk flow to satisfy the oxygen requirements of cells. To improve the liquid pumping capacity, the use of large D/T ratios has been suggested (Nienow, 1984). Prochem hydrofoil impellers, which produce axial flow, have been tested for use in high viscosity mycelial fermentations. Buckland et al. (1988a) demonstrated, in viscous mycelial fermentations at the pilot scale, that the replacement of standard radial flow Rushton impellers with larger diameter axial flow Prochem impellers, significantly improved the oxygen transfer efficiency. This improvement was attributed to the increase in size of the well-mixed, low viscosity cavern in the impeller region. The reader is also referred to the publications by Nienow (1990, 1998) and Nienow et al. (1995) for detailed discussions regarding the role of agitator design in fermentations. The effects of impeller geometry and type (including Scaba impellers) in high viscosity xanthan fermentations and fungal fermentations are discussed in Sections 18-3 and 18-4, respectively.

### 18-2.4 Mixing and Circulation Time Studies

It is generally recognized that the performance of bioreactors depends on the intensity of mixing of the gas and liquid phases. Therefore, extensive efforts have been devoted to understanding the mixing characteristics in bioreactors. Mixing in agitated vessels is dependent on both the levels of turbulence in the region of the agitator and in the remainder of the vessel, as well as the bulk turnover of the vessel contents. Einsele (1978) reported mixing times on the order of 160 s in aqueous, unaerated bioreactors of up to 100 $m^3$ in volume. He correlated the mixing times measured at different scales using the equation

$$t_m \propto V^{0.3} \tag{18-49}$$

Longer mixing times may result in aerated Newtonian and non-Newtonian systems. Einsele and Finn (1980) reported mixing times in aerated stirred tanks (0.02 and 0.35 $m^3$) using different aqueous solutions. From the pH response to a pulse input, it was shown that these mixing times increased with increasing gas holdup, and this was found to be more pronounced in higher viscosity solutions. They concluded that the mixing efficiency of a stirrer is adversely affected by the interaction between gas bubbles and the eddies that are generated by the stirrer under turbulent conditions. Using a flow follower technique, Bryant and Sadeghzadeh (1979) and Middleton (1979) observed an increase in the mean circulation time and the standard deviation under gassed conditions. Middleton (1979) indicated that this was expected in view of the decrease in the pumping capacity of the impellers caused by gas-filled cavities behind the impeller blades. This phenomenon was also reported by van Barneveld et al. (1987), who used a similar technique in a production scale 25 m³ fermenter. In contrast, Paca et al. (1976) observed shorter mixing times in aerated completely dispersed non-Newtonian systems than in unaerated ones. Under conditions of flooding, they reported higher mixing times than in unaerated systems. Correlations to estimate mixing times are given in Chapter 9. Although knowledge of the mixing and circulation times is potentially useful, these times do not on their own provide sufficient criteria with which the effects of inhomogeneities on microorganisms can be quantitatively explained. It is, in fact, the microenvironment experienced by the cells that determines the biological performance, not the partial view of the same system, provided by overall mixing times or oxygen transfer coefficients. In this regard, the concept of a circulation time distribution is very useful to describe the environment experienced by microorganisms.

*18-2.4.1 Circulation Time Distribution Models.* Circulation time is an important concept in the study of fluctuating environmental conditions because it provides an indication of the characteristic time interval during which a cell circulates through different regions of the reactor and hence possibly encounters different reaction conditions along the way. Clearly, consideration of a single circulation time in an agitated tank is a conceptual approximation. Upon leaving the impeller region, different elements of fluid will follow different paths in the vessel, giving rise to correspondingly different circulation times. An alternative method for characterizing the circulation in a stirred vessel is by the circulation time distribution (CTD). This is defined as the probability for each possible time interval that a fluid element takes to return to a fixed point, which is usually taken as the impeller region or in the case of substrate and pH addition as the feed (or addition) zone (Noorman et al., 1993; Larsson et al., 1996). A CTD is characterized by a mean circulation time and a standard deviation. Bryant and Sadeghzadeh (1979) described the measurement of circulation time distributions by means of a neutrally buoyant radio transmitter and a monitoring antenna around the impeller. Middleton (1979) used this technique to measure the CTD in 0.18, 0.60, and 1.80 m³ vessels with Rushton turbines using water under aerated and unaerated conditions. He showed that circulation times were lognormally distributed, with no relationship found between successive circulation times. Thus, in a fermenter, individual microorganisms will be subjected to oxygen depletion in a more-or-less random way. Under unaerated conditions, Middleton (1979) proposed the following equation to quantify the circulation time:

$$t_c = 0.5 V^{0.5} \frac{1}{N} \left(\frac{T}{D}\right)^3 \qquad (18\text{-}50)$$

For aerated systems, the mean circulation time and the standard deviation increased. However, correlations under these conditions were not reported. A lognormal CTD was also reported by Oosterhuis (1984) and van Barneveldt et al. (1987), who measured circulation times using a similar technique in a 25 m³ bioreactor using water. In mechanically agitated vessels, a very high degree of turbulence exists in the vicinity of the impeller. Cutter (1966) reports that up to 70% of the energy dissipation takes place in the impeller region. In this region, micromixing is complete (Bajpai and Reuss, 1982) and through it, the entire fluid in the vessel passes at a frequency dictated by the CTD. Away from the impeller, where the turbulence intensity is less, the mixing of fluid elements may range from complete micromixing to segregation. Using this concept, Bajpai and Reuss (1982) proposed, after Manning et al. (1965), a two-environment model which they named the *micro–macromixer model*. The volume of the micromixed region is very small compared to the macromixed region. The recirculating stream from the macromixed region is completely mixed with the incoming component, if any, and returned to the macromixed zone. In this manner the micromixed zone produces elements of age zero for the macromixed zone. This two-zone model was coupled with microbial kinetics to evaluate the performance of bioreactors. The CTD was described using the following equations:

$$f(t_c) = \frac{1}{\sqrt{2\pi}\sigma_l t_c} \exp\left[\frac{(\ln t_c - \mu_l)^2}{2\sigma_l^2}\right] \qquad (18\text{-}51)$$

$$\bar{t}_c = \exp\left(\mu_l + \frac{\sigma_l^2}{2}\right) \qquad (18\text{-}52)$$

$$\sigma_\theta^2 = \frac{\sigma^2}{\bar{t}_c^2} = \exp(\sigma_l^2) - 1 \qquad (18\text{-}53)$$

where $f(t_c)$ is the CTD, $t_c$ the range of circulation times, $\bar{t}_c$ the mean circulation time, $\sigma$ the standard deviation of the measured mean circulation time, $\sigma_\theta$ the normalized standard deviation with respect to the mean circulation time, and $\mu_l$ and $\sigma_l$ the mean and standard deviation of a lognormal distribution, respectively.

As an example, eqs. (18-51)–(18-53) have been used to calculate the CTD and the cumulative CTD for mean circulation times of 20, 40, and 120 s using a standard deviation of 8.9 s [as reported by Bajpai and Reuss (1982) for a 100 m³ fermenter] and are shown in Figure 18-3a. For $\bar{t}_c = 20$, 40, and 120 s, the circulation times are distributed in the ranges 0 to 90 s, 0 to 180 s, and 0 to 360 s, respectively. The cumulative CTD can be used with an estimate of the oxygen consumption time ($t_{oc}$) to determine the percentage of cells that may be subjected to conditions of oxygen starvation. The maximum oxygen consumption rate for a *Bacillus subtilis* culture was determined to be $4.73 \times 10^{-6}$ mol/g cell per second (Amanullah et al., 1993a,b). At a biomass concentration of 3 g/L and assuming zero-order kinetics, it would take $\approx 10$ s to reduce the DOT from 50% of air saturation to zero. As an example, this time ($t_{oc}$) is depicted by a dotted line in Figure 18-3a. Thus, the percentages of cells exposed to oxygen-depleted conditions for $\bar{t}_c = 20$, 40, and 120 s have been estimated as 66%, 88%, and 98%, respectively, from the intersection of this line with the respective cumulative CTD curves. Similar calculations to show the influence of increasing mean circulation times on the percentage of

SCALE-UP/SCALE-DOWN OF FERMENTATION PROCESSES  **1093**

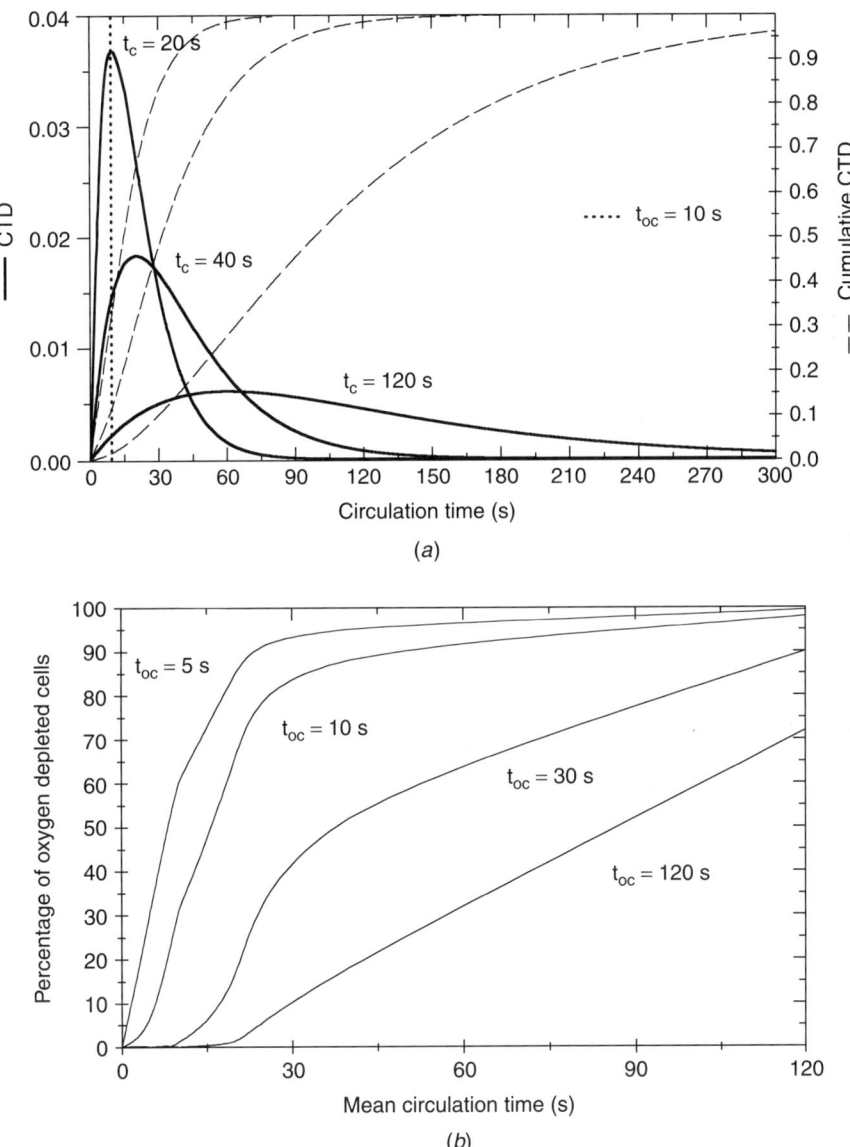

**Figure 18-3** (*a*) Calculated CTD [using eqs. (18-53)–(18-55)] and cumulative CTD at mean circulation times of 20, 40, and 120 s using a standard deviation of 8.9 s. Also shown as an example is a line of dots representing an oxygen consumption time of 10 s, which can be used to estimate the percent of oxygen depleted cells for a given circulation time (see Figure 18-3*b*). (*b*) Effect of mean circulation time on the percentage of cells subjected to oxygen depletion at different oxygen consumption times ($t_{oc}$).

cells exposed to oxygen-depleted conditions at different oxygen consumption times can be made, and the results are shown for $t_{oc} = 5$, 10, 30, and 120 s in Figure 18-3b. From this figure it is clear that as the mean circulation time increases, the percentage of cells exposed to oxygen-depleted conditions increases, and as the oxygen consumption time increases, the percentage of oxygen-deprived cells decreases. These calculations do not account for oxygen mass transfer, which would decrease the percentage of oxygen-deprived cells. Nevertheless, the concept of combining the CTD with oxygen uptake kinetics is very useful in analyzing the effects of nonhomogeneous DOT on cell metabolism.

### 18-2.5 Scale-down Approach

Scale-down is used to try to model physically at the laboratory scale the environmental conditions that microorganisms experience at the large scale. Oosterhuis and Kossen (1984) suggested that for a more realistic approach to scale-up, the rate-limiting step has to be determined first. The laboratory scale process is then designed by optimizing the rate-limiting step. The results of the optimization are then applied to the production scale, although this step has rarely been implemented. The most important requirement for experiments on the small scale is that they have to be representative of the conditions at the large scale. This obviously determines the possibilities and limits of the scale-down approach. A knowledge of the reaction kinetics and metabolic pathways is essential, and these have to be measured. Figure 18-4 shows the scale-down procedure as proposed by Oosterhuis and Kossen (1984). Scale-down strategies are based on actual or calculated measurements at the large scale and are designed on the individual characteristic features of the actual process (see Section 18-2.7.1 for details of scale-down studies).

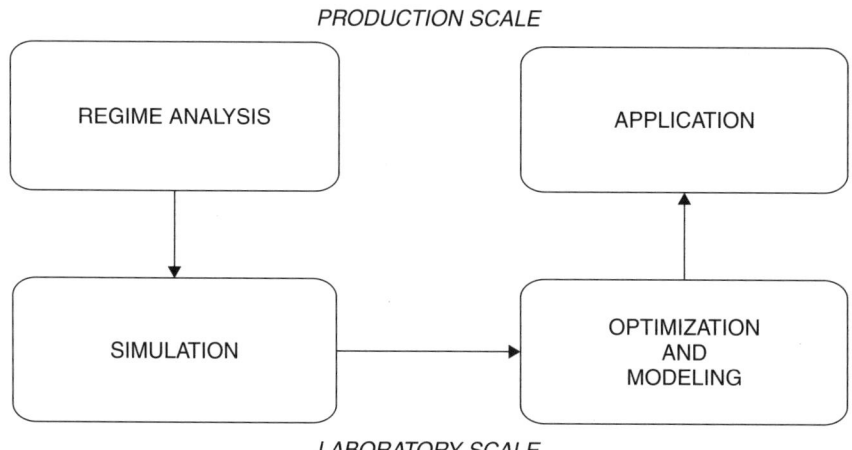

**Figure 18-4** Scale-down procedure. (From Oosterhuis and Kossen, 1984.)

## 18-2.6 Regime Analysis

The rate-limiting step in a bioprocess can be determined by carrying out a regime analysis. This analysis is based on a comparison of the characteristic times of various mechanisms in a process. Characteristic times can be defined as a ratio of capacity to flow. The comparison of times, in terms of orders of magnitude, can be made experimentally, or theoretically, qualitatively, or quantitatively. A low value of a characteristic time means a fast mechanism; a high value indicates a slow mechanism. Regime analysis can also be used to quantify the effects of changes from the well-mixed conditions at the small scale to the possible inhomogeneities arising at the large scale. It can also be used to determine whether there is a single rate-limiting mechanism (pure regime) or whether more than one mechanism (mixed regime) is responsible. Besides regime analysis, it is also important to use dimensional analysis and the principles of similarity to devise small scale experiments (Sweere et al., 1987). The use of regime analysis in conjunction with scaled-down optimization techniques has been reviewed by Sweere et al. (1987). Regime analysis of baker's yeast and *B. subtilis* fermentations can be found in Sweere et al. (1987) and Amanullah (1994). Oosterhuis (1984) conducted a regime analysis for the gluconic acid fermentation in a 25 $m^3$ fermenter. The results of the analysis are shown in Table 18-2, from which the following conclusions can be made:

1. The characteristic times for oxygen consumption and transfer to the liquid phase are of the same order of magnitude, and therefore oxygen limitation can occur. Also, the liquid circulation time is on the same order of magnitude, and hence oxygen gradients are likely to occur.

**Table 18-2** Regime Analysis of a 19 $m^3$ Gluconic Acid Fermentation

|  | Time (s) |
|---|---|
| *Transport Phenomena* | — |
| Oxygen transfer | 5.5 (noncoalescing) |
|  | 11.2 (coalescing) |
| Liquid circulation | 12.3 |
| Heat transfer | 330–650 |
| *Conversion* | — |
| Oxygen consumption | 16 (zero order) |
|  | 0.7 (first order) |
| Growth | $1.2 \times 10^4$ |
| Substrate consumption | $5.5 \times 10^4$ |
| Heat production | 350 |

*Source*: Oosterhuis (1984).

2. Growth and substrate consumption are unlikely to influence the performance of the process since the time constants for these processes are much larger compared to those for oxygen consumption and oxygen transfer.
3. From a comparison of the times for heat transfer and heat production, it is possible to say that heat transfer will not be a problem at this scale, and temperature gradients should not be present since the liquid circulation time is relatively small compared to the time constant for heat production.

### 18-2.7 Effects of Fluctuating Environmental Conditions on Microorganisms

Most of the studies in the literature are limited to investigating the effects of substrate and dissolved oxygen inhomogeneities on microorganisms, although more recently the effects of pH gradients have also been reported using *B. subtilis* as a model system (Amanullah et al., 2001b) and using GS-NSO myeloma cells (Osman et al., 2002). It should be pointed out that very few studies have been reported in animal cell culture systems, and therefore the review in the following sections has been limited to microbial systems. In investigating the influence of a changing environment on cells, a distinction has to be made between a single change (step signal or impulse) and continuous changes. The latter can be divided into periodic and nonperiodic. The response of microorganisms to forced variations in dissolved oxygen and substrate concentration has been used to study the effects of (1) periodic operation of both fed-batch and continuous culture fermentations as a method of improving culture performance, (2) transient conditions on the biochemistry of microorganisms in order to gain insight into microbial regulatory and control mechanisms, (3) determination of environmental fluctuations to which microorganisms will be exposed on scale-up, (4) verification of scale-down performance as an indicator of large scale performance, and (5) improvements to large scale fermentation processes.

Many of the inhomogeneities encountered in production scale fermenters are cyclic in nature and can be approximated by forced sine- or square-wave functions. Such periodic fluctuations lead to every point on the sine or square wave describing a change in pressure, dissolved oxygen, or substrate concentration as a function of time (Vardar and Lilly, 1982). Therefore, each point represents a different region in a mechanically agitated vessel or a discrete element of fluid in circulation through the micro- and macromixed regions of the vessel. Experimental small scale simulations of mixing and mass transfer limitations in large scale bioreactors have been made in several ways: by cyclic feeding of the limiting nutrient, by cyclic changes in the fermenter head pressure, by creating artificial dead zones, and by adding viscosity-enhancing agents. These studies are reviewed in the following sections and have been divided primarily into the effects of fluctuations in substrate and dissolved oxygen concentrations on microorganisms, although pH gradients are discussed in Section 18-2.11.

## 18-2.7.1 Scale-down Models to Simulate Substrate Inhomogeneities at the Large Scale.

An example of a microorganism that is particularly sensitive to glucose concentrations is *Saccharomyces cerevisiae*, which may alter its metabolism from oxidative to oxido-reductive, depending on the glucose level. A large number of investigations have been carried out using this microorganism, not only because of its commercial importance as baker's yeast, but also due to its sensitivity to the glucose effect. The primary goal in the baker's yeast process is to direct the glucose availability to biomass formation and prevent the formation of ethanol. Another industrial microorganism that is commonly used to produce recombinant proteins (Bylund et al., 2000) is *E. coli*. In this case, acetate formation due to overflow metabolism of glucose is undesirable. Thus, fed-batch strategies are commonly applied in industrial fermentations involving these microorganisms. Glucose feed concentrations of 500 to 600 g/L are generally used. However, the glucose saturation concentrations for *E. coli* and *S. cerevisiae* are 5 and 150 mg/L, respectively (Larsson et al., 1996). On a macroscale with respect to liquid-phase nutrients, it is reasonable to assume that higher concentrations of the substrate exist locally in the entry region of even well-mixed systems. Exposure of cells to this feed zone may affect their biological performance and the degree and duration of the perturbation will depend on the sensitivity of the microorganism to the change. The most comprehensive experimental evidence for such elevated concentrations in large bioreactors have been reported by the groups of Larsson and Enfors (Royal Institute of Technology, Stockholm, Sweden).

Larsson et al. (1996) measured spatial concentrations of glucose in fed-batch fermentations of *S. cerevisiae* at a scale of 30 m$^3$ (19.8 to 22 m$^3$ working volume) at the top (0.5 m below the liquid surface, where about 600 g/L glucose was added), middle and bottom of the bioreactor (close to the well-mixed region of the lowest impeller). Rapid sampling (0.15 s/sample) and inactivation allowed the determination of transient concentrations of glucose. At the top location glucose varied in the range of about 40 to 80 mg/L, whereas at the bottom location the level was relatively constant at about 22 mg/L. Given that the saturation constant is about 150 mg/L and Monod kinetics apply, these measurements suggest significant variations in cell metabolism, depending on the cellular spatial location in the bioreactor. Interestingly, the measured glucose variation was lower (about 16 to 36 mg/L) when glucose was fed near the bottom impeller.

The effects of glucose gradients in a 12 m$^3$ bioreactor on the production of recombinant protein by *E. coli* was also reported by Bylund et al. (1999). These authors measured spatial concentrations in the constant glucose feed (550 g/L) phase using the technique described by Larsson et al. (1996) except that the measurement location was altered such that samples were withdrawn 3 cm below the addition point and at 180° from the feed point. More significant variations in glucose levels were measured than those reported by Larsson et al. (1996); in the 180° position, the variation in glucose levels was in the range 70 to 4500 mg/L (Figure 18-5A). Of course, if it were possible to measure instantaneously from an infinitesimally small volume, the upper range in measured glucose levels would approach that in the feed solution. Such measurements also raise questions about feedback control strategies in large scale bioreactors based on point measurements using online probes.

A particularly interesting study of the physiological response of *E. coli* to glucose gradients in large scale bioreactors was reported by Enfors et al. (2001). Fed-batch cultivations were conducted at a 22 m³ scale (30 m³ bioreactor) equipped with either four radial flow Rushton turbines or four Scaba impellers in a radial/axial flow combination (see Section 18-4 for details of such impellers). Measurements of glucose concentration were made at three locations, including the glucose addition zone (top position), mid-liquid height, and in the plane of bottom impeller. The values of glucose measured were 57, 34, and 27 mg/L at the top, middle, and bottom positions, respectively. These differences were largely absent when the Scaba impellers were used. Spatial variations in glucose concentration with increasing distance from the feed point has been demonstrated in previous large scale studies (Larsson et al., 1996; Bylund

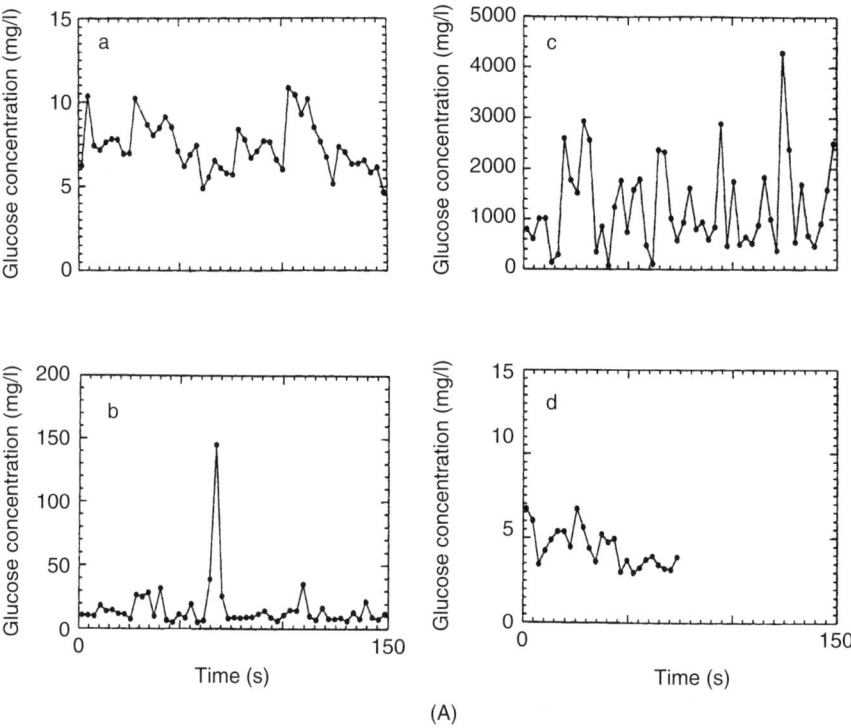

(A)

**Figure 18-5** (A) Glucose sampling with a frequency of 3 s in a 12 m³ *E. coli* fermentation. Glucose was fed to the top surface of the liquid at a concentration of 552 g/L and sampled at different locations: (*a*) 3 cm below feed point; (*b*) 30 cm below feed point; (*c*) 180° from feed point; (*d*) at the bottom of the bioreactor. (From Larsson et al., 1996.) (*Continued*)

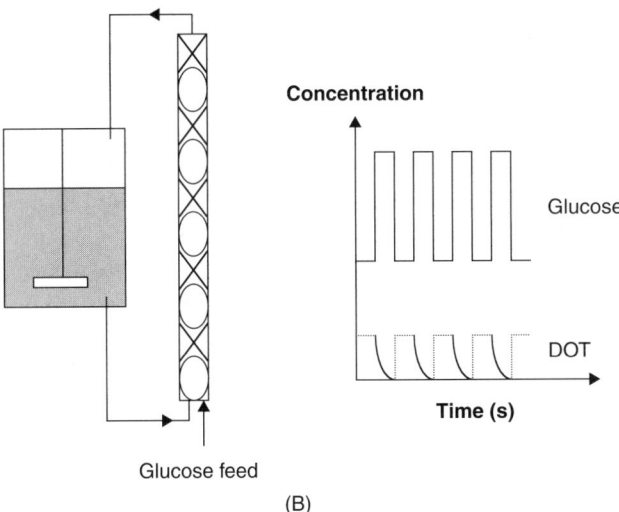

**Figure 18-5** (B) Scale-down model (STR + PFR) to simulate the effects of a high glucose concentration feed zone coupled to low dissolved oxygen (DOT) effects in large scale bioreactors. The scale-down model consists of an aerated STR and a PFR in series. The nonaerated PFR contains static mixers to promote plug flow and high oxygen mass transfer rates, DOT probe at the outlet, and a number of sampling points along its length. With glucose fed to the PFR it is possible to measure the physiological response of cultures subjected to high glucose levels at the addition point with low levels of DOT and low and high levels of glucose and DOT, respectively, in the bulk (represented by the STR). (From George et al., 1993.)

et al., 1998). The mixing time was measured between 1 and 2 min. Thus, cells circulating in the bioreactor encountered spatial variations in glucose concentrations. The consequence of this was formate accumulation, highlighting the effects of localized oxygen limitation in zones of high glucose concentrations. The reduced biomass yield compared to small scale cultivations (Bylund et al., 1998) was attributed to the repeated production/assimilation of acetate from overflow metabolism of glucose.

A two-compartment scale-down model, stirred tank reactor (STR) in combination with a plug flow reactor (PFR), was also used to model the presence of glucose gradients (Figure 18-5B). The PFR was fitted with static mixers to provide high oxygen transfer rates ($k_L a = 600$ to $1000 \text{ h}^{-1}$) as described by George et al. (1993). In this manner the cells were repeatedly exposed to high glucose levels with decreasing oxygen availability with increasing cell density. The mean residence times in the PFR and STR were 56 s and 9 min, respectively. Glucose was fed at the entrance of the PFR and samples were withdrawn at different locations in the PFR after 14, 28, 32, and 56 s. Acetate, lactate, and formate appeared in the oxygen-insufficient PFR whereas only formate accumulated in the oxygen-sufficient STR, indicating that acetate and lactate were readily assimilated there. Measurement of the mRNA levels of stress-induced genes showed increased levels with increasing residence time in the PFR, whereas very low levels were measured in the STR. Flow cytometric analyses of cells for viability and membrane potential from the scale-down model and the large scale bioreactor were in good agreement and significantly higher than when conducted in a STR without gradients in glucose levels. These results suggest that glucose heterogeneity in the large scale bioreactor was actually beneficial to the cells with respect to viability, although the biomass yield was lower.

Several other publications have described the use of two-compartment systems to investigate the influence of fluctuations in glucose concentration on a fed-batch baker's yeast production (Sweere et al., 1986, 1988c; Namdev et al., 1991; George et al., 1998) and *E. coli* fermentations (Hewitt et al., 1999; Bylund et al., 2000). The reader is referred to these studies for further details. In general, using baker's yeast, the studies showed that increasing circulation times caused a reduction in the biomass production and an increase in the product formation, especially ethanol. They concluded that the fluctuations in glucose concentrations at relatively rapid circulations was likely to have a distinct influence on the fed-batch production of baker's yeast.

### 18-2.7.2 Scale-down Simulations of Dissolved Oxygen Inhomogeneities at the Large Scale.
One of the most important aspects in the scale-up of any aerobic biochemical process is to maintain an adequate supply of oxygen to the microorganisms. As a result, many investigations have been devoted to the optimization of microbial growth and product formation with respect to dissolved oxygen tension concentrations. Several scale-down configurations (Figure 18-6) have been used, including well-mixed single-compartment (STR) and two-compartment systems (STR + STR and STR + PFR). In this section we review the studies conducted in scale-down models to simulate DOT gradients at large scales of operation.

*Well-Mixed Single-Compartment Model (STR).* Using such models, the dissolved oxygen can be fluctuated with a fixed frequency in a square- or sine-wave fashion by either varying the inlet gas composition or the fermenter head pressure to alter the liquid-phase dissolved oxygen concentration. A number of studies have implemented such strategies (Vardar and Lilly, 1982; Sokolov et al., 1983), and the reader is referred to these papers for further details. An easier setup to simulate dissolved oxygen gradients using a single well-mixed compartment would be to use timed pulsing of mixtures of nitrogen and oxygen (Oosterhuis, 1984; Sweere et al., 1988a,b; Namdev et al., 1991).

Namdev et al. (1991) simulated the fluctuating dissolved oxygen concentrations in large bioreactors in a 2 L vessel using a Monte Carlo approach. A lognormal distribution, described by a mean circulation of 20 s and a standard deviation of 8.9 s, was discretized into $n$ elements of equal probability, each with a corresponding circulation time. A uniform random number was then used to select a circulation time. Therefore, a random circulation time was selected

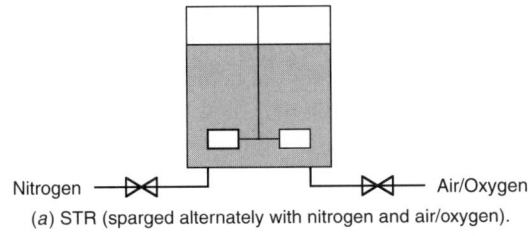
(a) STR (sparged alternately with nitrogen and air/oxygen).

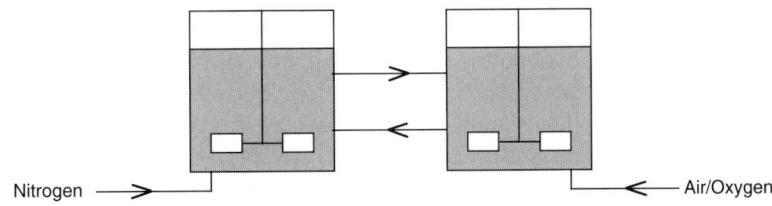
(b) STR + STR (one sparged with nitrogen and the other with air/oxygen to maintain low and high dissolved oxygen concentrations, respectively).

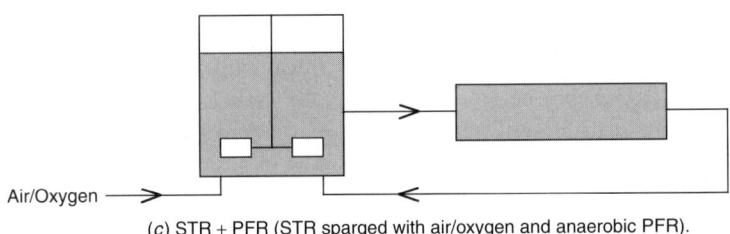
(c) STR + PFR (STR sparged with air/oxygen and anaerobic PFR).

**Figure 18-6** Different scale-down configurations to simulate dissolved oxygen gradients in large scale bioreactors.

within the bounds of the CTD. This method was used to control the aerated and nonaerated cycles to mimic the circulation time distribution of a production scale bioreactor. The results of the simulation were not sensitive to the order of the cycles since the circulation times used were much shorter than the growth rate of the cells. Due to the randomness of the circulation time selected, no single circulation time would be dominant, unlike the case for periodic oscillations. Using a culture of *S. cerevisiae* in a complex medium, they found that the biomass production decreased by 20% compared to experiments with continuous aeration. A 30% reduction in biomass level was found when the culture was subjected to periodic fluctuations, with 5 s of aeration and 15 s without aeration.

Yegneswaran et al. (1991) used a Monte Carlo method and CTD similar to those of Namdev et al. (1991) to investigate the effects of dissolved oxygen on a culture of antibiotic producing *Streptomyces clavuligerus*. They found that the yield of cephamycin C was suppressed by almost 44% due to the Monte Carlo simulation as compared to constant period cycling. One limitation in the studies of Namdev et al. (1991) and Yegneswaran et al. (1991) arises from the use of conditions producing relatively small $k_L a$ values to alter the liquid-phase dissolved oxygen level by varying the composition of oxygen in the gas phase. Therefore, only the effects of relatively slow fluctuations can be studied, since the lowest value of the fluctuating cycle time cannot be less than the time constant for oxygen transfer ($1/k_L a$) from the gas to the liquid phase. Another limitation of the methods of fluctuating head pressure or gas composition as a means of varying the liquid-phase dissolved oxygen concentration can be identified; only a mean circulation time is simulated and no information regarding the combined effects of poor oxygen transfer and liquid phase mixing is obtained. In practice, the microorganisms in production scale vessels will be subjected to a mean circulation time as well as a circulation time distribution. For more accurate scale-down modeling, the following should also be considered: (1) realistic values of the mean circulation time (typically, <60 s and preferably, <30 s), (2) the circulation time distribution in the model system resembles that measured on the production scale, and (c) the relative compartment volumes in the model should be correlated to those measured or calculated at the large scale.

*Two-Compartment Systems: Two Well-Mixed Stirred Tanks Model (STR + STR).* A two-compartment model, consisting of two well-mixed tanks (maximum working volumes of 0.6 and 1.6 L) with an exchange flow was presented by Oosterhuis et al. (1983, 1985) to model the scaling down of a circulation time distribution. This was based on experimental determinations of the local dissolved oxygen concentration in a production scale fermenter (Figure 18-2). From those results it could be concluded that it was possible to consider the reactor to consist of two parts: (1) high dissolved oxygen levels in the vicinity of the impeller, where maximum product formation can occur, and (2) dissolved oxygen concentrations close to the saturation constant of the microbial kinetics for oxygen in the other parts of the vessel. The exchange flow between the compartments was determined from radio pill flow follower experiments, which were also used to measure the CTD at the large scale. Close agreement was found between the experimental CTD measured in the production scale reactor and the CTD in the two-stirred-tanks system, as suggested by Levenspiel (1972). The mean circulation time and therefore the circulation time distribution could be varied by changing the liquid volumes in each vessel and/or by changing the circulation rate between the compartments. Air was sparged in to the smaller vessel, with agitator speed-controlled dissolved oxygen level, to simulate the small well-mixed highly oxygenated zone around the impeller. Nitrogen was sparged into the larger vessel to mimic the relatively poorly aerated areas away from the impeller. Such a model could therefore be used to model both the effects of gradients in dissolved oxygen and the bulk flow on microorganisms.

*Two-Compartment Model: Well-Mixed and Plug Flow Reactors Combination (STR + PFR).* Studies using a two-compartment model, consisting of a well-mixed stirred tank and a plug flow reactor to investigate the effects of glucose and dissolved oxygen gradients on cellular metabolism have been described

(Purgstaller and Moser, 1987; Larsson and Enfors, 1985, 1988; Amanullah et al., 1993a, 1993b, 2001).

It has been shown that various small scale models consisting of idealized reactor types can be used to simulate large scale fermentation processes, with respect to dissolved oxygen inhomogeneities. The reaction kinetic expressions, material balances on substrates, and products have to be formulated and solved in the context of the combined model network. The choice of the model configuration depends on (1) the system that has to be simulated, (2) knowledge of the hydrodynamics of the system, and (3) the equipment available and financial resources.

## 18-2.8 Required Characteristics of a Model Culture for Scale-down Studies

Two requirements can be stated for a reacting system to exhibit mixing sensitivity. First, there must be an inhomogeneous reactant distribution, such that the distribution is affected by the mixing intensity. Second, for the response to different mixing intensities to be appreciable, it is necessary for the characteristic reaction time to be less than the total time spent in regions where inhomogeneities exist. In other words, the Damkohler number, defined as the ratio of the system residence time to the characteristic reaction time, must be greater than unity (Fowler and Dunlop, 1989) (see also Chapter 13). The sensitivity of microorganisms to substrate concentrations can be used as a tool for studying mixing and transport effects in fermenters. However, sensitive instrumentation is required to detect instantaneous variations such as a membrane probe coupled to an online mass spectrometer to measure low boiling point volatiles in the liquid phase (Griot et al., 1987). The rapid secretion of low-molecular-weight products in response to substrate variations would be a desirable characteristic. Rapid growth and response would also be desirable.

*S. cerevisiae* possesses some of the desirable characteristics. It grows rapidly under aerobic conditions and responds rapidly to variations in glucose and oxygen (Furukawa et al., 1983). Einsele (1978) has shown that the response time of this organism to glucose pulses is approximately 4 s. Ethanol is produced under conditions of high glucose concentration and is independent of the oxygen concentration. However, it is also produced under low concentrations of glucose and oxygen. The interactive effect of glucose and oxygen make the interpretation quite difficult. The response is not reversible, and adaptation effects with respect to oxygen have been observed (Furukawa et al., 1983). Moes et al. (1985) reported the use of a *B. subtilis* culture with an oxygen-sensitive product distribution to characterize mixing and mass transfer in bioreactors. The claimed desirable characteristics of this culture included (1) extreme sensitivity with respect to oxygen supply and changes in reactor operating variables such as impeller speed and aeration rate, and (2) rapid and reversible response, allowing a number of investigations to be carried out within a single batch fermentation. However, in practice, the choice of the microorganism is dictated by the scale-up process of interest.

## 18-2.9 Use of *Bacillus subtilis* as an Oxygen- and pH-Sensitive Model Culture

In this *B. subtilis* strain (AJ 1992 from Ajinomoto, Japan) glucose is utilized to form pyruvate, which in the presence of excess oxygen is completely oxidized to carbon dioxide and water via the trichloroacetic and (TCA) and respiratory cycles. At low oxygen levels, pyruvate is used to produce acetoin, which in turn can be reduced by nicotinamide adenine dinucleotide (NADH) to form butanediol. Lactate is also formed from pyruvate at low levels of oxygen. Acetoin is produced primarily at oxygen levels above 150 parts per billion (ppb) and butanediol below 80 ppb. Moes et al. (1985) demonstrated the extreme sensitivity of the culture in the range 80 to 90 ppb. Using a complex medium, batch fermentations could be completed within 8 h with a biomass concentration between 2 to 3 g/L, using approximately 11 g/L of glucose. For oxygen concentrations above 100 ppb, typical values of acetoin and butanediol concentrations were in the range 0 to 3 and 0 to 0.5 g/L, respectively. At low oxygen concentrations (50 ppb), values of acetoin and butanediol were in the range 0 to 5 and 0 to 3 g/L, respectively. In batch fermentations, switching from one oxygen level to another caused one already accumulated product to be converted to the other in a reversible manner. The high rates of change of 0.5 to 1.0 g/L per hour enabled detection within 10 min.

Griot et al. (1986, 1987) showed that this strain changed the selectivity of excretion of metabolites in response to variations in dissolved oxygen levels in less than 1 s. The response time of the culture to a change in gas phase oxygen concentration was shown using mass spectrometry to be a total of 7.7 s. After subtracting the response of the gas–liquid transfer (3.2 s) and the acetoin detection (4 s), the response time of the culture was estimated to be of the order of 0.5 s. In addition to its sensitivity to dissolved oxygen levels, the production of the metabolites acetoin (Ac) and 2,3-butanediol (Bu) is sensitive to pH values between 6.5 and 7.2, with the total metabolite (Ac + Bu) concentration 3.5 times greater at pH 6.5 than at 7.2. Also, the acetic acid concentration was 0.56 g/L at pH 6.5, whereas its value diminished to zero at pH 7.2 (Amanullah et al., 2001b). In conclusion, the *B. subtilis* culture is a very useful tool to study the effects of mixing due to its unusual sensitivity to oxygen supply and pH. However, use of a well-defined medium is essential.

In the next two sections of this chapter we describe by way of examples, detailed practical designs of scale-down bioreactors for investigating the effects of dissolved oxygen and pH gradients and the results obtained. *B. subtilis* is employed as the model culture given its sensitivity to dissolved oxygen and pH.

## 18-2.10 Experimental Simulations of Dissolved Oxygen Gradients Using *Bacillus subtilis*

Given that significant oxygen gradients have been identified even at a 25 m$^3$ (19 m$^3$ working volume) scale (Oosterhuis et al., 1985), one approach has been to use two interconnected stirred tanks (STR + STR), one (well oxygenated) to represent the active (well mixed, oxygenated) zone and the other (with low oxygen levels), the quiescent zone. Both STR + STR and STR + PFR scale-down models are described here using an oxygen-sensitive culture of *B. subtilis* (see Section 18-2.9) to identify relationships between mixing and biological performance parameters and to compare the performance of each scale-down configuration. For successful scale-down, both mean and distribution of circulation times (CTDs) at the large scale have to be replicated within the scale-down configuration (Amanullah et al., 1993a; Amanullah, 1994). The CTD in the two stirred tanks scale-down configuration can be described by the following equation given by Levenspiel (1972) for a tanks-in-series model:

$$t_c C = e^{-t/t_c} \sum_{m=1}^{\infty} \frac{(t/t_c)^{mN-1}}{(mN-1)!} \qquad (18\text{-}54)$$

where C is the tracer concentration at time t, $t_c$ the mean circulation time, m the number of circulations, and N the number of tanks. For a single circulation or one complete recirculation (m = 1) in the STR + STR model, eq. (18-54) reduces to

$$C = \frac{e^{-t/t_c}}{t_c^2} t \qquad (18\text{-}55)$$

Oosterhuis et al. (1985) reported a mean circulation time of approximately 12 s in an unaerated 19 m³ fermenter using a radio pill flow follower. The experimental cumulative CTD associated with this mean is shown in Figure 18-7. This is the distribution of the residence time of the liquid outside the impeller region. If it is

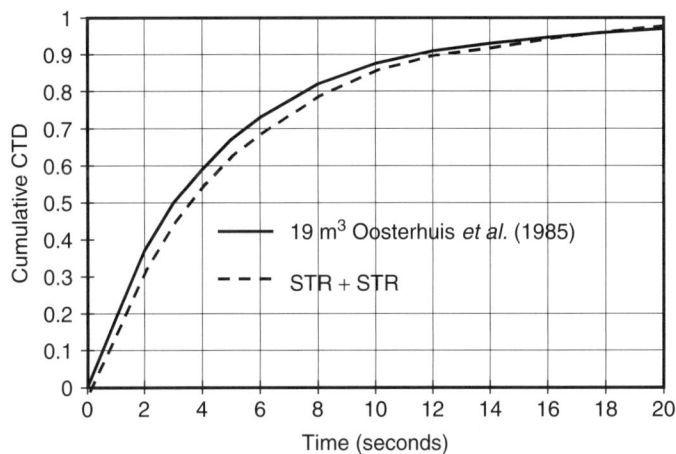

**Figure 18-7** Comparison of the experimental (Oosterhuis et al., 1985) and calculated cumulative CTDs for a single circulation in the STR + STR scale-down model (Amanullah et al., 1993a; Amanullah, 1994) and that at a scale of 19 m³. The mean circulation time in both cases is 12 s.

assumed that the residence time in the impeller region is relatively small and that the remainder of the vessel behaves like a well-mixed tank with a much longer residence time, eq. (18-55) can be used to describe the CTD. Figure 18-7 also shows the cumulative CTD calculated using eq. (18-55), with a mean circulation time of 12 s, for a single circulation in the STR + STR model. Comparison of the experimental and calculated cumulative CTDs shows that eq. (18-55) is fairly accurate in predicting the experimental CTD, and this is especially the case for the tail of the distributions. Therefore, the STR + STR model is well suited to study the effects of cell residence in the poorly oxygenated bulk regions, alternated by residence in the well-aerated impeller region.

Detailed descriptions of the scale-down models and their operation are provided in Amanullah et al. (1993a) and Amanullah (1994). Figure 18-8a depicts the scale-down STR + STR configuration in the laboratory. Two interconnected stirred vessels 6 L (2.8 or 4.8 L working volumes) and 2 L (1.2 L working volume) in capacity were used. Two variable speed pumps with a maximum estimated pumping capacity of 16 L/min, enabling a minimum mean circulation time of 15 s, were employed to provide the flow between the vessels. Dissolved oxygen tension (DOT) was measured and controlled at 5% ± 0.5% of air saturation in the smaller vessel by means of gas blending, keeping the total flow of gas constant at 1 vvm. Nitrogen was sparged at a rate of 0.5 vvm into the larger vessel to maintain a DOT level close to zero.

The mean circulation time ($t_c$) was varied in the range 15 to 300 s at $V_a/V_q$ of 0.25 and 0.43, where $V_a$ and $V_q$, are volumes of the active (small bioreactor with DOT at 5%) and quiescent (large bioreactor with DOT close to zero) zones, respectively. In the STR + PFR model (Figure 18-8b), the dissolved oxygen was maintained at 10% by gas blending in the STR. The plug flow volume was either 2 or 4 L, resulting in $V_a/V_q = 0.5$ and 0.25, respectively. The results from both the STR + STR and STR + PFR models were compared against a control batch fermentation with DOT = 10%. This control represented the ideal mixing situation in terms of oxygen supply and corresponded in effect to a zero-mean circulation time. The control experiment yielded maximum values of both final biomass concentration (6.44 g/L) and specific growth, rate ($\mu$) at 0.31 h$^{-1}$. Acetoin (Ac) production rate was also a maximum at 0.36 g/L per hour, while the 2,3-butanediol (Bu) production rate, as expected, was zero. The biological performance of the culture in the experiments conducted for 0 < $t_c$ < 300 s at $V_a/V_q = 0.25$ and 0.43 have been expressed in Figure 18-9a–c in terms of both maximum biomass ($x_{max}$) and metabolite concentrations, specific growth, and the Ac/Bu concentration ratio, which is dependent on the supply of dissolved oxygen. The values of $x_{max}$ and $\mu$ decreased with increasing mean circulation times, while the total metabolite concentration increased (Figure 18-9a and 18-9b). Although the percentage reduction in $x_{max}$ relative to the control was similar (10 to 15% for $t_c = 30$ s) for both ratios, the percentage reduction in $\mu$ for the 0.25 ratio was significantly higher than for the 0.43 ratio. This implies that although the potential for biomass formation remained unaffected at the different $V_a/V_q$ ratios, the fermentation time required to attain similar biomass concentrations was longer

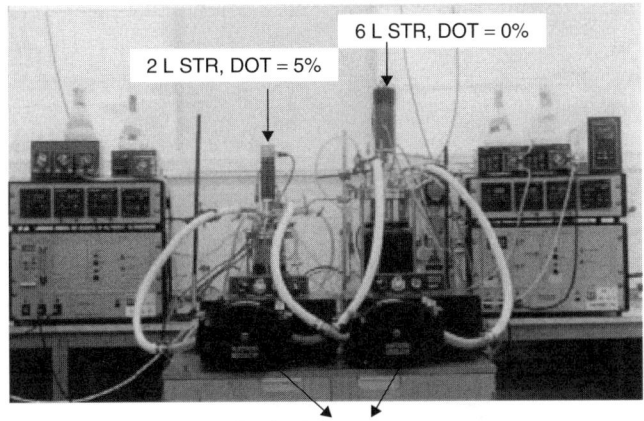

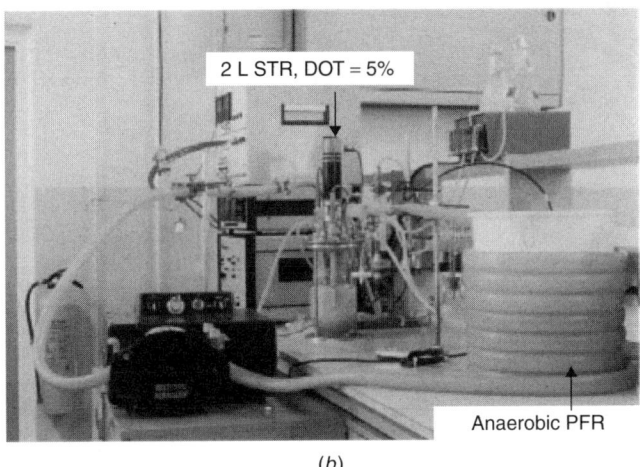

**Figure 18-8** (*a*) Laboratory scale-down model consisting of two interconnected stirred tanks (STR + STR) at different DOT levels to simulate dissolved oxygen levels at large scales of operation. The mean circulation time can be altered by manipulating the exchange flow rate via the recirculation pumps. (From Amanullah, 1994.) (*b*) Laboratory scale-down model consisting of an interconnected stirred tank and unaerated plug flow reactor (STR + PFR) to simulate dissolved oxygen levels at large scales of operation. The DOT level in the STR was controlled at 5% of air saturation. Different residence times in PFR could be imposed by altering the speed of the recirculation pump. (From Amanullah, 1994.)

at the 0.25 ratio. Figure 18-9*c* shows the Ac/Bu ratio as a function of the mean circulation time at $V_a/V_q = 0.25$ and 0.43 at a biomass concentration of 4 g/L. In both cases, the Ac/Bu ratio decreased sharply in the range $0 < t_c < 120$ s. In each case (increasing $t_c$ and lower $V_a/V_q$), the biological response of the culture

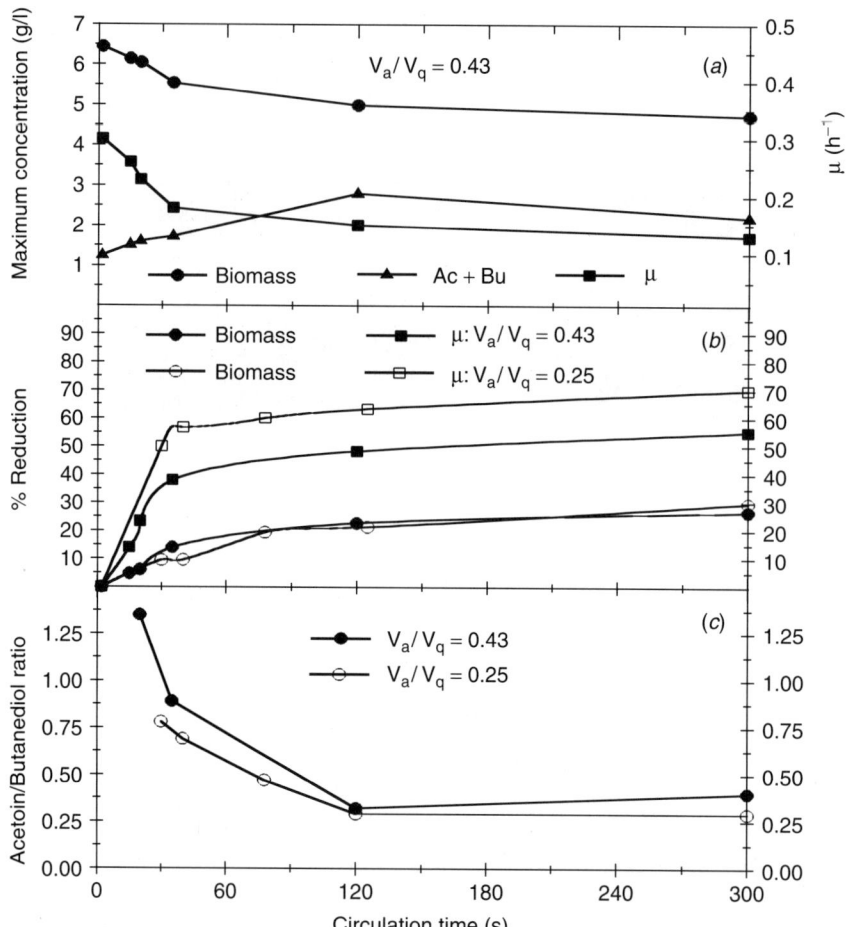

**Figure 18-9** (a) Maximum biomass, metabolites concentration, and specific growth rates as a function of mean circulation time. (b) Maximum biomass and specific growth rates expressed as percentage changes relative to performance under ideal conditions of oxygen supply for $V_a/V_q = 0.43$ and $0.25$. (c) Metabolite concentration ratios at a biomass concentration of 4 g/L at $V_a/V_q = 0.43$ and $0.25$ as a function of mean time circulation time. (From Amanullah, 1994.)

can be explained by an increase in the percentage of cells subjected to oxygen depletion with increasing mean circulation times (see Figure 18-3b).

The percentage reduction in $x_{max}$ for $V_a/V_p = 0.25$ for the different scale-down models is shown in Figure 18-10. For similar mean circulation times the percent reduction in maximum biomass concentration for the STR + STR model was significantly lower than for the STR + PFR model. One reason for this may due to the fact that in the STR + STR model there was always some oxygen available to the microorganisms in the quiescent zone due to the entrainment of

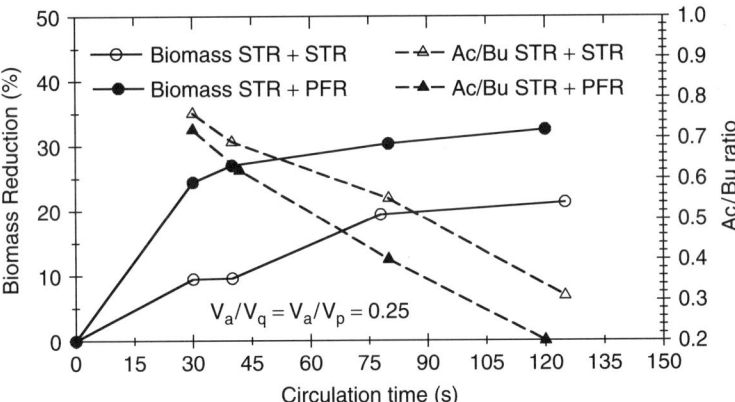

**Figure 18-10** Effects of mean circulation time on the percent reduction in maximum biomass concentration and product ratio, relative to ideal conditions of oxygen supply, for different scale-down models. (From Amanullah, 1994.)

air in the exchange flow. In contrast, the cells in the STR + PFR system were truly subjected to anaerobic conditions once the oxygen was consumed in the PFR. The Ac/Bu ratio at a specific biomass concentration of 4 g/L is shown in Figure 18-10 as a function of mean circulation time for the two-model configurations for $V_a/V_p = 0.25$. The Ac/Bu ratio for the STR + PFR model was $\approx 27\%$ lower than for the STR + STR model for $t_c \geq 80$ s.

The results presented also highlight the importance of the choice of the scale-down configuration when studying the impact of large scale dissolved oxygen inhomogeneities on microorganisms. The difference in biological performance between the two configurations can be explained in terms of the flow characteristics and oxygen availability in each system. The STR + STR model is more appropriate for use when the region outside the impeller zone, with a low dissolved oxygen concentration, behaves like a well-mixed tank characterized by mean circulation times with relatively large standard deviations. The STR + PFR model may be more suitable for situations where the motion of the fluid outside the impeller zone nearly resembles plug flow and may be characterized by relatively smaller standard deviations for a given mean circulation time. The results also suggest the fact that for successful scale-up or scale-down, apart from the CTD, both mean circulation times and relative compartment volumes have to be replicated at the different scales. In addition, use of the relative volume of the compartments or ratio of the residence times in each compartment is not suitable as a scale-up or scale-down criterion.

Significant changes in biological performance are likely to occur upon scale-up of this fermentation, due to the circulation of cells through oxygen-deprived regions. The performance can be enhanced if the microorganisms are circulated through the impeller region at a high enough frequency (small mean circulation times) such that the concentration of oxygen is kept above the critical value along all circulation paths or if the relative size of the well-mixed region is increased. However, increases in the flow capacity have to be balanced against cost of the higher power input. Although unicellular bacteria are generally thought to be insensitive to hydrodynamic stress in the range employed in bioreactors (Hewitt et al., 1998), the influence of increased agitation intensities may be important in other biological systems, especially mycelial fermentations (Jüsten et al., 1996; Amanullah et al., 2000), where increased hyphal fragmentation can occur.

## 18-2.11 Experimental Simulations of pH Gradients Using *Bacillus subtilis*

Studies simulating the effects of pH gradients on microbial growth and product formation are scarce, yet it is a fundamental parameter in the regulation of cellular metabolism, particularly in processes characterized by multiple end products. The first laboratory scale two-compartment system used to investigate the effects of pH fluctuations consequent of large scales of operation on microorganisms was reported by Amanullah et al. (2001b). *B. subtilis* was used as a model culture since in addition to its sensitivity to dissolved oxygen levels, production of the metabolites acetoin and 2,3-butanediol is sensitive to pH values between 6.5 and 7.2 (see Section 18-2.9).

The basis of industrial pH control (Figure 18-11a) is the point measurement by a pH probe (typically, near an impeller) and the subsequent on–off pulse injection of acid or base typically near the top surface in response to a deviation from the set point. Thus, the local pH values near the addition point may deviate from the bulk pH. In the study reported by Amanullah et al. (2001b), it was assumed that acid or base in production scale vessels was added at the broth surface, in response to a point measurement of a pH probe in the vicinity of the impeller region. This large scale situation can then be simulated experimentally by a stirred tank reactor + plug flow reactor (STR + PFR) configuration shown in Figure 18-11b. The STR represents the well-mixed impeller region, while the residence time in the PFR mimics the time spent by cells outside this region in the higher-pH zone. Figure 18-12 shows by way of example the transient pH probe response to a pulse injection of base to a 0.01 $M$ KH$_2$PO$_4$ buffer solution in a 4.4 m$^3$ bioreactor under unaerated conditions (Singh et al., 1986). Two pH probes were used, one near each of the two impellers used. The response is typical of that of a tracer injection at the liquid surface. Although mixing times give information about the rate of mixing of acid and/or base, it is not sufficient to characterize the mixing process because it does not provide any information on fluid segregation in the vessel for which the CTD is necessary to describe the microenvironment experienced by cells (Oosterhuis and Kossen, 1984; Amanullah et al., 1993a; Amanullah, 1994).

The experimental design of the scale-down experiments reported by Amanullah et al. (2001b) was based on the concepts proposed by Namdev et al. (1992) to evaluate the effects of glucose feed zone in fed-batch fermentations of *S. cerevisiae*. They represented the mixing process in a large scale fed-batch fermentation by a three-zone mixing model consisting of feed, bulk, and impeller

# SCALE-UP/SCALE-DOWN OF FERMENTATION PROCESSES 1111

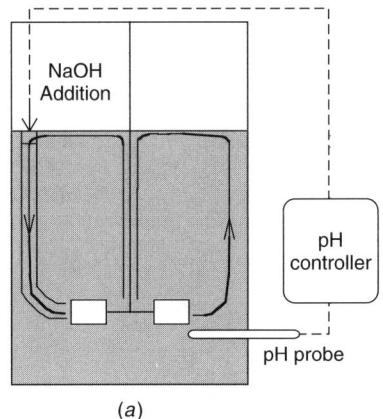

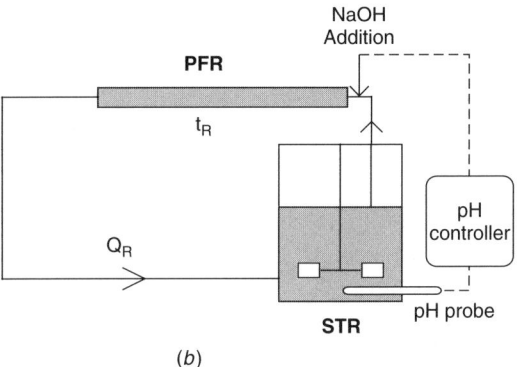

**Figure 18-11** (*a*) Typical pH control in a production scale bioreactor and (*b*) scaled-down model to simulate pH spatial fluctuations found in production scale bioreactors.

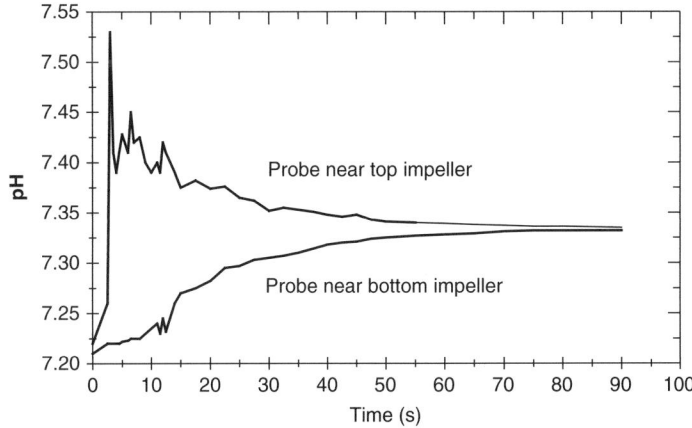

**Figure 18-12** Unaerated pH probe response in a 4.5 m$^3$ fermenter using a buffer solution. (From Singh et al., 1986.)

zones. Following the lognormal CTD suggested by Bajpai and Reuss (1982) for a 100 m$^3$ fermenter, Namdev et al. (1992) used a lognormal CTD based on the "network of zones" model of Mann et al. (1981) with a mean of 20 s and a standard deviation of 8.9 s to simulate glucose feeding in bioreactors of such scales. A recycle flow model was used which consisted of a stirred tank and a plug flow loop with a recirculation rate. The CTD of the recycle flow model is given by (Levenspiel, 1972)

$$F(t) = F \exp[-F(t - t_R)] \qquad (18\text{-}56)$$

where

$$F = \frac{Q_R}{V_{STR}} \qquad (18\text{-}57)$$

$$t_R = \frac{V_{PFR}}{Q_R} \qquad (18\text{-}58)$$

The recirculation rate, $Q_R$, determines the frequency of cells entering the PFR, F, from a given scale-down STR, $V_{STR}$ and $V_{PFR}$ are volumes of the STR and PFR, respectively, and $t_R$ is the residence time in the PFR. The residence time of a fluid element in the PFR is considered to be analogous to the time that it would spent in the acid or alkali addition zone in a production scale fermenter. The STR represents the well-mixed impeller region.

Since an estimate of the volume of the addition zone was not available, the volume of the loop was estimated as 5% of the total fermenter volume [similar to the volume fraction used by Namdev et al. (1992) and George et al. (1993)] at 50 mL. Using this PFR volume, the recirculation rate was varied to obtain different residence times in the PFR. For simulating pH fluctuations, the pH in the STR was controlled by the addition of 5 $M$ NaOH into the PFR via a mixing bulb which incorporated the incoming broth from the STR with the NaOH at the entrance of the PFR. The mixing bulb was essential to segregate the addition zone in compliance with the three-zone mixing model (zone 1: base addition (or feed) point; zone 2: circulation volume affected by the high pH; and zone 3: well-mixed impeller region). Since the culture was sensitive to dissolved oxygen, for each experiment with pH control by adding base into the PFR, equivalent experiments were conducted with pH control by addition of base into the STR, thus ensuring that any dissolved oxygen effects were common to both types of experiments. Residence times in the PFR was varied from 30 to 240 s. The results showed that without pH fluctuations in the PFR, there were no differences in performance between the batch STR and STR + PFR due to variations in dissolved oxygen or indeed any other parameters.

Since identical conditions (e.g., dissolved oxygen) were used for the STR + PFR controls with constant pH and STR + PFR scale-down model apart from the imposition of pH fluctuations in the latter case, the increase in acetic acid concentration, with a corresponding decrease of up to 27% in Ac + Bu concentration at a residence time of 240 s, must have been due to the increased exposure of cells to alkaline conditions at higher residence times in the PFR. Exposure of cells to the higher-pH environment did not affect the growth of the culture. The results from the scale-down studies clearly showed that cellular metabolism of this culture was affected by pH fluctuations at PFR residence times of 60 s or higher. It was proposed that these changes in metabolism may be linked to both the sensitivity of the acetoin and 2,3-butanediol forming enzymes to pH and to the inducing effects of dissociated acetate on the acetolactate synthase enzyme (Amanullah et al., 2001b).

It is well recognized that the optimum location for the addition and subsequent dispersion of any inlet feed, such as acid or base for pH control or glucose in fed-batch fermentations, is in the impeller region due to the prevailing high turbulence. (See Chapter 13 for further discussion of local mixing effects.) This would also reduce any overshoot from the set point, since the pH probe is generally in the close vicinity of this region. It is recommended to locate any pH feed near an impeller, perhaps employing several feeding points in order to eliminate or reduce pH fluctuations and their effect on microbial metabolism. A recent study of large scale free-suspension animal cell culture has shown such a change to be necessary and effective (Langheinrich and Nienow, 1999). Also, a comparison of the performance of large scale and scale-down experiments with respect to glucose feeding in *E. coli* fermentations for recombinant protein production also concluded that a two-feed-point glucose addition was superior to a single feed addition in preventing unwanted acetate formation at the large scale (Bylund et al., 1999).

## 18-3  POLYSACCHARIDE FERMENTATIONS

The large scale production of microbial polysaccharides exemplifies a fermentation industry with global markets representing hundreds of millions of dollars. One of the most challenging tasks in the fermentation industry today is the design of bioreactors for the production of rheologically complex polysaccharides at high concentrations of consistent high quality. Quality in this context refers to the molecular weight of the biopolymer, which in turn determines its viscosifying properties. These include commercially important polysaccharides such as xanthan, gellan gum, pullulan, alginate, curdlan, and glucan.

Traditionally, strain selection and more recently, genetic engineering have been used to potentially dramatically increase the maximum productivity and product concentration achievable in such processes. However, whether this potential is fully realized also depends on bulk mixing and oxygen mass transfer, which in turn are governed primarily by vessel design, impeller type, and fluid properties (rheological and chemical composition). These generic problems inherent in viscous fermentations, including those used to manufacture gellan gum (Dreveton et al., 1996), pullulan (Wecker and Onken, 1991), curdlan (Lee et al., 1999), and alginate (Peña et al., 2000) have been investigated and reported here using xanthan gum as a model polysaccharide fermentation system. For details of the other polysaccharide fermentations, the reader is referred to the aforementioned publications. Xanthan gum is an extracellular polysaccharide produced by *Xanthomonas campestris* and is commercially the most important bacterial polysaccharide. It has widespread applications as a viscosity-enhancing agent and stabilizer in the food, pharmaceutical, and petrochemical industries (Jeanes et al., 1976; Norton et al., 1981). Since the cost of downstream processing determines whether or not the manufacture of the gum is commercially feasible, a high product concentration (>25 g/L) is essential (Pace and Righelato, 1980; Galindo, 1994).

Ever since xanthan gum was reported by Rogovin et al. (1961), considerable research has been devoted to addressing the problems of poor bulk mixing and low oxygen transfer rates in xanthan gum fermentations (Moraine and Rogovin, 1973; Nienow, 1984; Funahashi et al., 1987a,b,c, 1988a,b; Nienow and Elson, 1988; Peters et al., 1989a,b, 1992; Galindo and Nienow, 1992, 1993; Herbst et al., 1992; Flores et al., 1994; Zhao et al., 1991, 1994). It is generally agreed in the literature that the process bottleneck in highly productive xanthan fermentations is related to these two parameters. However, the interpretation of experimental results in relation to these problems is difficult, due to the inability to separate the variable of oxygen transfer from that of vessel inhomogeneity. Accumulation of the extracellular gum also induces rheological complexities such that zones of significant motion (called a *cavern*) around impellers are formed (Figure 18-13), with essentially stagnant regions elsewhere (Nienow and Elson, 1988; Amanullah et al., 1998a,b). Cavern size is governed by the properties of the fluid, power input, and agitator design and is regarded as one of the limiting factors in the fermentation process. Thus, homogeneity of the broth is important to maintain optimal levels of dissolved oxygen, temperature, and pH and to prevent gradients in these parameters. In addition, oxygen transfer becomes increasingly difficult in these highly viscous broths.

### 18-3.1 Rheological Characterization of Xanthan Gum

During the course of the fermentation, the excretion of the polysaccharide increases the apparent viscosity of the broth by over three orders of magnitude. Initially, the broth is Newtonian and in turbulent flow. However, with increasing gum accumulation, it becomes increasingly viscous and non-Newtonian with Reynolds numbers in the transitional regime. The highly viscous and extremely shear thinning behavior of the gums, which also typify good-quality gums, can be characterized using the power law model with values of fluid consistency (K) and flow behavior (n) indices in the range 0 to 70 $N/m^2 \cdot s^n$ and 1 to 0.1, respectively. Typically, the flow behavior index remains constant (<0.2) at concentrations above 20 g/L. In addition, concentrations of xanthan in excess of 10 g/L generally possess a yield stress that can be obtained by fitting the Casson model (refer to Section 18-3.3) to data over the shear rate range 0.1 to 0.2 $s^{-1}$ (Amanullah et al., 1998b).

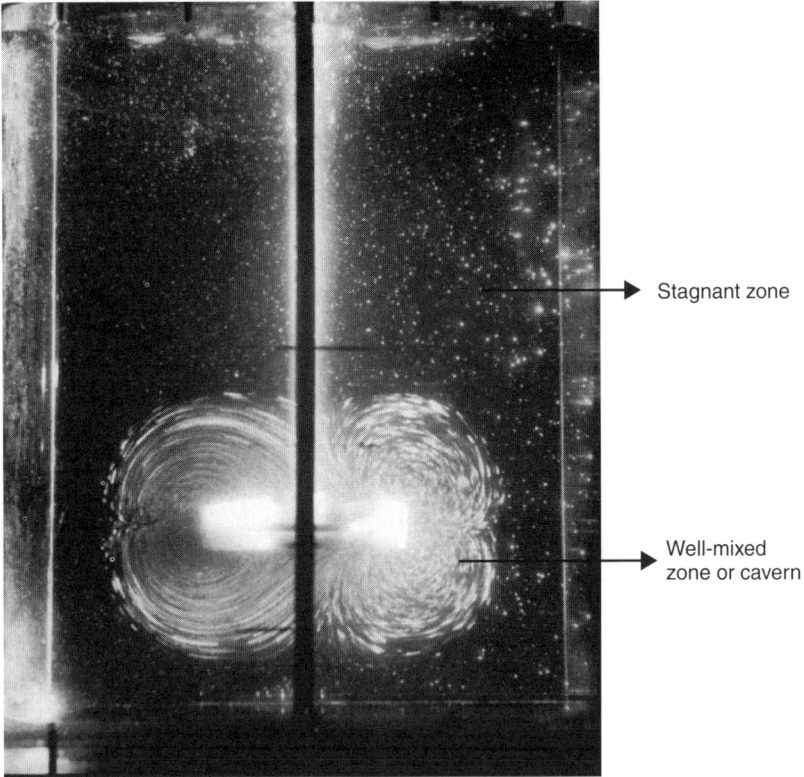

**Figure 18-13** Cavern formation in shear thinning fluids. A well-mixed region exists in regions of high shear rate, while stagnant regions form in regions of low shear rate.

## 18-3.2  Effects of Agitation Speed and Dissolved Oxygen in Xanthan Fermentations

*18-3.2.1  Agitation Speed.* Agitation speed affects both the extent of motion in xanthan fermentation broths because of their rheological complexity and the rate of oxygen transfer. The combination of these two effects causes the dissolved oxygen concentration and its spatial uniformity to change with agitator speed. Separating these complex interactions was achieved in the following way (Amanullah et al., 1998c). First, the influence of agitation speeds of 500 and 1000 rpm using three Rushton turbines (D/T = 0.5) at a 6 L scale was investigated at a constant nonlimiting dissolved oxygen concentration of 20% of air saturation using gas blending. Under these controlled dissolved oxygen conditions, the results demonstrated that the biological performance of the culture was independent of agitation speed (or shear stress) as long as broth homogeneity could be ensured. No difference in biological performance could be measured at different agitator speeds up to a xanthan concentration of 20 g/L. At higher gum concentrations, it was shown that the superior bulk mixing led to higher microbial oxygen uptake rates at 1000 rpm compared to 500 rpm, which in turn is responsible for enhanced performance at the higher speed. With the development of increasing rheological complexity lending to stagnant regions at xanthan concentrations greater than 20 g/L, it was shown that the superior bulk mixing achieved at 1000 rpm compared to 500 rpm, leading to an increased proportion of the cells in the fermenter to be metabolically active and hence higher microbial oxygen uptake rates, was responsible for the enhanced performance. Thus, for a given cavern size with equivalent dissolved oxygen levels, the specific xanthan production rate remained similar, independent of impeller speed.

The phenomenon of decreased specific xanthan production rate in the production phase is expected since it is partly growth associated. However, in many instances there is a continual decrease throughout the production phase, where the biomass concentration is approximately constant. Peters et al. (1989a) concluded that provided that oxygen limitation could be avoided (either by increased speed or by using oxygen enriched air), the specific xanthan production rate was not influenced by agitation speed or the shear stress related to it. It is important to point out that in that study, xanthan concentrations did not exceed 16 g/L, and as a result, stagnant zones did not develop in the moderately viscous broths. Their conclusions, however, were confirmed by Amanullah et al. (1998b) at much higher xanthan concentrations of 25 to 35 g/L, typical of those generally desired for commercial production of xanthan gum.

*18-3.2.2 Dissolved Oxygen.* The effects of varying dissolved oxygen were compared to a control in each case with an agitator speed of 1000 rpm to ensure full motion, but with a fixed nonlimiting dissolved oxygen of 20% of air saturation (Amanullah et al., 1998c). The specific oxygen uptake rate of the culture in the exponential phase, determined using steady-state gas analysis data, was found to be independent of dissolved oxygen above 6% of air saturation, whereas the specific growth rate of the culture was not influenced by dissolved oxygen even at levels as low as 3%, although a decrease in xanthan production rate could be measured. In the production phase, the critical oxygen level was determined to be 6 to 10%, so that below this value, both specific xanthan production rate as well as specific oxygen uptake rate decreased significantly. In addition, it was shown that the dynamic method of oxygen uptake determination was unsuitable even for moderately viscous xanthan broths, due to the presence of very small bubbles that act as an oxygen sink.

## 18-3.3  Prediction of Cavern Sizes in Xanthan Fermentations Using Yield Stress and Fluid Velocity Models

As discussed earlier, agitation of fluids with $n \leq \sim 0.3$ results in the formation of a cavern, a region of relatively rapid motion around the impeller (Wichterle and Wein, 1975) where high shear rates prevail with essentially stagnant regions elsewhere in the vessel. It is essential for the correct design and operation of bioreactors for highly viscous fermentations to determine the size of the region of motion as a function of the fluid rheology and agitation conditions. Two approaches have been employed for the estimation of cavern sizes. The first model [the Elson and Nienow (EN) model] is based on the concept of a yield stress that is defined as the minimum shear stress required to induce fluid motion and can be estimated using rheological models such as the Herschel–Bulkley or Casson equations applied at low shear rates.

The Herschel–Bulkley (HB) equation is given by

$$\tau_{HB} = (\tau_y)_{HB} + K_{HB}\dot{\gamma}^{n_{HB}} \tag{18-59}$$

the Casson equation is given by

$$\tau^{0.5} = (\tau_y)_C^{0.5} + K_C\dot{\gamma}^{0.5} \tag{18-60}$$

where $\tau$ is the fluid shear stress, $\tau_y$ the fluid yield stress, $\dot{\gamma}$ the fluid shear rate, and K and n are fluid consistency and flow behavior indices, respectively.

Solomon et al. (1981a) proposed a physical model to estimate cavern sizes based on a torque balance to predict its diameter, $D_c$. They assumed that the cavern was spherical, that the predominant motion at the cavern boundary was tangential in nature (applicable to radial flow impellers), and that the stress imparted by the impeller at the cavern boundary was equal to the fluid yield stress. This model was later modified by Elson et al. (1986) assuming the cavern to be a right circular cylinder with height, $H_c$, centered on the impeller to give

$$\left(\frac{D_c}{D}\right)^3 = \frac{1}{\pi^2[(H_c/D_c + \frac{1}{3}]}Po_t\left(\frac{\rho N^2 D^2}{\tau_y}\right) \tag{18-61}$$

where $Po_t$ is the power number in the turbulent regime and $H_c/D_c = 0.4$ for radial flow Rushton turbines. Equation (18-61) can be used to calculate the size of the cavern assuming that a yield stress can be determined for highly shear thinning fluids. However, Barnes and Walters (1985) point out that if sufficiently low shear stress and shear rates could be measured, these fluids would not exhibit a yield stress, but instead, a very high constant zero shear rate viscosity would be obtained. In addition, many highly shear thinning fluids for which $\sim\leq 0.3$ appear to indicate a yield stress when their flow curves are plotted on linear coordinates. Yet the rheological data could be modeled using a power law equation [eq. (18-62)] equally well:

$$\tau = K\dot{\gamma}^n \tag{18-62}$$

An alternative model was developed to address the problem of estimating fluid yield stress and is based on a fluid velocity approach (Amanullah et al., 1998a). This model considers the total momentum imparted by the impeller as the sum of both tangential and axial components and assumes a torus-shaped cavern. It combines torque and axial force measurements (for axial flow impellers) with the simple power law equation to predict the cavern diameter with the cavern boundary defined by a limiting velocity. The proposed new model is capable of predicting the measured cavern diameters for Re > 20 using both radial and axial flow impellers and is valid for sizes greater than the impeller diameter but less than the vessel diameter. This approach is also shown to be superior to the traditional EN yield stress model in extremely shear thinning fluids whose flow curve can be well fitted by the power law equation. Thus, the radius of a torus-shaped cavern ($r_c$) for a power law fluid is given by

$$r_c^{1-2/n} = v_o \left[ \left( \frac{2}{n} - 1 \right) \left( \frac{4\pi^2 K}{F} \right)^{1/n} \right] + b^{1-2/n} \quad (18\text{-}63)$$

where $r_c = D_c/4$, $b = T/4$, $v_o$ is the fluid velocity at the cavern boundary (estimated at $1 \times 10^{-3}$ m/s; Amanullah et al., 1998a), and F is the total force imparted by the impeller (radial and axial) defined by

$$F = \rho N^2 D^4 \sqrt{N_f^2 + \left( \frac{4Po}{3\pi} \right)^2} \quad (18\text{-}64)$$

Just as torque measurements can be made dimensionless through the power number, so the axial force measurements can be described in terms of a dimensionless axial force number, $N_f$, where

$$N_f = \frac{F_a}{\rho N^2 D^4} \quad (18\text{-}65)$$

The concept of an axial force number, $N_f$, is new. It has been shown that $N_f$ is a scale-independent parameter like Po for geometrically similar vessels (Amanullah et al., 1998a), although it would expected to depend on the Reynolds number and the impeller configuration. Axial force ($F_a$) measurements in a study by Underwood (1994) gave values too low to be detected for radial flow impellers ($N_f = 0$), while axial flow impellers gave significant and easily measured values. Therefore, for a given impeller $N_f$ can be determined as a function of Re and used in a scale-independent manner. Although eq. (18-63) is superior to (18-61) for estimating cavern diameters (Amanullah et al., 1998a) and can be used for both radial and axial flow impellers, it is more complex in its use. Consequently, eq. (18-61) has been used for the purposes of estimating cavern sizes in xanthan fermentations (Amanullah et al., 1998b). The yield stress model [eq. (18-61)] has been used successfully to predict cavern dimensions using model solutions of xanthan with a wide range of impeller designs (Elson et al., 1986; Nienow and Elson, 1988; Elson, 1988, 1990; Galindo and Nienow, 1992, 1993) in xanthan fermentations (Zhao et al., 1991, 1994) and in other yield stress fluids (Etchells et al., 1987). The gassed power consumption, $Po_g$, in eq. (18-61) can either be measured or estimated while the height/diameter ratio remains approximately constant at cavern diameters less than the vessel diameter. However, its value depends on agitator type. Once the cavern reaches the vessel wall, impeller type has little influence on the vertical expansion of the fluid for further increases in impeller speed (Elson, 1990). The rate at which the height of the cavern increases with impeller speed thereafter can be expressed as

$$H_c \propto N^p \qquad (18\text{-}66)$$

The values of $H_c/D_c$ and p are also impeller dependent (Amanullah et al., 1998b). For $N > N_w$, where $N_w$ is the agitator speed and $H_{cw}$ is the height of the cavern when $D_c = T$, the cavern height, $H_c$, can be calculated from

$$\frac{H_c}{H_{cw}} = \left(\frac{N}{N_w}\right)^p \qquad (18\text{-}67)$$

By definition, $H_{cw} = (H_c/D_c)T$ and $N_w$ can be obtained by rearranging eq. (18-61) and replacing $D_c$ with T

$$N_w = \frac{\pi}{D}\left\{\frac{[(H_c/D_c) + \frac{1}{3}](T/D)^3 \tau_y}{\rho Po_g}\right\}^{0.5} \qquad (18\text{-}68)$$

### 18-3.4 Influence of Impeller Type and Bulk Mixing on Xanthan Fermentation Performance

Amanullah et al. (1998b) compared the physical and biological performance of four pairs of impellers: a standard Rushton turbine (SRT, D/T = 0.33), a large-diameter Rushton turbine (LRT, D/T = 0.42), a Prochem Maxflo T (PMD, D/T = 0.44), and a Scaba 6SRGT (SRGT, D/T = 0.54) in a 150 L fermenter. The reasons for the choice of impellers studied and their characteristics are described in Amanullah et al. (1998a). Dissolved oxygen was controlled using agitation speed with maximum agitation speeds of up to 700 rpm with the SRT and PMD and 600 rpm with the LRT and SRGT (because of motor limitations). Throughout the LRT and SRGT fermentations, this method successfully maintained DOT levels above the 15% set point. However, the culture became oxygen limited (<10%) in the SRT and PMD fermentations. The impeller power consumption (and hence the gassed power number, $Po_g$) of the impellers was monitored using online torque measurements throughout each 150 L fermentation. As a result of their diameters and power characteristics, the specific power inputs in the stationary phase of the LRT and SRGT fermentations could be maintained sufficiently high to keep the DOT above 15% of saturation. The total energy requirement for agitation could be calculated by integrating the specific power input with respect to fermentation time. The lowest energy requirement was obtained using the PMD at 270 Wh/kg. This was approximately 14% lower than with the SRT. The energy consumption with the SRGT and LRT was 7 and 28% greater than that with the SRT.

The results of the duplicate fermentations with each pair of impellers did not vary by more than ±5%. Fifty g/L of glucose was added in a fed-batch mode to obtain high gum concentrations. Cell growth was limited by the supply of nitrogen. The highest concentration of gum was 34.8 g/L with the LRT, followed by 32.4 g/L using the SRGT with fermentations of 61 and 48 h, respectively. The highest overall productivity resulted with the SRGT at 0.68 g/L per hour. Fermentations using the SRT and PMD produced approximately similar quantities of gum at 29.7 and 29.0 g/L, respectively. On the basis of gum concentration and quality (as determined by measurement of viscosity), the best results were obtained with the LRT. The poorest quality of gum was obtained using the PMD. On the basis of energy consumption, the PMD was the most efficient. Xanthan productivity per unit energy input was 14% lower than with the SRT, which in turn was 11 and 1% lower than compared to the LRT and SRGT, respectively. Significantly, none of the criteria above favored the SRT. To elucidate the reasons for the very different performances obtained, a better understanding of the interaction between the bulk mixing and oxygen transfer was sought.

Equations (18-61), (18-66), (18-67), and (18-68) were used to estimate the cavern volume using each impeller. Since $N \geq N_W$ throughout the fermentation, the predicted diameter of caverns formed around each impeller was always greater or equivalent to the vessel diameter. Thus, adequate radial mixing could be achieved. However, mixing in the axial direction was poorer. The calculations demonstrated that the total height of the caverns was equivalent to the height of the broth for concentrations up to 16 g/L, ensuring complete homogenization. From 16 to 25.9 g/L, the caverns still interacted, although stagnant regions began to develop above the top and below the bottom impeller. At concentrations in excess of 27.3 g/L, the caverns did not interact and additional stagnant zones developed between the impellers. The calculated cavern volumes expressed as percentage of the total broth volume ($C_V/T_V$), plotted as a function of xanthan concentration for all four impellers, are shown in Figure 18-14. These results indicate that all impellers were equally effective in complete broth homogenization up to 16 to 18 g/L, although very different specific power inputs were necessary to achieve this. Differences in bulk blending were apparent only at higher concentrations; the volume of stagnant regions was greatest (45% at the end of the fermentation) with the SRT at a given gum concentration and least with the SRGT (19%).

The specific productivity in the stationary phase was correlated as a function of the well-mixed cavern region. The specific productivity decreased linearly from 0.36 to 0.04 g xanthan/g biomass per hour as the cavern volume decreased (Figure 18-15). It is proposed that the reduction in specific xanthan productivity was due to the oxygen limitation in the stagnant regions. This hypothesis was tested first, by plotting the measured oxygen transfer rates obtained with each impeller in the stationary phase as a function of the cavern volume. The linear relationship obtained suggested that oxygen transfer in these highly viscous fermentations predominantly occurs in the cavern. Therefore, with increasing fermentation time, the gas dispersion capability of all the impellers used was

POLYSACCHARIDE FERMENTATIONS **1121**

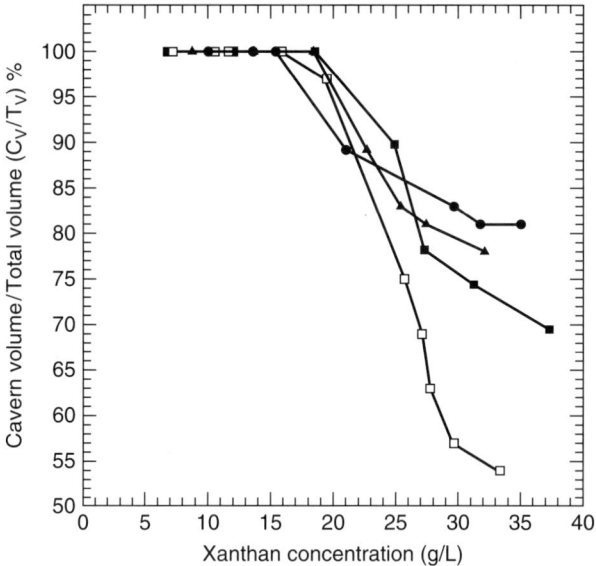

**Figure 18-14** Variation in the cavern volume (expressed as a percentage of the total broth volume) with xanthan concentration: (●) SRGT; (■) LRT; (▲) PMD; (□) SRT. (From Amanullah et al., 1998b.)

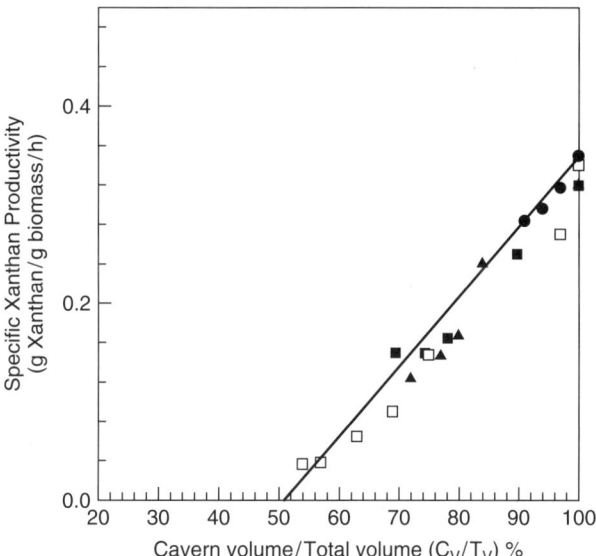

**Figure 18-15** Effect of the cavern volume (expressed as a percentage of the total broth volume) on the specific xanthan productivity: (●) SRGT; (■) LRT; (▲) PMD; (□) SRT. (From Amanullah et al., 1998b.)

severely restricted to the reducing cavern size due to the increasing viscous and non-Newtonian behavior of the broth. Furthermore, a linear correlation of the specific productivity as a function of specific oxygen uptake rate was also obtained. Similar results were reported by Peters et al. (1989a).

It is clear from the previous discussion that it is necessary to reduce the stagnant regions to a minimum and therefore increase oxygen availability in xanthan fermentations to achieve high productivity. In this respect, the agitator design has an important role. The generally enhanced biological performance obtained with the LRT and SRGT impellers can be attributed to their relatively superior bulk blending and oxygen transfer characteristics, which results at the expense of increased torque and energy consumption. Importantly, it should be recognized that a large torque agitation system for a given power input gives a larger cavern (Nienow and Elson, 1988). Thus, it is the combination of the largest D/T ratio and moderately high aerated power number for the Scaba SRGT and the high aerated power number and moderately large D/T ratio for the LRT that leads to them having similar and better performances than the SRT.

Solomon et al. (1981b) showed that pairs of large-diameter impellers were superior when compared to standard Rushton turbines for the mixing of yield stress fluids in terms of power consumption and energy costs. Zhao et al. (1994) demonstrated reduced operating costs and higher yields when using larger-diameter Rushton turbines in xanthan fermentations. Although larger caverns ensure motion in more of the fermenter, it may still be necessary to consider gross recirculation to prevent compartmentalization. Large solidity ratio axial flow impellers may be better in this respect (Nienow, 1990), especially as they also produce large $H_c/D_c$ ratios. To retrofit 0.33 D/T-ratio Rushton turbines with larger D/T systems with higher torque is more expensive than equal torque, speed, and power retrofitting in nonyield stress broths (Nienow, 1990). This factor, too, needs to be considered when selecting the agitation system, and it is important to do a proper economic assessment, allowing for the time value of money (Muskett and Nienow, 1987).

Literature data on the performance of novel impellers in fermentations are relatively scarce. Retrofitting large-diameter PMD impellers in place of standard Rushton turbines at constant power input in a range of bioreactor sizes has been reported to improve the oxygen transfer efficiency (attributed to enhanced bulk blending) and product yield in viscous shear-thinning mycelial fermentations (Gbewonyo et al., 1987; Buckland et al., 1988a,b). Thus, it is still possible that if larger D/T-ratio PMD impellers had been used in the study reported by Amanullah et al. (1998b), their performance might have matched that of the LRT and SRGT impellers. The results of that study also show that improved agitator performance can be used either to reduce operating costs significantly or to obtain enhanced productivity and product quality (at the expense of energy input) in xanthan fermentations. These findings further the strong grounds already established from a mixing viewpoint using model fluids for retrofitting the traditionally used standard Rushton turbines with large diameter impellers of similar designs. The latter can be used to improve both liquid pumping and gas-handling capacities (Nienow, 1990), thereby increasing the mass transfer potential. However, retrofitting of the standard with large diameter Rushton turbines in existing bioreactors may be difficult given the limitations of many motors and drive. An alternative is to use large diameter, low power number impellers such as the PMD and SRGT impellers, which can then be retrofitted for equal speed, torque, and hence power consumption.

## 18-3.5 Factors Affecting the Biopolymer Quality in Xanthan and Other Polysaccharide Fermentations

The quality of polysaccharide gums as determined by their mean molecular weight and molecular weight distribution is an important issue that in turn affects the rheological properties of the gum. A number of studies have reported that high DOT levels as well as good homogeneity of the broth result in xanthan gums of high molecular weight (Peters et al., 1989a,b). These studies demonstrated that as long as oxygen limitation could be overcome and homogeneous mixing conditions could be achieved, the hydrodynamics related to the use of different impellers or bioreactor design did not affect the quality of the gum produced. In a study reported by Amanullah et al. (1998a), mean molecular weight of xanthan using radial flow LRT and SRGT impellers was 8.8 and $9 \times 10^6$ kg/mol, respectively, while for the axial flow PMD impeller it was $8.1 \times 10^6$ kg/mol. The superior-quality gum obtained with the radial flow impellers was attributed to their enhanced bulk mixing and oxygen transfer characteristics.

Lawford and Rousseau (1991) reported that replacement of Rushton turbines with axial flow impellers under non-DOT-limiting conditions resulted in higher-quality (water-insoluble) curdlan. The authors suggested that the quality of the gum degraded at the higher shear rates imposed by the Rushton turbines, although bulk mixing characteristics were not analyzed, and it is difficult to ascertain whether local DOT limitations in the case of the Rushton turbines (lower cell viability was measured compared to the axial flow impellers) may have resulted in a lower-quality gum. Interestingly, the apparent viscosity of the gum increased to a maximum before decreasing in all fermentations. The authors speculated the role of endoglucanase activity with cell death associated in later stages of the fermentations. Dreveton et al. (1994) reported that the mean molecular weight of gellan gum could be increased twofold under homogeneous mixing conditions (obtained with helical ribbon impellers and enriched oxygen supply) compared to heterogeneous mixing conditions obtained with Rushton turbines. The authors claimed that these differences arose due to differences in shear rates obtained with these impellers, which in turn determined whether oxygen diffusional limitations were imposed on cells. However, since these results were obtained without separating dissolved oxygen effects, it is not possible to ascertain whether the differences in gum quality were not simply due to oxygen limitations in the poorly mixed regions using the Rushton turbines.

In other polysaccharide fermentations such as alginate, curdlan, and pullulan gums, the effects of agitation speed and dissolved oxygen are more complex. Peña et al. (2000) reported that at constant agitation speed of 300 rpm in a 1.5 L bioreactor, production of water-soluble alginate was enhanced at DOT levels of 5% compared to 0.5%. However, at DOT levels greater than 5%, the carbon source (sucrose) was utilized primarily for biomass formation rather than gum production. At a constant DOT of 3%, an increase in agitation speed to 700 rpm resulted in increased specific growth rate of the culture and alginate production, although the molecular weight of the polymer was nearly halved compared to 300 rpm. The authors stated that the measured alginate activity, controlled by the availability of oxygen to the cells (lower at lower speeds due to the formation of cell agglomerates), was responsible for degradation of the alginate. However, differences in alginase activity at the low and high agitation speeds were not reported, and in addition, higher leakage rates of alginase at higher agitation speeds cannot be ruled out either as the cause of the lower mean molecular weight of the alginate. On the other hand, Lee et al. (1999) found that the molecular weight of curdlan did not change with agitation speed in the range 300 to 700 rpm using a 5 L fermenter. Wecker and Onken (1991) showed that in the case of the water-soluble biopolymer, pullulan, production of the biopolymer by *Aureobasidium pullulans* was enhanced at 150 rpm compared to 500 rpm at a constant dissolved oxygen level of 100% in 6- and 50 L bioreactors. In addition, it was also shown that enhanced gum formation occurred at a DOT level of 50% compared to 100% at a constant agitation speed of 500 rpm. On the other hand, Gibbs and Seviour (1996) showed that pullulan production was reduced at high agitation speeds above 750 rpm using a 10 L bioreactor due to high dissolved oxygen. When operated at 1000 rpm at constant DOT, pullulan production was not affected.

## 18-4 MYCELIAL FERMENTATIONS

The industrial importance of filamentous fungi is illustrated by applications ranging across the production of antibiotics, organic acids, proteins, and food. The best known examples of the use of filamentous fungi are for production of penicillin by *Penicillium chrysogenum*, citric acid by *Aspergillus niger*, and recombinant proteins by *A. oryzae*. However, during the last two decades, filamentous fungi have been used increasingly as eukaryotic hosts for foreign gene expression, for which they have several attractions (Jeenes et al., 1991). First, due to their saprophytic life they are capable of secreting large quantities of proteins (van Brunt, 1986). Posttranscriptional modifications of proteins such as glycosylation are important capabilities offered by these hosts (Mackenzie et al., 1993). In addition, many species are generally regarded as safe by regulatory authorities. Despite the widespread industrial use and potential of fungal strains for secondary metabolite, organic acid, and heterologous protein production, relatively little is known about the influence of engineering variables such as agitation conditions upon the morphology of such organisms in submerged cultures. In many fungal fermentations, the high apparent viscosities and the non-Newtonian behavior of the broths necessitate the use of high agitation speeds to provide adequate mixing

MYCELIAL FERMENTATIONS     1125

and oxygen transfer. However, mycelial damage at high stirrer speeds (or power input) can limit the acceptable range of speeds, and consequently, the oxygen transfer capability and the volumetric productivity of the fermenter. The effects of hydrodynamic forces ("shear") on fungal physiology are poorly understood. An understanding of how agitation affects mycelial morphology and productivity ought to be valuable in optimizing the design and operation of large scale fungal fermentations for the production of secondary metabolites and recombinant proteins. Here the effects of agitation intensity on hyphal morphology and product formation in two fungal fermentations (*P. chrysogenum* and *A. oryzae*) are considered. *A. oryzae* was used to produce two proteins: α-amylase (homologous protein) and amyloglucosidase (heterologous protein).

Fungal morphology can be classified as dispersed or pelleted (Figure 18-16). The dispersed form is generally of greater importance, with the exception of some pelleted citric acid (*A. niger*) fermentations. Characterization of mycelial morphology is important for physiological and engineering studies of fungal fermentations and in the design and operation of such fermentations. The dispersed form of filamentous organisms in submerged cultures consists of branched and unbranched hyphae (freely dispersed) and clumps or aggregates (Tucker et al., 1992) and is most common in industrial filamentous fermentations. Although this classification does not have a physiological basis, it is nevertheless very useful when comparing relative changes in mycelial morphology. Morphological parameters of interest for the freely dispersed mycelia are mean total hyphal length, mean projected area, and the number of tips per hypha. Clump morphology can be quantified in terms of mean projected area (Tucker et al., 1992). The mean projected area of all elements was taken as a measure of the total

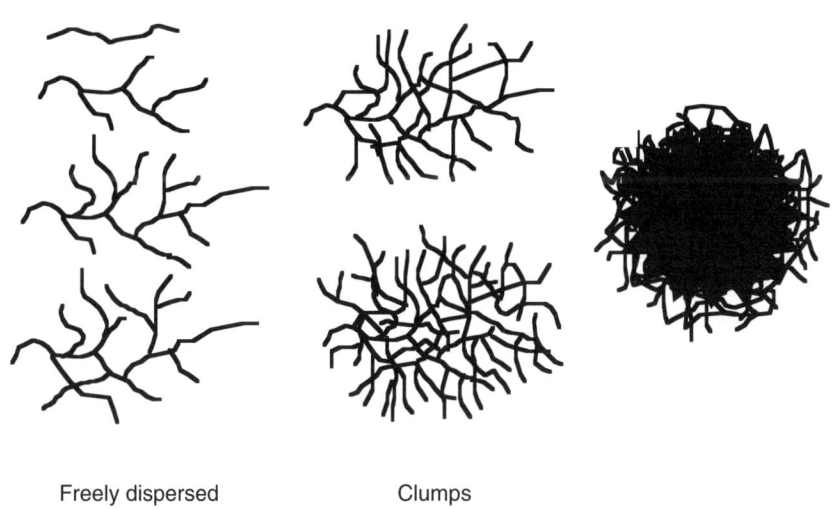

Figure 18-16  Classification of fungal morphology.

biomass (Packer and Thomas, 1990). Although the dispersed form of morphology is most commonly encountered in fungal fermentations, there are distinct benefits of operating with a pelleted morphology. Gbewonyo et al. (1992) demonstrated that in pilot scale Lovastatin fermentations using *Aspergillus terreus*, the pelleted form of morphology led to a fourfold lower apparent viscosity of the fungal broth compared to dispersed morphology fermentations. Although the oxygen uptake rate (and hence biomass concentration) was similar in the growth phase, significantly higher mass transfer and oxygen uptake rates were measured in the stationary phase, although overall product titers were similar. These results could therefore also be used to significantly lower power input (and hence operating costs) in pelleted fermentations [$k_L a = 77.4 P/V)^{0.6}$] and obtain similar mass transfer rates as dispersed morphology fermentations [$k_L a = 16.7(P/V)^{0.8}$].

Considering the importance of these complex morphologies on fermentation performance, and reports that changes in the morphology of *P. chrysogenum* can be caused by mechanical forces (Dion et al., 1954; Metz et al., 1981; van Suijdam and Metz, 1981; Smith et al., 1990; Nielsen, 1992; Makagiansar et al., 1993; Ayazi Shamlou et al., 1994; Jüsten et al., 1996, 1998a), the direct effect of agitation on morphology in submerged fermentations requires attention. As well as the total power input, the choice of impeller geometry determines the hydrodynamic forces that might affect the morphology (Jüsten et al., 1996; Amanullah et al., 1999) and differentiation (Jüsten et al., 1998a) of filamentous species, thereby influencing growth or production (König et al., 1981; Buckland et al., 1988a; Jüsten et al., 1998a; Amanullah et al., 1999). Other environmental factors, such as pH and spore concentration (in cases where spores are used as inoculum), also significantly influence mycelial morphology (Metz and Kossen, 1977). However, provided that these factors can be controlled and optimized, agitation-induced fragmentation, apart from growth, is considered to be one of the most important factors influencing mycelial morphology (especially in the design, operation, and scale-up of fungal fermentations). Although many studies have been conducted to investigate the effects of hydrodynamic forces on mycelial morphology and productivity (Dion et al., 1954; Ujcova et al., 1980; Metz et al., 1981; van Suijdam and Metz, 1981; Reuss, 1988; Smith et al., 1990; Makagiansar et al., 1993; Ayazi Shamlou et al., 1994; Nielsen et al., 1995), they generally suffer from two limitations. First, due to the lack of suitable methods for characterizing clumps (Tucker et al., 1992), only the freely dispersed form has been considered, although it may only account for only a small fraction of the biomass (Tucker et al., 1992; Jüsten et al., 1996). Second, it has not been possible to dissociate the influence of agitation from mass transfer effects. The dependence of product formation rates on impeller-generated fluid dynamic stresses has been observed for a wide variety of filamentous fungi (Ujcova et al., 1980; Vardar and Lilly, 1982; Smith et al., 1990; Braun and Vecht-Lifshitz, 1991; Märkl et al., 1991; Merchuk, 1991). There are also reports that mycelial fragmentation depends on the physiological state of the microorganisms (Smith et al., 1990; Paul et al., 1994).

## 18-4.1 Energy Dissipation/Circulation Function as a Correlator of Mycelial Fragmentation

Smith et al. (1990) and Makagiansar et al. (1993) proposed that the breakup frequency of mycelia would depend on $(P/D^3)(1/t_c)$, where P is the power input, D the impeller diameter, and $1/t_c$, the circulation frequency. This adaptation correlated the production rate and the morphology of the freely dispersed mycelia well at different scales up to 100 L and up to 1000 L, respectively. Unlike Smith et al. (1990) and Makagiansar et al. (1993), Jüsten et al. (1996) allowed for impeller designs other than Rushton turbines. Using well-established image analysis methodologies (Tucker et al., 1992; Paul and Thomas, 1998), Jüsten et al. (1996) were able to make quantitative measurements on the breakage of clumps and therefore to take into account the influence of realistic agitation conditions on the whole of the biomass. In off-line agitation studies it was demonstrated that *P. chrysogenum* morphological data using both radial and axial flow impellers of very different geometries and power numbers could be correlated with an energy dissipation/circulation function developed from the earlier work of Smith et al. (1990) and Makagiansar et al. (1993).

The percentage of clumps in a fungal fermentation depends not only on the strain, the specific set of operating conditions, but also on the physiological state of the microorganisms. For instance, the percentage of clumps in the rapid growth phase is much greater than in the fed-batch stage of such fermentations. The value of 80% is quoted as an example specifically for *P. chrysogenum* in the rapid growth phase of a fed-batch fermentation. Nevertheless, it is still clearly necessary to include clump measurements for a proper representation of the total biomass. The energy dissipation/circulation (EDC) function was defined by Jüsten et al. (1996) as $(P/kD^3 t_c)$, where P is the power input, D the impeller diameter, $t_c$ the mean circulation time, and k a geometric constant for a given impeller and is derived from a calculated impeller swept volume. This function arises from consideration of the energy dissipation in the impeller swept volume and the frequency of mycelial circulation through that volume. Although other correlating parameters, such as impeller tip speed and specific power input, were also considered, they were inferior to the energy dissipation/circulation function. The broader validity of these correlations was also verified in fragmentation studies at scales up to 180 L (Figure 18-17). The implications for the EDC function using tip speed and specific power input as scale-up criteria are interesting. Assuming that the flow number (Fl) and power number ($P_o$) are independent of scale (which is a reasonable assumption), for geometrically similar systems

$$\frac{P}{kD^3}\frac{1}{t_c} \propto N^4 D^2 \qquad (18\text{-}69)$$

At equal tip speed, $N \propto D^{-1}$ and therefore

$$\frac{P}{kD^3}\frac{1}{t_c} \propto D^{-2} \qquad (18\text{-}70)$$

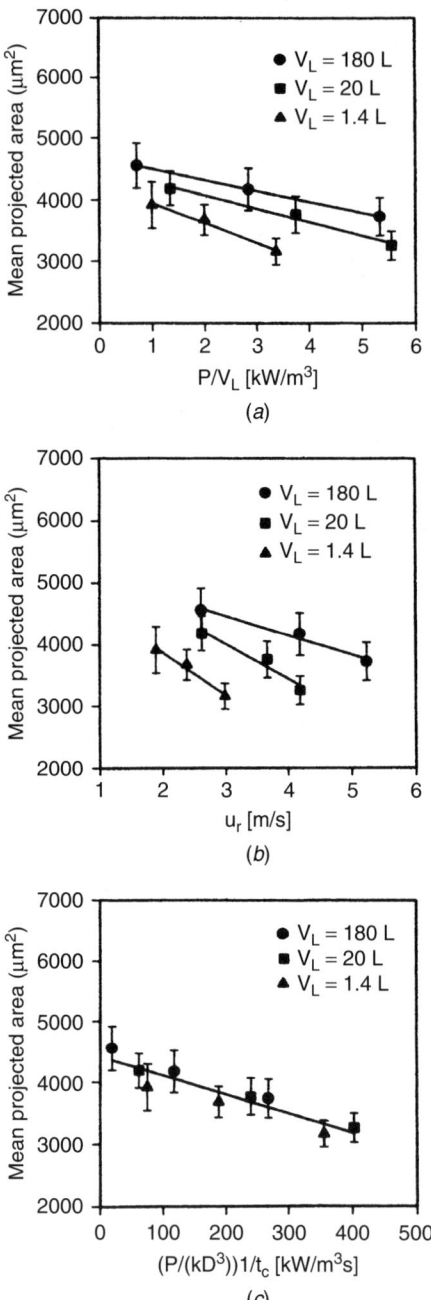

**Figure 18-17** Comparison of different scale-up criteria to correlate hyphal fragmentation in *P. chrysogenum*: (*a*) power per unit volume; (*b*) impeller tip speed; (*c*) the energy dissipation/circulation function. (From Jüsten et al., 1996.)

At equal P/V, $N \propto D^{-2/3}$ and therefore

$$\frac{P}{kD^3}\frac{1}{t_c} \propto D^{-2/3} \qquad (18\text{-}71)$$

Therefore, using both impeller tip speed and P/V as scale-up criteria, the value of the EDC function should decrease significantly on scale-up. This was confirmed in the study of Jüsten et al. (1996). Jüsten et al. (1998b) modeled the results of Jüsten et al. (1996) and suggested that clump fragmentation was the main cause of morphological changes and that the freely dispersed form was dominated by short fragments originating from clumps. It should be recognized that these studies were conducted in off-line vessels under nongrowing conditions, although it was subsequently shown that this function could also correlate the growth, morphology, vacuolation, and productivity in fed-batch penicillin fermentations (Jüsten et al., 1998a). It should also be noted that there is no fundamental understanding of how clumps might be broken by agitation.

The broader validity of the EDC function to correlate hyphal fragmentation of fungal cultures other than *P. chrysogenum* was reported for *Aspergillus oryzae* using different agitation intensities and different impellers under nongrowing conditions in stirred tanks (Amanullah et al., 2000), chemostats (Amanullah et al., 1999), and fed-batch cultures (Amanullah et al., 2002). The study by Amanullah et al. (2000) demonstrated that the EDC function could correlate hyphal fragmentation for both *P. chrysogenum* and *A. oryzae*. Samples for fragmentation studies were obtained from chemostat cultures. Details of the impeller type, geometry, and operating conditions are shown in Tables 18-3 and 18-4 for *A. oryzae* and *P. chrysogenum*, respectively.

**Table 18-3** Details of Agitator Type, Geometry, and Operating Conditions in Off-line Fragmentation Experiments Using *Aspergillus oryzae*

| Impeller | Impeller Diameter/Tank Diameter Ratio, D/T | Agitation Speed (rpm) | Power Number, Po[a] | Flow Number, Fl | Reynolds Number, Re |
|---|---|---|---|---|---|
| Rushton (radial flow) | 0.33 | 540 | 3.90 | 0.78 | 14 100 |
|  |  | 1200 |  |  | 31 400 |
|  |  | 1470 |  |  | 38 400 |
| Rushton (radial flow) | 0.65 | 290 | 4.2 | 0.87 | 29 400 |
|  |  | 330 |  |  | 33 500 |
| Prochem Maxflo T (axial flow) | 0.63 | 120 | 1.65 | 1.01 | 11 400 |
|  |  | 500 |  |  | 47 600 |
| Pitched blade (axial flow) | 0.40 | 1210 | 0.60 | 0.61 | 46 500 |
|  |  | 1500 |  |  | 57 600 |

[a] Off-bottom clearance: $0.25 \times T$.
*Source*: Amanullah et al. (2000).

**Table 18-4** Details of Agitator Type, Geometry, and Operating Conditions in Off-Line Fragmentation Experiments Using *Penicillium chrysogenum*

| Impeller | Symbol in Figures | Impeller Diameter/Tank Diameter Ratio, D/T | Speed Range (rpm) | Reynolds Number, Re[a] | Flow Number, Fl | Power Number, Po[b] |
|---|---|---|---|---|---|---|
| Paddle | ■ | 0.60 | 145–394 | 12 500–33 900 | 2.28 | 9.97 |
| (radial flow) | ● | 0.60 | 366–528 | 31 500–45 400 | 0.60 | 3.10 |
| Rushton turbine | ▼ | 0.65 | 238–407 | 24 000–41 100 | 0.87 | 4.52 |
| (radial flow) | ▲ | 0.40 | 548–1031 | 21 000–39 400 | 0.81 | 4.22 |
|  | ◆ | 0.33 | 767–1443 | 20 000–37 600 | 0.78 | 4.10 |
| Prochem Maxflow T | ⊙ | 0.63 | 340–639 | 32 200–60 600 | 1.01 | 1.90 |
| (axial flow) |  |  |  |  |  |  |
| Propeller | ◇ | 0.50 | 745–2021 | 44 500–120 700 | 0.60 | 0.55 |
| (axial flow) |  |  |  |  |  |  |
| Pitched blade | □ | 0.60 | 359–676 | 30 900–58 200 | 1.03 | 1.97 |
| (axial flow) | △ | 0.40 | 776–1461 | 29 700–55 900 | 0.91 | 1.48 |
|  | ▽ | 0.40 | 1032–1943 | 39 500–74 300 | 0.63 | 0.61 |
|  | ○ | 0.40 | 1201–2054 | 45 900–78 500 | 0.53 | 0.40 |
| Intermig set | ◌ | 0.65 | 423–795 | 42 700–80 300 | — | 0.81 |
| (radial and axial flow) |  |  |  |  |  |  |

[a] Viscosity = 0.001 Pa·s.
[b] Turbulent range with Reynolds number $> 10^4$ and no surface aeration. When surface aeration occurred, Po was reduced accordingly. The off-bottom clearance of the impellers was 0.7 × D.
*Source*: Jüsten et al. (1996); Amanullah et al. (2000).

Figure 18-18a shows the mean projected area of *A.oryzae* after 30 min of agitation normalized to a control (under nongrowing conditions). Details of impeller types and geometries are shown in Table 18-3. The control was taken as the mean of all the samples taken after dilution but just before the fragmentation tests were begun (i.e., 27 200 ± 7450 μm$^2$). Although the data are somewhat limited, the EDC is able to correlate the reduction in mean projected area for values greater than those used in the chemostat very well for both impeller types. Thus, this result supports the earlier work of Jüsten et al. (1996) with *P. chrysogenum*, using a greater range of impeller types and geometries (Table 18-4 and Figure 18-18b). For EDC values less than 90 kW/m$^3$ · s as used in the chemostat and marked in Figure 18-18a, fragmentation did not occur. This finding is consistent with the idea that the mycelia would be adapted to the agitation conditions in the chemostat. Figure 18-18b shows the data of Jüsten et al. (1996) with an indication of the EDC value (22 kW/m$^3$ · s) found in the chemostat. As before, the normalized data for all the impellers is correlated well by the EDC function for EDC values > (EDC)$_{chemostat}$. So, too, are the data for (EDC) < (EDC)$_{chemostat}$. However, even though the number of data at these low EDC values is much fewer than for EDC > (EDC)$_{chemostat}$, it is clear that there is a distinct change of slope, suggesting much less breakdown and is consistent with data for *A. oryzae*. Overall, the conclusion that can be drawn from these two examples is that off-line fragmentation, depending on the state of the mycelia, is either very low or zero if the EDC function is less than that used in the chemostat. If the slopes representing the reduction in clump size with increasing values of the EDC above that in the chemostat are compared, a difference is observed. For the *A. oryzae* (Figure 18-18*a*), the slope is −0.5, while for the *P. chrysogenum* (Figure 18-18*b*), it is −0.10.

The precise reasons for differences in fragmentation behavior for the two strains are unclear, but it would not be unreasonable to invoke differences in cell wall strength, clump size, and structure as possible causes. The implication of clumps in the fragmentation analysis is especially relevant since morphological distributional data showed that fragmentation seemed to be mainly of the clumps, with loss of small fragments gradually reducing clump size (Jüsten et al., 1998b). Clump rupture and fragmentation of freely dispersed hyphae did not appear to be of primary importance. Although the EDC function is successful in correlating hyphal fragmentation using different impeller types and geometries

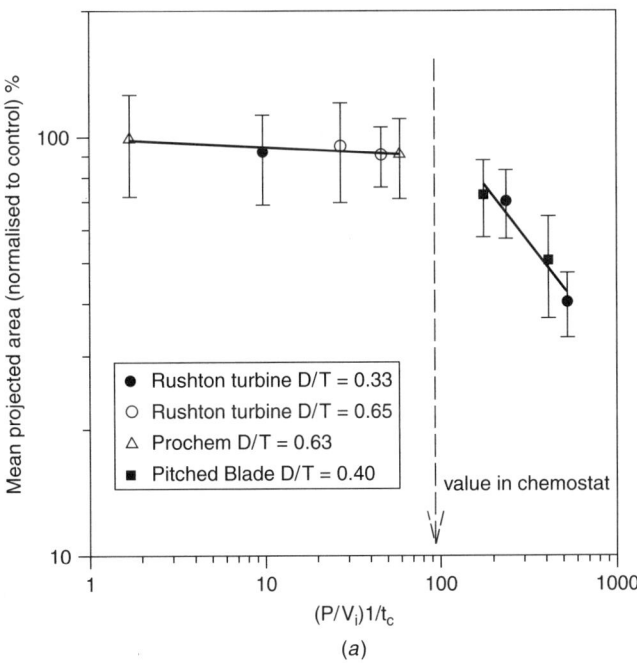

**Figure 18-18** (*a*) Variation in mean projected area at 30 min agitation time in an off-line vessel (nongrowing conditions) with the energy dissipation/circulation function using *Aspergillus oryzae*. Initial mean projected area (control) = 27 200 ± 7450 μm². The chemostat value of $(P/kD^3)(1/t_c)$ is shown by the arrow (Amanullah et al., 2000).

(*Continued*)

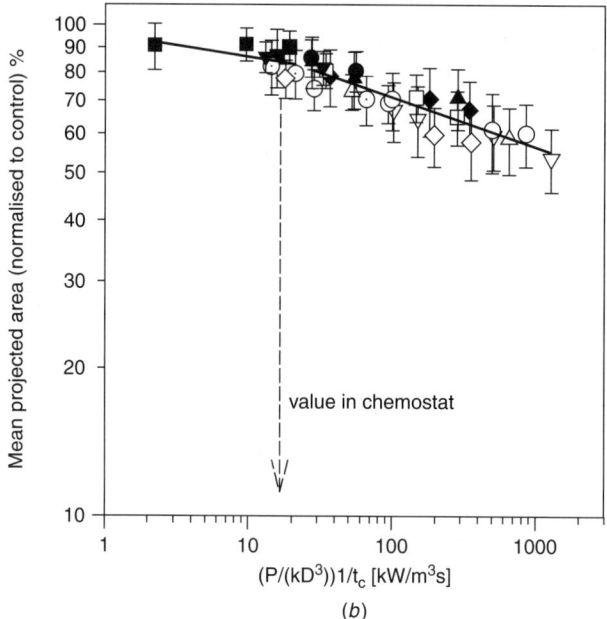

**Figure 18-18** (b) Variation of mean projected area at 30 min agitation time in an off-line vessel (nongrowing conditions) with the energy dissipation circulation function using *Penicillium chrysogenum* (Jüsten et al., 1996). Initial mean projected area (control) = 5620 ± 500 μm². Refer to Table 18-4 for symbols. The chemostat value of $(P/kD^3)(1/t_c)$ is indicated by the arrow (Amanullah et al., 2000).

as well as scale, it is not based on any fundamental understanding of the breakage process. It is also interesting to consider whether scale-dependent turbulent intermittency (short-term high-energy events), which can manifest itself in long-term turbulent mixing processes, has a role in the breakage process (Baldyga et al., 2001; also refer to Section 18.6.1). Perhaps the fragmentation experiments at different scales was not long enough (1 h) for intermittency to affect hyphal fragmentation.

### 18-4.2 Dynamics of Mycelial Aggregation

Although aggregation has been invoked as one possible way in which mycelia can increase their size, there is little evidence in the literature as to whether or not mycelia aggregate following a reduction in the agitation intensity (due to a reduction in agitator speed in a fermentation in response to a reduction in oxygen demand during a batch fermentation, for example, or when mycelia exit the high-energy dissipation impeller region in a large scale bioreactor). Given the importance of mycelial morphology, it is important to understand the factors, including aggregation, that influence it. Amanullah et al. (2001a) measured the dynamics of changes in mycelial morphology in response to a rapid and much reduced level of agitation intensity. Analysis of the transients of mycelial morphology in chemostat and batch cultures of A. *oryzae* under conditions of controlled, nonlimiting, dissolved oxygen tension, following a significant decrease in agitation speed, showed that a large and rapid increase in the mean projected area of the clumps plus freely dispersed mycelia occurred at a rate that cannot be explained by growth alone. Clearly, a physical mechanism must have been responsible for the rapid increase in mean projected area and for the changes in the freely dispersed morphology. It was suggested that a physical mycelial aggregation process with a time constant of minutes caused the initial changes. However, it appears that such aggregation only occurs in cultures with the availability of dissolved oxygen, as there was no significant change in the morphology in off-line experiments where the broth was sparged with nitrogen.

Vecht-Lifshitz et al. (1990) proposed that the aggregation of the filamentous bacterium *Streptomyces tendae* was caused by hydrophobicity, which itself was biologically regulated by the supply of oxygen, and it is possible that similar phenomena occur in fungi. These findings are important since it is possible that in a large scale fermentation, with long mean circulation times, mycelia aggregate rapidly outside the impeller swept volume after undergoing fragmentation within it. For instance, in a large scale aerated bioreactor operated with a viscous mycelial fermentation, a mean circulation time between 20 and 60 s would be reasonable, and assuming that the circulation time can be described by a lognormal distribution, they could be distributed in the range 0 to 240 s (i.e., a maximum of a few minutes) (see Section 18.2.4.1). This is of the same order of magnitude as the times measured for mycelial aggregation. Thus, mycelia may be repeatedly fragmented and aggregated as they circulate through a large bioreactor. By the same reasoning, mycelial aggregation may not be relevant in small bioreactors with very short circulation times.

### 18-4.3 Effects of Agitation Intensity on Hyphal Morphology and Product Formation

*18-4.3.1 Aspergillus oryzae.* The effects of agitation on fragmentation of a recombinant strain of A. *oryzae* and its consequential effects on recombinant protein production were investigated by Amanullah et al. (1999). Constant-mass 5.3 L chemostat cultures at a dilution rate of 0.05 $h^{-1}$ and a dissolved oxygen level of 75% of air saturation were conducted at 550, 700, and 1000 rpm. These agitation speeds were chosen to cover a range of specific power inputs (2.2 to 12 $kW/m^3$) from realistic industrial levels to much higher values. The use of a constant-mass chemostat linked to a gas blender allowed variation of agitation speed and hence gas hold-up without affecting the dilution rate (via gas hold-up effects) or the concentration of dissolved oxygen. The morphology of both the freely dispersed mycelia and clumps was characterized using image analysis. Statistical analysis showed that it was possible to obtain steady states with respect to morphology. The mean projected area at each steady state under growing conditions correlated well with the EDC function. Rapid changes in the hyphal mean projected area resulted in response to a speed change from 1000 rpm to 550 rpm (Figure 18-19). The steady-state mean projected area of the total biomass was found to increase significantly from $6100 \pm 1100$ μm$^2$ (mean ± standard error) at 1000 rpm to $16\,500 \pm 3800$ μm$^2$ at 550 rpm. Protein production (α-amylase and amyloglucosidase) was found to be independent of agitation speed in the range 550 to 1000 rpm (P/V = 2.2 and 12.6 kW/m$^3$, respectively) (Figure 18-20), although significant changes in mycelial morphology could be measured for similar changes in agitation conditions. This suggests that mycelial morphology in his strain does not directly affect protein production (at a constant dilution rate and therefore, specific growth rate). Although there is very limited use of continuous culture systems in industry, they are extremely useful research tools since they can give precise information on the influence of a single variable. However, it was important to verify the results in fed-batch fermentations at industrially realistic conditions of biomass concentration and specific power input.

To extend the findings of the chemostat study to realistic operating conditions, fed-batch fermentations of *A. oryzae* were conducted at biomass concentrations up to 34 g dry cell weight per liter and three speeds (525, 675, and 825 rpm) to give specific power inputs between 1 and 5 kW/m$^3$ (Amanullah et al., 2002) using two Rushton turbines (D/T = 0.5) in a 6 L bioreactor. Gas blending was used to control the dissolved oxygen level at 50% of air saturation except at the lowest speed, where it fell below 40% after 60 to 65 h. The effects of agitation intensity on growth, mycelial morphology, hyphal tip activity, and

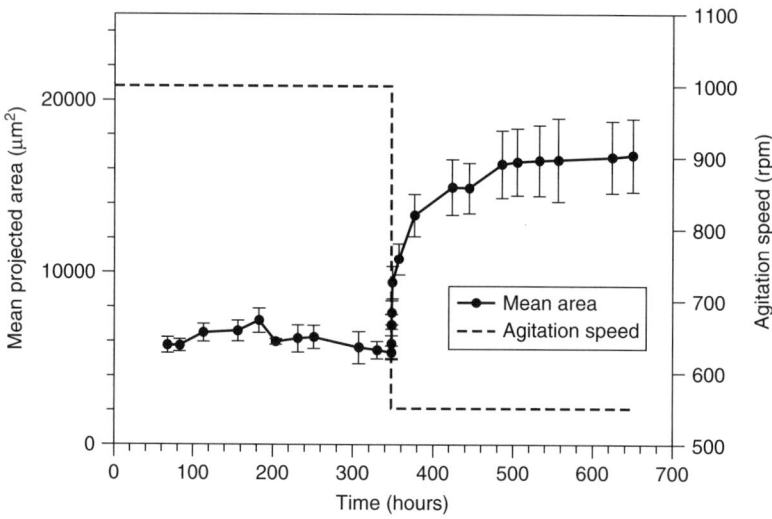

**Figure 18-19** Variation in mean projected area of the biomass with an agitation speed change from 1000 to 550 rpm using two Rushton turbines (D/T = 0.5) in a 5 L constant-mass chemostat at controlled dissolved oxygen of 75% of air saturation. (From Amanullah et al., 1999.)

MYCELIAL FERMENTATIONS    1135

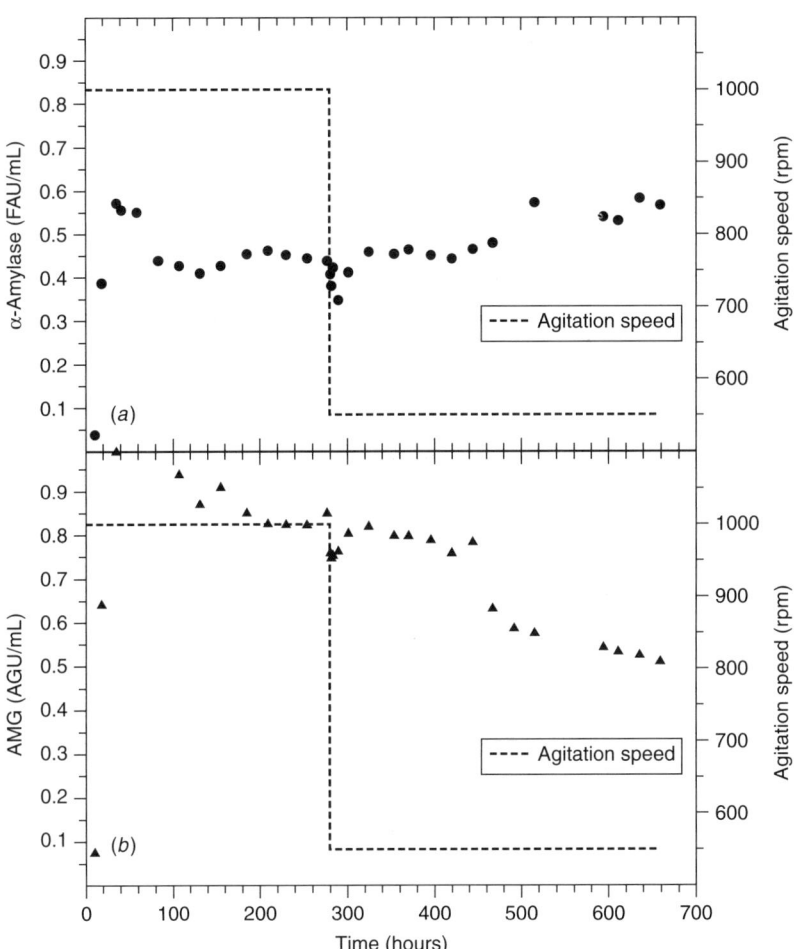

**Figure 18-20** Variation in (*a*) α-amylase and (*b*) AMG activities with an agitation speed change from 1000 to 550 rpm at 279 h in an agitation speed change from 1000 rpm to 550 rpm using two Rushton turbines (D/T = 0.5) in a 5 L constant-mass chemostat at a controlled dissolved oxygen level of 75% of air saturation. The data show that steady-state protein production remained independent of the changes in agitation speed. The decrease in the AMG activity at 450 h is thought to be related to a possible loss in the gene copy number for AMG. (From Amanullah et al., 1999.)

recombinant protein (amyloglucosidase) production in fed-batch cultures were investigated. In the batch phase of the fermentations, biomass concentration, specific growth rates, and AMG secretion increased with increasing agitation intensity. These early differences in specific growth rate were responsible for the dependence of biomass concentration on agitation intensity for the remainder of the fermentation, although the specific growth rate became independent. If in a fermentation, dissolved oxygen fell below about 40% due to inadequate oxygen transfer associated with enhanced viscosity, AMG production ceased. As with the chemostat cultures, even though mycelial morphology was significantly affected by changes in agitation intensity, enzyme titers (AGU/L) under conditions of substrate-limited growth and controlled dissolved oxygen of more than 50% did not follow these changes.

The practical implication of these results is that the agitation intensity in such fungal fermentations can be manipulated to meet process requirements in terms of dissolved oxygen levels and bulk mixing and possibly to control broth rheology by changing morphology without compromising recombinant protein production. Attempts were also made to repeat this study with higher biomass concentrations of up to 58 g/L when mycelial interactions should be pronounced. These biomass concentrations were achieved, but it became impossible to mix the broth adequately due to its highly shear thinning, viscous nature. This led to a much lower Reynolds number at the small scale and very long mixing times (Nienow, 1998), and possibly even complete stagnation due to yield stresses (Nienow, 1998; Amanullah et al., 1998a,b). A similar biomass and broth rheology on the large scale at equal specific power input would give relatively higher Reynolds numbers and better mixing. The problem of maintaining suitable scale-down conditions with respect to both fluid dynamics and biomass concentration is a particularly difficult one. Further work is still needed to resolve this problem.

### 18-4.3.2 Penicillium chrysogenum.
The influence of agitation conditions on growth, morphology, vacuolation, and productivity of *P. chrysogenum* in fed-batch 6 L fermentations was reported by Jüsten et al. (1998a). The results were compared using a standard Rushton impeller, a four-blade paddle, and a six-blade pitched blade impeller. Power inputs used ranged from 0.35 to 7.4 W/kg and the DOT was maintained above 40% of air saturation using gas blending. For a given impeller, the specific growth rate and biomass concentration in the batch phase increased with increasing agitation intensity, while the specific penicillin production rate decreased. These changes could be correlated to the EDC function. These results were in broad agreement with those reported by König et al. (1981), Vardar and Lilly (1982), Smith et al. (1990), and Makagiansar et al. (1993). The mean projected area also increased in the batch phase and remained relatively constant (dependent on agitation intensity) in the stationary phase. The proportion of vacillated regions also decreased with increasing agitation intensity possibly due to preferential fragmentation of the weaker vacuolated hyphal regions. This decrease is significant since penicillin synthesis is believed to be located in the vacuolated compartments. Clearly, the results obtained by the variation of agitation intensity in fed-batch cultures of *A. oryzae* (Amanullah et al., 2002) were different to those obtained with *P. chrysogenum*, where agitation intensity was found to strongly influence penicillin production (Jüsten et al., 1996, 1998a). For the latter case, it was suggested that the interrelationship is due to breakage of the relatively weaker vacuolated regions of hyphae, such regions being where penicillin synthesis is located. In contrast, although breakage still probably occurs at

the weaker vacuolated hyphal regions in *A. oryzae*, it does not affect AMG titer under conditions of substrate-limited growth. It is postulated that this lack of a relationship is because protein secretion in such strains appears to occur only at the hyphal tips (Wessels, 1990, 1993). Therefore, the results reported here are not generally applicable to all mycelial fermentations, and each system of interest should be considered individually.

### 18-4.4 Impeller Retrofitting in Large Scale Fungal Fermentations

Standard Rushton turbines of approximately one-third have traditionally been employed in fermentation processes. Bulk blending is also poor in highly shear thinning fungal broths, due to the tendency of Rushton turbines to compartmentalize. Buckland et al. (1988b) retrofitted two standard Rushton turbines (D/T = 0.33) with two low power number (Po = 1.1) Prochem hydrofoil impellers (D/T = 0.45) in 800 L in *Streptomyces avermitilis* fermentations. Two fermenters each equipped with either two Rushton or two Prochems were operated at 300 and 330 rpm with power draws of 4.1 and 2.7 kW, respectively. The values of $k_L a$, oxygen uptake rates and Avermectin titers in the two bioreactors were identical. Thus, the Prochem impellers were shown to be capable of providing the same oxygen transfer and product titers, while drawing about 50% less power. They attributed this to the increased well-mixed volumes generated by the Prochem impellers, leading to enhanced oxygen transfer rates. Such results have also been demonstrated in highly viscous xanthan fermentations (see Section 18.3.4). Buckland et al. (1988a) recommended that standard Rushton impellers could be replaced by low-power-number agitators (preferably axial flow) such that the retrofitting would result in similar operating speeds, torque, and power. In this manner, the same shaft, motor, and gearbox could be employed. Similar retrofitting of such impellers in 19 000 L penicillin batch fermentations (Buckland et al., 1988b) demonstrated that the use of Prochem impellers at 1.96 W/kg resulted in higher productivity than Rushton turbines drawing 2.35 W/kg. These results were again explained in terms of enhanced bulk mixing and oxygen mass transfer.

## 18-5 *ESCHERICHIA COLI* FERMENTATIONS

Protein production by recombinant technology has been the subject of much industrial interest. However, production has been limited to a few well-known overexpression systems such as *E. coli*, although knowledge of the process engineering variables on performance is still limited. The cultivation of *E. coli* in fed-batch mode to high cell densities is the preferred industrial method for increasing the volumetric productivity of bacterial-derived products such as nucleic acids (Elsworth et al., 1968), amino acids (Forberg and Haggstrom, 1987), and heterologous recombinant production (Risenberg and Schulz, 1991; Bylund et al., 2000). In such fed-batch *E. coli* cultivations, the carbon source (usually, glucose) is supplied continuously at a growth-limiting rate. This avoids problems associated with excessive oxygen demand, heat generation, and catabolite repression that occur in batch processes. Glucose is typically supplied at concentrations of up to 500 g/L. When cells grow at half their maximum specific growth rate, the concentration of the substrate is equal to the saturation constant of the Monod model, which for glucose is about 5 mg/L for *E. coli*. This creates a situation where the glucose concentration can vary theoretically in a large scale bioreactor from 500 g/L in the feed addition zone down to 5 mg/L elsewhere, with cells being exposed to very large concentration differences fluctuating in value in time and space. Mixing is critical in such situations to ensure that addition of the concentrated feed is dispersed as quickly as possible. However, the impact of intense mixing on bacterial physiology has not been widely reported.

### 18-5.1 Effects of Agitation Intensity in *E. coli* Fermentations

There are very few reports in the literature on the detrimental impact of fluid mechanical stress on bacteria and yeast in general. This is partly because bacteria are generally regarded as being significantly smaller than the Kolmogorov microscale of turbulence. For a detailed discussion and limitations of the Kolmogorov theory of isotropic turbulence, the reader is referred Section 18-6.1 (see also Chapter 2). The Kolmogorov microscale of turbulence, $\lambda_K$, is related to the local energy dissipation, $\varepsilon_T$, by the equation

$$\lambda_K = \left(\frac{\nu^3}{\varepsilon_T}\right)^{0.25} \tag{18-72}$$

where $\nu$ is the kinematic viscosity of the medium. Thus at 1 W/Kg in waterlike fluids, the value of $\lambda_K$ is 30 μm, while *E. coli* cells are 1 to 2 μm in length. This concept may also explain why hyphal fragmentation occurs in fungal fermentations since the size of hyphae are generally >50 μm. The only reported study that has investigated the effects of agitation intensity in *E. coli* fermentations was conducted by Hewitt et al. (1998). They cultivated *E. coli* in a chemostat equipped with two six-blade paddle impellers and varied the agitation speed of from 400 to 1200 rpm and back to 400 rpm. Dissolved oxygen was controlled independent of agitation speed using gas blending. This range in agitation speed covered a range in $\varepsilon_T$ from 1 to 30 W/kg, with $\lambda_K$ varying from 30 to 13.5 μm. Even if the maximum local energy dissipation value as found close to the impeller is used to estimate $\lambda_K$, it would still be about 6 μm, well above the size of *E. coli* cells. Physiological characteristics (cell viability and membrane potential as indicators of cell metabolism) of the cells were monitored using multiparameter flow cytometry. Neither biomass concentration, respiratory quotient, cell viability, nor membrane potential were affected in the range of agitation speeds tested. The only discernible change that could be detected via transmission electron microscopy was a stripping of the outer polysaccharide layer at the higher agitation speed, the physiological significance of which remains unclear.

Only one study (Toma et al., 1991) has reported growth and metabolic inhibition in *Brevibacterium flavum*, *S. cerevisiae*, and *Trichoderma reesei* due to turbulence and remains outside the main paradigm of cell–turbulence interactions and is difficult to explain, even considering local energy dissipation rates. Perhaps turbulent intermittency has a role to play here (see Section 18-6.1).

## 18-6 CELL CULTURE

The commercial use of animal and insect cells at scales greater than 10 000 L for the production of posttranslationally modified proteins in kilogram quantities and viral vaccines using recombinant DNA and cell fusion techniques has made cell culture a cornerstone of modern biotechnology. Cell culture is being used increasingly for the production of highly valuable biologicals such as viral vaccines, hormones, growth factors, enzymes, and monoclonal antibodies. The high cost of media can often result in the cell culture step becoming the most significant cost of the process. Hence, much work has been devoted to optimizing the culture media and feeding strategies for supporting high cell viabilities and productivity and product quality. However, given that animal cells are potentially more sensitive than microbial cells to agitation and aeration in stirred tank bioreactors, the proper design and operation of bioreactors in relation to agitation and aeration is critical to benefit fully from process optimization efforts. The perceived sensitivity of animal cells to hydrodynamic shear stresses has resulted in the use of low agitation intensities and aeration rates, which in turn can lead to inhomogeneities in dissolved oxygen and $CO_2$, but especially in pH at large scales of operation. The discussion in this section focuses on these issues. Although air-lift, hollow-fiber, and fixed bed (for anchorage-dependent cells) bioreactors are also used, the majority of industrial processes use stirred-tank bioreactors, and for that reason, the discussion here has been limited to stirred tank bioreactors. The issues related to agitation and aeration are summarized in this section, although several excellent reviews (Prokop and Bajpai, 1992; Aunins and Henzler, 1993; Wu, 1995; Joshi et al., 1996; Thomas and Zhang, 1998) on this subject have been published and the reader is referred to these for detailed discussions.

### 18-6.1 Shear Damage and Kolmogorov's Theory of Isotropic Turbulence

As Thomas and Zhang (1998) point out, the term *shear* must be one of the most abused words in the biochemical engineering literature and has gained a nearly colloquial usage to imply any hydrodynamic effect on biological materials. Even the term *shear damage* is problematic. It is often used to describe hydrodynamic, fluid mechanical, or interfacial damage. However, just because cell damage increases due to an increase in turbulence intensity, it does necessarily mean that the cell is being "sheared." Its use has not always resulted in an understanding of the underlying mechanisms, and this is symptomatic of many studies conducted to study shear sensitivity of biological materials, including fungal cultures and enzymes, but particularly cell and plant culture. Technically, *shear* refers to the relative motion of notional layers of liquid past one another due to a velocity gradient (or shear strain, dv/dy), and shear stress refers to the force per unit area acting tangentially to the surface of an object. The forces acting on a cell suspended in a fluid are the result of the dissipation of fluid kinetic energy to the external surface of the cell. The magnitude of these forces is a function of the fluid viscosity and velocity gradient immediately surrounding a cell and is given by

$$\tau = \mu \gamma = -\mu \left( \frac{dv}{dy} \right) \quad (18\text{-}73)$$

In laminar flow, the shear stress can be calculated relatively easily (Prokop and Bajpai, 1992). However, the situation is more complex in the turbulent flow regime. In this case the instantaneous velocity vector is the sum of the time-averaged velocity and a randomly fluctuating time-dependent velocity vector. The intensity of turbulence is directly related to the magnitude of the fluctuating velocity vector. Therefore, a cell suspended in laminar flow will experience an average shear stress that is independent of time, while in turbulent flow it can experience high shear stresses due to turbulence intermittency (Baldyga and Bourne, 1995; Baldyga et al., 2001). These authors point out that the local rate of energy dissipation displays fluctuations about its mean even under homogeneous flow conditions. These are intermittent (rare) fluctuations due to high energy events that are difficult to measure directly but could have a vital role in understanding cell–fluid interactions. Intermittency is especially important in cases where changes are irreversible and occur over long time frames (as is the case in cell culture systems). Consideration of intermittency also suggests that it is scale dependent, which may serve to undermine the conventional approach of using average energy dissipation rates for estimating cell damage or drop breakage in immiscible liquids. In fact, this was clearly shown in the highly significant paper of Baldyga et al. (2001), where difference in the steady-state Sauter mean drop diameter obtained at two different scales at equal mean energy dissipation rates could be explained by considering turbulence intermittency.

Almost equally abused is the Kolmogorov theory of isotropic turbulence (Kolmogorov, 1941a,b). According to this theory, kinetic energy at sufficiently large Reynolds numbers is imparted to the fluid by the rotating impeller blade and is initially transported by eddies (the largest eddy size is determined by the size of the impeller blade). The kinetic energy in the large eddies is rapidly cascaded to smaller eddies in an isotropic manner (i.e., directional information is lost), until finally most of the energy imparted by the impeller is transferred to the fluid via viscous dissipation of small eddies. These eddies are characterized by a length scale called the *Kolmogorov microscale* (see Section 18-5.1) and are often referred to as the *universal equilibrium range* of wavenumbers (inverse of the eddy size), where the energy carried by the eddies is independent of viscosity. The kinetic energy of turbulence is transferred from eddies to the cell surface to an extent determined by the relative size of eddies and the cells. Thus if the eddy size is significantly larger than the cell, negligible energy is transferred to the cell surface and the cell can be thought of as a freely rotating object in a relatively quiescent fluid bounded by the dimensions of the eddy. If the eddy size is on the same order as the cell, significant transfer of energy can take place, which can lead to damage.

Although estimates of the Kolmogorov microscale are useful in determining whether damage can occur, care should be taken in estimation of the stresses based on this theory. First, the turbulence intensity in a bioreactor is not homogeneous, with almost 100 fold higher local energy dissipation rates found in the impeller region (Zhou and Kresta, 1996). The maximum time-averaged stresses occur in the trailing vortices behind the impeller blades. Many studies do not consider this heterogeneity and have used an average energy dissipation rate, which may not reflect the local peak values. Second, Kolmogorov's theory does not shed light on the mechanisms of cell damage; it does not describe how the cell and a turbulent eddy may interact. It is possible that the presence of cells may affect the fluid flow field, and estimates of velocity and pressure fluctuations on the scale of the cell may not be accurate (Barresi, 1997), although at the typical 0.01 to 0.001 volume fractions, this is unlikely to occur. Also, Reynolds stresses, which arise from a consideration of momentum transport in turbulence are commonly used to estimate stresses on cells (Prokop and Bajpai, 1992), do not have obvious meaning since they include eddy motions of all sizes, whereas only eddies of sizes comparable to the Kolmogorov microscale may be relevant for cell–eddy interactions. Another problem in the application of Kolmogorov's theory is that related to turbulence intermittency as discussed earlier, and further work is required to determine how intermittency affects cell–fluid interactions.

## 18-6.2 Cell Damage Due to Agitation Intensity in Suspension Cell Cultures

Due to the lack of a cell wall, animal and insect cells have long been perceived to be shear sensitive. This perception has led to intense research over the past two decades to uncover the mechanisms responsible for cell damage in bioreactors, although some of the earliest efforts in this regard can be traced back to the 1960s for the BHK-based foot-and-mouth disease (FMDV) vaccine process (Capstick et al., 1965, 1967; Telling and Elsworth, 1965). The FMDV process developed by the Animal Virus Research Institute and the Wellcome Foundation (England) was groundbreaking in many respects, not least because it utilized stirred tank bioreactors at the 3000 to 8000 L scale for the cultivation of continuous tumorgenic animal cells in suspension. However, these early studies of cell damage were not conducted systematically, and hence the damage mechanisms were not readily identifiable. In fact, it was a similar 8000 L bioreactor that was characterized in the studies reported by Langheinrich et al. (1998) and Nienow et al. (1996) and shown to be unsatisfactory with respect to bulk mixing.

Cell damage can be characterized using a number of techniques, including, but not limited to, trypan blue exclusion test, fluorescein diacetate test,

LDH release, and morphological measurements using specific antibody labeling. Sublethal effects have been measured generally by metabolic parameters such as growth rate, product formation, enzymatic activity, protein synthesis rates, membrane activity, and mitochondrial activity, among others. One of the first systematic studies undertaken was by Oh et al. (1989, 1992) using hybridoma cells. They showed that even at relatively high specific power inputs, $P/\rho V$, compared to industrial practice [up to 0.25 W/kg (400 rpm) using Rushton turbines in baffled 1.4 L bioreactors], provided that surface air entrainment did not occur, hybridomas were insensitive to agitation [i.e., cell damage (defined here as cell membrane disruption and measured via trypan blue staining) did not occur]. They discussed this finding in relation to both the large scale vortices behind impeller blades and to the Kolmogorov eddy scale relative to the cell size (Nienow, 1998). They concluded that interactions between cells and isotropic turbulent eddies were unlikely to cause cell death. Considerable cell damage was measured, on the other hand, when air entrainment occurred or sparging was utilized (the relevance of this finding is discussed later in this section). Further evidence to support these findings was provided by the study of Kunas and Papoutsakis (1990), who grew hybridomas at agitation rates of 700 rpm in a 2 L bioreactor in the absence of an air–liquid interface in the headspace. The authors reported that the specific growth rate of the cells was indistinguishable from the control, which was operated at 60 rpm.

Thomas et al. (1994) showed that even when hybridoma cells were agitated at 1000 ($P/\rho V = 7$ W/kg) and 1500 rpm ($P/\rho V = 24$ W/kg) in a 2 L bioreactor, the viable cell concentration was only reduced by 20% and 40%, respectively, in the first 60 min. Thereafter, little change was measured. Even considering that the maximum energy dissipation rate in the impeller swept volume is typically 100 times greater than the mean energy dissipation rate (Zhou and Kresta, 1996), these translate to values significantly lower than those required ($10^5$ to $10^6$ W/kg) for total disruption of hybridoma cells (Zhang et al., 1993). It should be noted that deleterious effects other than cell disruption are possible, such as reduction in DNA synthesis and cell division (Oh et al., 1992) at high agitation rates (especially under sparged conditions), although cell membrane disruption by mechanical forces is by far the predominant issue.

The results reported by Oh et al. (1992) were also in good in agreement with the data reported by Kioukia et al. (1992), who also subsequently (Kioukia et al., 1996) used the same equipment to culture Sf9 insect cells under both virus-infected and virus-noninfected conditions. No difference in growth, infection kinetics, or recombinant protein expression could be measured in the range 100 to 400 rpm (i.e., up to 0.25 W/kg). However, the sensitivity to aeration was confirmed and was even greater for Sf9 cells than for hybridomas.

In all the studies above, we also need to be aware that the results must be viewed in proper context of cell concentration and culture duration. What may appear negligible in a 4 to 5 day culture at about $1 \times 10^6$ cells/mL may become significant in a modern industrial cell culture process at about $1 \times 10^7$ cells/mL for 14 to 21 days, especially using serum-free media (van der Pol and Tramper,

1998). However, in general the studies have shown that animal cell culture processes in stirred bioreactors can be operated at much higher mean specific energy dissipation rates without cell damage than had previously been thought possible.

Ideally, a mechanistic model is needed to better understand the cell–hydrodynamic interactions. However, bioreactor hydrodynamics are turbulent and complex, especially when sparged and have not yet been fully characterized, and given the difficulty of quantifying the magnitude of the shear stress on cells, it is difficult to compare the results obtained from different mixing systems. Therefore, a number of investigators have made use of relatively simpler laminar flow fields in capillary tubes and viscometers to study cell–hydrodynamic interactions. Many of these studies have drawn inconsistent conclusions, partly because of the range of different cell lines, culture history, mode of growth, and physical environment used, and it is difficult to predict from them whether damage can occur under particular hydrodynamic conditions other than those of the original experiments. Cell disruption was studied by Born et al. (1992) in a laminar flow cone and plate viscometer. Cell deformation was assumed to cause an increase in cell membrane tension, and cells would be disrupted if the intrinsic cell membrane tension was exceeded. A model was developed using micromanipulation measurements of the cell bursting membrane tension. Using these measurements, the model was compared to results from exposing animal cells to laminar shear stresses from 124 to 577 N/m$^2$ for 3 min and found to be able to predict successfully loss of viable cells with a maximum error of less than 30%.

In subsequent studies, Zhang et al. (1993) and Thomas et al. (1994) proposed a model to estimate cell disruption in turbulent capillary flows. The flow field in the capillary was described by a laminar sublayer close to the wall and a homogeneous turbulent region elsewhere consisting of eddies of different sizes. Energy exchange was assumed to occur between cells and eddies for eddies approaching or smaller than the size causing cell deformations, which in turn can cause an increase in the membrane tension. Cells from a holding flask were recirculated through the capillary and it was shown that the cell disruption was a first-order decay process (Zhang et al., 1993). Although the model only underestimated the experimental cell disruption by 15%, it lacked explanation of the independence of the specific lysis rate on the number of passes through the capillary. In a further study (Thomas et al., 1994), this problem was resolved by consideration of the localization of the energy dissipation in turbulent capillary flows to a small volume near the vessel wall. The mean specific energy dissipation rate, $\bar{\varepsilon}_T$, used in the study of Zhang et al. (1993) was on the order of $10^3$ to $10^4$ W/kg, while it was estimated that the $\bar{\varepsilon}_T$ value required to disrupt 95% of cells was $10^5$ to $10^6$ W/kg. Since the local $\varepsilon_T$ close to the capillary wall was estimated to be at least an order of magnitude higher than $\bar{\varepsilon}_T$, and the instantaneous $\varepsilon_T$ (due to the intermittency of turbulence) near the wall might even be another order of magnitude higher, it was suggested that nearly the entire cell population would be damaged close to the wall. The study of Thomas et al. (1994) employed a repeated single-pass exposure of the entire culture in a flask (the total volume of culture was collected in a second flask before repeating cell passage through the capillary) and demonstrated a deviation from first-order cell disruption kinetics, with a bias toward breakage of weaker cells first. Al-Rubeai et al. (1995) also suggested that the breakage of cells in turbulent capillary flow was cell-cycle dependent and demonstrated a bias toward the destruction of larger (and hence weaker) $G_2$ cells. The shearing studies described above have generally investigated short-term effects of agitation intensity, and it is possible in some cases that secondary deleterious effects can occur due to fatigue phenomena or reduced biosynthetic activity. For instance, Oh et al. (1989) reported that the specific metabolic activity (measured by the MTT assay) of the surviving cells under sparged conditions without Puronic F68 was higher throughout the duration of the culture than under unsparged conditions. Also, the specific MTT value increased with increasing agitation rates. It was suggested that the enhanced MTT levels were a stress response where the cells invoked the synthesis of enzymes involved in damage repair mechanisms. Under sparged conditions and the presence of Pluronic F68, although the MTT levels were higher at the early stages of the cultivation than under unsparged conditions, the differences became negligible with increasing cultivation time.

The general conclusion from the studies conducted in bioreactors and in simpler flow devices on the effects of agitation intensity in cell cultures is that in the absence of bubble entrainment, such cultures can be operated at much higher mean specific energy dissipation rates without cell damage than had previously been thought possible. For example, in the 8 m$^3$ bioreactors installed at the Wellcome Foundation as reported by Langheinrich et al. (1998), the maximum $\bar{\varepsilon}_T$ available was 0.01 W/kg.

## 18-6.3 Bubble-Induced Cell Damage in Sparged Suspension Cultures

The primary aim of aeration is to provide oxygen to cells. The oxygen requirements of animal and insect cells are low compared to bacteria, ranging from 0.1 to $1.6 \times 10^{-8}$ mg/cell per hour for animal cells to 0.64 to $2.4 \times 10^{-8}$ mg/cell per hour for insect cells (Aunins and Henzler, 1993). Furthermore, culture growth in many cell lines (e.g., CHO) is independent of dissolved oxygen across a range of 5 to 100% of air saturation (Oh et al., 1992). For insect cells the optimum value range of dissolved oxygen has been quoted at 40 to 60% of air saturation (Kloppinger et al., 1990). In most cases, it is not a major problem in meeting oxygen requirements. Headspace aeration and silicone tubing aeration is generally sufficient in laboratory, and in some cases, pilot scale equipment. However, as the scale increases, it becomes necessary to sparge air directly into the medium (especially for insect cells under infection conditions). Due to considerations of bubble-induced cell damage, particularly in the absence of surfactants for cell protection as well as $CO_2$ ventilation and pH control, aeration of cell cultures requires care.

After nearly two decades of research, it is now generally accepted that mechanical damage of freely suspended cells is due to bubble hydrodynamics, in particular the bubble-bursting phenomena at the headspace gas–liquid interface (Handa et al., 1987; Tramper et al., 1987; Handa-Corrigan et al., 1989; Oh et al., 1989,

1992; Chalmers and Bavarian, 1991; Kioukia et al., 1992; Meier et al., 1999). Four bubble-liquid-cell regions where cell damage can occur can be identified: (1) bubble formation at the sparger, (2) bubble coalescence and breakup in the impeller discharge, (3) bubble rise through the bioreactor, and (4) bubble bursting at the air–medium interface. Kilburn and Webb (1968) demonstrated the protective effects of Pluronic F68 in moderating cell damage. Tramper et al. (1987) suggested the concept of a killing volume that was associated with bubble frequency and independent of rise height, and therefore, cell damage was primarily as a result of bubble disengagement. Handa et al. (1987) and later Handa-Corrigan et al. (1989) formalized these concepts into an experimental framework and showed that cell damage occurs due to bubble bursting, and that the amount of damage was related to the bubble size and frequency and that cell damage could be reduced by the addition of Pluronic F68. Oh et al. (1992) extended and confirmed these findings and suggested that bubbles with diameters <5 mm were the most lethal and air sparging coupled with aeration led to increased damage compared to aeration alone. Less damage was found when air was sparged above the impeller compared to sparging below it. It was not possible to ascertain unequivocally whether the additional damage was associated with the trailing vortex bubble breakup per se, since this affected both bubble size and frequency, both of which are implicated in bubble bursting cell damage.

Tramper et al. (1987) estimated the shear stress generated by rising bubbles and found that it was considerably lower than that due to bubble bursting and negligible for cell damage. Bubble rupture at the gas–liquid interface involves several dynamic events (Newitt et al., 1954; Garcia-Briones and Chalmers, 1994; Boulton-Stone, 1995; Wu, 1995; Dey et al., 1997) which have been documented using high-speed video photography. For further details the reader is referred to these publications. However, in summary a bursting-bubble rupture starts at the thinnest apex of the film cap, where a hole is formed, followed by a rapid extension of the hole boundary. The receding bubble film is a very fast process, and for instance for a 2 µm liquid film, the receding velocity is estimated to be 8 m/s. The shear stress in the receding film has been estimated by Cherry and Hulle (1992) at around 95 N/m$^2$ in laminar flow and up to 300 N/m$^2$ in the boundary layer surrounding the bubble cavity (Chalmers and Bavarian, 1991). Following a cascade of events, two liquid jets are produced, one downward into the liquid and one upward over the bubble cavity. The velocity of the rising liquid jet has been estimated at 5 m/s for a 1.7 mm bubble (MacIntyre, 1972). Calculations by Kowalski and Thomas (1994) suggest that power dissipated during the ejection of the jet is in the region of 0.5 kW/mL ($5 \times 10^5$ W/kg), which itself can be a source of cell damage. The liquid jet eventually breaks into smaller liquid droplets, often at speeds up to 10 m/s. Wu and Gossen (1995) suggested that unless cells are concentrated into the bubble film cap before it ruptures, the contribution of this region to cell damage is likely to be small, as it constitutes a small proportion of the total bubble interface. Therefore, according to these results, the bubble crater collapse is the prime cause of cell death, although some caution should be applied in interpreting these results since Wu and Gossen

(1995) employed cell viability rather than viable cell concentration as an indicator of damage. Boulton-Stone (1995) developed a numerical model of the bubble-bursting process and was able to predict the bubble-bursting effects of a wide range of surface properties. The surface dilatational viscosity seemed particularly important. Dey et al. (1997) attempted to validate the model but could do so successfully only by setting the value of the surface dilatational viscosity to at least an order of magnitude higher than the value determined experimentally.

A foam layer can also develop at the gas–liquid interface at the surface in many cultures since most animal cell media contain chemicals that cause foaming. Cell damage measured in such cases have shown there to be a greater percentage of dead cells in the foam layer than in the bulk (Wu, 1995), probably attributable to liquid film drainage processes. It is likely that all the high-energy events associated with bubble bursting described above contribute to cell damage. From a practical viewpoint, an understanding of how the transport of cells to the headspace gas–liquid interface can be minimized can be used to reduce cell death.

The question of how cells are transported to the headspace gas–liquid interface has been addressed by various authors, including Tramper et al. (1987), Handa-Corrigan et al. (1989), Jöbses et al. (1991), Cherry and Hulle (1992), Garcia-Briones and Chalmers (1992), and Meier et al. (1997). Garcia-Briones and Chalmers (1992), who visualized cells attached to rising bubbles and suggested that the major cause of cell bubble attachment appears to be hydrophobic interaction via bubble–cell collision. This was in contrast to the observations made by Cherry and Hulle (1992) with insect cells and by Handa-Corrigan et al. (1989) with hybridoma cells—that few cells were attached to rising bubbles. The major cause of cell–bubble contact appears to be through hydrophobic interaction (Wu, 1995), although very little is known about such mechanisms with respect to animal and insect cultures. Knowledge of the mechanism of cell attachment to bubbles and the conditions under which such attachment occurs may help in minimizing cell damage due to bubble bursting.

### 18-6.4 Use of Surfactants to Reduce Cell Damage Due to Bubble Aeration in Suspension Culture

Surfactants such as the various members of the Pluronic family, especially Pluronic F68 (Kilburn and Webb, 1968), methylcellulose, polyethylene glycol, and serum, have long been known to decrease cell damage due to bubble aeration (Handa et al., 1987; Murhammer and Goochee, 1990). Papoutsakis (1991) reviewed the role of these media additives for protecting freely suspended cells and suggested that the all surfactants mentioned earlier imparted shear protective effects to differing degrees. However, a mechanistic understanding of their role in preventing cell damage was still lacking, and attempts to compare the results of using these additives in various studies was hampered by the diversity in equipment type, scale, operating conditions, and cell line. Michaels et al. (1991) suggested that the protective effects of surfactants can be both biological and physical in nature. A biological mechanism was taken to imply when addition of the surfactant directly resulted in an increase in the cell membrane strength. A physical mechanism implied that the cell resistance to shear remains unaltered, but the factors such as the medium properties that affect both the level and frequency of high-shear events (e.g., bubble bursting) change in a manner such that cell damage decreases. By far the most commonly used additive is Pluronic F68. Attempts to correlate cell death with media (containing Pluronic F68) properties such as surface tension (both static and dynamic) and viscosity have been unsuccessful (Dey et al., 1997). Dey et al. (1997) showed that bubble burst could be moderated by the addition of various concentrations of Pluronic F68 to basal medium: in particular, a decrease in the time of jet formation, the height and width of jets, and the number of subsequent liquid drops released.

The role of surfactants to reduce cell attachment as a means to reduce cell damage due to bubble bursting has been investigated by various researchers with inconsistent findings. Tramper et al. (1987), using a medium with 0.1% methylcellulose as surfactant, reported that cell death rate was directly proportional to gas flow rate and inversely proportional to culture volume in bubble columns, suggesting that the death rate is independent of the bubble residence time, a conclusion supported by the work of Handa-Corrigan et al. (1989), and Jöbses et al. (1991). On the other hand, others, including Cherry and Hulle (1992) and Garcia-Briones and Chalmers (1992), have clearly shown cell attachment to bubbles. Thus cell attachment to bubbles should result in a death rate that is dependent on bubble residence time, which in turn determines the number of attached cells. Meier et al. (1999) incorporated the foregoing findings into a single framework and concluded that cell death due to bubble bursting depends on cell attachment to bubbles. The discrepancy in the literature data was attributed to the differences in experimental setup (presence of surfactant in media, reactor geometry, and bubble size) and cell line.

Meier et al. (1999) concluded that insignificant cell attachment occurred in the presence of surfactants such as Pluronic F68 and that large scale bioreactors should be designed with a high aspect ratio, since bubble residence time is not important for cell damage mediated by bubble bursting. However, it is also important to remember that good homogenization is also required (see Section 18-6.6) and is much slower in vessels of high aspect ratio (Langheinrich et al., 1998), so this conclusion of Meier et al. (1999) may be inappropriate. On the other hand, in the absence of surfactants where bubble residence time is important, Meier et al. (1999) recommended a lower-aspect-ratio bioreactor, since this should lead to lower rates of cell attachment to bubbles and hence reduced cell damage.

Another mechanism by which surfactants such as Pluronic F68 can render a protective effect is biological in nature. Murhammer and Goochee (1990) and Goldblum et al. (1990) found that Pluronic F68 could be rapidly incorporated onto the membranes of insect cells and thus offer protection. Ramirez and Mutharasan (1990) used a fluorescence polarization method to demonstrate that Pluronic F68 decreased the membrane fluidity (and hence membrane strength) throughout a batch culture. Furthermore, Zhang et al. (1992) used a micromanipulation method to measure the strength of a hybridoma cell line from a continuous culture with and without added Pluronic F68. A significant increase in both the bursting membrane tension and mean elastic compressibility modulus (both indicators of strengthened cell membranes) was found in the presence of 0.5 g/L of Pluronic F68. It is likely that no single parameter is solely responsible for the protective effects offered by surfactants such as Pluronic F68, but rather, a combination of moderation of the bubble burst, reduced cell attachment to bubbles, and adsorption/incorporation into the cell membrane.

### 18-6.5 Cell Damage Due to Agitation Intensity in Microcarrier Cultures

Many animal cell lines such as VERO cannot be adapted to suspension culture, and such adherent cells are often grown on both conventional and macroporous microcarriers. For information on the effects of shear stress on anchorage-dependent cells, the reader is referred to the reviews by Prokop and Bajpai (1992) and Aunins and Henzler (1993). Microcarriers offer advantages of supporting high cell densities via large surface areas for growth, relatively easy cell-media separation, and scalability. Macroporous microcarriers provide some protection to adherent cells from agitation and aeration. However, to maximize this advantage, the internal space must be readily accessible to cells. Although agitation sufficient for microcarrier suspension may allow for adequate oxygen mass transfer, higher levels of agitation may be required if oxygenation is via surface aeration (especially at large scales of operation) or if high cell density cultures are desired. The main goal of successful microcarrier cultivation is to maximize cell density without detrimental effects of agitation, while providing adequate oxygen transfer and a homogeneous bioreactor environment. A major problem is that adherent cells can be removed from the microcarrier surfaces at steady laminar stresses in the range 0.5 to 10 N/m$^2$ accompanied by loss of viable cells and productivity (Aunins and Henzler, 1993). Unlike in suspension cultures at realistic power inputs (Oh et al., 1989, 1992), Croughan and Wang (1987) showed that cell damage in microcarrier cultures could be related to the Kolmogorov scale of turbulent eddy dissipation. In microcarrier suspension cultures, turbulent eddies in the viscous dissipation region are often intermediate in size between the cells and microcarriers. In this case, the high rate of specific energy dissipation as eddies interact with the surface of the microcarriers causes local transient shear rates to be sufficiently high to remove the cells from the microcarrier surface.

Croughan et al. (1989) analyzed selected published data on microcarrier cultivation of FS-4 cells in spinner flasks and showed that cell death occurred when the eddy size ($\approx$100 μm) calculated using the global energy input to the reactor was smaller than the average microcarrier diameter (185 μm) and proposed a model for cell death:

$$\frac{dx}{dt} = \begin{cases} \mu x & \text{for } \lambda_{\text{global}} > d_{\text{microcarrier}} \quad (18\text{-}74) \\ (\mu - k_d)x & \text{for } \lambda_{\text{global}} < d_{\text{microcarrier}} \quad (18\text{-}75) \end{cases}$$

where

$$k_d = K \left(\frac{\bar{\varepsilon}_T}{v^3}\right)^{0.75} \qquad (18\text{-}76)$$

and x is the cell concentration, $\mu$ the specific growth rate of the cells, $k_d$ the specific death rate constant, K a cell and bioreactor-dependent constant (Croughan et al., 1987), $\bar{\varepsilon}_T$ the mean specific energy dissipation rate in the bioreactor (= P/$\rho$V), and $v$ the medium kinematic viscosity. A note of caution may be appropriate when comparing the predictions of the models with experimental data given the inaccuracy of estimating the power input using a stirrer bar in the spinner flasks from literature correlations and the fact that the assumption of isotropic turbulence in a spinner of such scales at the Reynolds numbers employed was probably not valid. However, analysis of literature data showed that the dependency of $k_d$ on the average energy dissipation rate (slope = 0.75) was close to that suggested by eq. (18-76). From a scale-up viewpoint, provided that the turbulent eddy-cell model is valid, scale-up at equal energy dissipation rates should not lead to detrimental hydrodynamic effects. It should also be noted that data on the effects of aeration with respect to cell damage in microcarrier cultures are scarce (Aunins et al., 1986).

Croughan et al. (1988) also addressed the effects of cell–cell collisions and cell–impeller collisions with respect to cell damage. In a typical microcarrier bioreactor, the beads can collide with one another as well as with the impeller and other vessel internals if the collision forces are greater than the force required to remove the liquid film between them (Prokop and Bajpai, 1992). In this case, cell damage will be a function of collision frequency and the amount of energy transmitted during the collision event. They found that the rate of cell damage due to bead–bead collision, dc/dt $(\bar{\varepsilon}_T)^{3/4}C_b$, where $C_b$ is the microcarrier bead concentration. Design implications for bead–impeller collisions were also discussed by Croughan et al. (1988). The effects of cell–cell and cell–impeller collisions in suspension cultures is not significant compared to the hydrodynamic forces due to agitation and aeration and also because of dissipation of the collision energy due to cell-surface deformation.

A number of cell lines like BHK also exhibit aggregation behavior, and in such cases, Moreira et al. (1995) demonstrated that the aggregate size can be correlated to the Kolmogorov eddy scale using both $\bar{\varepsilon}_T$ and the local specific energy dissipation rates, although the latter was shown to be marginally superior in correlating the aggregate breakage. However, in contrast to microcarriers, breakage of cell aggregates at high agitation speeds led to smaller and more compact aggregates without resulting in cell damage. This advantage can be used to grow cells in aggregates as a natural immobilization system at higher energy dissipation rates where microcarriers cannot be used.

### 18-6.6 Physical and Chemical Environment

One of the consequences of the perceived shear sensitivity of animal cells to agitation intensity and operation at low agitation speeds is the inability to provide adequate liquid homogenization in large scale bioreactors. Typical values of energy dissipation rates are two orders below that used in bacterial fermentations. One of the very few mixing and aeration studies in large scale bioreactors under animal cell conditions, other than the one reported by Kiss et al. (1994), was reported by Nienow et al. (1996) and Langheinrich et al. (1998). These authors conducted comprehensive studies of homogenization in a commercial scale 8 m$^3$ bioreactor and in a geometrically similar scaled-down 0.61 m mixing tank. The 8 m$^3$ bioreactor was operated in industrial applications for cell culture growth at a maximum speed of 60 rpm with a single Rushton turbine with an unusually low D/T and clearance ratios (C/T) of 0.22 and an aeration rate via a pipe sparger of 0.005 vvm. Under such conditions the energy dissipation rates due to agitation and aeration were $11.5 \times 10^{-3}$ and $2 \times 10^{-3}$ W/kg, respectively. First, they showed that under such operating conditions the mixing time was >200 s for H/T = 1.3 and that the mixing time correlations derived in the literature for 1 < H/T ratio < 3 (Cooke et al., 1988) were capable of predicting the experimentally derived mixing times in the 8 m$^3$ bioreactor, even though the energy dissipation and aeration rates were two orders of magnitude lower than those used to derive the correlations. Second, they demonstrated that mixing at the top of the bioreactor was poor, due to the inability of the upper recirculation loop to reach the top of the bioreactor, and also that the aerated mixing time was significantly shorter than under unaerated conditions at the top of the bioreactor, due to the energy dissipated by bubble disengagement and convective flow. Suggested strategies for significant improvement in liquid blending arising from experimentation in the scaled-down tank included retrofitting the small diameter high power number Rushton with a large diameter low power number axial flow impeller such as an up-pumping Chemineer HE3 or a Prochem Maxflo T, the use of increased energy dissipation rates, and an increased impeller clearance ratio C/T of 0.5.

For caustic pH control, alkali is generally added at the liquid surface at the later stages of cell culture processes, and it is here that pH excursions are known to occur, especially if liquid blending is poor. Langheinrich and Nienow (1999) also used the geometrically scaled down 0.61 m tank to investigate possible pH excursions in the 8 m$^3$ bioreactor. At the operating conditions described earlier, they found significant pH excursions of up to 0.8 pH unit at the point of alkali addition in a buffered system using both pH probe measurements and flow visualization studies. Such excursions can be detrimental for cell viability and product formation (Brown and Birch, 1996; Osman et al., 2001, 2002). The pH variations were absent when alkali was added to the impeller region. They also found that an increase in the overall energy dissipation rate did not have a significant impact on the pH excursions given the substantial differences in the local energy dissipation rates at the liquid surface and the impeller region. The authors recommended that the pH excursions could, realistically, be avoided only with alkali added to the impeller region via a pipe. Traditionally, this option has not been implemented in industrial scale bioreactors due to "clean in place" (CIP) considerations.

One consequence of using low aeration rates to prevent bubble-associated cell damage is the possible accumulation of dissolved $CO_2$ with increasing hydrostatic pressure (or scale) to inhibitory or metabolism-altering concentrations, due to insufficient ventilation as a result of low air sparging rates or using oxygen-enriched air. This can be a major problem in large scale bioreactors, especially when operated at high cell densities. $CO_2$ production rates between 0.17 to $1.3 \times 10^{-8}$ mg/cell per hour have been reported for a range of cell lines (Aunins and Henzler, 1993). Kimura and Miller (1996) reported that CHO cells cultivated in laboratory scale bioreactors were moderately tolerant to dissolved $CO_2$ (up to 18% $CO_2$), while Gray et al. (1996) reported an inhibitory effect at 14% dissolved $CO_2$ in large scale cultures. $CO_2$ ventilation rate is affected by control strategy used to control dissolved oxygen (surface aeration, air sparging, partial pressure of oxygen in the inlet gas, and type of sparger) and therefore, a balance between oxygenation, $CO_2$ ventilation, and pH control needs to be made. One solution to this problem is to employ two spargers serving different functions; a sintered sparger to provide small bubbles to enhance oxygen mass transfer and a pipe sparger producing relatively larger bubbles for $CO_2$ ventilation while ensuring minimal cell damage. The inhibitory effects of $CO_2$ are evident at lower concentrations if the culture osmolality is >300 mOs, as may be the case in fed-batch cultures. Increased osmolality up to 400 mOs has also been shown to increase the specific antibody production rate, although this enhancement is cell-line dependent. Cyclic exposure of cells to increasing hydrostatic pressure with scale can also occur, although results reported by Tagaki et al. (1995) at constant pH, $pCO_2$, and $pO_2$ using hybridoma cultures suggested that this is unlikely to affect culture metabolism per se.

Despite advances in our understanding of cell damage due to bubble aeration and under realistic conditions of agitation, we still lack sufficient knowledge for an optimal a priori design. This is due to incomplete understanding of the complex two-phase hydrodynamics in bioreactors (e.g., bubble size distribution in large scale cell culture media), bubble–cell attachment mechanisms, effects of surfactants on media physical properties and due to the diversity of cell line, type of culture media, and operating conditions. However, a general guideline for design and operation of large scale bioreactors for freely suspended animal cells should probably consider the following (Nienow, 1997):

- Multiple impellers in high aspect ratio bioreactors with a large diameter radial flow Rushton turbine or Scaba (or up-pumping axial flow) impeller with clearance ratios between 0.33 and 0.5, good for gas dispersion, and an up-pumping or down-pumping large diameter axial flow impeller for liquid blending above it. Increased agitation intensity to ensure good liquid blending is also recommended.
- Design and operating conditions should be chosen to avoid pH excursions. "Clean in place" issues related to addition pipes for subsurface addition of concentrated reagents should be addressed.

- Oxygenation requirements should be balanced against $CO_2$ ventilation rates, perhaps employing different sparger types for each purpose if needed. Mean bubble sizes should generally be maintained above 5 mm, although this is not easy in practice.
- Pluronic at concentrations greater than 0.5 g/L should be used to minimize cell damage due to bubble aeration.

## 18-7 PLANT CELL CULTURES

Plants are recognized as an important source of natural compounds for use in the pharmaceutical and food industries. For example, the commercially important anticancer drug Taxol (Bristol-Myers Squibb) was originally isolated from the Pacific yew tree, *Taxus brevifolia*, requiring the bark of 1000 trees to produce 1 kg of Taxol (Kieran et al., 1997). It is currently produced via a semisynthetic route using taxane precursors extracted from trees. Plant cell cultures provide an alternative technology to produce such chemicals if the source plant is scarce, difficult to cultivate, or where chemical synthesis is challenging or not possible. The commercialization of plant cell culture has been limited due mainly to process economics, which in turn is governed by biological and engineering considerations. The success of industrial scale plant cell cultures depends on a number of factors, but is principally on the development of high-yielding cell lines and the ability to grow the plant cell suspension at large scales. The stirred tank bioreactor is still considered the most economically feasible bioreactor design for the cultivation of plant cells. In this section we summarize the key engineering factors that need to be considered from a mixing viewpoint. For detailed aspects of the impact of mixing in plant cell cultures, the reader is referred to selected publications and reviews by Leckie et al. (1991), Prokop and Bajpai (1992), Scragg (1992), Meijer et al. (1993, 1994), Shuler (1993), Namdev and Dunlop (1995), Joshi et al. (1996), Kieran et al. (1997), Doran (1999), and Rodríguez and Galindo (1999).

Although much of the know-how to design mixing systems for plant cell cultures can be borrowed from the fermentation and cell culture fields, a number of characteristics of plant cells make them unique. Plant cells are typically 10 to 100 μm in diameter and often grow in aggregates up to 500 to 2000 μm in size, due to the presence of hydrophobic glycan in the cell wall and incomplete cell separation after cell division. Due to their size and their rigid cellulosic cell wall, plant cells and their aggregates, similar to animal cells, are commonly considered to be sensitive to hydrodynamic shear stress. However, commercial production requires high cell densities, and coupled with the excretion of polysaccharides (unlike animal cells), often as a response to hydrodynamic stress (Rodríguez and Galindo, 1999), can result in rheological complexity, with the broth often exhibiting non-Newtonian behavior. Although the oxygen requirements of plant cells is low, with reported values in the range 1.5 to $6.3 \times 10^{-3}$ g $O_2$/kg · s (Doran, 1999), the ability to provide sufficient oxygen transfer is limited by considerations of shear sensitivity and rheology. Another requirement in plant cell cultures is cell suspension. Foaming is also a particular problem and in some cases can be severe (Zhong et al., 1992). In many cases the product of interest is intracellular, and cell disruption is required for downstream processing.

From a mixing viewpoint, only the effects of hydrodynamic shear stress and agitator design have been reported in the literature and to a much lesser extent and rigor compared to microbial, fungal, and animal cells. Hydrodynamic shear stress has been investigated both in bioreactors under growth conditions (using energy dissipation and tip speed as correlating factors) and under well-defined flow conditions in viscometric devices. Critical shear stress (using regrowth of cells as an indicator) in the range 50 to 200 N/m$^2$ have been reported (Kieran et al., 1997). Translation of such values to stirred tank conditions is difficult. Cell damage can be manifested as a lethal or sublethal effect. Lethal effects have been characterized by cell lysis, release of intracellular compounds, changes in aggregate size, and biomass yield. Prokop and Bajpai (1992) discussed the limitations of some of these measurements. For instance, since plant cells form large aggregates, cell number per unit volume is not an easily accessible parameter. Data on aggregate size distribution are scarce and often without statistical analysis. In addition, reports on biomass yields vary greatly among investigators, even using the same cell line and growth conditions. Namdev and Dunlop (1995) also cited the limitations of shear-related studies in plant cell culture and proposed a more systematic approach focusing on sublethal responses such as transmembrane activity, stress protein expression, osmoregulation, and aggregation.

Despite the difficulties in studying the effects of hydrodynamic shear stress in bioreactors, some useful conclusions have emerged. Doran (1999) reported a theoretical engineering analysis (cultivation of the culture was not performed) of impeller type and geometry for the application of plant cell culture in a 10 m$^3$ fermenter under continuous flow conditions. Mixing, mass transfer, impeller gas handling, and cell suspension characteristics of both Rushton turbines and axial flow impellers (both up- and down-pumping) demonstrated that upward-pumping axial flow impellers offered advantages compared to Rushton turbines for gas handling and cell suspension when the power input was restricted by cell damage considerations (Figure 18-21). Initial concerns of universal hydrodynamic damage to plant cells encouraged the widespread adoption of airlift bioreactors. As Meijer et al. (1993) point out, shear sensitivity of plant cells became established as a near axiom after quotation by many investigators. Meijer et al. (1993) conducted a through review of the effects of hydrodynamic stress on plant cells and concluded that although significant differences in shear tolerance existed between cell lines, plant cells were not as susceptible to hydrodynamic damage as had originally been thought. For instance, Leckie et al. (1991) demonstrated that using Rushton turbines and pitched blade impellers in 12 L fermenters, agitation speeds up to 300 rpm did not affect the growth of *Catharanthus roseus* and alkaloid production. In fact, culture growth was possible even up to 1000 rpm, although alkaloid production decreased. Similar conclusions were also reported by Scragg et al. (1988) and Meijer et al. (1994). Meijer et al. (1994) reported no difference in growth or product formation capabilities of *Catharanthus roseus*

**1154** MIXING IN THE FERMENTATION AND CELL CULTURE INDUSTRIES

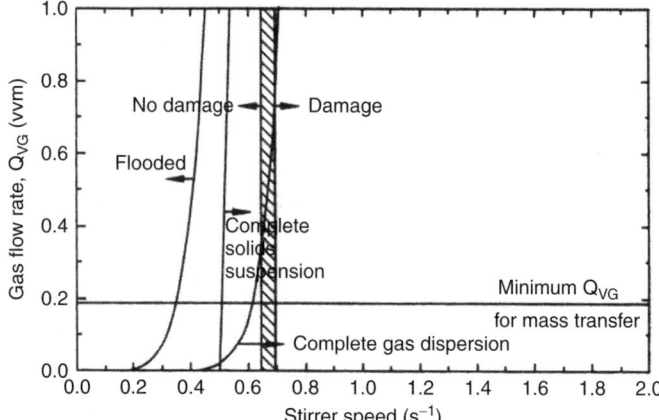

**Figure 18-21** Flow regime map for upward-pumping six-blade 45° pitched blade turbine with a D/T ratio of 0.40 using plant cell cultures in a 10 m³ bioreactor. The flow regime map shows that complete cell suspension occurs below the agitation speed required to damage cells independent of gas flow rate. Taking into account the minimum gas flow rate required for oxygen requirements, there is a small window of operating conditions where the impeller functions of complete gas dispersion, cell suspension, and oxygen mass transfer are met while avoiding cell damage. (From Doran, 1999.)

and *Nicotiana tabacum* up to agitation speeds of 1000 rpm using a Rushton turbine in a 12 L bioreactor, although cultures of *Cinchona robusta* and *Tabernaemontana divaricata* did exhibit detrimental effects at these agitation speeds. In general, plant cell damage mechanisms have been difficult to identify, due to the diversity of cell lines, aggregate morphologies, culture age, and cultivation history. Further systematic studies of oxygen mass transfer, nutrient availability, cell suspension, aggregation behavior, and cell damage in response to the intensity of mixing are required to provide a better understanding for successful scale-up. Furthermore, very little has been reported on the metabolism of plant cells, the effects of aeration as a cell-damaging parameter, and the influence of morphology on growth and product formation.

Interestingly, Meijer et al. (1993) and Kieran et al. (1997) also suggested that more effort should be devoted to establishing robust and reproducible cell lines than developing low shear bioreactors for fragile cell lines, which may not ultimately be robust and grow well. In some instances, plant cells can be adapted to withstand higher shear stresses by changing subculturing conditions and allowing the cells to proliferate for a longer time at low shear stresses in vitro (Drapeau et al., 1987).

## NOMENCLATURE

C          impeller clearance (m)
C          tracer concentration at time t (—)

# NOMENCLATURE

| | |
|---|---|
| $C_L^*$ | dissolved oxygen concentration at air saturation (% air saturation) |
| $C_L$ | dissolved oxygen concentration in the liquid phase (% air saturation) |
| CTD | circulation time distribution (−) |
| $C_b$ | microcarrier bead concentration |
| $C_v$ | cavern volume (m$^3$) |
| D | impeller diameter (m) |
| $D_c$ | cavern diameter (m) |
| DOT | dissolved oxygen tension (% air saturation) |
| Da | Damkoehler number—system residence time/characteristic reaction time (dimensionless) |
| $d_{microcarrier}$ | diameter of microcarrier beads (μm) |
| EDC | energy dissipation/circulation function $P/(kD^3 t_c)$ (kWm$^{-3}$s$^{-1}$) |
| F | total force imparted by the impeller, radial and axial (eq. 18-64) (Nm$^{-2}$) |
| F | $Q_R/V_{STR}$ (s$^{-1}$) |
| $F_a$ | axial force (eq. 18-65) (Nm$^{-2}$) |
| Fl | flow number $Q/ND^3$ (dimensionless) |
| $Fl_G$ | gassed flow number ($Q_G/ND^3$) (dimensionless) |
| Fr | Froude number ($N^2D/g$) (dimensionless) |
| F(t) | circulation time distribution (−) |
| g | gravitational constant (ms$^{-2}$) |
| H | liquid height (m) |
| $H_c$ | cavern height (m) |
| $H_{cw}$ | height of cavern when cavern diameter = vessel diameter (eq. 18-67) (m) |
| K | fluid consistency index (Nm$^{-2}$) |
| k | geometric constant for a given impeller (eq. 18-69) |
| $k_d$ | specific death rate constant of cells (h$^{-1}$) |
| $k_La$ | mass transfer coefficient (s$^{-1}$) |
| m | number of circulations (eq. 18-54) (−) |
| N | impeller rotational speed (rps) |
| N | number of tanks (−) |
| $N_F$ | minimum impeller speed to prevent flooding |
| $N_f$ | axial force number (eq. 18-65) (dimensionless) |
| $N_R$ | speed at which onset of gas recirculation occurs (rps) |
| $N_w$ | impeller speed at which cavern touches vessel wall (eq. 18-68) (rps) |
| n | flow behavior index (−) |
| OUR | oxygen uptake rate (mmol L$^{-1}$ h$^{-1}$) |
| OTR | oxygen transfer rate (mmol L$^{-1}$ h$^{-1}$) |
| P | fluid mixing power (W) |
| $P_g$ | gassed power (W) |
| Po | power number (dimensionless) |
| PFR | plug flow reactor |

| | |
|---|---|
| p | impeller dependent exponent defining cavern expansion (eq. 18-66) (−) |
| Q | impeller pumping capacity (m$^3$ s$^{-1}$) |
| $Q_{O2max}$ | maximum specific oxygen uptake rate (mmol/g biomass/s) |
| $Q_G$ | gas flow rate (m$^3$ s$^{-1}$) |
| $Q_R$ | gas recirculation rate |
| Re | Reynolds Number (ND$^2/\mu$) (dimensionless) |
| $r_c$ | radius of a torus shaped cavern (eq. 18-63) (m) |
| STR | stirred tank reactor |
| t | time (s) |
| T | vessel diameter (m) |
| $t_c$ | circulation time (s) |
| $\bar{t}_c$ | mean circulation time (s) |
| $t_{oc}$ | oxygen consumption time (s) |
| $t_m$ | mixing time (s) |
| $t_R$ | residence time in the PFR (s) |
| $U_T$ | impeller tip speed (ms$^{-1}$) |
| V | total volume of fluid in the vessel (m$^3$) |
| $V_a$ | volume of the active zone (m$^3$) |
| $V_i$ | impeller swept volume (m$^3$) |
| $V_o$ | fluid velocity at cavern boundary (ms$^{-1}$) |
| $V_q$ | volume of the quiescent zone (m$^3$) |
| $V_{PFR}$ | volume of PFR (m$^3$) |
| $V_{STR}$ | volume of STR (m$^3$) |
| $v_s$ | superficial gas velocity (ms$^{-1}$) |
| vvm | volumetric flow of gas per liquid volume per minute (min$^{-1}$) |
| w | impeller blade width (m) |
| x | cell concentration (g L$^{-1}$) |
| x | biomass concentration |
| $\dot{\gamma}$ | fluid shear rate (eq. 18-60) (s$^{-1}$) |
| ε | gas holdup (%) |
| $\varepsilon_T, \bar{\varepsilon}_T$ | rate of dissipation of turbulent kinetic energy (W kg$^{-1}$) |
| $\lambda_k$ | Kolmogorov scale (μm) |
| μ | fluid viscosity (Nm$^{-2}$s) |
| η | efficiency index (eq. 18-28) (−) |
| μ | specific growth rate (h$^{-1}$) |
| $\mu_l$ | mean of lognormal distribution (s) |
| ν | kinematic viscosity (m$^2$s$^{-1}$) |
| ρ | density (kg m$^{-3}$) |
| σ | surface tension (Jm$^{-2}$) |
| σ, $\sigma_\theta$ | standard and normalized deviations of mean circulation time (S, −) |
| τ | fluid shear stress (eq. 18-60) (Nm$^{-2}$) |
| $\tau_y$ | fluid yield stress (Nm$^{-2}$) |

*Subscripts*

c     Casson
CD   Complete dispersion
F     flooding
HB   Herschel–Bulkley
t     turbulent flow regime

## REFERENCES

Al-Rubeai, M., R. P. Singh, A. N. Emery, and Z. Zhang (1995). Cell cycle and cell size dependence of susceptibility to hydrodynamic forces, *Biotechnol. Bioeng.*, **46**(1), 88–92.

Amanullah, A. (1994). Scale down models of mixing performance in large scale bioreactors, Ph.D. dissertation, University of Birmingham, England.

Amanullah, A., A. Baba, C. M. McFarlane, A. N. Emery, and A. W. Nienow (1993a). Biological Models of Mixing Performance in Bioreactors, *Proc. 3rd International Conference on Bioreactor and Bioprocess Fluid Dynamics*, Cambridge University, Sept. 14–16, pp. 381–400.

Amanullah, A., A. W. Nienow, A. N. Emery, and C. M. McFarlane (1993b). The use of *Bacillus subtilis* as an oxygen sensitive culture to simulate dissolved oxygen cycling in large scale fermenters, *Trans. Inst. Chem. Eng., Part C*, **71**, Sept., pp. 206–208.

Amanullah, A., S. A. Hjorth, and A. W. Nienow (1998a). A new mathematical model to predict cavern diameter in highly shear thinning power law liquids using axial flow impellers, *Chem. Eng. Sci.*, **53**(3), 455–469.

Amanullah, A., L. Carreon-Serrano, B. Castro, E. Galindo, and A. W. Nienow (1998b). The influence of impeller type in pilot scale xanthan fermentations, *Biotechnol. Bioeng.*, **57**(1), 95–108.

Amanullah, A., B. Tuttiett, and A. W. Nienow (1998c). Agitator speed and dissolved oxygen effects in xanthan fermentations, *Biotechnol. Bioeng.*, **57**(2), 198–210.

Amanullah, A., R. Blair, C. R. Thomas, and A. W. Nienow (1999). Effects of agitation speed on mycelial morphology and protein production in chemostat cultures of recombinant Aspergillus oryzae, *Biotechnol. Bioeng.*, **62**(4), 434–446.

Amanullah, A., P. Jüsten, A. Davies, A. W. Nienow, and C. R. Thomas (2000). Agitation speed induced mycelial fragmentation in *Penicillium chrysogenum* and *Aspergillus oryzae* cultures, *Biochem. Eng. J.*, **5**, 109–114.

Amanullah, A., E. Leonildi, A. W. Nienow, and C. R. Thomas (2001a). Dynamics of mycelial aggregation in cultures of *Aspergillus oryzae*, *Bioprocess. Biosys. Eng.*, **24**, 101–107.

Amanullah, A., C. M. McFarlane, A. N. Emery, and A. W. Nienow (2001b). Experimental simulations of pH gradients in large scale bioreactors using scale down models, *Biotechnol. Bioeng.*, **73**(5), 390–399.

Amanullah, A., L. H. Christensen, K. Hansen, A. W. Nienow, and C. R. Thomas (2002). Dependence of morphology on agitation intensity in fed-batch cultures of *Aspergillus oryzae* and its implications for recombinant protein production, *Biotechnol. Bioeng.*, **77**(7), 815–826.

Aunins, J. G., and H. J. Henzler (1993). Aeration in cell cultures, in *Biotechnology*, 2nd ed., H. J. Rehm and G. Reed, eds., Vol. 3, *Bioprocessing*, G. Stephanopoulos, vol. ed., VCH, Weinheim, Germany.

Aunins, J. G., J. M. Goldstein, M. S. Croughan, and D. I. C. Wang (1986). Engineering developments in the homogenous culture of animal cells: oxygenation of reactors and scale-up, *Biotechnol. Bioeng. Symp.*, **17**, 699.

Ayazi Shamlou, P., H. Y. Makagiansar, A. P. Ison, and M. D. Lilly (1994). Turbulent breakage of filamentous micro-organisms in submerged culture in mechanically stirred bioreactors, *Chem. Eng. Sci.*, **49**(16), 2621–2631.

Bailey, J. E., and D. F. Ollis (1986). *Biochemical Engineering Fundamentals*, 2nd ed., McGraw-Hill, New York.

Bajpai, R. K., and M. Reuss (1982). Coupling of mixing and microbial kinetics for evaluating the performance of bioreactors, *Can. J. Chem. Eng.*, **60**, 384–392.

Baldyga, J., and J. R. Bourne (1995). Interpretation of turbulent mixing using fractals and multifractals, *Chem. Eng. Sci.*, **50**, 381–400.

Baldyga, J., J. R. Bourne, A. W. Pacek, A. Amanullah, and A. W. Nienow (2001). Effects of agitation and scale-up on drop size in turbulent dispersions: allowance for intermittency, *Chem. Eng. Sci.*, **56**, 3377–3385.

Barnes, H. A., and K. Walters (1985). The yield stress myth, *Rheol. Acta*, **24**(4), 323–326.

Barresi, A. A. (1997). Experimental investigation of interaction between turbulent liquid flow and solid particles and its effects on fast reactions, *Chem. Eng. Sci.*, **52**, 807–814.

Bartholomew, W. H. (1960). Scale-up of submersed fermentations, *Adv. App. Microbiol.*, **2**, 289–300.

Born, C., Z. Zhang, M. Al-Rubeai, and C. R. Thomas (1992). Estimation of disruption of animal cells by laminar shear stress, *Biotechnol. Bioeng.*, **40**, 1004–1010.

Boulton-Stone, J. M. (1995). The effect of surfactant on bursting gas bubbles, *J. Fluid Mech.*, **302**, 231–257.

Braun, S., and S. E. Vecht-Lifshitz (1991). Mycelial morphology and metabolite production, *Trends Biotechnol.*, **9**, 63–68.

Brown, M. E., and J. R. Birch (1996). The effect of pH on growth and productivity of cell lines producing monoclonal antibodies at large scale, *Cytotechnol.*, **17**, Suppl. 1, Abstr. 1.

Bryant, J. (1977). Characterisation of mixing in fermenters, *Adv. Biochem. Eng.*, **5**, 101–123.

Bryant, J., and S. Sadeghzadeh (1979). Circulation rates in stirred and aerated tanks, *Proc. 3rd European Conference on Mixing*, York, Yorkshire, England, Apr. 4–6, pp. 325–336.

Buckland, B. C., T. Brix, H. Fastert, K. Gbewonyo, G. Hunt, and D. Jain (1985). Fermentation exhaust gas analysis using mass spectrometry, *Bio/Technology*, **3**, 984–988.

Buckland, B. C., K. Gbewonyo, D. DiMasi, G. Hunt, G. Westerfield, and A. W. Nienow (1988a). Improved performance in viscous mycelial fermentations by agitator retrofitting, *Biotechnol. Bioeng.*, **31**, 737–742.

Buckland, B. C., K. Gbewonyo, D. Jain, K. Glazomitsky, G. Hunt, and S. W. Drew (1988b). Oxygen transfer efficiency of hydrofoil impellers in both 800L and 19000L fermenters, *Proc. 2nd International Conference on Bioreactor Fluid Dynamics*, R. King, ed., BHRA/Elsevier, London, pp. 1–15.

Bylund, F., E. Collet, S. O. Enfors, and G. Larsson (1998). Substrate gradient formation in the large-scale bioreactor lowers cell yield and increases by-product formation, *Bioprocess. Eng.*, **18**(3), 171–180.

Bylund, F., F. Guillard, S. O. Enfors, C. Tragardh, and G. Larsson (1999). Scale down of recombinant protein production: a comparative study of scaling performance, *Bioprocess. Eng.*, **20**(5), 377–389.

Bylund, F., A. Castan, R. Mikkola, A. Viede, and G. Larsson (2000). Influence of scale up on the quality of recombinant human growth hormone, *Biotechnol. Bioeng.*, **69**(2), 119–128.

Calderbank, P. H. (1958). Physical rate processes in industrial fermentation: 1. The interfacial area in gas–liquid contacting with mechanical agitation, *Trans. Inst. Chem. Eng.*, **36**, 443–463.

Capstick, P. B., A. J. Garland, W. G. Chapman, and R. C. Masters (1965). Production of foot-and-mouth disease virus antigen from BHK 21 clone 13 cells grown and infected in deep suspension cultures, *Nature*, **205**, 1135.

Capstick, P. B., A. J. Garland, W. G. Chapman, and R. C. Masters (1967). Factors affecting the production of foot-and-mouth disease virus in deep suspension cultures of BHK21 C13 cells, *J. Hyg. Camb.*, **645**, 273.

Carilli, A., E. B. Chain, G. Gualandi, and G. Morisi (1961). Continuous measurement of dissolved oxygen during fermentation in large fermenters, *Sci. Rep. Inst. Super Sanita*, **1**, 177–189.

Chalmers, J. J., and F. Bavarian (1991). Microscopic visualisation of insect cell–bubble interactions: II. The bubble film and bubble rupture, *Biotechnol. Prog.*, **7**, 151–158.

Charles, M. (1978). Technical aspects of the rheological properties of microbial cultures, *Adv. Biochem. Eng.*, **8**, 1–61.

Charles, M. (1985). Fermenter design and scale-up, in *Comprehensive Biotechnology*, Vol. 2, M. Moo-Young, ed., Pergamon Press, New York, pp. 57–75.

Cherry, R. S., and C. T. Hulle (1992). Cell death in the thin films of bursting bubbles, *Biotechnol. Prog.*, **8**, 11–18.

Cooke, M., J. C. Middleton, and J. Bush (1988). Mixing and mass transfer in filamentous fermentations, *Proc. 2nd International Conference on Bioreactor Fluid Dynamics*, R. King, ed., Elsevier, Amsterdam, pp. 37–64.

Croughan, M. S., and D. I. C. Wang (1987). Hydrodynamic effects on animal cells grown in microcarrier cultures, *Biotechnol. Bioeng.*, **29**, 130–141.

Croughan, M. S., J. F. Hamel, and D. I. C. Wang (1988). Effects of microcarrier concentration in animal cell cultures, *Biotechnol. Bioeng.*, **32**, 975–982.

Croughan, M. S., J. F. Hamel, and D. I. C. Wang (1989). Growth and death in microcarrier cultures, *Biotechnol. Bioeng.*, **33**, 731–744.

Cutter, L. A. (1966). Flow and turbulence in a stirred tank, *AIChE J.*, **12**, 35–45.

Dey, D., J. M. Boulton-Stone, A. N. Emery, and J. R. Blake (1997). Experimental comparisons with a numerical model of surfactant effects on the burst of a single bubble, *Chem. Eng. Sci.*, **52**(16), 2769–2783.

Dion, W. M., A. Carilli, G. Sermonti, and E. B. Chain (1954). The effect of mechanical agitation on the morphology of *Penicillium chrysogenum* Thom in stirred fermenters, *Rend. 1st Super. Sanita*, **17**, 187–205.

Doran, P. (1999). Design of mixing systems for plant cell suspension in stirred reactors, *Biotechnol. Prog.*, **15**, 319–335.

Drapeau, D., H. W. Blanch, C. R. Wilke (1987). Economic assessment of plant cell culture for the production of ajmalicine, *Biotechnol. Bioeng.* **30**, 946–953.

Dreveton, E., F. Monot, J. Lecourtier, D. Ballerini, and L. Choplin (1996). Influence of fermentation hydrodynamics on Gellan gum physico-chemical characteristics, *J. Ferment. Bioeng.*, **82**(3), 272–276.

Einsele, A. (1978). Scaling up bioreactors, *Proc. Biochem.*, **13**(7), 13–14.

Einsele, A., and R. K. Finn (1980). Influence of gas flow rates and gas holdup on blending efficiency in stirred tanks, *Ind. Eng. Chem. Process. Des. Dev.*, **19**, 600–603.

Elson, T. P. (1988). Mixing of fluids possessing a yield stress, *Proc. 6th European Conference on Mixing*, R. King, ed., BHRA, Cranfield, Bedfordshire, England, pp. 485–492.

Elson, T. P. (1990). The growth of caverns formed around rotating impellers during the mixing of a yield stress fluid, *Chem. Eng. Commun.*, **96**, 303–319.

Elson, T. P., D. J. Cheesman, and A. W. Nienow (1986). X-ray studies of cavern sizes and mixing performance with fluids possessing a yield stress, *Chem. Eng. Sci.*, **41**(10), 2555–2562.

Elsworth, R., G. Miller, A. Whitaker, D. Kitching, and P. Sayer (1968). Production of *E. coli* as a source of nucleic acids, *J. Appl. Chem.*, **17**, 157–166.

Enfors, S. O., M. Jahic, A. Rozkov, B. Xu, M. Hecker, B. Jürgen, E. Krüger, T. Schweder, G. Hamer, D. O'Beirne, N. Noisommit-Rizzi, M. Reuss, L. Boone, C. Hewitt, C. McFarlane, A. W. Nienow, T. Kovacs, C. Trägårdh, L. Uchs, J. Revteldt, P. C. Friberg, B. Hjertager, G. Blomsten, H. Skogman, S. Hjort, F. Hoeks, H. -Y. Lin, P. Neubauer, R. van der Lans, K. Luyben, P. Vrabel, and A. Manelius (2001). Physiological responses to mixing in large scale bioreactors, *J. Biotechnol.*, **85**, 175–185.

Etchells, A. W., W. N. Ford, and D. G. R. Short (1987). Mixing of Bingham plastics on an industrial scale, *Fluid Mixing 3, Inst. Chem. Eng. Symp. Ser.*, **108**, 1–10.

Flores, F., L. G. Torres, and E. Galindo (1994). Effect of the dissolved oxygen tension during the cultivation of *X. campestris* on the production and quality of xanthan gum, *J. Biotechnol.*, **34**, 165–173.

Forberg, C., and L. Haggstrom (1987). Effects of culture conditions on the production of phenylalanine from a plasmid-harboring *Escherichia coli* strain, *Appl. Microbiol. Biotechnol.*, **26**, 136–140.

Fowler, J. D., and E. H. Dunlop (1989). Effects of reactant heterogeneity and mixing on catabolite repression in cultures of *Saccharomyces cerevisiae*, *Biotechnol. Bioeng.*, **33**(8), 1039–1046.

Funahashi, H., H. Harada, H. Taguchi, and T. Yoshida (1987a). Circulation time distribution and volume of mixing regions in highly viscous xanthan gum solution in a stirred vessel, *J. Ferment. Technol.*, **20**, 277–282.

Funahashi, H., M. Machara, H. Taguchi, and T. Yoshida (1987b). Effect of glucose concentration on xanthan gum production by *Xanthomonas campestris*, *J. Chem. Eng. Jpn.*, **65**(6), 603–606.

Funahashi, H., M. Machara, H. Taguchi, and T. Yoshida (1987c). Effects of agitation by flat bladed turbine impeller on microbial production of xanthan gum, *J. Chem. Eng. Jpn.*, **20**(1), 16–22.

Funahashi, H., K. I. Hirai, T. Yoshida, and H. Taguchi (1988a). Mixing state of xanthan gum solution in aerated and agitated fermenter: effects of impeller size on volumes of mixed regions and circulation time distribution, *J. Ferment. Technol.*, **66**, 103–109.

Funahashi, H., K. I. Hirai, T. Yoshida, and H. Taguchi (1988b). Mechanistic analysis of xanthan gum production in a stirred tank, *J. Ferment. Technol.*, **3**, 355–364.

Furukawa, K., E. Heinzle, and I. J. Dunn (1983). Influence of oxygen on the growth of *Saccharomyces cerevisiae* in continuous culture, *Biotechnol. Bioeng.*, **25**(10), 2293–2317.

Galindo, E. (1994). Aspects of the Process for Xanthan Production, *Trans. Inst. Chem. Eng., Part C*, **72**, 227–237.

Galindo, E., and A. W. Nienow (1992). Mixing of highly viscous simulated xanthan fermentation broths with the Lightnin A315 impeller, *Biotechnol. Prog.*, **8**, 233–239.

Galindo, E., and A. W. Nienow (1993). Performance of the Scaba 6SRGT agitator in mixing of simulated xanthan gum broths, *Chem. Eng. Technol.*, **16**, 102–108.

Garcia-Briones, M., and J. J. Chalmers (1992). Cell–bubble interactions: mechanisms of suspended cell damage, *Ann. N.Y. Acad. Sci.*, **665**, 219–229.

Garcia-Briones, M., and J. J. Chalmers (1994). Flow parameters associated with hydrodynamic cell injury, *Biotechnol. Bioeng.*, **44**, 1089–1098.

Gbewonyo, K., D. DiMasi, and B. C. Buckland (1987). Characterization of oxygen transfer and power absorption of hydrofoil impellers in viscous mycelial fermentations, in *Biotechnology Processes, Mixing and Scale up*, C. S. Ho and J. J. Ulbrecht, eds., AIChE, New York, pp. 128–134.

Gbewonyo, K., G. Hunt, and B. C. Buckland (1992). Interactions of cell morphology and transport processes in the lovastatin fermentation, *Bioprocess. Eng.*, **8**, 1–7.

George, S., G. Larsson, and S. O. Enfors (1993). A scale-down 2-compartment reactor with controlled substrate oscillations: metabolic response of *Saccharomyces cerevisiae*, *Bioprocess. Eng.*, **9**(6), 249–257.

George, S., G. Larsson, and S. O. Enfors (1998). Comparison of Baker's yeast process performance in laboratory and production scale, *Bioprocess. Eng.*, **18**(2), 135–142.

Gibbs, P. A., and R. J. Seviour (1996). Does the agitation rate and/or oxygen saturation influence exopolysaccharide production by *Aureobasidium pullulans* in batch culture? *Appl. Microbiol. Biotechnol.*, **46**, 503–510.

Goldblum, S., Y. K. Bea, and J. Chalmers (1990). Protective effect of methylcellulose and other polymers on insect cells subjected to laminar stress, *Biotechnol. Prog.*, **6**, 383–390.

Gray, D. R., S. Shen, W. Howarth, D. Inlow, and B. L. Maiorella (1996). $CO_2$ in large-scale and high density CHO cell perfusion culture, *Cytotechnol.* **22**, 65–78.

Greaves, M., and K. A. H. Kobaccy (1981). Fluid Mixing 1, *Inst. Chem. Eng. Symp. Ser.*, **64**.

Griot, M., J. Moes, E. Heinzle, I. J. Dunn, and J. R. Bourne (1986). A microbial culture for the measurement of macro and micro mixing phenomena in biological reactors, *Proc. International Conference on Bioreactor Fluid Dynamics*, Cambridge, pp. 203–216.

Griot, M., E. Heinzle, I. J. Dunn, and J. R. Bourne (1987). Optimization of a MS-membrane probe for the measurement of acetoin and butanediol, in *Mass Spectrometry in Biotechnological Process Analysis and Control*, E. Heinzle and M. Reuss, eds., Plenum Press, New York, pp. 75–90.

Handa, A., A. N. Emery, and R. E. Spier (1987). On the evaluation of gas-liquid interfacial effects on hybridoma viability in bubble column bioreactors, *Develop. Biol. Standard.* **66**, 241–253.

Handa-Corrigan, A., A. N. Emery, and R. E. Spier (1989). Effects of gas-liquid interfaces on the growth of suspended mammalian cells: mechanisms of cell damage by bubbles, *Enz. Microb. Technol.* **11**, 230–235.

Harnby, N., M. F. Edwards, and A. W. Nienow, eds. (1997). *Mixing in the Process Industries.* 2nd Edition, Butterworths Heinemann, London.

Hempel, D. C., and H. Dziallas (1999). Scale-up, stirred tank reactors, in *Encyclopaedia of Bioprocess Technology: Fermentation, Biocatalysis and Bioseparation*, Vol. 3, M. C. Flickinger and S. W. Drew, eds., Wiley, New York, pp. 2314–2332.

Herbst, H., A. Schumpe, and W. Deckwer (1992). Xanthan production in stirred tank fermenters: oxygen transfer and scale up, *Chem. Eng. Technol.*, **15**, 425–434.

Hewitt, C. J., L. A. Boon, C. M. McFarlane, and A. W. Nienow (1998). The use of flow cytometry to study the impact of fluid mechanical stress on *Escherichia coli* W3110 during continuous cultivation in an agitated bioreactor, *Biotechnol. Bioeng.*, **59**(5), 612–620.

Hewitt C. J., G. Nebe-von-Caron, A. W. Nienow, and C. M. McFarlane (1999). The use of multi-parameter flow cytometry to compare the physiological response of *Escherichia coli* W3110 to glucose limitation during batch, fed-batch and continuous culture cultivation, *J. Biotechnol.*, **75**, 251–254.

Hickman, A. D., and A. W. Nienow (1986). *Proc. International Conference on Bioreactor Fluid Dynamics*, Cambridge, pp. 301–306.

Hughmark, G. A. (1980). Power requirements and interfacial area in gas-liquid turbine agitated systems, *Ind. Eng. Chem. Process. Des. Dev.*, **19**, 638–641.

Humphrey, A. E. (1977). Biochemical Engineering, in *Encyclopedia of Chemical Processing and Design*, Vol. 4, J. J. McKetta and W. A. Cunningham, eds., Marcel Dekker, New York, pp. 359–394.

Jeanes, A., P. Rogovin, M. C. Cadmus, R. W. Silman, and C. A. Knutson (1976). Procedures for culture maintenance and polysaccharide production, purification and analysis, *ARS-NC-51*, Agricultural Research Service, U.S. Department of Agriculture, North Central Region.

Jeenes, D. J., D. A. Mackenzie, I. N. Roberts, and D. B. Archer (1991). Heterologous protein production by filamentous fungi, *Biotechnol. Genet. Eng. Rev.*, **9**, 327–367.

Jöbses, I., D. Martens, and J. Tramper (1991). Lethal events during gas sparging in animal cell culture, *Biotechnol. Bioeng.*, **37**, 484–490.

Johnston, R. E., and M. W. Thring (1957). *Pilot Plants, Models and Scale-up Methods in Chemical Engineering*, McGraw-Hill, New York.

Joshi, J. B., C. B. Elias, and M. S. Patole (1996). Role of hydrodynamic shear in the cultivation of animal, plant and microbial cells, *Chem. Eng. J.*, **62**, 121–141.

Jüsten, P., G. C. Paul, A. W. Nienow, and C. R. Thomas (1996). Dependence of mycelial morphology on impeller type and agitation intensity, *Biotechnol. Bioeng.*, **52**, 634–648.

Jüsten, P., G. C. Paul, A. W. Nienow, and C. R. Thomas (1998a). Dependence of *Penicillium chrysogenum* growth, morphology, vacuolation and productivity in fed-batch fermentations on impeller type and agitation intensity, *Biotechnol. Bioeng.*, **59**, 762–775.

Jüsten, P., G. C. Paul, A. W. Nienow, and C. R. Thomas (1998b). A mathematical model for agitation induced fragmentation of *P. chrysogenum, Bioprocess. Eng.*, **18**, 7–16.

Kieran, P., P. MacLoughlin, and D. Malone (1997). Plant cell suspension cultures: some engineering considerations, *J. Biotechnol.*, **59**, 39–52.

Kilburn, D., and F. C. Webb (1968). The cultivation of animal cells at controlled dissolved oxygen partial pressure, *Biotechnol. Bioeng.*, **10**, 801–814.

Kimura, R., and W. M. Miller (1996). Glycosylation of CHO-derived recombinant tPA produced under elevated $pCO_2$, *Biotechnol. Prog.*, **13**, 311–317.

Kioukia, N., A. W. Nienow, A. N. Emery, and M. Al-Rubeai (1992). The impact of fluid dynamics on the biological performance of free suspension animal cell culture: further studies, *Food Bioprod. Process.*, **70**, 143–148.

Kioukia, N., A. W. Nienow, M. Al-Rubeai, and A. N. Emery (1996). Influence of agitation and sparging on the growth rate and infection of insect cells in bioreactors and a comparison of hybridoma culture, *Biotechnol. Prog.*, **12**, 779–785.

Kiss, R., M. Croughan, J. Trask, G. Polastri, M. Groenhout, A. Banka, S. Shurin, J. Paul, and H. Koning-Bastiaan (1994). Mixing time characterisation in large scale mammalian cell bioreactors, *AIChE Annual Meeting*, November 1994, San Francisco, CA.

Kloppinger, M., G. Fertig, E. Fraune, and H. G. Miltenburger (1990). Multi-stage production of *Autographa californica* nuclear polyhydrosis virus in insect cell bioreactors, *Cytotechnol.* **4**, 271–278.

Kolmogorov, A. (1941a). The local structure of turbulence in incompressible viscous fluid for very large Reynolds numbers, *Compt. Rend. (Dokl.) Acad. Sci. URSS*, **30**, 301–305.

Kolmogorov, A. (1941b). Dissipation of Energy in the Locally Isotropic Turbulence, *Compt. Rend. (Dokl.) Acad. Sci. URSS*, **32**, 16–18.

König, B., Ch. Seewald, and K. Schügerl (1981). Process engineering investigations of penicillin production, *Eur. J. Appl. Microbiol. Biotechnol.*, **12**, 205–211.

Kossen, N. W. F., and N. M. G. Oosterhuis (1985). Modelling and scaling up of bioreactors, in *Biotechnology*, Vol. 2, H. J. Rehm, and G. Reed, eds., VCH, Weinheim, Germany, pp. 572–605.

Kowalski, A. J., and N. H. Thomas (1994). Bursting of bubbles stabilised by surfactants for control of cell damage, in *Bubble Dynamics and Interface Phenomena*, J. R. Blake, J. M. Boulton-Stone, and N. H. Thomas, eds., Kluwer, Dordrecht, The Netherlands.

Kunas, K. T., and E. T. Papoutsakis (1990). Damage mechanisms of suspended animal cells in agitated bioreactors with and without bubble entrainment, *Biotechnol. Bioeng.*, **30**, 368–373.

Langheinrich, C., and A. W. Nienow (1999). Control of pH in large scale, free suspension animal cell bioreactors: alkali addition and pH excursions, *Biotechnol. Bioeng.*, **66**(3), 171–179.

Langheinrich, C., A. W. Nienow, N. C. Stevenson, A. N. Emery, T. M. Clayton, and N. K. H. Slater (1998). Liquid homogenisation studies in animal cell bioreactors of up to 8 $m^3$ in volume, *Trans. Inst. Chem. Eng., Part C*, **76**, 107–116.

Larsson, G., and S. O. Enfors (1985). Influence of oxygen starvation on the respiratory capacity of *Penicillium chrysogenum*, *Appl. Microbiol. Biotechnol.*, **21**, 228–233.

Larsson, G., and S. O. Enfors (1988). Studies of insufficient mixing in bioreactors: effects of limiting oxygen concentrations and short term oxygen starvation on *Penicillium chrysogenum*, *Bioprocess. Eng.*, **3**(3), 123–127.

Larsson, G., M. Tornkvist, E. S. Wernersson, C. Tragardh, H. Noorman, and S. O. Enfors (1996). Substrate gradients in bioreactors: origin and consequences, *Bioprocess. Eng.*, **14**(6), 281–289.

Lawford, H. G., and J. D. Rousseau (1991). Bioreactor design considerations in the production of high-quality microbial exopolysaccharide, *Appl. Biochem. Biotechnol.*, **28/29**, 667–684.

Leckie, F., H. Scragg, and K. Cliffe (1991). Effect of impeller design and speed on the large-scale cultivation of suspension cultures of *Catharanthus roseous*, *Enzyme Microb. Technol.*, **13**, 801–810.

Lee, I. Y., M. K. Kim, J. H. Lee, W. T. Seo, J. K. Jung, H. W. Lee, and Y. H. Park (1999). Influence of agitation speed on production of curdlan by *Agrobacterium species*, *Bioprocess. Eng.*, **20**, 283–287.

Levenspiel, O. (1972). *Chemical Reaction Engineering*, 2nd ed., Wiley, New York.

Luong, H. T., and B. Volesky (1979). Mechanical power requirements of gas–liquid agitated systems, *AIChE J.*, **25**, 893–895.

MacIntyre, F. (1972). Flow patterns in breaking bubbles, *J. Geo-Phys. Res.*, **77**, 5211–5228.

Mackenzie, D. A., D. J. Jeenes, N. J. Belshaw, and D. B. Archer (1993). Regulation of secreted protein production by filamentous fungi: recent development and perspectives, *J. Genet. Microbiol.*, **139**, 2295–2307.

Makagiansar, H. Y., P. A. Shamlou, C. R. Thomas, and M. D. Lilly (1993). The influence of mechanical forces on the morphology and penicillin production of *Penicillium chrysogenum*, *Bioprocess. Eng.*, **9**, 83–90.

Manfredini, R., V. Cavallera, L. Marini, and G. Donati (1983). Mixing and oxygen transfer in conventional stirred fermenters, *Biotechnol. Bioeng.*, **25**, 3115–3131.

Mann, R., P. P. Mavros, and J. C. Middleton (1981). A structured stochastic flow model for interpreting flow: follower data from a stirred vessel, *Trans. Inst. Chem. Eng.*, **59**, 271–278.

Manning, F. S., D. Wolf, and D. L. Keairns (1965). Model simulation of stirred tank reactors, *AIChE J.*, **11**(4), 723–727.

Märkl, H., R. Bronnenmeier, and B. Wittek (1991). The resistance of micro-organisms to hydrodynamic stress, *Int. Chem. Eng.*, **31**(2), 185.

Martin, T., C. M. McFarlane, and A. W. Nienow (1994). The influence of liquid properties and impeller type on bubble coalescence behaviour and mass transfer in sparged agitated bioreactors, *Proc. 8th European Conf. on Mixing*. Cambridge, Sept. 1994. I. ChemE., Rugby, pp. 57–64.

Meier, S. J., T. A. Hatton, and D. I. C. Wang (1999). Cell death from bursting bubbles: role of cell attachment to rising bubbles in sparged reactors, *Biotechnol. Bioeng.*, **62**(4), 468–478.

Meijer, J., H. ten Hoopen, K. Luyben, and R. Libbenga (1993). Effects of hydrodynamic stress on cultured plant cells: a literature survey, *Enzyme Microb. Technol.*, **15**, 234–238.

Meijer, J. J., H. J. G. ten Hoopen, Y. M. van Gameren, Ch. A. M. Luyben, and K. R. Libbenga (1994). Effects of hydrodynamic stress on the growth of plant cells in batch and continuous culture, *Enzyme Microb. Technol.*, **16**, 467–477.

Merchuk, J. C. (1991). Shear effects on suspended cells, *Adv. Biochem. Eng. Biotechnol.*, **44**, 66–95.

Metz, B. (1976). From pulp to pellet. Ph.D. dissertation, Delft Technical University, The Netherlands.

Metz, B., E. W. de Bruijn, and J. C. van Suijdam (1981). Method for quantitative representation of the morphology of molds, *Biotechnol. Bioeng.*, **23**, 149–163.

Metz, B., and N. W. F. Kossen (1977). Biotechnology review: the growth of molds in the form of pellets—a literature review, *Biotechnol. Bioeng.* **19**, 781–799.

Metzner, A. B., and R. E. Otto (1957). Agitation of non-Newtonian fluids, *AIChE J.*, **3**, 3–10.

Michaels, J. D., J. F. Peterson, L. V. McIntire, and E. T. Papoutsakis (1991). Protection mechanism of freely suspended animal cells (Crl 8018) from fluid mechanical injury: viscometric and bioreactor studies using serum, pluronic F68 and polyethylene glycol, *Biotechnol. Bioeng.* **38**, 169–180.

Middleton, J. C. (1979). Measurement of circulation in large mixing vessels, *Proc. 3rd European Conference on Mixing*, York, Yorkshire England, pp. 15–36.

Moes, J., M. Griot, J. Keller, E. Heinzle, I. J. Dunn, and J. R. Bourne (1985). A microbial culture with oxygen sensitive product distribution as a potential tool for characterising bioreactor oxygen transport, *Biotechnol. Bioeng.*, **27**, 482–489.

Moo-Young, M., and H. W. Blanch (1981). Design of biochemical reactors. Mass transfer criteria for simple and complex systems, *Adv. Biochem. Eng.*, **19**, 2–69.

Moraine, R. A., and P. Rogovin (1973). Kinetics of xanthan fermentation, *Biotechnol. Bioeng.*, **15**, 225–237.

Moreira, J. L., P. M. Alves, J. G. Aunins, and M. J. T. Carrondo (1995). Hydrodynamic effects on *Bhk* cells grown as suspended natural aggregates, *Biotechnol. Bioeng.*, **46**(4), 351–360.

Murhammer, D. W., and C. F. Goochee (1990). Structural features of non-ionic polyglycol polymer molecules responsible for the protective effect in sparged animal cell bioreactors, *Biotechnol. Prog.*, **6**, 142–148.

Muskett, M. J., and A. W. Nienow (1987). Capital and running costs: the economics of mixer selection, in Fluid Mixing III, *Inst. Chem. Eng. Symp. Ser.*, **108**, 33–48.

Nagata, S. (1975). *Mixing: Principles and Applications*, Halstead Press, New York.

Namdev, P. K., and E. H. Dunlop (1995). Shear sensitivity of plant cells in suspension, *Appl. Biochem. Biotechnol.*, **54**, 109–131.

Namdev, P. K., P. K. Yegneswaran, B. G. Thompson, and M. R. Gray (1991). Experimental simulation of large scale bioreactor environments using a Monte Carlo method, *Can. J. Chem. Eng.*, **69**, 513–519.

Namdev, P. K., B. G. Thompson, and M. R. Gray (1992). Effect of feed zone in fed-batch fermentations of *Saccharomyces cerevisiae*, *Biotechnol. Bioeng.*, **40**, 235–246.

Newitt, D. M., N. Dombrowski, and F. H. Knelman (1954). Liquid entrainment: 1. The mechanism of drop formation from gas or vapour bubbles, *Trans. Inst. Chem. Eng.*, **32**, 244–261.

Nielsen, J. (1992). Modelling the growth of filamentous fungi, *Adv. Biochem. Eng.*, **46**, 187–223.

Nielsen, J., C. L. Johansen, M. Jacobsen, P. Krabben, and J. Villadsen (1995). Pellet formation and fragmentation in submerged cultures of *Penicillium chrysogenum* and its relation to penicillin production, *Biotechnol. Prog.*, **11**, 93–98.

Nienow, A. W. (1984). Mixing studies on high viscosity fermentation processes: xanthan gums, in *World Biotech Report*, Vol. 1, Europe Online Publication, pp. 293–304.

Nienow, A. W. (1990). Agitators for mycelial fermentations, *Trends Biotechnol.*, **8**, 224–233.

Nienow, A. W. (1997). Large scale free suspension animal cell culture; cell fragility versus homogeneity, *Genet. Eng. Biotechnol.*, **17**, 111–113.

Nienow, A. W. (1998). Hydrodynamics of stirred bioreactors, *Appl. Mech. Rev.*, **51**, 3–32.

Nienow, A. W., and T. P. Elson (1988). Aspects of mixing in rheologically complex fluids, *Chem. Eng. Res. Des.*, **66**, 5–15.

Nienow, A. W., and J. J. Ulbrecht (1985). Gas–liquid mixing in high viscosity systems, in *Mixing of Liquids by Mechanical Agitation*, J. J. Ulbrecht and G. K. Patterson, eds., Gordon and Breach, New York, pp. 203–235.

Nienow, A. W., C. M. Chapman, and J. C. Middleton (1979). Gas recirculation rate through impeller cavities and surface aeration in sparged, agitated vessels, *Chem. Eng. J.*, **17**, 111–118.

Nienow, A. W., D. J. Wisdom, J. Solomon, V. Machon, and J. Vlcek (1983). The effect of rheological complexities on power consumption in an aerated, agitated vessel, *Chem. Eng. Commun.*, **19**, 273–293.

Nienow, A. W., M. M. C. G. Warmoeskerken, J. M. Smith, and M. Konno (1985). *Proc. 5th European Conference on Mixing*, Wurzburg, Germany, pp. 143–154.

Nienow, A. W., G. Hunt, and B. C. Buckland (1995). A fluid dynamic study using a simulated viscous shear thinning broth of the retrofitting of large agitated bioreactors, *Biotechnol. Bioeng.*, **49**, 15–19.

Nienow, A. W., C. Langheinrich, N. C. Stevenson, A. N. Emery, T. M. Clayton, and N. K. H. Slater (1996). Homogenisation and oxygen transfer in large agitated and sparged animal cell bioreactors: some implications for growth and production, *Cytotechnology*, **22**, 87–94.

Norton, C. J., D. O. Falk, and W. E. Luetzelschwab (1981). Xanthan biopolymer semipilot fermentation, *Soc. Petrol. Eng. J.*, **5**, Apr., pp. 205–217.

Oh, S. K. W., A. W. Nienow, M. Al-Rubeai, and A. N. Emery (1989). The effects of agitation intensity with and without continuous sparging on the growth and antibody production of hybridoma cells, *J. Biotechnol.*, **12**, 45–62.

Oh, S. K. W., A. W. Nienow, M. Al-Rubeai, and A. N. Emery (1992). Further studies on the culture of mouse hybridomas in an agitated bioreactor with and without continuous sparging, *J. Biotechnol.*, **22**, 245–270.

Oldshue, J. Y. (1966). Fermentation mixing scale-up techniques, *Biotechnol. Bioeng.*, **8**, 3–24.

Oldshue, J. Y. (1983). *Fluid Mixing Technology*, McGraw-Hill, New York, pp. 89–92, 192–215.

Oosterhuis, N. M. G. (1984). *Scale up of bioreactors: a scale down approach*, Ph.D. dissertation, Delft University of Technology, The Netherlands.

Oosterhuis, N. M. G., and N. W. F. Kossen (1983). Oxygen transfer in a production scale bioreactor, *Chem. Eng. Res. Des.*, **61**, 308–312.

Oosterhuis, N. M. G., and N. W. F. Kossen (1984). Dissolved oxygen concentration profiles in a production scale bioreactor, *Biotechnol. Bioeng.*, **26**, 546–550.

Oosterhuis, N. M. G., N. M. Groesbeek, A. P. C. Olivier, and N. W. F. Kossen (1983). Scale down aspects of the gluconic acid fermentation, *Biotechnol. Lett.*, **5**(3), 141–146.

Oosterhuis, N. M. G., N. W. F. Kossen, A. P. C. Olivier, and E. S. Schenk (1985). Scale down and optimisation studies of the gluconic acid fermentation by *Gluconobacter oxydans*, *Biotechnol. Bioeng.*, **27**, 711–720.

Osman, J. J., J. Birch, and J. Varley (2001). The response of GS-NSO myeloma cells to pH shifts and pH perturbations, *Biotechnol. Bioeng.*, **75**, 63–73.

Osman, J. J., J. Birch, and J. Varley (2002). The response of GS-NSO myeloma cells to single and multiple pH perturbations, *Biotechnol. Bioeng.*, **79**(4), 398–407.

Paca, J., P. Ettler, and V. Grègr (1976). Hydrodynamic behaviour and oxygen transfer rate in a pilot plant fermenter: 1. Influence of viscosity, *J. Appl. Chem. Biotechnol.*, **26**(6), 309–317.

Pace, G. W., and R. C. Righelato (1980). Production of extracellular microbial polysaccharides, *Adv. Biochem. Eng.*, **15**, 1–70.

Packer, H. L., and C. R. Thomas (1990). Morphological measurements on filamentous micro-organisms by fully automatic image analysis, *Biotechnol. Bioeng.*, **35**, 870–881.

Papoutsakis, E. T. (1991). Media additives for protecting freely suspended animal cells against agitation and aeration damage, *Tibtech*, **9**, 316–324.

Paul, G. C., and C. R. Thomas (1998). Characterisation of mycelial morphology using image analysis, *Adv. Biochem. Eng.*, **60**, 1–59.

Paul, G. C., C. A. Kent, and C. R. Thomas (1994). Hyphal vacuolation and fragmentation in *Penicillium chrysogenum*, *Biotechnol. Bioeng.*, **44**, 655–660.

Peña, C., M. A. Trujillo-Roldán, and E. Galindo (2000). Influence of dissolved oxygen tension and agitation speed on alginate production and its molecular weight in cultures of *Azotobacter vinelandii*, *Enzyme Microb. Technol.*, **27**, 390–398.

Peters, H. U., H. Herbst, P. G. M. Heselink, H. Lunsdorf, A. Schumpe, and W. D. Deckwer (1989a). The influence of agitation rate on xanthan production by *Xanthomonas campestris*. *Biotechnol. Bioeng.*, **34**, 1393–1397.

Peters, H. U., H. Herbst, I. H. Suh, A. Shumpe, and W. D. Deckwer (1989b). The influence of fermenter hydrodynamics on xanthan production by *Xanthomonas Campestris*, in *Biomedical and Biotechnological Advances in Industrial Polysaccharides*, V. Crescenzi et al., eds., Gordon & Breach, New York, pp. 275–281.

Peters, H. U., I. S. Suh, A. Schumpe, and W. D. Deckwer (1992). Modelling of batchwise xanthan production, *Can. J. Chem. Eng.*, **70**, 742–750.

Prokop, A., and R. Bajpai (1992). The sensitivity of biocatalysts to hydrodynamic shear stress, *Adv. Appl. Microbiol.*, **37**, 165–232.

Purgstaller, A., and A. Moser (1987). Mixing-modelling of a two compartment bioreactor for scale down approaches, *Chem. Biochem. Eng.*, **1**(4), 157–161.

Ramirez, O. T., and R. Mutharasan (1990). The role of the plasma membrane fluidity on the shear sensitivity of hybridomas grown under hydrodynamic stress, *Biotechnol. Bioeng.*, **36**, 911–920.

Reuss, M. (1988). Influence of mechanical stress on the growth of *Rhizopus nigricans* in stirred bioreactors, *Chem. Eng. Technol.*, **11**, 178–187.

Reuss, M., R. K. Bajpai, and K. Lenze (1980). Scale-up strategies based on the interactions of transport and reaction, *Paper F-7.2.1*, presented at the 6th International Fermentation Symposium, London, Ontario, Canada.

Revill, B. K. (1982). Pumping capacity of disc turbine agitators—a literature review, *Paper B1, Proc. 4th European Conference on Mixing*, Noordwijkerhout, The Netherlands.

Riley, G. L., K. G. Tucker, G. C. Paul, and C. R. Thomas (2000). Effect of biomass concentration and mycelial morphology on fermentation Broth Rheology, *Biotechnol. Bioeng.*, **68**(2), 160–172.

Risenberg, D., and V. Schulz (1991). High cell density cultivation of *E. coli* at controlled specific growth rates, *J. Biotechnol.*, **20**, 17–28.

Rodríguez-Monroy, M., and E. Galindo (1999). Broth rheology, growth and metabolite production of *Beta vulgaris* suspension culture: a comparative study between cultures grown in shake flasks and in a stirred tank, *Enzyme Microb. Technol.*, **24**, 687–693.

Roels, J. A. (1982). Mathematical models and the design of biochemical reactors, *J. Chem. Technol. Biotechnol.*, 59–72.

Rogovin, S. P., R. F. Anderson, and M. C. Cadmus (1961). Production of polysaccharide with *Xanthomonas campestris*, *J. Biochem. Microbiol. Technol. Eng.* **3**: 51–63.

Scragg, A. H., E. J. Allan, and F. Leckie (1988). Effect of shear on the viability of plant cell suspensions, *Enzym. Microb. Technol.* **10**, 361–367.

Scragg, A. H. (1992). Large scale plant cell culture: methods, applications and products, *Curr. Opin. Biotechnol.*, **3**, 105–109.

Shuler, M. L. (1993). Strategies for improving productivity in plant cell, tissue and organ culture in bioreactors in *Bioproducts and Bioprocesses*, Vol. 2, T. Yoshida and R. D. Tanner, eds., Springer-Verlag, Berlin.

Singh, V., W. Hensler, R. Fuchs, and A. Constantinides (1986). On-line determination of mixing parameters in fermentors using pH transients, *Proc. International Conference on Bioreactor Fluid Dynamics*, BHRA, Cambridge, pp. 231–256.

Smith, J. J., M. D. Lilly, and R. L. Fox (1990). Morphology and penicillin production of *Penicillium chrysogenum*, *Biotechnol. Bioeng.*, **35**, 1011–1023.

Smith, J. M., K. Van't Riet, J. C. Middleton (1978). Scale-up of agitated gas–liquid reactors for mass transfer, *Proc. 2nd European Conf. on Mixing*, Cambridge, 1977. BHRA Fluid Engineering, Cranfield, pp. F4-51 to F4-66.

Sokolov, D. P., S. A. Livova, and E. A. Sokolova (1983). Effect of cyclical changes in culturing conditions on the growth kinetices and physiological characteristics of yeasts, *Microbiologiya*, **52**, 909–916.

Solomon, J., T. P. Elson, A. W. Nienow, and G. W. Pace (1981a). Cavern sizes in agitated fluids with a yield stress, *Chem. Eng. Commun.*, **11**, 143–164.

Solomon, J., A. W. Nienow, and G. W. Pace (1981b). Flow patterns in aerated plastic and pseudoplastic viscoelastic fluids, in Fluid Mixing I, *Inst. Chem. Eng. Symp. Ser.*, **64**, A1–A13.

Steel, R., and W. D. Maxon (1966). Dissolved oxygen measurements in pilot and production scale novobiocin fermentations, *Biotechnol. Bioeng.*, **8**, 97–108.

Sweere, A. P. J., J. R. Mesters, K. Ch. A. M. Luyben, and N. W. F. Kossen (1986). Regime analysis of the Baker's yeast production, *Proc. International Conference on Bioreactor Fluid Dynamics*, BHRA, Cambridge, pp. 217–230.

Sweere, A. P. J., K. Ch. A. M. Luyben, and N. W. F. Kossen (1987). Regime analysis and scale down: tools to investigate the performance of bioreactors, *Enzyme Microb. Technol.*, **9**, 386–398.

Sweere, A. P. J., J. R. Mesters, L. Janse, K. Ch. A. M. Luyben, and N. W. F. Kossen (1988a). Experimental simulation of oxygen profiles and their influence of Baker's yeast production: I. One-fermenter system, *Biotechnol. Bioeng.*, **31**, 567–578.

Sweere, A. P. J., L. Janse, K. Ch. A. M. Luyben, and N. W. F. Kossen (1988b). Experimental simulation of oxygen profiles and their influence of Baker's yeast production: II. Two-fermenter system, *Biotechnol. Bioeng.*, **31**, 579–586.

Sweere, A. P. J., Y. A. Matla, J. Zandvliet, K. Ch. A. M. Luyben, and N. W. F. Kossen (1988c). Experimental simulation of glucose fluctuation, *Appl. Microbiol. Biotechnol.*, **28**, 109–115.

Tagaki, M., K. Ohara, and T. Yoshida (1995). Effect of hydrostatic pressure on hybridoma cell metabolism, *J. Ferment. Bioeng.*, **80**, 619–621.

Telling, R. C., and R. Elsworth (1965). Submerged culture of hamster kidney cells in a stainless steel vessel, *Biotechnol. Bioeng.*, **7**, 417.

Thomas, C. R., and Z. Zhang (1998). The effects of hydrodynamics on biological materials, in *Advances in Bioprocess Engineering*, Vol. II, E. Galindo and O. T. Ramírez, eds., Kluwer Academic, Norwell, MA, pp. 137–170.

Thomas, C. R., M. Al-Rubeai, and Z. Zhang (1994). Prediction of mechanical damage to animal cells in turbulence, *Cytotechnology*, **15**, 329–335.

Toma, M. K., M. P. Rukliska, J. J. Vanags, M. O. Zeltina, M. P. Leite, N. I. Galinina, U. E. Viesturs, and R. P. Tengerdy (1991). Inhibition of microbial growth and metabolism by excess turbulence, *Biotechnol. Bioeng.*, **38**, 552–556.

Tramper, J. J., D. Joustra, and J. M. Vlak (1987). Bioreactor design for growth of shear-sensitive insect cells, in *Plant and Animal Cell Cultures: Process Possibilities*, C. Webb and F. Mavituna, eds., Ellis Horwood, Chichester, West Sussex, England, pp. 125–136.

Trujillo-Roldán, M. A., C. Peña, O. T. Ramírez, and E. Galindo (2001). The effect of oscillating dissolved oxygen tension on the production of alginate by *Azotobacter vinelandi*, *Biotechnol. Prog.*, **17**, 1042–1048.

Tucker, K. G., T. Kelly, P. Delgrazia, and C. R. Thomas (1992). Fully automatic measurement of mycelial morphology by image analysis, *Biotechnol. Prog.*, **8**, 353–359.

Ujcova, E., Z. Fencl, M. Musilkova, and L. Seichert (1980). Dependence of release of nucleotides from fungi on fermenter turbine speed, *Biotechnol. Bioeng.*, **22**, 237–241.

Underwood, S. (1994). Mixing performance of the Prochem Maxflo T impeller in highly viscous model fluids, M.Sc. thesis, University of Birmingham, England.

Van Barneveld, J., W. Smit, N. M. G. Oosterhuis, and H. J. Pragt (1987). Measuring the liquid circulation time in a large gas–liquid contactor by means of a radio pill: 2. Circulation time distribution, *Ind. Eng. Chem. Process. Des. Dev.*, **26**, 2192–2195.

Van Brunt, J. (1986). Fungi: the perfect hosts? *Bio/Technology*, **4**, 1057–1062.

Van Suijdam, J. C., and B. Metz (1981). Influence of engineering variables upon the morphology of filamentous molds, *Biotechnol. Bioeng.*, **23**, 111–148.

Van der Pol, L., and J. Tramper (1998). Shear sensitivity of animal cells from a culture medium perspective, *Trends. Biotechnol.* **16**:8, 323–328.

Van't Riet, K. (1975). Turbine agitator hydrodynamics and dispersion performance, Ph.D. dissertation, University of Delft, The Netherlands.

Van't Riet, K. (1979). Review of measuring methods and results in nonviscous gas–liquid mass transfer in stirred vessels, *Ind. Eng. Chem. Process. Des. Dev.*, **18**(3), 357–363.

Vardar, F., and M. D. Lilly (1982). Effect of cycling dissolved oxygen concentrations on product formation in penicillin fermentations, *Eur. J. Appl. Microbiol. Biotechnol.*, **14**(4), 203–211.

Vecht-Lifshitz, S. E., S. Magdassi, and S. Braun (1990). Pellet formation and cellular aggregation in *Streptomyces tendae*, *Biotechnol. Bioeng.*, **35**, 890–896.

Wang, D. I. C., C. L. Cooney, A. L. Demain, P. Dunnill, A. E. Humphrey, and M. D. Lilly (1979). Aeration and agitation, Chapter 9 in *Fermentation and Enzyme Technology*, Wiley, New York, pp. 157–193.

Warmoeskerken, M. M. C. G. (1986). Gas-liquid dispersing characteristics of turbine agitators, Ph.D. dissertation, University of Delft, The Netherlands.

Warmoeskerken, M. M. C. G., and J. M. Smith (1985). *Proc. 5th European Conference on Mixing*, BHRA, Cranfield, Bedfordshire, England, pp. 127–142.

Wecker, A., and U. Onken (1991). Influence of dissolved oxygen concentration and shear rate on the production of pullulan by *Aureobasidium pullulans*, *Biotechnol. Lett.*, **13**(3), 155–160.

Wessels, J. G. H. (1990). Role of cell wall architecture in fungal tip growth generation, in *Tip Growth in Plant and Fungal Walls*, I. B. Heath, ed., Academic Press, San Diego, CA, pp. 1–29.

Wessels, J. G. H. (1993). Wall growth, protein excretion morphogenesis in fungi, *New Phytol.*, **123**, 397–413.

Wichterle, K., and O. Wein (1975). Agitation of concentrated suspensions, *Paper B4.6*, presented at CHISA '75, Prague, Czechoslovakia.

Wu, J. (1995). Mechanisms of animal cell damage associated with gas bubbles and cell protection by medium additives, *J. Biotechnol.*, **43**, 81–94.

Wu, J., and M. F. A. Gossen (1995). Evaluation of the killing volume of gas bubbles in sparged animal cell culture reactors, *Enzyme Microb. Technol.*, **14**, 980–983.

Yegneswaran, P. K., M. R. Gray, and B. G. Thompson (1991). Experimental simulation of dissolved oxygen fluctuations in large fermenters: effect on *Streptomyces clavuligerus*, *Biotechnol. Bioeng.*, **38**, 1203–1209.

Zhao, X., A. W. Nienow, S. Chatwin, C. A. Kent, and E. Galindo (1991). Improving xanthan fermentation performance by changing agitators, in *Proc. 7th European Conference on Mixing*, M. Bruxelmane and G. Froment, eds., Brugge, Belgium, pp. 277–283.

Zhao, X., H. Zongding, A. W. Nienow, C. A. Kent, and S. Chatwin (1994). Rheological characteristics, power consumption, mass and heat transfer during xanthan gum fermentation, *Chin. J. Chem. Eng.*, **2**(4), 198–209.

Zhang, Z., M. Al-Rubeai, and C. R. Thomas (1992). Effect of Pluronic F68 on the mechanical properties of mammalian cells, *Enzyme Microb. Technol.*, **14**, 980–983.

Zhang, Z., M. Al-Rubeai, and C. R. Thomas (1993). Estimation of disruption of animal cells by turbulent capillary flow, *Biotechnol. Bioeng.*, **42**(8), 987–993.

Zhong, J. J., T. Seki, S. Kinoshita, and T. Yoshida (1992). Effects of surfactants on cell growth and pigment production in suspension cultures of *Perilla frutescens*, *World J. Microbiol. Biotechnol.*, **8**, 106–109.

Zhou, G., and S. M. Kresta (1996). Impact of tank geometry on the maximum turbulence energy dissipation rate for impellers, *AIChE J.*, **42**, 2476–2490.

# CHAPTER 19

# Fluid Mixing Technology in the Petroleum Industry

RAMESH R. HEMRAJANI

ExxonMobil Research and Engineering Company

## 19-1 INTRODUCTION

Mixing applications in petroleum industry may be somewhat limited compared to chemical, pharmaceutical, and food manufacturing. In addition, refinery streams are less complex then specialty and fine chemicals in terms of fluid physical properties and process conditions. However, due to large volumes of petroleum streams to be mixed, mixing technology plays an important role in enhancing productivity and profitability. Of course, the petroleum companies that have large integrated chemicals operations use most aspects of mixing technology in reacting and nonreacting single- and multiphase systems. The refining processes involving mixing operations include making emulsion products for oil drilling, absorption of $CO_2$ from natural gas, crude oil–water homogenization for custody transfer, sludge suspension in crude oil storage tanks, desalting of crude oil, alkylation, caustic–oil contacting for neutralization, pH control, and more. Small enhancements in mixing efficiency can yield large benefits in reduced investment and operating costs.

A typical refinery flow scheme is described in Figure 19-1. Crude oil produced at onshore and offshore drilling fields is transported via pipelines and/or ships to the refineries. It is then stored in large (up to 1 Mbbl) tanks for several days. Sometimes a terminal with several large storage tanks is involved for temporary storage of crude oils and/or blending before pumping to refineries. The first step in crude processing consists of removing salt by emulsifying water and then demulsifying in an electrostatic field. Chemical aids are sometimes mixed in to improve water emulsification and demulsification. The oil is then processed in

---

*Handbook of Industrial Mixing: Science and Practice*, Edited by Edward L. Paul,
Victor A. Atiemo-Obeng, and Suzanne M. Kresta
ISBN 0-471-26919-0 Copyright © 2004 John Wiley & Sons, Inc.

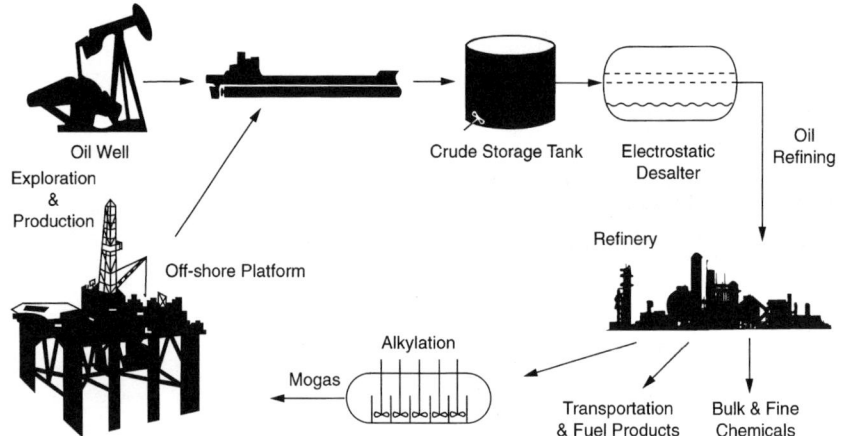

**Figure 19-1** Typical petroleum operations.

the refinery to make transportation/fuel products and feedstocks for bulk and fine chemicals. Part of the product stream is used to produce an alkylate by acid catalysis in mixing tanks. This alkylate is blended into the gasoline to increase its octane number.

Mixing plays important roles in many processing steps in the refinery's upstream and downstream areas listed below. The role of mixing technology in chemical operations is not discussed in this chapter, but it is covered in other chapters on single and multiphase systems.

*Mixing in upstream operations*

- Control fluid for oil drilling wells
- Gas treating on onshore and offshore platforms
- Oil/water homogenization in transfer lines

*Mixing in downstream operations*

- Sludge control in crude oil storage tanks
- Crude oil blending
- In-line mixing of water for desalting
- Solids suspension in bottoms slurry tanks
- Fuels and products blending tanks
- Acid/hydrocarbon mixing for alkylation
- Caustic/oil and water/oil contacting
- Acid or caustic mixing for pH control in wastewater treatment
- Caustic/gasoline emulsification for converting elemental sulfur

## 19-2 SHEAR-THICKENING FLUID FOR OIL DRILLING WELLS

Fluids with unique thickening properties are used during blowout and lost circulation of oil drilling wells. When pumped into pores surrounding the well, this fluid hardens and plugs the pores. The thickening time is required to be longer than 50 min to avoid premature stoppage of the flow. This fluid consists of a mixture of clay and a stabilized water-in-oil emulsion. The emulsion is prepared batchwise by dispersing water in oil in agitated tanks. Since this mixture is kept available nearby drilling wells, it must be highly stable to prevent settling out during long storage. The stability of the emulsion is achieved by forming a solid protective film at the interface between the water droplets and the continuous oil phase. The protective film is a product of a reaction between polyisobutylene succinic acid (PIBSA) and polyacrylic acid (Figure 19-2). These chemicals are added into the oil and water phases, respectively, before emulsification. During the mixing operation, the two reactants diffuse to the interface, where they react to form a stabilizing solid coating around the drops.

The mixing issues in designing a suitable mixing system for the manufacture of this shear-thickening emulsion are:

- Three volumes of water must be dispersed in one volume of oil, creating a potential for undesired phase inversion. Therefore, water must be added to the oil slowly with the mixer on.
- A narrow drop size distribution of the emulsion is desired for best shear thickening properties. The impeller selection, therefore, should be based on providing narrow distribution of shear.
- Formation of a thin protective coating around the dispersed drops is needed for emulsion to be stable. The emulsion stability is necessary for long shelf life at the oil field.

A laboratory study in two different-sized mixing tanks equipped with single pitched blade turbines lead to the following correlation for maximum drop size

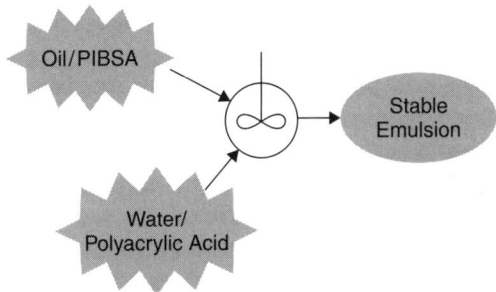

**Figure 19-2** Mixing of two phases to form a stable emulsion.

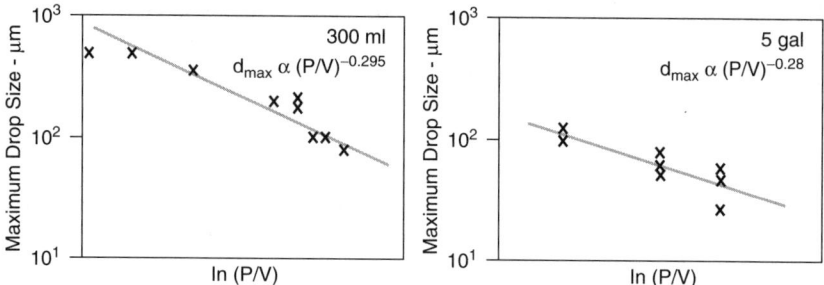

**Figure 19-3** Maximum drop size as a function of P/V: pitched blade turbine.

$d_{max}$ with mixing energy as power per unit volume P/V (Figure 19-3).

$$d_{max} = \left(\frac{P}{V}\right)^{-0.3}$$

The product analyses from these experiments indicated that the desired shear thickening properties are best achievable with maximum drop size of about 600 μm. A suitable mixing system consisting of a pitched blade turbine scaled up on the basis of constant power per unit volume has been commercially successful. Further improvements can be achieved by producing narrower distribution of drop sizes, perhaps through using hydrofoil impellers. The experiments with radial flow impellers yielded inferior product quality.

## 19-3 GAS TREATING FOR $CO_2$ REDUCTION

Natural gas often contains a high concentration (>2%) of $CO_2$, which makes it unsuitable for direct use as fuel gas. Conventional process for reducing this concentration to below 2% consists of absorption of $CO_2$ in an amine solution. The amine solution is regenerated and recycled back into the process. The absorption is carried out at pressures in excess of 100 bar and with a gas/liquid volumetric ratio of about 2:1. Packed bed contactors operated in countercurrent mode are commonly used for this purpose. Although these towers provide maximum driving force for mass transfer, they can be bulky and heavy with dimensions 5 m diameter by 25 m high, depending on the feed rates. They are, therefore, less suitable for use on offshore platforms. Static mixers can be the favored choice because of the following advantages:

- Plug flow conditions are available with
  - a large number of stages
  - good radial mixing
  - negligible axial mixing

- High mass transfer coefficient ($k_g a$ can be 10 to 20 times higher than in packed towers)
- Target product $CO_2$ concentration achievable in short residence time (1 or 2 s)
- Smaller in size and weight, with ease of installation on floating platforms
- Can be horizontal, vertical, or inclined (horizontal and inclined need to be within certain boundary conditions for successful application)
- Can easily handle foaming systems

Design guidelines for static mixers for this application are not available in the literature and therefore need to be developed through experimentation. Fundamental concepts of gas–liquid contacting in static mixers, discussed in Chapter 7, can be used to select an appropriate mixer type. Since no literature data are available for the specific fluids at process conditions, it is advised to conduct pilot plant tests at scale at actual pressures to quantify the effect of mixer pressure drop on mass transfer rate and to develop scale-up criteria.

## 19-4 HOMOGENIZATION OF WATER IN CRUDE OIL TRANSFER LINES

When crude oil is sampled for determination of water content, as a part of custody transfer, it is essential that the water be uniformly dispersed across the pipe cross-section. In the absence of good mixing, water can get stratified and flow near the pipe bottom and escape the sampler. While adequate mixing must be provided to create good dispersion; the resulting emulsion should not be stable, because water must later settle easily in storage tanks. Typical sizes of pipelines used for transferring oil range from 12 to 30 in. in diameter. The selection and design of an effective mixing system requires careful evaluation of various technologies for minimizing investment costs. Optimum mixing can add a high value for the refiner, as the cost of 0.1% sampling error can amount to $250 000 per medium-sized tanker.

The mixing system for adequate homogenization of water should be capable of handling changing flow rate because the ship pumps often operate at varying rates. The type and design of such a mixer depends on the length of the pipe, upflow/downflow sections, and pressure drop, creating elements such as bends and valves. The pipeline velocity requirements for adequate dispersion with different mixers and pipeline configurations can be obtained from the chart shown in Figure 19-4. This chart, based on field data, indicates that for less than 8 ft/s (2.5 m/s) oil velocity, some form of mixing is required. It can be used for deciding on the type of mixer needed based on oil velocity during normal and turndown rates.

When crude oil velocity is low and natural turbulence is inadequate even with pipe bends and valves, the following four types of mixers are commonly used:

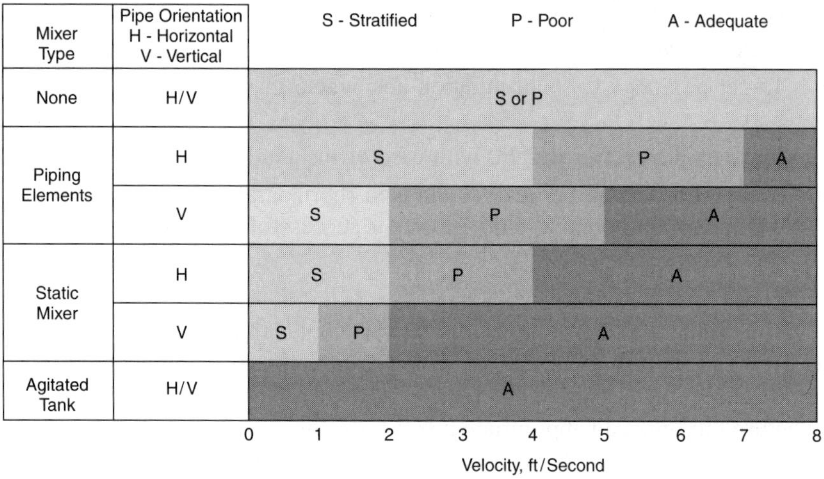

# 1176 FLUID MIXING TECHNOLOGY IN THE PETROLEUM INDUSTRY

| Mixer Type | Pipe Orientation H - Horizontal V - Vertical | | | | | | | | |
|---|---|---|---|---|---|---|---|---|---|
| None | H/V | \multicolumn{8}{l\|}{S or P} |
| Piping Elements | H | S | | | P | | A | | |
| | V | S | | P | | A | | | |
| Static Mixer | H | S | | P | | A | | | |
| | V | S | P | | A | | | | |
| Agitated Tank | H/V | A | | | | | | | |

S - Stratified    P - Poor    A - Adequate

Velocity, ft/Second: 0 1 2 3 4 5 6 7 8

**Figure 19-4** Pipeline velocity requirements with different mixers.

- Fixed geometry static mixers
- Variable geometry in-line mixers
- Rotary in-line blender
- Recirculating jet mixer

## 19-4.1 Fixed Geometry Static Mixers

The simplest type of fixed geometry in-line mixer is a distributor inside a pipe (Figure 19-5a), marketed by Neyrtec of France. This mixer is designed to create pressure drop in the two-phase flow through two perforated plates placed facing one another, and provide energy for dispersion. In addition, opposed jets are formed to create further shear, useful for breakup and homogenization of

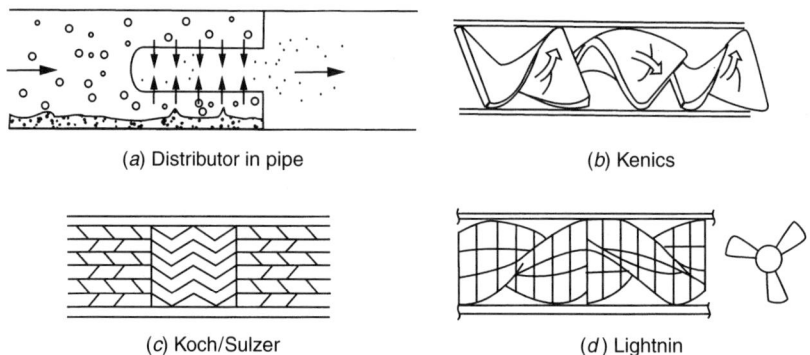

(a) Distributor in pipe    (b) Kenics

(c) Koch/Sulzer    (d) Lightnin

**Figure 19-5** Fixed geometry in-line mixers.

dispersed drops. Total energy per unit mass, $E_m$, is given by

$$E_m = Q \frac{\Delta P}{\rho V}$$

where Q is the flow rate, $\Delta P$ the pressure drop, $\rho$ the liquid density, and V the mixer volume.

The pressure drop, $\Delta P$, can be either measured for a given design or estimated based on distributor open area and physical properties of the emulsion. In this type of mixer the energy dissipation can be on the order of 0.4 kW/kg. The maximum drop size of dispersion, $d_{max}$, is related to $E_m$ by

$$d_{max} \propto E_m^{-0.4} \left(\frac{\rho}{\sigma}\right)^{-0.6}$$

where $\sigma$ is the interfacial tension.

There are a number of static mixers that can be used for liquid–liquid dispersion. Three of them—Kenics (Fasano and Ryan, 1988), Koch/Sulzer, and Lightnin—are shown in Figure 19-5b–d. These static mixers are compact pipe internals designed to provide radial mixing and turbulence to cause dispersion and to homogenize the emulsion in the pipe. The internals consist of a sequence of vanes that force the flowing fluid to change direction abruptly several times, thus imposing large shear forces on the fluid. The energy spent is delivered by the pump on the ship or at a terminal, and therefore no power supply is needed at the mixer location.

Literature data on dispersion in static mixers indicate that the Sauter mean diameter of dispersed drops can be correlated with Weber number and friction factor by

$$\frac{d_{32}}{D} \propto We^{-0.6} f^{-0.4}$$

where $d_{32}$ is the Sauter mean diameter, D the pipe diameter, We the Weber number ($= \rho v^2 D/\sigma$, where v is the velocity), and f is the friction factor.

If designed and operated at design pressure drop, these static mixers can produce a narrow drop size distribution. For example, 70% of dispersed volume can be within 20% of the mean drop diameter. However, these mixers work poorly at low velocities, because the pressure drop can fall below the necessary level. Also, at extremely high velocities, there is a danger of stable emulsion formation.

## 19-4.2  Variable Geometry In-line Mixer

The static mixer shown in Figure 19-6 overcomes the disadvantage of conventional static mixers by maintaining pressure drop and mixing energy constant over a wide range of flow rates. A servo system is used to vary the area of apertures as the flow rate varies. Creating opposing jets and turbulence in the mixer volume carries out mixing of dispersed aqueous drops. Although this mixer is

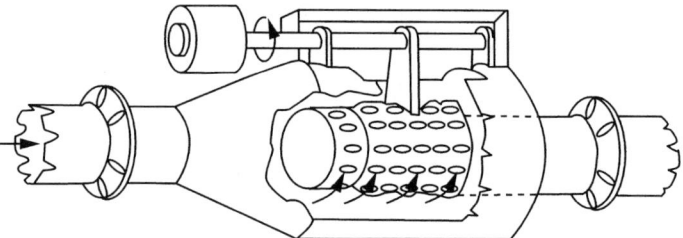

**Figure 19-6** Example of a variable geometry in-line mixer.

highly suitable for homogenization of water-in-oil in a wide range of flow rates, it is bulky and expensive.

### 19-4.3 Rotary In-line Blender

When crude oil velocity in large pipes is expected to be low, the small mixing tanks shown in Figure 19-7 are installed just upstream of the sampler. The design consists of at least two stages, each with an impeller, and internal baffles. The incoming flow is forced to pass through each stage with strategically designed baffles before exiting. The impellers can be radial flow or axial flow, and the mixer is sized to provide power per unit volume in the range 5 to 20 hp/kgal. This blender is installed in an oversized section of the pipe in order to increase residence time and reduce superficial velocity. More than two stages can be used to provide narrower residence time distribution. This mixer works best at low flow rates when mixing is most needed. When the flow rate is high, the mixer

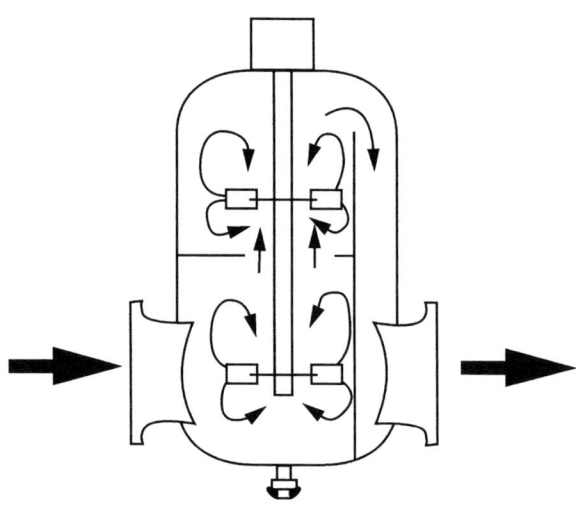

**Figure 19-7** Rotary in-line blender.

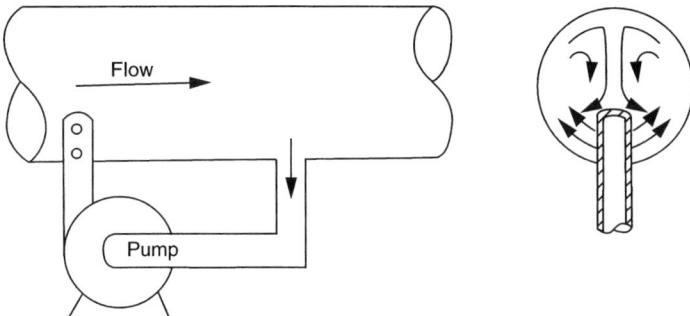

**Figure 19-8** Recirculating jet mixer.

can be turned off to prevent formation of a stable emulsion. It takes a significant head loss, however, and requires power supply at remote locations.

### 19-4.4 Recirculating Jet Mixer

The mixer shown in Figure 19-8 consists of a bypass loop with a pump to recirculate a portion of the flowing liquid, and reinject under pressures a series of jets which dissipate their energy through turbulent mixing with the main flow. In addition, these jets induce large scale vortices important for homogenization of dispersed drops. The jet nozzles, manufactured by Jiskoot, Inc. of Houston, are designed to provide velocities in the range 20 to 25 ft/s (6.5 to 8.3 m/s). Instead of generalized turbulence and shear in the whole fluid mass, the jet energy is concentrated locally near the bottom, where water concentration is expected to be highest. These jets are effective in dispersing settled water drops and evenly distributing over the complete cross-section of the pipe by twin helix rotation. The pump is generally sized at 7 bar above the line pressure, and its suction can be either downstream or upstream of the jets. The jet velocity can be varied independent of the pipe flow rate. Other than the jet nozzle, there are no protruding metal parts, and therefore the pressure loss is minimal. The pump is installed externally and can easily be repaired or replaced. When there is adequate velocity for mixing, the pump can be turned off to reduce additional energy input. The only disadvantage is that a power supply for the pump must be provided at remote locations.

## 19-5  SLUDGE CONTROL IN CRUDE OIL STORAGE TANKS

Crude oil almost always carries with it some amount of bottom sludge and water (BS&W) at a typical concentration of about 0.5 wt %. Sludge comprises a mixture of organic and inorganic products with water in the form of both types of stable emulsions, oil-in-water and water-in-oil. These products include waxes, asphaltenes, polymers, organic acids, salt, mud, sand, and corrosion products.

Because sludge is heavier than crude oil, it settles in storage vessels at terminals and refineries.

Excessive sludge accumulation can occur in tanks equipped with underpowered and/or improperly operated mixers. Low ambient temperatures can also cause reduced sludge dispersion and hydrocarbon solubility. Occasionally, crude oil tankers clean their ship tanks and pump high concentrations of sludge to storage tanks. Heavier and high sulfur crude oils carry higher than 0.5 wt % sludge. Once settled on the tank floor, the sludge hardens and cannot be removed by normal pumping.

Sludge settled on the tank floor can cause several problems such as:

- Loss of storage capacity or incorrect capacity assumptions
- Entrapment of settled water, which can occasionally leave as slugs and create operating problems in the downstream equipment, e.g.,
  - desalter overload, increased water carryover, and oil carryunder
  - crude preheat train fouling
  - pressure buildup in the pipestill
- Safety and environmental problems such as:
  - landing of floating roof on an uneven sludge layer
  - potential tank boilover due to trapped water in the event of fire
- Frequent need for off-stream tank cleaning, which is hazardous, expensive, time consuming, and requires sludge disposal
- Corrosion of the wall

Adequate mixing is therefore important for preventing settling of sludge on the tank floor and for suspending sludge that has hardened due to long-term accumulation. Conventional mixing systems used for this application are:

- Side-entering mixers
- Rotating submerged jet nozzles

### 19-5.1 Side-Entering Mixers

Side-entering mixers consist of a marine propeller or a hydrofoil impeller attached to a horizontal shaft and installed on the tank shell near the bottom. This mixer type generates a horizontal spiraling jet that provides the desired thrust to dislodge and entrain the sludge. The mixer shaft is positioned at a fixed angle of about $10°$ to the left of the tank diameter when the mixer is rotating clockwise looking from the motor side. If the mixer is installed facing the tank center, a vortex can be formed, resulting in sludge buildup in the tank center. The mixers can be designed

## SLUDGE CONTROL IN CRUDE OIL STORAGE TANKS

**Table 19-1** Guidelines for Selecting Number of Side Entering Mixers

| Tank diameter (m) | <30 | 30–45 | 45–60 | >60 |
|---|---|---|---|---|
| Number of mixers | 1 | 2 | 3 | 4 or 5 |

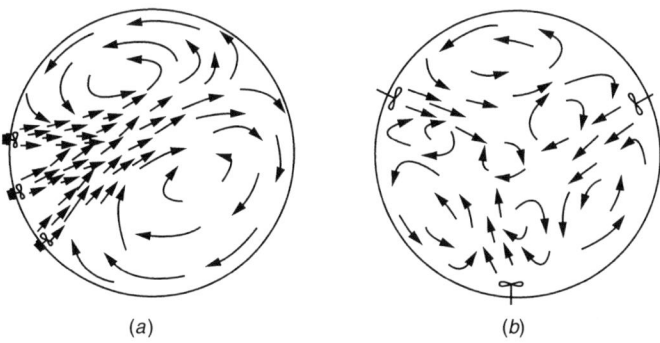

**Figure 19-9** Side-entering mixers.

with flexibility of changing angle from $-30°$ to $+30°$. These swivel angle mixers are more effective than a fixed-angle mixer for keeping the tank floor clean.

The mixer drives are generally limited to 60 hp for gear-driven and 100 hp for belt-driven systems, although larger mixers have occasionally been used. Since crude oil storage tanks can be very large, up to 300 ft in diameter, multiple mixers are commonly employed to provide adequate mixing energy. Guidelines for selecting appropriate number of mixers are given in Table 19-1. Multiple mixers generally are of equal size and can be installed in a clustered (Figure 19-9a) or distributed (Figure 19-9b) configuration. Field data indicate that the mixer performance with these two configurations are about the same if the total energy is adequate. It is recommended to install the tank outlet opposite the clustered mixers.

The side-entering mixers should be designed at a minimum of 0.4 hp/kbbl for light to medium crude oils. For heavy and high sludge containing crudes, higher mixing energy is needed to maintain sludge in suspension. This design guideline is based on sludge monitoring carried out for one year in two tanks equipped with mixers designed at 0.4 and 0.3 hp/kbbl. The data in Figure 19-10 clearly show that 0.4 hp/kbbl is required to minimize sludge accumulation. It should be noted that due to higher crude temperatures in summer months, the accumulated sludge volume is considerably reduced.

### 19-5.2 Rotating Submerged Jet Nozzle

A jet mixer can be designed to deliver a concentrated horizontal force on the tank floor to dislodge settled sludge. This jet must, however, be rotated to cover the entire floor. The jet can be energized by the oil flow during receipt or by pumping

**1182** FLUID MIXING TECHNOLOGY IN THE PETROLEUM INDUSTRY

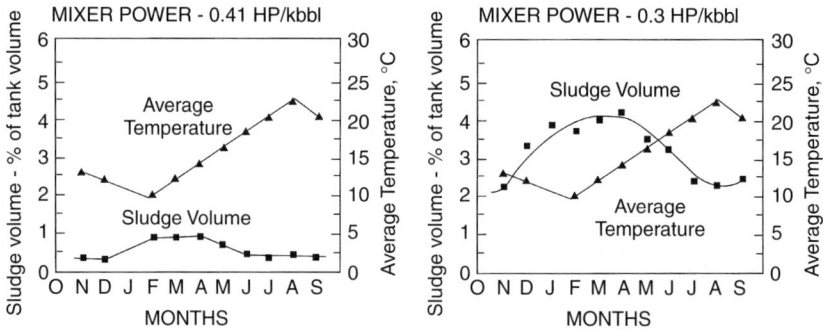

**Figure 19-10** Sludge measurements in two crude oil tanks at different mixer power.

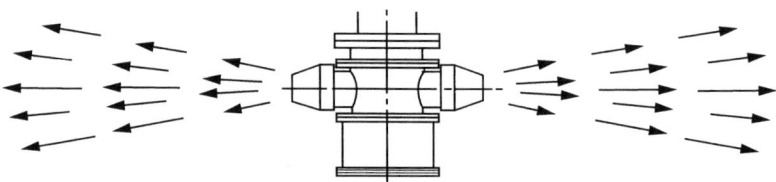

**Figure 19-11** Rotating submerged jet nozzles: P43 machine.

around the oil in the tank. Although a single rotating jet can be operated, a mixer with two diametrically opposite nozzles can have a better balance of forces on the mixer body. The latter design is marketed as the P43 machine by Sarp UK Ltd. and shown in Figure 19-11.

The P43 machine design is based on the required cleaning radius R, oil pumping rate, and available pressure drop. A portion of the flow is directed to an impeller inside the machine, which through a series of gears rotates the nozzles at rates from 1.5 to 3.5 deg/min. The P43 installation can be center mounted (CM) or shell mounted (SM). The CM system is cost-effective because only one machine is needed. In addition, the cleaning distance is equal to the tank radius and requires a lower pumping rate than do the SM machines. However, the tank must be fully cleaned and degassed before installation. The SM system can be installed safely, with the crude oil level lowered to below the manhole. Therefore, this configuration is selected when initial desludging is required followed by continued sludge control. The cleaning radius R of the SM system depends on the number of P43 machines, as shown in Figure 19-12. These machines can be operated sequentially for 8 h each, since sludge settling rates are quite slow.

The required pumping rate and pressure drop for effective sludge suspension can be calculated from

$$Q = 7.56 d^2 \left(\frac{R}{d}\right)^{1.01} \qquad \text{gal/min}$$

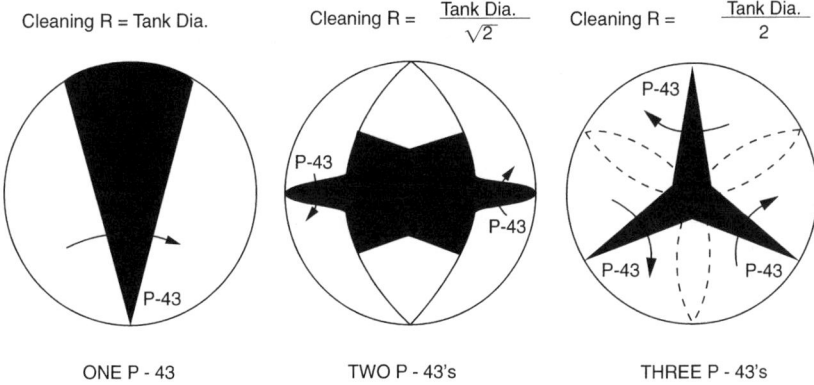

**Figure 19-12** Shell-mounted P43 configurations.

$$\Delta P = S\left(\frac{Q}{26.82d^2}\right)^2 \quad \text{psi}$$

where d is the nozzle diameter in inches, R the cleaning radius in feet, and S the specific gravity of the crude. Rotating submerged jet mixers are capable of both preventing sludge accumulation and suspending settled sludge, whereas side-entering mixers are only expected to prevent sludge settling. The guidelines discussed above are for normal crudes and not applicable to heavy crudes or slop oils.

## 19-6 DESALTING

*Desalting* is a process for removing salt from crude oil before sending it to the pipestill. This is done by first mixing a demulsifier in the pipe carrying crude oil, which is then followed by emulsifying fresh water using an in-line mixer (Figure 19-13). The water-in-oil emulsion is broken in an electrostatic separator to produce oil with negligible salt and water.

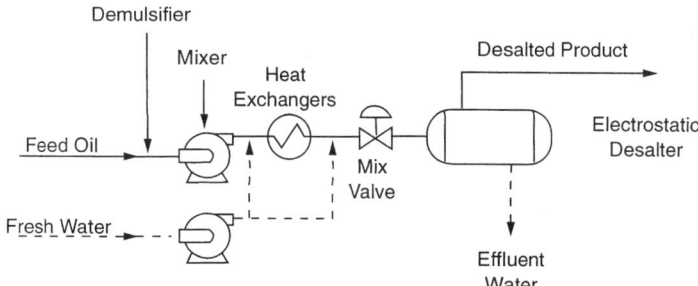

**Figure 19-13** Desalting process.

Optimum mixing of fresh water is desired for satisfactory desalting. Poor mixing can lead to carryover of salt in the crude, which causes:

- Fouling of heat exchangers
- Coking of furnace tubes
- Excessive corrosion of downstream equipment
- Poisoning of catalysts in downstream processes

Overmixing can result in the formation of a stable emulsion and poor separation. Water carried over and oil carried under can cause:

- Water flashing in the pipestill and hydrocarbon release into the atmosphere
- Poor product quality
- Loss of hydrocarbon value
- Effluent water contamination and disposal problems

Conventional in-line mixers used for mixing fresh water in oil include mixing valve, variable speed multistage mixing tank (Figure 19-7), and static mixers (Figure 19-5b–d). The first two mixer types are used more commonly because the mixing energy can be adjusted depending on the flow rate. This mixing energy must be optimized for adequate removal of salt without water carryover. The desalter operating data shown in Figure 19-14 indicate that as the mixing valve pressure drop is increased, salt concentration in the oil decreases rapidly and water carryover increases slowly. However, if pressure drop is increased to above a critical value, water carryover increases rapidly, which is then followed by increase in salt carryover. This performance plot depends on crude type, temperature, and demulsifier type. Continuous monitoring of BS&W in the

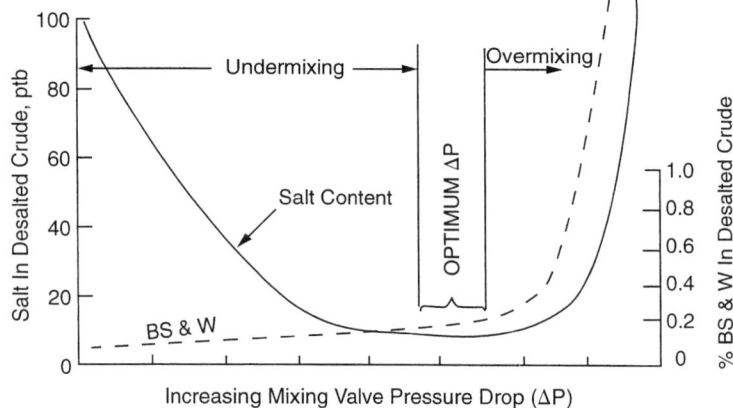

**Figure 19-14** Optimizing mixing valve pressure drop in desalting.

crude product, coupled with mix valve adjustment, is necessary to ensure high performance level.

## 19-7 ALKYLATION

Alkylation is a process for producing high-octane hydrocarbons with acid-catalyzed exothermic reactions of $C_4$ and $C_5$ olefins. The process can be based on HF or $H_2SO_4$. With the $H_2SO_4$ process, two configurations are used depending on the heat removal method, autorefrigeration, and indirect refrigeration (Figure 19-15). This process requires adequate contacting of hydrocarbons with concentrated sulfuric acid for a primary fast reaction and several parallel and consecutive secondary reactions. In the autorefrigerated system, heat is removed by boiling a hydrocarbon.

Good mixing in the reactor is important for producing a hydrocarbon-in-acid dispersion at minimum energy consumption and high mass transfer rate but short settling time. Excessive energy consumption can increase the load on refrigeration, and a stable emulsion can form and cause acid carryover and corrosion in the downstream equipment. The mixing system should be suitable for producing narrow distribution of drops. Such dispersion quality is desired for fast settling rates. Two types of impellers are used in commercial alkylation reactors, turbines, and hydrofoils. Both types result in nearly the same reaction yield when designed at 5.0 hp/kgal. However, the drop size distribution with hydrofoils is somewhat narrower than with turbines and can be better for minimizing acid carryover. Adequately designed static mixers are also used downstream of the reactor for neutralization of entrained acid by contacting with caustic.

## 19-8 OTHER APPLICATIONS

In addition to the applications discussed above, there are several other areas where mixing is important. They include suspension of catalyst fines in storage tanks for

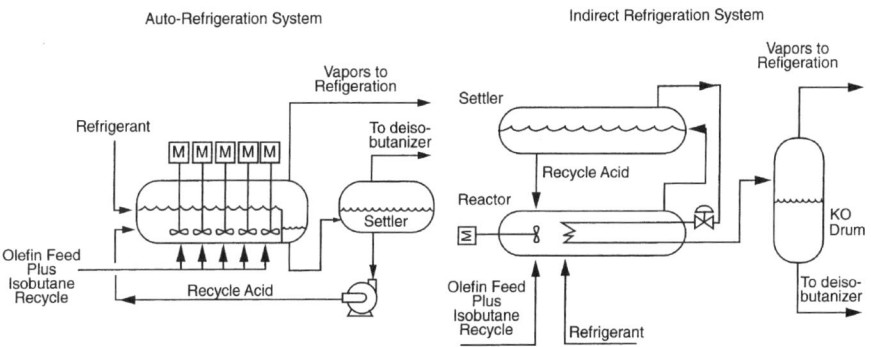

**Figure 19-15** Sulfuric acid catalyzed alkylation processes.

catalytic fractionator bottoms, product blending tanks, caustic treat/water wash of hydrocarbons for corrosion control, caustic contacting of gasoline for removal of elemental sulfur, and more. Although refineries have mixing systems in use, they are often designed on the basis of limited data and experience and can be inadequate. Therefore, design guidelines are needed for mixing systems to achieve good process performance and reliability. Other chapters in this book provide some of these design guidelines.

## NOMENCLATURE

| | |
|---|---|
| $\Delta P$ | pressure drop |
| d | drop size of dispersion |
| D | pipe diameter |
| $E_m$ | total energy per unit mass |
| f | friction factor |
| Q | volumetric flow rate |
| R | cleaning radius |
| S | specific gravity |
| v | velocity |
| V | mixer volume |
| We | Weber number $\dfrac{ev^2 D}{\sigma}$ |

*Greek Symbols*

| | |
|---|---|
| $\rho$ | liquid density |
| $\theta$ | interfacial tension |

## REFERENCES

Fasano, J. B., and D. C. Ryan (1988). Presentation at the AIChE 1988 Annual Meeting, Washington, DC.

JISKOOT, Inc., Houston, TX, technical literature.

Neyrtec, France, technical literature.

Sarp UK Ltd., private communications on P-43 jet mixers.

# CHAPTER 20

# Mixing in the Pulp and Paper Industry

CHAD P. J. BENNINGTON

*University of British Columbia*

## 20-1 INTRODUCTION

Since its invention in A.D. 105, paper use has steadily grown to become the world's preferred media for information dissemination and storage. Today, pulp and paper products cover a wide spectrum, from hygiene items, through a myriad of paper grades, to packaging materials. Pulp and paper production is a major and diversified world industry. In 2000, 188 million metric tons of pulp and 323 million metric tons of paper and paperboard were produced, with major production in Canada, the United States, and the Nordic countries (Pulp and Paper International, 2001). Paper is a renewable resource, and with proper stewardship, world demand can be met indefinitely. Part of this involves reuse of paper products. Recycling programs in many countries recover over 50% of their paper each year for inclusion in new products (FAO, 1999).

The demand for pulp and paper products continues to grow. Growth has typically been linked to gross national product levels, with developing countries using more paper per capita as standards of living rise. However, the emergence of new information technologies is changing the demand for paper. The success and acceptance of these technologies will determine the extent to which individual paper products will be affected. In the short term, demand for high-quality print-on-demand paper grades has increased. On longer time frames, demand for newsprint is forecast to decrease and that for packaging grades to increase.

The diverse range of paper products available necessitates a variety of pulp production processes. These range from mechanical to chemical processes, each producing pulps with unique attributes. These pulps are then blended to craft paper products having the desired properties for their end use.

---

*Handbook of Industrial Mixing: Science and Practice*, Edited by Edward L. Paul, Victor A. Atiemo-Obeng, and Suzanne M. Kresta
ISBN 0-471-26919-0 Copyright © 2004 John Wiley & Sons, Inc.

The most common chemical pulping process is the kraft process. Here, wood chips are treated to remove the lignin that binds the cellulose fiber to the wood matrix. Once released as individual fibers, residual lignin (which is also a chromophore) is progressively removed through a series of increasingly selective delignification and bleaching/brightening steps. Bleached kraft pulp can then be mixed with other pulps (and with minerals and additives) to produce any number of paper products, including the paper used to produce this book. The kraft process is the dominant chemical pulping process, due to the strength of the pulp it produces and the fact that an effective chemical recovery scheme exists. Process details may be found in a number of reference books, including Grace and Malcolm (1989), Smook (1992), Biermann (1996), and Gullichsen and Fogelholm (1999).

Mixing is an essential unit operation in all facets of pulp and paper manufacture. The range of mixing applications is extensive. These include examples from liquid–liquid to gas–solid–liquid mixing, from the blending of complex fluids to the mixing of reactive chemicals. For pulp fiber suspensions, mixing processes include blending pulp streams for papermaking, addition of wet-end and other chemical additives prior to sheet forming, consistency control ahead of most papermaking and pulp processing stages, and chemical contacting in pulp bleaching operations. Figure 20-1 is a typical process flow diagram for kraft pulp production. The NAMF (North American Mixing Forum) logo indicates those unit operations where mixing is important.

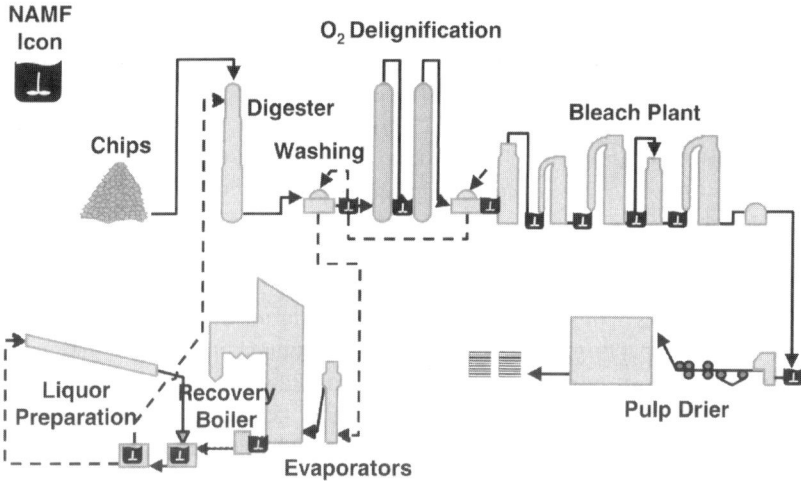

**Figure 20-1** Simplified overview of a typical kraft process showing chemical pulping, oxygen delignification, pulp bleaching, and chemical recovery systems. Important mixing operations are indicated by the NAMF mixing icon.

## 20-2 SELECTED MIXING APPLICATIONS IN PULP AND PAPER PROCESSES: NONFIBROUS SYSTEMS

Examples of mixing applications in pulp and paper processes are given below. They were selected to illustrate both the unique and diverse applications of mixing in pulp and paper technologies, and the critical role that mixing plays in pulp and paper manufacture. The examples have been classified under the traditional mixing disciplines and are not meant to be exhaustive.

### 20-2.1 Liquid–Liquid Mixing

Liquid–liquid mixing is common for chemical preparation in various pulp and paper processes. The dilution and makeup of bleaching solutions is done routinely using stirred vessels or static mixers. Applications of this type are discussed in other chapters.

One unique liquid–"liquid" mixing application is smelt dissolving. Molten smelt (molten salt at about 900°C) is added to a weak wash to form a green liquor (a green aqueous solution of $Na_2S$ and $Na_2CO_3$) in the recovery cycle. The smelt stream is "shattered" using steam jets to avoid large clumps of smelt entering the water and creating an explosion. The smelt solidifies on contact with the water and must be dissolved to produce the green liquor. Vigorous mixing is required to ensure solids suspension and dissolution. Stagnation zones must be avoided to prevent forming zones of high concentration that would lead to scaling problems (Frederick et al., 1990). The industry standard is to achieve this in baffled stirred tank reactors having a large fluid circulation (Holman et al., 1989; Ljungqvist and Theliander, 1995). Air entrainment, which would reoxidize the sodium sulfide generated by reduction in the furnace, must be avoided. Recent work has used computational fluid dynamics (CFD) and laboratory scale modeling to examine impeller configurations and mixing conditions to minimize the mixing time (Ljungqvist and Theliander, 1995).

### 20-2.2 Gas–Liquid Mixing

*20-2.2.1 White and Black Liquor Oxidation.* White liquor is an alkaline solution of sodium sulfide ($Na_2S$) and sodium hydroxide (NaOH) used to react with and remove lignin from wood chips during a kraft cook. Black liquor is the liquor separated from the pulp following the cook and contains the unreacted (residual) cooking chemicals and dissolved wood components. Both the white and black liquors contain $Na_2S$, which requires oxidation at certain points in the kraft process.

Black liquor is concentrated and incinerated to recover and regenerate the pulping chemicals and to utilize the heating value of the dissolved wood components. During liquor processing, older-style recovery furnaces use the sensible heat of the flue gas to evaporate water from the liquor. This is achieved in direct-contact evaporators. However, carbon dioxide in the flue gas reacts with sodium sulfide in the liquor (hydrolyzed to HS$^-$ under the alkaline processing conditions) to release hydrogen sulfide into the gas stream:

$$HS^- + CO_2 + H_2O \rightarrow HCO_3^- + H_2S \tag{20-1}$$

To prevent release of this pollutant, the liquor is oxidized to convert the Na$_2$S to thiosulfate (Na$_2$S$_2$O$_3$). The oxidation reactions can be written as

$$2HS^- + 2O_2 \rightarrow S_2O_3^{2-} + H_2O \tag{20-2}$$

$$S_2O_3^{2-} + 2O_2 \rightarrow SO_4^{2-} + H_2O \tag{20-3}$$

In black liquor oxidation, it is important to oxidize the sulfide but not to overoxidize the liquor and reduce its heating value. Consequently, target levels of sulfide in the strong black liquor going to the furnace are typically <0.1 g/L.

Industrial implementation of black liquor oxidation recognized the importance of effective gas–liquid mass transfer in the reaction. Sparged tank reactors are often used, with compartmentalized reaction zones to optimize reactor size given the reaction kinetics (Morgan and Murray, 1971; Shaw and Christie, 1984). Pipeline reactors can also be used to oxidize black liquor but gas–liquid mass transfer is critical and it may be necessary to increase the gas–liquid mass transfer by using static mixers in some applications.

White liquor is oxidized, so it may be used as a source of caustic in oxygen delignification and alkali extraction stages in the bleach plant. The degree of oxidation determines where the liquor can be used. Partially oxidized liquor can be used for oxygen delignification where reaction conditions do not oxidize the thiosulfate (S$_2$O$_3^{2-}$) remaining in the liquor. Fully oxidized liquor has all thiosulfate converted to sulfate. This liquor can be used in alkaline extraction stages reinforced using hydrogen peroxide. Under the reaction conditions of a peroxide-reinforced extraction stage, the peroxide would oxidize any thiosulfate remaining in the liquor and be wasted.

White liquor oxidation occurs under more severe reaction conditions than black liquor oxidation. Again, gas–liquid mass transfer is essential, and several reactor designs have been used for this purpose, including sparged stirred vessels, pipeline reactors (Thring et al., 1995), and buss loop reactors.

### 20-2.2.2 Generation of Alkaline Peroxymonosulfate.
Alkaline peroxymonosulfate (Na$_2$SO$_5$ or PMS) is a bleaching chemical that has been used successfully to augment delignification in the laboratory. In conjunction with oxygen, the addition of 1.0% PMS to an oxygen delignification system (on an active oxygen basis) increased delignification from 49% to 73% without reducing pulp strength (Bouchard et al., 2000). However, the commercial implementation of PMS requires an efficient and cost-effective strategy for its generation. One promising method of achieving this is the catalytic oxidation of sodium sulfite

with oxygen. Laboratory generation of PMS is readily achieved by this route, but typically at low yields (<20%) and low concentrations (<3.8 g/L).

The exact reaction mechanism for peroxymonosulfate generation using the copper-catalyzed sulfite oxidation is open to interpretation. The possible reaction mechanisms given in the literature can be simplified to the following reactions:

$$SO_3^{2-} + O_2 \xrightarrow{Cu^{2+}} SO_5^{2-} \tag{20-4}$$

$$SO_5^{2-} + SO_3^{2-} \rightarrow 2SO_4^{2-} \tag{20-5}$$

$$SO_5^{2-} \rightarrow 0.5O_2 + SO_4^{2-} \tag{20-6}$$

where PMS is produced by the catalytic oxidation of sulfite, but once produced can be reduced to sulfate by sulfite. In addition, peroxymonosulfate can decompose. If we ignore the decomposition reaction, which is slow, eqs. (20-4) and (20-5) represent a classic consecutive-competitive reaction scheme.

Studies conducted where PMS was generated under systematically varied mixing conditions (stirred tank reactor, semibatch reaction mode with sulfite maintained as the limiting reagent) showed the classic dependence of yield on sulfite feed time, chemical concentration, and mixing intensity. By adjusting mixing conditions, PMS yield was increased to 32.5% at 9.8 g/L and to 54% at 1.6 g/L (Shaharuzzaman and Bennington, 2001). The experimental results followed the E-model predictions of Baldyga and Bourne (1989) over the energy dissipation range studied, as shown in Figure 20-2. However, as power input

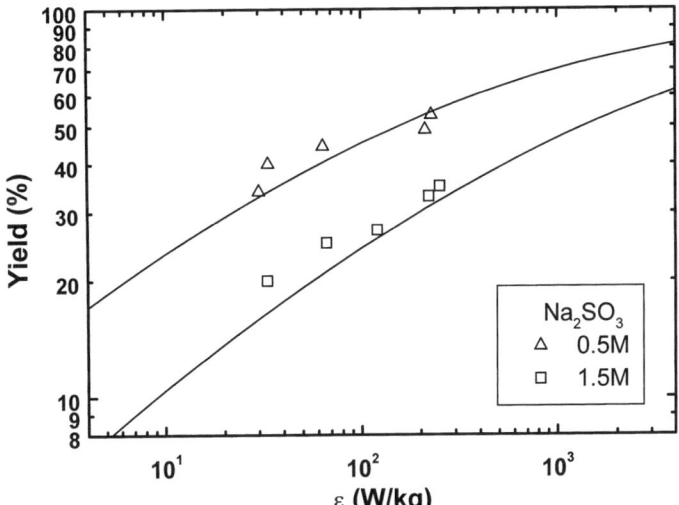

**Figure 20-2** Peroxymonosulfate yield versus average energy dissipation during reaction. The lines give micromixing predictions [with the E-model of Baldyga and Bourne (1989) using fitted rate constants].

increased, local gas–liquid mass transfer became insufficient to maintain the local oxygen concentration in excess and restricted the yield gain. Despite the improved generation efficiency, yields are still too low for economic implementation of peroxymonosulfate industrially. Further research may improve the economic feasibility of PMS use, while the results obtained demonstrate the role mixing can play improving generation of bleaching chemicals.

### 20-2.3  Solid–Liquid Mixing

Examples of solid–liquid mixing are found in kraft chemical recovery and in papermaking. In the recovery furnace, particulate material is removed from the boiler flue gases as dry solids and mixed with concentrated black liquor prior to reintroduction to the furnace. In coated papermaking production, coating preparation is critical for paper performance and uniformity. In chemical recovery, the recausticizing operation is an essential process for white liquor preparation and relies on effective solid–liquid mixing.

***20-2.3.1  Preparation of Coating Colors.*** Coating colors are concentrated slurries (of pigments, binders, and additives) applied to the surface of paper to enhance paper properties and print quality. The slurry concentration is as high as possible (often as high as 70% by mass) to limit the effect of moisture uptake (rewetting) on paper structure and to minimize the extent of subsequent paper drying required. Only as much water as required to achieve the desired flow properties of the coating color should be used.

The preparation of the coating formulation is extremely important (Robinson et al., 1997; Makinen, 1999; Thibault, 1999), as coating properties (especially their rheology) depend on the mixing applied. A wide range of specialized mixers that can operate over the wide range of process conditions encountered during coating preparation have been developed.

Several steps are used in coating preparation. The first involves preparing a pigment slurry from dry powder. This requires wetting the pigment. Two methods are used for this: forming a vortex in the mixer to draw the pigment into the slurry or using a rotor on the liquid surface to incorporate the dry pigment with the slurry. The power required for mixing increases significantly with increasing slurry solids content during this make-down period. Often, the slurry approaches the limit of fluid behavior. The slurry must then be homogenized, with any lumps and aggregates formed during powder addition removed by the erosion or fracture of the particles. This is achieved by applying shear during mixing, although dispersants can be added to aid this process. The power level falls as the suspension becomes homogenized. Several mixer types are used in this application, including the Cowles turbine, rotor–stator mixers (e.g., the Kady Mill), kneaders, and coaxial agitators. The coaxial agitators use two different impellers, each with its own motor, for mixing the dispersion. The central agitator is a high-speed turbine designed to mix and disperse the pigment. The outer agitator turns at low speed in proximity to the vessel wall to ensure that the entire suspension is well mixed. The two agitators can be rotated cocurrently (Ekato) or countercurrently (Cellier) (Thibault, 1999).

The effect of mixing (time and intensity) on the rheology of the pigment slurry is summarized by Robinson et al. (1997). Breakdown of the dispersion agglomerates requires imposition of shear on the slurry. Consequently, the impeller tip speed is critical for disperser operation and mixer design. As the maximum shear rate [ranging from 100 to $10^4$ s$^{-1}$ in dispersion operations (Makinen, 1999)] determines the ultimate size of the particles, extending the dispersion time cannot compensate for an inadequate shear level.

The second step in coating preparation is the coating formulation. The dispersed pigment is mixed with appropriate binders and cobinders (adhesives, such as starches and latexes) and other additives (optical whiteners and viscosity modifiers) to achieve the properties required for the coating. The quantity of materials and their order of addition greatly influences the ease of mixing and the properties and quality of the final color preparation (Makinen, 1999). The order in which the coating components are added determines their interaction with each other, and consequently, the size of the pigment particles and the rheology of the slurry. Certain components, if added in a particular order, rapidly form viscous mixtures that suddenly and dramatically increase the load on the mixer, placing unacceptable loads on it. Continued mixing may reduce the overall mixture viscosity as slurry aggregates are comminuted by shear during mixing. However, the final quality (rheology) of the color may not be an optimum. Once prepared, coating colors are kept in motion, by either mixing or recirculation, until applied to paper in a coating nip, where the shear rate can approach $10^6$ s$^{-1}$.

***20-2.3.2 Causticization.*** White liquor is prepared by mixing green liquor (largely sodium carbonate, $Na_2CO_3$) with calcium oxide (CaO). The calcium oxide slakes with water to form calcium hydroxide, $Ca(OH)_2$, which then reacts to produce sodium hydroxide:

$$CaO(s) + H_2O \rightarrow Ca(OH)_2(s) \qquad (20\text{-}7)$$

$$Ca(OH)_2(s) + Na_2CO_3(aq) \Leftrightarrow 2NaOH(aq) + CaCO_3(s) \qquad (20\text{-}8)$$

The reaction occurs in the solid pellets of CaO, governed by a shrinking core reaction mechanism. Consequently, particle size affects reaction progress, its yield (extent of unreacted material, referred to as *deadload*), and the ease of the subsequent separation of $CaCO_3$ from the system. The reaction is initiated in a slaker where calcium oxide is mixed into hot green liquor. The slaking reaction [eq. (20-7)] is strongly exothermic and completed in 10 to 15 min. The suspension is then passed through a series of causticizers (stirred tank reactors) to permit the equilibrium of eq. (20-8) to be established. A total residence time of 90 to 150 min is needed for this, and plug flow is desirable to limit short-circuiting of the solid phase. Short-circuiting would permit unreacted $Ca(OH)_2$ to reach the clarification stage, increasing system deadload and leading potentially to mud-settling problems. Typically, three or four stirred tanks in series are used, and the tanks can be compartmentalized to more closely approach plug flow. The causticizers use various impellers, with the goal of achieving well-mixed compartments and full suspension of the solids. The impeller tip speed is the most common criterion used by equipment manufacturers for both slaker and causticizer scale-up and is not varied during operation. Although increased mixing intensity can increase the rate of causticization slightly, it can dramatically affect the settling rate of the calcium carbonate particles, largely by reducing particle size (Dorris, 2000). Laboratory studies made by Dorris used a series of impellers to confirm that particle size decreased and filtration resistance increased with increased impeller tip speed. The absolute values depended on the specific impeller used. When the data were compared in terms of the average power dissipation in the vessel, filtration resistance depended only on the power input and the impeller/tank diameter ratio.

### 20-2.4 Gas–Solid–Liquid Mixing

*20-2.4.1 Polysulfide Generation.* Polysulfides, linear oligomers of sulfur ($S_nS^{-2}$, where n = 1 to 4), protect hemicelluloses during kraft cooking. This results in a significant increase in pulp yield if added during the cooking process. Polysulfides can be produced by the oxidation of sodium sulfide in the white liquor (hydrolyzed to $HS^-$) by oxygen in the presence of a catalyst (e.g., activated carbon or $MnO_2$). The reactions are complex and the mechanisms unknown. However, they can be represented by (Dorris and Uloth, 1994)

$$(2n+2)HS^- + nO_2 \longleftrightarrow 2S_nS^{2-} + 2(n-1)OH^- + 2H_2O \quad (20\text{-}9)$$

$$nS_nS^{2-} + H_2O \longleftrightarrow (n-1)S_nS^{2-} + HS^- + OH^- \quad (20\text{-}10)$$

with equilibrium established between polysulfide species of various chain lengths to give an average length, n, of 2.4 for typical reaction conditions (Dobson, 2001).

The polysulfide generated can be oxidized to thiosulfate:

$$2S_nS^{2-} + (n+2)O_2 + 2(n-1)OH^- \rightarrow (n+1)S_2O_3^{2-} + (n-1)H_2O \quad (20\text{-}11)$$

reducing the polysulfide yield.

The Paprilox® process catalytically oxidizes sulfide (present in the white liquor) to polysulfide during the causticization reaction (Uloth et al., 1996). Here, air or oxygen is sparged into one (usually, the last) causticizer, giving about 60 min for polysulfide generation. The impeller is used to create gas–liquid surface area, mix the gas throughout the vessel, and suspend the solids (the lime mud and added $MnO_2$ catalyze the oxidation). Mill implementation of the polysulfide process indicated sensitivity to mixing conditions (Uloth et al., 1996). Recent laboratory work (Dobson and Bennington, 2002) confirmed that mixing could alter the reaction rate, yield, and selectivity of polysulfide generation. For example, Figure 20-3 shows the effect of impeller speed on polysulfide selectivity

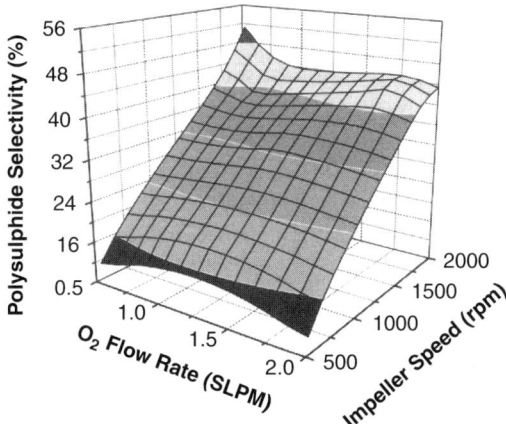

**Figure 20-3** Polysulfide selectivity (maximum selectivity achieved during 100 min of reaction) versus impeller speed and oxygen flow rate. Generation in a laboratory sparged stirred-tank reactor at T = 90°C using 2.0 g/L of Fisher $MnO_2$ catalyst. Selectivity is the fraction of polysulfide (as g S°/L) formed divided by the hydrosulfide ion consumed during oxidation (as g S/L). (From Dobson and Bennington, 2002.)

in a laboratory sparged-tank mixer. For fixed reaction conditions (reactant concentrations, temperature, etc.), increasing the impeller speed increased both the rate of polysulfide formation and the maximum polysulfide concentration reached during reaction. For the catalyst studied, the peak yield of polysulfide could be increased from 1.2 to 9.6 g/L (expressed as S°) as impeller speed increased from N = 500 to 2000 rpm. However, caution must be exercised when adjusting mixing to optimize oxidation in the mill. Any increased turbulence could increase the settling time of the lime mud and impair the clarification step that follows recausticizaton.

One final example of gas–liquid–solid mixing is the flotation process used in paper deinking operations. Ink is detached from fiber surfaces in the previous repulping operation, with ink redeposition minimized by creating an appropriate chemical environment. Air is used to collect the ink particles in the low consistency pulp suspension ($C_m < 0.01$) and float them to the surface where they are removed with the froth. Operating conditions are created to minimize entrapment of fiber in the rising bubble stream. The ink collection efficiency depends on the combined probability of a number of sequential events: collision between an ink particle and gas bubble, ink adhesion to the bubble, and maintenance of the ink–air bond. Mixing is critical for all these processes. The turbulence created in mixing determines the gas bubble size and creates interactions between the ink particles and bubbles that lead to collisions. The probability of adhesion depends on the magnitude of the inertial forces created in the flow. These forces must be strong enough to allow ink particles to penetrate the flow streamlines and approach the bubble surface, but it must not be so vigorous as to detach the ink particles once attached (Somasundaran and Zhang, 1998).

Industrial flotation cells use a variety of strategies to create bubble surface area and contact it with the suspension. These include mixing the gas (typically 10 to 20 vol %) with the suspension in some manner. Injector nozzles, turbines, or impellers can be used for this purpose. No well-defined criteria are published for designing gas/suspension mixing in flotation systems. However, all the mixing/contacting strategies attempt to create small gas bubbles and ensure efficient bubble/suspension contact in a well-mixed zone. This is followed by a quiescent zone in which the gas (and the attached ink) can readily separate from the suspension (McKinney, 1998).

## 20-3 PULP FIBER SUSPENSIONS

### 20-3.1 Pulp Suspension Mixing

Pulp suspensions must be mixed in many applications. For example, in papermaking many additives are mixed with pulp before sheet forming. These include a variety of wet-end chemicals, such as retention aids, antislimming agents, filler materials, and dyes. To be effective, these must be mixed uniformly throughout the suspension. Since wet-end processing is normally conducted at low suspension concentrations, mixing is readily accomplished with agitated chests. However, the unique rheology of the fiber suspensions, even at these low mass concentrations, creates challenges during mixing. As the suspension mass concentration increases, mixing becomes even more difficult. Consequently, characterizing and understanding suspension behavior is critical when designing pulp mixing systems.

### 20-3.2 Characterization of Pulp Suspensions

Pulp suspensions are composed of an array of wood components, including parenchyma cells, fibers, tracheids, and vessel elements, although the main components of interest are the fibers and tracheids. Individual fibers are small. A typical softwood tracheid (in wood) has a length of 3 to 7 mm, a width of 24 to 59 μm, and a cell wall thickness of 2 to 7 μm (Rydholm, 1965). After processing, the fiber length is reduced and the fiber lumen partially or fully collapsed. Following mechanical processing (using stoneground wood or thermomechanical pulp processes) the typical length-weighted fiber length, $l_w$, of a mill-produced hemlock–balsam pulp is 0.6 to 1.2 mm. The same fiber following chemical processing (e.g., the kraft process) would have a $l_w$ value of 1.9 to 2.4 mm (Bennington, 1988).

In suspension, fibers form coherent networks, which are interconnected systems with fibers in continuous contact with other fibers (see Figure 20-4). Forces exist at the fiber contact points, creating fiber aggregates (flocs) and giving the network mechanical strength. Before motion can be initiated within a suspension, whether to mix, transport, or disperse flocs, external forces sufficient to overcome the network forces must be applied.

**Figure 20-4** Photographs of fibers and suspensions: (*a*) photomicrograph of softwood trachieds ($l_w = 2.37$ mm); (*b*) light transmission through a $C_m = 0.005$ kraft pulp suspension showing mass nonuniformities (flocs). The line in the image is 5 cm in length; (*c*) hands hold a medium consistency ($C_m = 0.10$) kraft pulp suspension. See insert for a color representation of this figure.

Pulp suspensions are often characterized by two factors. The length-weighted fiber length is often used as a single parameter to characterize the average length of fibers within the suspension. The fiber length distribution is very broad (often extending two decades in length), and the length-weighted average provides greater weighting to the longer fibers, which disproportionately influence suspension behavior.

The mass concentration of the suspension, $C_m$, specifies the mass of fibers in the suspending medium, which is usually water. Thus,

$$C_m = \frac{m_f}{m_f + m_w} \qquad (20\text{-}12)$$

where $m_f$ and $m_w$ are the mass of fiber and water, respectively. This parameter is commonly used industrially (where it is referred to as the *consistency* of the suspension and often expressed as a percentage) to calculate production and chemical applications. Throughout this chapter, unless otherwise specified, the mass concentration is expressed as a fraction. Suspension rheology changes dramatically with mass concentration, varies widely among pulp and paper unit operations, and is typically categorized as being low ($C_m < 0.04$), medium ($0.08 \leq C_m \leq 0.16$), or high ($C_m > 0.20$) in pulping and bleaching operations.

The volume concentration of a pulp suspension, $C_v$, gives a better indication of suspension behavior, although it is not always easily calculated. Pulp fibers are hollow, and the lumen volume changes with the degree of mechanical action and the chemical treatment experienced during processing. The cell wall material also absorbs considerable water, causing fiber swelling and increasing fiber volume. Thus, calculation of the suspension volume concentration requires specification of at least two additional parameters: the quantity of water absorbed by the cell wall and the volume of the fiber lumen.

The degree of fiber collapse has only recently been measured (Jang and Seth, 1998). Fiber collapse, and hence lumen volume, has not been widely characterized as a function of fiber type or processing history and is presently a time-consuming measurement. However, the adsorption of water by the cell wall has been studied extensively. It can be measured by two tests, a solute exclusion test that yields the fiber saturation point and a water extraction test using a centrifugal field that yields the water retention value. Both tests can be used to estimate the amount of water responsible for fiber swelling, $X_w$, and agree over a range of test conditions (Scallan and Carles, 1972). Values for $X_w$ depend on pulp type and processing history, and for a typical kraft pulp can range from 0.8 to 2.0 kg water/kg fiber (Scallan and Carles, 1972). For a completely collapsed fiber (a reasonable approximation for a fully bleached kraft pulp) at low mass concentrations ($C_m \leq 0.04$ and where the gas trapped in the suspension is negligible), $C_v$ can be calculated using (Bennington et al., 1990)

$$C_v = \frac{1 + X_w(\rho_f/\rho_w)}{1 + [(1 - C_m)/C_m](\rho_f/\rho_w)} \quad (20\text{-}13)$$

where $\rho_f$ is the density of the cell wall material ($\rho_{cellulose} = 1500$ kg/m$^3$) and $\rho_w$ is the density of water. If gas is present in the suspension, which can occur at any suspension mass concentration but is common for $C_m > 0.08$, some measure of the gas volume must also be known to compute $C_v$. For example, if the bulk density of the suspension, $\rho_b$, is known,

$$C_v = C_m \left( \frac{1}{\rho_f} + \frac{X_w}{\rho_w} \right) \rho_b \quad (20\text{-}14)$$

Fiber suspensions are never uniform. They aggregate to form mass concentrations within the suspension, called *flocs* (Kerekes et al., 1985). These form even at low mass concentrations, as shown in Figure 20-4b. In various unit operations, fiber flocs must be disrupted and dispersed. The uniformity of a paper surface, called its *formation*, depends largely on floc dispersion during sheet formation in papermaking. Bleaching efficiency, particularly at higher mass concentrations, depends on floc dispersion during mixing with reactive chemicals. The latter can be difficult to achieve at typical bleaching processing conditions. Medium consistency pulp suspensions ($0.08 \leq C_m \leq 0.16$) are a case in point. As shown in Figure 20-4c, a $C_m = 0.10$ suspension, despite being 90% water by mass, displays solidlike behavior.

### 20-3.3 Suspension Yield Stress

Before motion can be initiated in a suspension, the yield stress must be overcome. Due to floc formation, the suspension first yields in low-strength regions between flocs. Thus, flow occurs before floc disruption. Suspension yield stress increases dramatically with concentration. At low mass concentrations ($C_m < 0.04$), where the gas content is negligible, the yield stress has been given traditionally by expressions of the form

$$\tau_y = aC_m^b \tag{20-15}$$

where $\tau_y$ is the suspension yield stress, $C_m$ the mass concentration, and a and b are parameters that depend on pulp type and processing history. Kerekes et al. (1985) found that the exponent, b, varied from 1.69 to 3.02. A more general correlation encompassing a wider range of pulps can be made using the volume concentration, $C_v$:

$$\tau_y = aC_v^b \tag{20-16}$$

which also accounts for the presence of gas within the suspension. Bennington et al. (1995) correlated an extensive set of experimental data to arrive at

$$\tau_y = 7.7 \times 10^5 C_m^{3.2}(1 - \phi_g)^{3.4} A^{0.6} \tag{20-17}$$

which expresses yield stress as a function of the mass concentration (expressed as a fraction), the void fraction of gas in suspension, $\phi_g$, and the aspect ratio of the fiber, A. The agreement between this equation (for $A = 70$) and extensive network strength data from the literature is given in Figure 20-5 (Bennington et al., 1995). The equation accounts for the observed dependency of suspension mass concentration, gas void fraction, and fiber aspect ratio displayed by a range of fiber networks, although a correlation developed for an individual pulp fiber would be more accurate.

Suspension yield stress must be considered in mixer design. Ideally, we want motion to exist throughout a mixing vessel, with no stagnation or dead zones present. As illustrated in Figure 20-6, this can be difficult to achieve. Here a $C_m = 0.02$ fully bleached kraft fiber suspension was agitated in a stirred vessel using a Rushton turbine. Dye was added to indicate regions of suspension motion. In Figure 20-6a the impeller was rotating at $N = 4$ rps. No motion was detected at the vessel wall. As impeller speed was increased to $N = 7$ rps, motion became apparent throughout most of the vessel volume, but stagnant regions existed in the

**1200** MIXING IN THE PULP AND PAPER INDUSTRY

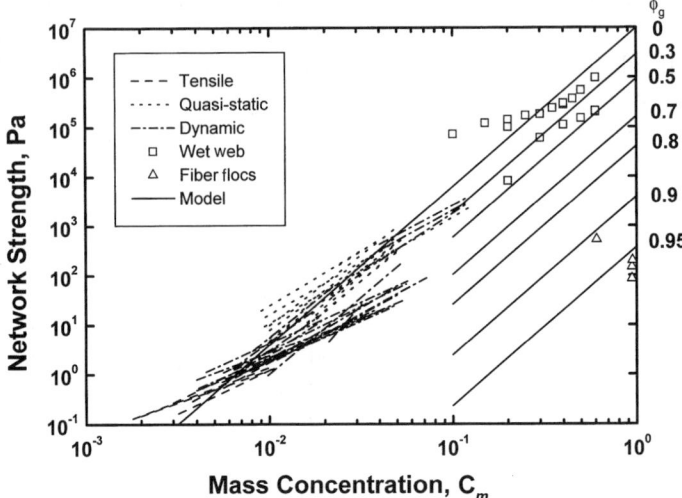

**Figure 20-5** Fiber network strength versus suspension mass concentration. Compilation of literature data. The predictions of eq. (20-17) for a fiber having an aspect ratio of $A = 70$ are given.

**Figure 20-6** Effect of yield stress on suspension motion in a stirred tank. $C_m = 0.02$ FBK suspension. The vessel is 30 cm in diameter with the suspension height set at 30 cm. A $D = 10$ cm diameter Rushton turbine was located 10 cm from the vessel floor. Impeller speeds are $N = (a)$ 4, $(b)$ 7, and $(c)$ 14 rps. The red dye shows regions of suspension motion. In image $(a)$ the cavern has not reached the vessel wall. See insert for a color representation of this figure.

low-shear regions at the wall. Increasing impeller speed to $N = 14$ rps produced motion throughout the vessel, but a wide range in local suspension velocity was apparent. The dramatic increase in network strength with mass concentration (Figure 20-5) means that the yield stress becomes more important for mixing as suspension concentration increases.

Yield stress fluids create caverns around impellers. The cavern boundary is delineated by the surface where the imposed shear force does not exceed the fluid yield stress. Pulp suspensions exhibit this behavior. The approach developed by Soloman et al. (1981) was used to characterize pulp suspension behavior in a laboratory mixer. A force balance at the cavern boundary (assumed cylindrical) gives its radial extent, $r_a$, as

$$r_a = \sqrt{\frac{T}{2\pi \tau_y L}} \quad (20\text{-}18)$$

where T is the measured torque, L the length of the cylindrical rotor, and $\tau_y$ the suspension yield stress. The cavern size measured for fully bleached kraft (FBK) pulp suspensions having mass concentrations from $C_m = 0.042$ to 0.095 is compared with predictions of eq. (20-18) in Figure 20-7 (Bennington, 1988). The results show that the equation underpredicts the cavern size. The accuracy with which the cavern radius could be measured and the accuracy of the yield stress determination are significant, as indicated by the error bars on the graph, and explain some of the discrepancy. The extent of suspension motion is important for operation of agitated chests but is difficult to predict.

### 20-3.4 Turbulent Behavior of Pulp Suspensions

Continued disruption of the fiber network is necessary to create and maintain suspension motion. This requires continuous application of force sufficient

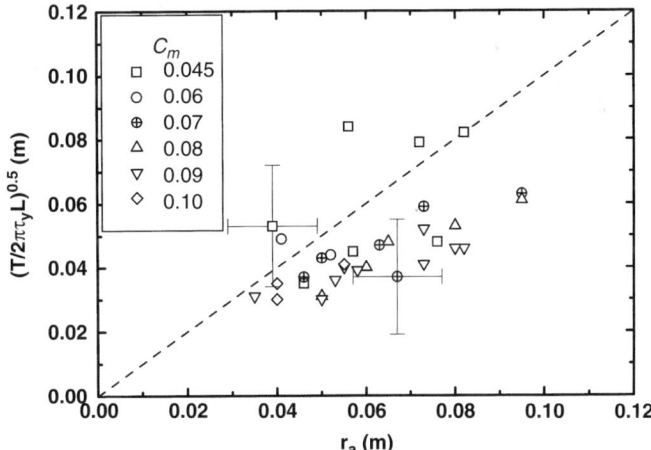

**Figure 20-7** Measured cavern radius versus that predicted using eq. (20-18) for tests made using a fully bleached kraft pulp. The test apparatus has a horizontal Couette geometry. The housing has a depth of 10 cm and a diameter of 22 cm. Six 1 cm baffles are spaced at 60° intervals around the periphery to limit slip at the vessel wall. The rotor is six-bladed and 10 cm in diameter. The blades protrude 1 cm from the rotor hub. The rotor is about 10 cm high and extends the full depth of the vessel.

to maintain network disruption with the consequent dissipation of energy. At medium mass concentrations, the solidlike properties of the suspension often required that motion be attained using positive-displacement types of pumps and mixers. In the early 1980s, Gullichsen and Harkonen (1981) showed that medium consistency suspensions behaved in a fluidlike manner provided that sufficient shear was applied to them. This corresponded to the onset of a turbulent flow regime. This led to development of centrifugal medium consistency devices, including mixers and pumps.

A typical plot of torque (shear stress) versus impeller rotational speed (shear rate) for a medium consistency pulp suspension is compared with that of water in Figure 20-8. The device used for the test had Couette geometry, with baffles on the vessel wall to prevent solid/suspension slip at the vessel periphery. For water, torque increased with the square of rotational speed, as expected for a Newtonian fluid mixed under turbulent conditions. For the pulp suspension, a significant torque was measured before rotation began. This was due to the network strength. Once the suspension yielded, a slip plane was created in the suspension immediately adjacent to the rotor. The torque fell dramatically. Increasing the rotor speed enlarged the region of active suspension motion as more and more flocs were entrained in the flow (illustrated in the inset to Figure 20-8). The flow followed streamlines around the rotor, and individual fiber flocs were observed in the flow. The torque increased as more suspension was brought into motion. When the flow reached the vessel wall, a transition to turbulence began. At this point (marked by a solid point on the graph), all suspension was in motion and the yield stress was exceeded everywhere in the vessel. Creation of this fluidized

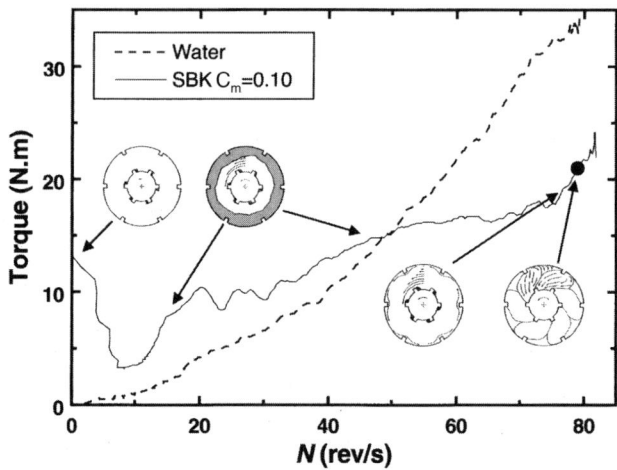

**Figure 20-8** Torque versus impeller speed for water and a $C_m = 0.10$ bleached kraft suspension measured in the Couette test apparatus described in the caption to Figure 20-7. The insert diagrams show the observed pulp suspension motion at points along the flow curve.

or fluidlike state became one criterion used in the design of these processing devices. Although individual flocs could be seen in the flow, they were dispersed as they moved into the rotor vicinity (Bennington, 1988).

The onset of fluidization marks the beginning of the transition to turbulence and is device dependent. This transition was studied under operating conditions typical of medium consistency pumps and mixers (Bennington et al., 1991; Bennington and Kerekes, 1996). The average energy dissipation measured at the onset of turbulence, $\varepsilon_F$ (in W/m$^3$), was correlated by the equation

$$\varepsilon_F = 4.1 \times 10^9 C_m^{2.5} (D/T)^{2.3} \tag{20-19}$$

where T is the housing (mixer) diameter, D the rotor diameter, and $C_m$ is expressed as a fraction. In the limit that $D/T \to 1$, a device-independent estimate of the power required to initiate turbulence in a pulp suspension was obtained. Industrial equipment usually operates at average dissipation levels lower than this value.

The measurement of the transition to turbulence and its comparison with power curves developed in the same device using Newtonian fluids allows the apparent suspension viscosity to be estimated. For medium consistency suspensions, the apparent viscosity is high. For example, for an FBK at $C_m = 0.10$, $\mu_a$ is approximately 2 Pa · s at the fluidization point (Bennington and Kerekes, 1996).

## 20-3.5 Turbulence Suppression in Pulp Suspensions

Creation of turbulence in pulp suspensions is needed to disperse fiber aggregates and to augment mass transfer within the suspension. Disruption of the fiber network requires considerable energy, and the fibers themselves modify the turbulence within the suspension. Experimental measurements for pulp suspensions have been limited to suspensions at low concentration due to interference between the fibers and invasive probes, or due to attenuation of optical signals by suspension opacity. Most studies have concluded that fibers suppress turbulence within the suspension, although some have reported turbulence enhancement. The studies indicate complex interaction between the fibers, fiber flocs, and the flow (Bennington and Mmbaga, 1996).

Mixing-sensitive chemical reactions have been used to probe liquid-phase turbulence within pulp suspensions (Bennington and Bourne, 1990; Bennington and Thangavel, 1993; Bennington and Mmbaga, 2001). As with previous studies, pulp fibers were found to attenuate liquid-phase turbulence under most mixing conditions. This was attributed to dissipation of energy by friction at fiber–fiber contact points. Under certain conditions, the local energy dissipation was slightly increased, probably due to local redistribution of energy within the mixer. Mapping this dissipation in a typical high-shear mixer showed modification of its distribution throughout the vessel volume when compared with water (Figure 20-9). However, the liquid-phase energy dissipation decreased exponentially with increasing suspension concentration.

**1204** MIXING IN THE PULP AND PAPER INDUSTRY

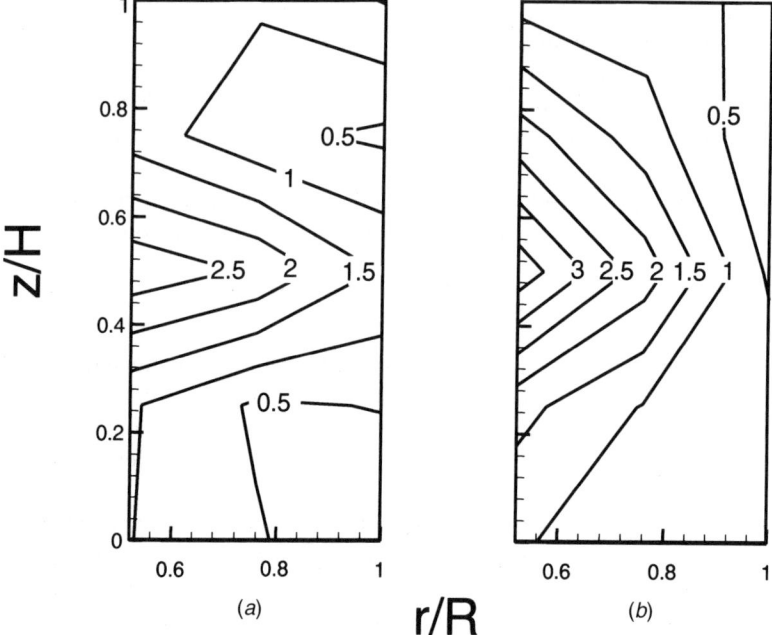

**Figure 20-9** Two dimensional map of the normalized local energy dissipation ($\varepsilon_{loc}/\varepsilon_{avg}$) measured in a plane midway between wall baffles in a medium-intensity mixer. The mixer is a concentric-cylinder device having a depth of 10 cm and a diameter of 19 cm. Four 1 cm baffles are spaced at 90° intervals around the vessel periphery. The rotor is four-bladed, 10 cm in diameter, with the blades protruding 2.5 cm from the rotor hub. The input ($\varepsilon_{in}$) and average ($\varepsilon_{avg}$) (measured) energy dissipations are given for two cases: (a) water ($\varepsilon_{in} = 101 \pm 5$ W/kg, $\varepsilon_{avg} = 120 \pm 13$ W/kg) and (b) a $C_m = 0.013$ FBK pulp suspension ($\varepsilon_{in} = 101 \pm 5$ W/kg, $\varepsilon_{avg} = 33 \pm 5$ W/kg). Note that the distribution map gives the local energy dissipation in the liquid phase only. Energy dissipated by the fibers is not measured. Tests were made at N = 17.3 rps.

### 20-3.6 Gas in Suspension

Gas is often present in pulp suspensions, reaching significant volumes in the medium consistency range (Dosch et al., 1986). For example, for a $C_m = 0.10$ kraft pulp, gas can occupy 10 to 20% of the suspension volume. This gas is readily separated from the suspension when subjected to centrifugal fields, such as those created by pumps and mixers. This provides a method of removing gas from suspensions (using centrifugal pumps) but creates difficulties when gases must be mixed into them. Gas, when present, can reduce the effective suspension density in the impeller vicinity, reducing both motion and energy dissipation in the suspension. This results in reduced gas–liquid mass transfer. Even when the gas volume in the suspension is small, gas can accumulate and be held up within the mixer.

The efficiency of gas mixing can be evaluated by measuring the rate of mass transfer from the gas to the liquid phase. This has been done in the laboratory using high-shear mixers similar to those used in industry (Bennington et al., 1997a; Rewatkar and Bennington, 2000). For an FBK fiber suspension in batch operation, the volumetric gas–liquid mass transfer coefficient, $k_L a$, was correlated with power dissipation per unit volume, $\varepsilon_v$, the gas void fraction in the mixer, $\phi_g$, and the suspension mass concentration (as a fraction), $C_m$. This gave

$$k_L a = 1.17 \times 10^{-4} \varepsilon_v^{1.0} \phi_g^{2.6} \exp(-38.6 C_m) \qquad (20\text{-}20)$$

Ninety-five percent confidence intervals were determined for each fitted parameter. For power dissipation this was $1.0 \pm 0.38$, for void fraction $2.6 \pm 0.65$, and for mass concentration $-38.6 \pm 11.0$. The dependence of $k_L a$ on power per unit volume is higher than that reported by other investigators, which typically have exponents ranging from 0.4 to 0.8 (Mann, 1983; Tatterson, 1991). Further, the dependence of $k_L a$ on the gas void fraction is significantly higher than found in stirred vessels, where proportionality is expected. Increasing the suspension mass concentration exponentially reduces gas–liquid mass transfer, with the reduction (compared with water) being an order of magnitude for a $C_m = 0.10$ suspension. This is shown in Figure 20-10, where $k_L a$ values measured for a representative set of operating conditions are plotted against suspension volume concentration. The reduction in $k_L a$ parallels the reduction in liquid-phase turbulence measured in the suspension (Mmbaga, 1999; Rewatkar and Bennington, 2000).

Gas–liquid mass transfer can dramatically affect certain bleaching reactions. For rapid reactions such as delignification with ozone, the rate of ozone mass

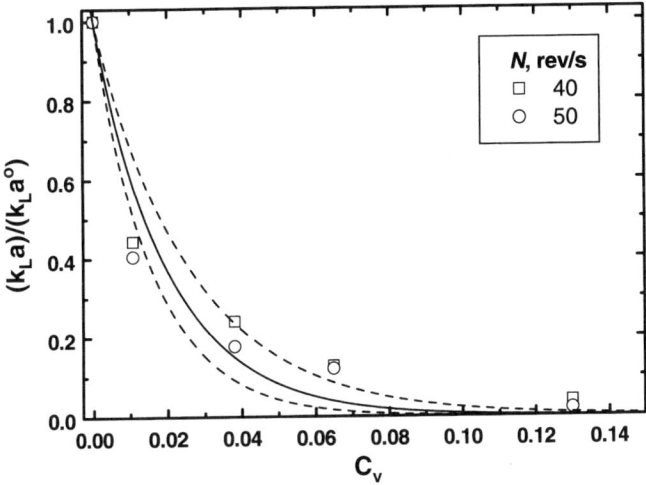

**Figure 20-10** Relative volumetric gas–liquid mass transfer coefficient ($k_L a$) versus suspension volumetric concentration for tests made in a high-shear laboratory mixer. Tests were made in a commercial laboratory pulp mixer (Quantum MK-IV).

transfer determines the effective reaction rate. Here, the extent of bleaching is determined by mass transfer achieved in the mixer (Bennington et al., 1997a).

## 20-4 SCALES OF MIXING IN PULP SUSPENSIONS

Mixing is often targeted at a particular scale, depending on the purpose of mixing (reaction, blending, etc.). In a liquid, mixing scales are determined by the scale of turbulence generated within them. In a pulp suspension, mixing scales are imposed by the fiber dimensions. A fiber scale can be defined as having dimensions between that of a fiber diameter and a floc length (typically, two to three fiber lengths). The macroscale encompasses scales larger than the floc dimension and the microscale encompasses scales below a fiber diameter, as summarized in Table 20-1.

As engineers, we often think of scales associated with processing equipment: the volume of a tank or mixer, the diameter of a section of piping, and so on. In many situations, a satisfactory process outcome also requires that good fiber scale and/or microscale mixing be attained. Accomplishing the required mixing and uniformity in a pulp suspension can be more difficult than in fluid systems due to the suspension rheology.

## 20-5 MACROSCALE MIXING/PULP BLENDING OPERATIONS

### 20-5.1 Homogenization and Blending

The agitation of pulp stock for homogenization and blending is the most common mixing operation in pulp and paper manufacture. The blend chest (a "well-mixed" stirred vessel) illustrates the importance of macroscale mixing in pulp and paper processes. It is the heart of the stock preparation system and is used to mix two or more pulp streams, often with wet-end chemicals, dyes, fillers or additives, as well as providing a uniform feed of stock to the paper machine. Mixing chests

**Table 20-1** Mixing Scales in Pulp Suspensions

| Mixing Scale | Scale Dimension Size (cm) [Volume (cm$^3$)][a] | Mixing Achieved by: | Physical Scale |
|---|---|---|---|
| Macroscale | >1.0 [0.5] | Bulk motion | Tanks, pipes, vessels |
| Fiber scale | 0.005–1.0 [7 × 10$^{-8}$–0.5] | Laminar and turbulent shear, diffusion | Fibers and flocs |
| Microscale | <0.005 [7 × 10$^{-8}$] | Diffusion aided by small scale fluid motion | Fiber diameter and fiber wall thickness |

[a] Assuming spherical geometry.

act as low-pass filters and ensure uniform mass flow ahead of many downstream operations in addition to the paper machine, including bleaching stages, washers, screens, and cleaners. In most cases, process control strategies are used to deal with long-term consistency fluctuations.

The economics of less-than-perfect macroscale mixing are difficult to assess, although an estimate can be made. From studies of paper machine variability, Bialkowski (1992) found that 55% of the variability in final paper quality was at frequencies due to flow instabilities and mixing deficiencies. These frequencies were higher than could be removed by process control strategies. It is difficult to relate paper machine efficiency (breaks, etc.) to this process variability. However, if we assume that the process variability is directly related to machine downtime, the economic impact of mixing nonuniformity can be estimated. Paper machine lost time averages approximately $10 000 per hour, and the total yearly operating efficiency of a paper machine is around 80 to 95% (personal communication, Pikulik, 2001). If only 1% of operating efficiency is lost due to mixing related inefficiencies, the cost for the average paper machine would be $800 000 per year! Other estimates have been made for improved consistency control and have yielded savings of similar magnitude (Jansson, 1999).

Another example of the importance of good macroscale mixing of pulp suspensions is consistency control ahead of a bleaching stage. Poor mixing can result in the over-or under-application of bleaching chemicals on pulp, which can impair pulp quality as well as increase chemical costs. A typical bleaching response is nonlinear, with increasing amounts of chemical required to obtain an increased bleaching response. If consistency control is poor ahead of a bleaching operation, some pulp must be overtreated to compensate for the periodic episodes of higher mass flow (a high consistency excursion). Because the frequencies of these disturbances are higher than can be eliminated by either feedforward or feedback control, overcharging is the only alternative to better mixing. Typically, the standard deviation of consistency measurement is 3 to 5% (Jansson, 1999). The increased chemical required to compensate for poor consistency control can be substantial and adds to the cost of bleaching the pulp. Just reducing the standard deviation from 5% to 3% (which could be done by measuring the consistency with a more accurate consistency meter) was estimated to be worth $250 000 per year in one chlorine dioxide bleaching application (Jansson, 1999). Further, the variability in fiber treatment can be propagated down the fiber line, affecting succeeding stages and contributing to nonoptimal strength delivery from the bleaching system.

The design of stock chests is concerned largely with selecting the power necessary to ensure complete motion throughout the chest volume. The impeller power for suspension agitation has been correlated with the apparent viscosity of the suspension, assuming Bingham plastic behavior. Thus,

$$\mu_a = \frac{\tau_y + \mu_p \gamma}{\gamma} \quad (20\text{-}21)$$

where $\tau_y$ is the yield stress of the suspension, $\mu_p$ the plastic viscosity, and $\gamma$ (du/dy) the shear stress. Using the Metzner–Otto correlation and assuming that the plastic viscosity is close to that of water (and thus an insignificant contribution to the apparent viscosity compared with the yield stress), the apparent viscosity can be expressed as

$$\mu_a = \frac{\tau_y}{10N} \quad (20\text{-}22)$$

which can then be used in calculation of the impeller Reynolds number. In practice, suspension behavior has been correlated with a stock parameter m (often used in correlations or to form a modified Reynolds number), given by

$$m = \sqrt{\frac{\tau_y}{\rho}} \quad (20\text{-}23)$$

The parameter m must be determined for each pulp stock [essentially by measuring the suspension yield stress, which is affected by many suspension parameters; see eq. (20-17)]. Correlations made using this procedure allow the behavior of different stocks to be correlated and for differences in impeller type and location to be expressed (Blasinski and Rzyski, 1972).

Gibbon and Attwood (1962) addressed the issue of scaling chest design, finding that impeller power scaled as

$$\frac{P_1}{P_2} = \left(\frac{D_1}{D_2}\right)^{2.75} \quad (20\text{-}24)$$

for equal process flow in geometrically identical chests. Here P is the impeller power and D the impeller diameter for chests 1 and 2. The measured scaling dependence of 2.75 lies between that of constant power per unit volume [one common method used for scaling pulp chest volume (an exponent of 3.0)] and that of constant tip speed (an exponent of 2.0). One still must measure the power consumption needed for adequate mixing at one chest scale. This is usually assessed visually as a condition of complete active suspension motion on the chest surface, which at best is a subjective measure. In addition, stagnant regions can exist below the suspension surface in vessel corners, as observed in a Plexiglas laboratory scale chest (Ein Mozaffari et al., 2001).

The dynamics of pulp chests in pulp processes have been modeled assuming ideal mixing and first-order behavior (Walker and Cholette, 1958; Reynolds et al., 1964). This allows chest design (essentially the chest volume) to be based on the anticipated disturbances and degree of attenuation required. However, these studies did not include allowances for nonideal suspension flow that can create bypassed regions and dead zones in the chest.

Industrial chests are known to behave nonideally. The response of a typical rectangular blend chest (L : W : D = 6.1 : 4.6 : 4.3 m, volume = 120 m$^3$) to a step change in consistency is shown in Figure 20-11. Based on these data, the Bode plot of the chest was constructed as shown in Figure 20-12. Significant departure from ideal behavior (the dashed curve) was measured between frequencies of

MACROSCALE MIXING/PULP BLENDING OPERATIONS  **1209**

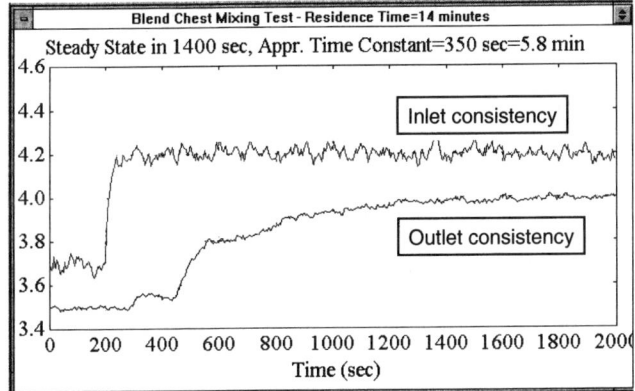

**Figure 20-11** Response of an industrial stock chest to a step change in mass concentration at t = 200 s. The mass concentration (in percent) is given on the y-axis.

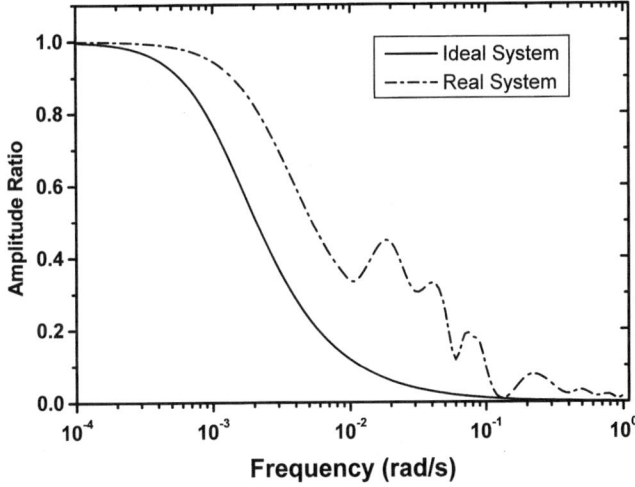

**Figure 20-12** Bode plot showing the amplitude ratio versus disturbance frequency for an industrial chest (Figure 20-11) and an ideal mixed chest of the same volume.

0.01 and 0.1 rad/s. This leads to problems in uniformity that cannot be removed by control strategies.

Currently, industrial chest design is based largely on proprietary criteria developed from accumulated experience (Oldshue, 1983; Yackel, 1998). One method uses the momentum required to agitate pulp stock within a chest. The momentum needed for surface motion of the pulp suspension has been correlated with a range of design variables, including the chest size and geometry, impeller location(s), the type of fiber and suspension mass concentration, and the desired retention time. Based on the required momentum, an appropriate impeller is selected. A similar approach is followed when designing limited agitation zones in larger vessels: for example, a low consistency extraction zone in a high-density pulp storage tower. These procedures are not linked directly to suspension rheology, although suspension rheology certainly dictated the correlations. Adjustments can be made depending on the pulp type used (through factors). However, these design procedures provide no indication of the degree of motion generated in the suspension. (Indeed, the original data upon which the correlations were developed are not given.)

Computational fluid dynamics permits solution of the entire suspension flow field, although considerable computer power is needed. CFD has been applied to pulp suspension agitation by Bakker and Fasano (1993), who treated the pulp suspension as a Bingham fluid. Calculations were first performed assuming turbulent flow. The turbulent shear rate was then compared with the shear rate required to disrupt the fiber network. If the shear was insufficient, calculations were redone for laminar flow. Calculated suspension velocities of less than 0.001 m/s were treated as being stationary. The simulation solutions corresponded to visual observations made in a laboratory chest, although the size of the agitated region was underpredicted. Turbulent motion was often limited to the immediate vicinity of the impeller.

Clearly, more work needs to be done in the area of macroscale mixing of fiber suspensions. Links are needed between suspension rheology, the desired process outcome (i.e., degree of suspension motion), and the chest design. The ability to design a blending system with known process dynamics is needed. This would also enable the synergism between process control and mixing to be fully exploited.

Pulp processing devices are also used to mix and simultaneously produce physical changes to the pulp. One example of this is in repulping operations.

### 20-5.2 Repulping

The repulping or reslushing of paper is an integral part of paper manufacture. *Broke and trim*, paper produced but not wound onto the reel, must be repulped and returned to the machine for reprocessing. In paper recycling, postconsumer paper products (often, old newspapers and old magazines) are the main source of recycled fiber. These must be repulped to liberate the individual fibers, and ink and other contaminants must be detached from the fiber surface and separated from the suspension. Repulpers must also blend the paper furnish thoroughly and mix chemicals and additives into the pulp.

Repulpers can be of a number of designs. Figure 20-13 shows a laboratory pulper modeled after a common industrial design. The repulper uses a triple-flighted helical rotor, a conical tub, and wall baffles to ensure complete motion in medium consistency suspensions. The laboratory unit operates on a batch size of 8 kg of pulp. Industrial units have volumes up to 160 m$^3$. Despite this disparity in size, the progress of deflaking and ink detachment can be modeled using a mixing-controlled mechanism based on the force transmitted to the fiber

**Figure 20-13** Photograph of the laboratory repulper showing the three-vaned helical rotor and wall baffles. The repulper tub has an inner diameter of 30 cm and a working volume of 8 L.

surface by the rotor (Bennington et al., 1998a; Bennington, 1999; Bennington and Wang, 2001).

If deflaking and ink detachment are caused by direct suspension–rotor interaction, the extent of deflaking should be proportional to the extent of this interaction. For a repulper where complete motion exists throughout the suspension, the rate of deflaking is given by the first-order expression

$$\frac{dF}{dC_R} = kF \qquad (20\text{-}25)$$

where F is the Tappi flake content (the mass of paper flakes retained on a 10-mesh screen following screening), k the rate constant (negative, as the flake content decreases during processing), and $C_R$ the contact area created between the fiber suspension and the rotor. $C_R$ is given by

$$C_R = SC_v \qquad (20\text{-}26)$$

where $C_v$ is the volume concentration of the suspension [see eq. (20-13)] and S is the area swept out by the rotor, given by

$$S = NBGt \qquad (20\text{-}27)$$

where N is the rotor speed, B the number of rotor vanes, G the surface area that a single rotor vane sweeps out during one revolution, and t the time. G is calculated knowing the rotor geometry and the height of suspension in contact with the rotor.

The rate of deflaking, k, will depend on two factors: the strength of the material being pulped and the force applied by the rotor to the suspension. Thus,

$$k = k_o \exp\left(-\frac{T_M}{KF_R}\right) \qquad (20\text{-}28)$$

where $k_o$ is the intrinsic rate constant, $T_M$ the strength of the material (the paper strength in the case of deflaking), $F_R$ the force per unit length of rotor vane, and K a proportionality constant. $F_R$ can be evaluated by measuring the torque on the rotor shaft:

$$F_R = \frac{T}{(D_m/2)BH_C} \qquad (20\text{-}29)$$

where T is the torque, $D_m$ the average rotor diameter, and $H_C$ the rotor height in contact with suspension. Alternatively, $F_R$ can be evaluated knowing the power number at the given operating conditions.

The use of these equations has allowed a wide range of data to be explained, as illustrated in Figure 20-14. Here a number of recycled paper materials of varying strength (wet-tensile strength) and suspension mass concentration were repulped in a 21 L laboratory repulper. The extent of flake removal is plotted against $kC_R$, which accounts for equipment geometry, operating conditions, and paper strength. Within experimental error, the data agree. This supports the interpretation of

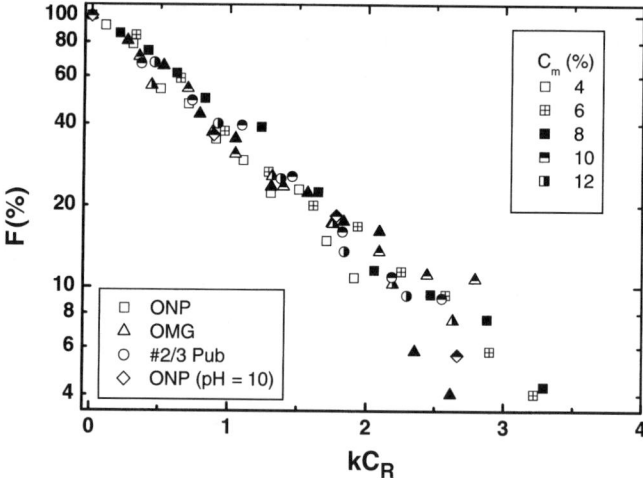

**Figure 20-14** Tappi flake content versus fiber–rotor contact area ($C_R$) [see eq. (20-25)] for four different recovered paper furnishes. The suspension mass concentration and paper type are parameters. ONP, OMG, and #2/3 PUB are old newspaper, old magazine, and a high-grade publishing furnish, respectively.

deflaking as being caused by physical interaction between the paper and the rotor. The model has been verified for varying rotor speeds and rotor designs, has been used successfully in the mill, and can be modified for different mass contents in the repulper (Bennington et al., 1998b).

Ink detachment is more difficult to model than deflaking. Ink is attached to fiber with a certain adhesive force. If this attachment force is greater than the force applied by the rotor, the ink will remain attached to the fiber. Ink can also be hidden in fiber interstices, and thus not be subjected to the rotor forces. Both these factors contribute to a floor level of ink that remains following repulping. Further, ink can be redeposited onto or inside the fiber once it has been detached from the fiber surface. The latter phenomenon, called *lumen loading*, is a function of mixing action, which increases with ink concentration and the duration and intensity of suspension treatment.

If one considers these factors, the observed detachment behavior can be modeled in a manner analogous to deflaking. This gives the ink concentration remaining on the fiber as a function of time, $C_I(t)$:

$$C_I(t) = C_I^f + (C_I^o - C_I^f)e^{k_I t} + k_R \int_{t=0}^{t} (C_I^o - C_I(t))\,dt \qquad (20\text{-}30)$$

where $C_I^f$ is the floor level of ink that cannot be removed and $C_I^o$ is the initial ink concentration. $k_I$ is the rate constant (again negative) for ink detachment, which can be expressed by an equation analogous to that of eq. (20-28). $k_R$ (positive) is the rate of ink redeposition, which can be measured experimentally. Thus, the ink concentration remaining on the fiber at any time depends on three factors: the ink that cannot be removed, the ink that can be removed (with ink removal following first-order kinetics similar to that for deflaking), and the amount of ink redeposited on the fiber (Bennington and Wang, 2001).

### 20-5.3 Lumen Loading

Most of the ink irreversibly redeposited in recycling operations is deposited inside the fiber lumen (Ben and Dorris, 1999) with the extent of redeposition affected by the mixing action imparted during suspension processing (Bennington and Wang, 2001). Lumen loading of ink is undesirable, as it cannot be removed subsequently. However, lumen loading can be used to enhance certain sheet properties. In papermaking, it is often desirable to load the sheet with filler materials to enhance their optical properties. However, filler that remains on exterior fiber surfaces interferes with bonding and reduces the physical strength properties of the paper. Processes to load the filler inside the fiber lumen allow the benefits of increased filler levels to be realized without compromising the strength properties of the sheet. Lumen loading is facilitated by the high-shear mixing of pulp in an excess of filler material. As shear increases, so does the level of mechanical action. This causes more filler to enter and remain in the lumen (Middleton and Scallan, 1993). No commercial application for intentional lumen loading of pulp fibers is currently operated.

## 20-6 MIXING IN PULP BLEACHING OPERATIONS

### 20-6.1 Pulp Bleaching Process

Pulp bleaching selectively removes unwanted components (chromophores) from the fiber wall using a series of increasingly selective chemicals and reaction conditions. One common bleaching sequence used to produce fully bleached kraft pulp is the OD(EO)DED sequence. (Each letter represents a distinct chemical used in a separate bleaching operation or stage). Typical reaction conditions for each stage in this sequence are given in Table 20-2, although conditions vary from mill to mill [see individual chapters in Dence and Reeve (1996)]. Other bleaching sequences and bleaching chemicals can also be used. For example, peroxide and ozone are increasingly incorporated in bleaching sequences, alone or in conjunction with other chemicals.

Bleaching reactions are complex, with many different chemical species involved in any given reaction. For example, hundreds of chlorinated products have been identified in extracts made from pulps bleached in a number of chlorinated sequences (Reeve and McKague, 1990). Although model compound studies are useful for elucidating chemical reactions and reaction mechanisms, the majority of bleaching studies measure an averaged bleaching response over all targeted compounds. As bleaching involves lignin removal (a heterogeneous polymer with a multitude of chemical structures and reactive sites) it is common to represent its removal using standardized tests. The kappa number (the percent lignin is approximately equal to 0.15 times the kappa number) is commonly used for this purpose (Tappi Test Methods, T236 cm-85). At low lignin concentrations (<0.5%) the kappa test is not accurate. Here, the progress of the bleaching reaction can be followed using brightness development (Tappi Test Methods, T452 om-98).

**Table 20-2** Typical Conditions for Bleaching Reactions: OD(EO)DED Sequence

| Stage Designation | Suspension Concentration ($C_m$) | Chemical | Chemical Charge (kg/t) | Reaction Temperature (°C) | System pH | System Pressure[a] (kPa) | Residence Time (h) |
|---|---|---|---|---|---|---|---|
| O | 0.08–0.14 | $O_2$ | 20–30 | 85–100 | 10–12 | 140–660 | 0.5–1.5 |
|   |   | NaOH | 25–30 |   |   |   |   |
|   |   | $MgSO_4$ | 0–2.5 |   |   |   |   |
| D | 0.03–0.04 | $ClO_2$ | 15 | 30–60 | 1.5–3 | 100 | 0.3–1.0 |
| EO | 0.10–0.12 | NaOH | 25 | 70 | 10.5 | 100–300 | 1.0 |
|   |   | $O_2$ | 5 |   |   |   |   |
| D | 0.10–0.12 | $ClO_2$ | 8 | 70 | 3.5 | 100 | 2.0 |
|   |   | NaOH | 5 |   |   |   |   |
| E | 0.10–0.12 | NaOH | 5 | 70 | 10.5 | 100 | 1.0 |
| D | 0.10–0.12 | $ClO_2$ | 5 | 70 | 4 | 100 | 3–5 |

*Source*: Based on data taken from Dence and Reeve (1996) and other sources.
[a] At top of tower. Pressure at the tower bottom will have hydrostatic head added, typically 300 kPa.

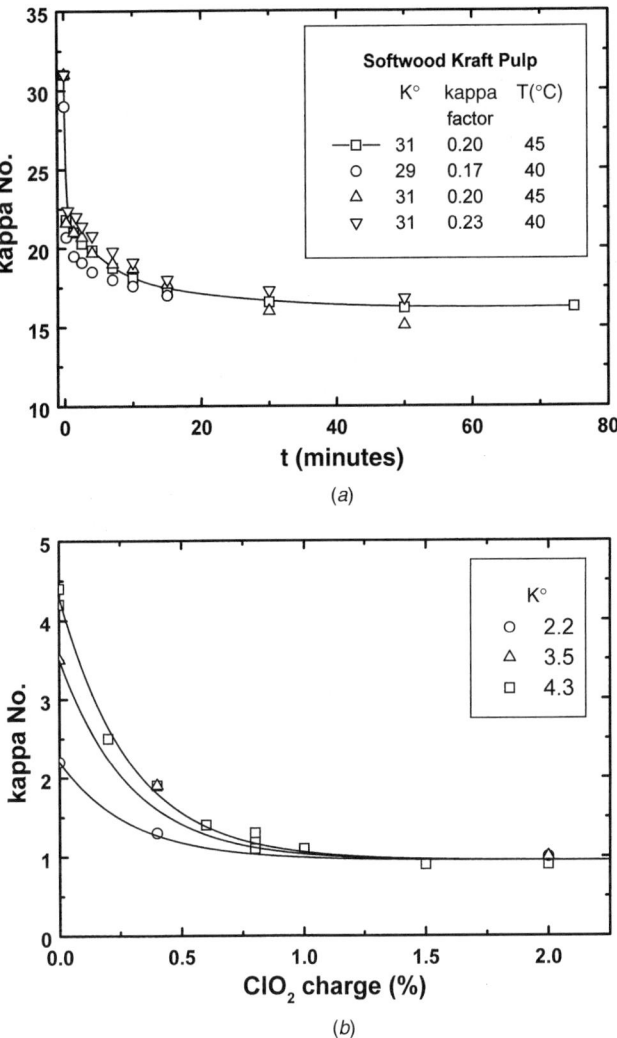

**Figure 20-15** Typical pulp bleaching results. (a) Kappa number versus reaction time for chlorine dioxide ($D_0$) delignification of a softwood kraft pulp ($C_m = 0.031$, initial pH ~ 3). (From Tessier and Savoie, 1997.) (b) Kappa number following chlorine dioxide ($D_1$) bleaching of a kraft softwood pulp. Pulp was bleached in the mill using a $DE_{OP}$ sequence and laboratory bleached ($D_1$) with hand mixing (at $C_m = 0.10$ and $T = 74°C$). (From Bennington et al., 2001.)

The progress of a typical bleaching reaction is illustrated in Figure 20-15. In Figure 20-15a, the reduction in kappa number is plotted against time for delignification with chlorine dioxide in the first delignification stage ($D_0$). The initial chemical reaction is rapid and the majority of the bleaching response occurs during it. The rapid reaction is followed by a falling-rate period, where reaction slows and eventually tapers off. This leaves residual lignin in the fiber. Increasing the chemical charge increases lignin removal (as shown for chlorine dioxide delignification in Figure 20-15b), but it is difficult to achieve complete removal in a single stage. The aggressive conditions required to delignify pulp completely in a single stage would impair pulp strength. Consequently, the residual lignin must be reactivated (using an extraction stage, for example) and removed in subsequent bleaching stages. All bleaching reactions display these characteristic attributes, which are reflected in the design of the typical bleaching stage, illustrated in Figure 20-16.

In a typical bleaching stage, pulp from the previous stage is washed to remove reacted and dissolved substances leached into the process liquor. The pulp is then prepared for reaction. Steam is used to raise the pulp to the reaction temperature and the process pH set. The bleaching chemical is then added, usually in a dedicated mixer, and the suspension passed into a tower for a period to ensure bleaching reaches completion. This time varies widely. For ozone delignification, it can be as short as 1 min. For chlorine dioxide brightening, it can be as long as 5 h. The pulp is then washed before being sent for further processing.

***20-6.1.1 Mixing and Chemical Reaction.*** Lignin and cellulose are contained within the fiber wall. The selectivity of a bleaching process (usually defined as the ratio of lignin removal to a measure of the concurrent damage to cellulose) is controlled primarily by the reaction chemistry and processing conditions (chemical charge, suspension mass concentration, temperature, pH). The extent of reaction (determined by chemical dosage and reaction time) can

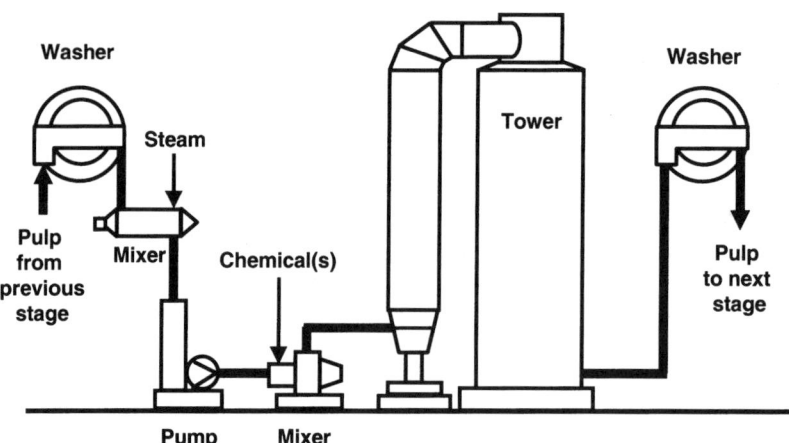

**Figure 20-16** A bleaching stage consists of prewashing, mixing for steam addition (temperature adjustment), mixing for chemical addition, pulp retention in a tower, and postwashing. The diagram is typical of a chlorine dioxide bleaching stage ($D_1$ or $D_2$) and shows only the pulp and chemical flows. Filtrate flows are not shown.

also be controlled to minimize pulp damage. Controlling these parameters permits the optimization of bleaching and pulp strength over an entire bleaching process.

In pulp passage through a mixer it is desirable to disrupt the fiber network (break apart the individual flocs) and commingle each fiber with the appropriate amount of bleaching chemical. Many mixers are designed to achieve this dispersed state. On exiting the mixer, the suspension rapidly reflocculates (for a $C_m = 0.10$ suspension in as little as $10^{-4}$ to $10^{-3}$ s). From this point onward, the reaction can be viewed as continuing in tiny individual reactors (the flocs) as they are conveyed through the remainder of the bleaching stage. Within the flocs, diffusion plays a large role in bringing chemical to regions where it is consumed. As the mixing quality attained on exiting the mixer can be essentially "locked in" for the remainder of the bleaching stage, mixture quality at the mixer exit must be good. Further, the rate at which this mixing occurs must be rapid compared with the net chemical reaction rate, or bleaching will occur under nonhomogeneous conditions.

As in aqueous systems, characteristic times for mixing ($\tau_M$) and chemical reaction ($\tau_R$) can be calculated and compared to determine whether mixing rate influences the outcome of a bleaching reaction. In the instantaneous ($\tau_M \gg \tau_R$) and fast regimes ($\tau_M \simeq \tau_R$), reaction occurs more rapidly, or at the same rate, as mixing. Here mixing can affect reaction. In the slow regime ($\tau_M \ll \tau_R$), mixing is complete before reaction begins and chemical kinetics govern the reaction outcome. As mixing occurs on a continuum of scales, time constants at appropriate scales must be compared with the bleaching reaction rate to see if mixing is important at that scale.

Mixing and reaction time constants are not readily calculated for pulp suspensions. Mixing rate is influenced by the complex suspension rheology. Bleaching rate is influenced by chemical diffusion into the fiber wall and can be controlled by mass transfer. Further, in laboratory experiments a net reaction rate is measured and may have been influenced by the mixing conditions during the test. Despite these concerns, a number of estimates can be made.

Mixing time constants can be estimated at a number of scales based on mixer design and operating conditions, chemical contacting strategy, and the level of turbulence generated in the suspension. A macroscale time constant may be based on the process time (the total residence time in the bleaching stage) or the mixer residence time. None of these time constants addresses the quality of mixing attained, which is discussed later.

The location of chemical injection may also be important for the bleaching reaction. Mesomixing time constants can be expressed as the characteristic time for turbulent dispersion of the chemical feed stream or for disintegration of large eddies into smaller ones. A number of methods for estimating these are given by Baldyga and Bourne (1999), and one can define others for fibrous suspensions. For example, the time required for dispersion of a floc could be used as a mesomixing time.

At the smallest scales, a micromixing time constant can be given by the reciprocal of the engulfment rate during mixing, $\tau_E$. This is given (Baldyga and

Bourne, 1989) as

$$\tau_E = 12.7 \left(\frac{v}{\varepsilon}\right)^{0.5} \quad (20\text{-}31)$$

The kinematic viscosity of suspension, $v$, can be estimated using an approximation suggested by Bennington and Kerekes (1996):

$$v = \frac{\tau_y^2}{\varepsilon \rho^2} \quad (20\text{-}32)$$

where $\varepsilon$ is the energy dissipation in W/kg and $\rho$ is the suspension density. The suspension yield stress can be estimated by an appropriate correlation. For a semibleached kraft pulp, for example, the yield stress was correlated using (Bennington et al., 1990)

$$\tau_y = 3.2 \times 10^6 C_m^{2.8} \quad (20\text{-}33)$$

where $\tau_y$ is in Pascal and $C_m$ is expressed as a fraction.

Estimates for mixing time constants are given in Table 20-3 for typical bleach mixers. Two estimates are made for $\tau_M$: the process mixing time constant, $\tau_P$ (assuming a 90% reduction in the incoming variability by a first-order process) and the micromixing time constant, $\tau_E$ [calculated using eqs. (20-31)–(20-33)].

Bleaching chemicals react with dissolved substances in the process liquor, and with chemical species both on the fiber surface and within the fiber walls. For reaction with compounds dissolved in the process liquor or located on the fiber surface, appropriate time constants can be estimated from model compound studies conducted in the aqueous phase. However, most bleaching studies measure a net or aggregate bleaching rate with the pulp.

This bleaching rate is determined by the bleaching chemical used and its form (liquid or gas), the nature of the target compounds (usually, a wide range of

**Table 20-3** Time Constants for Mixing in Pulp Bleaching Operations

| | Mixing Conditions | | | Mixing Time Constants | |
|---|---|---|---|---|---|
| Mixer | $C_m$ | Energy Dissipation, $\varepsilon$ (W/kg) | Mixer Residence Time (s) | $\tau_P$ (s) | $\tau_E$ (s) |
| Hand mixing | 0.03 | 20 | 30 | 13 | 0.43 |
| | 0.10 | 50 | 30 | 13 | 5.0 |
| Stirred tank | 0.03 | 0.4 | 300 | 130 | 21.3 |
| Static mixer | 0.03 | 30 | 4 | 1.7 | 0.3 |
| Peg mixer | 0.10 | 100 | 10 | 4.3 | 2.5 |
| High-shear | 0.03 | 4000 | 0.3 | 0.1 | 0.002 |
| | 0.10 | 4000 | 1 | 0.4 | 0.06 |

similar compounds), and their location in the suspension. Compounds dissolved in the liquid phase are readily accessible to the bleaching chemical. However, the target compounds are located within the fiber wall and diffusion may control the reaction rate. Reaction progress is followed using an aggregate measure of the compounds, including the kappa number (lignin content), brightness development (chromophore removal), or pulp viscosity (cellulose depolymerization or degradation). Further, bleaching studies (like those of Figure 20-15) are influenced by laboratory mixing conditions (and are often not characterized).

A range of kinetic expressions is reported in the literature for common bleaching chemistries (see, e.g., Dence and Reeve, 1996; Gullichsen and Fogelholm, 1999). The net bleaching rate can often be approximated by first-order kinetics over industrially relevant time scales (seconds to minutes); for example, for delignification

$$-\frac{dL}{dt} = k_{eff}[L] \qquad (20\text{-}34)$$

where L is a measure of lignin content (e.g., the kappa number) and $k_{eff}$ is the effective first-order rate constant. Typically, a rapid (or initial) and slow (or residual) rate constant can be determined for each bleaching reaction. The initial rate of reaction is appropriate for comparison with the rate of mixing. Reaction time constants, $\tau_R$, were given by the relaxation time

$$\tau_R = 1/k_{eff} \qquad (20\text{-}35)$$

Net reaction time constants were estimated by determining effective first-order rate constants using kinetic data found in the literature (Bennington et al., 1989). Calculated values were $\tau_R = 3.9 - 29$ s for chlorination, $\tau_R = 280$ s for delignification with chlorine dioxide, and $\tau_R = 1100$ s for caustic extraction (these include any mass transfer resistance in the system). If we compare these reaction time constants with the mixing time constants given in Table 20-3, we see that in most cases chemical reaction controls the bleaching outcome (i.e., $\tau_M < \tau_R$). Some exceptions are apparent and may explain past trends in mixer selection. For example, the industry moved away from using stirred vessels for pulp chlorination service in the 1960s. The unfavorable rate of mixing provided by the stirred tank reactors may have accounted for the shift toward in-line dynamic and static mixers in this application.

Mills are currently presented with the choice of using either high-shear or static mixers for low consistency delignification in the first bleaching stage (usually, with chlorine dioxide or with mixtures of chlorine dioxide and chlorine). In both cases, mixing is sufficiently rapid that either strategy can be used successfully.

High intensity mixers are now commonly used for chlorine dioxide brightening ($D_1$ and $D_2$) operations. Indeed, their introduction in the 1980s allowed bleaching efficiency achieved in the mill to meet or surpass that measured in laboratory studies. The high rate of mixing achieved by these mixers probably contributed to this success. However, this is not the complete story. Peg-type mixers are often found in $D_1$ and $D_2$ service and have $\tau_M < \tau_R$. The use of high-shear mixers in this application enabled significant chemical savings to be achieved. This is probably due to the low energy dissipation in the peg mixers, which is insufficient to rupture the fiber flocs. Consequently, all fiber is not effectively exposed to chemical, and mixing quality is not as good as is achieved in the high-shear mixers, where energy dissipation is an order of magnitude higher. Thus, the quality of the mixture exiting the mixer is important in bleaching. Retention towers are incorporated into the design of a bleaching stage to permit homogeneity to be attained (over longer time scales) as well as allowing slower bleaching reactions to reach completion.

### 20-6.1.2 Diffusion and Retention Towers.

The importance of contacting every fiber with bleaching chemical can be illustrated by computing a diffusive time constant for a fiber and comparing it with that of a fiber floc. The diffusive rate constant, $\tau_D$, can be expressed as

$$\tau_D = \frac{\alpha a^2}{D} \tag{20-36}$$

where D is the diffusion coefficient, a the appropriate size dimension of the particle, and $\alpha$ a constant determined by the geometry of the system and the criteria of mixedness chosen. D changes with the molecular weight of the diffusing substance. For bleaching chemicals (low-molecular-weight compounds) the chemical diffusion rate through water is a good approximation for the rate through a swollen fiber wall, although tortuosity, which effectively lengthens the diffusion path, has the effect of reducing D.

The time for diffusion (without reaction) to increase the average concentration within a fiber wall to one-half of its ultimate value, with the fiber treated as a fully collapsed slab of thickness equal to two fiber wall widths, gives $\alpha \simeq 0.19$. For a fiber wall thickness of $a = 4.0 \times 10^{-6}$ m and an effective diffusion coefficient of $D = 1.3 \times 10^{-9}$ m$^2$/s, $\tau_D = 0.0023$ s. The ozonation of pulp fibers follows a shrinking core mechanism, limited by the rate of ozone diffusion through the fiber wall. Exposed fibers are completely penetrated and reacted in seconds under typical operating conditions (Bennington et al., 1999).

A typical floc has a much longer diffusion path than an individual fiber. For a floc (where $2a \approx 4$ to 10 mm) using the same diffusion coefficient with $\alpha = 0.036$ [for the average concentration of a floc (treated as a sphere) to increase to half of its ultimate value], $\tau_D$ becomes 250 s. Tower residence times vary widely depending on the bleaching stage used. The residence time was often determined by the time needed for the bleaching reaction to reach an asymptotic level. These residence times are normally much larger than 250 s (in chlorine dioxide brightening, for example, they can be 3 to 5 h in duration). However, the ability of the retention tower to create homogeneity throughout the fiber suspension must be questioned.

The diffusion distance within fiber suspensions has been measured by several investigators. From an infinite reservoir of reacting chemical, diffusion/reaction distances of 3 to 6 mm were covered in 1 h at $C_m = 0.03$ and distances of 1 to 2 mm at $C_m = 0.10$ (Paterson and Kerekes, 1984). Similar results were found by Bennington (1988), who measured a diffusion distance of 5 mm for a nonreactive tracer through a $C_m = 0.03$ pulp suspension. However, diffusion distances are much shorter under typical bleaching conditions. Here, only a limited quantity of chemical is available for diffusion and the actual distance covered will depend on the amount of chemical available for diffusion and the stoichiometry of the bleaching reaction. If one makes reasonable assumptions for the chemical demand of pulp, and the size and concentration of bleaching chemical remaining in the segregated regions, penetration distances into the suspension can be estimated. For typical bleaching conditions this distance (to the point of complete chemical consumption) is small, only a fraction of a millimeter (Bennington, 1996).

Thus, an effective chemical contacting strategy for a pulp bleaching operation is to mix chemical to the fiber scale at a rate faster than the effective bleaching rate. The mixture quality exiting the mixer (and entering the retention tower) must also be of sufficient uniformity to optimize the desired pulp quality parameters. The residence time provided by the retention tower cannot be relied upon to create the degree of uniformity needed in all situations.

***20-6.1.3 Mixing Quality.*** The goal of pulp bleaching is to have each fiber react with the appropriate quantity of bleaching chemical. The typical bleaching response with chemical application is nonlinear, as shown in Figure 20-15*b*. Here, the effectiveness of delignification drops off after application of 0.5 to 0.8% $ClO_2$ on pulp. Chemical application in excess of this amount is not only wasted but permits strength damaging reactions to occur. Thus, an appropriate bleaching strategy must optimize the applied chemical charge.

When mixing is less than perfect, local clumps of fiber see a range of chemical application. Consequently, some fiber will be overtreated and some undertreated. The extent of the effect will depend on the degree of nonuniformity remaining in the suspension and the nature of the bleaching response curve. Homogenization must also be attained faster than the effective chemical reaction rate; otherwise, the degree of nonuniformity during contacting will also affect the bleaching outcome.

## 20-6.2 Mixing Equipment in Pulp Bleaching Objectives

Mixer design is dictated primarily by the rheology of the suspension, which in turn is largely determined by suspension mass concentration. Thus, certain mixers find predominant use in low, medium, or high consistency bleaching applications. Mixer technology has changed over the years. Certain mixers, such as the stirred tank, are no longer used for pulp bleaching. Other mixers have been developed in response to implementation of new bleaching technologies (e.g., medium consistency ozone bleaching). Descriptions of many mixers used in bleaching applications may be found in Perkins and Doane (1979) and Bennington (1996). A summary of important operational parameters for pulp mixers is given in Table 20-4.

**Table 20-4** Characterization of Mixers Used in Pulping Applications

| Mixer | Suspension Concentration, $C_m$ | Residence Time (s) | Power Dissipation (MW/m$^3$) | Energy Expenditure (MJ/t) |
|---|---|---|---|---|
| Laboratory mixing/mixers | | | | |
| Hand mixing (bag bleaching) | 0.03 | 180 | 0.02 | 120 |
| | 0.10 | 180 | 0.05 | 90 |
| High-shear | 0.05 | 10 | 0.45 | 63 |
| | 0.10 | 10 | 1.1–1.8 | 160–180 |
| Hobart | 0.28 | 30–240 | 0.08 | 11–86 |
| Fluffer | 0.25–0.40 | 40–200 | 0.12–0.16 | 50–20 |
| Industrial mixers | | | | |
| *Low consistency* | | | | |
| Agitated vessel | 0.02–0.04 | 150–400 | 0.0002–0.0006 | 5–9 |
| Static | 0.02–0.03 | 3–5 | 0.03 | 4 |
| High-shear/high intensity | 0.03–0.04 | 0.008–1 | 1–110 | 3–61 |
| *Medium consistency* | | | | |
| Static | 0.10 | 4 | 0.1 | 4 |
| Peg | 0.10–0.12 | 10–12 | 0.08–0.11 | 11–15 |
| High-shear/high intensity | 0.08–0.16 | 0.025–4 | 1–110 | 2–43 |
| Medium consistency pump | 0.08–0.11 | 0.3–0.5 | 4–5 | 13–18 |
| Valve and pipe expansion | 0.08–0.11 | 0.3–0.5 | 1–3 | 9–11 |
| *High consistency* | | | | |
| Schredder type | 0.20–0.50 | NM[a] | 30–70 | 22–90 |
| Kneader type | 0.25–0.45 | NM[a] | 86–170 | 70–700 |

[a] NM, not measured.

### 20-6.2.1 Low Consistency Applications ($C_m < 0.05$).

At low mass concentrations, the yield stress is relatively low and abundant free water exists in suspension. Creating motion is relatively easy and is accomplished routinely with agitators for mining stock from high density towers and for consistency control ahead of bleaching. In the past, stirred/agitated reactors of various designs were used for chemical contacting, but experience showed that increased energy dissipation (achieved at the expense of mixer residence time) improved bleaching and reduced chemical use (see Section 20-6.1.1). Although older mixers can still be found in many mill installations, modern practice is to use either static or higher intensity dynamic mixers in these applications.

Static mixers come in a wide range of designs, and a typical one is shown in Figure 20-17. The internal flow elements must allow unimpeded pulp passage to prevent fiber entrapment and plugging while providing mixing action. Residence times are on the order of seconds, but mixing in the axial direction is minimal. Thus, flow of chemical and pulp to the mixer must be uniform. Energy dissipation is approximately 0.03 MW/m$^3$ (calculated from the pressure drop across the mixer) and is provided by an external stock pump. Care must be taken to ensure that the mixer will provide the required mixing over the anticipated

MIXING IN PULP BLEACHING OPERATIONS **1223**

**Figure 20-17** Static mixer used in pulp suspension mixing applications. Shown is the Komax steam mixer used in medium consistency applications.

production range of the plant (turn down), which may fall to as low as 40 to 50% of the normal operating capacity. A typical installation will see two or three static mixers in series, particularly if more than one chemical is used in a bleaching stage. Dynamic high-shear mixers have also been used in low consistency applications. While the energy dissipation is high, the residence time is short, which necessitates precise control of both pulp and chemical flow to the mixer.

***20-6.2.2 Medium Consistency Applications*** $(0.08 \leq C_m \leq 0.16)$. Many methods have been used to mix chemicals into medium consistency pulp stock. One common approach is to make use of existing equipment to provide the needed chemical contacting. For example, for adjustment of suspension pH before a bleaching stage, one common method is to apply caustic solution across the width of the preceding washer drum discharge (often, >7 m wide) using a shower bar. The pulp is then discharged into a conveyor system and passes through a steam mixer before the bleaching chemical is added. The mixing obtained (through predistribution of chemical and its subsequent agitation in the conveyor and steam mixer) is usually sufficient for slow or nonreacting chemicals.

For rapidly reacting chemicals, turbulence generated in other process equipment can be utilized for mixing. Often, the energy dissipation is similar to that in a dedicated mixer and can be sufficient to disperse flocs. Medium consistency pumps subject pulp to turbulence in the pump approach, in the pump discharge, and across the flow control valve (Francis and Kerekes, 1992). Liquid chemicals (e.g., chlorine dioxide and hydrogen peroxide) can be added ahead of the pump or directly into the pump suction. For gas applications, sintered metal spargers can be used following the pump but before the flow control valve. Although many mills use these methods, the lack of a dedicated mixer can result in inconsistent mixing. For example, turbulence created across a discharge valve will vary depending on the valve opening (a function of valve position and production rate). Often, a dedicated mixer will improve mixing.

Dedicated medium consistency mixers include peg-type mixers (Figure 20-18a). Mixer residence times are around 10 to 12 s with energy dissipation typically about 0.1 MW/m$^3$. This energy dissipation is insufficient to disrupt individual fiber flocs; rather, pulp is broken into fiber clumps by the mixing action. Chemical is added ahead of or inside the mixer. Multiple injection ports are common for both chemical and steam mixing applications.

High-shear or high intensity mixers are commonly specified for medium consistency applications where the essential design criterion is disruption of fiber flocs during contacting with the bleaching chemical. This requires imposition of sufficient shear, which is often accomplished by creating turbulence within the suspension. Consequently, providing sufficient energy to "fluidize" the suspension is a common design criterion. The flow of pulp and chemical to the mixing zone must be uniform because the suspension rapidly reflocculates upon exiting the high-shear/turbulence zone of the mixer. This can be difficult to achieve, particularly as little backmixing occurs in these mixers. Residence times are typically less than 1 s, and the residence time in the high-shear mixing zone may be only a fraction of this time. A number of high intensity mixers are available, some of which are shown in Figure 20-18(b–e).

The installation of these high intensity/high-shear mixers in the early 1980s significantly improved chemical utilization in bleaching operations. However, the high power requirements needed by these mixers prompted development of high-shear mixers requiring less attached power. Since the energy dissipation required to create turbulence in the pulp suspension is fixed, power savings were obtained by reducing mixer volume (at the expense of mixer residence time). While one mixer was often adequate in a given bleaching application, two were often used or required.

Most mixer design has been accomplished using pilot trials and in-mill refinements that have resulted in generations of similar mixers having slightly different internals (location and method of chemical addition, rotor configuration, etc.). Recently, computational fluid dynamics has been used to help refine the design of one high-shear mixer. The original mill installation showed significant nonuniformity following the mixer (inferred from the nonuniform temperature profile measured around the pipe circumference at the mixer discharge). A commercial CFD code (Fluent) was used to model the existing mixer, which predicted the nonuniformity observed in the field (Figure 20-19). The model was then used to refine the mixer design to improve chemical distribution. Design changes included

MIXING IN PULP BLEACHING OPERATIONS **1225**

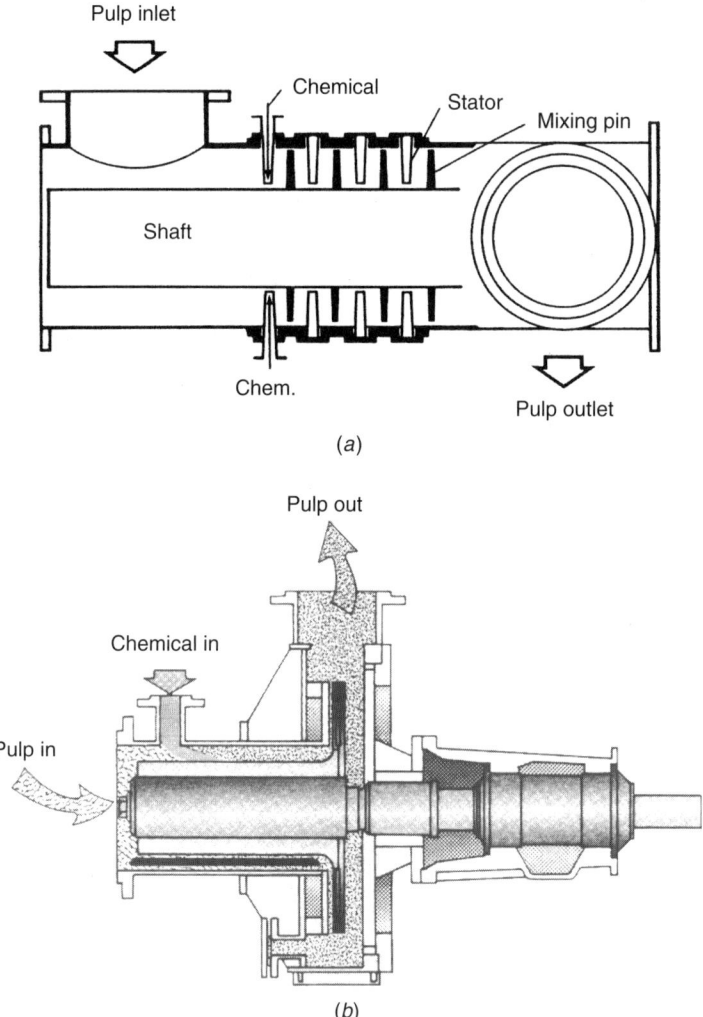

**Figure 20-18** Mixers used in medium consistency bleaching applications: (*a*) Beloit peg-type mixer; (*b*) Kamyr high-shear mixer (the first MC fluidizing mixer). (*Continued*)

enlargement of the predistribution vanes at the mixer entrance and use of an involute housing for the mixer chamber. These modifications were implemented in later mixer releases.

Gas mixing presents some unique challenges to mixer designers. For oxygen delignification applications, the typical oxygen volume needed could reach 20 to 30% of the total suspension volume. Rotary mixers tend to separate (de-mix) gas from the suspension. In addition, gas can be held up within the mixer, increasing the effective gas volume of the suspension and reducing the power applied to the suspension. This reduces gas–liquid mass transfer to the suspension. While many

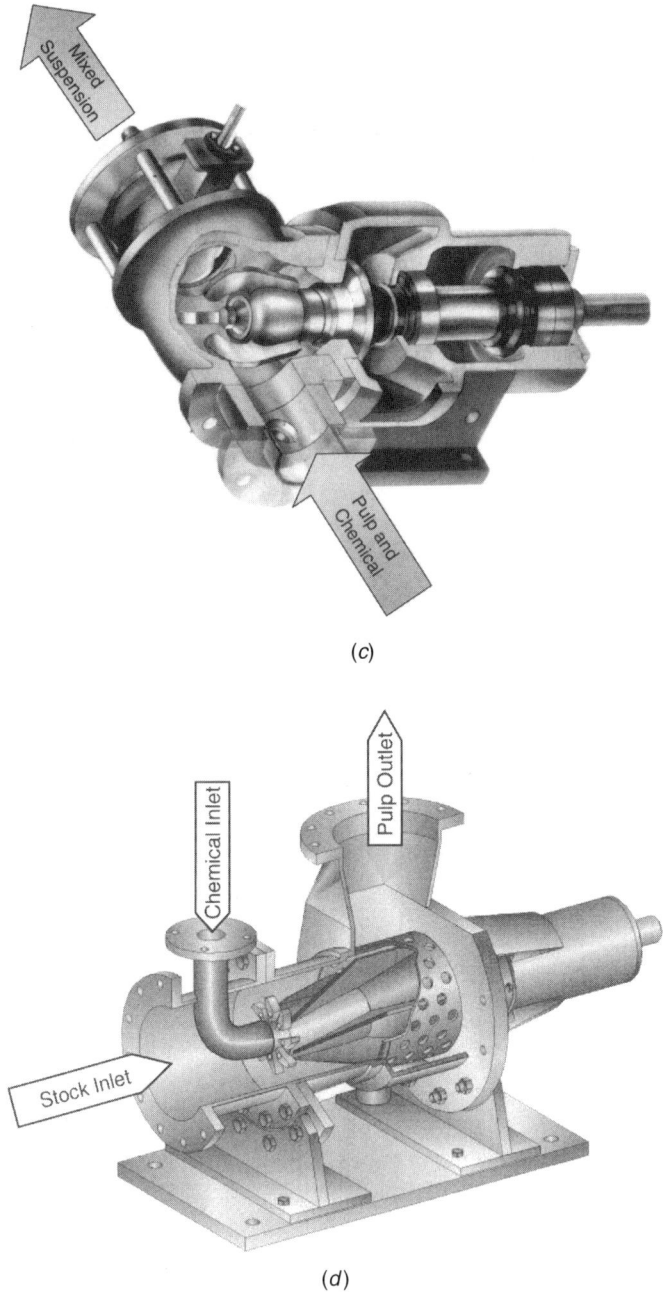

Figure 20-18 (c) Ahlmix (Ahlstrom); (d) Tri-phase mixer (GL&V).

MIXING IN PULP BLEACHING OPERATIONS **1227**

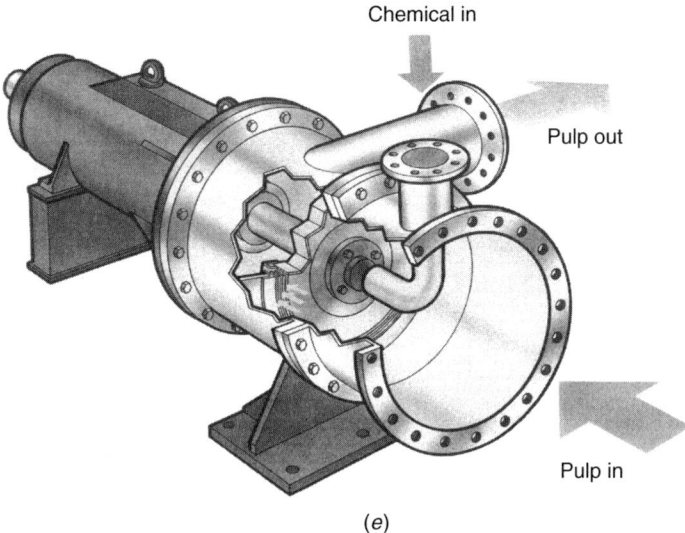

(e)

**Figure 20-18** (e) Sunds MC mixer.

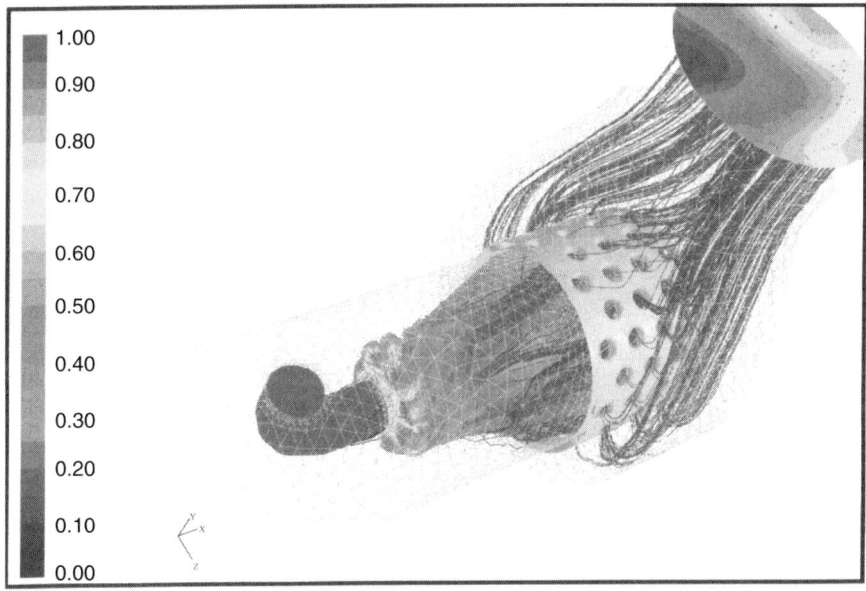

**Figure 20-19** CFD solution for flow in an early tri-phase mixer (GL&V) prior to design modifications. Note the spatial distribution of chemical (chlorine dioxide) in the pipe exiting the mixer. See insert for a color representation of this figure.

mixer manufacturers claim that their mixers were capable of handling gas void fractions up to 30% (Greenwood and Szopinski, 1992; Henricson, 1993; Miller et al., 1993), no operating data are published in the open literature. Frequently, mills operating at lower void fractions achieved better system performance.

The gassed power curve for a laboratory mixer operating in water and in a pulp suspension at mass concentrations up to $C_m = 0.16$ is given in Figure 20-20. While a high power draw was maintained in water up to a void fraction of $\phi_g = 40\%$, the power drawn for medium consistency pulp suspensions was reduced to 50% at a void fraction of only 10 to 13%. This is attributed to the yield stress of the suspension that inhibits flow back into the impeller region. Mixing and mass transfer are reduced with reduced power input, and the steep slope of the gassed power curve indicates that any variation in the gas content could dramatically affect the mixing achieved. Indeed, for this mixer design, a minor increase in gas content above 15% could prevent suspension motion entirely (Bennington, 1993; Smith and Bennington, 1995).

The efficiency of industrial high-shear mixers in gas mixing applications has not been measured directly. Despite this, the effectiveness of oxygen delignification processes (which rely on effective oxygen–pulp contacting) has been improved dramatically by implementing process changes designed to minimize gas volume during mixing (Bennington and Pineault, 1999). The magnitude of the improvements realized (delignification efficiencies were increased by 10 to 30 percentage points, more than doubling delignification in some cases) indicates the importance of mixing in these applications.

Medium consistency ozone delignification is practiced in only a limited number of mills worldwide (van Lierop et al., 1996). The amount of ozone required, plus the fact that it is generated at only 7 to 14 wt% in oxygen, means that gas volume fractions of 30% or greater are possible. Existing medium consistency

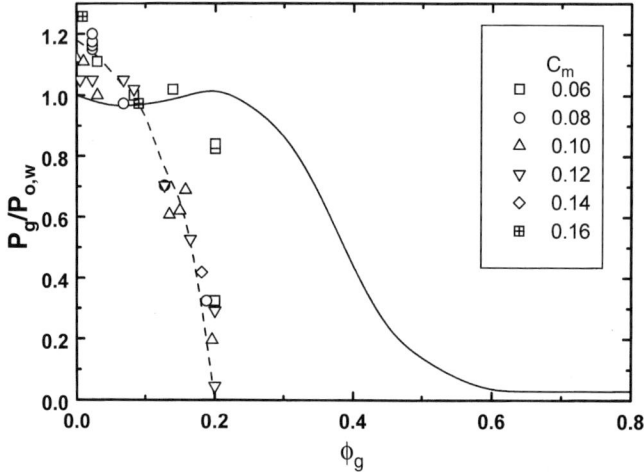

**Figure 20-20** Gassed power curve for laboratory pulp mixer (Quantum MK-IV).

mixers could not operate at these high gas contents. One manufacturer (Ahlstrom) designed a mixer to operate under these conditions. Here the flow within the mixer ensures continued gas-suspension contact to maximize the gassed power draw. Power dissipation in the suspension using these mixers is high, reaching 6.5 MW/m$^3$ for a treatment of 61 MJ/t pulp at void fractions close to 30%.

*20-6.2.3 High Consistency Applications* ($C_m > 0.20$). At high mass concentrations, little or no free water is present in suspension. The mixing strategy used is to expose as much damp fiber surface to chemical as possible. In high consistency oxygen delignification systems, fiber is "fluffed" to generate the fiber surface area using aggressive mixers designed to disrupt the high consistency flocs. These mixers are similar to refiners (see the plate pattern of the A.B. Nilsen mixer in Figure 20-21a) and apply tremendous power to the suspension. Chemical can be added during mixing (usually, for liquids), or the fluffed suspension can be transferred to a tower were contact with a gaseous chemical (e.g., oxygen) can be maintained.

Ozone delignification can also be achieved in a high consistency operation. Here, intimate contact with fiber must be achieved quickly. The Union Camp Corp. developed (White et al., 1993) a mixer/contactor based on a peg mixer design to create a dispersion of high consistency fiber in the gas phase (Figure 20-21b). Again, the goal was to create fiber surface area. Ozonation proceeds through a shrinking core process (Bennington et al., 1999). Fibers that are not exposed to ozone are not bleached. For liquid mixing, kneader-type mixers can be used. These squeeze the suspension to transfer liquid between regions in the suspension. Again, the high consistency of the suspension necessitates expenditure of considerable energy during the mixing process.

*20-6.2.4 Overview of Mixer Characteristics.* The wide range of mixers used in pulp bleaching applications can be characterized using a number of parameters. The residence time determines the extent of exposure to mixing action and places an upper limit on the macroscale mixing that might be achieved. The power dissipated during mixing determines if the suspension will be mixed under turbulent conditions, and if floc disruption is likely. Finally, the energy treatment gives an indication of the extent of mixing achieved in a given application. Table 20-4 summarizes these parameters for laboratory mixers (used for bench scale bleaching studies) and for a range of industrial mixers. Although these parameters are indicative of overall mixing performance, they do not account for the different efficiencies often observed between mixers. A number of observations can be made based on these data.

The first observation is that energy expenditure increases with suspension mass concentration. At low consistency, energy expenditure is typically 4 MJ/t. At medium consistency, this varies between 4 and 40 MJ/t. Although it is expected that mixing quality will improve with increased energy treatment, about 10 MJ/t is used by a wide range of mixers that achieve fluidization. For high consistency mixers, energy treatment rises significantly. The rate of mixing can be estimated

**1230** MIXING IN THE PULP AND PAPER INDUSTRY

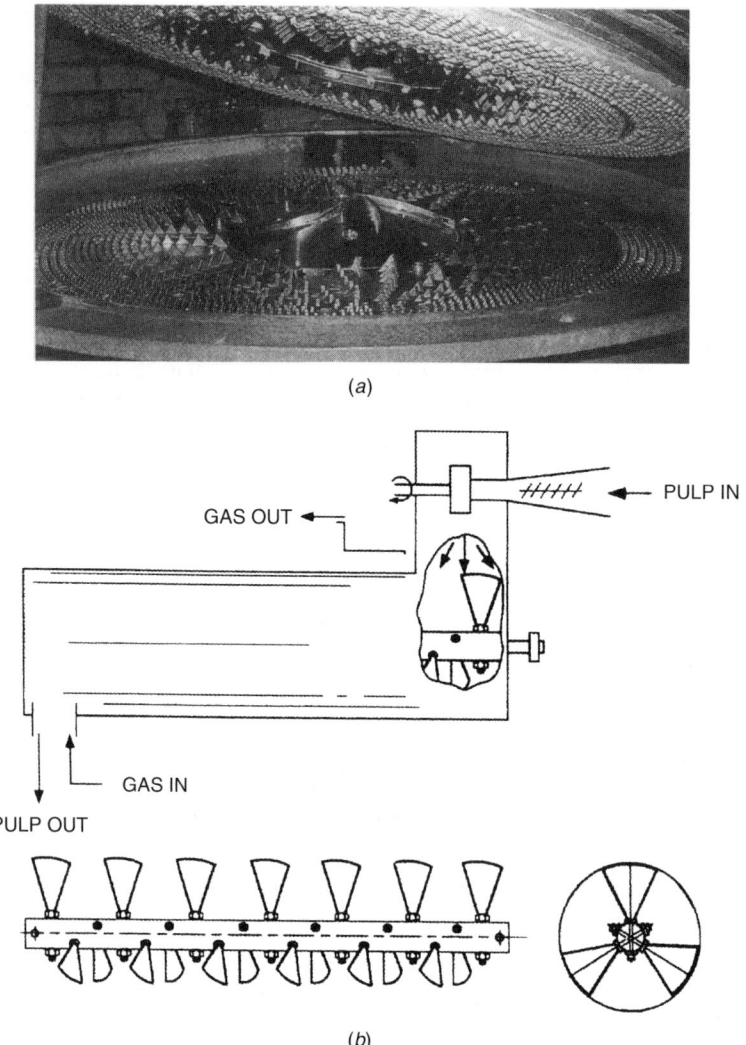

**Figure 20-21** Mixers used for high consistency bleaching applications: (*a*) Plate pattern of high consistency mixer (A.B. Nilsen); (*b*) Union Camp ozone mixer/reactor (White et al., 1993).

by the power dissipation per unit volume. In a given application, a mixer having a greater power will be more able to shear flocs apart and achieve a higher rate of fiber scale mixing.

In the past, laboratory bleaching was often represented as an ideal unlikely to be duplicated in the mill. This was based on pulp bleaching using slower-reacting, liquid-based bleaching chemicals mixed by hand. Here, the total energy treatment is typically an order of magnitude greater in the laboratory than in the mill (120 MJ/t versus 4 to 15 MJ/t). In the mill, low consistency bleaching was accomplished using agitated vessels or static mixers. Medium consistency mixing was largely accomplished using peg-type mixers. Mill adoption of medium consistency technologies (the high-shear/high intensity mixers) saw much higher mixing rates (power dissipation) as well as improved fiber scale mixing (floc dispersion). Mill results could now be as good or better than hand-mixed laboratory results. The mixing dependence of laboratory tests is particularly evident in reactions involving gases, where gas–liquid mass transfer is critical. Transferring laboratory results to the mill is even more difficult here.

*20-6.2.5 Incidental Mechanical Action Imparted during Mixing.* Mixing consumes energy and results in physical treatment of the pulp fiber. This can alter the physical properties of the pulp. For example, increased energy treatment reduces pulp drainage rate, increases bulk and opacity (for chemical pulps), increases tensile strength, and reduces tear strength. It is common to beat (mechanically refine) pulp to develop its strength properties prior to paper manufacture. Beating reduces tear strength and develops tensile strength, and a balance is sought between these two strength properties by the appropriate application of energy. Additional energy treatment during pulp production may alter the usual delivered pulp properties and affect its perceived strength.

In laboratory bleaching, energy treatment can be significantly higher than that experienced in an industrial setting. Consequently, pulp properties can be significantly different than achieved in the mill, and caution should be exercised when predicting certain pulp properties from these studies. In the mill, the addition of a single medium consistency mixer to a process would only add 3 to 30 MJ/t energy treatment to the pulp. The expected development in strength properties would be too small to be measured. The addition of six to eight of these units (to a bleaching sequence, for example) might increase fiber treatment by up to 200 MJ/t. This would change pulp strength properties by 5 to 10% and may be detectable. In any case, the intrinsic fiber strength, as measured by the zero-span tensile strength, is unaffected by the mechanical treatment (Bennington and Seth, 1989; Seth et al., 1993). Studies made by Bennington and Seth (1998) also showed that there was no synergistic degradation in pulp strength when intense mechanical mixing was applied during ozone bleaching.

### 20-6.3 Mixing Assessment in Pulp Suspensions

Mixing quality must be quantified to assess its impact on processes and products. Appropriate measurement techniques depend on a number of factors. For mill applications, the technique must not interfere with the process or change pulp properties in any way. More latitude is available for laboratory or pilot scale mixing assessment, although the presence of fibers complicates mixing measurement and assessment. Most mixing assessment techniques are tedious and time consuming to perform.

A number of indices can be used to quantify mixing in pulp suspensions. The coefficient of variation has been used by many investigators (Kolmodin, 1984;

Bergnor et al., 1985; Breed, 1985; Kuoppamaki, 1985; Bennington et al., 1997b) and a mixing index based on it, M, is given by

$$M_x = \frac{\sigma_x}{\bar{x}} \qquad (20\text{-}37)$$

where x is the measured property (usually, a tracer concentration). M is an appropriate mixing index as Gaussian residual chemical distributions have been measured following mixing (Bennington, 1996). The evaluation scale should also be specified, as M can vary with the assessment scale, as shown in Figure 20-22 (Bennington et al., 1997b).

Other indices can be used to quantify mixing quality. These include the charge deviation (Torregrossa, 1983), which assumes a rectangular distribution of chemical on pulp, and the intensity of segregation, $I_S$ (Paterson and Kerekes, 1985), which is analogous to the turbulent intensity. $I_S$ approaches M for large sample sizes, n ($M = \sqrt{n - 1/n}\, I_S$). In comparisons made in this chapter, n = 50, and M and $I_S$ can be considered equivalent. Other investigators (Atkinson and Partridge, 1966; Bennington et al., 2001) have used pulp quality parameters to quantify changes in mixing quality. These methods measure the result of mixing quality (the effect on some measured pulp property, i.e., kappa number, brightness, strength, etc.) rather than mixing uniformity directly.

### 20-6.3.1 Laboratory Techniques.
Mixing assessment at the laboratory and pilot scale have been made using pulp quality parameters and tracer distributions

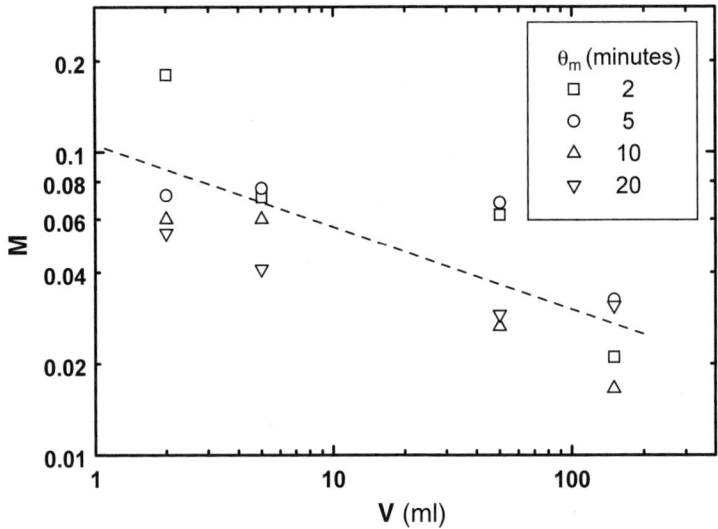

**Figure 20-22** Mixing quality, M, as a function of sample volume. Helical pulper (the mixer pictured in figure 20-13) with 8 L of softwood pulp ($C_m = 0.10$) operated at N = 6.7 rps. LiCl was used as the tracer. (From Bennington et al., 1997b.)

following mixing. A number of tracers have been used, including inert tracers (those that do not interfere with the pulp) (Breed, 1985; Bennington et al., 1997b; Kamal and Bennington, 2000), dyes (Francis and Kerekes, 1990), and radioactive tracers (Kuoppamaki et al. 1992; Kuoppamaki, 1985). These studies have confirmed that mixing quality measurably affects pulp properties and bleaching efficiency, that achieving uniformity becomes more difficult with increasing suspension mass concentration, and that mixing quality improves as the energy expended in mixing increases. Pilot tests have been used to optimize the location of chemical injection to a mixer (Breed, 1985) and confirm that the point of chemical addition is critical for attaining uniformity. Laboratory tests have also demonstrated the difficulties involved with assessing mixing in pulp suspensions. Suspension sampling is difficult, the quantification tests tedious, and the ability to obtain reproducible measurements problematic. It is common for the 95% confidence interval of a given mixing assessment to be ±30 to 60% of the measured value (Bennington et al., 1997b).

Predicting the effect of mixing on industrial scale bleaching can be difficult. Often, the pulp quality/economics predicted in the laboratory are difficult to attain in the mill. This leads to the notion that laboratory mixing is "ideal" and that mills that attain process results identical to those in the lab have achieved as good mixing as possible. Recent laboratory work has demonstrated the importance of mixing in certain bleaching stages. Tests measured the efficiency of oxygen delignification using a laboratory high-shear mixer and various mixing strategies (Berry et al., 2002). The extent of delignification varied widely. Delignification ranged from a low of 29% to a high of 50%, depending on both the intensity and duration of mixing. Similar results were found for ozone bleaching (Hurst, 1993) and are likely to occur in other bleaching chemistries. This raises the question: How do we scale-down industrial mixing so that we can use laboratory data to predict industrial performance reliably?

A further problem is the difference between spatial- and mass-based mixing assessments when made on the fiber scale. The distribution of tracer or a residual chemical through a suspension gives a spatial measure of mixing quality. It can be related to pulp quality only if the corresponding mass distribution is known. This is usually not known. Paterson and Kerekes (1986) measured scales of segregation from 2 to 4 mm in their mixing tests—the size of flocs within the suspension. How much does the measured mixing quality depend on the flocculated nature of the suspension, and how much on the distribution of chemical? Fortunately, when mixing is evaluated at scales greater than 5 to 10 $cm^3$, this problem does not arise because the mass distribution can be considered uniform at this scale.

*20-6.3.2 In-mill Mixing Assessment.* Mixing assessment in mill operations must be done in a manner that does not upset the process or compromise pulp quality. Past techniques have used mixing quality indices (i.e., pulp brightness, kappa number, etc.) (Atkinson and Partridge, 1966; Elliott and Farr, 1973; Abercrombie, 1986; Cameron, 1987), residual chemical profiles (Paterson and

Kerekes, 1986), inert tracers (Backlund et al., 1987; Robitaille, 1987), and temperature profiling (Torregrossa, 1983; Sinn, 1984; Robitaille, 1987; Rewatkar et al., 2001). Most measurements have been made on the macroscale (scales > 100 cm$^3$), although Paterson and Kerekes (1986) measured mixing quality on a fiber scale. The assessment techniques used are typically tedious, which limits the amount of data that can be collected. The suspension must also be sampled in a manner that avoids further mixing, which can often be difficult. The use of mixing quality indices requires that pulp samples be bleached in the laboratory for comparison with mill pulps.

The results of reported in-mill studies are included in Tables 20-5 and 20-6. The data show that mixing quality can vary substantially, regardless of the mixer used. Indeed, a given mixer will often perform differently in different mills or in different applications. Mixing quality also varies with process and operating conditions, although the time required to measure mixing has precluded most investigators from acquiring extensive sets of data. Paterson and Kerekes (1986) demonstrated this variability in one mill, where mixing quality was measured in 25 separate tests over a number of days. Even for ostensibly similar operating conditions, the fiber scale mixing quality varied significantly, from $M \simeq I_s = 0.00$ to $0.30$. The time interval between tests was too long (and varied) for the fluctuation frequencies to be identified. An online method of measuring mixing is needed to investigate this issue further.

Often, it is desirable to estimate mixing quality quickly. A common technique used for this purpose is temperature profiling. When a sufficient flow of a cold chemical stream is added to a sufficiently hot pulp stream (as in the D$_1$ stage), the spatial and temporal variation in temperature following mixing can be interpreted in terms of mixing quality. The assessment of mixing quality is qualitative (only

**Table 20-5** Fiber Scale Evaluation of Mixing Quality

| Mixer | Mill or Lab | Suspension Mass Concentration, $C_m$ | Tracer/ Technique Used | Sample Volume (cm$^3$) | Mixing Index, M or $I_s$ | Reference |
|---|---|---|---|---|---|---|
| CST | M | 0.03–0.04 | Res. Cl$^-$ | $7 \times 10^{-8}$ | 0.00–0.51 | Paterson and Kerekes (1986) |
|  | L | 0.01 | LiCl and | 0.0005–0.5 | 0.08–0.14 | Kamal and Bennington |
|  | L | 0.02 | 2-NSA | 0.0005–0.5 | 0.04–0.31 | (2000) |
| Static | M | 0.025 | Res. Cl$^-$ | $7 \times 10^{-8}$ | 0.0–0.11 | Paterson and Kerekes (1986) |
| Valve | L | 0.082 | Dye | Floc | 0.14–0.25 | Francis and Kerekes (1992) |
| Hobart mixer | L | 0.28 | Dye | Floc | 0.48–1.15 | Francis and Kerekes (1990) |
| Frotopulper | L | 0.26–0.27 | Dye | Floc | 0.27–0.45 | Francis and Kerekes (1990) |
| High shear | L | >0.25 | LiCl | Floc | 0.10–0.13 | Turnbull (private communication) |

**Table 20-6** Macroscale Evaluation of Mixing Quality

| Mixer | Mill or Lab | Suspension Mass Concentration, $C_m$ | Tracer/ Technique Used | Sample Volume (cm³) | Mixing Index, M | Reference |
|---|---|---|---|---|---|---|
| Tower | M | 0.095–0.12 | LiCl | 50 | 0.26–0.50 | Kolmodin (1984) |
|  | M | 0.089–0.122 | LiCl | 200 | 0.14–0.74 | Bergnor et al. (1985) |
|  | M | 0.108 | LiCl | Handful | 0.40 | Torregrossa (1983) |
| Repulper | L | 0.03 | LiCl | 50 | 0.008 | Bennington et al. |
|  | L | 0.08 | LiCl | 50 | 0.008 | (1997b) |
|  | L | 0.08 | LiCl | 2 | 0.05 |  |
| Static | M | Low | Ba-137 | 64 | 0.12 | Kuoppamaki (1985) |
|  | M | 0.13? | LiCl | Handful | 0.40–0.60 | Torregrossa (1983) |
|  | M | Low | Ba-137 | 15 000 | 0.005 | Kuoppamaki (1985) |
| Peg | M | 0.106 | LiCl | Handful | 0.20–0.40 | Torregrossa (1983) |
| High shear | L | 0.10 | LiCl | 2 | 0.03 | Bennington et al. |
|  | L | 0.10 | LiCL | 5 | 0.02 | (1997b) |
|  | M | 0.095–0.12 | LiCl | 50 | 0.02–0.05 | Kolmodin (1984) |
|  | M | 0.10 | LiCl | 200 | 0.06–0.08 | Bergnor et al. (1985) |
|  | L | 0.11–0.13 | LiCl | 300–400 | 0.06–0.53 | Breed (1985) |
|  | M | 0.106 | LiCl | Handful | 0.05–0.10 | Torregrossa (1983) |
|  | L | 0.09–0.11 | Ba-137 | ~500 (axial) | <0.05 | Kuoppamaki et al. (1992) |
|  | M | 0.08–0.11 | Temp | ~1400 (axial) | 0.10–0.40 | Rewatkar et al. (2001) |

the circumferential suspension temperature is measured, and temperature-based mixing indices do not allow estimation of potential chemical savings directly). Quantitative estimates can be made by making mass and energy balances on the added chemical (Rewatkar et al., 2001).

An example of temperature profiling is given in Figure 20-23. In a mill bleach plant, an older-style peg mixer used for chlorine dioxide mixing in the $D_1$ stage was replaced with a high-shear dynamic mixer to improve mixing quality. Following mixer replacement, chemical use unexpectedly increased 2.0 kg/t (18%). A series of thermograms taken of the process piping immediately following the new chlorine dioxide mixer showed that the cold chlorine dioxide stream channeled through the mixer. The temperature difference between the lower and upper sides of the pipe was 12°C. As the medium consistency pulp flowed as a plug after exiting the mixer, there was little dispersion or further mixing of the suspension. In this case, mixing was improved by changing the method of chemical addition ahead of the mixer. Following the modifications, chemical use decreased by 3.5 kg $ClO_2$/t, a savings of 1.5 kg $ClO_2$/t over the original peg mixer installation. This illustrates that mixing cannot be taken for granted—newer mixers do not necessarily guarantee better mixing.

Temperature profiling can also be used to examine variability in mixing quality with time (Rewatkar et al., 2001). As shown in Figure 20-24, the mixing index (based on the variability in measured temperature around the pipe circumference) changed significantly before and after the impeller. Following the mixer, mixing

**1236** MIXING IN THE PULP AND PAPER INDUSTRY

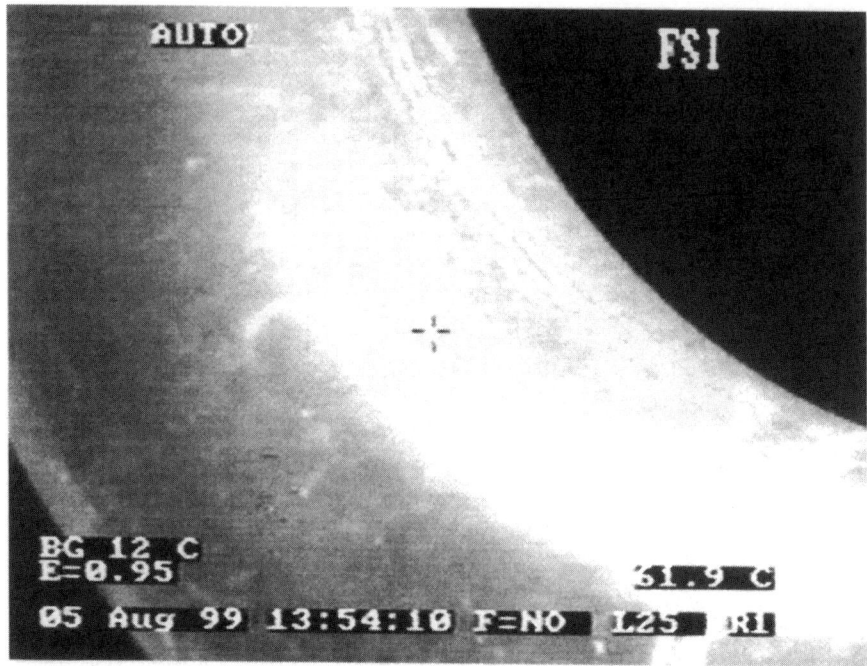

**Figure 20-23** Thermograph of exit piping following a high-shear chlorine dioxide mixer in $D_1$ service. Elbow immediately following mixer discharge. The temperature is 12°C higher along the top of the pipe. See insert for a color representation of this figure.

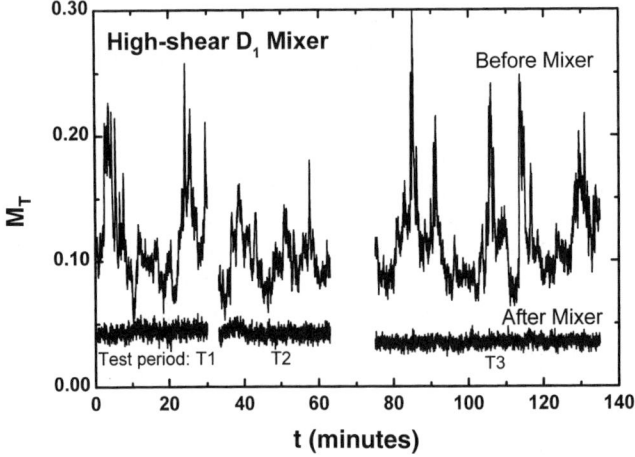

**Figure 20-24** Mixing index based on temperature variation ($M_T$) versus time for a high-shear chlorine dioxide mixer in $D_1$ service. Mixing quality is evaluated following chemical injection both before and following mixing.

MIXING IN PULP BLEACHING OPERATIONS  **1237**

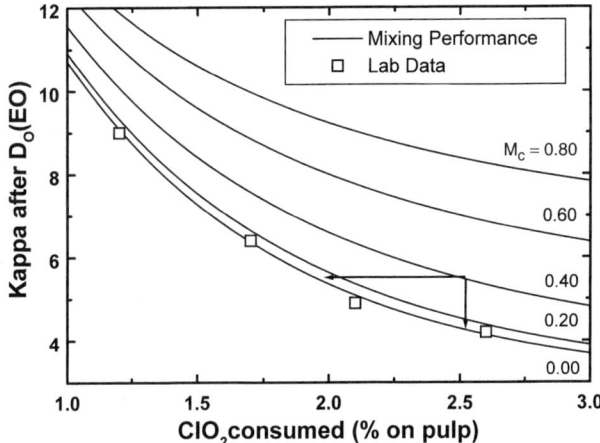

**Figure 20-25** Mixing performance curves generated for a softwood pulp having an initial kappa number of 27.6. Pulp was laboratory bleached using a D(EO) partial sequence. (Reaction conditions were: D: T = 48°C for 30 min at $C_m$ = 0.035 to an end pH of 1.9 to 2.5; EO: T = 70°C for 60 min at $C_m$ = 0.10 to an end pH of 11.5).

quality is improved and constant. Mass and energy balances can be used to convert the temperature-based index measured in Figure 20-24 to a chemical charge-based index that is more representative of system performance. For the $D_1$ mixer assessed in Figure 20-24, this would be M = 0.33, indicating that chemical savings are possible (see Figure 20-25).

### 20-6.4  Benefits of Improved Mixing

The benefits of improved mixing have been known for a long time. Following the introduction of medium consistency mixer technology in the 1980s, a CPPA survey documented chemical savings averaging 10 to 15% (Berry, 1990). Mills that took advantage of the improved mixing technology saw their capital investment returned in as little as three months. The situation is no different in mills today, and significant chemical savings can be achieved through improved mixing. However, mixing quality must be measured to know the degree to which it can be improved.

The chemical savings potential is readily demonstrated in Figure 20-25. Here, the mixing performance curves for a kraft pulp delignified using a $D_0$(EO) sequence (a chlorine dioxide stage followed by extractive oxidation stage) are given. The curve for "perfect" mixing (M = 0.0) was determined in the laboratory under "ideal" mixing conditions. The other curves were calculated with these data assuming a normal distribution of chemical throughout the suspension following mixing. The mixing index is defined using eq. (20-37). To illustrate the benefits of improved mixing, assume that an existing mixer achieves a mixing quality of M = 0.40 (which is typical of many installations; see Tables 20-5 and 20-6) and the process operates at the point indicated in the figure. Here an average chemical charge of 2.5% is needed to reduce the kappa number from 27.6 to 5.5. Improving mixing quality to M = 0.05 would allow the process to be moved anywhere on the corresponding performance curve (Note that the M = 0.05 curve is almost equivalent to the M = 0.0 curve.) This gives an engineer a number of options. He or she can choose to minimize chemical use in the $D_0$ stage, achieving the same exit kappa number using a chemical charge of only 1.6%. For a 1000 t/day mill, this would save about $1.4 million (Canadian) annually in chemical. Alternatively, the engineer can maintain the chemical charge and go into the subsequent bleaching stage at a lower kappa number. This is often the strategy of choice where the efficiency of a process using an inexpensive bleaching chemical is improved. Oxygen delignification is a good example. Here improved mixing has increased delignification efficiencies from 30% to the range 40 to 55+%. Chemical savings would be minimal in the oxygen stage (as oxygen is inexpensive), but by reducing the kappa number entering the Do stage, chemical savings in chlorine dioxide of up to $3 to $5 per metric ton could be achieved (Bennington and Pineault, 1999). This is equivalent to approximately 5 to 10% the bleaching cost of pulp. A combination of these strategies allows optimization of the overall bleaching sequence.

Improved mixing offers other benefits. Pulp strength and brightness ceilings can be improved, which would have beneficial marketing implications. In addition, synergism with other processes can be achieved, in particular, with process control systems used throughout the mill.

## 20-7 CONCLUSIONS

Mixing is a critical unit operation in many pulp and paper operations. A wide range of mixing strategies are employed throughout the industry. Some of these applications are common in other industries and are detailed in other chapters of this book. Some mixing applications are unique to pulp and paper production, particularly those involving pulp fiber suspensions. In either case, mixing affects the efficiency of many processes, directly and synergistically. Attaining optimal mixing conditions can improve product quality and reduce costs.

## NOMENCLATURE

| | |
|---|---|
| a, b | parameters |
| a | particle dimension (m) |
| A | fiber aspect ratio, $l_w/d$ |
| B | number of rotor vanes |
| $C_I$ | effective ink concentration (ppm) |
| $C_m$ | fiber mass concentration (as a fraction, unless specified otherwise) |
| $C_R$ | contact area between fiber and suspension (m$^2$) |
| $C_v$ | fiber volume concentration (as a fraction, unless specified otherwise) |

# NOMENCLATURE

| | |
|---|---|
| d | fiber diameter (m) |
| D | diffusion rate (m$^2$/s) |
| D | impeller/rotor diameter (m) |
| $D_m$ | mean impeller/rotor diameter (m) |
| F | Tappi flake content, fraction |
| $F_R$ | force applied per unit length of rotor (N/m) |
| G | rotor swept area (m$^2$/rev) |
| $H_C$ | height of rotor in contact with suspension (m) |
| $I_S$ | intensity of segregation |
| $k_{eff}$ | effective first-order reaction rate (s$^{-1}$) |
| k, $k_0$ | rate of deflaking (m$^{-2}$) |
| $k_I$ | rate of ink detachment (s$^{-1}$) |
| $k_L a$ | volumetric gas–liquid mass transfer coefficient (s$^{-1}$) |
| $k_R$ | rate of ink redeposition (s$^{-1}$) |
| K | proportionality constant |
| $l_w$ | length-weighted fiber length (m) |
| L | length of cylindrical rotor (m) |
| m | stock parameter [eq. (20-23)] (m/s) |
| $m_i$ | mass of component i (kg) |
| $M_x$ | mixing index of x [see eq. (20-37)] |
| n | sample size |
| N | rotor speed (rps) |
| P | impeller power (W) |
| $r_a$ | cavern radius (m) |
| S | rotor swept area (m$^2$) |
| t | time (s) |
| T | temperature (°C) |
| T | vessel diameter (m) |
| T | shaft torque (N · m) |
| $T_M$ | material strength (wet tensile strength) (N/m) |
| x | measured variable |
| $X_w$ | mass of water absorbed per mass of fiber material (kg/kg) |

## Greek Symbols

| | |
|---|---|
| α | constant determined by geometry and extent of diffusion |
| γ | shear rate (s$^{-1}$) |
| ε | energy dissipation (W/kg) |
| $\varepsilon_{avg}$ | average energy dissipation measured in the liquid phase, per unit mass (W/kg) |
| $\varepsilon_{in}$ | energy input per unit mass (W/kg) |
| $\varepsilon_{loc}$ | local energy dissipation measured in the liquid phase, per unit mass (W/kg) |
| $\varepsilon_v$ | energy dissipation per unit volume (W/m$^3$) |
| $\varepsilon_F$ | energy dissipation per unit volume at the point of fluidization (transition to turbulence) (W/m$^3$) |

$\mu_a$    apparent viscosity (Pa · s)
$\mu_p$    plastic viscosity (Pa · s)
$\nu$    kinematic viscosity (m$^2$/s)
$\rho_i$    density of component i (kg/m$^3$)
$\sigma_x$    standard deviation of variable x
$\tau_y$    yield stress (Pa)
$\tau_E$    micromixing time constant (s)
$\tau_M$    mixing time constant (s)
$\tau_D$    diffusion time constant (s)
$\tau_R$    reaction time constant (s)
$\phi_i$    volume fraction of component i

## Subscripts

f    fiber
g    gas
w    water

## Superscripts

0    initial
f    floor level

## REFERENCES

Abercrombie, D. A. (1986). C$_D$ and D$_1$ high intensity mixers reduce bleaching costs at Westar, *Proc. CPPA Spring Conference, Pacific Coast and Western Branches*, Jasper, Alberta, Canada, May 14–17.

Atkinson, E. S., and H. de V. Partridge (1966). Effects of mixing and degree of chlorination on quality and bleaching costs, *Tappi J.*, **49**(2), 66A–72A.

Backlund, B., E. Bergnor, P. Sandstom, and A. Teder (1987). The benefits of better mixing, *Pulp Paper Can.*, **88**(8), T279–T285.

Bakker, A., and J. B. Fasano (1993). A computational study of the flow pattern in an industrial paper pulp chest with a side entering impeller, in *Process Mixing: Chemical and Biochemical Applications*, Part II, G. B. Tatterson, R. V. Calibrese, and W. R. Penny, eds., *AIChE Symp. Ser.*, **293**(89), 118–124.

Baldyga, J., and J. R. Bourne (1989). Simplification of micromixing calculations: I. Derivation and application of new model, *Chem. Eng. J.*, **42**, 83–89.

Baldyga, J., and J. R. Bourne (1999). *Turbulent Mixing and Chemical Reactions*, Wiley, Chichester, West Sussex, England.

Ben, Y., and G. Dorris (1999). Irreversible ink deposition during repulping: II. ONP/OMG furnishes, *Proc. 5th Research Forum on Recycling*, PAPTAC, Ottawa, Ontario, Canada, Sept. 28–30, pp. 7–13.

Bennington, C. P. J. (1988). Mixing pulp suspensions, Ph.D. dissertation, University of British Columbia, Vancouver, British Columbia, Canada.

Bennington, C. P. J. (1993). Mixing gases into medium-consistency pulp suspensions using rotary devices, *Tappi J.*, **76**(7), 77–86.

Bennington, C. P. J. (1996). Mixing and mixers, in *Pulp Bleaching: Principles and Practice*, C. W. Dence and D. W. Reeve, eds., Tappi Press, Atlanta, GA, pp. 537–568.

Bennington, C. P. J. (1999). *Understanding defibering and ink detachment during repulping*, in *Paper Recycling Challenge*, Vol. III, *Process Technology*, M. R. Doshi and J. M. Dyer, eds., Doshi & Associates, Appleton, WI, pp. 268–282.

Bennington, C. P. J., and J. R. Bourne (1990). Effect of suspended fibres on macro-mixing and micro-mixing in a stirred tank reactor, *Chem. Eng. Commun.*, **92**, 183–197.

Bennington, C. P. J., and R. J. Kerekes (1996). Power requirements for pulp fluidization, *Tappi J.*, **79**(2), 253–258.

Bennington, C. P. J., and J. P. Mmbaga (1996). The use of mixing-sensitive chemical reactions for the study of mixing in pulp fiber suspensions, in *Mixed Flow Hydrodynamics: Advances in Engineering Fluid Mechanics*, N. P. Cheremisinoff, ed., Gulf Publishing, Houston, TX.

Bennington, C. P. J., and J. P. Mmbaga (2001). Fluid phase turbulence in pulp fibre suspensions, *Trans. 12th Fundamental Research Symposium*, Oxford, C. F. Baker, ed., Pulp and Paper Fundamental Research Society, Vol. 1, pp. 255–286.

Bennington, C. P. J., and I. Pineault (1999). Mass transfer in oxygen delignification systems: mill survey results, analysis and interpretation, *Pulp Paper Can.*, **100**(12), T395–T403 (123–131).

Bennington, C. P. J., and R. S. Seth (1989). Response of pulp fibres to mechanical treatment during MC fluidization, *Trans. 9th Fundamental Research Symposium*, Cambridge, Mechanical Engineering Publications, London, Vol. 1, pp. 87–103.

Bennington, C. P. J., and R. S. Seth (1998). Pulp strength and incidental mechanical treatment during kraft pulp bleaching, *Proc. International Pulp Bleaching Conference*, Helsinki, Finland, June 1–5, Book 1, pp. 167–173.

Bennington, C. P. J., and V. K. Thangavel (1993). The use of a mixing-sensitive chemical reaction for the study of pulp fibre suspension mixing, *Can. J. Chem. Eng.*, **71**, 667–675.

Bennington C. P. J., and M.-H. Wang (2001). A kinetic model of ink detachment in the repulper, *J. Pulp Paper Sci.*, **27**(10), 347–352.

Bennington, C. P. J., R. J. Kerekes, and J. R. Grace (1989). Mixing in pulp bleaching, *J. Pulp Paper Sci.*, **15**(5), J186–J195.

Bennington, C. P. J., R. J. Kerekes, and J. R. Grace (1990). The yield stress of fibre suspensions, *Can. J. Chem. Eng.*, **68**, 748–757.

Bennington, C. P. J., R. J. Kerekes, and J. R. Grace (1991). Motion of pulp fibre suspensions in rotary devices, *Can. J. Chem. Eng.*, **69**, 251–258.

Bennington, C. P. J., G. Azevedo, D. A. John, S. M. Birt, and B. H. Wolgast (1995). The yield-stress of medium- and high-consistency pulp fibre suspensions at high gas contents, *J. Pulp Paper. Sci.*, **21**(4), J111–J118.

Bennington, C. P. J., G. Owusu, and D. W. Francis (1997a). Gas–liquid mass transfer in pulp suspension mixing operations, *Can. J. Chem. Eng.*, **75**, 53–61.

Bennington, C. P. J., C. M. Peters, and R. MacLaren (1997b). Characterization of mixing quality in laboratory pulp mixers, *J. Pulp Paper Sci.*, **23**(9), J459–J465.

Bennington, C. P. J., O. S. Sui, and J. D. Smith (1998a). The effect of mechanical action on waste paper defibering and ink removal in repulping operations, *J. Pulp Paper Sci.*, **24**(11), 341–348.

Bennington, C. P. J., J. D. Smith, O. S. Sui, and M. -W. Wang (1998b). Characterization of repulper operation for newsprint deinking, *Proc. Tappi Pulping Conference*, Montreal, Quebec, Canada, Oct. 25–29, pp. 1083–1095.

Bennington, C. P. J., X. -Z. Zhang, and A. R. P. van Heiningen (1999). Effect of fibre-width distribution on ozone bleaching, *J. Pulp Paper Sci.*, **35**(4), 124–129.

Bennington, C. P. J., H. T. Yang, and G. Pageau (2001). Mill assessment of mixing quality using laboratory bleaching studies, *Pulp Paper Can.*, **102**(11), T305–T309.

Bergnor, E., B. Backlund, and A. Teder (1985). The Benefits of Better Mixing, *Proc. International Pulp Bleaching Conference*, Quebec, Canada, TAPPI/CPPA, pp. 233–241.

Berry, R. (1990). High-intensity mixers in chlorination and chlorine dioxide stages: survey results and evaluation, *Pulp Paper Can.*, **91**(4), T151–T159.

Berry, R., Z. H. Jiang, M. Faubert, B. van Lierop, and G. Sacciadis (2002). Recommendations from computer modelling for improving single stage oxygen delignification systems, *Proc. PAPTAC Annual Technical Conference*, Montreal, Quebec, Canada, pp. B151–B161.

Bialkowski, W. L. (1992). Newsprint variability and its impact on competitive position, *Pulp Paper Can.*, **93**(11), T299–T306.

Biermann, C. J. (1996). *Handbook of Pulping and Papermaking*, 2nd ed., Academic Press, San Diego, CA.

Blasinski, H., and E. Rzyski (1972). Mixing of non-Newtonian liquids: 2. Power consumption for fibrous suspensions, *Inzh. Chem.*, **2**(1), 169–182.

Bouchard, J., V. Magnotta, and R. Berry (2000). Improving oxygen delignification with peroxymonosulfate: the (OPx) process, *Proc. International Pulp Bleaching Conference*, Halifax, Nova Scotia, Canada, oral presentations, pp. 97–113.

Breed, D. B. (1985). Discovering the mechanisms of pulp mixing: a pilot approach to high shear mixing, *Proc. Medium Consistency Mixing Seminar*, Hollywood, FL, Nov. 7–8, Tappi Press, Atlanta, GA, pp. 33–37.

Cameron, M. (1987). Sunds high intensity mixer performance on $C_D$ and $D_1$ Stages, *Proc. CPPA Spring Conference, Pacific Coast and Western Branches*, Whistler, British Columbia, Canada, May 14–16.

Dence, C. W., and D. W. Reeve (1996). *Pulp Bleaching: Principles and Practice*, Tappi Press, Atlanta, GA.

Dobson, H. A. (2001). The mixing sensitivity of polysulfide generation, M.A.Sc. thesis, University of British Columbia, Vancouver, British Columbia, Canada.

Dobson, H. A., and C. P. J. Bennington (2002). The mixing sensitivity of polysulfide generation, *Can. J. Chem. Eng.*, **80**(2), 214–223.

Dorris, G. M. (2000). Effects of impeller type and mixing power on settling and filtering of lime mud, *J. Pulp Paper Sci.*, **26**(2), 47–53.

Dorris, G. M., and V. C. Uloth (1994). Analysis of oxidized white liquors: II. Potentiometric titrations for the determination of polysulfides and sulphoxy anions, *J. Pulp Paper Sci.*, **20**(9), J242–J248.

Dosch, J. B., K. M. Singh, and T. J. Stenuf (1986). Air content of medium and high consistency pulp slurries, *Proc. Tappi Engineering Conference*, Atlanta, GA, pp. 721–723.

Ein Mozaffari, F., G. A. Dumont, and C. P. J. Bennington (2001). Performance and design of agitated stock chests, *Pacwest Conference, Pulp and Paper Technical Association of Canada*, Whistler, British Columbia, Canada, May 16–19.

Elliott, R. G., and T. D. Farr (1973). Mill-scale evaluation of chlorine mixing, *Tappi J.*, **56**(11), 68–70.

FAO (1999). *Recovered Paper Data, 1997–1998*, Food and Agriculture Organization of the United Nations, Rome.

Francis, D. W., and R. J. Kerekes (1990). Measurement of mixing in high-consistency pulp suspensions, *J. Pulp Paper Sci.*, **16**(4), J130–J135.

Francis, D. W., and R. J. Kerekes (1992). Flow and mixing behaviour of low-consistency and medium-consistency fiber suspensions downstream of a valve, *Tappi J.*, **75**, 113–119.

Frederick, W. J., Jr., R. Krishnan, and R. J. Ayers (1990). Prissonite deposits in green liquor processing, *Tappi J.*, **72**(2), 135–140.

Gibbon, J. D., and D. Attwood (1962). The prediction of power requirements in the agitation of fibre suspensions, *Trans. Inst. Chem. Eng.*, **40**, 75–82.

Grace, T. M., and E. W. Malcolm, eds. (1989). *Pulp and Paper Manufacture*, Vol. 5, Alkaline Pulping Joint Textbook Committee, CPPA/TAPPI, Montreal, Quebec, Canada and Atlanta, GA.

Greenwood, B. F., and R. Szopinski (1992). Ozone Bleaching Technology '92, *Proc. Non-chlorine Bleaching Conference*, Miller Freeman, San Francisco.

Gullichsen, J., and C. J. Fogelholm, eds. (1999). *Chemical Pulping*, Books 6A and 6B, Fapet Oy, Helsinki, Finland.

Gullichsen, J., and E. Harkonen (1981). Medium consistency technology: I. Fundamental data, *Tappi J.*, **64**(6), 69–72.

Henricson, K. (1993). Modern bleaching technology, *Proc. Non-chlorine Bleaching Conference*, Miller Freeman, San Francisco.

Holman, K. L., R. P. Warrick, J. D. Farrelly, and T. D. McDonald, Jr. (1989). Analysis of mixing in a smelt dissolver, *Proc. Tappi Pulping Conference*, Seattle, WA, pp. 761–768.

Hurst, M. M. (1993). Effects of pulp consistency and mixing intensity on ozone bleaching, *Tappi J.*, **76**(4), 156–161.

Jang, H. F., and R. S. Seth (1998). Using confocal microscopy to characterize the collapse behaviour of fibers, *Tappi J.*, **81**(5), 167–174.

Jansson, I., ed. (1999). *Accurate Consistency: A Handbook on Accurate Consistency Measurement in Pulp and Paper Processing*, BTG Pulp and Paper Technology, Saffe, Sweden.

Kamal, N. R., and C. P. J. Bennington (2000). An on-line, in-situ, mixing assessment technique for pulp fibre suspensions, *J. Pulp Paper Sci.*, **26**(6), 214–220.

Kerekes, R. J., R. M. Soszynski, and P. A. Tam Doo (1985). The flocculation of pulp fibres, in *Trans. 8th Fundamental Research Symposium*, Oxford, V. Putnam, ed., Fundamental Research Committee, Oxford, Vol. 1, pp. 265–310.

Kolmodin, H. (1984). How to save costs by mixing chlorine dioxide and pulp homogeneously, *Svensk Papperstid.*, **87**(18), 8–14.

Kuoppamaki, R. (1985). The quality of mixing studied using a radiotracer technique, in *Medium Consistency Mixing Seminar*, Hollywood, FL, Nov. 7–8, Tappi Press, Atlanta, GA, pp. 13–17.

Kuoppamaki, R., O. Pikka, and K. Peltonen (1992). New high-intensity MC® mixer: direct measurement of mixing efficiency, *Proc. European Pulp and Paper Week, New Available Techniques and Current Trends*, SPCI/ATICELCA, Bologna, Italy, May 19–22, pp. 216–224.

Ljungqvist, M., and H. Theliander (1995). Mixing conditions in the smelt dissolver, *Proc. International Chemical Recovery Conference*, Toronto, Ontario, Canada, April 24–25, CPPA/TAPPI, pp. A283–A289.

Makinen, M. (1999). Coating color preparation, in *Pigment Coating and Surface Sizing of Paper*, E. Lehtinen, ed., Book 11 in the series Papermaking Science and Technology, Fapet Oy, Helsinki, Finland, pp. 319–388.

Mann, R. (1983). *Gas–Liquid Contacting in Mixing Vessels*, Institution of Chemical Engineers, Rugby, Warwickshire, England.

McKinney, R. (1998). Flotation deinking overview, in *Paper Recycling Challenge*, Vol. III, *Process Technology*, M. R. Doshi and J. M. Dyer, eds., Doshi & Associates, Appleton, WI, pp. 99–114.

Middleton, S. R., and A. M. Scallan (1993). Partial lumen loading, *Nordic Pulp Paper Res. J.*, **8**(1), 204–207, 231.

Miller, B., L. D. Shackford, and S. Minami (1993). Oxygen, peroxide and ozone process equipment, *Proc. Non-chlorine Bleaching Conference*, Miller Freeman, San Francisco.

Mmbaga, J. P. (1999). The use of mixing-sensitive chemical reactions to characterize mixing in the liquid phase of fibre suspensions, Ph.D. dissertation, University of British Columbia, Vancouver, British Columbia, Canada.

Morgan, J. P., and F. E. Murray (1971). Mass transfer and chemical reaction rate considerations in high efficiency black liquor oxidation designs, *Tappi J.*, **54**(9), 1500–1504.

Oldshue, J. Y. (1983). Pulp and paper agitation, in *Fluid Mixing Technology*, McGraw-Hill, New York.

Paterson, A. H. J., and R. J. Kerekes (1984). Fundamentals of mixing in pulp suspensions: diffusion of reacting chlorine, *Tappi J.*, **67**(5), 114–117.

Paterson, A. H. J., and R. J. Kerekes (1985). Fundamentals of mixing in pulp suspensions: measurement of microscale mixing of chlorine, *J. Pulp Paper Sci.*, **11**(4), J108–J113.

Paterson, A. H. J., and R. J. Kerekes (1986). Fundamentals of mixing in pulp suspensions: measurement of microscale mixing in mill chlorination mixers, *J. Pulp Paper Sci.*, **12**(3), J78–J83.

Perkins, J. K., and F. P. Doane, Jr. (1979). Pulp bleaching equipment, in *The Bleaching of Pulp*, R. P. Singh, ed., Tappi Press, Atlanta, GA.

Pulp and Paper International (2001). *World's Pulp, Paper Board Ind. Prod. Trade*, **43**(7); 8–9.

Reeve, D. W., and A. B. McKague (1990). Identification of chlorinated compounds in bleached pulp extracts, *EUCEPA 24th Conference Proc.*, Stockholm, Sweden, May 8–11, pp. 468–478.

Rewatkar, V. B., and C. P. J. Bennington (2000). Gas–liquid mass transfer in low- and medium-consistency pulp suspensions, *Can. J. Chem. Eng.*, **78**, 504–512.

Rewatkar, V. B., R. J. Kerekes, and C. P. J. Bennington (2001). The use of temperature profiling to measure mixing quality in pulp bleaching operations, *Proc. Annual Technical Meeting, PAPTAC*, Montreal, Quebec, Canada, pp. B135–B143.

Reynolds, E., J. D. Gibbon, and D. Attwood (1964). Smoothing quality variations in storage chests holding paper stock, *Trans. Inst. Chem. Eng.*, **42**, T13–T21.

Robinson, J. V., N. Millman and J. B. Whitley (1997). The dispersion of pigments for paper coating, Chapter 3 in *Pigments for Paper*, R. Hagemeyer, ed., Tappi Press, Atlanta, GA, pp. 21–53.

Robitaille, M. A. (1987). High intensity mixing at the chlorine dioxide stage, *Pulp Paper Can.*, **88**(4), T109–T111.

Rydholm, S. A. (1965). *Pulping Processes*, Wiley-Interscience, New York.

Scallan, A. M., and J. E. Carles (1972). The correlation of the water retention value with the fibre saturation point, *Svensk Papperstid.*, **75**, 699–703.

Seth, R. S., D. W. Francis, and C. P. J. Bennington (1993). The effect of mechanical treatment during medium stock concentration fluidization on pulp properties, *Appita*, **46**(1), 54–58.

Shaharuzzaman, M., and C. P. J. Bennington (2001). The effect of mixing on the generation of alkaline peroxymonosulfate, *Can J. Chem. Eng.*, **79**(4), 595–601.

Shaw, I. S. D., and R. D. Christie (1984). Low cost and energy efficient oxidizer (ECO), *Pulp Paper Can.*, **85**(3), T40–T43.

Sinn, S. (1984). State-of-the-art chlorine dioxide mixer installed at Weyerhaeuser, *Pulp Paper*, June, pp. 119–121.

Smith, R. J., and C. P. J. Bennington (1995). Mixing gases and pulp suspensions in a continuous laboratory mixer, *Appita*, **48**(6), 414–418.

Smook, G. A. (1992). *Handbook for Pulp and Paper Technologists*, 2nd ed., Angus Wilde Publishing, Vancouver, British Columbia, Canada.

Soloman, R., T. P. Elson, A. W. Nienow, and G. W. Pace (1981). Cavern sizes in agitated fluids with a yield stress, *Chem. Eng. Commun.*, **11**, 143–164.

Somasundaran, P. and L. Zhang (1998). *Fundamentals of flotation deinking*, in *Paper Recycling Challenge*, Vol. III, *Process Technology*, M. R. Doshi and J. M. Dyer, eds., Doshi & Associates, Appleton, WI, pp. 83–98.

Tatterson, G. B. (1991). *Fluid Mixing and Gas Dispersion in Agitated Tanks*, McGraw-Hill, New York.

Tessier, P., and M. Savoie (1997). Chlorine dioxide delignification kinetics and Eop extraction of softwood kraft pulp, *Can. J. Chem. Eng.*, **75**, 23–30.

Thibault, F. (1999). Analyse du procede de melange solide-liquide: application a la préparation des sauces de couchage du papier, Ph.D. dissertation, Université de Montreal, Montreal, Quebec, Canada.

Thring, R. W., V. W. Uloth, G. M. Dorris, T. S. Galvin, D. Hornsey, and J. R. Ayton (1995). White liquor oxidation in a pilot plant pipeline reactor, *Tappi J.*, **78**(1), 107–113.

Torregrossa, L. O. (1983). Effect of mixing efficiency on chlorine dioxide bleaching, *Proc. Tappi Pulping Conference*, Houston, TX, Oct. 24–26, pp. 635–641.

Uloth, V., G. Dorris, R. Thring, R. Hogikyan, J. Wearing, L. Tench and J. Ayton (1996). In situ production of polysulfide liquor in a kraft mill's causticizers, *Proc. Tappi Pulping Conference*, Tappi Press, Atlanta, GA, pp. 813–836.

Van Lierop, B., A. Skothos, and N. Liebergott (1996). Ozone delignification, in *Pulp Bleaching: Principles and Practice*, C. W. Dence and D. W. Reeve, eds., Tappi Press, Atlanta, GA, p. 335.

Walker, O. J., and A. Cholette (1958). Determination of the optimum size and efficiency of stock chests: I. The ideal chest, *Pulp Paper Can.*, Mar., pp. 113–117.

White, D. E., T. P. Gandek, M. A. Pikullin, and W. H. Friend (1993). Importance of reactor design in high-consistency ozone bleaching, *Pulp Paper Can.*, **94**(9), T242–T247.

Yackel, D. C. (1998). *Pulp and Paper Agitation: The History, Mechanics and Process*, Tappi Press, Atlanta, GA.

# CHAPTER 21

# Mechanical Design of Mixing Equipment

D. S. DICKEY
*MixTech, Inc.*

J. B. FASANO
*Chemineer, Inc.*

## 21-1 INTRODUCTION

Mixing equipment must be designed for mechanical and process operation. Although mixer design begins with a focus on process requirements, the mechanical design is essential for successful operation. Usually, a competent manufacturer of mixing equipment will take responsibility for the mechanical design. However, process conditions, such as impeller operation near a liquid surface, can impose severe mechanical loads. Similarly, the process environment will influence the selection of a motor enclosure. In many ways the process requirements can have a direct impact on the mechanical design. In other ways, such as the natural frequency of a mixer shaft, appropriate mechanical design must be determined by the equipment designer. Whatever the reason, knowledge of the mechanical requirements for a mixer will help guide the engineer toward a design that will meet both process and mechanical criteria.

The purpose of this chapter is to provide practical information about the mechanical design of mixing equipment. Therefore, descriptions, equations, and nomenclature will be given in both U.S. engineering units and metric units. Descriptions and equations using U.S. engineering units will follow common industrial practices used in the United States with design information for materials measured in inches and motors specified in horsepower. Descriptions using

---

*Handbook of Industrial Mixing: Science and Practice*, Edited by Edward L. Paul,
Victor A. Atiemo-Obeng, and Suzanne M. Kresta
ISBN 0-471-26919-0 Copyright © 2004 John Wiley & Sons, Inc.

metric units will reference materials commonly measured in millimeters (mm), while equations will do calculations in meters (m).

Metric units in equations will follow SI metric practice. To avoid confusion, values in the text that are also used in equations will use standard SI units even if more reasonable numeric values are possible with prefixes. Units for variables in U.S. engineering units (U.S. Eng.) are shown in brackets []. Units for variables in metric units (Metric) will be shown in braces {}. The nomenclature list shows both U.S. engineering units and metric units used in the equations. Care must be taken to use the correct units, since several equations contain dimensional constants. Results can be incorrect if the wrong units are used.

## 21-2 MECHANICAL FEATURES AND COMPONENTS OF MIXERS

Because of the diversity of fluid mixing applications and variety of vessels, many different styles of mixers are used in industrial applications. Mixer sizes include small fractional-horsepower portable mixers to huge 1000 hp plus mixers. Although normally viewed as a single piece of equipment, like a pump, the typical mixer is composed of several individual components, such as a motor, gear reducer, seal, shaft, impellers, and tank, which is often designed and purchased separately. Although highly customized for many applications, most mixers are a combination of standard components, sometimes with modifications, and often with unique characteristics, such as shaft length.

Generalizations, especially for mixers, can misrepresent individual situations, but some features are common to the largest number of mixers built worldwide. The most common motive force for a mixer is an electric motor, so a knowledge of standard motor characteristics is useful. Most mixers operate at or below typical motor speeds, so some type of speed reduction is common. Speed reduction can be accomplished with several different types of gears, usually in enclosed housings, or with belts and sheaves. Besides speed reduction, antifriction bearings are found in all types of rotating equipment. Some type of seal around the rotating shaft is required for closed-tank operation and the type depends on degree of seal required, operating pressure, and operating temperature.

The shaft for a mixer, especially a large one, involves significant mechanical design, partly because of the myriad of shaft lengths, impeller sizes, and operating speeds, and partly because both strength and rigidity are necessary for a successful design. The combination of custom process and mechanical design necessary for mixers is unique for chemical process equipment. Mechanical design does not end with the shaft, since strength and practical issues remain for the impeller.

Another part of mixer design is the tank in which the mixer is used, since tank dimensions influence mixer features, especially shaft length. Conversely, a mixer requires tank features, such as baffles, support strength, and other tank internals. Materials of construction, although most commonly metal alloys for mixers, depend on process chemistry and operational requirements.

Other mechanical features can be important in special-purpose mixers, such as high-shear mixers, dry-solids mixers, and static mixers. Without revealing trade secrets or emphasizing proprietary technology, elements of the same mechanical design considerations apply to special-purpose mixers. The primary mechanical emphasis in this chapter is on equipment discussed elsewhere in this book.

Each key element of the mechanical characteristics of mixers will be covered in this section. Although not comprehensive with respect to each topic, the equipment and design requirements discussed should cover most of the mixer types and applications. Even with the diversity of mixing equipment, features such as motors and materials of construction are mechanical considerations, common to all types of mixers.

### 21-2.1 Impeller-Type Mixing Equipment

Impeller-type mixing equipment represents the largest category of general purpose mixing equipment for fluid processing applications. From the process view of impeller-type equipment, an impeller, usually composed of blades mounted to a central hub and rotated by a drive shaft, pushes and moves the material to be mixed. The mixing action and the process results are primarily a result of this material, usually fluid, motion. The mechanical design of impeller-type mixing equipment is responsible for the process by which some form of energy, such as electricity, is converted into fluid motion. That fluid motion is ultimately dissipated as heat, hopefully after the process objectives are accomplished.

To present an organized understanding of mixing equipment, some common terminology is used to describe typical characteristics. Each category of equipment has some loosely defined limits, often with overlap to other categories, depending on features provided by different manufacturers of the equipment.

*21-2.1.1 Portable Mixers.* Portable mixers may or may not be truly "portable," depending on size and mounting. However, the term *portable mixer* most often refers to mixers with $\frac{1}{4}$ hp to 3 hp drives mounted with either a clamp or a bolted-swivel mount. Smaller mixers are usually considered laboratory or pilot-plant equipment and are not often used in industrial production processes. Most portable mixers operate at either motor speed, such as 1800 rpm (30 rps) or 1200 rpm (20 rps) with 60 Hz power, or with a single-reduction gear drive (approximately a 5:1 speed reduction) for 350 rpm (5.83 rps). Although details of impeller types vary, axial flow impellers, such as marine propellers or three-blade hydrofoil impellers, are used most often. A typical direct-drive portable mixer is shown in Figure 21-1 and a gear drive portable in Figure 21-2.

*21-2.1.2 Top-Entering Mixers.* The designation *top-entering mixers* has become accepted as a more restrictive term than the name would imply. Top-entering mixers are usually considered the equivalent of portable mixers with flange mountings, or perhaps larger mixers but with light-duty gear drives and motors less than 10 hp (7460 W). This designation is less of a true definition than an accepted industry practice used to describe basic mixer products.

By this definition, top-entering mixers have flange or pedestal mounts, compared with the clamp or swivel-plate mounts used on portables. Most top-entering

**1250** MECHANICAL DESIGN OF MIXING EQUIPMENT

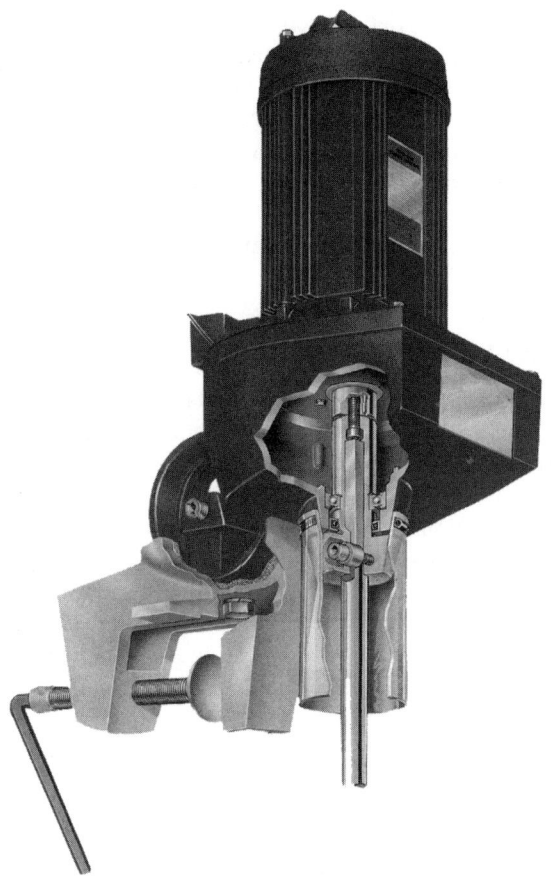

**Figure 21-1** Direct-drive portable mixer. (Courtesy of Lightnin.)

mixers are mounted on the vertical centerline of a tank with baffles, but may be off-center or off-center, angle mounted. Longer shafts and larger impellers cause more severe loads on top-entering mixers than portable mixers. A typical top-entering mixer is shown in Figure 21-3. Most top-entering mixers have an axial flow impeller, such as a hydrofoil impeller or sometimes a marine propeller. Typical seals for top-entering mixers are basic stuffing boxes or single, mechanical seals. For reasons of mechanical strength, sealing pressures are typically 30 psig (207 000 Pa) or less. For reasons of cost, single dry-running mechanical seals are common. More detail about different types of seals is given in Section 21-5.

**21-2.1.3 Turbine Mixers.** *Turbine mixer* is another industry designation that typically refers to more robust mixer designs that may have a variety of impeller and seal types and may have motors from 1 hp (746 W) to 1000 hp (746 000 W) or larger. The various sizes for turbine mixers are depicted in Figure 21-4. Turbine

# MECHANICAL FEATURES AND COMPONENTS OF MIXERS 1251

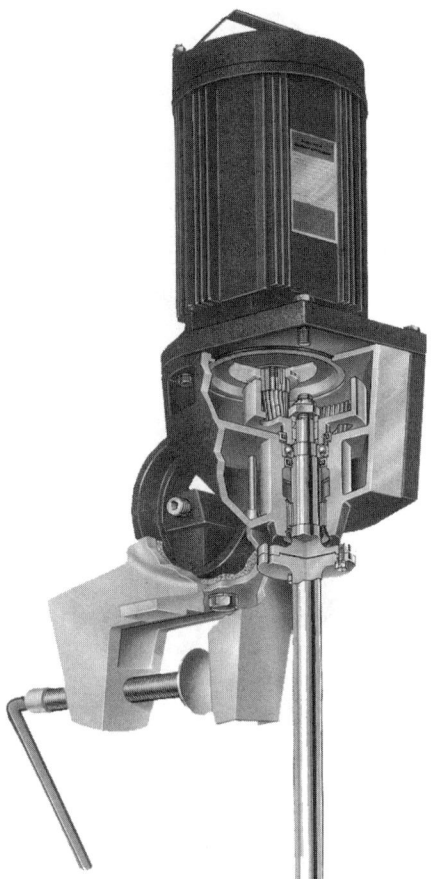

**Figure 21-2** Gear-drive portable mixer. (Courtesy of Lightnin.)

mixers are usually mounted vertically on the centerline of a cylindrical tank or rectangular basin or chest. The broader designation of turbine mixers may include top-entering mixers. Turbine mixer drives may be used with high viscosity, close-clearance impellers. Although none of these mixer designations are absolute and some equipment falls outside common or convenient terminology, knowing typical terminology can be helpful to understand the capabilities and limitations of different equipment.

Because of the broad use and versatile characteristics of turbine mixers, typical components are described at the beginning of this chapter. Essentially all turbine mixers have a motor, speed reducer, shaft, and impeller(s). Seals are used when containment is required. In this chapter we discuss motor and speed-reducer characteristics that commonly apply to turbine-style mixers. The shaft and impeller design characteristics are also typical for turbine mixers. A subset of these component characteristics and design procedures apply generally to other mixers.

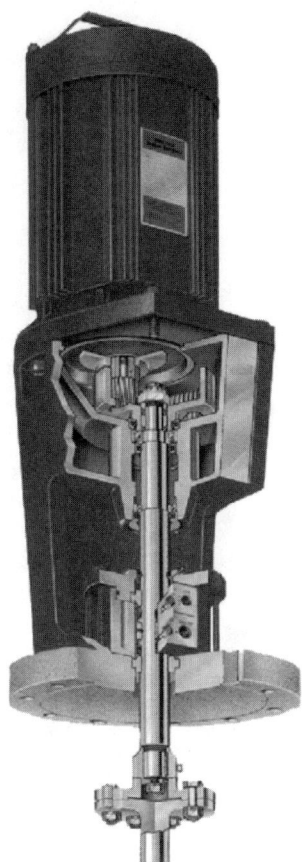

**Figure 21-3** Top-entering mixer with mechanical seal. (Courtesy of Lightnin.)

Obviously large, custom motors would never be applied to a portable mixer, but explosion-proof motors would.

***21-2.1.4 Side-Entering Mixers.*** *Side-entering mixers* are what the name implies, mixers that enter the tank or vessel from the side. For such mixers to mix the tank contents, they must be mounted below the liquid level. Consequently, they are most often mounted near the bottom to assure blending of the tank contents even at a low liquid level. The major disadvantage to side-entering mixers is a submerged shaft seal, which must operate in the process fluid. Process fluids may be lubricants, such as petroleum products, or abrasives, such as paper pulp and slurries. Many lubricant products require a positive seal, while abrasive products cause wear problems. The advantages of side-entering mixers are economic ones: lower initial cost, no mounting support on top of the tank, and simple speed reduction because of higher operating speeds than those of most turbine mixers. Many side-entering mixers use belt-drive, speed reductions, and

## MECHANICAL FEATURES AND COMPONENTS OF MIXERS 1253

**Figure 21-4** Different-sized turbine mixers and drives. (Courtesy of Chemineer.)

**Figure 21-5** Side-entering mixer with pillow-block bearings. (Courtesy of Chemineer.)

pillow-block bearings. A typical side-entering mixer is shown in Figure 21-5. Both belt drives and bearing types are discussed later in this chapter.

***21-2.1.5 Bottom-Entering Mixers.*** Bottom-entering mixers are usually the same basic drive arrangement as a turbine mixer, but mounted on the bottom of the tank. A bottom-entering mounting is shown in Figure 21-6. Most bottom-entering mixers have the disadvantage of a submerged seal without the

**Figure 21-6** Bottom-entering mixer. (*Chemical Engineering*, August 2, 1976, pp. 89–94.)

cost advantages of side-entering mixers. Bottom-entering mixers are used when process requirements or tank geometry makes top or side mounting impractical.

### 21-2.2 Other Types of Mixers

Although portable, top-entering, or turbine mixers account for the largest number of mixers built for the process industries, other common mixer categories with unique features are also important.

*21-2.2.1 High Viscosity Mixers*. While turbine mixers can handle low to moderate viscosities, high viscosity fluids [100 000 cP (100 Pa · s) and greater] usually require some type of close-clearance impeller design. The diameter of a typical turbine-style impeller is less than 70% of the tank diameter. Close-clearance impellers for high viscosity applications are 85 to 95% of the tank diameter. Some close-clearance impellers even have flexible scrapers, which are effectively 100% of the tank diameter.

Important mechanical features of high viscosity mixers are the low speed and high torque required to rotate large impellers in viscous fluids. Equally important, but more subtle, are requirements for the tank to have a very round cross-section. The tank must be round so that the clearance between the impeller and the wall remain nearly constant, and the shaft must be centered for the same reason. Shaft and impeller designs are primarily for strength and based on the hydraulic forces caused by viscous drag. Although high viscosity impellers can take many forms, two of the more common varieties are the helical-ribbon (Figure 21-7) and anchor-style (Figure 21-8) impellers.

*21-2.2.2 High-Shear Mixers*. High-shear mixers have many features opposite to those of high viscosity mixers. Typical high-shear mixers have small impellers, 10 to 20% of the tank diameter, and operate at high speeds, 1000 rpm (16.7 rps) to 3600 rpm (60 rps). To operate at high speeds, without requiring excessive power, high-shear impellers usually have small blades. The blades

# MECHANICAL FEATURES AND COMPONENTS OF MIXERS 1255

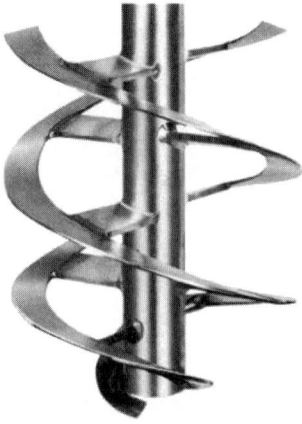

**Figure 21-7** Helical-ribbon impeller. (Courtesy of Chemineer.)

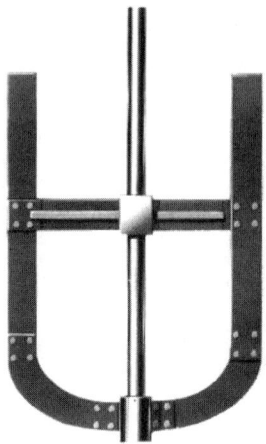

**Figure 21-8** Anchor impeller. (Courtesy of Lightnin.)

may appear as teeth on the edge of a disk or slots and holes in a rotating cylinder. A typical high-shear disk impeller is shown in Figure 21-9. The slotted-cylinder design is generally used for both a rotating and stationary element, called a *rotor–stator design*, as shown in Figure 21-10. Some high-shear mixing devices are used in-line, like pumps with high-shear blades inside a small housing, through which liquid flows or is pumped. Viscous fluids must be pumped through most in-line mixers. Such in-line style mixers or homogenizers still require some mechanical design, although with less emphasis on a long shaft support and more emphasis on tight tolerances.

A few high-shear mixing devices use impinging or interacting hydrodynamic flow to accomplish dispersion and mixing. These mixers operate more like

**1256** MECHANICAL DESIGN OF MIXING EQUIPMENT

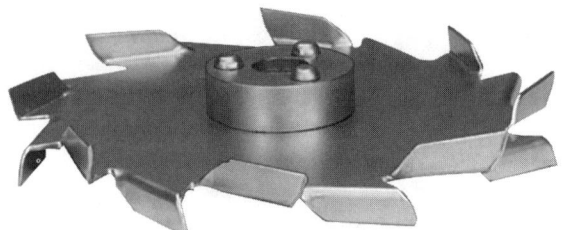

**Figure 21-9** High-shear impeller. (Courtesy of INDCO.)

**Figure 21-10** Rotor–stator high-shear impeller. (Courtesy of IKA Works.)

static mixers, with the mixing power provided by an external pump, often a high-pressure positive-displacement pump.

*21-2.2.3 Double-Motion Mixers.* As the name implies, double-motion mixers have a combination of mixer motions. Many double-motion mixers are a combination of a high viscosity, close-clearance mixer and high-shear mixer. The high viscosity part of the mixer provides bulk motion of the fluid(s), especially near the tank walls, and the high-shear mixer creates dispersion, often of two phases, either two liquids or a liquid and solids.

The double motion comes from two shafts with at least two impellers operating in the same tank. Other double-motion mixers have coaxial shafts with a close-clearance impeller and turbine impeller(s) operating at different speeds. Some mixers have shafts that move relative to the vessel, as in planetary motion mixers, where intermeshing impellers rotate on their own axis and move around the axis of the tank. Double-motion mixers provide a diversity of mixing actions selected to handle difficult or changing batch mixing requirements. The cost of more complicated equipment is offset by the ability to handle a wider range of mixing needs.

**21-2.2.4 Dry-Solids Mixers.** Dry-solids mixers are normally applied to flowable powdered materials. The action of the mixers can be categorized as (1) tumble mixers; (2) convective mixers, which use a ribbon, paddles, or blades to move material; (3) high-shear mixers, which create a crushing action like a mortar and pestle; (4) fluidized mixers, as in fluidized beds; and (5) hopper mixers, which use discharge and recirculating flow to cause mixing (Harnby et al., 1992). Although each type of dry solids mixer uses different equipment to accomplish the mixing action, the design methods discussed in this chapter for motors, drives, and even seals may apply.

**21-2.2.5 Static Mixers.** The mechanical design of static mixers, also called *motionless mixers*, is unique compared with other types of mixing equipment. Most other mixers involve some type of rotating equipment. Static mixers have no moving parts, and therefore design methods resemble those of piping and pressure vessels. The mixing elements of a static mixer can take many forms, but the most common is the twisted element style, shown in Figure 21-11. Most elements are merely inserted and fixed into a section of pipe, although some are designed to be removable for cleaning and others are sealed to the wall of the pipe. Design of the elements themselves is largely proprietary, although the pipe sections in which the elements fit are designed to piping standards for dimensions and end connections. Most static mixers are housed in the same size or one-size-larger pipe than the adjacent runs of piping and are the same material and schedule (wall thickness).

**21-2.2.6 Other Mixers.** A variety of other devices and methods can be used as mixers. Flow devices, such as jets and nozzles, can be used as mixers. Rising

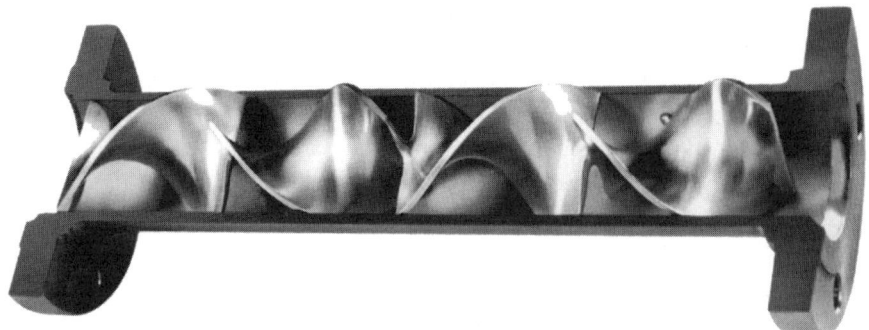

**Figure 21-11** Kenics static mixer. (Courtesy of Chemineer.)

**1258** MECHANICAL DESIGN OF MIXING EQUIPMENT

gas bubbles from injected air will cause mixing. Pulses of liquid or gas can create interesting flow patterns. It is beyond the scope of this book to provide mechanical design characteristics for such a diversity of equipment. However, within the scope of the equipment and methods described in this chapter, elements of many mixers and mixing systems can be designed or selected with an understanding of the basic requirements.

## 21-3 MOTORS

Motors are an essential part of most mixers, since a rotating shaft with an impeller is common. Electric motors are without doubt the primary source of rotating power for mixers. Air and hydraulic motors are used for some applications, especially where a combination of variable speed and explosion-proof performance are needed. Diesel engines are used occasionally where electric power is unavailable or unreliable.

### 21-3.1 Electric Motors

Electric motors take almost as many different forms as mixers. Motors can be classified by size, power source, enclosure, and even application. An essential part of any electric motor is the nameplate. Without the information found on a nameplate, most motors look like a cylindrical or rectangular housing with wires leading in and a rotating shaft coming out. Understanding the information on a motor nameplate will help identify an existing motor or specify a new motor. Although some information is unique to individual manufacturers, much of the information is essential for proper operation and application of a motor.

Some or all of the following information can be found on a typical motor nameplate:

- *Catalog number:* specific to the manufacturer.
- *Model number:* specific to the manufacturer.
- *Phase:* single, three, or direct current.
- *Type:* classification depends on the manufacturer.
- *NEMA (National Electrical Manufacturers Association) electrical design:* B, C, and D are most common and represent torque characteristics of the motor.
- *Duty:* most motors are rated for continuous operation, especially for mixers. However, motors for 15, 30, or 60 min duty are available.
- *Frequency (Hz):* electric frequency in cycles per second.
- *Speed (rpm):* revolutions per minute of shaft at full load.
- *Voltage:* single or multiple voltages, depending on winding(s).
- *Amperage (FLA):* full-load motor current.
- *Power (hp):* horsepower at rated full-load speed.
- *Frame size:* standard designation of dimensions.

- *Maximum ambient temperature (max. amb.) in Celsius (centigrade):* usually 40°C [104°F].
- *Insulation class:* standard insulation classes are B, F, and H, which establish the maximum safe operating temperature for the motor.
- *Enclosure:* indicates how the motor is protected and sealed from the surroundings.
- *Service factor:* a measure of continuous overload capacity.

A comprehensive description of manufacturer-specific information, such as catalog number, model number, type, and so on, can be found in the company's catalog. Many catalogs have a section of *engineering data* that may have more extensive tables of dimensions, enclosure features, and design calculations. Some manufactures even have separate technical data books (Leeson, 1994).

Because electric motors are used for an enormous range of applications and manufactured to many unique specifications. The full range of motor features cannot be covered in this book. The features most common for industrial mixer applications will be emphasized.

### 21-3.1.1 Phase.
Alternating current can be categorized as either single- or three-phase power. Single-phase power has a complete cycle of voltage from an alternating maximum positive value to a maximum negative value and back to the maximum positive value. Most household and office power in the United States is single phase. Three-phase power, commonly found in industrial environments, is carried by three conductors with three voltage cycles starting out of phase with one another. With three-phase power the voltage between two conductors never goes to zero, resulting in a smoother, more nearly constant voltage differential across motor windings. Most applications with motors 3 hp (2200 W) and larger use three-phase power.

### 21-3.1.2 Type.
Motor type depends on the manufacturer, power, and application. The most common motor type used on a mixer for single-phase power is a capacitor start motor. Capacitor start motors can be designed for both moderate (175% or less) and high (300% of full load) starting torque. Torque is the twisting force (moment of force), created by the motor and applied to the rotating shaft. Moderate torque is adequate for most mixer applications since impeller power is proportional to speed cubed in turbulent conditions, thus keeping starting torques low. Capacitor start motors use a start capacitor and a start switch. The start switch takes the capacitor and start winding out the electric circuit when the motor reaches approximately 75% of full-load speed. Split-phase motors can be used in light-duty applications, because of moderate to low (100 to 125%) starting torque and high starting current. Split-phase motors have no capacitor, only a start switch to drop out the start winding.

Three-phase motors have a high starting torque, high efficiency, and low current requirement. The torque characteristics are described by NEMA electrical design, which is discussed in the next section. Three-phase motors do not use a capacitor, switch, or relay for starting.

Other types of motors that may be encountered in mixer applications are gear motors, pony motors, and brake motors. Gear motors are composed of an electric motor with an attached gear reducer. Spur, helical, or worm gears can be used in single or multiple reductions to achieve a wide range of output speeds. Motor power, output speed, and output torque are all essential design variables.

A pony motor is a small gear motor used to turn a larger motor at slow speed and to provide additional starting torque. A pony motor or variable speed drive may be used to slowly start a mixer that could be embedded in settled solids. Pony motors are rarely used on mixers today because of available variable speed drives. Care must be exercised to match the output torque rating of the pony motor with the input torque rating for the mixer drive.

Brake motors have a fail-safe, stop-and-hold, spring-set brake on the back of the motor. When power fails, the brake sets and holds the motor and load. This feature is rarely needed on a mixer since the mixed fluid usually acts as its own brake.

**21-3.1.3 NEMA Electrical Design.** Three-phase motors are classified by electrical design type, B, C, or D, defined by NEMA. Design B motors provide normal (100 to 200%) starting torque at normal starting current and are suitable for most mixer applications. Design C motors provide high (200 to 250%) starting torque at normal starting current and may be used for special mixer applications, provided that the drive and shaft are not overloaded during startup. Design D motors have high (275%) starting torque with high slip at low starting current and are rarely used on mixers.

**21-3.1.4 Duty.** All motors used for mixer applications should be rated for *continuous* duty, since even batch runs may take more than the anticipated time should problems develop.

**21-3.1.5 Frequency.** The frequency of alternating current is measured in Hertz (cycles per second). Sixty-cycle (60 Hz) current is used throughout North America. Fifty-cycle (50 Hz) current is used in Europe and in many countries in Asia. The frequency of the current supplied affects the operating speed of an alternating current (AC) motor.

**21-3.1.6 Speed.** A typical AC motor is designed to operate within 2 to 3% of the synchronous speed. Synchronous speed depends on the number of poles in the winding:

U.S. Eng.

$$N \text{ [rpm]} = 120\frac{f}{p}$$

Metric

$$N \text{ \{rps\}} = 2\frac{f}{p}$$

(21-1)

where N is rotational speed [rpm] {rps}, f is frequency [Hz (cycles/s)] {Hz}, and p is the number of poles in the motor rotor. Typical motor speeds used for mixers with 60-cycle (60 Hz) power are 1800 rpm (30 rps) and 1200 rpm (20 rps), which correspond to four- and six-pole windings. Additional speeds occasionally encountered with mixers are 3600 rpm (60 rps) and 900 rpm (15 rps). Corresponding speeds for 50 cycle (50 Hz) power are 1500 rpm (25 rps), 1000 rpm (16.7 rps), 3000 rpm (50 rps), and 750 rpm (12.5 rps). Whether a mixer is designed to operate with 60 or 50 Hz power makes a major difference in the appropriate speed reduction for a mixer, since impeller power is a strong function of operating speed.

Multispeed motors can be built by using different connections to a single winding or with multiple windings. All single-winding two-speed motors have a 2 : 1 speed ratio, such as 1800/900 rpm (30/15 rps). Multiple winding motors can have two speeds, such as 1800/1200 rpm (30/20 rps). Multispeed electric motors have a large effect on mixer applications, because a 2 : 1-speed motor typically has an 8 : 1 effect on impeller power for turbulent conditions. Even a 3 : 2-speed motor has a 3.4 : 1 power effect for turbulent conditions. Multispeed motors can be applied when a viscosity change results in increased impeller power. However, motors are usually constant torque, so a 2 : 1 speed motor delivers only half the maximum power at the low speed. Multispeed motors have largely been replaced by variable speed (variable frequency) drives, because of the large power change with mixer speed.

*21-3.1.7 Voltage*. Like phase, voltage depends on the electrical supply to the location of the motor. Typical voltages in the United States are 125 and 230 V for single-phase power and 230 and 460 V for three-phase power. In Canada, 575 V, three-phase power is available in many industrial environments. Other low voltages, such as 200 and 208 V, can be found in certain facilities. Higher voltages, such as 2300 and 4160 V, are available in specific situations and may be needed for large motors. The higher the voltage, the lower the amperage and therefore the smaller the wire size and switching or starter capacity required for a given motor power.

*21-3.1.8 Amperage*. Amperage describes how much current is required to run a motor. A motor nameplate typically shows full-load amperage (FLA), which is the amount of current required when the motor is loaded to the rated power. Power or wattage of a motor is theoretically the product of voltage times amperage. However, motors are sized based on mechanical output. Because no motor is 100% efficient, the inefficiency is added to the theoretical power and reflected in the amperage required to operate a motor. Minimum efficiency standards for motors are established by the government to avoid unnecessary waste of energy. Motor manufacturers can offer higher-efficiency motors, which will waste less energy and therefore run cooler. High-efficiency motors are usually required when used with variable speed drives, such as variable frequency invertors, because of reduced cooling and efficiency at lower speeds.

**21-3.1.9 Power.** Power, in horsepower or kilowatts, is the primary criterion used to establish motor size. Commercially available motors, like those most often used on mixers in the United States, come in standard sizes, such as (in horsepower) 0.25, 0.33, 0.5, 0.75, 1, 2, 3, 5, 7.5, 10, 15, 20, 25, 30, 40, 50, 60, 75, 100, 125, 150, 200, and 250. Larger motors are nonstandard but typically follow similar increments of nominal power. Motors for international use are rated in kilowatts of power and roughly match these standard horsepower sizes, with some additions or exceptions.

Power alone does not describe a motor, especially with respect to physical size. The output torque, which is effectively power divided by speed, characterizes the frame and shaft dimensions. Thus, a 10 hp motor operating at 900 rpm will usually have the same dimensions as a 20 hp motor operating at 1800 rpm.

**21-3.1.10 Frame Size.** Frame size is set by standards such as those established by the National Electrical Manufacturers Association (NEMA). The frame size establishes critical dimensions for mounting and applying a motor. A motor, built by any manufacture, with a similar frame size should fit in the same application. Critical dimensions include shaft size, shaft location, overall size, and base or face dimensions and bolting patterns.

NEMA frame sizes begin with a number or number and letter combination, such as 56, 56C, 145T, or 213T. With two-digit frame numbers, such as 56, the distance from the mounting base to the centerline of the shaft is 56 divided by 16 in., or 3.5 in. The letter "C" following the number (e.g., 56C), indicates a "C face," which describes a bolting pattern and dimensions, such that the end of the motor can be bolted directly to the equipment it powers. C face motors are practical only when the size and weight of the motor can be supported from one end. Larger motors are "foot-mounted" so that a base on the side of the motor cylinder supports the motor from a central location.

Larger motors, starting with a three-digit number, such as 182, have a distance from the motor base to the shaft centerline of 4.5 in. which is 18 (the first two digits) divided by 4. The third digit in the number defines the distance from the motor centerline to the foot mounting holes. Similarly, a 213-frame motor has a shaft centerline $5\frac{1}{4}$ in. from the base: $21 \div 4 = 5.25$ in. The "T" designation indicates an integral horsepower motor with standard shaft dimensions, "D" indicates a flange mount, and "M" and "N" indicate flange mounts for oil burner applications. Motor part numbers often contain the frame size along with other letters or numbers indicating motor type or length. Complete tables of NEMA motor dimensions can often be found in the engineering section of motor catalogs, or catalog descriptions will define specific dimensions.

IEC (International Electrotechnical Commission) frame sizes serve the same purpose as NEMA sizes, by making motors with standard dimensions interchangeable. IEC frame sizes such as 63, 72, and 80 indicate that the shaft centerline is 63, 72, and 80 mm above the base, respectively. IEC motors are normally rated in kilowatts with standard increments similar to those for horsepower-rated motors. IEC and NEMA motors are not usually interchangeable without modifications to the support or motor, or both.

Table 21-1  NEMA Insulation Classes

| Class | Maximum Allowed Temperature | |
|---|---|---|
| A | 105°C | 221°F |
| B | 130°C | 266°F |
| F | 155°C | 311°F |
| H | 180°C | 356°F |

*21-3.1.11  Insulation Class.* Insulation systems are rated by NEMA for the maximum allowable temperature of the motor. Maximum allowable temperatures for different insulation classes are shown in Table 21-1.

*21-3.1.12  Enclosures.* Typical motor enclosures include open drip proof (ODP), totally enclosed nonventilated (TENV), totally enclosed fan cooled (TEFC), totally enclosed air over, totally enclosed hostile and severe environment, and explosion proof. Open motors are rarely used on mixers because splashed liquids or dust from dry powders can enter the motor. Totally enclosed motors are not airtight but are suitable for use in dirty or damp environments, but not hazardous locations. Nonventilated motors are usually small ($\frac{1}{3}$ hp or less) because cooling is not a problem. Fan-cooled and air over motors require moving air to provide sufficient cooling. Fan-cooled motors have their own fan on the end of the motor shaft to provide cooling. Hostile or severe environment motors have sealed housings sufficient to resist extremely moist or chemical environments, but not hazardous locations.

Explosion-proof motors meet standards set by independent testing organizations, such as Underwriters Laboratories (UL) or the Canadian Standards Association (CSA), for use in hazardous (explosive) locations. A location is considered hazardous if sufficient gas, vapor, or dust is present to cause an explosion. The National Electrical Code (NEC), published by the National Fire Protection Association, divides these locations into Divisions, Classes, and Groups according to the type of hazard present.

Division 1 motors are explosion proof, and Division 2 will not be an ignition source for an explosion. If explosion-proof motors are required, Division 1 is the correct specification. Division 2 usually applies only to large motors, generally used in an outdoor installation, where the atmosphere can become explosive only when a serious process failure occurs.

Class I is for flammable gases and vapors, Class II is for combustible dust, and Class III is for ignitable fibers and filings. Most explosion-proof motors are rated for Class I, with some for Classes I and II. Special motors are required for Class III.

Groups further define the materials for which a motor is designed to be explosion proof. Group D is for common flammable solvents and fuels, such as acetone, ammonia, benzene, butane, gasoline, hexane, methane, methanol, propylene, propane, styrene, and similar compounds. Many explosion-proof motors are rated Division 1, Class I, Group D. Group C motors provide additional protection for chemicals such as carbon monoxide, diethyl ether, ethylene, cyclopropane, isoprene, and others. Motors that satisfy Group C requirements also meet Group D requirements, so motors may be rated for Division 1, Class I, Groups C and D. No motors are rated as explosion-proof for Group B, which includes hydrogen, ethylene oxide, butadiene, propylene oxide, and other compounds, or Group A for acetylene.

Groups F, G, and E apply to dusts, powders, and fibers. Group F applies to carbon black, coke, and coal dust. Group G applies to flour, starch, grain, and nonconductive plastic or chemical dust. Group E applies to aluminum, magnesium, and other metal dusts. To handle some of these more severe explosion requirements, Division 1, Class I, Groups C and D, and Class II, Groups F and G explosion-proof motors are available. Other types of explosion ratings may be available by special design or rating.

***21-3.1.13 Service Factor.*** The service factor describes how much a motor can be overloaded without causing damage. Power requirements for mixer applications are difficult to estimate accurately, especially when process conditions change after the initial design. To reduce premature failure of electric motors, many have a service factor of 1.15, which means that the motor can be operated with a load 15% above the rated power without damage. While a 1.15 service factor may be common on many standard motors used for mixer applications, explosion-proof motors typically have a 1.0 service factor. This limitation is intended to reduce the possibility of the motor surface becoming hot enough to act as an ignition source.

***21-3.1.14 Wiring.*** For motors to operate with different supply voltages and to reverse direction of rotation, the wires for the internal windings are carried to the junction box. These wires can be connected to the power supply for the different voltages specified on the nameplate.

If single-phase motors are wound for multiple voltages, six or seven leads will be found in the junction box, as shown in Figure 21-12. With six leads and low-voltage operation, typically 125 V in the United States, three of the junction box leads are connected to each of the two line wires. With six leads and high-voltage operation, typically 230 V in the United States, three of the junction box leads are connected together and one or two of the other leads are connected to the two line wires. Details for the specific motor are often printed on the inside of the junction box lid.

Three-phase motors have three lines from the electric source connected to nine or more junction box wires. For low-voltage operation, typically 230 V in the United States, the line wires are connected to pairs of junction box leads, with additional leads interconnected, as shown in Figure 21-13. For high-voltage operation, typically 460 V in the United States and 575 V in Canada, each line wire connects to a single junction box lead, and pairs of the other leads are connected together. Additional leads may be present for thermostats and other motor features, such as heaters. Again, specific wiring is often shown inside the junction box cover.

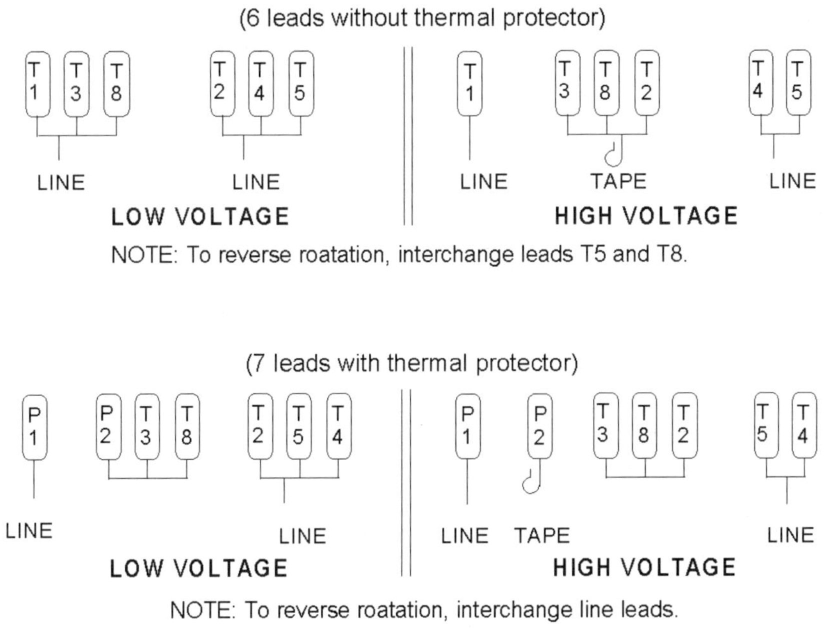

**Figure 21-12** Typical wiring for single-phase motors. (Courtesy of Chemineer.)

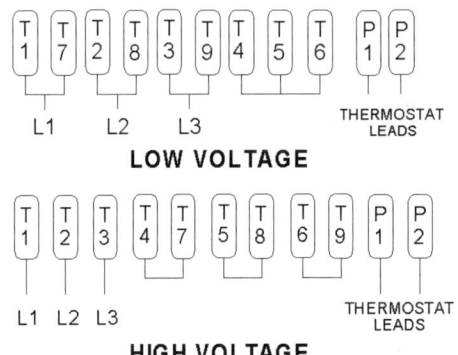

**Figure 21-13** Typical wiring for three-phase motors. (Courtesy of Chemineer.)

Single-voltage motors can have simplified wiring because the other connections are made internal to the motors. Portable mixers, especially those with single-phase motors, can be prewired with a suitable electric cord.

Electric motor wiring requirements go beyond simple connections to the correct motor lead. The NEC describes in detail many characteristics of proper wiring practices. References such as McPartland and McPartland (1990) provide

**1266** MECHANICAL DESIGN OF MIXING EQUIPMENT

additional background. However, a trained electrician familiar with motors, national and local codes, and facility requirements should install and wire any motor used in an industrial facility.

***21-3.1.15 Mounting.*** Most electric motors are mounted on feet or at the end on a face or flange. Face and flange mounted motors are usually 30 hp or less and may also have feet. The face or flange can be bolted directly to a gear reducer with a shaft coupling internal to the motor and drive assembly. The feet on a motor are on the side of the motor, so a foot-mounted motor typically has a horizontal shaft. Since many liquid mixers have vertical shafts, some type of right-angle drive is necessary to transition from a horizontal motor shaft to a vertical mixer shaft.

***21-3.1.16 Direct Current.*** Direct-current (DC) motors normally come with either an internal or external rectification system matched to the motor duty. The rectification of alternating current to direct current by solid-state electronics allows the control of motor speed by adjusting the applied voltage. Small DC motors, less than 5 hp (3700 W), are typically permanent magnet designs; larger motors are shunt wound.

Brushless DC motors can provide extremely accurate and efficient speed control. Digital feedback to the electronics provides the commutation required to set the motor speed. These added benefits come at a higher cost than typical DC or variable frequency AC drives.

Variable speed motors, whether AC, DC, or air driven, provide an added dimension to mixer operation. Many batch processes experience a range of liquid levels and fluid properties during a mixing operation. The ability to change mixer speed allows a mixer to be slowed to reduce splashing or mixing during part of the process and still permit more intense mixing at a higher speed during another part of the process.

While voltage control is the usual means of speed control with a DC motor, frequency control can be used to change the speed of an AC motor. A typical AC controller rectifies the incoming alternating current into a direct current, which is then converted back into a digitally controlled alternating frequency. The new alternating frequency can potentially be set between twice the normal frequency and one-tenth the normal frequency. The extremes of this range are rarely available simultaneously and are limited to use with lightly loaded motors. More practically, variable frequency controllers are used between the normal frequency and about one-fourth normal, which can provide a 64:1 power range for a typical mixer.

Limits to practical operation of motors controlled by variable frequency usually involve heat dissipation and noise. As a TEFC motor is slowed, the attached fan becomes less able to remove heat by blowing air across the outside of the motor. At low speeds (low frequency) the motor can also be very noisy. Another limit to motor design is usually torque, so as the speed is reduced, so is the output power. A variable speed motor at one-fourth speed can produce no more than one-fourth of the nameplate power. A variable speed motor must be rated for the desired speed range (frequency range shown on the nameplate) to be operated safely as an explosion-proof motor.

### 21-3.2 Air Motors

Air motors use compressed air to create rotational motion. Compressed air flows through the motor, turning a positive displacement rotor with vanes that extend to the wall of the housing. Sufficient air must be compressed to provide the required flow and pressure at the motor. Losses through piping, valves, filters, and flow meters must be considered.

Because an air motor is not electrically driven, the motor provides no direct source for sparks that could ignite an explosive atmosphere. Also, the expansion of air through the motor keeps it cool during operation. A simple valve in the air line provides speed adjustment. Air consumption increases as speed and air pressure are increased.

Besides providing variable speed and nonsparking drive, suitable for most hazardous locations, an air motor will not burn out when overloaded; instead, it slows with increased torque. Air motors are compact, portable, and lighter than comparable electric motors. Disadvantages are noise and inefficiency. A 1 hp air motor may require a 5 hp compressor for continuous operation.

### 21-3.3 Hydraulic Motors

Hydraulic motors provide some of the same features as air motors, except hydraulic fluid is circulated through the motor and back to a pump. A hydraulic pump can be operated in a safe location, while the motor can drive the mixer in a hazardous location. Hydraulic motors also provide high torque with variable speed. Sometimes multiple motors are driven from a single hydraulic supply system.

## 21-4 SPEED REDUCERS

Except for some portable mixers, high-shear mixers, and a few special mixers, most mixers operate below standard motor speeds. Typical motor speeds of 1800 or 1200 rpm (30 or 20 rps) are reduced to between 350 and 30 rpm (5.8 and 0.5 rps) for most mixer applications. Portable and side-entering mixers usually operate near the upper portion of this speed range from 420 to 170 rpm (7 to 2.8 rps). Turbine mixers operate in the middle range, from 125 to 37 rpm (2.1 to 0.6 rps), and high viscosity mixers operate from 45 to 20 rpm (0.75 to 0.33 rps) and slower.

In the upper portion of the speed range for mixers, a single speed reduction with either gears or belts is used. Gear reduction is used with most low-speed portable mixers, and belts are used with many side-entering mixers. Turbine mixers can use single-, double-, or triple-reduction enclosed gear drives. Sometimes a combination of gear and belt drives is used. Since most drives transmit essentially constant power, the reduced speed results in much higher torque. Torque is proportional to power divided by speed and represents the amount of turning force produced by a drive.

### 21-4.1 Gear Reducers

Gear reducers use a small rapidly turning toothed "gear" called a *pinion* to turn a larger gear with more teeth. How much speed reduction depends on the relative diameter of the pinion and gear, measured by the number of teeth on each. Thus a 5 : 1 gear reduction has five times as many teeth on the gear as on the pinion.

Although a nearly infinite number of gear ratios seem possible, practical limitations apply to different types of gears, and the American Gear Manufacturers Association (AGMA) recommends some nominal ratios for each type of gearing. Mixer drives using helical (parallel-shaft) gears or a combination of helical and spiral-bevel (right-angle) gears typically operate at 350, 230, 190, 155, 125, 100, 84, 68, 56, 45, 37, 30, 25, or 20 rpm (5.83, 3.83, 3.17, 2.58, 2.08, 1.67, 1.4, 1.3, 0.9, 0.75, 0.583, 0.5, 0.483, or 0.417 rps). Mixer drives using worm (right-angle) gears typically operate at 350, 233, 175, 146, 117, 88, 70, 58, 44, 35, 29, or 25 rpm (5.83, 3.88, 2.92, 2.43, 1.95, 1.47, 1.17, 0.967, 0.733, 0.583, 0.483, or 0.417 rps). Actual speeds can deviate from these nominal speeds by as much as 3 to 5%, depending on actual gear ratios and loaded motor speeds. When calculating impeller power, actual shaft speed should be determined by measurement (with a tachometer) or by calculation from actual gear ratio and motor speed under load.

Most gear reducers for mixer applications are enclosed to prevent the same potential contamination problems as those mentioned for motors. Also, open gearing poses safety hazards for operators. For the speed ranges required in mixers, one, two, or three gear reductions may be needed. Each reduction results in a successively lower speed. All of the reductions can be inside the same housing, or one reduction can be attached to the motor, with the other reduction or reductions inside the reducer housing.

A large diameter output shaft in the reducer is necessary to avoid deflections caused by hydraulic loads that could misalign gears. Typical mixers have a long overhung shaft, which is subjected to random hydraulic forces on the impellers. These forces cause seemingly small deflections, which can misalign the gears, resulting in rapid or premature wear. A large diameter shaft between the output bearings also increases the natural frequency of an overhung shaft.

Large bearings on the output shaft are also necessary to handle the loads transmitted by the mixer shaft. Output shaft bearings must handle radial loads caused by bending loads on the mixer shaft and axial loads. The bearings supporting the mixer shaft must handle loads that depend on the mixer application.

A dry-well seal is essentially a standpipe in the bottom of a gear reducer that surrounds the vertical output shaft. When the gear reducer is filled with lubricating oil, the normal oil level is below the top of the pipe. So even if the seal around the output shaft fails, the oil cannot leak out of the gear reducer.

The output shaft bearing near the bottom of the dry well is usually lubricated by grease, which is less likely to leak. This feature protects both the drive from the loss of lubricant and the process from oil contamination.

***21-4.1.1 Mixer Loads.*** The loads on a mixer are primarily those exerted by the impellers and transmitted by the shaft. The loads on a mixer shaft are depicted in Figure 21-14. For the mixer drive to turn the impellers, a torque must be applied to the mixer shaft. This twisting load contributes to the internal stresses in the shaft and must be considered when establishing the strength requirements for shaft design. Fluctuations in these loads are caused by the random motion of the fluid. Besides the torsion loads, bending loads are caused by the random hydraulic forces on the impeller(s). Bending loads can be large because of both the hydraulic forces and the length of the mixer shaft. The bending loads also contribute to the internal stresses in the shaft and must be considered in the shaft design.

Several axial loads are imposed on the mixer shaft and drive. First the weight (mass) of the impellers and shaft create a downward force. Then pressure forces with a closed tank will cause an upward force. The magnitude of the pressure force is the same as if the shaft were a piston (i.e., force equals pressure times the cross-sectional area of the shaft). The force will be downward with a vacuum, but because of the force limit of atmospheric pressure acting on a vacuum, the magnitude is rarely a problem. Finally, axial flow impellers can cause an axial thrust, usually upward. Many hydrofoil impellers create a measurable amount of axial thrust, often sufficient to counteract the weight of the impeller. Although

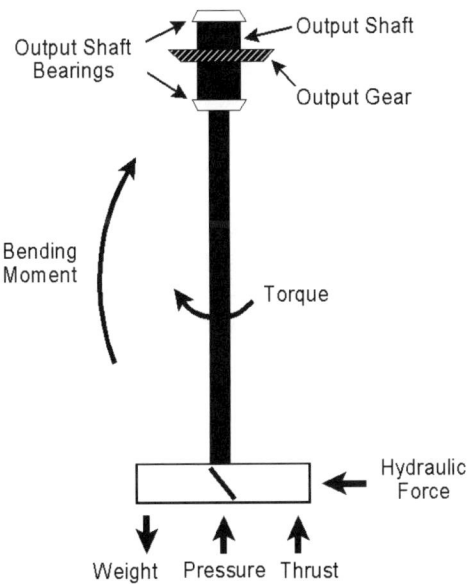

**Figure 21-14** Loads on a mixer shaft. (Courtesy of Chemineer.)

these axial forces are measurable, they are rarely significant to the design of the shaft, drive, or support. However, the ability of the axial thrust to lift an impeller must be considered when designing a means of attachment of the impeller to the shaft. Weight alone is not sufficient to hold some axial flow impellers down.

***21-4.1.2 Basic Configurations.*** Gear reducers are categorized according to the orientation of the input and output shafts, right-angle or parallel-shaft reducers. These different arrangements use different types of gearing. For mixer applications, both right-angle and parallel-shaft reducers have advantages and disadvantages. Right-angle reducers are typically shorter than parallel-shaft reducers, allowing them to fit better between floors and below roofs. Conversely, right-angle drives obstruct part of the top of the tank, which can make piping connections difficult. Mounting and adjusting foot-mounted motors may be easier with right-angle drives than with parallel-shaft reducers.

Parallel-shaft gear reducers use one or more sets of parallel-shaft gears, such as helical gears, to make the necessary speed reduction. Some parallel-shaft reducers have the motor stacked above the gear reducer to limit the overall diameter of the mixer drive system. Other parallel-shaft reducers have the motor mounted alongside the gear reducer to limit the overall height of the drive system. Generally, parallel-shaft reducers are easier than right-angle mixer drives to design and build. However, they do involve mounting and operating a vertical electric motor, which can cause additional problems with large motors.

In-line reducers are usually a variation on parallel-shaft reducers. A properly designed double-reduction reducer with two sets of gearing having the same center distance can be arranged so that the input and output shafts are not only parallel, but in line with one another. Compared with parallel shaft reducers, in-line reducers usually trade greater height for smaller diameter and centered weight. Other types of gearing, such as planetary gears, can make an in-line reducer. Whatever the basic configuration, well-designed gear reducers will provide good service in mixer applications.

Right-angle gear reducers must use at least one right-angle gear set, typically spiral bevel or worm gears. Both spiral-bevel and worm gears have unique advantages with respect to mixer applications. Spiral-bevel gears are some of the quietest and most efficient right-angle gears. Although less efficient than other gears, worm gears can make larger speed reductions in a single gear set. A single reduction usually means lower cost. However, lower efficiency can make heat dissipation more difficult.

***21-4.1.3 Gear Types.*** Although many different types of gearing are available, mixer applications usually warrant better quality for reliable, low-noise service. As a result, gears with curved profile teeth, such as helical and spiral-bevel gears, are used instead of similar gears with straight teeth. Some basic terminology of a gear mesh is shown in Figure 21-15. For operational purposes, gears must be adjusted so that tooth contact is made along the pitch circle with sufficient backlash to avoid contact on the back side of the teeth. Various sources provide

SPEED REDUCERS  1271

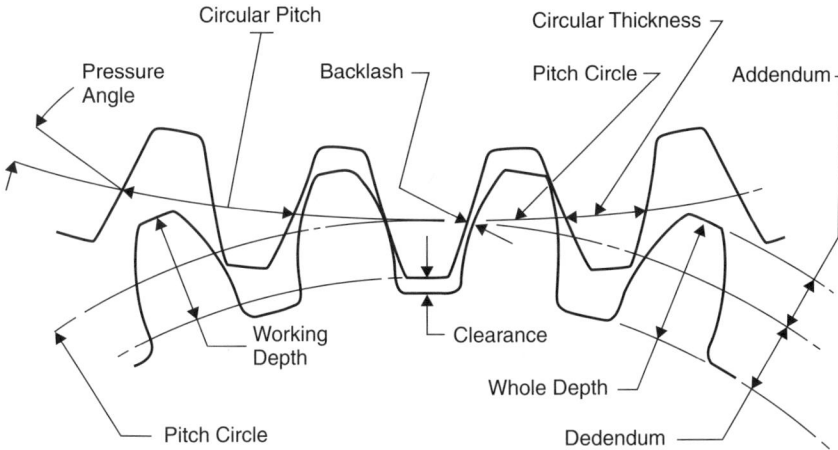

**Figure 21-15** Basic gear mesh. (Courtesy of Hamilton Gear.)

**Figure 21-16** Helical gear set. (Courtesy of Hamilton Gear.)

more detail about gear terminology and design, such as Baumeister et al. (1978). However, most mixer designs do not involve the actual selection of gearing, but instead, require a basic understanding of gear types and characteristics.

Helical gears (Figure 21-16) provide parallel-shaft gear reduction. The term *helical* comes from the fact that the teeth are cut along a helical path with respect to the axis of rotation. A typical spur gear is similar to this arrangement, except

**1272**  MECHANICAL DESIGN OF MIXING EQUIPMENT

**Figure 21-17**  Spiral-bevel gear set. (Courtesy of Hamilton Gear.)

that the teeth are straight and parallel to the axis of the shaft. While straight teeth are easier and cheaper to cut, they are noisier and more prone to wear than helical gears. The contact between straight teeth occurs across the entire face of the tooth simultaneously. Curved teeth, like those on helical gears, make a sweeping contact across the tooth surface for a more gradual contact transition. The gradual contact and transition between teeth with the helical gear make less noise than the full contact with a spur gear.

Spiral-bevel gears (Figure 21-17) are used to make a right-angle transition between the input and output shafts of a typical right-angle mixer drive. Similar to the helical gear, the curved shape of the spiral-bevel teeth makes gradual contact and less noise than bevel gears with straight teeth.

Worm gears (Figure 21-18) provide another means of making a right-angle transition. Because of a circular contact between the worm pinion and the gear, opportunities exist for curving either the gear teeth or both the pinion and gear teeth. The gear set shown in Figure 21-18 has curved (cupped) teeth on the gear, but a straight worm. Because a sliding contact is made between the worm and the gear, a worm gear reducer is quiet but less efficient, because of friction at the tooth contact. Special lubricants are used in worm gear reducers to reduce friction and tolerate higher temperatures. The biggest single advantage of a worm gear reducer is that large speed reductions can be made with a single gear set. A single set of gears make worm gear reducers less expensive than many other right-angle drives, requiring two or three gear reductions for the same output speed.

Beyond gear reducers made with similar straight-tooth gears, such as spur gears (parallel shaft) and bevel gears (right angle), planetary and other internal gear designs can be used for mixer drives. The gear sets shown in the previous figures are external gears with teeth around the outside of the gears. Internal gears with teeth on the inside of a ring can also be used. Planetary gears involve an internal gear with a small pinion (sun gear) surrounded by multiple small (planet) gears. Virtually any type of gear reduction, including the rear axle from an old pickup truck, can be used for a mixer drive, and probably has been used somewhere.

SPEED REDUCERS    **1273**

**Figure 21-18**   Worm gear set. (Courtesy of Hamilton Gear.)

*21-4.1.4 Bearing Types.* Another key mechanical element to a gear reducer used for a mixer drive are the rotating bearings. The typical components of an antifriction bearing are shown in Figure 21-19. The inner ring is usually fitted onto the machined surface of the rotating shaft. The outer ring is pressed into a machined opening in the reducer housing. Both the inner and outer rings are expected to remain stationary with respect to their mating components. All the rotation should take place on the rolling elements between the two bearing rings. The rolling elements can be balls or various shaped rollers, as discussed in the following sections.

The key dimensions of a bearing are bore (inside diameter, ID) and outside diameter (OD), which identify a bearing size. Most bore sizes are slightly smaller

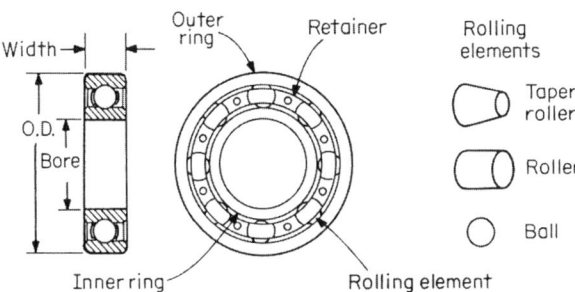

**Figure 21-19**   Bearing components.

**1274**   MECHANICAL DESIGN OF MIXING EQUIPMENT

**Figure 21-20**   Ball bearing (cross-sectional view). (Courtesy of NTN Bearing Corp.)

than standard bar stock dimensions, so that a minimal amount of machining is required to convert a piece of material into a drive shaft. The OD of the bearing must be sufficient to accommodate the ring and roller dimensions. Often, bearings with the same OD are available with different bore sizes for different applications.

Ball bearings use spherical balls as the rolling elements in the bearings (Figure 21-20). Ball bearings are the most common and inexpensive type of antifriction bearing. They provide good radial support to a rotating shaft, but only limited axial support. Ball bearings are good for motors, some high-speed drive shafts, and portable mixers. Large mixer output shafts and some internal gear-drive shafts transmit too much axial force to a bearing for satisfactory use of ball bearings.

Tapered-roller bearings use a tapered roller, as shown in Figure 21-21, instead of balls to support the rotating shaft. The tapered roller provides excellent load-carrying ability in the radial direction and considerable axial load capability against the taper of the roller. The tapered bearing carries an axial load in only one direction, so the orientation and location of these bearings are important. Besides the axial loads transmitted to the output shaft bearings by the mixer shaft, gearing transmits axial loads to internal shafts. The angled teeth on a helical gear set create an axial load on both the pinion and gear shafts as they transmit the rotating torque. A spiral-bevel gear set also creates an axial separating load on the shafts. Tapered-roller bearings are often used to support these axial loads in a mixer drive.

Spherical-roller bearings have two rows of rollers (Figure 21-22) to carry thrust loads in both directions. These bearings are rarely used as internal bearings, where load directions are known. However, they are used as output shaft bearings, especially in side-entering mixers, where fluctuating loads can occur.

SPEED REDUCERS **1275**

**Figure 21-21** Tapered-roller bearing (cross-sectional view). (Courtesy of The Timken Company.)

**Figure 21-22** Spherical bearing (cross-sectional view.) (Courtesy of NTN Bearing Corp.)

Pillow-block bearings provide both a roller bearing and a support housing. The housing can be used without a separate enclosure or machined opening in the mixer support. A pair of pillow-block bearings can be mounted directly on a support base for a simple mixer. Pillow-block bearings are often used for side-entering mixers.

*21-4.1.5 Reducer Ratings*. To provide a simple measure of drive capacity, a horsepower rating is used to describe gear capacity. Bearing ratings are given as

**Figure 21-23** Pillow-block bearing. (Courtesy of NTN Bearing Corp.)

time to failure for a percentage of the normal distribution of bearings. Because a simple power transmission rating also depends on fluctuations in load and number of hours of operation, ratings are usually converted into a service factor. A service factor of 1.0 normally defines the horsepower rating of a gear drive for a uniform load, operating 8 to 10 h per day. Since mixer applications are considered moderate shock loads, a 1.25 service factor is recommended. For heavy shock loads or moderate loads on mixers operating 24 h per day, a 1.5 service factor is needed for good gear life. With good maintenance and adequate service factors, mixer drives often last 10 years without major service, and some last 30 years or more.

The horsepower (wattage) ratings for gear drives involve several factors and are established by the AGMA. Gear reducer ratings take into account factors such as tooth shape, surface finish, and metal hardness. These factors, along with the housing characteristics, establish the wear, strength, and efficiency characteristics of the gearing. In effect, a gear reducer is rated for gear strength, gear wear, and heat dissipation, each represented by maximum horsepower. The smallest of the horsepower ratings is the nominal rating for the drive. Gearing for most mixer drives is designed to be wear limited rather than strength limited. Occasional process upsets can momentarily overload a gear drive. With a wear limit on the gears, the overload results in accelerated wear and is less likely to cause a gear to break (strength rating).

Like any mechanical component, bearings have a finite life and under load will fail with some expected variability or distribution. Bearing ratings are normally reported as number of hours to 10% failure, or $L_{10}$ life. The $L_{10}$ life for bearings may also be reported as a $B_{10}$ life. Typical bearing lives for mixer drives are long, 20 000 or 50 000 h. A mixer with severe service, or a long shaft, or high pressure usually means that the output shaft bearings have the shortest $L_{10}$ life. Because output shaft loads may affect the life of the mixer drive, each mixer may have a different life expectancy.

Power losses in a mixer drive may come from different sources and depend on the operating load. Although usually small compared with the rated load capacity, internal drive losses must be considered. Simply rotating the mixer drive will cause some friction in the gears and the bearings, plus some losses due to splashing of the lubricating oil. As the load on a drive increases, the loads on gears and bearings increase, reducing the efficiency from the no-load condition. Gear drive losses of 1 to 2% per reduction for helical and spiral-bevel gears are common. Losses for worm gears are higher: 4 to 10%, depending on the reduction ratio. These losses are based on the maximum drive capacity and may be a higher percentage of the power input if the drive is only partially loaded.

### 21-4.2 Belt Drives

Belt drives provide one of the simplest forms of speed reduction. A small diameter sheave (belt wheel) is attached to a high-speed shaft, typically from an electric motor (see Figure 21-24). The large sheave turns at the reduced shaft speed and can be attached directly to the mixer shaft. Typical speed reductions with a belt drive are limited to about 4:1. So one common use of belt drives is on side-entering mixers, where an output speed of about 300 rpm can be obtained from a 1200 rpm motor.

The most commonly used belts are basic V belts, but often, multiple belts are used on a sheave (Figure 21-25). Multiple belts are necessary to transmit the torque. Other belt types that can be used are ribbed V belts and cogged V belts. Ribbed V belts have multiple ridges that fit in multiple grooves on the sheaves, similar to multiple individual belts for greater torque capability. Cogged V belts have notches on the inside of the belt that fit teeth on the sheaves, to eliminate

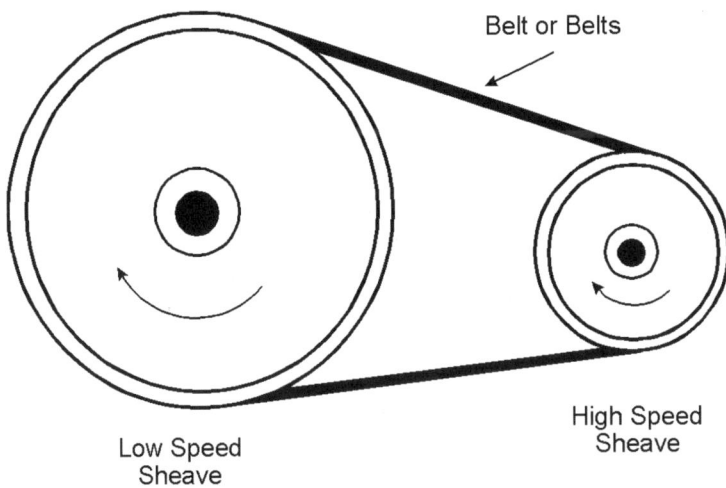

**Figure 21-24** Simple belt drive.

**1278** MECHANICAL DESIGN OF MIXING EQUIPMENT

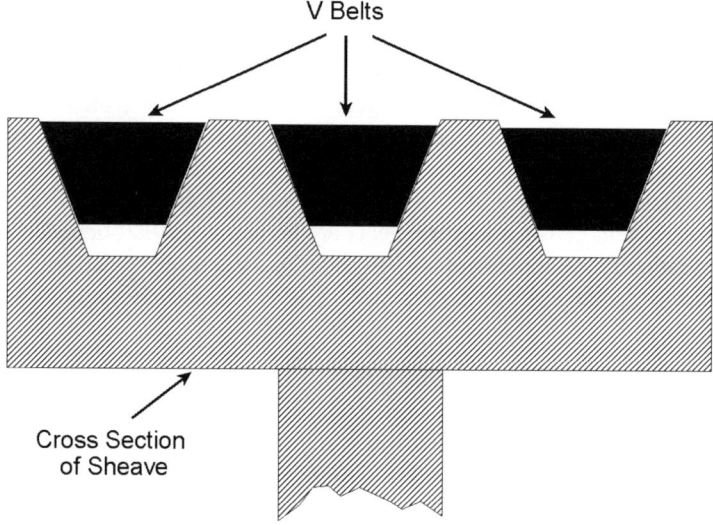

**Figure 21-25** Multiple V belt drive (cross-sectional view). (Courtesy of Chemineer.)

slippage. However, mixer applications take advantage of minor slippage to avoid equipment damage if an impeller stalls or strikes an object.

One advantage of a belt drive is the option of making minor changes in mixer speed. Gear reducers typically make significant changes is speed between standard reductions. Belt drives can have several different-sized sheaves that will fit the same shaft. Typically, the hubs and sheaves are sold separately, with the sheaves sized for speed change and the hubs sized for the shaft. By adjusting one or both sheave sizes, speed changes as small as 3% are possible. Because of a cubed relationship between speed and power, a 3% speed change will make almost a 10% change in power.

## 21-5 SHAFT SEALS

Shaft seals are necessary for tanks operating at elevated pressures, tanks containing hazardous, toxic, or noxious materials, and any mixer application where the shaft enters the tank below the liquid surface (i.e., side- and bottom-entering mixers). Several methods are available for sealing around the rotating mixer shaft. Although some methods are similar to seals used on pumps and other submerged equipment applications, the shaft deflection and runout for mixers are larger because of long shafts and large hydraulic loads. Thus, although similar in some ways to other applications, shaft seals for mixers are also unique.

### 21-5.1 Stuffing Box Seals

Perhaps the most versatile, yet simplest seals for a mixer are stuffing box seals. A stuffing box is essentially a housing around the shaft filled with a compression

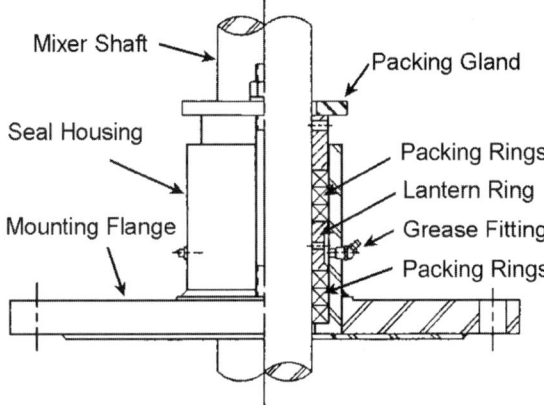

**Figure 21-26** Stuffing box shaft seal. (Courtesy of Chemineer.)

packing material to minimize leakage. The basic elements of a stuffing box seal are shown in Figure 21-26. The shaft enters the tank through an opening in the mounting flange, which is surrounded by a pipelike housing. Single or multiple rows of packing material are stacked in the housing and an adjustable plate and ring (gland and gland follower) compress the packing. As the packing rings are compressed, the material deforms and is pressed against the shaft and housing. Depending on the number of rings of packing, the shaft speed, tank pressure, and related parameters, some leakage will occur around the shaft. For seals above the liquid level the leakage will be vapor. For seals below the liquid level the leakage will be liquid, which can provide some lubrication, or possibly abrasion, depending on the tank contents.

Low pressure (less than about 30 psig [207 kPa]) applications of a stuffing box may require only one, two, or three rings of packing. Such applications can be as simple as a single ring of packing pressed into a machined opening in the face of the flange. The more rings, the more pressure capability and the less leakage.

At higher pressures (greater than about 30 psig [207 kPa]) more rings of packing are necessary. With more than three or four rings, the packing may deform enough that even compression is difficult. To help maintain uniform compression, a sleeve (lantern ring, Figure 21-26) is inserted in the housing between multiple layers of packing. The lantern ring serves two purposes, to keep the packing compressed evenly and to provide a means of lubricating the packing in contact with the shaft. With many rings of packing and even multiple lantern rings, high pressure capabilities can be obtained with a stuffing box, although some leakage will always occur.

All packing is made of a bulk material, such as braided fibers or metallic foils, which are coated or impregnated with a lubricant material. Some fibers used for braiding include acrylic, TFE (tetrafluoroethylene), Kevlar (aramid fiber), or graphite filament. Foil materials are aluminum or other alloys. The impregnating materials most commonly used are TFE and graphite. Some packings made of TFE or graphite fibers do not have a coating or impregnating (Crane, 1990).

Nearly all commercial packing materials are good for temperatures to 350°F [175°C] and many others will work to 500°F [260°C]. Some packings will work in applications where temperatures reach 1000°F [538°C] or higher. Because most materials used for packing are chemically inert, they resist a variety of acidic, caustic, oxidizing, reducing, or solvent materials. All packing material must be reviewed for chemical compatibility. Sometimes, fibers or graphite may contaminate the process.

Because stuffing box seals involve a rubbing contact with the rotating shaft, friction will cause heat and wear. To avoid problems and failures, the process of adjusting and maintaining stuffing box seals is important. Initial tightening of packing must be done gradually so that the rings of material deform uniformly and fit snugly around the shaft. Rapid or over-tightening can cause excessive friction and heat buildup. As the packing wears and compresses, the stuffing box must be tightened periodically to control leakage. When worn significantly, the packing must be replaced.

Because the shaft is also exposed to friction and wear in the region of the stuffing box, hardened coatings and sacrificial sleeves can be used to protect the shaft. In high speed or abrasive applications, protection of the shaft may be very important. Besides the maintenance cost of replacing a mixer shaft, a sudden failure may occur if shaft wear is sufficient to reduce the cross-sectional area and create a location of high stresses.

### 21-5.2 Mechanical Seals

Mechanical seals are an alternative and sophisticated means of sealing a mixer shaft. The rotating seal is formed between two seal elements, not the shaft and the seal as in a stuffing box. Mechanical seals allow for very tight control of the seal surfaces and materials. Such controls make possible high-pressure seals, even without external leakage.

The two working elements of a mechanical seal are called by various names. The part that rotates with the shaft may be called the *primary ring, rotating element*, or *washer*. The part that is stationary in the housing may be called the *mating ring, stationary element*, or *seat*. For convenience of nomenclature and understanding, we will use the terms *rotating* and *stationary* elements to be clear, even if they are not the most frequently used terms. One element, usually the rotating element, may be made of a wear (sacrificial) material, such as carbon. However, a hard material such as ceramic also can be used as the rotating element. The stationary element is usually a hard material such as tool steel or ceramic.

Seals are first described by the number of seal surfaces used to complete the seal assembly. Single seals have a single pair of seal elements, while double seals have two pairs. Further designations describe how the seals are mounted alone or together and how the seals move as wear takes place.

***21-5.2.1 Single Seals.*** Single seals are relatively simple (Figure 21-27) once the basic concept is understood. The stationary element, seal seat, is shown fitted into the mixer flange and sealed in place with an O-ring, which is a static seal. The rotating element is sealed against the shaft with another O-ring or wedge and pressed against the stationary element by a series of springs around the seal. Some seals have a single spring slightly larger than the shaft that holds the seal elements together.

A basic feature of mechanical seals is that the seal surface, between the rotating and stationary elements, is at right angles to the axis of rotation for the shaft. This arrangement allows the surfaces to be ground flat to within wavelengths of light and to remain smooth and flat even as the seal wears. The spring or springs hold the seal faces together and keep the seal closed, usually without outside adjustment.

Two other characteristics are commonly noted in mechanical seal terminology: the location of the seal and how the static seals move with seal wear. The seal shown in Figure 21-27 is called an *outside seal*, because the rotating seal components are located outside the tank. An *inside seal* would be inverted so that the rotating components would be inside the tank. All the component parts of an inside seal must be compatible with the process, but the pressure force tends to hold the seal faces closed. With an outside seal, the maximum containment pressure depends on the strength of the spring holding the seal closed. The seal in Figure 21-27 is also called a *pusher seal* because as the rotating element wears, it will push the O-ring between the seal element and the shaft. Sometimes, this motion may cause a leak to develop. A *nonpusher seal* uses a metal or elastomeric bellows to accommodate movement of the seal as wear occurs. The seal between the shaft and the rotating element, usually an O-ring, does not move once it is installed in a nonpusher seal.

Like split bearings, split seals can be replaced without sliding elements over the ends of the shaft. By reducing mixer disassembly for seal replacement, split seals have advantages. The rotating and stationary elements are usually carefully

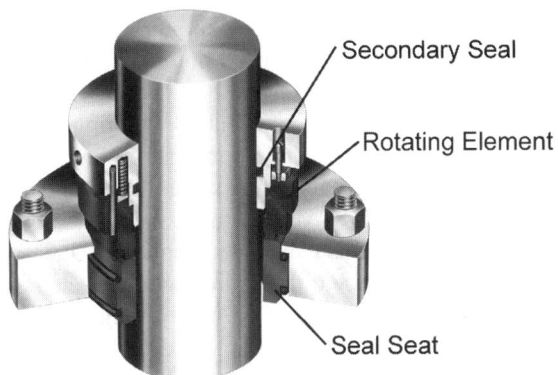

**Figure 21-27**  Single mechanical seal (cutaway view). (Courtesy of Flowserve.)

broken, so that reassembly puts the halves together at a labyrinth-like surface. Even O-rings must be made to fit around the shaft and be reconnected for a continuous loop. Other seal components, such as the spring assembly, may be bolted together.

**21-5.2.2 Double Seals.** Double seals, other than obviously having two sets of seal elements, can provide a positive seal that will not leak the tank contents into the surrounding environment. The way a nonleaking seal is formed is to pressurize a fluid between the seals such that the only direction of leakage is for seal fluid to leak outside into the surroundings or inside into the tank. This fluid is called a *barrier fluid*, because it forms a pressurized barrier between the tank contents and the surroundings. The barrier fluid also serves as a lubricant for the rotating seal surfaces.

The double mechanical seal, shown in Figure 21-28, is typical of seals used on mixers. The lower (inboard) seal is an outside seal, outside with respect to the tank contents, inside with respect to the seal fluid. The upper (outboard) seal is an inside seal, which keeps the seal fluid contained and under pressure. Pressure from the seal fluid helps keep the seal elements closed, and the differential pressure is held at some nearly constant value. Again, the barrier fluid pressure helps keep the seal elements closed. If a positive pressure is held in the tank and the barrier is at a higher pressure, the maximum pressure differential is always across the outboard seal. Because of the larger differential pressure, the outboard seal is more likely to wear first, so the first signs of leakage should be observed outside the tank. Maintenance of a double seal is mostly a matter of inspection. Inspection for leakage can be done by observing the seals visually or by checking the barrier fluid level or pressure. The barrier fluid also can be cooled externally to allow operation at higher process temperatures and to remove heat generated by friction at the seal faces.

Another common mixer feature seen in the double seal (Figure 21-28) is an additional bearing. The ball bearing between the double seals helps to reduce shaft deflections being transmitted to the seal elements. Typical pump seals are designed for 0.003 in. [0.08 mm] of runout. Mixer seals must tolerate 0.015 in. [0.38 mm] of runout. The larger the runout, the more rapid the seal wear. However, surface velocity for mixer seals may not be high, depending on shaft diameter and rotational speed.

Tandem seals are another seal arrangement for multiple seals. Tandem seals stack two or more seals together to handle portions of a high operating pressure. Much less common than double seals for mixer applications, tandem seals may be used with double seals to handle very high pressures.

Another characteristic of mechanical seals is balance. Unbalanced seals have the same shaft diameter for both the rotating and stationary elements, as shown on the left of Figure 21-29. The large area of pressure acting to close the unbalanced seal, shown as arrows at the bottom of the cross-section sketch, results in a large force on the seal faces. A balanced seal can reduce the pressure force acting to close the seal faces, by reducing the area on which the pressure acts, as

SHAFT SEALS **1283**

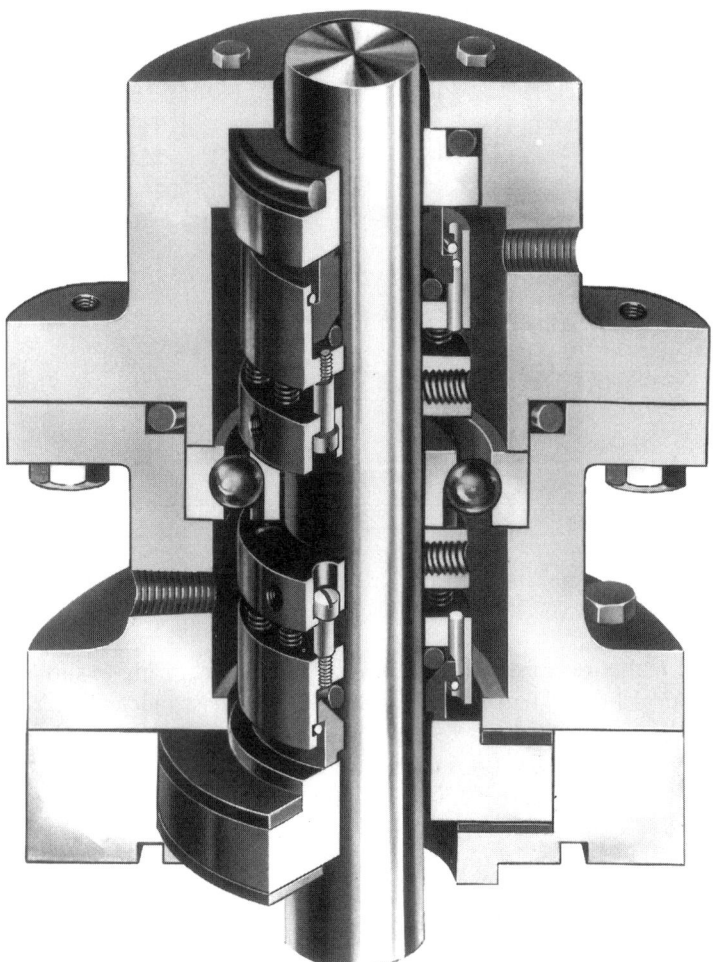

**Figure 21-28** Double mechanical seal (cutaway view). (Courtesy of Flowserve.)

shown on the right in Figure 21-29. To reduce the area a step is machined in the shaft so that the rotating element fits over a larger diameter shaft. A balanced seal arrangement requires additional machining operations, different-sized seal elements, and may be assembled from only one direction. However, because balanced seals will handle a higher operating pressure, some mixer applications require balanced seals.

Recent developments and adaptations of high-speed seal technology have brought gas barrier seals into mixer applications. With a gas such as nitrogen as the barrier fluid, process contamination is virtually eliminated, and a pressure monitor can be used to detect leaks. The problem is that gas is a poor lubricant for the seal surfaces. However, by machining swirls or pockets in the inside surface of the seal faces, gas pressure can be raised locally to lift the seal surfaces apart.

**1284** MECHANICAL DESIGN OF MIXING EQUIPMENT

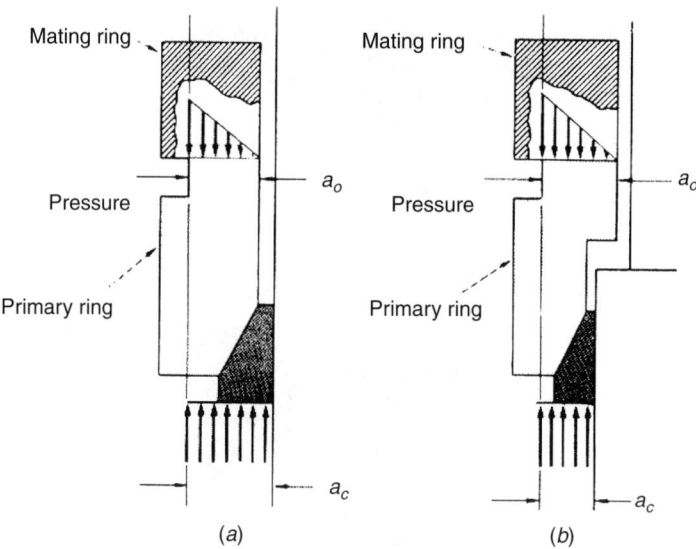

**Figure 21-29** (*a*) Unbalanced and (*b*) balanced mechanical seals.

This very small gap virtually eliminates surface contact. Without surface contact, the theoretical life of the seal faces is extended almost indefinitely. The reality is that gas barrier seals are more expensive initially than conventional lubricated seals, but may cost less to operate and maintain.

As mentioned in the general description of mechanical seals, the rotating (primary) element in a mechanical seal is typically made of various carbon formulations, tungsten carbide, or silicon carbide. The stationary (mating) element is usually tungsten carbide or silicon carbide. The hardware and springs inside the mechanical seal are usually 316 stainless steel but can be made of alloy materials such as Monel, 20 CB-3 stainless, Hastelloy B, or alloy C-276.

The real variety of materials comes in the elastomers used for the O-rings. The O-rings must be compatible with the process fluids and vapors and tolerant of the temperatures. Some standard and available options include: buna-N, fluoroelastomer, ethylene propylene copolymer, neoprene, Kalrez, silicone rubber, and FDA-approved materials. Instead of an O-ring seal between the rotating element and the shaft, some mechanical seals use a TFE wedge. With all the options and materials available, care must be exercised when ordering, replacing, and servicing mechanical seals.

**21-5.2.3  *Cartridge Seals*.** A cartridge seal places all of the mechanical seal components in a housing with the rotating element sealed against a sleeve. The sleeve slides over the shaft and the gap is sealed with O-rings. The advantage of a cartridge seal is ease of maintenance. Replacing a mechanical seal involves many small pieces and often breakable parts, such as graphite and ceramic elements. To replace a seal in place involves considerable maintenance skill, tools, and careful adjustments, often in uncomfortable or limited-access locations.

A cartridge seal can be removed as a major assembly and the individual seal parts replaced and adjusted on a bench in the maintenance facility. Once a double-seal cartridge is reassembled, the seal can be pressurized with fluid and checked for leaks before reinstalling the seal on the mixer. Cartridge seals may even be sent to the manufacturer or a service center for expert maintenance and assembly. Small cartridge seals could even be discarded.

### 21-5.3 Lip Seals

Lip seals are perhaps the simplest seals used in mixer service. The seal is formed by an elastomer material which fills the gap between the rotating shaft and the stationary flange. A typical lip seal is shown in Figure 21-30. The elastomer seal is held against the shaft by a small diameter spring. Although a lip seal can effectively seal the gap around the shaft, it cannot hold any appreciable pressure. Lip seals are typically used to keep dirt out of atmospheric tanks or to limit the free exchange of process vapors with the surroundings. Lip seals can also help hold elevated temperatures in a process tank. Lip seals are usually made of elastomeric materials similar to those used for O-rings in mechanical seals. The small spring is often made of stainless steel, to avoid attack by moisture and air.

### 21-5.4 Hydraulic Seals

Hydraulic seals are another type of simple seal used on mixers. Most hydraulic seals are used for vapor retention. A typical hydraulic seal is shown in Figure 21-31. An inverted cup is attached and sealed to the shaft. The cup runs inside a circular ring chamber welded to the mixer flange. The ring chamber

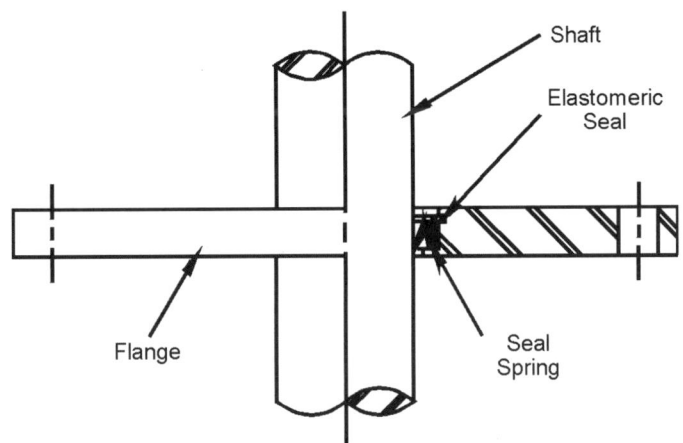

**Figure 21-30** Lip seal. (Courtesy of Chemineer.)

**1286** MECHANICAL DESIGN OF MIXING EQUIPMENT

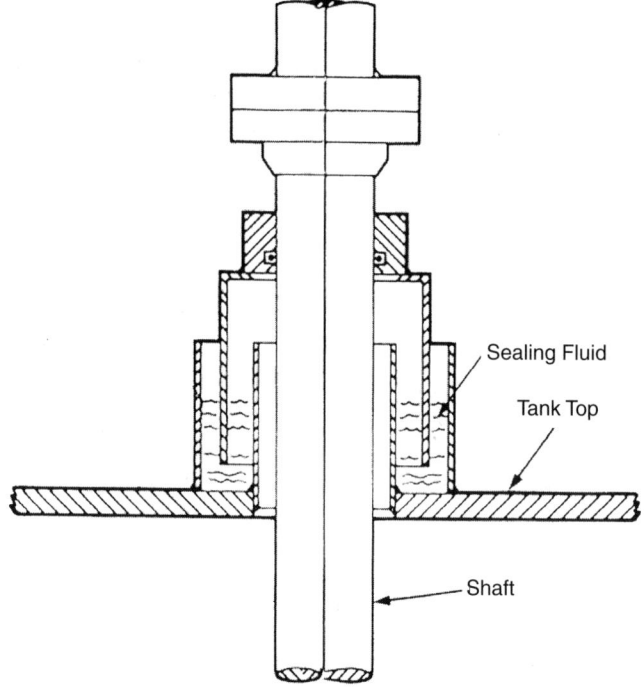

**Figure 21-31** Hydraulic seal. (Courtesy of Lightnin.)

is filled with a fluid, often simply water. The liquid forms a nearly frictionless barrier between the rotating shaft and the stationary flange.

The practical limit to a hydraulic seal is the hydrostatic head of the liquid. Effectively, the maximum pressure differential is only a few inches of water. The ability to maintain an effective vapor seal also depends on the solubility of the vapor in the seal liquid. To contain a vapor, the composition can be only slightly soluble in the seal fluid. Otherwise, it will dissolve in the seal fluid and revaporize outside the seal.

### 21-5.5 Magnetic Drives

Magnetic drives eliminate the problems of sealing a rotating shaft by using magnets to transmit torque from outside a vessel to inside the vessel. All the seals required for a magnetic drive are static seals and gaskets. A motor drives a rotating magnet outside a seal can, which turns a shaft inside the seal can by magnetic force. For small reactors, even at high pressures, the magnetic drives provide a simple and effective seal. The problems arise with larger drives. Magnets, even high-strength magnets, cannot transmit large torques, and the magnets need to be as close together as possible. To support the mixer shaft, the shaft bearings must be inside the vessel, which exposes them to the vapors or fluids from the process. This exposure can be a problem with corrosive materials. The expense of magnets is also a problem. Even small drives are expensive compared with typical mechanical seals. The biggest single advantage of a magnetic drive is the ability to handle high pressure without the leakage possibilities associated with a rotating seal.

## 21-6  SHAFT DESIGN

Shaft design must accommodate hydraulic and mechanical loads and must avoid vibration near the natural frequency. A typical overhung shaft arrangement with dimensional nomenclature is shown in Figure 21-32. Hydraulic loads on the shaft result from the torque required to turn the impeller(s) and random or systematic lateral hydraulic loads on the impeller(s). Other sections of the book describe methods for determining impeller power. Shaft design will use impeller power to calculate torque and hydraulic forces and thus size a shaft within allowable stress limits.

Natural frequency is the frequency of free vibration for the system. At the natural frequency an undamped system, one that continues to vibrate, with a single degree of freedom will oscillate after a momentary displacement. The operating speed of the shaft and impeller system must be sufficiently far from the system's natural frequency, often called the *critical speed*, to prevent undamped vibrations. If deflections caused by vibration become sufficiently large, the shaft could bend or break. Although torsional natural frequencies must be examined on very large mixers, in the following discussion we address only the lateral natural frequencies, which affect the design of all mixer shafts.

### 21-6.1  Designing an Appropriate Shaft

The steps necessary to design a mixer shaft first consider strength, then commercially available material, and finally, natural frequency. The following steps also consider alternatives if natural frequency problems are encountered.

1. Determine the material of construction and the allowable stresses for both combined shear and combined tensile (Table 21-3).
2. Calculate the minimum solid shaft size for an overhung shaft, which meets both shear [eq. (21-5)] and tensile [eq. (21-6)] stress limits. Then round up to the nearest $\frac{1}{2}$ in. increment (next-larger metric diameter) to obtain a size for commercially available bar stock.
3. For this standard shaft size, determine the natural frequency of the shaft and impeller system. If the system meets the natural frequency criterion, the design is complete.
4. If the shaft speed is near the natural frequency, increase the solid shaft size to the next $\frac{1}{2}$ in. larger, or metric equivalent. Then redo the critical speed calculation. Again, if the operating/critical speed ratio meets the natural

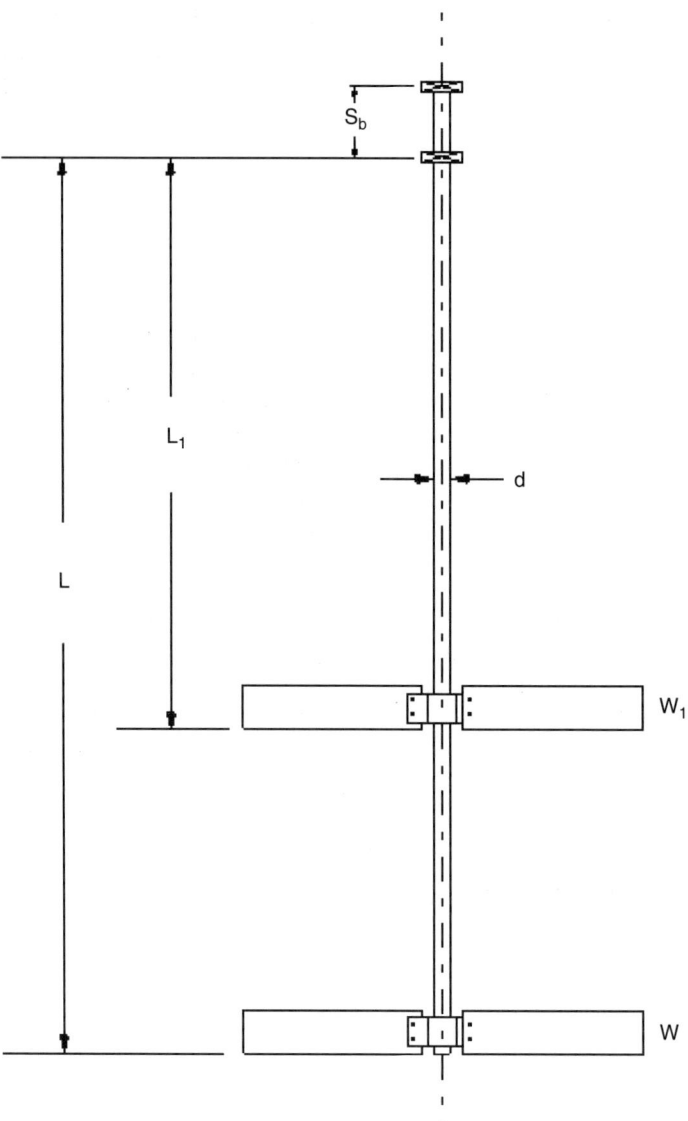

**Figure 21-32** Shaft and impeller schematic.

frequency criteria, design is complete. If necessary, repeat this process one additional $\frac{1}{2}$ in. increment. If more than 1 in. of additional diameter over the calculated diameter for strength is required to meet the natural frequency criterion, go to the next step.

5. Select a hollow shaft, usually a standard pipe size, that meets the mechanical stress requirements [eqs. (21-7) and (21-8)]. Compute the critical speed

and compare it with the natural frequency criteria. Typically, a hollow shaft can increase the critical speed about 20%. For strength the hollow shaft diameter is seldom more than twice the solid shaft diameter for equal strength. One drawback of hollow shafting is mounting impellers with adjustment for axial position.

6. If hollow shafting either cannot be made to work or is undesirable due to axial impeller adjustability, a foot (or steady) bearing will probably be required. Begin the design with the minimum shaft size for strength, and using the formulas for a shaft with a steady bearing to compute the critical speed (Section 21-6.4.6). Adjust the diameter upward in $\frac{1}{2}$ in. increments until the natural frequency criterion is met. Adding an inch or less to the minimum diameter for strength should satisfy natural frequency requirements.

In the following sections we present methods for calculating strength and natural frequency for different shaft types. All these methods make certain assumptions about the design. Other methods using other assumptions can be developed and used but usually give similar results.

### 21-6.2  Shaft Design for Strength

Computing shaft size for both allowable shear and tensile stress requires that the designer know the rotational speed of the mixer, plus the style, diameter, power, location, and service of each impeller. For overhung shafting the maximum torque will occur above the uppermost impeller. The maximum torque can be determined from the following equation:

U.S. Eng.

$$T_{Q(max)} = 63\,025 \frac{P}{N}$$

Metric

$$T_{Q(max)} = \frac{P}{2\pi N}$$

(21-2)

where $T_Q$ is torque [in.-lb$_f$] {N·m}, P is motor power [hp] {W}, and N is rotational speed [rpm] {rps}. To be sure that process upsets or changes do not exceed shaft design limits, the motor power is used instead of impeller power.

For design calculations, impeller power must be a calculated quantity, unless power has been measured on a previously built, identical mixer. Impeller power calculations based on empirical laboratory measurements can be used successfully for most mixer design. However, as a good design practice, total calculated impeller power should not be more than about 85 or 90% of motor power. Impeller power can be as little as 50% of motor power for a conservative design with uncertain process conditions.

In the following equation for bending moment, individual fractions of motor power are needed for each impeller, because the impellers are at different locations on the shaft. The following adjustment will give impeller power values that will sum to motor power:

$$P_i = P_{i_{calculated}} \frac{P_{motor}}{\sum_{i=1}^{n} P_{i_{calculated}}} \quad (21\text{-}3)$$

The maximum bending moment, $M_{max}$, for an overhung shaft is the sum of the products of the hydraulic forces and the distance from the individual impellers to the bottom bearing in the mixer drive (see Figure 21-32). The following expression computes an empirical hydraulic force related to the impeller torque acting as a load at a distance related to the impeller diameter.

U.S. Eng.

$$M_{max} = \sum_{i=1}^{n} \frac{19\,000 P_i L_i f_{H_i}}{ND_i}$$

Metric

$$M_{max} = \sum_{i=1}^{n} \frac{0.048 P_i L_i f_{H_i}}{ND_i} \quad (21\text{-}4)$$

where $M_{max}$ is the bending moment [in.-lb$_f$] {N · m}, $L_i$ the distance from the bottom drive bearing to the ith impeller location [in.] {m}, N the rotational speed [rpm] {rps}, and $D_i$ the diameter of the ith impeller [in.] {m}. The bending moment also depends on a hydraulic service factor, $f_H$, which is related to the impeller type and process operating conditions. Approximate hydraulic service factors for the various impellers and conditions can be found in Table 21-2.

Since the bending moment and the torque act simultaneously, these loads must be combined and resolved into a combined shear stress and a combined tensile stress acting on the shaft. The minimum shaft diameter for the allowable shear stress can be calculated as follows:

$$d_s = \left( \frac{16 \sqrt{T_{Q(max)}^2 + M_{max}^2}}{\pi \sigma_s} \right)^{1/3} \quad (21\text{-}5)$$

where $d_s$ is the minimum shaft diameter [in.] {m} for the shear stress limit, $\sigma_s$ [psi] {N/m$^2$}. Values for torque, $T_Q$, and bending moment, M, must be the appropriate values for the system of units [in.-lb$_f$] {N · m}. Some design shear stresses for common materials of construction are shown in Table 21-3.

# SHAFT DESIGN

**Table 21-2** Hydraulic Service Factors, $f_H$

| Condition | High Efficiency Impeller | 45° Pitched Four-Blade Impeller |
|---|---|---|
| Standard | 1.5 | 1.0 |
| Significant time at the liquid level | 2.5–3.5 | 2.0–3.0 |
| Operation in boiling systems | 2.0–3.0 | 1.5–2.5 |
| Operation in gas sparged systems | 2.5–3.5 | 2.0–3.0 |
| Large volume solid additions | 3.0–5.0 | 3.0–5.0 |
| Impacting of large solids | 5.0–7.0 | 5.0–7.0 |
| Startup in settled solids | 5.0–7.0 | 5.0–7.0 |
| Operation in a flow stream | 1.5–7.0 | 1.0–7.0 |

**Table 21-3** Allowable Stresses for Shaft and Blade Design

| Material | Shaft Design Tensile Stress [psi] | $\{N/m^2\} \times 10^6$ | Shaft Design Shear Stress [psi] | $\{N/m^2\} \times 10^6$ | Blade Design Stress [psi] | $\{N/m^2\} \times 10^6$ |
|---|---|---|---|---|---|---|
| Carbon steel | 9 000 | 62.1 | 5400 | 37.2 | 10 900 | 75.2 |
| Stainless steel 304 | 9 600 | 66.2 | 5800 | 40.0 | 11 600 | 80.0 |
| Stainless steel 304L | 8 400 | 57.9 | 5100 | 35.2 | 10 200 | 70.3 |
| Stainless steel 316 | 10 000 | 68.9 | 6000 | 41.4 | 12 100 | 83.4 |
| Stainless steel 316L | 8 700 | 60.0 | 5200 | 35.9 | 10 500 | 72.4 |
| Hastelloy C | 13 200 | 91.0 | 7900 | 54.5 | 15 900 | 109.6 |
| Hastelloy B | 14 300 | 98.6 | 8600 | 59.3 | 17 200 | 118.6 |
| Monel 400 | 9 200 | 63.4 | 5500 | 37.9 | 11 100 | 76.5 |
| Inconel 600 | 10 300 | 71.0 | 6200 | 42.7 | 12 400 | 85.5 |
| Nickel 200 | 7 300 | 50.3 | 4400 | 30.3 | 8 800 | 60.7 |
| Carpenter 20 | 11 100 | 76.5 | 6600 | 45.5 | 13 300 | 91.7 |

The minimum shaft diameter for the allowable tensile stress is calculated with a different equation:

$$d_t = \left[ \frac{16 \left( M_{max} + \sqrt{T_{Q(max)}^2 + M_{max}^2} \right)}{\pi \sigma_t} \right]^{1/3} \qquad (21\text{-}6)$$

where $d_t$ is the minimum shaft diameter [in.] {m} for the shear stress limit, $\sigma_s$ [psi] $\{N/m^2\}$. Values for torque, $T_Q$, and bending moment, M, must be the appropriate values for the system of units [in.-lb$_f$] {N · m}. Suggested tensile stresses for shaft design are shown in Table 21-3.

The minimum shaft diameter will be the greater of the two values calculated in eqs. (21-5) and (21-6). For practical purposes, most mixer shafts are made from bar stock, so standard sizes are usually available in $\frac{1}{2}$ or 1 in. increments or certain multiples of millimeters. For critical speed calculations, the next larger standard shaft diameter should be used.

Limits for shear and tensile stresses depend on shaft material, operating temperature, and chemical environment. Since nearly every chemical system is different, a review by a materials engineer should be made and appropriate allowable stresses established, especially for new or corrosive applications. Besides shaft strength, shaft straightness is important to avoid creating unnecessary loads and vibration. Typical shaft straightness for a mixer is 0.003 in./foot (0.25 mm/m).

### 21-6.3 Hollow Shaft

A hollow shaft, made from pipe, can increase the stiffness and reduce the weight (mass) of a mixer shaft in critical speed calculations. Such changes will increase the natural frequency and extend the allowable shaft length or operating speed. When determining the appropriate shaft size for the strength of a hollow shaft, begin with the dimensions for standard available pipe or tube. Then compute the shear and tensile stress values and compare them with the allowable values. The equations for combined shear and tensile limits in hollow shafts are, respectively,

$$\sigma_s = \frac{16\sqrt{T_{Q(max)}^2 + M_{max}^2}}{\pi} \frac{d_o}{d_o^4 - d_i^4} \quad (21\text{-}7)$$

where $\sigma_s$ is the shear stress [psi] {N/m$^2$}, $d_o$ the outside diameter [in.] {m}, and $d_i$ the inside diameter of the pipe [in.] {m}. Because nominal pipe dimensions have tolerances, the minimum wall thickness should be used to determine the inside diameter:

$$\sigma_t = \frac{16\left(M_{max} + \sqrt{T_{Q(max)}^2 + M_{max}^2}\right)}{\pi} \frac{d_o}{d_o^4 - d_i^4} \quad (21\text{-}8)$$

where $\sigma_t$ is tensile stress [psi] {N/m$^2$}. The smallest pipe dimensions that keep the shear stress, $\sigma_s$, and the tensile stress, $\sigma_t$, below allowable limits is probably a good start for further calculations.

In many tall tanks, designing a mixer with an overhung shaft is not economically practical because of shaft strength, or natural frequency, or both. Often, a lower bearing, called a *steady bearing* or *foot bearing*, is used to provide a more economical design. The steady bearing is typically attached to the bottom of the tank, as shown in Figure 21-33, or to the bottom nozzle. Such systems are effectively triple-bearing systems: two bearings in the gearbox, with the steady bearing

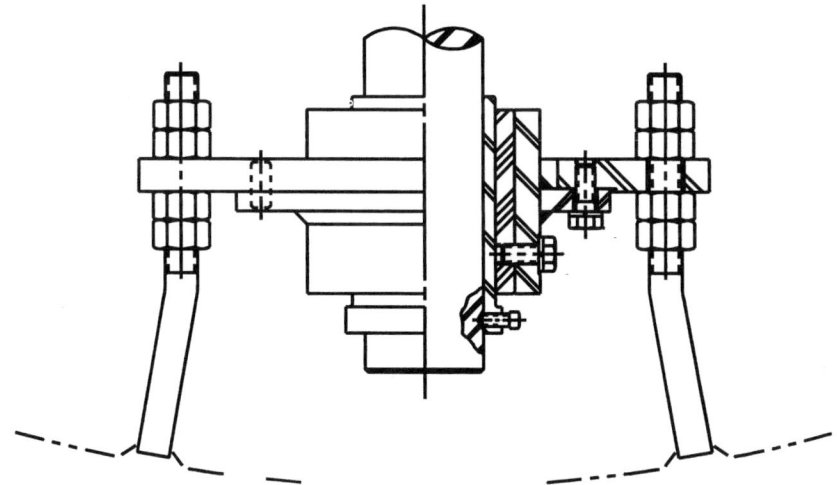

**Figure 21-33** Tripod steady bearing detail. (Courtesy of Chemineer.)

being the third bearing. Basic calculations for natural frequency typically consider only the lower drive bearing with the dimensions and nomenclature shown in Figure 21-34. Dynamic calculations can consider all three bearings.

The process of determining the location of maximum stress for a shaft using a steady bearing can be tedious. Usually, the location is just above the upper impeller. A conservative approach would be to assume prism supports at the ends of the shaft. Then apply hydraulic loads to create moments and sum them at the points shown in Figure 21-35 for the shaft and impeller system shown in Figure 21-34.

### 21-6.4 Natural Frequency

Natural frequency is a dynamic characteristic of a mechanical system. Of primary concern to mixer design is the first lateral natural frequency, which is the lowest frequency at which a shaft will vibrate as a function of length and mass. The first lateral natural frequency is analogous to the vibration of a tuning fork, except on a larger scale.

The concern about natural frequency is that an excitation such as mixer operating speed could cause undamped vibrations. Undamped vibrations occur when no resisting forces are present to diminish the amplitude of vibration. Such vibrations could result in sudden and catastrophic failure of the mixer shaft. The most dangerous conditions usually occur when the mixer is operated in air. Large mixers normally operate below the first natural frequency. Small portable mixers, which accelerate quickly, often operate above the first natural frequency. In either case, operating at or near the natural frequency must be avoided for both mechanical reliability and safety.

**1294** MECHANICAL DESIGN OF MIXING EQUIPMENT

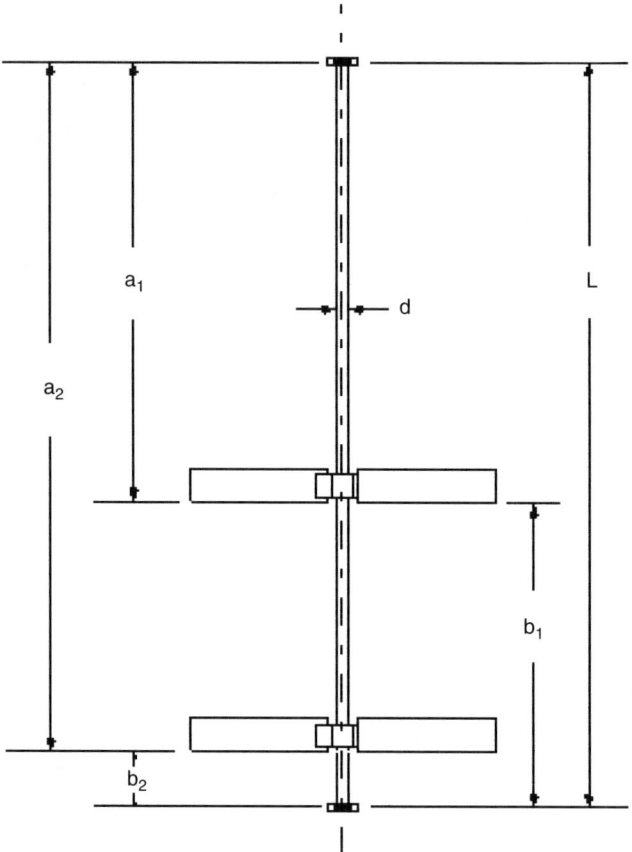

**Figure 21-34** Shaft and impellers with steady bearing.

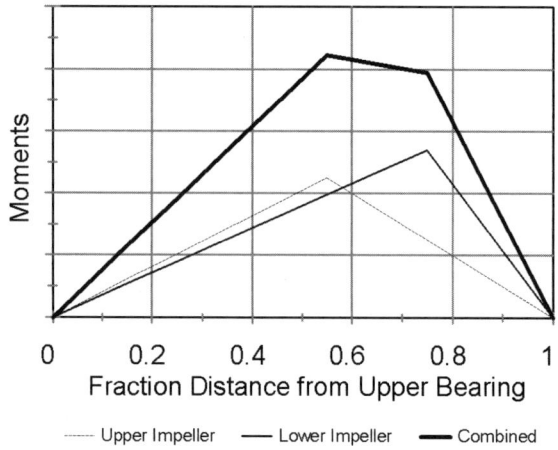

**Figure 21-35** Moment diagram for shaft and impeller system with steady bearing.

The standard vibration equation applies to a mixer shaft:

$$m\frac{d^2x}{dt^2} + c_v\frac{dx}{dt} + kx = f(t) \tag{21-9}$$

where m is the mass, $c_v$ the damping coefficient, k the effective spring constant for the system, and f(t) some type of forcing function. The forcing function for mixers can be approximated by a sine or cosine function. A mixer design must address several issues. The damping coefficient is seldom known to any degree of accuracy because it depends on the material being mixed, the type and number of impellers, and the size of the impeller compared with the shaft diameter. To simplify the solution, the effect of damping is generally represented as the ratio of the damping coefficient to the critical damping coefficient, $c_c$. The critical damping coefficient is the minimum value for the coefficient, $c_v$, in eq. (21-9), that results in nonperiodic motion:

$$\delta = c_v/c_c \quad \text{where} \quad c_c = 2\sqrt{km} \tag{21-10}$$

If more energy is added to a system than the amount dissipated through damping, the amplitude of vibration will increase. If the energy addition continues, the amplitude of vibration can exceed the deflection that will bend the shaft. The amplification factor depends on the proximity of the operating speed to the natural frequency. This relationship is shown in Figure 21-36.

Transmissibility is also often called the *force magnification factor*. Any force applied to a shaft under dynamic conditions will be amplified by this magnification factor. A side load of 100 units at rest, for a damping ratio of 0.1, will behave as a 257 unit side load when $N/N_c = 0.8$ and as a 388 unit side load when $N/N_c = 0.9$. Most mixer manufacturers use design stress limits based on an allowable approach to the first natural frequency, $N_c$. The worst-case scenario

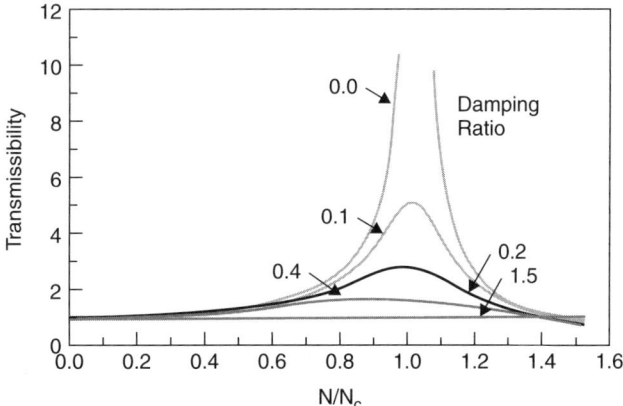

**Figure 21-36** Transmissibility for various damping ratios.

is to assume that no damping is present, $\delta = 0$. This assumption ensures that even if the mixer is operated in a vessel without liquid present (no damping), the shaft and impeller system will remain stable (i.e., will not cause deflections that could bend or break the shaft).

The other key design assumption is that the support stiffness is sufficiently large that the overall stiffness, k, is controlled only by the shaft stiffness. With a stiff support the natural frequency depends only on the shaft stiffness and associated mass. Mixer manufacturers generally assume that the mixer will be mounted on a structure where a small change in stiffness does not significantly affect the natural frequency.

Most structural engineers design primarily for strength. However, to mount mixers properly, stiffness (resistance to deflection) must be considered. Supports with adequate strength can experience noticeable deflection or movement with the dynamic load from a mixer. In most industrial applications stress levels are much less than allowable when appropriate stiffness is provided. High-pressure applications are the exception, where the structure to hold the pressure provides adequate stiffness.

The general rule used to design a mixer shaft and impeller systems is to keep operating speed 20% away from a critical speed:

$$0.8 N_c \not< N \not< 1.2 N_c \tag{21-11}$$

This rule applies to the first, second, and third natural frequencies. However, higher-order natural frequencies are seldom encountered in mixer applications.

Large mixers running at less than 150 rpm usually operate below the first critical speed. Small mixers operating above 250 rpm usually operate between first and second critical, $1.2 N_c$ to $0.8 N_{c2}$, where $N_{c2}$ is the second lateral natural frequency. Other frequencies, such as a blade-passing frequency, four times the operating speed for a four-blade impeller with four baffles, can cause mechanical excitations. Structural vibrations at certain fractions of operating speed can also contribute to natural frequency problems.

### 21-6.4.1 Using Stabilizers on Impellers to Improve Damping. 
Most mixing impellers do not need to be stabilized. The idea of stabilizers is to improve hydraulic damping, thus reducing deflection caused by imbalanced loads and consequently reducing stresses on the shaft. As a general rule of thumb, stabilizers are not required for impellers whose diameters exceed the shaft diameter by a factor of 10 or more. Shaft and impeller systems with impeller diameters greater than 10 times the shaft diameter have damping ratios of 0.4 or less, which is 40% of critical damping (Figure 21-36). Many are over critically damped. The only exception to this general rule is that mixed-flow impellers, pitched blade turbines, benefit from stabilizers when operating at or near the liquid level. Stabilizers on pitched blade turbines can reduce imbalanced hydraulic loads caused by splashing and surging on the liquid surface. A typical pitched blade turbine with stabilizers is shown in Figure 21-37. The stabilizers are the vertical fins mounted

SHAFT DESIGN **1297**

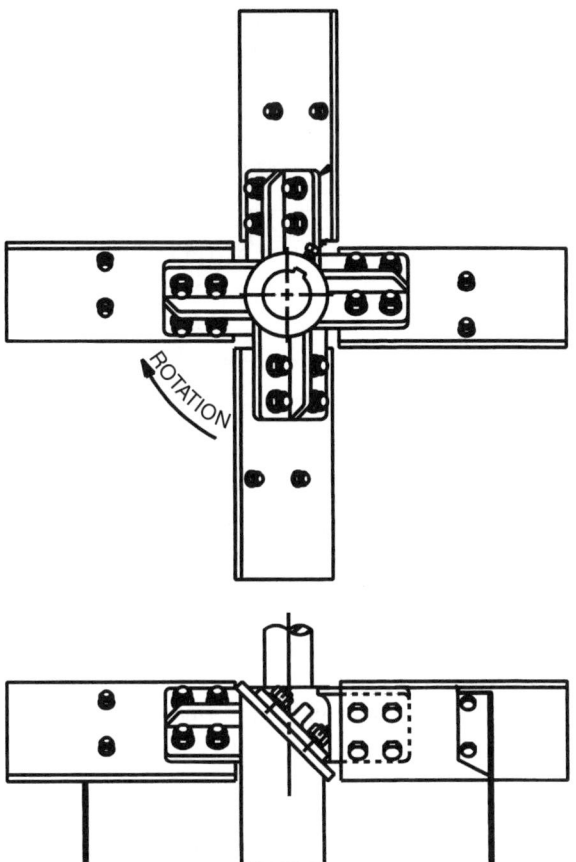

**Figure 21-37** Pitched blade turbine with bolted blades and stabilizers. (Courtesy of Chemineer.)

on the lower side of each blade. Shaft designs can tolerate operation at the liquid level for brief periods without the use of stabilizers, provided that appropriate hydraulic service factors are used in the bending moment calculation, eq. (21-4) and Table 21-2.

***21-6.4.2 Static Analysis for Natural Frequency of an Overhung Shaft.*** The elements that determine the lateral natural frequency are the magnitudes and locations of concentrated and distributed masses, the tensile modulus of elasticity of the material, and the moment of inertia of the shaft. Ramsey and Zoller (1976) presented the basic elements of natural frequency for a shaft and impeller system like the one shown in Figure 21-32. That method uses a lumped mass, static technique for computing the critical speed of a shaft and impeller system. The mass of the individual impellers and the distributed mass of the shaft is lumped into a single mass at the end of

**1298** MECHANICAL DESIGN OF MIXING EQUIPMENT

the shaft. The following equation estimates the first lateral natural frequency, or critical speed $N_c$ [rpm], of a top-entering mixer with a constant diameter overhung shaft:

U.S. Eng.

$$N_c = \frac{37.8 d^2 \sqrt{\frac{E_m}{\rho_m}}}{L\sqrt{L+S_b}\sqrt{W_e + \frac{wL}{4}}}$$

(21-12)

Metric

$$N_c = \frac{5.33 d^2 \sqrt{\frac{E_m}{\rho_m}}}{L\sqrt{L+S_b}\sqrt{W_e + \frac{wL}{4}}}$$

where $N_c$ is the critical speed [rpm] {rps}, d the shaft diameter [in.] {m}, $E_m$ the modulus of elasticity [psi] {N/m$^2$}, $\rho_m$ the density of the metal [lb$_m$/in$^3$] {kg/m$^3$} (see Table 21-4 for typical metal properties), L the shaft length [in.] {m}, $S_b$ the bearing spacing supporting the shaft [in.] {m}, $W_e$ the equivalent weight (mass) of the impellers [lb$_m$] {kg} at the bottom of the shaft, and w the specific weight (mass) of the shaft [lb$_m$/in.] {kg/m}.

$W_e$ in eq. (21-12) is the equivalent weight (mass) of each impeller resolved to the bottom of the shaft, which is defined as

$$W_e = \sum_{i=1}^{n} W_i \left(\frac{L_i}{L}\right)^3$$

(21-13)

where $W_i$ is the weight (mass) of the individual impellers [lb$_m$] {kg}, $L_i$ the shaft length to each impeller, and L the total shaft length [in.] {m}. A single impeller

**Table 21-4** Metal Properties for Natural Frequency Calculations

| Metal Type | Modulus of Elasticity, $E_m$ | | Density, $\rho_m$ | |
|---|---|---|---|---|
| | [psi] × 10$^6$ | {N/m$^2$} × 10$^{12}$ | [lb$_m$/in$^3$] | {kg/m$^3$} |
| Carbon steel | 29.8 | 0.205 | 0.283 | 7833 |
| Stainless steel 304/316 | 28.6 | 0.197 | 0.290 | 8027 |
| Hastelloy C | 30.9 | 0.213 | 0.323 | 8941 |
| Hastelloy B | 30.8 | 0.212 | 0.334 | 9245 |
| Monel 400 | 26.0 | 0.179 | 0.319 | 8830 |
| Inconel 600 | 31.0 | 0.214 | 0.304 | 8415 |
| Nickel 200 | 29.7 | 0.205 | 0.322 | 8913 |
| Carpenter 20 | 28.0 | 0.193 | 0.289 | 7999 |

at the bottom of the shaft results in an equivalent weight (mass) equal to the actual impeller weight (mass). In the following section we describe a method for estimating the weight of typical industrial impellers.

### 21-6.4.3 Estimating the Weight of Impellers.
Impeller weights, even for the same style of impeller, can vary from manufacturer to manufacturer, so the following weight calculations are only approximate. Small impellers, typically less than 24 in. in diameter, for portable and top-entering mixers (3 hp or less) are usually hydrofoil-style impellers. Most of these impellers are single-piece castings or welded fabrications. Some small impellers, less than 10 in. in diameter, can be marine propeller castings. Approximate weights for these typical impellers are shown in Table 21-5. Use only one method to estimate impeller weight, depending on whether the impellers are single-piece or bolted-blade designs.

The portable and small top-entering mixers that commonly use single-piece impellers often operate above first critical speed and may not have rigid mountings. Operation above first critical speed is not a problem if the shaft speed is more than 20% greater than first critical and the mixer accelerates quickly through first critical. Problems can develop when variable speed drives, electrical

**Table 21-5** Approximate Weights for Small Impellers

| Impeller Diameter [in.] | Propeller Weight [lb] | Hydrofoil Weight [lb] | Impeller Diameter {mm} | Propeller Weight {kg} | Hydrofoil Weight {kg} |
|---|---|---|---|---|---|
| 2.5 | 0.3 | 0.2 | 60 | 0.1 | 0.1 |
| 3.0 | 0.4 | 0.3 | 75 | 0.2 | 0.1 |
| 3.5 | 0.5 | 0.4 | 90 | 0.2 | 0.2 |
| 4.0 | 0.6 | 0.5 | 100 | 0.3 | 0.2 |
| 4.5 | 0.8 | 0.5 | 125 | 0.5 | 0.3 |
| 5.0 | 1.0 | 0.6 | 150 | 0.9 | 0.3 |
| 5.5 | 1.5 | 0.6 | 175 | 1.4 | 0.4 |
| 6.0 | 2.0 | 0.7 | 200 | 1.8 | 0.4 |
| 6.5 | 2.5 | 0.7 | 225 | 2.1 | 0.5 |
| 7.0 | 3.0 | 0.8 | 250 | 2.3 | 0.6 |
| 8.0 | 4.0 | 0.8 | 275 |  | 0.7 |
| 10 | 5.0 | 0.9 | 300 |  | 0.9 |
| 11 |  | 1.0 | 325 |  | 1.1 |
| 12 |  | 2.0 | 350 |  | 1.4 |
| 13 |  | 2.5 | 375 |  | 1.8 |
| 14 |  | 3.0 | 400 |  | 2.3 |
| 15 |  | 4.0 | 450 |  | 3.2 |
| 16 |  | 5.0 | 500 |  | 4.1 |
| 17 |  | 6.0 | 550 |  | 5.0 |
| 18 |  | 7.0 | 600 |  | 5.4 |
| 20 |  | 9.0 | 650 |  | 6.8 |

or air driven, are used. Variable speed drives may allow the mixer to operate, at least temporarily, at the critical speed, which could cause mechanical failure or even personal injury. Drives must be set or operators trained not to operate the mixer when large vibrations occur near first critical. Nonrigid mountings such as portable mixer clamps attached to the side of thin-walled tanks, will reduce the first natural frequency. The calculations assume rigid mountings, so actual critical speeds may be less than those calculated, depending on the mounting. Second natural frequencies are typically more than two or three times the first natural frequency and rarely cause problems for mixer shafts.

The two most common types of bolted-blade impellers used on turbine-style mixers today are the hydrofoil type (Figure 21-38), and the 45°C pitched blade type (Figure 21-37). These impellers, mostly larger than about 15 in. in diameter, are made of separate blades. The blades are metal plates, which can be shaped, rolled, or bent, bolted to a cast or fabricated hub. The hub is keyed and setscrewed to the shaft, as shown in Figure 21-39. To calculate the impeller weight, the weight of the hub found in Table 21-6 must be added to the weight, of the blades calculated by eq. (21-14) or (21-15).

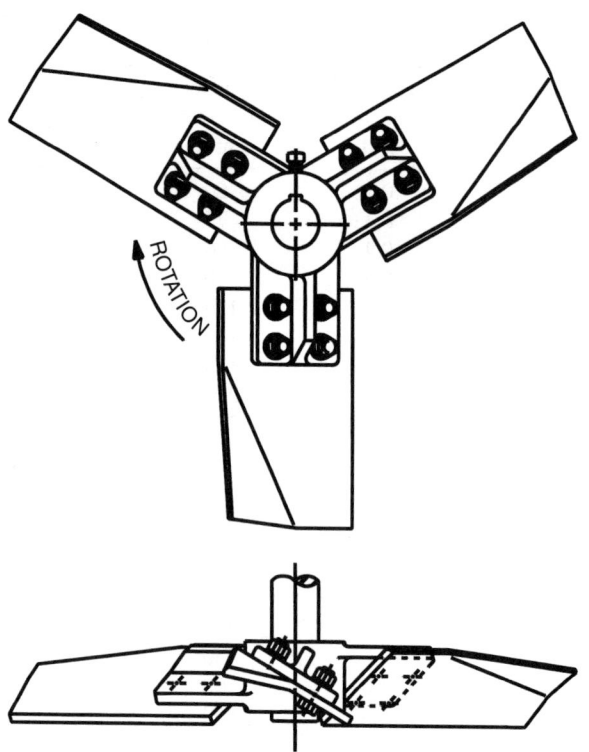

**Figure 21-38** Hydrofoil impeller with bolted blades. (Courtesy of Chemineer.)

SHAFT DESIGN **1301**

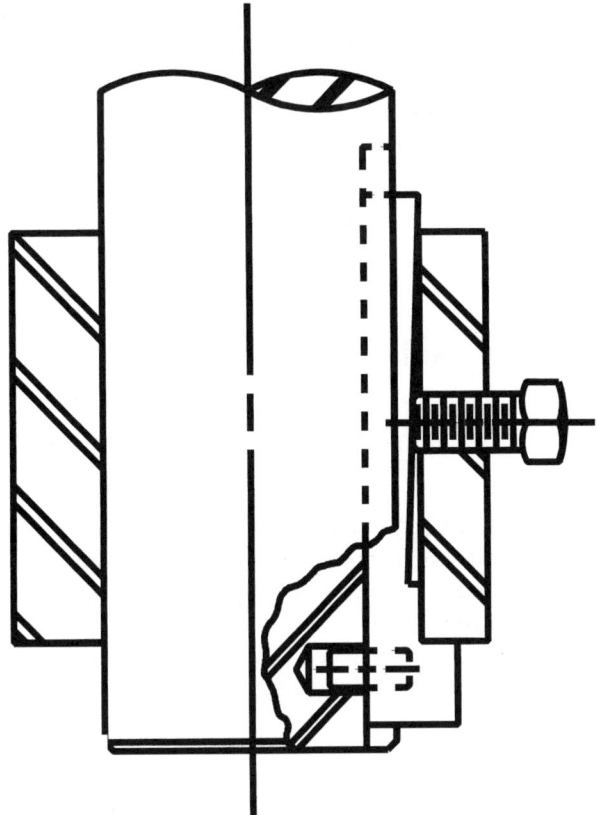

**Figure 21-39** Impeller hub with hook key detail. (Courtesy of Chemineer).

**Table 21-6** Impeller Hub Weights

| Shaft Diameter [in.] | Hydrofoil Hub Weight [lb] | 45° Pitched Four-Blade Hub Weight [lb] | Shaft Diameter {mm} | Hydrofoil Hub Weight {kg} | 45° Pitched Four-Blade Hub Weight {kg} |
|---|---|---|---|---|---|
| 1.5 | 12 | 10 | 40 | 6 | 5 |
| 2 | 24 | 20 | 50 | 11 | 9 |
| 2.5 | 30 | 25 | 60 | 13 | 11 |
| 3 | 48 | 40 | 75 | 21 | 18 |
| 3.5 | 60 | 50 | 90 | 28 | 13 |
| 4 | 73 | 60 | 100 | 32 | 27 |
| 4.5 | 91 | 75 | 110 | 39 | 32 |
| 5 | 121 | 100 | 120 | 47 | 39 |
| 5.5 | 151 | 125 | 140 | 69 | 57 |
| 6 | 182 | 150 | 160 | 91 | 75 |
| 7 | 242 | 200 | 180 | 112 | 93 |
| 8 | 302 | 250 | 200 | 134 | 111 |

Although hydrofoil impellers have only three blades compared with four blades for pitched blade turbines, hub weights for hydrofoil impellers are often greater because the shallow angle of the blades creates greater bending moments on the stub blades. The total weight of three hydrofoil blades for impellers from 15 to 90 in. in diameter can be estimated by the following equation:

U.S. Eng.

$$W_{b_{hydrofoil}} = \sqrt{\frac{0.50 D^3 P_i}{N}}$$

Metric

$$W_{b_{hydrofoil}} = \sqrt{\frac{0.14 D^3 P_i}{N}}$$

(21-14)

where $W_b$ is the weight (mass) of the blades [lb$_m$] {kg}, D the impeller diameter [in.] {m}, $P_i$ the power drawn by the impeller [hp] {W} (which is usually adjusted to a fraction of the motor horsepower to handle possible upset conditions), and N the rotational speed [rpm] {rps}.

The total weight of four 45° pitched blades for turbines from 15 to 90 in. in diameter can be estimated by the following equation:

U.S. Eng.

$$W_{b_{pitched}} = \sqrt{\frac{0.30 D^3 P_i}{N}}$$

Metric

$$W_{b_{pitched}} = \sqrt{\frac{0.084 D^3 P_i}{N}}$$

(21-15)

where $W_b$ is the weight (mass) of the blades [lb$_m$] {kg}, D the impeller diameter [inch] {m}, $P_i$ the power drawn by the impeller [hp] {W} (which is usually adjusted to a fraction of the motor horsepower to handle possible upset conditions), and N is the rotational speed [rpm] {rps}. Impeller power and speed enter the estimate because mechanical loads related to torque determine the blade thickness. Due to the wide variety of hydrofoil impellers, the accuracy of the estimated impeller weight is only about ±25% for narrow-blade hydrofoils (i.e., a blade width/impeller diameter ratios of about one-sixth). Wide-blade hydrofoil impeller blades can weigh two to three times the estimate from eq. (21-14). The weight for four-bladed 45° pitched turbines can be estimated more accurately because typical blade widths are about one-fifth the impeller diameter. The accuracy of the estimated pitched blade turbine weight is about ±15%. Adjustment for impellers with as few as two blades and as many as six blades can be made from these estimates.

***21-6.4.4 Shaft Couplings.*** Since most mixer shafts are long compared with the drive and impeller, the shafts are shipped separately and mounted at the time of installation. Large mixers may require several sections of shaft for shipping or installation. Wherever the shaft is attached to the drive or another section of shaft, a coupling is required. On small mixers the shaft coupling can be as simple as a collar with setscrews. On large mixers, shafts require stronger, more sophisticated couplings, to transmit loads and maintain alignment. A basic welded coupling is shown in Figure 21-40. If an impeller hub, shaft seal, or other assembly feature must be slid over the end of a shaft section, the coupling must be removable. A removable coupling with taper bore connections is shown in Figure 21-41. One reason for this focus on shaft couplings is that large couplings can add weight to the shaft. The additional weight will add to the equivalent weight and reduce the natural frequency. If couplings appear large with the potential for significant

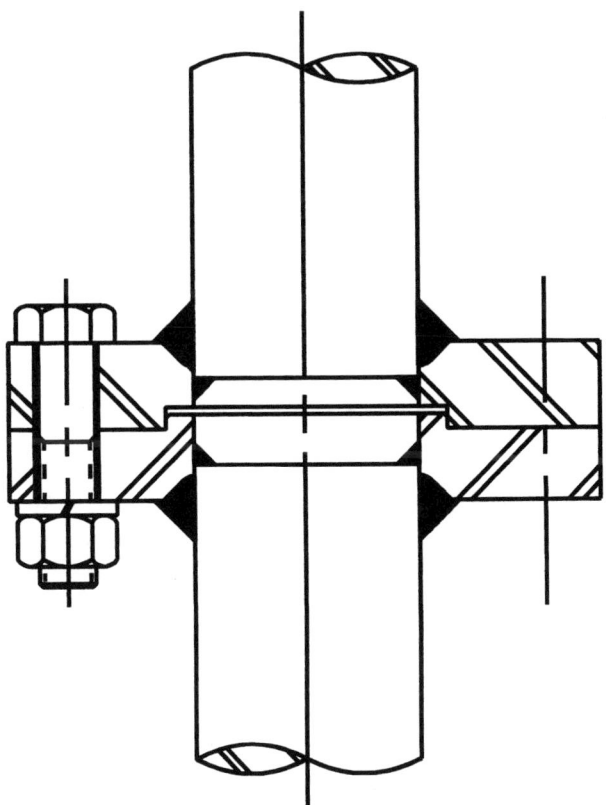

**Figure 21-40** Welded shaft coupling. (Courtesy of Chemineer.)

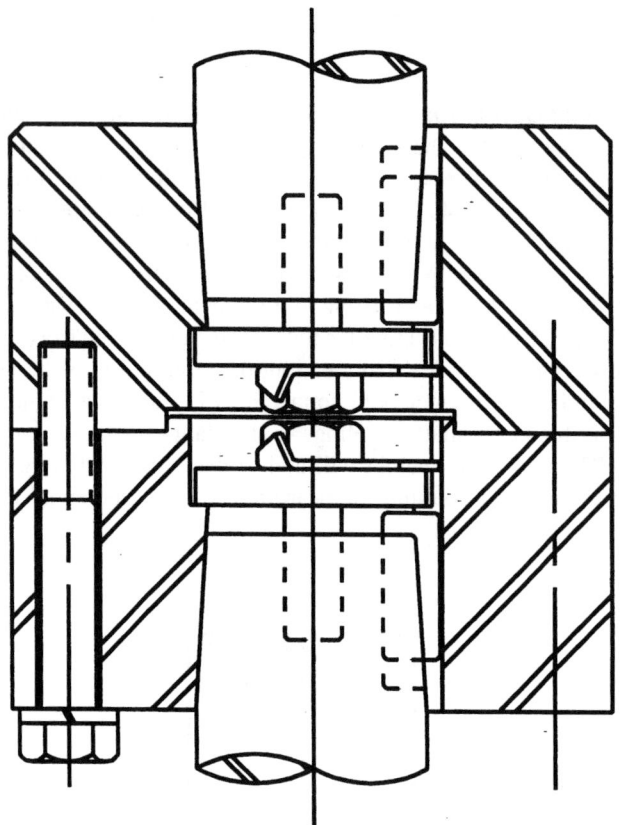

**Figure 21-41** Removable shaft coupling. (Courtesy of Chemineer.)

weight, they should be included as if they were an impeller in the equivalent weight calculations.

### 21-6.4.5 Static Analysis for Natural Frequency of a Steady Bearing Shaft. 
Although not an exact solution, the following equation can be used to estimate the first natural lateral frequency of a shaft and impeller system using a steady bearing with a stiffness over $5.00 \times 10^4$ lb$_f$/in. ($8.8 \times 10^{16}$ N/m). Remember that all these calculations for natural frequency assume that the mountings are rigid. The shaft, impellers, bearings, and dimensional nomenclature for a typical steady bearing design are shown in Figure 21-34.

To calculate natural frequency, $N_c$ [rpm] {rps}, for a shaft with a steady bearing, use shaft diameter, d [in.] {m}, shaft length, L [in.] {m}, shaft weight (mass), w [lb$_m$/in.] {kg/m}, equivalent weight (mass), $W_e$ [lb$_m$] {kg}, modulus of elasticity, $E_m$ [psi] {N/m$^2$}, material density, $\rho_m$ [lb$_m$/in.$^3$] {kg/m$^3$}, as shown in the following equations.

U.S. Eng.

$$X_1 = 2.44 \times 10^6 \frac{d^2}{L^2\sqrt{w}}$$

$$X_2 = 1.55 \times 10^6 \frac{d^2}{L^{3/2}\sqrt{W_e}}$$

$$N_c = 97.5 \times 10^{-6} \frac{X_1 X_2 \sqrt{\frac{E_m}{\rho_m}}}{\sqrt{X_1^2 + X_2^2}}$$

(21-16)

Metric

$$X_1 = 10.3 \times 10^6 \frac{d^2}{L^2\sqrt{w}}$$

$$X_2 = 6.55 \times 10^6 \frac{d^2}{L^{3/2}\sqrt{W_e}}$$

$$N_c = 3.25 \times 10^{-6} \frac{X_1 X_2 \sqrt{\frac{E_m}{\rho_m}}}{\sqrt{X_1^2 + X_2^2}}$$

The equivalent weight (mass), $W_e$, for this system is given by

$$W_e = \frac{16}{L^4} \sum_{i=1}^{n} (a_i b_i)^2 W_i \qquad (21\text{-}17)$$

Refer to Figure 21-34 for definitions of a, b, and L. If all the lengths are in the same units, the weights will be in the same units, since the lengths are essentially an adjustment factor.

**21-6.4.6 Static Analysis for Natural Frequency of a Pipe Shaft.** When a hollow shaft, such as a pipe or tube, is used, replace $d^2$ in eq. (21-16) or (21-17) by

$$\sqrt{d_o^4 - d_i^4} \qquad (21\text{-}18)$$

Be sure to use the same length dimensions in the respective calculations. Because of allowable tolerances on pipe dimensions, be sure to use the actual outside diameter and the maximum inside diameter or minimum wall thickness for design calculations.

**21-6.4.7 Dynamic Analysis for Natural Frequency.** Static models generally assume an infinitely stiff structure to which the mixer is mounted. When the support is not sufficiently rigid, the static model is compromised, and accurate predictions of natural frequency cannot be made. When a shaft and impeller system cannot be considered rigid in the operating frequency range, distributed properties need to be taken into account. In the following analysis, the shaft supports (drive bearings and steady bearings) are treated as *springs* to account for the lack of stiffness of the supports. To calculate natural frequency for a mixer shaft, the properties of a distributed and lumped system must be combined (Fasano et al., 1995). The transfer matrix method is an example of such a technique.

The transfer matrix (Pestel and Leckie, 1995) method can be used to calculate the critical speed and dynamic response of the shaft design. The matrix is composed of the mass and elastic characteristics of each span. The matrix is then multiplied by the deflection, slope, bending moment, and shear force at the position on one end of the span to calculate the deflection, slope, bending moment, and shear force at the position on the other end of the span. This calculation for each span is shown below in matrix form. Each span i has position i-1 on one end of the span and position i on the other end.

$$\begin{Bmatrix} -w_d \\ s \\ M \\ V \end{Bmatrix}_i = \begin{bmatrix} u_{11} & u_{12} & u_{13} & u_{14} \\ u_{21} & u_{22} & u_{23} & u_{24} \\ u_{31} & u_{32} & u_{33} & u_{34} \\ u_{41} & u_{42} & u_{43} & u_{44} \end{bmatrix} \cdot \begin{Bmatrix} -w_d \\ s \\ M \\ V \end{Bmatrix}_{i-1} \quad (21\text{-}19)$$

The U matrix, which is the transfer matrix, is unique for each span and is composed of the shaft properties of the span and lumped masses or spring supports at position i for span i. An example of a lumped mass is an impeller and an example of a spring is a bearing. This matrix is always a 4 × 4 square matrix no matter how many spans are used to describe the shaft because the matrix defines the characteristics of the individual spans, deflection, slope, bending moment, and shear force. The matrix calculations are done for each span, starting with span 1 and ending with span n (for n positions). Additionally, since the first position is always position 0 which is one of the two boundary conditions for the shaft, no span 0 exists and thus no transfer matrix for span 0.

The method for calculating the critical speed is first to multiply each transfer matrix for each span by the previous matrix. In other words, if n positions are present, the transfer matrix for the shaft design would be as follows:

$$U = U_n \cdot U_{n-1} \cdot U_{n-2} \cdots U_1 \quad (21\text{-}20)$$

The following general relationship now exists:

$$\begin{Bmatrix} -w_d \\ s \\ M \\ V \end{Bmatrix}_n = [U] \cdot \begin{Bmatrix} -w_d \\ s \\ M \\ V \end{Bmatrix}_0 \quad (21\text{-}21)$$

The shaft design is for the boundary conditions: deflection, slope, bending moment, and shear force at positions 0 and n, and the overall transfer matrix.

## SHAFT DESIGN

The advantage of using the transfer matrix method can be seen in eq. (21-21) because this matrix equation applies to any shaft design, whatever the number of spans, lumped masses, or spring supports. For computational purposes the bending moment and shear are zero at the boundaries. In this analysis, the bearings are assumed to behave as pinned joints and the shaft is assumed to extend some small distance above the upper bearing. Then the bending moment and shear are zero at both the upper and lower ends of the shaft. Rewriting eq. (21-21) reflecting the known boundary conditions yields:

$$\left\{ \begin{array}{c} -w_d \\ s \\ 0 \\ 0 \end{array} \right\}_n = [U] \cdot \left\{ \begin{array}{c} -w_d \\ s \\ 0 \\ 0 \end{array} \right\}_0 \quad (21\text{-}22)$$

The transfer matrix, U, is a function of frequency. So the equation involves four simultaneous equations and five unknowns: the deflection and slope at positions n and 0 and the frequency. Expanding eq. (21-22) yields the following equations:

$$0 = U_{31} \cdot (-w_{d0}) + U_{32} \cdot s_0 \quad (21\text{-}23)$$

$$0 = U_{41} \cdot (-w_{d0}) + U_{42} \cdot s_0 \quad (21\text{-}24)$$

Assuming that the deflection and slope at position 0 are nonzero for a nontrivial solution, the only way that eqs. (21-23) and (21-24) can be solved is if the following condition exists:

$$0 = U_{31} \cdot U_{42} - U_{32} \cdot U_{41} \quad (21\text{-}25)$$

The only unknown variable associated with eq. (21-25) is the frequency. The lowest frequency (greater than 0 rpm) that satisfies eq. (21-25) is the first critical speed.

The method for determining the deflection is very similar to calculating the critical speed. However, the hydraulic forces on the shaft are now taken into account. Also, the frequency is a known value. Hydraulic forces are determined based on the impeller torque. Because hydraulic forces used by mixer manufacturers already include the effect of dynamics, speed (frequency) is already included in the magnitude. Consequently, a forced response at the shaft speed would not be appropriate because the results would reflect the effect of frequency twice. Therefore, determining the forced response for the static condition is necessary (frequency = 0). From the bending moment for each position along with the torque from the impeller(s), the tensile and shear stresses can be calculated for each position. The static condition can only be calculated where the forcing frequency is effectively zero compared with the natural frequency, and such an analysis requires a 4 × 5 matrix.

## 21-7 IMPELLER FEATURES AND DESIGN

To most chemical engineers, impeller selection and design are driven primarily by the process requirements. In a high viscosity application a close-clearance impeller such as a helix or anchor may be the only practical means for achieving a uniform blend. In other, more general applications, such as solids suspension, a radial flow impeller could be used, although an axial flow impeller works better. The definition of "better" usually involves some evaluation of process performance with respect to mechanical design or operation. For example, an axial flow impeller often works better than a radial flow impeller for solids suspension, because less power is required. Less power means lower energy costs.

As for mechanical design, torque is often more important than power. Power requirements may dictate motor size and wiring requirements, but the primary consideration for power is operating costs associated with energy use. Torque, which is power divided by speed, influences the size of nearly all of the mechanical components of a mixer. To increase torque, a mixer drive must reduce the output shaft speed. To accomplish this speed reduction, a gear reducer may be used. The greater the speed reduction, the larger the gear reducer must be. Similarly, higher torque requires a larger shaft and thicker impeller blades. Higher torque is closely related to higher initial cost. So, often the better mixer for an application is the one that requires less torque and not necessarily less power, although sometimes less of both are possible.

For essentially all mixer applications, more torque or more power represents more intense mixing. A minimum level of intensity is necessary for any mixing requirement. Therefore, the mechanical design of an impeller extends beyond just blade thickness or hub strength, both of which are discussed later. If an impeller that produces more axial flow works better for fluid motion applications, such as liquid blending and solids suspension, a smaller blade angle with respect to the horizontal axis should be an advantage to impeller design. Solely on such a process basis, very small blade angles should produce the best impellers. However, the smaller the blade angle, the lower the power number, and therefore the larger the impeller diameter or the faster the rotational speed for the same power or torque input to the fluid.

Tank diameter places some firm limitations on impeller diameter. For axial flow to recirculate throughout the tank, an impeller cannot be larger than about 70% without obstructing the recirculation path. Half the tank cross-sectional area is in the inner 70.7% (the square root of one-half) of the tank diameter. Once the limit to impeller diameter is reached, increased speed is needed for more intense mixing. The maximum operating speed is limited by critical speed, but as blade angles become smaller, blade thicknesses increase for the same torque. Thicker blades mean greater mass and a lower critical speed. If a best impeller design ever exists, it must be an optimum based on impeller performance within the limits of mechanical design for impeller diameter, impeller weight (mass), and critical speed. Because of these mechanical limitations, a 45° pitched blade turbine may be a more cost-effective, "better" alternative to a hydrofoil impeller in certain intensely mixed axial flow applications.

With each new impeller concept comes the practical limitations of mechanical design. Whether a conscious design effort or an analysis based on previous practical experience, the process advantages of an impeller design must be weighed against the mechanical consequences. In the example of an axial flow impeller, a single blade should be most efficient because of small blade-tip wake interference, yet the mechanical imbalance would cause severe operating limitations. To reduce the induced loss caused by blade-tip vortices, the blade width should be narrow. However, a narrow blade has a low power number and therefore must have a large diameter or operate at a high speed. A shallow blade angle will increase the axial component of flow, but also increases blade thickness and decreases power number. Even blade camber, created by rolling or bending an airfoil-like cross-section into the blade, can transfer mechanical problems from the extension blade to the hub attachment. The consequence of these mechanical design considerations for an impeller is a variety of impeller design each with compromises and limitations designed to provide the best combination of process and mechanical performance under certain circumstances. Sometimes, maximum process flexibility or minimum cost can even override process performance as a determining factor.

### 21-7.1  Impeller Blade Thickness

Blade thickness is an obvious mechanical design consideration. The blades must be thick enough to handle fluctuating hydraulic loads without bending or breaking, and as thin as possible to conserve material and minimize weight (mass). Because of size and cost, the blades could even be the weakest link in a design. Breaking a blade would be less costly than breaking a shaft or gear reducer.

Designing the platelike extension blade for a pitched blade turbine (Figure 21-37), is a simple mechanical design task once the fluid forces are estimated. For turbulent conditions, the inertial forces dominate over the viscous forces, so pressure is the primary means of transferring fluid forces to mechanical objects. For design purposes, hydraulic forces corresponding to pressure are assumed to act normal to the blade. For transitional conditions, some combination of inertial pressure and viscous drag will act on the impeller blade. This combination of forces makes the pressure design of blades conservative for pitched blade turbines and equal for straight-blade (vertical-blade) turbines.

The commonly known or calculable force acting on the impeller blade is the force related to torque, which is horsepower, P [hp] {W}, divided by rotational speed, N [rpm] {rps}, divided by the number of blades, $n_b$, or the first term inside the parentheses in eq. (21-26). Because the pressure force acts normal to the blade and the torsional force must be horizontal for a vertical rotating shaft, a factor of the reciprocal of the sine of the blade angle enters the expression for blade thickness. The equivalent pressure force must act at some moment arm from the center of rotation, which would be the impeller radius, D/2 [in.] {m}, if the force acted at the blade tip. However, because pressure forces are lost around the tip of the blade, causing a vortex flow pattern, the effective force must act at a

**1310** MECHANICAL DESIGN OF MIXING EQUIPMENT

shorter moment arm, represented by a location fraction, $f_L$. For a typical pitched blade turbine, $f_L$ might be 0.8; for a narrower blade, it might be closer to 0.85. The point of the maximum design moment comes at the point of extension blade attachment to a stub blade or the central hub, the radius of which is represented by $D_S/2$ [in.]. Blade angle $\alpha$ also affects strength requirements.

The following calculation takes into account that the blade strength is partially provided by the width of the blade, W [in.] {m}, and number of blades, $n_b$. The thickness is also limited by an allowable stress, $\sigma_b$ [psi] {N/m$^2$}, suggested values of which are shown in Table 21-3.

U.S. Eng.

$$t = 615 \left\{ \frac{P}{Nn_b} \frac{f_L(D/2) - D_S/2}{\sin\alpha[f_L(D/2)]W\sigma_b} \right\}^{1/2}$$

Metric
(21-26)

$$t = 0.981 \left\{ \frac{P}{Nn_b} \frac{f_L(D/2) - D_S/2}{\sin\alpha[f_L(D/2)]W\sigma_b} \right\}^{1/2}$$

The coefficient in this blade thickness, t [in.] {m}, calculation returns the appropriate value with the units shown previously.

### 21-7.2 Impeller Hub Design

Calculations for the stub blades or welded attachment points of impeller blades can be done like calculations for the extension blade thickness. The details of welding, casting, or other methods of attachment become critical in the design. Conventional calculations for structural strength may be adequate, but for complicated geometry, finite element models can provide better design information.

Other features of hub design considered by mixer manufacturers include fastener selection to bolt blades to hubs and setscrew selection to handle keyed or unkeyed shaft attachment. Blade lengths must be uniform, because power is a function of impeller diameter to the fifth power in the turbulent range. Minor differences in blade length can create large imbalanced forces. The fit of a hub on a shaft must be tight to prevent dynamic forces from working the attachment loose. An impeller also needs to be balanced, usually statically for large mixers operating below 125 rpm, to keep the weight (mass) centered around the shaft.

## 21-8 TANKS AND MIXER SUPPORTS

The tanks and supports used with mixers are an integral part of the mechanical design process. Three primary loads must be considered in the design of a mixer support (Figure 21-42): vertical loads, such as equipment weight and pressure forces, torque, and bending moment. In addition, the torque and bending moment are dynamic loads. A major consideration in the design of a mixer support is

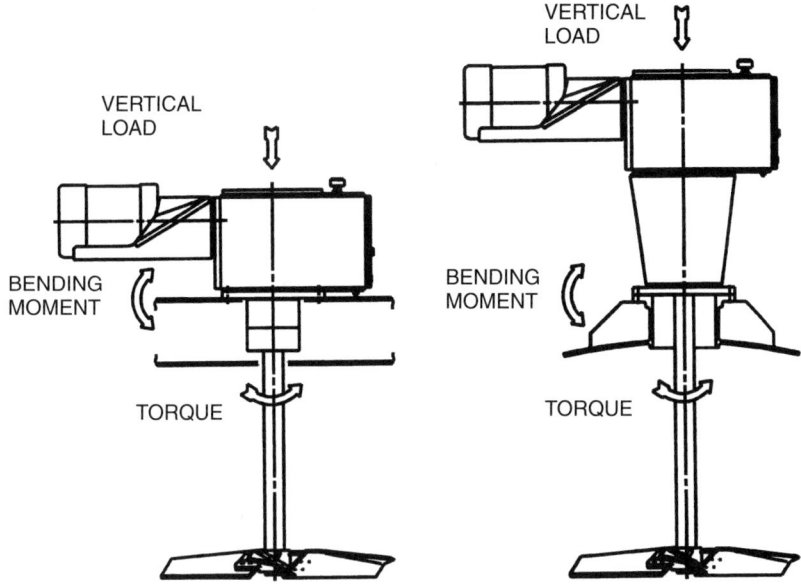

**Figure 21-42** Drive mounting loads: beams and nozzles. (Courtesy of Chemineer.)

the dynamic aspect of the loads, which can cause disturbing or even dangerous motion of the mixer and support structure. As a further reminder of the importance of a stiff mounting, the natural frequency of the mixer shaft will be reduced if the mount is not rigid. Three general categories of mixer mounting encompass most applications: beam mounting, nozzle mounting, and other structural mounting.

### 21-8.1  Beam Mounting

Mixers can be mounted on beams, as shown on the left in Figure 21-42, over both open and closed tanks. Mounting over an open tank is usually not as dimensionally critical as mounting over a closed tank, where seal alignment and thermal expansion may be significant factors. However, the concepts of all beam mounting are similar and cover a wide range of tank sizes, from a couple of feet to tens of feet.

A typical beam mounting arrangement is shown in Figure 21-43. The basic structure can be formed by two parallel beams mounted like a bridge over the tank. The dynamic nature of mixer loads becomes an immediate concern with respect to the appropriate structure for the support. If the only load considered in the design were the static weight of the mixer, relatively lightweight beams would be adequate. Deflections at the center of the beams could be $\frac{1}{8}$ or $\frac{1}{4}$ in. (3 to 6 mm) even on a short span without exceeding design stress limits. Even considering a static torque and bending moment, a beam structure could exceed good design practices for static loads and still demonstrate a measurable deflection

**1312** MECHANICAL DESIGN OF MIXING EQUIPMENT

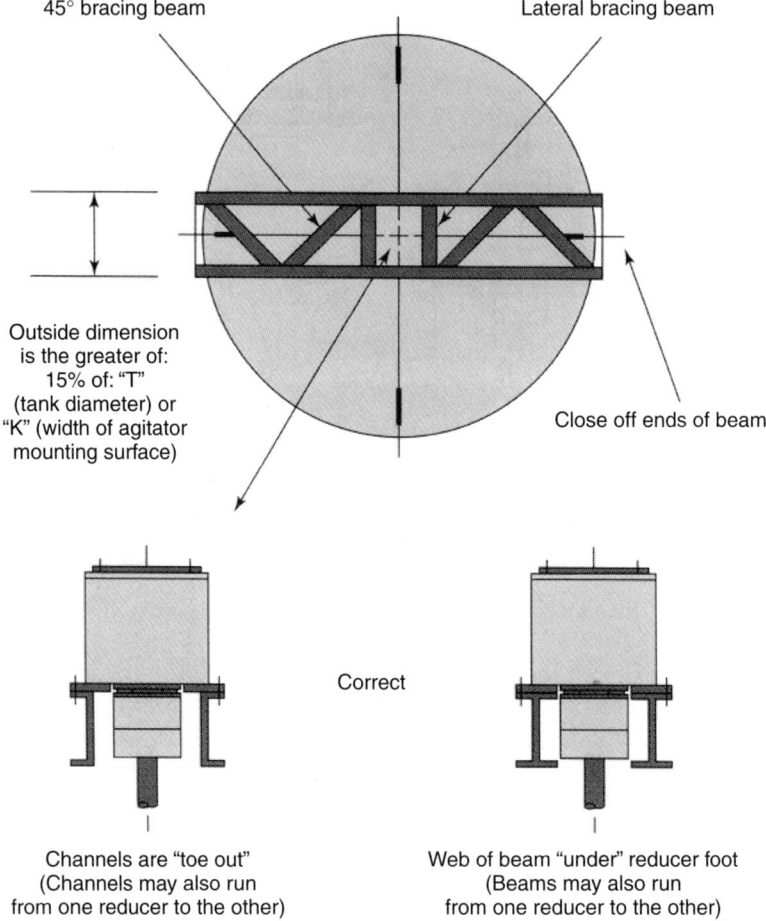

**Figure 21-43** Beam mounting for mixer drive. (Fasano et al., *Chemical Engineering Progress*, 1995 © AIChE.)

near the mounting location. When dynamic loads are considered acceptable static deflections become disturbing motions, similar to those experienced aboard a ship at sea. Such motions are disturbing to the operators of the mixer but can also accelerate damage to gears and bearings and even lead to catastrophic failure of a shaft or support component.

To handle dynamic loads, the structure must be stiff in all directions. Two parallel beams like the major ones shown in Figure 21-43 may be stiff along their length. However, without the cross bracing shown in Figure 21-43, the structure is flexible in the direction normal to the main beam lengths. End, lateral, and angled bracing strengthen the structure in all directions and a minimum spacing between the beams of 15% of the tank diameter or the width of the mixer drive provides the basis for a stiff support.

**Table 21-7**  Recommended Beam Sizes[a] [in. × lb/ft] for Mixer Mounting

| Mixer Torque [in.-lb$_f$] | \multicolumn{5}{c}{Vessel Diameter [ft]} | | | | |
|---|---|---|---|---|---|
| | 6 | 10 | 15 | 20 | 30 |
| 2 500 | C 5 × 6.7 | C 7 × 9.8 | W 12 × 14 | W 12 × 19 | W 18 × 35 |
| 4 000 | C 6 × 8.2 | W 8 × 10 | W 12 × 14 | W 14 × 22 | W 18 × 40 |
| 8 500 | C 7 × 9.8 | W 10 × 11.5 | W 12 × 19 | W 16 × 26 | W 21 × 49 |
| 17 000 | C 7 × 9.8 | W 12 × 14 | W 14 × 22 | W 18 × 35 | W 24 × 61 |
| 24 000 | W 8 × 10 | W 12 × 14 | W 16 × 26 | W 18 × 40 | W 24 × 76 |
| 33 000 | W 8 × 10 | W 12 × 16.5 | W 16 × 31 | W 21 × 44 | W 27 × 84 |
| 59 000 | | W 14 × 22 | W 18 × 35 | W 24 × 55 | W 30 × 99 |
| 87 000 | | W 14 × 22 | W 18 × 40 | W 24 × 61 | W 30 × 116 |
| 135 000 | | W 16 × 26 | W 21 × 44 | W 24 × 76 | W 33 × 130 |
| 225 000 | | W 18 × 31 | W 24 × 55 | W 27 × 84 | W 36 × 160 |
| 350 000 | | | W 24 × 61 | W 30 × 99 | W 36 × 194 |
| 525 000 | | | W 24 × 76 | W 30 × 116 | W 36 × 260 |

[a] C, American standard channel; W, wide-flange beam.

Recommended beam and channel sizes for different drive torques and beam spans are shown in Table 21-7. The torque values in the tables take into account typical hydraulic loads on the impellers and nominal shaft lengths for such mixers. If better information about axial, torsional, and bending loads are available, the support structure should be designed for deflections of 0.20 to 0.30 in./sec (5.0 to 8.0 mm/s). Structures designed to these standards typically give a satisfactory compromise between support structure cost and acceptable vibration levels (Fasano et al., 1995).

Also important is the correct use of channels or beams supporting a mixer, as shown in Figure 21-43. The web of the channel or beam should be under the drive so that more than just the flange carries the weight and loads. Additional lateral or longitudinal bracing may be required to support the drive adequately.

### 21-8.2 Nozzle Mounting

A typical nozzle mounting is shown on the right in Figure 21-42. Closed tanks typically have either a nozzle or a pad, shown in Figure 21-44, for mounting the mixer. If the vessel is designed to ASME Code pressure ratings only, the mixer support may not be adequate. For low pressure or nonpressurized tanks, the nozzle support may not be adequate to support even the weight of the mixer. Even when designed for vertical, bending moment, and torque loads as prescribed by the ASME Code, the dynamic nature of the loads can create undesirable motion. The problem of inadequate design strength to handle dynamic loads is often helped by the mixer manufacturers, who may provide design load values which are greater than the anticipated real loads.

Vertical forces downward on a mixer mounting come from the weight of the motor, drive, shaft, and impeller. Vertical forces upward are caused by pressure

**1314** MECHANICAL DESIGN OF MIXING EQUIPMENT

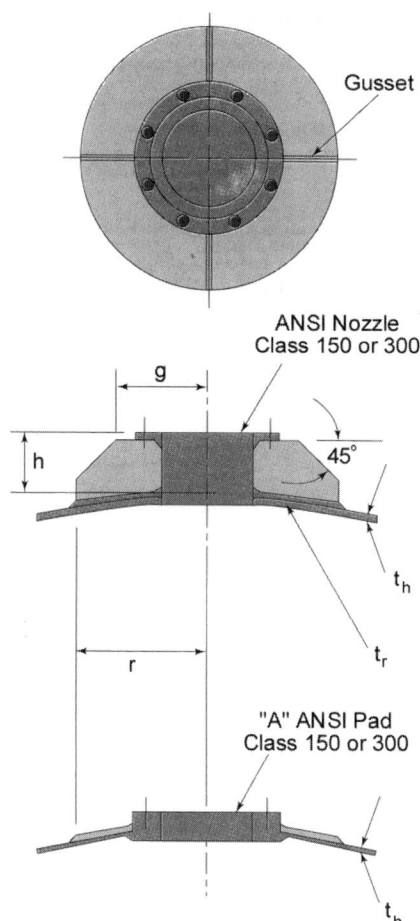

**Figure 21-44** Reinforcement for nozzles and pads. (Fasano et al., *Chemical Engineering Progress*, 1995 © AIChE.)

within the vessel, since the shaft passing through the seal transmits forces as if the shaft were a piston. An internal tank pressure of 100 psig (690 kPa gauge) will exert an upward force of 1257 lb$_f$ (5634 N) with a 4 in. (102 mm) diameter shaft, which has a 12.57 in$^2$ (0.008 m$^2$) cross-section. Because of pressure forces, a vertical downward load (Figure 21-42) will always have a minimum and maximum value, with the minimum sometimes being a significant upward force. To provide dynamic stiffness to the design calculations for a tank nozzle, the maximum drive torque may be multiplied by about 2.5 times for design values shown on a mixer drawing. Similarly, bending moments could be multiplied by three times or more, depending on the application and anticipated operating conditions. Failure to consider the dynamic nature of mixer loads can result in disturbing motion of the drive and shaft.

Dynamic stiffness and structural integrity can also be provided by adequate reinforcement of a nozzle or pad. Typical nozzle gussets with a backup reinforcing plate are shown in Figure 21-44. The nozzle diameter is usually set by a standard or available seal flange on the mixer. Suggested nozzle heights and gusset dimensions are provided in Table 21-8. Additional recommendations for minimum head thickness are shown in Tables 21-9 and 21-10. If the pressure requirements for the tank do not meet these minimum requirements, a reinforcement pad equal to the minimum head thickness is recommended.

A mounting pad may be welded into the top of a tank instead of using a nozzle. The pad eliminates the height of the nozzle and gusset requirements, as shown

Table 21-8    Nozzle and Pad Reinforcement Dimensions

| ANSI Nozzle/ Pad Size [in.] | Reinforcement Dimensions [in.] h | g | r | DIN Nozzle/ Pad Size {mm} | Reinforcement Dimensions {mm} h | g | r |
|---|---|---|---|---|---|---|---|
| 8  | 6  | 8.0  | 12 | 200 | 150 | 200 | 300 |
| 12 | 8  | 11.5 | 17 | 300 | 200 | 300 | 450 |
| 16 | 8  | 13.5 | 19 | 400 | 200 | 350 | 480 |
| 20 | 12 | 17.5 | 26 | 500 | 300 | 450 | 650 |
| 24 | 12 | 19.5 | 28 | 600 | 300 | 500 | 700 |
| 30 | 12 | 24.0 | 31 | 800 | 300 | 600 | 800 |

Table 21-9    Recommended Head Thickness [in.] for Nozzle-Mounted Mixers

| Mixer Torque [in.-lb$_f$] | ANSI Nozzle Size [in.] | Vessel Diameter [ft] 5 | 7 | 9 | 12 | 15 |
|---|---|---|---|---|---|---|
| 2 500   | 8  | 0.187 | 0.187 | 0.250 | 0.312 | 0.312 |
| 4 000   | 8  | 0.250 | 0.312 | 0.375 | 0.375 | 0.437 |
| 8 500   | 8  | 0.312 | 0.437 | 0.500 | 0.500 | 0.562 |
| 17 000  | 12 | 0.250 | 0.312 | 0.437 | 0.500 | 0.562 |
| 24 000  | 12 | 0.312 | 0.437 | 0.500 | 0.625 | 0.750 |
| 33 000  | 12 | 0.437 | 0.562 | 0.625 | 0.750 | 0.875 |
| 59 000  | 16 |       | 0.437 | 0.562 | 0.625 | 0.750 |
| 87 000  | 16 |       | 0.562 | 0.625 | 0.750 | 0.875 |
| 135 000 | 20 |       | 0.437 | 0.562 | 0.687 | 0.875 |
| 225 000 | 24 |       |       | 0.562 | 0.750 | 0.875 |
| 350 000 | 24 |       |       | 0.562 | 0.750 | 0.875 |
| 525 000 | 30 |       |       |       | 0.750 | 0.875 |
| 825 000 | 30 |       |       |       | 1.000 | 1.250 |

**1316** MECHANICAL DESIGN OF MIXING EQUIPMENT

**Table 21-10** Recommended Head Thickness {mm} for Nozzle-Mounted Mixers

| Mixer Torque {N · m} | DIN Nozzle Size {mm} | Vessel Diameter {mm} | | | | |
|---|---|---|---|---|---|---|
| | | 1400 | 2000 | 2800 | 3600 | 4600 |
| 300 | 200 | 6 | 6 | 8 | 8 | 8 |
| 400 | 200 | 8 | 8 | 10 | 10 | 12 |
| 900 | 200 | 8 | 12 | 14 | 14 | 16 |
| 1 900 | 300 | 8 | 8 | 12 | 14 | 16 |
| 2 700 | 300 | 8 | 12 | 14 | 16 | 20 |
| 3 700 | 300 | 12 | 16 | 16 | 20 | 25 |
| 6 600 | 400 | | 16 | 16 | 16 | 20 |
| 9 800 | 400 | | 12 | 16 | 20 | 25 |
| 15 000 | 500 | | | 16 | 18 | 25 |
| 25 000 | 600 | | | 16 | 20 | 25 |
| 39 000 | 600 | | | 16 | 20 | 25 |
| 59 000 | 750 | | | | 20 | 25 |
| 93 000 | 750 | | | | 28 | 32 |

**Table 21-11** Recommended Head Thickness [in.] for Pad-Mounted Mixers

| Mixer Torque [in.-lb$_f$] | ANSI Pad Size [in.] | Vessel Diameter [ft] | | | | |
|---|---|---|---|---|---|---|
| | | 5 | 7 | 9 | 12 | 15 |
| 2 500 | 8 | 0.125 | 0.125 | 0.125 | 0.125 | 0.187 |
| 4 000 | 8 | 0.125 | 0.187 | 0.187 | 0.250 | 0.312 |
| 8 500 | 8 | 0.187 | 0.250 | 0.250 | 0.312 | 0.375 |
| 17 000 | 12 | 0.187 | 0.187 | 0.250 | 0.250 | 0.375 |
| 24 000 | 12 | 0.250 | 0.250 | 0.312 | 0.375 | 0.437 |
| 33 000 | 12 | 0.250 | 0.312 | 0.375 | 0.437 | 0.500 |
| 59 000 | 16 | – | 0.250 | 0.312 | 0.375 | 0.437 |
| 87 000 | 16 | – | 0.312 | 0.375 | 0.437 | 0.500 |
| 135 000 | 20 | – | 0.312 | 0.375 | 0.437 | 0.500 |
| 225 000 | 24 | – | – | 0.437 | 0.437 | 0.562 |
| 350 000 | 24 | – | – | 0.375 | 0.437 | 0.562 |
| 525 000 | 30 | – | – | – | 0.500 | 0.625 |
| 825 000 | 30 | – | – | – | 0.625 | 0.750 |

in Figure 21-44. The bolt pattern for mounting the mixer on a pad is usually the same as a standard flange, but threaded holes are provided in the pad. Again, a minimum head thickness (Tables 21-11 and 21-12) is recommended for pad mounts. If the head thickness required for pressure is less than the minimum for mixing mounting, a reinforcing pad of the minimum head thickness and diameter, (Table 21-8) should be used.

**Table 21-12** Recommended Head Thickness {mm} for Pad-Mounted Mixers

| Mixer Torque {N · m} | DIN Pad Size {mm} | Vessel Diameter {mm} |  |  |  |  |
|---|---|---|---|---|---|---|
|  |  | 1400 | 2000 | 2800 | 3600 | 4600 |
| 300 | 200 | 4 | 4 | 4 | 4 | 6 |
| 400 | 200 | 4 | 6 | 6 | 8 | 8 |
| 900 | 200 | 6 | 8 | 8 | 8 | 10 |
| 1 900 | 300 | 6 | 6 | 8 | 8 | 10 |
| 2 700 | 300 | 8 | 8 | 8 | 10 | 12 |
| 3 700 | 300 | 8 | 8 | 10 | 12 | 14 |
| 6 600 | 400 | – | 8 | 8 | 10 | 12 |
| 9 800 | 400 | – | 8 | 10 | 12 | 14 |
| 15 000 | 500 | – | – | 10 | 12 | 14 |
| 25 000 | 600 | – | – | 12 | 12 | 16 |
| 39 000 | 600 | – | – | 10 | 12 | 16 |
| 59 000 | 750 | – | – | – | 14 | 16 |
| 93 000 | 750 | – | – | – | 16 | 20 |

### 21-8.3 Other Structural Support Mounting

The numerically largest category of other structural supports is clamp mounting for a portable mixer. Although the clamps are usually adequate to support the mixer, the tank wall may not be adequate. Most tanks, especially small storage tanks, are designed only for the weight of liquid contents. Dynamic loads, especially bending loads, can result in dangerous mounting situations. Poor mounting could result in the mixer falling into the tank or possibly cause personal injury. Some commercially available metal tanks have reinforced pads welded into the side specifically for mounting a mixer. Polymer and reinforced polymer tanks often lack sufficient rigidity to support a mixer. Without adequate support by the tank walls, the mixer should be attached to a strong external support beam or other mounting, often attached to a wall or other building support. Any time that a mixer exhibits significant motion relative to the surroundings, the situation should be considered dangerous and corrected immediately.

Many other types of support structures may be provided as an integral part of the manufactured mixing equipment. When designed by a competent mixer manufacturer, these structures should be adequate to handle all the mixer loads. Structures designed by the general equipment fabricators may or may not accommodate dynamic loads, especially ones that could occur under special circumstances, such as filling or emptying a tank.

Even if the mixer is mounted to an adequate support, if that support is independent of the tank, relative motion between the tank and mixer can cause safety problems. Large, high intensity mixers are often provided with a "change can" arrangement that allows the use of different tanks for batch processing. These tanks should be secured by a means that does not allow the tank to move relative to the mixer when in use.

## 21-9 WETTED MATERIALS OF CONSTRUCTION

Materials that have a proven successful history in the same or similar process will be the first choice. However, more economical materials should always be considered. Sometimes, "more economical" may mean lower cost materials with adequate resistance. In other cases, newer materials or materials with better resistance may be more economical because of expected life or other merits.

### 21-9.1 Selection Process

Selection of a material for a new process should involve three steps:

1. Screening of potential candidates
2. Selection of candidates and testing for fatigue strength
3. Final selection of the material based on an economic analysis

New alloys are usually slow to influence mixer design and manufacture. When new alloys become available in the marketplace, they are usually available in only plate and small diameter bar sizes. Finding bars of sufficient diameter and length for a mixer shaft can be difficult. If after several years the alloy has been a commercial success, quantities of the appropriate forms may become available. Material availability must be taken into account when specifying the materials of construction for a mixer.

The most difficult components to obtain are standard fasteners: bolts, nuts, and washers. Alloys must be well established in the CPI industry before fasteners made of new alloys become available. If fasteners in the desired alloy cannot be found, using an off-the-shelf fastener in a more-corrosion-resistant material will be more economical than making custom fasteners.

All alloys have ASTM, AISI, or DIN compositional ranges, which specify the percentage metal constituents. Alloy manufacturers today can hold the composition of an alloying ingredient close to the low-cost end of a range. This control provides for a higher degree of consistency but may not provide the corrosion resistance of an older material with the same alloy designation. For example, 317L stainless steel has an allowable molybdenum range of 3 to 4%. Today, the composition in 99 out of 100 pieces would contain 3 to 3.25% molybdenum. Twenty years ago, a mixer made out of 317L would, on average, have had a molybdenum content closer to 3.5%. This difference in composition may not seem significant, but molybdenum plays a major role in chloride corrosion resistance and could dramatically shorten the life of the alloy used in mixer service. Awareness of market realities can be important in selecting materials for a mixer.

The largest impact on materials has been in dual-phase stainless alloys. The two most popular dual-phase alloys for mixer service are Ferralium 255 (Bonar Langley Alloys Ltd. and Haynes International) and Alloy 2205 (AB Sandvik Steel). Both alloys are about 30% stronger than 316 stainless steel and approximately equivalent to 904L stainless steel in corrosion resistance. The high chromium and molybdenum content gives the newer alloys excellent pitting and crevice-corrosion resistance, particularly in chloride environments (Redmond, 1986a,b). However, duplex stainless steels have limited ductility and require special care in forming and welding.

Also, relatively new are the 6% molybdenum super-austenitic stainless steels, which possess the strength of duplex stainless steels and corrosion resistances about midway between the duplex stainless steels and the high-nickel-based alloys. Welding super-austenitic stainless is not a problem with molybdenum-enriched filler metals. Examples of high-molybdenum alloys are AL-6XN (Allegheny Ludlum Corp.) and 254 SMO (Avesta Sheffield, Inc.). Sorell (1994) and Kane (1993) present concise reviews of the various alloy groups.

### 21-9.2 Selecting Potential Candidates

A list of the most commonly used metals for mixers is shown in Table 21-13. Although this list is not exhaustive, only a few mixers require more exotic materials.

Most alloy manufacturers test their materials against certain standard solutions. These results are often reported in bulletins for the specific alloy. The results presented in these bulletins can be used to help select candidate materials. Metal failure usually begins on the surface of the material exposed to the chemical environment. Because changes in the surface contour usually act as stress risers, any chemical attack that pits the material can initiate failure long before general corrosion is a problem. Because metals become more anodic in high-stress areas, the pitting will be worst at areas of high-stress on the mixer shaft and impeller. This combination of effects makes the selection of alloys that minimize pitting essential.

One measurement of pitting resistance is the *pitting potential* in a standard solution such as 1 M sodium chloride. The higher the pitting potential or the higher the test temperature required to initiate pitting, the more resistant the alloy. The combined effects of various alloying elements have been related empirically through extensive laboratory testing in aqueous media. The pitting index, PI, of stainless steels and high nickel alloys in chloride-containing solutions can be

Table 21-13 Commonly Used Metals for Mixer Wetted Parts

| Alloy Group | Common Alloy Grades |
|---|---|
| Low carbon steels | AISI 1015 to AISI 1025 |
| Austenitic stainless steels | 304, 304L, 316, 316L, 317L, 20Cb3, 904L, 800, 825 |
| Duplex stainless steels (austenitic–ferritic) | 255, 2205 |
| Superaustenitic stainless steels | 254SMO, AL6XN, 3127hMo |
| High nickel alloys | G3, G30 600, 625, C276, C22, C4, B2, B3 |
| Other | Ti Gd.2, Zr 702 |

expressed as (Kane, 1993)

$$PI = \%Cr + 3.3 \times \%Mo + (X)\%Ni \qquad (21\text{-}27)$$

where $X = 0$ for ferritic stainless steels, $X = 16$ for duplex stainless steels, and $X = 30$ for austenitic stainless steels. A minimum PI has been established for satisfactory operation determined as a function of ppm Cl$^-$ and pH for operating temperatures of 120 to 150°F (50 to 65°C):

$$PI_{min} = 19.2 \text{ppm}^{0.151} \times 10^{-0.0453 \text{pH}} \qquad (21\text{-}28)$$

This minimum PI has been successful in establishing acceptable material candidates for systems with chloride ion present. Other halide ions can also pit stainless steels, but their natural presence is much less than for the chloride ion. Titanium alloys also have a high pitting potential in chloride ion solutions.

### 21-9.3 Corrosion–Fatigue

The chemical process industry requires that many pieces of equipment (including pumps, piping, tanks, mixers, and heat exchangers) be made from corrosion-resistant alloys. The costs of these metal alloys typically range from $2 to $25/lb$_m$ ($4 to $50/kg). The penalty for overdesign is high capital cost and for underdesign is failure. The unique nature of high-cycle corrosion–fatigue interaction requires an approach that defines allowable stresses for the expected life of the equipment. One method would be to generate data for the same number of fatigue load cycles as for the expected lifetime. Testing at the appropriate frequency for the expected lifetime of most equipment is not practical, since expected life can be 5 to 15 years or longer. Conversely, testing at higher frequencies will not produce the same life expectancy in corrosion–fatigue applications. Under the same loading conditions a lower frequency will typically produce a lower cycle count to failure than will a higher frequency. The need for a short-life-study technique to predict high-cycle fatigue life in various chemical environments is apparent.

*21-9.3.1 Background to Corrosion–Fatigue.* The term *corrosion–fatigue* describes the phenomenon of cracking in materials caused by the combined action of an applied cyclic stress and a corrosive environment. Corrosion–fatigue behavior is characterized by a shorter life, either in terms of cycles or time, than would be expected from fatigue or corrosion alone. Corrosion–fatigue is a recognized engineering problem and has caused major engineering structures and equipment to fail.

Although corrosion occurs strictly based on a time relationship, when combined with cyclic loading, the effects of environmental attack can be observed at low cycles. The co-joint chemical and mechanical mechanism has been shown by McAdam (1927) to be worse than both actions taken separately and sequentially. In aqueous systems, for example, highly stressed areas become more anodic and the rate of metal removal is more localized and severe than for unstressed materials in the same chemical environment.

Many examples in experimental testing referenced in the literature show that corrosion effects can be observed at cycles of 5000 to 100 000. Depending on the environment, a review of the literature suggests that corrosion–fatigue effects typically require exposure to the chemical environment for 100 to 1000 min. Effects for very aggressive environments can be observed in 100 min, while less aggressive environment could require 1000 min (17 h) of exposure. Environment C is more aggressive than environment B, which is more aggressive than environment A. Corrosion–fatigue cracks always nucleate at the surface unless near-surface defects act as stress concentration sites and facilitate subsurface cracking. Surface features that initiate corrosion–fatigue cracks depend on the alloy and environment conditions.

*21-9.3.2 Evaluating Corrosion–Fatigue.* Identifying allowable design stresses for high-cycle corrosion–fatigue environments has been hampered by the difficulties in evaluating the co-joint mechanism or by separating the effects of mechanical and chemical action. Every chemical system can have a different effect and, of course, an infinite variety of chemical systems exist.

*21-9.3.3 Significance of Corrosion–Fatigue.* In the United States alone, the chemical process industries (CPI) spend billions of dollars per year on equipment constructed of corrosion-resistant alloys. These alloys include the stainless steel, high nickel, and titanium groups and are designed with allowable stresses based on qualitative rules. The Bureau of Economic Analysis Division of the Department of Commerce estimated that the CPI (SIC 2800 Series) would spend $28 billion on capital equipment in 1999. High-cycle fatigue lives ($10^9$ to $10^{11}$ cycles) are the rule rather than the exception in the CPI, and failures occurring at low-cycle fatigue lives are seldom encountered. The ASME code design of vessels sets allowable stress limits as a function of tensile strength without regard for the chemical environment.

*21-9.3.4 Engineering Approach.* Curves of stress versus cycles are called *S-N curves.* Many techniques exist for approximating fatigue curves. Among them are the Collins method (Dowling, 1993), Juvinall method (Dowling, 1993), Shigley method (Dowling, 1993), Mitchell method (Dowling, 1993), Khan et al. (1995) method, and Wei and Harlow (1993) method. All these techniques are for an air environment and are not applicable for estimating high-cycle corrosion–fatigue life in liquid systems. Surprisingly little work has been reported concerning estimating high-cycle corrosion–fatigue for the chemical process industries, although much money is spent on equipment that should be designed for corrosion–fatigue.

Fasano (2000) has developed and demonstrated a technique for accurately extrapolating low-cycle ($10^3$ to $10^5$) corrosion–fatigue behavior to high-cycle ($10^7$ to $10^9$) corrosion–fatigue behavior. The technique includes a physical system for collecting low-cycle corrosion–fatigue data and a mathematical technique

**1322**  MECHANICAL DESIGN OF MIXING EQUIPMENT

for extrapolating the low-cycle data. Environmental effects on fatigue behavior were shown experimentally on stainless steel 316 at cycles from 5000 to 100 000. The technique was confirmed on many other alloys by using corrosion–fatigue data from many investigators.

### 21-9.3.5 Establishing Mean Life to Failure.
The mean cycles are determined from fitting the data at each stress level to a two-parameter Weibull distribution and computing the mean from the cumulative distribution equation for a probability of 50%. The two-parameter Weibull probability density function and cumulative distribution function as given by Madayag (1969) are

$$f(N) = \frac{qN^{q-1}}{N_a^q} \exp\left[-\left(\frac{N}{N_a}\right)^q\right] \qquad (21\text{-}29)$$

$$F(N) = 1 - \exp\left[-\left(\frac{N}{N_a}\right)^q\right] \qquad (21\text{-}30)$$

where N is the specimen life in number of cycles, $N_a$ is life of 36.8% of the population (36.8% = 1/e, e = 1.718), and q is the shape parameter for the Weibull distribution curve. The process used to adjust the fitting parameters $N_a$ and q is to adjust the parameters until a plot of

$$\log\log\left[\frac{1}{1-F(N)}\right] \quad \text{versus} \quad \log(N) \qquad (21\text{-}31)$$

produces a straight line (slope = q).

### 21-9.3.6 S-N Model and Evaluation of Goodness of Fit.
As reviewed previously, several different models for estimating high- and low-cycle fatigue behavior in air or inert environments exist. It could be argued that no inert industrial environments exist and even air environments will depend on whether the air is dry or has various amounts of humidity. Empirical and semiempirical approaches have been tried for singular environments with limited success. Environments posed in the chemical processing industry are so diverse and numerous that a catalog of S-N curves for every possible environment can never be developed. Environments must not only include the chemical constituents but the pressure and temperature as well. Running S-N studies for low-frequency devices to $10^9$ cycles is unreasonable in both time and money. A logical technique for estimating high-cycle corrosion–fatigue life is to use actual low-cycle data and develop a model that extrapolates to the high-cycle regime.

A two-term power law model was used for describing the S-N behavior. This model is similar to the low-cycle fatigue-based strain-life model:

$$S = aN^b + cN^d \qquad (21\text{-}32)$$

The model was also used to model the S-N corrosion–fatigue behavior from literature sources.

### 21-9.3.7 Testing the Method on Literature Data. 
Many examples of material and environment combinations exist in the literature. The recommended extrapolation technique was observed to work for a variety of alloys. The literature data are summarized by alloy group in Table 21-14. Table 21-15 summarizes the environments for the alloy groups, and Table 21-16 lists the alloys, the mechanical properties used for the model, the source of the mechanical property data if not from the corrosion–fatigue reference, and the corrosion–fatigue data references. Alloys spanned a considerable range.

### 21-9.3.8 Estimating Long-Life Behavior from Short-Life Behavior. 
Empirical expressions were developed to permit determination of the b and d exponents of the two-term power law model. The algorithm for d is

$$\text{If } P_d > -0.035, \quad d = 0.00$$
$$\text{If } P_d \leq -0.035, \quad d = 1.12 P_d + 0.0333 \tag{21-33}$$

where the parameter $P_d$ is defined as

$$P_d = \text{LLS}|_{10^5} \text{SR}^5 \left(\frac{\sigma_Y}{\sigma_T}\right)^{0.3} \tag{21-34}$$

SR is the ratio of the log-log slopes at $10^5$ cycles and $5 \times 10^4$ cycles:

$$\text{SR} = \frac{\text{LLS}|_{10^5}}{\text{LLS}|_{5 \times 10^4}} \tag{21-35}$$

The log-log slope, LLS, can be represented as

$$\text{LLS} = \frac{d(\log S)}{d(\log N)} \tag{21-36}$$

It is not required that low-cycle data be fitted to the two-term power model to obtain log-log slopes at $5 \times 10^4$ and $1 \times 10^5$. Any equation form can be used. Whichever mathematical form is used, however, it must be monotonically decreasing and should fit the data such that the $2\sigma$ error interval in predicting strength is within $\pm 4\%$ and the log-log slope ratio, SR, is less than 1.

The ratio $\sigma_Y/\sigma_T$ is the yield/tensile strength ratio. The algorithm for exponent b is

$$b = -0.0672 + 2.58 P_b + 1.85 P_b^2 \tag{21-37}$$

where the parameter $P_b$ is defined as

$$P_b = \text{LLS}|_{5 \times 10^4} \text{SR}^{-1.25} \left(\frac{\sigma_Y}{\sigma_T}\right)^{0.8} \tag{21-38}$$

**Table 21-14** Corrosion–Fatigue Data Available from the Literature

| Alloy Group | Group Abr. | Number of Data Sets | Alloy(s) | References |
|---|---|---|---|---|
| Low carbon steel | LCS | 1 | Low carbon steel | Dugdale (1962) |
| High strength carbon steels | HS-CS | 3 | AISI 4140, AISI 4620, AISI 6150 | Lee and Uhlig (1974), Baxa et al. (1978), Coburn (1984) |
| Cr stainless steels | 4xxSS | 3 | AISI 403, 13 Cr steel, 12.4 Cr steel | Spiedel (1977), Ebara et al. (1978), Sedriks (1979) |
| Cr–Ni–Mo stainless steel | 3xxSS | 3 | 316 SS, 21Cr–6Ni–2.4Mo | Huntington (1976), Amzallag et al. (1978), Larsson (1984) |
| High nickel alloys | Ni | 2 | Alloy 600, Udimet 720 | Walls et al. (1982), Korp and Olson (1987) |
| Zirconium alloys | Zr | 1 | Zircaloy 2 | Teeter and Hosbons (1977) |
| Titanium alloys | Ti | 1 | Ti–6Al–4V | Morton (1967) |
| Aluminum alloys | Al | 4 | Al 6061 T6, Al 7475 T7351, Al 7075 T6, Al 5.5Zn–2.5Mg–1.5Cu | Amax (1965), Gibala and Hehemann (1984), Goel (1986), Smith and Duquette (1986) |
| Copper alloys | Cu | 1 | IN 838 | Hahn and Duquette (1979) |

**Table 21-15** Literature Alloys and Environmental Systems

| Alloy Group | Group Abr. | No. | Alloy(s) | Dry Air | Ambient Air | Moist Air | Hot Air | Steam | Molten Salt | Na$_2$SO$_4$ Soln. | Chloride Solution | Acetic Acid Soln. |
|---|---|---|---|---|---|---|---|---|---|---|---|---|
| Low carbon steel | LCS | 1 | Low-carbon steel | 1 | – | – | – | – | – | – | 1 | – |
| High strength carbon steels | HS-CS | 3 | AISI 4140, AISI 4620, AISI 6150 | 1 | 1 | 1 | – | – | – | – | 3 | – |
| Cr stainless steels | 4xxSS | 3 | AISI 403, 13 Cr steel, 12.4 Cr steel | – | 3 | – | – | 1 | – | – | 2 | – |
| Cr–Ni–Mo stainless steel | 3xxSS | 3 | (2) 316 SS, 21Cr–6Ni–2.4Mo | – | 4 | – | – | – | – | – | 4 | 1 |
| High nickel alloys | Ni | 2 | Alloy 600, Udimet 720 | – | 1 | – | 1 | – | 1 | 1 | – | – |
| Zirconium alloys | Zr | 1 | Zircaloy 2 | – | 1 | – | – | – | – | – | 1 | – |
| Titanium alloys | Ti | 1 | Ti–6Al–4V | – | 1 | – | – | – | – | – | 1 | – |
| Aluminum alloys | Al | 4 | Al 6061 T6, Al 7475 T7351, Al 7075 T6, Al 5.5Zn–2.5Mg–1.5Cu | 1 | 3 | 1 | – | – | – | – | 3 | – |
| Copper alloys | Cu | 1 | IN 838 | – | 1 | – | – | – | – | – | 1 | – |

**Table 21-16** Actual or Typical Mechanical Properties for Referenced Corrosion Fatigue Data

| Alloy Group | Alloy | $\sigma_{yield}$ (ksi) | $\sigma_{uts}$ (ksi) | Fatigue Data Reference | Typical Properties Reference |
|---|---|---|---|---|---|
| LCS | Low carbon steel | 28.0 | 64.0 | Dugdale (1962) (yield est.) | — |
| HS-CS | AISI 4140 | 96 | 115 | Lee and Uhlig (1974) | — |
| HS-CS | AISI 4620 | 81.5 | 117 | Chiltum (1984) | *Materials Engineering Magazine* (1990, p. 39) |
| HS-CS | AISI 6150 | 132 | 155 | Baxa et al. (1978) | *Ryerson Stock List and Data Book* (1987–1989) |
| 4xxSS | AISI 403 | 119 | 164 | Sedriks (1979) | *Huntington Alloys Handbook* (1976) (interpolated on 23Rc actual) |
| 4xxSS | 13Cr steel | 91.4 | 115.3 | Spiedel (1977) | — |
| 4xxSS | 12.4Cr | 94.3 | 111 | Ebara et al. (1978) | — |
| 3xxSS | AISI 316SS | 42 | 84 | Spaehn (1984) | *Huntington Alloys Handbook* (1976) |
| 3xxSS | 21Cr, 6.2Ni, 2.4Mo | 63.1 | 94.1 | Amzallag et al. (1978) | — |
| Ni | Alloy 600 | 36 | 90 | Green (1987) | *International Nickel Bulletin* (1969) |
| Ni | Udimet 720 | 110.2 | 165.7 | Whitlow et al. (1982) (hypothetical data point provided; interpolated for 704°C) | — |
| Zr | Zircaloy 2 | 16.0 | 32.5 | Teeter and Hosbons (1977) | *Information and Data, Zirconium and Hafnium* (1965) |
| Ti | Ti–6Al–4V | 103 | 127 | Morton (1967) | — |
| Al | Al 6061-T6 | 40 | 45 | Tetelman and McEvily (1967) | *Materials Engineering Magazine* (1990, p. 82) |
| Al | Al 7475-T7351 | 59.0 | 70.0 | Lee (1986) | *Metallic Materials and Elements for Aerospace Vehicle Structures* (1988) |
| Al | Al 7075-T6 | 73 | 83 | Gibala and Hehemann (1984) | *Materials Engineering Magazine* (1990, p. 83) |
| Al | Al 7075-T6 | 72.5 | 82.7 | Smith and Duquette (1986) | — |
| Cu | IN838 | 34 | 58 | Hahn and Duquette (1979) | — |
| 3xxSS | 316 SS | 52.9 | 98.4 | Fasano (2000) | — |

The two-term power law model constants a and c can be determined through a linear least-squares fit.

***21-9.3.9 Model Development for Exponents.*** To estimate exponents b and d with only low-cycle fatigue data by this technique:

1. Find any reasonable model fit to the low-cycle data S(N).
2. Determine the log-log slopes $LLS|_{10^5}$ and $LLS|_{5 \times 10^4}$,
3. Determine the slope ratio, $SR = LLS|_{10^5}/LLS|_{5 \times 10^4}$,
4. Determine the ratio of yield to tensile strength, $\sigma_Y/\sigma_T$.
   (Actual data on specific material preferred, typical otherwise.)
5. Compute $P_b$ from eq. (21-37).
6. Compute b from eq. (21-36).
7. Compute $P_d$ from eq. (21-34).
8. Compute d from eq. (21-33).

***21-9.3.10 Probability Lower Bound Correction.*** As might be expected, the error interval becomes larger as the extrapolation becomes greater. This points out the need for replicate samples, experimental design that minimizes scatter, and stringent controls on testing parameters. The $2\sigma$ lower limit correction factor, $F_{LEB}$, to correct the model to the lower $2\sigma$ error interval is

$$F_{LEB} = 1 - [0.013 + 0.085 \log(\text{extrapolation ratio})] \quad (21\text{-}39)$$

Extrapolating from $10^5$ to $10^8$, for example, would result in a lower bound estimate of 0.732(model value). The ability to extrapolate is no better than the accuracy of the low-cycle data. This model assumes that the low-cycle data exhibit a $2\sigma$ error interval of no more than ±4%.

## 21-9.4 Coatings and Coverings

A number of different coatings can be applied to the process wetted parts of a mixer for chemical resistance.

***21-9.4.1 Coatings.*** Coatings are surface barriers, usually films 1/16 in. (1.5 mm) or less in thickness. The total thickness is generally built up through the application of several coats. The most commonly applied coatings are listed in the Table 21-17. Usually, the more fluorinated the carbon chain, the more chemically resistant the material is and the higher the temperature the coating can handle. Chlorine on the carbon chain also adds chemical resistance and temperature resistance, but generally to a lesser extent than fluorine. A coating thickness of 0.015 in. (0.4 mm) or greater is generally needed to ensure a surface without flaws or gaps, called '*holidays*.' The ability of a coating to resist corrosion

**Table 21-17** Commonly Applied Coatings

| Abbreviation | Chemical Name | Maximum Operating Range |
|---|---|---|
| FEP | Tetrafluoroethylene–propylene copolymer | −75 to 205°C |
| | | −100 to 400°F |
| PFA | Polyperfluoroalkoxy | −240 to 260°C |
| | | −400 to 500°F |
| PVDF | Polyvinylidene fluoride | −60 to 130°C |
| | | −80 to 265°F |
| ECTFE | Ethylene–chlorotrifluoroethylene copolymer | −75 to 150°C |
| | | −105 to 300°F |
| PVC | Polyvinyl chloride | −10 to 65°C |
| | | 15 to 150°F |
| Coal tar epoxy | Two-part catalyzed epoxy with solids | −5 to 65°C |
| | | 25 to 150°F |

depends on the tank contents and the temperature. The coating manufacturer or applicator should decide the acceptability of a coating for a given environment.

***21-9.4.2 Glass Coatings.*** Glass coatings, also called *linings*, are various proprietary formulations of glass fused at glass melt temperatures to a base of steel. Glass coatings are corrosion resistant to low-pH environments except solutions containing hydrofluoric acid (HF). High concentration of bases can also cause problems for most glass coatings. Standard glass formulations are often good to 500 to 600°F (260 to 315°C). One caution with glass is that most formulations will tolerate only about a 250°F (120°C) differential between the tank contents and the heat transfer medium. Processing procedures must be reviewed to ensure that such temperature differentials are not exceeded. Manufacturers of glass-lined equipment should be contacted for more definitive rules.

***21-9.4.3 Coverings.*** Coverings are normally applied as sheets of material and are generally about 3/16 in. (5 mm) to 1/4 in. (6.5 mm) thick. Some commonly applied coverings are described in Table 21-18. The elastomers used as coverings are generally cross-linked in autoclaves after the raw lining has been applied to the mixer shaft and impeller assembly. The degree of cross-linking will determine the hardness, which is usually measured as 'Shore hardness A.' The more highly cross-linked, the greater the corrosion resistance. Although a Shore A hardness greater than 120 is achievable, most hardness readings are 90 or less. A high hardness covering can become too brittle and microcrack, allowing the corroding material a path to the base metal.

Many coverings are used because of their ability to resist erosion when agitating abrasive materials. Maximum life in abrasive service has been observed when the Shore A hardness is within the range 35 to 45. To accommodate both corrosion and abrasion resistance, standard hardness values of these elastomers when applied to mixers are typically 35 to 70.

**Table 21-18** Commonly Applied Coverings

| Common Name | Chemical Name | Typical Maximum Temperature |
|---|---|---|
| Natural rubber | cis-1,4-Polyisoprene | 55°C, 130°F |
| Neoprene rubber | Polychloroprene | 85°C, 180°F |
| Butyl rubber | Isobutylene–isoprene copolymer | 85°C, 180°F |
| Chlorobutyl rubber | Chlorinated isobutylene–isoprene copolymer | 85°C, 180°F |
| Buna-N or nitrile rubber | Butadiene–acrylonitrile copolymer | 85°C, 180°F |
| Hypalon rubber | Chlorosulfonated polyethylene | 85°C, 180°F |

## NOMENCLATURE

| | |
|---|---|
| a | constant for two-term power model, eq. (21.32) (dimensionless) |
| $a_i$ | distance down to ith impeller [inch] {meter} |
| b | exponent in S-N two-term power model, eq. (21.32) (dimensionless) |
| $b_i$ | distance up to impeller ith [inch] {meter} |
| c | constant for two-term power model, eq. (21.32) (dimensionless) |
| $c_c$ | critical damping coefficient (dimensionless) |
| $c_v$ | damping coefficient (dimensionless) |
| d | exponent in S-N two-term power model, eq. (21.32) (dimensionless) |
| d | shaft diameter [inch] {meter} |
| $d_i$ | inside diameter of hollow shaft [inch] {meter} |
| $d_o$ | outside diameter of hollow shaft [inch] {meter} |
| $d_s$ | minimum shaft diameter required for shear [inch] {meter} |
| $d_t$ | minimum shaft diameter required for tensile [inch] {meter} |
| D | impeller diameter [inch] {meter} |
| $D_S$ | stub blade (on hub) diameter [inch] {meter} |
| $E_m$ | modulus of elasticity (tensile) [psi] {Pa} |
| $f_H$ | hydraulic service factor (dimensionless) |
| $f_L$ | location fraction for blade momentum (dimensionless) |
| f(t) | forcing function (dimensionless) |
| $F_{LEB}$ | lower 2σ bound correction factor, eq. (21.34) (dimensionless) |
| k | spring constant [lb$_f$/inch] {N/m} |
| L | shaft length [inch] {meter} |
| $L_i$ | shaft length to ith impeller [inch] {meter} |
| LLS | log-log slope (dimensionless) |
| $LLS_{10^5}$ | log-log slope at $1 \times 10^5$ cycles (dimensionless) |
| $LLS_{5 \times 10^4}$ | log-log slope at $5 \times 10^4$ cycles (dimensionless) |
| m | mass [lb$_m$] {kg} |
| M | bending moment [inch-lb$_f$] {N · m} |
| $M_{max}$ | maximum moment [inch-lb$_f$] {N · m} |
| n | number of impellers (dimensionless) |

**1330** MECHANICAL DESIGN OF MIXING EQUIPMENT

| | |
|---|---|
| $n_b$ | number of blades (dimensionless) |
| N | operating speed [rpm] {rps} |
| N | specimen life (cycles) in Weibull probability |
| $N_a$ | characteristic life (cycles) for 36.8% survival of the population, (36.8% = 1/e, e = 2.718) |
| $N_c$ | critical speed or natural frequency [rpm] {rps} |
| $P_d$ | parameter defined in (21.34) (dimensionless) |
| $P_i$ | impeller power for ith impeller [hp] {W} |
| q | shape parameter of the Weibull distribution curve (dimensionless) |
| R | stress ratio, maximum to minimum (dimensionless) |
| s | slope [radians] {radians} |
| S | fatigue strength [psi] {Pa} |
| $S_b$ | bearing span [inch] {meter} |
| $S_{LEB}$ | fatigue strength at the $2\sigma$ lower bound [psi] {Pa} |
| SR | log-log slope ratio, $LLS_{10^5}/LLS_{5\times10^4}$ (dimensionless) |
| t | time [minutes] {s} |
| $T_Q$ | torque [inch-$lb_f$] {N · m} |
| $T_{Q(max)}$ | maximum torque [inch-$lb_f$] {N · m} |
| V | shear force [$lb_f$] {N} |
| w | weight of shaft per unit length [$lb_m$/inch] {kg/m} |
| $w_d$ | deflection [inch] {meter} |
| W | impeller blade width [inch] {m} |
| $W_e$ | equivalent weight of impellers at point of calculation [$lb_f$] {N} |
| $W_{b,hydrofoil}$ | weight of hydrofoil blades [$lb_f$] {N} |
| $W_{b,pitched}$ | weight of 45° pitched blades [$lb_f$] {N} |
| $W_i$ | impeller weight of ith impeller [$lb_m$]{kg} |
| x | distance [inch] {meter} |
| $X_1$ | critical speed component (21.16) |
| $X_2$ | critical speed component (21.16) |

*Greek Symbols*

| | |
|---|---|
| $\delta$ | damping ratio (dimensionless) |
| $\rho_m$ | density of metal [$lb_m$/inch$^3$] {kg/m$^3$} |
| $\sigma_b$ | blade design stress [psi] {Pa} |
| $\sigma_s$ | combined shear stress [psi] {Pa} |
| $\sigma_y$ | yield stress [psi] {Pa} |
| $\sigma_t$ | combined tensile stress [psi] {Pa} |
| $\sigma_Y$ | 0.2% offset yield stress [psi] {Pa} |

## REFERENCES

Amzallag, C., P. Rabbe, and A. Desestret (1978). Corrosion fatigue behavior of some special stainless steels, *Corrosion Fatigue Technology*, ASTM STP 642, pp. 117–132.

Baumeister, T., E. A. Avallone, and T. Baumeister III, eds. (1978). *Mark's Standard Handbook for Mechanical Engineers*, 8th ed., McGraw-Hill, New York.

Baxa, M. S., Y. A. Chang, and L. H. Burck (1978). Effects of sodium chloride and shot peening on corrosion fatigue of AISI 6150 steel, *Met. Trans.*, **9A**, Aug., pp. 1141–1146.

Chiltum, J. F. (1984). Corrosion fatigue cracking of oil well sucker rods, in *Corrosion Source Book*, S. K. Coburn, ed., American Society for Metals, Metals Park, OH, pp. 378–382.

Crane (1990). *Engineered Sealing Products Catalog*, John Crane, International, Inc. Morton Grove, IL.

Dowling, N. E. (1993). *Mechanical Behavior of Materials*, Prentice Hall, Upper Saddle River, NJ, pp. 424–428.

Dugdale, D. S. (1962). Corrosion fatigue of sharply notched steel specimens, *Metallurgia*, **65**, 27–28.

Ebara, R., T. Kai, and K. Inoue (1978). Corrosion fatigue behavior of 13Cr stainless steel in sodium chloride aqueous solution and steam environment, in *Corrosion Fatigue Technology*, ASTM STP 642, pp. 155–168.

Fasano, J. B. (2000). *Development and generation of material behavior in a corrosion-fatigue environment*, D.E. dissertation, University of Dayton, Ohio, May.

Fasano, J. B., J. L. Miller, and S. A. Pasley (1995). Consider mechanical design of agitators, *Chem. Eng. Prog.*, Aug., pp. 60–71.

Gibala, R., and R. F. Hehemann (eds.) (1984). *Hydrogen Embrittlement and Stress Corrosion Cracking*, American Society for Metals, Metals Park, OH.

Green, S. J. (1987). Steam generator failure, in *Metals Handbook*, Vol. 13, 9th ed., L. J. Korb and D. L. Olson, co-chairs, p. 945.

Hahn, H. N., and D. J. Duquette (1979). The effect of heat treatment on the fatigue and corrosion fatigue behavior of a CuNiCr alloy, *Met. Trans.*, **10A**, Oct., pp. 1453–1460.

Harnby, N., M. F. Edwards, and A. W. Nienow (1992). *Mixing in the Process Industries*, 2nd ed., Butterworth-Heinemann, Boston.

*Huntington Alloys Handbook 10M 1-76-59* (1976). Published by Huntington Alloys, Huntington, WV, pp. 40–41.

*Information and Data, Zirconium and Hafnium* (1965). Published by Amax Specialty Metals, Akron, NY.

*International Nickel, Bulletin 12M 7-69 4463* (1969). Properties of some metals and alloys. Published by International Nickel, Inc., New York, NY.

Kane, R. D. (1993). Super stainless steels resist hostile environments, *Adv. Mater. Process.*, July, pp. 16–20.

Khan, Z., M. Younas, and G. Zuhair (1995). A fatigue-life prediction methodology for notched aluminum-magnesium alloy in Gulf seawater environment, *J. Mater. Eng. Perform.*, **4**(5), Oct., pp. 617–623.

Lee, E. U. (1986). Corrosion fatigue and stress corrosion cracking of 7454-T7351 aluminum alloy, in *Corrosion Cracking*, V. S. Goel, ed., American Society for Metals, Metals Park, OH, pp. 123–128.

Lee, H. H., and H. H. Uhlig (1974). Corrosion fatigue of Type 4140 high strength steel, in *Source Book in Failure Analysis*, American Society for Metals, Metals Park, OH, pp. 351–359.

Leeson (1994). *Practical Motor Basics*, Leeson Electric Corporation, Grafton, WI.

Madayag, A. F. (1969). *Metal Fatigue: Theory and Design*, Wiley, New York, pp. 155–156.

*Materials Engineering Magazine* (1990). Materials Selector (suppl.). Published by Penton.

*Metallic Materials and Elements for Aerospace Vehicle Structures* (1988). Military Handbook—5H, CD Version, Materials and Manufacturing Directorate's System Support Division Engineering and Design Data Section, Department of Defense, Dec.

McAdam, D. J. (1927). Corrosion fatigue of metals as affected by chemical composition, heat treatment and cold working, *Trans. Am. Soc. Steel Treat.*, **11**(5), 355–390.

McPartland, J. F., and B. J. McPartland (1990). *McGraw-Hill's National Electrical Code Handbook*, 20th ed., McGraw-Hill, New York.

Morton, A. G. S. (1967). Mechanical properties of thick plate Ti-6Al-4V alloy, *U.S. Navy Marine Engineering Laboratory Report 266/66*, Jan.

Pestel, E. C., and F. A. Leckie (1963). *Matrix Methods in Elastomechanics*, McGraw-Hill, New York.

Ramsey, W. D., and G. C. Zoller (1976). How the design of shafts, seals and impellers affects agitator performance, *Chem. Eng.*, Aug. 30, pp. 101–108.

Redmond, J. D. (1986a). Selecting second-generation duplex stainless steels: 1, *Chem. Eng.*, Oct. 27, pp. 153–155.

Redmond, J. D. (1986b). Selecting second-generation duplex stainless steels: 2, *Chem. Eng.*, Nov. 24, pp. 103–105.

*Ryerson Stock List and Data Book* (1987–1989 ed.). Published by Joseph T. Ryerson & Son, Inc., Cincinnati, OH.

Sedriks, A. J. (1979). *Corrosion of Stainless Steels*, Wiley, New York, p. 201.

Smith, E. F., III, and D. J. Duquette (1986). The effect of cathodic polarization on the corrosion fatigue behavior of a precipitation hardened aluminum alloy, *Met. Trans.*, **17A**, Feb., pp. 359–369.

Sorell, G. (1994). Choose the best alloy for incinerator heat exchangers, *Chem. Eng. Prog.*, Mar., pp. 49–60.

Spaehn, H. (1984). Electrochemical fundamentals of corrosion fatigue, in *Subcritical Crack Growth Due to Fatigue, Stress Corrosion and Creep*, L. H. Larsson, ed., Elsevier, London, pp. 275–304.

Spiedel, M. O. (1977). Corrosion fatigue in Fe–Ni–Cr alloys, *Proc. International Conference on Stress Corrosion and Hydrogen Embrittlement of Iron-Base Alloys*, Unieux-Firminy, France, June 1973, National Association of Corrosion Engineers, Houston, TX, pp. 1071–1094.

Teeter, M. C., and R. R. Hosbons (1977). Corrosion fatigue of annealed Zircaloy 2 in aqueous solutions at 575 K, *Fracture 1977*, Vol. 2, University of Waterloo Press, Waterloo, Ontario, Canada, pp. 887–896.

Tetelman, A. S., and A. J. McEvily (1967). *Fracture of Structural Materials*, Wiley, New York, p. 386.

Wei, R. P., and D. G. Harlow (1993). A mechanistically based probability approach for predicting corrosion and corrosion fatigue life, in *Durability and Structural Integrity of Airframes*, Vol. I, *Proc. 17th Symposium of the International Committee on Aeronautical Fatigue*, Stockholm, Sweden, June, pp. 347–366.

Whitlow, G. A., L. E. Willerts, and J. K. Tien (1982). Effects of frequency on the corrosion fatigue of Udimet 720 in a molten sulfate salt, in *Ultrasonic Fatigue*, J. M. Wells et al., eds., American Institute of Mining, Metallurgical and Petroleum Engineers, Warrendale, PA, pp. 321–322.

# CHAPTER 22

# Role of the Mixing Equipment Supplier

RONALD J. WEETMAN

*Lightnin*

## 22-1 INTRODUCTION

A qualified mixing equipment supplier can assist with process sizing, equipment selection, impeller options, capital versus operating cost evaluation, testing, reliability, and service. In this chapter we describe each type of support and its importance in the equipment selection process. Criteria for choosing a qualified supplier are identified.

Selection of a mixing configuration can be complex. The relationship between the open mixing impeller and its position in a mixing vessel is critical to process results. The proximity of the impeller or multiple impellers to the tank bottom, liquid surface, and tank wall can have a dramatic effect on the flow pattern and shear distribution in the tank. The placement of an inlet feed can affect the inlet dispersion and also greatly influence the yield.

To recommend a correct and optimum mixing configuration, the supplier must understand the most important process requirements and constraints of the application. This requires a close working relationship and exchange of critical application information between the supplier and the end user. Some process requirements suggest conflicting mixer characteristics. If the equipment supplier does not know what is really important in the application, he or she cannot provide the best possible solution. If the purchaser considers the details of the process to be proprietary information, he or she should ask the supplier to sign a nondisclosure agreement before providing this information.

When submitting a request for proposal (RFP), the purchaser should specify the liquids, solids, and/or gases involved in the process, the nominal and peak amounts of the constituents, and which is most important. For proprietary processes, detailed physical properties may be sufficient to describe materials. The

---

*Handbook of Industrial Mixing: Science and Practice,* Edited by Edward L. Paul,
Victor A. Atiemo-Obeng, and Suzanne M. Kresta
ISBN 0-471-26919-0 Copyright © 2004 John Wiley & Sons, Inc.

**1334** ROLE OF THE MIXING EQUIPMENT SUPPLIER

purchaser should also indicate how the constituents are introduced, whether it is a batch or continuous flow system, and which step in the process is rate controlling. In the case of a solids suspension application, the purchaser should specify the size and density of the particles, and whether off-bottom or full suspension is required. The RFP should also indicate whether it is a batch process requiring an outlet at the bottom of the tank, or a continuous process requiring a top overflow.

The purchaser should be sure to include any additional requirements that might affect the system configuration in the RFP. For example, if inlet flows, which are strong in terms of velocity or flow rate, are aimed at the impeller, the equipment supplier will need to calculate and allow for the additional forces on the mixing structure (Weetman and Salzman, 1981). The power and mechanical loads on the system are also sensitive to off-bottom clearance of the lower-mixing impeller. If the purchaser does not provide precise tank measurements, some variability of power, flow, and process may result (Oldshue, 1983).

## 22-2 VENDOR EXPERIENCE

The purchaser should seek a mixing equipment supplier with extensive experience in a variety of mixing applications, such as chemical process, fermentation, minerals, wastewater, and water treatment, even if the specific application is different. Comprehensive experience allows the supplier to adapt successful mixing configurations from other industries to the job at hand. The purchaser should request references and installation lists that show the supplier's depth and breadth of experience, including similar applications.

Experience in scale is also important. Mixing applications range from small processes requiring portable-size mixers to large processes involving millions of liters, such as those in the minerals processing industry. Considerations for mixer selection vary by scale, so it is essential for the equipment supplier to have experience in applications of all sizes, as well as proven success in process scale-up.

### 22-2.1 Equipment Selection and Sizing

Figure 22-1 outlines the steps in a typical sizing procedure. The customer usually specifies a particular process requirement, a tank diameter, and a volume. The process requirement may be solids suspension, blending, mass transfer, or combinations of the above. The mixer supplier uses this preliminary information to calculate process sizing and impeller type. From these calculations, the supplier selects a specific power requirement and impeller diameter, which in turn lead to the selection of the operating speed.

This preliminary analysis is simplistic, however, and does not take into account all necessary factors, including operating power, initial equipment cost, forces, erosion, dispersion, critical speed, and flow/shear balance. If the tank is similar to the one depicted in the middle of Figure 22-1, the impeller diameter may be

VENDOR EXPERIENCE **1335**

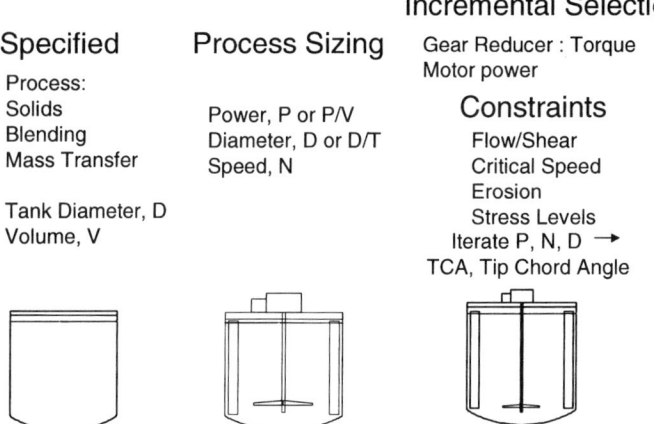

**Figure 22-1** Steps in a typical sizing procedure. First, specify the process and tank; proceed to process sizing; finish with incremental selection of mechanical aspects for the optimal configuration.

too large to provide optimum process sizing, or there may be a minimal power requirement for mass transfer in the process application. This configuration may not be efficient in terms of solids suspension, or it may create dead zones in blending applications.

After considering the process requirements, the supplier makes the incremental selection of a gear box or speed reducer. This is a critical part of the equipment specification, since the gear reducer is the largest cost of a mixing system. Gear reducers are manufactured with distinct operating speeds and torque capabilities. Although the process requirements may bring the selection close to the limit of one gearbox, a jump to the next gearbox may not be advisable because of certain ratings of particular gearboxes. The incremental values of the motors must also be considered.

Additional constraints pertaining to the flow to shear ratio and critical speed must be considered in the sizing procedure. The shaft of the mixer will vibrate at its natural frequency, much as a tuning fork vibrates when it is struck. This natural frequency is also called the *critical speed*. The supplier must be careful to avoid operating at the critical speed of the shaft and impeller.

Tip speed limits may be required to minimize erosion on an impeller, limit particle degradation, and control shear in the process. Other factors, such as mechanical stress limits on hardware, shafts, and other mechanical components, must also be considered. Normally, the equipment supplier will work with variables of power, speed, and impeller diameter to satisfy all the constraints and try to pick an optimum selection. In the case of hydrofoil impellers, the tip chord angle (TCA) can provide an additional parameter for optimization. The TCA is defined as the angle of the chord of the blade tip to the horizontal. The optimum selection is typically based on either operating efficiency or initial cost.

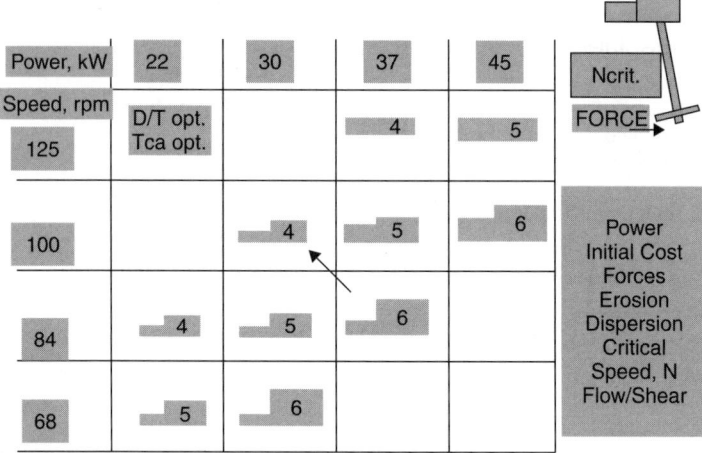

**Figure 22-2** Speed reducer selection: incremental selection of speed, motor power, and gearbox or speed reducer to optimize the process, the initial cost, and/or the operating costs.

Figure 22-2 addresses a number of issues affecting the selection of the optimum speed reducer for a process application. The chart specifies the proper speed reducer, indicated as sizes 4 to 6, at various impeller speeds. Since gear reducers are torque rated, they are positioned diagonally on the speed power chart. The power increments (in kW) on the top indicate the power of standard motors. The impeller speeds (in rpm) are indicated on the left of the chart. The figure in the upper right-hand corner shows the shaft and impeller reacting to the forces of the fluid on the impeller as a reminder that the critical speed, or natural frequency of the shaft, must be considered in the sizing procedure.

The arrow near the center of the chart indicates that it may be possible to reduce the gearbox size from 6 to 4 by creating a better ratio between the impeller diameter and the tank diameter (D/T). This can be accomplished either by switching from a pitched blade impeller to a more efficient hydrofoil impeller, or by changing the TCA of a hydrofoil impeller. An optimum D/T will reduce the power requirement for the process. This could reduce the impeller weight to achieve a corresponding increase in speed. The reduction in weight would reduce the shaft's natural frequency and permit an increase in speed. This demonstrates the optimum condition where the selection of a smaller gearbox size reduces the motor size and reduces the operating costs of the unit.

As an example, Figure 22-3 illustrates the very strong function of power versus the ratio of the impeller diameter to the tank diameter (D/T) for a solids suspension application. The optimum D/T occurs at 0.35 for this application. The ordinate shows the power versus the power at the optimum D/T. On the right side of the graph, the circled value indicates a power of 1.6 times the optimum at a D/T of 0.5. The large impeller cuts off the axial circulation loop,

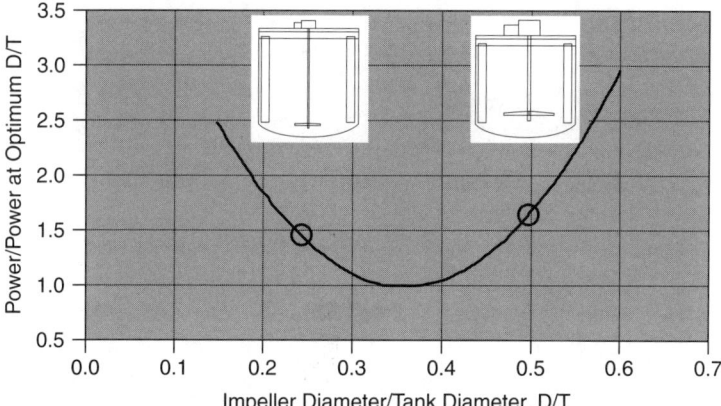

**Figure 22-3** Relative power versus D/T for solid suspension applications; there is a distinct minimum in power requirement at $D/T = 0.35$.

which requires a good sweeping motion around the blade tip in order to carry the entire flow up the sidewall of the tank. For turbulent flow, a very large D/T causes a recirculating flow, which results in an upflow beneath the impeller (see Figure 6-20). The larger D/T will also require a much larger gear reducer to deliver the same power. Changing the TCA of the impeller to produce the optimum impeller diameter could result in a lower power for the process and a smaller gear reducer.

The circled value on the left shows an increased power requirement of approximately 1.4 times the optimum when a D/T of 0.25 is selected. This is because the reduction in D/T causes a decrease in primary flow. These two examples of an important application demonstrate the significant difference between the power requirements for varying D/Ts. A power savings of up to 60% is achieved simply by changing the D/T to its optimum value of 0.35. This means that the sizing process needs to be a function of D/T. The effect of D/T on solid suspension is illustrated on the CD-ROM.

### 22-2.2 Scale-up

The purchaser should look for an equipment supplier who has good scale-up rules and experience in testing on different scales. One model test in the lab, shown in Figure 22-4, was scaled up in volume by a factor of 5000, from an 0.08 m$^3$ sample test tank to a 400 m$^3$ gold slurry production vessel, as shown in Figure 22-5. The process scale-up factor may be of the form: (power/volume)$^n$. The exponent on power/volume, n, may vary from 0.82 to 1.0. This gives power requirements that differ by a factor of 4.6 in the final full scale requirement: The required full scale power may be 50 or 230 kW, depending on the exponent used. The equipment supplier should also have many different tank sizes that can be

**1338** ROLE OF THE MIXING EQUIPMENT SUPPLIER

**Figure 22-4** Scale-model testing laboratory: testing and discussion between the customer and supplier clarify what is important in the process and allow the optimal design.

**Figure 22-5** Full scale gold slurry installation.

used to check scale-up and test full scale. Figure 22-6 shows a 3.2 m circulator impeller being tested in a 1900 m$^3$ test tank.

## 22-3 OPTIONS

### 22-3.1 Impeller Types

The figures below show examples of many types of impellers that are used for different applications. Availability of a broad range of impeller types permits the

OPTIONS  **1339**

**Figure 22-6**  Full scale testing: a 3.2 m impeller in a 1 900 000 L test tank.

equipment supplier to select an appropriate impeller and to optimize the process and the system cost (Weetman and Oldshue, 1988).

Impeller types range from high-flow/low-head impellers to high-shear impellers. Figure 22-7 shows a high-flow, low-head impeller that can be used for a wide variety of applications from mild blending to solid suspensions. This type

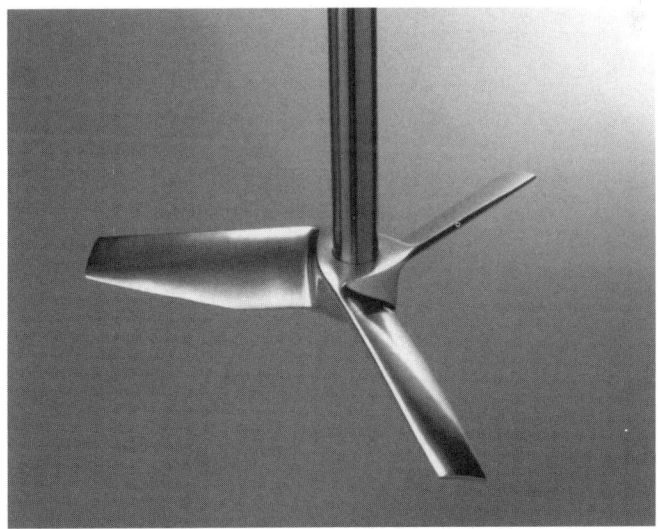

**Figure 22-7**  High efficiency flow impeller: $N_p = 0.3$.

of impeller has low operating costs and has a power number of approximately 0.3 for the turbulent regime. The power number is defined as

$$N_p = \frac{P}{\rho N^3 D^5}$$

The mixing power number is almost dimensionless but has units of revolutions raised to the $-\frac{1}{3}$ power.[1] The impeller in Figure 22-8 can handle three to five times more gas than the impeller shown in Figure 22-7 and is excellent for high viscosity and gas applications. Its power number is approximately 0.75. It obviously has more metal and weight than the same diameter impeller in Figure 22-7 and would therefore cost more than the other option. The weight is also an important consideration because it determines the maximum operating speed to keep the mixer running below its critical speed. The heavier the impeller, the lower the critical speed.

The impeller in Figure 22-9 is a pitched blade turbine that produces a mixed flow pattern. In other words, the discharge flow from the impeller has both a radial and an axial component. The high-angle 45° blade adds shear to the process due to the separated flow off the suction side of the blade. The pitched blade impeller creates less flow and more turbulence than the previous two impeller types. Its power number is approximately 1.3.

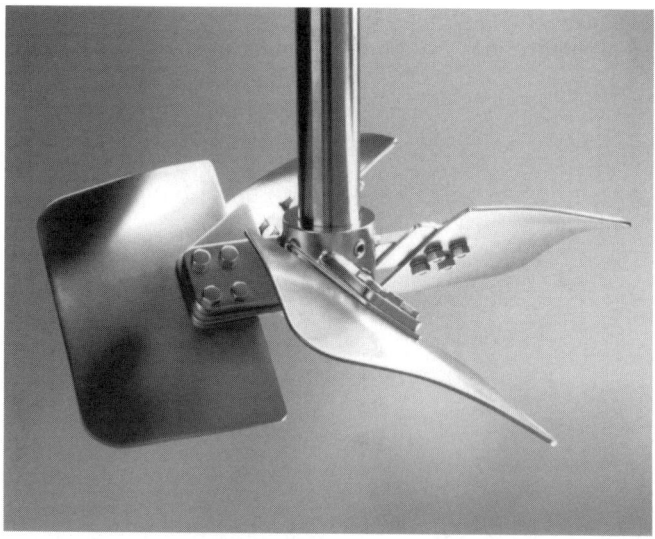

**Figure 22-8** High solidity hydrofoil impeller for high viscosity or gas applications: $N_p = 0.75$.

---

[1] A factor of $(2\pi)^{-3}$ would convert rotations to radians. This changes the power number by a constant factor, so for practical convenience N is always used in rps.

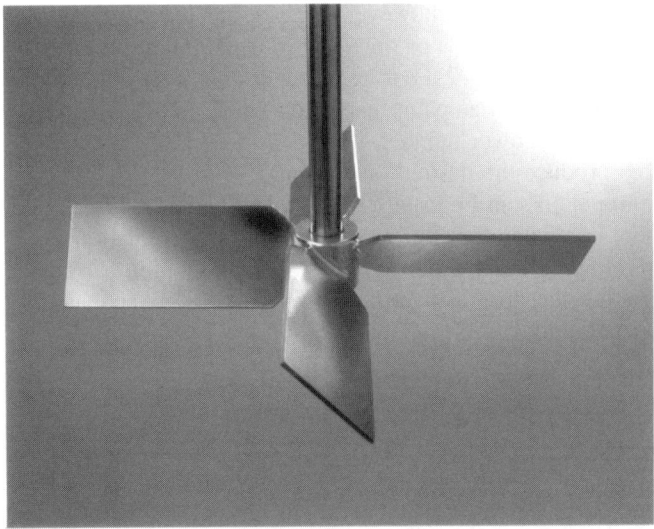

**Figure 22-9** Pitched blade impeller for mixed flow applications: $N_p = 1.3$.

**Figure 22-10** Rushton or radial flow disk impeller for intense turbulence: $N_p = 5.2$.

The fourth impeller type, shown in Figure 22-10, is the classic radial flow or Rushton impeller. It has an impeller disk with attached blades and produces a radial flow pattern and staging in a mixing tank. It has a power number of approximately 5.2. It generates high levels of turbulence but not a great deal of circulation or flow. It is relatively expensive to operate and does not provide

**1342** ROLE OF THE MIXING EQUIPMENT SUPPLIER

good top-to-bottom blending; however, applications requiring intense turbulence and the ability to disperse gas or other constituents might require this design.

The impeller in Figure 22-11 is referred to as a *high-shear impeller* and is used in applications such as paint dispersion where very high shear is required. It has a much lower power number, on the order of 0.45, but it is run at a very high speed to produce the desired dispersion.

Multiple impellers can be used as shown in Figure 22-12. The configuration shown in this computational fluid dynamics (CFD) simulation uses three high solidity, up-pumping axial flow impellers. The simulation shows the flow pattern by depicting the flow of neutral density particles (Weetman, 1998a). Other impeller types are discussed in Chapter 6.

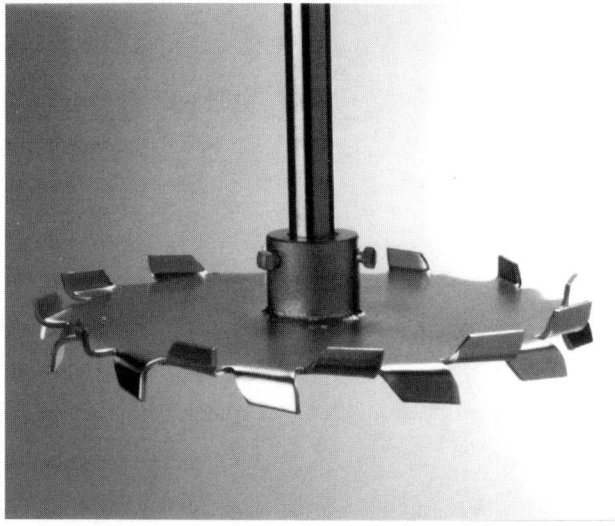

**Figure 22-11** High-shear impeller: used when a fine dispersion is required: $N_p = 0.45$.

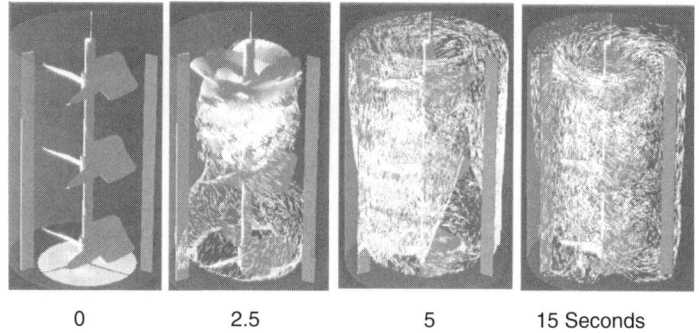

**Figure 22-12** Computational fluid dynamics plots of multiple up-pumping Lightnin A340 impellers: $N_p = 0.75$ for each impeller.

## 22-3.2 Capital versus Operating Costs: Torque versus Power

The trade-off of capital versus operating costs is a key consideration in the selection of a mixing system. As we saw in Figure 22-3, there is a large difference in power required for solid suspension processing as a function of D/T. For an impeller smaller than optimum, an increased impeller diameter would reduce operating costs (power). A larger impeller diameter usually means a larger shaft and drive, because of a greater torque requirement. The capital costs would therefore be much higher than for a smaller diameter or D/T ratio. An analysis of power costs is required to determine the best option. The cost of the impeller versus the process yield must also be considered. If a better mixing impeller or combination of impellers can yield better process results, a higher initial cost could create a better return on investment.

## 22-4 TESTING

There are many different types of testing that can be performed by a mixing equipment supplier. They range from customer sample testing to full scale witness testing. Tests may also include examination of the flow pattern with laser Doppler velocimetry (LDV) or computational fluid dynamics (CFD).

### 22-4.1 Customer Sample Testing

Customer sample testing, as shown in Figure 22-4, is usually done on a small scale (e.g., a 0.45 m diameter tank) with the impeller types under consideration for the mixing system. Once optimized on a small scale, the mixing results are scaled up to full-size equipment, according to the supplier's scale-up rules. Tremendous care is needed during the model testing. For example, if the same viscosity material is used in both sample and scale-up, the flow can be in the transitional regime on the small scale, but in the turbulent regime at full scale. Viscosity reduction must be included in the modeling to ensure comparable conditions in sample and full scale operation. Applications involving gas can also be difficult to model. On a small scale, the flow pattern can be overcome very easily with a small amount of gas, but this may not be the case at full scale.

Sometimes more elaborate process conditions must be tested, as shown in Figure 22-13. In this example, a high-pressure, high-temperature sulfide oxidation process is being tested in the laboratory to obtain process data for the full scale installation. The sample test is usually done as a pretest, followed by testing in the presence of the customer (see Figure 22-4). The pretest allows an opportunity to test options that can optimize the process and see the impact of changes in the mixer configuration.

Some testing requires the equipment supplier to have hazardous laboratory capabilities. In some cases, when it is not possible to handle the materials or duplicate the process in the supplier's test facilities, model tests can be done at the customer's site, using small scale impellers and tanks provided by the supplier.

**1344** ROLE OF THE MIXING EQUIPMENT SUPPLIER

**Figure 22-13** Laboratory test of high-temperature, high-pressure oxidation reactor.

### 22-4.2 Witness Testing

Witness testing at full scale is sometimes required. Figure 22-14 shows a witness test of a surface aeration impeller operating in a 15 m square tank with a 2000 m$^3$ (500 000 gal) capacity. The supplier is testing the oxygen uptake of the mixer for a waste treatment application. Witness tests are sometimes done at the actual installation after construction has been completed. Other tests performed might include flow velocity studies. For example, a purchase order for a large circulator precipitator might include a flow guarantee that the supplier would validate using an on-site flow velocity study.

**Figure 22-14** Aeration witness test in a 15 m square by 9 m deep tank.

### 22-4.3 Laser Doppler Velocimetry

Some equipment suppliers use laser Doppler velocimetry (LDV), as shown in Figure 22-15, or particle image velocimetry (PIV) to measure the velocity or flow field in a mixing vessel. These nonintrusive technologies can be used to develop new impeller designs or to measure equipment performance to optimize the mixing system configuration. Laser Doppler velocimetry (Weetman and Salzman, 1981) can also be used to measure the major flow patterns in the vessel and to identify low velocity regions, which are undesirable in most applications. Typical low velocity regions occur near the top of the tank (where feeds are often introduced) and underneath a large pitched blade impeller (if used) at the bottom of the tank.

Particle image velocimetry (PIV) is a newer technology that analyzes a velocity field in a spatial region or over the entire tank if the tank is small enough. It is especially useful for examining instantaneous spatial shear rates, but it is not as accurate as LDV for time-averaged measurements.

### 22-4.4 Computational Fluid Dynamics

Computational fluid dynamics (CFD) technology is an excellent tool for studying mixing applications (Weetman, 1994, 1998b). As computers are becoming faster and less expensive, CFD software is becoming more cost-effective. With the increasing availability of affordable hardware, the CFD software companies have put more options into their packages to provide more realistic modeling of the flow field in the mixing vessel. Figures 22-16 and 22-17 show a velocity field

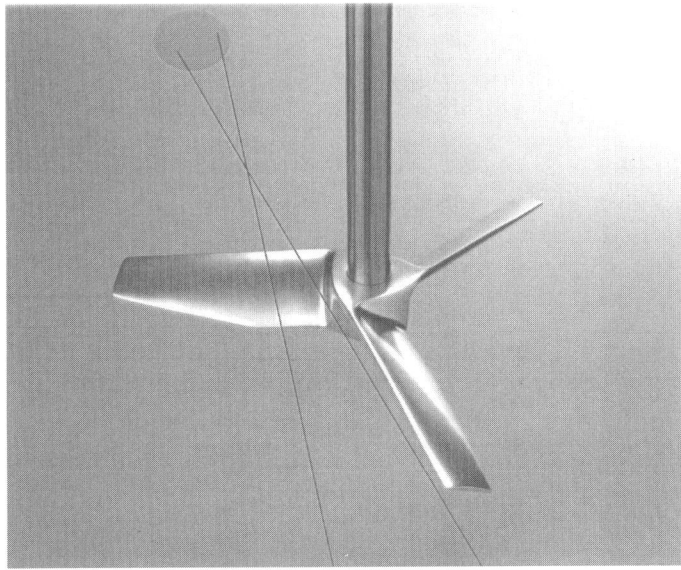

**Figure 22-15** Laser Doppler velocimeter measuring velocity near a mixing impeller.

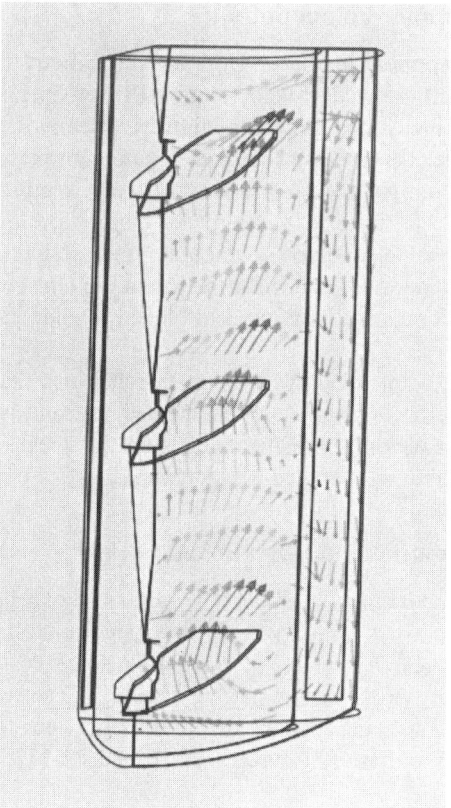

**Figure 22-16** CFD simulation of three up-pumping impellers.

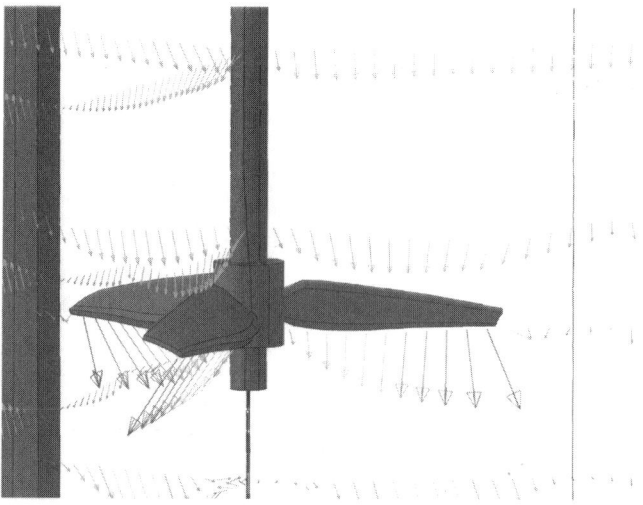

**Figure 22-17** CFD simulation of impeller: 3D using multiple-reference frame model.

near the impeller and demonstrate the capability of three dimensional modeling of the individual impeller blades. CFD modeling can be performed using velocity input from the outlet of the impeller or by modeling the impeller geometry as part of the simulation. The velocity output is usually obtained by laser Doppler velocimetry, as discussed in Section, 22-4.3, which is a quicker analysis than the 3D impeller geometry analysis. The former can be performed in minutes, whereas the 3D analysis may require days or weeks. The time required depends on the complexity of the impeller geometry and on whether the geometry has to be built. Preprocessors that convert mixer geometry to the general purpose CFD code can simplify tremendously the use of CFD. Multiphase or chemical reaction problems will take longer than simple models. CFD models are getting better each year, and it is now possible to calculate the power number or flow number with an accuracy of a few percent. CFD turbulence models are also getting better at predicting the turbulence qualities. An example of the use of CFD is the development of the circulator impeller shown in Figure 22-6, as discussed in Weetman (1994). See Chapter 5 for a full discussion of CFD in mixing applications.

## 22-5 MECHANICAL RELIABILITY

There are two major components to mechanical reliability: (1) knowing the fluid loads on the mixer from the mixing configuration, and (2) manufacturing techniques which ensure that the equipment can handle these dynamic loads. Dynamic fluid loads can cause fatigue and eventual failure of mixing system components, unless special manufacturing techniques, such as full penetration welds, have been used to prevent failure. It is essential to understand the physics of fluid loads to appreciate the importance of these manufacturing techniques to mechanical reliability. An experienced supplier knows that the loads on the mixing system and thus the installed dynamic equipment will have fewer mechanical problems. This is another reason why it is advantageous to give the supplier a process description instead of just a mechanical specification.

### 22-5.1 Applied Loads Due to Fluid Forces

The applied loads on a mixing vessel are illustrated in Figure 22-18. They consist of torque, bending, and thrust loads, which are transmitted from the mixing impeller to the shaft and drive. Because of the dynamic nature of the fluid flow, the blade loading can vary by as much as $\pm 40$ to $\pm 100\%$ (Oldshue, 1983). Mixing equipment suppliers use both model and full scale measurements to correlate these loads. Figure 22-19 shows a strain-gauged model impeller, which was used to test the loading from the blades. Fluid and mechanical loads can also be measured from the deflection of the gearbox. Equipment suppliers may also use fatigue testing, as shown in Figure 22-20, or computer modeling techniques such as finite element analysis (FEA) to determine the impact of load forces on their mixing systems.

**1348**  ROLE OF THE MIXING EQUIPMENT SUPPLIER

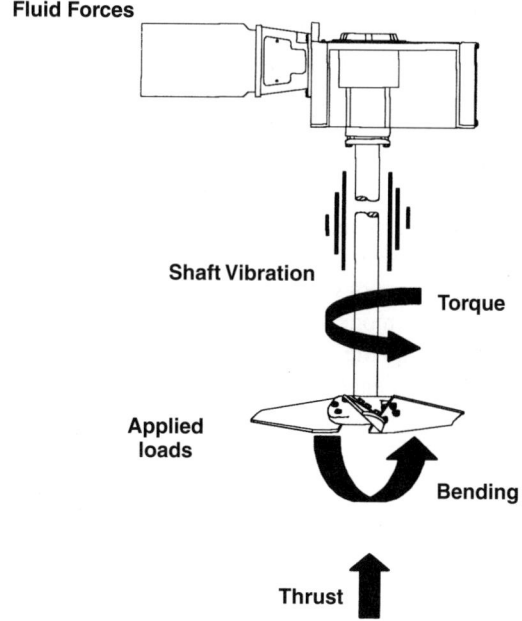

**Figure 22-18**  Applied loads on a mixer shaft due to fluid forces.

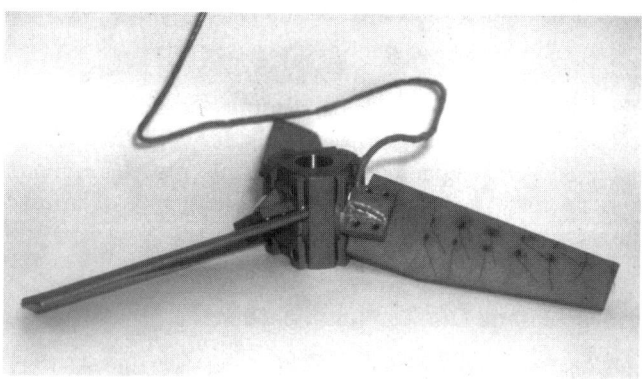

**Figure 22-19**  Impeller with strain gauges to measure blade bending loads.

### 22-5.2  Manufacturing Technologies

Large bending loads require special manufacturing techniques, such as heavy-duty welding and bolting, to ensure optimum performance and service life of the mixing system. Dealing with an equipment supplier who has an ISO 9001 registration or similar substantiation of a documented quality system is the purchaser's best assurance of consistent manufacturing procedures.

**Figure 22-20** Blade fatigue test stand to simulate bending loads on a composite impeller.

## 22-6 SERVICE

It is important to have a good service organization to provide support after the arrival of the mixing system (Figure 22-21).

### 22-6.1 Changing Process Requirements

Changes in customer requirements and/or unanticipated mechanical problems in the mixing system may require service after the sale. Process requirements may change, or the customer may discover that there were omissions in the specifications when the mixer was ordered. Mechanical problems could include an unanticipated frequency interaction between the mixer and the structure. Another problem might be additional loads on the mixer that were not understood when the system was configured. For example, gas distributions or liquid inlet flows might not be uniform, or the flow rates might exceed the initial specifications.

Regardless of the source of the problem, it is important for the equipment supplier's service organization to respond quickly. Supplier and purchaser need to discuss the requirements, the problem, and the accuracy of the installation

**Figure 22-21** Service engineer adjusting speed reducer.

to determine what modifications are necessary. Measurements such as strain or fatigue analysis can be very helpful in isolating and solving the problem. Since mixing is a very dynamic situation, vibration measurements may be needed to verify the calculated natural frequency of the entire mixer configuration, as well as other harmonic frequencies that might be present on the mixing structure and vessel. The shaft natural frequency (critical speed) and its high harmonics must be avoided throughout the system. Standard vibration instrumentation usually has a frequency range too high for mixing equipment, so the equipment supplier should have special instrumentation to conduct the analysis. See Chapter 21 for more details.

### 22-6.2 Aftermarket and Worldwide Support

Regular maintenance of the mixing system will maximize the life of the equipment. It is very desirable to have aftermarket support for preventive and predictive maintenance and service upgrades. This will enable the customer to budget and plan for needed maintenance procedures and parts. The service organization can check the equipment periodically against process requirements and suggest modifications or upgrades if the process has changed over time. It is not uncommon for mixing systems to be used for processes that were not anticipated in the initial offering, and a service engineer can assist in adapting or modifying the equipment for the new process. Equipment suppliers may need to offer worldwide support services. Global companies often want to have similar processes and equipment in multiple plants in various countries. Equipment ordered in one country may be shipped to and installed in another country. A worldwide service organization can provide better local support, installation, and long-term service.

## 22-7 KEY POINTS

1. The customer and equipment supplier need to work closely together to define the process requirements and constraints so that an optimum process and mixer configuration is selected.
2. The experience of the mixer supplier can be very important if scale-up rules and/or experience in pretesting is required.
3. In sizing a mixer, these are some of the items to consider:
   - Process requirements
   - Power/volume (P/V)
   - Impeller diameter/tank diameter ratio (D/T)
   - Impeller speed (N)
   - Tip speed ($\pi$ND)
   - Torque
   - Operating power
   - Flow/shear requirements
   - Critical shaft speed ($N_{crit}$)
   - Erosion/particle degradation
   - Mechanical stress levels and applied loads due to fluid forces
4. Scale-up experience at different scales is important as well as knowing what parameters are required to analyze for small scale testing.
5. The supplier should have a range of impeller types to choose from for the numerous processes that are encountered. Impellers required range from high flow, through high viscosity, gas controlled, mixed flow, radial flow, and high shear to specialty designs.
6. There is usually a trade-off between initial capital cost and operating cost. The most cost-effective option is usually the one that will produce the most yield for your process.
7. The supplier can be of great help in small scale testing and in process analysis to obtain the best solution for your mixing application.
8. The ability to test the process at full scale can be of great value.
9. Tools like laser Doppler velocimetry and computational fluid dynamics along with process testing are useful for analyzing mixing requirements.
10. Along with a process solution, mechanical reliability is a necessary part of a mixing solution. If the mixer is not reliable, substantial additional costs are incurred if the process is shut down even for a short time. The supplier needs to know the mechanical loads on the system and the customer needs to inform the supplier of conditions that might affect the loads on the system. This would include process changes, process upsets, and mixer configuration changes such as inlet flows and gas rate changes.
11. Aftermarket support for process changes or upgrades is important for the future and for global operations.

## REFERENCES

Oldshue, J. Y. (1983). *Fluid Mixing Technology*, McGraw-Hill, New York.

Weetman, R. J. (1994). Development of an erosion resistant mixing impeller for large scale solid suspension applications with CFD comparisons, presented at the 8th European Conference on Mixing, Cambridge.

Weetman, R. J. (1998a). Examining flow structures in a mixing vessel using computational fluid dynamics, presented at the AIChE Annual Meeting, Miami Beach, FL.

Weetman, R. J. (1998b). Automated sliding mesh CFD computations for fluidfoil impellers, presented at the 9th European Conference on Mixing, Paris.

Weetman, R. J., and J. Y. Oldshue (1988). Power, flow and shear characteristics of mixing impellers, presented at the 6th European Conference on Mixing, Pavia, Italy.

Weetman, R. J., and R. N. Salzman (1981). Impact of side flow on mixing impeller, *Chem. Eng. Prog.*, June.

# INDEX

**Acronyms**

| | |
|---|---|
| CFD | computational fluid dynamics |
| RTD | residence time distribution |
| LES | large eddy simulation |
| DNS | direct numerical simulation |
| RANS | Reynolds averaged Navier-Stokes equations |
| RSM | Reynolds stress model |
| CoV | coefficient of variation |
| LDA | laser Doppler anemometry |
| LDV | laser Doppler velocimetry |
| PIV | particle image velocimetry |

**Examples are identified in bold face type**
*Cross references are italicized*

Adsorption   545
Adsorption columns, residence time distribution   2
Aeration   353, 595, 1142
Agglomeration, solids   914, 921, 927, 1192
Agricultural chemicals   xxiii, **729**, 1027
Alkylation reaction   1185
Allowable stresses, table   1291
Alloys   1319
Anchor impeller   385, 872, 993, 1030, 1255
Animal cell culture   1139, 1150
Arrhenius model, reaction   268
Asymptotic directionality   110, **120**
Axial dispersion, pipeline mixing   418
Axial flow impeller   354, 361, 1032, 1049, 1296, 1308

*Bacillus subtilis*   1105, 1110
Backmixing, feed pipe   784, 823
Baffles   348
  2D-CFD   289
  glass-lined   351, 725, 1030, 1034
  heat transfer   876, 881
  liquid–liquid mixing   720, 724
  solids suspension   579
Batch mixing, viscous   1017, 1020, 1021
Batch reactor   780, 1029
  heat transfer   1029
  multipurpose plants   1029
  pharmaceuticals   1028
  residence time distribution   2
  scale-up   1035
Batchelor scale   49, 769, 775
Bearings   1273
  gear reducers   1273
  ratings   1275

---

*Handbook of Industrial Mixing: Science and Practice*, Edited by Edward L. Paul, Victor A. Atiemo-Obeng, and Suzanne M. Kresta
ISBN 0-471-26919-0   Copyright © 2004 John Wiley & Sons, Inc.

Belt drives   1277
Beta-PDF, reaction model   270, 853
Bimodal drop size distribution   644
Bin blender, solids   911
Biomass, non-Newtonian   1074
Biomass, yield stress   1074
Biopolymer, fermentation   1113
Bioreactors *see fermentation*
Black liquor, pulp   1189
Blend time   xiv, xl, 507
   CFD simulation   291, 301, 314, **315**, 843
   compartmentalization   606, 724
   confidence interval   511
   Corrsin   52, **57**, 510, 533
   definition   173, 764
   fermentation   1090
   gas–liquid mixing   590, 605, 1090
   impeller retrofit   512
   impeller selection   379, 511, 520, 529
   in pipes   **57**, 437, 438, 459
   jet mixing   534, **536, 537**
   laminar   527, **530**
   liquid–liquid mixing   724
   multiple impellers   513
   Newtonian, miscible   379, 508, **514**, 767
   rating problem   512, **515**
   shear-thinning fluids   518
   tall tanks   513
   transitional flow   510
   turbulent flow   509
   yield stress fluids   **525**
Blend time measurement   96, 164, 508
   colorimetric methods   167
   data analysis   172
   flow visualization   165
   indicators   168
   probe location   167
Blending, laminar   vii, 89, 527
Blending, petroleum   1171
Blending, solids *see powder blender*
Boiling   622
   cavitation number, table   614
   gas hold-up   622
   gassed power   614, **616**
   Smith number   614
   ungassed   612
Bottom-entering mixer   1253, 1278
Bottom port, stirred tank   376
Boundary layer   274
Bourne reactions   785, 822, 843, 846
   *see competitive-consecutive reactions*

   engulfment model   770, 776, 829, 834, 846, 1218
   parallel reactions   784, 786
   table   786
Boussinesq hypothesis, turbulence model   73, 262
Bubble size   632
Bubble size, measurement   199
Bypassing, residence time distribution   9

Calorimetry, reaction *see reaction calorimetry*
Capillary number , drop breakup   653
Causticization reaction   1193
Cavern formation   521, 530, 1114
   pulp   1200
   xanthan gum   1114, 1116
   yield stress fluids   521, **522, 526**, 1080
Cavitation number, boiling   614
Cavities, gas–liquid mixing   194, 598, 602, 608, 1081
CD6 impeller   331, 356, 597, 609, 614
Cell culture   xxiii, 1139
   animal cells   1139, 1150
   *see cell damage*
   equipment design   1151
   plant cells   1152
Cell damage   1139
   bacterial   1138
   characterization   1133, 1142
   circulation time distribution   1127
   due to aeration   1142, 1144, 1146
   due to bubble bursting   1142, 1144, 1146
   due to foam   1142, 1144, 1146
   due to turbulence   1076, 1127, 1139
   microcarrier cultures   1148
   plant cells   1152
   shear-sensitive cells   1140, 1176
   surfactants   1146
   suspension cultures   1141
Cell growth, model   69
Central differencing, CFD   278, 280
CFD   x, 257, 833
   blend time   291, 301, 314, **315**, 843, **845**
   circulation pattern   **80**
   equipment vendor role   1345
   extruder   **322**
   fermentation   **319**
   flow number   **294**, 314
   free surface   290, 300
   gas–liquid   290, 301, **320**, 604
   heat transfer   267

helical ribbon impeller  332
impeller design  **330**
laminar mixing  110, **119, 128, 332**
limitations and benefits  14, 81, 337, 856
macroinstability  **76**, 298
mass transfer  265
mesomixing  843
micromixing  270
Monte-Carlo coalescence-dispersion  848, **849**
multiphase flow  271, 290
non-Newtonian fluid  272, **321**
particle tracking  302
polymer processing  **322**
power number  **294**, 313
process objectives  302, 315
pulp and paper  1224
pulp chest  **321**
residence time distribution  14
scale-up  856
solid–liquid mixing  290, 301, **318**
start-up  297
static mixer  **325**, 414
tank geometry  258
time-dependent  284, 297
turbulence  291, 301
yield stress fluid  **321**
CFD, analysis of results  302
  animation  308
  helicity  311
  LES  315
  particle tracking  307
  path lines  306
  rate of deformation  311
  streamlines  305
  velocity vectors  304
  vorticity  310
CFD, numerical issues  273
  boundary layer  274
  central differencing  278, 280
  convergence  283
  differencing schemes table  280
  discretization schemes  277, 278
  finite difference method  281
  finite element method  281
  finite volume method  277
  grid generation  273, 303
  hardware  284, 336
  log-law of the wall  275
  mesh superposition  **323**
  numerical diffusion  284
  numerical methods  273
  parallel processing  284
  QUICK  279, 280
  SIMPLE algorithm  282
  structured grid  274
  *see turbulence model*
  underrelaxation  283
  unstructured grid  274
  upwind differencing  279, 280
  wall function  274
CFD, reaction  268, **316,** 833
  Bourne reactions  846
  closure equations  836
  closure models, table  834
  closure, beta-PDF  853
  combustion  407, 846
  engulfment model  846
  exothermic  854
  LDPE  **328**
  paired-interaction closure  850, **851**
  probability density function  37, 270, 853
CFD, stirred tank modeling  x, 285, 292
  baffles in 2D  289
  CPU time, figure  335
  impeller data  289
  impeller modeling  285
  multiple reference frame  292
  multiple shafts  **323**
  rotating reference frame  292
  sliding mesh  295
  snapshot model  300
Chaos *see laminar mixing*
Chemineer Greerco rotor-stator  483, 485, 487, 488, 489, 491
Chromatography column, residence time distribution  2
Circulation pattern  77, **80**, 203, 312, 366
Circulation time distribution  69
  cell damage  1127
  fermentation  1073, 1090, 1105
  liquid–liquid mixing  729
  measurement  70, 1102
Close-clearance impeller  385, 1254, 1308
Closure equations
  *see CFD, reaction*
  *see turbulence model*
Cloud height, solids suspension  562
$CO_2$ reduction reaction  1174
Coalescence  641, 679
  classification  685
  classification, table  686
  coagulum  707
  emulsion polymerization  707
  gas hold-up  621

**1356** INDEX

Coalescence *(Continued)*
  gas–liquid mixing  199
  *see liquid–liquid dispersion*
  pH  684
  practical implications  690, 696
  reaction  **738**
  scale-up  681, 685, 735
Coalescence, effect of
  electrolyte  707
  fine solids  718, **738**
  interfacial tension  683
  key variables  683, 690, 707
Coalescence, process of  682
  at an interface  687
  colliding drops  687
  collision frequency  642
  dispersion  713
  film drainage  642, 683, 687
  film thickness at rupture  689
  frequency  692
Coatings, equipment  1327
Coefficient of variation  410
  drop size distribution  645
  mixing quality  1231
  pipeline mixing  410, 416, 432
  pipeline mixing, figures  414, 416, 435
  pulp and paper  1231
  sample size effect  97
  static mixers  410, 414, 432, 435, 439, 455
Colloid mills  482
Combustion  407, 846
Compartmentalization, Rushton impeller  724
Compartment model  1099
Competitive-consecutive reaction  758, 764, 1039
  *see Bourne reactions*
  concentration effect  822
  peroxymonosulfate  1190
  scale-up  782, **785**, 787, 794, 827, 830
  selectivity  **758**, 782, 796, 806, 811
  simulation  843
Complete segregation  1, 5, 11
Computational fluid dynamics *see CFD*
Concentrated liquid–liquid dispersion  648, 663, 704, **732**
Concentration of reagents *see reaction*
Conductivity probes  178, 185, 187, 189
Constant temperature anemometry  203, 204, 205
Continuous mixing, viscous  1000, 1007, 1009, 1012
Continuous phase, definition  639

Continuous reactor *see reactor*
Convection, definition  259
Convective mixing, solids  912
Convergence, CFD  283
Conversion, definition  764
Conversion tables, units  lviii, lix
Corrosion  1319
Corrsin time scale  51, **57**, 510, 533, 763, 769, 771, 833, 836, 839, 849
Cosmetics and detergents  407
Costs
  capital vs operating  1343
  poor mixing  xxxv, 1237
  torque vs. power  1343
Couplings  1303
CoV *see coefficient of variation*
Critical micelle concentration  715
Critical speed, shaft design  1287, 1296, 1304, 1305, 1335
Crystallization  xxiii, xlviii, 1057
  crystal shape, table  1060
  growth  820, 858, 1028, 1061
  mass transfer  1061
  mixing  545, 820
  model  69
  nucleation  820, 858, 1059
  Ostwald ripening  717
  reactive  819, 1028, 1053
  regulatory control  1058
Crystallization equipment
  fluidized beds  1058
  impinging jets  1058
  multipurpose plants  1058
  scale-up  1056, 1057, **1054**, 1062, **1063**
CSTR  5, 778
  completely segregated model  5
  damping  9
  exponential distribution model  5
  first appearance time, table  7
  reactor volume, vs plug flow, figure  374
  residence time distribution  5, **12**, 15
  yield limits  12
Cumulative number frequency, drop size distribution  644
Cut diameter, drop size distribution  643

Damkoehler number  32, 756, 765, 823
  diffusion  259
  dispersed phase  639
  dispersive mixing  988
  distributive mixing  988
  drop size distribution, characteristics  643

drop size, mean  643
Eulerian frame of reference  62
fully isotropic turbulence  63
fully turbulent flow  63
granular solids  899
homogeneous turbulence  63
intensity of segregation  22
Lagrangian frame of reference  69
locally isotropic turbulence  63
macromixing  24, 765
mesomixing  24, 766
micromixing  24, 765, 768
mixing  xxxiv, 22
mixing and reaction  **32**, 756, 761, 777, 1039, 1043
reaction time  765
Reynolds number  394
rheology  148
Sauter mean diameter  643
scale of segregation  22
scale-up  1058
scale-up protocols  826
selectivity  764
turbulence  20, 45
turbulent shear flow  63
vorticity  310
vvm  1079
yield  764
Davies figure, drop size in equipment  500, 666
Definition
 blend time  173, 764
 continuous phase  639
 convection  259
 conversion  764
Deflaking, paper  1210
Deinking, lumen loading  1213
De-mixing, solids  887, 901, 906, 914, 926
Desalting, petroleum  1183
Design of equipment *see equipment specification*  429, 431
Desorption  545
Diagnostic chart  l
 dust explosion  980
 feed location  l
 gas–liquid reaction  li
 liquid–liquid extraction  lii
 liquid–liquid reaction  liii
 reaction  26
 slurry withdrawal  liv
 solid–liquid reaction  lv
 solids mixing  937, 938
 solids removal  liv

Differencing schemes, CFD  280
Diffusion *see mass transfer*
Dilute liquid–liquid dispersion, drop breakup  648, 659, **731**
Direct numerical simulation *see turbulence model, DNS*
Discretization schemes, CFD  277, 278
Dispersant  647
Dispersed phase, definition  639
Dispersed phase concentration, liquid–liquid mixing  647
Disperser operation, tip speed  1193
Dispersion
 floating solids  564
 solids  544
 viscous mixing  988
Dispersion and coalescence *see liquid–liquid dispersion*
Dissolving reagent  **796**, 819, 1044, 1052
Dissolving solids  545, 547, 556, **570**
Distillation column, residence time distribution  2
Distributive mixing, definition  988
DNS *see turbulence model*
Double cone blender, solids  908, 910
Double-motion impeller  1256
Draft tube  351, 724
Drawdown, floating solids  380, 564, 1192
Drop breakup  649
 capillary number  653
 coalescing system  676
 collision with walls  641
 dilute  659
 $d_{max}$, Davies figure  500, 666
 energy dissipation  658, 660
 Grace curve  654, 991
 intermittency  667
 laminar flow  651
 *see liquid–liquid dispersion*
 maximum stable drop size  657, **739**
 mechanisms  639, 641, 650, 713
 models  667
 power per volume  660, 666
 shear  641
 surfactant, role of  668
 time varying flow  650, 667
 turbulence  44, 650, 656, 658, 660
 viscosity ratio  654
 Weber number  658
Drop breakup, pipeline mixing  441
 drop size, laminar  447
 drop size, turbulent  443
 laminar de-mixing  461

**1358** INDEX

Drop breakup, pipeline mixing
  (*Continued*)
  large viscosity ratio **454**
  turbulent coalescence  461
  turbulent dispersion  **452, 454**
Drop size
  correlations, table  669
  drop size classification, table  646
  laminar pipe flow  447
  *see Sauter mean diameter*
  turbulent pipe flow  443
Drop size distribution  643, 662, 697
  bimodal  644
  control of  644
  CoV  645
  cumulative number frequency  644
  cut diameter  643
  Gaussian  644
  *see liquid–liquid dispersion*
  log-normal  644
  mean drop size  643
  normal  644
  population balance modeling  697
  Schwarz-Bezemer equation  645
  similarity  676
  time to equilibrium  668, 727
  transient  676
Drop suspension, off bottom  646
Dust explosion  979, 980
Dynamical systems theory *see laminar mixing*

Electric motors  1258
  amperage  1261
  direct current  1266
  duty  1260
  enclosures  1263
  explosion-proof  1263
  frame size  1262
  frequency  1260
  insulation class  1263
  mounting  1266
  NEMA design  1260
  phase  1259
  power  1262
  speed  1260
  type  1259
  variable speed  1266
  voltage  1261
  wiring  1264
Electrolyte, effect on coalescence  707
Emulsion  646
  critical micelle concentration  715
  desalting  1183

  impeller selection  1173
  latex paints  706
  *see liquid–liquid dispersion*
  oil drilling  1171
  overmixing  1184
  poor mixing  1184
  stability  706, 1173
  viscosity effects  706
  water in crude oil  1175
Emulsion polymerization  707, **739**
  heat transfer  705
  retreat curve impeller  707
Energy dissipation  47, 49, 53
  average *see power per volume*
  calculation  **55**, 781
  drop breakup  658, 660
  drop size, figure  500, 666
  effect of geometry  56, 660, 781
  impeller swept volume  **55**, 660
  maximum, table  781
  reaction and local  761, 763, 768, **773**,
    777, 783, 790, 809, 817, 833, 835,
    839, 851
  scale-up  53, **773**, 821, 829, 848, 1077
  viscous mixing  1022
Engulfment model  770, 776, 829, 834,
  846, 1218
Equipment  xi, xli
  *see baffles*
  cell culture  1151
  components  1248
  draft tube  351, 724
  extruders  464
  gas–liquid mixing, table  587,
    588, 801
  *see geometry*
  high-shear  1254
  high viscosity  1254, 1267, 1255, 1308
  *see impeller*
  *see liquid–liquid mixing equipment*
  *see materials*
  mechanical loads  1269
  mechanical seals  1280
  *see mixer*
  *see motor*
  mountings  1311, 1313, 1316, 1317
  non-Newtonian fluids  1222
  pipeline mixing  391, 422, 463
  *see powder blender*
  pulp and paper  1222, 1229
  rotor-stator  464, 504
  seals  1278
  *see stirred tank*
  shaft design  1287

solids mixing  934, 937, 938, 940, 965, 967, 977
static mixer  422
Equipment selection  378, 396, 573, 591, 719, 917, 933
  gas–liquid mixing  591
  gas–liquid mixing, table  586, 588, 801
  heat transfer, figure  993
  heat transfer, table  874
  impeller type  354, 579, 1249, 1308
  liquid–liquid mixing  719
  liquid–liquid mixing, table  500, 722
  materials of construction  1318
  solid–liquid mixing  548, 573, 577
  solids mixing  917, 924, 933, 951, 954
  static mixers  396, 429, 431, 1257
  static mixers, table  398, 405, 428, 430, 436
Equipment specification  1333, 1334
  CFD  1345
  laser Doppler velocimetry  1345
  loads due to fluid forces  1347
  mechanical reliability  1347
  motor power and speed  352, 1268
  process requirements  1333, 1349
  sample testing  1343
  scale-up  1338
  vendor role  1333, 1334
  vibration  1350
  welding  1348
  witness testing  1344
Equipment vendor
  importance of experience  1334
  role of  1351
*Escherichia coli*  1097, 1138
Eulerian frame of reference, definition  62
Evaporative cooling  873
Experimental methods  viii, 145
  blend time  96, 164, 508
  Bourne reactions  175
  cell damage  1143
  conductivity probe  178
  *see gas–liquid mixing*
  LDV  166, 202, 220
  liquid–liquid dispersion  187, 189
  mixing and flow visualization  145, 165
  mixing laboratory  146
  Monte-Carlo oxygen feed  1101
  Njd, liquid dispersion  189
  Njs, solid suspension  182, 184, 185
  phase Doppler anemometry  234
  PIV  166, 237

  power or torque  161
  probe size, laminar mixing  97
  *see reaction calorimetry*
  reactive mixing  175
  solid–liquid mixing  177, 182
  turbulence  71
  *see velocity measurement*
  *see visualization*
Explosion-proof motors  1263
External heat exchangers  873
Extraction, liquid–liquid troubleshooting chart  lii
Extruders  464, 1005
  heat transfer  993
  mixing enhancers  1008, 1019
  single-screw  1000, 1007, 1010, 1022
  twin-screw  1009, 1014, 1016, 1019, 1022
  twin-shaft  1020, 1021

Fatigue, tables  1324, 1325, 1326
Feed pipe  352, 375
  backmixing  375, 784, 823, 829, 848
  backmixing, table  376, 824
  injection velocity  783, 822, 829, 845, 848
  location  352, **773**, 775, 778, **782**, 817, 821, 829, 846, 851, 1113, 1233
  location, liquid–liquid mixing  726
  pipe diameter  868, 771, **773**, 782, 829
  reaction selectivity or yield  766, 771, **773**, 782, **783**, 789, 823, 828, 847, 853, 856, 1233
  troubleshooting chart  l
Fermentation  xxiii, 1071
  *Bacillus subtilis*  1105, 1110
  biopolymer  1113
  blend time  1090
  CFD  **319**
  circulation time distribution  1073, 1090, 1105
  compartment model  1099
  *Escherichia coli*  1097, 1138
  gassed power  **611**
  geometric similarity  1084
  impeller selection  1089, 1119
  mixing sensitivity  1103
  model culture  1103, 1105
  Monte-Carlo experiment  1101
  mycelia  1074, 1124, 1132
  mycelial aggregation  1132
  nutrient concentration  1074, 1096
  oxygen gradients  1073, 1105
  oxygen uptake rate  1073, 1087, 1095, 1100

**1360** INDEX

Fermentation *(Continued)*
  periodic fluctuations  1096
  pH  1074, 1096, 1110
  polysaccharide  1113
  product quality  1123
  rate limiting mechanism  1095
  retrofitting  1137
  *Saccharomyces cerevisiae*  1097, 1104
  scale-down experiment  1099
  scale-up  1071, 1076, 1094
  time scales  1095
  zone model  1099, 1112
Fiber length, pulp  1197
Film drainage, coalescence  642, 683, 687
Film thickness at rupture  689
Fine chemicals  xxiii, 407, 1027
Fine solids, effect of on liquid–liquid dispersion  647, 718, **738**
Finite difference method  281
Finite element method  281
Finite volume method  277
Fixed bed reactors, residence time distribution  2
Flat blade impeller  872, 1030
Floating solids drawdown  380, 564, 1192
Flotation, gas–liquid–solid mixing  **1195**
Flow characteristics of impellers  358
Flow follower, circulation time distribution  1102
Flow number, CFD  **294**, 314
Flow patterns  364, 393
  gas–liquid mixing  591, 599
  rotor-stator  493
  stirred tank  77, **80**, 203, 312
Flow regimes, pipeline mixing  393, 396
Flow visualization *see visualization*
Fluidized bed reactors, residence time distribution  2
Fluidized beds, crystallization  1058
Foam, cell damage by  1146
Food industry  408
Fourier number  509
Fractals  102, 136
Free settling velocity *see settling velocity*
Free surface, CFD  290, 300
Frequency shift, laser Doppler anemometry  217
Frequency spectrum  45
Friction factor
  derivation  395
  Reynolds number  394, 396, 431
Froude number, surface vortex  1078

Fully isotropic turbulence, definition  63
Fully turbulent flow, definition  63, **66**
Fundamentals, viscous mixing  987

Gas bubbles, liquid–liquid dispersion  719
Gas–gas pipeline mixing  **438**
Gas hold-up  196, 620
  boiling  622
  coalescence  621
  multiple impellers  **625**
  pulp  1204
  sparged, hot  623
  static mixers  620
  stirred tanks  620, **624**
  surfactants  621
Gas inducing mixers  382
Gas–liquid mixing  xvi, xli, 585
  aeration  595, 1142
  blend time  590, 605
  blend time, compartmentalization  606
  *see boiling*
  bubble size  199, 632
  cavities  194, 598, 602, 608, 1081
  cavity type  194
  CFD  290, 301, **320**, 604
  coalescence  199
  equipment classification  587
  equipment selection  591
  equipment table  588, 801
  flow patterns  591, 599
  *see gas hold-up*
  gas inducing mixers  382
  *see gassed power*
  gassing regime  194
  heat transfer  591
  hollow shaft  382
  impeller selection  597, 1084, 1119
  interfacial area  199
  mass transfer  590, 626, **628**
  mass transfer coefficient  196
  mass transfer, pulp  1205
  mixing mechanisms  589
  multiple impellers  597
  network of zones  198
  new developments  586
  operating regimes  599
  pipeline mixing  402, 457
  pulp and paper  1189, 1225
  recirculation  1082
  recommended tank geometry  596
  RTD  200
  scale-up  633
  *see sparged, hot*
  self-inducers  595

solids suspension 591
sparging 353, 595
stirred tank 382
surface aerators 595
tall tanks 597
Gas–liquid reaction **xlv**, 631, 800, 1045, 1189
  equipment performance values, table 588, 631, 801
  *see fermentation*
  gas as by-product 1046
  gas as reagent 1028, 1045, 1047
  mass transfer regimes, table 631
  scale-up **804**, 824, 831, 1049
  selectivity **806**
  tank geometry 801
  troubleshooting chart li
Gas–liquid–solid mixing 562
  flotation **1195**
  pulp and paper 1194
  reaction 1045, **1194**
Gassed power 196, 607, 1080
  boiling 614, **616**
  fermenter **611**
  measurement 196
  relative power demand 607
  relative power demand, table 609
  sparged 590, **610**, 1080
  sparged, hot 619
  static mixers 607
Gas sparger 353, 595
  heat transfer 874
Gas volume fraction *see gas hold-up*
Gas vs. liquid Schmidt number 49
Gear box *see gear reducer*
Gear reducers 1268, 1335
  bearings 1273
  configurations 1270
  gear types 1270
  mixer loads 1269
  ratings 1275
Geometric similarity
  fermentation 1084
  scale-up **58, 60**
Geometry 347
  baffles 348, 579, 876, 720, 1030
  energy dissipation 660
  *see equipment*
  gas–liquid mixing 382, 588, 596, 801
  heat transfer surfaces 871, 875, 881
  heat transfer, table 873, 874
  *see impeller*
  rotor-stator 482
  solid–liquid mixing 548, 577

static mixer 422, 426
  *see stirred tank*
Glass-lined equipment
  attachment 1030
  baffles 351, 725, 1030, 1034
  impellers 355, 1030
Glucose, feed location 1113
Grace curve, drop breakup 654, 991
Granular mixing 898
  definition of granular state 899
  dilation 900, 901
  double cone 908
  equipment 911
  fundamentals 898
  segregation 899, 906, 908, 914, 939
  V-blender 899, 908, 910
Gravity silo 953, 955, 974
Grid generation, CFD 273, 303

Hardware, CFD, 336
Heat and mass transfer, liquid–liquid reaction 717
Heat transfer xx, xliii, 869
  batch reactor 1029
  CFD 267
  emulsion polymerization 705
  equipment selection 874, 993
  evaporative 873
  external heat exchangers 873
  fundamentals 870
  gas sparging 874
  gas–liquid mixing 591
  geometry, table 873, 874
  impellers 872, 874, 881
  Nusselt number 878
  pipeline mixing 403, 465
  Prandtl number 878
  reaction calorimetry 869, 1036
  resistances 870
  static mixers, table 467
  stirred tank 352
  thermal conductivity 870
  viscous mixing **883**, 993
Heat transfer coefficient
  correlations 878, **882, 883**
  overall 870
  static mixers, table 467
  stirred tanks 869
  table 879
Heat transfer surface
  baffle pipes 876, 881
  internal coils 353, 871, 875, 881
  jackets 353, 871
  plate coil 353, 876, 881

**1362** INDEX

Helical ribbon impeller   332, 385, 872, **883**, 995, 1020, 1255
   CFD simulation   **332**
   heat transfer   **883**, 993
Helicity   311
Heterogeneous reaction   790, 796, 1044
   *see gas–liquid reaction*
   *see liquid–liquid reaction*
   *see reaction*
   scale-up   796, 830, 1049, 1051, 1056
   *see solid–liquid reaction*
   vs homogeneous   800
High solidity impeller   1340
High viscosity mixer   1254, 1267, 1255, 1308
High efficiency impeller   357, 365, 579, 1030, 1339
High intensity powder blender   953, 963
High-shear impeller   350, 358, 368, 1030, 1254, 1256, 1342
   glass-lined   1030
   pulp and paper   1234
High-shear powder blender   953, 963
Hindered settling   553, 554
Hollow blade impeller   331, 356, 597, 609, 614
Hollow shaft   382
Homogeneous reaction   785, 1038
   scale-up   773, **785**, 796, 827, 1042
   *see reaction*
   selectivity   756, 757, **785**, 796, 1038, 1042
   semibatch reactor   1043
Homogeneous turbulence, definition   63, **65**
Homogenization, water in crude oil   1175
Homogenizers, table   963
Hub design   1310
Hub weights, table   1301
Hydraulic seal   1285
Hydraulic service factors, table   1291
Hydrodynamics   302, 364, 489
   *see circulation pattern*
   *see circulation time distribution*
   *see flow*
   fluid-particle   549
   mixing and stratified flow   1176
   pipeline   364
   rotor-stator   489
   settling velocity   550
   transition region   489
   *see turbulence*
   turbulent   489
   solids suspension   548

Hydrofoil impeller   357, 365, 597, 609, 614, 872, 1030, 1300, 1339
IKA Rotor-Stator   484, 486, 492, 493, 494, 503
Image processing, PIV   239
Impeller   354, 1308, 1338
   anchor   385, 872, 994, 1030, 1255
   axial flow   350, 354, 361, 1032, 1049, 1296, 1308
   blade thickness   1308, 1309
   bottom-entering   1253, 1278
   clearance and spacing, table   372
   clearance, off-bottom   371
   close-clearance   385, 1254, 1308
   design and selection   579, 1308
   design using CFD   285, **330**
   double-motion   1256
   drop breakup   650
   emulsification   1173
   flat blade   872, 1030
   flow characteristics   358
   flow number   **294**, 314, 358, 491
   flow number, table   359
   geometric similarity   **58, 60**
   glass-lined   355, 1030
   heat transfer   872, 874, 881
   helical ribbon   332, 385, 872, **883**, 995, 1020, 1255
   high-shear   350, 358, 368, 1030, 1254, 1342
   high solidity   597, 609, 614, 1340
   high viscosity   1254, 1255, 1308
   high efficiency *see hydrofoil*
   hollow blade, CD6 or Smith   331, 356, 597, 609, 614
   hub design   1310
   hub weights, table   1301
   hydrofoil   357, 365, 597, 609, 614, 872, 1030, 1300, 1339
   Intermig   355, 681
   loop impeller   721
   low-shear *see hydrofoil* 1030
   Maxflow   357, 609, 614
   modeling in CFD   285
   multiple *see multiple impellers*
   multipurpose design   1032
   pitched blade   354, 597, 609, 614, 681, 872, 1030, 1341
   power number   **294**, 360, 387, 489, 609, 995, 1308, 1338
   power number, table   365
   pumping number *see flow number*
   radial flow   350, 356, 1032, 1049, 1308

retreat blade   355, 579, 1030
rotational speed *see tip speed*
Rushton turbine
  shear, shear rate   368
  side-entering   1180, 1252, 1278
  side-entering, table   365, 367, 368
  solids mixing   915, 941, 950
  spacing, table   371, 372
  stabilizers   872, 1296
  stirred tank   347, 385
  surface vortex   368, 380
  swept diameter   **55**
  tank diameter   1308
  tip chord angle   660, 1336
  torque   1289, 1307, 1309
  turbine, glass-lined   1030
  turbulent flow characteristics   358
  types   347, 385, 1249
  up-pumping   367
  variable speed   1032
  weight, table   1298, 1299, 1305
Impeller diameter
  gas dispersion   1084
  solids suspension   1336
Impeller location, phase inversion   714, 724
Impeller retrofit, blending   512
Impeller selection   354, 579, 1249, 1308
  blending   379, 511, 520, 529
  fermentation   1089, 1119
  gas dispersion   382, 597, 1119
  liquid–liquid mixing   381, 719
  solid–liquid mixing   380, 579
Impeller speed *see tip speed*
Impeller swept volume   **55**, 660
Impinging jets   399, 1058
In-line mixer   463
  mass transfer   630
  petroleum   1178
  pulp and paper   1222
  reactors   470, 778, **806**, 1044, 1058
Intensity of segregation
  Corrsin   52
  definition   22
Interfacial area, gas–liquid mixing   199
Interfacial area growth, laminar mixing   112, **124**
Interfacial tension
  coalescence   683
  liquid–liquid dispersion   188
Intermaterial area density, laminar mixing   114, **125**
Intermig   355, 681
Intermittency, drop breakup   667

Internal coil for heat transfer   871, 875, 881
Ion exchange   545
Islands of unmixedness, laminar mixing   95, 101, 107
Isotropy, turbulent   63, **66**

Jacket for heat transfer   871
Jet mixer   531
  blend time   534, **536, 537**
  sludge control   1179
  solids suspension   563
Just suspended speed, liquids   711
Just suspended speed, solids   182, 184, 547, 556, 558, **561**

Kappa number, pulp and paper   1215
Kenics mixer   **119**
k-ε turbulence model   73, 79, 263
Kinetics, mixing and reaction   1041
Kneading   996, 1001, 1018, 1021
Kolmogorov scale   48, 768

Lagrangian frame of reference, definition   69
Laminar mixing   vii, 89, 489, 987
  asymptotic directionality   110, **120**
  blend time   527, **530**
  *see blending*
  CFD   110, **119, 128, 332**
  chaos   94, 109
  characterization   92
  divergence of particle paths   92
  drop breakup   651
  fractals   102, 136
  interface growth   112, **124**
  intermaterial area density   114, **125**
  islands of unmixedness   95, 101, 107
  Kenics mixer   **119**
  material lines   112
  measures of, table   104
  mechanisms, summary   138, 987
  Metzner-Otto correlation   517, 530
  micromixing   776
  particle settling velocity   551, 552
  pipeline de-mixing   461
  pipeline mixing   401, 436, 440, 461
  Poincare sections   96
  power   384, 529
  probe size effect   97
  reaction   130
  repeating patterns   110
  scale of segregation   99
  similarity   110, **127, 130**, 137

**1364** INDEX

Laminar mixing *(Continued)*
  sine flow   103, 130
  static mixers   416, 432, 451, 461
  stirred tank   135, 383
  stretching and folding   90, 93, 989, 991
  stretching field   108, **123**
  striation thickness   99
  striation thickness distribution   116, **128**
  *see viscous mixing*
Large eddy simulation *see LES*
Laser Doppler anemometry *see LDV*
Laser Doppler velocimetry *see LDV*
Latex paints, emulsion   706
LDV   71, 166, 202, 207
  applications to mixing   224
  equipment specification   1345
  frequency shift   217
  fringe model   212
  measurements   220
  measuring volume   212
  optics   208, 216
  rotor-stator   493
  signal processing   218
  velocity bias   222
Length scales   24
  Batchelor scale   49, 50
  Kolmogorov scale   48, 50
  pulp   1206
  reaction   **24**
  reaction, table   50
LES
  CFD results   315
  macroinstabilities   **76, 333**
  turbulence model   74, 76, 265
Light sheet visualization   166
Liquid–liquid dispersion   649
  classification   640
  *see coalescence*
  concentrated   648, 663, 704
  dilute   648, 659
  dispersed phase concentration   647
  *see drop breakup*
  *see drop size*
  drop size classification, table   646
  drop size correlation, table   669
  drop size distribution   643
  *see emulsion*
  equipment selection   500, 504, 640, 666, 685, 715
  fine solids, effect of   647, 718
  gas bubbles, effect of   719
  interfacial tension   188
  just suspended speed   646, 711
  largest drop   **739**
  *see liquid–liquid mixing*
  Marangoni effect   648, 718
  mass transfer   648
  measurement of dispersed phase   187, 189
  measurement of Njd   189
  measurement, cleaning protocols   187
  *see phase inversion*
  pipeline mixing   443, 447, **452, 454**, 461
  power per volume   500, 660, 666
  rotor-stator   498, 665, 668
  *see Sauter mean diameter*
  scale-up   646, 730, **731, 732**, 735, 736
  scale-up, table   736
  solvent extraction   **659**
  static mixers   441, 665, 715, 1177
  stirred tanks   650, 681
  surfactant dose   **716**
  surfactant, stabilizer or dispersant   647, 668, 715, **716**
  volume fraction   647
Liquid–liquid dispersion, industrial   640, 646, 737
  emulsion polymerization   **739**
  pharmaceuticals   738
  pulp and paper   1189
  suspension polymerization   714, **738, 741**
Liquid–liquid extraction
  drop breakup   **452, 659**
  troubleshooting chart   lii
Liquid–liquid mixing   xvii, xlii, 639
  blend time   724
  circulation time distribution   729
  *see coalescence*
  dispersion and coalescence   713
  *see drop breakup*
  *see drop size*
  *see emulsion*
  *see liquid–liquid dispersion*
  *see liquid–liquid mixing equipment*
  *see liquid–liquid reaction*
  *see phase inversion*
Liquid–liquid mixing equipment
  baffles   720, 724
  coalescence   685
  equipment selection table   500, 666, 722
  feed location   726
  impeller location   724
  impeller selection   719

multiple impellers   721, **740**
pipeline mixing   402, 443
power   727
selection   504, 640, 715, 719
slurry drainage   721
static mixers   441
stirred tank design   381, 719
variable speed drive   721
Liquid–liquid reaction   xlii, **xliv, xlvii**, 809, 1050
  *see emulsion polymerization*
  heat and mass transfer   717
  *see heterogeneous reaction*
  nitration   640
  *see reaction*
  reactive extraction   **812**
  scale-up   **xliv, xlvii, 810, 812**, 825, 832, 1051
  *see solid–liquid reaction*
  *see suspension polymerization*
  selectivity   **xliv, xlvii**, 809, **810**, 1050
  troubleshooting chart   liii
Loads due to fluid forces-equipment specification   1347
Locally isotropic turbulence   63, **66**
Log-law of the wall   275
Log-normal drop size distribution   644
Loop impeller   721
Low-shear impeller   357, 365, 579, 1030, 1339

Macroinstability, CFD simulation   **76,** 298, **333**
Macromixing   1053
  definition   24, 765
  residence time distribution   1
Magnetic drives   1286
Marangoni effect, liquid–liquid dispersion   648, 718
Mass transfer   565, 626
  CFD   265
  coalescing   1086
  concentration at interface, figure   567, 631, 802, **808**
  crystallization   1061
  diffusion   259
  gas–liquid mixing   590, 626, **628**
  gas–liquid, pulp   1205
  in pipes   470
  in-line mixers   630
  liquid–liquid dispersion   648, 717
  noncoalescing   1086
  rate controlling steps   565
  reaction   790, 796
  scale-up   1086
  *see Schmidt number*
  solid–liquid mixing   182, 565, 567, **570**, 1220
  solid–liquid, impeller speed   569
  stirred tanks   627
Mass transfer coefficient, gas–liquid mixing   196
Materials of construction   1318
  alloys, table   1319
  coatings   1327
  corrosion   1319
  corrosion fatigue   1320, 1326
  density of   871
  failure   1322
  fatigue, tables   1324, 1325, 1326
  glass coatings   1328
  heat transfer properties, table   871
  laboratory   156
  life estimates   1323
  metal elasticity and density, table   1298
  properties, table   871
  seals   1279
  selection   1318
Maxflow impeller   357, 609, 614
Maximum stable drop size   657
Maximum viscosity ratio, drop breakup   654
Measurement *see experimental method*
Mechanical design   xxvi, 1247
  *see gear*
  *see impeller*
  *see materials*
  *see motor*
  rotor-stator   503
  *see seal*
  *see shaft*
  *see stirred tank*
Mechanical reliability, equipment specification   1347
Mechanical seals   1280
  balanced   1282, 1284
  barrier fluid   1282
  double   1282
  single   1281
  unbalanced   1282, 1284
Mechanism
  coalescence   682
  drop breakup   639, 641
  drop deformation   641
  gas–liquid mixing   589
  *see laminar mixing, turbulence*
  liquid–liquid dispersion   641

**1366** INDEX

Mechanism *(Continued)*
  reaction  31
  solids mixing  901, 904, 913, 919, 921, 945
  solids suspension  548
Mesh *see CFD, numerical issues*
Mesh superposition  **323**
Mesomixing  24, 765
  simulation  766, 843
  time scale  771, 1044
Metzner-Otto correlation
  constant, table  370
  laminar  368, 530
  shear-thinning fluids  517
Microcarrier cultures, cell damage  1148
Micromixing  24, 766
  CFD  270
  definition  24, 765, 768, 770
  laminar  776
  peroxymonosulfate  1190
  residence time distribution  1
  turbulent, high viscosity  776
Mixer  345, 396, 479, 909, 933, 994
  bottom-entering  1253, 1278
  double-motion  1256
  dry-solids  1257
  *see impeller*
  in-line  463
  mounting  1310
  portable  1249
  side-entering  1180, 1252
  top-entering  1249
Mixing
  batch reactors  780–781
  continuous reactors  782, 778
  cost of  xxxv
  crystallization  545, 820
  Damkoehler number  756, 761, 763, 765, 777
  definition  xxxiv, 22
  fundamentals  xxxix
  multipurpose plants  1028, 1058
  nomenclature  lv
  problems  xxxiv
  process development  xliv
  reaction model  38
  regulatory control  1027
  semibatch reactors  780, **785**
  turbulent  19
Mixing enhancers, extruders  1008, 1019
Mixing equipment
  *see equipment*
  *see impeller*
  *see stirred tank*

Mixing intensity, Reynolds number  418, 432
Mixing kinetics  1041
Mixing laboratory  146
  instrumentation  155
  lab scale pipes  160
  lab scale stirred tanks  156
  materials of construction  156
  safety  147
  scale of operations  154
  simulant fluids  152
Mixing quality, CoV  1231
Mixing sensitive reaction *see reaction*
Mixing time *see blend time*
Model
  cell growth  69
  *see CFD*
  coalescence frequency  692
  crystallization  69
  CSTR  6
  drop breakup  667
  network of zones  198
  population balance  697
  residence time distribution  6, 7, 9
  solids mixing  923
  turbulence  72, 261
  viscous mixing  1022
  wall jet  562
Model culture
  fermentation  1103, 1105
  xanthan gum  1113
Monte-Carlo coalescence-dispersion simulation  848
Monte-Carlo fermentation experiment  1101
Motionless mixer *see static mixer*
Motor  xxvi, 1258
  air  1267
  electric  1258
  gear reducers  1268, 1336
  hydraulic  1267
  power and speed table  352, 1268
  speed reducers  1267
  temperature  1266
Motor/gearbox  351
Mounting, electric motors  1266
Mountings, table  1313
Multiphase flow
  CFD  271, 290
  pipeline mixing  399, 441, 450, **452**
  pressure drop  451, 456, 457
Multiple impellers  372, 1342
  blending  513
  gas hold-up  **625**

INDEX    **1367**

gas–liquid mixing   597
glass-lined   1030
liquid–liquid dispersion   721, **740**
viscous mixing   1020
Multiple reference frame   292
Multiple shafts   **323**
Multipurpose vessel   1032
   batch reactor   1029
   crystallization   1058
   fine chemicals   1027
   impellers, variable speed   1032
   pharmaceuticals   1028
   regulatory control   1027
   scale-up   1028, 1032
   semibatch reactor   1027, 1032
Mycelia   1074, 1124, 1132

Natural frequency, shaft design   1287, 1293 1297, 1305
Network of zones   198
Newtonian fluids   149
Newton's law, settling velocity   552
Nitration reaction   640, **739**
Njd, liquid dispersion   189
Njs, just suspended speed
   liquid dispersion   711
   solids suspension   182, 184, 547, 556, 558, **561**
Nomenclature   lv
Noncoalescing liquid, mass transfer rate   1086
Non-Newtonian fluid
   biomass   1074
   CFD   272
   mixing equipment   1222
   rheology   150
   *see shear-thinning fluid*
   *see yield stress fluid*
Nucleation   820, 858, 1059
Numerical diffusion   284
Numerical methods *see CFD, numerical issues*
Nusselt number, heat transfer   878
Nutrient concentration, fermentation   1074, 1096

Oil and gas   xxv, 408, 1171
Oil drilling   1171, 1173
Operating regimes
   gas–liquid mixing   599
   liquid–liquid mixing   696, 736
   solid–liquid mixing   547, 556, 576
Optical probe   180, 187
Optics, LDV   216

Organic compounds   1027
Ostwald ripening
   crystals   717
   drops   717
Overmixing   1028, 1054, 1061,1063
   emulsion   1184
   viscous mixing   1022
Oxygen gradients, fermentation   1073, 1105
Oxygen uptake rate
   circulation time distribution   1091
   fermentation   1073, 1087, 1095, 1100
   xanthan gum   1114, 1116

Paddle blender, solids   941, 943, 945
Paired-interaction closure   850
Parallel processing   284
Parallel reaction   **759, 784, 816**, 823, 1040
   Bourne reactions   784, 786
Particle image velocimetry *see PIV*
Particle tracking
   CFD   302, 307
   circulation time distributions   70
   Lagrangian approach   69
Path lines, CFD results   306
PDF   37, 270, 853
Peclet number   278
Periodic fluctuations, fermentation   1096
Peroxymonosulfate reaction   1190
Petroleum processing   xxv, 1171
   alkylation   1185
   blending   1171
   desalting   1183
   sludge control   1179
   static mixers   408, 409, 1176
pH
   coalescence   684
   feed location for control   1113
   fermentation   1074, 1096, 1110
Pharmaceuticals   xxiii, 1027
   batch and semibatch   **738**, 1028
   continuous reactors   1035
Pharmaceuticals, multipurpose reactor design   1028
Phase Doppler anemometry   226
   effect of particle type, table   238
   equations   227
   implementation   233
   measuring modes, table   234
   optics   226, 233
Phase inversion   642, 707
   impeller location   714, 724
   *see liquid–liquid dispersion*
   oil drilling   1173

**1368** INDEX

Phase inversion *(Continued)*
  polymer processing  710
  suspension polymerization  **741**
Pilot plant *see scale-up*
Pipeline mixing  xii, 391
  applications  404, 407, 409, 410, 412
  applications, table  405
  blending, laminar  401, 436, 440
  blending, laminar thermal  440
  blending, turbulent  437, 438
  coefficient of variation  410, 416, 432, 435, 439, 455
  flow inverters  417, 422
  gas phase, continuous  **457**
  gas–gas  **438**
  gas–liquid  402
  heat transfer  403
  impinging jets  399
  laminar  401, 416, 432, 437
  liquid–liquid  402, 443
  mass transfer  470
  mixer types  391, 422, 463
  mixer types, table  398
  multiphase flow  399, 441, 450, 451
  operating regimes  392, 396
  reaction  403, 778
  residence time distribution  416, 470
  Reynolds experiments  393
  scale-up  **57**, 462
  solid–liquid  450
  Tee mixers  399, 419, 438
  transitional flow  459
Pitched blade impeller  354, 681, 597, 609, 614, 872, 1030, 1341
PIV  237
  applications in mixing  249
  compare with other methods  71, 166, 203, 207
  image processing  239, 246
  implementation  243
  principles  237
  seeding particles  243
  stereoscopic  247
Plant cells  1152
Plate coil for heat transfer  876, 881
Plough mixer  941, 942, 951, 1003
Plug flow reactor
  residence time distribution  5, **6**
  yield limits  12
Pneumatic mixer, solids  958
Poincare sections  96
Polymer processing  xxii, 987
  extruder  **322**, 464
  heat capacity of polymers  1006

phase inversion  710
static mixers  408
*see viscous mixing*
Polymerization, residence time distribution **12**
Polysaccharide fermentation  1113
Population balance models, drop size distribution  697
Portable mixer  1249
Powder blender
  comparison, tables  950, 951, 952, 954
  convective impellers  916
  gravity silo  953, 955, 974
  high intensity  963
  high-shear  953, 963
  homogenizers, table  963
  paddle  941, 943, 945
  plough  941, 942
  pneumatic  958
  power  943, 949, 971, 973
  ribbon  915, 945, 972
  screw  949,
  sigma blade  950, 999, 1001
  specifying  974, 1257
  speed  914
  testing  975
Powder blending  xxi, 887
  convective applications  912, 917, 941
  convective mechanisms  913
  granular convective mixing  911
  granular dispersive mixing  911
  scale-down  922
  scale-up  919, 968, 922
  scale-up of convective  917
  scale-up of tumbling  914
  segregation  888, 913, 926, 939
  *see solids mixing*
  time required  945, 971, 974
Power  358
  fermenter  **611**
  *see gassed power*
  impellers  384, 489, 1308, 1338
  laminar  384, 529
  liquid–liquid mixing  727
  measurement  155, 161
  motor speed  352, 1262, 1268
  rotor-stator  489
  scale-up *see power per volume*  376
  shear-thinning fluids  517
  solids mixing  943, 949, 971, 973
  solids suspension  580
  viscous mixing  992, 1001
Power number  360, 365, 384, 387, 489, 995, 1308, 1338

CFD  **294**, 313
  rotor stator  489
Power per volume
  drop breakup  660, 666
  *see energy dissipation*  **59**
  gas–liquid mixing  588, 801
  impeller swept volume  **55**
  liquid–liquid mixing  500, 666
  scale-up  53, **59**, 376, **773**, 821, 829, 848, 1077
Prandtl mixing length, turbulence model  73
Prandtl number, heat transfer  878
Precipitation  545, 819, 1053
Pressure drop
  multiphase flow  451, 456, 457
  static mixers, table  431
Probability density function  37, 137, 270, **837**, 853
Probe location, blend time measurement  167
Process development  xliv
Process intensification, pulp and paper  1224
Process objectives, CFD  302
Process requirements, equipment specification  1333, 1349
Pulp and paper  xxv, 1187
  black liquor  1189
  bleaching  1214, 1221
  caverns  1202
  CFD  1224
  costs of poor mixing  1237
  CoV  1231
  deflaking  1210
  deinking, lumen loading  1213
  gas hold-up  1204
  gas–liquid mass transfer  1205
  gas–liquid mixing  1189, 1225
  gas–liquid–solid mixing  1194
  high-shear mixing  1235
  kappa number  1215
  liquid–liquid mixing  1189
  mixing equipment  1222, 1229
  polysulfide generation  1194
  process intensification  1224
  *see pulp characterization*
  pulp yield  1194
  repulping  1210
  solids dissolution  1189
  static mixer  409
  white liquor  1190, 1193
  yield  1194

Pulp characterization  1195
  consistency  1197
  fiber length  1197
  high consistency  1229
  length scales  1206
  low consistency  1222
  medium consistency  1223
  rheology  1198, 1199, 1207
  yield stress  1199, 1207
Pulp chest, CFD  **321**
Pumping number  314, 358, 359, 491

QUICK, CFD  279, 280

Radial de-mixing, solids  906
Radial discharge rotor stator  482
Radial flow impeller  356, 1032, 1049, 1308
Radial mixing, pipeline flow  2, 409, 410, 412
RANS equations  78, 262
Rate controlling step, mass transfer  565
Rate of deformation, CFD results  311
Reaction  xliv, **xlv**, xix, 755, 1039
  Arrhenius model  268
  *see Bourne reaction*
  *see CFD, reaction*
  classification  794
  coalescence  **738**
  *see competitive-consecutive reaction*
  crystallization  819
  Damkoehler number  **32**, 756, 761, 777, 1039, 1043
  design flowsheet  26
  engulfment model  770, 776, 829, 834, 846, 1218
  exothermic  824, 854
  experiments  175
  fast  **33**
  *see feed pipe*
  *see gas–liquid reaction*
  gas–liquid–solid reaction  1045, **1194**
  heat and mass transfer  717
  heterogeneous  790, 795, **796**, 800, 830, 1044, 1049, 1051
  homogeneous  756, 773, **785**, **796**, 1038, 1042
  importance of diffusion  **28**
  kinetics  1041
  laminar mixing  130
  LDPE, CFD  **328**
  length scales, table  50
  *see liquid–liquid reaction*
  Magnussen model  268

**1370** INDEX

Reaction *(Continued)*
   mass transfer effects **28**, 790, **796**
   mixing mechanisms **25**
   mixing model 38
   mixing sensitive 26, **31**, 756, 1038, 1039, 1043
   modeling, *see CFD, reaction*
   parallel **759, 784, 816**, 823, 1040
   PDF 37, 137, 270, **837**, 853
   pipeline mixing 403, 778
   pulp and paper 1189, 1194
   rotor-stator 501
   scale-down 821
   *see scale-up*
   scale-up protocols 826
   scale-up, geometric similarity **60**
   scale-up, power per volume **59**
   *see selectivity*
   slow **33**
   solid-catalyzed 545
   *see solid–liquid reaction*
   supercritical **740**
   time and length scales **25**, 756, 794, 1217
   turbulence model 38
Reaction calorimetry 1036
   differential 1037
   heat transfer 1036
   integral 1036
Reactive crystallization 1053
Reactor
   *see batch reactor*
   continuous 782, 1035
   gas-sparged 800
   in-line flow 470
   pipeline 778
   *see semibatch reactor*
   static mixer 780, **785, 841**
   stirred tank, batch and semibatch 780, **785**
   Tee-mixers 779
Recirculation, gas–liquid 1082
Regulatory control
   agricultural 1027
   pharmaceuticals 1058
Relative power demand, gassed 607, 609
Repulping 1210
Residence time distribution vi, xxxvii, 1
   adsorption columns 2
   axial dispersion 2, 418
   axial mixing, temporal 2
   batch reactor 2
   bypassing 9, **1193**
   CFD 14
   chromatography columns 2
   complete segregation 1, 5, 11
   CSTR 5, **12**, 15, 174
   delta distribution 6
   diagnosis 9
   distillation columns 2
   fixed bed reactors 2
   fluidized bed reactors 2
   functions, table 3, 4
   gas phase 200
   macromixing 1
   maximum mixedness 1, 11
   micromixing 1
   models, hydrodynamic 6
   models, recycle 7
   models, side capacity 9
   moments, table 4
   nitration reaction **739**
   nth order reactions 11
   pipeline mixing 416, 470
   plant measurement 1234
   plug flow reactor (PFR) 5, 6
   polymerization **12**
   reaction yield 10, 12
   short circuiting 9, **1193**
   spatial mixing 1
   stagnancy 9
   temporal mixing 2, 9
   tracers 2, 9, 1234
   trickle bed reactors 2
   two-phase stirred tanks 2
   washout functions 3, 6, 9
Retreat blade impeller 359, 579, 1030
   emulsion polymerization 707
   glass-lined 1030
Reynolds experiment 393
Reynolds number 63, **66**, 261, 345
   critical 489
   definition 394
   friction factor correlation 394, 396, 431
   mixing intensity 418 432
   particle 551, **554**, 555
Reynolds stress model 74, 80, 263
Rheology 147
   Newtonian fluids 149
   non-Newtonian fluids 150
   pseudo-plastic fluids 151
   pulp 1198, 1199, 1207
   shear-thickening fluids 151
   shear-thinning fluids 150
   viscoelastic fluids 152
   xanthan gum 1114

INDEX     **1371**

Rheometers, torque measurement   155, 164
Ribbon blender   915, 945, 951, 972, 1003
Ross rotor-stator   482, 485, 490, 499
Rotating jet   1181
Rotating reference frame, CFD   292
Rotational speed *see tip speed*
Rotor-stators   xiii, 464, 479
   applications   480
   axial discharge   482
   characteristics   479
   Chemineer Greerco   483, 485, 487, 488, 489, 491
   colloid mills   482
   current knowledge   480
   design configurations   482
   drop breakup   498, 665, 668
   equipment suppliers   504
   geometry   482
   hydrodynamics   489, 492
   IKA   484, 486, 492, 493, 494, 503, 1256
   key questions   481
   mechanical design   503
   modes of operation   485
   multistage design   482, 483
   power draw   489
   power number vs. Reynolds number   489
   radial discharge   482
   reaction   501
   Ross   482, 485, 490, 499
   scale-up   496, 502
   selection, table   500
   shear rate range   479
   Silverson   480, 482, 483, 486, 490, 491, 499
   solids and powder dispersion   501
   tip speed   479
   velocity field   491
RSM, turbulence model   74, 80, 263
Rushton impeller   354, 361, 382, 597, 609, 614, 724, 872, **883**, 1020, 1250, 1341
   compartmentalization   724

*Saccharomyces cerevisiae*   1097, 1104
Safety   147, 977
Sample testing   1343
Sampling
   solid–liquid mixing   181
   solids mixing   891, 895, 931, 974
Sauter mean diameter
   correlations, table of   668
   definition   643
   drop breakup   658
Scale of segregation
   definition   22
   laminar mixing   99
Scale-down *see scale-up*
Scale-up
   batch and semibatch   1035
   blend time in pipes   **57**
   compare methods   377, 496, 502, 1088
   crystallization   1056, 1057, 1062
   Damkoehler number   826
   definition   1058
   energy dissipation   53, **773**, 821, 829, 848, 1077
   equipment vendor role   1338
   fermentation   1071, 1076, 1094
   gas–liquid mixing   633
   geometric similarity   **58, 60**
   liquid–liquid dispersion   646, 730, **731, 732**, 735, 737
   liquid–liquid mixing, table   736
   liquid–liquid, coalescing   681, 686, 735
   liquid–liquid, concentrated   **732**
   liquid–liquid, dilute, table   731, **732**
   mass transfer rate   1086
   pilot plant   154, 1035
   power per volume   53, **59**, 376, **773**, 821, 829, 848, 1077
   reaction, competitive-consecutive   782, **785**, 787, 794, 827, 830
   reaction, gas–liquid   **804**, 824, 831, 1049
   reaction, geometric similarity   **60**
   reaction, heterogeneous   796, 830, 1049, 1051, 1056
   reaction, homogeneous   **773, 796, 785**, 827, 1042
   reaction, liquid–liquid   **810, 812**, 825, 832, 1051
   reaction, mixing sensitive   **773, 785**, 821
   reaction, power per volume   **59**
   reaction, protocols   826
   reaction, solid–liquid   **796**, 825, 830, 1056
   rotor-stator   496, 502
   scale-down   154, 576, 821, 1099
   shear   1086
   simulation   856
   solid–liquid mixing   573
   solids mixing   914, 917, 919, 968, 922
   solids suspension   1194
   static mixers   462

**1372** INDEX

Scale-up *(Continued)*
   stirred tank, table 376, 377, 1088
   suspension polymerization **732**
   tip speed 1086
   torque 376
   turbulence 53, **57**
   turbulence, summary 61
Schmidt number
   blend time effects, table 52
   definition 24, 25
   gas vs. liquid 49
Schwarz-Bezemer equation, drop size distribution 645
Screw blender, powder 949,
Seals 1278
   barrier fluid 1282
   cartridge 1284
   hydraulic 1285
   lip 1285
   materials 1279
   mechanical 1280
   packing 1279
   shaft 1278
   stuffing box 1278
Seeding particles, particle image velocimetry 243
Segregation
   granular mixing 899, 906, 908, 914, 939
   powder blender 888, 914, 926, 939
Selectivity 764
   competitive-consecutive reactions **758**, 782, 796, **806**, 811
   CSTR, ideal yield 12
   enhancement 811
   expected yield, calculation 10, 760
   feed point 782, 829, 847
   feed rate, time 766, 771, **773**, 782, 789, 823, 828, 853, 856
   gas–liquid reactions 806
   heterogeneous reaction 790, 795, **796**
   homogeneous reaction 757, 785, **796**, 1038 1042
   liquid–liquid reactions 809, 810, 1050
   plug flow reactor (PFR), ideal yield 12, 758
   residence time distribution, limits 12, 758
   solid–liquid reaction **796**
Self-inducers, gas–liquid mixing 595
Semibatch reactor 780, **785**
   advantages 1029
   disadvantages 1029
   dissolving reagents 1044, 1052
   homogeneous reactions 1043
   multipurpose design 1032
   pharmaceuticals 1028
   scale-up 1035
Settling velocity 550
   correlations for 551
   drag coefficient, figure 551
   figure 552
   hindered settling 553, 554
   Newton's law 552
   particle orientation 553
   particle shape 553
   particle size 553
   solids concentration 553
   Stokes law 551, 552, **554**, 555
   turbulent vs laminar 551, 552, **555**
Shaft design xxvi, 1278, 1287
   allowable stresses, table 1291
   bending moment 1290, 1310
   couplings 1303
   critical speed 1287, 1296, 1335
   damping 1296
   design steps 1287
   diameter 1290
   dynamic analysis 1305
   failure 1293
   hydraulic service factors, table 1291
   metal properties, table 1298
   natural frequency 1287, 1293, 1297, 1305
   seal 1278
   solid–liquid mixing 580
   stabilizers 1296
   static analysis 1304
   steady bearing 1292, 1304
   strength 1289
   torque 1289, 1308, 1310
   vibrations 1293
   weight 1298, 1305
Shear 368
   cell damage 1076
   drop breakup 641
   impeller 368
   scale-up 1086
   solids mixing 919, 921
   viscous mixing 989
Shear flow **68**
   definition 63
Shear rate 368
   motionless mixers, table 433
   rotor-stator 479
   static mixers 432
Shear-sensitive cells 1140
Shear-thickening fluid, rheology 151

INDEX     **1373**

Shear thinning fluid   150
  blend time   518, 991
  Metzner-Otto correlation   517
  oil drilling   1173
  power   517
Short circuiting   **1193**
Side-entering mixer   1180, 1252, 1278
Sigma blade impeller   950, 998, 1001
Silverson rotor-stator   480, 482, 483, 486, 490, 491, 499
Similarity
  drop size distribution   676
  laminar mixing   110, **127, 130**, 137
SIMPLE algorithm   282
Simulation *see CFD*
Sine flow, laminar mixing   103, 130
Single-screw extruder   1000, 1007, 1010, 1022
Sinking solids   380
Sliding mesh   295
Slip velocity, solid–liquid mass transfer   568
Slow reaction   **33**
Sludge control   1179
Slurry drainage, liquid–liquid mixing   721
Slurry rheology, solid–liquid mixing   1193
Slurry withdrawal, troubleshooting chart   liv, 376
Smith impeller   331, 356, 597, 609, 614
Smith number, boiling   614
Snapshot model   300
Solid–liquid–gas reaction   804
Solid–liquid mass transfer   565
  correlations   569
  effect of impeller speed   182, 545, 556, 567, 568, **570**, 1220
  slip velocity   568
Solid–liquid mixing   xv, xl, 543
  aggregates, lumps   1192
  CFD   290, 301
  cloud height   562
  coating colors   **1192**
  complete suspension   556, 576
  dispersion or distribution   544, 556, 564
  dissolution   545, 547, 556, **570**
  dissolution, pulp and paper   1189
  drawdown   1192
  equipment selection   548, 573, 577
  floating solids   564
  flotation   1195
  key considerations   546
  measurement of Njs   182, 184, 185
  measurement, conductivity probes   178, 185
  measurement, optical probe   180
  measurement, sampling   181
  measurement, solids distribution   177
  measurement, tomography   182, 201
  measurement, visual observations   177, 183
  Njs, just suspended speed   547, 556, 558, **561**
  objectives   544
  operations requiring   544, 545, 557
  power   580
  scale-up   573
  slurry rheology   1193
  *see solids suspension*
  startup in settled solids   580
  turbulence   **61**
  wetting solids   1192
  *see Zwietering*
Solid–liquid mixing equipment
  baffles   579
  impellers   579
  jet mixers   563
  pipeline mixing   450
  rotor-stator   501
  shaft design   580
Solid–liquid reaction   **xlvi**, 818, 1052
  causticization   1193
  *see crystallization*
  dissolving reagents   **796**, 819
  precipitating products   819, 1053
  scale-up   **796**, 825, 830, 1056
  troubleshooting chart   lv
Solids mixing   887
  agglomeration   914, 921, 927
  axial de-mixing   907
  batch equipment   934, 937, 977
  bin blender   911
  characterization   888, 926
  continuous equipment   934, 938, 965, 967, 977
  convective blenders   912
  de-mixing   887, 901, 906, 914, 926
  double cone blender   910
  dust explosion   979, 980
  equipment selection   917, 924, 933, 951, 954
  equipment, classification   911, 940
  fundamentals   887, 936
  *see granular mixing*
  hard-particle methods   910
  ideal mixtures   888

Solids mixing *(Continued)*
  mechanisms, cohesive   904, 919, 921, 945
  mechanisms, free-flowing   901
  mixture types   926
  modeling   923
  ordered   889
  patterned   908
  *see powder*
  process safety   977
  radial de-mixing   906
  random   888, 926
  real mixtures   888
  sampling   891, 895, 931, 975
  scale of scrutiny   895, 932, 974
  shear   919, 921
  soft-particle methods   910
  statistical characterization   896, 928, 974
  textured mixtures   889
  tumbling   903, 909, 914, 917, 940
  V-blender   910
Solids removal, troubleshooting chart   liv
Solids suspension   **557**
  CFD   **318**
  cloud height   562
  factors affecting   549
  impeller diameter   1336
  jet mixed vessels   563
  mechanism   548
  Njs, just suspended speed   182, 184, 547, 556, 558, **561**
  operating regimes   547, 556, 576
  scale-up   1194
  sinking solids   380
  with gas dispersion   562
  *see Zwietering*
Sparged, hot   617
  gas hold-up   623
  gassed power   619
Sparged reactor   800
Sparging   353, 595
Spatial mixing, residence time distribution   1
Speed reducer   1267, 1335
Stability, emulsion   1173
Stabilizer, role of in liquid–liquid dispersion   647
Startup in settled solids   580
Static mixer   xii, 391, 1257
  applications   392, 397, 399, 401, 428, 430
  $CO_2$ reduction   1174

coefficient of variation   410, 414, 416, 432, 435, 439, 455
comparison, table   434, 436
de-mixing   461
density ratio   460
description   392, 422
downstream effects   460
drop breakup   665, 715
equipment   396, 422, 426, 1257
equipment selection, table   398, 405, 428, 430, 436
gas hold-up   620
gas-gas mixing   **438**
gassed power   607
heat transfer, table   467
inlet position   462
laminar   417, 432, **440**, 451, 461
laminar flow, CFD   **120, 325**, 413
liquid–liquid dispersion   441, **452**, 1177
orientation   460
petroleum processing   1176
*see pipeline mixing*
pressure drop, tables   431
reactor   780, **785, 841**
regimes, table   428
scale-up   429, 431, 462
shear rate   432
shear rate, table   433
turbulent, CFD   **326**
viscosity ratio   436, **454**
Weber number   1177
Steady bearing   1292, 1304
Stirred tank   xi, 345, 1310
  baffles   348
  bottom port   376
  CFD, geometry   285
  draft tubes   351, 383
  feed pipe backmixing, table   375, 376
  *see flow pattern*
  gas hold-up   620, **624**
  gas sparger   353, 595
  gas–liquid mixing, geometry   382, 596
  gas–liquid mixing, tall   513, 597
  geometric similarity   **58, 60**, 1084
  heat transfer   352, 871, 876
  heat transfer coefficient   869, 879
  *see impellers*
  impeller location, liquid–liquid mixing   724
  impeller, clearance and spacing   372
  inlets/outlets   352
  laminar mixing   135, 383
  liquid–liquid mixing   381, 719

INDEX     **1375**

mass transfer   627
motor power and speed, table   352
motor/gearbox   351
multistaged   372, 723
reactor   780, **785**
*see scale-up*
tank bottom shapes   351
tank diameter   1308
Stokes law, free settling velocity   551, 552, 554, 555
Strain gauges, torque   162
Streamlines   305
Stretching and folding   90, 93
Stretching field   108, **123**
Striation thickness   99
Striation thickness distribution   116, **128**
Structured grid   274
Stuffing box   1278
Surface aerators, gas–liquid mixing   595, 1142
Surface vortex
  Froude number   1078
  impellers   368, 380
Surfactant   647, 668, 715, **716**
  cell damage   1146
  dose   **716**
  gas hold-up   621
Suspension cultures, cell damage   1141
Suspension polymerization   545
  liquid–liquid dispersion   714, **738, 741**
  loop impeller   721
  phase inversion   **741**
  scale-up   **732**
Swept diameter, impeller   **55**

Tall tanks   513
Taylor's hypothesis, turbulence   42
Tee mixers
  pipeline mixing   419, 420, 438
  reactor   779
Temperature control, viscous mixing   1022
Temporal mixing, residence time distribution   2, 9
Terminal velocity *see settling velocity*
Thermal conductivity   870, 871
Time average Eulerian approach   62
Time scale
  Batchelor   50, 769, 775
  blending   52, 767
  Corrsin   51, 763, 769, 771, 833, 836, 839, 849
  fermentation   1095

Kolmogorov   50, 768
mesomixing   771
micromixing   770
mixing, reaction   24, 50, 756, 794
reaction   **25**, 756, 794, 1217
summary   775
table   50, 52
turbulent mixing   24, 50
Time to equilibrium, drop size distribution   676, 668, 727, 735
Time varying flow, drop breakup   650
Tip chord angle, impeller   660, 1336
Tip speed
  disperser operation   1193
  rotor-stator   479
  scale-up   580, 1086
  solids mixing   952
Tomography   182, 201
Top-entering mixer   1249
Torque
  cost due to   1343
  impeller   1289, 1308, 1307, 1309, 1310
  scale-up   376
  shaft design   1289, 1308, 1310
Torque measurement
  rheometers   155, 164
  strain gauges   162
  transducers   163
Tracer
  residence time distribution   2, 9
  washout   9
Trailing vortices, impeller blades   47
Transient drop size distribution   668, 676, 727, 735
Transitional flow   489
  blend time   510
Trickle bed reactor, residence time distribution   2
Troubleshooting chart   1
  dust explosion   980
  feed location   1
  gas–liquid reaction   li
  liquid–liquid extraction   lii
  liquid–liquid reaction   liii
  slurry withdrawal   liv
  solid–liquid reaction   lv
  solids mixing   937, 938
  solids removal   liv
Turbine impellers   354, 360, 382, 1030, 1250
Turbulence   vi, 19
  Batchelor scale   50
  cell damage   1076, 1127, 1139

**1376** INDEX

Turbulence *(Continued)*
  CFD  291, 301
  definition  20, 45
  drop break-up  44
  experimental methods  71
  frequency spectrum  **26**, 45
  fully isotropic  63
  fully turbulent  63, **66**
  intuitive framework  38
  Kolmogorov scale  50
  *see length scale*
  locally isotropic  63, **66**
  models, reaction *see CFD*  836
  pulp, interactions with  1201, 1203
  reaction model  38, 836
  scale-up  53, **57**
  solid–liquid  **61**
  statistical moments, table  223
  summary  81
  Taylor's hypothesis  42
  *see time scale*
  time and length scales, table  50, 52
  wavenumber  46
Turbulence model  72, 261
  Boussinesq hypothesis  73, 262
  DNS  74, 75
  k-ε  73, 79, 263
  LES  74, 76, 265
  Prandtl mixing length  73
  RANS equations  78, 262
  RSM  74, 80, 263
  table of  73, 266, 834
Turbulent dispersion, liquid–liquid pipeline  **452, 454**
Turbulent mixing  vi, 19
  blend time  509
  drop breakup  650, 656
  drop breakup, viscosity effect  658, 660
  impeller flow characteristics  358
  length scales  24, 50
  particle settling velocity  552
  time scales  24, 50
Twin-screw extruder  1009, 1014, 1016, 1019, 1022
Twin-shaft extruder  1020, 1021

Underrelaxation  283
Ungassed boiling  612
Unstructured grid  274
Up-pumping impellers  367
Upwind differencing  279, 280

Variable speed drive  721, 1266
V-blender, solids  899, 908, 910
Velocity measurement  203
  CFD results  304
  comparison of methods  203, 204, 206, 207
  rotor-stator  491
  streak photography  166
  transducers, table  204
Vendor role  xxvii, 1333
Vessel head thickness, tables  1315, 1316, 1317
Vessel materials *see materials of construction*
Vibrations  1293, 1350
Viscoelastic fluid rheology  152
Viscosity effects, emulsion  706
Viscous mixing  xxii, 987
  batch, change can  996
  batch, helical blade  995
  batch, single stirrer  994
  batch, special equipment  1017, 1020, 1021
  *see blending*
  challenges  987
  continuous  1000, 1007, 1009, 1012
  deformation  991
  elongation  989
  energy consumption  1022
  equipment  994, 1002
  extruder, CFD  **322**, 454
  fundamentals  987
  heat transfer  **883**, 993
  kneading  996, 1001, 1018, 1021
  *see laminar mixing*
  Metzner-Otto  517, 530
  model  1022
  overmixing  1022
  phase inversion  710
  power  992, 1001
  shear  989
  shear-thinning  991
  temperature control  1022
  Weber number  991
Visualization  145, 165, CD clips
  blending  165
  CFD animation  308
  light sheet  166
  solid–liquid mixing  177, 183
  streak photography  166
Vorticity
  CFD results  310
  definition  310
vvm, definition  1079

Wall function   274
Wall jet   562
Washout function, residence time
    distribution   3, 6, 9
Water in crude oil, homogenization   1175
Water treatment   409
Wavenumber of turbulence, definition   46
Weber number
    drop breakup   658
    static mixers   1177
    viscous mixing   991
Welding   1348
Wetting solids   1192
White liquor, pulp   1190, 1193
Witness testing   1344

Xanthan gum
    cavern formation   1114, 1116
    model culture   1113
    oxygen uptake rate   1114, 1115
    rheology   1114

Yield   764, 1042
    CSTR, ideal   12
    definition   764
    expected, calculation   10, 760
    plug flow reactor (PFR), ideal   12
    *see selectivity*
Yield stress fluid, biomass   1074
    blend time   525
    cavern formation   521, **522, 526**, 530,
        1080, 1114, 1200
    CFD   **321**
    pulp   1199, 1207

Zone model, fermentation   1099, 1112
Zwietering   557
    constant   558, 560, **561**
    constant, table   560
    correlation, Njs   547, 558
    criterion for solids suspension   556,
        557
    measurement of Njs   182, 184

**CUSTOMER NOTE: IF THIS BOOK IS ACCOMPANIED BY SOFTWARE, PLEASE READ THE FOLLOWING BEFORE OPENING THE PACKAGE.**

This software contains files to help you utilize the models described in the accompanying book. By opening the package, you are agreeing to be bound by the following agreement:

This software product is protected by copyright and all rights are reserved by the author and John Wiley & Sons, Inc. You are licensed to use this software on a single computer. Copying the software to another medium or format for use on a single computer does not violate the U.S. Copyright Law. Copying the software for any other purpose is a violation of the U.S. Copyright Law.

This software product is sold as is without warranty of any kind, either express or implied, including but not limited to the implied warranty of merchantability and fitness for a particular purpose. Neither Wiley nor its dealers or distributors assumes any liability of any alleged or actual damages arising from the use of or the inability to use this software. (Some states do not allow the exclusion of implied warranties, so the exclusion may not apply to you.)

**WILEY**